APPLICATION OF ACCELERATORS IN RESEARCH AND INDUSTRY

Application of Accelerators in Research and Industry
Sixteenth International Conference
CAARI 2000

Physics Department, University of North Texas
Denton, Texas 76203
http://orgs.unt.edu/CAARI

SUPPORTED BY:
US Department of Energy
National Science Foundation
University of North Texas

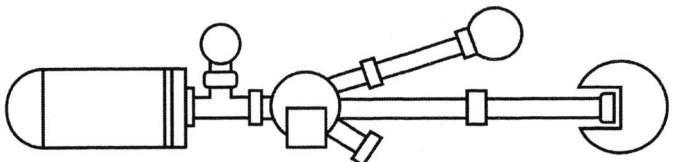

Previous Proceedings in the Series of Conferences on Application of Accelerators in Research and Industry

Year	Conference	Publisher	ISBN
1998	Fifteenth	AIP Conf. Proceedings Vol. 475	1-56396-825-8
1996	Fourteenth	AIP Conf. Proceedings Vol. 392	1-56396-652-2
1994	Thirteenth	North Holland, NIM B Vol. 99	

Other Related Titles from AIP Conference Proceedings

546 Beam Instrumentation Workshop 2000: Ninth Workshop
Edited by Kenneth D. Jacobs and R. Coles Sibley III, December 2000, 1-56396-975-0

538 Medical Physics: Fourth Mexican Symposium
Edited by J. J. Alvarado-Gil, J. G. Contreras, and G. Herrera Corral, October 2000, 1-56396-963-7

521 Synchrotron Radiation Instrumentation: Eleventh US National Conference
Edited by Piero Pianetta, John Arthur, and Sean Brennan, May 2000, 1-56396-941-6

520 Bates 25: Celebrating 25 Years of Beam to Experiment
Edited by T. W. Donnelly and W. Turchinetz, June 2000, 1-56396-949-1

451 Beam Instrumentation Workshop
Edited by Robert O. Hettel, Stephen R. Smith, and Jennifer D. Masek, December 1998, 1-56396-794-4

To learn more about these titles, or the AIP Conference Proceedings Series, please visit the webpage http://www.aip.org/catalog/aboutconf.html

APPLICATION OF ACCELERATORS IN RESEARCH AND INDUSTRY

Sixteenth International Conference

Denton, Texas 1-5 November 2000

EDITORS
J. L. Duggan
University of North Texas, Denton

I. L. Morgan
Advanced Molecular Imaging Systems Inc.

Melville, New York, 2001
AIP CONFERENCE PROCEEDINGS ■ VOLUME 576

Editors:

Jerome L. Duggan
University of North Texas
Department of Physics
P.O. Box 311427
Denton, TX 76203-1427
USA

E-mail: jduggan@unt.edu

I. Lon Morgan
Advanced Molecular Imaging Systems Inc.
2101 Shady Oaks Drive
Denton, TX 76205
USA

The articles on pp. 126-129, 261-264, 273-276, 362-365, 366-369, 559-562, 663-666, 828-832, 915-918, 935-938, 1053-1056, 1105-1108, 1113-1117, 1118-1121, 1152-1154, and 1155-1158 were authored by U.S. Government employees and are not covered by the below mentioned copyright.

Authorization to photocopy items for internal or personal use, beyond the free copying permitted under the 1978 U.S. Copyright Law (see statement below), is granted by the American Institute of Physics for users registered with the Copyright Clearance Center (CCC) Transactional Reporting Service, provided that the base fee of $18.00 per copy is paid directly to CCC, 222 Rosewood Drive, Danvers, MA 01923. For those organizations that have been granted a photocopy license by CCC, a separate system of payment has been arranged. The fee code for users of the Transactional Reporting Service is: 0-7354-0015-6/01/$18.00.

© 2001 American Institute of Physics

Individual readers of this volume and nonprofit libraries, acting for them, are permitted to make fair use of the material in it, such as copying an article for use in teaching or research. Permission is granted to quote from this volume in scientific work with the customary acknowledgment of the source. To reprint a figure, table, or other excerpt requires the consent of one of the original authors and notification to AIP. Republication or systematic or multiple reproduction of any material in this volume is permitted only under license from AIP. Address inquiries to Office of Rights and Permissions, Suite 1NO1, 2 Huntington Quadrangle, Melville, NY 11747-4502; phone: 516-576-2268; fax: 516-576-2450; e-mail: rights@aip.org.

L.C. Catalog Card No. 2001091237
ISBN 0-7354-0015-6
ISSN 0094-243X

Printed in the United States of America

CONTENTS

Preface xix
Dedication xxi
Organizational Committee xxix
Session Chairpersons xxxi
Contributors and Industrial Sponsors xxxiii
Photos xxxv

SECTION I — ATOMIC AND MOLECULAR PHYSICS

Stopping Cross Section and Charge Exchange Study on the $He^+ \to Ne$ System 3
    R. Cabrera-Trujillo, J. R. Sabin, E. Deumens, and Y. Öhrn

Determination of Stopping Power of Channeled α-Particles in SiO_2 in the Backscattering Geometry 7
    M. Kokkoris, S. Kossionides, R. Vlastou, X. Aslanoglou, R. Grötzschel, and T. Paradellis

Intriguing Results of Stopping Power Measurements with Light Ions Traversing Several Organic (Co) Polymer Targets 11
    L. E. Porter

Effect of Dechanneling on the Photon Emission by Ultra-Relativistic Positrons Channeling in a Periodically Bent Crystal 17
    A. V. Korol, A. V. Solov'yov, and W. Greiner

Measurement of the Nuclear Energy Loss Under Channeling 21
    H. Winter and A. Mertens

Stopping Powers of 2-10 MeV Si, P and S Ions in Ni, Cu and Ge Thin Films Using a Novel ERD-Based Technique 25
    M. Nigam, J. L. Duggan, M. El Bouanani, C. Yang, S. A. Datar, S. Matteson, and F. D. McDaniel

Effects of Binding Energy on Exchange Contributions to the Stopping of Electrons 29
    S. M. Cohen

Velocity Dependence of Electron Removal and Fragmentation of Water Molecules Caused by Fast Proton Impact 33
    A. M. Sayler, E. Wells, K. D. Carnes, and I. Ben-Itzhak

Transfer Ionization to Single Capture Ratio for Fast Multiply Charged Ions on He 36
    R. Ünal, P. Richard, H. Aliabadi, H. Tawara, C. L. Cocke, I. Ben-Itzhak, M. J. Singh, and A. T. Hasan

High Resolution X-Ray Emission Spectroscopy 40
    Y. Ito, A. M. Vlaicu, and T. Tochio

Intense Laser Field Studies of Positive Ions 44
    I. D. Williams, P. McKenna, B. Srigengan, I. M. G. Johnston, W. A. Bryan, J. H. Sanderson, A. El-Zein, T. R. J. Goodworth, W. R. Newell, P. F. Taday, and A. J. Langley

Polarization Study of the Extreme Ultraviolet (EUV) Emission from Helium Following Electron and Proton Impact 48
    H. Merabet, A. Siems, R. Bruch, M. Bailey, J. Hanni, H. C. Tseng, C. D. Lin, and A. G. Trigueiros

Understanding the Importance of Dynamic, Correlation, and Exchange Effects in New Low-Energy Photon Scattering Experiments 52
    J. P. J. Carney and R. H. Pratt

Consideration of the Continuum X-Ray Background in 50-300 keV Protons on Thick Targets of Ge, Ho and Au 56
    S. Cipolla, V. Horvat, and C. A. Quarles

Structures in the Energy Dependence of Classical and Quantum Bremsstrahlung 60
    A. Florescu, O. I. Obolensky, C. D. Shaffer, and R. H. Pratt

Polarizational Bremsstrahlung on Atoms and Ions: Relativistic and Non-Relativistic Cases 64
    A. V. Korol, A. G. Lyalin, O. I. Obolensky, I. A. Solovjev, and A. V. Solov'yov

Ionization Channel in Non-Perturbative Ion-Atom Collisions Dominated by Charge Exchange 68
    L. Kocbach and I. Ladadwa

Symmetry Trapping, Acceleration, and Charge Exchange Between Positive and Negative Ions in Production of Neutralized and Neutral Beams 72
    A. Y. Wong, G. Rosenthal, G. Paskalov, J. Chen, N. Hicks, D. Karfidov, and K. D. Kang

Charged Particle Trajectories in an Ideal Paracentric Hemispherical Deflection Analyser 76
    T. J. M. Zouros, E. P. Benis, and J. E. Schauer

Contributions of Negative Energy States (NES) to Magnetic Dipole Transitions in the Helium Isoelectronic Sequence, and for the Neutral Alkalis 80
    H. G. Berry and I. M. Savukov

Proton-Induced M X-Ray Production 84
    S. J. Cipolla

Non-Statistical Magnetic Substrate Populations Following Excitation of Helium by Electron and Proton Impact 89
    J. Hanni, H. Merabet, A. Siems, R. Bruch, M. Bailey, F. V. Fursa, I. Bray, K. Bartschat, H. C. Tseng, C. D. Lin, and A. G. Trigueiros

Projectile and Target Z-Scaling of Target K-Vacancy Production Cross Sections at 10 MeV/amu 93
    R. L. Watson, V. Horvat, and K. E. Zaharakis

Ionization of Noble Gases by Light Ion Impact: Scaling the Single Ionization Cross Section 96
    E. C. Montenegro, A. C. F. Santos, and G. M. Sigaud

Present Status of L-Shell X-Ray Production Cross Sections by Protons with Energies below 1 MeV 100
    J. Miranda and M. F. Lugo-Licona

Higher-Order Processes in Ion-Atom Collisions 104
    T. Mukoyama

Measured Energy and Angular Distributions of Sputtered Neutral Atoms from a Ga-In Eutectic Alloy Target 108
    A. W. Bigelow, S. L. Li, S. Matteson, and D. L. Weathers

Generalized Oscillator Strengths of the Rare-Gas Atoms 112
    Z. Chen and A. Z. Msezane

Fragmentation of H_2^{2+} Ions Following Impact of Slow Xe^{23+} and O^{5+} Ions with H_2: Projectile Charge and Velocity Dependence 114
    P. Sobocinski, J. Rangama, J.-Y. Chesnel, M. Tarisien, L. Adoui, A. Cassimi, X. Husson, and F. Fremont

Molecular Dynamics Simulations of Collisional Cooling and Ordering of Multiply Charged Ions in a Penning Trap 118
    J. P. Holder, D. A. Church, L. Gruber, H. E. DeWitt, B. R. Beck, and D. Schneider

Reactive Collisions of Hyperthermal Ions with Oxide Surfaces 122
    M. Maazouz, C. L. Quinteros, T. Tzvetkov, P. L. Maazouz, X. Qin, T. L. O. Barstis, and D. C. Jacobs

Electrostatic Trap for keV Ion Beams 126
    H. F. Krause, C. R. Vane, and S. Datz

Breakup and Recombination of Identical Bosons: He Dimer-Monomer Collisions 130
    J. H. Macek

Momentum Distributions in Fast Heavy Ion-Atom Collisions 133
    D. M. McSherry, S. F. C. O'Rourke, R. Moshammer, and J. Ullrich

Excitation Cross Sections of (1snp) $^1P^0$ (n=2-5) Levels of Helium by Fast Electron, Proton, and Molecular Hydrogen (H_2^+ and H_3^+) Impact 137
    M. Bailey, H. Merabet, R. Bruch, J. Hanni, S. Bliman, D. V. Fursa, I. Bray, K. Bartschat, H. C. Tseng, and C. D. Lin

Beyond the Shake Process 141
    T. Mukoyama and M. Uda

Mechanism of Secondary Ion Emission from an Al Sample under MeV Heavy Ion Bombardment 145
    J. Xue, S. Ninomiya, S. Gomi, and N. Imanishi

Stripping Energy Dependence of A B^{3+} ($1s^2\,^1S$, $1s2s\,^3S$) Beam Metastable Fraction 149
    M. Zamkov, H. Aliabadi, E. P. Benis, P. Richard, H. Tawara, and T. J. M. Zouros

Fragmentation of 50-100keV Molecular Hydrogen Ions in Collision with a Molecular Hydrogen Target 153
    C. McGrath, M. B. Shah, R. W. McCullough, P. C. E. McCartney, and J. W. McConkey

X-Ray Emission from Electron Capture by Highly-Charged Ions 157
    J. B. Greenwood, I. D. Williams, S. J. Smith, and A. Chutjian

Absolute Single Electron Loss in Collisions of Ar^+ with Various Atoms 161
    P. G. Reyes, H. Martínez, and F. Castillo

Recent Experiments on the Roles of Projectile Electrons in Ion-Atom Collisions 164
    D. Fregenal, J. Fiol, S. Suárez, G. Bernardi, P. Focke, and A. D. González

Ejected Electron Spectroscopy: Ionization in Ion-Atom Collisions 168
    D. M. McSherry, S. F. C. O'Rourke, C. McGrath, M. B. Shah, and D. S. F. Crothers

Zero-Degree Auger Electron Spectroscopy of Quasi-Free Electrons Scattered by Highly Charged Ions 172
    H. Aliabadi, R. Ünal, M. Zamkov, P. Richard, C. P. Bhalla, H. Tawara, M. Gealy, and A. T. Hasan

High-Resolution Photoelectron Spectroscopy in Atoms and Molecules 177
    N. Berrah, O. Nayandin, S. E. Conton-Rogan, E. Kukk, A. A. Wills, T. W. Gorczyca, G. Snell, C. N. Liu, J. D. Bozek, and M. Wiedenhoeft

Electron-Electron and Electron-Nuclear Interactions in Fast Neutral-Neutral Collisions 181
    R. D. DuBois

How Do Electrons Communicate About Time? 185
    A. L. Godunov, J. H. McGuire, S. G. Tolmanov, K. K. Shakov, R. Dörner, H. Schmidt-Böcking, and R. M. Dreizler

Photoelectron Spectroscopy and the Dipole Approximation 189
    O. A. Hemmers and D. W. Lindle

Measurement of Compton Scattered Electrons Using Monochromatic X-Rays 193
    W. A. Hollerman, N. A. Guardala, D. J. Land, G. A. Glass, and J. L. Price

Electrons from Neutral Projectiles and Targets: Peaks and Ramsauer-Townsend Structures 197
    D. H. Jakubaßa-Amundsen

Compton Profile of Multiply-Ionized Fluorine Atoms 201
    K. R. Karim and C. P. Bhalla

First Studies of State Selective Electron Capture from Atomic Hydrogen by State Prepared Doubly Charged Ions 205
    D. R. Gillen, D. M. Kearns, D. Voulot, R. W. McCullough, and H. B. Gilbody

Electron Capture and Loss Cross Sections for Neutral Projectiles Colliding with Atoms and Molecules 209
    J. M. Sanders, S. L. Varghese, and C. H. Fleming

EUV Photon Emission Spectroscopy for Diagnostics of Single and Multiple-Electron Capture Processes in 80 keV$Ar^{8+}+N_2$ Collisions 213
    A. Siems, H. Merabet, R. Bruch, V. Golovkina, G. Hinojosa, R. Phaneuf, J. Hanni, S. Bliman, and A. G. Trigueiros

Auger Decay of Triply Excited States of Li and Be^+ Projectiles Following Ion-Molecule Interactions 217
    R. Bruch, A. Siems, H. Merabet, J. Hanni, and S. Bliman

SECTION II — NUCLEAR PHYSICS

Spectrometers at RHIC 223
    K. Hagel

Tracking Chambers at RHIC ... 227
    J. Velkovska

First Results from the PHOBOS Experiment at RHIC ... 231
    A. H. Wuosmaa, B. B. Back, M. D. Baker, D. S. Barton, S. Basilev, B. D. Bates, R. Baum,
    R. R. Betts, A. Bialas, R. Bindel, W. Bogucki, A. Budzanowski, W. Busza, A. Carroll,
    M. Ceglia, Y. H. Chang, A. E. Chen, T. Coghen, C. Conner, W. Czyz, B. Dabrowski,
    M. P. Decowski, M. Despet, P. Fita, J. Fitch, M. Friedl, K. Galuska, R. Ganz, E. Garcia,
    N. George, J. Godlewski, C. Gomes, E. Griesmayer, K. Gulbrandsen, S. Gushue, J. Halik,
    C. Halliwell, P. Haridas, A. Hayes, G. A. Heintzelman, C. Henderson, R. Hollis, R. Holynski,
    B. Holzman, E. Johnson, J. Kane, J. Katzy, W. Kita, J. Kotula, H. Kraner, W. Kucewicz,
    P. Kulinich, C. Law, M. Lemler, J. Ligocki, W. T. Lin, S. Manly, D. McCleod, J. Michałowski,
    A. Mignerey, J. Mülmenstädt, M. Neal, R. Nouicer, A. Olszewski, R. Pak, I. C. Park, M. Patel,
    H. Pernegger, M. Plesko, C. Reed, L. P. Remsberg, M. Reuter, C. Roland, G. Roland, D. Ross,
    L. Rosenberg, J. Ryan, A. Sanzgiri, P. Sarin, P. Sawicki, J. Scaduto, J. Shea, J. Sinacore,
    W. Skulski, S. G. Steadman, G. S. F. Stephans, P. Steinberg, A. Straczek, M. Stodulski,
    M. Strek, Z. Stopa, A. Sukhanov, K. Surowiecka, J. L. Tang, R. Teng, A. Trzupek, C. Vale,
    G. J. van Nieuwenhuizen, R. Verdier, B. Wadsworth, F. L. H. Wolfs, B. Wosiek, K. Wozniak,
    A. H. Wuosmaa, B. Wyslouch, K. Zalewski, and P. Zychowski

The REX-ISOLDE Beam Diagnostic System ... 235
    P. Van den Bergh, M. Huyse, K. Krouglov, P. Van Duppen, and L. Weissman

Status Report on the ISAC Radioactive Ion Beam Facility ... 239
    P. Bricault

A Negative Surface Ionization Source for RIB Generation ... 244
    H. Zaim, Y. Liu, S. N. Murray, and G. D. Alton

Characterization and Fabrication of Target Materials for RIB Generation ... 250
    R. F. Welton, M. A. Janney, P. E. Mueller, W. K. Ortman, R. Rauniyar, D. W. Stracener,
    and C. L. Williams

RNB Production at SPIRAL/GANIL ... 254
    A. C. C. Villari, F. L. Pellemoine, C. Barué, G. Gaubert, S. Gibouin, Y. Huguet, P. Jarin,
    S. K. Rody, N. Lecesne, R. Leroy, M. Lewitowicz, C. Marry, L. Maunoury, J. Y. Pacquet,
    J. P. Rataud, M. G. Saint Laurent, C. Stodel, O. Bajeat, J. C. Angélique, and N. A. Orr

Radioactive Ion Beams at the HRIBF ... 257
    D. W. Stracener for the HRIBF Staff

Experiments with Radioactive Beams at ATLAS ... 261
    K. E. Rehm, I. Ahmad, J. Blackmon, F. Borasi, J. Caggiano, A. Chen, C. N. Davids,
    J. Greene, B. Harss, A. Heinz, D. Henderson, R. V. F. Janssens, C. L. Jiang, J. Nolen,
    R. C. Pardo, P. Parker, M. Paul, J. P. Schiffer, R. E. Segel, D. Seweryniak, R. H. Siemssen,
    M. S. Smith, J. Uusitalo, T. F. Wang, and I. Wiedenhöver

REX-ISOLDE—Post-Accelerated Radioactive Beams at CERN-ISOLDE ... 265
    T. Nilsson, J. Äystö, O. Forstner, H. L. Ravn, M. Oinonen, H. Simon, J. Cederkäll, L. Weissman,
    D. Habs, F. Ames, O. Kester, T. Sieber, H. Bongers, S. Emhofer, P. Reiter, P. G. Thirolf, G. Bollen,
    P. Schmidt, G. Huber, L. Liljeby, Ö. Skeppstedt, K. G. Rensfelt, F. Wenander, B. Jonson, G. Nyman,
    R. von Hahn, H. Podlech, R. Repnow, C. Gund, D. Schwalm, A. Schempp, K.-U. Kühnel, C. Welsch,
    U. Ratzinger, G. Walter, A. Huck, K. Kruglov, M. Huyse, P. van den Bergh, P. van Duppen,
    A. C. Shotter, A. N. Ostrowski, T. Davinson, P. J. Woods, I. Moukha, A. Richter, G. Schrieder,
    and the REX-ISOLDE Collaboration

The Production and Generation of Radioactive Beams at Louvain-la-Neuve ... 269
    M. Loiselet, M. Cogneau, J. M. Colson, M. Gaelens, N. Postiau,
    and G. Ryckewaert

Research with Radioactive Beams at ORNL ... 273
    J. F. Liang

Breeding 10^{10}/s Radioactive Nuclei in a Compact Plasma Focus Device ... 277
    J. S. Brzosko, K. Melzacki, C. Powell, M. Gai, R. H. France III, J. E. McDonald, G. D. Alton,
    F. E. Bertrand, and J. R. Beene

The Development of an ECRIS Charge State-Breeder for Generating RIB's ... 281
    T. Lamy, R. Geller, J. L. Bouly, N. Chauvin, J. C. Curdy, A. Lacoste, P. Sole, P. Sortais,
    T. Thuillier, and J. L. Vieux-Rochaz

Simulation of the Effusive-Flow of Reactive Gases in Tubular Transport Systems:
Radioactive Ion Beam Applications ... 285
    J.-C. Bilheux and G. D. Alton

Status of the New Los Alamos UCN Source 289
  K. Kirch, T. J. Bowles, B. Fillipone, P. Geltenbort, R. Hill, M. Hino, S. Hoedl, G. Hogan,
  T. M. Ito, T. Kawai, S. K. Lamoreaux, C.-Y. Liu, J. W. Martin, C. Morris, A. Pichlmaier,
  A. Saunders, S. Seestrom, A. Serebrov, D. Smith, B. Tipton, M. Utsuro, A. R. Young,
  and J. Yuan

**Measurements of the Electromagnetic and Spin Structure of Nucleons and Nuclei
with the Bates Large Acceptance Spectrometer Toroid** 293
  R. Alarcon

Tests of the Standard Model from Superallowed Fermi β-Decay Studies of ISAC 297
  G. C. Ball for the E823 Collaboration

Predicting Long-Lived, Neutron-Induced Activation of Concrete in a Cyclotron Vault 301
  L. R. Carroll

Sub-Coulomb Alpha Transfer Reactions 305
  C. R. Brune

Nuclear Physics with Fast Neutrons at LANSCE/WNR: GEANIE and FIGARO 309
  M. Devlin, N. Fotiades, G. D. Johns, R. O. Nelson, R. C. Haight, L. Zanini, J. A. Becker,
  L. A. Bernstein, P. E. Garrett, C. A. McGrath, D. P. McNabb, W. Younes, J. X. Saladin,
  and A. Aprahamian

Thick Target Yields for the Reaction ^{18}O(p,n)^{18}F above 16 MeV 313
  C. Gonzalez Lepera and S. R. Strangis

Utilization of the (n,n'γ) Reaction for Nuclear Structure Studies of ^{122,124,126}Te 315
  S. F. Hicks, J. R. Vanhoy, N. Warr, T. B. Brown, and S. W. Yates

Lifetime Measurements Using the Recoil Distance Method—Achievements and Perspectives 319
  R. Krücken

Cross Section Measurement of ^{11}B(p,α)^{8}Be and Its Application on Boron Characterization 323
  J. Liu, X. Lu, J. Jin, Q. M. Li, and W.-K. Chu

Cold Neutron Beam Studies of Parity-Violation in the n-α and n-p Systems 327
  D. M. Markoff

Effect of Symmetry Breaking on Transition Strength Distributions 331
  G. E. Mitchell and J. F. Shriner, Jr.

**Production of ^{90}Y by the ^{90}Zr(n,p)^{90}Y Reaction Using Neutrons Produced from a Variable
Energy Cyclotron** 335
  D. Necsoiu, I. L. Morgan, H. Hupf, J. Armbruster, D. Boyce, M. El Bouanani,
  and F. D. McDaniel

**Precision Measurement of the n+p→d+γ Cross Section for Testing Big-Bang
Nucleosynthesis Models** 339
  E. H. Seabury, R. C. Haight, J. M. O'Donnell, J. L. Ullmann, S. A. Wender, T. Akdogan,
  M. Chtangeev, W. A. Franklin, J. B. Hough, J. L. Matthews, and Y. Safkan

A Test of the Exponential Decay Law by Photo-Production of Nuclear Isomers 342
  D. P. Wells, J. F. Harmon, W. W. Scates, and R. Spaulding

Study of Reactions of Fast Neutrons with Nuclei with FIGARO at LANSCE 346
  L. Zanini, A. Aprahamian, M. B. Chadwick, M. Devlin, R. C. Haight, J. X. Saladin,
  and P. G. Young

Nuclear Astrophysics Measurements Using Neutrons 350
  P. Koehler

Nuclear Astrophysics with a Recoil Mass Separator 354
  F. Strieder

High Temperature rp-Process Nucleosynthesis of Mo and Ru, and β-Delayed Proton Emission 358
  R. N. Boyd

The Front-End Systems for the Spallation Neutron Source 362
  R. Keller

Extension of the Electronic Gamma-Ray Spectrum Catalogue Web Site 366
  R. J. Gehrke, J. W. Mandler, R. G. Helmer, and J. R. Davidson

SECTION III — ACCELERATOR MASS SPECTROMETRY

Transmission Properties for the Recombinator-Tandetron Configuration at the National Ocean Sciences Accelerator Mass Spectrometry Facility 373
    K. F. von Reden

Environmental Radiation Protection Studies Related to Nuclear Industries, Using AMS 377
    R. Hellborg, B. Erlandsson, M. Faarinen, H. Håkansson, K. Håkansson, M. Kiisk,
    C. E. Magnusson, P. Persson, G. Skog, K. Stenström, S. Mattsson, and C. Thornberg

Exotic Particle Searches Using the Purdue AMS Facility 382
    D. Javorsek II, D. Elmore, E. Fischbach, and T. Miller

A Structural Upgrade for the Iso Trace Tandetron Accelerator 386
    W. E. Kieser, X.-L. Zhao, A. E. Litherland, and K. H. Purser

Ion Preparation Systems for Atomic Isobar Reduction in Accelerator Mass Spectrometry 390
    A. E. Litherland, J. P. Doupe, I. Tomski, J. Krestow, X. L. Zhao, W. E. Kieser, and R. P. Beukens

Biokinetic and Dosimetric Investigations of ^{14}C-Labeled Substances in Man Using AMS 394
    S. Mattsson, M. Gunnarsson, S. L. Svegborn, B. Nosslin, L. E. Nilsson, O. Thorsson,
    S. Valind, M. Åberg, H. Östberg, R. Hellborg, K. Stenström, B. Erlandsson, M. Faarinen,
    M. Kiisk, C.-E. Magnusson, P. Persson, and G. Skog

Microbeam AMS Measurements of PGE and Au Trace and Osmium Isotopic Ratios 399
    S. H. Sie, D. A. Sims, and G. F. Suter

A 3 MV Heavy Element AMS System Using a Unique TOF Set-Up 403
    A. Gottdang, M. Klein, D. J. W. Mous, T. Kitamura, Y. Mizutani, T. Suzuki, T. Aramaki,
    O. Togawa, S. Kabuto, and K. Suto

Test of Positive Ion Beams from a Microwave Ion Source for AMS 407
    S.-W. Kim, R. J. Schneider, K. F. von Reden, J. M. Hayes, J. S. C. Wills, and W. G. E. Kern

SECTION IV — MATERIALS ANALYSIS WITH ION BEAMS, PIXE, RBS, ERD, NEUTRONS, AND NAA MICROBEAMS, OTHER TECHNIQUES

Practical Limitations of GUPIX 413
    J. L. Campbell and J. A. Maxwell

Applications of PIXE with 68 MeV Protons 417
    A. Denker and K. H. Maier

A Forensic Application of PIXE Analysis 421
    I. I. Kravchenko, F. E. Dunnam, H. A. Van Rinsvelt, M. W. Warren, and A. B. Falsetti

Study of the Elemental Composition of Yellow Pine Using Particle Induced X-Ray Emission (PIXE) 425
    C. Liao, W. A. Hollerman, and G. A. Glass

Investigation of Ancient Human Bone by Means of Ionoluminescence and μPIXE 428
    D. Spemann, S. Jankuhn, J. Vogt, and T. Butz

High Energy Ion Beam Analysis of Buried α-Fe$_2$O$_3$(0001)/α-Al$_2$O$_3$(0001) Interface 432
    S. Thevuthasan, V. Shutthanandan, E. M. Adams, and S. Maheswaran

Characterization of Multilayer Thin Film Optical Filters Using RBS 436
    R. Vlastou, E. Fokitis, M. Kokkoris, S. Kossionides, G. Koubouras, and R. Grötzschel

RBS and NRA of Cobalt Oxide Thin Films Prepared by The Sol-Gel Process 440
    E. Andrade, L. Huerta, E. Barrera, J. C. Pineda, E. P. Zavala, M. F. Rocha, and C. A. Vargas

Ion Beam Characterization of Advanced Metallization for ULSI Applications 443
    A. E. Bair, Y. Wang, J. W. Mayer, and T. L. Alford

Simultaneous Analysis of Multiple Elements by Combined Ion-Beam Methods 447
    W. Jiang, W. J. Weber, S. Thevuthasan, and V. Shutthanandan

Double Alignment Channeling at CIM — Alabama A&M University 451
    I. C. Muntele, C. I. Muntele, D. Ila, and R. L. Zimmerman

Investigation of Alkali Ion Exchange Processes in Waste Glasses Using Rutherford Backscattering Spectrometry and Nuclear Reaction Analysis 454
    V. Shutthanandan, S. Thevuthasan, D. R. Baer, E. M. Adams, S. Maheswaran,
    M. H. Engelhard, J. P. Icenhower, and B. P. McGrail

Ion Beam Analysis with Monolayer Depth Resolution Using the Electrostatic Spectrometer at the MPI Stuttgart ... 458
D. Plachke, G. Blohm, T. Fischer, A. Khellaf, O. Kruse, H. Stoll, and H. D. Carstanjen

RBS & HFS for Advanced Interconnect ... 463
M. D. Strathman

Hydrogen Analysis of Epitaxial Cu(111)/Nb(110) Multilayer Using MeV Ion Beams ... 466
S. Yamamoto and H. Naramoto

Advanced RBS Analysis of Thin Films in Micro-Electronics ... 470
B. Brijs, J. Deleu, C. Huygebaert, S. Nauwelaerts, K. Nakajima, K. Kimura, and W. Vandervorst

Strain Measurement of Semiconductor Multi-Layers by Ion Channeling, High Resolution XRD and Raman Spectroscopy ... 476
A. M. Siddiqui, S. V. S. Nageswara Rao, and A. P. Pathak

Depth Profiling Code for Analysing ERD-TOF Spectra ... 480
G. Mathot, G. Terwagne, and F. Bodart

Enhanced Hydrogen Detection and Depth Profiling System Using Coincidence Techniques ... 484
C. I. Muntele, D. Ila, and R. L. Zimmerman

Heavy Ion ERD of Nitrides with a Position-Sensitive Gas Ionization Detector ... 487
H. Timmers, T. D. M. Weijers, R. G. Elliman, and T. R. Ophel

Coded Aperture Fast Neutron Analysis: Latest Design Advances ... 491
R. Accorsi and R. C. Lanza

Methods for Quality Control/Quality Assurance of k_0-Assisted Neutron Activation Analysis ... 495
F. De Corte

Monitoring of D-T Accelerator Neutron Output in a PGNAA System Using Silicon Carbide Detectors ... 499
A. R. Dulloo, F. H. Ruddy, J. G. Seidel, and B. Petrović

The Preparation and Characterization of Synthetic Multi-Element Standards for Testing the Performance of k_0-NAA: the State of Affairs ... 504
M. Eguskiza, P. Robouch, F. De Corte, and S. Pommé

Applications of Nuclear Analytical Techniques to Environmental Studies ... 508
M. C. Freitas, A. M. G. Pacheco, A. P. Marques, L. I. C. Barros, and M. A. Reis

Metallic Pollutants in Mexico Valley ... 512
T. Martínez, J. Lartigue, P. Avila-Perez, M. Navarrete, G. Zarazúa, C. López, L. Cabrera, and A. Ramirez

Radiation Effects Microscopy ... 516
B. L. Doyle, G. Vizkelethy, K. M. Horn, D. S. Walsh, and P. E. Dodd

Report on the Acadiana Research Laboratory Nuclear Microprobe System ... 522
G. A. Glass, W. A. Hollerman, S. F. Hynes, J. Fournet, A. M. Bailey, and C. Liao

Latent Ion Tracks in Mica Studied with Scanning Force Microscopy in Air and in Vacuum ... 526
V. Hoffmann, J. H. Bremer, S. Bouffard, and N. Stolterfoht

Heavy Ion Microbeam Studies of Diffusion Time Resolved Charge Collection from p-n Junctions ... 531
B. N. Guo, M. El Bouanani, S. N. Renfrow, D. S. Walsh, B. L. Doyle, J. L. Duggan, and F. D. McDaniel

The Study of Phosphor Efficiency and Homogeneity Using a Nuclear Microprobe ... 535
C. Yang, B. L. Doyle, M. Nigam, M. El Bouanani, J. L. Duggan, and F. D. McDaniel

The Recent Progress of the High-Energy Heavy Ion Nuclear Microprobe at the University of North Texas ... 539
C. Yang, B. N. Guo, M. El Bouanani, M. Nigam, J. L. Duggan, and F. D. McDaniel

Dependence of Heavy Ion Induced Secondary Ion Emission on Electric Conductivity in MeV Energy Range ... 543
S. Ninomiya, S. Gomi, J. Xue, M. Imai, and N. Imanishi

Diffusion of TiN into Aluminum Films Measured by Soft X-Ray Spectroscopy ... 548
T. M. Schuler, D. L. Ederer, N. Ruzycki, G. Glass, W. A. Hollerman, A. Moewes, M. Kuhn, and T. A. Callcott

SECTION V — DETECTORS AND SPECTROMETERS

An Anti-Comption Suppression Ge-Telescope Detection System for Quality Control of Nuclear Waste Packages 555
    S. Agosteo, B. Chabalier, A. Foglio Para, U. Graf, N. Huot, T. Kekki, A. Ravazzani,
    P. Schillebeeckx, V. Tanner, and A. Tiitta

The Argonne Fragment Mass Analyzer and Measurements of Entry Distributions 559
    A. Heinz, T. L. Khoo, P. Reiter, I. Ahmad, P. Bhattacharyya, J. Caggiano, M. P. Carpenter,
    J. A. Cizewski, C. N. Davids, W. F. Henning, R. V. F. Janssens, G. D. Jones, R. Julin,
    F. G. Kondev, T. Lauritsen, C. J. Lister, D. Seweryniak, S. Siem, A. A. Sonzogni,
    J. Uusitalo, and I. Wiedenhöver

Neutron and Simultaneous Gamma Detection with $LiBaF_3$ Scintillator 563
    P. L. Reeder and S. M. Bowyer

Pulse-Height Spectrum Measurement Experiment for Code Benchmarking: Initial Results 567
    K. E. Sale, J. M. Hall, and C. M. Brown

SONTRAC—A Scintillating Plastic Fiber Tracking Detector for Neutron and Proton Imaging Spectroscopy 571
    J. M. Ryan, J. R. Macri, M. L. McConnell, and R. S. Miller

Absolute Calibration of the In-Beam Proton Polarimeter at IUCF 575
    E. J. Stephenson

Neutron Spectrometry in Neutron and Charged-Particle Mixed Fields with Phoswich Neutron Detector 579
    M. Takada, S. Taniguchi, T. Nakamura, and K. Fujitaka

The Gamma Ray Energy Tracking Array 583
    K. Vetter

Fractional Counts — The Simulation of Low Probability Events 587
    R. L. Coldwell, G. P. Lasche, and A. Jadczyk

Multielement Silicon Detectors for Registration of Charged Particles, X-Rays and Gamma-Radiation 591
    D. O. Frolov, O. S. Frolov, A. A. Sadovnickiy, O. F. Nimets, and V. A. Shevchenko

SECTION VI — ACCELERATOR TECHNOLOGY: ION SOURCES, NEW FACILITIES AND COMPONENTS

Application of ECR Ion Sources for Surface Modification of Materials 599
    A. Dobrosavljević, N. Bibić, and N. Nešković

New Ion Beam Development at Kansas State University 603
    C. W. Fehrenbach and M. P. Stockli

Solid State Pulsed Power Systems for Plasma Source Ion Implantation 607
    M. P. J. Gaudreau, J. A. Casey, M. A. Kempkes, J. M. Mulvaney, and T. J. Hawkey

Laser Ion Source for On-Line Production of Exotic Nuclei 611
    Y. Kudryavtsev, B. Bruyneel, J. Gentens, M. Huyse, P. Van den Bergh,
    and P. Van Duppen

Photocathode Electron Gun Applications in Research and Industry 615
    A. M. M. Todd, H. Bluem, M. D. Cole, J. R. Rathke, I. Ben-Zvi, T. Srinivasan-Rao,
    J. Schill, G. Neil, and C. Bohn

Emittance Studies of ARTEMIS — The New ECR Ion Source for the Coupled Cyclotron Facility at NSCL/MSU 619
    P. A. Zavodszky, H. Koivisto, D. Cole, and P. Miller

Spatial Distribution of Ion Species in a Source Plasma for Broad Beams 623
    N. Sakudo, K. Hayashi, Y. Nishiyama, K. Komatsu, J. Miyamoto, M. Yutani, and K. Awazu

The VERA Heavy Ion Program—Status and Prospects 627
    R. Golser, G. Federmann, W. Kutschera, A. Priller, P. Steier,
    and C. Vockenhuber

Accelerator System at the Wakasa-wan Energy Research Center 631
S. Hatori, Y. Ito, R. Ishigami, K. Yasuda, T. Inomata, T. Maruyama, K. Ikezawa, K. Takagi, K. Yamamoto, S. Fukuda, K. Kume, G. Kagiya, T. Hasegawa, M. Hatashita, M. Yamada, H. Yamada, M. Dote, N. Ohtani, S. Kakiuchi, Y. Tominaga, S. Fukumoto, and M. Kondo

The NSCL Coupled Cyclotron Project: Status and Latest News 635
T. Baumann

Recycling and Recommissioning a Used Biomedical Cyclotron 639
L. R. Carroll, F. Ramsey, J. Armbruster, and M. Montenero

The New IBA Laboratory to Be Installed at Universidad Autonoma de Madrid 643
A. Climent-Font, F. Agulló-López, O. Enguita, O. Espeso-Gil, G. García, and C. Pascual-Izarra

Radiation Effects Facilities, Dosimetry and Program at the Indiana University Cyclotron Facility 647
C. C. Foster, C. M. Berg, E. R. Hall, S. B. Klein, B. v. Przewoski, and K. M. Murray

A CW RFQ Injector for the IUCF Cyclotron 651
D. L. Friesel, V. Anferov, and R. W. Hamm

Electrostatic Focusing Accelerator Consisting of Multiple Coaxial Cylinders 655
A. D. Dymnikov and G. García

High-Intensity γ-Ray Source 659
J. H. Kelley, B. T. Crowley, V. N. Litvinenko, S. H. Park, I. V. Pinayev, E. C. Schreiber, W. Tornow, Y. Wu, and H. R. Weller

Conceptual Design of a 100 MW Electron Beam Accelerator Module for the National Hypersonic Wind Tunnel Program 663
L. X. Schneider

Laser-Cooling of Ions and Ion Acceleration in the RF-Quadrupole Ring Trap PALLAS 667
U. Schramm, T. Schätz, and D. Habs

The CERN-EU Radiation Facility for Dosimetry at Flight Altitude and in Space 671
A. Ferrari, A. Mitaroff, and M. Silari

Simulation of Cosmic Neutron Flux Using an Electron Linac 675
T. J. Collens, A. P. Tonchev, R. M. Jaber, K. C. Kennedy, and J. F. Harmon

RIKEN RI Beam Factory Project 679
Y. Yano, T. Katayama, A. Goto, and M. Kase

A Portable Neutron/Tunable X-Ray Source Based on Inertial Electrostatic Confinement 683
G. H. Miley

A High Intensity Radiation Effects Facility 687
V. H. Rotberg, O. Toader, and G. S. Was

Applications for the RFD Linac Structure 692
D. A. Swenson

Status Report of the Development of a Multi-mA Self-Extracted H^+-Beam Cyclotron 696
S. Lucas, W. Kleeven, M. Abs, E. Poncelet, and Y. Jongen

SECTION VII — SYNCHROTRON EXPERIMENTS AND FACILITIES

Threshold Photoelectron Spectroscopy Using Synchrotron Radiation 703
G. C. King, A. J. Yencha, and M. C. A. Lopes

Chemical State Analysis of Iron in Nerve Cells 707
S. Fujisawa, A. M. Ektessabi, and S. Yoshida

Recent Results in Photoionization of Atoms and Ions Using Undulator Radiation 711
F. J. Wuilleumier, D. Cubaynes, and J.-M. Bizau

Synchrotron Radiation Total Reflection for Rainwater Analysis 715
S. M. Simabuco and E. Matsumoto

Quantitative Analysis of Biomedical Samples Using Synchrotron Radiation Microbeams 720
A. Ektessabi, S. Shikine, and S. Yoshida

LIGA Spinnerets for Microfibers 724
B.-Y. Shew and Y. Cheng

The New Normal-Incidence-Monochromator Facility at CAMD 730
E. Morikawa, C. M. Evans, and J. D. Scott

The Development of the GCPCC Protein Crystallography Beamline at CAMD 734
    M. D. Miller, G. N. Phillips, Jr., M. A. White, R. O. Fox, and B. C. Craft, III

SECTION VIII — POSITRON EXPERIMENTS

An Intense, Compact Fourth-Generation Positron Source Based on Using
a 2 MeV Proton Accelerator. 741
    N. A. Guardala, J. P. Farrell, and V. Dudnikov
Slow Positron Beams — A Versatile Tool for Studying Ion Implantation Defect
Related Phenomena 745
    B. J. Sealy, A. P. Knights, R. M. Gwilliam, C. P. Burrows, and P. G. Coleman
Positron Annihilation Studies on Stable and Undercooled Metal Melts at the
Stuttgart Pelletron 749
    H. Stoll, A. Siegle, and J. Major
Low Energy Positrons at Semiconductor Surfaces 753
    N. G. Fazleev, J. L. Fry, and A. H. Weiss

SECTION IX — FUSION EXPERIMENTS

Performance Enhancement of Negative Ion Sources for the JT-60U Tokamak 759
    L. R. Grisham, M. Kuriyama, M. Kawai, T. Itoh, N. Umeda,
    and the JT-60U Team
Fusion Neutronics — Streaming, Shielding, Heating, Activation 763
    H. Freiesleben, D. Richter, K. Seidel, and S. Unholzer
Integrated Neutral Beam Measurements in a Tokamak Environment. 768
    D. M. Thomas

SECTION X — RADIATION PROCESSING: FACILITIES AND APPLICATIONS

A New Electron Accelerator Facility for Commercial and Educational Uses 775
    R. M. Uribe and C. Vargas-Aburto
Performance of the Electron Beam Fluidized Bed Process for Disinfection and Disinfestations
of Stored Products 779
    D. A. Cleghorn, S. V. Nablo, and D. N. Ferro
Treatments of Foods with High-Energy X Rays 783
    M. R. Cleland, J. Meissner, A. S. Herer, and E. W. Beers
A Sub-Picosecond Pulsed 5 MeV Electron Beam System. 787
    J. P. Farrell, K. Batchelor, I. Meshkovsky, I. Pavlishin, V. Lekomtsev, A. Dyublov, M. Inochkin,
    and T. Srinivasan-Rao

SECTION XI — MEDICAL APPLICATIONS:
POSITRON EMISSION TOMOGRAPHY AND TARGETS,
RADIOISOTOPE PRODUCTION, PARTICLE THERAPY BNCT MEASUREMENTS

Characterization of Neutron and Photon Sources from a 10.5 MeV Proton Beam
on [^{18}O] Enriched Water 793
    L. F. Miller, L. W. Townsend, and C. W. Alvord
Target and Accelerator Developments at CTI. 799
    C. W. Alvord, A. J. Mendez, and D. E. Wittner
GE PET Trace and Associated Systems, 4 Years Experience in Cambridge. 804
    J. C. Clark, F. I. Aigbirhio, P. Burke, and S. P. M. J. Downey
A Multi-Run Chemistry Module for the Production of [^{18}F] FDG. 808
    B. Sipe, M. Murphy, B. Best, S. Zigler, J. Lim, E. Dorman, T. Mangner, and M. Weichelt
The Plasma Separation Process as a Pre-Cursor for Large Scale Radioisotope Production 814
    N. R. Stevenson

Thermal Performance of CTI, Inc. Enriched Water Targets 817
    A. E. Ruggles and C. W. Alvord

On the Production of Radioactive Stents 824
    K. Schlösser and H. Schweickert

Proposal for a New High-Energy Isotope Production Facility at LANSCE 828
    E. J. Pitcher, M. W. Cappiello, and H. A. O'Brien

Computer Study of Isotope Production for Medical and Industrial Applications in High Power Accelerators 833
    S. G. Mashnik, W. B. Wilson, and K. A. Van Riper

Radio Indium and Gallium Labeled Porphyrins for Medical Imaging 837
    P. V. Kulkarni, D. Jain, and J. Narula

Non-Standard Isotope Production and Applications at Washington University 841
    T. J. McCarthy, D. W. McCarthy, R. Laforest, H. M. Bigott, F. Wüst, D. E. Reichert, M. R. Lewis, and M. J. Welch

Production of Ultra-Pure I-123 from the ^{123}Te(p,n)^{123}I Reaction 845
    H. B. Hupf, J. E. Beaver, J. M. Armbruster, and J. P. Pendola

Cyclotron Production and Potential Clinical Application of Iodine-124 Labeled Radiotracers 849
    R. Finn, J. Balatoni, P. Kothari, K. Pentlow, Y. Sheh, C. Lom, J. Dahl, W. Eckelman, P. Plascjak, H. R. Adams, and S. M. Larson

Industrial Production of ^{131}I by Neutron Irradiation and Melting of Sintered TeO$_2$ 853
    J. Alanis and M. Navarrete

The IBA State-Of-The-Art Proton Therapy System, Performances and Recent Results 857
    D. Prieels, B. Marchand, B. Bauvir, P. De Crock, G. Gevers, S. Schmidt, G. Andre, S. Ternier, and Y. Jongen

Proton Synchrotrons for Cancer Therapy 861
    G. B. Coutrakon

Ion Beam Therapy: Overview of the World Experience 865
    J. M. Sisterson

Progress of Particle Therapy in Japan 869
    F. Soga

Medical Applications of in vivo Neutron Inelastic Scattering and Neutron Activation Analysis: Technical Similarities to Detection of Explosives and Contraband 873
    J. J. Kehayias

Uranium Target for Electron Accelerator Based Neutron Source for BNCT 877
    A. P. Tonchev, F. Harmon, T. J. Collens, K. Kennedy, A. Sabourov, Y. D. Harker, D. W. Nigg, and J. L. Jones

Neutron Capture Therapy (NCT) Enhancement of Fast Neutron Radiotherapy: Application to Non-Small Cell Lung Cancer 881
    G. E. Laramore, K. J. Stelzer, R. Risler, J. L. Schwartz, J. J. Douglas, J. P. Einck, D. W. Nigg, C. A. Wemple, J. K. Hartwell, Y. D. Harker, P. R. Gavin, and M. F. Hawthorne

SECTION XII — ION IMPLANTATION: SEMICONDUCTORS, MATERIALS MODIFICATION, CLUSTERS, NANOTUBES, ORGANIC MATERIALS, SOURCES FOR IMPLANTATION

Fermi-Level Dependent Diffusion of Ion-Implanted Arsenic in Germanium 887
    T. Ahlgren, J. Likonen, S. Lehto, E. Vainonen-Ahlgren, and J. Keinonen

The Alternative Ion Implantation Approaches for Ultra-Shallow Junction 891
    W.-K. Chu, J. Liu, J. Jin, X. Lu, L. Shao, Q. Li, and P. Ling

Slicing Dielectric Crystals with Ions: A New Material Processing Technique for Electronic and Optoelectronic Materials Integration 896
    R. M. Osgood, Jr., A. M. Radojevic, M. Levy, and H. Bakhru

Recoil Implantation of Boron into Silicon by High Energy Silicon Ions 900
    L. Shao, X. M. Lu, X. M. Wang, I. Rusakova, G. Mount, L. H. Zhang, J. R. Liu, and W.-K. Chu

Low Energy Implantation of Boron with Decaborane Ions 904
    M. Sosnowski

Ultra Shallow Sb Doped Layer Formation in Si(001) by the Use of Recoil Implantation 908
  K. E. Daley, D. T. Vonk, and R. J. Culbertson
Low Energy Ion Beams for Surface Modification and Film Deposition 911
  J. W. Rabalais
Development of Corrosion-Resistant Metal Nitride Coatings via Ion Beam Assisted Deposition 915
  J. D. Demaree
Ion Beam Induced Pore Structure Changes in Porous Silicon 919
  F. Pászti, A. Manuaba, Z. E. Horváth, E. Szilágyi, and G. Battistig
Comparison Between Two Deposition Methods for Zirconia Film 924
  N. K. Huang and D. Z. Wang
Effects of Proton Irradiation on the Critical Current Densities of $YBa_2Cu_3O_{7-\delta}$ Thin Films 928
  C. Wang, H.-W. Seo, C.-J. Su, Y.-H. Wang, Q. Y. Chen, X. Lu, J. R. Liu, T. Johansen,
  and W. K. Chu
Measurement of Elastic Deformation of a Thin Foil by MeV-Energy Heavy Ion Irradiation 931
  H. Tsuchida, I. Katayama, S. C. Jeong, H. Ogawa, N. Sakamoto, and A. Itoh
In Situ Imaging of Highly Charged Ion Irradiated Mica 935
  L. P. Ratliff and J. D. Gillaspy
Functional Fabrication of MEMS by Ion Implantation 939
  S. Nakano and H. Ogiso
Experiments Using a 200 kV Implanter and a 5 MV Tandem Accelerator 943
  R. Ishigami, Y. Ito, K. Yasuda, and S. Hatori
Film Growth Using Mass-Separated Ion Beams 947
  H. Hofsäss, C. Ronning, and H. Feldermann
Diffusion and Roughening during Ion Beam Erosion of Graphite Surfaces 951
  S. Habenicht and K. P. Lieb
Films Formed by Hybrid Pulse Plasma Coating (HPPC) System 955
  K. Awazu, N. Sakudo, H. Yasui, E. Saji, K. Okazaki, Y. Hasegawa, N. Ikenaga, T. Sato,
  Y. Nambo, and K. Saitoh
**Chemical Functionalization and Modification of Carbon Nanotubes through
Ion Bombardment** 959
  B. Ni and S. B. Sinnott
DLC Film Formation by Ar Cluster Ion Beam Assisted Deposition 963
  T. Kitagawa, I. Yamada, J. Matsuo, A. Kirkpatrick, and G. H. Takaoka
Molecular Dynamics Simulations of Cluster Ion Implantation for Shallow Junction Formation 967
  T. Aoki, J. Matsuo, G. Takaoka, and I. Yamada
Defects and Nanocluster Engineering in MgO 971
  A. V. Fedorov, A. van Veen, M. A. van Huis, H. Schut, B. J. Kooi, J. T. De Hosson,
  and R. L. Zimmerman
Size-Specific Reactions of Transition Metal Clusters in Collision with Simple Molecules 975
  M. Ichihashi, T. Hanmura, R. T. Yadav, and T. Kondow
**Low Energy Cluster Beam Deposition: A Novel Approach to the Synthesis
of Nanostructured Materials** 979
  S. Iannotta and P. Milani
A Two-Step Annealing of Shallow Junctions Formed by GeB^- Cluster Ion Implantation of Si 983
  X. Lu, L. Shao, J. Jin, X. Wang, Q. Y. Chen, J. Liu, P. Ling, and W.-K. Chu
Nonlinear Effect of Carbon Cluster Induced Damage in Silicon 987
  Z. Xie, X. Wang, X. Lu, L. Shao, J. Liu, and W.-K. Chu
Development of Gas Cluster Ion Beam Equipment 991
  M. E. Mack
O_2 Cluster Ion Assisted Deposition for Tin Doped Indium Oxide (ITO) Films 995
  J. Matsuo, G. Takaoka, and I. Yamada
Small Clusters in Strong Laser Fields 999
  C. Siedschlag and J. M. Rost
STM Observation of a Si Surface Irradiated with a Single Ar Cluster Ion 1003
  T. Seki and G. H. Takaoka
The Extraction of Small Cluster Ions from Cesium Sputtering Ion Source 1007
  X. Wang, X. Lu, L. Shao, J. Liu, and W.-K. Chu

Spatial Control of Nanoparticle Structures Using Dynamic Processes under High Flux
Cu⁻ Implantation...1011
    N. Kishimoto, Y. Takeda, N. Umeda, N. Okubo, and C. G. Lee

Application of Ionizing Radiation for Nano-Cluster Engineering......................................1015
    D. Ila, R. L. Zimmerman, C. I. Muntele, D. B. Poker, and D. K. Hensley

Formation and Application of Metal Nanoclusters in SiC...1020
    R. L. Zimmerman, D. Ila, C. Muntele, I. Muntele, A. L. Evelyn, D. H. Hensley,
    and D. B. Poker

Application of Ion Beams for Polymeric Carbon Based Biomaterials................................1024
    A. L. Evelyn

A Negative Ion Beam Application for Improving Biocompatibility of Polystyrene Surface.....1028
    J. Ishikawa, H. Tsuji, H. Sato, H. Sasaki, and Y. Gotoh

Plasma Source Ion Implantation Technology for Engineering Surfaces of Materials..........1032
    E. H. Wilson, D. F. Lawrence, K. Sridharan, and P. W. Sandstrom

Resonance Ultrasonic Vibrations in Cz-Si Wafers as a Possible Diagnostic Technique
in Ion Implantation..1036
    Z. Y. Zhao, S. Ostapenko, R. Anundson, M. Tvinnereim, A. Belyaev, and M. Anthony

Ultrasonic Pulse from Fast Heavy-Ion Irradiation on Solids..1040
    T. Kambara, Y. Kanai, T. M. Kojima, Y. Nakai, A. Yoneda, Y. Yamazaki, and K. Kageyama

Ion Beam Enhanced Emission of Charged Particles from Hot Graphite..........................1044
    J. Lozano, Q. C. Kessel, E. Pollack, and W. W. Smith

High Power Photoconductive Semiconductor Switches Treated with Amorphic
Diamond Coatings...1047
    F. Davanloo, M. C. Iosif, T. Camase, C. B. Collins, and F. J. Agee

SECTION XIII — NONDESTRUCTIVE ANALYSIS: DETECTION OF DRUGS, NUCLEAR MATERIALS, EXPLOSIVES, CEMENT ANALYSIS

FIGARO: Detecting Nuclear Material Using High-Energy Gamma Rays from Oxygen......1053
    B. J. Micklich, D. L. Smith, T. N. Massey, D. Ingram, and A. Fessler

Relocatable Cargo X-Ray Inspection Systems Utilizing Compact Linacs........................1057
    W. W. Sapp, A. V. Mishin, W. L. Adams, J. Callerame, L. Grodzins, P. J. Rothschild,
    R. Schueller, and G. J. Smith

NELIS—An Illicit Drug Detection System...1061
    P. A. Dokhale, J. Csikai, P. C. Womble, and G. Vourvopoulos

A Commercial Elemental On-Line Coal Analyzer Using Pulsed Neutrons......................1065
    M. Belbot, G. Vouvopoulos, and J. Paschal

Detection of Explosives with the PELAN System..1069
    P. C. Womble, C. Campbell, G. Vourvopoulos, J. Paschal, Z. Gácsi, and S. Hui

A Portable System for Nuclear, Chemical Agent and Explosives Identification...............1073
    W. E. Parker, W. M. Buckley, S. A. Kreek, A. J. Caffrey, G. J. Mauger, A. D. Lavietes,
    and A. D. Dougan

Study of Cement Chemistry with Nuclear Resonant Reaction Analysis...........................1077
    J. S. Schweitzer, R. A. Livingston, C. Rolfs, H.-W. Becker, and S. Kubsky

A New On-Belt Elemental Analyser for the Cement Industry..1081
    B. D. Sowerby, C. S. Lim, J. R. Tickner, C. Manias, and D. Retallack

SECTION XIV — TOMOGRAPHY AND RADIOGRAPHY

The Detector Problem in Fast Neutron Radiography...1087
    J. I. W. Watterson, R. M. Ambrosi, and H. Rahmanian

Computed Tomography Investigation of Microgravity-Tested Sand Samples...................1091
    S. N. Batiste, K. A. Alshibli, M. R. Lankton, S. Sture, R. A. Swanson, and N. C. Costes

Cold Neutron and Monochromatic X-Ray Micro-Tomography...1095
    B. Masschaele, S. Baechler, P. Cauwels, J. Jolie, W. Mondelaers, and M. Dierick

Neutron Radiography Activity in the European Program Cost 524:
Neutron Imaging Techniques 1099
    P. Chirco, P. Bach, E. Lehmann, and M. Balasko

An Accelerator System for Neutron Radiography 1105
    B. Rusnak and J. Hall

Fast Neutron Resonance Radiography for Security Applications 1109
    G. Chen, R. C. Lanza, and J. Hall

Recent Results in the Development of Fast Neutron Imaging Techniques 1113
    J. Hall, F. Dietrich, C. Logan, and B. Rusnak

Development of Semiconductor Detectors for Fast Neutron Radiography 1118
    R. T. Klann, C. L. Fink, D. S. McGregor, and H. K. Gersch

SECTION XV — TEACHING UNDERGRADUATES WITH ACCELERATORS

Upper-Division Student Laboratory Experiments with a 2.5 MV Van de Graaff Accelerator 1125
    D. Bradbury, W. Dukes, E. Gerber, T. Terry, R. S. Peterson, and P. Sangsingkeow

PIXE and Moseley's Law 1128
    J. R. Huddle

Accelerator-Based Techniques for the Support of Senior-Level Undergraduate
Physics Laboratories 1132
    J. R. Williams, J. C. Clark, and T. Isaacs-Smith

Undergraduate Participation in the Crystal Ball Baryon Spectroscopy Program
at Brookhaven National Laboratory 1135
    M. E. Sadler and L. D. Isenhower

SECTION XVI — TARGETS FOR NUCLEAR RESEARCH

Tritium Target Manufacturing for Use in Accelerators 1141
    P. Bach, C. Monnin, M. Van Rompay, and A. Ballanger

Preparation of Actinide Targets by Electrodeposition for Heavy-Ion Studies and
Laserspectroscopic Investigations 1144
    K. Eberhardt, P. Thörle, A. Nähler, and N. Trautmann

Status of the Target Development for the Heavy Element Program 1148
    B. Kindler, S. Antalic, H.-G. Burkhard, P. Cagarda, D. Gembalies-Datz, W. Hartmann,
    S. Hofmann, J. Kojouharova, J. Klemm, B. Lommel, R. Mann, H.-J. Schött, and J. Steiner

Isotopic Germanium Targets for High Beam Current Applications at GAMMASPHERE 1152
    J. P. Greene and T. Lauritsen

Temperature Calculations of Heat Loads in Rotating Target Wheels Exposed
to High Beam Currents 1155
    J. P. Greene, R. Gabor, and J. Neubauer

APPENDIX

List of Participants 1161
Author Index 1189

PREFACE

The CAARI 2000: Sixteenth International Conference On The Application Of Accelerators In Research And Industry was held on the campus of the University of North Texas November 1-4, 2000. The major sponsors of the conference were the National Science Foundation, The US Department of Energy, and the University of North Texas. The conference is a topical conference of the American Physical Society sponsored through the Division of Nuclear Physics. An Industrial Exhibit Show composed of 50 companies that have products of interest to the accelerator community was held in parallel to the conference for the first two days. Approximately 850 accelerator scientists from 51 countries attended the conference. There were 510 invited papers that were distributed in 85 each, four hour sessions. There was a poster session Friday night with 112 posters on display. The conference was opened with four plenary speakers. The speakers were: Barney Doyle from Sandia, who gave an excellent overview of Radiation Effects Microscopy. The second speaker was Satoshi Ozaki from Brookhaven, who gave one of the first talks on the startup of the Relativistic Heavy Ion Collider (RHIC) at BNL. The third speaker was Leonard Feldman from Vanderbilt. He gave a beautiful treatment of Ultimate Ion Scattering Analysis and Semiconductor Interfaces. The final speaker was Graham Hubler from the Naval Research Lab. The talk he gave was Ion Beam Based Materials Studies: Applications and Spin-Offs: Ion Beams Can Lead You to the Strangest Places. This plenary session was probably the best that we have had in the 32 years that the conference has been in operation.

As has been the case in most of our previous conferences Accelerator Based Atomic Physics had the most sessions. There were 16 sessions that dealt directly with Atomic Physics. There were also three well attended sessions on Synchrotron Radiation. Many of the papers in these sessions also dealt with Atomic Physics. There were also 34 poster papers that dealt with Accelerator Based Atomic Physics. There were six sessions and 20 poster papers that dealt generally with Accelerator Technology. These papers covered such topics as New Accelerators, Beam Handling Systems, Ion Sources, Detectors, Spectrometers, Magnets, Control Programs, etc. Radioactive Beams and Nuclear Physics were the topics for eight sessions, which were very well attended. There were 13 sessions and 34 posters that dealt generally with ion beam analysis. These papers covered such topics as; Rutherford Backscattering Analysis, Particle Induced X-Ray Emission, Elastic Recoil Detectors, Nuclear Reaction Analysis, Accelerator Mass Spectrometry and Activation Analysis. Additional session and posters covered such topics as; Ion Implantation, Radiation Processing, Targetry, Detectors and Spectrometers, Energy Loss, Clusters, Free Electron Lasers, and Non Destructive Analysis. In addition to the sessions mentioned above, this year for the first time we ran a Medical Symposium in parallel with the conference. This symposium consisted of six, four session on the production and use of Medical Radioisotopes. The sessions were composed of 40 speakers, who are probably some of the best in the world over the subject matter covered. The key note speaker in the Medical Symposium was Dr. Henry Wagner from Johns Hopkins. Dr. Wagner is generally considered to be the "Father of Nuclear Medicine". His talk gave an overview of Molecular Medicine in the 21st Century. It was a distinct honor for us to have Dr. Wagner at our meeting.

The editors would like to thank the major sponsoring agencies, namely, The United States Department of Energy, The National Science foundation, and UNT for their continuing support of this conference series. Thanks are also due to the industrial sponsors. They not only helped financially but also provided two complete days of industrial exhibits that add greatly to the total conference experience that each participant enjoyed. We are also indebted to the program and advisory committees for the excellent slate of invited speakers whose presentations were given at the conference. Our gratitude also goes out to the 94 session chairpersons and co-chairpersons who not only helped to organize many of the sessions but also conducted the sessions at the conference. The editors now have about 500 referees that can be called on to help with the refereeing of the papers. We wish to especially thank the individuals that helped us referee papers for these proceedings. With faxes, email, and overnight mail we accomplished this task rapidly with the outstanding contribution of these referees.

We wish to thank the administrative staff, students, and professors at the University of North Texas for the monumental effort of helping us put the conference together and following through with the final publication of these proceedings.

The person, without question, who did the most administrative work on the conference was Margaret Hall. As most of you know this was Margaret's first conference and she was, so to speak, baptized by fire. She came through beautifully. Margaret was particularly interested in seeing to it that the participants were wined and dined and generally had a good time in Denton. From the participant point of view that is an admirable goal. We also wish to thank Margaret's staff Cindy Trimmier and Michaela Foucheux for doing an outstanding job. The barbeque dinner was a big hit. Thanks go to Lon Morgan for sponsoring it. We wish to thank all of the physics graduate students for helping with the slide projectors etc. Mohit Nigam and Dana Necsoiu organized the students and did the lions share of the work themselves. For this we especially thank them. We are indebted to Jonathan Reynolds of the UNT Center for Media Production, who took all of the pictures that appear in the front of these proceedings.

Finally, we wish to thank the UNT administration and faculty for the support that has been given this conference series since the first time it was held at UNT in 1974. It is not easy to bring 850 visitors onto a campus during the fall semester and provide classrooms, etc. for the meeting. Without the total support of the University, this would not be possible.

The next conference in this series will take place, as usual, on the campus of UNT, November 12-16, 2002.

DEDICATION

These proceedings are dedicated to three very active and productive scientists who have contributed not only to the methodology of ion beam techniques but also to the theoretical interpretation of the data. Each of these individuals has been very important to the accelerator community. It is therefore an honor to dedicate these proceedings to these outstanding individuals.

Jerome L Duggan

Dr. Sheldon Datz

In December 2000 Sheldon Datz capped his career with one of the highest honors in science-- the Enrico Fermi Award. This award is the U.S. government's oldest science and technology award dating back to 1956. He shares the award with Sidney Drell of the Stanford Linear Accelerator Center and Herbert York of the University of California. This award comes on the heels of the Davisson-Germer Prize awarded in 1998, one of the highest honors bestowed by the American Physical Society, when Sheldon was cited "For his broad contributions that have provided new understanding of the dynamics of atomic interactions with ions, electrons and photons at energies ranging from a fraction of a milli-electron volt to many trillion-electron volts."

He began his career in 1951 as a research chemist at Oak Ridge National Laboratory (ORNL), where he is now a Senior Corporate Fellow. He has published over 250 papers in peer-reviewed journals and has taught and lectured worldwide as a guest professor in a number of venues. In 1997, he was awarded Doctor Honoris Causis by Stockholm University. He is a Fellow of the American Physical Society and the American Association for the Advancement of Science.

Sheldon Datz earned his B.S. degree in chemistry in 1950 and his M.A. degree in physical chemistry in 1951 from Columbia University, and in 1960 he received his Ph.D. degree in physical chemistry from the University of Tennessee at Knoxville. In 1951, he began work on the application of molecular beam techniques to the study of chemically reactive collisions. In the first and two succeeding papers, Datz and Ellison Taylor described the experimental techniques and conditions necessary to study atom-molecule reactions in crossed beams and obtain detailed dynamical information. The famous Datz and Taylor papers of the 1950s and 1960s laid the foundation for an entirely new field of crossed molecular beam chemistry. In fact, two researchers who followed after the ORNL work won Nobel prizes. Sheldon organized several molecular dynamics conferences in the 1970's, even as his main interests shifted more toward high-energy atomic collision phenomena.

The availability of energetic multicharged ions brought new opportunities. In 1964, he initiated research on the effects of channeling energetic heavy ions in crystals that, with theoretical support from ORNL's Mark Robinson, led to the determination of crystal potentials and the states of ions penetrating solids. His experiments with the late Charles Moak and numerous others at ORNL resulted in i) the first demonstration that channeled ions interact with electrons in the crystal as a dense electron target, ii) the first demonstration that it is possible to can make quantitative measurements applicable to collisions in dense plasmas but with more monoenergetic collision conditions, and iii) the first demonstration of coherent excitation of ions by the periodic potential in a crystal lattice. In 1978, he helped to initiate work on radiation from channeled relativistic positrons and electrons, which yield line radiation in the X-ray region. In 1981, he initiated the first measurements of dielectronic recombination for multicharged ions. In 1990, he began studies of atomic collisions at ultrarelativistic energies at CERN and, in 1992, he initiated studies of molecular ion dissociative recombination in ion storage rings.

It has been a pleasure to collaborate with Sheldon and other long-term ORNL researchers such as Phillip Miller, Peter Dittner, and Charles Vane throughout the past 30 years. Sheldon's broad scientific interests and his ability to build long-lasting collaborative teams having superb interdisciplinary skills characterize the man first and foremost. As a catalyst, he has always encouraged his collaborators to move in new research directions with originality of approach and provided incisive advice to improve the scientific product. Sheldon has also worked to create a research and training environment in which young scientists could grow and flourish. Moreover, he has used his standing in the international community and gregarious nature to promote the exchange of information among scientists of all nations. Just as he brought molecular beams and people together in his early career, he is helping to bring atomic, molecular and optical physicists together for another major international conference that will take place in July 2001, 50 years after he started his illustrious career at ORNL. He officially "retired" from ORNL in November 2000.

For his past accomplishments, his present and future activities, and on behalf of his long list of collaborators, it is a pleasure to dedicate these proceedings to Sheldon Datz, a true friend to the scientific community.

Herbert F. Krause
ORNL, Physics Division
Oak Ridge, TN 37831-6372

John Reading

The topic represented by this conference, namely, applications of accelerators in Science and Industry, has been an important part of physics in the last half century. It owes its vitality to the ingenuity and insights of researchers in the many disciplines represented here. It is appropriate that we dedicate this volume of proceedings to one of those researchers. Those of you who know John Reading know how his presence livens up a program. Not only does he bring deep, fundamental insights to bear on accelerator-based physics, he does so with wit and enthusiasm. Points are driven home in a truly unique and memorable manner. He once referred to the semi-classical approximation, SCA, as the sometimes-curved approximation, which demonstrates the point. It is difficult to forget a lecture by John.

We can all appreciate these obvious aspects of John's contributions to this conference. To take a look back on John's career we note that he got his start in nuclear physics, but changed his area of concentration to atomic physics. We in the field of atomic physics can be eternally grateful for this change. John brought a fresh perspective to the field and wrote key papers on the many-body aspects of ion-atom collisions. His analysis of independent particle models is as relevant now as when he wrote the first papers on the subject. John's work in this area clearly defined what is actually computed with such models, and how such models should be compared with experiment.

Computation of cross sections for many electron processes in ion-atom collisions is the primary focus of his research. It is a subject that has challenged the best researchers in the field. Very early on, John concentrated upon developing techniques that directly addressed many-body problems. That these techniques correctly treated correlated electron motion was shown most dramatically by the analysis of cross sections for ionization of helium atoms by proton and anti-proton impact. The methods developed by John gave a full account of the unexpected behavior of these cross sections. Now that computational resources are readily available, these techniques are proving practical for a wide variety of atomic and molecular species.

John has always paid close attention to experimental observations and has been generous in sharing his work and ideas with his experimental colleagues. He has taken the lead in initiating workshops that fostered the confrontation of theory and experiment. The talks and discussions at these workshops were always at the highest scientific level, and John's naturally gregarious nature brought another dimension to the events. The social activities were delightful and provided another source of fond memories for the participants.

John's published work has stood the test of time and represents research in accelerator based atomic physics that has lasting value. Many of his insights are now part of the common knowledge that guides the field. It is with great pleasure that we recognize John's many contributions to the accelerator physics community by dedicating this volume to him.

Joe Macek, ORNL and Univ. of Tennessee
Pat Richard, Kansas State University

Joseph Tesmer

Joseph Tesmer is currently the supervisor of the Ion Beam Materials Laboratory at Los Alamos National Laboratory, and he is one of the finest experimental physicists available to the field of ion solid interactions today. Since 1985, he has transferred his rich nuclear physics expertise to the fields of materials science, ion beam analysis, and ion implantation. Joe received his Ph.D in nuclear physics from the University of Washington in 1971 and proceeded down the standard nuclear physics path at Purdue University as a Research Associate conducting basic research in low energy nuclear physics. In 1973 he moved on to the Medical Physics Section of the Department of Radiology at the University of Wisconsin. While there he worked on ion source and accelerator development of a high intensity neutron generator for cancer therapy.

In 1975 he joined the Physics Division of Los Alamos National Laboratrory where he expanded his horizons by making a series of fundamental measurements at somewhat higher energies (800 MeV protons using LAMPF/WNR). From 1977 until 1985 he was back in the realm of low energy nuclear physics with supervisory responsibility for the operation of the Los Alamos Van de Graaff/Ion Beam Facility. This historically important facility had a number of technological wonders such as polarized ion sources, a single-ended 7.5 MeV vertical accelerator that operated with either positive or negative beams and could be used as an injector to the FN tandem for three stage operation, pulsed beam operation for neutron time-of-flight spectroscopy, and a high resolution Q3D magnetic spectrograph. By the way, they also routinely accelerated tritium. It was the perfect environment to polish his skills as a complete experimentalist.

The ion beam community has been the fortunate recipient of his expertise since that time. Joe was and continues to be instrumental to all ion beam related materials work at Los Alamos. He took the lead in designing, building, and operating the Ion Beam Materials Laboratory. His low energy nuclear physics experience has served the ion beam analysis community very well by exploring the limits of light element detection in complex materials using resonant elastic backscattering. He is a reliable colleague known for his willingness to contribute the time and effort to organize conferences, sessions, and workshops. As one of the senior editors of the Handbook of Modern Ion Beam Materials Analysis his reputation and contribution to our community are secure. What you may not appreciate is the hard, behind-the-scenes work he has done on national accelerator standards and safety requirements, thankless but critical tasks.

Those of us who have had the pleasure of working with Joe in the lab, taking data, and figuring out what it means are lucky indeed. His enthusiasm and creativity are welcome additions to any experimental effort. His ability to fix whatever is broken is legendary.

For his contributions to the fields of ion implantation and ion beam materials analysis, for his enthusiastic service to the ion beam community, and for his friendship to all who know him, it is a pleasure to dedicate these Proceedings to Joseph Tesmer.

> Carl J. Maggiore
> Technanogy LLC
> Los Alamos Regional Commercializtion Center
> 101 DP Road, Los Alamos, NM 87544

ORGANIZATIONAL COMMITTEE

ADVISORY COMMITTEE

Bill **APPLETON**	Oak Ridge National Laboratory
Klaus **BETHGE**	JW Goethe University
Marshall **CLELAND**	Ion Beam Applications, S.A.
Geoff **DEARNALEY**	Southwest Research Institute
Robert L. (Cotton) **HANCE**	Motorola Incorporated
Homer B. **HUPF**	International Isotopes, Inc.
Carl **MAGGORIE**	Los Alamos National Laboratory
Klaus **MALMQUIST**	Lund, Sweden
Gregory A. **NORTON**	National Electrostatics Corporation
Themis **PARADELLIS**	Demoketres Greece
Robert D. **RATHMALL**	Axcelis Technologies, Inc
Patrick **RICHARD**	Kansas State University
Tom **TOMBRELLO**	Cal Tech
Vlado **VALKOVIC**	Croatia
Iso **YAMADA**	Kyoto, Japan

INDUSTRIAL SESSION

Chairman, Ira Lon MORGAN, Advanced Molecular Imaging Systems Inc.

Gerald D. ALTON	Oak Ridge National Laboratory
Joe E. BEAVER,	International Isotopes Incorporated
Frank CHMARA	Peabody Scientific
Cary N. DAVIDS	Argonne National Laboratory
Barney L. DOYLE	Sandia National Laboratories
Robert W. HAMM	AccSys Technology, Inc.
George M. KLODY	National Elecrostatics Corporation
Kenneth H. PURSER	Southern Cross
James F. ZIEGLER	IBM-Research

RESEARCH SESSION

Chairman, Jerome L. Duggan, University of North Texas

Klaus H. BETHGE	J. W. Goethe-Universität
Bert BRIJS	IMEC
H. Ken CARTER	Oak Ridge Associated Universities
Wei-Ken CHU	University of Houston
Sheldon DATZ	Oak Ridge National Laboratory
David L. EDERER	Tulane University
K. O. GROENVELD	J. W. Goethe Universität
Keith W. JONES	Brookhaven National Laboratory
Joseph MACEK	University of Nebraska
Floyd D. McDANIEL	University of North Texas
David J. PEGG	University of Tennessee
Friedel RAUCH	J.W. Goethe Universität
John F. READING	Texas A&M University
Patrick RICHARD	Kansas State University
William S. RODNEY	Georgetown University
Emile A. SCHWEIKERT	Texas A&M University
Stephen M. SHAFROTH	University of North Carolina
Soey H. SIE	CSIRO: Exploration & Mining
Joseph R. TESMER	Los Alamos National Laboratory
Vlado VALKOVIC	Analysis and Control Technologies Ltd.
S. L. VARGHESE,	University of South Alabama
George VOURVOPOULOS	Western Kentucky University
Richard L. WALTER	Duke University
Isao YAMADA	Kyoto University

Session Chairpersons

Gerald D Alton – Oak Ridge National Laboratory
Mark Anthony – University of South Florida
Lorenzo Avaldi – Consiglio Nazionale delle Ricerche

Hassaram Bakhru - University of Albany
Charles Barbour - Sandia National Laboratory
Paul Bergstrom – Lawrence Livermore National Laboratory
Adam Berryhill – PerkinElmer/ORTEC
Gerhard Brauer – Forschungszentrum Rossendorf
Bert Brijs - IMEC
Milos G Budnar – J Stefan Institute

Sam J Cipolla – Creighton University
Marshall R Cleland – Ion Beam Applications
Benjamin C Craft – Louisiana State University

William E Dance – Consultant/Neutron Radiology
Barney L Doyle – Sandia National Laboratories

Mohamed El Bouanani - University of North Texas

Alfredo Galindo-Uribarri – Oak Ridge National Laboratory
Gary A Glass – University of Louisiana at Lafayette
Tsahi Gozani – Ancore Corporation
Tom J Gray – Kansas State University

Marianne E Hamm – AccSys Technology Inc
Robert W Hamm – AccSys Technology Inc
Yale D Harker – Idaho National Engineering & Environmental laboratory
Frank J Harmon – Idaho State University
Tony Haynes – Oak Ridge National Laboratory
Richard D Hichwa – Pet Imaging Center
Wayne Holland - Oak Ridge National Laboratory
Homer B Hupf – International Isotopes Inc

Daryush Ila – Alabama A&M University

Dale C Jacobson – Lucent Technologies
Dennis James – Texas A&M University
Brant Johnson – Brookhaven National Laboratory
Keith W Jones – Brookhaven National Laboratory

David Knies – Naval Research Laboratory

William A Lanford – University at Albany
Richard Lanza – Massachusetts Institute of Technology
Larry Larson – SEMATECH

Keith B MacAdam – University of Kentucky
Joe Macek – University of Tennessee
Hans J Maier – University of Munich
Paul Mantica – Michigan Statte University
Daniel K Marble – Tarleton State University
Samuel B Matteson - University of North Texas
James Mayer – Arizona State University

Larry McIntyre – University of Arizona
Rahul Mehta – University of Central Arkansas
Marcus H Mendenhall – Vanderbilt University
Mitchell D Miller – Rice University
Gary E Mitchell – North Carolina State University

Michael Nastasi – Los Alamos National Laboratory
Boris Ni – University of Florida

Satoshi Ozaki - Brookhaven National Laboratory

Eugene Peterson – Los Alamos National Laboratory
Randolph Peterson – University of the South
Ronald A Phaneuf – University of Nevada-Reno
James M Potter – JP Accelerator Works Inc
Jack L Price – Naval Surface Warfare Center

Carroll A Quarles – Texas Christian University

Robert D Rathmell – Eaton Corporation
John F Reading – Texas A&M University
Mark L Roberts – The University of Georgia
James M Ryan – University of New Hampshire

John R Sabin – University of Florida
David Schlyer – Brookhaven National Laboratory
Emile A Schweikert – Texas A&M University
Carl Seidel – Consultant/Radioisotope & Radipharmaceutical
Ivan A Sellin – University of Tennessee
Stephen Shafroth – University of North Carolina-Chapel Hill
Soey H Sie – CSIRO Exploration and Mining
Janet Sisterson – Massachusetts General Hospital
Martin P Stockli – Kansas State University

Joseph R Tesmer – Los Alamos National Laboratory
Theva S Thevuthasan – Pacific Northwest National Laboratory
Samuel B Trickey – University of Florida

Vlado Valkovic – Ruder Boskovic Institute
S L Varghese – University of South Alabama
Vincent C Venezia - Lucent Technologies Bell Labs
Karl F Von Reden – Woods Hole Oceanographic Institution
George Vourvopoulos – Western Kentucky University

Kevin C Walter – Southwest Research Institute
Richard L Walter – Duke University
Rand L Watson – Texas A&M University
Duncan Weathers - University of North Texas
Steven A Wender – Los Alamos National Laboratory

Isao Yamada – Kyoto University
Yueh-Chung Yu – Academia Sinica

Robert L Zimmerman – Alabama A&M University

CONTRIBUTORS AND INDUSTRIAL SPONSORS

A&N Corporation
707 SW 19th Avenue
Williston FL 32696

AccelSoft Inc
10855 Sorrento Valley Rd
Suite 201
San Diego CA 92121

AccSys Technology Inc
1177A Quarry Lane
Pleasanton CA 94566

AMPTEK Inc
6 De Angelo Drive
Bedford MA 01730

Aptec-NRC Inc
125 Titus Avenue
Warrington PA 18976

AS&E High Energy Systems
3300 Keller Street # 101
Santa Clara CA 95054

Atomic Spectroscopy Instruments
3451 County Road 409
Taylor TX 76574

Austin Scientific
PO Box 18863
Austin TX 78760

Axcelis
108 Cherry Hill Drive
Beverly MA 01915

BICRON
12345 Kinsman Road
Newbury OH 44065

BOC Edwards
301 Ballardvale Street
Wilmington MA 01887

Charles Evans & Associates
810 Kifer Road
Sunnyvale CA 94086

CTI
810 Innovation Drive
Knoxville TN 37932

Dehnel Consulting Ltd
PO Box 201
Nelson BC V1L5P9 CANADA

Denton Economic Development
414 Parkway
Denton TX 76202

Department of Energy
19901 Germantown Rd
Germantown MD 20874

FusionStar
DaimlerChrysler Aerospace
Raumfahrt-Infrastruktur Postfach
D-29328 Fassberg GERMANY

Glassman High Voltage Inc
Route #22 East Salem Park
PO Box 551
Whitehouse Station NJ 08889

GMW Associates
955 Industrial Road
San Carlos CA 94070

High Voltage Engineering
Europa BV
PO Box 99
3800 AB Amersfoort,
Netherlands

Huntington Laboratories
1040 L'Avenida
Mountain View CA 94043

International Isotopes
3100 Jim Christal Road
Denton TX 76207

Ion Beam Applications
Chemin du Cyclotron 3
B 1348 Louvain-la-Neuve
BELGIUM

ISO-TEX Diagnostics Inc
1511 County Road 129
PO Box 909
Friendswood TX 77546

JP Accelerator Works Inc
2245 47th Street
Los Alamos NM 87544

Kimball Physics Inc
311 Kimball Hill Road
Wilton NH 03086

Kurt J Lesker Co
1515 Worthington Avenue
Clairton PA 15025

Linac Systems
1208 Marigold Drive NE
Albuquerque NM 87122

Los Alamos National Laboratory
PO Box 1663 MS P240
Los Alamos NM 87545

McAllister Technical Services
West 280 Prairie Avenue
Coeur d'Alene ID 83815

MF Physics
5074 List Drive
Colorado Springs CO 80919

Motorola Inc
PO Box 20922
Phoenix AZ 85036

National Electrostatics Corp
7540 Graber Road
PO Box 620310
Middleton WI 53562

North Star Research Corporation
4421 McLeod Road NE Suite A
Albuquerque NM 87109

Oak Ridge National Laboratory
Po Box 2008
Oak Ridge TN 38731

PerkinElmer/ORTEC
801 South Illinois Ave
Oak Ridge TN 37831

Potentials Inc
1704 Hydro Drive
Austin TX 78728

Princeton Scientific Corp
141 Snowden Lane
Princeton NJ 08540

SCIONIX
1720 Natchez Trace Blvd
Orlando FL 32818

SODERN
20 Avenue Descartes
F-94451 Lineil Brevannes
FRANCE

Spectrum Techniques
106 Union Valley Road
Oak Ridge TN 37830

Synergy Vacuum Inc
PO Box 16241
6310 Road, Suite A
Montrose CO 81401

Thero Reax & Thermo Eberline
504 Airport Road
Santa Fe NM 87505

Texas Instruments
PO Box 655214
Dallas TX 75265

Thermionics Vacuum Products
231 Otto Street
Pt Townsend WA 98368

Thomson Components & Tubes
40G Commerce Way
PO Box 540
Totowa NJ 07511

THT Net Mercury
13438 Floyd Circle
Dallas TX 75243

Vacuum Research Corporation
2419 Smallman Street
Pittsburgh PA 15222

Varian Medical Systems
PO Box 10022
Palo Alto CA 94303

Varian Vacuum Technologies
651 N Plano Road Suite 419
Richardson TX 75081

VAT Inc
500 West Cummings Park
Woburn MA 01801

Walker Scientific Inc
Rockdale Street
Worcester MA 01606

WIENER Plein & Baus Corp
300 E Auburn Ave
Springfield OH 45505

SECTION I

ATOMIC AND MOLECULAR PHYSICS

Stopping cross section and charge exchange study on the $He^+ \to Ne$ system

R Cabrera-Trujillo, J R Sabin, E Deumens and Y Öhrn

Departments of Physics and Chemistry, University of Florida, Gainesville FL, 32611-8435

Abstract. In this work we report calculations of the impact parameter dependent energy loss and charge exchange cross section of $^4He^+$ ions on atomic neon by taking into account the dynamics of both electrons and nuclei. The results are obtained for projectiles energies ranging from a few eV/amu up to 100 keV/amu. We also report the nuclear and electronic contribution to the total stopping cross section. Comparison of the present results for this channel with the available experimental data shows good agreement.

INTRODUCTION

The theoretical study of the energy loss processes that take place during the collision of an ion beam with a material target has been of interest since the beginning of the last century, beginning with the pioneering work of Bohr (2). Applications range from the understanding of the fundamental processes of the quantum world to radiotherapy, microelectronics and accelerators research.

When an ion beam penetrates a material target, several processes take place during the collision. One of those is the dynamical charge exchange between projectile and target which in the low- to intermediate projectile energies range is one of the most important effects in the slowing down of the projectile. In order to treat the problem of stopping and charge exchange during a collision properly, we employ a time dependent dynamical method that considers the dynamics of all the nuclei and electrons and allows for unrestricted charge and energy transfer between the projectile and target. This is the Electron-Nuclear Dynamics (END) approach (6).

In previous work (10, 18), we reported pilot calculations for helium ions colliding with neon targets within the END approach. In the present work, we continue with the analysis of the $He^+ \to Ne$ system to calculate the differential cross section for small angles, the charge exchange cross section, and the stopping cross section for several projectile energies.

ELECTRON-NUCLEAR DYNAMICS SURVEY

Our approach to analyze the energy loss and charge transfer processes is based on the application of the Time-Dependent Variational Principle (TDVP) to the Schrödinger equation (17), where the wave function is described in a coherent state representation. As the details of the END method have been reported elsewhere (6, 9, 7), we present here a resumé of the fundamental features of the theory.

We use a parametrization of the wave function in a coherent state manifold, which leads to a system of Hamilton's equations of motion (6). The variational wave function $|\xi\rangle$, is a molecular coherent state

$$|\xi\rangle = |z, \mathbf{R}, \mathbf{P}\rangle |\mathbf{R}, \mathbf{P}\rangle = |z\rangle |\phi\rangle \quad (1)$$

where $|z\rangle$ and $|\phi\rangle$ are the coupled electronic and nuclear wavefunctions, respectively.

The simplest level of the END approach employs a single spin unrestricted electronic determinant

$$|z\rangle = \det\{\varphi_i(x_j)\} \quad (2)$$

written in terms of the nonorthogonal spin orbitals φ_i

$$\varphi_i = u_i + \sum_{j=N+1}^{K} u_j z_{ji}, \quad i = 1, 2, \cdots, N \quad (3)$$

expressed in terms of a basis $\{u_j\}$ of atomic Gaussian type orbitals of rank K with complex coefficients $\{z_{ji}\}$. For the Gaussian type orbital we use

$$u_i = \sum_k c_k (x - R_x)^l (y - R_y)^m (z - R_z)^n \times \quad (4)$$

$$\exp[-\alpha_k (\mathbf{x} - \mathbf{R})^2 - i\mathbf{P} \cdot (\mathbf{x} - \mathbf{R})], \quad (5)$$

centered on the average positions $\mathbf{R}$ of the participating atomic nuclei and moving with a momentum $\mathbf{P}$. Here, c_k are the contraction coefficients and α_k are the exponents of the Gaussian basis set. This representation takes into account the momentum of the electron explicitly through the Electron Translation Factors (ETF) (5). The particular form of parametrization of $|z\rangle$ with complex, time dependent coefficients z_{ji} is due to Thouless (20) and is an example of so called generalized coherent states (15).

The nuclear part of the wavefunction is represented by localized Gaussians

$$|\phi\rangle = \prod_k \exp[-(\frac{X_k - R_k}{w})^2 + iP_k \cdot (X_k - R_k)] \quad (6)$$

or, in the narrow wave-packet limit ($w \to 0$), by classical trajectories ($\mathbf{R}_k, \mathbf{P}_k$).

Application of the TDVP then yields the dynamical equations (6)

$$\begin{pmatrix} 0 & -iC^* & -iC_R^* & 0 \\ iC & 0 & iC_R & -iC_P \\ iC_R^\dagger & -iC_R^T & C_{RR} & -I+C_{RP} \\ iC_P^\dagger & -iC_P^T & I+C_{RP} & C_{PP} \end{pmatrix} \times \begin{pmatrix} \dot{z} \\ \dot{z}^* \\ \dot{R} \\ \dot{P} \end{pmatrix} = \begin{pmatrix} \partial E/\partial z \\ \partial E/\partial z^* \\ \partial E/\partial R \\ \partial E/\partial P \end{pmatrix} \quad (7)$$

where the dot represent differentiation with respect to the time parameter and $E = \sum_k P_k^2/2M_k + \langle z|H_{el}|z\rangle/\langle z|z\rangle$ is the total energy of the system. H_{el} is the electronic Hamiltonian which contains the nuclear-nuclear repulsion potential energy. The nonadiabatic coupling terms between the electron and nuclear dynamics are given by

$$\mathbf{C} = \left.\frac{\partial^2 \ln S(z^*, \mathbf{R}, \mathbf{P}, z, \mathbf{R}', \mathbf{P}')}{\partial z^* \partial z}\right|_{\mathbf{R}'=\mathbf{R}, \mathbf{P}=\mathbf{P}'}, \quad (8)$$

$$\mathbf{C}_R = \left.\frac{\partial^2 \ln S(z^*, \mathbf{R}, \mathbf{P}, z, \mathbf{R}', \mathbf{P}')}{\partial z^* \partial \mathbf{R}'}\right|_{\mathbf{R}'=\mathbf{R}, \mathbf{P}=\mathbf{P}'}, \quad (9)$$

$$\mathbf{C}_{RR} = -2\,Im\left.\frac{\partial^2 \ln S(z^*, \mathbf{R}, \mathbf{P}, z, \mathbf{R}', \mathbf{P}')}{\partial \mathbf{R} \partial \mathbf{R}'}\right|_{\mathbf{R}'=\mathbf{R}, \mathbf{P}=\mathbf{P}'} \quad (10)$$

and similar definitions for $\mathbf{C}_{RP}$, $\mathbf{C}_P$, and $\mathbf{C}_{PP}$. Here, $S(z^*, \mathbf{R}, \mathbf{P}, z, \mathbf{R}', \mathbf{P}') = \langle z, \mathbf{R}', \mathbf{P}'|z, \mathbf{R}, \mathbf{P}\rangle$ is the overlap of determinantal states.

Solving the set of equations for $\{z, \mathbf{R}, \mathbf{P}\}$ as a function of the time t yields the evolving molecular state that describes the processes that takes place during the collision. For the purpose of discussing charge exchange, we make use of the Mulliken population analysis (16) of the projectile or target. From Eq. (3), the number of electrons in the system is

$$N = \sum_{\nu,\mu} P_{\mu\nu}\Delta_{\nu\mu} = \sum_{\mu}(\mathbf{P}\Delta)_{\mu\mu} = \mathrm{Tr}(\mathbf{P}\Delta) \quad (11)$$

where $P_{\nu\mu} = \sum_i^N z_{i\nu}^* z_{i\mu}$ and $\Delta_{\mu\nu}$ is atomic orbitals overlap matrix. It is possible to interpret $(\mathbf{P}\Delta)_{\mu\mu}$ as the number of electrons to be associated with the basis function u_μ. From this, $n_A = \sum_{\nu \in A}(\mathbf{P}\Delta)_{\nu\nu}$ is the number of electrons associated with atom A.

The END method has been implemented in the ENDyne program package (8).

THE $He^+ \to Ne$ SYSTEM

In previous work (10, 18), the energetics of the collision for 10 keV He^{q+} on Ne has been considered. Here, we present a more detailed analysis of the collision channel $He^+ \to Ne$ for a projectile energy range of a few eV/amu up to 100 keV/amu.

The electronic basis sets used consist of a [5s2p/3s2p] (21) for the He projectile and a [9s5p1d/3s2p1d] (11) for the Ne target, making a total of $K = 25$ for the rank K of the basis set.

In these calculations the projectile is initially at a distance of 30 a.u. from the target and the Ne is initially in its electronic ground state at the SCF level. The trajectory is continued 30 a.u, past the target. The impact parameter, b, considered in these calculations are in the range $\{0, 20\}$ a.u. in steps of 0.1 from 0.0 to 4.0 and 0.5 from 4.0 to 8.0 and in steps of 1.0 from 8.0 to 20.0. This gives a total of 60 fully dynamical trajectories, each with a total evolution distance of 60 a.u. for each projectile energy.

Differential cross section

In order to study the energy loss and the charge exchange for the collision of He^+ ions colliding with Ne, we first study the direct differential cross section. Since we have implemented the narrow width limit on the projectile wave packet, we describe it's nuclei as classical particles, thus requiring semiclassical corrections for the scattering process. We have implemented (3) the Schiff approximation (19) for small scattering angles which takes into account the quantum effects of the forward scattering. The advantage of using the Schiff approximation over some other semiclassical corrections (e.g. the Airy or uniform approximation) is that it includes all the terms of the Born series and treats the rainbow and glory angles in a single approach without requiring the separation of different scattering regions. The differential cross section is given by $d\sigma/d\Omega = k_f|f(\theta)|^2/k_i$, with

$$f(\theta) = ik_i \int_0^\infty \left[1 - \exp(2i\delta(b))\right] J_0(qb) b\, db, \quad (12)$$

where $d\delta(b)/db = k_i\Theta(b)/2$, with $\delta(b)$ the phase shift and $\Theta(b)$ the deflection function, such that the scattering angle $\theta = |\Theta|$. Here $q = 2k_i \sin(\theta/2)$, $J_0(x)$ s the Bessel

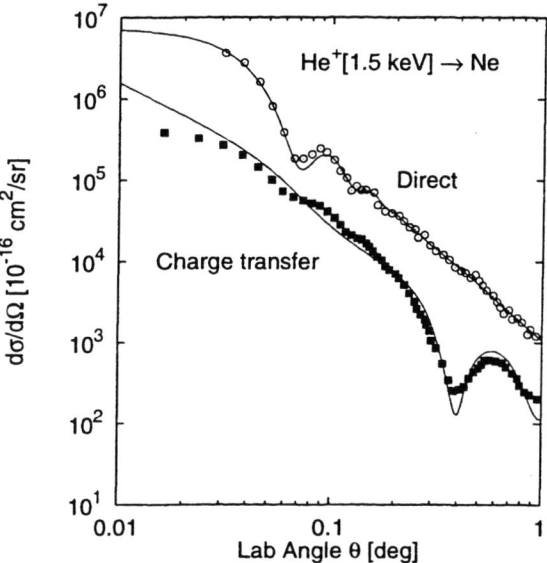

FIGURE 1. Differential cross section for direct scattering of $He^+ \to Ne$ at a projectile energy of 1.5 keV. The experimental points are from Johnson et al. (14).

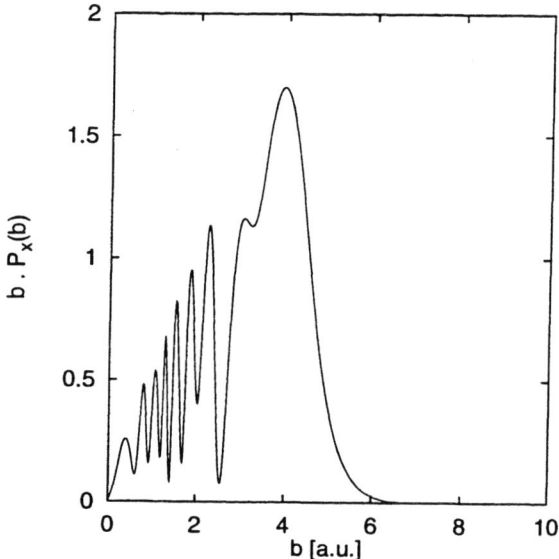

FIGURE 2. Transfer probability $P_x(b)$ times the impact parameter b as a function of the impact parameter for $He^+ \to Ne$ at a collision energy of 1.5 keV. Note the oscillatory behavior of the transfer probability in the small impact parameter region.

function of order zero and k_i and k_f the initial and final wave vector of the projectile, respectively.

In Fig. 1, we show the direct differential cross section for He^+ colliding at 1.5 keV with atomic Ne for small scattering angles and compare it with the experimental data of Johnson et al. (14). From this figure, we observe the interference pattern characteristic of the forward peak scattering when rainbow and glory angle are present (12).

Electron Charge Exchange

As the projectile departs from the target after the collision, it may capture or lose electrons.

From the Mulliken population analysis for each impact parameter dependent trajectory we calculate the probability of charge exchange, $P_x(b)$. In Fig. 2, we show $b \cdot P_x(b)$ for $He^+ + Ne \to He + Ne^+$ colliding at a projectile energy of 1.5 keV. In this figure, we observe that for small impact parameters ($b < 3.0$ a.u.) the probability for electron transfer from the target to the projectile shows a resonance pattern. Also, we note that for intermediate impact parameters ($3.0 < b < 6.0$ a.u.) the transfer probability is high. This is consequence of the high affinity of the He^+ ion to capture an electron in the vicinity of the target. In Fig. 1, we show the differential cross section for charge transfer as a function of the scattering angle and compare it with the experimental data of Johnson et al. (14). Contrary to the direct differential cross section, where use of semiclassical correction were made, the charge transfer results were obtained by means of the analysis of the classical trajectory without quantum correction (6). The strong oscillating pattern of Fig. 2, shows that semiclassical corrections are required. The effect of these can be observed in the small scattering angle region of Fig. 1, where the forward peak scattering produces a divergence in the classical approach. The implementation of the charge transfer within semiclassical corrections will be presented in a future publication.

Stopping cross section

The impact parameter dependent energy loss, $\Delta E(b)$, which we calculate as the difference between the initial and final kinetic energy of the projectile in the laboratory frame, can be obtained from the momentum **P** of the projectile at the end of the trajectory as a function of the impact parameter. Integrating the energy loss leads to the stopping cross section

$$S(v) = 2\pi \int_0^\infty \Delta E(b) \, b \, db \qquad (13)$$

In Fig. 3, we present the total, S_t, electronic, S_e, and nuclear, S_n, contributions to the stopping cross section as a function of the initial projectile velocity, for the case of He^+ colliding with atomic Ne.

In the same figure we compare with the experimental data of Chu and Powers (4), of Fukuda (13), and of

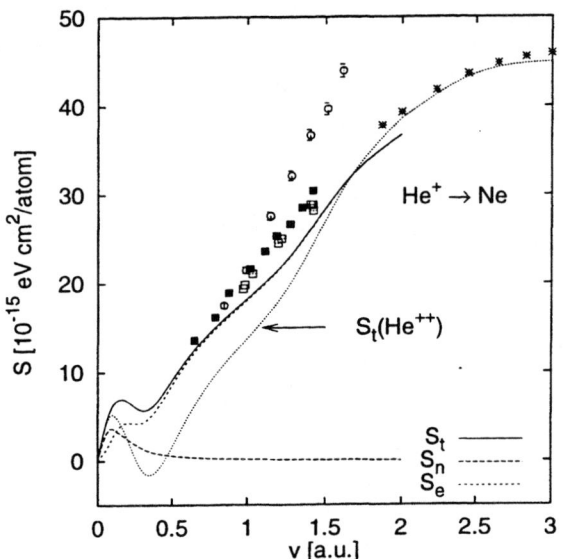

FIGURE 3. Stopping cross section for He^+ projectiles colliding with atomic neon. Continuum line: total stopping cross section. Long-dashed line: nuclear stopping cross section. Short-dashed line: electronic stopping cross section. The experimental data are from: (*) Chu and Powers (4); (o, full squares) Fukuda (13); (open squares) Baumgart et al. (1). Also for completeness we show the total cross section for He^{++} ions colliding with atomic neon (dotted line).

Baumgart et al. (1). Since the charge state of the incoming ion is not always unity, as we have seen from the charge transfer, it is necessary to calculate the contributions for He^{++} and for He^0 to obtain all the charge channels that contribute to the stopping cross section and be able to compare directly to the experiment. These studies will be reported elsewhere. The comparison of this single channel shows a good agreement with the experiment.

SUMMARY

In this work, we have calculated the charge exchange and differential cross sections, and the stopping cross section for He^+ ions colliding with atomic Ne by means of the Electron-Nuclear Dynamics approach, which includes the full dynamics of the electrons and nuclei. Quantum effects on the direct differential cross section have been taken in account by means of the Schiff approximation. We have found strong oscillations in the charge transfer probability. Finally, we obtain good results for the differential cross section, charge transfer and stopping cross section when compared with the experimental data.

ACKNOWLEDGMENTS

We thanks Prof. Darden Powers for the comments on Fukuda's work. This work is supported in part by CONACyT-Mexico to R.C.T., by NSF (grants No. CHE-9732902 and CHE-9974385), and by ONR (grants No. N0014-00-1-0197 and N0014-96-1-0707).

REFERENCES

1. Baumgart, H., Berg, H., Huttel, E., Pfaffe, E., Reiter, G., and Clausnitzer, G., *Nucl. Instr. and Meth.* **215**, (1983) 319.
2. Bohr, N., *Phil. Mag.* **30**, (1915) 581.
3. Cabrera-Trujillo, R., Sabin, J. R., Öhrn, Y., and Deumens, E., *Phys. Rev. A* **61**, (2000) 032719.
4. Chu, W. K., and Powers, D., *Phys. Rev. B* **4**, (1971) 10.
5. Delos, J. B., *Rev. Mod. Phys.* **53**, (1981) 287.
6. Deumens, E., Diz, A., Longo, R., and Öhrn, Y., *Rev. Mod. Phys.* **66**, (1994) 917.
7. Deumens, E., Diz, A., Taylor, H., and Öhrn, Y., *J. Chem. Phys.* **96**, (1992) 6820.
8. Deumens, E., Helgaker, T., Diz, A., Taylor, H., Oreiro, J., Morales, J. A., and Longo, R., *ENDyne version 2.7 Software for Electron Nuclear Dynamics*, Quantum Theory Project, University of Florida, 1998.
9. Deumens, E., and Öhrn, Y., *J. Phys. Chem.* **92**, (1988) 3181.
10. Diz, A. C., Öhrn, Y., and Sabin, J. R., *Nucl. Instr. and Meth.* **B96**, (1995) 633.
11. Dunning, T. H., and Hay, P. J., *In methods of electronic structure theory*, vol. 2, New York: Plenum Press, 1977.
12. Ford, K. W., and Wheeler, J. A., *Ann. Phys. (N. Y.)* **7**, (1959) 259.
13. Fukuda, A., *J. Phys. B: At. Mol. Opt. Phys.* **32**, (1999) 153.
14. Johnson, L. K., Gao, R. S., Hakes, C. L., Smith, K. A., and Stebbings, R. F., *Phys. Rev. A* **40**, (1989) 4920.
15. Klauder, J. R., and Skagerstman, B. S., *Coherent states, applications in physics and mathematical physics*, Singapure: World scientific, 1985.
16. Mulliken, R. S., *J. Chem. Phys.* **36**, (1962) 3428.
17. Öhrn, Y., Deumens, E., Diz, A., Longo, R., Oreiro, J., and Taylor, H., *Time-Dependent Quantum Molecular Dynamics*, New York: Plenum, 1992.
18. Sabin, J. R., *Adv. Quantum Chem.* **28**, (1997) 107.
19. Schiff, L. I., *Phys. Rev.* **103**, (1956) 443.
20. Thouless, D. J., *Nucl. Phys.* **21**, (1960) 225.
21. Woon, D. E., and Dunning, T. H., *J. Chem. Phys.* **100**, (1994) 2975.

Determination of Stopping Power of Channeled α-Particles in SiO$_2$ in the Backscattering Geometry

M. Kokkoris[a,1], S. Kossionides[a], R. Vlastou[b], X. Aslanoglou[c], R. Grötzschel[d], Th. Paradellis[a]

[a] *N. C. S. R. "Demokritos", Institute of Nuclear Physics, Tandem Accelerator, Laboratory for Material Analysis, Ag. Paraskevi 153 10, Athens, Greece.*
[b] *N. T. U. A., Department of Physics, Athens 157 80, Greece.*
[c] *University of Ioannina, Department of Physics, Ioannina, Greece.*
[d] *Institute of Ion Beam Physics and Material Analysis, Forschungszentrum Rossendorf, Dresden, Germany.*
[1] *Author to whom all correspondence should be addressed.*

Abstract. Energy spectra of α-particles channeling along the optical axis (c-axis) of a quartz crystal in the energy region E_α=3.0-3.5 MeV in the backscattering geometry were taken and analyzed. Computer simulations based on the assumption that the dechanneling of α-particles follows an exponential law are in good agreement with the measured spectra, yielding electronic stopping powers for the specific crystal orientation that vary between 17.82 and 16.12 eV/Å respectively for the energy interval into consideration.

INTRODUCTION

The behavior of charged particles channeled along a low index axis of a crystal, as far as their stopping power is concerned, is well documented [1, 2], with the energy loss being only a fraction of that corresponding to random direction of incidence. An accurate knowledge of α, the ratio of the energy loss for channeled versus randomly incident particles for different crystals is important for both basic physics studies as well as rapidly growing applications. Most of the experiments were carried out in the past in the transmission geometry. In these experiments the energy loss of channeled ions has been usually determined by measuring the final energy of the transmitted ions through thin crystals. Despite the many advantages of this method, it strongly depends on the preparation of homogeneous self-supported thin crystals, which are extremely difficult to manufacture in many cases. As an alternative, the use of the backscattering geometry has been proposed [3].

Following the pioneer works [4, 5] of the last decade, a new approach has been adopted recently, based on the use of a nuclear resonance as a marker for the range [6, 7], and on the assumption of an exponential rate of dechanneling of the incoming particles [8, 9]. This method led to the successful simulation of backscattering spectra in the systems p+^{28}Si <100> [9], p+^{28}Si <111> [10], and α+MgO [11]. In this approach, the dechanneling process is defined by only two parameters, the average ratio, α, of stopping powers in the channeling and random directions (over the energy range in which the particle travels inside the channel) and the mean channeling distance, λ. This technique allows *in situ* measurements and can be applied to several bulk single crystals (simple or compound) without any particular sample preparation, combining NRA and channeling.

In the present work, the channeling of α-particles in SiO$_2$ (quartz) along the optical axis (c-axis) is investigated. Backscattering spectra taken at the energy interval E_α=3.0-3.5 MeV, were simulated using the above model, and the values of α and λ were extracted. Although quartz is widely used as a substrate for implantations and depositions, mainly due to its excellent optical properties, there is a lack in bibliography concerning reported values for α. To the authors' best knowledge, this can be attributed to the fact that thin crystals, well polished, for transmission experiments and applications, were extremely difficult to manufacture in the past, as mentioned above.

EXPERIMENTAL SETUP

The experiments were performed in the Forschungszentrum, Rossendorf, Dresden, Germany, using the 3 MV TANDETRON Accelerator. Alpha particles were accelerated to energies E_α=3.0-3.5 MeV and were lead to a scattering chamber, which included a 4-motor goniometer system capable of determining the target orientation with an accuracy of 0.01°. The detection system consisted of a single Si surface barrier detector having an overall resolution of 12 keV for α-particles. The beam divergence was less than 0.07° and the vacuum pressure was kept constant during the measurements ($5 10^{-7}$ mbar). The carbon formation on the target's surface was negligible, verified by the absence of a strong carbon signal in the channeling spectra. The target used was a SiO_2 crystal wafer, with a diameter of 15 mm and a thickness of 0.4 mm, cut along the optical axis (c-axis) and very well polished. Before the measurements, a polar and an angular scan were performed, using 2.5 MeV α-particles in order to align the beam to the c-axis of the crystal. The sample showed an excellent crystalline behavior exhibiting a χ_{min} of the order of 4-6% (in an integrated region of 2 channels after the surface peak), as shown in the RBS/C spectra at E_α=2.9 MeV, presented in fig.1.

Spectra of α-particles backscattered at an angle Θ_{lab}=170° were taken within the energy interval, E_{lab}= 3.0-3.5 MeV, in steps of 100 keV, in both random and aligned angles of incidence, for the same accumulated charge of 20 μCb. The reaction $^{16}O(\alpha,\alpha)^{16}O$ exhibits two sharp resonances in that energy interval at energies E_{lab}= 3.04 MeV and 3.37 MeV [12, 13]. On the other hand, in the same energy interval, the elastic scattering cross section of α-particles in ^{28}Si does not deviate considerably from the pure Rutherford one, as reported in the past [12].

DATA ANALYSIS

The method used for the simulation of the energy spectra has recently been described in detail elsewhere [9]. The basic assumption is that beam particles dechannel exponentially,

$$N_d = N_o \cdot \left(1 - e^{-x/\lambda}\right) \quad (1)$$

where λ is considered to be energy independent (constant dechanneling rate). The part of the beam that has dechanneled, is considered to lose energy at a rate corresponding to the one calculated with the coefficients of Ziegler [14] in the random direction, while the part of the beam still channeled is considered to lose energy at a fraction of that rate,

$$S_{channel}(E) = \alpha \cdot S_R(E) \quad (2)$$

where α is also initially assumed to be energy independent and $S_R(E)$ is the specific energy loss in the random direction. The validity of this hypothesis is discussed in detail in the following section. For the calculations, the target was divided into slices of thickness dx = 10 μg/cm^2, and the beam was split into two components, a channeled and a dechanneled one. The evolution of each component was then followed throughout the target.

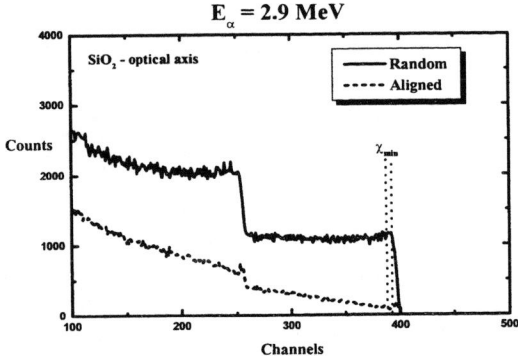

FIGURE 1. Plot of random and aligned spectra (RBS/C) in the backscattering geometry for the system α+SiO_2 at E_α=2.9 MeV. The χ_{min} value is determined by integrating 2 channels after the surface peak and by taking the corresponding ratio.

For the simulation of the experimental spectra, the accurate knowledge of the energy dependence of the elastic cross sections, for the reactions $^{28}Si(\alpha,\alpha)^{28}Si$ and mainly $^{16}O(\alpha,\alpha)^{16}O$, is mandatory. The excitation functions were thus obtained from the literature for the laboratory angle into consideration [12, 13]. The effect of the beam straggling was accounted for, using Bohr's equation [15], but not the multiple scattering effect, which is important at low energies. The excitation functions used, reproduced the spectra in the random direction within 5-15% accuracy, and were thus subsequently used for the simulation of the corresponding channeling spectra.

Therefore, given α and λ, simulated channeled spectra can be generated and compared to the experimental ones. The simulation program [9], written in standard FORTRAN, is machine-independent. The best possible fit was determined after χ^2 minimization, using the minimization code MINUIT developed at CERN. The output ASCII file can be displayed by any general-purpose graphic interface.

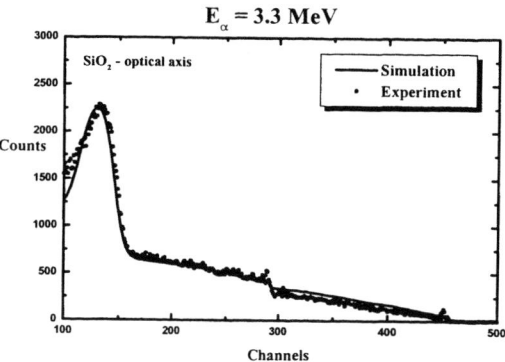

FIGURE 3. Experimental and simulation channeling spectra at E_α = 3.3 MeV. The determined values for α and λ are **0.82** and **3.9** μm respectively.

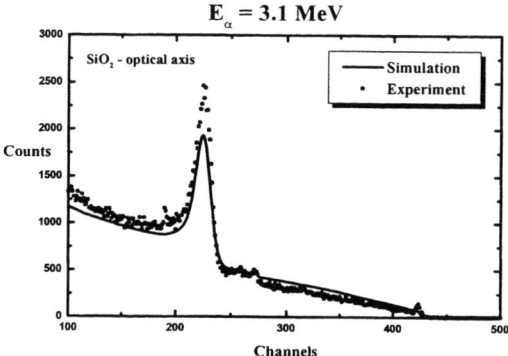

FIGURE 2. Experimental and simulation channeling spectra at E_α = 3.1 MeV. The determined values for α and λ are **0.80** and **3.7** μm respectively.

Experimental spectra, along with the simulations, in the aligned mode at energies E_{lab}= 3.1, 3.3 and 3.5 MeV are presented in figs. 2-4. The 3.04 MeV resonance appears clearly in the spectrum at E_α=3.1 MeV (fig. 2). At E_α = 3.3 MeV, the 3.04 MeV resonance is partly diffused (fig. 3). At E_α = 3.5 MeV, only the 3.37 MeV resonance is clearly visible (fig. 4).

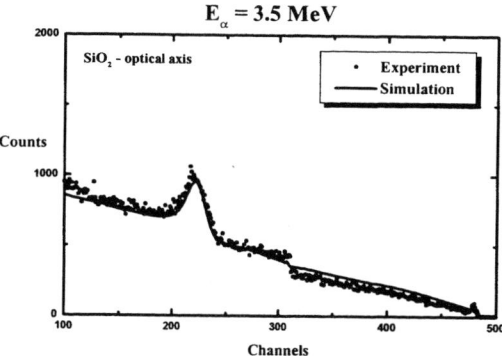

FIGURE 4. Experimental and simulation channeling spectra at E_α = 3.5 MeV. The determined values for α and λ are **0.86** and **3.6** μm respectively.

The reproduction of the experimental spectra is satisfactory in all cases. The discrepancies near the crystal surface could be attributed to a number of factors, such as, the existence of inaccuracies in the cross sections used from the bibliography [12, 13], as observed in the fits of the random spectra, and the fact that the initial dechanneling of the beam due to the distortion of the first few monolayers was considered to be of the order of 5% (constant). It should be mentioned here, that there are strong indications for the possible existence of a more complicated function, determining the dechanneling of α-particles in complex crystal structures, especially at small depths. As shown in the figures, the agreement between the simulated spectra and the experimental ones improves with the increase of E_α, that is when the resonance structure appears at greater depths. The final results for α and λ are presented in fig. 5.

DISCUSSION AND CONCLUSIONS

In the present work, simulations of backscattering energy spectra of α-particles channeled along the quartz optical axis, assuming an exponential rate of dechanneling, are presented. The data analyzed in the energy range, E_α=3.0-3.5 MeV, do not show any particular trend with regard to energy dependence of the parameters. The values of the parameters in this energy interval produce an average of λ=**3.7±0.4** μm and α=**0.81±0.05** for the axis into consideration, leading to channeling stopping powers that vary between 17.82 and 16.12 eV/Å for the energy interval into consideration.

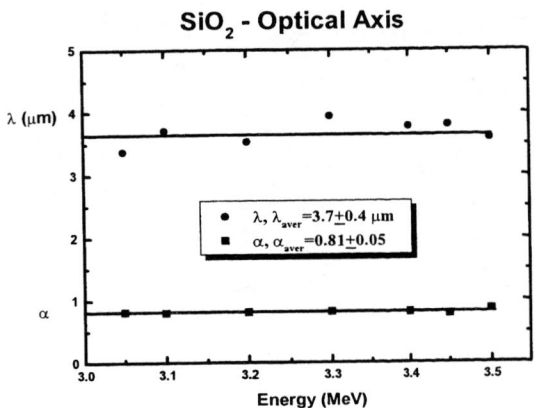

FIGURE 5. λ (in μm) and α as functions of E_α (in MeV) for the quartz optical axis (c-axis). The solid lines present correspond to the average determined values for these channeling parameters.

It should be noted here, that although α is energy depended and is expected to decline with energy, previous experiments did not show a consistent variation with energy in the range of $\cong 0.7$-1 MeV/amu [16, 17]. An analytical expression for α – if established [18] – could undoubtedly be incorporated in the simulation algorithm, but the quality of the simulated spectra in several crystals studied using this method, showed that this first-order approximation of a mean α value is valid for the energy interval into consideration.

It should be noted, however, that although these two parameters, λ and α, are in a way related, one can assume that α is inherent to the nature of the crystal channeling axis, while λ is much more sensitive, depending strongly not only on the quality of the investigated crystal axis or structure (e.g. concentration of defects), but also on the experimental conditions, such as, the initial divergence of the beam, the vacuum etc. It is thus the authors' belief, that a thorough study of the dechanneling and stopping powers of light beams in a variety of crystals is imperative before a complete understanding of the dechanneling process is accomplished.

ACKNOWLEDGMENTS

This work was done in the framewotk of the L.S.F. facility at Rossendorf, Dresden. It was also supported by the Greek National Scholarship Foundation.

REFERENCES

1. Lindhard J., *Mat. Fys. Medd.* **34**, no. 14 (1965).
2. Gemmel D. S., *Review of Modern Physics* **46**, 129-227 (1974).
3. Cembali F. and Zignani F., *Radiat. Eff.* **31**, 169-173 (1977).
4. dos Santos J. H. R., Grande P. L., Boudinov H., Behar M., Stoll R., Klatt Chr., Kalbitzer S., *Nucl. Instr. and Meth.* **B** 106, 51-54 (1995).
5. Kótai E., *Nucl. Instr. and Meth.* **B** 118, 43-46 (1996).
6. Hellborg R., *Physica Scripta* **4**, 75-82 (1971).
7. Vos M., Boerma D. O., Smolders P. J. M., *Nucl. Instr. and Meth.* **B** 30, 38-43 (1988).
8. Jack H. E., *Rad. Eff.* **13**, 101-114 (1972).
9. Aslanoglou X., Assimakopoulos P., Kokkoris M., Kossionides S., *Nucl. Instr. and Meth.* **B** 140, 294-302 (1998).
10. Aslanoglou X. A., Karydas A., Kokkoris M., Kossionides S., Paradellis Th., Souliotis G. and Vlastou R., *Nucl. Instr. and Meth.* **B** 161-163, 524-529 (2000).
11. Kokkoris M., Kossionides S., Aslanoglou X., Kaliambakos G., Paradellis Th., Harissopulos S., Gazis E.N., Vlastou R., Papadopoulos C.T., Grötzschel R., *Nucl. Instr. and Meth.* **B** 136-138, 137-140 (1998).
12. Cheng, H.-S., Shen, H., Tang, J. and Yang, F., *Nucl. Instr. and Meth.* **B** 83, 449-454 (1993).
13. Leavitt, J. A., McIntyre Jr., L. C., Ashbaugh, M. D., Oder, J. G., Lin, Z. and Dezfouly-Arjomandy, B. *Nucl. Instr. and Meth.* **B** 44, 260-264 (1990).
14. Ziegler J. F., Biersack J. P., and Littmark U., *The Stopping and Range of the Ions in Matter*, New York: Pergamon Press, (1980) Vol. 1.
15. Bohr N., *Mat. Fys. Medd.* **18**, no. 8 (1948).
16. Dias J. F., Azevedo G. de M., Behar M., Grande P. L., Klatt Chr., Kalbitzer S., *Nucl. Instr. and Meth.* **B** 148, 164-167 (1999).
17. Hetherington D. W., *Nucl. Instr. and Meth.* **B** 115, 319-322 (1996).
18. Simionescu A., Hobler G., Bogen S., Frey L., Ryssel H., *Nucl. Instr. and Meth.* **B** 106, 47-50 (1995).

Intriguing Results of Stopping Power Measurements with Light Ions Traversing Several Organic (Co)Polymer Targets

L. E. Porter

Washington State University
Pullman, WA 99164-1302

Abstract. Previously published measurements of stopping powers of four organic (co)polymers for protons, alpha particles, and ^{7}Li ions have been analyzed in terms of modified Bethe-Bloch theory. This procedure allows extraction of values of various parameters which constitute an integral part of the formalism, most notably the target mean excitation energy (I) and Barkas-effect parameter (b). Normally one can expect the extracted I-value to exceed by a few per cent the value based on the additivity assumption (I_B), whereas the extracted b-value should lie within the expected interval of about 1.3 - 1.5. Results of one series of measurements with the same experimental arrangement, utilizing thin target foils of formvar (a polyvinylformyl resin), polysulfone, kapton (a polymide), and vyns (a vinylchloride-vinylacetate copolymer), yielded results quite consistent with expectation except that the extracted I-value for formvar lay about 20% below the value of I_B. This salient anomaly will be examined in detail. Moreover, trends in extracted I- and b-values suggesting a dependence on projectile will be considered.

INTRODUCTION

The stopping power of matter for charged particles, a topic that has now been studied extensively for about a century, remains a fertile field for research both theoretical and experimental. One specific area of importance is the stopping power of various organic (co)polymers for light projectiles and heavy ions, partly because of the multitudinous applications of thin foils of these materials in experiments featuring ion beams from accelerators and partly because of possible improvement of understanding of energy loss processes for ions traversing target materials which manifest aggregation effects.

Results of a set of experiments conducted in recent years to measure accurately the stopping powers of four organic (co)polymer materials for light projectiles have been summarized in graphical form in Ref. [1], whereas corresponding data for ^{7}Li projectiles have been reported in Ref. [2]. The materials studied were formvar (a polyvinylformyl resin), polysulfone, kapton (a polyimide), and vyns (a vinylchloride-vinylacetate copolymer). Results of the experiments were originally characterized by comparison with widely employed models and parametrizations for stopping power [1,2]. When the author sought to analyze the results in terms of modified Bethe-Bloch stopping power theory as a continuation of a program spanning the previous three decades, a member of the experimental groups involved in the measurements graciously provided the detailed tabular data [3]. Subsequently the light projectile measurements have been analyzed in terms of modified Bethe-Bloch theory [4-7], as have more recently the ^{7}Li ion data [8]. The rather intriguing results of these analyses are the subject of this paper.

THEORY AND METHOD OF ANALYSIS

Modified Bethe-Bloch Theory

In a simple transmission measurement of the stopping power of a thin target the measured quantities are the projectile energy loss, ΔE, and the target areal density, ρ_A. When ΔE is expressed in MeV and ρ_A in g/cm^2, the stopping power, S, is given in units of MeV·cm^2/g by

$$S = \frac{\Delta E}{\rho_A}. \qquad (1)$$

In terms of Bethe Bloch theory [4-10] the stopping power of an elemental target of atomic number Z and atomic weight A for a projectile of atomic number z and velocity v = βc can be calculated in units of MeV•cm²/g from the expression,

$$S = \frac{0.30706}{\beta^2} z^2 \frac{Z}{A} L. \quad (2)$$

Thus the stopping power is written in terms of the stopping number per target electron, L, which depends on projectile velocity as well as on various target properties. Two modifications to Bethe-Bloch theory have been formulated by the addition of terms of higher order in z in the (dimensionless) L, and a third has been given by substitution of a velocity-dependent projectile atomic number z^* such that (z^*e) represents an effective charge to replace the basic charge (ze) when the projectile velocity has decreased to a value comparable to those of atomic electrons in the target. With the two modifications to L the stopping number is expressed as the sum of three terms:

$$L = L_o + \xi L_1 + L_2, \quad (3)$$

where L_o denotes the basic stopping number and L_1 and L_2 signify the terms of order higher than z^2 in projectile atomic number. In the absence of highly relativistic projectiles L_o is given by only four terms,

$$L_o = \ln\left(\frac{2mc^2\beta^2}{1-\beta^2}\right) - \beta^2 - \ln I - C/Z. \quad (4)$$

Here mc^2 denotes the rest mass energy of the electron, I represents the target mean excitation energy, and C signifies the sum of target shell corrections. Shell corrections can be calculated by the method of Bichsel [11], where the Walske K- and L-shell corrections [12,13] are used in conjunction with previously determined scaling factors applied to the L-shell correction to obtain the M- and N-shell corrections:

$$C = C_K(\beta^2) + V_L C_L(H_L\beta^2) + V_M C_L(H_M\beta^2) + V_N C_L(H_N\beta^2). \quad (5)$$

Here C_K and C_L represent the K- and L-shell corrections, respectively, whereas the V_i and H_i (with i = L, M, N) signify the scaling factors. The mean excitation energy, I, can be established from a fit to accurate stopping power measurements.

The correction terms in L were reviewed in the course of a random phase evaluation of the first of the two [14]. The first term, (zL_1), now called the Barkas-effect term [15], can be calculated with one of at least three existing formalisms [16-20]. In a comparison of these methods [21] it was shown that the first [16-18] furnishes a superior fit to measurements over the broadest energy interval. Hence this formalism is generally used by the author in the stopping number calculated for analyses. For that formalism,

$$L_1 = \frac{F(b/x^{1/2})}{Z^{1/2} x^{3/2}}, \quad (6)$$

where F denotes a function graphed in Ref. 16, x = $(18787)\beta^2/Z$, and b represents the only free parameter of the formalism. The appearance of ξ as an amplitude of the Barkas-effect term in Eq. (3) is an artifact of an earlier controversy over inclusion of close-collision contributions to the term [9,10]. In the aftermath of a study indicating that the controversy could not be resolved by mere fits to extant data [22], the value of ξ has generally been set at unity during data analyses.

The L_2 term, reintroduced by Lindhard [15], is called the Bloch term [23]. This term is given by

$$L_2(y) = \psi(1) - \mathrm{Re}[\psi(1+iy)], \quad (7)$$

where ψ represents the digamma function [24] and the fine structure constant, α, appears in the argument, y = zα/β.

The final modification to the Bethe-Bloch formula is incorporation of an effective charge factor to simulate the gain and loss of electrons by low-velocity projectiles. This technique enables the extension downward of the energy interval of applicability of the Bethe-Bloch formula. The form that is generally employed by the author, featured in an early study of heavy ion stopping powers [25], is written as (z^*e) = (γze) where

$$\gamma = 1 - \zeta e^{-\lambda v_r}. \quad (8)$$

The symbol v_r represents the ratio of projectile velocity in the laboratory frame (v) to the Thomas-Fermi velocity [$(e^2/h)z^{2/3}$], so that $v_r = \beta/\alpha z^{2/3}$. The effective charge parameters, ς and λ, presumably independent of both z and Z, are valid over the entire projectile-velocity interval. In fact, these parameters must be evaluated for any specified projectile-target combination [25]. The effective charge formalism was used in the current study for analysis of the stopping powers of all four target materials for ^{7}Li ions.

Clearly there are several parameters associated with the three terms in Eq. (3), namely, the mean excitation energy, I, the Barkas-effect parameters, b and ξ, and the scaling parameters for shell corrections, V_i and H_i. All of these parameters are independent of projectile velocity whereas the Bloch term clearly depends on projectile energy and charge but on no free parameters. Hence the process of fitting stopping power measurements in order to ascertain values of some subset of the above parameters ought to yield parameter values that are valid over the same projectile energy interval in which the modified Bethe-Bloch formula is valid.

Additivity Effects

The Bohr and Bethe theories originally applied strictly to pure monatomic targets in the gaseous state [26], whereas something as simple as an elemental target in the condensed state or a homonuclear diatomic molecule in the gaseous state will be subject to bonding effects. Fermi first studied the case of many atoms interacting simultaneously with a projectile and each other [27]. Yet it was Fano who explained the connection between Bethe and Fermi theories [26,28] and who subsequently described how the Bethe theory can be adapted to molecules and condensed matter [26]. Physical state effects and chemical bonding effects are both grouped together as aggregation effects. Compounds, alloys, and mixtures are clearly materials for which aggregation effects will pertain.

A first approximation in dealing with the reality of aggregation effects is to assume the linear additivity of stopping effects of the constituents of the material studied. This assumption, first advanced by Bragg [29], is appropriately known as "Bragg's rule" or the "additivity rule." The rule serves as a point of departure in the study of aggregation effects. The present investigation is a manifestation of continuing interest in the additivity problem.

Average parameter values can be calculated for materials evincing aggregation effects [17,26,30,31]. The Bloch term [23] possesses no target-dependent parameters, but the Barkas-effect term does require averaging [17,31]. Moreover, L_o contains target-dependent parameters, I and b, which must be assigned appropriate average values [26,30,31]. A problem inherent in the averaging procedure is the paucity of information about the correct parameter values for target constituents. For reasons of practicality mean excitation energies can generally be calculated from first principles only for low-Z targets, and aggregation effects influence the value of this parameter [26,30,31]. Shell correction scaling parameters have been established with some accuracy only for a few (elemental) target materials [11]. Hence the parameter most often chosen for a test of the additivity rule is the mean excitation energy, the (Bragg) average value of which, I_B, can be calculated from the expression [26,30],

$$\ln I_B = \sum_j n_j Z_j \ln I_j / \sum_j n_j Z_j. \quad (9)$$

In this formula n_j, Z_j, and I_j, respectively, represent the atomic concentration, atomic number, and mean excitation energy of the jth constituent of the target material.

One might expect deviations from the additivity rule to be greatest, with I-values exceeding the respective Bragg values, in cases of strong chemical bonds in low-Z compounds, where a large fraction of the total number of electrons participate in the bonding [26]. Furthermore, the chemical binding effect is greater at lower projectile velocities since the tightly bound inner shell electrons contribute less to the stopping than do the outer valence electrons [32].

The approach to additivity thus described is simply to approximate the stopping power (or stopping cross section) of a target possessing aggregation effects as a linear sum of the stopping powers (or stopping cross sections) of the atomic constituents with weight factors to reflect abundances [11,26,30,31]. An alternative approach, propounded by Neuwirth and Both [33] and Oddershede and Sabin [34], is to apply the additivity rule to molecular fragments such as bonds or functional groups, with the result of improving the accuracy of the additivity predictions [11].

PROCEDURE

A summary of target material information is displayed in Table 1. These data include the stoichiometry, average atomic number $(\bar{Z})$ and atomic weight $(\bar{A})$, shell correction scaling parameters (V_L and H_L), and the additivity-based mean excitation energy called the "Bragg value" (I_B). The method of analysis [4-10] was to fix all parameters except the mean excitation energy (I) and Barkas-effect parameter (b), and to search for the pair of values for I and b that provided the best fit to the data. Fits to the ^7Li ion data generally required a single charge-state parameter (λ) as well, thus

TABLE 1. Stoichiometry, average atomic number ($\bar{Z}$), average atomic weight ($\bar{A}$), logarithmic average mean excitation energy (I_b), and L-shell correction scaling parameters (V_L and H_L) assigned to the four organic (co)polymer targets.

Target	Stoichiometry	$\bar{Z}$	$\bar{A}$	V_L	H_L	I_B(eV)
Formvar	$H_{7.704}C_{5.558}O_2$	3.80	7.11	--	--	72.8
Polysulfone	$H_{22}C_{27}O_4S$	4.30	8.20	--	--	78.1
Kapton	$H_{10}C_{22}N_2O_5$	5.03	9.80	0.379	1.00	79.6
Vyns	$H_{33}C_{22}O_2Cl_9$	5.06	9.83	0.382	1.00	107.4

necessitating a three-parameter search. The figure of merit for any fit was the root-mean-square relative deviation of measured from calculated stopping power values over the selected projectile energy interval, given by

$$\sigma = \sqrt{\frac{1}{N} \sum_{i=1}^{N} \left(\frac{S_m - S_c}{\Delta S_m}\right)_i^2}, \qquad (10)$$

for measurements of stopping powers at N energies, with S_{ci} denoting the calculated stopping power at the the ith energy, and S_{mi} and ΔS_{mi} denoting the measured stopping power and statistical uncertainty in that quantity at the ith energy, respectively

RESULTS AND DISCUSSION

Analyses for stopping power measurements with light projectiles and kapton [4], vyns [5], and polysulfone [6] targets were completed and reported before the formvar data were considered. The latter analysis yielded anomalously low values of formvar mean excitation energy [7], some 20% below the additivity-based value (I_B), although the extracted values of Barkas-effect parameter (b) lay within or very close to the expected interval [35] of 1.3 - 1.5. The results for all four target materials are summarized in Table 2. Inspection of this display reveals a quite remarkable consistency with expectations from modified Bethe-Bloch theory, save for the aforementioned extracted value of mean excitation energy for formvar. The excellence of fits to the measurements achieved in these studies can be viewed in the original reports [4-7].

Initially the values of mean excitation energy characterizing formvar caused some consternation. Members of the experimental groups, led by J. Räisänen, joined the author in launching an investigation into the cause(s) of the anomalous results. However, all four target materials were studied with the same experimental apparatus and conditions, so that nearly any source of error applicable to the formvar case must also have applied to the other three target materials as well. Moreover, it was clear from a review of results of comparisons of the formvar stopping power measurements with predictions based on parametrization models [1] such as those featured in SRIM/TRIM [36, 37] that the measured values generally exceeded predictions for the formvar targets alone, thus indicating that the extracted I-values would lie considerably below expected values. Hence one might say that the direct source of anomalous values of mean excitation energy was anomalously high measured stopping powers. Nonetheless, several avenues of possible error were explored in an attempt to recheck the validity of the stopping power measurements, as described in Ref. [8]. These gambits included the possibility of an error in determining areal density (in which case even hygroscopicity was considered), and of an error in the reported stoichiometry of the target material utilized. The upshot of the investigation was that no source of error in the measurements could be found.

At this point the author and J. Räisänen decided to analyze the stopping power measurements for all four target materials and ^{7}Li projectiles [2] in order to check for consistency of results with those from light projectile studies. The case of the formvar target was obviously of special interest. Values of the two parameters, I and b, extracted from the ^{7}Li ion measurements are displayed in Table 3. The excellence of fits to the measurements achieved in this facet of the study can be seen in the original report [8]. In the cases of formvar and polysulfone the values of I and b proved to be consistent with those obtained from light projectile data, although in each case the value of b was more plausible when a single effective charge parameter (λ) was included in the modified Bethe-Bloch formula. In the case of kapton, the most plausible results emerged for a fit with only 60% of the data points, those at lower energies, and these parameter values were reasonably consistent with the light projectile values. The measurements with vyns yielded acceptable values over the full energy interval, but an improved fit was found for roughly 45% of the data points, those at higher energies. In the latter case, the best-fit values of b lay closer to those from the light projectile

TABLE 2. Results of fits to measurements with all four organic (co)polymer targets and proton and alpha particle projectiles, including the energy interval covered (ΔE), the mean excitation energy (I) and Bragg value (I_B), the Barkas-effect parameter (b), and the figure of merit (σ).

Target	Projectile	ΔE (MeV)	$I \pm \Delta I$ (eV)	I_B (eV)	$b \pm \Delta b$	σ
Formvar	Proton	0.67-1.75	58.8±2.1	72.8	1.39±0.45	0.29
	Alpha	1.79-3.02	58.6±4.3	72.8	1.62±0.32	0.40
Polysulfone	Proton	0.67-1.74	83.3±3.1	78.1	1.05±0.36	0.48
	Alpha	2.07-3.20	81.1±6.2	78.1	1.38±0.34	0.56
Kapton	Proton	0.56-1.54	80.6±2.2	79.6	1.46±0.31	0.46
	Alpha	1.84-3.19	77.8±4.6	79.6	1.87±0.35	0.39
Vyns	Proton	0.55-1.75	116.9±3.9	107.4	1.07±0.17	0.59
	Alpha	1.03-3.22	109.8±1.6	107.4	1.57±0.04	0.59

TABLE 3. Results of fits to measurements with all four organic (co)polymer targets and ^{7}Li ion projectiles, including the energy interval covered (ΔE), the mean excitation energy (I) and Bragg value (I_B), the Barkas-effect parameter (b), the effective charge parameter (λ), and the figure of merit (σ).

Target	ΔE (MeV)	I (eV)	I_B (eV)	b	λ	σ
Formvar	3.9-10.4	59.7	72.8	1.91	--	0.64
	3.9-10.4	61.6	72.8	1.55	1.82	0.63
Polysulfone	3.6-10.4	80.7	78.1	2.09	--	0.79
	3.6-10.4	83.1	78.1	1.71	1.94	0.78
Kapton	3.8-10.0	75.3	79.6	2.66	--	0.60
	3.8-10.0	76.2	79.6	2.51	2.91	0.60
	3.8-6.4	79.3	79.6	2.29	--	0.63
	3.8-6.4	82.1	79.6	1.95	2.11	0.63
Vyns	3.8-10.5	116.6	107.4	1.75	--	0.93
	3.8-10.5	117.6	107.4	1.71	6.87	0.93
	7.0-10.5	120.6	107.4	1.55	--	0.82
	7.0-10.5	122.2	107.4	1.49	2.58	0.82

analyses whereas the best-fit values of I rose about 6% above those from the light projectile analyses. Thus the data points for certain energy subintervals in the kapton and vyns measurements appear to deviate somewhat from the remainder of the measurements. However, in view of the uncertainties given for the two sets of measurements [1,2], results of the analyses of the ^{7}Li ion data can be considered supportive of the light projectile results.

TRENDS IN PARAMETER VALUES

In the course of reviewing results of analyses of the measurements with light projectiles, apparent trends in values of the parameters, I and b, emerged. That is, for a given target material the value of mean excitation energy extracted from the proton data exceeds that extracted from the alpha particle data, whereas the opposite trend prevails for values of the Barkas-effect parameter. A review of some previous studies revealed the same trends in the case of measurements with an alloy, havar [9]. When the ^{7}Li ion data were analyzed, results were checked for consistency with the observed trends. The ^{7}Li results were generally supportive. An initial report of this observation has been given [38], and a manuscript for a more detailed study is nearing completion [39].

SUMMARY

Recent measurements of the stopping powers of four organic (co)polymers for protons, alpha particles, and ^{7}Li ions have been analyzed in terms of modified Bethe-Bloch theory, in order to extract values of mean excitation energy and Barkas-effect parameter. The light projectile measurements, with a stated accuracy [1,3] of less than 1%, yielded results consistent with expectation (in the context of linear additivity) for target materials of polysulfone [6], kapton [4], and vyns [5], whereas the mean excitation energy for formvar [7] lay some 20% below the Bragg value of 72.8 eV. An exhaustive search for sources of error in the formvar measurements [7,8] yielded no plausible explanation for the anomalously low mean excitation energy characterizing both proton and alpha particle data. An explanation of the apparent anomaly may lie in the structural properties of the polyvinylformyl resin, formvar [40]. Results of two previous studies of chemical bonds in molecules containing carbon and hydrogen atoms [41,42] suggest the possibility of just such an explanation.

Corresponding stopping power measurements for the same target materials with ^{7}Li ion projectiles [2] were also analyzed. These data, of some 2% accuracy [2,3], yielded results that are reasonably corroborative of the light projectile results [8].

A review of results of the analyses conducted in the present study, as well as those of one previous study [9] in particular, revealed trends in values of mean excitation energy and Barkas-effect parameter extracted from the measurements. The trends suggest a dependence on projectile-z of both parameters. A possible explanation has been advanced [38,39]. Meanwhile, further study is in progress.

ACKNOWLEDGMENT

The author is deeply grateful to J. Räisänen, R. Cabrera-Trujillo, and J. R. Sabin for helpful guidance and advice.

REFERENCES

1. F. Munnik, A. J. M. Plompen, J. Räisänen, and U. Wätjen, *Applications of Accelerators in Research and Industry*, edited by J. L. Duggan and I. L. Morgan, (AIP Press, New York, 1997), pp. 1385-1389.
2. F. Munnik, K. Väkeväinen, J. Räisänen, and U. Wätjen, *J. Appl. Phys.* **86**, 3934 (1999).
3. J. Räisänen, private communication.
4. L. E. Porter, *Int. J. Quant. Chem.* **75**, 943 (1999).
5. L. E. Porter, *Nucl. Instrum. Methods B* **149**, 373 (1999).
6. L. E. Porter, *Nucl. Instrum. Methods B* **159**, 195 (1999).
7. L. E. Porter (in preparation).
8. L. E. Porter, J. Räisänen, and F. Munnik, (submitted for publication).
9. L. E. Porter, E. Rauhala, and J. Räisänen, *Phys. Rev. B* **49**, 11543 (1994).
10. L. E. Porter, *Phys Rev. A* **50**, 2397 (1994).
11. *Stopping Power and Ranges for Protons and Alpha Particles*, ICRU Report No. **49** (International Commission on Radiation Units and Measurements, Bethesda, MD, 1993).
12. C. Walske, *Phys. Rev.* **88**, 1283 (1952).
13. C. Walske, *Phys. Rev.* **101**, 940 (1956).
14. J. M. Pitarke, R. H. Ritchie, and P. M. Echenique, *Phys. Rev. B* **52**, 13883 (1995).
15. J. Lindhard, *Nucl. Instrum. Methods* 132, 1 (1976).
16. J. C. Ashley, R. H. Ritchie, and W. Brandt, *Phys. Rev. B* **5**, 2393 (1972).
17. J. C. Ashley, R. H. Ritchie, and W. Brandt, *Phys. Rev. A* **8**, 2402 (1973).
18. J. C. Ashley, *Phys. Rev. B* **9**, 334 (1974).
19. J. D. Jackson and R. L. McCarthy, *Phys. Rev. B* **6**, 4131 (1972).
20. S. H. Morgan and C. C. Sung, *Phys. Rev. A* **20**, 818 (1979).
21. L. E. Porter and H. Lin, *J. Appl. Phys.* **67**, 6613 (1990).
22. G. Basbas, *Nucl. Instrum. Methods B* **4**, 227 (1984).
23. F. Bloch, *Ann. Phys. (Leipzig)* **16**, 285 (1933).
24. M. Abramowitz and I. A. Stegun, Eds., *Handbook of Mathematical Functions*, (National Bureau of Standards, Washington DC, 1964), p. 259.
25. L. E. Porter, *Radiat. Res.* **110**, 1 (1987).
26. U. Fano, *Ann. Rev. Nucl. Sci.* **13**, 1 (1963).
27. E. Fermi, *Phys. Rev.* **57**, 485 (1940); see also E. Uehling, *Ann. Rev. Nucl. Sci.* **4**, 315 (1954).
28. U. Fano, *Phys. Rev.* **103**, 1202 (1956).
29. W. H. Bragg and R. Kleeman, *Philos. Mag.* **10**, 318 (1905).
30. W. H. Barkas and M. J. Berger, in *Studies in Penetration of Charged Particles in Matter*, National Academy of Sciences - National Research Council Publication No. 1133 (NAS-NRC, Washington, DC, 1967).
31. C. L. Shepard and L. E. Porter, *Phys. Rev. B* **12**, 1649 (1975).
32. D. Powers and H. G. Olson, *J. Chem. Phys.* **73**, 2271 (1980).
33. See, e.g., W. Neuwirth and G. Both, *Nucl. Instrum. Methods B* **12**, 67 (1985).
34. See, e.g., J. Oddershede and J. R. Sabin, *Nucl. Instrum. Methods B* **42**, 7 (1989).
35. R. H. Ritchie and W. Brandt, *Phys. Rev. A* **17**, 2102 (1978).
36. J. F. Ziegler, J. P. Biersack, and U. Littmark, *The Stopping and Range of Ions in Solids* (Pergamon Press, New York, 1985).
37. J. F. Ziegler, *TRIM 92 and SRIM 97 Computer Codes*.
38. L. E. Porter, in *Proceedings of the Twentieth Werner Brandt Workshop on Penetration Phenomena*, Feb. 9-12, 2000, University of Florida, Gainesville, FL. Edited by John R. Sabin (Feb. 2000).
39. L. E. Porter, in preparation.
40. R. Cabrera-Trujillo, private commuication.
41. J. Oddershede and J. R. Sabin, *Nucl. Instrum. Methods B* **42**, 7 (1989).
42. R. Cabrera-Trujillo, S. A. Cruz, and J. Soullard, *Nucl. Instrum. Methods B* **93**, 166 (1994).

Effect of dechanneling on the photon emission by ultra-relativistic positrons channeling in a periodically bent crystal

Andrei V. Korol[1,3], Andrey V. Solov'yov[2,3] and Walter Greiner[3]

[1] *Department of Physics, St.Petersburg State Maritime Technical University, St. Petersburg 198262, Russia*
[2] *A.F.Ioffe Physical-Technical Institute of the Academy of Sciences of Russia, St. Petersburg 194021, Russia*
[3] *Institut für Theoretische Physik der Johann Wolfgang Goethe-Universität, 60054 Frankfurt am Main, Germany*

Abstract. We investigate the influence of the dechanneling process on the parameters of undulator radiation generated by ultra-relativistic positrons channelling along a crystal plane, which is periodically bent.

INTRODUCTION

In this paper we proceed with the investigation of the properties of electromagnetic radiation accompanying ultra-relativistic positron planar channeling through a crystal which is periodically bent, see figure 1. This phenomenon was described recently in [1] and was called Acoustically Induced Radiation (AIR). The periodic pattern of crystal bendings (which can be achieved either through propagation of a high-amplitude transverse acoustic wave or by using static periodically strained crystalline structures [1,2]) gives rise to a new mechanism of electromagnetic emission of the undulator type, in addition to a well-known channelling radiation [3]. Provided certain conditions are met [1] the beam of particles, which enters the crystal at a small incident angle with respect to the curved crystallographic plane, penetrates further following the bendings of the channel. The related transverse oscillations of the beam particles (additional to the oscillations inside the channel due to the action of the interplanar force) become an effective source of spontaneous radiation of the undulator type due to constructive interference of the photons emitted from similar parts of the trajectory. It was demonstrated [1] that the system 'ultra-relativistic charged particle + periodically bent crystal' can produce undulator radiation of high intensity, monochromaticity and of a particular pattern of the spectral-angular distribution. It was also pointed out that this system leads, in addition to the spontaneous radiation, to the possibility of generating stimulated emission, similar to that in free-electron lasers, where the periodicity of a trajectory of a projectile is achieved by applying a spatially periodic magnetic field. The present study focuses on an accurate quantitative consideration of the in-

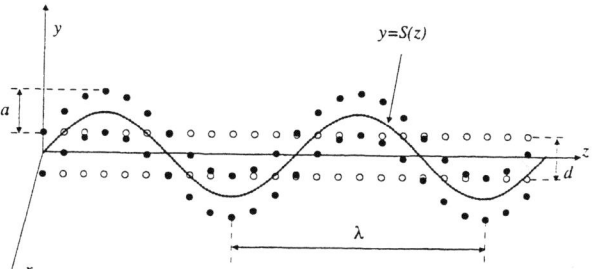

FIGURE 1. Schematic representation of the periodically bent crystallographic plane. The open circles mark the atoms of two neighbouring crystallographic planes (which are parallel to the xz- plane) in the initially linear crystal (d is the interplanar spacing). The filled circles denote the atoms when the crystal is bent. The z axis is directed along the linear channel centerline. The profile of the bent channel centerline (the dashed curve) is described by a periodic function $y = S(z)$, the amplitude, a, and the period, λ, of which satisfy $d \ll a \ll \lambda$.

fluence of the dechanneling process [4] on the parameters of the AIR. Dechanneling, i.e. a decrease in the volume density, $n(z)$, of the channeled particles with penetration distance z due to the multiple scattering of the projectiles from the target electrons and nuclei, is a parasitic effect which limits the crystal length L and, accordingly, on the number of the undulator periods N, which, in turn, defines the intensity of the AIR [1]. The factor, which plays a crucial role in obtaining the accurate data for the characteristics of the AIR, both spontaneous and stimulated, is a so-called dechanneling length L_d, which is the mean penetration distance covered by channeling particles. The quantity L_d depends on the projectile energy, the type of crystal and crystallographic plane, and on the parameters of the channel bending. In this connection the following problems were solved and are discussed briefly below: (i) A scheme for accurate quantitative treatment of the AIR in the presence of dechanneling is presented. As a result, a simple analytic expression is evaluated for the spectral-angular distribution of the AIR which contains L_d as a parameter; (ii) A simulation procedure of the dechanneling process for a positron in periodically bent crystals is outlined; (iii) The L_d values are calculated for positrons channeling in periodically bent crystals. The calculations have been performed for various parameters of the channel bending; (iv) Spectral-angular distributions of the AIR were calculated over broad ranges of the photon energies.

AIR IN THE PRESENCE OF DECHANNELING

There are two criteria essential when considering the properties of the AIR. The first one is a general criterion for a stable channeling of an ultra-relativistic particle of energy ε in a bent crystal (see e.g. [5]). It has clear physical meaning: the maximum centrifugal force must be less than that of the interplanar field. For a periodically bent crystal this condition reads

$$C = \varepsilon \left(R_{min} U'_{max} \right)^{-1} \ll 1, \qquad (1)$$

where U'_{max} is the maximum value of the interplanar force, R_{min} is the minimum curvature radius of the bent channel. The minimum curvature R_{min}^{-1} can be written in terms of the amplitude a and the period λ of the shape function $S(z)$ (see figure 1): $R_{min}^{-1} \sim a^{-1}(\lambda/2\pi)^2$. Thus, for a given crystal and crystallographic plane, the condition (1) establishes the ranges of projectile energies and the parameters a and λ for which channeling can occur.

The second criterion concerns the magnitude of the amplitude a of $S(z)$. Namely, we assume that a greatly exceeds the interplanar spacing d:

$$a \gg d \qquad (2)$$

When channeling through a periodically bent crystal the particle experiences two types of oscillations. Firstly, there are the channeling oscillations due to the action of the interplanar potential. The oscillations of the second type (the 'undulator oscillations') are related to the periodicity in the shape of the centerline of the channel. The characteristic frequencies of these oscillations, denoted as Ω_{ch} and Ω_u respectively, can be estimated as $\Omega_{ch} \sim (U'/dm\gamma)^{1/2}$ and $\Omega_u = (2\pi c/\lambda)$ (here $\gamma = \varepsilon/mc^2$ is the relativistic factor). These quantities define the characteristic frequencies, ω_{ch} and ω_u, of the photons emitted due to the channeling and undulator oscillations, $\omega_{ch} \sim \gamma^2 \Omega_{ch}$, $\omega_u \sim \gamma^2 \Omega_u$. Accounting for (1) and (2) one obtains $\omega_u^2/\omega_{ch}^2 \sim (d/a)C \ll 1$. Hence, the characteristic frequencies of the AIR and the ordinary channelling radiation are well separated [1], and, what is also very important, in the region $\omega \sim \omega_u$ there is no coupling of the two types of radiation [6].

If the dechanneling effect is neglected, then the spectral-angular distribution of AIR can be obtained by means of standard theory of the undulator radiation (e.g. [7]). Assuming that the shape function has the form $S(z) = a \sin kz$ (with $k = 2\pi/\lambda$) one writes

$$\frac{dE_N}{d\omega d\Omega_n} = \frac{e^2}{c} \frac{\omega^2}{\omega_0^2} \frac{D_N(\eta)}{4\pi^2} F(\omega, \gamma, \eta, p, \theta, \phi) \qquad (3)$$

where $\omega_0 = 2\pi c/\lambda$, θ and ϕ are the solid angles of the emission, p is the undulator parameter, and $\eta = (\omega/2\omega_0)\left[\gamma^{-2} + \theta^2 + p^2\gamma^{-2}/2\right]$. The factor $F(\omega, \gamma, \eta, p, \theta, \phi)$ is a smooth function of its argument, and its explicit form can be found in the literature (e.g. [7]).

The information on the specific pattern of the spectral-angular distribution of the undulator radiation is accumulated in the factor $D_N(\eta) = (\sin N\pi\eta/\sin\pi\eta)^2$ which defines the profile of the line of the emission in an ideal undulator. It has sharp maxima at the points $\eta = k = 1, 2, 3...$ where $D_N(k) = N^2$. The integer values $\eta = k$ define the characteristic frequencies ω_k (harmonics) of the undulator radiation for which $dE_N/d\omega d\Omega_n$ reaches the maxima:

$$\omega_k = 4\gamma^2 \omega_0 k, \left(2 + 2\gamma^2\vartheta^2 + p^2\right)^{-1} \qquad (4)$$

The natural width of each line of the emission can be estimated as $\Delta\omega' = (1/N)(\omega_k/k)$.

In an ideal undulator one may unrestrictedly increase the intensity of the radiation by considering larger values of the periods N. In reality, random scattering of the channeling positron by electrons and nuclei of the crystal leads to a gradual increase of the particle's energy associated with the channeling oscillations. As a result, the transverse energy at some distance from the entrance point exceeds the depth of the interplanar potential well, and the particle leaves the channel. Consequently, the volume density $n(z)$ of the channeled particles decreases with the penetration distance z. Although the exact explicit dependence of the channeled fraction of the particles $n(z)/n_0$ on z (n_0 is the volume density of the channeled particles at the entrance) can hardly be obtained by analyticall means, it has been argued (e.g. [5]) that in many cases the exponential decay law $n(z) = n_0 \exp(-z/L_d)$ is a good approximation. Basing on this relation it is possible to modify the spectral-angular distribution (3). Introducing the probability of the event, that after channeling through j periods of the undulator ($j = 1, 2, \ldots N-1$)

the particle dechannels within the subsequent $(j+1)$-th period, and taking into account that due to (2) the characteristics of the AIR are almost independent on the parameters of the fast channeling oscillations [1,5], one can construct the following quantity, which characterizes the spectral-angular distribution of the AIR formed in a crystal of length $L = N\lambda$ and accounts for the dechanneling:

$$\left\langle \frac{dE_N}{d\omega d\Omega_n} \right\rangle = \frac{e^2}{c} \frac{\omega^2}{\omega_0^2} \frac{\langle D_N(\eta) \rangle}{4\pi^2} F(\omega, \gamma, \eta, p, \theta, \phi) \quad (5)$$

$$\langle D_N(\eta) \rangle = e^{-\frac{N}{N_d}} D_N(\eta) + \left(e^{\frac{1}{N_d}} - 1 \right) \sum_{j=1}^{N} e^{-\frac{j}{N_d}} D_j(\eta) \quad (6)$$

where $N_d = L_d/\lambda$ is the number of undulator periods within L_d. The function $\langle D_N(\eta) \rangle$ has sharp maxima at the points $\eta = k$ corresponding to ω_k defined in (4). For a given N the maximum value $\langle D_N(k) \rangle$, which is the same for all k, is equal to

$$\langle D_N(k) \rangle = \begin{cases} N^2, & \text{for } N/N_d \ll 1 \\ 2N_d^2, & \text{for } N/N_d \gg 1 \end{cases} \quad (7)$$

The limit $N/N_d \ll 1$ corresponds to a thin crystal. In this case the number of channeling particles does not change noticeably on the scale L, so that the AIR intensity is proportional to N^2 as in an ideal undulator. For $L \sim L_d$ the probability of the particle to channel through the whole crystal length decreases, and $\langle D_N(k) \rangle$ starts to deviate from the N^2-law. In the limit of a very thick crystal, $L \gg L_d$, $\langle D_N(k) \rangle$ reaches its maximum value $\approx 2N_d^2$.

Expressions (5-6) describe the spectral-angular distribution of the AIR in the range $\omega \sim \omega_k \ll \omega_{ch}$. The dechanneling effect enters through the factor $\langle D_N(\eta) \rangle$ which depends on two parameters, N and L_d. The number of the undulator periods N can be varied by changing the length of the crystal L or/and the period λ.

NUMERICAL RESULTS

The approach which we developed to calculate the dechanneling lengths L_d of ultra-relativistic positron channeling through periodically bent crystal is based on a simulation of the projectile trajectories and dechanneling process. This is done by solving the three-dimensional equations of motion which account for (i) the interplanar potential, (ii) the centrifugal potential due to the crystal bending, (iii) the radiative damping force, (iv) the stochastic force due to random scattering of the projectiles by the lattice electrons and nuclei. To account for the random change in the direction of motion due to single collisions with target electrons we followed the procedure proposed by Biryukov [8] for heavy projectiles but having modified his formalism for the case of a light

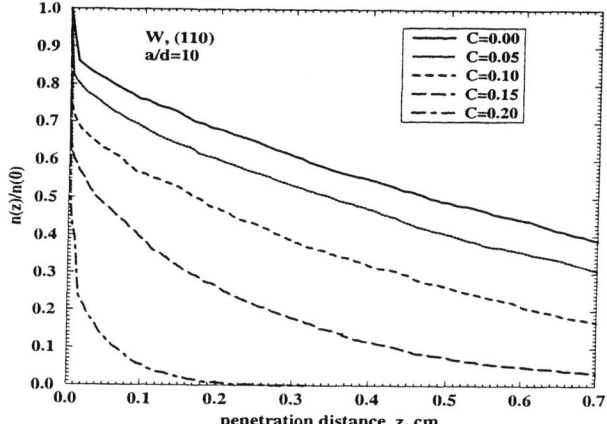

FIGURE 2. The calculated dependences $n(z)/n(0)$ for 5 GeV positrons channeling along (110) plane in the tungsten crystal for various values of the parameter C.

Table 1. Dechanneling lengths, L_d, the numbers of the undulator periods $N_d = L_d/\lambda$, the energies of the first harmonic $\hbar\omega_1$ at $\theta = 0°$ for 5 GeV positron channeling along the (110) planes in Si and W and for various values of the parameter C.

Crystal	d Å	C	λ μm	L_d cm	N_d	$\hbar\omega_1$ MeV
Si	1.92	0.05	100.9	0.430	39	1.38
		0.15	63.0	0.321	51	1.37
W	2.45	0.05	100.9	0.430	39	1.38
		0.15	63.0	0.321	51	1.37

projectile. Figure 2 represents the dependences $n(z)/n_0$ for 5 GeV positrons channeling along the (110) planes in W crystal. The ratio a/d equals 10. The curves in the figure refer to different values of C (eq. (1)). The value $C = 0$ stands for the case of a straight channel. The corresponding values of the spatial period λ of the shape function $S(z) = a\sin(2\pi z/\lambda)$ can be calculated as $\lambda = 2\pi\sqrt{(\varepsilon d/U'_{max})(a/d)C^{-1}}$. A detailed discussion of the obtained result will be given elsewhere. Here we only mention that the calculated values of the dechanneling lengths, summarized in table 1 along with the other data, are noticeably higher than those which one can obtain by applying the diffusion approach [1].

Figure 3 presents the spectral distributions (5) of the forward emission for a 5 GeV positron channeling in Si and W crystals. The spectra were calculated by using $C = 0.15$ for Si, and $C = 0.05$ for W. The number of the undulator periods varied from $N = 4N_d$ (figures 3(a,c) and the solid curves in figures 3(b,d)) to $N = N_d/2$ (the

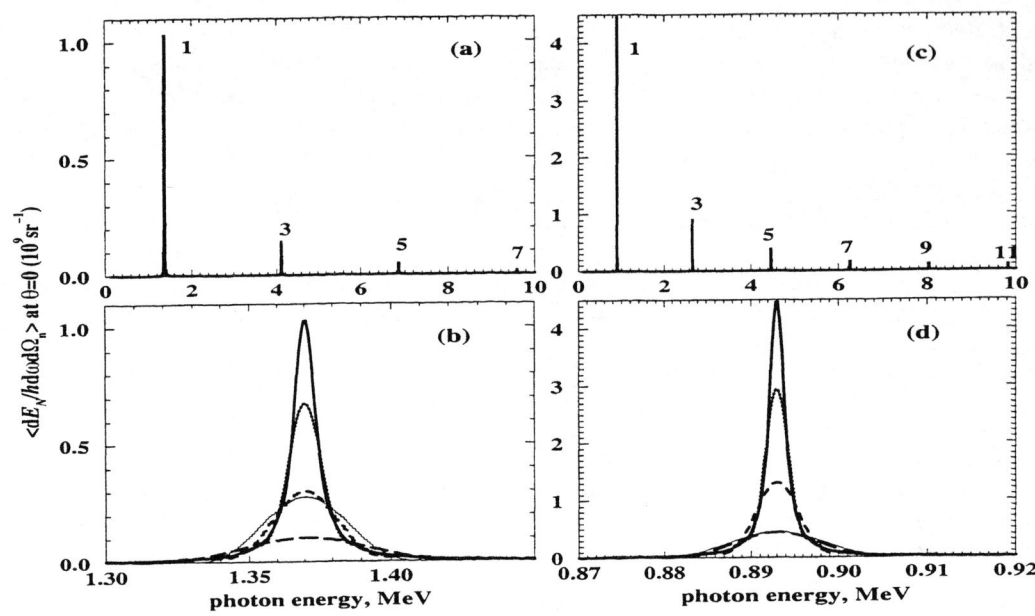

FIGURE 3. Spectral distribution (5) at $\theta = 0$ for 5 GeV positron channeling along periodically bent (110) planes in Si (figures (a) and (b)) and W (figures (c) and (d)) crystals. The upper figures (a) and (c) show $\langle dE_N/\hbar d\omega d\Omega_\mathbf{n}\rangle$ over the wide ranges of ω and correspond to $N = 4N_d$. The numbers enumerate the harmonics. The profiles of the first harmonic peaks (figures (b) and (d)) are plotted for $N = 4N_d$ (solid lines), $N = 2N_d$ (dotted lines), $N = N_d$ (dashed lines), $N = N_d/2$ (long-dashed lines). In both figures the thin solid line corresponds to the non-averaged spectrum (3) calculated for the number of undulator periods $N = N_d$.

long-dashed lines in figures 3(b,d)). The upper figures illustrate the spectral distribution over the wide range of the emitted photon energies. Each peak corresponds to the emission into the odd harmonics [7], whose energy is given by (4). The harmonics are well separated: the distance between two neighbouring peaks is 2.74 MeV for Si and 1.78 MeV for W, and the widths of the peaks are 8.7 and 2.5 keV, respectively.

Figures 3(b,d) exhibit the structure of the first-harmonic peaks. It is seen that for $N > N_d$ the intensity of the peaks is no longer proportional to N^2, as it is in the case without the dechanneling (eq. (3)). For both crystals, the intensities of the radiation for $N \rightarrow \infty$ exceeds those at $N = 4N_d$ by several per cent only. Thus, the solid lines correspond to almost maximal intensities which can be achieved for the used crystals, projectile energies and the parameters of the crystalline undulator.

ACKNOWLEDGEMENTS

The research was supported by DFG, GSI, and BMFT. AVK and AVS acknowledge the support by the Alexander von Humboldt Foundation.

REFERENCES

1. Korol, A.V., Solov'yov, A.V., and Greiner. W., *J.Phys.G.: Nucl. Part. Phys.* **24**, L45 (1998); *Int. J. Mod. Phys.* E **8**, 49 (1999); *ibid.* **9** 77 (2000).

2. Mikkelsen, U., and Uggerhøj, E., *Nucl. Inst. and Meth.* B **160**, 435 (2000).

3. Kumakhov, M. A., *Phys. Lett.* A**57** 17 (1976). *Nucl. Inst. and Meth.* B **160**, 435 (2000).

4. Lindhard., J., *Kong. Danske Vid. Selsk. mat.-fys. Medd.* **34**, 14 (1965).

5. Biruykov, V.M., Chesnokov, Y.A., and Kotov, V.I., *Crystal Channeling and its Application at High-Energy Accelerators*, Springer, 1996.

6. Krause, W., Korol, A. V., Solov'yov, A. V., and Greiner, W., *J.Phys.G.: Nucl. Part. Phys.* **26**, L87 (2000).

7. Baier, V. N., Katkov, V. M, and Strakhovenko, V. M., *High Energy Electromagnetic Processes in Oriented Single Crystals* World Scientific, Singapore, 1998.

8. Biryukov, V. M., *Phys. Rev.* E **51**, 3522 (1995).

Measurement of the nuclear energy loss under channeling

H. Winter and A. Mertens

Institut für Physik der Humboldt-Universität zu Berlin

Invalidenstr. 110, D-10115 Berlin, Germany

Abstract. Neutral Ne atoms with keV energies are scattered under channeling conditions, i.e. at a glancing angle of incidence, from a LiF(001) surface. By means of a time-of-flight method with a pulsed neutral beam we record energy distributions for scattered projectiles. For this specific system the small energy transferred to the crystal lattice ("nuclear energy loss") during channeling via binary collisions with large impact parameters dominates the dissipation of projectile energy. All other excitations of the solid can be brought to a negligible level.

INTRODUCTION

In Fig.1 we show a simple sketch of typical trajectories of fast atomic projectiles for the penetration between two atomic planes within the bulk of a monocrystalline sample ("planar channeling") [1-5]. This phenomenon, based on a steering of projectiles between atomic strings or atomic planes within the bulk of a crystal, is commonly called "channeling".

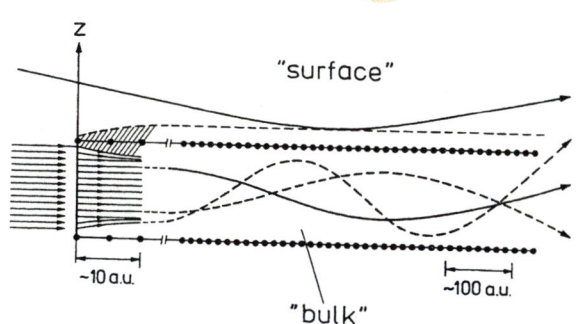

FIGURE 1. Sketch of scattering under channeling conditions between atomic planes in bulk and from surface.

Channeling effects are present if the direction of the incident beam of particles is close to a major crystal axis (low index directions). Then – in contrast to penetration into e.g. a polycrystalline sample – collisions of projectiles with target atoms proceed in a sequence of correlated small-angle-scattering events with relatively large impact parameters.

A large variety of different phenomena concerning atomic collisions in solids, as e.g. energy loss, resonant coherent excitation (RCE), or projectile ranges, have been studied in detail over the last decades. As a consequence, substantial details on ion-solid interactions could be cleared up by making using of this scattering regime.

An important aspect of collisions under channeling conditions is related to the feature that the steering of projectiles by the atoms of the crystal lattice proceeds in a sequence of small angle collisions. In contrast to large angle impact, these collisions result in a very small energy transfer of the projectile energy to lattice atoms ("nulcear energy loss"), so that, in general, this transfer can be neglected in comparison to the energy dissipated in electronic interactions. Indeed, it has been shown clearly that electronic stopping in the bulk of solids can be favourably studied under channeling. As a prominent example we mention here the so called "Z_1-oscillations", i.e. a periodic variation of electronic stopping powers in metal (and insulator) targets as a function of the projectile atomic number at low velocities ($v \leq v_o$ = Bohr velocity), which are particularly pronounced for data obtained under channeling [6-9].

Since the energy loss of projectiles under channeling is clearly dominated by electronic processes, there can be found to the best of our knowledge so far no experimental study on nuclear stopping in this regime. This can be understood from an estimate by using the simple formula for the energy E_1 of a projectile of mass M_1 and initial energy E_o in a binary elastic collision with a target atom of mass M_2, initially at rest. For an angle of scattering for projectiles θ_1 one finds from classical mechanics [4]

$$\frac{E_1}{E_o} = \frac{1}{1+(M_2/M_1)^2} \left(\cos\theta_1 + \sqrt{\left(\frac{M_2}{M_1}\right)^2 - \sin^2\theta_1} \right)^2 . \quad (1)$$

From this equation we deduce for angular deflections typical for low-energy channeling $\theta_1 < 2°$ and $M_1 \approx M_2$, an energy loss $\Delta E_1 = E_o - E_1 < 3 \cdot 10^{-3} E_o$. This number holds, however, for scattering by a single event. For a small angle collision sequence as met for channeling, this energy loss is further reduced, since for θ_1, achieved by a sequence in a number of n collisions $\theta_1 = \sum_i^n \theta_{1i}$, one obtains $\Delta E_1^{(n)} \approx \Delta E_1 / n$.

For channeling we have n >> 1 (typical estimate n ~ 10) so that the energy transferred to the steering lattice for the example given above is in the $10^{-4} E_o$ domain. It is evident that such a small energy loss can only be detected, if all other sources contributing to the projectile energy loss can be brought to a negligible level.

In our studies we made use of three features essential to resolve for the first time those small relative energy losses.

(1) Instead of transmission trough a thin foil, the projectiles are reflected from the topmost layer of crystal surface under a glancing angle of incidence ("(semi-planar) surface channeling" [5], see also Fig.1 [4]). This regime of scattering has the advantage that only a single reflection from an atomic plane takes place and low energy beams of some keV can be used.

(2) For a wide-band-gap insulator material (here: LiF), electronic excitations of valence band electrons are suppressed owing to the band gap. This is different from metal targets, where electronic excitations at the Fermi level are efficient and dominate stopping of atomic projectiles at low velocities.

(3) We use <u>neutral noble gas atoms</u> as projectiles with tightly bound ground-state electrons, in order to avoid stopping via charge transfer, electronic promotion, or long-range excitations.

EXPERIMENT AND RESULTS

In our experiments we scattered Ne^+ ions and Ne atoms with keV energies from a LiF(001) surface under grazing angles of incidence. The energy of scattered particles is obtained with a time-of-flight (TOF) setup, where a pulsed beam of keV neutral Ne atoms is produced via chopping of a Ne^+ ion beam by means of electric field plates and subsequent neutralization in a gas target operated with Ne atoms. Resonant charge transfer between Ne atoms in small angle scattering results only in a slight broadening of the energy width of the initial beam. The overall energy resolution of our TOF setup for the neutral projectile beam amounts to about 4 eV and thus allows us to resolve energy losses down to sub-eV energies. After collimation of the Ne° beam by sets of slits (forming components of differential pumping stages) to an angular spread of less than 1 mrad, the projectiles are scattered from a flat and clean LiF(001) surface at grazing angles of incidence ranging from $\Phi_{in} = 0.5°$ to 2°. The base pressure in the target chamber amounts to some 10^{-11} mbar and rises to about 10^{-10} mbar during the experiments. The target surface is prepared by cycles of grazing sputtering with 25 keV Ar^+ ions at a temperature of 300 °C, where LiF shows sufficient conductivity in order to avoid a macroscopic charging up of the sample, and subsequent annealing at temperatures up to 400 °C.

The energy of scattered atoms is measured with a TOF setup where the start-signal is provided from a micro-channelplate (MCP) that can be rotated in the scattering plane at a distance of 1.3 m behind the target. The digitally delayed signal from the beam chopper is used as the stop-signal for a Ortec Mod.567 time-to-amplitude converter (TAC) with an overall time resolution of about 0.2 ns.

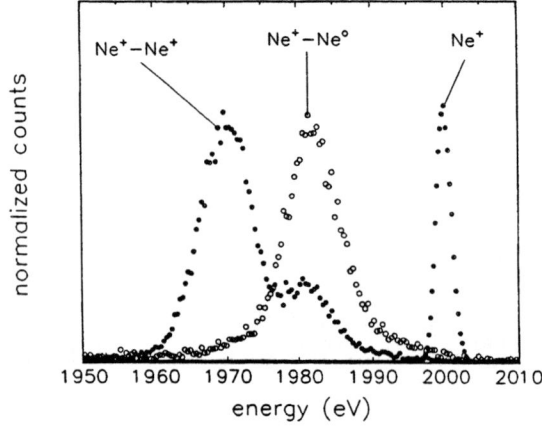

FIGURE 2. TOF spectra for 2 keV Ne^+ ions scattered from a LiF(001) surface under a grazing angle $\Phi_{in} = 1°$.

In Fig. 2 we show a TOF spectra converted to a projectile energy scale obtained for 2 keV Ne$^+$ projectiles. The full circles represent data for the complete scattered beam, wheras the open circles are data for the neutral fraction of the scattered beam separated by means of biased electric field plates behind the target. The data reveal an energy loss of about 30 eV for scattered ions and about half of this value for projectiles neutralized during the scattering. These energy losses exceed by one order of magnitude our estimates on the energy transfer to the crystal lattice given above. The origin for this dissipation of energy – not present for neutral projectiles – was attributed [10] to the excitation of optical phonons by long-range Coulomb forces between projectile ion and ions at lattice sites of the ionic crystal. The stopping for this problem can be calculated from dielectric response theory, where for the low velocities used here (v ~ 0.1 a.u.) excitations of the optical phonon band ($\hbar\omega$ ~ 38 meV) is the dominant mechanism. Note that owing to the broad band gap the (Auger) neutralization of projectiles is strongly suppressed and a dominant fraction of incoming ions survives the scattering from the surface. The energy loss shows a decrease with increasing angle which is reproduced by simple model calculations on a quantitative level (Fig.3). As we have said, it is clear that an energy loss observed for charged projectiles is much larger than the "nuclear energy loss"; furthermore we expect from eq.1 a monotonic increase of the nuclear stopping with angle.

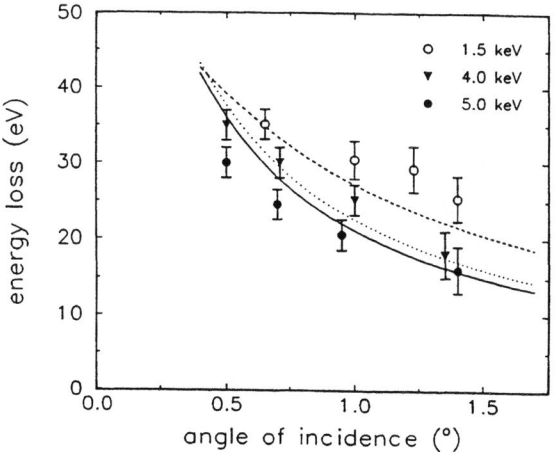

FIGURE 3. Energy loss for 1.5 keV, 4 keV, an 5 keV Ne$^+$ ions scattered from a LiF(001) surface as function of the angle of incidence. The curves represent results from calculations as outlined in the text.

In order eliminate energy dissipation effects caused by long-range response phenomena, we modified our setup in order to perform TOF-studies with a pulsed neutral beam of Ne atoms. The solid symbols in Fig. 4 represent data for scattering under Φ_{in} = 0.8° (solid circles) and 1.8° (solid squares). The spectra show a small but definite shift in energy from the spectrum for the direct beam (open circles) which serves as our reference with respect to the energy loss of scattered projectiles. Inspection of the energy scale reveals shifts of the spectra in the eV-domain and even lower. For recording TOF spectra of reflected atoms, the MCP detector has to be rotated by the scattering angle Φ_s from the direction of the incoming beam. For specular reflection we have Φ_s = 2 Φ_{in}. Since the axis of rotation for the detector differs from the position of the target surface, the actual distance between target and rotated detector is slightly reduced. This leads to corrections of the projectile energy which increase with Φ_s and amount here to about 50 % of the measured energy losses. In additional measurements with Ne$^+$ ions, deflected without surface scattering by an electric field at the target position, we could unambiguously demonstrate that the corrections in energy are in full agreement with slight reductions deduced from simple geometrical considerations.

From spectra as shown in Fig.4 we derive from the shift of the maxima of the data the <u>most probable</u> energy loss as displayed in Fig.5 for 3 keV (full diamonds) and 5 keV (full circles) Ne atoms as function of the angle of incidence Φ_{in}. The data reveal a monotonic increase with increasing angle and larger energy losses for the higher projectile energy. Note that the smallest energy losses are only about 0.5 eV and the highest about 3 eV (3 keV) and 6 eV (5 keV), i. e. a relative energy loss of only $\Delta E/E \approx 10^3$!

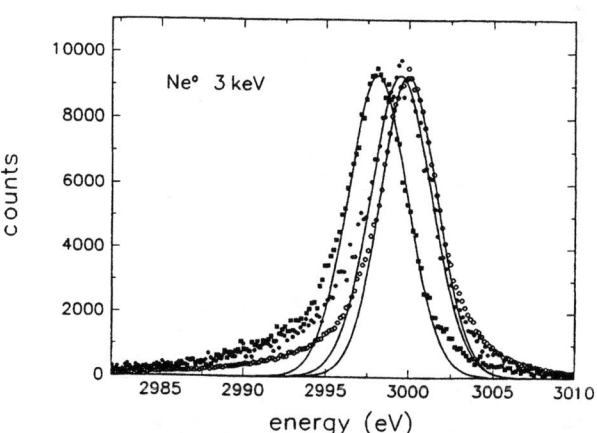

FIGURE 4. TOF spectra (converted to an energy scale) for 3 keV Ne° atoms scattered from a LiF(001) surface at a "random" azimuthal orientation. Open circles: direct beam without scattering, full circles: scattered beam for Φ_{in} = 0.8° and Φ_s = 1.6°, full squares: scattered beam for Φ_{in} = 1.8° and Φ_s = 3.6°. The solid curves represent best fits to a Gaussian lineshape.

The solid curves represent (most probable) energy losses for 3 keV and 5 keV Ne atoms obtained from computer simulations, where classical trajectories in front of an ideally terminated crystal surface (here LiF(001)) are derived from interatomic potentials with "universal" (ZBL) screening [11]. The energy transfer to nearest lattice atoms is calculated in small angle approximation from eq.1. Thermal vibrations of lattice atoms are taken into account by chosing a surface Debye-temperature of 250 K, and correlations between successive collisions are neglected [12]. The distances of closest approach for Ne projectiles are larger than about 3 a.u.

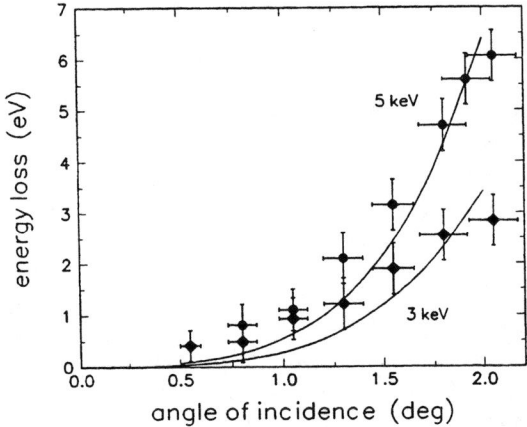

FIGURE 5. Most probable energy loss for Ne atoms scattered from a LiF(001) surface as a function of incidence angle. Full diamonds: projectile energy 3 keV, full circles: 5 keV. The solid curves represent results from our computer simulations (cf. text).

In view of the uncertainties in our experimental data, we find good agreement with the theoretical predictions and conclude the first demonstration of nuclear stopping under channeling conditions. The energy losses observed here are clearly smaller than the "mechanical energies" calculated by Ritchie and Manson [13] who considered the energy transfer to single lattice atoms instead of a collision sequence relevant for channeling conditions.

Furthermore our experiments show that under specific conditions keV projectiles can be scattered from the surface of a solid target without further energy loss, i.e. fully elastically. This means that simple conclusions with respect to suppression of electronic interaction channels owing to the wide band gap of the insulator target can indeed be observed in experiments at an extremely high level of sensitivity. The fact, that these ideal conditions can be experimentally realized for the dominant fraction of scattered projectiles may serve as an important demonstration, that for grazing collisions of fast atomic projectiles from carefully prepared monocrystalline surfaces the scattering process proceeds in specular reflection with very well defined trajectories.

ACKNOWLEDGMENTS

The assistance of K. Maass and S. Lederer in the preparation of the experiments and performance of computer simulations by T. Hecht and R. Pfandzelter are gratefully acknowledged. This work is supported by the Deutsche Forschungsgemeinschaft under contract Wi 1336/1

REFERENCES

[1] M. T. Robinson and O. S. Oen, Phys. Rev. 132, 2385 (1963).
[2] D. S. Gemmell, Rev. Mod. Phys. 46, 129 (1974).
[3] J. A. Davies, Phys. Scripta 28, 294 (1983).
[4] see e.g.: L. C. Feldman and J. W. Mayer, Fundamentals of Surface and Thin Film Analysis (Elsevier, Amsterdam, 1986).
[5] M. W. Thompson in: Channeling, D. W. Morgan (ed.), (Wiley, New York, 1973).
[6] J. H. Omrod, J. R. Macdonald, and H. E. Duckworth, Can. J. Phys. 43, 275 (1965).
[7] J. Bøttiger and F. B. Bason, Rad. Effects 2, 105 (1969).
[8] D. Ward, H. R. Andrews, I. V. Mitchell, W. N. Lennard, R. E. Walker, and N. Rud, Can. J. Phys. 57, 645 (1979).
[9] C. Auth, A. Mertens, and H. Winter, Nucl. Instr. Meth. B135, 302 (1998),
[10] A. G. Borisov, A. Mertens, H. Winter, and A. K. Kazansky, Phys. Rev. Lett. 83, 5378 (1999).
[11] J. F. Ziegler, J. P. Biersack, and U. Littmark, The Stopping and Range of Ions in Solids (Pergamon Press, New York, 1985).
[12] T. Hecht, R. Pfandzelter, and H. Winter, to be published.
[13] R. H. Ritchie and J. R. Manson, Phys. Rev. B49, 4881 (1994).

Stopping powers of 2-10 MeV Si, P and S ions in Ni, Cu and Ge thin films using a novel ERD-based technique

M. Nigam, J.L. Duggan, M.El Bouanani, C. Yang, S.A. Datar, S. Matteson and F.D. McDaniel

Ion Beam Modification and Analysis Laboratory, Dept. of Physics, University of North Texas, Denton TX 76203

Abstract. The stopping powers of 2-10 MeV Si, P and S ions in Ni, Cu and Ge thin films has been measured using a novel technique based on elastic recoil detection (ERD). This technique eliminates the need to recalibrate the silicon surface barrier detector while changing the incident ion species. Results have been compared to SRIM 2000 and other theoretical predictions and experimental measurements.

INTRODUCTION

The ability of ion beam analysis (IBA) to quantitatively determine the composition of thin films and provide surface characterization of thick targets is well known. IBA, combined with simulation programs like RUMP [1] and SIMNRA [2], makes an accurate and powerful tool. These simulation programs need accurate values of the stopping powers of ions in matter. These values are also of interest in radiobiology and radio therapy, and for ion-implantation technology, where shrinking feature sizes puts high demands on the accuracy of range calculations. The most common database used to determine stopping powers is the SRIM/TRIM program developed by J.F. Ziegler et al [3].

Some reports show that the SRIM codes for stopping powers sometimes differ significantly from experimentally determined values [4-7]. Since the TRIM code is a Monte Carlo simulation, its accuracy depends upon published experimental data. The experimental literature available for light ions is extensive, but this is not the case for heavy ions. Most experimental data are reported within an accuracy of 5% due to various parameter uncertainties that affect the measurements. One of the significant parameters that affect the accuracy of these experiments is the energy loss straggling in the foils. For heavy ions, the resolution of surface barrier detectors is worse than for light ions, and the radiation damage and pulse height defect are also larger.

In this study, a novel method based on elastic recoil detection (ERD) is used. The philosophy is to use an elastically recoiled lighter atom to indirectly measure the energy of the incoming heavy ion using a surface barrier detector. In this way, it is possible to reduce the damage in the surface barrier detector and improve the FWHM resolution of the detector. Using this technique, the stopping powers of Si, P and S ions have been measured in Ni, Cu and Ge targets. The results have been compared to the latest version of the TRIM code (SRIM-2000) and to other experimental data where available.

EXPERIMENTAL METHODS

High velocity heavy ion beams were produced by a 3 MV tandem pelletron accelerator (NEC Model 9SDH-2). The incident ion beam irradiated a $10\mu g/cm^2$ carbon film, and elastically recoiled carbon ions and reaction kinematics were then used to determine the energy of the incident ions. The energy loss measurements for the elemental films (Ni, Cu and Ge) were made in the foil transmission mode. The films were placed in and out of the path of the incident beam.

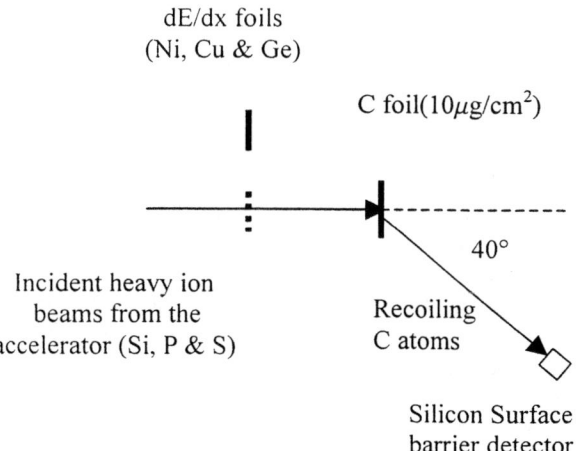

Figure 1: Experimental method for measuring dE/dx in elemental foils.

The energy loss was measured by estimating the shift in the energy of the recoiling C ions when the dE/dx foils were placed in and out of the path of the incident heavy ions. The experimental setup is shown in Fig1.

The energy of the pelletron was calibrated using a done using a 40μg/cm² LiF target foil and the reaction $F^{19}(p, \alpha\gamma)O^{16}$ that has proton resonance energies 0.8725, 1.3455 and 1.3732MeV. A thick Mica target was also used as a source of hydrogen for the reverse reaction $H^{1}(F^{19}, \alpha\gamma)O^{16}$ that has a higher resonance energy at 6.418MeV.

The foils were prepared by the vacuum evaporation technique. The Ni and Cu foils were self-supporting and the Ge films were deposited on 5μg/cm² carbon foils. The thickness of the foils was determined by Rutherford backscattering analysis using 2MeV α-particles from a Van de Graaff accelerator. The energy loss in the 5μg/cm² carbon foils was measured, and the data for Ge was corrected for the energy loss in the carbon backing foil.

The detector used for the energy loss measurements was an ORTEC Si surface barrier detector (Model # BU-012-050-100). The linearity of the detector electronics was checked using a precision pulse generator. For energy loss measurements in the Ni, Cu and Ge foils, the detector calibration was determined by irradiating the 10μg/cm² carbon foil with a heavy ion beam of known energy. The energy of the recoiling C particle was calculated using kinematics and plotted against the channel number that registered the particle in the multi-channel analyzer. The data points were then fitted using a straight line fit to determine the calibration curve. There are certain advantages to using this ERD-based technique. First, the detection system does not need to be re-calibrated while changing the ion species. The surface barrier detector needs to be re-calibrated for different ion species because of the pulse height defect associated with the interaction of heavy ions with surface barrier detectors [8]. Since the detected ion is always carbon, whatever the incident beam being (whether Si, P or S), the detector does not need to be re-calibrated. The damage done to the surface barrier detectors by heavy ions is dependent on the mass of the ion. Therefore using carbon ions instead of heavier ions reduces the damage to the detector. One also improves the FWHM resolution of the detector because carbon ions are lighter than the other ions used in this study. The resolution of the surface barrier detector is a function of the atomic number, mass and the energy of the incident ion.

DATA AND ERROR ANALYSIS

The stopping powers of Si, P and S ions were measured in Ni, Cu and Ge thin films. The stopping powers were reported at energy E-ΔE/2, where ΔE is the energy loss in the thin film. The energy E of the incident ion was changed by changing the terminal voltage and switching between charge states. For Ni, Cu and Ge thin films, the energy was corrected for the energy loss of the incident ion until it reached half the thickness of the thin carbon film used for scattering, and then for the energy loss of the recoiling carbon ion as it traversed the carbon film. The measurements were repeated two to three times in the same thin film (as in the case of the Ni films) at different spots. In the case of Cu and Ge thin films, the measurement was performed on two different films, and the average value of the stopping power was reported. The error in the measurements was estimated at approximately 5%. The uncertainties arose from the energy spread of the incident ion beam, error in the thickness measurement of the films, gain shifts in the electronics and the error in estimating the centroid of the peaks in the multi-channel analyzer. The uncertainty associated with the energy spread of the accelerator was system dependent and was measured and reported in the error. The thickness measurement of the foils (from RBS analysis) was accurate to approximately 2%. The error in the estimation of the centroid is much smaller than 1%. The spectra were fitted with a Pearson curve using PeakFit, and the mean value was compared to the mean value obtained from the multi-channel analyzer.

The stopping powers of energetic Si, P and S ions in thin films of Ni, Cu and Ge in the energy range of 2-10 MeV have been tabulated in Tables 1-3. The data have been compared to the predictions of SRIM-2000 [3] (the latest version of the TRIM code) and to other experimental results [9]. It has been assumed that SRIM-2000 stopping power values are normalized to other experimental data. Based on the above comparisons the following observations were made.

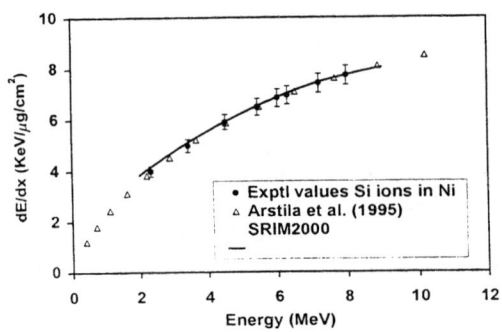

Figure 2: Stopping powers of Si ions in Ni

TABLE 1. Stopping power values for Si ions in Ni and Cu thin films

Ni			Cu		
E-dE/2 (MeV)	Exptl. (keV/μg/cm^2)	SRIM2000 (keV/μg/cm^2)	E-dE/2 (MeV)	Exptl. (keV/μg/cm^2)	SRIM2000 (keV/μg/cm^2)
2.34	3.98	4.13	2.35	3.74	3.59
3.44	4.97	5.15	3.43	4.79	4.58
4.51	5.90	5.94	4.52	5.58	5.32
5.45	6.48	6.52	5.46	6.02	5.84
6.02	6.83	6.84	6.03	6.42	6.12
6.31	6.93	6.98	6.32	6.55	6.25
7.22	7.41	7.39	7.23	6.97	6.61
8.00	7.73	7.69	8.00	7.36	6.90

TABLE 2. Stopping power values for P ions in Ni, Cu and Ge thin films

Ni			Cu			Ge		
E-dE/2 (MeV)	Exptl. (keV/μg/cm^2)	SRIM2000 (keV/μg/cm^2)	E-dE/2 (MeV)	Exptl. (keV/μg/cm^2)	SRIM2000 (keV/μg/cm^2)	E-dE/2 (MeV)	Exptl (keV/μg/cm^2)	SRIM2000 (keV/μg/cm^2)
2.37	3.94	4.13	2.38	3.77	3.57	2.57	3.86	3.83
2.76	4.32	4.54	2.77	4.15	3.97	2.98	4.28	4.17
3.49	4.88	5.20	3.50	4.73	4.61	3.73	4.73	4.72
3.53	5.02	5.24	3.54	4.79	4.65	3.78	4.74	4.75
4.08	5.29	5.66	4.08	5.12	5.07	4.33	5.29	5.09
4.63	5.83	6.06	4.66	5.60	5.45	4.92	5.62	5.41
5.22	6.08	6.44	5.22	5.90	5.79	5.51	5.89	5.71
5.40	6.31	6.55	5.42	6.14	5.89	5.71	6.10	5.80
6.91	7.34	7.37	6.92	7.05	6.61	7.28	6.67	6.44
7.30	7.50	7.54	7.31	7.29	6.76	7.69	6.76	6.58
8.07	7.81	7.88	8.08	7.63	7.06	8.47	7.05	6.84

TABLE 3. Stopping power values for S ions in Ni, Cu and Ge thin films

Ni			Cu			Ge		
E-dE/2 (MeV)	Exptl. (keV/μg/cm^2)	SRIM2000 (keV/μg/cm^2)	E-dE/2 (MeV)	Exptl. (keV/μg/cm^2)	SRIM2000 (keV/μg/cm^2)	E-dE/2 (MeV)	Exptl (keV/μg/cm^2)	SRIM2000 (keV/μg/cm^2)
2.37	3.89	4.36	2.37	3.73	3.78	2.56	4.02	4.07
2.75	4.40	4.76	2.77	4.11	4.18	2.96	4.54	4.39
3.15	4.73	5.13	3.15	4.56	4.52	3.70	4.88	4.90
3.45	5.02	5.38	3.47	4.68	4.78	4.33	5.29	5.26
4.05	5.54	5.82	4.06	5.28	5.22	4.39	5.50	5.29
4.10	5.63	5.86	4.12	5.35	5.26	4.89	5.64	5.55
4.60	5.84	6.20	4.62	5.49	5.58	5.49	6.24	5.84
4.61	6.10	6.21	4.63	5.84	5.59	5.80	6.37	5.98
5.16	6.57	6.56	5.18	6.25	5.91	6.39	6.86	6.26
5.36	6.60	6.68	5.37	6.26	6.01	7.25	7.29	6.69
5.44	6.89	6.73	5.46	6.49	6.06	7.63	7.34	6.87
6.01	7.30	7.06	6.04	6.82	6.35	8.43	7.67	7.22
6.010	7.25	7.12	6.12	6.85	6.40			
6.86	7.68	7.62	6.88	7.29	6.82			
7.22	7.97	7.83	7.25	7.49	7.02			
7.99	8.44	8.26	8.01	7.97	7.40			
8.37	8.48	8.44	8.39	8.00	7.57			

(i) For Si energetic ions in the energy range of 2-10 MeV in Ni thin films, the measured stopping power values were in overall good agreement with SRIM-2000 and with the data reported by Arstilla et al [9](Fig.2), the differences range from less than a percent to about 3.5%. For P and S ions in Ni absorbers in the energy range 2MeV to 5MeV, SRIM-2000 overestimated the experimental data by 3-12%. Above 5MeV, SRIM was in good agreement with the experimental data (differed by about 2%).

(ii) For stopping powers of Si ions through Cu in the energy range 2-10MeV, SRIM-2000 underestimated our data by about 4-6%. In the case of P and S ions in Cu absorbers, SRIM-2000 was in reasonable agreement (1-4%) with our data (Fig. 3)

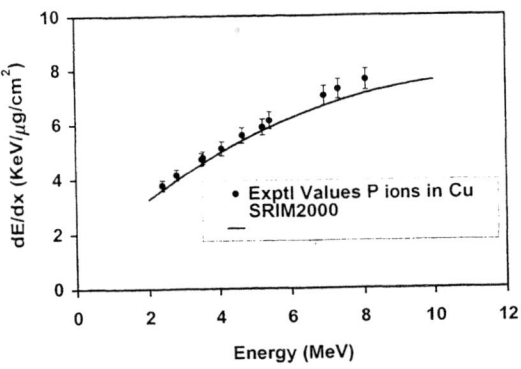

Figure 3: Stopping powers of P ions in Cu

for energies upto 5MeV. Above 5MeV, a distinct upward trend in the data was observed, and SRIM-2000 underestimated the data by as much as about 7.5%.

(iii) For Phosphorus ions in Ge thin films, SRIM-2000 and our data were in overall good agreement to within a few percent over the entire energy range of the measurements. However, for sulfur ions in Ge targets, SRIM-2000 was in overall good agreement with the experimental data until about 5MeV. Above 5MeV SRIM underestimated the experimental data by as much as 6-8%.

CONCLUSIONS

Stopping powers of 2-10MeV Si, P and S ions have been measured in thin films of Ni, Cu and Ge using a novel ERD-based technique. This technique eliminates the need to recalibrate the surface barrier detector while the incident ion species is changed, it reduces damage to the detector and allows better energy resolution in the acquired spectra. The data were compared to the predictions of SRIM-2000 and other experimental measurements. The following systematics were observed. For P and S ions in Ni absorbers, SRIM-2000 overestimated the data by as much as 10% for energies less than 5MeV. Above 5MeV SRIM predictions were in good agreement with the experimental data. For Si, P and S ions in Cu absorbers, SRIM-2000 was in good agreement with the experimental data upto about 5MeV, after which it underestimated the values by about 5-7%. For S ions in Ge absorbers, the predicted stopping power values by SRIM-2000 were in close agreement with experimental data below 5MeV. Above 5MeV SRIM underestimated the stopping power by 5-12%.

ACKNOWLEDGEMENTS

This work is supported in part by NSF, the State of Texas Advanced Technology Program, and the Robert A. Welch Foundation.

REFERENCES

[1] Doolittle, L.R., *NIM*B9, 344, (1985).
[2] Mayer, M., *SIMNRA Users Guide*, Technical Report IPP 9/113, Max-Planck Institut fur PlasmaPhysik, Garching, Germany, 1997.
[3] SRIM-2000, Ziegler, J.F., Biersack, J.P., Littmark, U. "The Stopping and Ranges of Ions in Matter", Pergamon Press, New York, 1985.
[4] Sharma, A., Kumar, S., Pathak, A.P., *Application of Accelerators in Research and Industry*, editors Duggan, J.L. and Morgan, I.L., AIP conference proceedings 475, New York, 1999, pp. 208-211.
[5] Shen Yixiong, Lu Xiting, Xia Zonghuang, Shen Dingyu, Jiang Dongxing, *NIM*B160 (2000) 11-15.
[6] Lennard, W.N., Massoumi, G.R., Simpson, T.W., Mitchell I.V., *NIM*B152, 370-376, (1999).
[7] Jiarui Liu, Zongshuang Zheng, Wei-Kan Chu, *NIM*B118, 24-28, (1996).
[8] Ortec AN-40, "Heavy-Ion Spectroscopy with Silicon Surface-Barrier Detectors.
[9] Arstila, K., Keinonen, J., Tikkanen, P., *NIM*B101, 321-326, (1995).

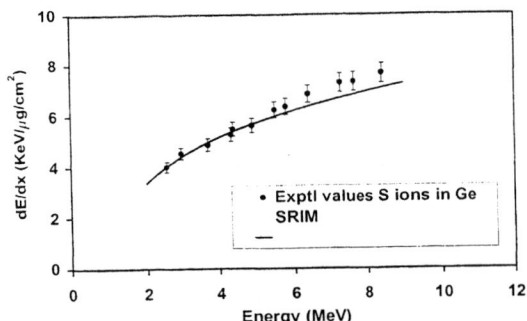

Figure 4: Stopping powers of S ions in Ge

Effects of Binding Energy on Exchange Contributions to the Stopping of Electrons

Scott M. Cohen

Department of Physics, Duquesne University, Pittsburgh, PA 15282-0321

Abstract. A theoretical treatment of the exchange contributions in Bethe's theory of the stopping of electrons is presented. The effects of the binding of the electron within the atom are systematically evaluated. Starting with the first Born approximation, an expansion in powers of the ratio of the electronic energy of the target atom to the incident electron's kinetic energy is obtained, and the leading contribution recovers Bethe's results. The approach involves the use of sum rules that are generalizations of the Bethe sum rule, specifically dealing with the contributions from electron exchange. Though the discussion is restricted to the stopping power here, the approach is quite general and may be applied to the calculation of the exchange contributions to other quantities related to the cross sections of electron-atom scattering. Our results indicate that there may be significant corrections when one accounts for the binding of the target electrons.

INTRODUCTION

In this talk, we will describe a recently developed extension of Bethe's theory [1-3] of the electronic stopping power for fast (but nonrelativistic) projectiles. The theory is extended, for the case of an incident electron, to include the effects of the binding energy of the target electrons in the treatment of the contributions due to exchange [4,5]. The analysis treats the target electrons as independent of each other, leaving to a later date the difficult question of the effects of their correlation.

One may recall [6] that Bethe's calculation of the stopping power involves the square of a transition matrix element of the Coulomb interaction between the target and projectile, connecting the initial and final states for the scattering system. When there are no exchange effects, this can be written in Fourier components as an integral over the wavevector $\vec{k}$, of

$$\langle f|e^{i\vec{k}\cdot(\hat{r}_t-\hat{r}_p)}|i\rangle/k^2 = \langle n|e^{i\vec{k}\cdot\hat{r}_t}|0\rangle \\ \times \langle \vec{p}'|e^{-i\vec{k}\cdot\hat{r}_p}|\vec{p}\rangle/k^2. \quad (1)$$

Here, $\hat{r}_t$ is the position operator of a target electron; $\hat{r}_p, \vec{p}, \vec{p}'$ are the position operator and the initial and final momenta of the projectile, respectively; while $|0\rangle$ and $|n\rangle$ are the initial and final states of the target. On the right-hand side of this equation, the projectile part leads to a delta function that requires $\hbar\vec{k} = \vec{q} = \vec{p} - \vec{p}'$, with $\vec{q}$ the momentum transferred to the target. This then leads to a great simplification of the analysis, in part due to the fact that the $\vec{k}$-integral now becomes trivial, giving

$$M_D = \int d^3k \langle f|e^{i\vec{k}\cdot(\hat{r}_t-\hat{r}_p)}|i\rangle/\hbar^2 k^2 \\ = \langle n|e^{i\vec{q}\cdot\hat{r}_t/\hbar}|0\rangle/q^2. \quad (2)$$

As a consequence, Bethe [1] was able to utilize closure to sum over the final states of the target, in the form of what is now well known as the Bethe sum rule [7]. He thus obtained a very compact result for the stopping power [1].

When the projectile is an electron, an accurate treatment must, of course, include the exchange contributions to account for the indistinguishability of the incident and target particles [3]. In this case, there is an "exchange" term of the form (we here omit the subscripts on the position operators)

$$M_E = \int d^3k \langle n|e^{i\vec{k}\cdot\vec{r}}|\vec{p}\rangle\langle\vec{p}'|e^{-i\vec{k}\cdot\vec{r}}|0\rangle/\eta^2 k^2, \quad (3)$$

which will appear in addition to the direct term, M_D, of Eq. (2). Here, the momentum-conserving delta function does not appear, and we do not have the consequent simplifications that occurred when there was no exchange.

Bethe was nonetheless able to derive a compact expression for the stopping power, including the effects of exchange [3]. He argued that, for the types of collisions being considered here, the exchange effect will be negligible when the momentum transferred between the two electrons is small. When the momentum transferred is large, on the other hand, the target electrons will behave very nearly as though they are free particles. Then the states $|0\rangle$ and $|n\rangle$ may be taken as momentum states, leading to simplifying delta functions in Eq. (3). He thus arrived at the following expression [3] for the stopping power, $-dE/ds$:

$$\left(-\frac{dE}{ds}\right) = \frac{2\pi e^4}{T} NZ \left(\ln\left(\frac{T}{I}\right) + \tfrac{1}{2}(1 - \ln 2) \right). \quad (4)$$

Here, $T = p^2/2m$ is the initial kinetic energy of the incident electron with m the electron mass; while N is the number density, Z the atomic number, and I the mean excitation energy, of the target atoms.

This result is beautiful in its simplicity, but it is only accurate when T is much larger than the binding energy of the electrons in the target material. As T decreases, the binding of the target electrons becomes more significant, and this expression must be supplemented through an analysis that takes that binding into account.

Recently, we have presented a new approach to this problem, which specifically accounts for the effects of the binding energy in the terms arising from exchange [4]. New sum rules have been derived, leading to an expression for the stopping power that is identical to Eq. (4), apart from the appearance of an additional term given by,

$$\Delta\left(-\frac{dE}{ds}\right) = \frac{7\pi e^4 N T_{tot}}{3T^2}, \quad (5)$$

with T_{tot} being the sum of the individual target electrons' kinetic energies, T_0. Additionally, we have calculated [5] the shell corrections [8,9], which correct for errors arising from approximations introduced to allow for the use of the sum rules mentioned previously. These corrections are of the same magnitude as those appearing in Eq. (5), above. Including both these corrections, we find for the stopping power [5],

$$\left(-\frac{dE}{ds}\right) = \frac{2\pi e^4}{T} NZ$$
$$\times \left(\ln\left(\frac{T}{I}\right) + \tfrac{1}{2}(1 - \ln 2) - \frac{T_{tot}}{ZT} \right), \quad (6)$$

which is accurate to $O(T^{-2})$. The magnitude of the correction we have found can be as much as several percent of Bethe's result, Eq. (4). We illustrate these results in Fig. 1, where we plot the fractional correction (ratio of the third term to the sum of the first two terms in the parentheses in Eq. (6)), as a function of the scaled incident kinetic energy, T/I, for a few low-Z elements. Values for T_{tot} have been taken from Huang, et.al. [10] and those for I are from Fano [6].

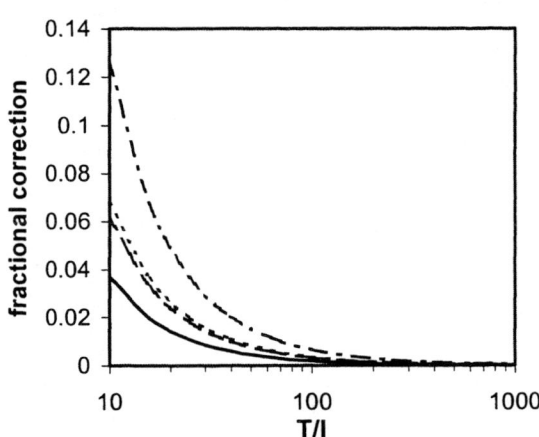

FIGURE 1. The fractional correction (ratio of the third term to the sum of the first two terms in the parentheses in Eq. (6)) plotted as a function of the scaled kinetic energy, T/I. The solid line is for H; dotted line, Li; dashed line, Be; and dot-dashed line, Al. Values for T_{tot} have been taken from Huang, et.al. [10] and those for I are from Fano [6].

As a specific example, we point out that when T is as large as twenty times the mean excitation energy (e.g., 3.3 keV for Al), these corrections can still be of the order of a few percent or more. The curve for Al indicates that these corrections will increase in importance as Z increases, perhaps dramatically.

SUM RULES FOR THE EXCHANGE CONTRIBUTIONS

The stopping power in Bethe's sum rule approximation may be written in terms of the differential cross-section, $d\sigma_n/dQ$, as [6,11]

$$-\frac{dE}{ds} = N \sum_{all} E_{n0} \int_{Q_{min}}^{Q_{max}} \frac{d\sigma_n}{dQ} dQ, \quad (7)$$

with E_{n0} the energy transfer, $Q_{min} = E_{n0}^2/4T$, $Q_{max} = (T + T_0)/2$, the summation is to cover energy transfers over the full range $0 < E_{n0} < \infty$, and the integration variable is related to the momentum transfer as $Q = q^2/2m$. The differential cross-section, neglecting correlations between the target electrons, may be written as

$$\frac{d\sigma_n}{dQ} = \frac{4\pi m^2 e^4}{T} \sum_{j=1}^{Z} \left(|M_D|^2 - \mathrm{Re}(M_E^* M_D) + |M_E|^2 \right) \quad (8)$$

In practice, the integration over Q is usually accomplished by introducing a cutoff value of $Q = Q_c$ to split the integral into two regions: a low-Q region within which a dipole approximation is introduced; and a high-Q region where the energy summation is performed before the Q-integration.

Our approach to handling the exchange parts begins with a momentum expansion of the initial target state [4]. Then, the exchange matrix element, Eq. (3), becomes

$$M_E = \int \frac{d^3k}{k^2} \int d^3P \langle n|e^{i\vec{k}\cdot\hat{r}}|\vec{p}\rangle$$
$$\times \langle \vec{p}'|e^{-i\vec{k}\cdot\hat{r}}|\vec{P}\rangle \langle \vec{P}|0\rangle$$
$$= \int \frac{d^3P}{(\vec{P}-\vec{p}')^2} \langle n|e^{-i(\vec{P}-\vec{p}')\cdot\hat{r}/\hbar}|\vec{p}\rangle \langle \vec{P}|0\rangle \quad (9)$$
$$= \int \frac{d^3P}{(\vec{P}-\vec{p}')^2} \langle n|e^{i\vec{q}\cdot\hat{r}/\hbar}|\vec{P}\rangle \langle \vec{P}|0\rangle.$$

This is to be used in Eq. (8) and then (7), with the aim of using closure to perform the sum over final states, $|n\rangle$, as Bethe has done. However, to proceed it is first necessary to account for the dependence of the outgoing momentum, $\vec{p}'$, on the final energy, E_n. In order to eliminate the energy-dependence by following the usual procedure [7] of using the identity, $E_n|n\rangle = \hat{H}|n\rangle$, with $\hat{H}$ the Hamiltonian of the target, we seek an expansion of the denominator in the integrand of Eq. (9).

We accomplish this expansion in essentially two steps. First we write,

$$2m(\vec{P}-\vec{p}')^{-2} = \big(T - Q + (Q - E_{n0}) + P^2/2m - \vec{P}\cdot\vec{p}'/m\big)^{-1}, \quad (10)$$

having used energy conservation in the form $p^2 - p'^2 = 2mE_{n0}$. Assuming the initial state is one of low energy, the factor of $\langle\vec{P}|0\rangle$ in Eq. (9) will ensure that only small momenta contribute significantly to M_E, allowing us to treat P as a small quantity. Furthermore, since the matrix element, $\langle n|e^{i\vec{q}\cdot\hat{r}/\hbar}|\vec{P}\rangle$, is significant only when $(Q - E_{n0})$ is small [12], we may treat the difference between these two energies as a small quantity as well. Hence, Eq. (10) may be expanded in powers of $(P^2 - 2\vec{P}\cdot\vec{p}')/2m(T-Q)$, and $(Q - E_{n0})/(T - Q)$.

The second step in expanding the denominator of Eq. (9) lies in the treatment of the terms containing $\vec{P}\cdot\vec{p}'$. Writing $\vec{p}'$ in terms of its components parallel and perpendicular to the incoming momentum, $\vec{p}$,

$$p^2\vec{p}' = \vec{p}(\vec{p}\cdot\vec{p}') + \vec{p}\times(\vec{p}'\times\vec{p})$$
$$= \vec{p}(p^2 - \vec{p}\cdot\vec{q}) - \vec{p}\times(\vec{q}\times\vec{p}), \quad (11)$$

and then using $\vec{p}\cdot\vec{q} = m(E_{n0} + Q)$ and $|\vec{q}\times\vec{p}| = \sqrt{p^2q^2 - (\vec{p}\cdot\vec{q})^2}$, we see that $\vec{P}\cdot\vec{p}'$ may also be written as a function of Q and $Q - E_{n0}$. These terms may therefore be expanded in powers of $(Q - E_{n0})/(T - Q)$, as well. Now the dependence of M_E on E_n can be rewritten as a power series and eliminated through the introduction of $\hat{H}$. This, in turn, allows the use of closure to obtain sum rules for the exchange contributions to the stopping power [4].

Within the sum rule approximation, then, we find a correction to Bethe's result for the stopping power, as given by Eq. (5).

SHELL CORRECTIONS

Following Bethe, we have in the foregoing discussion introduced certain approximations in order to implement the sum rule approach. The summation over energy transfer should, of course, be restricted by conservation principles to $E_{n0} < (T + T_0)/2 = Q_{max}$ (taking the outgoing electron with the highest energy to be the primary one, so that at most, only one-half of the total energy can be transferred). Furthermore, as discussed elsewhere [5,11,13], Q_{max} and Q_{min} are only approximations to the correct limiting values for the Q-integration. For the high-Q contribution, the region of the Q-E plane between $Q_c < Q < Q_{max}$, as indicated by the horizontal long-dashed lines in Fig. 2, has been included in our calculation. A correct approach should include that region which is enclosed by the full, curved line in this figure, and to the left of the vertical short-dashed line located at $E = Q_{max}$. As the figure illustrates, there are regions that have been improperly excluded, while others have been included incorrectly. Though there is some cancellation of these errors, the net effect is a "shell correction" of the same order of magnitude as the corrections found from the use of our new sum rules. Similarly, there are shell corrections from the low-Q region arising from the use of the dipole approximation, as well as from the use of the approximation, Q_{min}, as the lower limit in the integration over Q.

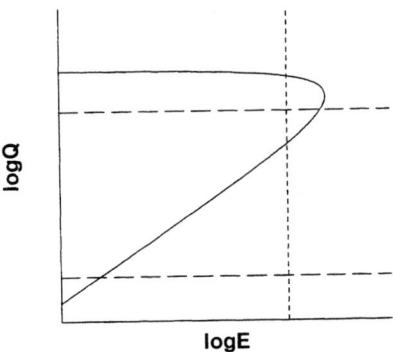

FIGURE 2. Illustration of the origin of the high-Q portion of the shell corrections. The region of the Q-E plane between the horizontal long-dashed lines has been included in our sum rule calculation. That region which is enclosed by the full, curved line in this figure, and to the left of the vertical short-dashed line, is the correct region to be included.

Following an approach presented by Kim and Inokuti [13], we have calculated the shell corrections to $O(T^{-2})$ [5]. Including these with the corrections previously found, one obtains the stopping power also to $O(T^{-2})$, as given in Eq. (6), above.

DISCUSSION

In the case of incident electrons, we have derived new sum rules for the exchange contributions to the stopping power. These sum rules take into account the effects of the binding energy of the target electrons. In addition, we have calculated the shell corrections, arriving at an analytical expression for the stopping power that is accurate to $O(T^{-2})$. We have seen that these corrections can be significant, and there is evidence that they may be quite large for high-Z atoms.

Our approach is general enough that it may also be applied to related problems, such as the calculation of the total inelastic cross-section; this work is in progress. It would be of interest to generalize our approach to include the case of relativistic incident electrons. This would serve as the starting point for a consideration of the stopping power of inner-shell electrons in heavy target atoms. Of course, as was mentioned above, our results neglect any correlations amongst the target electrons. It would be of considerable interest, as well, to look at the effects of these correlations on our results.

REFERENCES

1. Bethe, H.A., *Ann. Physik* **5,** 325 (1930).
2. Bethe, H.A., *Handbuch der Physik* **24/1**, 273 (1933).
3. Bethe, H.A., *Z. Physik* **76**, 294 (1932).
4. Cohen, S.M., *Phys. Rev. A* **61**, Article #022903 (2000).
5. Cohen, S.M., submitted to *Phys. Rev. A*.
6. Fano, U., *Ann. Rev. Nucl. Sci.* **13**, 1 (1963).
7. Jackiw, R., *Phys. Rev.* **157**, 1220 (1967); Bethe, H.A., and Jackiw, R.W., *Intermediate Quantum Mechanics*, 2nd Ed. New York: Benjamin, 1968, Chapter 11.
8. Walske, M.C., *Phys. Rev.* **88**, 1283 (1952); **101**, 940 (1956).
9. Bichsel, H., "Charged Particle-Matter Interactions," in *Atomic, Molecular, and Optical Physics Handbook*, edited by G.W.F. Drake, New York: AIP, 1996, p. 1032; *Phys. Rev. A* **46**, 5761 (1992).
10. Huang, K.N., Aoyagi, M., Chen, M.H., Crasemann, B., and Mark, H., *At. Data Nucl. Data Tables* **18**, 243 (1976).
11. Inokuti, M., *Rev. Mod. Phys.* **43**, 297 (1971).
12. When one or both of Q and E_{n0} are large, this statement follows from the same reasoning that leads to the conclusion that the Bethe surface is sharply peaked near $Q = E_{n0}$ (see Refs. [6] and [11]; see also, Rau, A.R.P., and Fano, U., *Phys. Rev.* **162**, 68 (1967)).
13. Kim, Y.-K., and Inokuti, M., *Phys. Rev. A* **3**, 665 (1971).

VELOCITY DEPENDENCE OF ELECTRON REMOVAL AND FRAGMENTATION OF WATER MOLECULES CAUSED BY FAST PROTON IMPACT

A.M. Sayler, E. Wells, K.D. Carnes, and I. Ben-Itzhak*

*James R. Macdonald Laboratory, Department of Physics,
Kansas State University, Manhattan, KS 66506-2604*

The yields of the dissociation products of water ionized by fast proton impact were measured relative to H_2O^+ using the coincidence time-of-flight method. The relative yields following single ionization were found to be independent of the collision velocity over the range of collision velocities measured and similar to the values reported for ionization by equal velocity electron impact, except for a higher rate of H_2^+ production. Double ionization of water leads predominantly to two dissociation channels, namely $H^+ + OH^+$ and $H^+ + O^+ + H$, which are equally likely and also independent of the collision velocity. The O^{2+} fragment was found to be always in coincidence with H^+, thus suggesting it is only a product of triple (or higher) ionization of water.

Electron impact ionization of water in the gas phase has been a topic of several studies (see, for example, Refs. [1–4]). Recently, Straub et al. [5] reported the absolute and relative cross sections of the dissociation products of water, as well as H_2O^+ formed by electron impact from the ionization threshold up to 1000 eV.

In this paper we describe the results of our coincidence time of flight measurements of single and double ionization of water by fast proton impact. This technique, in contrast to the mass spectrometry techniques commonly used in electron impact experiments, allows one to separate the double ionization channels from single ionization.

This is a compilation of background data taken over several years and only recently analyzed. The measurements were conducted using the same apparatus, experimental technique, and data analysis as described in our previous publications [6–12] and thus will be only briefly described here. A 1-8 MeV bunched beam of protons was directed through a target cell and collected afterwards in a Faraday cup. The target water vapor was the main remnant gas in our vacuum system. The recoil ions produced in the target cell were extracted and accelerated by uniform electric fields onto a microchannel plate detector of a time-of-flight spectrometer. The times of flight of the different recoil ions were recorded relative to a signal synchronized with the beam bunch, which was about 1 ns wide. Recoil ions produced in the same beam bunch were recorded in coincidence, thus separating single and double ionization events. The measured single ion and ion-pair events were corrected for random coincidences (recoil ions associated with the same beam bunch but from different molecules) and lost fragments (single ions measured from an ion-pair channel)(see details in Ref. [7]). The rate of O^+ from O_2 was also subtracted. The detector efficiency, used for these corrections, was measured to be 0.37 ± 0.03 [9]. The final charge state of the projectile was not determined in our measurements, and thus ionization and electron capture can both contribute to electron removal. However, electron capture by fast protons in the energy range under study is negligible in comparison to ionization, which is the dominant electron removal mechanism.

The products of singly ionized water were the molecular ion H_2O^+ and the fragment ions: OH^+, O^+, H_2^+ and H^+. Note that these ions are solely associated with the dissociation of the transient H_2O^+, and thus the missing fragments are neutrals. This is in contrast to previous measurements of mass spectra, in which a measured ion fragment can be associated with either neutrals or charged fragments. For example, we measure H^+ fragments either as single ions or as the ion pairs $H^+ + OH^+$ and $H^+ + O^+$. However, for fast proton impact (and most likely the same holds for fast electron impact) the ion-pair contribution to the single H^+ ion yield is small, and we can compare our single ion relative yields to the measured total ion production as we do in Fig. 1. At 4 MeV, for example, single H^+ ions are 95% of the total measured H^+ production. Thus, including the contribution of the ion-pair channels has only a small effect when comparing to the electron data. The same holds true for the OH^+ channel. In contrast, the O^+ singles are only 75% of the total O^+ production while the rest of the O^+ ions belong to the $H^+ + O^+$ ion pairs. It can be seen from Fig. 1 that the electron im-

---

*Corresponding author eMail: ibi@phys.ksu.edu

pact data for O^+ is higher by this factor than the proton impact data because it includes all O^+ ions while the latter is only O^+ singles. Also in Fig. 1, note that the relative yields of all single ions are constant over the velocity range studied. Furthermore, the branching ratios following single ionization of water are similar for electron and proton impact at the same collision velocity. An exception to this trend is the H_2^+ single ion which seems to be about a factor of 17 more likely for proton impact than for electron impact. It is important to note that this dissociation product was also the only channel for which an isotopic difference was observed when *Straub et al.* [5] compared H_2O and D_2O targets to each other. They reported that H_2^+ is about a factor of 2 more likely than D_2^+. The reason for these differences is not yet understood, and the magnitude of the difference between the proton impact data reported here and the electron data needs further investigation because of the relatively high level of carbohydrides in the background of our measurements. Clearly, C_mH_n molecules can also contribute to H_2^+ production, and it is not clear how much this background contributes to our measurements of H_2^+ fragments.

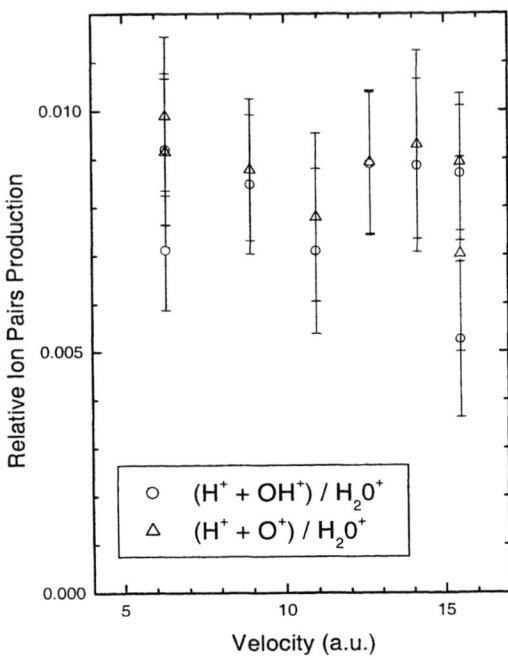

FIGURE 2. The yields of ion-pairs produced by fast proton impact relative to H_2O^+ as a function of collision velocity.

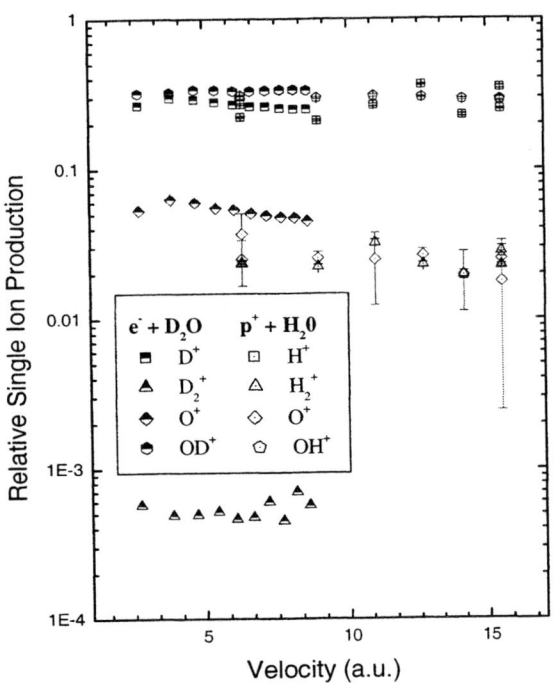

FIGURE 1. The yields of single ions produced by (1) fast proton impact - open symbols and (2) electron impact (Ref. [5]) - semi-filled symbols, relative to H_2O^+ as a function of velocity.

The O^{2+} final product is not shown in Fig. 1 because in our measurements these ions are part of the ion pair $H^+ + O^{2+}$, which is a product of triple ionization. The O^{2+} ions that appear in the singles spectrum are a result of the finite detection efficiency, i.e. cases where only one ion out of two is recorded. It is interesting to note that the O^{2+} ions seen in the electron impact ionization mass spectra have a high threshold, which is consistent with our observation that at least three electrons have to be ionized to produce this ion fragment. This channel was very small and, due to the low statistics of the measurement, it is not yet possible to determine the triple to single ionization ratio for water. In addition, when using the strong extraction fields needed to collect all ions, the typical time difference between two H^+ fragments is too small to distinguish the $H^+ + H^+$ ion-pair events from the $H^+ + H$ singles. This experimental problem limits our ability to determine the triple ionization rate. This problem might also affect the measurement of double ionization if the $H^+ + H^+ + O$ channel is important. To check if this is the case, we conducted a few measurements with low extraction field, thus allowing the detection of some of the $H^+ + H^+$ ion pairs, i.e. those dissociating along the spectrometer axis. However, this limits the fraction of ion pairs that can be detected, and we were only able to determine that this dissociation channel is smaller than the main two dissociation channels of doubly ionized water shown in Fig. 2. These two channels, $H^+ + OH^+$ and $H^+ + O^+ + H$, are equally likely within the accuracy of our measurement and also independent of the collision velocity.

The total double to single ionization ratio was evaluated from all the measured channels and is shown as a function of collision velocity in Fig. 3. This ratio is about 1% and constant over the range of measured velocities.

locity over the measured range. Finally, the measured O^{2+} fragments were solely associated with (at least) triple ionization of water, and they are not one of the dissociation products of H_2O^{2+}.

ACKNOWLEDGMENT

This work was supported by the Chemical Sciences, Geosciences and Biosciences Division, Office of Basic Energy Sciences, Office of Science, U.S. Department of Energy.

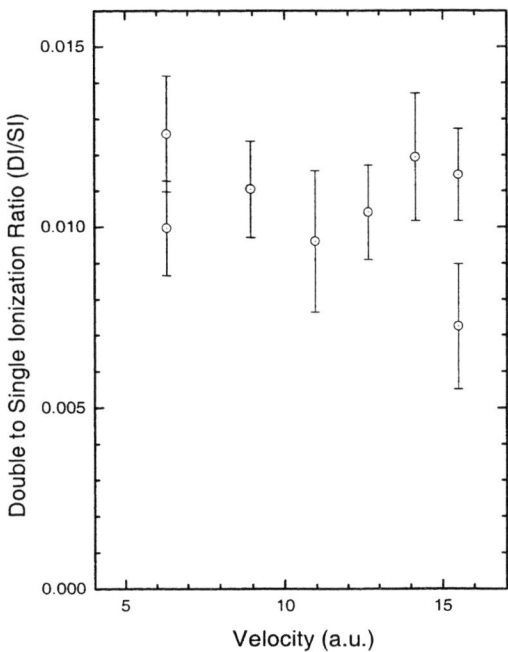

FIGURE 3. The ratio of double to single ionization of water for fast proton impact as a function of collision velocity.

To improve our understanding of the ionization and dissociation of water, further measurements are needed over a wider velocity range, with better statistics and a higher water to background ratio. In addition, we intend to image the position of each fragment and thus gain knowledge of the angular distribution and the kinetic energy release in the dissociation process.

In summary, double ionization of water by fast proton impact is about 1% relative to single ionization. The doubly charged transient water molecular ion dissociates mainly into two equally likely ion-pair channels, namely $H^+ + O^+ + H$ and $H^+ + OH^+$. The yields of single ions relative to the molecular ion H_2O^+ are similar to those measured for equal velocity electron impact except the H_2^+ channel, which is larger for proton impact by more than an order of magnitude. All the measured ratios were found to be independent of the collision ve-

[1] Märk, T.D., and Egger, F., Int. J. Mass Spectrom. Ion Phys. **20**, 89 (1976).
[2] Orient, O.J., and Srivastava, S.K. J. Phys. B **20**, 3923 (1987).
[3] Djuric, N.Lj., Cadez, I.M., and Kurepa, Int. J. Mass Spectrom. Ion Phys. **83**, R7 (1988).
[4] Rao, M.V.V.S., Iga, I., and Srivastava, S.K. J. Geophys. Res. **100**, 26 (1995).
[5] Straub, H.C., Lindsay, B.G., Smith, K.A, and Stebbings, R.F, J. Chem. Phys. **108**, 109 (1997).
[6] Ben-Itzhak, I., Ginther, S.G., Carnes, K.D., Nucl. Instrum. And Meth. B 66, 401 (1992).
[7] Ben-Itzhak, I., Ginther, S.G., Carnes, K.D., Phys. Rev. A 47, 2827 (1993).
[8] Ben-Itzhak, I., Carnes, K.D., Ginther, S.G., Johnson, D.T., Norris, P.J., and Weaver, O.L., Phys. Rev. A 47, 3748 (1993).
[9] Ben-Itzhak, I., Carnes, K.D., Ginther, S.G., Johnson, D.T., Norris, P.J., and Weaver, O.L., Nucl. Instrum. and Meth. B 79, 138 (1993).
[10] Ben-Itzhak, I., Carnes, K.D., Johnson, D.T., Norris, P.J., and Weaver, O.L., Phys. Rev. A 49, 881 (1994).
[11] Ben-Itzhak, I., Krishnamurthi, Vidhya, Carnes, K.D., Alibadi, H., Knudsen, H., Mikkelsen, U., *Nucl. Instr. and Meth.* B **99**, 104 (1995).
[12] Ben-Itzhak, I., Krishnamurthi, Vidhya, Carnes, K.D., Aliabadi, H., Knudsen, H., Mikkelsen, U., and Esry, B.D., *J. Phys.* B **29**, L21 (1996).

Transfer Ionization to Single Capture Ratio for Fast Multiply Charged Ions on He

R. Ünal, P. Richard, H. Aliabadi, H. Tawara, C.L. Cocke, I. Ben-Itzhak, M.J. Singh,[1] and A.T. Hasan[2]

J.R. Macdonald Laboratory, Kansas State University, Manhattan, KS 66506

Abstract. The charge state and energy dependences of Transfer Ionization (TI) and Single Capture (SC) processes are being investigated. The collision systems reported here are $O^{(4-8)+}$ ions interacting with Helium. The measurement is being made for beam energy 1 MeV/u using a supersonic He jet with two-stage collimation. A recoil ion momentum spectrometer is used to separate TI and SC by recording the longitudinal momentum transfer and time-of-flight of the recoil ions. A magnetic field is used to control the position of the recoil ions on the detector. The ratios of TI to SC are determined with high accuracy and are compared with previous results.[1,2] Higher Z-ions are under investigation.

INTRODUCTION

Over the last ten years several investigators have studied Single Capture, SC, and Transfer Ionization, TI, in ion-atom collisions for high velocity, highly-charged ions. Transfer Ionization is the removal of two electrons from He with one electron captured by the projectile while SC involves the removal of one electron of He by capture. Initial methods of determining TI and SC cross sections used time-of-flight of the recoil ions in coincidence with projectile ions in specific charge states. However, this method runs into difficulty when separating the contaminant processes of Single Ionization, SI, and Double Ionization, DI. It is known that the ionization of He by an impurity ion beam with one additional electron can simulate SC with a probability several orders of magnitude greater than true SC. A similar situation exists for TI, i.e., double ionization of He by an impurity ion beam with one additional electron.

EXPERIMENT

All measurements were performed in the J. R. Macdonald Laboratory at Kansas State University. Ion beams of interest were extracted from the EN tandem Van de Graaff accelerator, post-stripped when necessary, momentum analyzed, and the desired charge state directed to the collision area. The collision chamber is commissioned on the 15-degree port of a switching magnet, which allows the delivery of a beam with very little impurity. The ion beam is monitored with the beam profile monitor and the Faraday cups along its path. Monitoring the beam before and after the chamber was especially useful in determining its passage through the jet. The ion beam is well defined by two four-jaw slits (one before and one after the chamber) and three apertures where one (5 mm) is before the chamber and the other two are after the chamber (5 and 3 mm). The charge exchange beam is selected with a dipole magnet after the collision chamber.

---

[1] *Present address: Institute for Plasma Research, Bhat, Gandhinagar, India 382428*

[2] *Present address: American University of Sharjah, Physics Department, Sharjah UAE*

The target was provided using a supersonic He jet with a two-stage collimation as shown in Fig. 1.

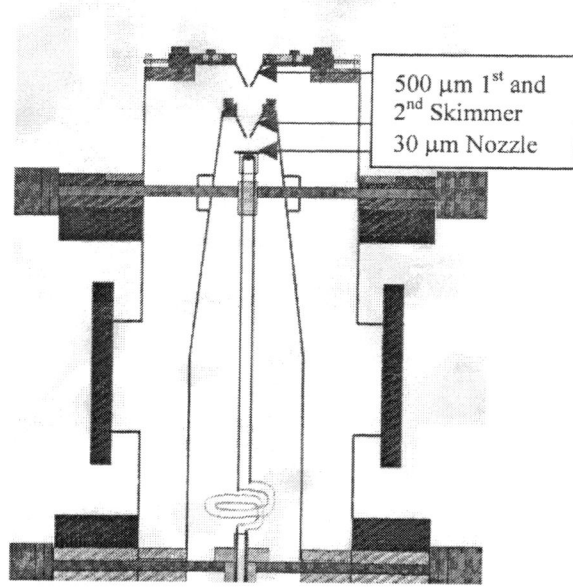

FIGURE 1. Two-stage gas jet assembly

At present the nozzle is located 6 mm from the tip of the first skimmer, and the distance between the tip of the second skimmer and the top of the first skimmer is 17 mm. The position of the first skimmer is fixed. The alignment of the aperture in the nozzle with respect to the first skimmer is done using two sets of micrometers. The position of the second skimmer can be adjusted with the set of four screws. The assembly allows one to laser-align the nozzle and the two skimmer apertures as a unit before mounting on the target chamber. The alignment of the assembly was initially tested using Nitrogen gas. It was found that the optimum flow of the gas was obtained at a gauge pressure of 15 lb/in^2. Above that pressure the properties of the jet appeared to be destroyed in that a Styrofoam ball placed above the second skimmer no longer reacted to the gas flow. The optimum pressure for He gas was found to be 28 lb/in^2. However, in vacuum the jet is found to behave completely different. It appears that even at a pressure of 55 lb/in^2 the jet behaves properly in that the rise in catcher pressure is linear with the increase in the driving pressure (gauge pressure on the gas bottle). The two-stage, geometrically cooled, supersonic He jet has significantly reduced background contribution to the spectrum compared to a single stage He jet.[4] In the case of a differentially pumped gas cell complex calculations based on assumptions for the correction due to the collisions with the contaminant beam led to corrections which were up to 50%.[3] The new setup allows one to make direct separation of contaminant processes in the experimental data using the longitudinal momentum spectra. Furthermore, this correction is much smaller (about 8.8%) yielding better over all precision.

Target recoil ions produced in collisions with the projectile were charge state analyzed using a time-of-flight spectrometer (Fig. 2).

FIGURE 2. 2-D Detector (e-side) and spectrometer.

The recoil ions were extracted by two static electric field regions, allowed to drift through a field free region, and detected by a chevron arrangement of two microchannel plates. Typical channel plate biases were 900 V across each plate. Recoil ion flight times are proportional to the square root of the ratio of the ion mass to charge. The dimensions of the spectrometer were designed to allow for time focusing of recoil ions produced at different positions in the extraction region due to the width of the beam, which was 2x2 mm at most. This condition was satisfied for the static electric field-free drift length that is twice the distance from the beam to the field free region in first order. Recoil ion flight times were 3.639 and 2.573 μs for He$^+$ and

He^{++} for an extraction voltage of 1500 V on the pusher plate and 852 V on the focusing plate. These plates set up fields of 66.4 V/cm and 232.8 V/cm.

Results and Discussion

Figure 3 shows a typical recoil-ion position spectrum of He^+ recoil ions produced by 1 MeV/u O^{5+} ions. Spectra are for recoil-ion O^{4+}-projectile coincidences. The beam direction is right to left in the figure. The left-most peak is associated with ionization. The right-most peak is the one corresponding to SC. The lower right figure indicates the longitudinal momentum delivered to the He^+ ions. Much higher resolution can be achieved by running the spectrometer at lower voltages. Figure 4 shows similar results for the He^{2+} recoil ions in coincidence with O^{4+}, for the same collision system.

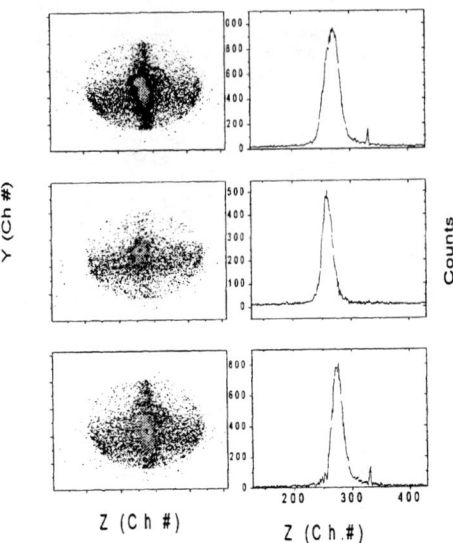

FIGURE 4. Similar spectrum as above for the case of He^{2+}.

Figure 5 shows the results measured in our COLTRIMS system for the TI/SC ratio for 1 MeV/u $O^{(4-8)+}$ + He compared with previous results taken from the literature.[1,2] The small error bars in the present work demonstrate the large improvement in the present set of measurements. The disagreement with the previous data is presumably due to beam contamination.

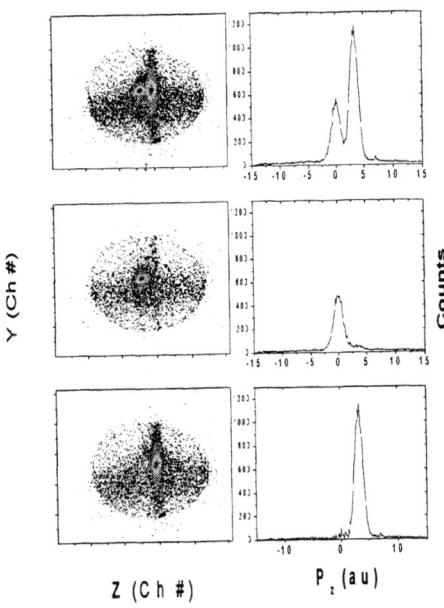

FIGURE 3. The 2-D recoil ion momentum spectrum of He^+ ions in coincidence with O^{4+} from 1 MeV/u O^{5+} + He collisions as observed on the position sensitive recoil-ion detector. In the left-most figures the ordinate is the momentum parallel to the beam (longitudinal momentum) and the abscissa is one component of the momentum perpendicular to the beam. The three spectra are for: total coincidences (top), random coincidences (middle) and the true SC coincidences (bottom). The right-most figures are the projections on to the longitudinal direction in a.u.

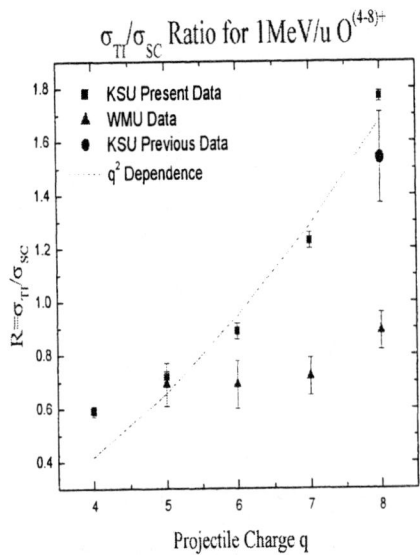

FIGURE 5. TI to SC ratio for 1 MeV/u O^{q+} on He.

CONCLUSIONS

We have measured TI and SC events for 1 MeV/u $O^{(4-8)+}$ on He collisions by COLTRIMS. Coincidences between charge changed projectiles of appropriate recoil longitudinal momentum yield accurate TI and SC ratios by eliminating contributions from SI and DI processes from impurity beam. The measured energy dependence of the ratio is compared to the previous measurements. The difference could be attributed to experimental technique, because small beam impurities can be a major source of error in this kind of measurement if not properly separated. We are planning to proceed with the measurement of TI/SC ratio in the range of q from 20 to 30 for energy range of 0.5 to 2.o MeV/u. As a next step, we will measure electron distributions associated with TI in a few selected collision systems and compare with CTMC calculations.

ACKNOWLEDGMENTS

This work was supported by the division of the Chemical Sciences, Geosciences and Biosciences, Office of Basic Energy Sciences, Office of Science, U.S. Department of Energy.

REFERENCES

1. J. L. Shinpaugh, J. M. Sanders, J. M. Hall, D. H. Lee, H. Schmidt-Böcking, T. N. Tipping, T. J. M. Zouros, and P. Richard, Phys. Rev. A **45**, 2922 (1992)

2. J. A. Tanis, M. W. Clark, R. Price, and R. E. Olson, Phys. Rev. A **36**, 1952 (1987)

3. J. L. Shinpaugh, J. M. Sanders, T. N. Tipping, D. H. Lee, T. J. M. Zouros, P. Richard, J. M. Hall, and H. Schmidt-Böcking, Nucl. Instrum. Methods Phys. Res. B **40/41**, 36 (1989)

4. E. C. Montenegro, K. L. Wong, W. Wu, P. Richard, I. Ben-Itzhak, C. L. Cocke, R. Moshammer, J. P. Giese, Y. D. Wang, and C. D. Lin, Phys. Rev. A **55**, 2009 (1997)

High Resolution X-ray Emission Spectroscopy

Yoshiaki ITO, Aurel M. VLAICU[#], and Tatsunori TOCHIO

Laboratory of Atomic & Molecular Physics, The Institute for Chemical Research, Kyoto University, Gokasho, Uji, Kyoto 611-0011 Japan
 # Division of Electronics and Applied Physics, Osaka Electro-Communication University, 18-8 Hatsu-cho, Osaka 572-8530 Japan

Abstract. A Johann-type single crystal spectrometer (2R=1500 mm) with a high resolution was equipped in BL01B1 in Spring-8. This has an excitation source over an energy range more than 3.1 keV for a core level investigation and material sciences. The experimental results with fluorescence x-ray spectra on copper and tungsten are first time presented here. The features in K and L emission spectra of copper and tungsten excited near threshold, and a clear relationship between them and spectral structures usually obtained in the laboratory, are mentioned. The monochromator use the Si double-crystal system, in which the energy resolution is within the width of Rocking curve in Bragg reflection. The spectrometer has a resolving power of $\Delta E/E = 2 \times 10^{-4}$ for Fe $K\alpha_1$ spectra. Si(100) crystal was used for the copper and Si(111) used for the tungsten, respectively.

INTRODUCTION

The third generation synchrotron gives interesting chances for material sciences. The high brightness of such sources made many difficult experiments much easier which no one could have carried out till now. A high resolution Johann-type X-ray spectrometer is just suitable for these experiments in this facility. Therefore, it was set in a bending BL01B1 at SPring-8 in October, 1999 and has been in operation for the past six months. The fluorescence end station was fabricated by a group of researchers and graduate students of Kyoto University and Osaka University. Here, we report on some initial experiments with the high energy x-ray fluorescence spectroscopy.

FLUORESCENCE X-RAY SPECTROMETER

BL01B1 consists of x-ray absorption spectrometer for XAFS (X-ray Absorption Fine Structure) which covers energy region from 4 to 90 keV, and a Johann-type single crystal monochromator observable in the emitted photon energy range more than 3.1 keV. Figure 1 shows the schematic overview of the spectrometer. There are three kinds of crystals, i.e., Si (100), Si(110), and Si(111), respectively. These crystals cover an energy range of 3.1 – 14.0 keV. The monochromator in BL01B1 uses the double-crystal

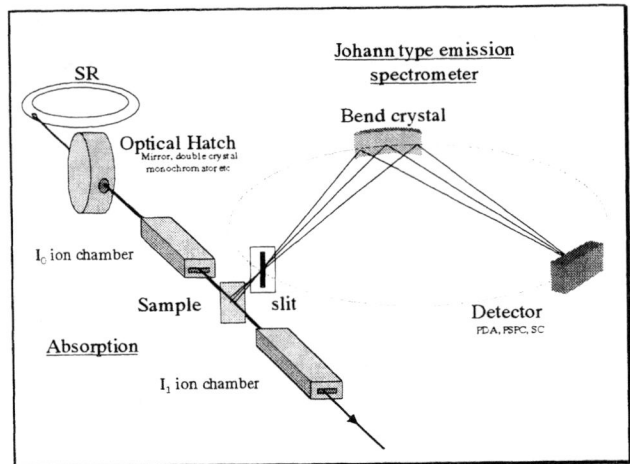

FIGURE 1. The schematic diagram of x-ray absorption and emission spectroscopy

system, in which can easily change the Bragg planes for the high energy, with the mirrors in order to reduce the harmonic components. The coherent radiation out of the monochromator is focussed onto the sample in the sample chamber of the spectrometer. The light then goes into the crystal housing in which three kinds of crystals are mounted. The radius of Rowland circle is 2R = 1500 mm. The optimal focusing condition can be met by moving the sample, crystal and detector to satisfy the Rowland geometry. For the fluorescence end station, a horizontal refocusing mirror is employed to narrow the spot size in that direction. The photon flux from the monochromator is estimated to be 10^{10} photons per second focused to a 10.0 x 1.0 mm^2 in size.

The high energy x-ray fluorescence end station is comprised of a Rowland circle single crystal spectrometer with a photon counting detector (SC, Scintillation Counter, PSPC, Position Sensitive Proportional Counter, or PDA, Photo-Diode Array) and a vacuum sample chamber. The emission spectrometer has a fixed entrance slit rigidly mounted on a flange in the sample chamber. The crystal chamber is separated from the sample chamber, where the three crystals are housed. The change of crystals is accomplished by a rotary feed through which activates the rotation of the carousel through spur-gear. Both scanning and data acquisition are automated through an IEEE-48 bus interfaced to a personal computer. The main properties of the emission spectrometer are listed in Table 1.

TABLE 1. X-ray Emission Spectrometer Characteristics

Crystal type	Johann-type
Crystals	Si (100, Si(110), Si(111)
2θ range	66 - 96
Energy range	3.1 –14.0 keV
Detector	SC, PSPC or PDA

EXPERIMENTAL RESULTS

X-ray fluorescence spectroscopy, an alternative method to photo-emission to study the electronic structure of materials, has been traditionally hampered by its weak spectral intensity. However, with the advent of intense monochromatic synchrotron radiation, such as the third generation SR, the investigation of the many interest systems has been executed. Especially, the tunability of the synchrotron radiation has an important key to the problem of understanding of the coupling of absorption and emission processes for the excitation near threshold.

We illustrate two examples of the experiments recently carried out in SPring-8 in order to show the capabilities of the fluorescence station.

Copper

It is generally believed that the multiple hole states may be considered to cause satellites in the x-ray emission spectra which result in the asymmetry of the diagram lines. Deutsch et al.[1] interpreted the $K\alpha_{1,2}$ emission spectra of copper in terms of the spectator satellites generated from the shake process. The aim of our experiment lies in elucidating the reason of the asymmetry with the attention to multi-electron processes of which the shake process is representative. In this sense, it is significant to measure the emission spectra with photon excitation energy is variable. Here we shall show three kinds of K emission spectra of copper, each of which was measured with photon excitation energy 8.982, 8.994, and 10.0 keV, respectively together with the x-ray absorption spectra as shown in Fig.2. After the Lorentzian fitting analysis of these spectra, the results were compared with those of the emission spectra of copper obtained using the double crystal x-ray spectrometer in the laboratory in Table 2. The full discussion of this work will be presented elsewhere.

FIGURE 2a. K absorption spectrum of copper.

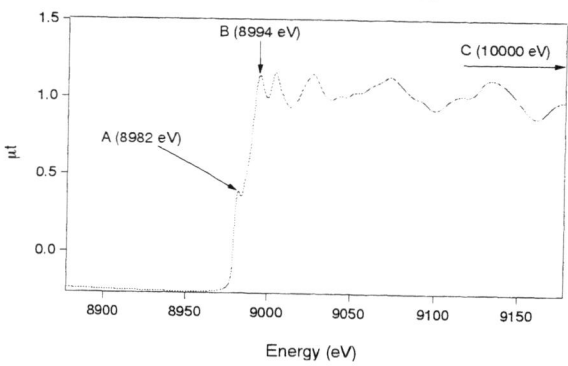

FIGURE 2b. $K\alpha_{1,2}$ emission spectra of copper at 8982 eV exciting photon energy.

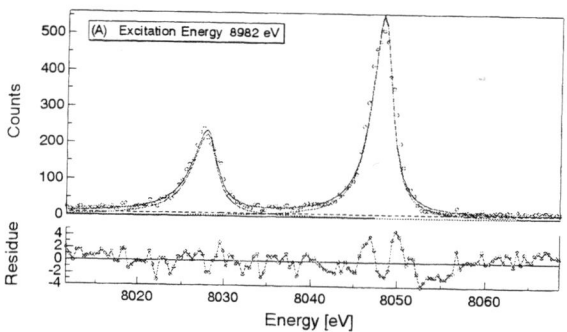

TABLE 2. FWHM and Asymmetry index of each spectrum obtained from the experiments.

Kα1			
Eex [eV]	Peak Position [eV]	FWHM [eV]	Asymmetry Index
8982	8048.3	2.72± 0.07	1.51± 0.07
8994	8047.8	3.02± 0.07	0.96± 0.05
10000	8047.9	3.74± 0.11	1.13± 0.07
Double crystal Spectrometer	8047.8	2.74± 0.01	1.17± 0.01

Kα2			
Eex [eV]	Peak Position	FWHM [eV]	Asymmetry Index
8982	8027.9	3.56±0.15	1.63±0.13
8994	8027.7	3.59±0.14	1.26±0.09
10000	8027.4	4.14±0.20	1.33±0.13
Double crystal Spectrometer	8027.6	3.23±0.01	1.22±0.01

Tungsten

It is well known that the satellites of the L shell emission lines corresponding to M spectator holes lead to lines that can be resolved well from the diagram lines, whereas those corresponding to N spectator holes almost coincide with the diagram lines. The widths of some L x-ray lines of $_{74}W$ were measured as a part of a program for compiling the L-series linewidths in heavy elements[2,3,4]. The disagreement between theory and experiment was suggested to be due to the large theoretical values of M- and N-subshells partial widths calculated by McGuire non-relativistically[5,6,7].

Among the L-shell emission lines of tungsten, the $L\beta_2$ $(L_3 - N_5)$ visible satellite is a special case which rises both a theoretical and experimental challenge in finding its origin. As we already mentioned, the wide separation of this satellite from its diagram line[8] (see Fig.3) suggests that the spectator hole involved in this transition is situated in the M-shell.

FIGURE 3. $WL\beta_{2,15}$ and visible satellite excited by electron bombardment reported by Vlaicu et al.[8]

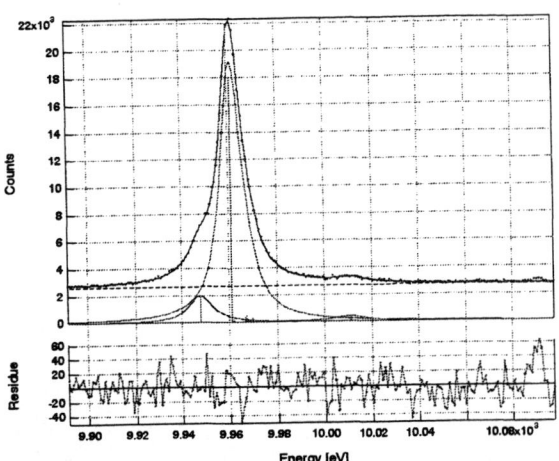

Salgueiro et al.[9] observed the satellite line in the high region of $L\beta_2$ $(L_3 - N_5)$ spectrum of tungsten, and concluded that this line was due to the $L_1 - L_3M_5$ Coster-Kronig transition. Their conclusion is based also on the experimental results of the dependence of the electron exciting energy on the relative intensity of satellite to the diagram, $Is(E)$. However, according to the work of Chen et al.[10] this transition is energetically forbidden for the atomic elements with the atomic number $50<Z<74$ and is allowed for $Z>75$.

Our recent results on this problem are the subject of the present work. The relative intensity of the satellite to the diagram line versus the exciting energy $Is(E)$ was first analyzed by using electron excitation, by taking into account also the dependence of the electron ionizing cross-section (see Fig.4).

FIGURE 4. $WL\beta_2$ relative intensity of satellite to diagram transition versus electron exciting energy reported by Vlaicu et al.[8]

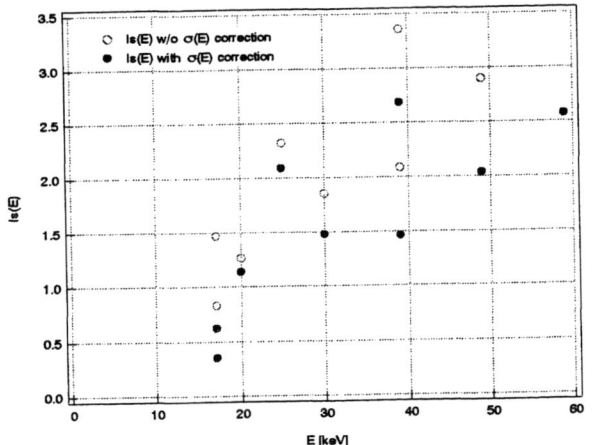

As the relative intensity of this satellite is below 2.5%, it is difficult to accurately reproduce the $Is(E)$ dependence for exciting energies below 20 keV. Therefore, this was recently investigated by synchrotron radiation from BL01B1 in Spring-8. The satellite is confirmed for the exciting energy of 12075 eV, just below the L_1 edge. This shows that the satellite exists even at an exciting energy on which the $L_1L_3M_5$ Coster-Kronig transition does not occur. Moreover, the intensity of the satellite does not increase drastically for exciting energies above the L_1 edge.

The emission lines of W $L\beta_{2,15}$, $L\beta_3$, and $L\beta_1$ where analyzed by a curved crystal spectrometer at various exciting energies situated between the L_2 and L_1 absorption edge and above the L_1 absorption edge (see Fig.5). The preliminary results show no clear threshold of the satellite intensity for exciting energies above the L_1 absorption edge, at which the supposed Coster-Kronig transitions should increase substantially the intensity of this satellite.

FIGURE 5. W L_3, L_2, L_1 absorption spectra (a) and the selected energies at where the spectra (b) of W $L\beta_{2,15}$, $L\beta_3$, and $L\beta_1$ are excited using the tunable synchrotron radiation.

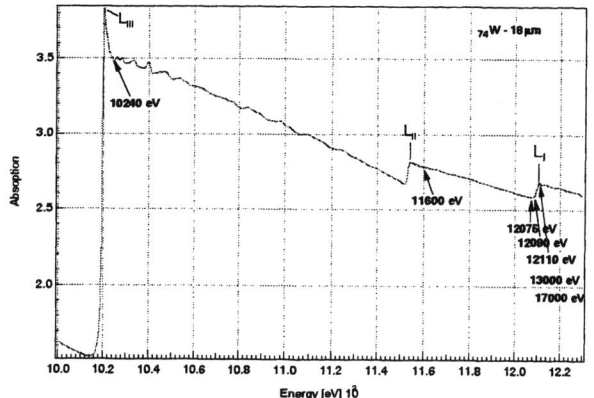 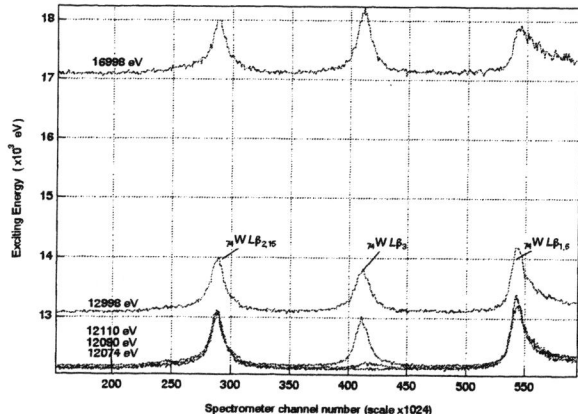

ACKNOWLEDGMENTS

The authors would like to express their thanks to Drs. S.Emura, Y.Nishihata, T.Uruga, J.Harada, T.Shoji, the graduate students and Y.Nakanishi in my laboratory for their kind cooperation to the spectrometer setup. We appreciate K.Tanno, M.Yasumoto, K.Imanishi, and T.Sasahara for their helping to make parts of the spectrometer.

REFERENCES

1. M.Deutsch etal., *Phys.Rev.***A52**,3661 (1995).
2. S.I.Salem andP.L.Lee, *Phys.Rev.***A10**, 2033 (1974).
3. S.I.Salem, S.L.Panossian, and R.A.Krause, *At.Data Nucl.Data Tables* **14**, 91(1974)
4. B.G.Gokhale, S.N.Shukla, and R.N.Srivastava, *Phys.Rev.***A28**, 858 (1983).
5. E.J.McGuire, *Phys.Rev.***A5**, 1043 (1972).
6. E.J.McGuire, *Phys.Rev.***A6**, 851 (1972).
7. E.J.McGuire, *Phys.Rev.***A10**, 13 (1974).
8. A.M.Vlaicu et at., *Phys.Rev.***A58**,3544(1998).
9. L.Salgueiro, M.L.Carvalho, and F.Parente, *J.de Phys. Colloque* **C9**, 48,609(1987).
10. M.H.Chen, B.Crasemann, K.N.Huang, M.Aoyagi, and H.Mark, *At.Data Nucl.Data Tables* **19**, 97(1977)

Intense Laser Field Studies of Positive Ions

I. D. Williams[1], P. McKenna[1], B. Srigengan[1], I. M. G. Johnston[1],
W. A. Bryan[2], J. H. Sanderson[2], A. El-Zein[2], T. R. J. Goodworth[2], W. R. Newell[2],
P. F. Taday[3] and A. J. Langley[3]

[1] *Physics Department, The Queen's University of Belfast, Belfast BT7 1NN, UK*
[2] *Department of Physics and Astronomy, University College London, London WC1E 6BT, UK*
[3] *Rutherford Appleton Laboratory, Chilton, Didcot, Oxon OX11 0XQ, UK*

There is considerable current interest in the study of atoms and molecules in very strong electric fields. The advent of short pulse high power lasers has enabled new experimental studies in this area leading to the discovery of new effects. We describe an experiment designed to extend these studies to targets of positive ions produced from conventional low energy accelerators. First results with H_2^+ molecular ions are presented, and noted differences with previous measurements using neutral H_2 molecules are discussed.

INTRODUCTION

Beams of positive ions produced from low energy accelerators are almost exclusively used as a probe of matter in the gaseous or solid state. With typical number densities seldom exceeding 10^6 cm^{-3}, the use of such beams as a target to be probed by some other beam has been severely limited. With the advent of new technology over the past two decades, the study of certain interactions involving the scattering of electrons (1) and ions (2) from ion beam targets have now been achieved. In this paper we describe a new experimental arrangement designed to use a beam of positive ions as a target for studies involving intense, ultrashort (femtosecond) pulses of laser radiation. First results from this instrument for the most fundamental molecular ion H_2^+ are also discussed.

BACKGROUND

A major development in atomic and molecular physics in the past decade has been the advances made in understanding the behaviour of matter in strong fields. Through the interaction of ultra-short pulses of intense radiation with atoms and molecules, new physical processes have been discovered. These include above threshold ionization and dissociation (3,4), Coulomb explosions (5), bond softening (6,7), and vibrational population trapping (8,9). Recently, clusters of atoms have been used as targets, with the resulting observations of coherent X-rays (10) and of high energy ions (11).

There is currently great interest in extending the study of intense ultra-short pulses of radiation to targets of ionized matter, in particular for positively charged molecular ions. Theory has predicted that molecular ions may be a far more efficient source of harmonic conversion than neutral atoms (12), and that isotope separation of HD^+ is possible in this way (13). The goal of controlling chemical dynamics via such interactions provides a strong motivating force behind these studies.

However, due to the inherent difficulties in working with pure ion beams as targets, experimental studies to date have all proceeded with targets of a precursor neutral molecule. Studies of this kind rely on the assumption that single ionization of the neutral molecule occurs as a distinct initial process at an early stage in the interaction with an intense laser pulse. Subsequently as the pulse achieves its maximum intensity, an interaction occurs with the product molecular ion, with the resulting heavy ionized fragments being extracted, analyzed and detected. Such an approach has clear drawbacks. Firstly, if we assume that ionization and fragmentation are sequential processes, the molecular ion will be produced with some range of internal energies, in particular a certain distribution of vibrational states. Most investigations seek to characterize the interaction as a function of field strength, *i.e.* of laser pulse intensity. However it is quite clear that as the pulse intensity varies, so too will the initial vibrational state distribution of the molecular ions. Secondly, it is not at all clear that a sequential model is always apt. For example, it has recently been shown that the dominant contribution to non-sequential, strong-field multiple-ionisation of Ne can be explained in terms of re-scattering of the first ionised electron (14). Clearly there is much to be gained by an experimental arrangement that permits the study of intense ultra-short laser pulses with pure beams of molecular ions.

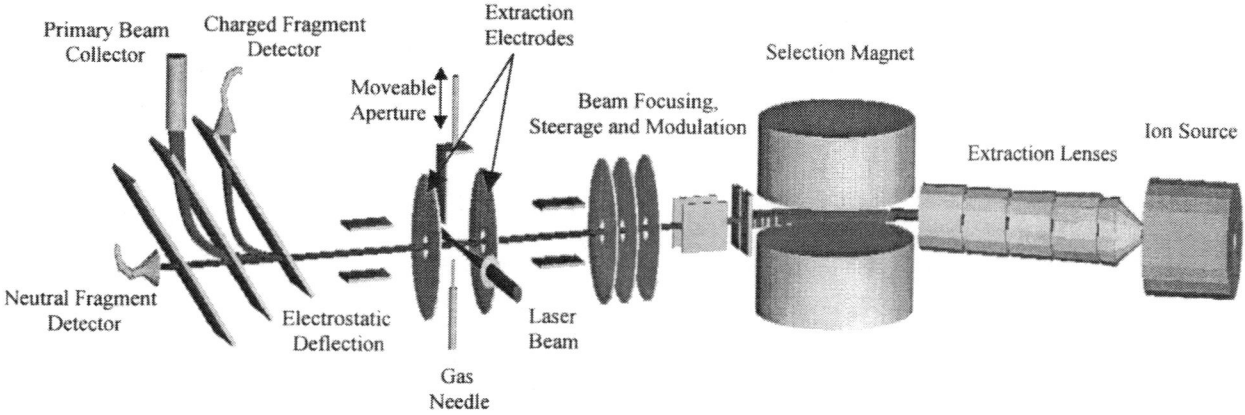

FIGURE 1. A schematic diagram of the experimental arrangement for the study of fragmentation of positive molecular ions by intense ultrashort laser pulses.

EXPERIMENTAL ARRANGEMENT

The Ti:Sapphire laser used in the investigation is housed at the Rutherford Appleton Laboratory. Operating at a fundamental wavelength of 790 nm it delivered 65 fs pulses of 30 mJ per pulse.

A schematic diagram of the ion beam arrangement is shown in Fig 1. A low energy (1 keV) beam of molecular ions is extracted from an oscillating electron ion source and momentum analyzed in a transverse magnetic field. The ion beam then passes through a differentially pumped region, prior to being focussed to a spot size of approximately 1 mm, as defined by a moveable slit and aperture assembly, at the point of interaction with the laser beam.

The laser beam enters and exits the vacuum system through high transmission ultra high vacuum glass windows. A lens of focal length 250 mm placed outside the vacuum system, focuses the 10 mm diameter laser beam to a spot size of approximately 0.02 mm over a confocal length of about 2 mm. Under these conditions the maximum intensity achievable at the interaction region is 5×10^{15} W cm^{-2}. This may be compared to the much higher intensities of approximately 10^{17} W cm^{-2} obtained with the same laser when focussed to a micron size spot by use of a shorter focal length mirror for studies with neutral molecules (15). The present study is designed to operate with a relatively large confocal volume, at the expense of pulse intensity, due to the low number densities present in the target ion beam. This also helps to minimize the so-called 'volume effect', which is due to the range of intensities present in the overlap with the target (16). This effect is also reduced by the well-defined finite size of the ion beam, by comparison to a diffusing neutral gas-jet. A systematic study of the volume effect, carried out by scanning the laser focus through the target, is discussed in detail elsewhere (17).

Following laser fragmentation of the molecular ions, the initial and product beams drift at the same velocity in a field free region before entering a 45° parallel plate electrostatic analyser. Here the primary beam and product ion beams are separated from each other by electrostatic deflection, whilst product neutral beams pass undeflected through the analyser. Fragmented ions are thus collected in an off-axis channel electron multiplier (cem) detector, whilst neutral fragments are detected in an on-axis cem. The primary ions are collected in a baffled Faraday cup. Great care is essential in this separation, since ratios of signal rates to primary beam rates of 10^{-13} - 10^{-14} must be distinguished. A high background rate due to fragmentation of the primary molecular ion beam in the background gas is unavoidable, but is minimised by operating at background gas pressures better than 10^{-9} mbar. The total rate from this background is minimised by modulating the ion beam using electrostatic deflection in the differentially pumped region. Further discrimination against the background is achieved by carrying out a time-of-flight analysis on the channel electron multiplier output.

A time-of flight analysis also permits the energy distribution of the fragmented neutral particles to be directly obtained, with due allowance being made for the laboratory to centre-of-mass kinematic transformation, along with the appropriate Jacobian correction. The product ions undergo additional energy analysis in a perpendicular electric field applied between the interaction region and the parallel plate analyser.

RESULTS

Measured energy spectra for both proton and H atom fragments from an H_2^+ molecular ion beam are shown in Fig 2, with the measurements carried out with linearly polarized laser pulses of intensity 3×10^{15} Wcm^{-2}. The intensity was calibrated by introducing Ar and Xe gas into the interaction region via a hypodermic needle, and extracting product ions into the time-of-flight spectrometer. Intensity values obtained by comparison with the threshold intensities of Augst et al. (18) agreed well with values calculated from the known laser parameters. The similarity between the normalized main peaks in the proton and neutral spectra of Figure 4 is immediately apparent. Clearly these peaks are both due to the (0,1) dissociation channel, leading to a proton and a H atom with equal kinetic energy.

resulting from bond softening (6,19-21), the first being due to a one photon absorption process, and the second due to three photon absorption with the re-emission of a single photon leading to a net two photon process. Indeed at similar intensities to the present study, the two photon process has been shown to be dominant. In the present work however, the (0,1) dissociation peak demonstrates a linear response with pulse intensity, suggesting dominance of the one photon absorption process.

The peak centred at ~ 2.0 eV in the proton spectrum, and absent in the neutral spectrum, is clearly due to the dissociative ionization, or Coulomb explosion, channel (1,1).

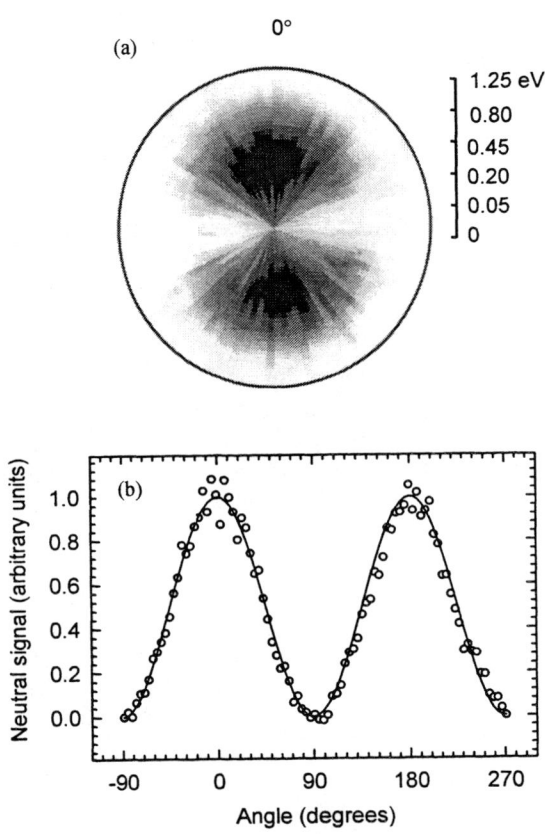

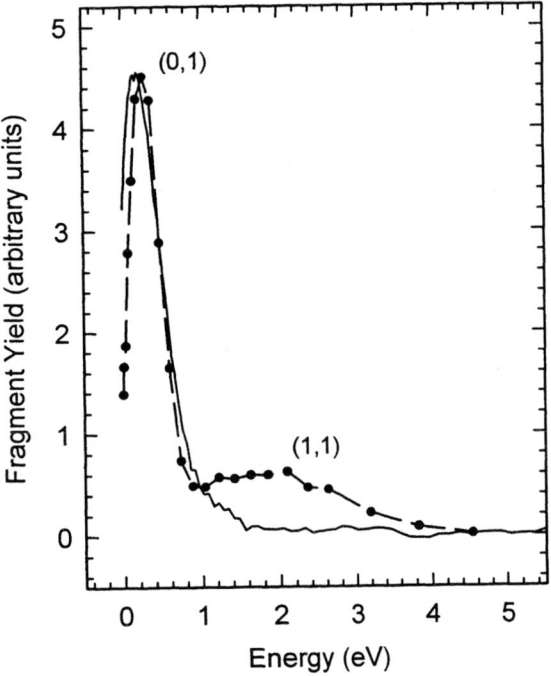

FIGURE 2. Kinetic energy spectra resulting from the fragmentation of H_2^+. H atom kinetic energy release spectrum (———); proton kinetic energy release spectrum (●) with dashed line drawn through the experimental points to guide the eye.

FIGURE 3. H atom yield as a function of angle, θ, with respect to the direction of polarization: a) 'momentum' map showing the energy and angular distribution of the H atoms; b) present measurements (○), $\cos^2 \theta$ function fitted to data (———).

The dissociation energy of 0.2 eV per particle derived from the neutral and ion results are in good mutual agreement, but differ markedly from previous results. Typical measurements in H_2 (19-21) show two proton peaks centred close to 0 eV and 0.5 eV respectively. These peaks have been explained as

In addition, by rotating the linear polarization of the laser pulse from 0°, where the polarization axis is parallel to the incident ion beam, through 360° in steps of 4°, we have measured the intensity of product H atoms as a function of angle, θ, with respect to the direction of polarization (see Fig 3). This indicates a $\cos^2 \theta$ distribution, in marked contrast to the $\cos^n \theta$ distribution ($n \geq 7$) observed for experiments with a primary H_2 neutral molecule (20). The higher the value of n the greater the degree of alignment of the molecular axis with respect to the laser polarization direction, and it has been shown to correspond to a reorientation of the molecular axis in the intense field. In the present case the value of n = 2 suggests that no reorientation of the H_2^+ molecular ion occurs in the dissociation process, the $\cos^2 \theta$ distribution merely mirroring the fall in $|\mathbf{E}|^2$ (where $\mathbf{E}$ is the laser electric field vector) in the direction of the detector. It is likely that the H_2^+ ions in the present study dissociate at an early stage on the rising edge of the laser pulse through the one photon process, allowing insufficient time for reorientation with the $\mathbf{E}$ vector to take place.

In experiments using H_2 as the primary species, the molecular ion dissociates predominantly via an initial three photon absorption process at these high laser intensities. It would thus appear that the neutral molecule itself aligns in the laser field prior to or during the initial ionization. This results in an oriented H_2^+ ion which is able to absorb three photons with high efficiency due to the alignment of the dipole moment with the field. The lack of alignment in the present study could explain the apparently low yield of dissociation products arising from initial three photon absorption.

The observed differences between the present fast-beam study of the dissociation of H_2^+ molecular ions, and previous studies with H_2 targets, provides compelling evidence that the neutral precursor plays an active role in the laser field. This is consistent with an emerging picture of the laser induced dynamics of neutral molecular species playing a significant role in the subsequent ion dynamics (21,22). Thus ion beam studies can not only provide a clearer understanding of molecular ion dynamics, but also by comparing with similar studies of the corresponding neutral species can improve understanding of the dynamics of the neutral molecule too.

OUTLOOK

Much of the understanding of fragmentation processes in intense laser fields has been due to the interpretation of new experimental measurements rather than *a priori* predictions. Experimental advances, such as the use of ion beams and detection of neutral products will further challenge the theoretical unfolding of the dissociation dynamics of both ionic and neutral species. It is thus anticipated that the use of ion beams will play a major role in the continued elucidation of this new and complex area of physics.

ACKNOWLEDGEMENTS

The financial support of EPSRC (UK) is gratefully acknowledged. PMcK and IMGJ are grateful to DENI (NI) and WAB, AEl-Z and TRJG are grateful to EPSRC (UK) for the award of postgraduate studentships.

REFERENCES

1. Williams, I. D., *Photonic, Electronic and Atomic Collisions*, Singapore: World Scientific, 1998, pp. 313-322.
2. Melchert F., *AIP Conf. Proc.* **295**, 574-584 (1993).
3. Giusti-Suzor, A., He, X., Atabek, O. and Mies, F. H., *Phys. Rev. Lett.* **64**, 515-518 (1990).
4. Zavriyev, A. *et al.*, *Phys. Rev. A* **42**, 5500-5513 (1990).
5. Codling, K., Frasinski, L. J. and Hatherly, P. A., *J. Phys. B: At. Mol. Opt. Phys.*, **22**, L321-L327 (1989).
6. Bucksbaum, P. H. *et al.*, *Phys. Rev. Lett.* **64**, 1883-1886 (1990).
7. Jolicord, G. and Atabek, O., *Phys. Rev. A* **46**, 5845-5855 (1992).
8. Giusti-Suzor, A. and Mies, F. H., *Phys. Rev. Lett.* **68**, 3869-3872 (1992); Yao, G. H. and Chu, S-I., *Chem. Phys. Lett.* **197**, 413-418 (1992).
9. Frasinski L. J. *et al.*, *Phys. Rev. Lett.* **83**, 3625-3628 (1999).
10. McPherson, A. *et al.*, *Nature* **370**, 631 (1994).
11. Hutchinson, H., *Science* **280**, 693 (1998).
12. Zuo, T. and Bandrauk, A. D., *Phys. Rev. A* **48**, 3837-3844 (1993).
13. Charron, E., Giusti-Suzor, A. and Mies, F. H., *Phys. Rev. Lett.* **75**, 2815-2818 (1995).
14. Moshammer R. *et al.*, *Phys. Rev. Lett.* **84**, 447-450 (2000).
15. Sanderson, J. H. *et al.*, *J. Phys. B: At. Mol. Opt. Phys.*, **31**, L59-L64 (1998).
16. Hansch P., Walker M. A. and VanWoerkom L. D. *Phys. Rev. A* **54** R2559-2562 (1996).
17. El-Zein A. *et al.*, *Physica Scripta* (2000) In course of publication.
18. Augst S., Strickland D., Meyerhofer D. D., Chin S. L. and Eberly J. H. *Phys. Rev. Lett.* **63**, 2212-2215 (1989).
19. Gibson G. N., Li M., Guo C. and Neira J. *Phys. Rev. Lett.* **79**, 2022-2025 (1997).
20. Walsh T. D. G., Ilkov F. A. and Chin S. L. *J. Phys. B: At. Mol. Opt. Phys.* **30**, 2167-75 (1997).
21. Thompson M. R. *et al.*, *J. Phys. B: At. Mol. Opt. Phys.* **30**, 5755-5772 (1997).
22. Larsen J. J., Sakai H., Safvan C. P., Wendt-Larsen I. and Stapelfeldt H. *J. Chem. Phys.* **111**, 7774-7781 (1999).
23. Bryan W. A. *et al.*, *J. Phys. B: At. Mol. Opt. Phys.* **33**, 745-766 (2000).

Polarization Study of the Extreme Ultraviolet (EUV) Emission From Helium Following Electron and Proton impact

H. Merabet[a], A. Siems[b], R. Bruch[a], M. Bailey[a], J. Hanni[a], H.C. Tseng[c], C.D. Lin[d], A. G. Trigueiros[b]

[a] *Department of Physics, University of Nevada Reno, Reno NV 89557 USA,*
[b] *Instituto de Física, Universidade Estadual de Campinas (Unicamp), 13083-970 Campinas, São Paulo, Brazil,*
[c] *Department of Physics, Chung Yuan Christian University, Chung Li, Taiwan 32023,*
[d] *Department of Physics, Kansas State University, Manhattan, KS 66506-2601 USA*

Abstract. A detailed investigation of excitation of He $(1s^2)$ ^1S to HeI $(1snp)$ $^1P^o$ (n=2-5) states and ionization-excitation of He $(1s^2)$ ^1S to HeII $(2p)$ $^2P^o$ and HeII $(3p)$ $^2P^o$ states following electron and proton impact on He is presented for a wide range of projectile velocities (2.2 a.u. < v < 6.9 a.u.). Specifically, new experimental data are presented on measurements of the degree of linear polarization for excitation and ionization-excitation of He following proton impact in the extreme ultraviolet (EUV) wavelength region. Furthermore, the proton experimental results are compared with theoretical polarization data using the first Born approximation and our recent atomic orbital close coupling (AOCC) calculations for the excitation process. A comprehensive comparison of experimental data for negatively and positively charged projectiles at equal impact velocities is given in order to elucidate differences in the collision mechanisms of two electron targets.

INTRODUCTION

Experimental and theoretical investigations of electron and proton impact on atoms are of considerable importance for our understanding of many-body collision dynamics. In particular, electronic processes in few-body atomic collision complexes such as helium are of great significance, both from a fundamental and an applied point of view. The helium atom is the second most abundant element in the universe and it is the simplest strongly bound two-electron system. For these reasons this atom is ideally suited for the study of collision dynamics and correlation effects (1-3) in ion-atom, few-electron processes leading to non-statistical population of magnetic substates. Such phenomena play an important role in the radiative emissions from singly charged ions such as He$^+$ which have been observed in solar spectra under solar-flare conditions (4). Hence, these studies of non-equilibrium, anisotropic, beam-like systems may provide a deeper understanding of electron and/or proton induced jets in solar flare and astrophysical investigation (5).

When a negatively or positively charged projectile collides with a helium atom at intermediate and high impact energies, the dominant collision mechanisms are excitation and simultaneous ionization-excitation of the target with subsequent detection of photons and/or ejected electrons (6). In cylindrical symmetry, the emitted radiation from line transitions between magnetic sublevels may be linearly polarized when observed at 90 degrees with respect to the quantization axis defined by the projectile beam. This is due to the non-isotropic, non-Maxwellian distribution of electron population between the sublevels (i.e., alignment). In turn, this alignment effect results in polarized line emission can be characterized by the degree of linear polarization P defined by,

$$P = \frac{I_\parallel - I_\perp}{I_\parallel + I_\perp} \quad (1)$$

where $I_\parallel$ is the intensity of radiation with electric field vectors oriented along the beam axis and $I_\perp$ is the intensity of radiation perpendicular to the quantization axis (7).

The present work represent a detailed experimental and theoretical investigation of excitation to HeI $(1snp)$ $^1P^o$ (n=2-5) states and ionization-excitation of He to HeII $(2p)$ $^2P^o$ and HeII $(3p)$ $^2P^o$ states following electron and proton impact for a wide range of projectile velocities (2.2 a.u. < v < 6.9 a.u.). Using our compact EUV polarimetry technique (8). Specifically, these states can be principally formed by the following reaction channels:

$$A + He\,(1s^2)\,^1S \rightarrow He\,(1snp)\,^1P^o + A$$
$$\hookrightarrow He\,(1s^2)\,^1S + h\nu, \quad (2)$$

$$A + He (1s^2)\,^1S \rightarrow He (np)\,^2P^o + A + e^-$$
$$\hookrightarrow He^+ (1s)\,^2S + h\nu, \quad (3)$$

where A is e^- or H^+, involving the emission of radiation with wavelengths from $\lambda=517$ to 584 Å for n=2 to 5 for HeI $(1snp)\,^1P^o$, the Lyman-α radiation for HeII $(2p)\,^2P^o$ at $\lambda=304$ Å and Lyman-β at 256 Å from HeII $(3p)\,^2P^o$, respectively.

In this study, intensive polarization measurements in the EUV range have been performed. The photon emission has been measured for proton impact on He from 121 keV to 1.19 MeV (2.2 a.u.< v <6.9 a.u.) and compared with previous electron impact polarization results at equal velocities. Moreover, a comparison with theoretical predictions, namely the first Born approximation (B1) (9) and our recent atomic orbital close-coupling (AOCC) calculations (10-11), is presented in the case of the excitation process. The optical device used for the polarization measurements is a multilayer mirror (MLM) polarimeter whose reflection characteristics have been optimized for the HeI and HeII emission in the EUV wavelength range (8).

RESULTS AND DISCUSSION

The apparatus used in this study is comprised of a 2 MV Van de Graaf accelerator or an electron gun, the EUV polarimeter, and the target chamber with the gas cell. For more details, a complete description of this experimental setup is given by Merabet and co-workers (8).

Collisional excitation/deexcitation, absorption processes, and repopulation due to secondary collisions may affect the observed degree of linear polarization. Consequently, a detailed pressure dependence of the emission from both excited and ionized-excited helium was obtained with our MLM polarimeter.

In Fig.1, we exhibit the degree of linear polarization as function of the target pressure for HeI and HeII states for electrons and protons. The polarization fraction exhibits a gradual change when the He gas pressure becomes greater than 1 mTorr for both electron and proton collisions for HeI (see Fig. 1 (a)). However, it remains constant up to 37 mTorr in the case of HeII proton impact while it also decreases after 1 mTorr for electrons (see Fig. 1(b)). Therefore, it is obvious that these two collision systems exhibit different collisional mechanisms. This is expected since protons are much heavier than electrons, thus involving smaller scattering angles during the collision processes. Therefore, a gas pressures of 30 mTorr and 1 mTorr were used for HeII measurements for proton and electron impact, respectively, whereas a gas pressure of 0.25 mTorr (electrons) and 1 mTorr (protons) was adopted for HeI case.

Before discussing our experimental results, we provide here the most relevant polarization expressions from Percival and Seaton (12) and their relations to the corresponding integral alignment parameter, A_0,

$$P(^1P^o) = \frac{\sigma(0) - \sigma(1)}{\sigma(0) + \sigma(1)} = \frac{3A_0}{A_0 - 2} \quad (4)$$

$$P(^2P^o) = \frac{3(\sigma(0) - \sigma(1))}{7\sigma(0) + 11\sigma(1)} = \frac{3A_0}{A_0 - 6} \quad (5)$$

where $\sigma(0)$ and $\sigma(1)$ are the magnetic subcross sections for the $M_L = 0$ and $M_L = \pm 1$ magnetic substates in LS coupling.

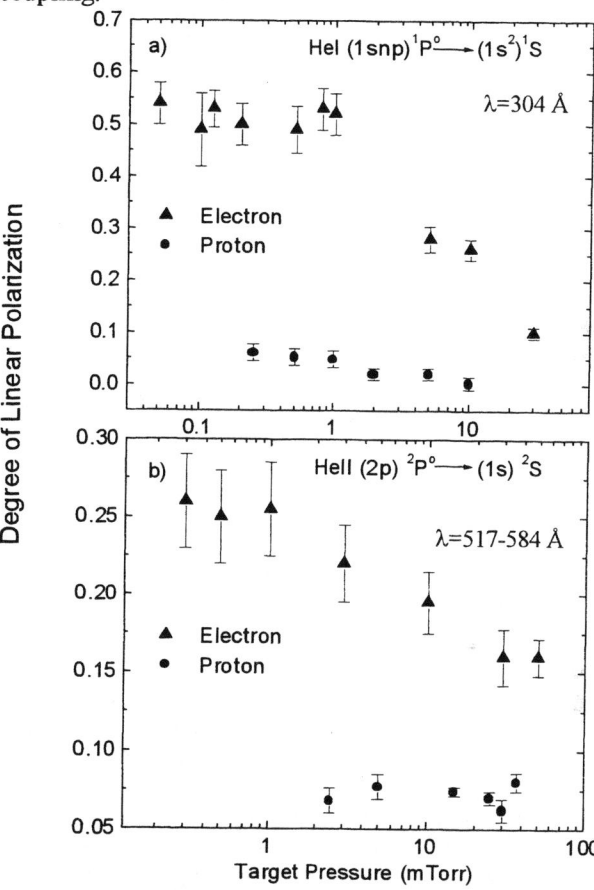

FIGURE 1: Gas target pressure dependence of HeI (a) and HeII (b) radiation following 60-eV-e^- + He and 220-keV-H^+ + He collision systems measured with the EUV-MLM polarimeter.

A. HeI $(1snp)\,^1P^o$ Polarization

Our HeI $(1snp)\,^1P^o$ polarization results in the energy range from 121 keV to 1.19 MeV proton impact are shown in Fig.2. The corresponding, previously obtained electron data are also depicted in this figure for comparison. Both sets of experimental data have not been corrected for cascade effects. It is evident from Fig. 2 that the degree of linear polarization behaves differently for negatively and positively charged particles at equal velocities. While the polarization fraction following electron impact is about 50% at a velocity of about 2 a.u., it is only 7% in the proton case. Furthermore, the two sets of experimental data exhibit a zero value of P at distinct impact velocities when comparing electron with proton projectiles at intermediate velocities. At higher velocities (7 a.u.), the electron and proton data still deviate. In Fig. 3, we have displayed the HeI proton experimental results along with B1 data and our theoretical calculations, based on the atomic-orbital close coupling (AOCC) method. As can be seen from this figure, neither the B1 nor the AOCC reproduce our experimental data.

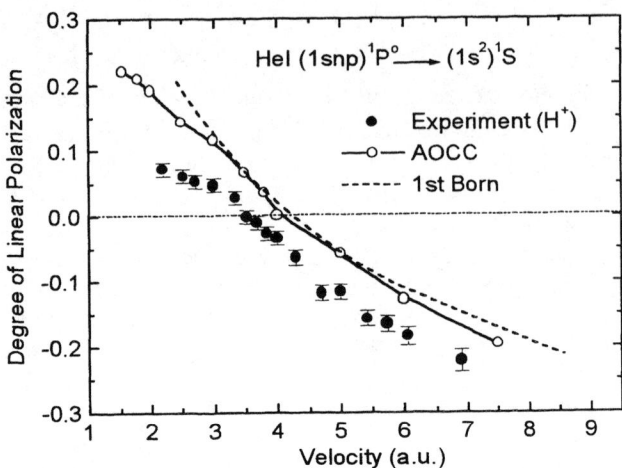

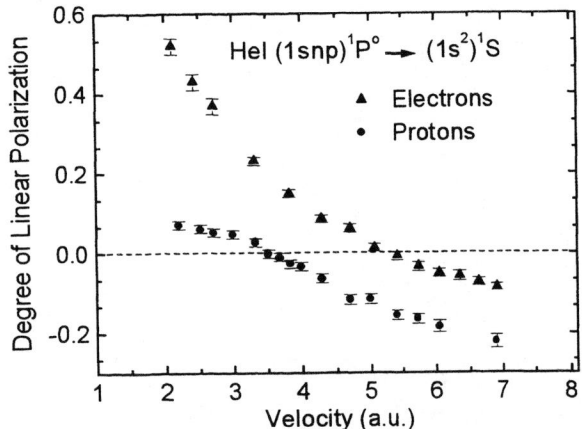

FIGURE 2. Polarization of HeI (1snp) $^1P^o \to (1s^2)$ ^1S radiation as a function of projectile velocity for electron and proton impact.

FIGURE 3. Polarization of HeI (1snp) $^1P^o \to (1s^2)$ ^1S radiation as a function of projectile velocity compared with AOCC and first Born predictions.

The $P=0$ occurs at different velocities for the AOCC method and the experiment. In addition, it is evident from Fig. 3 that the AOCC theory deviates at low energies from the first Born approximation and better approaches our experimental results. This deviation may be due to the limited excited target states used in AOCC calculations. At high impact velocities, up to 7 a.u. corresponding to 1.19 MeV proton impact, our AOCC results show a better agreement with experiment than B1 approximation. In summary, we have shown in this section that the degree of linear polarization strongly depends on the charge sign for electron and proton projectiles even for excitation of helium. These very interesting results may be used in the near future as a new diagnostic tool for solar flare studies and other plasma physics and astrophysical applications.

B. HeII (np) $^2P^o$ Polarization

The HeII (2p) $^2P^o$ and HeII (3p) $^2P^o$ polarization results, along with their integral alignment parameters A_0 are depicted in Fig. 4, for proton impact on helium. The HeII (2p) $^2P^o$ polarization results in this figure have not been corrected for cascade effects because the partial magnetic substate cross sections for the higher HeII

(nl) magnetic sub-states are not yet accurately known. However, due to the much shorter lifetimes of the excited HeII states in comparison with those of HeI, cascade effects associated with magnetic substate repopulation are expected to be smaller than for HeI (8).

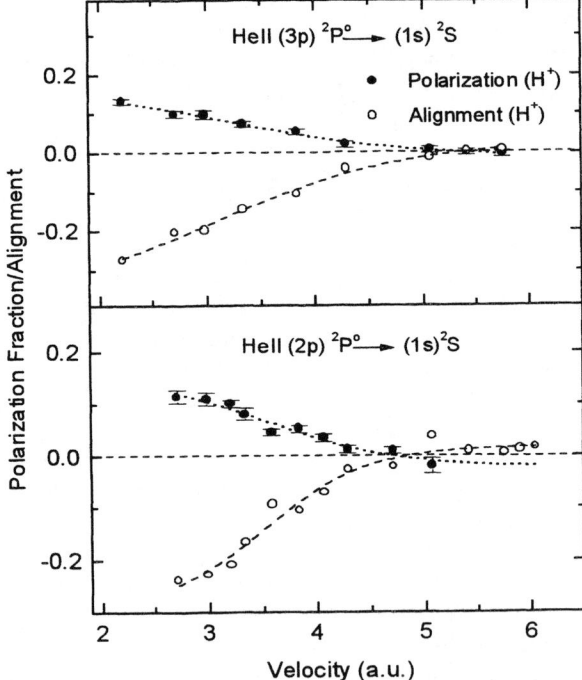

FIGURE 4: Polarization fraction P and integral alignment parameter A_0 of HeII (np) $^2P^o \to$ (1s) ^2S radiation as a function of proton impact velocity. The dashed lines are plotted to guide the eyes.

The HeII (3p) $^2P^o$ polarization results shown in Fig. 4 are slightly higher than those for the He (2p) $^2P^o$ level. Nevertheless, these two sets of data agree very well within our experimental uncertainties. Similar agreement was found in the case of electron impact on helium. This confirms the hypothesis that the degree of linear polarization is approximately independent of the principal quantum number n. In contrast, the HeII (2p) $^2P^o$ polarization data for proton and electron projectiles (see Fig. 5) indicate pronounced differences. These prominent deviations may be understood in terms of second order Born contributions and coherent superposition of first and second order scattering amplitudes. Indeed, using a semi-empirical model based on the first and second Born approximation, Bruch et al. (13) were able to decompose the He$^+$ (np) (n=2-5) cross section data, summed over M_L=-1,0 and 1, in leading order terms which are proportional to

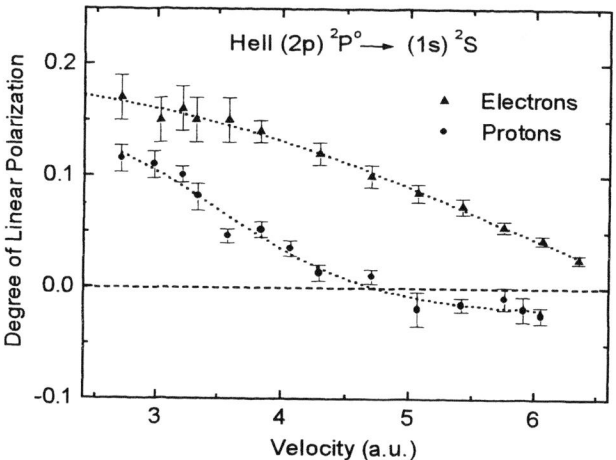

FIGURE 5. Polarization fraction of HeII (2p) $^2P^o \rightarrow$ (1s) ^2S emission as a function of electron and proton impact velocities. The dashed lines are plotted to guide the eyes.

Z_p^2, Z_p^3 and Z_p^4, where the Z_p is the projectile charge. In particular, these authors have assumed a strong Z_p^3 dependence, which may represent quantum interference effects. Such a strong Z_p^3 dependence may also affect the magnetic sub-level cross sections and, therefore, the observed degree of linear polarization following ionization plus excitation of helium by electron and proton projectiles. While both experimental polarization curves for electron and proton impact start with positive values and decrease with increasing impact velocity until reaching negative values, the proton data are substantially different from the electron impact data for all collision velocities.

In addition, the velocity dependence for HeI is completely different from the one for HeII states, i.e., the mechanisms of excitation and ionization-excitation channels exhibit completely different angular distributions of the EUV emission. To our knowledge, no accurate theoretical calculations for He Lyman-α and β polarization have been performed using sophisticated many-body approaches. The present study may serve as an important prototype test case for future, more advanced theoretical calculations including few-body dynamics such as ionization plus excitation of helium.

In summary, the HeI (1snp) $^1P^o$ (n=2-5) and HeII (np) $^2P^o$ results, presented for excitation and ionization plus excitation of helium, represent the first detailed polarization measurements of these states. We have provided evidence that both HeI and HeII EUV emissions associated with few body mechanisms lead to completely different polarization fractions for the proton and electron collision systems. These fundamentally different values could be used as a new diagnostic tool in solar flare analysis, where such positively and negatively charged projectiles are most abundant. Additional experiments are currently underway to investigate the degree of linear polarization for molecular ion projectiles (H_2^+ + He and H_3^+ + He), to elucidate these more complex processes where the role of multi-centered molecular projectile excitation and ionization-excitation will be explored.

ACKNOWLEDGMENTS

We would like to thank Dr. S. Fineschi from the Harvard-Smithsonian Center for Astrophysics for many interesting and fruitful suggestions. This project was supported in part by ACSPECT Corporation and Nevada Business and Science Foundation (NBSF), Reno Nevada.

REFERENCES

1. McGuire, J.H., *Electron Correlation Dynamics In Atomic Collisions*, (Cambridge University Press, 1997).
2. Stolterfoht, N., et al., *Phys. Rev. Lett* **57**, 74 (1986).
3. Chesnel, J., Merabet, H., Sulik, B., Fremont, F., Bedouet, C., Husson, X., Grether, M., and Stolterfoht, N., *Phys. Rev. A* **58**, 2935 (1998).
4. Kazantsev, S. A., Firstova, N.M., Bulatov, A.V., Petrashen, A.G., Henoux, J-C., *Optics and Spectroscopy* **78**, 729 (1995).
5. Kazantsev, S. A., Firstova, N.M., Petrashen, A.G., Henoux, J-C., *Optics and Spectroscopy* **82**, 714 (1997).
6. Bailey, M., Bruch, R., Rauscher, E., and Bliman, S., *J. Phys. B.* **28**, 2655 (1995).
7. Merabet, H., Bailey, M., Bruch, R., Fursa, D. V., and Bray, I., McConkey, J. W., and Hammond, P., *Phys. Rev. A* **60**, 1187 (1999).
8. Bailey, M., Merabet, H. and Bruch, R., *Applied Optics* **38**, No. 19, 4125 (1999).
9. Vriens, L. and Carrière, J.D., *Phys. Scp.* **49**, 517 (1970).
10. Fritsch, W., and Lin, C. D., *Phys. Rept.* **201**, 1 (1991).
11. Merabet, H., Bailey, M., Bruch, R., Hanni, J., Bliman, S., Fursa, D. V., Bray, I., Bartschat, K., Tseng, H.C., Lin, C.D., *Phys. Rev. A*, submitted.
12. Percival, I.C. and Seaton, M.J., *Philos. Trans. R. Soc. London, Ser. A* **251**, 113 (1958).
13. Bruch, R., Beigman, L., Rauscher, E. A., Fülling, S., McGuire, J. H., Träbert, E., and Heckmann, P. H., *J. Phys. B* **26**, L413 (1993).

Understanding the importance of dynamic, correlation, and exchange effects in new low-energy photon scattering experiments

J. P. J. Carney and R. H. Pratt

Department of Physics and Astronomy, University of Pittsburgh, Pittsburgh, Pennsylvania 15260

Abstract. We consider the current status of theory in photon scattering, and new approaches to both elastic and inelastic scattering, in the light of recent x-ray scattering measurements on neon and helium in the range 4-22 keV, obtained using the APS third-generation synchrotron. The precision of these experiments requires a detailed consideration of the effects of nonlocal exchange, electron correlations, and dynamic effects in making predictions. These effects can often be regarded as perturbations, though this assumption fails as threshold regions are approached. As well as considering procedures for making composite predictions including all significant effects, we also consider the approximations that are often employed in describing elastic (form-factor approximation) and inelastic (incoherent-scattering-factor and impulse approximations) scattering. Dynamic effects and nonlocal-exchange effects can be significant in elastic scattering. The incoherent-scattering-factor approximation works best for inelastic scattering at lower energies, and less so at higher energies. The converse occurs for impulse approximation.

INTRODUCTION

Recent x-ray scattering measurements on neon and helium in the range 4-22 keV, obtained using the APS third-generation synchrotron, have stimulated new theoretical efforts to understand the roles of electron correlation and exchange, of dynamics, and of relativity, in the dominant elastic and inelastic scattering processes (Rayleigh, Compton, and Raman scattering). Since no present calculation from first principles includes all of these effect intrinsically (corresponding to a full dynamic many-body-type calculation) one instead begins with calculations which include the dominant effects and handles additional effects perturbatively, as needed. An example would be to begin with relativistic S-matrix calculations using local exchange, and then to include perturbative estimates of the effects of using nonlocal exchange and electron correlations.

We are primarily concerned with elastic and inelastic scattering from low-Z atoms above (though not necessarily far above) threshold. Kinematic regimes can be found where any among the effects (exchange, correlations, dynamic) and processes (Rayleigh, Compton, Raman) are important. Anomalous dispersion (a dynamic effect beyond form-factor approximation) in elastic scatter-

ing is increasingly important in going to lower energies. Non-local exchange effects in elastic scattering can be significant above threshold, including in the high-momentum-transfer limit. The incoherent scattering factor appears to give a good approximation for inelastic scattering. For the Compton component corrections to impulse approximation can be significant, and the Raman component is large enough for helium that it cannot be treated as a perturbation on the Compton cross section.

First we will consider the current status of experiment and theory. We then consider possible new approaches to elastic and inelastic scattering, respectively. Finally we summarize our conclusions.

EXPERIMENT AND THEORY IN SCATTERING

Here we consider the current status and history of experiment and theory in describing scattering. In particular we mention three experiments on above-threshold scattering, and the theoretical efforts that complement them. The experiment of Chipman and Jennings in 1963 (1) was performed at the same time as new efforts in atomic structure calculations, using first Herman-Skillman and then Hartree-Fock wavefunctions (2).

Since that time there have been theoretical advances, producing relativistic variants (3, 4), correlated configuration-interaction incoherent scattering factors and form factors (5), and also relativistic S-matrix calculations for both elastic (6, 7) and inelastic scattering (8, 9). As was seen in 1998 with the experiment of Jung et al. (10), a composite theoretical treatment was needed, which encompassed knowledge of correlations, nonlocal exchange, and dynamic effects, as described in Ref. (11). Finally, new work (12) at lower momentum transfers spurred new advances, requiring a more detailed treatment. Here the detail of the treatment of the inelastic scattering component becomes an issue for low Z, and alternative approaches have been proposed, starting from the incoherent scattering factor (13, 14) or from the impulse approximation (11).

The earliest self-consistent models of the atom have given way to more sophisticated models involving nonlocal exchange, electron correlations, and full dynamic S-matrix calculations. Agreement is currently obtained by exploiting the full range of calculations, with dynamic effects estimated from S-matrix results (using local exchange within the independent particle approximation, which is usually sufficient for dynamic effects). Relativistic versions are also available. We note that the effects of correlation and nonlocal exchange are particularly well understood. The elastic case in particular is quite well understood at this point, except in the near vicinity of edges.

Conversely, the inelastic case is more complicated and has received less attention. Note that even neglecting relativity and $\boldsymbol{p} \cdot \boldsymbol{A}$ effects (which is actually reasonable in inelastic scattering, as opposed to elastic scattering, where $\boldsymbol{p} \cdot \boldsymbol{A}$ effects can be large) one does not have a simple expression for the inelastic scattering amplitude (as one does in the elastic case, namely the coherent form factor). Here instead further approximations are needed to obtain the incoherent scattering factor (describing inelastic scattering) or the Compton profile of the impulse approximation (describing Compton scattering). The A^2 expression for Raman scattering is simply a generalized form factor between (different) initial and final atomic states, but one has in principle an infinite sequence of final bound atomic states to consider.

Therefore in the inelastic case one has to consider the difference between the incoherent scattering factor result and the exact-A^2 result, when beginning from the incoherent scattering factor (assuming $\boldsymbol{p} \cdot \boldsymbol{A}$ effects are insignificant or else included perturbatively). Alternatively, starting from impulse approximation, one should consider the difference between this result and an exact-A^2 calculation for the Compton scattering contribution. Further, in obtaining the total inelastic scattering one should consider the Raman scattering contribution as well. Note that the incoherent scattering factor is an approximation to the total inelastic scattering (i.e. including both Compton and Raman scattering).

We will consider in more detail these two inelastic scattering approaches and the behavior of the corrections needed when starting with the incoherent scattering factor or with an impulse approximation result. We precede this however with a discussion of new approaches for elastic scattering, which are generally concerned with correlation effects in the anomalous scattering regime, as the above-threshold regime is already fairly well understood (until gamma-ray energies are reached).

ELASTIC SCATTERING APPROACHES

The treatment of elastic scattering is generally simpler than that of inelastic scattering. Within A^2 approximation the elastic scattering amplitude is given by the atomic form factor, which is easily evaluated for a given atomic wavefunction, and which has been tabulated using very sophisticated correlated wavefunctions. (The corresponding A^2 result for inelastic scattering involves a summation over allowed final states, and it only simplifies with further approximations.) We should mention that there are relativistic binding effects that lead to the use of the alternative modified form factor, but these are of little importance for low Z.

One of the more important above-threshold effects in elastic scattering is the nonlocal-exchange correction, particularly for low-Z elements. A substantial nonlocal-exchange effect was demonstrated in neon in the experiment of Jung et al. (10), and this effect has been theoretically investigated as a function of Z and momentum transfer for the low-Z atoms (15). In Ref. (15) it was seen that the correction can have a magnitude as large as 10% or more, and the effects of different higher-shell contributions are seen as one goes to higher Z. Although the magnitude of this effect tends to diminish overall with higher Z this happens fairly slowly, and even for high Z the high-momentum-transfer limit of this correction can still be a few percent. Correlation effects are seen to be much less important in elastic scattering in the above-threshold regime, except very close to threshold.

As one goes to lower energies, approaching the innermost thresholds and below, dynamic effects are becoming important (corresponding in neon to an $\approx$ 10% effect at 1 keV for which the K-shell binding energy ≈ 870 eV). As the anomalous dynamic contribution becomes large, so too do correlation effects in the anomalous amplitude. Consequently the composite approach described in Ref. (11) breaks down, since these effects can no longer be regarded as independent perturbations. In these situations a more sophisticated treatment is needed that properly incorporates correlation effects in a full dynamic calculation. Many-electron effects in the nonrelativistic approximation have been investigated for certain cases by Hopersky et al. (16, 17). Another approach utilized the optical theorem and dispersion relations to obtain correlation in scattering from correlation in photoeffect (18).

Another possibility, based on this second approach, is to proceed using the present relativistic S-matrix calculation, which as it stands performs a multipole expansion of the elastic scattering amplitudes, but uses only local exchange, with no account of electron correlations. However near and below threshold, where the subshell anomalous amplitude is large, is also the regime where the electric dipole anomalous amplitudes dominate the higher-multipole anomalous contributions. In photoeffect angular distributions and cross sections, quadrupole correlations have not been found important, except perhaps at threshold, even when the quadrupole amplitudes themselves are significant (19, 20), suggesting, through the optical theorem, that one only needs to include correlations in the anomalous amplitudes at the dipole level.

Therefore one would replace the (angle-independent) nonrelativistic nonretarded anomalous electric dipole amplitude components in the S-matrix calculations with better nonrelativistic nonretarded dipole amplitudes which include nonlocal exchange and electron correlations. These could be obtained utilizing dispersion relations and the optical theorem, to obtain the amplitude from photoeffect data which includes nonlocal exchange and electron correlations. Higher-multipole amplitudes from the usual S-matrix approach would be retained (as well as relativistic and retardation components of the dipole amplitudes), as is necessary to obtain the correct high-energy limit.

INELASTIC SCATTERING APPROACHES

As mentioned previously, the expression for the inelastic scattering cross section, even in the absence of relativistic and $\boldsymbol{p}\cdot\boldsymbol{A}$ effects, is not particularly simple, and one usually makes approximations to the exact A^2 result, leading to the incoherent scattering factor or the impulse approximation. One attraction of the incoherent scattering factor is that it is a function of momentum transfer alone (as is the coherent form factor). However the exact-A^2 inelastic scattering result that it approximates is not itself simply a function of momentum transfer, and hence the corrections δ_{GOS}, giving the difference between these results, encompass all the deviation from this behavior. These corrections have been investigated by Bonham (14), and are estimated from sum rules that apply to the generalized oscillator strength (GOS) that appears in the exact A^2 expression for the inelastic scattering cross section.

We focus on this exact-A^2 inelastic scattering cross section, corresponding to the nonrelativistic result for inelastic scattering in the absence of $\boldsymbol{p}\cdot\boldsymbol{A}$ effects (which we assume can be regarded as perturbations to the cross sections). We can of course separate the cross section into Compton and Raman contributions

$$\frac{d\sigma}{d\Omega}_{A^2}^{\text{inelastic}} = \frac{d\sigma}{d\Omega}_{A^2}^{\text{Compton}} + \frac{d\sigma}{d\Omega}_{A^2}^{\text{Raman}}.$$

Note that neither the exact-A^2 inelastic cross section, nor the separate Compton and Raman scattering cross sections, are strictly functions of momentum transfer only. In the incoherent scattering factor (ISF) approximation one obtains the ISF result for the total inelastic scattering by employing a closure approximation; the ISF is a function of momentum transfer K only, rather than energy and scattering angle separately. Therefore the ISF result is a strictly momentum-transfer-dependent approximation to the total inelastic scattering (Compton and Raman), to which one needs to add the corrections δ_{GOS} to get the total inelastic A^2 scattering:

$$\frac{d\sigma}{d\Omega}_{A^2}^{\text{inelastic}} = \frac{d\sigma}{d\Omega}_{\text{ISF}}^{\text{inelastic}}(K) + \frac{d\sigma}{d\Omega}_{\delta_{\text{GOS}}}^{\text{inelastic}}.$$

The corrections δ_{GOS} are manifestly not a function of momentum transfer alone, given that the exact A^2 result is not. As seen in Refs. (14, 12) these corrections are small at lower energies (still above threshold), indicating that the ISF works well. The corrections grow with energy, since in deriving the ISF one assumes that the final photon energy is equal to the incoming photon energy for the dominant contribution, which is less valid at higher energies.

Alternatively, we may consider the impulse approximation (IA), which approximates the Compton cross section, to which we need to add corrections δ_{IA} to get the exact A^2 result for the Compton cross section

$$\frac{d\sigma}{d\Omega}^{Compton}_{A^2} = \frac{d\sigma}{d\Omega}^{Compton}_{IA} + \frac{d\sigma}{d\Omega}^{Compton}_{\delta_{IA}}.$$

As seen in Refs. (11, 12), the corrections to impulse approximation δ_{IA} are small at high energies, being larger at lower energies (still above threshold), converse to the GOS corrections. One expects IA to work well at higher energies, where assumptions of free-particle kinematics are more valid. At the lower energies Raman scattering is also significant; it can be adequately described within A^2 approximation, although nonlocal-exchange effects and correlations in Raman scattering can be important (11, 12).

CONCLUSIONS

We have reviewed the status of theory and experiment in scattering, and we have seen that recent experiments (using third-generation synchrotrons) are probing the importance of the effects due to nonlocal exchange and electron correlations, and of dynamic effects (including anomalous scattering effects). These effects need to be included in making predictions. Nonlocal-exchange effects and electron correlations are well described in the literature (at least within the simpler form factor and incoherent scattering factor approximations), as are elastic anomalous scattering effects.

Less well understood and documented are (dynamic) corrections to the incoherent scattering factor and to impulse approximations in inelastic scattering, as well as the Raman scattering contributions to total inelastic scattering. Raman scattering is important at lower energy and for the lowest Z. Corrections to the incoherent scattering factor describing inelastic scattering are small at low energy. Conversely corrections to impulse approximation in Compton scattering are more important at lower energy.

ACKNOWLEDGMENTS

This work was supported in part by the National Science Foundation under Grant No. PHY-9970293.

REFERENCES

1. Chipman, D. R., and Jennings, L. D., *Phys. Rev.* **132**, (1963) 728–734.
2. Hubbell, J. H., Wm. J. Veigele, Briggs, E. A., Brown, R. T., Cromer, D. T., and Howerton, R. J., *J. Phys. Chem. Ref. Data* **4**, (1975) 471–538.
3. Hubbell, J. H., and Øverbø, I., *J. Phys. Chem. Ref. Data* **8**, (1979) 69–105.
4. Kahane, S., *At. Data Nucl. Data Tables* **68**, (1998) 323–347.
5. Wang, J., Esquivel, R. O., Smith, Jr., V. H., and Bunge, C. F., *Phys. Rev. A* **51**, (1995) 3812–3818.
6. Roy, S. C., Kissel, L., and Pratt, R. H., *Radiat. Phys. Chem.* **56**, (1999) 3–26.
7. Kane, P. P., Kissel, L., Pratt, R. H., and Roy, S. C., *Phys. Rep.* **140**, (1986) 75–159.
8. Bergstrom, Jr., P. M., and Pratt, R. H., *Radiat. Phys. Chem.* **50**, (1997) 3–29.
9. Bergstrom, Jr., P. M., Surić, T., Pisk, K., and Pratt, R. H., *Phys. Rev. A* **48**, (1993) 1134–1162.
10. Jung, M., Dunford, R. W., Gemmell, D. S., Kanter, E. P., Krässig, B., LeBrun, T. W., Southworth, S. H., Young, L., Carney, J. P. J., LaJohn, L., Pratt, R. H., and Bergstrom, Jr., P. M., *Phys. Rev. Lett.* **81**, (1998) 1596–1599.
11. Carney, J. P. J., and Pratt, R. H., *Phys. Rev. A* **62**, (2000) 012705.
12. Young, L., Dunford, R. W., Kanter, E. P., Krässig, B., Southworth, S. H., Bonham, R. A., Lykos, P., Morong, C., Timm, A., Carney, J. P. J., and Pratt, R. H., (unpublished).
13. Bonham, R. A., *Phys. Rev. A* **23**, (1981) 2950–2956.
14. Bonham, R. A., *J. Mol. Struct. (Theochem)* **527**, (2000) 103-111.
15. Carney, J. P. J., *Radiat. Phys. Chem.* **59**, (2000) 215–220.
16. Hopersky, A. N., Yavna, V. A., Novikov, S. A., and Chuvenkov, V. V., *J. Phys. B* **33**, (2000) L433–L438.
17. Hopersky, A. N., Yavna, V. A., and Popov, V. A., *J. Phys. B* **30**, (1997) 5131–5139.
18. Zhou, B., Kissel, L., and Pratt, R. H., *Phys. Rev. A* **45**, (1995) 6906–6909.
19. Amusia, M. Ya., Baltenkov, A. S., Felfli, Z., and Msezane, A. Z., *Phys. Rev. A* **59**, (1999) R2544–R2547.
20. Johnson, W. R., Derevianko, A., Cheng, K. T., Dolmatov, V. K., and Manson, S. T., *Phys. Rev. A* **59**, (1999) 3609–3613.

Consideration of the Continuum X-Ray Background in 50-300 keV Protons on Thick Targets of Ge, Ho and Au

Sam Cipolla*, Vladimir Horvat**, and C. A. Quarles***

*Creighton University, Omaha NE 68178
** Cyclotron Institute, Texas A&M University, College Station TX 77843-3366
*** Texas Christian University, Fort Worth TX 76129

Abstract. We consider the continuum bremsstrahlung spectrum from the bombardment of Ge, Ho and Au thick targets with protons from 50 to 300 keV. Two sources of the bremsstrahlung are considered. First is the bremsstrahlung from secondary electrons ejected from the target atoms in binary collisions with the projectile proton. The main contribution from secondary electrons is expected to come from collisions with inner-shell electrons that have significant momentum. The second process is polarization bremsstrahlung, the radiation from the target atom as it is dynamically polarized by the projectile charge. The data are compared with the secondary electron bremsstrahlung model developed here and with calculations of polarization bremsstrahlung.

INTRODUCTION

An extensive study of characteristic x-ray production cross sections has been carried out in the range of 50-300 keV proton bombardment on a wide variety of targets[1]. The x-ray spectra in all cases exhibit a significant continuum background. In the work reported here, we have focused our attention on a select set of targets (Ge, Ho, Au), and are trying to understand this continuum. Two sources of the x-ray continuum are being considered as the origin of the observed spectra. First is the bremsstrahlung from secondary electrons (SEB) ejected from the target atoms in binary collisions with the projectile. The main secondary electron effect is due to ejection of the nearly free valence electrons. Bremsstrahlung from these electrons has a kinematic cutoff at $4E_p m_e/M_p$. So for 300 keV protons, the cutoff is 0.65 keV, well below the observed continuum background. However, inner-shell electrons that have momentum in the direction of the projectile can be ejected with much higher energy and can contribute bremsstrahlung that extends well beyond the above quasi-free electron cutoff. In targets such as Ge, Ho, and Au, a significant contribution of secondary electron bremsstrahlung from K- and L-shell, and even M-shell electrons in the case of Au, can be expected.

A second process that may contribute is polarization bremsstrahlung. This process is radiation by the target atom electron cloud as it is polarized by the projectile charge. The kinematic endpoint for this process also extends to an energy much higher than the quasi-free electron cutoff mentioned above[2]; for example, for 300 keV protons, the endpoint is 33.5 keV for Ho and 38 keV for Au. Calculations of polarization for 1 MeV protons on Cu have been reported by Korol, Solovyov, Obolenski and Lyalin[3], and they have also provided calculations for Ge[4] for comparison with the data presented here. This calculation suggests that polarization bremsstrahlung may be an important contribution to the continuum x-ray spectrum, especially for low energy heavy projectiles.

EXPERIMENTAL SYSTEM

A 350-kV Cockcroft-Walton accelerator consisting of an RF ion source, a mass analyzer magnet, and a biased 1.5-mm beam collimator was employed to accelerate protons onto a thick target arranged on a vertical ladder oriented at 45° to the beam. Surrounding the ladder is a cylindrical copper screen that is biased with respect to the ladder to hold secondary electrons in the target. A current integrator connected to the target ladder is used to determine the number of protons hitting a target. A different spot on a target foil is used for each proton energy. Beam currents are corrected for leakage current from the target bias battery and for dead time (typically 2%) in the x-ray measurement. X-rays are measured with a Si(Li) detector equipped with an ultra-thin boron-nitride window. The detector axis is oriented at 45° to the target ladder and at 90° to the beam. To minimize dead-time effects from detection of very soft x-rays, one or two layers of 6-μm thick doubly-aluminized Mylar absorber are placed in front of the detector. The detector signals are processed for pile-up rejection and live-time correction. The detector efficiency responses were determined using the model equation described in ref. 5. Radioactive standards were used to determine the efficiencies above 1.84 keV (Si-K edge), while thick-target analysis along with ECPSSR K x-ray production cross sections were used to find the efficiency response below 1.84 keV.

CALCULATION OF SECONDARY ELECTRON BREMSSTRAHLUNG

The calculations of secondary electron bremsstrahlung (SEB) presented here are based on the following scenario. The projectile travels inside the target (positioned at 90° relative to the beam direction), gradually losing its energy with negligible angular straggling. (a) At some depth z inside the target the projectile collides with a target-atom electron. The electron emerges from the collision as a binary encounter (BE) electron with a kinetic energy E_e, traveling in the direction defined by the polar angle θ and the azimuthal angle ϕ. The positive y direction is chosen to be upward. It is assumed that the range of the BE electron is small compared with the diameter of the target and the target-to-detector distance, but not necessarily small compared with the target thickness. (b) The BE electron emits bremsstrahlung (BS) photons while inside the target. (c) The BS photons propagate through the target material on their way to the detector (positioned at 135° relative to the beam direction).

The number of detected BS photons with energy from the interval $(k, k+\Delta k)$ per beam particle $\Delta N_{det}^{BS} / N_p$ is then equal to

$$\frac{\Delta N_{det}^{BS}}{N_p} = \int dz \, P_c(z) \int dE_e \int d(\cos\theta)$$
$$\cdot \int d\phi \, P_b(z, E_e, \theta, \phi) \, P_a(z, E_e, \theta, \phi),$$

where P_a, P_b and P_c, respectively, are the (differential) probabilities for the events (a), (b), and (c) described above.

The probability P_a is given by the following expression:

$$P_a(z, E_e, \theta, \phi) = \frac{N_A \rho_2}{A_2} \frac{d^2\sigma_{BE}(z, E_e, \theta, \phi)}{dE_e d\Omega},$$

where $d^2\sigma_{BE} / (dE_e \, d\Omega)$ is the doubly differential cross section for the production of BE electrons in the collisions between projectile nuclei and target-atom electrons in the laboratory frame[6]. It was derived by transforming the corresponding expression (calculated in the impulse approximation)[7] from the center-of-mass frame. In the equation above, $d\Omega = d(\cos\theta)d\phi$ is the differential emission solid angle, while N_A is Avogadro's number, ρ_2 is the target density, and A_2 is the target-atom molar mass. The dependence of projectile velocity on depth inside the target was calculated using the method of Ziegler[8].

The number of BS photons with energy from the interval $(k, k+\Delta k)$ per BE electron (the probability P_b), was derived using the BS photon energy distribution of Storm[9], so that

$$P_b(z, E_e, \theta, \phi) = 2.76 \times 10^{-6} Z_2 [(E_e - E_e') \ln(1 + \frac{\Delta k}{k_0}) - \Delta k],$$

where $k + \Delta k \leq E_e$. The quantity $E_e'(z, E_e, \theta, \phi)$ is the energy of the BE electron eventually emerging from the back surface of the target, or zero if the BE electron is stopped inside the target. The calculations of the range of BE electrons and their energy at the surface of the target were based on the relativistic Bethe formula with the Bloch correction[6] assuming a straight-line path of the BE electron.

Finally,

$$P_c(z) = [1 - \exp(-\sqrt{2} \, \mu \, z)] \varepsilon,$$

where μ and ε are the attenuation coefficient and the detection efficiency (including the detector solid angle) for the BS photon at energy k.

RESULTS AND DISCUSSION

Results for 50 keV, 75 keV and 300 keV protons on a thick Ge target are shown in Figures 1-3. The SEB calculations are shown in the Figures as a solid line. It should be pointed out that the SEB calculations are done for a different angular arrangement than that of the experiment. The experiment looks at x-rays produced at 90° to the beam and the calculation is for 135°. However, since the assumption in the present SEB model is that the electron bremsstrahlung is isotropic, this difference is not expected to make any difference, except perhaps at low x-ray energy where differences in x-ray attenuation within the target may be more important.

The polarization bremsstrahlung calculations of Korol et al[4] are shown as dashed lines. The short dash line assumes that the proton projectile goes the full range in the target. The longer dash line is an effort to approximate the effect of proton energy loss in the

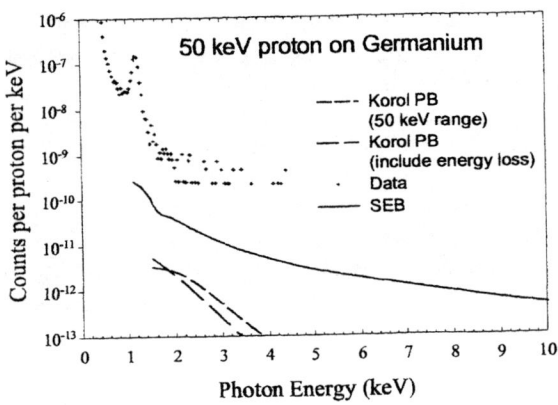

Figure 1

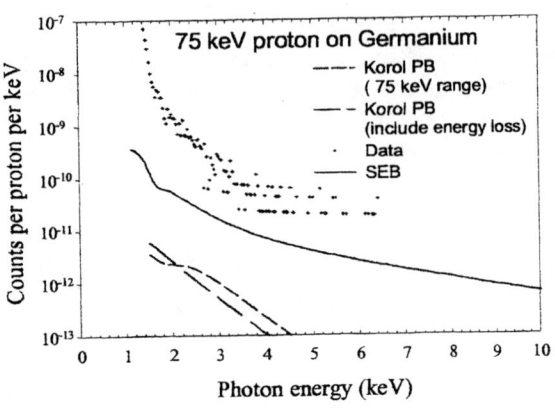

Figure 2

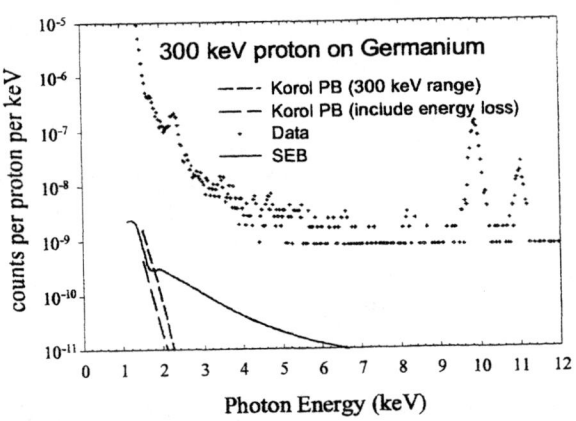

Figure 3

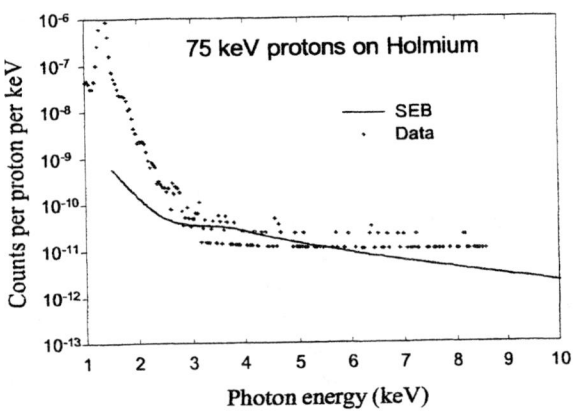

Figure 4

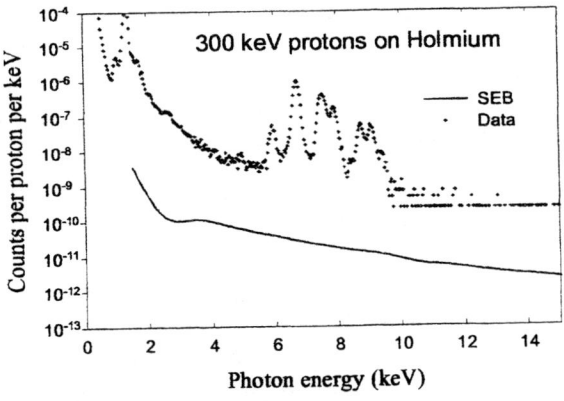

Figure 5

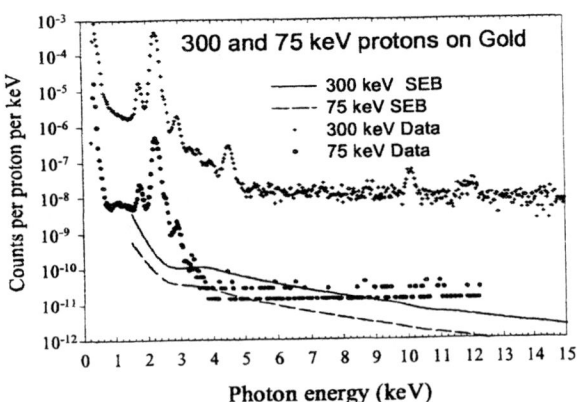

Figure 6

target. The bremsstrahlung was calculated for a coarse grid of successively lower proton energies in the target. Including energy loss reduces the contribution at photon energies below 2 keV for the 50 and 75 keV cases, but increases the contribution for photon energy below 2.5 keV at 300 keV. In Figure 1 the data follow the trend of the SEB calculation in both magnitude and energy dependence. Of course the statistical error is large (the lowest data points represent one count). The polarization bremsstrahlung alone under predicts the data and tends to fall off faster with photon energy than the trend of the data suggests. In Figure 2, the polarization bremsstrahlung is in somewhat better agreement with the data. SEB calculations tend to over predict the data. In Figure 3, at 300 keV the SEB calculation under predicts the data by about an order of magnitude. The polarization bremsstrahlung, if added in, would improve the agreement with the data.

In Figures 4 -6 data are shown for 75 and 300 keV protons on thick targets of Ho and Au. Polarization bremsstrahlung calculations for these targets are not yet available. In Figure 4, SEB significantly over predicts the data for 75 keV protons on Ho. In Figure 5, SEB for 300 keV protons on Ho is in better agreement with both the trend and magnitude of the continuum data (of course, the calculation is not concerned with the characteristic peaks). In Figure 6 for 75 keV and 300 keV protons on a thick Au target, the conclusions are similar to those for the Ho target. SEB over predicts the data at 75 keV and under predicts at 300 keV.

CONCLUSIONS

This work is a preliminary effort at both the observation of the bremsstrahlung and the development of an SEB model. We have concentrated on thick targets of higher atomic number with protons from 50 to 300 keV where there has been no previous data. Conclusions from the comparison of the SEB model with the data are mixed. The present SEB model tends to predict more continuum radiation than is observed at 75 keV and less than is observed at 300 keV. For 50 keV and 300 keV protons on Ge, the polarization bremsstrahlung seems to be an important additional contribution especially at lower photon energy.

Additional data with better statistics are needed. The SEB model can be improved by using a better approximation for the electron bremsstrahlung. New models for electron bremsstrahlung, such as that developed in ref. 10 or the Monte Carlo model discussed by Schiebl at this meeting[11], could be run for the targets and electron energy of interest to provide a numerical approximation to the SEB yield as input to the SEB model. It would also be interesting to have additional polarization bremsstrahlung calculations for higher Z targets. It appears from this preliminary effort that for low radiated photon energy, both polarization bremsstrahlung and SEB may be important, while at higher photon energy SEB is the dominant process.

ACKNOWLEDGMENTS

One of the authors (SC) wishes to thank the University of Nebraka-Lincoln for use of their accelerator in taking these data.

REFERENCES

1. S. Cipolla, Nucl. Inst. Meth. Phys. Res. A **422** (1999) 546; S. Cipolla, M. Dolezal, L. Casazza, AIP Conf. Proc. **392** (1997) 113; S. Cipolla, Nucl. Inst. Meth. Phys. Res. **B 99** (1995) 22.
2. K. Ishii, Nucl. Inst. Meth. Phys. Res. **B 99** (1995) 163.
3. A. Korol, A. Lyalin, Obolensky, and A. Solovyov, 21st ICPEAC Book of Abstracts, Sendai, Japan, p. 517 (1999).
4. Korol, A. Lyalin, Obolensky, and A. Solovyov, (private communication).
5. S. Cipolla and S. Watson, Nucl. Instr. & Meth. **B10/11**, 946 (1985).
6. V. Horvat and R. L. Watson, to be published.
7. D. H. Lee, P. Richard, T. J. M. Zouros, J. M. Sanders, J. L. Shinpaugh, and H. Hidimi, Phys. Rev. A **41**, 4816 (1990).
8. F. Ziegler (1996), *program SRIM* (private communication).
9. E. Storm, Phys. Rev. A**5** (1972) 2328.
10. M. Semaan and C. A. Quarles, X-ray Spectroscopy, to be published.
11. C. Schiebl, " Review of Sesame Program to Calculate Bremsstrahlung and Characteristic X-rays," presented at CAARI 2000.

Structures in the energy dependence of classical and quantum bremsstrahlung

A. Florescu[†‡], O. I. Obolensky[†¶], C. D. Shaffer[§], and R. H. Pratt[†]

[†] *Department of Physics and Astronomy, University of Pittsburgh, Pittsburgh, PA 15260, USA*
[‡] *Institute for Space Science, Bucharest, Romania*
[¶] *The A.F. Ioffe Physico-Technical Institute, St. Petersburgh 194021, Russia*
[§] *Department of Mathematics and Computer Science, Westminster College, New Wilmington, PA, USA*

Abstract. A study of the origin of structures in the energy dependence of electron bremsstrahlung is presented. There is a similarity between classical and quantum results. In the quantum case a connection between zeroes of dipole matrix elements and structures in the bremsstrahlung cross section and asymmetry parameter is established. In the classical case it is found that classical trapping, which results in a singular behavior of the scattering angle with impact parameter, is not directly responsible for the observed structures.

INTRODUCTION

We present here a study of the origin of structures in the energy dependence of the electron bremsstrahlung cross section $\frac{k d\sigma}{dk}$ and the asymmetry parameter a_2 for a given ratio of photon energy k and projectile electron energy T. The asymmetry parameter characterizes the angular distribution of radiation:

$$\frac{k d^2\sigma}{dk\, d\Omega_k} = \frac{1}{4\pi} \frac{k d\sigma}{dk}\left(1 + \frac{a_2}{2} P_2(\cos\theta_k)\right), \quad (1)$$

where Ω_k is the photon emission angle and P_2 is the Legendre polinomial.

We illustrate the structures for the case of an Al atom in Fig. 1 (for calculations performed in a quantum approach) and in Fig. 2 (for a classical approach). The structures were first seen in numerical calculations of angular distributions of electron bremsstrahlung in a screened atomic potential for Al within the framework of classical electrodynamics (1). Similar features were later observed in the corresponding quantum case for Eu (2). Here we show our quantum results for Al, which are qualitatively similar to the classical results. We have subsequently also seen similar structures in the energy dependence of the classical cross section (3), see Figs. 1 and 2. The correspondence between classical and quantum structures encourages the belief that they are real physical phenomena. (However modifications due to classical (4) or quantum (5) polarizational bremsstrahlung from the atomic structure have not yet been examined.)

It had initially been noted (1) that in the classical case these structures were developing as the electron en-

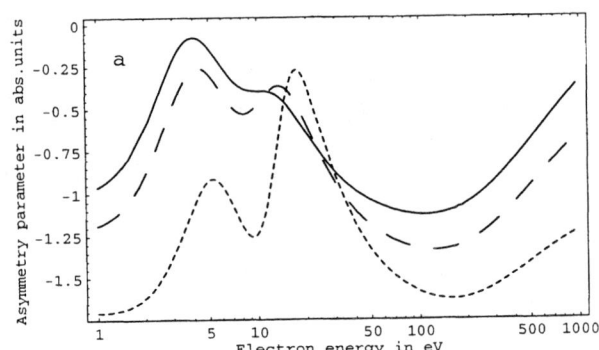

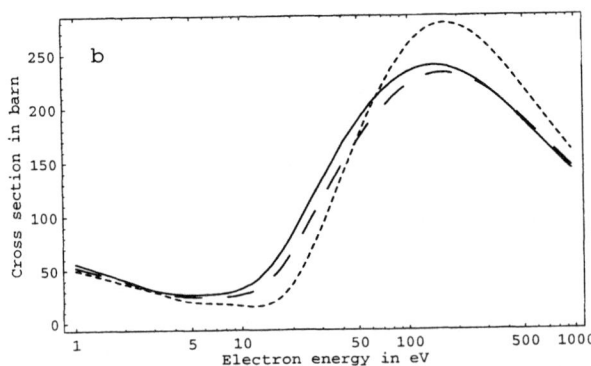

FIGURE 1. Quantum case: energy dependence of (a) the asymmetry parameter a_2 and (b) the cross section, for Al for different ratios of photon energy k and projectile electron energy T: solid line $k/T = 0.01$, dashed line $k/T = 0.2$, dotted line $k/T = 0.6$.

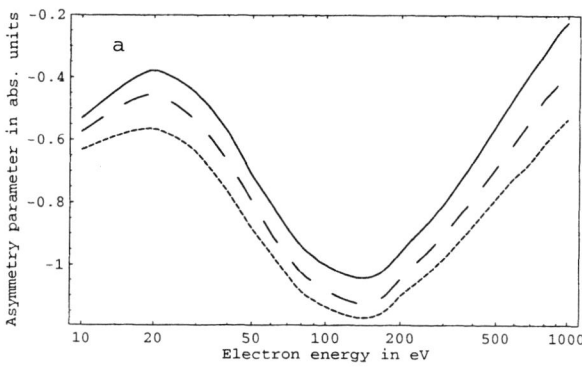

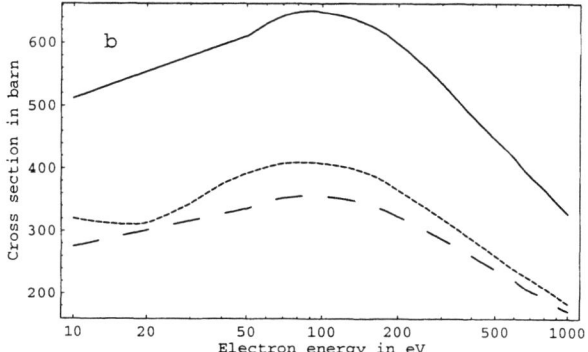

FIGURE 2. Classical case: energy dependence of (a) the asymmetry parameter a_2 and (b) the cross section, for Al for different ratios of photon energy k and projectile electron energy T: solid line $k/T = 0$, dashed line $k/T = 0.2$, dotted line $k/T = 0.4$.

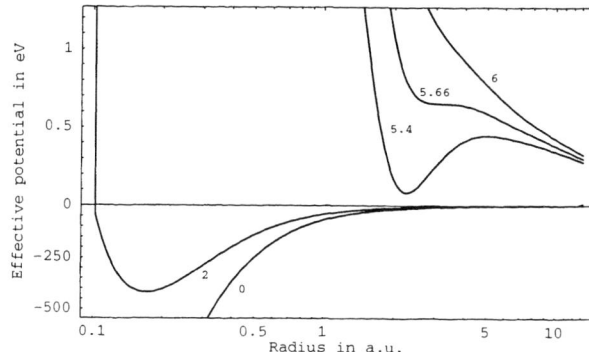

FIGURE 3. Effective potentials V_{eff} in Al for different values of Λ (see Eq. 2). The effective potential has an inflection point at $\Lambda = \Lambda_M = 5.66$ with a corresponding energy $T_M = 0.64$ eV. Note different scales for negative and positive values of the effective potential.

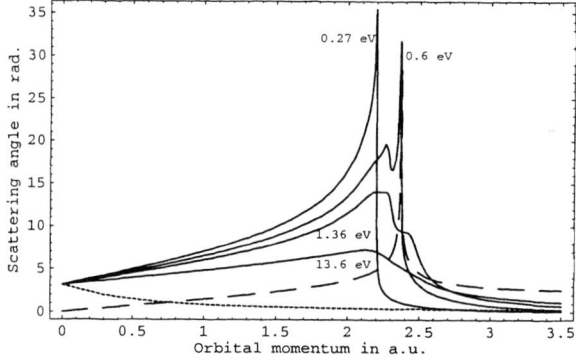

FIGURE 4. Scattering angle $\phi(T, L)$ as a function of L for several values of energy T (solid lines). Dashed line corresponds to Eq. 5, dotted line corresponds to the Coulomb point field.

ergy approached the maximum energy of classical trapping from above. Classical trapping occurs in a screened Coulomb potential when the projectile electron energy (at given angular momentum L) is equal to the height of the barrier in the effective potential

$$V_{\text{eff}}(L, r) = V(r) + \Lambda/(2mr^2), \qquad (2)$$

where $V(r)$ is the atomic potential, m is the projectile mass, $\Lambda = L(L+1)$ in the quantum case and $\Lambda = L^2$ in the classical case. (Effective potentials for various values of Λ are plotted in Fig. 3.) For such a trajectory the radial component of acceleration vanishes at the turning point of radial motion, and the incoming electron is trapped by the target. (The corresponding quantum phenomena is the presence of shape resonances in scattering.) For nearby energies and nearby angular momenta the electron will make many revolutions about the target before escaping. Thus, in a screened potential, the scattering angle $\phi(T, L)$, considered as a function of electron energy T and momentum L, has a singularity, changing fast for nearby energies and momenta, as shown in Fig. 4. At a certain value $\Lambda = \Lambda_M$ the effective potential has an in-

flection point (with a corresponding energy $T = T_M$) and then the outer barrier coalesces with the centrifugal barrier. For Al $\Lambda_M = 5.66$ and $T_M = 0.64$ eV. Trapping is not possible at energies higher than T_M. Since there is no outer barrier in the point Coulomb potential there is no trapping in the Coulomb case, and the scattering angle is a smooth function of L and T.

It had been supposed (1) that the classical structures were connected to the onset of trapping. Our present studies show that, to the contrary, the observed structures are not directly associated with the trapping or the resonances: The near-singular region of classical scattering does not contribute to the cross sections. Shape resonances do not affect the quantum asymmetry parameter (in fact resonances are not present at these energies), but the structures are rather associated with the presence of zeroes in dominant matrix elements. (Of course there are connections among these phenomena, just as the number of zeroes in a matrix element is connected with the number of bound states in a potential (6).)

In the next two sections we will describe the classical and then the quantum approaches, particularly focusing on the soft photon limit situation, simpler to describe and already illustrating the structure features.

CLASSICAL DESCRIPTION

A classical approach may be used for the description of bremsstrahlung processes at low projectile energies, both in the point Coulomb potential (7) and in screened atomic potentials (1, 3, 8). The motion of a beam of projectiles in the scattering potential is calculated in the frame of classical mechanics, and classical electrodynamics is used to obtain the radiation on electron orbits and the resulting radiation spectrum. Analytic results are available in the point Coulomb field case, for both the spectrum (9) and for the asymmetry parameter (7); both have been tabulated (7). Numerical results, as illustrated in Fig. 2, have been obtained for screened potentials (1, 3), obtaining structures (oscillations) which are not present in the Coulomb case. (In (1) the structures in the spectrum had not been observed because, as discussed in (3), the spectra were plotted with a scaling factor T/Z^2 which masked the behavior.)

We see from Fig. 2 that the main features of the structures are already present in the soft photon limit case, for which the description and procedure of calculation may be simplified. In this case the expressions for the spectrum and asymmetry parameter may be written as (we use the atomic system $e = \hbar = m = 1$)

$$\frac{k\,d\sigma}{dk} = \frac{8}{3c^3} \int_0^\infty L\,dL(1 - \cos\phi(T,L)), \qquad (3)$$

$$a_2 = \frac{\int_0^\infty L\,dL(1 - \cos\phi(T,L))(3\cos\phi(T,L) - 1)}{\int_0^\infty L\,dL(1 - \cos\phi(T,L))}. \qquad (4)$$

For energies near the maximum trapping energy T_M a singular behavior develops in L, as illustrated in Fig. 4. We have been able to obtain an expression for this behavior by making an expansion of $\phi(T,L)$ in Taylor series about the point $(T_M, L_M = \sqrt{\Lambda_M})$

$$\phi(T,L) \sim \left\{ \frac{\gamma}{2}\left[\frac{6}{\gamma}\left(\delta T - \frac{\delta L L_M}{mr_0^2}\right)\right]^{\frac{2}{3}} + \frac{2\delta L L_M}{mr_0^3} \right\}^{-\frac{1}{2}}, \qquad (5)$$

where $\delta T \equiv T - T_M$, $\delta L \equiv L - L_M$, r_0 is the inflection point in $V_{\text{eff}}(L_M, r)$, and

$$\gamma = -\left.\frac{\partial^3 V_{\text{eff}}(L,r)}{\partial r^3}\right|_{r=r_0, L=L_M}.$$

This expression well characterizes the singular behavior of the scattering angle and also the right wing of the peak, but it soon fails in the left wing (see Fig.).

One might suppose that this singular behavior is responsible for the observed structures or oscillations in the spectrum and asymmetry parameter. However, in fact, a singular behavior in $\phi(T,L)$ leads to rapid oscillations of the trigonometric functions of Eqs. 3,4, so that such a region does not contribute in the integration over impact parameters. One can see this analytically. (This result is contrary to the suggestion in (1) that oscillation in ϕ is unlikely to be averaged by the integration.)

Thus we conclude that the trapping singularity which occurs in screened potentials is not directly responsible for the observed oscillations.

QUANTUM DESCRIPTION

Our calculations indicate that the structures seen in classical bremsstrahlung are also can be found in the quantum case (Fig. 1). The similarity is only qualitative.

In a quantum calculation one obtains the spectrum and asymmetry parameter by summing contributions of dipole transition radial matrix elements of low initial and final angular momentum. The expressions for the spectrum is (2)

$$\frac{k\,d\sigma}{dk} = 16\pi^2 \frac{ck^4}{T} \sum_{l_1,l_2} l_> |<\nu_2||r||\nu_1>|^2, \qquad (6)$$

where l_1 and l_2 are the orbital momenta of the projectile electron in initial and final states, $l_> = \max(l_1, l_2)$, $<\nu_2||r||\nu_1>$ is the dipole matrix element between states $|\nu_1>$ and $<\nu_2|$. Quantum expression for the asymmetry parameter is rather complicated and we do not write it here.

Magnitudes of the terms in Eq. 6, which give the main contribution to the total cross section are shown in Fig. 5. The main features can again be understood from the soft photon case, on which we will now focus.

It is evident that within this formalism the structures in the spectrum (and similarly in the asymmetry parameter) are to be understood as resulting from zeroes in dominant radial matrix elements. In the spectrum at low energy the transition matrix elements $0 \to 1$ and $1 \to 0$ are dominant; we are denoting the matrix elements by the initial and final orbital momenta of the projectile electron. These then decrease to have zeroes, resulting in a minimum.

Note, that in the soft photon limit, $k/T \to 0$, matrix elements $0 \to 1$ and $1 \to 0$, $1 \to 2$ and $2 \to 1$ and so on are equal, while for larger k/T the behaviors of the pairs of matrix elements are more spread out, resulting in less pronounced structures. Comparing results for different k/T,

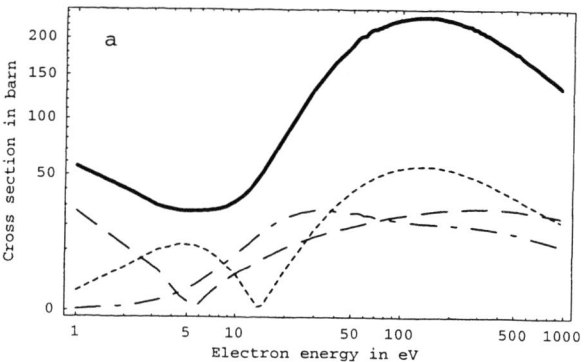

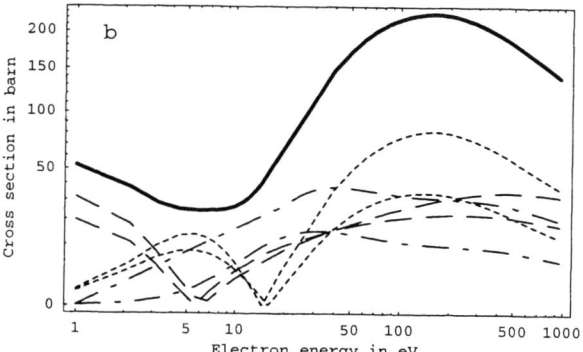

FIGURE 5. Partial contributions of dominant matrix elements to the total bremsstrahlung cross section (thick solid line) for the two values of ratio k/T: 0.01 (a) and 0.2 (b). Dashed lines correspond to $0 \to 1$ and $1 \to 0$ transitions, dotted lines correspond to $1 \to 2$ and $2 \to 1$ transitions and dashed-dotted lines correspond to $2 \to 3$ and $3 \to 2$ transitions. Note that for the soft photon limit $k/T \to 0$ matrix elements $\ell \to \ell'$ and $\ell' \to \ell$ are equal. For finite value of k/T the matrix elements are spread out; we do not label them separately here. Note also that positions of matrix elements zeroes depends on photon energy k.

one can also see that the positions of zeroes in matrix elements vary with the energies of the initial and final states; the trajectories of the positions of matrix element zeroes have been discussed in (6). Note also that from this viewpoint we would not expect further structures at still lower energies, unless the transition $0 \leftrightarrow 1$ had further zeroes – which it should not, based on our other work (6). This had not been clear in the numerical classical calculations, which failed for energies too near the maximum trapping energy T_M.

We see that the classical and quantum structures are qualitatively similar. Perhaps this reflects a connection between classical trapping and quantum bound states and resonances. It is known that there is a connection between numbers of bound states and numbers of matrix element zeroes (6). Evidently further study is needed.

ACKNOWLEDGMENTS

This work is supported by NSF grant 9970293 and by the Russian Foundation for Basic Research under the grant 99-02-18294-a.

REFERENCES

1. Kim, L., and Pratt, R. H., *Phys. Rev. A* **36**, 45-58 (1987).
2. Korol, A. V., Lyalin, A. G., and Solovyov, A. V., *J. Phys. B: At.Mol.Opt.Phys.* **28**, 4947-4962 (1995).
3. Shaffer, C. D., *Issues in Electron-Atom Bremsstrahlung*, Ph.D. thesis, University of Pittsburgh, Pittsburgh (1996).
4. Kogan, V. I., and Kukushkin, A. B., *Sov. Phys. — JETP* **60**, 665-675 (1984).
5. *Polarization Radiation*, edited by V. N. Tsitovich and I. M. Oiringel, Plenum, New York, 1993
6. Shaffer, C. D., Pratt, R. H., and Oh, S. D., *Phys. Rev. A* **56**, 3653-3658 (1997).
7. Florescu, V., and Costescu, A., *Rev. Roum. Phys.* **23**, 131 (1978).
8. Astapenko, V. A., Bureeva, L. A., and Lisitsa, V. S., *JETP* **90**, 434-446 (2000).
9. Kramers, H. A., *Phil. Mag.* **46**, 836-871 (1923).

Polarizational Bremsstrahlung on Atoms and Ions: Relativistic and Non-Relativistic Cases

A.V. Korol[1,5], A.G. Lyalin[2], O.I. Obolensky[3,4], I.A. Solovjev[5], A.V. Solov'yov[1,4]

[1]*Institut für Theoretische Physik der Universität Frankfurt am Main, Germany 60054*
[2]*Insitute of Physics, St Petersburg State University, St Petersburg, Russia 198904*
[3]*Department of Physics and Astronomy, University of Pittsburgh, Pittsburgh, USA PA 15260*
[4]*A.F.Ioffe Physical-Technical Institute, St Petersburg, Russia 194021*
[5]*Russian Maritime Technical University, St Petersburg, Russia 198262*

Abstract. Formalism for the full relativistic treatment of the spectral and angular distributions of the polarizational bremsstrahlung formed in collision of a charged particle with an atom is developed. A novel feature of the polarizational bremsstrahlung emitted in a fast collision of heavy charged projectile with an atom is described. The peculiarity in the cross section of the process originates from the kinematics of atomic electrons in the polarizational bremsstrahlung process and is similar to that occurs in inelastic scattering, where it is known as the Bethe ridge.

The polarizational bremsstrahlung (BrS) (for references see e.g. [1,2]) is one of the known sources of continuous spectrum radiation in collisions of charged particles with atoms.

The polarizational BrS arises as a result of the polarization of a target atom induced by the electric field of a projectile. In this process the projectile particle virtually excites (or ionizes) the target atom from its ground state. After the collision the atom remains in the ground state, but its virtual excitation leads to a photon emission.

This work is mainly focused on the study of spectral and angular distributions of the polarizational BrS as well as its dependence on the projectile velocity in the region of frequencies comparable with the ionization potentials of inner atomic shells. In this frequency range the polarizational BrS often dominates over other radiative mechanisms such as the ordinary BrS (see e.g. [3,4]), and radiative ionization (RI) or inelastic bremsstrahlung [5,6,7].

The ordinary BrS is produced by a charged projectile accelerated in the static electric field of a target. In collisions of electrons or positrons with atoms the intensities of the polarizational and the ordinary BrS are comparable in a wide range of photon frequencies [8,9]. In collisions of heavy projectiles with atoms the intensity of the ordinary BrS is suppressed by factor M_p^2 (M_p is a mass of the projectile) as compared to the intensity of the polarizational BrS [1].

The process of the radiative ionization is of importance in collisions of charged particles with atoms. It was demonstrated for fast non-relativistic and relativistic collisions [6,7] that atomic electrons radiate coherently in the polarizational BrS process and incoherently during the RI. This leads to the dominance of the polarizational BrS over the RI in the wide range of photon frequencies: $I < \omega < \mathrm{v} \cdot R_0^{-1}$.

Here I and R_0 are the ionization potential and the atomic radius, v is the collision velocity, which is $\mathrm{v} \gg 1$. Here we use the atomic system of units $\hbar = m_e = |e| = 1$. For fast heavy charged projectiles, beyond $I < \omega < \mathrm{v} \cdot R_0^{-1}$, the RI dominates over the polarizational BrS in the total photon emission spectrum [6,7]. An exception of this rule occurs in the region of sufficiently high photon frequencies. Indeed, the RI process has a threshold, which is equal to

$v^2/2$. However, the polarizational BrS of heavy projectiles on free atoms takes place at higher frequencies, up to $\omega \approx 2v^2$ and even above that, dominating in this region in the total photon emission spectrum. Therefore, the photon frequency range $v^2/2 \leq \omega \leq 2v^2$ is convenient for the observation of the polarizational BrS.

In this work both non-relativistic and relativistic situations are considered. Particular attention has been paid to the accounting for the relativism of inner-shell atomic electrons in the polarizational BrS process. In the frequency range considered the wavelength of the emitted photon becomes comparable with the radius of the electron inner-shell orbits. Therefore the dipole approximation is not sufficient for the correct description of the spectrum and higher multipole contributions to the photon emission cross section should be taken into account. To describe the scattering process we use relativistic distorted partial wave approximation in the field of the target atom with a `frozen' core. The opposite influence of the projectile on the atom which leads to the virtual atomic polarization and, thus, to the polarizational BrS, is treated as a perturbation.

The multipole expansion of the polarizational BrS cross section have been performed in both relativistic and non-relativistic cases, which generalize the results of earlier works where the polarizational BrS was treated in the dipole approximation. The double differential cross section of the polarizational BrS and the expression for the polarizational BrS spectrum have been expressed via the polarizabilities, which characterize the dynamic response of the target on the longitudinal, electric and magnetic L-pole components of the external field of projectile.

On the basis of these expansions a new algorithm is introduced for the calculation of the polarizational BrS cross section in relativistic case. It allows us to use for numerical calculations of the polarizational BrS cross sections the atomic relativistic wave functions obtained in various approximations. Here we point to the relativistic hydrogen-like approximation and Hartree-Fock-Dirac approximations as the most natural ones. In the latter case it is possible to take into account many-electron correlations, when calculating the atomic dynamical polarizabilities entering the problem. To describe the projectile–atom scattering in the process of polarizational BrS we use the Born approximation (for both non-relativistic and relativistic collisions), which is sufficient for the description of the polarizational BrS of heavy projectiles.

For the numerical calculation of the dynamic atomic polarizabilities the relativistic one-electron Green function method was used. Regular and irregular components of the wave functions from which the radial Green function is constructed were obtained by solving the Dirac equation with the atomic 'frozen'-core potential. Such an approach allows to reduce the summation over infinite number of the intermediate atomic states for solution of the differential equations. This method for the numerical calculation of the polarizational BrS cross section is particularly efficient for carrying out the slowly converging partial-wave series, which characterize the cross section of the process.

The spectral and angular distributions calculated on the basis of the non-relativistic and relativistic algorithms are compared with each other and with available experimental data [10]. Our calculations show that the mechanism of polarizational BrS plays the dominant role in a wide frequency range in heavy-particle–atom collisions both in relativistic and non-relativistic cases. In relativistic electron (positron) – atom collisions the polarizational and ordinary BrS can be well separated in the angular distribution.

A novel feature of the polarization BrS process predicted recently [11] was investigated in the region of frequencies comparable with the ionization potentials of inner atomic shells. It manifests itself as a powerful peak in the velocity dependence of the cross section arising under certain kinematical conditions. The physical nature of this peculiarity is similar to the one known from the theory of electron inelastic scattering on atoms as the Bethe ridge.

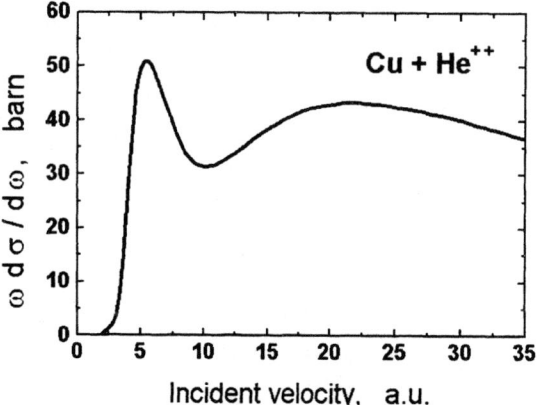

FIGURE 1. The dependence of the total BrS cross section on the projectile velocity V for the process $He^{++} + Cu \rightarrow He^{++} + Cu + \hbar\omega$ at $\hbar\omega = 30$ a.u. (ω is the photon frequency).

Figure 1 shows that a narrow extra maximum arises in the v-dependence of the BrS spectrum near $v=5$.

For projectile $e^{\pm}$ the v-dependence of the polarizational part of the BrS cross section has only a single maximum, origination of which has a clear explanation. At low collision velocities the induced dipole moment of the target alters slowly and thus the probability to emit high frequency photon becomes small. At high v, the interaction time of the projectile and the target decreases, which leads to the decrease of the polarizational BrS cross section. Therefore there is a value of v which provides the largest yield of radiation. Such a maximum manifests itself in heavy-particle – atom collisions as well. In figure 1 it is located near $v \approx 23$ a.u. The additional maximum, located at $v \approx 5$ a.u., is due to the process similar to the one responsible for the Bethe ridge in inelastic scattering of charged particles. In the Bethe type of a process the transferred momentum q is absorbed by a single electron outgoing from the target. In the polarization BrS process this electron becomes virtually ionized and radiates a photon being captured back by the target. Due to kinematic restrictions such a process is possible in collision involving heavy projectiles and forbidden in $e^{\pm}$- atom radiative collisions.

Let us conclude. In the present work we have developed the fully relativistic theory of the elastic bremsstrahlung process of charged particles on atoms accounting for both the ordinary and the polarizational BrS mechanisms and their interference. The details of these calculations, which are rather lengthy and cumbersome one can find in [12]. Our treatment generalizes all known approaches to the problem and can be considered as a unified theory for various limiting cases. Our formulae describe the angular and spectral distributions of photons emitted in various collision regimes ranging from the non-relativistic to ultra-relativistic, for non-relativistic (light) and relativistic (heavy) target atoms or ions.

Such calculations performed for Al, Cu, Ni, Ag, Xe targets will be an important step forward towards precise comparison theoretical results with the recent experimental data on the bremsstrahlung measured in collisions of 10–100 keV electrons with various solid, thin film and gaseous targets (see [13]). Experimental data show some indications on the significance of the polarizational BrS contributions [13] in the energy ranges in which the relativistic treatment of the process is necessary. Calculations for relativistic heavy-ion collisions are also of interest because of recent experimental efforts in this direction [16].

Our calculations also show that the Bethe peculiarity manifests itself in the velocity dependence of the polarizational BrS cross section. The peculiarity can be observed for frequencies lying below the ionization thresholds of the inner atomic shells. We have established relationship elucidating the connection between the photon frequency and the location of the peculiarity in the velocity dependence of the polarizational BrS cross section (see also [10]).

With slight modifications the formalism developed in our work can be applied for other colliding systems, where relativistic effects are important.

Let us note that if a relativistic projectile has some internal structure it is not necessary to develop the new formalism describing the polarizational BrS. Instead, the corresponding formulae can be obtained by using the Doppler and the aberration of light transformation similar with the method, which was developed in [17] for relativistic collisions of the non-relativistic light atoms.

For example, one can describe the bremsstrahlung arising in relativistic collisions involving nuclei. In this case the dynamic polarization of the colliders results in the photon emission via the polarizational BrS mechanism, and the main contribution comes out from the non-dipole radiation (quadrupole and higher).

We have not discussed the triply differential bremsstrahlung cross section in this paper. However let us note that it can be also elaborated using similar formalism and be a subject for the separate consideration. Also mention that the appropriate widths can be added to the resonant terms of the polarizabilities and the corresponding polarizational BrS cross section terms, which will incorporate the resonant X-ray photon emission in our calculation scheme.

Interesting problem, which deserves separate consideration, concerns the analysis of the asymptotic behaviour of the fully relativistic bremsstrahlung cross section in the region of high photon energies comparable with the rest mass of an electron. In this case the non-relativistic treatment of the atomic response is not adequate and one has to consider more sophisticated approaches when treating the the excitations of the atomic electrons into the negative energy continuum.

We left the processes of inelastic BrS beyond the scope of this work. The development of the fully relativistic formalism for the treatment these processes is the natural and straightforward continuation of the present research.

ACKNOWLEDGMENTS

We are grateful to C.A. Quarles for sending us preliminary experimental data [11]. Communication with him has stimulated our research. This work is partially supported by the Russian Foundation for Basic Research (grant 99-02-18294-a).

REFERENCES

1. Amusia, M.Ya., *etal.*, *Polarizational Bremsstrahlung of Particles and Atoms*, ed. Tsitovich, V.N., and Oiringel, I.M., Plenum, NY, 1993.

2. Korol, A.V., and Solov'yov, A.V., *J. Phys. B: At. Mol. Opt. Phys* **30,** 1105 (1997).

3. Akhiezer, A.I., Berestetskii, V.B., *Quantum Electrodynamics*, Nauka, Moscow, 1969.

4. Pratt, R.H., Tseng, H.K., Lee, C.M., and Kissel, L., *At. Data Nucl. Data Tables* **20**, 175 (1977).

5. Ozawa, K., Chang, J.H., Yamamoto, Y., Morita, S., Ishii, K., *Phys.Rev. A* **33**, 3018 (1986).

6. Amusia, M.Ya., Kuchiev, M.Yu., and Solov'yov, A.V., *Sov.Tech.Phys.Lett.* **11**, 577 (1985).

7. Amusia, M.Ya., Korol, A.V., Kuchiev, M.Yu., and Solov'yov, A.V., *Sov.Tech.Phys.Lett.* **12**, 290 (1986).

8. Korol, A.V., Lyalin, A.G., and Solov'yov, A.V., *J. Phys. B: At. Mol. Opt. Phys.* **30,** L115 (1997).

9. Korol, A.V., Lyalin, A.G., Obolensky, O.I., and Solov'yov, A.V., *JETP* **87,** 251 (1998).

10. Quarles, C.A., private communication (1999).

11. Korol, A.V., Lyalin, A.G., Obolensky, O.I., and Solov'yov, A.V., *J. Phys. B: At. Mol. Opt. Phys.* **33,** L179-186 (2000).

12. Korol, A.V., Obolensky, O.I., Solov'yov, A.V., and Solovjev, I.A., *J. Phys. B: At. Mol. Opt. Phys.*, to be submitted (2000).

13. Ambrose, V., Quarles, C.A., and Ambrose, R., *Nucl. Instr. Meth. B* **124**, 457 (1997).

14. Portillo, S., Quarles, C.A., McDaniel, F.D., Duggan, J.L., and El Bouanani, M., *21st International conference on the physics of electronic and atomic collisions*, Tu090, Sendai Japan, 1999, 574.

15. Quarles, C.A., and Portillo, S., *Applications of Accelerators in Research and Industry*, edited by J.L.Duggan and I.L.Morgan, AIP Press, 1999, 174-177.

16. Ludziejewski, *etal.*, *J. Phys. B: At. Mol. Opt. Phys* **31**, 2601 (1998).

17. Amusia, M.Ya., Kuchiev, M.Yu., and Solov'yov, A.V., *Sov. Phys. – JETP* **67,** 41 (1988).

Ionization channel in non-perturbative ion-atom collisions dominated by charge exchange

L. Kocbach and I. Ladadwa

Department of Physics, University of Bergen, Allégaten 55, N-5007 Bergen, Norway

Abstract. The development of the described method has been motivated by studies of the proton-atomic oxygen collision system, where the accidental closeness of energy levels makes the charge exchange quasi-resonant. In slow collisions the ionization channel must compete with the resonant exchange channel. The present theoretical approach is to add target continuum states, found by solving numerically the self-consistent field Schrödinger equation, to the set of basis states on target and projectile, as used in close coupling treatment of excitation and exchange. We have shown that the chosen theoretical approach is feasible and report here first model studies using the new computer code. It is found that the large population of the ionization channel at the lowest collision velocities found in a previous study is not seen in the present approach.

INTRODUCTION

One of the main motives for this work is the proton-atomic oxygen collision system, where the accidental closeness of energy levels makes the charge exchange quasi-resonant. The motivation to study this system, as well as our previous treatment of this system are described in ref. [1]. The results of the latter reference for ionization showed some unexpected features and the calculations have later been performed with a somewhat larger set of basis states, without a conclusive result [2]. In slow collisions the ionization channel must compete with the strongest resonant exchange channel. The situation is thus in many respects similar to the proton-hydrogen case, but it has different aspects, the most important is that a many electron system is studied and thus further approximations must be done to make the description of the many-electron problem feasible.

The present theoretical approach is to add to the set of basis states on target and projectile, as used in close coupling treatment of of excitation and exchange[1], a set of target continuum states, found by solving numerically Schrödinger equation with a potential obtained from the self-consistent field calculation.

Ionization in nonperturbative collisions has been studied in many works, some of them quite early. In some studies of Greiner's Group [3], [4] the continuum was taken into account by expansion over a set of real continuum states, while in the papers of Reading's Group (first in ref.[5]), as well as e.g. in L-shell ionization studies of Shingal *et al* [6], various forms of pseudostates are used. More recent use of pseudostates in typical two-centre collisions is very broad, a review can be found e.g. in [7]. The present theoretical treatment is thus an alternative or extension of previous methods, represented e.g. by the references [3], [5].

The studies of nonperturbative atomic collisions have become a large field where many innovative approaches have been developed. A very interesting set of ideas, named basis generator method (BGM) has been explored recently by Lüdde and collaborators [8]. Other approaches are based on purely numerical methods, e.g. direct numerical solution of Schrödinger equation on a mesh of spatially distributed points, e.g. in [9] and [10]. A special place has the approach of Sidky and Lin [11], [12], where the mesh is in the momentum space.

Here we explore a combination of two more traditional approaches, based on expansions in sets of certain eigenstates. We use a well established and tested coupled channels code, which uses atomic eigenstates (and pseudostates).

THE COUPLED CHANNEL CODES

The code which is the starting point of the present work is PSgatcc of J.P. Hansen and A. Dubois. This code is a recent version of a long series of codes starting from atcc (atomic coupled channels) originally written in 1988-89, later generalized to gatcc and more recently in the middle of 1990's fully included general pseudostates (PSgatcc). The one-electron wavefunction is expanded in a set of atomic or model eigenstates. These eigenstates are expressed as a linear combination of Slater orbitals on both target and projectile. Both direct and exchange matrix elements, are evaluated using Shakeshaft's method [13]. The basic routines of the original code have been published in [14]. The system of codes contains also a utility to plot the effective model molecular orbitals, which is useful for the discussion of collision dynamics.

Since the set of Slater type orbitals is larger than the set of atomic states, the positive energy solutions of the static problem can be used as pseudostates representing continuum, i.e. the ionization channel. A large number of studies has been performed using these codes, most recently e.g. [15]. and the paper [1] mentioned as one of the motives for this work.

CONTINUUM EXTENDED COUPLED CHANNEL CODE

The effective potential used to construct the bound states on the target for the program PSgatcc is used to obtain radial functions of the continuum states, by solving the radial Schrödinger equation. The evaluation of matrix elements is performed using the method described e.g. in [16]. For bound-bound coupling of states on one centre such matrix elements are already included, and they could in principle be evaluated in the same way. In fact, some of the earlier versions of gatcc used a set of routines similar to those included now. In the present PSgatcc the one centre bound-bound matrix elements are calculated in a different way, suitable for the fact that the expansion Slater orbitals are not eigenstates of the hamiltonian.

In the present version we do not include continuum-continuum coupling and we do not allow the bound states to change their 'instantaneous energies', i.e. we effectively work with the hamiltonian shown in Fig. 1. These

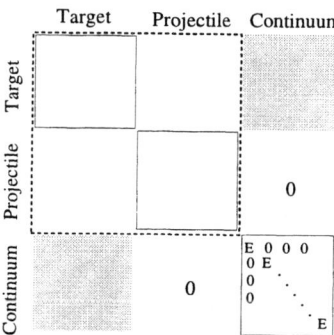

FIGURE 1. Schematic representation of the total hamiltonian. The part in dashed frame is evaluated by the original code, shaded area and the diagonals are new parts.

limitations might seem rather severe in light of other studies, e.g. [12] or [10], but for our purposes they in fact simulate important physical features (see the discussion). The coupling included is a sort of extension of perturbative ionization calculations. Perturbative ionization calculations have proven to be efficient even in many situations where the ionization channel dominates.

The inclusion of the continuum expansion in the coupled channel codes needs a method to discretize the continuum. This is implemented by normalizing the selected continuum states as explained in the next section.

NORMALIZATION OF THE CONTINUUM STATES

This procedure is discussed in a monograph of Bransden and McDowell [17], and applied e.g. in the paper of Schiwietz [18]. When we have a set of both continuum and discrete states, the expansion has a form

$$\Psi(\xi,t) = \sum_i c_i(t)\varphi_i(\xi) + \int_0^\infty b_\varepsilon(t) R_\varepsilon(\xi)\, d\varepsilon \quad (1)$$

where R_ε is a continuum state with energy ε and where ξ denotes the relevant electron coordinates. We need to consider only the second part of eq. (1), where our procedure should replace the integral by a summation. In [17] the discretization is introduced via so called eigendifferentials (there is a misprint obscuring the presentation in [17], therefore we discuss this in detail). For regular meshes of values ε_i the expression

$$\sum_i \left\{ \sqrt{\Delta\varepsilon}\, b_{\varepsilon_i}(t) \right\} \left\{ \frac{1}{\sqrt{\Delta\varepsilon}} \int_{\varepsilon_i - \frac{1}{2}\Delta\varepsilon}^{\varepsilon_i + \frac{1}{2}\Delta\varepsilon} R_\varepsilon(\xi)\, d\varepsilon \right\} \quad (2)$$

approaches the discussed integral as $\Delta\varepsilon$ approaches zero. Note that it is the expansion coefficient which is taken outside of the integrals. This expression is already in a form of summation

$$\sum_i \beta_i(t) \Phi_i(\xi) \quad (3)$$

where the functions $\Phi_i(\xi)$ are defined by the second bracket of eq. (2) and they are called eigendifferentials. The expansion coefficients β_i are related to the original b_{ε_i} in the first bracket of eq. (2).

It is easy to show that the eigendifferentials $\Phi_i(\xi)$ form an orthonormal set for any mesh of ε_i, when the continuum states are taken normalized on the energy scale

$$\int R_{\varepsilon'}(\xi) R_\varepsilon(\xi)\, d\xi = \delta(\varepsilon' - \varepsilon) \quad (4)$$

The eq. (2) is basis for the discretization, instead of a mesh with infinitesimal spacing a real mesh is chosen.

The fact that the set of $\Phi_i(\xi)$ is orthonormal for any energy mesh does not at all guarantee that it also is complete. We know, however, that as the mesh gets finer and finer, the expansion would converge to the original eq. (1). The coupled equations can thus be solved in a normal way when we use the fully discrete expansion implied by eq. (3). One needs to evaluate the matrix elements between all the states $\Phi_i(\xi)$ and $\varphi_i(\xi)$ of the coupling terms of the hamiltonian.

The total probability of the electron being in continuum is given by

$$\sum_i |\beta_i(t)|^2 = \sum_i |b_{\varepsilon_i}(t)|^2 \Delta\varepsilon \to \int_0^\infty |b_\varepsilon(t)|^2 d\varepsilon \quad (5)$$

and if differential quantities are needed, they can be based on the above relation.

We consider a very simple approximation, i.e. that the variation of wavefunction with energy in the second bracket of eq. (2) can be neglected. In this approximation, the eigendifferentials $\Phi_i(\xi)$ become

$$\Phi_i(\xi) \to \sqrt{\Delta\varepsilon} R_{\varepsilon_i}(\xi)$$

i.e. the energy-normalized continuum states with modified normalization states are used for the evaluation of the matrix elements. This method is adopted in the presented work.

The same method has also been used in the paper of Soff *et al* [3], who introduced it simply by the relation

$$\int_0^\infty d\varepsilon \to \sum_i (E_{i+1} - E_i)$$

The continuum states R_ε must be normalized on the energy scale, eq. (4). This is done according to Bethe's prescription, section 1.4 of [19].

COMPUTATIONAL ASPECTS

We use the possibility of the parallel execution by a small C-language code which performs in parallel multiple runs of the plain FORTRAN code without parallel modifications. This makes it possible to perform the same calculations also on any system with sufficient power but without parallelism. A large quantity of input and output data files are administered using standard Unix tools. The calculations are performed at the NOTUR site Parallab's Cray Origin 2000 for fast evaluation. However, it is possible to run the same code for example on any personal computer with Linux operating system, or any type of Unix workstation with standard FORTRAN compiler. The set of calculations for Fig. 3 would then take about one to two weeks of full time.

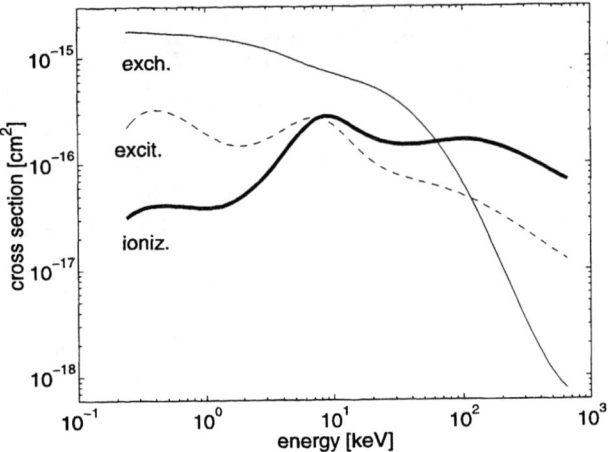

FIGURE 2. Comparison of cross sections for the three main channels in atomic O - H^+ collisions. Data from ref. [1].

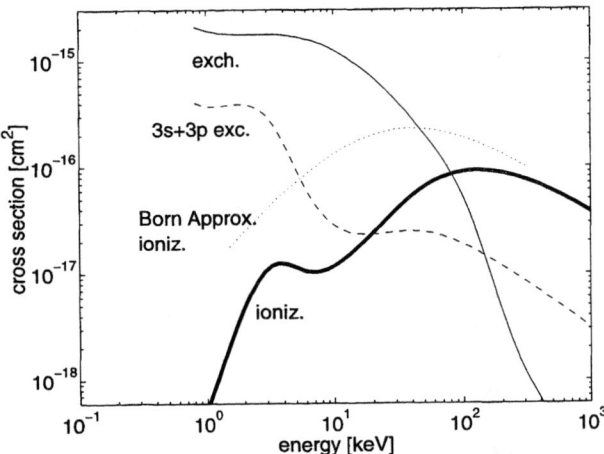

FIGURE 3. Comparison of present model cross sections for the three main channels in atomic O - H^+ collisions and Born Approximation for ionization.

The code is written so that it can be viewed by a html-browser with pointers to subroutines and places of interest. We hope to make it available at the address `http://www.fi.uib.no/AMOS/gatcc` which now contains additional material on this work.

DISCUSSION OF THE APPLICATION

It is well known (e.g. fig. 2.1 of [7]) that in proton-hydrogen collisions the exchange channel dominates and the ionization channel is much weaker than the excitation channel at low velocities. In the previous calculations on proton - atomic oxygen collisions [1] the situation was different, the ionization channel at low collision energies follows the cross sections for the excitation channel. In these calculations the ionization channel has been represented by the positive energy eigenstates (pseudostates), as in many other current works.

Fig. 2 prepared from results in [1] shows the competition of the three channels in atomic O - H^+ collisions. Fig. 3 shows the results of the present model calculations, which only add the probabilities of electron removal from the 2p-states without proper statistical treatment.

The two figures cannot be directly compared, since the present calculations are not modified by the statistical factors discussed in detail in [1], and also the size of the atomic basis has been reduced. Nevertheless, it is found that large population of the ionization channel at the lowest collision velocities found in the previous study is not seen in the present approach.

The reason for a spurious population of the ionization channel might be a sort of diving of the positive energy states, i.e. due to the necessarily limited basis some of the positive energy states get negative energies for small internuclear distances. This could be seen using the MO-tool mentioned in the description of the codes. If the basis were larger, other weekly bound states would keep the continuum states higher up. In our approach this is simulated by neglecting altogether the off-diagonal terms for the positive energy states.

CONCLUSION

We have modified a pseudostate-based coupled channel code to include a set of real continuum states obtained by solving electron Schrödinger equation with a realistic potential of the target atom. With this new code, which seems to be the first using such a combined approach, we have performed a set of model calculations for recently studied system proton on atomic oxygen. These calculations indicate that the previously obtained large cross sections for ionization are most probably spurious. The code has been shown functional and the results show a good agreement with perturbative calculations at high collision velocities. One of the above references [12] opens by a quote "to understand hydrogen is to understand all of physics" from ref. [20]. In accordance with this statement, we plan to study in the future the limitations of this model by applying it the proton-hydrogen case. Naturally, more extensive calculations on other suitable systems will also be performed in the near future.

ACKNOWLEDGMENTS

We would like to thank Prof. J.P. Hansen for numerous discussions and unpublished information on the PSgatcc code and for his interest in this project, and to Prof. J.M. Hansteen for comments on the manuscript. The research has been supported by NFR through a grant of computer time at Norwegian supercomputing facilities.

REFERENCES

1. Hamre, B., Hansen, J.P., and Kocbach, L., *J.Phys.B: At.Mol.Opt.Phys.* **32**, L127-L131 (1999).
2. Hansen, J.P., Kocbach, L., and Ladadwa, I., *XXI. ICPEAC, Abstracts of contributed papers*, ed. Y. Itikawa et al, Sendai, Japan, 1999, Volume II., p. 478
3. Soff, G., Reinhardt, J., Müller, B., and Greiner, W., *Z. Physik A* **294**, 137-147 (1980).
4. Mehler, G., Greiner, W., and Soff, G. *J.Phys.B: At.Mol.Opt.Phys.* **32**, 2787-2801 (1987).
5. Ford, A. L., Fitchard, E., and Reading, J.F., *Phys. Rev. A* **16**, 133-143 (1977).
6. Shingal, R., Malhi, N. B., and Gray, T. J., *J. Phys. B: At. Mol. Opt. Phys.* **25**, 2055-2063 (1992).
7. Fritsch, W., and Lin, C. D., *Phys. Reports* **202**, 1-97 (1991).
8. Kroneisen, O.J., Lüdde, H.J., Kirchner, T., and Dreizler, R.M., *J. Phys. A: Math.Gen.* **32**, 21411-2156 (1999).
9. Grün, N., Mühlhans, A., and Scheid, W., *J. Phys.B: At.Mol.Opt.Phys.* **15**, 4043-4061 (1982).
10. Wells, J. C., Schultz, D. R., Gavras, P., and Pindzola, M. S., *Phys. Rev. A* **54**, 593-604 (1996).
11. Sidky, E. Y., and Lin, C. D., *J. Phys. B: At. Mol. Opt. Phys.* **31**, 2949-2960 (1998).
12. Sidky, E. Y., Illescas, C., and Lin, C. D., *Phys. Rev. Lett.* **85**, 1634-1637 (2000).
13. Shakeshaft, R., *J. Phys. B: Atom. Molec. Phys.* **8**, 134-136 (1975).
14. Hansen, J.P., *Computer Phys. Comm.* **58**, 217-221 (1990); Hansen, J.P. and Dubois, A., *ibid.* **67**, 456-464 (1992).
15. Lundsgaard, M. F. V., Nielsen, S. E., Rudolph, H., and Hansen, J.P., *J.Phys. B: At. Mol. Opt. Phys.* **31**, 3215-3232 (1998).
16. Kocbach, L., *Z. Physik* **A 279**, 233-236 (1976).
17. Bransden B. H., and McDowell, M. R. C., *Charge Exchange and the Theory of Ion-Atom Collisions* Oxford: Clarendon Press, 1992.
18. Schiwietz, G., *Phys.Rev. A* **42**, 296-306 (1990).
19. Bethe, H. A., and Salpeter, E. E., *Quantum Mechanics of One- and Two-Electron Atoms* Berlin: Springer, 1957.
20. Kleppner, D., *Physics Today* **52**, 11 (1999).

Symmetry Trapping, Acceleration, and Charge Exchange Between Positive and Negative Ions in Production of Neutralized and Neutral Beams

A. Y. Wong, G. Rosenthal, G. Paskalov, J. Chen,
N. Hicks, D. Karfidov, K. D. Kang

University of California at Los Angeles

Abstract. This talk presents a coordinated experimental investigation and computer modeling of basic plasma processes in simultaneous trapping and acceleration of both positive and negative ions by propagating electrostatic structures such as RFQs. The concept of increasing the current through stacking modules will be discussed. Furthermore, the concept of enhancing the transfer of electrons between positive and negative ions through increasing the rate of encounters in a quadrupole will be presented. Applications to fusion research through the production of neutral and neutralized beams will be discussed.

INTRODUCTION

The overall objective of the research is to investigate a relatively new area of plasma physics in which the plasma is composed of positive and negative ions of equal mass (Symmetric Plasmas). The trapping, acceleration, and subsequent transfer of electrons between positive and negative ions form the area of concentration. The equations describing such plasmas are fully symmetric with respect to the two species and a 3-D computer code now exists to model the complete nonlinear process. Quantitative comparison between experiments, theory, and computer modeling in the phase space distribution of trapped particles will be made. The self-consistent fields and stability boundary will be investigated. A second objective is to investigate the scaling of the above process with decreasing size of the confining trap. For a given imposed wave potential the limiting trapped charge density should increase with decreasing cross section. Furthermore, the separation of charges of opposite signs by wave fields at small distances is to be demonstrated. The experimental verification will lead to understanding of plasmas confined in micro and nanostructures. A third objective is to correlate the basic findings with the design of tabletop neutral and neutralized ion beam sources. The efficient production of neutral and neutralized beams of controlled energies will have significant applications in space, the environment, plasma processing, and fusion energy research. Examples include ion propulsion where the exhaust must be efficiently neutralized, plasma processing where charge buildup on materials must be avoided and tabletop modules are desirable, and beam collision experiments where neutralized beams reduce space charge effects. The premise of this research is that radio frequency quadrupoles can create wave structures that simultaneously accelerate a beam of positive and negative ions and subsequently efficiently neutralize the beam.

Radio Frequency Quadrupoles

Radio frequency quadrupole (RFQ) accelerators are linear accelerators that have enjoyed considerable popularity because they can accelerate ions from typical ion sources of low energies (100 eV - 200 keV) up to a few MeV, with capture efficiencies approaching 80%. Other accelerators are not efficient below about 1 MeV because of difficulty in radially focusing at lower energies. The RFQ structure, on the other hand, automatically produces radial focusing as well as both acceleration and bunching. Over the past decade there have been many studies reported on RFQs, as well as many RFQs built and operated. The overall results of these studies and experiments can be summarized by stating that, for a large range of parameters, RFQs can be designed with great confidence using 3D codes, first developed at Los Alamos National Lab, to operate much as the design studies predict.

Neutral Beam Production

Present schemes of producing neutral beams require a separate neutralizer section where charge exchange between ions and neutrals occurs. Depending on the energy of the accelerated ions, the conversion efficiency of a high-energy ion beam to a high-energy neutral beam can be quite limited. For example, the neutralization efficiency for a H^+ or D^+

ion beam drops from ~50% to near zero for a beam energy increase from 100 keV to 1 MeV. Some recent Neutral Beam Injection (NBI) systems utilize negative ion beams in the high-energy range, because of higher efficiency for H^- or D^- ion neutralization compared to that for H^+ or D^+ ion neutralization. However, the efficient extraction of H^- using high DC voltage is difficult and the overall system dimensions are very large.

Description of Concept

The research program is to be in three phases that systematically progress from the known process of RFQ acceleration of a single species to RFQ trapping and acceleration of two oppositely charged ion species to neutralization of the energetic two species beam. The beam density will then be increased to quantify space charge effects. At each step we will measure and calculate the same quantifiable parameters such as distribution functions and electric fields so that there is no ambiguity in our conclusions. The plan emphasizes configurations in which the crucial physics can be delineated rather than pushing parameters.

Experimental Setup

The experimental system consists primarily of a beam source, a RFQ accelerator, a RFQ neutralizer, and a diagnostic assembly, as shown in Figure 1. The experiments will be carried out in a vacuum chamber that is 0.28 m in diameter and 1.20 m in length. The system is evacuated to an operating pressure of 1×10^{-6} Torr. The underlying goal is to inject positive and negative ions into a single RFQ accelerator, where they will be trapped, respectively, in each trough and crest of the RFQ traveling wave. The output of the RFQ accelerator will thus be an energetic beam of alternating bunches of positive and negative ions. This beam will then enter the neutralizer, which is designed to induce the bunches to oscillate axially, and the positive and negative ions in a given pair of adjacent bunches are driven to encounter each other repeatedly. This process enhances the efficiency of neutralization of the beam via electron transfer from negative to positive ions. Multiple diagnostics are used to provide confirmation of results; for example, ion velocities are measured by using energy analyzers, time of flight probes, and magnetic deflection. Well-tested diagnostic electron beams with minimum perturbations measure electric fields.

Beam Source

A total of three different beam sources will be used during the experiment. These consist of a double-plasma beam source, a one-beam ion source, and a two-beam ion source. The latter two types will be commercial units; two such sources will be acquired, one for H^+ ions and the other for H^- ions. The use of separate positive and negative ion sources allows us to accurately control the density and energy of both ion species individually.

RFQ Accelerator and Neutralizer

AccSys Technologies, Inc. supplied the RFQ used in all of our current experiments. This "prototype" accelerator is designed to be driven with 1.2 kV RF at 200 MHz, and to accelerate hydrogen ions from 1 keV to 20 keV. In general, when ions enter a RFQ, they are radially confined (subject to stability limitations that depend on the RFQ parameters and ion charge/mass ratio). As a perturbation, undulations are superimposed on the quadrupole pieces. The undulations on the vertical pieces are 180 degrees out of phase with those on the horizontal pieces, creating a wave that travels down the structure with a velocity dictated by the wavelength of the undulation and the driving frequency. The length of the undulations increases

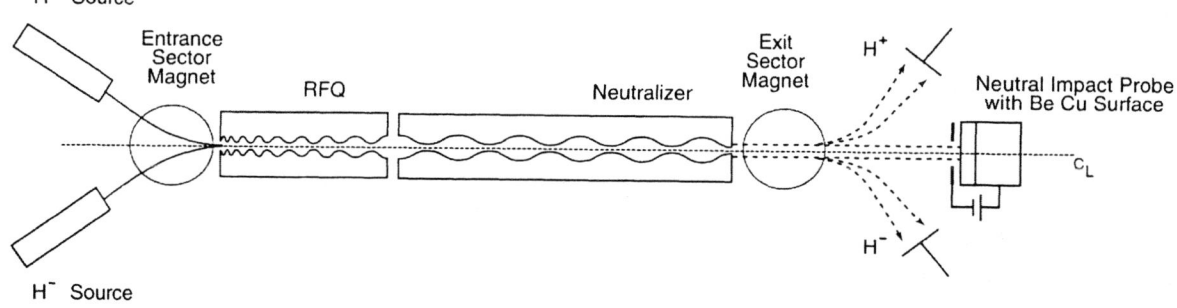

FIGURE 1. Experimental set up including ion source, RFQ accelerator, neutralizer, and a diagnostic assembly.

along the axis of the device. The end result is that ions entering the RFQ with velocity matching the initial wave velocity are trapped and accelerated by the wave. The neutralizer consists of a quadrupole system similar to the accelerator RFQ, and thus preserves radial confinement. Here, however, the undulation wavelength is constant and equal to twice the final wavelength of the accelerator, and the drive frequency is halved. This arrangement causes the ions to maintain their accelerated drift velocity, but also to oscillate axially. Positive and negative ion bunches are no longer located in wave potential extrema; instead, an axial electric field is imposed which causes repeated encounters between the bunches. This process was investigated in a simulation, as described below. The length of the neutralizer has to be sufficient for the required fraction of ions to neutralize before reaching the end of the system. The proposed neutralizer length is about 100 cm and is expected to yield 80% neutralization at 20 keV ion energy.

could increase the rate of encounters between them. A simple configuration is to apply an external field with spatial scale twice that of the acceleration process (and with frequency halved to maintain constant drift velocity). Thus a pair of previously separated positive and negative ion bunches will experience the same direction electric field at this stage. Based on the parameters of the prototype 20 keV RFQ design using H^+ and H^- ions, a frequency of 100 MHz is used for the external field. Since the ion bunches are at the same time moving at the ion drift velocity (v_D) of 1.96×10^6 m/s, the simulation program has computed the actual oscillation between the ion bunches at a rate of 42 MHz. This rate is found to be about 4 times the natural oscillation frequency ($\omega_{pi}/2\pi$) between H^+ and H^- ions, 10 MHz (where $\omega_{pi}^2 = 2\omega_p^2$, and $\omega_p^2 = 4\pi n e^2/M$ is the square of the ion plasma frequency). During each oscillation, the H^+ and H^- ion bunches will encounter each other along the axis two times. This is shown in the ion density vs. axial position plots of Figure 2.

Computer Modeling

Demonstration of Repeated Axial Encounters Due to External Field Imposed After Acceleration

A 1-D PIC simulation was performed to demonstrate that applying an external field after the process of accelerating the oppositely charged ions

Initial RFQ Accelerator Results

We are presently verifying the performance of our 20 keV prototype RFQ unit by injecting H^+ ions. So far, the best RF power matching is achieved for a

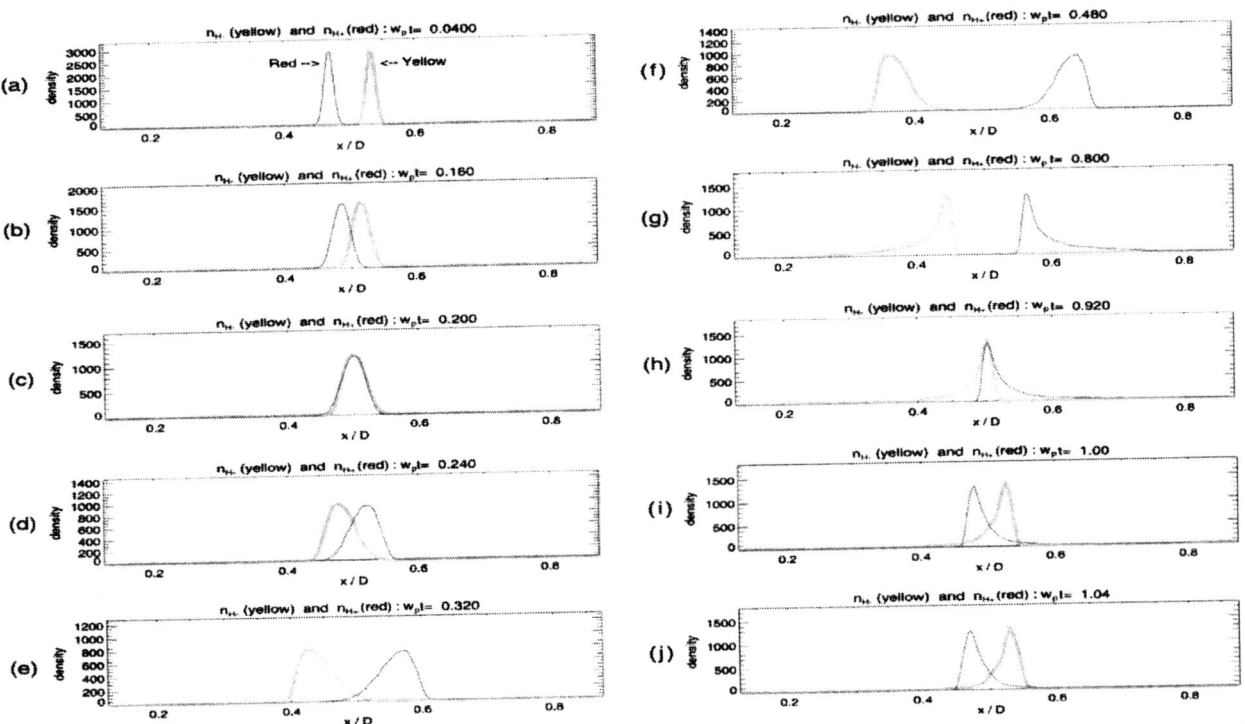

FIGURE 2. Time evolution of positive ion (red, thinner trace) and negative ion (yellow, thicker trace) densities vs. relative axial position during one RF oscillation in the neutralizer.

frequency of 120 MHz. The wavelength of the RFQ undulation at the entrance is ~2.2 mm. To match the ion input velocity to the initial phase velocity of the RFQ traveling wave, we must inject ions with energy in the neighborhood of 350-400 eV. This requirement is confirmed as shown in Figure 3.

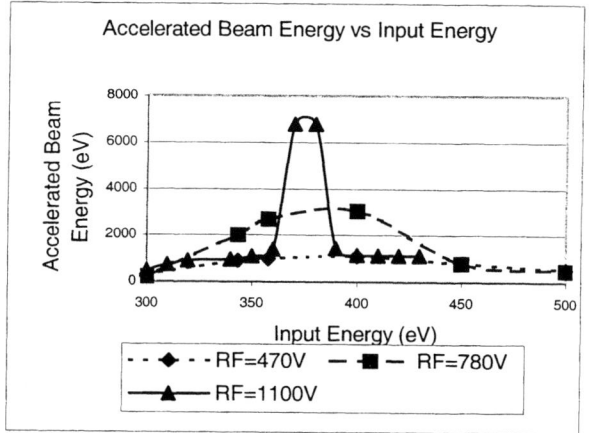

FIGURE 3. Energy of ions accelerated by RFQ vs. ion input energy. Data is shown for three different RF voltages.

The prototype unit is designed to impart a twentyfold energy increase to the ions it accelerates. Since we are injecting 380 eV ions, we hope to see ~7 keV ions exiting the RFQ. Figure 4 demonstrates this.

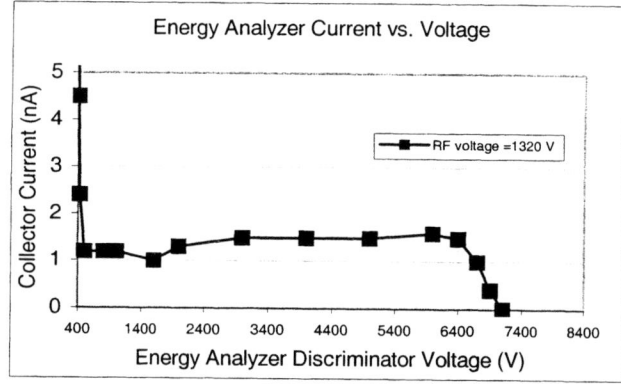

FIGURE 4. The energy analyzer current is shown as a function of the discriminator voltage. The drop in current as the voltage is raised from 6 kV to 7 kV indicates the presence of ions with energy 6-7 keV.

We also expect that the best ion trapping and acceleration should be achieved for an optimal RF voltage applied to the RFQ, as shown in Figure 5. This is a consequence of the stability limits of a quadrupole.

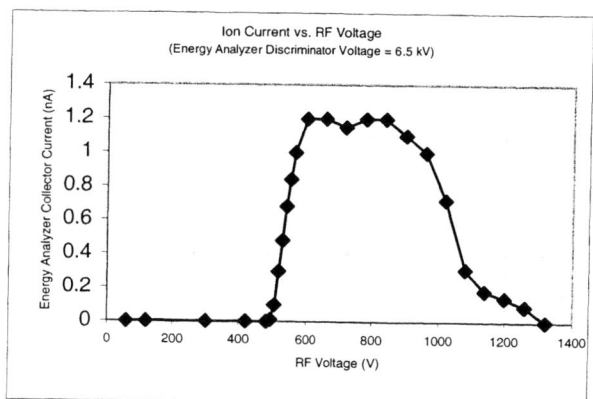

FIGURE 5. Energy analyzer current vs. RF voltage. The discriminator voltage is set to 6.5 kV.

Increasing Current by Source Stacking

Our proposed modular design allows us to multiply the number of accelerators to provide the desired size and current of the neutral beam while maintaining the optimal ion density of the small structure. An artist's sketch of a 3 x 3 RFQ design is shown in Figure 6. We recognize that more research is needed to optimize the number density of ions to be accelerated through the design of the accelerator geometry. The small size of our accelerator favors the containment of more ions per unit volume: the density increases with $1/r^2$ (r is the RFQ radius). The limiting density is obtained when the self-field of the trapped ions becomes comparable to the externally imposed field.

FIGURE 6. A conceptual design of a 3 x 3 RFQ array. Each "x" denotes a beam channel.

ACKNOWLEDGEMENTS

We wish to acknowledge the support by the US Department of Energy, and we thank Dr. R. Hamm and Dr. R. Wuerker for their assistance in the RFQ design.

Charged particle trajectories in an ideal paracentric hemispherical deflection analyser

T. J. M. Zouros,[*,a,b] E. P. Benis[a,b] and J.E. Schauer[a,c]

[a]*Dept. of Physics, University of Crete, P.O. Box 2208, 71003 Heraklion, Crete*
[b]*Institute of Electronic Structure and Laser, FORTH, Heraklion, Crete, GREECE*
[c]*Dept. of Physics, Concordia College, Moorhead, Minnesota, 56562, USA*

Abstract. We present an exact analytic solution for the trajectory of a charged particle moving in the ideal potential $\tilde{V}(r) = -k/r + c$ inside a hemispherical deflector analyser (HDA). Our treatment extends the known solutions to also include paracentric entry for which $R_0 \neq \bar{R} \equiv \frac{1}{2}(R_1 + R_2)$ and $\tilde{V}(R_0)$ is <u>not</u> necessarily zero, where R_0 is the centre of the HDA entry aperture. We also account for particle refraction at the potential boundary that cannot be neglected when $\tilde{V}(R_0) \neq 0$. A general 3-D vector treatment for calculating trajectories in a fixed frame is also described based on the conservation of the angular momentum and eccentricity vectors. These results find applications in modern hemispherical spectrographs incorporating large diameter position sensitive detectors (PSD) as for example in ESCA.

INTRODUCTION

Recently, we have reported on the use of a *paracentric* HDA with 2-D PSD (1) for high-resolution zero-degree Auger projectile electron spectroscopy. (2) A *paracentric* HDA has an *elliptical* central trajectory with particle entry at $R_0 < \bar{R} \equiv \frac{1}{2}(R_1 + R_2)$ and $\tilde{V}_0 \equiv \tilde{V}(R_0) \neq 0$, (1) unlike a *conventional* HDA having a *circular* central trajectory with $R_0 = \bar{R}$ and $\tilde{V}_0 = 0$. A typical paracentric HDA geometry is shown in Fig. 1.

The paracentric HDA has been shown (1) using charged particle optics program SIMION3D (3) to have superior energy resolution and larger acceptance energy window than that of a conventional HDA *without* the use of fringing field correctors. (4) The reason for this is not yet clear. By computing the 3-D trajectories of particles in an ideal and a <u>real</u> (simulated by SIMION) paracentric analyser we expect to get a better understanding of the specific ways by which fringing field effects modify the behavior of such analysers.

Paracentric HDAs have never been treated in the literature. However, very recently, a high-resolution tandem energy analyser incorporating an exit paracentric HDA (i.e. $R_0 = \bar{R}$ but $R_\pi < \bar{R}$) as the second stage has been briefly described.(5) We expect our general approach to be of particular interest to investigators using modern HDAs (for a recent review see (6)) with substantial interradial distances needed to accommodate large area PSDs (7) or second stages (8, 9) in which fringing fields (4, 10) and refraction at field boundaries

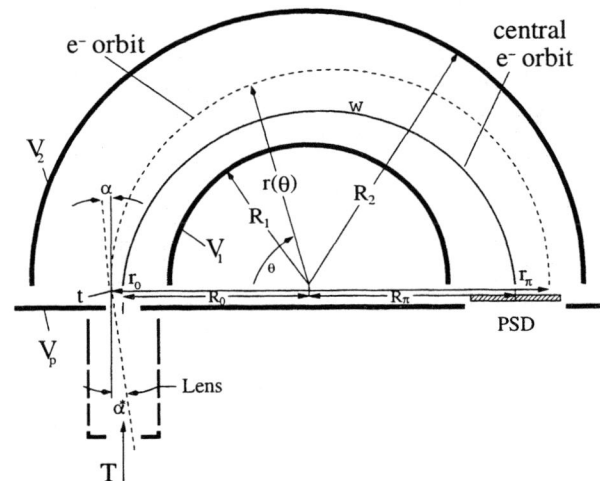

FIGURE 1. Schematic of paracentric HDA geometry. The charged particle initially enters the lens assembly with kinetic energy T and is then focused and decelerated by the lens and plate at potential V_p down to an energy t just prior to entering the interior region of the analyser (shaded area) with angle α^*. Upon entering at r_0 at potential $V(r_0)$, it is refracted to an angle α, follows the trajectory specified by $r(\theta)$ and exits at r_π after being deflected through an angle π. The centre of the entrance aperture is paracentric at $R_0 < \bar{R}$. Fixing the central trajectory ($\alpha = 0$) such that for $t = w$ and $r_0 = R_0$, $r_\pi = R_\pi$ fixes the analyser voltages.

---

* Corresponding author - tzouros@physics.uoc.gr

(8, 9) are important and may be used to advantage.(11) This is a work in progress and here we briefly present some of our first analytic results.

TRAJECTORY EQUATIONS

The classical, non-relativistic equations of motion for a particle of mass m and charge q in the potential $\tilde{V}(r)$ are given by:

$$m\ddot{\mathbf{r}} + q\boldsymbol{\nabla}\tilde{V}(r) = 0 \quad (1)$$

For the solution of $\mathbf{r}$ *inside* an ideal HDA we use $\tilde{V}(r) = -\frac{k}{r} + c$. The eccentricity vector $\boldsymbol{\epsilon}$ is given by:(12)

$$\boldsymbol{\epsilon} \equiv \frac{\dot{\mathbf{r}} \times \mathbf{L}}{qk} - \frac{\mathbf{r}}{r} \quad (2)$$

It is seen to be proportional to the Runge-Lenz vector $\mathbf{A} = qk\boldsymbol{\epsilon}$ (13) known to be *conserved* for motion in a $1/r$ potential. Clearly, $\boldsymbol{\epsilon}$ lies in the orbital plane since from Eq. 2 it is seen to be perpendicular to the angular momentum $\mathbf{L}$.

Taking the dot product of $m\mathbf{r}$ with Eq. 2 yields the scalar equation of motion for r:

$$r(\theta) \equiv r_\theta = r = \frac{p}{1 + \hat{\mathbf{r}} \cdot \boldsymbol{\epsilon}} = \frac{p}{1 + \epsilon \cos(\theta - \theta_\epsilon)} \quad (3)$$

Eq. 3 is seen to be the equation of a conic section in polar coordinates with the origin of the coordinate frame at the focus of the conic section. For $0 < \epsilon < 1$, the orbit is an ellipse with eccentricity $\epsilon = |\boldsymbol{\epsilon}|$ and *latus rectum* $p = L^2/(mqk)$ (13) with $L = mr_0v_0\cos\alpha$. The angle $\theta - \theta_\epsilon$ is just the angle between the two vectors $\mathbf{r}$ and $\boldsymbol{\epsilon}$. At entry, we have $\theta = \theta_0$ and $r_0 \equiv r(\theta = \theta_0)$. When $\theta = \theta_\epsilon$, it is seen that r is a minimum and thus $\boldsymbol{\epsilon}$ has the useful property that it *always points to periapse*.

Using Eq. 3 and specifying the *central* trajectory such that a particle with kinetic energy $t = w$ just prior to entry with $\alpha = 0$ and $r_0 = R_0$, exits at $r_\pi = R_\pi$, necessarily sets the values for $\tilde{V}(r)$ constants k and c:

$$qk = \gamma w R_0\left(1 + \frac{R_0}{R_\pi}\right) \qquad qc = w\left(1 + \gamma\frac{R_0}{R_\pi}\right) \quad (4)$$

where we have also defined $q\tilde{V}_0 \equiv (1-\gamma)w$ with γ a control parameter used to set the voltages of the HDA.

i. $\theta_\epsilon = 0$ vector form of the orbit

Clearly, the conserved vectors $\boldsymbol{\epsilon}$ and $\mathbf{L} \times \boldsymbol{\epsilon}$ are mutually perpendicular to $\mathbf{L}$ and therefore lie in the plane of the orbit. They can be used as a natural coordinate system of axes to describe the motion. Choosing to align the x-axis along the semi-major axis ($\theta_\epsilon = 0$) we may set: (14)

$$\mathbf{r}(t) = x(t)\frac{\boldsymbol{\epsilon}}{\epsilon} + y(t)\frac{\mathbf{L} \times \boldsymbol{\epsilon}}{L\epsilon} \quad (5)$$

with the focus of the ellipse (see Fig. 2) at $r = 0$ and

$$x(t) = a(\cos\zeta - \epsilon) \quad (6)$$
$$y(t) = a\sqrt{(1-\epsilon^2)}\sin\zeta \quad (7)$$
$$t = \sqrt{\frac{ma^3}{qk}}(\zeta - \epsilon\sin\zeta) \quad (8)$$

with the particle being at periapse at time $t = 0$ and $\zeta = \theta = 0$. The semi-major axis of the ellipse has length $a = (r_0 + r_\pi)/2 = p/(1-\epsilon^2)$ obtained directly from Eq 3.

The new angle ζ introduced above is known as the Kepler (14) or eccentric anomaly (12) and is related to the angle θ by:

$$\tan\frac{\zeta}{2} = \sqrt{\frac{1-\epsilon}{1+\epsilon}}\tan\frac{\theta}{2} \quad (9)$$

Eq. 9 is particularly useful as it avoids quadrant ambiguity since $\zeta/2$ is always in the same quadrant as $\theta/2$ (12). Using Eq. 5 it is straightforward to describe the 3-D tra-

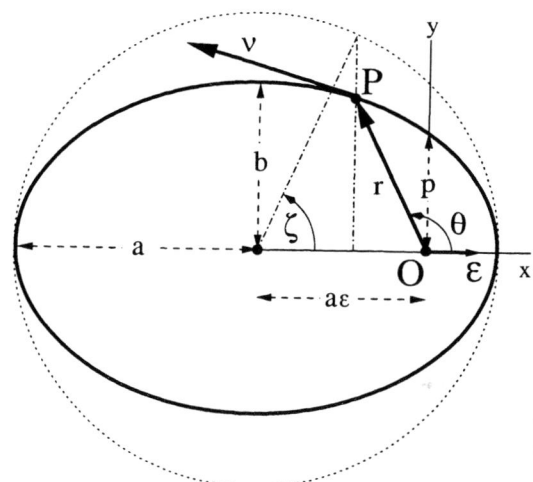

FIGURE 2. Elliptical particle orbit in the xy coordinate system showing the *true anomaly* θ and the *eccentric anomaly* ζ and *latus rectum* p. O is the center of attraction and focus of the ellipse. The eccentricity vector $\boldsymbol{\epsilon}$ is seen to start from O and point to periapse. It thus *always lies along the semimajor axis* of the ellipse.

jectory in any fixed coordinate system XYZ in which the initial components of $\mathbf{L}$ and $\boldsymbol{\epsilon}$ are known. In Fig. 3 we show such a 3-D plot made with the help of the software program *Mathematica*.

Eq. 8 is also very useful since it gives directly the time-of-flight (TOF) as a function of the eccentric anomaly ζ. Thus, for a particle entering the HDA at $t = t_0$ with θ_0, r_0 and v_0 and exiting after a deflection by π we have:

$$\text{TOF} = \sqrt{\frac{ma^3}{qk}}[\zeta_\pi - \zeta_0 - \epsilon(\sin\zeta_\pi - \sin\zeta_0)] \quad (10)$$

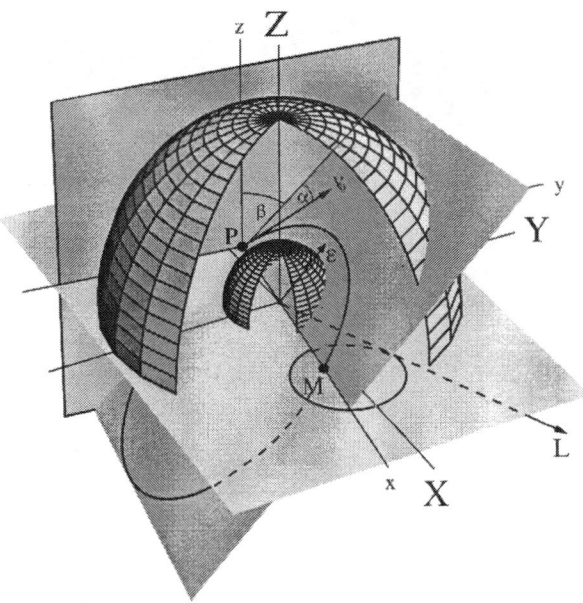

FIGURE 3. 3-D orbit in HDA: Charged particle enters at P and exits at M with $\alpha = -30°$, $\beta = -50°$, $\gamma = 1.5$, $\tau = 1.16$ and $w = 1000$ eV. **XYZ** is the fixed laboratory frame, while xyz is the relative reference frame traditionally used to describe the orbit in terms of angles α and β. The entry velocity $\mathbf{v}_0$, eccentricity ϵ and angular momentum **L** are also shown.

where $\zeta_0 \equiv \zeta(\theta_0)$ and $\zeta_\pi \equiv \zeta(\theta_0 + \pi)$. The orbit angle θ_0 is determined from Eq. 3 with $r = r_0, \theta = \theta_0$ and $\theta_\epsilon = 0$.

ii. $\theta_0 = 0$ scalar form of the orbit

Another useful form of the radial equation Eq. 3 is obtained by orienting our xy coordinate system so that $\mathbf{r}_0$ lies along the positive x-axis (i.e. $\theta_0 = 0$):

$$\frac{r_0}{r_\theta} = \frac{qk}{mv_0^2 r_0} \frac{(1 - \cos\theta)}{\cos^2\alpha} + \cos\theta - \tan\alpha \sin\theta \quad (11)$$

where we have also used the initial condition $\dot{r}_0 = v_0 \sin\alpha = -\frac{L}{mp}\epsilon \sin\theta_\epsilon$ obtained by evaluating $\dot{r}$ directly from Eq. 3 at entry. Eq. 11 is the well-known form introduced by Purcel (15) and discussed in more detail in Refs. (16, 17, 18, 7, 10). Eq. 11 also exhibits the well-known double focusing properties of the HDA, since for $\theta = \pi$ it is clear that $\partial r_\theta/\partial\alpha = 0$ at $\alpha = 0$. However, it does not include corrections for refraction discussed next.

REFRACTION CORRECTIONS

So far we have derived the trajectories in terms of initial conditions *within* the field of the analyser. However, right outside the analyser (just before entry) the potential is constant and thus changes discontinuously across the boundary at θ_0 (at the border of the shaded area in Fig. 1). This discontinuity can be represented mathematically by defining the step potential $\tilde{V}(r,\theta)$ in the orbital plane as:

$$\tilde{V}(r,\theta) = \tilde{V}(r)u(\theta - \theta_0) \quad (12)$$

where $u(\theta - \theta_0)$ is the unit step function.

It can be easily shown that the energy is conserved in going across the potential step and thus:

$$\Delta K^\star \equiv \frac{1}{2}mv_0^2 - \frac{1}{2}mv_0^{\star 2} = \frac{1}{2}mv_0^2 - t = -q\tilde{V}(r_0) \quad (13)$$

Furthermore, using the step potential $\tilde{V}(r,\theta)$ in the equation of motion for the θ coordinate we can show that:

$$\frac{dL}{dt} = -q\tilde{V}(r)\delta(\theta - \theta_0) \quad (14)$$

where $\delta(\theta - \theta_0) = du(\theta - \theta_0)/d\theta$ is the Dirac delta-function. After replacing d/dt with $\dot{\theta}d/d\theta$ we obtain:

$$\frac{d}{d\theta}(L^2) = -2mr^2q\tilde{V}(r)\delta(\theta - \theta_0) \quad (15)$$

with $L = mr^2\dot{\theta}$. Upon integrating across the boundary $\theta = \theta_0$ along a path of constant $r = r_0$ we obtain:

$$L^2 - L^{\star 2} = -2mr_0^2q\tilde{V}(r_0) \quad (16)$$

where the $\star$ tags parameters on the side where the particle is free. Thus, outside the analyser $L^\star = mr_0v_0^\star\cos\alpha^\star$, and inside the analyser $L = mr_0v_0\cos\alpha$ where $v_0, v_0^\star$ and $\alpha, \alpha^\star$ are the velocities and entry angles inside and outside the analyser, respectively. Using Eqs. 13 and 16 above it is straightforward to derive the law of refraction for charged particles, analogous to Snell's law for light:

$$v_{r_0}^\star = v_0^\star \sin\alpha^\star = v_0 \sin\alpha = v_{r_0} \quad (17)$$

or

$$\sin\alpha = \frac{\sin\alpha^\star}{\sqrt{1 + \frac{\Delta K^\star}{t}}} \quad (18)$$

From Eq. 17 it is seen that the radial velocity, $v_r = \dot{r}$, is continuous across the potential boundary(7), as opposed to the angular velocity, $v_\theta = \frac{L}{mr}$ which is not. The effect of refraction is shown in Fig. 4. The discontinuity effects across the sharp potential boundary at $\theta = \theta_0$, have not been given sufficient attention, especially in older publications, leading to some confusion in the literature. This has been primarily due to the fact that in *conventional* HDAs, the entry voltage $\tilde{V}_0 = 0$ and the entry slits are very narrow so that the approximation $\tilde{V}(r_0) \approx \tilde{V}_0 = 0$ is valid. Correct formuli for refraction can be found in Refs. (15, 19, 7, 8, 9).

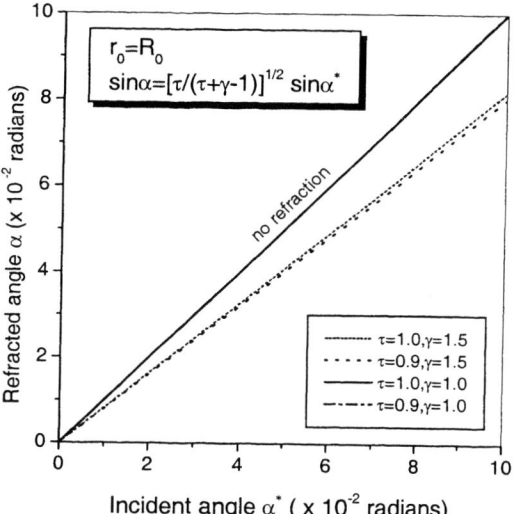

FIGURE 4. Relation between the entry angle α^* prior to refraction (angle of incidence) and angle α after refraction (angle of refraction) for two cases: (a) $\tilde{V}_0 = 0$ ($\gamma = 1$), $R_0 = \bar{R}$, (b) $\tilde{V}_0 = 0.5w$ ($\gamma = 1.5$), $R_0 = 0.8125\bar{R}$ (paracentric entry). w is the energy of the central trajectory or "tuning" energy of the HDA. In both cases, $w = 1000$ eV, $r_0 = R_0$ and $R_\pi = \bar{R}$. Clearly, the effect of refraction is non-negligible for paracentric entry and $\tilde{V}_0 \neq 0$ ($\gamma \neq 1$).

Using Eqs. 13, 16 and 18 we obtain the trajectory equation in terms of the entry angle α^* and velocity v_0^*:

$$\frac{r_0}{r_\theta} = \frac{qk}{mv_0^{*2} r_0} \frac{(1-\cos\theta)}{(1+\frac{\Delta K^*}{t\cos^2\alpha^*})\cos^2\alpha^*} + \cos\theta - \frac{\tan\alpha^* \sin\theta}{\sqrt{1+\frac{\Delta K^*}{t\cos^2\alpha^*}}} \quad (19)$$

where we have also introduced the reduced pass energy $\tau \equiv t/w$. Clearly, Eq. 19 also preserves the first-order focusing in α^* for $\theta = \pi$. Thus, following deflection by 180° and using Eq. 4 to rewrite potential constants k and c in terms of γ and the tuning energy w, the exit radial position r_π is given by the simple formula:

$$r_\pi = -r_0 + \frac{R_0 + R_\pi}{1 + \frac{R_\pi}{R_0 \gamma}(1 - \tau\cos^2\alpha^*)} \quad (20)$$

Eq. 20 is thus seen to extend the well known results for the exit point of a conventional HDA with $R_0 = R_\pi = \bar{R}$ and $\tilde{V}_0 = 0$ (e.g. see (7)) to those of the more general paracentric case, where $R_0 \neq R_\pi \neq \bar{R}$ and where $\tilde{V}_0$ might also be different from zero. The optical properties of the HDA (e.g. resolution, transmission, etc.) directly follow from Eq. 20 and are presented elsewhere (20).

ACKNOWLEDGMENTS

We acknowledge meaningful discussions with J. Erskine, D. Roy, E. Sidky, H. Wollnik, R. Woodard and M. I. Yavor. We thank Pat Richard and the J.R. Macdonald Laboratory at Kansas State University for their support. J.E.S. also thanks the Carl L. Bailey Centennial Scholarship of Concordia College for its support.

REFERENCES

1. E. P. Benis and T. J. M. Zouros, Nucl. Instrum. Methods Phys. Res. Sect. A **440**, 462 (2000).
2. T. J. M. Zouros and D. H. Lee, in *Accelerator-based atomic physics techniques and applications*, edited by S. M. Shafroth and J. C. Austin (American Institute of Physics Conference Series, New York, 1997), pp. 426–79.
3. D. A. Dahl, SIMION 3D v6.0, Idaho National Engineering Laboratory, Idaho Falls 1996.
4. D. Hu and K. Leung, Rev. Sci. Instrum. **66**, 2865 (1995) and references therein.
5. V. D. Belov and M. I. Yavor, Rev. Sci. Instrum. **71**, 1651 (2000).
6. D. Roy and D. Tremblay, Rep. Prog. Phys. **53**, 1621 (1990).
7. F. Hadjarab and J. Erskine, J. Electr. Spectr. and Rel. Phenom. **36**, 227 (1985).
8. A. Mann and F. Linder, J. Phys. E: Sci. Instrum. **21**, 805 (1988).
9. A. Baraldi and V. R. Dhanak, J. Electr. Spectr. and Rel. Phenom. **67**, 211 (1994).
10. P. Louette *et al.*, J. Electr. Spectr. and Rel. Phenom. **52**, 867 (1990).
11. S. C. Page and F. H. Read, Nucl. Instrum. Methods Phys. Res. Sect. A **363**, 249 (1995).
12. J. E. Prussing and B. A. Conway, *Orbital Mechanics* (Oxford University Press, Oxford, 1993).
13. D. L. Landau and E. M. Lifschitz, *Mechanics* (2nd edition Pergamon Press, Addison-Wesley Publishing Company, Inc., Reading, Massachusetts, 1969).
14. E. A. Solov'ev, Sov. Phys. JETP **55**, 1017 (1982).
15. E. Purcell, Phys. Rev. **54**, 818 (1938).
16. H. Wollnik, Nucl. Instr. & Meth. **34**, 213 (1965).
17. D. Roy and J.-D. Carette, Canadian J. of Phys. **49**, 2138 (1971).
18. M. E. Rudd, in *Low Energy Electron Spectrometry*, edited by K. D. Sevier (Wiley, New York, 1972), pp. 17–34.
19. R. E. Imhof, A. Adams, and G. King, J. Phys. E: Sci. Instrum. **9**, 138 (1976).
20. E. P. Benis and T. J. M. Zouros, Dept. of Physics, Univ. of Crete, internal report and to be published.

Contributions of negative energy states (NES) to magnetic dipole transitions in the helium isoelectronic sequence, and for the neutral alkalis

H. Gordon Berry and Igor M. Savukov

Department of Physics, University of Notre Dame, Notre Dame, IN 46556 USA

Abstract. Direct effects of negative energy states on atomic parameters are in general extremely weak and can be neglected. More precisely, it is found that the *no-pair* relativistic Hamiltonian is very precise for calculations of the energy and electric multipole operators. Only in certain cases of magnetic dipole transition rates is this approximation insufficient. We discuss the calculations of magnetic dipole transition rates in comparison with existing measurements, and the progress of our measurements in helium between two excited triplet states; we also discuss the NES effects and possible measurements in ns -> (n+1)s transitions in the alkalis.

NEGATIVE ENERGY STATES

It is well known (e.g Sucher in [1]) that the Dirac-Coulomb Hamiltonian has no true bound state eigenfunctions so that Brown and Ravenhall [2] proposed a *no-pair* Hamiltonian which leads to very accurate atomic energies and transition rates. No precise tests of the direct influence of negative energy states (NES) on atoms exist. Of course, there is no doubt that anti-particles exist; and quantum electrodynamic corrections (QED) such as the Lamb shift in atoms, also depend on the existence of virtual electron-positron pairs.

Our investigations are primarily questioning the form of the approximate Hamiltonian used in all atomic physics calculations, and the neglected NES terms, which can be significant for forbidden magnetic-dipole (M1) transition amplitudes. Such transitions have recently been studied in the helium-like isoelectronic sequence (eg. Refs. [3,4]), and we have begun our measurements on two M1 transitions. Negative energy states play an almost negligible role in all other multipole transition rates (as discussed in the references above). The largest effect in helium due to such negative energy states (NES) is in the M1 transition rate $1s2s\ ^3S - 1s3s\ ^3S$. It has been earlier pointed out [5] that the forbidden M1 transition rate from the ground state $1s^2\ ^1S - 1s2s\ ^3S$ might have some dependence on the NES; however, it was shown [4], that such NES effects are much less in transitions including a change of total spin. Our major program goals are to measure these two forbidden M1 transitions rates, to measure several other "forbidden" transitions from the n=2 states in helium and also to begin measurements of M1 transitions from the alkali $^2S_{1/2}$ ground states. Our calculations [6] indicate that a measurement of the 5s-6s M1 transition in rubidium is the most sensitive test of the contributions from the negative energy states of the correct relativistic Hamiltonian within the neutral alkali systems. Such measurements are also relevant to testing the theoretical atomic calculations underpinning the present very accurate atomic tests of the Standard Model in the electroweak interactions (PNC or parity non-conservation measurements) see for example the recent discussions by Derevianko [7].

LASER EXPERIMENTS IN HELIUM

The Grotrian energy level diagram (figure 1) shows the relevant transitions in the low-lying states of helium. The M1 transitions of interest are the $1s2s\ ^3S-1s3s\ ^3S$ transition at 427.7 nm and $1s^2\ ^1S - 1s2s\ ^3S$ transition in the ultraviolet at 62.5 nm. We focus on the transition at 427.7 nm which has never been observed (principally because of the normal electric dipole (E1) transition $1s2p\ ^3P-1s3s\ ^3S$ at 705 nm). An initial test measurement in progress is the two E1 photon transition rate between the 1s2s ^{3}S and 1s3s ^{3}S states. This involves, of course, the same un-doubled laser

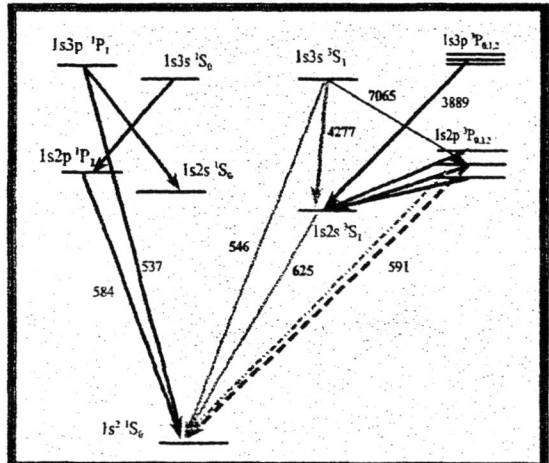

Figure 1. Lower levels of neutral helium showing some of the allowed E1 and forbidden M1 transitions.

frequency, and provides a helpful method of optimizing the experimental setup. Calculations are also in progress for the transition rates of the zero-order forbidden spin changing transitions between the singlet n=2, 3 S and P states and the triplet n=2, 3 P and S states. The same technology can be applied to measuring these transition rates.

The long helium absorption cell is shown in Figure 2. We observe the photo-emission following photo-absorption resonant laser absorption at 4277 Å from the 1s2s ^{3}S state to the 1s3s ^{3}S state. The absorption is followed by emission of the electric dipole transition to 1s2p ^{3}P at 7065 Å. We will measure this fluorescence intensity on and off resonance. The ratio of the two yields is proportional to the ratio of the M1 transition rate to the known E1 transition rates from the 1s3p ^{3}P state to the 1s2s and 1s3s ^{3}S states. This technique thus obviates the need for an absolute absorption measurement or a calibration of the fluorescence detection system.

The major difficulties in this experiment are the smallness of the M1 transition rate between the 1s2s and 1s3s ^{3}S states, and the fact that the lower state is not the ground state, and must be excited. The latter problem is solved by introducing a pulsed discharge in the cell to build up a population in the 1s2s ^{3}S$_1$ state. A typical low energy discharge at a pressure of 1 torr can generate population fractions of about 10^{-4}, and hence, population densities of the 1s2s ^{3}S state of the order of 10^{12} cm^{-3}. The helium cell is then probed by a laser at the resonance wavelength of 4277 Å. In a test 50 cm long helium cell, we have made measurements of the metastable densities (by absorption) and thus verified the expected population densities and the optimum discharge conditions for the experiment.

The 4277 Å input light is provided by a frequency-doubled CW Ti-Sapphire laser, pumped by an Argon Ion laser. The laser yields up to 1.8 watts single frequency (about 0.06 MHz width) in the range near 900 nm. Initial tests of the LiIO$_3$ doubling crystal system have been successful: we have built an extended cavity for a second Coherent Ti-sapphire broad band laser. This preliminary system has been used create 15 milliwatts of frequency-doubled intensity and to verify pumping in cesium (6s-7p transition) at the frequency doubled wavelength of 455 nm (laser wavelength of 910 nm).

The absorption cell system is the heart of this

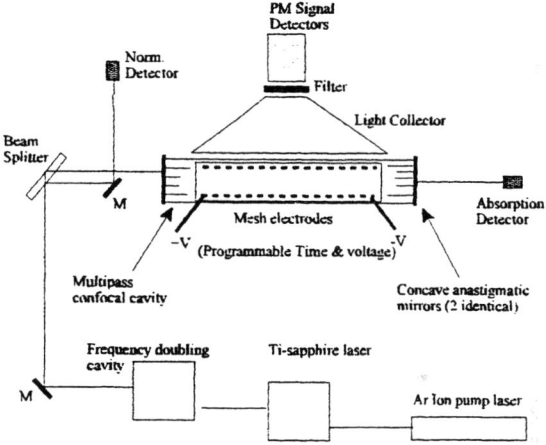

Figure 2. Experimental layout for helium absorption experiment

experiment: the pulsed discharge of a few hundred mAmp at a voltage of about 200 volts in a helium pressure of about 1 torr provides a hollow cathode glow between the two electrodes. At the end of each pulse, the fluorescence detection begins. The optical detection system is also a novel feature of the experiment: the emitted light is extremely weak (a few hundred photons per second), but is emitted more or less isotropically from a 50 cm long (depending on the cell length), 1 mm diameter pencil. It is much too expensive to surround the whole system directly with sensitive, low dark count phototubes. An obvious light collection system would be to place the pencil at one focus of an elliptical cylindrical mirror, and collect the light at the other focus - using fiber optics to change a long thin focus region to

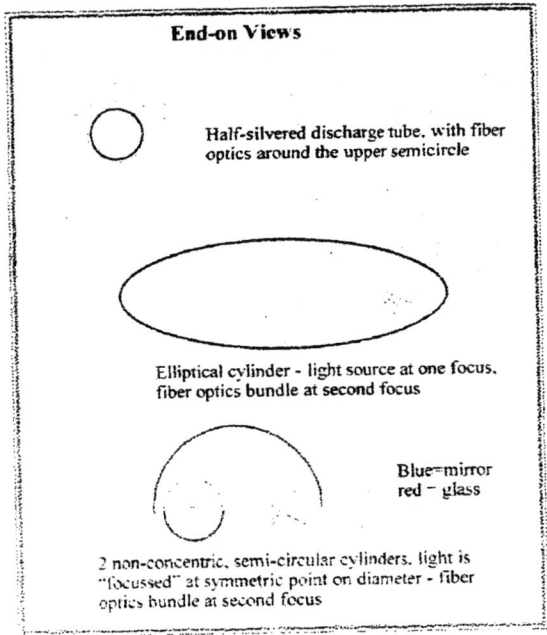

Figure 3. Some optics arrangements discussed in the text.

the circular geometry of the photo-tubes. Such techniques can achieve very high collection efficiency, even after accounting for losses at the entrance to the fiber optics and the efficiencies of the phototubes. However, a much cheaper and simpler system can be constructed using a circular cylinder in place of the elliptical one. Figure 3 shows schematically the systems mentioned above. The ribbon light source is placed off-center of the circular cylinder. An approximate focus appears on the same diameter on the opposite side of the center. Our calculations indicate almost no loss of focussing, using a 1 mm width source, and a 30 cm radius mirror. Initial tests of a 50 cm long mirror confirm these calculations. Several optical fiber geometries have also been considered (to be published). One filter/photomultiplier combination detects the 7065 Å transition 1s2p ^3P - 1s3s ^3S; another combination detects the 3889 Å 1s2s ^3S - 1s3p ^3P transition (for normalization).

The 4277 Å light excites directly the M1 transition, 2s ^3S - 3s ^3S. The 3s ^3S decays by E1 transition at 7065 Å. The yield $N_{res}(7065)$ is proportional to the input flux, the density of metastable 2s ^3S states, and the path-length in the helium cell. In addition, off-resonant excitation to the 3p ^3P state also takes place. Although this transition, 2s^3S - 3p ^3P at 3889 Å, is almost 200 Å off-resonance, its yield is still more than that from the M1 transition. The 3p ^3P state decays primarily directly back to the 2s^3S state, giving a fluorescence rate N(3889). But it also decays to the 3s ^3S state with a branching ratio of 0.114. Hence, this excitation mode also contributes to the 7065 Å fluorescence with a non-resonant contribution $N_{nonres}(7065)$. Since the E1 transition rates have been measured accurately, a measurement of the ratio of the yields of the 7065Å light on and off the M1 resonance gives the M1 transition rate. Similar contributions from higher n-states are too small to be appreciable.

[N_{tot}(7065) on resonance]/[N_{tot}(7065) off res.]
= [N_{res}(7065)+N_{nonres}(7065)]/[N_{nonres}(7065)]

We calculate that the M1 resonance rate in this ratio is 1/43 or about 2% of the total signal. With conservative assumptions of a laser power of 15 mW, 80 cm cell length, and density of metastable atoms of 10^{11} cm^{-3}, we calculate the following total numbers of photons per second: N(3889) = 6.5x10^5 s^{-1}, N_{nonres}(7065) = 8.3 x 10^4 s^{-1}, N_{res}(7065) = 2.0x10^3 s^{-1} Such fluorescence rates require a good collection efficiency, as discussed above.

OTHER HELIUM MEASUREMENTS

An initial test of the system will be the measurement of the 2 photon transition rate 2s ^3S - 3s ^3S, which uses the non-doubled initial wavelength of 2 x 4277 Å. in an internal cavity arrangement. We are also making an accurate calculation of this transition rate. The system can also be used to make the single-photon forbidden transition measurements from helium metastable states: we hope to make first observations of the transitions 2s ^1S - 3p ^3P at and 2s ^3S - 3p ^1P. These experiments will test ongoing calculations of these transition rates.

Finally, we have proposed to measure the M1 transition rate of the transition 1s^2 ^1S - 1s2s ^3S at 62.5 nm, by absorption of synchrotron radiation at the Advanced Light Source (ALS). There is a single existing measurement of this rate [8], which has an accuracy of 30%. It was earlier claimed that the transition is sensitive at the 25% level to effects from the negative energy states. However, further calculations have shown that this depends on the calculational technique. A more accurate measurement would be helpful to verify these results.

ALKALI M1 TRANSITIONS

Our recent theoretical calculations of M1 transition amplitudes in alkalis clearly demonstrate the important role of negative-energy state contributions [6]. We briefly discuss below the experimental possibilities to test these contributions. We focus on the transition from the ground state of each alkali ns $^2S_{1/2}$ (where n=2 for lithium, up to n=7 for francium) to the (n+1) $^2S_{1/2}$ state. Such transitions are forbidden M1 transitions and not allowed in a non-relativistic approximation.

The first order Dirac Hartree-Fock values of these transition rates dominate for low Z and become less significant for cesium and francium, due to larger second-order no-pair contributions. However, the values of NES contributions appear to be closely proportional to Z. In the case of cesium, the only measured transition in this sequence, the NES contributions are only a small fraction (4%). However, the rubidium case is the most surprising, where the no-pair contributions almost cancel, so that there is a strong dependence on the negative-energy corrections. Although such cancellation in second order may be a coincidence, and more accurate calculations may be

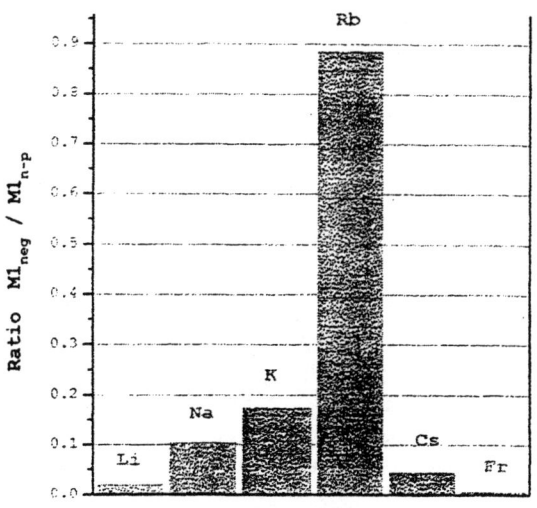

Figure 4. Ratio of the contribution due to the negative energy states to the total rate for the alkalis

necessary, the rubidium atom appears to be a most promising system to test NES contributions, while potassium may also provide a possible but difficult test.

We compare the NES fractional contributions in different alkaline atoms in Figure 4. In the light alkalis (Li, Na, K) the effect is proportional to Z and is maximal for K (17%). For heavy alkaline atoms such as Cs and Fr, it is small because of large no-pair contributions. Rubidium, in the middle, has a very large relative effect (89%) and is the most promising. If measurements in the alkalis reaches the precision achieved in Cs then all alkalis except Fr will be good candidates for testing NES theory. The accuracy of the calculations on the other hand can impose even more severe restrictions than experiment. The accuracy of our calculations as seen in the deviation from the experiment is 16% and, in principle, should become smaller for smaller Z.

SUMMARY

We have described an experimental arrangement for measuring absorption fluorescence for cases of very weak transition rates. We are applying these techniques to study magnetic dipole transitions in helium and the alkalis which are sensitive to the effects of negative energy states.

We are grateful for the help of Alex Vasilyev and Professor Carol Tanner in setting up this experiment. This work was supported in part by a grant from The Research Corporation.

REFERENCES

1. J. Sucher, *Phys. Rev.* **A 22**, 348 (1980)
2. G.E. Brown and D.E. Ravenhall, *Proc. Roy. Soc.* **A208**, 552 (1951)
3. J. Sapirstein, K.T. Cheng and M.H. Chen, *Phys. Rev.* **A 59**, 259 (1999).
4. A. Derevianko, I.M. Savukov, W.R.Johnson, and D.R. Plante, *Phys. Rev.* **A 58**, 4453 (1998).
5. P. Indelicato, *Phys. Rev. Lett.* **77**, 3323 (1996).
6. I.M. Savukov, A. Derevianko, H.G. Berry, and W.R.Johnson, *Phys. Rev. Lett.* **83**, 2914 (1999).
7. A. Derevianko, *Phys. Rev. Lett.* **85**, 1618 (2000).
8. J.R. Woodworth and H.W. Moos, *Phys. Rev.* **A 12**, 2455 (1975).

PROTON-INDUCED M X-RAY PRODUCTION

Sam J. Cipolla

Physics Department, Creighton University, Omaha NE 68178

Abstract. A brief survey is given of recent studies of M x-ray production from proton impact on heavy elements. Results of 50-300 keV proton-induced M x-rays on thick targets of Ho through Pb are presented and compared with ECPSSR theory and other predictions.

INTRODUCTION

The main emphasis in studying M x-ray production by proton impact has been the testing and refining of theoretical models of the ionization cross section. The ECPSSR (Perturbed Stationary State with Relativistic, Coulomb deflection, and Energy loss corrections) theory has been shown to describe the cross section fairly well at higher proton energies, but it increasingly under-predicts as the proton energy decreases. Since the cross section for M x-ray production from proton impact on heavy atoms is relatively high, this work also has potential for PIXE analysis of certain materials. Of course, basic and practical studies are related in that an accurate knowledge of the cross section is needed in application towards materials analysis. A general difficulty in the experimental determination of M-shell ionization is the complexity of the energy spectrum and the possible interference from contaminant low-energy x-rays. These obstacles can be largely overcome with the use of modern Si(Li) detectors and appropriate analytical techniques.

Merzbacher's group[1] pioneered a non-relativistic PWBA approach around 1958. From 1969-79, tables to facilitate calculation of M-shell ionization cross sections using the PWBA theory followed[2]. The semiclassical approximation (SCA) of Bang and Hansteen[3] paralleled the development of PWBA. During the 70s, Brandt's group developed a perturbed stationary state approach to achieve a better description of inner-shell ionization data over a broad projectile energy range, culminating in the ECPSSR theory of Brandt and Lapicki[4]. Liu and Cipolla[5] have made available a computer program to calculate PWBA and ECPSSR ionization and x-ray production cross sections. In 1983, Chen, et al[6] published relativistic PWBA calculations using binding-energy and Coulomb deflection corrections. Coster-Kronig factors and subshell fluorescence yields that are needed to obtain x-ray production cross sections have been non-relativistically calculated by McGuire[7] and relativistically by Chen, et al[8]. X-ray transition rates have come from the work of Bhalla[9].

Early proton-induced M-shell ionization experiments used proportional counters to measure x-rays; around the early 70s, Si(Li) detectors came into use, affording M sub-shell ionization to be studied. Lapicki[10] has presented a survey of M-shell work done up to 1989. Since then, Pajek's group[11] have made extensive measurements on Hf through Th using proton energies from 100 keV through 4 MeV, comparing the results with the PWBA, SCA, and ECPSSR theories. Sun, et al[12] have studied the rare-earths using 0.6-4.6 MeV protons, testing the PWBA and ECPSSR predictions. Cipolla, et al[13] has compared cross sections from 50-300 keV protons impacting elements ranging from Gd through Pb with the ECPSSR theory. Shokouhi, et al[14] tested the PWBA theory against their results from 2-6 MeV protons on Tb, Ho, Tm, and Lu. Verma's group has reported on 1-5 MeV protons striking Au, Pb, and Bi in comparison with PWBA and ECPSSR predictions[15].

The consensus of these studies is that ECPSSR theory represents measured x-ray production cross sections best, but not perfectly. Especially as the proton energy decreases, the ECPSSR theory as currently formulated increasingly under-predicts experimental data. Pajek, et al[16] have formulated reference cross sections for $M_{\alpha+\beta}$, M_γ, and $M_3O_{4,5}$ x-rays based on their extensive data base.

EXPERIMENTAL AND ANALYTICAL PROCEDURES

In our work, a 350-kV Cockcroft-Walton accelerator consisting of a RF ion source, a mass analyzer magnet, and a biased 1.5-mm beam collimator was employed to

TABLE 1. Measured x-ray production cross sections for some major peaks (quantities in parentheses are uncertainties).

E (keV)	M_ζ	M_α	M_β	M_γ	M_2N_4	M_1N_3
Ho						
100	1.87 (0.08)	7.20 (0.19)	4.56 (0.11)	0.137 (0.005)		0.0160 (0.0006)
150	5.74 (0.17)	16.4 (0.38)	13.6 (0.3)	0.458 (0.015)		0.0487 (0.0017)
200	10.4 (0.3)	26.1 (0.56)	28.0 (0.6)	0.968 (0.031)		0.0891 (0.0031)
250	17.0 (0.4)	38.9 (0.80)	45.5 (1.0)	1.78 (0.06)		0.136 (0.0048)
300	21.5 (0.5)	50.9 (1.0)	67.8 (1.6)	3.09 (0.10)		0.184 (0.0067)
Er						
100	4.07 (0.43)	11.4 (1.3)	5.3 (0.7)	0.224 (0.036)	0.0823 (0.0108)	0.0130 (0.0036)
150	8.96 (0.96)	21.9 (2.4)	14.4 (1.6)	0.590 (0.077)	0.208 (0.025)	0.0293 (0.0057)
200	20.8 (2.2)	44.6 (4.8)	31.8 (3.6)	1.72 (0.25)	0.532 (0.070)	0.0701 (0.0161)
250	360. (3.8)	60.8 (6.5)	71.6 (7.9)	3.32 (0.57)	1.09 (0.12)	
300	52.7 (5.6)	108 (12)	80.9 (9.0)	5.37 (0.63)	1.88 (0.21)	
Yb						
100	2.08 (0.22)	11.0 (1.2)	1.03 (0.19)	0.116 (0.016)	0.042 (0.008)	0.0089 (0.0017)
150	6.73 (0.70)	33.4 (3.5)	3.69 (0.61)	0.357 (0.043)	0.083 (0.011)	0.025 (0.004)
200	14.1 (1.4)	66.2 (6.9)	8.99 (1.33)	0.767 (0.087)	0.187 (0.023)	0.037 (0.005)
250	22.2 (2.2)	100 (10)	25.4 (3.7)	1.41 (0.15)	0.225 (0.025)	0.061 (0.007)
300	34.4 (3.5)	152 (15)	27.0 (3.3)	2.29 (0.24)	0.442 (0.048)	0.094 (0.011)
Hf						
100	1.11 (0.09)	6.39 (0.73)	1.25 (0.21)	0.140 (0.016)	0.0332 (0.0073)	0.0064 (0.0013)
150	3.83 (0.24)	21.5 (2.1)	4.96 (0.62)	0.494 (0.049)	0.134 (0.024)	0.026 (0.004)
200	8.43 (0.46)	45.9 (4.3)	9.51 (1.21)	1.17 (0.11)	0.303 (0.048)	0.060 (0.006)
250	13.3 (0.7)	77.9 (7.0)	17.9 (2.0)	1.87 (0.18)	0.542 (0.079)	0.091 (0.010)
300	20.6 (1.0)	121 (10)	29.3 (2.9)	3.12 (0.30)	0.884 (0.120)	0.137 (0.012)
W						
100	0.915 (0.079)	6.86 (0.58)	2.58 (0.27)	0.098 (0.008)	0.0254 (0.0021)	0.0073 (0.0018)
150	2.78 (0.22)	20.4 (1.6)	8.02 (0.78)	0.305 (0.025)	0.0842 (0.0068)	0.026 (0.005)
200	5.55 (0.45)	39.7 (3.1)	16.4 (1.5)	0.643 (0.054)	0.185 (0.015)	0.050 (0.007)
250	8.91 (0.74)	63.9 (5.0)	26.1 (2.4)	1.09 (0.10)	0.302 (0.026)	0.072 (0.010)
300	12.8 (1.1)	91.7 (7.2)	38.6 (3.5)	1.72 (0.16)	0.454 (0.040)	0.102 (0.015)
Au						
100	0.0305 (0.0113)	1.04 (0.36)	0.553 (0.185)	0.0216 (0.0062)	0.00530 (0.00208)	0.0002 (0.0001)
150	0.191 (0.084)	5.38 (1.64)	2.97 (0.91)	0.103 (0.023)	0.0305 (0.0090)	0.003 (0.001)
200	0.343 (0.095)	12.7 (3.0)	7.04 (1.63)	0.223 (0.038)	0.0687 (0.0152)	0.008 (0.003)
250	0.572 (0.130)	22.3 (4.2)	12.7 (2.4)	0.367 (0.053)	0.121 (0.023)	0.012 (0.003)
300	0.572 (0.130)	38.7 (7.1)	22.0 (4.1)	0.673 (0.093)	0.217 (0.038)	0.014 (0.004)
Pb						
100	0.0787 (0.0144)	1.31 (0.42)	0.606 (0.124)	0.027 (0.004)	0.00269 (0.00076)	
150	0.407 (0.067)	6.81 (2.10)	3.40 (0.68)	0.135 (0.017)	0.0195 (0.0046)	
200	1.05 (0.16)	18.7 (5.0)	8.96 (1.70)	0.330 (0.038)	0.0559 (0.0121)	
250	1.95 (0.27)	32.1 (8.9)	17.1 (3.1)	0.609 (0.067)	0.112 (0.023)	
300	3.10 (0.42)	51.5 (14.0)	28.2 (5.1)	1.01 (0.11)	0.198 (0.038)	

accelerate protons onto a thick target arranged on a vertical ladder oriented at 45° to the beam. Surrounding the ladder is a copper cylinder that is biased with respect to the ladder to hold secondary electrons in the target. A current integrator connected to the target ladder is used to determine the number of protons hitting a target. A different spot on a target foil is used for each proton energy. Beam currents are corrected for leakage current from the target bias battery and for dead time (typically 2%) in the x-ray measurement. X-rays are measured with a Si(Li) detector equipped with an ultra-thin boron-nitride window. The detector axis is oriented at 45° to the target ladder and at 90° to the beam. To minimize dead-time effects from detection of very soft x-rays, one or two layers of 6-μm thick doubly-aluminized Mylar absorber are placed in front of the detector. The detector signals are processed for pile-up rejection and live-time correction in conjunction with a multichannel analyzer.

DATA ANALYSIS AND RESULTS

X-ray peaks were fitted to a sum of gaussian functions plus a fixed linear background. The gaussian widths were fitted using FWHM = $2.355(A + B(E_x)^{1/2})$, where A and B are the fit parameters, and the x-ray energy E_x is determined from $E_x = a + bX + cX^2$, where X, a, b, c are fitted parameters, X being the channel number of the peak.

Following the spectral analysis, the x-ray production cross section for each x-ray transition at proton energy E_o was determined from

$$\sigma_x(E_o) = (\frac{1}{N\varepsilon})[\frac{dY}{dE_o}S(E_o) + \frac{\mu}{\rho}Y(E_o)] \quad (1)$$

where $Y(E_o)$ is the x-ray yield (x-rays/proton), dY/dE_o is the derivative of the yield at E_o, $S(E_o)$ is the proton stopping power (from TRIM[17]) for the target, μ/ρ is the mass absorption coefficient for x-rays (from XCOM[18]) in the target, N is the atom density of the target, and ε is the efficiency of the detector. The yield derivative was obtained by fitting the yield data for each peak to $A(E_o - B)^C$, with A, B, and C being the fitted parameters, and then differentiating. The detector efficiency responses were determined using the model equation developed by our group[19].

The M sub-shell x-ray production cross sections, σ_{Mi}, were determined from the M1N3, M2N4, M3N4,5(γ), M4N6(β) and M5N6,7(α) peak cross sections, σ_x, respectively, according to $\sigma_{Mi} = \sigma_x \Gamma_{Mi}/\Gamma_x$, where Γ_x is the x-ray peak transition rate to the M_i shell, and Γ_{Mi} is the total rate for the shell[9]. Figure 1 displays ratios of the total x-ray production cross sections, $\sigma_t = \Sigma\sigma_{Mi}$ to the ECPSSR values for 75-300 keV protons impacting Er, Ho, Yb, Hf, W, Au, and Pb as a function of the reduced projectile velocity, $\xi_i = 2v_1/\theta_i v_{2i}$, where v_1 is the projectile velocity, θ_i is the shell binding energy, and v_{2i} is the electron orbital velocity. Coster-Kronig factors and fluorescence yields came from McGuire[7]. The measured x-ray cross sections for the major peaks are presented in Table 1. Figure 2 shows the scaling of the $M_{\alpha,\beta}$ and M_γ cross sections with ξ, to compare with the empirical fit to similar data by Pajek, et al.[16]

DISCUSSION

The results shown in Fig. 1 are representative of those measured by other investigators. The ECPSSR theory increasingly under-predicts measured cross sections as the proton energy or reduced projectile velocity, starting at approximately $\xi_i = 0.6$. It also appears from Fig. 1 that ECPSSR over-predicts the cross sections of the lighter elements above $\xi_i = 0.6$. As the atomic number of the target element increases, the trend is for ECPSSR to increasingly under-predict the data. This trend is evidenced for these data in Fig. 3.

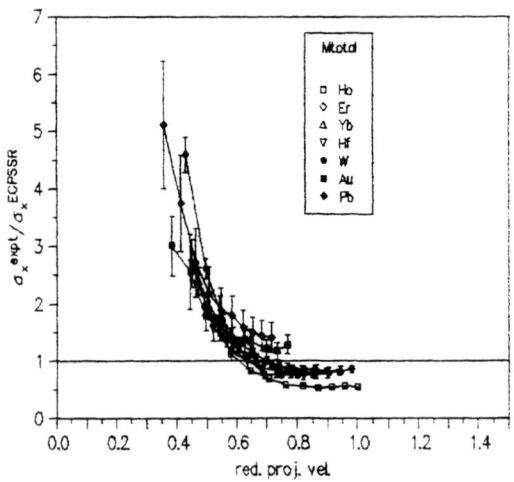

FIGURE 1. Ratio of measured total M x-ray production cross sections to ECPSSR values versus the reduced projectile velocity, $\xi_\tau = (2\xi_1 + 2\xi_2 + 4\xi_3 + 4\xi_4 + 6\xi_5)/18$.

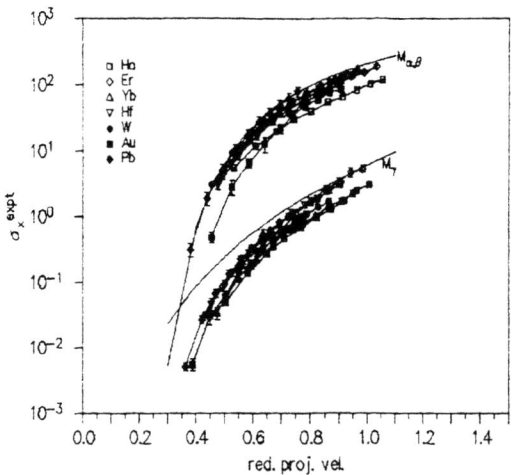

FIGURE 2. Measured $M_\alpha+M_\beta$ and M_γ x-ray production cross sections versus the reduced projectile velocity ($\xi_{\alpha,\beta} = (4\xi_4+6\xi_5)/10$, $\xi_\gamma = \xi_3$). Also shown (solid lines without data points) are the reference cross sections of Pajek, et al[16].

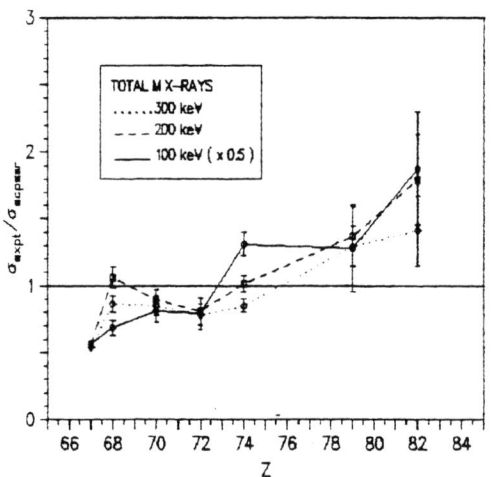

FIGURE 3. Ratios of measured total M x-ray production cross sections to ECPSSR values versus the atomic number of the target element for three representative proton bombardment energies.

The x-ray cross sections shown in Fig. 2 are seen to scale according to the reduced projectile velocity, as was shown by Pajek, et al[16] over a broader velocity range. They made an empirical polynomial fit to each cross section group for the purpose of application towards PIXE. Their fitted functions are shown in Fig. 2, which seem to match the data fairly well, except perhaps for the lower velocity range of the M_γ cross sections, a region that is beyond the fitted range of the polynomial.

CONCLUSION

The M sub-shell x-ray production cross sections for 75-300 keV protons on thick targets of Ho, Er, Yb, Hf, W, Au and Pb measured in this work are generally higher than ECPSSR predictions below $\xi_i = 0.6$. Above this region, there is better agreement for the outer-most subshells (which dominate the total cross section) as Z of the target atom decreases. It is also noticed that the x-ray cross sections scale with ξ. These trends are also noticed in L-shell ionization studies.

ACKNOWLEDGMENT

The author wishes to acknowledge the University of Nebraska-Lincoln for use of their laboratory where this research was conducted.

REFERENCES

1. E. Merzbacher and H. Lewis, Hand. d. Phys. 34 (1958) 166.
2. D. Johnson, G. Basbas, and F.D. McDaniel, At. Data Nucl. Data Tables 24 (1979) 1.
3. J. Bang and J. Hansteen, K. Dansk. Vidensk. Selsk. Mat. Fys. Medd 31 (1959) 13.
4. W. Brandt and G. Lapicki, Phys. Rev. A 23 (1981) 1717.
5. Z. Liu and S. Cipolla, Comp. Phys. Comm. 97 (1996) 315.
6. M.H. Chen and B. Crasemann, Phys. Rev. A 27 (1983) 2358.
7. E.J. McGuire, Phys. Rev. A 5 (1972) 1043.
8. M.H. Chen and B. Crasemann, Phys. Rev. A 21 (1980) 449; Phys. Rev. A 27 (1980); Phys. Rev. A 30 (1984) 170.
9. C.P. Bhalla, J. Phys. B 3 (1970) 916.
10. G. Lapicki, Abstr. Contr. Papers, XVI ICPEAC, ed. A. Dalgarno, R. Freund, M. Lubell, T. Lucatorto, New York, p. 619.
11. M. Pajek, A. Kobzev, G. Lapicki, Nucl. Instr. & Meth. Phys. Res. B 48 (1990) 87; M. Pajek, A. Kobzev, R. Sandrik, A. Skypnik, R. Ilkhamov, S. Khusmurodov, G. Lapicki, Phys. Rev. A 42 (1990) 261; A. Bienkowski, J. Braziewicz, T. Czyzewski, L. Jaskola, G. Lapicki, M. Pajek, Nucl. Instr. & Meth. Phys. Res. B 49 (1990) 19.
12. H. Sun, J. Kirchhoff, a. Azordegan, J. Duggan, F. McDaniel, R. Wheeler, R. Chaturvedi, G. Lapicki,

Nucl. Instr. & Meth. Phys. Res. B <u>79</u> (1993) 194.

13. S. Cipolla, *Nucl. Instr. & Meth. Phys. Res.* B <u>99</u> (1995) 22; S. Cipolla, P. Teeter, J. McClure, AIP Conf. Proc. CP475 (1999) 36; K. Welsh and S. Cipolla, AIP Conf. Proc. CP475 (1999) 23.
14. F. Shokouhi, S. Fazinic, I. Bogdanovic, M. Jaksic, V. Valkovic, H. Afarideh, *Nucl Instr. & Meth. Phys.Res.* B <u>109/110</u> (1996) 15.
15. J. Braich, P. Verma, D.P. Goyal, A. Mandal, B. Dhal, H. Padhi, H.R. Verma, *Nucl. Instr. & Meth. Phys.Res.* B <u>119</u> (1996) 317; J. Braich, P. Verma, H. Verma, *J. Phys.* B <u>30</u> (1997) 2359.
16. M. Pajek, M. Jaskola, T. Czyzewski, L. Glowacka, D. Banas, J. Braziewicz, W. Kretschmer, G. Lapicki, D. Trautmann, *Nucl. Instr. & Meth. Phys. Res.* B <u>150</u> (1999) 33.
17. J. Ziegler, TRIM, ver. 91.07, IBM, Yorktown Hts., NY, July 26, 1991.
18. M. Berger and J. Hubbell, XCOM, NIST, Gaithersburg, USA. Report NBSIR 87-3595 (July, 1987).
19. S. Cipolla and S. Watson, *Nucl. Instr. & Meth.* B <u>10/11</u> (1985) 946.

Non-statistical Magnetic Substate Populations Following Excitation of Helium by Electron and Proton Impact

J. Hanni[a], H. Merabet[a,*], A. Siems[b], R. Bruch[a], M. Bailey[a], D.V. Fursa[c], I. Bray[c], K. Bartschat[d], H.C. Tseng[e], C.D. Lin[f], A.G. Trigueiros[b]

[a]*Department of Physics, University of Nevada, Reno, NV 89557 USA*
[b]*Instituto de Física, Universidade Estadual de Campinas (Unicamp), 13083-970 Campinas, São Paulo, Brazil*
[c]*The Flinders, University of South Australia, GPO Box 2100, Adelaise 5001 Australia*
[d]*Department of Physics and Astronomy, Drake University, Des Moines, IA 50311, USA*
[e]*Department of Physics, Chung Yuan Christian University, Chung Li, Taiwan 32023*
[f]*Department of Physics, Kansas State University, Manhattan, KS 66506-2601 USA*

Abstract. The first experimental magnetic substate scattering angle-integrated cross sections following excitation of He $(1s^2)$ ^1S to He $(1snp)$ $^1P^o$ (n=2-5) in e– + He and H$^+$ + He collision systems have been determined using our differential cross sections and polarization fraction data in the extreme ultraviolet (EUV) range. The derived magnetic sublevel integrated cross sections, $\bar{\sigma}_0$ and $\bar{\sigma}_1$, for $M_L=0,\pm1$ have been studied over a wide range of projectile velocities: 1.4 to 8.5 a.u. and 1.4 to 7.5 a.u. for electron and proton impact on helium, respectively. In addition, the electron and proton collision data are compared with theoretical predictions using our improved first Born approximation (IFB), convergent close coupling (CCC) calculations for electron impact, and our recent atomic orbital close coupling (AOCC) calculations for protons. We have found that the electron results match very well the CCC predictions at all energies. However, the experimental proton data slightly deviate from the AOCC approach at intermediate velocities. Such findings are relevant to plasma and astrophysical applications.

INTRODUCTION

Two electron atomic systems are of fundamental importance to the investigation of complex many-body problems in physics (1). In the past, great effort has been devoted to the study of atomic and molecular collisions using helium atom as a target (2-5). Helium, is the simplest two-electron system, therefore it is well suited for studying the electron correlation effects (1,6-8). Previous research concerning excitation, ionization, double excitation and ionization-excitation of He has produced an extensive experimental and theoretical database necessary for achieving a deeper understanding of the collision dynamics in electron, proton, and multielectron ion impact on He at intermediate and higher energies (1-11).

We have measured differential excitation cross sections (DCS) $\bar{\sigma}$ for HeI $(1snp)$ $^1P^o$ states, n=2-5, in e– + He and H$^+$ + He collisions using EUV spectrometry (11). These absolute cross sections do not provide direct information about the magnetic substate populations, while linear polarization measurements yield only the magnetic sublevel scattering angle-integrated cross section ratios $\bar{\sigma}_0/\bar{\sigma}_1$. These ratios are directly related to the degree of linear polarization (11) as:

$$P = \frac{I_\parallel - I_\perp}{I_\parallel + I_\perp} = \frac{\bar{\sigma}_0 - \bar{\sigma}_1}{\bar{\sigma}_0 + \bar{\sigma}_1} = \frac{1-r}{1+r} \quad (1)$$

where $I_\parallel$ is the intensity of radiation with electric field vectors oriented along the beam axis and $I_\perp$ is the intensity of radiation with a transverse electric field vector (perpendicular to the plane formed by the incident projectile beam and the direction of observation); $r = \bar{\sigma}_0/\bar{\sigma}_1$, is the cross section ratio. Moreover, the differential cross section $\bar{\sigma}$ is the sum of the three magnetic sublevel angle-integrated cross sections,

$$\bar{\sigma} = \bar{\sigma}_0 + 2\bar{\sigma}_1 \quad (2)$$

Thus, by combining Eq. 1 and Eq. 2, the magnetic sublevel integrated cross sections can be obtained.

Another procedure has been employed, where the ratio of the magnetic sublevel (scattering) angle-differential cross section, σ_0, to the differential cross section, called the parameter $\lambda = \sigma_0/(\sigma_0 + 2\sigma_1)$, is determined experimentally using electron-photon coincidence techniques (12). The λ parameter, combined with available double differential cross

---

* Corresponding author: H. Merabet, email: hocine@physics.unr.edu

section ($\sigma = \sigma_0 + 2\sigma_1$) data, yields σ_0. This approach has been utilized by Chutjian et al. for the excitation of HeI (1s2p) $^1P^o$ level, at 60 and 80 eV electron impact energies (13), and HeI (1s2p) $^1P^o$ level at 80 and 100 eV (14); whereas Hummer et al. (15) have directly obtained the relative σ_0 for the excitation of the HeI (1s3p) $^1P^o$ state at 70 eV as a function of the scattering angle. Furthermore, Harris and co-workers (16) have measured triple differential cross sections for HeI (1s3p) $^1P^o$ for electron impact energy of 40 eV. In addition, Csanak et al. (17) have used the λ parameter procedure to extract HeI (1snp) $^1P^o$ (n=2-3) triply differential cross sections and have computed the corresponding theoretical integral values for n=2-6 with incident energy in the 25-500 eV range, using first-order many-body theory (FOMBT) (18). To our knowledge, no experimental data are available for HeI (1snp) $^1P^o$ n $\geq$ 4 levels.

In this work, we have combined measurements of two experimental techniques, namely EUV spectrometry and EUV polarimetry, to determine experimental cross section results for HeI (1snp) $^1P^o$ (n=2-5) excited states following electron and proton impact at a wide range of projectile velocities. These experimental data are compared with our IFB as well CCC results for electrons, and first Born (B1) (19) along with the AOCC predictions for protons, in an attempt to shed more light on the excitation processes of helium during the collision at intermediate and high energies. The CCC and AOCC methods have been described in detail by Fursa and Bray (20), and Fritsch et al. (21), respectively, whereas the IFB method is essentially similar to the first Born approximation, however with a better target description.

RESULTS AND DISCUSSION

The experimental setup used in this work consists of three main components: an EUV polarimeter; an electron gun or a 2 MV Van de Graaff accelerator, target cell and Faraday cup; and a 1.5 meter grazing incidence monochromator. A PC controlled data acquisition system has been used to operate the apparatus and to record the data. A detailed description of this experimental setup is given by Bailey et al. (7,22). In brief, the polarimeter utilizes a molybdenum-silicon (Mo/Si) MLM whose surface reflection has been used for radiation with a wavelengths $\lambda > 584$ Å at an incidence angle of 40° corresponding to the HeI (1snp) $^1P^o \rightarrow (1s^2)$ ^1S (n=2-5) transitions. It is assumed here that the degree of linear polarization does not depend on the principal quantum numbers n. The corresponding differential cross sections $\overline{\sigma}$ measurements have been conducted using a 1.5 m high resolution grazing incidence monochromator (7-8). These results have been put on an absolute scale by normalizing our data to the Bethe-Born cross section values (23), for electron and proton impact, at high velocities. This normalization procedure has been described in more details in Ref. (7). In this work, the obtained cross section data have also been corrected for alignment effects using (9),

$$\overline{\sigma}(\theta) = \overline{\sigma} \times \{1 + A_0 P_2 (\cos\theta)\} \quad (3)$$

where $\overline{\sigma}(\theta)$ is the measured cross section, $\theta=90°$ is the observation angle of the emitted photons, $\overline{\sigma}$ is the cross section for an isotropic distribution, $P_2(\cos\theta)$ is the second Legendre polynomial, and A_0 is the alignment parameter related to the degree of linear polarization by:

$$P(^1P^o) = \frac{3 A_0}{A_0 - 2} \quad (4)$$

In the following, we have analyzed the obtained cross sections for electron, and proton projectiles as a function of impact velocities for HeI (1s2p) $^1P^o$ states. The complete set of data for higher Rydberg states, with the quantum number n=3-5, are available and will be published in the near future. A preliminary comparison of both experimental and theoretical results is also given.

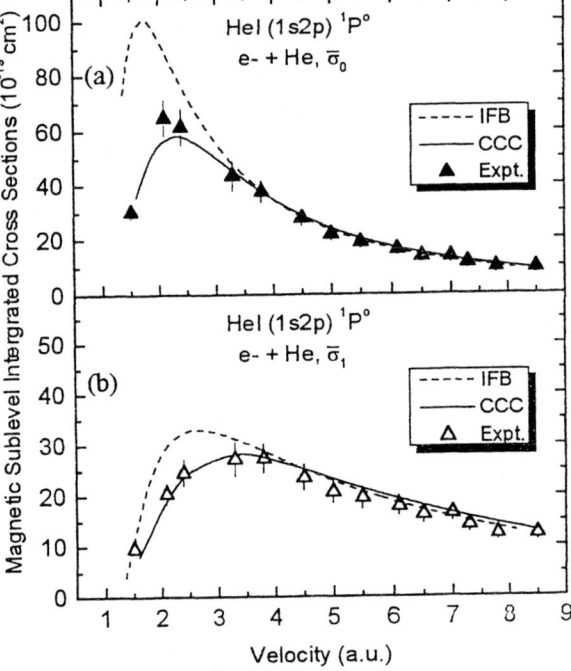

Figure 1. $\overline{\sigma}_0$ (a) and $\overline{\sigma}_1$ (b) for e– + He collisions, compared with CCC and IFB calculations.

A. Electron Impact on Helium

Using equations (1) and (2), we have derived $\overline{\sigma}_M$ following HeI (1snp) $^1P^o \rightarrow (1s^2)$ ^1S, (n=2-5), transitions in e– + He collisions at impact energies ranging from 30 to 980 eV (1.4 < v < 8.5 a.u.). As an example, we have shown in Fig. 1, $\bar{\sigma}_0$ and $\bar{\sigma}_1$ for HeI (1s2p) $^1P^o$ states along with our IFB and CCC predictions. We have used our polarization and DCS data for $v \geq 3$ a.u. The shown lower velocities magnetic substate cross sections $\bar{\sigma}_M$ for electrons have been extracted by combing the DCS of Westerveld et al (4) with our polarization results. As can be seen, both theories and experimental results are in excellent agreement in the high-energy range. At relatively lower energies, the IFB predictions diverge from the experimental data, while our CCC values coincide with the measured results. This confirms that the CCC is an excellent approach for describing the e-He scattering problem.

Furthermore, the $\bar{\sigma}_0$ cross sections are all equal to the theoretical predictions within experimental uncertainties. However, the $\bar{\sigma}_1$ cross sections slightly deviate from theory. Fig. 2 exhibit our $\bar{\sigma}_0$ compared with $\bar{\sigma}_1$ for electron impact. The velocity dependence of these cross sections reveals a strong anisotropy of the photon emission originating from the HeI (1s2p) $^1P^o$ states at v < 4 a.u.. From Fig. 2 it is also evident that the population of the sublevels with $M_L = 0$ is

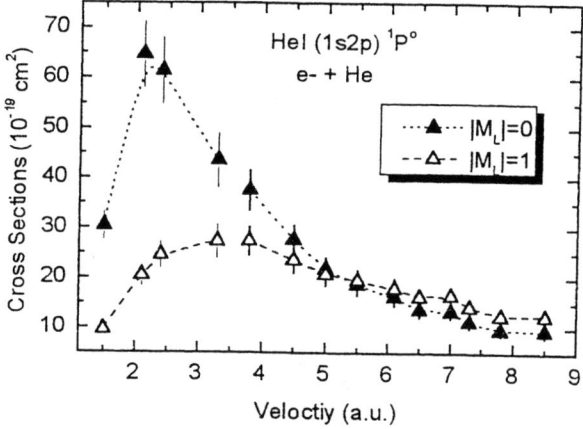

Figure 2. Comparison of MSICS σ_0 with σ_1 for electron impact on He. Fitted curves are provided to guide the eye.

greater than those with $M_L = \pm 1$ at impact velocities below 5 a.u. The cross sections $\bar{\sigma}_0$ and $\bar{\sigma}_1$ are equal at approximately v=5 a.u. For high electron velocities, the situation is reversed, i.e. $\bar{\sigma}_1$ is little larger than $\bar{\sigma}_0$, leading to a nearly isotropic photon emission.

B. Proton Impact on Helium

The cross sections $\bar{\sigma}_0$ and $\bar{\sigma}_1$ for HeI (1snp) $^1P^o$ states following proton impact have been extracted using the procedure outlined above at energies ranging from 50 to 1400 keV (1.4 < v < 7.5 a.u.). We have plotted in Fig. 3 the obtained experimental results in comparison with AOCC and B1 theoretical calculations corresponding to HeI (1s2p) $^1P^o \rightarrow$ (1s^2) ^1S transitions. In general, our experimental findings deviate from the B1 approximation at intermediate projectile velocities. The AOCC theory is somewhat closer to experiment than the first Born results, but it also partially fails to reproduce the measured values at lower velocities. Specifically, $\bar{\sigma}_1$ values match well the B1 and AOCC results in the studied velocity range. However, the 2 a.u. impact velocity point disagrees completely with both theories. We further note that in the case of $M_L = 0$, this lower impact velocity point is in excellent accord with the AOCC prediction and also in good agreement with the B1 approximation. In the high-energy limit, our experimental cross sections converge with both theories.

The AOCC calculations, presented in this paper, have been performed utilizing limited target states and they are, therefore, not expected to completely reproduce the experimental data in the intermediate energy range. A more accurate calculation, including extended target states may improve the convergence to the experimental data. Such more sophisticated calculations are now being performed.

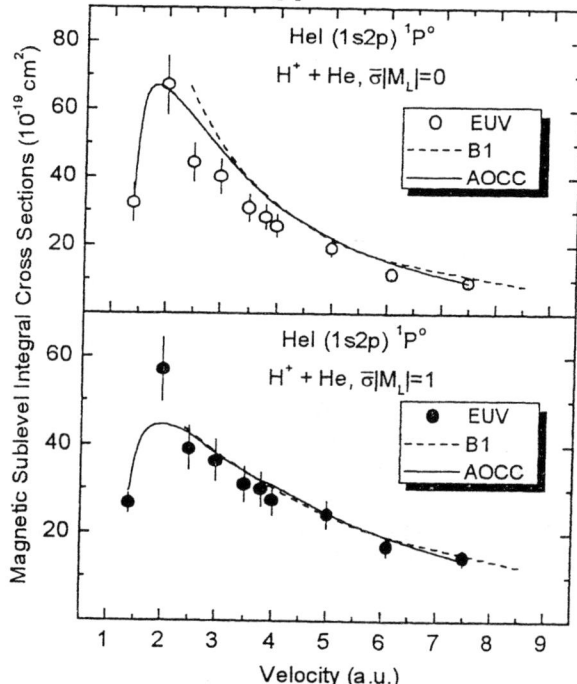

Figure 3. $\bar{\sigma}_0$ (a) and $\bar{\sigma}_1$ (b) for H$^+$ + He collisions, compared with CCC and IFB calculations.

Now let us compare the $\bar{\sigma}_M$ proton cross sections. Fig.4 shows these cross sections for $M_L = 0$ and $M_L = \pm 1$. In the vicinity of v = 3.5 a.u., $\bar{\sigma}_0$ is equal to $\bar{\sigma}_1$. At lower proton velocities, $\bar{\sigma}_0$ is slightly larger than

$\bar{\sigma}_1$, however the error bars are large. Above v = 4 a.u., $\bar{\sigma}_1$ overtakes $\bar{\sigma}_0$ and it clear that, for impact velocities higher than 5 a.u., an anisotropic population of the magnetic sublevel cross sections occurs.

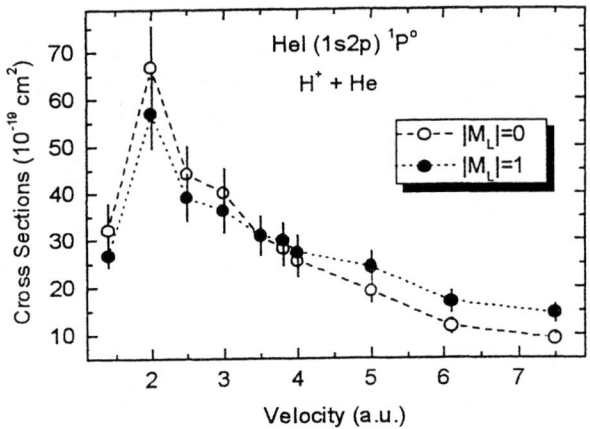

Figure 4. Comparison of $\bar{\sigma}_0$ with $\bar{\sigma}_1$ for proton impact on helium. Fitted curves have been provided to guide the eye.

It interesting to study the charge state dependence for electron and proton projectiles in order to gain a better understanding of the excitation process. This is the main focus of the following section.

C. Comparison of Electron and Proton Cross Section Results

We have depicted in Fig. 5, $\bar{\sigma}_0$ and $\bar{\sigma}_1$ for electron and proton impact. Fig. 5(a) compares $\bar{\sigma}_0$ cross sections. From 2.5 to 6 a.u., the electron and proton cross sections deviate, *i.e.*, the electron results are greater than the protons. For higher impact velocities, they converge, whereas at the lower velocities, they appear to be approximately equal. Fig. 5(b) indicates that the electron and proton cross sections are clearly unequal below 3.5 a.u., but they coincide at higher velocities. This excellent agreement in the high-energy limit is expected since both experiment and theoretical results agree very well for electrons and protons and confirm, therefore, that the excitation cross sections are independent of the charge-state and mass of the projectile.

In summary, excellent agreement between theory and experiment has been found for electron impact, although this is not quite the case for the H^+ + He collision system which requires a more accurate theoretical description to elucidate the detailed dynamics of the excitation process.

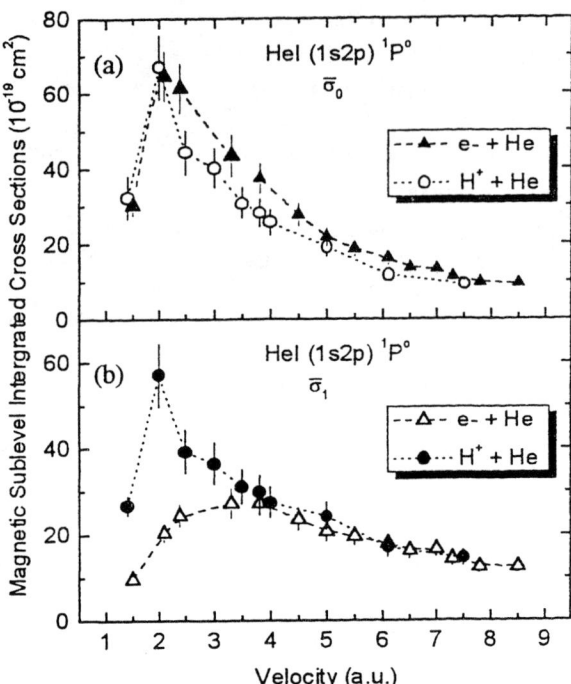

Figure 5: Comparison of for electron (a) and proton (b) impact on He. Fitted curves have been provided to guide the eye.

ACKNOWLEDGMENTS

This Project has been supported in part ACSPECT Corp. and the Nevada Business and Science Foundation, (NBSF) Reno, Nevada.

REFERENCES

1. McGuire, J. H., *Electron Correlation Dynamics In Atomic Collisions*, (Cambridge University Press, 1997).
2. van den Bos, J. *et al.*, Physica **40**, 357 (1968).
3. Hippler, R. et al., K.H., *J. Phys. B* **7**, 618 (1974).
4. Westerveld, W.B. *et al.*, *J. Phys. B.* **12**, 115 (1979).
5. Forand, J.L. *et al.*, *J. Phys. B:* **18**, 1409 (1985).
6. Bruch, R. *et al.*, *Encyc. Applied Phys.* **10**, 437 (1994).
7. Bailey, M. *et al.*, *J. Phys. B* **28**, 2655 (1995).
8. Merabet, H. *et al.*, *Phys. Rev. A* **60**, 1187 (1999).
9. Hammond, P. *et al.*, *Phys. Rev. A.* **40**, 1804 (1989).
10. Götz, A. *et al.*, *J. Phys. B.* **29**, 4699 (1996).
11. Merabet, H. *et al.*, this issue.
12. Blum, K. *et al.*, *Adv. At. Mol. Phys.* **19**, 187 (1983).
13. Chutjian, A. et al., *J. Phys. B* **8**, 2360 (1975).
14. Chutjian, A., *J. Phys. B* **9**, 1749 (1976).
15. Hummer, C.R. *et al.*, *Phys. Rev. A* **33**, 2995 (1986).
16. Harris, C. L., *Ph.D. Thesis*, University of Nevada, Reno (2000), unpublished.
17. Csanak, G. *et al.*, *Phys. Rev. A* **45**, 1625 (1992).
18. Csanak, G. *et al.*, *Phys. Rev. A* **3**, 1322 (1971).
19. L. Vriens and J.D. Carrière, *Phys. Scp.* **49**, 517 (1970).
20. Fursa, D.V. *et al.*, *J. Phys. B* **30**, 757 (1997).
21. Fritsch, W. and Lin, C.D., *Phys. Rept.* **201**, 1 (1991).
22. Bailey, M. *et al.*, *Appl. Opt.* **38**, 4125 (1999).
23. Kim, Y. K. *et al.*, *Phys. Rev.* **184**, 38 (1969).

Projectile and Target Z-scaling of Target K-vacancy Production Cross Sections at 10 MeV/amu

R. L. Watson, V. Horvat, and K. E. Zaharakis

*Cyclotron Institute and Department of Chemistry,
Texas A&M University, College Station, TX 77843, USA*

Abstract. Beams of 10 MeV/amu Ne, Ar, Cr, Kr, Xe and Bi ions have been employed to examine the dependence of target atom K-vacancy production cross sections on projectile and target atomic number. The collision systems studied ranged in (Z_1/Z_2) from 0.14 to 6.4 and in relative velocity (v_1/v_{2K}) from 0.31 to 1.87. Cross sections for near symmetric collision systems with $v_1/v_{2K} < 0.5$ were significantly larger than those expected from the systematic trends observed for asymmetric systems involving the same targets. A simple empirical scaling law was found to satisfactorily represent the overall Z_1 and Z_2 dependences of the K-vacancy production cross sections outside the regions of enhancement noted above.

INTRODUCTION

In two recent studies, attention was focused on the projectile atomic number (Z_1) dependence of Al and Cu target atom K-vacancy production cross sections in collision systems for which the ratio of projectile atomic number to target atomic number (Z_1/Z_2) ranged from 0.34 to 6.4 [1,2]. It was found that the cross sections for projectiles with $Z_1 > Z_2$ increase much more slowly as a function of Z_1 than predicted by theoretical (PWBA/ECPSSR) calculations and they appear to approach a saturation limit at high Z_1. In an attempt to develop a means of estimating K-vacancy production cross sections for heavy ion collisions, a simple scaling law was found that provided a reasonably good representation of the Z_1 dependences of the measured cross sections for both Al and Cu targets [2].

The object of the present work was to examine the general applicability of this scaling law and to investigate further the dependence of heavy ion K-vacancy production cross sections on Z_2. Using the same 10 MeV/amu beams as in the previous two studies, the investigation has been extended to much higher Z_2 by performing measurements for Mo, Ag, Sn, Sm, and Ta.

EXPERIMENTAL METHOD

The experimental details of the present measurements were nearly identical to those described in Ref. [2]. Thin metallic foils of Mo, Ag, Sn, and Ta, ranging in thickness from 0.79 mg/cm^2 to 2.67 mg/cm^2, and a Sm target prepared by vacuum evaporating 2.77 mg/cm^2 of metallic Sm onto a thin Al backing were mounted in a target wheel and positioned at a 45° angle relative to both the incident heavy ion beam and a Si(Li) detector. Absolute Kα x-ray production cross sections were determined for Mo, Ag, and Sn by measuring the K x-ray yields from these targets in coincidence with particle signals generated by a plastic scintillator detector mounted 4 cm behind the target position. In the cases of Sm and Ta, the cross sections were too small to allow direct particle counting, and so their x-ray yields were measured relative to those observed from Ag monitor foils mounted directly behind the Sm and Ta targets. The Ag monitor absolute x-ray yields per particle were determined in the same way as those for the other targets (i.e., by direct particle counting).

Based on previous target thickness dependence measurements [1,2], the thicknesses of the targets were chosen to give x-ray production cross sections for charge-equilibrated projectiles. Corrections for projectile energy loss have been neglected in this study, but checks performed using the Ag data obtained with the three targets Ag-only, Sm+Al+Ag, and Ag+Ta indicated that corrections to the cross sections for energy loss were less than 5% in the worst case.

Another possible source of error in the present measurements is the neglect of contributions to the K x-ray yields from secondary processes (e.g., photoionization and ionization by secondary electrons). However, based on the Z_2 dependence of x-ray production by secondary processes observed previously for Al and Cu targets [1,2], their contributions are expected to be of the order of 5% for Mo and less than 3% for the rest of the targets.

The measured $K\alpha$ x-ray yields per particle were converted to ionization cross sections using normal (single-vacancy) fluorescence yields. Although the fluorescence yields are undoubtably affected by multiple ionization, the degree of multiple ionization in these relatively high Z_2 targets is thought to be small enough that corrections can be neglected. Moreover, the values of the fluorescence yields are large enough that small changes do not introduce significant errors.

RESULTS AND DISCUSSION

The K-vacancy production cross sections obtained in the present experiments are listed in Table 1, along with the fluorescence yields used to convert them from x-ray production cross sections. Taking into account the errors associated with target thickness, detector efficiency, projectile energy loss, secondary x-ray production, fluorescence yield, and counting statistics, the overall uncertainty in the reported K-vacancy production cross sections is estimated to be ±12% for Mo, Ag, and Sn, and ±15% for Sm and Ta.

The K-vacancy production cross sections obtained in this work, together with those obtained previously for Cu (2), are shown in Fig. 1 plotted as a function of Z_1. The cross sections shown for protons ($Z_1 = 1$) are calculated (ECPSSR) values taken from Ref. [5]. The dashed line in each frame of this figure is drawn at the value of Z_1 that equals the target atomic number (Z_2) to delineate the region of symmetric collisions. In the cases of Ag, Sn, Sm, and Ta, it is evident that cross sections near the symmetric collision region (shown by the empty circle data points) are much larger than would be expected from the systematic trend of the cross sections in neighboring asymmetric collision regions (shown by the filled circle data points). Quadratic curves have been fit through the filled circle data points to emphasize this fact. Presumably, the enhanced cross sections for these near symmetric collision systems arise from the well established electron promotion mechanism associated with production cross sections determined in the previous studies [1,2] were crossings of quasimolecular orbitals [6]. The cross sections for Cu and Mo do not display

TABLE 1. Cross sections (kb) for K-vacancy production by 10 MeV/amu projectiles and target fluorescence yields ($\omega_{K\alpha}$). (The numbers in parentheses indicate powers of ten).

Target $\omega_{K\alpha}{}^a$		Projectile				
	Ne	Ar	Cr	Kr	Xe	Bi
Mo 0.64	8.5(0)	2.5(1)	4.3(1)	4.2(1)	5.7(1)	6.3(1)
Ag 0.68	3.0(0)	8.6(0)	1.2(1)	1.7(1)	3.3(1)	2.4(1)
Sn 0.70	1.8(0)	4.5(0)	5.9(0)	8.3(0)	2.1(1)	1.2(1)
Sm 0.73	2.3(-1)	4.3(-1)	4.5(-1)	6.7(-1)	3.0(0)	2.3(0)
Ta 0.75	4.2(-2)	8.4(-2)	9.5(-2)	1.3(-1)	5.0(-1)	1.2(0)

[a] The $K\alpha$ fluorescence yield ($\omega_{K\alpha}$) is equal to $\omega_K (1 + R)^{-1}$, where ω_K is the K-shell fluorescence yield (from Ref. [3]) and R is the ratio of the $K\beta$ and $K\alpha$ x-ray intensities (from Ref. [4]).

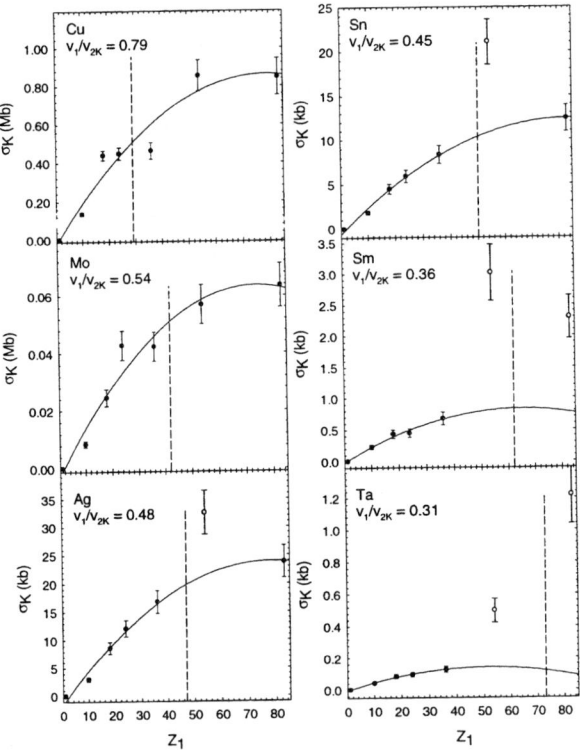

FIGURE 1. K-vacancy production cross sections for the indicated target elements as a function of projectile atomic number. See text for further information.

a similar enhancement in the region of symmetric collisions. This fact suggests that electron promotion does not contribute in a major way to K-vacancy production cross sections at relative velocities (v_1/v_{2K}) above 0.5.

The Z_1-dependence of the Cu and Al K-vacancy reasonably well represented by the empirical scaling law

$$\sigma(Z_1) = \sigma(1) Z_R^m,$$

where $\sigma(Z_1)$ is the K-vacancy production cross section for projectiles of atomic number Z_1, $\sigma(1)$ is the K-vacancy production cross section for protons, Z_R is the reduced atomic number defined by

$$Z_R = \frac{Z_1 Z_2}{Z_1 + Z_2},$$

and m is an exponent that slowly varies with Z_2. Fits of the above scaling law to the cross sections determined in the present work and to the cross sections for Cu and Al determined previously, are shown by the solid curves in Fig. 2. In making these fits, the enhanced cross sections for Ag, Sn, Sm, and Ta (shown by the open circle data points in Fig. 1) were excluded. Overall, the scaling law represents the general trend of most of the data reasonably well (with the exception of the enhanced cross section data points). The dependence of the scaling law exponent m is on Z_2 shown in Fig. 3. A third order polynomial (solid curve) fits the data very well and provides a means for estimating the exponent at other values of Z_2. The equation of this fitting function is

$$m = 2.158 + 4.091 \times 10^{-2} Z_2 - 1.431 \times 10^{-3} Z_2^2 + 1.046 \times 10^{-5} Z_2^3.$$

ACKNOWLEDGEMENTS

This work was supported by the Robert A. Welch Foundation. The authors thank Yong Peng for help with some of the experiments.

REFERENCES

[1] R. L. Watson, J. M. Blackadar, and V. Horvat, Phys. Rev. **A60**, 2959 (1999).
[2] R. L. Watson, V. Horvat, J. M. Blackadar, and K. E. Zaharakis, Phys. Rev. **A62** (in press).
[3] A. Langenberg and J. van Eck, J. Phys. B **12**, 1331 (1979).
[4] J. H. Scofield, Phys. Rev. **9**, 1041 (1974).
[5] D. D. Cohen and M. Harrigan, At. Data Nucl. Data Tables **33**, 255 (1985).
[6] R. Anholt, Rev. Mod. Phys. **57**, 995 (1985).

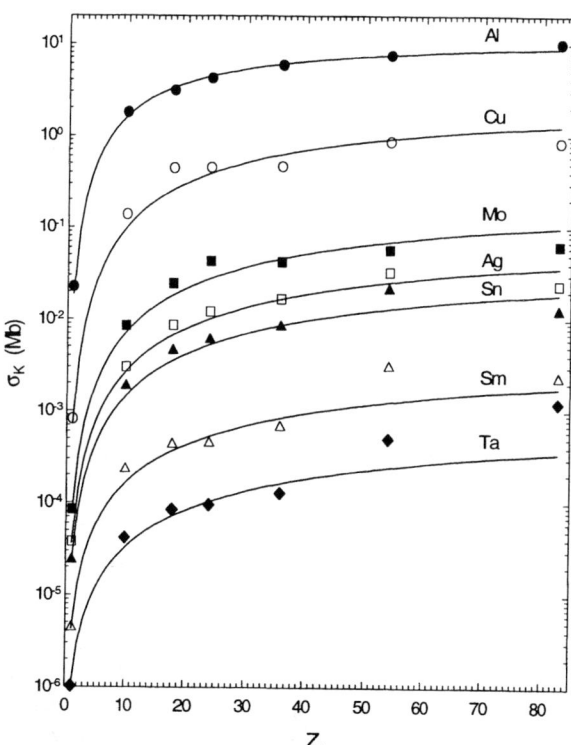

FIGURE 2. Fits of the scaling law to the K-vacancy production cross sections (solid curves).

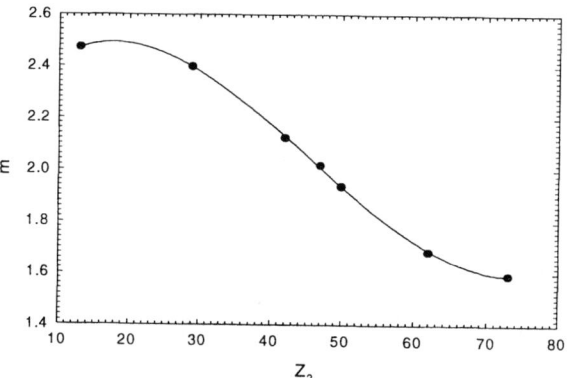

FIGURE 3. Dependence of the scaling law exponent on target atomic number. The curve shows a third order polynomial fit to the data.

IONIZATION OF NOBLE GASES BY LIGHT ION IMPACT: SCALING THE IONIZATION CROSS SECTION[1]

E. C. Montenegro[2], A. C. F. Santos[3] and G. M. Sigaud

Departamento de Física, Pontifícia Universidade Católica do Rio de Janeiro
Caixa Postal 38071, Rio de Janeiro, RJ 22452-970, Brazil.

Single ionization is the dominant channel for fast ion impact of multi-electron targets. This channel plays a major role in the modeling of several processes connected to the penetration of swift ions in solids, such as radiation damage, energy loss, sputtering, desorption, etc. This variety of applications makes having a scaling law for target ionization useful. The major difficulty in obtaining such a scaling lies in the fact that single ionization involves essentially the outermost electrons of the target, which have quite different wavefunctions if different atoms are considered. In this paper we used recent measurements of single-ionization of noble gases by He^+ ions together with previous data for proton and electron ionization to show that a consistent scaling law can be obtained for the single ionization of noble gases. A theoretical support of this scaling is also presented.

INTRODUCTION

Single ionization is the dominant channel for fast charged particle impact on multi-electron targets. Electron, proton or heavy-particle ionization plays a major role in the modeling of several processes connected to plasma and atmospheric physics as well as the penetration of swift ions in solids, such as radiation damage, energy loss, sputtering, desorption, etc. This variety of applications makes having a scaling law for target ionization useful so that a large number of systems can be modeled through the knowledge of a small number of parameters.

In the case of ionization by light charged particles, three parameters determine the main behavior of the cross section: the projectile charge and velocity, and the target atomic number. In the intermediate-to-high velocity regime and light projectiles, the behavior of the ionization cross section is conveniently described by first-order models, such as the Plane Wave Born Approximation (PWBA) [1]. Within this approximation the cross section scales with the projectile charge state as Z_p^2 and with the projectile velocity, v, as $\sim (\ln v^2)/v^2$ if the velocity is high enough.

The dependence of the ionization cross section with the target, on the other hand, is not that straightforward. As has been shown by several authors (see [2,3] and references therein) single ionization of multi-electron atoms involves mostly the outermost electrons which means that, when the periodic table of elements is scanned, electrons belonging to atomic outermost shells, with different quantum numbers, and described by quite different wavefunctions, are the main actors for the ionization process. Thus, it is not obvious that a scaling law for different atomic targets can be obtained.

Furthermore, although single ionization is the dominant process, multiple ionization is always present. If a simple scaling for each atomic shell is wished, inclusive cross section should be used. The inclusive cross section for each shell $n\ell$ is given by $\sigma_{n\ell} = \Sigma_q (q\, \sigma_{n\ell}^q)$ and relates the q-fold ionization of a particular shell, $\sigma_{n\ell}^q$, with the single ionization of the corresponding shell, $\sigma_{n\ell}$, in the single active electron limit [4,5]. The inclusive cross section $\sigma_{n\ell}$ can present a simple scaling but the same is not true for the exclusive cross section $\sigma_{n\ell}^q$ where the ionization probabilities are combined in a non-linear way through the binomial distribution [4,5]. Usually, experiments do not distinguish between the ionization from different shells and only the total inclusive cross section $\sigma = \Sigma\, \sigma_{n\ell}$ is available. However, because $(\sigma_{n\ell})_{\text{outer-shell}}$ dominates over the ionization of the other

---

[1]Work supported in part by CNPq, CAPES, FAPERJ and MCT (PRONEX).
[2]Corresponding Author. e-mail: ecmo@vdg.fis.puc-rio.br
[3]Present address: Instituto de Química, Universidade Federal do Rio de Janeiro, Brazil.

shells, the approximation $\sigma \approx (\sigma_{n\ell})_{outer-shell}$ can be used to obtain a scaling for the experimental total inclusive ionization cross section using parameters associated with the outermost atomic shell.

In this paper we show that a scaling law for the total inclusive ionization for light, single-charged projectiles impinging in noble gases can be achieved. This scaling can be useful not only for the modeling purposes mentioned earlier but also as a guide to check theoretical models for multiple ionization, which are presently quite scarce in the literature.

SHELL DEPENDENCE OF THE DIPOLE MATRIX ELEMENTS

It is well know that, in the high velocity limit, the ionization cross section is given by [1,6]:

$$\sigma_{n\ell} = \frac{8\pi Z_p^2 a_0^2 c_{n\ell} Z_{n\ell}}{I_{n\ell}/R_\infty} \frac{\ln(2mv^2/C_{n\ell})}{(v/v_0)^2} \quad (1)$$

In this equation, the constant $c_{n\ell}$ is the dipole matrix element between the $n\ell$ orbital and the continuum, summed over all the allowed continuum states, and is the factor which, together with the ionization energy $I_{n\ell}$, characterizes the initial electron state. The parameters a_0 and v_0 are the Bohr radius and velocity, respectively, and $Z_{n\ell}$ is the number of electrons in the $n\ell$ orbital. $C_{n\ell}$ is a parameter approximately proportional to $I_{n\ell}$.

With the exception of He, the outermost electrons of noble gases are n_p-type orbitals. An interesting result obtained by Bethe back in the thirties [1], but not very emphasized in the literature, is that the dipole matrix element $c_{n\ell}$ is pretty much constant (~ 0.13) for the n_p orbitals of atomic hydrogen. This result is the basis of our scaling. Indeed, if $c_{n\ell}$ is shell-independent for p electrons, it can be easily seen from Eq. (1) that the reduced cross section $I_{n\ell}^2 \sigma_{n\ell}/Z_{n\ell}$ is a function of $v^2/I_{n\ell}$.

The possibility to extend the above hydrogenic result to multi-electron atoms can be put forth with the help of PWBA calculations using more realistic representations of the atomic orbitals, such as the calculations of McGuire [7], for example. In that work, the ionization cross for several atomic sub-shells were calculated using atomic orbitals for multi-electron atoms based on Herman-Skillman potentials.

Unfortunately, if the parameterization is made as indicated above, the calculated cross sections for different atomic orbitals do not coalesce. However, if instead of considering $c_{n\ell}$ as a constant, we make $c_{n\ell} = (\delta_{n\ell} \times$ constant), where the parameter $\delta_{n\ell}$, of the order of unity, accounts for the small peculiarities of the dipole matrix element associated with the various atomic orbitals, an excellent coalescence is obtained among the cross sections for *all* n_p orbitals, and for the 1_s orbital as well, giving rise to an universal curve for this set of orbitals.

$\delta_{n\ell}$\shell	1s	2p	3p	4p	5p
Born fit	1	2/3	1.149	1.205	1.205
Exp. fit	1	2/3	1	1.21	1.33
$Z_{n\ell}$	2	6	6	6	6
$I_{n\ell}$ (a.u.)	1.8	1.59	1.15	1.09	0.92

Table 1. Parameters $\delta_{n\ell}$, $Z_{n\ell}$ and $I_{n\ell}$ used in this work. The table shows the parameter $\delta_{n\ell}$ selected through fits from the Born approximation as well as from the experimental data.

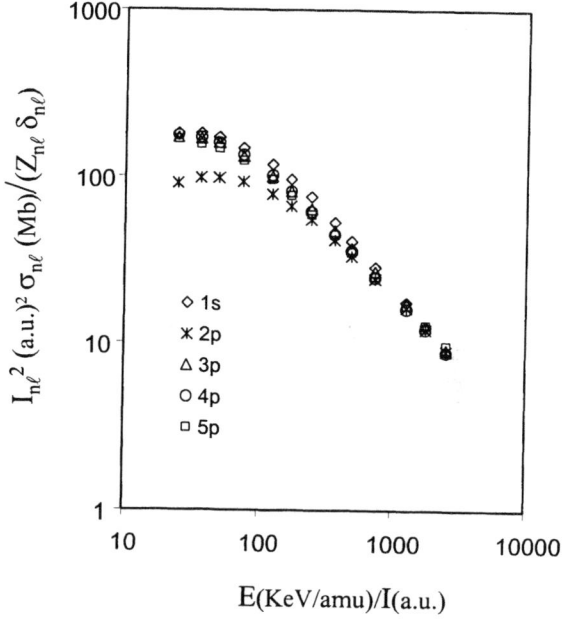

FIGURE 1. Scaled ionization cross sections for different atomic sub-shells obtained from the PWBA calculations of Ref. [7].

The second line in Table 1 (Born fit) gives the values of $\delta_{n\ell}$ used to obtain the curves shown in Fig. 1. The largest value of $\delta_{n\ell}$ is 1.5 for the 2_p shell, corresponding to the Ne case. The other values of $\delta_{n\ell}$ indicated in this table are quite close to one, indicating that the dependence of the n_p dipole matrix element on the principal quantum number, n, is, indeed, weak.

UNIVERSAL CURVE FOR SINGLE-CHARGED PROJECTILES IN NOBLE GASES

The above parameterization will be now used to arrange the experimental data for the total inclusive ionization cross section, or gross ionization, in noble gases by protons [8] and He+ [9] as well as the total ionization cross section by electrons [10] (= $\Sigma_{n\ell} \Sigma_q \sigma_{n\ell}^q$) in a universal curve. This is seen in Fig. 2 where a better coalescence of the experimental data is obtained with a slight change of the parameter $\delta_{n\ell}$ corresponding to the 5_p orbital (Xe case) as indicated in Table 1 (Exp. Fit). A very good universal behavior is obtained with parameters that are essentially the same as those given by the PWBA. The difference arising by using the total ionization cross section instead of the gross ionization for the electron case is not significant for our purposes.

Because the major contribution for the total inclusive ionization comes from the outermost atomic shells, the ionization energies used and the number of electrons are those corresponding to these shells. The values used are shown in Table 1.

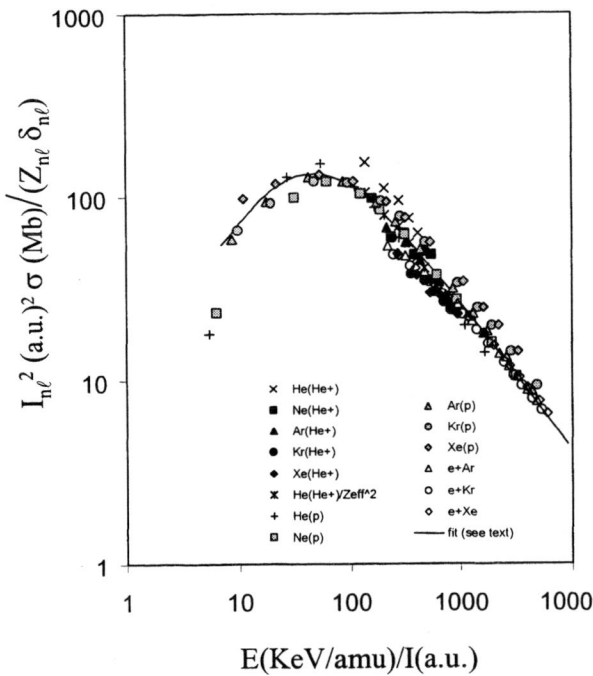

FIGURE 2. Universal curve for the inclusive ionization cross sections of noble gases by singly-charged ions. Except in the He(He+) case indicated explicitly, Z_p was taken equal to one.

With the parameter $\delta_{n\ell}$ included, the scaled cross section can be written as:

$$\frac{I_{n\ell}^2 (a.u)^2 \sigma_{n\ell} (Mb)}{Z_{n\ell} \delta_{n\ell}} = f(E(KeV/amu)/I(a.u.)) \quad (2)$$

where the universal function can be represented by the function

$$f(x) = \frac{A \ln(1 + Bx)}{x} - \frac{AB}{(1 + Cx)^4}, \quad (3)$$

where A=6150, B=0.07 and C=0.014.

DISCUSSION

Figure 2 shows three sets of data: two for bare projectiles and one for dressed projectiles. The comparison among these three sets in an universal plot allows us to identify some differences between the ionization dynamics of dressed and undressed projectiles.

For bare projectiles, the projectile charge is the same for all impact parameters while in the dressed case the projectile charge increases if the impact parameter decreases. Thus, for each projectile energy, dressed projectiles should have an effective charge which is between the ion charge and the nuclear charge. This reasoning explains why the set of measurements for the ionization of He by He+ lies clearly above the universal curve in Fig. 2. The effective charge in this case is larger than one, which is the effective charge used for all projectile-target combinations appearing in Fig. 2. Actually, if the experimental data is divided by an effective charge Z_{eff}^2, defined by the ratio $\sigma_{He^+}/\sigma_{proton}$, the He (He+) data promptly coalesce in the universal curve, as shown in Fig. 2. To determine Z_{eff}^2, the screened cross section σ_{He^+} is calculated within the PWBA along the same lines used to calculate the screening contribution of electron loss and using a form factor corresponding to one electron bound to a He^{2+} nucleus [11,12].

As mentioned above, for all other gases but He the experimental cross sections for He+ projectiles were plotted in Fig. 2 assuming $Z_p = 1$. The good agreement presented between the He+ data and the proton and electron data using this assumption indicates that distant collisions prevails over close collisions in the ionization by He+. It should be noted that the measurements of Ref. [9] were made assuming that He+ ions are present in the entrance and in the exit channel as well, i.e., correspond to ionization without stripping.

The restriction imposed by the measurements that the ionization is carried out without electron loss essentially eliminates the possibility of collisions with small impact parameters to participate in the ionization of the targets other than He. The reason for that is because the strength of the perturbing field *of a heavy target over the projectile* increases sharply towards a high value for decreasing inter-nuclear distances near the atomic radius. This strong field saturates the electron loss probability, which becomes nearly equal to unity for impact parameters smaller than the atomic radius [13,14]. Thus, only distant collisions are allowed if there is no electron loss, making the effective charge of the projectile for such allowed collisions to be approximately one.

Although the exclusion of close collisions for the He$^+$ case validates the use of $Z_p = 1$, it makes the cross section to be smaller than the corresponding bare case, where such a restriction does not appear, and all impact parameters contribute to the ionization cross section. This behavior can be seen in Fig. 2 where the experimental data for He$^+$ on Ar, Kr and Xe clearly lie below the data corresponding to bare projectiles. For example, the measured gross cross section for 3.5 MeV He$^+$ on Xe is 215 Mb [9] while Eqs. (2) and (3) give 257 Mb, i.e., about 20% difference, which can be partly attributed to the lack of contribution of close collisions. Although there are significant experimental uncertainties, which must be considered to properly determine the relative contributions between close and distant collisions from this analysis, the small deviation from the bare case clearly indicates that distant collisions play a major role in this collision regime.

CONCLUSIONS

In this paper, it was shown that the experimental total inclusive ionization cross sections by He$^+$, protons and electrons in gaseous targets from He to Xe can be reduced to an universal function which depends essentially on the parameters of the outermost atomic shells of the target. The proposed scaling has a clear physical basis and can be understood, within the PWBA, through the behavior of the dipole matrix elements involving the continuum states, which are quite insensitive to the initial n_p orbitals. Such a clear identification of the origin of the scaling is not present in some parameterizations available in the literature (see, for example, [15]), although a good coalescence for the various gases can be achieved.

Apart of the obvious usefulness of the availability of an universal behavior in simulation codes designed for a large variety of systems, such as those described in the Introduction, the comparison of the experimental data with theoretical models, which calculate exclusive cross sections, can be advantageously done with the present scaling.

Indeed, although there are several measurements of the inclusive ionization cross sections in noble gases, the same is not true for the exclusive cross sections. From the theoretical side, there are very few quantum mechanical calculations [16], which are able to give the exclusive cross sections $\sigma_{n\ell}^q$ needed to obtain the inclusive ones. The direct comparison of such calculations with the experiment is not conclusive in several cases due to the large discrepancies among the few experimental results available for the exclusive cross sections $\sigma_{n\ell}^q$. Because the inclusive cross section is a weighted sum of the exclusive ones, which are given by calculations, the proposed scaling can be a useful guide to verify the reliability of the theoretical calculations in cases where the experimental exclusive cross section is not available.

REFERENCES

1. Bethe, H., *Ann. Phys. (Leipzig)* **5**, 325, 1930.
2. DuBois, R. D., and Manson, S. T., *Phys. Rev.* **A35**, 2007 (1987).
3. Kim, Y. and Rudd, M. E., *Phys. Rev.* **A50**, 3954 (1994).
4. Sant'Anna, M. M., Montenegro, E. C., and McGuire, J. H., *Phys. Rev.* **A58**, 2148 (1998).
5. McGuire, J. H., *Electron Correlation Dynamics in Atomic Collisions*, New York: Cambridge University Press, 1997, ch. 4.
6. Mott, N. F., and Massey, H. S. W., *The Theory of Atomic Collisions*, London: Oxford University Press, 1965, p. 503.
7. McGuire, E. J., *Phys. Rev.* **A22**, 868 (1980).
8. Rudd, M. E., Kim, Y.-K., Madison, D. H., and Gallagher, J. W., *Rev. Mod. Phys.* **57**, 965 (1985).
9. Santos, A. C. F., Thesis (1999) unpublished.
10. Sorokin, A. A., Shmaenok, L. A., Bobashev, S. V., Möbus, B., Richter, M., and Ulm, G., *Phys. Rev.* **A61**, 22723 (2000).
11. Montenegro, E. C., Wong, K. L., Wu, W., Richard, P., Ben-Itzhak I., Cocke, C. L., Moshammer, R., Giese, J. P., Wang, Y. D. and Lin, C. D., *Phys. Rev.* **A55**, 2009 (1997).
12. Montenegro, E. C., Meyerhof, W. E., and McGuire, J. H., *Adv. Atomic Molecular Physics* **34**, 249-300 (1994).
13. Sant'Anna, M. M., Melo, W. S., Santos, A. C. F., Sigaud, G. M., and Montenegro, E. C., *Nucl. Instrum. Meth.* **B99**, 46 (1995).
14. Voitkiv, A. B., Sigaud, G. M., and Montenegro, E. C., *Phys. Rev.* **A59**, 2794 (1999).
15. Matsuo, T., Kohno, T., Makino, S., Mizutani, M., Tonuma, T., Kitagawa, A., Murakami, T., and Tawara, H., *Phys. Rev.* **A60**, 3000 (1999).
16. Kirchner, T., Lüdde, H. J., and Dreizler, R. M., *Phys. Rev.* **A61**, 012705 (2000).

Present Status of L-shell X-ray Production Cross Sections by Protons with Energies Below 1 MeV

J. Miranda and M. F. Lugo-Licona

Instituto de Física, Universidad Nacional Autónoma de México
Apartado Postal 20-364, 01000 México, D.F., MEXICO

Abstract. The use of low energy protons to induce the emission of X-rays has found application in many laboratories. However, the calculation of L-shell ionization and X-ray production cross sections does not seem to be very accurate in the energy range below 1 MeV, according to the available experimental data set. In the present work, a discussion is presented on the status of the L-shell X-ray production cross sections, for proton energies below 1 MeV. An evaluation of experimental results, together with the predictions given by the ECPSSR theory using two computer codes for numerical calculations and other semiempirical expressions, is given. This shows the need for performing more experiments in this field, as well as for the refining of theoretical models.

INTRODUCTION

The study of characteristic X-ray emission and inner-shell processes has attracted interest both from the point of view of basic research and the development of applications (1). In particular, this interaction of ion beams with atoms has as its main use the analytical technique Particle Induced X-ray Emission (PIXE) (2). Typically, protons with energies in the range 2 MeV to 3 MeV are used to obtain the best detection limits. However, many laboratories around the world have access only to particle accelerators providing proton beams with energies below 1 MeV (3). Moreover, most of the studies regarding X-ray emission by proton impact were carried out in the energy range above 1 MeV (4). On the other hand, K X-ray production has been widely studied, and even "reference" ionization cross sections were published several years ago (5). The situation for L and M X-rays is not at the same level, due to the complexity of the atomic transitions in these shells. Furthermore, the interaction of low energy protons with atoms to produce inner shell ionization and the subsequent X-ray emission in the L and M shells may require a different approach to understand other possible phenomena involved, such as electron capture and modification of electronic states. Therefore, a proper understanding of the ionization and X-ray production cross sections is required for the correct application of PIXE and the basic knowledge of the atomic processes and structure. The aim of this work, thus, is to show the present status of L-shell X-ray production cross sections when protons with energies below 1 MeV are used as the primary ionization agent. The theoretical models, atomic parameters, expressions used for practical calculations in PIXE quantitative analyses, and the existing experimental data set are discussed.

THEORETICAL MODELS

Although many efforts have been focused to develop theories describing the ionization process by ion impact, the most widely known is the ECPSSR model of Brandt and Lapicki (6). It is very useful to predict cross sections for proton energies above 1 MeV, as discussed by Orlic et al. (4) for the L-shell. However, disagreements are observed for projectile energies below 1 MeV.

Important contributions to improve the predictions of the ECPSSR theory have been followed by the group at the University of Naples with their United Atom (UA) approach, and subsequent refinements (7, 8). In the first case, a modification to the binding correction in the ECPSSR theory is done, leading to a better agreement, except for the L_1 ionization cross sections. Later on, they used the relativistic Dirac-Hartree-Slater (DHS) calculations made by Chen and Crasemann, and obtained the DHS-UA approximation (7). Recently, the group modified the coulomb deflection factor in the ECPSSR model (8), still using the DHS-UA procedure. In their papers, it is stated that a much better agreement is found for all the L subshells, and even for heavier projectiles, like ^{4}He.

ATOMIC PARAMETERS

The evaluation of the L-shell X-ray production cross sections from theory requires the knowledge of atomic parameters, like fluorescence yields, Coster-Kronig transition probabilities, and X-ray emission rates. There are several data sets that have been traditionally used. For fluorescence yields and Coster-Kronig probabilities, the tables published by Krause (9) are the most popular, although some authors (10) claim that the values given by Chen (11) give a better agreement with their experimental data. In this regard, there is still a discussion, as the Naples' group (8) preferred to use Krause's tables, which gave them the better results. On the other hand, the emission rates computed by Scofield (12) using Dirac-Fock approximations are widely used, although the DHS results by Scofield (13) provide better agreement with experiments. The latter values were improved by Campbell and Wang (14).

EXPRESSIONS FOR ANALYTICAL APPLICATIONS

The need for the evaluation of the X-ray production cross sections in PIXE analytical applications is limited when trying to use the best theoretical models. The reason is that the calculations are cumbersome, which increases the computing time in quantitative analysis. Therefore, semiempirical expressions have been developed to speed up the analyses. Orlic et al. (15) compiled the existing experimental data up to 1994, and calculated semiempirical expressions to the L_1, L_2, and L_3 ionization cross sections, and total L-shell ionization cross sections for light elements. Their expressions are based on the same reasoning as original fit presented by Johansson and Johansson in their PIXE review. Moreover, they also considered different atomic number ranges. On the other hand, a simple and quick approach developed by Smit (16)

allows the evaluation of the ECPSSR cross sections. Furthermore, Liu and Cipolla (17) formulated a computer code, known as ISICS, which is useful to calculate both the ECPSSR and PWBA cross sections, for the K, L, and M shells. However, the incorporation of this code in quantitative analysis programs is not straightforward. Anyway, the results provided by the Smit procedure and the ISICS code are essentially equivalent, except for the lowest $\xi_{L_2}^R$ interval, which, as will be seen later, is not covered by the experimental data. Figure 1 represents the ratio of the cross sections calculated by both procedures. The Smit algorithm gives higher cross sections, and little effect due to target atomic number is observed.

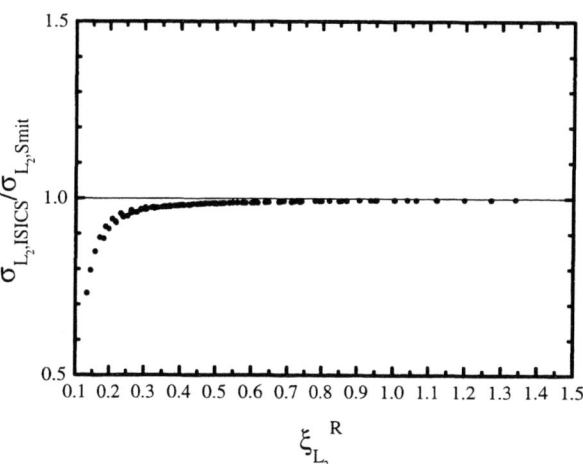

FIGURE 1. Ratio of L_2 ionization cross sections as calculated with the ISICS computer code (17), and the Smit procedure (16).

EXPERIMENTAL DATA SET

A compilation of experimental data was done, using as the main sources the tables published by Sokhi and Crumpton (18), and by Orlic et al. (4). Other papers published after those tables were also included (8, 10, 19-21). It must be emphasized that only X-ray production cross sections were considered, as most of the data were published in this way. Actually, during an experiment, this is the magnitude that is measured, although theoretical predictions are based on ionization cross sections. Table 2 displays the number of experimental points for each L X-ray line measured usually during experiments based on energy dispersive detection. By far, the most commonly reported measurements are for the L_α line, although a growing number of experiments covering the L_β and the L_γ lines is informed. In certain cases, separation within other L_β and L_γ transitions is also given, like $L_{\beta1}$ and $L_{\beta2}$. As the measurement of several lines is required for the determination of ionization cross sections starting from X-ray production cross sections, the number of useful data for this purpose is actually very limited. Also, the number of total L X-ray production cross sections for light elements ($26 \leq Z \leq 50$) is shown. As is well known, detectors have not enough resolution to separate the various lines in this atomic number interval, thus requiring the calculation of total cross sections.

TABLE 1. Number of published experimental points for the most commonly used L X-ray lines.

Line	Number of points
L_α	1184
L_β	199
L_γ	191
$L_{\beta1}$	558
$L_{\beta2}$	558
$L_{\gamma1}$	659
$L_{\gamma234}$	659
L_{Total}	428

Regarding the behavior of the data set according to the target atomic number, Table 2 displays the amount of data for several atomic number ranges, considering the L_α line, the most frequently reported. As can be seen, the heavy elements are not studied very often, due to the low cross sections that difficult the experiments. On the other hand, poor detection resolutions are the cause for not having many separate L_α data in lighter elements, a fact already mentioned. The available information for other lines is still scarce.

TABLE 2. Number of L_α X-ray production cross section measurements according to atomic number.

Z range	Number of points
40-49	133
50-59	238
60-69	406
70-79	259
80-92	149

Comparison Between Semiemprical Expressions and Experimental Data

In order to evaluate the goodness of the semiempirical expressions for calculating X-ray production cross sections, the data set was compared with the predictions of the Smit model, the ISICS code, and the Orlic at al. fits. In the first two cases, it is possible to use the reduced velocity scaling used by Rodríguez et al. (22). Fluorescence yields and Coster-Kronig probabilities by Krause (9), and the Dirac-Fock emission rates by Scofield (12) were used to go from ionization to X-ray production cross sections.

As a first step, the L_α line predictions are evaluated. Figure 2 presents the ratio S of experimental to theoretical cross sections, using the ECPSSR theory with the Smit procedure, while Figure 3 displays the same for the Orlic expressions. The reduced velocity parameter ξ_L^R is used. Very similar results are seen in both plots, showing increasing values of S for low ξ_L^R values, that is, low energies or high target atomic numbers.

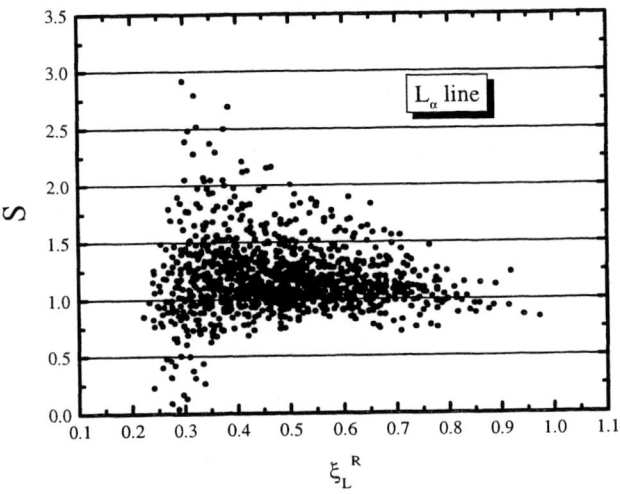

FIGURE 2. Ratio S of experimental to theoretical L_α X-ray production cross sections, following the Smit method.

As was observed in the past by several authors, the low energy range presents a much higher scattering in the ratios S. The first reasoning to explain this behavior is a dependence with target atomic number, which would make the scaling inappropriate. However, after looking at this possible dependence by plotting the data according to the atomic number ranges given in Table 2, no definitive conclusions can be extracted. Thus, the strong data scattering may be associated to differences in the experimental procedures rather than to target atomic numbers.

When the different approaches are used to compute all the remaining lines mentioned in Table 1, very similar dependence of S with the corresponding reduced velocity parameter is found in every case, namely, very close to unity for high values of ξ_L^R, and increasing dispersion for the low ξ_L^R range. Anyway, the data is very scarce for those lines. Thus, it is very difficult to draw any valid inference regarding the experiments or the expressions.

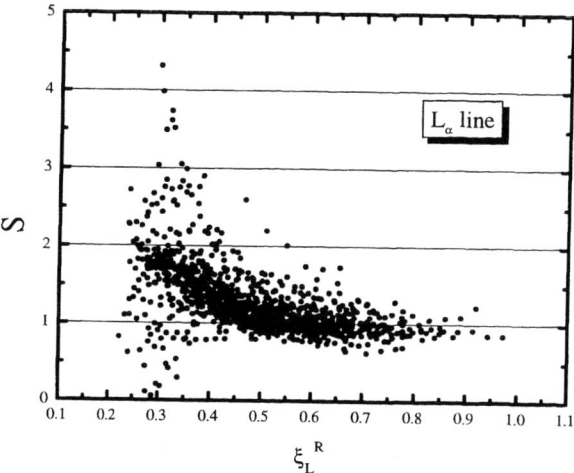

FIGURE 3. Ratio S of experimental to theoretical L_α X-ray production cross sections, using the Orlic et al. fits.

CONCLUSIONS

A short summary of the status of L-shell X-ray production cross sections, covering aspects about experimental, theoretical, and practical calculations, was given. It was shown that some improvements in the theories provide more accurate predictions, given that a proper data base of atomic parameters is used. The need to increase the experimental data set is emphasized, to solve the apparent inconsistencies at the lowest energies. Finally, it was demonstrated that it is possible to use one of several approaches to compute cross sections for analytical purposes, without significant differences, although there are still important errors in their predictions.

ACKNOWLEDGMENTS

Work supported by CONACYT (Project 25085-E) and the International Atomic Energy Agency (Contract 9944).

REFERENCES

1. Van Grieken, R.E. and Markowicz, A.A., *Handbook of X-ray Spectrometry*, New York: Marcel Dekker, 1992.
2. Johansson, S.A.E., Campbell, J.L., and Malmqvist, K., *Particle Induced X-ray Spectormetry (PIXE)*, New York: John Wiley, 1995.
3. Miranda, J., *Nucl. Instr.and Meth.* **B 118**, 346-351 (1996).
4. Orlic, I., Sow, C.H., and Tang, S., *At. Data and Nucl. Data Tables* **56**, 159-210 (1994).
5. Paul, H., and Sacher, J., *At. Data and Nucl. Data Tables* **42**, 105-156 (1989).
6. Brandt, W., and Lapicki, G., *Phys. Rev. A* **23**, 1717-1729 (1981).
7. De Cesare, N., Murolo, F., Perillo, E., Spadacini, G, and Vigilante, M., *Nucl. Instr. and Meth.* **B 84**, 295 (1994).
8. Balsamo, A., De Cesare, N., Murolo, F., Perillo, E., Spadaccini, G., and Vigilante, M. *J. Phys. B* **32**, 5699-5710 (1999).
9. Krause, M.O., *J. Phys. Chem. Ref. Data* **8**, 307-327 (1979).
10. Bogdanovic, I. and Fazinic, S., *J. Phys. B* **29**, 2021-2031 (1996).
11. Chen, M.H., Crasemann, B., and Mark, H., *Phys. Rev. A* **24**, 177 (1981).
12. Scofield, J.H., *At. Data and Nucl. Data Tables* **14**, 121-137 (1974).
13. Scofield, J. H., *Phys. Rev. A* **10**, 1507 (1974).
14. Campbell, J.L., and Wang, J.X., *At. Data and Nucl. Data Tables* **43**, 281-291 (1989).
15. Orlic, I., Sow, C.H., and Tang, S., *Int. J. PIXE* **4**, 217-230 (1994).
16. Smit, Z., *Nucl. Instr. and Meth.* **B 36**, 254-258 (1989).
17. Liu, Z., and Cipolla, S.J., *Comput. Phys. Comm.* **97**, 315-330 (1996).
18. Sokhi, R.S. and Crumpton, D., *At. Data and Nucl. Data Tables* **30**, 49-124 (1984).
19. Amirabadi, A., Afarideh, H., Haji-Sacid, S.M., Shokouhi, F., and Peyrovan, H., *J. Phys. B* **30**, 863-872 (1997).
20. Miranda, J., Ledesma, R., and De Lucio O.G., *Appl. Rad. Isotop.* **In press** (2000); and references cited there.
21. Cipolla, S.J, *AIP Conf. Proc.* **392**, 117-120 (1997).
22. Rodríguez-Fernández, L., Miranda, J., Oliver, A., Pegueros, J., And Cruz, F., *Nucl. Instr. and Meth.* **B 75**, 49-53 (1993).

Higher-Order Processes in Ion-Atom Collisions

Takeshi Mukoyama

Kansai Gaidai University, 16-1 Kitakatahoko-Cho, Hirakata, Osaka 573-1001, Japan

Abstract. The K- and L-shell ionization process of atoms by fast heavy charged particles has been studied with theories higher than the first Born approximation. The coupled-channel equation, the distortion approximation, and the coupled-subshell approximation are described. Comparison of theoretical results with the experimental data are shown.

INTRODUCTION

Inner-shell ionization by fast heavy charged particles is the fundamental process in atomic physics and has been extensively studied both theoretically and experimentally. It is well known that the K- and L-shell ionization processes by energetic ions can be well described within the framework of the first-order Born approximation, such as the plane-wave Born approximation (PWBA) (1) and the semi-classical approximation (SCA) with a straight-line trajectory (2). When nonrelativistic hydrogenic wave functions are used, the ionization cross sections based on this approximation can be expressed by a universal function as a function of the scaled binding energy of the target electron and of the scaled velocity of the projectile ion.

However, for low incident energy and highly charged ions the deviation of the experimental data from these simple theories becomes significant and the corrections for the effects, which are not included in the first-order Born approximation, such as the energy loss of the projectile during ionization, the increase in the binding energy of the target electron due to the presence of the projectile, the Coulomb deflection of the projectile by the target nucleus, polarization of the target electron cloud due to the projectile, and the relativistic motion of the target electron, should be taken into consideration.

Brandt and Lapicki (3) proposed the so-called ECPSSR theory, which takes into account these effects in the PWBA. The ECPSSR can fairly well predict the experimental results for K-shell ionization cross sections by light-ion impact (4, 5), except for low-energy projectiles and for light targets, and has been successfully used to estimate the theoretical values for K- and L-shell ionization cross sections by ions. The advantage of the ECPSSR consists in its convenience where the ionization cross section can be obtained by modifying the effective projectile velocity and binding-energy parameters in the universal PWBA tables and multiplying the correction factors for the Coulomb deflection and the energy-loss effects (6).

There have published a number of universal PWBA tables for K-, L-, and M-shell ionization and of the ECPSSR tables, as described in (6). The computer codes to calculate the ECPSSR cross sections are also available (7, 8). Several relativistic calculations based on the PWBA and SCA have also been performed (6).

It should be noted, however, that in the ECPSSR all the effects mentioned above are treated as correction factors under some simplifying assumptions (6). Furthermore, the ECPSSR is concerned only with the total cross sections. It is well known that there exists large discrepancy between theory and experiment for ionization probabilities as a function of impact parameter, especially in small impact parameter region.

In the case of L-shell ionization cross section, the L_1- and L_3-subshell ionization cross sections seem to be in reasonable agreement with the ECPSSR and the SCA, but the experimental L_2-shell cross sections are systematically larger than the theoretical predictions (9). Sarkadi and Mukoyama (10) found significant deviation of the L_3/L_2 subshell ionization cross section ratio for low-energy heavy ions on gold. Their result was confirmed by Sohki et al. (11) for different combinations of the projectile and targets. This discrepancy was later explained by Sarkadi and Mukoyama (12) due to the vacancy rearrangement between L subshells during L-shell ionization.

These facts suggest that in order to understand the inner-shell ionization process for ion-atom collisions in details theoretical models higher than the first-order Born approximation should be considered. In the present paper, higher-order theoretical models, such as the coupled-channel method, distortion approximation, and the coupled subshell method, are briefly described. The comparison of the theoretical values obtained with these models and the experimental data is made and the effect of higher-order processes is discussed.

THEORETICAL MODELS

We consider the system in which the target atom is located at the fixed origin and the projectile is moving along a classical trajectory with an impact parameter b. The time-dependent Schrödinger equation for the system is given by

$$\left(H - i \frac{\partial}{\partial t} \right) \Psi(r,t) = 0 , \qquad (1)$$

where r is the position vector of the target electron and the Hamiltonian is written as

$$H = H_0 + V(r, R) . \qquad (2)$$

The unperturbed Hamiltonian of the target atom is expressed as

$$H_0 = -\frac{1}{2} \nabla^2 + U(r) , \qquad (3)$$

and the time-dependent perturbation is given by

$$V(r, R) = -\frac{Z_1}{|R - r|} , \qquad (4)$$

where $U(r)$ is the atomic potential for the target electron and $R(t)$ represents the position vector of the projectile with the atomic number Z_1.

We expand the solution of Eq. (1) in terms of the wave function of the target atom as

$$\Psi(r,t) = \sum_k c_k(t) \, \psi_k(r) \, e^{-i\epsilon_k t} , \qquad (5)$$

where $c_k(t)$ is the expansion coefficient and $\psi_k(r)$ is the electron wave function of the target atom corresponding to the energy eigenvalue ϵ_k:

$$H_0 \, \psi_k(r) = \epsilon_k \, \psi_k(r) . \qquad (6)$$

Substituting Eq. (5) into Eq. (1), we obtain a set of the coupled equations:

$$i \frac{\partial c_k}{\partial t} = \sum_k c_k(t) \, V_{nk}(t) \, \exp\left[i \int^t (\epsilon_n - \epsilon_k) \, dt' \right] , \qquad (7)$$

where

$$V_{nk}(t) = -\int \psi_n^*(r) \frac{Z_1}{|r - R(t)|} \psi_k(r) \, dr . \qquad (8)$$

When the initial state is denoted by 0, the initial condition for Eq. (7) is given by

$$c_k(-\infty) = \delta_{k0} . \qquad (9)$$

and the electron transition probability from 0 to the state n can be obtained from

$$P_{0n}(b) = |c_n(+\infty)|^2 . \qquad (10)$$

To calculate the ionization cross sections, we have to solve the coupled-channel equation, Eq. (7). For ionization processes, the final state is in the continuum and it is tedious and time-consuming to solve Eq. (7) directly. In order to avoid this difficulty, some simplifying approximations have been often employed. The distortion approximation, proposed by Bates (13), is one of such approximations. In this approach, only two states corresponding to the initial and final states are retained and all the off-diagonal terms are neglected. Then Eq. (7) is decoupled and the ionization probability reduces to

$$P_{0n}(b) = \left| \int_{-\infty}^{\infty} V_{n0}(t) \exp[\,i\eta(t)\,] \, dt \right|^2 , \qquad (11)$$

where the phase shift is defined as

$$\eta(t) = \int_0^t \{[\epsilon_n + V_{nn}(t')] - [\epsilon_0 + V_{00}(t')]\} \, dt' . \qquad (12)$$

When we assume $V_{nn}(t) = V_{00}(t) = 0$, Eq. (12) is written by $\eta(t) = (\epsilon_n - \epsilon_0)t$ and Eq. (11) is equivalent to the first-order Born approximation. On the other hand, if we choose $V_{nn}(t) = 0$ and $V_{00}(t) = V_{00}(t_0)$, where t_0 is the time at the distance of the closest approach, $\eta(t)$ corresponds to the binding-energy correction in the ECPSSR.

In the case of the L-subshell ionization cross sections, the coupling effect between L subshells play a dominant role for low-energy heavy ions, as described above. Considering this fact, we can retain only the terms corresponding to L subshells in Eq. (7) and neglect all the contributions from other bound states (14). This approach is sometimes called the coupled-subshell approximation.

K-SHELL IONIZATION

With recent advance in high-speed and large-memory computers, ionization problems based on

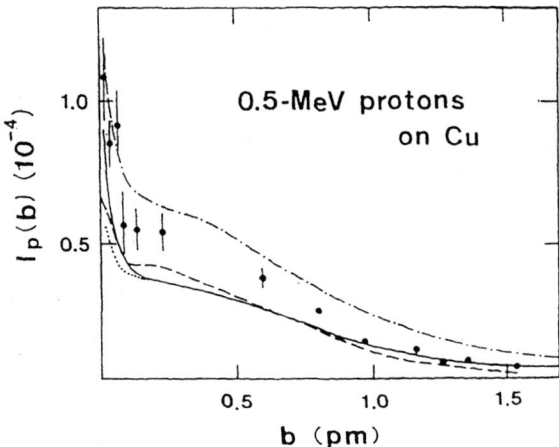

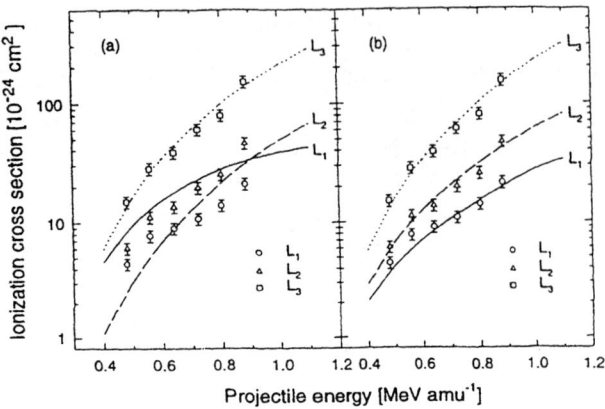

FIGURE 1. K-shell ionization probabilities for 0.5-MeV protons on Cu. The solid curve represents the distortion approximation, the dotted curve the distortion approximation with recoil effect, the dot-dashed curve the first-Born approximation, the dashed curve the coupled-channel calculations of Reading et al. (16), and the solid circles are the experimental data of Andersen et al. (23). From Ref. (21).

FIGURE 2. L-subshell ionization cross sections for B ions on Au. Theoretical: (a) the ECPSSR and (b) the coupled-subshell approximation. The experimental data are taken from Padhi et al. (30). From Ref. (30).

the models higher than the first-Born approximation can be performed with relative ease and a number of coupled-channel calculations have been performed. However, most of them are for simple systems, such as light ions on light targets, and calculations for medium and heavy elements are rather scarce.

Reading and coworkers (15, 16) carried out coupled-channel calculations using *pseudostates*, discrete states with positive energy eigenvalues. They expanded the time-dependent wave functions into a finite basis set of L^2 (square integrable) functions. Using these wave functions, the atomic Hamiltonian with the Hartree-Fock potential was diagonalized to obtain bound states and pseudostates. They calculated the K-shell ionization cross sections by protons including the second-order Born term. Their second-Born cross sections are smaller than the first Born ones for low-energy projectiles.

The coupled-channel calculations with the relativistic wave functions have been performed by Mehler et al. (17, 18, 19). They used the Dirac-Fock-Slater (DFS) wave functions for bound states and the continuum wave functions were expressed by the relativistic wave packet. The calculated K-shell ionization cross sections for protons on Cu and Ag are in agreement with the experimental values.

Mukoyama and Lin (20, 21, 22) calculated the K-shell ionization cross sections in the distortion ap-

proximation. They expanded the electron wave function in terms of Slater-type orbitals and diagonalized the atomic Hamiltonian for the Hartree-Fock-Slater potential. The obtained atomic states and pseudostates were used to calculate the ionization cross sections. Their result for ionization probabilities for 0.5-MeV proton on Cu (21) is shown in Fig.1 and compared with the experimental data of Andersen et al. (23). For comparison, the results of the first-order Born approximation and the coupled-channel method of Reading et al. (16) are also displayed. The values with the distortion approximation are in good agreement with the coupled-channel calculations. Agreement with the experimental values is also good, while the first-Born approximation gives systematically larger values.

The total K-shell ionization cross sections for protons on Cu in the distortion approximation are in good agreement with the nonrelativistic ECPSS values, ECPSSR without the relativistic correction, in the energy region between 0.5 and 2 MeV (21). The relativistic extension of the distortion approximation was also made by the use of the relativistic pseudostates for the DFS potential (20, 22, 24).

L-SHELL IONIZATION

In early calculations of the electron rearrangement process in L-shell ionization (12), the subshell coupling effect was estimated in the two-step model within the framework of the first-Born approximation. Sarkadi and Mukoyama (14, 25, 26, 27, 29) developed the coupled-subshell model with hydrogenic wave functions. The continuum wave function was approximated by that with zero kinetic energy. The calculated results for B ions on Au (29) are shown in Fig. 2 and compared with the experimental data of Padhi *et al.* (30). Figure 2 shows that the ECPSSR fails to predict the energy dependence of the experimental values and the coupled-subshell model can well reproduce it.

Mehler *et al.* (31) carried out the relativistic coupled-channel calculations for L subshells. On the other hand, Amundsen and Jakubaßa-Amundsen (32) and Legrand *et al.* (33) have performed the relativistic coupled-subshell calculations by the use of the hydrogenic Dirac wave functions.

SUMMARY

Higher-order theories for inner-shell ionization processes are described. For K-shell ionization, the distortion approximation gives the results in agreement with the coupled-channel method. In the case of L-subshell ionization, the coupled-subshell method can reasonably reproduce the experimental data.

REFERENCES

1. Merzbacher, E., and Lewis, H.W., in *Handbuch der Physik*, Vol. 34, ed., Flügge, S., Berlin: Springer, 1958, pp. 166–192.
2. Bang, J., and Hansteen, J.M., *K. Dan. Vidensk. Selsk. Mat. Fys. Med.* **31** (1959) No. 13.
3. Brandt, W., and Lapicki, G., *Phys. Rev. A* **23** (1981) 1717.
4. Paul, H., and Sacher, J., *At. Data Nucl. Data Tables* **42** (1989) 105.
5. Lapicki, G., *J. Phys. Chem. Ref. Data* **18** (1989) 111.
6. Mukoyama, T., *Internat. J. PIXE* **1** (1991) 209.
7. Mukoyama, T., and Sarkadi, L., *Bull. Inst. Chem. Res., Kyoto Univ.* **58** (1980) 60; **60** (1982) 67.
8. Liu, Z., and Cipolla, S.J., *Comput. Phys. Commun.* **97** (1996) 315.
9. Mukoyama, T., and Sarkadi, L., *Nucl. Instr. and Meth.* **190** (1981) 619; **205** (1983) 341; **211** (1983) 525.
10. Sarkadi, L., and Mukoyama, T., *J. Phys. B* **13** (1980) 2255.
11. Sohki, K.S., Crumpton, D., and Trautmann, D., *Nucl. Instr. and Meth. B* **42** (1989) 456.
12. Sarkadi, L., and Mukoyama, T., *J. Phys. B* **14** (1981) L255.
13. Bates, D.R., in *Atomic and Molecular Processes*, ed., Bates, D.R., New York: Academic, 1962, pp. 549–621.
14. Sarkadi, L., and Mukoyama, T., *Nucl. Instr. and Meth. B* **4** (1984) 296.
15. Reading, J.F., Ford, A.L., Becker, R.L., in *High-Energy Ion-Atom Collisions* eds., Berényi, D., and Hock, G., Budapest: Akadémiai Kiadó, 1982, pp. 33–52, and references cited therein.
16. Reading, J.F., Ford, A.L., Martir, M., and Becker, R.L., *Nucl. Instr. and Meth. B* **192** (1982) 1.
17. Mehler, G., de Reus, T., Müller, B., Reinhardt, J., Greiner, W., and Soff, G, *Nucl. Instr. and Meth. A* **240** (1985) 559.
18. Mehler, G., Greiner, W., and Soff, G, *J. Phys. B* **20** (1987) 2787.
19. Mehler, G., Müller, B., Greiner, W., and Soff, G, *Phys. Rev. A* **36** (1987) 1454.
20. Mukoyama, T., and Lin, C.-D., *Nucl. Instr. and Meth. A* **262** (1987) 15.
21. Mukoyama, T., and Lin, C.-D., *Phys. Rev. A* **40** (1989) 6886.
22. Mukoyama, T., and Lin, C.-D., in *Application of Accelerators in Research and Industry*, eds., Duggan, J.L., and Morgan, I.L., New York: AIP Press, 1997, pp. 101–104.
23. Andersen, J.U., Lægsgaar, E., Lund, M., and Moak, C.D., *Nucl. Instr. and Meth.* **176** (1976) 507.
24. Mukoyama, T., and Lin, C.-D., *Z. Physik D* **21** (1991) S237.
25. Sarkadi, L., and Mukoyama, T., *J. Phys. B* **20** (1987) L559.
26. Sarkadi, L., and Mukoyama, T., *Phys. Rev. A* **37** (1988) 4540.
27. Sarkadi, L., and Mukoyama, T., *Phys. Rev. A* **42** (1990) 3878.
28. Sarkadi, L., and Mukoyama, T., *Nucl. Instr. and Meth. B* **61** (1991) 167.
29. Sarkadi, L., Mukoyama, T., and Šmit, Ž., *J. Phys. B* **29** (1996) 2253.
30. Padhi, H.C., Dhal, B.B., Nandi, T., and Trautmann, D., *J. Phys. B* **38** (1995) L59.
31. Mehler, G., Reinhardt, J., Müller, B., Greiner, W., and Soff, G, *Z. Phys. D* **5** (1987) 143.
32. Amundsen, P.A., and Jakubaßa-Amundsen, *J. Phys. B* **21** (1988) L99.
33. Legrand, I.C., Zoran, V., Rörner, R., Schmidt-Böcking, H., Berinde, A., Fluerasu, F., and Ciortea, C., *J. Phys. B* **25** (1992) 189.

Measured Energy and Angular Distributions of Sputtered Neutral Atoms from a Ga-In Eutectic Alloy Target

A.W. Bigelow [1], S.L. Li, S. Matteson, D.L. Weathers

University of North Texas, Denton, Texas

The energy and angular distributions of ground-state neutral atoms sputtered from the surface of a liquid Ga-In eutectic alloy by normally-incident 25 keV Ar^+ have been measured using the technique of sputter-initiated resonance ionization spectroscopy (SIRIS). Details of the measurements and data analysis are presented. Differences between the distributions for the Ga and In atoms are discussed in the context of the extreme Gibbsian segregation of the alloy surface.

INTRODUCTION

Sputtering, a process of material erosion by particle bombardment, plays a significant role in many technologies, including integrated circuit manufacture and materials characterization techniques. In this study, the energy and angular distributions of neutral atoms sputtered from a multi-layered surface, a liquid Ga-In eutectic alloy, were measured. These results are useful for comparison with and possible further refinement of analytical and computational sputtering models. The reported distributions were measured using sputter-initiated resonance ionization spectroscopy (SIRIS), a technique for analyzing materials through selective resonance ionization of secondary neutral atoms.

Liquid Ga-In eutectic alloy is very well suited as a target to study the energy distribution of sputtered neutral atoms. Gibbsian segregation at the surface of the material results in an extreme concentration ratio gradient between the top two atomic layers. Previous studies have shown that the surface of the liquid Ga-In (16.5 at% In) eutectic is ≥94 at% In, and that this enhancement is almost exclusively in the top monolayer [1, 2]. The sputtering fraction from the first layer has been calculated to be 0.88 for a primary Argon ion beam energy of 25 keV [1]. Using the values from this study, one can estimate that the fraction of In sputtered from the top layer is >0.98. A corresponding estimate predicts that the fraction of Ga atoms sputtered from beneath the top monolayer is 0.75. This suggests, therefore, that the In energy distribution is dominated by atoms sputtered from the surface atomic layer. Likewise, any measurements of the energy distributions for Ga favor atoms sputtered from beneath the surface. The Ga-In liquid target also eliminates the dependence of sputtering on projectile fluence. With a liquid target, as sputtering continues, the surface layer and concentration gradient persist due to the high atomic mobility in the target.

EXPERIMENT

Components of the SIRIS machine at the University of North Texas (UNT) include a low-energy particle accelerator to provide a primary sputtering ion beam, an ultra-high vacuum target chamber, a tunable dye laser system for resonance ionization, and a position-sensitive detector. A time-of-flight detection method, initiated by a short pulse of primary ions on the sample and terminated by a resonance ionization laser pulse at the entrance to the detector, permits velocity measurements of the resonantly ionized particles. Fast timing electronics and data acquisition are handled through a virtual instrumentation program. Details of the

---

[1] Present address: Center for Radiological Research, Columbia University, New York, NY 10024, USA.

instrumentation are documented in a previous article [3]. The SIRIS apparatus cycles at 10 Hz. A schematic representation of the detector is shown in fig. 1. The sample orientation in the figure represents that used in this work. Individual photoions impinging on the detector's microchannel plates ultimately produce scintillations on a phosphor screen that are recorded by a charged coupled device (CCD) camera.

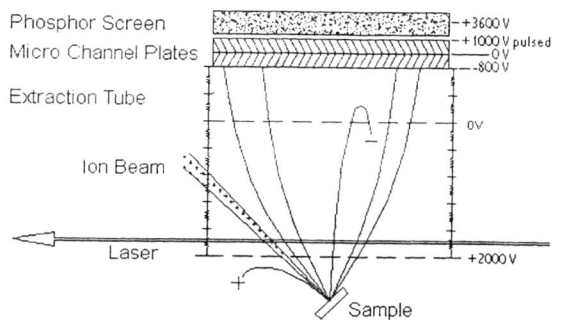

FIGURE 1. Schematic diagram of the interaction region and the position-sensitive detector. Dashed lines represent high-transmission grids.

The Ga-In eutectic alloy target material was obtained commercially. Appropriate handling of such material, which is liquid at room temperature, has been previously documented [4]. Two attributes of the liquid Ga-In eutectic alloy allowed for vertical mounting inside a vacuum system. In past experiments, the eutectic exhibited good wetting without dissolution on cobalt [5]. In addition, the eutectic has a low vapor pressure, rated as essentially zero by the manufacturer [6]; hence, it will not evaporate in a UHV environment. The sample was cleaned in a nitrogen back-filled glove bag attached to the sample introduction port. In this environment, the Ga oxide film layer that is known to form on the Ga-In eutectic [2] was swept off with a small wire brush.

To measure the energy distribution of sputtered neutral atoms, a series of different flight times for the atoms to reach the photoionization region was selected. Each time corresponded to a particular component of sputtered particle energy in the direction normal to the face of the detector. For Ga, the energy series comprised 1, 3, 5, 7.5, 10, 12.5, 15, 20, 25, 30, 35, and 40 eV. For In, the chosen energies were 1, 3, 4, 5, 6, 7.5, 10, 15, and 20 eV. Different energies were used for Ga and In because of differences in the energy distributions for the two different species. Data acquisition involved integrating a CCD image over many apparatus cycles to capture an overall intensity distribution built from successive individual ion impacts.

For each element, the primary sputtering ions were 25 keV Ar in a pulsed beam with pulse widths of 200 ns. The amplitudes of the ion pulses and the average laser power were recorded for data normalization. Typical instantaneous ion beam current was 100 nA and typical average laser power was 20 mW for Ga and 8 mW for In. In both cases, these laser power averages included a fundamental and a second harmonic wavelength. A BBO crystal used to double the dye laser output was measured to have 10% efficiency for second harmonic generation. Resonance ionization schemes exist for both Ga [7] and In [8]. The wavelengths chosen for Ga ionization were 287.4 nm for resonance and 574.8 nm for photoionization. For In, the chosen wavelengths were 303.9 nm for resonance and 607.8 nm for photoionization.

RESULTS

All ion impact positions on the CCD images were convolutions of particle energy and angle; image processing was required before final distributions of sputtered neutrals could be obtained. The first step in the processing of the raw data images was to extract the gray-scale pixel values representing particle impact intensity vs. position. The data were normalized with respect to the number of incident primary ions. The In data were also normalized to average laser power because the photoionization was linearly dependent on power in this case. A correction for spatial efficiency fluctuations across the microchannel plates and phosphor screen was required. Particle impact positions were then translated into initial ejection energies and angles. Sets of these data were expressed in terms of differential sputtering yields. The next figures illustrate the data processing steps for the 5 eV Ga measurement. They are followed by data set compilations and energy and angular sputtering distributions. Figure 2 depicts one of the images of acquired intensities at the 5 eV Ga setting.

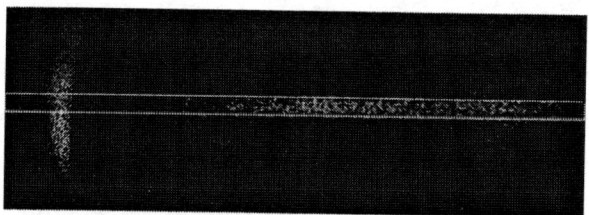

FIGURE 2. An image of SIRIS signal with the dynamic data exchange region of interest. Laser-induced stray-light background effects are responsible for the intensity arc on the left.

Figure 3 shows signal intensity averaged over three individual images. Such an average accounts for 18,000 apparatus cycles. All data were acquired and compiled in this fashion.

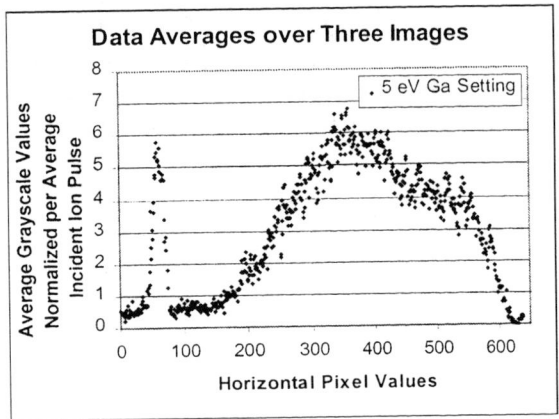

FIGURE 3. SIRIS signal averaged over three integrations with identical experimental parameters.

Following spatial efficiency correction, the modified image intensities for the 5 eV Ga setting are shown in fig. 4. These data were fit with a sixth order polynomial function. The sixth-order fit was used for subsequent deconvolution of the data.

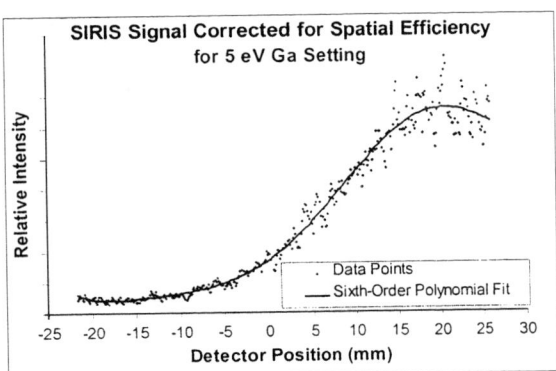

FIGURE 4. SIRIS data corrected for spatial efficiency of detector. The detector position axis range has been cropped to correspond to the angular emission range of interest: 0-85 degrees from sample normal.

Total kinetic energy for each sampled particle was calculated in a straightforward fashion once the ejection angles were known. To verify the correspondence between detector position and ejection angle, a series of images were acquired for which the range of ejection angles was truncated by a physical shield. These results were compared with computer simulations of particle trajectories through the detector and with analytical expressions for the particle trajectories. After the sputtering data were converted from impact positions to unique ejection angles and energies, the image intensity was converted to a differential sputtering yield with respect to energy and angle. The differential sputtering yield for the 5 eV Ga setting is plotted in fig 5.

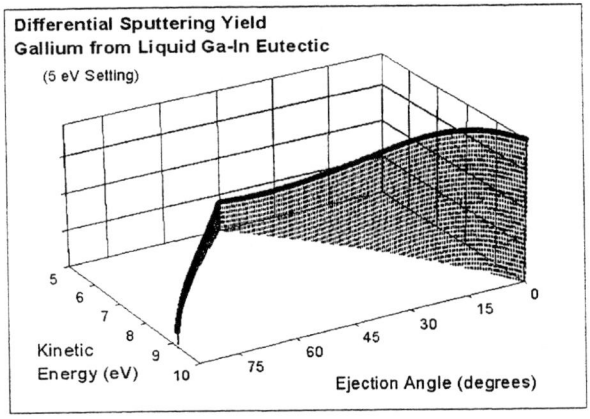

FIGURE 5. The differential sputtering yield with respect to energy and angle. This representation exhibits the image position deconvolution into angle and energy.

The differential sputtering yield, $\frac{dY}{dEd^2\Omega}(E,\theta)$, is shown in a three-dimensional surface plot in fig. 6. In contrast, the In data formed a similar plot but were broader in angle and narrower in energy.

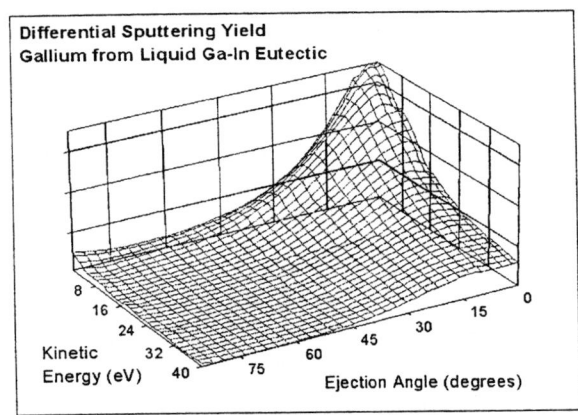

FIGURE 6. Grid plot of the differential sputtering yield with respect to energy and angle for Ga sputtered from liquid Ga-In eutectic alloy. The grid surface helps to visualize the energy peak in the distribution.

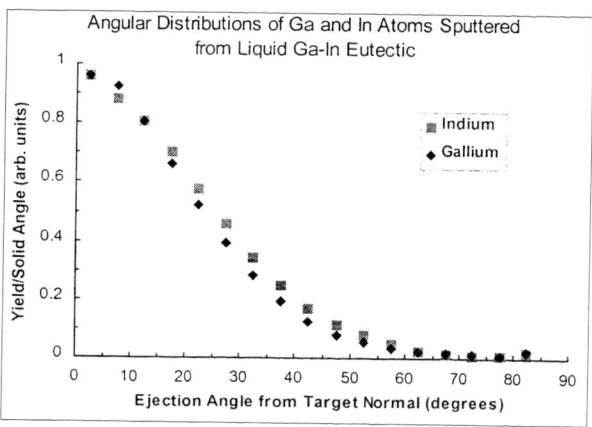

FIGURE 7. The angular distributions of sputtered Ga and In atoms from liquid Ga-In eutectic.

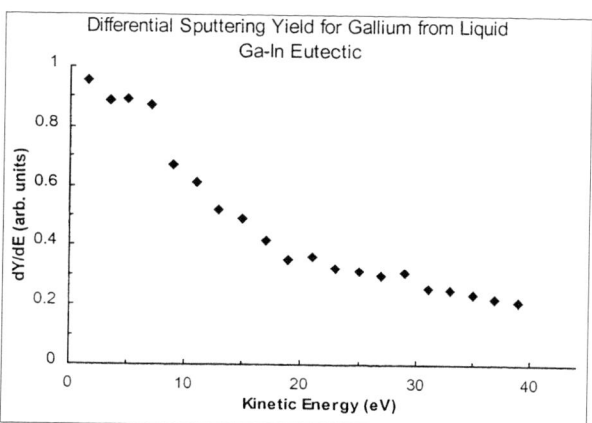

FIGURE 8. The energy distribution of sputtered Ga atoms from liquid Ga-In eutectic alloy.

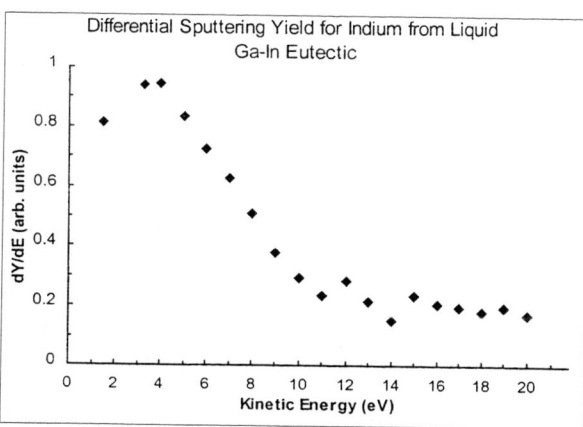

FIGURE 9. The energy distribution of sputtered In atoms from liquid Ga-In eutectic alloy.

Measured energy and angular distributions of sputtered neutral atoms from Ga-In eutectic alloy are displayed in fig. 7-9. For In, the trends are characteristic of sputtering from the top layer of a sample: the In yields are broader in angle and concentrated at lower energies than for Ga, which originates primarily from beneath the first monolayer. The angular distributions are consistent with the view that atoms originating from beneath the surface monolayer are less likely to escape in oblique directions because of a larger collision probability with the surface atoms in those directions. The energy distributions are consistent with the idea that the surface monolayer represents a barrier that prevents low-energy atoms from escaping from beneath the surface. Moreover, these results are in qualitative agreement with other angular measurements [1, 2] and with the results from molecular dynamics simulation [9]. Quantitative differences could be due in part to the coarseness of the sampling in the measurements reported here.

ACKNOWLEDGMENTS

This work was supported in part by the Texas Advanced Research Program under Grant No. 003594-068.

REFERENCES

1. K.M. Hubbard, R.A. Weller, D.L. Weathers, T.A. Tombrello, *Nucl. Inst. and Meth.* **B 36**, 395 (1989).

2. M.F. Dumke, T.A. Tombrello, R.A. Weller, R.M. Housley, and E.H. Cirlin, *Surface Sci.* **124**, 407 (1983).

3. A.W. Bigelow, S.L. Li, S. Matteson, D.L. Weathers, "Sputter-Initiated Resonance Ionization Spectroscopy at the University of North Texas," in *Appl. of Accel. in Res. and Ind.*, ed. by J.L. Duggan and I.L. Morgan, AIP Conf. Proc. 475, New York, 1999, pp. 569-572.

4. K.M. Hubbard and R.A. Weller, *Surface Sci.* **207**, 441 (1989).

5. T.B. Lill, W.F. Callaway, M.J. Pellin, and D.M. Gruen, *Phys. Rev. Lett.* **73** 12, 1719 (1994).

6. Alfa Aesar, Material Safety Data Sheet.

7. E.B. Saloman, *Spectrochimica Acta* **49B** 3, 251 (1994).

8. E.B. Saloman, *Spectrochimica Acta* **48B**, 1139 (1993).

9. M.H. Shapiro, K. Bengtson, Jr., and T.A. Tombrello, *Nucl. Inst. and Meth.* **B 103**, 2 (1995).

Generalized oscillator strengths of the rare-gas atoms

Zhifan Chen and Alfred Z. Msezane
Center for Theoretical Studies of Physical Systems
and Department of Physics,
Clark Atlanta University,
Atlanta, Georgia 30314, U. S. A.

ABSTRACT

The generalized oscillator strengths(GOSs) for the transitions from ground states to the excited states $np^5(n+1)s$ of Ne (n=2), Ar(n=3), Kr(n=4) and Xe(n=5) have been calculated as a function of the momentum transfer squared K^2 in the random phase approximation with exchange and in Hartree-Fock approximation. The positions of the characteristic minima and maxima in the GOSs have received particular attention in the evaluation. They are found to agree excellently with the experimental results. Exchange and correlation effects among the electrons in different subshells are found to influence the positions of the minima insignificantly.

I. INTRODUCTION

The generalized oscialltor strength (GOS) was first introduced by Bethe [1]. The existence of the minima and the maxima in the GOSs was first predicted and calculated by Bonham [2] for the Ar 3p-4s transition using Hartree-Fock wave functions published by Hartree and Hartree [3] and Knox [4]. Later Shimamura [5] obtained analytical forms of the GOSs for several excitation processes of some hydrogen-like atoms. Kim and Inokuti [6] measured and calculated GOSs for the excitation to the $5p^5(^2P_{\frac{3}{2}})6s$ from ground state of Xe. Their minimum occurs at a smaller value of K^2 in comparison with the experimental result. Ganas and Green [7] evaluated GOSs for the single-particle excitations of the rare gases which showed a complex nodal structure. Theoretical values of maxima and minima in the GOSs for the rare gases were also studied by Miller [8]. His position of the minimum for the Kr 4p - 5s transition is about 20% less than the experimental result of Wong et al [9]. Differential cross sections and GOSs for some rare gases have been measured by Li et al [10], Takayanagi et al [11], and Suzuki et al [12] at electron impact energies in the range 300-500 eV. Unfortunately these measurements did not cover the momentum transfer range where the first minima and maxima appear.

In this paper we report on the positions of the minima and maxima of the GOS for the transitions from ground states to the excited states $np^5(n+1)s$ of Ne(n=2), Ar(n=3), Kr(n=4) and Xe(n=5) obtained using the random phase approximation with exchange (RPAE) and the Hartree-Fock approximation (HFA).

II. THEORY

The generalized oscillator strength can be calculated in HFA by

$$f^l(K,w) = \frac{2(2l+1)N_l w}{(2l_i+1)K^2}|M_l|^2 \quad (1)$$

where N_l is the number of electrons in the excited state, l is the total orbital angular momentum of the electron-hole pair, which satisfies the triangular rule $|l_f + l_i| \geq l \geq |l_i - l_f|$ with l_i and l_f being respectively, the initial and final orbital angular momenta of the excited electron. The matrix element, M_l is calculated from

$$<\phi_f|M_l|\phi_i> = \sqrt{(2l_i+1)(2l_f+1)} \begin{pmatrix} l_f & l & l_i \\ 0 & 0 & 0 \end{pmatrix} \int_0^\infty P_i(r)P_f(r)j_l(Kr)dr \quad (2)$$

where $P_i(r)/r$ and $P_f(r)/r$ are respectively, the radial wave functions of the initial and final states of an electron in Hartree-Fock approximation.

In order to account for multi-electron correlation effects, the random phase approximation with exchange (RPAE) has also been used in the GOS calculations. The equation for the dipole matrix element in the RPAE [13] is

$$<\epsilon_f|M^{RPAE}|\epsilon_i> = <\epsilon_f|M_l|\epsilon_i> + (\sum_{\epsilon_3 \leq F, \epsilon_4 > F} - \sum_{\epsilon_3 > F, \epsilon_4 \leq F}) \frac{<\epsilon_4|M^{RPAE}|\epsilon_3><\epsilon_3\epsilon_f|U|\epsilon_4\epsilon_i>}{w - \epsilon_4 + \epsilon_3 + i\eta(1-2n_4)} \quad (3)$$

where ϵ_3 and ϵ_4 represent the virtual excitation states, F is the Fermi energy of the atom, and n_4 is the Fermi step : $n_4 = 1, \epsilon_4 \leq F$; $n_4 = 0, \epsilon_4 > F$. U is the combination of the direct and exchange Coulomb inter-electron potentials. The symbol $\sum$ denotes summation over discrete and integration over continuous states. The details of the RPAE method can be found in the book [13].

III. RESULTS AND DISCUSSION

The present calculations have used Hartree-Fock wave functions. Each channel has included three discrete excited states and seventeen continuum states. Since experiment may measure differential cross sections and GOSs to the excited states for example $4p^5(5s", 5s')$ of Kr, where the $5s"$ corresponds to the core angular momentum $J_c = \frac{3}{2}$ and $5s'$ to the $J_c = \frac{1}{2}$. In order to compare with the theoretical results the experimental energy of the $4p^5 5s$ level was obtained by the same method as that of Ganas and Green [7]. This method first subtracts the difference of the ionization limits of the $s"$ and s' systems from the energy of the $5s'$. Then it combines the resultant energy with that of the $5s"$ in the ratio $\frac{1}{3} : \frac{2}{3}$. The same procedure has been followed to obtain the experimental Ne $3s$, Ar $4s$, and Xe $6s$ excitation energies.

The positions of minima and maxima of the GOS for the transitions to excited states $np^5(n+1)s$ of the rare gases are listed in Table 1. The differences in the positions of the minima obtained in the RPAE and the Hartree-Fock calculations are less than 2.4%. Therefore, the positions of the minima can be obtained quite accurately using our numerical Hartree-Fock wave functions. Our calculations also show that exchange has very little effect on the positions of the minima. The difference between the positions of the minima with and without exchange effects is less than 2%. Exchange and correlation effects between the electrons in different subshells are found to influence the positions of the minima insignificantly. Exchange effects shift the positions of the minima to the smaller values of K.

IV. CONCLUSION

The GOSs have been calculated for the dipole allowed transitions $np^6 - np^5(n+1)s$ of Ne(n=2), Ar(n=3), Kr(n=4) and Xe(n=5) in the random phase approximation with exchange and in the Hartree-Fock approximation. The positions of these extrema agree very well with the experimental determinations. Correlation and exchange effects are found to have generally insignificant effect on the positions of the extrema in Ne, Ar, Kr and Xe atoms.

V. ACKNOWLEDGMENTS

Research was supported in part by the US DoE, Division of Chemical Sciences, Office of Basic Energy Sciences, Office of Energy Research and NSF.

[1] Bethe H., Ann. Physik **5**(1930) 325.
[2] Bonham R. A., J. Chem. Phys. **12**(1962) 3260.
[3] Hartree D. R. and Hartree W., Proc. Roy. Soc. (London) **A166**(1938) 450.
[4] Knox R. S., Phys. Rev. **110**(1958) 375.
[5] Shimamura I., J. Phys. Soc. Japan, **30**(1971) 824.
[6] Kim Y. K. and Inokuti M., Phys. Rev. Lett. **21**(1968) 1146.
[7] Ganas P. S. and Green A. E. S., Phys. Rev. A **4**(1971) 182.
[8] Miller K. J., J. Chem. Phys. **59**(1973) 5639.
[9] Wong T. C., Lee J. S., and Bonham R. A., Phys. Rev. A **11**(1975) 1963.
[10] Li G. P., Takayanagi T., Wakiya K., and Suzuki H., Phys. Rev. A **38**(1988) 1240.
[11] Takayanagi T., Li G. P., Wakiya K., and Suzuki H., Phys. Rev. A **41**(1990) 5948.
[12] Suzuki T. Y., Sakai Y., Min B. S., Takayanagi T., Wakiya K., and Suzuki H., Phys. Rev. A **43**(1991) 5867.
[13] Amusia M. Ya. and Chernysheva L. V., Computation of Atomic Processes: A Handbook for the ATOM Programs, IOP Publishing Inc. 1997, 111-115

TABLE I. Positions of the minima and maxima in the GOSs for excitation to $np^5 (n+1)s$ levels of Ne, Ar, Kr and Xe atoms

Atom	Authors	K for minima Expt.	K for minima Theory	K for maxima Expt.	K for maxima Theory
Ne	Wong	1.63± 0.05		2.72± 0.10	
	RPAE		1.608		2.56
	H-F		1.593		2.54
	Miller		2.04		3.39
Ar	Wong	1.12±0.05		1.72±0.10	
	RPAE		1.154		1.74
	HF		1.146		1.75
	Miller		2.70		3.01
Kr	Wong	1.02±0.05		1.58±0.10	
	Kim	0.95	1.10		
	RPAE		1.026		1.55
	H-F		1.051		1.57
	Miller		0.83		1.16
Xe	Wong	0.94±0.05		1.38±0.10	
	Kim	0.84			
	RPAE		0.8799		1.32
	H-F		0.8968		1.34
	Miller		0.99		1.16

Fragmentation Of H_2^{2+} Ions Following Impact Of Slow Xe^{23+} And O^{5+} Ions With H_2 : Projectile Charge And Velocity Dependence

P. Sobocinski, J. Rangama, J.-Y. Chesnel, M. Tarisien, L. Adoui, A. Cassimi, X. Husson and F. Frémont,

Centre Interdisciplinaire de Recherche Ions Lasers, Unité Mixte CEA-CNRS-ISMRA-Université de Caen Basse-Normandie, 6 Bd du Maréchal Juin, F-14050 Caen Cedex, France

Abstract. The energy distributions of H^+ fragments produced in $Xe^{23+} + H_2$ and $O^{5+} + H_2$ collisions at projectile velocities ranging from 0.1 to 0.6 a.u. have been investigated experimentally as a function of the detection angle. The analysis in projectile velocity shows that the projectile interacts significantly with each fragment after the collision. More precisely, the influence of the projectile is stronger when its velocity decreases.

INTRODUCTION

The study of collisions between slow multicharged ions and molecules has been of great interest for many years [1-4]. This study gives information on both the primary processes that occur during the collision and on the dynamics of the molecular dissociation after the removal of target electrons. In particular, much work has been devoted to simple molecular targets such as D_2 or H_2 [5-8].

At low impact velocities (< 1 a.u.), the electron capture is dominant, compared to ionization or excitation. Since the collision time is much shorter than the typical vibrational times in the molecule, the collision can be divided, in a first approach, into two independent steps [1]. The first step is the collision itself, where the transfer of one or several electrons from the target into excited states of the projectile occurs. With H_2 as a target,

$$A^{q+} + H_2 \rightarrow A^{(q-1)+*} + H_2^+ \quad (1)$$

for a single capture (SC), and

$$A^{q+} + H_2 \rightarrow A^{(q-2)+*} + H_2^{2+} \quad (2)$$

for a double capture (DC).

The second step is the fragmentation of the residual ionized target. It has been shown previously that the SC process (Relation 1) gives rise to an ionized target H_2^+ which is mainly in its ground state [8]. Hence, no dissociation of the target is expected. In contrast, the DC process leads to a fully ionized target H_2^{2+} that dissociates, giving rise to protons whose energy is of the order of 9.5 eV in the frame of the molecular center of mass.

A systematic study of the energy distribution of protons has been performed using Classical Trajectory Monte Carlo (CTMC) method [4] for the system $Xe^{54+} + H_2$ at projectile energies ranging from 1 eV/amu to 1 MeV/amu. According to the authors, the final energy of a proton has two sources : the recoil energy E_r transferred by the projectile to the center of mass of the molecule during the collision, and the fragment energy E_f originating from the Coulomb dissociation. At high projectile energies (1 MeV/amu), E_r is negligible compared to E_f. Consequently, the energy distribution consists in a sharp peak centered at ~9.5 eV (Fig. 6 of Ref [4]). At lower impact energies, E_r and E_f are of the same order of magnitude. Hence, both slow and fast protons are found, due to the competition between E_r and E_f.

Recently, it has been suggested that this two-step picture is too simple since the fragments, which are emitted in all directions, may also be influenced by the projectile [8,9]. An additional shift in the energy distributions of the fragments is thus expected. In particular, the energy of the fragments which are emitted in the forward (resp. backward) direction should decrease (resp. increase), due to the Coulomb interaction between the projectile and each fragment.

This effect is quite similar to the well-known post-collision interaction effect [10,11] which influences the energy spectra of Auger electrons emitted from the projectile in the field of the ionized target. Hence, the energy shift ΔE_{PCI} caused by the interaction between the projectile and a fragment (in the following, this interaction will be also referred to as *post-collision interaction*) is expected to depend on the projectile charge q_p and velocity v_p

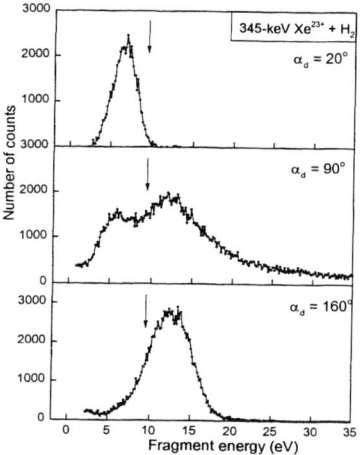
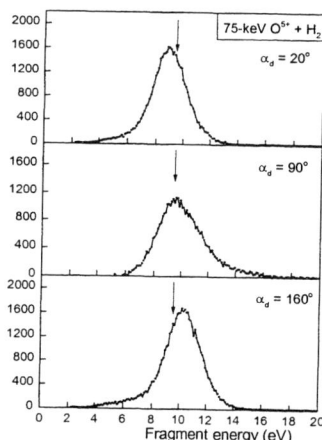

FIGURE 1 Energy distribution of the protons at detection angles of 20°, 90° and 160° following the fragmentation of ions after double capture in 345-keV Xe^{23+} + H_2 and 75-keV O^{5+} + H_2 collisions. The arrows, located at 9.5 eV, indicate the energy assuming a free fragmentation.

$$\Delta E_{PCI} \approx A(\theta)\frac{q_p}{v_p} \quad (3)$$

where $A(\theta)$ is a function of the emission angle θ of a fragment with respect to the beam direction. The aim of the present work is thus the analysis of the fragmentation of H_2 for two collision systems Xe^{23+} + H_2 and O^{5+} + H_2, whose projectile charges differ strongly. Also, to further enlighten the post-collision interaction, the collisions are investigated in a wide range of projectile velocities from ~ 0.1 a.u. to ~ 0.6 a.u. To separate the contributions of the different interactions, our experimental results are compared with model calculations.

EXPERIMENT AND CALCULATIONS

Charge Dependence

Figure 1 shows typical ion spectra for the systems Xe^{23+} + H_2 (left side) and O^{5+} + H_2 (right side) at about the same projectile velocity of ~ 0.4 a.u., and for detection angles of 20°, 90° and 160°. Details about the experimental method are given in Ref. [8]. The observed fragments in the present spectra originate only from the dissociation of H_2^{2+} ions after a DC process [8]. For both systems, it is seen that the mean energy of a fragment differs from the energy E_f of 9.5 eV (indicated by arrows). The difference with respect to E_f is even more pronounced for the system Xe^{23+} + H_2 than that for O^{5+} + H_2. In addition, two groups of peaks are observed at 90° for Xe^{23+} projectiles. This observation corroborates the theoretical predictions for slow and fast protons in the collision Xe^{54+} + H_2 at low impact velocities [4]. To interpret the energy shift observed in the spectra, mean energies of protons were calculated using two simple classical models [8]. In the first model, the problem is treated as two successive *two-body* problems. First, the residual target recoils with a velocity $\vec{v}_r$. Then, in a second step, the ion dissociates with a velocity $\vec{v}_f^{CM}$ in the frame of the molecular center of mass. Thus, the detected proton has a velocity defined by $\vec{v}_f = \vec{v}_f^{CM} + \vec{v}_r$. The velocity $\vec{v}_r$ is obtained using the momentum and energy conservation laws, which result in expressions for the longitudinal $p_{//}$ and transverse $p_\perp$ momenta of the recoiling target. The latter quantities depend on the inelastic energy transfer Q and the initial projectile momentum p_o that were derived from the over-barrier model [12]. The results for $p_{//}$ and $p_\perp$ are reported in Table 1 [13].

TABLE 1 Recoil energy E_r of the recoiling target ion in double (capture for the systems $O^{5+} + H_2$ and $Xe^{23+} + H_2$. The target recoil angle ψ_r is also given.

	θ (10^{-4}rd)	$p_{//}$ (a.u.)	$p_\perp$ (a.u.)	E_r (eV)	ψ_r (deg.)
Xe^{23+}	1.7	-8.6	13.2	0.92	123
O^{5+}	2.2	-2.2	2.8	0.05	128

From $p_{//}$ and $p_\perp$, the recoil energy E_r and the recoil angle Ψ_r were deduced, and utilized to determine the mean energy of the distribution as a function of the detection angle. It is noted that this first model does not take into account the post-collision interaction with the projectile. To take into account the post-collision interaction, we performed additional calculations using classical kinematics equations [8]. First, we assume a sudden double capture. Then, the problem to solve is a pure *three-body* problem involving Coulomb interactions between the three collision partners.

The comparison between the experimental mean energies (full circles and squares) and the calculation results (dashed curves and solid curves for the two- and three-body calculations, respectively) are presented in Figure 2 for both collision systems. The experimental values for the mean energies of the fragments were determined by fitting the spectra with Gaussian curves. For each collision system, both models agree with the experimental data (Fig. 2). While only one of the present models takes into account the post-collision interaction, the agreement between both models is remarkable. This shows that the present model calculations do not allow to evidence the projectile-fragment interaction *after* collisions at the velocity of ~ 0.4 a.u., even for the highest projectile charge of 23. Either the post-collision projectile interaction is negligible at this impact velocity or the models are too approximate to enlighten the post-collision interaction.

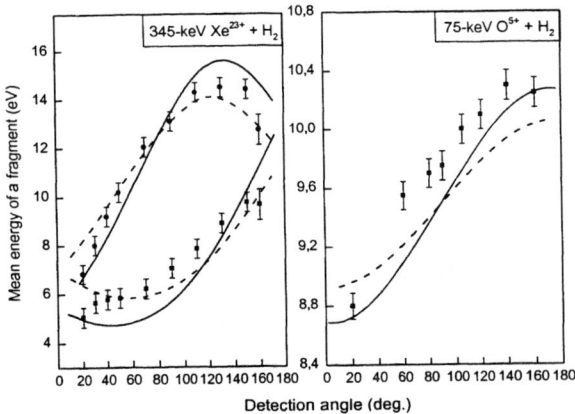

FIGURE 2. Mean energies of the protons following the fragmentation of H_2^{2+} ions after double capture in 345-keV $Xe^{23+} + H_2$ and 75-keV $O^{5+} + H_2$ collisions, as a function of the detection angle. Solid squares and circles, experimental values; – – –, two-body calculations; ———, three-body calculations.

This conclusion differs from that obtained in our previous work [8]. However, one has to notice that a small change for the collision parameters included in the models leads to significant variations for the results. It should be noted that the higher the projectile charge the higher the impact parameter at which the electron transitions occur. Hence, it is not obvious that an increase of the projectile charge results in an enhancement of the post-collision projectile-fragment interaction. On the other hand, it is likely that the post-collision interaction gains importance when decreasing the projectile velocity.

Velocity Dependence

Experiments were performed at lower impact velocities for $Xe^{23+} + H_2$ collisions. Differential cross sections for the detection of protons were determined from the ion spectra by integration of the energy distributions. The results for the system $Xe^{23+} + H_2$ are shown in Figure 3 for three projectile energies, as a function of the observation angle. When the impact energy decreases from 345 keV to 57.5 keV ($v_p = 0.16$ a.u.), significant changes occur for the detection cross sections. At the highest energy, the cross sections are found to be constant below $110°$, and slightly increase at backward angles. When decreasing the projectile energy, a strong anisotropy is observed. The fragments are preferentially emitted at backward angles. At the lowest energy, the number of detected ions at angles lower than $25°$ vanishes. This behaviour can be understood in term of interaction time. When decreasing the projectile energy, the interaction time between the projectile and a fragment increases. Hence, the protons that are emitted in forward direction are repulsed, due to the Coulomb interaction with the projectile.

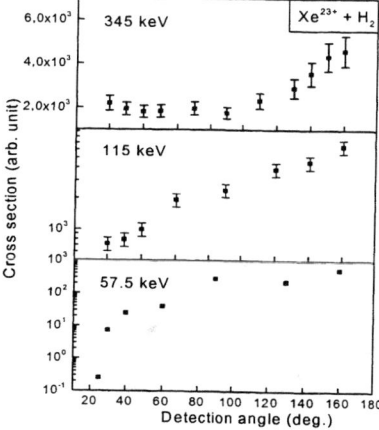

FIGURE 3. Differential cross sections for H^+-fragment emission following slow $Xe^{23+} + H_2$ collisions.

Summarizing, the fragmentation of H_2^{2+} ions following double capture in slow $Xe^{23+} + H_2$ and $O^{5+} + H_2$ collisions was studied. The emitted protons were detected as a function of the observation angle. Quantitatively, the two-step two-body model shows that, at a projectile velocity of ~ 0.4 a.u., the proton energies originate mainly from both the recoil energy of the center of mass of the ionized molecule during the collision, and the molecular dissociation energy (~ 9.5 eV).

To give evidence for the specific role of the projectile, experimental measurements were performed at lower impact energies. For the system $Xe^{23+} + H_2$, the analysis of the fragment emission cross sections reveals that the protons are repulsed in backward directions, due to the Coulomb forces induced by the projectile. Further experiments are planned to study the dynamics of this three-body system at very low impact velocities (< 0.1 a.u).

REFERENCES

1. Shah, M. B. and Gilbody, H. B., *J. Phys. B* **23**, 1491 (1990).
2. Folkerts, H. O., Hoekstra, R. and Morgenstern, R., *Phys. Rev. Lett.* **77**, 3339 (1996).
3. Tarisien, M., Adoui, L., Frémont, F., and Cassimi, A., *Phys. Scripta* **T80**, 182 (1999).
4. Wood, C. J., and Olson, R. E., *Phys. Rev. A* **59**, 1317 (1999).
5. Meng, L., Olson, R. E., Folkerts, H. O. and Hoekstra, R., *J. Phys. B* **27**, 2269 (1994).
6. Kravis, S., Saitoh, H., Okuno, K., Soejima, K., Kimura, M., Shimamura, I., Awaya, Y., Kaneko, Y., Oura, M., and Shimakura, N., *Phys. Rev. A* **52**, 1206 (1995).
7. Errea, L. F., Gorfinkiel, J. D., Macias, A., Mendez, L., and Riera, A., *J. Phys. B* **30**, 3855 (1997).
8. Frémont, F., Bedouet, C., Chesnel, J.-Y., Adoui, L., Cassimi, A., and Tarisien, M., *J. Phys. B*, **33**, L249 (2000).
9. DuBois, R. D., Schlathölter, T., Hadjar, O., Hoekstra, R., Morgenstern, R., Doudna, C. M., Feeler, R., and Olson, R. E., *Europhys. Lett.* **49**, 41 (2000).
10. Barker, R. B., and Berry, H. W., *Phys. Rev.* **151**, 14 (1966).
11. Arcuni, P. W., *Phys. Rev. A* **33**, 105 (1986).
12. Niehaus, A., *J. Phys. B* **19**, 2925 (1986).
13. After recheck of our previous calculations, we have detected errors in the Tables 1 and 2 in Ref. [8]. Compared to the previous estimations, the present values for θ and, thus, for E_r are more realistic.

Molecular Dynamics Simulations Of Collisional Cooling And Ordering of Multiply Charged Ions In A Penning Trap

J.P. Holder[1], D. A. Church[1], L. Gruber[2], H. E. DeWitt[2], B. R. Beck[2], and D. Schneider[2]

[1]*Physics Dept., Texas A&M University, College Station, TX 77843-4242*
[2]*Lawrence Livermore National Laboratory, Livermore CA 94550-0808*

Abstract. Molecular dynamics simulations are used to help design new experiments by modeling the cooling of small numbers of trapped multiply charged ions by Coulomb interactions with laser-cooled Be^+ ions. A Verlet algorithm is used to integrate the equations of motion of two species of point ions interacting in an ideal Penning trap. We use a time step short enough to follow the cyclotron motion of the ions. Axial and radial temperatures for each species are saved periodically. Direct heating and cooling of each species in the simulation can be performed by periodically rescaling velocities. Of interest are Fe^{11+} due to a EUV-optical double resonance for imaging and manipulating the ions, and Ca^{14+} since a ground state fine structure transition has a convenient wavelength in the tunable laser range.

INTRODUCTION

To study mixed non-neutral plasmas and to prepare for precision fine-structure and hyperfine-structure laser spectroscopy on ground term transitions of highly-charged ions, a beam pulse of multiply charged ions (MCIs) from the electron beam ion trap (EBIT) has been captured into a Penning ion trap (RETRAP) [1] previously filled with cold Be^+ ions. Typically, these ions have been Xe^{44+} due to ease of production and the goal of creating strongly coupled cold plasmas, but many other species and charge states can be created and extracted. The Be^+ ions were similarly captured from a metal vapor vacuum arc (MeVVA) source. These ions were initially cooled to near room temperature by tuning (via trap ring-endcap voltage) their axial oscillation frequency into resonance with a cryogenic parallel tuned circuit. The heating of the parallel circuit by currents induced by fluctuations in the centroid of the confined ion motion, together with ion-ion collisional coupling, initially cools the Be^+ ions at an exponential rate with a time constant of several minutes. Eventually, the cooling is dominated by a transverse laser beam de-tuned several GHz below the cycling resonance transition near 313 nm [2]. With decreased de-tuning of the cooling laser, the ions then rapidly cooled to below 1 K, as evidenced by a decrease in light scattering, and confirmed by probing the Be ions with a weak second laser tuned through a nearby non-cycling transition.

When the highly charged ions were subsequently captured, electrode switching and collisions with the multiply charged ions initially heated the Be^+ ions. This heating is associated with an increase in fluorescence rate, which quickly re-cools the Be^+ ions, and by collisonal coupling, cools the MCIs. As the mixed ion cloud cools, ions with different charge-to-mass ratio separate, leaving the cold highly charged Xe ions in the center of the trap and the Be^+ ions in an annulus that we image using scattered 313 nm light. A more detailed account of the procedure and results can be found in ref [3].

It is planned to directly image laser induced fluorescence from a magnetic dipole transition in the ground term of a suitable multiply-charged ion [4]. Of interest are Fe^{11+} due to a EUV-optical double resonance for imaging and manipulating the ions, and Ca^{14+} since a ground state fine structure transition ($2s^2 2p^2$ $^3P_0 - {}^3P_1$) has a convenient wavelength (~569 nm) in the tunable laser range. As a guide to future experiments oriented toward direct detection and imaging of the HCI, we have preformed molecular dynamics simulations of multiply-charged ions of potential interest cooled by Be^+ in a Penning trap.

SIMULATION

In preparation, a molecular dynamics code originally developed by E. L. Pollock, and used by H. DeWitt et al. to model strongly coupled plasmas [5] has been modified to exclusively calculate the interaction of two ion species confined in a Penning trap, with separate heating or cooling of each species. The code now runs on personal computers, using the GNU g77 compiler [6].

Simulations of the motion of 5 to 64 multiply charged ions (experimental ~100-500) and 64 to 200 Be$^+$ (experimental ~10^5) ions were performed using fields, potentials, ion masses, and ion charges correspond to those used in the experimental work. Initially, the ions were arranged in a body-centered cubic lattice, with random assignment of initial locations of each species. The center of mass was placed at the trap center for calculational ease. The ions were assigned pseudo-random initial velocities corresponding to an initial temperature, with the net linear momentum of the ensemble set to zero. The initial ion density, aspect ratio of the lattice, and ion rotation velocity were assigned.

The electrostatic force between each pair of particles was calculated using Coulomb's Law between point charges. The electrostatic trap potential was modeled as an ideal hyperbolic potential, with the voltage applied between ring and end caps of a trap of specified size characterized by a single parameter q_{trap}. For the 9.34 mm scale length trap, $q_{trap} = -2$ corresponds to about 540 V. This approximates the voltage at which the axial oscillation of a Be$^+$ ion is resonant with the 2.5 MHz tuned circuit. The value $q_{trap} = -0.039$ corresponds to a shallow axial well, used experimentally for lower ion plasma rotation frequencies, resulting in greater interaction stability between the laser beam and the plasma. Dynamic features associated with ion confinement measurements, such as r.f. excitations, thermal bath coupling to tuned circuits, and image charges, were not modeled.

The equation of motion of the ions was integrated in the lab frame using the Verlet method [7], based on the Stoermer formula $\mathbf{r}_i(t + \tau) = 2\mathbf{r}_i(t) - \mathbf{r}_i(t - \tau) + \tau^2 \mathbf{a}_i(t) + O(\tau^4)$. Here $\mathbf{a}_i$ is the acceleration of ion i due to the electric and magnetic forces of the trap, as well as the Coulomb forces. Since the trap symmetry axis is aligned with the uniform magnetic field B, this equation is simplest to apply in the z coordinate. For the x and y radial coordinate, the time-centered velocity $\mathbf{v}_i(t) = (\mathbf{r}_i(t + \tau) - \mathbf{r}_i(t - \tau))/2\tau + O(\tau^3)$ was used to evaluate the Lorentz force in the Stoermer formula and to get implicit expressions for $x(t + \tau)$ and $y(t + \tau)$ in terms of present acceleration, x and y position, and a single past time step of stored past x and y positions. The parameter τ was chosen to be a fraction of the period of the initial plasma oscillation, so that it is also small enough to follow the largest cyclotron frequency of the simulation. It was found that small values of τ (~30 – 300 ps compared to ~150 ns Be$^+$ cyclotron period) resulted in a stable simulation start.

The mean squared axial velocity component (proportional to axial temperature T_z of each species) is averaged for every other step. Every 1000 steps, an estimate of the rigid body rotation frequency was made, and in that reference frame, the mean squared radial velocity for each species (proportional to T_r) was calculated and recorded, along with the fixed frame kinetic energies and potential energies of the ions. When simulating cooling, if the running average axial temperature of an ion species exceeds the specified temperature by 3 %, the former z-axis positions of the species being "cooled" (proportional to axial velocity in a harmonic well) were re-scaled using a Taylor series expansion to approach the desired temperature. The other ion species could also be "heated" axially by a similar scaling, allowing a fixed amount of kinetic energy to be added on average to each ion. The temperature of the heated species can approach a steady state with the cooled species. Typically, the target temperature is lowered in steps as the system approaches quasi-steady state.

RESULTS

Major results of the simulation include observation of the onset of properties of a highly magnetized non-neutral plasma [8,9]. As ions are cooled in a magnetic field, the magnetic moment of the cyclotron motion eventually becomes an adiabatic invariant. This invariance occurs because the amplitudes of the Fourier components of a typical Coulomb collision become too low to excite the cyclotron motion. The invariance is broken in close, high velocity collisions. The multiply charged ions, with higher cyclotron frequencies, exhibited this effect at higher temperatures than the Be$^+$ ions, (ranging from 40 K to 2 K for Xe^{44+}) depending on density (or equivalently cloud rotation frequency), as the "radial cyclotron temperature" decoupled at an exponential rate from the "axial temperature". For example, a simulation of 186

Be$^+$ ions and 64 Fe^{11+} ion in a q_{trap} =-2.0 well and with a cold rigid body rotation frequency of 592 kHz, this temperature was about 2 Kelvin. A snapshot of this simulation is shown in Figures 1 and 2. The ions are separated, as discussed below. However, the radial and axial motions remain coupled through the ion cloud rotation. The Be$^+$ ions were not significantly magnetized for simulated temperatures > 1 Kelvin.

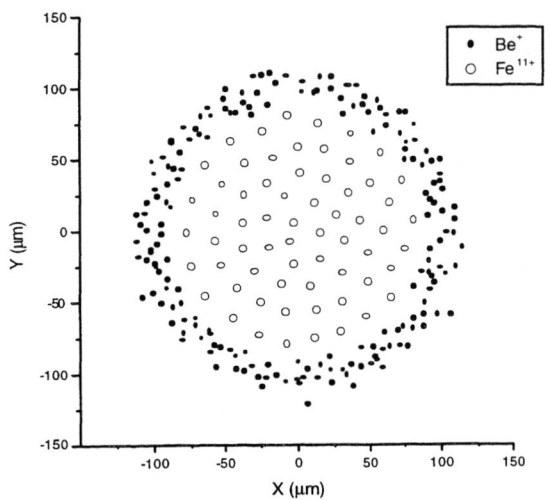

FIGURE 1. A "snapshot" x-y projection of the cloud of 186 Be$^+$ and 64 Fe^{11+} ions. The trap parameter is q =-2.0 corresponding to about 540 Volts applied to our trap. The Be$^+$ ions are axially cooled to 1 Kelvin. The rotation frequency of the cloud is approximately 592 kHz

Typically, the multiply charged ions and Be$^+$ ions centrifugally separate, as Figures 1 and 2 show. This centrifugal separation occurs when the plasma rotates as a rigid body. The "pseudopotentials" associated with the plasma rotation for each species then differ, because of the different mass-to-charge ratios, causing the ions with higher mass to charge ratio (in this case Be$^+$) to move radially outward and separate from the MCIs. The separation temperature was found to depend on the trap parameter q_{trap} and rotation frequency (bounds of which depend on q_{trap}) as expected, although the current code (no spacial averaging / small numbers of particles) and mode of simulation (step down of set temperature) does not track the separation in detail.

A particular case of 64 Al^{3+} ions and 64 Be$^+$ ions, for which the mass to charge ratios of each species are identical, was simulated in a q_{trap}= -0.5 trap. Here at milliKevin temperatures a charge separation is observed [9]. This is shown in Figure 3. Simultaneously ordering of the MCIs can be seen, first as shells, then, at lower temperatures, rings of ions are apparent. The shell surfaces then have the characteristic triangular structure of 2D Coulomb systems [7,10].

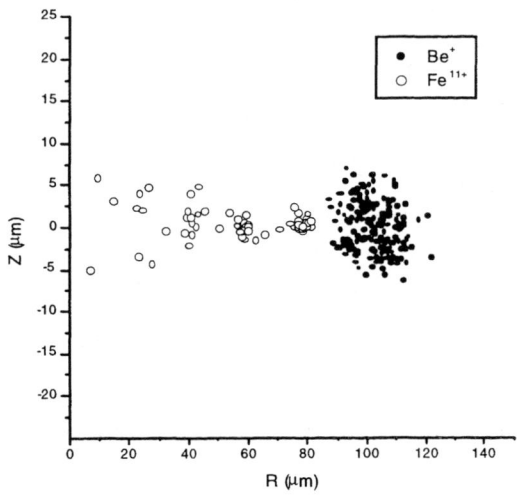

FIGURE 2. A "snapshot" cylindrical R-z projection of the ion cloud described in Figure 1.

Additional work involves the study of the effect of heating sources on the multiply charged ion temperatures given an experimentally measurable Be$^+$ temperature. From simulated motion and atomic data, spectroscopic lineshapes can be calculated and parameters for future experiments estimated. For example, Ca^{14+} has been caught and electronically cooled in our trap. Unfortunately, Be$^+$ laser cooling was not functioning at that time due to laser failure. Calculations are proceeding to model this system to optimize our procedure for a future experimental opportunity.

Further work is suggested by Figure 4, where the motion of the ordered ions might be used to quantum entangle the fine structure states of a few multiply charged ions using an appropriate laser. An ion state then could be read out using a EUV or x-ray electric dipole transition.

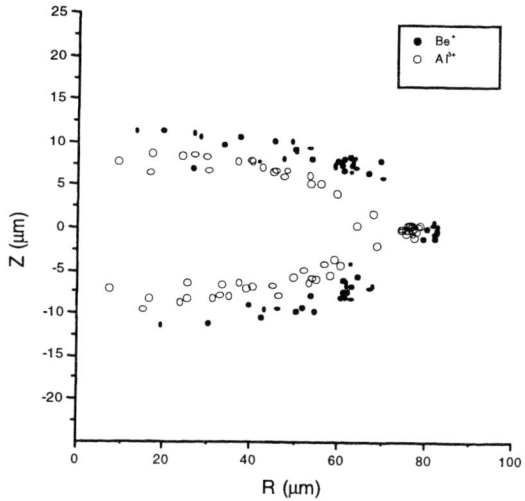

FIGURE 3. A "snapshot" cylindrical R-z projection of the ion cloud of 64 Be$^+$ and 64 Al^{3+} ions. The axial temperatures are ~2.5 milliKelvin while the rotational temperatures are ~640 milliKelvin. Trap parameter is q=-0.5 and ion cloud rotation frequency is ~163 kHz.

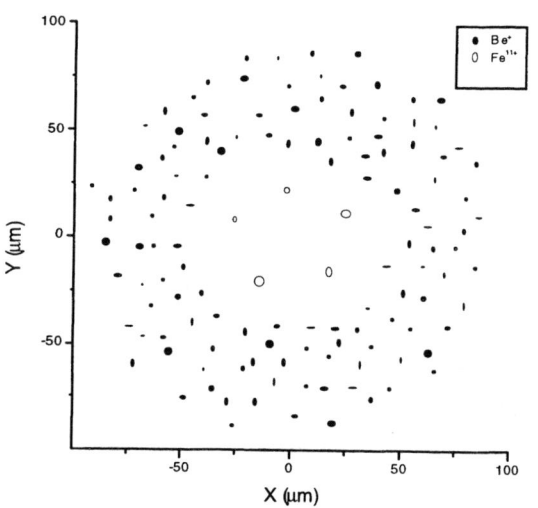

FIGURE 4. A "snapshot" x-y projection of the cloud of 123 Be$^+$ and 5 Fe^{11+} ions. The trap parameter is q =-2.0 corresponding to about 540 Volts applied to our trap. The Be$^+$ ions are axially cooled to 0.25 Kelvin. The rotation frequency of the cloud is approximately 544 kHz

ACKNOWLEDGMENTS

This work was Supported through NSF grant PHY-9876899, and State of Texas Advanced Research Program #010366-0018-1997.

Work was performed under the auspices of the U.S. Department of Energy by University of California Lawrence Livermore National Laboratory under contract No. W-7405-ENG-48 and supported in part by the U.S. Department of Energy, Basic Energy Sciences (Chemical Sciences).

REFERENCES

1. Schneider, D. *et al*, Rev. Sci. Instrum. **65**, 3472 (1994).

2. Brewer, L.R. *et al*, Phys. Rev. **A 57**, 859 (1988).

3. Gruber, L. *et al*, submitted to *PRL*

4. Church, D. A. *et al*, "RETRAP: An Ion Trap for Laser Spectroscopy of Highly-Charged Ions" in *Trapped Charged Particles and Fundamental Physics*, edited by D.H.E. Dubin and D. Schneider, AIP Conference Proceedings 457, New York: American Institute of Physics, 1999, pp. 235-241.

5. DeWitt, H.E., Slatterly, W.L. and Yang, J., "Monte Carlo Simulation of the OCP Freezing Transition," in *Strongly Coupled Plasma Physics*, edited by H. M. Van Horn and S. Ichimaru, Rochester: University of Rochester Press, 1993, pp. 425.

6. www.gnu.org

7. Verlet, L., Phys. Rev. **159**, 98 (1967).

8. Peurrung, A.J., Kouzes, R.T. and Barlow, S.E., Int. J. Mass Spectrom. Ion Processes **157/158**, 39 (1996).

9. Dubin, D. H. E. and O'Neil, T. M., RMP **71**, 87 (1999).

10. Hasse, R.W. and Schiffer, J.P., Ann. of Phys. **203**, 419 (1990).

Reactive Collisions of Hyperthermal Ions with Oxide Surfaces

M. Maazouz, C. L. Quinteros, T. Tzvetkov, P. L. Maazouz, X. Qin, T. L. O. Barstis,[*] and D. C. Jacobs

Dept. of Chemistry and Biochemistry, University of Notre Dame, Notre Dame, IN 46556

Abstract. Hyperthermal energy ions (NO^+ and O^+) are reacted with thin oxide films grown on Al(111) and Si(100), respectively. The scattered ionic products (NO_2^- and O_2^-, respectively) are detected with mass-, energy-, and angular-resolution. The translational energy distributions of the products are correlated with the collision energy, suggesting that the abstraction product emerges from the surface before it, or the incident ion, has had time to equilibrate. Thresholds for O-atom abstraction are 9 eV for NO^+ incident on O/Al(111) and 16 eV for O^+ on SiO_x. The data support an Eley-Rideal mechanism, in which the incident ion neutralizes along the inbound trajectory, abstracts an adsorbed oxygen atom through a direct collision, and scatters from the surface as a molecular anion.

INTRODUCTION

Hyperthermal energy ion/surface collisions can activate a number of reactive processes, e.g., electron transfer, fragmentation, atom abstraction, oxidation, and implantation. Scattering experiments on well-characterized surfaces offer an advantageous perspective for elucidating reaction mechanisms; ion beams allow one to measure how an incident ion's energy and approach geometry affect its reaction probability with the surface target. In the hyperthermal energy regime (10^0 - 10^2 eV), chemical barriers are easily surmounted without significant damage to the surface.

Surface reaction mechanisms are often categorized as falling into one of two limiting cases: Langmuir-Hinshelwood (LH) or Eley-Rideal (ER).[1] In the LH mechanism, both reactants are thermally equilibrated with the surface prior to reaction. In the ER mechanism, an incident gaseous particle reacts directly with a surface adsorbate without first trapping on the surface. In reality, a continuum exists between these two cases, and hot-atom precursors have been implicated for a variety of systems.[2] Deciphering the precise mechanism of a reaction requires a thorough study of the corresponding reaction dynamics.[3]

Two systems are presented here in which the incident ion abstracts an oxygen atom from a surface oxide layer.[4,5] The interaction of hyperthermal energy O^+ with SiO_x is relevant to the low-earth orbit environment, where energetic oxygen atoms and ions continuously bombard protective coatings such as SiO_2 on orbiting spacecraft.[6] Furthermore, oxygen plasma processing in the microelectronics industry is frequently used to etch and to deposit oxide films on silicon.[7] Despite these important technological applications, a fundamental understanding of how hyperthermal energy O^+ reacts with silicon oxide is lacking.

Many investigators have observed that hyperthermal molecular ions can abstract adsorbed species (predominantly hydrogen) from a surface; yet, the precise mechanism of such reactions is not definitively assigned.[8] Using state-selected $NO^+(X\ ^1\Sigma^+)$ ions, the abstraction dynamics on O/Al(111) are studied with precise control over the initial conditions of the reaction.

---

[*] Present address: Department of Chemistry and Physics, St. Mary's College, Notre Dame, IN 46556.

EXPERIMENTAL

The oxide surfaces are prepared *in situ* within an ultrahigh vacuum scattering chamber. Prior to introducing the NO$^+$ ions, the Al(111) surface is dosed with 750 L O$_2$ through a leak valve.[4] This results in a thin film containing chemisorbed and oxide phases.[9] An oxide layer is grown on Si(100) through 30 eV O$^+$ ion bombardment.[5] Prolonged exposure to the ion beam does not result in a thicker oxide film, because the incident oxygen ions erode the film at the same rate as oxygen deposition.[10]

A monoenergetic, mass-selected ion beam is targeted at each of the oxidized surfaces. The O$^+$(^{4}S) ions are formed in a Colutron plasma source;[5] whereas the NO$^+$(X $^1\Sigma^+$) ions are created in a state-specific manner at the intersection of a pulsed molecular beam and the frequency-doubled output of a Nd:YAG-pumped dye laser.[4] Scattered ionic products are detected with mass-, energy-, and angular-resolution.

RESULTS AND DISCUSSION

The abstraction of oxygen atoms from oxide surfaces is illustrated through two model reactions:

$$NO^+(X\ ^1\Sigma^+) + O/Al(111) \Longrightarrow NO_2^-$$

$$O^+(^4S) + SiO_x \Longrightarrow O_2^-$$

Following a discussion of the results for each reaction, the scattering dynamics for the two systems will be compared.

NO$^+$(X$^1\Sigma^+$) on O/Al(111)

Figure 1 presents the NO$_2^-$ yield as a function of NO$^+$(X $^1\Sigma^+$) collision energy for a Al(111) surface dosed with 750 L O$_2$. The NO$_2^-$ yield exhibits a threshold energy, 9 ± 1 eV, above which the yield increases monotonically with NO$^+$ translational energy. The observed threshold is consistent with thermodynamic estimates of the reaction barrier.[4] The NO$_2^-$ yield also scales with the total coverage of oxygen pre-dosed on the surface.

If a LH mechanism were operative, the incident NO would thermally equilibrate with the surface prior to forming NO$_2$. However, the velocity distribution of scattered NO$_2^-$ is nonthermal and cannot be described by a Maxwell-Boltzmann distribution at the surface

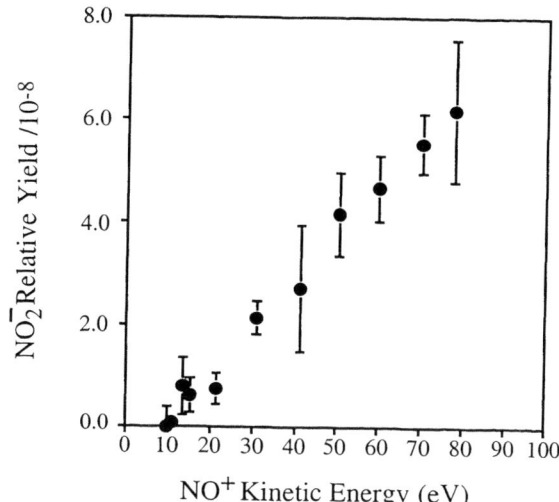

FIGURE 1. The collision-energy dependence to NO$_2^-$ emergence (From Ref. 4). The Al(111) surface was dosed with 750 L O$_2$ prior to NO$^+$ scattering at normal incidence.

temperature (See Fig. 2). Moreover, the mean translational energy of scattered NO$_2^-$ increases with NO$^+$ collision energy. This correlation implies that the reaction occurs via a direct collision between an incident NO molecule and an adsorbed oxygen atom. The data indicate that neither the incident NO molecule nor the scattered NO$_2^-$ product resides on the surface long enough to become thermally accommodated. It is proposed that NO$_2^-$ is

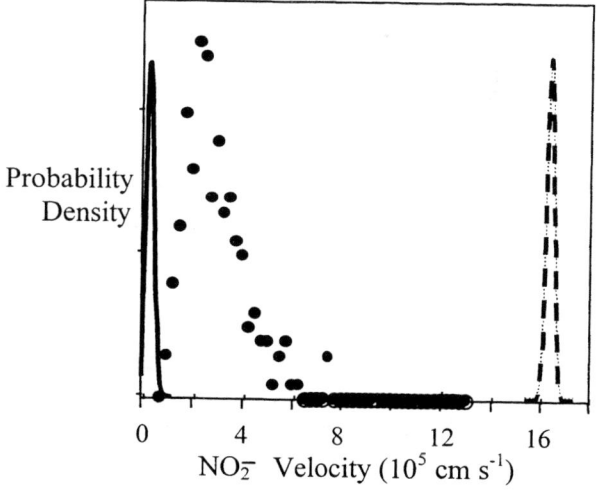

Figure 2. Velocity distribution of NO$_2^-$ produced from 40 eV NO$^+$ abstracting an oxygen atom from O/Al(111). For comparison, the plot shows the velocity distribution (dashed line) of incident NO$^+$, and the predicted Maxwell-Boltzmann distribution (solid curve) if NO$_2^-$ were thermally equilibrated with the surface at 300 K.

formed by a three-step mechanism: incident NO^+ is neutralized close to the surface; nascent NO impacts an adsorbed oxygen atom; and O^- is abstracted by NO to form NO_2^- via an ER mechanism.[4]

O^+ on Oxidized Si(100)

Figure 3 shows the collision-energy dependence to the O_2^- yield for O^+ scattering on a SiO_x surface. The threshold for O_2^- emergence occurs at 16 ± 1 eV.[5] Isotopic labeling experiments have helped to define the origin of each oxygen atom in the O_2^- product. An isotopically pure $Si^{16}O_x$ film was grown using a mass-selected $^{16}O^+$ beam. After a $^{16}O^+$ dose of 1×10^{16} ions/cm^2, scattering commenced with a mass-selected $^{18}O^+$ beam. For all $^{18}O^+$ incident energies (up to 60 eV) and scattering angles studied (45°-135°), > 96% of the scattered O_2^- appeared at a mass equal to 34 u. Therefore, O_2^- is produced when one oxygen atom from the incident ion beam (^{18}O) combines with one oxygen atom from the silicon oxide layer (^{16}O). The lack of signal corresponding to $^{32}O_2^-$ and $^{36}O_2^-$ precludes physical sputtering and projectile recombination, respectively, as significantly contributing to O_2^- formation.

A variety of remaining mechanisms can describe the association of a projectile and a surface oxygen atom to form O_2^-; yet most can be eliminated in consideration of the experimental evidence. In a LH mechanism, the O_2^- product would be expected to emerge with a translational energy distribution that is independent of the collision energy. However, Fig. 4 shows that the mean kinetic energy of scattered O_2^- correlates strongly with the incident O^+ kinetic energy. Therefore, the projectile (^{18}O) must be reacting before it has a chance to thermally equilibrate with the surface. Furthermore, the product O_2^- angular distribution is peaked near the specular angle and shows a correlation with the incident O^+ angle.[11] This is direct evidence for an abstraction reaction occurring in a single collision event (ER mechanism). If the O^+ ion were to undergo multiple bounces on the corrugated surface before abstracting the adsorbed oxygen atom, then the incident particle would lose memory of its initial direction, and the product angular distribution would be independent of the incident angle. It is conceivable however, that an angular correlation could persist if the incident oxygen atom underwent a single collision with the substrate and immediately abstracted an adsorbed oxygen atom on the outgoing trajectory.

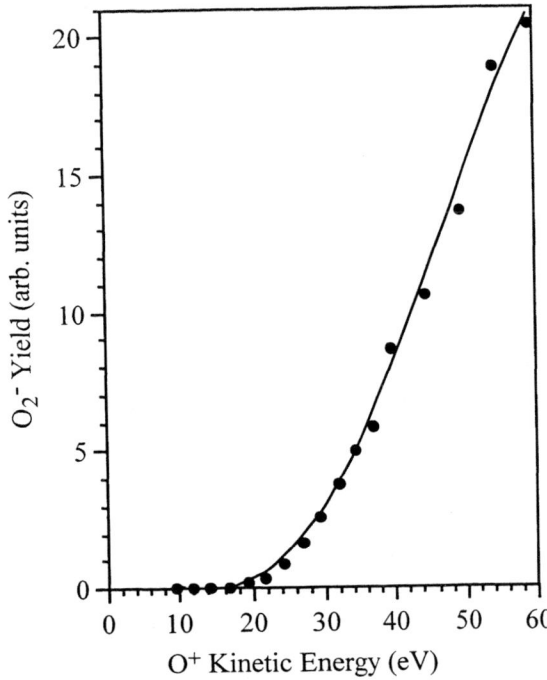

FIGURE 3. O_2^- yield as a function of O^+ collision energy. O^+ is incident at 45° on an oxidized Si(100) surface (From Ref. 5). The appearance threshold for O_2^- is 16 ± 1 eV. The curve is drawn to guide the eye.

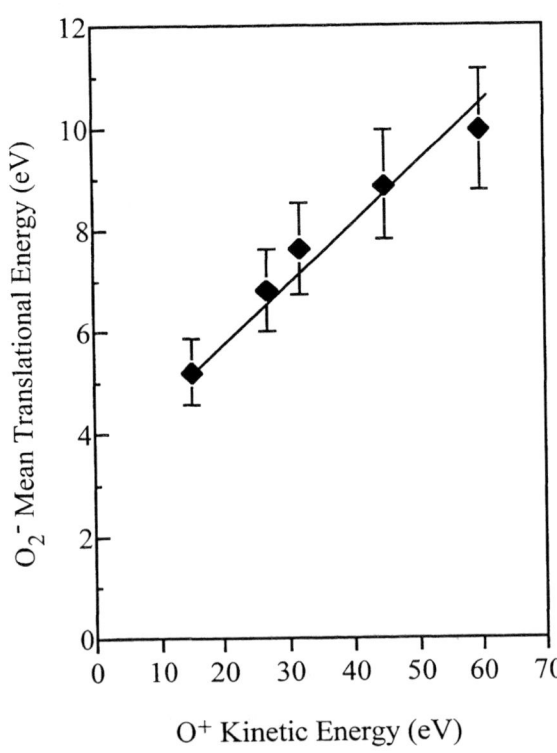

FIGURE 4. Mean translational energy of scattered O_2^- product versus O^+ kinetic energy (From Ref. 5). O^+ is incident at 45° on an oxidized Si(100) surface.

Comparison of Results

The two systems display a similar threshold behavior for abstraction. While the energetics for the two reactions are comparable, it is worth noting that the scattering of NO^+ on $O/Al(111)$ was performed at normal incidence, while O^+ was incident on SiO_x at 45°. Correspondingly, the former system exhibited a lower threshold than the latter.

In both systems, the mean translational energy of the product, $\langle E_{prod} \rangle$, shows a linear dependence on the collision energy, E_{coll}. A simple relation between the two quantities can be parameterized as:

$$\langle E_{prod} \rangle = F \cdot (E_{collision} - E_{threshold}) + E_o \quad (1)$$

where $E_{threshold}$ is the collision energy at which abstraction products first emerge, E_o is the mean translational energy of the products at threshold, and F is the fraction of excess incident energy that appears as translational energy in the products. Table 1 lists the values of these parameters for the two systems. At every incident energy, O_2^- emerged with greater kinetic energy than did NO_2^-. This may be attributed to three different effects. First, the scattering angle was ~90° and 180° in the two experiments, respectively. Second, the departing NO_2^- molecule is more massive than O_2^-. Binary collision theory would predict that both of these effects favor O_2^- as having the larger kinetic energy. Third, the final energy release is distributed into more degrees of freedom within the larger molecule. Consequently, NO_2^- should emerge with less translational energy and more internal energy than its O_2^- counterpart.

TABLE 1. Energetics of Abstraction

System	$E_{threshold}$ (eV)	E_o (eV)	F
$NO^+ + O/Al(111)$	9	1.1	0.023
$O^+ + SiO_x$	16	5.1	0.12

The two systems described here represent the first demonstration that hyperthermal ions can directly abstract oxygen atoms from an oxide surface. In both cases, the scattering dynamics strongly suggest that the oxygen atom is abstracted through an Eley-Rideal mechanism.

ACKNOWLEDGMENTS

Support from the National Science Foundation (CHE-9986374 and CHE96-15878) and the Air Force Office of Scientific Research (F49620-98-1-0029) are gratefully acknowledged.

REFERENCES

1. Weinberg, W. H., "Kinetics of Surface Reactions" in *Advances in Gas-Phase Photochemistry and Kinetics: Dynamics of Gas-Surface Interactions*, edited by C. T. Rettner and M. N. R. Ashfold, London: Royal Society of Chemistry, 1991, pp. 171 - 219.

2. Kratzer, P., *J. Chem. Phys.* **106**, 6752 – 6763 (1997).

3. Rettner, C. T. and Auerbach, D. J., *Science* **263**, 365 - 367 (1994).

4. Maazouz, M., Barstis, T. L. O., Maazouz, P. L., and Jacobs, D. C., *Phys. Rev. Letters* **84**, 1331 - 1334 (2000).

5. Quinteros, C. L., Tzvetkov, T., and Jacobs, D. C., *J. Chem. Phys.*, **113**, 5119 - 5122 (2000).

6. Murad, E., *Annu. Rev. Phys. Chem.* **49**, 73 - 98 (1998).

7. Herbots, N., Hellman, O. C., Ye, P., Wang, X., and Vancauwenberghe, O., "Chemical Reactions During Thin Film Synthesis By Low Energy Ions: Ion Beam Oxidation (IBO) and Ion Beam Nitridation (IBN)" in *Low Energy Ion-Surface Interactions*, edited by J. W. Rabalais, New York: John Wiley & Sons, 1994, pp. 387 - 480.

8. Wu, Q. and Hanley, L., *J. Phys. Chem.* **97**, 2677 - 6459 (1993).

9. Brune, H., Wintterlin, J., Trost, J., Ertl, G., Wiechers, J., and Behm, R. J., *J. Chem. Phys.* **99**, 2128 – 2148 (1993).

10. Todorov, S. S. and Fossum, E. R., *J. Vac. Sci. Technol.* B **6**, 466-469 (1988).

11. Quinteros, C. L., Tzvetkov, T., Qin, X., and Jacobs, D. C., (unpublished).

Electrostatic Trap for keV Ion Beams

H. F. Krause, C. R. Vane, and S. Datz

Physics Division, Oak Ridge National Laboratory, P. O. Box 2008, Oak Ridge, TN 37831 USA

Abstract. A reflecting ion beam electrostatic trap has been constructed and tested for the purpose of studying molecular ion physics. A trap of this type offers a large field-free region, wide mass range, and beam directionality that simplifies the detection of stored ions or breakup products. In our configuration, pulsed ions extracted from an ion source (5–10 μs) are accelerated to keV energies and magnetically analyzed before injection into the trap. After the ions enter the trap, high-voltage electrostatic entrance and exit mirrors and Einzel lenses (operated by fast switches) cause the ions to oscillate between the mirrors in nearly parallel trajectories. Neutralized ions, detected by a channel plate located beyond the exit mirror, indicate the time dependence of the stored ion population. Voltages and pulsed timing of the beam and trap are controlled by computer (38 parameters). The computer can quickly find the optimum trapping conditions in successive fills using the detected neutral yield. Numerous atomic and molecular ions (energy of 1.3–3.0 keV and m <100 amu) have been stored with a beam capture efficiency up to 2%. The storage lifetime, limited by neutralizing collisions with background gas, is typically 2–5 sec at a pressure of 2×10^{-10} Torr.

INTRODUCTION

A wide variety of techniques have been used to trap and study positive and negative ions [1]. Typically, very low-energy ions (E < 10 eV) are confined in a small volume using static electric and magnetic fields or time dependent fields. In most cases, the ions are born inside the trapping volume and there is no preferred direction of motion for the ions stored. These cold ion traps have led to major advances in physics and chemistry, such as in high-precision spectroscopy and mass measurements, fundamental constants and cold ion collision physics.

Another very successful approach to ion trapping has been the storage ring method that is based on accelerator technology [1-3]. Here, the ions are produced by an intense ion source and accelerated, and the beam is charge-state/mass analyzed before it is injected into the storage device. The large variety of positive and negative ion sources available (e.g., gas discharge, electron impact, sputter, ECR, electrospray ion sources) allows a much greater variety of ionic species to be studied than is possible in most cold ion traps. The directionality of the stored beam simplifies the detection of the stored species or decay products. Large medium- and high-energy storage rings (~30-m circumference), with the help of electron and laser-beam coolers, have produced major advances in the study of low-energy electron-ion collisions (e.g., dielectronic recombination and molecular ion fragmentation dynamics) and in nuclear physics.

A much smaller "table-top" electrostatic storage ring (6-m circumference), ELISA, which stores 25-keV atomic and molecular ions, has been developed recently [4]. It offers some of the basic features of the large storage rings mentioned above at a much lower price. It also offers a much larger mass range than is possible for the large magnetic rings because it is electrostatic.

The reflecting ion beam electrostatic trap (RIBET) described below, similar to [5-6], stores ions at low keV energies. A trap of this type, analogous to an optical resonator, offers many of the features of ELISA (e.g., a large field-free region, wide mass range, and beam directionality that simplifies the detection of stored ions and breakup products) in a much smaller and less expensive package. The large selection of ion sources available to storage rings can also be used with this trap. Its functionality is complementary to a storage ring like ELISA, and to low-energy ion traps where the ions have no preferred direction of motion.

ELECTROSTATIC TRAP ARRANGEMENT

The ion trap and the downstream detection chamber, shown in Fig. 1, operate as follows. Ions formed in a Colutron discharge ion source are pulse extracted, accelerated, focused, magnetically momentum analyzed, and then refocused before injection into the trap about 3 m from the ion source (E = 1.3–3.0 keV, $\delta E/E \sim 10^{-3}$). The ion beam is small and nearly parallel (1–2 mm dia., $d\vartheta \sim 0.3$ mrad) as it enters the trap and no entrance collimator slit is used. As an ion fill pulse is launched (width ~7 μs for projectile mass of 40 amu), high voltage is applied to the exit electrostatic mirror electrodes and its companion Einzel lens using a fast switch (shown at the right side of the trap). Incident ions pass through the grounded entrance mirror electrodes and its companion Einzel lens (left side) and the ions are

CP576, *Application of Accelerators in Research and Industry – Sixteenth Int'l. Conf.*, edited by J. L. Duggan and I. L. Morgan
2001 American Institute of Physics 0-7354-0015-6

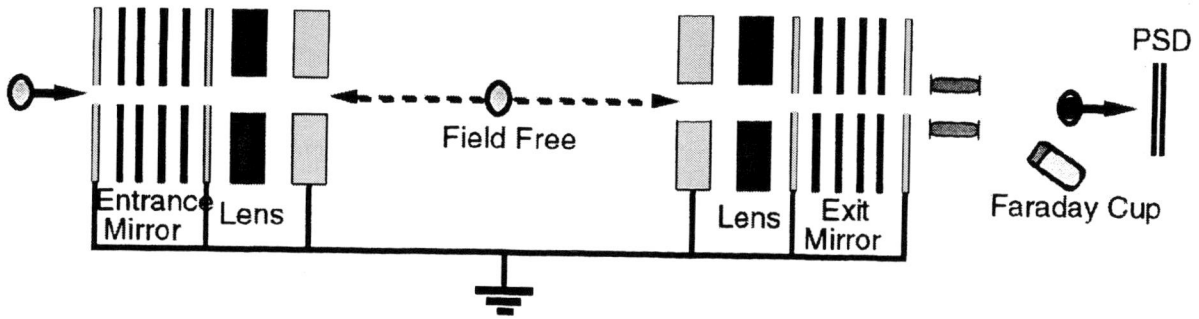

Figure 1. Sketch showing RIBET and the downstream detection chamber. High voltages are applied to mirror and lens elements that are shaded; text explains the voltage distribution.

reflected once from the exit mirror. As the leading edge of the reflected ion pulse approaches the entrance mirror, high voltages are applied to the Einzel lens and mirror electrodes that trap the ions. The ions oscillate back and forth between the mirrors until (i) ions undergo a charge-changing collision with residual gas and leave the trap or (ii) the confining potential on either electrostatic mirror is removed. Neutrals (keV energy) formed inside the trap leave through the ends; the neutrals traveling in the downstream direction are detected on a two-dimensional position sensitive detector (PSD). Electrostatic deflection plates and a Faraday cup, located between the trap and the channel plate, are used to measure the ion beam intensity and ensure that only exiting neutrals are measured throughout the trapping cycle.

An ion trap of this type, like any storage device, requires vacuum consistent with the desired storage time. The vacuum in our trap and adjacent differentially pumped chambers are maintained primarily by large Ti sublimation pumps with some assistance from ion pumps. Three differentially pumped stages between the ion source chamber and the ion trap chamber, consisting of turbomolecular, cryopump and another ion- and sublimation pumped stage, allow the trap vacuum to be maintained at or below 1×10^{-10} Torr range when the gas discharge source is used. Our trap chamber has been operated only at room temperature, but it could be cooled to much lower temperatures to prepare internally cold ions.

The voltages required to trap ions depend on the incident ion beam energy, and these voltages must be applied by fast high voltage switches. To trap 2.85-keV ions of any mass in our geometry (trap length, L = 38 cm) a maximum mirror voltage of 4.55 kV and trap Einzel lens voltages of 2.4 kV are used. The four mirror voltages at each end of the trap, sandwiched between grounded electrodes, decrease a factor of 2 stepping from outer to inner electrodes to provide a "soft bounce" for the stored ions. Because the voltages at the exit end are applied long before the ion fill pulse arrives at the trap (tens of μs), we drive all exit mirror electrodes through a 4×10^5 Ohm resistive divider circuit using one fast switch (Directed Energy, Inc., PVX-4130, 6 kV). At the entrance end, where higher speed is desirable, one fast switch drives the lowest three voltages through a resistive divider chain V_{max} = 2.22 kV; the measured rise time is about 2 μs using a total resistance chain of 2×10^5 Ohms and a short drive cable. The mirror element with the highest potential is driven by its own fast switch ($t_r \sim 0.2$ μs for a pure capacitive load). The entrance Einzel lens is also driven by a separate fast switch. Scaling experiments have shown that all voltages applied to the trap are approximately proportional to the ion energy in the 1.3–3.0-keV range (for masses below 100 amu). Ion energy scaling below 1.3 keV was not tested because the injected ion beam intensity and the detection efficiency for fast neutrals drops rapidly as the ion speed decreases below about 8×10^6 cm/s.

Operation of the trap is greatly simplified by computer control. All voltages (except ion source) and pulsed timing for the beam and trap are controlled by an Apple Macintosh computer through two CAMAC crates. The 38 control parameters consist of 24 digital-analog converters (DAC), 6 programmable pulser/latches and 8 programmable timing registers. The DACs control beam energy, beam focusing and x, y steering, the magnetic field for momentum analysis, and all high-voltage power supplies that control trap mirror and lens voltages. The programmable pulser/latches (LeCroy 4448) set the ion beam pulse length and timing delays for operating the trap mirrors and lenses and the latched signals control all fast high-voltage switches used for trapping, dumping, and the signals needed by other data acquisition computers. The CAMAC timing registers control the storage cycle length and any other devices (lasers, buncher etc.) that might be time-sequenced within a storage cycle. The control system software allows the user to vary any control parameter in a programmed manner in a series of successive trap fills, and it learns the preferred scan ranges for each parameter. This software allows the computer to quickly find the optimum trapping conditions in successive fills using the detected neutral yield (built-in optimizer software). Trapping parameters for any ion are stored and can be recalled so that ions and/or beam energy can be changed within seconds. The trap can also be monitored and operated from a remote location (using a telephone connection) using methods previously discussed [7].

TRAPPING PERFORMANCE

The storage lifetime for 2.85-keV Ar_{40}^+ ions (no long-lived excited electronic states) is shown for a hold time of 10 sec in Fig. 2. Here, the scaler signal from the PSD, read and reset every 10 ms, was accumulated for 200 storage cycles. In the simplest approximation we expect the stored ion loss rate to be proportional to the number of stored ions, $dN/dt = -RN$. If we assume that collisions between the stored ion and residual gas limit the lifetime, then the total ion loss rate is $R = \Sigma n\sigma v$, where n is the number density of residual gas in the trap (predominantly H_2 at this vacuum), σ is the effective total loss cross section (primarily charge exchange), and v is the ion speed; the sum is over all individual loss processes. We expect a time-dependent neutral signal of the form $N(t) = N_0 \exp[-Rt]$. The measured signal shown in Fig. 2 (small dark count subtracted) agrees with this expectation. In fact, the semi-log plot shows a linear loss in the 0–10 sec range where the stored ion intensity decreased a factor of 100. The storage lifetime, $\tau = 1.7$ sec for a trap pressure of 4.2×10^{-10} Torr, was shown to be inversely proportional to trap pressure, in agreement with the model (e.g., $\tau = 7.1$ sec at 1×10^{-10} Torr). The implied effective loss cross section derived from the lifetime, 4×10^{-15} cm^2, is the correct order of magnitude for charge exchange cross sections of 2.8-keV argon ions on H_2 or N_2. The effective loss rate in the mirrors involving slow ion charge-exchange collisions is probably smaller than for full energy ions because asymmetric charge exchange cross sections are much smaller and the effective path length in mirrors is much less than that experienced by the 2.8 keV ions. This would not be true in the case of a stored ion that can undergo symmetric or resonant charge exchange.

An estimate of the beam trapping fraction can also be derived using the data in Fig. 2. In the experiment, an 8 nA DC Ar^+ beam measured after the trap was pulsed for 7 μs to load the trap (~3.4×10^5 ions injected/fill) and about 1.64×10^3 neutrals/fill were detected on the PSD. Using an estimated detection efficiency of 0.8 for 2.8 keV Ar neutrals and recognizing that half of the fast neutrals that leave the trap are not traveling toward the detector, the estimated beam trapping efficiency is about 1.2%. This efficiency, dependent on the number of ions injected, could easily be in error by a factor of two. This estimate also ignores all low-velocity ions and neutrals lost in the end mirrors, which are undetectable. Our measurements indicate that the beam trapping fraction is reduced by another factor of 8-10 if the trap entrance Einzel lens is at full voltage when an ion pulse is injected (i.e., if this lens is not pulse synchronized with the entrance mirror).

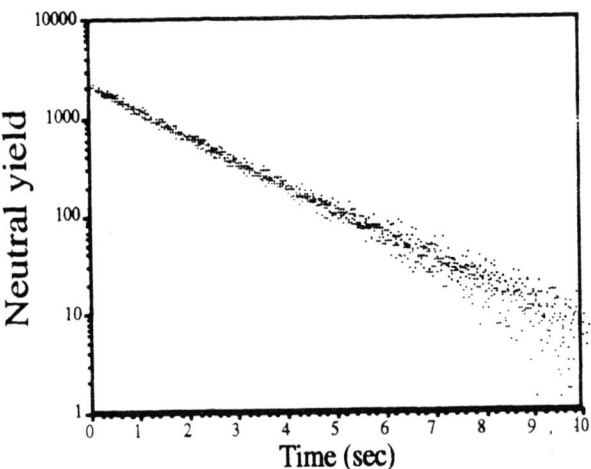

Figure 2. Stored ion population vs. storage time for 2.85 keV Ar^+ ions at a trap pressure of 4.2×10^{-10} Torr.

Similar long-term storage and loss results have been obtained for projectile ions of different masses and for beam energies in the range 1.3 -2.8 keV. These include: N^+, Ne^+, S^+, N_2^+, CO^+, Cl^+, CCl^+, S_2^+, and Cl_2^+. We have also stored Ar_{40}^{2+} without difficulty (same q/m and trapping conditions as Ne_{20}^+). Because the trap is electrostatic, projectiles of the same energy but different mass are trapped using essentially the same trapping voltages. But the timing of fast switches (ion beam pulse length and entrance trap door delay) must be adjusted for each mass to optimize trap filling at different ion speeds. Minor adjustments are usually made to maximize the trapped ion fraction in jumps between the lowest and highest masses tested (e.g., 20 and 70, respectively) and upstream of the trap to adjust for beams of different emittance, size, misalignment, etc. While the storage lifetime for different ions at fixed energy depends on the ion (because of varying charge exchange cross sections with the residual gas and different speeds), the beam trapping efficiency appears to be roughly mass independent in our tested mass range. Generally, the maximum storage lifetime for each tested ion was observed at the lowest projectile energy. This is probably because (i) each stored ion travels a shorter distance/second at the lowest speed and (ii) the cross section for charge loss in the case of cations decreases at lower energy when asymmetric charge exchange is dominant.

When a much more intense beam was trapped (e.g., 100 nA DC beam of Ar^+ pulsed for 7 µs), corresponding to about 4.2×10^6 ions/fill, a non-linear loss rate of stored ions was observed at the beginning of the storage cycle. The increased loss rate may be a consequence of space charge blow-up within the trapping volume, in the mirror regions or near focal regions in the field-free zone. Other pulse length experiments suggest that non-linear losses are minimal up to about 8×10^5 injected ions when those ions are uniformly distributed in time (longest injection pulse length possible). In fact, a major disadvantage of a short beam trap is the relatively short maximum filling time, especially for fast light ions; higher intensity ion beams of fast ions are required to maximize the fill.

Trapping efficiency vs. the voltage applied to trap Einzel lenses, which illustrates the "trapping window", is shown in Fig. 3. Here we see two regions where the efficiency is optimized. The minimum between the peaks corresponds approximately to the voltage where the incident beam is focused at the center of the trap (f = L/2, where L is the trap length). When the focus of each lens lies at the center of the trap, the next reflection produces a parallel beam in the mirror reflection. Parallel trajectories probably lead to higher losses when the ions are being decelerated in the mirrors. The efficiency drops to zero left of the lower voltage peak and to right of the upper peak where the lens voltages correspond approximately to a parallel beam and f = L/4, respectively. The maximum trapping efficiency occurs where the image of each lens is convergent in the mirror regions and where the electro-optic magnification does not lead to a collision with the trap's electrostatic elements.

The angular distribution of neutrals leaving the trap has also been measured on the PSD. The two-dimensional angular distribution (rotationally symmetric about the centroid) is nearly Gaussian with an angular width of 3.8 mrad (FWHM) when the injected beam is 1-mm dia. This width agrees with the trap Einzel lens geometry and trap focal conditions for the beam size used. Measurements also show that the angular distribution (shape or width) does not depend on the ion, storage time or the beam energy.

CONCLUSION

In summary, we have demonstrated that RIBET can be easily used to store cation beams of mass below

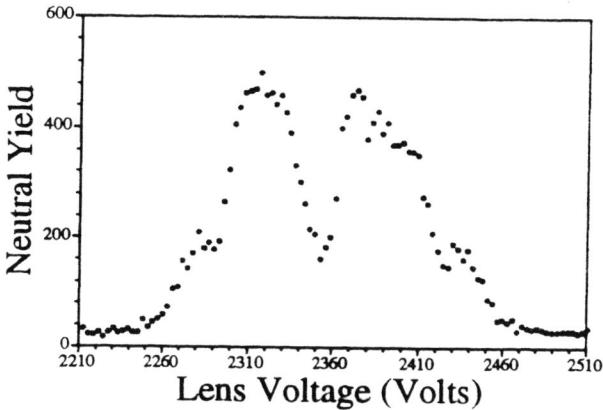

Figure 3. Trapped ion signal vs. the trap lens voltages for 2.85 keV Ar^+ ions. The trapping window is essentially identical for any mass ion at the same beam energy (see text).

about 100 amu throughout the 1.3–3.0-keV range. Part of this regime is well below the 4.2-keV ion beam energy used by Zajfman and co-workers [5-6] in a similar trap. Typical stored ion intensity in RIBET is orders of magnitude smaller than is possible in large storage rings, but so are the complexity and cost. These ion beam traps can be used for studying the lifetime for long-lived electronically excited states of atomic and molecular ions [8-10], cooling internal states of hot atomic and molecular ionic species having a permanent electric dipole moment for spectroscopic experiments, studying slow decay mechanisms of anion or cation cluster ions, and possibly as an ion source for internally relaxed ions for use in other experiments. The large field-free region (~25 cm) is a characteristic that may be especially important for some experiments.

Research sponsored by the U.S. Department of Energy, Office of Basic Energy Sciences, Division of Chemical Science, under Contract No. DE-AC05-00OR22725 with UT-Battelle, LLC.

REFERENCES

1. For overviews on ion trapping see full issues of Physica Scripta; **T22** (1988) and **T59** (1995).
2. Mokler, P.H. and Stohlker, Th., *Ad. At. Mol. Phys.* **36**, 297 (1997).
3. Andersen, L.H., Andersen, T., and Hvelplund, P., *Ad. At. Mol. Phys.* **38**, 155 (1997).
4. Møller, S. P., *Nucl. Instrum. Meth. Phys. Res.*, **A 394**, 281 (1997).
5. Zajfman, D., et al., *Phys. Rev. A* **55**, R1577.
6. Dahan, M., et al. *Rev. Sci. Instrum.* **69**, 76 (1998).
7. Krause, H. F. et al., *Nucl. Instrum. Meth. Phys. Res.* **B 124**, 128 (1997).
8. Wolf, A. et al., *Phys. Rev. A* **59**, 267 (1999)
9. Knoll, L., et. al. **Phys. Rev. A 60**, 1710 (1999)
10. Wester, R., et al., *J. Chem. Phys.* **110**, 11830, (1999).

Breakup and Recombination of Identical Bosons: He Dimer-Monomer Collisions

J. H. Macek [*]

Department of Physics and Astronomy, University of Tennessee, Knoxville, Tennessee 37996-1501
and
Physics Division, Oak Ridge National Laboratory, Oak Ridge, Tennessee

Abstract.
Three He atoms can react in vapor at temperatures of the order of a few degrees Kelvin to form a dimer and a free atom by three-body collisions. Conversely, the dimer may fragment by collisions with free He atoms. Cross sections and reaction rates for these processes have been computed at milliKelvin temperatures using the hyperspherical hidden crossing theory. At higher energies, of the order of 0.1 to 5 Kelvin, the impulse approximation has been used to estimate these processes. The computed fragmentation cross section is reported.

INTRODUCTION

In the theory of Bose condensates the two-body scattering length a is a fundamental parameter. If it is positive stable, homogeneous condensates can form, but if it is negative they cannot. If a is positive and large, then the bosons usually form a weakly bound dimer with binding energy $E_0 = 1/(2\mu a^2)$, where μ is the reduced mass and atomic units are used. The lifetime of the condensate may be set by the rate for dimer formation. One process that leads to dimer formation is the three-body reaction $B + B + B \rightarrow B + B_2$.

The rate for this process may be computed from the cross section for the inverse process $B + B_2 \rightarrow B + B + B$ and detailed balance. A prototype for this process is the fragmentation of the He dimer whose binding energy is 1.3 milliKelvin. The threshold region is important for temperatures in the millikelvin range. In this range the quantum theory of the full three-body system is needed for accurate calculations of the fragmentation S-matrix. Three different treatments have been given in the literature[1, 2, 3]. The hidden crossing theory of Ref. (1) gives an essentially closed form analytic result and is briefly reviewed in the next section.

At energies much higher than the dimer binding energy, the hidden crossing theory is no longer applicable and a high energy theory is needed. The Born approximation is the most widely known high-energy theory, but it requires that the collision energy be much greater than the interatomic potential. Since the He-He potential has a hard repulsive core, the Born approximation requires collision energies in the keV energy range, well outside of Kelvin range of interest for atomic beams. In this latter range the impulse approximation can be used since it does not require a weak interatomic potential. Rather, it is valid if the dimer binding is much less than the collision energy and the dimensions of the dimer are much larger than the range of the atom-atom interaction. These conditions are satisfied when the scattering length is large compared with the dimensions of the atoms. On that basis impulse approximation calculations for He-He$_2$ collisions are employed in the $1 - 10$ Kelvin energy range.

HIDDEN CROSSING THEORY

The hyperspherical close-coupling approximation[4] maps the full three-body problem onto two-body-like Schrödinger equations. Near the threshold for the fragmentation process, an analytic representation of the cross section has been found using the hidden crossing theory[1]. The hidden crossing theory gives

$$\sigma(E) = \frac{\pi}{k^2} \exp[-S(E)] \sin^2 \Delta(E) \qquad (1)$$

---

[*] This research is sponsored by the Division of Chemical Sciences, Office of Basic Energy Sciences, U.S. Department of Energy, under Contract No. DE-AC05-00OR22464 managed by UT-Battelle, LLC. Support by the National Science Foundation under grant number PHY997206 is also gratefully acknowledged.

where $S(E)$ and $\Delta(E)$ are the real and imaginary part of a WKB-like phase integral

$$\Delta(E) + iS(E)/2 = \int_{R_1}^{R_2} K(R) dR. \quad (2)$$

In the above equation, R is the hyper-radius, $K(R)$ is the local wave vector in the hyperspherical adiabatic representation, and the integral is taken along a contour in the complex plane connecting the zero of $K(R)$ in the dimer channel at R_1 with the zero in the fragmentation channel at R_2. This theory gives a fragmentation cross section proportional to a^4 times an oscillatory factor $\sin^2 \Delta(E)$.

The oscillatory factor plays an important role in setting the magnitude of the threshold cross section, but we may take it to be a constant for sufficiently small E. The remaining factors are proportional to the E^2 Wigner threshold factor and an a^4 scale factor. We find

$$\sigma(E) = \frac{4\pi\mu^2}{k^2} A E^2 a^4 \sin^2(s_0 \ln(a/R_0) + \Delta_0) \quad (3)$$

where $A = 0.167$, $s_0 = 1.006$, and R_0 and Δ_0 are constants.

The a^4 factor is not immediately obvious; rather an a^2 dependence is expected on the basis of an impulse approximation argument. In this approximation the cross section for dimer breakup is just double the cross section for elastic scattering of two He atoms, equal to $4\pi a^2$ at low energies. To reconcile these two different powers of a, note that it is the cross section averaged over an energy of the order of the binding energy E_0 of the dimer that should have an intuitive dependence upon a. One finds that $E^2 a^4$ averages to $1/E_0 \propto a^2$; thus the expected a^2 dependence is recovered.

IMPULSE APPROXIMATION

Of course, we note that the impulse approximation is only qualitatively correct in the threshold region, and one must rely upon more advanced three-body theories for quantitative results. At higher energies, for example, the 1-5 Kelvin range of interest for gases undergoing expansion(6), the impulse approximation is actually well justified. Because an energy of 1 Kelvin is 10,000 times larger than the dimer binding, this binding may be neglected in collisions with other He atoms. Also, the dimer is about 10 times larger than the nominal range of the He-He potential, so for atom wavelengths much less than the size of the dimer, one can consider that collisions with the dimer are effectively collisions with individual He atoms. If a collision takes place, the dimer breaks up with almost unit probability. In this case the impulse approximation cross section for breakup is just twice the He-He elastic

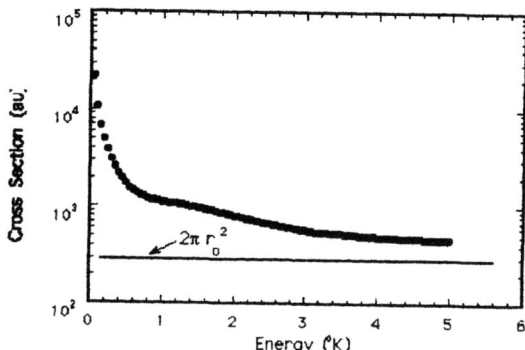

FIGURE 1. Cross section for the process $He + He_2 \rightarrow He + He + He$ in the impulse approximation. The horizontal line corresponds to the limit cross section for the hard core of the He-He potential with $r_0 = 4.5 au$.

scattering cross section at the same relative velocity. That is,

$$\sigma_{\text{breakup}}(E) = 2\sigma_{\text{elas}}^{(2)}(E). \quad (4)$$

We have computed $2\sigma_{\text{elas}}^{(2)}(E)$ by numerical solution of the Schrödinger equation for He-He scattering. We use a Morse potential chosen to fit the depth D, the radius $r_{\min}$ of the potential minimum, and the s-wave scattering length a from the standard LM2M2 potential of Aziz et al. The fit is very good over most of the range; however, the potential core is somewhat softer than for the Aziz potential. We find, for example, that the classical turning point for energies of 1 Kelvin is about 0.1 au smaller for the Morse potential than for the Aziz potential. This may affect cross sections by amounts of the order of 5%.

The Schrödinger equation was solved for the elastic scattering phase shifts in a partial wave expansion. Phase shifts for $\ell = 0, 1, 2$ and 3 were computed for energies between 0.05 and 5 Kelvin. The first three partial waves gave significant contributions, while the $\ell = 3$ contribution was negligible but was included for completeness. The computed breakup cross section is shown in Fig. 1.

Three features are apparent in Fig. 1. First there is the rapid rise toward low energies. This rise is mainly due to the s-wave component since it approaches $8\pi a^2$ with $a \approx 187 au$. A shoulder in the intermediate energy region near 1-2 Kelvin is a second noticeable feature. This is due to a top-of-barrier p-wave resonance. For this resonance, the phase shift increases to $\pi/4$, then begins to decrease with increasing energy giving the slight bump seen in the figure. A third feature is the flat region above 3 K. In this region the two-body elastic cross section has the same order of magnitude as πr_0^2 where r_0 is the distance at which the depth of the potential equals the incident energy. One expects that the cross section will decrease slowly from this value to πr_T^2 at high energies, where r_T is the range of the hard inner core. In the present case this range is equal to $r_T = 4.5 au$.

The cross section shown in Fig. 1 complements the low-energy hyperspherical and effective field theory calculations applicable at threshold. At the lowest energy ($E = 0.05K$) in the figure, the wavelength of Schrödinger wave for the He atom is of the order of the dimensions of the dimer and the impulse approximation ceases to be quantitatively accurate. Improved estimates could be obtained by adapting the impulse approximation amplitude given in the book by Mott and Massey(7) to the case of three identical particles. Using that amplitude would account for features, such as coherent scattering from the atoms in the dimer and energy shifts owing to the dimer binding, neglected in the simple expression Eq. (4). These corrections would be needed to connect with the threshold cross section. Even without these corrections, the cross section at 1.3×10^{-3} K from the hidden crossing theory neglecting the oscillatory factor is of the order of 10^4 au, in modest agreement with the value of $\sigma = 2^4$ au at $E = 50 \times 10^{-3}$ K from the impulse approximation.

SUMMARY

We have computed the cross section for $He + He_2 \rightarrow He + He + He$ in the impulse approximation applicable in the $1-5$ Kelvin energy range. Prominent features of the cross section are interpreted in terms of the expected behavior of the $\ell = 0$ and 1 partial waves.

REFERENCES

1. Nielsen, E., and Macek, J. H., *Phys. Rev. Lett.* **83**, 1566 (1999).
2. Esry, B. D., Burke, J. P. Jr., and Greene, C. H., *Phys. Rev. Lett.* **83**, 1751 (1999).
3. Bedaque, P. F., Hammer, H. W., and van Kolck, U., *Nuc. Phys. A* **646**, 444 (1999) and preprint.
4. Macek, J., *J. Phys. B* **1**, 831 (1968).
5. Aziz, R. A. and Staman, M. J., *J. Chem. Phys.* **94**, 8047 (1991).
6. Schöllkopf, W. and Toennies, J. P., *Science*, **266**, 1245 (1994).
7. Mott, N. F. and Massey, H. S. W., *Theory of Atomic Collisions*, Oxford, The Clarendon Press, 1965, pp 335 ff.

Momentum Distributions in Fast Heavy Ion-Atom Collisions

D.M. McSherry[a], S.F.C. O'Rourke[a],
R. Moshammer[b] and J. Ullrich[b]

[a] *School of Mathematics and Physics, The Queen's University of Belfast, N.Ireland, U.K.*
[b] *Universität Freiburg, Hermann-Herder Strasse 3, D-79104, Freiburg, Germany.*

Abstract. In recent years much progress has been made in understanding the dynamics of the three-body break-up in energetic ion-atom ionization processes. Much of this progress has been made due to the recent developments in efficient spectrometers combined with recoil ion momentum spectroscopy, which has allowed for targets heavier than helium to be considered. We discuss these experimental results in the context of our theoretical results achieved using the continuum-distorted-wave eikonal-initial-state approximation. Doubly differential cross sections, as well as transverse projectile momentum transfer shall be considered.

INTRODUCTION

Single ionization in fast heavy ion-atom collisions is a fundamental process which is important in any practical application based on the energy deposition in matter by fast ions. Such applications include track formation, bulk modifications and the radiation damage of biological and other materials. It is therefore of the utmost importance that an understanding of the three body dynamics of the situation is achieved. With the rapid developments in recoil ion momentum spectroscopy techniques (1), along with ejected electron spectroscopy it is now possible to experimentally produce kinematically complete investigations of single target ionization by fast heavy ion impact (2). This detailed information on the ionization mechanism can prove to be a stringent test for theory.

CALCULATIONAL DETAILS

For fast ion-atom collisons, longitudinal momentum and energy conservation require that

$$p_{R\parallel} = p_{p\parallel} - p_{e\parallel} = \frac{\Delta\varepsilon}{v} - k\cos\theta \quad (1)$$

where $p_{R\parallel}$ is the longitudinal recoil-ion momentum, $p_{p\parallel}$ is the longitudinal momentum transfer and $p_{e\parallel}$ is the longitudinal momentum of the ejected electron. The velocity of the projectile in the laboratory frame is given by $\underline{v}$, $\underline{k}$ is the momentum of the ejected electron relative to the target, and θ is the polar angle of $\underline{k}$ with respect to the polar axis $\underline{v}$. Also

$$\Delta\varepsilon = E_k - \varepsilon_i \quad (2)$$

where E_k is the continum energy of the emitted electron and ε_i is the binding energy of the target atom.

For equation (1) to be valid certain conditions apply namely:

1. the mass of the projectile ion must be much heavier than the electron, i.e. it must be very much greater than 1 a.u.,

2. the initial collision energy must be larger than $E_k - \varepsilon_i$,

3. the projectile scattering angle Θ must be small ($\Theta \ll 1$).

Normally in fast highly charged ion-atom collisions all of these conditions are satisfied.

Doubly differential cross sections

There are only very few experimental data sets of doubly differential cross sections for target heavier than helium (3, 4). It is only with the recent developments in entirely new and efficient electron spectrometers combined with recoil-ion momentum spectroscopy that heavier targets can be considered (1). It is for this reason we have considered the double differential cross section as a function of the longitudinal electron velocity $v_{e\parallel}$ for various transverse velocity $v_{e\perp}$ cuts. This can be derived from equation (1) noting that

$$E_k = \frac{1}{2}k^2 = \frac{1}{2}(p_{e\parallel}^2 + p_{e\perp}^2) \quad (3)$$

where $p_{e\perp}$ is the transverse momentum of the ejected electron. Hence

$$\frac{1}{2\pi v_{e\perp}} \frac{d^2Q}{dv_{e\parallel}dv_{e\perp}} = \frac{1}{2\pi} \int_0^{2\pi} \sigma(k) d\phi \qquad (4)$$

where $v_{e\perp}$ is the transverse velocity, ϕ is the azimuthal angle of $\underline{k}$ with respect to the collision plane and $\sigma(\underline{k})$ is the triply differential cross section achieved by integrating over $R_{if}(\underline{q})$, the scattering amplitude:

$$\sigma(k) = 2\pi a_0^2 \int_{\frac{\Delta\varepsilon}{v}}^{\infty} |R_{if}(\underline{q})|^2 q dq. \qquad (5)$$

The doubly differential cross sections are defined in this manner in order to correct for the increasing volume element with increasing $v_\perp$. Therefore the cross sections found will have the shape and dimensions of triply differential cross sections $dQ/d\underline{v}$, assuming azimuthal symmetry.

Transverse projectile momentum transfer

Experimentally it is now possible to measure the single differential cross sections as a function of the projectile scattering angle θ_P (5, 6). Measurements of this angle in fast ion-atom collisions have proved to be quite difficult because of the extremely small deflection of the projectile. Hence these experiments have concentrated on the ionization of helium by proton or deuteron impact.

Transverse projectile momentum transfer distributions can be obtained from the results of these measurements of the projectile scattering angle since $\eta \approx \mu v \theta_P$ for small values of θ_P.

The transverse projectile momentum transfer can be derived by integrating over the scattering amplitude $R_{if}(\underline{q})$ with respect to $\underline{k}$:

$$\frac{dQ}{p_{P\perp}} = 2\pi p_{P\perp} \int_0^\infty |R_{if}(\underline{q})|^2 d\underline{k} \qquad (6)$$

where

$$d\underline{k} = k^2 dk \sin\theta d\theta d\phi \qquad (7)$$

and $p_{P\perp} = \eta$, the transverse component of the change in relative momentum of the heavy particle.

The theory outlined above is independent of any theoretical approximations used to evaluate the scattering amplitude $R_{if}(\underline{q})$. We have chosen to use the continuum-distorted-wave eikonal-initial-state (CDW-EIS) approximation in our calculations which has proven quite successful in recent work describing ionization within the non-perturbation regime (7, 8). Its suitability arises from the fact that the ionized electron sees the Coulomb field from both the target and the projectile ion, the asymptotic boundary conditions are satisfied exactly and finally both the initial and final states are normalized. In this present work we have used the semi-classical straight line version of the impact parameter approximation with the inital bound states represented by Roothan Hartree Fock (RHF) wavefunctions. We have also adopted the independent electron approximation and assumed that the inactive electron remains in the RHF orbitals throughout the collison.

RESULTS

Doubly differential cross sections

In the following figures the experimental doubly differential cross sections as a function of the longitudinal electron velocity for different transverse velocity cuts are shown in comparision with the results for CDW-EIS calculations. Cylindrical coordinates in velocity space have been chosen and the experimental single ionization data have been normalized to the total CDW-EIS cross sections for each target.

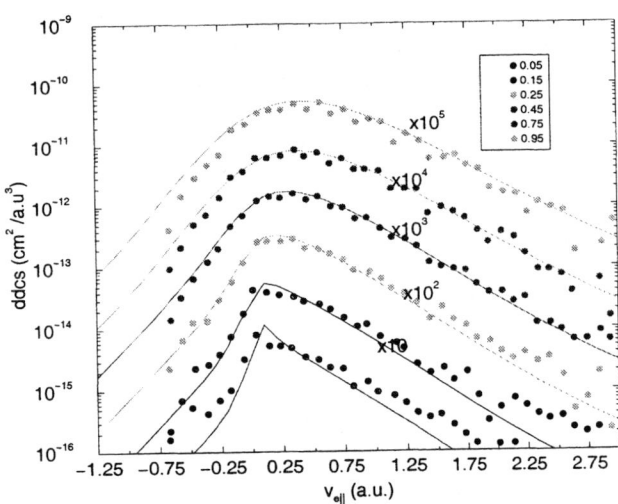

FIGURE 1. Double differential cross sections as a function of the longitudinal electron velocity for various transverse velocity cuts in singly ionizing 3.6MeV/u Au^{53+} on He. The experimental data is from Schmitt et al (9).

Considering the helium and neon figures first of all, there is good agreement with the theory and experiment in both the shape and absolute magnitude. The whole distribution including the maxima for the larger transverse momenta are shifted towards positive velocities and there is shown a strong forward-backward asymmetry. This asymmetry has been attributed to post collision interaction (PCI), i.e. the two centre effect. The emerging

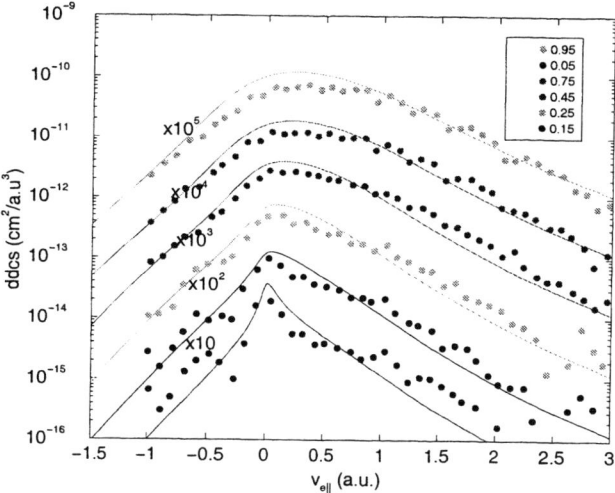

FIGURE 2. Double differential cross sections as a function of the longitudinal electron velocity for various transverse velocity cuts in singly ionizing 3.6MeV/u Au^{53+} on Ne. The experimental data is from Moshammer et al (1).

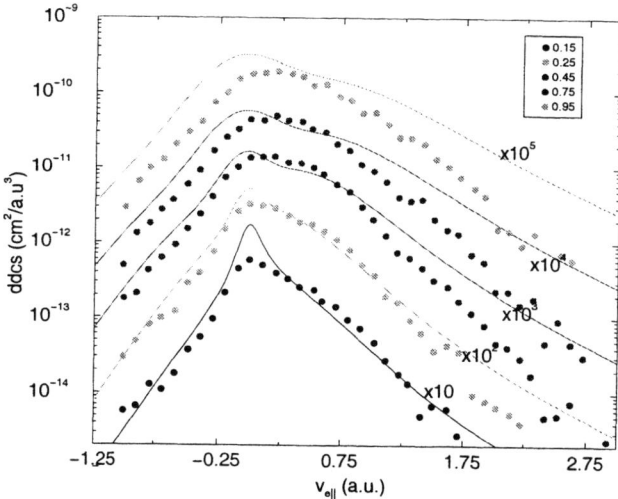

FIGURE 3. Double differential cross sections as a function of the longitudinal electron velocity for various transverse velocity cuts in singly ionizing 3.6MeV/u Au^{53+} on Ar. The experimental data is from Moshammer et al (1).

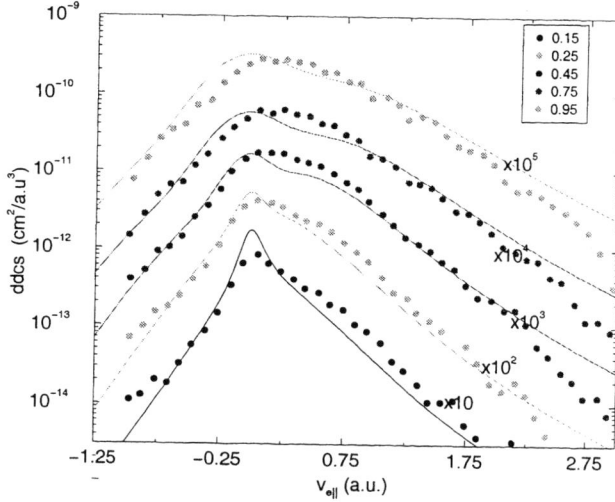

FIGURE 4. Double differential cross sections for electron emission due to single, double or triple ionization of Ar. The DDCS for the specified recoil-ion charge states are added according to their relative contribution to the total cross section. The experimental data is from Moshammer et al (1).

highly charged projectile drags the electrons in the forward direction. This PCI effect is possible to see here as the CDW-EIS approximation takes into account the long range interaction of the projectile and the target field.

Again for argon the same structures appear but here the cusps appear much narrower than for the lighter targets. Also a systematic discrepancy between experiment and theory seems to be appearing at the higher electron energies. In order to explain this discrepancy we need to return to one of the basic assumptions of CDW-EIS, i.e. we adopt the independent electron model and consider there to be only one active electron. In doing this we assume that the whole impact parameter range contributes only to single ionization. However, particularly at small impact parameters, single ionization can find itself in competition with double and mutiple ionization. By plotting the differential cross section irrespective of the degree of ionization this extra contribution can approximately be included. This considerably improves the agreement in the shape between experiment and theory over the entire range of electron energies.

Theoretical calculations for neon and argon were also carried out by Moshammer et al (1), using a generalized version of the CDW-EIS approximation which was obtained by solving stationary Schödinger equations with Hartree-Fock-Slater model potentials for the target atoms. In the case of argon the theoretical results using this model showed a shoulder effect at lower transverse velocity cuts. In our results this effect does not appear. This shoulder effect has also been analysed recently by Gulyás et al (10) using the CDW-EIS approximation with a target potential obtained from the optimised potential method (OPM). They concluded that this shoulder effect was an artifact and that the wavefuntion is sensitive to the inital state.

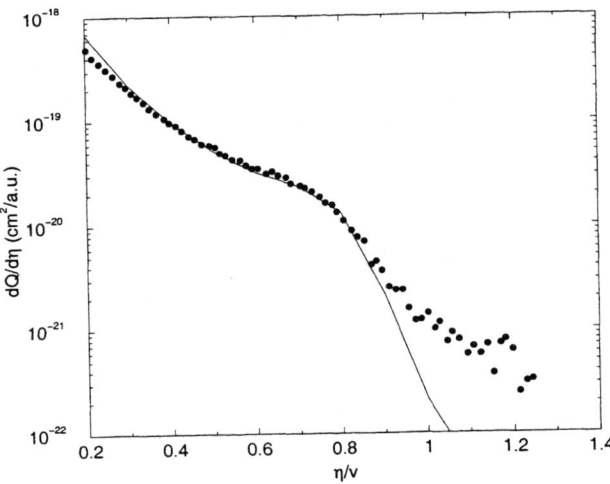

FIGURE 5. Transverse projectile momentum distribution for single ionization of He by protons at 3MeV. Experimental results are from Kamber et al (5)

Transverse projectile momentum transfer

In figure 5 we have the theoretical calculations of the transverse projectile momentum transfer for the single ionization of He by proton impact at 3MeV impact energy, compared to the experimental results of Kamber et al (5). The figure displays a distinctive shoulder effect at the proton scattering angle of 0.55 mrad in both the experimental and theoretical data. This has been attributed to binary collisions between the protons and the target electrons, in which the protons are scattered to near the maximum scattering angle for a proton, off a free electron. This maximum scattering angle is given by the ratio of the electron mass to that of the projectile. The recoiling electrons from such an event are seen in this binary encounter ridge. Considering higher impact energies this ridge appears to be slightly washed out by the momentum distribution of the target electrons however it does becomes stronger as the energy is decreased.

It is interesting to note that for larger angles the theoretical results fall off faster than the experimental results. Calculations by Rodríguez (11) show that this problem can be addressed by including an internuclear interaction to get the projectile deflection.

CONCLUSION

We have analyzed the doubly differential cross sections as a function of the longitudinal electron velocity for various transverse velocity cuts. There is satisfory agreement between the experimental and theoretical results with a strong forward-backward asymmetry being displayed. This asymetry has been attributed to the PCI effect.

We also examined the transverse projectile momentum transfer, finding again satisfory agreement between experimental and theoretical data. A shoulder effect was clearly seen which was attributed to dominant binary projectile-electron scattering at $\theta_P \leq 0.55$ mrad. It was also found that the transfer projectile momentum transfer is sensitive to internuclear interaction which confims the findings of Rodrigúez (11).

ACKNOWLEDGEMENTS

One of us (D.M. McS.) acknowledges the support of the Department of Higher and Further Education, Training and Employment (DHFETE) for the award of a Research Studentship. Also one of us (S.F.C. O'R) acknowledges support by a Royal Society University Research Fellowship.

REFERENCES

1. Moshammer, R., Fainstein, P.D., Schulz, M., Schmitt, W., Kollmus, H., Mann, R., Hagmann, S., and Ullrich, J., *Phys. Rev. Lett.* **83**, 4721-4724 (1999).
2. Stolterfoht, N., DuBois, R.D., and Rivarola, R.D., *Electron Emission in Heavy-Ion Atom Collisions*, edited by J.P. Toennies, Springer-Verlag, Berlin, 1997, Vol. 20.
3. Suárez, S., Garibotti, C., Bernardi, G., Focke, P., and Meckbach, W., *Phys. Rev. A* **48**, 4339-4339 (1993).
4. Stolterfoht, N., Chesnel, J.-Y., Grether, M., Skogvall, B., Frémont, F., Lecler, D., Hennecart, D., Hussion, X., Grandin, J.P., Sulik, B., Gulyás, L., and Tanis, J.A., *Phys. Rev. Lett.* **80**, 4649-4652 (1998).
5. Kamber, E.Y., Cocke, C.L., Cheng, S., McGuire, J.H., and Varghese, S.L., *J. Phys. B: At. Mol. Opt. Phys.* **21**, L455-459 (1988).
6. Kristensen, F.G., and Horsdal-Pederson, E., *J. Phys. B: At. Mol. Opt. Phys.* **23**, 4129-4149 (1990).
7. Crothers, D.S.F., and McCann, J.F., *J. Phys. B: At. Mol. Opt. Phys.* **16**, 3229-3242 (1983).
8. O'Rourke, S.F.C., Shimamura, I., and Crothers, D.S.F., *Proc. R. Soc. Lond. A* **452**, 175-184 (1996).
9. Schmitt, W., Moshammer, R., O'Rourke, S.F.C., Kollmus, H., Sarkadi, L., Mann, R., Hagmann, S., Olson, R.E., and Ullrich, J., *Phys. Rev. Lett.* **81**, 4337-4340 (1998).
10. Gulyás, L., Kirchner, T., Shirai, T., and Horbatsch, M., *Phys. Rev. A* **63**, 0022702 (2000).
11. Rodríguez, V.D., *Nucl. Instrum. Methods B.* **132**, 250-258 (1997).

Excitation Cross Sections of (1snp) $^1P^o$ (n = 2-5) Levels of Helium by Fast Electron, Proton, and Molecular Hydrogen (H_2^+ and H_3^+) Impact

M. Bailey, H. Merabet[*], R. Bruch, J. Hanni, S. Bliman[1], D. V. Fursa[2], I. Bray[2], K. Bartschat[3], H.C. Tseng[4], C.D. Lin[5]

Department of Physics, University of Nevada Reno, Reno, NV 89557 USA
[1]*Department de Physique, University de Marne la Vallée, 96166 Noisy LeGrand, France*
[2] *The Flinders, University of South Australia, GPO Box 2100, Adelaise 5001, Australia*
[3]*Department of Physics and Astronomy, Drake University, Des Moines, IA 50311, USA*
[4] *Department of Physics, Chung Yuan Christian University, Chung Li, 32023 Taiwan*
[5]*Department of Physics, Kansas State University, Manhattan, KS 66506-2601 USA*

Abstract. Experimental cross sections in the extreme ultraviolet (EUV) wavelength range for the excitation of helium following electron and H^+, H_2^+, and H_3^+ ion impact are presented for HeI (1snp) $^1P^o$ states with n = 2-5. These measurements extend over a large velocity range from 3.8-8.5 a.u. for electrons, 1.4-7.5 a.u. for protons, and 1.4-4.0 a.u. for H_2^+ and H_3^+ ions, and they represent the most complete data set obtained so far in the EUV. Furthermore, the convergent close coupling (CCC) and R-matrix with pseudo-states (RMPS) methods have been used here to predict excitation cross sections for the HeI (1snp) $^1P^o$ states following electron impact and the atomic-orbitals close coupling expansion (AOCC) for proton impact. In particular, our theoretical results are presented and compared with our EUV experimental cross sections for equal projectile velocities together with previous experimental results including cross sections derived from scaling procedures.

INTRODUCTION

Cross section measurements of processes involving helium by electron and ion impact are fundamental to the investigation of few electron interactions in atomic and ionic collision physics (1-3). The knowledge of such collision processes is not only important for the understanding of the collision dynamics but also for laboratory and astrophysical plasmas and helium based radiation diagnostics such as solar flare analysis (4). Such collision processes depend on a large parameter space which includes projectile velocity, projectile size, charge states, charge sign, mass, and structural complexity.

The processes investigated in this study are:

$$A + He (1s^2)\ ^1S \rightarrow He (1snp)\ ^1P^o + A$$
$$\hookrightarrow He (1s^2)\ ^1S + h\nu, \quad (1)$$
$$(A \equiv e^- \text{ or } H^+, n=2\text{-}5)$$

where the emitted EUV radiation is observed at 90° with respect to the projectile beam.

Theoretically, single electron excitation of helium by charged particle impact at high velocities (v > 3.8 a.u.) has been described in terms of the 1st Born approximation (B1) and the Bethe theory (5-6). In the intermediate energy range, where the cross sections exhibit their maximum values, all previous theoretical calculations are not in agreement with the measured cross sections for electron and proton impact. We have applied here three more sophisticated theoretical techniques namely the convergent close coupling (CCC) (7-8), R-matrix with pseudo-states (RMPS) (9-10), and atomic-orbitals close-coupling (AOCC) (11) methods to elucidate the excitation mechanisms of helium following electron and proton impact for a wide range of projectile velocities (1-9 a.u.). In this connection, we note that the CCC method used earlier by Fursa and Bray (7) for lower excited levels (1snp) $^1P^o$ (n=2 and 3) for the electron impact has been successfully extended here to include excitation to higher Rydberg states with n > 3. A similar method, namely AOCC, has also been utilized to derive excitation cross sections for HeI (1snp) $^1P^o$ states for the H^+ + He collision system (11). In addition we introduce new theoretical data for HeI (1s2p) $^1P^o$ levels following electron impact using the RMPS method for incident energies ranging from 30 up to 400 eV (i.e., $1.4 \leq v \leq 5.4$ a.u.).

Molecular projectiles (H_2^+ and H_3^+) have been utilized in an effort to improve the understanding of complex collision processes involving multi-scattering centers. In the following, we provide a detailed discussion and comparison of all data for e^- + He and

---

[*] Corresponding author: H. Merabet, email: hocine@physics.unr.edu

H^+ + He collision systems. We also focus on the molecular ion (H_2^+ and H_3^+) impact and compare them with the corresponding proton data.

RESULTS AND DISCUSSION

The experimental setup has been described in detail by Bailey et al. (12), therefore only a brief discussion is presented here. Positive ion beams (H^+, H_2^+ and H_3^+) produced by the University of Nevada Reno 2 MV Van de Graaff accelerator and electron beams created by an electron gun deliver projectile beams focused on an ultra-compact differentially pumped target cell. The EUV emission was observed at right angles to the projectile beam and analyzed with a 1.5 m grazing incidence monochromator in conjunction with a channeltron detector.

Cross sections for the excitation of He measured in this work were put on an absolute scale by normalizing our measured high velocity cross section data to the Bethe-Born cross section values for electron and proton impact velocities v > 3.8 a.u. This allowed us to determine the detection sensitivity for the wavelength region from 51 to 59 nm (HeI $(1snp)$ $^1P^o \rightarrow (1s^2)$ ^1S) and to effectively cross calibrate this region with the corresponding 22-31 nm range (HeII (np) $^2P^o \rightarrow (1s)$ ^2S). When instrumental uncertainties related to energy resolution of the Van de Graaff accelerator, absolute target pressure stability, polarization and charge normalization and cascade repopulation are combined, total experimental uncertainties ranging from 8% for the $(1s2p)$ $^1P^o$ to 18% for $(1s5p)$ $^1P^o$ are obtained.

A. e^- + He and H^+ + He Collisions

Our EUV results for electron and proton impact are plotted for the corresponding cross sections for HeI $(1snp)$ $^1P^o$ levels, n = 2-5, in Fig.1 and Fig.2. An overview of the cross sections for HeI $(1snp)$ $^1P^o$ states for e^- and H^+ are presented in Fig.3 as a function of projectile velocity (a.u.), where we also have included previously obtained results by other groups (13-18).

The EUV experimental and CCC, RMPS and B1 theoretical excitation cross sections for the e^- + He system are plotted in Fig. 1. The CCC calculations, shown in this figure, relate to HeI$(1snp)$ $^1P^o$ states with n=2-4 while the additional HeI $(1s5p)$ $^1P^o$ cross sections are estimated using an $1/n^3$ scaling law. Fig.1 includes the optical experimental data of Westerveld et al. (15) and Donaldson et al. (16), which are the recommended experimental data by de Heer and co-workers (19-20). We have found that the CCC results agree well with our EUV cross sections and the other recommended experimental data over the entire velocity range. Despite this favorable agreement, we note that overall the CCC results systematically overestimate the experimental cross sections by about 10%. This is may be due to the frozen-core target description used in the CCC model, which is somewhat problematic for the ground state of He. As expected, the B1 predictions do not agree with experiment in the low and intermediate velocity regime, but they coincide with our measurements for velocities of approximately 5 a.u. and above. We further note that the RMPS results for HeI $(1s2p)$ $^1P^o$ (see Fig. 1) and experiment agree well at all energies. Therefore, the apparent convergence of RMPS and B1 in the 3.8-5.5 a.u. velocity range, may lead to the conclusion that a sophisticated target description is more important than channel coupling in the high-energy regime.

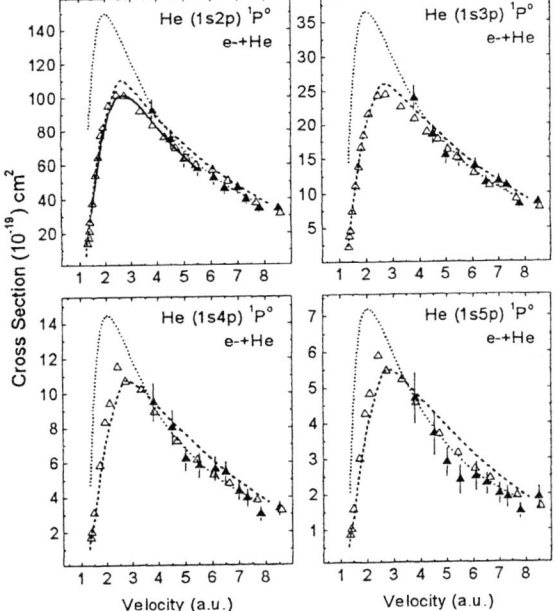

FIGURE 1. Experimental cross sections for $(1snp)$ $^1P^o$ states due to electron impact compared to recent theoretical results; Experiment; ▲, this work; ∆, de Heer et al. (20); Theory; - - - CCC calculation, this work, —— RMPS, this work; • • •, B1.

Now let us consider the H^+ + He collision system. In Fig. 2 we have exhibited our experimental results, the scaled data recommended by de Heer and co-workers (21), and the most prominent theoretical results for HeI $(1snp)$ $^1P^o \rightarrow (1s^2)$ ^1S (n = 2-5) emission (5,13) versus projectile velocity. The scaled data for proton impact stem from the earlier visible measurements of Van den Bos et al. (13), renormalized to agree with the UV measurements of Hippler et al. (14) for HeI $(1s3p)$ $^1P^o$ and HeI $(1s4p)$ $^1P^o$. In Fig.2, the general trend for all measured cross sections shows a moderately steep increase in cross section for small impact velocities, reaching a maximum around 2 a.u., followed by a gradual decrease with increasing projectile velocity toward the asymptotic high energy limit. We have also indicated in Fig.2 the predictions of our AOCC approach, together with other theoretical results. It is evident that all these calculations coincide with our experimental data in the higher velocity range (v > 2.5 a.u.), where the AOCC cross sections go nicely toward B1 results at higher energies. However we have observed slight deviations from the experimental EUV findings. Nonetheless, the agreement of the AOCC calculations with our EUV cross sections is reasonably good within the error bars. The AOCC calculations have been explicitly performed for the (1s2p) $^1P^o$ and (1s3p) $^1P^o$ states, whereas the data for the (1s4p) $^1P^o$ and (1s5p) $^1P^o$ states from the AOCC calculations are estimated using the $1/n^3$ scaling rule.

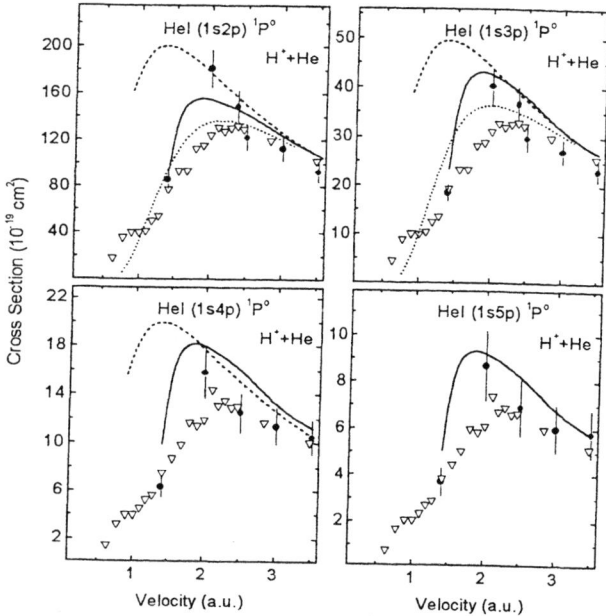

FIGURE 2. Theoretical cross sections for the excitation of (1snp) $^1P^o$ states in He following proton impact compared with experimental and scaled results. Experiment: ● EUV, this work, ∇ de Heer et al. (21), Theory: ---, 1st Born (6); —, AOCC calculations, this work; • • •, Van den Bos et al. (13).

In Fig. 2, results from earlier calculations and experiments have been shown as well. The theoretical data of van den Bos et al. (13) show good agreement with the scaled results from de Heer et al. (21) for excitation to the (1s2p) $^1P^o$ level. The corresponding (1s3p) $^1P^o$ calculations of van den Bos and co-workers also demonstrate reasonable agreement with the experimental data but deviate sharply from the visible results at lower impact velocities. In Fig. 3 we show a comparison of all the previously presented experimental data associated with electron and proton impact in conjunction with theoretical predictions from the first Born approximation. As can be seen in this figure, the proton induced cross sections are significantly larger with a maximum slightly shifted to lower velocities when compared to the corresponding electron data. From Fig. 3 it can be seen that in the high energy limit, the e^- and H^+ projectile results tend toward the first Born limit as expected. However significant differences between electron and proton impact occur at velocities below 3 a.u. for the (1snp) $^1P^o$ (n=2-5) states, well outside the range of the error bars. In particular, it is observed that the cross sections for the excitation of helium by proton impact are larger than those obtained for electron impact. Additionally, the peak in the cross section for electron excitation is slightly displaced towards higher velocities compared

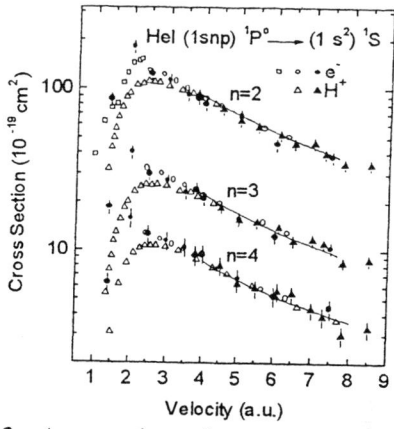

FIGURE 3. A comparison of proton and electron impact data as a function of projectile velocity. (1s2p) $^1P^o$ and (1s3p) $^1P^o$: ● H^+ and ▲ e^-, this work; □ H^+ Park and Schowengerdt (17); ○ H^+, Hippler and Schartner (14); Δ e^-, Westerveld et al. (15); (1s4p) $^1P^o$: (same as above except) Δ e^-, Donaldson et al. (16); solid lines = H^+ Bethe approximation (6), Kim and Inokuti (18).

to the proton results. In the lower velocity region, electron impact excitation results in a significant loss of the electron's incident energy. For proton impact such a loss is negligible. This may be one reason why electron impact excitation cross sections are smaller than proton impact excitation cross sections at equal incident velocities.

B. Comparison of H^+, H_2^+, and H_3^+ Results

We have examined our EUV cross sections for H_n^+ ion impact with helium by comparing them first to the semiempirical scaling model of Hasselkamp et al. (22). At high H_n^+ impact velocities, $v \geq 2$ *a.u.*, we have observed that for the cross section for excitation (σ^*) by H^+, H_2^+ and H_3^+ ions, the cross section difference for σ^* H_3^+ and H_2^+ is approximately equal to the difference between $\sigma(H_2^+)$ and $\sigma(H^+)$, i.e.

$$\sigma^*(H_3^+) - \sigma^*(H_2^+) \approx \sigma^*(H_2^+) - \sigma^*(H^+) \quad (3)$$

This expression can be rearranged to read

$$\sigma^*(H_3^+) \approx 2\sigma^*(H_2^+) - \sigma^*(H^+) \quad (4)$$

As a typical example we have plotted in Fig. 4 the cross sections for excitation of the (1s2p) $^1P^o$ state of helium by H_3^+, the prediction of Eq. (3) referred to as the "test" equation (23), and proportionally scaled cross sections from H^+ impact. To examine electron screening effects we have also plotted the ratios $\sigma^*(H_3^+)/3\sigma^*(H^+)$ and $\sigma^*(H_2^+)/2\sigma^*(H^+)$ for the transitions, HeI $(1snp)\ ^1P^o \to (1s^2)\ ^1S$, n = 2-5. A sample of typical error bars has been included in each figure. In the upper part of the figure for $v \geq 2$ a.u., we observe good agreement between the experimental data for $\sigma^*(H_3^+)$ and the "test" equation, $2\sigma^*(H_2^+) - \sigma^*(H^+)$, while the proportionally scaled cross sections, namely $3\sigma^*(H^+)$, are considerably larger than either of these quantities.

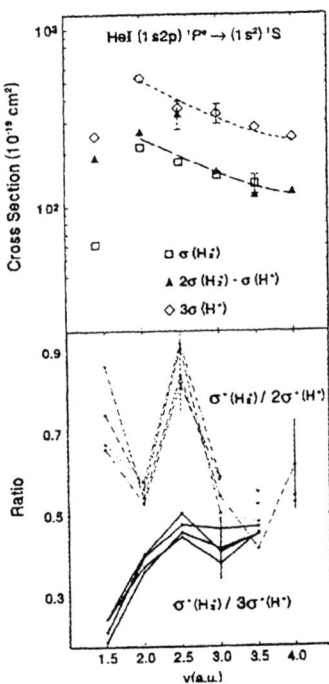

Figure 4. Cross-section results for $\sigma^*(H_3^+)$, □, in comparison with the "test equation", ▲, from the scaling model of Hasselkamp et al. (22) and scaled cross-sections, $3\sigma^*(H^+)$, ◊. The ratios of H_2^+ and H_3^+ cross-sections with proportionality scaled proton cross-sections for all the $(1snp)\ ^1P^o$ transitions measured in this work are presented in the lower part of the figure. Curves are provided to guide the eye.

The lower part of the figure shows the cross section ratios and clearly reveals contributions arising from dynamical effects, including screening, by the projectile electrons. It is obvious that a proportional scaling of 1:1 for $3\sigma^*(H^+)$ and $\sigma^*(H_3^+)$ is never attained. Interpreting excitation of He as a more distant collision phenomenon, screening by the projectile electrons is present; an under-proportionality is expected and in fact observed. At a projectile velocity of about v = 1.3 a.u. for H_3^+ the effective projectile electron energy is approximately equal to the excitation energies of the $(1snp)\ ^1P^o$ states and hence can be related directly to the energy defect for this projectile. However, for more distant collisions, the projectile electrons are not expected to contribute individually to the excitation process, and no contribution to the $\sigma^*(H_3^+)$ cross sections is apparent above this velocity, i.e. $\sigma^*(H_3^+)$ cross sections are consistently smaller than $3\sigma^*(H^+)$ results. Under-proportional scaling is also observed when cross section results for H_2^+, $\sigma^*(H_2^+)$, are compared with $2\sigma^*(H^+)$ though the effect is not as pronounced as in the case of H_3^+. This is expected since there is only one projectile electron to provide screening in this case.

By examining the ratios of $\sigma^*(H_n^+)$ to $n\sigma^*(H^+)$, the limitations associated with treating H_n^+ as a mixture of independent particle beams is clearly revealed. Since a more refined theoretical description of such complex multi-center collision processes does not currently exist, little can be said about contributions to the total scattering amplitude due to multi-center scattering and interference effects. However we have provided evidence that interstitial electrons due to molecular bonding play a significant role in these types of multi-center excitation processes.

ACKNOWLEDGMENTS

This work was supported, in part, by the National Science Foundation (KB), the Nevada Business and Science Foundation (NBSF) and ACSPECT Corporation.

REFERENCES

1. McGuire, J.H., *Electron Correlation Dynamics In Atomic Collisions*, (Cambridge University Press, 1997).
2. Stolterfoht, N. et al., *Phys. Rev. Lett* **57**, 74 (1986).
3. Fülling, S. et al., *Phys. Rev. Lett.* **68**, 3152 (1992).
4. Fineschi, S. et al. *Proceedings of the Workshop on Max 1991/SMM Solar Flares: Observations and Theory*, 124 (1991).
5. Bell, K. L. et al., *J. Phys. B.* **1**, 1037 (1968).
6. Inokuti, M., *Rev. Mod. Phys.* **43**, 297 (1971).
7. Fursa, D. V., and Bray, I., *Phys. Rev. A* **52**, 1279 (1995).
8. Fursa, D. V. et al., *J. Phys. B* **30**, 757 (1997).
9. Bartschat, K. et al., *J. Phys. B.* **29**, 2875 (1996).
10. Hudson, E. T. et al., *J. Phys. B.* **29**, 5513 (1996).
11. Fritsch, W. and Lin, C. D., *Phys. Rept*, **201**, 1 (1991).
12. Bailey, M. et al., *J. Phys. B.* **28**, 2655 (1995).
13. van den Bos, J. et al., *Physica* **44** 143 (1969).
14. Hippler, R. et al., K.-H., *J. Phys. B.* **7**, 618 (1974).
15. Westerveld, W. B., et al., *J. Phys. B.* **12** 115 (1979).
16. Donaldson, F.G., et al., J.W., *J. Phys. B.* **5**, 192 (1972).
17. Park, J. T. et al., *Phys. Rev.* **185**, 152 (1968).
18. Kim, Y. K., Inokuti, M., *Phys. Rev.* **184**, 38 (1969).
19. de Heer, F.J. et al., Joint European Torus publication (JET), 19 (1992).
20. de Heer, F. J., International Nuclear Data Committee Report IDC(NDS)-385 Vienna, Australia (1998).
21. de Heer, F. J., Hoekstra, R. and Summers, H. P., *Joint European Torus publication (JET)*, 47 (1992).
22. Hasselkamp, D. et al., *Z. Phys. D. Atoms, Molecules and Clusters* **6**, 269 (1987).
23. Alvarez, T. et al., *Phys. Rev. A* **14**, 602 (1976).

Beyond the Shake Process

Takeshi Mukoyama* and Masayuki Uda

Laboratory for Materials Science and Technology
Waseda University, Shinjuku-ku, Tokyo, 169-0051 Japan

Abstract. A new theoretical model for electron transition probability accompanying inner-shell vacancy production is presented. In this model, both the existence of the inner-shell vacancy and the Coulomb interaction between the ejected electron and atomic electrons are taken into consideration. The electron transition probability is expressed as the sum of the conventional shake process and multipole transitions due to the Coulomb interaction between electrons. The excitation probabilities for $2p$ electrons in Ne accompanying $1s$-shell photoionization are calculated and the contribution of the Coulomb interaction terms is discussed.

INTRODUCTION

It is well known that the atomic electrons have a small, but definite probability to be excited to a higher unoccupied state accompanying inner-shell ionization (1). This process is caused by the sudden change in the central potential due to creation of an inner-shell vacancy. The excitation probability can be treated with the shakeup model, i.e. imperfect wave function overlap between the initial and final states (2, 3). In this case, only the monopole transition is allowed for atomic electrons.

Recently we have developed a new model to estimate the electron excitation probability following inner-shell vacancy production (4). The change in the atomic potential due to the removal of the inner-shell electron is treated as the perturbation and the electron transition probability is obtained by the time-dependent perturbation theory. Although only the monopole transition is allowed, the transition matrix element is different from the simple overlap integral. The calculated probabilities for bound-bound transitions are in good agreement with the conventional shakeup probabilities.

However, both models are within the framework of the two-step approach, where only the existence of an inner-shell vacancy is taken in account and contributions from the ejected electron are not included. With recent advent of third-generation synchrotron radiation facilities, tunable photon beams become available easily and the electron transition probability during photoionization can be observed with high resolution as a function of incident photon energy. In order to understand the energy dependence of the multielectron excitation probability, it is important to develop theoretical models, which include the interaction between the outgoing electron and atomic electrons explicitly.

It is the purpose of the present paper to present a general model for electron transition accompanying inner-shell vacancy production beyond the shake process. This model is applied for the electron excitation probabilities accompanying photoionization.

THEORETICAL MODEL

We consider an electron in the atom and calculate its transition probability accompanying an inner-shell vacancy production. Its wave function, $\Psi(r,t)$, satisfies the time-dependent Schrödinger equation:

$$\left(\mathcal{H} - \frac{\partial}{\partial t}\right)\Psi(r,t) = 0 . \quad (1)$$

Throughout the present work atomic units are used. When an electron is ejected from the atom and the inner-shell vacancy is created at $t=0$ the Hamiltonian of the system can be written by

$$\mathcal{H} = \begin{cases} \boldsymbol{H}_0(r) & t < 0, \\ \boldsymbol{H}(r) + V(r,t) & t \geq 0 \end{cases}, \quad (2)$$

where $\boldsymbol{H}_0(r)$ is the Hamiltonian for the atomic ground state, $\boldsymbol{H}(r)$ is that for the ion with an inner-

---

* Permanent address: Kansai Gaidai University, 16-1 Kitakatahoko-Cho, Hirakata, Osaka 573-1001, Japan

shell vacancy, and $V(\mathbf{r},t)$ represents the interaction between the atomic electron and the ejected electron. The interaction Hamiltonian is given by

$$V(\mathbf{r},t) = \frac{1}{|\mathbf{R}(t) - \mathbf{r}|},$$

where $\mathbf{R}(t)$ is the position vector of the ejected electron. It is clear that $V(\mathbf{r},t) \to 0$ as $t \to \infty$.

For $t \geq 0$, we consider $V(\mathbf{r},t)$ as a time-dependent perturbation. The eigenfunction $\phi_n(\mathbf{r})$ and energy eigenvalue ϵ_n of $\mathbf{H}(\mathbf{r})$ are expressed as

$$\mathbf{H}(\mathbf{r}) \phi_n(\mathbf{r}) = \epsilon_n \phi_n(\mathbf{r}). \quad (3)$$

Using a set of these eigenfunctions, we expand the time-dependent wave functions $\Psi(\mathbf{r},t)$ and obtain a set of coupled equations

$$i \dot{a}_m(t) = \sum_n a_n(t) V_{mn}(t) e^{i\omega_{mn}t}, \quad (4)$$

where

$$V_{mn}(t) = \int \phi_m^*(\mathbf{r}) V(\mathbf{r},t) \phi_n(\mathbf{r}) d\mathbf{r}, \quad (5)$$

and $\omega_{mn} = \epsilon_m - \epsilon_n$.

Integrating Eq. (4) with respect to time from 0 to t, we obtain

$$a_m(t) = a_m(0) - i \sum_n \int_0^t a_n(t') V_{mn}(t') e^{i\omega_{mn}t'} dt'. \quad (6)$$

We assume that in the initial state the electron is in one of the eigenstates of the Hamiltonian $\mathbf{H}_0(\mathbf{r})$:

$$\mathbf{H}_0(\mathbf{r}) \phi_k^{(0)}(\mathbf{r}) = \epsilon_k^{(0)} \phi_k^{(0)}(\mathbf{r}). \quad (7)$$

The wave function at small t is expressed as

$$\phi_k^{(0)}(\mathbf{r}) e^{-i\epsilon_k^{(0)}t} = \sum_m a_m(t) \phi_m(\mathbf{r}) e^{-i\epsilon_m t}. \quad (8)$$

Using Eq. (8), the first term in Eq. (6) is written by

$$a_m(0) = <\phi_m(\mathbf{r})|\phi_k^{(0)}(\mathbf{r})>, \quad (9)$$

where the right hand side means the overlap integral between $\phi_m(\mathbf{r})$ and $\phi_k^{(0)}(\mathbf{r})$:

$$<\phi_m(\mathbf{r})|\phi_k^{(0)}(\mathbf{r})> = \int \phi_m^*(\mathbf{r}) \phi_k^{(0)}(\mathbf{r}) d\mathbf{r}. \quad (10)$$

Substituting Eq. (9) into the right-hand side of Eq. (6), we obtain

$$a_m(t) = <\phi_m(\mathbf{r})|\phi_k^{(0)}(\mathbf{r})> - i \int_0^t W_{mk}(t') e^{i\omega_{mk}t'} dt', \quad (11)$$

where $\omega_{mk} = \epsilon_m - \epsilon_k^{(0)}$ and

$$W_{mk}(t) = \int \phi_m^*(\mathbf{r}) V(\mathbf{r},t) \phi_k^{(0)}(\mathbf{r}) d\mathbf{r}. \quad (12)$$

The first term in Eq. (11) corresponds to the shake process, while the second term represents the transition due to the Coulomb interaction between the atomic electron and the ejected electron.

The first term is independent of time and the transition occurs at the moment of the vacancy creation. On the other hand, the second term is a function of time. This fact means that the electron transition due to the second term is possible only when the inner-shell vacancy exists. The probability that the inner-shell vacancy created at $t = 0$ survives at the time t is given by $\exp(-\Gamma t)$, where Γ is the level width of the inner-shell vacancy. Taking into this factor in the second term, Eq. (11) can be written as

$$\begin{aligned} a_m(t) &= <\phi_m(\mathbf{r})|\phi_k^{(0)}(\mathbf{r})> \\ &\quad -i \int_0^t W_{mk}(t') e^{(i\omega_{mk}-\Gamma/2)t'} dt' \end{aligned} \quad (13)$$

Then the electron transition probability from the state k to the state m accompanying the inner-shell vacancy production is given by

$$P_{mk} = |a_m(+\infty)|^2. \quad (14)$$

RESULTS AND DISCUSSION

In the present work, we confine ourselves to the case of bound-bound transitions. For such a case, the first term in Eq. (13) corresponds to the shakeup process. On the other hand, the second term means that the outgoing electron ejected from the primary ionization event kicks an atomic electron up to one of unoccupied bound states and can be called the *kickup* process.

In principle, the model described above is so general that it can be applied for all inner-shell ionization processes, such as photoionization, impact ionization by electron and other charged particles, and internal conversion in radioactive decay. It is also possible to apply this approach to the case of nuclear β decay, where there is no inner-shell vacancy, but nuclear charge increases by one. In this case, we have to use Eq. (11) instead of Eq. (13).

Here to test the validity of the present mode we consider the case where the $2p$-electron excitation probabilities accompanying K-shell photoionization

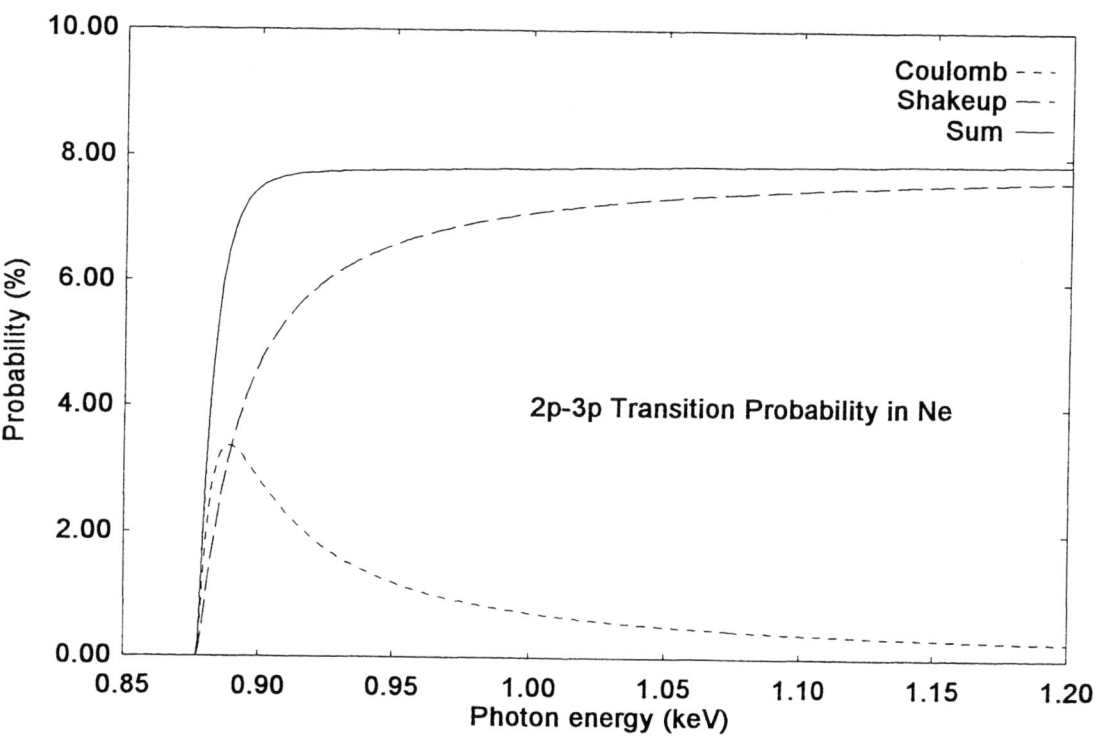

FIGURE 1. The $2p - 3p$ transition probability in Ne accompanying K-shell photoionization as a function of incident photon energy. The dashed curve represents the shakeup process, the dotted curve the Coulomb interaction term, and the solid curve the sum of two processes.

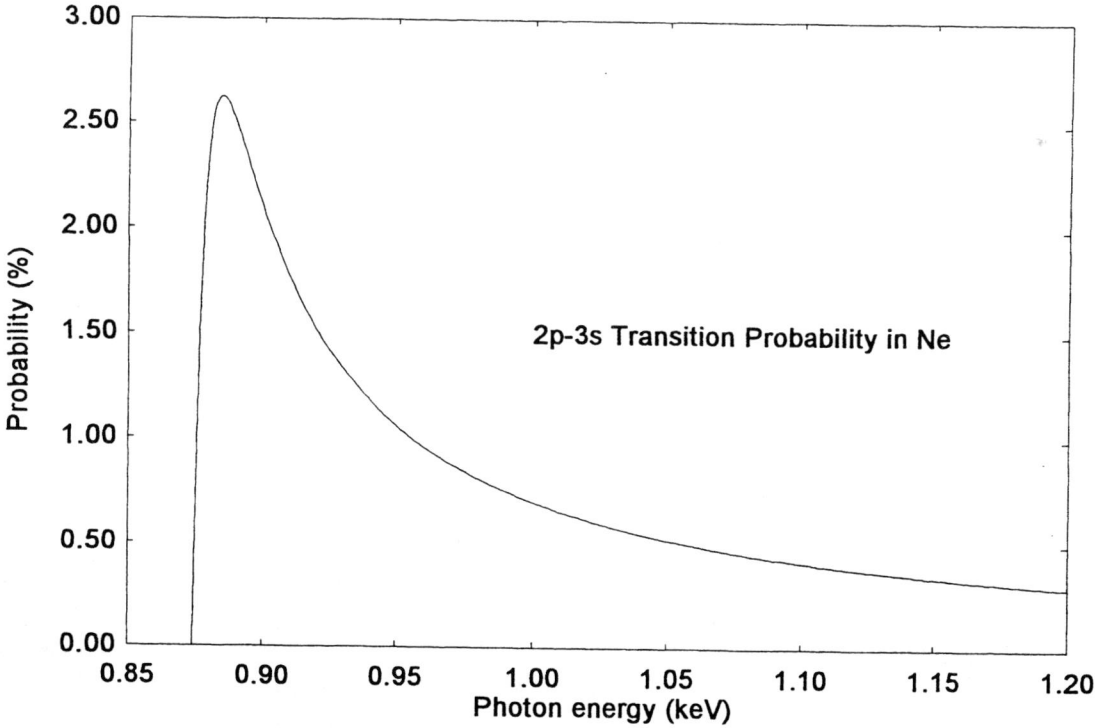

FIGURE 2. The $2p - 3s$ transition probability in Ne accompanying K-shell photoionization as a function of incident photon energy.

in Ne. When the incident energy of photon is E_p, the photoelectron ejected from $1s$ shell has the kinetic energy corresponding to $E_e = E_p + \epsilon_{1s}^{(0)}$, where $\epsilon_{1s}^{(0)}$ is the energy eigenvalue of the $1s$ state in the initial ground state. On the other hand, another electron makes a transition from the state k to m, a part of the kinetic energy of photoelectron is transferred to the atom and the final energy of the photoelectron escaped from the atom is given by $E'_e = E_e - \epsilon_m + \epsilon_k^{(0)}$. Then the electron transition probability as a function of photon energy is given by

$$P(E_p) = n_k \frac{\sigma(E_p, E'_e)}{\sigma(E_p, E_e)} P_{mk}, \qquad (15)$$

where n_k is the number of electrons in the state k and $\sigma(E_p, E_e)$ is the photoionization cross section for incident photon energy E_p and the ejected electron energy E_e (5).

For simplicity, we assume that the photoelectron is ejected from the center of the atom, which is located at the center of the coordinate system. This assumption can be justified because the mean radius of the $1s$ orbital is smaller than that of the $2p$ orbital. We choose the direction of emission of the photoelectron as the z axis. Then the position of the photoelectron at time t is given by $R = vt$ along the z axis, where v is the velocity of the photoelectron.

The atomic wave functions were calculated with the Hartree-Fock-Slater (HFS) method using the Herman-Skillman program (6). The wave functions for the final excited states were obtained by solving the Schrödinger equation in the HFS potential for the positive ion with an inner-shell vacancy.

The calculated results for the $2p-3p$ transition in Ne accompanying K-shell photoionization are shown in Fig. 1 as a function of incident photon energy. The dashed curve indicates the shakeup process, the dotted curve means the contribution from the Coulomb interaction between electrons (*kickup* process), and the solid curve represents the sum of two processes. Experimentally these two processes yield the same electron spectrum and cannot be distinguished. The shakeup probability at the asymptotic region is in good agreement with the calculated value of Mukoyama and Taniguchi (7). It is clear that the Coulomb interaction term is strong only in the low-energy region because the interaction time is longer for slower photoelectrons. Owing to the existence of this term, the curve for the excitation probability rises more steeply in the region near threshold than the curve for the shakeup process. Considering this fact, the contribution of the Coulomb interaction may be measured by observing the $2p-3p$ excitation probability near to the threshold energy.

In Fig. 2, the $2p-3s$ excitation probability in Ne as a function of incident photon energy. In the shake process, This excitation process is dipole transition and forbidden in the shakeup process due to the monopole selection rule. The excitation is possible only through the Coulomb interaction between two electrons. This fact suggests that the peak corresponding to the $2p-3s$ excitation be observed in photoelectron spectra. However, its energy position is a few eV lower than the $2p-3p$ excitation peak. This small energy difference makes it difficult to detect a weak peak of the former near to the strong shakeup peak.

CONCLUSION

In the present work, we have developed a new model for electron transition process accompanying an inner-shell vacancy production. The transition probability can be expressed as the sum of two processes; the conventional shake process and the Coulomb interaction between the ejected electron and atomic electrons. In the latter process, the outgoing electron kicks up other atomic electrons into higher energy states.

The present model was used for the $2p$-electron transitions in Ne following K-shell photoionization. Although the present work is limited to bound-bound transitions, it is interesting to apply this model also for bound-free transitions.

REFERENCES

1. Mukoyama, T., and Ito, Y., *Nucl. Instr. and Meth.* **B87** 26 (1994) and references cited therein.
2. Carlson, T.A., and Nestor, C.W., *Phys. Rev. A* **8** (1973) 2887.
3. Mukoyama, T., and Taniguchi, T., *Phys. Rev. A* **36** (1987) 693.
4. Mukoyama, T., and Uda, M., *Phys. Rev. A* **61** (2000) R030501.
5. Sauter, F., *Ann. Physik* **9** (1931) 217; **11** (1931) 454.
6. Herman, F., and Skillman, S., *Atomic Structure Calculations*, Englewood Cliffs, N.J.: Prentice-Hall, 1963.
7. Mukoyama, T., and Taniguchi, K., *Bull. Inst. Chem. Res., Kyoto Univ.* **70** (1992) 1.

Mechanism of Secondary Ion Emission from an Al Sample under MeV Heavy Ion Bombardment

J. Xue*, S. Ninomiya†, S. Gomi†, and N. Imanishi†

Quantum Science and Engineering Center, Faculty of Engineering, Kyoto University, Kyoto 611-0011, Japan
†*Department of Nuclear Engineering, Kyoto University, Kyoto 606-8501, Japan*

Abstract. Sputtering yields and kinetic energy distributions (KED) of Al atomic ions ejected from a pure aluminum sample under MeV silicon ion bombardment were simulated with the molecular dynamics (MD) method. Since the electronic energy loss (S_e) is much higher than the nuclear energy loss (S_n) when the incident ion energy is as high as several MeV, an S_e effect was also taken into consideration in the simulation. It was found that the simulated sputtering yield fits well with the experimental data and the electronic energy loss has a slight effect at incident ion energies higher than 4 MeV. The simulated secondary ion KED spectrum is a little lower in the peak energy and narrower in the peak width than the experimental.

INTRODUCTION

It is well known that an energetic ion loses its energy via two nearly independent processes in solid: nuclear energy loss (S_n) and electronic energy loss (S_e). When the incident ion energy is low (keV), the nuclear energy loss is dominant. The incident ion loses its energy mainly through the elastic collision with target atoms, some target atoms obtain enough energy to be ejected from the sample, and the contribution of S_e to the sputtering process could be neglected. This sputtering phenomenon (nuclear sputtering) can be explained well by the elastic collision theory.

When the incident ion energy is high, S_e becomes much higher than S_n, and its influence on the sample's sputtering behavior can not be neglected any longer. However, the mechanism of the electronic energy loss inducing sputtering (electronic sputtering) is still not clear, though some models were set up to explain it. Among these models, the thermal spike model [1,2] and the ionic model [3,4] are the most popular ones. The thermal spike model assumes that the electronic energy loss can be transferred to the lattice atoms through an electron-phonon coupling process. On the other hand, the ionic model assumes that the lattice atoms can obtain energy from the repulsive Coulomb force between the high-density excitations created by the electronic energy loss. No matter what the energy transfer mechanism is, it is sure that a part of the electronic energy loss can be transferred to the lattice atoms, which will influence the target's sputtering behavior. Since the energy that the target atoms can get depends on both the magnitude of S_e and the S_e-lattice transfer property, the electronic sputtering behavior depends also on both of them. It was found that [5-11] some materials, such as low temperature condensed-gas solids, organic materials [13] and oxides [6,7], have strong S_e-lattice energy transfer efficiencies; and some other materials, such as Al and Ni [12], have weak S_e-lattice energy transfer efficiencies. Therefore, the electronic sputtering effect in different materials should be different even if they have the same value of S_e.

For most MeV heavy ions, the values of S_e are higher than those of S_n when they travel in solid, and both S_n and S_e possibly influence the material's sputtering behavior in this energy region. It was found that the sputtering yield of Si^{4+} from an SiO_2 sample depends on both S_e and S_n [14]. In order to study the mechanism of the electronic sputtering process, it is necessary to investigate the S_e and S_n effects on the sputtering behavior of materials with different S_e-lattice transfer properties. For this purpose, Al, Al_2O_3, and SiO_2 samples that have different S_e-lattice transfer properties [6,7,12], have been or will be used in our sputtering measurement.

The Molecular Dynamics (MD) simulation method is a very powerful tool to investigate the ion sputtering process. Many MD simulations were carried out to simulate the sputtering process induced by low energy ions [15], and the simulated results fit well with the experimental data, including kinetic energy distributions (KED), angular distributions and sputtering yields of secondary ions [16,17]. For the electronic sputtering, a few MD simulations were done on weakly bonded materials, such as solid gas samples and biological samples. Though the S_e-lattice energy transfer mechanism is not clear, some simulated results are in agreement with the experimental results [18-20]. These works show that the MD simulation is suitable to study the electronic sputtering process, too. However, up to now, very few works have been done on the sputtering process of tight-bound-chemical materials irradiated with MeV heavy ions.

In this work, the MD method is used to simulate the sputtering process of pure Al bombarded with MeV silicon ions. Both S_e and S_n effects are taken into consideration in the simulation. The simulated sputtering yields and secondary ion KED spectra will be compared with our experimental results.

SIMULATION METHOD

The code used in the simulation was developed in our group on a PC computer. The average force method [21] was used to calculate the target atom movement. A Hookean like force was applied to the boundary atoms to keep the sample stable [15], except for those in the sample's free surface on which the incident ions hit.

The simulated Al sample contains $14\times14\times10 = 1960$ atoms organized in the FCC structure. The pair potential was used to describe the interaction between Al atoms in a combined form as given by:

$$V(r) = \begin{cases} Ae^{-Br} & r < R_1 \quad (1) \\ \text{cubic polynomial} & R_1 < r < R_2 \quad (2) \\ V_o\left(e^{-2\alpha(r-R_o)} - 2e^{-\alpha(r-R_o)}\right) & R_2 < r < R_3 \quad (3) \\ 0 & r > R_3 \quad (4) \end{cases}$$

where r is the distance between the pair of Al atoms; the values of A, B and α are 4285.1 eV, 0.2718 Å^{-1} and 1.21 Å^{-1}, respectively, which were adopted from Ref. [15]; V_o is the depth of the attractive well; R_0 is the aluminum crystal constant of the value 2.81 Å; R_1 and R_2 were chosen carefully to ensure the smooth connection between the potential (1) and (3); and R_3 is the cutoff distance for the potential. The Screened Coulomb Potential with ZBL parameters was used to describe the interaction between the incident silicon ion and the target atoms [22].

Nuclear energy loss could be taken into account in the simulation directly by the Coulomb force between the incident ion and the target atoms. As for the electronic energy loss, it is impossible to simulate the S_e-lattice energy transfer process because its mechanism is still not clear. In the simulation, we assume that the electronic energy transfers to the lattice atoms within a very short time, and the transferred energy is distributed only to the atoms within a cylindrical region around the ion track as given by the following equations:

$$E_{atom} \propto S_e \times R_{se} \times \frac{1}{r} \quad r < R_{cutoff} \quad (5)$$

$$E_{atom} = 0 \quad r \geq R_{cutoff} \quad (6)$$

where E_{atom} is the energy transferred to the lattice atom; S_e is the electronic stopping power, and the TRIM calculated values are used in the simulation; R_{se} is the S_e-lattice energy transfer coefficient, two arbitrary values of 0.025 and 0.1 were used in the simulation; r is the distance of the atom from the ion track; and R_{cutoff} is the assumed cylinder radius. Atoms' momenta are assumed to be distributed in random directions.

In the simulation, silicon ions with energies ranging from 100 keV to 6 MeV were incident on the Al <111> surface at an angle of 60° to the surface normal, and the incident ion number was 100 for every energy. The total simulation time was 5 ps, and then those atoms that went beyond 8 Å from sample's surface were picked out as sputtered atoms.

EXPERIMENTAL

Energy spectra of sputtered secondary ions were measured with a conventional time of flight (TOF) setup at Kyoto University. A silicon ion beam used in the experiment was produced with a 1.7 MV Cockcroft-Walton accelerator. A TOF ion detector was in the normal direction of the sample surface. The system vacuum was kept better than 4×10^{-6} Pa during the measurement. A surface barrier detector was placed at an angle of 60° to the incident beam to monitor the beam current intensity. A single crystalline Al sample was used in the measurement. An ion-beam-sputtering cleaning was applied to the sample surface for 30 minutes before every measurement. Other details of the experiment are described elsewhere [23,24].

RESULTS AND DISCUSSION

Secondary Ion Sputtering Yield

The simulated sputtering yields for $R_{se} = 0$, 0.025 and 0.1 are shown in Fig. 1 as a function of incident ion

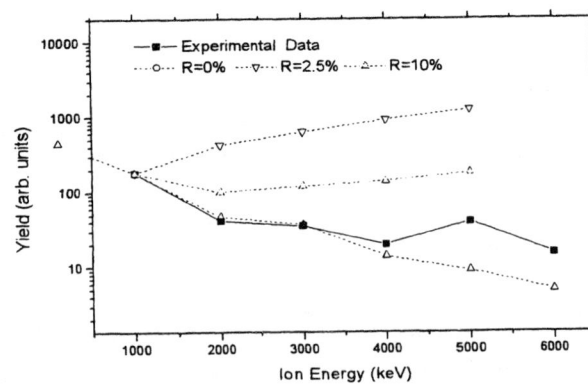

FIGURE 1. Comparison between the simulated and measured sputtering yields for the Al sample bombarded with silicon ions, shown as a function of the incident ion energy.

energy (E_{in}). The measured sputtering yields of Al bombarded with MeV silicon ions are also shown in Fig.1. All the simulation data are normalized to the experimental data at E_{in} = 1 MeV.

It can be seen that, if R_{se} = 0, which means only the nuclear sputtering is taken into account, the simulated yields fit well with the experimental data when E_{in} is lower than 4 MeV. However, when E_{in} is higher than 4 MeV, the simulated data become a little lower than the experimental.

The statistical uncertainty in the simulated yield has the highest value of 20% at E_{in} = 6 MeV and R_{se} = 0, and the uncertainty in the experimental yield is better than 1%. Though the statistical error of the simulation is high at high energies, it still can be seen that there exists an obvious deviation between the tendencies of the experimental and the simulated data. This deviation may come from the S_e effect: When E_{in} is low, S_e is also low, its contribution to the sputtering yield can be neglected, and the simulated yield of R_{se} = 0 fits the experiment well. However, when the incident energy becomes high, the S_e effect becomes comparable with that of S_n, and then it cannot be neglected any longer. The measured yield is a little higher than the simulated data for R_{se} = 0 at the high-energy region.

The simulated sputtering yields for R_{se} = 0.025 and 0.1 increase monotonically with the increase of E_{in}. This behavior disagrees with the experimental one. Even for R_{se} = 0.025, the simulated sputtering yield increases much faster than the measured ones. The above results show that the S_e-lattice energy transfer process is very weak in the Al material. This is in consistent with the theoretical analysis [12].

Secondary Ion KED

The KED spectrum obtained from the MD simulation contains all sputtered atoms, both the neutral and the charged. In order to compare it with the measured Al$^+$ KED, some corrections must be made. According to the tunneling model, the singly-charged secondary ion yield from metal can be written as: [25-28]

$$Y^+ \propto N_o(E) \exp(-\beta / v\cos(\phi)) \qquad (7)$$

where $N_o(E)$ is the kinetic energy distribution for all the sputtered atoms; β is a parameter which characterizes the interaction between the departing ions and the surface, its value depends on the energy of the secondary ions [27]. However, when the secondary ion energy is lower than 50 eV, its value for Al$^+$ from Al is nearly a constant and could be estimated to be 1.1×10^4 m/s [27]. v is the secondary ion velocity and ϕ is the angle between v and the sample surface normal direction, and ϕ = 0 in our experimental arrangement.

By using Eq. (7), the Al$^+$ KED spectrum can be obtained from the simulated KED spectrum of the secondary ions. The simulated secondary particle KED spectra for an Al sample bombarded with 1 MeV Si$^+$ ions,

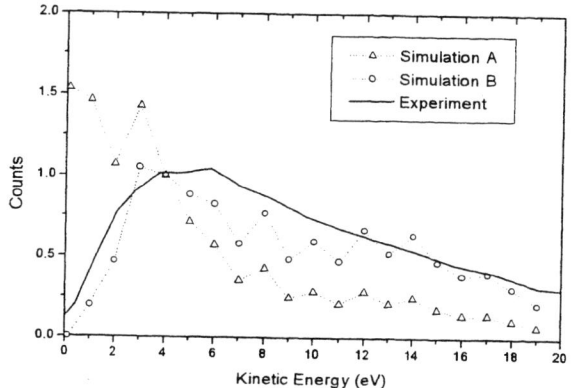

FIGURE.2. The simulated and measured secondary ion kinetic energy distribution spectra for the Al sample bombarded with 1 MeV Si$^+$ ions. A: Simulated KED spectrum of all the sputtered atoms; B: Simulated Al$^+$ KED spectrum obtained with Eq. (7)

for both the total and Al$^+$ atoms, are shown in Fig. 2. The experimental Al$^+$ KED spectrum is also shown in Fig. 2 for comparison. All the spectra are normalized to the experiment data at E_k = 4 eV. It can be seen that the Al$^+$ KED spectrum of the MD simulation fits roughly with the experimental one. Compared with the experimental result, the peak energy of the simulated spectrum is a little low, and the simulated peak width is narrow. Both the simulated and measured results show that the shape of the secondary ion KED spectrum does not depend significantly on the incident ion energy in the range from 1 to 6 MeV.

In our simulation, the pair wise potential was used, because it is simple and allows a faster simulation speed. If the embedded-atom type (EAM) potential is used, it could be expected that the peak energy of the KED spectrum of the sputtered atoms increases, and the energy peak could become broader [16,29]. The former potential may be the reason for the above discrepancy between the simulated spectrum and the experimental.

The above results indicate that nuclear sputtering is still the dominated process for the Al sample irradiated with MeV silicon ions. At the ion energies higher than 4 MeV, S_e shows a slight influence on the sputtering yield, though S_e is much higher than S_n. Since SiO_2 and Al_2O_3 have stronger S_e-lattice transfer efficiencies compared with Al, it could be hoped that electronic sputtering effect could be more significant than in Al.

SUMMARY

Both the MD simulation and measurement for an Al sample sputtered with MeV silicon ions have been carried out. It was found that, if only S_n is concerned, the simulated sputtering yields fit well with the experimental ones at the incident ion energies lower than 4 MeV. By using the tunnel model, the Al^+ KED spectrum was derived from the simulated KED spectrum of all the secondary ions. The simulated Al^+ KED spectrum fits roughly with the measured spectrum, though its peak energy is a little low compared with the measured spectrum.

These results show that, when Al is irradiated with silicon ions at the energy range from 1 to 6 MeV, the nuclear sputtering is still the dominant process, and the electronic sputtering effect is very slight, which means that the S_e-lattice energy transfer process of Al is very weak.

ACKNOWLEDGEMENTS

This work was done with the Experimental System for Ion Beam Analysis at Kyoto University. We thank A. Itoh, M. Imai, K. Yoshida, and K. Norizawa for their useful advice and technical support during the experiments. This work has been supported in part by a Grant-in-Aid for Scientific Research from the Ministry of Education, Science, Sports and Culture of Japan.

REFERENCES

1. Toulemonde, M., Dufour C. and Paumier E., *Phys. Rev. B* **46**, 14362 (1992).
2. Meftah, A., Brisard, F., Costantini, J.M., Dooryhee, D., Hage-Ali, M., Hervieu, M., Stoquert, P., Studer, F., and Toulemonde, M., *Phys. Rev. B* **49**, 12457 (1994).
3. Dunlop A., Lesueur D. and Dural J., *Nucl. Instr. Meth. B* **42**, 182 (1989).
4. Dunlop A., Lesueur D., Morillo J., Dural J., Sphor R. and Vetter J. *Nucl. Instr. Meth. B* **48**, 419 (1990).
5. Toulemonde, M., Bouffard, S., Studer, E., *Nucl. Instr. Meth. B* **91**, 108 (1994).
6. Meftah, A., Brisard, F., Costantini, J.M., Dooryhee, E., Hage-Ali, M., Hervieu, M., Stoquert, J.P., Studer F., and Toulemonde, F., *Phys. Rev. B* **48**, 920(1993).
7. Canut, A., Benyagoub, A., Marest, G., Meftah, A., Moncoffre, N., Ramos, S.S.M., Studer, F., Thevenard, P., Toulenmonde, P., *Phys. Rev. B* **51**,12194(1995).
8. Itoh, N., and Stoneham, A.M., *Nucl. Instr. Meth. B* **146**, 362(1998).
9. Toulemonde, M., Costantini, J.M., Dufour, C.H., Meftah, A., Paumier, E., and Studer, F., *Nucl. Instr. Meth. B* **116**, 37(1996).
10. Toulemonde, M., Dufour, Ch., Wang, Z.G., and Paumier, E., *Nucl. Instr. Meth. B* **112**, 25(1996).
11. Meftah, A., Djebara, M., Khalfaoui, N., Stoquert, J.P., Studer, P., and Toulemonde, M., *Nucl. Instr. Meth. B* **122**, 470(1997).
12. Wang, Z.G., Dufour,, C., Paumier, E., and Toulemonde, M., *J.Phys. Matter.* **6**, 6733(1994).
13. Runne, M., Becker, J., Lassch,W., Varding, D., Zimmer, G., Liu, M., Johnson, R.E., *Nucl. Instr. Meth. B* **82**, 301(1993).
14. Kyoh, S., Takakuwa, K., Umezawa, M., Itoh, A. and Imanishi, N., *Phys. Rev. A* **51**, 554(1995).
15. Eckstein, W., *Computer Simulation of Ion-Solid Interaction*, Springer-Verlag(1991).
16. Garrison, B.J. et.al. *Phys. Rev. B* **36**, 3516(1987).
17. Garrison, B.J., Nicholas Winograd, Deaven, D.M., Reimann, C.T., *Phys. Rev. B* **37**,7197(1988).
18. Sundaqvist, B.U.R and Karlsson, B.R., *Phys. Rev. B* **42**, 1895(1990).
19. Bringa, E.M., Johnson, R.E., *Nucl. Instr. Meth. B* **143**, 513(1998).
20. Bringa, E.M., Johnson, R.E., Dutkiewicz, L., *Nucl. Instr. Meth. B* **152**, 267(1999).
21. Harrison, D.E., Gay, Jr. W.L., Effron, H.M., *J. Math Phys.* **10**, 1179 (1969).
22. Ziegier, J.F., Biersack, J.P., Littmark, U., *The stopping and Range of Ions in Solids*, Pergamon, New York, 1985.
23. Imanishi, N., Shimizu, A., Ohta, H., and Itoh, A., *Applications of Accelerators in Research and Industry*, edited by J.L. Duggan and I.L. Morgen, 396(1999).
24. Ninomiya, S., Gomi, S., Xue, J., Imai, M., and Imanishi, N., This proceeding.
25. Yu, M.L. and Lang, N.D., *Nucl. Instr. Meth. B* **14**, 403(1986).
26. Garrison, B.J., *Surf. Sci.* **167**, 225(1986).
27. Garrett, R.F., Macdonald, R.J. and O`Connor, D.J., *Nucl. Instr. Meth.* **218**, 333(1983).
28. Vasile, M.J., *Surf. Sci.* **115**, 1141(1982).
29. Heinrich G. and Urbassek, H.M., *Nucl. Instr. Meth. B* **69**, 232(1992).

Stripping Energy Dependence Of A B^{3+} ($1s^2\ ^1S$, $1s2s\ ^3S$) Beam Metastable Fraction

M. Zamkov, H. Aliadadi, E.P. Benis*, P. Richard, H. Tawara, T..J.M Zouros*

*James R Macdonald Laboratory, Department of Physics,
Kansas State University, Manhattan, KS 66502-2604*
**Dept. of Physics, University of Crete, P.O. Box 2208, 71003 Heraklion, Crete, Greece*

Abstract. The fraction of metastable ions in the B^{3+} ($1s^2\ ^1S$, $1s2s\ ^3S$) beam was measured as a function of the stripping energy in the range of 0.85 to 9 MeV. This was done by comparing the electron yield of the two final states formed in the collision of B^{3+} on hydrogen; the $1s2s2p\ ^4P$ formed by capture from the metastable $1s2s\ ^3S$ component of the beam and $1s2p^2\ ^2D$ formed by resonant transfer excitation from the ground-state $1s^2\ ^1S$ component of the beam. The double differential cross sections for $1s2s2p\ ^4P$ and $1s2p^2\ ^2D$ lines were then calculated. The stripping energy was converted to the ratio of projectile velocity to the K-shell electron velocity, v_p/v_k, which allows using our findings for other He-like ions. It was found that at $v_p/v_k = 0.3$ no metastable ions are produced. The maximum of the metastable fraction was observed when the stripping foil-projectile relative velocity was near the K-shell electron velocity. The collisionally produced electrons emitted in the forward direction were detected with the new paracentric hemispherical spectrograph[1]. The detailed calculation of the metastable beam fraction is presented.

INTRODUCTION

The analysis of various transitions in metastable ion-atom collisions requires quantitative information on the fraction of metastable ions in the beam. The production of metastable ions was found to be dependent on a number of parameters such as the density of the stripping gas or the thickness of the foil, an atomic number of the element used for the stripper, the stripping energy (i.e. the energy the beam was obtained at). He-like ions can be produced in the long-lived $1s2s\ ^3S$ metastable state. The fraction of He-like ions in this state was previously measured as a function of the stripping energy for a number of various projectiles[2]. In present work, we report the first measurement of the metastable fraction for B^{3+}($1s^2\ ^1S$, $1s2s\ ^3S$) ions performed for a wide range of stripping energies, including a previously uninvestigated low energy region. The electron-ion double differential scattering cross sections were obtained using fast ion-atom collisions. High resolution zero degree Auger spectra were measured using the new paracentric hemispherical spectrograph[1].

COMPUTATION AND EXPERIMENTAL METHOD

As a result of a collision of B^{3+} ($1s^2\ ^1S$, $1s2s\ ^3S$) with target gas (H_2), doubly excited states of B^{2+} are formed. The ^4P state is produced by capture from metastable ions and can be used as an indication of a metastable component in the beam. While the ^2D state, formed by Resonant Transfer Excitation from the ground state, will show the ratio of projectiles in the ground state.

For the given collision parameters, the measured electron yields at zero degrees, for these two states are given by

$$Z(^4P) = N_m\ n\ \sigma_{Capture}\ \xi_{4P}\ \eta\ \Delta\Omega$$
$$Z(^4P) = N_g\ n\ \sigma_{RTE}\ \xi_{2D}\ \eta\ \Delta\Omega \qquad (1)$$

where N_g and N_m are the number of projectile ions in the ground and metastable states, respectively, n is the target number density, ξ is the Auger yield and $\Delta\Omega$

is the effective solid angle. The RTE double differential cross section was calculated using the LS-coupling scheme[3], and the total capture cross section was obtained from the Schlachter scaling rule[4].

The metastable beam fraction is expressed by:

$$F = \frac{N_m}{N_n + N_g} = \left(\frac{N_m}{N_g} + 1\right)^{-1}$$
$$= \left(\frac{Z(^2D)\,\sigma_{Capture}\,\xi_{4P}}{Z(^4P)\,\sigma_{RTE}\,\xi_{2D}} + 1\right)^{-1} \quad (2)$$

The lifetime of the metastable ^4P state is long compared to the time of flight of the ion in the gas cell, so the deexcitation behavior along the projectile trajectory should be taken into account. The fraction R of ^4P ions that decay inside of the gas cell can be found by integrating over the gas cell and is given by:

$$R = \sum a_J \xi_{4P,J}\left(1 - \frac{L_J}{L_{cell}}\left(1 - e^{-\frac{L_{cell}}{L_J}}\right)\right) \quad (3)$$

here, L_{cell} is the length of the gas cell.

$L_J = t_{4P,J}\,v_{projectile}$, where $t_{4P,J}$ is a lifetime of the ^4P_J state, v_p – projectile velocity, $\xi_{4P,J}$ is an Auger yield of the ^4P_J state.

So, finally, for the metastable beam fraction we have the following equation:

$$F = \left(\frac{Z(^2D)\,\sigma_{Capture}\,R}{Z(^4P)\,\sigma_{RTE}\,\xi_{2D}} + 1\right)^{-1} \quad (4)$$

The experiments were performed at the Macdonald Laboratory at Kansas State University, using a 7 MV EN tandem Van de Graaff accelerator. Boron ions were produced either directly in the gas stripper inside of the tandem or after it in the 5 g/cm^2 carbon foil post-stripper to achieve the desired charge state of 3+. A zero degree Auger spectrometer was used to analyze the energy of all the collisionally produced electrons. The target consisted of H_2 molecules in a 7.3 cm long differentially pumped gas cell. The pressure of the target gas was in the range of 10-20 mTorr. The electrons were recorded in the single collision regime. To obtain high resolution Auger electron spectra, the electrons were decelerated.

According to equation (4), only ^4P electrons produced inside of the gas cell should be included in the metastable fraction calculation. Therefore a small voltage (20-30 V) was applied to the gas cell to separate electrons produced outside of it.

RESULTS

Figure 1 shows 2 examples of Auger spectra obtained by the collision of B^{3+} ions on H_2. The Auger electrons at 157 eV and 166.5 eV attributed to the $1s2s2p\ ^4P$ to $1s^2\ ^1S$ and the $1s2p^2\ ^2D$ to $1s^2\ ^1S$ Auger decay were observed. To normalize the spectra, the binary encounter peak enhancement factors were measured for $B^{(2-5)+}$ on H_2. After subtraction of the binary encounter contribution, the double differential cross sections for $1s2s2p\ ^4P$ and $1s2p^2\ ^2D$ lines were obtained.

The ratio R_0 of ^4P electrons produced inside of the gas cell to those produced outside of it was found for a number of beam energies and then extrapolated for the entire beam energy range, used in the experiment. In order to exclude the ^4P electrons produced outside of the gas cell from the spectra, the ^4P yield was multiplied by R_0.

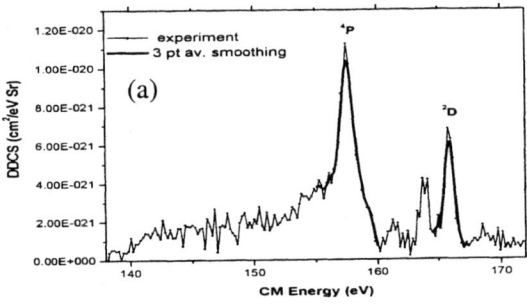

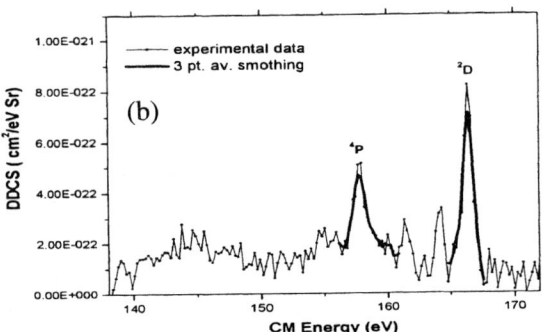

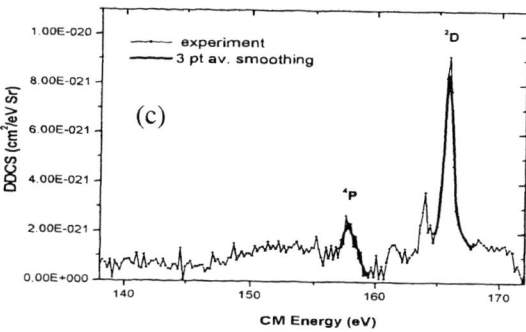

FIGURE 1. B^{3+} on H_2 Auger zero degree spectra. To obtain the B^{3+} ($1s^2$ ^1S, $1s2s$ ^3S) beam the B^{2+} was stripped at a) projectile to K-shell velocity ratio equal to 0.87 b) projectile to K-shell velocity ratio equal to 0.51 c) projectile to K-shell velocity ratio equal to 0.44. The $1s2s2p$ ^4P is formed by capture from the metastable $1s2s$ ^3S component of the beam and $1s2p^2$ ^2D is formed by resonant transfer excitation from the ground-state $1s^2$ ^1S component of the beam.

The metastable beam fraction has been calculated using equation (4) for each stripping energy and plotted as a function of the ratio of the projectile velocity to the K-shell electron velocity (Fig. 2)

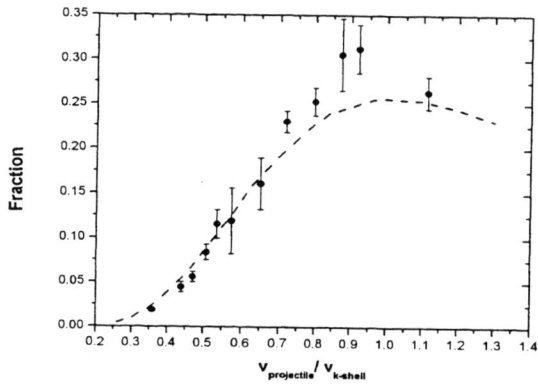

FIGURE 2. The metastable fraction $F = B^{3+}$ ($1s2s$ ^3S)/B^{3+} ($1s^2$ ^1S, $1s2s$ ^3S) versus the ratio of the projectile velocity to the K-shell electron velocity. Circles – experimental data, dashed line - scaled K-shell vacancy production cross section

For comparison the scaled K-shell vacancy production cross section, the primary production mechanism of metastable B^{3+} ($1s2s$ ^3S) by stripping from lower charged beam, is given.

CONCLUSION

High resolution zero degree Auger spectra are measured for 3.5-11.6 MeV B^{3+} projectiles incident on hydrogen. These results were used to calculate the fraction of $1s2s$ ^3S metastable ions in the B^{3+} beam. This was done by comparing the electron yield of the two final states formed in the collision, namely, the $1s2s2p$ ^4P formed by capture from the metastable $1s2s$ ^3S component of the beam and $1s2p^2$ ^2D formed by resonant transfer excitation from the ground-state $1s^2$ ^1S component of the beam. The double differential cross sections for $1s2s2p$ ^4P and $1s2p^2$ ^2D lines were then found by subtracting the binary encounter contribution. For that purpose the binary encounter peak enhancement factors were measured for $B^{(2-5)+}$ on H_2.

The metastable fraction of the B^{3+} beam was calculated as a function of both the stripping energy and the ratio of the projectile velocity to the K-shell electron velocity. The first measurement of the metatsable fraction of He-like ions was taken for the projectile to K-shell electron velocity ratio less than 0.5. It was demonstrated that there are almost no metastable ions produced if the ratio of projectile velocity to 1s electron velocity is less then 0.3. This can be explained by the relatively low K-shell vacancy production cross section at these low velocities. The maximum of the metastable fraction was observed when the projectile-stripping foil relative velocity was near the 1s orbital velocity.

The low velocity B^{3+} beam was obtained by stripping from B^{1-} using the Ne gas stripper inside of the Tandem accelerator. The high velocity B^{3+} beam (ratio of the projectile velocity to the K-shell electron velocity grater than 0.68) was obtained from the primary B^{2+} beam by stripping in the Carbon foil. The difference in the projectile production mechanism may result in the difference in metastable fraction. That partly explains the fact that the K shell vacancy production scaled cross section fits low velocity metastable fraction very well, but deviates that of high velocity projectiles. This dependence is the goal of future experiments.

ACKNOWLEDGEMENT

This work was supported by the division of Chemical Sciences, Office of Basic Energy Sciences, Office of Energy Research, U.S. Department of Energy.

REFERENCES

1. E.P. Benis, K. Zaharakis, M.M. Voultsidou, T.J.M. Zouros, M. Stokli, P.Richard, S. Hagmann *Nuclear Insr. In Phys. Res. B 146* (1998) 120-125

2. T.R. Dillingham, eight references in a doctor's dissertation pp52-54 (1983).

3. T.J.M. Zouros, C.P. Bhalla, D.H. Lee, and P. Richard, *Phys.Rev.* A 42, 678 (1990)

4. A.S. Schlachter, J.W. Sterns, W.G. Graham, K.H. Berkner, R.V. Pyle, and J.A. Tanis, *Phys. Rev. A27*, 3372 (1983)

Fragmentation of 50 – 100keV Molecular Hydrogen Ions in Collision with a Molecular Hydrogen Target

C. McGrath[§], M. B. Shah[§], R. W. McCullough[§], P. C. E. McCartney[†] and J. W. McConkey[‡]

[§]The School of Mathematics and Physics, The Queen's University of Belfast, Belfast, BT7 1NN, N. Ireland
[†]Jet Propulsion Laboratory, NASA, Pasadena, California 91109, U. S. A.
[‡]School of Physical Sciences, University of Windsor, Windsor, Ontario, N9B 3P4, Canada,

Abstract. Dissociation of the molecular hydrogen ion, H_2^+ in the energy range 50 – 100keV in collision with a H_2 target has been investigated. Production of one or two fast protons formed from the projectile break-up was distinguished by use of a Si barrier energy detector. Various projectile fragmentation channel cross-sections were determined by coincidence counting techniques between the target ions (separated by time-of-flight analysis) and one or two fast protons. Such data is of importance in the understanding of astrophysical and high temperature laboratory plasmas.

INTRODUCTION

The most fundamental collision involving a molecular ion and a molecule is the H_2^+ ion on the H_2 molecule. A clear understanding of the processes that can occur in this collision is essential if one is to study more complex molecular ion/molecule collision systems. These processes are also of great importance in the diagnostics of high temperature laboratory and astrophysical plasmas, such as supplementary heating by neutral beam injection in tokamak fusion devices.

This collision has been studied by a number of groups, as can be seen from the data compilations of Barnett[1] and Phelps[2]. The H_2^+ energy range investigated in our experiment was 50 to 100keV. The twenty reaction channels possible for H_2^+ colliding on H_2 are shown in Fig. 1. As can be noted from Fig. 1, we have not included the production of negative ions, as this has been shown to have a very small cross-section[3], and likewise for H_3^+ formation[2], in this energy range.

The pre-1960 work carried out by Damodaran[4], Barnett[5] and Fedorenko et al[6] was later extended by Sweetman[7], Schmidt[8], Guidini[9] and McClure[10], but none of the above experimentalists considered the outcome of the target. To quote Sweetman[7], "*we are not concerned with the fate of the slow ions produced in the collision process and*

Target H_2	+	**Projectile** H_2^+	→
Target product(s)		Projectile product(s)	
H_2	+	$\underline{H}^+ + \underline{H}$	(1)
H_2	+	$\underline{H}^+ + \underline{H}^+ + e$	(2)
$H + H$	+	$\underline{H_2}^+$	(3)
$H + H$	+	$\underline{H}^+ + \underline{H}$	(4)
$H + H$	+	$\underline{H}^+ + \underline{H}^+ + e$	(5)
H_2^+	+	$\underline{H_2}$	(6)
$H_2^+ + e$	+	$\underline{H_2}^+$	(7)
H_2^+	+	$\underline{H} + \underline{H}$	(8)
$H_2^+ + e$	+	$\underline{H} + \underline{H}^+$	(9)
$H_2^+ + e$	+	$\underline{H}^+ + \underline{H}^+ + e$	(10)
$H + H^+$	+	$\underline{H_2}$	(11)
$H + H^+ + e$	+	$\underline{H_2}^+$	(12)
$H + H^+$	+	$\underline{H} + \underline{H}$	(13)
$H + H^+ + e$	+	$\underline{H} + \underline{H}^+$	(14)
$H + H^+ + e$	+	$\underline{H}^+ + \underline{H}^+ + e$	(15)
$H^+ + H^+ + e$	+	$\underline{H_2}$	(16)
$H^+ + H^+ + 2e$	+	$\underline{H_2}^+$	(17)
$H^+ + H^+ + e$	+	$\underline{H} + \underline{H}$	(18)
$H^+ + H^+ + 2e$	+	$\underline{H} + \underline{H}^+$	(19)
$H^+ + H^+ + 2e$	+	$\underline{H}^+ + \underline{H}^+ + e$	(20)

FIGURE 1. Reaction channels available to the $\underline{H_2}^+$ on H_2 collision. Underlining signifies a fast particle.

CP576, *Application of Accelerators in Research and Industry – Sixteenth Int'l. Conf.*, edited by J. L. Duggan and I. L. Morgan *merely regard them as donating or extracting electrons to maintain charge balance*". These later authors were, however, able to separate various projectile channels. We have considered the fate of both the target and the projectile in coincidence and non-coincidence with each other, hence identifying specific channels in Fig. 1. This was achieved through the use of time-of-flight separation of the target ions produced and energy analysis of the projectile products. This work is part of a complete study to determine cross-sections for all the processes available in the H_2^+ - H_2 and H_2^+ - H collision systems[11].

EXPERIMENTAL APPROACH

The experimental arrangement is shown in Fig. 2. The basic experimental approach is very similar to our previous work with H_2^+ ions on an atomic hydrogen target[11]. A pulsed H_2^+ beam from an accelerator intersects a molecular hydrogen beam at right angles. Slow ions resulting from the interaction are extracted using a pulsed extraction field and detected by a channel electron multiplier. The different slow species can be identified by their time-of-flight to this detector. Deflection plates subsequent to the interaction region allow separation of the projectile beam into neutral and residual ion species. The primary fast H_2^+ ion beam is collected in a Faraday cup. Fast H^+ fragments are deflected into a Si barrier detector, which has an output signal proportional to the energy deposited by the impinging particles. If the incident beam was 80keV, fragmentation could produce one or two fast, 40keV H^+ fragments, which impinge on the active area of the detector. Pulse height analysis of the output of this detector allows clear separation of these two possibilities and provides non-coincidence measurements of the production of one or two fast H^+, yielding the summed cross-sections for the families of channels (1), (4), (9), (14) and (19), $\sigma_{1,4,9,14,19}$ and (2), (5), (10), (15), and (20), $\sigma_{2,5,10,15,20}$, respectively. This terminology, referring to Fig. 1, will be used throughout.

The novel feature of our experimental arrangement is that the output of the energy detector can be recorded in coincidence with H^+ or H_2^+ slow products. A time window derived from the channeltron output corresponding to the arrival times of the H^+ or H_2^+ ions is used to gate the detector. Hence, measurements of a slow H_2^+ formed in coincidence with one or two fast H^+ give directly σ_9 and σ_{10}, respectively, and measurements of slow protons in coincidence with one or two fast H^+ give σ_{14} and σ_{15}, respectively. The processes (19) and (20) are assumed to be very small in this energy range. Subtraction of σ_9 and σ_{14} from $\sigma_{1,4,9,14,19}$, and σ_{10} and σ_{15} from $\sigma_{2,5,10,15,20}$, yield $\sigma_{1,4}$ and $\sigma_{2,5}$, respectively.

All measurements, both coincidence and non-coincidence, are recorded with both H_2 gas present and not present. This enables us to remove contributions due to any fast H_2^+ ions colliding with the background gas. As can be expected, this is negligible in coincidence mode. Outputs from the current and pressure monitors are routed through voltage-to-frequency converters and counted, minimizing errors

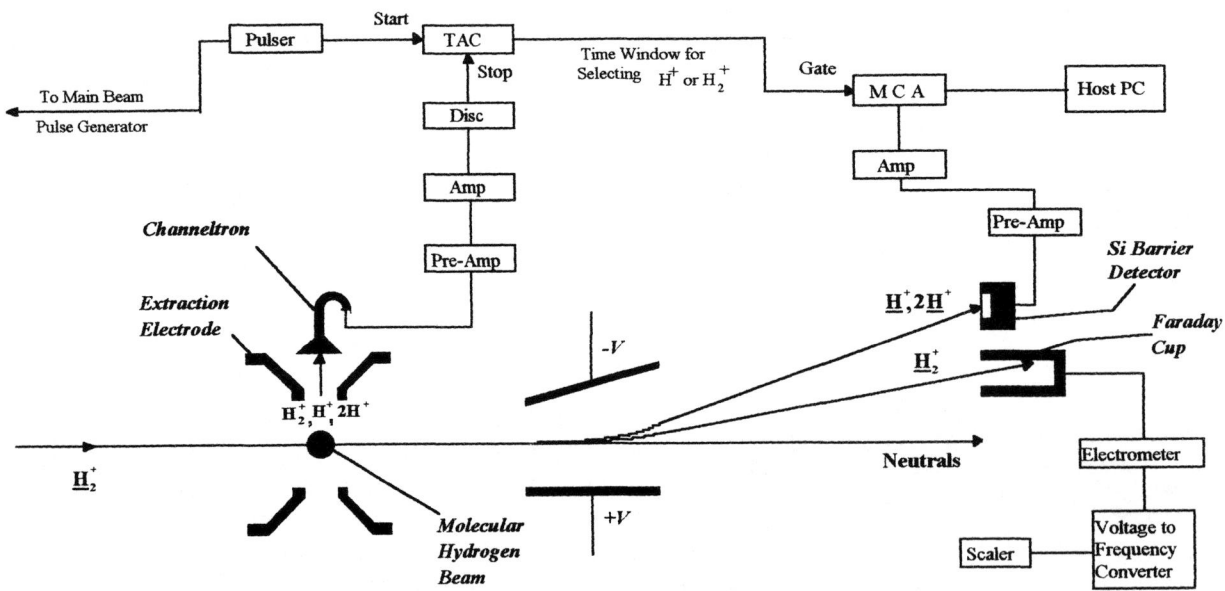

FIGURE 2. Schematic of experimental apparatus, including signal recovery set-up.

from this source. A typical operating pressure for this experiment is ~ 8 x 10^{-7} torr with H_2 present and a typical ion beam current is ~ 1 pA.

Normalization Considerations

Normalization in this experiment took place in two stages, namely non-coincidence and coincidence normalization. This is necessary because the fast H_2^+ ions "see" essentially different target thickness. In non-coincidence mode, fragmentation into one or two fast H^+ can occur at any position along the beam in the interaction region, whereas in coincidence mode, the target thickness is defined by the combination of the extraction field and the detection of the slow ions by the channeltron.

Non-coincidence Normalization

McClure[10] measured the total production of fast protons for this collision system. His cross-section is therefore equal to $\sigma_{1,4,9,14} + 2\sigma_{2,5,10,15}$. We calculate our fast proton production and normalize it to his value. Then we can separate the total proton production cross-section into $\sigma_{1,4,9,14}$ and $\sigma_{2,5,10,15}$.

Coincidence Normalization

To quantify the effective target thickness using coincidence mode, we measure slow H_2^+ production for both H_2^+ and H^+ projectiles at 100keV, using the same target extraction fields. Thus, both projectiles see the same target thickness. The production of H_2^+ for H^+ impact can proceed in two ways:

$$\underline{H^+} + H_2 \rightarrow \underline{H} + H_2^+ \qquad (21)$$
$$\underline{H^+} + H_2 \rightarrow \underline{H^+} + H_2^+ + e \qquad (22)$$

Stier and Barnett[12] measured the straight forward charge-exchange cross-section, σ_{21} and Shah and Gilbody[13] measured the non-dissociative ionization cross-section, σ_{22}. We used these two cross-sections to normalize our H_2^+ impact measurements.

RESULTS AND DISCUSSION

Fig. 3 shows the cross-sections of the individual channels and families of channels that we have been able to identify with the present experimental set-up. Error bars are not included for clarity but our data are accurate to ±10%. This is the first time individual channels have been presented.

The dominant cross-section in the lower end of our energy range is $\sigma_{1,4}$, where the projectile simply fragments into H^+ and H and the target either remains unchanged or dissociates symmetrically into two H atoms. Between 70keV and 80keV $\sigma_{1,4}$ falls more steeply and falls below σ_9. Channels 1 and 4 (and likewise 2 and 5) are extremely difficult to individually separate as detection of low energy neutrals is required. It is important to note, that excitation and subsequent de-excitation of either or both, the target and projectile products can also occur in the collision. Large numbers of UV photons were detected by our slow ion channeltron. Cross-section, σ_9, although not increasing significantly, becomes the largest at about a projectile energy of 75keV. Next in order of magnitude, $\sigma_{2,5}$ increases as for σ_9, and at the upper limit of our energy range investigated approaches $\sigma_{1,4}$ where there is only a measured 20% difference in the H_2^+ fragmenting into two or one fast protons. The cross-sections, σ_{14} and σ_{10} are an order of magnitude lower than $\sigma_{1,4}$, σ_9 and $\sigma_{2,5}$ and are of similar magnitude to each other within the energy range investigated. However while σ_{14} exhibits structure σ_{10} appears to follow the same trend as $\sigma_{2,5}$. As could be expected, σ_{15} is the smallest and also displays similar structure to σ_{14}. The magnitudes of the above cross-sections can be understood qualitatively in terms of the impulse excitation of the H_2^+ ion and the relevant energy defects for the different processes, i.e., smaller

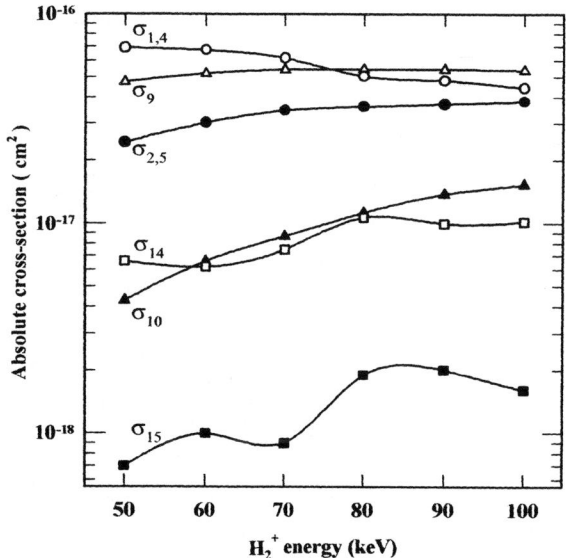

FIGURE 3. Cross-sections of the various fast proton channels and families of channels, when H_2^+ in the energy range 50-100keV collides with H_2, revealed by this experiment.

energy defects give rise to larger cross-sections.

It has been assumed in the past (Williams et al.[14]) that the collisional destruction of H_2^+ in this energy range proceeds primarily via the following three paths:

$$\underline{H_2^+} + H_2 \rightarrow \underline{H_2} \ldots \quad (23)$$
$$\underline{H_2^+} + H_2 \rightarrow \underline{H^+} + \underline{H} + \ldots \quad (24)$$
$$\underline{H_2^+} + H_2 \rightarrow \underline{H} + \underline{H} + \ldots \quad (25)$$

As can be seen from Fig. 3, the break up of the projectile into two fast protons is also important, (i.e., cross-sections $\sigma_{2,5}$ and σ_{10}) particularly at high energies.

Although, this is the first time these cross-sections have been measured, cross-sections for the production of one or two fast protons without knowing the fate of the target, i.e. our non-coincidence data, do exist. Comparison with such results is shown in Fig. 4. The data of Pivovar et al[15] is well above our energy range but is included for completeness. The data of Sweetman[7] also covers a higher energy range, but at the common energy of 100keV, both our cross-sections are in extremely good agreement. However, the work of Guidini[9], covering the energy range 25-250keV, does not provide good agreement. A reason for this is possibly due to the high operating pressures used by Guidini[9]. This would increase the probability of secondary collisions, thus artificially enhancing the production of two protons and lowering the production one proton.

It is important to note that the initial vibrational state of the H_2^+ projectile is not known. It has been highlighted in the past[16] that this factor can influence dissociation channels.

In conclusion, we have been able to identify, for the first time, individual channels when a fast H_2^+ molecular ion collides with a H_2 molecule. Our non-coincidence data compare extremely well to those of Sweetman[7] but raise questions regarding Guidini's[9] results. Work is already in progress to identify the remaining channels of the H_2^+-H_2 collision system.

ACKNOWLEDGMENTS

The authors wish to acknowledge financial assistance from UK EPSRC. JWMcC thanks NSERC, Canada for financial support, whilst a visiting professor in Belfast. CMcG is indebted to the Department of Education, Northern Ireland for the award of a research studentship. We also thank Mr. Christof Hanke and Mr. Didier Voulot for translating some of the manuscripts.

REFERENCES

1. Barnett, C. F., *Atomic Data for Fusion* **1**, ORNL-6086, Oak Ridge National Laboratory, TN (1990).
2. Phelps, A. V., *J. Chem Phys Ref. Data* **19**, 653 (1990).
3. Williams, J. F. and Dunbar, D. N. F., *Phys. Rev.* **149**, 62 (1966).
4. Damodaran, K. K., *Proc. Roy. Soc. (London)* **239**, 382 (1957).
5. Barnett, C. F., *Proc. 2nd UN Intern. Conf. Peaceful Uses Atomic Energy, Geneva*, **32**, 398 (1958).
6. Fedorenko, N. V. et al., *Soviet Physics JETP* **36**, 267 (1959).
7. Sweetman, D. R., *Proc. Roy. Soc. (London)* **A256**, 416 (1960).
8. Schmid, A., *Z. Physik* **161**, 550 (1961).
9. Guidini, J., *Proc. 3rd Intern. Conf. Physics Electronic and Atomic Collisions, London*, 751 (1964).
10. McClure, G. W., *Phys. Rev.* **130**, 1852 (1963).
11. McCartney, P. C. E. et al., *J. Phys. B: At. Mol. Opt. Phys.*, **32**, 5103 (1999).
12. Stier, P. M. and Barnett, C. F., *Phys. Rev.*, **103**, 896 (1956).
13. Shah, M. B. and Gilbody, H. B., *J. Phys. B: At. Mol. Opt. Phys.*, **15**, 3441 (1982).
14. Williams, I. D. et al., *J. Phys. B: At. Mol. Opt. Phys.*, **17**, 811 (1984).
15. Pivovar, L. I. et al., *Soviet Physics JETP* **13**, 23 (1961).
16. Lindsay B. G. et al., *J. Phys B: At. Mol. Opt. Phys.*, **21**, 2593 (1988).

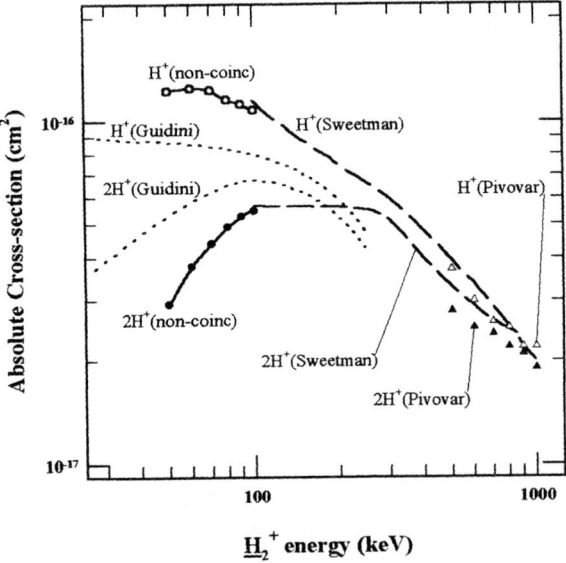

FIGURE 4. Comparison of our non-coincidence data for the production of one fast proton and two fast protons with others; Guidini[9], Sweetman[7] and Pivovar et al[15].

X-ray Emission From Electron Capture By Highly-Charged Ions

J. B. Greenwood[1], I. D. Williams[1], S. J. Smith[2] and A. Chutjian[2]

[1] Department of Pure and Applied Physics, Queen's University Belfast, Belfast BT7 1NN, UK
[2] Jet Propulsion Laboratory, California Institute of Technology, Pasadena, CA 91109

Abstract

Recently, electron capture by highly-charged ions has been recognized as the major source of X-ray emission from comets. It has been proposed that highly-charged ions of minor solar wind species such as C, N, O, and Ne emit X-rays following electron capture from cometary atoms and molecules such as H, CO, OH, H_2O, and CO_2. In our work we have measured accurate cross sections for single and multiple electron capture by solar wind ions in collision with various gas targets. We have also observed X-ray Lyman transitions from H and He-like ions formed in the capture process. The importance of multiple transfer, autoionization and the l-distribution of initial capture states will be discussed in light of our results.

INTRODUCTION

Although electron capture in ion collisions with atoms or molecules has been studied for many years, interest in these collisions continues unabated, particularly at low collision energies (<< 1 a.u.). The need for total and state-selective cross sections has been recognized through recent discoveries in astrophysics, and in the quest for controlled nuclear fusion.

Heavy solar wind ions in highly-charged states are now being acknowledged as an important source of X-rays in the solar system, through electron capture with cometary, atmospheric and interstellar gases [1-11]. Electron capture also yields visible, UV and X-ray photons from bodies as diverse as Jupiter's Aurora [12-14], the interstellar medium, HII regions, and nebulae[15,16].

In magnetically-confined fusion devices, present efforts are focussed on understanding the divertor region which is needed to provide continuous operation of an ignited plasma. Gas is injected into the divertor to cool hot impurity ions before their removal from the system. Electron capture is thus an important process in the physics of tokamaks and as a diagnostic of plasma parameters.[17,18]

Comprehensive charge-transfer data are needed for analyzing these interactions and the light emitted from the collisions. Although simple target systems, like H, H_2 and He, have been investigated for a number of projectile ions[19,20], there is still need for further study, particularly at low energies. In more complex targets, like H_2O, CO and CO_2 important to comets, studies have only recently been initiated.

If detailed studies of the line intensities observed from fusion or astrophysical plasmas are undertaken to characterize ion or neutral densities, then electron capture needs to be well understood. While state-selective cross sections are essential, consideration of autoionizing multiple capture and anisotropic emission of photons due to magnetic sublevel population can also be important.

We have designed an experiment to measure total cross sections for single and multiple electron capture, providing absolute results which can be used by modelers and other experimentalists. Installation of an X-ray detector has also allowed us to measure emission cross sections for soft X-ray transitions.

These cross sections can be applied directly to the analysis of line strengths observed in X-ray spectroscopy of plasmas. Total cross sections have been obtained for collisions of C, N, O and Ne ions with various gases. X-ray spectra of Lyman-like transitions following one electron capture have been measured.

EXPERIMENTAL TECHNIQUE

The experimental setup has been described in more detail elsewhere[3,21], but a brief description follows. The ions of interest are produced from a *Caprice* type Electron Cyclotron Resonance ion source [22], capable of producing fully-stripped ions of C, N, O and Ne. The ions are accelerated to an energy of $7q$ keV (q is the ion charge state) and the desired ion is selected by a double-focussing 90° bending magnet. An electrostatic switcher directs the beam into the charge exchange beamline. There, the beam is collimated before entering a collision cell containing the target gas. On exiting, the ion beam current is measured in a deep Faraday cup. The pressure in the cell is determined by a temperature-stabilized capacitance manometer, while separation of different final ion charge states is achieved by a series of retarding potential apertures in front of the Faraday cup. A high-purity Ge X-ray detector is located at 90° to the ion beam direction and views the interactions through a 2 mm aperture in the cell wall. A 7.5 µm Be window separates the detector from the vacuum chamber, and blocks any photons of energies less than 500 eV.

RESULTS

Total cross sections for single and double charge exchange are shown in Figure 1 (and will be reported in detail elsewhere [21]). There, single-exchange values are compared, as a function of ion charge state, to predictions of the classical over-barrier model [23]. It can be seen that although this model gives a good estimate of the cross section, the discontinuities caused by filling of the next highest n-level are not reproduced. This suggests that one cannot assume that capture occurs into a unique n-level.

Double-exchange cross sections are also shown in Figure 1. For a CO_2 target it can be seen that these cross sections are a significant proportion of single exchange, typically 20-30%. However for He, with the exception of $q = 4$ where single exchange is very low, the double-exchange cross sections are relatively much smaller. As He has only two electrons, double exchange proceeds only by capture of two electrons which subsequently radiatively stabilize. While autoionization is often the preferred mode of decay in a multiply-excited ion, double-exchange cross sections in CO_2 are more significant as the process of multiple capture followed by emission of one or more Auger electrons is available.

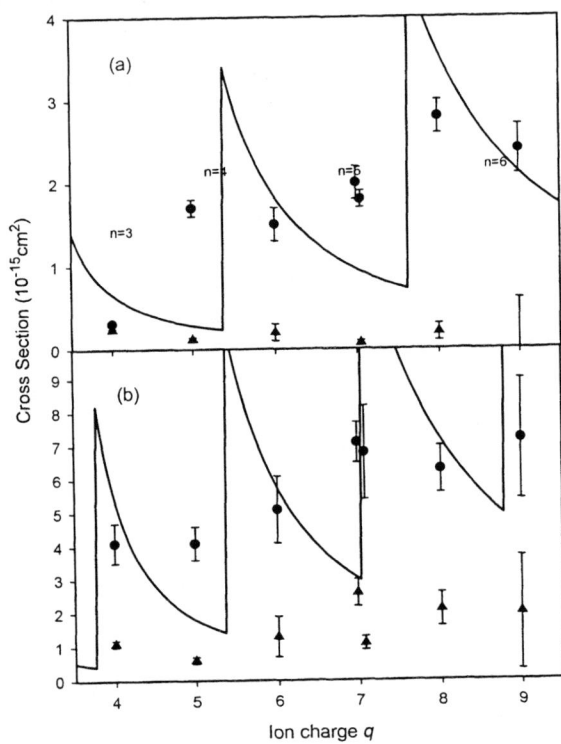

FIGURE 1. Single $\sigma_{q,q-1}$ λ, and double $\sigma_{q,q-2}$ σ, charge-exchange cross sections as a function of ion charge state q, in collision with (a) He and (b) CO_2. Comparison is made to the predictions of the classical over-barrier model $\sigma_{q,q-1}$ (solid line).

X-ray spectra for collisions of bare and H-like O and Ne ions in He are shown in Figure 2. These spectra are uncorrected for transmission of the detector's Be window. The observed peaks represent transitions to the ground state of the $q-1$ ion from np levels and have been fitted by Gaussian profiles representing the energy resolution of the detector. It can be seen that when the transmission of the Be window is taken into account, the lowest-energy transition Ly α ($2p - 1s$) is the dominant one, having been populated entirely by cascades from higher levels. Collisions of Ne^{10+} demonstrate the existence of a Ne^{8+} transition generated from radiative stabilization of double capture (the small low energy peak marked A in Figure 2).

For O^{7+} and Ne^{9+}, capture of a single electron can occur into triplet or singlet levels of the He-like ion. However, transitions to the ground state (ones energetic enough to be transmitted by the Be window) can only be observed from short-lived singlet levels, with the exception of the intercombination transition $1s2p\ ^3P_1 - 1s^2\ ^1S_0$. Decays of the long-lived, metastable triplet states $1s2s\ ^3S_1$ and $1s2p\ ^3P_{0,2}$ proceed outside the viewing angle of the detector.

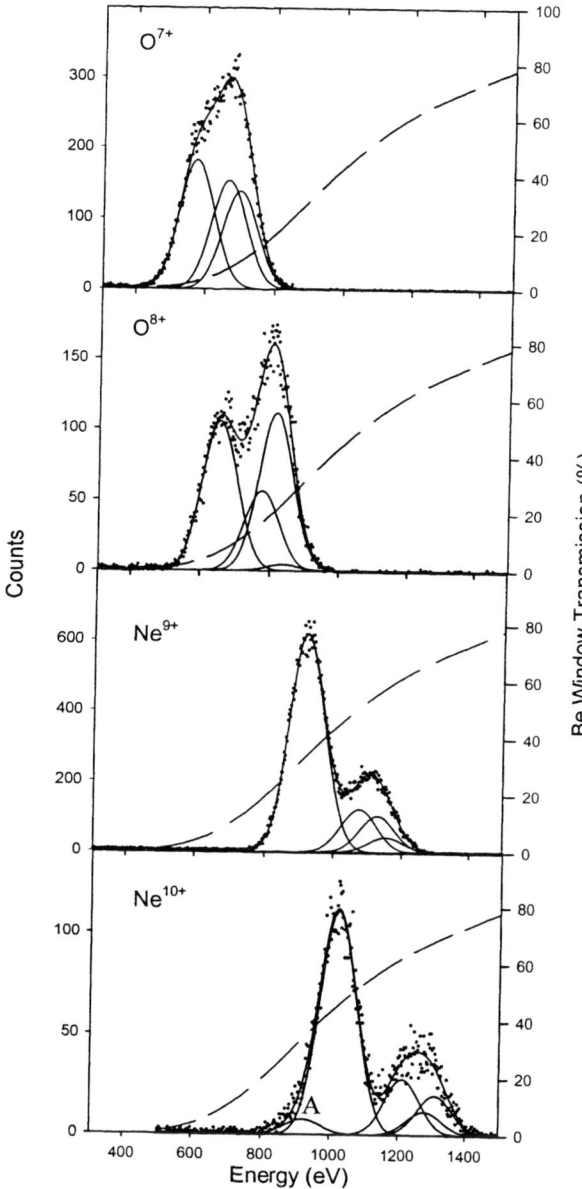

FIGURE 2. X-ray emission spectra obtained for collisions in He, fitted to the known transition energies of the product ions by a Gaussian profile. Data are uncorrected for transmission of the Be window (given as the long dashed line).

From analysis of the branching ratios for H-like ions produced from single capture by O^{8+} and Ne^{10+}, it is clear that capture into states with higher values of l tends to result in population of the $2p$ level, thus strengthening the Ly α emission intensity. However, our results indicate that the relative intensity of the Ly α lines is not as large as that expected from the assumption that the l levels are statistically populated.

Using an analysis similar to that used by Vernhet et al. [24], the average value of the initial angular momentum state $<l>$ for collisions of Ne^{10+} in He is 2.1, compared to the statistical average of 2.8 for capture into $n=5$. This result agrees with the predictions of Burgdörfer et al. [25] who extended the over-barrier model to include a centrifugal barrier term.

CONCLUSION

Measurements of total cross sections for single and double charge exchange in collisions of charge states of C, N, O and Ne in He and CO_2 are reported at a collision energy of $7q$ keV. The limitations of the classical over-barrier model in accurately determining cross sections, and the significant contribution of double exchange in collisions involving many-electron targets have been highlighted. X-ray emissions from these collisions have been observed. The dominant transition is $2p - 1s$, the strength of which is determined from the distribution within initial capture states nl. Analysis has shown that a statistical distribution within l-states would give stronger Ly α transitions than are observed. This suggests that the collision velocity is too low to give the captured electron enough angular momentum to statistically populate f and g states.

ACKNOWLEDGMENTS

This work was carried out at the Jet Propulsion Laboratory/Caltech, and was funded by the JPL Director's Discretionary Fund and NASA. JBG acknowledges the support of the National Research Council through the Associateships Program and the Queen's University Belfast Research Fund. IDW acknowledges support of both EPSRC and the QUB Research Fund.

REFERENCES

1. Lisse C. M., Dennerl K., Englhauser J., Harden M., Marshall F. E., Mumma M. J., Petre R., Pye J. P., Ricketts M. J., Schmitt J., Trumper J., and West R. G., *Science* **274**, 205-209 (1996).

2. Cravens T. E., *Geophys. Res. Lett.* **24**, 105-108 (1997).

3. Greenwood J. B., Chutjian A. and Smith S. J., *Ap. J.* **529**, 605-609 (2000).

4. Greenwood J. B., Williams I. D., Smith S. J. and Chutjian A., *Ap. J.* **533**, L175-L178 (2000).

5. Cravens T. E., *Adv. Space Res.* **26**, 1443-1451 (2000).

6. Cravens T. E., *Ap. J.* **532**, L153-L156 (2000).

7. Häberli R. M., Gombosi T. I., De Zeeuw D. L., Combi M. R. and Powell K. G., *Science* **276**, 939-942 (1997).

8. Kharchenko V. and Dalgarno A., *J. Geophys. Res.* **105**, 18351-18359 (2000).

9. Krasnopolsky V. A., Mumma M. J. and Abbot M. J., *Icarus* **146**, 152-160 (2000).

10. Lisse C. M., Christian D., Dennerl K., Englhauser J., Trumper J., Desch M., Marshall F. E., Petre R. and Snowden S., *Icarus* **141**, 316-330 (1999).

11. Wegmann R. Schmidt H. U., Lisse C. M., Dennerl K. and Englhauser J., *Planet. Space Sci.* **46**, 603-612 (1998).

12. Cravens T. E., Howell E., Waite J. H. and Gladstone G. R., *J. Geophys. Res.* **100**, 17153-17161 (1995).

13. Kharchenko V., Lui W. H.. and Dalgarno A., *J. Geophys. Res.* **103**, 26687-26698 (1998).

14. Lui W. H. and Schultz D. R., *Ap. J.* **526**, 538-543 (1999).

15. Heil T. G., *NIM* **B23**, 222-225 (1987).

16. Kingdon J. B. and Ferland G. J., *Ap. J.* **516**, L107-L109 (1999).

17. Isler R. C., *Plasma Phys. Controll. Fusion* **36**, 171-208 (1994).

18. Anderson H., von Hellerman M., Hoekstra R., Horton L. D., Howman A. C., Konig R. W. T., Martin R., Olsen, R. E. and Summers H. P., *Plasma Phys. Controll. Fusion* **42**, 781-806 (2000).

19. Janev R. K., Phaneuf R. A. and Hunter H. T., *At. Data, Nucl. Data Tables* **40**, 249-281 (1988).

20. Wu W. K., Huber B. A. and Wiesemann K., *At. Data, Nucl. Data Tables* **40**, 57-200 (1988).

21. Greenwood J. B., Williams I. D., Smith S. J. and Chutjian A., *Phys. Rev. A*, submitted (2000)

22. Liao C., Smith S. J., Chutjian A. and Hitz D., *Phys. Scr.* **T73**, 382-383 (1997).

23. Ryufuku H., Sasaki K., Watanabe T., *Phys. Rev. A* **21**, 745-750 (1980).

24. Vernhet D., Chetioui A., Rozet J. P., Stephan C., Wohrer K., Touati A., Politis M. F., Bouisset P., Hitz D. and Dousson S., *J. Phys. B*, **21**, 3949-3968 (1988).

25. Burgdörfer J., Morgenstern R. and Niehaus A., *J. Phys. B* **19**, L507-L513 (1986).

ABSOLUTE SINGLE ELECTRON LOSS IN COLLISIONS OF Ar$^+$ WITH VARIOUS ATOMS

P. G. Reyes[a], H. Martínez[b] and F. Castillo[c]

[a] Facultad de Ciencias, UNAM, Apartado Postal 70-542, 04510, México, D. F.
[b] Centro de Ciencias Físicas, UNAM, Apartado Postal 48-3, 62151, Cuernavaca, Morelos, México.
[c] Instituto de Ciencias Nucleares, Apartado Postal 70-542, 04510, México, D. F.

Absolute differential and total cross sections for single electron loss were measured for Ar$^+$ ions on various atoms in the energy range of 1.5 to 5.0 keV. The laboratory angular scan for the distributions ranged from -2.5 to 2.5 degrees. The measured differential cross sections have been integrated over the experimental angular range providing absolute total cross sections. The behavior of the total electron loss cross sections with the target atomic number, Z_t, shows different dependences as the collision energy increases. In all cases it displays a saturation as Z_t increases.

I.- INTRODUCTION

Single electron loss in ions colliding with atoms is one of the most typical phenomena accompanying the passage of fast atomic particles through matter. Because of its fundamental nature, the knowledge of the behavior of the cross section for that process is essential in many applied areas, such as atomic physics [1], plasma physics [2], chemical physics [3] material science [4], radiotherapy and radiation research [4] and energy loss information in organic material, such as human tissue. Until now, there are few systems dealing with projectiles carrying more than one electron, for which experimental data and theoretical calculations are known. Many of the experiments and calculations have been restricted to the simplest case of He$^+$ colliding with Ne, Ar, Kr and Xe atoms and even these often restricted to the high velocity regime [5]. Recently, theoretical calculations of the cross sections for the loss of an electron by a fast He$^+$ and C^{3+} particles colliding with H and H$_2$ in the 0.5 MeV$\leq$E$\leq$3.5 MeV energy range [6, 7], were made in a wide range of velocities for two very simple cases: loss of an electron by a simple charged helium ion colliding with heavy atomic targets [5, 8, 9], and loss of an electron by hydrogenlike ions with atomic number Z_p ($2 \leq Z_p \leq 7$) and with velocity v ($0.3 v_0 \leq v/Z_p \leq 6 v_0$; v_0 is the Bohr velocity) colliding with N$_2$, O$_2$, Ne and Ar [9, 10].

This lack of information was one of the reasons, for which we decided to measure the single electron loss cross sections of Ar$^+$ on various atoms, in order to obtain a consistent set of results. The Ar$^+$ projectiles had energies ranging from 1.5 to 5.0 keV.

II.- EXPERIMENT

The experimental arrangement has been described in detail elsewhere [11, 12], so only a brief survey of the experimental setup will be given here. A schematic diagram of the apparatus is shown in Fig. 1. Ar$^+$ ions were formed in a colutron-type ion source and accelerated in the energy range of 0.3 to 5.0 keV. The selected ion beam was velocity analyzed with a Wien filter, passed through cylindrical plates to deviate it by 10^0, and through a series of collimators before entering the gas target cell. The cell is a cylinder of 2.5 cm in length and diameter, with a 1-mm entrance aperture, and a 2-mm wide, 6-mm long exit aperture. All apertures and slits had knife edges. The target cell was located at the center of a rotatable, computer controlled vacuum chamber that moved the whole detector assembly which was located 47 cm away from the target cell. A precision stepping motor ensured a high repeatability in the positioning of the chamber over a large series of measurements. The detector chamber housed a Harrower-type parallel plate electrostatic analyzer, located at 45^0 with respect to the incoming beam direction, with two channel electron multipliers (CEMs). The Ar^0 atoms formed by electron capture passed straight through the analyzer through a 1-cm orifice on its rear plate, and impinged on a CEM so that the neutral counting rate could be measured. Separation of charged particles occurred inside the analyzer, which was set to detect the Ar^{2+} species with a second CEM. To measure the angular distributions, a 0.36-mm diameter pinhole was located at the entrance of the analyzer. This geometry permitted the measurement of neutral atoms, the directions of which make an angle of up to $\pm 7^0$ with respect to the incoming beam direction. Path lengths and apertures gave an overall angular resolution of the system of $0.1°$. The target thickness was 10^{13} atoms/cm^2 in order to ensure a single collision regime. Absolute gas pressures in the cell were measured with a capacitance manometer. The absolute differential cross section was calculated from the relation:

$$\frac{d\sigma}{d\Omega} = \frac{I(\theta)}{nLI_0}$$

where I_0 is the number of Ar$^+$ ions incident per second on the target, n, the number of atoms per volume unit, L, the effective length of the scattering chamber, and $I(\theta)$, the number of Ar^{2+} counts per unit solid angle per second detected at laboratory angle (θ) with respect to the incident beam direction. The total cross section was derived by integrating the differential cross section over the solid angle $d\Omega$:

$$\sigma = 2\pi \int_0^\pi \frac{d\sigma}{d\Omega} sin(\vartheta) d\theta$$

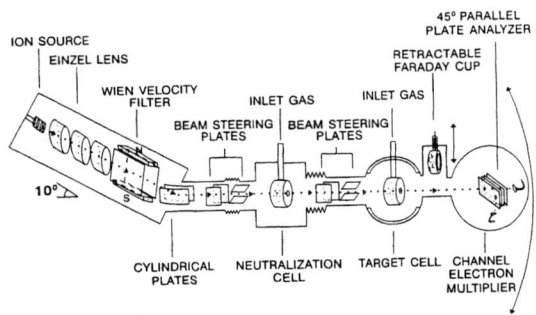

Figure 1. Schematic diagram of the apparatus

Extreme care was taken when the absolute differential cross section was measured. The reported value of the angular distributions was obtained by measuring it with and without gas in the target cell with the same steady beam in order to eliminate the counting rate due to ionization of the Ar$^+$ beam on the slits and those arising from background distributions. The Ar$^+$ beam intensity was measured before and after each angular scan. Measurements not agreeing to within 5% were discarded. Angular distributions were measured on both sides of the forward direction to assure they were symmetric. The estimated rms error is 15%, while the cross sections were reproducible to within 10% from day to day.

III.- RESULTS AND DISCUSSION

Characteristic angular distributions for single electron loss of Ar$^+$ in various atoms are shown in Fig. 2 at the projectile energy of 3.0 keV (the estimated error is of the size of the data point). The behavior of the differential cross sections is qualitatively identical for each of the targets studied in this work. All curves plotted in Fig. 1 display a tendency to decrease in the differential cross section with increasing angle. For Ne as a target, the data show slight structures, while for Ar, Kr and Xe they display a constant behavior between 0.5 and 1.7 degrees. The measured differential cross sections have been integrated numerically over the observed angular range ($0 \leq \theta \leq 2.5$ degrees) to yield the total cross sections σ_{10}.

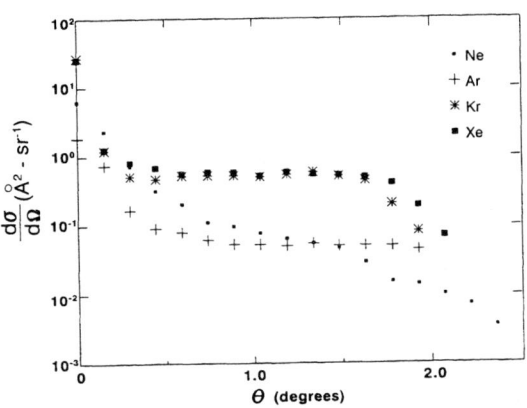

Figure 2. Measured absolute differential cross sections for 3 keV

The cross sections obtained for each colliding system exhibit an increasing behavior as a function of the incident energy; we only show the Ar$^+$-Ne system in Fig 3.

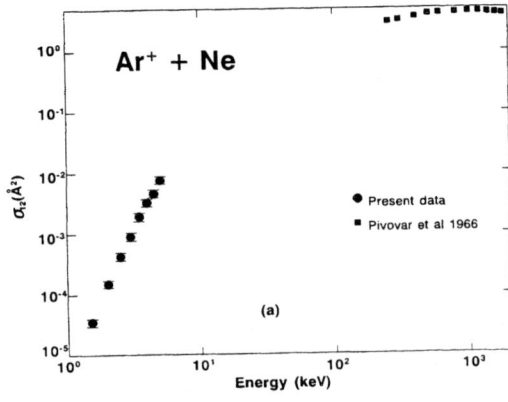

Figure 3. Calculated total cross sections for single electron loss of Ar$^+$ ions in Ne atoms

Fig. 4 and table I show the dependence with Z_t (the target charge) of the electron loss cross sections for 1.5, 3.0 and 5.0 keV Ar^+ ions colliding with various atoms. For all the energies studied, the electron loss cross section has a smooth Z_t dependence. At 1.5 keV, the electron loss cross section decreases faster than at 3.0 keV, while at 5.0 keV it has a smooth Z_t dependence. At all energies there is a saturation as Z_t increases.

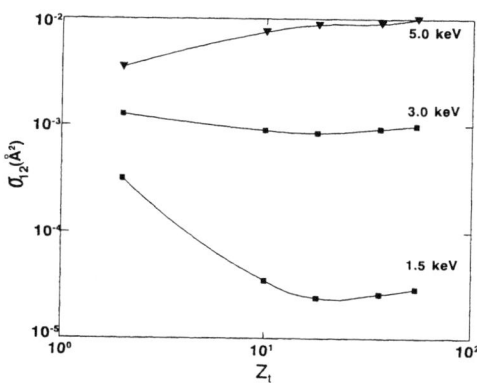

Figure 4. Total single electron loss cross sections

It thus appears desirable that a detailed theoretical analysis be carried out to further check this behavior. The results of the present work can be summarized as follows:

a) Differential and total cross sections for single electron capture of Ar^+ on various atoms are reported.

b) The total cross sections for single electron capture show an increasing behavior.

c) The single electron loss cross sections as a function of Z_t show a strong energy dependence, changing from a decreasing to an increasing behavior as the collision energy increases.

d) A striking feature of the results is the experimental evidence of a saturation in the single electron loss cross sections as the nuclear charge of the target increases.

ACKNOWLEDGMENTS

We are grateful to B. E. Fuentes for helpful suggestions and comments. The authors wish to thank Mr. Anselmo González and Mr. Jose Rangel G. for their technical assistance. This work was supported by the DGAPA-IN100392 and CONACyT 32175-E.

REFERENCES

[1] S. Datz, G. W. F. Drake, T. F. Gallagher, H. Kleinpoppen and G. Zu Putitz, Reviews of Modern Physics 71, S223 (1999).

[2] R. K. Janev, in Atomic and Molecular Processes in Fusion Edge Plasmas, Plenum Press, New York, 1995.

[3] D. Herschbach, Reviews of Modern Physics 71, S411 (1999).

[4] W. Möller, in Final Programme and abstract of the 14th International Conference on Ion Beam Analysis and European Conference on Accelerators in Applied Research and Technology, 1999.

[5] A. B. Voitkiv, G. M. Sigaud, E. C. Montenegro, Phys. Rev. A59, 2794 (1999).

[7] M. M. Sant'Anna, W. S. Melo, A. C. F. Santos, M. G. Sigaud, E. C. Montenegro, M. B. Shah and W. E. Meyerhof, Phys. Rev. A58, 1204 (1998).

[8] E. C. Montenegro, W. E. Meyerhof and J. H. McGuire, Adv. At. Mol. Opt. Phys. 34, 249 (1994).

[9] T. Kaneko, Phys. Rev. A32, 2175 (1985).

[10] T. Kaneko, Phys. Rev. A34, 1779 (1986).

[11] H. Martínez and P. G. Reyes, Phys. Rev. A59, 2504 (1999).

[12] H. Martínez, J. of Phys. B: At. Mol. Opt. Phys. 32, 189 (1999).

[13] L. I. Pivovar, M. T. Novikov and A. S. Dolgov, Sov. Phys. JETP 23, 357 (1966).

Recent Experiments on the Roles of Projectile Electrons in Ion-Atom Collisions

D. Fregenal, J. Fiol, S. Suárez *, G. Bernardi *, P. Focke * and A.D. González *

Centro Atómico Bariloche and Instituto Balseiro, 8400 Bariloche, Río Negro. Argentina
*Consejo Nacional de Investigaciones Científicas y Técnicas, Argentina
gonzalez@cab.cnea.gov.ar

Abstract. By measuring recoil-ion–electron, projectile-electron, and triple coincidences, multiple ionization differential cross sections and probabilities for impact of H, He, H$^+$, He$^+$ and He^{++} colliding with He, Ne, Ar and Kr have been obtained. Here we concentrate on differential measurements of multiple ionization of heavy targets produced by impact of neutral and charged H and He projectiles. In the case of H$^+$ impact, previous experiments have shown a large enhancement of electron emission associated with the ratio of Ar^{2+}/Ar$^+$, with a maximum at an electron velocity close to the projectile velocity for electron emission angles around 90 degrees. Both, double ionization (DI) and transfer ionization (TI) are responsible for the enhancement, which was also observed for Ne and Kr targets. New results demonstrate that, for 100 keV H$^+$+Ar collisions, the contribution to the enhancement due to DI is larger than TI. Neutral beams produce no effect of this sort. A simultaneous projectile and target ionization contribution is discussed.

INTRODUCTION

In our laboratory, many studies have dealt with multiple ionization and capture processes with various projectiles and targets. We have measured differential cross sections and probabilities in collisions of protons, H, He, He$^+$ and He^{++} with He, Ne, Ar and Kr targets, in the energy range from 10 keV/amu to 250 keV/amu. Basically two complementary setups have been used. In one of them, the emitted electrons are energy and angle analyzed with a cylindrical mirror spectrometer, while the outgoing projectile is charge selected and detected in coincidence [1-4]. The other setup uses a recoil-ion and electron time-of-flight coincidence technique [5-7]. Afterwards, we have added a particle detector at the end of the line in order to perform triple coincidences between electrons, recoil ions and projectiles.

The motivation of these research lines are mainly: i) to study the relative importance of different collision channels in conditions of simultaneous competition; ii) to investigate the effects of electron correlation in multiple electron emission; iii) to establish the influence of electron capture in ionization of target electrons; iv) to determine the role of projectile electrons in these processes.

Multiple electron emission has been investigated by many authors, and it is today one of the most exciting subjects in ion-atom collisions. The many-body problem, so far not solvable in a closed form, obliged to find alternative strategies to understand the data. Even the simplest multi-electron processes, namely Double ionization (DI) and Transfer Ionization (TI), involving only two active electrons, require electron correlation to explain the experimental results [8,9].

For single capture, an asymptotically high energy dependence of v^{-11} for the cross section was determined. In contrast, for TI a dependence of v$^{-7.5}$ was obtained experimentally from high energies down to a proton beam energy as low as 300 keV, in agreement with a theoretical approach [8]. This finding encouraged us to direct part of our research to study the mechanisms that produces TI for beam energies lower than 300keV/amu.

We have extended the study to Double Transfer Ionization (DTI), in which two electrons are captured by the projectile with simultaneous emission of electrons into the continuum. Doubly differential cross sections of DTI were investigated for He^{2+} colliding on Ar, as a function of electron ejection angles and energy [4]. The poor statistics for the measurements at 90 degrees with respect to the beam axis, did not allow so far to establish if a Thomas-like peak would appear in DTI. Nevertheless, the results for the forward direction showed a sharp cusp peak associated with DTI, narrower than the observed in total electron emission. Following a model based on resonance phenomenon for emerging neutral projectiles, cusp electrons in DTI should be related to capture processes on metastable states of the outgoing He atom. A large relative contribution of DTI to the binary encounter electron region was also observed. This is in agreement with previous data obtained by measuring recoil ions in coincidence with integrated electrons, for which a large contribution of Ar^{3+} recoils were observed [7, 10].

RECOIL-ELECTRON COINCIDENCES

Figure 1 shows a scheme of the recoil-electron time-of-flight coincidence spectrometer. Integrated differential cross sections in electron energy above a threshold in a solid angle around 90° emission are obtained. Though the data are integrated and lack the details of the high resolution measurements, the advantage of this setup is to measure with an experimental time short enough to obtain results with a variety of targets, projectile structures and energies.

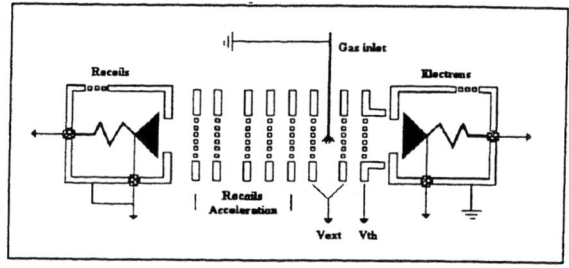

Figure 1: Experimental setup for recoil-electron coincidence measurements. Vext is the recoil-electron extraction electrostatic potential and Vth is a barrier potential. Only electrons above this threshold will reach the channeltron detector. The beam impinges perpendicular to the drawing.

In Figure 2 integrated differential cross sections (IDCS) for multiple ionization at 90 degrees emission as a function of voltage threshold are plotted for 50 keV protons and neutral H colliding with a Kr target. Due to the detection technique, the singly charged recoil ion in coincidence with the electron corresponds to Single Ionization (SI) of the target for proton impact. For higher ionization degrees the data account for the sum of multiple pure ionization and capture with simultaneous ionization. Note the marked difference in the shape of the curves for H^+ and H projectiles, for the various ionization channels. In the case of a neutral beam as H, the electrons in coincidence with singly charged recoils account for the sum of target ionization and simultaneous ionization of the target and projectile, not distinguished in the present work.

We have obtained IDCS for a variety of targets and impact energies [5-7]: 10, 25, 50, 120 and 210 keV H^+, H colliding with Ne, Ar and Kr; 10, 25, 35, 50 keV/amu He, He^+ and He^{++} colliding with Ne, Ar and Kr; and 100 keV/amu He^{++} + Ar.

To the details allowed by the present setup, the systematic comparison of the data give: a) no significant difference as a function of target structure and number of target electrons; b) for an Ar target the integrated cross sections for single ionization reproduce very well previous results by Rudd et al [11]; for all targets investigated, Ne, Ar and Kr, the IDCS curves as a function of threshold potential show sharp changes in slope for proton impact and are smother for neutral H as projectiles.

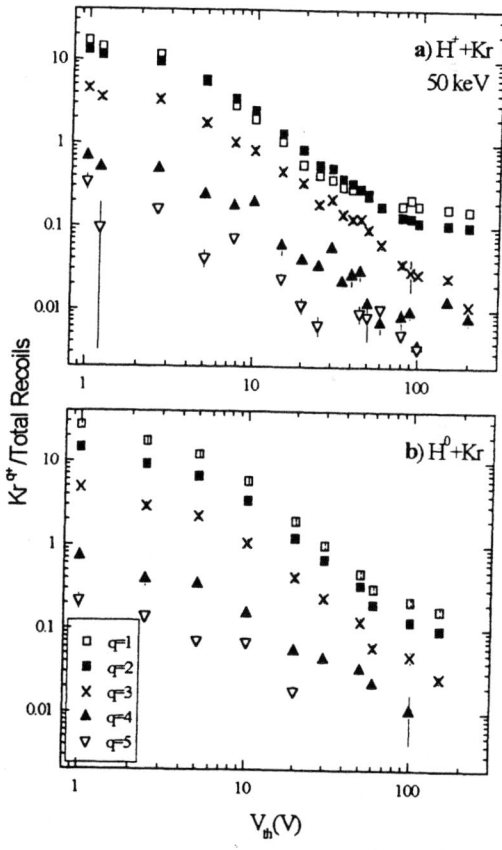

Figure 2: Differential cross sections integrated in electron energies above the threshold value of the x-axis, for the multiple ionization of Kr by 50 keV H^+ and H impact. The normalization is to the total non-coincident recoil ions detected. The vertical scale is plotted in arbitrary units.

From data on IDCS as in Fig. 2, we have studied the ratio of doubly charged recoils compared to singly charged, shown in Figure 3 for different H^+ impact energies colliding with Ar. For proton impact, Ar^+ data account only for SI, and due to the fact that the beam charge is not selected in this experiment, Ar^{2+} data account for the sum of DI+TI processes. In Fig. 3 the enhancement observed for all impact energies has a maximum at an electron threshold corresponding closely to the beam velocity. These enhancements mean that for all the beam energies shown in Fig. 3, an emission of a larger number of electrons at 90° with velocities similar to that of the beam occur for Ar^{2+}.

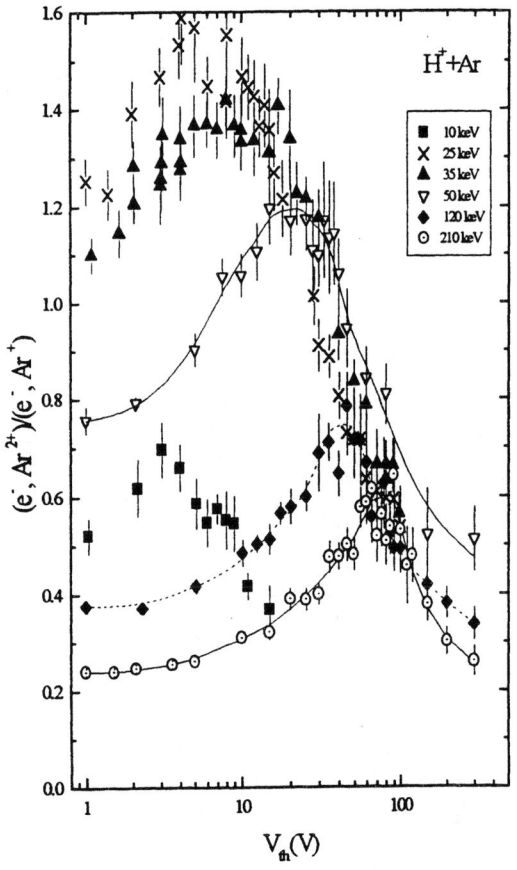

Figure 3: Differential cross section ratio Ar^{2+}/Ar^+ for 10, 25, 35, 50, 120 and 210 keV H^+ +Ar collisions. The curves for 50, 120 and 210 keV are only drawn to guide the eye and make clearer the various plots. The symbol (e^-, Ar^{q+}) means electron-recoil ion coincidence.

Similar enhancements were obtained for Ne and Kr as targets, and with He^+ and He^{++} as projectiles [6,7]. By differentiating the curves in Fig. 2, differential cross sections as a function of electron energy were obtained. The electron energy corresponding to the maxima were found to match very closely the beam velocity [5].

In Fig. 3, different ratio values at the maximum of the enhancement for each energy are observed. This is due to the beam dependence of the differential cross section. Data available so far are only for Total cross sections (TCS) [12], however the behavior is similar to the observed in Fig. 3.

In Fig. 2b) the IDCS are plotted for neutral H colliding with Kr. The curves corresponding to different degrees of ionization are quite parallel, in contrast to the curves in Fig. 2a). When the ratios of <u>multiple</u> ionization with respect to the singly charged recoil ion are studied for neutral beam impact, functions with no enhancements are obtained, suggesting a different collisional behavior from that observed in Fig. 3 for proton impact [5-7]. TCS for the production of Ar^+ in H + Ar collisions show a significant large contribution of simultaneous projectile and target ionization in the range 25 to 200 keV projectiles energy [13]. At 50 keV simultaneous projectile and target ionization accounts for 85% of the production of Ar^+, being the values 83% and 90% at 100 and 200 keV respectively. Even though TCS do not have the same energy dependence as differential cross sections, we can infer from this result that a strong projectile ionization might mask the effect of target ionization for neutral projectiles in the present range of impact energies.

TRIPLE-COINCIDENCE MEASUREMENT

In order to study the relative contribution of DI and TI to the enhancements observed, we have recently setup a triple coincidence experiment in which electrons integrated in energy above a threshold, recoil-ions and charge selected projectiles are analyzed. We have investigated the collision 100 keV H^+ + Ar, separating the contribution of charge transfer and pure ionization for the production of doubly charged Ar ions. DI is obtained selecting H^+ and TI by selecting neutral H as the outgoing projectiles, respectively. In Figure 4, a data for the ratio Ar^{2+}/Ar^+ are plotted. DI gives the larger contribution to the enhancement of the ratio. TI also shows an enhancement in the same position as DI, however it contributes with a much smaller value. The sum DI+TI (filled squares) gives a similar result to the data plotted in Fig. 3 for 120 keV impact energy (open squares). From this measurement, done so far for one beam velocity, we can not derive a definite conclusion on which channel is responsible for all the enhancements observed in Fig. 3. It might reflect that

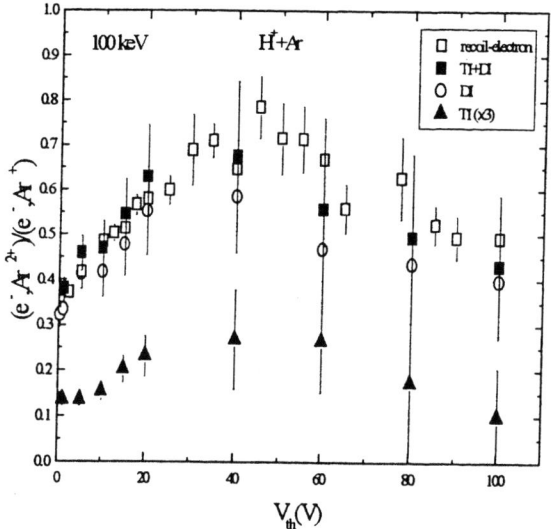

Figure 4: Differential cross section ratio Ar^{2+}/Ar^+ for 100keV H^+ +Ar collisions. Open squares are the data plotted for 120 keV impact in Fig. 3. Open circles and filled triangles are data obtained with a triple coincidence technique. Open circles correspond to DI and filled triangles to TI. The sum of the results of the triple coincidence are plotted in filled squares.

the contribution of DI is larger than TI for the particular 100 keV impact energy investigated. To study the beam energy dependence of the separate contributions of these two mechanisms will bring much understanding of two-fold ionization of heavy targets.

CONCLUSION

In conclusion, we have studied multiple electron emission at intermediate beam velocities, for a wide range of energies and a variety of targets and projectiles [1-7]. New interesting features have been obtained, which require both theoretical studies and further detailed measurements to be understood in deep. The projectile electrons seem to play an important role in the dynamical effects observed, since important effects like the beam velocity dependent enhancements in Figures 3 and 4 do not appear for neutral beams. Further experiments and calculations will be of great interest, to understand the underline mechanisms to produce such noticeable results.

ACKNOWLEDGEMENTS

The work presented here was funded partially by a grant from Agencia de Promoción Científica y Tecnológica, Argentina, through the FONCYT program, and partially by a grant (PEI 118/98) from Consejo Nacional de Investigaciones Científicas y Técnicas (CONICET).

REFERENCES

1. P. Focke, G. Bernardi, D. Fregenal, R.O. Barrachina and W. Meckbach, J. Phys. B **31**, 289 (1998)
2. P. Focke, W. Meckbach, G. Bernardi and N. Stolterfoht, J. Phys. B **31**, 3893 (1998)
3. G. Bernardi, P. Focke, A.D. González, S. Suárez and D. Fregenal, J. Phys B **32**, L451 (1999)
4. D. Fregenal, J. Fiol, G. Bernardi, S. Suárez, P. Focke, A.D. González, A. Muthig, T. Jalowy, K.O. Groeneveld and H. Luna, Phys. Rev. A **62**, 012703 (2000)
5. A.D. González, D. Fregenal, S. Suárez, H. Wolf and W. Wolff, J. Phys. B **31**, L257 (1998)
6. D. Fregenal, S. Suárez, G. Bernardi and A.D. González, J. Phys. B **33**, 3345 (2000)
7. A.D. González, D. Fregenal and S. Suárez, Nucl. Inst. Meth. B **132**, 236 (1997)
8. V. Mergel et al, Phys. Rev. Lett. **79**, 749 (1997).
9. J.F. Reading, T. Bronk and A.L. Ford, J. Phys. B. **29**, 6075 (1996).
10. P. Moretto-Capelle, D. Bordenave-Montesquieu, A. Bordenave-Montesquieu and M. Benhenni, J. Phys. B **31**, L423 (1998)
11. M.E. Rudd, L.H. Toburen, N. Stolterfoht, Atomic Data and Nuclear Data Tables **23**, 405 (1979)
12. R.D. Dubois and S.T. Manson, Phys. Rev. A **35**, 2007 (1987)
13. D. Fregenal, Ph.D. Thesis, Instituto Balseiro and Universidad de Cuyo (2000)

Ejected Electron Spectroscopy: Ionization in Ion-Atom Collisions

D.M. McSherry, S.F.C. O'Rourke, C. McGrath,
M.B. Shah and D.S.F. Crothers

School of Mathematics and Physics, The Queen's University of Belfast, N.Ireland, U.K.

Abstract. The continuum-distorted-wave eikonal-initial-state (CDW-EIS) approximation has had much success describing the single ionization of a bare atom by ion impact within the non-perturbative regime [Proc. R. Soc. Lond. A. **452** 175-184 (1996)]. This model is first order in a distorted wave series incorporating an eikonal phase distortion in the initial state and a continuum-distorted-wave in the final state. The latter of these accounts for the ejected electron travelling in the presence of two long ranged Coulomb potentials. The single ionization of multielectron atoms shall be discussed, with particular emphasis on heavier targets such as neon and argon. A comparison will be made with other available theoretical models and experimental data.

INTRODUCTION

In the single ionization of an atom by a bare ion as described by the collision process:

$$P^{Z_P+} + T^{(Z_T-1)+} \longrightarrow P^{Z_P+} + T^{Z_T+} + e^-$$

the ion causes the atom to break up into an ionic core, consisting of the projectile ion and the recoil ion, and one electron. From a theoretical point of view the main difficulty here is the representation of this final state where the ejected electron travels in the presence of the two Coulomb potentials. Therefore we adopt the CDW-EIS model which incorporates an eikonal phase distortion in the inital state and a continuum distorted wave in the final state, accounting for the long ranged two centered potential. This approximation is particularly suitable since the asymptotic Coulomb boundary conditions are satisfied exactly and both the initial and final states are normalized.

The aim of this paper is to examine the single ionization of multielectron atoms by proton impact. We have carried out both the measurement and calculations for various ion-atom collisions. The notation first introduced by Crothers and McCann (1) shall be followed, and atomic units shall be used throughout unless otherwise stated.

CALCULATIONAL DETAILS

The triply differential cross section can be found from

$$\sigma(\underline{k}) = \int d\underline{b} |a_{if}(\underline{b})|^2, \quad (1)$$

where $a_{if}(\underline{b})$ is the transition amplitude, $\underline{k}$ is the momentum of the ejected electron in the laboratory frame and $\underline{b}$ is the impact parameter. The first of these in post form, correct to first order, is found from

$$a_{if}(\underline{b}) = -i \int_{-\infty}^{\infty} dt \langle x_i^+ | H_e - i\frac{d}{dt_r} | \varepsilon_f^- \rangle, \quad (2)$$

where x_i^+ and ε_f^- represent both the initial and final wavefunctions and the Hamiltonian is defined to be

$$H_e = -\frac{1}{2}\nabla_r^2 - \frac{Z_T}{r_T} - \frac{Z_P}{r_P}. \quad (3)$$

Here Z_T and Z_P are the charges of the target and the projectile and $\underline{r_T}, \underline{r_P}$ and $\underline{r}$ are the position vectors of the electron relative to the target nucleus, the projectile nucleus and their midpoint respectively. The internuclear potential $\frac{Z_T Z_P}{R}$ has been removed by a phase transformation (2).

In this present work we have used the semi-classical straight line version of the impact parameter approximation with the inital bound states represented by Roothan Hartree Fock (RHF) wavefunctions. We have adopted the independent electron approximation and considered there to be only one active electron with the others remaining in the RHF orbitals throughout the collision thus defining the effective potential. It is in this effective potential the active electron evolves.

Integrating over the triply differential cross section with respect to ϕ the double differential cross section can be achieved:

$$\frac{dQ}{dE_k d(\cos\theta)} = k \int_0^{2\pi} \sigma(\underline{k})d\phi \qquad (4)$$

Here ϕ is the azimuthal angle of $\underline{k}$ with respect to the collision plane, and θ is the electron emission angle with respect to the incident beam direction.

The single differential $\frac{dQ}{dE_k}$ and total Q cross sections can then be achieved by successive integrations.

EXPERIMENTAL SETUP

A schematic diagram of the apparatus used for the DDCS measurements is shown in figure 1 and a detailed description has been given previously (3). A momentum analysed, 100keV proton beam, 0.5mm in diameter and 1nA intensity, was passed through the 10mm long differentially pumped target cell, where the pressure of the target gas was kept low enough (typically about 10^{-4} torr) to ensure single collision conditions. Differential pumping maintained the pressure in the surrounding region (of the main chamber) at about 10^{-8} torr. The ion beam and electrons emerging from the target cell at zero degrees with respect to the ion beam travelled 50mm before entering a large hemispherical electrostatic energy analyser. The full acceptance angle of the analyser was $1.5 \pm 0.2°$. Screening from the Earth's magnetic field was provided by a double layer of 2 mm thick mu-metal. The fast ions, which were not deflected significantly in the electron spectrometer, passed through an aperture in the side of the spectrometer and were collected in a Faraday cup which recorded the projectile beam intensity. The electron energy spectra were recorded using the hemispherical analyser in conjunction with a channel electron multiplier in pulse counting mode. The energy resolution of the analyser ($\Delta E/E$) was 0.012 (FWHM) and the analyser was operated at a pass energy of 14eV.

The electron distribution was recorded when the target gas was introduced into the target cell. This was then carefully corrected by subtracting the corresponding distribution obtained when the main chamber was flooded with gas (carefully introduced through a needle valve) to a pressure equal to the value attained when the gas was introduced into the target cell. Passage of the fast primary ion beam through the background gas within the hemispherical analyser inevitably gave rise to a large number of electrons that were then detected. However, the application of additional pumping to the hemispherical analyser greatly reduced this background signal. Much attention was also paid to the collimation of the projectile beam.

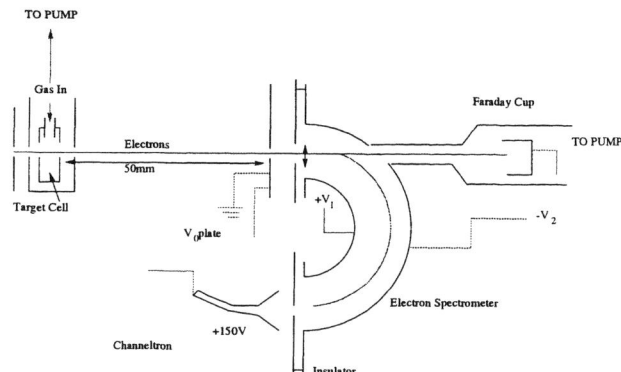

FIGURE 1. Experimental setup

One major improvement over other previous investigations is the inclusion of the target cell in our work compared to the use of an effusive target gas from a needle tip. In the present work a well defined and concentrated target geometry is obtained compared to the very diffused and diluted target geometry of others.

COMPARISON BETWEEN THEORY AND RESULTS

He and H_2

In the figures below the experimentally double differential cross sections for electron emission at zero degrees

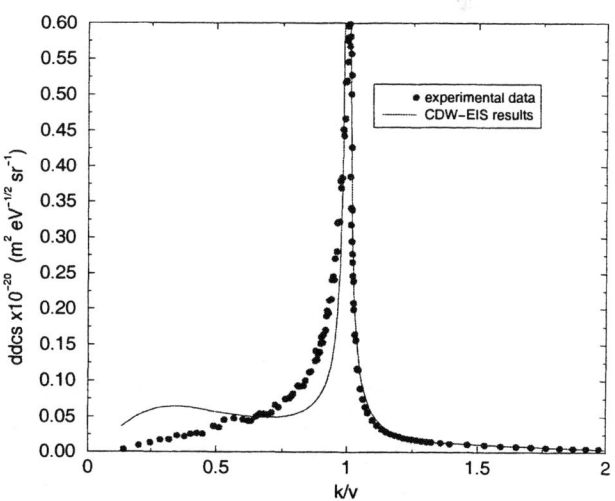

FIGURE 2. Measured double differential cross section for electron emission at zero degrees in collisions of 100keV protons with H_2 compared to CDW-EIS predictions

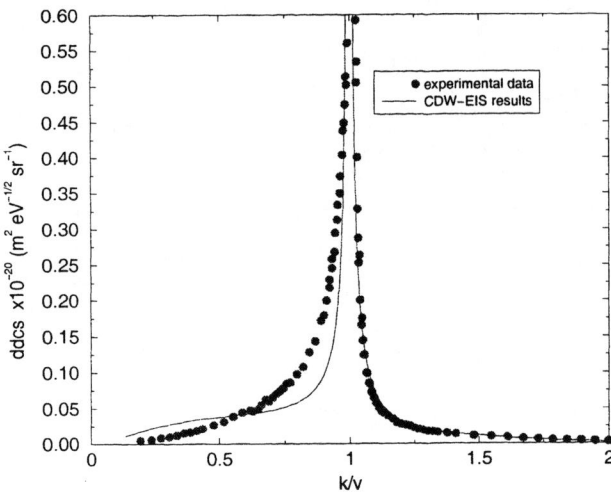

FIGURE 3. Measured double differential cross section for electron emission at zero degrees in collisions of 100keV protons with He compared to CDW-EIS predictions

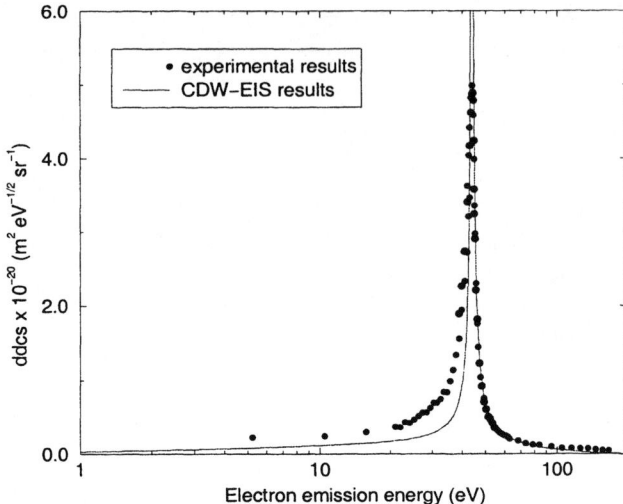

FIGURE 4. Measured double differential cross section for electron emission at zero degrees in collisions of 80keV proton with Ne compared to CDW-EIS predictions

for 100keV protons on He and H_2 are compared with theoretical predictions. Uncertainities associated with the experimental results vary from ±1% near the peak to about ±15% near the extreme wings of the distribution. These results were then scaled to provide a best fit with the CDW-EIS calculations.

In both cases there is satisfactory agreement between experimental and theoretical results with excellent agreement for electrons with velocities greater than v, where v is the velocity of the projectile. For lower energy electrons the eikonal description of the inital state may have its limitations especially for lower impact parameters.

The whole spectrum is dominated by the ECC peak which in accordance with the theory, extends much higher than the experimental data. The cusp itself shows the usual asymmetry with cross sections after the peak falling off more steeply than those before it. The binary peak is negligibly small at this projectile energy because the projectile velocity is of the same magnitude as the velocity spread associated with the target electron.

Ne

In figure 4 the experimental double differential cross sections for electron emission at zero degrees of a neon target in collision with 80keV protons are compared to theoretical predictions. Again we have quite good agreement between experimental results and the CDW-EIS predictions even for this low collision energy. The experimental data has been normalized to the theoretical predictions but the qualitative agreement is quite clear.

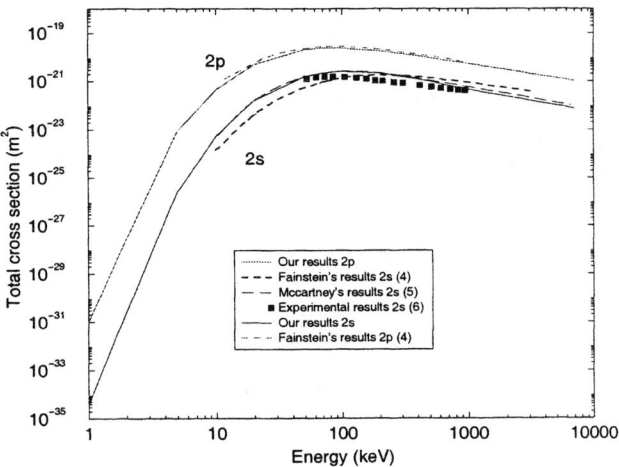

FIGURE 5. Total cross sections for neon by proton impact

Also in figure 5 the total cross sections for neon by proton impact are considered. As can be seen the electron emission from the 2p subshell will give the main contribution to the total cross section. We have been unable to reproduce the results of Fainstein *et al* (4) concerning the 2s subshell, and our results tend to be similar to McCartney *et al* (5). This result seems to reproduce the trend of the experimental data more satisfactorily.

Ar

In considering the case of argon we calculated the double differential cross sections for electron emission at various degrees in collision with 1MeV protons. Overall the theoretical results are in good agreement with experimental data except at higher energies for 90 and 125 degrees, where the theoretical predictions fall off quickly compared to experiment. It has been suggested from the work of Madison (8) and Manson *et al* (9) that this underestimation of the doubly differential cross sections at large emission angles could be due to the non-orthogonality of the inital and final states. Indeed in our analytical approximation our inital bound states are represented by RHF wavefunctions written as a linear combination of Slater type orbitals. The continuum states on the otherhand, are described by a hydrogenic wavefunction with an effective charge chosen from the energy of the inital bound state. Hence there is an non-orthogonality which exists due to the states corresponding to different potentials.

The double differential cross section for electron emission at various degrees in collions of 1MeV proton with argon was also considered by Gulyás *et al* (10). In this the CDW-EIS approximation was also used, however a completely numerically bound and continuum wavefuntion was obtained as solutions to a model potential. This generalized CDW-EIS significantly improved the theoretical predictions for higher energies at large emission angles of 90 and 125 degrees yet below 10eV the theoretical predictions were much higher than experiment.

The question of describing the targets states and whether or not the non-orthogonality of the target states in our present analytical solution really makes a significant difference, requires futher research. Yet the qualitative agreement between the CDW-EIS model and experiments leads to the conclusion that it does represents the physics of the process correctly.

SUMMARY

To conclude we have examined the single ionization of multielectron atoms by proton impact by both experimental measurement and theoretical calculation using the CDW-EIS approximation. There is satisfactory agreement between experimental and theoretical results in all cases considered, leading to the conclusion that the CDW-EIS model does indeed model the physics of the ionization processes considered here correctly.

ACKNOWLEDGEMENTS

This work is part of a large programme supported by the U.K. Engineering and Physical Sciences Council. Two of us (D.M. McS. and C. McG.) are also indebted to the Department of Higher and Further Education, Training and Employment (DHFETE) for the award of a Research Studentship. One of us (S.F.C. O'R) acknowledges support by a Royal Society University Research Fellowship.

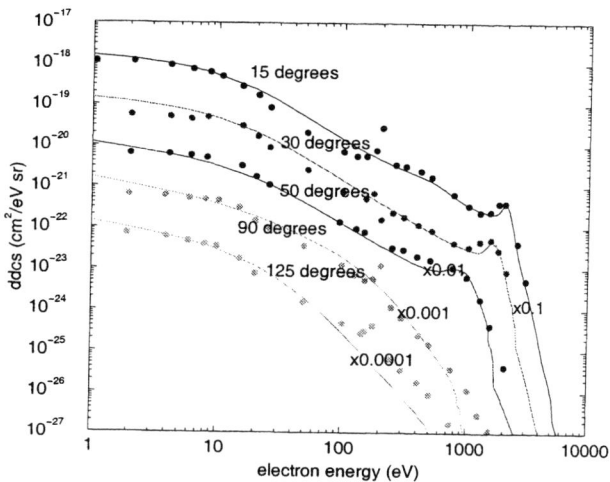

FIGURE 6. Measured double differential cross section for electron emission at various degrees in collisions of 1MeV protons with Ar (7) compared to CDW-EIS predictions.

REFERENCES

1. Crothers, D.S.F., and McCann, J.F., *J. Phys. B: At. Mol. Opt. Phys.* **16**, 3229-3242 (1983).
2. Bransden, B.H., and McDowell, M.R.C., *Charge Exchange and the Theory of Ion-Atom Collisions* Oxford:Clarendon, 1992, p74.
3. Nesbitt, B.S., Shah, M.B., O'Rourke, S.F.C., McGrath, C.,Geddes, J. and Crothers, D.S.F., *J. Phys. B: At. Mol. Opt. Phys.* **33**, 637-651 (2000).
4. Fainstein, P.D., and Rivarola, R.D., *Phys. Lett. A* **150**, 23-26 (1990).
5. McCartney, M., and Crothers, D.S.F., *J. Phys. B: At. Mol. Opt. Phys.* **26**, 4561-4574 (1993).
6. Eckhardt, M., and Schartner, K.H., *Z. Phys. A* **312**, 321-328 (1983).
7. Rudd, M.E., Tobarin, L.H., and Stolterfoht, N., *At. Data Nucl. Data Tables* **23**, 405-442 (1974).
8. Madison, D.H., *Phys. Rev. A* **8**, 2449-2455 (1973).
9. Manson, S.T., Toburen, L.H., Madison, D.H., and Stolterfoht, N., *Phys. Rev. A* **12**, 60-79 (1975).
10. Gulyás, L., Fainstein, P.D., and Salin, A., *J. Phys. B: At. Mol. Opt. Phys.* **28**, 245-257 (1995).

Zero-Degree Auger Electron Spectroscopy of Quasi-Free Electrons Scattered by Highly Charged Ions

Habib Aliabadi, Ridvan Unal, Mikhail Zamkov, Patrick Richard, Chander P Bhalla, Hiro Tawara

J R Macdonald Laboratory, Department of Physics, Kansas State University, Manhattan KS 66506

Mark Gealy

Physics Department, Concordia College, Morehead MN

Asad T. Hasan

American University of Sharjah, Physics Department, Sharjah UAE

Abstract. Ion-atom collisions can be used to determine singly differential electron scattering cross-sections for the direct and resonant, elastic and inelastic electron-ion collisions. A new high-resolution tandem parallel-plate electron spectrometer system capable of sub eV resolution has been designed and constructed. This spectrometer has been used to analyze electrons up to 12 keV in energy. An estimated analysis limit of 21 keV has been determined, which will allow the measurement of the $2p^2$ resonances up to Z=22. We will report the results of direct and resonant elastic scattering studies for hydrogen-like F, Mg and Si ions.

INTRODUCTION

Ion-atom collisions are an indispensable tool in determining resonant elastic and resonant inelastic electron-ion (e-ion) scattering cross sections [1]. Such cross sections are necessary for design and implementation of nuclear fusion reactors and for evaluation of theoretical treatments of electron ion interactions [2,3]. Previous experimental studies of ion-atom collisions have measured e-ion cross sections in the range of Z=6-9 [4]. We expect to improve our understanding of e-ion interactions by determining such cross sections in heavier elements in the range of Z=12-14. In the present work we have determined the resonantly scattered $2p^2$ ^1D cross-section for hydrogen-like Mg so as to explore the possible limits of the theoretical treatment of such interactions, pioneered by our group.

EXPERIMENTAL APPARATUS

Figure 1 shows the gas cell and high-voltage two-stage parallel-plate electron spectrometer that we have designed for the purpose of our experiments. The gas cell is 12 cm in length and doubly differentially pumped. The entrance aperture of the cell is 4 mm in diameter and the exit aperture is 2 mm in diameter. With these apertures this gas cell is capable of maintaining pressures up to 100 mTorr while the chamber pressure remains below 3 μTorr. If necessary, the gas cell or its apertures can float from ground up to 3000 volts. All areas of the scattering chamber are kept under vacuum with turbo molecular pumps.

The electron spectrometer is comprised of two sets of electron deflectors positioned in tandem [5]. The primary ion beam of enters the first stage of the spectrometer at a 45° angle through a 4 mm slit and exits its first stage through a 4 mm slit in the back plate. The deflected electrons exit the first stage through a 3 mm slit. These electrons then enter the second stage through a 2.4 mm slit and are detected by a channel electron multiplier (CEM) after passing through a 1.2 mm exit slit. The CEM is mounted on the second stage, and it can be made to maintain voltages so as to detect electrons at energies higher or lower than their exit energy. Another CEM is mounted on the back plate of the second stage. This CEM is used to determine the spectrometer constant of the first stage independent of the second stage. The spectrometer constant for each state is determined by deflecting an electron beam of known energy through the stage and maximizing the number of detected electrons detected by the CEM for that stage. The spectrometer constants for the first and second stage are found to be $f_1=0.60$ and $f_2=0.63$. The flight path of the electrons from the center of the spectrometer to the CEM is 30 cm, and the average solid angle subtended to the gas cell of the spectrometer is estimated to be 10^{-4} sr.

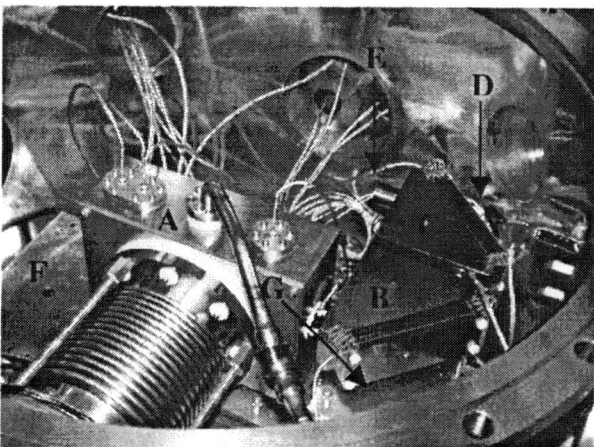

FIGURE 1. This is the present spectrometer system placed inside a scattering chamber. A: Differentially pumped chamber houses the gas cell as well as a set of deflector plates located in front of the gas cell. The deflector plates are used to minimize the contribution of electrons scattered by slits and apertures ahead of the gas cell. B: The first parallel plate analyzer separates the electrons scattered at zero degrees from projectile ions. C: The second parallel plate analyzer deflects the electrons for detection by CEM. D: CEM mounted on the back plate of the second stage. E: CEM mounted at the exit of the second stage. F: Three-point adjustable mounting table for the gas cell. G: Three-point adjustable mounting table for the spectrometer. The primary ion beam enters the chamber from the left and passes through the first stage of the spectrometer (B). In high-resolution mode the second stage is placed under high voltage to retard electrons prior to entrance into the second stage.

The gas cell assembly and the high-resolution electron spectrometer are mounted on three-point adjustable tables. The apertures are easily aligned to better than 0.01 of an inch with respect to the ions. Also, the gas cell assembly may be placed after the first stage of the spectrometer. This allows for electron scattering studies at 180 degrees in the laboratory frame.

RESULTS

The performance of the spectrometer has been tested with a number of ion beams of different energies colliding with molecular hydrogen at various target pressures. Figure 2 shows a typical zero-degree electron spectrum. The cusp peak is due to electrons moving at the same velocity as the projectiles. The binary encounter electrons (BEe) move at twice the speed of the projectiles. The projectile energy is selected such that the quasi-free target electrons collide with the projectiles at the energy resonant to the $2p^2$ state of F^{7+}. This resonant peak is expected to occur on top of the BEe peak. Indeed at a high resolution, the resonant elastically scattered (RES) electrons' peak appears superimposed on BEe peak (Fig. 3).

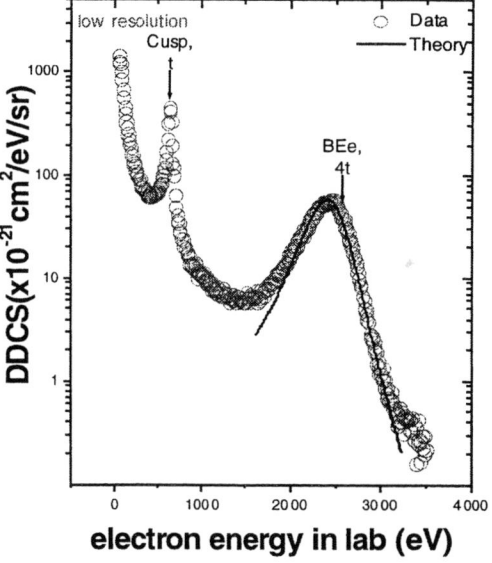

Figure 2. Zero-degree electron spectrum produced in collisions of 22MeV F^{8+} ions with H_2 target.

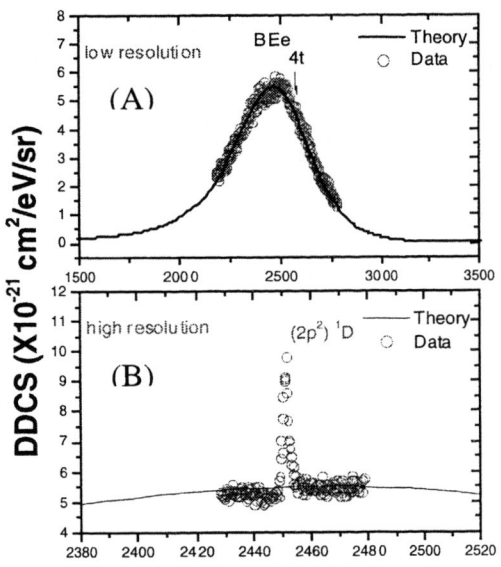

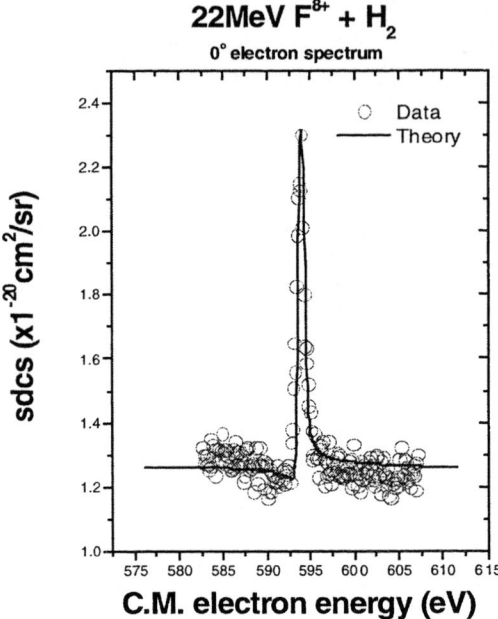

Figure 3. Electron spectra in low- (A) and high-resolution (B) in 22MeV $F^{8+} + H_2$ collisions. The RES electrons are visible in (B). In high-resolution mode the electrons are decelerated to 50 eV pass energy. The solid lines are calculated with the ESM.

Figure 4 shows this peak transformed to the projectile frame with the help of the elastic scattering model (ESM) [3]. The theoretical curve is the calculation of Bhalla *et al.* [4]. The observed full width at half maximum of 1.3 eV is more than twice as large as the 0.5 eV kinematic broadening of the peak. The kinematic broadening is due to the 1.7 degree acceptance angle of the spectrometer at the exit slit. This spectrum was taken with electron energies retarded to 50 eV prior to entrance to the second stage.

Figure 4. Single differential cross section of RES electron. The peak corresponds to $(2p^2)$ ^1D state of F^{7+}. The solid line is the R-matrix calculation of Bhalla et al [3].

Figure 5 shows an electron spectrum of 70 MeV hydrogen-like Si^{13+} ions colliding with hydrogen at 60 mTorr target. Despite the weakness of the beam the BEe peak is visible.

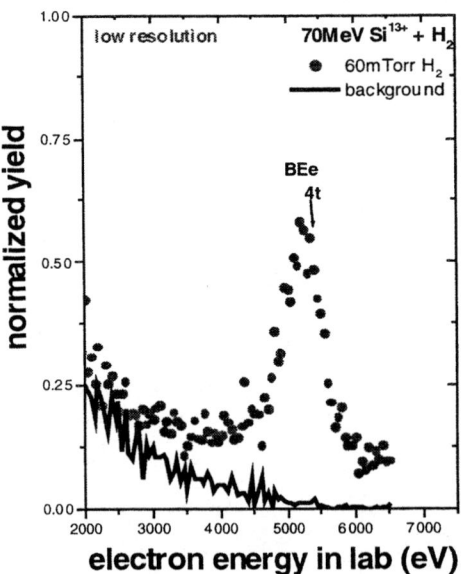

Figure 5. BEe peak due to collision of 70MeV Si^{13+} ions with H_2 target.

We performed similar experiments with hydrogen-like magnesium ions (Mg^{11+}) (see Fig. 6). In this case we were also able to see the $(2p^2)$ ^1D peak at a high resolution (see Fig. 7). The electrons in the high-resolution mode of the spectrometer were decelerated to 500 eV prior to the entrance to the second stage. Figure 8 shows the same peak transformed to the projectile frame with the help of ESM.

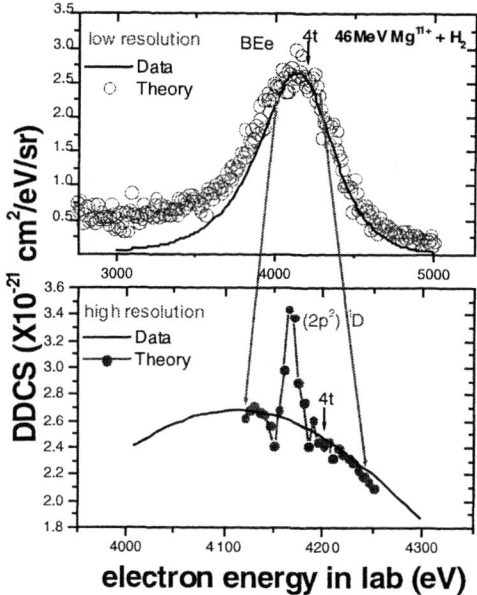

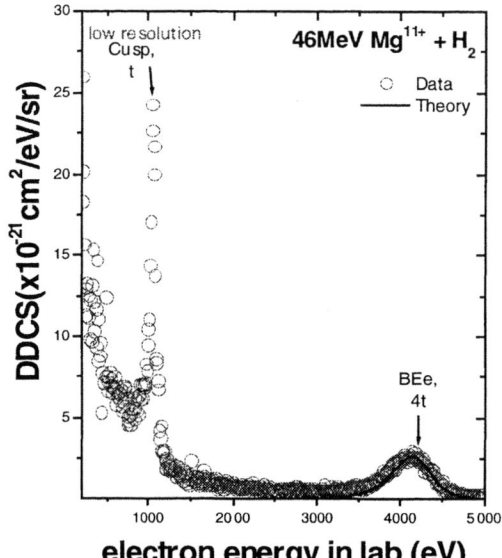

Figure 6. Zero-degree electron spectrum of 46MeV Mg^{11+} ions colliding with hydrogen target. The solid line corresponding the BEe peak is calculated based on the ESM.

Figure 7. Expanded electron spectrum from 46MeV Mg^{11+}+H_2 collisions. At 500eV pass energy the $(2p^2)$ ^1D RTEA peak appears on top of the BEe peak.

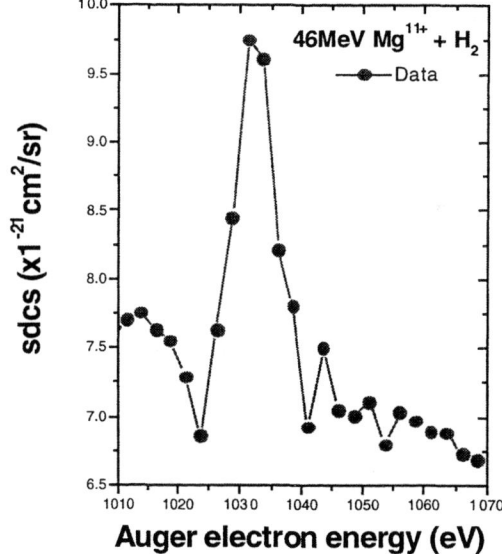

Figure 8. Single differential cross section of RES electrons. The peak corresponds to $(2p^2)$ ^1D state of Mg^{10+}.

CONCLUSION

A new high resolution zero degree electron spectrometer capable of analyzing energetic electrons has been designed and built. A resonant transfer and excitation followed by an Auger decay (RTEA) peak in 22 MeV F^{8+} colliding with H_2 was resolved at a pass energy of 50 eV. In the projectile frame this peak has only 0.65 eV FWHM and matches well with the theoretical calculation. A BEe peak in collisions of Si^{13+} with hydrogen molecules due to the direct elastic scattering of quasi-free target electrons was observed. The $2p^2$ ^1D RTEA peak of 46 MeV Mg^{11+} colliding with H_2 has also been resolved by decelerating electrons by a factor of eight from about 4000 eV to a passing energy of 500 eV through the second stage. The resolution of the spectrometer is determined to be 2.6% of the pass energy of the electrons in the second stage.

ACKNOWLEDGMENTS

We would like to acknowledge Mr. E. P. Benis., Dr. M. J. Singh and Professor T. J. M. Zouros for helpful discussion during the design and implementation of the spectrometer. This work is supported by the Office of Chemical Sciences, Geosciences, and Biosciences, Office of Basic Energy Sciences, Office of Sciences, U.S. Department of Energy.

REFERENCES

1. P. Richard et al., Physica Scripta T80, 87 (1999).
2. R. K. Janev et al., At. Data and Nucl. Data Tables 40, 249 (1988)
3. C. P. Bhalla, Phys. Rev. Lett. 64, 1103 (1990).
4. G. Toth et al., to be published.
5. D. H. Lee et al., Nucl. Instrum. Methods Phys. Res. B 40/41, 1229 (1989)

High-Resolution Photoelectron Spectroscopy in Atoms and Molecules

N. Berrah[a], O. Nayandin[a], S. E. Conton-Rogan[a], E. Kukk[a,d], A. A. Wills[a,c], T. W. Gorczyca[a], G. Snell[a,b], Chien-Nan Liu[c], J. D. Bozek[b], M. Wiedenhoeft[a]

[a]*Department of Physics, Western Michigan University, Kalamazoo, MI 49008*
[b]*Lawrence Berkeley National Laboratory, University of California – Berkeley, CA 94720*
[c]*Max-Planck-Institut fur Phusik komplexer Systeme, Nothnitzer Str. 38, D-01187 Dresden, Germany*
[d]*Department of Physical Sciences. Oulu University, Linnanmaa, Oulu FIN-90570 FINLAND*

Abstract. New features are revealed by critically combining high photon resolution from the Advances Light Source and differential photoelectron spectroscopic techniques. Two LS-forbidden doubly-excited resonances have been observed in the $3p^{-1}_{3/2,1/2}$ partial cross-sections of Ar which exhibit mirroring profiles, resulting in complete cancellation in the total photoionization cross-section as was predicted by Liu and Starace [Phys. Rev. A 59, R1731 (1999)]. We will also discuss recent preliminary measurements on the vibrational structure and partial rates of resonant Auger decay of the N 1s→2π core excitation in nitric oxide.

INTRODUCTION

Recently, there has been renewed interest in photoexcitation and photoionization experiments for the investigation of the structure and dynamics of atoms and molecules due to the rapid technological development of high-resolution spectrometers coupled with third-generation synchrotron radiation sources [1]. Technological advances have permitted the improvements of instruments, the development of spectroscopy and new methods for the quantum control of atomic and molecular processes. These continued improvements allow ever more difficult atomic and molecular experiments to be performed with a level of detail that is unprecedented. The aim of this paper is to report highly detailed studies on an atom, Ar, and a molecule, NO. The work on Ar consists of measurements of the low-lying resonances in the Ar $3p^{-1}_{3/2,1/2}$ continua in the spectral range 26.4 eV $\leq h\nu \leq$ 29.4 eV. Two new doubly exited, predominantly triplet resonances are observed in Ar which belong to a class of *mirroring* resonances [2] giving equal and opposite resonant contributions to the individual partial cross sections. As a result, their net contribution to the total cross section vanishes, which is why they are not observable in photoabsorption experiments [3,4,5]. These resonances have not been detected in previous measurements of the partial cross sections [6,7] since the achievable resolution was insufficient for resolving the rather narrow features (< 3 meV FHWM) that can now be seen.

This paper will also describe recent high-resolution measurements of resonant Auger decay of the 2π excitations at the nitrogen 1s edge of NO.

MIRRORING DOUBLY-EXCITED RESONANCES IN AR

The experiment was performed at the Advanced Light Source on the high flux and high resolution Atomic and Molecular undulator beamline 10.0.1. A photon resolution of 3 meV was achieved, close to the theoretical 10,000 resolving power available at this beamline. Two complementary experimental setups were used to measure partial differential cross sections. In the first one, a spectrometer with a hemispherical analyzer fitted with an ISL (Integrated

Sensors Ltd), position sensitive detector [8] was employed to record photoelectron spectra at the magic angle ($\theta=54.7^0$), giving the partial cross sections and the corresponding branching ratio. In the second one, two time-of-flight analyzers, at $\theta=0^0$ and $\theta=90^0$ with respect to the light polarization axis, were used to determine asymmetry parameters [7].

In both experiments, the data were collected with a two-dimensional acquisition technique [9, 10], consisting of a systematic accumulation of photoelectron spectra at many, close and equally spaced photon energies, from which constant ionic state (CIS) spectra were extracted. The spectra were corrected for variations in incident light intensity. The electron transmission efficiency curves were slowly varying functions of energy and were assumed constant over the spectral range corresponding to the separation of the photoelectron lines (177 meV). The photon energy scale was calibrated using the positions of the $3s^{-1}$ np window resonances by comparison with photoabsorption work [3, 4].

The CIS spectra of the two final states $3p^{-1}_{3/2,1/2}$, are presented in Fig 1 for all three angles. At $\theta=54.7^0$, their sum is proportional to the total ionization cross-section.

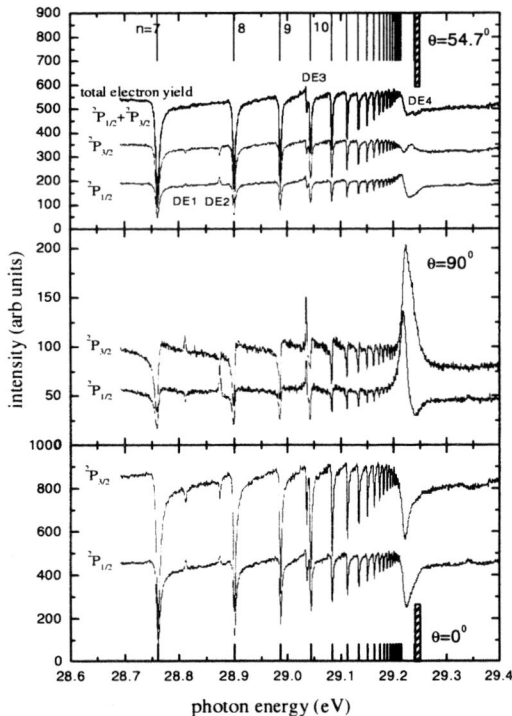

FIGURE 1. Partial differential cross sections at $\theta = 54.7^0$, $\theta=90^0$, $\theta=0^0$.

In addition to the indicated $3s^{-1}np(^1P_1)$ Rydberg series, the two previously reported low-lying doubly excited resonances DE$_3$ and DE$_4$, assigned respectively as $3s^23p^44s(^2P_{3/2})4p(^1P_1)$ and $3s^23p^44s(^2P_{1/2})4p(^1P_1)$ [3], are clearly seen. While there are no other observable resonances in the *total* cross section, the *partial* cross sections reveal two new resonances, DE$_1$ and DE$_2$, with mirroring profiles between the $3s^{-1}7p$ and $3s^{-1}8p$ members. Although similar mirroring behavior has been reported in partial photoionization cross sections of atoms [11,12,13], and very recently in photodetachment of negative ions [14], those observations were considered coincidental until the analytic developments of Liu and Starace [2] showed that this mirroring behavior is guaranteed under certain conditions.

The important implication of mirroring behavior is that the intrinsic interference effects involved in the resonant photo fragmentation processes due to two indistinguishable quantum paths are not negligible, even though the symmetric resonance profiles suggest otherwise. For example, in studies of resonant Auger processes, a two-step sequential model has been commonly used in describing the symmetric resonance profiles observed in the total cross section. However, recent theoretical and experimental measurements show asymmetric resonance profiles in the partial cross sections as well as mirroring behaviors among different partial cross sections [15]. Therefore, using a two-step sequential model to describe the autoionization resonances in photofragmentation processes does not give a correct picture, even if the resonance profile in the total cross section is Lorentzian. The present results illustrate that intrinsic interference effects involved in resonant photofragmentation processes are not negligible, even if the symmetrical resonance profiles in the total cross section suggest a small interference effect.

RESONANT AUGER DECAY OF 2π EXCITATIONS AT THE NITROGEN $1S$ EDGE OF NO

An accurate interpretation of molecular resonant Auger spectrum requires a thorough understanding of the core-excitation process and a characterization of the numerous final ionic states of the decay process. Recently this approach has been successfully employed in modeling high resolution resonant Auger spectra of carbon monoxide [16], for example. Among the diatomic molecules, nitric oxide offers an interesting case due to its open-shell configuration with several differently coupled core-excited electronic states.

In this work, motivated by recently reported *ab initio* calculations [17], we report high-resolution measurements of resonant Auger decay of the 2π excitations at the nitrogen 1s edge of NO. In the theoretical work, the excitation of N and O 1s core electrons to the 2π valence orbital and the subsequent Auger decay was investigated [17]. However, no detailed comparison with the experiment could be made, since the vibrational and electronic structure was unresolved in the available experimental spectra [18, 19]. In the presented work, Auger electron spectra are reported at several photon energies across the $1s \rightarrow 2\pi$ absorption profile, probing different $\Lambda\Sigma$-coupled electronic states of the N $1s^{-1}2\pi^2$ core-excited configuration.

The experiment was performed at the ALS beamline 8.0.1. X-ray radiation generated by a 5 cm period undulator was monochromatized by a spherical grating monochromator (SGM) using a grating with 925 lines/mm groove density. A spectrometer based on a Scienta SES-200 hemispherical energy analyzer was used to measure the electron spectra [9]. The target gas was introduced into the a gas cell with differentially pumped openings for the photon beam. The spectrometer was operated in constant (40 eV) pass energy.

Electron emission spectra measured at several photon energies across the N $1s \rightarrow 2\pi$ resonance are shown in Fig. 2. The binding energy range of the electron spectra in the figure covers final ionic states with single vacancies in the 2π, 5σ and 1π orbitals, states that can be created both by direct valence photoionization and by resonant Auger decay of the core-excited state. The upper most spectrum (V) in Fig 2 was measured at a photon energy of 380 eV, well below the resonance structure in the absorption spectrum, and is due only to direct valence ionization. Spectrum (V) was measured with lower photon and electron energy resolution to compensate for the much smaller cross-section of the direct valence photoionization at these high photon energies. The cross-section for non-resonant valence photoionization is, according to these measurements, at least an order of magnitude less than that of any of the resonant spectra shown in the figure. Significant changes in the relative intensities of the electronic bands and their vibrational progressions can be seen at the resonant photon energies (A-G). Vibrationally resolved final states in the experimental electron spectra provide a stringent test of the quality of calculated potential energy curves for the core-excited and lowest ionic states. Further comparison of the experimental results with theoretically calculated partial Auger decay rates to several final ionic states will test the theory at the level of individual electronic transitions.

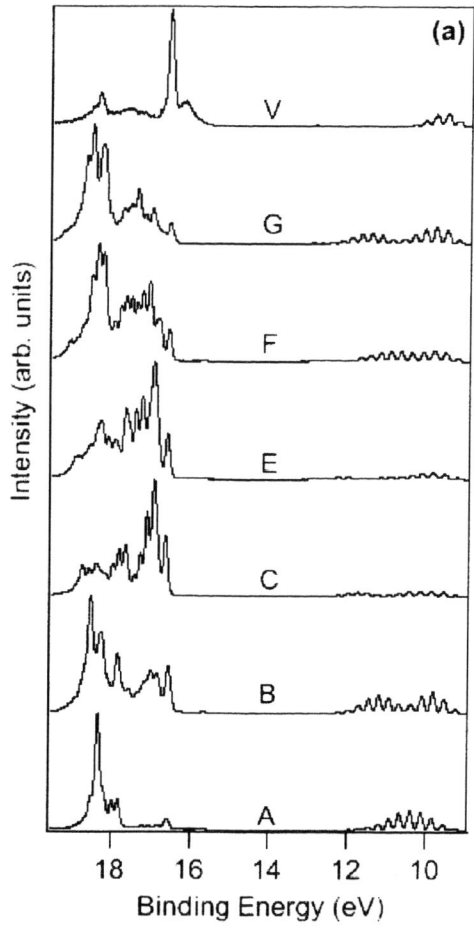

FIGURE 2. Experimental Auger electron spectra of the decay of the N $1s^{-1}2\pi^2$ core-excited states. A, B, C, E, F, G, V spectra are measured at the corresponding excitation energies 399.38 eV, 399.56 eV, 399.67 eV, 399.95 eV, 400.02 eV, 400.17 eV, 380 eV.

The Auger decay spectra of the N 1s core-excited states of NO are strongly influenced by lifetime vibrational interference effects as well as the Auger resonant Raman effect, both resulting in line shape distortions and peak position shift [20]. The interference effects can, in addition, drastically alter the intensity distribution between the vibrational peaks in the spectra. The experimental finding compared to the modeled spectra will provide better understanding of these effects.

CONCLUSIONS

Detailed high-resolution work allowed: 1) two new mirroring resonances in the low-energy photoionization spectra of Ar, that are undetectable in the total photoionization cross-section, to be observed, and 2) vibrationally resolved resonant Auger electron spectra from the decay of the N $1s^{-1}2\pi^2$ core-excited states to be measured. The latter gave the opportunity to compare with the theoretical calculations to better understand the molecular core-excitation process and a characterization of the numerous final ionic states of the decay process. These experimental results were possible due to continued improvements and rapid technological development of high-resolution spectrometers coupled with third-generation synchrotron radiation sources

ACKNOWLEDGMENTS

This work was supported by the Department of Energy, Office of Science, Basic Energy Science, Division of Chemical Science, under contracts No.DE-FG02-92ER14299.

REFERENCES

1. Cheuk-Yiu Ng, *Photoionization and Photodetachment*, Publisher City: World Scientific, New Jersey, 2000, part 2, pp. 1203-1289 and references therein.

2. Liu, C. N., Starace, A. F. *Phys. Rev. A* **59**, R1731 (1999).

3. Madden, R. P., Ederer, D. L., and Codling, K., Phys. Rev. **177**, 136 (1969).

4. Baig, M., A., and Ohno, Z. Phy. **3**, 369 (1986).

5. N. Berrah, B. Langer, J. D. Bozek, T. W. Gorczyca, O. Hemmers, D. W. Lindle, and O. Toader, J. Phys. B **29**, 5351 (1996).

6. Codling, K., West, J. B., Parr, A. C., Dehmer, J. L., and Stockbauer, R., J. *Phys. B* **13**, L639 (1980).

7. Svensson, A., Krauss, M. O.,and Carlson, T. A., *J. Phys. B* **20**, L271 (1987).

8. Hatfield, J. V., Burke, S. A., Comer, J., Currell, F., Goldfinch, J., York, T. A., and Hicks, P. J., Rev. Sci. Instrum. **63**, 235 (1992).

9. Berrrah, N., Langer, B., Wills, A. A., Kukk, E., Bozek, J. D., Farhat, A., and Gorczyca, T. W., J. Electr. Spectr. And Relat. Phenom., **101**, 1 (1999).

10. Sokell, E., Wills, A. A., Cubric, D., Currell, F. J., Comer, J., and Hammond, P., J. Electron. Spectros. Realt. Phenom. **94**, 107 (1998).

11. Samson, J. A. R., and Cairns, R. B., *Phys. Rev.* **173**, 80 (1968).

12. Krause, M. O., Cerrina, F, Fahlman, A., and Carlson, T. A., *Phys. Rev. Lett.* **51**, 2093 (1983).

13. Flemming, M. G., Wu, J. Z., Caldwell, C. D., and Krause, M. O., *Phys. Rev. A* **44**, 1733 (1991).

14. Haeffler, G., Kiyan, I. Yu., Hanstorp, D., and pegg, D. J., *Phys. Rev. A* **57**, 2216 (1998).

15. Gorczyca, T. W., and Robicheaux, F., *Phys. Rev. A* **60**, 1216 (1999).

16. Kukk, E., Bozek, J. D., Cheng, W.-T., Fink, R. F., Wills, A. A., and Berra, N., *J. Chem. Phys.* **111**, 9642 (1999).

17. Fink, R., *J. Chem. Phys.* **106**, 4038 (1997).

18. Carroll, T. X., and Thomas, *J. Chem. Phys.* **97**, 894 (1992).

19. Carroll, T. X., Coville, M., Morin, P., Thomas, T. D., *J. Chem. Phys.* **101**, 998 (1994).

20. Kukk, E., Aksela, S., and Aksela, H., *Phys. Rev. A.* **53**, 3271 (1996).

Electron-electron and Electron-Nuclear Interactions in Fast Neutral-Neutral Collisions

R.D. DuBois

University of Missouri-Rolla
Rolla, MO 65401 USA

Abstract. Contributions to the ionization cross sections by electron-electron and electron-nuclear interactions are discussed for fast neutral-neutral and ion-atom collisions. The first Born approximation is used to illustrate the relative importance of e-e and e-n processes and how they scale with projectile net charge and number of projectile electrons. New data He – He and He – Ne ionizing collisions is combined with existing atom impact data to compare with the theoretical expectations.

INTRODUCTION

Inelastic atomic interactions occurring in nature often involve dressed particles (meaning partially stripped ions) and complex targets. When compared to fast bare ion impact on simple targets where single electron removal processes dominate and the first Born approximation has been shown to be quite good for predicting total and differential cross sections, interactions involving dressed projectiles are less well understood. This is because electrons bound to both centers participate in various ways. First, for dressed ion-atom collisions both the target or the projectile can be ionized during the collision. Second, electrons bound to each center partially screen the nuclear charge; hence the coulomb forces are altered. A further complication is that the screening depends on the impact parameter and therefore can be different for single, double, triple, etc. electron removal. Third, direct interactions between projectile and target electrons can occur. These lead to excitation or ionization of electrons on both collision partners. Fourth, depending on the impact energy and the number and binding energies of the projectile and target electrons, the possibility of multiple electron removal from either, or both, collision partners, exists. Overall, this makes modeling and understanding interactions between dressed particles very difficult.

A particularly interesting subset of dressed particle interactions are those between fast atoms. Here there is no long range Coulomb force, i.e., the screening on both centers is maximized, so one would expect that the various cross sections are much smaller than those for ion impact. On the other hand, for atom impact the number of weakly bound electrons on each center is maximized. Therefore, according to items 3 and 4 above one would expect larger cross sections than for ion impact.

If the Born approximation is used to model dressed ion-atom collisions, two terms must be calculated for ionization of each collision partner. The first involves ionization of one of the collision partners by the partially screened nucleus of the other. This is sometimes referred to as the electron-nuclear (e-n) interaction. The second term involves direct interactions between projectile and target electrons and is sometimes called electron-electron (e-e) interactions. Screening/antiscreening or singly/doubly inelastic processes are also terms that are used. Of importance is that both involve a <u>single</u> interaction and therefore can be equally important in contributing to the total ionization cross section of either the projectile or the target. Also of importance is that e-e interactions lead to excitation or ionization of both collision partners whereas e-n interactions only lead to ionization of one partner. See, for example, Ref. 1 for more details.

According to the Born approximation, the doubly differential cross sections for the e-n and e-e processes are given by:

$$d^2\sigma(\varepsilon) = \int_{K\min} A(\varepsilon,K) \left[|Z - N F(K)|^2 \right] dK$$

e-n interactions (1)

$$d^2\sigma(\varepsilon) = \int_{K'\min} A(\varepsilon,K) \left[N - N |F(K)|^2 \right] dK$$

e-e interactions (2)

Here $d^2\sigma(\varepsilon)$ is the differential cross section for liberating an electron of energy ε, $A(\varepsilon,K)$ contains all information about the particle from which the electron is liberated, N and Z are the number of bound electrons and nuclear charge of the collision partner, and $F(K)$ is the single electron form factor as a function of momentum transfer for the initial state of the collision partner, i.e., $F(K) = \langle i | e^{iKr} | i \rangle$. Note that by definition, $1 < F < 0$ for $0 < K < \infty$ which corresponds to large and small impact parameter collisions respectively. The above formulae apply for ionization of either the target or the projectile with the results being applicable within the appropriate frame, i.e., target ionization is calculated in the laboratory frame whereas projectile ionization is calculated in the moving projectile frame and the results must then be transformed to the laboratory frame of reference.

As seen, the differences in these two expressions are a) differences in the bracketed terms and b) slight differences in the lower limits of integration. Another thing to keep in mind is that the e-e process has an impact energy threshold given by $\frac{M}{m}\Delta E$ where M and m are the projectile and electron masses and ΔE is the energy required to ionize/excite both collision partners. For impact energies far above threshold, differences in the bracketed terms are the most important factors in determining the differential cross sections. As pointed out some years back,[2] for positive ion impact the ratio of the bracketed terms [ ]$_{e-e}$ / [ ]$_{e-n}$ ranges from 0 to N/Z^2 as the impact parameter ranges from infinity to zero and has a maximum value of $N/[Z^2 - N^2]$ at $F(K) = N/Z$. Thus for He$^+$ impact, the maximum e-e/e-n ratio is 0.33, for Li$^+$ and C$^+$ impact it is 0.4 and 0.56 respectively, but for doubly and triply charged ion impact it is smaller, e.g., 0.125 and 0.2 for Li^{2+} and C^{2+} and 0.11 for C^{3+}. In sharp contrast, for fast atom impact where $N = Z$, the [ ]$_{e-e}$ / [ ]$_{e-n}$ ratios range from infinity to Z^{-1} and e-e processes always dominate for $F(K) > (Z-1)/(Z+1)$.

Total cross sections for e-e and e-n interactions can be obtained by integrating the above formulae. Since total cross sections are dominated by small momentum transfer, again by looking at the bracketed terms as $K \rightarrow 0$ we can estimate how the processes scale for different projectiles. Thus, we expect the e-n cross sections to scale as $[Z - N F]^2$ and the e-e processes to scale as $N(1 - F^2)$. $[Z - N F]$ is commonly noted as Z_{eff} and is obtained by comparing cross sections for dressed ion impact with those for fully stripped ion impact. Thus, e-n processes should scale as Z_{eff}^2. In contrast, e-e processes should scale with the number of bound projectile electrons since $1 - F^2 \rightarrow 2\varepsilon$ as $F \rightarrow 1$, where $\varepsilon \ll 1$.

As shown, the relative importance of e-e processes is maximum for atom impact. For this reason, several studies have been done in the past[2-5] to investigate how e-e processes influence total and differential ionization cross sections and how they scale for different collision systems. It is the purpose of this paper to extend these studies by presenting new data for fast atom-atom collisions and to combine these new data with existing data to demonstrate the relative importance of e-e versus e-n processes for both fast atom and ion impact ionization of helium.

RESULTS:

Using similar techniques to those discussed in Ref. 6, cross sections for He atom impact on helium and neon were measured between 0.1 and 2 MeV. The neon data are shown in Fig. 1 where the initial and final charge states of the projectile are the superscripts and the final target charge state is the subscript. Included in the figure are cross sections for total electron, σ_-, and target ion, σ_+, production which were obtained by adding the appropriate individual cross sections. Also included are total single electron loss cross sections, σ^{01}, taken from Ref. 7. These were obtained by using their quoted metastable fraction (roughly 25%) and cross sections for ground state and metastable atom impact. By subtracting the measured values for σ_q^{01}, cross sections for single projectile loss with no ionization of the target, σ_0^{01}, were obtained. Included in the figure are thresholds for removing various numbers of electrons from both centers. Note the scale change on the y-axis for the double loss cross sections.

As seen, multiple electron removal processes play an important role at all impact energies. In addition, processes where the projectile and the target are both ionized are comparable to processes where only the target is ionized. Also note that the cross sections for

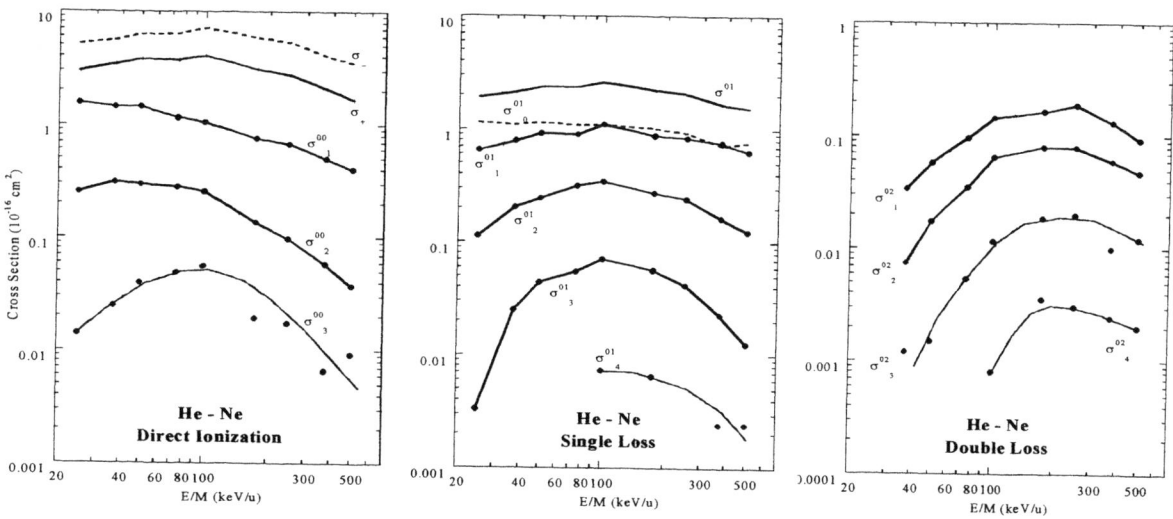

FIGURE 1. Cross sections for projectile and target ionization in He – Ne collisions.

ionizing the target or the projectile via a e-n process, σ_1^{00} and σ_0^{01}, are rather similar at higher impact energies, as expected because of the similar ionization potentials. The higher values for σ_0^{01} might also be associated with larger loss cross sections for metastable He impact.[7] Both the σ_1^{00} and σ_0^{01} cross sections are associated with an e-n process and are seen to have nearly the same magnitude as the e-e process, σ_1^{01}, where both centers are ionized. Regarding the σ_1^{01} cross sections, it is also possible that both centers can be ionized via a second order e-n process, i.e., where the target is ionized by the screened projectile nucleus and the projectile is ionized by the screened target nucleus. But Montenegro et al.[8] have shown that such second order processes can be neglected at higher impact energies.

Using these data and those from previous studies,[5,6] scaling dependences and the relative importance of e-e versus e-n processes can be investigated for ionization of helium by fast atoms ranging from hydrogen to neon. The neon impact data are obtained by reversing the collision, i.e., viewing helium as the target rather than the projectile. In Figure 2 cross sections for single electron removal from helium by various neutral and singly charged ion impact at 0.5 MeV/u are shown. For neutral impact, e-n cross sections leading to target and projectile ionization as well as e-e cross sections leading to ionization of both are shown. The first thing to note is that even though no long range coulomb forces exist for atom-atom collisions, the target ionization e-n cross sections are comparable to those for singly charged ion impact. This is because the screening by the bound electrons is not complete for neutral particle impact. The second thing to note is that the e-e cross sections do not scale linearly with the number of bound projectile electrons (which is equal to Z for neutral impact) but rather is roughly in the ratio of 1: 2.5: 3.5: 3.5: 7 for H, He, Li, C and Ne. This is attributed to the fact that the impact energy is not sufficiently high such that all projectile electrons can be considered to be loosely bound, see Figure 1. Figure 2 implies that all projectile electrons contribute to the e-e cross section for H, He and Li impact but only the n = 2 shell contributes for C and Ne impact.

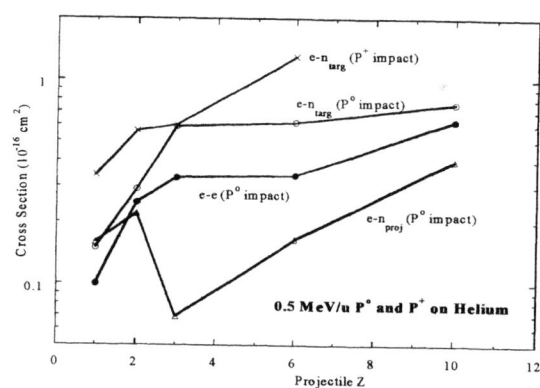

Figure 2. e-e and e-n cross sections for 0.5 MeV/u neutral and singly charged ion impact ionization of helium.

Using the first Born formulation, rough estimates of the relative importance of e-e versus e-n processes were made for neutral particle and ion impact. These estimates are compared to experimental data in Figure 3. Using our present data and that from references 5,6, (e-e)/(e-n) ratios for both single and double ionization of helium were calculated. For ion impact, He^+ impact data were taken from Ref. 9 and Li and C data were taken from the study of Sanders et al.[5] As can be seen, the relative importance of e-e processes is maximum for neutral particle impact and systematically decreases with increasing projectile charge state. Within accuracy of the data, no major differences for the various projectiles is seen. (The low data point at q = 2 is for Li^{2+} impact where the final 1s electron is tightly bound with respect to our reduced projectile velocity of 250 eV.) For double target ionization, the ratios are larger but have a similar projectile and charge state dependence. Included in the figure are the maximum values expected from "theory" for ion impact, i.e., $N / [Z^2 - N^2]$, which are in reasonable agreement with our experimentally derived ratios for q = 1, 2.

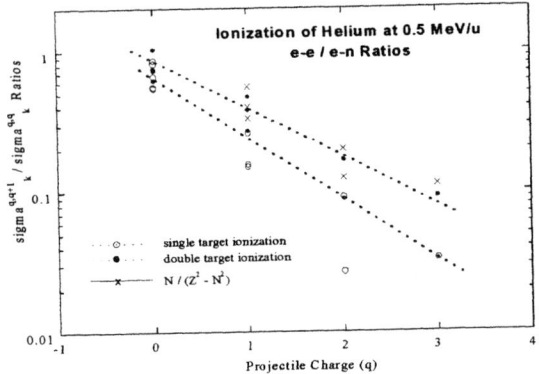

FIGURE 3. e-e / e-n ratios for 0.5 MeV/u neutral and ion impact ionization of helium.

SUMMARY:

In summary we have presented new data for target and projectile ionization in atom-atom collisions. These data were combined with existing data to demonstrate the relative importance of e-e and e-n ionization processes and to show how these scale with projectile charge (for neutral particle ion impact) and with number of projectile electrons (for fast neutral impact). Deviations from the predicted scaling behavior for e-e process were attributed to the impact energy which was not large compared to the binding energies of various inner shells. Therefore, for heavier atom impact only outer shell electrons contributed to the e-e cross sections.

ACKNOWLEDGMENTS

The support of the National Science Foundation, Grant No. PHY9732150, during the preparation of this paper is gratefully acknowledged.

REFERENCES

1. R.D. DuBois and S.T. Manson, Phys. Rev. A 42, 1222 (1990).

2. R.D. DuBois and S.T. Manson, Nucl. Inst. Meth. B 86, 161 (1994).

3. S.T. Manson and R.D. DuBois, Phys. Rev. A 46, R6773 (1992).

4. R.D. DuBois and S.T. Manson, Nucl. Inst. Meth. B 79, 93 (1993).

5. J. M. Sanders et al., XIX ICPEAC, book of abstracts (1995).

6. R. D. DuBois and À. Kövèr, Phys. Rev. A 40, 3605 (1989).

7. E. Horsdal Pedersen et al., J. Phys. B 13, 1167 (1980).

8. E.C. Montenegro et al., Phys. Rev. A 48, 4259 (1993).

9. R.D. DuBois, Phys. Rev. A 39, 4440 (1989).

How do electrons communicate about time?

A.L. Godunov[a] J.H. McGuire[a] S.G. Tolmanov[a] Kh. Kh. Shakov[a] R. Dörner[b]
H. Schmidt-Böcking[c] R.M. Dreizler[d]

[a]*Department of Physics, Tulane University, New Orleans, LA 70118-5698*
[b]*Fakultät für Physik, Universität Freiburg, 79104 Freiburg, Germany*
[c]*Institut für Kernphysik, Universität Frankfurt, 60486 Frankfurt, Germany*
[d]*Institut für Theoretische Physik, Universität Frankfurt, 60054 Frankfurt, Germany*

Abstract. We show that time correlation between electrons requires that the Dyson time ordering operator, T, differs from its uncorrelated value *and* spatial electron-electron correlation be present. In this paper we decompose T into an uncorrelated term, T_{unc}, plus a correlated term, $T_{cor} = T - T_{unc}$, which leads to time correlation in time dependent external interactions. Effects of time correlation between electrons can be observed. Two examples are presented. In transfer ionization the time correlation operator incoherently changes the shape of an electron-electron Thomas peak. In double excitation the influence of T_{cor} in amplitudes for coherently interfering pathways changes resonance intensities and profiles.

Understanding how electrons communicate about time requires ideas about both correlation and time. The mechanism for electrons to interact is the electron-electron Coulomb interaction, which is the source of spatial electron correlation (1). Without this spatial correlation the electrons are independent and cannot communicate. Time is often regarded as a parameter common to both the Schrödinger-wave and Newtonian-particle equations. However, the way in which time operates is quite different in the wave and particle limits. In the quantum wave limit of broad delocalized wavepackets, operators for time are difficult to define (2), as reflected in Pauli's remark (3) that it is "impossible to find a self adjoint (local) time operator conjugate to any Hamiltonian with a bound spectrum (such as an atom)". Fortunately the mathematics of quantum mechanics is straightforward. The concept of time correlation has been used in non-equilibrium statistical quantum mechanics, where it is similar to spatial correlation (4, 5).

In this paper for the first time we specifically address the question of how time correlation between electrons affects cross sections for two electron transitions. The key conceptual tools of this paper are temporal correlation of the external interactions and spatial correlation between electrons. Both are required for time correlation between electrons. We give two examples in which time correlation between electrons plays an observable role in atomic reaction cross sections. The first case is a kinematic peak in a reaction in which electron transfer and ionization both occur. In this case time correlated and time uncorrelated amplitudes add incoherently. The second case is double electron excitation, where coherent reaction pathways interfere. In the second case time correlation between electrons produces a large observable effect on both the shape and intensity of a double excitation resonance.

In general, time correlation in many-body systems is basic to understanding timing among subsystems, cause and effect, dynamic control, and information processing. Transmission of information in multi-electron quantum systems depends on how electrons are correlated in time. Control of reaction pathways in chemical and biological reactions (6, 7, 8), application of fast atomic switching (9), time dependence in multi-electron quantum computing and quantum communication (10, 11) and in general dynamics of nanostructures (12, 13, 14) all rely on understanding time correlation in multi-electron systems.

Time dependence is imposed on a quantum system (2, 15) by an external time dependent interaction, $V_I(t)$. The general expression for the probability amplitude, $a_{fi}(t) = <f|U_I(t,t_i)|i>$, for the transition of one or more electrons from $|i>$ at time t_i to $|f>$ at time t may be described most conveniently in the interaction representation (5, 16) using the evolution operator, $U_I(t,t')$, which satisfies,

$$i \, \partial U_I(t,t')/\partial t = V_I(t) U_I(t,t') \ , \qquad (1)$$

with the initial condition $\lim_{t \to -\infty} U_I(t, -\infty) = \hat{I}$. The formal solution for the evolution operator may be expressed as a time ordered exponential (16, 17),

$$U_I(t, t_i) = T \exp\{-i \int_{t_i}^{t} V_I(t') dt'\} = \qquad (2)$$

$$\sum_{k=0}^{\infty} \frac{(-i)^k}{k!} \int_{t_i}^{t} \ldots \int_{t_i}^{t} T[V_I(t_1) \ldots V_I(t_k)] dt_1 \ldots dt_k,$$

where T is the Dyson time ordering operator,

$$T[V_I(t_1) V_I(t_2) \ldots V_I(t_k)] \qquad (3)$$
$$\equiv \sum_{P(1,2,\ldots,k)} \theta(t_1 - t_2) \theta(t_2 - t_3) \ldots \theta(t_{k-1} - t_k)$$
$$\times V_I(t_1) V_I(t_2) \ldots V_I(t_k).$$

Here $\theta(t - t')$ is the Heavyside step function. The sum above is taken over all possible permutations, P, of the parameters $1, 2, \ldots, k$. The Dyson time ordering operator, T, imposes ordering of the $V_I(t_j)$ interactions in time to enforce causality in the time evolution of the system (16). Here $V_I(t) = \sum_j^N V_{Ij}(t)$ is implicitly summed over electrons.

We seek correlation in time between the $V_I(t_j)$'s, which provide (5) the time dependence to the quantum wave amplitudes, $a_{fi}(t)$, via Eq.(2). Requiring that correlation in time be independent of the mathematical form of $V_I(t)$, we use the only time dependent term available other than V_I, namely the time ordering operator, T. All time dependence in T arises from the $\theta(t_i - t_j)$ terms in Eq.(3). Thus, time correlation may be removed by replacing all $\theta(t_i - t_j)$ by a constant. Then $T[V_I(t_1) V_I(t_2) \ldots V_I(t_k)]$ is a simple product of $V_I(t_j)$ and is therefore uncorrelated in time. Hence there can be no time correlation in U_I. Therefore, we now separate the T operator into two terms,

$$T = T_{unc} + (T - T_{unc}) \equiv T_{unc} + T_{cor}. \qquad (4)$$

where T_{unc} is the uncorrelated part of T, and $T_{cor} \equiv T - T_{unc}$, acting on $V_I(t_1) \ldots V_I(t_k)$, is our time correlation operator. In first order in V_I there is no time correlation. In second order one has,

$$T(V_I(t) V_I(t')) = \theta(t - t') V_I(t) V_I(t') \qquad (5)$$
$$+ \theta(t' - t) V_I(t') V_I(t)$$

where,

$$T_{unc}(V_I(t) V_I(t')) = \frac{1}{2}(V_I(t) V_I(t') + V_I(t') V_I(t)) \qquad (6)$$

whence it is easily shown that,

$$T_{cor}(V_I(t) V_I(t')) = \frac{1}{2} \, \text{sign}(t - t') \, [V_I(t), V_I(t')]. \qquad (7)$$

Calculations using $T \simeq T_{unc}$ correspond to an independent time approximation (18), where the $V_I(t_j)$ interactions are not correlated in time. In second order a two step process is reduced to two independent one step processes (19, 20).

Since entanglement is conceptually and mathematically similar to electron correlation (5, 21), our time correlation operator, T_{cor}, may also be regarded as a time entanglement operator. Observable effects due to T_{cor} occur in both single (22, 23) and multiple electron transitions (5). However, in this paper we consider only the effects of time correlation between electrons, i.e. how electrons communicate with one another about time.

In multiple electron transitions correlation in time between electrons generally requires spatial electron-electron correlation in addition to time ordering (5, 24). Physically this is obvious. In the uncorrelated independent electron approximation without exchange, the probability is represented as a product of single electron probabilities, namely, $P(t) = |a_{fi}(t)|^2 = \Pi_j |<f_j|U_{Ij}(t, t_i)|i_j>|^2 = \Pi_j P_j(t)$. In this limit there is no mechanism for time correlation between transitions of different electrons. Without spatial electron correlation phase information between electrons is lost. Only when spatial electron correlation is included can T_{cor} cause time correlation between different electron transition amplitudes.

In calculations presented in this paper electron exchange is included. Nevertheless, we note that it is conceptually convenient to neglect exchange. This simplifies the meaning of 'an electron' and 'an electron transition' and also it allows one to regard electrons as distinguishable. Inclusion of exchange is mathematically straightforward, but adds complexity both conceptually and technically. In fast atomic collisions the effects of exchange are often small.

In two examples below we have evaluated the effects of the T_{cor} operator in calculations through second order in $V_I(t)$ by separating the second order term in U_I into parts corresponding to the T_{unc} and T_{cor} parts of T. Calculations of cross sections are done with and without the T_{cor} time correlation terms.

As a first example let us consider a resonant reaction in which both electron transfer and ionization occurs, namely the purely second order electron-electron Thomas peak in transfer ionization (25). In this two step example a positively charged particle first interacts with an electron in an atomic target. Then the target electron rescatters from a second target electron such that it travels out of the collision with the projectile. Because of the presence of electron correlation the probability for this reaction cannot be written as a product of two independent electron transitions (transfer and ionization). The cross section for this peak is shown in figure 1. The effect of sequencing of the two interactions is carried by the T_{cor} term in U_I. The effects of T_{unc} and of the time correlation operator T_{cor} add incoherently since the corresponding matrix elements differ by a factor of i (27). The node in the contribution from time correlation at the center of the resonance is typical of anomalous dispersion (26), which connects the correlated contribution to the uncorrelated contribution and forces the correlated contribution to zero at the center of our resonance. This peak has been studied in detail experimentally (28, 29, 30).

sees a strong effect from the time correlation term on both the shape and the intensity of $(2p^2)^1D$ and $(2s2p)^1P$ resonance spectrum. In the $(2p^2)^1D$ resonance time correlation changes the resonance shape from a window-type to a nearly asymmetric resonance profile. At the same time the intensity of the $(2s2p)^1P$ resonance increases by a factor of three. The effect of time correlation varies with both scattering angle, θ_f, and emission angle, θ_e. Calculations for double electron excitation by fast ion impact also show clear effects due to time correlation. These resonances have been studied experimentally using high-resolution spectroscopy for both electron (32) and ion (31) impact.

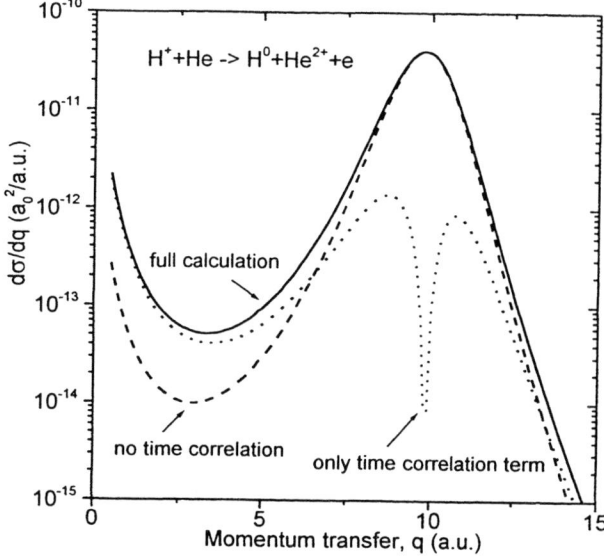

FIGURE 1. Cross section for transfer ionization as a function of the momentum transfer, q, in 2.5 MeV proton helium collisions in the vicinity of the electron-electron Thomas peak showing the effects of time correlation. Full curve, full second order calculation including both T_{cor} and T_{unc} terms of Eq.(4); long dash, approximate calculation using only the uncorrelated time term, T_{unc}; short dash, approximate calculation using only the correlated time term, T_{cor}. In this case the effects of T_{cor} and T_{unc} add incoherently as explained in the text.

A second example is double electron excitation (31). In figure 2 we present calculations of the electron emission spectrum in the region of the $(2p^2)^1D$ and $(2s2p)^1P$ resonances of helium excited by 200 eV electron impact. Unlike the previous example, there is interference between reaction pathways in this case, namely direct single ionization and single ionization proceeding through the double excitation resonance. The effect of time correlation is amplified when the relative phase between competing pathways is close to $(2n+1)\pi$. In figure 2 one

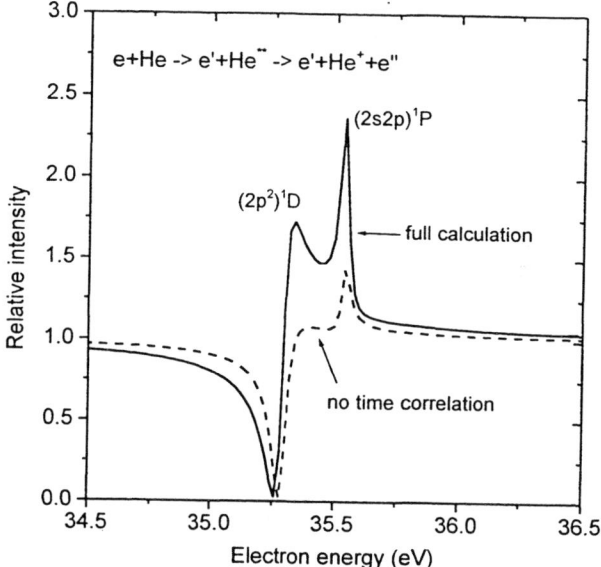

FIGURE 2. Effect of time-ordering on the autoionizing $(2p^2)^1D$ and $(2s2p)^1P$ resonances of helium in electron emission spectrum excited by 200 eV electron impact. The electron angle of emission is $60°$ and the projectile scattering angle is $30°$. Full curve, full second order calculation including the T_{cor} term of Eq.(4); broken curve, approximate calculation using only the uncorrelated time term, T_{unc}. In this case the effects of T_{cor} and T_{unc} are partially coherent. The cross section is normalized to the background of direct ionization.

In summary, we have considered time correlation between electrons in fast two electron transitions. Time dependence enters via an external $V_I(t)$. Time correlation among the $V_I(t_j)$'s in the time evolution of the system is carried by the Dyson time ordering operator, T, which may be decomposed into an uncorrelated term, T_{unc}, plus a time correlation term, $T_{cor} = T - T_{unc}$. Interaction between electrons occurs via the two-body electron-electron correlation interactions, i.e., $1/r_{ij}$ Coulomb interactions, sometimes modified by mean field potentials. This interaction produces spatial electron correlation. When correlation in time is combined with spatial correlation, electrons are connected in time, as well as in space. This gives time correlation between electrons, which governs time sequencing in multi-electron quantum systems. Effects of time correlation (entanglement in time) are observable. Two examples were given, one with and one without interfering pathways to a final state. Our approach applies to impact of ions, electrons and photons (including multi-photon effects) on atoms. Extension past second order in V_I, and also to more complex (e.g., nanoscale) systems, both appear feasible.

We thank P. Ivanov, V. Mergel, B. Shore and A. Goodman for stimulating discussions. This work was supported by the Division of Chemical Sciences, Office of Science, U.S. Department of Energy. R.D. acknowledges support by the Heisenberg Program der DFG.

REFERENCES

1. C. Froese Fischer, *Atomic, Molecular and Optical Physics Reference Book*, (G.W.F. Drake, ed.), AIP Press, NY, Chapter 21, (1996).

2. J.S. Briggs and J.M. Rost, Eur. Phys. J. D **10**, 311 (2000); M. Murao, M.B. Plenio, S. Popescu, V. Vendral and P.L. Knight, Phys. Rev. A **57**, R4075 (1999).

3. W. Pauli, *Encyclopedia of Physics*, edited by S. Flugge, Vol. 5/1, 60 (Springer, Berlin, 1958);

4. R. Balescu, *Equilibrium and non-equilibrium Statistical Mechanics*, John Wiley, NY, 1975), Chap. 21, Sec. 1. A simple product form is uncorrelated with this definition of correlation.

5. J.H. McGuire, *Electron Correlation Dynamics in Atomic Collisions*, (Cambridge University Press, 1997).

6. C. Winstead and V. McKoy, Adv. At., Mol., Opt. Phys. **43**, in press.

7. L.R. LeClair and J.W. McConkey, J. Phys. B **27**, 4039 (1995).

8. L.J. Dubé and P. Després, *The Physics of Electronic and Atomic Collisions*, ed. Y. Itikawa, AIP Conf. Proceedings, March 2000.

9. S. Gao, M. Persson, and B.I. Lundqvist, Solid State Communication **84**, 271 (1992). (1993).

10. C.H. Bennett, D.P. DiVincenzo, J.A. Smolin and W.K. Wootten, Phys. Rev. A **54**, 3824 (1996).

11. Colin P. Williams, *Quantum Computing and Quantum Communications*, (Springer, NY, 1999).

12. E.J. Heller, M.F. Crommie, C.P. Lutz and D.M. Eigler, *Invited Talks of ICPEAC XIX*, Whistler, Canada, AIP Conf. Proc. **360**, 3 (1995).

13. J. Gao and J.B. Delos, Phys. Rev. A **56**, 356 (1997).

14. M.T. Frey, F.B. Dunning, C.O. Reinhold, S. Yoshida, and J. Burgdörfer, Phys. Rev. A **59**, 1434 (1999).

15. J.H. McGuire and O.L. Weaver, Phys. Rev. **A34**, 2473 (1986).

16. M.L. Goldberger and K.M. Watson *Collision Theory*, (Wiley, NY, 1964).

17. A.L. Fetter and J.D. Walecka, *Quantum Theory of Many-Particle Systems*. McGraw Hill, San Francisco, CA, (1971).

18. A.L. Godunov et al., in preparation.

19. G.R. Satchler, *Direct Nuclear Reactions*, (Oxford University Press, 1983), p. 300.

20. P.K. Bibdak and R.D. Koshel, Phys. Rev. C **6**, 506 (1972).

21. R. Grobe, K. Rzazewski and J.H. Eberly, J. Phys. B. **27**, L503 (1994).

22. H.Z. Zhao, Z.H. Lu and J.E. Thomas, Phys. Rev. Let. **79**, 613 (1997).

23. L. Mandel and E. Wolf, *Optical Coherence and Quantum Optics*, (Cambridge University Press, 1995), Sections 4.3.1, 4.6.3 and 8.2.

24. L. Nagy, J.H.McGuire, L. Vegh, B. Sulik, and N. Stolterfoht, J. Phys. B **30**, 1939 (1997); N. Stolterfoht, Phys. Rev. A **48**, 2980 (1993).

25. S.G. Tolmanov and J.H. McGuire, Phys. Rev. A, **62**, 032711 (2000)

26. J. H. McGuire and O. L. Weaver, J. Phys. **B17**, L583 (1984).

27. J.H. McGuire, A.L.Godunov, S.G.Tolmanov, H.Schmidt-Böcking, R.Dörner, V.Mergel, R.Dreizler and B.W.Shore, Intl. J. Mass Spectrometry **192**, 65 (1999).

28. J. Palinkas et al., R. Schuch, H. Cederquist and O. Gustafsson, Phys. Rev. Let. **22**, 2464 (1989).

29. V. Mergel, R. Dörner, M. Achler, Kh. Khayyat, S. Lencinas, J. Euler, O. Jagutzki, S. Nüttgens, M. Unverzagt, L. Spielberger, W. Wu, R. Ali, J. Ullrich, H. Cederquist, A. Salin, C.J. Wood, R.E. Olson, Dž. Belkić, C.L. Cocke, and H. Schmidt-Böcking, Phys. Rev. Let. **79**, 387 (1997); V. Mergel, Ph.D thesis, Universität Frankfurt (1996).

30. R. Schuch, private communication.

31. A.L. Godunov, V. A. Schipakov, P. Moretto-Capelle, D. Bordenave-Montesquieu, M. Benhenni, A. Bordenave-Montesquieu J. Phys. B: At. Mol. Opt. Phys., **30**, 5451-5477 (1997).

32. J. Lower and E. Weigold, J. Phys. B: At. Mol. Opt. Phys., **23** 2819-2845 (1990)

Photoelectron Spectroscopy and the Dipole Approximation

O. A. Hemmers and D. W. Lindle

Department of Chemistry, University of Nevada, Las Vegas, Las Vegas, NV 89154-4003, USA

Abstract. Over the past three decades, the dipole approximation has facilitated a basic understanding of the photoionization process in atoms and molecules. Advances in gas-phase photoemission experiments using synchrotron radiation have recently highlighted nondipole effects at relatively low photon energies while probing the limits of the dipole approximation. Breakdowns in this approximation are manifested primarily as deviations from dipolar angular distributions of photoelectrons. Detailed new results demonstrate nondipolar angular-distribution effects are easily observable in atomic gases at energies well below 1 keV, and, in molecules, a previously unexpected phenomenon greatly enhances the breakdown of the dipole approximation just above the core-level ionization threshold.

INTRODUCTION

Although breakdowns in the dipole approximation in the soft-X-ray photon energy range (hν ≤ 5 keV) were first observed 30 years ago and have been studied theoretically for many years, their significance at low photon energies has remained generally unappreciated within the broader photoemission community. Ultraviolet (UV) and X-ray photoelectron spectroscopy (PES) is a common technique for studying matter of all kinds.

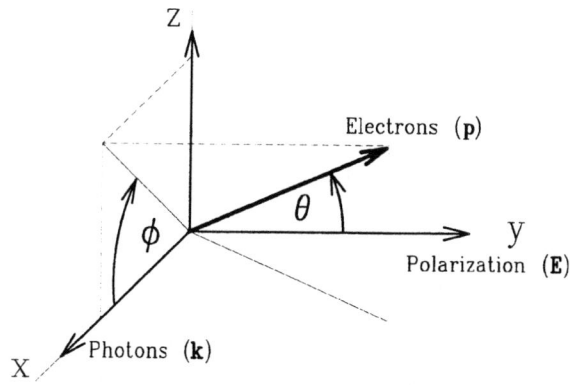

FIGURE 1. Geometry applicable to photoelectron angular-distribution measurements using polarized light. The polar angle θ is measured between the photon polarization vector **E** and the momentum vector **p** of the photoelectron. The azimuthal angle φ is defined by the photon propagation vector **k** and the projection of **p** into the x–z plane.

The power of PES stems from its ability to probe directly, via the measurement of electron kinetic energies, orbital and band structure in valence and core levels in a wide variety of samples: atoms, molecules, clusters, solids, surfaces and adsorbates.

The technique is even more powerful in an angle-resolved mode, where photoelectrons are distinguished not only by their kinetic energies but by their directions of emission as well. Probability of electron ejection as a function of angle is an excellent probe of quantum-mechanical channels available to a photoemission process because it is sensitive to phase differences among these channels. As a result, angle-resolved photoemission has been used successfully for many years to provide stringent tests of our understanding of basic physical processes underlying gas-phase and solid-state interactions with radiation, and also a tool to probe physical and chemical structure in solids and surfaces.

THE DIPOLE APPROXIMATION

One mainstay in the application of angle-resolved PES is the well-known dipole approximation (DA) for photon interactions, which leads to easily characterized and quantified behavior as a function of electron ejection angle. The electric-dipole ($E1$) approximation assumes the electromagnetic field of the photon beam, $\exp(i\mathbf{k}\mathbf{r})$, expressed as a Taylor-series expansion, $1+i\mathbf{k}\mathbf{r}+\ldots$, can be truncated to unity. In this simplification, all higher-order interactions, are neglected [1].

In the UV and far-UV photon-energy ranges, the DA for photoionization is grounded in solid physical reasoning based on two qualitative arguments: (1) photoelectron velocities following UV photoemission are extremely small compared to the speed of light, rendering relativistic effects unlikely, and (2) the wavelength of UV light (e.g., He I radiation) is much larger than the orbitals from which electrons are ejected, mitigating higher-order effects in the photon interaction. At the other extreme, in the hard-X-ray range (h$\nu\geq$5 keV), the breakdown of the DA essentially becomes complete, requiring the use of the full Taylor-series expansion for the photon interaction. Somewhere between the UV and hard-X-ray ranges, it is clear effects due to interactions beyond the DA must eventually become important.

BEYOND THE DIPOLE APPROXIMATION – FIRST ORDER CORRECTIONS

The first hints of low-photon-energy deviations from the DA in angle-resolved photoemission were provided by Krause [2] and Wuilleumier and Krause [3] in measurements on rare gases using unpolarized Mg and Al Kα X-rays. Recently, more extensive measurements, focusing on noble-gas core levels (Ar K and Kr L) and tunable photon energies between 2 keV and 5 keV, began to investigate nondipole effects in photoelectron angular distributions in more detail [4,5]. Probing the limits of the DA at lower energies (0.15 keV $\leq$ hν $\leq$ 1.2 keV), other experiments measured nondipolar angular distributions in Ne 2s and 2p valence photoionization [6-8]. Even at the lowest energy studied, nondipole effects are observable, bringing into question the usual assumption of the DA in soft-x-ray applications of PES. Finally, nondipole effects have been observed in molecules; in N1s photoemission from N$_2$, deviations from dipolar angular distributions peak just 60 eV above threshold, due to an entirely new physical phenomenon [9]. At the peak, measured relative photoemission intensities as a function of angle vary by as much as 100% compared to DA expectations. At very high photon energies, where the DA is completely invalid, it is necessary to include the exact expression, $exp(i\mathbf{k}\mathbf{r})$, for the electromagnetic field of the photon beam. For soft-x-ray photoionization, in contrast, the first-order correction to the DA is expected to be dominant [10,11]. The first step beyond the DA is to truncate the expansion of $exp(i\mathbf{k}\mathbf{r})$ after the second term, $1+i\mathbf{k}\mathbf{r}$, which amounts to including E2 and M1 interactions. These higher-order amplitudes contribute through cross terms with the E1 dipole amplitude (order $\mathbf{k}$, or $O(k)$). They contribute only to odd multipoles, leaving σ and β unaffected, but leading to forward/backward asymmetries in the photoejection probability with respect to the photon propagation vector. In addition, M1 interactions vanish in a nonrelativistic treatment in which core relaxation is unimportant, and are generally considered to be much less significant than E2 interactions for soft-x-ray photoionization [10,11]. In the soft-x-ray range, cross terms yield the largest deviations from the DA. At a level of approximation in which nondipole effects are due only to first-order E2-E1 (and the weaker M1-E1) cross terms, two new nondipolar angular-distribution parameters, in addition to the dipole parameter β, are required. We adopt the parameterization used by Cooper [11] for the differential cross section in the first-order nondipole approximation,

$$\frac{d\sigma}{d\Omega} = \left(\frac{\sigma}{4\pi}\right)\left[\begin{array}{l}1+\frac{\beta}{2}\left(3\cos^2\theta-1\right)\\+\left(\delta+\gamma\cos^2\theta\right)\sin\theta\cos\phi\end{array}\right] \quad (1)$$

which is valid for 100% linearly polarized light. The angles θ and ϕ are described in Fig. 1. As with σ and β, δ and γ depend on subshell and photon energy. Equation (1) makes it clear nondipolar angular-distribution patterns for photoelectrons exhibit forward/backward asymmetry, relative to the photon propagation direction ($\mathbf{k}$), due to the presence of cos (ϕ) in the last term.

Enhanced probability in the forward direction corresponds to the classical notion of momentum transfer from the photon to the ionized electron. To illustrate graphically the extent to which nondipole effects can modify photoelectron angular distributions, Figure 2 shows δ and γ in the horizontal x-y plane,

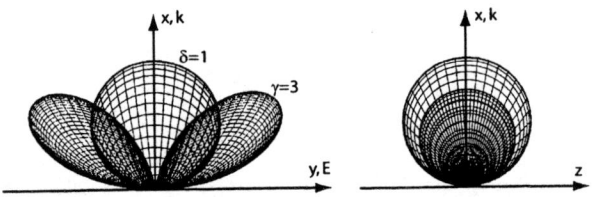

FIGURE 2. Forward/backward asymmetries of the nondipolar angular-distribution patterns are shown for the γ and δ parameters. The shapes of the patterns stay the same for all values of γ and δ and there are in general no upper or lower limits unlike for β (-1 to 2).

containing the photon propagation (**k**) and polarization (**E**) vectors and in a side view (x-z plane) just containing the photon propagation vector. As under the DA, photoejection at the magic angle (θ_m) is independent of β, and also is independent of δ and γ in the dipole plane (see Fig. 2). Thus, one can refer to "magic directions," for the combination of angles θ_m=54.74°, 125.26°, 234.74°, 305.26° and ϕ=90°; only in these (four) directions will the probability of photoejection depend solely on σ.

BEYOND THE DIPOLE APPROXIMATION – SECOND ORDER CORRECTIONS

The second step beyond the DA is to truncate the expansion of $exp(i\mathbf{kr})$ after the third term $1+i\mathbf{kr}+0.5(i\mathbf{kr})^2$, which amounts to including pure electric-quadrupole (E2) and magnetic-dipole (M1) interactions (order $\mathbf{k}^2$, or $O(k^2)$) and E3 and M2 cross terms. The even multipoles in the spherical-harmonic expansion, affect directly both σ and β. Four new parameters ($\Delta\beta$, η, ξ, μ) have to be included in the differential cross section, which arise from interferences between E1-E3, E1-M2, E2-E2, E2-M1, M1-M1 as well as from retardation corrections to E1-E1. Three of the new parameters satisfy the constraint $\eta+\xi+\mu=0$ and their angular-distribution patterns are shown in Fig.3.

All parameters depend on subshell and photon energy and contribute intensities even to the above mentioned "magic directions". The differential cross section including the second order corrections is given by Derevianko et al. [8] as,

$$\frac{d\sigma}{d\Omega} = \left(\frac{\sigma}{4\pi}\right) \begin{bmatrix} 1+(\beta+\Delta\beta)\,P_2(\cos\theta) \\ +(\delta+\gamma\cos^2\theta)\sin\theta\cos\phi \\ +\eta P_2(\cos\theta)\cos 2\phi \\ +\xi(1+\cos 2\phi)P_4(\cos\theta) \\ +\mu\cos 2\phi \end{bmatrix} \quad (2)$$

with $P_2(\cos\theta) = 0.5(3\cos^2\theta-1)$ and $P_4(\cos\theta) = 1/8(35\cos^4\theta-30\cos^2\theta+3)$ and for 100% linear polarized light. The angles are the same as in Eq. (1).

The experiments were performed at the Advanced Light Source with four electron analyzers mounted in a chamber which can rotate about the photon beam [12]. At the nominal angular position of the apparatus, two analyzers are at θ_m and $\theta=0°$ in the plane perpendicular to the photon beam ($\phi=90°$), which we

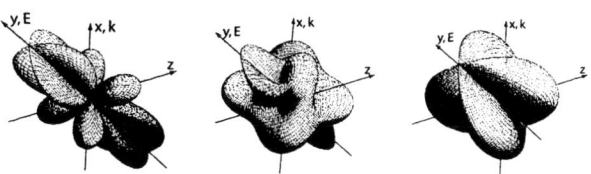

FIGURE 3. The angular-distribution patterns correspond from left to right to the parameters η, ξ and μ. The photon propagation vector is in x-direction and the polarization vector is in y-direction. As for δ and γ the shapes of the patterns stay the same for all values of η, ξ and μ and there are in general no upper or lower limits.

refer to as the dipole plane because first-order corrections vanish, while two more analyzers are positioned on the forward 35.3° cone with respect to the photon beam. At the nominal position, these two "nondipole" analyzers are at (θ_m, $\phi=0°$) and ($\theta=90°$, $\phi=35.3°$). Photoemission intensities in the two magic-angle analyzers are independent of β and can differ only because of nondipole effects. While the magic angle is no longer strictly valid when second-order effects are included, calculations show they can be unimportant in certain geometries.

We present experimental results for Ne γ_{2s} and ζ_{2p}, assuming the validity of Eq. (1), for comparison with $O(k)$ and $O(k^2)$ calculations. The first data set is based on angle-resolved photoemission intensities from the two magic-angle analyzers. Figure 4 compiles old [6] and new values for γ_{2s} and ζ_{2p} (open squares) determined using this geometry. The solid curves represent $O(k)$ calculations [10,11,13], which agree well with the 2s results but disagree with the 2p results above 800 eV.

For the magic-angle geometry, Eq. (2) and calculated values for $\Delta\beta$, η, ξ, and μ [14] can be used to estimate $O(k^2)$ influences on the experimental determination of γ_{2s} and ζ_{2p}. Measured values of ζ_{2p} will be perturbed by second-order effects as follows:

$$\zeta(k^2) \approx \frac{\gamma+3\delta+\sqrt{54}(\mu-7\xi/18)}{1-\mu} \quad (3)$$

Effective values for ζ_{2p} (and similarly γ_{2s}) have been determined, yielding the dotted curves in Fig. 4. We find excellent agreement for γ_{2s} and clearly improved agreement for ζ_{2p}. The second-order effects thus included account for much of the difference between first-order theory (solid curve) and experiment for ζ_{2p}.

To confirm this unexpected finding, new measurements in a different geometry were performed by rotating the apparatus to ten different angular

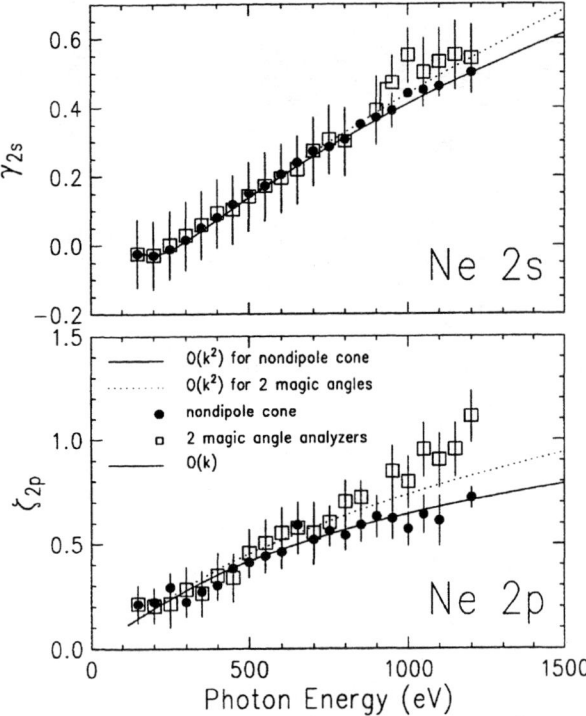

FIGURE 4. Experimental and theoretical values of γ_{2s} and ζ_{2p} for neon determined under different geometrical conditions: (1) open squares and dotted curves relate to the magic-angle geometry; (2) solid circles and curves relate to the nondipole-cone geometry. Both dotted and solid curves include effects up to $O(k^2)$. The solid curves also represent first-order theory, independent of geometry.

positions about the photon beam, yielding 20 angle-resolved intensities for Ne $2s$ and $2p$ photoemission at different angles θ and ϕ around the 35.3° nondipole cone. Here, influences of the $O(k^2)$ parameters are superimposed on intensity variations due to the dipole β and the $O(k)$ δ and γ parameters. But for both γ_{2s} and ζ_{2p}, our calculations predict effects due to η, μ, and ξ also mostly cancel in the nondipole-cone geometry, yielding the solid curves in Fig. 4. Furthermore, small residual effects around this cone are similar in sign and magnitude for $2s$ and $2p$, which is relevant because $2s/2p$ intensity ratios are the raw input for data analysis. Assuming no influence of second-order effects in the nondipole cone, we modeled the measured ratios around this cone using Eq. (1) to derive values for γ_{2s} and ζ_{2p}. These results (solid circles in Fig. 4) agree extremely well with $O(k)$ calculations [10,11,13], confirming our prediction of near cancellation of $O(k^2)$ effects in this geometry.

In conclusion, the experimental study of breakdowns in the DA for soft-x-ray photoemission from gas-phase targets has experienced a resurgence in the past few years. Although DA breakdowns, and their effects on photoelectron angular distributions, have been predicted for some time, their magnitude and potential significance have remained relatively unappreciated within this community.

ACKNOWLEDGMENTS

We would like to thank our colleagues P. Glans, H. Wang, S.B. Whitfield, R. Wehlitz, I.A. Sellin, for their support during the experiments and we thank A. Derevianko and W.R. Johnson for their collaboration and development of the theoretical framework. This research is funded by the NSF (PHY-9303915), the DOE EPSCOR, Research Corporation, and The Petroleum Research Fund. The Advanced Light Source is supported by the DOE under Contract No. DE-AC03-76SF00098.

REFERENCES

1. Bethe, H.A. and Salpeter, E.E. *Quantum Mechanics of One- and Two-Electron Atoms* Springer-Verlag, Berlin, 1957.
2. Krause, M.O., Phys. Rev. **177**, 151 (1969).
3. Wuilleumier, F.J. and Krause, M.O., Phys. Rev. A **10**, 242 (1974).
4. Krässig, B., Jung, M., Gemmell, D.S., Kanter, E.P., LeBrun, T., Southworth, S.H., and Young, L., Phys. Rev. Lett. **75**, 4736 (1995).
5. Jung, M., Krässig, B., Gemmell, D.S., Kanter, E.P., LeBrun, T., Southworth, S.H., and Young, L., Phys. Rev. A **54**, 2127 (1996).
6. Hemmers, O., Glans, P., Hansen, D.L., Wang, H., Whitfield, S.B., Lindle, D.W., Wehlitz, R., Levin, J.C., Sellin, I.A., Perera, R.C.C., Dias, E.W.B., Charkraborty, H.S., Deshmukh, P.C., and Manson, S.T., J. Phys. B **30**, L727 (1997).
7. Lindle, D.W. and Hemmers, O., J. Electron Spectrosc. Relat. Phenom., **100**, 297 (1999).
8. Derevianko, A., Hemmers, O., Oblad, S., Glans, P., Wang, H., Whitfield, S.B., Wehlitz, R., Selling, I.A., Johnson, W.R., and Lindle, D.W., Phys. Rev. Lett., **84**, 2116 (2000).
9. Hemmers, O., Wang, H., Lindle, D.W., Focke, P., Sellin, I.A., Arce, J.C., Sheehy, J.A., and Langhoff, P.W., submitted to Phys. Rev. Lett.
10. Bechler, A. and Pratt, R.H., Phys. Rev. A **39**, 1774 (1989); **42**, 6400 (1990).
11. Cooper, J.W., Phys. Rev. A **42**, 6942 (1990); **45**, 3362 (1992); **47**, 1841 (1993).
12. Hemmers, O., Whitfield, S.B., Glans, P., Wang, H., Lindle, D.W., Wehlitz, R., and Sellin, I.A., Rev. Sci. Instrum., **69**, 3809 (1998).
13. Johnson, W.R., Derevianko, A., Cheng, K.T., Dolmatov, V.K., and Manson, S.T., Phys. Rev. A **59**, 3609 (1999).
14. Derevianko, A., Johnson, W.R., and Cheng, K.T., At. Data Nucl. Data Tables **73**, 153 (1999).

Measurement of Compton Scattered Electrons Using Monochromatic X-Rays

W.A Hollerman[*], N.A. Guardala[†], D.J. Land[†], G.A. Glass[*] and J.L. Price[†]

[*] *Acadiana Research Laboratory, University of Louisiana at Lafayette, Lafayette, Louisiana 70504, USA*
[†] *Radiation Technology Branch, Naval Surface Warfare Center/Carderock Division, West Bethesda, Maryland 20817, USA*

Abstract: The technique of zero degree electron spectroscopy, first developed in experiments involving ion-atom collisions, has been applied to study the distribution of inelastically scattered electrons to the K-shell ionization limit. Focused x-rays, generated at the National Synchrotron Light Source (NSLS), were used to produced ejected electrons from a thin 20 nm carbon foil at both zero and 180 degrees. Electrons were energy analyzed using a spherical sector electrostatic analyzer coupled to a charge multiplier detector. At zero degrees, both Compton and K-shell photoelectrons are emitted. The emission of Compton-recoil electrons is suppressed at the highest electron energies, where the infrared divergence is greatest. Comparison of the zero and 180 degree spectra could provide experimental evidence of the infrared divergence, due to the large enhancement of inelastically-scattered recoil electrons observed at zero degrees.

BACKGROUND

Compton scattering is a quantum process that involves the inelastic scattering of a photon by a bound atomic electron. The orbital motion of atomic electrons broadens the scattered photon spectrum. Usually the Compton process is studied by measuring the energy spectrum of the inelastically-scattered photon. The measured spectrum of recoil electrons exhibits a similar, but not identical, broadened emission peak.

Compton scattering has recently attracted significant attention from both the experimental and theoretical points of view [ref. 1-3]. The areas of relevance related to inelastic x-ray scattering from 10 to 100 keV are quite diverse. These processes have applications in many areas of applied and basic science, especially physics that involves a more complete understanding of quantum electrodynamics and the deeper insights into the role of structure in modeling scattering from bound target electrons [ref. 4].

The authors measured the Compton-recoil spectrum, first in the energy region of the so-called Compton profile and more recently in the regime of largest momentum transfer to the bound electrons. The latter spectral region is anticipated to be most significantly altered in both shape and intensity due to manifestation of the infrared divergence, which results in an enhanced yield of low-energy, or "soft" photons. This behavior in cross section for Compton scattering has not been successfully studied by photon spectroscopy due to inherent problems with unambiguously measuring low-energy photons, or the "soft" photon portion of the scattered x-ray spectrum.

It seems reasonable to investigate the emission probability of "hard" electrons, (particle component of the atomic Compton process) so as to observe the distinct rise in the inelastic scattering cross section. Hard electrons are inelastically scattered with an energy that is 10-20 times larger than the energy for recoil electrons at the Compton profile peak. For example, the recoil-electron energy for zero degree emission is approximately 400 eV at the so-called "Compton profile" region, based on a photon energy of 11.3 keV

Spectroscopy can be used to determine electron emission to the kinematic limit for the production of K-shell photoelectrons. A merging of the photoionization and inelastic scattering processes is possible and can lead to interference effects. Interference between the smoothly varying continuum and an isolated resonance is a well-known phenomenon in both nuclear and atomic physics [ref. 5]. In this system, the smoothly varying continuum can be identified with the inelastically scattered electron amplitude. Conversely, the isolated resonance peak corresponds to the carbon K-shell photoelectron line. It is assumed that the cross section for soft photon emission will be related, though not necessarily identical, to the cross section for hard inelastic electron emission.

A spectrum measured at 180 degrees contains only background electrons produced through the interaction of the incoming photon beam and K-shell photoelectrons. The zero degree spectrum contains components observed in the 180 degree spectrum and the kinematically allowed Compton recoil electrons. The difference between the zero and 180 degree spectra should reflect the rise in the cross section, due to the stronger influence of both the infrared divergence and the presence of the p • A term in the interaction Hamiltonian [ref. 6-7]. In the low-momentum transfer regime, this part of the Hamiltonian can be neglected and only A^2 is used. It should be noted that the p•A term for atomic photoionization is also the most significant part of the interaction Hamiltonian.

EXPERIMENTAL SETUP

A spherical sector double focusing electrostatic spectrometer has been used in previous studies of ion-atom collisions [ref. 8-10]. These measurements were performed at the X26C and X14A beam lines located at the National Synchrotron Light Source (NSLS).

The energy of the x-ray beam was found to be 11.25 keV, using the K-shell photoelectron spectrum obtained at 90 degrees from a 20 nm carbon foil. When positioned at zero degrees, the x-ray beam impinges on the carbon foil and then passes through the body of the spectrometer. At 180 degrees, the photon beam passes through the sectors of the spectrometer before hitting the carbon foil. The zero degree spectrum contains Compton-recoil electrons and inelastically-scattered electrons. This results in ejected electrons identical in kinetic energy to the K-shell photoelectrons. These photoelectrons are produced by K-shell photoionization. The 180 degree data contains K-shell photoelectrons and inelastically-scattered recoil electrons. It was assumed that the differential cross section for emission of the K-shell photoelectrons is identical at both zero and 180 degrees [ref. 11]. The emission probability for the highest energy recoil electrons (infrared divergence) is expected to be suppressed at 180 degrees relative to forward angle emission.

The spectrometer constant for this detector/analyzer was measured to be 0.44 [ref. 12]. Electrons were detected using a channel electron multiplier. An argon gas-filled ionization chamber (IC) was positioned outside the vacuum chamber downstream of the spectrometer. This detector was used to monitor the x-ray beam for differences in photon fluence during each measurement. The output current of the IC was digitized and used to advance each channel of a multi-channel scalar (MCS). Each electron energy bin was accumulated for the same number of incoming photons through the target foil. The measured electron yield was normalized for spectrometer resolution and detector efficiency [ref. 13].

The energy resolution for the carbon K-shell photoelectron peak was measured to be between 2.5% and 3.0% at an emission angle of 90 degrees. At this angle, the photoelectron emission amplitude should be far more pronounced than the probability for recoil electron emission. The background inside the vacuum chamber was measured after each run by removing the thin carbon foil from the beam path and placing an identical empty aluminum target holder in its place. Data was collected in the same fashion as when the foil was in place. In the vast majority of cases, the contribution from background counts for all electron energies accounted to less than 5% of the total accumulated counts. For the largest spectrometer potentials, the background contribution was typically less than 1% of the total counts. The background was well behaved and predictable between scans, which resulted in a spectrum that was monotonically decreasing with increasing spectrometer potential.

ANALYSIS AND DISCUSSION

FIGURE 1 shows spectral results collected at zero and 180 degrees over the range of 8.5 to 11.7 keV at the NSLS X26C beam line in April 1998. Notice the zero degree emission probability rises sharply around 9.2 keV and continues to the onset of the large photopeak with a energy of about 10.7 keV. This increase is due to both the K-shell photoionization peak and L-shell Compton-scattered electrons, which are identical for photons above the K-shell binding energy [ref. 14]. The corresponding 180 degree emission yield does not exhibit this behavior for the ejected electrons. This difference is believed to be a manifestation of the infrared divergence for Compton scattering from carbon.

FIGURE 2 shows results collected from 10.2 to 11.3 keV at the NSLS X14A beam line in November 1999. The emission probability at zero degrees in FIGURE 2 is significantly larger (roughly 3 times more intense at the photoelectron peak centroid) than the corresponding value measured at 180 degrees. This is further evidence of the infrared divergence for Compton scattering. We believe that additional data should be collected in order to quantify these results.

Photoionization is conventionally considered to be a totally non-radiative process. The term photoabsorbtion is often used interchangeably with photoionization for bound atomic states. However, theoretical studies [ref. 15-16] involving the radiative branch opens the possibility for coherence between the two photon-induced ionization processes.

We believe it is reasonable to assume the initial bound K-shell electrons, incoming synchrotron produced x-rays, and the final electron states can be classified as indistinguishable. Our observations show that this coherence tends to be much more pronounced at zero than at 180 degrees. It is possible that the observation of non-dipole photoionization amplitudes from the carbon K-shell is due to the extreme forward emission probability at zero degrees. These amplitudes should be identical to the non-dipole nature of Compton scattering, regardless of which portion of the Hamiltonian is making the largest contribution and whether the infrared divergence is playing a role in the scattering.

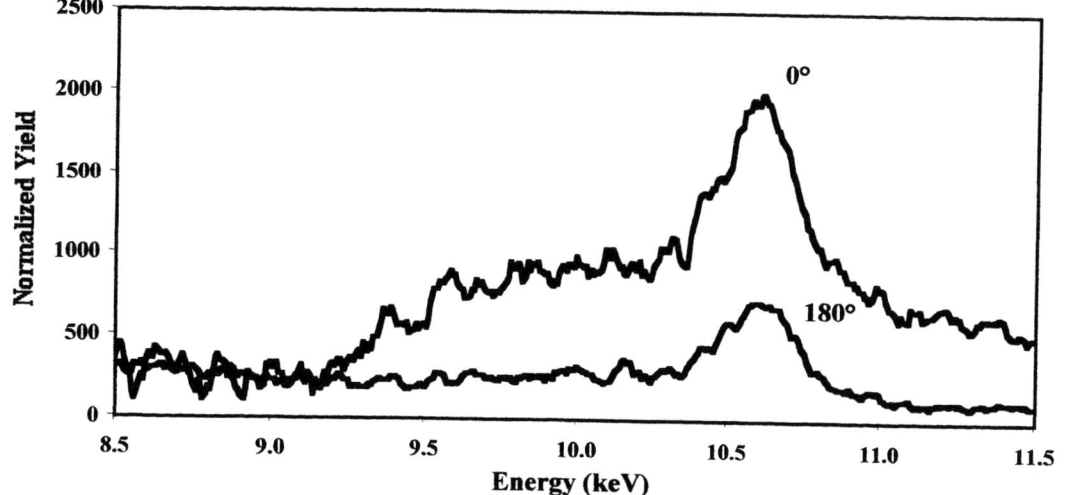

FIGURE 1. April 1998 NSLS X26C Spectral Results

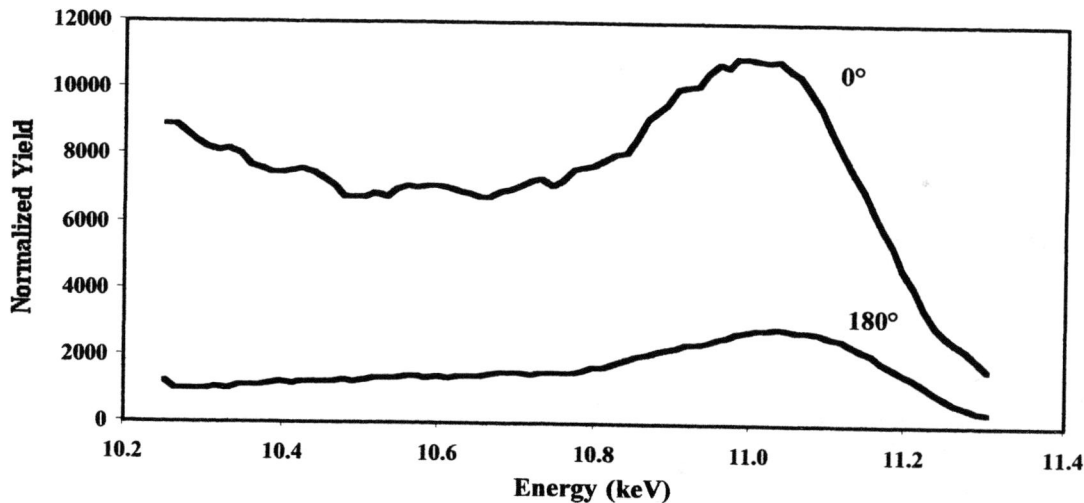

FIGURE 2. November 1999 NSLS X14A Spectral Results

CONCLUSION

The technique of electron spectroscopy has been applied to study the energy distribution of electrons inelastically scattered to the K-shell ionization limit. Focused 11.25 keV x-rays were used to produce ejected electrons from a thin 20 nm carbon foil at both zero and 180 degrees at NSLS in 1998 and 1999. These preliminary results could provide experimental evidence of the infrared divergence in the atomic Compton scattering process for carbon. Additional measurements, using larger x-ray fluences, are necessary to further quantify this phenomenon.

ACKNOWLEDGEMENTS

The U.S. Department of Energy and the Louisiana Education Quality Support Fund supported this research under contract numbers DOE/LEQSF (1993-95)-03 and DE-FC02-91ER75669. The authors would like to thank the staff of the National Synchrotron Light Source for their assistance during this research.

REFERENCES

[1] P.M. Bergstrom and R.H. Pratt, Radiat. Phys. Chem., 50, 23 (1997).

[2] D.L. Anastassopoulous, G.D. Priftis, C. Toprakeioglu, and A. A. Vradis, Phys. Rev. Lett., 81, 830 (1998).

[3] L. Rosenberg, Phys. Rev., A52, 364 (1995).

[4] C. Rocchi, and F. Sachetti, Phys. Rev., B51, 81, (1995).

[5] U. Fano, Phys. Rev., 124, 1866, (1961).

[6] P. M., Bergstrom, National Institute of Standards and Technology (NIST), private communication.

[7] Z. Kalman, T. Suric, K. Pisk, and R. H. Pratt, Phys. Rev., A57, 2683, (1998).

[8] T.T. Shy, Calibration of a Spherical Sector Spectrometer, Master's thesis, University of Southwestern Louisiana, 1987, unpublished.

[9] S.D. Berry et al., Targets, Phys. Rev. A31, 1392 (1985).

[10] K.D. Sevier, *Low Energy Electron Spectroscopy*, Wiley-Interscience, New York, 1972.

[11] P.H. Pratt, A. Ron, and H.K. Tseng, Rev. Mod. Phys., 45, 273 (1973).

[12] N.A. Guardala, J. L. Price, D. J. Land, D. G. Simons, D. H. Lee, B. M. Johnson, G. A. Glass, and J. G. Brennan, Nucl. Inst. Meth., A347, 504, (1994).

[13] G. Pashman et al., Rev. of Sci. Instr., 41, 1706 (1970).

[14] P.M. Bergstrom T. Suric, K. Pisk, and R.H. Pratt, Phys. Rev., A48, 1134 (1993).

[15] J. McEnnan and M. Gavrilla, Phys. Rev., A15, 1537, (1972).

[16] D.J. Botto and M. Gavrilla, Phys. Rev., A26, 1237 (1982).

Electrons from neutral projectiles and targets: Peaks and Ramsauer-Townsend structures

D.H.Jakubaßa-Amundsen

Physics Section, University of Munich. 85748 Garching, Germany

Abstract. The interrelation between the binary encounter peak and the electron loss peak on one hand, and the soft electron peak and the electron loss cusp on the other hand is discussed both theoretically and experimentally. Particular examples concern the Ramsauer-Townsend structures in the binary encounter peak from 5.88 MeV/amu U^{29+} + Ne and 0.1 MeV/amu Ne^0 + He, He^0 + Ne collisions where absolute experimental data are compared with the electron impact approximation. Measurements and strong potential second Born calculations for the soft electron peak from 0.1 MeV/amu Ne^0 + He are also presented. The peak asymmetry is studied as a function of collision velocity and projectile species, and marked deviations from the behaviour of the soft electron peak induced by charged particle impact are found.

The characteristic features in the electron spectra from energetic atom-atom collisions are the soft electron (SE) peak and the binary encounter (BE) peak from target ionisation, as well as the electron loss to continuum (ELC) cusp and the electron loss (EL) peak from projectile ionisation. All these features are very sensitive to strong perturbing potentials, and an accurate description requires nonperturbative treatments.

The discovery of Ramsauer-Townsend structures in the angular-dependent singly differential EL cross section from collisions with heavy targets dates back to the late seventies [1]. About ten years later, structures at certain angles were also observed in the BE peak from collisions with heavy partly stripped projectiles [2]. Figs.1 and 2 show new experimental data from U^{29+} + Ne collisions which in contrast to the earlier experiments were measured on an absolute scale [3]. These structures are described in terms of quasielastic electron scattering [5]: The doubly differential cross section for BE electron emission results from a convolution of the cross section $d\sigma_e/d\Omega$ for elastic electron scattering from the projectile field, with the momentum distribution φ_i^T of the bound target electron (of energy E_i^T),

$$\frac{d^2\sigma}{dE_f d\Omega_f} = \frac{k_f}{v} \int d\mathbf{q}\, \delta(E_f - E_i^T + \mathbf{q}\mathbf{v} - \mathbf{k}_f\mathbf{v} + v^2) \cdot \frac{d\sigma_e}{d\Omega}(k,\theta) \left|\varphi_i^T(\mathbf{q}+\mathbf{v})\right|^2 \quad (1)$$

Atomic units are used, and the energy, momentum and emission angle of the outgoing electron are denoted by E_f, $\mathbf{k}_f$ and ϑ_f, respectively, while $\mathbf{v}$ is the collision velocity. The Ramsauer-Townsend structures leading to a vanishing BE peak around $\vartheta_f = 30°$ (Fig.1) are, however, not well described by theory which only shows a broadening of the BE peak around 30°. (In addition, the data give evidence for another peak located around $E_f = 2v^2$ for all ϑ_f which may arise from a rescattering of the BE electrons [3]).

It was soon understood that these BE peak structures were indeed Ramsauer-Townsend structures as viewed in the target frame of reference, and the quasielastic scattering model (1) is just the transformed electron impact approximation (EIA) put forth for the interpretation of the EL peak structures [6] (for a review, see [7]). The interrelation between the doubly differential cross sections for electron emission from a $P \to T$ collision (in the target frame of reference) and from a $T \to P$ collision with reversed collision velocity (in the projectile frame of reference; denoted by primed quantities) is simply

$$\frac{d^2\sigma}{dE'_f d\Omega'_f} = \frac{k'_f}{k_f} \frac{d^2\sigma}{dE_f d\Omega_f} \quad (2)$$

with $\mathbf{k}'_f = \mathbf{k}_f - \mathbf{v}$. Subsequently, a search for structures in the (doubly differential) EL peak was carried out [8], but no peak splitting could be observed. On the other hand, for these collision systems the agreement between experiment and theory is much better than in case of the BE peak investigations, although an important contribution from a simultaneous ionisation of projectile and target is present due to the loosely bound electrons of the neutral perturber atom [7].

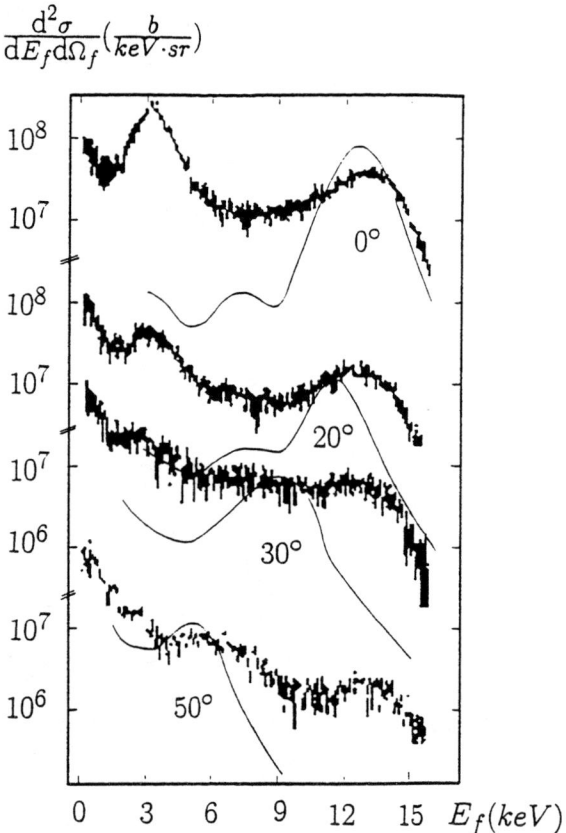

Fig. 1. Doubly differential cross section for electron emission from Ne bombarded with 5.88 MeV/amu U^{29+} as a function of electron energy E_f, at emission angles $\vartheta_f = 0°, 20°, 30°, 50°$. Histograms, experimental data from Bechthold et al [3, 4];
—, EIA theory (using Hartree-Fock wavefunctions for the Ne target, a Herman-Skillman potential for U^{29+} and including exchange).

An immediate comparison between the BE peak structures and those observed in EL-experiments is made possible by considering the (energy-integrated) singly differential cross section. One can obtain this cross section in the projectile frame from the one measured in the target frame by means of [3, 7]

$$\frac{d\sigma}{d\Omega'_f} = \int_{E_{min}}^{E_{max}} dE_f \frac{|\mathbf{k}_f - \mathbf{v}|}{k_f^2} (k_f - v\cos\vartheta_f) \cdot \frac{d^2\sigma}{dE_f d\Omega_f} \quad (3)$$

where E_{min} has to be chosen such that the integration regime covers the whole BE peak region with a nonnegative integrand. In addition, one may use the approximation $\vartheta'_f \approx 2\vartheta_f$, valid near the BE peak maximum.

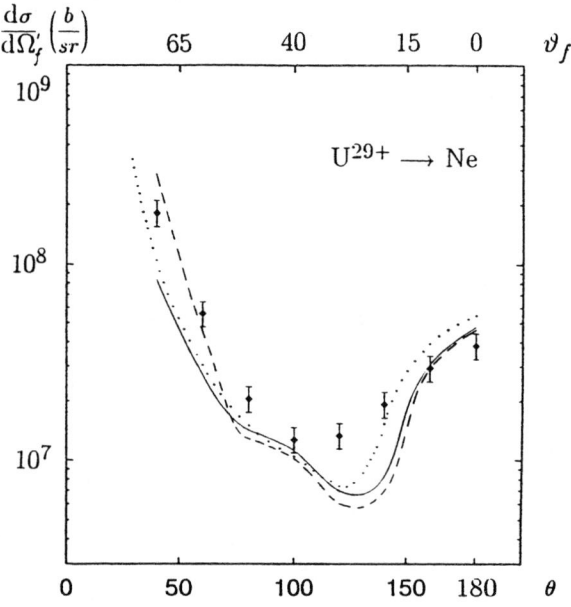

Fig. 2. Projectile-frame singly differential cross section for BE electron emission from 5.88 MeV/amu $U^{29+} \div$ Ne collisions. ◇, experimental data [3]; ——, EIA results obtained in the same way; - - - -, EIA results obtained from (3); ····, elastic electron scattering from a U^{29+} potential, multiplied by the number of target electrons (= 10) (taken from [3]).

Defining θ to be the scattering angle in the projectile frame of reference (see (1)) and keeping in mind that $\mathbf{v}$ has to be replaced by $-\mathbf{v}$ when doing the kinematical transformation, one has approximately

$$\theta = \pi - \vartheta'_f = \pi - 2\vartheta_f. \quad (4)$$

Using $k_f \approx 2v\cos\vartheta_f$, formula (3) may be even further simplified to [3]

$$\frac{d\sigma}{d\Omega'_f} \approx \frac{1}{4\cos\vartheta_f} \frac{d\sigma}{d\Omega_f} \quad (5)$$

where $d\sigma/d\Omega_f$ is the (energy integrated) BE peak yield. The result for U^{29+} + Ne using (5) with the integration limits $E_{min,max} = 2v^2 \cos^2 \vartheta_f \mp 2.5$ keV, both experimentally and theoretically, is shown in Fig.2. From a calculation using (3) instead it is evident that (5) with the restricted integration limits breaks down for $\vartheta_f < 60°$.

In order to test the independence of the Ramsauer-Townsend structures from the chosen frame of reference, detailed BE peak measurements on 0.1 MeV/amu $Ne^0 \to$ He were carried out, frame-transformed via the exact formulae (2) and (3) and compared to equivelocity EL experiments on the reversed collision system, $He^0 \to$ Ne. Agreement within error bars was found both for the singly differential and doubly differential cross sections except at very small electron energies where stray fields complicate the detection [9]. These measurements provide the first experimental proof of the validity of the 'method of inverse kinematics' in the case of neutral collision partners.

Inverse kinematics also relate the soft electron peak to the electron loss cusp, both features originating from low-energy electrons in the parent-atom reference frame. A theory well suited for the emission of low-energy electrons is the strong potential second Born (SB2) approximation, successfully applied to calculations on the electron loss cusp induced by neutral perturbers [10]. In that case, one has to fully account for the parent atom field in the electronic final state, while the influence of the perturber potential can safely be neglected (this is, however, not the case for charged perturbers which require a continuum-distorted-wave-type theory [11, 12]). The SB2 cross section for target electron emission is given by [12, 13]

$$\frac{d^2\sigma^{SB2}}{dE_f d\Omega_f} = \frac{k_f}{v} \int d\mathbf{q}'\, \delta(E_f - E_i^T + \mathbf{q}'\mathbf{v}) \quad (6)$$

$$\cdot \left| \int d\mathbf{k}\, \varphi_f^{*T}(\mathbf{k}+\mathbf{v})\, f(\mathbf{q}'+\mathbf{k},\mathbf{k})\, \varphi_i^T(\mathbf{q}'+\mathbf{k}+\mathbf{v}) \right|^2$$

which differs from the EIA cross section (1) by the additional weighting factor φ_f^{*T}, the Fourier transform of the final target continuum eigenstate.
$f(\mathbf{p},\mathbf{k})$ is the (off-shell) amplitude for electron scattering from the projectile field with change of the momentum from $\mathbf{p}$ to $\mathbf{k}$. It may, similarly as in the EIA theory, be replaced by the elastic scattering amplitude for some average momentum [13]. To (6), the cross section for the simultaneous projectile and target ionisation has to be added incoherently [10, 12].

In Fig.3 the spectra of SE electrons from 0.1 MeV/amu Ne^0 on He are shown for forward ($\vartheta_f = 0°$) and backward ($\vartheta_f = 180°$) emission. Comparison is made with experimental 0° data [9, 14] and satisfactory agreement is found up to $E_f \approx 35$ eV where experiment shows the onset of the ELC cusp. From Fig.3 where also results for C^0 and Ar^0 impact are included, it can be seen that there is not necessarily an increase of intensity with increasing projectile nuclear charge, in contrast to the behaviour for charged projectiles (see e.g. [15]). Also the SE peak asymmetry does not show a systematic increase with perturber nuclear charge as would be expected from charged perturbers. Table 1 depicts the 0°/180° SE peak asymmetry for small E_f as obtained from the singly inelastic (SI) contribution (6). For the simultaneous projectile and target ionisation (the doubly inelastic (DI) contribution) a first-order Born treatment is sufficient which does not show any asymmetry for $E_f \to 0$. Hence inclusion of the DI term reduces the asymmetry considerably.

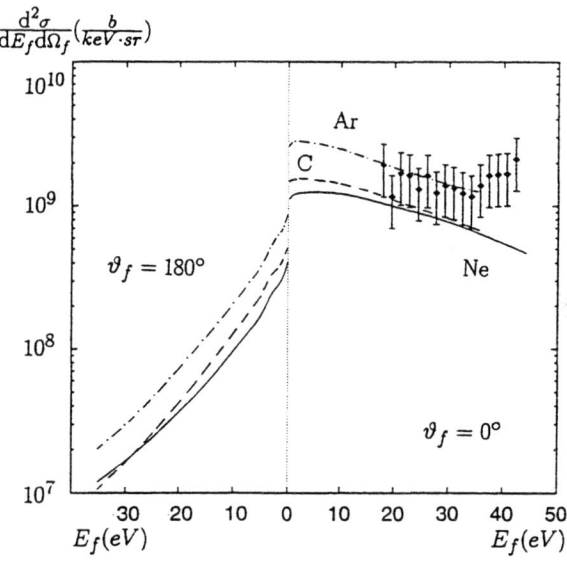

Fig. 3. Doubly differential cross section for soft electron emission in collision of 0.1 MeV/amu C^0, Ne^0, Ar^0 with He as a function of E_f at $\vartheta_f = 0°$ (right-hand side) and 180° (left-hand side). ◇, experimental data for Ne^0 from Jalowy et al [9, 14]. Theory: EIA (SI + DI) for —— Ne^0, - - - - C^0, — · — · — Ar^0.

	SI	SI + DI
C	36.11	2.90
Ne	15.35	2.85
Ar	15.00	3.02

Table 1. Asymmetry $d^2\sigma(0°)/d^2\sigma(180°)$ at $E_f = 0.136$ eV for 0.1 MeV/amu C^0, Ne^0, Ar^0 + He collisions.

A study of the velocity dependence of the asymmetry for the test case Ne^0 + He shows a smooth decrease of the total (SI + DI) asymmetry with v from v_0 onward where $v_0 \sim 1.5$ a.u. is close to the limit of validity of the SB2 theory (the orbiting velocity of a He electron). If, however, an experiment would be designed which allows to separate the SI contribution, theory would predict a maximum of the (SI) SE peak asymmetry near $v = 2.4$ a.u.

The complete angular distribution of the soft electrons from 0.1 MeV/amu Ne^0 + He collisions is shown in Fig.4. A comparison with measurements for $E_f = 20$ eV indicates that SB2 is able to describe the experimental data at least up to $\vartheta_f \sim 120°$. When the collision velocity is varied, a change in the angular pattern is observed (see Fig.4): For small velocities, the electrons are predominantly ejected into the direction of the projectile (due to the effect of electron-projectile attraction at low v), while at the higher velocities the electrons are preferrably ejected perpendicular to the beam direction (following the ridge between projectile and target).

Such a behaviour is in accord with the findings for charged projectiles as was recently investigated experimentally for electrons with $E_f \leq 20$ eV from 0.2 - 1.3 MeV p + He colisions [16]. Further experiments with heavy neutral projectiles are highly desirable to support these theoretical conjectures on the angular dependence of the soft electrons.

Acknowledgments

I should like to thank T.Jalowy and Th. Weber for discussions and for the communication of unpublished experimental data. I am also very grateful to the members of the Mathematical Institute of the LMU Munich for their support.

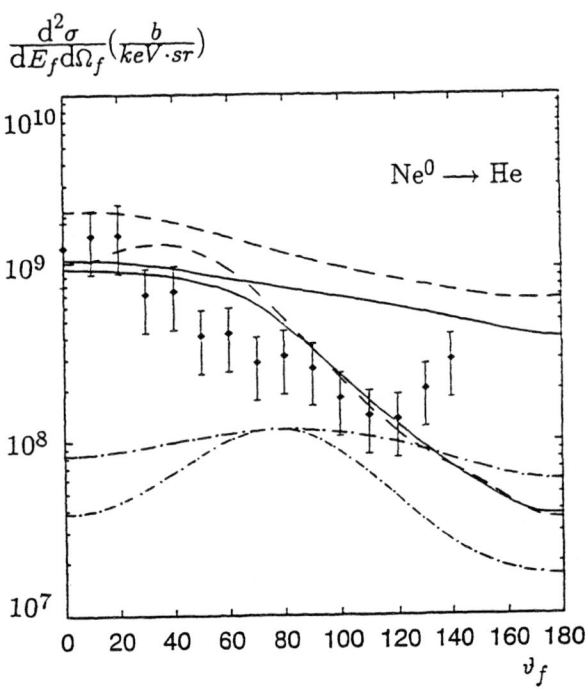

Fig. 4. Doubly differential cross section for 0.136 eV (upper curves) and 20 eV (lower curves) electrons emitted in Ne^0 + He collisions as a function of emission angle ϑ_f at 0.05 MeV/amu ($v = 1.415$ a.u., - - -), 0.1 MeV/amu ($v = 2$ a.u., ———) and 1 MeV/amu ($v = 6.328$ a.u., – · – · –). ◊, experimental data for $v = 2$ a.u. and $E_f = 20$ eV [9, 14]. Theory: EIA (SI + DI).

References

1. Duncan M.M. et al, *Phys.Rev.A* **19**, 49 (1979)
2. Reinhold C.O. et al, *Phys.Rev.Lett.* **66**, 1842 (1991)
3. Bechthold U. et al, *Phys.Rev.A* **58**, 1971 (1998)
4. Bechthold U., Thesis, University of Frankfurt (1996)
5. Wang J. et al, *Phys.Rev.A* **44**, 7243 (1991)
6. Jakubaßa D.H., *J.Phys.B* **13** 2099 (1980)
7. Lucas M.W. et al, *Int.J.Mod.Phys.A* **12**, 305 (1997)
8. Kuzel M. et al, *J.Phys.B* **27**, 1993 (1994)
9. Jalow T. et al, *Phys.Rev.A* **61**, 022714 (2B000)
10. Jakubaßa-Amundsen D.H., *J.Phys.B* **26**, 2853 (1993)
11. Tribedi L.C. et al, *J.Phys.B* **31**, L369 (1998)
12. Jakubaßa-Amundsen D.H., *Eur.Phys.J.D* **10**, 319 (2000)
13. Hartley et al, *J.Phys.B* **20**, 3811 (1987)
14. Jalowy T., Private Communication
15. Pedersen J.O.P. et al, *J.Phys.B* **24**, 4001 (1991)
16. Weber Th. et al, submitted to *J.Phys.B* (2001)

COMPTON PROFILE OF MULTIPLY-IONIZED FLUORINE ATOMS

K. R. Karim,* and C. P. Bhalla**

*Department of Physics, Illinois State University, Normal, IL 61790-4560, USA
**Department of Physics, Kansas State University, Manhattan, KS 66506, USA

We have calculated Compton profiles of neutral and ionized fluorine atoms using impulse approximation. The radial single-electron wave functions were obtained from Hartree-Fock atomic model. The Compton profiles of individual orbitals and the total profiles are presented for fluorine atoms with electronic configurations $1s^2 2s^2 2p^n$, with $n = 5$, 3, and 1. The results from the present calculation are compared with those available in the literature for neutral fluorine. The theoretical values of Compton profiles presented here could be used to interpret cross sections of recent ion-atom collision experiments involving multiply-ionized fluorine atoms. The dependence of the Compton profiles on the degree of ionization is presented.

INTRODUCTION

X-rays scattered from moving electrons bound in an atom reflect the momentum distribution of the electrons. The Doppler broadened profile of the scattered x-rays is known as the Compton profile. A knowledge of the Compton profile is needed to determine the doubly differential electron production cross section (DDCS) in ion-atom collision experiments. Though theoretical values of Compton profile for neutral atoms are available in the literature [1-2], no calculations have been reported, to the best of our knowledge, on ionized atoms.

If $\chi(p)$ is the wave function of the atomic electrons in the momentum space then the Compton profile in impulse approximation can be written as [1-4]

$$J_{nl}(q) = \frac{1}{2} \int_q^\infty |\chi_{nl}(p)|^2 p\,dp. \quad (1)$$

The $\chi_{nl}(p)$ can be obtained from the radial wave funtion R_{nl} by the Fourier transformation

$$\chi_{nl}(p) = \sqrt{\frac{2}{\pi}} \int_0^\infty R_{nl}(r) j_l(pr) r^2 dr, \quad (2)$$

where $j_l(pr)$ are the spherical Bessel functions of the first kind.

If the differential cross section $\frac{\partial\sigma}{\partial\Omega}$ is known for electron-ion collision, then the DDCS of ion-atom collision in the projectile frame can be expressed as [5-6]

$$\frac{\partial^2 \sigma}{\partial\Omega\partial\epsilon} = \left(\frac{\partial\sigma(\theta)}{\partial\Omega}\right)\left(\frac{J(q)}{q + V_p}\right) \quad (3)$$

where V_p is the velocity of the projectile and $q = \sqrt{2}\sqrt{\epsilon + E} - V_p$, and ϵ is the energy of the incident electron and E is the ionization energy of the target electron. Recently there has been several ion-atom collision experiments involving variously ionized fluorine atoms. The purpose of this paper is to report on a calculation of Compton profile of selected multiply-ionized fluorine atoms. These theoretical data can be used to determine the DDCS in the rest frame of the target atoms.

RESULTS AND DISCUSSION

The single-electron radial wave functions were obtained by using Hartree-Fock atomic model [7]. The Compton profile $J(q)$ for each orbital was then obtained by using equation (1). We have verified that the Compton profiles reported here satisfy the condition

$$2\int_0^\infty J_{nl}(q)dq = 1.$$

In Table 1 and Table 2 we list, respectively, the Compton profile of F^{2+} and F^{4+} ions with electronic configurations $1s^2 2s^2 2p^n$, with $n = 3$, and 1. The values of Compton profile for neutral fluorine have been reported by Biggs et al [1]. There is excellent agreement between our calculation and the values reported by Biggs et al [1] except at large q where the value of $J(q)$ is negligible. For example, the value of the total $J(q)$ reported by Biggs et al for neutral fluorine atoms at $q = 0, 1, 2, 3$, and 4 are, respectively, 2.75, 1.71, 0.615, 0.279, and 0.163 (in a.u.). The corresponding values from our calculations are 2.745, 1.710, 0.6168, 0.2786, and 0.1625, respectively.

Table 1. The values of q (in units of $me^2/\hbar$) and $J_{nl}(q)$ and total $J(q)$ (in units of $\hbar^2/me^2$) for F^{2+} ions with electronic configuration $1s^2 2s^2 2p^3$.

q	$1s^2$	$2s^2$	$2p^3$	TOTAL
0.0	.1001E+00	.4512E+00	.2468E+00	.1843E+01
0.1	.1001E+00	.4476E+00	.2468E+00	.1836E+01
0.2	.9994E-01	.4370E+00	.2467E+00	.1814E+01
0.3	.9972E-01	.4200E+00	.2464E+00	.1779E+01
0.4	.9942E-01	.3974E+00	.2455E+00	.1730E+01
0.5	.9903E-01	.3704E+00	.2438E+00	.1670E+01
0.6	.9856E-01	.3401E+00	.2409E+00	.1600E+01
0.7	.9801E-01	.3079E+00	.2367E+00	.1522E+01
0.8	.9738E-01	.2750E+00	.2310E+00	.1438E+01
0.9	.9667E-01	.2425E+00	.2238E+00	.1350E+01
1.0	.9589E-01	.2112E+00	.2152E+00	.1260E+01
1.2	.9412E-01	.1548E+00	.1946E+00	.1082E+01
1.4	.9208E-01	.1090E+00	.1712E+00	.9157E+00
1.6	.8980E-01	.7414E-01	.1472E+00	.7695E+00
1.8	.8732E-01	.4917E-01	.1242E+00	.6455E+00
2.0	.8466E-01	.3214E-01	.1032E+00	.5433E+00
2.5	.7742E-01	.1196E-01	.6242E-01	.3660E+00
3.0	.6969E-01	.6949E-02	.3679E-01	.2637E+00
4.0	.5426E-01	.6237E-02	.1285E-01	.1596E+00
5.0	.4056E-01	.5388E-02	.4769E-02	.1062E+00
6.0	.2949E-01	.3964E-02	.1908E-02	.7264E-01
7.0	.2110E-01	.2707E-02	.8210E-03	.5007E-01
8.0	.1497E-01	.1798E-02	.3774E-03	.3467E-01
10.0	.7536E-02	.7926E-03	.9477E-04	.1694E-01
15.0	.1532E-02	.1289E-03	.6283E-05	.3341E-02
20.0	.3943E-03	.2975E-04	.8095E-06	.8505E-03
30.0	.4596E-04	.3159E-05	.4050E-07	.9837E-04
40.0	.8705E-05	.5778E-06	.4526E-08	.1858E-04
50.0	.2194E-05	.1439E-06	.1035E-08	.4678E-05
75.0	.1211E-06	.7757E-08	.2813E-10	.2578E-06

In Fig. 1 (A) we compare the total Compton profiles of F, F^{2+}, and F^{4+} ions. In Fig. 1 (B) we have plotted the Compton profiles the 2p-electron of F, F^{2+}, and F^{4+} ions. Fig. 1 (C) and 1 (D) shows the Compton profiles of F^{4+} and F^{2+}, respectively. It can be seen that the

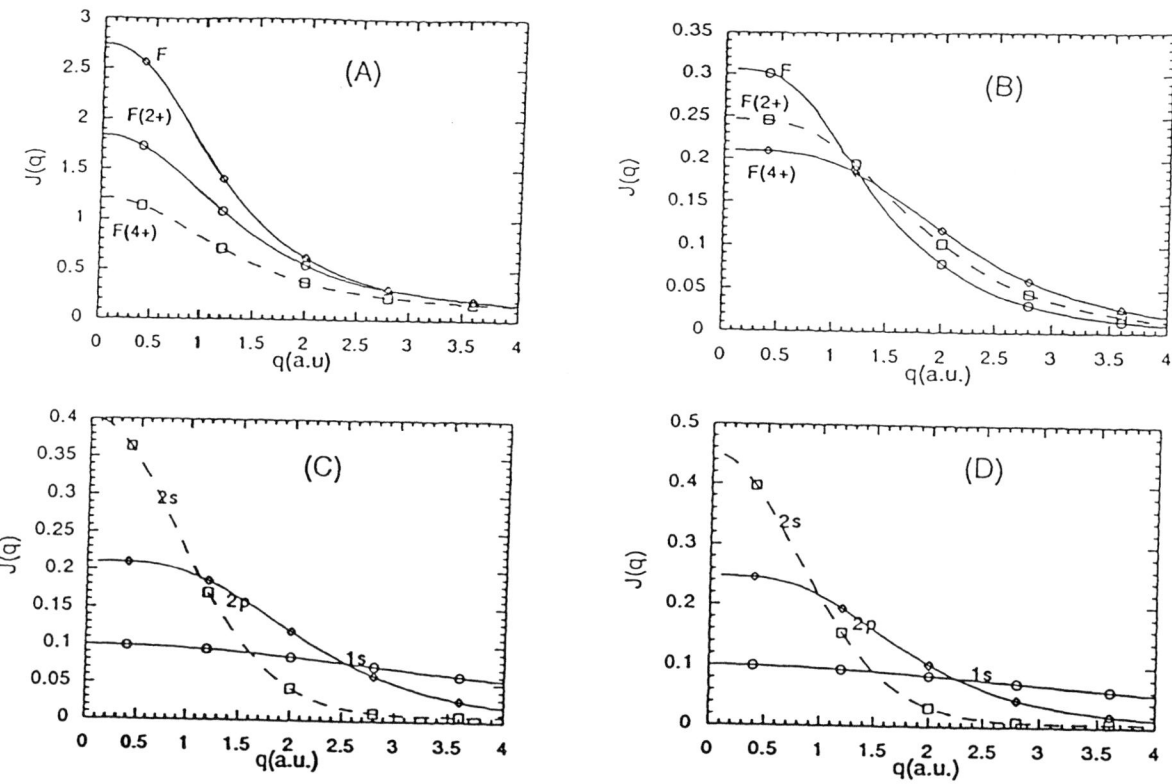

Fig. 1. The total Compton profile (a.u.) as a function of momentum (a.u.). (A) The total Compton profile: the circle, the square, and the diamond represent, respectively, the F, F^{2+}, and F^{4+} ions; (B) the Compton profile of 2p-electron: the circle, the square, and the diamond represent, respectively, the F, F^{2+}, and F^{4+} ions; (C) the Compton profile of F^{4+} ions; (D) the Compton profile of F^{2+} ions.

Table 2. The values of q (in units of $me^2/\hbar$) and $J_{nl}(q)$ and total $J(q)$ (in units of $\hbar^2/me^2$) for F^{4+} ions with electronic configuration $1s^2 2s^2 2p^1$.

q	$1s^2$	$2s^2$	$2p^1$	TOTAL
0.0	.9953E-01	.4020E+00	.2100E+00	.1213E+01
0.1	.9949E-01	.3995E+00	.2100E+00	.1208E+01
0.2	.9937E-01	.3920E+00	.2100E+00	.1193E+01
0.3	.9916E-01	.3799E+00	.2098E+00	.1168E+01
0.4	.9886E-01	.3636E+00	.2095E+00	.1134E+01
0.5	.9849E-01	.3438E+00	.2088E+00	.1093E+01
0.6	.9803E-01	.3211E+00	.2076E+00	.1046E+01
0.7	.9750E-01	.2965E+00	.2057E+00	.9937E+00
0.8	.9689E-01	.2706E+00	.2031E+00	.9381E+00
0.9	.9620E-01	.2442E+00	.1997E+00	.8806E+00
1.0	.9544E-01	.2181E+00	.1954E+00	.8224E+00
1.1	.9461E-01	.1927E+00	.1902E+00	.7648E+00
1.2	.9371E-01	.1685E+00	.1841E+00	.7087E+00
1.4	.9173E-01	.1254E+00	.1699E+00	.6041E+00

Table 2. (Continued)

q	$1s^2$	$2s^2$	$2p^1$	TOTAL
1.6	.8951E-01	.9010E-01	.1535E+00	.5127E+00
1.8	.8708E-01	.6290E-01	.1360E+00	.4360E+00
2.0	.8447E-01	.4300E-01	.1185E+00	.3734E+00
2.5	.7734E-01	.1666E-01	.7913E-01	.2671E+00
3.0	.6971E-01	.8763E-02	.5000E-01	.2069E+00
3.5	.6196E-01	.7317E-02	.3074E-01	.1693E+00
4.0	.5440E-01	.7250E-02	.1875E-01	.1420E+00
4.5	.4727E-01	.7036E-02	.1147E-01	.1201E+00
5.0	.4073E-01	.6470E-02	.7093E-02	.1015E+00
6.0	.2965E-01	.4844E-02	.2834E-02	.7182E-01
7.0	.2122E-01	.3327E-02	.1210E-02	.5031E-01
8.0	.1506E-01	.2211E-02	.5515E-03	.3510E-01
9.0	.1067E-01	.1460E-02	.2671E-03	.2453E-01
10.0	.7582E-02	.9700E-03	.1366E-03	.1724E-01
15.0	.1540E-02	.1561E-03	.8895E-05	.3400E-02
20.0	.3960E-03	.3586E-04	.1140E-05	.8649E-03
30.0	.4615E-04	.3794E-05	.5600E-07	.9994E-04
40.0	.8740E-05	.6927E-06	.6269E-08	.1887E-04
50.0	.2202E-05	.1727E-06	.1552E-08	.4752E-05
75.0	.1216E-06	.9232E-08	.4216E-10	.2617E-06

value of $J(q)$ for the 2p-electron falls off more slowly as the degree of ionization increases. This trend is similar for the total $J(q)$ and the $J(q)$ for the 2s and 1s electrons. Thus it is more likely to find an electron with larger momentum component as the atom is progressively ionized.

Acknowledgment

This work was supported by the Division of Chemical Sciences, Geosciences and Biosciences Division, Office of Basic Energy Sciences, Office of Science, U.S. Department of Energy, Chandra X-Ray Observatory Center Award Number EL9-1006A made by Smithsonian Astrophysical Observatory for and on behalf of NASA, and in part by the University Research Grant Office of Illinois State University.

References

[1] F. Biggs, L.B. Mendelsohn, and J. B. Mann, Atomic Data Nuclear Data Tables **16**, 201 (1975)

[2] A. Harmalkar, P. V. Panat, and D. G. Kanhere, J. Phys. B. **13**, 3075 (1980)

[3] M. Cooper, Advances in Physics, **20** 453 (1971)

[4] M. Cooper, X-ray and Inner-Shell Process page 18, edited by Dunford et al (American Institute of Physics, New York,1999)

[5] S. R. Grabbe, G. Toth, C. P. Bhalla, and P. Richard, Nucl. Instr. And Methods in Physics Research **B 124**, 347 (1997)

[6] G. Toth, S. Grabbe, P. Richard, and C. P. Bhalla, Phys. Rev. A **54**, R4613 (1996)

[7] C. Froese-Fischer, Comput. Phys. Commun. **64**, 369 (1991)

First Studies Of State Selective Electron Capture From Atomic Hydrogen By State Prepared Doubly Charged Ions

D.R. Gillen, D.M. Kearns, D. Voulot, R. W. McCullough and H.B. Gilbody

The Atomic and Molecular Physics Research Division, The School of Mathematics and Physics, Queen's University Belfast, Belfast, BT7 1NN, N. Ireland

Abstract. Electron Capture by multiply charged ions in atomic hydrogen and nitrogen is an important process in planetary atmospheres and in many astrophysical and technological plasmas. Excited state formation and decay results in photon emission in the optical, UV and X-ray regions. In these studies doubly charged ions of carbon or nitrogen and oxygen have been prepared in either ground or metastable states prior to colliding with atomic hydrogen and nitrogen target enabling, for the first time, true state electron capture processes to be investigated. Comparison between present results and existing theoretical calculations have been made. Although the comparison is good for the major reaction channels there is disagreement between experimental results and theory for the minor reaction channels.

INTRODUCTION

The formation of excited states during electron capture collisions by low energy (<25keV amu^{-1}) multiply charged ions is an important process in astrophysical environments and in technological plasmas. For example, electron capture by solar wind ions interacting with cometary gases [1] and planetary atmospheres [2] results in X-ray and extreme UV emissions. The thermal and ionisation balance of astrophysical plasmas is influenced by one-electron capture processes of the type,

$$X^{q+} + H \rightarrow X^{(q-1)+}(n, l) + H^+$$

involving a wide variety of partially stripped ions [3]. These processes are also important in fusion plasmas [4]. It is well known that, at impact velocities (v < 1 au), the electron capture process results in the formation of a relatively small number of excited product states.

There have been a large number of experimental measurements and theoretical calculations of one electron capture by partially stripped ions in atomic hydrogen resulting in both total and state selective cross sections (see for example [5] [6] [7]). In our laboratory we have used the technique of translational energy spectroscopy (TES) to study state selective electron capture (SSEC) by a wide variety of partially ionised species in atomic hydrogen [8]. In the TES technique the difference ΔT between the kinetic energy T_1 of the primary X^{q+} ions and the kinetic energy T_2 of the fast forward-scattered $X^{(q-1)+}$ products is measured. This difference can be expressed as $\Delta T = (T_2 - T_1) = (\Delta E - \Delta K)$ where ΔK is the target recoil energy. Provided that $\Delta E/T_1 \ll 1$ and the scattering is confined to small angles then $\Delta T \approx \Delta E$. Thus, within the limitations of the available energy resolution, an analysis of the $X^{(q-1)+}$ excited product yields in the energy change spectrum enables relative cross sections for the main excited product channels to be determined. However, detailed comparison of experiment and theory has been hindered by the presence of ions in metastable states in the primary ion beams used in the experiments. These metastable ions may have significantly different capture probabilities than their ground state counterparts. In our TES study of SSEC by 2 - 8 keV C^{2+} in H [9] product channels associated with C^{2+} (^3P) metastable ions accounted for more than 60% of the total C^+ product ion yield. Such TES studies indicate that many previous measurements of total electron capture cross sections carried out with ion beams of unknown metastable content in different laboratories should be treated with caution.

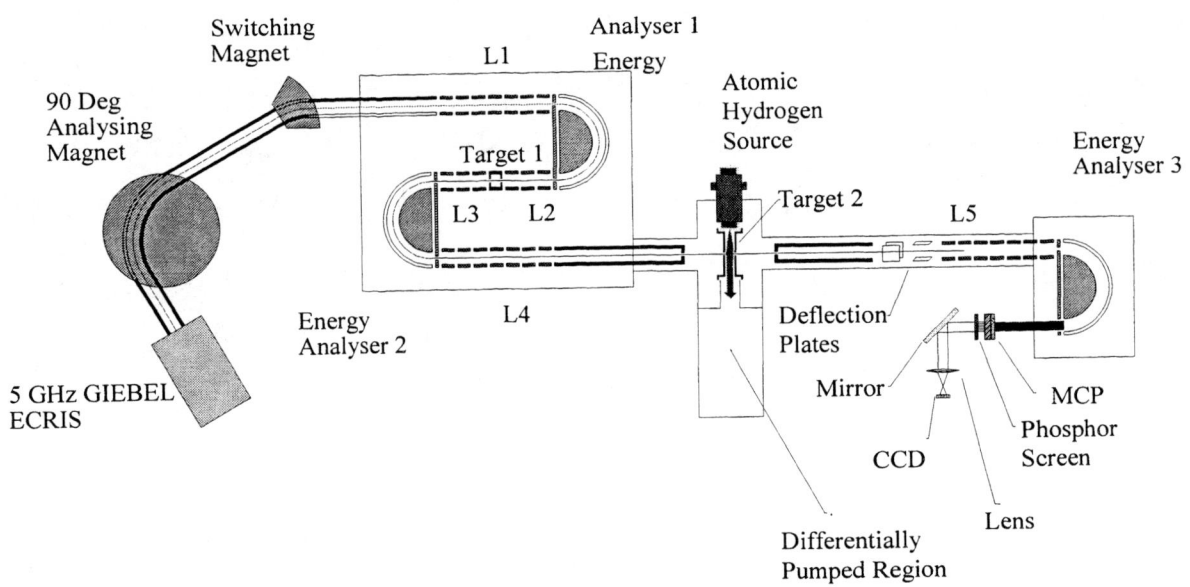

FIGURE 1. Schematic of experimental apparatus used in DTES experiment

EXPERIMENTAL APPROACH

A full description of the DTES experiment has been given previously[10] and only an outline will be given here. The technique used for producing a pure ground or metastable beam will be explained by using the case of N^{3+} as an example. High purity nitrogen gas is fed into a 5 GHz ECR plasma source, see Fig. 1, and a 12 keV beam of N^{3+} is extracted focussed and momentum analysed by a double focusing magnet. After focusing and deceleration by an electrostatic lens system L1, the beam passes through an electrostatic energy analyser EA1 at an energy of 180 eV. The emerging energy resolved beam with an energy spread corresponding to about 1eV FWHM, was accelerated by lens L2 to an energy of 3 keV and enters the first target cell T1. This cell is used to prepare the state selected beam of pure ground or metastable state N^{2+} (n,l) ions by filling T1 with He and selecting the appropriate forward scattered N^{2+} (n,l) ion. These product ions were decelerated by lens systems L3 and L4 before entering the hemispherical analyser EA2. By applying an appropriate biasing voltage to EA2, the N^{3+} ions could be rejected while allowing the passage of the N^{2+} ions. By employing a scanning voltage the translational energy change spectrum obtained from EA2 could be displayed. The excited N^{2+} product ions from the main observed channel

$$N^{3+}\ 1s^22s2p(^3P) + He \rightarrow N^{2+}\ 2s2p^2(^2P) + He^+ (^2S) \quad (2)$$

were selected. This channel, with an energy defect of 13.06 eV, corresponds to peak C in Fig. 2. These excited N^{2+} ions are known to decay rapidly to the ground state so that, within the minimum beam flight

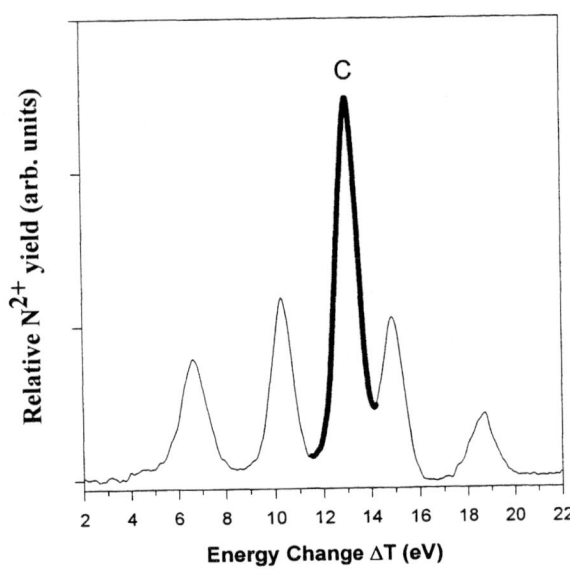

Figure 2. Energy Change spectrum of N^{2+} ions formed in one electron capture by 12 keV N^{3+} ions in collisions with helium

time to the second target gas cell, all the ions were in the $N^{2+}\ 2s^22p\ (^2P^\circ)$ state.

TABLE 1. Main C^+ states formed in one electron capture by 1.5 keV, 3.0 keV and 6.0 keV C^{2+} $(2s^2)$ ^1S ground state ions in H(1s). Percentages of the total yield for individual product channels obtained from the experiment (E) are compared with theoretical values (T) from reference [11].

C^+ Product State	Energy Defect eV	Percentage of Total Yield					
		1.5 keV		3.0 keV		6.0 keV	
		E	T	E	T	E	T
$(2s^22p)$ $^2P^o$	10.78	-	0.0	0.4	0.4	1.3	3.5
$(2s2p^2)$ ^2D	1.49	100	83.4	96.6	78.6	98.7	70.5
$(2s2p^2)$ ^2S	-1.18	-	12.7	-	15.6	-	16.3
$(2s2p^2)$ ^2P	-2.94	-	0.7	-	1.3	-	1.4
$(2s^23s)$ ^2S	-3.67	-	1.3	-	2.1	-	3.4
$(2s^23p)$ $^2P^o$	-5.55	-	2.0	-	3.0	-	4.9

The pure beam of $^2P^o$ ground state ions emerging from EA2 were accelerated and focused by lens system L4 into the main target region T2. A μ-wave driven atomic hydrogen atom source was connected to T2 and fed with hydrogen. After the charge transfer collision the forward scattered ions were decelerated and focused into energy analyser EA3. The energy analysed product N^+ ions from EA3 were counted by a computer controlled position sensitive detector.

RESULTS AND DISCUSSION

In the present work we have studied SSEC at collision energies in the range 800 eV – 6 keV involving state prepared ground state ions of C^{2+} and N^{2+} in atomic hydrogen.

Studies involving C^{2+} ^1S ions

Energy change spectra for one electron capture by ground state C^{2+} ^1S ions in H at 1.5keV, 3keV and 6keV have been obtained. Possible reaction channels are listed in table 1 together with a comparison of the percentage contributions of observed channels with recent close coupling calculations [11]. It is clear from this comparison that whilst experiment and theory agree that the $C^+(2s2p^2)$ ^2D state is the dominant state produced in the one electron capture process there are considerable differences for the other C^+ product states. At 1.5 keV the experimental measurements indicate a negligible contribution from the $(2s^22p)$ $^2P^o$ state whereas theory predicts a contribution of 35%. At 3keV the agreement between theory and experiment is better although there are differences of more than a factor of two for the minor capture channels. At the higher energy of 6 keV the agreement is poor except for the dominant $C^{2+}(2s2p^2)$ ^2D product state and the $(2s^22p)$ $^2P^o$ state. Clearly there is a need for further SSEC measurements on this collision system at other energies and for experimental measurements of the partial cross sections.

Studies involving N^{2+} $^2P^o$ ground state ions

Our previous TES measurements [12] for N^{2+} in H were found to independent of the primary beam content and these cross sections are now well established. For this reason it was possible to normalise our results on previously measured cross sections. These relative contributions of the channels for the capture into specific states have been determined and are shown in Fig. 3. The present results are in excellent agreement with our previous TES measurements obtained with N^{2+} mixed beams [12]. A small contribution from the channel $N^+(2s^22p3p)$ ^3P, with an energy defect of −5.17 eV, was identified by Wilkie et al [12] for collision energies above 6 keV. This channel is not observed in our spectra, even though the statistics and resolution are superior in the present measurements. Calculations for capture into the $(2s2p^3)$ $^3D^o$ channel [13] are also in good agreement with our measurements. A recent calculation considering capture into both the $(2s2p^3)$ $^3D^o$ and $(2s2p^3)$ $^3P^o$ states was performed [14]. There is again good agreement with measurements for the $(2s2p^3)$ $^3D^o$ state, however their results for the $(2s2p^3)$ $^3P^o$ state are one order of magnitude lower than those published by Wilkie et al [12] and also the present measurements.

CONCLUSIONS

These DTES measurements represent the first detailed study of the one-electron capture processes in atomic hydrogen in which the initial state of the projectile ion has been uniquely defined. The results enable for the first time a detailed comparison with theoretical

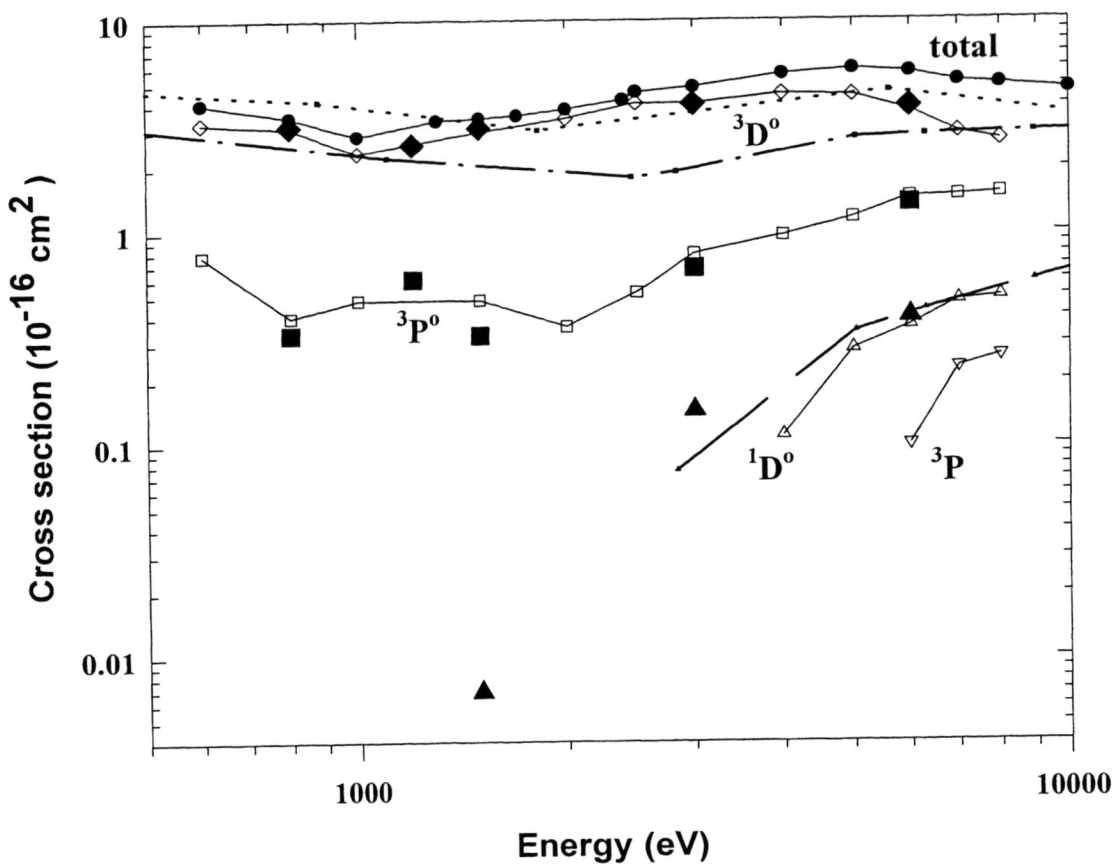

Figure 3. Cross sections for one electron capture into specified states of N^+ in N^{2+}- H(1s) collisions in relation to total cross section σ_{21} for capture into all final states. Experiment: ●, Total; ◊, N^+ $2s2p^3$ ($^3D^o$); □ N^+ $2s2p^3$ ($^3P^o$); Δ, N^+ $2s2p^3$ ($^1D^o$); ∇, N^+ $2s^22p3p$ (^3P); [12]. ◆, N^+ ($^3D^o$); ■, N^+ ($^3P^o$); ▲, N^+ ($^1D^o$); present DTES work. Theory: ----- N^+ ($^3D^o$) [13]; — — — N^+ ($^3D^o$); —— —— N^+ ($^3P^o$); [14].

models and demonstrate the need for more effort in this field. DTES studies are in progress using state prepared beams of metastable ions.

ACKNOWLEDGMENTS

This research was part of a Programme Grant supported by the UK Engineering and Physical Sciences Research Council and forms part of LEIF 2000, an EC Framework V Thematic Network on Low Energy Ion Facilities. The support Department of Education for Northern Ireland and the European Social Fund for the award of Research Studentships is also acknowledged.

REFERENCES

1. Cravens, T. E., *Geophys. Res. Lett.* **24**, 105 (1997)
2. Liu, W. and Schultz, D. R., *Astro. Phys. J.* **526**, 538 (1999)
3. Kingdon, J. B. and Ferland, G. J., *Astro.Phys. J.* **516** L107 (1999)
4. Janev, R.K., IAEA report INDC(NDS)-277 (1993)
5. Phaneuf, R. A., Meyer, F. W. and McKnight, R. H., *Phys. Rev. A* **17** 534 (1978)
6. Havener, C. C., *Accelerator-Based Atomic Physics Techniques and Applications,* edited by S M Shafroth and J C Austin, AIP, New York, 1997 pp. 117
7. Hoekstra, R., Beijers J. P. M., Schlatmann, A. R., Morgenstern, R. and de Heer, F. J., *Phys. Rev. A* **41**, 4800 (1990)
8. Gilbody, H. B., *Advances in Atomic, Molecular Physics and Optical Physics,* edited by B Bederson and H Walther, Academic Press, New York 1994, **32**, pp. 149.
9. McCullough, R. W., Wilkie, F. G. and Gilbody, H. B., *J. Phys. B: At. Mol. Opt. Phys.* **17**, 1373 (1984)
10. McCullough, R. W., *Int. J. Mass Spec* **192**, 141 (1999)
11. Errea, L. F. *et al*, *J. Phys. B: At. Mol. Opt. Phys.* **33**, 1369 (2000)
12. Wilkie, F. G. *et al*, *J. Phys. B: At. Mol. Phys.***18**, 479 (1985)
13. Bienstock, S., Dalgarno, A. and Heil, T., *Phys Rev A* **33**, 2078 (1986)
14. Herrero, B. *et al*, *J. Phys. B: At. Mol. Opt. Phys.***28**, 711 (1995)

Electron capture and loss cross sections for neutral projectiles colliding with atoms and molecules

J.M. Sanders, S.L. Varghese, and C.H. Fleming

Department of Physics, University of South Alabama, Mobile, AL

Abstract. Collisions of neutral projectiles with atoms and molecules arise in a number of situations: in the magnetosphere of the earth, in heating or diagnostics of fusion plasma, in neutral-beam injection into accelerators, and when ion beams come to a stop in solids. Scaling relations for capture cross sections will be presented for collisions involving He targets. Recent results for electron loss in collisions of neutrals with hydrocarbon molecules will also be discussed. It is found that the additivity rule holds for these molecular targets.

INTRODUCTION

Intermediate and high-energy collisions of neutral projectiles with atoms and molecules occur in several situations of interest. In the magnetosphere of the earth, high-energy ions become collisionally neutralized and then are free to leave the magnetosphere. The flux of neutral atoms can therefore serve to image the ion currents where they were created (1). The details of the transport of the neutrals out of the magnetosphere requires a knowledge of collision processes that may re-ionize the neutral atom. In plasma physics, high-energy neutral beams are employed in heating and in obtaining diagnostics of fusion plasma (2). Neutral-beam injection into tandem Van de Graaff accelerators, the work-horses of accelerator mass spectroscopy, has been proposed as an effective means of extending the use of the accelerators to species which do not easily form a negative ion (3). Loss cross sections are of particular relevance to this use of neutral beams. Of course, when an ion comes to a stop in a solid, its final charge state is neutral, so neutral collisions will be relevant to the very broad application of ion beam interactions with solids.

Even the simplest atom-atom collision (H+H) is a multibody problem involving four particles. Theoretical work on the collisions is therefore quite difficult and is also rather sparse. The experimental database is also incomplete even for total cross sections. For hydrogen and helium projectiles a fair amount of work has been done. Of heavier projectiles, only carbon has had total cross sections measured over the full range of energies from a few keV to the MeV range (4). Work has been done in electron emission in neutral-neutral collisions for both light and heavy projectiles (e.g. (5, 6)), and studies have been made of ionization of the target (7, 8).

In the next two sections of this paper, we will discuss scaling relations for electron capture in collisions of neutral projectiles with helium targets and electron loss cross sections for hydrogen atoms colliding with hydrocarbon molecules.

ELECTRON CAPTURE SCALING RELATIONS

Systematic studies of the dependence of the cross sections can lead to simple empirical or semi-empirical scaling rules (for an example, see (9)). These scaling rules, in turn, are very useful to the experimenter as well as the theorist in providing easily-calculated estimates of cross sections for new collision systems or at energies where measurements have not yet been made. Scaling relations for cross sections are often helpful when more detailed theoretical treatments are non-existent or prohibitively difficult. They are also useful when only a ready approximation is required, for example, in making estimates of counting rates for planned experiments.

For capture by neutral projectiles, the variables that will affect the cross section are the projectile velocity, target atomic number, and the electron affinity of the projectile. The cross section for electron capture by a neutral projectile rises rapidly as a function of projectile energy, reaches a maximum, and then falls rapidly for high energies. A relation which gives the position of the maximum of the cross section is the Massey adiabatic parameter

$$\frac{a|\Delta E|}{hv_{max}} \approx 1 \qquad (1)$$

where ΔE is the energy difference between the initial and final states, v_{max} is the projectile velocity at which the maximum occurs, and a is an empirically determined length scale which typically ranges from 3 to 12 Bohr radii (10). For complex target atoms, the target may be left in one of a variety of exited states, so there will be several possible ΔE's. This has the effect of producing double peaks, or shoulders on the main peak. In order to avoid these complications, we will limit ourselves to a simpler target, helium.

Fogel (11) proposed that at energies less than the maximum, the cross section would scale as

$$\sigma_c = \sigma_0 e^{-k\frac{a|\Delta E|}{h v}}. \quad (2)$$

While this expression works well at low velocities, it is certainly not suited to predicting cross sections at or above the cross section maximum. Further, the constant ka has to be determined for each collision system, and it is not obvious how the cross sections might scale with parameters of interest such as the electron affinity of the projectile.

Limiting the collisions to He targets, we have obtained a simple scaling for the electron capture cross sections in terms of the projectile energy and the electron affinity of the projectile. Figure 1 shows electron capture cross sections for H, C, O, F, and Cl atoms colliding with He at energies ranging from 1 keV/u to 25 keV/u. These energies are below the cross section maximum. The common curve is given by

$$\sigma_c = (8.43 \times 10^{-19} \text{cm}^2)(E/N) A_e^{1.66} \quad (3)$$

where E/N is the energy in keV divided by the mass number and A_e is the electron affinity in eV. As can be seen in Fig. 1, the data are described reasonably well by Eq. 3 to within a factor of two. Equation 3, like Eq. 2, is limited to energies below the maximum in the cross section. Insufficient data are available to provide similar scaling relations at energies near or above the cross section maximum.

ELECTRON LOSS

As discussed in the Introduction, electron loss from neutral projectiles is a process that is of interest in a number of fields. We describe measurements investigating electron loss from atomic hydrogen projectiles colliding with molecular hydrogen (H_2), methane (CH_4), acetylene (C_2H_2), ethylene (C_2H_4), ethane (C_2H_6), and propane (C_3H_8) in the 60- to 120-keV energy range. The primary purpose of these measurements was to test the simple Bragg additivity rule for molecular cross sections.

In its simplest form, the Bragg additivity rule states that the cross section for a collision between a projectile and a molecular target is simply the sum of the cross

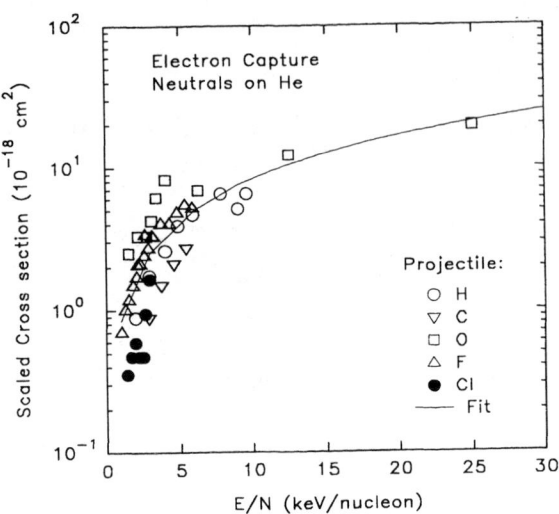

FIGURE 1. Electron loss capture sections for H, C, O, F, and Cl atoms colliding with helium. The cross sections have been scaled by dividing by $A_e^{1.66}$. The curve is given by Eq. 3 again scaled by dividing by $A_e^{1.66}$. The data are from Hird et al. (12, 13) and the tabulations of Wu (14, 15).

sections for collisions between the projectile and the constituent atoms of the molecule (16). If additivity holds, the cross section for loss in a collision with a C_nH_m molecule would be

$$\sigma_l^{C_nH_m} = n\sigma_l^C + m\sigma_l^H \quad (4)$$

where σ_l^C and σ_l^H are the cross sections for loss in a collision with a carbon and hydrogen atom, respectively. Equation 4 describes a plane where the independent variables are n and m. If the measured loss cross sections are well-described by such a plane, we may conclude that additivity holds for these collision partners and in this energy range.

Experiment

Cross sections were obtained from charge state fractions resulting from bombardment of the target gases with 60- to 120-keV protons. The protons were accelerated by a 150 kV Cockcroft-Walton accelerator and directed into a differentially-pumped gas cell containing the target gas. The gas cell was 3.49 ± 0.37 cm long with the uncertainty reflecting the end-correction for gas streaming from the entrance and exit apertures. The pressure in the gas cell was monitored by a capacitance manometer. After leaving the gas cell, the projectiles passed between a pair of electrostatic deflector plates which steered H^+ projectiles onto a surface barrier detector, while the H^0 projectiles continued undeflected to a second surface barrier detector. Comparison of the response of these detectors to both the charged and neutral beams indicated that they had equal efficiencies for both projectile species. Signals from the two detectors were amplified and pulse-height analyzed by multichannel analyzers. Total counts of the two charge states were used to compute the charge-state fractions.

The charge-state fractions were measured as a function of gas cell pressure over a wide range of pressures, from very low, single-collision conditions, to very high, equilibrium conditions. The resulting data were then fit using a two-charge state model to obtain cross sections (17):

$$f^0 = \frac{1}{1+R} + \left(\frac{R}{1+R} - f_0^0\right)\exp[-\sigma_c(1+R)x] \quad (5)$$

where f^0 is the charge-state fraction for the neutral species, f_0^0 is the background neutral fraction, R is the ratio of the cross section for loss to the cross section for capture, σ_c is the cross section for capture by H^+, and x is the areal number density of the target. When performing the non-linear least-squares fit, f_0^0, σ_c, and R were the fitting parameters. The fit gave σ_c directly, and $\sigma_l = R\sigma_c$. This two-state model does not include any contributions from H^-, since cross sections involving the negative ion are at least an order of magnitude smaller than the corresponding charge states for H^+ and H^0 (18, 19).

The uncertainties of the fitting parameters σ_c and R provided the relative statistical uncertainties for the cross sections, and absolute uncertainties were obtained by taking the uncertainty in the gas cell length and gas cell pressure into account. In addition, each cross section measurement was repeated at least once, and a weighted average of the separate measurements was used in all subsequent analyses.

Results

For each projectile energy, a plane given by Eq. 4 was fit to the measured loss cross sections for the six targets. The independent variables in the fit were n and m, while the adjustable parameters resulting from the fit were the two effective cross sections σ_l^C and σ_l^H. The effective cross sections are given in Table 1.

In order to display the three dimensional dataset and fitted plane on a two dimensional page, we rotate the fitted plane about the vertical axis until it is perpendicular to the plane of the paper. The data points are then projected onto the plane of the paper. The result of this projection is that a point $(n,m,\sigma_l^{C_nH_m})$ is mapped to a point $(D,\sigma_l^{C_nH_m})$ where D is given by

$$D = n\cos\theta + m\sin\theta \quad (6)$$

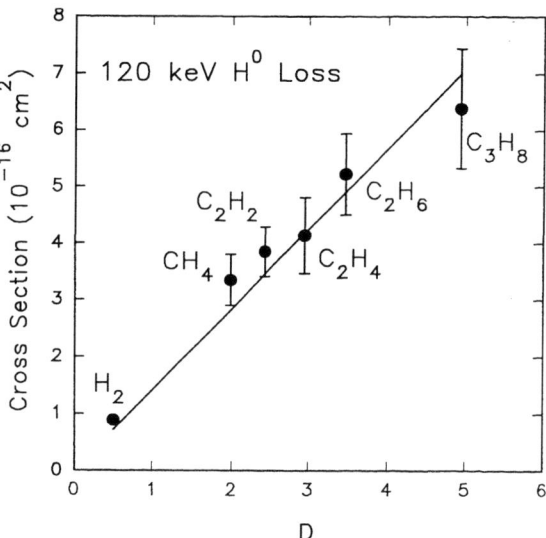

FIGURE 2. Electron loss cross sections for 120 keV H^0 colliding with hydrogen and hydrocarbons. The line presents the fitted plane. The view has been rotated about the vertical axis, so the fitted plane is perpendicular to the plane of the paper.

where $\theta = \tan^{-1}(\sigma_l^H/\sigma_l^C)$. By plotting in this fashion, any deviations of the data from the fitted plane can be easily seen. The result of such a plot of the data for 120 keV H^0 is shown in Figure 2 and similar fits and graphs were obtained at the other energies. As can be seen in the figure, the data are well described by the fitted plane, from which we can conclude that the simple additivity rule holds for these collisions.

This conclusion is in accord with expectations based on the "exit effect" model which Bissinger et al. developed to account for deviations from additivity in electron capture by protons (16). In that model, additivity breaks down when the projectile suffers a second collision with an atom on its way out of the molecule. For example, in the case of electron capture by protons, the proton may capture an electron but then lose it in a collision with another atom as it leaves the molecule. The effect is large

Table 1. Effective loss cross sections for H^0 colliding with C and H atoms.

Energy (keV)	σ_l^C (10^{-16}cm^2)	σ_l^H (10^{-16}cm^2)
60	1.67 ± 0.61	0.44 ± 0.24
75	1.79 ± 0.28	0.37 ± 0.11
90	1.66 ± 0.38	0.41 ± 0.15
105	1.36 ± 0.42	0.36 ± 0.17
120	1.37 ± 0.30	0.36 ± 0.12

in this example, because the loss cross section is much larger than the capture cross section. In the present case of electron loss from H^0 colliding with molecules, the exit effect would occur if the H^0 were to lose an electron but then capture another in a second collision on its way out of the molecule. However, since the cross section for capture is rather smaller than the loss cross section, it would be expected that capture within the molecule would not be sufficiently probable to cause deviations from additivity.

CONCLUSION

We have reported a simple empirical scaling rule for electron capture by low-energy (less than 25 keV/u) neutral projectiles colliding with He. The cross sections scale linearly with the projectile energy and as a power of the electron affinity of the projectile. More data in the 50- to 500-keV/u energy range will be needed, if scaling rules are to be developed for higher energies.

We have also reported measurements of electron loss in collisions of H^0 with molecular hydrogen and hydrocarbon gases. It was found that the loss cross sections obeyed the simple additivity rule, and this allowed the extraction of effective cross sections for electron loss in collisions of H^0 with hydrogen and carbon atoms.

ACKNOWLEDGMENT

This work was supported in part by the University of South Alabama Research Council and the University of South Alabama University Committee on Undergraduate Research.

REFERENCES

1. Gruntman, M., *Rev. Sci. Instrum.* **68**, 3617–3656 (1997).
2. Hutchinson, I. H., *Principles of Plasma Diagnostics*, New York: Cambridge University Press, 1987.
3. Litherland, A. E., and Kilius, L. R., *Nucl. Instrum. Meth. Phys. Res. B* **123**, 18–21 (1997).
4. Sanders, J. M., Varghese, S. L., Datz, S., Deveney, E., Krause, H. F., Shinpaugh, J. L., Vane, C. R., and DuBois, R. D., in *Application of Accelerators in Research and Industry: Proceedings of the Fourteenth International Conference*, edited by J. L. Duggan, and I. L. Morgan, Woodbury, NY: AIP, 1997 pp. 93–96. AIP Conf. Proc. No. 392.
5. DuBois, R. D., and Kövèr, A., *Phys. Rev. A* **40**, 3605–3612 (1989).
6. Heil, O., DuBois, R. D., Maier, R., Kuzel, M., and Groeneveld, K.-O., *Phys. Rev. A* **45**, 2850–2858 (1992).
7. González, A. D., Fregenal, D., Suárez, S., Wolff, W., and Wolf, H., *J. Phys. B* **31**, L257–L263 (1998).
8. Fregenal, D., Suárez, S., Bernardi, G., and González, A. D., *J. Phys. B* **33**, 3345–3362 (2000).
9. Schlachter, A. S., Stearns, J., Graham, W. G., Berkner, K. H., Pyle, R. V., and Tanis, J., *Phys. Rev. A* **27**, 3372–3374 (1983).
10. Hasted, J. B., *Physics of Atomic Collisions*, second edn., New York: American Elsevier Publishing, 1972.
11. Fogel, Y. M., *Sov. Phys.–Usp.* **3**, 390–416 (1960).
12. Hird, B., Rahman, F., and Orakzai, M. W., *Can. J. Phys.* **66**, 972–977 (1988).
13. Hird, B., Rahman, F., and Orakzai, M. W., *Phys. Rev. A* **37**, 4620–4624 (1988).
14. Wu, W. K., Huber, B. A., and Wiesemann, K., *At. Data Nucl. Data Tables* **40**, 57 (1988).
15. Wu, W. K., Huber, B. A., and Wiesemann, K., *At. Data Nucl. Data Tables* **42**, 157–185 (1989).
16. Bissinger, G., Joyce, J. M., Lapicki, G., Laubert, R., and Varghese, S. L., *Phys. Rev. Lett.* **49**, 318–322 (1982).
17. McDaniel, E. W., Mitchell, J. B. A., and Rudd, M. E., *Atomic Collisions: Heavy Particle Projectiles*, New York: John Wiley & Sons, 1993.
18. Stier, P. M., and Barnett, C. F., *Phys. Rev.* **103**, 896–907 (1956).
19. Toburen, L. H., Nakai, M. Y., and Langley, R. A., *Phys. Rev.* **171**, 114–122 (1968).

EUV Photon Emission Spectroscopy For Diagnostics Of Single And Multiple-Electron Capture Processes In 80 keV Ar^{8+} + N_2 Collisions

A. Siems[$,#], H. Merabet[#*], R. Bruch[#], V. Golovkina[#], G. Hinojosa[#], R. Phaneuf[#], J. Hanni[#], S. Bliman[&], and A.G. Trigueiros[$]

[#] *Department of Physics, University of Nevada, Reno, NV 89557 USA*
[$] *Instituto de Fisica, Universidade Estadual de Campinas, 13083-970 Campinas, Sao Paulo, Brazil*
[&] *Department de Physique, Universitéde Marne la Vallé, 93166 Noisy LeGrand, France*

Abstract. We present the first EUV emission spectra following single and multiple-electron capture for the 80 keV Ar^{8+} + N_2 multi charged ion-molecular collision system. Our analysis provides evidence of single and double capture processes leading to radiative deexcitation of Ar^{7+*} (nl) n=3-6 and Ar^{6+*} (3lnl') n=3-5 projectile states. Furthermore we have identified numerous N^{q+} (q=2,3) ionic target lines in the EUV spectral range (10-80 nm), owing to multiple ionization of the N_2 molecule and consecutive dissociation plus excitation of the fragments. The subsequent photon emission arises from high lying states of the target such as N^{2+*} ($1s^22s2lnl'$) ^{2,4}L, n=2-10 and N^{3+*} ($1s^22lnl'$) ^{1,3}L, n=2-5 with excitation energies up to 52 eV and 78 eV, respectively. The N^{2+} and N^{3+} excited dissociated fragments are due to multiply electron capture processes involving up to six electrons.

INTRODUCTION

EUV photon emission spectrometry of one- and two-electron processes in fast ion-atom and electron-atom collisions with few electron targets such as helium has attracted considerable interest over the past decade [1,2]. Recently, slow collisions involving highly charged ions have been studied including the Ar^{8+}+He system [3-5]. Furthermore spectroscopic investigations of ion-molecule systems such as Ar^{8+}+H_2 and Ar^{8+}+N_2 have been performed to elucidate multiple electron charge transfer, molecular dissociation, and excitation dynamics [6-9].

In this work we present new data obtained from Ar^{8+}+N_2 single collisions utilizing EUV spectroscopy. This collision system is of particular interest due to the many electron molecular target, which allows us to study multiple electron capture reactions and subsequent fragmentation and excitation of highly charged N_2^{q+} molecular complexes. These highly charged N_2^{q+} molecular target ions are unstable and rapidly dissociate via Coulomb explosion processes leading to highly excited ionic nitrogen states. EUV photoemission spectroscopy is the ideal tool for selectively analyzing these excited ionic N^{q+} states. These states provide detailed collisional information about new dissociation channels following multiple electron capture processes with many electron molecular targets. In a recent publication Bliman et al. [9] have addressed the importance of Non-Frank-Condon dissociation mechanisms caused by large space and time dependent electric field effects which may lead to large intermolecular interactions.

EXPERIMENTAL SETUP

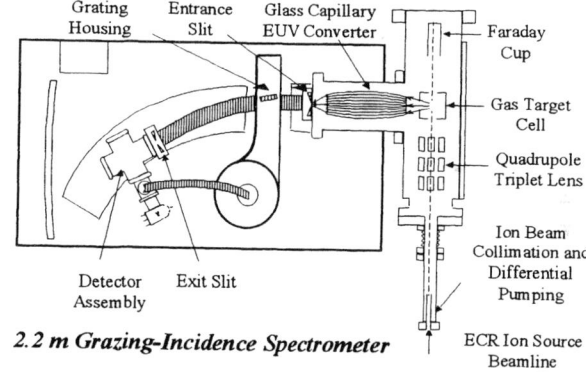

FIGURE 1. Experimental set-up for EUV spectroscopy of highly charged ion-molecule collisions.

In order to obtain a deeper understanding of the many-electron correlation effects and molecular fragmentation in Ar^{8+} + N_2 collisions, we have measured EUV spectra over a large wavelength range from 10 to 80 nm at a projectile energy of 80 keV (v = 0.29 a.u.). The experiments have been performed at the University of Nevada, Reno, 14-Ghz Electron Cyclotron Resonance (ECR) multi-charged ion source facility.

---

* Corresponding author: H. Merabet, email: hocine@physics.unr.edu

CP576, *Application of Accelerators in Research and Industry – Sixteenth Int'l. Conf.*, edited by J. L. Duggan and I. L. Morgan
© 2001 American Institute of Physics 0-7354-0015-6/01/$18.00

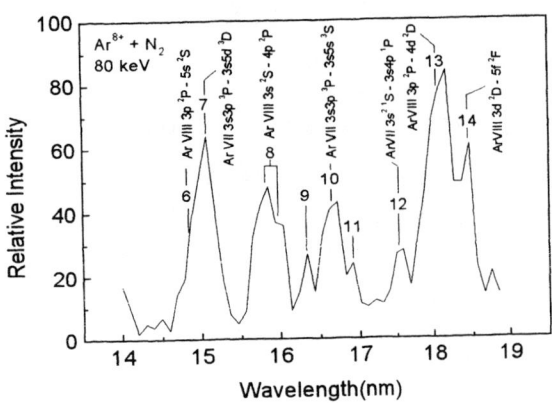

FIGURE 2. EUV Spectrum for $Ar^{8+}+N_2$ collisions in the wavelength range from about 14-19 nm. The dominant spectral features arise from single and double electron capture processes.

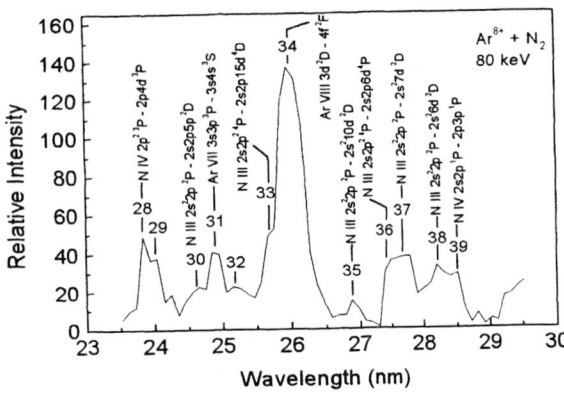

FIGURE 3. EUV Spectrum for $Ar^{8+}+N_2$ collisions in the wavelength range from about 23-29 nm. The dominant spectral features arise from ionic projectile and target lines.

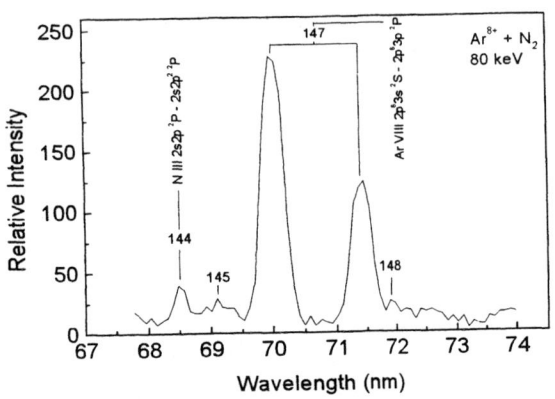

FIGURE 4. EUV Spectrum for 80 keV $Ar^{8+}+N_2$ collisions in the wavelength range from about 68-74 nm. The dominant spectral features arise from ionized projectile and target lines.

As can be seen in Fig. 1, the experimental apparatus is composed of a 2.2. m grazing incidence monochromator (Mc Pherson Model 247), a target chamber with a differentially pumped gas cell, collimator system, quadrupole lens for focusing and steering of the ion beam, and a Faraday cup for ion beam normalization. The target cell is operated at a pressure of about 0.7 mTorr and the background pressure of the vacuum chamber is approximately 5×10^{-7} Torr. The gas pressure in the target cell is controlled via a capitance manometer (Barocel 655). The emitted photons are observed under an angle of 90^0 with respect to the incident ion-beam direction and the measured line intensities have been substantially enhanced through the use of a sophisticated Glass Capillary Converter (GCC) system [10-11]. This GCC device collects, guides, and focuses the emitted radiation onto the entrance slit of the 2.2 m grazing incidence monochromator equipped with a 600 lines/mm grating. Data acquisition and control is accomplished using a PC-CAMAC computer system, and the photon counts are normalized with respect to the amount of charge collected in the Faraday cup. The slit width of the grazing incidence monochromator has been set to be 400 μm and the short wavelength radiation is detected by a channel electron multiplier (CEM). We have relatively calibrated our obtained spectra by dividing by the established efficiency curve for the monochromator [12]. Here we present an overview of the experimental results and identification of the most prominent target and projectile lines.

RESULTS AND DISCUSSION

In Fig. 2, 3, and 4 we exhibit some of our measured characteristic EUV spectra following single $Ar^{8+} + N_2$ collisions. We have provided evidence for more than 150 spectral lines; many of them have been identified as projectile lines arising from Ar^{7+} and Ar^{6+} ions due to single and double electron capture processes. For the identification of the target and projectile line structures we also utilized the comprehensive EUV line tabulations of Kelly [13]. Especially, we have identified numerous ionic nitrogen lines originating from $N^{2+}*$ and $N^{3+}*$ states stabilizing by radiative decay. We also established correlation diagrams for $Ar^{8+} + N$ single and double capture processes indicated in Fig. 5 and 6a and b, respectively. It is important to note that have we used a simplified collisional model replacing the N_2 potentials by the atomic nitrogen curves and assumed that the cross sections are approximately half the cross section for N_2 projectile states. Using this approach, we predicted that for single electron capture, the $Ar^{7+}*$(nl) states with n=3-5 are most abundantly populated. Furthermore we have performed extensive calculations of energy levels, transition wavelengths, and transition probabilities for $Ar^{7+}*$(nl), n=3-6 and $Ar^{6+}*$(nln'l'), n,n'=3-5 projectile states using the Cowan atomic structure code [14]. In addition we have utilized correlation diagrams for the single and double capture processes based on the classical overbarrier model to receive more detailed information about the complex $Ar^{8+} + N_2$ system. It is evident from the identification of the observed line spectra, that the prominent single and double electron capture processes:

$$Ar^{8+} + N_2 \rightarrow Ar^{7+*} (nl, n=3-6) + N_2^+ \qquad (1)$$

$$Ar^{8+} + N_2 \rightarrow Ar^{6+*} (3lnl, n=3-5) + N_2^{2+} \qquad (2)$$

are the main channels that stabilize via radiative transitions, whereas the higher lying doubly excited Ar^{6+*} (nlnl') states are expected to decay predominantly via autoionization. These results will be discussed in detail elsewhere [15].

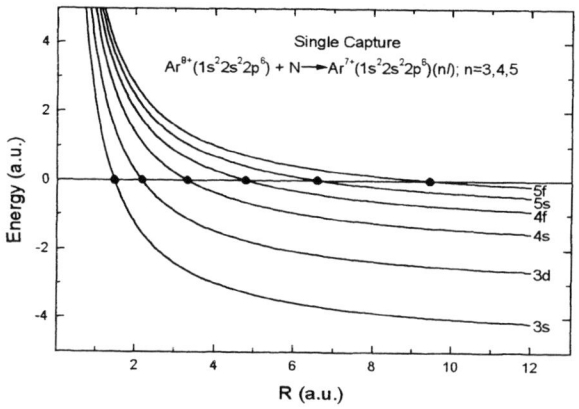

FIGURE 5. Correlation diagram for $Ar^{8+} + N$ single electron capture processes using the classical over-barrier model.

In this work we have focused mainly on the fragmentation of multi-charged N_2^{q+}, q=1-6 molecular complexes and excitation of the consecutive dissociation fragments. In Table 1 we have summarized the main EUV transitions originating from N^{2+*} and N^{3+*} ions in the 10-80 nm range. According to Remscheid et al.[7] such nitrogen ions generated in slow $Ar^{8+} + N_2$ collisions are most likely produced by :

$$N_2^{q+} \rightarrow N^{r+} + N^{s+} \qquad (3),$$

where in the dissociation of N_2^{q+} the symmetric distribution of the charge to both fragments is preferred, i.e. that N^{2+}, N^{4+}, and N^{6+} fragments most likely have the same charge, whereas for odd charge states such as N^{3+} and N^{5+} they differ mainly by one charge unit. In our experiment we have observed the EUV- photon emission from such ionic nitrogen fragments, arising from the highly excited target states N^{2+*} $(1s^22s2lnl')$ $^{2,4}L$, n=2-10 and N^{3+*} $(1s^22lnl')$ $^{1,3}L$, n=2-5 with excitation energies up to 52 and 78 eV, respectively (see Table 1). For example N^{2+*} and N^{3+*} excited nitrogen fragments may result from four and six electron capture mechanisms. The importance of non-Franck-Condon dissociation processes in such highly charged ion-molecule collisions has been suggested by Bliman et al [9]. In particular they have estimated that the electric space and time-field variation for $Ar^{8+} + H_2$ collision system is of the order of 1.5×10^{25} V/ms leading to dissociation processes deviating from the Frank-Condon Principle.

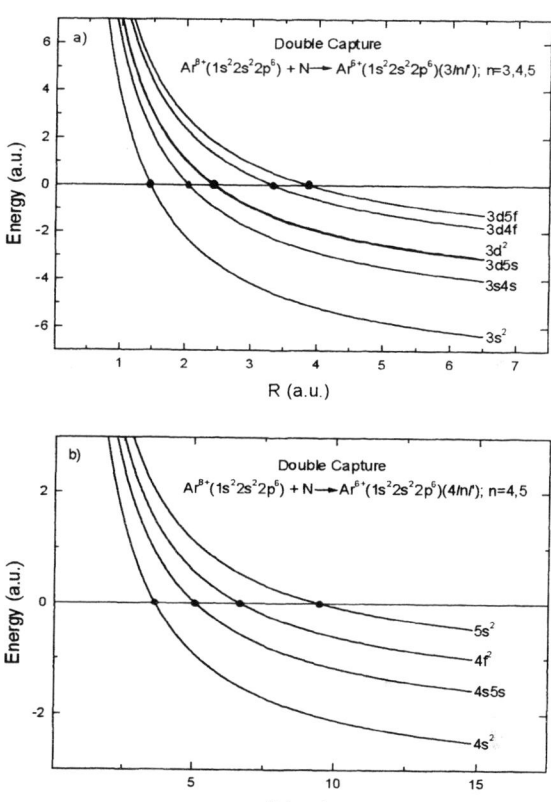

FIGURE 6a and b. Correlation diagrams for $Ar^{8+} + N$ double electron capture processes using the classical over-barrier model.

When comparing the relative line intensities for N^{2+*} and N^{3+*} optical transitions, we have found that 60% of the identified target lines stem from N^{2+*} excited states whereas 40% originate from N^{3+*} emission lines in our wavelength region. From Table 1, it is obvious that the N^{2+*} $(2s2p3d, 2s^23d, 2s2p4d, 2s2p^2)$ configurations are most abundantly populated in 3,4, and 5 electron capture processes. Furthermore, for the N^{3+*} fragmentation channels, the configurations N^{3+*} (2s2p, 2s4p, 2s4s, 2p4p, 2p5s, 2s4s) dominate in the measured spectral range.

CONCLUSIONS

We have studied single and multiple electron capture processes in slow 80 keV collisions between Ar^{8+} and molecular nitrogen using EUV spectrometry. Exotic electron capture processes with up to six electrons in one collision could explain the observed highly excited ionic nitrogen Rydberg states of the type N^{2+*} $(1s^22s2lnl')$ $^{2,4}L$, n=2-10 and N^{3+*} $(1s^22lnl')$ $^{1,3}L$, n=2-5. Thus, capture of four electrons may lead to the production of N^{2+*} excited ions. If five electrons are captured, this may lead to the creation of N^{2+*} and N^{3+*} excited states. Whereas, in the six electron capture reaction, N^{3+*} excited ions are most likely created. These findings are consistent with our measured relative target line intensities. These measurements will be complimented in the near future by target ion TOF spectrometry.

ACKNOWLEDGMENTS

This Project has been partially supported by ACSPECT Corp., Reno, Nevada, and by CNPq (Conselho National de Desenvolvimento Cientifico e Technológico), Brazil.

TABLE 1: Observed main ionic target lines owing to fragmentation plus excitation following multiple electron capture processes in 80 keV Ar^{8+} + N_2 collisions.

Peak Position	Relative Intensity	Charge State	Optical Transition	Wavelength[a] λ (nm)	Wavelength[b] λ (nm)	Energy (eV)[c]
30	25	N^{2+}	$2s^22p\ ^2P_{3/2} - 2s2p5p\ ^2D_{5/2}$	24.63	24.6249	50.374
33	56	N^{2+}	$2s2p^2\ ^4P_{5/2} - 2s2p15d\ ^4D_{7/2}$	25.70	25.7502	55.260
35	17	N^{2+}	$2s^22p\ ^2P_{3/2} - 2s^210d\ ^2D_{5/2}$	26.89	26.8473	46.206
36	37	N^{2+}	$2s2p^2\ ^4P_{5/2} - 2s2p6d\ ^4P_{5/2}$	27.44	27.4374	52.299
37	46	N^{2+}	$2s^22p\ ^2P_{3/2} - 2s^27d\ ^2D_{5/2}$	27.70	27.6326	44.894
38	40	N^{2+}	$2s^22p\ ^2P_{3/2} - 2s^26d\ ^2D_{5/2}$	28.22	28.2209	43.959
42	26	N^{2+}	$2s^22p\ ^2P_{3/2} - 2s^25s\ ^2S_{1/2}$	29.98	29.9818	41.378
48	19	N^{2+}	$2s^22p\ ^2P_{3/2} - 2s^24d\ ^2D_{5/2}$	31.50	31.4850	39.402
53	16	N^{2+}	$2s^22p\ ^2P_{3/2} - 2s^24s\ ^2S_{1/2}$	33.20	33.2333	37.332
61	92	N^{2+}	$2s2p^2\ ^4P - 2s2p3d\ ^4P$	35.79	35.8578	41.688
62	112	N^{2+}	$2s2p^2\ ^2P_{3/2} - 2s2p3d\ ^2P_{3/2}$	36.06	36.1288	52.420
65	109	N^{2+}	$2s^22p\ ^2P_{3/2} - 2s^23d\ ^2D_{5/2}$	37.28	37.4441	33.136
68	77	N^{2+}	$2s2p^2\ ^2S_{1/2} - 2s2p4d\ ^2P_{3/2}$	39.35	38.7483	48.243
74	94	N^{2+}	$2s2p^2\ ^2P_{1/2} - 2s2p4d\ ^2P_{1/2}$	40.91	41.1056	48.252
144	200	N^{2+}	$2s^22p\ ^2P_{3/2} - 2s2p^2\ ^2P_{3/2}$	68.43	68.5816	18.101
154	116	N^{2+}	$2s2p^2\ ^4P_{5/2} - 2p^3\ ^4S_{3/2}$	77.14	77.2385	23.162
19	38	N^{3+}	$2s2p\ ^3P_2 - 2s5s\ ^3S_1$	21.04	20.9471	67.552
22	52	N^{3+}	$2s2p\ ^1P_1 - 2p4p\ ^1S_0$	21.73	21.7218	73.287
24	49	N^{3+}	$2p^2\ ^3P_2 - 2p5s\ ^3P_2$	22.22	22.1789	77.695
28	52	N^{3+}	$2s2p\ ^3P_2 - 2s4s\ ^3S_1$	23.80	23.7991	60.460
39	36	N^{3+}	$2s2p\ ^1P_1 - 2p3p\ ^1P_1$	28.53	28.5561	59.626
49	25	N^{3+}	$2p^2\ ^1S_0 - 2s5p\ ^1P_1$	31.87	31.7596	68.224
50	27	N^{3+}	$2s2p\ ^3P_2 - 2s3s\ ^3S_1$	32.26	32.2722	46.780
64	57	N^{3+}	$2p^2\ ^1S_0 - 2s4p\ ^1P_1$	36.78	36.8108	62.868
153	194	N^{3+}	$2s^2\ ^1S_0 - 2s2p\ ^1P_1$	76.54	76.5148	16.205

[a]This work. [b]Kelly data [13]. [c]Excitation energies are derived from data of Kelly [13].

REFERENCES

1. McGuire, J.H., *Electron Correlation Dynamics in Atomic Collisions*, Cambridge University Press (1997).
2. Fuelling, S., et al. *Phys. Rev. Lett.* **68** 3152 (1992).
3. Bliman, S., et al., *Phys. Rev A*, **46** 1321 (1992).
4. Zou, Y., et al., *Phys. Scr.* **T73** 79 (1997).
5. Kambara, T., et al., *J. Phys. B* **31** L909 (1998).
6. Meyer, F.W., et al., *Nucl. Instrum. Methods Phys. Res. B* **24/25** 106 (1987).
7. Remscheid, A., et al., *J.Phys. B* **29** 515 (1996).
8. Shiromaru, H., et al, *Phys. Scr.* **T80** 110 (1999).
9. Bliman, S., et al., *Phys. Rev. A.*, **60** 1 (1999).
10. Bruch, R., et al., *Surf. Interface Anal.*, **27** 236 (1999).
11. Bruch, R., et al., *Rev. Sci. Instrum.*, **68** 1091 (1997).
12. Fülling, S., *Ph.D. Thesis*, UNR, Reno, NV USA (1991).
13. Kelly, R.L., *Atomic and Ionic Lines Below 2000 Å, H through Ar* (1982).
14. Cowan, R.D., *The Theory of Atomic Structure and Spectra*, Univ. of California Press (1981).
15. Siems, A., *Ph.D. Thesis*, State University of Campinas Brazil (2000).

Auger Decay of Triply Excited States of Li and Be⁺ Projectiles Following Ion-Molecule Interactions.

R. Bruch[a], A. Siems[a,b], H. Merabet[a*], J. Hanni[a], S. Bliman[c]

[a]*Department of Physics, University of Nevada, Reno, NV 89557 USA*
[b]*Instituto de Física, Universidade Estadual de Campinas (Unicamp), 13083-970 Campinas, São Paulo, Brazil*
[c]*Department de Physique, Université de Marne la Vallée, 93166 Noisy LeGrand, France*

Abstract. In this study, we present high resolution electron spectra and new identifications of numerous triply excited states of the Li-isoelectronic sequence produced by intermediate velocity Li^+ and Be^+ projectiles interacting with CH_4 molecules under single collision conditions. In particular, we show here as typical examples high resolution lithium and beryllium projectile Auger electron spectra following double K-shell excitation in 200 keV Li^+ + CH_4 and 300 keV Be^+ + CH_4 collisions. For the beryllium case, the impact velocity is 1.15 a.u. Many new lines appearing in these spectra have been assigned to triply excited Li and Be^+ and doubly excited Li^+ and Be^{2+} states. Only recently due to the pioneering theoretical predictions by Chung and coworkers (1), such collisional line structures can now be unambiguously identified. We have not only identified the lowest lying, most prominent $2s^22p$, $2s2p^2$ and $2p^3$ triply excited states (2), but also new excitation and deexcitation channels of higher lying $2l2l'nl'$ hollow atom/ion Rydberg states following ion-molecule collisions.

INTRODUCTION

Triply excited states in Li and Be^+, also referred to as "hollow atoms," represent fundamental highly-correlated few-body dynamical systems (3-10). In earlier studies of foil excitation of Li, such states were observed by Auger-electron projectile spectroscopy, however the resolution was limited due to kinematic effects (4). Using projectile beam-gas excitation, higher resolution electron spectra from double K-shell vacancy states in Li and Be^+ were obtained (5-7). Furthermore, Simons et al. (8) were the first to calculate the Auger and radiative decay rates of the $2s^22p$ level in Li using many-body perturbation theory. The first comprehensive description of hollow atoms in Li and Be^+ was presented by Rødbro et al. (5-7). Auger decay channels of triply excited three-electron systems using beam-gas spectroscopy were reviewed by Bruch and Chung (5). Extensive calculations of even and odd parity triply excited Li-like states were provided by Chung et al. (1, 9-10,18), and Conneely and Lipsky (3). In addition, optical emission from the (2p2p2p) $^4S°$ states in three-electron systems has been studied both experimentally and theoretically for He^-, Li, Be^+, B^{2+}, and C^{3+} (11-12).

Recently, photoelectron and photoion spectrometry studies of the triply excited states of Li have been conducted using synchrotron radiation sources (13-17). The selectivity of such photo-excitation techniques populate only $^2P°$ symmetry states from the ground state of Li $(1s^22s)$ ^2S. States of different symmetry and parity have been examined by means of laser excited $(1s^22p)$ $^2P°$ states from which synchrotron radiation has produced even-parity triply excited states (17). Photoexcitation by synchrotron radiation has proven to be a very successful method to study energies, widths, and branching ratios of such states in Li.

In this study, we provide a new examination of our double-hole K-shell vacancy Li^+, Li, Be^{2+} and Be^+ high-resolution electron spectra originating from Li^+ and Be^+ ion beams interacting with He and CH_4 under single-collision conditions. When compared to synchrotron radiation experiments, we have observed numerous additional new transitions arising from both doublet and quartet states of even and odd parity not previously discovered. Several of the triply excited Be^+ results presented here are the first reported in the literature. These results are compared with theoretical predictions of Li-like triply excited hollow atoms (9).

EXPERIMENTAL METHOD

The ejected electron spectra of highly excited resonance states of Li and Be were studied by using the projectile-electron spectroscopy method. High resolution was obtained by selecting a small observation angle with respect to the beam axis.

---

* Corresponding author: H. Merabet, email: hocine@physics.unr.edu

The apparatus used in this study consisted of a 600 keV heavy-ion accelerator, equipped with a universal ion source of the Nielsen type, and target chamber with a high resolution electron spectrometer, which can be continuously rotated between 0° and 150°. The excitation of the observed hollow atoms took place in a differentially pumped gas cell. Before entering the gas cell, the ion beam was carefully collimated using different apertures. The pressure in the differentially pumped gas cell was kept low enough (typically 10^{-3} torr) to ensure single-collision conditions. For normalization purposes, the ion beam was collected in a Faraday cup. The details of the experimental setup were described by Rødbro et al. (6-7). Since projectile electrons are ejected from fast moving particles, the laboratory energies E were transformed to energies E_o in the source-particle frame.

FORMATION OF TRIPLY EXCITED Li-LIKE STATES IN ION-ATOM AND ION-MOLECULE COLLISIONS

The projectile Auger electron spectra of Li and Be were studied for incident beam energies from 100 to 500 keV excited in single collisions with He and CH_4. Double-hole vacancy states were weakly excited in collisions with helium, whereas strong lines appeared when CH_4 was used as the target gas. Therefore, we primarily focused on molecular targets such as CH_4 to obtain high excitation probabilities for helium- and lithium-like doubly core-excited states (hollow atoms). The formation of triply excited doublet and quartet projectile states involves both direct excitation, electron capture, and or electron exchange mechanisms. For instance, Li $(2l2l'nl'')$ ^2L states may be formed by two-electron double K-shell excitation processes plus electron capture. On the other hand, Be^+ $(2l2l'nl'')$ $^{2,4}L$ states may be formed by double K-shell excitation plus additional outer shell excitation and/or electron exchange.

RESULTS AND DISCUSSION

Fig. 1, shows the high-energy portion of the Li electron spectrum obtained in 200 keV Li^+ + CH_4 collisions in the energy range from 66 to 93 eV. The observed spectral features (peaks 18-48) and the additional satellite lines (labeled a-l) are associated with doubly core-excited states and their decay in Li and Li^+. Several triply excited resonance channels are indicated in this figure and the line identification is summarized in Table I. In this table, we have identified resonance states and decay channels of eighteen lines originating from triply excited states. Especially, we have seen numerous decay channels of $(2l2l'nl')$ ^2L resonances with n = 2, 3, and 4. It is

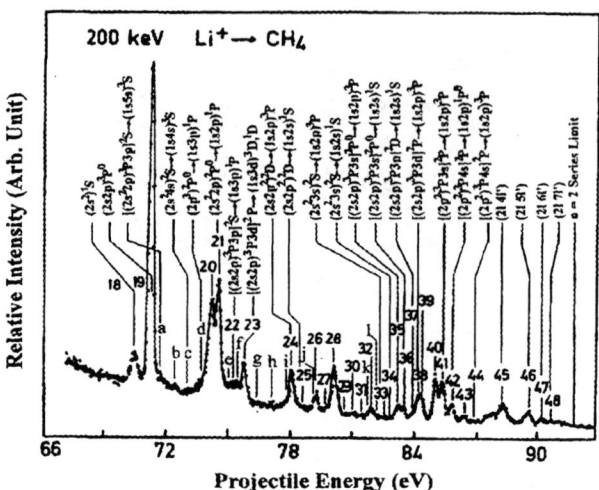

Figure 1. High-resolution projectile Auger electron spectrum for the 200 keV Li^+ + CH_4 collision system observed at an angle of 6.4°. The structures shown correspond to doubly core-excited Li^+ and triply excited Li states (hollow atoms).

interesting to note that both even and odd parity triply excited states contribute to the spectrum.

In Fig. 2, we have schematically shown the Auger decay path of Li-like triply excited states. As can be seen, some of the $(2l2l'nl')$ states can not only decay to the lowest $(1s2l)$ $^{3,1}L$ continua, but also to higher lying continua such as $(1s3l)$ $^{3,1}L$ and $(1s4l)$ $^{3,1}L$. The Auger

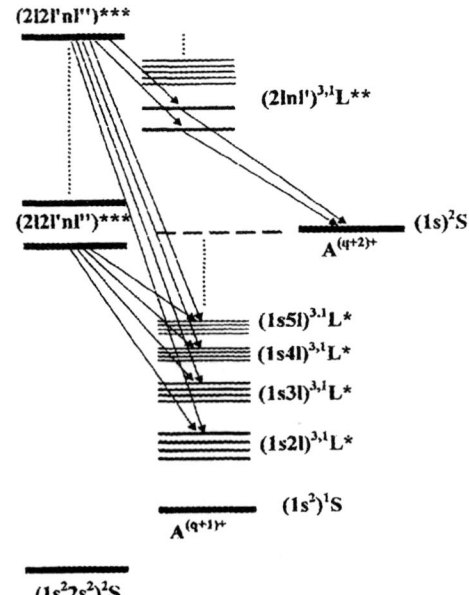

Figure 2. Schematic of different Auger decay channels of Li-like triply excited states.

rates of such triply excited series have been examined theoretically by Verbockhaven and Hansen (19); in particular, they have found that the Auger rates of triply excited Auger series behave rather differently when compared to doubly excited series. For example, the Auger decay rates for $(2l2l'nl'')$ Rydberg series with n ≥ 2 are expected to be nearly independent of n, while typical $(2lnl')$ decay rates decrease with increasing values of n. As demonstrated in Fig. 1, the Li$^+$ $(2lnl')$ doubly excited states for n = 4-7 agree with this predicted trend. Furthermore, all the observed transitions from triply excited resonances are in close agreement with the predictions of Chung and coworkers (1,9,10,18). In this work, we have also observed Auger transitions to Li$^+$ Rydberg states including (1s3p) $^1P^o$, (1s3d) $^{3,1}D$, (1s4s) ^3S, and (1s5s) ^3S. According to Fig. 2, the higher lying $(2lnl'nl'')$ Rydberg states can also decay to doubly excited $(2lnl')$ $^{3,1}L$ manifolds, which in turn can decay further to the Li^{2+} (1s) ^2S ground state. Hence, some of the measured contributions of the doubly excited Li$^+$ $(2lnl')$ states (see Fig. 1) may arise from the decay of such triply excited states.

The high energy portion of the 300 keV Be$^+$ + CH$_4$ spectrum, in the projectile energy range from 120 to 160 eV is illustrated in Fig. 3. The observed spectral features (peaks 30-62) are attributed to Be$^+$ and Be^{2+} states with double K-shell vacancies. In Table II, fifteen lines, labeled 32-48, are associated with triply excited resonances. They are identified via resonance

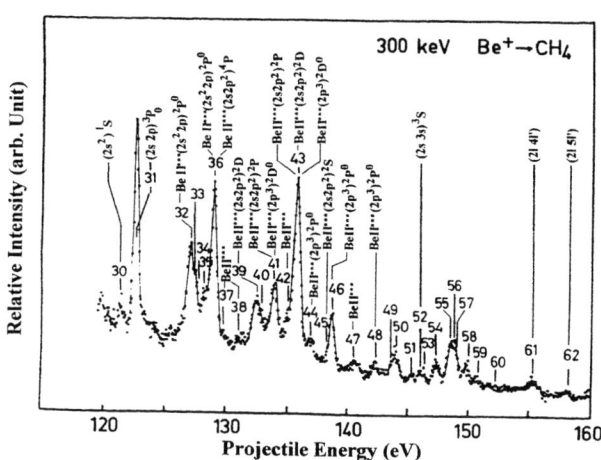

Figure 3. High-resolution projectile Auger electron spectrum for 300 keV Be$^+$ + CH$_4$ collisions observed at an angle of 6.4°. The structures shown correspond to doubly core-excited Be^{2+} and triply excited Be$^+$ states (hollow atoms).

Table I. Observed Auger transitions of triply excited $2l2l'nl''$ (n=2-4) states in Li following Li$^+$ + CH$_4$ collisions.

Line Number	Resonance State	Decay Channel	Theoretical Prediction (eV)
A	[(2s2p) ^{3}P 3p] ^{2}S	(1s5s) ^{3}S	71.800$^\alpha$
B	(2s^{s}4s) ^{2}S	(1s4s) ^{3}S	72.957$^\alpha$
d	(2p^3) ^{2}P^o	(1s3p) ^{1}P^o	73.730$^\beta$
21	(2s^{2}2p) ^{2}P^o	(1s2p) ^{1}P^o	74.660$^\gamma$
22	[(2s2p) ^{3}P 3p] ^{2}S	(1s3p) ^{1}P^o	75.453$^\alpha$
23	[(2s2p) ^{3}P 3d] ^{2}P^o	(1s3d) ^{3}D	75.970$^\gamma$
23	[(2s2p) ^{3}P 3d] ^{2}P^o	(1s3d) ^{1}D	75.970$^\gamma$
24	(2s2p^2) ^{2}D	(1s2p) ^{3}P^o	78.140$^\delta$
25	(2s2p^2) ^{2}D	(1s2s) ^{1}S	78.498$^\delta$
33	(2s^{2}3s) ^{2}S	(1s2p) ^{3}P^o	82.667$^\delta$
34	(2s^{2}3s) ^{2}S	(1s2s) ^{1}S	83.025$^\delta$
35	[(2s2p) ^{3}P 3s] ^{2}P^o	(1s2p) ^{3}P^o	83.170$^\gamma$
36	[(2s2p) ^{3}P 3s] ^{2}P^o	(1s2s) ^{1}S	83.530$^\gamma$
37	[(2s2p) ^{3}P 3p] ^{2}D	(1s2s) ^{1}S	83.892$^\alpha$
38	[(2s2p) ^{3}P 3d] ^{2}P^o	(1s2p) ^{3}P^o	84.270$^\gamma$
41	[(2p^2) ^{3}P 3s] ^{2}P	(1s2p) ^{3}P^o	85.453$^\alpha$
42	[(2p^2) ^{3}P 4s] ^{2}P	(1s2p) ^{1}P^o	85.910$^\alpha$
44	[(2p^2) ^{3}P 4s] ^{2}P	(1s2p) ^{3}P^o	86.850$^\alpha$

$^\alpha$Zhang and Chung (10), $^\beta$Kiernan et al. 13), $^\gamma$Chung and Gou (9), $^\delta$Chung (1).

Table II. Observed Auger transitions of triply excited $2l2l'2l''$ states in Be^+ following $Be^+ + CH_4$ collisions.

Line Number	Resonance State	Decay Channel	Peak Position (eV)	Theoretical Prediction (eV)*	Branching Ratio (%)$^\alpha$
32	$(2s^22p)\ ^2P^o$	$(1s2p)\ ^1P^o$	126.9 ± 0.1	126.896	16
36	$(2s^22p)\ ^2P^o$	$(1s2s)\ ^1S$	128.9 ± 0.1	128.913	11
36	$(2s2p^2)\ ^4P$	$(1s2p)\ ^3P^o$	128.9 ± 0.1	128.992	100
39	$(2s2p^2)\ ^2D$	$(1s2p)\ ^3P^o$	132.4 ± 0.1	132.514	34
40	$(2s2p^2)\ ^2D$	$(1s2s)\ ^1S$	132.8 ± 0.1	132.783	12
41	$(2s2p^2)\ ^2P$	$(1s2p)\ ^1P^o$	134.0 ± 0.1	133.913	35
41	$(2p^3)\ ^4D^o$	$(1s2p)\ ^1P^o$	134.0 ± 0.1	134.132	21
42	$(2s2p^2)\ ^2S$	$(1s2s)\ ^1S$	135.4 ± 0.1	135.477	20
43	$(2s2p^2)\ ^2P$	$(1s2p)\ ^3P^o$	135.8 ± 0.1	135.660	65
43	$(2s2p^2)\ ^2D$	$(1s2s)\ ^3S$	135.8 ± 0.1	135.839	53
43	$(2p^3)\ ^2D^o$	$(1s2p)\ ^3P^o$	135.8 ± 0.1	135.879	79
44	$(2p^3)\ ^2P^o$	$(1s2p)\ ^1P^o$	137.0 ± 0.1	137.094	14
45	$(2s2p^2)\ ^2S$	$(1s2s)\ ^3S$	138.5 ± 0.1	138.532	80
46	$(2p^3)\ ^2P^o$	$(1s2p)\ ^3P^o$	138.8 ± 0.1	138.841	82
48	$(2p^3)\ ^2P^o$	$(1s2s)\ ^3S$	142.4 ± 0.2	142.166	1

$^\alpha$Gou and Chung (9).

states, decay channels, and branching ratios. As can be seen from this table, the experimental Auger peak positions of the observed triply excited ($2l2l'2l'$) states are in excellent agreement with predictions from Gou and Chung (9). We note that the peak 36 may be composed of transitions arising from Be^+ $(2s^22p)\ ^2P^o$ and the Be^+ $(2s2p^2)\ ^4P$ states, where the quartet state decays to the $[(1s2p)\ ^3P^o\ \epsilon p]\ ^4P$ continuum.

CONCLUSION

In this work, we have presented an experimental study of triply excited states of three electron systems formed by intermediate energy impact of Li^+ and Be^+ projectiles on CH_4. We have demonstrated that the triply excited states are populated in ion-molecule collisions under single collision conditions. Using high resolution electron spectroscopy individual lines arising from Li and Be^+ so called "hollow atoms" with two K-shell vacancies were resolved and identified. Theoretically, excitation energies, probabilities for Auger decay, and branching ratios have been derived from the literature and compared with our experimental results. Excellent agreement has been found between our experimental results and predicted non-radiative transition energies.

ACKNOWLEDGMENTS

This project has been partially supported by CNPq (Conselho National de Desenvolvimento Cientificico e Technológico), Brazil.

REFERENCES

1. Chung, K.T., *Phys. Rev. A* **25** 1596 (1982).
2. Safronova, U.I. and Bruch, R, *Phys. Scrpt.* **57** 519 (1998).
3. Conneely, M.J and Lipsky, L., *Phys. Rev. A* **61** 2506 (2000).
4. Bruch, R. *et al.*, *Phys. Rev. A* **12** 1808 (1975).
5. Bruch, R. and Chung, K.T., *Comments on At. Mol. Phys. D* **14** 117 (1984).
6. Rødbro, M. *et al.*, *J. Phys B* **12** 2413 (1979).
7. Rødbro, M. *et al.*, *J. Phys. B* **10** L275 (1977).
8. Simons, R.L. *et al.*, *Phys. Rev. A* **19** 682 (1979).
9. Gou, B.C. and Chung, K.T., *J. Phys. B* **29** 6103 (1996).
10. Zhang, Y. and Chung, K.T., *Phys. Rev. A* **58** 1098 (1998).
11. Agentoft, M. *et al.*, *J. Phys. B* **17** L433 (1984).
12. Bruch, R., *et al*, private communication (2000).
13. Kiernan, L.M. *et al.*, *J. Phys. B* **28** L161 (1995).
14. Azuma, Y. *et al.*, *Phys. Rev. Lett.* **79** 2419 (1997).
15. Wehlitz, R. *et al.*, *Phys. Rev. A* **60** 1050 (1999).
16. Diehl, S. *et al.*, *Phys. Rev. Lett.* **84** 1677 (2000).
17. Madsen, L.B., *et al.*, *Phys. Rev. Lett.* **85** 42 (2000).
18. Chung, K.T. and Gou, B.C., *Phys. Rev. A* **53** 2189 (1996).
19. Verbockhaven, G. and Hansen, J.E., *Phys. Rev. Lett.* **84** 2810 (2000).

SECTION II

NUCLEAR PHYSICS

Spectrometers at RHIC

K. Hagel

Cyclotron Institute, Texas A & M University, College Station, Texas 77843 USA

Abstract. A review of various spectrometers at RHIC is presented with an emphasis on how different physics goals influenced their design. These physics goals have in turn be influenced by the historical development of the field. Particular attention will be paid to the BRAHMS experiment which seeks the maximum possible acceptance in transverse and longitudinal momenta while accepting a small azimuthal coverage. The very wide range in momentum drives all aspects of the design. The performance of the detector and its physics capabilities will be presented.

INTRODUCTION

First collisions were observed at the Relativistic Heavy Ion Collider (RHIC) at Brookhaven National Laboratory on June 13, 2000. This momentous event marked the end of a nearly 10 year process of bringing RHIC from a concept on paper to a functioning accelerator. It also marked the beginning of a hopefully long fruitful period of physics which will dramatically increase our understanding of the nature of the universe.

There are four experiments at RHIC each of which address in various ways the problem of understanding the behavior of nuclear matter as it was in the first moments after the big bang. The goal of the RHIC experiments is to study reaction mechanisms of relativistic heavy ions at RHIC energies and and the properties of highly excited nuclear matter formed in these reactions. Another goal is to search for evidence of the Quark Gluon Plasma (QGP). Two of the experiments are global in scope and aim to measure a large fraction of the products produced in the Relativistic Heavy Ion reaction and completely reconstruct the event. The other two experiments are smaller in scope, but aim to make very detailed measurements of small parts of the reaction.

In these proceedings we discuss spectrometers at RHIC with a particular emphasis on BRAHMS, one of the "small" experiments. Various components of the spectrometers are discussed elsewhere in these proceedings, so we concentrate on how these components work together.

RELAVISTIC HEAVY ION EXPERIMENTS WITH SPECTROMETERS

There is a long history of the use of spectrometers in relativistic heavy ion experiments from which a large body of data has been acquired. Heavy ion experiments can be either global where an attempt is made to measure as much of the remnants of the reaction as possible. This comes with a cost. The momentum distributions of reaction products becomes larger and larger as the angle becomes smaller and smaller. This causes an exponential increase in the cost of building detectors sensitive to these high momenta products. Global detectors therefore usually have limited acceptance at forward angles or small η. Another approach is to make inclusive measurements which measure a small fraction of the reaction products in great detail, that is over the entire rapidity and transverse momentum distribution. This is typically accomplished with spectrometers. There is some overlap between the two philosophies in that the global experiments usually have one or a few detectors that have some limited capabilities in detecting particles at high rapidities and/or high transverse momentum. In addition, the inclusive experiments typically have some "global" detectors which provide some global information on the reaction albeit not with the high degree of detail of the global experiments.

Some of the inclusive experiments include the E802/859/866[1] series of experiments at the AGS. This series of experiments employed a 25msr spectrometer which was capable of particle identification between 0.5 and 4.7 GeV/c. The experiment was designed to survey semi-inclusive hadron production of nucleus-nucleus collisions using relativistic heavy ion beams ranging from 14.6 GeV/u ^{16}O and ^{28}Si to ^{197}Au. In addition to the spectrometer, there was a target multiplicity array used for event characterization. The BRAHMS experiment has its roots in this series of experiments.

Another inclusive experiment which employed a spectrometer was the NA44[2] experiment at CERN. NA44 was a focusing spectrometer that used superconducting quadropole and conventional dipole magnets to produce a magnified image of the target in the spectrometer. This gave it excellent momentum resolution. It used cerenkov detectors, calorimeters and time of flight TOF to identify particles while a silicon pad array functioned as a global detector to measure centrality.

An example of a global experiment was the NA49[3] experiment at CERN. It consists of a wide acceptance magnetic spectrometer combining momentum measurement and particle identification in a set of large Time Projection Chambers (TPC's) with the aim of measuring the complete range of hadronic interactions up to the highest particle densities produced in central Pb + Pb collisions.

One of the large experiments at RHIC is the Solenoidal Tracker At RHIC (STAR). STAR is a large acceptance multi-purpose spectrometer that detects all charged particles produced in the pseudo-rapidity interval -2<η<2 with a Time Projection Chamber (TPC). The solenoidal nature of the magnet allows a measurement in 4π as the solenoidal magnet causes the particles to curve in circles rather than being bent out of the way as in a dipole magnet.

The other large experiment at RHIC is PHENIX. This experiment measures leptons which necessitates completely different technology. The PHENIX detector system is also a multi-purpose spectrometer which has a tracking system that consists of drift chambers, pad chambers and time expansion chambers. These tracking detectors are designed to locate all charged particles, measure the particle momenta and identify which of the tracks are electrons. The PHENIX Muon Arms allow good statistics measurements of production of Drell-Yan, J/ψ, ψ', and ϒ through their decay into di-muon pairs.

One of the small experiments at RHIC is PHOBOS. PHOBOS consists of many silicon detectors surrounding the interaction region. With these detectors the PHOBOS group hopes to identify unusual events, fluctuations in the number of particles and angular distribution. In order to obtain more detailed information about these particular events the PHOBOS detector also has two high quality spectrometers which will study, in detail, 1% of the produced particles. This detector system can measure particles down to very low p_t.

The other small experiment at RHIC is the Broad RAnge Hadron Magnetic Spectrometer (BRAHMS). The experimental goals of RHIC physics are accomplished by measuring and identifying charged hadrons over a wide range of rapidity and transverse momenta. The BRAHMS experiment is discussed in some detail in the following sections.

BRAHMS

BRAHMS consists of two magnetic spectrometer arms and four global detectors. In the following we first describe the components of the detector. Then the physics capabilities of the detector will be discussed.

BRAHMS Overview

A perspective view of BRAHMS is shown in figure 1 where we see the two spectrometers, namely the Forward Spectrometer (FS) to the right of and near the beampipe and the Mid Rapidity Spectrometer (MRS) to the left of the beampipe near 90°. The FS covers angles between 2.3° and 30° and is sensitive to hadrons having rapidities up to y=4 while the MRS covers angles between 30° and 95° and will probe particles at y ≈ 0.

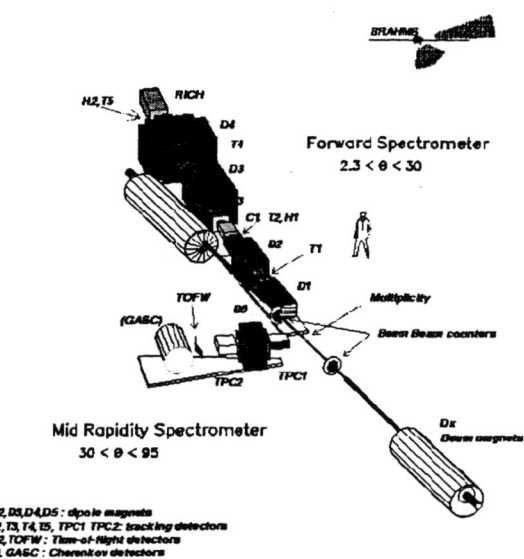

FIGURE 1. A depiction of the BRAHMS experiment.

BRAHMS has its foundation in five dipole magnets. Dipole magnets in a momentum spectrometer are the analog of lenses in an optical spectrometer. In figure 1 we note the five dipole magnets D1, D2, D3 and D4 in the FS and D5 in the MRS. These magnets have the purpose of bending the particles between the various tracking stations and allowing a measurement of the momentum. Different field settings allow a wide range of momenta to be studied.

Along with the magnets are a number of tracking stations. There are four TPC's, T1 and T2 in the FS and TPC1 and TPC2 in the MRS. In the FS, T1 and T2 with magnet D2 in between allow tracking and momentum determination in a low momentum mode. In the MRS, tracking is accomplished with TPC1 and TPC2 with D5 in between.

In addition to the TPC's used for tracking, the FS also has three drift chamber stations (DC's), T3, T4 and T5 each separated by magnets D3 and D4. The three DC tracking stations are composed of three submodules each. The most forward DC, T3 experiences the highest particle density (~.007 particles/cm^2 for central collisions at 2.2°.) while T4 and T5 have 1/4-1/7 of that particle density.

Each module in T3 has 10 planes to accommodate the higher particle density whereas T4 and T5 have only 8 planes each.

In a high momentum mode, D1 and D2 are used for sweeping and the momentum determination is made using T2, T3, T4, T5 with D3 between T3 - T4 and D4 between T4 - T5.

Three plastic scintillator walls, H1 and H2 in the FS and TOFW in MRS are used along with the tracking stations for particle identification (PID). H1 can separate pions from kaons (π/K) up to 3.3 GeV/c and kaons from protons (K/p) up to 5.7 GeV/c. H2 is able to separate π/K up to 5.8 GeV/c and K/p up to 8.5 GeV/c.

Two other detectors enable to extend the PID capabilities to higher momentum. C1, a Cerenkov detector, extends the π/K identification to 9 GeV/c and a Ring Imaging Cerenkov detector[4] (RICH), when used along with H2, extends the π/K/p identification up to 25 GeV/c.

BRAHMS employs a series of global detectors to enable event characterization. Around the interaction vertex (IV) is a global multiplicity array used to estimate event by event the centrality. The multiplicity array is made up of two components. There is an array of plastic scintillator tiles which provide a measure of the charged particle multiplicity. Simulations indicate that the multiplicity can be determined to 8% for semi-central Au-Au collisions and to 4% for central Au-Au collisions.

There is also an array of silicon strip detectors in the multiplicity array. These detectors will provide some redundancy with the tiles, but the strips provide segmentation to measure dN/dη.

Further away from the IV near the beampipe are Beam-Beam counters (BB) which are made up of Cerenkov radiators. The beam-beam counters are located 220 cm from the nominal IV. They are constructed from 2 types of Cerenkov radiators glued to photomultiplier tubes. The diameters of the radiators are 19 and 51mm, respectively.

The beam-beam counters provide a start time and a level 0 trigger. The 50ps time resolution allows to determine the vertex position to ~2cm. These detectors also provide multiplicity information at high η.

Behind DX (not shown in figure 1) on both sides of the IV are Zero Degree Calorimeters[5] (ZDC's) which detect neutrons emitted in a small cone around the beam. The ZDC's measure the total energy (ZDC resolution is 28% for 1 neutron at 65 GeV) of neutrons in a cone of θ<2mr from the beam axis. The ZDC coincidence of the two beam directions can provide for a minimum bias selection making it useful as an event trigger as well as a luminosity monitor[6]. In addition, the neutron multiplicity is also known to be correlated with the event geometry[7] and can therefore be used along with the multiplicity array and the beam-beam counters to estimate the centrality of the collision.

There are identical sets of ZDC's in all of the RHIC experiments. They provide luminosity measurements for the RHIC machine group as well as neutron measurements for BRAHMS. Since the ZDC's provide some centrality information, having identical detectors in all four of the RHIC experiments provides redundancy which will enable results from the different experiments to be cross checked.

BRAHMS Physics Capabilities

Figure 2 shows the acceptance of BRAHMS for the different spectrometer settings. Region I and II in the figure show the regions accessible by the FS and region III shows the region accessible to the MRS. It should be noted that region III can be measured simultaneously with the first two regions.

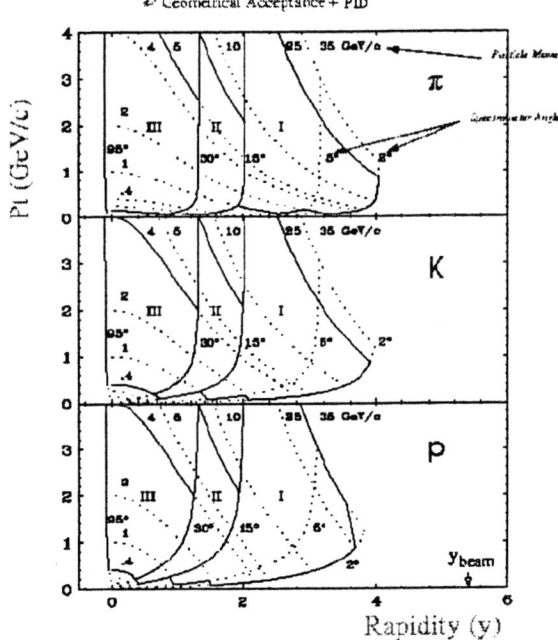

Region I : With the full Forward Arm
Region II : With the D1 - D2 (Forward Arm) Complex alone
Region III : With the Mid-Rapidity Arm

FIGURE 2. Acceptance of the BRAHMS experiment

We note in the figure that the rapidity-p_t range accessible to BRAHMS is very large. Distributions for all of the rapidity can be measured as a function of all of the variables accessible by the global detectors. The rapidity distributions we can measure extend over the whole range of rapidity expected in the reactions at RHIC energies. This makes an immediate comparison to the models possible and will provide constraints to the models.

Although as noted in the introduction BRAHMS can measure only a small fraction of the reaction in any given event, the wide range of rapidity and p_t accessible for measurement makes it possible to obtain many interesting quantities on an average basis. Such quantities include π/K ratios which give a measure of strangeness, one of the signatures of the QGP. A measurement of lambda production is possible in the MRS. HBT measurements over a broad y-pt range can be performed utilizing the FS as well as the MRS.

In summary we have described the BRAHMS detector system at RHIC and shown what the capabilities of this detector system are.

ACKNOWLEDGMENTS

The heroic efforts of the RHIC team to commission the accelerator and deliver the beam are gratefully acknowledged. This work was supported by the United States Department of Energy grants #DE-FE05-86ER40256, by the Danish Natural Science Research Council, by the Norwegian Natural Science Research Council and by the Polish ministry of Scientific Research.

REFERENCES

1. T. Abbott, et al., Nucl.. Instrum. Meth. A**290**, 41 (1990).

2. I. G. Bearden, et al., Phys. Rev. Lett.. **85**, 2681 (2000).

3. S. Afanasiev et al., Nucl.. Instrum. Meth. A**430**, 210 (1999).

4. R. Debbe et al., Nucl.. Instrum. Meth. A**371**, 327 (1996).

5. C. Adler et al., submitted to Nucl.. Instrum. Meth. (2000), nucl-ex/0008005.

6. A. J. Baltz, C. Chasman, and S. N. White, Nucl.. Instrum. Meth. A**417**, 1 (1998).

7. H. Appelshauser et al., Eur. Phys. J. A**2**, 383 (1998).

Tracking Chambers At RHIC

Julia Velkovska

SUNY Stony Brook

Abstract. The extremely high multiplicity environment of ultra-relativistic heavy ion collisions imposes stringent performance requirements on the tracking detectors used in the four RHIC experiments. This paper will emphasize the tracking detectors of the PHENIX experiment.
The PHENIX central arm spectrometers utilize three different types of tracking chambers, each with unique detector and electronics design. Two multi-layer focusing drift chambers provide high resolution charged particle momentum measurement with excellent two-track resolution. Three layers of pad chambers with unique pixel-pad readout system aid the tracking by supplying three-dimensional coordinates along the charged particle trajectories in the field-free region. Time-expansion chambers are used for tracking, as well as particle identification, by multiple dE/dx measurements along each track. The detector designs and their performance in the RHIC2000 run are presented.

INTRODUCTION

During the first run of the Relativistic Heavy Ion Collider (RHIC) in summer 2000, the PHENIX experiment operated successfully and recorded more than 5 million events from ultra-relativistic Au-Au collisions at sqrt(s) = 130 GeV/c. This paper will review the design and performance of the tracking chambers of PHENIX that were instrumented for this run.

The two central arms, shown in Fig. 1 consist of alternating layers of tracking and particle identification (PID) detectors. Particle ID is done using Ring Imaging Cherenkov detectors (RICH), Time-Expansion-Chamber (TEC), Time-Of-Flight counters (TOF) and electromagnetic calorimeters (EMCAL). Tracking is done in two stages: inner tracking, performed with drift chamber (DC) and pad chamber (PC1), and outer tracking, performed with TEC and pad chamber 3 (PC3). The magnetic field is axial and drops almost to zero at the entrance of the drift chamber. The momentum of the charged particles crossing the detector is determined by the bend in the magnetic field, while the sign of the bend angle provides information about the charge.

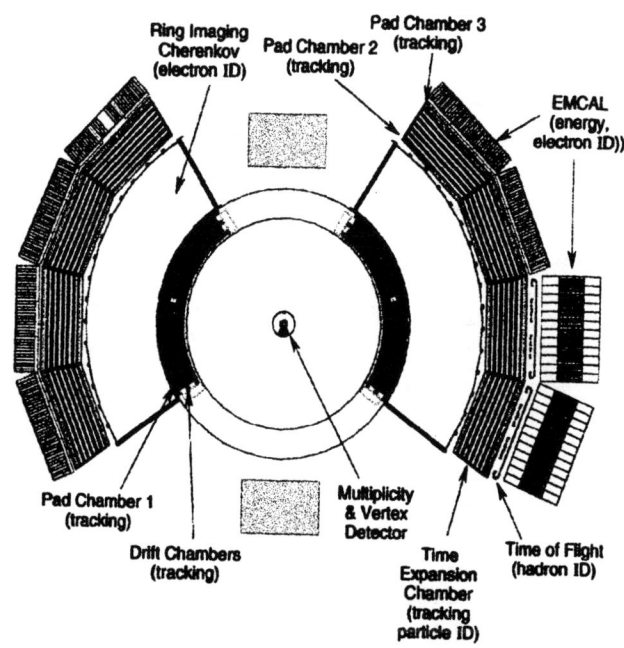

FIGURE 1. PHENIX central arms in a cross-section perpendicular to the beam axis.

DRIFT CHAMBER

Two cylindrical drift chambers covering 90 degrees in azimuth each provide tracking for charged particles with momentum > 200 MeV/c. These innermost tracking detectors located in the radial tracking region between 2.02 and 2.46 m set the momentum and mass resolution of PHENIX. Forty radial layers of sense wires are arranged in eighty cells along ϕ. To reduce ambiguity in the z coordinate of the track, the wires are split in the middle and read out from both ends. A total of 12800 readout channels were instrumented to handle the enormous track density in a relativistic heavy ion collision

Drift chamber electronics

The large number of readout channels requires state-of-the art electronics. All DC electronics is entirely contained within the titanium frame of the detector and all digitization is done onboard. Each plane of 40 readout channels is connected to one ASD8/TMC card shown in the picture. Eight channels of amplifier/shaper/discriminator are contained in the 1cm x 1cm ASD8 chip. After discriminator with a typical threshold of 6 fC, the signals are passed to a 10 channel Time-Memory-Cell (TMC), which is a multi-hit time to digital converter with 0.8 ns least count and a ring buffer memory that allows storing 6.4 µs history. The latency provision is common to the electronics of all tracking detectors in PHENIX, since they do not participate in the trigger decision, but need to be able to retrieve the information correlated with the trigger at a later time without interrupting data taking. The communication with the trigger system and the data collection modules is carried out over 1Gbps optical fibers by the Front-End-Modules (FEM). The photograph in Fig. 2 shows one "keystone" of electronics, which provides readout for 160 channels.

Electrode configuration

The PHENIX drift chamber utilizes a novel focusing/protected field shape. The wire configuration and the electric field lines around the sense wires are shown in Fig 3. In addition to the sense and field wires common to all drift chambers, the basic cell of the PHENIX drift chamber contains two channel wires and a guard wire. The channel wires control the drift space, by collecting the ionization drifting towards the sense wire along the longest field lines. This configuration minimizes the charge collection time and provides excellent two-track resolution. The guard wires collect all the charge coming from the side opposite to the channel wires. This makes the sense wires only sensitive to one side and eliminates the left-right ambiguity present in conventional drift chambers. This feature is especially valuable in the high track density environment of RHIC and tremendously eases the pattern recognition.

FIGURE 2. One "keystone" of DC electronics: 4 ASD8/TMC cards connected to one FEM. Optical fibers carry the trigger signal to the FEM and transmit the data out of the FEM to a data collection module.

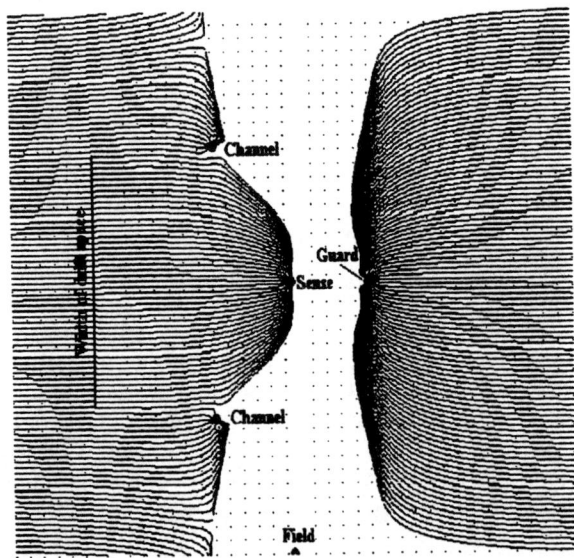

FIGURE 3. Electrode configuration and electric field lines in the basic drift chamber cell.

Drift chamber performance

The drift chamber performance is measured by the single point resolution, double track resolution and single wire efficiency.

The single point resolution was measured by constructing the residual distributions of the hits and the fitted tracks. Resolution of 140 µm was achieved. Double track resolution of 1.76mm was determined from the average width of the timing signal (34 ns) from hits that belong to reconstructed tracks, multiplied by the constant drift velocity in the gas (54 µm /ns). During the RHIC2000 run, the drift chamber operated with single wire efficiency between 90% and 95%, which due to the large number of measurements along the track (i.e. not all hits are necessary for track reconstruction) translates into tracking efficiency between 95% and 98%.

PAD CHAMBERS

Two layers of pad chambers: PC1 and PC3 located behind the DC and TEC, respectively, were installed for the RHIC2000 run. The pad chambers provide three-dimensional points along the charged particle trajectories and are crucial for determining the z-coordinate of the tracks.

Read-out Electronics And Pixel-Pad Configuration

A charged particle traversing the gas of the pad chamber creates an avalanche on the anode wire, which induces a charge cluster on the conductive cathode pad plane. After pre-amplification, the signals are discriminated with typical threshold of 2fC and digitized onboard, providing a binary yes/no output. The pad chamber read-out electronics is located in the space traversed by the particles, hence low mass requirements have to be met in order to reduce background. This is achieved by using bare Si chips wire-bonded to kapton substrates and "capped" with epoxy. A more detailed description of the pad chamber electronics can be found in reference 1.

The PHENIX pad chambers have unique pixel-pad configuration [2] shown in Fig.4. In conventional cathode readout wire chamber, the size of the pads is comparable to the size of the charge cluster induced on the cathode plane. However, in this configuration, a single dead channel may cause a particle to be missed completely. To avoid this problem, each pad in the PHENIX pad chambers is divided into three pixels. At least two pixels are required to fire in order to reconstruct a hit. The efficiency is greatly improved, since any one pixel signal can be missed. In addition, by imposing the coincidence requirement, the device becomes practically noise-free. The short-come of such configuration is that if all pixels are read-out separately, one needs three times more electronics, compared to conventional designs. A clever solution was to connect nine pixels arranged in a "staircase" to one electronics channel, preserving the separate readout for all three pixels on a pad. That way, the necessary electronics is reduced by a factor of three.

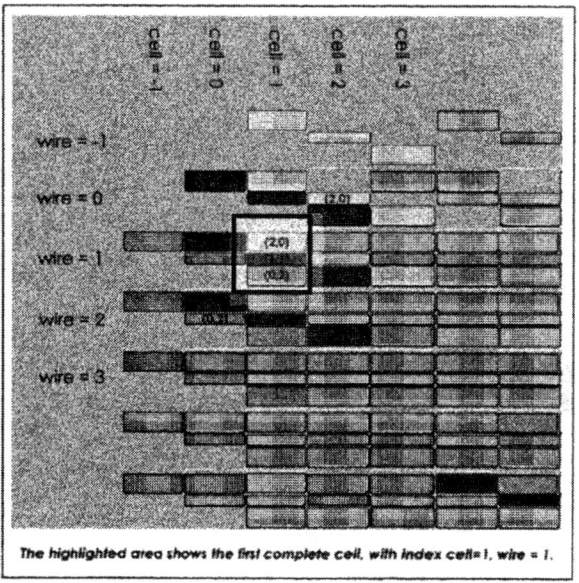

FIGURE 4. Pad chamber pixel-pad configuration: The different shades show groups of 9 pixels connected to one read-out channel. Three adjacent pixels form a cell.

Pad Chamber Performance

The pad chamber efficiency as a function of the high voltage and threshold setting were studied extensively with cosmic rays. It was determined that the device operates with efficiency >99% when the average size of the charged cluster is >1.3 cells. In beam, during the RHIC2000 run, the measured average cluster size was 1.54 cells. The efficiency of the pad chambers was also measured independently by matching PC hits to DC tracks and found to be >99%. Noise-free operation with optimal efficiency was achieved.

TIME-EXPANSION CHAMBER

The time expansion chamber is used for tracking as well as for particle ID. 24 TEC chambers arranged in 4 sectors containing a stack of 6 chambers each, were installed on the PHENIX East carriage. The active area covers π/2 in azimuth and pseudorapidity +/- 0.35. Two sectors were instrumented for the RHIC2000 run. All 4 sectors will be instrumented for RHIC2001 run.

The anode/cathode configuration and the electric field lines in the TEC are shown in the figure below.

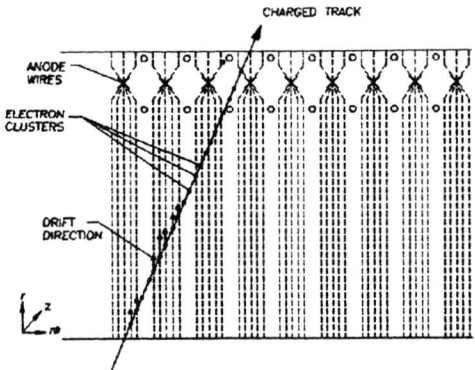

FIGURE 5. TEC anode/cathode configuration and the electric field lines along a charged track.

Unlike in the drift chamber, the electric field lines in the TEC are primarily along the track direction. The wire signals are sampled at 40 MHz using 5-bit non-linear flash ADC. This allows for multiple measurements along each track. By mapping the charge distribution with a large number of samples, one obtains excellent dE/dx measurement, which is used for PID. Tracking is two-dimensional. PC3 is used for reconstruction of the z-coordinate of the tracks. Single point resolution of 250 μm and two-track separation of 5 mm were achieved. Additional information on the design and in-beam testing of the TEC can be found in ref. 3.

TRACKING SOFTWARE

Both DC and TEC use combinatorial Hough transform for pattern recognition in the track bend plane. Tracks are defined by two angles φ and α, as shown in Fig.6. Combinations of hits that belong to the same track have the same angles φ and α, hence they form a peak in the Hough space (φ,α). The z-coordinate of the track is determined from matching hits in DC/PC1 and TEC/PC3. A more detailed description of the pattern recognition algorithm can be found in ref. 4. After the pattern recognition, the tracks are handled by a track model based on field-integral look-up table. The momentum and the path length to the PID detectors are predicted.

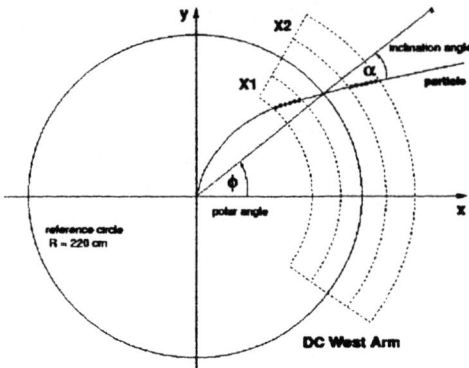

FIGURE 6. The Hough space in the XY plane of the track.

CONCLUSIONS

The PHENIX tracking system is a diverse array of detector technologies and techniques. All tracking chambers worked within design specifications during the first RHIC run. First physics results for inclusive and identified charged particle transverse momentum distributions, charged particle multiplicity distributions were obtained (for details see the presentations of A. Franz and J. Lajoie "First results from the PHENIX experiment").

ACKNOWLEDGEMENT

The author acknowledges the contribution of all members of the PHENIX collaboration.

REFERENCES

1. J. Barrette et al, "The pixel readout system for the PHENIX pad chambers", *Nuclear Physics A661*, 1999, pp. 665c-668c.

2. L. Carlen et al., "Two-dimensional pixel readout of wire chambers", *NIM A396*, 1997, pp. 310-319

3. K. Barish et al., "The PHENIX Time Expansion Chamber", *Nuclear Physics A661*, 1999, pp. 669c-672c.

4. S. C. Johnson et al., "Three-dimensional track finding in the PHENIX drift chamber by a combinatorial Hough transform method", *Proceedings of CHEP*, 1998

First Results From the PHOBOS Experiment at RHIC

A. H. Wuosmaa[1], for the PHOBOS Collaboration: B. B. Back[1], M. D. Baker[2], D. S. Barton[2], S. Basilev[5], B. D. Bates[5], R. Baum[8], R. R. Betts[1,7], A. Białas[4], R. Bindel[8], W. Bogucki[3], A. Budzanowski[3], W. Busza[5], A. Carroll[2], M. Ceglia[2], Y.-H. Chang[6], A. E. Chen[6], T. Coghen[3], C. Conner[7], W. Czyz[4], B. Dabrowski[3], M. P. Decowski[5], M. Despet[3], P. Fita[5], J. Fitch[5], M. Friedl[5], K. Galuska[3], R. Ganz[7], E. Garcia[8], N. George[1], J. Godlewski[3], C. Gomes[5], E. Griesmayer[5], K. Gulbrandsen[5], S. Gushue[2], J. Halik[3], C. Halliwell[7], P. Haridas[5], A. Hayes[9], G. A. Heintzelman[2], C. Henderson[5], R. Hollis[7], R. Hołynski[3], B. Holzman[7], E. Johnson[9], J. Kane[5], J. Katzy[5,7], W. Kita[3], J. Kotuła[3], H. Kraner[2], W. Kucewicz[7], P. Kulinich[5], C. Law[5], M. Lemler[3], J. Ligocki[3], W. T. Lin[6], S. Manly[9,10], D. McCleod[7], J. Michałowski[3], A. Mignerey[8], J. Mülmenstädt[5], M. Neal[5], R. Nouicer[7], A. Olszewski[2,3], R. Pak[2], I. C. Park[9], M. Patel[5], H. Pernegger[5], M. Plesko[5], C. Reed[5], L. P. Remsberg[2], M. Reuter[7], C. Roland[5], G. Roland[5], D. Ross[5], L. Rosenberg[5], J. Ryan[5], A. Sanzgiri[10], P. Sarin[5], P. Sawicki[3], J. Scaduto[2], J. Shea[8], J. Sinacore[2], W. Skulski[9], S. G. Steadman[5], G. S. F. Stephans[5], P. Steinberg[2], A. Straczek[3], M. Stodulski[3], M. Strek[3], Z. Stopa[3], A. Sukhanov[2], K. Surowiecka[5], J.-L. Tang[6], R. Teng[9], A. Trzupek[3], C. Vale[5], G. J. van Nieuwenhuizen[5], R. Verdier[5], B. Wadsworth[5], F. L. H. Wolfs[9], B. Wosiek[3], K. Wozniak[3], A. H. Wuosmaa[1], B. Wysłouch[5], K. Zalewski[4], P. Zychowski[3]

[1]*Physics Division, Argonne National Laboratory, Argonne IL, 60439-4843*
[2]*Chemistry and C-A Departments, Brookhaven National Laboratory, Upton NY 11973-5000*
[3]*Institute of Nuclear Physics, Krakow, Poland*
[4]*Department of Physics, Jagellonian University Krakow, Poland*
[5]*Laboratory for Nuclear Science, Massachusetts Institute of Technology, Cambridge MA 02139-4307*
[6]*Department of Physics, National Central University, Chung-Li, Taiwan*
[7]*Department of Physics, University of Illinois at Chicago, Chicago IL 60607-7059*
[8]*Department of Chemistry, University of Maryland, College Park MD 20742*
[9]*Deparment of Physics and Astronomy, University of Rochester, Rochester NY 14627*
[10]*Department of Physics, Yale University, New Haven CT 06520*

Abstract. The PHOBOS experiment at RHIC has measured the charged-particle density $dN/d\eta$ at mid-rapidity for central Au+Au collisions at center of mass energies of $\sqrt{s_{NN}}$ =56, and 130 GeV. We deduce that $dN/d\eta$=408±12(stat) ±30(syst) and 555±12(stat)±35(syst) for collision energies of 56 GeV and 130 GeV, respectively. These numbers suggest energy densities that are some 70% higher than have been achieved in any heavy-ion collisions previously studied, and also 25-40% higher than nucleon-nucleon collisions at comparable center of mass energies.

INTRODUCTION

In June of 2000, the Relativistic Heavy Ion Collider (RHIC) produced its first collisions of Au ions at energies of $\sqrt{s_{NN}}$ =56 and 130 GeV, the highest energies at which heavy ions have yet been collided in the laboratory. Of particular interest are global characteristics of the collisions that provide information about the thermodynamic properties of the early evolutionary phases of the expanding hot system. Chief among these are the values of the energy, and entropy density a few fm/c after the collision. Bjorken[1] has suggested that these quantities may be directly related to the density of charged particles emitted at mid rapidity. In order to probe these conditions, it is therefore interesting to study the charged-particle density at mid-rapidity, as well as its scaling with collision energy and the number of interacting nucleons. A comparison of the charged particle density per participant nucleon with data from proton-proton or proton-antiproton scattering may yield information about the collective effects or rescattering. The early PHOBOS measurements attempt to address these questions.

EXPERIMENTAL DETAILS

The PHOBOS experiment consists primarily of a large collection of silicon pad detectors. In the full experimental configuration used to take data in 2000, the setup consisted of a large acceptance (~4π) multiplicity detector, a small-acceptance two-arm silicon pad tracking spectrometer within a 2T magnetic field, plastic-scintillator and Cerenkov-radiator trigger detectors, two plastic-scintillator time-of-flight walls, and two zero-degree calorimeters (ZDCs). A more detailed description of the experimental setup may be found in References 2 and 3.

In the initial phases of collider operations at RHIC, there was some, ultimately unfounded, concern about radiation damage to the silicon sensors and delicate readout electronics in PHOBOS, in the case of possible large background radiation fields. Consequently, for the first few weeks of operations a small subset of the silicon pad detectors was installed. This collection included 13 of the total 92 elements of the central multiplicity detector, six of a total of 24 of the central vertex tracking detectors, and six planes of tracking spectrometer. The plastic-scintillator trigger detectors, as well as the ZDCs were also installed. In addition, the magnet was not energized for this commissioning period. A schematic diagram of the experimental setup for this initial stage appears in Fig. 1.

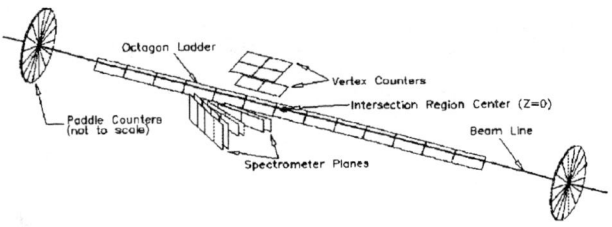

FIGURE 1. Schematic diagram of the commissioning run setup.

DATA ANALYSIS

The first step in data analysis concerns the identification of collision events. This identification was achieved using timing data from the PHOBOS trigger detectors. Particles produced by collisions in the nominal interaction region at the center of the experiment were detected in the two plastic scintillator trigger arrays. The difference in time-of-arrival for these particles at the trigger counters was approximately zero, whereas events produced by background interactions upstream of the experiment traversed one, then the other trigger array, yielding very different timing characteristics. This timing measurement provided a very clean separation between collisions and background. The contribution to the sample of collision data from double beam-gas interactions was calculated from the observed rate of single beam-gas interactions, and was found to be negligible.

The centrality of the collision events was determined from a measurement of the total energy deposited in the two paddle-counter arrays (see Fig. 1). To avoid contributions to this energy signal from low-energy particles that are not minimum ionizing, and consequently deposit more energy, the paddles with the four highest of the 16 energy signals were taken out of the analysis, and the remaining data were used to determine the centrality. Figure 2 shows the resulting paddle energy distribution obtained at a center-of-mass collision energy of 130 GeV. The shaded region denotes the 6% of events that deposit the most energy. These are the most central collisions included in this analysis.

To determine the number of participating nucleons for this event sample, events from the HIJING event generator[4,5] were processed with a detailed simulation of the response of the trigger counters using the GEANT simulation package. The same 6% energy cut was applied to the HIJING Monte-Carlo data, and the corresponding number of participating nucleons from HIJING was determined to be $<N_{part}> = 330 \pm 4(stat) \pm^{10}_{15} (syst)$ and $<N_{part}> = 343 \pm 4(stat) \pm^{7}_{14} (syst)$ at $\sqrt{s_{NN}}$ =56 and 130 GeV, respectively.

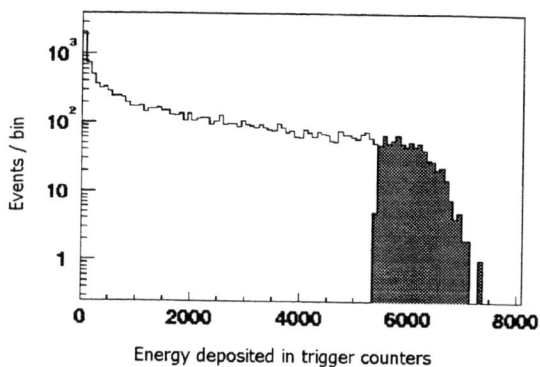

FIGURE 2. Energy spectrum measured in the trigger counters. The shaded region approximately corresponds to the 6% most central collisions.

In order to determine the pseudorapidity η for each particle, it is first necessary to identify the collision vertex. In PHOBOS this may be accomplished in a number of different ways. For the analysis presented here, vertices were identified using found tracks in the silicon tracking spectrometer. Here, tracks are identified with a road-following algorithm connecting hits in at least four of the six silicon spectrometer planes. The resulting tracks are then analyzed to determine if they project back to the same point in space to within approximately 1 mm. The average number of reconstructed tracks at the lower energy was approximately 13, and at the higher energy 18, for central collisions. Figure 3 shows a typical event obtained at a center of mass energy of $\sqrt{s_{NN}}$ =56 GeV.

After identification of the collision vertex, the hits in the first two planes of the tracking spectrometer, as well as those in the two planes of the vertex detector, are analyzed to reconstruct "tracklets." A tracklet is a straight-line track emanating from the collision vertex, which connects to two points in first two layers of the tracking spectrometer, or the two layers of the vertex detector. As the tracklets are required to originate at the identified vertex, the number of identified tracklets is insensitive to backgrounds arising from secondary particles produced, for example, in the beam pipe.

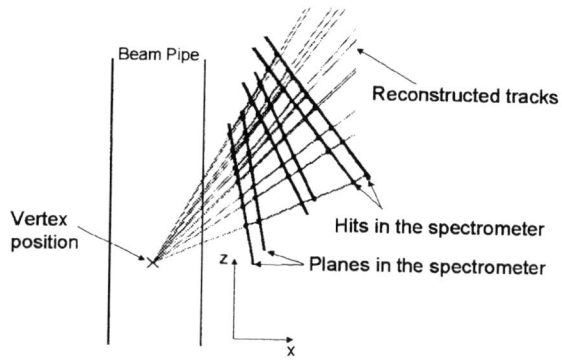

FIGURE 3. Reconstructed tracks for an Au+Au collision at $\sqrt{s_{NN}}$ =56 GeV. The thick lines represent planes of the tracking spectrometer, and the thin lines denote reconstructed tracks.

In order to deduce the pseudorapidity density from the measured number of tracklets, a number of corrections were applied to the data to account for the acceptance of the spectrometer and vertex detectors, inefficiencies of tracklet finding, false tracks from noise hits, as well as secondary particles and the products of weak decays of strange particles. These corrections were obtained from analyses of the response of the detector using the event generator HIJING and the simulation code GEANT. For each event, the Monte-Carlo data were analyzed in precisely the same manner as real events. The number of tracklets per event for the Monte-Carlo sample was compared to the actual number of particles emitted in to the spectrometer or vertex acceptance (for |η|<1), and the corresponding correction factor was then applied to the real data. This procedure was also carried out for events simulated using the VENUS event generator. The correction factors obtained from the HIJING and VENUS event sets differed by less than 5%, demonstrating that this procedure produces a reliable measure of the charged-particle density.

RESULTS AND DISCUSSION

From this analysis, we obtain the values $dN/d\eta|_{|\eta|<1}$= 408±12(stat) ±30(syst) for $\sqrt{s_{NN}}$ =56 GeV, and 555±12(stat)±35(syst) at $\sqrt{s_{NN}}$ =130 GeV. To compare our results with those of nucleon-nucleon scattering at similar center of mass energies, we divide our results by the approximate number of participating nucleon-nucleon pairs $<0.5N_{part}>$. We determine $dN/d\eta|_{|\eta|<1}/<0.5N_{part}> = 2.47 \pm 0.10(stat) \pm 0.25(syst.)$ at 56 GeV, and $3.24 \pm 0.10(stat) \pm 0.25(syst.)$ at 130 GeV. Figure 4 shows the PHOBOS data points, as well as the results from the SPS Pb+Pb collisions[7] at $\sqrt{s_{NN}} = 17.8$ GeV, and from $p\bar{p}$ collisions from CERN[8]. We observe that the central pseudo-rapidity density for charged particles is approximately 70% higher at 130 GeV than for the SPS energy, suggesting a correspondingly higher central energy density for these collisions. Also, the number of charged particles produced per participant nucleon is approximately 40% larger for Au+Au collisions than for nucleon-nucleon collisions at similar bombarding energies, ruling out simple superposition models, such as the wounded nucleon model[9]. Also, the increase in the normalized particle densities of approximately 30% between $\sqrt{s_{NN}} = 56$ and 130 GeV is dramatically larger than for proton-antiproton collisions.

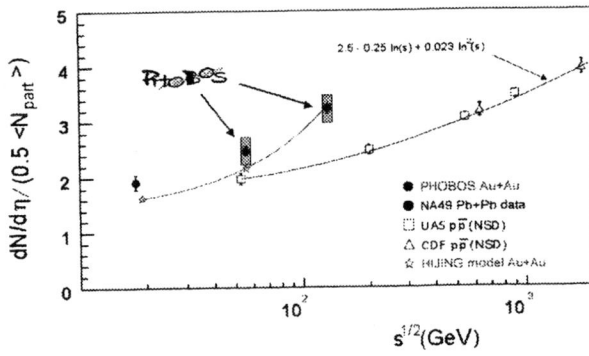

FIGURE 4. Charged particle density $dN/d\eta$ ($|\eta|<1$) divided by number of participant pairs, plotted versus center of mass collision energy.

Our current results are in good agreement with the predictions[5] of the code HIJING, which includes the effects of particle production via hard-scattering processes, as seen in Fig. 4. This calculation also predicts the dependence of $dN/d\eta$ on N_{part}, as well as on η. An experimental determination of this behavior can readily distinguish between HIJING-like models, and saturation models employing different pQCD cutoff schemes (e.g. Ref. 10). Also, while the effects of jet quenching in HIJING are modest at the highest energy studied so far, the results at $\sqrt{s_{NN}} = 200$ GeV show a marked sensitivity to this aspect of the calculation. It will be extremely interesting to follow the evolution of the data to even higher energies.

With the conclusion of the physics run in September 2000, PHOBOS has accumulated a data set consisting of more than 3.5M Au+Au collisions at $\sqrt{s_{NN}} = 130$ GeV, using the fully instrumented experimental setup. These data are currently under analysis and will shed additional light on the issues examined in the early running phase.

ACKNOWLEDGMENTS

We gratefully acknowledge the support of the personnel of the RHIC facility, the C-A and Chemistry Departments at BNL. We thank Fermilab and CERN for their assistance in silicon detector assembly. This work was supported in part by U.S. Department of Energy grants DE-AC02-98CH10886, DE-FG02-93ER40802, DE-FC02-94ER40818, DE-FG02-94ER40865, DE-FG02-99ER41099, W-31-109-ENG-38, and National Science Foundation Grants 9603486, 9722606, and 0072204. The Polish groups were partially supported by KBN grant 2 P03B 04916. The NCU group was partially supported by the NSC of Taiwan under contract NSC 89-2112-M-0088-024.

REFERENCES

1. Bjorken, J. D., *Phys. Rev.* **D27**, 140-151 (1983).

2. Back, B. B. *et al.*, *Phys. Rev. Lett.*, (in press) and hep-ex/0007036.

3. G. S. F. Stephans, "How Strange is PHOBOS" in *Proceedings of the Vth International Conference on Strangeness in Quark Matter*, to be published in *Journal of Physics G*.

4. Gyulassy, M., and Wang, X.-N., *Phys. Rev.* **D44**, 3501 (1991).

5. Wang X.N, and Gyulassy, M., *Phys. Rev. Lett. (submitted)* and nucl-th/0008014.

6. Werner, K, *et al.*, *Phys. Rep.* **232**, 87 (1993).

7. Bachler, J. *et al.*, *Nucl. Phys.* **A661**, 45 (1999).

8. Abe, F., *et al.*, *Phys. Rev.* **D41**, 2330 (1990).

9. Bialas, A, Bleszynski, M. and Czyz, W., *Nulcear. Phys.* **B111**, 461 (1976).

10. Eskola, K. J., *et al.*, *Nucl. Phys.* **B570**, 379 (2000).

The REX-ISOLDE Beam Diagnostic System

P. Van den Bergh [a], M. Huyse [a], K. Krouglov [a], P. Van Duppen [a], L. Weissman [b]

[a] *Instituut voor Kern- en Stralingsfysica, Katholieke Universiteit Leuven, Celestijnenlaan 200 D, B-3001 Leuven, Belgium*
[b] *CERN, CH-1211 Geneva 23, Switzerland*

Abstract. A beam diagnostic system was developed for the REX-ISOLDE post-accelerator. This system shows the position of the beam and its intensity distribution for ion beams in a wide dynamic range of both intensity and energy. The design of this system is discussed here as well as the results of the various tests that were performed on it.

INTRODUCTION

The REX-ISOLDE project at CERN [1] aims at accelerating radioactive ions from the ISOLDE mass separator using a new accelerator concept. The ions delivered by ISOLDE will undergo several manipulations, including trapping, charge breeding and several stages of acceleration, from 60 keV up to 2.2 MeV/u. Beam diagnostic units, required after every section of REX-ISOLDE had to be developed that meet the following requirements:

1. monitor the position of the ion beam with a spatial resolution of 1 mm or better;
2. show the intensity distribution of the ion beam;
3. work in a wide dynamic range in energy, from 60 keV up to the final 2.2 MeV/u;
4. work in a wide dynamic range in beam intensity from a few particles per second (pps), up to 10^{10} pps;
5. the beam diagnostic system also must be compatible with the time structure of the REX-ISOLDE beams.

The majority of existing beam diagnostic systems cannot satisfy these requirements [2]. Grid based profiler systems demand high beam intensities and many electronic channels, which makes them expensive. Other systems such as avalanche chambers and position sensitive silicon detectors have a higher sensitivity but are not applicable for low energy beams. Position sensitive electron multipliers, such as micro channel plates (MCP) and micro sphere plates (MSP) have very high sensitivities, for electrons as well for ions that are directly implanted in the detector. These detectors can also be used to monitor the secondary electrons that are created by the impact of the ion beam on a target.

Considering all the above listed possibilities for beam diagnostic systems and the requirements implied by the REX-ISOLDE postaccelerator, we have developed universal beam diagnostic units, each containing a Faraday cup, a stepping motor driven collimator-wheel and a beam profiler [2].

THE CONCEPT OF THE BEAM PROFILER

A schematic drawing of the low intensity beam profiler is shown in Fig. 1. The beam impinges on an Al plate, which is mounted at 45° relative to the beam axis. The Al plate can be introduced in the beam path by a pneumatic feedthrough. Because of the impact of the ion beam, secondary electrons are emitted from the Al plate. The number of secondary electrons emitted depends on the electronic stopping power of the projectile in the medium, but is generally in the order of 1 to 1000 electrons per ion [3,4]. The majority of

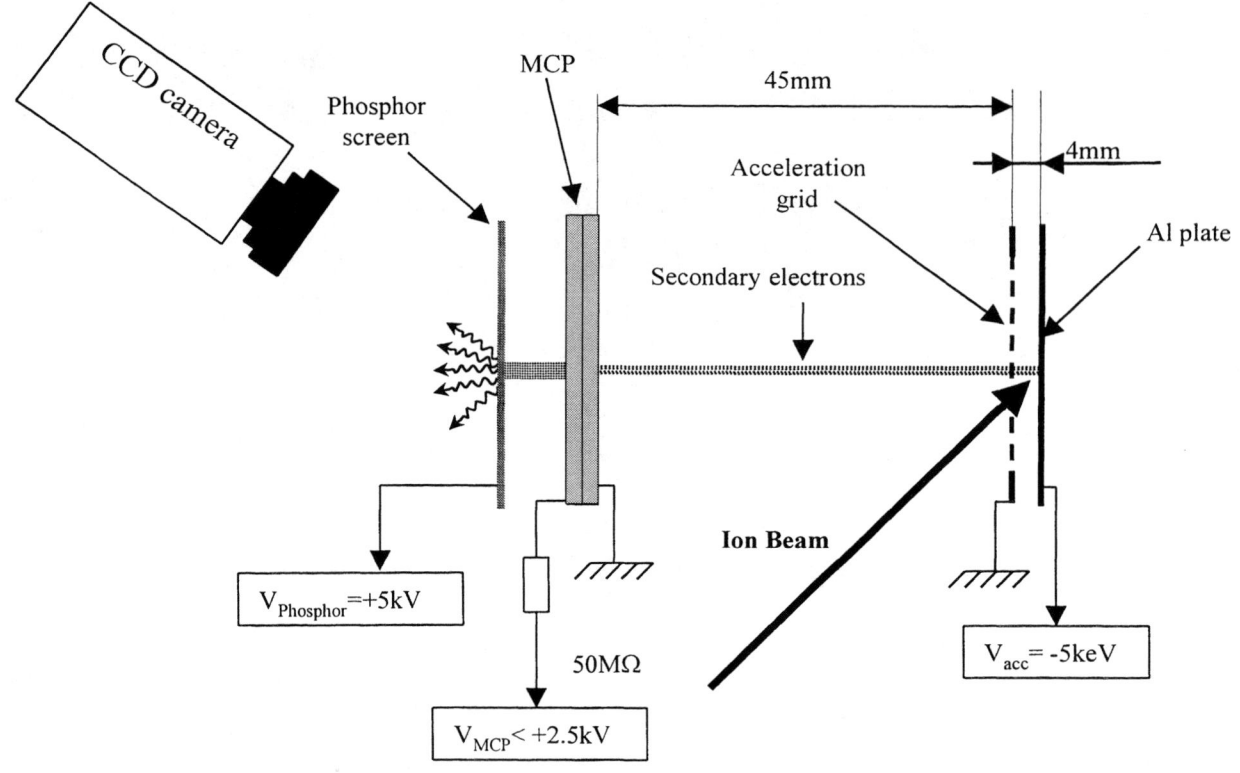

FIGURE 1. A schematic drawing of the beam profiler.

the electrons have an energy in the order of 1 eV [5]. A transparent grid electrode, mounted in front of the Al plate creates an electric field that accelerates the electrons to 5 keV. Beyond the grid, the electrons enter a field free zone and travel towards the entrance of the MCP detector. Because of their low initial energy, the position information of the electrons is preserved and their position on the MCP's entrance plane reflects the position of impact of the ion on the plate. A similar technique was used by Shapiro et al. [6,7].

The MCP detector consists of 2 MCPs in the so-called chevron configuration. The gain of the detector, defined by the number of electrons emerging on the exit divided by the number of electrons on the entrance, can be changed in a wide range by varying the voltage across the MCP's and can be as high as 10^7.

The cascade of electrons is then accelerated towards a phosphor screen where a 2 dimensional image of the ion beam becomes visible. A slow phosphor was used (P20), with a decaytime of 4 ms to 10 % but a considerably longer decay-time of 55 ms to 1% of the initial intensity. Therefore, also very short ion beam bunches can be observed. The image on the phosphor screen is observed by a standard CCD camera and captured by a frame-grabber. The image-data can then be processed by software and displayed on a monitor or distributed to operator consoles.

TEST RESULTS

Experimental Setup

Based on the configuration explained in the previous section, we developed a prototype beam-profiler in order to test if such a system can satisfy the REX-ISOLDE requirements. The initial design, however, used ring-electrodes to create the acceleration field between the plate and the detector. This was changed later to a grid, as in Fig. 1, to improve the resolution. During some of the tests that are reported here, a MSP detector was used instead of a MCP detector and a few different cameras were also used. Although different configurations were tested, the final design uses the components that produced the best results, the results obtained with earlier configurations are still valid. In the final configuration, a double MCP detector with an effective diameter of 25 mm in chevron configuration was used, equipped with a P20 phosphor screen. The camera is a standard 1/3" CCD camera with a sensitivity of 0.1 Lux. The images are processed by a standard personal computer with a commercial frame-grabber. Up to 20 frames per second can be displayed if the system is operated in stand-alone mode and the image is displayed directly on the computer monitor.

Sensitivity And Energy Range

To show that the profiler can handle ion beams in a wide energy and intensity range, it was tested at different facilities, including the IMBL pelletron in Leuven, the LISOL online isotope separator in Louvain-la-Neuve and the cyclotron at CRC, Louvain-la-Neuve. Some of these tests are listed in Table 1. In all of these tests images of the ion beam, showing its position, and intensity profiles (x and y projections) were obtained. Examples of beam images can be found in [2].

TABLE 1. Some tests with stable beams

Place / facility	Ion Beam	Intensity
IMBL	1 MeV ^{4}He	10^{10} pps
LISOL	50 keV ^{40}Ar	10^6 pps
LISOL	50 keV ^{219}Rn	10^3 pps
CRC	459 MeV ^{129}Xe	5 pps/cm^2

Resolution

To measure the spatial resolution, the image from a beam passing through a 1 mm collimator was observed. Profile analysis showed that the resolution (FWHM of the beam profiles) was better than 1 mm. The image of a beam passing through 2 collimators, 1 mm apart and each with a diameter of 0.3 mm showed that the two spots were clearly resolved. These measurements were done with beams of different energy and intensity, showing that the resolution is independent of the properties of the ion beam.

Besides these measurements, several other tests were done to study the properties of the system. More information can be found in [2].

THE CONTROL SYSTEM

Hardware

In total, 6 beam-diagnostic units were built in Leuven, each of them equipped with a beam profiler, a Faraday cup and a collimator-wheel. Next to this, 3 beam-diagnostic systems where developed by ISOLDE for the UHV beam-lines, where the ion beam is directly implanted in a MCP. To deliver the necessary supply voltages and control signals, 9 electronic control units were designed and built by our institute's electronic workshop. These units communicate with the control computer via an RS485 network. A video-multiplexer, connected to the same RS485 network, selects the video-signal from 1 of the nine cameras and sends it to a frame-grabber inside the control computer. The signals from the Faraday cups are connected to an electrometer via a commercial low-level current multiplexer. Both instruments communicate with the control computer using a IEEE-488 bus.

Software

The control computer, that has access to all functions of the nine beam-diagnostic units and related hardware is integrated into the REX-ISOLDE control system. This is a client-server based control system, where all the hardware is controlled by personal computers that or connected to the CERN ethernet network. Client programs (operator interfaces) can run anywhere on this network and acces these control computers, the so-called FEC's (Front End Computer), using TCP/IP or UDP datagrams. The control computer for the beam-diagnostics is configured as a FEC, so it can be used and accessed in a similar way as all other REX-ISOLDE equipment.

The client program allows users to control and read all necessary hardware parameters (MCP high voltage, Faraday cup position, electrometer reading, ...) and acquires and displays the camera-images and the X- and Y-profiles. Because a full image from the camera takes more than 400 kbytes of data, transmitting the raw images from the FEC to the client programs will be an important load for the already busy CERN network, considerably reducing the refresh time. Therefore, the images are preprocessed by the control computer to reduce the data-size. In a first step, reducing the resolution and the pixeldepth very significantly decreases the amount of data. The resolution of the camera-image, defined by the CCD camera and the frame-grabber, is about 0.05 mm. The resolution of the image on the phosphor screen, however, is defined by the initial energy of the secondary electrons and is about 1 mm. This means that we can safely decrease the number of pixels that are taken into account without losing information. The pixel-depth (the number of bits per pixel) is reduced from 8 bit to 4 bit for the image, for the calculation of the profiles, a full depth of 8 bit is used. After this, the resulting image is compared against the camera noise and a data-package is composed, containing only the areas where the signal is above the noise level. This data-package is then sent to the clients where it is decoded and where the image and the profiles are drawn on the screen. The size of the data-packages that are sent over the network has been reduced to about 1 kbyte only for normal beam images, with a maximum of 18 kbyte for full images without suppression of dark areas. This makes it possible to update the images on the screen every 0.5 s. A planned upgrade of the network in the near future will probably allow even faster update times.

Other features of the software include averaging, and rejection of dark images (usefull for pulsed beams).

CONCLUSION

We have developed a simple and cost effective beam diagnostic system, based on observation of secondary electrons created by the impact of an ion beam on an aluminum plate. Tests have shown that this system performs well over a wide range of beam intensity, from 1 nA down to 5 pps. The performance is also independent of the energy of the ions over a wide range. The spatial resolution is about 1 mm. Software has been developed that integrates the complete system into the REX-ISOLDE control system. It can be operated from anywhere without causing too much network traffic.

ACKNOWLEDGMENTS

We would like to express our gratitude to Mr. J. Gentens for his help in operating the LISOL separator, Mrs. S. Hogg for help in operating the Leuven pelletron and all members of the technical staff of our institute for the design and production of the mechanical and electronic components. We also want to thank the whole LLN cyclotron crew for providing us with high-energy beams, O. Kester, T. Sieber and K. Rudoph for their support during our tests in Munich and R. Repnow and the Heidelberg MPI accelerator crew for their help during the tests with pulsed beams. We also thank the ISOLDE team for the support and help during the installation in CERN. This work is supported by the Inter-University Attraction Poles (IUAP) Research Program, the Fund for Scientific Research - Flanders (FWO) and the Research Fund K.U. Leuven (GOA). M. Huyse is a Research Director of the FWO.

REFERENCES

1. D. Habs et al., *Nucl. Phys. A* **616**, 214-234 (1997).

2. K. Kruglov et al., K., *Nucl. Instr. And Meth. A* **441**, 595-604 (2000).

3. E. J. Sternglass., *Phys. Rev.* **108**, 1 (1957).

4. H. Rothard eta al., *Nucl. Instr. And Meth. B* **48**, 616 (1990).

5. B. L. Henke, J.P. Knauer, K. Premaratne, *J. Appl. Phys.* **52(3)**, 1509 (1981).

6. D. Shapira, T.A. Lewis, L.D. Hulett., *Nucl.Instr. And Meth. A* **454**, 409 (2000).

7. D. Shapira, T.A. Lewis, L.D. Hulett and T. Ciao, *Nucl.Instr. And Meth. A* **454**, 409 (2000).

Status Report on the ISAC Radioactive Ion Beam Facility

Pierre Bricault

TRIUMF, 4004 Wesbrook Mall, Vancouver, BC, Canada, V6T 2A3

Abstract. The first phase of the ISAC radioactive ion beam facility at TRIUMF is well under way. The ISAC facility includes: a new building with 5000 m^2 of floor space, a beam line with adequate shielding for up to 100 μA protons at 500 MeV from the TRIUMF H$^-$ cyclotron. Due to the relatively low intensities of some of the radioactive species, continuous (cw) operation of the accelerator is required. The linear accelerator is composed of a 36 MHz split-ring RFQ and five interdigital H-type structures, operating at 106 MHz. The first radioactive ion beams were produced in the fall of 1998 and now the experimental program has started. A novel approach for the target station permits the operation of the ISOL target up to 100 μA. During the fall 1999, a prototype target was tested at 100 μA for temperature distribution. Last April we received funding for the second phase of ISAC. This funding includes an extension of the accelerator which will be able to deliver RIB up to mass 150 at energies up to 6.5 MeV/nucleon and the construction of necessary components to equip the second target station.

INTRODUCTION

The TRIUMF's ISAC is a second-generation Radioactive Ion Beams (RIB) facility, it uses the 500 MeV – 100 μA primary proton beam extracted from the TRIUMF H$^-$ cyclotron. The radioactive nuclei are produced by interaction of the proton beam on a thick target material. The atoms are released into an ion source and then the RIB is selected using the isotope separation on line (ISOL) method. A new beam line has been built to transport the 100 μA proton beam to two target stations. An experimental program has started using the low energy beam. First, we will give a brief description of the new concept for the target station that allows high power beam on target. Secondly, we will give a description of the mass separator, the accelerator for the ISAC-I and ISAC-II phases and the experimental areas. References[1,2] give a more exhaustive description of the shielding and target station concept used at ISAC.

TARGET STATION

The target station contains proton beam monitoring equipment, production target and ion source, a beam dump, and the front-end heavy ion beam optics. A strategy has been adopted in which the target station is contained in a heavily shielded building connected directly to a hot cell facility. This approach is based on the successful experience at TRIUMF of vertically servicing and remote handling of modular components embedded in a close-packed radiation shield, coupled with the requirement for quick access to the production target and of containment of any mobile activity. Careful design of both the modular components and the remote-handling systems was carried out to ensure the operational viability of this system. The ISAC target-handling concept and the ISAC target facility are based on the twenty years of experience at operating meson factories. The meson production target and beam stop areas of these facilities have power dissipation and radiation levels similar to, or even greater than, those expected at ISAC. Meson factory experience shows that the correct approach to handle components in high-current and thick-target areas is to place them in tightly shielded canyons. All the components are accessed vertically. The repair and services are carried out in a dedicated hot cell.

Three important factors not encountered in meson factory targets have to be addressed. These are: the containment of large amounts of mobile radioactivity; the high voltage required for beam extraction; and quick routine replacement of the short-lived target systems. In the present design, these issues are solved by placing the target in a sealed self-contained module, which can be transferred directly to the hot cell facility for maintenance. The main guidelines for radiation protection are the same as for nuclear industry.

The target stations are located in a sealed building serviced by an overhead crane. The target maintenance facility includes a hot cell, assembly area, decontamination facilities and a radioactive storage area. The target area is sufficiently shielded so that the building is accessible during operation at the maximum proton beam intensities.

Beam-line elements near the target are installed inside a large T-shaped vacuum chamber surrounded by close-packed iron shielding. This general design eliminates the air activation problem associated with high current target areas by removing all the air from the surrounding area. The design breaks naturally into modules; an entrance module containing the necessary primary beam diagnostics, an entrance collimator and a pump port; a beam dump module containing the water-cooled copper beam dump; a target module containing the target/ion source, extraction electrodes and first beam steering elements and heavy ion beam diagnostics; and two exit modules containing the optics and the necessary beam diagnostics for the transport of heavy ion beams.

The vacuum design seeks to eliminate the need for radiation-hard vacuum connections at the beam level by using a single vessel approach. The front-end components, with their integral shields, are inserted vertically into the T-shaped single large vacuum vessel. Most vacuum connections are situated where elastomer seals may be used. Only two beam-level connections exist; one at the proton beam entrance and one at the heavy ion beam exit.

The target stations are shielded by approximately 2 m of steel placed close to the targets. Outside this steel shielding, the operating radiation fields are sufficiently low enough that radiation damage to equipment is not a concern. The steel shielding is surrounded by an additional 2 to 4 m of concrete, which provides the required personnel protection during operation. To service the targets, shielding above the target station is removed giving access to the services at the top of the steel shielding plugs. Residual radiation fields at this level are low enough to allow hands-on servicing.

MASS SEPARATOR SYSTEM

The front end of the mass separator includes an electrostatic triplet followed by a doublet. The ion optics calculations were performed up to the third order. The mass separator will handle beams from mass 6 to 238 AMU, and source extraction voltages between 12 and 60 kV. Preliminary mass selection is achieved using a ± 60° pre-separator magnet. The pre-separator is followed by three matching-sections that allow enough flexibility to adapt the beam in order to obtain the same mass dispersion from either target station. The mass separator magnet including the entrance and exit matching sections are on a high voltage platform in order to allow reduction of the cross contamination and ease the magnet tuning

The optics components following the ion source are suspended at the bottom of the exit modules shielding plugs and are composed of an electrostatic triplet and a doublet, which prepare a parallel beam for the pre-separator magnet. The role of the pre-separator is to act as a cleaning stage in order to limit the contamination in the rest of the mass separator.

The mass separator magnet is the former Chalk River mass separator[3]. We modified the entrance and exit arms in order to adapt the magnet to our needs. The magnet is equipped with so-called α and ß coils[4]. The α coil allows correction of the magnet index and the ß coil allows adjustment of the second order correction provided by the pole face curvature. The magnet is on a high voltage platform that will allow rejection of the cross contamination. Such cross contamination comes from ions of different masses that have the same momentum, for example, due to collision with the residual gas. After acceleration, these ions will have different energies and momentum. Consequently, better separation becomes possible.

100 µA TARGET TEST

In collaboration with a group from Amparo Corporation[5] the ISAC facility was used to thermally test a target designed to operate with 100 µA proton beam at 500 MeV. Since the experiment was only a thermal test, no attempt to extract ion beams was made. The target was instrumented with an array of thermocouples to record the target temperature profiles under various conditions of heating by the proton beam. The 15 cm long prototype target was fabricated from 1200 Molybdenum foils and spacers that were diffusion bonded together. The final machining was done using an electrical discharge (EDM) to a specific shape. The target design goal was to provide nearly constant temperature at the center of the 0.95 cm radius target material. In principle the target density has to vary along the length to make the linear energy deposition profile approximately constant by compensating with target mass for beam fluence reduction through the target. The target shaping that resulted from thermal analysis incorporated two longitudinal fins to which the cooling lines were attached, with each fin having a machined thermal constriction with a widening taper along the target length. The experiment was performed last December with the incident proton beam current varied from 1.0 µA to 100 µA, back down to 10 µA, and cycled up again. The target required only 10 minutes to reach thermal equilibrium. The resulting temperature profiles and the predicted profiles do not agree perfectly. However, the deviation is less than 10 %. Nevertheless, this experiment confirms that it is possible to use appropriate computer analysis to develop reasonable target designs for producing radioactive ion beams under conditions of high power beam heating.

ISAC ACCELERATOR COMPLEX

The accelerator complex comprises an RFQ [4] to accelerate beams of q/A ≥ 1/30 from 2 keV/u to 150 keV/u and a LINAC (DTL) to accelerate ions of q/A ≥ 1/6 to a final energy between 0.15 MeV/u to 1.5 MeV/u. Both LINACs are required to operate in continuous mode (cw) to preserve beam intensity.

RFQ

The ISAC RFQ is an 8 m long, 4-rod split-ring structure operating at 35 MHz in CW mode[6,7]. The rods are supported by 19 rings spaced 0.4 m apart. The rings are unique in the RF surfaces and have been structurally decoupled from the mechanical support structure to improve dynamic stability. The quadrature positioning of the electrodes is crucial. A three dimensional theodolite inspection alignment method was used and the results are within the stringent specifications, ± 0.08 mm. The longitudinal field variation was also measured and found to be within ± 1 %, using the standard bead pull method. Signal level measurements gave a frequency of 35.4 MHz and a Q value of 8700 and a resonant shunt impedance of 283 kΩ*m. At the beginning of operation, we experienced a rapid growth of dark currents associated with field emission. Careful cleaning procedures and high power pulsing reduced the growth rate of dark currents by two order of magnitude. With no dark current present, the power requirement to reach an inter-electrode voltage of 75 kV is 78 kW. To reduce the longitudinal emittance growth in the RFQ the gentle buncher section was removed and replaced by an external four harmonics buncher. This has the advantage of a smaller longitudinal emittance and a 2 m shorter RFQ. Beam dynamic tests were performed using $^{14}N^+$ and $^{14}N_2^+$ beam coming from an off-line ion source. The RFQ beam capture efficiency at the nominal voltage is 80% in the bunched case (with only three harmonics) and 25% for the unbunched case, in good agreement with PARMTEQ predictions.

Drift Tube Linear Accelerator

Due to the requirement of continuous energy variability and preservation of the time structure, the DTL structure has been configured as a separated function DTL[8,9]. Five independently phased IH tanks operating at $\phi_s = 0°$ provide the main acceleration. Longitudinal focussing is provided by three independently phased, split-ring resonator structures positioned before the second, third and fourth IH tanks. Quadrupole triplets placed after each of the four IH tanks maintain transverse focussing. Full power tests are under way. The RF amplifier was operated in self excited mode delivering 3.6 kW and 87.5 kVolts to the drift tube gap. Stable operation for more than 100 hours was achieved. The last four DTL-IH tanks are built from a forged mild steel cylinder, 2.54 cm thick. The first tank is 26 cm long and 96 cm in diameter. The last four are 70 cm in diameter by 50, 77, 90 and 98 cm long. The two lids are made from 25 mm thick copper. The installation and alignment of the tanks will proceed during the summer and we are planning to have our first accelerated beam through the whole LINAC at the end of this year.

ISAC-II Accelerator

In the ISAC-II facility, the mass range is increased to 150 AMU and the energy is increased to 6.5 MeV/nucleon. In order to accept higher mass, the RIB will be charge boosted by either an ECR charge state booster or a low frequency RFQ and gas stripper combination[10]. In order to minimize the total accelerating voltage the beam will be stripped at around 400 keV/nucleon. The resulting beam will be accelerated by a series of super-conducting structures. The output energy will vary from 15 MeV/nucleon for light ions to 6.5 MeV/nucleon for the heaviest. The pre-stripper LINAC will be a room temperature IH structure capable of accelerating ions with a q/A ≥ 1/30 from 150 keV/nucleon to 400 keV/nucleon. The post-stripper LINAC will be a two-gap super-conducting quarter wave resonator because of the high velocity acceptance and proven stability.

ISAC EXPERIMENTAL AREAS

TRINAT facility

Laser trapping of neutral atoms is a very rapidly growing field. The TRINAT facility is aimed to study symmetry properties of the Standard Model using isotopically pure atoms confined in space by magneto-optical forces. TRINAT focuses on β–ν correlation in $^{38m}K(0^+ \rightarrow 0^+)$ transition.

Lifetime measurement

Precise measurements of the intensities for super-allowed Fermi $0^+ \rightarrow 0^+$ β decays provides a demanding test for the CVC hypothesis. This station is equipped with a fast transport tape system (5 m/s) a very efficient 4π gas counter and two 80% HPGe counters[11].

LTNO

The Low Temperature Nuclear Orientation setup at ISAC is operated in collaboration with the Oregon State University, University of British Columbia, Michigan State University, Georgia Institute of Technology, and TRIUMF. The cryogenic system was tested at 10 mK.

β-NMR

We developed at ISAC a β-decay nuclear magnetic resonance spectrometer (β-NMR) which will accept an intense beam of low energy, highly polarized radioactive ions produced by the ISAC facility. Although the main application is condensed matter physics, the polarizer itself may also have applications in nuclear physics.

Polarizer

Ions emerging from an ISOL target are unpolarized. A fast collinear polarization technique is used to generate high nuclear polarization in all the alkali metal atoms such as Li. This is possible since the ground state of a neutral alkali atom can be excited with visible or near visible lasers. Firstly, the Li^+ ion beam is neutralized by passing through a Na vapour cell. The neutral Li beam drifts 1.7 m in a small longitudinal magnetic holding field of 1 mT while being pumped with circularly polarized laser light brought in along the beam axis. The resulting nuclear polarization is longitudinal with respect to the beam axis.

DRAGON

The DRAGON facility aims to measure radiative capture reactions. It uses the inverse kinematics to measure capture rates, in which the beam is the heavy nucleus and the target is Hydrogen or Helium. The main components of the DRAGON facility are a windowless gas target; a two-stage electromagnetic mass separator; a detector facility for the reaction products. The beam suppression is expected to be of the order of 10^{-12}. A BGO gamma array to be positioned around the gas target will complete the system.

REFERENCES

1 P. Bricault *et al*, Proc. of the 15th Int. Conf. on Cyclotron and their Application 1998, Caen, France, Ed. Eric Baron and Marcel Lieuvin, p. 347.
2 P. Bricault M. Dombsky, P. Schmor and G. Stanford, Nucl. Instr. and Method B126 (1997) p. 231-235.
3 H. Schmeing, J. C. Hardy, E. Hagberg, W. L. Perry, J.S. Wills, J. Camplan and B. Rosenbaum, Nucl. Instr. and Methods, 186 (1981) 47-59.
4 J. Camplan and R. Meunier, Nucl. Instr. and Methods, 186 (1981) 445-452.
5 W. Talbert, private communication and RNB2000, to be published.
6 G. Stanford, P. Bricault, G. Dutto, R. Laxdal, D. Pearce, R. Poirier, R. Roper, Proc. XIX Int. LINAC Conf., ANL, Chicago, USA, p. 980.
7 R. Poirier, P. Bricault, K. Fong,. A. Mitra, and W. Uzat, Proc. LINAC2000, SLAC, Monterey, USA, to be published.
8 R. E. Laxdal and P. Bricault, Proc. Of the XVIII Inter. Linear Accelerator Conf. 1996, Genève, Suisse, p. 435

9. R. E. Laxdal, P.G. Bricault, T. Reis, D.V.Gorelov, Proc. of the 1997 Particle Accelerator Conf. Vancouver, Canada, p. 1195
10. P. Bricault, Proc. XX Int. LINAC Conf., SLAC, Monterey, USA, to be published.
11. G. Ball, "Tests of the Standard Model from superallowed Fermi β-decay studies at ISAC" this conference.

A Negative Surface Ionization Source for RIB Generation

H. Zaim*, Y. Liu, S. N. Murray and G.D. Alton

Oak Ridge National Laboratory, P.O. Box 2008, Oak Ridge, TN 37831-6368, USA
* Ph.D. Student, Université de Versailles-Saint-Quentin, Versailles, France

Abstract. An efficient negative surface ionization source has been designed, fabricated, and initial tests begun for potential on-line use in generating radioactive ion beams of members of the group VIIA elements (F, Cl, Br, I, and At) for the Holifield Radioactive Ion Beam Facility research program. The source utilizes direct-surface ionization to form negative-ion beams resulting from interactions between highly electronegative atoms or molecules and a spherical-sector LaB_6 surface ionizer maintained at ~1722 °C. Despite its widely publicized propensity for being easily poisoned, no evidences of this effect were experienced during testing of the source. The source has been extensively evaluated off-line in terms of ionization efficiency for generating beams of Br^- by feeding $AlBr_3$ vapor at low feed rates into the source. The results of initial testing indicate that the source is reliable, stable and easy to operate, with nominal efficiencies of 15 % for Br^- beam generation when account is taken of the fractional thermal dissociation of the $AlBr_3$ carrier molecule. The design features and principles of operation of the source are described and initial performance, operational parameter and beam quality (emittance) data are presented in this article.

INTRODUCTION

Ion sources based on the surface ionization principle are generally characterized by a high degree of ion beam purity (chemical selectivity), thermal energy spreads (~2 kT << 1 eV)), and limited range of species capability. The ionization efficiency can be high or low, depending on the electrochemical character of the species in relation to the work function of the ionizing surface. Because of the fact that the surface-ionization process is highly chemically selective, it can be used to great advantage for radioactive ion beam (RIB) applications to eliminate isobaric contaminants that may compromise experimental results with these beams. Experimental methods and techniques for negative ion production by surface ionization have been reviewed by Kawano et al. [1-3].

Surface ionization has not been utilized frequently for generation of negative ion beams – principally due to the lack of chemically stable low-work-function materials for use as ionizers. Unfortunately, few chemically stable materials are available for this purpose in contradistinction to its positive ionization complement where several high-work-function metals may be chosen. LaB_6 is usually used for negative surface ionization because of its relatively low work function (φ: 2.3 to 3.2 eV [4–8]) and ready availability, despite its widely publicized propensity for poisoning [9,10]. The poisoning mechanism appears when LaB_6 interacts with residual gases in the vacuum system, usually under high flow rate conditions or higher than optimum pressure conditions. The effect raises the work function of the LaB_6 surface, thereby reducing the probability of ionizing electronegative atoms as they evaporate from the surface. Under high flow-rate conditions, the poisoning process also affects the reliability of operation of sources equipped with this material through time varying fluctuations of ion beam intensity caused by variations in work function [10]. A raising of the work function causes an exponential diminution of the probability for negative ion formation and, consequently, a reduction in intensity of extracted negative ion beams.

Despite this problem, sources based on the use of LaB_6 ionizers have been described in the literature [10-12], including their use at ISOL facilities for negative ion generation of high-electron-affinity radioactive species [12]. A LaB_6 surface ionizer was also chosen for use with the source described in this article because the poisoning effects were not expected to be pronounced at existing flow-rates that characterize on-line source operation at ISOL based RIB facilities.

THEORY OF NEGATIVE ION FORMATION

The process of direct-surface ionization is statistical in nature, and therefore statistical and thermodynamic arguments can be used to determine the degrees of positive or negative ion formation. For thermodynamic equilibrium processes, the ratio of ions to neutrals that leave an ideal surface can be predicted from Langmuir-

Saha surface ionization theory appropriate for either positive or negative ion formation. The form of the Langmuir-Saha equation for the probability of negative-ion formation of neutral particles of electron affinity, E_A, interacting with a hot metal surface at temperature, T, and low work function, P_i, is given by

$$P_i = \frac{\omega_-}{\omega_0}\left(\frac{1-r_-}{1-r_0}\right)\exp\left(\frac{E_A - \phi}{kT}\right) \\ \times \left[1 + \frac{\omega_-}{\omega_0}\left(\frac{1-r_-}{1-r_0}\right)\exp\left(\frac{E_A - \phi}{kT}\right)\right]^{-1} \quad (1)$$

where r_- and r_0 are the reflection coefficients of the negative or neutral particle at the surface and ω_- and ω_0 are statistical weighting factors for the negative ion or neutral atom, respectively. ω_- and ω_0 are related to the total spin of the respective species given by

$$\omega = 2\sum_i s_i + 1$$

where s_i is the spin of the electron. From Eq. 1, it is evident that negative ion yields can be enhanced by lowering the work function, ϕ, or increasing the surface temperature, T, for elements where $E_A \leq \phi$.

DESCRIPTION OF THE SOURCE

The salient design features of the negative surface ionization source are schematically illustrated in Fig. 1 which shows a cross sectional side view of the target reservoir, the vapor transport tube, ionization region and extraction electrode system of the source. The target material reservoir is positioned within the inner diameter of a series-connected, resistively-heated, Ta tube designed to reach temperatures exceeding 2000 °C. The vapor transport tube is typically heated resistively to ~ 1400-2000 °C by passing a current through the tubular structure. In order to provide halogen atoms for evaluating the efficiency of the source, AlB_3 vapor was fed into the target material reservoir from the vapor feed system described below.

According to Eq. 1, electronegative species that strike and are subsequently evaporated from the spherical-geometry LaB_6 ionizer surface have a finite probability of being negatively ionized, these species are then extracted by applying a positive voltage to the extraction electrode.

Ionizer Design

The spherical geometry ionizer (spherical radius: 2.5 mm; diameter: 4.3 mm) is machined from a solid LaB_6 rod and pressed into a 6 mm diameter Ta holder with 0.63 mm deep slots machined in the outer periphery through which vapor flows from the transport tube into the ionization region of the source. The temperature distribution along the vapor transport system and at the LaB_6 surface ionizer were computed as function of the heating current by use of the thermal transport code ANSYS [13]. These results were used to design the transport tube so that the tube can be operated at temperatures up to ~ 2000 °C while achieving a maximum temperature at the surface of the ionizer of ≤ 1740 °C, the thermal dissociation temperature of LaB_6 (calculated by use of the chemical reaction code described in Ref. 14). This effect was achieved by adjusting the thickness of the transport tube and the interface position of the return current bus relative to the position of the LaB_6 ionizer holder so that an optimum temperature of ~1722 °C could be maintained during normal operation of the source.

Ion Optics

Figure 2 displays the ion optics of the ion extraction system of the source. As noted, the spherical geometry electrode system is designed to focus the beam through a small aperture (φ: 0.41mm). The perveance, P_C, for space charge limited flow of Br^- through the electrode system is $P_C = 1.0 \times 10^{-9}$ [A/{V_{ex}}$^{3/2}$] as calculated by use of the computer code described in Ref. 15. Of course, this figure of merit has no practical meaning at the very low flow rates used in these experiments or that will be present during on-line operation of the source where space charge effects are negligible.

The Vapor Feed System and Vapor Transport

The system (see, e.g., Fig. 3) consists of a reservoir, located external to the main vacuum system of the source, for holding a relatively volatile halogen feed material (e.g., $AlBr_3$ or $AlCl_3$). The material is fed through a small diameter transfer tube of effective radius, a, and length, l, that is designed to conductance limit the flow rate of vapor from the reservoir to the source. The transfer tube is isolated from the heat-sink effects of the vacuum chamber wall by means of a thermally isolated vacuum feed-through. Feed material, placed in the reservoir, is raised to temperature, T_R, where it is vaporized, creating a pressure gradient, $(p_R - p_{IS}) / l = \Delta p / l$ along the tube due to differences in pressure in the reservoir, p_R, and in the ionization region of the source, p_{IS}. The reservoir is heated by thermal conduction from the independently heated transfer tube with temperature control maintained by adjusting the transfer tube temperature.

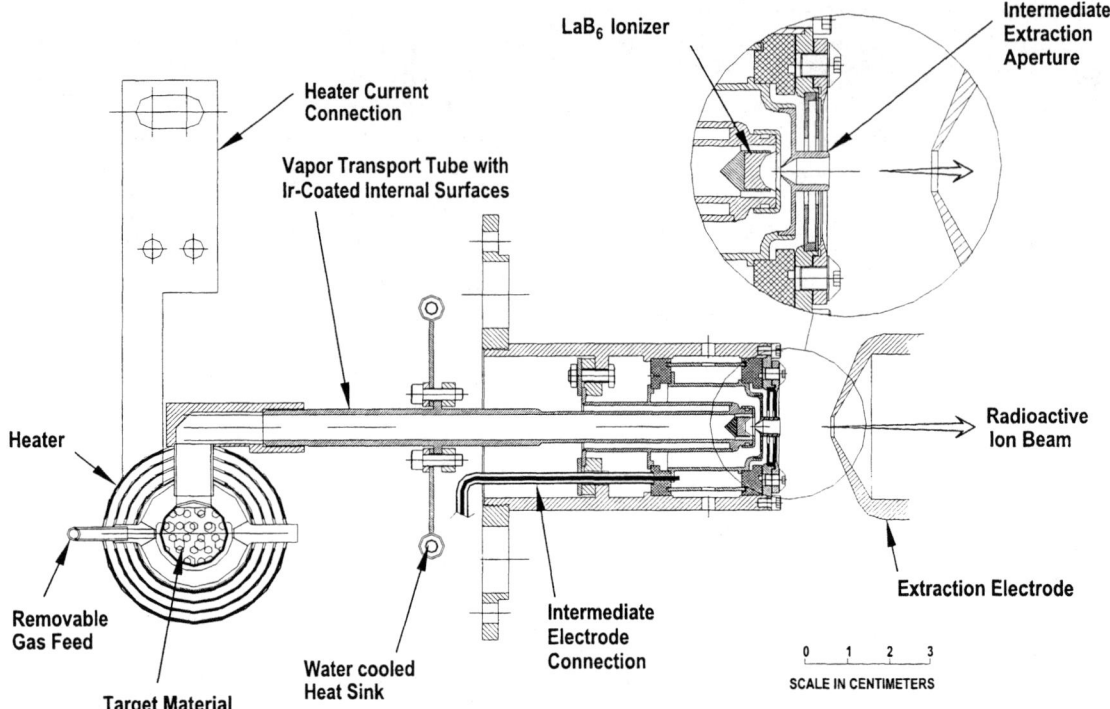

FIGURE 1. Schematic drawing of the negative ionization source equipped with a spherical geometry ionizer.

Vapor pressure

The vapor pressure of the $AlBr_3$ feed material, used to investigate the ionization efficiency of the source, was computed by use of the thermo-chemistry computer code, ThermoCalc [16]. During all measurements, the reservoir was operated between 20 and 22°C, correlating to a range of vapor pressures between 3.5×10^{-1} Pa and 4.6×10^{-1} Pa.

Flow rate of Br atoms into the ion source

The rate of flow of halogen molecules through the transfer tube into the source can be estimated from the familiar relation (see, e. g., Ref. 17), given by

$$dN_M/dt = \{2\pi a^3/3k_BT\}v_M\Delta p/l \qquad (2)$$

In Eq. 2, Δp is the pressure drop across the transport tube of length, l, and effective radius, a; T is the average temperature from the reservoir to the ionization volume of the source and ; v_M is the average velocity of an $AlBr_3$ molecule in transit from the

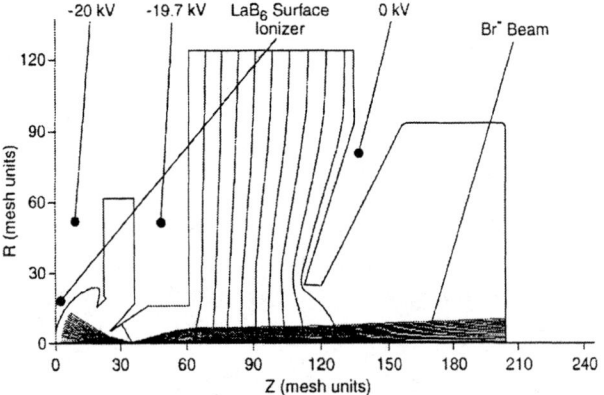

FIGURE 2. Ion optics of the spherical-geometry surface-ionization source for extraction of Br^-. 1 Mesh Unit = 0.1 mm.

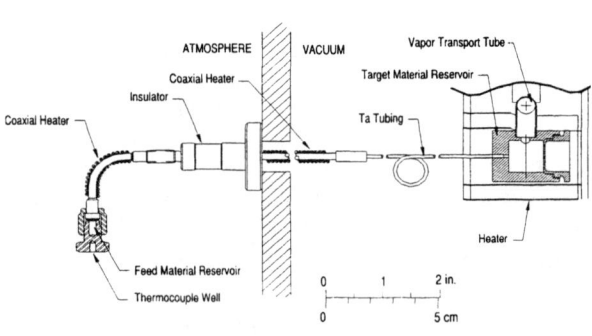

FIGURE 3. Vapor transport system used to feed $AlBr_3$ vapor into the source

feed material reservoir to the ionizer volume of the source. At the operational temperature of the vapor transfer tube, $AlBr_3$ thermally dissociates. Therefore, the number of halogen atoms entering the source per unit time, dN_A/dt, is equal to

$$dN_A/dt = 3F_D dN_M/dt = F_D\{2\pi a^3/k_B T\}v_M \Delta p/l \quad (3)$$

where F_D is the dissociation fraction for releasing atomic Br atoms into the source from the $AlBr_3$ molecule during transit through the hot vapor transport tube and the factor, 3, is the number of Br atoms in the $AlBr_3$ molecule. F_D can be estimated by computing the equilibrium composition of the molecule as a function of temperature, T. The equilibrium composition versus temperature of $AlBr_3$ in a Ta tube is displayed in Fig. 4 as computed with the chemical reaction and equilibrium code described in Ref. 14. As noted the equilibrium dissociation fraction for $AlBr_3$ is 0.66 at $T \geq 1500\ ^0C$. For optimum efficiency, the LaB_6 ionizer operates at ~1722 ^0C.

The rate of neutral Br atoms striking the LaB_6 ionizer

The number of neutral Br atoms, dN_{IS}/dt, striking the ionizer per unit time can be expressed as

$$\begin{aligned}dN_{IS}/dt &= \{dN_A/dAdt\}A_{IS} \\ &= \{F_D\{2\pi a^3/k_B T\}v_A \Delta p/l\}A_{IS}/A_T \\ &= n_A v_A\{A_{IS}\}/4 \end{aligned} \quad (4)$$

where v_A is the average velocity of a Br atom within the volume surrounding the ionizer surface at temperature T; n_A is the number of particles per unit volume in the region of the ionizer, A_{IS} is the effective area of ion extraction from the ionizer surface, determined from ion optics data (see Fig. 2) and A_T is the total surface area of the ionization chamber.

The negative ion beam intensity

Since the negative ion beam intensity is proportional to the number of neutral Br atoms striking the ionizer surface per unit time, the Br^- intensity can be estimated by multiplying Eq. 4 by the probability of negative ion formation given by Eq. 1.

Operational Parameters

In order to optimize the performance of the source, the dependence of Br^- beam intensity on the following parameters must be known: 1) Br^- beam intensity versus feed material reservoir temperature; 2) Br^- beam intensity versus vapor transport tube current (ionizer temperature); and 3) Br^- beam intensity versus extraction voltage.

Br^- beam intensity versus feed material reservoir temperature

The feed material reservoir temperature was held essentially constant during all measurements (22 ^0C) correlating to an equivalent flow rate of $AlBr_3$ into the source of ~ 3.33 μA at a vapor transport tube current of 368 A.

Br^- beam intensity versus vapor transport tube temperature

Onsets are observed in the $^{79}Br^-$ beam intensity versus vapor transport tube current (ionizer temperature), shown in Fig. 5, for data taken after the *First Day* and *Second Day* of source operation that we attribute to the dissociation of $AlBr_3$ ($AlBr_3$ = $AlBr$ + $2Br$). The surface temperature of the LaB_6 ionizer, measured with an optical pyrometer, is also shown.

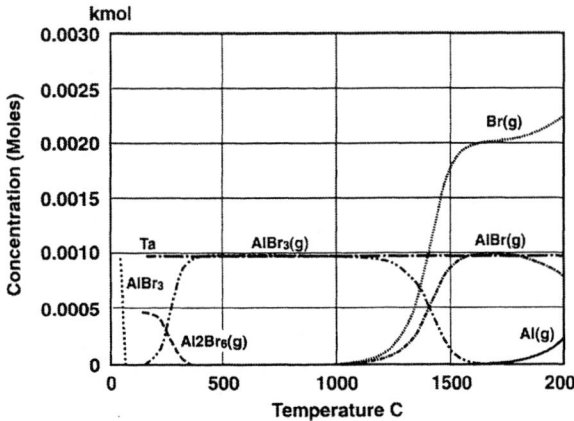

FIGURE 4. Equilibrium composition of $AlBr_3$ in a Ta tube.

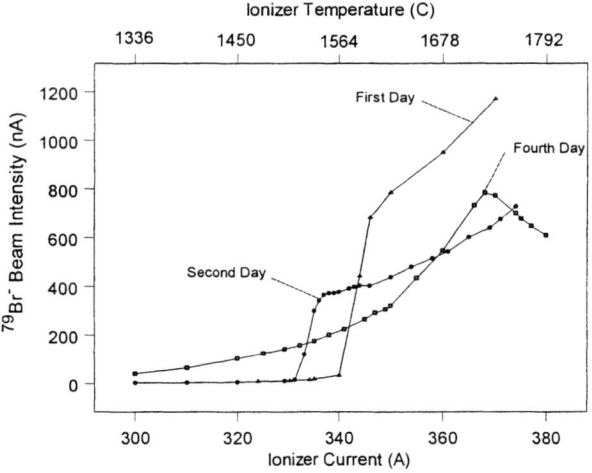

FIGURE 5. $^{79}Br^-$ ion beam current and ionizer temperature versus ionizer current.

However, after more than two days of operation, the $^{79}Br^-$ intensity versus ionizer current relation reverted to the curve designated as *Fourth Day*. As noted, the beam intensity increases monotonically with vapor transport tube current until it reaches an optimum value at ~ 368 A, beyond which the intensity drops. (This current correlates to a temperature of ~ 1722 ^0C at the surface of the LaB_6 ionizer.) The decrease in intensity beyond the optimum value for the *Fourth Day* of operation is presumably attributable to the onset of thermal dissociation of LaB_6 (~1740 ^0C).

Br^- beam intensity versus extraction voltage ΔV_{ex}

Figure 6 displays a typical Br^- ion beam intensity versus extraction voltage curve. As noted, the extracted ion current increases with extraction voltage, ΔV_{ex}, until it reaches a constant value at ~ 150 V, suggesting that the Br ions are extracted as fast as they are evaporated from the hot LaB_6 surface.

Source Performance: Experimental Results

The negative surface ionization source (Fig. 1) was characterized by feeding $AlBr_3$ vapor from a feed material reservoir located external to the main vacuum system, as displayed in Fig. 3. All efficiency measurements were made with the feed material reservoir temperature set at 22^0C and the transport tube current fixed at 368 A, correlating to an ionizer temperature of ~ 1722 ^0C.

Mass spectrum

The mass spectrum obtained while feeding $AlBr_3$ in to the source is shown in Fig. 7. The selective nature of the surface ionization process is clearly demonstrated

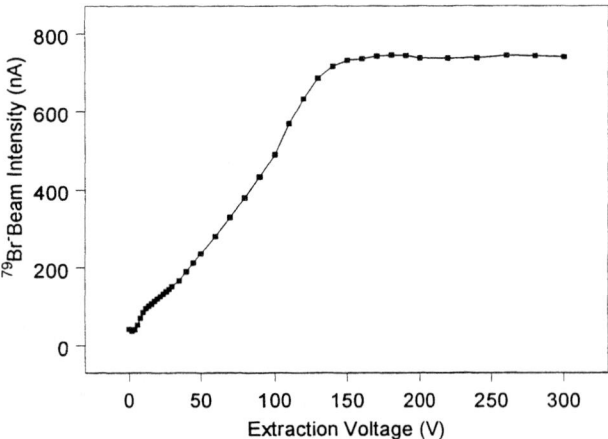

FIGURE 6. Br^- ion beam intensity versus extraction voltage.

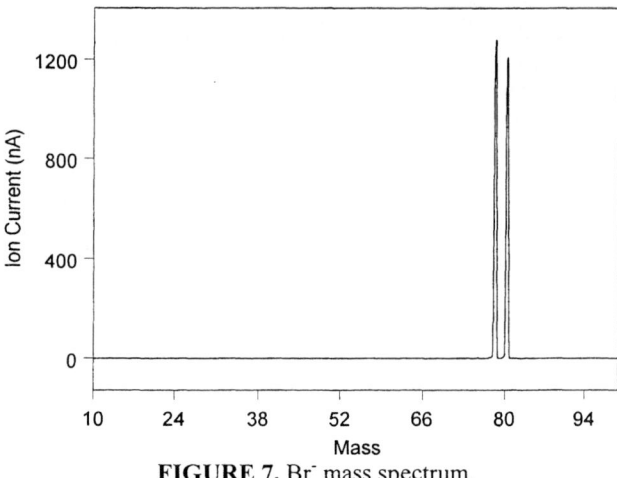

FIGURE 7. Br^- mass spectrum.

by the cleanliness of the spectrum. The only masses with significant intensity are the two isotopes of interest (i.e., ^{79}Br and ^{81}Br).

Emittance Data

Isobaric contamination problems can cause serious difficulties in interpretation and analysis of on-line experimental data and consequently, seriously compromise experimental results. In such cases, high quality beams (low energy spread, low-emittance) are very important, making possible mass resolution of such contaminants with existing magnetic isobar separation systems. Fig. 8 displays a typical emittance diagram for a 20 keV Br^- beam generated with the source. As noted the emittance is quite small.

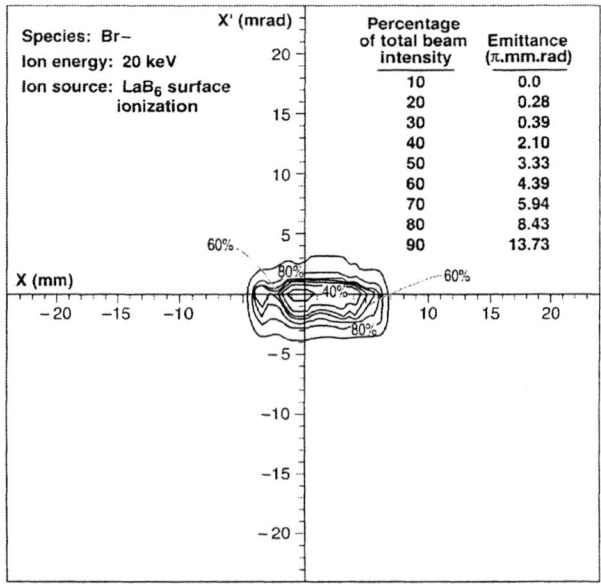

FIGURE 8. A typical emittance plot.

Br⁻ ionization efficiency estimates

According to the results of thermal equilibrium composition calculations, displayed in Fig. 4, the dissociation fraction of $AlBr_3$ is 0.66 at T= 1722 ^{0}C, the operational temperature of the ionizer. Thus, two of three of the Br atoms are available for ionization. The feed material reservoir was operated at 22 ^{0}C for all efficiency measurements, correlating to a vapor pressure of $p_R \cong 0.46$ Pa. The halogen atom flow rate into the ionization region of the source was calculated from Eq. 3, yielding an equivalent flow-rate of atomic Br into the source of ~ 10 μA. The overall efficiency for negative ion formation is then computed from the following relation:

$$\eta \cong I^-(Br)/\{dN_A/dt\} \qquad (7)$$

Mass analyzed ^{79}Br⁻ intensities, ranging 0.8 and 1.5 μA, were typical of the currents recorded during source evaluation, corresponding to ~ 10 to 17 % total efficiency. Thus, the over all efficiency for generating beams of Br⁻ is quite high (~24.3% for the *First Day* of operation), as required for on-line applications. Fig. 9 displays efficiency versus temperature of the LaB_6 ionizer.

CONCLUSIONS

The surface ionization source has proved to be a stable, reliable, versatile, and efficient means for generating beams of highly electronegative species. These characteristics and flexibility make it a viable candidate for use in several research and applied science applications, including RIB generation. Off-line tests, the source has demonstrated that it can be used to efficiency ionize atomic species such as Br. The source was found to be easy to operate with no evidence of poisoning effects which plague more traditional negative surface ionization sources equipped with LaB_6 ionizers that operate under higher flow rate conditions [10]. Preliminary estimates suggest that source can be used to ionize all halogens except F because it is transported in strongly-bound, molecular form. Mass analyzed ^{79}Br⁻ intensities, ranging between 0.5 and 0.75 μA, were typical of the currents recorded during source evaluation, corresponding to ~ 10 to 15 % total efficiency.

ACKNOWLEDGEMENTS

Research sponsored by the Oak Ridge National Laboratory, managed by UT-Battelle for the U.S. Department of Energy under contract number DE-AC05-00OR22725.

REFERENCES

1. H. Kawano and F. M. Page, Inst. J. Mass. Spectr. and Ion Phys. **50** (1983) 1.
2. H. Kawano, Y. Hidaka and F. M. Page, Int. J. Mass Spectr. and Ion Phys. **50** (1983)35.
3. H. Kawano, Y. Hidaka, M. Suga, and F. M. Page, Int. J. Mass Spectr. and Ion Phys. **50** (1983)77.
4. J. M. Lafferty, J. Appl. Phys, **22** 299 (1951).
5. V. S. Fomenko, Emission Properties of Materials, JPRS-56579, (NTIS, U.S. Dept. Comm., Springfield, VA 1972).
6. H. Ahmed and A. M. Broers, J. Appl. Phys, **43** 2185 (1972).
7. S. Hosoki, S. Yamamoto, K. Hayakawa, and H. Okano, Jpn. J. Appl. Phys. **Suppl. 2, Part 1**, 285 (1974).
8. H. Yamauchi, K. Takagi, I. Yuito and U. Kawabe, Appl. Phys. Lett. **29** (1976) 638.
9. A. Avdienko and M.D. Malev, Vacuum **27** (1977) 583.
10. G.D. Alton, M.T. Johnson and G.D. Mills, Nucl. Instr. and Meth. **A 328** (1993) 154.
11. N. Kashihira, E. Vietske and G. Zellermann, Rev. Sci. Instr. **48** (1977) 150.
12. B. Vosicki, T. Bornstad, L.C. Carraz, J. Heinemeyer and H.L. Ravn, Nucl. Intr. and Meth. **186** (1981) 307.
13. ANSYS is a finite element code marketed by ANSYS, Inc., Cannonsburg, PA, 15317, USA.
14. HSC Chemistry is a chemical reaction and chemical equilibrium software package marketed by Outokumpu Research Oy, Pori, Finland.
15. PBGUNS is an electron/ion optics simulation code, developed by Thunderbird Simulations, Garland, TX, USA.
16. ThermoCalc is a thermodynamic-equilibrium, phase diagram computer software package developed by the Royal Institute of Technology, Stockholm, Sweden.
17. *The Kinetic Theory of Gases*, R. D. Present, McGraw-Hill Book Company (1958), Chap. 4, p. 63.

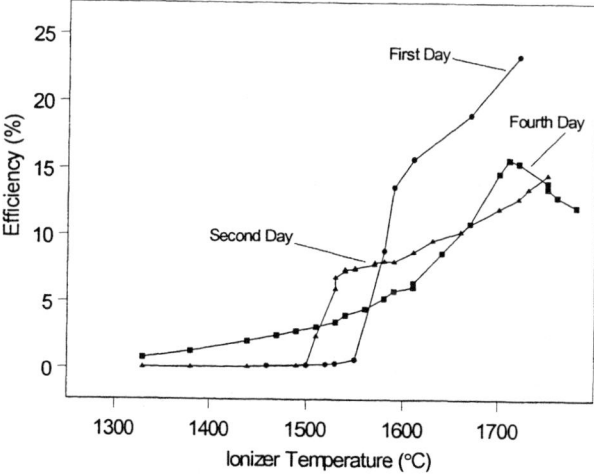

FIGURE 9. Efficiency versus surface ionizer temperature.

Characterization and Fabrication of Target Materials for RIB Generation

R.F. Welton,[a] M.A. Janney,[b] P.E. Mueller,[a] W.K Ortman,[c] R. Rauniyar,[c] D.W. Stracener[a] and C.L. Williams[c]

[a] *Physics Division, Oak Ridge National Laboratory, Oak Ridge, TN.*
[b] *Metals and Ceramics Division, Oak Ridge National Laboratory, Oak Ridge, TN.*
[c] *Oak Ridge Institute for Science and Engineering, Oak Ridge, TN.*

Abstract. This report discusses two techniques developed at the Oak Ridge National Laboratory (ORNL) that are employed for the fabrication and characterization of targets used in the production of Radioactive Ion Beams (RIBs). First, our method of in-house fabrication of uranium carbide targets is discussed. We have found that remarkably uniform coatings of UC_2 can be formed on the microstructure of porous C matrices. The technique has been used to form UC_2 layers on highly thermally conductive graphitic foams. Targets fabricated in this fashion have been tested under low-intensity proton bombardment and yields of selected radioactive species are reported. This report also describes an off-line test stand for the investigation of effusive and diffusive transport in RIB target/ion sources. Permeation rates of gases and vapors passing through a high temperature membrane or through an effusive channel constructed from the material under investigation are recorded. Diffusion coefficients and adsorption enthalpies, which characterize the interaction of RIB species with materials of the target/ion source, are extracted from the time profile of the recorded data. Examples of diffusion, effusion and conductance measurements are provided.

INTRODUCTION

The Holifield Radioactive Ion Beam Facility (HRIBF) located at ORNL is devoted to the production of Radioactive Ion Beams (RIBs) for nuclear structure and nuclear astrophysics research. Radioactive ions are produced by directing light-ion beams accelerated by the k=100 Oak Ridge Isochronous Cyclotron (ORIC) onto thick, hot, refractory targets. The radioactive atoms diffusing and effusing from the target material are ionized and injected into the 25 MV tandem accelerator [1, 2]. The most significant losses of RIB particles occur within the target/ion source, a single, high temperature enclosure, which houses the target material, transfer line and ion source [3]. This report describes two techniques developed jointly by the Metals and Ceramics (M&C) Division and the HRIBF at ORNL with the goal of improving our understanding of these loss processes and improving the thermal properties of uranium targets.

UC FOAM TARGETS

Performing nuclear structure experiments with RIBs of neutron-rich nuclei, at energies around the coulomb barrier, has been a long-standing goal of the physics community [4]. Intense, low energy (keV) neutron-rich beams produced from U targets have been available at CERN-ISOLDE for some time. Targets employed there have been either a U impregnated graphite cloth or pressed pellets of UC [5]. The HRIBF has recently delivered neutron-rich isotopes of Ag (500 MeV), Sn (300 MeV) and Te (300 MeV) for nuclear structure experiments using targets of Reticulated Vitreous Carbon (RVC) coated with a layer of UC_2 [1].

In order to improve the thermal properties of the HRIBF target and facilitate procurement, we have developed an in-house technique to form layers of UC_2 on arbitrary C matrices. We have chosen a highly thermally conductive graphite foam as a host matrix for U as opposed to the glassy (RVC) low thermal conductivity matrix employed in the first-generation

HRIBF targets. The thermal conductivity of the graphitic material is ~100 greater than the conductivity of the glassy material [6]. Thermal calculations show that use of this host material will virtually eliminate large temperature gradients that exist across the RVC due to beam-heating effects. In addition, eliminating the high peak temperatures in the center of the material will also allow the application of significantly greater proton beam intensities.

ORNL Graphitic Foams

The M&C Division at ORNL has recently developed a highly thermally conductive mesophase pitch-based graphitic foam for thermal management applications like automobile radiators [6]. The material has an open, interconnected, microcellular structure and is available in densities ranging from 10-50% of that of solid graphite with mean pore sizes of ~300 μm and specific surface areas larger than 4 m^2/g. The cell wall consists of an aligned graphitic structure resulting in large thermal conductivities, greater than 180 W/mK at 25 C. Fig. 1 shows a scanning electron micrograph (SEM) of this type of material.

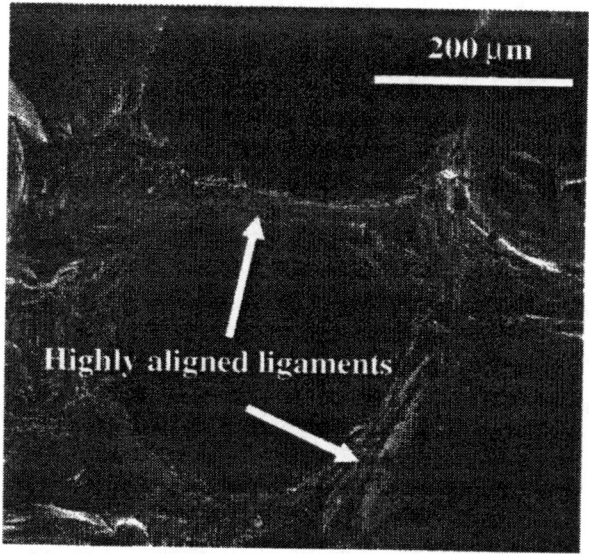

Figure 1. SEM of the ORNL graphitic foam.

Target Fabrication

Our fabrication technique is similar to that used at CERN-ISOLDE where aqueous uranyl nitrate ($UO_2(NO_3)_2$) is applied to the graphite cloth, dried, reapplied, repeated for a number of cycles, until a sufficient amount of $UO_2(NO_3)_2$ is deposited. Once the desired amount of $UO_2(NO_3)_2$ is in place, the target is fired to 600 C to remove the nitrate, leaving a layer of UO_2 on the carbon fibers. During operation, the target is brought to 1900-2000 C and conversion to the carbide takes place [7]

$$UO_2 + 4C \rightarrow UC_2 + 2CO. \qquad (1)$$

Our approach is similar with the exception that the uranyl nitrate hexahydrate ($UO_2(NO_3)_2 \cdot 6H_2O$) is melted directly into the C host with no additional water added. The C foam is cut with a cork punch to fit just inside a glass vessel. Uranyl nitrate granules are then added to the vessel above the C foam and allowed to melt into (infiltrate) the foam as the temperature is raised through the melting point of uranyl nitrate to ~80 C. A vacuum desiccator then is used to draw air pockets out of the interior of the target. The vessel is then opened to air and dehydrated at 70 C for several hours. The target is then removed from the glassware, inserted into a graphite crucible, and fired to 600 C in an Ar tube furnace for several hours, effectively removing the nitrate. If more U is desired, these infiltration steps are repeated. Once a sufficient quantity of U has been deposited, the target is then removed and fired at 1900 C (~12 hours) in the vacuum furnace to insure conversion to UC_2 (Eq. 1).

We have found this process to be quite efficient, allowing ~70% of the space within the foam to be filled with $UO_2(NO_3)_2 \cdot 6H_2O$ during each infiltration step (using a vacuum desiccator). Conversion to UO_2 involves a ~7.5 fold decrease in molar volume that allows a large quantity of UO_2 to build up in the interior space of the foam by subsequent infiltration and firing steps. Microscopy of the final UC_2 targets revealed quite uniform structure with UC_2 evenly distributed through the interior of the target body. Mass analysis revealed that ~4 μm layers of UC_2 could be formed as a result of each uranyl nitrate infiltration step, and the effects of subsequent infiltration steps on coating thickness were approximately additive. The maximum U:C weight ratio seems to be 6:1 which can be achieved by 3 infiltration steps on the 20% dense foam. Analysis also shows that ~ 3.8 g/cm^3 of U can be achieved by 10 infiltration steps of 50% dense foams with an approximate thermal conductivity of 12 W/mK at room temperature.

RIB Yields from Target

A single on-line test of this target material has been performed to date. The target was fashioned by applying 3 infiltration steps, described above, to a cylinder (len.=15 mm; dia.=15 mm) of 23% dense ORNL foam. This resulted in a target with a total thickness of 2.3 g/cm^2; U:C weight ratio of ~3.7:1; UC$_2$ layer thickness of ~8 μm; overall porosity ~76% and estimated thermal conductivity of 20 W/mK.

The target was bombarded with a low-intensity beam of 10 nA of 30 MeV protons at the UNISOR mass separator located in the HRIBF. Since this low-beam intensity produced no beam heating effects, this was a test of the release properties of the target and not the thermal properties. The target was close coupled to our standard Electron Beam Plasma Ion Source (EBPIS) [2] and detection was accomplished using the UNISOR Ge(Li) detector and tape system. The target material was maintained at a constant temperature of 1800 C during these experiments. The table below shows an abbreviated list of measured yield and cumulative production rate for each species [1]. The yields have been scaled to ions/s/μA to facilitate extrapolation to yields expected during use with the cyclotron (1-20 μA).

TABLE 1. Yields of selected isotopes produced from the UC target and cumulative production rates. Yields and production rates are listed in the order of masses shown in column 2.

Element	Selected mass	Cumulative production rate for each listed mass 10^5 ions/s/μA	RIB Yield for each listed mass 10^5 ions/s/μA
Ge	78, 80, 82	119, 106, 28	8, 0.7, 0.1
As	79, 81, 82	222, 326, 287	4, 1.5, 0.13
Se	83	689	2.1
Br	84, 87, 88	1080, 1090, 672	19, 26, 11
Kr	87, 89, 92	1730, 2090, 362	30, 29, 2.6
Rb	89, 91, 93	2480, 4120, 1910	45, 27, 6.3
Sr	91, 93, 96	3980, 5750, 1680	59, 7.4, 0.1
Ag	114, 117, 121	6790, 5160, 937	99, 65, 15
Cd	119, 121, 123	4400, 4130, 2130	77, 37, 9.6
In	119, 122, 127	4390, 6090, 1500	57, 68, 24
Sn	127, 130, 132	6050, 1320, 997	214, 13, 2.2
Sb	127, 129, 134	5640, 7430, 229	326, 231, .9
Te	129, 134, 136	5380, 3530, 456	63, 203, 8.8
I	130, 134, 138	306, 6950, 787	70, 231, 7.4
Xe	135, 137, 138	7110, 7160, 5200	108, 38, 53
Cs	138	6670	15

The target/ion source efficiency can be obtained by dividing column 4 by column 3 in the table above. These values are species dependent and ranged from less than 1% to ~20% and decreased sharply with decreasing half-life for a given element. Overall, the yields were similar to those of the earlier generation HRIBF UC-RVC target [1] and comparable to efficiencies obtained at CERN-ISOLDE [5]. Thus, we have shown that this target has suitable release properties and we will next investigate the target's thermal properties under high-intensity bombardment with 1-20 μA from the cyclotron. We anticipate higher overall yields will be achieved during high intensity bombardment as a result of the superior thermal characteristics of this material.

EFFUSION AND DIFFUSION MEASUREMENT TECHNIQUES AND INSTRUMENTATION

Our goal was to develop off-line, stand alone, instrumentation capable of directly measuring effusive and diffusive transport of RIB particles occurring in the target/ion source. Losses of RIB particles incurred in these steps dominate all other loss processes and can be characterized by the following quantities: the diffusion coefficient (cm^2/s), the adsorption enthalpy (kJ/mole) and the vacuum conductance (cm^3/s).

The measurement apparatus consists of two vacuum chambers, each capable of operation under UHV conditions. The chambers are connected by an interface consisting of either a thin membrane or an effusive passage constructed from the material under investigation. The interface can be heated to 2100 C and gases or vapors to be studied are introduced into one chamber while permeation across the interface is monitored by a sensitive mass spectrometer (~10^{-14} Torr) located within the other chamber.

Measurements are made by suddenly exposing one side of the interface with the gas or vapor and recording the rate of permeation through the interface as a function of time. Both diffusion through a membrane and effusion through a narrow passage can be described by the familiar diffusion equation in one dimension

$$\frac{\partial C}{\partial t} = D \frac{\partial^2 C}{\partial x^2} \quad (2)$$

where C(x,t) is the concentration of the species under investigation and D is the coefficient of diffusion. After applying the correct initial and boundary conditions Eq. 2 can be solved and the time profile of the permeation can be determined [8]

$$J = -D\frac{\partial C(x,t)}{\partial x}\bigg|_{x=L} = D\frac{C_1}{L}\left(1 - 2\sum_{n=1}^{\infty}(-1)^{n+1} Exp\left(-\frac{n^2 t}{\alpha}\right)\right) \quad (3)$$

here $\alpha = L^2/\pi^2 D$ where L is the thickness of the membrane or length of the effusive channel.

In order to evaluate the ability of this instrument to measure diffusion coefficients, adsorption enthalpies and vacuum conductance, we initially investigated several well-known systems. The diffusion coefficient of oxygen through stabilized zirconia was measured over a temperature from 500-1000 C. Fig. 2 shows an example of the oxygen permeation spectra. The diffusion coefficient was determined by fitting Eq. 3 to the oxygen ion signal from the mass spectrometer as a function of time.

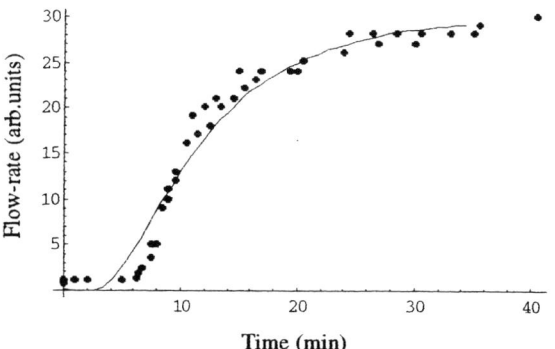

FIGURE 2. Diffusion spectra of O through a 1 mm thick stabilized zirconia (ZrO_2/Y_2O_3) membrane at 830 C.

Excellent agreement was found between measured values of D ($\sim 2 \times 10^{-6}$ cm^2/s) for this system and those found in the literature [9].

Particle-surface sticking times were also measured to determine enthalpies of adsorption in a fashion similar to the diffusion measurements. The Br/Ta system was first studied because previous measurements can be found in the literature [10]. The membrane was removed and replaced with a long 1 m x 0.8 mm Ta tube, coiled to fit within the heater. Time spectra, similar to that shown in Fig. 2, was taken and fit using Eq. 3. The resulting D was then related to a sticking time τ using the Clausing relation [8]

$$D = \frac{4}{3} \frac{r^2}{2r/\langle v \rangle + \tau} \quad (4).$$

where r is tube radius and v is the RMS gas velocity. Once τ has been determined the Frenkel relation is employed to obtain the adsorption enthalpy [3]. Again, good agreement was found between the literature value, $\Delta H = 2.7$ eV and our measurements $\Delta H = 2.5$ eV at 1800 C. Ar and Xe permeation spectra also agreed very closely with Eqs. 2 and 4 with $\tau \sim 0$.

Conductance measurements were also performed using this apparatus by recording evacuation curves of He, Ne, Ar, Kr and Xe flowing through a known geometrical conductance. Excellent agreement was found between measured and calculated values for the conductance.

Thus, we have shown that in the temperature range applicable to RIB target/ion source operation, we can effectively measure diffusion, adsorption and conductance phenomena. The next step will be to integrate this capability into our RIB beam program allowing optimal selection of wall materials, coatings, target materials, vapor transport species, target microstructure, etc.

ACKNOWLEDGMENTS

The authors gratefully acknowledge the numerous contributions of the HRIBF and Metals and Ceramics Division staff. Oak Ridge National Laboratory is managed by UT-Battelle, LLC under contract number DE-AC05-00OR22725 with the US department of energy.

REFERENCES

1. D.W. Stracener, these proceedings.
2. R.L. Auble, Proceedings of the Fifteenth International Conference on the Application of Accelerators to Research and Industry, Denton, Texas, Nov. 4-6 1998, AIP Press, Woodbury, New York.
3. G.D. Alton, J.R. Beene and Y.Liu, Nucl. Instrum. and Meth. A 438 (1999) 438.
4. "Scientific Opportunities with an Advanced ISOL Facility", A White Paper, Nov. 1997, Columbus, Ohio, USA.
5. L.C. Carraz, I.R. Haldorsen, H.L. Ravn, M. Skarestad and L. Westgaard, Nucl. Instrum. and Meth. 148 (1978) 217.
6. J. Klett, Composites in Manufacturing (USA), vol. 15, no. 4, pp. 1-5, 1999.
7. H. Ravn, A private communication.
8. R. M. Barrer, "Diffusion in and Through Solids", Syndics of the Cambridge university press, London (1951).
9. H. Solmon, J. Chaumont, C. Dolin and C. Monty, Cer. Trans. 24 (1991) 175.
10. G.Bolbach, J.C. Blais, A. Brunot and A. Marilier, Surface Science 90 (1979) 65.

RNB production at SPIRAL/GANIL

A.C.C. Villari[1], F. Landré Pellemoine[1], C. Barué[1], G. Gaubert[1], S. Gibouin[1],
Y. Huguet[1], P. Jardin[1], S. Kandri Rody[1], N. Lecesne[1], R. Leroy[1], M. Lewitowicz[1],
C. Marry[1], L. Maunoury[1], J.Y. Pacquet[1], J.P. Rataud[1], M.G. Saint Laurent[1],
C. Stodel[1], O. Bajeat[2], J.C. Angélique[3], N.A. Orr[3]

[1]GANIL, B.P. 5027 14076 Caen Cedex 5, France
[2]IPN, B.P. 1, 91406, Orsay Cedex, France
[3]LPC-ISMRa, Bld. Marechal Juin, 14050 Caen

Abstract. We present the on-line production system and tests for the Phase-I of SPIRAL/GANIL (Radioactive Ion Production System with Acceleration on-Line) where exotic multicharged noble gas ion beams have been obtained with the ECRIS (Electron Cyclotron Resonance Ion Source) NANOGAN-III. A new ECRIS (MONO1000) for monocharged radioactive ions is also presented, together with the diffusion properties of different graphite targets for He and Ar isotopes.

The SPIRAL facility (1) is based on the ISOL (Isotopic Separator On Line) technique for production and separation of Radioactive Ion Beams (RIB). The primary heavy-ion beam accelerated by the GANIL facility bombards a production target located inside a well shielded cave. The radioactive atoms produced by nuclear reactions diffuse out of the target, which is at high temperature (~ 2300K) and pass though a transfer tube to reach a permanent magnet ECRIS (Electron Cyclotron Resonance Ion Source) NANOGAN-III. The low-energy RIB is selected by a relatively low resolution separator (R=250) and accelerated by the compact cyclotron CIME. The energy range of the RIB varies from 1.7A to 25A MeV. Concerning the production of Radioactive Nuclear Beam, different systems have been developped (2) by GANIL for rare gaz and helium beams (NANOGAN-III), for alkali (MONOLITHE) and for the 1+/n+ project (MONO1000).

The NANOGAN-III (3) (Figure 1) is a compact permanent magnet ECRIS developed for the first phase of the SPIRAL project. The magnetic circuit consists in a sextupolar magnetic structure for radial confinement superposed by two axially and one radially magnetised permanent magnet rings. This ion source has been designed for operation with a 10GHz transmitter. Its power consumption is of 200W when tuned for the best charge state distribution performance. NANOGAN-III is an evolution of the preceeding model NANOGAN-II which worked at 14.5GHz. The new version, even working at lower frequency, has several advantages. In order to lower the magnetic field in the center of the plasma chamber, one of the permanent magnet rings has been removed. This decreased the cost and the weight of the ECRIS to 4/5 of the preceeding one. The transmission of the beam through the separator has been improved, which could mean that the emittance of NANOGAN-III is smaller than the preceding version. The ion source is linked to a carbon target by a cold and short transfer tube. This allows efficient production of noble gas elements with reasonable suppression of condensable contaminants.

The ionisation efficiency - all charge states - of Ar ions has been measured to be better than 95% when NANOGAN-III is tuned in order to maximise the 8+ charge state. (see Figure 1). In this same figure, we compare the charge state distribution of ^{40}Ar leak with and without beam on the target. The comparison between the off and on-line distributions attests that the ion source is almost insensitive to the heating and to the presence of the beam on the target.

At GANIL, the development of ECRIS for monocharged production is intrinsically related to the 1+/n+ project, i.e. the reinjection of the beam into another ion source for charge multiplication before the injection into the accelerator. This important condition fixes not only the required efficiency but also the characteristics of the beam provided by the 1+ ECRIS which its cost is cheaper than a multicharged ion source. If one consider that the life time of the production system is limited, a smaller and cheaper ion source is needed inside the production cave - very close to the irradiation point. The MONO1000 ECRIS (4) (Figure 2) has been designed in order to have a large magnetised chamber

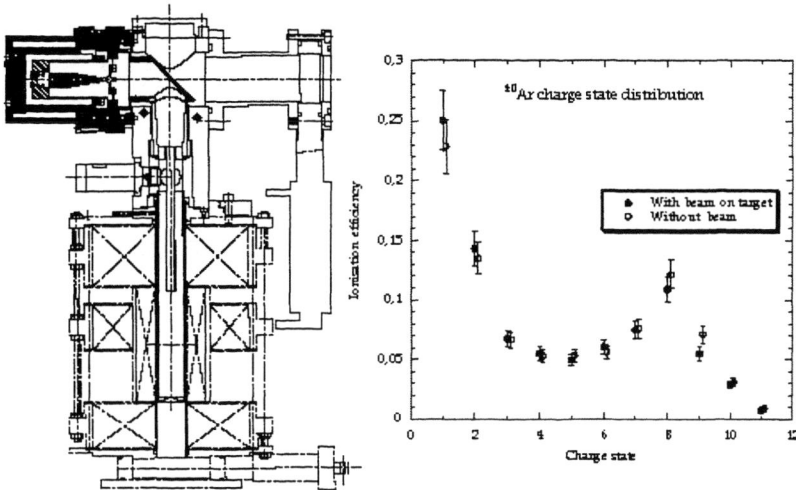

FIGURE 1. NANOGAN-III system and charge state distribution of ^{40}Ar with and without beam in target

of approximately 1 liter of volume, allowing one to place a target and/or a heating internal wall system with external cooling. The magnetic structure is made with two permanent magnets rings (total weight of 22.2 kg), which allows to create a closed 2000 Gauss surface at the wall of the plasma chamber. The plasma electrode is located in a 1800 Gauss magnetic field. It has to be pointed out that no specific radial structure is used - the magnetic field is on revolution symmetry - and that the magnetic field in the extraction area presents a cylindrical geometry. During the first commissioning tests, the 2.45 Ghz microwave was injected into the large diameter (90 mm) cylindrical plasma chamber through a coaxial transition ended by an antenna.

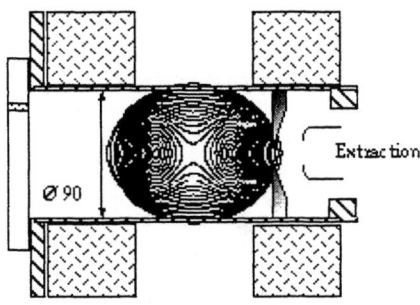

FIGURE 2. The MONO1000 ion source

The first tests of this ion source have been made with Argon through a calibrated leak and Helium as support gas. The measured ionisation efficiencies for ^{40}Ar at 1+ and 2+ states were respectively 90% and 9%, inside a root mean squared emmitance of 27 π mm mrad at $E = 14 keV$. Moreover, ^{32}S beam production has been also studied injecting a calibrated quantity of the gaseous molecule of SO_2 in the source. This test revealed an overall efficiency of ionisation for the compounds of ^{32}S of around 95% It is clear that several different molecules are still present on the spectrum of the extracted ions. The ratio between the yield of this molecules and the ^{32}S (1+) beam varies strongly with the microwave power. Presently we are limited to around 20W due to our microwave coupling. A new coupling is being developed which could allow one to deliver one order of magnitude more power in the ion source.

The efficient production of radioactive nuclear beams by ISOL method is intrinsically related to the efficiency of the ion source and to the release delay time of the target. The carbon target has been chosen due to its excellent release properties, low atomic number and high sublimation temperature. A special conical design has been choosen to distribute uniformly the power density of the beam (due to the Bragg peak) (5). We present in figure 4 the diffusion-effusion efficiency of carbon targets with different grain microstructure, i.e. 1μm, 4μm and 15μm.

The results (6) show an important enhancement of the diffusion effusion efficiency for the 1μm microstructure. We also observed an enhancement of a factor 2 for the production for ^{35}Ar and a factor 10 for ^{32}Ar. The diffusion effusion efficiency of ^6He in the 1μm microstructure target has been also mesured as a function of the target temperature. The obtained efficiencies are presented in figure 4.The most important observation is the fact that the ^6He diffusion effusion efficiency reaches 100% at reasonably low temperatures, around 1600K.

We also developed a carbon target for helium production which is devided in two parts, due to the long range of ^8He in C (figure 5). The first part (production target) is only heated by the ^{13}C primary beam ; the second part

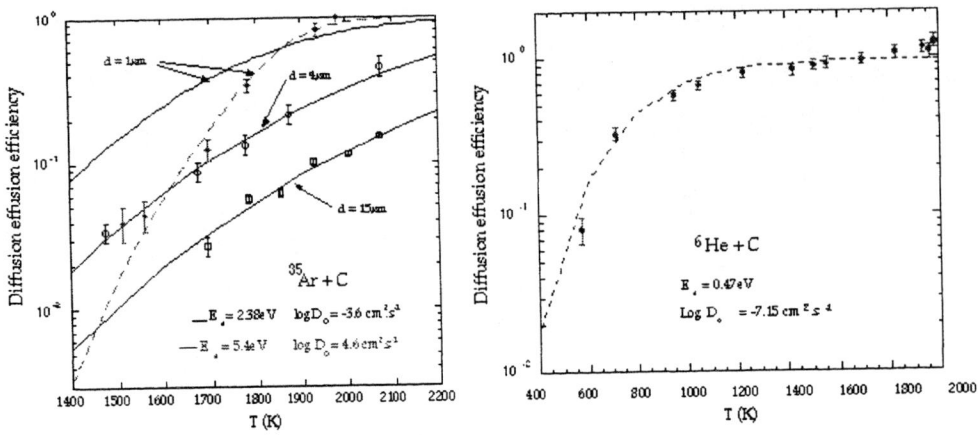

FIGURE 3. Diffusion-effusion efficiency as a function of the target temperature for ^{35}Ar and ^{6}He

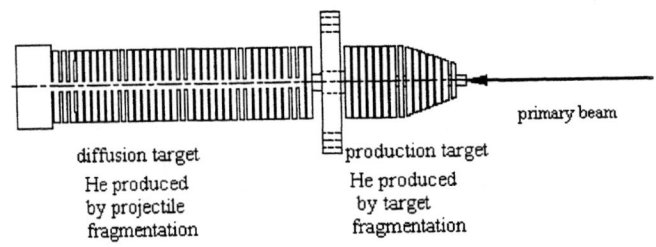

FIGURE 4. Target for helium production

(diffusion target), stops the fragmentation products, and is heated by an electric current through the axis.

Yields of different isotopes have been measured at the SIRa on-line test bench. For exemple, we obtained yields of $5.4 \ 10^8$ and $1.48 \ 10^6$ for $^6He^+$ and $^8He^+$, $2.7 \ 10^3$ and $5.4 \ 10^8$ for $^{32}Ar^{7+}$ and for $^{35}Ar^8$, $4.49 \ 10^4$ $1.95 \ 10^7$ and $6.76 \ 10^8$ respectively for $^{73}Kr^{15+}$, $^{75}Kr^{15+}$ and $^{77}Kr^{15+}$ (7) normalised to 1pµA.

For the future at GANIL (8), a new project is developped for the production of neutron-rich beams by fission of Uranium induced by neutrons or photons : SPIRAL Phase-II. To study the optimum conditions of this method, different experiments and simulations are being performed. The energy range and the angular distribution of the neutrons produced at different deuteron incident energies on different converters has been already measured. Cross sections of fast neutron induced fission and the isotopic yields of several elements diffused out of a thick target are now available. Moreover, the photofission is also investigated in order to choose between fission induced by fast neutrons and photofission. Simulations are also being made in order to be able to predict the production rate for neutron-rich nuclei.

REFERENCES

1. *The SPIRAL Radioactive Beam Facility*, GANIL Report **R-94-02** Caen(1994).
2. A.C.C. Villari, *Nucl. Instr. Meth. Phys. Res.*, **B126** (1997) 35.
3. L. Maunoury et al., *18th Int. Work. ECR Ion Sources*, Texas USA (1997).
4. A.C.C. Villari, *Ion source development for RNB production at SPIRAL/GANIL*, Proceedings of RNB 2000, to be published in Nucl. Phys. A.
5. R. Lichtenthäller and al., *A simulation of the temperature distribution in the SPIRAL target*, NIM **B 140** (1998) 415-425.
6. F. Landré-Pellemoine and al., *Recent results at SIRa test bench : Diffusion properties of Carbon graphite and Boron carbide targets*, Proceedings of RNB 2000, to be published in Nucl. Phys. A.
7. L. Maunoury, *Production de faisceau radioactifs multichargés pour SPIRAL : Etudes et réalisation du premier ensemble cible-source*, GANIL PhD **T-98-01** Caen (1998).
8. M.G. Saint Laurent and collaborators, *SPIRAL Phase-II contract number ERBFMGECT980100*, GANIL Report, April 2000.

Radioactive Ion Beams at the HRIBF

D. W. Stracener for the HRIBF Staff

Oak Ridge National Laboratory, Oak Ridge, Tennessee, USA.

Abstract. The Holifield Radioactive Ion Beam Facility (HRIBF) is now delivering beams of radioactive ions for nuclear physics experiments. The radioactive ions are produced using the Isotope-Separator-On-Line (ISOL) technique and then injected into a 25 MV tandem electrostatic accelerator and accelerated to energies of 0.1 – 10 MeV per nucleon for light nuclei and up to 5 MeV per nucleon for nuclei heavier than 100 amu. In the past year, the facility has provided almost 1200 hours of ^{17}F and ^{18}F beams for research with intensities up to 3×10^6 ions/s on target. Also, the facility has provided beams of neutron-rich radioactive ions using proton-induced fission in a highly permeable uranium carbide target. The first such beam delivered to an experiment was ^{118}Ag with intensities up to 2×10^6 ions/s on target. Details of the techniques used to produce these beams and other beam development projects will be discussed.

INTRODUCTION

The HRIBF, at the Oak Ridge National Laboratory, utilizes the ISOL technique to produce ion beams of short-lived radioactive nuclei. The nuclei of interest are produced by bombarding a thick target with high intensity beams of light ions. The radioactive atoms diffuse out of the target matrix, are transported to an ion source, ionized, extracted at 40 keV, and then mass separated. The HRIBF is unique in that these low energy beams are then injected into a second accelerator and accelerated up to energies of a few MeV per nucleon and delivered to targets for nuclear physics experiments. A diagram of the ISOL process as implemented at the HRIBF is shown in Figure 1. Brief descriptions of the driver accelerator, the Radioactive Ion Beam (RIB) Injector, and the post-accelerator systems are given in the next section. Details of the development of ^{17,18}F beams and neutron-rich beams are also presented in this paper.

ACCELERATOR SYSTEMS

The driver accelerator for the facility is the Oak Ridge Isochronous Cyclotron (ORIC), which was first commissioned in 1962. The ORIC is a variable energy cyclotron (k=100) and in the present configuration has delivered high-intensity beams of protons at 42 MeV, deuterons at 49 MeV, and 85 MeV for ^{4}He beams. The maximum beam intensity delivered has been limited by the ability of the RIB production target to dissipate the power deposited in the target. Proton and deuteron beams with intensities up to 12 µA have been delivered to RIB production targets. The reliability of the cyclotron has greatly improved due to a number of improvements made over the last couple of years. These include the replacement of several aging power supplies (one of which was built in 1938), improved extraction efficiency (1), and a change in the design of

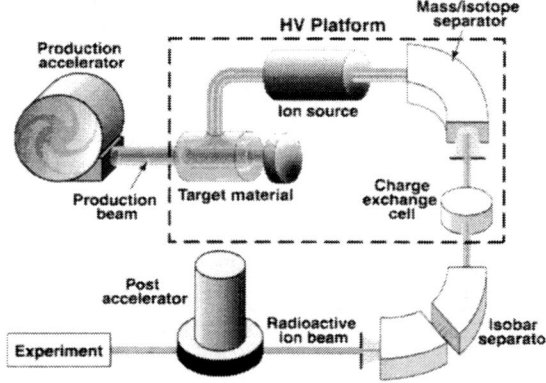

FIGURE 1. Shown are the various elements of the HRIBF for the production and acceleration of RIB's. (not to scale)

---

* Managed by UT-Battelle, LLC, for the U.S. Department of Energy under contract DE-AC05-00OR22725.

the ORIC ion source, which improved the operational lifetime from 300 hours to more than 800 hours. Recently, the cyclotron operated nearly continuously for a three-month period.

The post-accelerator at the HRIBF is a folded-geometry 25-MV tandem electrostatic accelerator that has been in routine operation since 1982 and has provided more than 75 different beams for research. The tandem has operated with terminal potentials up to 25.5 MV (highest in the world), and as low as 1 MV with excellent reliability. Negative ions, injected into the tandem, are stripped in the terminal and the positive ions are then accelerated down the high energy side of the tandem. For light nuclei (A<100), single-stripping is enough to allow for acceleration up to ~5 MeV per nucleon, but for heavy ions (A>100) a second stripping is required to achieve this energy. Double-stripping has successfully been used with radioactive ion beams at intensities of 10^6 ions/second. Often, the radioactive ion beams contain several isobars that are accelerated along with the beam of interest. One technique used to separate isobars of light ions is to fully strip the ions after acceleration in the tandem. This technique was used recently to separate ^{17}F ions from the large ^{17}O contamination.

These two accelerators are linked together by the RIB Injector that is comprised of the production target, the ion source, initial mass separation, and acceleration of negatively charged ions to 200 keV for injection into the tandem. The main component is the target/ion source (TIS) and its enclosure. Two TIS systems have been used in the past year for production of ^{17,18}F beams from a hafnium oxide target, and neutron-rich beams from proton-induced fission of uranium. Both of these have been quite successful and have provided beams of sufficient quality, intensity, and duration to facilitate the completion of several experiments as shown in Table 1. The success of these two systems was due to recent improvements and modifications to the RIB Injector. The TIS enclosure has been modified to increase its reliability and lifetime in the high-temperature and high-radiation environment present on the RIB Injector. A new Cs-vapor charge exchange cell has been installed, which gives longer operational lifetimes, better reliability, and better transmission than the previous version. A tape system has been installed after the isobar-separator magnet to allow for analysis of the intensity and purity of the beam. Since several isobars may be present in the beam, target/ion source parameters and beam-tuning parameters will be adjusted to optimize the purity of the beam.

^{17,18}F BEAM DEVELOPMENT

The target/ion source system used to produce the fluorine beams consisted of a Kinetic Ejection Negative Ion Source (2) coupled to a target of hafnium oxide fibers (3). A SEM of these HfO_2 fibers is shown in Figure 2. The radioactive fluorine isotopes were produced in the oxide target via the $^{16}O(d,n)^{17}F$ and the $^{16}O(\alpha,pn)^{18}F$ reactions, using 45 MeV deuterons and 85 MeV α-particles. The intensity of the beam on the production target was limited to less than 3 μA since previous tests with high intensity beams from ORIC had shown that the target rapidly deteriorated at higher beam currents. The low-density (1.15 g/cm^3) hafnia-fiber target consisted of four rolls, each having a diameter of 15 cm and a length, along the beam axis, of 1 cm. The rolls were separated along the beam axis by ~1 cm to allow heat to radiate out to the walls of the target holder. Previous tests had shown that the yield of ^{17}F was enhanced if a small amount of Al_2O_3 was present in the target holder, so a liner of alumina fiber material was placed around the hafnia cylinders. This TIS and its enclosure operated without failure for 1420 hours (59 days) and was working when it was removed. The target was irradiated for more than 3400 μA-hours with deposited power per unit length of 52 W/g/cm^2. As shown in Table 1, seven experiments (4,5) were recently completed with radioactive fluorine beams.

TABLE 1. Radioactive Ion Beams Delivered to Experiments During the Last Year

RIB Species	Beam Energy (MeV)	Beam Intensity (ions/second)	# of hours beam-on-target	# of Completed Experiments
^{17}F	10 – 68	3 x 10^6	1060	5
^{18}F	10 – 14	6 x 10^5	120	2
^{117}Ag	460	1 x 10^6	72	detector tests
^{118}Ag	500	2 x 10^6	240	1
^{126}Sn	300	> 10^5	4	low-intensity diagnostics tests
^{134}Te	300	1 x 10^6	8	

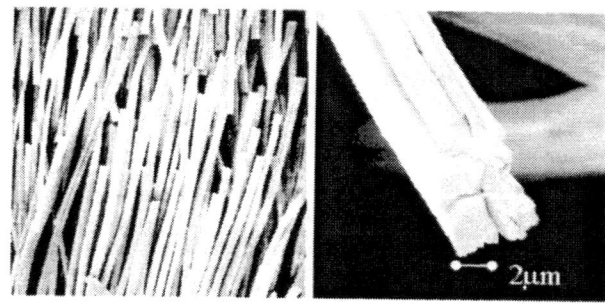

FIGURE 2. These SEM's show the HfO_2 target material used in the production of radioactive fluorine beams.

NEUTRON-RICH BEAMS

At the HRIBF, the focus of the various beam development projects for the last few years has been on proton-rich isotopes of As, Ga, Cu, Ni, and F (1,3,6). After tests of a uranium carbide target at the UNISOR Facility (7) using low-intensity proton beams from the tandem, administrative approval was recently granted to use a fissionable target on the RIB Injector. The ion source used was an Electron-Beam-Plasma Ion Source (EBPIS) (8) and was operated for more than 1200 hours (50 days) without a failure.

The target, supplied by Babcock and Wilcox, McDermott Technologies of Lynchburg, VA, is a low density highly permeable graphite matrix coated with a thin layer of uranium carbide (9). Scanning electron micrographs of the reticulated vitreous carbon (RVC) matrix before and after coating are shown in Figure 3. The fibers of the RVC matrix are ~60 μm thick and the material has a density of $0.06\ g/cm^3$. After the coating process, the surface layer was analyzed and found to be ~12 μm thick and to contain an atomic ratio of U:C of 1:1.85. The density of the coated matrix was 1.34 g/cm^3 with an overall U:C mass ratio of 6.6:1. The target material was delivered as disks that were 15 mm in diameter and 2 mm thick. Nine of these disks were used for the target on the RIB Injector, which resulted in a uranium target thickness of 2.1 g/cm^2. The target thickness was chosen to allow a 40 MeV proton beam to exit with 15 MeV since the cross-section for proton-induced fission drops off rapidly for proton energies less than 18 MeV. Also, since less power is deposited in the target, the operating temperatures should be lower, resulting in longer operational lifetimes. No significant decrease in yields was observed during a period of 12 days during which there was an average proton beam intensity of 10 μA on target. The average beam power deposited per unit length in this target was 125 $W/g/cm^2$. Not only did the ion source and the target perform above expectations, but the enclosure also endured the thermally and radiologically harsh environment. More than 5000 μA-hours were logged on this TIS and its enclosure before a vacuum leak began to develop. However, this problem was slow to develop and the system was used for several days after the leak was first noticed.

A survey of positive ion yields from this TIS was made using a tape system and a γ-ray detector that were located just off the RIB Injector Platform. At least 130 isotopes of 20 different elements were observed between ^{78}Ga and ^{144}La. Not all elements in that range were observed since some are quite refractory (e.g. Zr-Rh) and will not be released from the target. Some, observed as positive ions, will not be available as accelerated beams because either they do not form negative ions (e.g. noble gases) or the charge exchange efficiencies are very small. A table of the measured yields can be viewed on-line on the HRIBF Website (10). After the initial survey of yields, several beams were tuned through the tandem and accelerated to experimental stations (see Table 1).

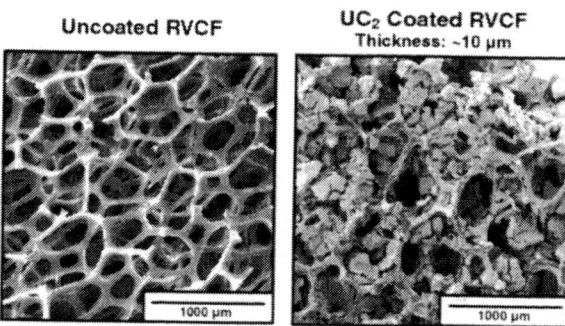

FIGURE 3. SEM's of the graphite matrix used for the uranium carbide targets, before and after coating.

OTHER BEAM DEVELOPMENT

While the main effort of the facility has been focused on the aforementioned development projects, other targets and ion sources are being developed and tested. An ion source for long-lived RIB's has been tested extensively off-line and is ready for installation on the RIB Injector. This source (6), called the batch-mode source, is a multi-sample, Cs-sputter type source that produces negative ions directly. It is similar to the source that has been used for almost 20 years in the stable-ion injector for the tandem with the added feature of a rotating target wheel with up to eight different target positions. The sputter target is initially aligned with the ORIC beamline and the radioactive species of interest is produced in nuclear reactions using intense beams of light ions. This target is then rotated 90° into the sputter region where the negative ions are formed and extracted to form the beam. While this target is being used, the next target is irradiated with the production beam. The frequency at which the target wheel is rotated depends on the half-life of the isotope of interest. This ion source will initially be used on the RIB Injector with four graphite pellets to produce beams of ^{11}C ($T_{1/2}$ = 20.3 m) and four CaF$_2$ pellets to produce beams of ^{18}F ($T_{1/2}$ = 110 m). After these runs, nickel pellets will be inserted to produce ^{56}Ni beams.

Interest in neutron-rich isotopes of bromine is strong in the nuclear structure community and the release efficiency of bromine from the uranium carbide target is quite good. The measured positive-ion yields from this target coupled to an EBP ion source are ~10^8 ions/s but the charge exchange efficiency of bromine in a Cs-vapor cell is quite low (~1%). To overcome this limitation, a negative surface ionization source, using a LaB$_6$ surface, has recently been tested and optimized for the direct production of negative ions of bromine (11). This source has an efficiency for Br (~10%) that is comparable to the efficiency of the EBPIS but with the advantage that the ions are already negatively charged and thus no charge exchange is needed. With the negative surface ionization source, the intensities of bromine beams delivered to experiments should be two orders of magnitude higher. This will soon be tested on-line at the UNISOR Facility (7).

A positive surface ionization source using a hot tantalum or tungsten surface is being developed and tested for use with electropositive elements such as rubidium and strontium. These two elements, in particular, have high production rates in the uranium carbide target and have high release efficiencies. This type of source has been shown to have a very high efficiency (up to 90%) for the Group I elements and good efficiencies for many other elements (12). While the charge exchange efficiency for Rb is low (0.3%), this source should result in at least one order of magnitude improvement in the extracted beam intensities measured from the EBPIS and should deliver beam-on-target intensities of 10^6 ions/s for Rb isotopes with half-lives greater than one second.

Another type of uranium carbide target, built on a carbon matrix with a high thermal conductivity, has been tested recently. Details of this target are given in Ref. 13. Other targets under development include SiC (powder and fibers) for ^{25}Al beams and CeS powder for beams of ^{33,34}Cl.

ACKNOWLEDGMENTS

The author wishes to acknowledge that this paper is possible due to the efforts of the entire HRIBF Staff. Any successes realized in the last few months are the result of several years of hard work by many individuals. Research at the Oak Ridge National Laboratory is supported by the U.S. Department of Energy under contract DE-AC05-00OR22725 with UT-Battelle, LLC.

REFERENCES

1. Auble, R. L., in *Application of Accelerators in Research and Industry*, edited by J. L. Duggan and I. L. Morgan, AIP Conference Proceedings 475, New York: American Institute of Physics, 1999, pp. 292-295.

2. Alton, G.D., Liu, Y., Williams, C., and Murray, S.N., *Nucl. Instr. and Meth.* **B 170**, 515-522 (2000).

3. Welton, R.F., and the HRIBF Staff, accepted for publication in *Proceedings, 5th International Conference on Radioactive Nuclear Beams (RNB-5)*, Divonne, France, April 3-8, 2000, Nucl. Phys. A.

4. Bardayan, D.W., et al., *Phys. Rev. Lett.* **83**, 45-48 (2000).

5. Liang, J.F., et al., *Phys. Lett.* B, in press.

6. Mills, G.D., Alton, G.D., Haynes, D.L., and Beene, J.R., *Physics Division Progress Rpt. ORNL-6957, Sept. 1998.* (www.phy.ornl.gov/progress/hribf/randd/hri031.pdf).

7. Stracener, D.W., Carter, H.K., Kormicki, J., Poland A.H., Reed, C.A., Welton, R.F., and Williams, C.L., *Physics Division Progress Rpt. ORNL-6957, Sept. 1998.* (www.phy.ornl.gov/progress/hribf/facility/hri013.pdf).

8. Carter, H.K., et al., *Nucl. Instr. Meth.* **B 126**, 166 (1997).

9. Alton, G. D. and Moeller, H. H., *Physics Division Progress Report ORNL-6957, September, 1998.*

10. The tables of RIB yields can be found on the HRIBF Website at www.phy.ornl.gov/hribf/users/beams/.

11. Zaim, H., Liu, Y., Murray, S.N., and Alton, G.D., in these proceedings.

12. Bjørnstad, T., Hagebø, E., Hoff, P., Jonsson, O.C., Kugler, E., Ravn, H.L., Sundell, S., Vosicki, B., and the ISOLDE Collaboration, *Physica Scripta* **34**, 578-590 (1986).

13. Welton, R.F., et al., in these proceedings.

Experiments with Radioactive Beams at ATLAS

K. E. Rehm[a], I. Ahmad[a], J. Blackmon[b], F. Borasi[c], J. Caggiano[a], A. Chen[d], C. N. Davids[a], J. Greene[a], B. Harss[a], A. Heinz[a], D. Henderson[a], R. V. F. Janssens[a], C. L. Jiang[a], J. Nolen[a], R. C. Pardo[a], P. Parker[d], M. Paul[e], J. P. Schiffer[a], R. E. Segel[c], D. Seweryniak[a], R. H. Siemssen[a], M. S. Smith[b], J. Uusitalo[a], T. F. Wang[f], I. Wiedenhöver[a]

[a] *Argonne National Laboratory, Argonne, IL 60439*
[b] *Physics Division, Oak Ridge National Laboratory, Oak Ridge, Tennessee 37831*
[c] *Northwestern University, Evanston, Illinois 60208,*
[d] *Wright Nuclear Structure Laboratory, Yale University, New Haven, Connecticut 06520,*
[e] *Hebrew University, Jerusalem, Israel,*
[f] *Lawrence Livermore National Laboratory, Livermore, California 94550*

Abstract.
Various beams of short- and long-lived radioactive nuclei have recently been produced at the ATLAS accelerator at Argonne National Laboratory, using either the so-called In-Flight or the Two-Accelerator method. The production techniques, as well as recent results with ^{44}Ti ($T_{1/2}$=60y) and ^{17}F ($T_{1/2}$=64s) beams, which are of interest to nucleosynthesis in supernovae and X-ray bursts, are discussed.

INTRODUCTION

The availability of beams of unstable nuclei at various facilities worldwide has allowed to investigate questions in several areas of nuclear physics that previously could not be addressed. For example, nuclei at the neutron or proton drip line exhibit new structures, such as skins and halos, which have been shown to strongly influence other reaction channels such as transfer or fusion. The effects of neutron-proton pairing can best be studied in heavier N=Z nuclei, which beyond ^{40}Ca are β-unstable. In nuclear astrophysics it has been known for some time that a large fraction of the elements above A~20 is produced in explosive nucleosythesis where the reactions occur on such a rapid time scale that unstable nuclei produced in these processes do not have time for β-decay, but rather continue to react with protons, neutrons or α particles. Nature has no difficulties producing these unstable, short-lived nuclei in the stellar furnaces. In the laboratory, however, it was only during the last decade that some of these important reaction rates could be studied.

BEAM PRODUCTION

The majority of the radioactive beams are presently produced either via the isotope-separation- online (ISOL) technique or with the projectile-fragmentation method. In the former the radioactive material is produced with a high-current driver accelerator or at a reactor. At high enough temperatures some of the nuclei effuse from the target material, are ionized and accelerated with a post accelerator. In the latter technique a high-energy heavy-ion beam is fragmented in a thin target, and the fragments, after electromagnetic selection are used directly in the experiment. Because of the production technique, the ISOL beams show better beam qualities compared to fragmentation beams. On the other hand the effusion from the target is for many nuclei a relatively slow process and, thus, short-lived nuclei are more efficiently produced with the fragmentation technique.

At the ATLAS accelerator at Argonne National Laboratory we have used modifications to the two techniques mentioned above. For longer-lived isotopes ($T_{1/2}$ ≥2h) the irradiated material can be extracted from the production accelerator and converted into a suitable chemical form allowing to produce beams of nuclei that, because of their chemical properties, are difficult to extract using conventional ISOL techniques. Examples for these radioactive beams produced by the two-accelerator method are ^{18}F or ^{56}Ni.

For shorter lived nuclei we have used the in-flight method to produce beams of e.g. ^{17}F ($T_{1/2}$ = 64s) or ^{25}Al ($T_{1/2}$ = 7.18s). In this technique a high intensity

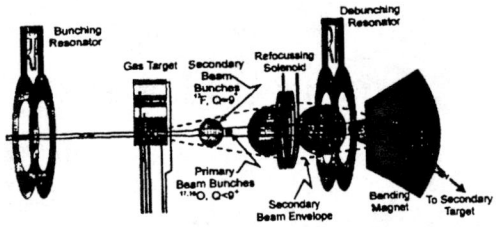

FIGURE 1. Schematic of the experimental setup used to produce short-lived radioactive beams via the in-flight technique.

(d,n) or (^{3}He,n) reactions secondary beams of unstable nuclei which are separated from the primary beam and transported onto a secondary target that is to be studied. An example for this technique is the production of a ^{17}F beam via the d(^{16}O,^{17}F)n reaction. In this case a few ppm of the primary ^{16}O beam particles are converted into ^{17}F in the deuterium gas cell and transported through a 12m long beamline/separator system onto target. With a 100 pnA primary ^{16}O beam a ^{17}F intensity of 2×10^6 has been achieved. The experimental setup is shown schematically in Fig.1.

A superconducting bunching resonator, located 10 m upstream from the production target provides a time focus of the primary beam at the gas cell, minimizing the longitudinal emittance of the secondary beam. The ^{17}F particles produced via the inverse (d,n) reaction are emitted within a cone with an opening angle of a few degrees. A superconducting solenoid located after the target is used to capture the particles within this cone and to focus them through a 22° bending magnet that separates the secondary ^{17}F^{9+} particles from the primary ^{16}O^{8+} beam. The selection of the particles according to their magnetic rigidity results in a suppression of the primary beam by a factor of $\sim 3\times 10^{-7}$. The debunching resonator located about 3 m after the production target can be used to improve the energy spread of the secondary beam. By making use of the energy-time correlation and choosing the RF phase of this resonator appropriately the energy resolution of the secondary beam has been improved by a factor of 3 (see Fig.2).

The 400 keV energy spread achieved for the ^{17}F beam translates into a 23 keV spread in the c.m. system for a study of the p(^{17}F,^{14}O)α system.

The beams produced with these two techniques at the ATLAS accelerator, including their energies and intensities, are summarized in Table.1.

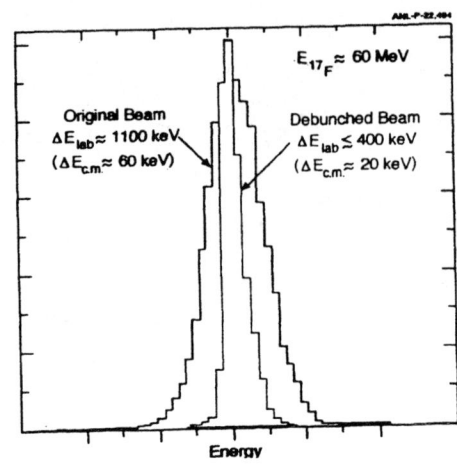

FIGURE 2. Effects of the debunching resonator on the energy distribution of a 60 MeV beam of ^{17}F

Details for some of the ion beams produced with the two-accelerator or the in-flight technique can be found in Refs.(1, 2, 3).

ION SOURCE AND ACCELERATOR

For the production of radioactive beams via the two-accelerator method a negative sputter source SNICS(4), dedicated to radioactive material, has been installed at the Tandem accelerator which is one of the two injectors of the superconducting heavy ion accelerator ATLAS. It has quite large efficiencies for certain elements(e.g. $\sim 1\%$ for fluorine) and it's compact geometry makes decontamination after a run much simpler. For elements with small electron affinity (e.g. Ti$^-$) molecules (e.g. TiO$^-$) have been used.

The beam intensities of the radioactive beams extracted from the ion source are usually too weak to either stabilize the terminal voltage of the tandem accelerator or the timing of the bunched beam which is required for injection into the superconducting RF accelerator. For this reason a wedge-shaped Bicron Corporation plastic scintillator was mounted on a photomultiplier and installed in the vertical plane behind the tandem 90° analyzing magnet. Timing signals from this scintillator were used to stabilize the time-of-arrival of the beam pulse by adjusting the phase of the pre-tandem buncher(5).

Another problem that had to be addressed was the tuning of the RF accelerator and the beam transport system. In some cases the stable isobar which is present in the source material can be used (e.g. ^{18}O for a ^{18}F beam).

Table 1. Radioactive ion beams produced at ATLAS

Beam	$T_{1/2}$	Production Method	I_{source} [sec^{-1}]	I_{target} [sec^{-1}]	E/A [MeV/u]
^{18}F	110m	two-accelerator	5×10^7	3×10^6	0.6
^{56}Ni	6.1d	two-accelerator	2×10^7	6×10^4	5
^{56}Co	77d	two-accelerator	1×10^8	3×10^5	5
^{44}Ti	60y	two-accelerator	2×10^7	5×10^5	2-7
^{17}F	65s	in-flight	5×10^7	3×10^6	3-6
^{21}Na	22.5s	in-flight	2×10^7	5×10^5	5
^{25}Al	7.2s	in-flight	1×10^7	2×10^5	5

In other cases, however, the intensity of this beam could overwhelm the detector system. For this reason the whole accelerator was tuned with a pilot beam which has the same magnetic rigidity and velocity as the beam of interest. For 250 MeV ^{56}Ni^{10+}, a pilot beam of 125 MeV ^{28}Si^{5+} was used while for ^{44}Ti^{8+} the pilot beam was ^{66}Zn^{12+}.

EXPERIMENTAL RESULTS

In the following results from two recent experiments are discussed in more detail.

Study of the ^{17}F(p,α)^{14}O Reaction

The relatively long half-life of the nucleus ^{14}O ($T_{1/2}$=70.6s), produced via the ^{13}N(p,γ)^{14}O reaction limits the energy production in the hot CNO cycle. ^{14}O is, therefore, considered a bottleneck which is only broken when, at higher temperatures, breakout via the ^{14}O(α,p)^{17}F reaction starts to become possible. Because of the difficulties with a direct measurement of the ^{14}O(α,p)^{17}F reaction, which requires a low energy ^{14}O beam and a He gas target, the inverse reaction ^{17}F(p,α)^{14}O was studied to get information about the properties of excited states in ^{18}Ne.

In the experiment CH$_2$ targets with thicknesses of 100 or 500 μg/cm^2 were used. The energy and the scattering angle of the outgoing particles (p and ^{14}O) were measured in coincidence with two position sensitive double-sided annular silicon strip detectors. The measurement of the four quantities θ_α, E_α, θ_{14_O} and E_{14_O} allowed for a clean identification of the ^{17}F(p,α)^{14}O reaction with a detection efficiency of about 65%. The measured cross sections are shown in Fig.3.

The circles represent measurements with a thin (100μg/cm^2) target, while the crosses are the results from

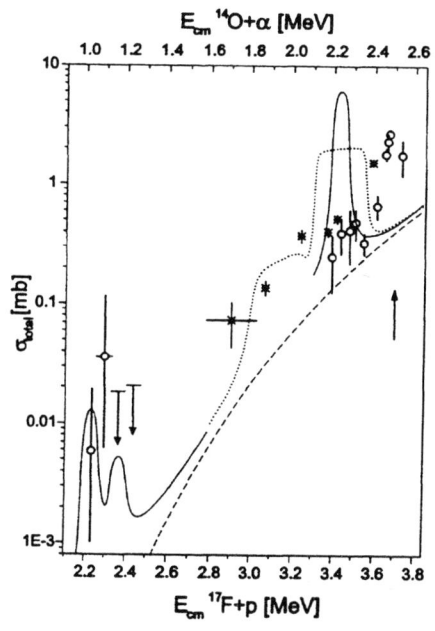

FIGURE 3. Cross sections measured for the ^{17}F(p,α)^{14}O reaction. The symbols are explained in the text

a 500 μg/cm^2 target. The horizontal bars, shown for one point only, indicate the energy interval covered by the target thickness. The dotted line is the expected thick target yield using parameters from Ref.(6), while the solid line is that for the thinner target. The dashed line represents the direct component estimated in Ref.(7). The difference between the observed and the expected cross sections is due to incorrect spin-assignments for particle-unbound states in ^{18}Ne and to the omission of higher-lying states in Ref.(6).

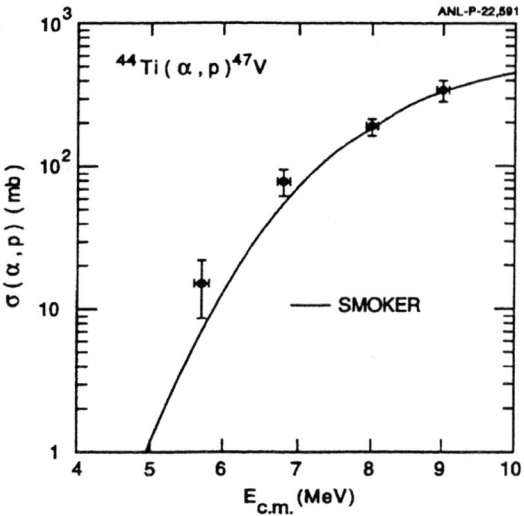

FIGURE 4. Measured excitation function for the ^{44}Ti$(\alpha,p)^{47}$V reaction. The solid line corresponds to a calculation done with the code SMOKER (see text for details).

Study of the ^{44}Ti$(\alpha,p)^{47}$V Reaction

Nuclei of ^{44}Ti are produced in the last stages of a supernova event in the so-called alpha-rich freeze-out(8). The recent observation of γ rays associated with the decay of ^{44}Ti from Cassiopeia A showed that the ^{44}Ti afterglow can be used to locate individual supernovae remnants. With the launch of the next-generation gamma-ray observatory INTEGRAL, new supernovae remnants are likely to be discovered. The amount of ^{44}Ti generated in a supernova is governed by a subtle interplay between the nuclear reactions that produce it and those that destroy it. For the latter the ^{44}Ti$(\alpha,p)^{47}$V reaction has been shown(9) to have the strongest influence.

This reactions was studied with the recently developed ^{44}Ti beam at ATLAS. The material was produced via the ^{45}Sc$(p,2n)$ reaction using a 50 MeV, 20 μA proton beam from the injector of Argonne's Intense Pulsed Neutron Source. After a 70 hour long irradiation, 1.3 μg of ^{44}Ti were produced which was chemically separated from the Sc material. The ^{44}Ti$(\alpha,p)^{47}$V reaction was studied in inverse kinematics using a ^{4}He gas target and the Fragment Mass Analyzer for separating the ^{47}V particles from the incident beam. Details of the experimental setup can be found in Ref.(10). Figure 4 shows the measured cross section for the ^{44}Ti$(\alpha,p)^{47}$V reaction in comparison with results from the statistical model code SMOKER (11).

While at the higher energies good agreement with the theoretical predictions is observed the falloff towards lower energies is slower than predicted resulting in cross sections that are about a factor of two larger than the SMOKER predictions. This translates to an astrophysical reaction rate which is higher by at least a factor of two than earlier theoretical predictions (12). This higher rate results in a reduction of the amount of ^{44}Ti produced in supernovae explosions. However, changes in other reaction rates which have not been measured so far, could effect the ^{44}Ti yield as well.

4. ACKNOWLEDMENT

This work was supported by the U. S. Department of Energy, Nuclear Physics Division, under Contract No. W-31-109-ENG-38, Grant No. DE-FG02-98ER41086, the National Science Foundation, and by a University of Chicago/Argonne National Laboratory Collaborative Grant.

REFERENCES

1. A. Roberts et al., Nucl. Instr. and Meth. **B103**, (1995)523
2. K. E. Rehm et al., Nucl. Instr. and Meth. **A449**, (2000)208
3. B. Harss et al., Rev. Sci. Instr. **71**, (2000)380
4. National Electrostatics Corporation, Graber Road, Box 310, Middleton, WI 53562
5. G. P. Zinkann et al., 8th International Conference on Heavy Ion Accelerator Technology, Argonne National Laboratory, Argonne, 1998, AIP Conference Proceedings Vol 473, 1999, p.279
6. K. I. Hahn et al., Phys. Rev. **C94**, (1996)1999
7. C. Funck et al., Z. Phys. **A332**, (1989)109
8. David Arnett, *Supernovae and Nucleosynthesis* (Princeton University Press, Princeton, NJ, 1996
9. L. S. The et al., Astrophys. J. **504**, (1998)500
10. K. E. Rehm et al., Phys. Rev. Lett. **84**, (2000)1651
11. T. Rauscher et al., Phys. Rev **C56**, (1997)1613
12. S. Woosley et al., Astrophys. J. Suppl**101**, (1995)181

REX-ISOLDE - POST-ACCELERATED RADIOACTIVE BEAMS AT CERN-ISOLDE

T. Nilsson[1], J. Äystö[1], O. Forstner[1], H.L. Ravn[1], M. Oinonen[1], H. Simon[1,12],
J. Cederkäll[1], L. Weissman[1], D. Habs[2], F. Ames[2], O. Kester[2], T. Sieber[2],
H. Bongers[2], S. Emhofer[2], P. Reiter[2], P.G. Thirolf[2], G. Bollen[2], P. Schmidt[3],
G. Huber[3], L. Liljeby[4], Ö. Skeppstedt[4], K.G. Rensfelt[4], F. Wenander[5], B. Jonson[5],
G. Nyman[5], R. von Hahn[6], H. Podlech[6], R. Repnow[6], C. Gund[6], D. Schwalm[6],
A. Schempp[7], K.-U. Kühnel[7], C. Welsch[7], U. Ratzinger[8], G. Walter[9], A. Huck[9],
K. Kruglov[10], M. Huyse[10], P. van den Bergh[10], P. van Duppen[10], A.C. Shotter[11],
A.N. Ostrowski[11], T. Davinson[11], P. J. Woods[11], I. Moukha[12], A. Richter[12],
G. Schrieder[12], and the REX-ISOLDE collaboration

[1] CERN, CH-1211 Geneva 23, Switzerland
[2] Sektion Physik, LMU München, D-85748 Garching, Germany
[3] Johannes-Gutenberg Universität, D-55099 Mainz, Germany
[4] Manne Siegbahn Laboratory, S-10405 Stockholm, Sweden
[5] Chalmers University of Technology, S-41296 Göteborg, Sweden
[6] Max-Planck-Institut für Kernphysik, D-69117 Heidelberg, Germany
[7] Institut für Angewandte Physik, J.W. Goethe Universität, D-60325 Frankfurt, Germany
[8] GSI, D-64220 Darmstadt, Germany
[9] Universite Louis Pasteur, Strasbourg, France
[10] Instituut voor Kern- en Stralingsfysica, K.U. Leuven, B-3001 Leuven, Belgium
[11] University of Edinburgh, GB-Edinburgh EH9 3JZ, UK
[12] TU-Darmstadt, D-64289 Darmstadt, Germany

Abstract.
The ISOLDE RIB-facility at CERN has today been producing a vast range of radioactive beams since more than 30 years. The low-energy beams of ISOLDE will be complemented by a post-accelerator, REX-ISOLDE, currently being assembled. In order to convert the pseudo-DC, singly-charged beam from the ISOLDE mass separators into a cooled and bunched beam at higher charge states, a novel scheme of trapping, cooling and charge-state breeding has been devised, using a linear Penning trap and an Electron Beam Ion Source (EBIS). This allows for subsequent acceleration by a short, cost-effective LINAC consisting of an RFQ, an IH-structure and three seven-gap resonators, reaching 0.8 - 2.2 MeV/u. The installation of REX-ISOLDE is well underway and the first post-accelerated radioactive beams are expected to be obtained during late 2000.

Introduction

Radioactive Ion Beam (RIB) Facilities is presently a topic attracting interest from a large part of the nuclear physics community. A number of new or upgraded facilities are currently being planned, built or commissioned world-wide, and in a longer perspective, NUPECC (Europe) and DOE (USA) independently have recommended the construction of a "second-generation" RIB facility where radioactive beam intensities 100 times higher than at the current facilities can be produced.

The production of RIB is done according to two different methods; one is to let an energetic heavy-ion beam impinge on a thin production target and by subsequent in-flight ($B\rho - \Delta E - B\rho$) separation of the fragments select the wanted isotopes (1). The other method, which is the one used at ISOLDE (2), is to let a driver beam (protons, neutrons or heavy ions) impinge on a thick production target where a huge amount of unstable nuclides are created, and from this target volume extract, ionise and separate the wanted isotopes.

The driver beam of the ISOLDE facility is a pulsed

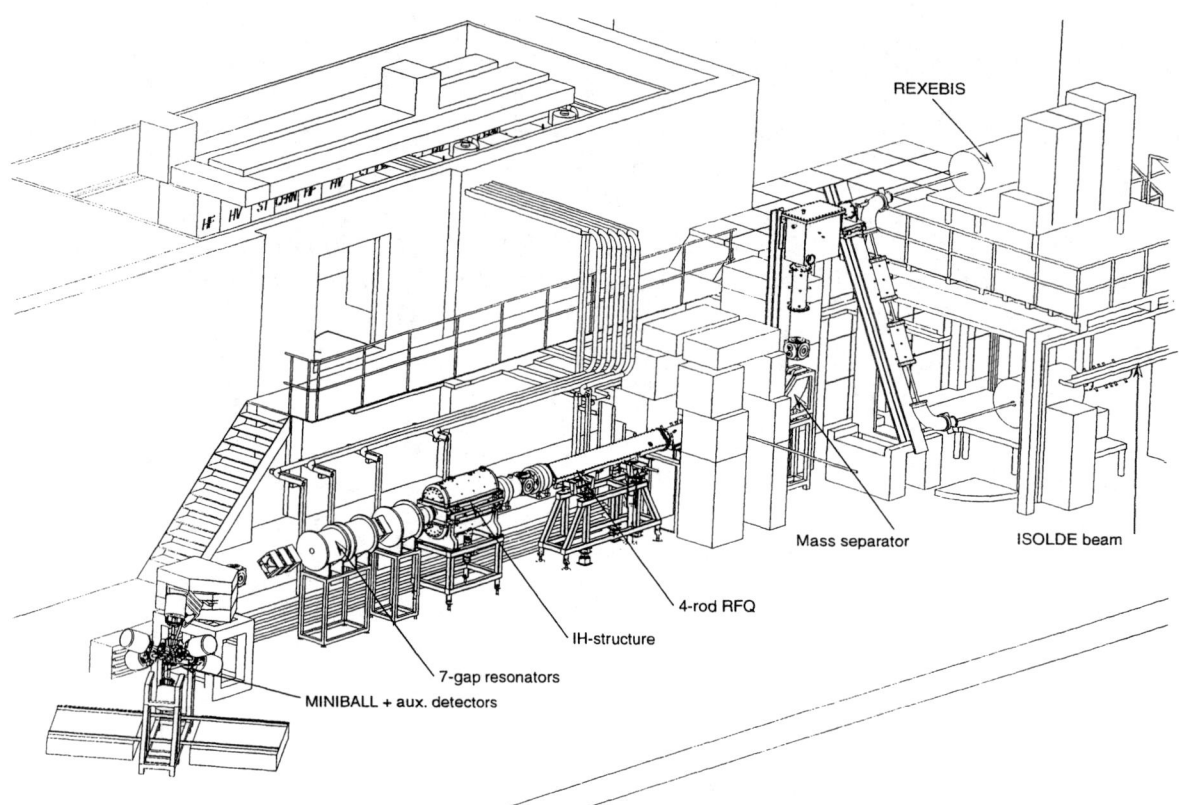

FIGURE 1. Schematic view of the REX-ISOLDE installation in the ISOLDE experimental hall.

proton beam from the PS-Booster synchrotron of 1.0 or 1.4 GeV energy with a repetition rate of typical 1/2.4 s^{-1}. The average current exceeds 2 μA. The high-energetic protons induce spallation, fragmentation and fission, giving rise to a large number of radioactive species. ISOLDE regularly uses a large variety of target materials (metal foils, oxides and carbides) and 1$^+$-ion sources (surface, plasma and laser ionisation) to produce more than 600 isotopes of >60 elements. A constant development of targets with respect to new materials with release of the often short-lived species of interest (down to a few ms) is intensely pursued. The Resonant Ionisation Laser Ion Source (RILIS) (3, 4) has been intensely and successfully used at ISOLDE in recent years to improve the ionisation efficiency, enhance selectivity and to ionise elements not otherwise achievable with the traditional ISOLDE ion sources.

With this as a starting point, it is clear that the prerequisites for building a post-accelerator at ISOLDE are optimal, apart from the proliferation and intensities of the available radioactive beams, it also benefits from the large know-how in targetry and RIB handling and the existence of CERN infrastructures.

The REX-ISOLDE post-accelerator

Until now, ISOLDE has been limited to study radioactive nuclei at very low energies only, 60keV. The REX-ISOLDE (5, 6) experiment will change this situation dramatically; almost any nuclide currently produced at ISOLDE will be possible to accelerate to an energy in the range 0.8 - 2.2 MeV/u. This opens up a completely new field of experiments with RIB with A<50, with an energy range tailored for Coulomb excitation and few-nucleon transfer reactions, being well-proven spectroscopic tools for stable beams. An overview picture of REX-ISOLDE in its final state is shown in 1.

REXTRAP - REXEBIS Cooling, bunching and charge breeding

The RIB from the ISOLDE separator is a continuous beam, with the intensity varying in time after the proton impact as a function of the release characteristics and the half-life of the nuclide in question. To transform this singly-charged, "pseudo-DC" beam from the ISOLDE separator to a multiple-charged (q/A > 1/4.5) pulsed

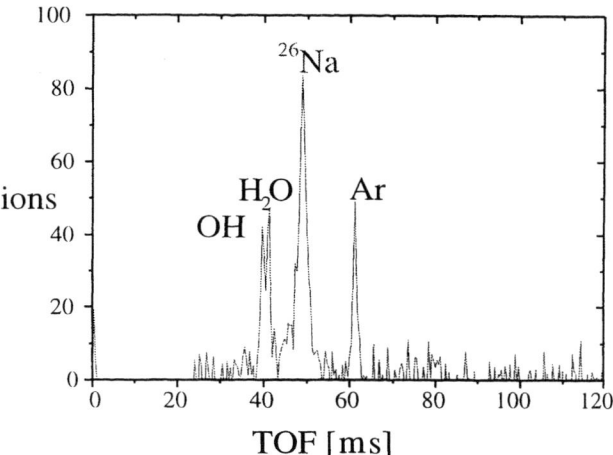

FIGURE 2. TOF distribution of captured, cooled and bunched ^{26}Na ($T_{1/2}$ = 1.07 s) ejected from REXTRAP.

beam suitable for acceleration with acceptable losses requires novel techniques. Bunching of the radioactive beam is required in the chosen acceleration scheme because the charge-state breeding in the REXEBIS requires a typical time of about 20 ms and the LINAC operates with a duty factor of 10%. The basic technique applied to accumulate and bunch the beam is to continuously inject the radioactive ions into a large Penning trap, REXTRAP, where they are stopped by collisions with the atoms of a buffer gas, accumulated and finally extracted as pulses. REXTRAP consists of a 1 m long cylindrical trap structure filled with buffer gas in a 3 T magnetic field. It is mounted on a high voltage platform close to 60 kV in order to retard the ions from 60 keV to some eV suitable for injection into the trap. After the ions have passed the potential barrier at the entrance of the trap, final deceleration is done by friction in the Ar buffer gas. The ions can be fully captured if their energy loss during a single oscillation in the trap is larger than the energy spread of the ISOLDE beam after deceleration, after which they can not escape the entrance barrier. Typical cooling times down to room temperature are a few ms. By this cooling the emittance of the extracted beam is considerably improved. The trapping and cooling are expected to reach almost 100% efficiency. Recent tests with stable and radioactive beams have shown capture efficiencies up to 27% for ion beams from ISOLDE and 45% from an auxiliary ion source. In fig 2 a TOF distribution of captured, cooled and bunched ^{26}Na ($T_{1/2}$ = 1.07 s) ejected from REXTRAP is shown.

Mass separator and LINAC

The bunched beam is then transported to REXEBIS (Electron Beam Ion Source). An EBIS makes use of a dense electron beam that is focused to a high current density by a strong magnetic field of a solenoid. The electron beam forms a radial potential well for the ions while the longitudinal confinement is performed by electric potentials applied to cylindrical electrodes surrounding the electron beam. Trapped low-charged ions undergo stepwise ionisation via electron impact collisions until the ions are extracted by changing the longitudinal potential distribution. The centroid of the charge distribution is determined by the product of the confinement time and the electron beam current density. In the REXEBIS a superconducting solenoid creates a magnetic field of 2 T, with a homogeneity of about 0.25% along the confinement length of 0.8 m which compresses the 0.5 A, 5 keV electron beam into a current density larger than 200 A/cm^2. The isotopes used at REX-ISOLDE will require breeding times between 5 and 20 ms to reach a charge-to-mass ratio larger than 1/4.5. In order to decrease the injection energy into the RFQ to 5 keV/u the platform voltage has to be switched from 60 kV (injection potential) down to 20 kV, and rapidly back again to accept the next bunch of ions.

The yield of the radioactive isotopes from ISOLDE can be several orders of magnitude lower than the amount of residual gas ions from C, N, O and Ar coming out of the EBIS. Therefore, a mass separator similar to a Nier-spectrometer will be employed with a q/A-resolution of about 1/150 which is sufficient to select the highly charged rare radioactive ions from rest-gas contaminants.

In the first stage of REX-ISOLDE LINAC the ions are accelerated from 5 keV/u to 300 keV/u by a 4-rod RFQ. In order to match the beam from the RFQ into the acceptances of the IH-structure a section consisting of two magnetic quadrupole triplet lenses and a rebuncher is required.

The Interdigital-H-type (IH)-structure is an efficient drift-tube structure, similar to structures like the GSI HLI-IH-structure or 'tank 1' of the lead LINAC at CERN. A new feature of the REX-ISOLDE-IH resonator is the possibility to vary the final energy between 1.1 and 1.2 MeV/u by adjusting the gap voltage distribution via two capacitive plungers and by adjusting the rf-power fed into the resonator. The lower final energy of the IH-structure is important for deceleration of the ions down to 0.8 MeV/u, since the deceleration from 1.2 MeV/u down to 0.8 MeV/u through the 7-gap resonators would perform a non-acceptable phase spread at the target.

The high-energy section (0.8 - 2.2 MeV/u) of the REX-ISOLDE LINAC consists of three 7-gap resonators similar to those built for the high-current injector at MPI

Heidelberg. These spiral resonators are designed and optimised for synchronous velocities of β=5.4%, 6.0% and 6.6%. Each resonator has a single resonance structure, which consists of a copper half shell and three arms attached to both sides of the shell. Each arm consists of two hollow profiles, surrounding the drift tubes and carrying the cooling water. All LINAC structures are operated at 101.28 MHz.

Planned experiments at REX-ISOLDE

The REX-ISOLDE pilot experiments (7, 8) aims to probe whether the magic numbers N=20 and N=28 are conserved when going to very neutron-rich nuclei by using Coulomb excitation and transfer reaction experiments. Other experiments have hinted a weakening of the nuclear shell structure in the region of ^{32}Mg (9) and below ^{48}Ca (10, 11), with deduced sizable deformations. The main detector system to be used here is the state-of-the-art gamma-detector array MINIBALL (12) currently being assembled by a large European collaboration. To facilitate Doppler corrections, charged particles will be detected in a high-granularity, segmented silicon detector with "compact-disc" geometry, the CD. The residual nuclei will be detected in a PPAC (Parallel Plate Avalanche Counter) downstreams from the reaction target.

A large number of other experiments will also be using REX-ISOLDE. Two further experiments (13, 14) are already approved; both concerned with investigating the unbound sub-systems of halo nuclei by one-neutron pick-up reactions and elastic resonance scattering. E.g. the unbound sub-system ^{10}Li of the halo nucleus ^{11}Li will be studied by the ^{9}Li(d,p)^{10}Li reaction and through the IAS in ^{10}Be by the ^{9}Li(p,p')^{9}Li reaction. Other experiments are foreseen in the short-term future concerning the structure of nuclei along the N=Z line, proton radioactivity and nuclear astrophysics.

The bunched and cooled beams from REXTRAP are also well suited for other experiments without subsequent acceleration. In one approved experiment (15), the cooled beam will be extracted and stored in an electromagnetic trap, combined with a retardation spectrometer. The latter is used to measure the recoil momentum of the daughter nuclei after beta decay, allowing studies of β-ν correlations. This could give information on possible scalar current components in the nuclear beta decay, meaning "new physics" beyond the Standard Model. Furthermore, the higher available implantation energy gives rise to new possibilities to use radioactive methods to probe the bulk properties of condensed matters.

Outlook

The REX-ISOLDE post-accelerator will keep ISOLDE in a unique position in the world, with a unprecedented choice of radioactive beams for low-energy reactions. Together with the continuous beam development, this assures many years to come with ISOLDE research at the forefront of physics. The techniques used in REX-ISOLDE are also highly interesting for future, second-generation ISOL-facilities, such studied in the recently started EURISOL (16) project. It is foreseen that after the commissioning phase and the first generation of pilot experiments, REX-ISOLDE will be integrated into the CERN accelerator structure.

REFERENCES

1. Geissel, H., Münzenberg, G., and Riisager, K.: *Annu. Rev. Nucl. Part. Sci.*, **45**, (1995) 163.
2. Kugler, E., et al.: *Nucl. Instr. Meth.*, **B70**, (1992) 41.
3. Lettry, J., et al.: *Rev. Sci. Instrum.*, **69** (2), (1998) 761–763.
4. Köster, U., et al.: *Nucl. Instr. and Meth.*, **A160** (4), (2000) 528–535.
5. Habs, D., et al.: *Nucl. Instr. and Meth.*, **B139**, (1998) 128–135.
6. Habs, D., et al.: *Hyperfine Interactions*, **ISOLDE Laboratory portrait, acc. for publ.**
7. REX-ISOLDE collaboration: Radioactive beam experiments at isolde: Coulomb excitation and neutron transfer reactions of exotic nuclei, CERN/ISC 94-25 P64, November 1994.
8. Scheit, H., et al.: Investigation of the single particle structure of the neutron-rich sodium isotopes $^{27-31}$Na, CERN-ISTC-99-20 ISC-P-114, 1999.
9. Motobayashi, T., et al.: *Phys. Lett.*, **B346**, (1995) 9–14.
10. Glasmacher, T., et al.: *Phys. Lett.*, **B395** (3-4), (1996) 163–168.
11. Scheit, H., et al.: *Phys. Rev. Lett.*, **77** (19), (1996) 3967–3970.
12. Habs, D., et al.: *Z.Phys.*, **A358**, (1997) 161–162.
13. Axelsson, L., et al.: Study of the unbound nuclei ^{10}Li and ^{7}He at REX ISOLDE, CERN/ISC 98-11 ISC-P-100, 1998.
14. Axelsson, L., et al.: Investigations of neutron-rich nuclei at the dripline through their analogue states: The cases of ^{10}Li-^{10}Be (T=2) and ^{17}C-^{17}N (T=5/2), CERN/ISC 98-23 ISC-P-105, 1998.
15. Beck, D., et al.: Search for new physics in beta-neutrino correlations using trapped ions and a retardation spectrometer, CERN/ISC 99-13 ISC-P-111, 1999.
16. EU-network EURISOL: proposal to the EU, No. HPRI-1999-50016, 2000.

The Production and Generation of Radioactive Beams at Louvain-la-Neuve

M. Loiselet, M. Cogneau, J.M. Colson, M. Gaelens, N. Postiau, G. Ryckewaert

Centre de Recherches du Cyclotron, Université catholique de Louvain
Chemin du Cyclotron, 2
B-1348 Louvain-la-Neuve, Belgium

Abstract. Recent development work on various aspects of the generation of Radioactive Ion Beams at Louvain-la-Neuve is reported. The performances of CYCLONE44, the new isochronous cyclotron designed for the combined acceleration and isobaric separation of RIB at low energy are discussed. The results obtained with a new target dedicated to the production of ^{15}O, and able to stand 9 kW of 30 MeV proton beam power are presented. The present status of the research and development efforts aiming at the acceleration of an intense ^{7}Be beam is given.

INTRODUCTION

The Louvain-la-Neuve facility is providing post-accelerated radioactive beams on a routine basis since the beginning of the nineties. It uses two cyclotrons coupled by an ion source to produce, ionize and selectively accelerate the radioactive ions. The intense 30 MeV proton beam of the first machine, CYCLONE30, is used to produce unstable elements in dedicated targets. After extraction from the target, these are ionized to the requested charge in an ECR source before being accelerated in the second cyclotron, CYCLONE, which is also used as a high resolution mass separator, to remove (or considerably reduce) any isobaric contaminant (fig. 1).

During the last years, while continuing to deliver these exotic beams to a relatively large community of European physicists, most of the efforts have been put in the design and construction of a new post-accelerator, CYCLONE44. The present status of this new machine, which started to operate in 1999, is given in the first part of this paper. In the second part, we report on the latest target and ion source developments aiming at the generation of ^{15}O and ^{7}Be beams.

CYCLONE44, A CYCLOTRON FOR LOW ENERGY RADIOACTIVE BEAMS

CYCLONE44 is a compact cyclotron dedicated to the acceleration *and* isobaric separation of unstable ions. It covers an energy domain between 0.2 and 0.8 MeV per nucleon, and is used for experiments in nuclear astrophysics. It will later be extended to 1.7 MeV/amu. In order to provide pure radioactive beams in the presence of isobaric contaminants that are sometimes many orders of magnitude larger, it is designed as a high efficiency accelerator and a radiofrequency mass spectrometer with a high resolving power.

To assure the requested mass separation ($\Delta m/m \approx 10^{-4}$), it has been shown (1,2) that the number of turns inside the machine has to be rather large. For this reason, it has been equipped with two acceleration electrodes (so-called "dee") with a non-conventional shape (fig. 2). At small radius, the dee angle is close to the optimum value (30° in harmonic 6 mode) to have the maximum energy gain per turn in the first few turns. Outside this region, the dee angle is reduced to decrease the energy gain per turn, and so increase the number of turns it takes for a particle to reach the extraction zone.

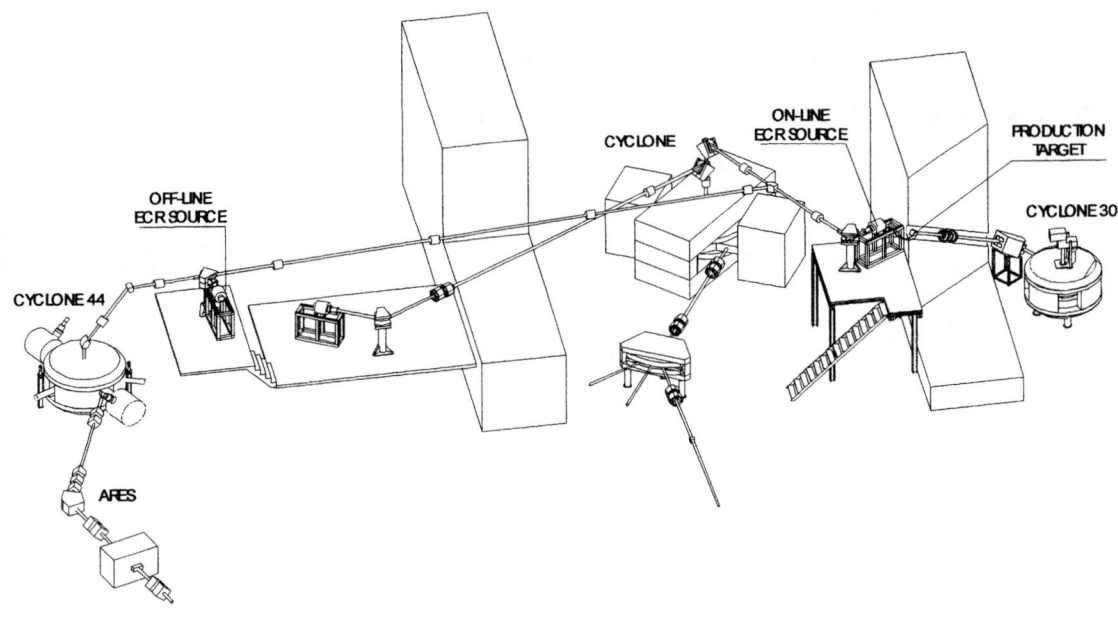

FIGURE 1. The Louvain-la-Neuve radioactive beam facility

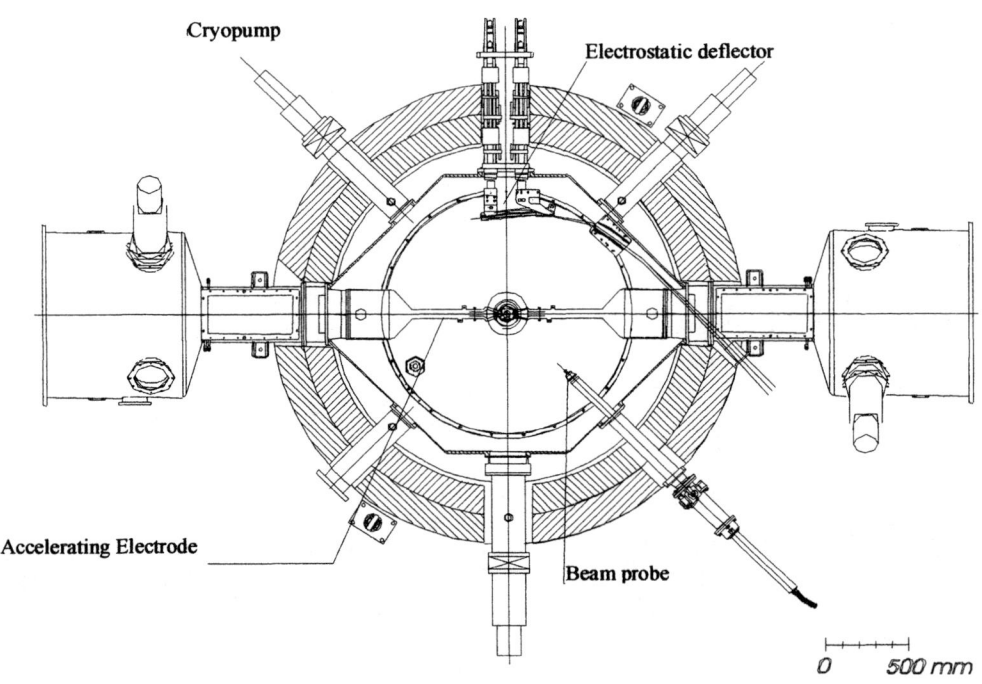

FIGURE 2. Median plane view of CYCLONE44, with the non-conventional accelerating electrodes

The ions can be injected from an off-line ECR source, which was used for the commissioning, or from the on-line ECR source for the acceleration of the unstable elements (fig. 1). At the end of 1999, CYCLONE30 and CYCLONE44 were connected, and a first radioactive beam ($^{19}Ne^{3+}$) was accelerated and separated from its isobaric contaminant $^{19}F^{3+}$, which has a relative mass difference ($\Delta m/m$) of $1.8\ 10^{-4}$ with respect to $^{19}Ne^{3+}$. An overall transmission efficiency (from the analyzing magnet after the on-line source to the exit of the cyclotron) of 7 % was obtained, and the isobaric suppression factor for ^{19}F was measured to be $2\ 10^3$. The beam loss measurements have shown that the poor transmission efficiency of the injection line (50%) could be improved by minor mechanical modifications. However, both the isobaric suppression factor of the cyclotron, which has to reach a value of 10^5 for ($\Delta m/m$)=$1.0\ 10^{-4}$, and the acceleration efficiency have to be further increased. After detailed measurements and a careful reanalysis of the numerical simulations, it appears that the debunching effect induced by the geometry of the acceleration electrodes spoils the mass resolving power of the cyclotron. Therefore, new acceleration electrodes, with a conventional dee-shape, and a new central region are presently under design. Meanwhile, CYCLONE44 is accelerating stable beams for the development of the recoil separator ARES, which will be used for experiments in nuclear astrophysics.

DEVELOPMENTS OF NEW RADIOACTIVE ION BEAMS

In parallel with the commissioning of CYCLONE44, the beam development work has been concentrated mainly on the production of ^{15}O and ^{7}Be beams with CYCLONE30 and CYCLONE. The typical intensities of the various beams available after acceleration and isobaric separation in CYCLONE are given in table 1.

$^{15}O^{2+}$ Beam

As it was described in the previous conference of this series (3), ^{15}O is produced in a LiF target through the $^{19}F(p,\alpha n)$ reaction with a production yield at 30 MeV of $1.6\ 10^{-3}$ ^{15}O per incident proton. In order to enhance the ^{15}O extraction from the target material as $C^{15}O$ or $C^{15}O_2$ molecules, the target is made of a graphite matrix (diameter: 50 mm, height: 17 mm) in which 48 holes of 5 mm diameter are filled with compressed LiF powder (see ref 3 for details).

The target assembly has been intensively tested with the 30 MeV proton beam up to 300 µA, which corresponds to a beam power of 9 kW totally dissipated in the target. The ^{15}O extraction efficiency has been measured at different proton beam intensities by condensing the released gases in a cryogenic trap, and monitoring the activity in the trap. The result of these measurements is given in fig. 3. It shows that the extraction efficiency, which is of the order of 6% at 200 µA, increases to 24% at 300 µA. This corresponds to respectively $4.5\ 10^{10}$ and $2.8\ 10^{11}$ ^{15}O atoms extracted from the target per second. We define the extraction efficiency as the ratio between the measured number of ^{15}O condensed in the trap and the number of ^{15}O produced inside the target, as calculated from the production yield.

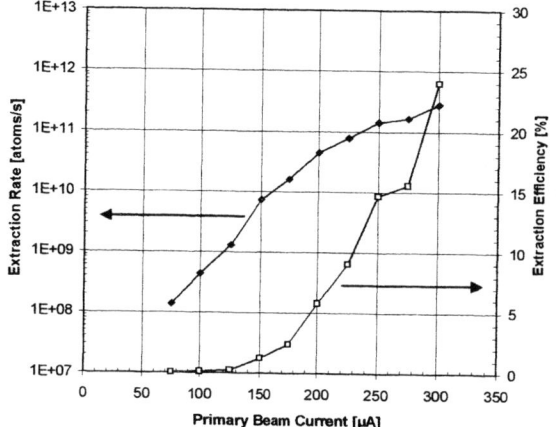

FIGURE 3. Extraction of ^{15}O as a function of the proton beam intensity. The left-side scale (logarithmic) shows the number of extracted atoms per second, while the right-side scale (linear) shows the extraction efficiency.

A first preliminary measurement of the ionization efficiency of the ECR source, under on-line conditions, was done by using a calibrated leak of $^{17}O_2$. With 210 µA on target, the pressure at the injection side of the source was $2\ 10^{-5}$ mbar and the ionization efficiency for 2^+ ions was measured to be 5%. A systematic measurement of this quantity, at various proton beam intensities, and with an outgassed target, has still to be made. The intensity of the $^{15}O^{2+}$ beam after acceleration and isobaric separation in CYCLONE is $6\ 10^7$ pps, with a primary beam of 250 µA.

$^7Be^{1+}$ Beam

The main interest for the ^7Be beam in nuclear astrophysics is the study of the $^7Be(p,\gamma)^8B$ reaction which is of importance for the "solar neutrino problem". To allow the measurement of this reaction at 1MeV/A, the requested beam intensity after acceleration in CYCLONE should be larger than 10^8 pps.

The long half-life of ^7Be (53.3d) allows us to develop a scheme completely different from the ones previously used in our laboratory. ^7Be is produced in a metallic lithium target with 27.5 MeV protons. To avoid the problems of the chemical reactivity of liquid lithium, the proton beam intensity is limited to 25μA to keep the lithium in the solid phase. After irradiation, ^7Be is chemically separated from the lithium and is deposited on a sputtering electrode which is introduced in the plasma chamber of the ECR ion source (4,5,6). Because of the long half-life of ^7Be and the toxicity of Be, special care has been taken to avoid any contamination. The sputtering and ionization of Be has been studied in a dedicated plasma chamber, with a thin layer of ^9Be (200 μg/cm^2) on the sputtering electrode, mixed with ^7Be for monitoring purposes. An ionization efficiency of the order of a 2-4 % was measured for $^9Be^{1+}$. This value, combined with the acceleration efficiency of CYCLONE, requires that at least 10^{17} atoms of ^7Be have to be deposited on the sputtering electrode if the beam has to be delivered during 3 days. For safety reasons, several test runs have been done with less than 10^{16} atoms of ^7Be on the sputtering head, and a pure beam of $2\ 10^7$ pps of $^7Be^{1+}$ at 7 MeV has been accelerated with CYCLONE during several hours. The reliability, reproducibility and safety of each step of the process (i.e. production, chemical extraction and ionization) have to be further improved before starting an experiment with 10^{17} atoms of ^7Be on the sputtering head.

ACKNOWLEDGMENTS

This paper presents results of research funded by the Belgian Program on Interuniversity Pole of Attraction (PAI 4/48) initiated by the Belgian State, Federal Services of Scientific, Technical and Cultural Affairs, and by the Interuniversity Institute for Nuclear Sciences (IISN- contract N°4.4502.95). M.G. is a Scientific Research Worker of the F.N.R.S.

TABLE 1. Typical intensities available for the various radioactive ions accelerated by CYCLONE (Nov. 2000)

	T1/2	q	Intensity [pps]	Energy range [MeV]
^6He	0.8 s	1+	$9\cdot10^6$	5.3-18
		2+	$3\cdot10^5$	30-73
^7Be	53 d	1+	$2\cdot10^7$	5.3-15.7
^{11}C	20 m	1+	$1\cdot10^7$	6.2-10
^{13}N	10 m	1+	$4\cdot10^8$	7.3-8.5
		2+	$3\cdot10^8$	11-34
		3+	$1\cdot10^8$	45-70
^{15}O	2 m	2+	$6\cdot10^7$	10-29
^{18}F	110 m	2+	$1\cdot10^6$	11-24
^{18}Ne	1.7 s	2+	$6\cdot10^6$	11-24
		3+	$4\cdot10^6$	24-36,45-55
^{19}Ne	17 s	2+	$2\cdot10^9$	11-23
		3+	$1.5\cdot10^9$	23-38,45-50
		4+	$8\cdot10^8$	60-93
^{35}Ar	1.8 s	3+	$2\cdot10^6$	20-28
		5+	$1\cdot10^5$	50-79

REFERENCES

1. M. Loiselet et al., *Proc. of the 15th Int. Conf. on Cyclotrons and their Applications*, edited by E. Baron and M. Lieuvin, Institute of Physics Publishing, 1998, pp. 305-310.

2. C. Barué et al., *Proc. of the 6th European Particle Accelerator Conf*, Institute of Physics Publishing, 1998, pp. 544-546.

3. M. Gaelens, M.Loiselet, G.Ryckewaert, *Proc. of the 15th Int. Conf. on Applications of Accelerators in Research and Industry*, AIP Conf. Proc. 475, 1999, pp. 305-308.

4. C. Barué et al., *Rev. of Sc. Instr.* **69**, pp. 764 -766 (1998).

5. B.W. Filippone, M. Wahlgren, *NIM* **A243**, pp. 41-44 (1986).

6. A. Y. Zyuzin et al., *NIM* **A438**, pp. 109-115 (1999).

Research with Radioactive Beams at ORNL

J. F. Liang

Physics Division, Oak Ridge National Laboratory, Oak Ridge, TN 37831

Abstract. Radioactive ^{17}F and ^{18}F beams have been produced at the HRIBF using the ISOL technique. The beams were postaccelerated by the 25 MV tandem accelerator. Experiments were carried out to search for F(p,p) resonances, measure F(p,α) reaction rates, study two-proton decay from ^{18}Ne produced by ^{17}F+p, and study the reaction mechanisms in ^{17}F + ^{208}Pb.

INTRODUCTION

The Holifield Radioactive Ion Beam Facility (HRIBF) employs the Isotope Separator On-Line (ISOL) method to produce radioactive beams. Light-ion beams from the Oak Ridge Isochronous Cyclotron (ORIC) bombard a close-coupled target-ion source. The radioactive ions are extracted, mass analyzed and subsequently accelerated by the 25 MV tandem accelerator for experiments. The ^{17}F beam was produced by the ^{16}O(d,n) reaction using a 42 MeV proton beam bombarding a refractory fibrous hafnium oxide target. The ^{18}F beam was made with the same target material but with an 85 MeV ^{4}He beam by the ^{16}O(α,pn) reaction. The highest intensity achieved for experiments is 10^6 pure ^{17}F per second and 2×10^5 ^{18}F per second with some ^{18}O contaminations.

The facility houses three experimental end stations: Recoil Mass Separator (RMS), Daresbury Recoil Separator (DRS), Enge Split-Pole Spectrograph, and a variety of detectors and electronics. This report presents several experiments performed with the radioactive ^{17}F and ^{18}F beams. Other measurements at the HRIBF include observation of the 3^+ state in ^{18}Ne in ^{1}H(^{17}F,p)^{17}F which was published(1) and the two-proton decay from ^{18}Ne populated by the ^{17}F+p reaction which will be presented elsewhere in this conference(2).

MEASUREMENT OF ^{1}H(^{17}F,α)^{14}O CROSS SECTION

In x-ray bursts and novae, ^{12}C captures protons through ^{12}C(p,γ)^{13}N(p,γ)^{14}O to produce ^{14}O. Further proton capture by ^{14}O is inhibited because ^{15}F is particle unstable. Since the lifetime of ^{14}O is long ($t_{1/2}$ = 71 s), a significant accumulation of ^{14}O can build up. However, in x-ray bursts where the temperature is higher, ^{14}O(α,p)^{17}F can take place leading to the production of heavier elements. Measuring the reaction rate of ^{14}O(α,p)^{17}F is important for understanding the conditions under which the ^{14}O β-decay is bypassed.

The time-reversed reaction ^{17}F(p,α)^{14}O was measured at 21 beam energies between E(^{17}F) = 40.1 – 68.2 MeV corresponding to E* = 6.1 and 7.7 MeV, the excitation energy in ^{18}Ne, which is the region relevant to reactions occurring in x-ray bursts and novae. The accelerated ^{17}F beam was stripped to the 9$^+$ charge state to eliminate ^{17}O contamination and incident on a thin polypropylene target. The α-particles were detected in a large annular silicon detector array (SIDAR) covering angles from 10.5° to 25.8°. The recoiling ^{14}O ions were detected in coincidence with the α-particles in a smaller annular silicon strip detector spanning θ$_{lab}$ = 3.4° – 6.7°. The excitation function is shown in Fig. 1 and compared to calculations by Hahn et al.(3).

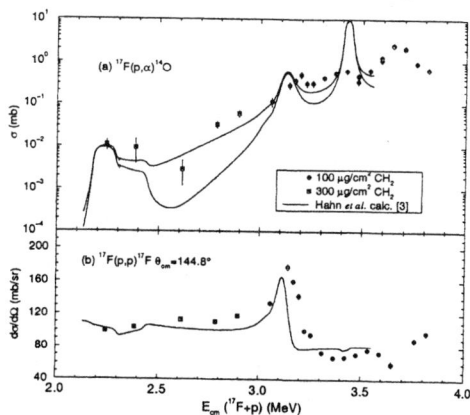

FIGURE 1. The ^{17}F(p,α)^{14}O excitation function and ^{17}F(p,p)^{17}F differential scattering cross section at θ$_{cm}$ = 144.8°.

From the measured ^{1}H(^{17}F,α)^{14}O cross sections and other recent data(3, 4), the reaction rates for ^{14}O(α,p)^{17}F$_{g.s.}$ can be determined with reasonable uncertainty (~ 30 %). The reaction rates for ^{14}O(α,p)^{17}F* which populates the first excited state of ^{17}F are still unknown. This reaction may be significant and would increase the ^{14}O(α,p)^{17}F reaction rate. Further measurements are planned to constrain the uncertainties in the ^{14}O(α,p)^{17}F reaction rates.

MEASUREMENT OF ^{1}H(^{18}F,p)^{18}F AND ^{1}H(^{18}F,α)^{15}O EXCITATION FUNCTIONS

Large quantities of proton rich nuclei are produced in nova explosions. Because ^{18}F has a relatively long lifetime ($t_{1/2}$ = 109.8 m) and large abundance, its positron decay produces the majority of γ-rays observable during the first few hours after the explosion. The observation of γ-rays from nova ejecta would provide information for testing nova models. The destruction of ^{18}F by the ^{18}F(p,α)^{15}O reaction will reduce the amount of ^{18}F available to decay. Determining whether γ-ray observations are feasible would rely on knowing a more precise value of the ^{18}F(p,α)^{15}O stellar reaction rate. This reaction is thought to be dominated by a resonance near $E^* \simeq 7.08$ MeV in ^{19}Ne(5). Moreover, the competition between ^{18}F(p,α)^{15}O and ^{18}F(p,γ)^{19}Ne in part determines the production of heavy-elements in x-ray bursts. Therefore, it is important to measure the stellar reaction rates of ^{18}F(p,α)^{15}O and ^{18}F(p,γ)^{19}Ne.

The experimental setup is shown in Fig. 2. Both the ^{1}H(^{18}F,p)^{18}F and ^{1}H(^{18}F,α)^{15}O reactions were measured simultaneously with a thin polypropylene target bombarded by ^{18}F. In the ^{1}H(^{18}F,p)^{18}F reaction, the proton was detected by the SIDAR in coincidence with ^{18}F which was detected in the gas ionization chamber. E and ΔE information was provided by the gas ionization chamber for particle identification. Since the beam was contaminated by ^{18}O (^{18}F/^{18}O ~ 0.1), coincidence measurement was necessary to discriminate against events originated from ^{18}O+p. For the ^{1}H(^{18}F,α)^{15}O reaction, both ^{15}O and α-particles were detected in SIDAR. Because of the differences in Q-value, the ^{1}H(^{18}F,α)^{15}O events were distinguished from the ^{1}H(^{18}O,α)^{15}N events by reconstructing the total energy of the recoils. The yield was measured at 15 energies to obtain the excitation functions.

The measured excitation functions for ^{1}H(^{18}F,p)^{18}F and ^{1}H(^{18}F,α)^{15}O are shown in Fig. 3. A simultaneous fit to the ^{1}H(^{18}F,p)^{18}F and ^{1}H(^{18}F,α)^{15}O excitation functions was performed to extract the resonance properties. For the ^{1}H(^{18}F,p)^{18}F data, the Breit-Wigner methodology and R-matrix analysis were used. The ^{1}H(^{18}F,α)^{15}O data were fitted with the standard formula for an isolated resonance:

$$\frac{d\sigma}{d\Omega} = \frac{1}{4}\frac{\lambda^2\omega}{4\pi^2}\frac{\Gamma_p(E)\Gamma_\alpha(E)}{(E-E_r)^2+(\Gamma(E)/2)^2},$$

where ω is the statistical factor and $\Gamma = \Gamma_\alpha + \Gamma_p$. The resonance energy, width, and proton partial width adopted from this measurement and analysis are E_r = 664.7±1.6 keV, Γ = 39.0±1.6 keV, and Γ_p/Γ = 0.39 ± 0.02, respectively. Fig. 4 presents the temperature dependence of stellar reaction rate based on the measured resonance properties. This work removes the uncertainty in the ^{1}H(^{18}F,α)^{15}O rate at high energies and resolves the discrepancies in the previous measurements(5, 6, 7). However, the rate at lower temperature is still uncertain because of the uncertain properties of lower energy states in ^{19}Ne. Further measurements with ^{18}F beams are planned to address these uncertainties.

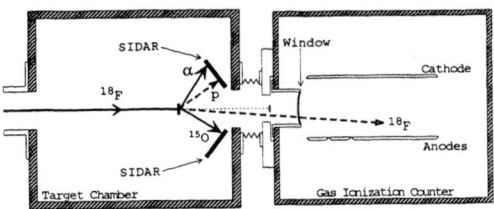

FIGURE 2. Experimental setup for measuring ^{1}H(^{18}F,p)^{18}F and ^{1}H(^{18}F,α)^{15}O reactions.

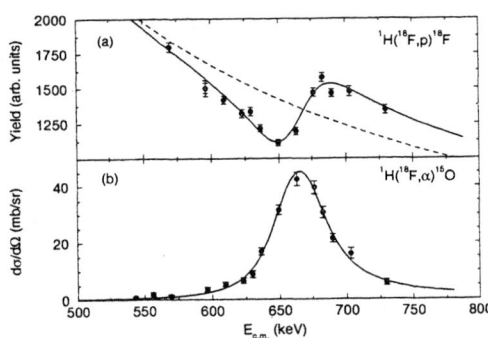

FIGURE 3. (a) The normalized proton yield as a function of the average center of mass energy in the target. The solid curve shows the R-matrix fit and the dashed curve shows the expected yield if there is no resonance. (b) The ^{1}H(^{18}F,α)^{15}O excitation function.

MEASUREMENT OF ^{17}F+^{208}PB REACTION

^{17}F is a proton drip-line nucleus and its valence proton is bound by 0.6 MeV. The first excited state, E^* =

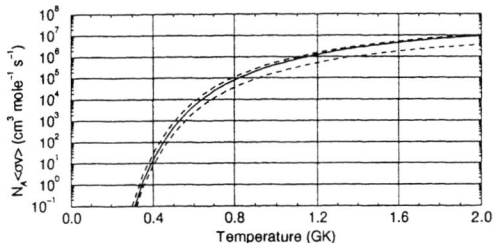

FIGURE 4. The ^{1}H(^{18}F,α)^{15}O rate contribution from from the $\frac{3}{2}^+$ resonance as a function of the stellar temperature. The dashed curves are limits of the rate from the previously measured properties of this resonance.

0.495 MeV and $J^\pi = \frac{1}{2}^+$, of ^{17}F is expected to have an extended rms radius(8). It is predicted that reactions involving weakly bound nuclei may be influenced by the breakup reaction channel. Fusion excitation functions measured with ^{17}F+^{208}Pb found no enhancement in fusion at energies near the Coulomb barrier(9).

The breakup of ^{17}F into ^{16}O and p was measured by bombarding a ^{208}Pb target with a 170 MeV ^{17}F beam. The breakup products were measured in coincidence by a large area double-sided silicon strip detector (DSSD) spanning angles from $\theta_{lab} \sim 30° - 60°$. The upper limit of the breakup cross section is found to be 6.6 mb/sr which is about a factor of 4 less than theoretical predictions. This small cross section can have little influence on fusion(10).

In a separate experiment, a 900 mm^2 thin silicon surface barrier detector (SBD) was placed in front of the DSSD to measure the energy loss of the reaction products. Since the energy resolution was not very good, the angular distribution of elastic scattering was obtained with contributions from inelastic scattering, as shown in Fig. 5. Optical model fits to the elastic and inelastic scattering data were performed with the Wood-Saxon potential. The results are shown by the solid curve in Fig. 5 and the potential parameters are listed in Table 1.

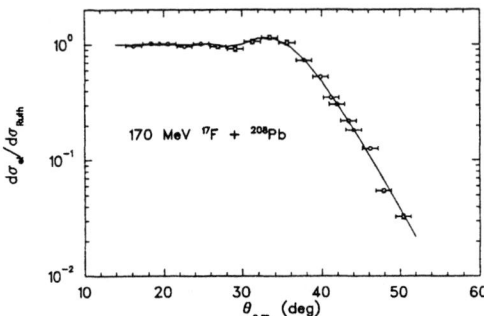

FIGURE 5. Angular distribution of ^{17}F+^{208}Pb elastic scattering. The optical potential fit to the data is shown by the solid curve.

Table 1. Optical model potential parameters obtained from fitting the ^{17}F+^{208}Pb elastic and inelastic scattering data. V and W is the depth of real and imaginary part of the potential, respectively.

V	W	$r_o = r_{io}$	$a = a_i$
40	73	1.29	0.5

Angular distributions of the reaction products with atomic number (Z) different from fluorine are shown in Fig. 6. The angular distributions of N (Z=7) and C (Z=6) have a peak near the grazing angle. This shows that these are products originated from transfer reactions. The shape of the angular distribution of O (Z=8) is similar to that of N and C but the peak is shifted slightly to forward angles. As shown in Ref (10), the angular distribution of ^{17}F breakup is predicted to have a peak at 34°. The oxygen produced in ^{17}F+^{208}Pb can be attributed to transfer and breakup reactions.

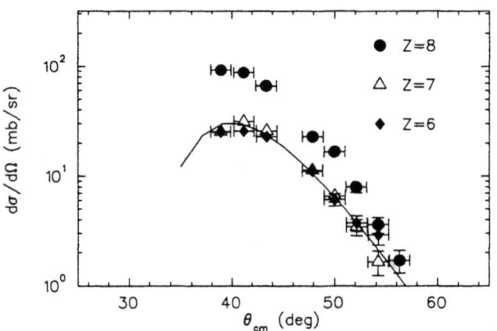

FIGURE 6. Angular distribution of ^{17}F+^{208}Pb transfer reactions. The solid curve is for one-proton stripping calculated by DWBA and the solid curve is for the DWBA calculation shifted by 5°.

One step DWBA calculations for one-proton stripping were carried out with the code PTOLEMY(11) using the potential parameters obtained from the elastic and inelastic scattering data. As shown in Fig. 6, the shape of the angular distribution predicted by DWBA (solid curve) agrees with the data for N and C when the results of calculations are shifted backward by 5°. This may be due to the multistep processes involved in the transfer of several nucleons at the high reaction energy. On the other hand, if the ^{17}F is polarized while approaching the target such that the valence proton is pushed away from the target and shielded by the ^{16}O core, a shorter reaction distance is required for transfer reactions to take place. This would result in the grazing peak shifting backward. However, a detailed model calculation is required to estimate the polarizability of ^{17}F.

The cross section for ^{17}F breakup is measured to be small and the influence of breakup on fusion is found to be small. The first excited state in ^{17}F is below the breakup threshold and can be excited from the ground state by an E2 process with a large B(E2) value(12). Further data analysis and experiments will be used to study the couplings to this excitation channel.

SUMMARY AND OUTLOOK

The resonance properties measured with the radioactive ^{17}F and ^{18}F beams produced at HRIBF have resolved discrepancies seen in previous measurements. As a result, the uncertainty in stellar reaction rates is reduced significantly. The polarization of ^{17}F by the ^{208}Pb target might have shifted the grazing peak in the transfer reactions. Further experimental and theoretical work are necessary to study the reaction mechanisms.

Neutron rich radioactive ions have been produced at HRIBF with a uranium carbide targets, recently. Yields of some accelerated neutron rich beams have been measured and extrapolated to the unmeasured species. Experiments using neutron rich beams to study nuclear structure and reaction dynamics are underway. In the near future, a batch mode source will be installed to produce "long-lived" proton rich nuclei such as ^{56}Ni.

ACKNOWLEDGMENTS

The author would like to thank D. W. Bardayan and J. C. Blackmon for providing their data and discussing their results. Thanks also go to J. R. Beene, D. J. Dean, H. Esbensen, J. Gomez del Campo, M. L. Halbert, G. R. Satchler, D. Shapira, R. L. Varner, and C. Y. Wong for helpful discussions. This work would not be possible without the hard work of the HRIBF staff. Research at the Oak Ridge National Laboratory is supported by the U.S. Department of Energy under contract DE-AC05-00OR22725 with UT-Battelle, LLC.

REFERENCES

1. D. W. Bardayan et al., *Phys. Rev. Lett.* **83**, 45 (1999).
2. A. Galindo-Uribarri et al., this conference.
3. K. I. Hahn et al., *Phys. Rev. C* **54**, 1999 (1996).
4. B. Harss et al., *Phys. Rev. Lett.* **82**, 3964 (1999).
5. S. Utku et al., *Phys. Rev. C* **57**, 2731 (1998).
6. R. Coszach et al., *Phys. Lett. B* **353**, 184 (1995).
7. K. E. Rehm et al., *Phys Rev C* **52**, R460 (1995); **53**, 1950 (1996).
8. R. Morlock et al., *Phys. Rev. Lett.* **79**, 3837 (1997).
9. K. E. Rehm et al., *Phys. Rev. Lett.* **81**, 3341 (1998).
10. J. F. Liang et al., *Phys. Lett. B* (in press).
11. M. H. Macfarlane and S. C. Piper, PTOLEMY: *A Program for Heavy-ion Direct-reaction Calculations*, ANL-76-11 Rev. 1, (1978).
12. B. A. Brown, A. Arima, and J. B. McGrory, *Nucl. Phys.* **A277**, 77 (1977).

Breeding 10^{10}/s Radioactive Nuclei in a Compact Plasma Focus Device [1]

Jan S. Brzosko *, Krzysztof Melzacki *, Charles Powell *, Moshe Gai †, Ralph H. France III †, James E. McDonald †, Gerald D. Alton ¶, Fred E. Bertrand ¶, and James R. Beene ¶

* *DIANA-HiTech, Detection and Instrumentation for Advanced Neutron Applications LLC. 152 Harrison Ave, Jersey City NJ 07304-1906, USA.*

† *Department of Physics, University of Connecticut, Storrs, CT 06269-3046, USA.*

¶ *Oak Ridge National Laboratory, Oak Ridge TN 37831-6369, USA.*

Abstract. In the early 90's, it was discovered that a Plasma Focus (PF) system self-creates a plasma-target in which high energy-threshold nuclear-reactions occur at high reaction rates. Short life radioisotopes (SLR)s such as ^{18}F, ^{17}F, ^{15}O, ^{14}O, ^{13}N have been generated (10^6 - 10^8 per pulse) with a PF-machine using 7 kJ energy storage to produce the plasmas. β^- radioactivity from the SLRs is measured with rugged, Geiger counters inserted into the PF-chamber, and a specific SLR is identified by its half-life. The PF chamber (before discharge) is filled with a mixture of gases that constitutes the latter plasma-target – beam system, e.g., the elements required to produce specific SLRs through nuclear reactions. In this paper, arguments are presented showing that a modest sized PF-machine, using a 50 -75 kJ fast capacitor-bank, when operated at pulse frequencies of 1-10 Hz can produce $\geq 10^9$ SLRs/pulse. This paper reports the results of testing a PF as a breeder of SLRs with dual applications for: (*i*) Secondary Radioactive Nuclear Beams ion-sources (Z < 35), and (*ii*) as a breeder of radioisotopes for biomedicine (Z ≤ 10) and/or PET imaging.

1. INTRODUCTION

The Plasma Focus (PF) was independently discovered by J. Mather [1] and N.V.Filippov [2] in the late 50's. Since then, many laboratories have studied the plasma-focus phenomenon and its remarkable capabilities for producing short pulses (1 ns – 300 ns) of X-rays, neutrons and fast ions depending on PF mode of operation. PF operation can be briefly described as a five stage process (see Fig. 1) with the appearance of high-energy ions and nuclear reactions in occurring in the last stage.

FIGURE 1. Conceptual drawing of the plasma focus header indicating the sequence of plasma sheath positions. The central electrode diameter is $\phi \approx 50$ mm, the chamber is filled with gas mixture at p = 0.1 – 7 Torr. Numbers refer to plasma development stages: 1 – the plasma sheath is formed, 2 - the plasma sheath moves toward the anode nozzle (v ≈ 10^5 /s), 3 - the sheath arrives at the end of anode and rearranges itself into a cylinder with a conical opening, 4 - the plasma is compressed at the axis (10^{25} ions/m^3), 5 - the plasma column quickly develops instabilities associated with high-energy acceleration, high nuclear reactivity and X-ray emission.

Measurements of fast ion spectra have been carried out elsewhere and with the use of various methods ranging from time-of-flight, filters, faraday cups, Thomson spectrometers to track detectors and blisters method.

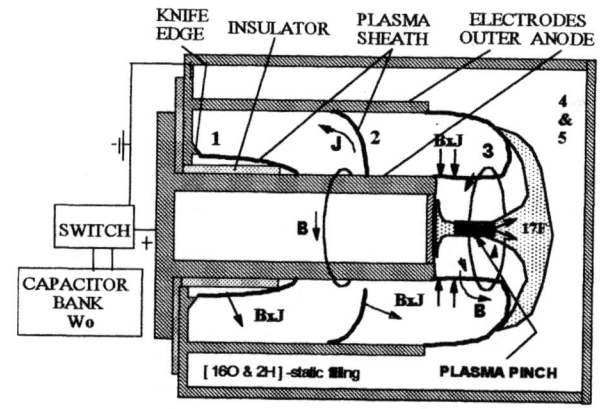

Small (ϕ: 20 - ϕ: 300 μm) plasma domains are created. The domains have above solid state densities, kT ≥ 3 keV and magnetic fields sufficient to trap ions of 5 MeV/nucleon.

---

[1] Work supported in part by USA Department of Energy, contract #DE-FG02-00ER82988 (2000).

All these studies [3] suggest the following empirical formula ($E_i \geq 0.1$ MeV):

$$d\Phi_i/dE_i \propto E_i^{-m} \quad (2 \leq m \leq 3) \quad (1)$$

Similar (as in Eq.1) spectral dependence for the trapped and reacting in plasma ions was drawn by unfolding of the D(d,n) neutron spectra. The data provide evidence that the fast beam yields (Φ_i) as well as reaction yields in plasma (Y_p) and/or on external targets (Y_t) scale with the square of the energy (W) stored in the capacitor bank [4], i.e.

$$\Phi_i \propto W^2 ; \quad Y_p \propto W^2 ; \quad Y_t \propto W^2 \quad (2)$$

In addition to direct ion-beam observation, the activation methods have also been broadly used. In such approaches energy threshold reactions leading to formation of radioactive species in externally positioned solid targets (exposed to D^+ irradiation) have been used as PF-diagnostics. Under these conditions, radioisotopes such as ^{11}C, ^{11}N, ^{15}O, ^{17}F, ^{28}Al, ^{66}Cu have been observed and are reported elsewhere.

The scope of this paper is to evaluate/demonstrate of potential of the PF device as a breeder of radioactive isotopes that can be utilized for the production of exotic and/or radioactive ion beams for radioactive ion beam applications and/or for medical imaging (PET) and biomedicine applications.

2. NUCLEAR REACTIONS INDUCED IN THE MAGNETIZED PLASMA-TARGET BY FAST IONS TRAPPED IN THE TARGET

In this section, the experimental evidence and discussion is oriented to the situation where fast ions are trapped in domains of magnetized plasma. Such fast ions can induce nuclear reactions by collisions with nuclei in the background plasma. To produce a selected nuclear reaction and as a result, produce a specific radioisotope, one has to create a dense, magnetized plasma composed of two isotopes with different atomic numbers and an appreciable population of high-energy (fast) ions. The cross-section for the production of the radioisotope of interest (reaction product) has to be sufficiently large ($\sigma \geq 100$ mb) at the ion-projectile energy of 0.5 MeV/nucleon.

Only one experimental group has achieved high efficiency (yield per kJ of stored energy) for the production of short-lived radioisotopes in a plasma environment [5]. The experiments reported here were carried out with a Mather-type PF device with a capacitor bank energy of W= 7 kJ (V = 18 kV) and at a total gas mixture pressure of $p \approx 5$ Torr. Gas mixtures were composed of low-Z (LZ) isotopes (H, D, ^{3}He) mixed with high-Z (HZ) isotopes.

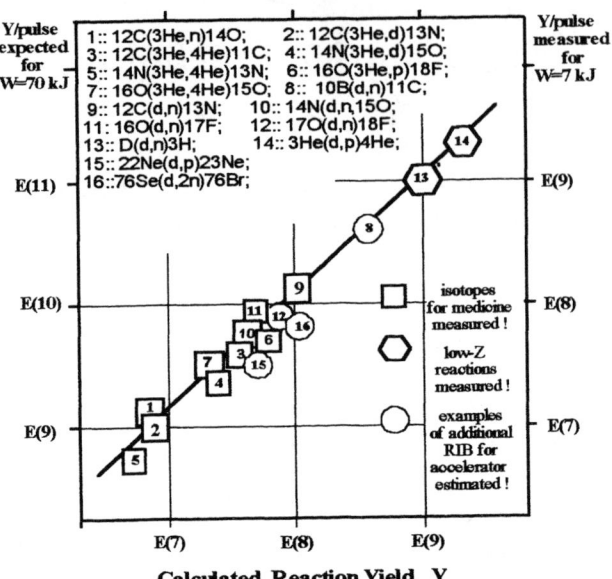

FIGURE 2. Compilation of experimental data measured with a PF (W= 7 kJ) filled with a mixture of HZ and LZ gases. The box sizes are representative of standard deviations (vertically) obtained by averaging many experimental points (discharges) and model uncertainties (horizontal). For convenience experimental data are adjusted (see Eq. 3 and 4) to $n_{HZ}/n_{LZ} \approx \rho_{HZ}/\rho_{LZ} \approx 1$.

A summary of these experiments is shown in Fig.2. From Fig.2 it is clear that a broad range of radioisotopes, with large yield, can be produced in a PF device. A relatively small PF-machine, operated at a discharge power of W= 7 kJ, can produce $\times 10^6$ to 5×10^8 radio-nuclei per pulse, while with a medium size machine, operated at discharge power of, W = 70 kJ, one can expect a hundred times higher yields.

The semi-empirical model assumes: (i) the instantaneous creation of a magnetized plasma domain, containing fast trapped-ions, (ii) a domain stability period, $\tau \approx 0.5$-5 ns, and (iii) an instantaneous domain destabilization, causing fast ions to be ejected out of the pinch zone. For briefness, and demonstration of trends, a simplified formula (instead of full theory [5]) for (HZ-LZ) radioisotope production is used here:

$$Y_p = n_{HZ} \times \tau \times \Phi_p^\circ \times \int_0^{E_m} (d\Phi_p/dE) \times \sigma(E) \times (2E/M)^{1/2} \times dE \quad (3)$$

In the Eq. 3, $d\Phi_p/dE$, (see Eq.1) is normalized to unity in the range, $E_0 \leq E \leq E_m$; Φ_p° is the total number of the projectile ions that are accelerated and trapped in the plasma; n_{HZ} is the numeric density of plasma nuclei that serve as target during radioisotope production; $\sigma(E)$ is the reaction cross section and M is the mass of the LZ ion. Eq.3 neglects stopping powers of fast ions in the plasma domain and considers only LZ as projectiles inducing nuclear reactions. Eq.3 simplifies further if one assumes that: (*i*) the density of HZ particles in the plasma target, n_{HZ}, is proportional to the atomic density, ρ_{HZ}, of the HZ in the filling gas (constant plasma compression), *i.e.*, $n_{HZ} \propto \rho_{HZ}$; (*ii*) the number of accelerated LZ ions is proportional to the abundance of such ions in the plasma target, *i.e.*, $\Phi_p \propto n_{LZ} \propto \rho_{LZ}$. As a result, Eq.3 converts to the following form:

$$Y_p \propto \tau \times \rho_{HZ} \times \rho_{LZ} \times \int_0^{E_m} \sigma(E) \times E^{-m+1/2} \times dE \quad (4)$$

The theoretical values (in Fig.2) are calculated by use of Eq. 3. The beam yield in the plasma was assumed to be the same as those activating the external targets. The $n \times \tau$ product was taken from Ref. 13. Eq. 4 agrees well with experimental results (see Fig.2).

3. EXPERIMENTAL TESTS OF RADIOISOTOPES PRODUCTION IN THE PF-PLASMA

Radioactivity (decay curve) is measured by use of a thin-wall, cylindrical Geiger-Muller counters (GM) having walls (cathode) made of high resistivity metal alloy (no Al or Cu). GM counters were used because they can survive an electromagnetic shock and jets of plasma emitted during PF operation. In the course of experiments, various geometrical set-ups were used [5], one of which is shown in Fig.3.

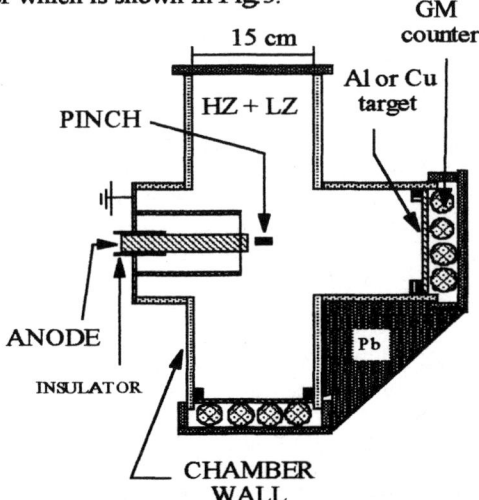

FIGURE 3. An experimental set-up used for simultaneous measurement of radioactivity induced in the plasma and on the target foils (Al or Cu). Each set of the four GMs is connected in parallel to a multi-scaler counter and counting begins 10 s after the PF-discharge is completed. The chamber walls are clad with Ta.

The identification of isomers is based on their lifetimes ($T_{1/2}$) and reaction yields are estimated from initial radioactivity of the component (associated with the identified radioisotope) and the efficiency (solid angle included) of the detection system.

Plasma origin of the radioisotopes breeding is supported by three observations: *(A)* For PF discharges with a chamber Ta-clad and filled with a mixture of deuterium and one HZ component (^{12}C, or ^{14}N and/or ^{16}O) only one $T_{1/2}$, as expected from (d,n) reaction product, is observed; no radioactivity occurs for pure D_2 fillings. *(B)* Whenever the PF-chamber is filled with different relative compositions of HZ and LZ gases while keeping the total atomic density constant – the resulting radioactivity changes proportionally to the product of $\rho_{HZ} \times \rho_{LZ}$ as expected from Eq.4. *(C)* In experiments with Al or Cu external targets and the chamber filled with HZ and LZ gases, two radioisotopes are produced: one <u>in</u> the plasma and one <u>on</u> the target. Fig.4 demonstrates complete disappearance of the plasma radioactivity when the chamber is instantly evacuated during measurements of the decay curve.

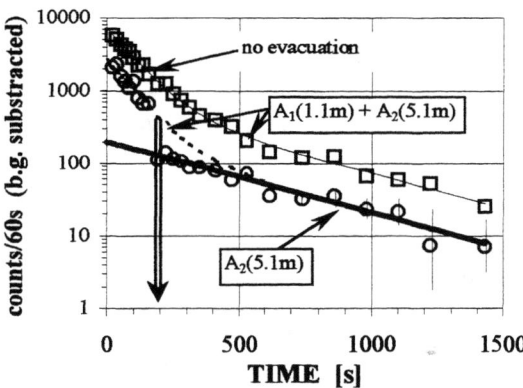

FIGURE 4. An example of the decay curves each composed of two half-lives: $T_{1/2} = 1.1$ min (^{17}F) and 5.1 min (^{66}Cu). Three min after the discharge ($\rho_{HZ}/\rho_{LZ} \approx 0.1$), the chamber was pumped-out (radioactive gas evacuation) after which the plasma-induced radioactivity ($T_{1/2} = 1.1$ min) disappeared. A decay curve, measured after another discharge and without gas evacuation, is shown for comparison.

4. PLASMA FOCUS AS AN EFFICIENT BREEDER OF SHORT LIFE RADIOISOTOPES FOR ACCELERATORS AND BIOMEDICINE

PF devices have the capability of producing radio-nuclei <u>on</u> external targets and <u>in</u> the plasma. This second application is addressed here. Data, such as shown in Fig.4, replicate results where the external target is assumed to be solid ^{16}O, accounting for differences in the ion stopping power, cross-sections, efficiency and considering a $\rho_O/\rho_D \approx 1$ plasma. In such cases, the production of ^{17}F radionuclei <u>in</u> the plasma system would be 45 times larger than production of ^{17}F in a solid oxygen target. This example demonstrates the superiority of a plasma breeder over a solid-target.

The number of radionuclei in the chamber, N, at the end of the breeding cycle (Δt) in various PF-operating conditions can be roughly estimated from the Y_p data shown in Fig. 2 ($W_o = 7$ kJ) and Eqs. 2 ÷ 4.

$$N \equiv Y_p \times [\rho_{HZ} \times \rho_{LZ}/(\rho_{HZ}+\rho_{LZ})^2] \times (W/W_o)^2 \times f \times (1-e^{-\lambda \times \Delta t}) \quad (5)$$

where: W is the capacitor bank energy, f is the discharge repetition frequency, $\lambda \equiv \ln 2 / T_{1/2}$.

Some examples of expected breeding capability are shown in Fig. 5. Chosen activities reflects typical biomedical and/or accelerator demands. Between biomedical urgent market demands one can refer to compounds labeled with ON, very recently recognized by the 1998 Noble prize committee for its importance. Other examples include the use of ^{18}F-radio tracers as Fluoro-Deoxy Glucose (FDG) in oncology, cardiology and neurology [6]. Applications for short-life radio-nuclei have been well documented in the literature but have been used sparingly because cyclotron breeders are costly and radiation must be handled with heavy shields.

Secondary Radioactive Nuclear Beams provide one of the most powerful tools for studying Nuclear Physics and Astrophysics phenomena [7, 8]. The Holifield Radioactive Ion Beam Facility (HRIBF) [9], Oak Ridge National Laboratory (ORNL) is the first US-Isotope Separator On-line (ISOL) facility. In addition to present capabilities, some experiments will require higher intensities, than ISOL can provide, specially for studies of nuclear structure of short-life radioisotopes with A ≥ 20. Their short life-times make it difficult to generate useful beam in of isotopes with lifetime less than several seconds at ISOL type facilities. For these isotopes, the recently proposed method of producing radioisotopes in a PF type ion source could overcome these handicaps and be substituted for the equivalent ISOL method.

5. SUMMARY

The data presented in this paper consistently show the remarkable ability to produce intense levels of radioisotopes (including those with short half-lives). The existing mathematical models constitute reliable tools for extrapolation of experimental data to other radioisotopes as well as for scaling plasma focus breeders.

Radioisotopes produced in a PF device are embedded in the low-pressure (up to 5 Torr) gas of reacting components and are ready to be transported as ions and/or as neutral atoms to an accelerator or to an ion source.

As already noted the proposed PF-breeding method could be utilized for the production of radioisotopes needed in nuclear as well as in biomedical applications. In fact, the needs in both fields overlap, as the same isotopes are used for different applications.

6. REFERENCES

1. Mather, J.W, Physics of Fluids **8**, 336-341 (1965).
2. Filippov, N.V., Filippova, T.I., and Vinogradov, V.P., Nuclear Fusion Suppl. "Dense and High Temperature Plasma in Non-Cilindrical Z-Pinch" in *1st Int. Conf. on Plasma Fusion Physics*, Salzburg-1962, Publ. by IAEA as Nuclear Fusion Suppl., 1962, vol.2, pp577-587.
3. Nardi V., Bortolotti A., Brzosko J.S., Esper M., Luo C.M., Pedrielli F., Powell C., and Zeng D., *IEEE Trans. On Plasma Science* **16**, 368-373 (1988).
4. Brzosko, J.S., Degnan, J., Filippov, N.V., Freeman, B.L., Kiuttu, J., Mather, J.W., "Comments on the Feasibility of Achieving Scientific Break-Even with a Plasma Focus Machine" in *Current Trends in International Fusion Research* edited by E.Panarella, Plenum Press, New York, 1997, pp. 11-32.
5. Brzosko J.S., and Nardi V., Physics Lett. **A155**, 162-168, (1991), Physics Lett. **A192**, 250-257 (1994), Phys. Plasmas **2**, 1259-1269 (1995); see also Brzosko J.S. private communication, 1999, review paper to be send for publication.
6. Garg P.K., Garg S., Bigner D.D., and Zalutsky M.R., Cancer Research. **52** 5054 (1992), Nuclear Med. Biol. **21** 21-28 (1994).
7. " The Isospin Laboratory, Research Opportunities with Radioactive Beams", Report LALP, 91-51, 1991.
8. Gai M., Nuclear Phys. **A570** 87c (1994).
9. Alton G.D., and Beene J.R., Journal of Physics, **G24** 1347 (1998).

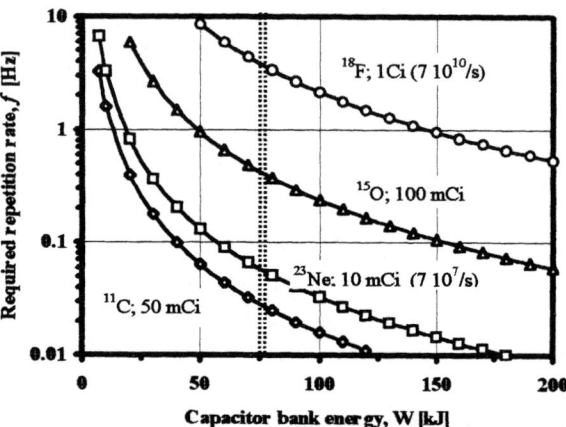

FIGURE 5. Expected relation between discharge repetition rate and capacitor bank energy for obtaining useful radioactivity of specific radioisotopes after evacuation from the discharge chamber. The time of PF operation is equal to the half-life of the particular radioisotope, *i.e.* the radioactivity is equal to $\frac{1}{2}$ of the radio-nucleus production rate (given in brackets). The assumed mixture of fill gas is $\rho_{HZ}/\rho_{LZ} = 1$.

The Development Of An ECRIS Charge-State Breeder For Generating RIB's

T. Lamy, R. Geller, J.L. Bouly, N. Chauvin, J.C. Curdy, A. Lacoste, P. Sole, P. Sortais, T. Thuillier, J.L. Vieux-Rochaz

Institut des Sciences Nucléaires UJF-IN2P3-CNRS, 53 Av. des Martyrs, 38026 GRENOBLE cedex France

Abstract. ECRIS have proven their ability to fulfill the requirements of all types of accelerators[1]. Initially developed for radioactive ion beam production, the ECR charge breeder shows that the beam injection of a primary beam inside an ECR ion source is a very general process for beam production and better understanding of the ECRIS physics. A new dedicated ECRIS PHOENIX has been specifically built to improve a means for the already very satisfying results obtained with the MINIMAFIOS source adapted to the so called "1+/n+" method. In this paper, we will review the latest results obtained on the ISN Grenoble test bench for the production of c.w. or pulsed metallic ion beams. The results are already very promising, thanks to an improved and versatile magnetic field configuration. The efficiencies of specific charges have been improved and the overall breeding time has been decreased by factors of 3 to 9 (i.e. : 25 ms to obtain Ag19+).

INTRODUCTION

Many present or future accelerator facilities, dedicated to the production of Radioactive Ion Beams (RIB's), are interested in a fast and efficient charge breeding method [2] following initial 1+ production. Depending on the type of accelerator, it is possible to use one of two ECRIS 1+ → n+ schemes: c.w. mode or pulsed mode. These schemes must lead to the highest efficiency for a specific charge-state. However, while the c.w. mode must be as fast as possible, the pulsed mode must be able to trap the multi-charged ions, in a time that depends on the duty cycle of the post-accelerator. Moreover, the number of ions available for the accelerator strongly depends on the time required for the charge breeding process when using short life isotopes. In this paper, after a short explanation of experimental methods used to measure, efficiencies and charge-breeding times, we will compare the results obtained with the MINIMAFIOS ECR source and the new dedicated PHOENIX ECR booster source, using the same R.F. frequency (10 GHz). The results obtained with a new element (Sn) will also be presented.

EXPERIMENTAL METHOD

The RIB is simulated by a low energy (i.e. : 20 keV) 1+ stable ion beam, produced by any type of 1+ RIB source (ECR, hollow cathode, surface ionization). This primary beam is characterized after passing through a 90° magnetic separator, by measuring the emittance and current (Figure 1). The beam is then decelerated and injected inside a n+ ECR ion source. A slight variation of the n+ source voltage (ΔV) [3,4], with respect to the 1+ voltage permits tuning the final energy 1+ ions and therefore optimizing their capture by the n+ plasma (Figure 2).

Continuous Mode

The charge breeding efficiency is given by :

$$\eta = \frac{1}{n}\frac{I_{n+}}{I_{1+}} \qquad (1)$$

where I_{n+} and I_{1+} are the intensities of the multi-charged and mono-charged ions measured in the Faraday cups. These evaluations are performed by the pulsing the 1+ beam with deflection plates and then measuring the pulsed n+ beam having the same mass. The charge breeding time is measured by the same procedure. We measure the time to reach 90 % of maximum intensity for a specific n+ beam with respect to the instant of injection of the 1+ beam.

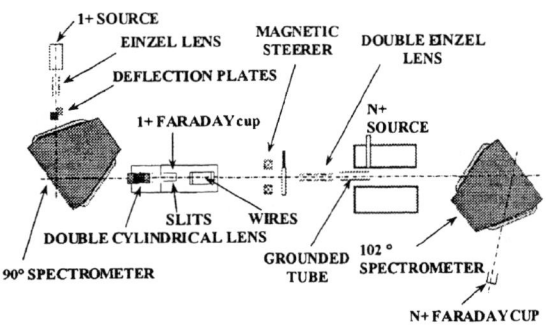

FIGURE 1. Experimental setup for the 1+ → n+ line.

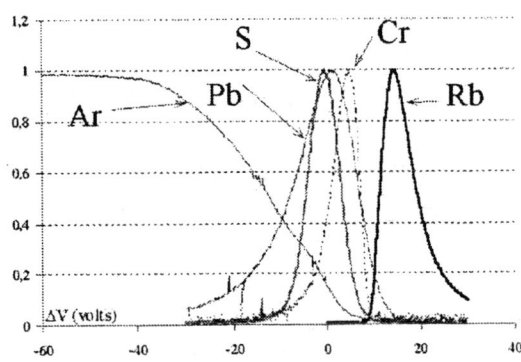

FIGURE 2. Normalized variation of the 1+ ion capture voltage, ΔV, for different elements.

Pulsed Mode (ECRIT)

For pulsed accelerators (synchrotrons, linacs), it is necessary to bunch the n+ beam due to the continuous production of RIB's. For this situation, the physical process incorporates a new stage: capture and multiple ionization like in the c.w. mode, and trapping of the n+ ions corresponding to the cycle of the accelerator. These n+ trapped ions are extracted by use of the afterglow method [1,5]. The ECRIT (Electron Cyclotron Resonance Ion Trap) system that we have developed fulfills these requirements, and permits minimizing radioactive ions losses [6,7]. The main additional parameters for this mode are the trapping time and the charge capacity of the n+ ion source.

RESULTS OBTAINED WITH MINIMAFIOS

Three groups of elements have been defined due to their physical interaction properties with respect to surfaces and/or to their production device. The rare gases are produced by an ECR source and are reflected after neutralization close to a cold surface. The alkali metals, produced by surface ionization stick to cold surfaces. Metallic ions usually produced in ECRIS also stick to cold surfaces. Many elements have been extensively studied for years with the MINIMAFIOS ECR ion source [2]. Let us recall the main results obtained from these studies.

Efficiency Yields

High efficiency yields have been obtained for the most abundant charge state of each element (Table 1). The global efficiency η_G is defined as

$$\eta_G = \sum_n \eta_{n+} \qquad (2)$$

(which is proportional to the global capture efficiency) thus about 50% for rare gases and in the range of 13 to 35 % for metallic and alkali ions.

TABLE 1. Efficiency yields obtained with MINIMAFIOS.

1+ Ion	N+ ion	η %
^{84}Kr	Kr^{11+}	9
^{40}Ar	Ar^{8+}	10.4
^{22}Ne	Ne^{4+}	9
^{107}Ag	Ag^{17+}	3.2
^{64}Zn	Zn^{11+}	3
^{26}Mg	Mg^{5+}	2.8
^{85}Rb	Rb^{15+}	5.5

Charge Breeding Time

The charge breeding time (τ_{cbt}) strongly depends on the tuning of the ECR source. The shortest reproducible breeding times obtained are summarized in Table 2 and are plotted in Figure 3. One can clearly see from the latter figure, the mass and charge dependence of the breeding time. This dependence is completely consistent with the highest efficiency obtained from the classical afterglow extraction method, which is more efficient for heavy highly charged ions. Finally, what we call the charge breeding time τ_{cbt} is in fact an image of the trapping time of the ECR source.

TABLE 2. Charge breeding time obtained with MINIMAFIOS.

N+ ion	τ_{cbt} (ms)
N^{4+}	30
Ne^{4+}	50
Mg^{5+}	35
Ar^{8+}	80
Zn^{11+}	100
Kr^{11+}	180
Rb^{15+}	225
Xe^{16+}	220
Ag^{17+}	225

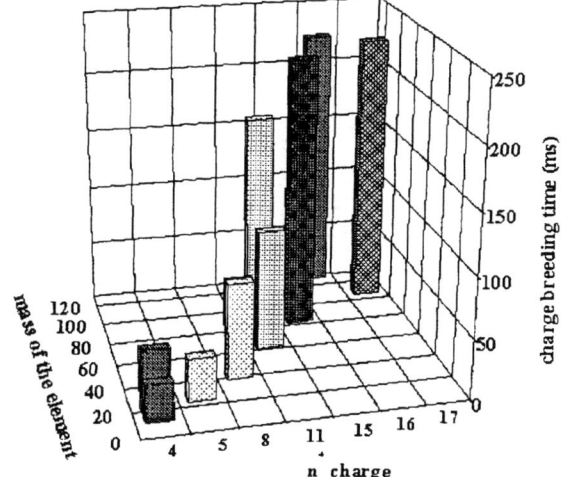

FIGURE 3. Mass and charge dependence of the breeding time (MINIMAFIOS).

THE PHOENIX CHARGE STATE BREEDER

The PHOENIX source (Figure 4) was built to overcome the limitations of the 18 year old MINIMAFIOS source and to allow extensive R&D studies of the breeding process. The R.F. input geometry has been kept unchanged, and the high voltage insulation has been increased to permit extraction at 60 kV. The magnetic field configuration has been designed for injection of frequencies from 10 to 18 GHz (hexapole 1.1 T). The completely flexible design is so as to permit exploring many operational parameters (including those for the MINIMAFIOS ECRIS).

Charge Breeding With The Phoenix Source

Four elements have already been studied with the PHOENIX source : Ar, Kr, Ag, Sn.

Argon, Krypton And Silver Results

For Argon, the efficiency $\eta(Ar^{1+} \rightarrow Ar^{9+})$ is 11.9 %. The global efficiency is $\eta_G = 56$ % and the charge breeding time is $\tau_{cbt} = 25$ ms.

For Krypton, we obtain 10.3 % for $\eta(Kr^{1+} \rightarrow Kr^{14+})$, a charge breeding time $\tau_{cbt} = 60$ ms and a global efficiency of $\eta_G = 65$ %.

A typical Ag spectra is shown Figure 5. The efficiency is $\eta(Ag^{1+} \rightarrow Ag^{19+}) = 3.9$ %, the global efficiency is $\eta_G = 23.5$ % and the charge breeding time, shown Figure 6, is $\tau_{cbt} = 25$ ms.

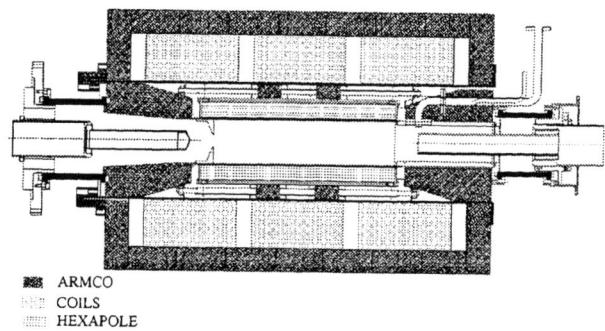

FIGURE 4. Cross-section of the PHOENIX source.

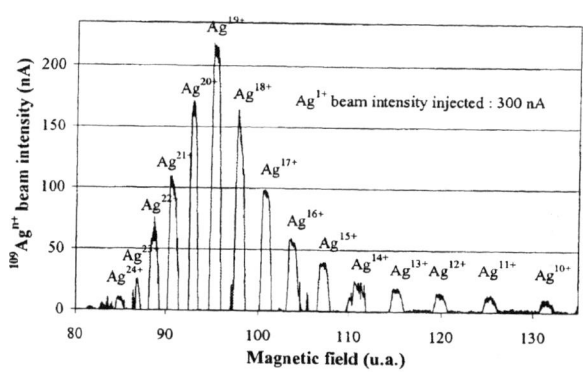

FIGURE 5. Charge breeding with the PHOENIX source: Silver.

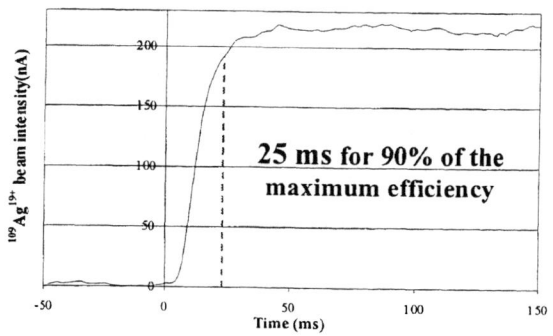

FIGURE 6. Charge breeding time with the PHOENIX source: Silver.

A New Element : Tin

First charge breeding experiments with Sn have been performed with the PHOENIX source. A 1 µA $^{120}Sn^{1+}$ has been produced and injected into the PHOENIX ECRIS, the n+ CSD is peaked on charge-state 21+, as a preliminary result we obtained a maximum efficiency of $\eta(Sn^{1+} \rightarrow Sn^{22+}) = 4\%$ and the global efficiency yield is almost the same as for Ag i.e. $\eta_G = 25\%$ (Figure 7). The charge breeding time $\tau_{cbt} \leq$ 20 ms for Sn^{19+} (Figure 8) is very impressive. This result was obtained with a R.F. power of 580 W.

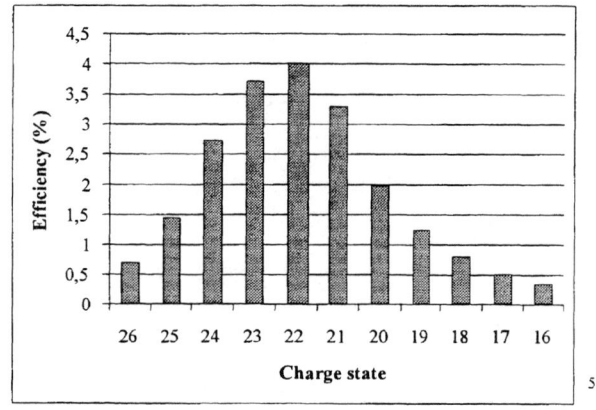

FIGURE 7. PHOENIX charge breeding efficiency for Sn.

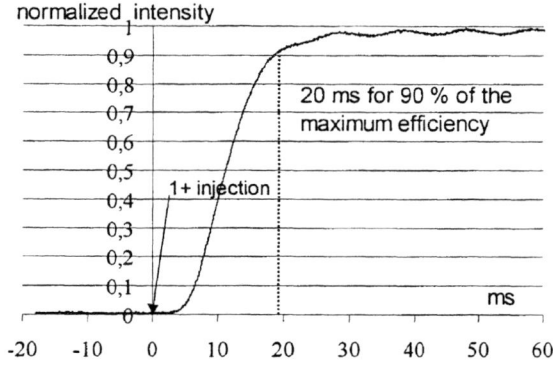

FIGURE 8. Charge breeding time for Sn^{19+}.

PHOENIX - MINIMAFIOS COMPARISON

To summarize the results of this paper, it is important to point out the improvements already obtained with the ECRIS PHOENIX charge state breeder (with the same R.F. : 10 GHz). The global efficiency yields have been maintained with respect to the MINIMAFIOS yields. The highest efficiency for a specific n+ beam has been increased due to a steeper CSD as it's clearly observed for Silver (Figure 9). Moreover the average CSD has been shifted to a few higher charge states. Due to the much greater range of tunings, the charge breeding times have been reduced by a factor 3 to 10 which is a very promising result for efficient RIB production.

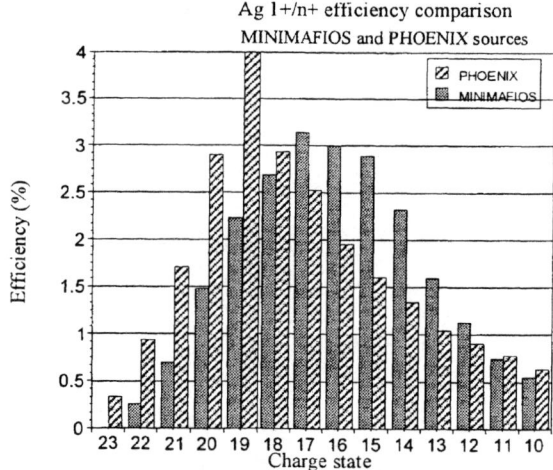

FIGURE 9. Silver efficiency yield improvement with PHOENIX.

REFERENCES

1. R.Geller, *Electron Cyclotron Resonance Ion Sources and ECR plasmas*, Bristol and Philadelphia: Institute of Physics Publishing, 1996.

2. N.Chauvin : *La transformation d'état de charge $1^+ \rightarrow n^+$ pour l'accélération des ions radioactifs*, Thèse de Doctorat Université Joseph Fourier Grenoble, Juillet 2000.

3. T.Lamy et al., *Rev. Sci. Instrum* **69**, nb.3 1322-1326 (1998).

4. P. Sortais et al., "Ecris As Ion Source And Charge Breeder" in *Nuclear Physics A*, to be published., RNB2000 Conference Proceedings.

5. P. Sortais : *Rev. Sci. Instrum.* **63**, 2801-2805 (1992).

6. N. Chauvin et al, *Nucl. Inst. Meth. A* **419 (1)**, 185-188 (1998).

7. N. Chauvin et al., "The 1+ n+ charge breeding method for the production of radioactive and stable continuous/pulsed multi-charged ion beams", Proceedings of the 14th International Workshop on ECR Sources, ECRIS99 CERN, Geneva, 1999

Simulation of the Effusive-Flow of Reactive Gases in Tubular Transport Systems: Radioactive Ion Beam Applications

J.-C. Bilheux[*] and G.D. Alton

Oak Ridge National Laboratory, P.O. Box 2008
Oak Ridge, TN 37831-6368

[*]*Ph.D. Student, University of Versailles, Versailles, France.*

Abstract. An analytical formula has been developed that accurately reproduces Monte Carlo simulations for the effusive-flow transport of chemically active species through tubular transport systems under ideal conditions, independent of species, tube material, and operational temperature. Through its use, the choice of materials of construction for a given transport system can be made on a relative basis that minimizes effusive-flow times, independent of system geometry and size. In this report, we describe the formula and compare results derived by its use with those determined by use of Monte-Carlo techniques for noble gases as well as a variety of chemically active species.

INTRODUCTION

Excessively long delay times due to diffusion from solid or liquid target materials followed by their effusive-flow transport to the ion source are the principal means whereby short-lived radioactive species are lost between initial formation and acceleration at radioactive ion beam (RIB) facilities based on the ISOL technique [1]. Due to the low production rates and short lifetimes of species of research interest, RIB intensities are at a premium at such facilities and consequently, time delays due to each of these processes must be minimized. Since the residence times of chemically active species depend strongly on the enthalpies of adsorption of the species/surface combination as well as the operational temperatures used system during transport, it is desirable to choose low enthalpy of adsorption, highly refractory materials for constructing the transport system. We have developed an analytical formula for estimating transport times of species through tubular geometry transport systems. The results derived by its use can be used on a relative basis to select materials of construction that will minimize the transport times for any species for which enthalpies of adsorption are known, independent of size and geometry of the system. The equation accurately replicates results derived by use of Monte-Carlo statistical techniques.

EFFUSIVE FLOW

In the case of a straight tube at constant temperature T and low pressure, transport through the system is referred to as free-molecular diffusion or Knudsen-flow. Under these conditions, a gas molecule collides many times with the walls of the tube before it encounters another molecule. Therefore, molecule-molecule collisions are negligible.

For an ideal gas in a tube of radius, a, and length, l, at low pressure, the steady-state flow rate, dN/dt, for particles of average velocity υ, flowing through the tube under a density gradient along the tube of dn/dz, is given by the following relation [2]:

$$\frac{dN}{dt} = -\left\{\frac{2\pi a^3}{3}\right\}\upsilon\frac{dn}{dz} = -\left\{\frac{2\pi a^3}{3k_B T}\right\}\upsilon\frac{dp}{dz} \quad (1)$$

where n is the particle density, υ is the Maxwellian velocity and k_B is Boltzmann's constant.

Time delays associated with adsorption that are excessively long in relation to the lifetime of the radioactive species can result in significant losses of beam intensity in an ISOL facility [1]. Therefore, it is desirable to minimize residence times of atoms/molecules on surfaces in the target/ion source.

The residence time, τ_{ad}, for a particle on a surface at temperature, T, is given by the Frenkel Equation (see, e.g., Ref [3]):

$$\tau_{ad} = \tau_0 \cdot \exp\{-H_{ad}/k_B T\} \quad (2)$$

where H_{ad} is the heat of adsorption or enthalpy required to evaporate the atom or molecule from the surface, k_B is Boltzmann's constant, T is the absolute temperature and τ_0 is the time required for a single lattice vibration.

The desorption rate of atoms dN/dt in thermal equilibrium with a surface at temperature T, can be expressed by [4]:

$$\frac{dN}{dt} = P(T) A_0 Q_i k_B T \cdot \exp[-H_{ad}/k_B T]/\sigma N_0 h \Lambda^2 \quad (3)$$

where $P(T)$ is the temperature-dependent probability that the particle will stick to the surface, Q_i is the internal partition function of the gas phase molecule, $\Lambda = (h^2/2\pi M k_B T)^{1/2}$ is the thermal wavelength, A_0 is the area of the surface having N_0 adsorption sites. Thus, the rate of desorption is governed by the enthalpy of adsorption, H_{ad}, and the temperature, T, of the surface.

COMPUTER SIMULATIONS

The Monte-Carlo simulation program, *Effuse* [5], tracks particles by randomly choosing starting parameters (radial position and angle of injection) at entrance to the tube and starting angles at positions of de-sorption along the tube. The code accounts for residence times that particles spend on the walls of the transport system during passage. Knudsen-flow conditions are assumed, i.e., the mean free path, λ, is considered to be large with respect to the radius, r, of the tube and therefore, collisions between particles are negligible. Consequently, each particle is assumed to travel in a straight line between collisions with the wall at temperature, T, at an average Maxwellian velocity, v. After each particle is tracked through the system, the code records the average distance traveled per particle, L, and the average number of bounces, N_b, that the particle makes during transit through the tube of radius, a, and length, l. The code typically tracks 10,000 particles through a given system, starting at $t=0s$ and ending when all particles have been transferred through the system.

DEVELOPMENT OF THE EFFUSIVE-FLOW FORMULA

The number of particles, N, radioactive particles with lifetime, $\tau_{1/2}$, remaining in a tube at time, t, after evacuation can be deduced by solving the time dependent form of Eq. 1, resulting in the following expression:

$$N = N_0 \exp\{-\lambda t\} \exp\{-t/\tau_C\} \quad (4)$$

where N_0 is the number of particles in the volume at time $t=0s$, $\lambda = 0.693/\tau_{1/2}$; H_{ad} is the enthalpy of adsorption; τ_0 is a constant with value taken as $\tau_0 = 3.4 \cdot 10^{-15}$s; L is the average distance that a particle travels during transit through the tube; and τ_C is the characteristic transport time given by:

$$\tau_C = \tfrac{3}{4}[N_b \tau_0 \exp\{-H_{ad}/k_B T\} + L/v] \quad (5)$$

The factor ¾ was found by fitting Eq. 2 to Monte Carlo results derived for transport of identical particles through identical tubular systems under the same conditions.

Average distance traveled per particle, L

The average distance, L (cm), traveled by a particle during transit through a tubular system is assumed to be independent of tube material, species and temperature, T, and only a function of the length, l, dimensions of the tube and radius, a, of the tube. A formula for L was found by use of *Mathematica* [6] to fit to Monte Carlo results derived by use of *Effuse*. The resulting formula is given by:

$$L(l,a) = l[-0.86 \times \log_n(a) + 3.5505] + l^{2.31} \times 2.40 \cdot 10^{-3} + 3.5a - 10 \quad (6)$$

where $\log_n(a)$ is the natural logarithm of the tube of radius a. Results derived by use of this formula are found to be in very good agreement with those obtained from *Effuse* for the following range of tube dimensions: $5 \text{ cm} \leq l \leq 100 \text{ cm}$ and $0 \text{ cm} \leq a \leq 2.5 \text{ cm}$.

Number of wall collisions, N_b

A linear relation between the number of bounces, N_b, and the average distance traveled per particle during transit through a tubular system, $L(l,a)$, is found from

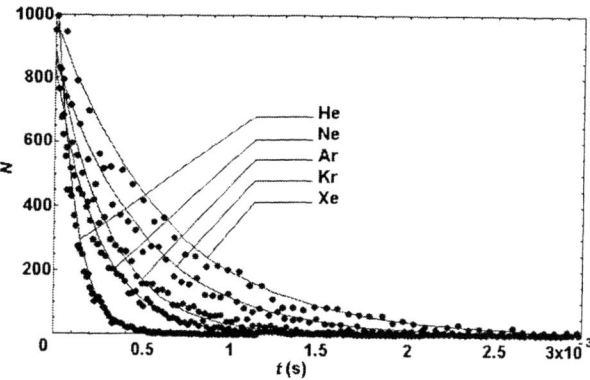

Figure 1. Time distribution of noble gas elements flowing through Nb, Ta, Re and Ir tubes as computed by use of *Effuse* (•) and Eq. 2 (—), at 2273 K. Tube length, l: 12.62 cm and tube radius, a: 0.293 cm.

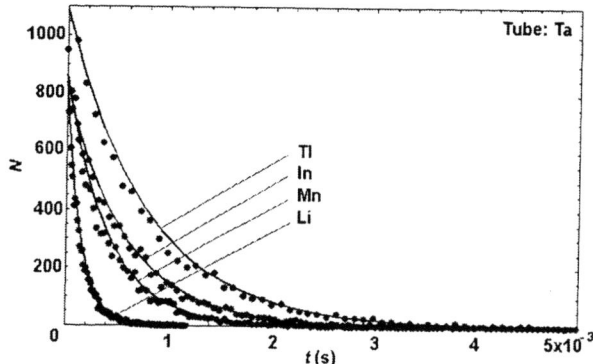

Figure 2. Time distribution of Li, Mn, In and Tl flowing through a transport tubes made of Ta as computed by use of Effuse (•) and Eq. 2 (—), at 2273 K. Tube length, l: 12.62 cm and tube radius, a: 0.293 cm.

Effuse calculations and after fitting with Mathematica and using the expression of $L(l,a)$, the expression of N_b can be written:

$$N_b = \left[-0.1019 - \frac{0.103}{a}\right] \times l \times \log_n(a) + 0.4207 \times l + 0.4253 \times \frac{l}{a} + 0.4148 \times a + \frac{0.8}{a} - 0.195 + \left[11.85 + \frac{12}{a}\right] \times l^{2.31} \times 2.40.10^{-5} \quad (7)$$

Again, results derived by use of this formula are found to be in excellent agreement with those of *Effuse* over the following range of tube dimensions: 5 cm ≤ l ≤ 50 cm and 0 cm ≤ a ≤ 2.0 cm.

Values of L and N_b, determined, respectively, from Eqs. 6 and 7, can then be used in Eq. 4, to calculate the number of particles, N, remaining in a tube of specified dimensions at time, t, for any atomic or molecular species for which enthalpies of adsorption, H_{ad}, are known.

SIMULATION RESULTS

Noble gas elements

For these elements, residence times on the walls are negligible because of their very low enthalpies of adsorption, H_{ad}, and, consequently, the velocity dependent term, L/v in Eq. 5 dominates. As noted in Fig. 1, the time dependent flow characteristics of these elements, derived by use of Eq. 5, are found to be in good agreement with *Effuse* values.

Electropositive stable elements

Many of the less refractory and more electropositive elements move through a given transport system, maintained very quickly with characteristic times approaching those for noble gas elements of comparable mass when the transport tube is operated at high temperature. For these elements, the characteristic transit times are less sensitive to the materials of which the transport systems are made, as noted in Fig. 2 at the assumed operating temperature (i. e., T = 2273 K).

Electronegative stable elements

On the other hand, the characteristic times for more electronegative species are very sensitive to the choice of the materials of construction of the transport system and consequently vary widely. The transit times for these elements can be reasonably short, moderate or prohibitively long (see, e.g., Table 1).

Selected radioactive isotopes

Delay times due to adsorption/desorption can have a dramatic influence on the rate at which radioactive particles arrive at the ion source if the species has a lifetime less than the characteristic transport time for the stable isotope of the element in question. The importance of selecting the most appropriate material for manufacture of the vapor transport system for selected radioactive isotopes in various tubes is illustrated in figure 3. As noted in figure 3, Nb is a better choice for the transport of ^{94}Pd whereas either Ir or Re would be a viable choice for the transport of ^{66}As.

Table 1. Characteristic effusive flow times, τ_C, of various electronegative elements through various tube materials. Tube length, l: 12.62 cm and tube radius, a: 0.293 cm. Temperature = 2273 K.

Characteristic effusive flow times, τ_c (s)				
Elements	Tube materials			
	Ta	Re	Ir	Nb
B	1070	60	118	454
C	10^9	10^6	10^6	10^9
N	10^9	421	143	10^9
O	10^{10}	81	10^{-4}	10^{10}
Si	1017	16	48	206
Ge	2	10^{-2}	10^{-1}	10^{-1}
Pd	21	10^{-3}	10^{-4}	5

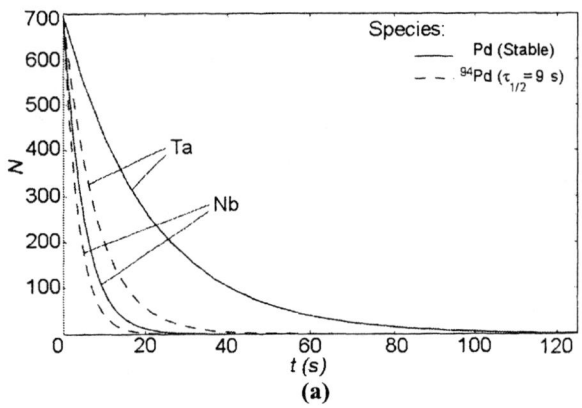

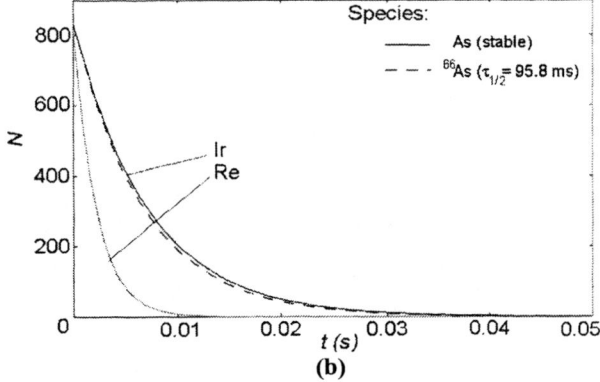

Figure 3. Time distribution a) for stable Pd and ^{94}Pd ($\tau_{1/2}$=9s) flowing through Nb and Ta tubes and b) for As and ^{66}As ($\tau_{1/2}$=95.8ms) flowing through Re and Ir tubes. Tube length, l: 12.62 cm and tube radius, a: 0.293 cm.

DISCUSSION

The utility of the formula lies in the fact that it can be used as a replacement for Monte-Carlo techniques to compute the time dependence for transport of a wide variety of elemental/molecular species through simple tubular systems under ideal conditions, on an absolute basis. Moreover, the equation permits separation of time dependence aspects of the diffusion-release and effusive-flow processes. Another utilitarian aspect of the equation is that it permits the choice of materials of tube construction that minimize the time required for transport of a given species through arbitrary geometry transport systems, on a relative basis.

ACKNOWLEDGEMENTS

Research at the Oak Ridge National Laboratory is supported by the U.S. Department of Energy under contract DE-AC05-00OR22725 with UT-Battelle, LLC.

REFERENCES

1. Alton, G.D., Beene, J.R., and Liu, Y., *Nucl. Instrum. and Meth. A*, **438**, 190 (1999).
2. Present, R.D., *Kinetic Theory of Gases*, Mc Graw-Hill Book Company, New York, 1958.
3. Adamson, A.W., *Physical Chemistry of Surfaces*, third Edition, John Wiley and Sons, New York, 1976.
4. Kevan S.D., *Applications of Accelerators in Research and Industry*, **CP392**, 425 (1997).
5. *Effuse* is a Monte-Carlo code, written by J. Dellwo and G.D. Alton, developed for calculating the time dependent transport of gaseous elemental/molecular species through simple tubular systems.
6. Wolfram S., Mathematica 3.0, Wolfram Research, Champaign, Il 61820, USA.

Status of the New Los Alamos UCN Source

K. Kirch,[a] T. J. Bowles,[a] B. Fillipone,[b] P. Geltenbort,[c] R. Hill,[a] M. Hino,[f]
S. Hoedl,[e] G. Hogan,[a] T. M. Ito,[b] T. Kawai,[f] S. K. Lamoreaux,[a] C.-Y. Liu,[e]
J. W. Martin,[b] C. Morris,[a] A. Pichlmaier,[a] A. Saunders,[a] S. Seestrom,[a]
A. Serebrov,[d] D. Smith,[e] B. Tipton,[b] M. Utsuro,[f] A. R. Young,[e] J. Yuan[b]

[a]*Los Alamos National Laboratory,* [b]*California Institute of Technology,* [c]*Institut Laue-Langevin,* [d]*Petersburg Nuclear Physics Institute,* [e]*Princeton University,* [f]*University of Kyoto*

Abstract. Ultra-cold neutrons (UCN) have been produced at reactors over the last 30 years. Although very successful, experiments often suffer from low UCN statistics - as limited by the traditional production scheme. A new type of UCN source has been developed at Los Alamos. The source combines a spallation target, a cold neutron flux trap, and a solid deuterium converter for the down-scattering of cold neutrons into the ultra-cold regime. The breakthroughs of the last year include the theoretical understanding of the production mechanism and its experimental verification. The new technique is capable of delivering orders of magnitude higher densities of UCN. The highest UCN density ever stored in an experiment is reported.

INTRODUCTION

Ultra-Cold Neutrons (UCN) are defined as neutrons which are totally reflected from certain materials at all angles of incidence. They were first considered theoretically in 1959 by Zel'dovich (1) or maybe even earlier by Fermi and first experimentally observed in 1968/69 (2, 3). Typical UCN kinetic energies are a few hundred neV, corresponding to velocities below about 8 m/s. UCN can be stored in material traps ("UCN bottles"), using the total reflection from material surfaces. Due to the neutron magnetic moment, magnetic fields of the order of several Tesla provide potential energies of the same order as UCN kinetic energies. Also the change in the gravitational potential for height differences of a few meters is of the order of the kinetic energy. As a consequence UCN can be confined in material traps, magnetic traps, and combined gravitational traps.

Storage of UCN is a very important feature for a variety of fundamental experiments. It allows the measurement of the neutron lifetime from a well defined sample and leads to ingenious experiments searching for an electric dipole moment (EDM) of the neutron (see e.g. (4)). Although UCN offer greatly improved sensitivity and systematics compared to cold neutron based experiments, one main limitation of these experiments is the low UCN statistics. Typically UCN densities of 1-10 cm^{-3} are obtained. Sources with increased UCN density of the order of 1000 cm^{-3} may lead to much improved measurements of the neutron lifetime and the neutron EDM. Moreover, neutron decay studies may become possible with UCN. There is already a proposal to measure the neutron decay correlation "A" using the new Los Alamos UCN source. Accurate measurements of the shape of the neutron decay β-spectrum and the investigation of radiative neutron decay also come into reach with high densities.

PRODUCTION OF ULTRA COLD NEUTRONS

Traditionally, UCN are produced at reactors. The principal difficulty of UCN production is that their energy region is far out in the tail of a thermal Maxwell distribution, leading to suppression factors of the order of 10^{-9}. Tricks to increase the UCN yield from a cold neutron source include vertical extraction (3), using gravity to decelerate neutrons of higher velocities into the UCN regime; and mechanical deceleration (5), using collisions of faster neutrons with a moving scatterer. The fast neutrons can more easily penetrate windows and can be transported over longer distances with few reflections. The distances for UCN transport to the experiments can thus be short and UCN losses small.

Alternative UCN production schemes have been proposed and demonstrated, such as conversion of cold neutrons into UCN in superfluid helium (6, 7) or other suitable cold converters as, e.g., solid deuterium (8, 9, 10, 11).

THE LOS ALAMOS PROTOTYPE UCN SOURCE

Our design closely follows the original Pokotilovski idea for an UCN source operated at a pulsed neutron source (12); our principle modifications include production of neutrons on a spallation target (see also (13)), a cold neutron flux trap to maximize the flux of cold neutrons (30 - 40 K) at the position of a solid deuterium converter, and vertical UCN extraction from the deuterium to account for the material potential step.

The downscattering rate of the cold neutrons into UCN is almost independent of the converter temperature. Because the maximum achievable density from a superthermal source is given by the product of the downscattering rate and the lifetime of the UCN in the converter (6), the most important issue for the production of UCN with a solid deuterium converter is to maximize the neutron lifetime in the deuterium. A short lifetime hinders the extraction of the UCN from the converter.

There are four contributions to the lifetime of UCN in solid D_2 and the different contributions add up reciprocally: Nuclear absorption of neutrons on deuterons results in 150 ms lifetime. A SD_2 crystal temperature of 4-5 K gives 150 ms lifetime due to phonon induced upscattering. A contamination of the D_2 with 0.2% H_2 results in 150 ms lifetime. A para-D_2 fraction of about 1% again gives 150 ms lifetime. Although there was strong first evidence for a better UCN production from ortho deuterium in a Gatchina reactor experiment (11), the detailed mechanisms involved have been clarified only recently (14).

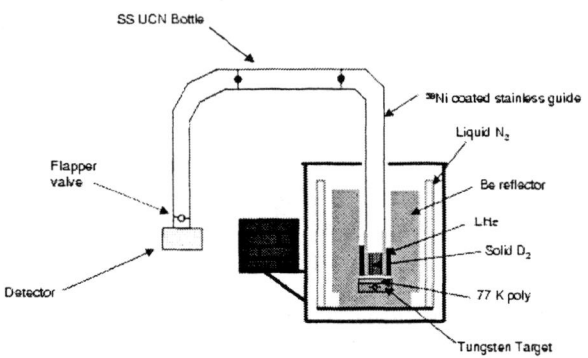

FIGURE 1. The Los Alamos prototype source. Details are explained in the text.

Figure 1 shows schematically the Los Alamos prototype source lay out. The design presents a compromise between the important aspects mentioned above, a simple geometry, and easily obtainable materials. An 800 MeV proton beam from the Los Alamos Neutron Science Center (LANSCE) accelerator is used to produce neutrons

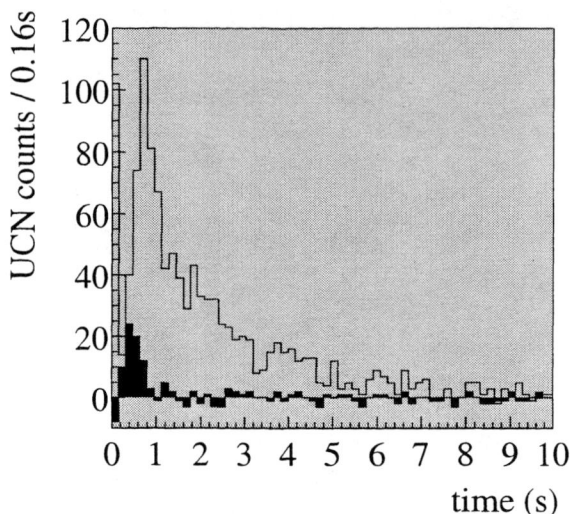

FIGURE 2. Background subtracted arrival time spectra for UCN at the detector. t = 0 corresponds to the time of the proton pulse. The bottle valves (see fig. 1) are open. The black spectrum is obtained with the flapper valve in front of the detector closed. The valve is a 0.25 mm thick Al plate with a 150 nm layer of ^{58}Ni. The flapper valve is open for the other spectrum.

from a tungsten spallation target. At this energy and in the specific geometry used the neutron production amounts to about 18 per proton, 12 from direct spallation, 6 from secondary (n,xn)-reactions. The spallation neutrons with 1-2 MeV average energy are moderated and trapped by an assembly of polyethylene (PE) moderators and beryllium reflectors. Part of the PE is at liquid nitrogen temperature, and part is integrated into the liquid helium cryostat. The liquid helium cryostat holds the solid deuterium (SD_2) converter. The SD_2, which has to be predominantly in the ortho molecular state, sits in a windowless ^{58}Ni coated stainless steel guide tube of 80 mm diameter. UCN which leave the SD_2 can travel along the guide tube, might be stored in a UCN bottle, and can be detected in a ^{3}He wire chamber at the other end of the guide tube system. An additional flapper valve in front of the detector allows further tests; it can e.g. hold a thin ^{58}Ni foil or different size apertures. Figure 2 shows results from a run with such a nickel foil. UCN are totally reflected from the high ^{58}Ni potential, but they would pass through the substrate Al with small losses. There are nearly no counts when the flapper valve is closed, which demonstrates that the signal when this valve is open is due to UCN.

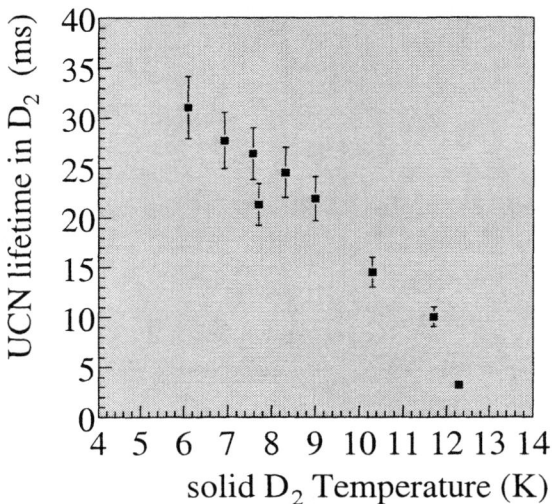

FIGURE 3. UCN lifetime in SD$_2$ as a function of the converter's temperature. The fraction of para-D$_2$ is about 2%.

EXPERIMENTAL RESULTS

Deuterium lifetime measurements

As described above, the production of UCN from solid deuterium is strongly dependent on the lifetime of UCN in SD$_2$. A measurement of this lifetime was done by storing UCN in the volume between the SD$_2$ and the nearer bottle valve (see Fig. 1). By opening the valve after different times and counting the surviving neutrons, one can measure the lifetime of the UCN in the storage volume which includes the deuterium. The extraction of the UCN lifetime in the SD$_2$ is accomplished by fitting Monte Carlo transport calculations with adjustable lifetime to the data. Figure 3 shows the results for the SD$_2$ lifetime as a function of the SD$_2$ temperature for deuterium with a 2% para fraction and 0.2% hydrogen contamination. These data were taken with 20 cm^3 SD$_2$ in the cryostat. The expectation for the SD$_2$ lifetime for the 2% para deuterium at the lowest temperatures of about 6 K and taking into account the 0.2% hydrogen contamination is about 30 ms, which is in very good agreement with the data. Figure 4 shows the UCN lifetime in the SD$_2$ as measured for different para-D$_2$ fractions. The SD$_2$ temperature was about 6 K again. The fall off of UCN lifetime for larger para fractions is explained by the upscattering of UCN on para-D$_2$, taking along the 7 meV transition energy of the para to ortho deexcitation (14). The result underlines the difficulty of building a high density UCN source using unconverted D$_2$ (33% para, 67% ortho).

Volume scans

Figure 5 shows the dependence of the UCN production on the SD$_2$ volume, which essentially is the dependence on the SD$_2$ thickness. The measured data are shown together with Monte Carlo calculations based on the SD$_2$ model as proposed in (14). There is a simple qualitative explanation of the data: when increasing the SD$_2$ volume the UCN production goes up linearly. Although this is true, also the absorption of UCN (including upscattering) is increased. At a certain thickness the SD$_2$ is no longer transparent for UCN (the elastic mean free path is about 8 cm, typical inelastic interaction lengths are about 20 cm).

High proton current runs

In principle, the UCN production rate on solid deuterium is proportional to the incident cold neutron flux. Thus, by increasing the proton beam current, the cold neutron flux and the UCN production can be increased. The fundamental limitation of the method is warming of the SD$_2$ because of the heat load. Because calculation of the thermodynamic properties of the SD$_2$ converter in the cryostat is a difficult task, it is of interest to experimentally test the UCN production limits. Figure 6 shows the UCN density after 0.5 s storage in the bottle of Fig. 1 as a function of the proton beam charge. Each point repre-

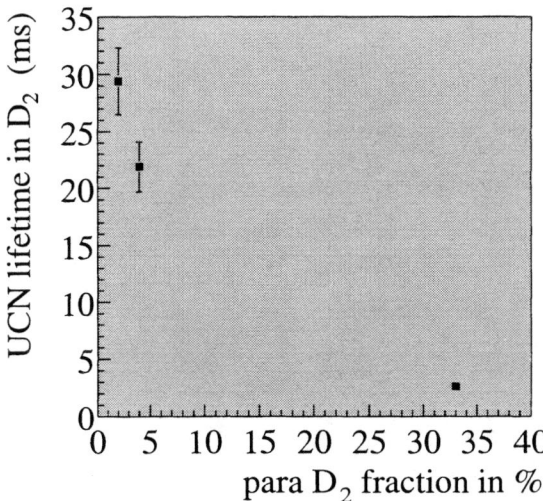

FIGURE 4. UCN lifetime in SD$_2$ as a function of the fraction of para-D$_2$ in the converter. The temperature is about 6K.

sents data taken after a beam pulse in which the protons hit the target distributed with the same time structure over 1 s. There is some evidence for a nonlinear behaviour in the slight fall off at high beam charges. However, no drastic decrease in UCN production has been found up to 100 μC. The highest density measured in this experiment was 98 ± 5 cm^{-3} in a 3.6 l stainless steel bottle. The highest UCN density previously obtained was 41 cm^{-3} (15). This value was measured in a stainless steel bottle at the ILL high flux research reactor. The fact that this small prototype source has set a new density record underlines the great potential of the new methods.

FUTURE DEVELOPMENTS

The successful test of the Los Alamos prototype UCN source leads the way for the construction of a dedicated SD$_2$-based UCN source at LANSCE. Modeling suggests that the new source could run with about 4 μA average proton beam current and feed a 40 l storage volume to an average UCN density of 300 cm^{-3}. According to the current plans, the storage volume will continously feed the experimental volume of a neutron decay experiment.

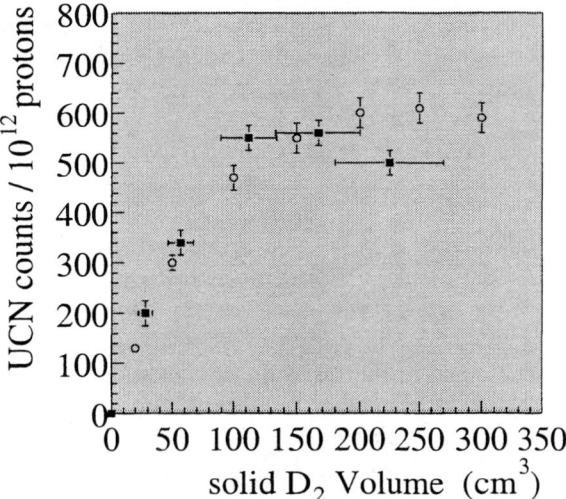

FIGURE 5. The influence of the SD$_2$ volume (surface area = 50 cm^2) on UCN production. The vertical axis gives the number of UCN detected in our detector (see Fig. 1) normalized for the number of protons (10^{12} protons correspond to 0.16 μC). The black squares show the results from measurements with 2% para D$_2$ and 0.4% HD contamination. The horizontal error bars are systematic 20% errors on the volume measurements and correlated for all points. The open circles are results of Monte Carlo calculations for the given geometry and SD$_2$ properties.

ACKNOWLEDGEMENTS

We would like to thank M. Anaya, L. Marek, R. Mortensen, and W. Teasdale for their technical support and the LANSCE accelerator team for providing the proton beam.

REFERENCES

1. Zel'dovich, Ya. B., *Sov. Phys. JETP* **9**, 1389 (1959)
2. Luschikov, V. I., et al., *JETP Lett.* **9**, 23 (1969)
3. Steyerl, A., *Phys. Lett.* **B29**, 33 (1969)
4. Ignatovich, V. K., *Ultracold Neutrons*, Clarendon Press, Oxford, 1990; Golub, R., Richardson, D. J., Lamoreaux, S. K., *Ultra-cold Neutrons*, Adam Hilger, Bistol, 1991
5. Steyerl, A., *Nucl. Instr. Meth.* **125**, 461 (1975)
6. Golub, R., Pendlebury, J. M., *Phys. Lett.* **62A**, 337 (1977)
7. Huffman, P. R., et al., *Nature* **403**, 62 (2000)
8. Golub, R., and Böning, K., *Z. Phys.* **B51**, 95 (1983)
9. Yu, Z.-Ch., et al., *Z. Phys.* **B62**, 137 (1986)
10. Serebrov, A. P., et al., *JETP Lett.* **59**, 757 (1994)
11. Serebrov, A. P., et al., *JETP Lett.* **62**, 785 (1995)
12. Pokotilovski, Yu. N., *Nucl. Instr. and Meth.* **A 356**, 412 (1995)
13. Serebrov, A. P., et al., *JETP Lett.* **66**, 802 (1997)
14. Liu, C.-Y., et al., *Phys. Rev.* **B62**, R3581 (2000)
15. Steyerl, A., et al., *Phys. Lett.* **A116**, 347 (1986)

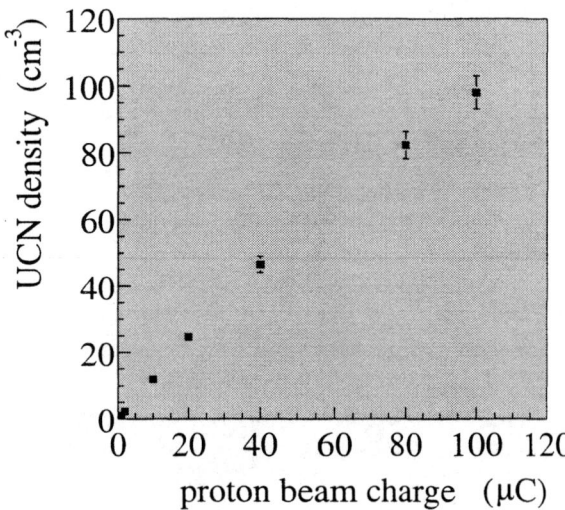

FIGURE 6. Density of stored UCN as a function of the incident proton beam charge.

Measurements of the Electromagnetic and Spin Structure of Nucleons and Nuclei with the Bates Large Acceptance Spectrometer Toroid

R. Alarcon[*,#,1]

Department of Physics and Astronomy, Arizona State University, Tempe, AZ 85287-1504
#*MIT-Bates Linear Accelerator Center, Middleton, MA 01949-2846*

Abstract. The Bates Large Acceptance Spectrometer Toroid (BLAST) is a detector designed to study the spin-dependent electromagnetic response of few-body nuclei at momentum transfers up to 1 $(GeV/c)^2$ at the MIT/Bates Linear Accelerator Center's South Hall Ring (SHR). The BLAST detector consists of an eight-sector copper coil array producing a toroidal magnetic field, instrumented with two opposing wedge-shaped sectors of wire chambers, scintillation detectors, Cerenkov counters, neutron detectors, a lead-glass forward calorimeter, and recoil detectors. The ability of BLAST to carry out multi-particle detection over a large solid angle from polarized internal targets will provide an unprecedented and unique opportunity to study simultaneously the spin structure of the few-body nuclear ground states, the reaction mechanisms, and the nucleon form factors. Presently, BLAST is under construction and it is on schedule to be done in summer 2001. A status of the project is presented as well as highlights of the scientific program.

INTRODUCTION

The Bates Large Acceptance Spectrometer Toroid (BLAST) is a detector designed to study in a comprehensive and precise way the spin-dependent electromagnetic response of few-body nuclei at momentum transfers up to 1 $(GeV/c)^2$ at the MIT-Bates South Hall Ring. The BLAST scientific program is focussed on the study of these systems in terms of nucleon structure, the ground state few body structure built from the nucleon-nucleon interaction and the nature of the interaction of the virtual photon for $Q^2 \leq 1$ $(GeV/c)^2$. A major consideration in the design of BLAST has been the realization that these aspects of the study of the electromagnetic response are interrelated in a complicated way and can only be unambiguously separated by a broad study of the few body systems. In addition, both the choice of few body systems and the relatively low momentum transfers should allow accurate comparison with theoretical calculations.

To accomplish its scientific goals, BLAST will utilize the latest technology available in the form of polarized electron scattering from pure, polarized internal gas targets. The newly commissioned Bates South Hall Ring (SHR) will deliver longitudinally polarized electrons at the location of the BLAST detector. The internal targets are highly polarized, $\geq$ 50 %, do not have any dilution from non-polarized species, are rapidly reversible, can be oriented with low magnetic fields and the resulting luminosity is well matched to that of a large acceptance detector. Further, thin walled target cells allow straightforward detection of low energy recoil particles which will allow complete reconstruction of the final state in the electrodisintegration of few body systems. The internal target technique has now been successfully implemented at several laboratories (Novosibirsk, IUCF, NIKHEF, DESY) and it is widely recognized that this arrangement can provide the lowest combined statistical and systematic error. The ability with BLAST to carry out multiparticle detection over a large solid angle from polarized internal targets will provide an unprecedented and unique opportunity to study simultaneously the spin structure of the few-body nuclear ground states, the reaction mechanism, and nucleon form factors.

---

[1] On behalf of the BLAST Collaboration

THE BLAST PROJECT

The BLAST detector consists of an eight-sector copper coil array producing a toroidal magnetic field, instrumented with two opposing wedge-shaped sectors of wire chambers, scintillation detectors, Cerenkov counters, neutron detectors, a lead glass calorimeter and recoil detectors (see Fig. 1) [1]. The open geometry maximizes acceptance while allowing good momentum and angular resolution, and with a luminosity capability that is matched to the densities of the polarized internal targets.

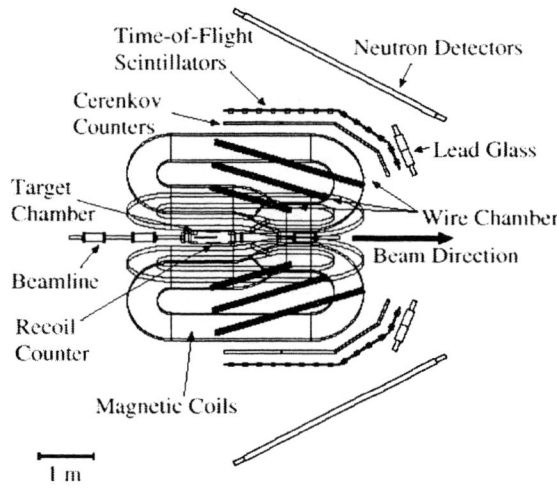

FIGURE 1. Top view of BLAST showing the eight coils and the two instrumented sectors. The tracking drift chambers are located between the coils followed by the Cerenkov detectors, the timing scintillators and the neutron detector array. Updated information can be found at http://mitbates.mit.edu/blast/.

The design emphasizes proven technology, commercial electronics, and existing data acquisition system software to achieve low cost and a short implementation time. Clear upgrade possibilities exist so that the detector can evolve to match developing physics priorities.

Construction of BLAST, by an international collaboration of physicists from 11 institutions, is well underway. The coils have been constructed and, with the appropriate mechanical supports, will be installed in the SHR by early 2001. A polarized proton and deuteron target has been installed and it will be tested in the BLAST configuration as soon as the coils are in operation.

Testing of the first set of detectors with a 569 MeV electron beam was carried out at the BLAST configuration in March 2000. At present the South Hall Ring delivers high currents (100-200 mA) of unpolarized electron beams through gas targets of the required density, with lifetimes of several minutes. Polarized internal beams are available with the operation of the Siberian snake magnets to ensure longitudinal polarization at the position of the BLAST targets. The installation and commissioning of a laser Compton back-scattering electron polarimeter was carried out in March 2000. It is planned that this instrument will be taking data with a polarized beam by the end of 2000. Commissioning of the full detector is scheduled to begin by fall 2001.

SCIENTIFIC PROGRAM

BLAST will provide precise information on nucleon form factors for momentum transfers up to about 1 $(GeV/c)^2$. In particular, BLAST will provide data on the neutron magnetic and electric form factors with both deuteron and ^{3}He targets. These are fundamental quantities that are essential to any description of electromagnetic scattering from nuclei. Because of the significantly larger binding energy of the neutron in ^{3}He compared to that in deuterium, it is possible that the neutron charge distribution in ^{3}He may be modified from its free value. BLAST has the necessary precision to probe for such an effect.

BLAST will carry out a precise measurement of the spin-dependent momentum distribution in few-body nuclei in order to understand the spin structure of few-body systems in terms of the successful theoretical framework which has been developed primarily for unpolarized scattering. Effects such as final-state interactions, meson exchange currents, and the off-shell nature of the bound nucleon can be studied over the broad kinematics range provided by the BLAST detector. In addition, the spin-dependent momentum distributions are used as input for calculations of spin-dependent scattering in the deep-inelastic region where polarized deuterium and ^{3}He are used to determine the neutron spin structure at the quark level.

With a tensor polarized deuterium target BLAST will provide precise data from elastic e-d scattering (T_{20}), particularly in the region of the first minimum of the charge form factor of the deuteron. In addition, data will be obtained in the same experiment for the exclusive scattering channels.

BLAST will carry out measurements of spin-dependent charged pion electroproduction on few-body systems from threshold to beyond the Δ-resonance. Such studies are important for understanding the role of the nucleon resonance in few-body systems. BLAST allows reconstruction of the resonance from its π-nucleon decay channel. In this way, for example, the presence of pre-existing Δ-components in the ^{3}He ground state can be studied.

With a polarized proton target and a polarized electron beam, BLAST will study the N → Δ transition to isolate components beyond the dominant M1 transition and will provide additional precise data on the proton form factors. In addition, at lower energies, an experiment is planned to extract the charge radius of the proton.

There is also a rich program of physics to be pursued with unpolarized targets and light nuclei. Examples are the study of multinucleon processes which are well suited for a large acceptance detector like BLAST, measurements of the neutron momentum distributions in nuclei, and the detection of heavy recoils in processes of astrophysical significance.

The Charge and Magnetic Form Factors of the Neutron

How charges and currents are distributed in the nucleon is central to answer fundamental questions about the quark structure of everyday matter. A centerpiece of the BLAST scientific program is measurement of the neutron electric $G_E^n(Q^2)$ and magnetic $G_M^n(Q^2)$ form factors up to about 1 (GeV/c)^2. The expected precision and accuracy will be comparable to that of the proton form factors $G_E^p(Q^2)$ and $G_M^p(Q^2)$.

Precise determination of the neutron charge distribution has been an elusive goal in electromagnetic nuclear physics over the last several decades. However, Bates with its BLAST detector and high intensity stored beam can deliver the definite data at long distances. $G_E^n(Q^2)$ provides information on the sea quark distribution in the nucleon as the charge of the valence quarks of the neutron sum to zero [2]. The neutron form factors G_E^n and G_M^n are important for testing modern theoretical calculations of few-body nuclei based on hadronic degrees of freedom. These calculations are needed for interpretation of all Coulomb and electric multipoles in nuclei. Finally, high quality data on the neutron form factors is needed for the extraction of the nucleon strange quark content from parity-violating electron scattering experiments [3].

Polarization data on the charge form factor of the neutron G_E^n [4-7] are summarized in Fig. 2 together with the projected results from BLAST. At Bates BLAST will measure $G_E^n(Q^2)$ from 0.1 to 0.8 (GeV/c)^2 with high precision both on deuterium and ^{3}He within the same apparatus. With overall uncertainties ≤ 5 % it should be possible to carry out a high precision comparison of G_E^n in the deuteron and ^{3}He. This will allow the most precise search for the predicted modification of the neutron's pion cloud in the nuclear medium. In addition, BLAST will address the reaction mechanism effects by the simultaneous measurement of several reaction channels over a broad kinematic range and for various orientations of the target polarization direction. BLAST will use its symmetric capability to perform a simultaneous determination of $G_M^n(Q^2)$ at identical kinematics to $G_E^n(Q^2)$, both in the case of deuterium and ^{3}He.

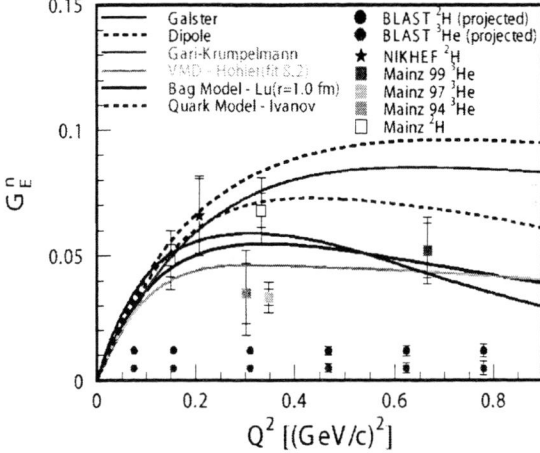

FIGURE 2. Values for G_E^n as extracted from spin-dependent electron scattering. Also shown are the BLAST projections from polarized deuterium and ^{3}He. The curves correspond to various theoretical projections.

Structure of the Deuteron

The deuteron, as a two-nucleon system, is often used as a benchmark to test nuclear theory because exact calculations can be performed in both non-relativistic and relativistic models. Precise descriptions of the nucleon-nucleon interaction, investigation of the role of non-nucleonic degrees of freeedom in nuclei, the role of relativity in nuclear structure are key issues where the deuteron is the best testing ground. It is also one of the few nuclear systems for which predictions based on QCD have been made and tested.

The electromagnetic structure of the spin-1 deuteron is described by the charge monopole G_C, charge quadrupole G_Q, and magnetic dipole G_M form factors. The unit spin of the deuteron requires a spin measurement for separating the G_C and G_Q charge distributions and thus allowing the separation of all multipole form factors for the simplest nuclear system. G_C is expected in most models to pass through zero as a result of a node in the S-state wave function that comes from the repulsive nature of the NN interaction at short distances.

Fig. 3 shows the world data [8-11] for the observable T_{20}. With BLAST, polarization data for elastic scattering of comparable quality to that available for the unpolarized response functions will be obtained, allowing for a much more precise comparison among the theoretical models than has been possible up to now. The data projected with BLAST will be particularly sensitive to the position of the first minimum of the charge monopole form factor.

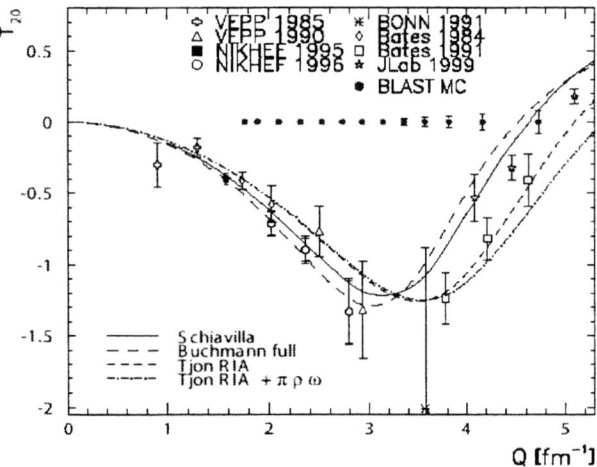

FIGURE 3. Data and theoretical predictions for T_{20} as a function of momentum transfer. The curves represent various theoretical models. Also indicated are the expected data (solid circles) with BLAST. The first minimum of G_C occurs in the region between 4.0 and 4.5 fm^{-1}.

Structure of ^{3}He

The ^{3}He nucleus has several properties that make the study of its spin particularly interesting. The three body system is unique in that, although it is relatively tightly bound, essentially exact Faddeev solutions in non-relativistic approximation of the ground state have been obtained from a variety of two-nucleon potentials. In addition, unlike a heavy nucleus where the total spin is usually determined by only a few valence nucleons, the spin of ^{3}He involves all the nucleons in the nucleus.

Spin-dependent electron scattering from a polarized ^{3}He internal target with BLAST is a unique program in medium energy nuclear physics. In a single experiment BLAST can simultaneously determine in ^{3}He the ground state spin-dependent spectral function, the nucleon form factors, study the reaction mechanism, the neutron resonance structure, the Δ-components in the ^{3}He wave function at the level of about 0.2 %, and the elastic and threshold region.

The nuclear structure information is contained in the spin-dependent spectral function [12]. Quasielastic spin-dependent knockout of the constituent nucleons of ^{3}He with good resolution in the energy and momentum of the initial state nucleon offers the most direct experimental approach to constrain the spectral function. With BLAST data will be obtained simultaneously for a wide range of the momentum transfer Q^2. In the same experiment, in addition to measuring the spin dependent momentum distributions of the neutron and the proton in ^{3}He, BLAST can provide clean separation of the two-body and three-body contributions to the proton momentum distribution and determination of the presence of the D-state at high missing momenta.

REFERENCES

1. Alarcon, R., *Progress in Particle and Nuclear Physics* **44**, 253-272 (2000).
2. Gorski A. Z., Grummer F., and Goeke K., *Phys. Lett. B* **278**, 24 (1992).
3. Aniol K., et al., *Phys. Rev. Lett.* **82**, 1096 (1999).
4. Passchier I., et al., *Phys. Rev. Lett.* **82**, 4988 (1999).
5. Meyerhoff M., et al., *Phys. Lett. B* **278**, 24 (1992).
6. Ostrik M., et al., *Phys. Rev. Lett.* **83**, 276 (1999).
7. Rohe D., et al., *Phys. Rev. Lett.* **83** (1999) 4257.
8. The I., et al., *Phys. Rev. Lett.* **67**, 173 (1991); Garcon M., et al., *Phys. Rev. C* **49**, 2516 (1994), and references therein.
9. Ferro-Luzzi M., et al., *Phys. Rev. Lett.* **77**, 2630 (1996).
10. Bouwhuis M., et al., *Phys. Rev. Lett.* **82**, 687 (1999).
11. Abbott D., et al., *Phys. Rev. Lett.* **84**, 5053 (2000).
12. Milner R., et al., *Phys. Lett. B* **379**, 67 (1996).

Tests of the Standard Model from Superallowed Fermi β-Decay Studies at ISAC

G. C. Ball*

TRIUMF, 4004 Wesbrook Mall, Vancouver, B. C. Canada V6T 2A3,

Abstract. Precise measurements of the intensities for superallowed Fermi $0^+ \rightarrow 0^+$ β-decays have provided a powerful test of the CVC hypothesis at the level of 3×10^{-4} and also led to a result in disagreement with unitarity (at the 98% confidence level) for the CKM matrix. Since this result would have profound implications for the minimal Standard Model it is essential to address possible "trivial" explanations for this apparent non-unitarity, such as uncertainties in the theoretical isospin symmetry-breaking correction. At ISAC we are pursuing a program to study the $T_z = 0$ (odd-odd) emitters with $A \geq 62$ where this correction is predicted to be large (>1%) and readily tested. The nucleus ^{74}Rb was chosen for the first experiments. A Nb foil target was bombarded with a 10 μA beam of 500 MeV protons to produce ~ 4000 ^{74}Rb ions/s. Sufficient data were obtained in these measurements to improve the precision in the previously measured half-life of ^{74}Rb by a factor of sixteen; the result is $t_{1/2} = 64.761 \pm 0.031$ ms. High precision branching ratio measurements for the β-decay of ^{74}Rb to excited $(0,1)^+$ states in ^{74}Kr are in progress. The results of these first experiments at ISAC together with the prospects for the future are presented

INTRODUCTION

Precise measurements of the intensities for superallowed Fermi $0^+ \rightarrow 0^+$ β-decays provide demanding and fundamental tests of the properties of the weak interaction. In particular, since the axial vector decay strength is zero for such decays the intensities are directly related to the weak vector coupling constant, G_V. According to the conserved vector current (CVC) hypothesis, the measured ft values for Fermi decays of $0^+, T = 1$ analog states are given by:

$$ft(1+\delta_R)(1-\delta_C) \equiv \mathcal{F}t = \frac{K}{2G_V^2(1+\Delta_R)} = \text{constant}, \quad (1)$$

where f is the statistical rate function, t is the partial half-life for the transition, Δ_R and δ_R are the nucleus-independent and nucleus-dependent parts of the radiative correction and δ_C is the isospin symmetry-breaking (Coulomb) correction.

Presently, nine transitions have been determined with sufficient precision (2, 3, 4) to confirm the CVC hypothesis at the level of 3×10^{-4}. The average value of $\mathcal{F}t$ is $3072.3 \pm 0.9 \pm 1.1$, where the first error is the statistical error of the fit and the second is an error related to the systematic difference between two calculations of δ_C (2).

These data together with the muon lifetime also provide the most accurate value for the up-down quark mixing matrix element of the Cabibbo-Kobayashi-Maskawa (CKM) matrix, V_{ud}. The current status of V_{ud} is reviewed by Towner and Hardy in ref. (2). In particular the value of V_{ud} is given by:

$$V_{ud}^2 = \frac{K}{2G_F^2(1+\Delta_R)\overline{\mathcal{F}t}}, \quad (2)$$

where G_F is the weak coupling constant from muon decay: $G_V = G_F V_{ud}$, and $\overline{\mathcal{F}t}$ is the average $\mathcal{F}t$ from the nine precision Fermi superallowed β decays measured so far. The result obtained is $|V_{ud}| = 0.9740 \pm 0.0005$. It is important to note that the error associated with $|V_{ud}|$ is not predominantly experimental in origin; the largest uncertainties come from Δ_R (± 0.0004) and δ_C (± 0.0003).

The unitarity test as it relates to the elements in the first row of the CKM matrix,

$$V_{ud}^2 + V_{us}^2 + V_{ub}^2 = 1, \quad (3)$$

can be examined using the recommended (5) values for V_{us} and V_{ub} (i.e. 0.2196 ± 0.0023 determined from K_{e3} decay and 0.0036 ± 0.0010, respectively). The result becomes:

$$\sum_i V_{ui}^2 = 0.9968 \pm 0.0014, \quad (4)$$

which fails to meet unity by 2.2 standard deviations.

---

* For the E823 collaboration(1)

The value of V_{ud} can also be determined from the decay of the free neutron and pion beta decay. These decays have the advantage over nuclear decay that there are no nuclear-structure-dependent corrections. However, at the present time the values for $|V_{ud}|$ obtained from measurements of these decays are much less precise than those from superallowed Fermi β-decay (2). Ultimately, they will also be limited by the theoretical uncertainty in Δ_R.

Since the failure in the unitarity of the CKM matrix would imply physics beyond the minimal Standard Model, it is important to re-examine possible explanations for the apparent discrepancy such as the calculated nuclear-structure-dependent corrections. To restore unitarity, the calculated radiative corrections for all nine nuclear transitions would have to shift downwards by 0.3% (nearly one-quarter of their present value). This seems unlikely since the leading order terms in these corrections are well founded. Although smaller than the radiative corrections, the Coulomb corrections are clearly sensitive to nuclear structure because the Coulomb and charge-dependent nuclear forces destroy isospin symmetry between the initial and final states in superallowed beta-decay. Two consequences result: there are different degrees of configuration mixing in the two states and their radial wave functions differ because their binding energies at not identical.

Methods to calculate δ_C have been developed by Towner et al (6) and Ormand and Brown (7). The values of δ_C predicted by the two methods are in reasonable relative agreement for the nine well-known transitions with $A \leq 54$ but differ absolutely by $\sim 0.07\%$. This difference is much less than the 0.3% required to resolve the unitarity problem. For the heavier fp-shell nuclei both theoretical methods predict much larger and also very different values for the Coulomb corrections (7). At ISAC a program has been initiated to measure the half-lives and branching ratios for the odd-odd, $T_z = 0$ nuclei with $A \geq 62$. These data together with accurate Q_{EC} values, will provide a critical test of these theoretical calulations for isospin-symmetry breaking.

EXPERIMENT AND RESULTS

The ISAC radioactive Beam Facility

A new second-generation radioactive beam facility, ISAC, is being built at TRIUMF (8) to provide intense beams of both low-energy (<60 keV) and accelerated radioactive ions up to 1.5 MeV/u. The facility began delivering low-energy beam for experiments early in Dec. 1998; the accelerated beams at full energy will be available for experiments in early 2001. ISAC produces radioactive beams by the ISOL method using very intense beams of protons at 500 MeV to bombard thick production targets. The facility has been constructed to handle proton beam intensities on target of up to 100 μA. The target is coupled to the ion source via a small transfer tube. Several types of ion sources will be available; the first operation has used a surface ionization source to produce beams of alkali elements. Development of a compact microwave ECR ion source to produce beams of volatile elements is near completion. Plans to develop a laser ion source are being formulated.

The radioactive ion beams are mass analyzed by a low-resolution pre-separator followed by a high acceptance mass analyzer operating in the low resolution mode with a resolving power of $M/\Delta M \cong 1000$. This will be increased by a factor of five in the future when the mass analyzer is operational in the high-resolution mode.

ISAC is ideally suited to precision nuclear beta-decay studies and it should be possible to study the series of odd-odd, $T_z = 0$ nuclei from ^{62}Ga to ^{86}Tc. The nucleus ^{74}Rb was chosen for the first experiments since it is the easiest to produce with the surface-ionization source. A Nb foil target (11.5 g/cm^2) was used to produce ^{74}Rb. At proton beam currents above 3 μA the yields of radioactive Rb isotopes were found to increase nonlinearly (9). The yield of ^{74}Rb obtained at a beam current of 10 μA was ~ 440 ions $\mu A^{-1} s^{-1}$; comparable to the yield previously observed at ISOLDE (10). During a one day test run when the beam current was increased to 20 μA, the yields of all Rb isotopes increased by a factor of four. The Nb target was bombarded for several weeks at a beam current of 10 μA with no noticeable decrease in yield.

High-Precision β-decay Half-life Measurement for ^{74}Rb

The experiment was carried out at ISAC using a technique that was developed by members of this collaboration in previous high-precision β-decay half-life measurements (11). Although the measurement is simple in principle, great care must be taken to achieve the required precision ($\sim 0.05\%$). The low-energy (29 keV) radioactive ion beam from ISAC was implanted into a 25 mm wide aluminized mylar tape of a fast tape-transport system. After a collection period of ~ 4 half-lives, the ISAC beam was interrupted and the samples were moved out of the vacuum chamber through two stages of differential pumping and positioned in a 4π continuous-gas-flow proportional counter. Several improvements were made to optimize the experimental setup for short-lived (~ 100 ms) radioactive isotopes such as ^{74}Rb. With these changes it was possible to begin the counting period 100 ms after the collection period. After multiscaling the signals from the 4π counter for about 25 half-lives, the data were stored and the cycle was repeated, continuously.

A Stanford Research DS335 function generator provided a 100kHz $\pm$ 0.2 Hz time standard for the experiment. An instantly retriggerable gate generator was used to give a well-defined, non-extendable dead time that was significantly longer than any series dead time preceding it. Systematic errors introduced by the measurement techniques were determined by making on-line measurements under a variety of different conditions. Sample purity was monitored by measuring the gamma-ray spectrum using a 40% HPGe detector located close to the 4π counter.

A total of 38 runs each consisting of $\sim 4 \times 10^5$ events were carried out. The data obtained from one run are shown in Fig 1. A small ($\sim$2%) long-lived background was observed in the decay curves that was identified in the gamma-ray spectra as ^{74}Ga ($t_{1/2} = 8.12$m). The data from each run was analyzed using a method described elsewhere (11). After prescreening the data to eliminate low-statistics and noisy samples, the remaining data were dead-time-corrected, summed and then fitted with a single short half-life, intensity and long-lived background. No evidence of any systematic error was observed in the complete data set. In addition, limits on possible short-lived contaminants determined from the γ-ray spectra were found to decrease the fitted ^{74}Rb half-life by less than 0.01%. The resulting value for the half-life of ^{74}Rb was 64.761 $\pm$ 0.031 ms (see ref (12) for more details). This value is a factor of sixteen more precise than the only previous measurement by D'Auria et al. (13) who obtained a value of 64.9 $\pm$ 0.5 ms.

High-Precision Branching Ratio Measurements for ^{74}Rb

Search for β-Delayed γ emission

An experiment to search for allowed transitions in the β-decay of ^{74}Rb to excited $(0,1)^+$ states in ^{74}Kr was carried out using a technique similar to that of Hagberg et al. (14). The same experimental apparatus described in the previous section was used to collect and move the ^{74}Rb samples out of the vacuum chamber but in this case the samples on the transport tape were positioned between two thin plastic scintillator paddles each backed by a large ($\sim$80%) HPGe detector mounted collinearly. The background in the HPGe detectors resulting from positrons was reduced by accepting only those β-γ coincidence events in which a positron was detected in the scintillator paddle located on the opposite side of the tape to

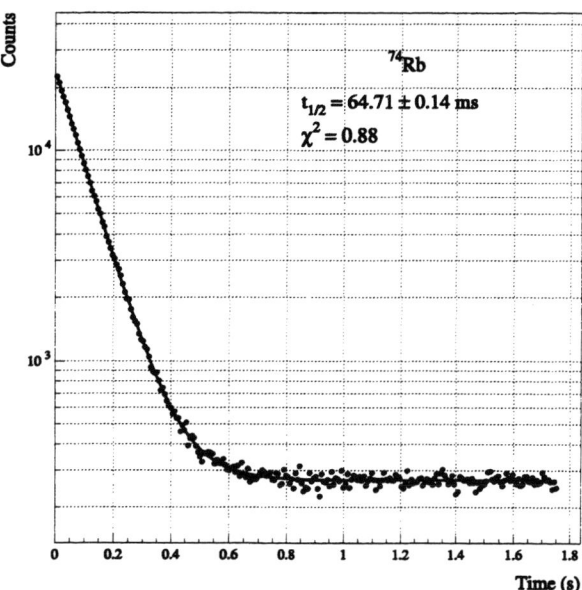

FIGURE 1. Typical decay curve obtained for ^{74}Rb. The dwell time was 7 ms per channel (see text for details).

the HPGe detector. The β-γ coincidence events were time stamped and recorded event by event. The positrons detected in each scintillator paddles were also multiscaled. After a measurement time of 0.5 s the cycle was repeated. A total of $\sim$15 M positrons were detected resulting in $\sim$1 M β-γ coincidence events.

The analysis of these data is in progress. Weak Gamow-Teller/Fermi decays to one or more high-lying levels in ^{74}Kr were observed by their depopulation through the known first excited 2^+ level at 456 keV. However, preliminary results indicate that the superallowed branch is the dominant transition (> 99%), similar to those observed previously for odd-odd $T_z = 0$ superallowed decays.

Search for the non-analog transition to the 0_2^+ state in ^{74}Kr

The determination of the transition strengths for non-analog $0^+ \rightarrow 0^+$ decays provide a critical test of the model predictions for the isospin mixing component of the Coulomb correction for superallowed β-decays. Recently, in-beam experiments (15, 16) have revealed the existence of a low-lying, isomeric 0_2^+ level in ^{74}Kr at 508 keV which decays primarily by an electric monopole transition to the ground state of ^{74}Kr. An experiment to search for the β-decay of ^{74}Rb to this 0_2^+ level has also been carried out at ISAC. In this measurement the ^{74}Rb atoms were implanted into the tape of a moving tape transport system that operated in vacuum. The collection point was viewed by two large plastic scintillation counters to detect positrons, three LN-cooled Si(Li) diodes for the detection of conversion electrons and an ($\sim 80\%$) HPGe detector for gamma-rays. Both Si(Li)-plastic and Ge-plastic coincidence data were recorded, The tape was moved every 5 s to reduce the background from the decay of ^{74}Kr and ^{74}Ga. A weak transition was observed in the Si(Li) spectra at 495 keV corresponding to the decay of the 508 keV level in ^{74}Kr. From an analysis of these it should be possible to measure the lifetime for the decay of this level and the strength of the non-analog transition.

CONCLUSIONS

The present half-life and branching ratio measurements for the superallowed β-decay of ^{74}Rb demonstrate the potential of the new second-generation radioactive-beam facility ISAC to deliver the beams required to extend precision superallowed β-decay studies to heavier short-lived $T_z = 0$ emitters. In particular, these are the first precision measurements of any of the key decay properties for an odd-odd $N = Z$ nucleus with A > 54. A precise measurement of the Q_{EC}-value for the β decay of ^{74}Rb should be possible in the near future using Penning traps such as the Canadian Penning Trap (17).

ACKNOWLEDGMENTS

The authors would like to thank the entire TRIUMF-ISAC team whose hard work and dedication led to the availability of high-quality radioactive beams from the ISAC facility ahead of schedule. TRIUMF is funded by a contribution from the National Research Council of Canada.

REFERENCES

1. The E823 Collaboration: G.C. Ball, J.A. Behr, S. Bishop, G. Boisvert, P. Bricault, J. Cerny, J.M. D'Auria, M. Dombsky, J.C. Hardy, V. Iacob, J.R. Leslie, T. Lindner, J.A. Macdonald, H.-B. Mak, D.M. Moltz, A. Piechaczek, J. Powell, G. Savard, I.S. Towner, J. Wood and E.F. Zganjar

2. Towner, I.S., and Hardy, J.C., *Proc. of the V Int. WEIN Symposium: Physics Beyond the Standard Model, Santa Fe, NM, June 1998*, edited by P. Herczeg, C.M. Hoffman and H.V. Klapdor-Kleingrothaus, World Scientific, Singapore, 1999, pp. 338-359.

3. Hardy, J.C., Towner, I.S, Koslowsky, V.T., Hagberg, E., and Schmeing, H., *Nucl. Phys.* **A509**, 429-460 (1990).

4. Towner, I.S., Hagberg, E., Savard, G., Hardy, J.C., Koslowsky, V.T., Galindo-Uribarri, A., and Radford, D.C., *Proc. of the IV Int. WEIN, Osaka, Japan 1995*, edited by H. Ejiri, T. Kishimoto and T. Sato, World Scientific, Singapore, 1995, pp. 313-318.

5. Groom, D.E., et al., *Eur. Phys. Journal* **C15**, 1 (2000).

6. Towner, I.S., Hardy, J.C., and Harvey, M., *Nucl. Phys.* **A284**, 269-281 (1977).

7. Ormond, W.E., and Brown, B.A., *Phys. Rev.* **C52**, 2455-2459 (1995).

8. Ball, G.C., Baartman, R., Behr, J., Bricault, P., Buchmann, L., D'Auria, J.M., Delheij, P., Dombsky, M., Dutto, G., Hutcheon, D., Jackson, K.P., Kiefl, R., Laxdal, R., Levy, P., Poutissou, J.-M., Schmor, P., Standford, G., and Ruth, T., *Proc. of the Int. Conf. on the Nucleus: New Physics for the New Millennium, Faure, South Africa, Jan. 1999*, edited by F.D. Smit, R. Lindsay and S.V. Fortsch (Kluwer Academic/Plenum Publishers, New York, 1999) pp. 69-76.

9. Dombsky, M., Bricault, P., Schmor, P., Hodges, T., and Hurst, A., *Proc of the V Int. Conf on Radioactive Nuclear Beams, Divonne, France, April 2000* (to be published).

10. Jokinen, A., Oinonen, M., Aysto, J., Baumann, P., Didierjean, F., Hoff, P., Huck, A., Knipper, A., Marguier, G., Novikov, Yu.N., Popov, A.V., Ramdhane, M., Seliverstov, D.M., Van Duppen, P., Walter G., and the ISOLDE collaboration, *Z. Physik* **A355**, 227-230 (1997).

11. Koslowsky, V.T., Hagberg, E., Hardy, J.C., Savard, G., Schmeing, H., Sharma, K.S., and Sun, X.J., *Nucl. Instr. Meth.* **A401**, 289-298 (1997).

12. Ball, G.C., et al., to be published.

13. D'Auria, J.M., Carraz, L.C., Hansen, P.G., Jonson, B., Mattsson, S., Ravn, H.L., Skarestad, M., Westgaard L., and the ISOLDE collaboration, *Phys. Lett.* **66B**, 233-235 (1977).

14. Hagberg, E., Alexander, T.K., Neeson, I., Koslowsky, V.T., Ball, G.C., Dyck, G.R., Forster, J.S., Hardy, J.C., Leslie, J.R., Mak, H.-B, Schmeing H., and Towner, I.S., *Nucl. Phys.* **A571**, 555-568 (1994).

15. Chandler, C., Regan, P.H., Pearson, C.J., Blank, B., Bruce, A.M., Catford, W.N., Curtis, N., Czajkowshi, S., Gellelty, W., Grzywacz, R., Janas, Z., Lewitowicz, M., Marchand, C., Orr, N.A., Page, R.D., Petrovici, A., Reed, A.T., Saint-Laurent, M.G., Vincent, S.M., Wadsworth, R., Warner, D.D., and Winfield, J.S., *Phys. Rev.* **C56**, R2924-R2928 (1997)

16. Becker, F., Korten, W., Hannachi, F., Paris, P., Buforn, N., Chandler, C., Houry, M., Hubel, H., Jansen, A., Le Coz, Y., Liang, C.F., Lopez-Martens, A., Lucas, R., Mergel, E., Regan, P.H., Schonwasser, G., Theisen, Ch., *Eur. Phys. Journal* **A4**, 103-105 (1999)

17. Savard, G., Barber, R.C., Beeching, D., Buchinger, F., Crawford, J.E., Feng, X., Gulick, S., Hagberg, E., Hardy, J.C., Koslowsky, V.T., Lee, J.K.P., Moore, R.B., Sharma, K.S., and Watson, M., *Nucl. Phys.* **A626**, 353c-356c (1997).

Predicting Long-Lived, Neutron-Induced Activation of Concrete in a Cyclotron Vault

L. R. Carroll

Carroll & Ramsey Associates

Abstract

Many elements in concrete can become activated by neutrons in a cyclotron vault, but only a few of the activation products are long-lived. The most prominent of these are Eu-152, Eu-154, Co-60, and Cs-134 which build up over time from (n,γ) reactions in trace amounts of stable Europium, Cobalt, and Cesium that are normally present in concrete in concentrations of a few parts per million, or less, by weight.

A retrospective analysis of data taken in connection with a previous decommissioning of a cyclotron vault, coupled with independent published data, gives us an estimate of the concentrations of these elements in concrete. With that estimate as a benchmark, we then employ a Monte Carlo Radiation Transport Code to estimate the long-term activation profile in concrete for arbitrary irradiation conditions.

Introduction

Now that Positron Emission Tomography (PET) has been approved as a diagnostic imaging modality by the U.S. Food and Drug Administration, and now that reimbursement for clinical PET procedures by government health agencies and third-party insurers is generally available, a large number of cyclotrons are being installed and commissioned world-wide for commercial-scale production of short-lived positron-emitting isotopes, particularly ^{18}F.

Many of these cyclotrons are compact, low-energy (11-12 MeV proton energy) systems which come equipped with their own built-in radiation shielding. Others, however, are higher-energy (16-18 MeV proton-energy) accelerators which are intended for installation in a concrete vault for radiation protection.

Long-term activation of the concrete in the walls of the accelerator vault may become a liability when -- after many years of operation -- the time finally comes to decommission the facility.

There are many elements in normal concrete that become activated when irradiated by neutrons from a cyclotron target. Fortunately, only a few of the resulting radio-isotopes are long-lived. These are identified in Table 1 [1]. These radioactive species build up over time from neutron capture (n,γ) reactions in trace amounts of stable Europium, Cobalt, and Cesium which are present in concrete in concentrations of a few parts per million or less by weight.

Isotope	Reaction	Half-life	Principal γ's MeV (%)
^{152}Eu	^{151}Eu(n,γ)^{152}Eu	13.4 y	0.122 (37%), 0.344 (27%) 0.779 (14%), 0.96 (15%) 1.087 (12%), 1.11 (14%) 1.408 (22%)
^{154}Eu	^{153}Eu(n,γ)^{154}Eu	8.5 y	0.12 (38%), 0.72 (21%) 1.00 (31%), 1.278 (37%)
^{60}Co	^{59}Co(n,γ)^{60}Co	5.27 y	1.17 (100%), 1.33 (100%)
^{134}Cs	^{133}Cs(n,γ)^{134}Cs	2.065 y	0.57 (23%), 0.605 (98%) 0.796 (99%)

Table 1:
Long-lived activation Products in Concrete

To calculate the build-up of activation in a cyclotron vault, we first need a benchmark -- an estimate of the concentrations of these trace elements. We then incorporate that information in the formula for the concrete used in a neutron transport program. Using that formula, we can then compute the activation-reaction profiles in concrete for arbitrary irradiation conditions.

Initial Benchmark

In 1981 a Model CP-42 H$^-$ ion cyclotron, designed and built by the Cyclotron Corporation of Berkeley, CA, underwent high-current endurance testing in the factory per customer contract specification. Beam conditions during the test irradiation were: 42 MeV protons on a Cu target; 200µA for ~500 hours; target located ~18 inches from surface of the vault wall. The test resulted in long-lasting thermal-neutron-induced activation of concrete in the region of the vault wall closest to the target.

This volume of concrete was assayed in 1992 and finally removed and shipped to a rad-waste site for disposal in 1996. Data from the 1992 assay, reported in a 1997 article [4] is shown in Figure 1.

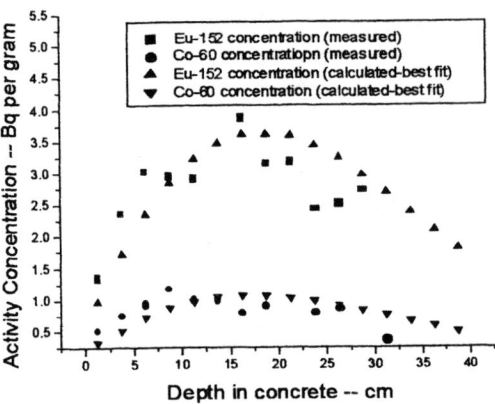

Figure 1: ^{152}Eu and ^{60}Co concentrations as a function of depth in concrete 11 years post-irradiation.

The fluence and spectrum of forward-directed neutrons resulting from 42 MeV proton bombardment of copper are derived from experimental data by others. The next step is to insert the (now known) fluence and spectrum of neutrons from the original target irradiation into our radiation transport code **TART98** [3] which, in turn, provides a tally of neutron-induced activation reactions.

It is then a simple matter to vary the concentrations of trace elements in the concrete formula used in our computer model, and repeat the calculations until the average (within the first 30 cm of concrete depth) in the computed result is forced to match the average (after correcting for isotope decay) which was actually measured in 1992. See Fig. 1.

The measured data are made to match the TART98 simulation results when concentrations of 0.294 parts per million by weight for natural (stable) Europium and 2.54 parts per million by weight for natural Cobalt are chosen. We entered these parameters (concentrations are expressed as "atomic fractions" per TART98 input data format) into the concrete recipe in the simulation program (See table 2), and repeated the TART98 calculations for three different neutron spectra, as shown in Fig. 2 below[1]:

Element	Symbol	Atomic Fraction
Hydrogen	1001	0.1047
Oxygen	8016	0.584
Magnesium	12000	0.0157
Aluminum	13027	0.0317
Silicon	14000	0.2115
Calcium	20000	0.0479
Iron	26000	0.01
Europium	63000	3.7×10^{-8}
Cobalt	27059	0.826×10^{-6}

Table 2: Material definition for "normal concrete" used in TART98 simulations.

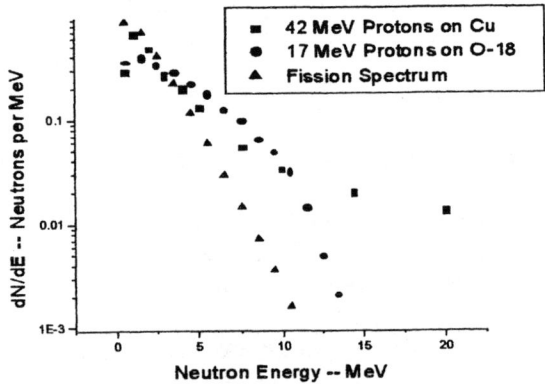

Figure 2: TART98 Neutron Spectra

The three target conditions are:
1) neutrons from 42 MeV protons on copper,
2) neutrons from 17 MeV protons on $H_2^{18}O$,
3) a fission neutron spectrum.

The spectrum for 42 MeV protons on Copper, obtained from [2], is truncated at 20 MeV since TART98, like other commonly-used neutron transport codes, currently accepts neutron energies up to 20 MeV. This introduces only a slight error, however, since the great majority of source neutrons are less than 10 MeV, and only a very small percentage of the total exceeds 20 MeV.

---

[1]For comparison, the concentration by weight of these elements, averaged over the Earth's crust, are 2.1 ppm for Europium and 29 ppm for Cobalt. See, e.g., ref [6].

The "thick target" neutron spectrum for 17 MeV protons on $H_2^{18}O$ is constructed from a weighted sum of "thin target" spectra which, in turn, are computed using the computer code **ALICE-91** [5].

All spectra shown here are normalized so that the area under their respective curves = 1.0. Note that there is relatively little difference in the shape of the neutron spectrum for 42 MeV protons on Copper, versus 17 MeV protons on ^{18}O-water. This is reflected, in turn, by the "per-neutron" activation profiles shown in Fig 3, in which the result for Ep = 17 MeV on $H_2^{18}O$ is almost indistinguishable from that for Ep = 42 MeV on copper.

The activation-reaction profile from fission-neutrons, which have a substantially lower mean energy, does peak more toward the surface of the concrete, since lower-energy neutrons do not penetrate as far before thermalizing and finally being absorbed.

The shape of the spectrum governs the depth-profile of activation but -- more importantly -- it is *the number of neutrons* in concrete which determines the overall average level of activation. The number of possible reaction channels increases very rapidly with increasing energy. For example, in the reaction $^{18}O(p,n)^{18}F$ at Ep = 17 MeV ALICE-91 predicts almost four times as many neutrons produced per second as ^{18}F atoms per second. Experimentally-measured neutron production data are not always available; when in doubt, one is well-advised to use a code such as ALICE-91 to obtain as accurate an estimate as possible of total neutron yield.

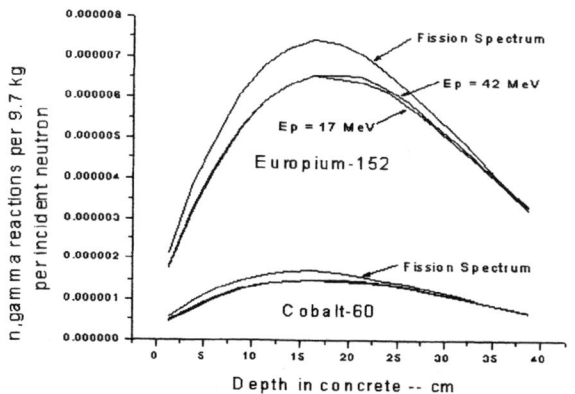

Figure 3. TART98 Activation Profile per incident neutron

Comparison with Other Data

The data in Table 3 below, excerpted from [1], shows the results of an assay of concrete in an accelerator vault in 1985 as reported by Phillips, et al, at the University of Colorado. The data were recorded ~one year post-shutdown, after operating a cyclotron there for 21 years. When back-corrected for isotope decay, our levels of ^{152}Eu and ^{60}Co activation are almost 2 orders of magnitude higher than the corresponding data in Table 3, due to a much higher average neutron flux in the samples tested. Further, after back-correcting to 1-year post-irradiation to better match the Colorado conditions, our ratio of concentration of ^{152}Eu / ^{60}Co is ~1.51.

Our data in Fig. 1 were obtained from relatively small (~10 gm) samples of concrete dust recovered at various depths from small-diameter (3/4") bore-holes. In our assay, (taken 11 years post-irradiation) activity concentrations of ^{134}Cs and ^{154}Eu were too low for reliable quantitation. The data in Table 3, in contrast, were taken only 1 year post-shut-down using much larger (10 cm x 20 cm x 40 cm) concrete slabs.

Isotope	Activity
^{60}Co	110 Bq kg^{-1}
^{134}Cs	37 Bq kg^{-1}
^{152}Eu	89 Bq kg^{-1}
^{154}Eu	11 Bq kg^{-1}
^{40}K	300 Bq kg^{-1}

Table 3: Long-term activation products[2] in concrete in the walls of another cyclotron vault.

It shouldn't be too surprising that trace elements which are present in concentrations of fractional parts -- to a few parts -- per million in concrete should vary by 50% to 100% or more from sample to sample.

Notwithstanding these potential sources of error, we shall endeavor to estimate the build-up of activation in concrete which is in close proximity to a PET cyclotron target during many years of commercial service. Back-projecting our data from Fig. 1 for ^{152}Eu and ^{60}Co, and scaling in proportion to the data from the University of Colorado (Table 3) for ^{154}Eu and ^{134}Cs, we present Figure 4.

Assumptions underlying the plots in Fig. 4 are:
1) 80,000 µA-hours per year proton bombardment of ^{18}O water target at Ep = 17 MeV.
2) 18" - 30" distance from target to the surface of a concrete wall or floor.

---

[2] Naturally-occurring (and ubiquitous!) ^{40}K ($t_{1/2}$ = 1.3 x 10^9 y) is not an activation product; it is included here for comparison only.

3) No additional shielding between target and concrete.
4) The concrete in our sample is reasonably (within the limitations discussed earlier) representative of all concrete.

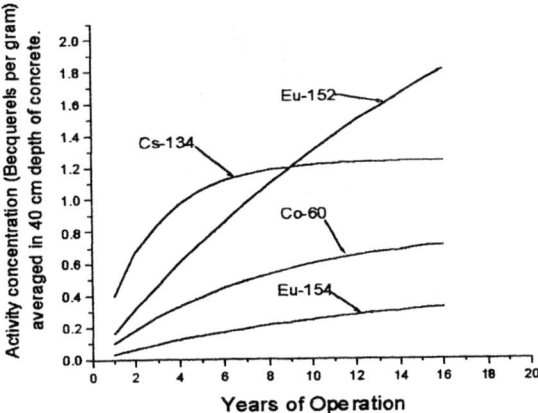

Figure 4. Build-up of long-term, neutron-induced activity in concrete close to the target.

If the same bombardment conditions are maintained for many years (several half-lives of the longest-lived product) the average activity concentrations in the concrete nearest the target eventually level off at the "saturation activity concentrations", which are shown in table 4:

Isotope	Average Conc.	Total in 6 ft. dia. x 1.5 ft. deep plug.
^{152}Eu	3.22 Bq gm^{-1}	~ 0.24 mCi
^{154}Eu	0.447	~ 0.032 mCi
^{60}Co	0.814	~ 0.06 mCi
^{134}Cs	1.24	~ 0.0925 mCi

Table 4: Saturation Activities close to the target

The build-up of radioactivity averaged over the interior of the vault as a whole -- away from the target -- is obviously much lower than that in the hottest region near the target. Moreover, as shown in Table 3, concrete already contains radioactive (but naturally-occurring) ^{40}K ($t_{1/2}$ = 1.3 x 10^9 years) at a concentration of ~0.3 Becquerel / gram. Thus, areas of concrete which are far from the target should maintain concentrations of induced activity at or below that of natural radioactivity.

Conclusions and recommendations

Cyclotron-based production of radioisotopes on a commercial scale can result in significant activation of concrete. Unless preventive measures are taken to reduce the neutron flux which is incident on the wall or floor nearest the target, there will be a gradual build-up of radioactivity which may eventually require remediation by removal of concrete and disposal as rad-waste up to a depth (typically) of ~18 inches and approximately 2 - 3 ft. radius around the target(s). Beyond that radius, the activity concentration should be low enough to qualify for disposal as rubble in an ordinary landfill -- if that is necessary to reduce the ambient dose-rate in the room to a normal background level.

To reduce this build-up of radioactivity in the concrete, the area immediately surrounding the target should -- at the outset -- be stacked with neutron-absorbing material such as borated polyethylene (2 - 5% boron by weight) or equivalent.

For example, for a medium-energy (Ep = 16 to 18 MeV) PET cyclotron, one foot of borated polyethylene will reduce the number of neutrons reaching the concrete wall by a factor of ~30, so that the "hot spot" activity concentration will be maintained at or below the level of naturally-occurring ^{40}K, and the induced activity away from the hottest spot will be 1-2 orders of magnitude lower.

REFERENCES

1. A. B. Philips, et al: "Residual Radioactivity in a Cyclotron and its Surroundings", *Health Physics* Vol 51, No. 3 (September) 337-342, 1986.
2. Thomas Marshall Amos, Jr., *Neutron Yields from Proton Bombardment of Thick Targets*, Ph.D. Thesis in Nuclear Physics, Michigan State University, 1972. University Microfilms International, Ann Arbor MI #73-12,658.
3. *TART98 – A Coupled Neutron-Photon Monte Carlo Transport Code* by D.E. Cullen, Lawrence Livermore National Laboratory. Available from RSICC, Oak Ridge National Laboratories.
4. J. E. Cehn and L. R. Carroll; "Fast Neutron Activation and Ultimate Decommissioning of a Cyclotron Vault: 1981-1996" in *Health Physics of Radiation-Generating Machines; Proceedings of the 30th Midyear Topical Meeting of the Health Physics Society*, San Jose, CA, January, 1997.
5. M. Blann, *PSR-146 / ALICE-91 Code Package*, Available from RSICC, Oak Ridge National Laboratories.
6. N.N. Greenwood and A. Earnshaw; *Chemistry of the Elements*, Pergamon Press, 1990. ISBN 0-08-022057-6.

Sub-Coulomb Alpha Transfer Reactions

C. R. Brune

University of North Carolina, Chapel Hill, 27599-3255, U.S.A. and Triangle Universities Nuclear Laboratory, Durham, NC, 27708-0308, U.S.A.

Abstract. The α transfer reactions $X(^6\text{Li},d)$ and $X(^7\text{Li},t)$ can be used to probe the degree of α clustering in levels of the $(X+\alpha)$ compound nucleus. There are many cases in nuclear astrophysics where this information is crucial for predicting the the low-energy $X+\alpha$ cross section which in turn determines the reaction rate. A major advantage of using sub-Coulomb energies is that the results are largely independent of the assumed nuclear potential parameters. We have recently applied this technique to the $^{12}\text{C}(\alpha,\gamma)^{16}\text{O}$ reaction, which is crucial for the understanding of He burning in massive stars. The $^{12}\text{C}(^6\text{Li},d)^{16}\text{O}$ and $^{12}\text{C}(^7\text{Li},t)^{16}\text{O}$ reactions populating the bound 2^+ and 1^- states of ^{16}O have been measured at energies below the Coulomb barrier. The cross sections can be determined by detecting the charged particles produced, or in some cases via the de-excitation γ rays from the populated excited states.

INTRODUCTION

The rates of many α-induced reactions of relavance to astrophysics are highly dependent upon the properties of nuclear states near the α-separation threshold. Levels below the threshold act as sub-threshold resonances which enhance the low-energy cross section, while levels above the threshold appear as narrow Breit-Wigner resonances. In many cases the spectroscopic information needed to compute the level's effect upon the reaction rate is known, with the exception of the partial width in the α particle channel (reduced α width). Astrophysically-important reactions in this category include $^{12}\text{C}(\alpha,\gamma)^{16}\text{O}$, $^{13}\text{C}(\alpha,n)^{16}\text{O}$, and $^{22}\text{Ne}(\alpha,n)^{25}\text{Mg}$. The major difficulty with a direct determination of the α widths is the very small cross sections at the low energies where the resonances impact the cross section.

It has long been known that reduced α widths can in principle be determined from α transfer reactions. However, the results reported previously using these methods have been subject to rather large uncertainties, attributed mainly to uncertainties in the potential parameters and the influence of compound-nuclear contributions to the reaction cross section. The Li-induced α transfer reactions at very low energies offer several attractive features, which have not been previously investigated. For levels near the α separation threshold, the slightly negative Q-values mean that the outgoing deuterons or tritons will also have energies below the Coulomb barrier. As discussed in Ref. (1), distorted-wave Born approximation (DWBA) calculations under these conditions are determined mainly by Coulomb potentials, and have very little dependence on nuclear potential parameters. The calculated cross sections are thus essentially model-independent, except for the absolute normalization (which contains the reduced α width information).

The first reaction we have studied via this technique is $^{12}\text{C}(\alpha,\gamma)^{16}\text{O}$. This helium burning reaction is a very important process in massive stars, as the fusion rate greatly affects the resulting ratio of ^{12}C to ^{16}O, the subsequent nucleosynthesis of heavier elements, and the final fate of the star (i.e., whether it becomes a black hole or neutron star) (2, 3). The cross section for this reaction at the energies required for stellar helium burning ($E_{c.m.} \approx 0.3$ MeV) is far too small for direct measurement using presently available techniques.

The extrapolation of the measured cross sections ($E \geq 1$ MeV) to lower energies is complicated by the presence of two states located 45 keV ($J^\pi = 1^-$) and 245 keV ($J^\pi = 2^+$) below the $^{12}\text{C}+\alpha$ threshold. The cross section at astrophysical energies arises largely from the high-energy tails of these states, but their properties are only weakly constrained by cross-section measurements at higher energies. The γ-ray widths of these levels are known, but there is considerable uncertainty in their reduced α widths. Very little information exists for the reduced α width of the 2^+ state. Reviews of the present status of the $^{12}\text{C}(\alpha,\gamma)^{16}\text{O}$ problem can be found in Refs. (4, 5).

This paper will focus on the experimental challenges of sub-Coulomb α transfer measurements, and review the results obtained so far. Our first measurements of $^{12}\text{C}(^6\text{Li},d)^{16}\text{O}$ and $^{12}\text{C}(^7\text{Li},t)^{16}\text{O}$ have recently been reported in Ref. (6).

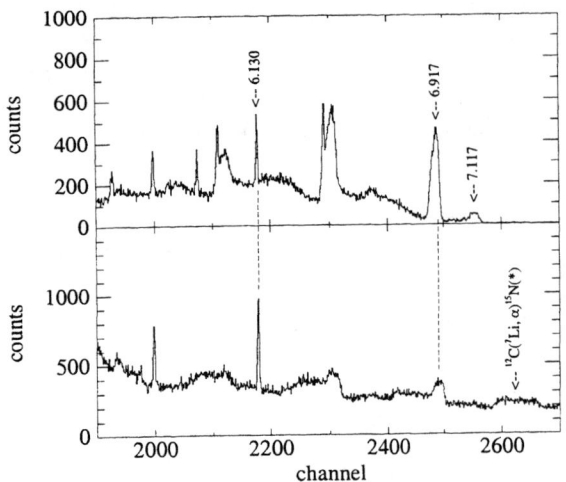

FIGURE 1. γ-ray spectra obtained at 31°, for ^{12}C + ^{6}Li at $E_{c.m.}$ = 2.4 MeV (upper panel), and ^{12}C + ^{7}Li at $E_{c.m.}$ = 3.1 MeV (lower panel).

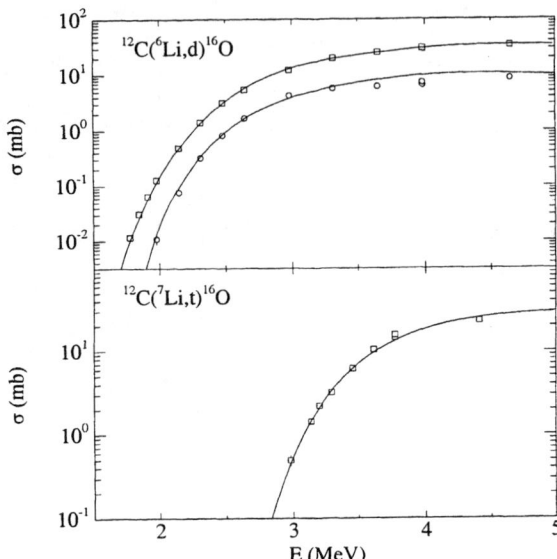

FIGURE 2. Total cross sections versus center-of-mass energy measured using ^{6}Li (upper panel) and ^{7}Li (lower panel) beams for the 6.92-MeV 2^+ state of ^{16}O (squares) and the 7.12-MeV 1^- state (circles, ^{6}Li beam only). The solid curves are DWBA calculations normalized to the data.

EXPERIMENTAL APPROACHES

The traditional approach to studying transfer reactions has been to utilize the heavier nucleus as the target and the lighter nucleus as the projectile (normal kinematics). The light reaction products are then detected, for example with a magnetic spectrograph. This method presents some difficulties when applied to sub-Coulomb α transfer reactions. For normal kinematics, the differential cross section for sub-Coulomb transfer reactions is peaked at backward angles. This angular region is the most interesting to measure, but the energies of the light particles are extremely low, e.g. < 500 keV for ^{12}C(^{6}Li,d)^{16}O and ^{12}C(^{7}Li,t)^{16}O reactions leading to the ^{16}O bound states near E_x = 7 MeV. These deuteron or triton energies are too low to detect the reaction products in the presence of the intense background of elastically-scattered particles.

γ-ray detection methods

One approach which bypasses the difficulty of directly detecting the reaction products is to detect γ rays from the residual nuclei. This technique is applicable if the states of interest have a significant probability for γ decay. Since the 6.92- and 7.12-MeV states of ^{16}O are bound and only decay by γ-ray emission, this method is well suited for measuring ^{12}C(^{6}Li,d)^{16}O and ^{12}C(^{7}Li,t)^{16}O. Our first measurements were conducted by bombarding ^{12}C targets with ^{6,7}Li beams covering $2.7 \leq E[^{6}\text{Li}] \leq 7.0$ MeV and $4.75 \leq E[^{7}\text{Li}] \leq 7.0$ MeV, supplied by the Caltech 3-MV Pelletron accelerator. De-excitation γ-rays were detected using high-purity Ge detectors placed at angles of 31° and 110°. Sample γ-ray spectra are shown in Fig. 1. The ^{12}C(^{7}Li,α)^{15}N reaction populates several bound states between 7 and 10 MeV excitation energy in ^{15}N, producing considerable background in the region of interest. These states are not populated with the ^{6}Li beam, giving rise to significantly cleaner spectra. The 7.12-MeV γ ray from the ^{12}C(^{7}Li,t)^{16}O reaction is not positively identified.

It is very efficient to determine the total cross section for populating the 6.92- and 7.12-MeV ^{16}O by applying Gaussian quadrature to the measured γ-ray yields. The Gaussian quadrature using the two detection angles of 31° and 110° [zeros of $P_4(\cos\theta)$] is exact for any decaying level with spin $J \leq \frac{7}{2}$. This result relies upon two facts: that γ-ray angular distributions must be symmetric about $\theta = 90°$ (ignoring very small center-of-mass motion effects), and that the maximum order of the Legendre polynomial present in the angular distribution is given by twice the spin of the decaying state. Finally note the branching ratios of 6.92- and 7.12-MeV states to the ground state of ^{16}O are essentially 100%. The total cross sections thus obtained for the bound 2^+ and 1^- states are shown in Fig. 2.

In these initial measurements we were unable to detect the γ decay of the 7.12-MeV 1^- state for the ^{12}C(^{7}Li,t)^{16}O reaction, because this transition lies in a high-background region of the γ-ray spectrum. Recently we made additional measurements of the ^{12}C + ^{7}Li re-

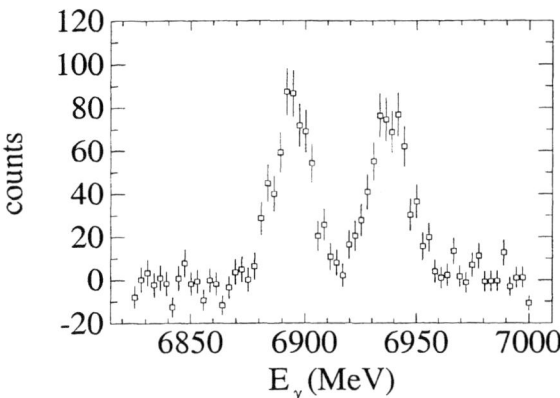

FIGURE 3. Background-subtracted lineshape for the 6.917 MeV → 0 transition observed at $\theta_\gamma = 90°$ for $^{12}C(^6Li,d)^{16}O$ at $E_{c.m.} = 2.3$ MeV.

action at Notre Dame University. The setup was similar to the previous measurements, but utilized Compton-suppressed Ge detectors detectors, which made it possible to resolve the decay of the 7.12-MeV state. This measurement, when compared to measurements of $(^6Li,d)$ to the same state, will provide a consistency test of the the reduced α widths determined from sub-Coulomb α reactions

γ-particle correlations

The analysis of measured γ-ray lineshapes offer another interesting possibility. The correlation between the emitted charged particle and the γ decay of the ^{16}O nucleus can provide a very conclusive signature of direct α transfer. For example, consider the $^{12}C(^6Li,d)^{16}O$ reaction. Since the recoil direction of the ^{16}O nucleus is directly related to the direction of the emitted deuteron, certain γ-d angular correlations can be inferred from the Doppler-shift distribution observed in the γ-ray spectrum. Note also that the lifetimes of the 6.92- and 7.12-MeV states are sufficiently short (∼ 10 fs) that slowing down effects are negligible, while the lifetime of the 6.13-MeV state is so long (27 ps) that it nearly always decays at rest. In addition the effects of detector resolution and finite detector size are small compared to the Doppler spread. The correlations are particularly straightforward to analyze for $\theta_\gamma = 90°$ and $0°$. The spectrum with considerable structure is shown in Fig. 3.

Charged-particle methods

We noted previously that charged-particle detection for sub-Coulomb transfer reactions in normal kinematics is very difficult. However, since many reactions of interest cannot be investigated by the γ-ray detection approach, other charged-particle detection methods have been considered. Inverse kinematics has been found to have several advantages. The light reaction products are now emitted at forward angles, with significantly greater energies. The higher energy in particular allows several possibilities for improved detection. We have carried out a test measurement of the $^6Li(^{12}C,d)^{16}O$ reaction at Triangle Universities Nuclear Laboratory. The target was ^6LiF backed by a Ni foil. The foil was chosen to be of sufficient thickness to stop the incident ^{12}C beam and yet allow the deuterons to pass through with minimal energy loss. The deuterons were detected with an $E - \Delta E$ Si detector telescope at forward angles. The test indicates that the deuterons can be detected with sufficient resolution to separate deuteron groups from the 6.92- and 7.12-MeV states.

ANALYSIS

The reduced α width information is extracted from the magnitude of the experimental cross sections measured at sub-Coulomb energies. The sub-Coulomb α transfers on ^{12}C have been analyzed using the finite-range DWBA code FRESCO (7). It is found that the calculated cross sections at sub-Coulomb energies are sensitive only to the asymptotic part of the α particle bound-state wavefunctions, while other factors relating to the bound states such as the geometry of the binding potentials, or the numbers of radial nodes, are unimportant. The calculations describe the energy dependence of the data very well (see Fig. 2). The normalization of the theoretical cross sections to the experimental data then gives the product of two normalization constants, one from the Li ion which contributed the α particle and one from the ^{16}O final state. The normalization constants for the Li ions appear to be reasonably well determined (6). The R-matrix reduced widths γ_i^2 can be calculated from the normalization constant C_i via

$$C_i^2 = \frac{2\mu a}{\hbar^2 W^2(a)} \left(\frac{\gamma_i^2}{1 + \gamma_i^2 \frac{dS}{dE}} \right), \quad (1)$$

where μ is the reduced mass, a is the channel radius, and $S(E)$ is the shift function (8). Note that for α-unbound states the analysis is somewhat modified, as it is the α partial width Γ_α which is determined rather than the normalization constant.

Once the reduced α widths have been determined, the impact on astrophysical reaction of interest must be studied. Ideally the approach adopted would take into account all of the available data, e.g. direct measurements, elastic scattering, known excitation energies and partial widths, β-delayed particle spectra, etc... We have analyzed the $^{12}C(\alpha,\gamma)^{16}$ reaction using R-matrix theory (9), the reduced α widths determined from our sub-Coulomb α transfer studies, and other experimental input data. The details of this procedure and our results for the low-energy $E1$ and $E2$ multipolarities are given in Ref. (6). The results for the $E2$ multipolarity are much more accurate than attained by any previously available method.

CONCLUSIONS

Sub-Coulomb α transfer reactions are a useful tool for indirectly studying several α-induced processes in astrophysics. This technique provides a means for obtaining needed reduced α width information which cannot be easily obtained by other methods. Sub-Coulomb transfer reactions have the advantage over transfer reactions at higher energies that the model dependence due to uncertainties in optical and binding potentials is minimized. We have given an overview of some experimental possibilities available for sub-Coulomb α transfer reactions. Several useful experimental techniques are identified which we plan to pursue vigorously in the future.

We have measured the $^{12}C(^6Li,d)^{16}O$ and $^{12}C(^7Li,t)^{16}O$ reactions in order to determine the reduced α widths of the bound 2^+ and 1^- states of ^{16}O. These reduced α widths of these levels are crucial for predicting the low-energy $^{12}C(\alpha,\gamma)^{16}$ cross section. The resulting improved understanding of the $^{12}C + \alpha$ fusion rate in turn reduces uncertainties in supernova nucleosynthesis calculations, and allows for better study of convective mixing processes.

ACKNOWLEDGMENTS

The experimental work discussed in this paper has been carried out in collaboration with physicists from the University of North Carolina, Chapel Hill and Triangle Universities Nuclear Laboratory (A. Danner, B.M. Fisher, W.H. Geist, H.J. Karwowski, D.S. Leonard, X. Lu, and K.D. Veal); Caltech (R.W. Kavanagh); and the University of Notre Dame (J. Daly, R. Detwiler, J. Görres, P. Tischhauser, and M. Wiescher). It is a pleasure to thank all of my collaborators. This research was supported in part by the U.S. Department of Energy, Grant No. DE-FG02-97ER41041, and the National Science Foundation.

REFERENCES

1. R. Bass, *Nuclear Reactions with Heavy Ions* (Springer-Verlag, Berlin, 1980), pp. 180-188.
2. T. A. Weaver and S. E. Woosley, Phys. Rep. **227**, 65 (1993).
3. M. Hashimoto, Prog. Theor. Phys. **94**, 663 (1996).
4. L. Buchmann *et al.*, Phys. Rev. C **54**, 393 (1996).
5. G. Wallerstein *et al.*, Rev. Mod. Phys. **69**, 1022 (1997).
6. C. R. Brune, W. H. Geist, R. W. Kavanagh, and K. D. Veal, Phys. Rev. Lett. **83**, 4025 (1999).
7. I. J. Thompson, Comp. Phys. Comm. **7**, 167 (1988); private communication.
8. A. M. Mukhamedzhanov and R. E. Tribble, Phys. Rev. C **59**, 3418 (1999).
9. F. C. Barker and T. Kajino, Aust. J. Phys. **44**, 369 (1991).

Nuclear physics with fast neutrons at LANSCE/WNR: GEANIE and FIGARO

M. Devlin,[1] N. Fotiades,[1] G. D. Johns,[1] R. O. Nelson,[1] R. C. Haight,[1] L. Zanini,[1]
J. A. Becker,[2] L. A. Bernstein,[2] P. E. Garrett,[2] C. A. McGrath,[2] D. P. McNabb,[2]
W. Younes,[2] J. X. Saladin,[3] A. Aprahamian[4]

[1] *Los Alamos National Laboratory, Los Alamos, NM 87545*
[2] *Lawrence Livermore National Laboratory, Livermore, CA 94551*
[3] *University of Pittsburgh, Pittsburgh, PA 15260*
[4] *University of Notre Dame, Notre Dame, IN 46556*

Abstract. GEANIE is an array of 26 HpGe detectors used to study nuclear reaction dynamics and structure following reactions with high-energy ($1 < E_n < 200$ MeV) neutrons, for both basic and applied research projects. Studies have included the measurement of *(n,xn)* partial cross sections as a function of E_n for a variety of nuclei, particularly actinides. More recently, studies of *n*-induced fission-fragment distributions and nuclear structure in the actinide region have been started. A second beam line and experimental station (FIGARO) have been set up to complement and extend this program. Research conducted on this second beam line includes the use of conversion electron spectroscopy to explore nuclear structure, using the University of Pittsburgh ICEBall II array.

INTRODUCTION

Spectroscopic studies of reactions between nuclei and fast neutrons can provide information on both nuclear structure and nuclear reactions. Of particular applied interest are the measurements of both partial and total fast neutron-induced reaction cross sections with a variety of nuclei, especially actinides, for the DoE stockpile stewardship program as well as other applications.

There is also considerable basic physics interest in such reactions. Since *(n,xnypzα)* reactions preferentially populate relatively low spin states, detailed studies of these reactions can provide information on both the reaction mechanism and the nuclear structure involved. Such studies are useful for improving detailed reaction models, including the description of nuclear level densities. In this paper we will present the current status and recent results of the program at LANSCE/WNR to measure the properties of neutron-induced reactions by spectroscopic means.

LANSCE/WNR

The Weapons Neutron Research (WNR) facility, part of the Los Alamos Neutron Science CEnter (LANSCE), provides a "white" spectrum of neutrons from spallation of 800 MeV protons on a thick tungsten target. Six beam lines are available for a variety of defense, industrial and basic science measurements, with the experimental target positions typically 20-30 m from the spallation target. The proton beam structure consists of numerous micropulses 1.8 μs apart, contained in macropulses typically 625 μs wide, at a rate of up to 120 Hz. The overall duty cycle is 6-8%. The pulsed nature of the beam allows the time-of-flight determination of the incident neutron energy on an event-by-event basis. Figure 1 shows the neutron flux spectrum at WNR for the GEANIE flight path.

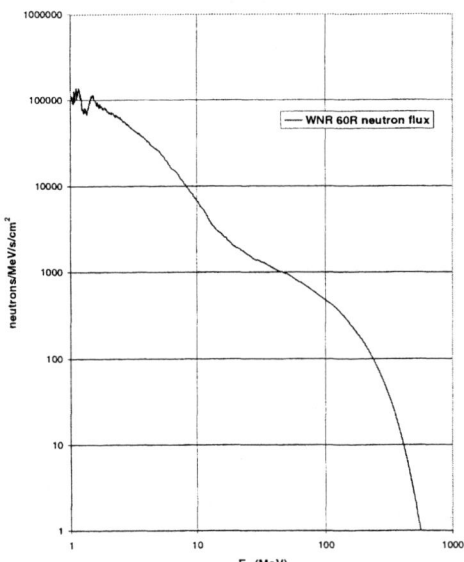

FIGURE 1. The neutron flux spectrum at WNR for the GEANIE flight path. Other flight paths have similar neutron flux spectra.

The GEANIE array consists of 26 high-purity germanium detectors for γ-ray spectroscopy; 11 of these are LEPS detectors for low-energy γ-rays (up to 1 MeV), and twenty of these, including all the LEPS, are Compton suppressed. It is based on the HERA array from LBNL, with the addition of the LEPS detectors.

A second beam line is currently being reconfigured to expand this program and relieve the beam time pressure on GEANIE. It can accommodate both Ge and neutron detectors, and has recently been used to explore the use of conversion electron detection at WNR. FIGARO is described in another submission to this conference[1], though a description of the electron detection experiment will be presented below.

^{238}U(n,xn) RESULTS FROM GEANIE

The GEANIE array has typically been used in "singles" mode to accurately measure yields of individual γ-rays, and hence to determine partial γ-ray cross sections. The partial cross sections of individual γ-rays are determined as a function of incident neutron energy by measuring the γ-ray yield, corrected for computer deadtime, attenuation in the target, internal conversion, and the absolute efficiency of the array. The incident neutron flux is monitored using ^{235}U and ^{238}U fission foils[2].

The extraction of γ-ray yields from "dense" spectra, in which literally thousands of discrete lines can be identified, can of course be problematic. An elaborate and versatile peak-fitting routine has been developed for this purpose[3], and the resulting fits to the data give reliable yields for all of the cases studied. Nonetheless, doublets and triplets, etc., need to be considered on a case by case basis; the excitation functions provided by the incident neutron spectrum at WNR allow the separation of doublets from different reaction channels.

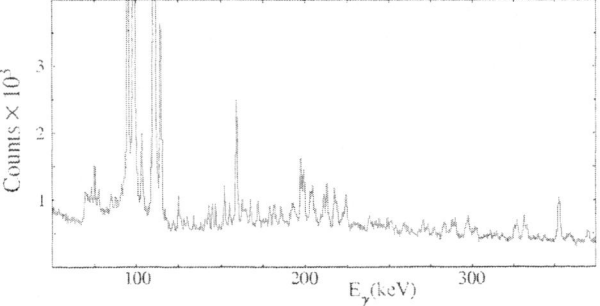

FIGURE 2. Total γ-ray spectrum for $n + ^{238}$U, for $1 < E_n < 100$ MeV. Note the density of γ-ray lines.

Figure 2 shows a sample γ-ray spectrum, and figure 3 shows some partial cross section results from reactions with ^{238}U as a function of incident neutron energy. The data in the top panel are ground-band transitions in ^{238}U, populated by (n,n') reactions. The second, third and fourth panels are lighter U isotopes populated in (n,xn) reactions, and the bottom panel contains a sample transition in ^{100}Zr, a fission fragment. The accompanying lines are predictions from the pre-equilibrium plus Hauser-Feshbach reaction code GNASH,[4,5] which includes a description of the γ-ray decay of the residual nuclei. The detailed comparison of GNASH with data of this type has improved the nuclear structure input to GNASH, resulting in better fits to individual γ-ray transitions.

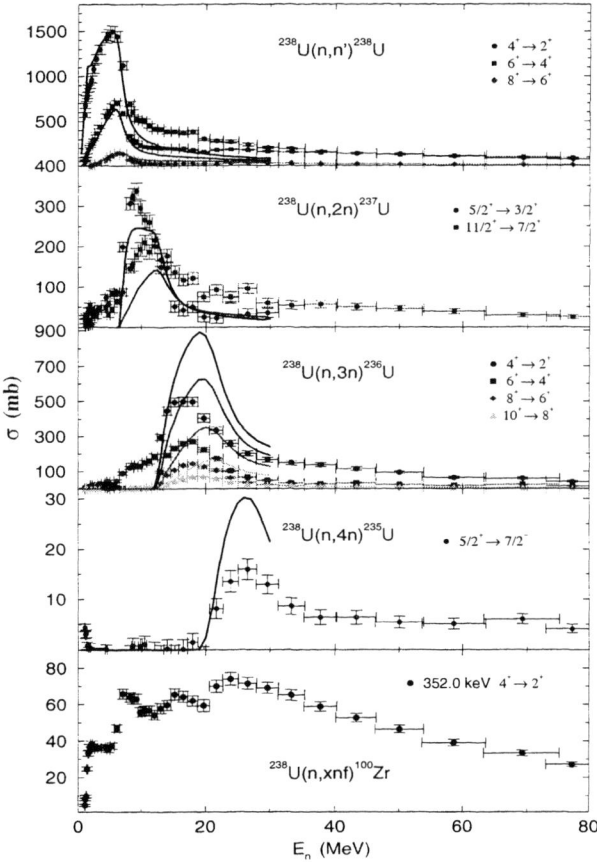

FIGURE 3. Partial cross sections measured at GEANIE. Shown are various transitions from the *(n,n')* [top], *(n,xn)* [middle three], and one fission fragment line from *(n,xnf)* [bottom]. Solid lines are GNASH calculations[5] for the same transitions. In addition to γ-ray background, contaminant lines are subtracted from the yields, where known.

^{239}Pu$(n,2n)^{238}$Pu CROSS SECTION

The ^{239}Pu$(n,2n)^{238}$Pu cross section has been one focus of our research. Previous measurements have given inconsistent results below 14 MeV, resulting in a large uncertainty in the evaluated cross section. The combination of WNR and GEANIE using the technique of measuring partial γ-ray cross sections has been applied to this problem.

Figure 4 shows a preliminary result for the ^{239}Pu$(n,2n\gamma)$ partial cross section, for the 6^+ to 4^+ transition in ^{238}Pu (top, left), and for the sum of five independent transitions (bottom, left). The resulting total cross sections, using GNASH calculations of the proportion which feeds through the measured transitions, are shown on the right. These results indicate that using as many independent transitions as possible provides a better estimate of the total cross section. The ^{239}Pu$(n,2n)^{238}$Pu cross section has now been accurately measured as a function of incident neutron energy from threshold to over 50 MeV.

OTHER PROJECTS AND FUTURE DIRECTIONS

In addition to the examples presented above, GEANIE has been used for a variety of other research projects. These include studies of states populated in spallation neutron-induced reactions with ^{92}Mo[ref. 6], ^{235}U(n,xn) partial cross sections[7], the distribution of fission fragments for *n*-induced fission on ^{235}U[ref. 7], and ^{238}U[ref. 8] and $(n,2n)$ partial cross section measurements on a variety of isotopes of interest for applications such as stockpile stewardship and Accelerator Transmutation of Waste (ATW). In addition, future experiments will explore actinide spectroscopy via (n,xn) reactions and *n*-induced fission fragment distributions.

CONVERSION ELECTRON SPECTROSCOPY AT WNR

Internal conversion represents an increasing contribution to the decay of nuclear states as atomic number increases. Conversion electron spectroscopy is therefore an appealing technique to study actinide structure. In typical, charged-particle induced reactions, the profuse production of atomic electrons and x-rays supplies a large background for conversion electron detectors. For neutron-induced reactions, however, this background should be considerably reduced. Moreover, *E*0 transitions can only be observed via electron detection, and such observations have provided significant information on β bands[9] and 0$^-$ bands[10] in U isotopes.

In order to test the feasibility of conversion electron detection at WNR, we set up the University of Pittsburgh ICEBall II detector array[11] on the FIGARO beam line to look for *E*0 and other converted transitions following neutron reactions with ^{238}U. ICEBall II consists of six SiLi detectors with mini-orange focusing magnets; it has a peak efficiency of 14% at 400 keV. Despite the reduced background from the target, we encountered a significant background for scattered neutrons.

Improved shielding was successful at reducing this background, and further improvements in shielding are planned. Analysis of the data is in progress, though to date only continuous, not discrete, electrons from the ^{238}U target have been identified.

ACKNOWLEDGMENTS

This research was supported by the U.S. DoE contracts W-7405-ENG-36 (LANL) and W-7405-ENG-48 (LLNL). Additional support was provided by the US NSF (Notre Dame) and the University of Pittsburgh.

REFERENCES

1. Zanini, L., et al., these proceedings.
2. Wender, S.A., et al., *Nucl. Instr. and Meth.* **A336**, 226 (1993).
3. Younes, W., to be published.
4. Young, P.G., Arthur, E.D., and Chadwick, M.B., LANL Report LA-12343-MS (1992).
5. Chadwick, M.B., personal communication.
6. Garrett, P.E., et al., *Phys. Rev.* **C62**, (2000) 014307; Garrett, P.E., et al., *Phys. Rev.* **C,** in press.
7. Younes, W., et al., to be published.
8. Fotiades, N., et al., to be published; Ethvignot, T., et al., to be published.
9. Janssens, R. V. F., et al., Phys. Lett **90B** (1980) 209; W. Z. Venema, et. al., Phys. Lett. **156B** (1985) 163.
10. Zeyen, P., et al., Z. Phys. **A328** (1987) 399.
11. Saladin, J.X., et al., to be published; see also Metlay, M.P., Saladin, J.X., Lee, I.-Y., and Dietzsch, O., *Nucl. Instr. and Meth.* **A336**, 162 (1993).

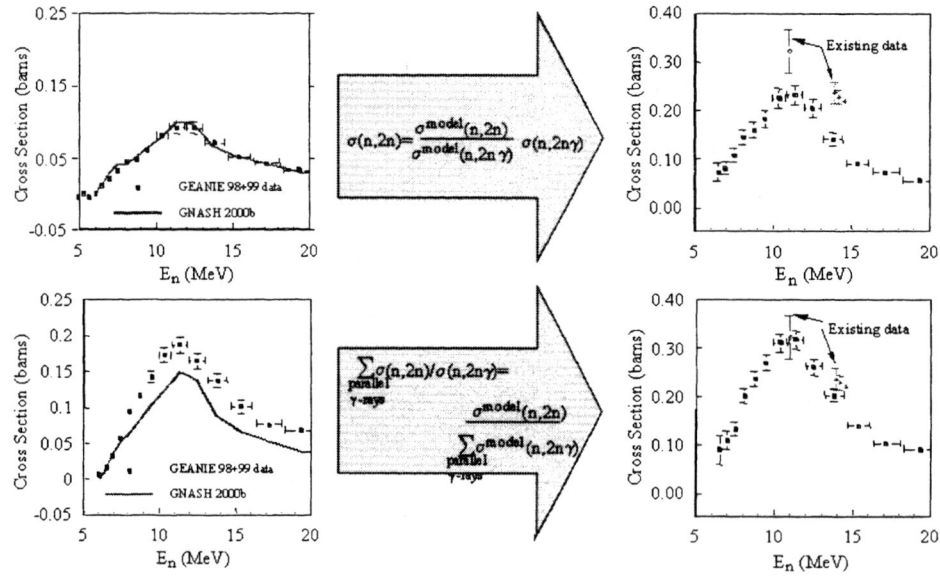

FIGURE 4. Preliminary partial ^{239}Pu(n,2n) cross sections extracted for the 6$^+$ to 4$^+$ transition in ^{238}Pu (top left) and a sum of five independent transitions in ^{238}Pu (bottom left). Extracted total cross sections for the transitions (right top and bottom) using GNASH calculations.

Thick Target Yields For The Reaction $^{18}O(p,n)^{18}F$ above 16 MeV

Carlos Gonzalez Lepera[a] and S. Roberto Strangis[b]

[a]*Cyclotron Facility, University of Texas Health Science Center, 6431 Fannin, Houston, Texas 77030 USA*
[b]*Cyclotron Facility, Centro Atómico Ezeiza, Comisión Nacional de Energía Atómica, Buenos Aires, Argentina*

Abstract. Two high-pressure targets were built to investigate the potential benefits of higher than typical (11-16 MeV) beam energy for production of ^{18}F using 95% enriched ^{18}O-Water. Saturation yields for this reaction between 19 and 25 MeV are presented and compared with available data at lower energies.

INTRODUCTION

The recent approval of ^{18}F-labeled glucose [FDG] by the USA Food and Drug Administration as a safe and effective radiopharmaceutical for human use has considerably increased the demand for this product. Regional distribution centers with multi-Curie production capacity can supply extended areas covering a high percentage of the population, therefore reducing the need for dedicated in-house cyclotrons. Previously available thick target yields for the reaction $^{18}O(p,n)^{18}F$ were obtained using $^{18}O_2$ from threshold up to 14.7 MeV [1]. In practice, most centers produce ^{18}F irradiating enriched water. A high-pressure target was built to investigate the potential benefits of higher than typical (11-16 MeV) beam energy for production of ^{18}F using 95% enriched ^{18}O-Water under routine production conditions.

TARGET DESIGN

Several modifications were introduced to the target design previously described in reference [2]. Figure 1 shows a schematic of the target assembly and Table 1 presents characteristics for both targets. The target body was machined from high purity titanium plate approximately 9.5 mm thick and fitted with a 32 μm thick front titanium foil. A 0.5 mm aluminum plate compressed the titanium foil against the target giving the necessary strength to support high-pressure operation while degrading the beam energy to the desired value. Two targets were fabricated and installed on two separate but otherwise similar beam lines at the Scanditronix MC40 cyclotron of the University of Texas Houston. Bulging under pressure of the front foil increases the actual depth by approximately 0.5 mm on each target. No gaskets were used to achieve a leak tight seal between target body and front foil. With gas pressure applied to the target, bolts were adjusted in a rotating cross pattern until a leak tight seal was obtained as verified with a digital pressure gauge. Targets were subsequently tested with He gas to pressures up to 68 bars. Deionized chilled water supplied at a temperature of 11°C and a flow rate of 3.5 liters/min was circulated on the back of the targets. Chilled He gas circulating on the front foil provided appropriate cooling. Target loading and transfer of the product was accomplished using two low dead-volume check valves (V3 and V4 - Fig.1) as described in [2]. This method has been in use by one of the authors for several years and ensures reliable loading and recovery of target volume with an uncertainty of ± 3%.

TABLE 1. Targets Characteristics.

Irradiation Energy	19.4 MeV	24.2 MeV
Volume	1.8 cm³	2.5 cm³
Water Fill Volume	1.4 cm³	1.9 cm³
Depth	4.6 mm	7 mm
Beam Collimation	14 mm	14 mm

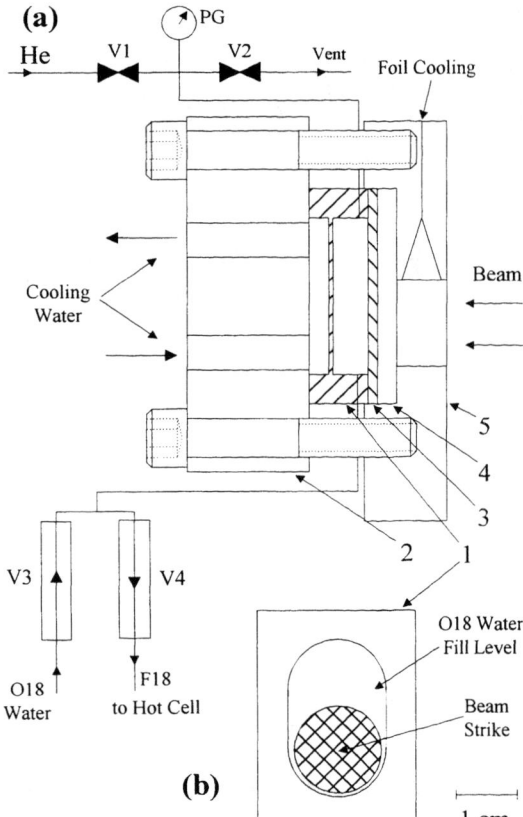

FIGURE 1. (a): Schematic of high pressure target showing (1) titanium target; (2) water cooling flange; (3) titanium target foil; (4) energy degrader; (5) He cooling flange. (b): Ti target front view shown to scale.

TABLE 2. Saturation Yield as a Function of Beam Energy.

Beam Energy on Target [MeV]	Saturation Yield [mCi/µA]	Number of Runs
19.4 ± 0.3	275 ± 13	20
24.2 ± 0.3	291 ± 13	10

The targets were irradiated with beam currents between 20 to 35 µA at 19.4 MeV and 10 to 20 µA at 24.2 MeV. More than 70% of the runs at 19.4 MeV were conducted at approximately 20 µA for 1 hr obtaining on average 62.9 GBq (1.7 Ci) of ^{18}F at EOB. Subsequent conversion into FDG showed average decay corrected yields of 62%. No saturation yield dependence on beam current was observed for any of these targets. The results of Table 2 show a significant increase on the saturation yield as previously reported [2] from 200 mCi/µA at 18 MeV to 275 mCi/µA at 19.4 MeV. However, increasing the energy from 19.4 to 24.2 MeV does not result in a significant increase in saturation yield. Considerable amounts of ^{17}F were observed at the energies listed on Table 2, particularly at 24.2 MeV.

CONCLUSIONS

Considering beam power dissipated on the target as the limiting factor to the maximum amount of ^{18}F produced at any given energy per unit of time and using the data from Table 2 and reference [2] we observe that the ratio of saturation yield to beam energy presents a maximum around 20 MeV for the reaction and conditions described in this work. We conclude that maximum ^{18}F production per unit of bombardment time under these conditions takes place at incident proton energies near 20 MeV.

ACKNOWLEDGMENTS

We wish to thank the personnel of the Cyclotron Facility in Houston for their support and help to conduct this work.

REFERENCES

1. Ruth, T. J., and Wolf, A. P., *Radiochimica Acta* **26**, 21-24 (1979).

2. Gonzalez Lepera, C. E., and Dembowski, B., *Appl. Radiat. Isot.* **48**, 613-617 (1997).

RESULTS

Previous to each irradiation, targets were pressurized with He gas between 35.4 and 36.7 bars (520-540 psi) and subsequently loaded with ^{18}O water. Valve V3 is preset by the manufacturer to open at 250 psi and V4 opens at 750 psi. Both check valves are rated to withstand backpressures of 1000 psi. The incoming beam energy from the cyclotron was fixed at 27.2 MeV and using aluminum plates degraded to 19.4 MeV on one target and 24.2 MeV on the other. Since our main goal was to establish experimental results under actual production conditions, we decided not to rinse the target after each irradiation to include residual ^{18}F left inside the target. However, to estimate the magnitude of this effect, the target was rinsed only during a few runs obtaining after two rinses, no more than 3% of the amount of ^{18}F originally recovered. This correction was not included on our experimental values.

Utilization of the (n,n′γ) Reaction for Nuclear Structure Studies of ^{122,124,126}Te

S. F. Hicks[1], J. R. Vanhoy[2], N. Warr[3], T. B. Brown[3], and S. W. Yates[3]

[1]Department of Physics, University of Dallas, Irving, TX 75062 USA
[2]Department of Physics, United States Naval Academy, Annapolis, MD 21402 USA
[3]Department of Physics and Astronomy, University of Kentucky, Lexington, KY 40502 USA

Abstract. The collective properties of levels to 3.3 MeV excitation in ^{122,124,126}Te have been investigated using the (n,n′γ) reaction. Measurements were made at the neutron scattering facilities at the University of Kentucky. Gamma-ray angular distributions, Doppler shifts, excitation functions, and γ-γ coincidences were measured. Level spins, parities, lifetimes, branching ratios, and multipole-mixing ratios have been deduced. The experimental methods used in this study and examples of nuclear structure questions in ^{122,124,126}Te that have been investigated are discussed.

INTRODUCTION

The use of γ-ray detection following inelastic neutron scattering (INS) offers many unique advantages for probing low-lying nuclear structure. Levels can be examined very near threshold as there is no Coulomb barrier; the (n,n′γ) reaction is in general non-selective exciting all levels with spins J≤6; and the energy resolution available with γ-ray detection is excellent. Additionally, level lifetimes in the fs to ps range can be determined using the Doppler-shift attenuation method following INS.

The stable ^{122,124,126}Te nuclei have been investigated using neutron scattering techniques to examine the role of collective and few-particle excitations in low-lying levels. Of particular interest in these nuclei are multiphonon excitations, states of mixed neutron-proton symmetry, and 4p-2h excitations across the Z=50 major proton shell. The advantage of using the (n,n′γ) reaction mechanism in this study is that in addition to determining the level scheme to approximately 3.3 MeV excitation, the ability to deduce electromagnetic decay probabilities offers a very stringent test for model comparisons. Very diverse nuclear models can sometimes reproduce the level energies and spins, but electromagnetic transition rates are a much more sensitive test of the calculated wave function. The techniques used in this investigation as well as some results are presented.

Experimental Techniques

Measurements were made using the neutron scattering and γ-ray detection facilities at the University of Kentucky 7 MV Van de Graaff Laboratory. The $^3H(p,n)^3$He reaction was used as a source of nearly monoenergetic neutrons. Isotopically enriched samples, with enrichments of ≥ 94%, were used in the study of all three nuclei. For singles measurements, a Compton-suppressed n-type HpGe detector was used with a relative efficiency of 51% and an energy resolution of about 2.1 keV at 1.33 MeV. A BGO annulus detector surrounding the main detector was used for Compton suppression. For coincidence measurements, four such detectors were used without Compton suppression. All detectors were monitored for gain stability using radioactive sources ^{56}Co and ^{152}Eu. The singles mode was used for γ-ray angular distributions, excitation functions, and Dopper shifts. The coincidence mode was used for the (n,n′γγ) measurements. Each of these will be briefly discussed below.

Gamma-Gamma Coincidence Measurements

Gamma-gamma coincidence measurements are useful for making unambiguous placements of γ rays in the level scheme and for resolving doublets. The experimental facilities and techniques for analyzing coincidence spectra are detailed in Ref. [1]. An example coincidence spectrum that has only new assignments labeled with γ-ray energies is shown in Fig. 1.

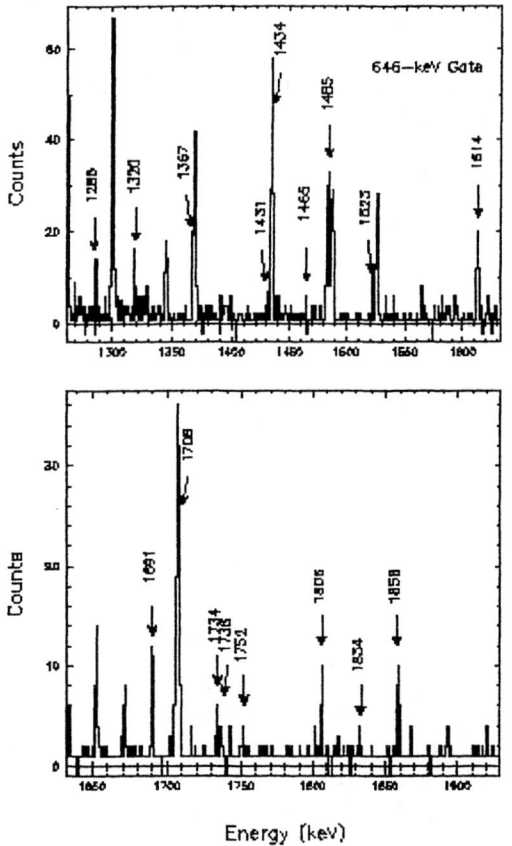

FIGURE 1. Coincidence spectrum from a gate placed on the 646-keV transition from the 1249-keV level in ^{124}Te. Only new γ-ray placements are noted.

Angular Distributions and Excitation Functions

During the INS process the reaction takes place predominantly through the compound nuclear mechanism, which results in the excited nucleus being aligned. The shape of the angular distribution of the emitted γ rays reflects this alignment and can be used to determine the multipolarity of the radiation, including electric quadrupole and magnetic dipole mixing, and the spin of the initial level. Compound nucleus model calculations are compared to the measured angular distributions in making these assessments. Gamma-ray branching ratios are also determined from these measurements by considering the angle-integrated production cross sections for each transition from a given level.

Gamma-ray excitation functions are used to establish new levels, place γ-ray transitions in the level scheme and to evaluate level spins and branching ratios determined from the angular distributions. The threshold of the excitation functions is used to assign transitions to levels, and the shape as a function of incident neutron energy is compared to theoretical statistical model calculations of the relative production cross sections to assess the spin and branching ratios. Some examples of these comparisons are shown in Fig. 2 for three different level spins.

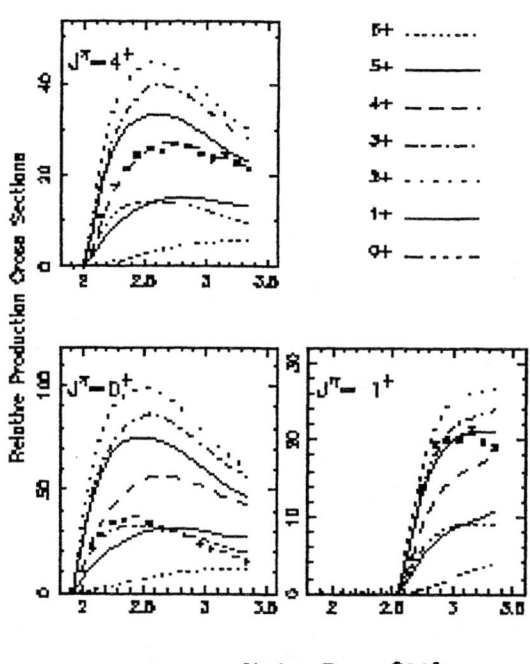

FIGURE 2. Calculated relative production cross sections are compared to measured γ-ray excitation functions for levels of $J^\pi = 0^+$, 1^+ and 4^+. The excellent agreement between theory and experiment indicates that both the branching ratios and level spins are well determined for the transitions. The levels shown have energies of 1940, 2592, and 2040 keV, respectively, for the three different spins.

Doppler Shifts

Level lifetimes were extracted using the Doppler-shift attenuation method following INS. The shifted γ-ray energy $E_\gamma(\theta)$, as a function of scattering angle θ, is given by the following expression,

$$E_\gamma(\theta) = E_o[1 + F_{exp}(\tau)\frac{v_{cm}}{c}\cos(\theta)], \quad (1)$$

where E_o is the un-shifted γ-ray energy, $F_{exp}(\tau)$ is the experimental Doppler-shift attenuation factor, and v_{cm} is the velocity of the center of mass in the collision between the neutron and nucleus. Lifetimes are determined by comparing experimental and theoretical $F(\tau)$ values using a procedure described in Ref. [2]. Some examples of Doppler shifts measured in ^{122,124}Te are shown in Fig. 3.

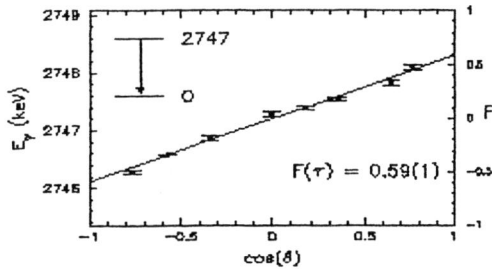

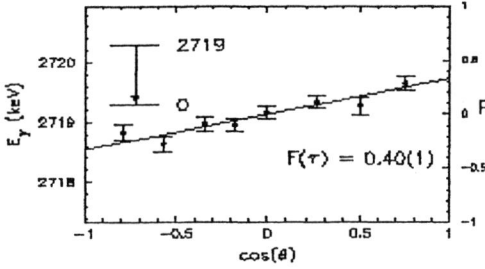

FIGURE 3. Doppler shifts for ground-state transitions from spin 1 states in ^{124}Te (2727 keV) and ^{122}Te (2719 keV). The resulting lifetimes are 50(1) fs and 86(5) fs, respectively.

Nuclear Structure Interests in ^{122,124,126}Te

The level systematics for all stable even-even Te nuclei is shown in Fig. 4. The low-mass Te nuclei are thought to be excellent examples of vibrational nuclei in which the excited levels result from the coherent collective vibrations of neutrons and protons about a spherical equilibrium shape, i.e., normal collective excitations. This is evinced in the figure by the nearly parabolic behavior of the 2_1^+, 2_2^+, and 4_1^+ levels and by the fact that the energy of the 2_2^+ and 4_1^+ states are almost exactly twice the energy of the 2_2^+ level, as expected for multi-quadrupole phonon states in a spherical vibrator. For ^{120,122}Te, even the 0_2^+ state appears almost degenerate with the 2_2^+ and 4_1^+ levels as it should in a harmonic picture. Away from these two nuclei, however, the 0_2^+-level energy acts rather anomalously, and the flatness of the 6_1^+-state energy for masses A>120 is indicative of a few-particle excitation rather than collectivity [3]. In the nearby Cd and Sn nuclei, such anomalous behavior of the 0^+ state is attributed to quasiparticle 2p-4h and 2p-2h intruder excitations across the Z=50 closed shell, respectively, that result in an additional low-lying 0^+ excitation near the two-quadrupole phonon energy. (See Ref. [4] and references therein.) In the Te nuclei, the identification of such states has remained elusive.

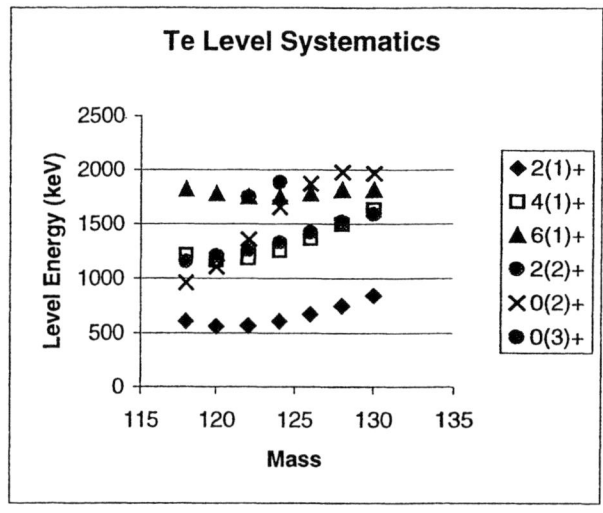

FIGURE 4. Energies of excited levels in the eight stable Te nuclei. The 2_1^+, 2_2^+, and 4_1^+ levels behave much like expected from a classic vibrational picture. The 0_2^+ level, which would be a 2-phonon state in a vibrational model, exhibits an anomalous behavior. Level energies were obtained from the National Nuclear Data Center for nuclei other than ^{122,124,126}Te.

Another type of low-lying excitation of interest in the Te nuclei is collective excitations in which the neutrons and protons vibrate incoherently. The lowest such state is expected to be a 2^+ level at about 2 MeV excitation in these nuclei. Such states of mixed neutron-proton symmetry have very distinct decay characteristics. One predicted characteristic is a strong magnetic dipole transition to the 2_1^+ state.

Considerable progress has been made in identifying and characterizing the low-lying collective states in ^{122,124,126}Te via these (n,n'γ) measurements. Lifetimes were found for numerous adopted and new levels in these three nuclei from which absolute electromagnetic transition probabilities have been determined and used for model comparisons.

Results

States of Mixed Neutron-Proton Symmetry

The determined absolute transition rates for low-lying 2^+ levels have been used to look for states exhibiting mixed proton-neutron symmetry. The magnetic dipole and electric quadrupole rates found for the first five 2^+ states in ^{122,124}Te are shown in Fig. 5. None of these levels exhibit the large $B(M1)$ rates predicted for states of mixed neutron-proton symmetry.

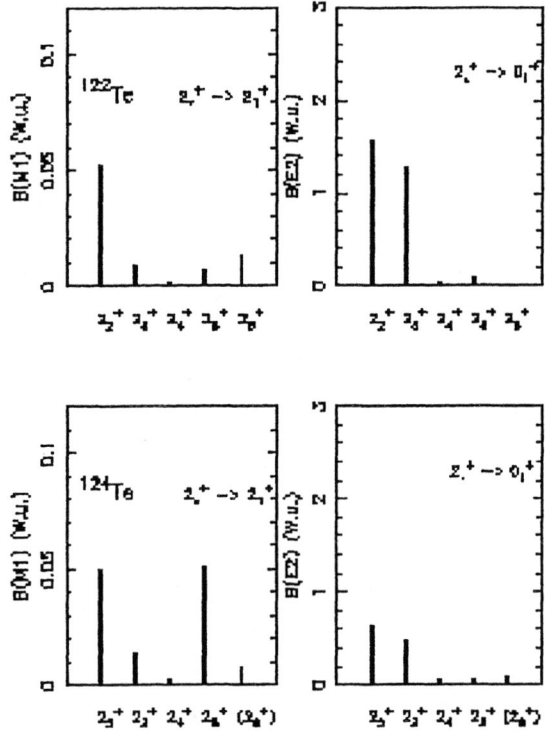

FIGURE 5. Electric quadrupole $B(E2)$ and magnetic dipole $B(M1)$ transition probabilities for low-lying 2^+ levels in ^{122,124}Te.

Normal Collective Excitations and Intruder States

Comparisons between experimental absolute transition probabilities measured in this work and model calculations indicate that the low-lying states in these nuclei consist of highly mixed configurations. Only when the investigation is extended to high spin states not observed in these (n,n'γ) measurements is there any convincing evidence of isolated intruder excitations.

Knowledge of normal collective excitations, including multi-quadrupole phonon excitations and quadrupole-octupole coupled (QOC) phonon states has been extended. New candidates for these collective states have been found based on the observed decays into low-lying levels. For example, the 2592-keV level in ^{122}Te has been proposed as the 1^- member of the QOC quintet in this nucleus in Ref. [5]. Both the ground-state decay and the decay of this level into the 2_1^+ are observed in this work to be fast electric dipole transitions typically seen in QOC 1^- state decays. Other candidates for this quintet have been found and will be reported more extensively elsewhere.

In conclusion, γ-ray detection following inelastic neutron scattering has proven to be quite useful in the investigation of collective excitations in ^{122,124,126}Te. The ability to measure lifetimes within the fs to ps range for all spins J≤6 makes this reaction mechanism quite unique for investigating nuclear structure questions.

ACKNOWLEDGMENTS

This work was supported in part by the National Science Foundation.

REFERENCES

1. McGrath, C. A., Villani, M. F., Garrett, P. E., and Yates, S. W., Nucl. Instrum. Methods A **421**, 458 (1999).

2. Belgya, T., Molnár, G., and Yates, S. W., Nucl. Phys. A**607**, 43 (1996).

3. Lee, C. S., Cizewski, J. A., Barker, D., Tanczyn, R., Kumbartzki, G., Szczepanski, J., Gan, J. W., Dorsett, H., Henry, R. G., and Farris, L. P., Nucl. Phys. A **528**, 381 (1991).

4. Rikovska, J., Stone, N. J., Walker, P. M., Walters, W. B., Nucl. Phys. A **505**, 145 (1989).

5. Schwengner, R., et al., Nucl. Phys. A **620**, 277 (1997).

Lifetime measurements using the recoil distance method - Achievements and Perspectives.

R. Krücken[*]

A. W. Wright Nuclear Structure Laboratory, Yale University, New Haven CT 06520

Abstract. The recoil distance method (RDM) for measuring pico-second nuclear level lifetimes and its use in nuclear structure studies is reviewed and perspectives for the future are presented. High precision measurements in the mass-130 region, studies of multi-phonon states in rare earth nuclei, the investigation of shape coexistence and the recently discovered phenomenon of 'magnetic rotation' are reviewed. Prospects for lifetime measurements in exotic regions of nuclei such as the measurement of lifetimes in neutron rich nuclei populated via spontaneous and heavy-ion induced fission are discussed. Other prospects include the use of the RDM technique in conjunction with recoil separators. The relevance of these techniques for experiments with radioactive ion beams will be discussed.

INTRODUCTION

The knowledge of matrix elements for transitions between excited nuclear levels adds important insight into the structure of nuclei. While relative transition matrix elements can be determined by measuring gamma-ray branching ratios, absolute transition matrix elements are only accessible by measuring the lifetimes or the Coulomb excitation cross sections of the nuclear levels. For example, in deformed nuclei, the reduced transition probability B(E2) for the $2_1^+ \rightarrow 0_1^+$ ground state transition provides a measure of the charge quadrupole moment and thus the ground state deformation of an even-even nucleus. Other examples for the importance of transition matrix elements include the possibility to identify collective excitation modes on the basis of the magnitude and spin dependence of B(E2) values, or their sensitivity to the mixing of different configurations, in particular in the case of shape coexistence.

The precision achievable in lifetime experiments has dramatically increased with the availability of large, highly efficient γ-ray multi-detector arrays. These instruments have made exotic, weakly populated nuclear excitations, such as superdeformed (SD) bands accessible for lifetime measurements. In particular for such exotic nuclear excitations lifetime measurements have proven to be essential for a sound understanding of the excitation mechanism.

This contribution will review some recent achievements that have been made in measuring nuclear level lifetimes in the picosecond range, which is a critical lifetime range for low-lying collective excitations, as well as some levels in more exotic nuclear excitations. The most common method used is the so-called recoil distance method (RDM), also called the recoil distance Doppler-shift (RDDS) technique. Both terms are used interchangeably in the course of this article.

THE RECOIL DISTANCE METHOD

The RDM is the standard method to measure lifetimes of excited nuclear states in the picosecond range. The method uses a target chamber, called a plunger device, that contains a stretched target- and stopper-foil which are mounted parallel to each other at a variable distance. Excited states in the nuclei of interest are populated in the target foil. The nuclei then recoil with a velocity of a few percent the speed of light in the direction of the stopper foil where they are stopped. The distances are chosen such that the flight time is on the order of the effective lifetime of the levels of interest and vary typically from about 10 μm to several mm. The lifetime of a level of interest is determined from the changing intensities of fully Doppler-shifted and stopped (or slowed) γ-ray components detected by the surrounding γ-ray detectors when varying the target-to-stopper distance.

In recent years new plunger devices were developed for the use with large multi-detector gamma-ray arrays. The New Yale Plunger Device (N.Y.P.D.)(1) is one such device, which follows the design of the most recent

---

[*] In part supported by US DOE under Grant No. DE-FG02-91ER-40609

Cologne plunger. The N.Y.P.D. fits into the center of arrays such as Gammasphere and the SPEEDY array (see Fig. 1) and can take advantage of the high efficiency of these arrays.

Additionally there has been a significant improvement of the reliability of lifetimes from RDM experiments due to the analysis of coincidence data with the so-called Differential Decay-Curve Method (DDCM) (2), which extracts a value for the level lifetime for each target-to-stopper distance directly from observables. The DDCM with coincidence gates from above the level of interest eliminates various systematic uncertainties such as feeding and side-feeding times as well as the deorientation effect. At the same time the method provides a consistency check of the analysis.

RECENT RESULTS

The resolving power of modern γ-ray multi-detector arrays has enabled the application of the RDM to excited levels that are populated only with a few percent of the total fusion cross section. At the same time lifetimes for states along the yrast line can be measured with unprecedented relative accuracy of a few percent.

Decay out of superdeformed bands

The sudden disappearance of the intensity at the bottom of SD bands has been a puzzling and intensely investigated problem. Only recently, major breakthroughs have been accomplished with the first observations of discrete linking transitions between the SD bands in ^{194}Hg (3, 4) and ^{194}Pb (5, 6) and the respective near yrast normal deformed (ND) levels. These observations have for the first time enabled the determination of the excitation energies, spins and parity of superdeformed states in the mass-190 region. It was also possible to measure lifetimes at the bottom of some SD bands in the mass-190 region using the RDDS technique (7, 8, 9).

The mechanism leading to the decay out of the SD bands can be understood in a simple mixing picture (9, 10), where the lowest observed SD states mix with their nearest neighboring normal deformed (ND) states with the same spin and parity. The transition quadrupole moments at the bottom of the SD bands did not show a significant reduction compared to the SD quadrupole moments at higher spin. This simple finding was the first experimental indication that the mixing between SD and ND states has to be very weak. In the case of ^{194}Pb many direct transitions between the lowest members of the SD band and the near yrast states were observed (6) and multipolarities of pure E1 and mixed E2/M1 character were determined. B(E1) values of about 8×10^{-6} W.u. and upper limits for B(E2) and B(M1) values of 5×10^{-2} W.u. and 5×10^{-4} W.u., respectively, were determined for the pure ND states at the excitation energy of the SD states, which are consistent with a statistical decay (8).

Lifetimes in mass-130 nuclei

The lifetime of the first excited 2^+ level of an even-even nucleus is a measure for the deformation of the nucleus in its ground state. Thus the ratio of the transition quadrupole moments for transitions between states in the ground band and that of the $2^+ \rightarrow 0^+$ ground state transition is an excellent measure for possible changes in the deformation within the ground band. However, small changes in the deformation can only be detected if the lifetimes are measured with very high precision. Such accurate measurements have recently been performed for the first time using large γ-ray detector arrays.

Table 1 shows the relative transition quadrupole moments within the ground state bands of several Xe, Ba and Nd nuclei (11) measured with GASP and the Cologne

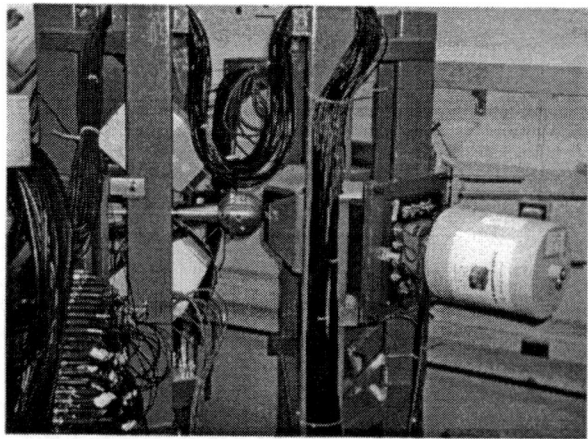

FIGURE 1. The Yale SPEEDY array.

Table 1. Relative transition quadrupole moments $R_I = Q_t(I^+ \rightarrow (I-2)^+)/Q_t(2^+ \rightarrow 0^+)$ for several even-even A≈130 nuclei.

	R_4	R_6	R_8	R_{10}
^{122}Xe	0.96 (2)	0.99 (4)		
^{126}Xe	1.00 (3)	0.93 (6)	0.8 (1)	
^{124}Ba	1.04 (2)	1.00 (3)	0.99 (6)	
^{126}Ba	0.99 (2)	1.11 (3)	1.20 (5)	1.13 (7)
^{128}Ba	1.00 (3)	0.98 (6)	0.9 (1)	0.7 (4)
^{132}Nd	0.99 (2)	1.01 (2)		

plunger at the Laboratory Nazionali di Legnaro, Italy. The quadrupole moments are divided by the quadrupole moment for the $2^+ \rightarrow 0^+$ ground state transition. The results in ^{126}Ba are most impressive since, for example, the lifetime of the 4^+ level was measured to be 8.59 ± 0.18 ps (12) which represents an uncertainty of only 2% including systematic uncertainties. The ratios of the Q_t values are remarkably constant for most levels in the ground state bands below the crossing of the $\pi(h_{11/2})$ or $\nu(h_{11/2})$ intruder bands. An exception is the band in ^{126}Ba that shows clear deviations for the 6^+ and 8^+ levels. This dynamic shape effect is attributed to the underlying shell structure (12).

The example of ^{126}Ba demonstrates that high precision lifetime measurements can help to investigate nuclear structure beyond the standard approach of simple collective models. More such measurements need to be performed in order to carry out critical tests of collective models.

Shears bands in the Pb region

Very regular rotational level sequences connected by strong magnetic dipole (M1) transitions in the neutron deficient Pb isotopes (13) have been observed in recent years. The band head of these M1 bands is generated by the perpendicular coupling of $h_{9/2}$ and $i_{13/2}$ protons to $i_{13/2}$ neutron holes. This leads to a large perpendicular component $\vec{\mu}_\perp$ of the magnetic moment with respect to the total angular momentum vector $\vec{J}$, which in turn leads to enhanced M1 transitions between the rotational levels.

Spin is generated in these M1 bands by the gradual alignment of the particle and hole angular momenta $\vec{j}_\pi$ and $\vec{j}_\nu$ with the total angular momentum vector $\vec{J}$. This is reminiscent to the closing of the blades of a pair of shears, leading to the name 'shears mechanism'. This behavior arises naturally in the framework of the tilted-axis-cranking (TAC) model (14). The closing of the particle and hole spin-vectors with respect to the total angular momentum vector leads to a specific large drop of B(M1) values with increasing spin, since $|\vec{\mu}_\perp|$ is decreasing with decreasing opening angle and B(M1)$\propto |\vec{\mu}_\perp|^2$.

Many of the features of the shears bands have been described by the shears mechanism as well as models that did not involve this new concept. Lifetime experiments were carried out at Gammasphere using the DSAM technique and RDDS technique in $^{193-199}$Pb (15, 16) and ^{198}Pb (17), respectively. The lifetime experiments have established the characteristic decrease of the B(M1) values with increasing spin. These key observations have provided the essential support for the concept of magnetic rotation as the underlying mechanism for the shears

bands. This is a very impressive example for the importance of lifetime measurements.

FUTURE PROSPECTS

The examples given in the previous section have highlighted that important insights can be gained by the study of absolute nuclear transition matrix elements. Since major thrusts of current and future nuclear structure research are aiming at the study of nuclei far from stability, it is useful to take a look at the prospective areas in which the measurement of nuclear level lifetimes can make an impact in accordance with these goals.

Use of channel selection devices

Nuclei near the N=Z line and the proton dripline are currently intensely investigated. However, spectroscopic experiments are only possible by using auxiliary channel selection devices in combination with large gamma-ray detector arrays. Near the N=Z line it is essential to detect light charged particles as well as neutrons in order to pick out individual weak reaction channels from the many open reaction channels. For heavier nuclei near the proton dripline the weak fusion reaction channels have to compete with a large fission background. Here it is essential to detect the fusion products directly by means of recoil spectrometers, such as the Argonne FMA or the Oak Ridge RMS, or using gas-filled recoil separators such as RITU in Jyväskylä, the BGS in Berkeley, or SASSYER at Yale.

First experiments have been performed in neutron deficient Sn isotopes using the Cologne plunger, the GASP spectrometer and its silicon ball ISIS. In a first combination of a plunger device with a recoil detector the N.Y.P.D. has recently been combined with Gammasphere and the FMA at Argonne to measure lifetimes in ^{188}Pb in order to study the phenomenon of shape coexistence in this nucleus. Future experiments of this type will be possible at Yale with the Small Angle Separator System at Yale for Evaporation Residues (SASSYER).

Lifetimes in neutron-rich nuclei

A currently much utilized technique to study neutron-rich nuclei is the spectroscopy of prompt or β-delayed γ-rays of fission fragments. Lifetimes of low lying states have in the past mostly been measured using β-delayed γ-rays and electronic time techniques but in a few cases the RDDS technique has also been employed (18).

Neutron-rich nuclei exhibit a variety of structural phenomena, which include shape coexistence, strong octupole correlations, the existence of low-lying intruder states, signs of triaxiality and γ-softness as well as vibrational excitations. However, very little lifetime information, which would help to classify the structure in these nuclei, is available.

Recently a RDM experiment in this region has been performed using the N.Y.P.D. and SPEEDY in conjunction with a ^{252}Cf source. This technique was already used once by Mamane et al. (18) using only one Germanium detector and a thin ^{252}Cf fission source. Fig. 2 shows RDM spectra for the $6^+ \rightarrow 4^+$ transition from which a lifetime of 29 ± 6 ps was extracted.

Lifetimes in "rare isotopes"

With the construction of second generation radioactive beam facilities such as the U.S. Rare Isotope Accelerator beams of exotic neutron rich isotopes will become available. One way to study these exotic nuclei is by Coulomb exciting them on a stable fixed target and performing spectroscopy using the emitted gamma radiation. From this type of data one can extract information about the level scheme as well as the transition and intrinsic matrix elements from the excitation cross sections. The reliability of the matrix elements is, however, sometimes dependent on the knowledge of all transitions going to and from a certain level as well as assumptions about the signs of the matrix elements involved. This can particularly become a problem when multi-step Coulomb excitation plays a role. Here lifetime measurements with the RDDS technique can provide an alternative and model independent way to determine absolute transition matrix elements.

Acknowledgements

Important contributions by W.-T. Chou, R.M. Clark, J.R. Cooper, A. Dewald and P. von Brentano are gratefully acknowledged.

REFERENCES

1. R. Krücken, J. Res. Natl. Inst. Stan. **105**, 53 (2000).
2. A. Dewald et al., Z. Phys. A **334**, 163 (1989).
3. T.L. Khoo et al., Phys. Rev. Lett **76**, 1583 (1996).
4. G. Hackman et al., Phys. Rev. Lett **79**, 4100 (1997).
5. A. Lopez-Martens et al., Phys. Lett **B 380**, 18 (1996).
6. K. Hauschild et al., Phys. Rev. **C55**, 2819 (1997).
7. R. Kühn et al., Phys. Rev. **C55**, R1002 (1997).
8. R. Krücken et al., Phys. Rev. **C55**, R1625 (1997) and references therein.
9. R. Krücken et al., Phys. Rev. **C54**, 1182 (1996) and references therein.
10. E. Vigezzi et al., Phys. Lett. **B 249**, 163 (1990).
11. P. von Brentano et al., Abstract to the Symposium "New Spectroscopy and Nuclear Structure 1997", Copenhagen 1997.
12. A. Dewald et al., Phys. Rev. **C54**, R2119 (1996).
13. R.M. Clark et al., Nucl. Phys. **A562**, 121 (1993), and references therein.
14. S. Frauendorf, Nucl. Phys. **A 557**, 259c (1993).
15. R.M. Clark et al., Phys. Rev. Lett. **78**, 1868 (1997).
16. R.M. Clark et al., Phys. Lett. B **440**, 251 (1998).
17. R. Krücken et al., Phys. Rev. **C58**, R1876 (1998).
18. G. Mamane et al., Nucl. Phys. **A 454**, 213 (1986).

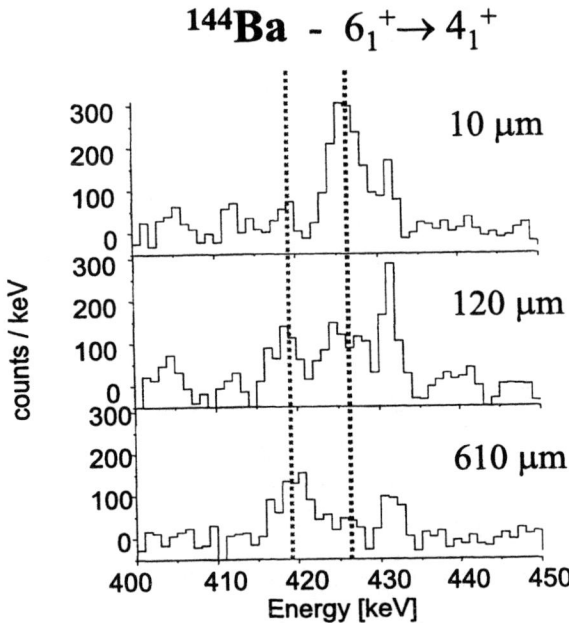

FIGURE 2. RDM spectra for the $6^+ \rightarrow 4^+$ transition in ^{144}Ba from which a lifetime of 29 ± 6 ps was determined.

Cross Section Measurement of ^{11}B(p,α) ^{8}Be and its Application on Boron Characterization

Jiarui Liu, Xinming Lu, Jianyue Jin, Q. M. Li and Wei-Kan Chu

Department of Physics, and Texas Center for Superconductivity, University of Houston, Houston, Texas 77204

Abstract. Existing problems in the published cross section data of the ^{11}B(p,α) ^{8}Be reaction and the consequences in material analysis will be reviewed. The existing cross section data on this reaction shows errors of up to 30% and inconsistency as high as 50%. New experimental data of the cross sections with the standard deviation better than 3% are presented. Application of NRA/Channeling for both the chemical concentration profiles and the carrier profiles of B implanted into silicon will be introduced. We used NRA/Channeling combined with anodic stripping to get the chemical and carrier depth profiles simultaneously. ^{11}B(p, α) ^{8}Be reaction at incident beam energy of 660 keV was used to probe ^{11}B atoms. Channeling technique was applied to measure the lattice location of the ^{11}B atoms. Anodic oxidation stripping was used to get differential depth information. This technique can deliver consistent and reliable chemical depth profiles of ^{11}B as well as the carrier depth profiles. The differential NRA/Channeling technique is essential in studies of ion implantation, TED, activation and deactivation, and defect engineering application in ultra-shallow junction research.

INTRODUCTION

It is well known that the ^{11}B(p, α)^{8}Be reaction has some sharp resonance and shows a large cross section at 660 keV, but unfortunately it is not widely used in B characterization in silicon. This is due to two reasons. First the large resonance at 660 keV has a continuos spectrum without depth information in the energy spectrum. Second, the cross section of this reaction has never been measured with good accuracy for ion beam analysis. In this paper, we present our recent result of the cross section with the experimental error less than 3%. Then we demonstrate the combination of Nuclear Reaction Analysis (NRA) with channeling and the anodic oxidation stripping for the depth profiling of both total B concentration and substitutional boron concentration simultaneously. Among many analytical techniques for depth profiling, SIMS is the principal diagnostic technique for characterization of shallow dopants in semiconductor structures. Since semiconductor devices require low implant concentration, dynamic SIMS is an ideal technique to do the implant depth profiling. Dynamic SIMS shows excellent elemental concentration sensitivity (<ppm), depth resolution of 2-3 nm and reasonable lateral spatial resolution. However, in the near surface region, a steady-state equilibrium of SIMS processes has not yet been reached due to the chemical effects of sputtering ions, O_2^+ or Cs^+. The thickness of this pre-equilibrium region is typically about 20 to 30 nm, depending on sputter conditions. The lack of a steady-state equilibrium in SIMS processes at the top surface region seriously affects both the depth scale and the concentration quantification, so the data within this top layer (20-30 nm) is unreliable. The pre-equilibrium region is becoming the principal obstacle for SIMS application for ultra-shallow implants. Moreover, SIMS provides the chemical concentration only without information on carriers. Different electrical measurement techniques are being used for carrier profiling. [1] In this article we focus on the NRA of Boron dopant. Combination of NRA with anodic oxidation stripping can provide very good depth resolution. Combination of NRA/channeling with anodic oxidation stripping can provide the

substitutional boron profile along with the chemical profile.

CROSS SECTION OF ^{11}B(p, α) ^{8}Be REACTION

Nuclear reaction with charged particles is often used for depth profiling by energy analysis of the reaction products or by the reaction resonance. Different approaches for boron depth profiling in semiconductors have been studied based on this reaction, such as profiling by low energy resonance at 165 keV,[2,3] profiling by energy analysis at 2.62 MeV[4] and by spectrum fitting at broad resonance at 660 keV.[5] All these approaches are based on out-of-date erroneous cross section data or by using B-standard samples.

The reaction ^{11}B(p,α)^{8}Be at low proton energy shows two reaction channels:

$$^{11}B + p \rightarrow \alpha_o + {}^8Be + Q(8.586 \text{ MeV})$$
$$\rightarrow \alpha_1 + {}^8Be^* + Q(5.65 \text{ MeV})$$
$$\rightarrow \alpha_{12} + \alpha_{12}(Q=3.028 \text{ MeV})$$

The reaction product α-particles belong to three groups. The first group is from mono-energetic α_o-particles in the first reaction channel with the ^{8}Be at ground state. The second reaction channel with α_1-particles shows a very sharp resonance at the incident proton energy as low as 163 keV ($\Gamma = 0.2$ keV, $\sigma = 0.2$ mb).[2] The resonance analysis with tilted samples can provide a depth resolution of 10nm. The reaction channel with α_1 is more attractive due to the huge reaction cross section at 660 keV. The differential reaction cross section at this broad resonance is in the order of 0.1 b/Sr, which is larger than the Rutherford cross section (0.066b/Sr). The sensitivity of the total amount of B by this nuclear reaction analysis is about 0.01 ppm, if interference from Nitrogen is negligible.[7] This high sensitivity is very attractive for different applications in material analysis, such as total amount of B (dosage) and uniformity measurement, NRA/Channeling for lattice location determination and B depth profiling combining with stripping.[8,9] Unfortunately, analysis with α_1-particles is complicated by overlapping with the third group of α-particles – the continuous spectrum of α_{12} –particles from B* break-up.

The existing cross section data on this reaction shows a very large error up to 30% and inconsistency as high as 50%. The first measurement of the cross section of the ^{11}B(p, α_1)^{8}Be* reaction was published by Beckman et al. about 50 years ago.[10] The integral cross section of 600 mb was obtained just from the differential cross section for 97° times 4π. The authors claimed the experimental error was 30%, but they also recognized the discrepancy between the proton backscattering yield and the Rutherford cross section was as large as 50%.

Later, in 1972, Ligeon et al. measured the cross section of the ^{11}B(p, α_1)^{8}Be* reaction at 150°, which is useful for nuclear reaction analysis.[11] Due to the complexity of the α-spectrum consisting of α_1 and α_{12}, the authors obtained the cross section under a "convention". Under this convention, all the α-particles must be taken whose energy is more than the spectrum minimum. If the first assumption by Beckman et al. is roughly right, then Legion's cross section under "convention" is roughly twice that of the real nuclear reaction cross section. Then after the 1970's, following the *Ion Beam Handbook for Material Analysis* which quoted this cross section under convention as a real nuclear reaction cross section, the Legion's data became a popular cross section of this reaction.[12] In fact, the "spectrum minimum" in this convention depends on many experimental factors and is difficult to control. Recently, Lin et al. published the accurate integral cross section of this reaction channel at the proton energy of 667 and 1370 keV with the experimental error of 8%.[13] The integral cross section at 667 keV was 894 mb and is about 50% higher than Beckman's measurement [10]. Unfortunately, Lin et al. did not present any differential cross section for this incident beam energy. For repeatable ion beam analysis we measured this cross section under detail spectrum analysis to get an accurate differential cross section. For more convenient practical applications, a cross-section under a repeatable convention was obtained.

The cross section measurements were performed at the 5SDH-2 pelletron tandem accelerator at the Texas Center for Superconductivity, University of Houston. Five improvements in experimental technique are adapted to the cross section measurement of the ^{11}B(p,α)^{8}Be reaction. Here, we just mention the main improvements briefly and then give the details about the spectra analysis and how to define the cross section under "convention". The improvements are as follow:

1. Self-supported thin ^{11}B target versus BO$_2$ on a substarte in the early days.
2. High-resolution silicon barrier detector without front absorber.
3. Normalization by Rutherford cross section on ^{11}B.
4. Accurate spectra analysis:
5. Cross section under convention using the sharp peak of α_0 as a benchmark.

In this measurement, we obtained the ratio of the differential reaction cross section to the Rutherford cross section with the absolute identical set-up first.

The cross section is obtained as: $d\sigma = d\sigma_R N_b / N_R$, where $d\sigma_R$ is the Rutherford cross section with good accuracy, N_b is the count of the reaction yield and N_R is the count of Ruthsrford backscattering. Because the Rutherford cross section has been measured very accurately[14], the reaction cross section can been obtained with good accuracy too.

The final experimental error for this (p,α_1) reaction cross section is estimated to be 3.3%. The cross section of (p,α_1) reaction is shown in Fig. 1.

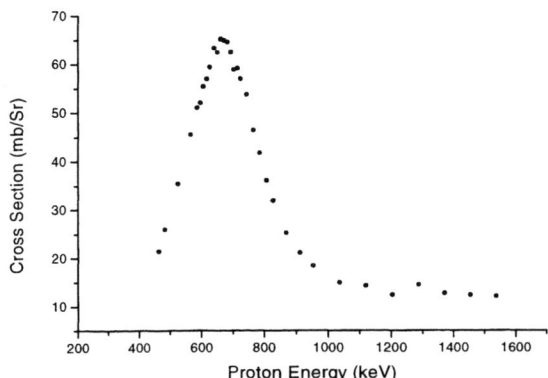

FIGURE 1. Cross section of $^{11}B(p,\alpha_1)^8Be^*$ reaction as a function of incident proton energy for the detector at 150°.

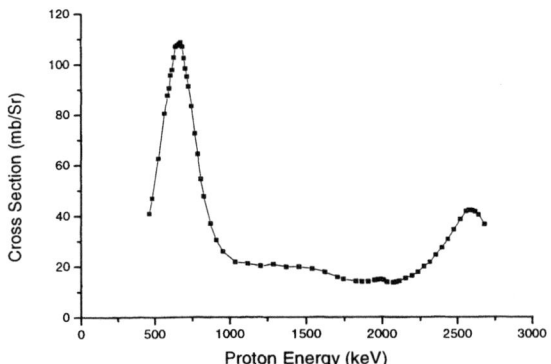

FIGURE 2. Cross section of $^{11}B(p, \alpha_1)^8Be^*$ reaction under $E_{min}/E_{\alpha o} = 0.398$ convention as a function of incident proton energy for 150°.

The cross section under convention is a practical way to take boron measurement from the continuos α_1-spectrum. Take the integral count of the α-particle above a threshold at the spectrum minimum E_{min} as the reaction yield for the reaction cross section under convention. The energy threshold E_{min} is well repeatable, if it is defined relative to the sharp peak of the α_o. The cross section is given as a cross section under convention of a ratio of $E_{min}/E\alpha_o$. The cross section of $E_{min}/E\alpha_o = 0.398$ convention is shown in Fig.2. The experimental error is 2.4% due to the accurate counting of the reaction yield.

PROFILING: DIFFERENTIAL NRA WITH ANODIC OXIDATION STRIPPING

Depth profiling of boron by NRA is obtained by successive layer stripping and measurement of the yield of the $^{11}B(p,\alpha_1)^8Be^*$ reaction. Well controlled thicknesses of silicon are removed using anodic oxidation, followed by oxide etch in 5% HF. Anodic oxidation stripping is being used in differential Hall Effect profiling and differential four-point-probe with anodic oxidation stripping.[8,9] The anodic oxidation was performed in a standard solution of ethylene glycol and 0.04N KNO_3.[15] A trace of water (0.1-1%) was present and actually found necessary for best oxide characteristics. The measured amount of KNO_3 was added to the bottled ethylene glycol which was stoppered and mixed just before usage. Anodization was carried out by immersing the silicon wafer installed in the electrode assembly in the fresh prepared ethylene glucol-KNO_3 solution at a temperature of 20-25°C. A constant current source with current density ranging from 1-5 mA/cm² is used. The oxide layer thickness was monitored with a digital voltmeter. The oxide layer thickness is close to a linear function of the voltage across the film, but it has to be calibrated. The calibration of oxide layer thickness was carried out by glancing angle RBS/channeling with the depth resolution of about 20A. The depth resolution of this profiling depends mainly on the reproducibility of the anodization process and the calibration. Fig. 3 shows RBS/channeling spectra of 2 independent anodized oxide films. The reproducibility is within one channel or 10A, so we believe a depth

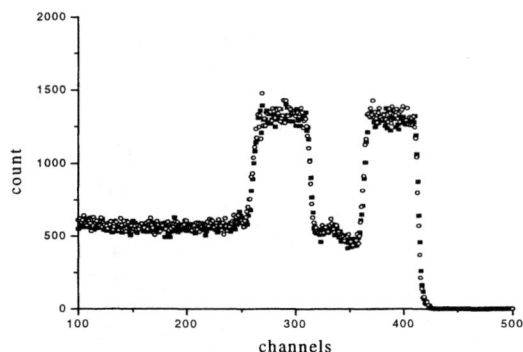

FIGURE.3. RBS/Channeling spectra of 2 independent oxide films with the same anodization parameters. The spectra show good reproducibility of the anodic oxidation stripping.

resolution can be obtained with well-controlled process. NRA analysis was conducted after each layer was removed and then the differential reaction yield will give the depth profile of B atoms. The combination of NRA with Channeling provides both the chemical profile (total amount of B) and the carrier

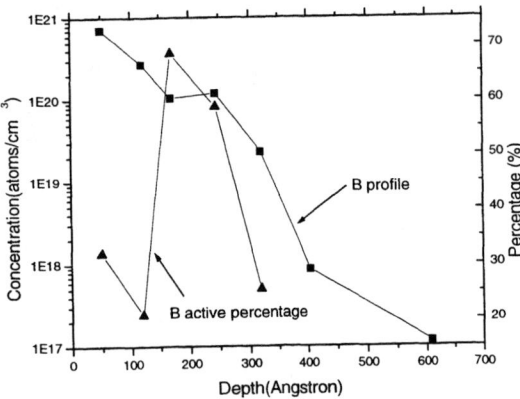

FIGURE.4. Depth profiles of total implanted boron (solid square) and activated boron (solid triangle). The sample was boron implanted silicon at 1 keV and annealed at 1000°C/10s.

depth profile (activated amount of B atoms). The yield of the random NRA spectrum gives the chemical depth profile of boron. The difference between the random and the aligned spectra gives the carrier depth profile. Fig.4 shows the depth profiles of total boron dopant and the profile of activated boron atoms. Combination of NRA with Channeling and oxidation stripping provides depth profiling of total B dopant and the activated boron profile.

CONCLUSION

Nuclear reaction of $^{11}B(p,\alpha_1)^8Be^*$ is being used for boron implant dosage and uniformity measurement. The best sensitivity is $10^{13}/cm^2$ obtained by using an annular detector with a large solid angle. The cross sections of the $^{11}B(p,\alpha_1)^8Be^*$ reaction at the broad resonance at 660 keV were measured with high accuracy. The experimental error for the cross section under convention was 2.4%. The real cross section of the $^{11}B(p,\alpha_1)^8Be^*$ was obtained for the first time in the energy range 0.4-1.6 MeV with the experimental error of 3.3%.

ACKNOWLEDGEMENT

This project was supported by The State of Texas through the Texas Center for Superconductivity, partially by the 003652-797 ARP and the National Science Foundation through the Materials Research Science and Engineering Center.

REFERENCES

1. Ishida, E., and Felch, S. B., *J. Vac. Sci. Technol.* **B14,** 397 (1996).

2. Scanlon, P. J., Farrell, G., Ridgway, M. C., and Valizadeh, R.., *Nucl. Instr. and Meth* **B16,** 479-482 (1986).

3. Moncoffre, N., Millard, N., Jaffrezic, H., Tousset, J., *Nucl. Instr. and Meth* **B45,** 81-85 (1990).

4. Lu, Xiting., Xie, Yuan., Zhen, Zongshuan., Jian, Wei-lin., and Liu, Jiarui., *Nucl. Instr. and Meth* **B43,** 565-569(1989).

5. Liao, Changgeng., Wang, J. H. Yonggiang, Zhen, Zhihao., *Nucl. Instr. and Meth ,* 97-101(1995).

6. J. R. Liu and W. K. Chu, submitted to *Nucl. Instr. and Meth B*

7. Lappalainen, R., Raisanen, J., and Anttila, A., *Nucl. Instr. and Meth ,* 55-59(1985).

8. Huang, M. B., and Mitchell., I. M., *J. App. Phys.* **85,** 174 (1999).

9. Iho, K. , Taniguchi, K. , Ohmi, T., *Jpn. J. Appl. Phys.* **37,** 4277(1998).

10. Beckman, O., Huus, T., Zupancic. C., *Phys. Rev.* **91,** 606 (1953).

11. Ligeon, E., Bontemps. A., *J. Radioanal. Chem.* **12,** 335-351(1972)

12. Mayer, J. W., Rimini. E.., *Ion Beam Handbook for Material Analysis*. New York: Academic Press, In. 1977.

13. Lin Erh-kan. Wang Chang-wan, Yuan Jian, Liu Xiao-dong, Li Cheng bo, Sun Zu-xun, Zhang Pei-hua, Chen Jin_xiang, Yang Qi-xiang, Wang Jian-yuon, Gong Ling-hua. *Chinese Physics Lett.* **15,** 796(1998).

14. Liu, J. R.., Zheng, Z. S. , Chu. Wei-Kan., *Nucl. Instr. and Meth* **B108,** 1-6(1996).

15. Duffek, E.F., Benjamini, E. A., and Mylroie, C., *Electrochemical Technology* 3, 75(1965).

Cold Neutron Beam Studies of Parity-Violation in the n-α and n-p Systems

D. M. Markoff*

Department of Physics, North Carolina State University, Box 8202, Raleigh, NC 27695 USA
Triangle Universities Nuclear Laboratory, Box 90308, Durham, NC 27708 USA

Abstract. Long wavelength neutrons ($\lambda > 1$Å) in a cold neutron beam provide a valuable probe to study the strong and weak nuclear forces in hadronic systems, where the description is complicated by the quark structure of the particles. As a consequence of parity-violation (PV) arising from the weak interaction, the low-energy neutron transverse spin-polarization vector rotates as the neutrons traverse a medium. The magnitude of the PV spin-rotation observable in the n-α system provides important new data to determine the strength of the neutron-nucleus weak interaction. Measurement of the spin-rotation in the bare neutron-proton system with a parahydrogen target, will provide important constraints on the weak nucleon-nucleon (NN) interaction including the neutral current contribution, and will increase our understanding of the strong NN interaction. This paper will review the recent spin-rotation measurement in a liquid helium target, and the proposed measurement in a parahydrogen target.

INTRODUCTION

In contrast to the leptonic sector, the hadronic weak interaction is difficult to study because of the presence of the strong interaction. Both the strong and the electromagnetic interactions must be suppressed by a symmetry principle in any experimental study of the hadronic weak interaction. Parity-violating observables in nuclear systems are the only accessible means to study the neutral current contribution to the hadronic weak interaction since the neutral current Z^0 exchange is forbidden in high-energy measurements of flavor-changing weak decays. Theoretically, the description of the low-energy nucleon-nucleon (NN) weak interaction is limited by our ability to describe the underlying strong interaction in the regime of non-perturbative QCD.

In the latest comprehensive review paper on the subject of nuclear parity-violation (PV), the authors emphasized that the weak NN interaction amplitudes are not well known, and that further experimental studies of these parameters are needed (1). Specifically, experiments in few-nucleon and two-nucleon systems enable the extraction of the weak amplitudes independent of nuclear calculational models. Measurements in the bare neutron-proton system are the only practical way to test the weak neutral-current quark sector of the Standard Model, are a foundation upon which to understand multi-nucleon effects on the weak hadronic interaction, and will provide a useful benchmark for testing calculational techniques for the strong interaction at low energies.

Recent developments of cold neutron beams has advanced the study of neutron induced PV reactions. A review of neutron experiments is provided in the proceedings of the recent workshop on Fundamental Physics With Pulsed Neutron Beams (2). This paper will focus on the neutron spin rotation experiment, by reviewing the recent (n-α) measurement at a reactor beam line and the proposed (n-p) measurement that will take advantage of the properties of a pulsed cold neutron beam available at the Oak Ridge Spallation Neutron Source (SNS).

SPIN-ROTATION OBSERVABLE

In 1964, Michel first proposed that the weak interaction could produce a measurable effect with neutrons that is analogous to the observed optical rotation of polarized photons propagating through a "handed" medium (3). Sub-thermal neutrons that propagate through a crystal of spin-zero nuclei experience a parity-violating rotation of the spin polarization vector. The Coulomb, magnetic, and spin-dependent nuclear interactions do not contribute to this effect.

The PV neutron spin-rotation, φ_{PNC}, is linearly dependent upon the target length, ℓ, and the density, ρ and is given by (3)

$$\varphi_{PNC} = 4\pi \rho \ell f_{PNC}. \quad (1)$$

---

* Supported in part by DOE Grant No. DE-FG02-97ER41042.

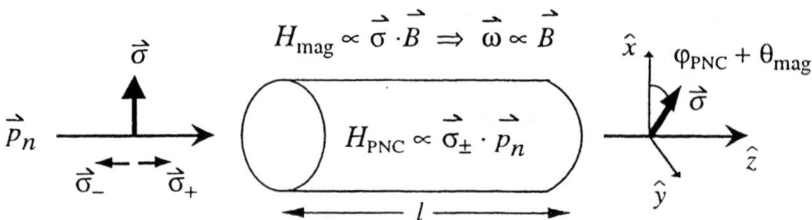

FIGURE 1. Simplified representation of an experiment for measuring φ_{PNC}.

where f_{PNC} is the parity nonconserving (PNC) part of the forward scattering amplitude that is dependent upon the weak interaction matrix element. A positive phase difference (when $f_{PNC}(+\text{helicity}) > f_{PNC}(-\text{helicity})$) induces a rotation about the neutron momentum that follows the sense of a right handed screw. Note that the PNC spin rotation is independent of the neutron velocity or energy.

This effective neutron "optical rotation" was confirmed in the first spin-rotation experiments carried out in the tin isotopes (4) and was found to be on the order of 10^{-5} radians. Enhancements in heavy nuclei is absent for the few nucleon systems. Calculations for the n-α system predict the spin-rotation magnitude between $(0-1.5) \times 10^{-6}$ rad/m, where the range arises from uncertainties in the weak interaction amplitudes (5). Theoretical analysis of the neutron spin rotation through a parahydrogen target predicts $(0-9) \times 10^{-7}$ rad/m (6).

SPIN-ROTATION APPARATUS

The experiment uses a beam of transversely polarized cold neutrons, with energies in the 10^{-3} eV range ($\lambda \sim 5\text{\AA}$). Previous spin-rotation measurements were performed at cold neutron beam lines at reactor facilities, namely at the Institut Laue Langevin (ILL) in Grenoble, France and at the National Institute of Standards and Technology (NIST) in Gaithersburg, MD. With Beryllium filters, the neutron wavelengths range from $\lambda > 4.7\text{\AA}$ with the peak at about $5\text{\AA}$. To maximize the signal, (refer to Equation 1) solid tin and lanthanum targets were used and a cryogenic liquid helium target was used for the recent $\vec{n} + \alpha$ measurement. A cryogenic liquid parahydrogen target, where the proton spins are anti-aligned, is proposed for the n-p measurement. A schematic representation of the experiment is given in Figure 1.

A polarimeter for measuring the spin-rotation of transversely polarized neutrons through solid targets was developed by Heckel et al. (7) This polarimeter consisted of a target region located between a crossed polarizer and analyzer pair. The target region is divided into two target positions separated by the "π-coil". The π-coil magnetic field axis is aligned along $\hat{x}$, the direction of the initial neutron polarization where $\hat{z}$ is the neutron momentum direction. The π-coil magnetic field strength is set so that the neutron spin precesses about $\hat{x}$ by 180 degrees as it passes through the coil. The π-coil thereby reverses the sign of the upstream spin-rotations in the xy plane relative to rotations downstream of the coil.

As the target is moved between the front and back targets positions, the π-coil causes the PNC signal to be modulated against a nearly constant background rotation. In the ideal case, the difference in measured rotations between the front and back target configurations, eliminates the background rotation and leaves twice the PNC signal.

The analyzer magnetic field is flipped for analyzing in the $\pm \hat{y}$ direction at a rate of approximately 1 Hz. For neutron counts $N_\pm$ corresponding to the $\pm \hat{y}$ analyzing direction, the total rotation angle, φ, of the spin vector about the momentum axis in the ideal case is given by

$$\sin \varphi = \frac{(N_+ - N_-)}{(N_+ + N_-)} \quad (2)$$

where a positive rotation is given by a right-hand rotation about the neutron momentum direction, $\vec{p}_n = p_n \hat{z}$.

To reduce background magnetic field induced rotations, the target is contained within a magnetic field-free region. Two layers of μ-metal magnetic shielding and field compensation solenoid coils were used to minimize the axial magnetic fields. Field guide coils preserve the polarization direction as the neutrons pass into or out of the field-free region.

To reach a rotation sensitivity on the order of 10^{-8} radians, we improve the background subtraction by measuring the PNC signal from both the front and back target positions simultaneously. The apparatus is divided into two side-by-side sections in which each half effectively constitutes a separate experiment run under the same conditions. A simultaneous comparison of the counts measured in the two sides decreases the signal's sensitivity to neutron count-rate noise caused by beam fluctuations.

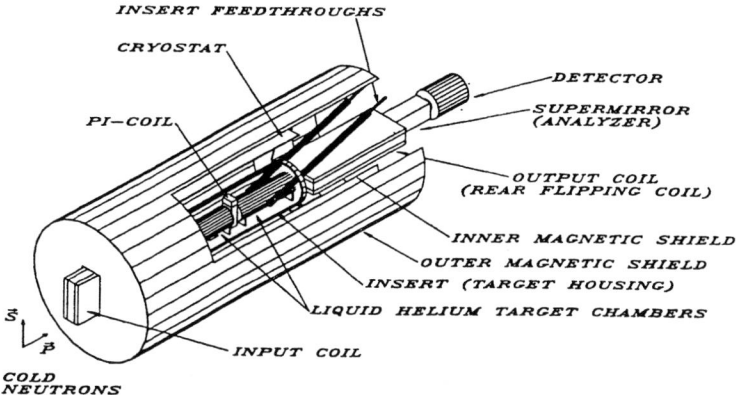

FIGURE 2. Schematic diagram of the neutron spin rotation apparatus used with a cryogenic liquid helium target.

SYSTEMATICS

The split-beam polarimeter with a two target system was designed to separate constant background rotations from the modulated PNC signal. The method was limited to the extent that the two target states produced the same background rotations. Parity-conserving changes in the spin-rotation that are dependent upon the target position, will mimic the PNC signal. The basic design of the apparatus was dominated by the challenge of reducing to the level of 5×10^{-8} rad (for the n-α experiment), the difference in background rotations measured for the two target positions.

Target-induced changes in local magnetic fields and target-dependent neutron scattering effects are the two primary sources of spin-rotations that will mimic the PNC signal. The diamagnetic properties of the target material alters the magnetic field in the chamber region. In addition, the effective index of refraction for neutrons traveling in the liquid target is less than that for the gaseous state reducing the neutron velocity in the full target compared to the "empty" chamber. The neutron spin therefore feels the magnetic field for a different length of time, thus changing the integral background rotation. The material properties set the maximum allowable magnetic field to assure that the effect of these target position dependent rotations integrated over the length of the chamber is less than the desired systematic limits.

In the recent liquid helium experiment, the fields were on the order of 40nT, producing a background rotation of ~ 2 mrad. Magnetic field induced target-dependent rotations were below the desired systematic limit.

A change in the neutron beam profile for target in or out of the chambers will mimic a PNC signal. Neutron scattering effects may change the beam divergence or alter the apparatus acceptance for different neutron wavelengths. This may induce a background count-rate asymmetry, or alter the background rotation since the detected neutrons will have traveled different paths depending on the target position. Scattered neutrons are therefore removed from the system and magnetic fields are reduced to a level so that the change in rotation due to a change in path length for the fraction of detected neutrons is below the systematic limit. In the n-α experiment, the neutron asymmetry was below 10^{-4}.

PROPOSED EXPERIMENT AT THE SNS

All previous spin-rotation measurements were taken at reactor cold-neutron beam lines. While the fundamental design of the split-beam spin-rotation polarimeter would be employed, the use of a spallation neutron source pulsed beam requires a few apparatus design modifications and introduces some advantages to the management of systematics.

A measurement of the neutron spin-rotation in a parahydrogen target has been proposed (11) for the future SNS facility at the Oak Ridge National Laboratory.

Neutron polarization

To take advantage of the full spectrum of cold-neutrons from the moderated pulsed source requires the use of a broad band ^{3}He spin polarizer. In previous measurements, a supermirror polarizer and analyzer were used. The supermirrors, based on spin-dependent critical angles of reflection, are not designed to accommodate the wide range of neutron wavelengths ($\lambda > 1$ Å) available in a spallation source cold-neutron beam. The development of ^{3}He spin polarizers and analyzers has progressed and promising increases in the figure of merit, $P_n^2 T_n$, can be achieved, where P_n is the neutron polarization and T_n is the neutron transmission (8).

An additional feature of the ^{3}He spin filter polarizer is the ability to flip the direction of the initial neutron spin. This provides an additional systematic check not previously available using supermirror polarizers.

Neutron time-of-flight

A pulsed neutron beam allows for a straightforward time-of-flight determination of the neutrons' velocity as they are detected. This is a powerful tool for eliminating background rotations and obtaining high sensitivity to the true PNC spin-rotation. The non-constant background rotations originate from Larmor precession around stray fields in the apparatus and change linearly as a function of the time spent in the presence of these fields. Therefore, magnetic field induced background rotations, not completely canceled in the analysis subtraction scheme, will exhibit an inverse velocity dependence. Measurement of the velocity dependence of the experimental signal will strongly separate the velocity-independent PNC rotation from residual background rotations.

As discussed above in the Systematics section, false signals from target-dependent PC rotations will vary with neutron velocity including those from neutron scattering effects. For example, the coherent scattering in liquid helium decreases as a decreasing function of energy. In liquid parahydrogen, neutron scattering exhibits a more complicated dependence on energy than has been measured. The true PNC signal, therefore must be independent of velocity or time as measured from a pulsed beam, and analysis as a function of time provides a limit on the false signal contribution.

To obtain time-of-flight information, a fast response detector (such as a scintillator) is required. Previously a segmented ^{3}He ionization chamber integrated the beam with a coarse, 4 group velocity separation dependent upon a model of the neutron spectrum (9). Using a pulsed beam, time-of-flight determination of the neutron velocity is independent of knowing the beam spectrum.

π-coil configuration

When using a reactor based, polychromatic neutron beam, a range of neutron velocities is present in the apparatus at any given time. The π-coil current must be tuned so that neutrons of some average velocity receive exactly a 180 degree rotation. Neutrons of other velocities receive larger or smaller rotations and their associated spin-rotation signal is washed out. At a pulsed spallation neutron source, neutrons of different velocities pass through the apparatus at different times with respect to the initial proton pulse. The π-coil can be ramped to produce a nearly 180 degree rotation for every neutron as it enters the apparatus. This will reduce the net statistical uncertainty per neutron by a factor of almost two.

STATUS OF SPIN-ROTATION MEASUREMENTS

The recent neutron spin-rotation measurement in a liquid helium target gives (10) $\varphi(n,\alpha) = (8.0 \pm 14_{(stat)} \pm 2.2_{(syst)}) \times 10^{-7}$ rad/m. Although this result is statistically limited, it is the most sensitive spin rotation measurement to date. The systematic limits of the apparatus were not reached at this measurement. Plans are underway to repeat the experiment in 2001 (with an improved apparatus for increased reliability and decreased downtime) at the NIST Cold Neutron Research Facility.

The systematic checks and control of false contributions to the PNC target dependent signal will be greatly improved by moving the spin-rotation apparatus to a spallation neutron source beam line. This will allow for an achievable goal of systematic contributions less than $1 \cdot 10^{-8}$.

REFERENCES

1. W.C. Haxton and E. M. Henley, *Symmetries and Fundamental Interactions in Nuclei* Chapter 2, pages 17-66, World Scientific, 1995.
2. *Proceedings of the Workshop on Fundamental Physics with Pulsed Neutron Beams* North Carolina, June 2000.
3. F.C. Michel, *Phys. Rev.* **133**, B329, (1964).
4. M. Forte, *et al. Phys. Rev. Lett.* **45**, 2088, (1980).
5. V.F. Dmitriev *et al. Phys. Lett. B* **125B**, 1, (1983).
6. Y. Avishai and P. Grange, *J. Phys. G: Nucl. Phys.* **10** L263 (1984)
7. B.R. Heckel *et al., Phys. Lett. B* **119B**, 298 (1982).
8. F. Tasset *et al. Physica B* **180**, 896, (1992).
9. S.D. Penn *et al.*, Nucl. Inst. Meth. (1999).
10. D.M. Markoff *at al., Bull. Am. Phys. Soc.* **42**, (1997).
11. D.M. Markoff and F.E. Wietfeldt, *Proceedings of the Workshop on Fundamental Physics with Pulsed Neutron Beams*, North Carolina, June 2000.

EFFECT OF SYMMETRY BREAKING ON TRANSITION STRENGTH DISTRIBUTIONS

G. E. Mitchell[a] and J. F. Shriner, Jr.[b]

[a] *North Carolina State University, Raleigh, North Carolina 27695*
and Triangle Universities Nuclear Laboratory, Durham, North Carolina 27708
[b] *Tennessee Technological University, Cookeville, Tennessee 38505*

Abstract. The quantum numbers of over 100 states in ^{30}P have been determined from the ground state to 8 MeV. Previous measurements had provided complete spectroscopy in ^{26}Al. For these $N = Z =$ odd nuclei, states of isospin $T = 0$ and $T = 1$ coexist at all energies. These spectra provide a unique opportunity to test the effect of symmetry breaking (of the approximate symmetry isospin) on the level statistics and on the transition strength distributions. The level statistics are strongly affected by the small symmetry breaking and the transition strength distributions differ from the Porter-Thomas distribution.

INTRODUCTION AND MOTIVATION

After its introduction by Wigner (1), Random Matrix Theory (RMT) (2) was used primarily to describe compound nuclear resonances. There are now many applications of RMT: the most recent review (3) lists over 800 references! Since Bohigas *et al.* (4) suggested that quantum systems whose classical analogs were chaotic should show eigenvalue fluctuations that agree with RMT, level statistics have been used as "signatures of chaos."

There have been few nuclear tests of RMT. The reason is that the standard measures used to analyze the level statistics are very sensitive. The spectrum must be pure (few or no misassigned quantum numbers) and complete (few or no missing levels). Few nuclear data sets satisfy these stringent criteria. Complete spectroscopy was first obtained for ^{26}Al (5, 6). We have recently completed a new experimental study of the nuclide ^{30}P (7).

Consider the approximate symmetry isospin. The nuclides ^{26}Al and ^{30}P have the unusual feature that the $T = 0$ and $T = 1$ states coexist at all energies. For ^{26}Al the level statistics (8, 9) are consistent with the prediction (10, 11) that a small symmetry breaking (on average isospin symmetry is broken by about 3% in ^{26}Al) can have a large effect. Guhr and Weidenmüller (12) analyzed the effect of isospin symmetry breaking on the eigenvalue distribution. Our new results for ^{30}P (13) agree with the ^{26}Al results. There have been no other experimental tests in a quantum system. However, there have been tests in analog systems: acoustic resonances in quartz blocks (14) and electromagnetic resonances in superconducting microwave billiards (15). In both cases the results were consistent with theory and with our earlier experiment.

RMT also predicts the distribution of the transition matrix elements. According to RMT, the strength distribution should be a Porter-Thomas (PT) distribution. Until 1999 there were no theoretical predictions for the effect of symmetry breaking on the transition distribution. Our analysis of the ^{26}Al transition data (16) indicated that the strength distributions deviate appreciably from Porter-Thomas. Recent theoretical analyses (17, 18) indicate that symmetry breaking changes the transition strength distribution from the Porter-Thomas. Our new ^{30}P transition distributions (13) also deviate from the PT distribution.

GENERAL BACKGROUND

Random matrix theory predicts that reduced width amplitudes will follow a Gaussian distribution. The corresponding distribution for the reduced widths is the Porter-Thomas distribution

$$P(y) = \frac{1}{\sqrt{2\pi y}} e^{-y/2}, \quad (1)$$

where $y = \gamma^2 / \langle \gamma^2 \rangle$, γ^2 is the reduced width, and $\langle \gamma^2 \rangle$ is the average reduced width. For electromagnetic transitions, the appropriate strength measurement is the reduced transition probability B. We define the dimensionless strength parameter y as $y = B(XL\Delta T)/<B(XL\Delta T)>$, where B is the reduced transtion probability, X is the electromagnetic radiation character (electric or magnetic), L is the multipolarity of the transition, and ΔT represents the change of isospin between initial and final states.

The reduced matrix elements range over several orders of magnitude. It is therefore convenient to change to a new variable $z = log_{10}(y)$. The Porter-Thomas distribution in terms of z is

$$P(z) = \ln 10 \sqrt{\frac{y}{2\pi}} e^{-y/2}. \qquad (2)$$

EXPERIMENTAL METHODS

The level schemes were obtained via a variety of experimental measurements, which we illustrate by considering ^{30}P. The key reactions for the spectroscopic studies are the (p,p) and (p,γ) reactions. Our group (19) performed a high resolution study of the ^{29}Si(p,p_0) and ^{29}Si(p,p_1) reactions and identified 66 resonances.

One crucial issue is whether all of the resonances are observed. The beam energy resolution is about 200 eV, corresponding to a quality factor (or Q value) of about 10^4. Therefore we observed almost all of the resonances. We performed another high resolution study (20) of the capture reaction with a γ-ray detector that had poor γ-ray energy resolution, but very high efficiency. We observed every resonance seen in the scattering measurements plus several new ^{30}P resonances.

We measured the ^{29}Si(p,γ) reaction with very good beam energy resolution and good detector resolution. These experimental results were reported by Wallace et al. (21) and Vavrina et al. (22). Spectra were measured for about 50 resonances. Since these resonances have different J values (0 to 5), different parities, and different structure, the cumulative probability of missing any bound state is very small. Therefore the spectrum is complete.

Next we used information about the transition strengths. Endt (23) tabulated experimental values of γ-ray strengths and classified them by mass region, electromagnetic radiation type XL, and as isoscalar or isovector (IS or IV). He adopted an empirical Recommended Upper Limit (RUL) for each type of transition in a given mass region. Application of these RUL's places additional restrictions on the quantum numbers of the initial and final states. Multipole mixing often occurs in the γ-ray transitions. The relative amount of each multipole can be determined from the γ-ray angular distributions. We measured 230 angular distributions of primary and secondary γ-ray transitions from 32 resonances [7] and determined many multipole mixing ratios.

At this stage we had determined almost all of the J and π values, but T was uncertain for a number of states. We then used information from the parent nucleus ^{30}Si: all $T = 1$ states in ^{30}P must be analogs of states in ^{30}Si. We used this correspondence to aid in the assignment of T to states with known J and π. We also compared shell model calculations (24) for ^{30}P with our experimental spectrum; the agreement for the positive parity states is very good. For a few states with known positive parity and a J-value ambiguity, we used the shell model calculations to assign J.

At this stage we had complete assignments for almost all states. Our final determination of the ^{30}P level scheme is that there are 107 states between 0 and 8 MeV with spins $J = 0 - 5$. We have J, π, and T assignments for 103 of these states, with only 4 states with unknown J value. A total of 472 $B(E1)$, $B(M1)$, and $B(E2)$ values are known.

EXPERIMENTAL RESULTS

For the transitions the quantities of interest are the reduced transition matrix elements. To determine the reduced matrix elements, one needs the energy E, angular momentum J, parity π, and isospin T for each level, the γ-ray decay width Γ_γ, the branching ratios to each final state, and the multipole mixing ratios for transitions with competing decay modes. We define a transition sequence as a set of transitions with the same electromagnetic radiation type X (electric or magnetic), multipolarity L, and isospin change ($\Delta T = 0$ or $\Delta T = 1$). We then convert to reduced matrix elements.

To put results in terms of the dimensionless quantity z requires a determination of $\langle B \rangle$ as a function of energy for each transtion sequence. We defined the local average for a given initial and final energy as the average over all B values in that transition sequence whose initial and final energies were both within N_t levels of the transition in question. We chose $N_t = 15$, as it was large enough to provide a meaningful average but small enough not to remove any local energy dependence.

The experimental results for the overall reduced transition probability distributions in ^{30}P are shown in Fig. 1, and those for ^{26}Al are shown in Fig. 2. In both cases, the maximum is shifted from the value $z = 0$ expected for the Porter-Thomas distribution.

ANALYSIS

Since isospin is approximately conserved, one might expect a GOE distribution when the states are separated by their isospin values and close to a two-GOE distribution when isospin is ignored. The experimental level statistics do not agree with this simple assumption: within the experimental uncertainties, isospin appears irrelevant.

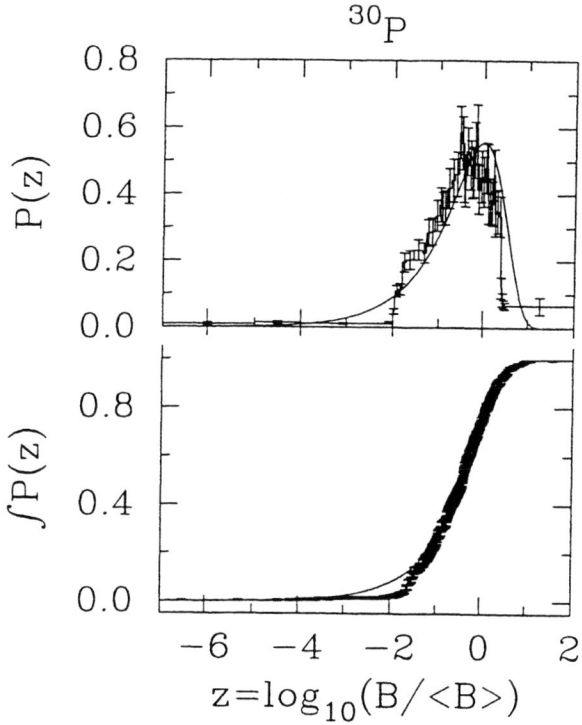

FIGURE 1. Probability density and distribution functions for reduced transition probabilities in ^{30}P. The smooth curve shows a Porter-Thomas distribution.

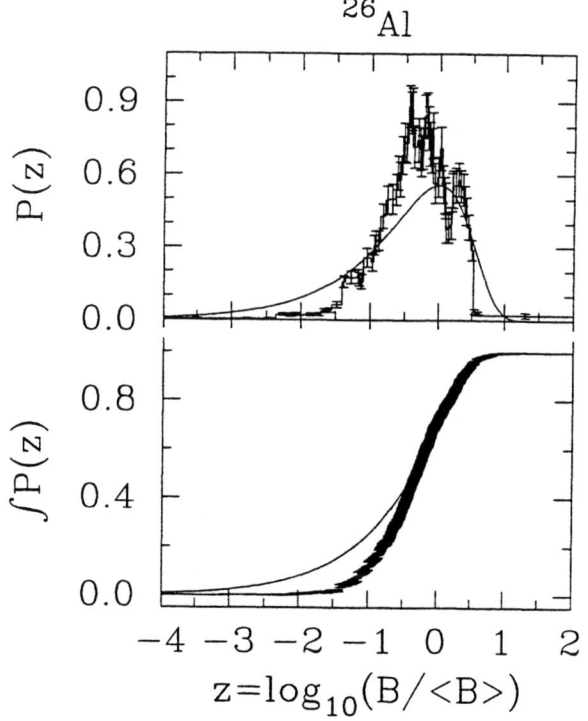

FIGURE 2. Probability density and distribution functions for reduced transition probabilities in ^{26}Al. The smooth curve shows a Porter-Thomas distribution.

This is what Dyson (10) and Pandey (11) predicted for the effect on level statistics when a discrete symmetry is broken. Consider a Hamiltonian of the form

$$H = H_{sc} + \alpha H_{sb}, \qquad (3)$$

where H_{sc} conserves the symmetry and H_{sb} breaks the symmetry. The term H_{sc} is block diagonal in the symmetry in question (here isospin) and H_{sb} has only off-diagonal elements (here Coulomb). The two submatrices in H_{sc} are random matrices characterized by the isospins $T = 0$ and $T = 1$. The NNSD for each of these blocks is a GOE distribution. After the symmetry breaking and rediagonalizing, the new NNSD is clearly different. The naive expectation is a small change due to the smallness of α (about 3% for ^{26}Al). However, the change in the distribution is governed by both the magnitude of the symmetry breaking (α) *and* by the average separation between the levels (D). The relevant parameter for the new NNSD is $\lambda = \alpha/D$. For D small enough, a very small symmetry breaking can have a large effect. The ^{26}Al data provided the first experimental test of this prediction. Guhr and Weidenmüller (12) performed a detailed analysis of the ^{26}Al data and concluded that our data are consistent with the RMT predictions.

The experiments with analog systems by Ellegaard *et al.* (14) and by Alt *et al.* (15) confirmed our results with much better statistics. Agreement with RMT was excellent. Our new measurements for ^{30}P are in agreement with all other results. Therefore it appears that the effect of symmetry breaking on the level statistics is well understood.

Our results are the first and only experimental tests for the transition strength distributions. The experimental transition distributions are not Porter-Thomas. This result seem surprising, since the usual argument is that once the wavefunctions are sufficiently complicated, the central limit theorem should apply. Therefore the components of the wavefunction should be Gaussian distributed, and transitions involving such complicated states should also be Gaussian distributed.

Barbosa, Guhr, and Harney (17) first considered the effect of symmetry breaking on the transition matrix elements in a random matrix formalism. They started from the same Hamiltonian as used for the level statistics problem. The symmetry breaking changes both the eigenvalues and the wavefunctions. The matrix elements of the operator in question are calculated with these new (perturbed) wavefunctions. Barbosa *et al.* express the operator as a combination of two terms

$$O = O_{sc} + \beta O_{sb}, \qquad (4)$$

where O_{sc} and O_{sb} are the symmetry conserving and symmetry breaking parts of the operator, and β is a measure of the symmetry breaking. The electromagnetic operator obviously breaks isospin symmetry. With this ansatz, they calculated the distribution of transition strengths and obtained a deviation from the PT distribution. Hussein and Pato (18) use a somewhat different approach, but also consider the effect on the transition distribution of symmetry breaking in both the Hamiltonian and the operator. They also find that the distribution is modified from the PT distribution by symmetry breaking in either the Hamiltonian or the operator. Since there are no other experimental data, and our data have limited statistics, it would be very interesting to examine these issues in other quantum systems or in analog systems.

SUMMARY

With a variety of experimental methods, the quantum numbers of essentially every state in ^{30}P have been determined from the ground state to 8 MeV. Previous measurements had provided complete spectroscopy for ^{26}Al over a similar energy region. These $N = Z =$ odd nuclei have the property that states of isospin $T = 0$ and $T = 1$ coexist at all energies. Since the energy E, spin J, parity π, and isospin T are known for every level, these spectra test the effect of symmetry breaking on the level statistics and on the transition strength distributions. The level statistics are strongly affected by the small symmetry breaking, in agreement with the RMT predictions. Measurements in analog systems also agree very well with the RMT predictions for the level statistics.

We also determined the electromagnetic transition strengths in ^{26}Al and ^{30}P. To obtain the transition strengths one needs not only the quantum numbers of the levels, but also transition properties such as branching ratios, multipole mixing ratios, etc. In ^{26}Al and ^{30}P the distribution of the normalized reduced matrix elements differs from the Porter-Thomas distribution. Recent RMT calculations predict a deviation from the PT distribution.

ACKNOWLEDGMENTS

This work was supported in part by the U.S. Department of Energy, Office of High Energy and Nuclear Physics, under grants No. DE-FG02-97-ER41042 and DE-FG02-96ER40990. We thank Y. Alhassid, C. I. Barbosa, B. A. Brown, T. Guhr, H. L. Harney, M. Hussein, W. E. Ormand, M. P. Pato, H. A. Weidenmüller, and V. Zelevinsky for valuable discussions. We thank A. A. Adams, C. A. Grossmann, M. A. LaBonte, J. D. Shriner, G. A. Vavrina, and P. M. Wallace for their valuable contributions to the measurements and analysis.

REFERENCES

1. Wigner, E. P., Oak Ridge National Laboratory Report ORNL-2309, 59 (1957); in *Statistical Theories of Spectra: Fluctuations*, edited by C. E. Porter, Academic Press, New York, 1965, p. 199.
2. Brody, T. A. *et al.*, Rev. Mod. Phys. **53**, 385 (1981).
3. Guhr, T., Müller-Groeling, A., and Weidenmüller, H. A., Phys. Rep. **299**, 189 (1998).
4. Bohigas, O., Giannoni, M. J., and Schmit, C., Phys. Rev. Lett. **52**, 1 (1984).
5. Endt, P. M., de Wit, P., and Alderliesten, C., Nucl. Phys. A **459**, 61 (1986).
6. Endt, P. M., de Wit, P., and Alderliesten, C., Nucl. Phys. A **476**, 333 (1988).
7. Grossmann, C. A. *et al.*, Phys. Rev. C **62**, 024323 (2000).
8. Mitchell, G. E. *et al.*, Phys. Rev. Lett. **61**, 1473 (1988).
9. Shriner, J. F., Jr. *et al.*, Z. Phys. A **335**, 393 (1990).
10. Dyson, F. J., J. Math. Phys. **3**, 1191 (1962).
11. Pandey, A., Ann. Phys. (NY) **134**, 110 (1981).
12. Guhr, T. and Weidenmüller, H. A., Ann. Phys. (NY) **199**, 412 (1990).
13. Shriner, J. F., Jr., Mitchell, G. E., and Grossmann, C. A., to be published.
14. Ellegaard, C. *et al.*, Phys. Rev. Lett. **77**, 4918 (1996).
15. Alt, H. *et al.*, Phys. Rev. Lett. **81**, 4847 (1998).
16. Adams, A. A., Mitchell, G. E., and Shriner, J. F., Jr., Phys. Lett. B **422**, 13 (1998).
17. Barbosa, C. I., Guhr, T., and Harney, H. L., Phys. Rev. E (to be published).
18. Hussein, M. S. and Pato, M. P., Phys. Rev. Lett. **84**, 3783 (2000).
19. Nelson, R. O. *et al.*, Phys. Rev. C **27**, 930 (1983).
20. Frankle, S. C. *et al.*, Phys. Rev. C **45**, 2746 (1992).
21. P. M. Wallace *et al.*, Phys. Rev. C **54**, 2916 (1996).
22. G. A. Vavrina *et al.*, Phys. Rev. C **55**, 1119 (1997).
23. P. M. Endt, At. Data Nucl. Data Tables **55**, 171 (1993).
24. W. E. Ormand (private communication).

Production of ^{90}Y by the ^{90}Zr (n, p)^{90}Y Reaction using Neutrons Produced from a Variable Energy Cyclotron

D. Necsoiu[a], I.L. Morgan[b], H. Hupf[b], J. Armbruster[b], D. Boyce[b], M. El Bouanani[a], F.D. McDaniel[a]

[a]*Ion Beam Modification and Analysis Laboratory, Department of Physics, University of North Texas, Denton, Texas 76203-11427*
[b]*International Isotopes, Inc., 2101 Shady Oaks, Denton, 76201*

Abstract. Traditionally, ^{90}Y obtained from a ^{90}Sr/^{90}Y generator contains a small concentration of ^{90}Sr (range: 1μCi ^{90}Sr/100mCi to 1 Ci ^{90}Y) and due to the 28.78 y half-life of ^{90}Sr, special waste handling and storage is required. In this study, the medical isotope ^{90}Y has been produced by an alternate method using the ^{90}Zr(n,p)^{90}Y reaction. Neutrons for the activation process were produced using ^{nat}Rh(p,xn) reaction with a 27 MeV proton beam from a cyclotron. Since ^{90}Y is a pure beta emitter, the gamma rays from the ^{90}Zr(n,2n)^{89}Zr reaction were used to quantify the incident neutron flux on the ^{90}Zr sample. Experimental results of the neutron production and ^{90}Y activity are presented.

INTRODUCTION

Most radioisotopes are currently obtained from bombarding enriched stable isotopes with protons from an accelerator or by neutron bombardment in nuclear reactors. The production of radioisotopes using charged particle accelerators is gaining increasing importance due to improved handling of the radioactive waste and the lower cost of isotope fabrication compared with the production of radioisotopes using nuclear reactors.

Radioisotopes are essential for a wide variety of applications in medicine, where they are used for diagnosis, as well as, therapy for illnesses. Radioimmunotherapy, which is used for cancer therapy, is an in vivo treatment technique which involves attachment of a radionuclide to a monoclonal antibody or a smaller protein fragment that is targeted at a particular type of tumor cell [1]. Yttrium-90 (^{90}Y), a beta emitter with a short half-life, is a good candidate for several medical applications, including the radiolabeling of antibodies for tumor therapy and treatment of liver malignancy using radiolabeled particles. The physical data for ^{90}Y are listed in table 1. The monoclonal antibody, with the attached ^{90}Y radionuclide, is injected into the blood stream and absorbed by the tumor. The 2.28 MeV beta emission from the radionuclide kills the tumor cells.

TABLE 1. ^{90}Y physical data [2]

Half-Life	$T_{1/2}$ = 64.10 h
Decay Mode	β^-
Main Decay Energy	E_β = 2.28 MeV
Branching Ratio	I_β = 99.9885 % to the ground state of stable $^{90}_{40}$Zr

Traditionally, ^{90}Y is obtained from the decay of Strontium-90 (^{90}Sr) according to the following scheme [3]:

$$^{90}_{38}\text{Sr} \xrightarrow[28.78\text{ y}]{\beta^- \, (0.546 \text{ MeV})} {}^{90}_{39}\text{Y} \xrightarrow[64.10\text{ h}]{\beta^- \, (2.28 \text{ MeV})} {}^{90}_{40}\text{Zr (stable)}$$

In this decay process, since ^{90}Y contains some low concentration of ^{90}Sr (range: 1μCi ^{90}Sr/100mCi to 1Ci ^{90}Y), a chemical separation is needed in order to be safely employed in clinical applications. Furthermore, due to the long half-life of ^{90}Sr (28.78 y), special waste handling and storage is required.

In order to eliminate the influence of ^{90}Sr and to produce a carrier free isotope at a lower production cost, the present work focuses on the production of ^{90}Y using the ^{90}Zr(n,p)^{90}Y nuclear reaction. Neutrons for activation of ^{90}Zr, were produced using the $^{nat}_{45}$Rh (p,xn) reaction with a 27 MeV proton beam from the cyclotron. In the ^{90}Zr neutron activation process, neutrons initiated a number of nuclear reactions including the ^{90}Zr(n,p)^{90}Y and the ^{90}Zr(n,2n)^{89}Zr reactions.

The ^{89}Zr decays to ^{89}Y with a half-life of 78.41 h and 99.87% of the time results in the emission of a 908.96 keV gamma ray from the first excited state to the ground state of ^{89}Y. Since ^{90}Y decays 100% by beta decay with no observable gamma ray emission, the activity of ^{90}Y was determined from the calculated number of neutrons per second from the measured ^{89}Zr activity.

The main advantage of this technique for ^{90}Y production is that the production of neutrons is essentially free. Therefore, the neutron-produced isotopes using a charged particle accelerator are much less expensive to produce if they can be "piggybacked" at the same time with the production of a proton rich isotope.

EXPERIMENTAL PROCEDURE

The experimental arrangement is show in Fig. 1. A 27 MeV proton beam from the CP42 cyclotron at International Isotope Inc., Denton, Texas, bombarded a thick ^{nat}Rh disk target (thickness 2.81 mm, 25.31 mm in diameter) attached to a water-cooled Faraday cup located at the end of the beam line.

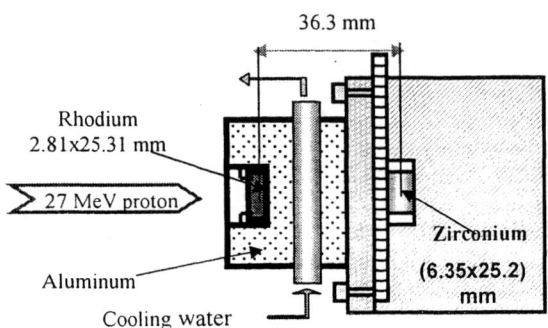

FIGURE 1. Rhodium – Zirconium Target

Neutrons produced by the ^{103}Rh(p,xn) reaction are used to bombard a ^{nat}Zr cylinder of 6.35 mm diameter and 25.2 mm length. The solid angle of the Zr cylinder as seen by the neutron source from Rh, was calculated to be 0.123 sr. The range of 27 MeV incident protons was calculated to be 1.10 mm, using the Stopping and Range of Ions in Matter computational model (SRIM2000) [4]. The beam current was 10μA and the Rh disk was irradiated for 2.5 h for a total of 25 μA.h. The integrated charge on the target was measured using a current integrator.

Measurements were made of all gamma rays from the activation of the Rh and Zr targets for a number of days after irradiation using a high purity germanium n-type detector (EG&G Ortec) connected with a DSPEC (Digital Gamma-Ray Spectrometer) unit. The data were processed by the gamma ray spectrum analysis software, Gamma-Vision 32 [5].

RESULTS AND DISCUSSION

Because the ^{90}Y decays 100% by beta decay to ^{90}Zr (stable), with no observable gamma emission, a second reaction from the same parent, ^{90}Zr, with a measurable gamma ray emission was used to determine the activation of ^{90}Y. The measured activity for the gamma ray emission for the ^{90}Zr(n,2n)^{89}Zr reaction was used to determine the number of neutrons per second (n/s) striking the Zr sample. The ^{90}Zr(n,2n)^{89}Zr reaction was chosen due to the large (n,2n) cross section as shown in fig.2 [6-9].

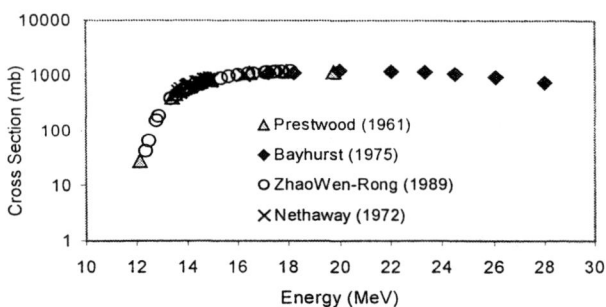

FIGURE 2. The ^{90}Zr(n,2n)^{89}Zr cross section

The decay scheme shows that ^{89}Zr decays by electron capture 100 % of the time, to the excited states of ^{89}Y. By measuring the 908.96 keV gamma ray from the decay of ^{89}Y from the first excited state to the ground state, the activity of ^{89}Zr produced in the (n,2n) reaction was calculated using Gamma Vision software. Figure 3 shows the gamma ray spectrum from the Zr cylinder. The spectrum was taking after 87.15 h from the end of beam (EOB) irradiation. The activity of the Zr cylinder was counted two times per day during the two weeks from the EOB date. The activity of the ^{89}Zr determined from the 908.96 keV gamma-ray peak was calculated and plotted in fig. 4.

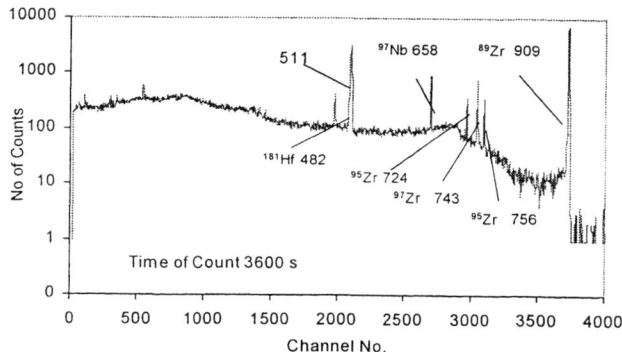

FIGURE 3. Gamma Spectrum for ^{nat}Zr

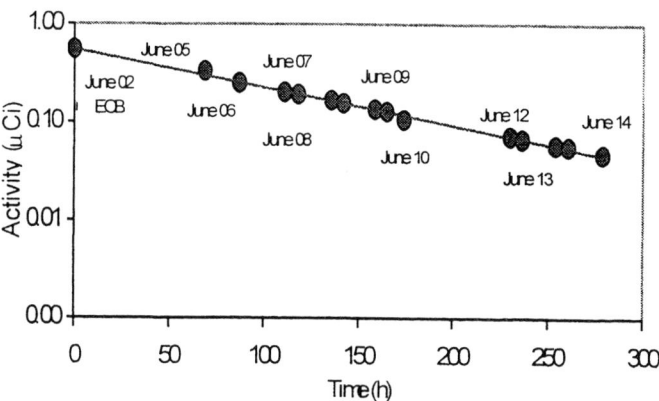

FIGURE 4. Decay activity for ^{89}Zr (908.96 keV)

Using the Gamma Vision software, the time corrected activity (EOB corrected) for ^{89}Zr at 908.96 keV was determined to be 0.55 µCi for 25 µA-h of proton beam. From the measured time corrected activity of the ^{89}Zr and using the ^{90}Zr(n,2n)^{89}Zr cross section, the number of neutrons per second (n/s) and the total number of neutrons incident on the Zr cylinder during the irradiation time was determined.

The number of neutrons per second, N(n/s), incident on the Zr cylinder was calculated using the following formula [10]:

$$N\left(\frac{n}{s}\right) = \frac{3.7(10^{10})\left(\frac{dis}{sCi}\right)A(Ci)}{nd\left(\frac{atoms}{cm^2}\right)\sigma\left(\frac{cm^2}{atoms}\right)(1-e^{-\lambda t})f} \quad (1)$$

The A(Ci) is the ^{89}Zr activity, nd(atoms/cm^2) is the Zr cylinder thickness, σ (cm^2/atoms) is the average cross section for ^{90}Zr(n,2n)^{89}Zr reaction, f is the ^{90}Zr fraction, t (s) irradiation time of 2.5 h and λ (s^{-1}) ^{89}Zr radioactive decay constant. In table 2 the results for neutron production determined from the activation of the Zr cylinder are listed:

TABLE 2. Calculation of the fast neutron flux on the Zr cylinder by activation

Total no. of protons on Rh target(p)	5.625E+17
Total no. of fast neutrons from Rh target with E_n>12 MeV (N)	5.625E+15
(N/p)$_{theory\ for\ Rh}$ [11]	~1E-02
Measured fast neutrons flux on Zr cylinder for E_n>12 MeV (n/sec)	8.43E+07

The total number of neutrons (N) and the number of neutrons / second (n/sec) are the neutrons that have enough energy to cause activation of the ^{89}Zr from the ^{90}Zr(n,2n)^{89}Zr reaction. The calculated threshold for the ^{90}Zr(n,p)^{90}Y reaction is 1.51 MeV. The reaction cross section reaches a maximum value of ~ 48 mb, around 12 MeV as shown in fig. 5 [12-14]. The calculated threshold for the ^{90}Zr(n,2n)^{89}Zr is 12.11 MeV and the reaction cross section reaches a maximum value of ~ 1000 mb around 14 MeV. The ^{90}Y activity is calculated using the number of neutrons per second determined from ^{89}Zr measured activity as shown in formula (2):

$$A(Ci) = \frac{nd\left(\frac{atoms}{cm^2}\right)\sigma\left(\frac{cm^2}{atoms}\right)N\left(\frac{n}{s}\right)(1-e^{-\lambda t})f}{3.7(10^{10})\left(\frac{dis}{sCi}\right)} \quad (2)$$

where N(n/s) is the number of neutrons per second incident on the Zr cylinder, nd(atoms/cm^2) is the Zr cylinder thickness, σ (cm^2/atoms) is the average cross section ^{90}Zr(n,p)^{90}Y reaction, f is the ^{90}Zr fraction, and λ (s^{-1}) ^{90}Y decay constant. The calculated value of ^{90}Y activity for an irradiation time of 2.5 h and 10 µA current beam was determined to be 0.032 µCi.

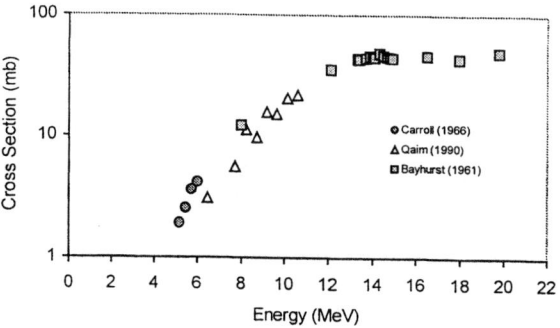

FIGURE 5. The ^{90}Zr(n,p)^{90}Y cross section

The ^{90}Y activity was produced by the (p,xn) reaction with ^{nat}Zr. The efficiency of this method is the fact that neutron rich isotopes may be produced at the same time as proton rich isotopes. Using a special target design, neutrons that are created essentially free, during the production of the proton rich isotopes, can activate a second target for production of the neutron rich isotopes. Therefore, in the same production time, one can obtain two medical isotopes at the same production cost. For example, neutrons can be produced during the production of Tl-201 from the ^{203}Tl(p,xn) reaction and from the ^{nat}Cu(p,xn) reaction. A Zr plate may be added to a slot in the target holder and located just below the center of the copper target plate to maximize the neutron flux on the Zr. Neutrons that are produced from (p,n), (p,2n), (p,3n), etc. reactions with the thallium and copper are produced free in a "piggyback" process with the production of thallium or any other product. The thallium will produce neutrons as the protons lose a portion of their energy energy (depending of the thickness of the thallium), and also the copper will produce neutrons, as the protons lose energy down to 10 MeV where the reaction threshold occurs. Several studies showed the neutrons are produced mainly in the forward direction [15-16], which is the direction of the Zr target material to be neutron activated. The activated Zr or any other element can be then transported in the same target holder in the existing target transfer system to the existing hot cells for chemical processing.

CONCLUSIONS

We have demonstrate an alternative method of fast neutron activation for the production of ^{90}Y that does not involve a ^{90}Sr generator with its waste handling and storage problems. The rate of production of ^{90}Y can be improved with an improved geometry, i.e. decreasing the distance between Zr target and the Rh neutron source. Also, by increasing the proton beam current to 250µA, the irradiation time to 25 hours, the solid angle by a factor of 50, the mass and the isotope purity by a factor of 5, the neutron yield will increase by a factor of 6.25E+05. The great advantage of this technique is the wide selection of fast neutron nuclear reactions available to produce a variety of medical radioisotopes, based on (p, xn) reactions.

The increasing market demand for ^{90}Y due to more specific delivery of radiation of beta particles at higher energy and dose rates, greater safety to health care workers, and lower cost, makes ^{90}Y a very good choice for production using neutron reactions from charge particle accelerators.

ACKNOWLEDGEMENTS

This work was supported by International Isotopes Inc., Robert A. Welch Foundation, National Science Foundation, and The Coordinating Board of the State of Texas.

REFERENCES

1. Coursey, B.M., and Nath, R., *Phys. Today*, April 2000, 25-30.
2. Browne, E., Dairiki, J.M., and Doebler, R.E., *Table of Isotopes, seventh edition*, John Wiley & Sons, Inc., 1978, pp. 343
3. Dietz, M.L., and Horwitz E.P., *Appl. Radiat. Isot.* **43**, 9, 1093-1101 (1992).
4. Ziegler, J.F. "Stopping Power and Range of Ions in Matter," *Instruction Manual, version 96.xx* (1999).
5. EG&G Ortec, *Gamma Vision software*
6. Prestwood, R.J., and Bayhurst, B.P., *Phys.Rev.*, **121**, 1438 (1961).
7. Bayhurst, B.P., Gilmore, J.S., Prestwood, R.J., Wilhelmy, J.B., Jarmie, N., Erkkila, B.H., Hardekopf, R.A., *Phys.Rev. C*, **12**, 451 (1975).
8. Zhao Wen-Rong, Lu Han-Lin, and Fan Pei-Guo, *Rept. INDC(CPR)*, August 18 (1989).
9. Nethaway, D.R., *Nucl.Phys.A*, **190**, 835 (1972).
10. Ehmann, W.D. and Vance, D., *Radiochemistry and Nuclear Methods of Analysis*, John Wiley & Sons, Inc., 1991, pp. 141
11. Patterson, H.W and Thomas, R.H., *Accelerator Health Physics*, Academic Press NY and London, 1973, pp. 128.
12. Caroll, E.E, and Stooksberry, R.W., *J.Nuclear Science Eng.*, **25**, 285 (1996)
13. Qaim, S.M., and Stoecklin, G., *Phys.Rev. C*, **42**, 363 (1990).
14. Bayhurst, B.P., and Prestwood, R.J., *J. Inorg. Nucl. Chem.*, 23, 173 (1961)
15. Nakamura, T., Masahiko, F., and Kazuo, S., *Nucl. Science Eng.*, **83**, 444-458, (1983).
16. Chadwick, M.B., Young, P.G., Chiba, S., Frankle, S.C., Hale, G.M., Hughes H.G., Koning, A.J., Little, R.C., MacFarlane, R.E., Prael, R.E., and Waters, L.S., *Nucl. Science Eng.*, **131**, 293-328 (1999).

Precision Measurement Of The n+p→d+γ Cross Section For Testing Big–Bang Nucleosynthesis Models

E.H. Seabury[1], R.C. Haight[1], J.M. O'Donnell[1], J.L. Ullmann[1], S.A. Wender[1],
T. Akdogan[2], M. Chtangeev[2], W.A. Franklin[2], J.B. Hough[2], J.L. Matthews[2],
Y. Safkan[2]

[1]*Los Alamos National Laboratory, Los Alamos NM*
[2]*Massachusetts Institute of Technology, Cambridge MA*

Abstract. The radiative neutron capture process plays an important role in Big Bang Nucleosynthesis (BBN) models. Recent measurements of the deuterium abundance in high-redshift hydrogen clouds have allowed a precise comparison of the light elemental abundances with BBN-model predictions. These measurements have also demonstrated the need for more precise cross section data. In particular there are very few measurements of the np?d? cross section in the energy region of interest to BBN, none of which cover the important energy range of 50-400keV. Measurements of this cross section are bing undertaken at the WNR facility of the Los Alamos Neutron Science Center (LANSCE).

INTRODUCTION

The radiative neutron capture cross section of hydrogen plays an important role in testing the predictions of Big Bang Nucleosynthesis (BBN) models. The deuterium abundance or D/H ratio has been described as a "cosmic baryometer"[1], i.e. a means of determining the upper limit to the primeval baryon density. This is because only big bang processes contribute to the production of deuterium whereas astrophysical processes tend to destroy it. The D/H ratio depends on both the baryon density and the neutron capture cross section of hydrogen.

Recently, measurements of deuterium abundances in high-redshift hydrogen clouds have been made to an accuracy of about 10% [2,3], allowing a precise comparison of the light elemental abundances with BBN-model predictions. The uncertainty in these measurements is also expected to decrease significantly in the next few years[1]. However, there is an almost complete lack of data[4] for the n(p,d)γ reaction at energies of interest to BBN. Only three sets of measurements have been performed[5], none of which covers the important energy range of 50-400keV. This cross section is therefore a limiting factor in the precision of calculations and BBN-model development.

Measurements of this cross section are being undertaken at the WNR facility of the Los Alamos Neutron Science Center (LANSCE), with the goal of obtaining results in the energy region of interest with an uncertainty of approximately 5%.

EXPERIMENTAL ARRANGEMENT

The experimental arrangement is very simple. It consists of a germanium detector, polyethylene target and shielding for the detector as shown in Figure 1. Neutrons are produced at the WNR facility by impinging a pulsed 800MeV proton beam onto a tungsten target. Neutron energies are obtained from the resulting white spectrum neutron beam by using time-of-flight techniques. The detector is located approximately 10m from the neutron production target. The proton beam consisted of pulses 15ns wide and 3.6μs apart. The neutron flux was monitored by using a fission chamber with a ^{235}U fission foil.

RESULTS

The polyethylene target provides the ability to perform a relative measurement of the np-capture cross section by comparing with the well-known inelastic neutron scattering cross section of ^{12}C. The ^{12}C(n,n') cross section was obtained by generating gamma-energy versus neutron energy matrices and gating on the neutron energy. The area under the 4.4MeV gamma peak in the resulting spectrum was then obtained. A sample gate on neutron energies between 5 and 6 MeV is shown in Figure 2.

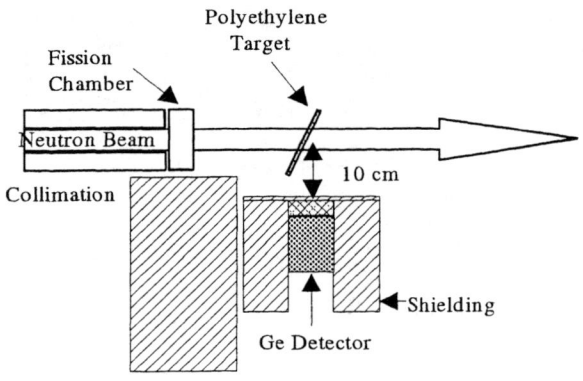

Figure 1. Experimental Arrangement

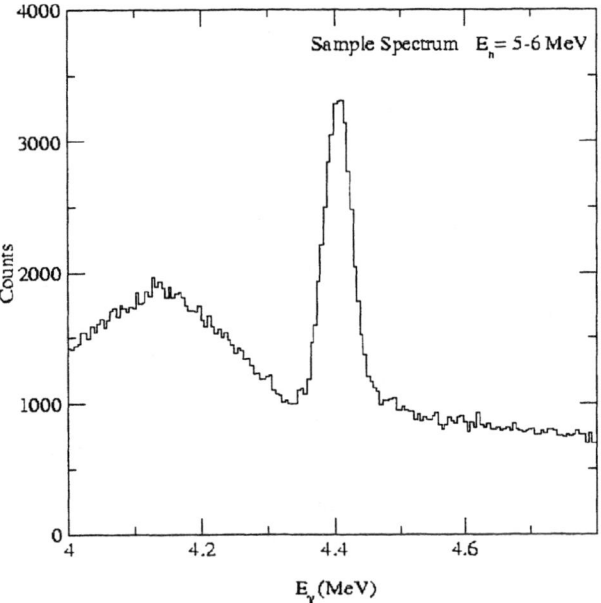

Figure 2. Sample Gamma Spectrum, E_n 5-6 MeV

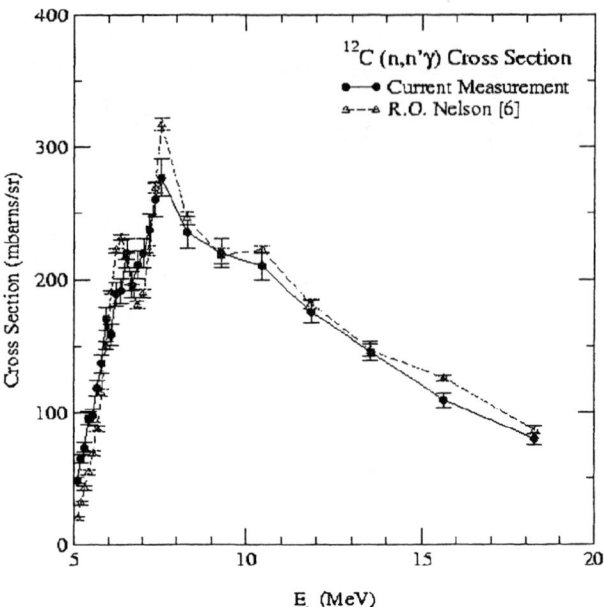

Figure 3. ^{12}C (n,n'γ) Cross Section Comparison

Using the area under the peak and the relative neutron flux obtained from the fission chamber allows a relative cross section measurement to be obtained as a function of neutron energy. The relative cross section is shown in Figure 3 where it is compared with previous measurements performed at WNR on a different flight path[6]. A relative normalization has been performed to account for the efficiencies of the fission chamber and germanium detector, and the previous data set has been rebinned and convoluted to match our energy binning and time-resolution. As can be seen in the figure, there is good general agreement between the two independent measurements.

There are currently limited statistics in the data set. Because the np-capture cross section is predicted to be approximately 50 microbarns at 200keV[7] (compared with 0.26 barns at 6.5 MeV for carbon inelastic scattering) we are currently unable to extract the cross section from the data. The count rate for the reaction should be able to be increased by increasing the beam intensity and using a more efficient germanium detector.

CONCLUSION

The np-capture cross section is of significant importance to studies of big bang nucleosynthesis. The measurements described here are the first to cover the energy region of interest to BBN and with further statistics, will provide cross section data with uncertainties on the order of 5%.

ACKNOWLEDGMENTS

This work is supported in part by the US Department of Energy under Contract W-7405-ENG-36.

REFERENCES

1. D.N. Schramm and M.S. Turner, *Rev. Mod. Phys* **70**, 303 (1998).

2. S. Burles and D. Tytler, *Astophys. J.* **499**, 699 (1998).

3. S. Burles and D. Tytler, *Astrophys. J.* **507**, 732 (1998).

4. S. Burles, K.M. Nollett, J.W. Truran, M.S. Turner, *Phys. Rev. Lett.* **82**, 4176 (1999).

5. Y. Nagi, T.S. Suzuki, T. Kikuchi, T. Shima, T. Kii, H. Sato, *Phys. Rev. C* **56**, 3173 (1997).

6. R.O. Nelson (private communication).

7. ENDF/B-VI Data Files.

A Test of the Exponential Decay Law by Photo-Production of Nuclear Isomers

Douglas P. Wells*, J. Frank Harmon*, Wade W. Scates* and Randy Spaulding*

Idaho State University, Idaho Accelerator Center, Campus Box 8263, Pocatello, ID 83204

Abstract: Modern tests of grand unification theories and the standard model spend considerable experimental effort in pursuit of rare decays. A common feature of these experiments is that they involve extremely rare decay processes and probe regions of the systems' decay curves which are very short compared to their mean lifetimes. A potential complication to interpretations of such experiments is the approximate nature of the exponential decay law for quasi-stationary states. We use the decay of the isomeric nuclear states $^{91}Zr(\gamma, n)^{90m}Zr$ ($t_{1/2} = 0.8$ s) and $^{137}Ba(\gamma, n)^{136m}Ba$ ($t_{1/2} = 0.3$ s) in the short time limit to search for predicted deviations from the exponential decay law. These experiments address the short-time electromagnetic decays of nuclei with half-lives of order a few seconds, and explore the as-yet untapped electromagnetic sector for short-time ($t_{min}/t_{1/2} \approx 10^{-8}$) violations of the exponential decay law. Isomeric states are photo-populated with bremsstrahlung beams from ISU's 30 MeV pulsed electron linac.

Introduction

Modern tests of grand unification theories have recently spent considerable experimental effort in pursuit of proton decay and neutrinoless double beta decay. These experiments are designed to test theories of the fundamental forces and grand unification schemes. A common feature of these experiments is that they probe regions of the systems' decay curves which are very short compared to their expected mean lifetimes. A potential complication to interpretations of such experiments is the approximate nature of the exponential decay law for quasi-stationary states, which has been acknowledged by numerous authors [1, 2, 3, 4]. The first observation of non-exponential decay by Wilkinson et al. in 1997 [5], in a non-nuclear system, lends additional credence to the importance of this phenomena in interpreting rare decays.

Merzbacher has noted that [3], "the exponential decay law can be derived only as an approximate, and not a rigorous, result of quantum mechanics and ... it holds only if the decay process is essentially independent of the manner in which the decaying state was formed and of the particular details of the incident wave packet." Kalfin [6] showed that quantum mechanics predicts that the exponential decay law <u>cannot</u> hold in the short and long time regime and that the decay rate must approach zero as t → 0. Here short and long mean with respect to the half life of the system when measured from the time of preparation of the unstable state. Deviations from the exponential law can be traced to a number of approximations made in its derivation. These include (1) the assertion that the initial level is well separated in energy from other discrete states, (2) that the final group of states is truly continuous, (3) that the spectrum is flat and unbounded (the so-called Weisskopf-Wigner (WW) approximation) and (4), that the continuum couples only to the initial state or, equivalently, that the products of the decay are stable.

It should be noted that several authors [7, 8] have pointed out that an unstable state, when monitored for its existence at sufficiently small time intervals, should live longer than one that is not periodically observed. This effect is sometimes called the quantum Zeno effect and is intimately related to the deviation from the exponential decay law at small times in that the measurement of the existence of the particle amounts to a collapse of its wave-function and a return to the beginning of its decay curve.

Thus in addition to testing a basic prediction of quantum mechanics and fundamental conservation laws, and shedding light on tests of grand unification theories, the observation of non-exponential decay offers the possibility of slowing down the decay rate of radio-nuclides. One obvious application of this is the production of new medical or industrial radio-nuclides through intermediate states that would ordinarily be too short-lived to serve as a production intermediary.

Much of the early theoretical and experimental work focused upon the long-time limit. No deviation from the exponential decay "law" in this regime has ever been observed. For the remainder of this paper, we focus upon the short time regime.

Experimental Tests to Date

It is a well known experimentally that the exponential decay law is valid to a very high degree of accuracy in the intermediate time domain where t is not too different from $t_{1/2}$, and both are measured from the time of preparation of the unstable system. Since it is also known from fundamental theoretical considerations that the decay is non-exponential for a short time after preparation, it follows that there must be some intermediate time in which the transition from non-exponential to exponential decay occurs. The interval of non-exponential decay $[0, t_{ne}]$ depends on the energy spectrum of the system, on the decay mechanism, and on the details of the preparation of the unstable particle. To the extent that no firm statements concerning the region of non-exponential decay have been made, the numerous theoretical investigations exploring decay in this time regime have been inconclusive. Thus the current theoretical limits lead to the conclusion that non-exponential effects could occur at any time as long as t < < $t_{1/2}$. Despite this, there have been few attempts to test the exponential decay law at short times.

In nuclei, short-time non-exponential decay has yet to be observed. Norman et al. [9] conducted experiments that studied the weak decays of ^{56}Mn down to 0.3 half lives and ^{60}Co down to around 10^{-4} half lives and found no deviations. In the experiments on ^{56}Mn and ^{60}Co, the isotopes were produced in a reactor over a very short time interval, and their subsequent beta decay curves were observed. More recently, Norman et al. performed a weak-decay experiment on ^{40}K [10], where the decay rate of a newly prepared sample was compared to that of naturally occurring ^{40}K and found to be the same to an accuracy of ±11%. This experiment pushed the observational time-scale $t_{min}/t_{1/2}$ down to the 10^{-10} level.

The first short-time positive result (in a non-nuclear system) was reported in 1997 by Wilkinson et al. [5]. They observed short-time departures from exponential "decay" due to quantum tunneling of a system of ultra-cold sodium atoms that were trapped in an optical potential. Surprisingly, they found violations of exponential decay for time scales as long as $t / t_{1/2} \approx 10^{-2}$ - 10^{-1}. This lends credence to the theoretical work that suggests that the time-scale for violations from exponential decay is sensitive to the form of the interaction or the preparation of the unstable system [2, 4]. The discovery of this phenomena in a non-nuclear system lends new urgency to the discovery and exploration of the systematics of non-exponential decay in nuclei.

It may be that deviations from exponential decay are too small in nuclear or particle systems to be detected with current experimental techniques. However, the recent demonstration of this effect in trapped sodium atoms and the absence of experimental limits in a variety of nuclear and particle systems leaves open the question of the importance of this effect in interpreting rare decays. Further, the observation of non-exponential decay may open the door to new applications, such as the production of new medical or industrial isotopes. We are extending the weak-decay studies of Norman et al. to the electromagnetic sector using nuclear isomeric decays.

Experimental Methods

We are investigating the short-time electromagnetic decays of isomeric nuclei with half-lives of order a few seconds. These experiments explore the as-yet untapped electromagnetic sector for short-time violations of the exponential decay law.

We are using (γ, n) reactions to produce short-lived (~ seconds) isomeric nuclei. The subsequent purely electromagnetic isomeric transitions (IT) will be measured as a function of time after irradiation. Isomeric states will be populated with high-intensity pulsed bremsstrahlung photon beams. ISU's 30 MeV Electron LINAC is capable of producing pulse widths from 10 pico-seconds to 10 micro-seconds. A bremsstrahlung photon beam (maximum energy = 30 MeV) will be produced with an electron beam of 5 nano-second width (200 nano-Coulombs per pulse) incident upon a thick tungsten target (20 g/cm^2). This yields a photon flux per pulse above 8 MeV (nominal neutron threshold) of approximately 10^{11} photons/cm^2. The energy spread of this flux will completely span the predominantly (γ, n) giant dipole resonance of these nuclei, with typical peak cross sections of a few hundred mb.

Decay gammas will be detected with high-efficiency germanium detectors. Detectors will be initially shielded with 10 cm of lead, with a 7.5 cm diameter collimator. Additional shielding will include a NaI(Tl) annular Compton-suppression shield and borated paraffin. We will count from t = 10 ns to t = 20 half lives, where each time decade (10 ns to 100 ns, for example) will be routed to its own energy spectrum. Events will also be recorded individually for later "play-back" and analysis. These pulsed time/energy yields will be repeated until adequate statistics are obtained, typically about one week's irradiation. Pulses will be separated by 20 half-lives to ensure that essentially all of the activation from the previous pulse has decayed.

The targets will be of natural isotopic composition and approximately 10 g/cm^2. The nuclear reactions will be ^{91}Zr(γ, n)^{90m}Zr ($t_{1/2}$ = 0.8 s), ^{137}Ba(γ, n)^{136m}Ba ($t_{1/2}$ = 0.3 s).

To estimate the yield of isomeric nuclei, the number of isomeric nuclei (N) produced in a sample over the duration of the experiment is:

$$N = n_p \, n_t \, b_I \, a_I \int \sigma(E) \, \phi(E) \, dE, \quad (1)$$

where n_p is the number of irradiation pulses, n_t = is the number of target nuclei, b_I is the branching ratio of the giant resonance excited state to the isomeric state, a_I is the natural isotopic abundance of the target nuclei, $\sigma(E)$ is the cross section of the activation reaction, and $\phi(E)$ is the photon flux from 200 nC of incident electron beam. The b_I were taken from Gangrskii et al. [11], while the a_I were taken from the "Chart of the Nuclides". Photon fluxes were calculated with Sandia's "ACCEPT" code and the giant dipole resonance cross sections were taken from Berman and Fultz [12].

Since we will count until essentially all of the induced activity has decayed, the total experimental yield is then:

$$Y_{tot} = \epsilon_\gamma(E) \, I_\gamma \, N \quad (2)$$

where $\epsilon_\gamma(E)$ is the detector efficiency (including solid angle and other geometric effects) for the associated isomeric transition gamma energy and I_γ is the relative intensity (branching ratio) of that transition. The yield for a given time decade is, if we assume exponential decay:

$$Y_{\Delta t} = \lambda \, \Delta t \, Y_{tot} \quad (3)$$

where λ is the decay constant of the reaction product; and Δt is the time width of the decade in question. This leads to a few hundred counts in the shortest time-decade (10 ns to 100 ns) in a one-week run, with commensurately larger yields for each longer time-decade.

In order to estimate sensitivities to non-exponential decay effects, some simple assumptions about the shape of the decay rate curve were made. It is known that at times comparable to the mean lifetime the curve is essentially exponential, and the rate is expected to be zero at t=0. For the purposes of investigating sensitivities, the decay curve was parametrized as a linearly increasing decay rate from t = 0 to t = t_{ne}, and exponential thereafter:

$$R(t) = a_1 \, t \text{ for } t < t_{ne} \quad (4)$$

and

$$R(t) = a_2 \, e^{-wt} \text{ for } t > t_{ne} \quad (5)$$

where t_{ne} is a variable parameter which characterizes the transition time between the exponential and non-exponential regions, w is the commonly accepted decay constant, and a_1 and a_2 are determined by normalization and continuity requirements.

For the short lifetime experiments ($t_{1/2}$ ~ seconds), the expected yield in a given decade of the decay (if the decay is exponential) is: $Y_{\Delta t} = \lambda \, \Delta t \, Y_{tot}$. On the other hand, if the decay follows the simple linear model above from one end of the time-decade to the other, the yield will be approximately: $Y_{\Delta t} = 0.5 \, \lambda \, \Delta t \, Y_{tot}$. Thus the experimental signature of non-exponential decay in this simple model is a yield of half of that expected from exponential decay.

Each increase in time decade has approximately a factor of 10 more counts, as long as the time-scale is still less than the half-life. Thus for the ^{91}Zr(γ, n)^{90m}Zr and ^{137}Ba(γ, n)^{136m}Ba reactions, it should be relatively easy to observe the onset of non-exponential decay on time scales greater than 10 ns if background rates are relatively small in comparison. Note that because we are comparing different parts of the nearly complete decay curve in these experiments, we are not sensitive to the usual systematic errors associated with absolute photo-nuclear cross sections, photon flux, target thickness, etc. The predominant uncertainties in these experiments will be the counting uncertainties.

The background rates from natural background is completely negligible when compared to the expected rates from these experiments. Other backgrounds stem from neutron activation of the detector's environment, from neutron capture in the germanium detector, and from photon-induced activation of materials other than the target isotope. Our estimates of these rates in the energy region of interest are based on an extrapolation of our experience on μs time scales. In general, these background rates are relatively small when compared to the expected rates.

Of greater concern are pileup and dead-time due to the lower-energy portion of these beam-related background rates. This can "mis-route" useful events, add background counts to the useful part of the spectrum, and "lose" useful counts from the spectrum. Without *a priori* knowledge of these effects to our spectra on the nanosecond time-scale, we assumed a nominal absolute efficiency (intrinsic efficiency times solid angle) of 0.01 throughout these estimates. This allows us room to optimize our electronics and passive and active neutron/gamma shielding to minimize pileup, deadtime and background contributions to the spectrum. We expect a final efficiency somewhat better than this nominal value.

Summary

We are pursuing an experiment that will search for the much discussed, but never observed effect of non-exponential decay in nuclei. Non-exponential decay has recently been observed (for the first time) in ultra-cold sodium atoms trapped in an optical potential; however, the absence of any observation of this effect in nuclei or particles and its major ramifications for rare decays and new applications makes these measurements extremely important.

Acknowledgements

*Supported in part by the U.S. Department of Energy under DOE Idaho Field Office Contract DE-AC07-99ID13727.

References

1. Jacob, R., R.G. Sachs, Phys. Rev., **121**, no. 1, (1961) 350.
2. Goldberger, M.L., K.M. Watson, Phys. Rev., **136**, no. 5B, (1964) 1472.
3. Merzbacher, E., *Quantum Mechanics*, second edition, 1970.
4. Peres, A., Ann. of Phys., **129** (1980) 33.
5. Wilkinson, S.R., et al., Nature **387**, 575 (1997).
6. Khalfin, L.A., JETP Lett., **8**, (1968) 65.
7. Ekstein, H., A.J.F. Siegert, Ann. of Phys. **68**, (1971) 509.
8. Sudbery, A., Ann. of Phys. **157** (1984) 512.
9. Norman, E., S.B. Gazes, S.G. Crane, D.A Bennett, Phys. Rev. Lett. **60**, no. 22, (1988) 2246.
10. Norman, E., et al. Phys. Lett. **B 357** (1995) 521.
11. Gangrskii, Y.P., A.P. Tonchev and N.P. Balabanov, Phys. Part. Nucl. **27**, 428 (1996).
12. Berman, B.L., and S.C. Fultz, Rev. Mod. Phys. 47, 713 (1975).

Study of reactions of fast neutrons with nuclei with FIGARO at LANSCE

L. Zanini[1], A. Aprahamian[2], M. B. Chadwick[1], M. Devlin[1], R. C. Haight[1], J. X. Saladin[3], P. G. Young[1]

[1]*Los Alamos National Laboratory, Los Alamos, NM 87545*
[2]*University of Notre Dame, Notre Dame, IN 46556*
[3]*University of Pittsburgh, Pittsburgh, PA 15260*

Abstract. We present the new FIGARO facility under construction at the Weapons Neutron Research facility at LANSCE, for the measurement of γ rays and neutrons from interactions of fast neutrons with nuclei. A first measurement was performed in 1999 with ^{59}Co, using a single germanium detector. Prompt γ rays from reactions with neutrons in the energy range from 1 to 200 MeV have been measured. Preliminary cross sections for γ transitions from (n,n'), $(n,2n)$, $(n,3n)$, (n,p), (n,np), $(n,2np)$ and (n,α) reactions have been obtained. The results have been compared with existing data, and with calculations using the code GNASH. They indicate that with FIGARO high quality data can be obtained in the considered energy range.

INTRODUCTION

From the study of reactions of fast neutrons with nuclei in which γ rays are produced, it is possible to obtain information on the nuclear level densities (1). In the neutron energy range available at the Weapons Neutron Research (WNR) facility at LANSCE (1 to several hundreds MeV), neutrons produce reactions with nuclei, leaving the residual nucleus in an excited state, which may decay by γ emission. The energies of the emitted γ's are a signature of the residual nucleus, and therefore of the type of the reaction taking place. Information on the level density at high excitation energies can be obtained from the population of low-lying excited states. Further information can be obtained if one or more of the emitted neutrons are detected: in this case an $n-\gamma$ coincidence measurement allows not only to identify the type of reaction but also to determine the excitation energy of the residual nucleus. Using this technique it is also of great interest to investigate the level densities near the neutron separation energy. This kind of measurement would provide information on the angular momentum distribution of the level density.

In 1999 we have begun setting up the FIGARO (Fast neutron-Induced GAmma Ray Observer) experiment at WNR, for the study of this type of reactions. Besides measurements of nuclear level densities for a broad range of excitation energies, FIGARO can be used for other types of applications, like for instance the study of transitions where the internal conversion coefficient is high,

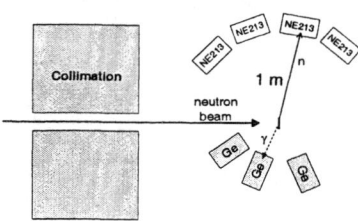

FIGURE 1. Schematic view of the FIGARO experimental setup.

by means of detection of conversion electrons, in order to extract information on nuclear structure, or measurements of interest for waste transmutation.

In its final configuration, the basic components of FIGARO will be (Fig. 1): *i*) an array of three-four high purity germanium detectors, for the detection of γ rays; *ii*) a similar number of neutron detectors, which will be NE213 liquid scintillators because of their capability to perform pulse shape discrimination and separate the γ-ray background; *iii*) according to the needs of a particular experiment, different types of detectors will be installed. An example of an alternative detector is constituted by the

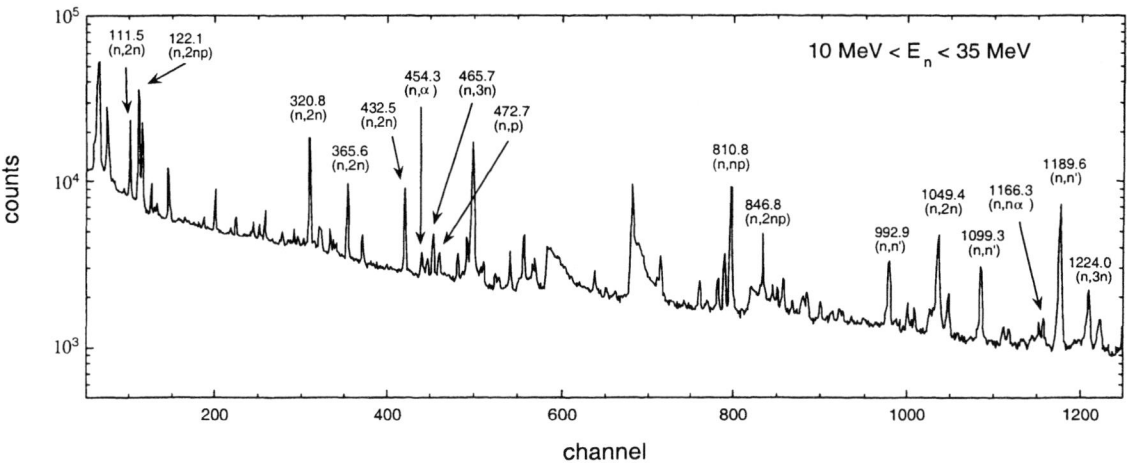

FIGURE 2. Part of γ-ray spectrum from interactions of neutrons in the energy range 10-35 MeV with ^{59}Co, obtained from a 24 hours measurement with FIGARO. Energies of γ rays from different reactions are indicated in keV.

ICEBALL II array (2), from the University of Pittsburgh, for the measurement of conversion electrons (3).

MEASUREMENT WITH COBALT

Experimental Setup

We began to build the facility in 1999. The first goal was to evaluate the possibilities for this kind of measurement at the experimental station chosen for FIGARO. The experiment was prepared on the flight path at 30° with respect to the incoming proton beam, at a distance of about 22 m from the neutron source.

Collimation of the neutron beam is very important for this type of experiment. High energy neutrons, above a few MeV, have great penetrating power and they induce reactions that make lower energy neutrons and γ rays. Present-day germanium detectors with large volumes are very sensitive to these background radiations. Because these detectors have no directional sensitivity (until tracking detectors are available), background radiations incident from all directions are indistinguishable from γ rays from neutron interactions in the sample under study. For these reasons, we concentrated on the design of beam collimators and of shielding to reduce radiations from the collimators from reaching the detectors.

The beam line consists of an adjustable shutter, pre-collimators at 9 and 15 m from the source, a defining collimator at 19 m followed by clean-up collimators. Shielding of iron or copper surrounds each of these collimators.

Lead is used as the last of the clean-up collimators. The defining collimator for the experiments reported here had a diameter of 2 cm. Inserts are now available to reduce the beam diameter further. Care was taken to further shield the γ-ray detector from backgrounds produced at neighboring flight paths.

For this measurement we used one germanium detector, of about 25 % efficiency. The detector was placed at about 15 cm from the sample, at an angle of 130° with respect to the beam line. A fission chamber (4) was placed in the beam at a farther distance, of about 24 m from the neutron source, for the measurement of the neutron flux.

Measurements were performed with ^{59}Co. This sample was chosen to support previous studies of $(n,x\alpha)$ reactions performed at WNR (5); moreover, we could compare with previous data from a similar measurement performed in Oak Ridge (6).

The sample consisted of monoisotopic ^{59}Co, cut in ten square tiles of area 6.5 cm^2 each, and of 0.05 cm thickness. They were arranged forming samples with one, two or five layers: by measuring with samples of different thicknesses we could evaluate the multiple scattering inside the sample (1), which can be important at high incident neutron energies.

The accelerator was running at 120 Hz repetition frequency, with an average current of 6 μA. The proton beam had a time structure of macropulses, about 625 μs long, composed by micropulses spaced of 1.8 μs, each micropulse being about 1 ns long. In this configuration, at the FIGARO flight station neutron energies above 1.5 MeV could be measured.

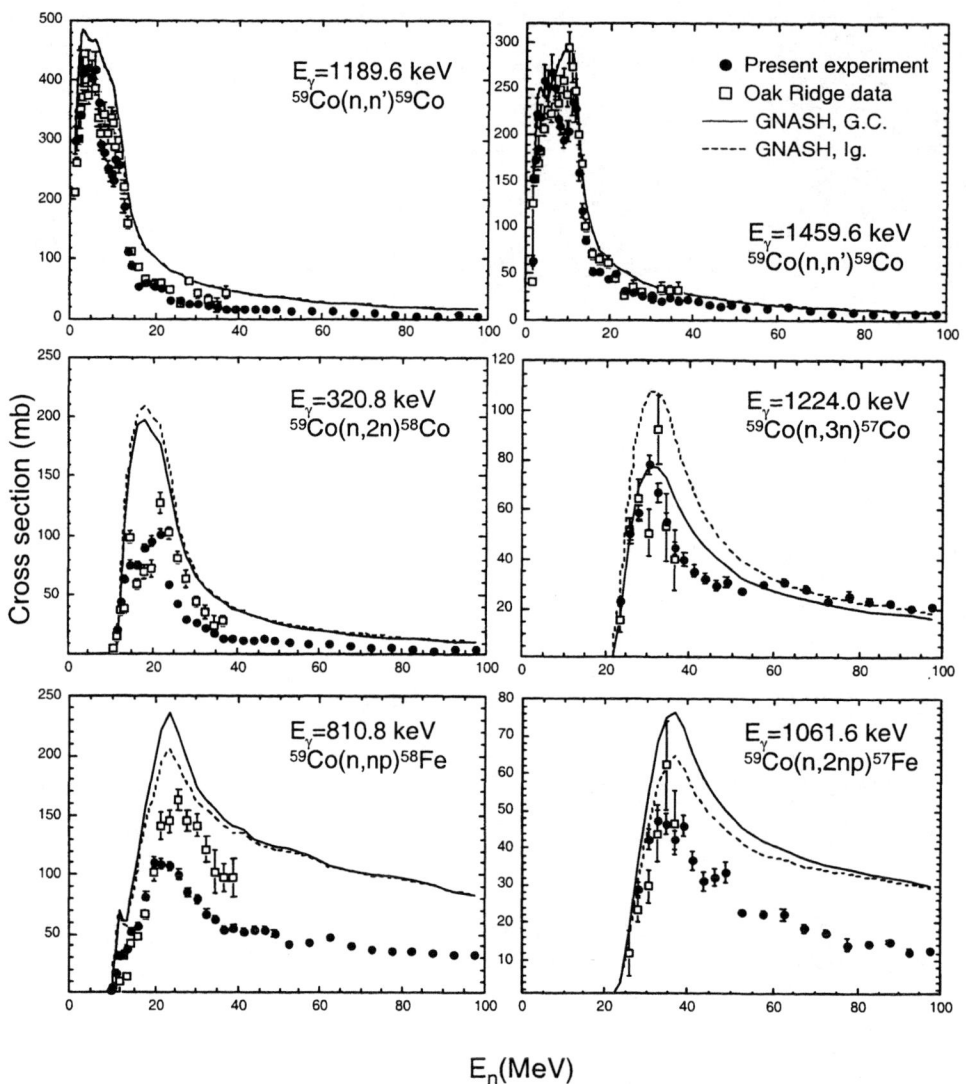

FIGURE 3. Cross sections for γ rays of indicated energies, from different ^{59}Co$(n,xnp\,\gamma)$ reactions, for incoming neutron energies from 1 to 100 MeV. Data from the present measurement are indicated with black dots; they are normalized to the data from Ref. (6), indicated with open squares. Statistical uncertainties only are shown. Results from GNASH calculations are represented with full and dashed lines, using the Gilbert-Cameron and Ignatyuk level density models, respectively.

Data were taken with the cobalt samples in the different configurations. The data shown in this paper refer to a 24 hours measurement only. Measurements were also taken with the sample out to measure the background. The electronics setup consisted of standard NIM and CAMAC modules. Data were collected using the VMS based data acquisition system XSYS (7). The standard fast/slow coincidence logic was applied to measure the time and amplitude of the signals from the germanium detector, as well as the fission chamber signals.

Data Analysis and Results

In Fig. 2 part of the γ-ray spectrum measured by the germanium detector, for energies of incoming neutrons in the range from 10 to 35 MeV, is shown. Several lines corresponding to different reactions are observed; the stronger ones and the corresponding reactions are indicated in the figure.

The data were sorted out in selected neutron energy bins from the two-dimensional spectra representing the pulse height spectrum versus the neutron time-of-flight. The widths of the neutron energy bins were different, increasing with the neutron energy, from 1.5 MeV to 200 MeV. The neutron energy was determined by time-of-flight, having as a starting time reference the time T_0 determined by the γ flash emitted in the spallation process inside the neutron source. A correction for the dependence of the T_0 on the time-of-flight, which was due to a walk effect, was applied.

The obtained spectra were analyzed with the fitting program XGAM (8), which allowed to fit simultaneously the entire γ spectrum from 0.1 to 3 MeV.

The cross section for a specific γ for energies of incident neutrons in the interval ΔE_n was obtained from the formula

$$\sigma(E_\gamma, E_n) = \frac{\text{counts}_{\gamma, \Delta E_n}}{\varepsilon_{\text{det}} \times (\text{N. neutrons}_{\Delta E_n}) \times (\frac{\text{N. atoms}}{\text{cm}^2})_{\text{sample}}}. \quad (1)$$

where ε_{det} is the germanium detector full-peak efficiency. The number of neutrons in the energy range ΔE_n within the solid angle determined by the collimator was measured by the flux monitor. A correction for the dead time was also applied.

Given the uncertainty in the knowledge of the exact amount of ^{235}U deposit in the fission chamber at our disposal, and of the uniformity of its distribution, we normalized our data to the Oak Ridge data, using for normalization the cross section for the 1189.6 keV line. For future measurements, absolute cross sections will be determined by using a well characterized flux monitor, and by comparing with reference cross sections.

We obtained the excitation functions for several γ rays. In Fig. 3 six excitation functions for five different reactions are shown. Data from Ref. (6) are also shown. As can be seen, the statistical uncertainties for the present measurement are considerably lower. For absolute measurements we will have to take into account also systematic uncertainties. The main contribution to the systematic uncertainties comes from the uncertainty in the amount of ^{235}U in the fission chamber. At neutron energies above 30 MeV, the uncertainty in the ^{235}U fission cross section becomes significant.

Cross sections for the same γ rays were calculated using the reaction code GNASH (9), which includes sequential Hauser-Feshbach calculations and pre-equilibrium processes. Additionally, direct (n,n') contribution to low-lying states in ^{59}Co were calculated using DWBA theory. For the γ decay part, the $E1$ γ-ray strength function model of Kopecky-Uhl (10) was used. Lorentzian models for the $M1$ and $E2$ strength functions were also used. Two models of level densities were tested: the Gilbert-Cameron (11) and the model by Ignatyuk (12). The results of the calculations with the two level density models are shown in Fig. 3 for six transitions.

For most γ rays both models are generally higher than the data. For the 1189.6 keV transition, direct reaction contributions were included for the excitation of the initial level, and at higher incident neutron energies the calculated results are dominated by the direct contribution. Discrepancies between theory and experiment at higher energies could therefore be reduced through future DWBA calculations, perhaps using a smaller deformation parameter for this transition. For the 1459.6 keV transition, which also comes from the (n,n') reaction, the agreement is better also at higher energies. In other cases the disagreement may originate from incomplete knowledge of level schemes, of spins and parities of low-lying levels, and of the branching ratios for γ transitions.

An absolute measurement with ^{59}Co, as well as with other isotopes, using the complete FIGARO setup, will be soon performed; since, for most of the transitions, GNASH calculations using different level density models give different results (see Fig. 3), an absolute measurement will allow a better comparison with calculations, and the determination of the best level density model. Furthermore, the measurement in coincidence of the emitted neutrons and γ's will constitute a powerful tool for the study of nuclear level densities.

The preliminary results shown in this paper demonstrate that precise measurements of γ-ray production cross sections are possible at FIGARO, in the neutron energy range from 1 to hundreds of MeV.

REFERENCES

1. H. Vonach *et al.*, Phys. Rev. **C 50**, 1952 (1994).
2. M. P. Metlay, J. X. Saladin, I. Y. Lee, O. Dietzsch, Nucl. Instrum. Methods **A 336**, 162 (1993).
3. M. Devlin *et al.*, these proceedings.
4. S. A. Wender *et al.*, Nucl. Instrum. Methods **A 336**, 226 (1993).
5. S. M. Grimes *et al.*, Nucl. Sci. Eng. **124**, 271 (1996).
6. T. M. Slusarchyk, ORNL/TM-11404 (1989).
7. N. R. Yoder, Internal Report, Indiana University (1991).
8. W. Younes, private communication, 1999.
9. P. G. Young, E. D. Arthur, M. B. Chadwick, Report LA-12343-MS (1992).
10. J. Kopecky and M. Uhl, Phys. Rev. **C 41**, 1941 (1990).
11. A. Gilbert and A. G. W. Cameron, Can. J. Phys. **43**, 1446 (1965).
12. A. V. Ignatyuk, G. N. Smirenkin, A. S. Tishin, Yad. Fiz. **21**, 485 (1975) [Sov. J. Nucl. Phys. **21**, 255 (1975)].

Nuclear Astrophysics Measurements Using Neutrons

Paul Koehler

Physics Division, Oak Ridge National Laboratory, Oak Ridge TN 37831-6354

Abstract. Recently, there has been a resurgence of interest in experimental neutron nuclear astrophysics. New precision abundance determinations, recent changes and improvements in stellar models, and the realization of new ways in which neutron experiments could yield vital nuclear astrophysics data have all contributed to the need for new experiments. New (n,γ) and (n,α) cross-section data are needed for both stable and radioactive nuclides across a broader range of energies and with much higher precision than previously available. With the increase in precision of the (n,γ) data, new inelastic neutron scattering data are needed so that the stellar enhancements of reaction rates measured in the laboratory can be determined. Recent improvements in experimental techniques have made it possible to begin to fulfill these needs. In addition, the extremely high fluxes at the new spallation neutron sources should enable a large expansion in (n,γ) cross-section measurements on relatively short-lived radioactive nuclides as well as on stable nuclides having very small natural abundances.

INTRODUCTION

Recent improvements in elemental and isotopic abundance measurements, both in meteorites and stars, have provided a rich store of new data with which to test and improve astrophysical models. At the same time, advances in the size and speed of computers have made possible more realistic models of stars and explosive astrophysical objects such as novae and supernovae, and the nucleosynthesis that occurs inside them. These advances have led to the need for more and better determinations of the rates for the nuclear reactions that form the backbone of the networks determining the nucleosynthesis output from these astrophysical environments. The data needs are many and varied.

For elements heavier than iron, nucleosynthesis is dominated by an interplay between chains of (n,γ) reactions and β decays. In the slow-neutron-capture process (s process) [1], neutron captures occur on a time scale that is slow compared to β decay. As a result, the nucleosynthesis flow follows the valley of β stability and the resulting isotopic abundances are roughly proportional to the (n,γ) reaction rates. Because the s process proceeds mainly through stable nuclides, many of the (n,γ) reaction rates have been measured. However, the uncertainties on many of the previously measured rates are too large to be useful, so they need to be remeasured. Also, the most recent and successful s-process models indicate that most of the neutron exposure during the s process occurs at much lower temperatures (kT = 6 - 8 keV) than previously thought (kT = 30 keV). Many of the old (n,γ) measurements were made over an energy range that was too small to determine the reaction rates at the lower temperatures needed by the new models, so the data were extrapolated to obtain the rates. Recent studies [2] have shown that such extrapolations are often not reliable. Therefore, new measurements are needed to lower neutron energies, especially for the so-called s-only isotopes (isotopes produced solely or at least predominantly by the s process) and for isotopes having precise abundance ratios measured in meteoric grains.

The p process [3] is the name given to the mechanism by which the low-abundance, proton-rich isotopes of the intermediate to heavy elements were synthesized. Little is know about the details, but is seems certain that the p isotopes originated in a high-temperature environment (perhaps in supernovae or in the later burning stages in massive stars) where seed nuclides built up by a previous s process were "photo-eroded" via (γ,n), (γ,α), and (γ,p) reactions towards more proton-rich isotopes. At present, p-process calculations must rely on theoretical rates for most of the reactions in the nucleosynthesis flow. The largest nuclear physics uncertainties in the *p* process and other high-temperature environments, are the rates for reactions involving α particles. Direct (γ,α) measurements are typically extremely difficult (even via the inverse reaction) and the rates currently are poorly constrained by theory. The main difficulty in the theory is that the α-nucleus potential is poorly constrained. Recent work [4] has shown that a series of low-energy (n,α) measurements, across a wide range of masses, appears to be the best means of constraining the α-nucleus potential and thus improving the calculation of these rates.

The reaction rates inside the thermal plasma of a star can be significantly different from the rates measured in the laboratory due to reactions involving thermally populated excited states. These stellar enhancement effects cannot be directly measured, but can be determined by measuring neutron inelastic cross sections to the same levels populated in the stellar environment. The enhancement of the stellar reaction rates due to this effect can be as large as 30% and the enhancements calculated by various nuclear statistical models can differ by substantial amounts. Such large and uncertain effects are particularly troublesome for the *s*-only isotopes that are the main calibration points for *s*-process models. Five of the thirty *s*-only isotopes are calculated to have enhancements in excess of 10%. In contrast, current techniques can determine the laboratory rates to 1-3% accuracy and isotopic abundances can often be measured with per mil accuracy. There are about 25 other nuclides along the *s*-process path with calculated enhancements larger than 10%. A program of (n,n') measurements for these isotopes is clearly needed to determine the enhancements for *s*-only isotopes and to improve the statistical models so that reliable enhancements can be calculated for other nuclides of interest.

Finally, there are several radioisotopes along the *s*-process path with half lives long enough that there is a branching in the reaction flow due to the competition between β decay and neutron capture. If the (n,γ) reaction rates could be measured for these nuclides, then they would provide a natural handle on the dynamics of the *s*-process environment. Because the radioactive samples needed for these measurements are difficult and expensive to produce, and because backgrounds from the decays of such samples can be large, techniques that require the minimum sample sizes are favored. In general, activation methods require the smallest amount of sample and the least stringent sample purity requirements. However, the activation method works only for a small subset of the nuclides for which measurements are needed and the reaction rate is obtained at only one or two temperatures. Because of their extremely high fluxes, measurements at spallation sources require much less sample material than previous time-of-flight techniques and hence offer the possibility of making measurements on a fairly wide range of radioactive samples. For the same reason, spallation sources offer the possibility of measuring the reaction rates for stable isotopes having very small natural abundance. Such measurements are needed for many isotopes involved in the *p* process and would also be very useful for improving nuclear models which must still be relied upon for calculating the rates for the many reactions that remain unmeasurable.

SOME RECENT EXPERIMENTS

Recent measurements [2] have demonstrated that contributions to the astrophysical reaction rates at kT=6-8 keV due to resonances below the low-energy cutoff of previous experiments can be very large (up to 70%). Furthermore, these measurements have shown that the extrapolations used to account for the contribution of resonances below the measured range in even the most recent high-precision measurements have typically been in error by factors of 2 to 3 times the estimated uncertainties. Therefore, although the reaction rates at the canonical *s*-process temperature of 30 keV may be determined to a precision of 3%, without low energy measurements the rates at kT=6-8 keV are much less precise. These results demonstrate the importance of cross section measurements in the E_n=100 eV to 10 keV energy range, especially for *s*-only isotopes where high precision is require. So far low energy measurements have been made for only a few of the 25 *s*-only isotopes.

Isotopic abundance ratios have been measured with per mil accuracy for several elements found within microscopic grains of SiC and other refractory materials recovered from certain primitive meteorites. The measured ratios for many elements match general expectations for grains formed in the atmospheres of Asymptotic Giant Branch (AGB) stars inside of which the *s* process had occurred. However, detailed *s*-process calculations failed to reproduce the measured ratios, so it was suggested that the input (n,γ) cross sections were in error. Subsequently, new and improved (n,γ) measurements were made and it was demonstrated that the old data were incorrect. For example, with new precision ^{142,144}Nd(n,γ) measurements [5] it was demonstrated for the first time that stellar s-process calculations were in precise agreement with the meteorite data. In addition, it was shown that calculations employing the "classical" s-process model overproduced the s-only isotope ^{142}Nd, providing some of the first evidence that this simplified model was no longer adequate to explain the observations. Similarly, the meteoric abundance ratios for isotopes of barium were not in good agreement with the s-process model calculations until new precision measurements were made. However, although new precision reaction rate measurements from different laboratories are in agreement to within the experimental uncertainties, a significant discrepancy still exists between the s-process calculations and the meteorite measurements for ^{137}Ba [6]. At present, this discrepancy is not understood, but it appears as if the problem lies in the model and can no longer be attributed to errors in the reaction rates. There are several other elements for which precise isotope ratios have been measured. More precision (n,γ) reaction rate measurements are needed to take advantage of the opportunities presented by the meteoric data for testing and improving s-process models.

First measurements of the ^{143}Nd and ^{147}Sm(n,α) cross sections [4] were made across the range of energies needed for astrophysics applications. Previous measurements of this type were limited to energies below a few keV (which is too small of an energy range to be useful for comparison to statistical models) due to overload problems in the detectors and associated electronics resulting from the γ flash at the start of each neutron pulse. In the new experiments, this problem was overcome by employing a compensated ion chamber (CIC) as the detector. Although a CIC can have poorer pulse-height resolution than, for example, a gridded ion chamber, it allows the γ-flash background to be reduced to the point where measurements are possible to much higher energies (500 keV in the cases of ^{143}Nd and ^{147}Sm). The resulting cross sections in the unresolved region are compared to statistical model calculations in Fig. 1.

As can be seen, the older calculation of Holmes et al. [7] is much closer to the data than the newer NON-SMOKER [8] or MOST [9] calculations, which differ from the data by about a factor of 3 in opposite directions. The better agreement of the older model may be due to a fortuitous cancellation of effects due to the neutron and α potentials. The newer models employ a neutron potential that is known to be more reliable in this mass region. The sensitivity of the calculated cross sections to the α potential and level densities employed in the model was studied [4]. It was found that differences of about a factor of 30 could be accounted for in the variation of the potential alone. The different level density prescriptions changed the cross section by a factor of about 1.5, by far smaller than the effect of the α potential. More (n,α) data across as wide a range of masses and energies as possible are needed to constrain the several parameters thought to be necessary to define the global α potential needed for astrophysics applications. Counting rate estimates based on these experiments indicate that as many as 30 measurements should be possible across the mass range from S to Hf.

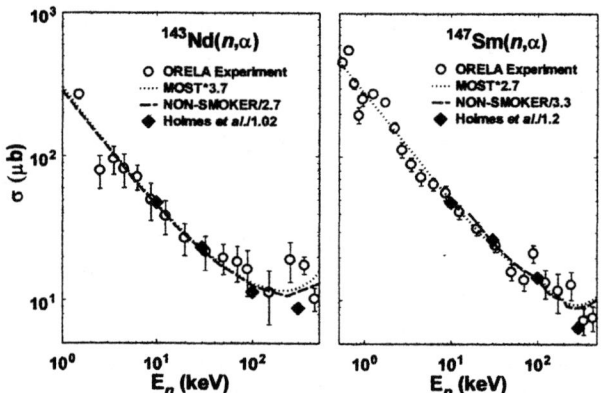

FIGURE 1. Cross sections for the ^{143}Nd and ^{147}Sm (n,α) reactions in the unresolved region. Shown are the measurements of the present work (circles with error bars), and calculations by Holmes et al. (diamonds), as well as calculations using the newer statistical model codes NON-SMOKER (long-dashed curves), and MOST (dotted curves). Note that the theoretical calculations have been normalized by the factors given in the legends.

The stellar enhancement factor (SEF) for one nuclide along the s-process path (^{187}Os) was determined [10] through neutron inelastic scattering measurements several years ago. This measurement was particularly difficult because the excitation energy of the first excited state in ^{187}Os is only 9.8 keV; hence, it was not possible to detect the de-excitation γ rays directly and it was difficult to resolve the inelastically scattered neutrons from the larger elastic group. However, by using a clever technique exploiting the excellent time-of-flight resolution available at the Oak Ridge Electron Linear Accelerator (ORELA) facility, it was possible to measure this cross section. There are 4 s-only isotopes (^{154}Gd, ^{160}Dy, ^{170}Yd, and ^{176}Hf) in addition to ^{187}O calculated to have SEFs greater than 10%, and there are substantial differences between the SEFs calculated by different statistical models [7,8,11]. Measurements should be easier for these 4 s-only isotopes than for ^{187}Os because both the excitation energies and the natural abundances are higher. Given advances in detector technology since the ^{187}Os experiment, it may even be possible to measure the de-excitation γ rays directly. There are about 25 other nuclides along the s-process path calculated to have SEFs larger than 10%. First excited state energies range from 8.4 keV (^{169}Tm) to 100.106 keV (^{182}W). For those with the smallest excitation energies it appears as if a technique similar to that used in the ^{187}Os experiment will be necessary. However, it should be possible to use a flight path about a factor of 4 shorter (and hence have a higher counting rate or use smaller samples) and still obtain resolution sufficient to separate the elastic and inelastic groups.

First proof-of-principle experiments [12] demonstrating the ability to make (n,γ) measurements across a fairly wide range of energies of interest to astrophysics on very small (mg) samples of stable isotopes were performed some years ago at the Manuel Lujan Jr. Neutron Scattering Center (MLNSC). The first (n,γ) measurement on a relatively short-lived nuclide (^{171}Tm) was recently reported [13] using the same apparatus. Unfortunately, subsequent activation measurements [14] on the same nuclides resulted in cross sections substantially smaller than in the time-of-flight measurements. These differences are probably due to background and/or baseline shift problems in the time-of-flight measurements. Current plans call for constructing a new detector on a longer and better shielded flight path at the MLNSC. These improvements, together with the implementation of an automated sample changer should allow these problems to be overcome in future experiments.

Two new spallation sources currently under construction should allow even more measurements on small samples to be made. Recently, a study [15] was done to compare the suitability of different white neutron sources for measurements on very small samples. With the advent of the Spallation Neutron Source (SNS) [16] in Oak Ridge, and the TOF facility [17] at CERN it should be feasible to determine reaction rates for several s-process branching point isotopes, for nuclides of interest to the p process, and for studies of freeze-out conditions during the end of the r process.

ACKNOWLEDGMENTS

ORNL is managed by UT-Batelle, LLC for the U.S. Department of Energy under Contract No. DE-AC05-00OR22725.

REFERENCES

1. Käppeler, F., *Progress in Particle and Nucl. Phys.*, **43**, 419 (1999).

2. Koehler, P. E. et al., *Phys. Rev. C* **54**, 1463 (1996).

3. Rayet, M. et al., *Astron. Astrophys.*, 298, 517 (1995).

4. Gledenov, Yu. M., Koehler, P. E., Andrzejewski, J., Guber, K. H., and Rauscher, T., *Phys. Rev. C* **62**, 042801(R) (2000).

5. Guber, K. H., Spencer, R. R., Koehler, P. E., and Winters, R. R., *Phys. Rev. Lett.* **78**, 2704 (1997).

6. Koehler, P. E. et al., *Phys. Rev. C* **57**, R1558 (1998).

7. Holmes, J. A., Woosley, S. E., Fowler, W. A., and Zimmerman, B. A., *At. Data Nucl. Data Tables* **18**, 305 (1976).

8. Bao, Z. Y., et al., *At. Data Nucl. Data Tables* **75**, 1 (2000).

9. Goriely, S., "Improved Predictions of Neutron Capture Rates and Their Impact on the s- and r-process Nucleosynthesis" in *Nuclei in the Cosmos V*, edited by N. Prantzos and S. Harissopulos, AIP Paris: Editions Frontieres, 1998, p. 314.

10. Macklin, R. L., Winters, R. R., Hill, N. W., and Harvey, J. A., *Astrophys. J.* **274**, 408 (1983).

11. Harris, M. J., *Astrophys. Space Sci.* **77**, 357 (1981).

12. Koehler, P. E. and Käppeler, F., "Measurements of (n,γ) Cross Sections for Very Small Stable and Radioactive Samples of Interest to the s- and p-process" in *International Conference on Nuclear Data for Science and Technology*, edited by Edited by J. K. Dickens, La Grange Park: American Nuclear Society, 1994, p. 179.

13. Ullmann, J. L., et al., "Neutron Capture Measurements on Unstable Nuclei at LANSCE" in *Proceedings of the Fifteenth International Conference on the Application of Accelerators in Research and Industry*, edited by J. L. Duggan and I. L. Morgan, New York: American Institute of Physics, 1999, p. 251.

14. Jaag, S., *Nucl. Phys.* **A621**, 251c (1997); Käppeler, F., private communication (2000).

15. Koehler, P. E., accepted for publication in *Nucl. Instr. and Meth.*

16. Olsen, D. et al., Oak Ridge National Laboratory Report SNS 100000000PL0001R00 (1999).

17. Rubbia, C. et al., European Laboratory for Particle Physics Reports CERN/LHC/98-02 (EET) and CERN/LHC/98-02 (EET)-Add. 1, (1998).

Nuclear Astrophysics with a Recoil Mass Separator

Frank Strieder

*Institut für Experimentalphysik III**
Ruhr-Universität Bochum, D-44780 Bochum, Germany

Abstract. Radiative capture reactions, like (α,γ)- and (p,γ)-reactions, are of great importance for the understanding of the different burning phases in stars. In most cases laboratory studies of some key reactions are very difficult due to the low cross section at the relevant Gamow energy where the stellar burning occurs. A new approach to measure these capture cross sections involves a two-sided differentially pumped gas target, a recoil mass separator, and a ΔE-E detector telescope (allowing for particle identification) as detection system. This combination allows a direct measurement of the produced recoils in inverse kinematics. The direct observation of the recoils requires an efficient recoil mass separator to filter out the incident beam particles from the recoils. The recoil separator must not only have a high filtering power but also a high transmission of the recoils (for the selected charge state) between the gas target chamber and the ΔE-E telescope. The feasibility of the separation of projectiles and recoils with a mass difference of 1 amu to 1 part in 10^{11} or more has been demonstrated in test experiments.

INTRODUCTION

The quest of nuclear astrophysics is to understand how chemical elements that make up our world are formed in the cosmos. Nucleosynthesis started in the Big Bang, and continues in the various stages of stars. Chemical elements are released into the Universe following explosive collapse of a star. While a great deal is now known due to the pioneering efforts of Hans Bethe, Carl F. von Weizsäcker, and William Fowler, nevertheless major questions are still present (1).

The capture reaction ^{12}C(α,γ)^{16}O is a key reaction of nuclear astrophysics due to its influence on several areas of stellar evolution. This reaction takes place in the helium burning of Red Giants and determines not only the nucleosynthesis of elements up to the iron region but also the subsequent evolution of massive stars, the dynamics of a supernova, and the kind of remnant after a supernova explosion. For these reasons, the cross section should be known with a precision of at least 10 %. In spite of tremendous experimental efforts in measuring the cross section with standard techniques over nearly 30 years, one is still far from this goal.

Originally, the cross section at the relevant Gamow energy $E_0 = 300$ keV was assumed to be dominated by the capture process into the ^{16}O ground state with an E1 amplitude arising from the low-energy tail of a broad $J^{\pi} = 1^-$ resonance at $E_R = 2.42$ MeV and the high-energy tail of a $J^{\pi} = 1^-$ subthreshold resonance at $E_R = -45$ keV including interference effects between both E1 sources. The first ^{12}C(α,γ)^{16}O measurements (2, 3) revealed a constructive interference at energies between the two resonances. Studies in inverse kinematics (4) suggested an E2 capture amplitude of the same order as the E1 capture amplitude at E_0, where the major E2 source is a $J^{\pi} = 2^+$ subthreshold resonance at $E_R = -245$ keV. Subsequent measurements of γ_0 angular distributions (5, 6, 7) supported this suggestion. The γ_0 capture transitions were studied also in coincidence with the ^{16}O recoil nuclides (8). The analysis and extrapolation of the dominant E1 and E2 γ_0 capture transitions depend critically on the reduced α widths of the two subthreshold resonances. Values for these widths were inferred from elastic-scattering data (9), the β-delayed α spectrum of ^{16}N (10), and α-transfer reactions. However, a close inspection of the available data for the E1 multipole shows that at energies below and above the $E_R = 2.42$ MeV resonance the various data sets have systematic differences. There are essentially only 3 data sets (5, 6, 7), where the E1 multipole was observed at $\Theta_{\gamma} = 90°$ in far geometry. The other data sets (3, 4, 8) were obtained in close geometry observing the sum of E1 and E2 multipoles and corrected for the contribution of the E2 multipole, being thus in a way model-dependent. Furthermore, some data sets

---

* The reported project is supported by the Deutsche Forschungsgemeinschaft (Ro 429/35-1) and the Istituto Nazionale di Fisica Nucleare.

(using an extended gas target) did not include the interference effects between E1 and E2 multipoles and may thus be too high (or too low) at energies far below (or far above) the 2.42 MeV resonance; the effects are absent for a solid target. Note that measurements above $E_{cm} = 3.0$ MeV are as important as low-energy measurements, because they fix the E1 background amplitude. However, such measurements are hampered by the background due to the $^{13}C(\alpha,n)^{16}O$ reaction (normal kinematics) or due to the $^{12}C + ^{12}C$ fusion reactions (inverse kinematics). In inverse kinematics, one could observe the γ_0 capture transition in coincidence with the ^{16}O recoils, which can emerge from a windowless gas target. Since there is always a charge state representing about 50 % of all ^{16}O recoils produced, such coincidence measurements would have a reduction in γ-ray efficiency of only 50 % but the use of the recoil mass separator would essentially eliminate the background signals.

Improved data of the ratio $\sigma_{E2}(E)/\sigma_{E1}(E)$ may be obtained from measurements of angular distributions using a crystal ball. If the crystal ball could be combined with the recoil separator, background-free data could be obtained. Such setup would also allow clear observation of cascade transitions via the $E_x = 6.92$ and 7.12 MeV excited states over a wide range of energies. One may also pose the question, whether the capture into the ^{16}O ground state could not also proceed by a monopole (E0). In those nuclides, where monopoles have been observed they were always of the same order as quadrupoles. Since the E2 quadrupole plays a non-negligible role in $^{12}C(\alpha,\gamma)^{16}O$, one may suggest also a non-negligible role of an E0 monopole. The existance of such a monopole may be tested experimentally by the observation of the ^{16}O recoils in a recoil separator, i.e. through a comparison of the ^{16}O flux with the γ_0 - ^{16}O coincidence flux.

THE ERNA-PROJECT: $^{12}C(\alpha,\gamma)^{16}O$

On the basis of the experiences of an experiment at the Accelerator Mass Spectrometry facility of the University of Naples, Italy, aiming in the direct measurement (11, 12) of the capture reaction $^7Be(p,\gamma)^8B$ a new sophisticated recoil mass separator for other reactions involving stable nuclei is in preparation. This separator, called ERNA (European Recoil separator for Nuclear Astrophysics) is being installed at the 4 MV Dynamitron tandem accelerator of the Ruhr-Universität Bochum, Germany, and designed mainly for an improved measurement of the key reaction $^{12}C(\alpha,\gamma)^{16}O$. In this approach, the reaction is initiated in inverted kinematics, $^4He(^{12}C,\gamma)^{16}O$, i.e. a ^{12}C beam is guided into a windowless ^4He jet gas target and the kinematically forward-directed ^{16}O recoils are detected downstream on the beam line. The direct observation of the ^{16}O recoils requires an efficient recoil separator to filter out the intense ^{12}C beam particles from the ^{16}O recoils: the number of ^{16}O recoils per incident ^{12}C projectile is 1×10^{-18} for a cross section $\sigma = 1$ pb and a target density $n(^4He) = 1 \times 10^{18}$ atoms/cm^2. The recoil separator must also filter out beam contaminants, small-angle elastic scattering products, and background events from multiple scattering processes leading to a degraded tail of the projectiles. If the filtering of the separator is sufficiently effective, the ^{16}O recoils can be counted directly in a ΔE-E telescope placed in the beam line at the end of the recoil separator, where the telescope allows for particle identification. It is expected that the high detection efficiency of the ^{16}O recoils and the negligible contribution of cosmic-ray events in the ΔE-E coincidences allows a measurement of the $^4He(^{12}C,\gamma)^{16}O$ cross section to as low as $E_{cm} = 0.7$ MeV ($\sigma \approx 1$pb).

Beam contaminants

Since the ^{12}C projectiles and the ^{16}O recoils have essentially the same momentum and since the ^{12}C ion beam emerging from the accelerator passes a momentum filter (analysing magnet), a nearly complete elimination of any ^{16}O beam contaminant in the ^{12}C ion beam incident on the ^4He gas target is of utmost importance for the new approach: the ^{16}O beam contaminant and the ^{16}O recoils cannot be distinguished in the recoil separator, since both have the same momentum.

In a first ERNA report (13), a Wien filter and a ΔE-E telescope were used to investigate the ^{16}O beam contamination accompanying a momentum-filtered ^{12}C beam. The resulting identification matrix (Fig. 1) shows indeed the presence of a contaminant ^{16}O beam. The energy of the contaminant ^{16}O beam is about 3/4 the energy of the ^{12}C incident beam, as expected from the momentum filter for equal charge states. Furthermore, the leaky ^{12}C beam has about the same velocity as that of the contaminant ^{16}O beam, as expected from the action of the Wien filter (velocity) filter, and represents a velocity-filtered section of the degraded tail of the ^{12}C beam. The intensity ratio of ^{16}O to ^{12}C was found to be $P_0 = 6 \times 10^{-10}$, i.e. much higher than the intensity ratio 1×10^{-18} between the ^{16}O recoils and ^{12}C projectiles at $E_{cm} = 0.7$ MeV, or even 5×10^{-14} at $E_{cm} = 2.45$ MeV with maximum radiative capture cross section $\sigma \approx 50$ nb. If the injection magnet after the ion source of the accelerator is set at mass 16 (rather than at mass 12 for the ^{12}C ion beam), the telescope shows the dominant presence of ^{16}O ions, at the same point in the matrix. Thus, the main source of the ^{16}O beam contamination lies in the ion source setup aris-

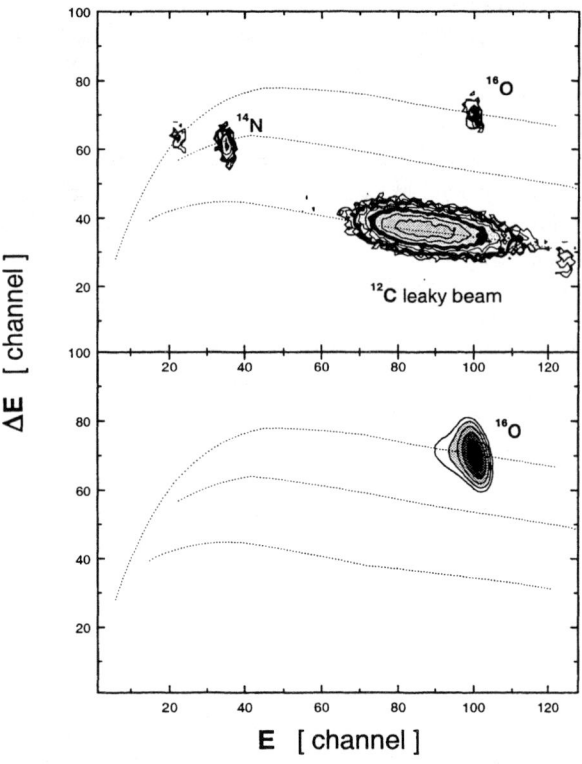

FIGURE 1. The ΔE-E identification matrix for a $^{12}C^{3+}$ ion beam is shown with the injection magnet of the accelerator (Fig. 2) set at mass 12 (upper part) and at mass 16 (lower part): the contaminant ^{16}O beam appears at the same point in the matrix.

ing from the finite mass resolution of the injection magnet.

The data (Fig. 1) indicate that the ^{12}C ion beam intensity can be suppressed by a single Wien filter to about 4×10^{-8}. Since the studies of $^4He(^{12}C,\gamma)^{16}O$ using the recoil separator ERNA include the combination of a Wien filter, a momentum filter, and another Wien filter (see below), the above results suggest that the needed suppression factor of 1×10^{-15} can be achieved with ERNA. The ^{16}O beam purification of a momentum-filtered $^{12}C^{3+}$ ion beam (from the 4 MV Dynamitron tandem) using a single Wien filter is about $6 \times 10^{-10} \times 4 \times 10^{-8} = 2 \times 10^{-17}$, where we have assumed an ^{16}O degraded tail identical to that of the ^{12}C ion beam. This purification is not quite sufficient for a measurement with $\sigma \approx 1pb$ using a single Wien filter. For this reason, one Wien filter will be installed - in the final setup of the ERNA project - before the analysing magnet and a second Wien filter will be placed between the analysing magnet and the jet gas target (Fig. 2), where this setup should provide a sufficient ^{16}O beam purification for the ERNA aims.

Ion beam optics

Although the ^{16}O recoils - produced in the ^4He jet gas target via the reaction $^4He(^{12}C,\gamma)^{16}O$ - are kinematically forward-directed, the emission of the capture γ-rays (energy E_γ) leads to an emission cone of half-angle $\Theta = \arctan(E_\gamma/pc)$, where p is the momentum of the ^{16}O recoils and c the velocity of light. Associated with the γ-ray emission there is also a spread $\Delta p/p$ in momentum. For example, at $E_{cm} = 0.7$ MeV ($E_\gamma = 7.9$ MeV) one finds $\Theta = 1.8°$ and $\Delta p/p = 6.2$ %. At $E_{cm} = 0.7$ MeV, the cone has reached a diameter of 3.1 cm at a 0.5 m distance from the jet gas target. Thus, shortly after the jet gas target there must be a focusing element followed by filter-elements and other focusing elements up to the site of the telescope, where all elements must have an angle acceptance of at least $\Theta = 1.8°$ and a momentum acceptance of at least $\Delta p/p = 6.2$ %, in order to transport the ^{16}O recoils with 100 % transmission to the telescope. This requirement demand a compact design of the jet gas target system involving several pumping stages, where the present technical plan involves an extension of 35 cm on both sides of the jet gas target.

The calculations of ion beam optics for the recoil separator ERNA were performed up to third order using the program COSY INFINITY (15). They showed that the above requirements can be fulfilled with the setup indicated in Fig. 2. After the jet gas target, the separator will consist sequentially of the following elements: (i) a magnetic quadrupole triplet (MQT; length 105 cm, inner diameter 10.6 cm), (ii) a Wien filter (WF3, DANFYSIK) containing two electrostatic plates (12 cm width, 50 cm effective length, 7 cm gap), the maximum electric field is ±40 kV with an magnetic field of 0.135 T, (iii) a magnetic quadrupole singlet (MQS1, length 5.2 cm, inner diameter 10.6 cm), (iv) a 60° dipole magnet (radius of curvature: 40 cm, 7.6 cm gap, max. B-field: 1.3 T), (v) a magnetic quadrupole doublet (MQD, length 50.8 cm, inner diameter 10.6 cm), (vi) a second Wien filter (WF4, electrical plates: 6.3 cm width, 57.8 cm effective length, 7 cm gap), (vii) a magnetic quadrupole singlet (MQS2, length 5.2 cm, inner diameter 7.7 cm), and (viii) the ΔE-E telescope (13). The calculations were performed assuming a parallel ^{12}C ion beam of 3 mm diameter and a pointlike ^4He jet gas target with a density of 1×10^{18} atoms/cm². To calculate the emittance of the ^{16}O recoils, the energy loss of the ^{12}C projectiles and the ^{16}O recoils in the gas target as well as the recoil from the capture γ-rays to the ^{16}O ground state were taken into account; the effects of angle and en-

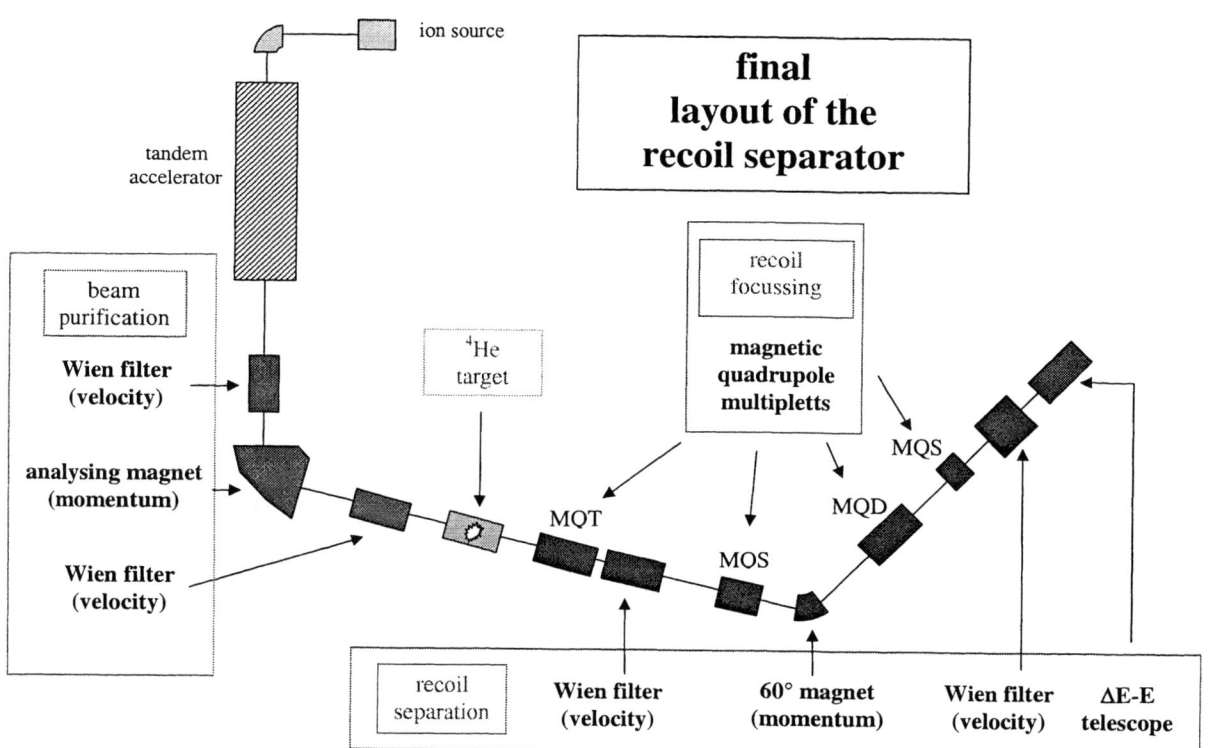

FIGURE 2. Schematic diagram of the complete ERNA setup (MQT = magnetic quadrupole triplet, MQD = magnetic quadrupole doublet, MQS = magnetic quadrupole singlet)

ergy straggling in the target have also been included. The calculations were performed over the planned energy region, $E_{cm} = 0.7$ to 5.0 MeV, where at each energy the most probable charge state was selected for the ^{16}O recoils. The trajectories emitted within the entire emission cone in the jet gas target are 100 % transmitted to the telescope, where they enter the telescope within a diameter of about 30 mm and are nearly parallel.

Conclusions

Since the setup used for the measurement of the beam suppression factor (14) is nearly identical to that in the final ERNA layout, a successful direct measurement of the cross section of the reaction ^{12}C$(\alpha,\gamma)^{16}$O with a recoil separator seems to be feasible. First results of this experiment are expected for summer 2001.

REFERENCES

1. C.E. Rolfs, W.S. Rodney, Cauldrons in the Cosmos (University of Chicago Press, 1988).
2. R.J. Jaszczak et al., *Phys. Rev.* **C2**, 2452 (1970).
3. P. Dyer, C.A. Barnes, *Nucl. Phys.* **A233**, 495 (1974).
4. K.U. Kettner et al., *Z. Phys.* **A308**, 73 (1982).
5. A. Redder et al., *Nucl. Phys.* **A462**, 385 (1987).
6. J.M.L. Ouellet et al., *Phys. Rev.* **C54**, 1982 (1996).
7. G. Roters et al., *Eur. Phys. J.* **A6**, 451 (1999).
8. R.M. Kremer et al., *Phys. Rev. Lett.* **60**, 1475 (1988).
9. R. Plaga et al., *Nucl. Phys.* **A465**, 291 (1987).
10. R.E. Azuma et al., *Phys. Rev.* **C50**, 1194 (1994).
11. L. Gialanella et al., *Nucl. Instr. Meth.* **A376**, 174 (1996).
12. L. Gialanella et al., *Eur. Phys. J.* **A7**, 303 (2000).
13. D. Rogalla et al., *Nucl. Instr. Meth.* **A437**, 266 (1999).
14. D. Rogalla et al., *Eur. Phys. J.* **A6**, 471 (1999).
15. M. Berz: Computational aspects of design and simulation (COSY INFINITY), *Nucl. Instr. Meth.* **A298**, 473 (1990).

High Temperature rp-Process Nucleosynthesis of Mo and Ru, and β-Delayed Proton Emission

R.N. Boyd

Department of Physics, Department of Astronomy
The Ohio State University, Columbus, OH 43210, USA

Abstract. The trajectory of the high-temperature rp-process passes through nuclei close to the proton drip line. Some of these can decay by β-delayed proton emission, which may shift the final mass distribution produced by this form of stellar burning from that of its progenitors. This will also apply to other processes that can synthesize very proton-rich nuclides, including the γ-process and the repetitive rp-process, or $(rp)^2$-process. The very proton-rich nuclides above mass 74 u that cannot be synthesized by either the r- or s-processes can serve as excellent tests of the theoretical models of the rp-, $(rp)^2$-, and γ-processes, especially in the mass region 70 to 100 u, where these processes may operate concurrently.

INTRODUCTION

The astrophysical rp-process occurs in high-temperature (T) H-rich environments. It does not take place in the quasi-static burning stages through which a massive star evolves, since the required combination of T and H-richness does not occur. Rather, the rp-process is thought to occur in situations in which a compact star—a white dwarf or a neutron star—accretes matter from a less evolved companion. The accreted matter comes from the periphery of the star, so will be H-rich. It will also be hot, as it will have fallen into a deep gravitational potential well as it approaches the surface of the compact star. Thus the requisite rp-process conditions are fulfilled. Another site may occur in the shock wave of a supernova, as the energy deposited in the pre-existing nuclei as the shock wave passes can provide the requisite T, while the required H-density is supplied by (γ,p) reactions on the pre-existing nuclei or by mixing with material from H-rich zones. In any event, the rp-process occurs rapidly, in tens of seconds or less, and produces a large amount of energy. If the accretion is onto a white dwarf, T will be 200-400 million K, and the result may be a nova. If accretion occurs onto a neutron star, T will exceed 1 billion K, and the result may be an x-ray burst. General descriptions of the nuclear astrophysics of this process have been given by Wallace and Woosley, [1], Champagne and Wiescher [2], and Schatz et al. [3].

The γ-process also occurs in a very hot environment, often concurrently with the rp-process, with T exceeding 2 billion K. Heavier nuclides undergo photonuclear reactions, emitting α-particles and neutrons. This process also drives its seeds toward the proton drip line, so its progenitors are subject to many of the same considerations as in the rp-process. The γ-process is described in Woosley and Howard [4, 5].

SPECIFICS OF THE rp-PROCESS

Because the rp-process involves a rapid succession of proton captures, it will drive its seed nuclei to near the proton drip line. Occasional (rapid) β^+-decays will allow the progression from light seed nuclei to the heavier ones. Occasional (p,α) reactions may close cycles, but the general progression from light to heavy nuclides is expected to proceed none the less. The reactions relevant to the rp-process are indicated in Fig. 1.

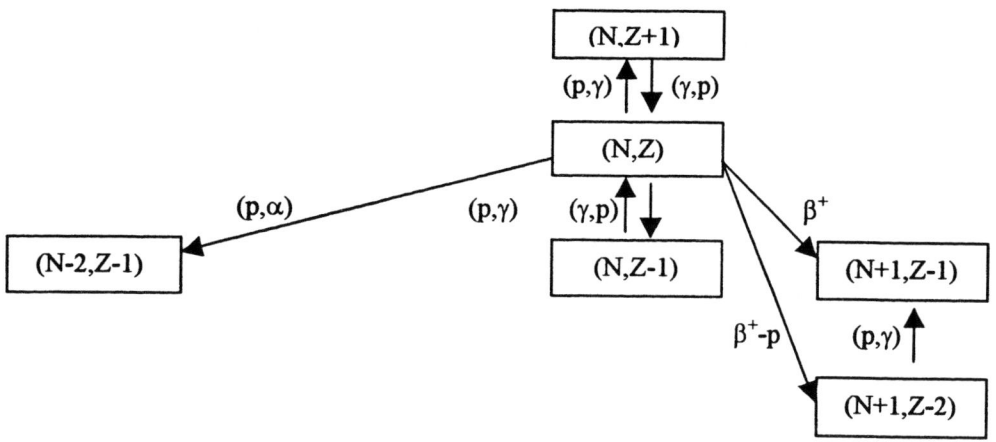

FIGURE 1. Various reactions that can occur on nucleus (N,Z) in rp-processing. If (γ,p) dominates over (p,γ), and both β⁺ and β⁺-p decays are slow, nucleus (N,Z) will become a waiting point.

If the rp-process persists for a long time it will drive its seed nuclei to nuclides at which further (p,γ) reactions are prohibited by negative Q-values, or inhibited by small positive Q-values. For nucleus (N,Z) in Fig. 1, then, (N,Z+1)(γ,p) would dominate over (N,Z)(p,γ) due to the high-energy photons in the environment. If the nucleus (N,Z) β⁺-decayed rapidly, the resulting nucleus, (N+1,Z-1) would quickly undergo a (p,γ) reaction, and the rp-process would continue unabated. If, however, the half-life were long, (N,A)'s abundance would build up during rp-processing. After the high T subsided, it would β⁺-decay to the stable products of the rp-process, producing a large abundance.

Of course, β⁺-p emission could affect the trajectory of the rp-process, but only if the nuclide from which the β⁺-p emission occurred was long-lived. Then abundance would build up at that nucleus, some of those decays would occur after the rp-process conditions ended and β⁺-p decay would shift the resulting abundances.

SIGNATURES OF rp-PROCESSING

Nuclides that might provide definitive signatures of rp-processing probably do not exist, although the p-process nuclei (see section by Boyd in [6]) do come close. They lie to the proton-rich side of stability, and are blocked by stable isobars from synthesis by the r- or s-processes, which synthesize most nuclei heavier than Fe. Thus, the p-nuclides must be created by processes that pass along the proton-rich side of stability; one such process is the rp-process. Of particular interest are the Mo and Ru p-nuclides, because of their very high abundances. While p-nuclear abundances in this mass region are typically 1% of the elemental abundances, those for ^{92}Mo, ^{94}Mo, ^{96}Ru, and ^{98}Ru are large: 14.8%, 9.3%, 5.5% and 1.9% respectively [7]. These have posed a longstanding problem for nuclear astrophysics.

OTHER PROCESSES THAT SYNTHESIZE THE p-NUCLIDES

There are other processes, e.g., the γ-process [4,5], that produce p-nuclides. It processes abundant high-mass nuclei, e.g., Pb, at high T via (γ,n), (γ,p), and (γ,α) reactions to lighter nuclei. The Coulomb barrier drives the γ-process trajectory to the proton-rich nuclei. The nuclei that are synthesized during the period of high T will decay back to stability after T decreases. This process is inevitable in high T environments, so always accompanies the rp-process. Some of the stable nuclei it ultimately produces will be p-nuclei.

An adaptation of the rp-process is the repetitive-rp-process, or (rp)²-process [8]. This occurs under the same conditions as the rp-process, but alternates rp-processing bursts and quiescent periods. The very proton-rich nuclides synthesized in each of the processing bursts have time to decay back to stability before the next processing burst begins, thus circumventing the rp-process waiting points. This scenario could result from accretion onto a neutron star in which thermonuclear bursts occurred as the accreted matter reached a critical value, then would subside until enough matter built up to produce a subsequent burst.

SUCCESSES AND FAILURES OF rp-PROCESS MODELS

The γ-process [9] explains the abundances of most of the heavy p-nuclides, those heavier than 100 u, quite well, but the rp-process appears to be essential for fitting the lower mass p-nuclides, Se to Sr. However, no theory has successfully reproduced the abundances of the Mo and Ru p-nuclides without some downsides. One fairly successful description arises from the rp- plus γ-process with high T (T = 1.5×10^9 K) and densities (10^6 g cm^{-3}), and very long processing times (100 s) [3]. These conditions are extreme, but they might occur in the accretion of matter onto a neutron star. However, its deep gravitational well would allow very little matter to escape into the interstellar medium, necessitating a huge enhancement of the Mo and Ru abundances from rp-processing. Such enhancements are achieved in the model of Schatz et al. [3]: $\sim 10^7$. The downside of that model, though, was the production of large amounts of ^{80}Kr, a nuclide that is made in the s-process, so is potentially ultimately overproduced.

The (rp)^2-process also can achieve large enhancements of the Mo and Ru p-nuclides, but it does not require such long processing times or high densities; ~1 s and ~10^4 g cm^{-3} are sufficient. However, it also produces some non-p-nuclides, e.g., ^{93}Nb, with abundance enhancements comparable to those of the Mo and Ru p-nuclides. Since ^{93}Nb is made in the s-process, this might present the same sort of problem that ^{80}Kr does for the rp-process.

THE EFFECTS OF β-DELAYED PROTON EMISSION

The path of the rp-process [3] through the 90 to 100 u nuclides is shown in Fig. 2. A similar picture occurs in the (rp)^2-process. The abundances and decay modes of the progenitor nuclei determine the resulting rp-process abundances. The progenitor nuclei for ^{92}Mo, ^{94}Mo, ^{96}Ru, and ^{98}Ru are ^{92}Pd, ^{94}Pd, ^{96}Cd, and ^{98}Cd respectively, all waiting points along the high-T rp-process path. The progenitor of ^{93}Nb is ^{93}Pd. If it decayed by β$^+$-p a significant fraction of the time, it would produce ^{92}Mo, thus solving a problem in the (rp)^2-process of overproducing ^{93}Nb. It would also enhance the production of ^{92}Mo, as the 60 s half-life of ^{93}Pd would produce a large ^{92}Mo abundance. Energetics suggest that β$^+$-p emission is very likely for ^{93}Pd, but it has not been measured. Because the other nearby progenitors are all even-even nuclides, their β$^+$-p emission would be unlikely.

A similar situation exists for ^{97}Cd, although its half-life, 3 s, is much shorter than that of ^{93}Pd. If it decayed by β$^+$ emission, it would populate ^{97}Mo. It is known [10] to decay somewhat by β$^+$-p, so some ^{97}Cd abundance will end up as ^{96}Ru, benefiting both the rp- and (rp)^2-processes. However, attention must alsobe given to potential isomers in ^{93}Pd and ^{97}Cd.

CONCLUSIONS

Although the differences between the rp- and (rp)^2-processes are not large, it may be possible to test their predictions. The rp-process appears to require long processing times to reach the mass 90 to 100 u nuclides. However, it then terminates with a Sn-Sb-Te cycle (Schatz [11]) just above mass 100 u, as the sequence of reactions converts four protons to an α-particle as in the CNO cycle. The (rp)^2-process, by contrast, involves much shorter times, and may not even get to the proton drip line. In either event, it can circumvent the Sn-Sb-Te cycle by resetting the nuclei through which each successive rp-process passes back to stability. This could lead to an interesting test of these models; any observation of nuclei heavier than those to which the progenitors decay in the rp-process, e.g., In or Ag, in the x-ray data would strongly suggest that the rp-process had passed beyond the Sn-Sb-Te cycle, *i.e.*, that its seeds had been processed in a *previous* rp-process.

For β$^+$-p emission to affect the rp- and (rp)^2-processes, it must involve nuclei that are long-lived with respect to the time scale of the rp-process, ~10 s, and they must lie along the proton drip line. These two conditions are rarely simultaneously fulfilled until the nuclei that lie along the rp-process trajectory attain masses in the 90 to 100 u region.

However, the possibility that progenitor nuclides such as ^{93}Pd and ^{97}Cd do decay by β$^+$-p

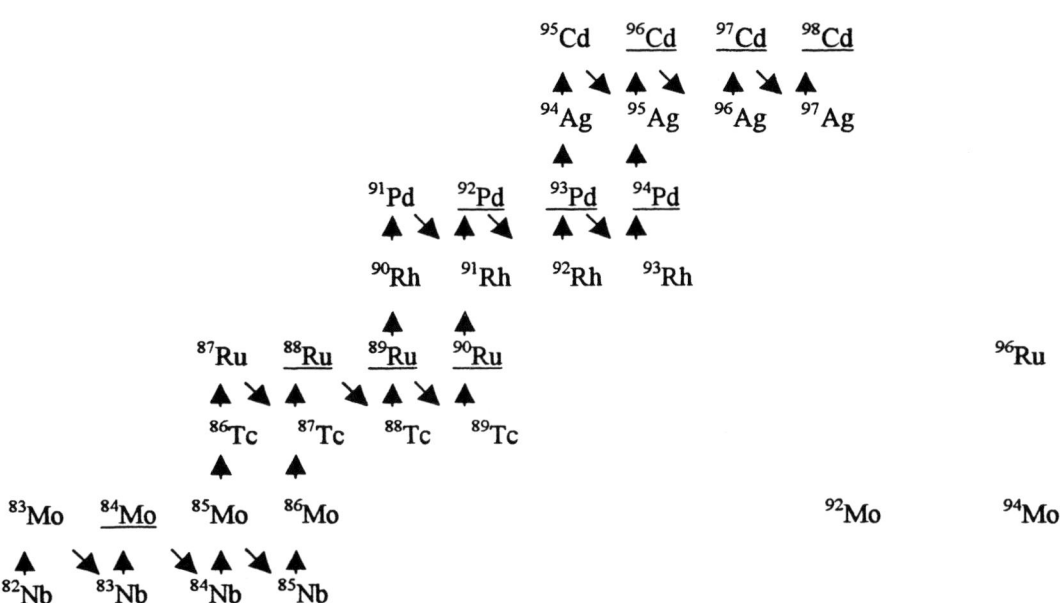

FIGURE 2. Nuclides along the primary trajectory of the rp-process that populate the Mo and Ru p-nuclides, three of which are shown (lower right). The underlined nuclides are waiting points, in that they have half-lives of about 1 s or longer (although some of the half-lives have not been measured) and (p,γ) Q values less than 1 MeV. Upward arrows represent (p,γ) reactions, whereas arrows pointing downward and to the right represent β$^+$ decays.

emission may impact the nucleosynthesis that is performed by the rp- and (rp)^2-processes, and may even help to solve some of the difficulties with current models of the synthesis of the Mo and Ru p-nuclides. Further data on β-delayed proton emission probabilities of the key nuclei will help greatly to resolve these issues.

ACKNOWLEDGEMENTS

The support of the National Science Foundation through grant PHY9901241 is gratefully acknowledged.

REFERENCES

1. Wallace, R.K. and Woosley, S.E., Astrophys. J. Suppl. Series 45, 389 (1981)
2. Champagne, A.E. and Wiescher, M., Ann. Rev. Nucl. Part. Sci. 42, 39 (1992)
3. Schatz, H., Aprahamian, A., Gorres, J., Wiescher, M., Rauscher, T., Rembges, J.F., Thielemann, F.-K., Pfeiffer, B., Moller, P., Kratz, K.-L., Herndl, H., Brown, B.A., and Rebel, H., Phys. Reports 294, 167 (1998)
4. Woosley, S.E. and Howard, W.M., Astrophys. J. Suppl. 36, 285 (1978)
5. Woosley, S.E. and Howard, W.M., Astrophys. J. 354, L21 (1990)
6. Wallerstein, G., Iben, I., Jr., Parker, P., Boesgaard, A.M., Hale, G.M., Champagne, A.E., Barnes, C.A., Kappeler, F., Smith, V.V., Hoffman, R.D., Timmes, F.X., Sneden, C., Boyd, R.N., Meyer, B.S., and Lambert, D.L., Rev. Mod. Physics 69, 995 (1997)
7. Anders, E. and Grevesse, N., Geochim. Cosmochim. Acta 53, 197 (1989)
8. Boyd, R.N., Hencheck, M., and Meyer, B.S., *Origin of Matter and Evolution of Galaxies 97*, ed. by S. Kubono, T. Kajino, K.I. Nomoto, and I. Tanihata, (World Scientific, Singapore) 350 (1998)
9. Howard, W.M., Meyer, B.S., and Woosley, S.E., Astrophys. J. 373, L5 (1991)
10. Schmidt, K., et al., J., Phys. A624, 185 (1997)
11. Schatz, H., private communication (2000)

The Front-End Systems for the Spallation Neutron Source.*

Roderich Keller

Ernest Orlando Lawrence Berkeley National Laboratory, Berkeley, CA 94720, USA

Abstract. The Front-End Systems (FES) of the Spallation Neutron Source (SNS) project are being built by Berkeley Lab and will deliver a 52-mA H⁻ ion beam at 2.5 MeV energy to the subsequent Drift-Tube Linac, to be built by Los Alamos National Laboratory. The FES comprise a volume-production, cesium-enhanced Ion Source, an electrostatic Low-Energy Beam Transport (LEBT), an RFQ accelerator, and a Medium-Energy Beam Transport (MEBT) that includes rebuncher cavities, magnetic quadrupoles, and beam diagnostics. The macro-pulse duty factor is 6%, and the macro pulses have to be chopped into a mini-pulse structure with a time scale of hundreds of ns, to reduce beam losses and component activation during extraction from the SNS Accumulator Ring. Delivery of the entire FES to the main SNS facility in Oak Ridge is planned for April 2002. This paper discusses the design features and status of the major FES subsystems with special emphasis on Ion Source and LEBT for which first experimental results have been obtained. After successful completion, the FES could be viewed as a prototype of a high-current, high duty-factor injector for other accelerator projects or, without the elaborate MEBT, as an independent 2.5-MeV accelerator for various applications.

INTRODUCTION

The Spallation Neutron Source (SNS) project [1] is presently in the third year of its construction phase. The project is being carried out under a collaboration agreement among six U. S. National Laboratories, led by ORNL whose responsibilities include the project leadership. LBNL is building the front end, mainly consisting of the ion source, low-energy beam-transport section (LEBT), RFQ accelerator, and medium-energy beam-transport section (MEBT). Some subsystems of the front end, i.e. the rf power system for the RFQ and the MEBT chopper structures with their power supplies, will be supplied by LANL.

The SNS accelerator systems aim at delivering intense proton-beam pulses of less than 1-μs duration to the spallation target at 60-Hz repetition frequency and with an average power of about 2 MW. The 1-ms long H⁻ macro pulses that are accelerated in the linac to about 1-GeV energy have to be chopped into 'mini pulses' of 645-ns duration, with 300-ns pauses, to avoid activation of accumulator-ring components during the extraction kicker rise time.

Chopping is performed in the front end by two separate chopper systems located in LEBT and MEBT, respectively. The LEBT chopper removes most of the beam power during the mini-pulse gaps, and the MEBT chopper reduces the rise and fall time of the transported beam.

The main requirements for the SNS Front-End Systems are listed in Table 1. The front end will be assembled and commissioned at the Integrated Testing Facility at Berkeley Lab before shipment to ORNL.

Many of the FES design features, such as the chopping system, the ion-generator particularities associated with the production of a negative ion beam, and the entire MEBT are typical for pre-injectors designated to serve a high-energy accumulator ring. Other FES elements, however, such as the multi-cusp ion generator, the purely electrostatic LEBT with two einzel lenses, and the RFQ lend themselves to all kinds of applications of high-current, high-brightness ion beams with high duty factor. After its completion, the SNS front end will therefore be an example of a new gen-

TABLE 1. FES Key Performance Parameters

Ion species	H⁻
Output energy (MeV)	2.5
H⁻ peak current:	
MEBT output (mA)	52
Nominal ion-source output (mA)	65
Output normalized transv. rms emittance (π mm mrad)	0.27
Output normalized longit. rms emittance (π MeV deg)	0.126
Macro pulse length (ms)	1
Duty factor (%)	6
Repetition rate (Hertz)	60
Chopper:	
Rise, fall time (ns)	10
Off/on beam-current ratio	10^{-4}

*Work supported by the Director, Office of Science, Office of Basic Energy Sciences, of the US Department of Energy under Contr. No. DE-AC03-76SF00098.

eration of compact and efficient accelerators for the MeV energy range.

BEAM GENERATION

Earlier in its history, the SNS project had as its goal 1-MW of average beam power on the spallation target and required only 28 mA of beam current to be injected into the Drift Tube Linac (DTL) being built by LANL. Under the conservative assumption of 20% beam loss in the RFQ, this translates into the need for a 35-mA beam at the LEBT exit. An 'R&D' ion source was built first to demonstrate this current capability at 6% duty factor and to investigate the cesium enhancement and electron separation processes [2]. To provide a beam for injection into the first of four RFQ modules at 65 keV energy as early as possible, a 35-mA 'startup' ion-source/LEBT system was produced as a next step.

In the meantime, the SNS project goal was changed to 2-MW average power with a 52-mA output-current requirement for the FES—not quite twice as much as the earlier requirement because of an increase in the circumference of the accumulator ring that resulted in a slightly higher mini-pulse duty factor. Design and construction of the final 65-mA 'production' ion-source/LEBT system, consistent with the 52-mA requirement, were started in parallel with the fabrication of the startup system.

A schematic of the startup system is shown in Figure 1.

Ion Source

The H$^-$ ion-source is derived from the SSC model which had demonstrated beam currents in the 100-mA range at very short duty factor [3]. The plasma generator utilizes a 2-MHz, rf-driven discharge, confined by a multi-cusp magnet configuration. A magnetic dipole filter separates the main plasma from a smaller H$^-$ production region where low-energy electrons generate copious amounts of negative ions. A heated collar, equipped with eight cesium dispensers, surrounds this H$^-$ production chamber.

The outlet plate of the plasma generator contains another dipole-magnet configuration that creates a deflecting field across the extraction gap in order to separate extracted electrons from the ion beam and steer them towards a 'dumping' electrode attached to the outlet plate. Because this dumping field steers the ion beam as well, the entire plasma generator is tilted at an adjustable angle with respect to the LEBT axis to compensate for the resulting kink.

LEBT Design

The LEBT structure has to serve five main purposes, i.e., beam formation, 2-parameter matching into the RFQ, steering in angle and transverse offset, pre-chopping, and gas pumping. An earlier proton-LEBT design [4] had proven the viability of the purely electrostatic matching approach, and a similar configuration with two einzel lenses was chosen for the SNS startup LEBT, see Fig. 1 [5]. The second one of these lenses is split into four quadrants which can be biased with d.c. and pulsed voltages to provide angular steering as well as pre-chopping [6]. Transverse offset correction is achieved by moving the ion source and LEBT together with respect to the RFQ.

The last LEBT electrode is part of the RFQ entrance wall, and on its upstream side it carries a diagnostic electrode made again from four insulated quadrants. To create the mini pulse structure, chopping voltages of ±2.5-kV amplitude and 300-ns duration are applied to opposing pairs of lens quadrants in a rotating pattern, directing the chopped beam alternatively towards each of the four separation zones between the diagnostic-electrode quadrants. In this way, any parts of the beam that are not intercepted by the diagnostic electrode are prevented from hitting the RFQ vanes themselves whose accurate shapes could otherwise gradually be eroded by sputtering.

The LEBT-electrode shapes were optimized by simulating proton beams, using the 2-d code IGUN [7] in a novel way that allows introduction of finite ion temperatures into the calculation without experiencing unrealistic deformations of the plasma meniscus [8]. In essence, the plasma section of the problem is calculated with zero ion temperatures, and finite angles, corresponding to the assumed ion temperature, are added

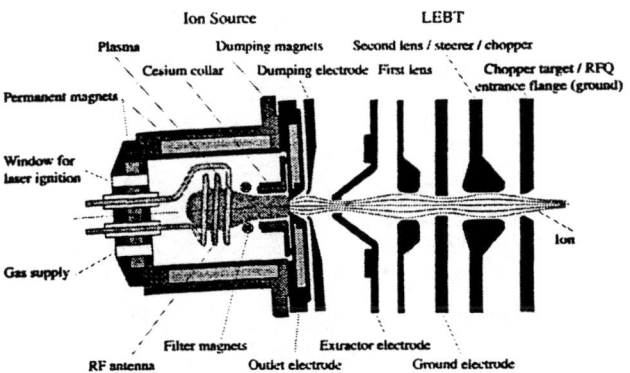

FIGURE 1. Schematic of the 35-mA startup ion-source and LEBT. Note that the actual filter and electron-dumping magnetic fields are oriented orthogonally to the illustration plane. The width of the ion beam is greatly exaggerated in this schematic to emphasize the focusing action of the double-lens system.

to all trajectories on the equipotential surface just 200 V below meniscus potential. This method was validated by comparing a few measured emittances with simulation results.

Another problem arises from the fact that copious amounts of electrons are extracted together with the desired negative hydrogen ions and deflected towards the dumping electrode as mentioned above. Following a strategy that was developed earlier [9], these electrons are accounted for in our simulations by assuming an increased proton current value between the plasma meniscus and the equipotential surface 2 kV farther downstream. Depending on the condition how the ion source is operated, with or without cesium enhancement, two different equivalent proton-current factors were found to best match simulations and measured beam envelopes. The values are 1.4 and 7.1, respectively, for these two cases.

The electrode shapes developed for the 65-mA production LEBT are shown in Fig. 2.

Ion Source and LEBT Status

Startup ion source and LEBT are presently being commissioned at Berkeley Lab in the SNS-FES Integrated Testing Facility, and beam pulse-currents up to 42 mA have been obtained at 6% duty factor and transported through the LEBT [10], so far at energies up to 55 keV. 35 kW of peak rf power is needed to create the necessary plasma density for these results. The vertical normalized rms emittance of a 35-mA beam was measured by an Allison scanner as $0.17\ \pi$ mm mrad. All electrons are removed from the ion beam at low energy and deposited on the outlet or dumping electrodes, as verified by operating the ion source with helium.

THE RFQ ACCELERATOR

The RFQ design [11] is derived from an earlier RFQ that has run for many years in the AGS injector at BNL [12]. The SNS RFQ is 3.72-m long overall and consists of four modules built as composite structures with an outer GlidCop shell and four oxygen-free copper vanes [13]. The RFQ will accelerate the H$^-$ beam from 0.065 to 2.5 MeV with an expected transmission far better than 80% [14]. Peak surface fields reach 1.85 Kilpatrick, and the total rf power consumption is 800 kW during pulses. Water-cooled π-mode stabilizers, following the JHC layout [15], separate unwanted dipole modes from the quadrupole frequency. Static frequency tuning is achieved by 20 stub tuners per module, and dynamic tuning by adjusting the temperature difference between vane tips and the outer walls of the modules.

Figure 3 shows the assembled first module before the final brazing operation. This module has now been completed and installed at the Integrated Testing Facility, and all ancillary components have been attached, such as tuners, rf pickup probes, vacuum manifolds, and cooling lines. The resonance frequency with tuners at nominal positions is very close to the design frequency of 402.5 MHz, and the field flatness is better than ±1% peak-to-peak.

The module has been conditioned to the full nominal power amplitude at 3% duty factor, operating either

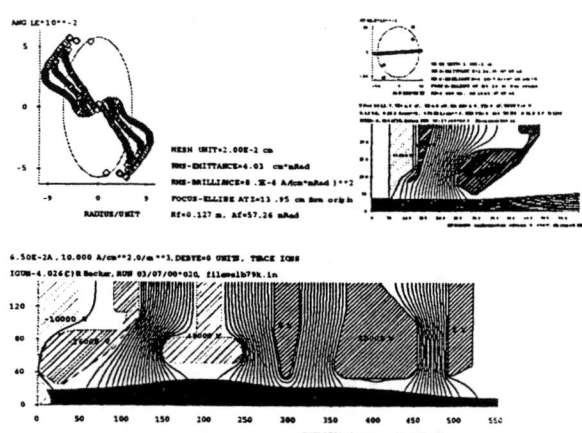

FIGURE 2. IGUN simulation of the 65-mA production LEBT, split into two major parts, i.e., extraction gap (top right) and main LEBT (bottom). The calculated transverse normalized rms emittance in the exit plane (see top left) amounts to $0.1\ \pi$ mm mrad, leaving a considerable safety margin for the emittance of the actual beam.

FIGURE 3. End-on view of the assembled RFQ Alpha Module prior to the final brazing operation. The upstream ends of the four vanes are seen at the center, with π-mode stabilizers penetrating the vanes horizontally and vertically.

at 60 Hz/0.5 ms or at 30 Hz/1.0 ms, with only one out of two designated power couplers installed so far. The dynamic tuning procedure was tested and works well; the calibration factor is very close to the calculated value of 33 kHz per °C.

MEBT

The MEBT [16], shown in Fig. 4, is 3.67-m long and has three main functions, i.e., matching the beam from the RFQ exit plane into the MEBT chopper plane, cleanup chopping, and matching the remaining particles into the Drift Tube Linac that is being built by LANL. Matching in both transverse and in the longitudinal direction is provided by 14 quadrupole magnets, arranged in three families, and four rebuncher cavities. The beam dynamics features of the SNS MEBT are discussed in detail elsewhere [14]; the hardest challenge for the lattice design is the limitation of the transverse emittance growth to 20%.

During the 10-ns rise and fall of the chopper pulses, the ion beam will be only partially chopped, and to get these parts of the beam back on axis an anti-chopper with the same functional characteristics as the chopper is inserted downstream of the chopper target.

The most critical component in the MEBT is the chopper target because of the extreme power density, reaching a peak level of 300 kW/cm^2 for a few ns duration. The target consists of a tilted, directly cooled TZM (molybdenum alloy) brazement with numerous cooling channels. The MEBT will also contain diagnostic elements [17] such as beam-position monitors, wire scanners to measure beam profiles, two fast current transformers, and an in-line emittance device.

All MEBT elements are grouped on three rafts that can be individually aligned. At present, all major components, including power supplies, are being fabricated or on order, and two chopper-targets are being built.

REFERENCES

[1] B.R. Appleton, J.R. Alonso, J.E. Cleaves, T.A. Gabriel, W.T. Weng, A.J. Jason, R.A. Hardekopf, D.P. Gurd, R.A. Gough, R. Keller, B.S. Brown, R.K. Crawford, "Status Report on the Spallation Neutron Source (SNS) Project," EPAC '98, Stockholm (1998).

[2] M.A. Leitner, D.W. Cheng, S.K. Mukherjee, J.Greer, P.K. Scott, M.D. Williams, K.N. Leung, R. Keller, R.A. Gough, "High-Current, High-Duty-Factor Experiments with the RF Driven H- Ion Source for the Spallation Neutron Source," PAC '99, New York (1999).

[3] K. Saadatmand, G. Arbique, J. Hebert, R. Valicenti, and K.N. Leung, Rev. Sci. Instr. 67 (3), p.1318-1320 (1996).

[4] J.W. Staples, M.D. Hoff, and C.F. Chan, "All-electrostatic Split LEBT Test Results," Linac '96, (1996).

[5] D.W. Cheng, M.D. Hoff, K.D. Kennedy, M.A. Leitner, J.W. Staples, M.D. Williams, K.N. Leung, R. Keller, and R.A. Gough, "Design of the Prototype Low Energy Beam Transport Line for the Spallation Neutron Source," PAC '99, New York (1999).

[6] J.W. Staples, J.J. Ayers, D.W. Cheng, J.B.Greer, M.D.Hoff, and A.Ratti, "The SNS Four-Phase LEBT Chopper," PAC '99, New York (1999).

[7] R. Becker, "New Features in the Simulation of Ion Extraction with IGUN," EPAC '98, Stockholm (1998).

[8] J. Reijonen, R. Thomae, and R. Keller, "Evolution of the LEBT Layout for SNS," Linac'00, Monterey (2000).

[9] M.A. Leitner, D.C. Wutte, and K.N. Leung, "2D Simulation and Optimization of the Volume H- Ion Source Extraction System for the Spallation Neutron Source Accelerator," Proc. Int. Conf. on Charged Particle-Beam Optics, Delft, Netherlands (1998).

[10] R. Thomae, P. Bach, R. Gough, J. Greer, R. Keller, and K. N. Leung, "Measurements on the H- Ion Source and Low Energy Beam Transport Section for the SNS Front-End Systems," Linac 2000, Monterey (2000).

[11] A. Ratti, R. DiGennaro, R. A. Gough, M. Hoff, R. Keller, K. Kennedy R. MacGill, J. Staples, S. Virostek, R. Yourd., "The Design of a High Current, High Duty Factor RFQ for the SNS," EPAC 2000, Vienna (2000).

[12] R. Gough, J. Staples, J. Tanabe, D. Yee, D. Howard, C. Curtis, and K. Prelec, "Design of an RFQ-based H- Injector for the BNL/ FNAL 200-MeV Proton Linacs, Linac '86, Stanford (1986).

[13] A. Ratti, C. Fong, M. Fong, R. MacGill, R. Gough, J. Staples, M. Hoff, R. Keller, S. Virostek, R. Yourd, "Conceptual Design of the SNS RFQ, " Linac '98, Argonne (1998).

[14] J. W. Staples, "SNS RFQ and MEBT Work at LBNL," ICFA Beam Dynamics Newsletter **21**, April 2000.

[15] A. Ueno, et al., "Beam Test of the Pre-injector and the 3-MeV H- RFQ with a New Field Stabilizer PISL," Linac '96, Geneva (1996).

[16] J. Staples, D. Oshatz, and T. Saleh, " Design of the SNS MEBT," Linac 2000, Monterey (2000).

[17] L. Doolittle, T. Goulding, D. Oshatz, A. Ratti, and J. Staples, "SNS Front-End Diagnostics," Linac 2000, Monterey (2000).

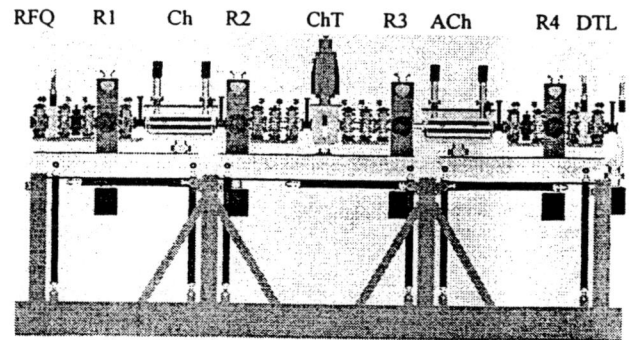

FIGURE 4. MEBT layout. The main elements are 14 quadrupoles, 4 rebuncher cavities, R, the chopper, Ch, and anti-chopper, ACh, structures, and the chopper target, ChT. Five ion pumps are connected to the beam line.

Extension of the Electronic Gamma-ray Spectrum Catalogue Web Site

R. J. Gehrke[†], J. W. Mandler[†], R. G. Helmer[‡], and J. R. Davidson[†]

[†]Idaho National Engineering and Environmental Laboratory, P.O. Box 1625, Idaho Falls, ID 83415-2114
[‡]Idaho Accelerator Center, Idaho State University, Pocatello, ID 83209

Abstract. The electronic version of the γ-Ray Spectrum Catalogue, at the Web Site http://id.inel.gov/gamma/, has been extended to include additional radionuclides measured with modern Ge detectors. The set of data for each nuclide includes a new spectral plot, a complete decay scheme, and a table of γ-ray energies and intensities downloaded from the Evaluated Nuclear Structure Data File (ENSDF). Each γ-ray is color coded in the same way in each of these three presentations. γ rays from daughter activities are identified in the spectrum by labeling the peak with the isotope with which it belongs. X-rays, artifact peaks (e.g., sum peaks), and contaminant radionuclides are distinctly colored. For each available nuclide, any available spectra from the earlier NaI(Tl) and Ge(Li) Catalogues have been included. The date of all of the downloaded ENSDF data is also recorded to provide a pedigree. Actinide decay chains allow hot links to other members of the decay chain. Links to other user information resources have been included.

INTRODUCTION

Extensive use of the CD version of the γ-Ray Spectrum Catalogue[1-4] since its release in March 1999 has demonstrated its usefulness and also its potential to serve an increasingly larger community of users of radionuclides and gamma-ray spectrometrists. The rapid access to spectral data via a CD is tempered by the limitations of its material being dated.

With high-speed Internet connections becoming available even to the individual user, the time is ripe to modify and improve on the structure of the present Web-based gamma-ray spectrum catalogues. This Web-based version is able to take advantage of frequent updates, links to additional or supporting information, new references, and new material.

With this intention the Electronic γ-Ray Spectrum Catalogues at http://id.inel.gov/gamma are being expanded with new data. This Web Site is being designed to be fast and easy to use with hot links to various other nuclear data resources, and references.

γ-RAY CATALOGUES

The γ-Ray Spectrum Catalogues on the INEEL γ-Ray Spectrometry Center Web site are in Adobe Acrobat (PDF) form and are color-coded for ease of use and to correlate between the presentations. At the present time three sets of data are available on the Web Site:

1. radionuclide spectra acquired with NaI(Tl) from 1964[5]

2. radionuclide spectra acquired with Ge(Li) from 1974[6]

3. new radionuclide spectra acquired with modern Ge detectors.

The NaI(Tl) γ-Ray Spectrum Catalogue[5] has been scanned (but not the higher quality digitized data) and placed on the Web Site with the addition, in color, of the associated decay schemes. The energies and intensities in the NaI(Tl) Catalogue were entered earlier and are not as current as those in the Ge(Li) or Ge Catalogues. Over the past two years a significant number of previously missing actinide data have been and continue to be, added to the Ge Catalogue on this

Web Site. A few other spectra from other radionuclides have also been acquired. Presently, all of these Ge spectra are listed in the Actinide Catalogue. This Catalogue is formatted as a series of single nuclei data sets to facilitate rapid access via the Internet. Some earlier NaI(Tl) and Ge(Li) spectra of the actinides and their progeny have been updated and included with these new nuclide data sets.

The color coding for the NaI(Tl), Ge(Li) and Actinide Catalogues is based on the most current, highest resolution spectral data available at the time of their creation. Those γ-ray peaks whose maximum height extends a factor of ten above the spectral continuum are colored red and the weaker γ-ray peaks are blue in spectra, decay schemes, and tables. γ rays from this radionuclide but are not observed in the γ-ray spectrum are labeled black in the decay scheme and in the table of γ ray energies and intensities. Peaks not representing the full energy of a γ ray from this radionuclide are color-coded gold (e.g., x rays, escape peaks, annihilation radiation, coincidence sum peaks, background peaks, and interfering radionuclides). The energy labels associated with the full-energy peaks from the decay of progeny are colored brown and the nuclide is identified. As an example, Figures 1-4 show portions of the data set for the decay of ^{239}Pu.

In this paper, due to the lack of color print, the different size and type peaks are distinguished by the size of their font (the peaks in the most intense category having the largest font size). Peaks not representing the full-energy of a γ ray emitted from the radionuclide peaks appear gray.

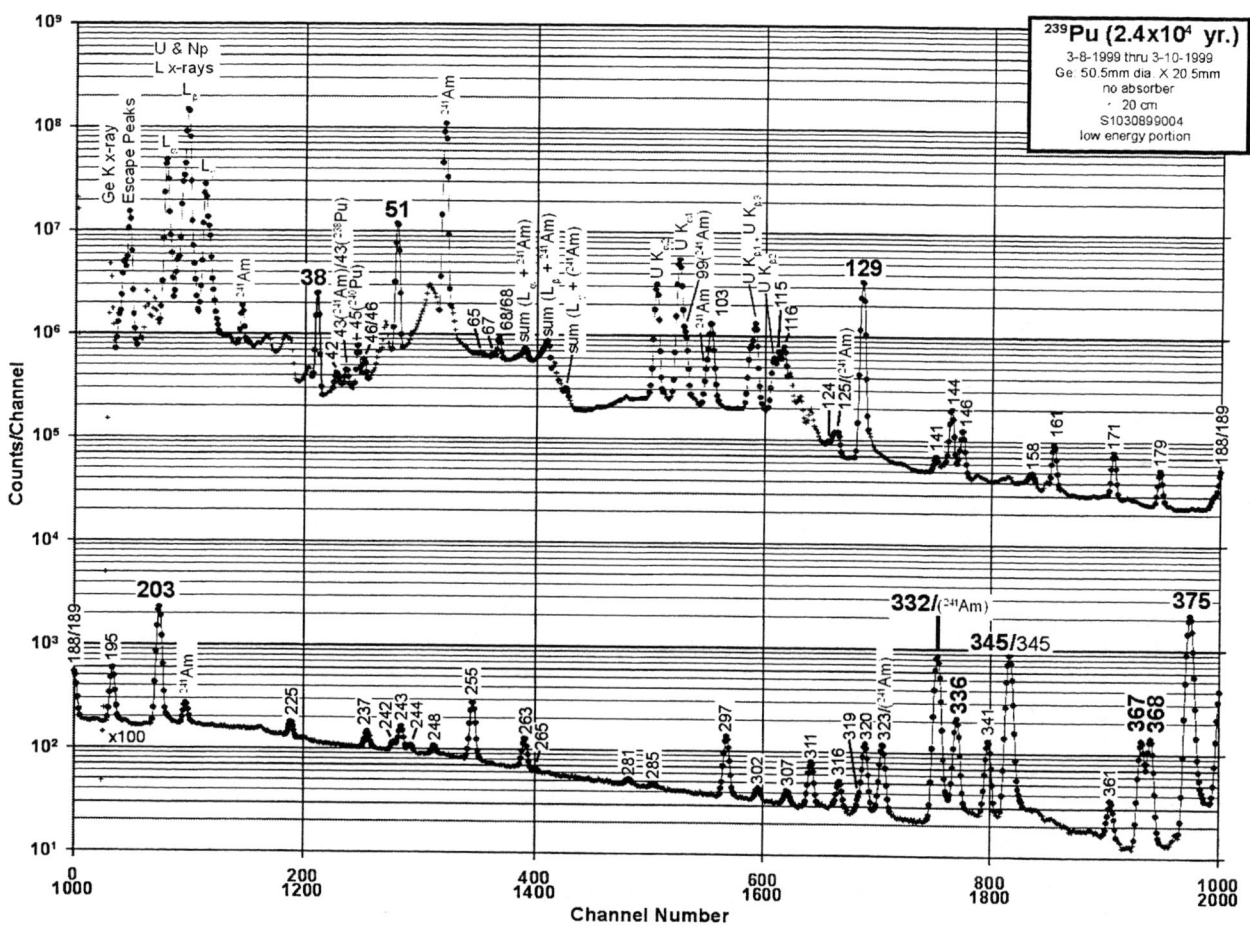

FIGURE 1. The low-energy portion of the ^{239}Pu spectrum as it appears in the new Catalogue. The L and K x rays are identified by their component symbol, γ-rays are labeled by energy truncated to keV. Ge K x-ray escape peaks are labeled "Ge escape peaks." Coincidence and random sum peaks are labeled "sum" followed, in parenthesis, by the energies of the component peaks. Decay schemes, and tables of the γ-ray energies and emission probabilities are provided for each member (i.e., parent and progeny) of the decay chain observed in the γ-ray spectrum.

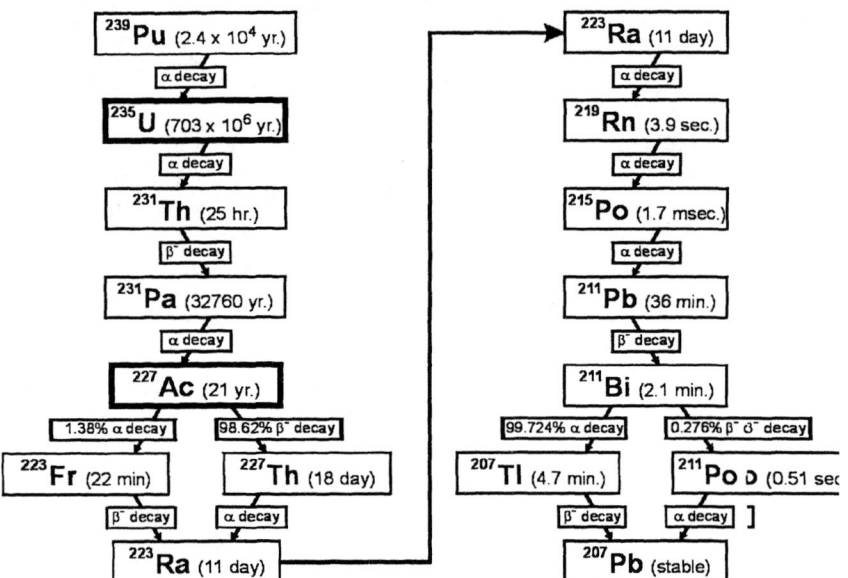

FIGURE 2. The decay chain of ^{239}Pu. The bold box, magenta in the Web catalogue, around some progeny nuclides indicate that the corresponding data set is available. Selecting this box with a mouse will transfer the user directly to that data set.

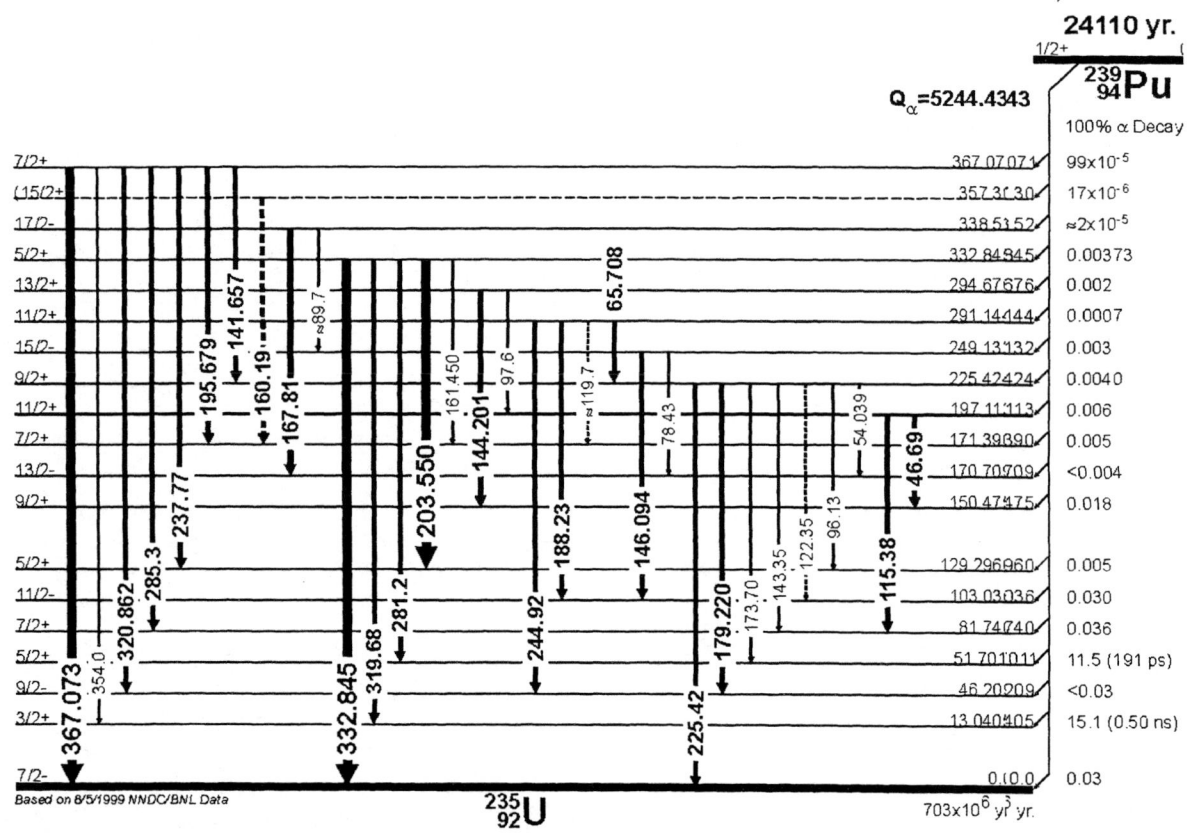

FIGURE 3. A portion, one page of six, of the decay scheme of ^{239}Pu that populates levels in ^{235}U. In the case of ^{239}Pu, the decay schemes for the progeny are not included in the data set since the γ-ray peaks from the progeny are not observed in the acquired spectra, due to the extremely long half-life (7.03×10^8 y) of ^{235}U. The boldness of the transition arrows and the font sizes of the corresponding energy labels indicate the relative intensities of the γ rays. A dotted line for the transition arrow shows γ-ray transitions whose placement is uncertain. The Evaluated Nuclear Structure Data File (ENSDF)[7] citation gives the reference for the downloaded data.

GAMMA-RAY ENERGIES AI

Nuclide: 239Pu E_γ σE_γ I_γ σI_γ Levels- from ENSDF

E_γ (keV)	σE_γ	I_γ	σI_γ	Level	
0.076 8	0.000 5			0.076 8	α
12.965	0.003			13.040 5	
30.04	0.02	0.000 217	0.000 006	81.740	α
38.661	0.002	0.010 5	0.000 2	51.701 1	α
40.41	0.05	0.000 162	0.000 016		α
42.06	0.03	0.000 165	0.000 005	171.390	α
46.21	0.05	0.000 737	0.000 074	46.209	α
46.69		0.000 058	0.000 004	197.113	α
47.56	calculated	0.000 056	calculated	129.296 0	α
51.624	0.001	0.027 1	0.000 5	51.701 1	α
54.039	0.008	0.000 197	0.000 003	225.424	α
56.828	0.003	0.001 130	0.000 025	103.036	α
65.708	0.030	0.000 045 6	0.000 001 4	291.144	α
67.674	0.012	0.000 164	0.000 003	170.709	α
68.696	0.006	0.000 3	0.000 1	81.740	α
68.74	calculated	0.000 110	0.000 060	150.475	α
74.96	0.10	0.000 038	0.000 006		
77.592	0.014	0.000 410	0.000 020	129.296 0	α
78.43	0.02	0.000 141	0.000 006	249.132	α
89.7	≈	0.000 002		338.52	α
89.73	0.04	0.000 030	0.000 006	171.390	α
96.13	0.05	0.000 022 3	0.000 004 0	225.424	α
97.6	0.3			294.876	α
98.780	0.020	0.001 220	0.000 040	150.475	α

FIGURE 4. The table format adopted for the γ-ray energies and emission probabilities. The figure represents a portion of the data set for the decay of ^{239}Pu. The same color and font size coding is used in the tables as in the decay schemes.

FUTURE DEVELOPMENTS

R. L. Heath who died in 1997 held the original vision. His vision was to develop the INEEL Gamma-Ray Spectrometry Center Web Site beyond a compilation of the NaI(Tl) and Ge γ-Ray Spectrum Catalogues and into a world resource for γ-ray spectrometry.

In line with this vision, our first goal is to add spectra and data for missing radionuclides of importance to the various disciplines, such as nuclear medicine, basic science and engineering applications. In addition, a major goal is to expand the Catalogue to include prompt neutron capture and inelastic scattering γ-ray spectra measured with large volume HPGe and thin-window detectors.

A second goal is to fully integrate spectra from different detectors into each nuclide's data set, eventually replacing current separate Catalogues with one integrated Catalogue access via a periodic table of elements interface. The Actinide Catalogue on the Web Site is the beginning of this effort. In this way the user can quickly access the appropriate isotope data and visualize spectral expectations when using a specific detector. As an example, a large Ge detector spectrum may be used by someone using a NaI(Tl) detector to ascertain the presence of lines that do not show in the NaI(Tl) spectrum as separate lines.

Further, there is a need to make available current nuclear data and information on methodology to both the experienced spectrometrists and the novice practitioner who has been trained in another area of science or in an engineering discipline. We wish to develop a tutorial section of the Web Site to illustrate various aspects of γ-ray spectrometry.

These additions will occur as funding and time permit.

ACKNOWLEDGEMENTS

This work was supported by the US Department of Emery, Assistant Secretary for Environmental Management, under DOE Idaho Operations Office Contract DE-AC-07-99ID13727 BBWI.

REFERENCES

1. Gamma-ray Spectrum Catalogue, Ge(Li) & Si(Li) Gamma-ray Spectra; CD ROM edition, Technical Coordinators: Dr. R. G. Helmer, R. J. Gehrke, and J. R. Davidson; INEEL, PO Box 1625, Idaho Falls, ID 83415-2114; published March 1999.

2. Heath, R. L., J. Radioanal. Nucl. Chem. 233, 81-89 (1998).

3. Helmer, R. G., Gehrke, R. J., Davidson, J. R., and Mandler, J. W., J. Radioanal. Nucl. Chem. 243, 109-117 (2000).

4. Gehrke, R. J., Davidson, J. R., Taylor, P. J., Helmer, R. G., and Mandler, J. W., J. Radioanal. Nucl. Chem accepted for publication.

5. Heath, R. L., AEC report IDO-16880 –1 & -2, (1964).

6. Heath, R. L., AEC report ANC-1000-2 (1974).

7. Evaluated Nuclear Structure Data File (ENSDF) database maintained at the National Nuclear Data Center (NNDC) by Brookhaven National Laboratory.

SECTION III

ACCELERATOR MASS SPECTROMETRY

Transmission Properties For The Recombinator - Tandetron Configuration At The National Ocean Sciences Accelerator Mass Spectrometry Facility

Karl F. von Reden

NOSAMS Facility, Woods Hole Oceanographic Institution, Woods Hole, MA 02543
E-mail: kvonreden@whoi.edu

Abstract. After a year of full operation, the newly installed 134-cathode ion source (MC-SNICS, NEC Corp.) has proven to be compatible with the AMS system and in fact the results are as good or even better than those obtained with the predecessor device. Preliminary data[1] were shown at the 8th International Conference on AMS, Vienna, Austria, 1999. At this point the operational parameters are better understood and progress was made in optimizing the extracted beam for highest transmission through the system. Control software was written for the AMS system to facilitate the analysis of the ion beam optics at the device level. Transmission profiles are presented in a detailed comparison of the two NOSAMS injectors.

INTRODUCTION

One of the challenges of accelerator mass spectrometry is the requirement that ion beams of the isotopes of interest be transmitted through the system with very high stability and reproducibility over long time periods (at least several hours). Whether sequential or simultaneous injection is used, the method relies on the assumption that the isotopic ratio measured on well-known "standard" reference samples can be used to estimate the relative abundance of the isotope of interest in an unknown sample. The number of standards needed in a sequence of analyses depends on the stability of the system over time. For reasons of throughput economy it is important to minimize the required number of standards. The overall system stability is determined by the degree to which fluctuations in the beam optics devices (e.g. magnets, lenses, accelerator) around their tuning optimum lead to changes in the absolute and relative intensities of the transmitted ion beams. A desirable situation is that at least the relative intensities remain constant over a range of fluctuation much larger than the intrinsic regulation stability of the power supplies involved. This is already difficult to achieve with two equally abundant isotopic species. In the case of carbon we are dealing with three isotopes whose natural abundance spread is at least 12 orders of magnitude. This paper investigates the transmission properties of the NOSAMS system in light of the replacement of one of its original ion sources with an MC-SNICS model by National Electrostatics Corporation.

THE NOSAMS "RECOMBINATOR" INJECTORS

The two standard methods currently in use in AMS, simultaneous (dc) injection and sequential injection (rapidly alternating the injected beams) have proven to be equivalent in achieving very accurate isotope ratios. Both methods have to allow either spatial or temporal separation of the ion beams during the transmission through major sections of the system. The NOSAMS injectors are an implementation of the simultaneous injection method in a mirror symmetric 4-magnet 180° inflection layout[2]. The discussion of transmission properties will concentrate on this section of the NOSAMS system because here the ion beams are separated by as much as 40 mm. All other devices in our system displayed transmission profiles with large overlaps in their flat-topped region making them easily tunable.

Layout and Calculations

Figure 1 shows the layout of the second NOSAMS recombinator as it was calculated with code boundaries, designed in this way to correct for second order aberrations. The layout[4] reflects a reversal of the order in which the concave and convex boundaries are arranged from the original design[2]. This fact does not seem to change the calculated behavior significantly as long as complete symmetry is maintained. The 40 keV A=12,13,14 ion beams are first dispersed and travel RAYTRACE[3]. The field of magnet #4 can be reversed to accept beams from the identical recombinator #1 (to the left). The pole faces are shown together with the alternating convex and concave effective field parallel between magnets 2 and 3 where a rectangular aperture stops ions of masses outside of this range (e.g., NH). Magnets 3 and 4 refocus the beams for subsequent injection into the accelerator. It should be mentioned that for vertical focusing and displacement correction slot lenses[5] and steerers are positioned between magnets 1 and 2 as well as 3 and 4 (not shown here), creating an intermediate focus in the symmetry plane of the recombinator.

One of the most probable alignment errors in this system is a rotation of one or more of the magnets around their vertical axes. This would have the largest effect in magnets 2 and 3. In the figure the result of a 10° rotation is superimposed for magnet 3 and the insert (vertically compressed by a factor 10) shows how that affects the ion beams at the exit of the recombinator (past magnet 4). The solid lines depict the solution with good alignment yielding about 2 mm waist size for incoming rays of up to 12 mrad divergence angles. The dashed lines correspond to the massive misalignment of the third magnet. An increased waist size of about 4 mm is calculated with a distribution favoring ^{12}C rays on the left and ^{14}C rays on the right side of the midpoint in the insert of fig.1.

Scaling the fields of all four magnets at the same time will not significantly change the lateral position of the exit beam. However, changes in the fields of a subset of the magnets will produce an exit beam of an angle different from 180°. At NOSAMS, a common power supply powers the first three magnets and a

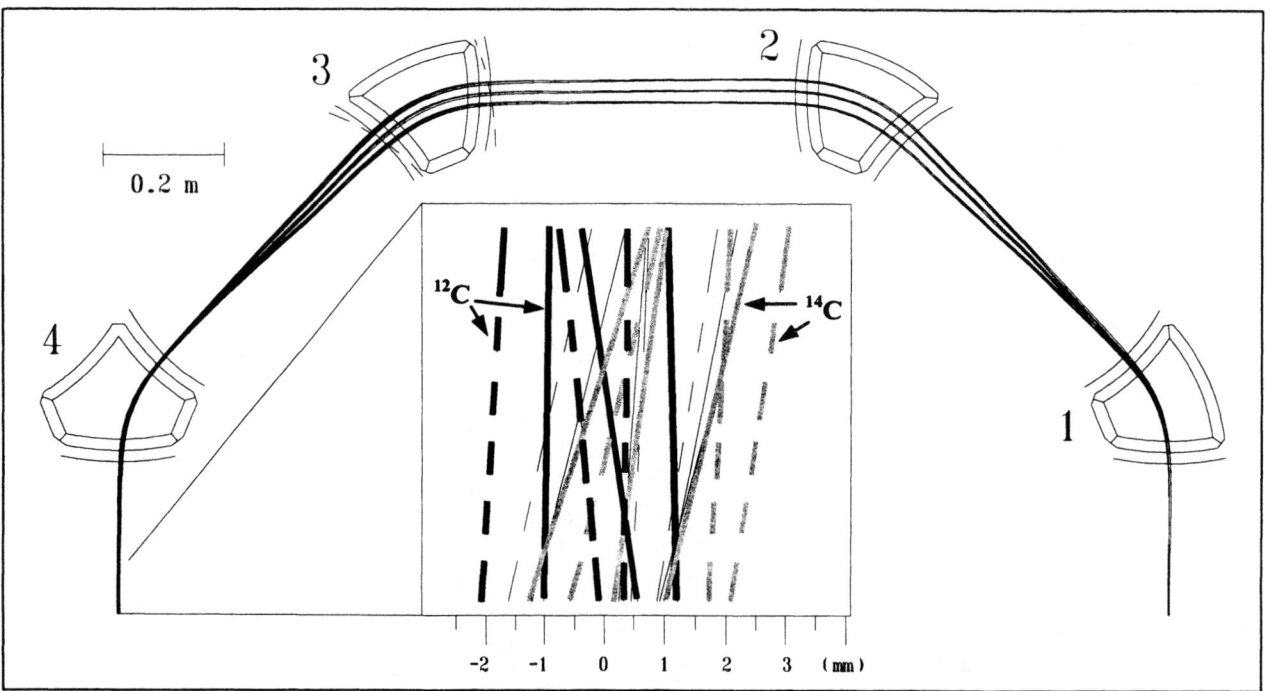

FIGURE 1. NOSAMS recombinator layout with RAYTRACE calculations of the three carbon isotope beams. The effect of a misaligned magnet (3) is also shown. See text for details.

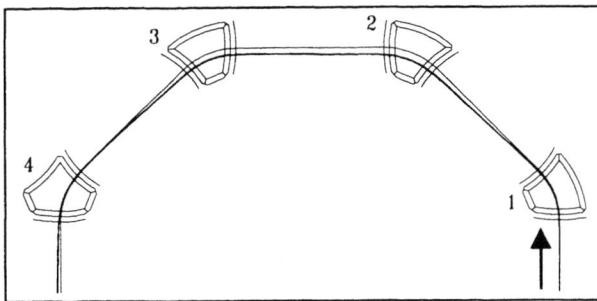

FIGURE 2. The effect of raising the field in magnets 1-3 by 5% over the normal setting for optimal tune. The thin line is the layout ray and the darker lines are the rays for A=13. For dimensions see figure 1.

separate bipolar supply controls the fourth magnet. Lowering the fields in the first three magnets is equivalent to raising the field in the fourth and vice versa, with respect to the deviation of the outgoing beam from 180°. Figure 2 shows the case of ^{13}C with the fields in magnets 1 – 3 raised by 5%. The field in magnet 4 will have to be raised as well to straighten the exit beam. The measured transmission profiles should be judged in light of these calculations.

Transmission Profile Measurements

The system was first carefully tuned to achieve maximal transmission for the three ion beams. Here the ten years of operating the NOSAMS system helped to arrive at a reasonable set of "book values" for the tuning parameters. Then a computer routine was used to slowly lower the setting for one device at a time until in each case the predominant ^{12}C beam current, measured at the high-energy end of the system after the main analyzing magnet, dropped to less than 60% of its maximum value. For a given device the routine now stepwise raised the setting in five-second intervals until the current dropped below 60% on the high side of the maximum. The step time interval was chosen to allow the ^{14}C detector to accumulate reasonable counting statistics at about 120 cps. Figures 3 and 4 show the measured profiles for (a) the linked set of recombinator magnets 1-3 and (b) the fourth magnet, the switching magnet shared by the two injectors. The vertical axis "Intensity" is offset by 1000 for (b) and denotes the number of counts for ^{14}C and the beam current in nA for ^{12}C and ^{13}C. Note that in our system the ^{12}C beam is chopped by a factor of about 90 before acceleration. The transmission profiles very closely resemble the expected behavior of a beam with a gaussian distribution swept over a rectangular slit. The width of the profiles is between 1 and 2% of the magnet current setting, considerably wider than the <0.1% stability of the observed magnetic fields.

Aberrations in the two injectors (possibly in part from misalignment) lead to small shifts of the profiles relative to each other. This makes it more difficult to find regions where all ion beam currents remain constant under variation of the field setting.

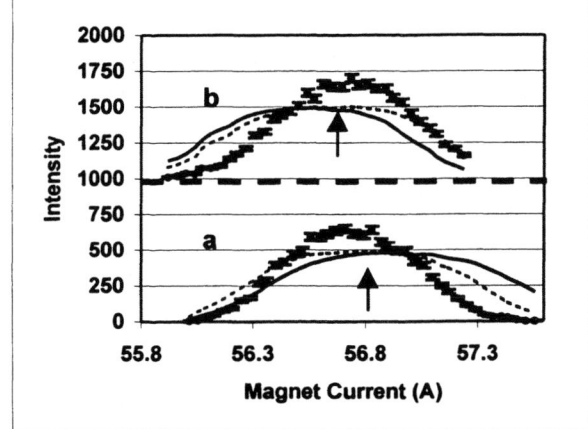

FIGURE 3. Measured transmission profiles for (a) Recombinator #1 magnets 1-3 (linked) and (b) the fourth magnet. ^{12}C: solid line; ^{13}C: dashed line; ^{14}C: filled circles, the error bars reflect the counting statistics.

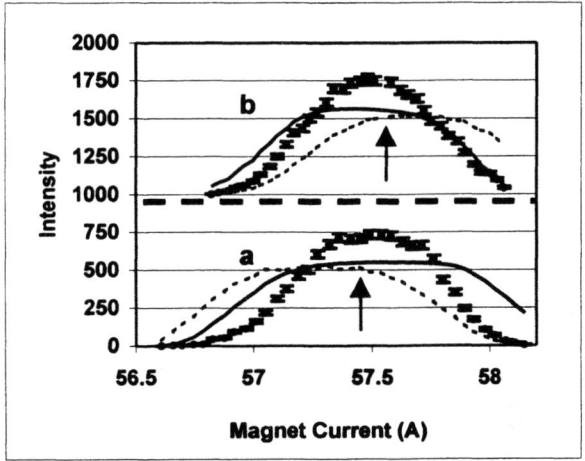

FIGURE 4. Transmission profiles for Recombinator #2. The data reflect ion beams extracted from the new MC/SNICS ion source. See figure 3 for a description of the curves.

Attempts[6] to correct the aberrations with field clamps were unsuccessful, in the past. One drawback of the manufacturer's design[4] is the very compact assembly of elements in the injector with almost no possibility of simple alterations in the layout. The two recombinators display a slightly different behavior with respect to the three isotopic ion beams: #1 has a "normal" mass dependence in the shifts of the profiles, whereas #2 has a good alignment of the ^{12}C and ^{14}C profiles and only the ^{13}C profile is shifted. For both it is possible to reliably tune the system (the arrows indicate the optimal settings). Note that in both recombinators the effect described in figure 2 leads to a mirror-reflected distribution of the individual profiles for magnets 1-3 and the switching magnet (4).

Conclusions

The measured profiles indicate that transmission of the three carbon ion beams is subject to small aberrations in the recombinator and that future alterations along the lines of reference 7 would be desirable. A chance to address this matter will come up when we will install the newly developed microwave ion source[8]. Meanwhile, stable tunes have been obtained consistently over the last ten years with an accuracy of the results matching any other AMS facility around. Even without corrective measures the profiles in figure 4 are very encouraging because they show that a near perfect tune can be obtained for ^{12}C and ^{14}C while ^{13}C is somewhat on the edge of the tune. This is acceptable because the important result is the ratio of $^{14}C/^{12}C$ while the ^{13}C beam is primarily used for monitoring and control purposes (e.g., stabilization feedback for the accelerator terminal voltage). All NOSAMS samples are separately analyzed for ^{13}C on a stable isotope ratio mass spectrometer. None of the calculations so far helped to explain the somewhat unexpected behavior of the ^{13}C beam relative to the others but it will allow us to run the new MC/SNICS ion source on recombinator #2 with even higher reproducibility than injector #1.

ACKNOWLEDGMENTS

This work is supported by the National Science Foundation under Cooperative Agreement OC-9807266.

REFERENCES

1. von Reden, K.F., Bellino, M., Long, P., Schneider, R.J., Loger, R., "Installation of a 134-sample MC-SNICS ion source at NOSAMS and first results", *Nuclear Instruments and Methods in Physics Research* B **172**, 247-251 (2000).

2. Litherland, A.E., Kilius, L.R., "A recombinator for radiocarbon accelerator mass spectrometry", *Nuclear Instruments and Methods in Physics Research* B **52**, 375-377 (1990).

3. Enge, H.A., Kowalski, S., "Raytrace" computer code, Laboratory of Nuclear Science, MIT, 1990, unpublished.

4. Purser, K.H., Smick, T.H., Purser, R.K., "A precision ^{14}C accelerator mass spectrometer", *Nuclear Instruments and Methods in Physics Research* B **52**, 263-268 (1990).

5. Litherland, A.E., Kilius, L.R., US Patent 487207 (1990).

6. Schneider, R.J., von Reden, K.F., "Corrections for aberrations in the Woods Hole recombinators", 30th Symposium of Northeastern Accelerator Personnel, Woods Hole, 1996, Proceedings edited by K.F. von Reden and R.J. Schneider, World Scientific 1998, 123-127.

7. Mous, D.J.W., Gottdang, A., van der Plicht, J., "Status of the first HVEE ^{14}C AMS in Groningen", *Nuclear Instruments and Methods in Physics Research* B **92**, 12-15 (1994).

8. Kim, S-W. Schneider, R.J., von Reden, K.F., Hayes, J.M., Wills, J.S.C., Kern, W.G.E., "Test of positive ion beams from a microwave ion source for AMS", these proceedings.

Environmental radiation protection studies related to nuclear industries, using AMS

Ragnar Hellborg, Bengt Erlandsson, Mikko Faarinen, Helena Håkansson, Kjell Håkansson, Madis Kiisk, Carl-Erik Magnusson, Per Persson, Göran Skog and Kristina Stenström

Department of Physics, Lund University, Sölvegatan 14, SE-223 62 LUND, Sweden

Sören Mattsson and Charlotte Thornberg

Department of Radiation Physics, Malmö University Hospital/Lund University, SE-205 02 MALMÖ, Sweden

Abstract. ^{14}C is produced in nuclear reactors during normal operation and part of it is continuously released into the environment. Because of the biological importance of carbon and the long physical half-life of ^{14}C it is of interest to study these releases. The ^{14}C activity concentrations in the air and vegetation around some Swedish as well as foreign nuclear facilities have been measured by accelerator mass spectrometry (AMS). ^{59}Ni is produced by neutron activation in the stainless steel close to the core of a nuclear reactor. The ^{59}Ni levels have been measured in order to be able to classify the different parts of the reactor with respect to their content of long-lived radionuclides before final storage. The technique used to measure ^{59}Ni at a small accelerator such as the Lund facility has been developed over the past few years and material from the Swedish nuclear industry has been analysed.

INTRODUCTION

Accelerator mass spectrometry (AMS) is one of the applications of electrostatic accelerators which has been used to great advantage in different fields of scientific endeavour. The many steps from ion source to detector via an accelerator and several analysing units have been found to lead to a strong reduction of the background.

In Lund the AMS system is used for radiocarbon dating, for studies of releases into the environment of reactor-produced ^{14}C by analysis of air and plant material in the vicinity of nuclear installations, for classification of nuclear waste by detecting ^{59}Ni, for studies of ^{14}C-labelled pharmaceuticals used in clinical nuclear medicine, and – in the near future – for detection of ^{10}Be and ^{26}Al for environmental and biomedical studies. In this report some recent AMS studies related to nuclear industries are high-lighted.

DEVELOPMENT OF THE LUND AMS SYSTEM

The Lund AMS system is built up around a 3 MV Pelletron tandem accelerator. Since the end of the 80's this accelerator has been extensively remodelled to become a high quality AMS facility [1]. The most recent modification is the installation of a new gas stripper including terminal pumping. The very limited space and the absence of electrical power in the high voltage terminal, required an unusual design for the installation and powering of the two turbomolecular pumps [2,3]. A new injector with a spherical electrostatic analyser and a 90° magnet is at the moment being installed [4]. This most recent development will give a better mass and energy resolution, especially for heavy atoms.

APPLICATIONS AT THE LUND AMS FACILITY RELATED TO NUCLEAR INDUSTRIES

^{14}C emission from nuclear industries and its effect on the ^{14}C levels in the environment

^{14}C is one of the radionuclides which are produced by neutron-induced reactions in all types of nuclear reactors. The continuous release through the ventilation system leads to an increase in the ^{14}C specific activity of the atmosphere and, hence, to an increased radiation exposure of man. Because of the biological importance of carbon it is of interest to study its local and global distribution and especially its pathways to man and the resulting activity concentration levels. Of all radionuclides released in routine operation by the nuclear power industry, ^{14}C is likely to produce the largest collective dose to man [5]. Extensive investigations of the ^{14}C releases from some Swedish nuclear power plants have been performed during a time period of more than ten years [6,7,8,9,10]. During recent years also the release from foreign nuclear facilities –in Estonia, Romania, Canada, Germany and England– have been studied.

TABLE 1. Comparison between ^{14}C in annual tree rings of pine near the Barsebäck nuclear power plant and the content in rush at a "clean air" site at Måryd.

Year	Barsebäck Mean (Bq/kgC)	Måryd (Bq/kgC)	Barsebäck Excess activity (Bq/kgC)
1981	295±3	287±2	8±4
1982	281±5	283±2	-2±5
1983	275±3	279±2	-4±4
1984	277±3	276±2	1±4
1985	275±3	272±2	3±4
1986	273±3	269±2	4±4
1987	273±3	266±2	7±4
1988	266±4	262±2	4±4
1989	264±5	262±2	2±5
1990	265±4	261±2	4±4
1991	260±3	260±2	0±4
1992	262±3	258±2	4±4
1993	258±3	254±2	4±4
1994	256±3	254±2	2±4
1995	252±2	252±2	0±4
1996	252±2	251±2	1±4

The ^{14}C content in annual tree rings of pine (*Pinus*), located a few km in the direction frequently down-wind of the Barsebäck nuclear power plant, was measured. The power plant, which is located in South Sweden, consists of two boiling light water reactors (BWR) 600 MW$_{el}$ each. The mean ^{14}C specific activity (given in Bq per kg carbon) obtained from two different trees at 3 and 1 km from the power plant, are in Table 1 compared to the activity in samples of rush (*Juncus*) from the "clean air" site at Måryd (located 10 km east of Lund). The Måryd samples have been measured at the conventional ^{14}C laboratory in Lund. By subtracting the Måryd activity from the mean Barsebäck activity the excess activity at Barsebäck was obtained. There might be a small excess of ^{14}C for some of the years, but the effect is small and rarely higher than the uncertainty of the measurements. The mean excess specific activity could be around 2 Bq/kg.

Activity concentrations of ^{14}C in willow leaves (*Salix viminalis*) and air samples taken at different distances up to 9 km from the Barsebäck power plant also showed - with few exceptions - values very close to the Måryd data. A few of the air samples showed significantly lower specific activities than Måryd, probably due to the influence of the combustion of fossil fuels.

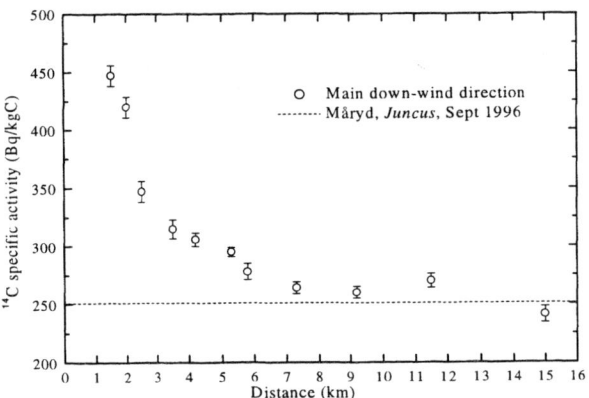

Figure 1 The ^{14}C content in grass samples collected at various distances in the NNE direction from the Sellafield nuclear facility. The distances are relative to the Thorp reprocessing facility.

At the Forsmark nuclear power plant (115 km north of Stockholm) three BWR 970, 970 and 1155 MW$_{el}$ resp., the activity concentrations of ^{14}C in the yearly rings –taken in 1992– from a pine at 2.5 km in the direction frequently down-wind of the plant and in sallow (*Salix caprea*) have been measured. The tree ring sample shows an excess ^{14}C specific activity of 25±5 Bq/kgC, *i.e.* an elevation of 11±2% above the background. The sallow samples collected at 1.5 and 3.2 km show excess activities of 28±4 Bq/kgC (11±2% above background) and 6±4 Bq/kg C (2±2% above background), resp.

The fuel reprocessing plant at Sellafield in northwest England is known to release substantial amounts of ^{14}C. In Fig. 1 the ^{14}C content in grass samples collected in September 1996 at various distances from the Sellafield plant is presented. The highest activity, found at the sample site closest to the facility was 447±9 Bq/kgC.

TABLE 2. Local excess ^{14}C specific activity due to releases from various nuclear installations and the related effective dose to the most exposed individual[1]. Values within () according to ref 11.

Site	Excess ^{14}C specific activity (Bq/kg C)	Effective dose rate (μSv/year)
Barsebäck	2-4	0.1-0.2
Forsmark	6-30	0.3-0.7
Sellafield	200	11
Pickering	4600 (12000)	300 (7000)
Natural production	226	12

The Pickering nuclear plant station in Canada is equipped with 8 units, each 540 MW$_{el}$, of Canadian deuterium uranium (CANDU) pressurized heavy-water reactors. According to ref. 5 the normalised ^{14}CO$_2$ release during operation is 4800 GBq (GW$_{el}$ a)$^{-1}$ (for pressurised light water reactors, PWR, the corresponding figure is 120 GBq (GW$_{el}$ a)$^{-1}$, and for BWR 450 GBq (GW$_{el}$ a)$^{-1}$). Four grass samples were collected in April 1998 at various distances from the plant. The highest activity measured, at about one kilometre north-east of the plant, showed a specific activity of 4820±100 Bq/kgC, *i.e.* about 20 times the natural ^{14}C level. At 700-800 m east, at 1.5-2 km north-east and at 100 km a specific activity of 511±7, 1530±50 and 268±5 Bq/kgC, resp., were found.

At Paldiski submarine training centre in Estonia two PWR (full power 70 MW and 90 MW) were in operation during 1968-1989 and 1983-1989, resp. Tree ring samples from the vicinity of Paldiski were collected in the spring of 1998. The samples show no significant excess ^{14}C activity.

From the excess ^{14}C specific activity within 3 km from the different types of nuclear facilities reported above, the effective dose to man has been calculated and is presented in Table 2. The absorbed dose, in mGy, to fat and to bone marrow may be 2-3 times higher than the effective dose presented in the table. The results of this investigation show that the effective dose to man due to the releases of ^{14}C from the Swedish BWR at Barsebäck and Formark is very low, also when compared to other nuclear installations.

It has to be stressed that the local collective doses due to ^{14}C releases from reactors represent only a small fraction of the total collective dose commitment. The main significance of ^{14}C lies in the fact that it enters the carbon cycle and its resulting global dispersion, leading to long-term irradiation.

For details on these measurements, see ref.10.

^{59}Ni in wastes from the nuclear industry

In nuclear waste management ^{59}Ni is an important radionuclide. It is produced by neutron activation in the stainless steel shielding surrounding the fuel. The total activity concentration of ^{59}Ni, as well as of other long-lived radionuclides, has to be established in the planning of the final disposal. Because ^{59}Ni decays only via electron capture and has a very long physical half-life (7.6×10^4 years), it is difficult to measure the radiation emitted in its decay. AMS is in this case advantageous. However, for small tandem accelerators, such as the Pelletron in Lund, the common detection techniques of energy or energy loss are not able to distinguish atomic isobars of heavy elements such as Ni. One way to eliminate this problem is to combine AMS with the detection of characteristic projectile X-rays emitted when the ions pass a suitable target. In Lund this technique was introduced some years ago [12] and a method to extract nickel chemically from stainless steel has been developed [13,14,15].

The chemical purification process is performed in two steps. In the first step – developed during the initial stage of the nickel project in Lund – the nickel is extracted from the stainless steel by dissolution in HCl and precipitation with dimethylglyoxime [13]. The lower limit obtained in the purification from cobalt is found to be set by contamination within the process. Tools and beakers of plastic and glass have to be used and recently the distilled water used in this step has been replaced by distilled water deionised down to the 10^{-9} level. This later improvements gives – with three standard deviations- a detection limit of 12 Bq/gNi (Bq per gram nickel). This corresponds to a ratio ^{59}Ni/Ni of 4×10^{-9}. For samples with ^{59}Ni content lower than this limit a second and more sophisticated step has recently been introduced which uses the reaction between nickel and carbon monoxide to form gaseous nickel carbonyl (Ni(CO)$_4$) [16]. This second step is expected to decreased the detection limit to 1 Bq/gNi.

Samples from the nuclear industry have activities up to about 10^8 Bq per gram steel, the main

---

[1] The "most exposed individual" is a person living within 1-3 km from the source of the releases and producing all his food locally.

radioisotope in the material being ^{60}Co. The residual products from the chemical steps contain most of the activity and are returned to the nuclear industry.

The new gas-stripper with terminal pumping, mentioned above, is used routinely for the Ni experiment and has resulted in less energy spread, a more stable beam current and a higher beam transmission than when the old gas stripper or the foil stripper were used.

TABLE 3. ^{59}Ni activities in samples taken from different positions in the BWR no. 1 at the Oskarshamn nuclear power plant on the east coast; m.t.=moderator tank; s.s.=steam separator; g.r.=guiding rod.

Sample no.	Sample detail	Activity MBq/(gNi)
1	m. t.	3.2±0.8
2	m. t.	2.5±0.7
3	m. t.	4.5±1.1
4	m. t.	1.8±0.5
5	m. t.	2.2±0.6
6	g. r. for the m. t. head	0.30±0.08
7	m. t.	2.3±0.6
8	m. t.	3.1±0.8
9	Rods between s. s.	0.0038±0.0010
10	g. r. for the m. t. head	0.0063±0.0016
11	m. t.	4.2±1.0
12	Flange joint between s. s. and m. t. head	0.064±0.016
13	Rods between s. s.	0.012±0.003

Improvements to decrease the detection limit has also been done by remodelling the target-detector set-up. The detector solid angle has been increased by combining target and exit window into one foil. The lead shield around the detector has been thickened and the X-ray yield for nickel has been optimised using optimal beam energy and target material [17]. The excellent resolution of the Ge-detector (145 eV FWHM at 5.9 keV) separates the K_α peaks of Co and Ni completely. The Co K_β peak, however, influences the Ni K_α peak, but as the ratio of the K_α and K_β yields for Co is known, this influence can be subtracted and it is possible to tolerate a certain level of Co in the samples. To normalise the measurements, X-rays detected from the stable isotope ^{61}Ni (with a natural abundance of 1.14%) are used. This simplifies the normalisation compared to using the beam current of the stable isotope since neither the X-ray production rate nor the detection efficiency for the two different Ni isotopes need to be considered.

The detection limit obtained by AMS has been compared with radiometric measurements made with the same Ge-detector. As ^{59}Ni decays via electron capture a characteristic X-ray from the daughter isotope ^{59}Co will be emitted and is detected. For a sample containing few other radioisotopes, *e.g.* steel samples, a detection limit corresponding to 20 Bq/gNi is obtained for the radiometric method. This is comparable with the AMS limit obtained for samples that have passed the first chemical step. However, for a sample containing other radioisotopes, such as a nickel sample extracted from recirculating water (see below), the detection limit for the radiometric method is much worse.

A number of steel samples obtained from the Swedish nuclear industry have been analysed by the Lund AMS system. In Table 3 some of the recent results of rather active steel samples are presented. Samples of recirculating water from the PWR reactors at the Ringhals nuclear power plant on the west coast have also been analysed. The activities obtained was preliminary found to be 15, 20 and 26 kBq/L (kBq per litre recirculating water) for the reactors no. 2, 3 and 4 at Ringhals, resp. The measured values agree with expected values. Up to our knowledge this is the first reported measurement of ^{59}Ni in recirculating water.

SUMMARY

The power of AMS for extremely sensitive radioisotope measurements has been demonstrated over the past ten to fifteen years. In this article a few examples of AMS experiments carried out at the AMS facility in Lund, especially on samples related to environmental studies, are described.

REFERENCES

1. Hellborg, R., Håkansson, K. and Skog, G., "The modification of an over 20-year old Pelletron to function as a high quality AMS facility" in *Proc. of the "31st Symp. of Northeastern Accelerator Personnel"* 13-15 October, 1997, Jülich Germany, p. 165-179.

2. Håkansson, K. and Hellborg, R., "A new Design of Terminal Pumping in the Lund Pelletron Tandem" in *Proc. of the 8th Int Conf on Heavy-Ion Acc. Techn.*, Argonne Nat Lab USA, 5-9 October, 1998, p. 94-99, (AIP conference Proceedings no. 473).

3. Håkansson, K., Hellborg, R., Uthas, S. and Olsson, F., "Improvements of the terminal pumping in the Lund Pelletron Tandem", to be published in the *Proc. of the "1999 Symposium of Northeastern Accelerator Personnel"*, Knoxville, Tennessee, 25-28 October 1999.

4. Hellborg, R., Bazhal, S., Faarinen, M. and Magnusson, C.-E., "Calculations for the optics for a new injector of the Lund Pelletron tandem", to be published in the *Proc. of the "1999 Symp. of Northeastern Accelerator Personnel"*, Knoxville, Tennessee, 25-28 October 1999.

5. UNSCEAR (United Nations Scientific Committee on the Effects of Atomic Radiation Sources) 1993 sources and effects of ionising radiation. Report to the General Assembly.UN, NY.

6. Stenström, K., Erlandsson, B., Hellborg, R., Skog, G., Wiebert, A., Vesanen, R., Alpsten, M. and Bjurman, B., *Journal of Radioanalytical and Nuclear Chemistry* **198:1**, 203-213 (1995).

7. Stenström, K., Erlandsson, B., Hellborg, R., Wiebert, A. and Skog, G., *Radioactivity and Radiochemistry* **7:1**, 32-36 (1996).

8. Stenström, K., Erlandsson, B., Hellborg, R., Wiebert, A. and Skog G., *Nucl. Instr. and Meth.* **B113**, 474-476 (1996).

9. Stenström, K., Skog, G., Thornberg, C., Erlandsson, B., Hellborg, R., Mattsson, S. and Persson, P., ^{14}C levels in the vicinity of two Swedish nuclear power plants and at two "clean air" sites in southernmost Sweden, *Radiocarbon*, **40:1**, 433-438 (1998).

10. Stenström, K., Erlandsson, B., Mattsson, S., Thornberg, C., Hellborg, R., Kiisk, M. and Persson, P., Skog, G.. ^{14}C emission from Swedish nuclear power plants and its effect on the ^{14}C levels in the environment, LUNFD6/(NFFR-3079)/1-44/(2000) Lab report, Department of Nuclear Physics, Lund University.

11. Milton, G.M., Kramer, S.J., Brown, R.M., Repta, C.J.W., Kimg, K.J. and Rau, R.R., *Radiocarbon* **37:2**, 485-496 (1995).

12. Wiebert, A., Persson, P., Elfman, M., Erlandsson, B., Hellborg, R., Kristiansson, P., Stenström, K. and Skog, G., *Nucl. Instr. Meth.* **B109/110**, 175-178 (1996).

13. Persson, P., Erlandsson, B., Freimann, K., Hellborg, R., Larsson, R., Persson, J., Skog, G. and Stenström, K., Internal Report LUNFD6/(NFFR-3074)/1-31/(1999).

14. Persson, P., Erlandsson, B., Freimann, K., Hellborg, R., Larsson, R., Persson, J., Skog, G. and Stenström, K., *Nucl. Instr. Meth.* **B 160**, 510-514 (2000).

15. Persson, P., Kiisk, M., Erlandsson, B., Faarinen, M., Hellborg, R., Skog, G. and Stenström K., *Nucl Instr Meth* **B 172**, 190-194 (2000).

16. McAninch, J.E., Hainsworth, L.J., Marchetti, A.A., Leivers, M.R., Jones, P.R., Dunlop, A.E., Mauthe, R., Vogel, J.S., Proctor, I.D. and Straume T., *Nucl. Instr. Meth.* **B123**, 137-143 (1997).

17. Kiisk, M., Persson, P., Hellborg, R., Šmit, Ž, Erlandsson, B., Faarinen, M., Stenström, K. and Skog, G., *Nucl. Instr. Meth.* **B 172**, 184-189 (2000).

Exotic Particle Searches using the Purdue AMS Facility

D. Javorsek II*, D. Elmore*, E. Fischbach*, and T. Miller*

Department of Physics, Purdue University, West Lafayette, IN 47907

Abstract. Two exotic particle searches are being performed using the Accelerator Mass Spectrometer (AMS) at the Purdue Rare Isotope Measurement Laboratory (PRIME Lab). Recent theoretical developments allow for the possibility of small violations of the symmetrization postulate, which may lead in turn to detectable violations of the Pauli exclusion principle. We report the results of a new experimental search for paronic (Pauli-violating) Be, denoted by Be', in samples where Be' retention would be highest. Our limits represent an improvement by a factor of approximately 300 over a previous search for Be'. There are also several recent cosmological motivations for strongly interacting massive particles (SIMPs). We present results from our current search for anomalous heavy isotopes of Au in samples of Australian and laboratory gold with a limit on SIMP abundance ratios as low as 10^{-12}. This experiment provides significant constraints on the existence of such particles in high Z nuclei.

INTRODUCTION

The standard model of particle physics is one of the twentieth century's most significant achievements, but few if any physicists would say that it is the last word in the understanding of fundamental processes. At PRIME Lab (7), the Purdue Accelerator Mass Spectrometer (AMS) facility, we are involved in two projects which test for new physics beyond the standard model.

PARONIC BERYLLIUM

The Pauli Exclusion Principle (PEP) has played a central role in quantum mechanics since its formulation in 1925 (22). Notwithstanding its many successful predictions, the PEP remains somewhat enigmatic, particularly with respect to the question of whether small deviations from it are possible. Recently this question has become the subject of renewed investigation both theoretically (3, 9, 10, 11, 12) and experimentally (2, 5, 8, 14, 20, 26, 25). One of the most interesting conclusions to emerge from the theoretical efforts is what Greenberg and Mohapatra (GM) (11) term the "surprising rigidity" of the PEP. By this GM refer to the apparent impossibility of formulating a consistent local field theory of small PEP violations within the framework of a positive-metric Hilbert space.

Since the PEP makes clear predictions which have direct experimental implications, tests of the PEP are possible even without a fully consistent theoretical framework. However, in the absence of such a framework comparisons among different experiments are not straightforward, even though the interpretation of individual experiments may be clear, as has been emphasized recently by Baron, Mohapatra, and Teplitz (3).

We are using AMS to search for paronic (Pauli-violating) beryllium, Be'. Here Be' is Be in which the PEP-forbidden electronic configuration $1s^4$ replaces the usual configuration $1s^2 2s^2$. In contrast to previous experiments in which the PEP is violated by a single additional fermion in a forbidden state, a search for Be' probes for a PEP violation by two electrons. We also note that in Be' *every electron* resides in the ground state of the atom manifesting behavior consistent with bosonic particles. Although such a violation of two fermions may appear at first sight to be suppressed relative to that for a single fermion, in the absence of a consistent fundamental theory of PEP violation one cannot exclude the possibility that the reverse may be true.

In this experiment we assume that Be' behaves as an inert gas, and we search for the presence of this gas trapped in Be metal, Be ore, natural gas, and air. Each gas sample is admitted into the ion source through a tube attached to the back of a specially designed sample holder (cathode) containing an Sm_2O_3 surface. When the gas exits the cathode it is exposed to a 5 keV Cs^+ ion beam. Any Be' in the sample is first converted to Be through collisions with the energetic Cs^+ ions and is then combined with oxygen from the Sm_2O_3 surface to form the BeO^- ions which enter the accelerator.

Before running the unknown samples it was necessary to calibrate the sensitivity of the AMS to Be concentration and determine the background inherent in the system. This was done by running seven standard samples of known Be concentration along with a blank made from

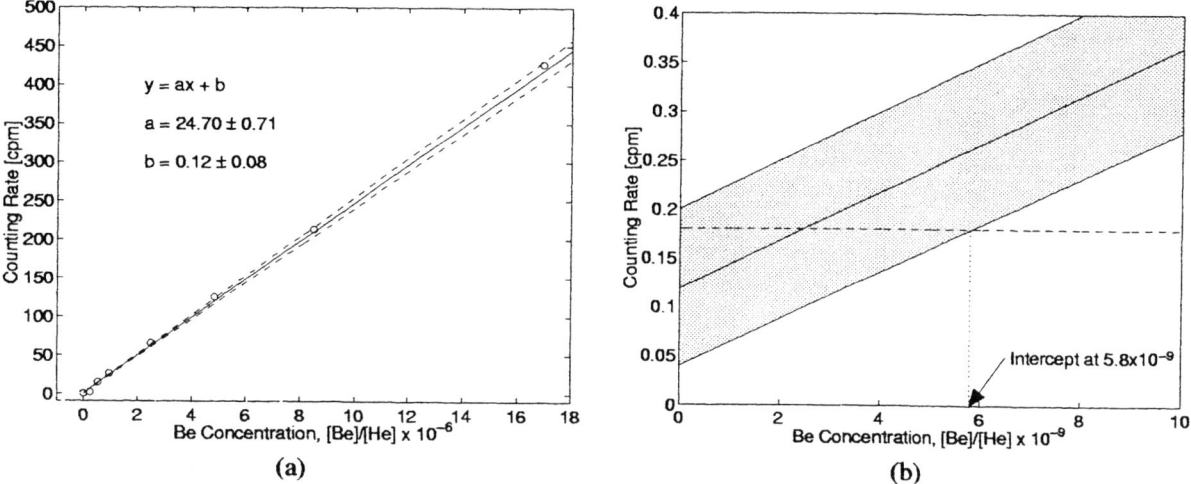

FIGURE 1. (a) Calibration of the AMS for gaseous Be. The open circles represent the experimental data for the blank and the seven Be Standards. The solid line is the result of the least-squares fit shown in the figure and the 1σ errors are represented by the dashed lines. (b) The shaded region is an enlarged version of the lower end of (a) with the 1σ errors included. The horizontal dashed line is the measured counting rate for Be, which is presumed to arise from the presence of Be$'$ in the samples. The limits on [Be$'$]/[He] for the beryl and gas field samples are identical, as shown in Table 1, and are obtained from the intersection of the horizontal line and the 1σ band as shown. The limits from the other samples are obtained in a similar manner.

pure He. The gaseous standard samples were prepared by mixing Be(C$_5$HF$_6$O$_2$)$_2$ with He to produce mixtures with known Be/He concentrations. (In this paper all concentrations are quoted as ratios of the number of Be or Be$'$ atoms to the number of He atoms in the sample.) The calibration of the AMS is provided in Figure 1a. As expected the Be counting rate in the detector varied linearly with the concentration. The blank sample, with no Be, provided us with a measure of the background counting rate.

In contrast to Ref. (8), which set a limit on Be$'$ in air, we have started from samples where the probability of Be$'$ retention is highest. We tested for the presence of Be$'$ in a pure metal sample produced in 1967. We also looked for Be$'$ in beryl which has a nominal chemical formula 3BeO·Al$_2$O$_3$·6SiO$_2$. Both of these samples were obtained by liberating and collecting any Be$'$ gas which may have been trapped inside. The gas field sample containing a significant amount of He (and possibly Be$'$) was obtained from a natural gas field in Texas. The assumption behind the use of the gas field sample is that a geological region capable of trapping He might also trap Be$'$, which may arise from outgassing of the Earth's crust or through direct cosmic ray production. In addition to the gas field sample, we also ran a sample of laboratory air in order to compare to previous experiments (8) (for a more in depth discussion of the samples see Ref (13)).

Since the Be metal, ore, and gas field samples were prepared in the same way as the standards, with similar He partial pressures, a comparison of the Be counting rates in the unknown samples and the standards determined the Be/He ratio in each sample.

Thus, after calibrating the AMS with the standards and blank, the unknown samples were analyzed. The intersection of the horizontal band corresponding to the Be counting rate in the unknown sample with the calibration line then determined the minimum detectable Be concentration, as illustrated in Figure 1b. No signal for Be above background was detected in any of the unknown samples (Be metal, ore, gas field, and laboratory air). This translates into a 1σ limit on the concentration of Be$'$ relative to He in each of the samples, as shown in Table 1. This gives a limit on the Be$'$ concentration, ρ', in air of $\rho' < 3 \times 10^{-14}$, which improves on the limit in Ref. (8) by a factor of almost 300. However, the more significant results are those from the Be metal, beryl, and gas field sample, since these were obtained from sources where Be$'$ was more likely to be found than in air.

In the absence of a rigorous fundamental theory of PEP violation, it is unclear how to convert the results from these specific samples into meaningful limits on the PEP-violating parameter $\beta^2/2$ often quoted in the literature (11). For this reason we list only the quantities directly measured by our experiment in Table 1, and defer the interpretation of these results to a future complete theory.

Table 1. Summary of unknown sample concentrations. The first column gives the Be$'$ concentration and the second column expresses the Be$'$ concentration relative to Be or air. All results are at the 1σ level and assume that the probability of converting Be$'$ to Be in the ion source is unity.

Sample	Be$'$ Concentration, [Be$'$]/[He]	Final Be$'$ Concentration
Be Metal	$< 2 \times 10^{-8}$	[Be$'$]/[Be] $< 9 \times 10^{-12}$
Beryl	$< 6 \times 10^{-9}$	[Be$'$]/[Be] $< 1 \times 10^{-11}$
Gas Field	$< 6 \times 10^{-9}$	—
Air	$< 5 \times 10^{-9}$	$\rho' =$ [Be$'$]/[air] $< 3 \times 10^{-14}$

STRONGLY INTERACTING MASSIVE PARTICLES

With the introduction of Big Bang cosmologies (6, 27) it was possible to predict the formation of new heavy (10 GeV to 100 TeV) stable particles during the early moments of the universe. Recently superheavy particles have been introduced as an explanation of a wide array of cosmological problems and inconsistent results seen in astrophysics. One method of searching for such particles would be to identify anomalous isotopes of known heavy nuclei (like gold, $A = 197$) which can result if a neutral stable strongly interacting massive elementary particle (SIMP) binds with known nuclei.

There are both particle physics and cosmological motivations for the existence of SIMPs. These include the dark matter problem (18, 19), predictions of gauge mediated supersymmetry (SUSY) models (4, 15, 23, 24), and predictions of models that explain perplexing ultra high energy cosmic ray (UHECR) events (1, 16).

Each of the above theories raises the possibility of SIMPs existing in nature, and each predicts a range for the SIMP mass, M_X. If in the process of cosmological evolution SIMPs formed the gravitational potential well and became halo dark matter, Mohapatra and Teplitz (18) predict $M_X \leq 100$ GeV. Raby (24) has proposed a class of SUSY models in which the gluino is the lightest supersymmetric particle and predicts a mass range from a few GeV to 100 GeV. Finally, Albuquerque, Farrar, and Kolb (1) show the SIMP mass must be below 50 GeV after detailed analysis of air showers produced by UHECRs (particles with energies above the Greisen-Zatsepin-Kuzmin cutoff, 5×10^{19} eV). The predicted range for M_X made by each of the above theories is summarized in Table 2.

Calculations performed by Mohapatra and Teplitz (18) show that binding is more likely in massive high Z nuclei than in low Z nuclei. As mentioned above, identifying massive isotopes of gold would be one method of searching for the existence of SIMPs. Mohapatra, Olness, Stroynowski, and Teplitz (17) have shown the fraction of gold nuclei which may have captured SIMPs is within the

Table 2. Summary of predicted SIMP masses M_X by theory.

Theory	Predicted SIMP Masses
Dark Matter	$M_X \leq 100$ GeV
LSP	6.3 GeV $< M_X <$ 100 GeV
UHECR	10 GeV $< M_X <$ 50 GeV

sensitivity range of accelerator mass spectrometry (AMS) (17). Since AMS permits us to select for a particular combination of A and Z, a sensitive search for anomalous gold nuclei was performed.

In our experiment the gold sample is introduced to a cesium sputter source where it is negatively ionized and accelerated through the lower end of the tandem. Eight of the atom's electrons are then stripped off as it passes through argon gas at the accelerator terminal. The remaining positively charged atom (+7) accelerates to 22.92 MeV, and we use a series of magnetic and electrostatic fields downstream to select for the particular isotope mass. We then proceed to change (7) the injector magnetic field, terminal voltage, and electrostatic analyzer in two amu (where 1 amu ≈ 0.94 GeV/c^2) steps to select SIMP masses ranging from 2.8–144 GeV/c^2.

The scan for massive isotopes of Au was performed on lab gold and gold extracted from Laverton in the Leanora District of western Australia. The Australian gold was discovered within 15 cm of the surface using a metal detector. These samples are likely to have the longest near-surface residency of any gold on Earth, perhaps on the order of 50 million years (21).

The scans performed on both the lab and western Australia gold did not detect the presence of any heavy isotopes, and hence no evidence for SIMPs was found in the indicated mass range. Figure 2 shows the limit set on the existence of SIMPs in Au as a function of the SIMP mass. This limit is not the same for all masses because the efficiency of stripping eight electrons from the particle decreases with velocity. The X/Au ratios for both the lab

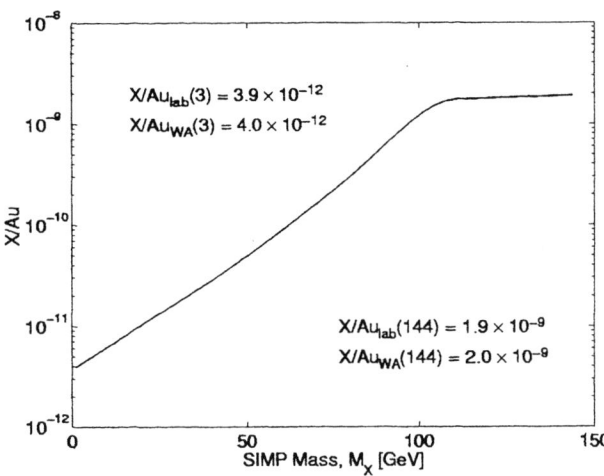

FIGURE 2. Fraction of anomalous nuclei, Au with SIMP, to regular gold nuclei, X/Au, as a function of SIMP mass M_X in GeV.

Table 3. Summary of X/Au ratios for the predicted SIMP masses M_X in GeV.

Sample	$M_X = 50$ GeV	$M_X = 100$ GeV
Lab	$X/Au < 4.8 \times 10^{-11}$	$X/Au < 1.1 \times 10^{-9}$
W. Aus.	$X/Au < 4.9 \times 10^{-11}$	$X/Au < 1.1 \times 10^{-9}$

gold and western Australia gold at each of the predicted masses is provided in Table 3.

ACKNOWLEDGEMENTS

The authors are deeply indebted to Professor Don Gaines and Dr. Dovas Saulys for providing us with the sample of $Be(C_5HF_6O_2)_2$; to Jim Marder and the Brush Wellman Corporation for the samples of Be metal and Be ore; to Bill Moore of the U.S. Bureau of Mines for providing us with the Bush Dome gas field sample and for determining its chemical composition; to Mike Bourgeois, Darren Hillegonds, Henry Rohrs and Stephen Vogt for sample preparation; to Mike Stohler for AMS assistance and data analysis; to Dr. Doug Oliver for his analysis of Australian gold and Dr. Vigdor Teplitz for helpful discussions. This work was supported in part by the U.S. Department of Energy under Contract No. DE-AC02-76ER01428, and National Science Foundation Grant No. 9809983-EAR.

REFERENCES

1. Albuquerque, I. F. M., Farrar, G. R., and Kolb, E. W., *Physical Review D* **59** (015021), (1998) 1–12.
2. Barabash, A. S., Kornoukhov, V. N., Tsipenyuk, Y. M., and Chapyzhnikov, B. A., *JETP Letters* **68** (2), (1998) 112–116.
3. Baron, E., Mohapatra, R. N., and Teplitz, V. L., *Physical Review D* **59** (036003), (1999) 1–4.
4. Chacko, Z., Dutta, B., Mohapatra, R. N., and Nandi, S., *Physical Review D* **56** (9), (1997) 5466–5474.
5. Deilamian, K., Gillaspy, J. D., and Kelleher, D. E., *Physical Review Letters* **74** (24), (1995) 4787–4790.
6. Dover, C. B., Gaisser, T. K., and Steigman, G., *Physical Review Letters* **42**, (1979) 1117–1120.
7. Elmore, D., et. al., *Nuclear Instruments and Methods in Physics Research B* **123**, (1997) 69–72.
8. Fischbach, E., Kirsten, T., and Schaeffer, O. A., *Physical Review Letters* **20** (18), (1968) 1012–1014.
9. Greenberg, O., and Hilborn, R., *Physical Review Letters* **83**, (1999) 4460–4463.
10. Greenberg, O. W., and Mohapatra, R. N., *Physical Review Letters* **59** (22), (1987) 2507–2510.
11. Greenberg, O. W., and Mohapatra, R. N., *Physical Review Letters* **62** (7), (1989) 712–714.
12. Ignat'ev, A. Y., and Kuz'min, V. A., *Sov. J. Nucl. Phys.* **46** (3), (1987) 444–446.
13. Javorsek II, D., et. al., *Physical Review Letters* **85** (13), (2000) 2701–2704.
14. Logan, B. A., and Ljubičić, *Physical Review C* **20** (5), (1979) 1957–1958.
15. Mohapatra, R. N., and Nandi, S., *Physical Review Letters* **79** (2), (1997) 181–184.
16. Mohapatra, R. N., and Nussinov, S., *Physical Review D* **57** (3), (1998) 1940–1946.
17. Mohapatra, R. N., Olness, F., Stroynowski, R., and Teplitz, V. L., *Physical Review D* **60** (115013), (1999) 1–5.
18. Mohapatra, R. N., and Teplitz, V. L., *Physical Review Letters* **81** (15), (1998) 3079–3082.
19. Nardi, E., and Roulet, E., *Physics Letters B* **245** (1), (1990) 105–110.
20. Nolte, E., et. al., *Nuclear Instruments and Methods in Physics Research B* **52**, (1990) 563–567.
21. Oliver, D. (Private Communication).
22. Pauli, W., *Z. Phys.* **31**, (1925) 765–783.
23. Raby, S., *Physical Review D* **56** (5), (1997) 2852–2860.
24. Raby, S., *Physics Letters B* **422**, (1998) 158–162.
25. Ramberg, E., and Snow, G. A., *Physics Letters B* **238**, (1990) 438–441.
26. Reines, F., and Sobel, H. W., *Physical Review Letters* **32** (17), (1974) 954.
27. Wolfram, S., *Physics Letters B* **82** (1), (1979) 65–68.

A Structural Upgrade for the IsoTrace Tandetron Accelerator

W. E. KIESER, X-L. Zhao, A. E. Litherland and K. H. Purser*

IsoTrace Laboratory, University of Toronto, Toronto, Canada, and
* Krytec Corporation, Peabody, Massachusetts, U.S.A.

Abstract. As a result of the failure of most of the glass accelerator tube insulators, the IsoTrace model 4130 Tandetron accelerator has been upgraded by re-building the tubes with new glass insulators, improving the alignment of the tube / terminal vacuum housing assembly and by adding a support insulator for the terminal. The engineering rationale, installation procedures and results of this upgrade are discussed.

1. INTRODUCTION

After eighteen years of nearly continuous operation, the accelerating tubes of the IsoTrace model 4130 (nominally 3 MV) Tandetron accelerator were found to have deteriorated significantly in an unusual fashion. Most of the glass insulators had developed many 1 - 2 millimeter sized cracks inside the body of the glass itself and for one insulator, these cracks resulted in a complete shearing of the glass. This deterioration could not be linked to any specific event or condition, such as radiation damage, impact, or excessive vibration, although eight years earlier a compression rod underneath the low energy tube (cf. section 2) had spontaneously released (but with no noticeable damage to the tubes).. As the glass and titanium tubes form an integral part of the column structure in these accelerators, it was suspected that the structural design of the column might have resulted in excessive stress on many of the glass insulators. Following an analysis of this design (sections 2 and 5) and bench testing of elements of the column structure, (sections 4 and 5), the structure was rebuilt incorporating two modifications:

a) the measurement of the tilts of the end flanges of each of the rebuilt accelerator tubes and the corresponding adjustment of their matching flanges (section 4), and

b) the installation of a terminal support insulator[1] and the removal of the tension bars (sections 5 and 6). The accelerator has been operated now for six months since this upgrade; results are summarized in section 7.

2. ACCELERATOR COLUMN DESIGN

A schematic of the model 4130 Tandetron accelerator, as well as some discussion of the column structural elements is given in ref [2]. As originally built, the model 4130 accelerator column could be viewed as consisting of two overlapping bridges:

a) one exterior bridge, made up of vertical Lucite plates in the areas of high gradient and square steel tubing in the terminal region. This supports the equipotential rings, column resistor plates, the terminal alternator and its drive shaft and the stripper gas reservoir and control system.

b) one interior bridge, consisting of the accelerator tubes and the terminal vacuum housing with tension bars placed parallel to the longitudinal axis of the tubes, 140 mm below that axis. In the tube section these consisted of 32 mm diameter delrin rods held at a tension of 7500 N by a spring assembly. In the terminal region, continuity of this tension was maintained by two 12 mm diameter stainless steel rods under similar tension.

Tensions for the delrin rods were specified by the manufacturer based on the calculated sag due to gravity of the ~2 m long tubes. Alignment in the vertical plane was assisted by selecting the rotational orientation of the tubes which provided the largest wedging (cf. section 4) in the downward direction (i.e., with the shortest length of the tube on the bottom). The terminal tension bars, were adjusted so that the accelerator tube entrance and exit and the stripper canal were as co-linear as possible. On the IsoTrace machine, once the tank was pressurized, the deflection at the terminal was approximately 4 mm. Once these alignment functions were completed, a steel tab on the terminal vacuum housing was attached to a bulkhead on the Lucite column (exterior) bridge to prevent longitudinal movement of the accelerator tube bridge (especially in the event of a vacuum accident). Unfortunately, this connection also couples any vibration from the terminal alternator drive and any lateral movement of the exterior bridge system into the interior tube bridge system.

3. THE ACCELERATOR TUBE FAILURE

At the beginning of the 1999 July in-tank maintenance the tubes showed a significant increase in the number of 1 - 2 mm bubble or disc-like structures in the body of the glass insulators. These had been observed in previous tank openings, but only in a few locations. Several days later, a sharp crack was heard and the accelerator vacuum deteriorated to 10^{-4} mbar. A vertical crack in the centre of one side of the second insulator from the terminal on the low energy tube. had opened slightly While efforts were underway to obtain a non-conducting epoxy to seal this crack, it increased in length and eventually migrated all the way around the insulator. At this point the pumps could no longer maintain a vacuum.

The tension underneath the low energy tube was reduced to 3200 N and the column structure was carefully removed. To support the tube when the tension rods were fully released, and to protect workers in the event that sudden glass failures might occur, a polyethylene tarpaulin was placed around the tube in a U-shaped configuration. Once the tension rod was removed, the tarpaulin was used to lift the tube out of the exterior column assembly.

Outside the tank, the ~2 mm long "bubbles" seen in transmitted light inside the tank, were actually cracks. These were found in both tubes, with a number density which ranged from ~30 "bubbles" per glass at the terminal end and to 0 at the ground end of the tubes. the failure pattern in the one broken insulator could be approximated by the intersection of a vertical cylinder (centre away from the terminal) with the glass electrode, possibly indicating some lateral stress on the glass.

Both low and high energy tubes were shipped to High Voltage Engineering Europa B.V. for rebuilding. In the process of heating the tubes to release the vinyl acetate used to bond the glass to the titanium most of the glass insulators disintegrated; hence the tubes were re-built with entirely new glass.

4. PROCEDURE FOR ALIGNING RE-BUILT TUBES

For a tube of length ~2 m, composed 23 mm thick glass sections and 2 mm thick titanium electrodes bonded with vinyl acetate, it is highly unlikely that the end flanges of the tube will be exactly perpendicular to the longitudinal axis of the tube. In the tube bridge of the 4130 Tandetron, the tube flanges are mated directly to other rigidly held flanges, with no interconnecting bellows, as would be the case in a machine with a supporting column structure. Thus any tilt in the end flanges of the tube would result in a lateral displacement of the tube assembly where it joins the terminal. For example, a tilt of 1 mr would result in a 2 mm misalignment for a 2 m long tube. An attempt to force this displacement into alignment might cause excessive stress on a number of insulators.

When the tubes were received, they were each placed on an optical bench and an alignment telescope was set on the longitudinal axis of the tube. Adapter flanges were machined for each end of the tube which held a front surface mirror optical target parallel to the mating surface of the flange to better than 0.05 mr, checked by rotating the adapter around the tube flange. The alignment telescope was then used in autocollimation mode [3] to measure the tilt and its azimuthal axis of the end flange. The values measured for each tube are given in table 1.

TABLE 1. Tilts measured for the rebuilt tube ends		
Tube / End	Tilt (mr)	Rotation of Tilt* (from vertical)
Low Energy / Ground	2.97	197°
Low Energy / Terminal	0.85	21°
High Energy / Terminal	1.46	18°
High Energy / Ground	0.87	186°

* the azimuthal angle of the tilt axis is as seen from the low energy end of the accelerator. The vertical direction for each tube was selected so that the tilts at the terminal end were as close to downwards as allowed by the flange bolt pattern (8 holes).

To match the terminal end tilts, each end flange of the terminal vacuum housing was re-machined. The entire housing was aligned longitudinally in a lathe, supported by plates machined to engage the alignment edges. These plates were in turn supported by a four jaw chuck and a live tailstock. The chuck end was then offset to produce the appropriate tilt and azimuth at the tailstock end, and the mating surface and O-ring groove at this end were then re-machined.

The ground end of the tubes are attached to tube extension flanges which are supported in the tank wall ~ 170 mm away from the tube end. The mating surface of the low energy tube extension was also re-machined, in this case (as it was significantly shorter than the terminal housing) using a numerically controlled mill. After machining, each joint was tested on the optical bench and the resulting assembly was found to be straight within the accuracy of the alignment telescope (± 0.05 mm).this end were then re-machined.

5. PROPERTIES OF THE ACCELERATING COLUMN

A number of useful formulas for estimating the stress in beams with various cross sections and modes of support are given by Roark[4]. These were used to calculate maxima for the vertical deflection, the bending moment of the tube and the longitudinal stress for two cases: with and without support at the terminal. The tubes and vacuum housing were approximated by a hollow circular cross section (Table 1, case 11 of ref 4), using the Young's modulus for glass, but including the titanium in the distributed weight. The formulas of table 3 in ref. 4 were used to calculate the above parameters for such a beam with a uniformly distributed load (its own weight), using the following support configurations. The case of the tube extensions in the tank bulkhead was treated as "simple support", i.e., vertically supported but free to tilt (true within the limits of the clearances of tube extension in the bulkhead, which in practice are not reached). In the case of an unsupported terminal, the terminal end of the tube was considered as "guided", i.e., fixed in angle, but not in position. For the supported terminal case, the terminal end of the tube was considered "fixed" i.e., in both angle and position. These formulae show that the maximum vertical deflection varies as ℓ^4 (where ℓ is the unsupported length) and the maximum stress varies as ℓ^2, indicating the advantage of a supported terminal. Using a Young's modulus for glass of 69 GPa, the estimates in table 2 were obtained. For the unsupported terminal case, the weight of the terminal was taken into account by superimposing a calculation for a point load equal to half its weight at the end of a tube with zero mass, on the calculation with the equally distributed weight. The supported terminal calculation was also used to estimate the vertical force transmitted through the support insulator to the tank wall: 1200 N.

Before final installation, the tube and terminal vacuum housing system was assembled on the optical bench, supported at each end and at the terminal. In order to measure the force required at the terminal, the yoke assembly included a compression spring with spring constant 63.5 N/mm.. Optical targets were located at each end of the terminal vacuum housing and at the exterior ends of the tubes, by flanges machined to hold them concentric with the alignment surfaces in each location. With the exterior ends of the tubes on the optical axis of the alignment telescope, the tension on the terminal support needed to bring the terminal targets into alignment was 1110 N.

A test of the maximum displacement of the unsupported terminal was carried out. As a precaution longitudinal tension cables were installed under the tubes in the same location as the delrin tension rods in the original design. (These also proved to be a useful safeguard when transferring the overall column assembly into the tank). With 4,450 N on each cable, the terminal support tension was gradually released. The deflection soon exceeded the measurement range of the alignment telescope, so it was estimated by noting where the horizontal axis intersected the markings on the alignment target in the terminal. The unsupported displacement was approximately 2.7 mm.

TABLE 2. Estimates of Accelerating Column Properties

Estimated Property	Terminal Supported	Terminal Unsupported	Units
Maximum vertical displacement	0.009	0.6	mm
location of max. displacement	0.9 m from tank wall	at terminal	
Maximum Bending Moment	159	1205	N-m
Max longitudinal stress	325	2458	kPa

This measurement drew our attention to the effect of neglecting the vinyl acetate in the displacement calculations. This material has a considerably lower Young's modulus of 600 MPa [5] and if each joint used a layer 0.03 mm thick, the aggregate thickness in the low energy tube would be 5 mm. This is consistent with the greater elasticity of the tubes shown by the unsupported terminal measurements..

6. INSTALLATION OF THE COLUMN SUPPORT INSULATOR

The support insulator was attached to the tank through an existing access port on top of the tank near the centre of the terminal. To adjust the applied force, the same tension spring used in the optical bench test (section 5) was installed above the tank, enclosed in an extension flange. The lower end of the insulator was attached to a yoke consisting of a stainless steel bar supporting a stainless steel band around the terminal. The upper terminal cover plate was modified by cutting a hole to allow the insulator to pass through it; the electrical gradient near this hole was shaped by welding a ring, numerically machined with a quarter circular cross section which followed the cylindrical shape of the cover plate.

The individual tube electrodes are electrically connected to the planes of the resistor divider chain (stainless steel plates supported by the two Lucite plates of the exterior column, to which the column resistors are attached) by a stainless steel spring. The force exerted by this spring averaged 8.8 N. These increase the downward force on the tube to 2.3 times the force exerted by gravity; hence a greater force on the support insulator (and the greater need for the support insulator). The final adjustment of the support tension was accomplished by viewing the stripper canal through an alignment telescope set on the longitudinal axis of the accelerator. Vertical alignment was achieved to within the precision of the alignment telescope at that distance (~0.1 mm). However, during the re-pressurization of the tank, it was noticed that the stripper canal had deflected horizontally by 0.4 mm. The support yoke has been designed to accommodate some lateral adjustment and this will be attempted during a future maintenance shutdown.

7. CONCLUSIONS

Since the upgrade, the accelerator has operated continuously for approximately 6 months. The voltage stability, especially for heavy element analysis (129I) has significantly improved over the conditions shortly before the break occurred. During initial voltage conditioning, a voltage dependent yield of X-rays was observed at one location in the low energy end indicating perhaps either the presence of a dust particle or a rough spot on an electrode. These have gradually decreased with the continued operation of the accelerator.

ACKNOWLEDGEMENTS:

The authors are grateful to High Voltage Engineering Europa B.V. for their rapid rebuilding of the tubes and the staff of the University of Toronto Physics Department workshop and Peterborough Precision Machining, who were asked to do a number of unconventional jobs. We also wish to thank Messrs. Z. Zlicic and Z. Fang of the IsoTrace technical staff who quickly disassembled and accurately re-assembled the accelerator system. The University of Toronto provided part of the funds required for this upgrade. The IsoTrace authors particularly wish to thank their AMS clients for their forbearance during the time the accelerator was out of service.

REFERENCES

1. manufactured by Potential Corp., Cambridge, Massachusetts, U.S.A.
2. Kieser, W. E., Kilius, L. R., Nadeau, M-J., Perez, J. and Litherland, A. E., *Nucl. Instr. and Methods*, **B45**, 570-574 (1990).).
3. Kissam, P., Optical Tooling for Precise Manufacture and Alignment, McGraw-Hill, New York, 1962, pp. 103-107.
4. Roark, R. J. and Young, W. C., *Roark's Formulas for Stress and Strain (6th ed.)*, McGraw-Hill, New York, 1989.
5. *Private communication*: J. Fox, McGean Rohco Inc., Cleveland, Ohio, U.S.A.

Ion Preparation Systems for Atomic Isobar Reduction in Accelerator Mass Spectrometry

A. E. Litherland, J. P. Doupe, I. Tomski, J. Krestow, X-L. Zhao, W.E. Kieser and R.P.Beukens

IsoTrace Laboratory, University of Toronto, Toronto, M5S 1A7, Canada

Abstract. Methods for carrying out isobar reduction, at or near the ion source, are reviewed. These range from charge changing and electron transfer reactions at keV energies to ion reactions with molecules at a few eV. All of these methods can emphasize particular isobars by many orders of magnitude. These additions to the original ion source are part of a more complicated ion source and ion beam preparation system, prior to a tandem accelerator, which must be an integrated whole. The purity of ion source materials, the sputtering beam and the nature of the residual gas are also reviewed.

1. INTRODUCTION

A key event in the introduction of AMS for ^{14}C studies was the realization that an existing negative ion sputter ion source, tandem accelerator and peripherals, used for nuclear physics studies, might be used to solve both the mass spectrometric atomic and molecular isobar problems. Previously the study of the positive ions of the ^{14}C isotope had been plagued by the ubiquitous ^{14}N ion and the 14 amu hydride ions of carbon. These effectively obscured the ^{14}C ions because in nature ^{14}C/C is only $\sim 10^{-12}$. The N$^+$ problem was also very evident when AMS using a cyclotron was re-introduced [1]. However, once the instability of the N$^-$ ion was finally established [2] and the molecular isobars destroyed during tandem acceleration, ^{14}C detection and measurement followed quickly. This was followed by similar tests, which showed that the same was true for the pairs ^{26}Al & ^{26}Mg [3] and ^{129}I & ^{129}Xe [4]. Surprisingly only one other similar pair, ^{202}Pb & ^{202}Hg [5], has been found.

Since the beginning of AMS other methods for separating additional isobaric pairs, such as stable ^{36}S$^-$ and ^{36}Cl$^-$, have been developed using the differing rate of energy loss of the isobars in matter after acceleration and molecular ions from the ion source; such as ^{41}CaF$_3^-$, which discriminates against ^{41}K. These processes have been quite successful [6]. However, the isobar reduction methods usually require higher energy, and so larger accelerators, to finally separate the atomic isobars adequately. The difficulty of removing intense beams of atomic isobars after tandem acceleration suggests that their reduction before acceleration would be well worthwhile.

In this paper, we will consider aspects of the ion beam preparation system before the tandem accelerator that are relevant to the goal of reducing atomic isobar interferences. This system may consist of a positive or negative ion source, followed by charge changing or even a chemical reaction region, to reduce, or effectively eliminate in some cases, the atomic isobars. It was first discussed in some detail about six years ago [8].

Since the beginning of AMS the range of ^{14}C variation studied has increased [6] from $\leq 10^4$ to $>10^6$ [7]. At the IsoTrace Laboratory an effort is being made to increase this still further, possibly to 10^8. This introduces some additional problems for the AMS sputter ion source and the ion preparation system as a whole and these will also be mentioned briefly.

2. SPUTTER ION SOURCES AND THE AMS PREPARATION SYSTEM

The sputter ion source was introduced [9] using a keV Kr$^+$ beam to sputter neutral atoms from a surface, the work function of which had been lowered by the deposition of Cs vapor. The neutral atoms then became negative by electron tunneling as they left the surface. Later these two functions were combined when Cs$^+$ ion beams were used to sputter a surface [10], with the implanted Cs ions diffusing to the surface to lower the work function.

There are now two main types of Cs^+ negative ion sputter sources. In the earlier version the Cs^+ ion gun was well separated from the sputter target by transfer optics. A later version locates the Cs^+ generator much closer to the sputter target and Cs vapor is present to lower the work function in addition to the implanted Cs^+ ions. The second arrangement in general gives higher currents but it is not clear whether or not this feature is intrinsic to the design. An advantage of the first arrangement is discussed below.

(a) Cesium Ion Beam Purity

A disadvantage of the higher current arrangement is that the Cs^+ beam cannot then be analyzed for impurities prior to the sputtering. Measurements have shown [11] a remarkable variety of atomic and molecular ions accompanying the Cs^+ ion beam at the ppm to ppb level. These implant in the target and then become potential background for low-level measurements. For the measurement of rare atoms, < ppt, in materials such as ultra-pure silicon, this can be a disadvantage, so a Cs^+ gun with transfer optics and both electric and magnetic analysis is to be preferred if very low levels of atoms are to be studied. For example, the NASA Genesis Project [12] will use an ultra-pure silicon collector, among others, to collect atoms from the solar wind. Consequently a prior knowledge of the actual level of all impurities in the collector material is required. This is one of many challenges for analysis during the project. The re-introduction of Neutral Injection AMS [13], with an appropriately designed ion injection system may be relevant here.

(b) The Purity of the Sputter Ion Source Materials

This problem is of concern when very rare isotopes are to be studied, for example from old carbon with $^{14}C/C < 10^{-16}$. When Al is used to support the carbon targets in the ion source for ^{14}C dating it must contain little or no radiocarbon. A similar restriction must apply to the iron or other catalyst used for making the targets for the sputter ion source [6]. Fortunately it appears that not only is carbon at a low level in Al [14] but it is also probably not contemporary due to the use of petroleum coke in the industrial apparatus electrolyzing the Al [15]. At 100 ppb, such carbon if contemporary would contribute at the level of $^{14}C/C <$ 10^{-19} and so the actual level is probably much less. However, as the level of radiocarbon in natural gas has already been shown to be $< 10^{-18}$ [7] there is clearly need for further study. Radiocarbon can be observed while sputtering the purest Al alloy available but at the very low level of ~ 0.0025 pMC, or $^{14}C/C \sim 3\times 10^{-17}$. This may be due to ^{14}C in gases in the ion source, which adsorb onto the target and then are sputtered. Indeed the problem of distinguishing between residual gas effects and the purity of ion source materials is a formidable one requiring further study.

The target material inserted into a sputter ion source can in some cases be purified in situ. BeO target material has been observed to lose boron if it is heated in a vacuum to $\sim 1700^\circ C$ [16]. For AMS work the target support material must also be depleted in boron and this too can be accomplished [17] by heating to high temperature in a vacuum.

(c) Negative Ion Sputter Sources

These have been used [18] in an exploratory way for AMS. Such ion sources are sources of positive ions so that for normal tandem operation these have to be converted into negative ions. This requires at least two charge changes at energies of a few keV so that the multiple scattering of the ions causes the beam quality to deteriorate. The conversion to neutral can usually be carried out with large cross sections, and so low scattering, but the cross sections for conversion to negative are in general smaller so that more scattering centers are needed. A solution to this dilemma has been to employ two vapors in sequence [19] to minimize the scattering. Another solution is to employ Neutral Injection AMS [13] as the resonant neutralization cross sections are large. Negative Ion Sputter Ion Sources also have the advantage of being able to sputter insulating materials with little surface charging [20].

(d) Molecular Ions and Sputter Tails

Negative ion sputter sources produce copious molecular ions also, but the use of a tandem results in their complete destruction during charge changing in the stripping canal. However, the molecular fragments themselves, after further acceleration in the tandem, can also create problems for the detection of the rare isobar unless sequences of pairs of electric and magnetic analyzers, arranged appropriately, are used to remove them. The fragments can then be removed [21] completely, unless they have the same E/Q and

M/Q ratios as the rare ion. Such values of E/Q and M/Q should not be chosen. If other atomic isobars remain after this process then, depending upon the ratio of the rare to common isotope being studied, they must at present be handled by high-energy methods [6].

In low-energy mass spectrometry any fragmenting molecular ions can produce peaks [22] superposed on a continuous tail in the ion energy spectrum, below the main atomic ion peak. Above the peak the sputter tail proper extends and this tail, although mainly associated with atomic ions, can also be present for some molecular ions. This broad spectrum of ions for a single mass creates potential problems, which by adding electric analyzers before and after the magnetic analyzer, are readily solved. If electric analyzers are not used then some of the abundant isotope will be injected into the tandem, with only a small difference in energy from the rare isotope, especially for the heavier elements. Then the large additional energy added by the tandem makes it difficult to resolve these beams in energy from the rare ions by the high-energy electric analyzers and also in the final ionization detector. This problem is made even worse by the fact that the tandem itself also produces an energy peak on top of a continuum [23] due to the presence of gas in the acceleration tubes. As a result, for the optimum analysis of the rare isotope, the low energy ion beam preparation system should remove as many of the abundant isotopes as possible. The injection of all isotopes for simultaneous measurement is clearly not suitable for measurements of the largest isotope ratios, unless more than one magnet is used for each of the high-energy ions. This is because each high-energy magnet only attenuates ions with the same E/Q [23] because of scattering in the residual gas of the magnet.

3. GAS ION SOURCES FOR AMS

A CO_2 gas hybrid sputter ion source has been developed for radiocarbon dating and described extensively by Bronk-Ramsey and Hedges [24]. In this ion source a Cs^+ beam bombards a Ti target upon which the CO_2 is adsorbed. Good quality carbon negative ion beams can be obtained and the memory effect controlled by changing and ion-beam-cleaning the Ti targets for each new gas sample.

A gas ion source based upon the generation of C^+ ions by microwave ionization of CO_2 is being developed [25] with control of the memory effect. Conversion to negative will be necessary for the elimination of the N^+ ions and tandem acceleration.

As a result of this experience gas or vapor ion sources for AMS appear to be very promising in general provided the purity of the sputtering ion beam and the surroundings can be controlled. It has been noticed [17], for example, that because S_2I_2 is thermodynamically unstable with respect to other iodine and sulfur-bearing phases at normal temperatures, it dissociates and cannot accompany the gas ICl. Consequently, the gas ICl may be a good starting point for analyzing ^{36}Cl by a gas ion source provided the gases H_2S and SO_2 can be excluded sufficiently. Low sulfur content materials for the ion source must be identified for this to be exploited

4. CHARGE TRANSFER ION SOURCES

These ion source systems have been discussed previously [13] and so only recent developments will be discussed. An example of such an ion source system, which can discriminate between isobars, has been known for some time. It can be derived from the cross sections given by Allison [26]. If a low-intensity low-energy positive ion beam of tritons is available and the discrimination against the accompanying $^3He^+$ ions is required, it can be seen from the cross sections [26] that the passage through helium gas could be advantageous. The cross sections σ_{10} for T^+ and $^3He^+$ at 2.5 keV are different for the two ions by a factor of about 300 because He^+ is resonant with He. Consequently an isobar attenuation of 10^6 is possible by first of all selecting the mass 3 beam and charge changing in helium. The actual attenuation possible is ultimately controlled by the loss due to scattering at this energy.

Many other examples can be given of this selection process for isobars at the ion source and cross sections for charge changing for each pair must be measured [27].

A combination gas and charge transfer ion source can, in principle, be used for the gas ICl, especially as the transfer of an electron to the ICl from a negative ion has a large cross section [17]. Several excited states of ICl^- have been observed to fission into $I + Cl^-$ [28]. However, the negative ions created this way inevitably have an energy spread, which must be allowed for in the ion preparation system. The ICl^- or Cl^- ions are also created at rest so that ideally they must be created also in a static hyperbolic potential [29] or the equivalent. Then scattered ions cannot stray far from the beam axis and can be extracted from the collision region.

The problem of scattering has been solved in a related ingenious manner by the introduction of the Dynamic Reaction Cell (DRC) [30]. In this device, positive ions of low eV energies are extracted from an Inductively Coupled Plasma (ICP) argon torch and sent through an appropriate gas in a radio frequency quadrupole electric field. In the DRC both resonant transfer reactions and chemical reactions can be used. For example, the copious $^{40}Ar^+$ ions from the torch, in contrast with the $^{40}Ca^+$ ions, readily react with a molecule such as NH_3 to form Ar neutrals or ArH^+ ions. The $^{40}Ar^+$ ions are reduced by a factor of about 10^9 so that $^{40}Ca^+$ can then be observed at ppt levels. By adjusting the parameters of the DRC the scattering of the ions at this low energy can be controlled and losses are much reduced. This is similar to the use of a static hyperbolic potential mentioned above in that the scattered ions are effectively in a potential well.

ACKNOWLEDGMENTS

We wish to thank Drs R. R. Raghavan, G. C. Wilson and Prof. J. C. Rucklidge for many valuable discussions and the Natural Sciences and Engineering Research Council and the University of Toronto for their financial support.

REFERENCES

1. Muller, R. A., *Science* **196**, 489-494 (1977).
2. Purser, K.H. et al., *Rev. de Phys. Appl.* **12**, 1487-1497 (1977).
3. Kilius, L. R. et al., *Nature* **282**, 488-489 (1979).
4. Elmore, D. et al., *Nature* **277**, 22 (1979).
5. Nadeau, M-J. et al., *Nucl. Instr. and Meth.* **B123**, 521-526 (1997).
6. Tuniz, C., Bird, J. R., Fink, D. and Herzog, G. F., *Accelerator Mass Spectrometry*, Boca Raton, Boston, London, New York and Washington, D.C.: CRC Press, 1998, pp 371.
7. Beukens, R. P., *Nucl. Instr. and Meth.* **B79**, 620-623 (1993).
8. Litherland, A. E., *Nucl. Instr. and Meth.* **B92**, 207-212 (1994)
9. Hortig, G., *IEEE Trans. Nucl. Sci.* **NS-16**, 38 (1969).
10. Middleton, R., *Nucl. Instr. and Meth.* **122**, 35-43 (1974) and references therein.
11. Matteson S. et al., *Nucl. Instr. and Meth.* **B52**, 327-333 (1990).
12. http://www.gps.caltech.edu/genesis/genesis3.html
13. Litherland, A. E. and Kilius, L. R., *Nucl. Instr. and Meth.* **B123**, 18-21 (1997).
14. Fedoroff, M., Loos-Neskovic, C. and Revel, G. *J., Radioanal. Chem.* **38**, 107-113 (1977).
15. Kirk-Othmer, *Encyclopedia of Chemical Technology; Aluminum and Aluminum Alloys,* **2**, 184-251 (1992).
16. Belshaw, N. S., O'Nions, R. K. and von Blanckenburg, F., *Int. J. Mass Spectr. and Ion Proc.* **142**, 55-67 (1995).
17. Doupe, J. P., private communication.
18. Aardsma, G., *Nucl. Instr. and Meth.* **B113**, 170 (1985).
19. Schuessler, H., private communication from Stracenor, D.
20. Kilius, L. R. et al., *Nucl. Instr. and Meth.* **B123**, 5-9 (1997).
21. Kilius, L. R. et al., *Nucl. Instr. and Meth.* **B123**, 10-17 (1997).
22. Kilius, L. R., Rucklidge, J. C. and Litherland, A. E., Nucl. Instr. and Meth. **B31**, 433-441 (1988).
23. Zhao, X-L., private communication
24. Bronk Ramsey, C. and Hedges, R. E. M., *Nucl. Instr. and Meth.* **B123**, 539-545 (1998).
25. Schneider, R.J. et al., *Nucl. Instr. and Meth.* **B123**, 546-549 (1997).
26. Allison, S. K., *Revs. Modern Phys.* **30**, 1137-1168 (1958).
27. Tomski, I., private communication.
28. Le Coat, Y., Guillotin, J-P. and Bouby, L. *J. Phys. B: At. Mol. Opt. Phys.* **24**, 3285-3294 (1991).
29. Rudenberg, R., *Journal of the Franklin Institute,* **246**, 311-408 (1948).
30. Tanner, S. D., and Baranov, V. I. *J. Anal. At. Spectrom.* **20**, 45-52 (1999).

Biokinetic And Dosimetric Investigations Of ^{14}C-Labeled Substances In Man Using AMS

Sören Mattsson, Mikael Gunnarsson, Sigrid Leide Svegborn, Bertil Nosslin, Lars-Erik Nilsson, Ola Thorsson, Sven Valind, Magnus Åberg and Henrik Östberg

Departments of Radiation Physics, Clinical Physiology and Plastic Surgery, Lund University, Malmö University Hospital, SE-205 02 MALMÖ, Sweden

Ragnar Hellborg, Kristina Stenström, Bengt Erlandsson, Mikko Faarinen, Madis Kiisk, Carl-Erik Magnusson, Per Persson and Göran Skog

Department of Physics, Lund University, Sölvegatan 14, SE-223 62 LUND, Sweden

Abstract. Up to now, radiation dose estimates from radiopharmaceuticals, labeled with pure β-emitting radionuclides e.g. ^{14}C or ^{3}H have been very uncertain. Using accelerator mass spectrometry (AMS) we have derived new and improved data for ^{14}C-triolein and ^{14}C-urea and are currently running a program related to the biokinetics and dosimetry of ^{14}C-glycocholic acid and ^{14}C-xylose. The results of our investigations have made it possible to widen the indications for the clinical use of the ^{14}C-urea test for *Helicobacter pylori* infection in children. The use of ultra-low activities, which is possible with AMS (down to 1/1000 of that used for liquid scintillation counting), has opened the possibility for metabolic investigations on children as well as on other sensitive patient groups like new-borns, and pregnant or breast-feeding women. Using the full potential of AMS, new ^{14}C -labeled drugs could be tested on humans at a much earlier stage than today, avoiding uncertain extrapolations from animal models.

INTRODUCTION

Radioactive tracers have been a primary diagnostic tool in medicine and in medical and pharmacological research for over 60 years. To trace organic molecules, ^{14}C and ^{3}H have been used during a long time. The long-lived ^{14}C (half-life 5 730 years) is today complemented with the short-lived ^{11}C (half-life 20 min.). The later is used for imaging and quantitative measurements in positron emission tomography (PET). The time periods available for studies with ^{11}C are however too short for many biomedical applications.

The long physical half-life of ^{14}C creates two difficulties in the current clinical use: 1) A risk for high absorbed doses if even a small part of the compound is efficiently retained in the body (1). 2) Unfavorable conditions for decay counting. One day of decay counting detects only 0.000033% of ^{14}C present in the sample, while accelerator mass spectrometry (AMS) in a short time detects about 1 % of the ^{14}C-atoms in a small sample. AMS has been shown to quantify 10^{-18} mole or 10 nBq from milligram-sized samples. AMS brings at least three advantages to biochemical tracing: high sensitivity enabling use of sub-toxic chemical amounts of a substance, small sample size for painless biopsies, measurement on highly specific biochemical separations or subcellular fractions, and reduction of radiation exposure. Our program started in 1994 and has been focused on measurements with AMS of the biokinetics and especially the long-term retention of ^{14}C from clinically used radiopharmaceuticals in volunteering patients and healthy volunteers at activity levels, which earlier had not been possible to follow using liquid scintillation counting (LSC). Our current program has also exposed the potential of AMS for further biomedical research. The studies conducted so far have been related to four so called "breath tests". The ^{14}C-labelled compound is ingested and metabolized, resulting in the end-product $^{14}CO_2$ which is exhaled and easily collected for measurement. We have used the AMS technique to study the long-term retention of ^{14}C in connection with clinical tests for the presence of *Helicobacter pylori* in the stomach with ^{14}C-urea, fat malabsorption using ^{14}C-labelled triolein, and for demonstration of abnormal intestinal bacterial flora using ^{14}C-glycocholic acid and ^{14}C-xylose, where the former also can be used to reveal changes in the bile salt metabolism.

AIMS OF THE PROJECT

- to study the long-term biokinetics, in volunteering patients and healthy volunteers, of ^{14}C from ^{14}C-labelled radiopharmaceuticals in current use

- to estimate the absorbed dose to various tissues and organs and the effective dose

- to determine the minimal administered activity and minimum amount of sample material needed for a clinical study in patients of various age.

MATERIALS AND METHODS

^{14}C-Urea

The urea breath test is used for diagnosis of infection of *Helicobacter pylori (Hp)* in the stomach. The bacteria produce urease, which metabolizes urea with the production of CO_2, which is exhaled. Intact ^{14}C-urea leaves the body with the urine and the degraded urea is exhaled as $^{14}CO_2$. For persons with *Hp* infection, the amount of $^{14}CO_2$ exhaled is higher than for persons without this infection. The long-term biokinetics and dosimetry of ^{14}C-urea has been investigated in nine adults and fifteen children (2,3,4); most of them volunteering patients. After an overnight fast, the adult persons were given 110 kBq ^{14}C-urea orally. Eight children (7-15 years) got 55 kBq and seven children (3-6 years) 440 Bq.

^{14}C-Triolein

The ^{14}C-triolein breath test is used for investigation of fat malabsorption of the gastro-intestinal tract. A $^{14}CO_2$ exhalation larger than 3.5% of the administered activity within 6 hours is considered normal; a lower value indicates fat malabsorption. The long-term biokinetics and dosimetry of ^{14}C have been investigated extensively in four adults. After an overnight fast three healthy volunteers were given 74 kBq ^{14}C-triolein in a fat meal. Another healthy volunteer was given 1/50 of the standard activity, or 1.4 kBq (5,6,7). One of the volunteers who got 74 kBq was followed with measurements of the exhalation of $^{14}CO_2$ up to 4.5 years after administration (8). From the same person it was also possible to take biopsies from body fat, muscles and bone 4.5 years (8) and body fat and muscle biopsies 6 years after administration. Two new healthy volunteers have recently entered the study and follow an extended sampling program with measurements on expired air (together with spirometry), urine, feces and biopsies. To follow the incorporation of ^{14}C in the skeleton via the bicarbonate cycle, we plan to study a third healthy volunteer with bone biopsies before and after the administration.

^{14}C-Glycocholic Acid

^{14}C-glycocholic acid ("bile acid") breath test is used to investigate changes in the metabolism of bile acid indicating abnormal bacterial growth or reduced resorption of bile acids in the small intestine. Nine individuals are presently under study. After an overnight fast 6 volunteering patients and 3 healthy volunteers were given 200 kBq ^{14}C-glycocholic acid per os together with a portion of gruel. In seven persons, expired air and urine have been collected up to one year after administration (9). In four of them, feces has also been sampled at regular intervals.

^{14}C-Xylose

^{14}C-xylose is used for investigations of an abnormal intestinal bacterial flora (9). The advantage of using xylose instead of glycocholic acid is that xylose is absorbed primarily in the proximal half of the small intestine and very little reaches the bacterias in the colon, thus avoiding the problem of false positive results that the bile-acid breath test suffers from. After an overnight fast, nine volunteers got 74 kBq of ^{14}C-xylose together with 1 gram of non-radioactive xylose. As $^{14}CO_2$ equilibrates with the CO_2/bicarbonate pools in the body, there is an interest to assess the concentration of ^{14}C in trabecular and cortical bone in connection with all these type of tests. Especially the cortical bone is supposed to have a long biological half-time (20-30 years) and an uptake there could generate a considerable dose contribution. The various patient studies carried out so far are summarized in Table 1. All studies were approved by the Ethical Committee at Lund University and by the Radiation Protection Committee at Malmö University Hospital.

RESULTS AND DISCUSSION

^{14}C-Urea

Of the 24 subjects, who took part in the investigation, 21 were found to be *Hp* negative and 3 *Hp* positive. In all *Hp* negative subjects older than 7 years, a majority of the given ^{14}C was excreted through the kidney-bladder system (88%±4%), most likely as unchanged ^{14}C-urea. A minor part was exhaled as $^{14}CO_2$; in adults 4.6%±0.6% and in children 7-15 years 2.6%±0.3%. A maximum of $^{14}CO_2$ in exhaled air occurred within an hour. It was assumed that the fecal elimination was insignificant. The highest organ dose was received by the urinary bladder wall: in adults 0.15 mGy/MBq and in children 7-15 years: 0.36-0.14 mGy/MBq (2) and in younger children 3-6 years: 0.58–0.41 mGy/MBq (4). Compared to other radiopharma-ceuticals, the absorbed dose and effective dose is low. With the low activity administered (25-110 kBq), producing effective doses as low as 2 µSv, there is no reason for restrictions on even repeated screening investigations with ^{14}C-urea in whole families, including children, at least down to the age of 3 years. We therefore do not see the rationale in investing in equipment for ^{13}C and mass spectrometry (MS) as an alternative to ^{14}C measurements with LSC. The studies in children were made possible as we could use only 1/250 (440 Bq) or 1/50 (2.2 kBq) of the activity given to adults.

TABLE 1. Summary of studies on volunteering patients and healthy volunteers carried out within the current program

^{14}C-compound	Investigation	Age group	No of individuals	Effective dose µSv/MBq	Publication
Urea, 110 kBq Urea, 0.44 - 55 kBq	*Helicobacter pylori*	Adults Children (3-15 yr)	9 15	19 19 (15 yr) 25 (10 yr) 41 (7 yr) 40-80 (6-3 yr)(p) 190 (6-3 yr, HP+) (p)	(2,4) (2,4)
Triolein, 1.5-74 kBq	Fat malabsorption	Adults	4 (+3)	≤2 000	
Glycocholic acid, 200 kBq	Bacterial overgrowth/bile salts malabsorption	Adults (26-73 yr)	9	≈1 000 (p)	(9)
Xylose, 74 kBq	Abnormal intestinal bacterial flora	Adult	9	≈300 (p)	(9)

(p) = preliminary figure; (HP+) = *Helicobacter pylori* positive; number of individuals in () refer to planned investigations

^{14}C-Labelled Triolein

Earlier estimates indicated that about 30% of the given amount of ^{14}C was exhaled rapidly (within 24 h) (5,6). This value was based on an endogenous CO_2 exhalation of 9 mmol per hour and kg body weight. A recent critical analysis has shown that this value might have been as much as a factor of two higher which means around 60% was exhaled rapidly, while the remaining 40% was assumed to have a slow turnover. In total 90% was exhaled and 5-10% was assumed to be excreted via feces and some via urine. In the healthy volunteer, who was followed several years after administration, it was repeatedly demonstrated that provocation in the form of fasting significantly increased the exhalation of $^{14}CO_2$. The amount of "extra" $^{14}CO_2$ exhaled had a half-time of 400 days (8). The results of the measurements of the fat biopsies indicate that the initial amount of ^{14}C in the adipose tissue was 4.3% of the administered activity. The dose calculations give a high value for the effective dose per unit administered activity (2100 µSv/MBq). The cortical bone was found to have a specific activity of

0.266±0.004 Bq/g C and the trabecular bone 0.260±0.004 compared with an estimated atmospheric background value of 0.253 Bq/gC. To estimate a fraction of ^{14}C-bicarbonate in the skeleton, there is a need to determine the bone background value before the study. Due to slow turn-over of carbon in bone, the ^{14}C/carbon-ratio may not be the same in bone and atmosphere, because the atmospheric ^{14}C level had a maximum in 1963 (the "bomb-peak") and since then has been continuously decreasing. Therefore it is advisable to try to get an individual background bone sample before the administration of the ^{14}C-compound. However, the present result gives an estimate of the upper level of the absorbed dose contribution.

^{14}C-Glycocholic Acid

The results from 1 year studies of 7 patients indicate that about 90% of the administered activity is exhaled. After one year there is no significant extra exhalation. 6% of the administered activity is excreted in feces within 3 days, 3% in urine within one month. Healthy volunteers will get a higher dose than patients with changed metabolism of bile acid or bacterial overgrowth due to a more efficient re-circulation of bile acids in healthy persons.

^{14}C-Xylose

Results for four volunteering patients, show that 65% of the administered activity is found in urine and 35% in expired air within three months. Fecal samples from two individuals show no increased values of ^{14}C. Pathological individuals show a higher exhalation and lower urinary excretion than healthy volunteers.

Possibilities To Use Ultra-Low Activity And/Or To Reduce The Amount Of Sample Material Taken When Using AMS

We have shown that it is possible to reduce the administered activity compared to the standard amount given when ordinary LSC technique is used, both for ^{14}C-triolein test and for ^{14}C-urea used in small children. A prerequisite for the clinical use and thus our biokinetic studies of ^{14}C-urea on children 3-6 years was that we could lower the administered activity to 0.4 % of the usual activity of ^{14}C-urea. Our results indicate a possibility for a further reduction down to 0.4 ‰ (!) of the activity used in connection with liquid scintillation counting. In one ^{14}C-triolein study, the activity was reduced to 1/50 of the normal value. The results indicate possibilities for a reduction to 1 ‰ of the standard activity. This means that AMS has a great potential in the field of metabolism studies and related areas. In particular, it will enable the administration to humans of very low activities, e.g. 10 Bq of ^{14}C. This will, for most substances lead to effective doses of less than 1 µSv, which is so low that in many countries authorization via radiation protection authorities is not required.

SUMMARY AND FUTURE PLANS

AMS has enabled improvements of biokinetic and dosimetric information for ^{14}C-triolein and ^{14}C-urea. The data are subject to continuous enlargement and revision. We are currently also running a program to collect biokinetic data for ^{14}C-glycocholic acid and ^{14}C-xylose. The results of our investigations have already made it possible to widen the indications for the clinical use of the ^{14}C-urea test for *Helicobacter pylori* infection in the stomach. The use of ultra-low activities, which is possible with AMS, opens the possibility for metabolic investigations on healthy volunteers and on sensitive groups of patients like new-borns, children and pregnant or breast-feeding women. Using the full potential of AMS, new ^{14}C-labelled drugs could be tested on humans at a much earlier stage than today, avoiding uncertain extrapolations from animal models. This can be done since the sensitivity of the technique is high enough to use sub-toxic chemical amounts of a substance. The use of small amounts of material opens the possibility to analyze small tissue biopsies, some microliter of blood, scraping, highly specific biochemical separations and subcellular fractions including purified DNA (10, 11). This also means that frequent samples may be taken for detailed temporal records of the absorption, distribution, metabolism, and elimination of the labeled compound.

ACKNOWLEDGMENTS

This study was supported by the Swedish Medical Research Council (14X-011272).

REFERENCES

1. Mattsson, S., Leide-Svegbörn, S., Stenström, K., Erlandsson, B., Hellborg, R., Nilsson, L.-E., Nosslin, B. and Skog, G., "Improved absorbed dose estimates for

^{14}C-labelled pharmaceuticals" in *Proc. 6th International Symposium on Radiopharmaceutical Dosimetry*, edited by A S-Stelson et al., Oak Ridge Associated Universities, Oak Ridge, Tenn., USA, pp. 613-619 (1999).

2. Leide-Svegborn, S., Stenström, K., Olofsson, M., Mattsson, S., Nilsson, L.-E., Nosslin, B., Pau, K., Johansson, L., Erlandsson, B., Hellborg, R. and Skog, G., *Eur J Nucl Med* **26,** 573-80 (1999).

3. Leide-Svegborn, S.: "Radiation exposure of the patient in diagnostic nuclear medicine. Experimental studies of the biokinetics of ^{111}In-DTPA-D-Phe1-octreotide, ^{99m}Tc-MIBI, ^{14}C-triolein and ^{14}C-urea and development of dosimetric models", *Doctoral Diss.,* ISBN 91-628-3491-6, Malmö, 1999

4. Gunnarsson, M., Leide-Svegborn, S., Stenström, K., Mattsson, S., Nilsson, L.-E., Nosslin, B., Hellborg, R. and Skog, G: Estimates of absorbed dose from a ^{14}C-urea breath test in children aged 3-7 years, Manuscript, 2000

5. Stenström, K., Leide-Svegborn, S., Erlandsson, B., Hellborg, R., Mattsson, S., Nilsson, L.-E., Nosslin, B., Skog, G. and Wiebert, A., *J Appl Radiat Isotopes* **47:4,** 417-422 (1996)

6. Stenström, K.: "New applications of ^{14}C measurements at the Lund AMS facility", Doctoral Dissertation, ISBN 91-628-1816-3, Lund, 1995

7. Stenström, K., Leide-Svegborn, S., Erlandsson, B., Hellborg, R., Mattsson, S., Nilsson L-E,., Nosslin, B. and Skog, G., *Nucl Instr Meth B* **123** 245-248 (1997)

8. Gunnarsson, M., Mattsson, S., Stenström, K., Leide-Svegborn, S., Erlandsson, B., Faarinen, M., Hellborg, R., Kiisk, M., Nilsson, L.-E, Nosslin, B., Persson, P., Skog, G. and Åberg, M., *Nucl Instr Meth B,* **172:1**, 942-946 (2000).

9. Gunnarsson, M., Leide-Svegborn, S., Stenström, K., Mattsson, S., Nosslin, B., Nilsson, L.-E., Hellborg, R., Skog, G. and Östberg, H., Biokinetics and radiation dosimetry in patients undergoing ^{14}C-glycocholic-acid and ^{14}C-xylose breath tests, Manuscript, 2000

10. Cupid, B. C. And Garner, R. C.: "Accelerator mass spectrometry – A new tool for drug metabolism studies", in *Drug metabolism: Towards the Next Millennium* edited by N Gooderham, pp 175-187, 1998

11. Vogel, J S, *Nucl Phys News,* **10:1**, 8-13, 2000

Microbeam AMS Measurements of PGE and Au Trace and Osmium Isotopic Ratios

S.H. Sie, D.A. Sims and G.F. Suter

*Heavy Ion Analytical Facility, CSIRO Division of Exploration and Mining,
P.O. Box 136, North Ryde, NSW 2113, Australia*

The microbeam Accelerator Mass Spectrometry (AMS) system, AUSTRALIS (AMS for Ultra Sensitive TRAce eLement and Isotopic Studies) at CSIRO Heavy Ion Analytical Facility (HIAF) has been used to measure trace levels of precious metals, the platinum group elements (PGE) and Au, from an assortment of geological and meteoritic samples with spatial resolution of 30 micrometres. For Au, detection sensitivity as low as sub parts per billion has been obtained, which will be of great benefit in studies of ore deposit mineralogy and mineral processing. We have also demonstrated the facility for osmium isotope measurements in meteorite samples, opening up the possibility of widespread use of the Re-Os system in exploration programs.

INTRODUCTION

The ability to detect low levels of precious metals such as PGE (platinum group elements) and Au distributed in carrier minerals is desirable in the minerals industry, which enables selective mining and optimisation of minerals beneficiation process. The PGE abundance pattern itself carries important geological information that provides powerful tools for tracing the evolution of magma into rocks and subsequent alterations. One of the osmium isotopes (^{187}Os) can be augmented by the presence of radioactive ^{187}Re, and with their distinct geochemical behaviour, the ratios can be used for dating sulfide ores as well for differentiating crustal/mantle sources [1]. These elements are often concentrated in distinct mineral phases, and partition of these elements in coexisting minerals precipitating from the same magma provides additional information. In typical geological samples, the carrier minerals often occur as inter-grown microscopic constituents, necessitating an in-situ microanalytical approach.

Microbeam AMS offers the prospect of an ideal tool for such application, by providing the means to counter the problem of mass interference encountered in SIMS (secondary ion mass spectrometry) using the ion microprobe. The AUSTRALIS system recently commissioned at CSIRO is designed for such geological applications [2-6]. The use of negative ions in the first stage works in favour for AMS, due to the high ionisation efficiency for PGE, especially the heavy PGE, which include Os. The large effect on Os isotope ratio do not require the high precision of better than 1 permil typically required for other isotope system (e. Pb, S isotopes) for geochronology or tracer applications [5,6]. This factor mitigates the counting statistics practical limitations considering that the Os content of most samples to be analysed are typically less than parts-per-million weight.

The present paper describes the result of tests of the system for PGE elements and Os isotopes.

AUSTRALIS

Main features that distinguishes AUSTRALIS from most AMS systems are the microbeam Cs source [3], and a fast sequential isotope detection system at the high energy end (high energy bouncer) [4]. The sample chamber is designed to facilitate microanalysis of geological samples, incorporating a high magnification zoom microscope based sample viewing system in the reflected geometry, and a three-axis microstage sample mount for precise positioning of samples. The ion source produces a 30 micrometer diameter Cs$^+$ beam routinely, striking the sample at 45° incidence and the negative secondary ion beam is extracted at normal angle. An electron flood gun, mounted at the opposite 45° angle, can be used to compensate charge build-up during analysis of insulating samples. The secondary ions pass through a 45° bend spherical electrostatic analyser (ESA), and a 90° bend analysing magnet. The magnet box is insulated to allow rapid sequential injection of the different isotopes at a fixed magnetic field by applying modulating voltages, comprising the low energy 'bouncer' system. At the focal point of the magnet the analysed ion beam can be monitored in either a Faraday cup or an ETP electron multiplier acting as an ion counter. Coupled to a fast pulse amplifier, the ion counter can be driven as high as ~10 MHz, giving a good overlap with current measurements where the noise level is at ~ 0.1 pA.

At the high energy side the beam enters a 90° bend, 1.3m mean radius, 2.5cm gap magnet, corrected to the 2nd order, with a beam product of 140 MeV.amu, operated in the double focusing, unity magnification mode. The high energy bouncer consists of two pairs of electrostatic deflector plates at the entrance and exit ports of the magnet box, deflecting the beam in the orbit plane. The magnet is followed by two 3 m mean radius, 22.5° bend spherical ESA's, separated by an electrostatic quadrupole doublet that focuses the beam into the detector chamber. The ions can be detected in a Faraday cup, a Frisch gridded gas proportional counter or an ETP counter.

The high energy bouncer makes it possible to switch the isotopes at rates faster than the beam fluctuations and achieve high precision in the measurements. The bouncer can be driven as fast as 1 ms per isotope, and tests on Pb isotopes demonstrated that better than permil precision can be achieved at bouncing rate < 10 ms per isotope. In this bouncer system, the exit beam is restored to the main axis and therefore the subsequent beam transport system does not require any correction, at least in the central region of the dynamic range. For mass region of ~ 190, the transmission efficiency stays within 5% in a range of ~ +/- 4 amu of the mass of the central trajectory (neutral bouncer setting). Instrumental fractionation effects which include a significant component due to the mass dependence of stripping efficiency, is corrected against measurements on standards.

RESULTS

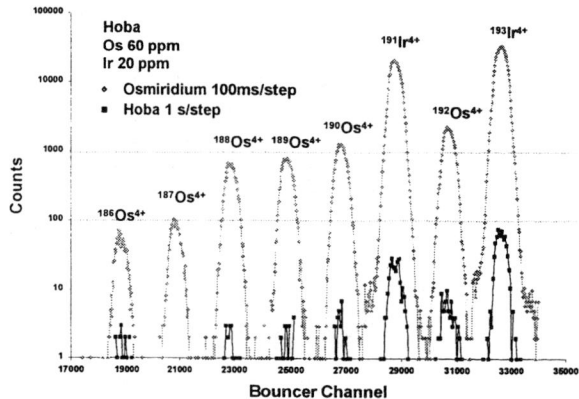

Figure 2. Composite mass spectrum from osmiridium and an iron meteorite (HOBA) for the 4+ ions obtained for 1.5 MV terminal voltage, at the high energy end using the ion counter, for the corresponding injected isotopes. The high energy magnet is set at a fixed field to pass $^{193}Ir^{4+}$ from $^{193}Ir^-$. Since the bouncer system is limited to scan over 8 isotopes, the full range of the Os-Au isotopes were measured in two passes, with overlap on Ir.

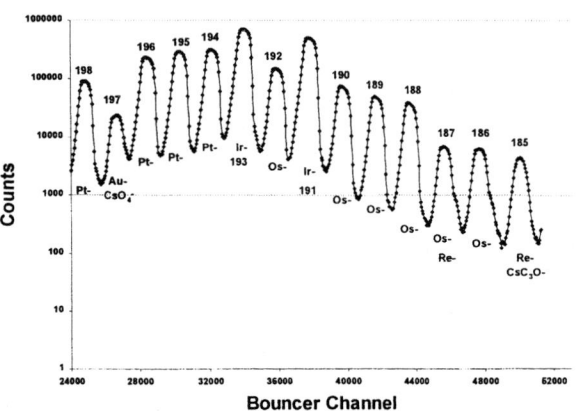

Figure 1. The low energy mass spectrum from an osmiridium sample in the mass region of interest covering the Re, Os, Ir, Au and Pt isotopes.

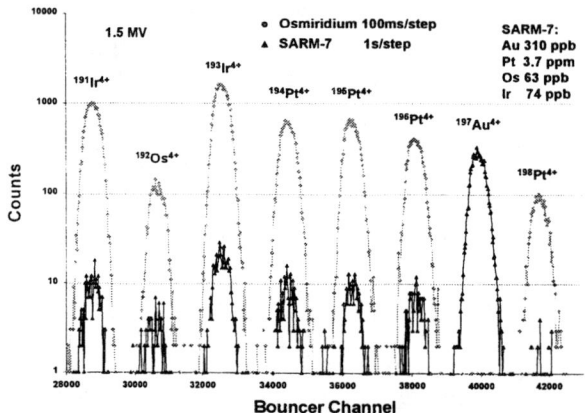

Figure 3. Composite mass spectrum obtained from a NiS fire-assay bead prepared from SARM-7, a standard for PGE trace element measurements, superimposed on the osmiridium spectrum.

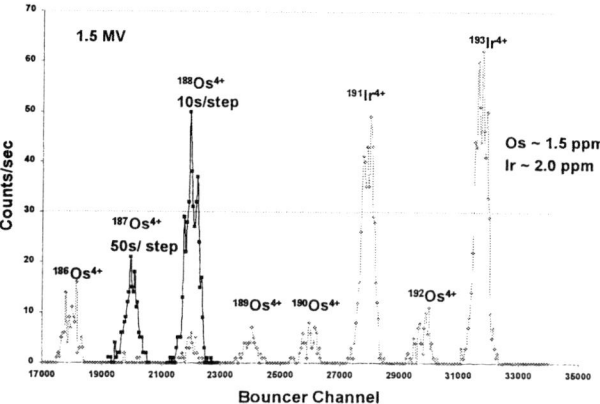

Figure 4. Composite mass spectrum obtained from the Canyon Diablo meteorite showing the feasibility of measurements of the important $^{187}Os/^{188}Os$ ratio using a microbeam.

One favourable factor in PGE and Au measurements is the fact that these elements are generally prolific negative ion producers which result in high sensitivity for their detection. For this work we have measured only the heavy PGE's, namely Os, Ir and Pt. To calibrate the mass spectrometers we have used samples of osmiridium, naturally occurring mineral alloy of Ir and Os with significant Pt content. The setting for Au can be calibrated separately, or interpolated using the Pt isotopes. For in-situ Os isotopes measurements, AMS offers two advantages: elimination of possible hydride interference (e.g. ^{187}Os by ^{186}OsH and ^{188}Os by ^{187}OsH), and suppression of ^{187}Re isobaric interference due to the property that Os^- is more prolific than Re^-.

Figure 1 shows the low energy mass spectrum from an osmiridium sample (45% Os, 49% Ir and 8% Pt) in the mass region of interest covering the Re, Os, Ir, Au and Pt isotopes. The mass 185 and mass 197 peaks contain contributions due to the primary beam, namely the $^{133}Cs^{12}C_3O^-$ and $^{133}CsO_4^-$ ions respectively, and any hydride ions would be unresolved. The spectrum was obtained with an intrinsic mass resolution of ~1000 and with the image slits widened to achieve flat topped transmission. Figure 2 shows the composite mass spectrum for the 4+ ions obtained for 1.5 MV terminal voltage, at the high energy end using the ion counter, for the corresponding injected isotopes. The high energy magnet is set at a fixed field to pass $^{193}Ir^{4+}$ from $^{193}Ir^-$. Since the bouncer system is limited to scan over 8 isotopes, the full range of the Os-Au isotopes were measured in two parts, with overlap on Ir. The second part of the spectrum is shown in figure 3. In both figures, the osmiridium spectra data points are shown in gray.

The high efficiency resulting from not requiring high resolution enables measurements of isotopes at low concentration levels. This is illustrated by the measurement on an iron meteorite (HOBA) which contains PGE at several ppm levels [7], shown superimposed in figure 2. Incidentally, the osmiridium spectrum shown in figures 2 and 3 was obtained using much lower beam intensity in order to keep the maximum count rate in the detector below 10^6 counts per second and minimise dead time correction.

The highest sensitivity is obtained in the detection of Au. Figure 3 shows a mass spectrum obtained a NiS fire-assay bead prepared from SARM-7 standard for PGE trace element measurements, superimposed on the osmiridium spectrum. The bead only results in a 1.1 pre-concentration, but provides a good conducting matrix for the analysis. The Au peak is very prominent, reflecting the high ionisation yield of Au. In this sample the expected Au concentration is 310 ppb, and the high count rate obtained demonstrate the high sensitivity for Au detection at sub-ppb level.

The feasibility of $^{187}Os/^{188}Os$ ratio measurements for low level samples is demonstrated in figure 4, showing the composite mass spectrum from the Canyon Diablo meteorite. The Os content of this meteorite has been reported in literature at ~ 1.5 ppm. All data have been obtained using a 30 micrometer Cs beam, and using the 4+ ions. It is possible to use the 3+ ions for the mass range covered by these elements, which have magnetic rigidity just below the maximum power of the present magnet. A tenfold yield was obtained over the 4+ ions, which facilitates detection of lower levels of concentration. An application envisaged is dating sulfide deposits through routine measurements in chalcopyrite ($CuFeS_2$).

CONCLUSION

AUSTRALIS, a microbeam AMS system designed for in-situ geological applications has been used to measure PGE and Au in mineral and meteoritic samples with sensitivity as low as parts-per-billion concentration levels, with 30 micrometer spatial resolution. We have also demonstrated the feasibility of Os isotopic ratio measurements in samples containing low level osmium in the ppm range.

The system paves the way for applications in isotope geology previously not feasible by the in-situ method.

ACKNOWLEDGEMENT

We thank the Australian Museum, Sydney for the osmiridium samples and the ISOTRACE group, University of Toronto for the meteorite samples.

REFERENCES

1. Shirey, S.B. and Walker, R.J., Annu. Rev. Earth Planet. Sci. **26**, 423-500(1998)
2. Sie, S.H., Niklaus, T.R. and Suter G.F., Nucl. Instr. Meth. **B123**, 112-121(1997)
3. Sie, S.H, Niklaus, T.R., Suter, G.F. and Bruhn, F., Rev. Sci. Instr. **69**, 1353-1358 (1998)
4. Sie, S.H., Sims, D.A., Suter, G.F., Cripps, G.C., Bruhn, F. and Niklaus, T.R., Nucl. Instr. Meth. **B172** (2000) in press
5. Sie, S.H., Sims, D.A., Niklaus, T.R. and Suter, G.F., Nucl. Instr. Meth. **B172**(2000) in press
6. Sie, S.H., Sims, D.A., Bruhn, F., Niklaus, T.R. and Suter, G.F., AIP Conf. Proc. **475**, 648-651 (1999).
7. Wilson, G.C., Rucklidge, J.C., Kilius, L.R., Ding, G.J. and Cresswell, G.C., Nucl. Instr. Meth. **B123**, 583-588 (1997)

A 3 MV Heavy Element AMS System Using a Unique TOF Set-up

A. Gottdang[1], M. Klein[1], D.J.W. Mous[1], T. Kitamura[2], Y. Mizutani[2], T. Suzuki[2], T. Aramaki[2], O. Togawa[2], S. Kabuto[3] and K. Suto[3]

1) High Voltage Engineering Europa B.V., P.O. Box 99, 3800 AB Amersfoort, The Netherlands
2) Japan Atomic Energy Research Institute (JAERI), 4-24 Minato-machi, Mutsu, Aomori, 035-0064, Japan
3) Japan Marine Sience Foundation (MSF), Mutsu, Japan

Abstract. A heavy element AMS system, based on a 3 MV Tandetron, has been put into operation at JAERI, Mutsu, Japan. The system uses sequential injection, designed for cycle frequencies of up to 1000 Hz. The high-energy section is unique in that the identification of the isotopes of interest is done in two successive steps, each using a separate foil combined with energy discrimination. This method allows for the detection of elements that suffer from problematic isobar interference like ^{36}Cl and ^{41}Ca. In that case the foils are chosen to be relatively thick in order to achieve the required energy dispersion. In order to cope with the large scattering caused by the foils, the applied TOF has a unique design that features the acceptance of extremely high divergent beams of up to 80 mrad.
During the acceptance tests the precision was shown to be ~1.1 % for ^{129}I measurements. The background was found to be below 10^{-13}.

INTRODUCTION

It is well known within the AMS community that the analysis of some elements like ^{36}Cl and ^{41}Ca is hampered by the interference of isobars with lower nuclear charge Z. In the case of ^{36}Cl the ^{36}S content in the sample is suppressed by natural abundance and sample pre-treatment by factors of 10^4 and 10^6 respectively [1]. So a further reduction by the system of 10^5 is needed to achieve a desired background level of 10^{-15}.

The general approach to achieve such reduction is the use of beam energies in excess of 50 MeV [2], [3] combined with a multi-anode detector. Such energies correspond to a terminal voltage of at least 5 MV.

It was suggested that with a so-called two-foil arrangement the needed sensitivity can be achieved at terminal voltages as low as 3 MV [4]. This method uses two successive steps of isobar suppression. After acceleration, the ions are mass analyzed and pass through a first thin foil. This preferentially reduces the energy of ions with higher nuclear charge. With the electrostatic analyzer (ESA) that follows, an energy window is set that allows only ions of wanted energy and charge state to pass a slit system. Calculations

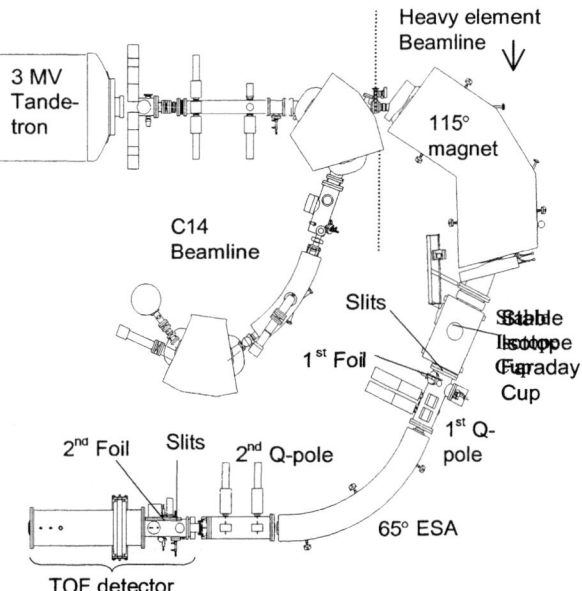

FIGURE 1. The high energy section.

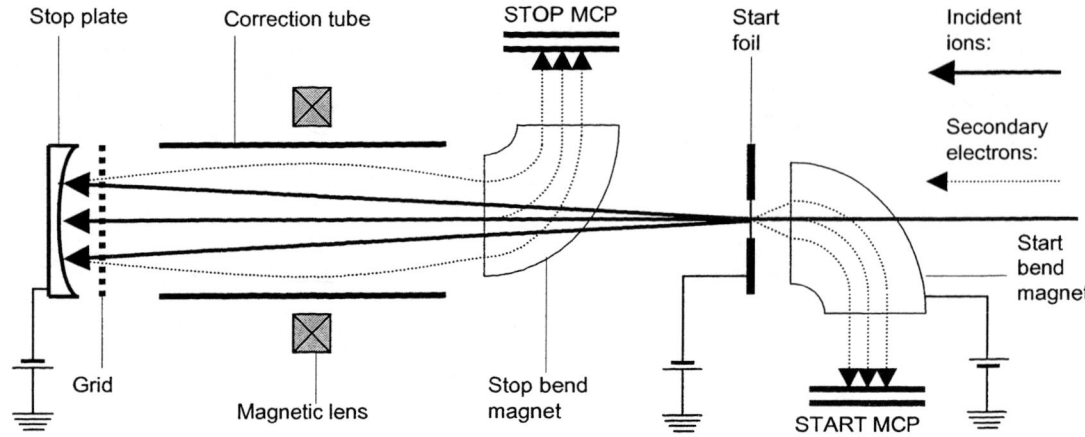

FIGURE 2. The principal layout of the TOF detector.

have shown that by this first step a background reduction of 2000 can be achieved. The remaining ions pass a second foil located inside a time-of-flight (TOF) detector. By means of the resulting second energy shift the TOF detector discriminates between the ^{36}Cl ions and their isobars ^{36}S. With the combination of these two steps an overall sensitivity of 10^{-15} is achieved. However, the ion beam is highly divergent because of the scattering in the foils. The acceptance of the beam line and the detector must be sufficient to cope with these divergent beams.

HVEE has recently finalized an AMS system at the Japanese Atomic Energy Research Institute (JAERI), Mutsu, Japan, that consists of a dedicated ^{14}C AMS system using a 3 MV Tandetron accelerator, extended with a sequential injector and a high energy section for the analysis of heavy elements, which is designed to meet the requirements of the two-foil arrangement. Up to three different isotopes can be inserted with the achromatic injector at cycle frequencies as high as 1000 Hz. For a more detailed description of the system see [4] and [5].

BEAM TRANSPORT IN THE HIGH ENERGY SECTION

Figure 1 shows the high-energy section of the system. Ions entering the high-energy section are mass analyzed by a higher order corrected 115° magnet having a radius of 1200 mm and a calculated mass resolution of 2000. A Faraday cup is located at the image of the magnet to measure the current of the stable isotope. It is equipped with slit stabilization. Another slit system is mounted in the path of the radioisotope. Behind the slits provisions are made to insert the first foil at the image of the magnet. In case of ^{36}Cl measurements, 2 μm thick mylar foil is used. The following ESA has a radius of 1700 mm, an angle of 65° and a calculated resolution of 1000. A slit system is located at the image of the ESA inside the TOF detector housing. Two pairs of electrostatic quadrupole lenses, before and after the ESA, ensure proper focussing of the beam, which is highly divergent because of scattering in the first foil. ESA and quadrupoles are designed to accept a beam with a divergence of up to 40 mrad. A Faraday cup and an aperture can be inserted at the entrance of the detector for tuning purposes.

THE TIME-OF-FLIGHT DETECTOR

Calculations have shown that an energy resolution of 6 $^0/_{00}$ is required for the time of flight detector, in order to achieve sufficient background reduction in the second stage of the two-foil arrangement for an overall background level of 10^{-15}. This can be converted to 1 ns time resolution, given a flight path of 1.5 m. The system must accept the highly divergent trajectories of up to 80 mrad, which result form the scattering in the second foil. A unique TOF detector was developed to meet these demands. Figure 2 shows the layout of the detector, of which figure 3 shows a photograph. The positions of start foil and stop plate inside the housing are indicated. Ions enter from the right side.

The incoming ions pass the start foil, which is mounted on an isolated frame with an inner diameter of 20 mm. The secondary electrons emitted from this foil are accelerated by a frame potential of -2400 V. They are directed by a permanent magnet onto a micro channel plate (MCP) for the generation of the start

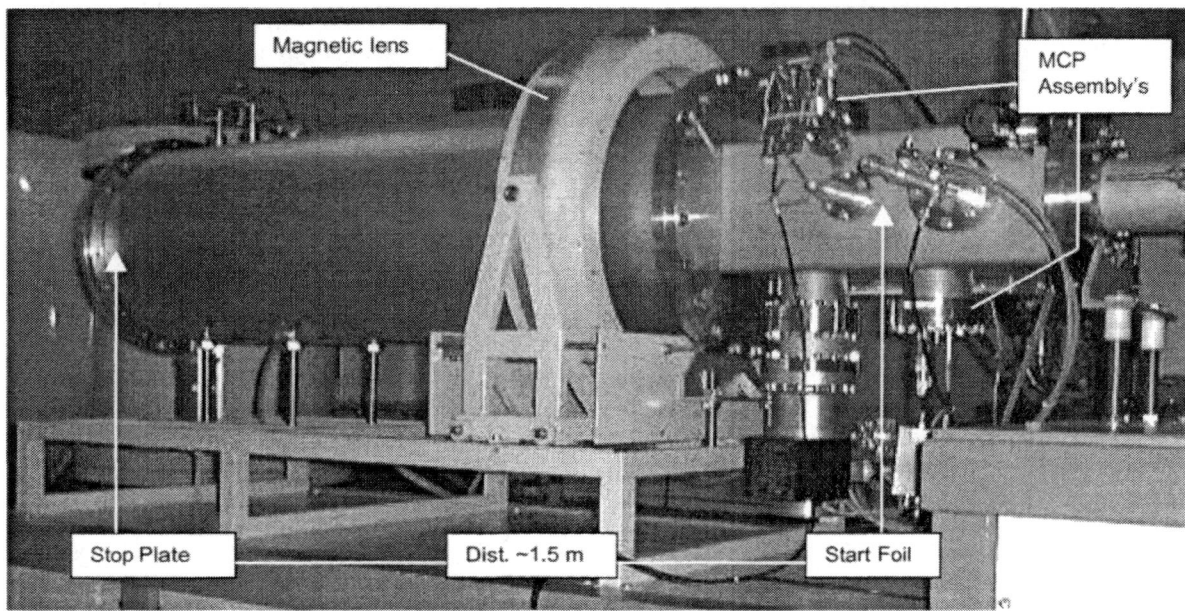

FIGURE 3. The time-of-flight detector.

signal. The magnet assembly is set on a potential of 27 kV for proper transport of the electrons to the MCP. The applied MCP's are manufactured by Comstock (model CP-640C/50F).

Because a foil is impractical for stop detection when an angular acceptance of 80 mrad is needed, a different design is implemented. The ions are stopped on a spherical aluminum stop plate with a diameter of 400 mm located 1.5 m behind the start foil. The emitted secondary electrons are accelerated backwards to the stop detection by a plate potential of 8.6 kV. A current controlled stop bend magnet directs these electrons 90° upward to the stop-MCP, where the stop signal is created. A magnetic lens is implemented to focus the electrons on the MCP. This lens is placed outside of the detector housing and can be adjusted in position and rotation as well as in field strength, for optimal detection efficiency.

Detailed calculations on ion and electron trajectories using a 5th order Runge-Kutta trajectory solver have indicated that a time resolution of 1 ns can be achieved. The ions that have passed the foil will be distributed in different charge states, which have different transit times between plate and grid. The mentioned calculations take this into account, as well as the influence of the angular distribution of the ions. A correction tube can be set on a positive potential to minimize the influence of the different charge states.

The MCP signals are amplified by 1 GHz preamplifiers Ortec model 9306, discriminated by pico timing discriminators Ortec model 9307, and their time difference is measured with the time to amplitude converter Ortec model 566.

PERFORMANCE

The following measurements were made during the final acceptance tests of the system on Iodine. Because ^{129}I has no problematic isobar interference, no foil aside the detector start foil was needed. The terminal voltage was set to 2.5 MV and the charge state 5+ was selected. The current of ^{127}I in the stable isotope Faraday cup was 100 to 200 nA DC, depending on the target. The ion transmission of the ^{127}I through the accelerator was ca. 5 %. For the precision measurements injection times for ^{129}I and ^{127}I were chosen to be 8 ms and 2 ms per cycle respectively.

Figure 4 shows scans on the alignment of the high-energy section. The current in the stable isotope Faraday cup as well as the current on the deflector cone is plotted as a function of the voltage of the last Y-steerer in front of the accelerator. The transmission to the cup is limited only by the geometry of the stripper channel.

For the current measurement on the deflector cone, the ^{127}I current was drastically reduced by decreasing the target voltage of the ion source while keeping the total extraction energy constant. The ions passed the start foil before hitting the cone. The TOF detector voltages were switched off during this measurement. The width of the peak is limited by the diameter of the foil holder, but it still shows clearly a flat top tuned area. Both scans are normalized for comparison.

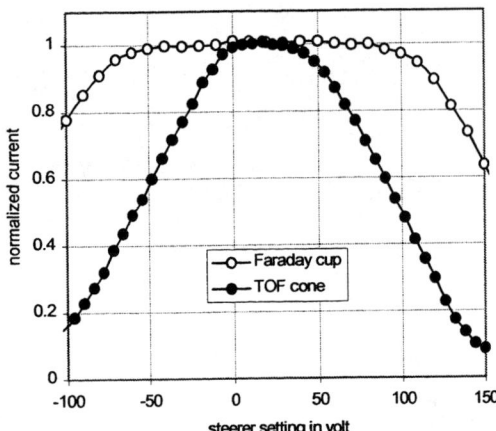

FIGURE 4. Current of the stable isotope Faraday cup and the TOF cone during a scan of the last Y steerer in front of the accelerator.

Figure 5 shows a ^{129}I- spectrum of the TOF detector. Assuming that the energy straggling in the start foil causes the low energy tail at higher channel numbers, the time resolution of the detector was determined to be ~ 1 ns. This is in close agreement to the design value.

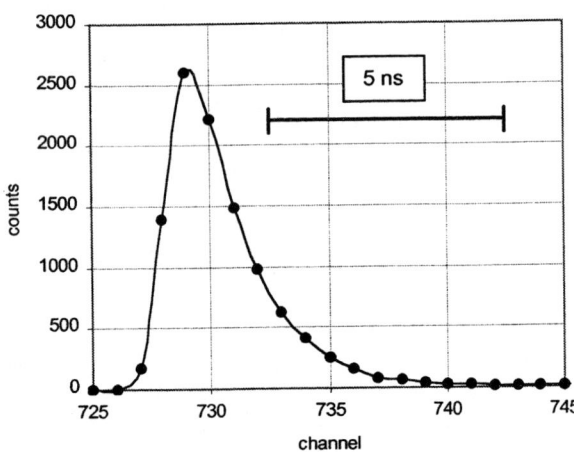

FIGURE 5. ^{129}I in the TOF spectrum. One channel corresponds to 0.5 ns.

The efficiency of the TOF detector was measured to be about 70%, which is probably dominated by the efficiency for electron detection of the MCP's.

Table 1 presents the results of the precision measurements, which were made on occasion of the acceptance tests for the heavy element section. The samples were made from AgI powder provided by Isotrace Laboratory, Toronto, Canada, which had a ^{129}I content of about $1.1 \cdot 10^{-10}$. Before any data were taken, the targets were sputter cleaned for 5 to 10 minutes. Each day 4 samples were measured repeatedly in several cycles. The results show that a typical precision on ^{129}I/^{127}I of 1.1 % was achieved. In addition measurements were performed at the background level. These measurements yielded a background level below 10^{-13}, but it was anticipated that this number reflected real ^{129}I atoms present in the sample.

TABLE 1. Precision measurements during acceptance for iodine samples. Average normalized to $1.1 \cdot 10^{-10}$

#	July 17, 2000		July 18, 2000	
	^{129}I/^{127}I (x 10^{-10})	Rel. std. dev.	^{129}I/^{127}I (x 10^{-10})	Rel. std. dev.
1	1.0778	1.40 %	1.0912	2.15 %
2	1.1015	1.35 %	1.0967	0.73 %
3	1.1001	1.44 %	1.1087	1.79 %
4	1.1207	1.83 %	1.1033	0.83 %
Av.	1.1	1.50 %	1.1	1.38 %
Relative standard deviation:				
	1.59 %		0.69 %	
Statistical uncertainty				
	0.58 %		0.63 %	

CONCLUSION

A heavy element AMS system has successfully been put into operation at JAERI, Mutsu, Japan. It features a sequential injection system and a high-energy section, which applies isobar suppression in two successive steps. The ions are identified by a time-of-flight detector with an angular acceptance of 80 mrad and a time resolution of ~1 ns. The average on the precision of the two succeeding days was typically ca. 1.1 %. The background level was determined to be below 10^{-13}.

REFERENCES

1. Synal, H.A., Beer, J., Bonani, G., Lukaczyk, Ch. and Suter, M., *Nucl.Instr. and Meth. B 92* (1994) 79-84
2. Elmore, d., et al. *Nucl. Instr.and Meth. B 92* (1994) 65-68
3. Knie, K., et al., *Nucl. Instr. and Meth. B 123* (1997) 128-131
4. Gottdang, A., Mous, D.J.W., *Nucl. Instr. and Meth. B 123* (1997) 163-166
5. Gottdang, A., Mous, D.J.W., "*Appl. of Accelerators in Res. And Ind.*", AIP Conference Proceedings 475, New York: American Institute of Physics, 1999, pp. 652-656.

Test of Positive Ion Beams from a Microwave Ion Source for AMS

S-W. Kim[a], R.J. Schneider[a], K.F. von Reden[a], J.M. Hayes[a], J.S.C. Wills[b], W.G.E. Kern[c]

[a]*Woods Hole Oceanographic Institution, Woods Hole, MA 02543, USA*
[b]*Chalk River Laboratories, AECL, Chalk River, Ontario, Canada, K0J 1J0*
[c]*University of Massachusetts Dartmouth, North Dartmouth, MA 02747, USA*

Abstract. A test facility has been constructed to evaluate high-current positive ion beams from small gaseous samples for AMS applications. The flow of gas into a compact microwave ion source is regulated by a restrictor constructed from vitreous-silica capillary tubing. A double-focusing spectrometer magnet is utilized to isolate carbon ions derived from CO_2 samples from other products of the plasma discharge, including argon ions produced from the carrier gas. Dual Faraday cups have been constructed to measure the $^{13}C/^{12}C$ isotope ratio by collecting $^{13}C^+$ and $^{12}C^+$ ions simultaneously. When CO_2 gas is injected, the molecules fragment to yield C^+, O^+, CO^+, O_2^+, and CO_2^+ ions. The relative abundances vary by less than 5 % as the extraction voltage is changed. However, the abundances of atomic ions relative to molecular ions increase as the RF power of the microwave is increased. The overall efficiency of ($^{12}C^+$ ions detected)-to-(C atoms injected) is ~ 1.7 %. This includes the beam transport efficiency of ~ 30 % between the ion source and the collector. With a continuous flow of pure CO_2 and a mixture of CO_2 and argon (16 % CO_2 by mole), the microwave ion source produces $^{13}C/^{12}C$ ratio that is stable to within 0.1 % for over two hours after it reaches thermal equilibrium. It takes approximately one hour for the ion source to reach thermal equilibrium. With pulses of CO_2 entrained in the argon carrier gas, the ratio of integrated carbon-13 and carbon-12 current peaks varied by 1.5 % over the period of 22 hours. The next step will be to construct an efficient charge-exchange cell and produce negative ions for injection into the accelerator mass spectrometer.

INTRODUCTION

The Chalk River compact microwave ion source, which was described at AMS-7 in Tucson [1], at the Radiocarbon Conference in Groningen [2], and at AMS-8 in Vienna [3], is being further developed at the National Ocean Sciences AMS (NOSAMS) Facility in Woods Hole Oceanographic Institution. A detailed description of its construction can be found in reference [4]. The use of a gas ion source can open further applications for AMS, specifically, those in which the preparation of the graphite sputter targets needed for a cesium sputter source would be impractical or impossible. These would include very small samples (below a few micromoles of carbon), gas-profiling where changes in the isotopic ratio of an evolved gas are of interest, and also studies where large numbers of small gas samples may be collected, such as in atmospheric trace gas research.

Highly efficient sputter ion sources that can accept gaseous inputs are now available commercially [5, 6] and are very useful for the analysis of small samples. However, the fact that the surface concentration of CO_2 affects the target work function limits the flow rate of CO_2 and thus the maximum current attainable. In circumstances useful for routine measurement, reported maximum currents are approximately 12 μA [6]. This limitation on the sample flow rate makes sputter gas sources unsuitable for continuous-flow systems in which the analytes flow directly and continuously from a gas chromatograph to an ion source. In such cases, the peak flow rate of the CO_2 sample would substantially exceed the limit of the sputter gas source, and most of each sample would be wasted. Meanwhile, the microwave ion source can produce a C^+ current of more than 1 mA. With a suitable charge-exchange cell, which has a typical efficiency of about 10 % for producing negative carbon ions [7], the microwave ion source could generate more than 100 μA of negative carbon ions.

Such a source would be useful for the analysis not only of chromatographic effluents but also of any gaseous material, for example CO_2 produced from carbonates. In addition, the operation and maintenance of the microwave ion source are much simpler than those of a sputter gas source. These reasons make the microwave ion source an attractive candidate for a gas ion source for AMS applications.

TEST STAND

The layout of the test stand is shown in Figure 1. A tapered waveguide directs 2.45 GHz microwaves into the plasma chamber through an aluminum nitride window. The waveguide has a d.c. isolator section so that the tapered adapter leading up to the chamber is held at the ion source potential (+ 24 kV). The water-cooled copper plasma chamber has an inner diameter of 5 cm and a length of 2 cm. The sample gas enters the chamber through a tube adjacent to the microwave window. The other end of the tube is connected to a silica capillary tube from the gas bottle. The required magnetic field, having a constant axial field of about 900 G, is provided by a permanent magnet assembly. The extraction system is an accel-decel triode structure with an acceleration gap of 5 mm (variable from 2.5 to 10 mm) and a deceleration gap of 2 mm (fixed). The extraction aperture has a diameter of 2.5 mm (variable up to 5 mm). A 90-degree, double focusing spectrometer magnet with 30-cm bending radius is used to analyze the ion beams. Dual Faraday cups are used to collect carbon-12 and carbon-13 ions simultaneously, as well as to collect other ion species.

MICROWAVE ION SOURCE

The microwave ion source utilizes the absorbed microwave power to heat the electrons, resulting in excitation and ionization of particles in the plasma chamber. A flat, axial magnetic field is superimposed to improve the microwave heating of the plasma electrons. Because we need singly charged ions, the source is not operated at the electron cyclotron resonance (ECR), which would create highly energetic electrons and result in high charge state ions. In general, the extractable ion current density is proportional to the product of electron density and the square root of electron temperature [8]. Therefore, in order to achieve higher ion beam current, either or both of these parameters must be raised by increasing the absorbed microwave power. Because the impedance of a high-density plasma changes with magnetic field, gas pressure, and the degree of

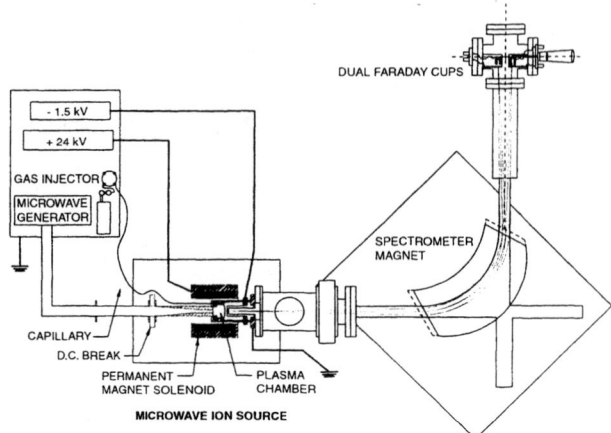

FIGURE 1: Test stand for the compact microwave ion source at the NOSAMS Facility. For scale, the permanent magnet solenoid which surrounds the ion source has a 10 cm bore. The spectrometer magnet has a 30 cm bending radius.

ionization, a tuner (not shown) is placed between the microwave generator and ion source to prevent high reflection of microwave power due to mismatch of the plasma impedance. The ions drift along the axial magnetic field until a fraction of them reaches the extraction region. The accel electrode in the middle of the triode extraction system (at -1.5 kV) limits the backstreaming of electrons into the discharge chamber and preserves the space charge compensation of the extracted beam. For a given geometry and extraction field strength, the plasma density must be matched to generate a beam with the desired properties (e.g. low emittance, aberration). Therefore, although the extractable ion current density is highest at the maximum achievable plasma density, it is not always desirable to inject the highest microwave power available.

GAS SAMPLE INJECTION SYSTEM

Schematic of the gas-sample injection system is shown in Figure 2. Operation of the system is explained in the caption. The ion source can be operated either with a continuous flow of sample gas or pulses of sample gas entrained in a carrier gas.

In order to introduce small gas samples without drastically changing the plasma conditions, an argon carrier gas which readily sustains a plasma discharge is used to deliver samples to the plasma chamber. A laminar flow into the ion source is maintained by the use of vitreous-silica capillary tubing. The flow rate into the plasma chamber is kept at a constant level by keeping the constant pressure drop across the capillary. This is accomplished by an open split device in which one end of the capillary is kept at

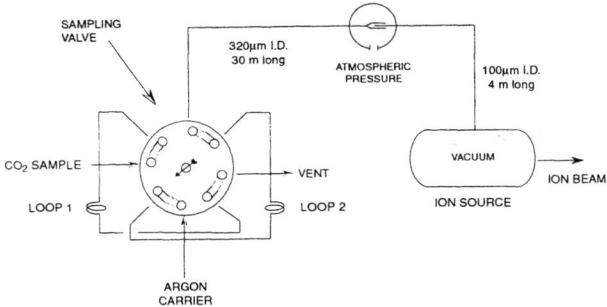

FIGURE 2: The gas sample injector for the test stand. As shown, the sample valve connects adjacent pairs of ports. CO_2 always flows to the vent by way of Loop 1 or Loop 2 (108μl). The Ar carrier always flows to the ion source by way of Loop 2 or Loop1. Toggling the sample valve places a loop full of sample gas into the Ar stream. The 30-m capillary is used to produce chromatograph-like peaks by allowing longitudinal diffusion as the peak passes through the capillary (retention time ≈ 5 min). A constant pressure drop and voltage gradient are maintained across the 100 μm capillary, which leads to the ion source.

atmospheric pressure and the other at vacuum. A silica capillary also provides good electrical isolation between the ion source at + 24 kV and ground.

RESULTS

The microwave ion source was operated with a flow rate of 0.08 standard cc/min. (calibrated for argon), which provided a stable operation. For this flow rate and with the given extraction geometry, the beam-matching is achieved at the microwave power of 160 W and the extraction voltage of + 24 kV with respect to ground.

When CO_2 gas is injected, the molecules are fragmented and ionized by the electron bombardment to yield C^+, O^+, CO^+, O_2^+, and CO_2^+ ions. Figure 3 shows an ion mass spectrum obtained when the source was operated on pure CO_2. The relative abundance of $^{12}C^+$ ions was typically around 9 %. The relative abundances varied by less than 5 % as the extraction voltage was changed. However, the abundances of atomic ions relative to molecular ions increased as the microwave power was increased (i.e. from 8.5 to 13.5 % for $^{12}C^+$ ions when the microwave power was increased from 82 to 250 W).

The overall efficiency for producing positive carbon-12 ions from carbon dioxide molecules was calculated from the collected $^{12}C^+$ ion current and the flow rate of the CO_2 molecules into the source when the source was operated with a continuous flow of CO_2. The best efficiency to date is 1.7 %.

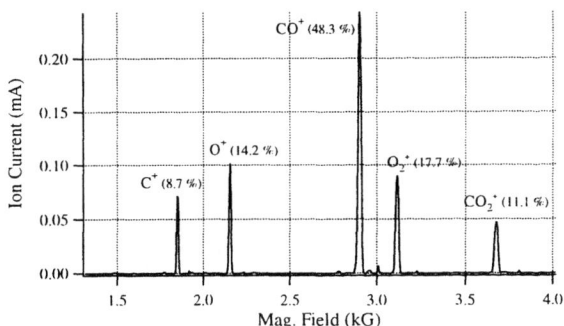

FIGURE 3: Analyzed positive ion spectrum from the microwave ion source. A flow rate of about 0.12 cc/min. of CO_2 was used.

This includes the detection efficiency of about 30 %, which reflects losses during the ion transport from the ion source to the collector. The detection efficiency was approximated by measuring the ratio of the collected argon ion current to the total current extracted from the source with a continuous flow of argon. We hope to improve the efficiency by utilizing an electrostatic or magnetic lens to focus the ion beams. Also, since there are several different extraction geometries available, they will be investigated for best transmission characteristics.

The dual Faraday cups were constructed to measure the ratio of carbon-13 to carbon-12. Good stability in the measurement could imply that production of the carbon-14 beams would also be stable. With a continuous flow of pure CO_2 and with a continuous flow of a mixture of CO_2 and argon (16 % CO_2 by volume), the microwave ion source produced $^{13}C/^{12}C$ ratio that was stable to within 0.1 % for over two hours after it reached thermal equilibrium. It took approximately one hour for the ion source to reach thermal equilibrium.

When a pulse of CO_2 gas (5.7 μmole) was injected into the stream of argon carrier gas, the ratio of integrated carbon-13 and carbon-12 current peaks for each pulse was calculated using the data processing method described by Ricci et al. [9]. This process was repeated for many pulses over a long period of time to investigate the long-term stability of the obtained isotope ratio in the pulse mode. Figure 4 shows the $^{12}C^+$ and $^{13}C^+$ ion current peaks resulting from the injection of a CO_2 pulse entrained in the argon carrier gas into the ion source. It took approximately 5 min. for the current peaks to appear after an injection of CO_2 into the carrier gas stream.

The result of the ratio measurements is shown in Figure 5. The CO_2 pulses were injected into the argon carrier gas stream once every 25 minutes (so that the

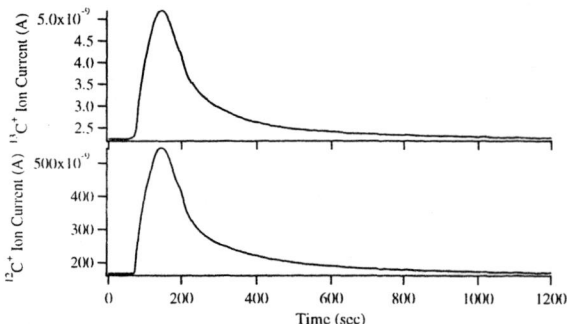

Figure 4: Positive carbon-12 and carbon-13 ion peaks resulting from small amounts of CO_2 gas injected into the microwave ion source. The background levels were 167 nA for $^{12}C^+$ current and 2.24 nA for $^{13}C^+$ current. Peak currents are low due to an aperture of 1 cm diameter which has been placed at the entrance to the 90-degree spectrometer magnet in order to shape the ion beams entering the magnet. Without this aperture, the two Faraday cups would have measured different currents for the same ion beam because the beam size was larger than the entrance area of the magnet.

carbon ion currents come down to the baseline level before the next current peaks start), and the data were taken over the period of 22 hours. The isotope ratio varied most at the beginning of the measurements but then stabilized. But, the ratio started to increase near the end of the experiment. The relative standard deviation of the obtained ratios is 1.5 %. We are hoping to achieve a similar stability for $^{14}C/^{12}C$ ratio measurements when we install the microwave ion source on the AMS system.

CONCLUSION

Positive ion beams from the microwave ion source were investigated for AMS applications. It is found that a stable C^+ ion beam can be produced from CO_2 gas. The overall efficiency of producing positive carbon ions from CO_2 molecules was approximately 1.7 %, including the detection efficiency of about 30 %. The microwave ion source produced stable $^{13}C/^{12}C$ ratios for both a continuous flow and pulses of CO_2, which is promising for the final goal of measuring the radiocarbon isotope ratio. After optimizing the ion beams, the next step will be to construct an efficient charge-exchange cell and produce negative ions for injection into the accelerator mass spectrometer.

ACKNOWLEDGEMENT

We thank the Isotrace Laboratory of the University of Toronto for making the ion source available to us

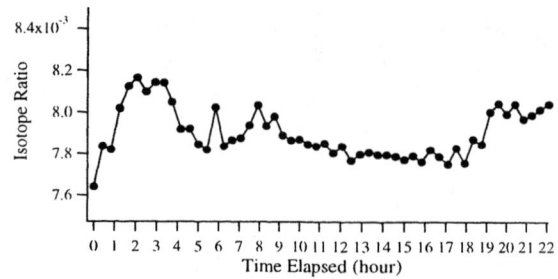

FIGURE 5: Ratios of integrated positive carbon-13 to carbon-12 ion peaks resulting from small amounts of CO_2 gas injected into the microwave ion source. The time between two pulses was 25 minutes.

and Argonne National Laboratory for loaning us the spectrometer magnet. This work is partially supported by a Senior Technical Staff Award to one of us (RJS) by Woods Hole Oceanographic Institution. The National Ocean Sciences AMS (NOSAMS) Facility is supported by the National Science Foundation, under Co-operative Agreement OCE-9807266.

REFERENCE

1. Schneider, R. J., von Reden, K. F., Wills, J. S. C., Diamond, W. T., Lewis, R., Savard, G., and Schmeing, H., *Nucl. Instr. and Meth.* B **123**, 554-557 (1997).

2. Schneider, R. J., Hayes, J. M., von Reden, K. F., McNichol, A. P., Eglinton, T. I., and Wills, J. S. C., *Radiocarbon* **40**, 95-102 (1998).

3. Schneider, R. J., von Reden, K. F., Hayes, J. M., Wills, J. S. C., Kern, W. G. E., and Kim, S-W., *Nucl. Instr. and Meth.* B **172**, 254-258 (2000).

4. Wills, J. S. C., Lewis, R., Diserens, J., Schmeing, H., and Taylor, T., *Rev. Sci. Instr.* **69**, 65-68 (1998).

5. Ferry, J. A., Loger, R. L., Norton, G. A., and Raatz, J. E., *Nucl. Instr. and Meth.* A **382**, 316-320 (1996).

6. Ramsey, C. B., and Hedges, R. E. M., *Nuc. Instr. and Meth* B **123**, 539-545 (1997).

7. Tykesson, P., "The Production of Negative Heavy Ion Beams Through Charge Exchange Processes" in *Symposium of Northeastern Accelerator Personnel*, Oak Ridge, TN, 1978.

8. Sakudo, N., "Microwave Ion Sources," Ch. 11 of *The Physics And Technology Of Ion Sources*, Edited by Brown, I. G., New York: John Wiley & Sons, 1989, pp. 229-244.

9. Ricci, M. P., Merritt, D. A., Freeman, K. H., and Hayes, J. M., *Org. Geochem.* **21**, 561-571 (1994).

SECTION IV

MATERIALS ANALYSIS WITH ION BEAMS, PIXE, RBS, ERD, NEUTRONS, AND NAA MICROBEAMS, OTHER TECHNIQUES

Practical Limitations of GUPIX

J. L. Campbell and J.A. Maxwell

Guelph-Waterloo Physics Institute, University of Guelph, Guelph, Ontario, Canada N1G 2W1

Some pitfalls and limitations of the GUPIX software are examined. An improper choice of options by the user can result in erroneous outcomes. Imperfections in the atomic physics database limit accuracy of analysis, especially when L X-rays are concerned. The very simple background approach used in GUPIX worsens precision for very small peaks. Improvements in progress are sketched.

INTRODUCTION

While GUPIX [1-3] now offers much versatility (thin, thick, layered specimens; excitation by H, D, He beams) and sophisticated options (invisible elements, oxide constituents, etc), there are pitfalls for the unwary or inexperienced user, and limitations in the database and fitting procedure that users need to bear in mind.

ERRONEOUS OUTCOMES

Although the channel (c)-versus-energy (E) calibration of an energy-dispersive X-ray spectrometer is usually linear, a quadratic term was added at an early stage in the development of GUPIX because of reports of possible very small departures from linearity in pulse processors. The peak-width (σ) - versus- $E^{0.5}$ relation is linear, reflecting the role of Fano statistics and system noise. There are thus 5 parameters p_1 - p_5 to be determined If the spectrometer system is operated at fixed settings, then these could be determined just once, and held fixed in the fitting software. However, it was part of the design philosophy to define all five as possible variables of the non-linear fit, in order to provide flexibility and avert small misfits arising from day-to-day fluctuations. We expected p_3 to be rarely needed and, if needed, to be small. The GUPIX documentation advises fixing p_3 at zero value unless it is indubitably necessary to do otherwise; it advises determining the constant c in the relation $p_5 = c\, p_2^2$, and then fixing p5 accordingly

Good estimates of the p1 - p_5 must be input by the user if GUPIX is to converge. The simplest case of GUPIX producing erroneous energy calibration parameters is with blanks inserted at intervals in a specimen suite. The resulting detection limits will also be wrong. If peaks in the blanks are few and weak, then only p_1 and p_2 should vary: if they are non-existent all five parameters must be fixed.

The adoption of a simple top-hat filter convolution to remove continuous background, another element of the early GUPIX design philosophy, was intended to avoid the complexities of multi-parameter analytical backgrounds where the determination of starting estimates turned out in practice to be a difficult issue. The top-hat has a central positive lobe whose width is set at the value of peak FWHM that prevails at the centre of the spectrum, and symmetric, negative outer lobes each of half that width. Obviously the top-hat filter will be most reliable when the background curvature is low enough to preclude significant departure from linearity over several peak widths. Caution is needed when the continuum has significant curvature.

An aerosol spectrum, collected by a GUPIX user using a 2% aperture "funny" absorber, provides an example of concurrent pitfalls involving p_3 and background removal. The lower limit of the region of fit was defined to include the bremsstrahlung hump to 0.6 keV, enabling the inclusion of the elements Na, Mg, Al, Si in the fit element list. The user found poor fits until he introduced a finite p_3 parameter. The fit thus achieved is shown in the upper panel of Fig 1; it is quite good, but a fitted FWHM of 220 eV at the spectrum centre and high peak areas for Al and Si must cast doubt upon it.

We set p_3 to zero and curtailed the region of fit to start at 1.7 keV, thus excluding the steep downward trend of the continuum, and we removed Na, Mg, Al and Si from the element list. The fit (lower panel) was better but, more importantly, the central FWM was at the expected 160 eV. It is notable that the erroneous initial fit had overestimated the sulfur concentration by 100% and that of iron by 30%.

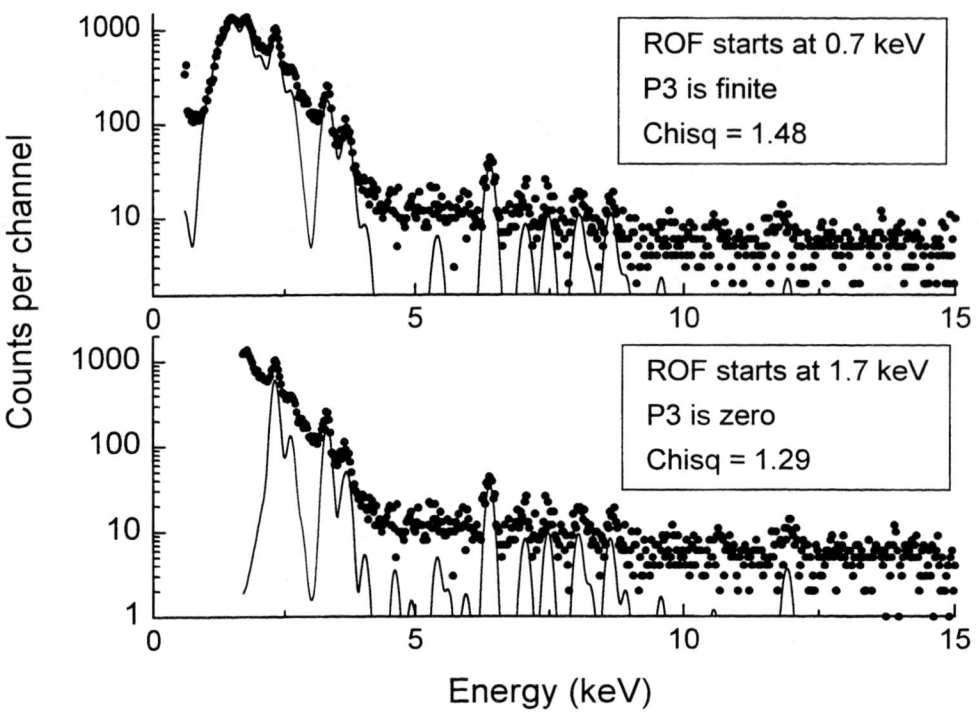

FIGURE 1. GUPIX fits to an aerosol spectrum, exemplifying problems due to background curvature

This leaves the question of whether the light elements can be quantified. We fitted a short region (0-.9 - 2.7 keV) that included only the K lines of Na through S, and observed that the peak width parameters changed significantly and the sulfur concentration increased by 16% We therefore fixed all 5 parameters at the previously determined values and as a result the sulfur concentration was similar to that determined from the main fit, and the Al and Si concentrations did not appear anomalous. This example thus demonstrates various pitfalls concerning the calibration parameters.

IONIZATION CROSS-SECTIONS

The high accuracy with which various of the necessary atomic physics quantities are known enables GUPIX to take a "fundamental parameters" approach, in which the only spectrometer variable is an instrumental constant H [3] that is effectively the detector solid angle multiplied by a smooth function with low dependence upon X-ray energy that represents any imperfections in the database and the detector description.

The upper panel of Fig 2 shows (squares) the quantity H determined with 3 MeV protons and thick

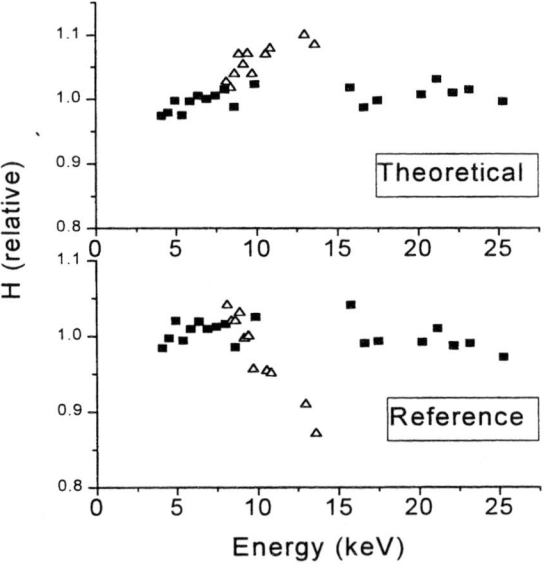

FIGURE 2. H-values for K X-rays (squares) and L X-rays (triangles): from [3] with permission of Elsevier Science

single-element standards, assuming the ECPSSR-DHS cross-sections which employ Dirac-Hartree-Slater atomic wave functions. It is indeed constant. But the corresponding H values (triangles) for L X-rays of heavy elements depart systematically from this trend. Orlic [4] has created a set of reference L subshell cross-sections by compiling measured values, applying statistical selection criteria, computing the average of selected values, and determining the ratio of these averages to the predictions of the basic ECPSSR-H model with hydrogenic wavefunctions. This is an approach that was applied earlier to the K-shell case by Paul and Sacher [5]. We therefore used these K and L reference cross-sections to create the lower panel of Fig 2; our own variation upon Orlic's interpolation scheme is described elsewhere [3]. We observe little difference to the K shell data, but an L-shell discrepancy persists, albeit with altered shape.

This shows the pitfall of adopting a uniform H for both K and L X-rays. Until the L shell cross-section situation is improved, GUPIX users should measure H for the L-shell cases pertinent to their application and should input the values as an ascii file into GUPIX. Despite the plethora of measured cross-section data, more effort is required to generate highly refined L subshell data for the simplest proton and helium cases. Such an effort needs to be supported by equally sophisticated attempts to improve our knowledge of L subshell fluorescence and Coster-Kronig yields; it is remarkable that most cross-section experimenters still have to rely upon a 26-year old empirical fit [6] to these quantities.

INACCURATE ERROR ESTIMATES

Despite the simplicity of the top-hat filter for background removal, many tests of GUPIX on standard reference materials have demonstrated accuracy in determining concentrations. The principal drawback of the top-hat approach may well be that it generates larger error estimates than would a polynomial approach. This is illustrated via a suite of pentlandite (FeNiS) mineral specimens containing varying tellurium (Te) concentrations. Fig. 3 shows a typical PIXE spectrum.

Fig. 4 shows, for different Te concentrations, the significant reductions in the percentage uncertainty in the Te Kα peak area (or concentration) as the extent of the top-hat exterior lobe is increased by various factors relative to the default value. The default filter of constant dimensions, optimized for the centre of the spectrum, significantly degrades experimental uncertainty in cases of small peak intensity. In addition for the cases of 360 and 180 peak areas, the default filter generates areas that are 10-15% too low.

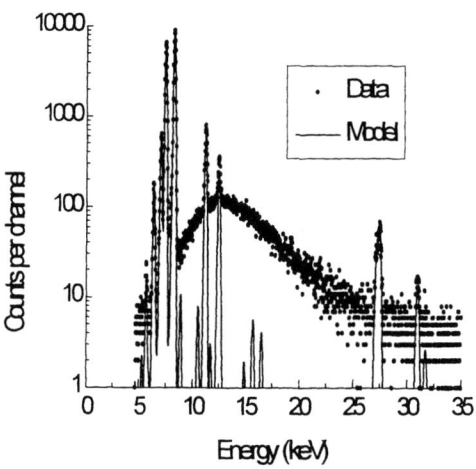

FIGURE 3. Pentlandite spectrum showing Te K peak at 27.4 keV

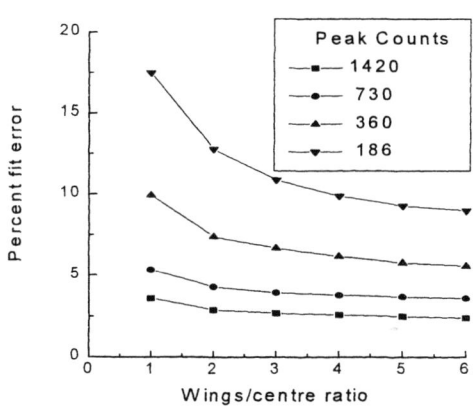

FIGURE 4. Peak area error versus relative dimensions of top-hat filter

We have developed a version of GUPIX (VDF, for variable digital filter) in which the top-hat width is not fixed at the centre-spectrum value but instead increases with channel number so that it is close to McCarthy's optimum recipe [7] at every channel. But this goes only part-way to deal with the above problem. Therefore, a version will be released wherein the user can adopt the standard-recipe VDF in regions of clustered peaks (usually at low energies), a narrower filter in regions of high spectrum curvature, and a filter with three times wider outer lobes than the VDF recipe in regions of low curvature and low peak intensity. This approach will markedly reduce the problems sketched here.

X-RAY DATABASE ISSUES

An element's K X-ray spectrum contains diagram lines and less intense non-diagram lines, plus spectrum artefacts such as the escape peak and low-energy tailing features. PIXE analysts generally assume that the energies and relative intensities of the diagram lines are accurately known, they accept that there is uncertainty in the small radiative Auger contributions, and they make varying attempts to determine low-energy tailing contributions and include them in GUPIX fitting. Scofield's Dirac-Fock theory [8], with known exceptions for the 3d transition elements, does provide the relative intensities with high accuracy. But, as an independent-particle model approach, it does not account for the $K\beta'$ satellite [9] in the important $21<Z<29$ region. PIXE analysts usually ignore this feature. Its causes are debated but must include the multiplet effect caused by the coupling of the 3p vacancy to the open 3d shell. Around $Z=25$ the satellite lies nearly 15 eV below the $K\beta1$ energy. Literature values of its relative intensity display considerable scatter but range up to 25-30%, and also show a dependence upon valence state. This explains in part why energy and width calibrations in GUPIX that are dominated by elements such as iron can be sufficiently erroneous as to cause peaks in the region above 20 keV to be grossly misfitted. The early work of Salem et al [10] on iron gave the the intensity ratio of the satellite to the diagram line as 34% and its shift 11.7 eV, while Holzer et al [11] recently reported a multiplet fitting approach that has a 39% component lying 14 eV below the generally used $K\beta1$ energy. The omission of this feature from the X-ray intensity database causes an error of ~4 eV in the energy assigned to the $K\beta1$ line. Consider a pyrite (iron sulfide) sample with traces of silver - a typical micro-PIXE mineralogical sample. With Fe $K\alpha$ and $K\beta1$ as the predominant determinants of the energy calibration parameters, then GUPIX could construct the silver K lines several tens of eV distant from their correct position in the pulse height spectrum. We are attempting now to refine the rather limited existing database on $K\beta1$ and to incorporate this into GUPIX. For the moment, weighting schemes available for the least-squares fit procedure in GUPIX provide means to avert such errors.

DETECTOR LINESHAPE

Finally, there is the ever-present issue of the low-energy adjuncts to any given peak, arising from X-ray escape and from partial escape of Auger and photo-electrons, from hot electron diffusion, and from fabrication problems. If a description of these adjuncts is not incorporated into the lineshape file in GUPIX, then there is a well-known pitfall. Elements in the fitting list whose peaks overlap "undescribed" tails can be accorded finite peak areas and concentrations as GUPIX attempts to fit the tail by the only means available to it. For example, we [12] showed elsewhere in PIXE analysis of zircons that the strong Zr tail could masquerade as a significant yttrium presence, in a context where yttrium concentration was one of the important parameters to be determined

Erroneous conclusions can be averted by paying critical attention to the results and by refraining from the inclusion in the fit list of elements that overlap with strong tails. But GUPIX will be a more powerful tool if the tails can be determined. If this task is attempted using K X-ray spectra of pure elements, the challenge of the poorly known radiative Auger features that overlap the $K\alpha$ and $K\beta$ tail features of interest immediately arises. If these features are simply considered part of the tails, then the tail estimates will be exaggerated and will not be applicable to L or M X-rays.

We are developing a simple package to assist GUPIX users in estimating these tail features via PIXE K X-ray spectra. It has of necessity to make various simplifying assumptions, especially as to the shape of underlying background, but it appears to provide a reasonable approximation to the tail features that is surely superior to neglecting them.

ACKNOWLEDGEMENTS

This work was supported by the Natural Sciences and Engineering Research Council of Canada.

REFERENCES

1. Maxwell, J.A., Campbell, J.L. and Teesdale, W.J., *Nucl. Instrum. Methods* **B43**, 218-230 (1989).
2. Maxwell, J.A., Teesdale, W.J. and Campbell, J.L., *Nucl. Instrum. Methods* **B95**, 407-421 (1995).
3. Campbell, J.L., Hopman, T.L., Maxwell, J.A. and Nejedly, Z., *Nucl. Instrum. Methods* **B170**, 193-204 (2000).
4. Orlic, I., private communication.
5. Paul, H. and Sacher, J., *Atomic Data Nucl. Data Tables* **42**, 105-156 (1989).
6. M.O. Krause, *J.Phys. Chem. Ref. Data* **8**, 307-327 (1979).
7. McCarthy, J.J., *Scanning Electron Microscopy* **1980/II**, 259-270 (1980).
8. Scofield, J.H. *Phys. Rev.* **A9**, 1041-1049 (1974).
9. Ekstig, B., Kallne, E., Noreland, E. and Manne, R. *Physica Scripta* **2**, 38-44 (1970).
10. Salem, S.I., Hockney, G.M. and Lee, P.L., *Phys. Rev.* **A13**, 330-333 (1976).
11. Holzer, G., Fritsch, M., Deutsch, M., Hartwig, J. and Forster, E., *Phys. Rev.* **A56**, 4554-4568 (1997).
12. Halden, N.M., Hawthorne, F.C., Campbell, J.L., Teesdale, W.J. and Maxwell, J.A., *Can. Mineral.* **31**, 637-647 (1993).

Applications Of PIXE With 68 MeV Protons

A. Denker, K. H. Maier

Ionenstrahllabor, Hahn-Meitner-Institut, Glienickerstr. 100, D 14109 Berlin, Germany

Abstract. The usual PIXE analysis with 2 to 4 MeV protons is limited to thin layers. The range of 68 MeV protons is large compared to the absorption length of X-rays, and 68 MeV protons produce abundantly K X-rays of heavy elements that are weakly absorbed. Therefore PIXE with high-energy protons is particularly suited for the analysis of thick objects. Simultaneously radiation damage is low and K X-ray spectra are more easily evaluated than the complicated L X-ray spectra. The composition, thickness and sequence of pigments in the layers of paintings has been measured. The results are compared with those from conventional "destructive" methods. The elementary composition of ancient metal objects below a thick cover of patina has been be analyzed.

INTRODUCTION

Unique or highly priced objects d'art may often only be examined by a truly non-destructive analysis. For analysis close to the surface, XRF or PIXE with 2 to 4 MeV protons are available (1,2). Both methods yield elemental compositions up to 100 µm of material. High-energy PIXE using 68 MeV protons permits non-destructive analysis up to several mm (3) due to the large range of high-energy protons, and the high X-ray production cross section at these energies, particularly for K X-rays of heavy elements (4,5) which are less absorbed than the L X-rays. Applications of high-energy PIXE to paint mock-ups representing ancient painting techniques, consisting of paint layers of several hundred micron thickness as well as the bulk analysis of metal objects are presented.

EXPERIMENTAL SET-UP

The 68 MeV proton beam from the ISL cyclotron leaves the vacuum of the beam-line through a 30 µm thin Kapton foil at the end of a 7 cm long and 1 cm diameter aluminum tube. The distance from the exit-foil to the target in air is 11 cm. The energy loss of the protons in the Kapton foil and the air is about 150 keV. The beam is focused to a diameter of 1 mm on a luminescent, artificial ruby. After focussing, the intensity of the beam is reduced to typically 0.5 to 1 pA, using slits far upstream the beam-line. The exact size can be controlled by measuring the X-rays from a thin wire, e.g. 100 µm tantalum, that is moved across the beam. Difficulties with sample charging does not occur in normal atmosphere.

Two detectors are mounted at 135° with respect to the beam axis: A 12.5 mm^2 Si(Li) detector, 155 eV resolution at 5.9 keV, measures the L X-rays of the heavy and K X-rays from the light elements. A 300 mm^2 HPGe, 180 eV resolution at 5.9 keV, measures the K X-rays of heavy elements and γ-rays eventually occurring from nuclear reactions. Both detectors are shielded with two 2 cm thick tungsten blocks and a 5 mm thick aluminum sheet against radiation from the exit foil.

The objects to be measured are mounted on an x-y table that can support 50 kg with a positioning precision of 0.1 mm. A laser beam can be inflected onto the beam axis marking the beam spot on the object. The objects and the beam spot are viewed by a color TV camera, and its images can be stored on video tape to document the measured spots.

The complete system is remote controlled by a PC, allowing an automatic scanning or mapping of the object. After the experiment, the line intensities of the different X-ray lines are determined by using the evaluation code QXAS/AXIL (6).

APPLICATIONS

Paintings

At the Kunsthistorisches Museum in Vienna, 12 paint test mock-ups, simulating traditional painting techniques, have been prepared (7) to study what can be learned by high-energy PIXE on paintings. After the PIXE examination small specimens were cut out and their cross-section perpendicular to the surface has been investigated under a microscope. The data from PIXE were compared to the results from this cross-section analysis which is a well established method. Besides merely detecting the presence of the characteristic elements of the pigments, the sequence of paint layers has been estimated from the PIXE data, as follows: The 68 MeV proton beam looses only about 1 MeV in the paint layers. Hence, the X-ray production cross section is independent of depth. The various X-rays from one element, like K, Lα, or Lβ, are absorbed to a different degree on their way out of the sample. Therefore the measured intensity ratio gives a qualitative information on the depth of the emitting layer. The composition of the absorbing layers is not known; therefore only a nominal depth, namely the thickness of a hypothetical $CaCO_3$ layer is evaluated, reproducing the measured absorption. Since the sequencing of paint layers is of greater concern than the layer thickness, which is strongly dependant on the brush stroke, an exact determination of the layer thickness is not required.

The irradiation time was 200 s for all measured points. To check that the method is truly non-destructive, the radiation dose was increased by a factor of 400 for one mock-up. This beam spot was investigated using optical, Fourier-Transformed-Infra-Red, and Scanning-Electron microscopy. No difference between surrounding and irradiated material could be observed.

The results from the most simple paint test mock-ups, composed of successive homogeneous paint layers containing only one pigment each, have been evaluated before and the results have been published (7). All pigments could be detected applying high-energy PIXE by observing the X-ray lines of at least one element characteristic for each pigment, except ultramarine, carbon black and organic dyes, as these pigments do not contain any element heavier than sulfur. In addition, the sequence of the paint layers in these mock-ups was shown to correspond with the results obtained from cross-section analysis. But the paint layering structure in most real paintings is significantly more complex than the above cited examples. In the following two painted mock-ups with different complexity will be discussed.

Mock-up 2/III

This paint mock-up (17th century Northern painting technique on wood panel) represents a structure with four layers, and all layers vary in thickness and contain mixtures of different pigments. The characteristic elements are given in brackets in the following. The ground layer is composed of a mixture of chalk (Ca) and lead-white (Pb). It is covered by a paint layer of lead-white (Pb), carbon-black(/), and umber-paint (Fe, Mn), followed by a verdigris (Cu)/carbon-black (/) glaze. The painting is coated with a mastic resin (varnish)(/).

PIXE analysis was performed at two positions: Position 1 includes the verdigris/carbon black glaze layer, whereas on position 3 the glaze is absent. The resulting PIXE spectra, shown in Fig. 1, clearly distinguish between the two positions, showing no evidence of Cu for position 3. Interestingly at position 3, Ca X-rays have been detected where in fact they should have been masked by the overlying paint layers. This suggests that either the surface at the selected position must have been somewhat abraded providing a channel for the X-rays to exit, or that Ca is present in an overlying paint layer. In both positions, the elements characteristic for the paint, could be detected: The umber of the paint layer can be identified by detecting Mn together with Fe lines. The Pb lines, slightly differing in intensity at the two positions, originate from the ground and paint layer.

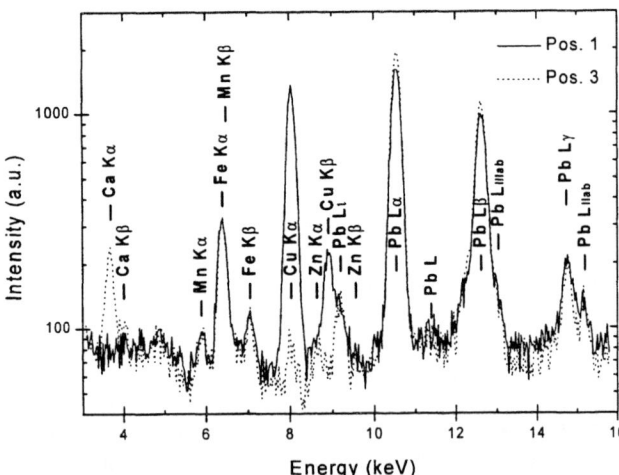

FIGURE 1. High-energy PIXE spectra of mock-up 2/III. The solid line represents the spectrum measured on the complete sequence (ground, paint layer, glaze and varnish; position 1) and the dotted line at position 3 where no glaze is applied.

The evaluation of the intensity ratios of the Kα/Kβ lines show that at position 1 Cu is clearly present in the upper layer in a nominal depth of 14 μm and that for Fe and Pb the relative depths are similar of around 74 μm. This strongly suggests that Fe is located at the same or slightly lower depth as Pb (a Pb layer on top of the Fe would influence strongly the absorption– altering the intensity ratio of the Fe lines in the spectra). For Mn and Ca no depth information could be obtained, as the statistic of the Kβ lines is not sufficient. Providing that adequate knowledge concerning historical painting technique within a certain period (17th century) and location (Northern Europe) is at hand, the presence of a chalk ground, followed by an umber/lead white paint layer and a verdigris glaze can be derived. The addition of different amounts of carbon black to the glaze can be seen directly on the painting surface without further investigation. The cross-section analysis confirmed the derived sequence. For this mock-up, high-energy PIXE revealed the correct sequence of paint.

Mock-up 3/II

This mock-up, the most complex one, involves a preparation technique used by Rembrandt and Claude Lorraine. To one part of a copper support a lead-white (Pb) ground was applied. The entire surface including the bare copper support (position 1) and the ground (position 2) was then covered with a thin tin foil. This was followed by a paint layer containing azurite (Cu), smalt (Co), lead-white (Pb), and lead-tin-yellow (Pb, Sn). The uppermost layers consisted of highlight applications with lead-white/lead-tin-yellow (Pb, Sn) and a mastic resin (/).

X-rays from all characteristic elements are observed in Fig 2. Only for Cu, Sn, and Pb are the lines distinct enough to be able to calculate the nominal depth of the elements from the intensity ratios. At both positions, the following sequence was obtained from the intensity ratios: Pb in a nominal depth of 140/150 μm (pos. 1) and 98/68 μm (pos. 2), followed by Cu in a depth of 207 μm (pos. 1) and 245 μm (pos. 2), and Sn at 600 μm respectively 980 μm. For Pb two depths have been given, one derived from the Lα/Lβ intensity ratio and the other from Lα/Kα. For position 1, Pb has correctly be found being on top (in highlight and paint layer), and Cu being at a larger depth (being in the paint layer). But for Sn, the presence in the top layers is masked by the strong signal from the metal foil and therefore one must recognize that in such a circumstance, usable PIXE results can only be achieved by already having a somewhat clear idea of the layering sequence, and possible pigment content, i.e. through microscopic investigation of the paint layers where the various strata are exposed, and through comparison with traditional practice as described in the appropriate literature of the time. The same result is obtained for position 2, but as the nominal depth of Pb at position 2 differs clearly, when the Lα/Kα or Lα/Lβ ratio is evaluated, this indicates that at least two layers containing Pb are present at significantly different depths.

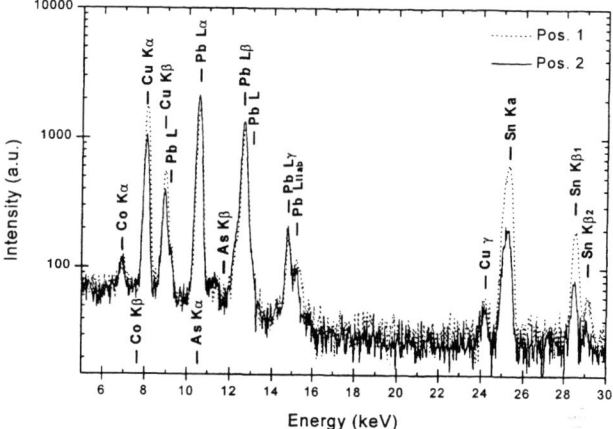

FIGURE 2. High-energy PIXE spectra of mock-up 3/II at two positions.

Italian Plaquettes

The Sculpture Collection of the "Staatliche Museen" in Berlin possesses around 1000 Italian plaquettes. They form one of the worlds largest collection of this type of small scale reliefs, although some of them have been badly damaged or even disappeared in the second world war. Fig. 3 shows an example. Plaquettes had been very fashionable as collectors items during the 15th and 16th centuries. The present measurements aim to determine the composition of the copper alloys, in order to establish a metallurgically correct designation and classification of the Berlin plaquettes. The color of the plaquettes ranges from red through brown to nearly black, some are covered with patina or plated with a different metal. Therefore a surface analysis is not sufficient. On the other hand the pieces are also too valuable to remove specimens. In this case, high-energy PIXE is an adequate, fast and non-destructive analytical method.

765 plaquettes have been measured, some on front and back, or at several points, resulting in more than 1000 spectra. As one could see already during the measurements, the color of the surface gives no hint to the composition in the bulk. The X-ray spectra were evaluated in a rather qualitative way: The intensity of the measured Kα lines was determined and compared to those from brasses and bronzes of known composition to

take the absorption of the X-rays in the plaquettes into account. The error in the determination of the concentration of the main components, that are of interest for the classification, proved to be smaller than 5%. It was assumed, that no light elements like Al that cannot be detected due to the strong absorption of their low energy lines, were used in these old metal objects.

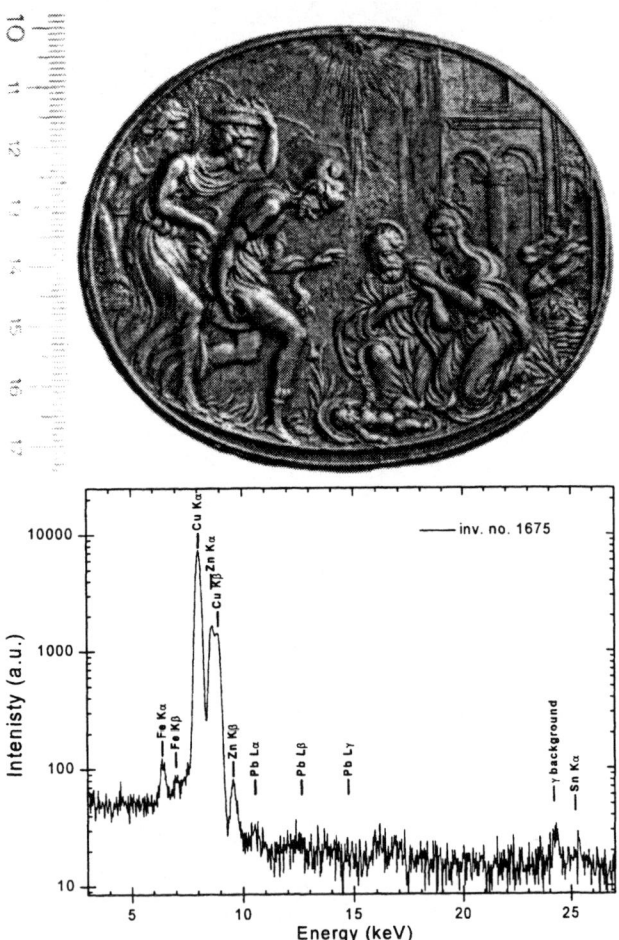

FIGURE 3. . Gian Giacomo Caraglio, The Adoration of the Shepherds, 1540-50. Brass, 74.4 x 90.3 mm. Berlin, Sculpture Collection, inv. no.1675, zinc brass, photograph and spectrum.

So far, art historians had assumed that most of the objects are made of regular tin-bronzes. However, the results of these measurements shows, that only 9 % of the investigated objects are made of bronze; the others belonging to various alloys, even a few iron and two silver objects have been found.

CONCLUSIONS

The analysis of thick paint layers and the plaquettes has shown the capabilities and limitations of high-energy PIXE. Information can be gained on the elementary composition of samples deep below the surface, particularly for heavy elements; this is a fairly unique feature. These measurements are fast and completely non-destructive. Therefore extensive surveys can be performed on very precious objects. However, a quantitative evaluation of the X-ray spectra is usually not possible, because of the absorption of X-rays in the sample, of generally unknown composition, that is inherent to the study of thick objects. The different absorption of the various lines from one element, leading to changes in the intensity ratios of these lines, permits nevertheless a qualitative determination of the depth distribution. Art historians might use these properties to complement the information from other examinations.

ACKNOWLEDGMENTS

A part of this work was performed within the European COST G1 framework "Ion beam analysis applied in art and archaeometry". The authors are indebted to M. Griesser, H. Musner, Kunsthistorisches Museum Vienna, for the fruitful collaboration with the paint mock-ups. M. Schreiner, D. Jembrih, Academy of Fine Arts and R. Erlach, S. Fischer, University of Applied Arts, are thanked for the investigations using FTIR and SEM.

REFERENCES

1. Campbell, I. L., and Johansson, S. A. E., *A novel technique for elemental analysis*, New York: J. Wiley and Sons, 1988

2. Schreiner, M., Mantler, M., Weber, F., Ebner, R., Mairinger, F., *Advances in X-ray Analysis* **35**, 1157-1163 (1992)

3. Denker, A., and Maier, K. H., *Nucl. Instr. Meth.* **B 150**, 118-123 (1999)

4. Paul, H., and Sacher, J., *At. Data and Nucl. Data Tables* **42**, 106-156 (1989)

5. Pineda, C. A., and Peisach, M., *Nucl. Instr. Meth.* **A 299**, 618-623 (1990)

6. Bernasconi, G. A.., *QXAS User Manuals*, IAEA Laboratories Seibersdorf, 1993

7. Denker, A., Griesser, M., Maier, K. H., and Muser, H., "Investigation of Paint Test Samples by High Energy 68 MeV PIXE" in *ART '99 - 6th Int. Conf. on Non-Destructive Testing and Microanalysis for the Diagnostics and Conservation of the Cultural and Environmental Heritage*, edited by M Marabelli et al., Proceedings., Rome, Italian Society for Non-Destructive Testing, 1999, pp. 983-997

A FORENSIC APPLICATION OF PIXE ANALYSIS

*I.I. Kravchenko, *F.E. Dunnam, and *H.A. Van Rinsvelt,
**M.W. Warren, and **A.B. Falsetti

*P.O.Box 118440, Department of Physics, University of Florida, Gainesville, FL32611
** Department of Anthropology and C. A. Pound Human Identification Laboratory
University of Florida, Gainesville, Florida 32611

Abstract. PIXE measurements were performed on various calcareous materials including identified bone residues, human cremains, and samples of disputed origin. In a forensic application, the elemental analysis suggests that the origin of a sample suspectly classified as human cremains can tentatively be identified as a mixture of sandy soil and dolomitic limestone.

INTRODUCTION

The incidence of cremation as a method of disposal of the dead is increasing. Nearly 25% of deaths are cremated in the United States (1). As the incidence of commercial cremation increases, there has been an associated increase in civil litigation involving the cremation industry. Legal issues usually involve improper disposal of cremated remains (cremains), questioned identity, and commingling of more than one individual. Since cremated remains are chiefly calcined skeletal fragments, forensic anthropologists are usually the experts consulted.

Commercial cremation involves the incineration of the body at temperatures reaching 600 degrees centigrade. After incineration, only the calcined, inorganic component of the skeleton remains. A pulverizer, or processor then further reduces these bone fragments, until they consist of extremely small fragments suitable for scattering or inurnment. Older types of processors render cremains to particles less than 4 mm. In these cases, artifactual or bony evidence of identity may be found among even thoroughly processed cremains. However, newer cremation retorts and processors render cremains to a fine ash. In these cases, little is found among the cremains that offer evidence as to the identity of the deceased. The newest types of rotary-blade processors may render cremains to a point where it is difficult for the anthropologist to determine if the ash is bone or some other material. In these cases, elemental analysis of the cremains may provide evidence that the cremains are indeed bone.

Case Report

During July of 2000, an urn reportedly containing the cremains of an elderly female were presented to the faculty and staff of the C.A. Pound Human Identification Laboratory at the University of Florida. The cremains were a source of ill will among the decedent's children, both of whom wanted custody of the remains of their mother. The decedent's son, the legal executor of the decedent's estate, suspected that the remains given him by his sister were not the mortal remains of his mother.

On inspection, the urn was filled with an appropriate volume of fine-grained material. Microscopic examination revealed a homogenous mixture of particulate matter, but no bone structure was noted. Instead, a large percentage of dense, globular semi-transparent material was present. This material was radiographically opaque – more so than bone. Although the fragments under investigation were extremely small, the absence of bony structure led us to the preliminary conclusion that the material was not the product of human cremation, but a combination of sandy soil or other material. The extreme fragmentation of the cremains precluded a firm, final conclusion. This case offered an outstanding opportunity to test the application of PIXE elemental analysis in a forensic context

EXPERIMENTAL

PIXE spectra were obtained with 2.5 MeV protons from the 4MV Van de Graaff accelerator at the University of Florida. The beam was defined by a series of graphite collimators and defocused in such a way that its diameter on the target was approximately 5 mm. The beam intensity was kept below 20 nA in order to avoid overheating and charge build-up on the sample. The X-rays were detected in a 30 mm^2 x 3 mm thick Kevex Si(Li) detector located inside the vacuum chamber, in the horizontal plane, and making an angle of 135° with the incident beam direction. The solid angle of the detector was defined by means of a high purity aluminum collimator and data were obtained with a "funny" filter: a 660 μm thick Mylar absorber with an axial aperture of 1% area inserted between the sample and the detector. A Kevex 4525P amplifier/pulse processor was used in conjunction with the detector. The resolution obtained with the detector-pulse processor combination is 180 eV for the Mn K$_\alpha$ line. Having passed through the target, the beam is dumped in a Faraday cup for integration. The preset charge was typically up to 20 μC.

Kimfoil™ was chosen as backing for the PIXE targets. To prepare good quality thin targets where charge build-up and matrix effects are minimized, a spin-drop technique was utilized. Fixation of the sample was achieved with a 1% solution of polystyrene in benzene where some amount of powdered sample (typically 0.5 gram per 4 ml of the solvent) was suspended. In the next step, this liquid sample suspension was pipetted in small increments onto the Kimfoil backing rotated in a spin-coater. The spinning velocity is adjusted in such a way that the solvent is thrown off while at the same time the sample particles are left on the backing surface. Excessive amount of polystyrene residue produces visible charge build-up and related bremstrahlung background. If the target is properly prepared, no charging effects are observed. PIXE runs on blank targets (polystyrene without sample material on Kimfoil substrate) revealed no contaminants other than a trace amount of Zn.

Analyses of the PIXE spectra were carried out using Pixfit [2]. The sample targets contained on the surface some grains of bone material which would violate thin target assumption and cause an error in concentration evaluation due to self-absorption that is difficult to evaluate. Therefore, only qualitative results were used in this particular case.

In order to develop familiarity with the PIXE spectra from calcareous substances of this type as well as to establish the reliability of our target preparation methods, a variety of target materials was examined. Included were soil samples, soil amendments [e.g. limerock, dolomitic limestone, bone meal] and other related materials.

RESULTS AND DISCUSSION

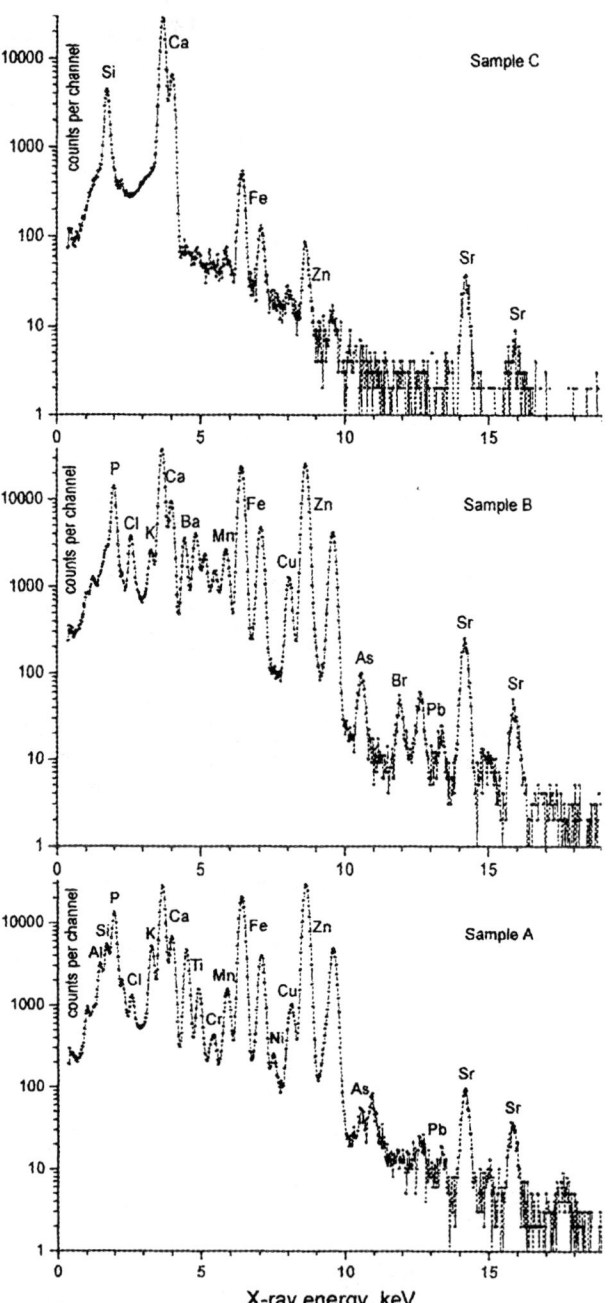

FIGURE 1. PIXE spectra of the samples studied.

TABLE 1. Relative elemental composition of the investigated samples (average error is +/- 10%).

Element	Sample A Counts/total beam charge	Sample B Counts/total beam charge	Sample C Counts/total beam charge
Al	584	0	68
As	13	13	0
Br	2	12	0
Ca	11316	11155	11848
Cl	125	683	14
Cr	67	184	7
Cu	310	420	7
Fe	8968	7568	228
K	1852	574	171
Mn	586	611	15
Na	84	0	0
Ni	52	0	0
P	4084	3305	79
Rb	6	3	0
Se	8	0	0
Si	1415	412	1301
Sr	53	577	17
S	254	80	20
Ti	1758	129	0
V	380	0	0
Zn	12973	8377	35

Three samples were the primary subjects of this investigation. Sample A represents contemporary cremation done with the newest types of retort (furnace). Sample B was taken from cremains incinerated, "hand processed" and stored in a brass urn since 1962. Sample C represents the disputed material described in the introduction. A number of targets was bombarded and the averaged PIXE spectra for the materials analyzed in this study are presented in Figure 1. The elemental composition obtained from the spectra is summarized in Table I. A striking similarity is observed in the chemical composition of samples A and B while sample C has a distinguishably different elemental pattern. The latter is apparently composed mostly of calcium and silicon, the transition metals are at residual concentration levels, and such trace elements as bromine and rubidium are completely absent. The difficulty of human cremains identification lies in the fact that their chemical composition varies and depends on many factors such as the level of trace elements in different individuals, alteration of the element ratios due to the incineration process, contamination of the cremated remains during the collection and storage procedures, and even the medical history of the deceased. For example, sample A contains an elevated level of silicon, which was most likely introduced into the cremains as a result of scraping the refractory material of the furnace liner. We also want to point out some other curious peculiarities of the studied human cremains. Compared to sample A, sample B contains sixty times as much barium. This estimation is derived from the BaKα line (not shown in fig.1). We can only speculate that this

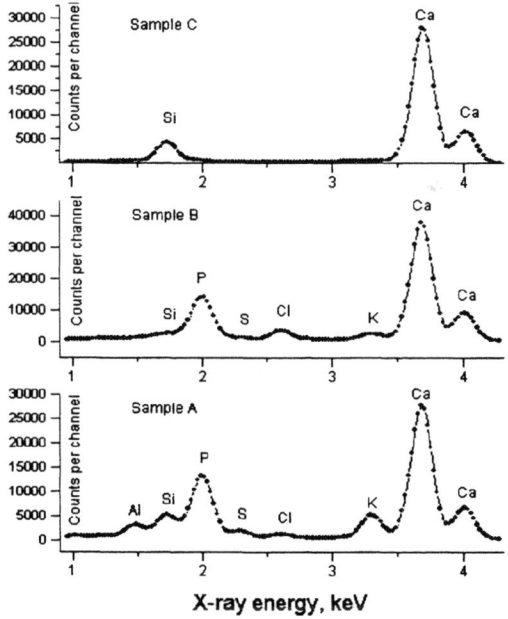

FIGURE 2. PIXE spectra of the low X-ray energy region

elevated Ba concentration may be related to a medical procedure that the deceased person had undergone prior to death. On the other hand, sample A has a much higher level of titanium: a close examination of the sample material revealed a small (~10mm) object subsequently identified as a partially oxidized titanium vascular clip that has most likely contaminated the sample.

It is obvious that the key element that would be a definite indication of real human cremains is phosphorus, a major constituent of bones. The phosphorus content in the identified cremains is much higher than in sample C where this element is at a residual level (see table I and fig.2). If we assume that the incineration of bone does not significantly change the P/Ca concentration ratio then this ratio would be also an additional indicator of skeletal remains. Indeed, a rough estimation of the P/Ca concentration ratio for samples A and B (around 0.4) is close to the bone ratio (around 0.44) provided by the Report of the Task group on Reference Man [3] confirming the suspicion that sample C does not represent human cremains. Additional PIXE test runs of dolomite, sand, limestone, and soil samples lead us to the tentative conclusion that sample C can be identified as dolomitic limestone with an admixture of sand.

ACKNOWLEDGMENTS

The authors would like to thank Dr. Robert Coldwell for valuable consultations concerning Pixfit software applications.

REFERENCES

1. Cremation Association of North America "Statistics for cremation rates in North America" http://www.cremationassociation.org/docs/state98.pdf

2. Coldwell R.L., H.A.Van Rinsvelt "Pixfit-A Special Analysis Program for PIXE" in *Application of Accelerators in Research and Industry-1997*, edited by J.L.Dugan and I.L.Morgan, AIP Conference Proceedings 475, New York: American Institute of Physics, 1997, pp. 555-558.

3. Report of the Task Group on Reference Man in *International Communications on Radiological* Protection, Publ.23, Oxford, Pergamon Press, 1975.

Study of the Elemental Composition of Yellow Pine Using Particle Induced X-Ray Emission (PIXE)

Changgeng Liao, William A. Hollerman. Gary A. Glass, and Richard Greco

Acadiana Research Laboratory, University of Louisiana at Lafayette, Lafayette, Louisiana 70504, USA

Abstract: It has been found that metals in woody tissue will influence the growth rate of trees. Utilization of particle induced x-ray emission (PIXE) for determination of trace element levels in biological systems is rather extensive. In this work, three rings taken from a 50-year-old Yellow Pine tree were PIXE analyzed with minimal sample preparation. Eight elements (potassium, calcium, titanium, chromium, manganese, iron, copper, and zinc) were studied during the PIXE analysis. The relationship between relative yield and tree ring thickness is presented for each detected element. These results show that wood taken from the oldest ring has the widest ring thickness and possesses the largest quantities of all the tested elements. Calcium appears to have the largest relative yield of all the tested elements and is roughly proportional to ring thickness. Heavier elements, such as lead and mercury, were not detected in any of the Yellow Pine samples.

BACKGROUND

There has been considerable interest in assessing environmental pollution using tree ring analysis. This method provides a historical snapshot of pollutant level near the tree. Since 1970, it has been used to determine the history of pollution levels [ref. 1-6]. This research has shown that metals existing in woody tissue will influence the growth rate of trees [ref. 7]. Utilization of particle induced x-ray emission (PIXE) to determine trace element yields in biological systems is rather extensive. However, most of the research on tree rings is focussed on assessing the pollution, not on investigating the elements that influence tree growth.

In this research, three tree rings of various years were analyzed using the PIXE technique for the desired eight elements, potassium, calcium, titanium, chromium, manganese, iron, copper, and zinc. The relationship between relative yield and tree ring thickness for these eight elements were studied for the selected rings.

EXPERIMENTAL PROCEDURE

The PIXE analysis completed in this research used the 5SDH-2 tandem Pelletron accelerator at the Acadiana Research Laboratory (ARL), which is located on the campus of the University of Louisiana at Lafayette. The experimental arrangement used in this study was described in a previous paper [ref. 8]. A beam of 3 MeV protons was used for the PIXE irradiation sequence. Two sets of collimators upstream of the target provided a measured beam diameter of 2 mm to each tested sample. X-rays emitted from the sample were detected using a Si(Li) detector with an active area 30 mm^2 and energy resolution of 180 eV (measured at 5.9 keV). The detector was positioned about 20 cm from the target, which makes an angle of 20 degrees relative to the incident proton direction. A 250 µm thick carbon foil was placed in front of the detector to reduce the intensity of low energy x-rays and bremsstrahlung photons. PIXE spectra were normalized to a total integrated beam charge of 10 µC. Beam current was maintained at less than 5 nA throughout each irradiation sequence.

Samples were taken from wood cut from a 50-year-old Yellow Pine (subgenus pinus) tree that was approximately 50 cm in diameter when it was cut down in 1995. The sample tree was located near the headwaters of the Upper East Fork Popular Creek at the Oak Ridge Y-12 Plant in Tennessee [ref. 9-10]. Each sample had no pre-analysis preparation, except for the addition of a 50 nm aluminum film evaporated on the surface to increase electrical conductivity. Individual samples were attached to an aluminum card (slightly larger than the ring segment) with a small amount of epoxy to minimize contamination. Each card was clipped to an aluminum sample holder inside the ARL macroscopic irradiation chamber for PIXE analysis.

It should be noted that the authors did not physically take the sample slice from the cut Yellow Pine tree in 1995. The selected slice was used in other analyses at the Oak Ridge Y-12 Plant [ref. 10]. Since this work involved pollutant levels in the parts per billion (ppb) range, the authors assumed that the Yellow Pine sample had minimal contamination and was suitable for this research.

A plot of the thickness for each tree ring as a function of year is shown in FIGURE 1. Each sample had a volume of 3,000 mm^3 and was cut along the axial direction of the ring. The incident proton beam was directed along the radial direction of the ring.

RESULTS AND CONCLUSIONS

The relative x-ray yield data for the eight tested elements is shown in TABLE 2. The relative yield for an element is proportional to the number of counts under the K_α x-ray peak divided by the equivalent value for the calcium K_α x-ray peak in sample A (1947). The relative yield was normalized for production cross section and fluorescence yield.

These results show that sample A has the widest ring thickness and possesses larger quantities of all the tested elements. Calcium appears to have the largest relative yield of all the tested elements and is roughly proportional to ring thickness.

Heavier elements, such as lead and mercury, were not detected in any of the Yellow Pine samples.

Earlier atomic fluorescence research determined the mercury concentration in the Yellow Pine tree reached a maximum of 285 ppb [ref. 9-10]. This data was obtained using the Environmental Protection Agency (EPA) Method 1631 sampling protocol. Mercury was used at the Oak Ridge Y-12 Plant as part of the lithium isotope separation and purification process. During this period, a total of 330 metric tons of mercury were lost to the environment near the sample tree [ref. 9-10]. Currently, the mercury concentration in the Upper East Fork Popular Creek (near where the sample Yellow pine tree was growing) is less than 1 ppb [ref. 9-10]. The PIXE detection limit for mercury in a similar wood sample was determined to be less than 8 ppm at the start of this analysis. Neutron activation analysis for several Yellow Pine samples will be completed in the near future to correlate these results.

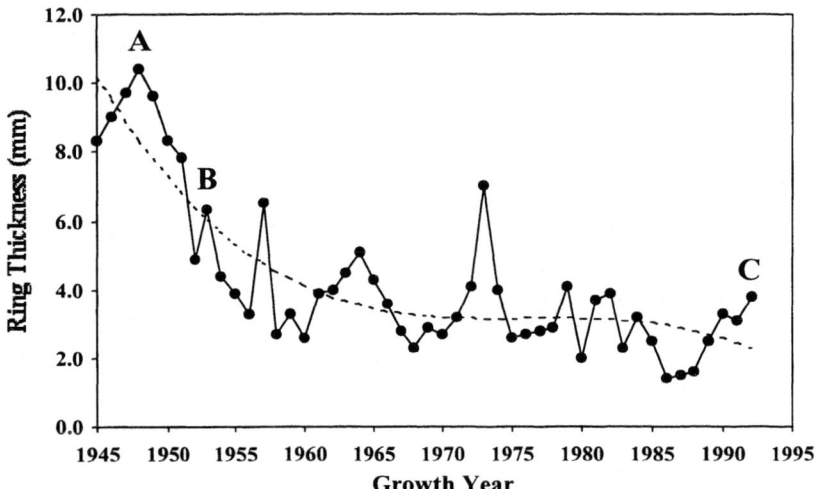

FIGURE 1. Plot of Yellow Pine Ring Thickness Versus Year

TABLE 2. Yellow Pine Relative X-Ray Yield

Ring Sample	A	B	C
Year	1947	1953	1992
Thickness	9.8 mm	6.5 mm	3.8 mm
Element	**Relative Yield**	**Relative Yield**	**Relative Yield**
Potassium	0.11	0.45	0.08
Calcium	1.00	0.64	0.59
Titanium	0.05	0.00	0.00
Chromium	0.03	0.00	0.00
Manganese	0.08	0.08	0.05
Iron	0.32	0.07	0.09
Copper	≤ 0.01	≤ 0.01	0.00
Zinc	0.03	0.00	0.00

All relative yields are normalized to the calcium K_α x-ray peak.

This research could provide insight into the growth patterns for Yellow Pine. However, there are many factors affecting tree growth, including temperature, rainfall, degree of sunlight, availability of nutrients, and anthropogenic activity.

A natural tree ring sample can be directly analyzed without chemical or physical treatment, thereby avoiding loss of material or contamination during preparation of the sample. Trees analyzed with PIXE can help determine the history of pollutants and can also give information on elements present during its life cycle. PIXE provides a good method to determine the composition of tree rings. However, analytical sensitivity and precision must be improved to make PIXE more useful for heavy elements. Additional research is needed to fully quantify the relationship between ring growth and elemental composition.

ACKNOWLEDGEMENTS

This research was supported by the U.S. Department of Energy and the Louisiana Education Quality Support Fund under contracts DOE/LEQSF (1993-95)-03 and DE-FC02-91ER75669. We would also like to thank Changgeng Liao, for his assistance during his stay at the Acadiana Research Laboratory, Dr. Liao is currently at the Department of Modern Physics, Lanzhou University, Lanzhou, Gansu, 73000, Peoples Republic of China.

REFERENCES

[1] W.U Ault, R.G. Senechal and W.E. Erlebach, Isotop. Sci. Technol., 4, 305 (1970).

[2] P. S. Szpoa, E. A. McGuinness, Jr. and J. O. Pierce, Wood Sci. Technol., 6, 72 (1973).

[3] N. I. Ward, R. R. Brooks and R. D. Reeves, Environ. Polllut. 6, 149 (1974).

[4] V. Valkovic, *X-Ray Spectroscopy in Environmental Sciences*, CRC press, 291 (1989).

[5] C.F. Base III and S.B. Mclaughlin, Science, 224, 494 (1984).

[6] J.R. McClenahen, J.P. Vimmerstedt and A.J. Scherzer, Can. J. Forest Res., 19, 880, (1989).

[7] C. W. Berish and H. L. Ragsdale, Can. J. Forest Res. 15, 477 (1985).

[8] Gary A. Glass, Karl H. Hasenstein and His-Tsung Chang, Nucl. Instr. And Meth., B79, 393 (1993).

[9] W. Hollerman, L. Holland, D. Ila, J. Hensley, G. Southworth, T. Klasson, P. Taylor, J. Johnston, and R. Turner, Journal of Hazardous Materials, 68 (3), 193-204 (1999).

[10] R. Turner, Frontier Geosciences, Seattle, WA, personal commmunication, 1998.

Investigation of ancient human bone by means of ionoluminescence and μPIXE

D. Spemann*, St. Jankuhn, J. Vogt, and T. Butz

Nukleare Festkörperphysik, Fakultät für Physik und Geowissenschaften, Universität Leipzig, Linnéstraße 5, 04103 Leipzig, Germany

Abstract. We studied diagenetic alterations on ancient human bones by means of ionoluminescence and μPIXE. It was found that diagenetically altered regions show an orange-red luminescence activated by Mn^{2+} ions. In order to study the incorporation of Mn into the bone mineral in more detail, μPIXE measurements were performed in the periosteal region of a bone cross section from which maps of elemental distributions could be obtained. The Mn distribution is characterized by a rather uniform Mn concentration of 130 μg/g over the scanned area and additional small areas with concentrations up to 110 mg/g. These areas are mainly located at holes and pores in the bone mineral indicating that they act as suitable pathways for the incorporation of Mn. Furthermore a line scan was made which showed an enhanced concentration of Mn, Fe, and Zn in the area of the periosteal surface which shows that this region is strongly influenced by diagenetic alterations.

INTRODUCTION

One of the possibilities to obtain information about bone mineral content, diseases like osteoporosis, nutrition, and living conditions of our ancestors is to determine the elemental content of the bone mineral especially in the WARD's triangle by ion beam analysis. In order to compare these data with those about the recent population, one must consider diagenetic alterations (4). Since the ionoluminescence method (IL) is able to provide information about the chemical form of elements (speciation) (5) and allows the detection of Mn and rare earth elements in host minerals like apatite with a detection limit of a few μg/g (3, 6), we used IL as well as spatially resolved particle-induced X-ray emission (μPIXE) in order to study diagenetic alterations of ancient human bone.

SAMPLE

The ancient bone investigated in this study originates from a female Merovingian individual from Essingen (Rheinland-Pfalz, Germany) und was dated to the 6th–8th century AD (4).

---

* Corresponding author
email: spemann@physik.uni-leipzig.de

The sample was prepared as follows: A cross section with a thickness of 1 mm was cut with a diamond saw from the femoral shaft at about 120 mm apart from the proximal epiphysis, then cleaned in an ultrasonic bath with extrapure water (17 MΩ) for 2×15 min, dried in an oven for 12 h at 50°C and finally vacuum-dried for 1 h.

EXPERIMENTAL

The IL measurements were carried out at the external ion beam facility of the 2 MV VAN DE GRAAFF accelerator using a 1.3 MeV proton beam with a beam diameter of 250 μm on the sample and a current of 0.5 nA. The bone cross section was radially scanned from the periosteal surface to the endosteal surface. A JARRELL-ASH MONOSPEC 27 (SCIENTIFIC INSTRUMENTS INC.) grating spectrometer fitted with a PELTIER-cooled SOLITON SCCD-array was used to analyse the light emitted from the sample. A detailed description of the IL-setup can be found in (8).

The μPIXE measurements were performed at LIPSION, the high energy ion nanoprobe at Leipzig (1). A 2 MeV proton beam was used with a beam spot diameter of 1 μm at a beam current of 70 pA. For each scan a total charge of 200 pC was collected. The X-ray spectra were recorded using a EG&G ORTEC HPGe X-ray detector (IGLET-X series) with an energy resolution $\Delta E_X = 148$ eV at 5.9 keV and an active area of 95 mm^2. A 60 μm PE

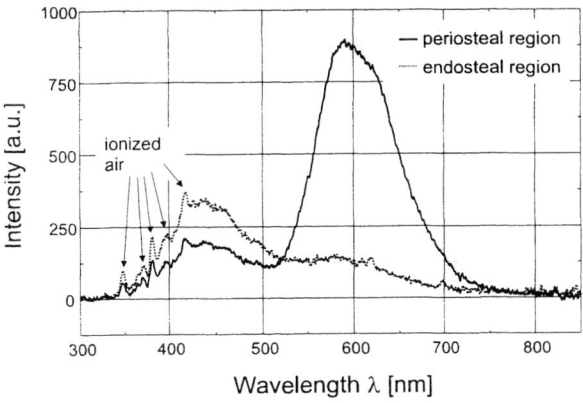

FIGURE 1. IL spectra from the periosteal and the endosteal region of the bone cross section.

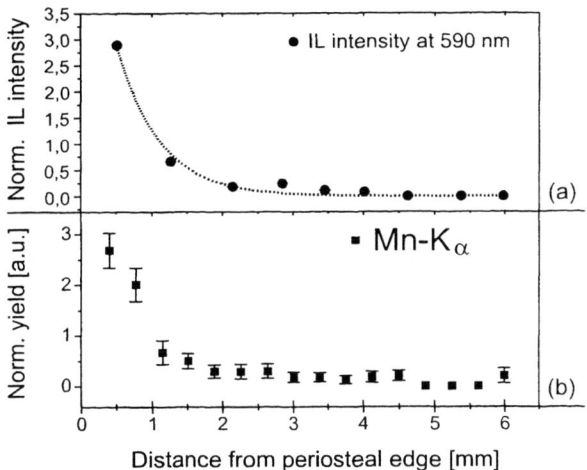

FIGURE 2. (a) Normalized IL intensity of the luminescence band centered at 590 nm. (b) Manganese distribution (normalized to calcium content).

foil was used for proton stopping. In order to suppress the P-K-X-rays a aluminum funny filter with a thickness of 30 μm was used additionally.

The PIXE spectra were analyzed with the standard peak-fit program-code ORIGIN 5.0 (Microcal Software). The elemental concentrations were calculated from the extracted peak areas using YIELD (2).

RESULTS AND DISCUSSION

With increasing distance from the periosteal surface, the luminescence of the ancient bone sample changes from orange-red to blue. The corresponding IL spectra consist of two broad luminescence bands centered at 440 nm (FWHM: 95 nm) and 590 nm (FWHM: 100 nm), respectively (Fig. 1) (8). The intensity of the blue luminescence band at 440 nm is nearly constant over the bone cross section, contrary to the orange-red luminescence band at 590 nm, which rapidly decreases in intensity with increasing distance from the periosteal surface (Fig. 2(a)). As has been shown previously (8), the latter one is most likely caused by Mn^{2+} ions which are well known luminescence activators in apatite mineral (6). The good correspondence between the IL intensity and the Mn content which has been evaluated from the same area of the sample using μPIXE, indicates that Mn is mainly present in the form of Mn^{2+} ions (Fig. 2) incorporated into the bone mineral due to the ion exchange process $Ca^{2+} \longleftrightarrow Mn^{2+}$ (8). Since Mn will not be incorporated in bone tissue by biological processes during the lifetime of an individual (7), the presence of Mn in the periosteal region of ancient bones results from diagenetic processes which occurred during the time that the bone was buried in the soil.

In order to study the incorporation of Mn into the bone mineral in more detail, μPIXE measurements were performed from which maps of elemental distributions could be obtained. Figure 3 shows the elemental distributions of Ca, Mn, and Fe in the periosteal region of the bone (scan size 750×750 μm^2). The bone mineral $Ca_5(PO_4)_3OH$ is represented by the Ca-map, because Ca is one of the main components. The elemental distribution shown in the Mn-map is characterized by a rather uniform Mn concentration of 130 μg/g over the scanned area and additional small areas with concentrations exceeding 4.5 mg/g (shown as white spots in the map). The fact that these areas with high Mn concentration are mainly located at holes and pores in the bone mineral (see Ca-map for comparison) indicates that they act as suitable pathways for the incorporation of Mn. The Fe-map not only shows small areas with high elemental concentration similar to those in the Mn-map, but also a remarkable enrichment in Fe content at the periosteal surface. In Fig. 4 a secondary electron image (SE) and the elemental distributions of Ca and Mn are shown (scan size 84×84 μm^2) which were taken from the scan indicated by the white square in Fig. 3. As can be seen from the Mn-map, most of the Mn is located at a pore in the bone mineral as indicated by the dashed shape in the SE image and the Ca-map. In these areas the Mn concentrations amounts up to 110 mg/g, reaching the same concentration level as phosphorus, another main component of bone mineral. The SE image shows clearly that the bone mineral located around the pore is very porous, which obviously supports the incorporation of Mn. However, it is very unlikely that all Mn is incorporated into the bone mineral by ion exchange processes at these high concentrations. Contrary to μPIXE, quantitative IL or cathodolumines-

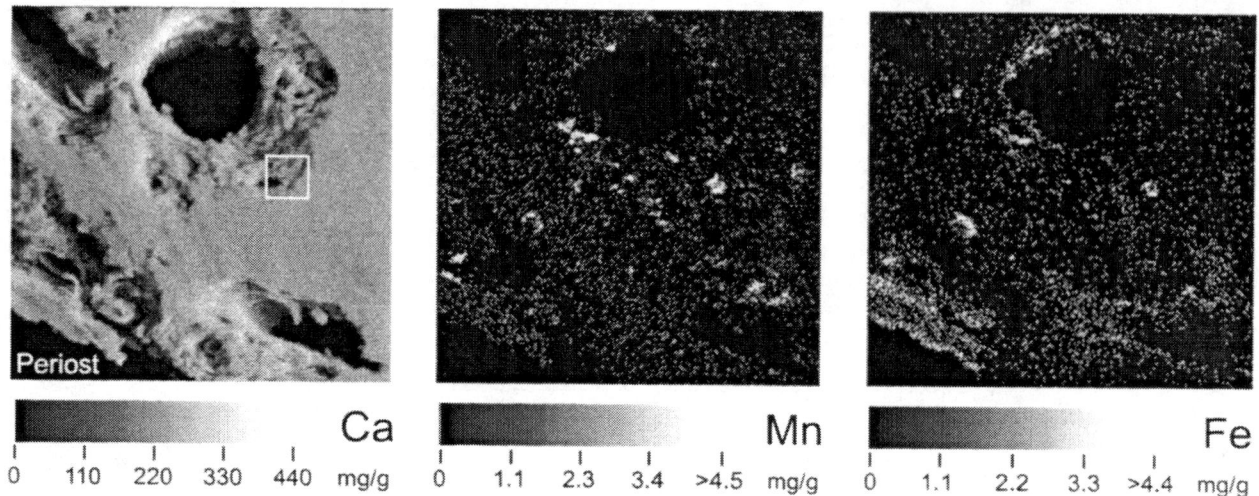

FIGURE 3. Elemental maps of Ca, Mn, and Fe from the periosteal region (scan size $750 \times 750\ \mu m^2$). The bone mineral is represented by the Ca-map. The Mn-map shows spot-like areas with concentrations exceeding 4.5 mg/g which are mainly located at holes and pores in the bone mineral. The Fe-map shows a remarkable enrichment in elemental content at the periosteal surface (lower left corner).

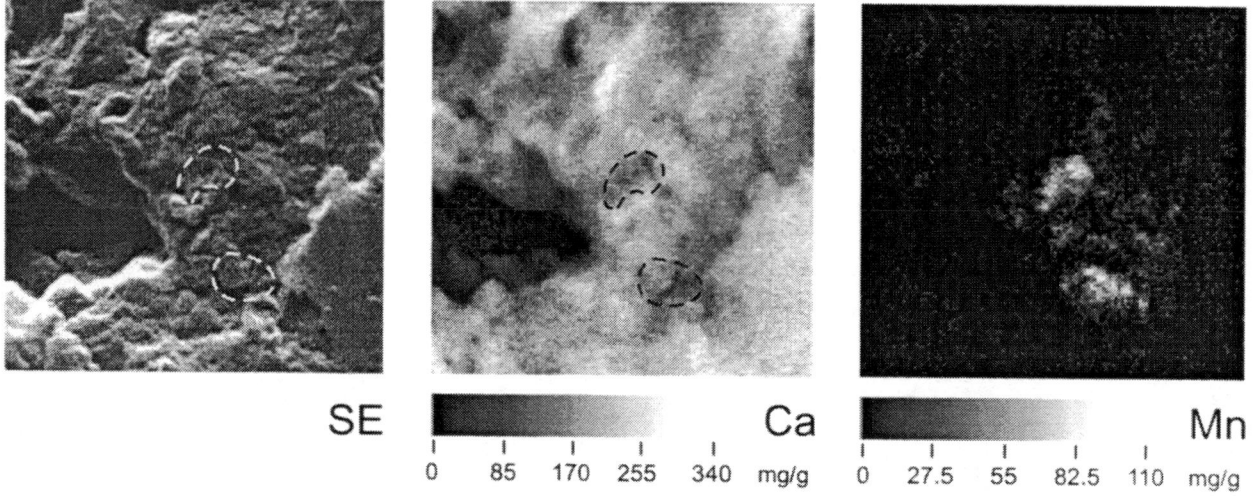

FIGURE 4. SE image and elemental maps of Ca and Mn from the region indicated by the white square in Fig. 3 (scan size $84 \times 84\ \mu m^2$). The Mn-map shows the elemental distribution within an area of high Mn concentration. As can be seen from the SE image, this area is dominated by cracks and pores in the bone mineral.

cence measurements (CL) would therefore underestimate the amount of incorporated Mn.

In order to obtain elemental concentration profiles from the periosteal region of the ancient bone, a line scan was performed as indicated in the upper part of Fig. 5. Besides the porous surface, the scan area consisted of compact bone mineral, where Mn was not present at concentration levels comparable to those shown in Fig. 4. As can be seen from the elemental concentration profiles of Mn, Fe, and Zn, the porous surface area is strongly influenced by diagenetic alterations yielding enhanced concentrations of those elements. In the compact bone mineral the concentrations of Mn and Zn slowly decreases with increasing distance from the periosteal edge. In addition the profiles of Mn, Fe, and Zn show several areas in the compact bone mineral with enhanced concentrations. In the case of Fe and Zn these areas correspond well in lateral position and width which shows that these local enrichments probably result from the same geochemical and biochemical processes. However, the enrichments of Mn are more numerous and narrower in width, which indicates that the incorporation of Mn is due to diagenetic processes being possibly even specific for that element.

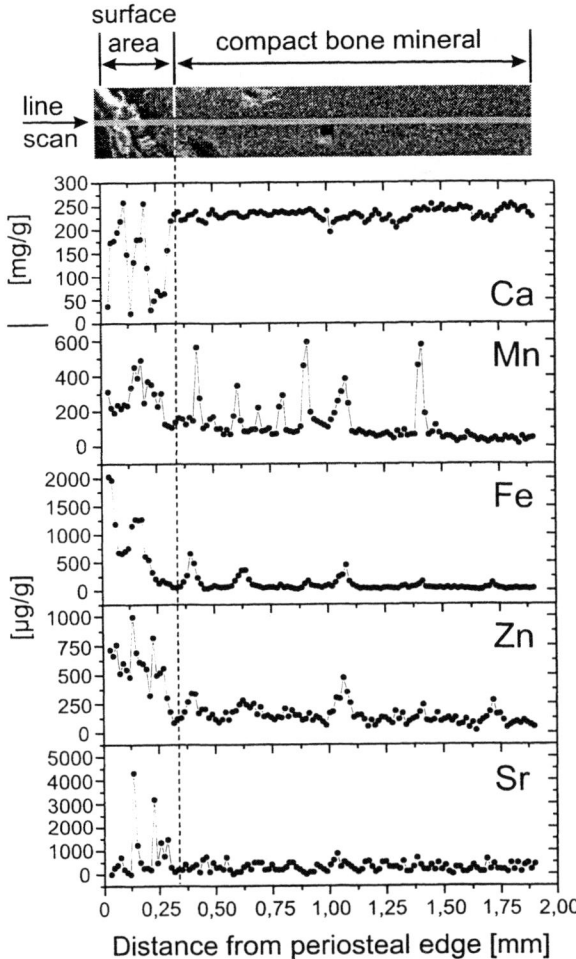

FIGURE 5. Elemental distributions of Ca, Mn, Fe, Zn, and Sr obtained from the line scan indicated in the upper part. The areas in the compact bone mineral with a high Mn concentration as well as the enrichment in Fe content at the periosteal surface are clearly visible.

CONCLUSIONS

For a qualitative and quantitative interpretation of the data about the chemical composition of ancient bone mineral diagenetic alterations have to be considered.

The bone investigated in this study was characterized by a remarkable uptake of Mn, Fe, and Zn, however, it was restricted to a depth of 2 mm at maximum from the periosteal surface.

Because of its high sensitivity the IL method is very suitable for the detection of Mn in bone mineral. This method allows to reveal diagenetically altered regions in bone within a few seconds. Since quantitative information about the Mn content from IL or CL measurements can only be obtained for concentrations up to 10 mg/g due to selfquenching processes (6), the determination of the elemental concentrations by a method like μPIXE is always necessary in order to accurately quantify diagenetic alterations of bone mineral.

ACKNOWLEDGEMENTS

This work was supported by the Deutsche Forschungsgemeinschaft (DFG), Innovationskolleg INK 24 B1/1 (Germany). The financial support of D. S. by the DFG under grant SP 656/1-1 is gratefully acknowledged.

Furthermore the authors wish to thank J. HAMMERL (Inst. für Anthropologie und Humangenetik, Universität Frankfurt a. M.) for providing us with the bone samples.

REFERENCES

1. Butz, T., Flagmeyer, R.-H., Heitmann, J., Jamieson, D. N., Legge, G. J. F., Lehmann, D., Reibetanz, U., Reinert, T., Saint, A., Spemann, D., Szymanski, R., Tröger, W., Vogt, J., and Zhu, J., *Nucl. Instr. and Meth.* **B 161-163**, (2000) 323.
2. Frey, H., Vogt, J., and Otto, G., *J. Radioanal. Nucl. Chem.* **99** (1), (1986) 193.
3. Homman, N. P.-O., Yang, C., and Malmqvist, K. G., *Nucl. Instr. and Meth.* **A 353**, (1994) 610.
4. Jankuhn, S., Butz, T., Flagmeyer, R.-H., Reinert, T., Vogt, J., Barckhausen, B., Hammerl, J., Protsch von Zieten, R., Grambole, D., Herrmann, F., and Bethge, K., *Nucl. Instr. and Meth.* **B 136-138**, (1998) 329.
5. Malmqvist, K. G., Elfman, M., Remond, G., and Yang, C., *Nucl. Instr. and Meth.* **B 109-110**, (1996) 227.
6. Marshall, D. J., *Cathodoluminescence of Geological Materials*, Boston: Unwin Hyman, 1988.
7. Int. Commission on Radiological Protection No. 23, *Report of the Task Group on Reference Man*, Oxford: Pergamon Press, 1975.
8. Spemann, D., Jankuhn, S., Vogt, J., and Butz, T., *Nucl. Instr. and Meth.* **B 161-163**, (2000) 867.

High Energy Ion Beam Analysis of Buried α-Fe$_2$O$_3$(0001)/α-Al$_2$O$_3$(0001) Interface

S. Thevuthasan, V. Shutthanandan, and E.M. Adams
Environmental Molecular Sciences Laboratory, Pacific Northwest National Laboratory, Richland, WA 99352

S. Maheswaran
School of Science, University of Western Sydney, Nepean, Kingswood, NSW 2747, Australia

Abstract: We have investigated the disordering at the buried interface of α-Fe$_2$O$_3$(0001)/α-Al$_2$O$_3$(0001) interface using Rutherford backscattering spectrometry (RBS) and channeling techniques. Although expitaxially-grown α-Fe$_2$O$_3$(0001)/α-Al$_2$O$_3$(0001) thin films exhibit about 1.5-2.5% of minimum yield, disordering at the interface is visible due to the misfit dislocations because of the lattice mismatch between the substrate and the film. Theoretical simulations of surface and interface peak areas were simulated using VEGAS program and compared with the experimental data.

INTRODUCTION

Synthesis of model oxides as thin films on various oxide and metal substrates to obtain high quality surfaces is a growing interest in the scientific community. High quality surfaces are important for the investigation of physical and chemical properties of these film surfaces. In most cases, the single crystal oxides purchased from the commercial vendors still exhibit large amount defects in the material. In some cases, especially in the case of iron oxides, high quality single crystals are not commercially available. In addition, high quality iron oxide films have applications in several areas including heterogeneous catalysis, magnetic thin films, surface geochemistry, corrosion and integrated microwave devices [1-6]. As such, the growth of high crystalline quality iron oxide thin films is of increasing interest. Recently, several high-quality well-oriented single crystal iron oxide films with various stoichiometries have been synthesized using Molecular Beam Epitaxial (MBE) growth and the structural properties have been analyzed by various surface and bulk sensitive techniques[7-12]. Most of these studies are surface related and it has been shown that high-quality well-ordered surfaces can be obtained in these films. On the other hand, the bulk related studies are mostly limited to x-ray diffraction (XRD) studies, and the crystalline quality, impurity concentrations and locations, dislocations and disordering at the interfaces, elemental in and out diffusion at the interfaces are not known in some cases. Recently, we have investigated the Al diffusion at the α-Fe$_2$O$_3$(0001)/α-Al$_2$O$_3$(0001) interface [13] and Mg diffusion at the γ-Fe$_2$O$_3$(001)/MgO(001) and Fe$_3$O$_4$(001)/MgO(001) interfaces [14-15] using Rutherford backscattering spectrometry (RBS) and channeling (RBS/C) techniques. Al diffusion into the α-Fe$_2$O$_3$(0001) film was observed after heating the sample to 970 K using RBS [13]. On the other hand, Mg diffusion into the γ-Fe$_2$O$_3$(0001) and Fe$_3$O$_4$(0001) films starts at much lower temperatures of about 870 K [15]. In fact, if the films were grown around 770 K small amount of Mg would be incorporated in the film during the growth [10]. Annealing the γ-Fe$_2$O$_3$(001) and Fe$_3$O$_4$(0001) films in 2.0x10^{-6} Torr of oxygen at temperatures up to 970 K enhances Mg out diffusion into the films and increases the film thickness depending on temperature. The magnetite film thickness reach a limiting value at 870 K anneal while the maghemite film thickness did not maximize after annealing at 970 K. After the annealing at 970 K, both films produced a compound with composition close to magnesioferrite (MgFe$_2$O$_4$). On the other hand, when γ-Fe$_2$O$_3$(001) film was annealed in vacuum at 970 K, in addition to the Mg diffusion into the film, Fe diffusion into the MgO substrate was observed [14].

The goal of the present work is to investigate the disordering at the α-Fe$_2$O$_3$(0001)/ α-Al$_2$O$_3$(0001) interface. Three different thick epitaxially grown high quality films were used in this work. A systematic study of the interface disordering using these three films is currently in progress. This paper describes the RBS and channeling measurements along channeling and random geometries along with some computer simulations from the investigation of interface disordering due to misfit dislocations at the interface using a 70 nm thick film.

EXPERIMENTAL

The samples were grown in the molecular beam epitaxial (MBE) facility at the Environmental Molecular Sciences Laboratory (EMSL) at PNNL using the procedures described elsewhere [9]. High quality iron oxide films with low minimum yields (ratio of aligned to random yields) can be routinely grown in the MBE facility. After growth, the samples were carefully removed from the MBE system and introduced into the channeling end station at the accelerator facility. The details of the accelerator facility and the end stations are described elsewhere [16]. The samples were heated to 200-250°C to desorb hydrocarbons from the surface. The samples were mounted on a molybdenum backing plate using Ta clips and a conventional alumel-chromel thermocouple was attached to the backing plate close to the sample for temperature measurements. The standard dose of helium ions for one spectrum was 4.4×10^{15} ions/cm^2. The backscattering spectrum was collected using a silicon surface barrier detector at a scattering angle of 150°. The primary energy of the ions was 2.04 MeV and the incident ion beam was directed along the normal to the sample surface. Theoretical simulations of surface peak area were simulated using VEGAS program. This computer program uses Monte Carlo calculations and a Moliere screened potential to simulate the ion scattering interactions and it is described elsewhere [17]. The root mean square vibrational amplitudes are calculated using the Debye temperature, reported for Fe_3O_4 [9] and the vibrations are assumed isotropic.

RESULTS AND DISCUSSION

Aligned and random RBS spectra for the 70 nm thick film are shown in Fig. 1. A small energy window (ΔE = 733-743 channels) near the surface region was used to calculate the minimum yield (χ_{min}) for Fe. Another small energy window (ΔE = 485 - 505 channels) near the aluminum surface peak region for the substrate was used to calculate the minimum yield for Al. The minimum yield is the ratio of the yield in the channeling geometry to that for a random, non-channeling geometry. For the film, χ_{min} is calculated to be 1.8±0.3%. In general the film has high crystalline quality and the minimum yield is accordingly very low. The minimum yield for the substrate Al is 11.9±1.3%. There are five peaks visible in the aligned spectrum. The first peak at the high-energy side is the surface peak at the front of the film (Fe – SP). The second peak is attributed to some Fe atoms visible to the ion beam at the interface (back surface of the film – Fe - IP). Apparently, the ion beam sees some Al atoms (substrate surface) at the interface as indicated by the third peak (Al – IP). The fourth peak is related to the backscattered ion contribution due to the surface oxygen atoms (O – SP) of the film and the fifth peak is due to the visibility of oxygen to the ion beam at the interface (O – IP). Since the Fe, O and Al atoms are visible to ion beam at the interface, there must be some disordering of the iron oxide film

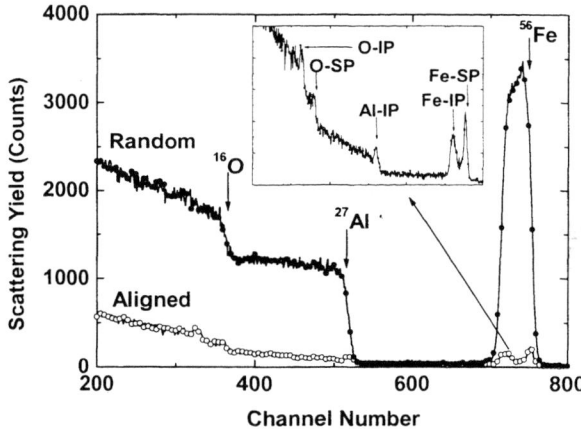

Fig. 1: Aligned and random RBS spectra from epitaxially grown 70 nm thick α-Fe$_2$O$_3$(0001) film on α-Al$_2$O$_3$(0001) substrate. Incident energy of the He$^+$ was 2.04 MeV and the scattering was 150°. Magnified version of the aligned spectrum is presented in the insert.

at the interface or there must be substrate-film mixing present at the interface. Although mixing of the substrate and the film is possible at the interface, no evidence for mixing has been observed in the random spectrum within the experimental resolution and the uncertainties. It could be that the lattice mismatch between the substrate and the film is too large [5.4%] to allow a continuous channeling trajectory through the interface between the two crystal structures. A transmission electron microscopy (TEM) and selected-area diffraction study [18] reported that there are two distinct types of dislocation arrays present at the interface, one of perfect lattice dislocations and the other of partial dislocations of the corundum structure common to the hematite film and substrate. In a recent study [19], Chambers and co-workers have investigated the α-Cr$_2$O$_3$(0001)/α-Al$_2$O$_3$(0001) interface using TEM and identified misfit dislocations at the interface. The in-plane lattice parameters (a-axis) for bulk α-Al$_2$O$_3$(0001), α-

$Cr_2O_3(0001)$ and $\alpha\text{-}Fe_2O_3(0001)$ are 0.476, 0.492, and 0.503 nm respectively. As such, the in-plane lattice mismatch is 3.36% and 5.80% for $\alpha\text{-}Cr_2O_3(0001)$ and $\alpha\text{-}Fe_2O_3(0001)$ on $\alpha\text{-}Al_2O_3(0001)$ respectively. Since all three structures are corundum structures and the lattice mismatch for both for $\alpha\text{-}Cr_2O_3(0001)$ and $\alpha\text{-}Fe_2O_3(0001)$ on $\alpha\text{-}Al_2O_3(0001)$ are comparable, misfit dislocation is expected at the $\alpha\text{-}Fe_2O_3(0001)/\alpha\text{-}Al_2O_3(0001)$ interface as in the case of for $\alpha\text{-}Cr_2O_3(0001)/\alpha\text{-}Al_2O_3(0001)$ interface. The present ion scattering results are consistent with the misfit dislocations at the interface, so that Al and O atoms from the substrate and Fe and O atoms from the film are visible to the channeled He^+ ion beam.

The experimental peak area of the front Fe – SP is determined to be 3.9 atoms/row and this area is approximately same as the area of the back Fe surface peak (4.3 atoms/row). The simulated surface peak area using VEGAS code was 3.8 atoms/row and there is a good agreement between the experimental and theoretical surface peak areas. A single Fe-layer terminated bulk like hematite structure with appropriate surface relaxations [20] was used in this simulation. Vibrational amplitudes are crucial for these simulations and isotropic root mean square vibrational amplitude of 0.0068 nm (determined from the Debye temperature of Fe_3O_4) was used for both Fe and O atoms. A cluster with misfit dislocations at the interface was used in the simulations of the Fe -IP area. Although the VEGAS code is not large enough to accommodate larger clusters, the simulations with smaller clusters indicate the possibility for an existence of an interface with misfit dislocation for the $\alpha\text{-}Fe_2O_3(0001)/\alpha\text{-}Al_2O_3(0001)$ system.

We show in Fig. 2 the normalized Fe – SP, Fe - IP and Al - IP angular yield curves with respect to the [0001] direction for the film and the substrate respectively. The energy regions used to calculate the minimum yield for Al (substrate) along with the energy regions for Fe - SP (ΔE = 743-775 channels) and Fe - IP (ΔE = 704-733 channels) were used to extract the angular yield curves. The Fe - SP, Fe - IP and Al angular yields are normalized to the maximum Fe - SP (40091 counts), Fe - IP (72997 counts) and Al (23278 counts) yields in the random geometry for the respective energy regions. During the whole angular scan, aligned spectra were collected at the beginning and at the end of the angular scan to check for sample damage. These two aligned scans were essentially identical within the experimental uncertainties and, as a

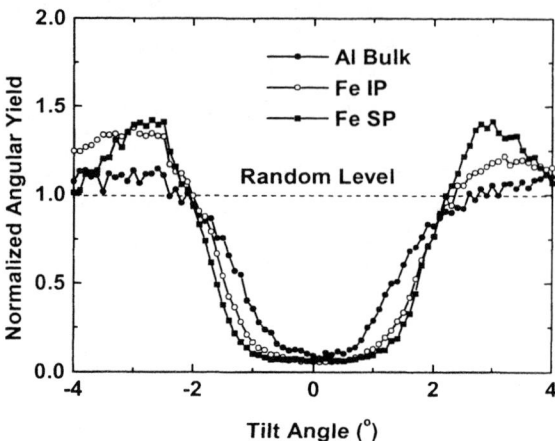

Fig. 2: Normalized angular yield curves for Fe - SP, Fe - IP, and Al are presented as a function of tilt angle. Incident energy of the He^+ was 2.04 MeV and the scattering was 150°.

result, the sample damage is negligible. The full width at half minimum (FWHM) for Fe - SP is 3.68°, for Fe - IP peak is 3.49° and for Al is 2.89°. Similar angular yields curves for Al - SP and Al from a clean sapphire (0001) substrate showed 3.12° and 2.91° FWHMs respectively and the difference between these two FWHMs is about 0.21°. The difference between the FWHMs for Fe - SP and Fe - IP is about 0.19° and this value and the difference between the FWHMs from the clean sapphire substrate (0.21°) are comparable. This indicates that there is no significant disordering for a large depth into the film from the interface and it is consistent with the disordering generated by the misfit dislocations in a few layers at the interface.

CONCLUSIONS

Rutherford backscattering and channeling techniques were used to determine the crystalline quality of epitaxially grown $\alpha\text{-}Fe_2O_3(0001)$ film on $Al_2O_3(0001)$ and disordering at the interface. The minimum yield obtained from the aligned and random spectra for these films are 1.8±0.3% for this film. Although the film shows high crystalline quality, it appears that some disordering due to misfit dislocations exists at this interface. The simulated Fe surface peak area is in good agreement with the front Fe surface peak (Fe - SP) from the experiment. The results from preliminary simulations using clusters with misfit dislocations at the interface are consistent with the experimental observations. Systematic experimental investigations and detail simulations are currently under progress.

ACKNOWLEDGMENTS

The authors gratefully acknowledge Dr. S. A. Chambers and Dr. S.I. Yi for providing samples for this study. The authors also gratefully acknowledge Prof. R.J. Smith from Montana State University for helpful suggestions. Pacific Northwest National Laboratory is a multi-program national laboratory operated for the U.S. Department of Energy by Battelle Memorial Institute under contract No. DE-AC06-76RLO 1830. The authors gratefully acknowledge partial support from the US Department of Energy, Offices of Basic Energy Sciences, and Biological and Environmental Research - Environmental Management Science Program.

REFERENCES

1. J.W. Geus, Appl. Catl. 25 (1986) 313.
2. H.H. Kung, *Transmission Metal Oxides: Surface Chemistry and Catalysis*(Elsevier, New York, 1989).
3. T. Fujii, M. Takano, R. Katano, and Y. Bando, J. Appl. Phys. 66 (1989) 3168.
4. T.D. Waite, Rev. Mineral, 23 (1990) 559.
5. R.K. Wild, in *Surface Analysis: Techniques and Applications, Special Publication 84*, edited by D.R. Randell and W. Neagle (Royal Society of Chemistry, London, 1990).
6. D. M. Lind, S.D. Berry, G. Chern, H. Mathias, and L.R. Testardi, Phys. Rev. B 45 (1992) 1838.
7. J.F. Anderson, M. Kuhn, U. Diebold, K. Shaw, P. Stoyanov, and D. Lind, Phys. Rev. B 56 (1997) 1134.
8. J.M. Gaines, P.J.H. Bolemen, J.T. Kohlhepp, C.W.T. Bulle-Lieuwma, R.M. Wolf, A. Reinders, R.M. Jungblut, P.A.A. van der Heijden, J.T.W.M. van Eemeren, J. aan de Stegge, W. J.M. de Jonge, Surf. Sci. 373 (1997) 85.
9. Y.J. Kim, Y. Gao, and S.A. Chambers, Surf. Sci. 371 (1997) 358.
10. Y. Gao, Y.J. Kim, S. Thevuthasan, P. Lubitz, and S.A. Chambers, J. Appl. Phys. 81 (1997) 3253.
11. Y. Gao, Y.J. Kim, S.A. Chambers, and G. Bai, J. Vac. Sci. Technol. A 15 (1997) 332.
12. Y. Gao and S.A. Chambers, J. Cryst. Growth 174 (1997) 446.
13. S. Thevuthasan, D.E. McCready, W. Jiang, S.I. Yi, and S.A. Chambers, Applications of Accelerators in Research and Industry, ed. J.L. Duggan and I.L. Morgan, American Institute of Physics (1999), pp 508-511.
14. S. Thevuthasan, W. Jiang, D.E. McCready, and S.A. Chambers, Surf. and Interface Anal. 27(1998) pp 194-198.
15. S. Thevuthasan, D.E.McCready, W. Jiang, S.I Yi, S. Maheswaran, K.D. Keefer, and S.A. Chambers, Nucl. Intsr. Meth. B 161-163 (2000) pp. 510-514.
16. S. Thevuthasan, C.H.F. Peden, M.H. Engelhard, D.R. Baer, G.S. Herman, W. Jiang, Y. Liang, and W.J. Weber, Nucl. Instr. Meth. A 420 (1999) pp 81-89.
17. J. W. Frenken, R. M. Tromp, and J. F. Van der Veen, Nucl. Instrum. Methods B 17 (1986) 334.
18. I.M. Anderson, L.A. Tietz, and C.B. Carter, Mat. Res. Soc. Symp. Proc. 238 (1992) 807.
19. S.A. Chambers, Y. Liang, and Y. Gao Phys. Rev. B, 61 (2000) pp. 13223-13229.
20. S. Thevuthasan, Y.J. Kim, S.I. Yi, S.A. Chambers, J. Morais, R. Denecke, C.S. Fadley, P. Liu, T. Kendelewicz, and G.E. Brown Jr. Surf. Sci. 425 (1999) 276.

Characterization of Multilayer Thin Film Optical Filters Using RBS

R. Vlastou[1,a], E. Fokitis[a], M. Kokkoris[b], S. Kossionides[b], G. Koubouras[b] and R. Grötzschel[c]

[a] Department of Physics, National Technical University of Athens, GR-157 80, Greece
[b] Institute of Nuclear Physics, NRCPS "Demokritos", GR-153 10, Greece
c Institute of Ion Beam Physics and Material Analysis, Forschungszentrum Rossendorf, Dresden, Germany

Abstract. The composition and thickness of wide band multilayer optical filters, especially designed for the Fluorescence Detector of the Pierre Auger Project, were measured using the RBS (Rutherford Backscattering Spectroscopy) method. The filters were made of 6 pair thin film layers of ZrO_2/SiO_2, deposited on UV glass or absorption filters. The depth profile of the ZrO_2 layers have been measured directly, while reliable results for the SiO_2 layers could also be extracted. Relative thickness and density of the individual layers have been deduced for the samples, as prepared by electron beam deposition technique as well as after thermal annealing.

INTRODUCTION

RBS (Rutherford Backscattering Spectroscopy) has been used during the development of multilayer thin film optical filters, especially designed for the fluorescence detector of the Pierre AUGER Project [1]. These filters must be able to select the UV light from the nitrogen atmospheric fluorescence which is emitted from the Extensive Air Showers induced in the atmosphere by the extremely high energy cosmic ray events. For the study of these events, two different type of detectors will be used : a) The Ground Detector array consisting of water Cherenkov tanks, aiming to detect the muons which arrive in the ground from the shower development in the atmosphere and b) the Fluorescence Detectors consisting of spherical mirrors, photomultipliers and optical filters placed in front of them to detect the air fluorescence induced by the ionization and excitation of nitrogen from the Extensive Air Showers. The fluorescence emitted by the nitrogen atoms, molecules and ions lies in the UV region 300-410 nm. Thus the optical filters should have high transmittance (higher than 85%) in the range 300-410nm (pass band) and very low one (less than 2%) in the visible region 410-650nm (stop band), in order to improve the signal to noise ratio and contribute efficiently to the trigger logic adapted in the Fluorescence Detector for a valid detection of an extremely energetic cosmic ray event. [2]. In addition to the optical properties, the uniformity, the mechanical stability and the cost for the mass production of the filters, must be acceptable.

The required properties of the optical filters led to specific production techniques, such as the dielectric multilayer thin film deposition on UV glass, using high-low refractive index combinations of UV-transparent materials. Several combinations have been tried for the design such as Al_2O_3, Sc_2O_3, ZrO_2, WO_3 for high index and MgF_2, SiO_2 for low index layers. The appropriate software has been developed for the design of the filters [3], to calculate the optical thickness and combination of layers required for an optimal performance. After the production of a series of specific filters, the RBS method has been used, between other characterization techniques, to provide information concerning the relative thickness, reproducibility and homogeneity of the individual layers as well as possible deviations from the desired stoichiometry [4]. As a result, the combination of ZrO_2/SiO_2 has proven to have the best performance concerning stoichiometry, packing density of SiO_2, optical properties and cost. Filters, made of 6 thin film layers of ZrO_2/SiO_2, deposited on UV glass or absorption filters, were thus produced for further tests concerning inhomogeneities, thickness variation and reproducibility, environmental behavior and packing density. The results of this investigation will be presented in the present work.

RBS EXPERIMENTS AND RESULTS

For the RBS measurements, a 2.9 MeV alpha beam was used, supplied by the tandem T11/25 accelerator at the NRCPS "Demokritos" or by the 3 MV tandetron accelerator at the Forschungszentrum Rossendorf. The detector was Si surface barrier, positioned at 166^0 with respect to the beam direction in both laboratories.

The RBS spectra were analyzed by utilizing the computer code RUMP [5]. Typical results for a sample consisting of 6 pair layers of ZrO_2/SiO_2 on absorption filter, is shown in Fig 1. It is seen that a good fit to the data could be achieved and reliable results could be extracted (with an uncertainty up to 10%). The peaks of the heavy element Zr can be clearly separated by the ^{4}He beam and can thus be used to determine the relative thickness of the ZrO_2 films (by using the nominal density of the bulk material) as well as the stoichiometry. The peaks of Si corresponding to the layers of SiO_2 are weak and can only be analyzed by considering the additional information coming from the 'valleys' between the Zr peaks, which arise from the energy loss of the beam passing through the SiO_2 layers. In the samples containing SiO_2 as the first deposited layer on the substrate, and ZrO_2 on the surface, this first SiO_2 layer could not be separated from the Si of the substrate and its thickness is given with an uncertainty up to ~100%. In the samples containing ZrO_2 as the first deposited layer on the substrate, and SiO_2 at the surface, there exists no 'valley' for this surface layer which, however, could be indirecty analyzed from the energy shift of the Zr signal. This type of filters is more favourable for the implementation of the RBS technique. Additionally, the presence of SiO_2 at the surface seems to provide protection against humidity, as will be discussed in detail later in the text. In Fig. 1 the small peaks in front of the surface peak are due to the heavier isotopes of Zr. The high energy signal from the backing corresponds to the presence of Zn and trace quantities of Ca in the absorption filter.

In order to acquire accurate results, the composition of the substrate had to be well determined. Thus, RBS spectra were taken for the substrates of the multilayer films at E_α = 3.05 MeV, where the well known ^{16}O resonance [6] at E_α = 3.04 MeV can provide, with great accuracy, the Si/O ratio present in the UV glass or the absorption filter. A typical RBS spectrum for the substrate absorption filter, corresponding to the multilayer structure shown in Fig. 1 is presented in Fig. 2, along with the simulation. In the case of ZrO_2 and SiO_2 this resonance could not provide additional information concerning the stoichiometry of oxygen, due to the overlapping of oxygen concentration between the layers.

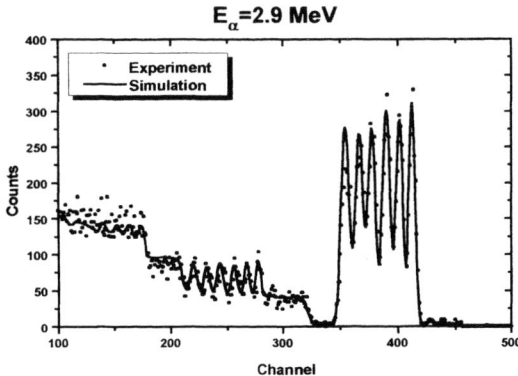

FIGURE 1. RBS spectrum of 6 layer pairs of ZrO_2/SiO_2 films on absorption filter taken with 2.9 MeV ^{4}He^{++}. The 6 strong peaks of Zr correspond to the 6 layers of ZrO_2 and the next 6 weak peaks of Si to the 6 layers of SiO_2. The small peaks at about channels 420 to 450 are due to the heavier isotopes of Zr. The step around channel 320 corresponds to the presence of Zn in the absorption filter. The solid line represents simulation by the RUMP code.

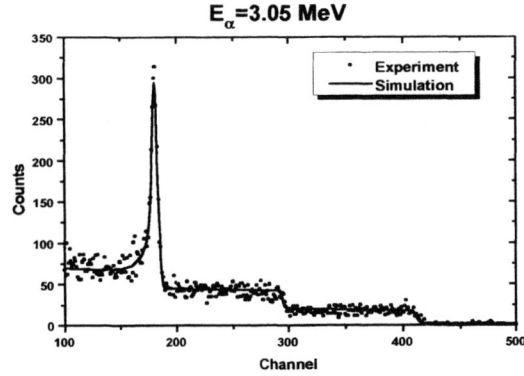

FIGURE 2. RBS spectrum of the absorption filter on which the 6 layer pairs of ZrO_2/SiO_2, presented in Fig. 1, have been deposited. The spectrum has been taken at E_α = 3.05 MeV, where the reaction $^{16}O(\alpha,\alpha)^{16}O$ exhibits a strong resonance. The solid line represents simulation by the RUMP code with the implementation of the nuclear resonance data [6].

The stoichiometry of the thin films in both ZrO_2 and SiO_2 layers in all the samples were found to follow the desired composition, in agreement with the results of refs [4] and [7].

The optimum calculated thicknesses [3] of each of the high and low index materials, were the same (quarterwave) for each material. However, the thickness of the deposited layers were found to vary, due to the instability of the quartz crystal monitor during the deposition process. From the analysis of the RBS spectra, the relative thicknesses of the individual layers exhibit a variation from the average value of the order of 16% for ZrO_2 and 10% for SiO_2. To test the homogeneity of each sample, several measurements were taken at different points ant analyzed by RBS. In addition, the reproducibility of the layer thickness and thus the reliability of the production method were tested with the RBS method.

During the deposition of ZrO_2 a reduced packing density of the films can be produced which may affect the film properties [4, 7]. For the SiO_2, on the other hand, the packing density of the deposited layer turns out to be close to that of the bulk material [4, 7]. The packing density p is defined as the ratio of the density of the film material p_f to the bulk material p_b. Since the same mass of the material in a film with reduced packing density, fills a larger volume than a dense film, the geometrical and optical film thickness are affected, as well as the hardness, stresses and stability of the film. The refractive index n of the film is also influenced by the lower packing density since the voids of the film with refractive index n=1.0, tend to absorb water vapors with n=1.33 when exposed to air and humidity [7].

The thickness of the layers was monitored by a quartz crystal monitor, during the deposition. After the preparation of the samples the total thickness was also measured with a Tally step Profilometer as well as with ellipsometry. Using the RBS relative thicknesses (by using the density of the bulk material), in conjunction with the other thickness measurements and the fact that the packing density of SiO_2 p=1, the packing density of ZrO_2 layers could be deduced and the mean value was found to be p=0.7±0.1.

In order to investigate the influence of thermal treatment to the packing density of ZrO_2, samples were examined with RBS before and after thermal annealing. Several samples, consisting of 6 pair layers of ZrO_2/SiO_2 with either ZrO_2 or SiO_2 at the surface, were annealed at 250-300°C for 24 or 36 hours in regular atmospheric conditions. Within the experimental errors, the packing density of ZrO_2 did not seem to be affected by the annealing. Only an interdiffusion of Zr in SiO_2 layers has been observed and is illustrated in Fig. 3, where a typical RBS spectrum of a sample before and after annealing is presented.

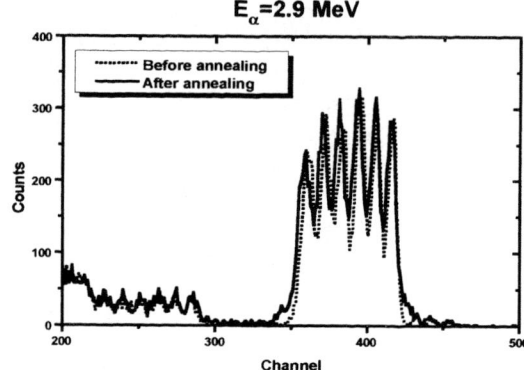

FIGURE 3. Typical RBS spectra of an optical filter consisting of 6 layer pairs of ZrO_2/SiO_2 on UV glass, measured with 2.9 MeV $^4He^{++}$ ions, before and after annealing. The interdiffusion of Zr in SiO_2 layers is evident.

Finally, in order to test the environmental behavior of the filters, 6 pair layer samples with SiO_2 at the surface were exposed to 95% humidity for 3 weeks and no adsorption of water vapor has been observed. This is probably due to the fact that the surface layer of SiO_2, having a packing density of p=1, provides a protection coating against humidity.

SUMMARY

RBS has been used for the characterization of thin film optical filters consisting of 6 pair layers of ZrO_2/SiO_2 on UV glass or absorption filters. The relative thickness and the stoichiometry of the individual layers, the thickness variation of the layers, the homogeneity of the films, as well as the packing density of ZrO_2 after the film deposition have been deduced. The influence of the thermal annealing to the packing density of ZrO_2, turned out to be insignificant. Further, exposure to humidity was found to leave the samples unaffected.

ACKNOWLEDGEMENTS

It is our pleasure to thank A.Braem of Optics Laboratory, CERN, for preparing the optical multilayer filters and for useful discussions. Our warmest thanks to G.Polysos and E.Drakaki of NTUA Physics Department, who made the tests of the filters against humidity and the ellipsometry measurements, respectively. Part of this work was done in the framework of the LSF facility at Rossendorf, Dresden.

REFERENCES

1. AUGER Collaboration, "Design report of the Pierre AUGER Project", March 1997.

2. Elbert J., in *Proceedings of Tokyo Conference on Techniques for the Study of Extremely High Energy Cosmic Rays*, edited by M.Nagano, 1993, 232-243.

3. Fokitis E., Maltezos S., and Papantonopoulos E., in J.H.E.P. Euroconference on the Standard Model and beyond, PRHEP-corfou98/045, 1998.

4. Vlastou R., Fokitis E., Maltezos S., Kalliabakos G., Kokkoris M., and Kossionides S., Nucl. Instr. and Meth. **B** 161-163, 590-594 (2000).

5. Doolittle L. R., Nucl. Instr. and Meth. B 9, 344-352 (1985).

6. Cheng, H.-S., Shen, H., Tang, J. and Yang, F., *Nucl. Instr. and Meth.* **B** 83, 449-454 (1993).

7. Ritter E., Applied Optics 15 No10, 2318-2327 (1976).

RBS and NRA of Cobalt Oxide Thin Films Prepared by The Sol-Gel Process

E. Andrade[a,1], L. Huerta[a], E. Barrera[b], J.C. Pineda[a], E.P. Zavala[a], M.F.Rocha[c], C.A. Vargas[d]

[a] *Instituto de Física, UNAM, Apartado Postal 20-364, México D.F. 01000, México*
[b] *Departamento de Ingeniería de Procesos e Hidráulica, CBI, UAM-I, Universidad Autónoma Metropolitana Iztapalapa, A.P. 55-534, C.P. 09340, México D.F., México.*
[c] *Escuela Superior de Ingeniería Mecánica y Eléctrica, IPN, C.P. 07738, México D.F., México.*
[d] *Laboratorio de Fenómenos Críticos y Fluidos Complejos, Depto. Ciencias Básicas, UAM-A, A.P. 16-600, , México, D.F. 02011., México*

Abstract. This work presents a study of cobalt oxide thin films produced by the sol-gel process on aluminum and glass substrates. These films have been analyzed using two ion beam analysis (IBA) techniques: a) a standard RBS ^{4}He 2 MeV and b) nuclear reaction analysis (NRA) using a 1 MeV deuterium beam. The $^{12}C(d,p_o)^{13}C$ nuclear reaction provides information that carbon is incorporated into the film structure, which could be associated to the sinterization film process. Other film measurements such as optical properties, XRD, and SEM were performed in order to complement the IBA analysis. The results show that cobalt oxide film coatings prepared by this technique have good optical properties as solar absorbers and potential uses in solar energy applications.

INTRODUCTION

Some cobalt oxide coatings as black cobalt (Co_3O_4) have shown good optical properties. This material has high values of solar absorptance α_s (0.70-0.95) and low values of emittance ε (0.1-0.2). These film properties make these materials very attractive to be used in solar energy applications due to the good thermal efficiency as solar energy collectors [1,2].

Cobalt oxide thin film materials can be obtained by a variety of techniques, for example, physical vapor deposition (PVD), chemical vapor deposition (CVD), and thermal vacuum evaporation, [3,4,].

The sol-gel technique is a very attractive process because is non vacuum, low cost and simple, and it has been used to prepare several oxide materials. However, to the authors knowledge, this method has not been used to prepare cobalt oxide coating.

In this study, a sol-gel process was used to prepare cobalt oxide films using aluminum and glass as substrates. The atomic film composition of the cobalt oxide thin films and the incorporation of some impurities must be known in order to correlate to the sol-gel production parameters such as dipping cycles and sintering process. We report here the atomic concentration of the prepared thin films and its correlation with the sol-gel deposition parameters. The atomic film composition was obtained using two IBA techniques: a) a standard RBS 2 MeV ^{4}He and b) nuclear reaction analysis (NRA) using a 1 MeV deuterium beam. Other materials characterization (XRD, SEM, etc) were also performed to complement the IBA measurements.

EXPERIMENTAL DETAILS

The cobalt oxide surfaces were prepared on aluminum frames (60mmx60mmx1mm) and microscopic Corning glass slides (76 mm x 26mm x 1 mm) as substrates. These substrates were selected in order to measure the optical properties of the films.

The sol-gel method has been described elsewhere [5], however in this section, a short description is given. We used cobalt nitride (0.1M) as precursor compound and it

---

1.Corresponding author Tel.: 52 5 622 5055; fax: 52 5 622 50 46;
e-mail: andrade@fenix.ifisicacu.unam.mx

was dissolved in distilled water. This solution was mixed with an ammonium hydroxide solution (2 M) in order to precipitate the cobalt hydroxide. This cobalt hydroxide was dissolved in acetic acid (0.01 M) in order to obtain the cobalt hydroxide acetate. The following chemical reactions are expected to occur during the sinterization process:

$Co(OH)_2 + C_2H_4O_2 \rightarrow CO(OH)(C_2H_3O_2) + H_2O$
$CO(OH)(C_2H_3O_2) + H_2O \rightarrow Co_3O_4 + CO_2$.

The substrates were dipped into the cobalt solutions and they are withdrawn with a constant speed of 1.5mm /s. The gel adhered to the substrate was allowed to dry at room temperature and it was converted into a solid film by heating the samples in an oven in air at 400 ^{0}C, during 60 min. The sample film thickness was expected to increase with the number of dipping cycles. Samples with 3, 5 and 7 dipping cycles were prepared.

The film optical measurements: the transmittance and reflectivity spectra were obtained using a Varian 5E spectrophotometer. The emittance, ε_s, of the samples, was determined at room temperature using an emisometer model AE Devices and Services Co. A Scanning Electron Microscopy, model Carl Zeiss DSM 940, was used to determine the morphology of the coatings. A Siemens X-ray diffractometer, model D500 was used to obtain the crystal properties.

The IBA facilities at the University of Mexico [6], based on a vertical single ended 5.5 MV Van de Graaff accelerator, were used to obtain the composition of the cobalt oxide films A combination of α-RBS and d-NRA is often used to obtain a better information of films such as metallic oxides. In general, the RBS is effective at measuring the content of heavy elements but it is less sensitive to light elements. NRA using a deuteron beam to induce oxygen or carbon (d, p) reactions is more sensitive to obtain the concentration of these elements. A conventional 2 MeV ^{4}He$^+$ RBS and a ^{2}H$^+$ 1 MeV NRA techniques with the ion beams at normal incidence to the sample and the detector set at $\theta = 165°$ were used to analyze the films. A surface barrier detector subtending a solid angle of 1.8msr and standard electronics were used to obtain the particle energy spectra.

No absorbing foil was used in front of the detector, in order to detect all the particles produced when the ^{2}H$^+$ beam was used to bombard the samples. A low beam current ($\cong$ 5 nA) was set in order to avoid pile-up pulses in the energy spectra.

EXPERIMENTAL RESULTS AND DISCUSSION

Fig. 1a and b show typical energy spectra from one of the cobalt oxide films on aluminum substrate bombarded with a ^{2}H$^+$ 1 MeV beam. The cobalt oxide was prepared using 7 dipping cycles Fig. 1a corresponds to the low-energy part of the spectrum arising from elastically scattered ^{2}H$^+$ particles and fig. 1b shows the high high-energy part containing the peaks from nuclear reaction (NR) products.

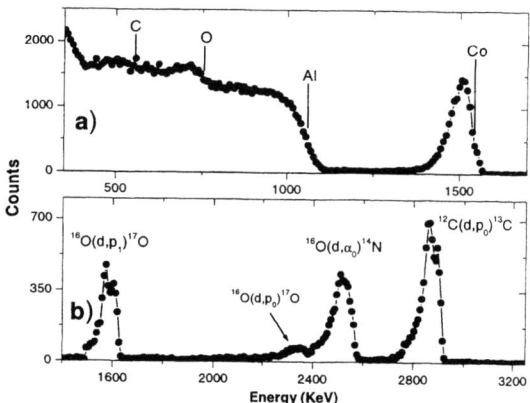

FIGURE 1. A typical charge particle spectrum from the 1 Mev ^{2}H$^+$ bombardment of a thin film cobalt oxide on aluminum substrate. The sample was prepared by sol-gel process using 7 dipping cycles. a) It is the ^{2}H$^+$ elastically region of the spectrum. b) It is the high energy region of the spectrum and the peaks arising from carbon and oxygen NR are indicated.

Using kinematic equations, Q-values, excitation energies and energy calibration, the peaks in the spectrum were identified as arising from $^{12}C(d,p_o)^{13}C$ (Q=2.7220 MeV), $^{16}O(d,p_o)^{17}O$ (Q=1.92 MeV), $^{16}O(d,p_1)^{17}O$ (Q= 1.0460 MeV), and $^{16}O(d,\alpha)^{14}N$ (Q=3.1100 MeV) nuclear reactions. At this bombardment energy the Coulomb barrier for the Al nuclei is high and the elastic cross section is about 4 orders of magnitude bigger than possible $^{27}Al(d, p)$ and $^{27}Al(d, \alpha)$ reactions, and their contribution to the spectrum yield is negligible.

The analysis of particle spectra were performed using the SIMNRA program [7] to obtain areal density (atoms/cm^2) and the composition of the cobalt oxide. This information can be obtained by simulating the elastic region of the spectrum. However, the ^{2}H$^+$ backscattered from the O and C nuclei are seen in the spectrum as weak peaks and they overlap the Al substrate yield. The errors in the determination of the concentration of O and C are large ($\approx$ 20%) due to the background Al yield. The SIMNRA software has a very extensive nuclear reaction library that includes O and C data. By fitting simultaneously the elastically and NR spectrum regions, a more accurate determination of the O and C ($\approx$ 7%) was possible.

The SIMNRA code was also used to analyze the $^4He^+$ RBS spectra. The limitation in the accuracy for obtaining O and C concentrations in the films are similar to those mentioned above. However, the advantage of using $^4He^+$ in comparison with $^2H^+$ is the better atomic depth profile resolution.

This SIMNRA fitting to the spectrum in Fig. 1a and 1b was perform with 2 layers on the Al substrate, with thickness given in monolayer units ML (10^{15} atoms/cm^2) and the atomic composition as follows: layer I: 500 ML, Co_3O_4, layer II: 350 ML, $Al_1 Co_{0.14} O_{0.53} C_{0.09}$. Layer I is the film with stoichiometry Co_3O_4, and layer II is an interface layer that shows that Co, O, C diffused in to the Al.

The IBA analysis of the samples provides very useful information related with the sol-gel process, and it is summarized as follows: 1) The cobalt oxide films produced with 3 and 5 dipping cycles show a high carbon concentration from 20% -30%. 2) In the films prepared with 7 dipping cycles no carbon is incorporated, however carbon is incorporated into the interface layer. 3) The cobalt oxide film thicknesses are almost constant ($\approx 500 \times 10^{15}$ atoms /cm^2) for the 3, 5 and 7 dipping cycles. 4) All the films prepared with 3,5 dipping cycles have an interface layer with C, O and Co diffused into the substrate.

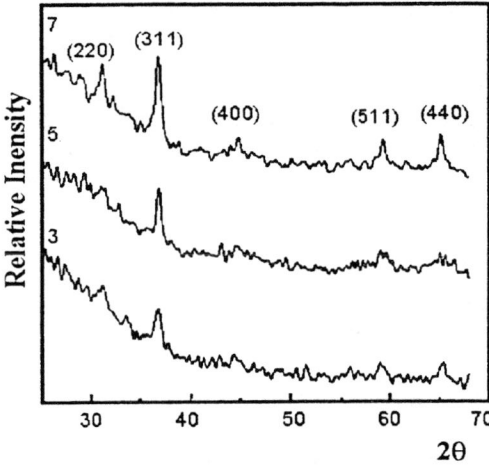

FIGURE 2. X-ray diffraction spectra of cobalt oxide coatings on glass for 3, 5 and 7 dipping cycles. The comparison of these XDR spectra with and standard sample indicate that: Co_3O_4 have been produced in the films.

Fig.2 shows a typical XRD spectrum, measured for the cobalt oxide coating over glass substrate prepared with 3, 5 and 7 dipping cycles. The comparison of these XRD spectra with standard Co_3O_4 shows that this phase is present in all the cobalt oxide films. However, the XRD spectrum that corresponds to 7 dipping cycles shows that the signal is stronger in the Co_3O_4 compared with 3 and 5 dipping cycles.

From the optical characterization of the cobalt oxide film and using the Duffie and Beckmann method [5] an absorptance $\alpha = 0.71 \pm 0.01$ and emittance $\varepsilon = 0.10 \pm 0.002$ were obtained. These values were measured in the light spectrum region λ: 200 – 2500nm. The values for α and ε were not seen to change with the number of dipping cycles.

The SEM measurements of the cobalt oxide films show a flat microstructure that did not change with the number of dipping cycles. This may be the cause of the optical parameters (α and ε) were almost constant with the dipping cycles.

CONCLUSIONS

Cobalt oxide films were prepared by sol-gel process. The IBA methods were very useful for correlating the properties of the films with the preparation procedure. The $^2H^+$ NRA of the films shows that carbon is incorporated into the films when 3 and 5 dipping cycles were used, and with 7 dipping cycles carbon is not incorporated in the film, but it is incorporated in the interface layer. The XRD spectrum intensity for samples prepare with 7 cycles show that the Co_3O_4 phase signals are stronger in comparison when the samples prepared with 3 or 5 cycles. The optical properties of the films did not change with dipping cycles used and the absorbance and emittance values are similar are those obtained with other methods.

ACKNOWLEDGMENTS

This work supported by DGAPA-UNAM project IN108798, and CONACYT No.400200-5-1776PA. We also to thank Arcadio Huerta for their help in the final revision.

REFERENCES

1. G.A. Niklasson and C.G. Granqvist, Review Surfaces for Selective Absorption of Solar Energy, Journal of Materials Science, 18, 3475-3534, (1983).
2. C.G. Granqvist, "Materials Science for Solar Energy Conversion System". Pergamon Press. Oxford, (1991).
3. Barrera C.E., Viveros G. T, Montoya A. and Ruiz M, sent to be published.
4. C.J. Brinker and G.W. Sherer, Sol-gel Science, Academic Press INC, London (1990. B)
5. J. A. Duffie and W.A: Beckman, Solar Engineering of Thermal Processes, John Wiley&Sons, New York, (1980).
6. E.Andrade, Nucl. Instr. And Meth. B 57 (1991) 799.
7. M. Mayer, SIMRA User Guide, Technical Report IPP 9/113, Max-Plank Institut Für Plasmaphysik, Garching, Germany.

Ion Beam Characterization of Advanced Metallization for ULSI Applications

A. E. Bair, Y. Wang, J. W. Mayer, and T. L. Alford

Department of Chemical and Materials Engineering
NSF Center for Low Power Electronics
Arizona State University, Tempe, AZ 85287-6006

Abstract. The self-encapsulation kinetics of Ag/Al and Ag/Ti bilayers were studied as part of the effort to introduce Ag as an alternative metallization scheme for future ultra large-scale integrated (ULSI) technologies. The amount of segregated Al was monitored by Rutherford backscattering spectrometry (RBS), which included the use of nitrogen and oxygen elastic resonance reactions. Ion beam analysis was used to show both the transport of Al and Ti through the Ag layer, as well as the formation of Al oxynitride and TiN surface layers. These techniques were used to confirm the self-passivating characteristics of $Al_xO_yN_z$ diffusion barriers formed by Ag/Al and Ag/Ti bilayers annealed between 400~725 °C.

INTRODUCTION

The future ULSI technologies will shrink the device sizes in order to achieve higher speeds and higher component density [1]. This tendency will make the *RC* delay (inherent with the metallization schemes) much more significant than before since thinner and narrower multiple metal layers will be employed for future electronic devices. This is why Ag with the lowest resistivity among all the metals has been explored as an alternative for future metallization schemes [2-6]. At the same time, this scaling tendency will also generate more stringent requirements on other issues, such as robust diffusion barriers and better thermal management. Previous studies on self-encapsulation, or holistic metallization, of Ag/Al and Ag/Ti bilayers provided good approaches for the above questions since these processes have achieved the lower resistivity (~ 1.75 $\mu\Omega$-cm) of the as-processed Ag layer and formed thin $Al_xO_yN_z$ and TiN diffusion barriers with good thermal stability [4-6]. The physical process involved in the encapsulation process is similar to the so-called surface segregation phenomena [7], which have been noticed for a long time, but had not been intensively studied until the establishment of surface sensitive analytical equipment a few decades ago. The segregation phenomena at the grain boundary also plays an important role in improving the electromigration resistance of the conventionally used Al-Cu (0.5wt%) alloy films [8]. Segregation is involved in many thin film phenomena. However, it is not a well understood process due to its complexity. Most of the work on this subject was restricted to merely predict the segregating component and the body of experimental data is rather limited. Therefore, a better understanding of segregation, including the encapsulation process, will definitely benefit the application of thin alloy films in electronic devices in the future [6].

The process of self-encapsulation of Ag/Al and Ag/Ti bilayers in ammonia ambient occurs when Al or Ti atoms diffused through Ag layers and reacted with both NH_3 and the residual O_2 in the ambient to form thin $Al_xO_yN_z$ or TiN diffusion barriers at the surface. Characterization of this process is enhanced by the use of ion beam analysis. The quantification of low Z elements is typically problematic due to the dependence of the scattering cross section σ on Z^2 when scattering occurs under Rutherford conditions [9]. The use of elastic nuclear resonances, including $^{14}N(\alpha, \alpha)^{14}N$ and $^{16}O(\alpha, \alpha)^{16}O$ provide useful means of enhancing the sensitivity toward light elements using the same experimental setup as for Rutherford backscattering. This gives the ability to study the transport of both the relatively heavy Al and Ti, and the formation of oxides and nitrides on the surface simultaneously.

EXPERIMENTAL

Bilayers of Ag/Al and Ag/Ti were deposited sequentially by electron-beam evaporation without breaking vacuum. The base and operation pressure were ~10^{-7} and 10^{-6} Torr, respectively. For the Ag/Al bilayers, self-encapsulation was performed in an AST rapid thermal annealer (RTA) with a flowing gas mixture of N_2 and 1% NH_3 for different times at 500 and 725 °C, respectively. The Ag/Ti samples were annealed at temperatures ranging from 400 to 600 °C for various times (15, 30, 60, and 120 min) in a Lindberg quartz-tube furnace with flowing electronic grade ammonia.

These films were characterized by backscattering measurements made using a General Ionex tandem Cockroft-Walton accelerator. The vacuum in the backscattering chamber was 1×10^{-7} Torr. The detector scattering angle was $170.0 \pm 0.3°$. The detector solid angle, Ω, was $1.07 \pm 0.05 \times 10^{-3}$ steradians. The total incident ion fluence was 2.5×10^{14} He^{++}. The incident ions used were 3.7 MeV He^{++} for characterizing the nitrogen content using the $^{14}N(\alpha, \alpha)^{14}N$ reaction. To characterize the oxygen content of the films, an incident ion beam of 3.04 MeV He^{++}, the energy of the $^{16}O(\alpha, \alpha)^{16}O$ reaction, was used. The oxygen and nitrogen yields were obtained from each spectra by integrating the area under the peaks after subtracting the silicon substrate signal with a linear background fit. Beam energy calibration was made using the $^{16}O(\alpha, \alpha)^{16}O$ elastic resonance reaction in a method based on Scott and Paine [10]. Each spectrum was additionally calibrated with the aid of an Americium marker which emits 5.486 MeV alpha particles. The KeV/channel conversion and the energy offset were determined by aligning the front edges of the Ag and Al or Ti signals with the aid of the simulation program RUMP [11]. With the combination of these methods the beam energy and detected energy scales were calibrated to within 5 keV. The scattering cross sections of the $^{14}N(\alpha, \alpha)^{14}N$ and $^{16}O(\alpha, \alpha)^{16}O$ elastic resonance reactions were incorporated into RUMP to allow simulations of these elements. Careful calibration of the energy of the incident ions is critical with the use of elastic resonance techniques due to the rapid change in the scattering cross section with energy.

RESULTS AND DISCUSSION

Both the schematic diagrams of sample configuration before and after the encapsulation are shown in Fig. 1 and can be used for reference throughout this paper for both bilayers. It has been shown that during the high temperature anneals (>500 °C) of Ag/Al bilayers, Al atoms did diffuse through the silver layer. There they reacted with either NH_3 or residual O_2 and formed a thin layer of Al-oxynitride on the surface [4].

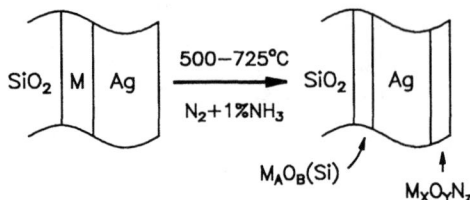

FIGURE 1. Schematic diagram of the prior to and after the self-encapsulation process.

Figure 2 shows the RBS spectra from the as-deposited and an annealed sample, which were processed with RTA for 25 minutes at 500 and 725 °C. The formation of the Al-oxynitride is confirmed by the use of the $^{14}N(\alpha, \alpha)^{14}N$ elastic resonance which shows the evolution of the nitrogen peak. The energy of the leading edge of the peak confirms that this film is on the surface of the sample. Figure 2 also displays the time evolution of the surface Al peak after various anneal times at 500 and 725 °C. Table I lists the normalized backscattering yield of the surface Al peaks as a function of anneal times, obtained from two bilayer samples, Ag(100nm) and Ag(200nm). Inspection of Table I reveals that at 725 °C the height of the Al peaks from both sets of samples increased with the anneal time; but the Al peaks remained constant for anneal times longer than 10~15 min. The peak heights after the 500 °C anneals increased much slower than in the previous case.

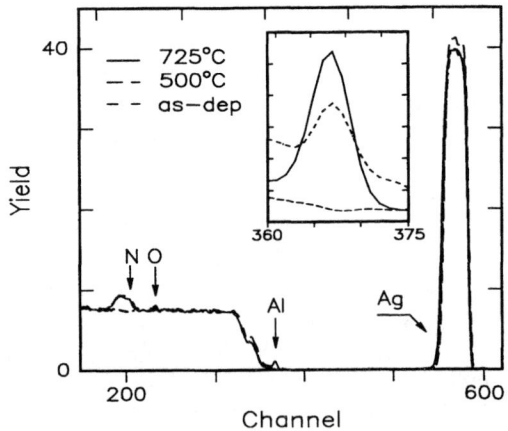

FIGURE 2. RBS spectra from the as-deposited sample and after rapid thermal annealing at 500 and 725°C.

Table I. Normalized surface Al backscattering yield of Ag(100nm) and Ag(200nm) bilayer samples and the corresponding anneal times at 725 °C and 500 °C, respectively.

Thick	Temp.	Normalized Backscattering Yield Time (min)							
(nm)	(°C)	0	3	5	10	15	20	25	30
100	500	0		0.410	0.431	0.468	0.498	0.536	0.578
100	725	0		0.915	1.048	1.106	1.097	1.097	1.097
200	725	0	0.856	1.061	1.086		1.099		1.099

The amount of surface Al atoms after the above anneals was determined by using the RUMP simulation software. These values were used to validate the proposed encapsulation model in Wang et. al. [6].

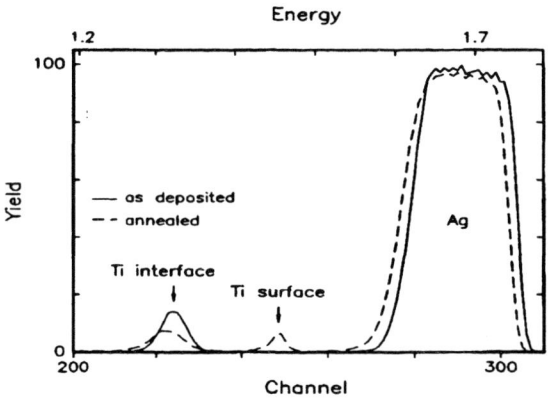

FIGURE 3. RBS spectra from the as-deposited sample and after a 25 minute anneal at 400 °C in flowing NH$_3$.

The RBS spectra of the as-deposited Ag (100 nm)/Ti (50 nm) bilayers are compared with those nitrided at 400 °C in Fig. 3. A surface segregation of Ti (peak labeled "surface Ti") was observed in the RBS spectrum after annealing the sample. The nitride layer formed at the surface has a thickness of 3-6 nm, as determine by RUMP simulation. It is noted that the Ti signal in the silver layer goes to the background level, which implies that the residual Ti in the Ag layer is less than the RBS detection limit (~1 at.%). This demonstrates the power of ion beam analysis in studying the process of self encapsulation. In one spectrum it is shown that Ti has transported through the Ag layer to form a surface film. It also shows that no more than a trace quantity remains in the silver.

After the nitridation reaction at temperatures of 400 – 600 °C, all the samples have a yellow gold color. This color is found only for compounds close to the stoichiometric composition of TiN{12}. This was verified by fitting the experimental data using a RUMP simulation, which included the cross section for the nitrogen resonance reaction.

CONCLUSIONS

It has been shown that ion beam analysis provides the ability to study the transport of heavy ions through thin films while simultaneously showing the formation of oxide and nitride films. This technique, when combined with other methods such as transmission electron microscopy, van der Pauw four-point probe and x-ray diffraction make it possible to study complex segregation phenomena and help advance ULSI technology.

ACKNOWLEDGEMENTS

The work was partially supported by The National Science Foundation, (L. Hess, Grant No. DMR-9624493), to whom the authors are greatly indebted. This work was carried out at the National Science Foundation's State/Industry/University Cooperative Research Centers' (NSF-S/I/UCRC) Center for Low Power Electronics (CLPE). CLPE is supported by the NSF (Grant #EEC-9523338), the State of Arizona, and the following companies and foundations: Burr-

Brown, Inc., Conexant, Gain Technology, Intel Corporation, Medtronic Microelectronics Center, Microchip Technology, Motorola, Inc., The Motorola Foundation, Raytheon, Texas Instruments and Western Design Center.

REFERENCES

1. T. Seidel and B. Zhao, Mat. Res. Soc. Symp. Proc., **427**, 3 (1996).
2. T. L. Alford, D. Adams, T. Laursen and B. Manfred Ullrich, Appl. Phys. Lett. **68**, 23 (1996).
3. T. L. Alford, J. Li, J.W. Mayer, and S.-Q. Wang, Thin Solid Films **262**, (1995).
4. Y. Wang and T. L. Alford, Appl. Phys. Lett. **74**, 52 (1999).
5. Y.L. Zou, T.L. Alford, Yuxiao Zeng, F. Deng, S.S. Lau, T. Laursen, A.I. Amali, and B.M. Ullrich, J. Appl. Phys. **82** (7), 1997.
6. Y. Wang, T.L. Alford, J.W. Mayer, J. Appl. Phys. **86** 10, 1999.
7. *Interfacial Segregation*, edited by W. C. Johnson and J. M. Blakely (American Society for Metals, Metal Park, Ohio, 1977), p39.
8. K.-N. Tu, James W. Mayer and L. C. Feldman, *Electronic Thin Film Science for Electrical Engineers and Materials Scientists*, (Macmillan Publishing Company, New York, 1992).
9. S.W. Russell, T.E. Levine, A.E. Bair, T.L. Alford, Nucl. Instr. And Meth. **B118** (1996).
10. D.M. Scott, B.M. Paine, Nucl. Inst. And Meth. **218** (1983).
11. L. R. Doolittle, Nucl. Inst. Meth. Res. **B9**, 344 (1985).

Simultaneous Analysis of Multiple Elements by Combined Ion-Beam Methods

W. Jiang, W.J. Weber, S. Thevuthasan, and V. Shutthanandan

Pacific Northwest National Laboratory, Richland, WA 99352

Abstract. Rutherford backscattering spectrometry (RBS) and nuclear reaction analysis (NRA) in channeling geometry have been combined to study the accumulation and recovery of Au^{2+}-induced disorder on both sublattices in 6H-SiC. Conventional He^+ RBS/channeling is used to analyze the Ga disorder and Au profiles in Au^{2+}-implanted GaN and the disorder on both the Sr and Ti sublattices in Au^{2+}-implanted $SrTiO_3$. Results on the disorder accumulation in these materials, disorder recovery in 6H-SiC, as well as the mobility of Au implants in GaN are presented and discussed.

INTRODUCTION

Rutherford backscattering spectrometry in a channeling geometry (RBS/C) has been well established since 1960s. This method has been often used to characterize implantation damage in single crystals [1-4], strain at superlattice interfaces [5], and impurity locations [6]. Complementary to the local examination of defect types by electron microscopy, RBS/C provides a quantitative and atom-identifiable depth profiling of atomic disorder in single-crystal materials. In addition, a combination of various ion-beam methods may result in simultaneous analysis of multiple elements in solids [3,7].

Both silicon carbide (SiC) and gallium nitride (GaN) are wide bandgap semiconductor materials with outstanding properties that make them promising candidates in fabrication of advanced electronic and optoelectronic devices. SiC has also been proposed as a structural component for fusion and fission reactors. Strontium titanate ($SrTiO_3$) with the perovskite structure represents a rich class of materials with applications ranging from electronic devices to immobilization of nuclear wastes. A fundamental understanding of disorder accumulation and recovery in these materials is important in using ion-implantation techniques for device fabrication and in predicting performance in nuclear applications. This paper summarizes some of our recent results on damage accumulation and recovery, as well as on Au mobility, in 6H-SiC, GaN and $SrTiO_3$.

EXPERIMENTAL PROCEDURES

Ion irradiation and in-situ RBS/C analysis have been performed with a 3.4 MV tandem accelerator within Environmental Molecular Sciences Laboratory (EMSL) at Pacific Northwest National Laboratory (PNNL). Study samples were 6H-SiC single-crystal wafers cut along (0001) plane, wurtzite GaN single-crystal films (~2.0 µm thick) epitaxially grown on c-plane sapphire substrates, and <100>-oriented $SrTiO_3$ single-crystal wafers. Irradiation experiments on these samples were carried out at low (~180 K) and room temperatures using 1 or 2 MeV Au^{2+} ions over a range of fluences. Samples were titled to 30° or 60° to produce shallow damage profiles that could be readily analyzed by ion-beam methods. In all cases, weak beam fluxes on the order of 0.01 $Au^{2+}/nm^2/sec$ were applied to avoid high-dose rate effects on damage production. For low-temperature irradiation, the implanted samples were maintained at or below those temperatures during the interim technical procedures for switching Au^{2+} to analyzing beams. In general, these procedures are considered necessary to minimize thermal recovery of defects and thermal diffusion of implants in the low-temperature-implanted specimens. In situ RBS/C measurements along <0001> or <100>-axial directions were conducted at the low temperatures using 0.94 MeV D^+ or 2.0 MeV He^+ beams at a scattering angle of 150°. Simultaneous determination of disorder on both the Si and C sublattices has been achieved based on the $^{28}Si(d,d)^{28}Si$ RBS and

$^{12}C(d,p)^{13}C$ NRA. Damage in the investigated depth region during the ion-beam analysis was insignificant. Isochronal anneals (20 min) were performed in vacuum from 300 to 870 K with a step increment of 150 K. Similar ion-beam methods were utilized at a temperature below and at 300 K for annealing cycles at and above room temperature, respectively, to minimize self-defect recombination during analysis.

RESULTS AND DISCUSSION

A series of in-situ 0.94 MeV D^+ RBS and NRA channeling spectra for 6H-SiC irradiated with 2.0 MeV Au^{2+} at 170 K to various ion fluences are shown in Fig. 1. Also included in the figure are random-equivalent and <0001>-aligned spectra from a virgin area. From Fig. 1, the Si damage peaks for the irradiated specimens are readily measurable from the D^+ RBS/C. Although the C damage peaks also appear in the RBS/C spectra, accurate analysis of disorder on the C sublattice is not straightforward because of the low C scattering yield and spectrum overlap. In addition to the RBS/C spectra, $^{12}C(d,p)^{13}C$ NRA/C spectra for the C disorder profiles are well resolvable in a background-free region. This condition allows simultaneous analysis of disorder on both the Si and C sublattices from one measurement.

The accumulated disorder at the damage peak for both the Si and C sublattices at 170 and 300 K is shown in Fig. 2 as a function of dose in units of displacements per atom (dpa). The dpa dose is an equivalent ion fluence converted using SRIM97 [8]. Complete amorphization on both the Si and C sublattices corresponds to a relative disorder of 1.0. The solid lines in Fig. 2 are data fits using the direct-impact / defect-stimulated (DI/DS) model for amorphization [9]. This sigmoidal-like dependence of relative disorder on dose has also been observed for 6H-SiC irradiated with other ion species at or below room temperature [1-3,10].

From Fig. 2, the 6H-SiC becomes fully amorphous on both the Si and C sublattices at ~0.24 dpa under Au^{2+} irradiation at 170 K. The experimental data indicate that the C disorder is slightly higher at low doses. This behavior is most likely associated with a smaller displacement energy on the C sublattice, which is consistent with molecular dynamics simulations [11] and other experimental measurements [7]. At higher doses, the residual disorder on the Si sublattice appears to be in excess of that on the C sublattice, suggesting that at higher doses a higher dynamic recovery rate may be occurring on the C sublattice during the ion irradiation. The results in Fig. 2 show that the dose to achieve the fully amorphous state increases to ~0.35 dpa at 300 K. This decrease in rate of disordering is primarily attributed to a higher dynamic recovery rate at the higher irradiation temperature (300 K). As with the irradiation at low temperatures, the results at 300 K also show a higher disordering rate on the C sublattice at low damage levels and a higher degree of disorder retained on the Si sublattice at higher damage levels.

The results from the isochronal annealing of 6H-SiC irradiated at 170 K with Au^{2+} to selected ion fluences are shown in Fig. 3 and exhibit similar recovery behavior on the Si and C sublattices. Significant recovery processes on both sublattices occur below room temperature (Stage A), between 420 and 570 K (Stage

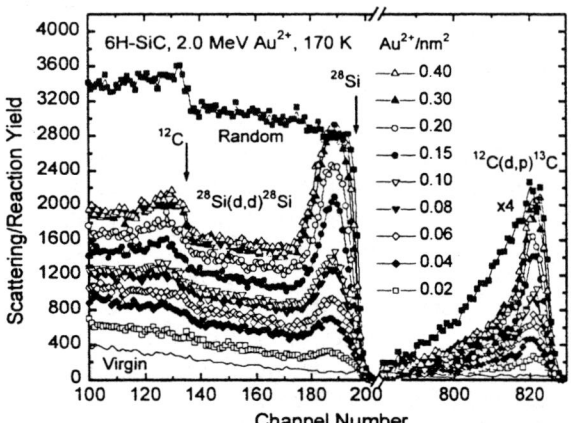

Figure 1. A sequence of in-situ 0.94 MeV D^+ RBS and NRA channeling spectra for <0001>-oriented 6H-SiC wafers irradiated 60° off surface normal at 170 K with 2.0 MeV Au^{2+} ions. Also included are random and aligned spectra from a virgin area.

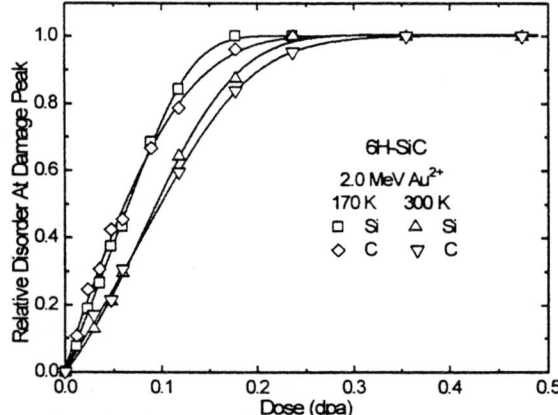

Figure 2. Relative disorder on the Si and C sublattices as a function of dose (dpa) at the damage peak for 6H-SiC irradiated with 2.0 MeV Au^{2+} at 170 K and 300 K. Solid lines are the data fits using the direct-impact / defect-stimulated model [9].

B), and above 570 K (Stage C). The activation energies for the recovery processes associated with stages A and C on the Si sublattice have been determined to be 0.3 ± 0.15 eV and 1.5 ± 0.3 eV [12], respectively, and stage B is expected to have a value on the order of 1.3 ± 0.25 eV [10]. Additional experiments are planned to determine the activation energies for all the observed recovery stages on the Si and C sublattices.

The dependence of relative Ga disorder (at both damage peak and surface) in GaN on the Au^{2+} ion dose is shown in Fig. 4. Full amorphization corresponds to 1.0 on the vertical scale. The solid lines in Fig. 4 are sigmoidal fits to the data, and the dotted lines are a simple spline fit to the data. This sigmoidal dependence of amorphous fraction on dose is consistent with the DI/DS model for amorphization [9]. The data in Fig. 4 suggests that at both 180 and 300 K amorphization at the surface occurs more readily than at the damage peak. Similar behavior in Au implanted GaN has also been reported [13]. This result might be attributed to the mobile defects diffusing into the surface region and the surface acting as a large sink for some types of defects. Clearly, more work is needed to understand this behavior. Again from Fig. 4, the surface region disorders more rapidly at 180 K than at 300 K due to the lower dynamic annealing rate at 180 K. The damage accumulation behavior in the damage peak at 300 K exhibits markedly different responses; there is a rapid accumulation of disorder below a dose of ~7 dpa, followed by an intermediate saturation stage at a disorder level of ~0.6 for doses between ~7 and ~25 dpa, and a rapid amorphization process at higher doses. This behavior has been interpreted as due to the formation of nucleation sites for amorphization [13]. Note that the full amorphization at the damage peak at 300 K arises from the extension of the surface amorphous layer [4]. Also shown in Fig. 4 for comparison are the results [14] for O^+ irradiated GaN at ~200 K. The relative disorder on the Ga sublattice from the O^+ implantation exhibits similar saturation over a comparable dose range as the sample implanted with Au^{2+} at 300 K, thus supporting the existence of the stable state. Recent TEM studies by others [13] have showed that this state constitutes defect clusters and planar defects.

Under the experimental conditions, Au spectra are well resolvable from the GaN spectra [4]. The Au distributions, determined from the RBS analysis, together with the results predicted by SRIM97 simulations are shown in Fig. 5 for different implantation fluences. The calculated curves are normalized to the same integrated Au content as the Au spectra. The as-implanted Au profile in Fig. 5 is in reasonable agreement with the SRIM97 calculations at fluences below 20 Au^{2+}/nm^2. At an ion fluence of 36 Au^{2+}/nm^2 and higher, there is a discrepancy between the experimental and calculated profiles. Because the SRIM97 simulation is a static calculation, and it does not allow for implant diffusion in host materials, the difference of the profiles may be attributed to the room temperature migration of Au atoms in GaN at higher Au concentrations. The data in Fig. 5 further suggests that the implanted Au atoms tend to diffuse into the surface region where the amorphous state of GaN is present [4]; at higher fluences, the fraction of Au migrating to the surface increases.

The accumulation of relative disorder on both the Sr and Ti sublattices in $SrTiO_3$ irradiated at 170 and 300 K with 1.0 MeV Au^{2+} ions is shown in Fig. 6 as a function of dose [15]. The relative disorder shows a sigmoidal dependence on dose with an amorphization dose of ~1.0 dpa at 300 K and of ~0.8 dpa at 170 K. The disorder on the Sr sublattice appears to show a slightly lower level than on the Ti sublattice at both 170 and 300 K. The curves shifted to higher doses because of the faster dynamic recovery processes during implantation at 300 K. Below a dose of 0.3 dpa,

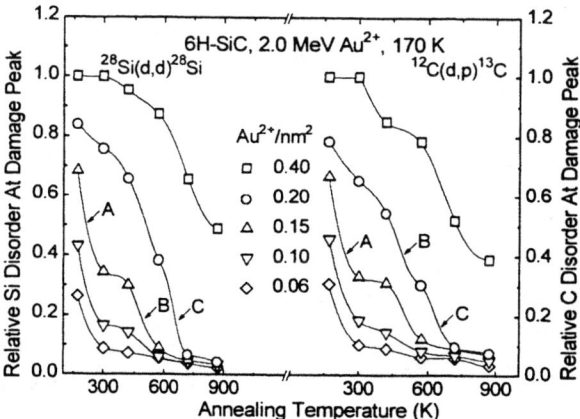

Figure 3. Isochronal recovery (20 min) of relative disorder on the Si and C sublattices in Au^{2+} irradiated 6H-SiC as a function of annealing temperature.

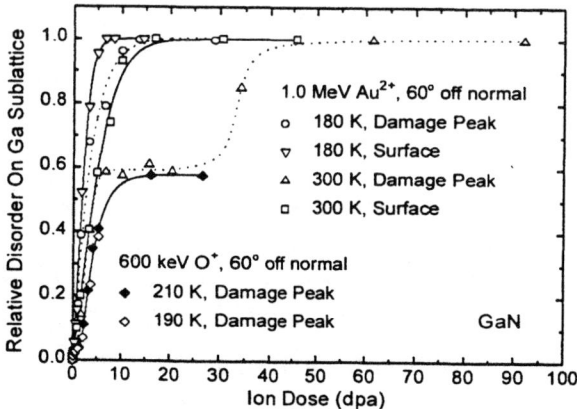

Figure 4. Relative Ga disorder as a function of dose (dpa) at the damage peak and surface for GaN irradiated with Au^{2+} at 180 and 300 K, and with O^+ at 190 and 210 K.

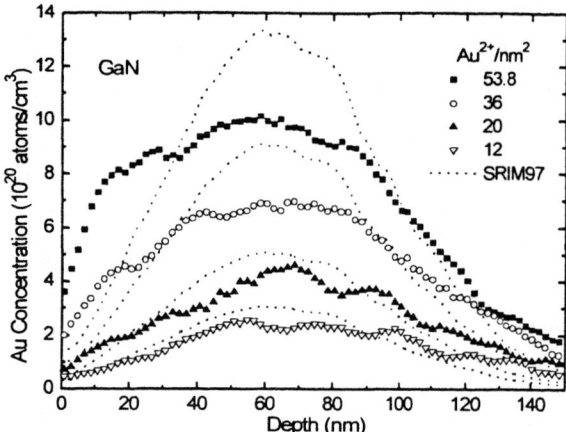

Figure 5. Depth profiles of Au atoms implanted into GaN at room temperature to different ion fluences. Also included are the profiles predicted by SRIM97.

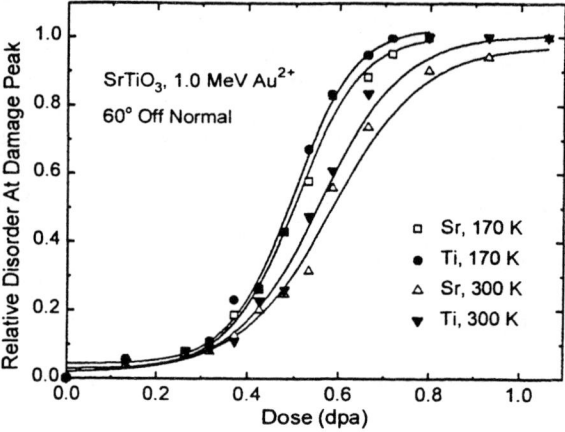

Figure 6. Relative disorder on the Sr and Ti sublattices as a function of dose (dpa) at the damage peak for $SrTiO_3$ irradiated with 1.0 MeV Au^{2+} at 170 K and 300 K.

the disordering rate on both the Sr and Ti sublattices is relatively low, probably due to a critical disorder level, above which defect-assisted amorphization processes become important.

SUMMARY

Disorder accumulation and recovery in Au^{2+}-implanted 6H-SiC have been studied using a combination of RBS/C and NRA/C. Disorder on the C sublattice is higher than on the Si sublattice at low doses, which is consistent with the smaller displacement energy on the C sublattice. Three similar recovery stages are observed for both the Si and C sublattices. Results show that the accumulation of Ga disorder in GaN at the surface is faster than in the bulk, and there exists an intermediate state in GaN implanted with Au^{2+} at room temperature. The implanted Au atoms in GaN exhibit diffusive behavior at 300 K and high doses. The disorder accumulation on both the Sr and Ti sublattices in $SrTiO_3$ shows a sigmoidal dependence on Au^{2+} dose with a slightly higher rate of disordering on the Ti sublattice.

ACKNOWLEDGMENTS

This study was supported by the Division of Materials Sciences, Office of Basic Energy Sciences, U.S. Department of Energy. Support for the accelerator facilities within the Environmental Molecular Sciences Laboratory (EMSL) was provided by the Office of Biological and Environmental Research, U.S. Department of Energy. The Pacific Northwest National Laboratory is operated by Battelle Memorial Institute for the U.S. Department of Energy under Contract DE-AC 06-76RLO 1830.

REFERENCES

1. Weber, W.J., Yu, N., Wang, L.M., and Hess, N.J., *J. Nucl. Mater.*, **244**, 258-265 (1997).
2. Weber, W.J., Yu, N., and Wang, L.M., *J. Nucl. Mater.* **253**, 53-59 (1998).
3. Jiang, W., Thevuthasan, S., Weber, W.J., and Grötzschel, R., *Nucl. Instr. and Meth. B* **161-163**, 501-504 (2000).
4. Jiang, W., Weber, W.J., and Thevuthasan, S., *J. Appl. Phys.* **87**, 7671-7678 (2000).
5. Jiang, W., Thevuthasan, S., Weber, W.J., *Appl. Phys. Lett.* **74**, 3501-3503 (1999).
6. Kobayashi, H., and Gibson, W.M., *Appl. Phys. Lett.* **74**, 2355-2357 (1999).
7. Nashiyama, I., Nishijima, T., Sakuma, E., Yoshida, S., *Nucl. Instr. and Meth. B* **33**, 599-602 (1988).
8. Ziegler, J.F., Biersack, J.P., Littmark, L., *The Stopping and Range of Ions in Solids*, Pergamon, New York, 1985.
9. Weber, W.J., *Nucl. Instr. and Meth. B* **166-167**, 98-106 (2000).
10. Weber, W.J., Jiang, W., Thevuthasan, S., *Nucl. Instr. and Meth. B*, 2000, submitted.
11. Devanathan, R., Weber, W.J., *J. Nucl. Mater.* **278**, 258-265 (2000).
12. Weber, W.J., Jiang, W., Thevuthasan, S., *Nucl. Instr. and Meth., B* **166-167**, 410-414 (2000).
13. Kucheyev, S.O., Williams, J.S., Jagadish, C., Zou, J., and Li, G., *Phys. Rev. B* **62**, 7510-7522 (2000).
14. Jiang, W., Weber, W.J., Thevuthasan, S., Exarhos, G.J., and Bozlee, B.J., *Mater. Res. Soc. Symp. Proc.* **537**, G6.15.1 (1999).
15. Thevuthasan, S., Jiang, W., Shutthanandan, V., Weber, W.J., *J. Nucl. Mater.* (2000), in press.

Double alignment channeling at CIM – Alabama A&M University

I. C. Muntele, C. I. Muntele, D. Ila, R. L. Zimmerman

Center for Irradiation of Materials, Alabama A&M University, P.O. Box 1447, Normal, AL 35762-1447

Abstract. Using a 5SDH-2 tandem accelerator, at the Center for Irradiation of Materials of Alabama A&M University we have developed facilities for ion beam based studies of materials, such as PIXE, RBS/Channeling, NRA/Hydrogen Profiling. The RBS/Channeling setup (experimental chamber and attached software for data acquisition and analysis), provided by NEC, has a goniometer with 4 degrees of freedom, two linear upon the x and z axes, and two rotational upon the theta (around the z axis) and phi (upon the y axis – beam axis). We designed an addition to this setup to allow us to perform double alignment channeling. A description of the setup, as well as double alignment channeling results are presented in this paper.

INTRODUCTION

Usually, the scattering of ions on target materials result in implantation and, with a smaller yield, backscattering. In a highly organized material, i.e. crystalline structure, there are planar and axial elements of symmetry. The alignment of the incident beam with the axes or planes of symmetry results in the phenomena of scattering at small angles along the elements of symmetry. The trajectory of the ion along the channel is dictated by a series of small angle scatters due to the coulombian interaction with the atoms bordering the channel. Due to the decreased electron density in the channels, the scattered ion experiences less inelastic collisions on the electrons of the target atoms and therefore the energy loss is decreased. In this condition the penetration depth of the ion is increased compared with the penetration depth in amorphous targets.

The presence of impurities in the host lattice at interstitial sites makes the channeled ions to backscatter. If the impurity is centered on the channel (see Fig. 1) and its mass is much higher than the mass of the incident particle, the incident ion may be backscattered at high angles on the channel. Placing the detector close to the incident direction the particle scattered under certain angles can be detected. If both the incident beam and the detector are aligned with the crystal channels, double alignment channeling occurs (Fig.2). The advantage of using double alignment channeling instead of simple channeling arise from the fact that the background given by the host lattice scattering is highly decreased. The yield of scattered ions is increased in the presence of small amounts of point defects or dislocations in the crystal.

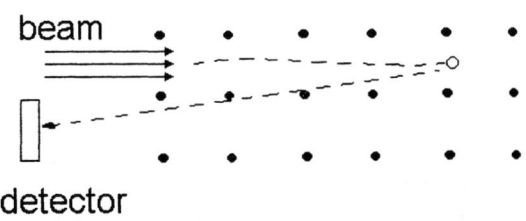

FIGURE 1. Single alignment channeling

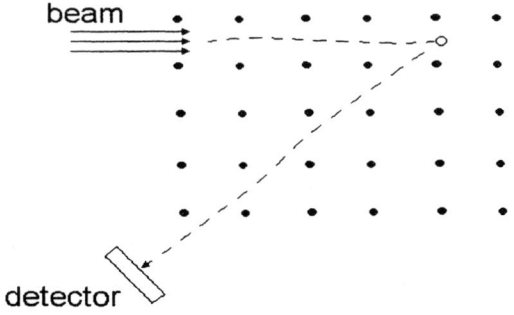

FIGURE 2. Double alignment channeling.

The experimental conditions required in order to achieve double channeling spectra are:

- Well collimated beams in 0.3-2.5 MeV range;
- Beam divergence of 0.1° or less (apertures at least 1-2 m apart);
- Beam at target ~ 1mm diameter;
- For intense micrometer sized beams the ion damage causes serious problems;
- For defect studies the sample should be cooled;
- Ideally, the sample should be surrounded by a cold shield of liquid N_2, negatively biased from the sample;
- For surface studies, UHV chamber;
- Goniometer is a crucial part; Angular precision of 0.01° is desirable for the sample tilt angle.

Most applications are with 1-2 MeV $^4He^+$ ($^1H^+$ for more depth). Too heavy ions imply more damage with little increase in resolution. Above 2 MeV 4He+, cross sections increase by a factor of 10 because of the non-Rutherford behavior ((α,α) channels on O, N, C). Although non-Rutherford BS, X-rays, or nuclear reactions are possible, usually RBS is the best choice for channeling studies of crystal perfection [1].

EXPERIMENTAL SETUP

The Center for Irradiation of Materials at Alabama A&M University has an NEC 2 MeV Pelletron - 5SDH-2 Accelerator with two ion sources (alphatross and sputtering) and a RC43 RBS chamber. In the initial phase of the experiment the chamber had a goniometer provided by NEC. We introduced a modification which allows us to modify the position of the detector (Fig. 3). This addition allows us to control the angular positioning of the silicon surface barrier (SSB) detector with respect to the incident beam's axis. This arrangement allows us to align the incident beam with the crystal orientation of the target as well as to align the detector with back scattered particles which exit along the channeling axis of the target crystal.

To calibrate the values on the goniometer, with respect of the direction of the incident beam, we use an alignment laser placed at the analyzing magnet along beam's axis.

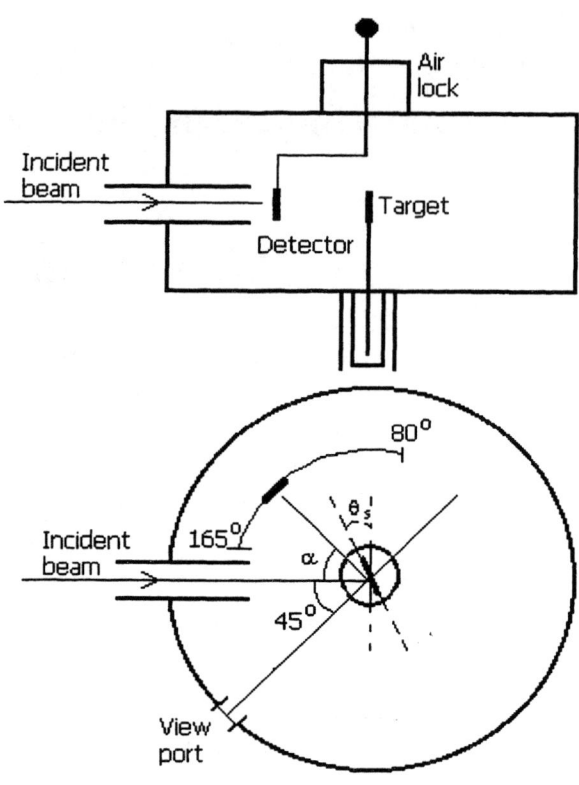

FIGURE 3 Cross sectional and top view of the modified RC43 RBS chamber

A CCD camera is placed on the viewport, at 45 degrees from the beamline. When the laser spot can be seen reflected from the sample, then the theta position of the goniometer is established to be 22.5 degrees. The next step is to calibrate the position α of the detector. This is done by catching its image on the sample. In that moment, the position of the detector is given by relation 1 (see also Figure 3).

$$\frac{\beta - \alpha}{2} = \theta_s \quad (1)$$

where β=45 deg. is the angle between the axis of the incident beam (alignment laser) and the viewport, α is the angle between the detector and the axis of the incident beam, and θ_s is the tilt angle of the sample from the normal to the incident beam.

Once the detector has its scale calibrated, the alignment of crystal axes begin with performing regular channeling, for aligning the incoming particles along the desired channel (see Fig. 4b). Once this is set, the sample remains locked in position, and the detector's angle is varied such that the outgoing channel is detected (see Fig. 4c).

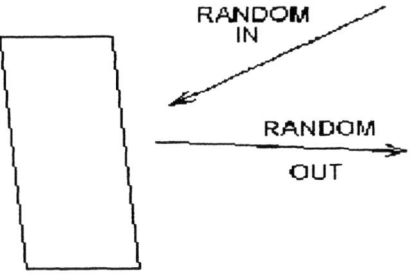

FIGURE 4a. Random RBS

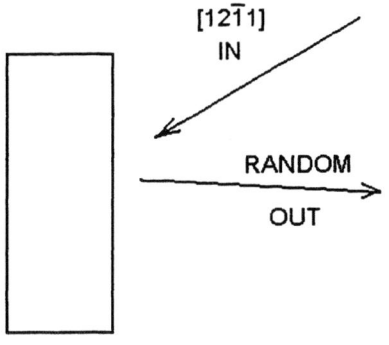

FIGURE 4b. Single channeling

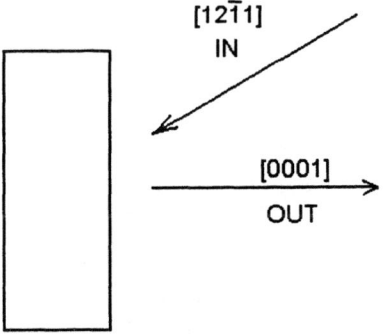

FIGURE 4c. Double alignment channeling

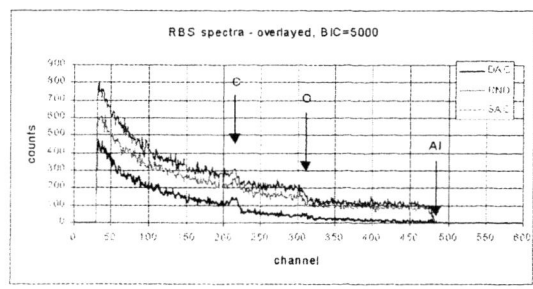

FIGURE 5. RBS spectra taken on sapphire. The carbon signal comes from the carbon tape used to hold the sample on the sample holder.

The RBS spectrum taken in the single alignment channeling configuration appears slightly smaller than the random RBS spectrum, while the RBS spectrum taken at the double alignment channeling location shows a very low baseline.

Further improvements aim at a software controllable position of the detector, and integration of it in the already available software for RBS/channeling existing at CIM.

REFERENCES

1. *Handbook of Modern Ion Beam Materials Analisys,* Material Research Society, Pittsburgh, Pennsylvania, 1995
2. *Backscattering Spectroscopy,* Wei-Kan Chu, James W. Mayer, Marc-A. Nicolet, Academic Press Inc., 1978
3. *Ion Implantation and Annealing of Crystalline Oxides,* C. W. White, C. J. McHargue, P. S. Sklad, L. A. Boatner, G. C. Farlow, Material Science Reports 4 (1989) 41-146

Investigation of Alkali Ion Exchange Processes in Waste Glasses Using Rutherford Backscattering Spectrometry and Nuclear Reaction Analysis

V. Shutthanandan[a], S. Thevuthasan[a], D. R. Baer[a] E. M. Adams[a], S. Maheswaran[b], M. H. Engelhard[a], J. P. Icenhower[a], and B. P. McGrail[a]

[a]*Pacific Northwest National Laboratory, Richland, WA 99352, USA*
[b]*School of Science, University of Western Sydney-Nepean, Kingswood, NSW 2747, Australia*

Abstract. A series of dissolution experiments using isotopic labeled ($D_2^{18}O$) aqueous solution were carried out to investigate the ion exchange mechanism in Na_2O-Al_2O_3-SiO_2 glasses with fixed Na_2O and variable Al_2O_3 concentrations. The sodium removal and the deuterium and oxygen uptake in the glass coupons were measured using ion beam methods such as Rutherford backscattering spectrometry (RBS) and nuclear reaction analysis (NRA). Both RBS and NRA experimental spectra were simulated using the SIMNRA simulation code with a thin layer approximation. Na, D, and ^{18}O concentrations as a function of depth in reacted and non-reacted glasses were determined using simulated spectra. On the basis of the depth distributions of these elements, three different zones (reaction, transition and diffusion zones) were identified in both samples.

INTRODUCTION

Recent performance assessment calculations of a disposal system for low-activity waste glass at the Hanford Site, Washington, show that a Na ion-exchange reaction can effectively increase the radionuclide release rate by a factor of over 1000 [1]. Because of this, the Na ion-exchange reaction is a major factor that currently limits waste loading. The discovery of the significance of ion exchange to long-term radionuclide release rates requires simulation of the coupled processes of glass dissolution, mass transport, and chemical reactions in a complex disposal system [1]. This observation stresses the importance of understanding and minimizing (through the formulation of new glasses) alkali ion exchange.

Although recent glass interaction work has not been focused on the ion exchange process, it was the focus of numerous earlier studies. Ion exchange (in which an H^+ or H_3O^+ ion exchanges for an alkali ion (M^+) in the glass, thereby generating a hydrated layer on the glass surface) was, in fact, the primary process involved in traditional ideas of glass "leaching." The overall chemical reaction describing the process could be written as:

$$\equiv Si\text{-}O\text{-}M + H^+ \rightarrow \equiv Si\text{-}OH + M^+ \quad (1)$$

or $\equiv Si\text{-}O\text{-}M + H_3O^+ \rightarrow \equiv Si\text{-}OH + M^+ + H_2O.$ (2)

Rana and Douglas [2,3] were among the first to report on this mechanism. Because reactions (1) and (2) produce a characteristically different ratios of H/M in the hydrated layer, surface analytical techniques have been used to map these elemental distributions. These experiments provide clues as to whether reaction (1), (2), or another reaction is rate controlling. This paper describes a series of ion beam measurements related to the release of Na from Na_2O-Al_2O_3-SiO_2 glass exposed to an aqueous solution. The current work quantifies the removal of sodium and the uptake of oxygen and hydrogen in a glass that contains approximately 10 mole% Al_2O_3 using Rutherford backscattering spectrometry (RBS) and nuclear reaction analysis (NRA). Detailed descriptions of related solution and glass structure measurements will be reported in a separate publication [4].

EXPERIMENTAL

The dimensions of the glass coupons used in this study are ≈1cm x 1cm x 1.5mm and the chemical compositions of the coupons are approximately 14.6 at% of Si, 5.9 at% of Al, 17.9 at% of Na, 55.2 at% of O and 0.014 at % of Mo. Glass reaction experiments were carried out using a single pass flow-through (SPFT) apparatus [4]. Solution compositions were made up of 0.05 m THAM (tris hydroxymethyl aminomethane), an organic buffer, $D_2^{18}O$ and dissolved silicon. Subsequent to silicon addition, the solutions were pH-adjusted to the desired value. Because of cost, $D_2^{18}O$ was mixed with $D_2^{16}O$. The ratio of these concentrations was used to obtain the effective ^{18}O uptake amount.

Following exposure to solution for 1, 2, 3, 4, and 5 days, the coupons were carefully removed from the SPFT reactors and carbon coated to avoid charging during the ion beam experiments. The ion beam experiments were carried out in the accelerator facility at the Environmental Molecular Sciences Laboratory [5].

Sodium concentration in the reacted glass was determined by RBS measurements using 2.04 MeV helium ion beam at normal incidence. The backscattering spectrum was collected using a silicon surface barrier detector at a scattering angle of 135°. The D(d,p)T nuclear reaction was used to measure the D uptake into the glass samples. D nuclear reaction was collected using 0.650 MeV d^+. A surface barrier detector at the scattering angle of 170° was used with aluminized Mylar foil to reduce the backscattered d^+ ions and the alpha particle yields generated during the ion scattering and nuclear reactions respectively. Similarly, the total number of ^{18}O uptake was determined using the $^{18}O(p,\alpha)^{15}N$ nuclear reaction. The surface barrier detector at 135° scattering angle was used to collect the alpha particles generated during this nuclear reaction. The energy of the incident protons was 0.750 MeV and the energy of the alpha particles generated during the nuclear reaction was 3.25 MeV.

Theoretical simulations of RBS, D NRA and ^{18}O NRA experimental spectra were performed to extract the Na, D, and ^{18}O depth profiles using SIMNRA program [6]. In these simulations the near surface regions were divided into thin layers with the thickness ranging from 0.05 μm to 0.2 μm.

RESULTS AND DISCUSSION

An experimental RBS spectrum along with the simulated spectrum using SIMNRA from a 3 day reacted glass coupon is shown in Fig.1. The solid line in figure 1 indicates the SIMNRA results of the experimental spectrum. In the simulations, the near surface region of the coupon is divided into several thin layers (ten 80 nm thick layers near the surface and five 420 nm thick layers deeper into material) with different elemental compositions. Also the concentrations of all the elements were kept the same except those for Na, H, ^{18}O and D in these simulations. It is clear from the figure 1 that the simulated spectra agree very well with the experimental spectra.

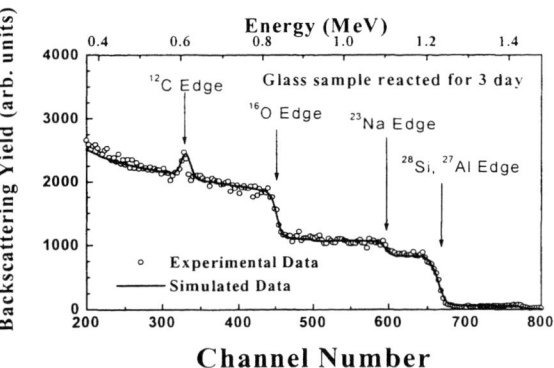

Figure 1: Experimental and simulated RBS spectra for a glass coupon that was exposed to the solution for 3 days. He^+ ions with 2.04 MeV were used for these measurements. The scattering angle was 135°.

Since the RBS results reflect the glass compositions near the surface region, the concentrations obtained from the RBS using blank glass were used in the simulations of experimental RBS, D and ^{18}O NRA spectra on reacted glass samples.

The Na depletion profile (*concentration of Na in the blank sample - concentration of Na in the reacted sample* versus *depth*) determined for a 3 day reacted glass using the simulation is presented in Fig. 2. The filled circles in the figure represents the average Na concentration from each layer and the corresponding X-value represents the midpoint of the layer thickness. The smooth thick curve represents the spline fit of these points. The profile for the reacted glass shows that approximately half of the Na is removed from the outer portion of the glass. These results suggest a relatively slow variation of Na more deep into the material and a partial removal of Na from the actual outer surface.

Fig. 3 shows a typical experimental spectrum measured during the D(d,p)T nuclear reaction along with the simulated spectrum from a glass coupon that was exposed to the solution for 3 days. The first (near channel number 675) peak is due to the proton (p_0) yield from the $^{16}O(d, p_0)^{17}O$ oxygen nuclear reaction and the second (near channel 760) peak is due to the proton yield from the D(d,p)T nuclear reaction. The solid line in Fig. 3 indicates the results from the simulations. As in the case of RBS, the simulations were performed using SIMNRA by dividing the near surface region of the sample into several thin layers. In these simulations, only the deuterium concentrations in different layers were varied while the concentrations of all the other elements that were determined from the RBS results from reacted glass coupons, were kept constant. Appropriate scattering cross sections relevant to this nuclear reaction were also used in these simulations [7]. Using these simulations it is possible to determine deuterium concentration profiles as a function of depth for each of the glass samples and the depth profile obtained for the 3 day reacted glass is shown in Fig 2 (Filled squares).

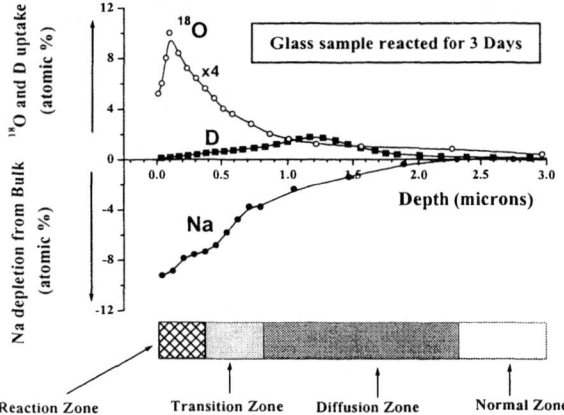

Figure 2. Na Depletion profile, D uptake profile and ^{18}O uptake profile for 3 days reacted glass sample are presented as a function of depth.

A typical ^{18}O(p,α)^{15}N nuclear reaction data from a glass coupon that was exposed to the solution for 3 day is shown with the simulated spectrum in Fig. 4. Here again the solid line in Fig. 4 represents the results from the simulations. These simulations were also performed using the SIMNRA with thin layer approximation. In these simulations, only ^{18}O concentration was varied in different layers while the concentrations of other elements in the corresponding layers were kept constant. Appropriate cross sections for this nuclear reaction were also used in these simulations [7]. ^{18}O depth profile obtained for a 3 day reacted glass sample using this simulation is also shown in Fig. 2 (open circles).

As shown in Fig. 2, a significant amount of Na has been removed from the outer portion of the glass. The Na depth profiles in the outer part of the glass include approximately 9 atomic % of Na in the first few simulated layers before the concentration gradually increases towards the bulk value. A notable feature of this result was the extent to which Na was retained in the outer or surface portions of the glass. Because RBS is not as surface sensitive as other methods, some XPS data were collected to confirm this observation. The results from the XPS depth profile measurements in the first 150 nm depth show about 10 atomic % of Na (consistent with the RBS results) in the surface.

All of the glass specimens appear to have nearly the same amount of Na in the outer surface reaction layer. It seems that the Na composition of the reaction layers quickly reached a constant "steady state" composition. There appears to be a very fast exodus of

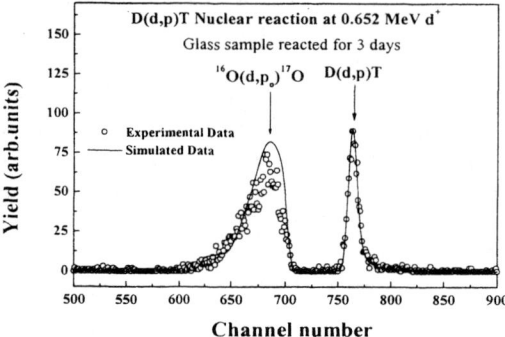

Figure 3. The NRA spectrum from the D(d,p)T nuclear reaction along with the simulated spectrum for a glass coupon that was exposed to the solution for 3 day is presented. The incident ions were 0.65 MeV d$^+$. The scattering angle was 170°.

some Na from the near surface region of the glass upon solution exposure. After this initial rapid loss, the additional Na loss appears to be at a significantly slower rate.

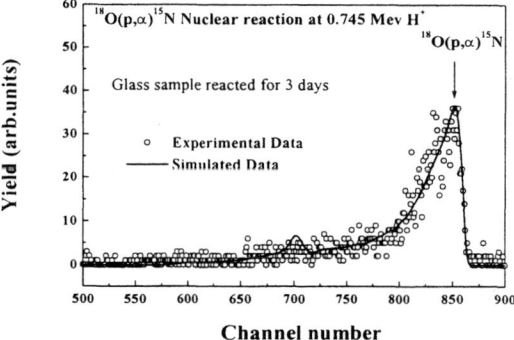

Figure 4. The NRA spectrum from the ^{18}O(p,α)^{15}N nuclear reaction along with the simulated spectrum for a glass coupon that was exposed to the solution for 3 day is presented. The incident ions were 0.75 MeV H$^+$. The scattering angle was 135°

Since Na can be lost due to ion exchange or reaction induced breakup of the matrix, the D and ^{18}O data must be used to examine glass reaction layer formation. It is clear from figure 2 that when the glass samples were exposed to the solution, both D and ^{18}O were incorporated into the glass samples. There is a general increase in amounts of ^{18}O and D incorporated in the glass with increasing exposure. A few results are readily apparent from figure 2. First, The amount of Na loss roughly mirrors the oxygen uptake, both in shape and magnitude. Second, the deuterium incorporated into the reacted glass is at a significantly different depth than the ^{18}O. Although the depth of the maximum ^{18}O concentration occurs around 0.25 µm, the peak of the D concentration appears near one micron depth. The deuterium accumulates in the region where the Na concentration is increasing toward the bulk value.

Based upon Na depletion profile, D uptake profile, and the ^{18}O uptake profile shown in fig. 2 it appears that outer regions of the glass exposed to the saturated solution can be divided into roughly three zones as a function of depth. A region containing the largest amount of ^{18}O near the surface involves the reaction of the labeled water with the silica backbone of the glass. This region is identified as the **reaction zone**. If the thickness is defined as the depth in which the ^{18}O drops to 50% of the maximum value, the reacted layer thickness would have been somewhat less than 0.5 µm. The D concentration in this region is very low. The dominant process visible in this region is the breaking and reforming of Si-O-Si bonds. The net result will be the inclusion of ^{18}O in the glass but the retention of little D as shown in Fig. 2. A significant amount but less than half, of the Na is removed from this region.

The region closest to the undisturbed glass region into the glass contains significantly less ^{18}O, the maximum amount of D and an increasing amount of Na. We believe that the dominant process in this region is ion diffusion and this region is identified as **diffusion Zone**. The ratio of ^{18}O uptake to D in this region is in the range of 0.3 to 0.5. This ratio is generally quite consistent with H_3O^+ but some H^+ and H_2O may also be present. Measurements on several specimens demonstrate very reproducible total amount of ^{18}O uptake, but the amounts of D in the samples are somewhat variable. On an apparently random basis some specimens have significantly larger uptake or retention of D in comparison to other specimens run under supposedly identical times. The exact nature of the glass structure, effects of cutting and polishing and possibly small variations in glass composition may vary somewhat and this may influence the variability in the D measurements. We attribute the consistency of the ^{18}O as due to consistent formation of the reaction layer.

The region between the reaction and diffusion zones is identified as **transition Zone** in the cartoon picture. Although there is less ^{18}O (which drops off with increasing depth) still there is readily measurable amounts of ^{18}O associated with the glass reaction process. The amount of D (associated with ion exchange related diffusion) is slowly increasing. The amount of Na loss decreases slowly, to a significant extent following related to the extent of the ^{18}O uptake.

CONCLUSION

The types of measurements achieved by the ion beam techniques provide the sample context for understanding solution measurements of ion exchange rates. The Na release in relation to the glass reaction layers is confirmed. Assuming diffusion limited ion exchange process it is also possible to use solution data to calculate a depth of Na depletion. Such calculations are usually based upon the total removal of Na. The ion beam work shows that less than half of the Na is removed even in the surface reaction layer so that such a calculation would severely underestimate that depth into the glass from which some material is removed.

ACKNOWLEDGMENT

Pacific Northwest National Laboratory is a multi-program national laboratory operated for the U.S. Department of Energy (DOE) by Battelle Memorial Institute under contract No. DE-AC06-76RLO 1830. The authors gratefully acknowledge partial support from the US DOE, Environmental Management Science Program and the Office of biological and Environmental Research.

REFERENCES

[1]. J. K. Bates, C. R. Bradley, E. C. Buck, J. C. Cunnane, W. L. Ebert, X. Feng, J. J. Mazer, D. J. Wronkiewicz, J. Sproull, W. L. Bourcier, B. P. McGrail, and M. K. Altenhofen. **DOE-EM-0177**, U.S. Department of Energy, Office of Waste Management, Washington, D.C (1994).

[2]. M. A. Rana and R. W. Douglas. *Phys. Chem.Glasses* **2**(6):(1961) 179.

[3]. M. A. Rana and R. W. Douglas. *Phys. Chem. Glasses* **2** (6) (1961) 196.

[4]. B. P. McGrail, J. P. Icenhower, D. K. Shuch, J. G. Darab, D. R. Baer, S. Thevuthasan, V. Shutthanandan, M. H. Englehard, Submitted to *J. Non-Cryst. Solids* (2000).

[5]. S. Thevuthasan, C.H.F. Peden, M.H. Enghard, D.R.Baer, G.S. Herman, W. JiangY. Liang and W.J. Weber, Nucl. Instr. Meth. A 420 (1999) 81.

[6]. *SIMNRA User's Guide*, ed. by M. Mayer, Max-Planck-Institut fur Plasmaphysik, Germany (1997).

[7]. *Handbook of Modern Ion Beam Materials Analysis*, ed. by J.R. Tesmer, M. Nastasi, Materials Research Society (1997).

Ion Beam Analysis with Monolayer Depth Resolution Using the Electrostatic Spectrometer at the MPI Stuttgart

D. Plachke, G. Blohm, Th. Fischer, A. Khellaf, O. Kruse, H. Stoll and H.D. Carstanjen

Max-Planck-Institut für Metallforschung, Stuttgart, Heisenbergstr. 1, D 70569 Stuttgart, Germany

Abstract. At the Pelletron of the Max Planck institute for metal research in Stuttgart an electrostatic spectrometer is operated which is used for high-resolution ion beam analysis using MeV ion beams. In high-resolution RBS and ERD experiments the instrument was shown to provide energy resolutions in the 1 keV range, thus resulting in monolayer depth resolution. The present contribution lines out the main features of the instrument and gives various examples for applications. These include: RBS studies on the surface segregation in Cu_3Au samples and on Stranski-Krastanov growth of ZnO during hetero-epitaxy on silicon, ERD studies on low-energy boron implants in Si as well as on how to resolve single monolayers of oxygen on copper surfaces and of carbon layers in graphite.

INTRODUCTION

During the last twenty years the analysis of surfaces and near-surface layers has gained an increasing interest. While for the surface itself there exists a variety of efficient and well established analysis techniques such as LEED, AES, XPS, STM and AFM, there is only a small number of methods available for the analysis of near-surface layers with thicknesses of 1 to 10 nm. Among these are SIMS and related techniques, transmission electron microscopy, X-ray surface diffraction, RHEED and ellipsometry the perhaps most powerful. All these methods have their merits, but also their disadvantages. SIMS is destructive and has only a limited depth resolution (generally 2 to 3 nm, but in best cases 1 nm), the sample preparation for TEM is rather complicated and tiresome, X-ray diffraction techniques and RHEED require very flat surfaces. Both of these as well as ellipsometry are not element-specific. So other techniques which allow element-specific depth profiling with depth resolutions in the monolayer range are highly welcome. Such techniques are high resolution RBS and ERD (they are called HRBS and HERDA in the following) which are the subject of this contribution.

In principle RBS and ERD allow depth profiling of even very thin surface layers (below 1 nm) with reasonable depth resolutions (of the order of atomic monolayers), since the mean energy loss of the particles in a material per unit length, which provides the depth scale, is – though being a statistical quantity – a well defined function of depth. The limiting fact, indeed, has been the limited energy resolution of the commonly-used silicon surface barrier detectors which for standard detectors is of the order of 14 keV for light ions such as protons and alphas. An estimate for 1 MeV alphas in Au shows that the depth resolution obtainable for normal incidence is of the order of 10 nm. For grazing incidence (e.g. 12°) this improves by about a factor of 5, yielding a depth resolution of 2 nm which, however, is still off from a monolayer resolution by a factor of 10.

In Stuttgart our solution to this problem was to set up an electrostatic spectrometer for MeV particles [1] which allows depth profiling by RBS and ERD with monolayer depth resolution [2]. A short description of the instrument is given in section 2. At first it was used for high resolution RBS purposes only. Recently various techniques were adopted to also allow it to be used for high resolution, background-free ERD measurements. The principles of this technique are included in section 2. Examples for HRBS experiments are given in section 3 and for experiments using HERDA in section 4. Particular emphasis is made on examples with monolayer depth resolution.

THE HIGH RESOLUTION ELECTROSTATIC SPECTROMETER

The spectrometer [1] consists essentially of three parts (see Fig. 1): i) an electrostatic lens system (four quadrupole lenses and one hexapole), ii) the analyzer

CP576, *Application of Accelerators in Research and Industry – Sixteenth Int'l. Conf.*, edited by J. L. Duggan and I. L. Morgan
© 2001 American Institute of Physics 0-7354-0015-6/01/$18.00

(a 100° cylindrical sector field of 700 mm radius and 19.8 mm gap width) and iii) a 1-dimensional position-sensitive detector. The lens system focuses particles emitted parallel to the optical axis of the instrument onto the entrance slit of the analyzer. This design in particular allows for extended samples and beam spots without spoiling the energy resolution of the instrument due to kinematic errors. With a maximum voltage of ± 60 kV applied to the capacitor plates of the analyzer, it allows to analyze 2 MeV particles of charge state one. Since these numbers do also not depend on the particle's mass, heavy particles of the same energy can be analyzed as well.

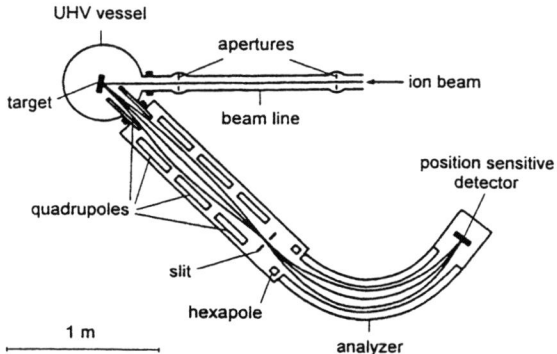

FIGURE 1. Schematic drawing of the electrostatic spectrometer and the scattering chamber set up at the Pelletron accelerator of the Max Planck Institut für Metallforschung, Stuttgart.

The simultaneously available energy window amounts to 2.8 % of the analyzed energy; the energy resolution of the instrument is better than $3 \cdot 10^{-4}$. For the detection of the particles, 1-dimensional position-sensitive detectors are used. With silicon surface barrier detectors (lengths: 10 to 30 mm) energy resolutions in the range 1.4 to 2.0 keV were obtained for 1 MeV ions, depending on the detector size. In addition they provide rough energy discrimination which is useful for separating different groups of ions or different charge states. Channelplates yield the by far better spatial and hence energy resolution. Here energy resolutions better than 1 keV were obtained, but no energy discrimination is possible. If this is needed, as in the case of HERDA experiments, the primary ion beam is chopped and the particle time of flight through the spectrometer is measured in coincidence with the particle's position on the detector (Fig. 6, see e.g. [3,4,5]). In this way background-free HERDA spectra of e.g. oxygen at metal surfaces are obtained (see Fig. 8). In case of HERDA analysis of hydrogen and deuterium a thin foil of appropriate thickness is placed directly in front of the position-sensitive detector which discriminates backscattered particles of the primary ion beam [6].

HIGH RESOLUTION RUTHERFORD BACKERSCATTERING SPECTROSCOPY (HRBS)

Fig. 2 shows as a first example the high energy edge of a HRBS spectrum of a clean Cu sample [7]. In the figure two distinct steps are visible. They are due to the two Cu isotopes ^{63}Cu and ^{65}Cu which are present in Cu with concentrations of 69% and 31%, respectively. From the scattering kinematics (1 MeV α-particles, scattering angle: 75°) the distance of the two steps is calculated to 2.7 keV which serves as energy reference. From the fit curve to the data as calculated by RUMP an energy resolution of 0.658 keV is obtained which corresponds to a relative resolution of 0.072 %. It is the best resolution we were able to obtain so far in ion beam analysis. The spectrum was obtained by counting He ions of charge state 2 by use of a position-sensitive channelplate. The background seen at high energies arises from a small amount of singly charged He ions of half the energy which were scattered deep in the bulk of the sample. No pulsed beam was used in this experiment. Under the conditions of the experiment the energy resolution of 0.658 keV corresponds to a depth resolution of 0.3 nm in Cu. Fig. 3 shows the result of 1 MeV Ne ion backscattering (again charge state 2) by Au under the same experimental conditions [7]. In this case the energy resolution is only 0.78 keV (0.09 %), but due to the far bigger stopping power of Ne this corresponds to a depth resolution of exactly 0.1 nm. Also the influence of different slit settings can be seen clearly.

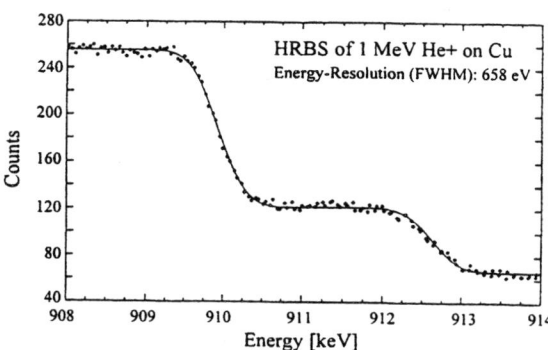

FIGURE 2. High-energy edge of a high-resolution RBS spectrum as obtained with the electrostatic spectrometer shown in Fig. 1 from a clean Cu sample (1 MeV He ions, channelplate detector) [7]. Due to the high energy resolution of the analyzer the edge is resolved into two substages corresponding to scattering by ^{63}Cu (69 %) and ^{65}Cu (31 %). The energy resolution obtained in this experiment is 0.658 keV.

The next example is dealing with the investigation of ordering phenomena at the (100)-surface of Cu_3Au crystals by HRBS [8]. Above 390 °C Cu_3Au is in a disordered state, i.e. Cu and Au atoms are distributed at random over the lattice sites. Below this temperature it starts ordering with pure (100)-Cu layers interlaced by CuAu layers. If the sample is terminated by a (100) surface the order starts with a Au-rich layer at the surface. Fig. 4 shows the result of a HRBS analysis of an ordered (100)-oriented single crystal using 3 MeV N-ions (scattering angle: 38°, incidence angle to surface: 7°). The spectrum clearly shows a first maximum in the scattering by the Au atoms at the surface and a second maximum at about 11.5 keV lower energies which is exactly the position of the second Au-containing layer, at a distance of one lattice constant (= 0.377 nm) below the surface.

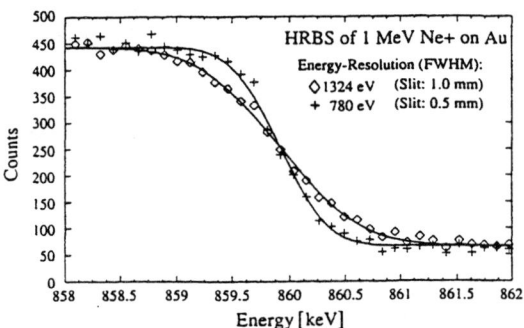

FIGURE 3. High-energy part of HRBS spectra from a clean Au sample [7] (1 MeV Ne ions, channel plate detector). The two spectra correspond to two different settings of the entrance slit of the analyzer. Due to the excellent spatial resolution of the channel plates an energy resolution of 0.78 keV could be obtained for a slit of 0.5 mm which corresponds to a depth resolution of 0.10 nm in Au.

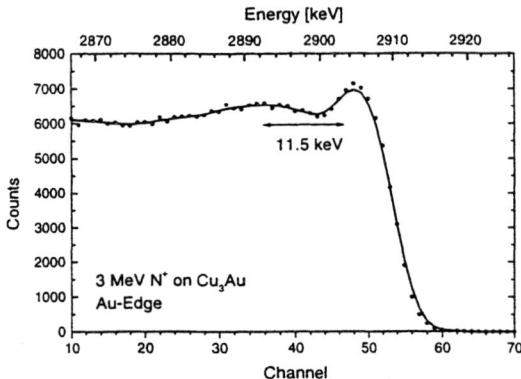

FIGURE 4. High energy part of an HRBS spectrum from an ordered (100)-oriented Cu_3Au crystal [8]. The two maxima correspond to Au-rich layers at the surface and at a distance of one lattice constant (= 0.377 nm) from the surface.

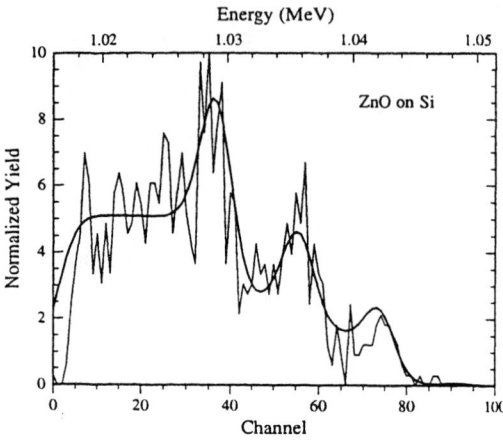

FIGURE 5. HRBS spectrum of a ZnO layer on silicon grown from an aqueous solution. Due to the three Zn isotopes ^{64}Zn, ^{66}Zn, and ^{68}Zn with higher abundance the spectrum consists of three partial spectra. The shapes of the spectra are indicative of Stranski-Krastanov growth.

The last example of this paragraph is dealing with the analysis of ZnO layers on Si surfaces deposited out of aqueous solutions. The samples have been prepared by T. Fuchs, MPI für Metallforschung, Stuttgart. By scanning electron microscopy it was observed that during deposition small isolated crystallites of ZnO grow at the Si surface which cover about 4 % of the total surface. The question now was if there in addition existed a thin layer of ZnO covering the Si surface according to a Stranski-Krastanov growth. Since the ZnO crystallites stick out considerably (to a height of several micrometer) no STM techniques could be applied for analysis. Also AES failed since the electron beam probably destroyed the water containing layers. Fig. 5 shows the high-energy edge of the Zn HRBS spectrum obtained from such a sample. The spectrum exhibits a 3-step structure according to the three isotopes of natural Zn (^{64}Zn, ^{66}Zn and ^{68}Zn with the abundances of 49 %, 28 % and 18.5 %, respectively). As outlined in detail in ref. [1], such spectra represent (besides a convolution with the resolution function and, here, the superposition of three partial spectra because of the three Zn isotopes) directly the thickness distribution of the layer if the spectrum is taken in backward direction. Here we used 1.25 MeV He ions and a scattering angle of 135° which is an almost backward direction. It also explains the low count rates seen in the experiment. The height of the spectra (when compared with a bulk sample consisting of the same material) is a measure of the coverage of the surface by the deposited material. So in the present case the peaks at the energy edges of the steps correspond to thin homogeneous layers of ZnO completely covering the Si substrate, while the lower parts of the steps extending to the left arise from the crystallites which cover the surface to a small fraction. A detailed analysis using a simulation with RUMP (smooth lines in Fig. 5) yields that the thickness of the homogenous layer is almost exactly 2 monolayers (0.5 nm) in this particular case, while the coverage by crystallites is about 4 %.

HIGH RESOLUTION ELASTIC RECOIL DETECTION ANALYSIS (HERDA) OF BORON, CARBON AND OXYGEN

As mentioned before, if one is to analyze light atoms like carbon, nitrogen or oxygen by ERD, one has to discriminate the particles under consideration against the backscattered particles (e.g. Ne or Ar) of the primary beam as well as against unwanted recoil atoms or particles of other charge states. This can not be done by a stopper foil since the ranges of these particles in such a foil are very similar. In our case it is done by measuring the particle energy in coincidence with the particle time of flight through the spectrometer using a chopped particle beam [2,3]. Fig. 6 shows the result of such a 2-parameter measurement in a 3-dimensional presentation [4,5]. The sample is oxidized aluminium. As a primary beam 1.5 MeV Ar ions were used. As the figure shows, the backscattered Ar ions as well as the recoil Al atoms are well separated from the recoil O atoms.

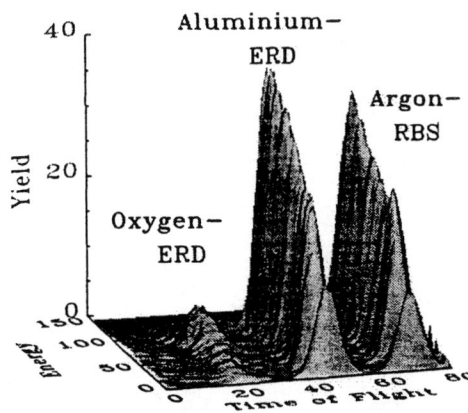

FIGURE 6. Intensity versus time of flight and energy plot of particles ejected from an oxidized aluminum sample during a HERDA experiment (1 channel corresponds to 13 ns and 0.26 keV, respectively). The backscattered Ar^{2+} ions are well separated from the Al^{2+} and O^{2+} HERDA particles. Primary beam: 1.5 MeV Ar^+ ions.

Fig. 7 shows as a first example the depth profile of 2 keV boron ($4 \cdot 10^{15}$ at/cm²) implanted into Si. For this experiment a 3 MeV Ne ion beam was used (scattering angle: 43°). The mean range of the B atoms well agrees with the mean range calculated by SRIM2000 (9.9 nm). However the range straggling found here is considerably bigger. For comparison HERDA measurements by A. Bergmaier and G. Dollinger in Munich are shown (indicated by 'HERDA-M' in Fig. 7) which are in good agreement with our measurements. For their experiments the authors used 170 MeV iodine ions (scattering angle: 15°).

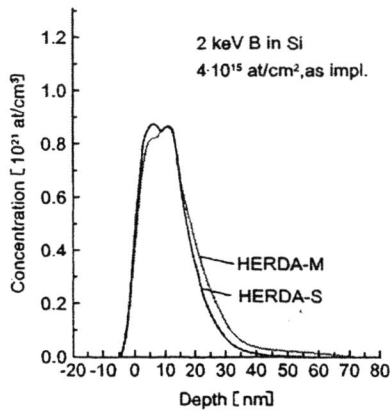

FIGURE 7. Depth profiles of 2 keV boron implanted in Si as obtained from HERDA experiments in Stuttgart and by the group of G. Dollinger in Munich (indicated by 'S' and 'M', respectively).

Fig. 8 shows the analysis of a thin layer (less than 1 monolayer) of the oxygen isotope ^{18}O on a clean (111) surface of a Cu single crystal. The HERDA spectrum was obtained by a 1.3 MeV Ar beam and exhibits a sharp peak with a width of 1.48 keV (FWHM). For the condition of the present experiment (scattering angle: 38°, incidence angle: 19° to surface) this corresponds to a depth resolution of 0.17 nm in Cu (distance of (111) planes: 0.19 nm) and is the best resolution so far obtained by our group in HERDA experiments.

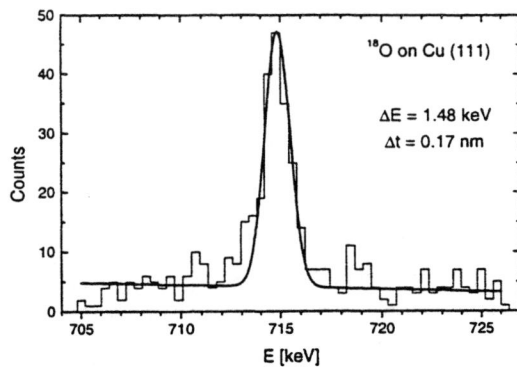

FIGURE 8. HERDA spectrum of a thin layer of ^{18}O on a clean (111)-Cu surface. The energy width of the peak of 1.48 keV corresponds to a depth resolution of 0.17 nm in Cu.

Fig. 9 finally shows the analysis of the (0001) surface of a sample of highly-oriented pyrolytic graphite. For this experiment a beam of 1.3 MeV Ar ions was used with a scattering angle of 38° and an exit angle of 3.8° to the surface. In this case recoil C ions of charge state 2 were recorded. The HERDA spectrum clearly shows well pronounced oscillations. They are due to the three first monolayers of the graphite sample which are well resolved in this experiment. The distance between two adjacent maxima corresponds to the distance between two neighboured (0001) planes which is 0.34 nm in this material. For this experiment the sample was annealed at 1000°C for 1 hour and kept at this temperature during the measurement in order to allow the annealing of defects produced by the analysis. We would like to point out that the existence of these oscillations is due to the local separation of scattering and stopping: while scattering (here the recoil of the C atoms) takes place only at the positions of the carbon nuclei, i.e. essentially at the lattice planes, stopping due to interaction with electrons is occurring also between the planes, thus shifting recoils from the second layer in energy by an amount ΔE with respect to the recoils from the first layer.

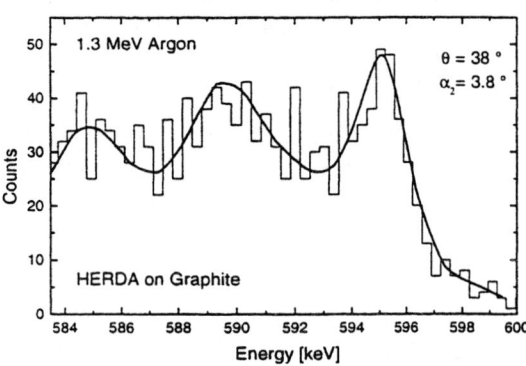

FIGURE 9. HERDA spectrum of a highly-oriented pyrolytic graphite crystal with [0001] surface orientation. The first three (0001) lattice planes are clearly visible.

ACKNOWLEDGMENTS

The author would like to thank M. Bechtel and the staff of the Pelletron accelerator at the MPI für Metallforschung in Stuttgart for kind cooperation.

REFERENCES

1. Enders, Th., Rilli, M., and Carstanjen, H.D., *Nucl. Instr. Methods* **B64**, 817 (1992)

2. Carstanjen, H.D., *Nucl. Instr. Methods* **B136-138**, 1183 – 1190 (1998)

3. Plieninger, R., Diploma-Thesis, Univ. Stuttgart 1994

4. Jamecsny, S., Plachke, D., and Carstanjen, H.D., Proceedings of the 14th Int. Conf. on the Application of Accelerators in Research and Industry, Denton, Texas 1996, edited by J.L. Dugan and I.L. Morgan, New York: AIP Press, pp. 723 – 726

5. Jamecsny, S., Kruse, O., Plachke, D., and Carstanjen, H.D., in *Microscopy of Oxidation III*, edited by S.B. Newcomb and J.A. Little, London: The Institute of Materials, 1997, pp. 369-381

6. Kruse, O., and Carstanjen, H.D., *Nucl. Instr. Methods* **B89**, 191 (1994)

7. Fischer, Th., Diploma-Thesis, Univ. Stuttgart 1996

8. Blohm, G., Diploma-Thesis, Univ. Stuttgart 1999

RBS & HFS for Advanced Interconnect

Michael D. Strathman

Thin Film Analysis, Inc.
250 Santa Ana Court
Sunnyvale, CA 94085
408-238-6351

Abstract. This paper will address the application of RBS and HFS to issues relating to existing and advanced interconnect technology. In particular the paper will address issues relating to the need for monitoring the stoichiometry of bi-elemental systems such as silicides, titanium nitrides, copper doped aluminum, and silicon nitrides. The use of a stopping power correction for improving the RBS precision to better than 1% will be presented. New dielectric films present analytical challenges to measure the elemental composition using conventional RBS techniques. Several methods for improving the analysis of Low K films, including the uses of channeling the sample substrate, will be presented. .

INTRODUCTION

Interconnect systems used for semiconductor devices continue to develop into new and complex films for both conductors and dielectrics. Many of these new films are covered in the "International Technology Roadmap"[1] including barriers, Cu, Low K and High K dielectrics. These films often require new analytical approaches to fully characterize the films.

STOPPING POWER CORRECTION

Many metallic thin film systems do not always conform to tabulated stopping power predictions[2] As shown in figure 1, this effect is particularly evident in the WSi bi metallic system. Figure 2 shows that when a slight correction to the stopping power is used, an increase of 4% in this case, the theoretical curve fits the experimental data both in the W region as well as in the Si region. Note that because the W signal is fitted in each case, the measured W atoms./cm2 remains constant, however the Si atoms/cm2 drops slightly to reflect the better fit to the experimental data (see table 1).

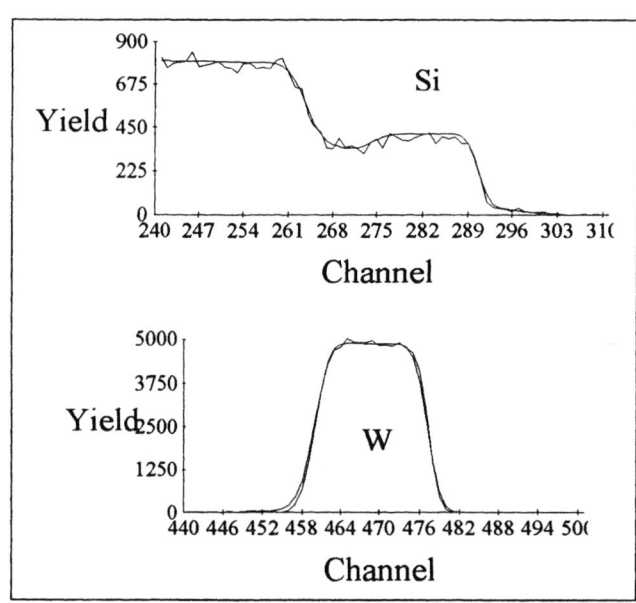

FIGURE 1. This figure shows the Si and W regions of the RBS data with no stopping power correction. Notice that although there is a good fit in the W region, the black Theoretical data is well above the Silicon data in the WSi

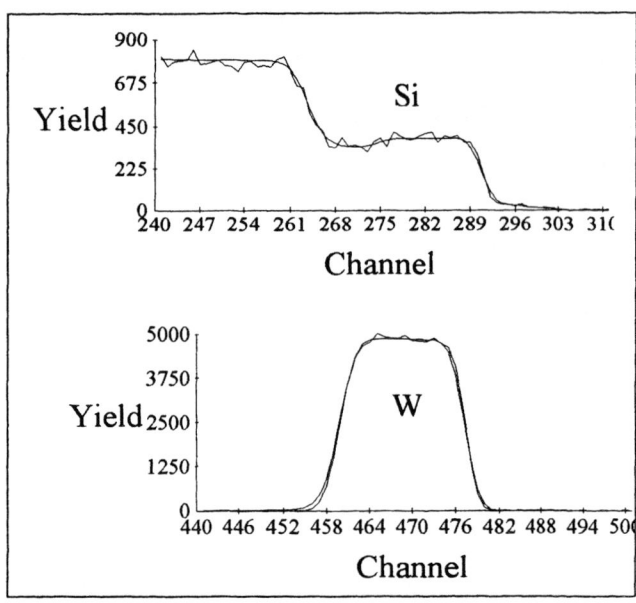

FIGURE 2. This figure shows the Si and W regions of the RBS Data with a 1.04 Stopping Power Correction. Notice the improved fit to the Silicon Data in the WSi region.

I have observed many pure elemental systems such as Cu that require stopping corrections as large as 10% to obtain a good fit to the experimental data. Many bi metallic systems have stopping power correction factors that fall in the 4-8% range. Some deposition parameters seem to effect the required stopping power correction, I have seen WSi systems with no stopping power correction to those having as much as 8% correction required. The correction value appears to be somewhat correlated to the grain structure of the films (with columnar structures requiring more correction) however clearly more investigation into this parameter and its physical explanation is required.

A bimetallic sample has been analyzed on a weekly basis for the last four years, and during this time no change in the stoichiometry or stopping power correction has been observed. Clearly the stopping power correction is stable as a function of time in some systems.

TABLE 1. Fitting Parameters used in figures #1 and #2

Stopping Power Correction	Tungsten Atoms/cm2	Silicon Atoms/cm2	Stoichiometry
No Correction	1.84E17	4.3E17	2.33
1.04 (+4%)	1.84E17	4.0E17	2.17

MEASUREMENT LOW K FILMS

Low K films are usually made up of silicon and elements lighter than silicon. These films can contain hydrogen, carbon, nitrogen, oxygen, fluorine as well as silicon.[3] This presents a challenge for conventional RBS. As shown in Figure 3 if the low K dielectric is very thick then all of the signals overlap, making it very difficult to deconvolute the data.

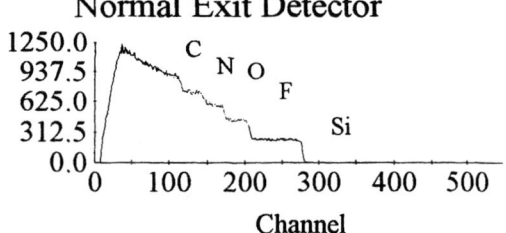

FIGURE 3. The simulated RBS data from a thick low K dielectric containing C,N, O,F, and Si.

However by utilizing one experimental design improvement and two analytical improvements, low K films can be analyzed. The first improvement is in the design of the film which is to be characterized. If the film is made to be less than about 2000 angstroms two very important things happen. First the signals from the individual elements do not overlap, and second the film is now thin enough so that the underlying substrate can be channeled. By channeling the substrate, the first analytical improvement, the signal to noise ratio in the region for the light elements is greatly improved. Next by using the grazing exit detector, which is located near a 110 backscattering angle, the scattering cross sections of the light elements are increased relative to the silicon substrate.[4]

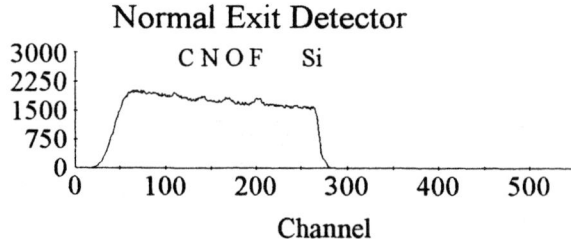

FIGURE 4. The simulated RBS data from a thin low K dielectric containing C,N, O,F, and Si showing the separation of the peaks but poor signal to noise.

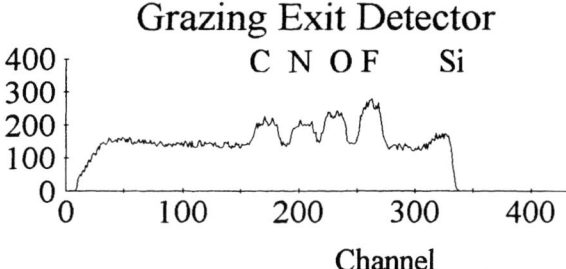

FIGURE 5. The simulated RBS data from a thin (2000 angstrom) low K dielectric containing C,N, O,F, and Si showing the separation of the peaks, and improved signal to noise by channeling the substrate.

It has been found experimentally that for films less than 2000 angstroms thick that the yield from the silicon substrate can be reduced routinely by 80% and greater in some cases. This allows the following detection limits and precision to be obtained for low K films.

Table 2. Typical Error and Detection Limits for LowK films < 2000 angstroms.

Element	Error Bar	Detection Limit
H	2%	1%
C	6%	4%
N	5%	3%
O	3%	3%
F	2%	2%
Si	1.5%	1%

CONCLUSION

Careful design of the thin film system to be analyzed and optimization of the experimental acquisition parameters are required to obtain the maximum information from many of the new IC interconnect materials.

ACKNOWLEDGMENTS

The author would like to acknowledge the time, lab space, and technical discussions with Peiching Ling and Harry Kawayoshi of Advanced Material Engineering Research, Inc. (AMER) have provided to make this work possible.

REFERENCES

1. http://public.itrs.net/Files/1999_SIA_Roadmap/Home.htm

2. Chu, Mayer, and Nicolet, *Backscattering Spectrometry*, Academic Press, 1978 pp. 364-365.

3. H. Kudo, S. Takeishi, R. Shinohara and M. Yamada, DUMIC Conference, Santa Clara, CA; ISMIC-222D, p. 85, 1997

4. Chu, Mayer, and Nicolet, *Backscattering Spectrometry*, Academic Press, 1978 p. 30, Fig 2.5.

Hydrogen Analysis of Epitaxial Cu(111)/Nb(110) Multilayer using MeV Ion Beams

S. Yamamoto, H. Naramoto*

*Department of Materials Development, * Advanced Research Science Center, JAERI, 1233 Watanuki, Takasaki, Gunma, 370-1292, Japan*

Abstract. High-quality epitaxial Nb films and Cu/Nb multilayers were prepared on α-Al_2O_3 substrate with different orientations by electron beam evaporation. The characterization of these films was made with RBS/channeling and XRD. The lattice locations of hydrogen atoms in Nb(100), (110) and (111) films on r-, a-, and c-plane α-Al_2O_3 substrates were successfully performed by ^{15}N-NRA/channeling. It was confirmed that the hydrogen atoms occupied the tetrahedral interstitial sites of bcc lattice even in hetero-epitaxial Nb layer on α-Al_2O_3 substrate. The detailed hydrogen profiling in hydrogen charged multilayer films was also realized by ^{15}N-NRA. In hydrogen charged epitaxial Cu/Nb multilayer less than 20 at. %, the hydrogen concentration in Nb(110) layers between two Cu(111) layers is remarkably low in comparison with Nb(110) layer on the α-Al_2O_3 a-plane substrate.

INTRODUCTION

The growth of high-quality epitaxial Nb films on single crystal α-Al_2O_3 substrate has attracted much interest for both applied and fundamental reasons [1-3]. Recently, the influence of hydrogen introduction into single crystal Nb films and/or multilayers has been studied using XRD and ion beam analysis [4-10]. In these studies, the attention was paid for the depth-sensitive measurement of the hydrogen concentration in thin films, and several ion beam techniques were employed to obtain the detailed hydrogen profile in thin films and multilayers consistently. Especially, the excellent depth resolution has been realized by ^{15}N-NRA technique with the very narrow resonant nuclear reactions at 6.385 MeV [11].

Fcc Cu and bcc Nb have large misfit (10.7 %) even in between the close-packed Nb(110) and Cu(111) planes, and these elements are immiscible with each other even at higher temperatures. This combination is supposed to be suitable for tailoring the multilayer structure with the sharp interface for the study of hydrogen interaction with metallic thin films and multilayers [12-14]. In these studies, it is required to prepare high-quality epitaxial multilayer samples to avoid the influence by possible lattice strains.

In the present study, we have explored the suitable condition for epitaxial Cu(111)/Nb(110) multilayer on the α-Al_2O_3 a-plane substrate by electron beam evaporation under UHV condition. The structural analysis of multilayer films using RBS/channeling technique and XRD evidenced the high-quality thin films comparable to well-treated bulky crystals. The lattice location of hydrogen atoms in Nb(100), (110) and (111) films on r-, a-, and c-plane α-Al_2O_3 substrates by ^{15}N-NRA/channeling confirmed the tetrahedral interstital site occupation even in this Nb crystals. The hydrogen profiling in hydrogen charged multilayer films was made also by ^{15}N-NRA, which suggests the possible strain effect on the hydrogen distribution.

EXPERIMENT

The epitaxial Nb films and Cu/Nb multilayers were grown on the α-Al_2O_3 substrates with four different orientations using the electron beam evaporation technique [14]. The vacuum in a growth chamber was maintained around 5×10^{-8} torr with the aid of a Ti-sublimation pump and shroud cooled with liquid nitrogen. The thickness of each layer was monitored with quartz oscillators. The Nb and Cu films were deposited at the same rate of about 0.2 nm/s onto the α-Al$_2$O$_3$ substrate kept at 750°C for Nb and less than 200°C for Cu. For the hydrogen charging, samples were transferred to a hydrogenation vessel connected to a evaporation chamber and heated at 150°C for 30~60 min in H$_2$ (99.9999% purity) atmosphere purified after passing through a liquid nitrogen cold trap.

The ion beam analyses were performed using 3 MV tandem and 3 MV single stage accelerators at JAERI/Takasaki. ^{4}He$^+$ ions with the broad energy range of 2.0 to 2.7 MeV were employed to analyse the quality of epitaxial layer and crystallographic relationship of each layers with RBS/channeling technique. Backscattered particles were detected by a standard surface barrier detector at 165° to the incident beam. The crystallographic orientations between Nb films and substrates were determined by x-ray diffraction analysis using a high-resolution diffractometer (X'Pert-MRD, Philips). The rocking curve measurements were made using a Ge(220) asymmetric 4-crystal monocromater in the incident beam to select the Cu-Kα radiation.

The lattice locations of hydrogen atoms in Nb films on the α-Al$_2$O$_3$ substrates were performed using ^{1}H(^{15}N, αγ)^{12}C resonant nuclear reaction (^{15}N-NRA) at 6.385 MeV combined with channeling. The hydrogen depth profile in the Cu/Nb multilayers was determined by ^{15}N-NRA. The yield of the characteristic γ-rays of the resonant nuclear reactions was measured as a function of incident ^{15}N ion energy with a 3 inch NaI(Tl) detector placed just behind samples. To avoid the loss of hydrogen during the ion beam analysis, comparison was made between the γ-ray yields measured by increasing and decreasing modes of incident energy at the step of 40 keV.

RESULTS AND DISCUSSION

The crystal quality of grown Nb films on the α-Al$_2$O$_3$ substrate was examined using RBS/channeling. Fig. 1 illustrates 2 MeV ^{4}He$^+$ RBS spectra from the epitaxial Nb film on the α-Al$_2$O$_3$ c-plane measured under random and <111> aligned condition. The Nb film was prepared at 750°C with 100 nm thickness. The measured χ_{min} value for the <111> axis is about 0.03 at the depth of about 20 nm just behind the surface peak of the Nb film and the crystal quality of the Nb film is high comparable to that in bulky Nb single crystals. In the <111> aligned spectra of Fig. 1, there appear two peaks at both sides of the Nb layer. The peak at the higher energy side corresponds to the surface peak, and the peak at the lower energy is supposed to be the crystallographic imperfections in the interface layer adjacent to the α-Al$_2$O$_3$ c-plane. The similar results were obtained in Nb films deposited at 750°C on the a-, r- and m-planes α-Al$_2$O$_3$. The detailed XRD studies conclude that Nb(110), Nb(111), Nb(100) and Nb(211) films are grown on the a-, c-, r- and m-planes of the α-Al$_2$O$_3$ substrates, respectively. The crystal quality of the Nb films is dependent on the substrate temperature when deposition. The χ_{min} values of the Nb films deposited at the temperature in the rang of 600 to 800°C with 100 nm thickness were measured by RBS/channeling. Fig. 2 summarizes the χ_{min} values of the Nb films as a

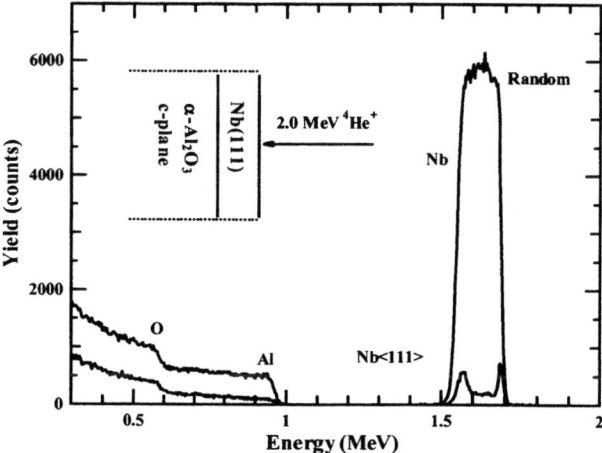

FIGURE 1. 2.0 MeV ^{4}He$^+$ RBS/channeling spectra from the deposited epitaxial Nb(111) film with 100 nm on the α-Al$_2$O$_3$ c-plane substrate. The aligned spectrum was taken with the beam directed along the <111> axis of the Nb film.

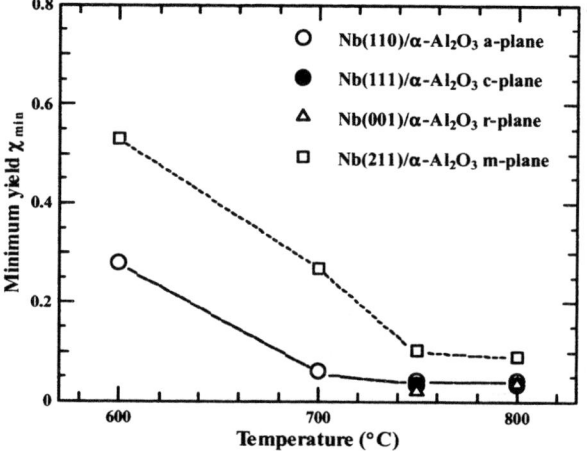

FIGURE 2. χ_{min}, relative of the axial channeling minimum yields for the Nb films on the four kind of α-Al$_2$O$_3$ substrates are plotted as a function of substrate temperature during deposition.

function of the substrate temperature. It can be shown that the α-Al_2O_3 substrates should be heated at temperature higher than 750°C for the growth of high-quality epitaxial Nb films. The FWHM of Nb(110) rocking curves in the Nb film on the α-Al_2O_3 a-plane was 0.0061°, which suggests the epitaxial Nb films with the high-quality again in an entire volume.

[15]N-NRA at 6.385 MeV combined with channeling technique is promising for the lattice-location of hydrogen atoms in epitaxial thin films because the incident ions do not induce the lattice defects in the analyzing region as a result of transmission through the relevant layer. The lattice locations of hydrogen atoms were performed under exactly the same condition in Nb(100), (110) and (111) films on r-, a-, and c-plane α-Al_2O_3 substrates by ^{15}N-NRA/channeling. The good quality Nb single crystalline films were prepared at 750°C on the c-, a- and r-plane α-Al_2O_3 and then charged with hydrogen into up to about 2.4 at. %. The Fig. 3 shows the angular scans for hydrogen and niobium around the Nb<111> axis in the α-Al_2O_3 c-plane substrate. The yield for hydrogen was obtained by detecting γ-ray yields from the nuclear reaction ^1H (^{15}N, $\alpha\gamma$) ^{12}C at 6.385 MeV. The peaking with open circles is for hydrogen and the dip with triangles is for the Nb layer. Further series of angular scan measurements around the Nb<100> axis of the Nb(100) film on the α-Al_2O_3 r-plane and the Nb<110> axis of the Nb(110) film on the α-Al_2O_3 a-plane suggest that the hydrogen atoms occupy the tetrahedral interstitial sites in bcc lattice even in epitaxial Nb layer on α-Al_2O_3 substrate.

The high-quality epitaixal Cu/Nb multilayers are realized by the combination between Cu(111) and Nb(110) layers. Fig. 4 shows the random and the aligned spectra from the Nb(140 nm)/Cu(42 nm)/Nb(48 nm) on the α-Al_2O_3 a-plane substrate. In this multilayer the deposition was made at 750 °C for the Nb layer on α-Al_2O_3, 200 °C for the Cu layer and 500°C for the top Nb layer. The results of XRD analysis confirm that the stacking of crystallographic planes is in the order of Nb(110)/Cu(111)/Nb(110) on the α-Al_2O_3 a-plane substrate and that the in-plane orientation relationship of Cu[011]//Nb[001] is kept between Cu and Nb layers. The minimum yield of the top Nb layer on the Cu layer is about 0.08, which suggests the high-quality crystal growth even at the second period in Cu/Nb bi-layer stacking. As a result of systematic RBS/channeling examination of the substrate temperature dependency on the high-quality growth of Nb and Cu, it has become clear that the substrate temperatures for the high-quality eptitaxial growth are 200 °C for the Cu layer on Nb layer, and 500°C for the Nb layer on Cu layer.

In the epitaxial Cu/Nb multilayer sample with the low hydrogen concentration, the hydrogen atoms were incorporated differently depending on the layer positions. Fig. 5 shows a typical hydrogen distribution in the epitaxial Cu(111)/Nb(110) multilayered sample composed of three Cu(50 nm)/Nb(50 nm) periods on the α-Al_2O_3 a-plane substrate. For hydrogen charging, a sample was kept at 150°C for 30 min under H_2

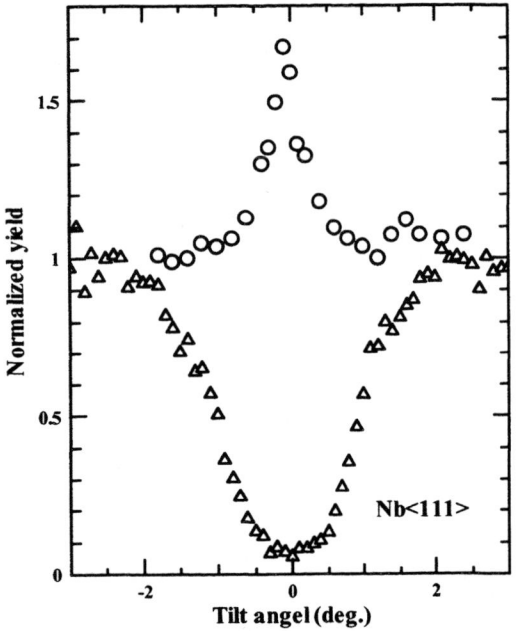

FIGURE 3. Normalized angular scans for hydrogen and Nb around Nb<111> axis in Nb film on the α-Al_2O_3 c-plane. Open circles for hydrogen were calculated from γ-ray yields of ^1H (^{15}N, $\alpha\gamma$) ^{12}C nuclear reactions and triangles correspond to the normalized backscattering yields from Nb film.

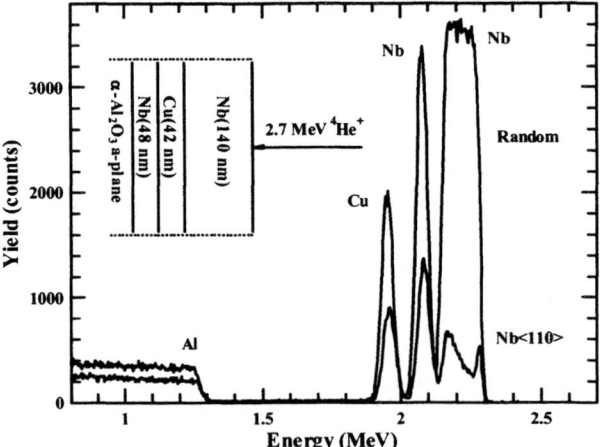

FIGURE 4. RBS/channeling spectra of the 2.7 MeV $^4He^+$ from the Cu(111)/Nb(110) layer on the α-Al_2O_3 a-plane substrate.

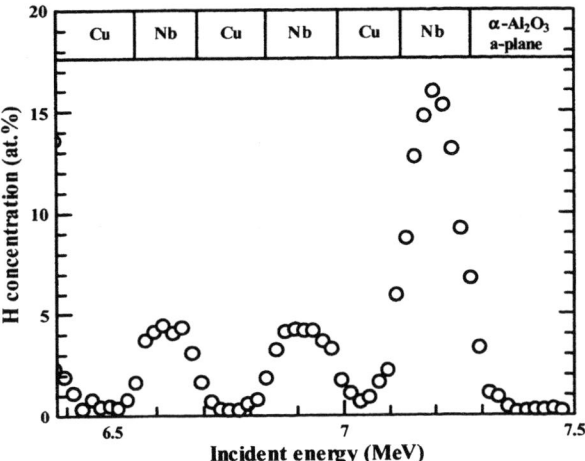

FIGURE 5. The result of ^{15}N-NRA analysis of the epitaxial Cu(111)/Nb(110) multilayer composed of three Cu(111)/Nb(110) on the α-Al$_2$O$_3$ a-plane substrate. The thicknesses of Cu and Nb layers are 50 nm each.

atmosphere. The hydrogen concentration was profiled from the top to the substrate by increasing the incident energy of ^{15}N ions. The result shows that the hydrogen atoms are transported into the interface between the Nb layer and the α-Al$_2$O$_3$ substrate through Cu and Nb layers during the hydrogen-charging process. The energy straggling analysis of incident ions concludes that there is no hydrogen incorporation in the Cu layers and the α-Al$_2$O$_3$ substrate.

The layer-dependent hydrogen incorporation was observed only in the epitaxial multilayer samples with the low hydrogen concentration. The low hydrogen concentration in the Nb(110) layers between the Cu(111) layers may result from the interface strain due to large misfit (10.7 %) between Nb(110) and Cu(111) compared with that(1.8 %) between Nb(110) and α-Al$_2$O$_3$ a-plane substrate. The misfit between Cu(111) and Nb(110), along Nb[001] direction is 10.7% and along Nb[111] direction is 22.7%. The large misfit between Nb(110) layer and Cu(111) layers generate the strain at the interface so that it suppresses the hydride formation in the inserted Nb(110) layers.

CONCLUSION

The high-quality epitaxial Nb film and Cu/Nb multilayer are formed on the α-Al$_2$O$_3$ substrate by the electron beam evaporation technique. The characterization of multilayer films was made using RBS/channeling and XRD. The lattice locations of hydrogen atoms in Nb(100), (110) and (111) films on r-, a-, and c-plane α-Al$_2$O$_3$ substrates were performed by ^{15}N-NRA/channeling. It is suggested that hydrogen atoms occupy the tetrahedral interstitial sites of the bcc lattice even in hetero-epitaxial Nb layer on α-Al$_2$O$_3$ substrate. The hydrogen profiling of H-charged multilayers was made with ^{15}N-NRA. It is found that the hydrogen incorporation in the inserted Nb(110) layer between the Cu(111) layers and the Nb(110) layer on the α-Al$_2$O$_3$ substrate are different in low hydrogen concentration region. The low hydrogen concentration in the Nb(110) layer between Cu(111) layers is probably due to the interfacial strain caused by large misfit between Nb(110) and Cu(111) crystallographic plane.

REFERENCES

1. S. Yamamoto, H. Naramoto, K. Narumi, B. Tsuchiya, Y. Aoki, H. Kudo, Nucl. Instr. and Meth. B **134**, 400-404 (1998).
2. B. Wölfing, K. Theis-Bröhl, C. Sutter and H. Zabel, J. Phys.: Condens. Matter **11**, 2669-2678 (1999).
3. E.J. Grier, M.L. Jenkins, A.K. Petford-Long, R.C.C. Ward, M.R. Wells, Thin Solid Films **358**, 94-98 (2000).
4. P.F. Miceli, H. Zabel, J.A. Dura and C.P. Flynn, J. Mat. Res., **6**, 964-968 (1991).
5. J. Steiger, S. Blässer, and A. Weidinger, Phys. Rev. B **49**, 5570-5574 (1994).
6. G. Song, M. Geitz, A. Abromeit, and H. Zabel, Phys. Rev. B **54**,14093-14101 (1996).
7. F. Klose, Ch. Rehm, D. Nagengast, H. Maletta, and A. Weidinger, Phys. Rev. Lett. **78**, 1150-1153 (1997).
8. G. Song, A. Remhof, K. Theis-Brohl, and H. Zabel, Phys. Rev. Lett. **79**, 5062-5065 (1997).
9. N.M. Jisrawi, M.W. Ruckman, T.R. Thurston, G. Reisfeld, M. Weinert, M. Strongin, M. Gurvitch, Phys. Rev. B **58**, 6585-6590 (1998).
10. Ch. Rehm, H. Fritzsche, H. Maletta, F. Klose, Phys. Rev. B **59**, 3142-3152 (1999).
11. W.A. Lanford, Nucl.Instr. and Meth. B**66**, 65-82 (1992).
12. S. Yamamoto, H. Naramoto, Y. Aoki, J. Alloy and Comp. **253-254**, 66-69 (1997).
13. S. Yamamoto, H. Naramoto, Nucl.Instr. and Meth. B **161-163**, 605-608 (2000).
14. S. Yamamoto, H. Naramoto, B. Tuchiya, K. Narumi, Y. Aoki, Thin Solid Films **335**, 85-89 (1998).

Advanced RBS Analysis of Thin Films in Micro-Electronics

B. Brijs[a], J. Deleu[a], C. Huyghebaert, S. Nauwelaerts[a], K. Nakajima[c], K. Kimura[c] and W. Vandervorst[ab]

[a]IMEC, Kapeldreef 75, B-3001 Leuven, Belgium
[b]KULeuven, INSYS, Kard. Mercierlaan 92, B-3001 leuven, Belgium
[c]KYOTO University, Yoshida-honmachi, Sakyo-ku, Kyoto 606-8501, Japan

Abstract. Rutherford Backscattering Spectrometry (RBS) is well known for quantitative compositional analysis of thin films. As the semiconductor industry heads towards shrinking device dimensions and new materials, there is a need for accurate and reliable characterization of very thin films. With a careful selection of RBS analysis conditions and instrumentation, a depth resolution down to 1 nm can be obtained. Mainly the reduction in analysis energy and improved detector resolution contribute to this enhancement. At the same time it is demonstrated that the sensitivity of RBS improves to below the 1% level. The effect of reduction in analysis energy will be illustrated in the study on the segregation process of Cu on a Al(Cu) surface. Furthermore, it will be demonstrated that the use of dedicated spectrometers to improve detector resolution leads to extreme depth resolutions allowing the analysis of SiGe layers less than 10 nm thick and the analysis of thin oxynitrides which are only 2.5 nm thick.

INTRODUCTION

Rutherford Backscattering Spectrometry (RBS) is widely applied to quantitative compositional analysis of thin films in microelectronics. The most well-known applications are directed at multi-elemental films (SiGe, CoSi, TiN, SiN, etc...) deposited on Si or oxide substrates. As the semiconductor industry heads towards shrinking device dimensions and new materials, there is a need for accurate and reliable characterization of very thin films. However in the past conventional RBS has primarily been used for thicker films. The typical depth resolution, 15 nm, is not sufficient for accurate material analysis in micro electronics. Whereas conventional RBS analysis is performed at 2 MeV, a reduction in energy can be quite advantageous. First, we will demonstrate that lowering the beam energy in conventional RBS extends the depth resolution and sensitivity of the technique. This methodology will be used to study the Cu segregation in a Al(Cu) layer. By an appropriate choice of analysis conditions and a high resolution spectrometer, depth resolution can be improved below 10 nm as will be illustrated in the analysis of thin SiGe layers. Improving the scattering geometry will transform the above methodology to a real surface technique as will be demonstrated by the analysis on ultra thin oxynitrides.

EXPERIMENTAL

In order to demonstrate the use of RBS for the characterisation of thin films in micro electronics three kind of thin films were analysed :
a) Al films with 0.21 atomic % Cu, 1 µm thick, were sputter deposited at 500°C by an Applied Material's Endura system. Next, etching of Al was performed with a LAM TCP9600, by using a Cl_2/BCl_3 (1/1) gas mixture. After etching, the wafer was transferred in situ

to the Down Stream Quartz (DSQ) chamber, where the remaining resist was stripped in an $H_2O/CF_4(8/1)$ plasma at a paddle temperature of about 190°C. After the DSQ treatment, the wafer was rinsed in the water rinse station. After the in situ strip, a wet chemistry (Microstrip 2001) was used to remove the remaining residues. Finally wafers were further treated in O_2 plasma at a temperature of 170°C for 65 min.

Parallel lines with 1 μm spacing were patterned by standard lithography using IX845 photoresist. To investigate the etched surface, blanket films were directly etched without the photoresist to approximately 50% of the original thickness (500nm).

RBS measurements on these samples were carried out in an RBS-400 endstation (CE&A), which is installed on a 6SDH Pelletron accelerator. In the present measurements, the accelerator beam has been decreased (800 and 500 keV) to improve the sensitivity as well the depth resolution. In the case of parallel lines, samples were analysed in a tilted (45°) and non tilted geometry. The detector was installed at 17° with respect to the beam. The accelerator beam was focused to a diameter of 0.5 mm by a NEC RBS lens

b) A 10 nm $Si_{0.7}Ge_{0.3}$ layer was grown in an Epsilon-One epi-reactor, built by ASM U.S.A. It is essentially a horizontal, single wafer, load locked reactor, with a lamp heated graphite susceptor in a rectangular quartz tube, generally operated at atmospheric pressure, but equipped with a dry pump for reduced pressure operation. High Resolution Rutherford Backscattering Spectrometry (H-RBS) has been used to characterise the SiGe layers. H-RBS differs from conventional RBS by its detection system. H-RBS makes use of a high-resolution energy spectrometer consisting of a 90° magnet (300 mm radius, 1200 mm dispersion) in combination with a position sensitive (dx = 0.13 mm) detector instead of the solid state detector in the classical RBS set-up. For better depth resolutions, sub MeV He ions are most suitable and in this measurement the beam energy has been lowered to 300 keV. The SiGe layers were measured under the following parameters, scatter angle 100°, exit angle 30° and total beam dose 3.5 μC.

c) Thin nitrided oxides are grown by a combination of O_2 and NO oxidation steps in a vertical furnace (ASM A400) on top of a Si wafer. The thickness of the oxynitride film (2.5 nm) was measured with a single-wavelength ellipsometer assuming a fixed refractive index of 1.465. H-RBS has been used to analyse this material. To enhance the depth resolution, the exit angle was decreased to 13° while the beam energy was 350 keV. The scatter angle of the magnetic spectrometer is fixed at 100°.

RESULTS AND DISCUSSION

Depth resolution in RBS

In order to understand the determining factor for depth resolution in RBS, a review of the underlying physics is necessary. As the high-energy particles penetrate in a layer, they loose energy as the result of interactions with elements therein. This means that a particle that backscatters from an element at some depth in a sample will have a measurable smaller energy than a particle that backscatters from the same element on the sample surface. The amount of energy a primary He ion loses per distance traversed in a sample depends on its incident energy, the elements in the sample and the scatter geometry. The ratio of the projectile energy after a collision (E_1) to the projectile before a collision (E_0) is defined as the kinematic factor K. The energy difference ΔE between ions scattered at the surface and ions backscattered from a collision at a depth x in the target can be written as (1,2)

$$\Delta E = KE_0 - E_1 \quad [1]$$

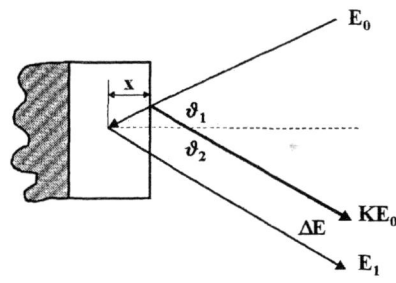

Fig. 1 *Experimental basic backscattering geometry*

The energy loss ΔE depends on the experimental conditions and can be written as

$$\Delta E = [S]x \quad [2]$$

with S the energy loss factor defined as

$$[S] = \left[K\left(\frac{dE}{dx}\right)_{in}\frac{1}{\cos\theta_1} + \left(\frac{dE}{dx}\right)_{out}\frac{1}{\cos\theta_2}\right] \quad [3]$$

with ϑ_1 the incident angle measured from the surface plane (fig. 1) and ϑ_2 the exit angle. dE/dx is the stopping power cross section on the inward and outward path. Depth resolution is defined by the ability to sense compositional changes with depth or variations in impurity distributions with depth. It can be approximated by the lowest resolvable energy difference δE assigned to the smallest resolvable depth interval δx:

$$\delta x = \frac{\delta E}{[S]} \quad [4]$$

From this equation it is clear that a better depth resolution is obtained by reducing the lowest resolvable energy difference δE. δE is determined by several contributions such as the detector resolution, energy spread in the incident beam and straggling by traversing the sample. Since straggling is proportional to $\sqrt{x}$ (x the path length inside the sample) and independent of the energy the only way to make δE as small as possible is to improve the detector resolution. From equation (4) is also clear that δx can be reduced by making the energy loss factor S as large as possible. The latter can be done by lowering the beam energy thereby increasing dE/dx and by choosing more grazing in and exit configuration. The reduction in energy is primarily advantageous as it increases at the same time the scattering cross section and thus the sensitivity.

Cu segregation in Al

In this first example the Cu segregation in an Al layer after dry etching and the subsequent cleaning is studied. The samples show Cu concentration variation between 0.1 and 1 atomic percent. To enhance the depth resolution the beam energy was decreased from the conventional 2 MeV to 0.5 MeV. This enabled the lowering of the average detection limit to 0.1 atomic percent for Cu on two etched blanket films (samples F5 and F4b) (4). Fig.2 shows the Cu spectrum obtained with RBS. When comparing spectra of the films after different treatments (F5 and F4B) a clearly higher surface Cu surface peak for F5 is detected as well as a shift to lower energy of this peak for F4b. The latter can be associated to the formation of a native oxide of different thickness (3.7 nm for F5, 2.5 nm for F4b) whereas the difference in Cu concentration at the surface is due

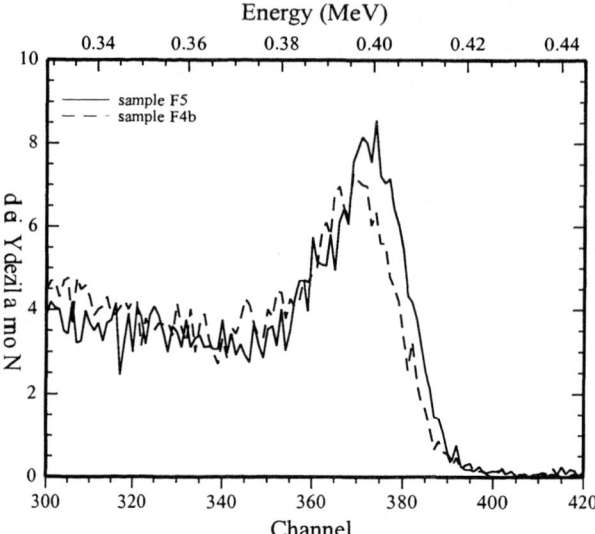

Fig. 2 *500 keV 4He$^+$ backscattering spectra of etched Al(Cu) films F5 and F4b. Only the channels belonging to Cu are presented*

to the different treatment. Based on the RBS measurements a Cu peak concentration of 0.8% with respect to the bulk value of about 0.3% is found. The Cu is distributed over a range of about 40 nm, with a peak concentration located at about 15 nm below the surface.

The next example studies the Cu segregation at the sidewalls of etched lines (0.8 μm wide and equally spaced by 1 μm) and annealed at different temperatures. In order to overcome the limited lateral resolution of conventional RBS (0.5 to 1 mm) the patterned sample was tilted by 45° to extract similar information as for the etched blanket films. Two experiments were performed : the first with the probing beam at normal incidence (fig. 3.a) and a second with the incident He$^+$ ions under 45° to the substrate normal and perpendicular to the line length direction (fig. 3.b). The analysis under 0° directly measures the surface composition of the Al(Cu) layer although with a reduced intensity (compared to the blanket film) as no Al(Cu) is present in between the lines. Under normal incidence the probing beam will not scatter at the 'surface' of the side walls and the surface layer can be seen as a region composed of 40% Al(Cu) and 60 % Si. The measurements under 45° samples the top layer as well as the side wall composition with roughly the same contribution from both layers.

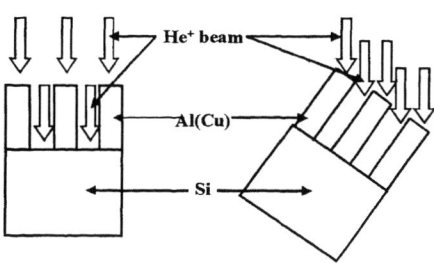

Fig. 3 Schematic view of the sample layout. In fig. 3a (left), surface scattering occurs as well on the top of the Al(Cu) layer and the Si substrate. In fig.3b (right), where the sample normal is tilted over 45°, surface scattering occurs on the top as well on the side walls of the Al(Cu) layer

.Fig. 4.a shows the 800 keV RBS results with the sample at 0° and fig. 4.b with the sample at 45°. The strong surface peak in the tilted spectra (almost absent in the 0° measurement) clearly demonstrates that a strong Cu segregation occurs on the sidewalls of the etched lines. The latter difference remains even after annealing at 250° C for 20 minutes. Annealing at 250° C for 20 min in N_2 slightly widens the range of Cu distribution. At 350°C, the Cu segregation totally disappears. The data in fig. 4.a (0° experiment) indicate no enrichment of Cu occurs in the top layer.

The latter can be explained by the fact that it was protected by photoresist during dry etching. The side wall surface is however created by the dry etching process. The above results clearly indicate that dry etching is mainly responsible for the Cu enrichment. This example clearly demonstrates that an optimised geometry between probing beam and sample can lead to characterisation with conventional RBS.

The analysis of thin SiGe layers

In the previous examples, sensitivity and depth resolution could be improved by decreasing the probing beam energy. Under these circumstances, the near surface modifications of the Al(Cu) layer due to etching and cleaning could be studied. In the next example we will study the analysis of thin (10nm) SiGe layers. In order to improve further on the depth resolution a detector set up based on a magnetic spectrometer was used leading to a

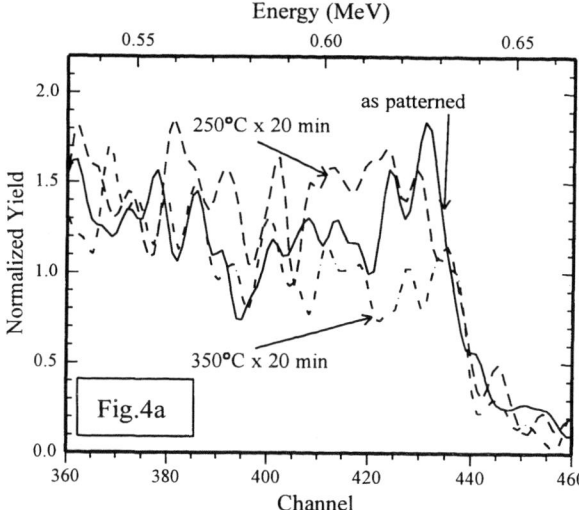

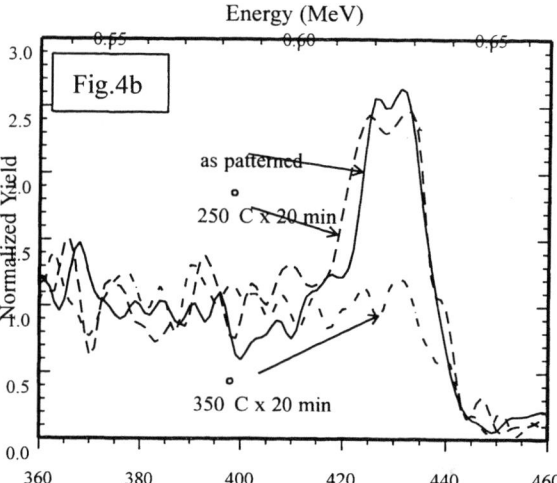

Fig.4 800 keV RBS spectra of ..0.8 µm wide and 1 µm spacing lines taken with the incident ions parallel to the sample normal and 45° to the sample normal. Only the channels belonging to Cu are presented

resolution of 1 keV. When applied to the analysis of SiGe and with the typical analysis conditions for this system (scatter angle 100°, exit angle 30°, 0.3 MeV He+) one could expect a depth resolution of better than 1 nm based on equation (4). However experimental data in fig. 5 show an apparent detector resolution of 3 keV (taken as the 16-84% on the leading edge). The latter is related to the presence of multiple isotopes of Ge with masses from 72.6 to 75.9 a.m.u. The kinematic factor differs for every isotope and as a consequence the RBS spectrum will be similar to a compound. The isotopic steps (69.9, 71.92, 72.63, 72.92, 73.92, 75.92) for Ge at the high-energy side of the spectrum can not be resolved, and will give rise to an apparently poor detector resolution. The RBS spectrum shown on fig. 5 was measured with

High Resolution RBS with a system resolution of 1 keV. Only the Ge profile is demonstrated. Experimental conditions were 0.3 MeV He+ ions, scatter angle 100°, exit angle 30°, total beam dose 3.5 μC. The measured system resolution on Ge was 3 keV induced by the kinematic factor for the different Ge isotopes which ranges from 0.8917 to 0.8998. This gives rise to a ΔE of 2.43 keV between the surface energy of Ge_{70} and Ge_{76}.

The lowest resolvable energy difference δE due to isotopic broadening under these measuring conditions is increased from 1 keV to 3 keV. As a consequence, the depth resolution is decreased with almost a factor of three. While for a mono isotopic element the depth resolution would be 0.7 nm, for Ge (due to isotopic broadening) the depth resolution is limited to 2.1 nm.

From this Ge spectrum, a Ge content of 33% and a thickness of 10-11 nm can be extracted. This Ge content is in excellent agreement with a conventional

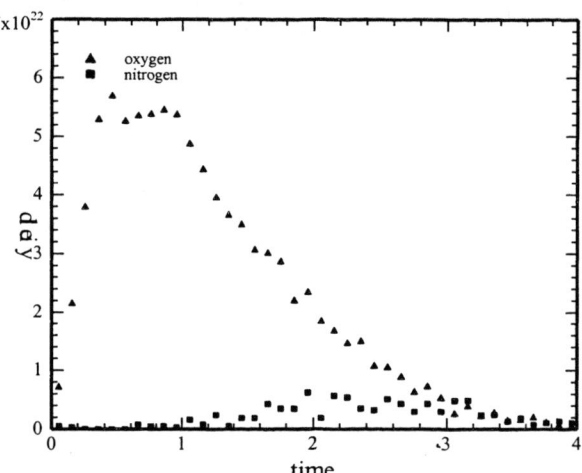

Fig. 6 *LE-SIMS spectra of a 2.5 nm oxynitride layer. Only the oxygen and the nitrogen profiles are drawn*

First, it is hard to quantify the oxygen content at the real surface, as the transient region in SIMS limits quantification. Moreover one can not exclude that these layers could also be contaminated with hydrocarbons or water. Secondly, a very long tail (≅ 1-2 nm/decade) on the oxygen profile is noticed, which would suggest a very broad SiO_2/Si interface. This is rather hard to believe since the SiO_2/Si interface is expected to be very sharp. Whereas the nitrogen profile is situated between the SiO_2 and the Si substrate, it would be nice to exclude the existence of possible artefacts and confirm the location of the nitrogen. The same sample was therefore analysed with H-RBS as shown in fig. 7. While the sensitivity of H-RBS is much lower than the sensitivity of SIMS in general, complementary information could be obtained from H-RBS.

In this measurement, the oxygen concentration could be better profiled leading to a nominal concentration of 66% over a region of 1 nm. H-RBS also reveals the same large interface. Straggling of the beam in this low energy region (350 keV) might be a possible explanation. The similarity with the interface width measured with SIMS may however also suggest a rough interface in the sample itself despite the fact that the surface roughness as deduced from AFM measurements is less than 1 nm.

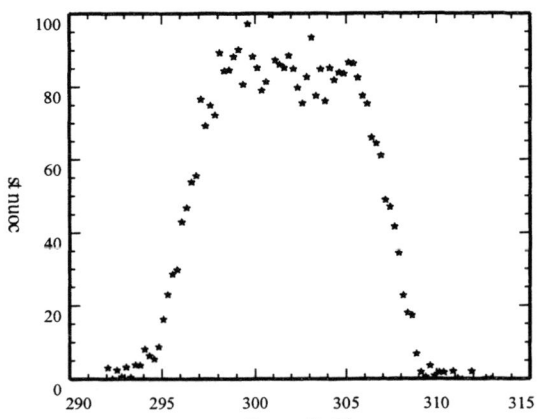

Fig. 5 *Ge yield of a 10 nm Si0.7Ge0.3 layer measured with H-RBS*

RBS measurement on a second, 48 nm thick sample, grown under identical conditions.

The analysis of 2.5 nm oxynitrydes

With future advanced semiconductor technologies and the scaling down of structure dimension, new gate dielectric will replace the SiO_2. The physical dimension of these layers will be limited to a few nanometer. While low energy SIMS is expected to be the ideal tool to characterise these layers, sputtering artefacts can not be excluded. H-RBS, less sensitive than SIMS, is expected to be a non destructive technique(7) meaning that the layer of interest will not be modified during the measurement. Due to the extreme depth resolution, H-RBS can profile even the near-surface layers which causes severe problems for SIMS Fig. 6 illustrates a low energy SIMS measurement obtained with an ATOMIKA 4500 of a 2.5 nm oxynitride. In this graph only the oxygen and nitrogen are presented.

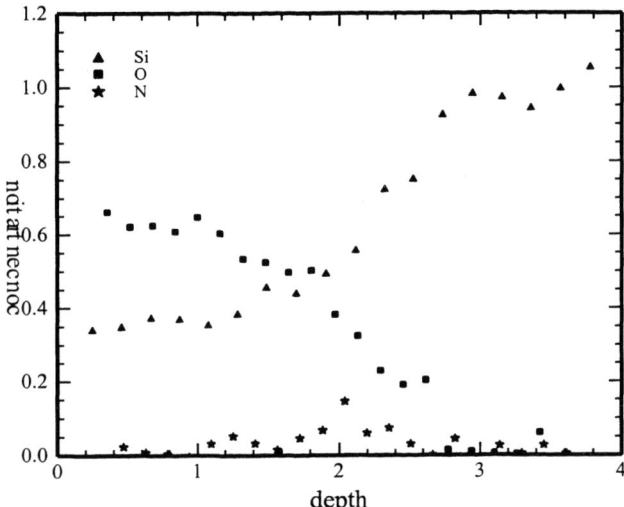

Fig. 7 *H-RBS spectrum of a 2.5 nm oxynitryde obtained with H-RBS*

While RBS is not sensitive to light elements on top of a heavy substrate, H-RBS could demonstrate the location of the N profile. Here too, the N profile was situated at the end of the SiO_2 layer in front of the interface in complete agreement with the low energy SIMS profile.

DISCUSSION

Rutherford Backscattering Spectrometry is an interesting asset to the quantitative determination of absolute concentrations of multi elemental thin films. By lowering the energy of the probing beam from the conventional 2 MeV to 0.5 MeV, the depth resolution could be improved while at the same time the sensitivity increases from a few atomic % to better than 0.5 atomic %. Under these circumstances we studied the segregation process of Cu in an Al(Cu) surface on blanket films as well as on patterned lines. Using optimised geometry it could be demonstrated that Cu segregation primarily occurs on the sidewalls of the lines due to the interaction with the etching plasma. Furthermore, using a high-resolution magnetic spectrometer in combination with the grazing exit geometry depth resolutions down to 1 nm can be obtained. This methodology has been successfully used to determine the stoichiometry of a 10 nm SiGe layer and a 2.5 nm thin oxynitryde film. The non destructive character of RBS in combination with extreme detector resolution have led to a comparative study with low energy SIMS characterisation of ultra thin oxynitrides confirming that the nitrogen located at the interface with the Si substrate was formed during the treatment of the sample and was not induced by sputter artefacts.

REFERENCES

1. W. Chu, J. Mayer and M. Nicolet, Backscattering Spectroscopy, Academic Press (1978)

2. J. Tesmer and M. Nastasi, Handbook of modern ion beam materials analysis, Materials Research Society (1995)

3. H. Li, K. Maex, B. Brijs, T. Connard, W. Vandervorst, M. Baklanov, W. Boullart and L. Froyen, Mat. Res. Soc. Symp. Proc. Vol. **516**, 77-82 (1998)

4. H. Li, M. Baklanov, W. Boullart, T. Connard, B. Brijs, K. Maex and L. Froyen (submitted to Electrochem. Soc.)

5. K. Kimura and M. Mannami, Nucl. Instrum. And Methods B, **113**, 270 (1966)

6. M. Caymax, R. Loo, B. Brijs, W. Vandervorst, D. Howard, K. Kimura, Mat. Res. Soc. Symp. Proc. Vol. **533**, 339-344 (1998)

7. Brijs, B.; Deleu, J.; Conard, T.; De Witte, H.; Vandervorst, W.; Nakajima, K.; Kimura, K.; Genchev, I.; Bermaier, A.; Goergens, L.; Neumaier, P.;Dollinger, G., and Döbeli, M. Nuclear Instruments and Methods B. Vol. 161-163:429-434; 2000.

Strain Measurements of Semiconductor Multi-layers by Ion Channeling, High Resolution XRD and Raman Spectroscopy

Azher M. Siddiqui, S. V. S. Nageswara Rao and Anand P. Pathak[*]

*School of Physics, University of Hyderabad,
Central University P. O., Hyderabad - 500 046, India*

The ion-beam channeling technique has been used, along with other characterization techniques, to characterize the epilayer and interface of $In_{0.1}Ga_{0.9}As/GaAs$ superlattice structures grown by Organo Metallic Vapour Phase Epitaxy (*OMVPE*). Strain produced n the epilayer and at the interface due to the lattice mismatch between $In_{0.1}Ga_{0.9}As$ and GaAs substrate can be directly measured by carrying out ion-channeling in the off-normal channeling axes. The energy of the probe beam plays a very important role in the determination of strain, the critical angle for channeling being directly related with it. At low incident energies, the critical angle is comparable to the angular misalignment of the axes, giving rise to ambiguous measurements in the strain values. We discuss here the Beam Steering effect occurring at low energy channeling and compare the channeling results with High Resolution XRD and Raman Spectroscopy.

INTRODUCTION

Strained-Layer Superlattices (SLS) consist of alternating layers of two materials of similar crystal structures with lattice constants which have a mismatch of ~1 % (1). Layers are grown sufficiently thin or the mismatch is kept sufficiently small (commensurate) to avoid the generation of misfit dislocations which deteriorate the performance of the devices. Typical layer thicknesses for such mismatches are limited to a few hundred angstroms. These structures have alternating compressive and tensile strain and thus provide new physics and device possibilities. The other main advantage of such structures over the conventional lattice matched systems is that different band-gap ranges are achievable through the freedom in choosing the composition thereby enhancing the electronic and optoelectronic properties of the materials manyfold.

Rutherford Backscattering Spectrometry (RBS/Channeling) is a powerful tool to determine the thickness, composition, defect densities and strain (2-5). When the incident ion beam is directed along a high-symmetry crystal direction, it undergoes a correlated series of small angle scatterings, a phenomenon known as *channeling*. The ranges of channeled particles is anamolously large and close impact parameters like Rutherford

---

• Corresponding Author: Fax +91-40-3010181/3010227/3010120

Email: appsp@uohyd.ernet.in.

Backscattering Spectrometry (RBS) or Inner Shell excitation yield are reduced drastically, indicating the alignment of the beam with axes or planes (6). χ_{min} which is the ratio of the backscattered particles when aligned (A) to a crystallographic axis to that in the random (R) condition (i.e., $\chi_{min} = Y_A/Y_R$) is a measure of the crystalline quality of the sample. The use of ion channeling in conjunction with RBS provides a measure of the crystalline quality as a function of depth. A strain measurement by ion-channeling technique in multi-layered structures is based on the tetragonal distortions induced in the layers which results in the shift in the channeling dip. The tetragonal distortion in the epilayer ε_t is evaluated by the formula

$$\varepsilon_t = \frac{\Delta\theta}{\sin\theta\cos\theta} \qquad (1)$$

where θ is the angle between the [100] and [110] directions in the substrate and Δθ is the angular difference in the substrate and epilayer.

With a view to point the important role of the incident energy in the determination of strain using channeling experiments, we have carried out ion channeling on $In_{0.1}Ga_{0.9}As/GaAs$ superlattice structures grown by Organo Metallic Vapour Phase Epitaxy (*OMVPE*) using 1.2 MeV He^+ and 3 MeV He^{++} beam. When the angular misalignment due to the lattice strain is comparable to or smaller than the channeling critical angle, angular scan measurements about the off-normal axes often provide ambiguous measurements. This is because some of the channeled ions can be steered into the misaligned channels after crossing the interface (5,7). This *beam steering effect* can also alter the symmetry of angular scan profiles of deep layers. The results of 3 MeV He beam channeling could be compared straightforwardly with the other complimentary techniques. The critical layer thickness for $In_{0.1}Ga_{0.9}As$ material lies between 100 Å - 300 Å, (8,9). The sample was thus grown within the critical layer thickness. High Resolution X-Ray Diffraction (*HRXRD*) is another versatile and non-destructive tool to characterize heteroepitaxial structures, and complementary to other characterization techniques. It is capable of detecting strains with a sensitivity of about 10^{-5} (10) and provides information on the interface structure with monolayer precision. It gives a Fourier transform of a crystal volume of typically 1mm x 1 mm x (5-50) μm. The data are X-ray intensity distributed in the vicinity of a reciprocal lattice point (or a Bragg peak), which is integrated over the direction normal to the diffraction plane by a detector wide-open in that direction. A high angular resolution, and thus a high strain sensitivity is achieved by monochromating and/or collimating the incident X-ray beam, the procedure for which is given in ref. 10.

Differentiating the Bragg condition, we get

$$\frac{\Delta d}{d} = \varepsilon_\perp = -\cot\theta_B \Delta\theta \qquad (2)$$

Let θ_{BS} be the Bragg angle for the substrate and θ_{BL}, that for the epilayer. For GaAs samples and materials grown on them, (004), (224), ($\bar{2}24$), (115) and ($\bar{1}15$) are the directions in which the Bragg equation is satisfied. If the thickness of the epilayer exceeds the critical thickness, the layer relaxes giving rise to misfit dislocations; the tetragonal symmetry is cancelled and the unit cells of the layer assume a cubic symmetry. In the case of fully strained layer, $\varepsilon_\parallel$ is zero and when the layer partially or fully relaxes, $\varepsilon_\parallel$ is no longer zero. Thus by carrying out HRXRD, one can establish whether the sample has relaxed or not *and* also quantify the strain if the system has not relaxed (11,12).

Raman Spectroscopy also allows one to determine this strain. The quantitative measurement of strain is obtained from the phonon wave number shift between commensurate and incommensurate layers, governed by the formula

$$\varepsilon_p = \frac{1}{\beta} \frac{\Delta v}{v} \qquad (3)$$

where β is a constant which depends on the material, growth direction and conditions and phonon type mode. v_o is the wave number of the strain relaxed material for the same composition, which is evaluated from figure 3 of ref 13. v is the LO phonon wave number of $In_xGa_{1-x}As/GaAs$ structure (13,14). Δv is the wave number shift. Constant *et-al* (13) determined the internal strain in the pseudomorphic $In_xGa_{1-x}As/GaAs$ structures over a wide range of composition x. Comparison of 1.2 MeV channeling results with HRXRD and Raman Spectroscopy reveals that there is an ambiguity in the strain values determined by channeling at this energy.

EXPERIMENTAL

$GaAs/In_{0.1}Ga_{0.9}As/GaAs/GaAs$(substrate) strained layer structure is grown by Organometallic Vapour Phase Epitaxy. The details of the experimental set-up of this facility are given in ref. 15. Ion channeling work on this sample has been carried out at the Center of Irradiation of Materials, Alabama A & M University, at 3.058 MeV He^{++} beam at a scattering angle of 170° and a solid angle of 0.2249 msr, the total FWHM is about 35-40 keV. The goniometer of this set-up has four degrees of freedom; x (horizontal transversal), z (vertical transversal) with a precision of 0.0025mm each, θ rotation (about the z-axis) of 0.01° precision and a φ rotation (about the beam direction) of 0.001° precision. A pair of collimators of 1.5 mm and 2 mm collimates the beam respectively.

RESULTS AND DISCUSSION

From the random spectrum, the thickness of top GaAs layer is found to be 250 A°, that of the InGaAs epilayer as 350 A° and the composition of In to be 0.1 The spectrum in the channeling direction suggests that the sample is of good crystalline quality. The tetragonal distortion is thus evaluated and all the results are tabulated in the following table. X-ray measurements of (004) reflection along four azimuth directions give mean separation between the GaAs Bragg peak and the $In_{0.1}Ga_{0.9}As$ Bragg peak $\Delta\Omega$ such that $\varepsilon_\perp$ and $\varepsilon_\parallel$ are evaluated directly. SLS's of $In_xGa_{1-x}As/GaAs$ have been studied by Raman Spectroscopy rigorously (14). v_o is found to be 288.3/cm^{-1} and v, the wavenumberr of the InGaAs layer, to be 291.353/cm^{-1}. β value for the indium content of 10% has been found by least square fit of the values given in table 3 of reference (13).

It comes out to be 1.6602. Strain value is then evaluated using equation (3). An earlier channeling study at 1.2 MeV He beam reveals that ε_t is 0.17% indicating that the sample has relaxed (16); contradicting thereby the experiments of HRXRD and Raman Spectroscopy. *A careful study shows therefore, that the incident energy in ion beam channeling plays a very significant role in the determination of strain in these semiconductor multi-layered structures, which are technologically very important.*

Sl. No	Characterization Technique.	Strain Value (%)
1.	Ion Channeling (3 MeV)	1
2.	HRXRD	0.96
3.	Raman Spectroscopy	1
4.	Ion Channeling (1.2 MeV) Ref. 16.	0.17

Table 1: The values of strain for various samples using Ion Channeling, XRD and Raman Spectroscopy.

ACKNOWLEDGEMENTS

AMS and SVSNR are thankful to the CSIR for providing Senior Research Fellowship. This work is partially supported by an UFUP project of NSC, New Delhi (India). We are very grateful to Prof B M Arora of TIFR, Mumbai, Dr. Keshav Murthy of IGCAR Kalpakkam, Dr. V N Kulkarni of IIT Kanpur, Dr. Eric Williams of Alabama A & M University for extending necessary laboratory facilities to carry out some parts of this work. We also thank them for very useful and valuable suggestions.

REFERENCES

1. Osbourn G. C., *J. Appl. Phys.*, 53, 1586, 1982.
2. Mayer J. W., Zeigler J. F., Chang L. L., Tsu R. and Esaki L., *J.Appl. Phys.*, 44, 2322, 1973.
3. Saris F. W., Chu W. K., Chang C. A., Ludeke R. and Esaki L., *Appl. Phys. Lett.*, 37, 931, 1980.
4. Picraux S. T., Dawson L. R., Osbourn G. C. and Chu W. K., *Nucl. Ins. and Meth.*, 218, 57, 1983.
5. Chu W. K., Pan C. K. and Chang C. A., *Phys. Rev. B*, 28, 4033, 1983.
6. Feldman L. C., Mayer J. W. and Picraux S. T., *Materials Analysis by Ion Channeling*, New York, Academic Press, 1982.
7. Hashimoto Shin, Feng Y.Q.and Gibson W.M , *Nucl. Inst. And Meth, B* 13, 45, 1986.
8. Coleman James J. *Strained-layer Quantum Well Heterostructure Lasers*, Chap 8, ed: Zory Peter J, Academic Press, New York (1993).
9. Matthews J. W. and Blakeslee A. E., *J. Crystal.Growth*, 27, 118, 1974.
10. Wie C. R., *Mat. Sci. Engg. R*, 13, 1, 1994.
11. Multani J.S. and Sandhu J.S., *J. Electr. Soc.* 126, 1086, 1979.
12. Bartels W.J., *J. Vac. Sc. Tech*, B1, 338, 1983.
13. Constant M., Matrullo N., Lorriaux A. and Boussekey L., *J. Raman Spect.*, 27, 225, 1996.
14. Emura S., Gonda S., Matsui Y. and Hayashi H., *Phys. Rev. B*, 38, 3280, 1998.
15. Siddiqui Azher M., Pathak Anand P., Sundaravel B., Das Amal K., Sekar K., Dev B. N. and Arora B. M., *Nucl. Inst. Meth. B*, 142, 389, 1998.
16. Pathak Anand P, Nageswara Rao S.V.S. and Siddiqui Azher M., *Nucl. Instr. and Meth. B*, 161-163, 488, 2000.

Depth profiling code for analysing ERD-TOF spectra

G. MATHOT, G. TERWAGNE and F. BODART

Facultés Universitaires Notre-Dame de la Paix, Laboratoire d'Analyses par Réactions Nucléaires, rue de Bruxelles 61, B-5000 Namur, Belgium

A computer program calculating depth profiles of light elements in surface layer of various materials from experimental ERD-TOF spectra has been developed. The program, which is able to identify the recoil particles, makes multi-element profiling by sorting the spectra by mass. The interactive spectrum synthesis compare the real recoils spectra with simulated spectra of the assume target. The program is also able to calculate the atomic concentration ratios without any a priori assumption of the composition of an unknown target. The stopping power used in the analysis package respect the Alegria [1] format and can be easily upgraded and modified by the user. It can be calculated for any particle target combination and beam energy between 100keV and 15 MeV. The calculation takes also into account for the straggling, the energy loss in the carbon foils of the start and the stop detectors and the entry window of the particle detector.

1. Introduction

Elastic recoil detection time of flight system is a simple and fast method for quantitative multi light elements depth concentration profiling. We have developed a computer package, which is able to identify the recoil particles, to sort the spectra by mass and to calculate depth concentration profiles of light elements in surface layer of various materials from experimental ERD-TOF spectra.

2. Mass identification

These spectra are build up by measuring in coincidence the total energy E of the each detected particles and the time needed by particles to pass between two time detectors (T_1 & T_2) separated by a length L as shown Fig 1. The time signal is due to electrons emission when the recoil atom passes through a thin carbon foil (10 µg/cm²). The time signal T_1 is delayed in order to start the coincidence when an event is observed in the energy detector.

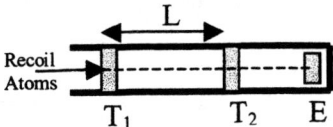

Fig. 1: TOF telescope system

A typical spectrum is plotted in Fig 2. It's the result of $Si_vC_wN_xO_yH_z$ ERD-TOF analyse by ^{35}Cl incident ions of 9.3 MeV. The recoil atom signals are separated. This effect is due to the mass dependant velocity of the same energy detected recoil atoms:

$$v = \sqrt{\frac{2E}{M}}$$

Where, v is the velocity, M the mass and E the energy of the recoil atom.

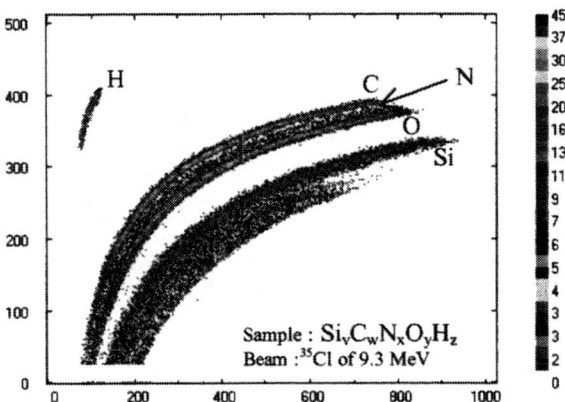

Fig. 2: ERD-TOF coincidence spectra. Horizontal and vertical scale is proportional to the energy and flight time respectively

The mass of the detected particle can be calculated using:

$$M = 2E(\frac{T}{L})^2$$

Where T is the time required by the particle to pass between two time detectors separated by a length L and E is the measured energy of the particle. This transformation gives one yield-energy spectrum for each recoil atom detected, as shown Fig. 3.

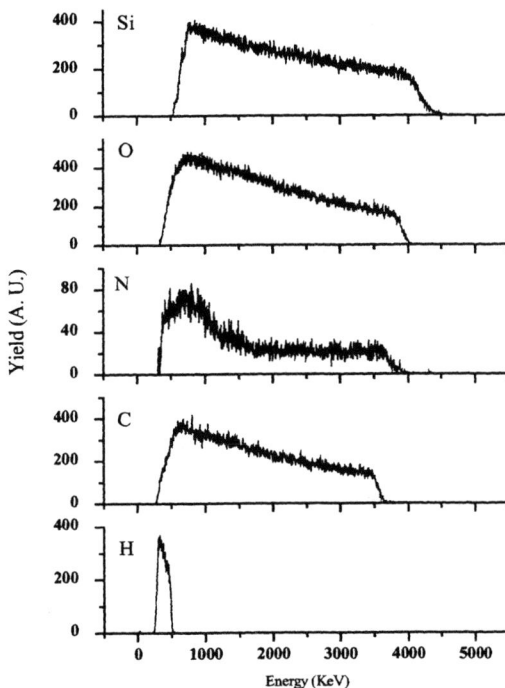

Fig. 3: Mass separated yield energy spectra

These mass separated spectra leads to quantitative results by using Interactive spectrum synthesis method or spectra scaling [2].

3 Interactive spectrum synthesis

In this method, theoretical ERD spectra are calculated and compared with the experimental data. The calculation takes into account for the straggling, the energy loss in the carbon foils of the start and the stop detectors and the entry window of the particle detector

The first step is to describe the sample as a stack of sub-layers. Each one is characterised by its thickness and its uniform composition. The concentration depth profile is determined by trials and errors, changing sub-layers specifications.

The simulated spectrum is made up of contributions from each sub-layer in the sample. Any such contribution will be name a brick.

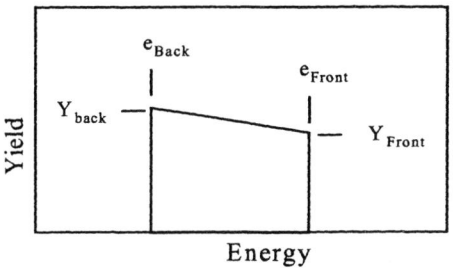

Fig. 4. Single Brick and notation

To determine the location in energy of a brick, we compute the energy lost by the beam and the recoil particle in the inward path and the outward path respectively. In both case the particles lose energy according to the differential equation:

$$\frac{1}{N}\frac{dE}{dx} = -\varepsilon_p(E)$$

Where N is the atomic density (at/cm³), dx is the path length into the target, in 10^{15} atoms/cm², dE the energy loss and $\varepsilon_p(E)$ the stopping cross sections in which the subscript p refer to the particle loosing energy. to calculate the stopping cross sections, the analysis program uses fits parameters in respect with the Alegria [1] standard. The parameters of the spline functions used are fits of TRIM92 [3] data but can be easily upgraded and modified by the user.

Using the Bragg rule for compound, energy loss can be calculated for any particle target combination and beam energy between 100 keV and 15 MeV.

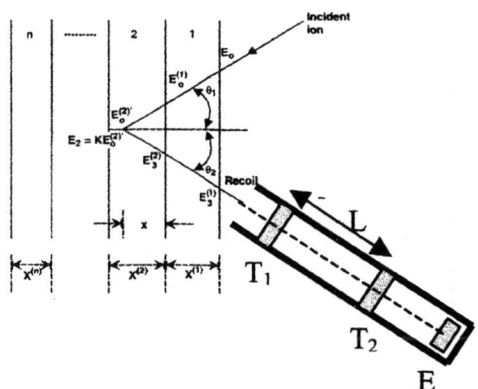

Fig. 5. Slab notation and experimental set-up

The location in energy of each brick is determined from the calculation above and the energy transfer in the scattering event. The program uses elastic scattering [4] in which the energy of the recoil after collision is proportional to the energy of the beam before collision. The kinematics factor of the collision is define by:

$$K = \frac{E_r}{E_p} = \frac{4M_p M_r}{(M_p + M_r)^2} \cos^2(\phi)$$

Where M_p and M_r are the masses of the incident and recoil particle respectively and ϕ the scattering angle.

The height of the front and the back edges of the brick are evaluated from the equation giving the yield from nuclide r for the recoil atoms detected in energy channel E_d [2]. This yield is given by:

$$y^r(E_d) = \frac{QN_r(x)\sigma_r(E_0',\phi)\Omega \partial E_d}{\cos\Theta_1 dE_d/dx}$$

Where Q is the incident projectile fluence, $N_r(x)$ is the atomic number density of the recoil atom at depth x, Ω is the detector solid angle, ∂E_d is the width of one channel, dE_d/dx is a correcting factor taking into account the dispersion of the scattered particles during the outward path and $\sigma(E_0',\phi)$ is the recoil differential cross section in laboratory system.

The cross section used by the program is calculated using Rutherford law. This cross section is governed by coulomb scattering and can be expressed in laboratory reference by:

$$\sigma(E_0',\phi) = \frac{[Z_p Z_r e^2 (M_p + M_r)]^2}{[2M_r E_0']^2 \cos^3\phi}$$

Here $(e^2/E_0')^2 = 0.020731$ barns for $E_0' = 1$ MeV

3.1 Straggling

When the incident particles penetrate the material, statistical collisions slow down the ions. Straggling effect leads to an energy spread for the incident and recoils atoms, which can be a limited factor in the depth resolution of ERD-TOF depth profile technique.

The Bohr model [2] takes into account for the energy transfer from the ion to the electron of the target. In the limit of high ions velocity, the energy distribution is gaussian and the straggling in keV is given by:

$$\sigma_{Bohr} = \sqrt{0.26 Z_1^2 Z_2 \Delta x}$$

Where Z_1, Z_2 are the incident beam and the target atomic number respectively and Δx the layer thickness in 10^{18} at/cm².

Expression for σ_B describes the amount of rounding for each brick's back and front edge. For calculating the energy dependant rounding, the program replaces the yield brick by two triangles [5] as shown Fig. 6. This front and the back triangles are separately convolved with gaussians of with σ_{Back} and σ_{Front} respectively.

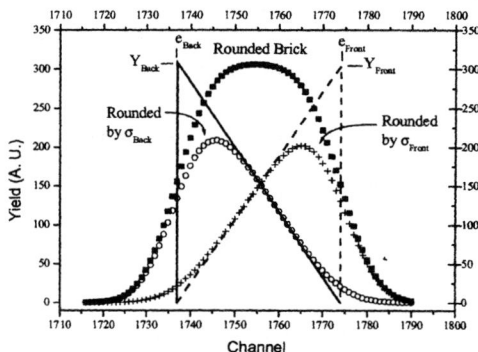

Fig. 6.: Example of rounding effect of a brick due to straggling.

The convolution of a unit triangle with a gaussian of width σ can be written as

$$R(x,\sigma) = \int_0^1 (1-t) \frac{1}{\sigma\sqrt{\pi}} e^{-(\frac{x-t}{\sigma})^2} dt$$

The rounded brick is describe by

$$y_r(E) = y_{Back} R(\frac{E - e_{Back}}{e_{Front} - e_{Back}}, \frac{\sigma_{Back}}{e_{Front} - e_{Back}})$$
$$+ y_{Front} R(\frac{e_{Front} - E}{e_{Front} - e_{Back}}, \frac{\sigma_{Back}}{e_{Front} - e_{Back}})$$

This rounding way is an approximation that gives good result assuming that σ_{Back} is nearly equal to σ_{Front}.

3.2 Example

This method has been applied to the $Si_vC_wN_xO_yH_z$ spectra. The comparison between the computed and measured spectra is shown Fig. 7. The concentrations obtained are:

Element	Concentration (at. %)
H	3
C	29
N	4
O	33
Si	31

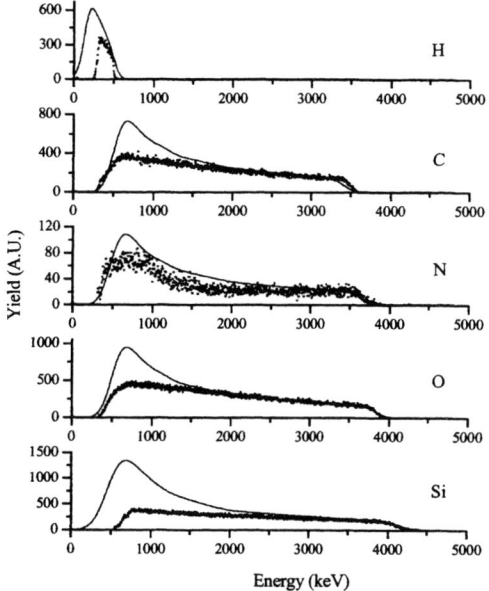

Fig 7. Comparison between experimental (dots) and simulated spectra (solid line).

At low energies, stopping power uncertainties leads to bad reaction cross-sections and yield calculations

4 Scaling spectrum

Another method to obtain quantitative result is the scaling spectrum method. It takes into account for the energy loss in the carbon foils of the start and the stop detectors and the entry window of the particle detector. The straggling effect is neglected.

The above analytical equations describing ERD spectrum can be use to calculate composition profile without any a priori assumption of the composition of an unknown target. In this approach the program creates a table containing the sample depth, $\sigma(E_0^n)$, E_d and dE_d/dx as a function of slab layer. The spectra are then converted into depth profile by interpolating values in the table and scaling the yield in each channel. Mathematically, the scaling is given by:

$$N_r^n(x) = \frac{\cos\Theta_1}{Q\Omega\partial E_d} \frac{dE_d/dx|_n}{\sigma_r(E_0^n,\phi)} Y_r^n(E_d)$$

To take into account the variation of the energy loss with concentration depth profile, the program creates a new table based on the known specification of the target after the first iteration and rescale the spectra.

5. Conclusion

We have presented algorithm for ERD-TOF qualitative and quantitative analyses. The methods deal with the basic physical concept of the stopping power, Rutherford cross-section and Bohr straggling. Non Rutherford Cross section, Geometrical straggling due to the beam spot size will be implemented in the future.

6. Acknowledgements

The authors would like to thanks F. Schiettekatte for his help concerning the calculation and the fits of stopping cross sections.

7. References

[1] F. Schiettekatte, A. Chevarier, N. Chevarier, A. Plantier, G.G. Ross, *Nucl. Instrum. and Meth.*, B118(1996)307.

[2] J.R. Tesmer, M. Nastasi, J. Charles Barbour, C. J. Maggiore, J. W. Mayer, *MRS, Pittsburgh, 1995*

[3] J.F. Ziegler, J.P. Biersack, and U. Littmark. *Pergamon Press, New York, 1985.*

[4] J. Tirira, Y. Serruys and P. Trocellier, *Plenum Publishing Corporation, New York,1996.*

[5] Lawrence R. Doolittle, *Nucl. Instrum. and Meth.*, B9(1985)344

Enhanced hydrogen detection and depth profiling system using coincidence techniques

C. I. Muntele, D. Ila, R. L. Zimmerman

Center for Irradiation of Materials, Alabama A&M University, P. O. Box 1447, Normal, AL 35762-1447

Abstract. Hydrogen is perhaps the most common elemental contaminant that appears in thin film materials, and can have dramatic effects on material's properties. Most modern analytical techniques are incapable of hydrogen detection. Because of this, special ion beam techniques based on resonant nuclear reactions have been developed to fill this gap in material analysis [1]. The yield of the resulting gamma-rays is proportional to the hydrogen content within a very narrow depth region which satisfies the resonant energy "window". Such setups have been developed at SUNY at Albany, and GSI in Germany. The setup developed at the Center for Irradiation of Materials of Alabama A&M University has an active protection against the parasitic radiation, that can reduce the detecting threshold to .1 ppm. The full description of the assembly, as well as experimental results on profiling hydrogen amounts in different materials is provided below.

INTRODUCTION

Over the past twenty years or so, hundreds of papers have been published on the topic of hydrogen concentration profiling using MeV ion beams. The materials analyzed varied from semiconductor materials used in electronics (silicon, diamond, silicon dioxide etc.) to metals and lunar glasses. The reason for which this subject gained that much attention is that hydrogen is the most common elemental contaminant whose presence into materials can drastically change their electrical, chemical and mechanical properties. Besides, other modern analytical techniques cannot identify it.

Although in principle any proton-induced nuclear reaction can be used in reversed kinematics for performing hydrogen profiling, considerations of available equipment, as well as target requirements, reduce drastically the number of useful reactions that can be explored. Apparently, the favorite nuclear reaction is the one introduced by Lanford et al. [2], using ^{15}N, as having the best combination of depth resolution and sensitivity. The reaction is

$$^{15}N + ^{1}H \rightarrow ^{16}O^* \rightarrow ^{12}C^* + ^{4}He$$

then

$$^{12}C^* \rightarrow ^{12}C + \gamma$$

The reaction diagram is shown detailed in Fig. 1.

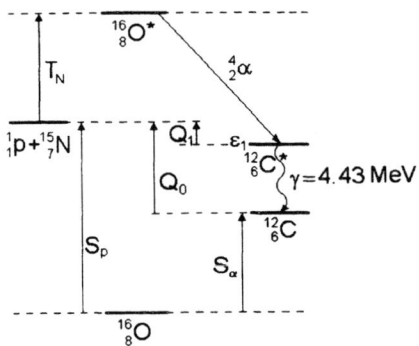

FIGURE 1. Schematic diagram of the nitrogen-hydrogen nuclear reaction.

The values shown in the above diagram are for the resonance energy when ^{15}N is the projectile:

$$T_N = 6.385 MeV$$
$$S_p = (m_p + m_N - m_O) \cdot c^2 = 12.11 MeV$$
$$S_\alpha = (m_\alpha + m_C - m_O) \cdot c^2 = 7.17 MeV$$
$$\varepsilon_1 = 4.439 MeV$$

The resulting gamma rays are directly measured with a scintillation detector, and their yield is proportional to the hydrogen concentration at the specific location where the nuclear reaction took place. Since the resonance energy of the incoming ^{15}N ion is 6.385±0.005 MeV, the depth window is

~75 Å in solid materials. The profiling is done by varying the ion energy; to see the surface hydrogen concentration, the beam should have the resonance energy. By increasing the energy, one can "scan" deeper under the surface of the target. The depth at which the nuclear reaction takes place is derived considering the distance on which the incoming ion loses the difference between its initial and resonant energy. However, as the depth increases, energy straggling will alter the resolution. Also, the energy cannot be increased arbitrarily; there is an upper limit (13.35 MeV), unless one wants to open the next resonant channel, in which case the analysis of the spectrum gets more complicated.

CIM EXPERIMENTAL SETUP

Usual sensitivities for hydrogen profiling analysis lay in the range of hundreds of ppm (10^{18} at/cm^3). In order to push the detection limit to lower levels, one should pay attention to reducing the background due to cosmic radiation and decays of natural radioisotopes, as well as parasitic reactions due to deuterium and tritium, where those elements are present in the sample. Solutions would be increasing the detection efficiency by using BGO rather than NaI(Tl) scintillators, shielding the detector against cosmic rays, and increasing the analysis time. The increase of the beam current is not recommended for some samples, since the hydrogen content can be altered due to sample overheating.

Rather than shielding the detector for reducing the cosmic background, we have chosen to use coincidence techniques for active rejection of the parasitic signal. A schematic drawing of our setup is presented in Fig. 2.

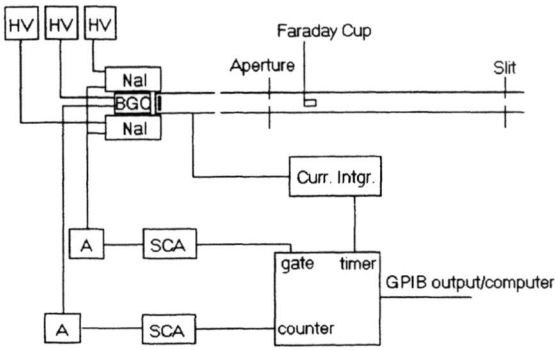

FIGURE 2. CIM NRA geometry and coincidence electronics arrangement.

Our setup uses a BGO detector surrounded by four NaI connected in pairs of two. The identification of an event (nuclear reaction on hydrogen) is based on the first escape peak of the 4.43 MeV (that is 3.92 MeV) being registered by the BGO while the 511 keV gamma escaped is registered by one of the NaI detectors. The electronics setup is very stable and requires only a daily calibration with isotopic calibration sources. Once the calibration is verified, the setup has an option for acquisition only, and energy tuning via terminal potential, or to be passed on automatic computer control for data acquisition, as well as target fine biasing in the range +30kV to –30kV for ion energy fine tuning, without any requirement for the tandem terminal voltage modification.

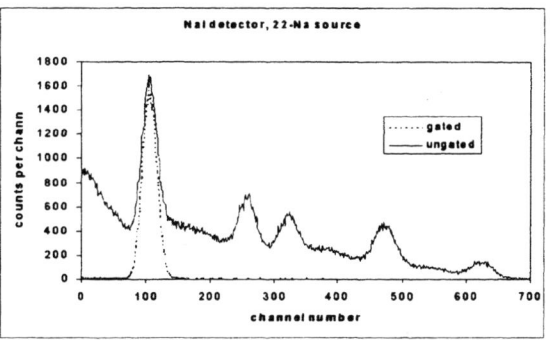

FIGURE 3. Calibration plots for the NaI(Tl) detectors, before and after the SCA gating.

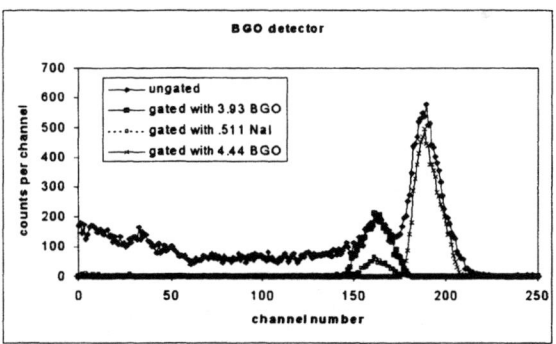

FIGURE 4. Calibration plots for the BGO detector, before and after gating.

In figures 3 and 4 one can see the raw signals at the outputs of each detector's amplifier, and then, overlapped, the trimmed signals going out of the SCA's. Using this configuration, the average values for background measurements are of 36 counts in 10 hours. Using lead shielding around detectors, lower values can be attained. Practically, the only limit in pushing down the sensitivity is the accuracy in defining beam's energy, i. e. apertures, slits, terminal potential etc.

For producing the ^{15}N ions, we use either cathodes of titanium provided by NEC, or homemade cathodes of potassium cyanide, both of them having as constituent enriched ^{15}N. For defining the beam, we use a pair of slits placed just after the analyzing magnet, and another aperture placed 3 m apart from them. Common settings for raw measurements are 4 mm opening at slits, and 4 mm diameter aperture. For precision measurements we close the slits to 1 mm, and use a 2 mm diameter aperture. Current values on targets are between 0.5 and 100 nA, depending upon the target. The target chamber is insulated from the rest of the beam line, so that no electron emission suppression is necessary and the current integrator picks up the charge from the whole assembly. For calibration purposes we use polyethylene foils of known density, and very low currents (0.5 nA) in order to ensure that hydrogen does not migrate away from the sample. Repetitive measurements on the same foil have shown similar profiles, within statistic fluctuation.

EXPERIMENTAL RESULTS

In Figure 5 are shown typical results obtained during this type of analysis. One can clearly see the water content at the surfaces of all the samples.

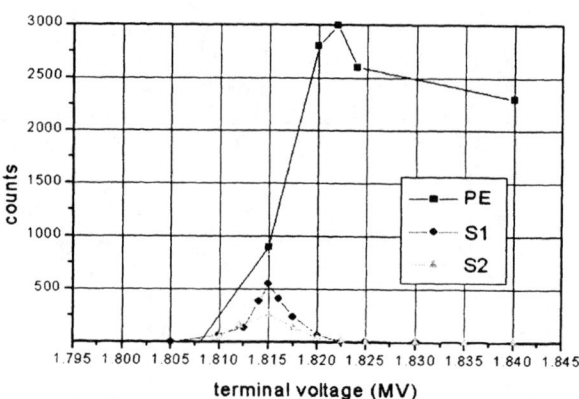

FIGURE 5. Results from a polyethylene calibration sample, along with two other samples. The numbers on the vertical scale are for 2×10^{-5} coulomb set on the current integrator.

REFERENCES

1. *Handbook of Modern Ion Beam Materials Analysis*, Materials Research Society, Pittsburgh, Pennsylvania, 1995, Chapter 8.
2. Lanford, W. A., Trautvetter, H. P., Ziegler, J. F., Keller, J. *Appl. Phys. Lett.* **28**, 566, (1976).

Heavy Ion ERD of Nitrides with a Position-Sensitive Gas Ionization Detector

H. TIMMERS[1,2], T.D.M. WEIJERS[1,2], R.G. ELLIMAN[1], T.R. OPHEL[2]

Departments of Electronic Materials Engineering[1] and Nuclear Physics[2], Research School of Physical Sciences and Engineering, Australian National University, ACT 0200, Canberra, Australia

A gas ionization detector with novel design features has been developed for the compositional depth-profiling of materials with Elastic Recoil Detection (ERD) using very heavy incident beams. The detector features a large solid angle and thus has high detection efficiency. The detection of the ion position using a saw-tooth ΔE electrode within the anode is energy and species independent and enables the correction of kinematic energy broadening. The energy information is obtained from a single grid-electrode, which considerably simplifies data analysis. All chemical elements, including hydrogen, can be detected simultaneously with similar sensitivity. While it is versatile and applicable to many materials, the technique has unique capabilities when applied to thin films containing carbon, nitrogen or oxygen in combination with heavier elements and hydrogen. In contrast to other techniques, heavy ion ERD resolves all elements in a single measurement and the experimental uncertainty is limited only by counting statistics. However, the desorption of some elements, such as nitrogen or hydrogen, during the analysis can be significant. For the GaN films studied, the nitrogen and hydrogen desorption rates were found to be linear with dose. This allowed accurate extrapolations to zero dose retaining the precision of the measurement. In contrast, no nitrogen desorption occurred for nitrided steel samples enabling nitrogen depth-profiles to be extracted.

INTRODUCTION

The compositional depth-profile of the near-surface region of materials and thin films can be measured with Elastic Recoil Detection (ERD) analysis using very heavy incident ion beams, such as 200 MeV ^{197}Au [1,2]. For elements heavier than hydrogen the recoil yields increase smoothly with atomic number Z and ions as heavy as the projectile have sufficient energy so that they can be detected. As a consequence the technique is sensitive to all chemical elements. It is most sensitive to hydrogen, since protons have a charge to mass ratio of one, leading to a more than threefold increase of the recoil cross-section, when compared to other light elements. The simultaneous detection, distinction and direct quantification of light elements such as oxygen, nitrogen, carbon and hydrogen in combination with heavier elements is a unique capability of the technique.

Gas ionization detectors are well suited for the detection of the recoil ions, because they are element dispersive and do not suffer radiation damage from heavy ion bombardment. In addition, large entrance windows and detection solid angles of several millisteradian are possible, since position sensitivity can be incorporated to allow the correction of kinematic energy broadening over the acceptance angle [3]. Energy resolutions of less than 1 % have been obtained for lighter ions such as oxygen. As a consequence optimum depth resolution can be achieved, if the energy broadening caused by multiple scattering and energy straggling of the ions in the sample exceeds 1 %. This is typically the case beyond a depth of about 30–50 nm [4]. The probing depth of the technique is of the order of 1 µm.

Information about the compositional depth-profile of the sample is reflected in the particle and energy spectra of the recoil ions. The extraction of this information for the complete range of ions spanning hydrogen to the heaviest elements poses a challenge to the design of gas ionization detectors. The stopping powers differ greatly, so that, for example, hydrogen cannot be stopped and thus no total energy signal be obtained when the detector operation is optimized for heavy ion detection. Distortions of the electric field at the detector entrance due to the presence of a large window can produce a species and energy dependent deficit of the cathode signal. The use of the cathode for position detection consequently becomes complex and requires detailed calibrations [5]. The presence of the window can also influence the collection of the electrons by the anode, causing an undesirable height-dependence of the ΔE signal. For ERD analysis the total energy signal from the detector is crucial, since it contains the depth information. It is therefore preferable to obtain it from a single electrode rather than combining signals from two or more electrodes, thus obviating the need for relative calibrations and avoiding a reduction in resolution. This is not necessarily compatible with the need to also obtain energy loss and position information.

The possibility of large detection solid angles ensuring efficient ion detection is a distinct advantage of gas ionization detectors, when they are compared to alternative detection systems. For certain materials a significant desorption of some elements during the analysis can, however, not be avoided. In particular hydrogen and nitrogen are prone to ion beam induced desorption. It is thus essential that in such cases all elements are detected simultaneously, as efficiently as possible and as a function of the incident projectile dose, so that the initial composition can be precisely extrapolated. Published data on the shape of the desorption curves, the dependence of the process on the material, the incident beam and energy are rather limited [6]. It may be expected that the choice of projectile beam and energy can be optimized for particular applications, so that desorption rates are reduced.

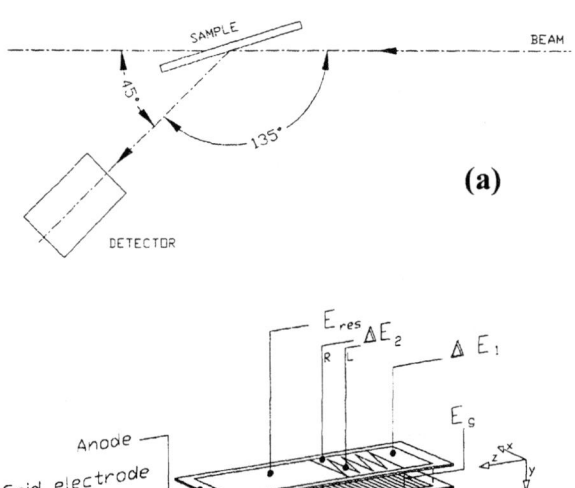

FIGURE 1 (a) Detection geometry for ERD analysis. (b) The design of the gas ionization detector. The guard wires are only shown for one side.

DETECTOR DESIGN

A gas ionization detector with novel design features has been developed for the specific demands of ERD analysis with heavy ion beams. It is located inside a large scattering chamber (radius 1 m) at the 14UD Pelletron accelerator facility of the Australian National University. A typical detection geometry used for ERD analysis is shown in Fig.1(a). The location of the detector relative to the beam direction can be varied and the sample can be rotated about an axis perpendicular to the detection plane. Recoil ions enter the detector through a 0.5 μm thick Mylar window, which is supported by a rectilinear grid of tungsten wires. The wires are in electrical contact with the window support frame, which is at a fixed electrical bias. For the measurements discussed here the detector was positioned at a scattering angle of 45°, 278 mm from the sample with a solid angle of 3.5 ± 0.05 msr. The angle between sample normal and beam was 67.5°. Propane was passed through the detector at a constant pressure of 70 mbar. The projectile beam was ^{197}Au at an energy of 200 MeV. It was collimated before the sample using two pairs of slits with the dimensions of 0.5 mm x 3 mm and 1 mm x 4 mm, respectively. The slits were 200 mm apart.

The detector, Fig. 1(b), which is discussed in more detail elsewhere [7,8], features an undivided grid electrode for a direct measurement of the total ion energy thus obviating the need for relative calibrations of the anode signals. The drifting electrons induce a signal on the grid electrode after they have passed through the Frisch grid, however, after they have moved through the electrode their drift towards the anode causes an immediate decrease of that signal. In contrast to the anode signals the decay time of the signal is thus short and comparable to the rise time. For 28.6 MeV ^{16}O ions the grid signal was found to have a resolution of 1.6 %. This may be compared with a resolution of 0.9 % obtained under the same conditions for the combined anode signal, however, for most applications this reduction in resolution is found to be insignificant.

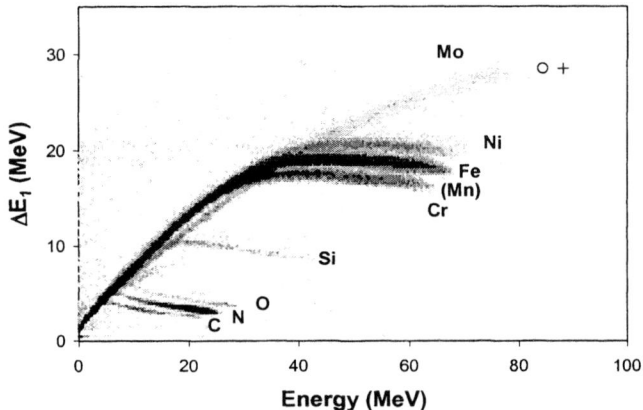

FIGURE 2 The ΔE_1 signal versus total energy as measured for elastic recoils from a nitrided steel sample. For Mo the magnitude of the pulse height deficit (o) and the energy loss in the detector window (+) are indicated.

The anode of the detector consists of three sections, labeled in Fig. 1(b) as ΔE_1, ΔE_2 and E_{res}, respectively. Combining the signal from the short ΔE_1 electrode with that from the grid electrode E_g allows the separation of lighter and heavier elements with acceptable resolution at a single gas pressure, in contrast to other detector designs, which require measurements at two different gas pressures. This is shown in Fig. 2 for the broad range of ions from a nitrided steel sample.

It was found [5,7] that the response of the short ΔE_1 electrode adjacent to the entrance window depends sensitively on the electric field in the entrance region, since electrons generated underneath the ΔE_1 electrode may be collected by the ΔE_2 electrode resulting in a height dependence of the signal. A careful placement of the window and appropriate bias on the window assembly ensure, however, that the geometric division between the ΔE_1 and the ΔE_2 electrodes is reflected by the electric field. This is assisted by the truncation of the first part of the cathode, which has the same length as the combined ΔE_2 and E_{res} electrodes. When necessary, a considerably better separation of the lighter elements than in Fig. 2 can be obtained by plotting the sum ($\Delta E_1 + \Delta E_2$) versus E_{res}.

The initial use of a sawtooth geometry within the cathode for the detection of the horizontal ion position proved to be unsatisfactory, since the response is both ion and energy dependent [5]. It has been shown [7] that this is due to field modifications caused by the presence of the window assembly in the entrance region. The detector shown in Fig. 1(b) features a subdivision of the ΔE_2 section of the anode, again using a sawtooth geometry, which allows an essentially species- and energy-independent detection of the horizontal ion position with an angular resolution of about 0.18°. For a detector angle of 45°, the vertical position of the ion is almost irrelevant. It can be obtained, however, with sufficient resolution from the ratio $C/(\Delta E_2 + E_{res})$, which has a near-quadratic dependence on height.

At typical gas pressures used for heavy ion detection, protons recoiling from the sample are not stopped in the detector. They can, however, be separated from the tail of low energy heavy ions, because the relationship between their E_g and E_{res} signals is unique. These two signals are also separately amplified with gains sufficiently high to isolate the protons. This is illustrated in the inset of Fig. 3. This technique has been found to be ideally suited for the determination of the hydrogen content of films deposited on hydrogen-poor substrates. The hydrogen yield, measured with the same solid angle, can directly be related to the yields for the heavy ions. Furthermore, when the electrode response is calibrated using a sample with uniform hydrogen content, this technique can be extended to determine the hydrogen content of other materials.

The capabilities of the novel detector design, when used for the ERD analysis of GaN films and nitrided steel, have been investigated.

ANALYSIS OF GaN FILMS

GaN films are being developed for applications in optoelectronics, UV detectors and microwave power switches. Fig. 3 shows the result of the ERD analysis of a GaN film, which was deposited on a silicon substrate at low temperatures by laser-induced, plasma assisted chemical vapour deposition. All elements in the film can be distinguished and are separated from the substrate signal.

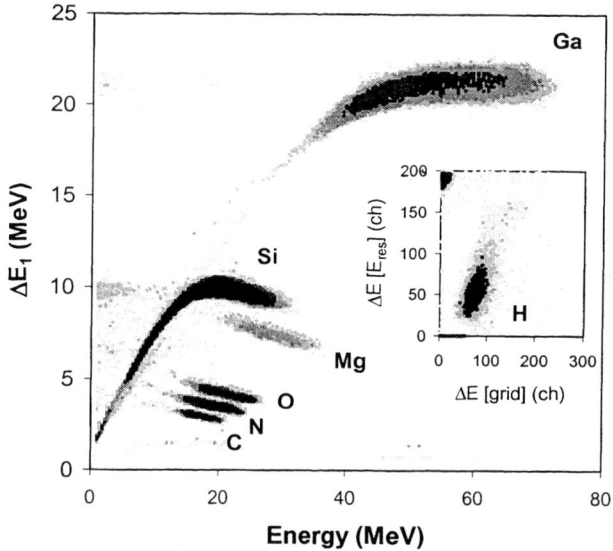

FIGURE 3 The ΔE_1 versus energy spectrum measured for ions recoiling from a GaN film on silicon. The inset shows the hydrogen signal for the same sample.

Assuming uniform composition and no desorption during the analysis, the stochiometry can directly be obtained by dividing the individual yields by the recoil cross sections. The experimental precision is excellent, being limited only by counting statistics. While it was found that, with few exceptions, the films analysed are uniform with depth, all films suffered a significant desorption of nitrogen and hydrogen during the ERD analysis. Fig. 4 shows that the stoichiometry gradually changes with fluence, which is caused by the desorption of nitrogen and, to a lesser extent, of hydrogen. The incident ion dose was determined by adapting simulations to the energy spectra extracted from Fig. 3 using the code RUMP [9]. The kinematic energy broadening over the acceptance angle of ~4° was corrected. The uncertainty of the total fluence, which is of the order of 20%, arises mainly from the measurement of the beam spot area on the sample.

The linearity of the change in stoichiometry enables extrapolations to zero fluence which retain the precision of the measurement. The relative uncertainties of the extrapolated atomic fractions are better than 1% for Ga and of the order of 2-3% for the other elements.

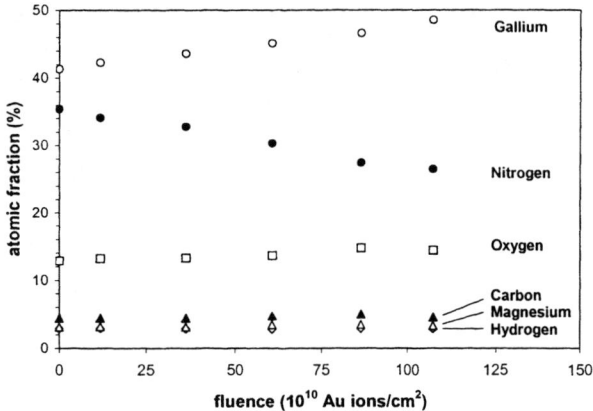

FIGURE 4 The atomic fractions obtained from the data in Fig. 3 as a function of the fluence of incident projectile ions. The stoichiometry indicated at zero fluence is a linear extrapolation.

NITROGEN DIFFUSION IN STEEL

RF-plasma nitriding of steel to increase surface hardness and wear resistance is being developed as an alternative to commercial techniques. Several samples of nitrided stainless steel surfaces were analyzed with heavy ion ERD. A typical result is shown in Fig. 2. All elements in the steel are resolved, with the exception of Mn, which is nominally present at a level of ~1.4 at-%, but cannot be distinguished from the stronger Cr and Fe signals. Energy spectra have been extracted and kinematic energy broadening has been corrected. The spectra have been calibrated using the high energy edges of the C, N and O signals. The energy loss in the detector window and the pulse height deficit effect [10], which is significant for heavier ions, have been taken into account. Fig. 2 gives an indication of the magnitude of these two effects for Mo. The composition of the steel obtained by adapting a RUMP simulation to the spectra is in agreement with the expected stoichiometry.

The inset of Fig. 5 shows that, in contrast to the results for the GaN films, any nitrogen desorption during analysis of the nitrided steel samples is negligible. The extracted nitrogen depth-profile, which is shown in Fig. 5, is thus reliable. It has been obtained by dividing the measured energy spectrum for nitrogen by that expected theoretically from a RUMP simulation assuming constant nitrogen concentration. The unexpected shape of the diffusion profile, which is not yet fully understood, is in agreement with that obtained from SIMS measurements.

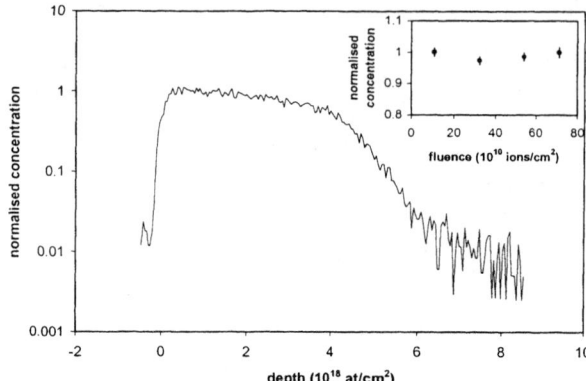

FIGURE 5 The nitrogen depth-profile extracted from the data in Fig. 1. The inset shows that the nitrogen content of the sample is constant with fluence.

SUMMARY

A position-sensitive gas ionization detector with novel design features has been developed for the demands of elastic recoil detection with very heavy ion beams. It is ideally suited for the compositional depth-profiling of materials, which contain carbon, nitrogen, and oxygen, in combination with hydrogen and heavy elements. It has been shown that the stoichiometry of GaN films can be determined with excellent precision, if nitrogen and hydrogen desorption during the measurement is quantified. In contrast, any nitrogen desorption from nitrided steel has been found to be negligible. This has enabled the measurement of nitrogen diffusion profiles in steel.

ACKNOWLEDGEMENTS

The authors would like to acknowledge K.S.A. Butcher and M.P. Fewell for their contributions to this work.

REFERENCES

1. Stoquert, J.P. et al., Nucl. Instr. and Meth. B 44 (1989) 184.
2. Siegele, R. et al., J. Appl. Phys. 76 (8) (1994) 4524.
3. Assmann, W. et al., Nucl. Instr. and Meth. B 85 (1994) 726.
4. Elliman, R.G. et al., Nucl. Instr. and Meth. B 136-138 (1998) 649.
5. Ophel, T.R. et al., Nucl. Instr. and Meth. A 423 (1999) 381.
6. Walker, S.R. et al., Nucl. Instr. and Meth. B 136-138 (1998) 707.
7. Timmers, H. et al., Nucl. Instr. and Meth. B 161-163 (1998) 707.
8. Timmers, H. et al., Nucl. Instr. and Meth. A 447 (2000) 19.
9. Doolittle, L.R., Nucl. Instr. and Meth. B 9 (1985) 344.
10. Weijers, T.D.M. et al., to be published.

Coded Aperture Fast Neutron Analysis: Latest Design Advances

Roberto Accorsi and Richard C. Lanza

Department of Nuclear Engineering, Massachusetts Institute of Technology, Cambridge, MA 02139, USA

Abstract. Past studies have showed that materials of concern like explosives or narcotics can be identified in bulk from their atomic composition. Fast Neutron Analysis (FNA) is a nuclear method capable of providing this information even when considerable penetration is needed. Unfortunately, the cross sections of the nuclear phenomena and the solid angles involved are typically small, so that it is difficult to obtain high signal-to-noise ratios in short inspection times. CAFNA© aims at combining the compound specificity of FNA with the potentially high SNR of Coded Apertures, an imaging method successfully used in far-field 2D applications. The transition to a near-field, 3D and high-energy problem prevents a straightforward application of Coded Apertures and demands a thorough optimization of the system. In this paper, the considerations involved in the design of a practical CAFNA system for contraband inspection, its conclusions and an estimate of the performance of such a system are presented as the evolution of the ideas presented in previous expositions of the CAFNA concept.

INTRODUCTION

CAFNA© is a nuclear technique for the determination of the spatial distribution of the elemental density of materials. As such, it shares the major advantage of nuclear techniques in identifying materials by their elemental composition, not simply by their density as with x-ray techniques [1]. As in Fast Neutron Analysis (FNA), in CAFNA fast neutrons are used as the probing radiation. The gamma rays emitted from elements such as C, O, N or Cl and H after inelastic scattering or capture in the container are detected. The central difference between CAFNA and other neutron techniques is the method of imaging the gamma rays. Coded Apertures (CA) were developed mainly for astronomy applications in which photons have energies too high (> 70 keV) to be imaged with refraction or diffraction devices. At the same time, higher efficiency is achieved by replacing the classic pinhole / collimator optics with a multiple pinhole aperture, for an open fraction as high as 50%.

We have been analyzing the performance of coded apertures through simulation and experiment. In the following, after an overview of the principles of coded aperture imaging, we explain the design procedure of a small cargo inspection system, and compare its performance to that of a pinhole system. Some experimental results that support the principles underpinning the design close the paper.

HOW DOES CAFNA WORK?

Coded aperture imaging (for overviews see [2] and [3]) employs straight-line ray optics. Gamma rays from the inspected object are passed through an aperture (or mask), whose shadow is cast on a radiation detector (Figure 1). The projection **P** so produced is obtained in terms of the activity distribution **S** and the aperture **A** with the correlation:

$$\mathbf{P} = \mathbf{S} \times \mathbf{A} \qquad (1)$$

From knowledge of the aperture and of the detected pattern, the distribution of the source can be reconstructed. The choice of **A** is critical to the performance of the system, but patterns for which image reconstructions are artifact-free, such as the Uniformly Redundant Arrays (URAs), have been devised [4]. The relevant property of the patterns is the existence of a decoding pattern **G** such that the periodic correlation $\mathbf{A} \otimes \mathbf{G}$ is a δ function. For example, for URAs, **G** = **A** and **A** ⊗ **A** = δ. If **G** exists, a perfect image of **S** can be reconstructed by taking the periodic correlation of **P** with **A**. In fact:

P ⊗ **G** = (**S** × **A**) ⊗ **G** = **S** ∗ (**A** ⊗ **G**) = **S** ⊗ δ = **S**   (2)

This operation is carried out by digital processing and takes about 1 second on a personal computer. In the literature, a number of families of suitable arrays can be found [3].

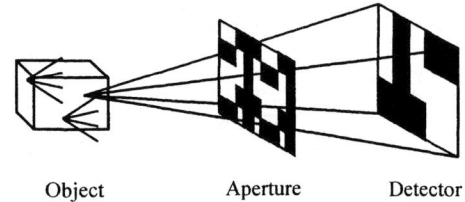

FIGURE 1. Basic concept of coded-mask imaging.

Figure 2 shows how such a system may be implemented for container inspection. The entire inspected object is simultaneously probed by a flood beam of neutrons from a simple, low-cost, source like a standard commercial sealed D-T tube of the type used in the well logging industry. Since the emission location of the ensuing gamma rays (typically 1.5-11 MeV) is imaged directly through the coded aperture, neutron scattering does not affect the image and all gamma rays can be used. Furthermore, not having to tag the neutrons in space and time allows the use of the entire output of the source without compromising resolution or sensitivity. By using 14 MeV neutrons, depth penetration is high, and more uniform penetration can be achieved by arranging multiple sources around the object. This solution would also increase flux uniformity and provide enhanced reliability through redundancy. Similarly, multiple coded aperture detectors give increased three-dimensional information and efficiency. The system is scalable in size in the sense that the number of sources and detectors can be suited to the inspection volume: CAFNA can be tailored to work with small objects, such as luggage, as well as with cargo containers.

These characteristics compare to those of another nuclear technique recently developed, PFNA, which mechanically scans the container with a narrow beam of 8 MeV neutrons and determines the position of the material along the beam line by means of precision timing of the arrival of the gamma rays. With current technology, this appears to place a limit of ~ 5 cm on spatial resolution. The technique also requires a large, expensive accelerator, complex mechanical motion, and is not scalable in size.

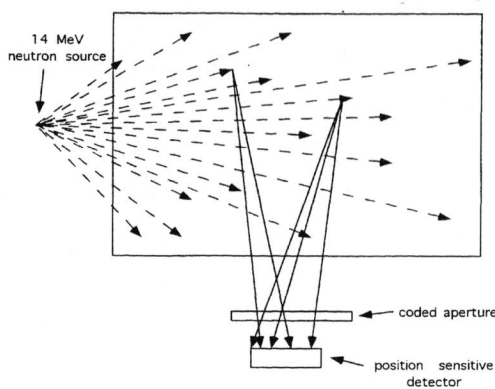

FIGURE 2. Operating principle of a CAFNA system.

SYSTEM DESIGN

The relationship between the Field of View (FoV) and the resolution λ of a coded aperture system is:

$$FoV = \lambda \sqrt{N_T} \quad (3)$$

with N_T total number of positions (open or closed) in the mask (assumed square). A realistic problem in contraband detection could be that of finding some 50 kg of material in a cargo of size 2×2 m. Assuming a density of 1.2 g/cm^3, we are looking for a 35×35×35 cm^3 volume: a system resolution of 20 cm should be sufficient. From the equation for the FoV, we see that the mask should have 10×10 elements. Since URAs of this dimension do not exist, we will use an 11×11 MURA pattern, which has 60 open positions, with a little gain in FoV at constant resolution. The condition that the projection of the mask pattern be sampled correctly at the detector is that each mask position be sampled twice:

$$mp \, m = 2 \, dp \quad (4)$$

where *dp* is the size of a detector pixel and we have defined the magnification coefficient *m* as the ratio of the projection on the detector of a mask hole to the size *mp* of the hole itself. This can be used in conjunction with the equation that links these parameters to the resolution:

$$\lambda = mp \frac{m}{m-1} \quad (5)$$

The detector currently available is a 8×8 square of 64 10×10×10 cm NaI(Tl) scintillators, which provide the necessary energy resolution (a better discussion of detector design can be found in [5]), which leads to

$mp = 10$ cm and $m = 2$, which, as can it be worked out from the definition of m, means that the mask should be midway in between object and detector. Note that the object-to-detector distance has not been fixed and, thus, can be optimized in some other way. The trade-off involved is between counts, higher rates being achieved for shorter distances, and near-field artifacts, which worsen with decreasing distance.

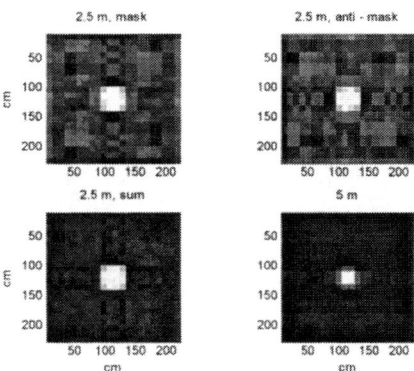

FIGURE 3. Artifact reduction: simulation.

We have recently developed a technique to reduce artifacts in pictures taken at short object-to-mask distance. The discussion of this method is outside the scope of this paper: here we just present a sample application relevant to this discussion. In Figure 3 are shown the results of three simulations, all having infinitely good statistics. A sphere of diameter 25 cm was placed at the center of a 2×2×2 m container. The two top pictures were obtained with an object-to-detector distance of 2.5 m. Since the mask-to-detector distance was 1.25 m, the mask was as close as practical to the object. While the top left picture was taken with the above-mentioned MURA pattern, the top right picture was taken after inverting the open and closed pixels of the pattern. Artifacts change sign and, when the two pictures are summed, artifacts cancel out (bottom left). The resulting picture has the same quality of that obtained with an object-to-detector distance of 5 m, but has a Signal-to-Noise-Ratio (SNR) advantage of about 4 (not simulated in this example).

The effect of mask thickness and penetration has also been simulated. At 5 MeV, penetration is considerable (1% for 9.7 cm of lead). However, the mask pixel is 10 cm wide, and the mask thickness does not lead to collimation artifacts.

Finally, it should be noted that the projection of the mask measures 2.2×2.2 m, but our detector is only 0.8×0.8 m. While in a real system a larger detector can be built, we can also think of scanning the projection with our smaller detector or, better yet, move the mask so that the different parts of the projection are, in turn, shifted onto the detector. This would more be practical, because the mask, lighter and not wired, would have to be moved by smaller shifts, and advantageous, because incidence angles at the detector would be lower, reducing near-field artifacts.

SYSTEM PERFORMANCE

The sensitivity advantages realized in CAFNA depend on several factors: the aperture pattern, its transmission and transparence to radiation, background and the appearance of the object. Let S_i be the number of counts due to the source in the ith pixel of the reconstructed image (of total intensity $I_T = \Sigma_i S_i$) and B the number of counts at all image pixels due to background (assumed uniform). For half-open URAs, the ratio of the SNR of the coded aperture system to that of the pinhole (the SNR advantage) is:

$$\frac{SNR_{URA}}{SNR_P} = \sqrt{N} \sqrt{\frac{\psi_i + \xi}{1 + 2\xi}} \qquad (6)$$

where $\psi_i = S_i / I_T$, $\xi = B / I_T$ and N is the number of pinholes in the coded aperture (each of the same size as the pinhole). To gain some insight in the meaning of eq. (6) we will proceed by particular cases. If the object is a point and there is no background, $\psi = 1$ and $\xi = 0$. The SNR advantage is maximum and equals $\sqrt{N}$. In our case $N = 60$ and the maximum advantage is 7.75. The second factor at the RHS of eq. (6) is the reduction of this maximum advantage. The most unfavorable case for coded apertures is when this factor is minimum, i.e. for $\xi = 0$ and ψ minimum. ψ is the lower the more sparse over the image pixels is the activity. It is minimum (excluding the case in which we are interested in points whose intensity is below average) in a uniform object, for which $\psi = 1/N_T$ at all reconstruction points. Since the array is 50% open $\psi = 1/2N$: the advantage is reduced to $1/\sqrt{2}$, i.e. the pinhole is actually favored. This is actually the case in some low-background applications, such as in medical imaging. In a CAFNA application, however, the background due to both gamma rays and neutrons is typically very high: some preliminary experiments performed at our irradiation facility have suggested $\xi = 30$. In this situation, ψ is irrelevant because eq. (6) reduces to $\sqrt{N/2}$, in our case a SNR advantage of 5.5. This figure should not be underestimated because it corresponds to a 30-fold reduction in time or activity, quantities proportional to the square of the SNR.

EXPERIMENTAL RESULTS

Our preliminary experiments aimed at verifying our ability to determine the composition of samples by FNA, and the predictions of our simulations. The neutron source used was a MF Physics D-T tube model A-325. It was run at 10kHz, with a duty cycle of 20%, with a nominal output of 3×10^7 n/s. Data were acquired for 5 minutes in a 16 µs gate delayed by a few µs with respect to the pulse signal. The detectors used were seven $10\times10\times10$ cm NaI(Tl) connected to a standard multi-channel analyzer.

TABLE 1. Example of Fast Neutron Analysis results.

Sucrose sample	Total mass (g)	Molar ratio C / O
Measured	864 ± 28.7	1.13 ± 0.07
Actual	808 ± 7.09	1.09
Difference	+ 7 %	+ 4 %

The first set of experiments aimed at determining the molar ratio of water and carbon in sucrose ($C_{12}H_{22}O_{11}$). For this experiment one detector only was used. Three spectra were acquired, respectively, from a water, a graphite and a sucrose sample. To determine the molar ratio, we found with a least-mean-square fit the best coefficients to combine the water and the oxygen spectra and obtain the sucrose spectrum. From the coefficients, after background and dead time correction and sample mass normalization, we produced Table 1, were the molar ratio is seen to be within one standard deviation of the expected value.

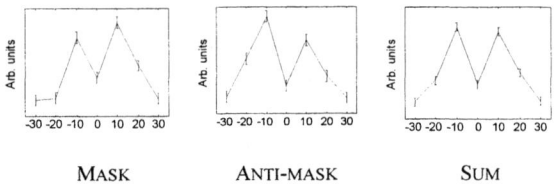

FIGURE 4. Artifact reduction: 1d experiment. The graphs show graphite mass as a function of position (cm).

The second set of experiments aimed at verifying the formulae for the FoV and resolution and the performance of the artifact reduction method. A 1d experiment was carried out: two ~8×8×8 cm cubes of graphite were placed, 20 cm apart, 50 cm from the detector. The coded aperture was designed to image with 10 cm resolution over a FoV of 70 cm. As explained above, two images were taken and the result added (Figure 4). First, two peaks, each at the correct position, are seen. Second, as predicted, the two blocks were resolved. Third, after artifact correction, the peaks have, within one standard deviation, the same height, confirming the effectiveness of artifact correction. Finally, comparison with the results obtained from a pinhole experiment was in agreement with the predictions of eq. (6) applicable to this case.

CONCLUSIONS

The results to date of simulations and experiments have shown that near-field coded aperture imaging is feasible. Since the advantage over pinhole systems depends critically on the background level, future investigations will aim at verifying that the preliminary findings on background are also applicable to realistic 2d and 3d situations. While difficulties are not expected for 2d applications, extension to 3d seems more challenging. Nevertheless, the possibility and the need of developing a manageable nuclear technique for security applications are a strong incentive for further investigation.

ACKNOWLEDGMENTS

This material is based upon work supported by the Office of National Drug Control Policy under Contract DAAD07-98-C-0117 and by the Federal Aviation Administration through Grant 93G-053.

REFERENCES

1. Bell, C. J., Krauss, R. A., and Lanza, R. C., "Analysis of complex targets using fast neutrons" in *Substance Detection Systems*, edited by L. Myers et al., SPIE Proceedings 2092, Bellingham: International Society for Optical Engineering, 1994, pp. 514-524.

2. Caroli, E., Stephen, J. B., Di Cocco, G., Natalucci, L., and Spizzichino, A., *Space Science Reviews* **45**, 349-403 (1987).

3. Skinner, G. K., *Nuclear Instruments and Methods in Physics Research* **221**, 33-40 (1984).

4. Fenimore, E. E., and Cannon, T. M., *Applied Optics* **17**, 337-347 (1978).

5. Lanza, R. C., Accorsi, R., and Chen, G., "CAFNA, Coded Aperture Fast Neutron Analysis: application to contraband and explosive detection" in *Third International Topical Meeting on Nuclear Applications of Accelerator Technology*, ANS Conference Proceedings, LaGrange Park: American Nuclear Society, 1999, pp. 147-154.

Methods for Quality Control/Quality Assurance of k_0-Assisted Neutron Activation Analysis

Frans De Corte

Laboratory of Analytical Chemistry, Institute for Nuclear Sciences, Gent University, Proeftuinstraat 86, B-9000 Gent, Belgium

Abstract. A survey is given of methods that were developed and applied for controlling and assuring the quality of k_0-assisted neutron activation analysis. These methods are: 1. Parallel but independent determination, in several laboratories, of the fundamental nuclear data, leading to the detection and elimination of systematic uncertainties; 2. Identification and quantification of all sources of uncertainty related to every step and parameter involved, and the application of error propagation theory to make an uncertainty budget of the analytical result; 3. Overall accuracy control via the analysis of various reference materials; 4. Detailed study of the traceability, and its enhancement by the introduction of neutron flux monitor alloys with certified composition; and 5. Currently, the development of synthetic multi-element standards ("SMELS") for testing the performance of the k_0-standardization when implemented in a laboratory.

CONCEPT AND PRACTICE OF k_0-NAA

For a detailed outline of the fundamentals and the practice of the k_0-standardization method, reference is made to former work [1]. Let it be recalled here that in k_0-NAA the concentration (ρ) of an analyte element ("a") is obtained from co-irradiation of the sample with a monitor ("m") in a calibrated reactor neutron spectrum, followed by measuring both on an efficiency-calibrated Ge detector. In the commonly used Høgdahl formalism [2], the basic equation is

$$\rho_a = \frac{(N_p/WSDCt_c)_a}{(N_p/wSDCt_c)_m} \frac{1}{k_{0,m}(a)} \frac{f+Q_{0,m}(\alpha)}{f+Q_{0,a}(\alpha)} \frac{\varepsilon_{p,m}}{\varepsilon_{p,a}} \quad (1)$$

with:

$$Q_0(\alpha) = \left(\frac{Q_0 - 0.429}{(\overline{E_r})^\alpha} + \frac{0.429}{(2\alpha+1)(E_{Cd})^\alpha}\right)(1eV)^\alpha \quad (2)$$

The $k_{0,m}(a)$ factor – of analyte "a" versus monitor "m" – is an experimentally measured compound nuclear constant defined as:

$$k_{0,m}(a) = (M_m \theta_a \sigma_{0,a} \gamma_a)/(M_a \theta_m \sigma_{0,m} \gamma_m) \quad (3)$$

with M the atomic weight, θ the isotopic abundance, σ_0 the 2200 ms^{-1} (n,γ) cross section and γ the gamma-ray emission probability. Furthermore, in Equations (1) and (2): N_p is the net number of counts in the full-energy photopeak (corrected for pulse losses); W is the sample mass; w is the mass of the co-irradiated monitor; S, D and C are the saturation, decay and counting factors, respectively; t_c is the counting time; f is the thermal-to-epithermal neutron flux ratio; $Q_0(\alpha)$ [= $I_0(\alpha)/\sigma_0$] is the ratio of the resonance integral (I_0) to the thermal cross section (σ_0), obtained from correction of Q_0 (in fact, of I_0) for a non-ideal epithermal neutron flux distribution, approximated by a $1/E^{1+\alpha}$ shape; $\overline{E}_r$ is the effective resonance energy; E_{Cd} (=0.55 eV) is the effective Cd cut-off energy; and ε_p is the full-energy peak detection efficiency. It should be noted that, in the methodology developed, the k_0-factors are expressed versus Au as the monitor [^{197}Au(n,γ)^{198}Au; E_γ = 411.8 keV].

Since its launching in the mid-1970s [3], the k_0-standardization of neutron activation analysis became widely accepted as a valuable analytical tool. This is not only evident from the more than 50 university, industrial and governmental NAA laboratories worldwide employing the k_0-methodology, but also from the continuing series of "International k_0 Users Workshops" organized by the Institute for Nuclear

Sciences, Gent in 1992, by the Jožef Stefan Institute, Ljubljana in 1996, and jointly by the Institute for Reference Materials and Measurements, Geel and the Nuclear Study Center, Mol in 2001. One of the main reasons for the spread of k_0-NAA is that – throughout its development – indefatigable care was exercised to make the analysis protocol at the same time user-friendly, generally applicable and – above all - reliable. In this context of reliability, a survey is given in the present paper of methods that were developed and applied for the quality control/quality assurance of k_0-NAA.

NUCLEAR DATA LIBRARY

The k_0-factor itself is linearly present in the expression for the calculation of the concentration [Equation (1)], and hence its accuracy is of the utmost importance for the quality of the analytical output. Therefore, it was decided to perform the determination of k_0's via parallel, but independent measurements in varying experimental conditions. This work was done at the analytical laboratories of the Institute for Nuclear Sciences (INW), Gent and the Central Research Institute for Physics (KFKI), Budapest (although, in a later stage, other institutes were involved as well). Thus, in both institutes different preparations were made of standards and monitors to be co-irradiated. Next, the irradiations were performed in at least two positions (with different neutron thermalization) of their respective reactors, namely the "swimming pool"-type reactor THETIS in Gent (water/graphite moderator, graphite reflector) and the "water boiler"-type reactor WWR-M in Budapest (light water moderator, beryllium reflector). Finally, the countings in both institutes were done on several independently calibrated Ge detectors, followed by net peak integration based on different gamma-ray spectrum analysis codes. In each reactor channel, the experiments were carried out in 3-5-fold. By following this protocol, it was possible to detect and eliminate systematic errors. By finally making the grand mean of the INW and KFKI values, it is believed that a randomization of the uncertainty (as a rule better than ±2%, on the average ~ ±1%) was achieved. Up to the present, the thus obtained "k_0-data library" contains k_0-factors for the relevant gamma-lines of about 130 analytically interesting radionuclides, the measurement and evaluation of which - following the above outlined principles - is perhaps best described and illustrated in Reference [4]. As an example, the experimental k_0-factor determination for ^{24}Na is shown in Table 1. The monitor used was an Al-0.1%Au alloyed wire, the composition of which was checked versus a home-made Au standard. The relevant neutron spectrum characteristics were: f=34 and α=0.016 in channel "Mila"; f=18 and α=-0.007 in channel "Csöpi"; f=25 and α=0.015 in channel 3; and f=71 and α=0.084 in channel 15.

In addition to the k_0-factor, attention was also paid to the other nuclear data figuring in Equation 1, such as Q_0 and $\bar{E}_r$, required for the conversion of Q_0 to $Q_0(\alpha)$. However, in view of the herewith-associated error reduction factors towards the analytical results, less stringent criteria were handled for the evaluation of these data.

UNCERTAINTY BUDGET

The main parameters and steps that are determining the uncertainty on the k_0-standardization are related to the nuclear data – as outlined above –, and to the calibration of the irradiation facility and the Ge-detector. For both calibrations, not only were user-friendly and accurate procedures worked out, but a large effort was also spent to identify all sources of uncertainties and to evaluate their propagation towards the analytical result.

The neutron spectrum parameters to be entered in Equation 1 are the thermal-to-epithermal neutron flux f and the epithermal spectrum shape factor α. Traditionally, measurement of these parameters involves the use of Cd-covered monitors (or of both Cd-covered and bare monitors, to obtain Cd-ratios), and these methods were optimized for being implemented in k_0-NAA [5]. Innovative, however, in the development of the k_0-method, was the elaboration of procedures based on the irradiation of bare monitors

TABLE 1. Measurement of k_0-factors for ^{24}Na.

Sample preparation	Isotope formed	E_γ, keV	Measured $k_{0,Au}$-factor (relative uncertainty, %)				Grand mean
			KFKI, WWR-M reactor		INW, THETIS reactor		
			Ch. "Mila"	Ch. "Csöpi"	Ch. 3	Ch. 15	
KFKI: 1mg NaCl on Al-foil; pellet 6.4mm diam. × 0.2mm. INW: Na$_2$CO$_3$ powder in PE tube; 20mg (Ch.3); 50mg (Ch.15).	^{24}Na	1368.6	4.71E-2 (2.2)	4.74E-2 (2.4)	4.61E-2 (2.5)	4.65E-2 (1.8)	4.68E-2 (0.6)
		2754.0	4.60E-2 (1.6)	4.74E-2 (1.6)	4.56E-2 (1.8)	4.59E-2 (0.5)	4.62e-2 (0.9)

(minimum three for α-measurement, minimum two for f-measurement), which can thus simply be co-irradiated with the sample to be analyzed

In the case of α-monitoring, it was found from both theoretical and practical considerations that the best sets for the "bare triple-monitor"-method are ^{197}Au-^{94}Zr-^{96}Zr or ^{238}U-^{94}Zr-^{96}Zr. Although the uncertainty on the thus obtained result for α [from ~ ±20% (relative) for α ≅ 0.1 to ±80% for α ≅ 0.01) is rather poor compared with the "Cd-covered multi-monitor"- and (especially) the "Cd-ratio for multi-monitor"-methods, it is quite acceptable in view of the considerable error reduction factor towards the analytical result (see below).

Once α known, f can be obtained from the "bare bi-isotopic monitor"-method:

$$f = \frac{\frac{k_{0,Au}(1)}{k_{0,Au}(2)} \frac{\varepsilon_{p,1}}{\varepsilon_{p,2}} Q_{0,1}(\alpha) - \frac{A_{sp,1}}{A_{sp,2}} Q_{0,2}(\alpha)}{\frac{A_{sp,1}}{A_{sp,2}} - \frac{k_{0,Au}(1)}{k_{0,Au}(2)} \frac{\varepsilon_{p,1}}{\varepsilon_{p,2}}} \quad (4)$$

where it was found that ^{94}Zr-^{96}Zr is the best couple giving the most accurate results [6].

Thus, from the above it follows that co-irradiation of a Zr-foil and a dilute Al-Au or Al-$^{(238)}$U alloy yields acceptably accurate and precise values for both the α and f parameters. The nuclear data are: Q_0 = 5.05 and $\overline{E}_r$ = 6260 eV for ^{94}Zr(n,γ); Q_0 = 248 and $\overline{E}_r$ = 338 eV for ^{96}Zr(n,γ); Q_0 = 15.7 and $\overline{E}_r$ = 5.65 eV for ^{197}Au(n,γ); Q_0 = 103.4 and $\overline{E}_r$ = 16.9 eV for ^{238}U(n,γ).

Propagation of the uncertainty on f and α towards the analytical result was also examined in detail and, as an example, the error propagation factor s(α→ρ) for Equation 1, with the input parameter f obtained from Equation 4, is given by (1 eV terms omitted):

$$s(\alpha \to \rho) = \alpha \times$$
$$\left\{ \frac{q_{0,m}(\alpha) \ln \overline{E}_{r,m}}{f + Q_{0,m}(\alpha)} - \frac{q_{0,a}(\alpha) \ln \overline{E}_{r,a}}{f + Q_{0,a}(\alpha)} + \frac{Q_{0,m}(\alpha) - Q_{0,a}(\alpha)}{Q_{0,1}(\alpha) - Q_{0,2}(\alpha)} \times \right.$$
$$\left. \frac{[f + Q_{0,1}(\alpha)]q_{0,2}(\alpha) \ln \overline{E}_{r,2} - [f + Q_{0,2}(\alpha)]q_{0,1}(\alpha) \ln \overline{E}_{r,1}}{[f + Q_{0,m}(\alpha)][f + Q_{0,a}(\alpha)]} \right\}$$

with:
$$q_0(\alpha) = (Q_0 - 0.429)/(\overline{E}_r)^\alpha \quad (5)$$

This shows, for instance, that the uncertainty on ρ (induced by α) decreases with lower α, higher f, a small spread in Q_0-values for analyte and monitor, and a large spread in Q_0-values for both f-monitors.

As to the detector calibration, Equation 1 shows that it is the uncertainty on the ε_p-ratio – rather than on the absolute value of ε_p – which is linearly transferred to the analytical result. Evidently, ε_p refers to the geometric sample-detector configuration on hand, and a procedure was worked out [7] for the accurate conversion of the experimentally measured values of "$\varepsilon_{p,ref}$" (for point sources at large distance from the detector) to "$\varepsilon_{p,geo}$" (for voluminous sources and/or close-in counting geometries). The detection efficiency is also involved in the correction for true-coincidence effects: ε_p in case of summing-in, ε_t – the total detection efficiency – in case of summing-out. In the terminology of our work, the latter is defined as $\varepsilon_p/(P/T)$, and user-friendly, reliable procedures were worked out for the measurement of the peak-to-total ratio P/T. Here, all these parameters, as well as the decay scheme data involved, undergo an appreciable uncertainty reduction towards the analytical result.

From these considerations, and taking into account correlation, it was possible to evaluate the uncertainty on the analytical result induced by the k_0-standardization. It was concluded that, for average experimental irradiation and counting conditions, the total uncertainty is of the order of 3.5 % (see Table 2).

EXPERIMENTAL ACCURACY CHECK

In addition to the evaluation of the uncertainty, in the course of the years this estimation was experimentally verified by analyzing a variety of reference materials, such as "standard" SRMs (NIST), "certified" CRMs (IRMM/BCR), USGS materials, etc. Here again, this work was performed cooperatively at the INW, Gent and the KFKI, Budapest, and it could indeed be proven that in practice the accuracy of the k_0-standardization is of the order of 3-4% (see e.g. [8]).

Next to this, in an indirect way the accuracy of the k_0-method followed from the results of participation in European certification campaigns. Among the most recent [9], Figure 1 gives a comparison of our results with the certified values for the polymeric reference material BCR CRM-680. The error bars are ±2s stand-

TABLE 2. k_0-standardization: uncertainty budget.

Uncertainty on analysis result	
Induced by	Contribution
k_0	~ 1 %
Q_0	~ 1 %
α	~ 1.5 %
f	~ 1 %
ε_p (meas. & convers.)	~ 2 %
coincidence correction	~ 1.5 %
Total (quadratic Σ)	~ 3.5 %

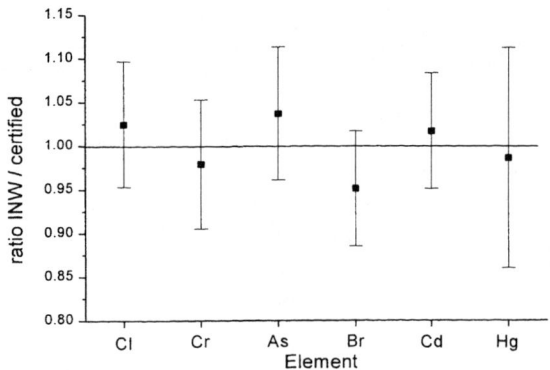

FIGURE 1. INW results in the certification of the polymeric reference material BCR CRM-680.

ard deviations, obtained from quadratic combination of the INW and the certified data. Here again, it is shown that with respect to its accuracy, k_0-NAA is quite satisfactory for certification purposes.

TRACEABILITY

The concern for the traceability of the k_0-method was triggered in the 1980s by a controversy on the acceptability of k_0-NAA for certification purposes. To this aim, the method was examined in detail [10], by unraveling the procedure into its basic steps and parameters, with special emphasis on the three main topics: the co-irradiated gold monitor, the nuclear data, and the methodology of calibrating the irradiation facility and the Ge-detector. It was concluded that k_0-NAA provides adequate traceability, as well for routine analysis as for certification work.

As to the gold monitor, it goes without saying that the introduction, in 1991 [11], of the IRMM-530 Al-0.1%Au reference material for the k_0-standardization of NAA, was significantly contributing to the traceability of the method. The present development of an IRMM Zr-Au-Lu alloy (enabling the simultaneous monitoring of α, f and of the neutron temperature T_n), will further add to the traceability of the k_0-methodology.

As a result of the above efforts, k_0-assisted NAA became accepted for European certification analysis, and is nowadays frequently used in CRM certification rounds. This is illustrated by a contribution on the topic: "k_0-NAA, a valuable tool for RM producers", presented at the recent BERM-8 symposium [12].

THE "SMELS" PROJECT

To demonstrate the quality of implementation of k_0-NAA in a laboratory, synthetic multi-element standards ("SMELS") are being developed, consisting of an inert material doped with a variety of analytically interesting elements. As outlined earlier [13], three types of standards will be made available, namely for analysis via short-, medium- and long-lived radionuclides. The work, done in cooperation with IRMM,Geel, involves the preparation of the spiked matrix (phenol-formaldehyde resin), the control of its resistance for irradiation in high neutron fluxes, homogeneity control of the dopant elements, and the final concentration characterization [14].

CONCLUSION

As a result of the various QC/QA-actions undertaken in the course of its development, k_0-NAA grew from a mere theoretical concept in the mid 1970s to a fully-operational and reliable analytical tool at present.

ACKNOWLEDGMENTS

The financial support of the Fund for Scientific Research-Flanders is gratefully acknowledged.

REFERENCES

1. De Corte, F., Simonits, A., De Wispelaere, A., and Hoste., J., *J. Radioanal. Nucl. Chem.* **113**, 145-161 (1987).
2. Høgdahl, O.T., "Neutron Absorption in Pile Neutron Activation Analysis", *Report MMPP-226-1* (Dec. 1962).
3. Simonits, A., De Corte, F., and Hoste, J., *J. Radioanal. Chem.* **24**, 31-46 (1975).
4. De Corte, F., Simonits, A., De Wispelaere, A., and Elek., A., *J. Radioanal. Nucl. Chem.* **133**, 3-41 (1989).
6. De Corte, Sordo-El Hammami, K., Moens, L., Simonits, A., De Wispelaere, A., and Hoste, J., *J. Radioanal. Chem.* **62**, 209-255 (1981).
6. Simonits, A., De Corte, F., and Hoste, J., *J. Radioanal. Chem.* **31**, 467-486 (1976).
7. Moens, L., De Donder, J., Lin Xilei, De Corte, F., De Wispelaere, A., Simonits, A., and Hoste, J., *Nucl. Instr. Methods* **187**, 451-472 (1981).
8. De Corte, F., Demeter, A., Lin Xilei, Moens, L., Simonits, A., De Wispelaere A., and Hoste, J., *Isotopenpraxis* **20**, 223-226 (1984).
9. Lamberty, A., (IRMM, 2000), *private communication*.
10. De Corte, F., *J. Trace Microprobe Techn.* **5**, 115-133 (1987).
11. Ingelbrecht, C., Peetermans, F., De Corte, F., De Wispelaere, A., Vandecasteele, C., Courtijn, E., and D'Hondt, P., *Nucl. Instr. Methods A* **303**, 119-122 (1991).
12. Robouch, P., presented at *BERM-8*, Bethesda, MD, USA (Sept. 2000).
13. De Corte., F., De Wispelaere, A., Kramer, G.N., Robouch, P., and Simonits, A., *Nucl. Instr. Methods A* **422**, 891-894 (1991).
14. Eguskiza., M., Robouch., P., and De Corte, F., presented at *CAARI-2000*, Denton, TX, USA (Nov. 2000).

Monitoring Of D-T Accelerator Neutron Output In A PGNAA System Using Silicon Carbide Detectors

Abdul R. Dulloo, Frank H. Ruddy, John G. Seidel, and Bojan Petrović

Science & Technology Department, Westinghouse Electric Company
1330 Beulah Road, Pittsburgh, PA 15235, USA

Abstract. Silicon carbide (SiC) detectors are being employed to monitor the neutron output of the D-T accelerator in a pulsed Prompt Gamma Neutron Activation Analysis (PGNAA) system. Detection of the source neutrons relies on energetic neutron reactions in the detector material. Experimental testing has been performed to confirm that the detector response is caused by fast neutrons from the accelerator source. Modeling calculations have also been carried out to provide additional verification. Use of the SiC detectors in the PGNAA system is expected to assist in evaluating system performance as well as ensuring accurate data interpretation and analysis.

INTRODUCTION

Monitoring the output intensity of D-T neutron generators is needed in order to verify generator performance and to assist in data interpretation and analysis for applications employing these generators. Examples of such applications include Prompt Gamma Neutron Activation Analysis (PGNAA) and well logging.

Present monitoring methods rely on detectors that contain a constituent element which produces a short-lived radioactive nuclide upon neutron activation. The intensity of the photon radiation emitted during the subsequent decay of the radioactive nuclide is then related to neutron source strength. Typically, photon scintillation detectors such as thallium-activated sodium iodide or cerium-activated yttrium orthosilicate are used. In the former case, the product of the ^{24}Na (n, α) ^{20}F reaction is monitored, whereas that of the ^{16}O (n, p) ^{16}N reaction is used in the latter. The disadvantages of these detectors are that they are generally cumbersome and are subject to interference from background photon radiation. Furthermore, because the half life of the activated nuclide can be up to 11 s, there is a corresponding delay in the detector response to changes in neutron source strength.

In order to overcome the deficiencies described above, we propose a novel method, based on the use of silicon carbide (SiC) detectors, to measure the neutron output of a D-T source. SiC semiconductor diodes are being developed for neutron and high-intensity gamma-ray radiation detection [1 - 5]. By virtue of the wider band gap energy of SiC, these detectors are less sensitive to temperature than conventional silicon semiconductor detectors, and are also more resistant to radiation damage. The nuclear detection properties of miniature SiC detectors have been demonstrated for charged particles, thermal neutrons (through the juxtaposition of a ^{6}LiF layer), and gamma radiation.

The proposed method to monitor D-T generator neutron output with SiC detectors relies on energetic neutron reactions - such as ^{28}Si (n, p) ^{28}Al, ^{28}Si (n, α) ^{25}Mg, ^{28}Si (n, n') ^{28}Si, ^{12}C (n, α) ^{9}Be, ^{12}C (n, n') ^{12}C and ^{12}C (n, n') 3α - that occur in the detector material. The recoil charged-particle products of these reactions produce ionization in the active volume of the SiC detector, and the resultant charge pulses are collected and counted with conventional signal-processing equipment. The count rate is used to determine the neutron fluence rate incident on the detector, which is in turn related to the neutron output of the D-T source.

The main advantages of this method are as follows:

- There is no delay in the detector response to changes in neutron source strength since the signal-producing charged particles are emitted within a negligible time of neutron capture. The detector response is not dependent on activation and decay.

- Charge collection is relatively fast in the SiC detector. Typical charge-induced signal rise times are 10-20 ns long, allowing counting during neutron source pulses in applications where pulsing on a microsecond time scale is utilized.

- The small size of a SiC diode allows for flexibility in configuring the detector to match a particular application.

- The SiC detector can function at high temperatures [1, 4], and is relatively insensitive to thermal neutrons and to gamma radiation [1, 3].

- The SiC detector is designed so that radiation damage to the active region of the diode is minimized [5].

In this paper, the performance of a SiC detector used to monitor the fast-neutron output of a pulsed D-T source in a PGNAA system is described. We also present the results of preliminary modeling calculations that were performed to study the energy spectrum of neutrons that are incident on the SiC detector for the case of a D-T neutron generator and to evaluate the effect of surrounding materials on the neutron energy spectrum.

The proposed method can be applied to neutron-based applications that employ D-T neutron generators. Examples of such applications include PGNAA, downhole logging of oil and gas wells, nondestructive assay of transuranic waste containers, nuclear safeguards measurements, and industrial process monitoring.

EXPERIMENTAL TEST RESULTS

Two 2.5-mm diameter SiC diodes, connected in parallel and mounted on a connecting box, were utilized to monitor the output of a D-T neutron generator employed in a PGNAA system. A negative voltage bias of 80 volts was applied to the detectors. The position of the detectors relative to the D-T source is shown in Figure 1. The generator (MF Physics, Model A320) and connecting box are located inside a polyethylene cavity.

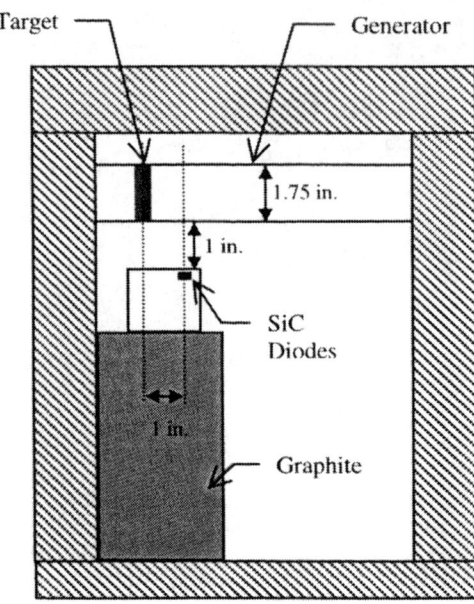

FIGURE 1. Location of SiC Detectors Relative to Neutron Generator Target.

The output of the connecting box is connected to a charge-sensitive preamplifier, which in turn is connected to the fast-spectroscopy amplifier of a pulsed PGNAA signal-processing system. Data collection from the diodes is correlated with pulsing of the neutron generator via a timing module. The pulse timing scheme employed in the experimental tests, illustrated in Figure 2, contains four time groups (1 – 4). Typical values of the frequency and neutron pulse width are 1112 Hz (899 μs) and 90 μs, respectively.

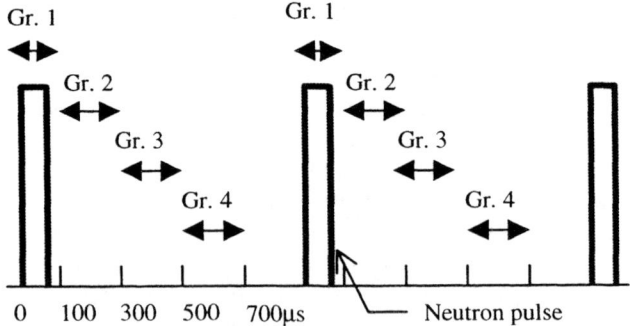

FIGURE 2. Neutron Pulsing Scheme Employed in Experimental Tests. Time scale is in μs.

The time-dependent spectra acquired from the SiC diodes during an 1800-s run (No. SIC13) are shown in Figure 3. Although each group's spectrum contains

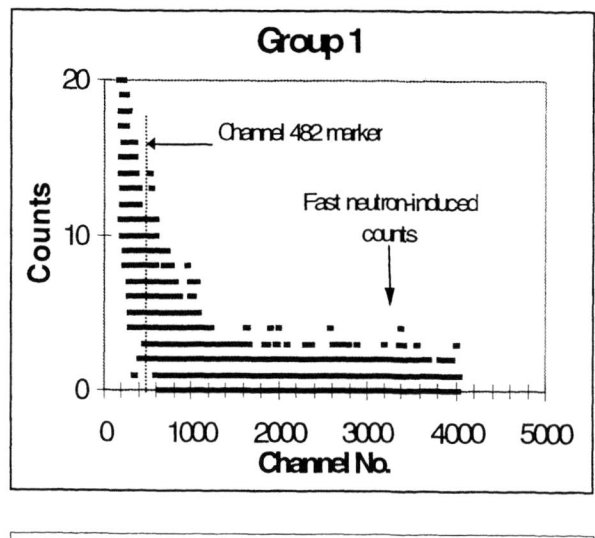
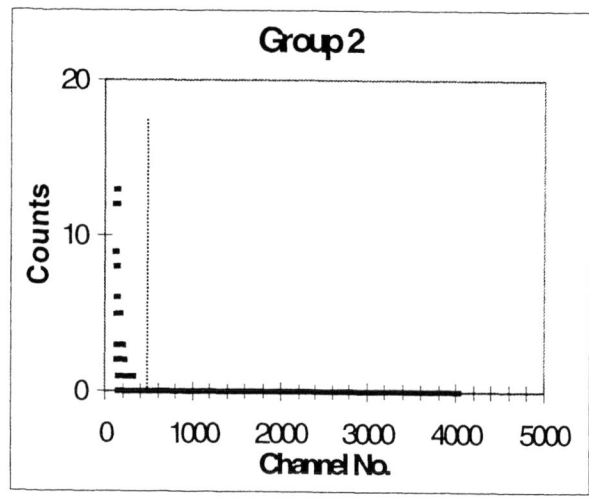
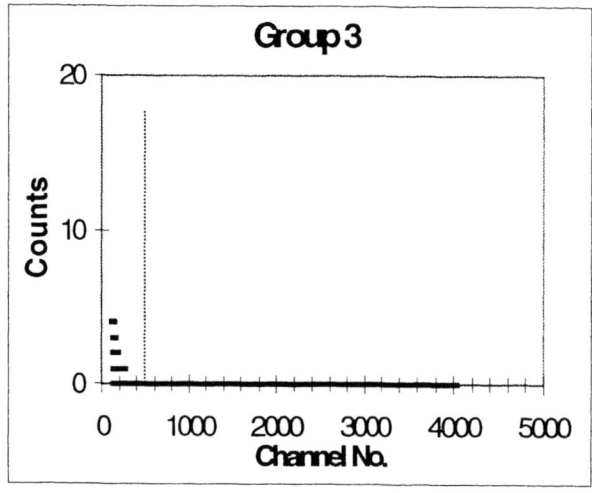
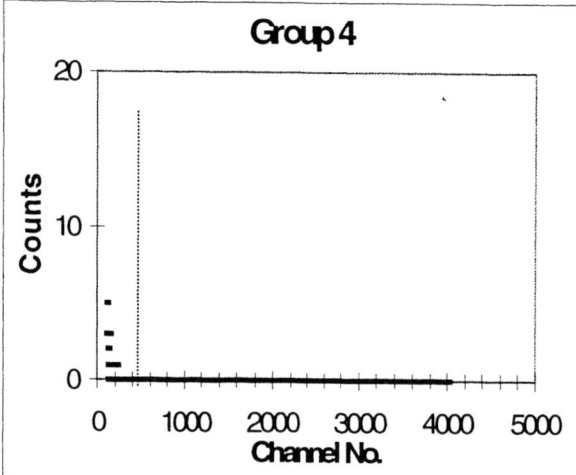

FIGURE 3. SiC Detector Signal Acquired from an 1800-s Run. Counts past channel 482 are only observed in Group 1.

8192 channels, only data in the first 4000 channels are shown. Whereas no counts above channel 482 are observed in the spectra of groups 2, 3 and 4, the spectrum of group 1 (which is concurrent with the 14-MeV neutron pulses) contains counts beyond channel 482. Monte Carlo simulations [6] have shown that the fast neutrons emitted by the source in Group 1 are thermalized within 10 to 20 μs and, consequently, fast neutron-induced events in the SiC detectors should occur mainly in Group 1. The counts observed in the spectrum of Group 1 beyond channel 482 are therefore attributed to charge pulses generated primarily by fast neutron interactions with the Si and C nuclides of the detector material. Furthermore, PGNAA measurements [7] and the Monte Carlo simulations indicate that a significant flux of thermal neutrons exists in groups 2 and 3. Since no count is observed in the 482-8192 channel region of these two groups in run SIC13, it can be concluded that thermal neutrons do not produce counts in the SiC detectors beyond channel 482. This result confirms the expectation that thermal neutron-induced reactions will generally produce insufficient ionization in the detector to produce a measurable signal. The SiC detectors are therefore insensitive to thermal neutrons.

Replicate measurements confirmed good reproducibility and consistency of the detector signal induced by the fast source neutrons. A passive run was employed to further confirm that the counts observed in the active run spectrum are induced by neutrons from the generator.

MODELING RESULTS

Based on the experimental results and discussion presented in the preceding sections, the response of a "bare" SiC detector to a 14-MeV neutron source is caused primarily by fast neutrons incident on the detector. In order to accurately monitor the source output, the detector response should be elicited by neutrons coming directly from the source. If a significant fraction of the detector response is caused by neutrons that have first scattered from materials outside of the generator, then the composition of surrounding materials will influence the response. This latter situation would be undesirable. Therefore, two key issues that need to be addressed are the energy distribution of the fast neutrons primarily responsible for the detector response, and the effect of the surrounding environment on this neutron energy distribution.

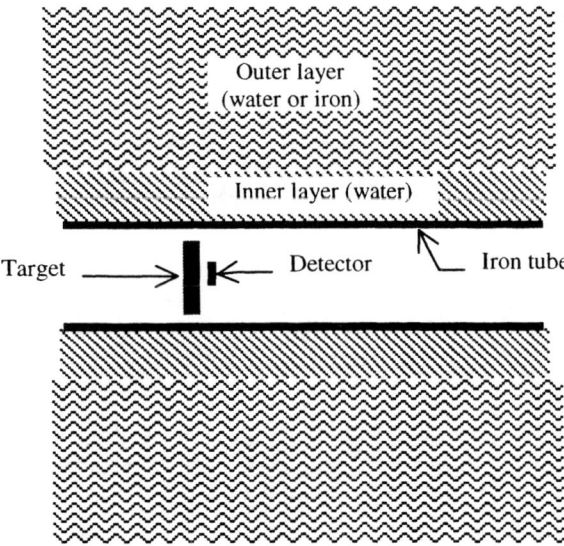

FIGURE 4. Geometrical Model Employed in MCNP Calculations.

Monte Carlo calculations using the MCNP code [8] were performed to study these issues. The geometrical model employed consists of a neutron generator, a SiC detector and a two-layer surrounding medium, as shown in Figure 4. The generator consists of a target (neutron source), which has a radius of 0.75 cm and a thickness of 0.4 cm, located inside an iron tube of inner radius 2.0 cm and outer radius 2.4 cm. The detector is located inside the generator tube at a distance of 0.4 cm from the target. The surrounding material was modeled as two concentric layers. An inner water layer is 1.6-cm thick. The outer layer material (reflector) was either water or iron. Time-dependent Monte Carlo simulations were performed in order to obtain the neutron energy distribution at the detector location using pointwise cross sections. We assumed that an instantaneous neutron burst occurred at time zero, and that 14.01-MeV monoenergetic neutrons were emitted isotropically. The time-dependent neutron energy spectrum was tallied over the first μs, and then the next 10, 100, and 1000 μs. In such a configuration where the detector is close to the source (i.e., D-T neutron generator target), most of the flux is expected to come directly as uncollided neutrons from the source in the first μs. In this particular case, this fraction is ~92%. Another ~2% is contributed by neutrons that had undergone small-angle scattering, leading to similar energies (13-14 MeV), and another ~5% accounted for neutrons of all other energies in the first μs. The remaining ~1%, integrated over the subsequent 1110-μs time after the burst, are low-energy neutrons. The surrounding environment (external to the generator tube) causes a negligible perturbation on the neutron energy distribution at the detector location in the 0 – 1111 μs time interval.

The above results and discussion demonstrate that the energy spectrum of the neutron flux incident on a SiC detector is essentially determined by the target detector configuration within the time interval considered. For a configuration where the detector is located close to the source, the absolute flux level recorded by the detector is proportional to the total source output. Furthermore, perturbation in the SiC detector response due to the change in the surroundings is small. These features indicate that the use of SiC detectors should be accurate for on-line monitoring of a pulsed D-T neutron generator's output/intensity, or for monitoring of other fast neutron sources, as long as the detector is in reasonably close proximity to the source.

CONCLUSIONS

A novel method to monitor the fast neutron output of a D-T accelerator source has been proposed. This method employs SiC diodes as fast-neutron detectors. Preliminary measurements and Monte Carlo numerical simulations carried out to study this method have established the following conclusions:

- Charge pulses above a characteristic amplitude are generated in the SiC detector only in the presence of fast neutrons from the D-T source.

- The bare SiC detector is insensitive to thermal neutrons.

- A SiC detector located in close proximity to the target of a D-T neutron generator is exposed mainly to uncollided source neutrons in the first few hundred microseconds after neutron pulsing. The detector response in this interval is essentially proportional to the source output.

The proposed method can be used to monitor the D-T source output in applications where pulsed D-T sources are employed, such as PGNAA, oil and gas well logging, process monitoring, nuclear safeguards monitoring, and nondestructive assay of transuranic waste.

REFERENCES

1. Ruddy, F. H., Dulloo, A. R., Seidel, J. G., Seshadri, S., and Rowland, L. B., *IEEE Trans. Nuclear Sci.* **45**, 536-541 (1998).

2. Dulloo, A. R., Ruddy, F. H., Seidel, J. G., Adams, J. M., Nico, J. S., and Gilliam, D. M., *Nucl. Instr. and Meth. A* **422**, 47-48 (1999).

3. Dulloo, A. R., Ruddy, F. H., Seidel,, J. G., Davison, C., Flinchbaugh, T., and Daubenspeck, T., *IEEE Trans. Nuclear Sci.* **46**, 275-279 (1999).

4. Babcock R. V., and Chang, H. C., "SiC Neutron Detectors for High-Temperature Operation" in *Proceedings of the 1962 Symposium on Neutron Detection, Dosimetry and Standardization.*, IAEA STI/PUP/69 1, 1963, pp. 613-622.

5. U.S. Patent No. 5,726,453, "Radiation Resistant Solid State Neutron Detector," assigned to the Westinghouse Electric Company (1998).

6. Petrović, B., Haghighat, A., Congedo, T. V., and Dulloo, A. R., "Development and Validation of Transport Theory Methodology For Accurate Modeling and Simulation of PGNAA System For Non-Destructive Mixed Waste Characterization" in *Proceedings of 6th Nondestructive Assay Waste Characterization Conference-1998*, Department of Energy/INEEL Proceedings, Salt Lake City, 1998, pp. 605-627.

7. Dulloo, A. R., Ruddy, F. H., Congedo, T. V., Seidel, J. G., and Gehrke, R. J., Nucl. Technol. **123**, 103-112 (1998).

8. *MCNP - A General Monte Carlo N-Particle Transport Code, Version 4B*, edited by J. F. Briesmeister, Los Alamos National Laboratory Report, LA-12625-M, Version 4B (1997).

The Preparation and Characterization of Synthetic Multi-Element Standards for Testing the Performance of k_0-NAA: the State of Affairs

M. Eguskiza[a], P. Robouch[a], F. De Corte[b] and S. Pommé[c]

[a] *European Commission, Institute for Reference Materials and Measurements, Retieseweg, B-2440 Geel, Belgium*
[b] *Lab. Anal. Chem. Inst. Nucl. Sci., Gent University., Proeftuinstraat 86, B-9000 Gent, Belgium*
[c] *SCK·CEN, Belgian Nuclear Research Centre, Boeretang 200, B-2400 Mol, Belgium*

Abstract. As part of the general QC/QA management of the k_0-standardization, three types of 'SMELS' (synthetic multi-element standards) are being developed to check the performance of k_0-NAA when implemented in a laboratory. Phenol-formaldehyde resin is proposed in this paper as the matrix of choice. Based on previous work, this material could be prepared with sufficient purity so as to make interferences with the spiking elements negligible. The matrix was found to be adequately resistant towards high neutron flux irradiation. A selection was made of compounds, soluble in ethyl alcohol, for spiking the prepolymer ethanolic solution with the elements of interest. Finally, some prototypes were prepared and a satisfactory homogeneity of the spiked elements was found for 50 mg sample intake.

INTRODUCTION

Among the strategies developed for QC/QA of the k_0-standardization of NAA [1], the generation of a dedicated tool for demonstrating its performance after its implementation in a laboratory, is considered to be of great practical importance. To this aim, a research project was started for the development of synthetic multi-element standards (SMELS) with well-characterized elemental concentrations, as a collaboration between the IRMM (Institute for Reference Materials and Measurements, Geel, Belgium) and the INW (Institute for Nuclear Sciences, University of Gent, Belgium). Three types of standards are envisaged, each consisting of an inert pure matrix doped with a group of elements according to the half-lives of the radionuclides formed: short-lived (type I), medium-lived (type II) and long-lived (type III). In an introductory paper of De Corte et al. [2], an account was given of the selection of the spiking elements with their corresponding radionuclides and concentrations, following some basic criteria: overall well-balanced spectra, coverage of a wide range of half-lives and absence of interferences on the analytical gamma lines. In the choice of the matrix some crucial requirements have to be met as well. It should be composed of "inert" elements that do not give rise to interfering neutron activation and do not lead to appreciable neutron shielding, it should be sufficiently pure, so as to avoid any reaction or spectral interferences arising from the impurities, it should stand irradiation in high-neutron flux irradiation facilities, and finally, it should offer the possibility to be spiked homogeneously. In the present paper, a survey is presented of the state of affairs in the development of the SMELS, based on a bakelite matrix. The topics highlighted are: preparation of the matrix, investigation of its behavior in an intense neutron field, spiking procedure, interferences caused by matrix impurities and assessment of the homogeneity of doping.

CHOICE OF MATRIX AND DOPANTS

Among the various matrices for the production of synthetic NAA multi-element standards that were formerly described in the literature [2-10], in the SMELS project phenol formaldehyde resin (bakelite) was selected as the matrix of choice. The reason for choosing this material, that was already studied in the

1970s [4], is that it is likely to fulfill all of the above mentioned matrix requirements. The selected elements and target mass fractions (mg/kg) for each of the types were: type I, 8 elements: Au (100), Cl (5500), Cs (900), Cu (4000), I (150), La (270), Mn (100) and V (50); type II, 10 elements: As (100), Au (5), Br (175), Ce (17000), Mo (5400), Pr (1300), Sb (200), Th (4000), Yb (220), Zn (7500); type III, 15 elements: Au (1), Co (20), Cr (80), Cs (20), Fe (8000), In (400), Sb (50), Sc (1), Se (120), Sr (8000), Th (25), Tm (20), Yb (20), Zn (600), Zr (6000).

EXPERIMENTAL

Matrix Preparation

Resol, a prepolymer of phenol formaldehyde resin (bakelite) was synthesized by mixing 94 g of phenol with 123 g of formaldehyde in aqueous solution (37% by weight) and 40 g NH_4OH as catalyst [11]. The reagents were stirred and heated at 70°C for 2 hours at pH=8. After finishing the reaction, an upper aqueous layer was formed above a lower solid phase containing the resol. The water was poured out, leaving the resol in the reactor vessel, where it was dissolved in acetone. The thus prepared resol had to be purified and the remaining initial reagents and secondary products had to be removed. To this aim, the resol acetone solution was added to water where the resol precipitated free of impurities. Then, it was filtrated using a filtration funnel equipped with a net filter, which consists of Nylon PA 6,6 (41 μm pore). Finally, the resol was dried at 70°C and crushed into a powder which was the starting material of the spiking step.

Spiking Procedure

The resol solubility in ethanol dictated that, in order to arrive at a homogeneous doping, also the compounds had to be soluble in ethanol. The absence of elements (like the Cl^-) that could interfere and the purity of the reagents were other factors to be taken in account. Therefore, analytical grade or high-purity compounds were selected. From each compound, an ethanolic solution was prepared with ~10 times higher concentration than the target concentration and therefore small spiking volumes of ~1 ml could be used. After adding the elements one by one to the resol solution, the ethanol was removed on a boiling water bath; the solid residual was dried at 100°C, crushed in a melamine mortar and dried until constant weight.

RESULTS

Matrix Characterization

The organic composition of the matrix was studied by proton and carbon nuclear magnetic resonance spectroscopy (^{1}H- and ^{13}C-NMR), proving a fairly complete resol formation. The stability of the resol at high fluxes was studied at the Nuclear Physics Institute in Řež, Czech Republic. Two samples were irradiated for up to 20 hours in reactor LVR-15 at a thermal neutron flux of $8*10^{13}$ $cm^{-2}s^{-1}$. No significant weight loss was observed. Visual observation revealed no change in the structure of the irradiated powder, but a slight color change was seen, from brown to black.

The content of elemental impurities in the unspiked matrix was determined by k_0-NAA [12] at reactor Thetis (Gent), via short- and long-time analysis. For the analysis via short-lived radionuclides, four 150 mg samples were irradiated for 5 minutes in channel 9 at a thermal neutron flux of $1.9*10^{12}$ $cm^{-2}s^{-1}$. The co-irradiated monitor was a certified IRMM-530 Al-0.1%Au foil. Two spectra were collected for each sample on an HPGe-detector (relative efficiency 38%, FWHM 1.77 keV), the first one after 2 minutes decay time, with 5 minutes counting time, and the second one after 7 minutes, with 30 minutes counting time. As to the analysis via long-lived radionuclides, two 150 mg samples were irradiated for 7 hours in channel 3 at a thermal neutron flux of $1.2*10^{12}$ cm^{-2} s^{-1}, with the same Au monitor as mentioned above. Three spectra were collected on the HPGe-detector for each sample: after 1 hour decay with 30 minutes counting time, after 1 day with 15 hours counting time, and after 1 week with 24 hours counting time. The mass fractions (ρ), expressed as indicative values or detection limits of the elemental impurities found in the resol are shown in Table 1. Knowing the concentrations of the impurities in the resol, to analyze their influence was needed, i.e. their interference on the spiking elements. The interferences were divided in two categories: *type a*, from elements that were planned to be present in the standards and *type b*, from other elements. Note that an interference from Cl is of *type a* for the type I standard, but of *type b* for types II and III. In the *type a* interferences, only the content of the impurity and of the target concentration in the standard had to be compared. It was generally found that, as shown in Figure 1, the impurity concentrations in the matrix amounted at the maximum to 0.2 % of the target concentrations, and were mostly much lower. As to *type b*, all possible (n, p), (n, α), (n, n') and (n, 2n) activation reaction interferences on the impurity elements were quantitatively examined, as well as all possible spectral interferences arising from gamma lines (with an energy close to an analytical gamma-line), emitted by the radionuclides formed by (n,γ) and the above mentioned threshold reactions. These *type b* interferences were always found to be at the maximum 0.1% (and usually far below this).

TABLE 1. Resol matrix impurity content. Indicative values or detection limits.

Element	ρ, mg/kg	Element	ρ, mg/kg
Al	4.5E+00	La	2.6E-03
As	< 3E-03	Mg	< 2E+04
Au	1.0E-03	Mn	8.0E-02
Ba	1.6E-01	Mo	< 4E+00
Br	1.1E-01	Na	4E+00
Ce	< 6E-02	Pr	< 1E-02
Cl	6.0E+00	Sb	< 3E-02
Co	< 3E-02	Sc	< 8E-04
Cr	4.4E-02	Se	< 9E-02
Cs	3.1E-02	Sm	7.8E-04
Cu	< 1E-01	Sr	< 8E-02
Dy	< 1E-04	Th	1.3 E-02
Eu	< 5E-03	Tm	< 1E-02
Fe	< 6E+00	V	< 3E-03
Gd	< 2E-01	W	< 2E-03
I	< 1E-02	Yb	1.0E-02
In	< 5E-04	Zn	7.6E-01
K	1.1E+00	Zr	< 6E-01

Homogeneity of doping

Prototypes of SMELS types I, II and III were prepared as described above, and the homogeneity of the spiked elements at 50 mg sample intake was examined. The irradiations were performed in the BR1 reactor at the SCK·CEN, Mol: for type I in the fast pneumatic irradiation facility S84 (thermal neutron flux of $1.1*10^{11}$ cm^{-2}s^{-1}), for types II and III in channel Y4 (thermal neutron flux of $3.85*10^{11}$ cm^{-2}s^{-1}). For all spiking elements in all three types, the homogeneity, expressed as relative standard deviation (RSD, %), n=6, was found to be better than 1.5%. As an example, the results for type II are shown in Table 2, where a comparison is also made with the observed counting statistics of the measured gamma-lines. Six samples of 50 mg were irradiated for 7 hours together with a neutron flux Au monitor. After 1 day decay time, the samples were measured for 20 minutes on an HPGe-detector (relative efficiency 40%, FWHM 1.78 keV). A spectrum collected from a type II sample is shown in Figure 2. An overall well-balanced spectrum appearance can be observed, with the areas of the most prominent analytical gamma-peaks of the 10 spiking elements being very similar.

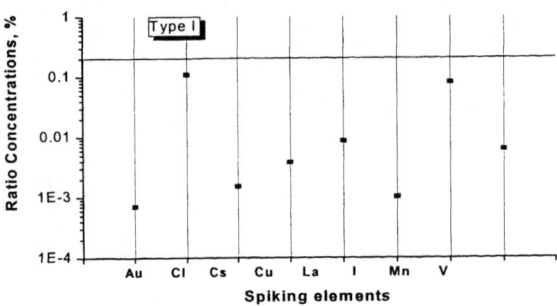
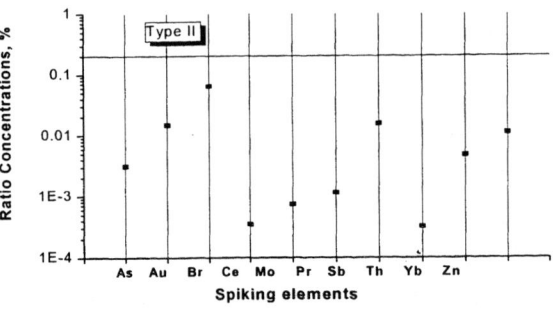
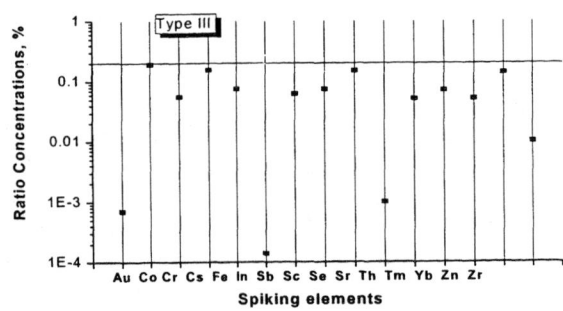

FIGURE 1. Interferences (in %, logarithmic scale) from impurity elements on spiking elements in the SMELS type I, II and III.

TABLE 2. Test of spiking homogeneity (expressed as relative standard deviation, n=6), for SMELS type II.

Element	Measured isotope	E_γ, keV	Count. stat., %	RSD (%)
As	^{76}As	559.1	0.47	0.97
Au	^{198}Au	411.8	0.5	0.52
Br	^{82}Br	554.3	0.57	0.95
Ce	^{143}Ce	293.3	0.29	1.29
Mo	^{99m}Tc	140.5	0.42	1.55
Pr	^{142}Pr	1575.6	0.55	1.12
Sb	^{122}Sb	564.1	0.43	0.72
Th	^{233}Pa	312.2	0.3	1.15
Yb	^{175}Yb	396.3	0.64	1.09
Zn	^{69m}Zn	438.6	0.61	0.56

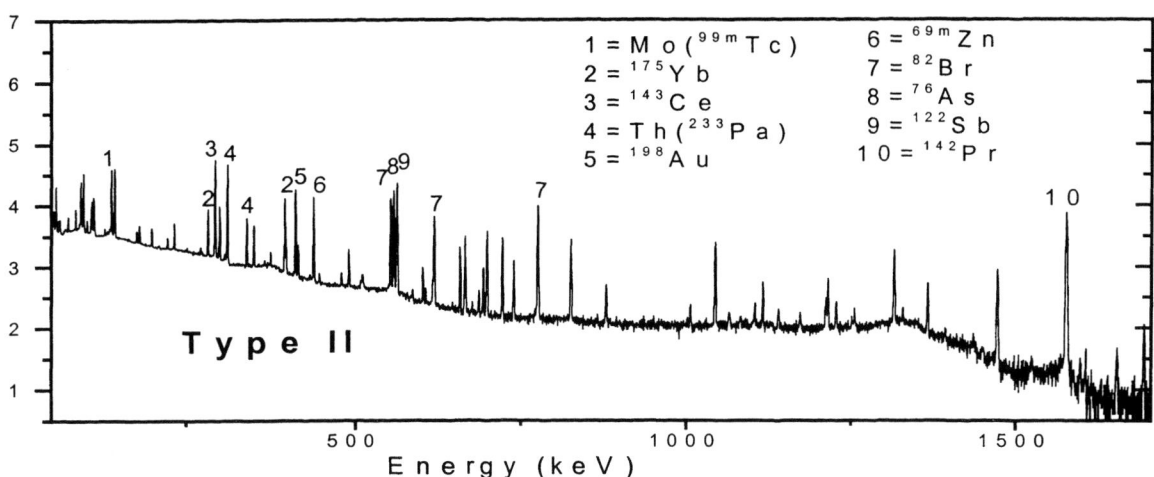

FIGURE 2. Gamma-ray spectrum of SMELS type II. The most prominent analytical gamma lines are numbered.

CONCLUSION

The results obtained here show that, in the development of the SMELS for k_0-NAA, the important problem has been solved of the preparation of a stable and interference-free resol matrix that can be homogeneously doped. This allows to foresee the finalization of the project by the year 2001.

ACKNOWLEDGMENTS

The authors would like to thank Mrs. Inge De Witte, Prof. Eric Goethals and Dr. Jan Kučera for their valuable help. This work was performed in the frame of the IRMM - SCK·CEN collaboration on k_0-NAA. The financial support of the FWO-VL greatly acknowledged (FDC).

REFERENCES

1. De Corte, F., *Proc. 16th International Conference on the Application of Accelerators in Research and Industry CAARI 2000*, edited by J. L. Duggan and I. L. Morgan, AIP Conference Proceedings, in press.

2. De Corte, F., De Wispelaere A., Kramer, G.N, Robouch, P., and Simonits, A., *Nucl. Instr. Meth.* **A422**, 891-895 (1999).

3. Lindstrom, R. M., *Nuclear Analytical Methods in Standards Certification*, IAEA meeting, Oak Ridge, Tennessee, USA, 3-7 October (1986).

4. Mosulishvili, L. M, Kolomiytsev, M. A., Dundua V. Y., Shonia, N. I., and Danilova, O. A., *J. Radioanal. Chem.* **26**, 175-188 (1975).

5. Iwata, Y., and Suzuki, N., *J. Radioanal.Nucl. Chem.* **233**, 49-53 (1998).

6. Barnes, L., Garner, E. L., Gramlich, J. W., Moore, L. J., Murphy, T. J., Machlan, L. A., Shields, W. R., Tatsumoto, M., and Knight, R. J., *Anal. Chem.* **45**, 880-885 (1973).

7. Anderson, D. H., Murphy, J. J., and White, W. W., *Anal. Chem.* **48**, 116-117 (1976).

8. Mitchell, J. W., Blitzer, L. D., Kometani, T. Y., Gills, T., and Clark, L. Jr., *J. Radioanal. Chem.* **39**, 335-342 (1977).

9. Date, A. R., *Analyst* **103**, 84-92 (1978).

10. Kayasth, S. R., Iyer, R. K., and Sankar Das, M., *J. Radioanal. Chem.* **59**, 373-379 (1980).

11. Sorenson W. R., and Campbell T. W., *Preparative methods of polymer chemistry*, John Wiley & Sons, 1968, pp. 455-456.

12. De Corte, F., Simonits, A., De Wispelaere, A., and Hoste, J., *J. Radioanal. Nucl. Chem.* **113**, 145-161 (1987).

Applications of Nuclear Analytical Techniques to Environmental Studies

M.C. Freitas[1], A.M.G. Pacheco[2], A.P. Marques[1], L.I.C. Barros[2], and M.A. Reis[1]

[1]*DEA-ITN – Nuclear and Technological Institute, Estrada Nacional 10, 2686-953 Sacavém, Portugal*
[2]*CVRM-IST – Technical University of Lisbon, Av. Rovisco Pais 1, 1049-001 Lisboa, Portugal*

Abstract. A few examples of application of nuclear-analytical techniques to biological monitors – natives and transplants – are given herein. *Parmelia sulcata* Taylor transplants were set up in a heavily industrialised area of Portugal – the Setúbal peninsula, about 50 km south of Lisbon – where indigenous lichens are rare. The whole area was 10x15 km around an oil-fired power station, and a 2.5x2.5 km grid was used. In north-western Portugal, native thalli of the same epiphytes (*Parmelia* spp., mostly *Parmelia sulcata* Taylor) and bark from olive trees (*Olea europaea*) were sampled across an area of 50x50 km, using a 10x10 km grid. This area is densely populated and features a blend of rural, urban-industrial and coastal environments, together with the country's second-largest metro area (Porto). All biomonitors have been analysed by INAA and PIXE. Results were put through nonparametric tests and factor analysis for trend significance and emission sources, respectively.

INTRODUCTION

Biological monitoring of airborne contaminants by means of vegetable organisms (or parts of them) has been increasingly used, especially in Europe, as a complement or even an alternative to classical, instrumental methods of studying the deposition of airborne substances to the terrestrial environment. On the other hand, and apart from any particular search for an element pattern or specific pathway, biomonitoring at large is likely to gain an enormous advantage from the use of multielement analytical techniques, particularly concerning the screening of potential biomonitors and/or source apportionment.

The usefulness and adequacy of the combined use of biological monitors and nuclear-analytical techniques have been shown through a number of studies [1-4] and, more recently, there is every indication that tree bark can be analysed in the same way as lower epiphytes (or aerosols, for that matter) and for many of the same elements. As an air-pollution monitor, bark has been used far less than lichens, bryophytes or non-lichenised fungi [5]: bark studies are truly scarce and mostly related to environmental acidification [6-9]. Nevertheless, biomonitoring with bark could have several advantages that should not be overlooked, meaning an availability of biological material year-round; an easier identification and sampling when compared to lichens or bryophytes, and an ubiquity of some genera that makes it feasible to survey extensive areas without endangering any species or putting them in short supply. This paper is thus aimed not only at presenting and discussing results from epiphytic lichens, but also at briefly looking into the ability of bark as an air-pollution monitor.

EXPERIMENTAL

Transplants of the lichen *Parmelia sulcata* Taylor were enclosed into nylon bags and suspended in the Setúbal region. In each of the 47 exposure sites, two sets of four samples each were displayed. Care was taken: i) in covering the samples with a polythene shelter to prevent the leaching of elements off the lichen; ii) in building a hanging device that could rotate according to the wind direction; iii) in orienting one set towards the wind and the other set against the wind. Transplants facing the wind (F-set) and opposing the wind (T-set) were all removed after 3 months.

Olive-tree (*Olea europaea*) bark and native *Parmelia* spp. (*Parmelia sulcata* Taylor, mostly) were sampled at a height of 1-2 m above ground from an average of five trees at each of the 28 collection sites in north-western Portugal. Samples within each site were combined into a single, nominal one prior to laboratory work. Local-variation aspects have been studied and given elsewhere [3]. Lichen transplants, indigenous lichens and tree bark were analysed by means of two multielement techniques: instrumental neutron activation analysis (INAA), using the Portuguese Nuclear Research Reactor, and proton-induced X-ray emission (PIXE), using the Van de Graaff accelerator.

Following standard procedures, pelletised materials were put through INAA and/or PIXE, and assessed for their element contents.

RESULTS AND DISCUSSION

Transplanted Species

Figure 1 shows the transplant sites in the Setúbal peninsula, as well as the approximate location of some relevant features of the area. Lichen elemental contents were determined in 10 sub-samples before exposure, corresponding to an intrinsic (0-month) background. Statistical data obtained for a 3-month exposure are presented in Table 1. No significant difference between F- and T-set results is apparent, though an evident increase – relative to their initial (0-month) values – is observed for As and Zn in both sets. Through previous publications [2,4,10], these elements were already pointed out as coming from oil-fired and coal-fired (remote, about 50 km south) power plants, and from a Portland-cement factory (local).

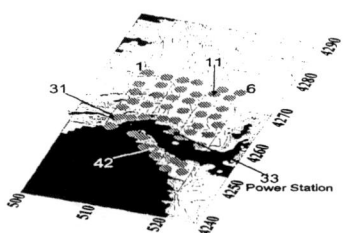

FIGURE 1. Transplant sites (●) and other relevant features in the Setúbal area (11: agricultural land; 31: cement factory; 33: power plant; 42: sandy shore).

The element contents obtained by INAA and PIXE were put through factor analysis [2,4]: the results for both sets are given in Tables 2 and 3 (only values greater than 0.67 were included).

In Table 2 and for the F-set of transplants, 5 factors were found. Factor 1 relates to an industrial-urban contribution. Factor 2 is clearly a soil component. Factor 3 is assigned to sea spray with equal weights of Na and Cl, and some significant contribution of S from marine origin. Factor 4 may be connected with land-use, agricultural activities, as shown before [10]. Factor 5 singles out Ca, pointing out the cement processing. A clear response to exogenous factors is thus given by the transplants facing the wind after a relatively short exposure time.

For the T-set in Table 3, factor analysis is indicating a rather different behaviour, disclosing the actual influence of the wind on the element uptake/release. Now, factor 1, related to soil sources, does include a larger matrix of elements, hence showing how the wind contributes to cleaning up the F-set transplants from soil particles. Factor 2 appears associated with the sea spray to a larger extent than before, as suggested by the significant weight of other marine elements – Mg, Br.

Factor 3 is the industrial-urban one. The remaining factors differ from those found for the F-set: factor 4 is surely related to the physiology of the organisms, while factor 5 probably points to waste incineration or non-ferrous metallurgy. The transplants opposing the wind seem to give a slower response and yield a broader spectrum of significant factors. All in all, as discussed before [10], F-type transplants are more likely to indicate local sources, whereas the T- type transplants could be more sensitive to remote ones.

TABLE 1. Concentration (in $\mu g\ g^{-1}$) of selected elements in *P. sulcata* transplants after a 3-month exposure. The 0-month exposure corresponds to initial (lichen background) values.

0-month	Ca	V	Zn	As	Hg
Nr. samples	10	10	10	10	10
Mean	7090	3.44	49.4	0.790	0.160
StD	1410	0.704	9.30	0.160	0.050
F-set	Ca	V	Zn	As	Hg
Nr. samples	39	39	39	39	39
Mean	7430	3.56	80.7	1.35	0.143
StD	3050	1.44	100	1.93	0.0323
Median	6830	3.04	60.4	0.955	0.137
T-set	Ca	V	Zn	As	Hg
Nr. samples	39	39	39	39	39
Mean	6960	3.84	75.8	1.24	0.164
StD	1670	1.15	79.8	0.456	0.0426
Median	6610	3.63	55.7	1.16	0.157

TABLE 2. Results of factor analysis on the F-set of *P. sulcata* transplants after a 3-month exposure.

Element	Factor 1	Factor 2	Factor 3	Factor 4	Factor 5
As				0.75	
Ca					0.74
Ce		0.93			
Cl			0.94		
Co		0.87			
Cu	0.81				
Fe		0.91			
Mn	0.83				
Na			0.94		
Ni	0.90				
Pb	0.92				
S			0.75		
Sc		0.96			
Ti	0.91				

Indigenous Species

Figure 2 outlines an area in the Portuguese north-west where native lichens of the *Parmelia* genus (mostly, *Parmelia sulcata* Taylor) and olive-tree (*Olea europaea*) bark were sampled. Statistical data for Na and Cl (marine influence) and As and V (industrial influence) can be seen in Table 4. Generally speaking, element contents are larger in the lichen thalli than in the tree bark, although the differentiation (spread) between maximum and minimum levels is alike.

TABLE 3. Results of factor analysis on the T-set of *P. sulcata* transplants after a 3-month exposure.

Element	Factor 1	Factor 2	Factor 3	Factor 4	Factor 5
Al	0.84				
Br		0.84			
Ce	0.96				
Cl		0.96			
Co	0.82				
Cr	0.83				
Fe	0.91				
Hg	0.80				
K					0.87
La	0.90				
Mg		0.72			
Mn			0.70		
Na		0.94			
Ni			0.74		
P					0.81
Pb			0.86		
S		0.79			
Sc	0.95				
Se	0.87				
Si	0.84				
Ti	0.71				
Zn					0.85

TABLE 4. Concentration (in µg g^{-1}) of selected elements in *Olea europaea* bark and *Parmelia* spp thalli. from the same sampling area.

Bark	Na	Cl	As	V
Nr. samples	28	28	26	28
Mean	387	425	0.978	8.08
StD	178	194	0.991	9.19
Median	371	386	0.652	4.61
Max.	744	1220	4.94	33.2
Min.	99.2	185	0.159	1.81
Lichen	**Na**	**Cl**	**As**	**V**
Nr. samples	28	28	28	28
Mean	768	1290	2.63	12.9
StD	550	488	1.86	9.13
Median	624	1230	1.84	11.1
Max.	2700	2890	8.51	42.7
Min.	244	427	0.749	3.44

TABLE 5. Results of factor analysis for lichen samples from north-western Portugal.

Element	Factor 1	Factor 2	Factor 3	Factor 4
Al		0.78		
Cl	0.67			
Fe	0.88			
K	0.88			
Mg				0.82
Na	0.89			
Ni			0.77	
Pb			0.73	
Rb	0.79			
Sc	0.84			
Ti		0.92		
V		0.75		
Zn				0.76

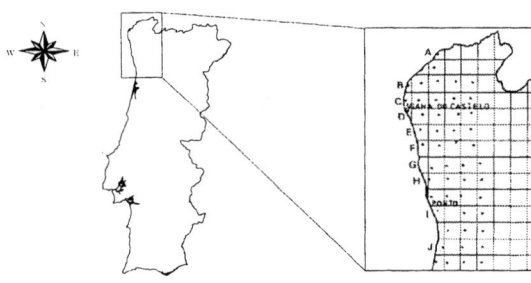

FIGURE 2. Sampling area for the biomonitoring of airborne contaminants through olive-tree bark and indigenous lichens of the *Parmelia* genus.

Results from factor analysis are shown in Tables 5 and 6. Except for the factor 3 in *Parmelia* spp. and tree bark, there seems to be a dissimilar elemental loading. Lichen splits factor 1 of the bark into two – factors 1 and 2. Different elemental contributions appear in factor 4 for the lichen (Mg and Zn) and in factor 2 for the bark (K and As).

Despite the former divergence in emission-source identification, element signals from lichen and bark are quite consistent in what concerns spatial trends across the sampling area. As an example, Figure 3 compares arsenic and vanadium concentrations in bark and lichen samples from every collection site: such elements are usually associated with emissions from coal-fired and oil-fired power plants.

TABLE 6. Results of factor analysis for bark samples from north-western Portugal.

Element	Factor 1	Factor 2	Factor 3
Al	0.86		
As		0.74	
Fe	0.92		
K		0.70	
Mn			
Na	0.81		
Ni			0.68
Pb			0.77
Rb	0.84		
Sc	0.90		
Ti	0.84		
V	0.61		

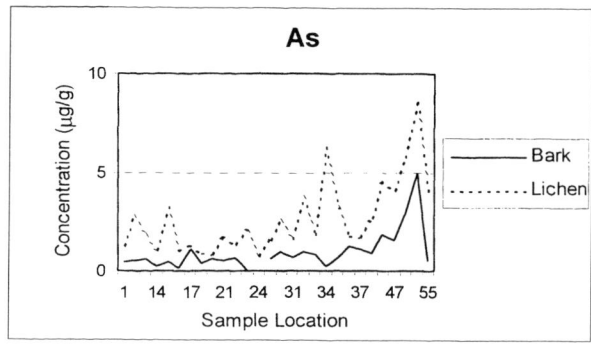

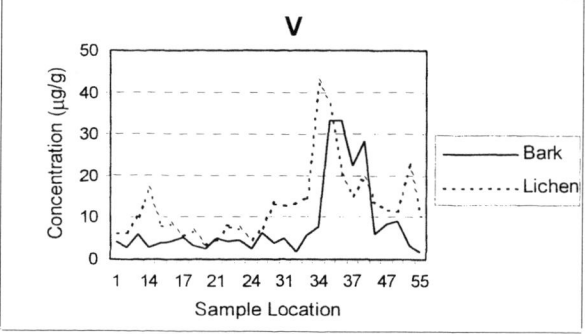

FIGURE 3. Arsenic and vanadium levels in *Olea europaea* bark and *Parmelia* spp. thalli across the study area.

Table 7 lists the pairwise strength of association between field variables, as measured by the Spearman (R_S) and Kendall (R_K) rank-order correlation coefficients. Values of the gamma (G) statistic, another nonparametric test that explicitly accounts for tied variates, are included as well. Every coefficient and the corresponding significance (p-level) was computed to the 6th-decimal place, then rounded off to the nearest thousandth. Other than an overall (visual) impression of close agreement, the numerical (statistical) results are highly significant. Correlations for both elements are still holding well beyond the .02 level, whatever measure of association (R_S, R_K, G) is considered. Statistically speaking, this means that an accidental pattern of such significance and consistency would hardly occur. Bark signals appear weaker than lichens', as expected, yet they are satisfactory to the extent that consistent indication is always much more an asset than signal magnitude.

TABLE 7. Nonparametric statistics and associate probability levels for the concentrations of As and V in *Olea europaea* bark and *Parmelia* spp. thalli.

	R_S (p-level)	R_K (p-level)	G (p-level)
$[As]_{BARK}$ vs $[As]_{LICHEN}$	0.487 (.012)	0.370 (.008)	0.370 (.008)
$[V]_{BARK}$ vs $[V]_{LICHEN}$	0.446 (.017)	0.332 (.013)	0.332 (.013)

CONCLUSIONS

In this work, a few examples of air monitoring with transplanted and native organisms were presented to illustrate the use of nuclear techniques – INAA and PIXE – for multielement analysis. Lichen transplants were found to respond quickly and effectively to the elemental emission sources, which they reflect, though their position relative to the prevailing winds may turn into differences in source indication. The response of indigenous lichens appeared somewhat less clear. However, some biological variability and a more complex and diverse sampling area may account for such an outcome. Studies with olive-tree bark have shown that it could be an interesting alternative to common epiphytes, at least in southern Europe. Generally speaking, and other than their analytical quality, the multielement capability of nuclear techniques seems most relevant in environmental terms, especially in what concerns biomonitoring.

ACKNOWLEDGEMENTS

The authors are indebted to Mrs. Isabel Dionísio (DEA-ITN) for helping with sample preparation. Thanks are also due to the International Atomic Energy Agency (Austria) and to the Ministry of the Environment (Portugal) for financial support. Helpful comments by one Reviewer were truly appreciated.

REFERENCES

1. Freitas, M.C., Reis, M.A., Alves, L.C., Wolterbeek, H.Th., Verdurg, T., and Gouveia, M.A., *ANST/Nuclear Methods in Environmental Research* **74**, 117-118 (1996).
2. Reis, M.A., Alves, L.C., Wolterbeek, H.Th., Verdurg, T., Freitas, M.C., and Gouveia, M.A., *Nucl. Inst. Meth. Phys. Res.* **B109/110**, 493-497 (1996).
3. Freitas, M.C. and Nobre, A.S., *J. Radioanal. Nucl. Chem. (Articles)* **217**, 17-20 (1997).
4. Freitas, M.C., Reis, M.A., Alves, L.C., Wolterbeek, H.Th., Verdurg, T., and Gouveia, M.A., *J. Radioanal. Nucl. Chem. (Articles)* **217**, 21-30 (1997).
5. Nimis, P.L., "Air Quality Indicators and Indices: The Use of Plants as Bioindicators for Monitoring Air Pollution", in *EUR 13060 EN Report*, edited by A.G. Colombo and G. Premazzi, EEC, Luxembourg, 1990, pp. 93-126.
6. Staxang, B., *Oikos* **20**, 224-230 (1969).
7. O'Hare, G.P., *J. Biogeogr.* **1**, 135-146 (1974).
8. Grodzinska, K., *Water, Air, and Soil Pollut.* **7**, 3-7 (1978).
9. Grodzinska, K., "Monitoring of Air Pollutants by Mosses and Tree Bark", in *Monitoring of Air Pollutants by Plants – Methods and Problems*, edited by L. Steubing and H.-J. Jäger, W. Junk Publishers, The Hague, 1982, pp. 33-42.
10. Freitas, M.C., Reis, M.A., Marques, A.P., Wolterbeek, H.Th., *J. Radioanal. Nucl. Chem. (Articles)* **244**, 109-113 (2000).

Metallic Pollutants in Mexico Valley

Trinidad Martínez[*], Juan Lartigue[*], Pedro Avila-Perez[**], Manuel Navarrete[*], Graciela Zarazúa[**], Carmen López[**], Luis Cabrera[*], Alejandro Ramirez[*]

*National University of Mexico, Chemistry Faculty, Bldg. D, C.U. (04510), Mexico City, Mexico.
** National Institute of Nuclear Research (ININ), Ocoyoacac, 50045, Mexico.

Abstract. Pollution has reached critical levels in the Metropolitan Zone of Mexico Valley. Concerned about, the Faculty of Chemistry has been performing environmental studies since 1995. This work presents the distribution and evolution of metallic pollutants in the Metropolitan zone of Mexico Valley. Samples consisted in aerosol filters (classified as total solid particles and respirable particles) as well as dry fallout. Samples were collected in several areas of the Mexico Valley, in different seasons along successive years. Metallic elements were determined by Instrumental Neutron Activation Analysis (INAA), X-ray fluorescence, and others techniques. Simultaneously, total solid particles (TSP) and respirable particles (RP) were determined by gravimetry. Elemental analysis of samples and matrix correlation allow us to establish some contaminant sources as well as relationship between concentration and relevant parameters.

INTRODUCTION

Pollution is one of the most important problems around the world. Thousand of million of world inhabitants suffer health problems related to industry and atmospheric pollutants [1]. Some of the most dangerous contaminants to health are: Total suspended particles (TSP), Respirable particles (RP) lesser than 10 µm (coarse, PM10, and fine PM2.5), Hydrocarbons, Sulphur dioxide, Nitrogen oxides, Ozone, Lead and other heavy metals (Cd, As, Mn, Ni, Zn, etc.). Big Cities, like Mexico, are the most affected by this type of pollution. In the Metropolitan Zone of Mexico Valley (MZMV) pollution has reached critical levels. It is caused by: its topography and meteorology (which led to an extremely poor atmospheric mixing, mainly in the winter); its great population (roughly 20 million inhabitants); its accelerated increase in gasoline consumption from 16 million liters per day (mld) in 1989 to 40 mld in 1996, this one due to an increase in the number of motor vehicles in the zone at an annual rate of 10%. At present time, the zone holds up 3.3 million vehicles (45% of them older than 10 years). Besides, one fifth of the population travels by car and only 55.1% of travels are made by collective transport. Only 13.4% and 0.6 % of travels are made by metro and trolley respectively [2]. The zone concentrates also 41% of the national industrial park as well as two old power plants (thermoelectric).

As a reaction to this problem, an official program has been adopted to decrease present pollution levels. One of the actions was to take inventory of fixed and ambulatory contaminants sources. It showed that 75.5%, from more than 4 million tons of pollutants, correspond to transportation. Other steps were taken, such as fuel conversion from highly leaded gasoline to a low leaded one, and the introduction of low sulphur content diesel in 1986. As well, an improved diesel was introduced in 1993 and a Lead-free gasoline was adopted in 1996. Additionally, regulations to reduce motorized traffic are in force as a part of stricter environmental laws [2].

Concerned about, the Faculty of Chemistry has been performing environmental studies [3-6] since 1995. This work presents the distribution and evolution of metallic pollutants in the Metropolitan Zone of Mexico Valley.

EXPERIMENTAL

Surveys were carried out from 1995 to 1999 in several zones covering the MZMV and different seasons along the year. Air samples were obtained utilizing two types of samplers: a WLM-1A with an air sampler pump, 0.13 l min^{-1} flow rate on a millipore filter of 0.8 µm during 24 or 120 h periods; and a high volume sampler (General Metal Gord, FMG, with an average flow rate of 1.5 m^3 min^{-1} during 24 h [7].

Nuclear analytical techniques such as Instrumental Neutron Activation Analysis (INAA) and X-ray fluorescence (XRF) were utilized for analysis. In the 1995 and 1997 sampling periods, whole filters were analyzed using ^{239}Pu as an excitation source in a detection system consisting of a Si Li detector coupled to a Norland Inotech 5400 channel analyzer (resolution 180 eV for K_α Mn of 5900 eV). Quantification was carried out utilizing the AXIL program [8], based on standard reference materials. In the 1999 sampling period a small section of the 7"x 9" filter (deposit area) was analyzed in a total reflection X-ray fluorescence spectrometer Model TX-2000 utilizing traditional 45° geometry with an IAEA external standard. Aerosol particles were collected in glass fiber for TSP and quartz filters for respirable particles (PM10). The method used to prepare samples for irradiation include cutting of pieces from the millipore filter using a bone cutting instrument and weighting them. Irradiations were performed in the Nuclear Reactor Triga Mark III, 1 MW(th), at the National Institute of Nuclear Research (ININ) with fluxes of 0.9×10^{13} n cm^{-2} s^{-1} at the SIFCA position for long irradiation times and 1.3×10^{13} n cm^{-2} s^{-1} at the SINCA position for short irradiation times. Gamma spectra were taken utilizing a HP Ge detector (with 1.8 KeV resolution for 1,332 KeV peak of ^{60}Co source) coupled to a PC with a Master program.

Metallic elements determined by these two analytical methods were: Ca, K, Na, Ti, Cr, Cd, Co, Mn, Fe, Ni, Cu, Zn, Pb, As, Br, Cl, Hg, Mn, K, and V.

Total suspended and respirable particless were determined by gravimetry.

Dry deposit samples were collected during two months in two periods in 1999 utilizing an automatic sampler model ASP78100, placed one-meter height. After being dried, an aliquot of 0.1 g of each sample was submitted to three types of acidic digestion. One for X-ray fluorescence determination (with Ga as an internal standard); one for Hg Atomic Absorption determination in an Hiranuma Spectrometer; and one for As determination in a Plasma Emission spectrometer SPS-1200 AR Seiko, with an hydride generator Model THG-1200. A quality control program with internal, external standards and duplicate measurements was established.

RESULTS AND DISCUSSION

Regarding weight abundance, metallic elements in air samples can be distributed in three classes: major elements (Ca, Fe, Zn, Co, Cl, Pb) with a concentration higher than 1 μg m^{-3}, including Xalostoc monitoring station in November 1999; in all others samples from the 1999 period, Pb is included between minor elements (K, Na, Ti, Cu) with a concentration between 0.1 and 1μg m^{-3}; finally, trace elements (V, Cr, Mg, Mn, Se, Ni, As) with a concentration between 0.01 and 0.1μg m^{-3}. Low enrichment factors [9,10,110] (taking Si as reference), have been shown by K, Ca, Ti, Fe, Rb, Sr, which are soil related elements and have a significant contribution to the atmospheric pollution. Besides, their concentrations are similar to those reported previously for the same City from 1974 to 1990 [12, 9]. Very high Ca and high K concentration in air samples are well correlated with salty soils conditions with negligible moisture content. Such values are also similar to soil and dry fallout concentration [6, 9]. Iron concentration is higher than those reported for some U.S. Cities [13]. It seems that these natural elements are associated with large particulate matter (TSP) since PM10 concentration represents around 35% of TSP.

Elements with a high enrichment factor (Cl, V, Ni, Cu, Zn, As, Cd, Hg) in airborne particulate matter and also S in dry fallout [6], are anthropogenic from several sources.

V, Ni, Cu, Zn, Br, Cl and Pb were found in airborne particular matter in a higher concentration than in soil [9], but lower to Theshold Level Values (TLV) for eight-hour occupational exposure [14]. These concentrations range from 5 to 300 higher than the soil concentration. In dry fallout samples [6] these elements have shown to be 10 to 100 times more abundant than in soil, exception made of Br (its concentration is similar to that of soil: 8 μg g^{-1}).

Figure 1 shows the total average contaminant concentration (autumn 1999) for TSP and PM!0 as well as contaminant concentration by each month. As it can be seen, November shows the highest levels of contaminants because thermal inversions are frequently present. The most contaminated zone is the Northeast (Xalostoc monitoring station). Lead concentration in November in that station was 6.9 μg m^{-3} in TSP and 5.5 μg m^{-3} in PM10. Averaging the three months of this sampling period according to the Mexican Norm [15], the resulting concentration for Xalostoc was higher than the maximum average value of WHO [16], EPA [17], and Mexican Norm (1.5 μg m^{-3}) [15]. With the only already mentioned exception (Northeast) lead concentration in other zones (NW, Center, SE and SW) and the total average concentrations for all zones in the airborne sampling period (autumn) are lower than those reported previously [3,4,5,9] and the Mexican Norm.

If we compare total averages obtained from 1980 to 1999, Figure 2, it is evident that the Lead problem has been decreasing to date, though it is still present.

Lead dry fallout concentration was in average 230 μg g^{-1}, higher than most of the Mexican agricultural soils

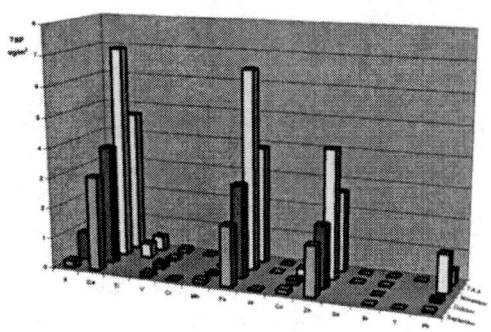 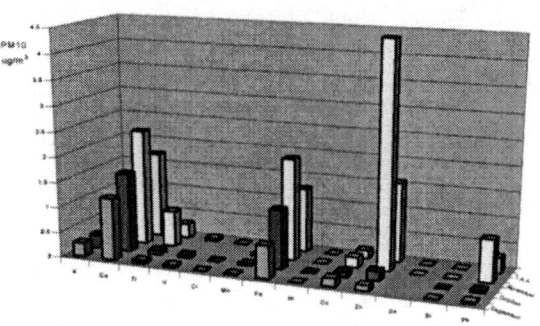

FIGURE 1. 1999 Monthly and Total Autumn Average (T.A.A) Metals Concentration in TSP and PM10.

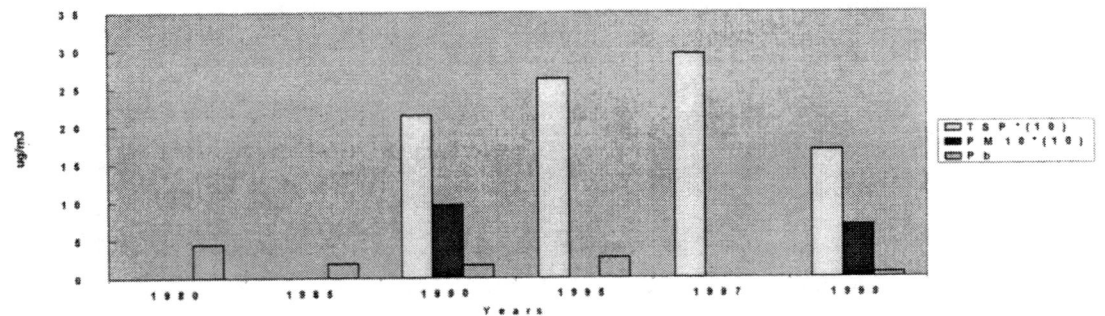

FIGURE 2. Evolution from 1990 to 1999 of TSP, PM10 and Lead Concentration ($\mu g/m^3$) in Mexico Valley.

($6\mu g\ g^{-1}$), similar to those obtained in dust samples [18]. So, Mexico City inhabitants may be significantly exposed to this element as a result of environmental retention (e.g. soil and dust) of previous release, the legacy of past domestic uses (e.g. pipes and paints) and remission of soils. Its quantification is a critical subject because Pb is a major toxic element, antagonist to central nervous system, inhibitor of macrophague alveolar activity [19], with effects on intelligence and behavior of young children. It has been found that 68 % of lead was bounded to respirable particles (PM10) though another author mentions 90% [20].

Correlation coefficients were established between metallic elements concentration in airborne and dry fallout samples, that suggest grouping and therefore common sources [10, 11]: natural (soils, e.g. Ca, K, Ti, Mn), mixed (e.g. Cr, Cd, Hg), and anthropogenic (automotive, Pb, Cu, Zn, Br, fuels, e.g. Ni, V, etc).

The total average of TSP and PM10 were 167.63± 93.7 and 68.9±43.9 $\mu g\ m^{-3}$ respectively; both lower than previously reported values [4, 5, 9], but still above the U.S and Mexican Norms [2] and lower to IMECA 100 (Metropolitan Index Air Quality) [21]. Only Xalostoc and Virgencitas monitoring stations have shown values above IMECA100. The decrease in TSP is based mainly on large particles (probably because sampling period was at the end of the rainy season).

It was reported [19] that fine particles PM2.5 represents more than 70% of PM10; it means that around 50 $\mu g\ m^{-3}$ could be secondary or photochemical particles.

CONCLUSIONS

Results show that Particles, Pb, Zn and Cu are the main environmental pollutants in the MZMV. Fortunately Lead problem have diminished to date with the only exception at Northern (Xalostoc, monitoring stantion) probably because it is the industrial zone and there are contribution of local sources; Lead still remain as a good labeler of traffic-derived pollutants. Elemental analysis and matrix correlation coefficient stand as useful tools to evaluate anthropogenic and natural particulate matter as well as to define its influence on health.

REFERENCES

1. Moore, C., "La calidad del aire urbano". *Series: Documento Verde* **1**, edited by Servicio cultural e Informativo de los Estados Unidos, 1994, pp. 1-14.
2. INEGI, SEMARNAP. "Estadísticas del Medio Ambiente Natural, Asentamientos y Actividades Humanas" in *Estadísticas del Medio Ambiente. Informe de la Situación*

General en Materia de Equilibrio Ecológico y Protección al Ambiente, 1995-1996, edited by Instituto Nacional de Estadística, Geografía e Informática, México, 1998, pp. 119-248.
3. Martinez, T. et al. *J. Radioanal. Nucl. Chem,.* **216**, 1, 37-39 (1997).
4. Martinez, T., et al. *IAEA TEC-DOC,* **1152**, 209-214 (2000).
5. Martinez, T., et al., *J. Radioanal. Nucl. Chem,* **244**, 1, 127-131 (2000).
6 Martinez, T., et al., "X-ray fluorescence analysis of dry deposit samples in Mexico City" Presented in *Fifth International Conference on Methods and Applications of Radioanalyitical Chemistry-MARC V,* 2000.
7. NOM-CCAM-002-ECOL. Diario Oficial, 18 octubre (1993).
8. QXAS, Users Manual: X-Rays Analysis System, edited by IAEA, Physics Section PCI Laboratory Seibersdorf, 1993.
9. Barbiaux, M., *Characterization of respirable particulate matter in Mexico City.* Thesis for Degree. Graduate College of the University of Chicago, 1990.
10. Miranda, J., et al. *Atmósfera* **5**, 95-108 (1992).
11. Person, A., et al. *Pollution Atmosphérique,* Juillet-Septembre, 75-88 (1993).
12. Navarrete, M., et al. *Radiochem. Radioanal. Lett,* **19** 163-170 (1974).
13. Thompson, R. J., "Collection and Analysis of Airborne metallic elements" in *Ultratrace Metals Analysis in Biological Sciences and Environment,* edited by Risby, T. H., American Chemical Society. Washington, D.C., 1979, pp. 54-72.
14. American Conference of Governmental Industrial Hygienists (ACGIH): *Theshold Limit Values (TLV) and Biological Exposure Indices* for 1985-*1986.* ACGHI, Cincinnati, Ohio, (1985).
15. NOM-026-SA 1. Criterios `para evaluar la calidad del aire con respecto al plomo. Diario Oficial, 23 Diciembre, 1994.
16. WHO. Air quality guidelines for Europe. Report **23** Copenhagen, 1987, pp. 426.
17. EPA-600/8-83/028, F. Air Quality Criteria for Lead, **1** (1986).
18. Albert, L. A., and Badillo, F. *Review of Environmental Contamination and Toxicology.* **117**, 1-46 (1991).
19. Fortoul, T., and Barrios, R., "Metales y partículas. Daño al aparato respiratorio" in *Memorias sobre Salud y Ambiente en la Ciudad de México,* edited by México, 1989, pp. 227-231.
20. Melgar, M., and Ruiz, M. E. *Tecno-Fórmula Ambiental* **1**, No. 2 Abril/Mayo, 10-11 (2000).
21. Indice Mexicano de la Calidad del Aire (IMECA). Diario Oficial, 29 noviembre (1982).

Radiation Effects Microscopy*

B.L.Doyle, G.Vizkelethy**, K.M.Horn, D.S.Walsh, and P.E.Dodd

Sandia National Laboratories, Albuquerque, NM, USA

Abstract. Nuclear microscopy is usually associated with the use of highly focused MeV ions to measure microscopically the composition of solids. Another use of such focused ions emerged 10 years ago with the introduction of radiation effects microscopy or REM. With REM one exploits the charge deposited by each ion, or the effect of this charge. The power of this new technique stems from the reproducible, and very well understood, linear charge density produced by ions in semiconductors, coupled with the capability of a nuclear microprobe to provide individual ions to a specimen with high spatial resolution. Several techniques form the bases for REM, and these can be categorized under the headings: ion beam induced charge collection (IBICC) and single event effects (SEE) imaging, this paper reviews these techniques, and gives examples of their use in studying charge transport, and the effects of this transport, in semiconductors and integrated circuits.

1. Charge introduction and transport

The passage of energetic ions through semiconductor materials results in the creation of electron-hole pairs. Radiation effects microscopy or REM, exploits both the reproducibility and high state of knowledge of this charge deposition process in semiconductors to form the basis for a new nuclear microscopy of charge transport and radiation effects in semiconductors and devices.

If this charge is deposited in a region of no bias, the charges can either spontaneously recombine or begin to diffuse. The diffusive charge that reaches a reverse biased collection junction, and doesn't suffer trapping or recombination, will induce a charge detectable with standard nuclear counting electronics.

If this charge, or a fraction thereof, is deposited in a region which experiences an electric field, then it will drift through this potential and also be detectable.

If this charge is introduced into a node of an integrated circuit (IC), it can, if of sufficient magnitude and duration, cause the transient or permanent failure of a device.

The utility of REM stems from coupling the extremely well understood physics of charge deposition, drift, diffusion and collection, with the extraordinary capability of a conventional nuclear microprobe to provide a stream of individual ions to a specific position on a semiconductor or device, with sub micron resolution (recently 41 nm resolution has been reported) [1]. The modeling of this charge collection processes is equally important to the REM measurements themselves so that the results are interpretable in terms of the details of either materials or circuit electrical performance. When REM is used to study semiconductors, electrical properties such as diffusivity, mobility, carrier lifetimes, carrier collection lengths, and the effects of bulk and surface defects can be determined on a microscopic scale. In the case of ICs, REM has become vital for validating predictive design simulation codes where it is used to experimentally replicate the conditions that are simulated by three dimensional charge transport calculations of a circuit's response to a single ion strike at a specific circuit location.

This paper reviews radiation microscopy techniques used in studying semiconductors and integrated circuits, also provided are examples of both REM and theoretical calculations used to model these microscopic electrical measurements. These examples are selected in order of the application of REM to semiconductors with increasingly complex electric field configurations, from unbiased to constant fields to ICs.

2. Nuclear Microprobes and Radiation Effects Microscopy

Over the last decade, focused, ion microbeams have been used for radiation testing at labs in Japan, Europe, Africa, Australia and the United States. A conceptual rendering of a microprobe-based test system is illustrated in Figure 1. An energetic beam of ions is magnetically focused to a sub-micron-sized spot at the surface of a sample. The focused ion beam can be scanned across a region of the sample or directed to a specific spot. The use of ion fluences of several hundred ions/s or less permits the effect of each ion to be measured and analyzed individually in addition to limiting the displacement damage and/or oxide charging which is also caused by these ions.

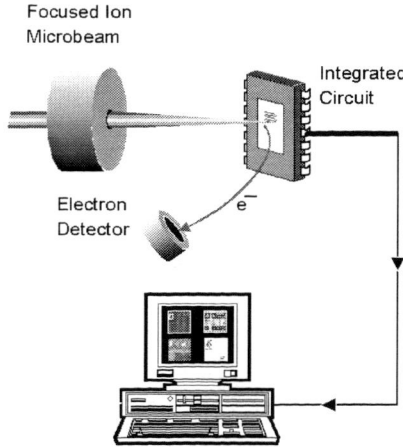

Figure 1. Schematic drawing of experimental setup for Radiation Effects Microscopy.

We have published descriptions of the operation of the REM system elsewhere [2] and shall only briefly summarize it here. Once the ion beam has been focused, the size of the scan is calibrated by imaging TEM (transmission electron microscope) grids of known dimensions and pitch. The full charge generation of the incident ions is then measured in a fully depleted silicon p-i-n diode whose depletion depth exceeds the range of the incident ion. This measurement is used to calibrate the signal electronics (charge sensitive pre-amp, amplifier and digitizer) for subsequent, quantitative, ion beam induced charge collection (IBICC) measurements. In order to measure a circuit's functional response to the incident ions, two computers act in tandem to record the single event effect image (SEE imaging). One computer controls the positioning of the focused ion beam and the dwell time of the beam at each pixel of the x-y scan while a digital parametric analyzer exercises the target circuit and 'notifies' the first computer whenever a change of logic state is detected. The first computer then records the X and Y position of the beam, and the occurrence of the SEE malfunction.

2.1 Ion Beam Induced Charge Collection - IBICC

IBICC involves the highly controlled deposition of charge by microfocused MeV ions coupled with the subsequent collection of what remains of this charge after being transported though a semiconductor or integrated circuit. Measurements of charge collection using apertured ion exposures were previously done in the 1980's and 90's [3,4,5]. In these measurements, a charge sensitive pre-amplifier is connected to the V_{DD} or V_{SS} pins of an integrated circuit. The network of metalizations that biases the circuit structures of the device are in this way also used to collect charge from the structures. Breese et al. first applied scanned, focused ion microbeams to the imaging of charge collection within integrated circuits [6], using this same methodology. An example of the IBICC analysis of a Sandia TA670 16k SRAM is shown in Figure 2. It is usual to perform a median filter on the IBICC "spectrum" at each point of the scan and plot this median pixelized to form the image. In Figure 2 lighter pixels correspond to higher charge collection.

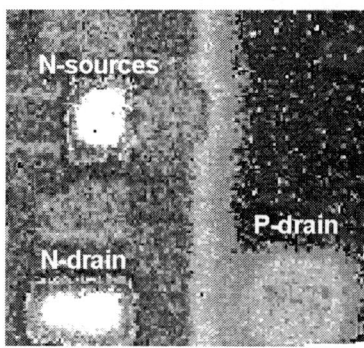

Figure 2. IBICC image of two memory cells in a TA670 16k SRAM designed and manufactured at Sandia.

The use of IBICC to study semiconductor radiation detectors was started by Jaksic [7] and Manfredotti [8]. IBICC constitutes the most common form of radiation microscopy by directly viewing charge collection magnitudes within a device in almost real-time.

Other variations of this technique include the recording of the current transient caused by each ion, and this is called Time Resolved IBICC or TRIBICC [9]. TRIBICC can either be made with extremely high bandwidth (e.g. 75 GHz***check this***) linear amplifiers and transient digitizers to measure the current transient in ICs, or by digitizing the output of a charge sensitive preamp used in the case of lower bandwidth measurements for studying bulk semiconductors. Yet another variation involves an IBICC measurement of a sample with multiple charge collection points where the timing signals of the preamps are used for a type of "charge arrival time difference" measurement. Since this has been applied to measure diffusivities in bulk semiconductors, it has been called Diffusion TRIBICC or DTRIBICC [10]. IBICC and its variations are now staring to become a routine tool to nuclear microscopists.

2.2 Single Event Effects (SEE) Imaging

The charges collected by IBICC also represent the underlying cause of single event effects (SEE) in ICs. A single ion can potentially deposit enough charge at a sensitive node within a circuit to cause a change of logic state (single event upset - SEU), an inability to change state (single event latch up - SEL), or even a destructive power short (single event burnout - SEB). In the case of SEU, the total area of the circuit over which an ion strike can cause upset, measured in units of cm^2, is called the upset cross-section. Linear Energy Transfer, (LET), measured in units of MeV/mg/cm^2, is the amount of energy deposited by the incident ion per unit track length. A circuit is 'hardened' to radiation by reducing its upset cross section or increasing the threshold LET at which an incident ion can cause SEE, or both. Three-dimensional charge transport simulations of circuit response to ion strikes are used to evaluate and predict the radiation hardness of circuit designs. Experimental measurements of upset cross sections and threshold LETs for single event effects are used to verify radiation hardness specifications.

In October of 1990, the nuclear microprobe was first used to directly image single event upsets (SEU) in SRAMs fabricated with 1.25 micron technology [2]. It has since been used to measure and image upsets in DRAMS, EEPROMS, buffers, shift registers, and other semiconductor devices. In the field of radiation effects, researchers had previously used apertured systems to localize the exposure of integrated circuits (IC) to ionizing radiation [11,12,13]. By focusing rather than aperturing the incident ions, higher ion influences are obtained than with apertured systems and it is possible to perform two-dimensional scans quickly and with flexible control of the scan area. An SEU image for the same TA670 SRAM discussed above is shown in Figure 3 together with the circuit layout.

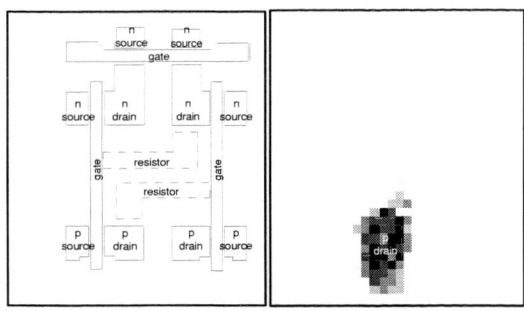

Figure 3. The circuit layout for a single memory cell of the TA670 16k SRAM is shown next to the corresponding SEU image recorded using 30 MeV Cu ions.

Upset cross sections can also be measured directly from the upset-image, rather than inferred from the statistics of whole-die exposures. By processing the data livetime and navigating on the die using GDS-II mask design files, the experimenter can directly image and identify circuit structures susceptible to upset - thus the whole system acts as a sort of radiation microscope to "view" upsets and even the underlying charge collection occurring at different circuit structures.

2.3 Combined IBICC and SEU-imaging

When combined with SEE-Imaging, these two techniques yield complementary information: (1) how much charge is collected at a specific site and (2) whether that charge causes a circuit malfunction. In the charge collection image shown in Figure 2, the n-drain and n-sources exhibit the highest charge collection within the memory cell, while from Figure 3 it is seen that it is the p-drain, which actually causes circuit upset when exposed to 30 MeV Cu ions.

3. Examples and Theory

As shown in the preceding sections, energetic particles incident on an electronic device create electron-hole pairs as they transit the device. Modeling the flow, trapping and collection of this charge provides an analysis of the individual materials charge transport parameters that leads to, say, a certain IBICC signal, TRIBICC transient or a single event upset.

The Shockley-Ramo (S-R) theorem is central to understanding the source and magnitude of signals generated by the charge generation, transport and collection in semiconductors and even ICs:

$$i = \frac{1}{V_0} \int_{Volume} \mathbf{E}_1 \cdot \mathbf{j} \, d^3 x \qquad (1)$$

where i is the current measured at the collection electrode, V_0 is the volume of the device, $\mathbf{E}_1$ is the electric field due to the applied bias, and $\mathbf{j}$ is the current.

This theorem says that the charge induced on the electrodes which bias a semiconductor, either as a capacitor or a reversed bias pn junction, is caused by the ion-induced electrons and holes as they drift through the region of electric field. An excellent review of this theorem and its ramifications to IBICC has recently been given by Vittonne [14]. Since it is this induced charge or the transient change of this charge that provides all of the various IBICC signals, it is impossible to overstate the importance of understanding the Shockley-Ramo theorem and applying it to the system being analyzed. This is true whether the sample is a capacitor-like detector, PIN diode, or an SRAM cell. We now look at three separate cases of modeling charge transport: field free, constant fields and in ICs. The reader is referred to the references given in these examples for the complex details and discussion merited by these experiments.

3.1 Field Free Transport

The stable electron-hole pairs that are created in regions with no electric field will diffuse in the semiconductor according to Fick's Law which must be solved with boundary conditions consistent with the source (the ion strike) and sinks (the collection points) of the problem. It is clear from the S-R equation that the sink collection points better involve electric fields, or nothing will be measured by the IBICC experiment. In the simplest case where no charge traps exist in the bulk of a semiconductor, it is straightforward to show that the distance x through which the charge diffuses after the ion strike is just $\sqrt{D \cdot t}$ where D is the diffusion coefficient. Using two np reversed diode strips (these are the fielded regions required by the S-R theorem to induce charge) to detect the arrival of this charge, Guo et al [10] have shown that the DTRIBICC signal, which is just the difference in arrival time of the charge to these strip, can be expressed as:

$$\Delta t = \frac{2 \cdot d \cdot (d - |x|)}{D} \quad (2)$$

where the strips are separated by d, and the ion strikes at a distance x away from one of the diodes. A plot of this Δt measurement is shown in Figure 4, and the straight-line fit of Δt vs. x gives the diffusion coefficient in this case to $D = 18.5$ cm²/s.

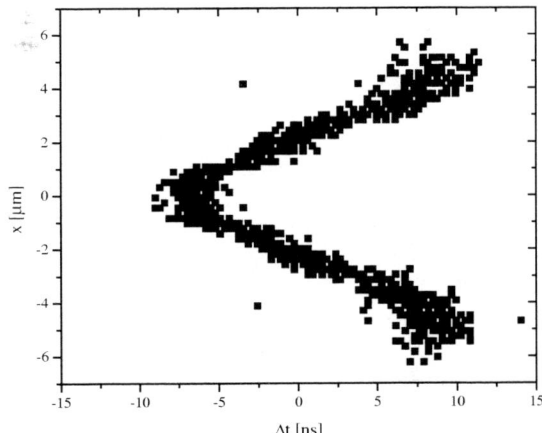

Figure 4. REM experiment using DTRIBICC to measure the diffusion coefficient of ion induced charge produced between strip-like diodes. There are actually three diodes in this sample, an inner and two outer ones that are connected.

The mobility of the diffusing charge, which in this case was electrons, can be obtained using the Einstein relation:

$$\frac{D}{\mu} = \frac{kT}{e} \quad (3)$$

and is found to be 714 cm²/Vs, which is quite close to what was expected for this sample (~700 cm²/Vs). This simple example shows just one case how IBICC can be used to microscopically determine electronic properties of semiconductors.

3.2 Constant fields

The situation of charge transport in semiconductors that are biased with a constant field becomes a little more complicated, but here, again, the S-R theorem provides the solution. Vizkelethy et al [15] have used both frontal (ions striking the front and back electrodes) and lateral (ions striking between the electrodes) to determine $\mu\tau$ product and charge collection lengths of Cadmium Zinc Telluride - CZT capacitor structure. In this case the S-R theory reduces to the two-carrier Hecht equation [16]:

$$q(x) = Ne \cdot \left[\frac{\lambda_e}{d}\left(1 - e^{-\frac{d-x}{\lambda_e}}\right) + \frac{\lambda_h}{d}\left(1 - e^{-\frac{x}{\lambda_h}}\right) \right] \quad (4)$$

for lateral IBICC, and the one carrier Hecht equation for frontal IBICC:

$$q = Ne \frac{\lambda_e}{d}\left(1 - e^{-\frac{d}{\lambda_e}}\right) \quad (5)$$

where q is the measured charge, Ne is the total created charge, λ_e and λ_h are the electron and hole drift lengths ($\lambda = \mu \cdot \tau \cdot E_1$), d is the thickness of the detector, and x the distance from the cathode of the point where the charge is generated.

Using both lateral and frontal IBICC and TRIBICC, it possible to microscopically measure the $\mu\tau$ product, the carrier extraction lengths, the carrier lifetimes, the mobilities of both the holes and the electrons, and in some cases, even variations in the electric fields, particularly near contacts. As an example of this, we plot both μ and τ maps for electrons in CZT in Figure 5.

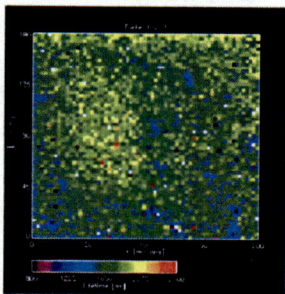
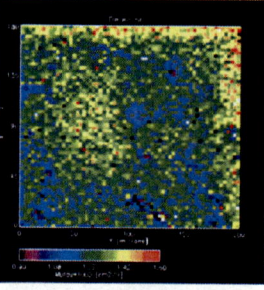

Figure 5. Mobility and lifetime maps of electron transport in CZT. The product of these two parameters is proportional to the carrier extraction length, which is the key property of a detector.

3.3 Integrated Circuits

Computer modeling using a code like DAVINCI [17] of the three dimensional charge transport that occurs within the device during passage of an energetic ion through the circuit helps designers to understand the complex interplay among device design, operation, and the introduction of charge into the device at arbitrary locations due to irradiation. The detailed calculation of charge transport provided by DAVINCI within a semiconductor structure is performed using finite-element charge transport calculations that solve Poisson's equation and the continuity equation for each volume element comprising the modeled circuit structure. Using these results, full-scale circuit simulations are performed in order to determine the effect of a specific ion strike at a specific circuit location on overall circuit operation. The S-R theory can be used to qualitatively interpret the main features in IBICC experiments on ICs.

The occurrence of simulated upset in the SRAM cell is determined by introducing a time-dependent current transient at the site of the ion strike into a circuit-level simulation of the SRAM cell. (Recall that the time-resolved current transient is calculated by a finite element calculation within a confined region of the circuit, e.g. drain, source). Both the magnitude and duration of the collected charge transient are important in determining whether circuit malfunction occurs. For any given memory cell, the magnitude of the collected charge must be greater than a critical amount, (Q_c), and persist for a time longer than the characteristic feedback time of the memory cell.

Simulations indicate that single event upset occurs in the n-off drain of the TA788 memory cell at a threshold LET of approximately 11.5 MeV/mg/cm^2. This result has been confirmed by broad beam upset testing of the device. An example of combined IBICC and SEU imaging, both experimental and theoretical, is shown in Figure 6.

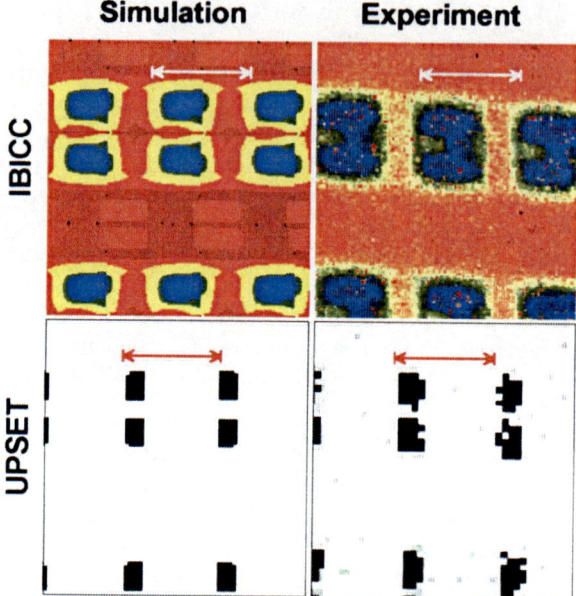

Figure 6. 35 MeV Cl SEU (lower) and 20 MeV C IBICC (upper) images with DAVINCI theory (left) of charge collection and upset of Sandia TA788 SRAM.

A measured upset image from the TA788 is shown in the lower right of Figure 6. The dark pixels in the upset images reflect the occurrence of 0-to-1 logic-state transitions in the memory cells during 35 MeV Cl irradiation. The corresponding simulated upset image is shown in the lower left of Figure 6. Since the Cl ions have an LET of approximately 17 MeV/mg/cm^2, which is above the threshold, one would expect these ions to upset this IC. In Figure 6 simulated charge collection image is shown in the upper left panel and the experimentally measured IBICC image is shown in the upper right. This data was taken with C ions which only have an LET of 5 MeV/mg/cm^2, and therefore the IC is not seen to upset but remains fixed in its initial state. This is born out by the model calculations as well. The highly localized and periodic clusters of upset sites in the measured upset image agree well with the prediction.

4. Conclusion

The use of scanned, focused ion microbeams for high spatial resolution measurements of electronic transport parameters and radiation testing of ICs has proved to be a useful tool to study semiconductors and the simulation verification and design evaluation of integrated circuits. This paper has briefly described the radiation effects microscopy techniques used to measure the radiation sensitivity of integrated circuits, as well as measuring the mobility and carrier lifetime in semiconductors such as CZT being developed for room temperature gamma ray detectors. We also showed comparisons between three dimensional charge transport simulations of charge collection and ion microbeam experimental measurements of a radiation hardened SRAM designed and manufactured at Sandia. While more and more microbeam groups are using REM, the tool clearly has yet to reach its full potential.

---

* Sandia is a multiprogram laboratory operated by Sandia Corporation, a Lockheed Martin Company, for the United States Department of Energy under Contract DE-AC04-94AL85000

** On leave from Idaho State University, Pocatello, ID, USA

1. T. Butz, R.-H. Flagmeyer, J. Vogt, D. Lehmann, St. Jankuhn, T. Reinert, and D. Spemann, "First results of the Leipzig nuclear nanoprobe LIPSION in biomedical research" in these Proceedings

2. K.M. Horn, B.L. Doyle, D.S. Walsh and F.W. Sexton, "Application of the Nuclear Microprobe to the Imaging of Single Event Upsets in Integrated Circuits", Scanning Microscopy, Vol. 5, No. 4, 1991, pp. 969-976

3. A.R. Knudson and A.B. Campbell, "Charge collection Measurements for Energetic Ions in Silicon", IEEE Trans. Nucl. Sci., vol. NS-29, 1982, pp.2067-2071

4. P.J. McNulty, W.J. Beauvais, D.R. Roth, J.E. Lynch, A.R. Knudson, and W.J. Stapor, "Microbeam analysis of MOS circuits", RADECS 91: First European Conf. On Radiation Effects on Devices and Systems 1991, pp.435-439

5. T.J. Aton, J.A. Seitchik, S.D. Hantz and H. Shichijo, "Accurate measurements of small charges collected on junctions from alpha particle strikes using an accelerator-produced microbeam", Proc. Intl. Reliability Physics Symp., 1995, pp.303-310

6. M.B.H. Breese, P.J.C. King, G.W. Grime, and F. Watt, "Microcircuit imaging using an ion-beam induced charge", J.Appl. Phys., vol. 72, no. 6, 1992, pp. 2097-2104

7. M. Jaksic, S. Fazinic, T. Tadic, M. Bogovac, I. Bogdanovic, and Z. Pastuoviz, "IBIC study of charge collection properties in Si(Li) detectors", Nucl. Instr. Meth., B138, 1998, pp. 1327-1332

8. C. Manfredotti, F. Fizotti, P. Polesello, P.P. Trapani, E. Vittone, M. Jaksic, S. Fazinic, and I. Bogdanovic, "Investigation on the electric field profile in CdTe by ion beam induced current", Nucl. Instr. Meth., A380, 1996, pp. 136-140

9. H. Schöne, D.S. Walsh, F.W. Sexton, B.L. Doyle, P.E. Dodd, J.F. Aurand, N. Wing, N., "Time-resolved ion beam induced charge collection (TRIBICC) in micro-electronics", Nucl. Instr. and Meth. B158, 1999, pp. 424-431

10. B.N. Guo, M. El Bouanani, S.N. Renfrow, D.S. Walsh, B.L. Doyle, E.B. Smith, R.C. Baumann, J.L. Duggan, and F.D. McDaniel, "Heavy ion microbeam studies of diffusion time resolved charge collection from p-n junctions", in these Proceedings

11. A.B. Campbell and A.R. Knudson, "Use of an ion microbeam to study single event upsets in microcircuits," IEEE Trans. Nucl. Sci., vol. NS-28, 1981, pp. 4017-4021

12. F.J. Henley and W.G. Oldham, "Soft error studies using a scanning source", Proc. 20th IEEE Reliability Physics Symp., 1982, pp.88-91

13. D.F. Heidel, U.H. Bapst, K.A. Jenkins, L.M. Geppert, and T.H. Zabel, "Ion microbeam radiation system", IEEE Trans. Nucl. Sci., vol. 40, 1993, pp.127-134

14. E. Vittone, F. Fizzotti, A. Lo Giudice, C. Paolini, and C. Manfredotti,"Theory of ion beam induced charge collection in detectors based on the extended Shockley-Ramo theorem", Nucl. Instr. Meth., B161-163, 2000, pp. 446-451

15. G. Vizkelethy, B.L. Doyle, D.S. Walsh, and R.B. James, "Nuclear Microprobe Studies of the Electronic Transport Properties of Cadmium Zinc Telluride (CZT) Radiation Detectors" in Hard X-Ray, Gamma-Ray, and Neutron Physics II, Ralph B. James, Richard C. Schirato, Editors, Proceedings of SPIE Vol. 4141, 2000, pp. 178-185

16. K. Hecht, Z. Physik, 77, 1932, p 235

17. DAVINCI 3.0 (Technology Modeling Associates, Inc., 1994)

Report on the Acadiana Research Laboratory Nuclear Microprobe System

Gary A. Glass, William A. Hollerman, Shelly F. Hynes, Justin Fournet,
Alan M. Bailey and Changgeng Liao*

Acadiana Research Laboratory, University of Louisiana at Lafayette, Lafayette, Louisiana
* Department of Modern Physics, Lanzhou University, Lanzhou, Gansu, 73000, Peoples Republic of China

Abstract. The Acadiana Research Laboratory of the University of Louisiana at Lafayette provides high energy ion beams for materials research. Major components of the ion beam systems include a National Electrostatics Corporation (NEC) 1.7 MV tandem Pelletron accelerator system with both SNICS and RF ion sources and a Varian CF-4 200 kV implanter. The NEC Pelletron has three operational beamlines that provide a wide range of capabilities for materials modification and analysis, including such techniques as PIXE, PIGE, RBS, RFS, TOF-ERDA and ion implantation. An Oxford Microbeams Ltd. microprobe system was recently declared operational with the attainment of a 1.5 μm x 2.0 μm beam spot size. Microprobe techniques presently available include μPIXE, μRBS and scanning transmission ion microscopy (STIM).

INTRODUCTION

The physical facility at Acadiana Research Laboratory (ARL) of the University of Louisiana at Lafayette (UL Lafayette) was designed for research in atomic and low energy nuclear physics, surface science, and materials science. The physical facility at ARL is the only laboratory in the state capable of supporting research in atomic physics involving ion-atom collisions, surface science, surface chemistry, and materials science utilizing ion beam technology. A 1.7 MV tandem Pelletron® accelerator (National Electrostatics Corporation), and a 200 kV ion implanter are presently housed in the target room, but development of the present laboratory began two decades ago. The present-day laboratory facility is located approximately 1 mile from the main campus of UL Lafayette.

ACADIANA RESEARCH LABORATORY HISTORY

In the mid-1970's A Nationals Electrostatics Model JN 3 MV Van de Graaff accelerator was obtained from NASA/Houston. In addition, a complete inventory of wood and metal machine shop equipment and supplies was acquired from the NASA/Michaud facility in New Orleans. After construction of a 5,000 ft^2 laboratory facility in 1978, the JN accelerator was installed to provide a rudimentary PIXE analysis capability. Later during 1978, a National Electrostatics Model KN 1 MV Van de Graaff accelerator was acquired from the University of Virginia. In 1982, a 4,600 ft^2 specially shielded area was added to the ARL complex and both the KN and JN accelerators were installed in that area. The KN was devoted primarily for use as an instructional tool while operation of the JN focused primarily on PIXE analysis. Once the accelerators were moved into the new area, the original building was partitioned into areas for use as metal and wood machine shops as well as storage. In 1986, the first of several proposals was submitted to fund the development of a complete ion beam research facility including (1) a high energy accelerator/implanter, (2) a low energy implanter, and (3) a high energy ion microbeam system. However, due of the large costs involved in acquiring all of the necessary component systems, the acquisition of each major equipment component has proceeded through funding of individual systems. The 1.7 MV tandem Pelletron® accelerator system, with RF (radiofrequency) and SNICS (source of negative ions by cesium sputtering) negative ion sources, along with vacuum hardware and detectors for an experimental beamline was acquired with two 1990 Louisiana Board of Regents (BOR) enhancement grants. The Pelletron system began operations began in August 1991 and replaced the 1 MV Van de Graaff accelerator in the target room. A

Varian CF-4 200 kV ion implanter system was installed in April 1993. This implanter system has a capacity for 100 mm diameter wafers. In 1996, funding was obtained to acquire an Oxford Microbeams, Ltd. nuclear microprobe system to occupy a dedicated beamline on the ARL 1.7 MV tandem Pelletron accelerator system. A schematic representation of the ion accelerator systems at ARL is shown in Figure 1.

Figure 2. ARL Nuclear Microprobe Beamline

Determination of ion beam dimensions is accomplished by scanning the beam across a metal (typically copper or gold) grid with a predetermined repeat distance. Both PIXE and STIM were used to provide images of the grid and corresponding determination of the beam size in the vertical and horizontal directions. Examples of a typical map obtained with a copper grid having a 12.5 μm repeat distance and a grid wire size of 7.5 μm is shown in Figure 3. [2]

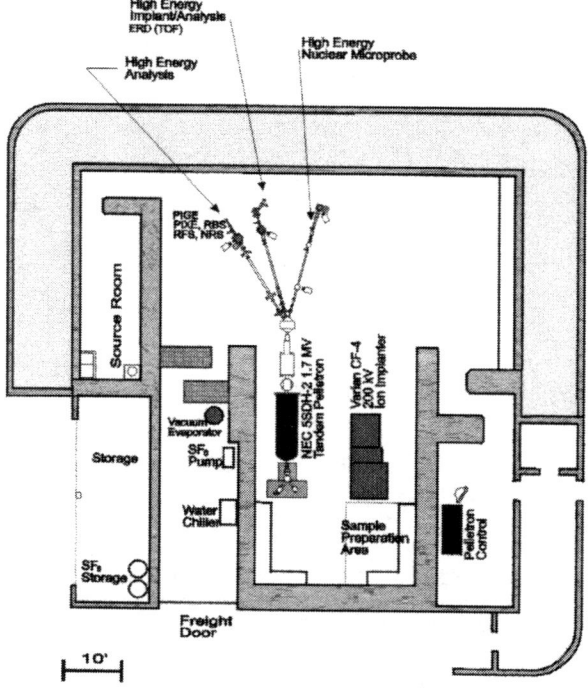

Figure 1. Acadiana Research Laboratory Accelerator Area

MICROPROBE SYSTEM

The microprobe system was funded in 1996 by the Louisiana Board of Regents and provides the capability to undertake basic research in many different areas of physics, chemistry and engineering. The small size (<1.5 x 1.5 μm) proton beam obtained with the microprobe system will allow the study of a wide variety of surface phenomena with very high spatial resolution with good sensitivity for multi-elemental analysis, particularly when compared with other microbeam techniques. [1] The microprobe system is designed to greatly expand the micro-analytical capabilities provided by atomic force (AFM), scanning tunneling (STM), and scanning electron microscope (SEM) systems presently at the at the university. Figure 2 shows a photograph of the ARL microprobe beamline and endstation.

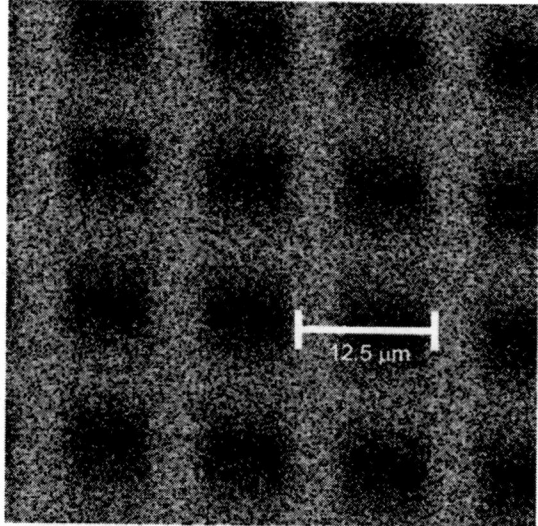

Figure 3. Cu-K$_\alpha$ PIXE Map of Copper Grid

At present, the nuclear microprobe system at ARL has PIXE, RBS and STIM capability with a readily attainable beam size of 1.5 μm x 1.5 μm and beam currents up to 100 pA with a proton energy of 2 MeV. The following are presented as examples of some of the work now in progress.

μPIXE MAPPING OF ELEMENTAL CONTENT IN A FOSSIL SECTION

A geological sample was collected from a shallow (11.8 m) core of Louisiana coastal sediments located about 30 km inland from the present coast. The sample itself consists of a vertebra on which was precipitated a thick layer of carbonate cement (See Figure 4). Compositions of such cements provide information on early chemical processes in the sediment. [3,4] This, in turn, contributes to our understanding of early fossilization and lithification of sediments. Hence, chemical examination of the sample was undertaken. Determination of elemental distributions within the sample requires microanalytical techniques.

μPIXE examination using the nuclear microprobe system at ARL has provided encouraging early results. Figure 5 shows maps and some profiles for Ca, P, Fe, Mn, and S. Areas of high Ca and P counts correspond to the bone while areas of high Fe around the bone indicate that the cement is Fe carbonate with impurities. One of the impurities is Mn, which tends to be higher next to the bone, indicating early Mn-rich solutions. Also present are areas with both high Fe and S. These correspond to pyrite crystals in the sample. The study of trace metal distributions in such samples is continuing.

Figure 6 is a presentation of a line scan taken across the iron and calcium carbonate boundary. The interesting feature in this line scan is the presence of manganese at the interface.

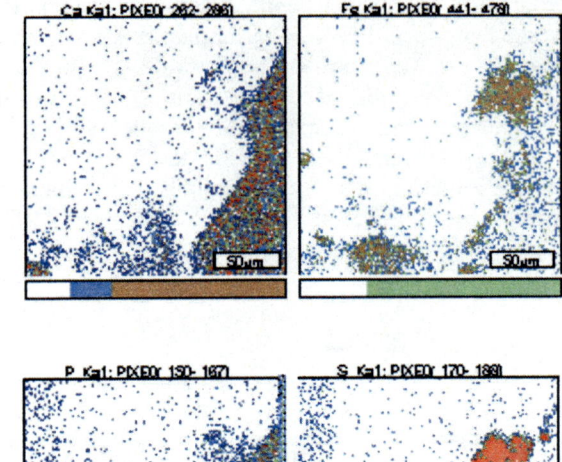

Figure 5. Elemental Map of Fossil Section Near Calcium and Iron Carbonate Boundary

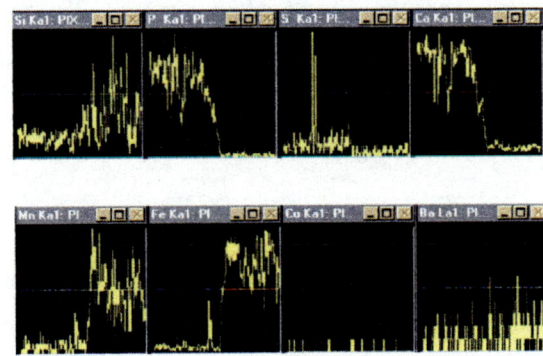

Figure 6. PIXE Line Scan across Boundary between Calcium and Iron Carbonate Regions

STUDY OF NEW FLUOR MATERIALS

There is worldwide interest in the use of fluor materials that emit visible light when exposed to ionizing radiation. Typically, fluors are used as components in high performance electromagnetic calorimeters, down-hole oil well loggers, temperature sensors for equipment with high speed moving parts, and beam positioning systems for large particle accelerators. A candidate fluor should have a large fluorescence efficiency, small reduction in output as a function of exposure, intense visible fluorescence spectrum, large material density, and small prompt fluorescence decay time. Over the last few years, a

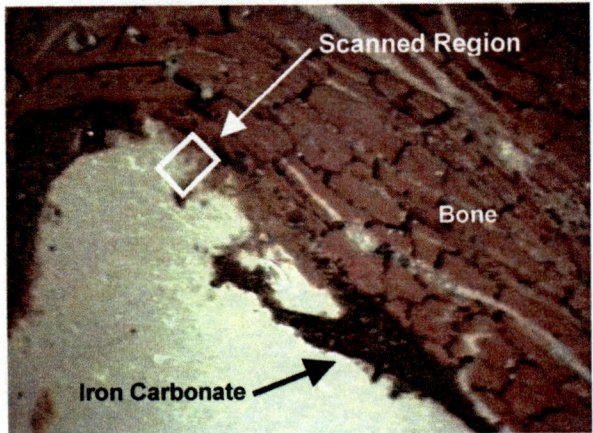

Figure 4. Bright Field Optical Image of Fossil Section (x50)

number of new fluor materials have been developed. These new fluors, such as $YAlO_3$:Ce, $YSiO_5$:Ce, $GdSiO_5$:Ce, Y_2O_2S doped with Tb, Pr, and Eu, and Gd_2O_2S doped with Tb, Pr, and Eu, show a great deal of promise for use in many applications. [5]

ARL has recently begun a study of new scintillators and other fluorescent materials. One of the tasks for this new research program is the determination of the extent of radiation damage in fluor materials to see if enhancements in their characteristics can occur as a result of exposure to protons, alphas, and selected heavy ions. The prompt fluorescence decay time and half brightness dose will be measured for each fluor sample. However, the nuclear microprobe provides a unique opportunity to study individual fluor particles. Figure 7 shows a PIXE map of S, Eu and Y in fluor particles on the surface of a fluor material consisting of Y_2O_2S doped with Eu. The fluor particle is clearly seen in the 2-dimensional maps. Since the silicon signal originates only from the glass substrate, it is believed that the reduction of the silicon signal indicates the presence of relatively thick particles. By focusing the beam on an individual fluor particle, it will be possible to measure fluorescence and optical properties of the individual particles including light frequency response, attenuation effects within the fluor material, overall amplitude and time response, and radiation damage effects.

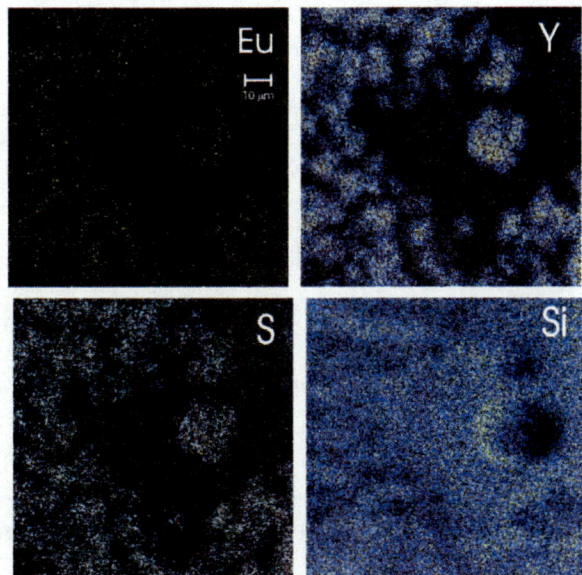

Figure 7. µPIXE Maps of Eu-doped Y_2O_2S Fluor Material and Glass Substrate

CONCLUSIONS

The Acadiana Research Laboratory nuclear microprobe system will provide the capability to study surfaces, coatings, and interfaces between substrates and thin coatings with spatial resolutions eventually less than 1 µm. The utilization of the nondestructive elemental micro-analysis capabilities provided by the microprobe offer advantages for researchers in metallurgy, mineralogy, geology, geochemistry, environmental sciences, archaeology, semiconductors, microbiology, plant sciences, biology, and medicine. Further refinements of the system are in progress.

ACKNOWLEDGMENTS

Supported in part by the U.S. Department of Energy and the Louisiana Board of Regents Support Fund under contracts DOE/LEQSF (1993-95)-03, DE-FC02-91ER75669 and LEQSF(1996-97)-ENH-TR-75.

REFERENCES

1. Watt, Frank and Grime, Geoff W., "The High Energy Ion Microprobe," in *Particle-Induced X-ray Emission Spectrometry (PIXE)*, edited by S.A.E. Johansson, J.L. Campbell, K.G. Malmqvist, Publisher John Wiley & Sons, Inc., 1995, pp. 101-165.

2. Hynes, S.F., Hollerman, W.A., Pastore, J.J. and Glass, G.A., "Comparison of Measured and Calculated Beam Spot Sizes for the Acadiana Research Laboratory Nuclear Microprobe," in *Sixteenth International Conference on the Application of Accelerators in Research and Industry – 2000*, AIP Conference Abstracts (2000).

3. Bailey, A.M., Roberts, H.H., and Blackson, J.H., 1998, Early diagenetic minerals and variables influencing their distribution in two long borings (>40m), Mississippi River delta plain: Journal of Sedimentary Research, v. 68, (1998) pp. 185-197.

4. Bailey, A. M. and Roberts, H. H., "Controls on minor element compositions of early diagenetic siderites and dolomites in the Mississippi River delta plain," J. Conf. Abs., v. 5, (Tenth Goldschmidt Conference, Oxford, UK), (2000) p. 175.

5. Hollerman, W.A., Glass, G.A. and Allison, S.A., "Survey of Recent Results for New Fluor Materials," Materials Research Society Symposium Proceedings, 560 (1999) pp. 335-340.

Latent Ion Tracks In Mica Studied With Scanning Force Microscopy In Air And In Vacuum

V. Hoffmann[1], J.H. Bremer[1], S. Bouffard[2], and N. Stolterfoht[1]

[1] *Hahn-Meitner Institut GmbH, Glienicker Strasse 100, D-14109 Berlin, Germany and*
[2] *Centre Interdisciplinaire de Recherche Ions Lasers, Unite Mixte CEA-CNRS-ISMRA, F-14050 Caen Cedex, France*

Abstract. Latent ion tracks in mica have been studied using scanning force microscopy (SFM) on freshly cleaved samples in air and under vacuum conditions. The tracks have been produced by irradiating mica with 2.8 MeV/u uranium ions. The topography images were recorded simultaneously both in forward and backward scanning direction. When scanning in air a strong friction force on the latent ion tracks complicates the analysis of their topography. Since we measured in vacuum these hillocks can be seen due to a drastically reduced friction. Thus, we observed hillocks of about 0.5 nm height.

INTRODUCTION

The miniaturization of technological devices successively requires smaller structures. In microelectronics, structures below a micrometer are commonly used already. The material modification with fast heavy ions allows for the production of single structures in the nanometer range. Latent ion tracks are investigated intensively in the past decades, gaining new attention since it is possible to image single ion tracks with scanning force microscopy SFM [1-7] and transmission electron microscopy TEM [8-10].

Along its trajectory through the material, the ion transfers kinetic energy to the electrons of the solid and, thereby heating it up. If the deposited energy density is high enough to melt the material, an amorphous cylinder is formed along the ion track. The diameter of the latent track is a characteristic quantity for the ion solid interaction.

Muscovite mica is very suitable for the investigation of latent ion tracks. It is a layered crystalline material. Thin mica sheets with an atomically flat and clean surface can be produced by cleavage. On uranium irradiated muscovite mica, amorphous tracks with diameters around 10 nm have been reported from small angle x-ray scattering [11] and TEM [9] measurements. The same diameter follows from theoretical model calculations based on the energy dissipation by secondary electrons [14]. However, the material transport during the ion-solid interaction by a thermal spike and the tensions inside the latent ion tracks formed far away from equilibrium are not well understood up to now.

Whereas those methods yield information about the atomic order/disorder in the bulk, SFM images the topography and, the force between the mica surface and the tip. Hillocks with a diameter around 20 nm and a height in the order of 1-2 nm have been found on mica, when the sample was not recleaved after irradiation [12,13]. These hillocks have been interpreted as residual material of the melted ion tracks, which tend to expand at the surface because they have a lower density than the crystalline mica lattice.

In order to get information from the interior of the latent tracks the mica sample are recleaved after irradiation, producing a new surface within the former bulk. Recently, SFM measurements at ambient conditions on these surfaces showed a higher friction on the tracks referred to the crystalline mica lattice [15]. Latent ion tracks appeared as hillocks and craters, depending on the scanning direction [2]. The images are inverted due to the strong friction force, which influences the topographic z(x,y) signal in contact mode SFM, as will be discussed below. The larger friction coefficient of the track surface is attributed to adsorbates such as water and hydrocarbons, which are present at ambient conditions.

In this work we report on contact mode SFM measurements on mica samples, which were recleaved after irradiation. We compared the results taken in air and under high vacuum conditions, in order to obtain a better insight to the role of adsorbates on SFM imaging. The aim is to clarify the topographic structure of latent ion tracks on recleaved muscovite mica.

EXPERIMENTAL

The mica samples were irradiated at GANIL cyclotron accelerator (Caen) with 2.8 MeV/u U^{55+} ions, with a dose amounting to $4\cdot10^{-10}$ ions/cm^2. Prior to irradiation, the samples were cleaved to produce a clean surface.

SFM measurements were performed in contact mode with an Omicron SFM/STM system connected to the UHV chamber at the ECR source of the Ionenstrahl-Labor ISL at Hahn-Meitner Institut (Berlin). Prior to SFM measurements the mica surface was freshly produced recleaving the sample with a tape. Mica exhibits areas up to a µm^2 with an atomically flat surface. This is an essential condition to visualize a possible topographic structure of nm range originated by latent ion tracks.

First, we measured the surface in air, and subsequently in high vacuum, at a pressure of 10^{-8} mbar. Finally, we exposed the surface several hours to ambient conditions and repeated the SFM measurements. Topographic and lateral force signals were recorded simultaneously both, in forward and backward scanning direction ± x (trace and retrace). The fast scanning direction x is perpendicular to the long axis of the cantilever. We used Si_3N_4 cantilevers with a bending spring constant of 0.09 N/m. The load force was set as low as possible, so that stable imaging was still achieved. It ranged from 15 to 35 nN, depending on the experimental conditions.

RESULTS AND DISCUSSION

In Fig. 1a and 1b topographic SFM images in air (500 x 500 nm^2) are shown in forward and backward scanning direction, respectively. The line scans correspond to the arrows in the topview images. Left-right is the fast scanning direction, which is perpendicular to the cantilever axis.

Here the base width is referred to as the diameter of the ion tracks. In order to determine the diameter d of a single ion track we measured the base width in the trace and the retrace images, d_t and d_r, respectively. This was done for the fast x and as well as for slow scanning direction y. The average of these four base width values was used as diameter of one ion track d = $(d_{xt} + d_{xr} + d_{yt} + d_{yr})/4$. This procedure was repeated for several tens of tracks.

The mean diameter amounts to 20 ± 5 nm, which is in fair agreement with related data reported from other groups [13], ranging from 10 to 35 nm for uranium irradiated mica. The discrepancy between results from different groups is likely to be due to differences in determining the diameter of the tracks. It might also be an effect of the resolution of the cantilever, which depends on its curvature. The SFM image results from the convolution of surface structure and cantilever. Sharp structures are smeared out and appear larger than they are [16]. This effect is considerable when the surface structures are about as small as the radius of the cantilever, which is the case for latent ion tracks. Lateral extensions of surface structures in SFM images therefore have to be interpreted as an upper limit.

The topographic profiles in forward and backward scanning direction in Fig. 1 appear to be inverted, showing hillocks and craters with apparent height and depth values of up to roughly ±10 nm, respectively. A higher friction on the ion tracks than on the surrounding mica lattice can explain this. Using the SFM in contact mode, the friction force at the contact area produces a distortion of the cantilever, which deviates the reflected laser beam from the center of the detector. In the method of scanning force microscopy the position of the laser spot on the detector is used as feedback signal. Therefore, a deviation due to an additional friction force on the track is superimposed to the topographic and the lateral force signals. Note that the friction primarily causes a torsion or bending of the cantilever, while scanning perpendicular or parallel to its long axis, respectively. Therefore, in our experimental setup the friction should not influence the topographic signal. Nevertheless, the friction is transmitted through the contact area of the cantilever tip. The orientation of the edges of the contact area

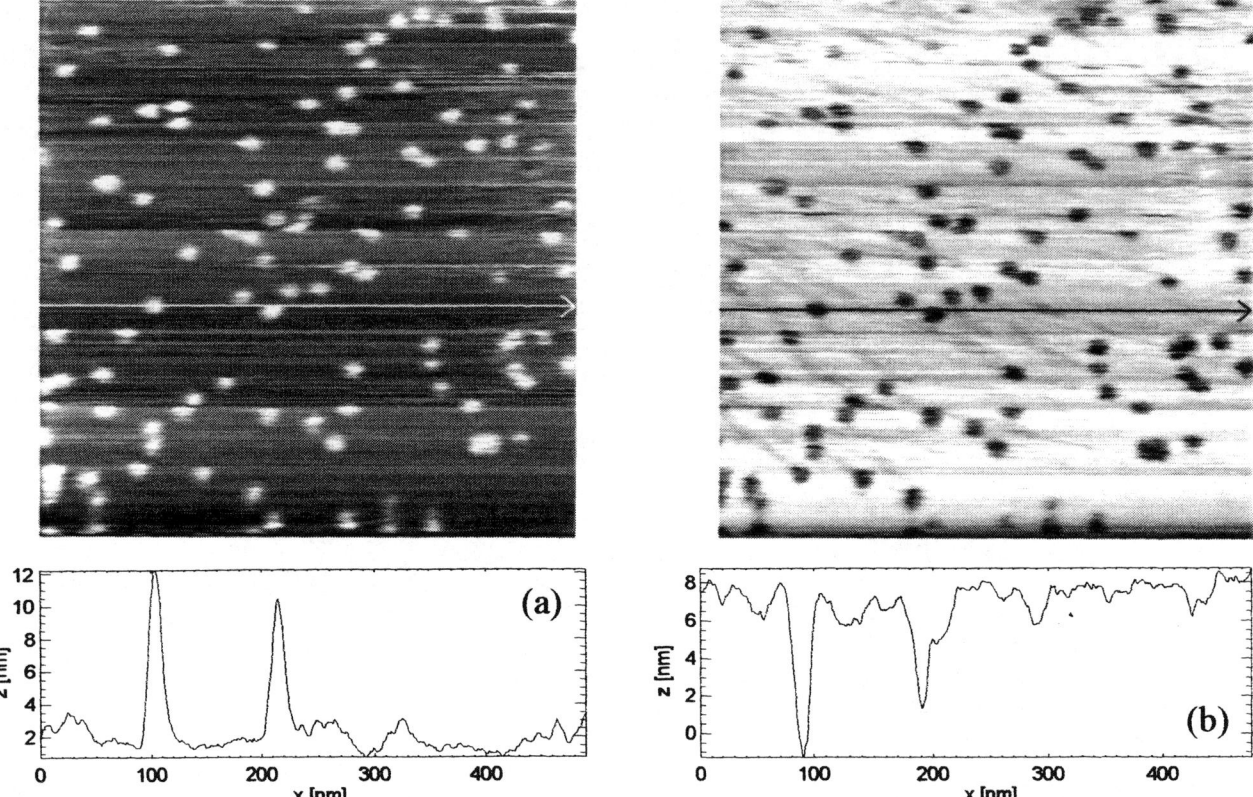

FIGURE 1. Topographic AFM images of a 500 x 500 nm² area of an irradiated mica surface measured in air. The data in (a) and (b) are obtained scanning from left to right and from right to left, respectively. The line profiles in the lower frames correspond to the arrows in the upper frames.

with respect to the scanning direction determines the distribution of the lateral friction force to the components F_x and F_y. An asymmetry concerning shape and orientation of the contact area with respect to the scanning direction results in a contribution of both force components. Then the friction induces both, torsion and bending of the cantilever.

The strong influence of the friction force on the topographic signal with apparent height and depth values of roughly ±10 nm complicates the analysis of the surface shape from the SFM data. If there were real topographic hillocks or craters on the surface, their height or depth, respectively, is small compared to 10 nm. Therefore it is not possible to conclude from our SFM measurements in air whether the cleaved mica surface exhibits hillocks or craters at the track sites, or whether it is completely flat. The high friction on the tracks in air is mainly attributed to the accumulation of water and hydrocarbons. To obtain SFM data with less adsorbates on the mica surface we evacuated the system to a pressure of 10^{-8} mbar.

In Fig. 2a and 2b topographic SFM images of a mica surface (300 x 300 nm²) in vacuum are shown in trace and retrace, respectively. The line profiles correspond to the arrows in the topview images. Again, left-right is the fast scanning direction. The pictures show round ion tracks with an average diameter of 20 ± 2 nm, which coincides with the one measured at ambient conditions. Contrary to the SFM measurements in air, under vacuum conditions the topographic $z(x,y)$ signals of trace and retrace show the same surface morphology. This is due to the low friction force in the vacuum. The lateral forces on the mica surface are about a factor 20 smaller than scanning in air, thereby drastically reducing the influence of the friction on the topographic image.

From bottom to top, which is the slow scanning direction y, the surface of the tracks seems to consist of a part above and a part below the undisturbed mica surface. We interpret this hillock-crater appearance of the tracks as due to the influence on the measurement by the slope grad $z(x,y)$ of a topographical hillock. At the contact area the surface produces a horizontal force component $F_H = F_N$ grad $z(x,y)$ directed to the

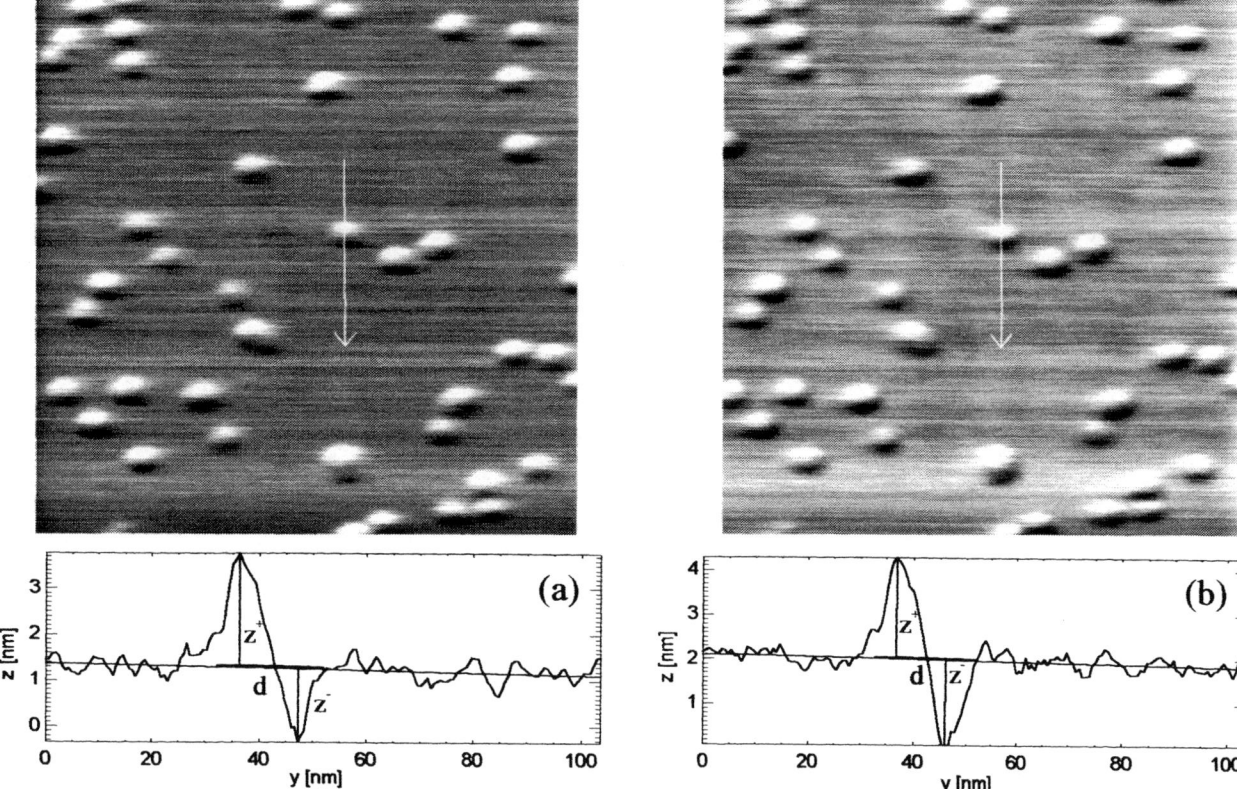

FIGURE 2. Topographic AFM images of a 300 x 300 nm² area of an irradiated mica surface measured in vacuum. The data in (a) and (b) are obtained scanning from left to right and from right to left, respectively. The line profiles in the lower frames correspond to the arrows in the upper frames. z^+ and z^- are the maximum and minimum of the track structure, respectively, and d is the diameter.

cantilever tip to compensate the applied vertical load force F_N. With the length of the tip l_T a torque $l_T F_H$ is applied at the end of the cantilever. As can readily be shown, this produces a bending and torsion of the cantilever, depending on the position of the tip on the hillock. On the way to the top of a hillock in the slow scanning direction y, F_H bends the cantilever, as a crater would do, whereas on the way down the bending (z signal) appears as a hillock. The influence of a topographical crater would result opposite signals. Our estimations concerning the influence of the topographical slope grad z(x,y) of a hillock suggest that apparent z values in the order of ±1 nm as shown in the line scans in Fig. 2 are consistent with the experimental conditions.

In first approximation, we determined the extreme values z^+ and z^- of a measured track structure z(x,y) in the trace z_t as well as in the retrace z_r image. Then we averaged these values in order to obtain the topographic height $h = (z_t^+ + z_t^- + z_r^+ + z_r^-)/4$ of one track. This procedure was repeated for several tens of track structures. Thus, we concluded that the latent ion tracks form hillocks with a mean height of roughly 0.5 nm. We note that this elevation of 0.5 nm corresponds to about the thickness of two atomic layers, whereas the diameter of these hillocks is up to factor 40 larger.

The formation of hillocks on recleaved mica surfaces suggests that there remains a pressure inside the latent ion tracks. At a freshly produced surface, the pressure could relax with a vertical expansion in a surface near region, and thereby forming new hillocks. Nevertheless, with a pressure of 10^{-8} mbar during the measurements it can not be disregarded that the detected elevation of 0.5 nm might also be due to residual adsorbates at the ion track surface.

CONCLUSIONS

We compared contact mode SFM measurements performed on muscovite mice surfaces under ambient and high vacuum conditions. The samples have been recleaved after the irradiation with 2.8 MeV/u U ions. The same diameter of the latent ion tracks follows from both experiments. However, the friction force on the cleaved mica surface is by roughly a factor 20 larger in air, than in vacuum. The convolution of the friction force with the recorded z(x,y) signal at ambient conditions does not allow for the resolution of topographic structures in the sub-nm range. Under high vacuum conditions the latent ion tracks show elevations with a height of roughly 0.5 nm. In order to clarify the origin of the hillocks formed on freshly cleaved mica samples further SFM measurements are planed.

ACKNOWLEDGEMENTS

The authors would like to thank R. Neumann from the Gesellschaft für Schwerionenphysik (GSI) and R.C. Birtcher from the Argonne National Laboratory for fruitful discussions.

REFERENCES

1. Thibaudau, F., Cousty, J., Balanzat, E., and Bouffard, S., Phys. Rev. Lett. **67**, 1582 (1991).

2. Bouffard, S., Cousty, J., Pennec, Y., and Thibaudau, F., Radiat. Eff. Def. Sol. **126**, 225 (1993).

3. Schneider, D., Briere, M.A., Clark, M.W., McDonald, J., Biersack, J., and Siekhaus, W., Surf. Sci. **294**, 403 (1993).

4. Parks, D.C., Stöckli, M.P., Bell, E.W., Ratliff, L.P., Schmieder, R.W., Serpa, F.G., and Gillaspy, J.D., Nucl. Instr. and Meth. in Phys. Res. B **134**, 46 (1998).

5. Neumann R., Nucl. Instr. and Meth. in Phys. Res. B **151**, 42 (1999).

6. Döbeli, M., Ames, F., Musil, C.R., Scandella, L., Suter, M., and Synal, H.A., Nucl. Instr. and Meth. in Phys. Res. B **143**, 503 (1998).

7. Ackermann, J., Angert, N., Neumann, R., Trautmann, C., Dischner, M., Hagen, T., and Sedlacek, M., Nucl. Instr. and Meth. in Phys. Res. B **107**, 181 (1996).

8. Albrecht, D., Armbruster, P., Spohr, R., Roth, M., Schaupert, K., and Stuhrmann, H., Appl. Phys. A **37**, 37 (1985).

9. Vetter, J., Scholz, R., Dobrev, D., and Nistor, L., Nucl. Instr. and Meth. in Phys. Res. B **141**, 747 (1998).

10. Birtcher, R.C., and Donnelly, S.E., Phys. Rev. Lett. **77(21)**, 4374 (1996).

11. Chailley, V., Dooryhee, E., Bouffard, S., Balanzat, E., and Levalois, M., Nucl. Instr. and Meth. in Phys. Res. B **91**, 162 (1994).

12. Ackermann, J., Grafström, S., Hagen, T., Kowalski, J., Neumann, R., and ,Sedlacek, M., *Micro/Nanotribology and its Applications*, edited by B. Bhushan, Kluwer Academic Publishers, Netherlands, 1997,pp. 261.

13. Barlo Daya, D.D.D., Hallen, A., Hakansson, P., Sunqvist, B.U.R., and Reimann, C.T., Nucl. Instr. and Meth. in Phys. Res. B **103**, 454 (1995).

14. Tombrello, T.A., Nucl. Instr. and Meth. in Phys. Res B **94**, 424 (1994).

15. Hagen, T., Grafström, S., Ackermann, J., Neumann, R., Trautmann, C., Vetter, J., and Angert, N., J. Vac. Sci. Technol. B **12(3)**, 1555 (1994).

16. Ackermann, J., Müller, A., Neumann, R., and Wang, Y., Appl. Phys. A **66**, 1151 (1998).

Heavy Ion Microbeam Studies of Diffusion Time Resolved Charge Collection from p-n Junctions

B.N. Guo,[a*] M. El Bouanani,[a] S.N. Renfrow,[b**] D.S. Walsh,[b] B.L. Doyle,[b]
J.L. Duggan,[a] and F.D. McDaniel[a]

[a]*Ion Beam Modification and Analysis Lab, Department of Physics, University of North Texas, Denton, TX 76203*
[b]*Ion Beam Materials Research Lab, Sandia National Laboratories, MS 1056, PO Box 5800, Albuquerque, NM 87185*

Abstract. The knowledge of (diffusion, drift, and funneling assisted) charge collection within electronic devices is essential to design radiation hardened Integrated Circuits (ICs). In the present work, diffusion time resolved charge collection studies were performed on stripe-like junctions using 12 MeV carbon and 28 MeV silicon microbeams and MEDICI simulation calculations. The relative average arrival time of the diffused charge on the junctions was measured along with the amount of charge collection by the junctions. The average arrival time of the diffused charge is related to the first moment (or the average time) of the arrival carrier density on the junction. The experimental results and MEDICI (a 2D-device simulator) calculations support this interpretation. These results show the importance of the diffusive charge collection by junctions, which is especially significant in accounting for Single Event Upsets (SEUs) and Multiple Bit Upset (MBUs) in digital devices.

INTRODUCTION

Due to intrinsic properties of semiconductor devices, ionizing radiation can induce undesired excess carrier generation and migration within microelectronic components. Technology trends, such as reduced operating voltage and device area scaling, will result in a higher susceptibility to ionizing radiation induced effects, which are of great concern for the reliability of future devices [1].

The knowledge of (diffusion, drift, and funneling assisted) charge collection dynamics is essential to design radiation hardened devices. A new technique, Diffusion Time Resolved Ion Beam Induced Charge Collection (DTRIBICC) [2], is proposed to measure the average arrival time for an ion track in addition to the charge collection by junctions using a multiple parameter data acquisition system.

As IC design moves to a smaller scale, it is found that models based on the diffusion mechanism can interpret some experimental results better than those based on the drift and funneling assisted mechanism [3,4]. The time duration of charge collection by the charge collection node has been proposed to estimate the speed of the diffusive charge collection process [5]. If the transient current is $I(t)$ through the node caused by the induced charge at the source point x, then the total charge collection and the statistical average arrival time (or the first moment of the transient current), respectively, are

$$Q(\tau) = \int_0^\tau I(t)dt \quad \tau \to \infty \qquad (1)$$

$$T(\tau) = \frac{1}{Q(\tau)} \int_0^\tau tI(t)dt \quad \tau \to \infty. \qquad (2)$$

Along the ion track, a series of source points can be defined. Therefore, the average arrival time can be used to estimate the speed of charge collection for the ion track. The diffusion time from a specific striking spot to the junction was recorded using a Time-To-Amplitude Converter (TAC) as a new parameter along with collected charge.

In this paper, we present experimental and model simulation results of the average arrival time on the

---

*Present address: Varian Semiconductor Equipment Associates, 35 Dory Road, Gloucester, MA 01930. Email: baonian.guo@vsea.com
**Present address: Mission Research Corporation, 6703 Odyssey Drive, Suite 101, Huntsville, AL 35806. Email: s.renfrow@mrchsv.com

specially designed stripe-like junction test structure. IBICC measurements from the ring-gate-inner diodes and large diode on the same specially designed ICs were previously reported elsewhere [6,7].

EXPERIMENTAL DETAILS

The experiments, conducted at the Ion Beam Materials Research Laboratory at Sandia National Laboratories, employed 12 MeV C and 28 MeV Si microbeams with ~ 1 μm beam spot size to scan junction test structures. The test structure is the outer-inner diodes. The cross sectional view of the stripe-like portion is shown in Fig. 1. Phosphorus and Arsenic are used to form $n+$, and the Boron doped p-substrate has a $p+$ retrograde boron well.

Three data collection channels were used to measure the charge collection from the outer and inner diodes, and the relative average arrival time. Two separated 4 V reverse biases were applied to the outer and inner diodes, and the p-substrate was grounded. The timing outputs from the two preamplifiers were fed into two Constant Fraction Discriminators (CFDs). The two fast timing outputs from CFDs were then fed into a TAC as the start (from outer junction) and stop (from inner junction) signals. The TAC signals were digitized and recorded along with striking spot coordinates and collected charge from the two junctions. The TAC range was set at 100 ns with 48 ns offset inserted into the stop signal channel. The time scale was calibrated with an ns delay box. The total charge collected by a pin diode was used to calibrate the charge collection electronics. The counting rate of charge collection was ~400 counts per second.

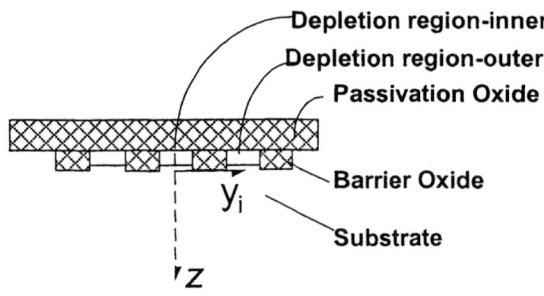

FIGURE 1. The cross sectional view of outer-inner diode test structure for the DTRIBICC measurements. The 2 μm wide stripes are separated by 2 μm wide oxide barriers.

The relative charge arrival time approach is an advantage [8], if it is assumed the charge is collected by the junction centers. If the ion strikes at a distance y_i from the center of the inner junction, and penetrates a distance z into the sample (Fig.1), it is straightforward to show that this arrival time difference, Δt, is just:

$$\Delta t = \frac{(d-|y_i|)^2 + z^2}{D} - \frac{|y_i|^2 + z^2}{D} = \frac{d(d-2|y_i|)}{D} \quad (3)$$

where d is the separation between the junction centers (4 μm) and D is the diffusivity of minority carriers (electrons in the p-type substrate). Note that the time difference is not a function of z, and therefore charge that diffuses from any point along the trajectory of the ion will diffuse to these diodes with the same Δt.

Computer simulation results based on the MEDICI code are also provided for comparison [9]. The junction doping profiles were incorporated into MEDICI simulations. A localized electron-hole pair generation based on SRIM calculation along the ion track is incorporated into the MEDICI codes to simulate the induced carrier transport in the device [10].

RESULTS AND DISCUSSIONS

Figs. 2 and 3 are the DTRIBICC measurements on the outer-inner test structure using 12 MeV C and 28 MeV Si microbeams, respectively. Figs. 2-A and 3-A are the charge collection by the inner and outer junctions. Figs. 2-B and 3-B are the relative arrival times. MEDICI simulation results are also plotted in Figs. 2 and 3.

The transient currents vs. time can be obtained from the MEDICI simulations for the outer and inner junctions. The amount of charge and the average arrival time for the outer and inner junctions can be calculated using Eqs. 1 and 2. Since simulation results show that carriers arriving later than 1~2 μs can be ignored, the values at $\tau=2$ μs are used for the charge collection and the average arrival time [2]. The simulated relative arrival time has been adjusted with the 48 ns offset. Due to the structural symmetry along the inner junction, only half of test structure is simulated.

The charge collection and relative arrival time in Figs. 2 and 3 indicate:

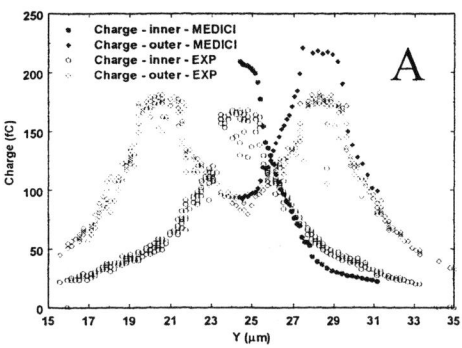
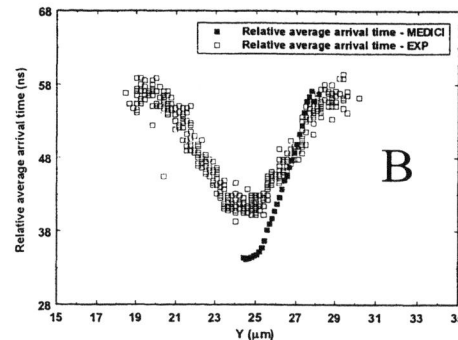

FIGURE 2. DTRIBICC experimental and MEDICI simulation results using a 12 MeV carbon microbeam.

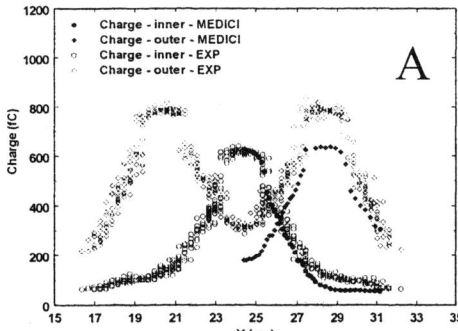
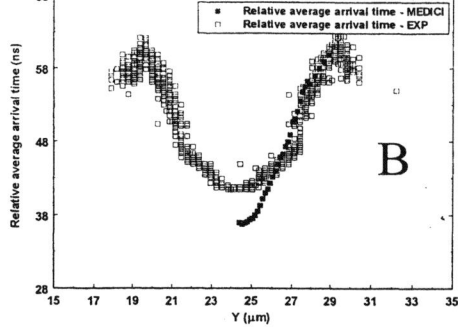

FIGURE 3. DTRIBICC experimental and MEDICI simulation results using a 28 MeV silicon microbeam.

When ions strike outside of and far away from the outer junction, the charge is collected by the outer junction through the diffusion process and no charge is collected by the inner junction. As the beam strikes closer to the outer junction, the inner junction begins to pick up some charge through diffusion. Once the stop timing signals begin to be triggered, non-zero TAC signals are registered.

The large increases at the peaks in Figs. 2-A and 3-A indicate the transition from the outside (barrier oxide) to inside of the junctions. When ions directly strike the junctions, the struck junction will collect most of the charge through the funneling effect, and the other junction shares charge through diffusion. It also indicates charge collection through diffusion is not as effective a mechanism to collect charge compared with directly striking the junction, where the charge funneling is the main charge collection mechanism.

When the ions strike between the outer and inner junctions, the induced charge is shared between the two junctions and the relative arrival time should follow Eq. 3. When ions strike in the middle of the outer and inner junctions (i.e. $|y_i|=d/2$ in Eq. 3), the TAC will record the relative arrival time as 48 ns which corresponds to $\Delta t=0$ in Eq. 3. If the data in Fig. 2-B is fit to the straight line(s) derived in Eq. 3, the diffusivity, D, of electrons in silicon is determined to be 18.5 cm2/s, which corresponds to an electron mobility of 714 cm^2/V-s. A 28 MeV Si ion has a decreasing LET (Linear Energy Transfer) along the ion track compared with a relative constant LET for a 12 MeV C ion [6], it is expected that the measured relative average times are different for different ion species. Eq. 3 should be modified to account for ion with varying LET along its penetration track. In Fig. 4, the collected charge by the outer junction is plotted against the collected charge by the inner junction. This charge sharing also indicates the ion striking spots, which is very similar to a Position Sensitive Detector. Since the distance between the outer and inner junctions is only 2 μm, the position can be inferred based on the charge sharing with a potential ~0.1 μm resolution.

As shown in Figs. 2, and 3, it is clear that the charge collection and the relative arrival time predicted by the MEDICI simulations are in agreement with the experimental results. We believe the differences between the experimental and simulation results are due to limitations of the 2D MEDICI code [9]. To better predict and compare with experimental results, 3D codes such as DAVINCI should be utilized to simulate correct ion track and test structure [11].

Since the experimental results were recorded in list mode, an off-line analysis of the collected data shows that ion induced damage did not affect the charge collected by the junctions at the accumulated dose ~ 10 ions/μm^2. The relative arrival times were measured using two CFDs, the uncertainty of the CFD determined the uncertainty of the measurement. The CFDs were triggered by the rising edges of the preamplifiers. The timing uncertainty also was affected by the capacitance of the junctions. The estimated uncertainty was about 0.5 ns. Still, the timing measurements provide useful information to determine the charge collection dynamics.

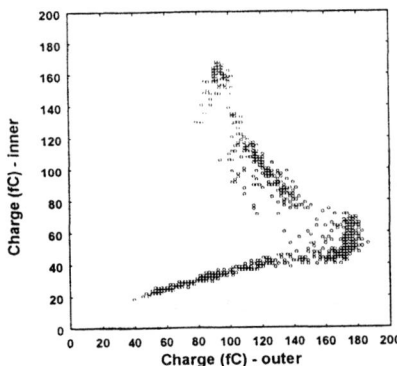

FIGURE 4. Charge collection from the outer junction is plotted against that from the inner junction.

CONCLUSIONS

The fundamental knowledge of charge collection dynamics in semiconductor devices due to ionizing radiation is essential for device radiation hardness assurance. The experimental results show the importance of diffused charge and charge sharing between adjacent junctions. In particular, the effect of charge sharing between adjacent memory nodes can result in MBUs. Since the MBUs are extremely difficult to diagnose and correct, the charge sharing is especially significant to account for MBUs in ICs. In addition, DTRIBICC experimental results and MEDICI calculations support the interpretation that the average arrival time of the diffused charge is related to the first moment (or the average time) of the arrival carrier density on the junction. DTRIBICC, using relative arrival timing, represents an important new single-ion radiation effects microscopy for studying charge sharing and measuring basic electrical properties such as charge carrier diffusivity, mobility, and lifetime in ICs or detectors. DTRIBICC forms the basis for a new type of Position Sensitive Detector for MeV ions with a resolution approaching 0.1 μm, which could clearly be improved by optimizing the design of the p-n diode stripe structure. The fact that very little charge was collected on the inner diode when ions struck outside the outer diodes suggests that sensitive junctions could be shielded from distant ion strikes by using a perimeter-type diode structure. Such a perimeter shield could potentially ameliorate MBU effects.

ACKNOWLEDGEMENTS

Work supported by the NSF, the State of Texas-Advanced Technology Program, and the Robert A. Welch Foundation. Sandia is a multiprogram laboratory operated by Sandia Corporation, a Lockheed Martin Company, for the United States Department of Energy under contract DE-AC04-94AL85000. The authors would like to thank T.J. Aton, E.B.Smith, and R.C. Baumann of Texas Instruments Inc for assistance with samples and for many helpful discussions.

REFERENCES

1. Semiconductor Industry Association (SIA), The National Technology Roadmap for Semiconductors, (1997).
2. B.N. Guo, Ph.D. dissertation, University of North Texas, 2000.
3. E.C. Smith, E.G. Stassinopoulos, K. LaBel, and C.M. Seidlick, IEEE Trans. on Nucl. Sci. **NS-42**, No. 6, (1995), 1772.
4. L.D. Edmonds, IEEE Trans. on Nucl. Sci. **NS-38**, No. 6, (1996), 3207.
5. L.D. Edmonds, IEEE Trans. on Nucl. Sci. **NS-43**, No. 4, (1996), 2347.
6. B.N. Guo, S.N. Renfrow, B.L. Doyle, D.S. Walsh, T. J. Aton, M. El Bouanani, J.L. Duggan, and F.D. McDaniel, CP475, Application of Accelerators in Research and Industry, edited by J.L. Duggan, I.L. Morgan, AIP press, New York, (1999), 1121.
7. F.D. McDaniel, B.N. Guo, S.N. Renfrow, M. El Bouanani, J.L. Duggan, B.L. Doyle, D.S. Walsh, and T.J. Aton, Nucl. Instr. and Meth. **B158** (1999), 264.
8. B.L. Doyle, private communications.
9. MEDICI/DAVINCI device simulation tools, Avant! Corporation, 46871 Bayside Parkway, Fremont, CA 94538, http://www.avanticorp.com/.
10. J.F. Ziegler and J.P. Biersack, SRIM-96: The Stopping and Range of Ions in Matter, (1996), http://www.research.ibm.com/ionbeams/.
11. P.E. Dodd, IEEE Trans. on Nucl. Sci. **NS-43**, No. 2, (1996), 561.

The study of phosphor efficiency and homogeneity using a nuclear microprobe

C. Yang [a], B. L. Doyle [b], M. Nigam [a], M. El Bouanani [a],
J. L. Duggan [a], and F. D. McDaniel [a]

[a] *Ion Beam Modification and Analysis Laboratory, Department of Physics*
University of North Texas, Denton, Texas 76203, USA

[b] *Ion Beam Materials Research Laboratory, MS 1056, PO Box 5800*
Sandia National Laboratories, Albuquerque, NM 87185, USA

Abstract. Ion Beam Induced Luminescence (IBIL) and Ion Beam Induced Charge Collection (IBICC) have been used to study the efficiency and the homogeneity of the luminescence emission in phosphors. The IBIL imaging was made by using sharply focused ion beams or broad/partially-focused ion beams. Samples were examined to reveal possible distributed crystal-defects that may lead to the inhomogeneity of the luminescence emission. The purpose of the study is to search for suitable thin films that have high homogeneity of luminescence emission and large IBIL efficiency under heavy ion excitation. These films may be placed as a thin layer on the top of microelectronic devices to be analyzed with Ion Photon Emission Microscopy (IPEM). The emission yields were found to be low for organic materials, due to saturation of the light output dependence on the energy deposition of heavy ions. The emission yield of a typical Bicron plastic scintillator is about 70 photons/ion/micron. Inorganic materials may have higher IBIL yields under high-energy and heavy-ion excitation, but the challenging problem is the inhomogeneity of the IBIL emissions. The IBIL image techniques are applied in the investigation of the homogeneity of a GaN epitaxial thin film, a zircon single crystal and a thin layer coated by Thiogallate (EuII) ceramic.

Introduction

The luminescence property of various materials has been the subject of research interests for a long time. However, the knowledge of IBIL excited by high-energy, heavy-ions in organic and inorganic materials is still very limited. The information on the IBIL yields from thin film phosphors is crucial to the success of a new nuclear emission microscopy: Ion Photon Emission Microscopy (IPEM), currently under development at Sandia National laboratory (SNL) [1]. The IPEM can be a very promising new tool for performing single ion effects microscopies without beam focusing, and potentially with a radioactive source instead of an accelerator. These thin films must be deposited or placed on the surface of microelectronic devices to enhance the production of IBIL. The IBIL photons are then projected with a high magnification lens system onto a position sensitive detector (PSD) that is sensitive to single photons. Using the IPEM technique, the efficient generation, transmission and detection of these photons is required to determine the arrival position of each ion that strikes the sample. Therefore, the utility of IPEM hinges on finding an optimum luminescent layer.

Plastic phosphors have several advantages over inorganic ceramic materials. They usually have a fast decay time (a few ns to tens of ns); they are clear and smooth (reducing or even eliminating light scattering); they are easily made thin and self-supporting; and such plastic scintillators are already commercially available. The drawback is that the emission yield is relatively low, due to saturation of the light output dependence on the energy deposition of heavy ions [2]. IBIL and IBICC may be applied together to quantify the phosphors efficiency [3].

Inorganic materials may have higher IBIL yields under high-energy, heavy-ion excitation, but they have common problems of inhomogeneity of luminescence emission. Therefore, it is necessary to check the IBIL homogeneity prior to a further quantification analysis of the luminescence efficiency. In this paper IBIL image techniques were applied to the investigation of the homogeneity of the luminescence emission in a GaN epitaxial thin film, zircon single crystal and Thiogallate (EuII) ceramic.

Instrumentation

Measurements were performed on the UNT heavy-ion microprobe beam line in the Ion Beam Modification and Analysis Laboratory (IBMAL). A microscope, (SM-OM40, JEOL USA, Inc.) which was originally designed for scanning electron microscopy (SEM), is installed in the nuclear microscope chamber. The microscope allows the observation of micrographic images with a magnification of 300 and a viewing field 650 microns in diameter. The optical microscope's objective lens is a long retractable arm. The objective lens, which is a type of refracting lens, has a hole drilled through its center that allows the ion beam to pass during nuclear microprobe experiments. The optical microscope has an easy working distance of 5 mm and numeral aperture (N.A.) of about 0.3.

Due to the large magnification, the microscope is ideal for viewing very small features (approximately one μm resolution). A CCD camera or a luminescence detector such as photo-multiplier tube (PMT) for Ionoluminescence applications can easily replace the monocular eyepiece of the microscope. In this work, a PMT-based integrated photon-counting head (Hamamatsu Model H5920-01) is used for a single photon counting detector. This luminescence detector has an output pulse height of 3.3 volts and 25 ns pulse width. The detector has the highest sensitivity at about 400 nm wavelength with about 16% quantum efficiency. The dark current of the photon counting head is about 1-2 cps, after being stored in a dark room more than 24 hrs.

In this study, a PIN-diode (Hamamatsu, model S1223) is used to detect and count the incoming particles, which excite the IBIL photons. Thick and thin plastic phosphors can be placed on the top surfaces of the S1223 PIN-diodes. For the thick phosphors, the ion beam is stopped completely, and only IBIL signals are detected. In the uncovered area of the pin-diode, only IBICC signals are produced and detected. One ion hitting the uncovered pin-diode area produces one IBICC event signal. For the thin phosphors, the ion beam passes through the phosphors into the pin-diode and both IBICC and IBIL signals are detected. For each pixel in the scanned region, the IBICC event and the IBIL photons are simultaneously detected for each incoming ion striking the phosphor. The experiments to determine the detected photon/ion efficiency is similar to an IBICC measurement on the PIN diode, where in addition, the IBIL signal is measured at the same time. The combination of the IBICC and IBIL information gives the quantitative information of the thick phosphor efficiency in the term of *photons/ions*.

Two types of IBIL imaging methods are applied in this study. One employs a broad (non-focused) ion beam. The IBIL image is captured through a video system attached to an optical microscope. In the second type of operation, a focused ion beam is used and the IBIL image is simultaneously obtained with other techniques in nuclear microscopy, such as IBICC, PIXE and RBS. This approach utilizes the full power of nuclear microscopy.

The measurement of the IBIL efficiency of organic scintillation materials

The nature of the luminescence in organic materials is characteristic of the molecular structure, because the interaction of inter-molecules is much weaker than that of intra-molecules. The organic phosphors are commonly aromatic hydrocarbon compounds containing linked or condensed benzene-ring structures. The luminescence can be explained with free valence electron transitions in the energy scheme of π-molecular orbitals [4].

The analysis with a scanned alpha particle beam was carried out for thick samples mounted on the PIN-diodes. The scans were made to include 1) an area for IBICC analysis of the PIN-diode surface that remained uncovered and 2) part of the sample area covered by the thick plastic that produced an IBIL signal. A typical IBIL and IBICC image for such a thick plastic phosphor is displayed in Figure 1. Here, a 6.0 MeV

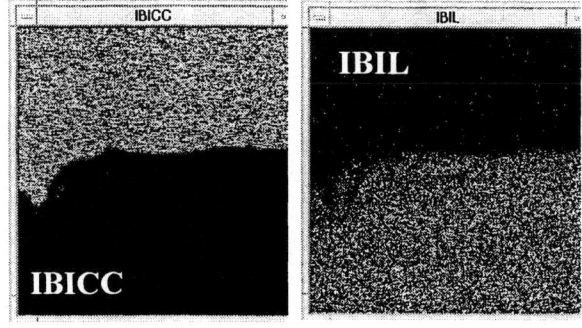

Figure 1. IBICC and IBIL images of a thick phosphor over a pin-diode. A 6.0 MeV alpha particle beam is used in the study. The quantitative information of *ions/pixel* is extracted from the IBICC image area where the PIN-diode is not covered by the thick sample. Similarly, the quantitative information on *photons/pixel* is extracted from the IBIL image area where the PIN-diode is covered by the thick sample. The combination of the IBICC and IBIL information gives the quantitative information of the phosphor efficiency in *photons/ions*.

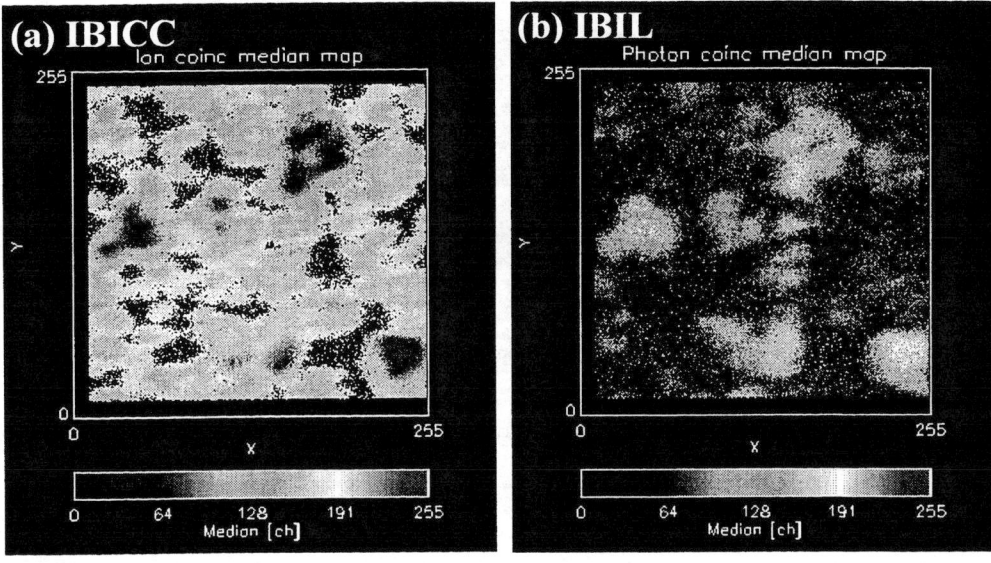

Figure 2. (a) IBICC and (b) IBIL images over the layer made by coating the inorganic ceramic Thiogallate (EuII) phosphor on a pin-diode. A 20 MeV carbon beam is used in the study. It is evident that the phosphor grains dominate the IBIL photon distribution in the sample. (data from Sandia National Laboratory).

alpha particle beam was scanned across both the covered and uncovered areas. The quantitative information of *ions/pixel* is extracted from the IBICC image area where the PIN-diode is not covered by the thick plastic sample. Similarly, the quantitative information on *photons/pixel* is extracted from the IBIL image area where the PIN-diode is covered by the thick plastic sample. Therefore, by dividing the *photons/pixel* from the IBIL measurement by the *ions/pixel* from the IBICC measurement, the IBIL efficiency in *photons/ion* can be determined. The investigation of the plastic scintillation materials, such as the Bicron samples: BC 400, BC 404, BC 408 and BC 430, indicates the relative low IBIL efficiency ~ 70 photons/ion/micron [3]. The image of IBIL from the organic material usually shows a homogenous distribution of the IBIL yield. The high homogeneity of the IBIL emission is essential for the successful IPEM application.

The investigation of the homogeneity of the IBIL emission in inorganic phosphors

The mechanism of luminescence in inorganic materials can be explained with semiconductor band structure theory [5] or crystal field theory [6]. It may be relatively easy to make thin films by coating a device with a fine powder of efficient luminescent materials; but it can not be used for the IPEM applications if there is a variation in mass density. The variation of the mass density in the thin film can lead to a large variation of the IBIL homogeneity. Figure 2 displays the IBICC (Figure 2a) and IBIL (Figure 2b) images over the layer made by coating the inorganic ceramic Thiogallate (EuII) phosphor on a pin-diode. A 20 MeV carbon beam is scanned over the sample. It is evident that the phosphor grains dominate the IBIL photon distribution in the sample. A further effort will be made to fabricate this type of thin film phosphor with a large improvement of the homogeneity.

The other causes of the IBIL variation in samples may not be associated with the variation of the mass density, but instead, may be related to internal crystal defects [7], even though the crystal may appear to be very smooth. Figure 3a reveals the line dislocation defects in the IBIL image produced with a broad 2 MeV alpha particle beam excitation in zircon, which is a large-band-gap crystal. The IBIL image was captured with a video system attached to the optical microscope. The size of the "broad beam" applied in this analysis is about 200 μm by 160 μm. The structure of the crystal defects that may occur in various forms is usually invisible when investigated by an optical microscope with reflected light. The broad beam IBIL image method can be easily used to screen the inorganic crystals for homogeneous thin films for the IPEM applications.

In Figure 3b, GaN (Mg doped), made by an epitaxial growth on an aluminum oxide substrate, yields a bright green and homogeneous IBIL emission distribution. Here, a 2.0 MeV alpha particle beam with a partially focused beam size of 100 μm is used. The high homogeneity of the luminescence emission

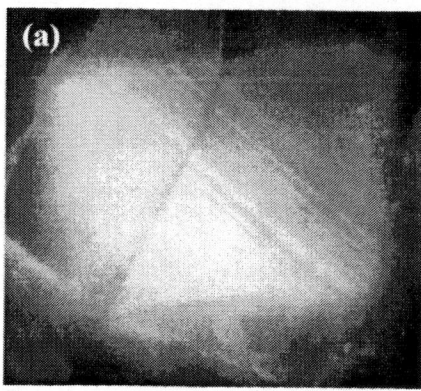 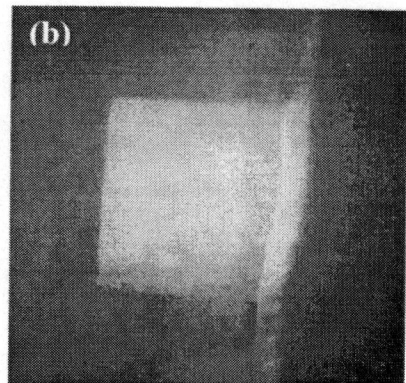

Figure 3. (a) The crystal dislocation defects in a zircon mineral are revealed by the IBIL image using partially focused beam (200μm x 160 μm); (b) Smooth IBIL distribution in an epitaxial thin film made of GaN (Mg-doped). The thin film yields a bright green IBIL emission. Both IBIL images are excited with a 2 MeV alpha particle beam.

distribution, as revealed in the IBIL image, makes it a good candidate thin film for the IPEM application.

Conclusions

Plastic phosphors have the advantage of being relatively easy to produce homogeneous luminescent layers over some inorganic materials. The disadvantage is that the emission yield is relatively low. The challenging problem of inorganic phosphors for the IPEM application is the inhomogeneous distribution of the IBIL emission due to crystal defects or variation of mass density. IBIL and IBICC measurements with the nuclear microprobe can be used to quantify the IBIL efficiency and study the homogeneity of the IBIL emission in the search for the candidate films suitable for the IPEM application.

Acknowledgements

This work is supported in part by NSF, the State of Texas Advanced Technology Program, and the Robert A. Welch Foundation. Sandia is a multiprogram laboratory operated by Sandia Corporation, a Lockheed Martin Company, for the United States Department of Energy under Contract DE-AC04-94AL85000.

References

1. B. L. Doyle, D. S. Walsh, S. N. Renfrow, G. Vizkelethy, T. Schenkel and A.V. Hamza, Proceedings of the 7th International Conference on Nuclear Microprobe Technology and Applications, Bordeaux, France, September 10-15, 2000, to be published in Nucl. Instrum. and Methods. B 2001.
2. J.B. Birks, Proc. Phys. Soc. A64, 874 (1951).
3. C. Yang, B. L. Doyle, P. Rossi, M. Nigam, M. El Bouanani, J. L. Duggan and F. D. McDaniel, Proceedings of the 7th International Conference on Nuclear Microprobe Technology and Applications, Bordeaux, France, September 10-15, 2000, to be published in Nucl. Instrum. and Methods. B 2001.
4. W.R.Leo, Techniques for Nuclear and Particle Physics Experiments (Springer-Verlag, 1987) p159.
5. B.G. Yacobi and D.B. Holt, Cathodoluminescence Microscopy of Inorganic Solids (Plenum Press, New York and London, 1999) p25.
6. B. Henderson and G.F. Imbusch, Optical Spectroscopy of Inorganic Solids (Clarendon Press, Oxford, 1989).
7. P.D. Townsend and J.C. Kelly, Colour Centres and Imperfections in Insulators and Semiconductors (Sussex University Press, 1973).

The recent progress of the high-energy heavy ion nuclear microprobe at the University of North Texas

C. Yang, B.N. Guo, M. El Bouanani, M. Nigam, J. L. Duggan, and F. D. McDaniel

Ion Beam Modification and Analysis Laboratory, Department of Physics, University of North Texas, Denton, Texas 76203-1427

Abstract. The paper reports the recent progress of a high-energy, heavy ion nuclear microprobe facility established at the University of North Texas. The microprobe system is installed on a 3MV NEC 9SDH-2 Pelletron tandem accelerator. A high demagnification factor (~60) has been achieved with the system, using a probe-forming lens system (from MARC, Melbourne, Australia) with the new Russian quadruplet configuration. The spatial resolution of 2-3 μm has been achieved for 4.0 MeV carbon ions or 9.0 MeV alpha particles with a beam current of ~ 50-100 pA. Better spatial resolution (approaching one μm) is achievable when an extremely low beam current (100-2000 ions/sec) is used in the applications of IBICC and IBIL. Applications of the analytical techniques with the nuclear microprobe are outlined and discussed.

Introduction

The major components of a high–energy, heavy ion, focused microprobe were purchased from MARC in Melbourne, Australia and the beam line, pumping system and analysis chamber were constructed at UNT. The microprobe began operating in September 1999. The nuclear microprobe beam line shares the 3.0 MV tandem accelerator with five beam lines for other applications. Currently, the major research projects involve Ion Beam Induced Charge Collection (IBICC), Ion Beam Induced Luminescence (IBIL) and Single Event Upset (SEU) techniques, which require a low beam current and a high spatial resolution (~ few μm).

The ion sources and the accelerator

Two types of ion sources are used with the Nuclear Microprobe on the tandem accelerator. A source of negative ions by cesium sputtering (NEC SNICS) provides a wide variety of ion species (H, Li, B, C, Si and etc.). The second ion source, (NEC Alphatross) is used mainly for the production of negative helium ions. The Alphatross ion source provides much higher beam brightness and reproducibility for the microbeam operation. However, due to the electron-stripping process in the tandem accelerator, the beam brightness achieved is relatively low, about one or two orders of magnitude lower than the typical values found in single-ended accelerator [1].

The major components of the microprobe system

Figure 1 displays the layout of the focusing microprobe system. The microprobe beam line hardware consists of object and aperture slits; probe-forming lenses with a computerized focusing control system, a computerized beam scan unit, and a target chamber. The magnetic probe-forming lens system is the Russian-Quadruplet configuration, which gives a relatively high demagnification factor (~ 60). The advantage of higher demagnification means a larger image on the objective slits without sacrificing the spatial resolution; therefore, a larger beam current can be obtained.

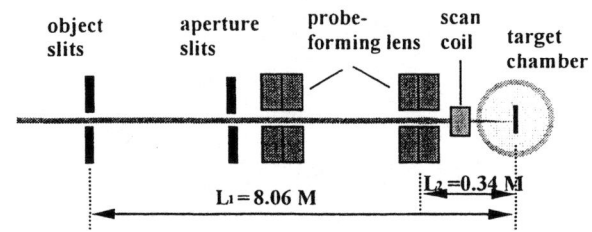

Figure 1. The layout of the microbeam line

The multiparty boxes for both the object and aperture slits were purchased from MARC. The beam viewers in the slits boxes are made of quartz plate and are not designed for heavy ions such as Si and C. The radiation damage to the quartz viewers by the heavy ion beams can lead to a rapid reduction of the efficiency of the luminescence emission. An aluminum oxide ceramic (doped with chromium) is a good replacement for the quartz plate. Under any ion beam strike, the Al_2O_3 (Cr doped) yields a very strong reddish light (~630 nm). This wavelength falls into a sensitive wavelength region of common CCD cameras for remote monitoring the ion beam shape and intensity. The new beam-viewer makes it possible to remotely view the beam shape and intensity at the location of the object and aperture slits. Even if the beam current is as low as 50 pA, the CCD camera can still easily pickup a clear picture of the beam spot for normal light illumination conditions of the room.

A high-quality optical microscope is used to examine and localize very small features in targets prior to the nuclear microprobe analysis. An optical microscope (JEOL SM-OM40) has been installed in the nuclear microprobe chamber in a target front viewing position. This microscope was originally designed for scanning electron microscopy (SEM) for the observation of micrographic images with a very high magnification (x300) and a large viewing field (650 μm in diameter). The optical microscope is equipped with an object lens in a retractable long arm. The object lens has a refracting lens with a holed reflection mirror. When viewing a sample perpendicularly, an ion beam goes through the hole in the reflection mirror. The monocular eyepiece of the microscope can be easily replaced by a CCD camera for a video picture, or be replaced by a luminescence detector such as photo-multiplier tube for ion luminescence applications.

The microprobe imaging and Estimation of the spatial resolution

When minimizing the microbeam spot on a target (optimizing the spatial resolution), a thin quartz glass can be used if there is a large enough beam current to generate a strong luminescence. For the case of low beam currents, scintillation materials, such as CsI(Tl) and BC480, may be used. These materials have higher luminescence efficiency and their IBIL wavelength is in the blue to green color range, which is also the sensitive range for human eyes. However, in order to capture the microscopic image of focused beam spots by a normal CCD camera (best for the reddish light), one may need to use the IBIL beam viewing materials that luminesce in the red region.

Figure 2 shows a sharply focused 4.0 MeV carbon ion beam spot captured by a CCD camera mounted on an optical microscope with a very high spatial resolution. A thin section of an emerald mineral is used as the beam-viewing target. A 1000 mesh (1000 lines per inch) gold grid is mounted at the top surface of the target for scaling purposes. The size of the beam spot is estimated to be ~ 2-3 μm at a beam current of 50-100 pA.

The minimum beam size of 2 μm has also been measured for 9.0 MeV alpha particles, which has a higher beam rigidity. Figure 3 displays an IBIL image from a test sample made of a luminescent GaN(Mg) film with a 1000 mesh gold grid mounted on its top surface. The measurement was made at a beam current intensity of about 2000 ions/sec. Figure 3 demonstrates the potential of the imaging technique for the study of the microstructures in targets with a high spatial resolution.

Due to poor mechanical rigidity of the homemade chamber, there is a noticeable mechanical vibration (on the order of one μm) observed when

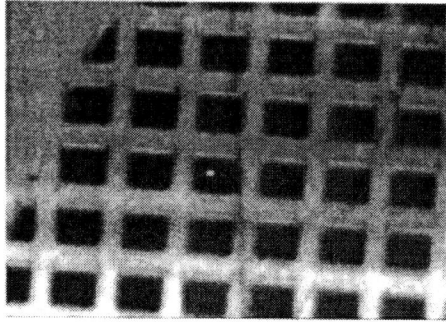

Figure 2. Microscopic image of sharply focused beam spot on an emerald mineral. A 1000 mesh grid is over-laid on the surface of the emerald crystal. The IBIL is bright reddish light.

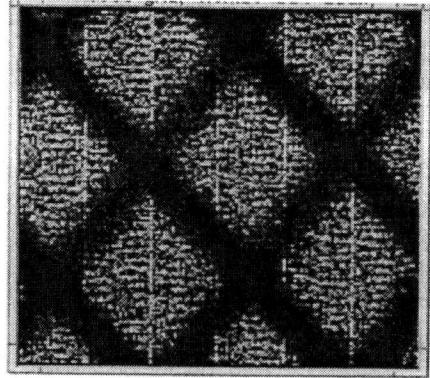

Figure 3. A sharp IBIL image of a 1000 mesh gold grid on a GaN(Mg) thin film.

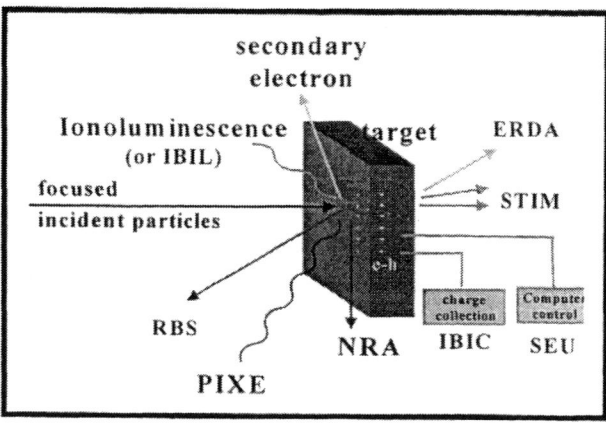

Figure 4. The analytical methods can be used for imaging in the nuclear microprobe.

someone walks near the nuclear microprobe beam line. In order to achieve sub-micrometer beam resolution in the future, a compact new chamber with a higher mechanical rigidity is being developed.

Microprobe applications

In general, all of the analytical techniques normally found in the Ion Beam Applications (IBA) field can be used with the nuclear microprobe, as shown in Figure 4. These analytical techniques also include the conventional methods: micro-PIXE, -RBS, -NRA and -ERDA. Because the beam brightness is relatively low, high spatial resolution can only be obtained at the cost of a large reduction in beam current intensity. For high spatial resolution measurements, there may not be enough beam current to achieve good statistics in the experimental analysis. Therefore, the nuclear microprobe has advantages for those applications when low beam current is required. The nuclear microprobe at UNT is currently used mainly for microelectronics with the analytical techniques of IBICC, SEU and IBIL. In general, these techniques require a low beam current of a few thousands ions per second.

In studying the microelectronic device response to ionizing radiation, SEU imaging and IBICC are two complementary techniques. SEU imaging directly reveals those weak nodes within a device, where upsets or malfunctions are caused when exposed to energetic heavy ions, but it does not provide quantitative information concerning the amount of charge, which produced the upsets. IBICC, on the other hand, provides quantitative measurement of the amount of charge collected by these circuit nodes when struck by ions, but does not reflect the effect of the ion strike on the circuit's operation. The combination of these two techniques allows quantitative investigation of the upset process within a circuit. Time Resolved IBICC (TRIBICC) [2] has been used to measure the transient charge collection on MOSFET structures using a high bandwidth data acquisition system and a microbeam. TRIBICC and Diffusion TRIBICC (DTRIBICC) [3] provide approaches to experimentally verify computer simulation results of charge diffusion and collection.

The IBIL technique can be combined simultaneously with IBICC or applied on its own for the characterization of large band gap materials and micro-electronic devices. Since IBIL can be simultaneously combined with IBICC at extremely low beam currents, high spatial resolution can be readily achievable, as indicated in Figure 3. By the combined use of IBIL and IBICC, it is possible to characterize the phosphor's efficiency quantitatively and study the problems of the luminescence inhomogeneity distributions in thin films [4]. IBIL and IBICC can also be simultaneously applied in the characterization of properties of large band-gap materials and devices [5].

The Technique of Ion Beam Induced Luminescence (IBIL) has also been developed with two technical options for IBIL imaging. One employs a sharply focused ion beam and a beam scan over a selected area to be imaged. The IBIL image can be simultaneously obtained with other techniques in nuclear microscopy, such as IBICC, PIXE and RBS. In this option, the imaging was reconstructed from saved data-events by using dedicated software including on-line data acquisition and off-line data sorting utilities developed for the nuclear microprobe. This approach

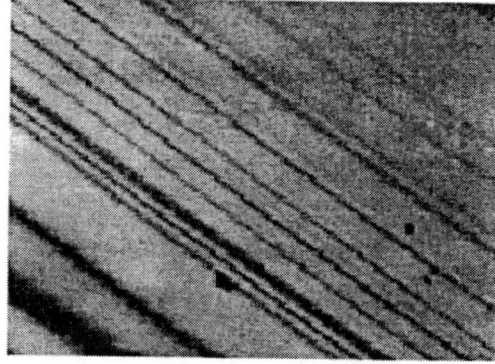

Figure 5. The internal crystal defects are revealed by the IBIL imaging technique in a single crystal zircon. A 2.0 MeV alpha beam is applied in the study. The IBIL image is directly captured through a video system attached to a high-resolution optical microscope. The image area is about 200 by 200 μm.

utilizes the full power of nuclear microscopy. In the second type of operation, a broad or partially focused ion beam is applied over an interesting area in a sample and the IBIL image is captured through a video system attached to an optical microscope.

Figure 5 displays an IBIL image excited with a 2.0 MeV alpha particle beam in a single crystal zircon. The internal crystal defects are clearly revealed. The IBIL image is directly captured through a video system attached to a high-resolution optical microscope. The image area is about 200 µm by 200 µm. The resolution of the image is determined by the quality of the optical microscope, which in this case, is about one µm.

Conclusions

A high-energy, heavy ion nuclear microprobe facility has been constructed at UNT with components from MARC. The facility is used for the analytical techniques IBCC, IBIL and SEU for the applications of failure analysis of microelectronic devices and characterization of materials. Currently, spatial resolution of 2 µm has been routinely obtained. Better spatial resolution should be obtained in the future when a compact target chamber with a higher mechanical rigidity is installed.

Acknowledgements

This work is supported in part by NSF, the State of Texas Advanced Technology Program, and the Robert A. Welch Foundation.

References

[1] R. Szymanski and D. Jamieson, Nucl. Instrum. and Methods **B130** (1997) 80-85.

[2] H. Schone, D.S. Walsh, F.W. Sexton, B.L. Doyle, P.E. Dodd, J.F. Aurand, R.S. Flores and N. Wing, Nucl. Instrum. and Methods **B158** (1999), 424.

[3] B.N. Guo, M.El Bouanani, S.N. Renfrow, M.Nigam, D.S.Walsh, B.L. Doyle, J.L. Duggan and F.D.McDaniel, Proceedings of the 7th International Conference on Nuclear Microprobe Technology and Applications, Bordeaux, France, September 10-15, 2000. To be published in Nucl. Instrum. and Methods.

[4] C. Yang, B. L. Doyle, P. Rossi, M. Nigam, M. El Bouanani, J. L. Duggan and F. D. McDaniel, Proceedings of the 7th International Conference on Nuclear Microprobe Technology and Applications, Bordeaux, France, September 10-15, 2000. To be published in Nucl. Instrum. And Methods.

[5] C. Yang, A. Bettiol, D. Jamieson, X. Hua, J.C.H. Phang, D.S.H. Chan, F. Watt, and T. Osipowicz, Nucl. Instrum. and Meth. **B158** (1999) 481-486.

Dependence of Heavy Ion Induced Secondary Ion Emission on Electric Conductivity in MeV Energy Range

S. Ninomiya*, S. Gomi*, J. Xue[†], M. Imai*, and N. Imanishi*

Department of Nuclear Engineering, Kyoto University, Kyoto 606-8501, Japan
[†]*Quantum Science and Engineering, Kyoto University, Kyoto 611-0011, Japan*

Abstract. We have measured mass and kinetic energy distributions of secondary ions emitted from Al, Si, Al_2O_3, and SiO_2 targets bombarded by MeV Si ions, where the electronic stopping power dominates over the nuclear stopping power. The obtained emission energy distributions of atomic ions from the conductor and the semiconductor are very broad and have exponentially decaying tails at the high-energy side in contradiction to those of singly charged atomic and cluster ions from the insulators. The most probable and mean energies of atomic ions are proportional to their electric charge irrespective of the target conductivity. The observed facts can be explained by the simple combined mechanism of the simultaneous recoiling and ionization by the projectile and the Coulomb repulsion by the short-lived ionized track region.

INTRODUCTION

The electronic sputtering has been widely studied using laser, electrons, highly charged heavy ions, and high-energy heavy ions for various target materials. Most of studies are concerned with sputtering yields including secondary ions, but emission velocity and kinetic energy distributions have been scarcely measured except for secondary hydrogen and cluster ions from organic compounds, which gave insight into some aspects of the formation mechanism of the secondary ions [1-3]. For example, the energy distribution of hydrogen ions was explained by the microscopic charges in the ion track [3], and that of very large organic ions emitted in directions far from the target surface normal are in agreement with the value obtained by the pressure-pulse model [1].

We have measured yields and emission energies of secondary ions emitted from an SiO_2 tightly bound chemical compound and found that cluster ions are formed with a well defined low emission energy and their yields are characterized by a cluster-size dependent power function of the electronic stopping power, supporting the ion track model [4]. The aim of the present work is to extend the measurement of emission energies of individual secondary ions to various target species conductive to insulating and to reveal the dynamic mechanism of secondary ion emission in an MeV energy range.

EXPERIMENTAL

Measurements of the emission energies of secondary cluster ions were carried out in a similar way described previously using a conventional time-of-flight (TOF) technique at the Kyoto University 1.7-MV tandem Cockcroft-Walton accelerator facilities [4]. A heavy ion beam was chopped every 100 μs to a width of 50 ns and was incident at an angle of 60° with respect to the target surface normal. The resulting secondary ions were detected with a channel electron multiplier (Ceratron) [5] set on the axis of the surface normal. The flight time was analyzed with a multi-stop time-to-amplitude converter. Data were taken typically in a vacuum of 2×10^{-6} Pa for Al, Si, Al_2O_3, and SiO_2 targets in an energy range of 1 to 5 MeV for Si projectiles. The front surface of each target was purified by bombarding with intense continuous beam for 30 min before each 90-min data taking. The insulating targets were grown on conductive specimens and were thin enough for the ions to penetrate through the insulating layer. This made certain that the target was free from the macroscopic electrical charging-up as verified in previous experiments [4].

RESULT AND DISCUSSION

Axial Emission Energies

Figure 1 shows typical examples of the TOF spectra in a low mass region for the Al, Si, Al_2O_3, and

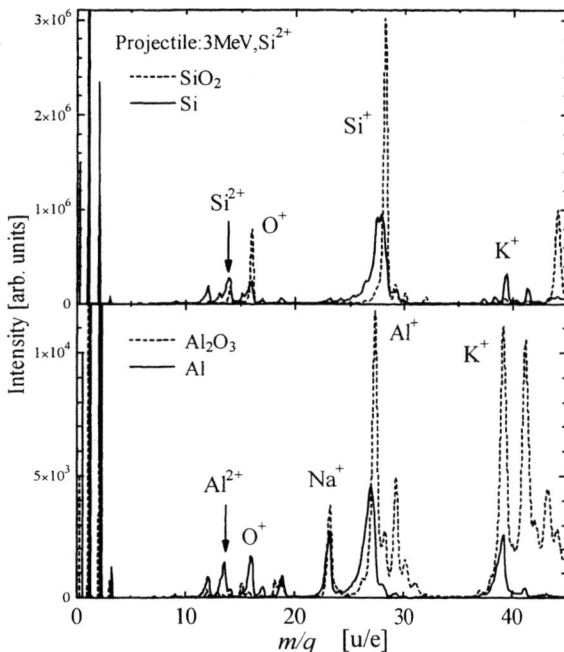

Figure 1. Examples of mass spectra of secondary ions in a low mass region for the Al, Si, Al$_2$O$_3$, and SiO$_2$ targets. The intensities are shown relative to a projectile fluence.

SiO$_2$ targets bombarded by 3 MeV Si ions. Singly and multiply charged atomic ions of the constituting elements were identified in the spectra. In addition to atomic species, cluster ions were observed in the case of the insulating targets though they are not shown in Fig. 1. Hydrogen, F, C, Na, K, and hydrocarbons originated from contaminants on each target surface. On the mechanism of the cluster ions we previously discussed basing on the yield and emission-energy dependence on species, energy, and stopping power of incident particles [4], and here we focus on the secondary atomic ion formation process. Though the obtained mass spectra resemble with one another at first glance, it should be noted that multiply charged Al and O ions from the Al$_2$O$_3$ target are missing in the spectrum in contrast with the dominance of those species in the SiO$_2$ spectrum.

Axial emission energy distributions of O^{q+}, Al^{q+}, and Si^{q+} secondary ions (qe: electric charge) produced from the targets are shown in Fig. 2. The kinetic energy distribution depends strongly on the target species. In the case of Al and Si, it is strongly asymmetric extending to a high-energy side with an exponentially decaying component and broadens gradually with increasing electric charge of secondary ions. It is wider for Si than for Al. The distributions of the singly charged ions from the Al$_2$O$_3$ and SiO$_2$ targets have symmetric shapes except for tiny tails on

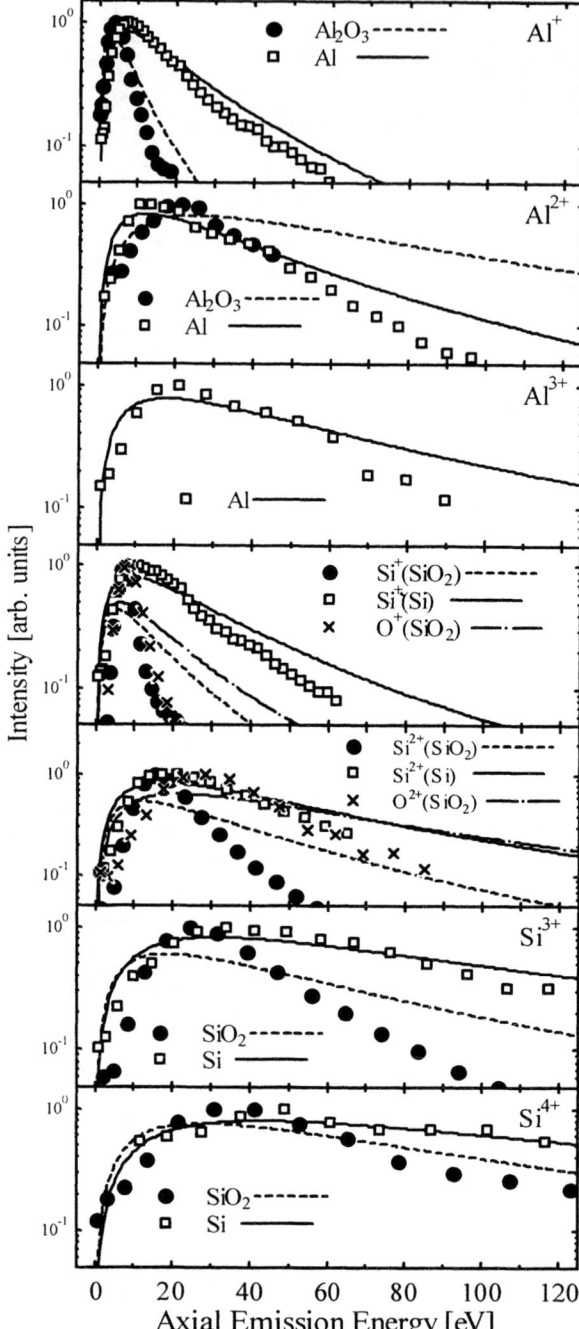

Figure 2. Axial emission energy distributions of secondary O^{q+}, Al^{q+}, and Si^{q+} ions (qe: electric charge of secondary ions) for the Al, Si, Al$_2$O$_3$, and SiO$_2$ targets bombarded by 3 MeV Si ions. The lines show the results calculated with Thompson-Sigmund model.

the high-energy side and are very narrow compared with those for the conductive targets. Any tails were not observed in the case of the secondary cluster ions. The kinetic energy of the doubly charged ions of O^{2+},

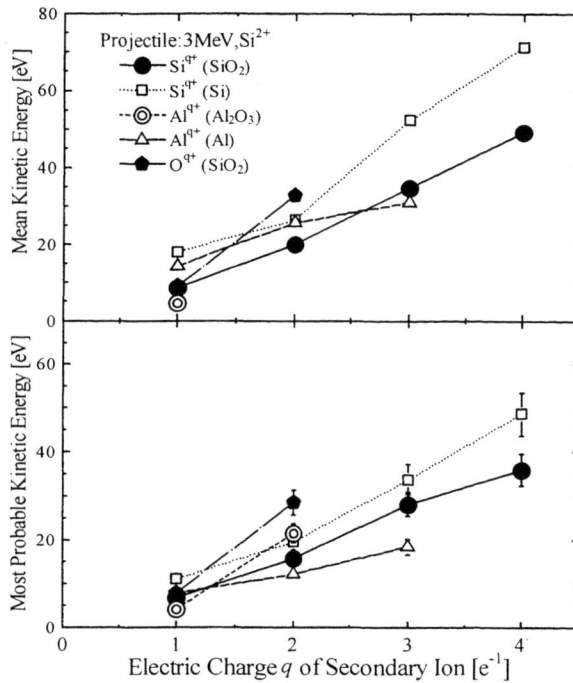

Figure 3. The most probable and mean energies of the axial emission energy distributions of O^{q+}, Al^{q+}, and Si^{q+} ions are plotted vs. the electric charge qe of the secondary ions for the Al, Si, Al_2O_3, and SiO_2 targets bombarded by 3 MeV Si ions. The lines are drawn to guide the eye.

Al^{2+}, and Si^{2+} from those insulating targets increases abruptly from the value of the singly charged ions. The solid curves in Fig. 2 show the Thompson-Sigmund calculation of the linear cascade model successfully applied in sputtering in a low energy region [6,7]. Though the rough reproduction of the shapes between the measured and calculated distributions could be achieved, the obtained fitting parameter corresponding to the planar surface barrier potential was unreasonably high. Therefore it is hard to apply the Thompson-Sigmund model to the emission process of atomic ions in the MeV energy range.

The resultant most probable and the mean energies are plotted in Fig. 3 as a function of the electric charge of secondary ions. The obtained values are high compared with those of cluster ions and the impurities such as Na, K and hydrocarbon ions, whose values are around 4-6 eV. The most probable kinetic energy increases overall with increasing the electric charge. The value is the highest for the Si^{q+} ion species from the Si target. The kinetic energy of Si ions from the SiO_2 target increases with the electric charge of the secondary ions at a same rate as for the Si target but the increment of the kinetic energy of the Al ions from the Al target is rather weak. Therefore, though the most probable kinetic energy is lower for the single-charged Si ions from the SiO_2 target than that for the Al ions from Al, the value becomes higher for the multiple-charged ions from the SiO_2 target than those for Al. In the case of O ions from SiO_2 the most probable value of O^{2+} jumps three times from the value of O^+. The value of Al^+ from the Al_2O_3 targets is the lowest among the values obtained in the present experiment. O^{q+} and Al^{q+} ions of $q \geq 2$ were hardly produced from the Al_2O_3 target.

The general trend of the mean kinetic energy is similar to that of the most probable kinetic energy except for a few facts. That is, owing to the asymmetric nature of the kinetic energy distribution for the conductive materials, the mean energy of Al ions from Al becomes higher than those of the atomic ions from the insulators.

Secondary Atomic Ion Yields

Ratios of respective atomic ion yields to those of doubly charged ions are plotted in Fig. 4 as a function of incident particle energy. Emission mechanism of singly charged ions depends so much on the projectile-target system. Therefore, the yields of doubly charged ions were used as the references. For comparison, the ratios of some clusters are also shown. As shown in our previous work the cluster ion yields

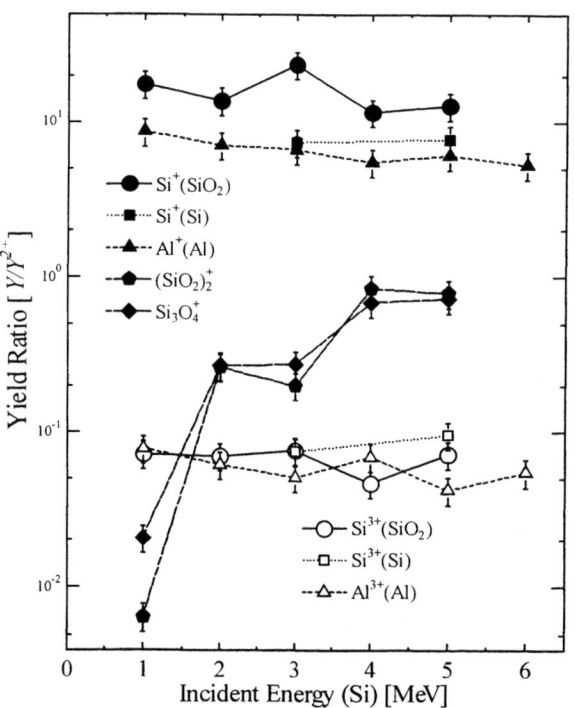

Figure 4. Yield ratios of secondary ions relative to the yield of the corresponding $q=2$ ions are plotted as a function of incident energy for the Al, Si, and SiO_2 targets.

nonlinearly depend on the projectile energy and they are described well by power functions with exponents depending on the cluster size [4]. However, the yields of the atomic ions show weak dependence on the incident energy, and the yield ratios between ions with $q=3$ and 2 are independent not only of the incident energy but also of the target species. The ratios for $q=1$ to 2 have similar independence as far as the conductive targets concerned. The yield of Si^+ from SiO_2 is enhanced when compared with the conductive targets. It is pointed out that, in the cases of alkali halides and SiO_2 irradiated with laser, electrons, and highly charged ions, sputtering results from the formation and successive decay of electronic defects such as self-trapped excitons and self-trapped holes [8-10]. Then, Si^+ ions could be formed from SiO_2 via this process and could result in the symmetric narrow distribution with a very low energy compared with the singly charged ions from the conductive targets as shown in Fig. 2. The same emission process as that for the conductive targets probably causes the small asymmetric component in the distribution.

Emission Process of Atomic Ions

The important observed facts are that the emission energies of the secondary atomic ions from the conductive targets are very high compared with those of the clusters and impurity ions, their distributions have strongly asymmetric shapes exponentially decreasing at the high-energy side, and their most probable and mean energies are proportional to their electric charge. These facts cannot be explained by the linear cascade model of Thompson-Sigmund without using unreasonable surface potential energies nor by a single collision between a projectile and a target atom resulting in ionization and emission because of the present detection angle of 120°. It was found in previous measurements that H ions produced from organic solids at the MeV energy region have high axial emission energies compared with the values for other impurity atoms and parent molecular ions [3]. This phenomenon was explained by the microscopic charges in the ion track. That is, when the fast ion interacts with the solid, secondary electrons are ejected from the ion track region, and the resultant very short-lived positively charged spot then accelerates the ejected ions. This feature has been clearly observed in the present experiment as the linear dependence of the emission energy on the electric charge of the atomic ions. The microscopic charging up in the track region is neutralized by electrons drifting with velocities around 10^6 m·s^{-1} near track region and stands for a very short period of about 10^{-15}-10^{-14} s. Then, kinetic energies given to stationary atoms by the repulsive Coulomb force before the neutralization of the track region is too low for the atoms to override the potential barrier. However, target atoms on the surface recoiled by a projectile at an angle close to 90° can get kinetic energies of a few eV and their M shell electrons are ionized at the same time. These moving ionized target atoms are accelerated more to the backward direction by the repulsive Coulomb force from other ionized target atoms in the track region and can get out the surface within 10^{-15}-10^{-14} s. The Coulomb repulsive energy is estimated to be about (10 to 15)×q eV on an average when the nearest lying atoms are assumed to be doubly ionized. It is expected then that the exponentially decaying component in the higher energy site of the emission energy distribution shows the neutralization life of the ionized track region. That is, the fact that the exponentially decaying tail extends to a higher energy in the Si target than in the Al target probably mean that the neutralization proceeds faster in the conductive Al target than in the semi-conductive Si target. This simple mechanism initiated by the simultaneous recoiling and ionization is supported by the independence of the yield ratios on incident energy and target species as shown in Fig. 4. That is, in the conductive materials the nuclear collision still plays an important role in the emission process of secondary ions in the MeV energy region and the yield of secondary atomic ions depends both on the nuclear and stopping powers as described in our previous report [11].

SUMMARY

The axial emission energy distribution has been measured for O^{q+}, Al^{q+}, and Si^{q+} secondary ions produced from the Al, Si, Al_2O_3, and SiO_2 targets in the electronic collision dominant MeV energy region. The emission energies of secondary atomic ions from conductive targets are very high compared with those of the clusters and impurity ions. Their distributions are strongly asymmetric extending to a high-energy side with an exponentially decaying component, and their most probable and mean energies are proportional to their electric charge. The observed facts can be explained by the simple combined mechanism of the simultaneous recoiling and ionization by the projectile and the Coulomb repulsion by the short-lived ionized track region.

ACKNOWLEGEMENTS

This work was done with the Experimental System for Ion Beam Analysis at Kyoto University. We thank A. Itoh, K. Yoshida, and K. Norizawa for their useful advice and technical supports during the experiments.

It has been supported in part by a Grant-in-Aid for Scientific Research from the Ministry of Education, Science, Sports and Culture of Japan.

REFERENCES

1. Fenyö, D., Sundqvist, B.U.R., Karlsson, B.R., and Johnson, R.E., *Phys. Rev. B* **42**, 1895 (1990).
2. Papaléo, R.M., Brinkmalm, G., Fenyö, D., Eriksson, J., Kammer, H.-F., Demirev, P., Håkansson, P., and Sundqvist, B.U.R., *Nucl. Instr. Meth. B*, **91**, 667 (1994).
3. Wien, K., Koch, Ch., and Tan, Nguyen van, *Nucl. Inst. Meth. B*, **100**, 322 (1995).
4. Imanishi, N., Kyoh, S., Takakuwa, K., Umezawa, M., Akahane, Y., Imai, M., and Itoh, A., Proc. of Int. Conf. on Accelerator Application in Research and Industry, Denton, 1997, p. 507; Imanishi, N., Kyoh, S., Shimizu, A., Imai, M., and Itoh, A., *Nucl. Instr. Meth. B* **135**, 424 (1998); Imanishi, N., Shimizu, A., Ohta, H., and Itoh, A., Proc. of Int. Conf. on Accelerator Application in Research and Industry, Denton, 1999, p. 396; Imanishi, N., Ohta, H., Ninomiya, S., and Itoh, A., *Nucl. Instr. Meth. B* **164-165**, 803 (2000).
5. Murata Mgf. Co., Ltd. in Japan.
6. Sigmund, P., *Sputtering by Particle Bombardment I*, Berlin, Springer, 1981, p. 9.
7. Thompson, M.W., *Philos. Mag.* **18**, 377 (1968).
8. Menzen, D., and Gomer, R., *J. Chem. Phys.* **41**, 3311; Redhead, P., *Can. J. Phys.* **42**, 886 (1964).
9. Itoh, N., *Nucl. Instr. Meth. B* **122**, 405 (1997).
10. Schenkel, T., Newman, M.W., Niedermayr, T.R., Machicoane, G.A., McDonald, J.W., Barnes, A.V., Hamza, A.V., Banks, J.C., Doyle, B.L., and Wu, K.J., *Nucl. Instr. Meth. B* **161-163**, 65 (2000).
11. Kyoh, S., Takakuwa, K., Sakura, M., Umezawa, M., Itoh, A., and Imanishi, N., *Phys. Rev. A* **51**, 554 (1995).

Diffusion Of TiN Into Aluminum Films Measured By Soft X-ray Spectroscopy

T. M. Schuler[a], D. L. Ederer, and N. Ruzycki
Department of Physics, Tulane University, New Orleans, Louisiana 70118

G. Glass and W. A. Hollerman
Acadiana Research Laboratory, University of Louisiana at Lafayette, P.O. Box 44210, Lafayette, Louisiana 70504

A. Moewes
Center For Advanced Microstructures and Devices, CAMD at Louisiana State University, 6980 Jefferson Hwy., Baton Rouge, Louisiana 70803[b]

M. Kuhn
Digital Equipment Corporation, 77 Reed Rd., Hudson, Massachusetts 01749[c]

T. A. Callcott
Physics and Astronomy Department, University of Tennessee, Knoxville, TN 37996

Abstract. Understanding the atomic bonding properties at the interface between thin films is crucial to a number of key modern technical devices, including integrated circuits, magnetic disk read/write heads, batteries, and solar cells. Semi-conducting materials such as titanium nitride (TiN_x) are widely used in the manufacturing of modern electronics, requiring a wealth of information about its electronic structure. We present data from soft x-ray emission and absorption experiments involving a sample consisting of a 40 nm TiN layer on top of an aluminum film 550 nm thick. Soft x-ray emission spectroscopy (XES) and near-edge x-ray absorption fine structure (NEXAFS) spectroscopy are tools that provide a non-destructive, atomic site-specific probe of the interface, where the electronic structure of the material can be mapped out element by element. From these measurements, we show that the Ti and the N diffuse into the Al film to form an equivalent material depth of about 4.5 nm, and the NEXAF structure reveals that the nitrogen has probably formed AlN, and the Ti has also diffused to form a titanium-aluminum compound.

INTRODUCTION

The study of interactions between thin films has become a matter of increasing importance over the past few years as the demand for innovative materials has intensified greatly, most notably in microcircuit industry[1]. The manner in which thin films react with each other is an important consideration for the use of thin films in devices. The study presented here concerns the reactions on the interface between titanium nitride thin films and a film composed of 99% aluminum and 1% copper.

Titanium nitride (TiN_x) is a unique material because of its unusual combination of properties. Its basic structure is that of a face-centered cubic lattice. The properties attributed to TiN_x consist of a high level of hardness, high thermal conductivity, an immunity to wear and corrosion, chemical inertness, and a resistance to atomic diffusion within its matrix. These properties suggest a partially filled band and a chemical bond consisting of metallic, covalent, and ionic character simultaneously[2]. The properties of TiN_x also lead to a great number of applications that include more efficient UHV equipment[3], and photo-

---

[a] email address: tschule@tulane.edu
[b] current address: Department of Physics and Engineering Physics, University of Saskatchewan, Saskatoon, SK, S7N 5E2, Canada
[c] current address: 2501 229th Ave., Hillsboro, OR 97124

thermal conversion of solar energy[4]. It is also used for diffusion barriers in microcircuits[5], which is interesting because our data shows that TiN itself seems to diffuse.

X-ray emission spectroscopy (XES) and x-ray absorption spectroscopy (XAS) are useful tools in measuring the electronic states of both bulk and thin film systems. Using these spectroscopic methods, one can examine the properties of the thin film/substrate interface in a non-destructive manner, and they can provide a probe of the interface that is atomic site-specific[6]. This permits the material's electronic structure to be studied in a manner that maps out the different components present in the material element by element[7]. The electronic structure of titanium, along with a variety of titanium compounds, has been extensively studied[2,7,8] by means of soft x-ray emission and absorption spectroscopy.

In this work we present data from XAS and XES measurements on titanium nitride thin films and their interaction with an aluminum/copper intermediate layer. We also use a model of the x-ray fluorescence intensity as a function of the exciting photon energy across the titanium $L_{2,3}$ absorption edge in TiN to discuss migration of Ti and N into the substrate. Our main point of interest in performing these experiments was whether a significant change in composition occurred in any of the materials at the interface, which could be attributed to either the preparation temperature or the presence of the other layers of the sample.

EXPERIMENT

The soft x-ray fluorescence and emission measurements were made at Beamline 8.0 of the Advanced Light Source located at Lawrence Berkeley National Laboratory, which is an undulator beamline equipped with a spherical grating monochromator as described by Jia et al.[9]. The resolving power for this monochromator was set at $E/\Delta E = 500$, and the Rowland circle grating spectrometer located in the fluorescence end station provides a resolving power of about 300 at an energy of 400 eV. A more detailed description of this type of experimental process is provided by Jia, et al.[9].

A diagram of the sample used is shown in Fig. 1. Measurements were taken on samples annealed at temperatures of 410°C and 450°C with the TiN over-layer present, then the over-layer was removed and the measurements were repeated in an attempt to reach the TiN sub-layer. The removal of the over-layer was accomplished by a selective wet chemical etch. Both the Al(Cu) and TiN films were sputter-deposited using an Al(Cu) or Ti target. In the case of the TiN films, nitrogen was introduced during the sputter deposition process.

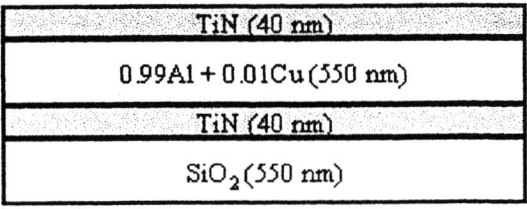

FIGURE 1. Diagram of the sample used. Measurements were also performed on samples with the TiN over-layer etched away.

The samples were energy calibrated at different photon excitation energies with hexagonal boron nitride, to calibrate the data taken at the N K-edge, and TiO_2, to calibrate the data taken at the Ti $L_{2,3}$-edge, using the previously reported measurements of Tegeler et al.[10] and Jimenez-Mier et al.[11]. All of the data was normalized by dividing the fluorescence emission intensity by the intensity of the incoming radiation, derived from the emitted current of a gold mesh.

RESULTS

If we set our fluorescence spectrometer to the N K-edge emission at 400 eV, we will also record Ti emission from the $2p^5 3s^2 V$ level to the $2p^6 3s^1 V$ level. The overlap of these Ti transitions with the N K-edge emission produces the structure at the Ti $L_{2,3}$-edge due to the excitation of a titanium 2p core electron, as shown in Fig. 2.

We modeled the obtained fluorescence intensity from the titanium nitride cross section, by using transmission data calculated by the Center For X-Ray Optics (CXRO) at Lawrence Berkeley National Laboratory[12] as a starting point. We then applied the equation for the intensity developed by Jaklevic, et al.[13]. In our model we took the fluorescence yield of titanium to be one-tenth of that of nitrogen. This model, shown in Fig. 3, provides an excellent qualitative match to our data, as can be seen by comparing the titanium and nitrogen peaks in Fig. 3 with the data plots shown in Fig. 2.

Now that we have a reasonable model that accounts for the major spectral features, we can point out a few three TiN samples. In plot (a), we see the absorption

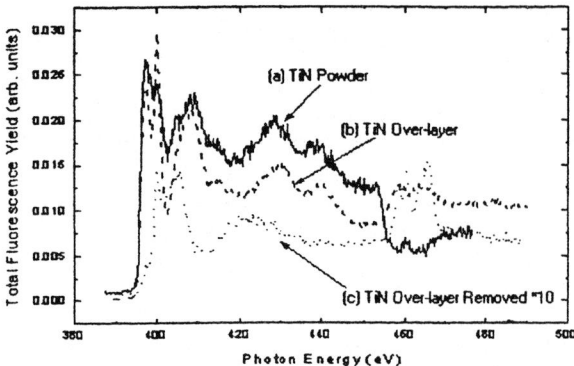

FIGURE 2. NEXAFS fluorescence measurements for fine structure of bulk TiN powder, plot (b) is that of our layered sample with the TiN over-layer present, and plot (c) shows our layered sample with the TiN over-layer etched away. The intensity of plot (c) has been multiplied by a factor of ten to allow for comparison with the other plots.

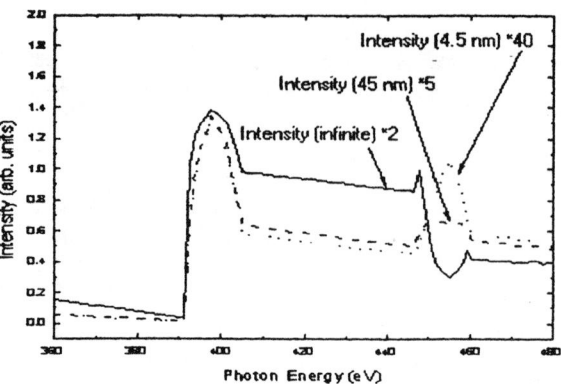

FIGURE 3. Adjusted model of fluorescence intensity of TiN sample accounting for self-absorption in titanium, assuming $Y_{Ti} = Y_N / 10$.

features within the data. First of all, the near edge x-ray absorption fine structure (NEXAFS) for bulk TiN matches very well with the thin film of TiN (Fig. 2). Secondly, both N and Ti remain after the TiN over-layer is removed, when one should not observe either element. Thirdly, the NEXAFS structure at the N K-edge of the etched sample is very different than that of bulk TiN. Later we will show that the nitrogen K-edge NEXAFS structure is very closely matched to that of AlN. Finally, we are led to the conclusion from our model comparison in Fig. 3 to Fig. 2(c), that the N and Ti diffuse into the Al to an extent that could be equivalent to a film of thickness 4.5 nm.

The soft x-ray emission spectra obtained from our sample with the TiN over-layer present, show very little change between the bulk TiN sample and that of the thin film over-layer at the two different temperatures. The similarity of these thin film over-layer spectra implies that the presence of the substrate has very little, if any, influence on the TiN layer presiding above it. Changes occur in the spectra at both the N K and Ti $L_{2,3}$-edges when the over-layer is removed, as shown in Figs. 4(a) and 4(b). The emission at the N K-edge closely resembles that of the TiN bulk powder. The major difference occurs at the Ti $L_{2,3}$-edge. (Fig. 4(b)).

In an attempt to explain the variation from TiN at the Ti L-edge, we compare our data to previous measurements on hexagonal aluminum nitride as reported by Lawniczak-Jablonska et al.[14] shown in Fig. 4. There is convincing agreement between the AlN data and the data from our sample with the TiN over-layer removed, shown in Fig. 4(b) which is a strong indication that the nitrogen from the TiN layer has diffused into the aluminum to form an AlN material.

Indirect evidence also suggests that titanium has diffused into the aluminum. Measurements taken at the Ti $L_{2,3}$-edge confirm the presence of titanium in the etched sample. The signal from the etched sample is weaker than that of the TiN over-layer by about a factor of five, whereas we would expect this signal to be reduced by a factor of 25 if the emission were coming from the TiN sub-layer. Therefore, we conclude that all of the emission measured at the Ti $L_{2,3}$-edge of the etched sample must be coming from the aluminum intermediate layer, implying that titanium has diffused into this layer as well as the nitrogen. It is also possible that the etchant may not have completely removed the TiN film.

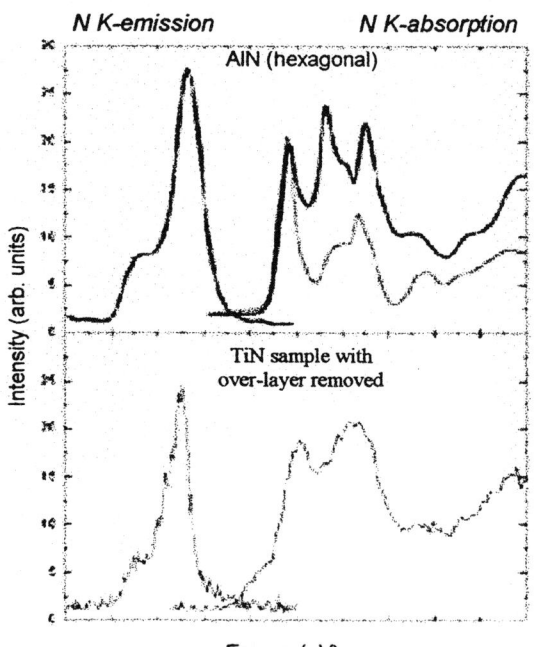

FIGURE 4. Our measurements performed on the sample with the TiN over-layer (bottom plot), compared with previously measured emission and absorption data (top plot) performed on hexagonal AlN adapted from Ref. 15. The N K-edge emission peaks (left) compare very well with the AlN measurements at that energy. The zero energy point on the x-scale is aligned with the maximum of the SXE emission.

CONCLUSIONS

From our data analysis, we conclude that on our multi-layer sample the TiN thin film over-layer has electronic properties that are very similar to those of bulk TiN.

We also found that when the TiN over-layer was removed, the aluminum layer below it changed in a manner closely resembling that of aluminum nitride. From this we conclude that the nitrogen from the TiN layers had diffused into the intermediate layer, which has been substantiated by RBS measurements. XES measurements indicate that titanium from the over-layer has also diffused into the aluminum. RBS data showed that after the TiN over-layer had been removed from the samples, a very thin layer of titanium remained after etching. Our data on the etched sample also suggested that emission from the TiN sub-layer was probably not being detected.

ACKNOWLEDGMENTS

This work was partially supported by the National Science Foundation Grant No. DMR-9801804, a DOE-EPSCor Cluster Research Grant No. DOE-LEQSF (1993-1995)-03, and the Science Alliance Center of Excellence Grant from the University of Tennessee. The Advanced Light Source is supported by the Office of Basic Energy Sciences, U.S. Department of Energy under contract No. DE-AC03-765f00098.

REFERENCES

1. W. Monch, *Semiconductor Surfaces and Interfaces* (Springer, New York, 1993)

2. D. W. Fischer and W. L. Baun, J. Appl. Phys. **39** (1968), 4757.

3. M. Minato and Y. Itoh, J. Vac. Sci. Technol. A **13** (1995), 540.

4. Y. Claesson, M. Georgson, A. Roos, and C. -G. Ribbing, Solar Energy Mater. **20** (1990), 455.

5. M. Eizenberg, MRS Bull. **20** (1995), 38.

6. D. L. Ederer, J. A. Carlisle, J. Jimenez, J. J. Jia, K. Osborn, T. A. Callcott, R. C. C. Perera, J. H. Underwood, L. J. Terminello, A. Asfaw, and F. J. Himpsel, J. Vac. Sci.Technol. A **14** (1996), 859.

7. L. D. Finkelstein, E. Z. Kurmaev, M. A. Korotin, A. Moewes, B. Schneider, S. M. Butorin, J-H. Guo, J. Nordren, D. Hartmann, M. Neumann, and D. L. Ederer, Phys. Rev. B **60** (1999), 2212.

8. J. Barth, F. Gerken, and C. Kunz, Phys. Rev. B **31** (1985), 2022.

9. J. J. Jia, T. A. Callcott, J. Yurkas, A. W. Ellis, F. J. Himpsel, M. G. Samant, J. Stohr, D. L. Ederer, J. A. Carlisle, E. A. Hudson, L. J. Terminello, D. K. Shuh, and R. C. C. Perera, Rev. Sci. Instrum. **66** (1995), 1394.

10. E. Tegeler, N. Kosuch, G. Weich, and A. Faessler, Phys. Stat. Sol. (b) **84** (1977), 561.

11. J. Jimenez-Meier, J. van Ek, D. L. Ederer, T. A. Callcott, J. J. Jis, J. Carlisle, L. Terminello, A. Asfaw, and R. C. C. Perera, Phys. Rev. B **59** (1999), 2649.

12. B. L. Henke, E. M. Gullikson, and J. C. Davis. *X-ray interactions: photoabsorption, scattering, transmission, and reflection at E = 50 – 30,000 eV, Z = 1-92*, Atomic Data and Nuclear Tables. **54**(2) (1993), 181.

13. J. Jaklevic, J. A. Kirby, M. P. Klein, A. S. Robertson, G. S. Brown, and P. Eisenberger, Solid State Commun. **23** (1997), 679.

14. K. Lawniczak-Jablonska, T. Suski, I. Gorczyca, N. E. Christensen, K. E. Attenkofer, E. M. Gullikson, J. H. Underwood, D. L. Ederer, R. C. C. Perera, and Z. Liliental Weber, Phys. Rev. B **61** (2000), 16623.

SECTION V

DETECTORS AND SPECTROMETERS

An Anti-Compton Suppression Ge-Telescope Detection System for Quality Control of Nuclear Waste Packages

S. Agosteo[a], B. Chabalier[b], A. Foglio Para[a], U. Graf[c], N. Huot[b], T. Kekki[d], A. Ravazzani[c], P. Schillebeeckx[c], V. Tanner[d], A. Tiitta[d].

[a] Dipartimento di Ingegneria Nucleare, Politecnico di Milano, via Ponzio 34/3, 20133 Milano, Italy,
[b] Commisariat a l'Energie Atomique, Departement d'Enterposage et de Stockage de Dechets, Centre d'Etudes de Cadarache, 13108 Saint Paul lez Durance, France, [c] EU, JRC Ispra, Institute for Systems, Informatics and Safety, 21020 Ispra, Italy, [d] VTT Chemical Technology, P.O. Box 1404, FIN-02044 VTT, Finland.

Abstract. An anti-Compton suppression system is studied for the quality control of radioactive waste packages by non-destructive assay. The main objective is the reduction of the detection limit of actinides in the packages. The optimization of a final device is based on Monte Carlo simulations (MCNP and FLUKA) validated by experiments using a prototype consisting of a Ge-telescope detector surrounded by a NaI detector. The validation reveals that most of the discrepancies between experimental and simulated data are due to an incomplete description of the experimental conditions. After fine-tuning of the input file the uncertainties on the simulated full-energy peak efficiency are reduced to less than 5 %. Also the total detector response for mono-energetic photons and real waste, including the photon interactions within the drum, can be simulated satisfactorily.

INTRODUCTION

The characterization of waste drums for radionuclide identification and quantification is often based on the detection of photons spontaneously emitted by the material of interest [1]. A dominant problem for the analysis of waste packages containing actinides is the identification of full-energy peaks of low energy (from Pu and Am) superimposed on a high background due to the presence of fission products and activated fuel cladding material. The minimum detectable activity for such waste packages can be improved by optimizing the detection system and the procedures of measurement and analysis. As recommended in Ref. [2], such an optimization requires an analysis of the complete measurement process.

Sources creating the background continuum can be grouped into two classes: ambient and source-induced radiation. The former can be largely reduced by passive shielding and by the use of spectrometers fabricated with low-background materials. The source-induced background is originated from Compton scattering of higher energy photons in the materials of the measurement set-up (detector, source matrix, etc.). The reduction of the component scattered in the detector can be obtained by active Compton suppression methods [3,4]. Different principles for the reduction of the Compton continuum exist [5,6], in this work the so-called anti-Compton spectrometer was used.

Such a spectrometer was studied for the characterization of waste packages containing plutonium. The gamma rays emitted by the actinides of interest are in the low energy region (50-500 keV) and require a high spectral resolution. The background radiation is mainly due to photons following the decay of ^{137}Cs and ^{60}Co, which undergo Compton scattering not only in the detector but also in the drum and in the environment. A strong collimating geometry is required for the detector to avoid dead time effects caused by this intense photon field. As already mentioned, background reduction using a Compton Suppression Spectrometer (CSS) will only be effective if the Compton scattering in the detector is the dominating contribution. Therefore, before any optimization study, the various background contributions have to be identified.

The present work is in the framework of a project aiming at improving the minimum detectable activity of waste packages [7]. The influence of the main background contributions on the response of an existing CSS will be discussed here. The CSS response functions to isotopic point sources and waste packages were calculated with the Monte Carlo method using the MCNP [8] and FLUKA [9] codes and the results were validated experimentally.

DETECTOR DESIGN

The CSS used in this work is a simplified version of that described in ref. [10]. The system (Fig. 1) consists of a Ge-telescope detector, a NaI guard detector and a passive Pb-shield, including Cd and Cu liners (not covering the central collimator hole). The primary role of the liners is to reduce dead time effects and accidental coincidences due to the detection of lead x-rays in the guard detector. Liners covering the collimator hole are only useful in very intense photon fields of low-energy.

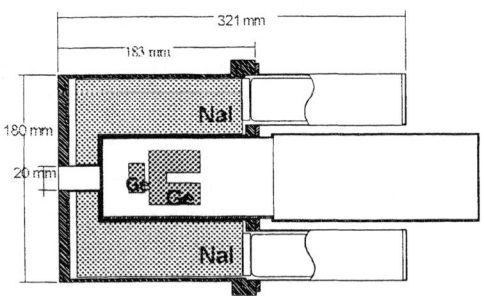

FIGURE 1. The lay-out of the detector system.

The Ge-telescope is made of two closely packed Ge-detectors, a planar n-type (Canberra GL0515R, diameter 25.5 mm, and 15 mm thick) and a coaxial p-type detector (Canberra GC1518, diameter 47 mm, length 46 mm) mounted in one cryostat housing (Canberra 7600/S). The NaI guard detector together with the coaxial detector acts as a Compton suppressor for the planar detector. Similarly the guard detector and the planar detector act as a Compton suppressor for the coaxial.

When the Monte Carlo technique is applied properly, is a very powerful tool for simulating the detector response. One of the most determining factors is the correctness of the input geometry reflecting the measurements conditions. Therefore, the supplier's design specifications were verified by transmission tomographic measurements. A comparison of the geometry based on the tomograph with that provided by the supplier showed significant differences [7]: a larger distance between the end cap and the planar detector and a difference in the mounting structure of the planar detector. A possible displacement (1-2 mm) of both the planar and the coaxial crystals from the central axis of the cylindrical housing was observed by a two-dimensional scanning measurements with a ^{137}Cs point source.

MEASUREMENTS

Experimental response functions of the detector system were obtained for quasi-monoenergetic photon beams from ^{99m}Tc, ^{51}Cr, ^{137}Cs, ^{65}Zn. ^{51}Cr and ^{65}Zn were produced in the thermal neutron field at the reactor of VTT (Finland). The Compton suppression factors are shown in Fig. 2 for ^{137}Cs.

The measurements emphasized that: (a) the background continuum between the Compton edge and the full energy peak in the spectra of the coaxial detector is caused mainly by forward scattered events in the planar detector; (b) the region between the Compton edge and the full energy peak for the planar detector is hard to suppress. This continuum originates from scattering in the inactive material and from multiple scattering events in the detector with a low energy escaping photon. These are hard to suppress when inactive material is present between the central and the guard detector; (c) the suppression factor as a function of angle for the coaxial detector does not depend strongly on the energy of the incoming photon and therefore also not on that of the scattered one. The observed low suppression factor for the coaxial detector is not caused by a limited thickness of the guard detector but is due to scattering in inactive material; (d) the suppression factor for a fixed angle in the planar detector increases with energy of the primary and consequently of the scattered photon. An increase in energy of the scattered photon increases the probability that it reaches the guard detector without absorption in the inactive material; (e) the effect of the coaxial detector as a guard detector for forward scattering events in the planar one is visible on the high suppression factor for the planar detector in the low energy region

The main limitations of the suppression factors of the CSS under study are due mainly to the presence of inactive material. For the coaxial detector the use of the planar as a veto is not effective: scattering events are firstly created and afterwards mostly suppressed. The coaxial detector is effective to reduce the events forward scattered in the planar.

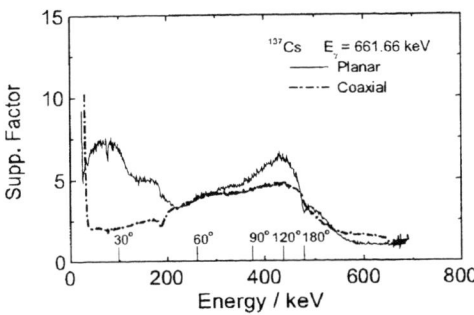

FIGURE 2. The measured suppression factors for ^{137}Cs.

Further measurements were performed to evaluate the influence of the Compton scattering on the environment surrounding the measurement set-up [7]. A reduction of the suppression factor was observed. A shield covering externally the guard detector must be used to reduce the strong influence of events scattered in the environment. When a waste drum is analyzed, this passive shield also contributes to reduce the influence of photons reaching the detector after having scattered inside the matrix.

MONTE CARLO SIMULATIONS

As already mentioned, the Monte Carlo simulations were performed with the MCNP and FLUKA codes. The geometry deduced from the tomographic measurements was reproduced strictly. For the simulation of coincident or anticoincident spectra the FLUKA code has the necessary options available in the tally DETECT. Such options are not available in the MCNP code in its original version. Therefore, a special routine was implemented at CEA Cadarache [7]. The response functions of the CSS in the available coincidence modes were calculated both for the point isotropic sources used in the measurements described above and for a waste drum. A comparison between the experimental and simulated (both with MCNP and FLUKA) detection efficiency of the full-energy peak is shown in Fig. 3. The data refer to isotropic point sources at 1 m from the cryostat endcap. No dead layer was considered for the planar detector in these MCNP simulations, while FLUKA results refer also to a dead layer 0.5 mm thick with a diameter of 8.5 mm.

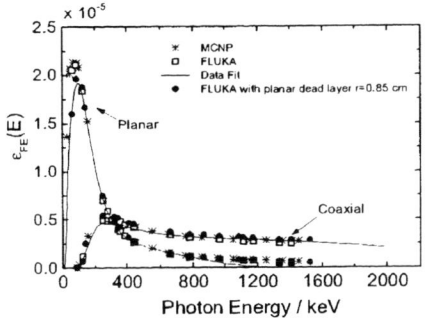

FIGURE 3. Experimental and simulated full-energy efficiency.

The agreement of the MCNP with the FLUKA results (no planar dead layer) is satisfactory. The discrepancy with the experimental data in the low energy region is reduced strongly when the dead layer of the planar detector is considered. As its dimensions were not known accurately, they were assessed with a set of simulations aiming at finding the agreement with the experimental data at low energies by varying the dead layer diameter, while keeping its thickness fixed at 0.5 mm (provided by the supplier).

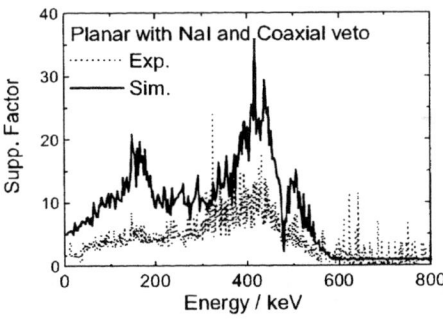

FIGURE 4. Measured and simulated suppression factors of the planar detector for a ^{137}Cs point source at 1 m from the cryostat end cap.

A comparison between simulated and experimental suppression factors for a ^{137}Cs point source at 1 m from the cryostat endcap is shown in Fig. 4. The simulated data do not account for any resolution broadening or incomplete charge collection effects. Besides a systematic underestimation of the Compton continuum, the simulations reproduce all the features of the experimental response. For the planar detector, the largest discrepancies are observed when the veto signal of the coaxial detector is also activated. This underestimation of the Compton continuum reflects in an overestimation of the suppression factor and indicates the presence of inactive material between the planar and the coaxial detectors neglected in the input file. Additional inactive material, such as an increased dead layer thickness, would also explain the discrepancies for the coaxial detectors shown in Fig. 5, which reports the influence of the ambient scattering and the dead layer thickness of the coaxial detector on the suppression factor, as results from FLUKA simulations. The performance of a not properly shielded CSS is reduced largely by the presence of a scattering wall at 50 cm from the spectrometer. Generally, these evaluations were repeated for different energies and interpolated by dedicated procedures.

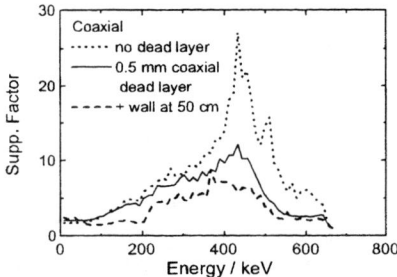

FIGURE 5. The influence of the thickness of the coaxial dead layer and the presence of scattering material on the suppression factor for a ^{137}Cs point source.

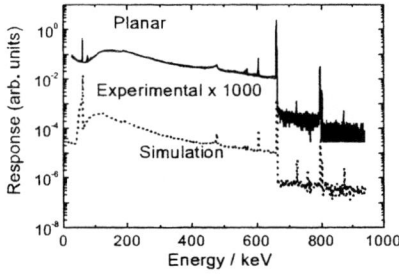

FIGURE 6. Experimental and simulated response (planar detector) for a bituminized waste drum.

A direct Monte Carlo simulation of the complete detector response for a waste drum measurement is very time consuming. To reduce CPU time the calculations were split into two parts. In a first stage the photon energy distribution in front of the detector was calculated with MCNP. Subsequently, this distribution was used as a source file for FLUKA and the response of the detection system was calculated. Such a procedure does not preserve the photon direction, but at large distances as in the present case, the parallel beam approximation provides accurate results as shown in Fig. 6. The figure shows the experimental and simulated response for a bituminized drum containing ^{241}Am, ^{134}Cs, ^{137}Cs and ^{154}Eu. The same agreement holds for the coaxial detector.

CONCLUSIONS

The measurements and simulations here discussed were very useful guidelines for the design of an optimized CSS for the characterization of waste packages [7]. In particular, it has been emphasized that: (a) a dual system (central + guard detector) provides better experimental data than a Ge-telescope; (b) the presence of inactive material in the detector should be reduced as much as possible; (c) the system should be shielded against environmental scattering effects.

ACKNOWLEDGEMENTS

This work was supported by the European Union under contract F14W CT 96 0037.

REFERENCES

1. Birkhoff, G., *Monitoring of Plutonium Contaminated Solid Waste Drums: A Technical Guide to Design and Analysis of Monitroing Systems*, EUR 10026, 1985, pp. 129-220.
2. Currie, L. A., and Austin, K., *Analyt. Chem. Letters* **40**, 586-593 (1968).
3. Verplancke, J., *Nucl. Instr. Meth.* **A312** 174-182 (1992).
4. Paulus, T.J., Keyser R.M., *Nucl. Instr. Meth.* **A286** 364-368 (1990).
5. Aspacher, B., Coldwell, R.L., Rester, A.C., *Nucl. Instr. Meth.* **A330** 243-253 (1993).
6. Palms, J.M., Wood, R.E., Puckett, O.H., *IEEE Trans. Nucl. Sci.* **NS-15** 397- (1968).
7. Schillebeeckx, P., Ravazzani, A., Graf, U., Tomin, R.J., Foglio Para, A., Agosteo, S., Tiitta, A., Kekki, T., Tanner, V., Chabalier, B., Artaud, J.L., Huot, N., *Quality Control of Nuclear Waste Packages with a Compton Suppression and Ge-telescope Detection System*, JRC Ispra Tech. Note I.00.78, 2000.
8. Briestmeister, J.F., *MCNP-A General Monte Carlo N-Particle Transport Code–version 4A*, LA-1262-M, 1993.
9. Fassò, A., Ferrari, A., Ranft, J., Sala, P.R., "New Developments in FLUKA Modeling of Hadronic and EM Interactions" in Proc. 3rd Workshop on Simulating Accelerator Radiation Environments (SARE3), edited by H.Hirayama, KEK Proceedings 97-5, 1997, pp. 32-44.
10. Verplancke, J., Schoenmaeckers, W., Hesselink, W.H.A., Hacquebord, A., Penninga, J., Stolk, A., *IEEE Trans. Nucl. Sci.* **NS-33** 340-342 (1986).

The Argonne Fragment Mass Analyzer and Measurements of Entry Distributions

A. Heinz[a], T.L. Khoo[a], P. Reiter[b], I. Ahmad[a], P. Bhattacharyya[c], J. Caggiano[a], M.P. Carpenter[a], J.A. Cizewski[d], C.N. Davids[a], W.F. Henning[a], R.V.F. Janssens[a], G.D. Jones[e], R. Julin[f], F.G. Kondev[a], T. Lauritsen[a], C.J. Lister[a], D. Seweryniak[a], S. Siem[a], A.A. Sonzogni[a], J. Uusitalo[f], I. Wiedenhöver[a]

[a] *Argonne National Laboratory, Physics Division, 9700 South Cass Avenue, Argonne, Il 60439, USA,*
[b] *Ludwig Maximilian University Munich, Sektion Physik, Am Coulombwall 11, D-85748 Garching, Germany,*
[c] *Purdue University, 1396 PHYS, W. Lafayette, IN 47907, USA,*
[d] *Department of Physics and Astronomy, Rutgers University, New Brunswick, New Jersey 08903, USA,*
[e] *Department of Physics, University of Liverpool, Liverpool L69 7ZE, England,*
[f] *Department of Physics, University of Jyväskylä, Finland*

Abstract. The Argonne Fragment Mass Analyzer (FMA) is designed to separate and identify evaporation residues according to their mass-to-charge ratio. The FMA in combination with GAMMASPHERE - an array of 100 compton-suppressed germanium detectors - allowed for a number of very interesting in-beam gamma-spectroscopy studies, as the FMA provided a very clean trigger on evaporation residues to obtain extremely background-free gamma-spectra. This setup was used to measure the total gamma-energy and multiplicity after particle evaporation - the so-called entry distribution - by exploiting the calorimetric properties of GAMMASPHERE, using its germanium detectors as well as its BGO shields for a maximum gamma efficiency. The entry distribution can be used to estimate the height of the fission barrier as a function of angular momentum. This method is especially favorable for unstable nuclei, for which the fission barrier is otherwise very difficult to measure. Here, the entry distributions of ^{220}Th at beam energies of 206 MeV and 219.5 MeV in the ^{176}Yb(^{48}Ca,4n) reaction are presented. The results are compared to a previous measurement.

INTRODUCTION

The Argonne Fragment Mass Analyzer (FMA) (1) at the ATLAS facility of Argonne National Laboratory is designed to separate and identify nuclear reaction products according to their mass-to-charge ratio. In the last two years GAMMASPHERE, an array consisting of 100 compton-suppressed germanium detectors, was installed at the target position of the FMA. This allowed for high-precision in-beam gamma spectroscopy. The FMA provided a clean trigger on the evaporation residues of interest, allowing GAMMASPHERE to detect almost background-free spectra of unstable isotopes, which were produced with cross sections down to several micro barn.

In this work, this setup was used to measure the multiplicity and the total energy emitted by gamma radiation, in order to determine the hight of the fission barrier of an unstable nucleus. The height of the fission barrier is an important property of heavy nuclei, controlling their stability against spontaneous fission. In particular the lifetimes and decay modes of super-heavy elements are determined by the height and shape of the fission barrier. This quantity for a wide range of nuclei is obviously important for the understanding of the structure of heavy nuclei.

The fission barrier plays also an important role in the production cross sections of heavy nuclei in fusion-evaporation and in fragmentation-type reactions. Moreover, the renewed interest in the production of neutron-rich nuclei via in-flight fission of a relativistic primary ^{238}U beam requires also a knowledge of the fission barriers of nuclei, which might fission after abrasion and ablation of some nucleons.

Experimentally, the fission barrier is often very difficult to determine. High-precision experiments to determine the effective height of the fission barrier require stable or long-lived targets for measurements of the fission probability as a function of the excitation energy. Fission induced by neutrons or light-charged particles is very well suited for this method, as the formation of the compound nucleus in these types of reactions is very well understood. Unfortunately, this is not the case for heavy-ion induced reactions.

One method to determine the height of the fission barrier of proton-rich light actinides and preactinides has been proposed by Grewe et al. (2), who used electromagnetic excitation of relativistic secondary beams. Here, the measured fission cross section was matched with a model calculation of the excitation process, with the fission barrier as the only free parameter. The agreement with earlier measurements using conventional techniques is very good. The disadvantage of this method is, besides its model-dependence, the fact that it is not useful for isotopes which cannot be produced in fragmentation reactions.

Another attempt is to extract the height of the fission barrier from beta-delayed fission (see e.g. (3)). Here a beta-unstable nucleus populates states in its daughter nucleus, which are above the fission barrier. This method possesses two drawbacks: first it is limited to isotopes that populate the above-mentioned states after beta-decay. A second complication is that the extraction of the fission barrier height requires a detailed knowledge of this population, which is then again model-dependent.

A new alternative approach to estimate the height of the fission barrier of ^{254}No was proposed by Reiter et al. (4). ^{254}No nuclei produced by the ^{208}Pb(^{48}Ca,2n) reaction were identified in a recoil separator. The recoil decay tagging technique confirmed that the mass identification is sufficient to separate the nobelium gamma-rays from a large background. The calorimetric properties of GAMMASPHERE were used to measure the total gamma-ray sum energy and multiplicity. The entry distribution (the two-dimensional distribution in spin and excitation energy after neutron evaporation) was obtained by correcting for the detector response. This method is applicable when the saddle-point energy (the fission barrier height plus the yrast energy) is below the neutron-separation energy. As the fission probability is expected to be dominant over the much slower deexcitation by gamma emission, the entry distribution is limited by the fission barrier, thus allowing an estimate of its height, which was in the case of ^{254}No at least 5 MeV.

One major advantage of this method is the fact that it is the only way to study the spin-dependence of the fission barrier. For ^{254}No, the data show the fission barrier to be remarkably stable against spin, which is attributed to the fact that the ground-state shell correction is predominantly responsible for the fission barrier.

To test this entry-distribution method for the determination of the fission barrier height, a measurement on a nucleus whose fission barrier has been determined with a different experimental technique is necessary. Unfortunately, nuclei which were studied with neutron or light-charged particle emission are not accessible for fusion-evaporation reactions. The ^{220}Th system was chosen, however, because this nucleus has been studied by Grewe et al. (2), using electromagnetic excitation, as described above. As theoretical expectations for the height of the fission barrier of this nucleus do not agree (see discussion in ref. (2) and references therein) it is also interesting from the theoretical point of view. The fission barrier of ^{220}Th is expected to show a much stronger spin dependence than for ^{254}No, as it is derived largely from a liquid-drop term that decreases with spin.

EXPERIMENT AND DATA ANALYSIS

The ATLAS accelerator at Argonne National Laboratory provided a ^{48}Ca beam, which impinged on a ^{176}Yb target with a thickness of 0.81 mg/cm^2. For the measurement, beam energies of 206 MeV and 219.5 MeV were chosen. The lower beam energy is centered at the maximum production cross section; the higher beam energy was used to test whether additional angular momentum could be brought into the system. Gamma radiation was detected using GAMMASPHERE, an array of 101 BGO-shielded germanium detectors. The recoils were selected and identified with the Argonne Fragment Mass Analyzer (FMA), using a position-sensitive parallel-grid avalanche counter (PGAC) to measure the mass-over-charge spectrum for identification of the residues. This identification was confirmed by implanting the recoils into a double-sided silicon strip detector (DSSD) and measuring the subsequent alpha-decay. Due to the short half-life of ^{220}Th (9.7 μs) the use of the recoil decay tagging (RDT) technique was not possible, as it would reduce the efficiency too much. In the case of the lower beam energy this is not a problem because ^{220}Th was the most dominant channel. For the higher beam energy the production rate of ^{219}Th is about as high as that of ^{220}Th, but mass identification was sufficient to separate the two products. Later, an RDT condition will be used to test whether the tails of the ^{219}Th peak in the focal-plane spectrum have an influence on the entry-distribution of ^{220}Th at all. Scattered beam particles at the focal plane were excluded by measuring the energy loss of the recoils in the DSSD as a function of the time-of-flight between the PGAC and the DSSD. In this spectrum, recoils and scattered beam particles were clearly separated. Using the above-mentioned conditions and additional conditions on the time-of-flight between the accelerator RF and the firing germanium or BGO detectors to suppress random events, the measured gamma sum-energies and the module multiplicity for ^{220}Th residues was obtained. A module is defined as one germanium detector of GAMMASPHERE with its surrounding BGO shields.

Unfolding and conversion from multiplicity to spin

The measured gamma sum-energy as well as the multiplicity have to be corrected for the response of GAMMASPHERE in order to obtain the total gamma energy and multiplicity. The sum-energy and multiplicity responses of GAMMASPHERE to 898-keV photons was measured with a ^{88}Y source, using an event-mixing technique (5). The response was used to unfold the measured data using a Monte-Carlo simulation procedure (6). The final energy-multiplicity distribution was also corrected for the multiplicity dependence of the trigger efficiency, as the trigger of GAMMASPHERE required a minimum of two detectors firing in coincidence.

To convert multiplicity into spin (I) we used the relation:

$$I = \Delta I * (m - m_{stat}) + \Delta I_{stat} * m_{stat} + I_{elec}$$

Here, ΔI_{stat} is the spin carried away by a statistical gamma ray, m_{stat} is the multiplicity of statistical gamma rays, ΔI and m are the respective quantities for non-statistical gamma rays. For the calculation the values $\Delta I_{stat} = 0.5$, $m_{stat} = 4$ and $\Delta I = 1.75$ were chosen. I_{elec}, the spin carried by conversion electrons, is, for the moment, taken to be zero, but needs to be added later (after the electron contribution has been determined).

RESULTS AND DISCUSSION

The measured entry distributions of ^{220}Th for the two beam energies are shown in figure 1. The two distributions look very similar, which is remarkable. The larger amount of fluctuations for the higher beam energy is due to the fact that the statistics for the entry distribution is about one order of magnitude lower than that for $E_{Beam} = 206$ MeV. It is significant that the maximum spin appears to be about $20\hbar$ in both cases, which suggests that this is the maximum angular momentum that this nucleus can withstand. The entry distribution does not follow the yrast line as might be expected, but is tilted with respect to it. This is similar to the results obtained for ^{254}No (4). This behavior is not understood, as the phase space for low energies and low angular momenta is expected to be small. Reiter et al. (4) proposed the emission of pre-compound neutrons as an explanation.

Concerning the excitation energy, the entry distribution does not reach below the yrast line, as it is expected. The saddle point energy shown in figure 1 is the sum of the fission barrier and the yrast energy (9). This saddle-point energy is calculated as the sum of the ground-state shell effect (7) and a spin-dependent liquid-drop component (8). The entry distribution reaches well beyond the boundary given by the saddle-point energy, which indicates a fission barrier higher than the expected value.

The neutron separation energy shown is the sum of the neutron-separation energy at spin zero (7) and the yrast energy. The neutron separation energy is not a hard cut-off in excitation energy either, which is expected, as the neutrons carry kinetic energy and the phase-space close to the threshold is very small.

It is important to note that in ^{220}Th the fission barrier and the neutron separation energy are expected to be much closer than in the ^{254}No case. Thus, the upper limit of the entry distribution might be determined by either of the two.

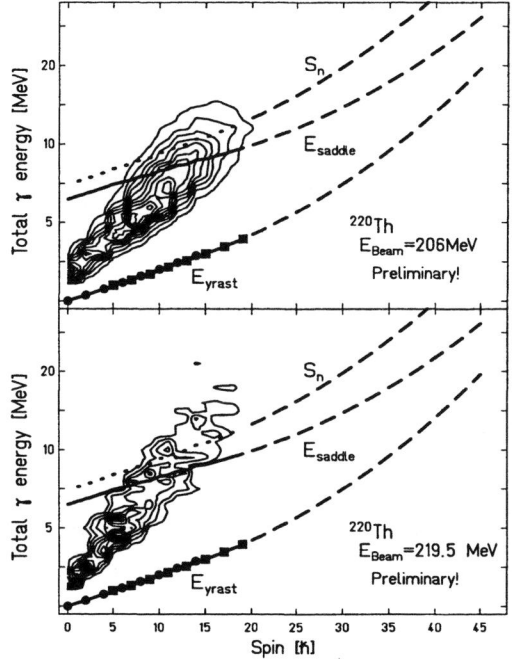

FIGURE 1. Preliminary entry distributions of ^{220}Th at a beam energy of 206 MeV and 219.5 MeV. In the figure, the yrast line, the neutron-separation energy S_n and the saddle-point energy E_{saddle} are shown. The saddle point energy is defined as $E_{saddle}(I) = E_{yrast}(I) + B_f(I)$, with $B_f(I)$ being the fission barrier at a given angular momentum I. $B_f(I)$ is calculated as the sum of a liquid drop component (8) and the ground-state shell effect (7). The dashed lines are extrapolations. The yrast line data have been taken from reference (9). The neutron separation energy shown is calculated according to $S_n(I) = S_n(I = 0) + E_{yrast}(I)$. $S_n(I = 0)$ is calculated by using masses from reference (7). The lowest cut in both figures corresponds to a limit of 20 in yield (arbitrary units).

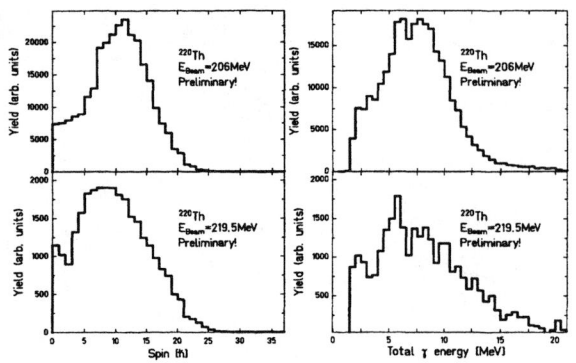

FIGURE 2. Projections of the entry distributions (preliminary data) shown in figure 1 with respect to spin and total gamma-energy.

In figure 2 projections of the entry distributions of figure 1 on the spin and energy axis are shown. The cutoff in spin at $20\hbar$ is again revealed. The excitation energy distribution drops less sharply for the higher beam energy and shows a rather slow decrease, which reflects the extra excitation energy in ^{220}Th at the higher beam energy.

The maximum excitation energy, which is the beam energy in the center-of-mass system plus the Q-value, is 14.9 MeV in the case of $E_{Beam} = 206$ MeV and 25.1 MeV at $E_{Beam} = 219.5$ MeV. The maximum excitation energies were calculated for the center of the target. The maximum excitation energy of the entry distribution at $E_{Beam} = 206$ MeV and at $E_{Beam} = 219.5$ MeV are comparable, which means that the average kinetic energy of the neutrons is significantly higher in the latter case as expected due to the higher excitation energy.

To obtain an estimate for the fission barrier the half-maximum values at a given spin were used. The data suggest a fission barrier of at least 10 MeV at a spin of $15\hbar$, which is closer to the value of 9 MeV calculated by Pashkevich (10) for spin zero.

Further analysis is in progress to confirm the multiplicity-spin conversion and to compare our results with a statistical model calculation. A subtraction of random events will be done as well. For the future it would be very interesting to study the entry distribution of ^{220}Th by using different projectile-target combination in order to gain a deeper understanding of the reaction mechanism.

ACKNOWLEDGEMENTS

This work has been supported by the U.S. Department of Energy under contract Nos. W-31-109-ENG-38 and DEFG05-88ER40411 and by the National Science Foundation.

REFERENCES

1. C.N. Davids *et al.*, Nucl. Instr. Meth. **B70** (1992) 358.
2. A. Grewe *et al.*, Nuc. Phys **A614** (1997) 400.
3. H.L. Hall and D.C. Hoffman, Annu. Rev. Nucl. Part. Sci. **42** (1992) 147.
4. P. Reiter *et al.*, Phys. Rev. Lett. **84** (2000) 3542.
5. M. Jääskeläinen et al., Nucl. Instrum. Methods Phys. Res. **204** (1983) 385.
6. Ph. Benet, Ph.D. thesis, L'Universite Louis Pasteur d Strasbourg, CRN/PN 88-29, 1998.
7. P. Möller *et al.*, At. Data Nuc. Data Tab. **59** (1995) 185
8. A.J. Sierk, Phys. Rev. **C33** (1986) 2039.
9. B. Schwarz, Ph.D. Thesis, Univ. Heidelberg, 1998.
10. V.V. Pashkevich, Int. School Seminar On Heavy Ion Physics, ed. E.V. Ivashkevich and B. Koliesova, (INP Dubna, Dubna 1983), p. 119.

Neutron and Simultaneous Gamma Detection With LiBaF$_3$ Scintillator

P. L. Reeder and S. M. Bowyer

Pacific Northwest National Laboratory
Richland, WA 99352

Abstract. Pulse shape discrimination techniques using the scintillator LiBaF$_3$ allow very clean separation of densely ionizing radiation (protons, tritons, alphas, etc.) from less densely ionizing radiation (electrons). The pulse shape discrimination is based on the presence of sub-nanosecond core-valence luminescence for gammas and electrons. We use a pulse discrimination based on integration of the signal over a short time interval compared to integration over a long time interval. Thermal neutron capture events, fast neutron capture events, and fast neutron elastic scattering events on Li have distinctive loci in a two-dimensional plot of the short gate pulse heights versus total gate pulse heights. Applications of this scintillator include measurements of fast to thermal neutron ratios, fast neutron spectroscopy, and alpha contamination, all in the presence of significant gamma radiation.

INTRODUCTION

The detection of neutrons in the presence of significant gamma radiation is often required in arms control, material accountability, and nuclear smuggling scenarios as well as in basic nuclear research. The new scintillator material LiBaF$_3$ offers the possibility of measuring neutron count rates and energy spectra while measuring gamma count rates and spectra simultaneously using a single detector.

The scintillation properties of LiBaF$_3$ have been reported by researchers at Delft University of Technology and by others.[1,2,3,4,5] This scintillator has both core valence luminescence (CV) and self-trapped-exciton luminescence (STE) under gamma irradiation whereas only the STE luminescence is present under neutron or alpha irradiation. Because of the high energy of the CV photons, the luminescence from LiBaF$_3$ must be detected by a quartz window photomultiplier tube. Relatively simple pulse shape analysis techniques can be used to obtain excellent neutron/gamma discrimination.

In a previous paper, we reported some preliminary results comparing light output as a function of dopant concentrations.[6] In this paper, we present examples of pulse height spectra for gamma, neutron, and alpha radiations.

EXPERIMENTAL

The results reported here are based on a crystal obtained from AC Materials[7] that was 1.9-cm diam. by 0.95-cm high having a volume of 2.7 cm^3. The crystal was drawn from a melt containing 1 mol% Rb and 1 mol% Ce. The concentration of dopants in this crystal has not been measured, but from our previous work we expect the Rb concentration to be about 0.5 mol% and the Ce concentration to be about 0.005 mol%. Although the Ce concentration is very small, it plays an essential role in the pulse shape analysis for neutron/gamma discrimination.[6]

The data acquisition system consists of a multi-input Charge-to-Digital Converter (QDC) interfaced to a multiparameter data acquisition computer. The PMT signal is sent to two separate inputs to the QDC. Each input is gated separately to record the amount of charge in portions of the pulse. A "short" gate of 60 ns starting 20 ns before the pulse records the fast component of the scintillation light due to the CV and Ce^{+3} fluorescence. The combined short and long components are recorded by a "total" gate of about 1.4 µs.

A two-dimensional array of the "short" signal versus the "total" signal gives excellent separation of the neutron induced events from the gamma induced events. A region of interest that includes only neutron events can be defined for the two-dimensional array. Events that fall within the neutron region have their "total" amplitude recorded in a separate one-dimensional histogram. Likewise, events in the gamma region have their "total" amplitude recorded in a separate histogram. The two regions are defined such that there is no overlap. Neutron capture events in ^{6}Li give pulses which fall well within the neutron region.

Gamma Pulse Height Spectra

An example of the two-dimensional histogram of "short" versus "total" pulse heights is shown in Figure 1 for a ^{137}Cs source. The photopeak for the 662-keV gamma from ^{137}Cs shows up as a distinct island beyond the Compton continuum in the region labeled "Gamma". A weak distribution to the left of the "Gamma" region is attributed to pulses originating directly from the photocathode of the PMT rather than the scintillator.

The pulses within the "Gamma" region can be collapsed to the "total" axis to give a one-dimensional histogram of the gamma pulse height as shown in Figure 2. The Compton continuum is rather large relative to the photopeak because of the small size of the crystal. The resolution of the photopeak is about 17% FWHM. Spectra of other gamma sources having photopeaks between 123 keV and 834 keV were measured. The calibration curve of photopeak energy versus pulse height was linear.

Thermal Neutron Spectra

Pulse height spectra were obtained for ^{252}Cf and ^{239}PuBe neutron sources with various shields and moderators around the LiBaF$_3$ scintillator. This

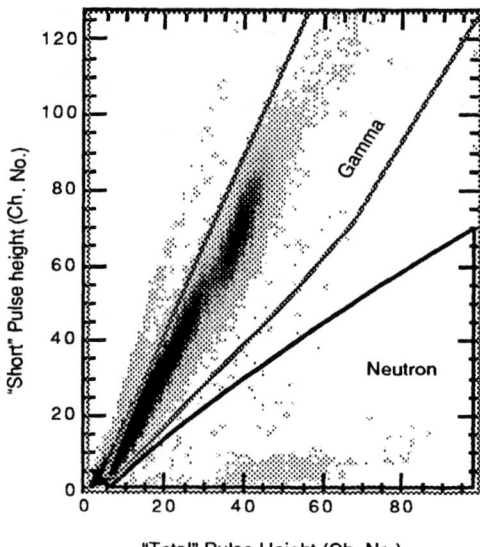

FIGURE 1. Two-dimensional histogram for ^{137}Cs source. Gamma and neutron regions are outlined. Note location of 662-keV photopeak at XY channels (39,69).

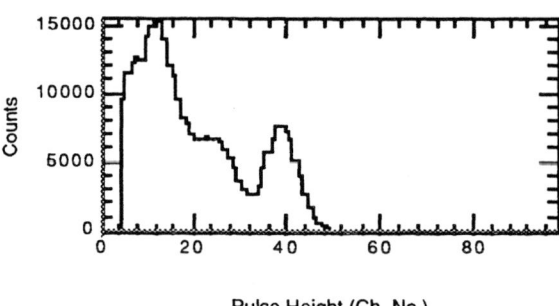

FIGURE 2. Pulse height spectrum of gammas from ^{137}Cs in LiBaF$_3$ scintillator. Photopeak of 662-keV gamma is in channel 39.

allowed interpretation of the data in terms of thermal neutron response and fast neutron response. Capture of thermal neutrons in ^{6}Li produces a triton and an alpha with a total kinetic energy of 4.79 MeV.

The two-dimensional histogram for a ^{252}Cf source is shown in Figure 3. The scintillator and PMT were surrounded by polyethylene to thermalize the neutron spectrum. In addition, a 1.27-cm thick lead shield was placed between the source and the detector to reduce the number of gamma rays reaching the scintillator.

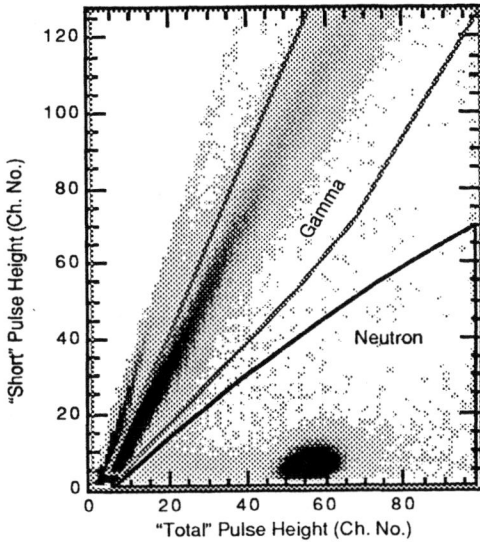

FIGURE 3. Two-dimensional histogram for ^{252}Cf source with neutron moderator. Note location of thermal neutron peak at XY channels (56,6).

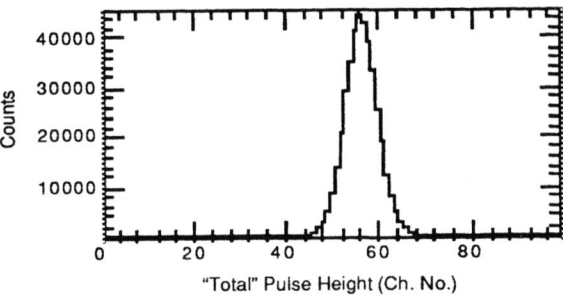

FIGURE 4. Pulse height spectrum for thermal neutrons from ^{252}Cf source.

The one-dimensional histogram of the "total" pulse height for neutrons shows a distinct peak as shown in Figure 4. The peak is centered at channel 56.3 and has a FWHM of 8.2 channels corresponding to an energy FWHM of 0.59 MeV and a resolution of 12.2%. Although the alpha and triton share 4.79 MeV of kinetic energy, the observed pulse height corresponds to only 0.96-MeV electron equivalent energy (based on the gamma calibration curve).

Background

A few counts always appeared in the neutron region even if no source of neutrons was present. To determine whether this background was due to internal alpha activity or to thermal neutrons from the surroundings, the pulse height spectra were measured with two different sets of shielding and moderation. One background was taken with the same polyethylene moderator as for the thermal neutron measurement discussed above. A second background was taken without the polyethylene moderator and with the scintillator and PMT covered with a cadmium shield to remove thermal neutrons from external sources. The neutron spectra for these two shielding conditions are shown in Figure 5.

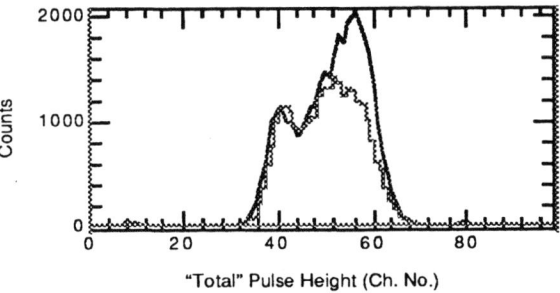

Figure 5. Pulse height spectra in neutron region for backgrounds taken with different shields. The smooth curve is for the measurement with the polyethylene moderator. The histogram curve (light gray) is for the measurement with the Cd shield and no polyethylene.

The difference between the two background pulse height spectra is equivalent to the thermal neutron spectrum shown in Figure 4. and is thus due to thermal neutrons produced externally to the scintillator. The majority of the background pulses are due to an internal radioactivity. Because Ba is a major constituent of the scintillator and is frequently contaminated with Ra, we assume that the internal radioactivity is due to alphas from decay of ^{226}Ra and its daughters. The lowest energy peak in Figure 5 is due to 4.78-MeV alphas from ^{226}Ra. The broad peak around channel 54 is a mixture of 5.49-MeV alphas from ^{222}Rn and 6.00-MeV alphas from ^{218}Po.

Fast Neutron Capture and Elastic Scattering on Li

Although the cross sections for fast neutron capture on ^{6}Li are much lower than the thermal neutron cross section, the kinetic energy of the fast neutron will add to the reaction Q value and give pulse heights greater than the pulse height due to thermal neutron capture. Scattering of fast neutrons on ^{7}Li can produce a recoil Li ion with a maximum energy of 0.438 times the neutron energy. Such pulses should appear at pulse heights below the thermal neutron peak.

In Figure 6, the pulse height spectrum in the neutron region is shown for a PuBe source. The scintillator and PMT were surrounded by cadmium to minimize the contribution of thermal neutrons. The peaks from channels 35-65 are due to the internal alpha contamination as shown by the normalized background spectrum also shown in Figure 6.

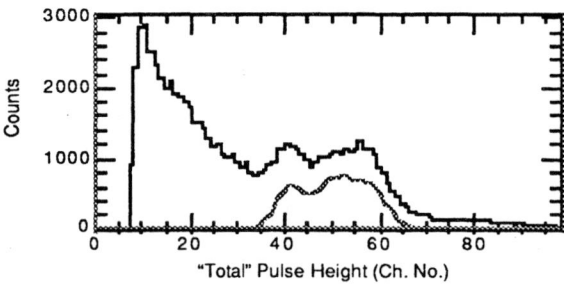

FIGURE 6. Upper histogram is pulse height spectrum for PuBe source with Cd shield. Lower curve is normalized background spectrum.

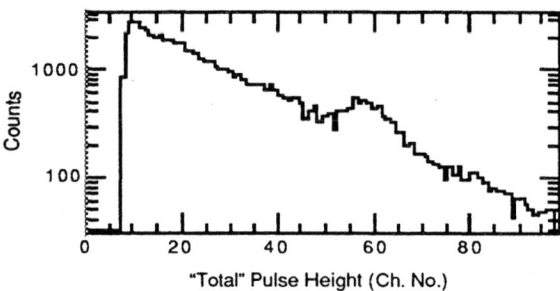

FIGURE 7. Net pulse height spectrum for PuBe source after subtraction of background. Thermal neutrons were eliminated by Cd shield.

The net spectrum after subtracting the background is shown in Figure 7. The PuBe source has a spectrum of neutrons with energies up to 10 MeV. It is thus possible to have Li ion recoils with energies up to the energy of the thermal neutron peak. The cross section for elastic scattering is higher than for fast neutron capture so the net spectrum is dominated by the elastic scattering at lower pulse heights. The excess of counts with pulse heights above the thermal peak is attributed to the fast neutron capture reaction on ^{6}Li. These conclusions have been verified by a Time-of-Flight experiment reported elsewhere using a tagged ^{252}Cf source.[8]

CONCLUSIONS

The Ce-doped LiBaF$_3$ scintillator shows excellent discrimination between gamma induced events and charged particle induced events. Pulse height spectra for both types of reactions can be obtained simultaneously from a single detector. Thermal neutrons can easily be identified based on their capture in ^{6}Li. It is also possible to detect fast neutrons and obtain a low resolution neutron energy spectrum based on the ^{6}Li capture reaction. Fast neutrons can also be detected with higher yields by their elastic scattering on ^{6,7}Li. However, the elastic scattering reaction is more difficult to convert to a neutron energy spectrum.

REFERENCES

[1] Knitel, M. J., Dorenbos, P., de Haas, J. T. M., and van Eijk, C. W. E., *Nucl. Instr. Meth. Phys. Res. A* **374**, 197 (1996).

[2] Combes, C. M., Dorenbos, P., van Eijk, C. W. E., Gesland, J. Y., and Rodnyi, P. A., *J. Luminescence* **72-74**, 753 (1997).

[3] Combes, C. M., Dorenbos, P., Hollander, R. W., and van Eijk, *Nucl. Instr. Meth. Phys. Res. A* **416**, 364 (1998).

[4] Gektin, A., Shiran, N., Voloshinovski, A., Voronova, V., and Zimmerer, G., *IEEE Trans. Nucl. Sci.* **45**, 505 (1998).

[5] Baldochi, S. L., Shimamura, K., Nakano, K., Mujilatu, N., *J. Crystal Growth* **200**, 521 (1999).

[6] Reeder, P. L. and Bowyer, S. M., Proceedings of "Methods and Applications of Radioanalytical Chemistry - MARC V", Kailua-Kona, HI, April 9-14, 2000, to be published in *J. Radioanal. Nucl. Chem.*

[7] AC Materials, 2721 Forsyth Road, Suite 264, Winter Park, FL 32792.

[8] Reeder, P. L. and Bowyer, S. M., Proceedings of IEEE Nuclear Science Symposium, Lyon, France, Oct. 15-20, 2000.

Pulse-Height Spectrum Measurement Experiment for Code Benchmarking: Initial Results[*]

K.E. Sale, J.M. Hall and C.M. Brown

Lawrence Livermore National Laboratory, P.O. Box 808, Livermore, CA 94551-9900

Abstract. We have completed a set of gamma-ray pulse-height benchmark experiments using a high-purity germanium detector to measure absolute counting rate spectra from ^{60}Co, ^{137}Cs and ^{57}Co isotopic sources. The detector was carefully shielded and collimated so that the geometry of the system was well known. The measured absolute pulse-height spectrum counting rates are compared to energy-deposit spectra calculated using the Monte Carlo radiation transport code COG [1, 2]. We present here a small subset of our results. The agreement between the calculated and measured spectra and known sources of discrepancies will be discussed.

INTRODUCTION

Monte Carlo radiation transport codes are in wide use in research and industry throughout the world. Although the codes have been well tested in some specific applications, the tests have generally been against integral data, *e.g.* dose, total energy deposit or activation. In this paper we present initial results for a photon transport benchmark experiment testing the capability of a code to predict absolute differential results, specifically pulse-height counting rate spectra in a high-purity germanium detector. No scaling factors have been employed to help the code predictions match the experimental data.

THE EXPERIMENT

In our design of the experiment we wanted to create a geometry that was realistic and yet simple enough that it could be modeled accurately and completely. The main tactics we used were to minimize the mass of supporting structures and carefully collimate both the detector and source. A second goal was to provide data that would test the quality of code predictions over a wide range of materials (*i.e.* range of atomic number) and a wide range of penetration depth. The materials used for collimation around the source ranged from carbon (Z=6) to tungsten (Z=74). Material thickness ranged from essentially zero to over 500 mean free paths through the thickest parts of the source collimator. The third goal was to perform a test with no "adjustments" or normalization factors. To this end, we used sources with known, NIST traceable strengths. A photograph of the experimental set-up is shown in Figure 1. Four different source collimators were used. They were made of graphite, aluminum, iron and HeviMet (a tungsten alloy). The source holder and collimator could be rotated keeping the source position fixed. Pulse-height spectrum counting rates were measured using ^{60}Co, ^{137}Cs and ^{57}Co isotopic gamma-ray sources mounted in the throat of the collimator which could be rotated to angles of 0°, 15°, 35°, and 90° relative to the source-detector axis (48 histograms in all plus background measurements).

At 0° most of the gamma flux was unscattered with the exception of a contribution from small angle scattering from the walls of the collimators. This case tested the quality of the physics simulation within the bulk of the germanium detector. At 15° the source was just obscured from the detector position with a small part of the throat of the collimator still visible. This arrangement was meant to be sensitive to small angle scattering. The 35° setting was chosen to maximize the amount of material between the source and the detector, thereby testing the accuracy of the predictions in a deep penetration situation. At 90° the gamma flux at the detector was a mixture of both direct and highly-scattered photons.

---

[*] Work performed under the auspices of the U.S. Department of Energy by the Lawrence Livermore National Laboratory under Contract W-7405-Eng-28

Figure 1. Photograph of the experimental apparatus. From upper left to right, the alignment scope, LN$_2$ dewar, steel detector collimator, graphite source collimator and plastic source holder are visible. Each piece is mounted on a low mass aluminum support.

Data acquisition was done using a standard MCA system (Canberra *InSpector* Multi-Channel Analyzer run with Canberra's Genie-2000 software) into 4096 channels. We used the same system gain for the ^{60}Co (E$_\gamma$= 1.1732, 1.3325 MeV) and ^{137}Cs (E$_\gamma$= 0.66166 MeV) sources and a higher gain for the ^{57}Co (E$_\gamma$= 0.12206, 0.13627 MeV) source. We also measured background spectra with each source collimator at each system gain. These spectra were scaled to the same live time as the source spectra and subtracted from the measured data to produce the spectra used to compare with code predictions.

THE PREDICTIONS

The pulse-height spectrum predictions were made using the Monte Carlo radiation transport code COG [1, 2]. The model for the germanium detector used as-built dimensions for the specific detector we used (a 70% relative efficiency high purity Germanium detector from Ortec). It has been our experience that it is nearly impossible to over emphasize the importance of using a sufficiently accurate and complete geometric model to achieving correct results. COG-generated pictures of the geometry model are shown in Figure 2.

In our simulations we used the NIST traceable source strengths from the manufacturer of the sources and the known source ages to make absolute predictions of the energy deposit spectrum counting rates (counts per second per MeV of energy deposited). No scaling factor was used to compare the predictions to the measured data.

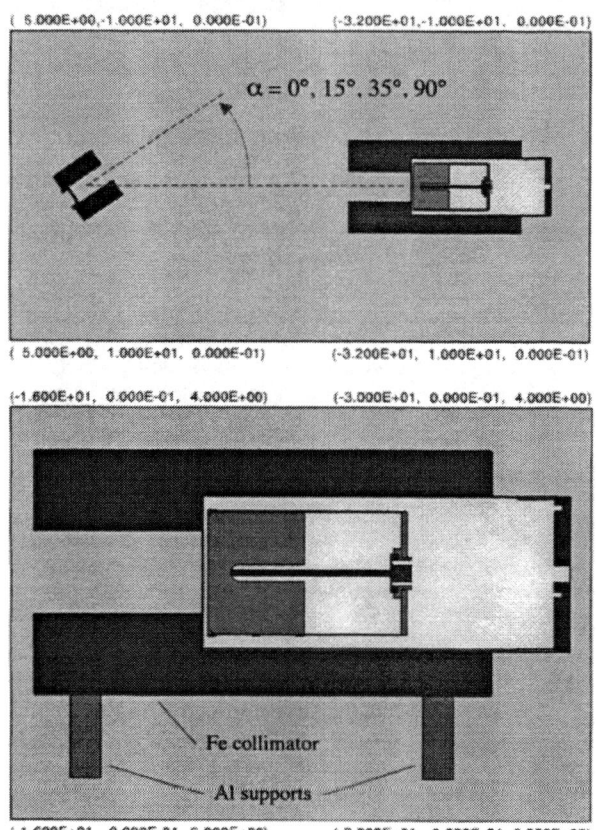

Figure 2. Cross sectional pictures of the geometry model generated using the COG Monte Carlo transport code.

The detector used in this experiment was well collimated so that secondary electrons generated in the Ge crystal were unlikely to escape; therefore, in this case, tracking of charged particles was unnecessary. Any charged particles generated were treated with the zero-range approximation and the energy they would have carried deposited at the interaction site. The scattered photons were tracked using cross sections derived from the EPDL data base [3].

The Monte Carlo simulations were done as follows. For each source an energy grid was chosen with very narrow bins around the photopeaks and other expected features in the spectrum and much wider bins over the rest of the spectrum, which was expected to be more or less featureless. In this way, good statistical convergence could be achieved everywhere along with good energy resolution where it was needed.

The final step in generating the predicted pulse-height spectra was to convolve the energy-deposit histograms with a detector response function. We chose to use a Gaussian peak shape with an energy-dependant width inferred from the observed widths of about a dozen peaks in the measured background spectra. The fact that the energy grid used in the calculations was different from that used in the measurements *and* was non-uniform made the convolution with the experimental (electronic) peak shape slightly more complicated.

THE RESULTS

Due to the space constraint for this paper we have chosen to show only the data and predictions for the ^{60}Co spectrum with a iron source collimator rotated to 15° and a ^{57}Co spectrum with an aluminum collimator rotated to 15°. These two cases are neither the easiest nor the most difficult to predict. The agreement between the COG simulation and the experimental data is excellent as seen in Figure 3. The overall agreement (in the full data set) between experiment and code predictions is excellent. There is a slight systematic discrepancy between the ^{137}Cs measurements and predictions, possibly due to an inaccurate source strength. Also, the count rates were so low for the ^{57}Co source mounted in the higher Z collimators that backgrounds dominated the experimental data making it difficult to assess the quality of the predictions.

REFERENCES

1. T. Wilcox and E. Lent, *COG - A Particle Transport Code Designed to Solve the Boltzman Equation for Deep-Penetration (Shielding) Problems, Vol. 1 Users Manual*, LLNL Rept. # M-221-1 (1989); see also R. Buck and E. Lent, "COG: A New, High-Resolution Code for Modeling Radiation Transport," *LLNL Energy and Technology Review* (June 1993); additional information on COG may be obtained at URL http://www-phys.llnl.gov/N_Div/COG/.
2. T. Wilcox and E. Lent, *COG - A Particle Transport Code Designed to Solve the Boltzman Equation for Deep-Penetration (Shielding) Problems, Vol. 4: Benchmark Problems*, LLNL Rept. # M-221-4 (1989).
3. D. Cullen, L. Kissel and J. Hubbell, "EPDL97: the Evaluated Photon Data Library, '97 Version," Lawrence Livermore National Laboratory, UCRL--50400, Vol. 6, Rev. 5, September 1997.

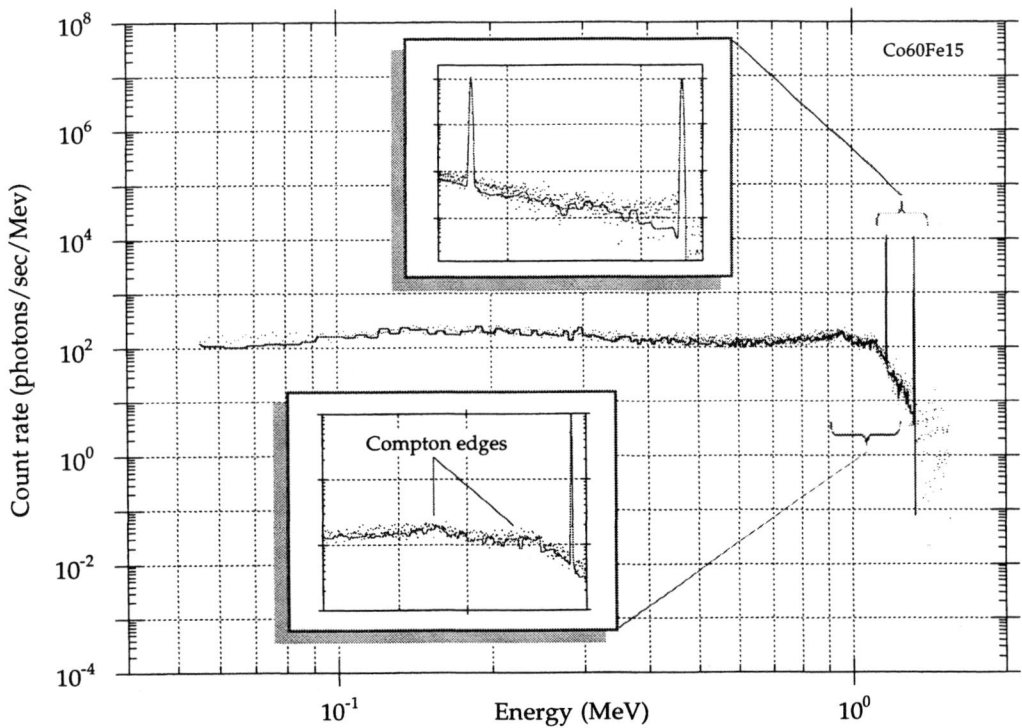

(Figure 3a) ^{60}Co source mounted in the iron collimator at 15°.

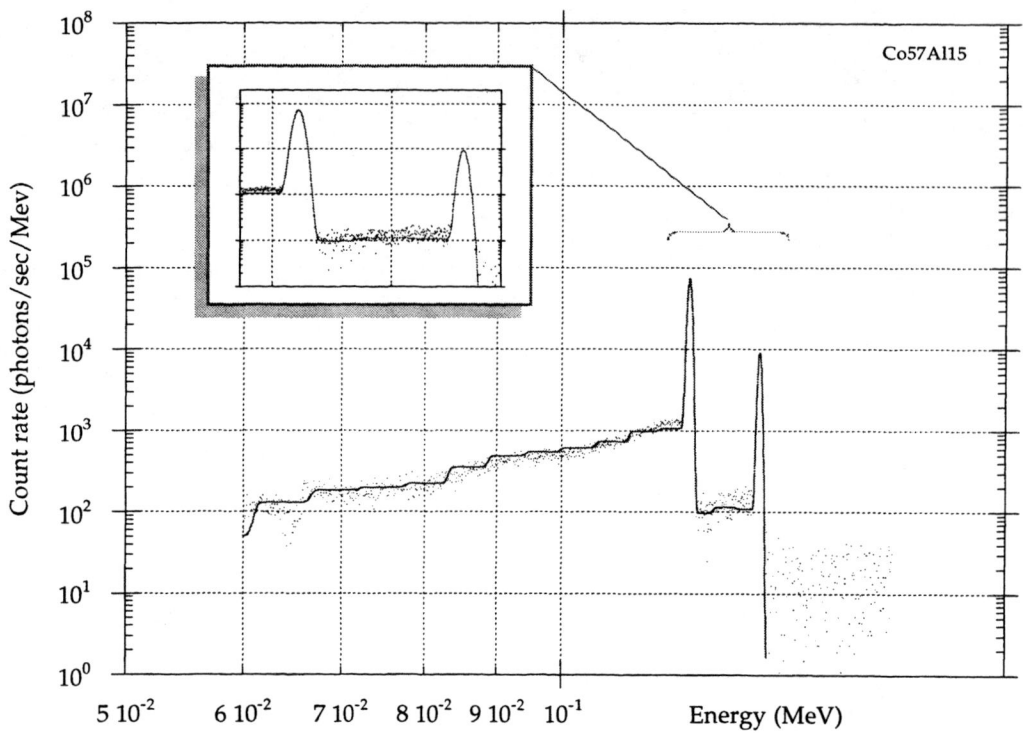

(Figure 3b) ^{57}Co source mounted in the aluminum collimator at 15°.

Figure 3. Measurements and predictions of pulse-height histograms for ^{60}Co source mounted in the iron collimator at 15° (upper plot) and ^{57}Co source mounted in the aluminum collimator at 15° with particular features highlighted in the insets.

SONTRAC – A Scintillating Plastic Fiber Tracking Detector For Neutron And Proton Imaging Spectroscopy

James M. Ryan, John R. Macri, Mark L. McConnell, Richard S. Miller

Space Science Center, Institute for the Study of Earth, Oceans, and Space
University of New Hampshire, Durham, NH 03824

Abstract. SONTRAC (SOlar Neutron TRACking imager and spectrometer) is a conceptual instrument intended to measure the energy and incident direction of 20–150 MeV neutrons produced in solar flares. The intense neutron background in a low-Earth orbit requires that imaging techniques be employed to maximize an instrument's signal-to-noise ratio. The instrument is comprised of mutually perpendicular, alternating layers of parallel, scintillating, plastic fibers that are viewed by optoelectronic devices. Two stereoscopic views of recoil proton tracks are necessary to determine the incident neutron's direction and energy. The instrument can also be used as a powerful energetic proton imager. Data from a fully functional 3-d prototype are presented. Early results indicate that the instrument's neutron energy resolution is approximately 10% with the neutron incident direction determined to within a few degrees.

INTRODUCTION

We are developing an instrument capable of simultaneously performing neutron/proton imaging and spectroscopy. Originally conceived and developed for space-based astrophysics investigations, the SONTRAC (Solar Neutron TRACking telescope) instrument concept has other applications ranging from medical therapy to environmental radiation monitoring. In the SONTRAC concept a bundle of tightly-packed scintillating fibers tracks recoil protons from neutron scattering. The detection and measurement of double scatter events within the fiber bundle permit a determination of the incident energy and direction of each neutron in the 20–250 MeV energy range. Using a scintillating fiber detector built up from orthogonal layers of fiber planes, a full 3-dimensional image of recoil proton tracks can be recorded and reconstructed. This technique provides excellent energy and angular resolution in a compact and efficient package for neutron measurements and represents the next generation of space-based neutron telescopes and spectrometers. We have successfully demonstrated the performance of a small 3-dimensional SONTRAC prototype, describe that instrument here, and present preliminary performance characteristics.

MOTIVATION

Besides solar flare neutrons, the technology that will be developed for SONTRAC will benefit any application where directional sensitivity to fast neutrons and/or neutron spectroscopy is important The SONTRAC instrument concept provides a number of unique capabilities for the detection and measurement of neutrons, including:

- Simultaneous determination of energy and direction of each incident neutron

- Angular resolution varying from <20° at 20 MeV to ~1° at 250 MeV

- Energy resolution better than 10% over the energy range of the telescope

- Ability to distinguish minimum ionizing and highly ionizing particles

- Significant background rejection capabilities

- Portability, large effective area and compactness

Technical Status Summary

We summarize the current SONTRAC development status as follows:

- We have an operational 5 × 5 × 5 cm three-dimensional science model detector

- There is sufficient signal to detect and track minimum ionizing particles.

- Event-by-event detection and measurement of track length (energy) track direction and particle type has been achieved.

- We have developed track recognition algorithms and used them to analyze proton beam data for characterization of energy and angular resolution capabilities up to 67.5 MeV.

The SONTRAC Concept

The SONTRAC detector can unambiguously reproduce the energy and direction of each incident neutron. The approach is based on the non-relativistic elastic double scatter of neutrons off ambient hydrogen. SONTRAC is based on an earlier concept investigated at Case Western Reserve University (Frye et al. 1985, 1987; Pendleton et al. 1988) and developed, to the level of a small prototype, with NASA Supporting Research & Technology (SR&T) funds by the University of New Hampshire.

Basic Principles

The double-scatter of a non-relativistic neutron in a solid block of plastic scintillator is illustrated in Figure 1 (top). Neutrons interact in plastic scintillator either by elastically scattering from hydrogen (n-p) or by interacting with carbon (n-C). The n-p events are the most useful. For non-relativistic scattering

$$\sin^2 \phi_{n'} = \cos^2 \phi_{p'} = \frac{E_{p'}}{E_{n'} + E_{p'}} = \frac{E_{p'}}{E_n}$$

where, E_n is the incident neutron energy; $E_{n'}$ and $E_{p'}$ are the scattered neutron and proton energies, respectively; $\phi_{n'}$ and $\phi_{p'}$ are the neutron and proton scatter angles, respectively. The kinematics of nonrelativistic elastic scattering implies that the scattered neutron and proton momenta are mutually orthogonal. Relativistic corrections at higher energies are necessary.

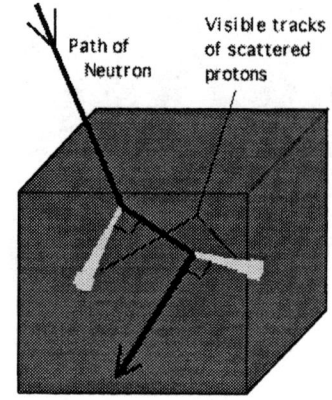

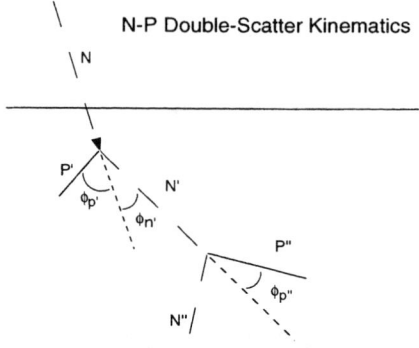

FIGURE 1. Schematic of non-relativistic double scatter neutron event in a block of plastic scintillator.

If the incident direction of a given neutron is known, then the measurement of the energy and direction of a recoil proton in a single scattering is sufficient to determine the incident neutron energy. In particular, if the incident direction is known, then $E_{n'}$ is determined and the neutron energy is,

$$E_n = \frac{E_{p'}}{\cos^2 \phi_{p'}}$$

An unambiguous approach, however, is provided by double-scatter events (Figure 1, bottom). If both recoil protons in a double scatter event can be measured, then the energy and incident direction of the neutron are uniquely determined. A system that can measure the parameters of both recoil proton tracks in three dimensions therefore provides the information that is necessary and sufficient to unambiguously determine the incident neutron energy and direction. The angular and energy resolutions depend on the ability to measure the energy and direction of the recoil protons. Because there are no formal restrictions on instrumental field-of-view double scatter events can be used to measure the neutron intensity from an extended sources without significant loss of sensitivity. Double scatter events are also preferred, because they allow for a more complete separation of the source signal from the background.

Scintillating Fiber Detector

We use orthogonal plastic scintillator fiber bundles for measuring neutrons in the 20 to 250 MeV range by recording the image of the recoil proton tracks induced by neutron interactions. The scintillating fibers serve as the source of ambient proton scattering centers, and as light pipes for image readout. The basic instrument consists of a closely packed bundle of square cross section, scintillating fibers. A fiber pitch of 300 μm (250 μm active fiber size) was selected so that a 10 MeV proton traverses several fibers before stopping. The fibers are arranged in stacked planes with the fibers in each plane orthogonal to those in the planes above and below. This alternating orientation of fiber planes provides a stereoscopic view of recoil proton tracks and allows the reconstruction of these tracks in three dimensions. These ionizing tracks in turn provide the information necessary to determine the arrival direction and energy of incident neutrons.

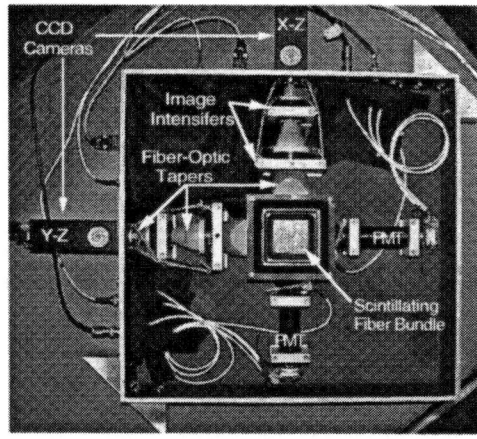

FIGURE 2. SONTRAC SM assembly.

The fibers in each dimension are coupled to imaging electronics (image-intensified CCD cameras, II-CCD). The intensity of each fiber's scintillation light combined with the location of the fiber bundle within the bundle provides the information for reconstruction. As expected, the ionization track length provides a sensitive measure of the recoil proton energy. The Bragg peak, corresponding to the stronger ionization near the end of a proton track, identifies the particle direction.

Early designs of this detector concept were studied with Monte Carlo simulations (Frye et al., 1985, 1987; Pendleton et al., 1988). The concept suffered at the time from a lack of sufficiently mature technology and existed in simulations only. Applicable technologies (small fibers, optoelectronics) have since become commercially available.

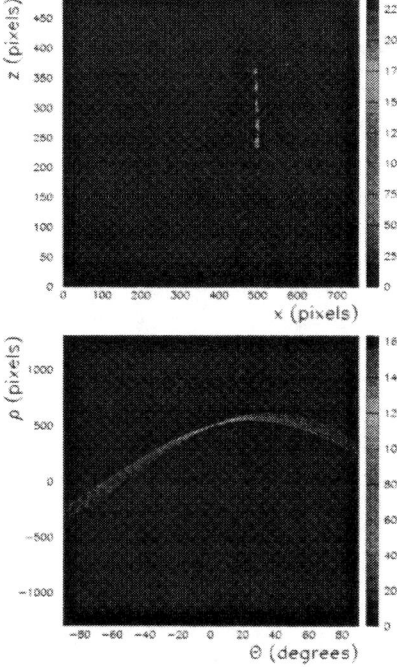

Figure 3. (top) Calibration image showing a normal incidence proton. (bottom) Hough Transform of image.

SONTRAC Science Model

We have assembled and tested a three-dimensional science model (SM), developed track recognition and data analysis tools, and used these to begin characterizing the performance parameters important for space- and ground-based applications.

The SONTRAC science model is a true 3-dimensional version of the SONTRAC instrument concept. The SM fiber bundle is a 5 × 5 × 5 cm scintillating plastic fiber tracking detector with appropriately sized optoelectronic readout components. Figure 2 is a photograph of the assembled SONTRAC SM. The plastic scintillating fiber detector bundle (center) as well as the light guides, fiber optic tapers, image intensifiers, and associated power supplies are contained within a single light-tight enclosure (cover removed in photo). The CCD cameras are mounted externally. The physical size of the optoelectronics required for readout, relative to the active detector area (fiber bundle) is illustrated in the photo. Much of the volume, mass, and cost of the assembly is associated with the fiber optic tapers and photomultiplier tubes (PMT). Future designs will eliminate the PMTs and will have alternative readout schemes to replace the II-CCD chains.

The SM assembly can be rotated through a ±45° range of incident angles for exposure to calibrated beams. External equipment, not shown, includes NIM electronics and a CAMAC ADC for processing the PMT signals and performing the trigger logic, and the computer for control and data acquisition. The SM is self-triggered; a 2-fold coincidence of the PMTs forms the trigger for image acquisition. The 5 cm size of the SM fiber bundle permits measurements to be made from 10-70 MeV, a range starting below the nominal threshold up to energies where neutron double scatter events are no longer contained.

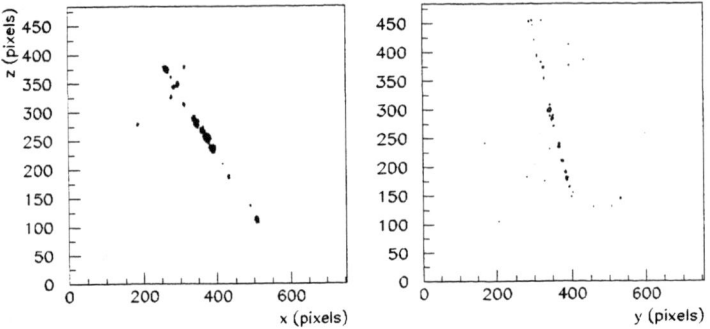

FIGURE 4. Raw CCD images showing orthogonal views of a cosmic ray muon track detected with SONTRAC science model.

Preliminary Calibration Results

Proton data was acquired with the SM at Crocker Nuclear Laboratory (U.C. Davis) and used to develop image-processing software for event analysis. The Hough Transform (HT) (Duda, et al. 1972; Sklansky 1978) is an effective technique for track reconstruction while simultaneously providing a large degree of data compression. Although other analysis techniques have been applied to the SM images (e.g. chi-square line fit) the HT is faster, and in this application more accurate. The HT also provides information useful for determining event quality. Image reconstruction using the HT effectively compresses the data from roughly 1 megabyte per image to a tuple of a few bytes composed of data such as θ, r, *starting points, length,* and *intensity*. A data example (single projection of incident proton) from the SM is shown in Figure 3; Figure 4 shows two orthogonal projections of an incident muon event.

Initial results from the proton calibration effort yield a 1σ angular resolution of 1.5°. Similar results were obtained for data taken ~17° off-axis. Taking track length to be proportional to energy gives an energy resolution in one projection of ~7.5% (1σ) at 67.5 MeV. Calibration of the instrument response to incident protons (and neutrons) at various energies is planned in the near future.

ACKNOWLEDGMENTS

We have built a fully functional 3-dimensional SONTRAC prototype and performed initial calibrations to determine instrument response. The excellent results from these efforts suggest that larger, more sensitive detectors based on this instrument concept are possible, with unique and useful application to a variety of science disciplines including medical imaging and nuclear physics research.

REFERENCES

1. Frye, G.M., et al., 20th ICRC, 4, 392, 1987.

2. Frye, G.M., et al., 19th ICRC, 5, 498, 1985

3. Pendleton et al., Workshop on Scint. Fiber Devel., Fermilab, 1093, 1988

4. Duda, Richard O. and Hart, Peter E., Commun. Ass. Comput. Mach., vol. 15, pp 11-15, Jan 1972.

5. Sklansky, Jack, *On the Hough Technique for Curve Detection*, IEEE Transactions on Computers, vol. C-27, No. 10, Oct. 1978.

Absolute Calibration of the In-beam Proton Polarimeter at IUCF

E. J. Stephenson

Indiana University Cyclotron Facility
2401 Milo B. Sampson Lane, Bloomington, IN 47408 USA

Abstract. This contribution covers the absolute calibration to better than 1% of the in-beam proton polarimeters at IUCF. The polarimeters observe p+d elastic scattering from thin CD_2 targets at $\theta_{lab}(d)=42.6°$. The calibration standard was double scattering using a ^{12}C first target and the focal plane polarimeter of the K600 magnetic spectrometer. This procedure gave the beam polarization and the analyzing powers from both targets. Systematic errors are discussed.

INTRODUCTION

Within the last few years, the requirements for a precise knowledge of the proton beam polarization at IUCF have extended from 200 MeV down to 80 MeV at the same time that typical beam currents have risen. This required improvements in the design of our in-beam proton polarimeters to handle the higher rate, and a recalibration over a larger energy range. For this, we needed a new polarization standard to replace that used originally and reported by Wells.[1] Here we discuss a refinement of double scattering for targets with unequal analyzing powers so that we could take advantage of the high efficiency of the focal plane polarimeter mounted on the K600 magnetic spectrometer. By initiating the scattering with polarized beam, it becomes possible to acquire enough information to determine independently the average beam polarization and the analyzing power of each target. This method is sensitive to systematic errors arising from instrumental asymmetries in the second scattering, and the consequences of this will be discussed.

POLARIMETER IMPROVEMENTS

The first polarimeter of this type was constructed to observed p+d scattering from a CD_2 target using two plastic scintillators to simultaneously record the proton and deuteron signals. Separation of the elastic events from breakup protons rested on measurements of the relative time of flight between the two scintillators as well as deposited energy. To help with this separation, thick stopping scintillators were used for the deuterons. Because of the large light output, these scintillators had unstable gains for beam currents above 50 nA.

A sketch to scale of one detector plane in the new polarimeter is shown in Fig. 1. The large deuteron scintillators were replaced with thinner ones, and absorbers were placed in front of both the proton and deuteron detectors to reduce the rate of breakup protons. The thickness of the deuteron detector absorber was chosen so that the elastic deuterons reach the end of their range close to the back of the scintillator, thus raising the recorded pulse height for the deuterons of interest. In this case, both scintillator signals became too broad for background-free peak sums, and the number of events was determined from the relative time peak for the arrival of the proton and deuteron.

REFINEMENTS TO DOUBLE SCATTERING

The previous calibration[1] made use of an A=1 point in elastic proton scattering from ^{12}C confirmed by Wissink.[2] This was only useful for calibrations between 175 and 200 MeV where the deviations from A=1 are small. At other intermediate energies, double scattering points have been used.[3]

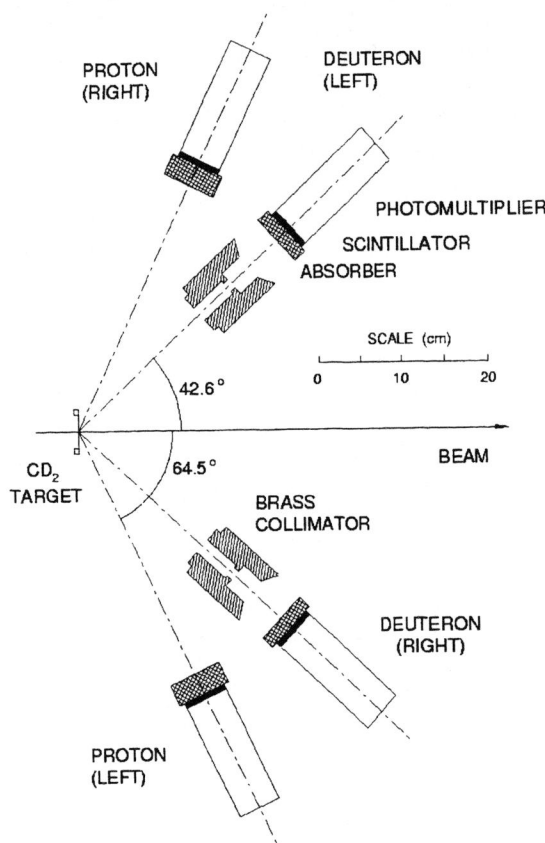

FIGURE 1. Layout to scale of the new polarimeter.

Traditional double scattering starts with an unpolarized beam. The protons scatter twice from two identical targets. The first polarizes the protons, the second creates a left/right asymmetry based on this induced polarization and the analyzing power of the target. If the scattering geometry is the same for the two targets (beam spot size, beam energy, scattering angle, acceptance) and the targets have spin=0, then the analyzing power and induced polarization are equal and the asymmetry at the second target is just $\varepsilon = AP = A^2$, which becomes known without reference to another standard. This scheme is hard to realize in practice because the scattered protons have less energy than the beam, and the tight geometrical collimation usually produces small double scattering count rates.

It would be desirable to use a more efficient double scattering polarimeter such as that installed at the focal plane of the K600 magnetic spectrometer.[2] But this violates the rule that the two double scattering analyzers need to be similar. Doing the double scattering experiment twice with beam polarized up or down with respect to the scattering plane recovers enough information that the system is again solvable.

To see how this works, consider the schematic layout in Fig. 2 that also incorporates detection after the first scattering.

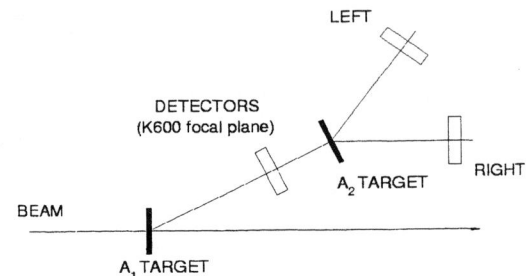

FIGURE 2. Schematic double scattering layout.

If the count rate at the focal plane detectors is C and the count rates in the two rear detectors are L and R, then there are three asymmetries that are available.

$$\varepsilon_1 = \frac{C_+ - C_-}{C_+ + C_-} \quad \varepsilon_+ = \frac{L_+ - R_+}{L_+ + R_+} \quad \varepsilon_R = \frac{L_- - R_-}{L_- + R_-} \quad (1)$$

These asymmetries can be combined to form

$$X = \varepsilon_+(1 + \varepsilon_1) + \varepsilon_-(1 - \varepsilon_1)$$
$$Y = \varepsilon_+(1 + \varepsilon_1) - \varepsilon_-(1 - \varepsilon_1) \quad (2)$$

from which it is possible to obtain the magnitudes of the average beam polarization and the two analyzing powers as

$$p^2 = \frac{Y\varepsilon_1}{X} \quad A_1^2 = \frac{X\varepsilon_1}{Y} \quad A_2^2 = \frac{XY}{4\varepsilon_1} \quad (3)$$

Calibrations were run at several energies between 100 and 200 MeV using ^{12}C and ^{90}Zr targets in order to maintain a large value of A_1. It was also possible to make a calibration at 70.9 MeV using a previously established analyzing power value by Eversheim.[4]

SYSTEMATIC ERRORS

Comparison to Other Methods

The asymmetries ε_+ and ε_- depend on the absence of any instrumental contribution in order to remain accurate. These cannot be recast, for example, as cross ratios in which solid angle and efficiency differences between left and right are cancelled.

Another absolute calibration has been reported by Clajus[5] in which these two asymmetries were replaced with

$$\varepsilon_L = \frac{L_+ - L_-}{L_+ + L_-} \quad \varepsilon_R = \frac{R_+ - R_-}{R_+ + R_-} \quad (4)$$

where the only remaining effect is the appropriate normalization of the spin up and down run lengths. As in Eqs. (1)-(3), it is possible to form the magnitudes of p, A_1, and A_2. However, the calculation of these quantities involves differences that become statistically imprecise when the value of A_2 gets small, as it does in our case. Figure 3 shows the ratio of the statistical error in the polarization for this method compared to the one described in the previous section. Despite the freedom from instrumental asymmetries, the errors using this procedure were larger than when we explicitly considered systematics, as described next.

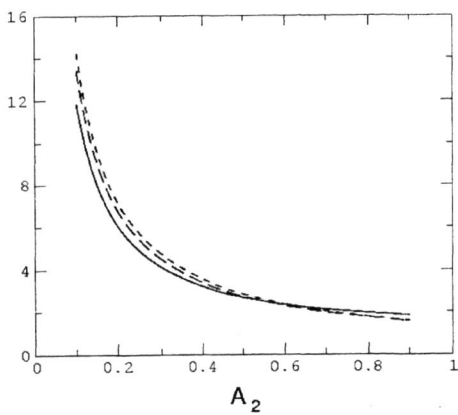

FIGURE 3. Ratio of the statistical errors using asymmetries from Eq. (4) rather than Eq. (1). The curves represent A_1=0.8 (solid), 0.5 (long dash) and 0.3 (short dash).

Measuring Systematic Errors

Corrections to the results of Eq. (3) can be derived to account for the difference in the magnitudes of p_+ and p_- are known from an external source. In this case such information comes from another upstream polarimeter that intercepts the beam between the two IUCF cyclotrons. The difference is parametrized as

$$\Delta = \frac{p_+ + p_-}{p_+ - p_-} \quad (5)$$

a small number. Δ was usually negative and smaller in magnitude than 0.01.

If measurements are made deliberately where A_1 is small and the instrumental asymmetry is parametrized as a change to the efficiency of the 'left' rate as

$$L \leftarrow L(1+\phi) \quad (6)$$

then ϕ may estimated from the first-order relation

$$\varepsilon_+ + \varepsilon_- = (1 - p^2 A_2^2)\phi + 2A_2(\Delta + [1-p^2]A_1)$$

Runs of this sort were made throughout the calibration experiment. They yielded 13 estimates of ϕ. These values did not appear to be correlated in any simple way with the running conditions. So it was decided to average these estimates and to assume that this was a systematic error that applied to each calibration measurement. The average value was

$$\phi_{rms} = 0.033.$$

This was large enough that it became the primary error (exceeding statistics) for several of the data points.

RESULTS

Figure 4 shows the values of A_2, the analyzing power for the focal plane polarimeter.

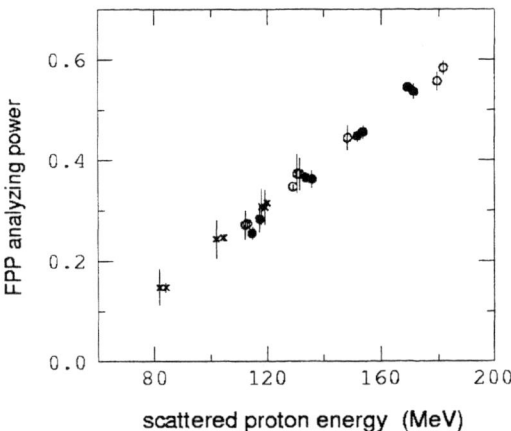

FIGURE 4. Measurements of the analyzing power for the focal plane polarimeter on the K600. The thickness of the carbon analyzer was 1.27 cm (x points), 2.54 cm (open points), and 5.08 cm (dots).

When the values of the focal plane polarimeter are plotted as a function of the energy at the center of the carbon analyzer target during the measurements, as they are in Fig. 4, then the values of A_2 form a smooth trend. At the lower end of the energy range, the small size of this analyzing power coupled with the rms average value for ϕ caused the errors in the calibration to rise to 5% of their value. At the upper end of the energy range, the calibration errors were limited by statistics and were less than 1%.

Data for the in-beam polarimeters was recorded simultaneously with the double scattering data from the K600. The beam line contained two such polarimeter, separated by a 45° bending magnet that made it possible to infer all three polarization components of the beam from measurements of the normal and sideways components at each polarimeter. It was assumed that the normal component of the beam polarization remained the same in each polarimeter as well as in the K600 double scattering. Once the average beam polarization was known, the analyzing power of the p+d polarimeter was inferred from the measured (cross ratio) asymmetry. These results are shown in Fig. 5.

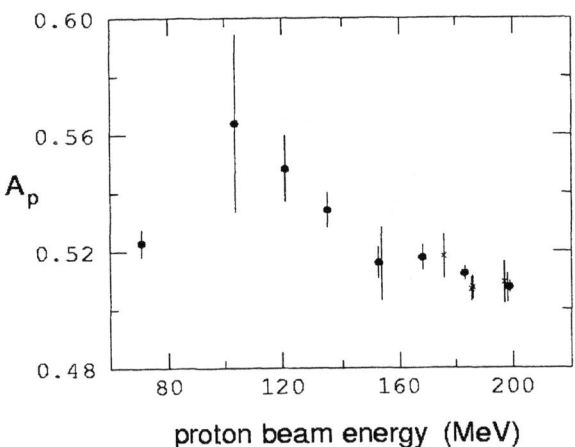

FIGURE 5. Values for the analyzing power of the in-beam polarimeter as a function of the beam energy. The values are positive, in contrast to the p+d elastic analyzing power, because the values are referred to the observation of the recoil deuteron.

The values of this calibration, shown by the dots in Fig. 5, agree well with the original calibration of Wells[1], shown by the x values. In the region above 150 MeV, there is an improvement in the precision of the results. The value at 70.9 MeV has again a small error since it is calibrated against the experiment of Eversheim[4] and not our double scattering standard.

These values have been published[6] in conjunction with three-body Faddeev calculations of the analyzing power in the p+d elastic scattering channel. Over this energy region they are more positive than the calculations based only on two-body interactions among the nucleons. Inclusion of the Tuscon-Melbourne three-body force shifts the analyzing power calculations in the right direction, but the change is too large and results in a similar difference between theory and experiment with the opposite sign. In particular, the three-body calculations confirm the rollover between 70 and 100 MeV and would suggest that there is no inconsistency between the calibration values at these two energies.

These results illustrate that it is possible to use double scattering techniques to establish precise standards for the measurement of the polarization of an intermediate-energy proton beam. Efficient double scattering polarimeters may be used whose analyzing powers differ from that of the primary target provided data is taken with the beam polarization up and down with respect to the scattering plane.

ACKNOWLEDGEMENTS

The assistance of A.D. Bacher, T.C. Black, Seonho Choi, W.A. Franklin, K. Jiang, S.W. Wissink, X. Yang, and C. Yu from IUCF, C. Hautala, M. Plarczyk, and J. Rapaport from Ohio University, and B.D. Anderson, A.R. Baldwin, and D. Prout from Kent State University with this experiment is acknowledged. This work was supported in part by the National Science Foundation under grant NSF-PHY-9602872.

REFERENCES

1. Wells, S.P. *et al.*, *Nucl. Instrum. Methods* **A325**, 205 (1993)

2. Wissink, S.W., in *Spin and Isospin in Nuclear Interactions,* eds. Wissink, S.W., Goodman, C.D., and Walker, G.E. (Plenum, New York, 1991) p. 253.

3. Hoistad, B. *et al.*, *Nucl. Phys.* **A119**, 290 (1968).

4. Eversheim, P.D., *et al.*, *Phys. Lett.* **B234**, 253 (1990).

5. Clajus, M. *et al.*, *Nucl. Instrum. Methods* **A281**, 17 (1989).

6. Stephenson, E.J., *et al.*, *Phys. Rev. C* **60**, 061001 (1999).

Neutron spectrometry in neutron and charged-particle mixed fields with phoswich neutron detector

Masashi Takada[a], Shingo Taniguchi[b], Takashi Nakamura[c], and Kazunobu Fujitaka[a]

[a] National Institute of Radiological Sciences, Chiba 263-8555, Japan
[b] Japan Synchrotron Radiation Research Institute, Hyogo 679-5198, Japan
[c] Quantum Science and Energy Engineering, Tohoku University, Sendai 980-8579, Japan

Abstract. We have developed the phoswich neutron detector consisting of the NE115 and NE213 scintillators which can distinguish neutron and photon events from charged-particle events. The performance of distinguishing photon and neutron events from charged-particle events was investigated in two neutron and charged-particle mixed fields in NIRS, which gave satisfactory results. In these radiation environments, photon and neutron energy spectra were obtained by the unfolding technique and the evaluated response functions. The proton energy spectra were simultaneously obtained by using the relation between the proton light yield and the incident proton energy.

INTRODUCTION

The radiation environments inside aircrafts and human spacecrafts are highly complex due to the coexistence of charged and neutral particles. High-energy neutrons in these radiation environments contribute a significant fraction of the total dose equivalent, however only insufficient data have ever been acquired. A flight neutron spectrometer capable of properly distinguishing neutrons and charged-particles is necessary to estimate the equivalent dose of neutron in these environments.

In this study, we have developed the NE213-NE115 coupled phoswich neutron detector which can distinguish both neutron and photon events from charged-particle events and potentially measure high energy neutrons upto 130MeV. This phoswich detector is an extension of the previously developed phoswich detectors(1, 2), that can measure neutrons upto 70MeV.

DETECTOR

The developed phoswich neutron detector consists of an organic liquid scintillator, NE213 (133mm diameter by 133mm long) surrounded by a slow plastic scintillator, NE115 (15mm thickness), as shown in Fig.1. In order to protect the NE115 plastic scintillator from the NE213 organic liquid scintillator, the inner NE213 scintillator is encapsulated in the glass cell of 3mm thickness. Both scintillators are coupled to the same photomultiplier tube.

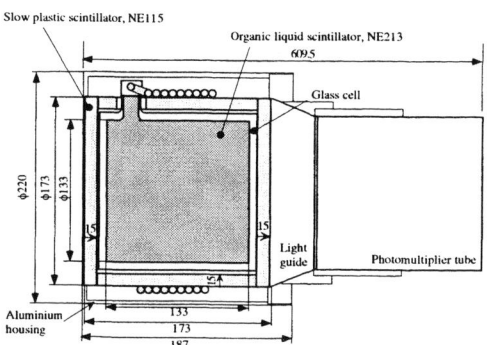

FIGURE 1. Cross-sectional view of the NE213-NE115 coupled phoswich detector.

The decay time of the scintillator light outputs produced by photon is 3.7nsec and that by neutrons is about 30nsec(3). The scintillator length of 133mm corresponds to the range of 130MeV proton and 25MeV electron. The NE115 plastic scintillator has the 225nsec decay time of light outputs(3) which is much slower than that of the NE213. The thickness of 15mm for NE115 was selected for distinguishing neutron events from proton events completely.

PARTICLE IDENTIFICATION

By using the different pulse shapes from the phoswich detector, three particles (photon, neutron, and proton) can

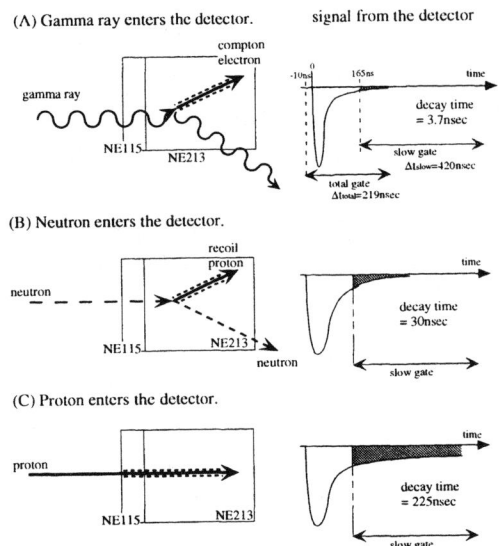

FIGURE 2. Schematic models of signals produced by the phoswich neutron detector. (A), (B), and (C) are the cases that photon, neutron, and proton enter into the detector, respectively. The right sketches are the time relation between the pulse from the detector and the gate inputs to two ADCs. Two light outputs (total and slow components) are obtained by integrating the signal over the relevant time periods.

be separated, as shown in Fig.2. The three cases shown in Fig.2 are described below:

(A) A compton electron scattered by a photon loses its energy only in the NE213 scintillator, and its signal from the detector has the only fast component (the decay time of a few nsec).

(B) A proton recoiled elastically by a neutron also loses its energy only in the NE213 scintillator, and its signal has the only fast component, but has the slower decay time of about 30nsec than that of photon.

(C) A proton incident to the detector loses its energy in both of the scintillators, and its signal becomes a sum of fast and slow components (the decay times of about 30nsec and 225nsec, respectively).

These three particles are detected separetely by the use of a pulse shape discrimination technique based on the standard CAMAC charge integration ADCs. The charge integration of the signal is carried out during the time period specified by a gate pulse (total and slow gates) as shown in the right side of Fig.2. The total and slow components can be obtained by a total gate pulse adjusted at the peak of the signal and by a wider delayed slow gate set over the long tail of the signal, respectively.

PERFORMANCE

In order to investigate the performance of the phoswich detector, as the particle spectrometer, two radiation environments, a neutron and proton mixed field and a neutron and heavy-charged-particle mixed field were formed at two accelerator facilities.

Experiment in a neutron and proton mixed field

The radiation environment consisting of photon, neutron, and proton was formed at the cyclotron facility(4) in the National Institute of Radiological Sciences (NIRS). This neutron-proton mixed field was produced from a 2mm-thick Be metal target injected by 70MeV protons through the ^{9}Be(p,nx) and ^{9}Be(p,px) reactions. The phoswich neutron detector was set at 45deg with respect to the beam direction and 2.8m downstream from the target.

Particle identification

The two-dimensional distribution of the slow versus the total components measured with the phoswich detector in a neutron and proton mixed field is shown in Fig.3. In Fig.3, the symbols (A) shows electrons scattered in the NE213 scintillator by gamma rays, the symbol (B) shows protons produced by elastic collision with hydrogen, H(n,n)H and the ^{12}C(n,xp) reaction with carbon in the NE213 scintillator, and the symbol (C) shows α particles produced by the ^{12}C(n,α) and ^{12}C(n,n'α) reactions. The symbols (D) and (E) show the external protons which were stopped in the front NE115 scintillator due to its low energy, and those which passed only along the sidewall of the NE115 scintillator, and the external protons stopped in the NE213 scintillator after crossing the NE115 scintillator. It is clearly seen from Fig.3 that the photon events (A), the neutron events (B and C), and the proton events (D and E) can be separated each other.

Particle spectra

From the pulse heights of photon, neutron, and proton obtained by selecting each region of the interest in Fig.3, the photon and neutron energy spectra were obtained by using the unfolding technique with the FERDOU code(5) coupled with our evaluated photon and neutron response functions(6). The proton energy spectrum was obtained by using the relation between the proton light yield and

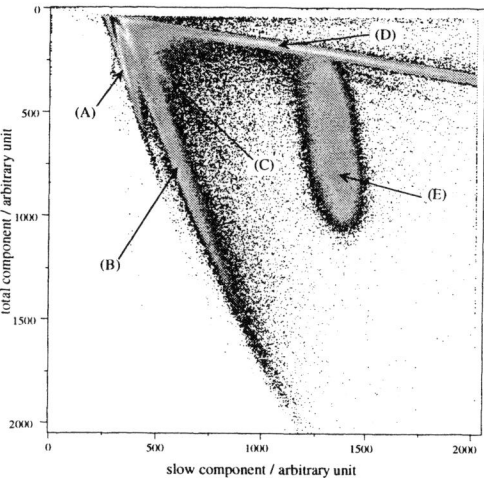

FIGURE 3. Two dimensional distribution of the slow versus the total component measured with the phoswich detector in a neutron and proton mixed field produced from a thin Be target injected by 70MeV protons. The mark (A) represents the photon events, (B)-(C) the neutron events, and (D)-(E) the proton evnets, respectively (See text in detail).

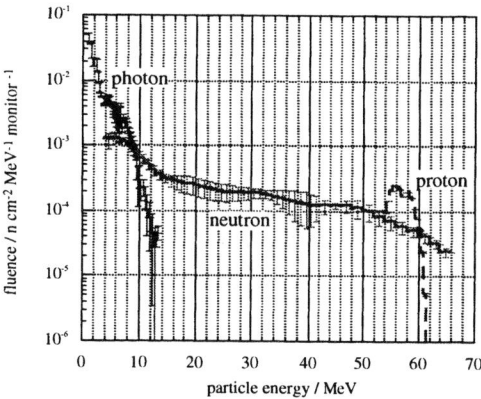

FIGURE 4. Photon, neutron, and proton energy spectra in a neutron and proton mixed field produced from a thin Be target injected by 70MeV protons.

Experiment in a neutron and heavy-charged-particle mixed field

The radiation environment consisting of photon, neutron, proton, helium and heavier particles was formed at the Heavy Ion Medical Accelerator in Chiba (HIMAC)(7) in NIRS. This neutron and heavy-charged-particle mixed field was composed of secondary particles produced from a full-stopping-length carbon target by the bombardment of 100MeV/nucleon carbon ion. The phoswich neutron detector was placed at θ=0,15, and 30deg with respect to the beam direction and 6.1m downstream from the target.

Particle identification

The two dimensional distribution of the total versus the slow components obtained by the bombardment of 100MeV/nucleon carbon ion with a full-stopping-length carbon at 30deg with respect to the beam direction is shown in Fig.5, as an example.

In Fig.5, the symbols (A), (B), and (C) represent the electron events produced by photons, the proton events produced by the H(n,n)H and the ^{12}C(n,px) reactions, and the α particle events produced by the ^{12}C(n,αx) reactions, respectively. The symbol (D) indicates the events that external protons completely stopped in the NE213 scintillator after crossing the NE115 scintillator. The symbol (E) indicates the events that protons stopped in the back side of the glass cell after crossing the front NE115 and the NE213 scintillators. The symbol (F) indicates the events that protons stopped in the back side of the NE115 scintillator after crossing the front NE115 and the NE213 scintillators. The symbol (G) indicates the protons that passed through both sides of the NE115 scintillator and the NE213 scintillator. The symbols (H) and (I) show the events that the external deuteron and triton stopped in the NE213 scintillator, respectively. The photon and neutron events can be clearly distinguished from the charged-particle events.

Particle spectra

From the pulse height spectra of the thus-discriminated particle events, the energy spectra of photon and neutron energy spectra were obtained by using the unfolding method coupled with the response functions of this detector. The proton energy spectra were obtained by using the relation between proton energy and light yield. The neutron spectra at 0,15, and 30deg from a carbon target are shown in Fig.6 comparing with the neutron spectra measured with the BC501A scintillator using the TOF method(8), as an example. The neutron spectra extend over 100MeV and become

the incident proton energy(6). Fig.4 shows these particle energy spectra in a neutron and proton mixed field. The neutron spectrum in the energy range from 4MeV to 66MeV and the photon spectrum in the energy range from 1MeV to 12MeV were obtained. The proton energy spectrum has a peak at about 56MeV.

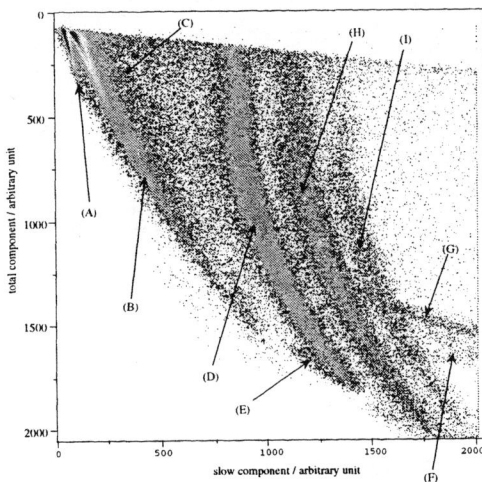

FIGURE 5. Two dimensional distribution of the total versus the low components for 100MeV/nucleon carbon ions incidence on a 14.45mm thick carbon at θ=30deg. The components marked (A), (B) and (C), (D)-(G), (H), and (I) are identified as photon, neutron, proton, deuteron, and triton, respectively (See text in detail).

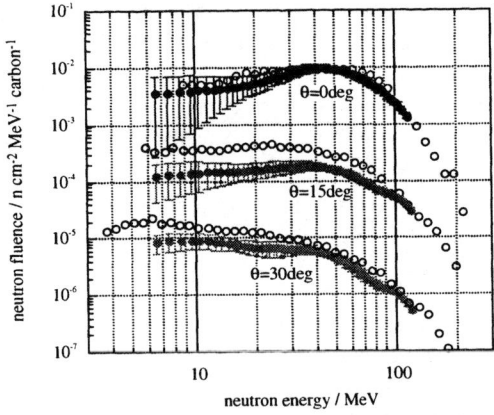

FIGURE 6. Neutron energy spectra (solid circle) at 0, 15, and 30deg from a carbon target bombardment by 100MeV/nucleon carbon ion, comparing with neutron spectra measured with the TOF method (open circle) (8).

softer with the emission angle. The absolute values also decrease with the emission angle. The agreement between two experiments is rather good, especially at 0deg.

CONCLUSIONS

A new kind of phoswich neutron detector combined with the NE213 and NE115 scintillators was developed in order to distinguish high energy neutrons from charged particles in the neutron and charged-particle mixed field. Using this phoswich detector, the performance of distinguishing photon and neutron events from charged-particle events gave satisfactory results. Neutron and photon spectra in neutron and charged-particle mixed fields were obtained by the unfolding technique and our evaluated response functions. Proton energy spectra were simultaneously obtained by using the relation between the proton light yield and the incident proton energy. It was also found that the NE213-NE115 coupled phoswich detector has a sufficiently good quality to obtain the neutron, photon, and proton energy spectra separately in neutron and charged-particle mixed fields which is usually realized in aircrafts and spacecrafts. By using this NE115-NE213 coupled phoswich detector, the photon and neutron energy spectra can be obtained by distinguishing neutral particles in neutron and charged-particle mixed fields.

ACKNOWLEDGMENTS

The authors are very grateful to Dr.T.Yamada, Dr.T.Honma for the cyclotron operation and also to Dr.Y.Uchihori for his helpful assistances. The authors are very grateful to the staff members for the cyclotron and HIMAC operation during the experiments. This work was financially supported by a Grant-in-Aid for Scientific Research from the Japanese Ministry of Education and Culture. This work was also achieved at Research Project with Heavy Ions at NIRS-HIMAC.

REFERENCES

1. Takada, M., Taniguchi, S., Uwamino, Y., and Nakamura, T., *Nucl. Instr. and Meth.* **A379** 293-306 (1996).
2. Takada, M., Taniguchi, Nakamura, T., and Fujitaka, K., *IEEE Trans. Nucl. Sci.* **NS-45(3)** 888-893 (1998).
3. Knoll, G.F., *Radiation Detection and Measurement* John Wiley & Sons, Inc., New York, 1989, pp.220-227.
4. Ogawa, H., Yamada, Y., Kumamoto, Y., Sato, Y., and Hiramoto, T., *IEEE Trans. Ncul. Sci.* **NS-26 (2)** 1988-1991 (1979).
5. Shin, K., Uwamino, Y., and Hyodo, T., *Nucl. Technol.* **53** 78-85 (1981).
6. Takada, M., Taniguchi, Nakamura, T., Nakao, N., Uwamino, Y., Shibata, T., and Fujitaka, K., *Nucl. Instr. and Meth.* (submitted).
7. Hirao, Y., et al., *NIRS-M-89/HIMAC-001* (1989).
8. Kurosawa, T., Nakao, N., Nakamura, T., Uwamino, Y., Shibata, T., Nakanishi, N., Fukumura, A., and Murakami, A., *Nucl Sci. Eng.* **132** 30-37 (1999).

The Gamma Ray Energy Tracking Array

K. Vetter

Nuclear Science Division, Lawrence Berkeley National Laboratory, Berkeley, California 94720

Abstract. Gamma-ray tracking is a new concept for the detection of γ radiation. One proposed implementation of this concept, called GRETA for Gamma Ray Energy Tracking Array, aims at an improvement in nuclear physics and is based on an array of highly segmented HPGe detectors. We have developed new techniques to determine three-dimensional positions and energies of interactions based on pulse-shape analysis in a two-dimensionally segmented Ge detector and algorithms which use these informations to reconstruct the scattering sequence of γ rays, even if many γ rays hit the array at the same time. Such a detector will have a high efficiency and a good peak-to-background ratio, an excellent Doppler-shift correction and high count rate capability, as well as a high polarization sensitivity. However, the concept will not only improve the sensitivity for γ rays in nuclear physics but large potential gain is also possible in other areas, such as γ-ray imaging used in astrophysics or medicine. Only recently we have shown the proof-of-principle of the proposed concept based on the measured position resolution of better than 1 mm in three dimensions in a 36-fold segmented Ge detector at an γ-ray energy of 374 keV.

INTRODUCTION

Many new facets of the nucleus have been discovered and explored over the last several years due to the construction and operation of large γ-ray arrays such as Gammasphere or Euroball. These arrays, consisting of approximately 100 modules of Compton suppressed Ge detectors, have a total peak efficiency of about 0.1 (for a 1.3 MeV γ ray) and a peak-to-total ratio (P/T) of about 0.6. These instruments provided a factor of about 100 improvement over previous detector systems in the ability to resolve weak features in a complex spectrum.

To improve this performance, the efficiency and/or the P/T haveto be increased. However, the highest efficiency that can be reasonably achieved for a Compton suppressed Ge detector array is limited to about 0.15. This is partly due to the scattered γ rays escaping from the Ge detector (a 1.3 MeV γ ray deposits full energy in a 7cm-by-7cm detector only 20% of the time) and partly due to the solid angle lost to the Compton shield (about 50% for Gammasphere).

An alternative approach consists of a closed shell of Ge detectors which allows to almost completely cover the entire solid angle, and by adding the signal from neighboring detectors, the escaped energy is recovered and much higher efficiency can be achieved. However, for events with many coincident γ rays, such as long cascades in the decay of nuclear high-spin states, the summing of two γ rays hitting neighboring detectors reduces the efficiency and increases the background. In order to reduce this summing, a large number of detectors, of the order of 1000, is required. The cost of such a detector array will be prohibitive.

GRETA CONCEPT

To circumvent the limitations associated with the previously mentioned detector systems we have developed a new concept which is based on the determination of the positions and energies for every interaction of each γ ray. This novel technique of γ-ray tracking and its implementation called GRETA (Gamma-Ray Tracking Array) aims at the identification and separation of individual γ rays and is based on highly segmented Ge detector elements in combination with pulse-shape analysis to determine the location and energy of every interaction of each γray . This information will be used to "track" all interactions of each γ ray by using the energy-angle relation given by the Compton scattering formula or particular pattern of the pair production process for higher γ-ray energies.

The tracking will not only enable to identify and separate multiple, coincident γ rays, but in addition is able to distinguish full-energy events in the Ge crystal from partial-energy events, thereby improving considerably the response function (peak-to-total ratio). Furthermore, it allows the determination of the locations of the two first interactions of the scattering sequence. The localization of the first interaction point in a detector defines the angle of emission of that γ ray from a source of known location relative to the detector and therefore allows to correct for the Doppler shift of γ-rays emitted in flight. With the identification and position of the first two interactions it is possible to determine the linear polarization of a γ ray and thereby define its electric or magnetic character. Due to the close packing of the Ge crystals and the use of the "add-back" feature of GRETA, the efficiency for detecting full-energy γ rays will be much higher than previous detector systems, especially for high-energy γ rays. The most impressive gain, however, to the currently existing arrays will be for experiments using fusion reactions to study of high-spin states in the nucleus. These reactions are related to the emission of many simultaneous γ rays (20-30) and a gain in sensitivity per γ ray will result in a large sensitivity increase for the whole event.

In the following we present some of the key aspect of the development to show the proof-of-principle for GRETA.

Segmented Detector

The ability to manufacture coaxial Ge detectors with a high degree of two-dimensional segmentation is an essential component of our approach towards γ-ray tracking. The combination of segmentation and pulse-shape processing of segment signals provides the energies and positions of interaction points which are used as input for tracking algorithms to identify and separate individual γ rays and to determine the time sequence of the interactions.

Since GRETA focuses on the implementation of γ-ray tracking for a 4π γ-ray detector array, many of these Ge detectors have to be closely packed to maximize the solid angle coverage (1). Several packaging schemes are possible, such as a spherical shell, or barrel-, or cube-like arrangements. The prototype detectors we obtained are designed to fit into a spherical shell of about 100 tapered hexagonal and pentagonal detectors, very similar to the Gammasphere geometry (2). A 36-fold segmented prototype detector which was built by Eurisys Mesures. It consists of a 9 cm long, closed-ended HP-Ge n-type crystal with a tapered hexagonal shape (3)and a maximum diameter of 7 cm. The outer electrode is divided into 36 parts by 6 longitudinal and 5 transverse segmentation lines. The 37 FETs for the 36 segments and the central channel are located and cooled in the same vacuum as the crystal. Cold FETs provide low noise which is important for optimizing the energy and position resolution. An average energy resolution of 1.14 keV and 1.94 keV was measured at a γ-ray energy of 60 keV and 1332 keV, respectively. A total integrated noise of about 4 keV was measured up to a frequency of 40 MHz. This low noise is not only important for energy, time or position resolution but also for trigger purposes (3).

Pulse-shape analysis

Pulse-shape analysis in a two-dimensional segmented detector allows to determine the position in three dimensions with an accuracy far better than the segmentation size. This is achieved by not only measuring the signal of the charge collecting electrode with the net charge signals but also by analyzing the transient signals of neighboring segments which display temporary image charges. We studied the position sensitivity as a function of the location and the direction in the crystal, and as a function of the energy, as well as the limiting factors. This was accomplished by measurements as well as pulse-shape calculations (5).

We extracted a three-dimensional position sensitivity of about 0.2 mm along the electrical field lines, which is very close to the drift direction of the charge carrier,and about 0.5 mm in the complementary directions, away from boundary lines. These values were obtained at a γ-ray energy of 374 keV. The position sensitivity reflects the relative change of the measured signals for the different positions in terms of the noise, which is main uncertainty in the signals. It measures the minimum distance between interactions that produce distinguishable signals. The obtained sensitivity is remarkable considering the fact that the size of the segment is about $2 \times 2 cm^2$ implying an improvement of about a factor of 100. However, it neglects the absolute position of the detector relative to the source or effects, such as the range of the Compton electron or the intrinsic momentum of the Compton electron. A position resolution was measured in a two-dimensional collimation system and decomposing the measured set of signals with calculated ones. Generally, the fit reproduced the measured data exceptionally well. Based on the shape of purely calculated and fitted position spectra we were able to obtain a position resolution of 0.5 mm to 0.9 mm in all three dimensions. The position for the direction with the worst resolution was off by about 2 mm which can be explained by crystal orientation effects which were not taken into account in the calculations. While the dependence of the magnitude of the drift velocity from the crystal orientation have been taken into account, the change in the direction of the charge carrier as a function of electrical field and crystal orientation direction have not yet been taken into account. This is due to the missing theoretical description of the hole mobility properties.

While the three-dimensional position sensitivity obtained above represent an essential step to enable γ-tracking, it only reflects the accuracy of locating single interactions. In order to determine locations and energies of multiple interactions in multiple segments, we developed algorithms to decompose signals into their individual components. For example, a 1.3 MeV γ ray interacts on average 4 times (two interactions in two segments) before it is fully stopped in one crystal. However, the goal of these decomposition algorithms is not only to optimize the position resolution, but also to perform the decomposition in real time.

We have explored three approaches, so far: an adaptive grid method, a method which allows to solve systems of equations using generalized matrix-inverses by using the singular value decomposition, and wavelet transformations to reduce the large amount of parameters. Work is also pursued using artificial neural networks (ANN), genetic algorithms (GA) or wavelet transformations. All except the ANN approach have a set of "basis" signals in common which are used for determining the number, amplitudes and positions of interactions. The aim of all of these approaches is to minimize the error between the measured signals and signals taken from the basis.

Tracking Algorithm

We have developed tracking algorithms to associate the interactions we obtained in the above described way with a certain γ ray (6). The goals of this algorithms are to identify interactions belonging to a given γ ray and to resolve the tracks of multiple, coincident γ rays, to distinguish between γ rays which only left a partial energy and γ rays which deposited their full energy in the detector system and to determine the first and second interactions. The first interaction is required for a proper Doppler correction, the second in connection with the first interaction is used for determining the linear polarization.

Most of our efforts have been focussed on the treatment of Compton scattering since this is the dominant interaction process for γ-ray energies between 150 keV and 5 MeV in germanium. Below 150 keV the photo-electrical effect and above 5 MeV the pair production process dominates. Recently, we developed also an algorithm to identify and recover γ rays interacting by the pair production process by means of tracking.

The current Compton-tracking algorithm consists of three steps. Cluster identification is the first step of the algorithm. The interaction points within a given angular separation as viewed from the target are grouped into a cluster. In the second step, each cluster is evaluated by tracking to determine whether it contains all the interaction points belonging to a single γ ray. The tracking algorithm uses the angle-energy relation of Compton scattering to determine the most likely scattering sequence from the position and energy of the interaction points. If the interaction points had infinite position and energy resolution, the tracking would be exact and the properly identified full-energy clusters will show no deviation from the scattering formula ($\chi^2 = 0$). Wrongly identified clusters or partial-energy clusters will deviate from the formula and the separation of the good and bad clusters would be easy. However, in reality, with finite position and energy resolution, the good clusters will also have a non-zero χ^2 and they cannot be separated cleanly from the bad clusters. This causes a lower efficiency and poorer P/T ratio. In the third step, we try to recover some of the wrongly identified γ rays, e.g. by splitting or adding clusters of interactions. This simulation was carried out for a number of different conditions such as the multiplicity and energies of the γ rays as well as position resolution of the detector. Assuming a position resolution of 1 mm which appears to be feasible based on the previous results and an angle parameter of 8 degrees an efficiency of 33% and a peak-to-total of 72% can be achieved. This has to compared with Gammasphere which has an efficiency of about 8% and a peak-to-total of about 50% under the same conditions, which implies a gain of four in efficiency and 1.5 in peak-to-total for each of 25 emitted γ rays (6).

In addition to the possible improvements discussed above another advantage of GRETA is the photo-peak efficiency for high energy γ-rays (e.g. $E_\gamma \geq 10$ MeV). Above the threshold of 1.022 MeV, the probability of pair production increases as energy increases. At 10 MeV, this probability is about 60 % and therefore the pair-production events need to be identified with a high efficiency. Only recently we have developed a first version of an algorithm to identify and recover pair-production processes.

The algorithm was tested using simulated data of interaction points generated by GEANT3. The detector consisted of a shell of Ge with 21 cm outer radius and 12 cm inner radius, which is a simplified model of a possible GRETA configuration. In each event, a 10 MeV γ-ray was launched from the center in coincidence with several 1 MeV γ-rays, which simulate background γ-rays. A preliminary result gives a tracking efficiency of $\varepsilon_t \sim 0.5$ at 10 MeV, assuming an energy resolution of 2 keV and a position resolution of 3 mm. Note that the photo-peak efficiency for pair events is the product of ε_t and full absorption efficiency ε_a (~ 0.7 at 10 MeV) for the Ge shell.

POTENTIAL CAPABILITIES

The estimates of GRETA's final capabilities come from simulations using a spherical-shell geometry built out of 110 irregular hexagons and 10 regular pentagons, similar to the Gammasphere geometry. We considered other, cube-like geometries which resulted in comparable results. The advantage of a spherical geometry is the symmetric arrangement of Ge detectors distributing the total count rate equally over all detectors and the increasing space which is available behind the detectors with increasing radius, which will be useful to mount the large number of preamplifiers. For one emitted γ ray, we obtain a full-energy efficiency of 0.55 and 0.12 for γ-ray energies of 1.33 MeV and 15 MeV, respectively. Furthermore, we obtain a P/T ratio of 0.85 at 1.33 MeV. This high value is mainly due to tracking. As mentioned above, even assuming a γ-ray multiplicity of 25 we still obtain an efficiency of 0.33 and a P/T ratio of 0.72 for a γ-ray energy of 1.33 MeV.

A variety of physics can be addressed with the new capabilities of GRETA. Tracking can determine the scattering sequence and the linear polarization of a γ ray and thereby define its electric or magnetic character, import ant information in most nuclear structure studies. The localization of the first interaction point with GRETA can define the angle of emission of a γ ray to better than one degree, and thus eliminate Doppler broadening for nuclei having v/c less than 10% and greatly reduce it for higher velocities. As an example consider the study of neutron-rich light nuclei, where the production of the interesting near-dripline nuclei is by fragmentation reactions resulting in product recoil velocities of v/c = 30%, or higher. For a 1 MeV γ ray emitted at 90° to the recoil direction, the contribution to the FWHM of the γ-ray peak due to the Doppler broadening would be 39 keV with a standard Gammasphere detector, but only 3.7 keV with GRETA. Clearly this improvement can have a large effect on the extracted physics, both in detecting weak γ rays and in separating close-lying peaks. Localization is important for many experiments, especially ones using inverse-reaction kinematics and heavy-ion Coulomb-excitation studies.

The very high efficiency for high-energy γ rays, together with excellent energy resolution, will open up new studies of giant resonances in nuclei. For example, it will be possible to tag the giant-resonance γ rays with known low-lying γ rays to define the process being studied. As pointed out above, the new concept in detection of γ radiation described here will not only increase the sensitivity for nuclear physics applications but will enhance the efficiency in other areas, such as γ-ray imaging, too (e.g. Compton camera). Here, the tracking of the path of a γ ray is used to measure the direction of an incoming γ ray to determine the location of γ-ray sources in the environment or space. Due to the symmetry of the Compton effect each event only defines a cone of possible direction of the incoming γ ray. However, using multiple events the intersection of the cones determine the location of the γ-ray source. For GRETA and applications where the location of the source is known, all the above discussed properties of the γ-ray tracking procedure is available on the event-by-event basis.

We have performed Monte-Carlo simulations and measurements with our 36-fold segmented GRETA detector to determine the capability of this kind of Ge detector (7). Other crystal geometries, such as segmented planar detectors have been used earlier, already, to employ γ-ray tracking. However, due to their arrangement the efficiency of these devices were very small. Using a segmented coaxial geometry will allow to increase the efficiency by about one order of magnitude.

CONCLUSIONS

A γ-ray energy tracking array which is based on highly segmented Ge detector elements would provide far better efficiency, peak-to-back ground, counting rate and localization than any existing γ-ray array. These new capabilities will have a major impact in many areas of physics. Recent progress in detector segmentation, signal processing and γ-ray tracking has provided the proof-of-principle for the proposed concept.

ACKNOWLEDGEMENTS

The work on GRETA is embedded in the Nuclear Structure Group of the LBNL and therefore relies on many contributions from the group members. The Compton-tracking code was mainly developed by Greg Schmid while he was at LBNL; the pair-tracking algorithm was implemented by Takashi Teranishi during his stay at LBNL. Results shown in this talk, particularly regarding the imaging, are based on a collaboration with Greg Schmid and Dean Beckedahl from LLNL and Jerry Blair from Bechtel Nevada. Eurisys Mesures manufactured the prototype detector. This work was supported by the Director, Office of Energy Research, Division of Nuclear Physics of the Office of High Energy and Nuclear Physics of the U.S. Department of Energy under contract No. DE-AC03-76SF0093.

REFERENCES

1. M.A. Deleplanque, et al., Nucl. Instr. and Meth. A 430 (1999) 292
2. I Y. Lee, Nucl. Phys. A 520 (1990) 641c
3. K. Vetter, et al., Nucl. Instr. and Meth. A 452 (2000) 105
4. M.R. Maier, et al., submitted to the Proceedings of the IEEE Symposium on Nuclear Science, Seattle Oct. 1999
5. K. Vetter, et al., Nucl. Instr. and Meth. A 452 (2000) 223
6. G.J. Schmid, et al., Nucl. Instr. and Meth. A 430 (1999) 69
7. G.J. Schmid, et al., submitted to Nucl. Instr. and Meth.

Fractional Counts - The Simulation of Low Probability Events

R.L. Coldwell

Department of Physics, University of Florida, Gainesville, FL 32611

G. P. Lasche, and A. Jadczyk

Constellation Technology Corporation, 7887 Bryan Dairy Road, Suite 100, Largo, FL 33777

Abstract. The code RobSim has been added to RobWin[1]. It simulates spectra resulting from gamma rays striking an array of detectors made up of different components. These are frequently used to set coincidence and anti-coincidence windows that decide if individual events are part of the signal. The first problem addressed is the construction of the detector. Then owing to the statistical nature of the responses of these elements there is a random nature in the response that can be taken into account by including fractional counts in the output spectrum. This somewhat complicates the error analysis, as Poisson statistics are no longer applicable.

INTRODUCTION

The code began life as BSIMUL. It was built to aid in the construction of the BGO shielded GRAD[2] detector that was eventually used in an Antarctic balloon borne observation of supernova 1987A[3]. In its present incarnation the numerically intensive part is written in Fortran while the user interface is written in C.

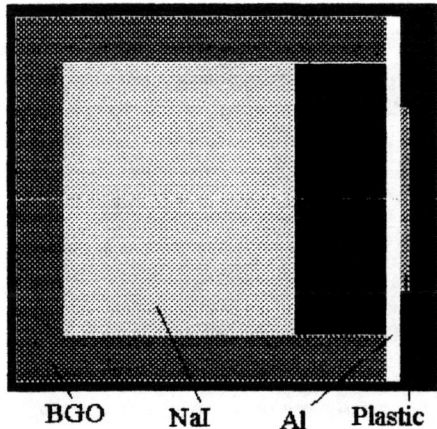

FIGURE 1. Sample detector similar to that used by the NEAR-X and Gamma Ray group at NASA Goddard.

The detector is described by cylindrical coordinates z and rho. In figure 1 z is measured from the left and rho from the dashed line in the center. There is a repetition in that rho values less than zero in figure 1 are simply repeats of those greater than zero.

The simulation works by assigning materials and detector properties to a series of zones. There are five zones in figure 1, which is the detector as drawn on the screen. The user can add zones and can change the sizes and the material independently in each zone. In addition there can be a detector associated with each zone with a different energy-dependent response function for each zone. Spectra are constructed as shown for the anti-coincident NaI spectrum below. The term En refers to the energy in zone n absorbed in zone n from an input photon while it is in the detector.

NAIAC.SP
E3 if not (10 < E1+E2 < 10000) or (50 < E4 < 10500)

The first line is the name of the file in which the spectrum will be placed. The second line tells the code to add the probability of energy absorbed in NaI, zone 3, to the spectrum only if the energies in BGO, zones 1 and 2, and that in the plastic, zone 4 are "essentially" zero. The code spreads the energies according to the input full width at half-maximum and input photo-peak response function. The above set of directions with 50000 input photons produce the results shown in figure 2 in 2-5 minutes on a laptop computer with a 366 MHz Celeron processor.

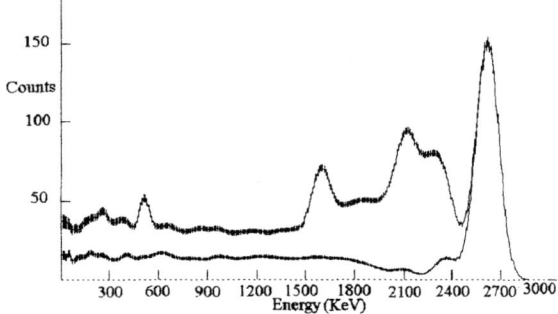

FIGURE 2. Input photon is 2614 KeV. Top spectrum is NaI alone. Bottom spectrum is NaI in anti-coincidence with any event in plastic or BGO.

COMPUTATIONAL METHOD

The code follows photons through matter. These photons have 3 possible interactions. They can be completely absorbed in a single encounter, they can Compton scatter and they can engage in pair production. The experimental cross sections for these three reactions are stored for most materials. In addition it is possible to use information about the density and charge dependence to deduce the relationship of non-stored cross sections to stored ones[4] -- this last has been implemented only for HgI_2 at this time. Electrons are not followed, they are absorbed with the photons.

The detector in figure 1 is defined by a set of lines in rho and z. A subroutine returns the material present at any value of (rho,z). The basic method is to calculate the distance to the nearest defining line in rho and in z. If the material on both sides of the line is the same, the distance is extended to the next line. Stored experimental cross sections are then used to find the probability of interaction. A random number generator[5] determines if, where, and how the photon interacts. If the interaction is photo absorption, the photon tracking stops. If the interaction is a Compton scattering, its direction is changed randomly in accordance with the Klein-Nishina formula[6]. Pair production emits both a pair of 511 KeV photons and a series of bremsstrahlung[7] photons. The raw and anti-coincidence spectra are given in figure 2 for 2614 KeV input photons. The same spectra for 6 MeV photons are shown in figure 3.

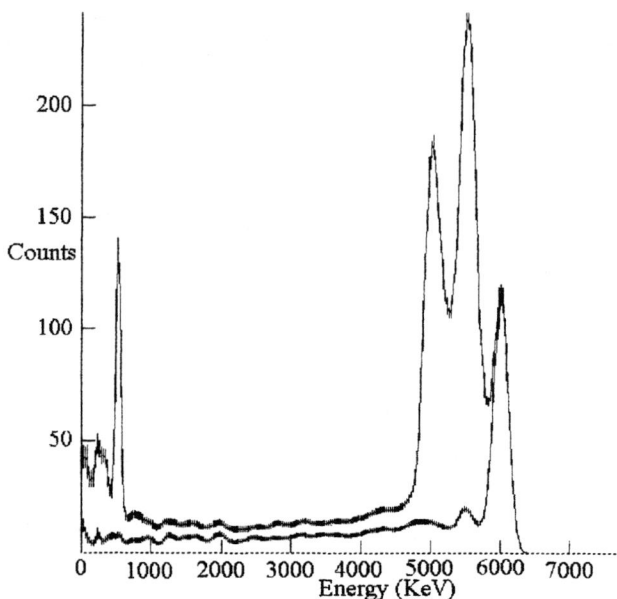

FIGURE 3. Input photon is 6000 KeV. Top spectrum is NaI alone. Bottom spectrum is NaI in anti-coincidence with any event in plastic or BGO.

The largest peak in this spectrum is the single escape peak in which all the incoming energy except that in one of the 511 KeV gamma rays produced by a pair production event is absorbed in the NaI in zone 3. The anti-coincidence spectrum is still a very clean representation of the case in which all of the energy is absorbed in the NaI, but in the presence of background from other sources, it is not as easily found as is the first escape peak.

FRACTIONAL COUNTS

The response function of each detector is used to calculate the probability of a detector reporting a count in a given channel. The sums of these probabilities over many input photons are the spectra plotted in figures 2 and 3. This could have been accomplished, however, by reporting the energy in the single channel corresponding to it, followed by a convolution with the appropriate photo-response function. The earlier versions of the code did exactly this, although the convolution step was frequently not made.

Fractional counts begin to be worthwhile in simulating the attempts made to suppress the photo-peak in figure 3. The following method for isolating peaks from high-energy lines is due to the NASA Goddard X and Gamma- Ray group[8]. The lines defining the spectrum are

```
NaiCoBgo.sp    2
//NaI in coincidence with 511 or 1022 MeV in BGO
//in anti-coincidence with a plastic front shield
//Coin window for BGO is  491 KeV to 531 KeV
//and from 1002 KeV to 1042
E3 if (491 < E1 + E2 < 531) or (1002 < E1 + E2 <
1042) not (10 < E4 < 10000)
```

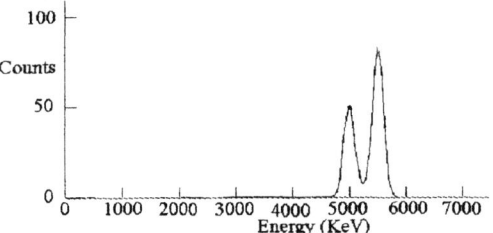

FIGURE 4. Input photon is 6000 KeV. Spectrum is NaI in coincidence with a 511 or 1022 KeV photon in the BGO and nothing detected in the plastic.

The spectrum reported in figure 4 is

$$S_k(i) = \sum P_{k,i} \qquad (1)$$

The k refers to the type of spectrum or equivalently to the set of zones in which energy E has been absorbed. The subscript i refers to energy of channel i. The probability P includes the spread due to the photo-response of the NaI and also that in the other zones. That is, the probability is calculated that the event would be vetoed or accepted for every event. An energy higher than the window around the 511 KeV could be accepted owing to the fact that the detector in charge of vetoing responds as though it were in the proper range, or it could be vetoed even though it is in the correct range. This is a straightforward calculation from the photo-response functions in each zone on an event by event basis, but it is not one that would be easy to make after the events are summed. If there were an energy for which a series of unlikely events need to conspire such as two or more detectors responding at the edge of their usual response functions, this event will be seen as a small value of P_k for many photons in equation 1. It is of course possible to randomly select a single channel in which to report a 1. In this case it will take 10000 photons to see a single channel with an average probability of 10^{-4}.

The photo-response function of the BGO used to generate figure 4 was the same as that of the NaI, which is approximately that of either peak in figure 4. Actually BGO is usually not that energy sensitive. Quadrupling the width of the BGO response function leaves "essentially" the same spectrum but divides the strength of the lines by about 4. This is plotted on a logarithmic scale in figure 5.

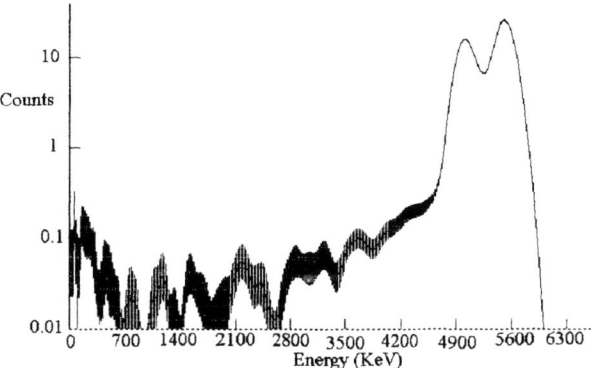

FIGURE 5. Input photon is 6000 KeV. Spectrum is NaI in coincidence with a 511 or 1022 KeV photon in the BGO -- wide response function -- and nothing detected in the plastic.

The vertical lines in figure 5 are the error estimates in the determination of the spectrum. They are the error of the entire curve and not that of the individual points.

STANDARD DEVIATION OF THE SPECTRUM

Include zero as a perfectly acceptable probability. Define P_j to be the probability reported for channel i of the k'th spectrum after the j'th photon has been stopped or exited the system. Then denote the average value of the i'th channel in the k'th spectrum as $<P>$. The standard deviation estimate begins with the definition

$$\langle P \rangle = \frac{1}{N} \sum_{j=1}^{N} P_j \qquad (2)$$

where j is a sum over the N photons tracked by RobSim.

The variance in this value is defined by

$$\langle \{P - \langle P \rangle\}^2 \rangle = \frac{1}{N} \sum_{j=1, k=1}^{N} (P_j - \langle P \rangle)(P_k - \langle P \rangle) \qquad (3)$$

Owing to the fact that the P's are totally uncorrelated, it is assumed that the off diagonal terms average to zero so that

$$\frac{1}{N} \sum_{j=1, k \neq j}^{N} (P_j - \langle P \rangle)(P_k - \langle P \rangle) = 0 \qquad (4)$$

This means that

$$\langle \{P-\langle P\rangle\}^2\rangle = \frac{1}{N}\sum_{j=1}^{N}(P_j-\langle P\rangle)(P_j-\langle P\rangle)$$
$$= \frac{1}{N}\sum_{j=1}^{N}(P_j^2 - 2P_j\langle P\rangle + \langle P\rangle^2) \quad (5)$$
$$= \{\langle P^2\rangle - 2\langle P\rangle\langle P\rangle + \langle P\rangle^2\} = \{\langle P^2\rangle - \langle P\rangle^2\}$$

The standard deviation in <P> is the square root of this quantity divided by the number of observations N

$$\sigma_P = \frac{1}{\sqrt{N}}\sqrt{\langle P^2\rangle - \langle P\rangle^2} \quad (6)$$

The spectrum of course is N times <P>. Thus the standard deviation in the spectrum is

$$\sigma_S(i) = \sqrt{N}\sqrt{\left(\frac{1}{N}S_2(i)\right) - \left(\frac{1}{N}S(i)\right)^2} \quad (7)$$

where S(i) and $S_2(i)$ are the sums of the P's and P^2's respectively.

To see how the usual Poisson statistics come from this, imagine that there is a channel that receives a single count M<<N times. In this case

$$\sigma_S(i) = \sqrt{N}\sqrt{\left(\frac{1}{N}M\right) - \left(\frac{1}{N}M\right)^2} \quad (8)$$

Since by assumption M/N << 1, the second term is much less than the first, this becomes

$$\sigma_S(i) = \sqrt{N}\sqrt{\left(\frac{1}{N}M\right)} = \sqrt{M} \quad (9)$$

The vertical lines at each point figure 5 extend from one standard deviation below the value of the spectrum to one above. A second simulation of the spectrum made with different random numbers has a 65% probability of finding the j'th channel within this vertical line. The adjacent points are highly correlated with this point. If the j'th spectral value is above the line in the second calculation, all values with approximately the same j will also be above the line and vice versa. If the standard deviations were multiplied by the square root of the full-width at half maximum, then the visual estimate based on adjacent points would seem better, but there would be other problems.

CONCLUSION

The code RobSim estimates a detector response function $P(E,E')$. Typically this is calculated by repeatedly allowing a photon of energy E' to enter the system of zones. The code allows a user to imagine detectors connected to each of a number of various zones and to use these to find the "best" possible response function. An attempt has been made to make this process as fast and intuitive as possible.

REFERENCES

[1] R.L. Coldwell, "Robust Fitting of Spectra to Splines with Variable Knots", CP475, *Application of Accelerators in Research and Industry*, edited by J.L. Duggan and I.L Morgan AIP Press, New York (1999)

[2] A.C. Rester, R.B. Piercey, G. Eichhorn, R.L. Coldwell, J. McKisson, D.W. Ely, H.M. Mann and D.A. Jenkins, "The GRAD Gamma-Ray Spectrometer", IEEE Trans. on Nucl. Sci. **33**, 732 (1986).

[3] A.C. Rester, R.L. Coldwell, F.E. Dunnam, G. Eichhorn, J.I. Trombka, R. Starr and G.P. Lasche, "Gamma-Ray Observations on Supernova 1987A from Antarctica", Ap. J. **342**, L71 (1989).

[4] Robley D. Evans, *The Atomic Nucleus*, McGraw Hill, 1955 pp.686-715

[5] R.L. Coldwell, "Correlational Defects in the Standard IBM 360 Random Number Generator and the Classical Ideal Gas Correlation Function",, J. Computational Phys. **14**, 223(1974)

[6] E.U. Condon, "X Rays",in *Handbook of Physics*, ed E.U. Condon and Hugh Odishaw, McGraw-Hill (1958)pp 7-128-7-129

[7] R.L. Coldwell, M.W. Katoot, and P.S. Haskins, "Interactions of Multi-Mev Gamma Rays with Matter", pp. 125-131,*High Energy Radiation Background in Space*, AIP Conference Proceedings #186, Edited by A.C. Rester and J.I. Trombka (AIP, New York, 1989)

[8] L.G. Evans, R. Starr, J.I. Trombka, T. McClanahan, S.H. Bailey, I. Mikheeva, J. Bhangoo, J. Bruckner, and J.O. Goldsten, "Calibration of the NEAR Gamma-Ray Spectrometer", **Icarus** (in press)

Multielement Silicon Detectors for Registration of Charged Particles, x-Rays and Gamma-radiation[*]

D. O. Frolov [1], O. S. Frolov [1,4], A. A. Sadovnichiy [1,4], O. F. Nimets [2], V. A. Shevchenko [3,4]

[1]State Enterprise R&D Institute of Microdevices, Kiev, 03136 Ukraine.
[2]Institute of Nuclear Research, Kiev, Ukraine
[3]Tarasa Shevchenko National Kiev University, Physics Department, Kiev, 01033, Ukraine.
[4]Polinom&K, Kiev, Ukraine

Abstract. A novel type multi-sectional detectors consisting of separate sections has been developed. Signals from each section pass through a shaping amplifier channel and then are mixed in a special manner. This principle allows reduction of detector's and amplifier's electrical noise in many times, and due to this allows one to develop detectors and spectrometers for various types of radiation with characteristics unreachable at present using traditional methods. For example, it's supposed to obtain energy resolution for alpha-particles 30 keV with active area of the detector 40 cm^2. It's possible to create semiconductor spectrometers and radiometers with high resolution and the area of detectors of hundreds of cm^2. Possibilities and limitations of this principle had examined and a series of multi-element detectors and an alpha- spectrometer based on them has been developed.

INTRODUCTION

In detection of ionizing radiation large area of the detector is often required when intensity of the registered particles is low and time given to collect a certain number of signals is limited. However, increase of the detector active area is associated with significant difficulties. As the area of a silicon detector grows, the detector's leakage current and noise associated with it grow.

Besides, electrical capacity of the detector and noise of a charge-sensitive preamplifier, associated with it grow. Increase of noise leads to deterioration of energy resolution and to increase of the minimum detectable energy of a particle. The described fact causes the most serious limitations on sphere of application of silicon radiation detectors.

In the given activity the way of overcoming of this difficulty and application of this way for mining α, β, γ - spectrometer of a new type with the improved characteristics is described.

1. MULTISECTIONAL PRINCIPLE OF DETECTION UNIT DESIGN

We have proposed an idea of multi-sectional detectors which allows one to bypass the mentioned dependency of noise on the area of a detector[1-4]. This idea lies in the following. The detector is divided into **n** independent parts (sections) with area **s** of each section. Each section has its own amplification channel. Signals from all channels are applied to a special mixer [2] which passes them to one bus and then are applied to a counter or an amplitude analyzer. The mixer is designed in such a way that it transmits noise only from one channel, in which the signal passes at the moment. So, we have a detector with the area n×s but the noise of the detector and its preamplifier are corresponded to the area **s**.

We have produced such multi-sectional detection modules. A detector with the area of 8 cm^2 and

---

[*] Work supported by The Sciense and Technology Center in Ukraine, grant № 1578.

consisted of 8 sections of 1 cm² each had complete energy resolution for α-particles of 25 keV while the world best samples of ordinary type reached 32 keV. A detector with the area of 24 cm², consisted also of 8 sections had the resolution of 44 keV (the world best samples of ordinary type have 70 keV [5]). Resolution of the whole detector is only 5 - 10 % worse then resolution of one section with the area of 3 cm² (see Table 1).

TABLE 1. Energy resolution of the multi-sectional detector S= 8x3 cm².

Energy resolutions (keV).	S=24 cm² (8 detectors placed in parallel)	S=8x3 cm² (multi-sectional)	S=24 cm² (single) ORTEC [5]
Total	145	43.7	69
α-source	6	6	6
Preamp	89	15.5	41
Detector	114	40.4	55

Here, the quality of used silicon diodes was not very high. Thus, implementation of the proposed method will allow one to achieve record value of the main parameter of the detector using silicon crystals of not the best quality and correspondingly of low cost.

We believe that the proposed method is universal and applicable for creation of spectrometers and radiometers of various purpose with detectors of large area and high energy resolution. The following devices can be developed:

1) α-spectrometer with energy resolution of 35 keV with the detector area of 30 cm² and 50 keV with the area of 100 cm², which is significantly better then the current level.

2) β-radiometer with silicon detector having active area of 20 - 100 cm².

3) γ- spectrometer with energy resolution of about 7% for ^{137}Cs with the detector area up to 1000 cm².

4) X-ray spectrometer with high energy resolution. The detector by the common area of 9mm² consists of 8 units. The series application of a principle of a multisection system will allow to reduce a self noise of the detector and in 6-8 times to increase speed of registration.

5) α, β, γ- spectrometer with a two-layer detector (silicon, scintillator + photomultiplier).

The outcomes of experiments are below described which allow to estimate capabilities and limitations of this device.

2. α, β, γ - SPECTROMETER

Structurally detector represents thin silicon detector arranged sequentially concerning a bundle of particles and behind it - detector such as phoswich (see Fig. 1).

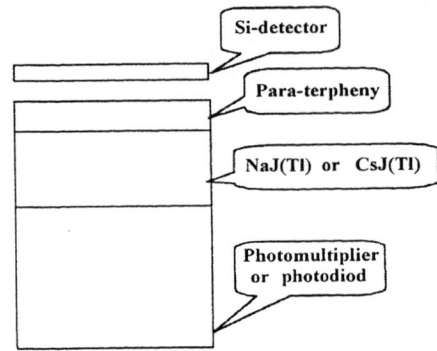

FIGURE 1. Structure of the α, β, γ detector.

Diameter of an active area depends on assignment of the device and can be selected up to 100 mm. For measurement of α-spectra the first silicon detector from high purity of silicon by depth from 100 up to 300 microns will be used only. The measurement of β-spectra is made with the help of silicon and scintillation detectors. The electrons lose energy either in a silicon detector, or in silicon and scintillation simultaneously. The fact of presence of synchronized signals in the first and second detectors testifies to that, what is it signal of an electron. In this case energy of an electron is evaluated as the sum of losses in silicon and scintillator. There are two channels of an analog-digital converter processing signals of the semiconducting detector and photomultiplier. The signals from a silicon detector are divided on α- and β-signals on energy or by more composite mode (see below). The signals of coincidence of both channels fall into to electrons and tot (with definite factors). The signals from a photomultiplier fall into to γ-quanta.

The absorption of γ-quanta descends mainly in a heavy scintillator such as NaJ (Tl) or CsJ (Tl) and in significant of a smaller degree - in a plastic scintillator, as which will use para-terphenyl more often. However, it will not put to considerable errors in a γ-spectrum, as the efficiency of a plastic scintillator is only 1-2 %.

For mocking-up α, β, γ-spectrometer the silicon detector manufactured which is a matrix of diodes arranged on one slice of silicon. The silicon slices n-type as with specific resistance 5 kOhm×cm and thickness 300 μm were used. The area of one die backer is made 0,5 cm², with the fissile area of all detector - 16 cm². The outcomes of experiments on the detector consisting of 8 sections on 2 cm² are below adduced which is represented to us optimum.

2.1. α- spectrometry

The main regime parameters, on which the energy resolution of the α-detector - bias voltage on the detector depends and τ of the shaper (the condition, that $\tau = \tau_{dif} = \tau_{int}$) is received.

TABLE 2. The full energy resolution and position of a peak ^{238}Pu for the detector with S = 2 cm^2 at different U and τ (position of a peak at U = 40 V is accepted for 1).

τ, μs	Energy resolution, keV		Peak position	
	U = 40 V	U = 0 V	U = 40V	U = 0
12,8	30	53	1	0.9983
3,2	24,5	70	1	0.9964
0,1	32	119	1	0.3274

For our detectors the relation of a noise to bias voltage has very smoothly varying minimum and the optimum bias voltage on the detector lies in area 20-40 V. However for registration of electrons rather interesting the mode is a low voltage and small τ. The relation of a noise from τ also has a minimum lying in range 3-15 μS. In the table 2 the data about the energy resolution and position of a peak are adduced at different bias voltage and τ.

2.2. β - spectrometry

In offered α, β, γ-detector at a spectrometry of electrons the signals of a silicon and scintillation detector will be used. The main problems are correct separation of an α-spectrum and spectrum of losses of electrons, electrical noise and supression of a γ-background in a β-channel. Below these problems surveyed explicitly.

2.2.1. Condition of a congregating of charges in a silicon detector at effect of charged particles

Let's consider the factors determining value of registered energy losses of electrons in a silicon detector. The bias voltage on the detector changes depth of a space charge region in silicon [6]:

$$d_f = |2\varepsilon_{Si}(U + U_b + 2kT/q)/qN_D|^{1/2}, \quad (1)$$

where ε_{Si} - permittivity of silicon, q - charge of an electron, N_D - concentration of the donors in a substrate of silicon, U_b - applied bias voltage, U - contact potential difference on p-n transition. In the elementary unidimensional and fixed case at homogeneous rate of a carrier generation a full current of hole

$$I = I_h + I_{dif} \sim d_f + L_D, \quad (2)$$

Where L_D - length of diffusion of hole in a substrate of silicon. In an actual case a full signal charge is equal:

$$Q = Q_f + Q_D, \quad (3)$$

where Q_f - field component of a charge, it is derivated by hole generated in a space charge region, Q_D - diffusive component. Here it is necessary to enter an effective length of diffusion of hole $L_{D\,eff}$, which are depended on the shape and arrangement of a packet of hole after process of ionization. Then

$$Q_f \sim d_f; \quad Q_D \sim L_{D\,eff}. \quad (4)$$

For an estimation $L_{D\,eff}$ it is possible to offer a following way. The back side of the detector is irradiated with α-particles. If $(d_{Si} - d_f)$ there is more run of α-particles, the signal is determined only by diffusive flow of hole and $Q = Q_D$. In a Fig. 2 the relations from U and τ of amplitudes of signals are

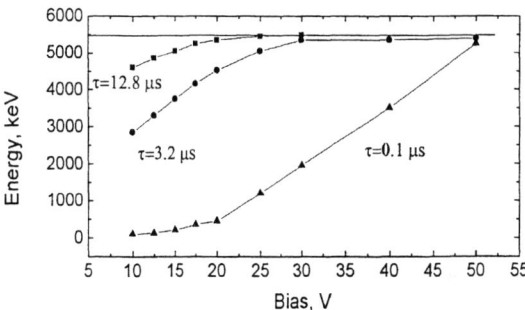

FIGURE 2. The relations from U and τ of amplitudes of signals are adduced at irradiation of a back side.

adduced at irradiation of a back side, reduced to E_α. The values calculated ground these data, of effective lengths of diffusion are equal (for bias voltage range 30-50V) are listed in table 3.

TABLE 3. Relation of length of diffusion from τ.

τ, μs	$L_{D\,eff}$, μm
12,8	550
3,2	200
0,1	60

It is necessary to mark very high $L_{D\,eff}$ for range of times τ from 3 μs up to 13 μs, which often will be used in spectrometric practice. Apparently, it testifies first of all to excellence of initial silicon (corporation Wacker) and master schedule of manufacturing of

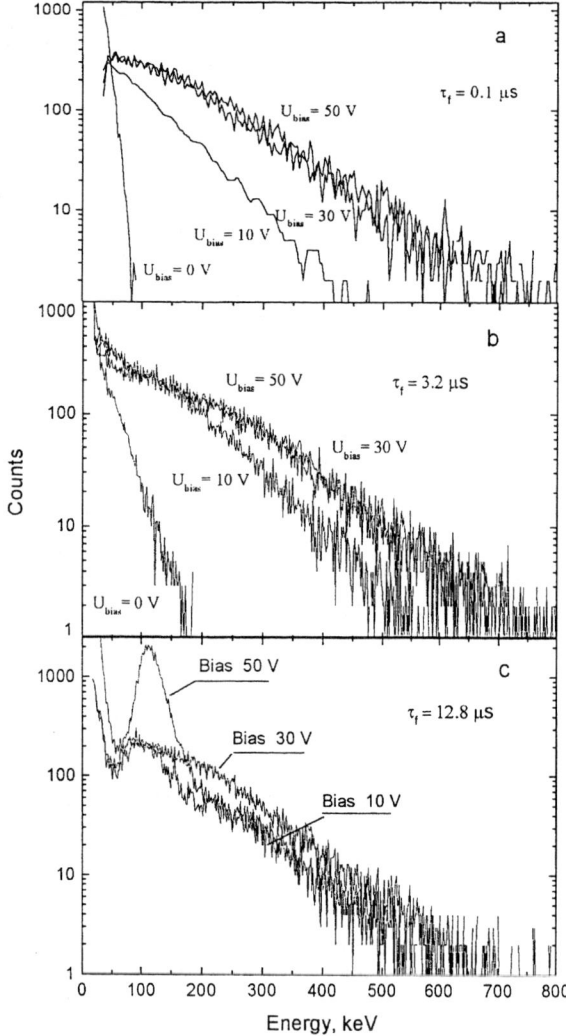

FIGURE 3. The spectrum of losses of electrons from Y+Sr sourse.

detectors that provides a high life time of hole. The selection of bias voltage on the detector and τ of formation of an impulse allows over a wide range to vary a depletion layer of silicon in which the ionization hole will derivate a signal charge.

2.2.2. Spectra of losses of electrons

The measurements of losses of electrons were conducted at different U and τ shown on a Fig. 3. The irradiation was made by electrons from a source (Sr + Y) at normal falling of particles. In the table 4 are adduced of maximum energy of losses of electrons at different U and τ. From the table follows that by selection U and τ it is possible considerably to change maximum energy of a spectrum of losses of electrons. So at U = 0, E_{max} = 190 keV in a case τ = 3.2 μs and

TABLE 4. Maximum energy of losses of electrons in a silicon detector depending on U and τ.

τ, μs	Losses, keV	
	U = 50V	U = 0
12,8	700	338
3,2	787	188
0,1	770	100

E_{max} = 100 keV in a case τ = 0.1μs. This circumstance can be utilised for correct separation α and β of spectra (see section 2.2.3).

2.2.3. Separation of spectra of α-particles and electrons

The presence of bumper zone between borders of a spectrum of losses of electrons and α-spectrum from 800 keV up to 2 MeV confidently allows to conduct discrimination of signals in a silicon detector from α-particles and electrons. For correct α-β separation the bias voltage on a silicon detector decreases. The depth of an active zone drops approximately up to 30 microns. Thus in a spectrum of α-particles the energy resolution is some aggravated. At the same time spectrum of losses of electrons is essentially reduced on energy.

2.2.4. Electric noises at registration of electrons

The silion detectors differ by rather high leakage currents and according to a high electrical noise. In our device this difficulty is overcome usage of the multisection detector. In another way of noise reduction of the detector is the selection of regime parameters U and τ. The decreasing U results in decreasing a leakage current ($I \sim U^{1/2}$) and amplitude of a noise ($U_d \sim I^{1/2}$). From here

$$U_d \sim V^{1/4}, \qquad (5)$$

The noise charge sensitive of the preamplifier depends on capacity of the detector:

$$U_{pa} = U_o + k(C_d + C_{pa}), \qquad (6)$$

where U_o - initial noise (at capacity of the detector C_d = 0), k - declination of relation of a noise from capacity in eV/pF, C_{pa} - input capacitance of the preamplifier. As a rule, $C_d \gg C_{pa}$, therefore in these cases

$$U_{pa} = U_o + kC_d. \qquad (7)$$

As $C_d \sim U^{-1/2}$, then $\quad U_{pa} = U_o + kU^{-1/2}. \qquad (8)$

The expressions (5) and (8) display, that with increase U the detector noise grows, and the noise of the preamplifier drops. It results that there is a minimum in relation of a full noise $U = (U_d^2 + U_{pa}^2)^{1/2}$ from U.

In many appendices of a β-spectrometry and β-radiometry a pacing factor limiting minimum measured activity (MDA), is the γ-background [7]. The detector, offered us, has low sensitivity to γ-radiation, at first, because of low efficiency of absorption and dissipation of γ-quanta by a thin slice of silicon. At depth of a slice 300 μm the efficiency makes 3×10^{-3} [8]. The spectra of losses of γ-quanta ^{60}Co on one unit of this multisection detector were measured. The area of a unit S = 2 cm^2. The outcomes are adduced in a Fig.4. It is possible to receive a factor of supression of a γ-background about $5 \times 10^{-4} - 5 \times 10^{-5}$.

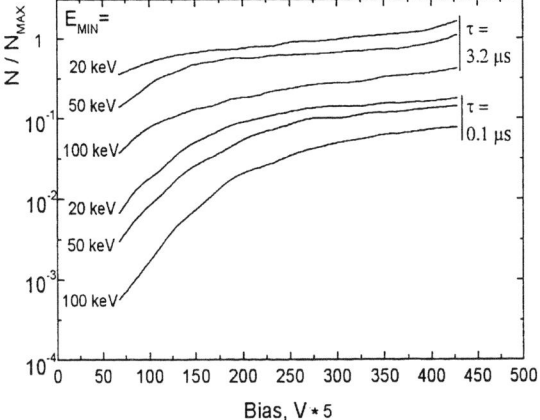

FIGURE 4. Reduced number of signals, which energy exceeds selected values E_{min} =20, 50, 100 keV, depending on U and τ. Time of measurement 30 min.

3. SUMMARY

In the given activity the principle of a multisection silicon detector is offered and tested.

The outcomes testify that controlling parameters, it is possible to receive at the area 20cm^2 the energy resolution on α- particles 25-30 keV and MDA=10^{-1} Вк/kg and less on β-particles.

Let's put two examples of selection of a mode of measurements pursuant to the main purpose of measurements.

1. If the main purpose of measurements is the measurement of minimum activity of β-radiators (for example, ^{90}Sr), it is necessary to select small τ and small V. It allows to measure β-activity up to 0,1Bq/kg. Thus the energy resolution in α-spectra will be about 35keV (at the area of the detector 18 cm^2).

2. If it is required to receive the maximum energy resolution in α-spectra, it is possible to increase τ and V. Thus is a little aggravated MDA on β. It makes about 1Bq/kg, that is connected to some increase of an electrical noise.

ACKNOWLEDGMENT

The authors would like to express their thanks to Dr. I.N.Kadenko for help on this work.

Development and research of silicon matrixes were made under the grant STCU № 1578.

REFERENCES

1. Nimets O. F., Frolov O. S., Shevchenko V. A., Gavrilenko V.I., Sadovnichiy A. A.,"The multy-element detection unit," Positive solution about issue of the patent for invention from 04 Feb. 1999 under the patent application of Ukraine №99020641.
2. Shevchenko V. A.," The analogue summator for multy-element spectrometers of a nuclear radiation," patent application of Ukraine №30082, 18.12.1997.
3. Nimets O. F. , Frolov O. S., Shevchenko V. A., Gavrilenko V.I., Sadovnichiy A. A., "The multy-element detection unit for research of parameters of an ionizing radiation," in *The Collected of scientific works of Institute of Nuclear Researches of a NAS of Ukraine.*, Kiev: NAS Ukr. Press, 1999, page 121 - 123.
4. Nimets O. F., Frolov O. S., Shevchenko V. A., "Alpha spectrometer with a large illumination power," in *The Kiev university as center of national confessor, science and culture*, Kiev: Kiev Univ. Press, 1999., page 57- 60.
5. EG&G ORTEC 97/98, Detector and Instruments for Nuclear Spectroscopy. Charged-Particle Detectors, 1998, p.1.8 -1.16.
6. Sze S.M., *Physics of Semiconductor Devises*, 1981, part 2.3.1.
7. Atom Complex Devices, Catalog, Kiev: ACD Press, 1998, p.10.
8. Akimov Yu.K., et. al. *Semiconducting detectors in experimental physics*, Moskow: Energoatomizdat, 1989, p. 72.

SECTION VI

ACCELERATOR TECHNOLOGY:
ION SOURCES, NEW FACILITIES AND COMPONENTS

Application of ECR Ion Sources for Surface Modification of Materials

Aleksandar Dobrosavljević, Nataša Bibić, and Nebojša Nešković

Vinča Institute of Nuclear Sciences, P. O. Box 522, 11001 Belgrade, Yugoslavia

Abstract. Electron cyclotron resonance ion sources (ECRIS) have been used primarily as injectors for cyclotrons and linear accelerators. Significant improvement of their performance in the last two decades and substantial reduction of their price and running cost have opened the new fields of their application. In this paper we describe the applications of ECRIS in the field of surface modification of materials, which represent a relatively novel approach. The major advantages of these sources when compared to other sources applied in this field are: production of multiply charged ions, wide range of ion species obtained from gaseous and solid substances, and uninterrupted stable operation in a long period of time. New types of low power consumption ECR ion sources, with the magnetic structure made entirely of permanent magnets, are well suited for the applications using high voltage platforms, e.g., for ion implantation.

INTRODUCTION

The first practical applications of the electron cyclotron resonance ion sources (ECRIS) were in the field of accelerator techniques, where the benefits of multiply charged ions could justify the relatively high price and high power consumption of these sources. Since then, the ECRIS have been widely used as cyclotron heavy ion injectors. The use of multiply charged ions in cyclotrons is important because the extracted beam energy per nucleon is proportional to $(Q/A)^2$, where Q is the charge state and A is the mass number of the accelerated ion. It is obvious that cyclotrons can benefit tremendously from accelerating high charge state ions.

There was also considerable interest in using multiply charged ion beams at low energies for different investigations related to the ion beam interaction with solids. Unfortunately, wide application of the ECRIS as a stand-alone machine was mostly restricted by its high price. A good opportunity for such research was found on the multipurpose cyclotron installations that accelerate light as well as heavy ions. During acceleration of light ions, the ECRIS heavy ion injector is free for other applications at low energies.

Recently, new types of low power and cheap ECRIS were developed. Their magnetic structure is completely made of permanent magnets and their power consumption is reduced even below 100 W, enabling easy installation on high voltage platforms. This kind of ECRIS is optimized for production of high currents of low or singly charged ions (few mA). They are well suited for applications in the field of ion implantation and surface modification of materials.

SURFACE MODIFICATION OF MATERIALS BY ION BEAMS

Ion beam interaction with solids is of practical interest for such fields as ion beam processing of materials, surface modification and characterisation of materials, and reactor technology. By using a well defined beam of ions, with energies in the range from a few tens to a few hundreds of keV, ion implantation allows the introduction of any element into any solid resulting in the formation of surface alloys. By this means, important modifications of various physical, mechanical, and chemical properties have been developed. An interesting aspect is that the ion beams deposit high energy densities over a very short period of time, thus it is possible to overrule thermodynamic forces and produce new solid phases such as extended solid solutions of different metals or amorphous metallic alloys [1]. These materials have some unique properties thus they are interesting for wide range of applications.

A typical scheme of an experimental installation for surface modification of materials by ion beams is presented in Fig. 1. It is basically an ion implanter with additional capabilities regarding the sample treatment and analysis. After the ion source (IS), there should be an analyzing magnet used to separate the required ions (AM). If the energy of the ions obtained directly from the ion source is not sufficient, some kind of beam postacceleration (PA) has to be introduced. Usually, the whole ion source together with AM is lifted to the high voltage platform and an acceleration tube with several electrodes is used. Afterwards, some kind of

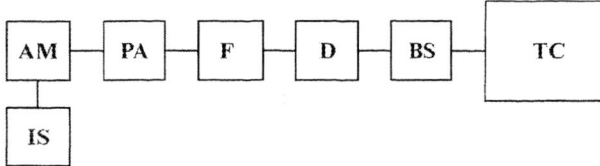

FIGURE 1. The scheme of an experimental channel for surface modification of materials by ion beams: IS - ion source, AM - analyzing magnet, PA - ion post acceleration, F - beam focusing elements, D - beam diagnostic elements, BS - beam scanning system, TC - target chamber.

ion beam focusing ellements (F) have to be installed (e.g., triplet of quadrupole magnets, or electrostatic quadrupols). Beam diagnostic ellements like Faraday cups and beam profile monitors (D) are installed along the beam line at certain locations. In order to spread the beam over the large sample surfaces, an electromagnetic or electrostatic two-dimensional beam scanning system (BS) is installed in front of the target chamber (TC). The samples that are exposed to the ion bombardment are installed inside the TC on a sample holder and can be positioned and rotated regarding the incomming beam. Some additional systems are often located inside the target chamber, for instance, a system for the evaporation of materials over the sample surface, a sample heating/cooling unit, sample neutralisation assembly, etc. Also, additional devices for analysis and measurement are usually required, e.g., for residual gas analysis (RGA), secondary ion analysis (quad SIMS), thin layer deposition rate and thickness measurement, sample temperature measurement. Finally, the vacuum system is an important factor of the whole installation. A good vacuum, free of hydrocarbons is required to obtain good beam transport and avoid sample contamination. The vacuum should be better than 10^{-4} Pa. For that purpose, a combination of turbomolecular and cryo pumps is the best choice.

WHAT KIND OF REQUIREMENTS SHOULD ECRIS FULFILL

Ion implantation and surface modification of materials represent a highly demanding field of ECRIS application. The major requirements are generally related to obtaining high currents (high doses) and a wide energy range (from several keV up to several MeV), providing uniform doses over the sample surface, and producing of different ion species from gases and solids. Of course, the total price of the ECR ion source is another important aspect that can restrict its field of applications.

Sample irradiation dose

The major parameter for the sample irradiation is fluence or dose, defining the total number of ions per unit area that are implanted into the sample. Generally, doses over 10^{17} ions/cm^2 require high currents from ECRIS (over 1 mA) in order to perform the sample irradiation in several hours, not days. This is important in case of commercial applications where the irradiation time and the electrical power consumption define the price of product. Tribological and corrosion protection applications require typical doses of 10^{17} ions/cm^2; silicon-on-insulator (SOI) technology for the formation of micromechanical structures requires implantation of oxygen ions to a doses of 10^{18} ions/cm^2. On the other hand, there are other applications like ion beam mixing, microstructural changes etc. requiring doses in the range of 10^{13}-10^{16} ions/cm^2 that can be easily achieved.

The modern, high performance ECRIS is capable of producing high yields of high charge state ions. The ion beams that are widely used for the surface modification of materials can reach several hundreds of eμA, even crossing the 1 emA limit (e.g. 500 eμA of N^{5+}, 1000 eμA of O^{6+}, 1000 eμA Ar^{8+}). The new generation of specially designed ECR sources can provide even higher currents of low or singly charged ions.

The ion beam energy

The desired penetration depth of the ions that are used usually defines the required beam energy. In general, the ions produced in an ion source are accelerated by an electrostatic field. The ion acceleration can be performed by the extraction voltage of the ion source alone, or by introducing some additional acceleration system if the higher energies are needed (e.g., high voltage platform, Van de Graaff generator). The use of multiply charged ions is often the most convenient way to extend the energy range to higher energies.

The energy of an ion extracted from ECRIS is defined by the extraction voltage (usually V_{ex} = 10-25 kV) and the ion charge state ($E = Q \cdot V_{ex}$). Thus, we can increase the beam energy by choosing the higher charge state ions (e.g., for 25 kV extraction voltage, Ar^{4+} ions can reach 100 keV and Ar^{8+} ions 200 keV, while Xe^{20+} ions can reach 500 keV). Unfortunately, detailed analyses show that for some important applications we can hardly reach the required energy level. For example, C and N ions, which are important for improvement of tribological properties of stainless steel, or O ions, which are important for semiconductor technologies, can hardly cover the basic 100–200 keV range, even taking into account the highest charge states with reasonable ion yields (C^{4+}, N^{5+}, O^{6+}) and maximal extraction voltage. Improving the surface hardness of polymers is another case, where the best results could be obtained with C, N, O, Ne, or Ar ion beams with energies over 500 keV. These energies could not be obtained from ECRIS without additional post acceleration.

Also, one has to be aware that use of the high charge state ions for obtaining the higher energies might have some disadvantages. Usually, the particle current I_p (pμA) decreases by increasing the charge state of ions. For example, if we obtain the electrical current I_e = 200 eμA for Ar^{4+} and the same current for Ar^{8+} ions (measured on the Faraday cup), the energy of Ar^{8+} beam will be two times higher for the same extraction voltage but in return, the particle current will be two times lower ($I_p = I_e/Q$), doubling the sample irradiation time (to obtain the same dose).

The ion energy can be increased by applying some kind of beam post-acceleration: (i) One possibility is lifting the ECRIS on the high voltage platform. This solution was too expensive and complicated until the invention of the low power ECR sources, made completely of the permanent magnets. They can be easily used on conventional implanter platforms (100-200 kV), providing high-current beams of singly or low charged ions. (ii) It is possible to introduce the rf accelerator (e.g., RFQ) but this solution is expensive and gives low acceleration efficiency (bunched beam). (iii) Lifting the sample inside the interaction chamber to a negative bias voltage (e.g., -100 kV) is another possibility, especially attractive in the case of high power ECRIS.

Dose uniformity

Often it is important to obtain a highly uniform dose over a relatively large sample surface (e.g., 20×20 cm^2). This can be achieved using the beam scanning system. Usually, the beam scanning system is based on electrostatic deflection plates for horizontal and vertical beam scanning. It enables uniform irradiation of the sample surface inside the interaction chamber, achieving dose homogeneity within 1 %. Spreading the beam over a large surface decreases the equivalent particle current density, and thus significantly increases the irradiation time.

Production of different ion species

Experimental or applied programs in the field of surface modification of materials require a variety of ion species. The most commonly used ion beams that are produced from gases are He, N, O, Ne, Ar, and Xe. In the case of solid substances, ions are produced either from gaseous compounds, or by evaporating the pure elements or oxides inside a miniature oven. For example, B ions can be obtained from BF_3 gas, C ions from CO_2 gas, and Si ions from SiH_4 gas, while Zn or Pb ions are obtained using miniature oven. There is so called MIVOC technique that uses special metalloorganic volatile compounds such as ferrocene ($C_{10}H_{10}Fe$) for obtaining ions of Fe, or $Cr(CO)_5$ for ions of Cr. Most of them are toxic or flammable, so special care has to be taken in this cases. Generally, poisonous or chemically aggressive species are not recommended for use (e.g., F, P, Cl, As) because they can harm the people or damage the expensive equipment (e.g., cryogenic pump may collect on its cold cap a significant amount of poisonous material that could be released in atmosphere during the regeneration process).

AN EXAMPLE OF A CHANNEL FOR MODIFICATION OF MATERIALS BY ION BEAMS FROM AN ECRIS

TESLA Accelerator Installation is a multipurpose cyclotron facility that can accelerate light and heavy ions. It is still under construction in the Vinča Institute of Nuclear Sciences in Belgrade. A part of this installation that is functioning represents successful application of ECRIS in the field of surface modification of materials. The mVINIS Ion Source serves as an injector of the VINCY Cyclotron, but it works also as a stand-alone machine, connected to the low energy experimental channel for modification of materials (L3A) [2,3]. A schematic layout of mVINIS connected to the L3A channel is given in Fig. 2.

In general, the L3A channel offers two kinds of experimental work: (a) using only the target (interaction) chamber as a stand-alone device, and (b) using the complete configuration, employing beams from mVINIS.

(a) The target chamber offers thin film deposition using an electron gun for evaporation of solid materials. An additional argon gun can be used to assist the deposition process (IBAD), improving the coating properties. Another type of experiments is related to the formation of hard coating materials using reactive deposition (TiN or CrN layers).

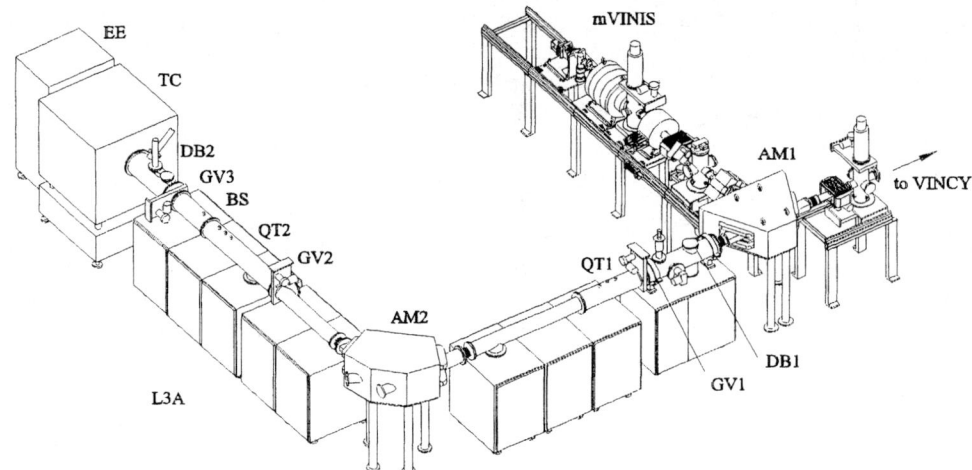

FIGURE 2. Layout of the complete facility including the mVINIS Ion Source and L3A channel: AM1, AM2 - analysing magnets; QT1, QT2 - electrostatic quadrupole triplets; DB1, DB2 - diagnostic boxes; BS - beam scanning system; GV1, GV2, GV3 - gate valves, TC - target chamber, EE - electronics rack.

(b) Ion beams obtained from mVINIS are used to irradiate samples inside the target chamber, broadening the spectrum of possible experiments. These experiments include direct ion implantation, ion beam mixing, and ion beam assisted deposition. The ion irradiation induces compositional and microstructural changes in materials, improving their physical, chemical and mechanical properties.

CONCLUSIONS

The electron cyclotron resonance ion sources entered just recently into the field of ion implantation and surface modification of materials. Significant improvement of the performance of the ECRIS over the last two decades, as well as the reduction of their price, and their running cost through the significant reduction of their electric power consumption made them attractive for applied research or industrial applications in the low energy range (below 1 MeV).

A high performance ECRIS, which is capable of producing high charge state ions, is still an expensive machine (over 1 million $) and a high electric power consumer (over 100 kW). Therefore, its use on conventional implantation installations, especially for commercial use, cannot be approved. It may be used in this field only as a multipurpose ion source, operating primarily as an accelerator injector and partially as a stand-alone machine for low energy applications in the field of surface modification of materials and other related research fields. If the energy of extracted ions is not sufficient, it seems that the most convenient solution is to bias the sample inside the target chamber to the negative potential.

Several compact, reliable and cheap ECRIS have been developed recently by Pantechnik (NANOGAN, MICROGAN, PICOGAN). Their magnetic structure is completely made of permanent magnets and they require 10-100 W of microwave power. The MICROGAN source is well suited for production of medium currents of multiply charged ions (500 eμA of Ar^{4+}, or 32 eμA of Ar^{8+}) or high currents of low or singly charged ions (1-3 mA of Ar^{1+}, P^{1+}, B^{1+}). This kind of compact and low power consuming ECRIS can be easily mounted on high voltage platforms and thus enter into the field of ion implantation and surface modification of materials, replacing the standard Freeman ion sources [4].

The major advantages of the electron cyclotron resonance ion sources relative to other ion sources are: production of high charge state ions, wide range of ion species produced from solid and gaseous materials, and uninterrupted stable operation over a long period of time (no filaments or other consumable parts). The beams of multiple charged ions from an ECRIS can provide the most versatile and sophisticated family of methods for tailoring of the surface properties of all classes of materials such as semiconductors, metals, polymers, glassy carbon, ceramics.

REFERENCES

1. Riviere, J. P., *Nucl. Instr. and Meth.* B **68**, 361-368 (1992).
2. Dobrosavljević, A., et al., *Rev. Sci. Instrum.* **71**, 915-917 (2000).
3. Dobrosavljević, A., et al., *Rev. Sci. Instrum.* **71**, 786-788 (2000).
4. Bieth, C., et al., *Rev. Sci. Instrum.* **71**, 899-901 (2000).

New Ion Beam Development at Kansas State University

C. W. Fehrenbach and M. P. Stockli[*]

Kansas State University, Manhattan, KS 66506

Abstract. New experiments planned for the Low Energy Ion Collisions Facility in the J. R. Macdonald Laboratory at Kansas State University are requiring high fluxes of low-charge ions. We have begun construction of new ion source facilities to supplement the Electron Beam Ion Source in the LEICF area to try to better meet these needs. This initiative includes a new accelerator platform with multiple ion-source capabilities for delivering low-charged ions to existing beam lines. We are also working on portable ion sources that can be used for setup work on experiments or allow experiments to run independently of the main ion accelerator. In this talk I will review the ion-beam requirements for the experiments in the LEICF and describe the planned facility improvements.

INTRODUCTION

The Low Energy Ion Collisions Facility in the J. R. Macdonald Laboratory at Kansas State University has been primarily centered around experiments using the Cryogenic Electron Beam Ion Source. The CryEBIS was designed to deliver highly-charged ions at low energies to experimental beam lines. It has been quite valuable both in facilitating pioneering studies with multiply-charged ions at low energies [1-3] and as a calibration source for other groups setting up detectors for multiply-charged ions [4].

Many recent proposals for the LEICF, however, have required projectiles of low charge. It becomes problematic for the CryEBIS to deliver heavy ions of low charge because the length of our trap region is such that even without confinement, the drift time of ions through the trap will result in considerable ionization of the outer shell of heavy ions. This can be seen, for example, in the yield curves for Ar, in which Ar^{7+} is dominant even with no confinement time [5]. Production of low charge states generally requires that high seed-gas pressures be fed into the source so that adequate beam can be produced at the low efficiency. This, in turn, causes problems when the source is tuned again for high charge states in the same or in succeeding experiments. The high residual gas pressure left in the trap increases the rate for highly-charged ions in the trap to capture electrons from neutrals which drift into the electron beam.

Because of this difficulty of providing projectiles of low charge to experiments in the low energy facility, the Macdonald lab has begun construction of new ion-source systems to provide low-charged beams to users. The biggest effort in this upgrade is to build a new accelerator platform. This platform will be able to hold several ion sources to produce low to moderately charged beams. The new platform can be biased to 250 kV. The ion beam from this facility will be merged into the common beam line from the present CryEBIS source, so that experiments can make use of charge states from one up to the limit of the CryEBIS.

Additionally, we have made provisions for other experiments to run on the ECR source which is set up for the ion-ion collision studies in the LEICF. We are also setting up stand-alone ion sources for various uses in the lab, either to aid experimenters to set up their apparatus in advance of beam time on the accelerator facilities, or so that experiments which do not need the full capabilities of the accelerators can run independently.

---

[*] present address: Oak Ridge National Laboratory, Oak Ridge, TN 37380

LOW CHARGED ION PLATFORM

The primary goal in adding a new accelerator platform to the LEICF is to relieve the CryEBIS of having to produce low-charge-state ions. This way the CryEBIS can be kept conditioned for the production of high-charge states. In addition, we can do development work on the CryEBIS while experiments are running with the low-charge-state source. The beams from CryEBIS and from the new platform will be merged into a single switching magnet. The scheme is shown in Fig. 1. With this arrangement a single experimental beam line can use beams from either source, thus allowing it to take advantage of the full range of charge states that we will be able to deliver without the interaction chamber having to be moved from one facility to another.

The new platform is designed to have multiple ion sources mounted on it so that the source which best matches the users requirements can be selected without having to vent the vacuum system. Switching between sources is accomplished by using an analyzing magnet with multiple entry ports. This is shown in Fig. 1.

Commercial ion sources are planned for all of the new platform. We already have a hot cathode type ion-gauge source (VG Microtech model EX-05) which can produce singly charged ions from gas samples. We also have Penning Ion Gauge sources from Physicon, Inc., which can produce charge states up to 3+ from gas sources or from solid sputter cathodes [6]. We are also trying to obtain funding to purchase an ECR source, the SuperNanogan from PanTechnik, Inc. This source will produce charges up to 14+ in Ar, and 24+ in Xe. With this source functional on the new platform, the CryEBIS can be dedicated to producing only highly-ionized species.

ION-ION ECR

Other changes in the LEIC area are being made in addition to the new accelerator platform being added to the CryEBIS beamline area. These are also indicated in Fig. 1. A new beam line has been added to the ECR source on the ion-ion facility to run a series of experiments by a group from Colorado State University. These measurements will focus on determining the angular momentum distributions in product states following electron capture by multiply charged ions.

These new investigations help to underscore the value of adding high-intensity low-charge state ion sources to augment the CryEBIS in the LEICF. This same group was able to successfully complete a series of experiments on the CryEBIS which looked at target energy dependence for electron-capture into a particular final state [2]. Their technique used laser spectroscopy to analyze the collision products. The CryEBIS was able to deliver enough C^{3+} that they could get reasonable signal-to-noise to detect Rydberg states with the l levels not resolved by the laser.

In the new experiments the group wants to look at the distributions of l states, so they have changed to a species with a greater dipole polarizability so that the l levels will be resolved in their laser signals, with the amplitude of the signal correspondingly decreased. The CryEBIS could no longer deliver adequate flux of ions to make the measurements possible. Moving of the experiment to the ECR source, where it shares source time with the ion-ion measurements, has resulted in a hundred-fold increase in S/N in a factor of three less measurement time than what they were using on the EBIS. This series of experiments is expected to keep running in this location as an ongoing effort.

STAND-ALONE ION SOURCES

By stand-alone sources I mean an ion source which does not use an accelerator column, the ion beam energy is determined by the extraction potential applied to the source. Small stand-alone sources are playing an increasingly prominent role in JRM lab operations.

One of the first uses of such a source in a major research effort was on the ion-ion collisions setup. A small PIG source produces singly charged ions which collide with an ion beam from an ECR source. The PIG source produces much greater fluxes of singly-charged ions than is possible with the EBIS. It also has the advantage that the experiment does not have to share source time with other experiments.

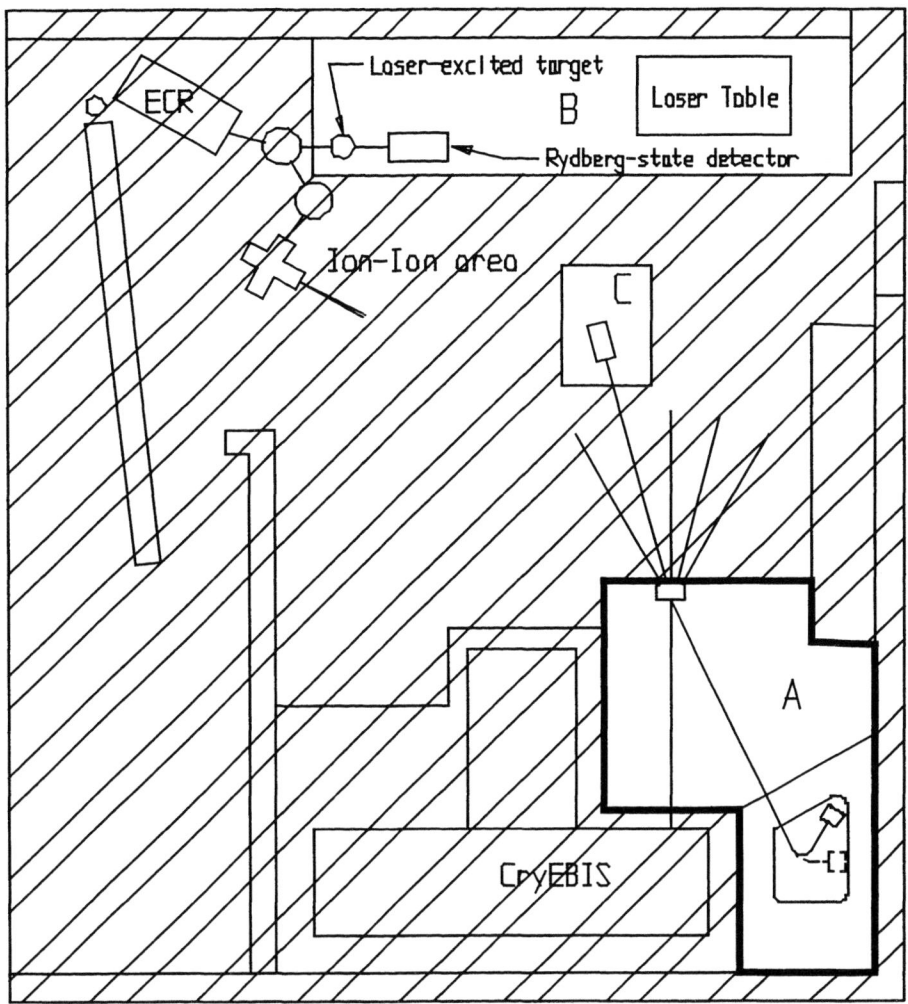

FIGURE 1. Floor plan of the Low Energy Ion Collisions Facility in the J. R. Macdonald Laboratory. The areas which are not cross-hatched indicate changes to the lab. The area marked 'A' shows where the new accelerator platform will go and how the ion beam is merged into existing beam lines. Two ion source locations are shown on the platform (with one dashed.) The platform can accommodate three different ion sources. Area B is where an outside group from Colorado State University has set up an experiment to study the interaction of ions with a laser-excited target. The ion beam comes from the ECR on the Ion-Ion collisions facility. Area C shows a stand-alone ion source mounted at the end of a beam line used for a TRIMS experiment. This ion source provides higher fluxes of protons than is possible with the EBIS. The EBIS system can still be used to deliver beams with energies above 20 keV.

Similar considerations have led to the installation of a stand-alone source on one of the CryEBIS beam lines. The experiment is trying to do momentum imaging of the target ion when atomic hydrogen is ionized by a proton. The target gas is produced from H_2 by an rf dissociator which gives much lower target densities than typical COLTRIMS jets. Getting reasonable signal rates, then, means having to increase the projectile flux. A commercial PIG source (Physicson, Inc. model CGAS 221) has been installed on the interaction region in an interesting double-ended configuration which allows beam to be delivered to the target from either the stand-alone source, or from the CryEBIS, which is diametrically positioned with respect to the PIG source. The CryEBIS will be used if projectile energies are needed that are greater than what the PIG source can deliver. Currently the PIG source is limited to 10 kV extraction voltage by the isolation transformer in the power supplies to the discharge. The physical voltage isolation of the vacuum vessel and power supplies is limited to 20 kV.

The idea of running a beam from a stand-alone source "backward" through a COLTRIMS apparatus originated as a way of making more efficient use of beam time on the main accelerators. A group setting up a COLTRIMS run on the LINAC facility in the JRM lab typically has to use the first day of their beam time to align their target jet and get their electronics tuned up. By using a stand-alone ion source on the beam-dump side of their apparatus, they can do this alignment off line.

To facilitate this alignment procedure, we have begun building low-cost ion sources for this purpose in the Macdonald lab. Our first prototype is the cold-cathode source shown in Fig. 2. The entire source is built on a single 2.75 inch conflat vacuum flange. The extraction system fits into a stock commercial hv isolator, also with 2.75 in c.f. flanges. With the small size, standard vacuum components and availability of low cost dc-to-dc converters to power the discharge, this proves to be a very inexpensive source to set up. The source produces much more ion beam than is needed for COLTRIMS measurements, so we are planning to refine the design with a smaller aperture and greater attention to gas sealing, so as to reduce the gas load in the extraction region.

This cold-cathode source is getting an interesting application in a TRIMS experiment which is looking at the dissociation of HD. In this case, the investigators want to use a D^+ beam to compare with their results using protons [6]. Because of the relatively low gas throughput needed by this source, it is quite attractive to operate it with an expensive source gas, like D_2, compared to producing the beam with the diode source on the Tandem Van de Graff accelerator.

One final not on ion-source development at the JRM laboratory. We are working on developing a source for Th^{3+} for experiments on atomic spectroscopy of Fr-like ions. This is another example of a beam which our CryEBIS would have great difficulty in producing at the flux levels needed for this experiment. We are instead trying to use a PIG ion source (Physicon model CC 2.21) with a means of obtaining the sample by sputtering from a metal cathode [7]. With such a system we can make relatively efficient use of the radioactive sample, and the waste material remains well localized for clean up and disposal.

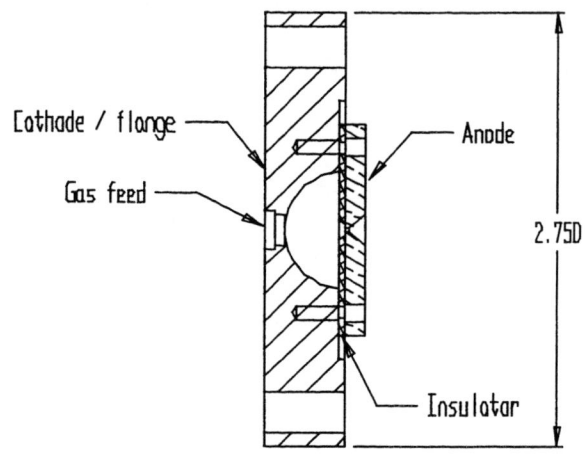

FIGURE 2. Cold Cathode Ion Source made in the Macdonald lab. The source is shown full size in the original document.

ACKNOWLEDGMENTS

Support for this work was provided by the Division of Chemical Sciences, Office of Basic Energy Sciences, Office of Energy Research, U.S. Department of Energy.

REFERENCES

1. M-T Huang, et al, *J. Phys. B*, **30**, 2425-2442 (1997).

2. D. S. Fisher, et al, *Phys. Rev. Lett.* **81**, 1817-1820 (1998).

3. E. Y. Camber, M. A. Abdallah, C. L. Cocke and M. Stockli, *Phys. Rev. A*, **60**, 2907-2920 (1999).

4. W. Mroz, D. Fry, M. P. Stockli and S. Winecki, *Nucl. Instrum. Methods Phys. Res. A*, **A 437** 335 (1999).

5. Stockli, Martin P. "The Operation of Electron Beam Ion Sources for Atomic Physics", in *Accelerator-Based Atomic Physics Techniques and Applications*, edited by Stephen M. Shafroth and James C. Austin., publisher, American Institute of Physics, New York, 1998, pp. 67-116.

6. I. Ben-Itzhak, et al., *Phys. Rev. Lett.* **85**, 58-61 (2000).

7. H. Baumann and K. Bethage., *Nucl. Instr. and Meth.* **122**, 517-525 (1974).

SOLID STATE PULSED POWER SYSTEMS FOR PLASMA SOURCE ION IMPLANTATION

Dr. Marcel P.J. Gaudreau, P.E., Dr. J. A. Casey, M. A. Kempkes, J.M. Mulvaney and T. J. Hawkey

Diversified Technologies, Inc. 35 Wiggins Avenue, Bedford, MA 01730 USA

Abstract. Diversified Technologies, Inc. (DTI) has developed and patented solid state technology to provide fast, high power switches capable of both opening and closing. These switches are modular, and can be combined in series and/or in parallel to provide a wide range of voltage, current, and power handling capabilities for Plasma Immersion Ion Implantation (PIII).

INTRODUCTION

One of the key requirements for PIII is the delivery of fast, tightly regulated, high voltage, high current, pulsed power to the plasma implantation chamber. This requires a nearly ideal high voltage switch capable of operating at short pulsewidths, fast rise and fall times, and high pulse repetition rates. Older, tube-based technologies such as gridded vacuum tubes, thyratrons, and pulse forming networks (PFNs) are limited in providing the ideal switching required. Solid-state high power systems are now available to allow PIII to be a cost-effective, commercially viable process, ready for widespread commercialization.

PIII REQUIREMENTS

PIII's unique requirement, in comparison to established processes like PVD, is high voltage pulsed power for implantion. Though current may vary by an order of magnitude, very tight voltage control must be maintained during a pulse.

There are two obstacles to the efficient application of constant voltage-accelerating pulses in PIII applications. First, the peripheral passive circuitry plays a significant role in pulse shaping and power loss. Second, non-ideal switching elements can significantly degrade pulse shape, efficiency, system complexity, and cost of a PIII system.

Regardless of the switch characteristics, the passive elements of the PIII system should be optimized to minimize cable and feedthrough capacitance; minimize inductance (especially for high current systems); and allow for large peak currents many times the average pulse current.

SOLID STATE MODULATOR BASICS

To use semiconductor devices for high voltage switching, many devices must be cascaded in series.

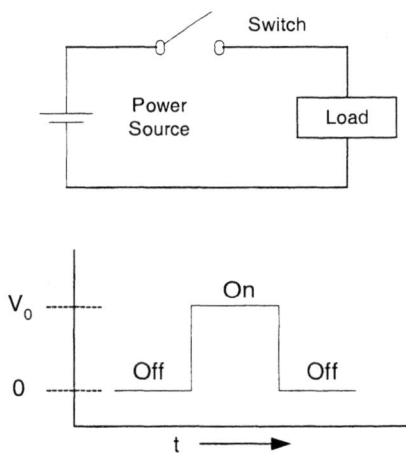

FIGURE 1. Ideal Circuit (top) and Ideal Pulse (bottom)

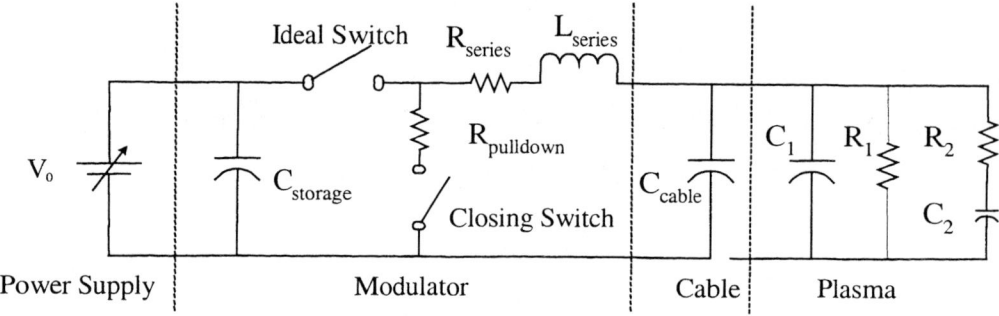

FIGURE 2. Real World PIII Circuit Model

This concept is a two edged sword: one gains the flexibility of a modular design, with no inherent limit to voltage handling, but also faces the formidable task of ensuring that the load is shared equally between devices so that no single device sees harmful or destructive voltages. In addition, the series circuit must be able to accommodate repetitive high voltage and high opening current operation, and non-repetitive fault extremes, such as opening under very large arc-currents without damage to the switch or the load. DTI has patented advanced snubbing and synchronization techniques to solve these problems.

The Ideal Pulse Power Circuit

The top drawing in FIGURE 1 shows the "ideal" pulse power circuit for PIII, powered by an ideal voltage source, with unlimited current and constant voltage. It contains an ideal switch having the following characteristics:

- the switch is both an opening and closing switch
- the switching time is fractions of a microsecond
- it has nearly zero series impedance when closed and infinite impedance when open

With such an ideal circuit, the plasma accelerating voltage will be a perfect square pulse, as shown in the bottom drawing in FIGURE 1. This ideal voltage pulse has a minimal rise and fall time and a flat top, independent of load, current and repetition rate.

PULSED POWER IN A REALISTIC CIRCUIT

FIGURE 2 shows a simplified, but much more realistic PIII circuit. This circuit retains the ideal switch, as in Figure 1, but adds several "real world" factors: a variable DC power supply with finite current capability, a storage capacitor for achieving high peak current, and series and parallel (pulldown) output impedances. The series impedance is inherent in any circuit, as is the cable capacitance between the power source and the plasma. The pulldown resistor is added to provide a discharge path for this cable capacitance when the switch is off. Finally, real world plasma loads are complex so the plasma load is represented as a complex RC circuit.

FIGURE 3 and FIGURE 4 show typical voltage and current pulses seen by the load in this model. Several critical observations are apparent from this simple model. First, even with an ideal switch, the pulse shape is no longer a perfect square wave. The pulse top can be kept nearly flat by choosing a large enough value of storage capacitor, such that the voltage droop, ΔV, from the pulse current is small. Note that longer pulsewidths or higher pulse currents will require larger storage capacitors.

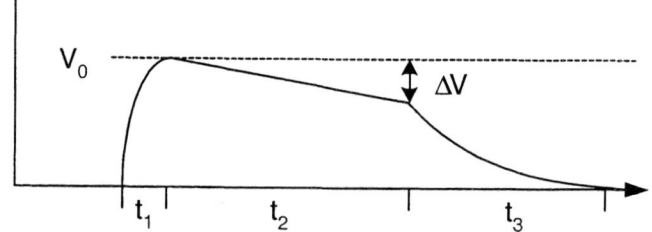

FIGURE 3. Voltage Waveform for the Figure 2 Circuit

The fall time of the pulse (t_3) is given by the RC constant of the cable and plasma capacitance and the pulldown and plasma impedance. In all cases, it is desirable to keep the series impedance as low as possible. A tradeoff must be made, however, with the pulldown resistor value. A low resistance speeds fall time, but also shunts power from the load when the

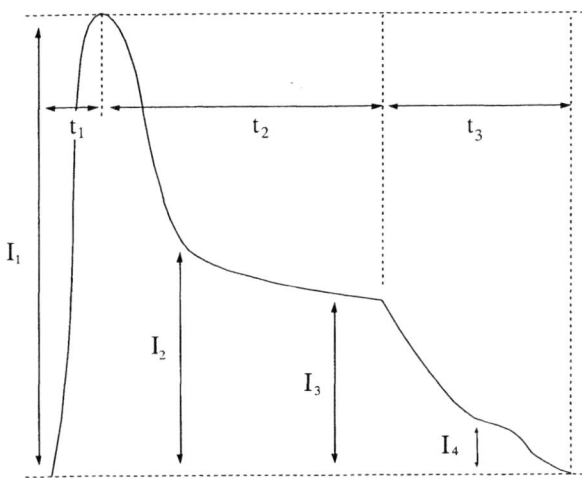

FIGURE 4. Current Waveform for PIII Circuit

switch is closed. Balancing the competing desires for fast fall time and power efficiency is a critical factor in selecting the value of pulldown resistor. Alternately, a second switch can be placed in series with the pulldown resistor, which closes briefly to discharge the plasma after the main switch has opened at the completion of the pulse.

FIGURE 4 shows the typical current drawn from the pulsed power source during a voltage pulse: I_1 is the initial charging current. I_3 is the typical pulse current normally considered in sizing PIII average power and dose rate systems. A key here is to note that the magnitude of the peak current (I_1), while relatively short in duration, will often be many times this average current (I_3). I_2 is the transition current.

Three major conclusions can be drawn from this model. First, the stability of the voltage pulse is critical over a wide range of current levels during the pulse. The switch, power supply, and storage capacitance must have the ability to support very fast dI/dt and high peak current at the beginning of the pulse, while maintaining V_0. This requires low series equivalent resistance and inductance. Second, while the plasma characteristics are a function of the process, the PIII system should be designed to minimize external capacitance.

Primarily, this can be achieved by minimizing the cable length from the switch to the plasma, and by careful design of the feedthrough for minimum capacitance. Minimizing this capacitance will reduce the peak current required from the pulse power system, and improve both the rise and fall times of the pulse. At high current levels, cable inductance will dominate this series impedance, and the circuit must be designed to minimize this inductance.

Finally, note that all of these effects are seen with an ideal switch model. The result is that, even with an ideal switch, it is not possible to achieve perfect PIII pulses. The objective is to minimize the effects of each of the previously discussed factors on the pulse.

HIGH VOLTAGE SWITCHING OPTIONS

Recently discussed PIII switching requirements include:

- High Voltage 1-100kV
- Low switch impedance
- Very fast voltage rise and fall times (typically less than 1µS).
- High peak current handling (10-3000A)
- Very high dI/dt capability (100 - 2000 A/µS)

Three technologies are in use today to address these requirements: (1) vacuum electron switch tubes (e.g., tetrodes); (2) pulse forming networks with thyratrons; and (3) the relatively new approach of solid state switching developed by DTI, which comes closest to providing the 'ideal' characteristics demanded by PIII processes.

Solid state modulators feature insulated gate bipolar transistor (IGBT) switches in series and parallel configurations that allow nearly arbitrary high voltages (1kV-200kV) and currents (10A-5kA). They operate as both opening and closing switches, providing extensive flexibility in pulsewidth, and very fast fault protection.

A solid state switch is very nearly ideal. Current through the switch has very minimal impact on output voltage. Switch times are typically 500 nS, independent of voltage. When the switch is closed, the voltage drop across it is very small so the switch adds very low series impedance into the PIII system. The DC power supply voltage required with these switches is virtually the same as the implant voltage required - a 100 kV process requires only a 100.3 kV power supply. For high current low voltage, commercial-scale PIII processes such as DLC, the most cost effective solution uses high current (1200+A), high voltage (3.3kV+) IGBTs in series.

PIII AT LOS ALAMOS NATIONAL LABORATORY

DTI's HVPM 20-2000 solid state modulator (FIGURE 5) uses series connected, high current IGBTs, to provide up to 40 MW of peak power at 20 kV. At Los Alamos the modulator was used to demonstrate diamond-like coating (DLC), high voltage PSII processes, and high voltage (50 – 100 kV) semiconductor implantation systems.

FIGURE 6 shows typical current and voltage pulses in the PIII processing of 1000 automotive engine pistons at LANL. The flattop of the voltage pulse, during which the current varies by a factor of three, demonstrates the nearly ideal performance of the solid state modulator used in the processing.

For very high voltage (20-200kV) implantation, either a directly coupled power supply and modulator;

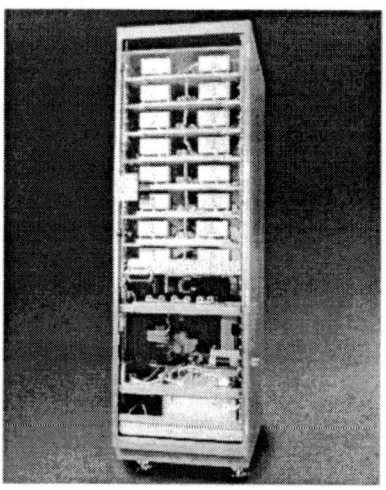

FIGURE 5. PowerMod™ 20-2000, 40MW solid state modulator

or a lower voltage power supply / modulator with a step-up pulse transformer can be utilized. The required peak current determines the optimal approach. The transformer coupled approach is generally limited to peak currents below 100A, while direct coupled systems can support thousands of amperes. In either configuration, cable capacitance can be a major source of inefficiency. For example, the power lost to repetitively charge and discharge the cable capacitance of six feet of coaxial cable between the modulator and PIII chamber at 100kV and 5kHz is found using:

$$P = 2fE = fCV^2 \approx 9kW$$

It is critical that the DC power supply is capable of providing the power required by the PIII process itself,

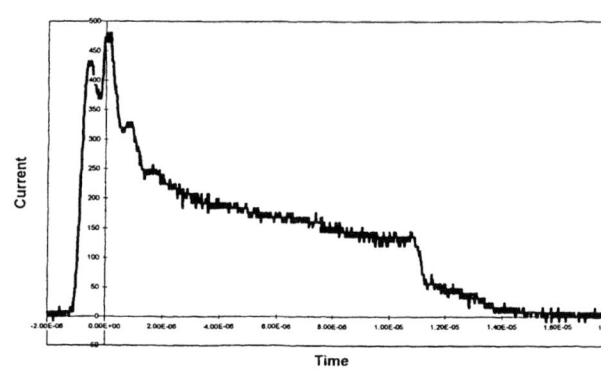

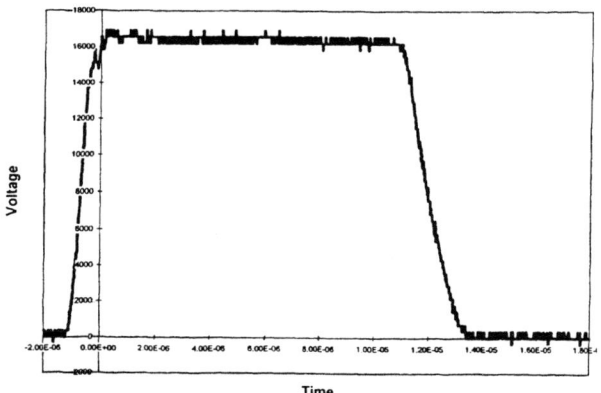

FIGURE 6. Upper Trace - Varying Current (475A peak) of 16.5kV, 12μS Pulse into Plasma Load; Lower Trace- Constant Voltage of 16.5kV, 12μS into Varying Plasma load (Source: LANL)

plus the additional power dissipation associated with charging the power cable, feedthrough, and plasma.

CONCLUSIONS

Very tight voltage control must be maintained during a pulse in which the current may vary by an order of magnitude.

Loss of efficiency occurs as a result of peripheral passive circuitry. Non-ideal switching elements can significantly degrade pulse shape, efficiency, system complexity, and cost of a PIII system.

Solid state pulse modulators offer the following advantages: (1) short pulse to DC flexibility; (2) high power efficiency resulting in lower power and cooling costs; (3) fast risetime and fast opening for arc protection; (4) a modular design which can be scaled to specific PIII systems and processes; (5) very high reliability; (6) small footprint, and (7), no x-ray emissions.

Laser Ion Source for On-Line Production of Exotic Nuclei

Yu. Kudryavtsev, B. Bruyneel, J. Gentens, M. Huyse,

P. Van den Bergh, P. Van Duppen

Instituut voor Kern-en Stralingsfysica University of Leuven
Celestijnenlaan 200 D, B-3001 Leuven, Belgium

Abstract. An on-line laser ion source is used at the Leuven Isotope Separator On-Line (LISOL) for the production of pure beams of exotic nuclei. The operational principal of the ion source is based on the element-selective multistep laser resonance ionization of nuclear reaction products thermalized and neutralized in a high-pressure noble gas. A number of improvements have been carried out to improve the ion source parameters.

INTRODUCTION

Modern studies of nuclei far from the valley of stability at on-line mass separators require the development the target-ion source system that can deliver exotic nuclei with high efficiency, high selectivity and short delay time. Such nuclei are produced in nuclear reactions in very small quantities and usually are overwhelmed by much more abundant isobars. Laser resonant ionization can provide a very efficient and highly selective way to ionize the exotic atoms only. An on-line laser ion source has been developed at the Leuven Isotope Separator On-Line (LISOL) for the production of purified beams of exotic nuclei [1-6]. The laser ion source allowed to collect nuclear spectroscopic data for exotic nuclei produced in light ion-induced fusion reactions: $^{54-55}$Ni [7], $^{54-55}$Co, in proton-induced fission reactions: $^{68-74}$Ni [8], $^{67-70}$Co [9-11], $^{70-77}$Cu, $^{110-114}$Rh and in heavy ion-induced fusion reactions: $^{91-95}$Rh, ^{98}Rh, $^{90-91}$Ru, $^{42-43}$Ti. A gas cell filled with noble gas can be used as an effective stopping and storage place for short-lived isotopes from a fragmentation mass separator. Chemical reaction of trace elements with small amount of impurity gases or with buffer gas itself is of prime importance. The influence of these interactions can be studied in the laser ion source.

PRINCIPLE OF OPERATION

The operational principle of the laser ion source is based on the element-selective multistep laser resonance ionization of nuclear reaction products thermalized and neutralized in a high-pressure noble

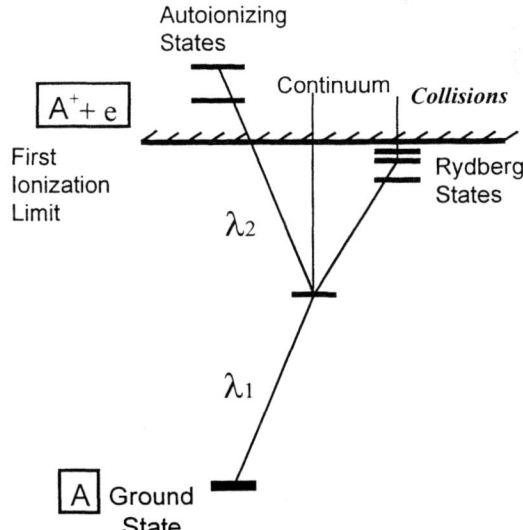

FIGURE 1. Schemes of two-step laser ionization

gas. Fig. 1 shows a scheme of atomic levels and possible ways for a two-step two-color ionization. The atoms thermalized in the ground state are excited by the first step laser λ_1 to an intermediate level. Then three ways are possible. The second step laser λ_2 ionizes the excited atoms through the continuum or an autoionizing state or excites the atoms in a high-lying Rydberg state and then they are ionized via collisions with the buffer gas atoms. Of the three ways, the ionization through the Rydberg or autoionizing states is preferable since the cross section for these states is 10-100 times higher than for the continuum. With commercially available lasers it is possible to ionize about 80% of the elements of the periodic table using this two-color two-step scheme.

DESCRIPTION OF THE ION SOURCE

Figure 2 shows a general view of the Laser Ion Source (LIS) and its coupling to the mass separator through the SextuPole Ion Guide (SPIG) [5]. The laser ion source has been optimized for on-line work. The nuclear reaction products recoil out of the target and are thermalized in the He or Ar buffer gas. They then move in the buffer gas flow towards the exit hole of 0.5 mm in diameter. The inner diameter of the ion source is equal to 5 cm. The body of the cell is made of stainless steal and is electro-polished to reduce the level of contamination. The configuration of the inner part of the gas cell can be changed by putting different inserts inside the cell. This allows to optimize the volume and the shape of the stopping volume for different types of nuclear reactions. The argon gas at pressure of 500 mbar was used for fission reaction and for heavy-ion-induced fusion reaction. For light ion-

TABLE 1. The laser wavelengths used for the two-step ionization of Co, Cu, Ni, Rh, Ru and Ti atoms.

Element	First step (nm)	Second step (nm)
Co	230.903	481.90
Cu	244.164	441.60
Ni	232.003	537.84
Rh	232.258	572.55
Ru	228.538	553.09
Ti	395.821	339.35

induced fusion reaction, helium gas at the same pressure was used. The laser optical system consists of two tunable dye lasers pumped by two time-synchronized XeCl (308 nm) excimer lasers. The laser pulse length equals to 15 ns. The dye laser bandwidth equals to 0.15 cm^{-1}. To get the UV light the frequency of the first step laser radiation is doubled in the second harmonic generator. The dye laser beams are directed to the ion source located at the distance of 15 m at a small angle where two laser beams are overlapped. The maximum laser pulse repetition rate is equal to 400 Hz. The laser light can enter the cell longitudinally as well as transversely. In the longitudinal case the laser beams ionize neutral atoms along the axis of the cell. In the transverse case, a prism reflects the laser light and only atoms next to the exit hole are ionized. The dye laser frequencies of both lasers are tuned to ionize the selected element through an autoionizing state. Table 1 shows the laser wavelengths, which were used for two-step ionization of different elements at on-line conditions. Ionization schemes through autoionizing levels have been developed also for the following elements: Cr, Mn, Cu, Fe, Mo, Cd, Hf, Ta, W, Re, Pt, and Pb.

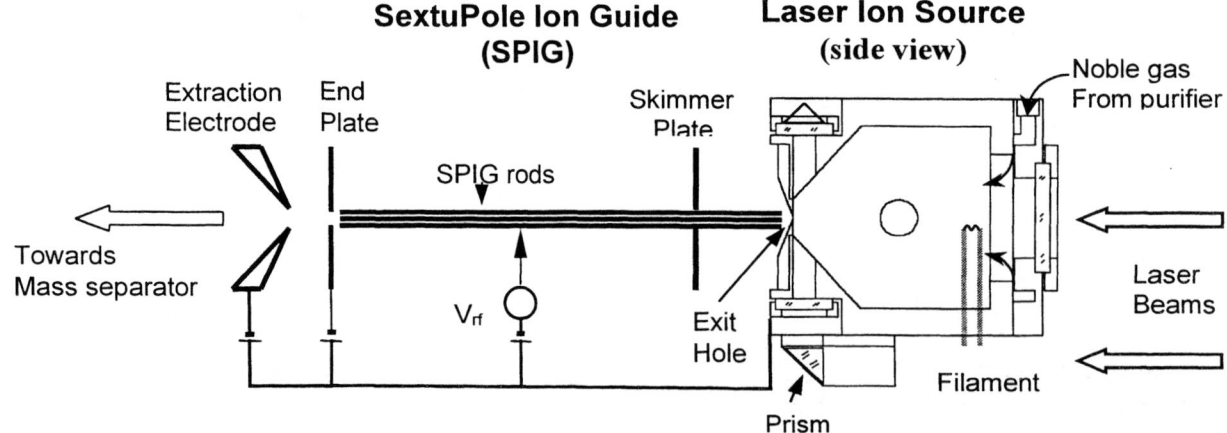

FIGURE 2. A view of the Laser Ion Source together with The SextuPole Ion Guide (SPIG)

The ions leaving the gas cell are captured by the SPIG and transported towards the extraction electrode (figure 2). The SPIG combines static and RF electrical fields to transports ions from the high-pressure zone of the gas cell to the high-vacuum zone of the mass separator. The introduction of the SPIG led to the higher stopping efficiency of the gas cell for energetic nuclear reaction products, since the buffer gas pressure inside the ion source can be increased without affecting the high vacuum in the accelerating stage of the mass separator. The ion beam quality is drastically improved due to the cooling capacity of the SPIG and finally the ion guide allowed us to study the processes of molecular ion formation and ion neutralization inside the ion source [12] as these molecular ions can be transported to the mass separator without applying a DC electrical field and consequently without destruction. The SPIG consists of 6 rods (124 mm long and a diameter of 1.5 mm) cylindrically mounted on a sextupole structure with an inner diameter of 3 mm. The distance between the SPIG rods and the ion source is equal to 2 mm. An oscillating voltage V_{rf} with fixed frequency of 4.7 MHz and variable peak-to-peak amplitude of 0-500 V is applied to the rods with every rod in antiphase to the neighboring rods. The buffer gas is pumped out efficiently through the gaps between the rods while the ions are confined and transported to the extraction electrode with the gas jet velocity. The skimmer plate separates the high-vacuum chamber of the separator and low-vacuum part around the gas jet The main acceleration of ions takes place only near the extraction electrode where a relative high vacuum (10^{-5} Torr) is achieved. The travel time of ions through the SPIG and mass separator was measured to be 180 μs for helium and 300 μs for argon.

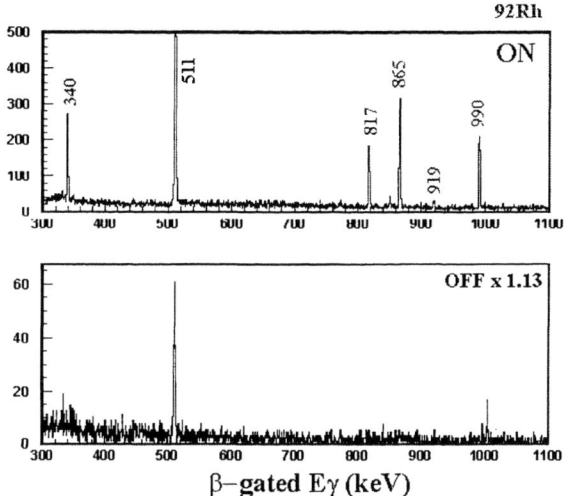

FIGURE 3. β-gated γ-spectra of rhodium-92 with lasers tuned on resonance (top panel) and without lasers (bottom panel). The bottom spectrum was multiplied by 1.13 to account for the difference in beam dose between the two spectra. The γ-rays indicated with their energy in keV and are identified as belonging to ^{92}Rh.

ON-LINE PARAMETERS

Different types of nuclear reactions have been used to produce and to study nuclei far from stability. The mass separated radioactive isotopes were implanted in a movable tape. Two high efficiency germanium γ-ray detectors and thin plastic detectors for β particles were used to measure γ-spectra. The next two pictures illustrate the possibilities of the ion source. Neutron-

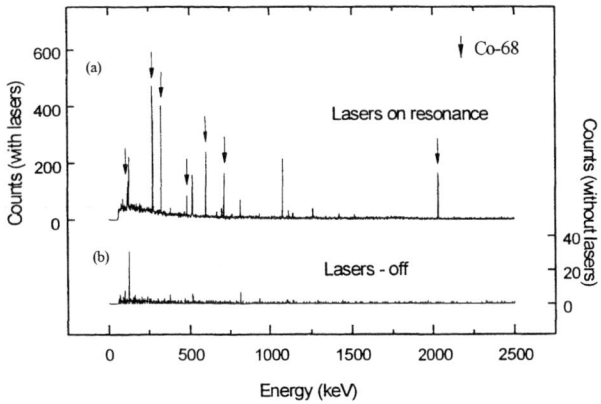 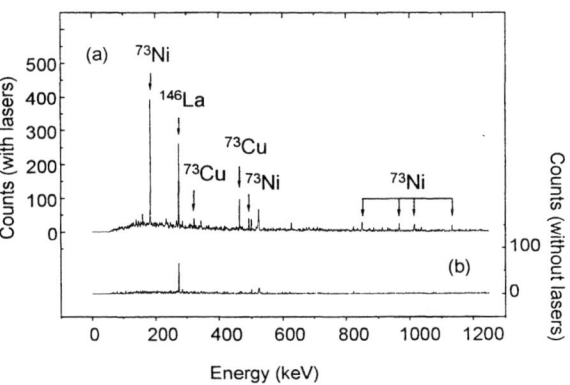

FIGURE 4 β-gated γ-spectra of cobalt-68 (left panel) and nickel-73 (right panel) with lasers tuned on resonance (a) and off resonance (b).

TABLE 2. Parameters of the laser ion source

Nuclear reaction	Reaction products	Efficiency,%[a]	Selectivity [b]
Light ion-induced fusion	^{54,55}Ni, ^{54}Co	5	300
Proton-induced fission of ^{238}U	$^{68-74}$Ni, $^{66-70}$Co, $^{110-114}$Rh	0.2	50-80
Heavy ion-induced fusion	$^{91-95}$Rh, ^{98}Rh, $^{90-91}$Ru, $^{42-43}$Ti	0.04	80

(a) [ions/s in mass separated beam] / [atoms/s recoiling out the target] (%),

(b) [ions/s in mass separated beam (lasers on resonance)] / [ions in mass separated beam (lasers off resonance)]

deficient nuclei in the N=Z region were produced in the ^{58}Ni(^{36}Ar^{10+}, 1p1n) reaction at incident beam energy of 130 MeV in the middle of the target. Figure 3 shows β-gated γ-spectra at mass 92 with lasers tuned on-resonance (top panel) and without lasers (bottom panel). The experimental production rate of ^{92}Rh was 11 at/μC. A comparison between the two spectra in figure 3 shows new γ lines belonging to the decay of ^{92}Rh. The half-life value obtained from the TDC spectrum gated by the most intense lines (817, 865 and 990 keV) equals 5.6±0.3 sec.

The ion source was used successfully for the study of nuclear decay properties of neutron-rich $^{68-74}$Ni [8] and $^{66-70}$Co isotopes [9-11]. New nuclear spectroscopic information has been obtained. The right panel in figure 4 shows the β-gated γ-spectrum on mass 73 with lasers tuned on-(a) and off-resonance (b) with nickel atoms. The γ-lines at 166 keV and 479 keV are present only in the on-resonance spectrum. The observed Cu-activity in the on-resonance spectrum is produced by the decay of ^{73}Ni. The measured production cross section of ^{73}Ni is equal to 3 μbarn while the total fission cross section is equal to 2 barn. The line at 258 keV, observed as well in the on- as in the off-resonance spectra, is due to the double charged ^{146}La^{++} ions produced with a very high cross section. The half-lives as well as the experimental reaction cross sections have been measured for the chain of nickel isotopes (A=68-74). The left panel in the figure 4 shows a β-gated γ-spectrum on mass 68 with lasers tuned on resonance with cobalt atoms (a) and off resonance (b). The arrays indicate peaks that can be associated to the β-decay of ^{68}Co. The comparison of these spectra clearly shows the laser enhancement of Co production.

The parameters of the laser ion source for different types of nuclear reactions are summarized in table 2.

ACKNOWLEDGMENTS

This work is supported by the Inter-University Attraction Poles and the GOA Research Programs, by the Funding for Scientific Research - Flanders (FWO) and by the EXOTRAPS and the EURISOL projects.

REFERENCES

1. Van Duppen, P., et al., Hyperfine. Interactions. **74**, 193-204 (1992).
2. L.Vermeeren, et al., Phys. Rev. Lett. **73**, 1935-1938 (1994).
3. Kudryavtsev, Yu., et al., Nucl. Instr. and Meth. in Phys. Research, **B114**, 350-365 (1996).
4. Vermeeren, L., et al., Nucl. Instr. and Meth. in Phys. Research, **B126**, 81-84 (1997).
5. Van den Bergh, P., et al., Nucl. Instr. and Meth. in Phys. Research, **B126**, 194-197 (1997).
6. Kudryavtsev, Yu., et al., Rev. of Scientific Instrum. **69**, 738-740 (1998).
7. Reusen, I., et al., Phys. Rev. **C59**, 2416-2421 (1999).
8. Franchoo, S., et al., Phys. Rev. Lett. **81**, 3100-3103 (1998).
9. Weissman, L., et al., Phys. Rev. **C 59**, 2004-2008 (1999).
10. Mueller, W. F., at al., Phys. Rev. Lett., **83**, 3613-3616 (1999).
11. Mueller W.F. et al., Phys. Rev. **C 61** 05438 (2000).
12. Kudryavtsev, Yu., et al., Nucl. Instr. and Meth. in Phys. Research (to be published).

Photocathode Electron Gun Applications in Research and Industry

A. M. M. Todd[a], H. Bluem[a], M. D. Cole[a], J. R. Rathke[a], I. Ben-Zvi[b],
T. Srinivasan-Rao[b], J. Schill[b], G. Neil[c] and C. Bohn[d]

[a] *Advanced Energy Systems, 29 Airpark Road, Princeton, New Jersey 08540, USA*
[b] *Brookhaven National Laboratory, Upton, New York 11973, USA*
[c] *Thomas Jefferson National Accelerator Facility, 12000 Jefferson Avenue, Newport News, Virginia 23606, USA*
[d] *Fermi National Accelerator Laboratory, P.O. Box 500, Batavia, Illinois 60510, USA*

Abstract. Bright, radio frequency photocathode electron guns are the choice for most high-performance research accelerator sources. Until recently, their complexity, cost and average beam power had restricted their application in industry. Several novel photocathode electron sources and their performance are described. These include a fully superconducting, compact source for quasi-continuous applications that is being developed in collaboration with Brookhaven National Laboratory and a modification of the Jefferson Laboratory injector for high average power free-electron laser applications. Normal-conducting, high-brightness sources for intense, low-repetition-rate research and commercial applications utilizing metal or robust, high quantum efficiency photocathodes, are discussed.

INTRODUCTION

Because of their ability to produce very bright beams of electrons, radio frequency (RF), photocathode electron guns are the source of choice for most high-performance research accelerator systems. However, due to economic, reliability and average-power limitations, photocathode RF guns have not successfully penetrated industrial applications to date.

Economic photocathode operation at high average power can best be achieved using fully superconducting RF[1] or coupled DC guns[2], which are just now being developed. The choice between high quantum efficiency (QE), which reduces the economic impact and complexity of the drive laser, and durability of the photocathode is inevitably a trade off between robust, relatively low-QE, metal cathodes and high-QE semiconductor cathode materials that are highly vacuum sensitive.

We describe our ongoing efforts, in collaboration with several institutions, to develop solutions to these issues. In order to produce high average power, a fully superconducting RF gun is being developed in collaboration with Brookhaven National Laboratory (BNL) that utilizes the niobium (Nb) itself as the cathode material. A modification of the Jefferson Laboratory DC photocathode gun coupled to a superconducting RF (SRF) accelerator is currently under design. The robustness of a cesium telluride (Cs_2Te) cathode coated with a protective layer[3] has been tested in collaboration with Los Alamos, as a solution for room-temperature guns. We also describe high-brightness, normal-conducting injector systems for intense, low-repetition-rate, commercial and research applications such as pulsed radiolysis[4], x-ray production[5] and fundamental accelerator research[6].

FULLY-SUPERCONDUCTING RF GUN

We are developing an SRF photocathode injector using the niobium itself as the photoemitter[1]. Using niobium avoids the complications involved in introducing foreign materials into the interior of the superconducting cavity. The main problem with this concept is the low QE of Nb. To overcome this, the Schottky effect can be used to increase the practical QE.

A series of successful measurements at BNL characterized the QE of Nb under the influence of the Schottky effect. The tests were performed at 6.5 MV/m, but the results obtained can be extrapolated to the higher gradient level, approaching 100 MV/m, at which the gun will be operated.

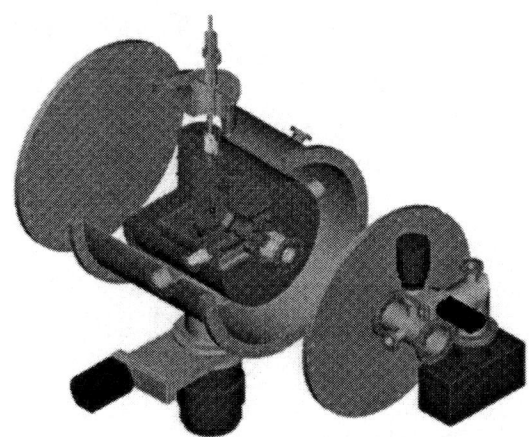

FIGURE 1. Exploded view of the fully superconducting electron gun, shown in blue with the cathode knob in red.

The QE of Nb was measured using two test samples that had undergone different surface preparation. The first cathode was chemically etched using a standard treatment for superconducting cavities. This heavy etch removed approximately 150 μm of material and resulted in a surface that was visibly rough to the eye, as opposed to the mirror-polished surfaces desired for test cathodes. The 100 MV/m extrapolated QE of ~ 2×10^{-5} for this sample was disappointing. A second identical Nb cathode, mechanically polished after chemical etching using a series of diamond polishing compounds, was cleaned ultrasonically and installed in vacuum with minimum exposure to ambient air. The system was baked for twelve hours before the QE was measured. Extrapolation of the data to the 100 MV/m gradient level yielded a very encouraging QE of 3.745×10^{-3}.

The design of the superconducting gun cavity, shown in Figure 1, is based on the traditional elliptically-shaped cavity in wide use today. The cavity is terminated with a flat end-wall on which is located the small knob that will serve as the photocathode. The intent is to enhance the cathode surface fields using the knob while keeping the peak electric and magnetic fields within an achievable range.

The limiting factor chosen for the design is the peak magnetic field that would quench the cavity, rather than the maximum electric field. Given a 1 kG maximum H field at the cathode base, the E field at the center of the cathode is 131 MV/m, with a peak electric field at the cathode edge of 201 MV/m. The peak field on the aperture is 37 MV/m and the E_0T is 16 MV/m. Since the cavity will be produced to permit high fields on a limited surface area, we believe it will be able to operate at peak fields higher than those traditionally achieved.

Table 1 summarizes the system operating parameters and the resulting performance predictions for two operating temperature regimes. The QE measurements for the mechanically-polished sample, the present cavity design and 1W of average drive laser power are assumed. The gun fabrication and testing is scheduled for completion in 2001.

TABLE 1. Projected output beam voltage, current, and power together with RF power requirements for two helium temperature ranges as a function of cathode field gradient

Cathode Gradient (MV/m)	Beam Voltage (MV)	2° K RF Power (W)	4.2° K RF Power (W)	4.2° K LHe Boil Off (L/Hour)	Current from 1 W Laser (mA)	Beam Power (kW)
50	1.5	1.84	17.44	24.15	0.483	0.725
80	2.4	2.94	44.66	61.83	0.679	1.629
100	3.0	3.68	69.78	96.61	0.804	2.411
120	3.6	4.42	100.48	139.12	0.926	3.332

COUPLED DC AND SRF GUN

This approach to efficient high average power electron injectors is derived from the present Jefferson Laboratory system, which consists of a high-voltage DC Gun directly coupled to a cryomodule with a sequence of superconducting accelerator cavities[2]. The difficulty in this case is obtaining sufficient initial DC acceleration to minimize emittance growth and then getting high capture efficiency in the accelerator. We are currently designing a modified DC gun and 500 MHz superconducting RF injector system in collaboration with Jefferson Laboratory. Fabrication is scheduled to begin in 2001. This concept has a larger footprint than the fully superconducting RF gun but should be capable of approaching the 100 mA average current level with a gallium arsenide cathode, which is greater than the postulated practical limit for Nb cathodes.

HIGH-QE CATHODE GUN

A drawback of high-QE photocathode materials is their sensitivity to residual gas which means they are difficult to use and even more difficult to transport. Overcoating these cathodes with a protective layer has been shown to be effective in increasing their durability[3]. The feasibility of utilizing this type of cathode in a commercial RF gun has been examined. The goal was to determine whether they could be produced in one location and then transferred to a photocathode gun at a remote site without the complication and expense of a UHV load lock system.

To perform these tests, an overcoated Cs_2Te cathode was obtained from Los Alamos. The cathode was shipped in an evacuated container and transferred to the test chamber. This transfer took place within a nitrogen-purged glove box. Other than the nitrogen purge, no attempt was made to control the atmosphere within the box. The cathode was quickly transferred to the test chamber and re-evacuated. The total time outside of vacuum was about 20 minutes.

Cathode QE testing was performed at BNL. A nanosecond pulse length, quadrupled Nd:YAG laser provided the excitation of the cathode. During testing, the laser energy was monitored using a calibrated portion of the signal that was picked off from the main beam. To draw the electrons from the cathode, a positive voltage was placed on the anode. This voltage was kept at roughly 200V. The charge from the cathode was measured directly after passing through a calibrated charge amplifier by using an oscilloscope. The QE was calculated from the measured laser energy and the cathode charge.

The original QE, measured at LANL, had been 2%. During subsequent testing in the test chamber, the QE was measured to be up to 0.2%. Since these types of overcoated cathodes are predominantly susceptible to water vapor[3], this level of decrease was to be expected. A partial pressure of water as low as 10^{-4} is enough to defeat the protective layer in a few minutes. Given the simple transfer method employed, the 0.2% obtained is not unexpected. Heating the cathode is a proven method for rejuvenation[7,8] and should have restored the QE to the 1-2% level. Unfortunately, a proper rejuvenation test could not be performed and thus the overcoated cathode results presently remain inconclusive.

INTENSE LOW-REPETITION-RATE APPLICATIONS

A number of applications, some of them commercial, of intense, bright, low-repetition-rate photocathode guns are emerging. These include pulse radiolysis sources for Chemistry research, such as the Laser Electron Accelerator Facility (LEAF) at BNL, for which Advanced Energy Systems delivered the accelerator system shown in Figure 2. The design specifications for this system called for the delivery at 9 MeV, 10 Hz of 10 nC electron bunches with a pulse length of less than 5 psec on one target which utilized an achromatic bend for compression, and greater than 20 nC with a pulse length of less than 30 psec on a straight-ahead target. The photocathode gun is a 3.5 cell S-band system with a magnesium cathode. The LEAF accelerator has been operating successfully for over two years. Similar S- and L-band systems have been designed for other customers[4].

FIGURE 2. BNL 9 MeV LEAF photocathode gun and beamline designed and built by Advanced Energy Systems.

Widespread interest is developing in the use of Compton backscattering of a high-power laser off an intense, bright electron beam as a tunable source of monochromatic x-rays[5]. Applications for such systems include medicine and protein crystallography. One of several medical advantages of tunable, monochromatic x-rays is the ability to obtain improved images with significantly reduced radiation dose. This is particularly attractive for applications such as mammography screening. Under contract to MXISystems of Nashville, Tennessee, Advanced Energy Systems has delivered a photocathode electron accelerator for a prototype x-ray system.

Recently, we have completed the design for the Fermi National Accelerator Laboratory (FNAL) of an injector system for an Engineering Test Facility (ETF) for the Next Linear Collider (NLC). The beam envelopes as calculated by the TRACE code[9] are shown in Figure 3. The system has to deliver a 357 MHz train of a hundred spherical, 0.5 mm rms, 3.5 nC bunches with an rms energy spread of less than 0.2% and normalized transverse emittance of less than 10π mm-mrad. We propose to utilize three SLAC S-band structures to deliver the 132 MeV beam.

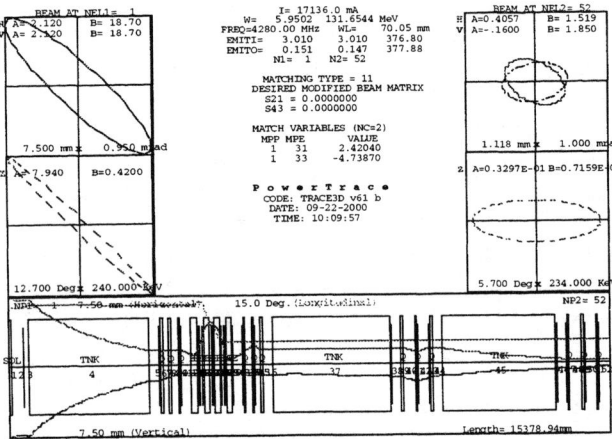

FIGURE 3. Standard TRACE-3D[9] beam envelope output for a proposed 132 MeV NLC test stand injector.

SUMMARY

The use of photocathode electron guns for commercial as well as research applications continues to expand rapidly. The feasibility of operating a superconducting RF gun with niobium as the photo-emitter has been demonstrated. This type of gun will enable the generation of high-average-power, high-brightness beams in a compact, relatively simple structure. DC guns directly coupled to SRF cavities are also being developed as high-power FEL injectors. We have begun to study the use of robust, high-quantum-efficiency photocathodes in a commercial setting and hope to prove that transfer of these cathodes in a simple purged atmosphere, followed by rejuvenation with heat, will restore the quantum efficiency and lead to practical, economic, high-current guns. Finally, we have used three examples of the use of low-repetition-rate, intense, photoinjectors for research beamlines and commercial x-ray systems which illustrate their growth in number and varied application.

ACKNOWLEDGMENTS

This work was performed under US Department of Energy (DOE) Small Business Innovative Research (SBIR) Grants DE-FG02-99ER82723 (High QE Cathode Gun), DE-FG02-99ER82724 (Fully-Superconducting Gun) and United States Army Strategic Missile and Defense Command SBIR DASG60-00-M-0134 (Coupled DC and SRF Gun). In order, the specific projects listed under Intense, Low-Repetition Rate Applications were performed under contract to BNL, MXISystems Inc. and FNAL. BNL, Jefferson Laboratory and FNAL personnel were respectively supported under DOE contracts DE-AC02-98CH10886, DE-AC05-84ER40150 and DE-AC02-76CH03000.

REFERENCES

1. Bluem, H., et al. "Photocathode Electron Source Development at Advanced Energy Systems," to appear in *Proceedings EPAC 2000*, Vienna, Austria, June 2000.

2. Kehne, D., "Experimental Results from a DC Photocathode Electron Gun for FEL," *Proceedings FEL 97 Conference*, Beijing, China, August 1997.

3. E. Shefer, "Coated Photocathodes for Visible Photon Imaging with Gaseous Photomultipliers," *Nucl. Instr. Meth. A* **433**, 502 (1999).

4. Todd, A. M. M., et al. "Picosecond and Subpicosecond High Charge Electron Linacs," *Proceedings of the XIX International Linac Conference*, Chicago, Illinois, ANL-98/28, **1** 397 (1998).

5. See Monochromatic X-Ray Imaging Systems at "http://www.mxisystems.com".

6. Bharadway, V., et al. "The NLC Injector System," *Proceedings of the 1999 Particle Accelerator Conference*, IEEE 99CH36366 **5** 3447 (1999).

7. Kong, S. H. et al., "Fabrication and Characterization of Cesium Telluride Photocathodes: A Promising Electron Source for the Los Alamos Advanced FEL," *Nucl. Instr. Meth. A* **358**, 276 (1995).

8. Michelato, P., et al., "Characterization of Cs_2Te Photoemissive Film: Formation, Spectral Responses and Pollution," *Nucl. Instr. Meth. A* **393**, 464 (1997).

9. Crandall, K. R., and Rusthoi, D. P., "TRACE-3D Documentation, Third Edition," *Los Alamos National Laboratory Report* LA-UR-97-886 May (1997).

Emittance Studies Of ARTEMIS – The New ECR Ion Source For The Coupled Cyclotron Facility At NSCL/MSU

Peter A. Zavodszky[1], Hannu Koivisto[2], Dallas Cole[1] and Peter Miller[1]

[1]Michigan State University, National Superconducting Cyclotron Laboratory, East Lansing, MI 48824, USA
[2]University of Jyväskylä, Department of Physics, Jyväskylä, Finland

Abstract. In order to match the acceptance of the coupled cyclotrons as well as the beam intensity requirements, the emittance and the brightness of the new Electron Cyclotron Resonance ion source built for the Coupled Cyclotron Project at NSCL/MSU (**A**dvanced **R**oom **TEM**perature **I**on **S**ource - ARTEMIS), was optimized for several ion source parameters: microwave power, bias disk voltage, plasma aperture – puller aperture gap distance, puller electrode voltage, gas pressure, and axial magnetic field intensity. The emittance was measured with a simple setup composed from a screen with parallel slits and a NEC rotary wire scanner connected to a digital oscilloscope and a PC. The present study was done only for ions produced from gases, a radial insertion micro-oven capable reaching 1500°C is under construction.

INTRODUCTION

The upgrading and coupling of the present K500 and K1200 cyclotrons at the National Superconducting Cyclotron Laboratory at Michigan State University (NSCL/MSU), will greatly improve the performance of the facility [1]. Before the upgrade, highly charged ion beams of relatively low intensity obtained by direct acceleration in the K1200 cyclotron have been used for production of rare isotopes by fragmentation. In the Coupled Cyclotron (CC)-operation an intense, medium-charged ion beam will be accelerated to about 10-15 MeV/u energy in the K500 cyclotron, then after stripping to a high charge state, further accelerated in the K1200 cyclotron up to energies of 200 MeV/u. The maximum primary beam intensities will reach 6×10^{12} ions/s, an improvement by a factor of 100. The CC-operation mode will also increase the available energy of mid mass and heavy ions.

From the existing three ECR ion sources at NSCL only the superconducting ECR ion source (SC-ECRIS) met the requirements concerning the CC-operation. This ion source operates in the high magnetic field mode at 6.4 GHz microwave frequency. The old room temperature ECR ion source (RT-ECRIS) and the ComPact ECR (CP-ECRIS) were decommissioned. In order to assure increased reliability of the CC-operation and to minimize the time necessary to switch ion beams, we decided to design and construct a new high performance ECR ion source: ARTEMIS (Advanced Room TEMperature Ion Source). Table 1 shows the requirements for the new accelerator facility for initial beams to be developed.

Due to the tight schedule of the upgrade, there was no available time to develop a new ion source. Instead, a proven design developed at Lawrence Berkeley National Laboratory was considered for possible use at the NSCL. The AECR-U [2] at Berkeley was tested in order to find out how it meets the requirements shown in Table 1. As a result of the tests, the AECR-U source was chosen as a starting point for adapting the design to fulfill our requirements. The main modifications to meet the NSCL's requirements are the following: the source has vertical orientation, the magnetic field in the injection end of the plasma chamber was increased and the insulation was improved to allow the use of bias voltages up to 30 kV, necessary for CC-operation. The injection iron was redesigned, one double pancake solenoid was added to the injection side and small modifications to the shape of the yoke were made. The radius of the coil was also increased slightly. By running 1000 Amps in the injection solenoid and by lowering slightly the extraction current, the 2.6 T field will allow operation at both 14 and 18 GHz [3].

TABLE 1. The initial ion beams to be developed and the requirements for the CC-operation.

ECR ion	Q1	ECR intensity [pμA]	ECR voltage [kV]	K500 final energy [MeV/A]	Stripping efficiency [%]	Q2	Extracted beam intensity [pnA]	K1200 final energy [MeV/A]	Beam power [W]
$^{16}O^{3+}$	3	50	22.6	12.5	100	8	1000.0	140	2240
$^{18}O^{3+}$	3	50	25.4	12.5	100	8	1000.0	140	2520
$^{36}Ar^{7+}$	7	30	24.1	13.8	75	18	450.0	155	2511
$^{40}Ar^{7+}$	7	30	26.8	13.8	73	18	438.0	155	2715
$^{48}Ca^{8+}$	8	50	25.4	12.5	61	19	610.0	140	4099
$^{68}Zn^{12+}$	12	8.33	24.0	12.5	40	29	66.7	140	653
$^{70}Zn^{12+}$	12	8.33	24.7	12.5	40	29	66.7	140	635
$^{78}Kr^{14+}$	14	8.57	23.6	12.5	42	34	72.0	140	786
$^{86}Kr^{14+}$	14	8.57	26.0	12.5	42	34	72.0	140	867
$^{112}Sn^{19+}$	19	2.63	24.9	12.5	34	45	17.9	140	281
$^{136}Xe^{21+}$	21	3.33	23.5	10.7	3	50	2.0	120	33
$^{136}Xe^{19+}$	19	4.21	21.6	8.9	27	46	22.7	100	309

Details of the mechanical design and the design and the construction of the magnetic structure as well as the results of the magnet mapping are given elsewhere [4].

FIRST EXPERIMENTAL RESULTS

Charge State Distributions

The ion source was first optimized for the benchmark CC-operation beam, O^{3+}. A value of 300 pμA was obtained which exceeds the CC-operation requirements by 500%.

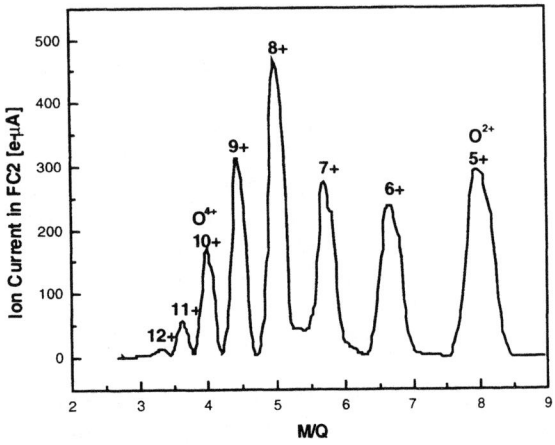

FIGURE 1. Charge state distributions for argon ions measured in a Faraday cup after a 90° analyzing magnet. The source was tuned for Ar^{7+}. Oxygen was used as mixing gas.

During the same run we obtained 170 pμA of O^{6+} with 1500 W microwave power, helium mixing gas and 25 kV extraction voltage.

Figure 1 shows a charge state distribution of Ar ions, optimized for Ar^{7+} production The acceleration voltage was 25 kV, the diameter of the hole in the plasma electrode was 8 mm, Oxygen was used as mixing gas, the microwave power was 800 W, and the bias disc voltage was –60 V. The 40 pμA current of Ar^{7+} already exceeds the 30 pμA required for the CC-operation.

Emittance Measurements

One of the CC-operation performance assumptions is that 50% of the intensities listed in Table 1 will be transported from the ion sources to the cyclotron match point (3 m below the cyclotron median plan) within a transverse emittance of $\varepsilon_x, \varepsilon_y \leq 75\pi$ mm.mrad. In order to optimize the brightness of ARTEMIS to match this intensity and emittance requirements, we performed a systematic study using a simple beam emittance scanner. The experimental setup is shown schematically on Figure 2. The scanner is composed from a set of parallel slits of 0.254 mm width, spaced 3.81 mm from each other (2.54 mm between the two central slits), a National Electrostatics rotary beam profile monitor connected to a digital oscilloscope and a PC connected via a serial port to the oscilloscope. The distance between the slits and wire scanner is 250 mm. At the end of the test beam line the ion beam was stopped on a scintillator made from KBr on an Al plate placed 45° relative to the beam direction. This scintillator could be viewed by a video camera, connected to a PC with a video capture card.

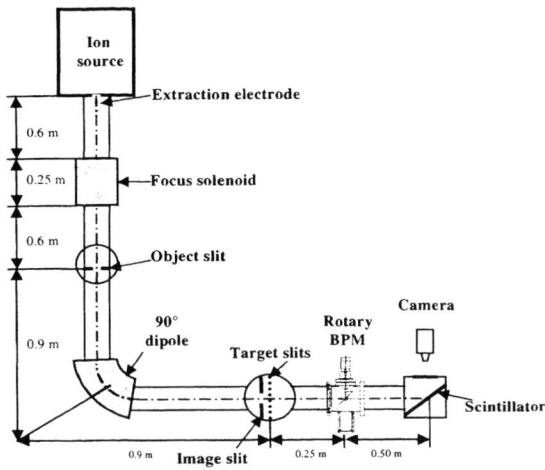

FIGURE 2. Experimental setup used to study the transverse emittance of ARTEMIS.

Knowing the distance between the slit system and the wire scanner and the geometry of the slits, we were able to determine the emittance values by fitting the base of the measured spectra with an ellipse. A typical measured slit image, transformed in 3D (x – distance from the center of the beam pipe, x' – the angular deviation from the beam pipe axis, z – measured ion current) is shown in Figure 3.

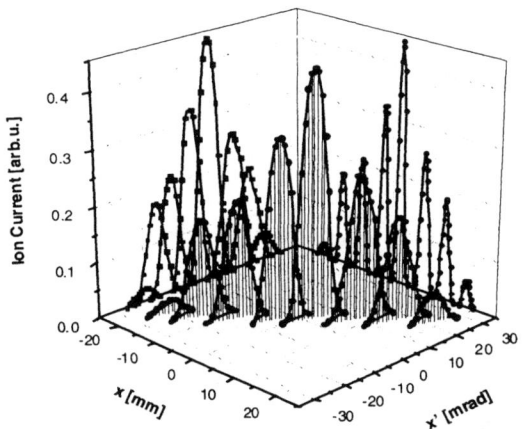

FIGURE 3. Typical ion current distribution in the x-x' plane. The measured spectrum is the projection on the x-Ion current plane.

Several source parameters such as, gas pressure, microwave power, bias disk voltage, axial magnetic field structure, distance between the plasma electrode and puller electrode, and puller electrode voltage were varied for a fixed 20 kV extraction voltage, with a 4 mm diameter extraction hole in the plasma electrode, selecting O^{3+} ions for measurement. The transverse emittance is equal to the area closed by the ellipse shown in Figure 4.

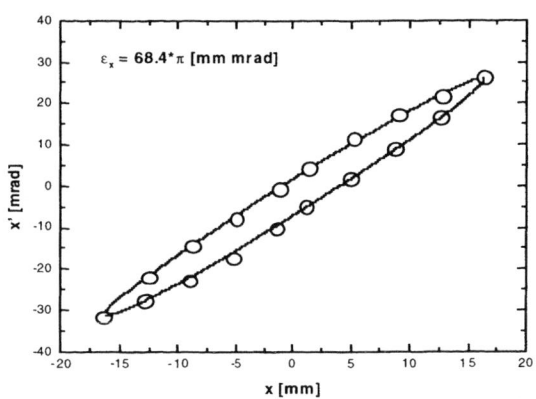

FIGURE 4. Transverse emittance ellipse of an O^{3+} ion beam extracted at 20 kV from ARTEMIS.

The emittance values determined from the video image of the ions passed through the slit system were systematically lower by 30-40% than those obtained with the wire scanner. One of the possible explanations for this discrepancy is the low sensitivity of the used scintillator at the very low ion beam intensities at the tails of the slit images. For this reason the scintillator images were used only to qualitatively monitor the ion beam behavior.

Figures 5, 6 and 7 show the ion beam intensity, emittance and brightness, respectively, as a function of microwave power and bias disc voltage. The microwave power was varied between 30 and 1000 W, the bias disc voltage was varied between 0 and –120 Volts. At each bias disc voltage value there is a remarkable drop in the ion current and emittance around 150-200 W. This causes a maximum in the brightness, which is the ratio between the ion current intensity divided with the square of the transverse emittance. The two transverse emittances are assumed equal; this would be rigorously true only for a cylindrically symmetric ion source and transport system. The hexapolar magnetic field and the magnetic focusing lens cause a coupling in the two transverse directions, but for the sake of simplicity we will neglect this effect.

The cause of the pronounced minimum in the ion beam intensity and transverse emittance around 150 W at each bias disc voltage value is not completely understood. Similar behavior was observed in the case of laser ion sources by a group at ITEP Moscow [5]. Their explanation for this behavior was that at certain ion source parameter combinations the optimum extraction conditions were met, causing a smaller emittance, i.e., better quality ion beam. Another possibility would be that the plasma has two distinctive modes as a function of the microwave power coupled to the plasma, producing the observed behavior of the extracted current intensity and emittance. This was observed qualitatively in the scintillator images of the ion beam passing through the slit system. To further check this explanation, we will monitor the plasma brightness and volume with a CCD camera using one of the available radial ports on the plasma chamber as well as the X-ray spectra.

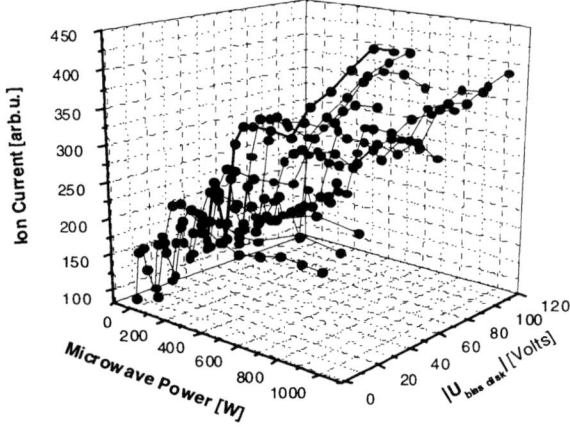

FIGURE 5. 60 keV O^{3+} ion beam intensity in function of microwave power and bias disc voltage.

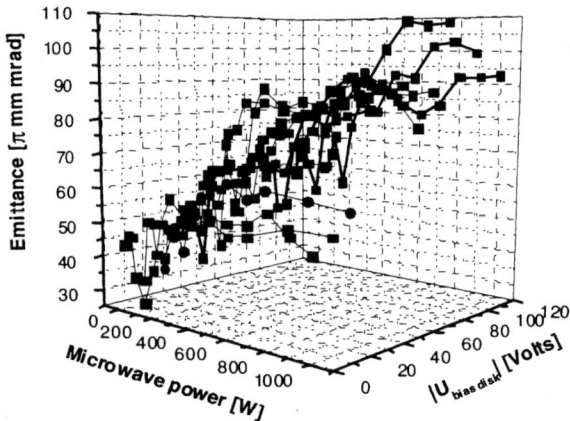

FIGURE 6. 60 keV O^{3+} ion beam transverse emittance in function of microwave power and bias disc voltage.

Although the ion beam intensity has an increasing trend with the increasing bias disc voltage, the emittance of the beam is also increasing. The resulting effect in the ion beam brightness is a decreasing trend with the increasing bias disc voltage. This justifies our approach that to match the CC-operation requirements it is not enough to optimize the intensity of the extracted current from our ECR ion sources, it is equally important to study the behavior of the emittance in function of the source parameters and to do an optimization in ion beam brightness.

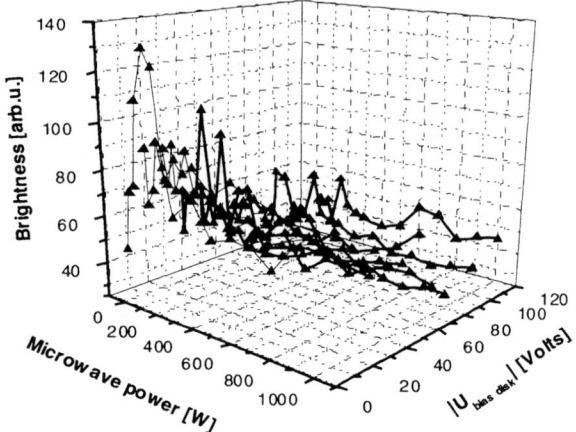

FIGURE 7. 60 keV O^{3+} ion beam brightness in function of microwave power and bias disc voltage.

In order to develop the metal ion beams listed in Table 1, a radial insertion micro oven capable reaching 1500°C was constructed.

ACKNOWLEDGMENTS

This work has been supported by National Science Foundation under grant PHY-95-28844.

REFERENCES

1. The K500⊗K1200 A Coupled Cyclotron Facility at the National Superconducting Cyclotron Laboratory Michigan State University, MSUCL-939, July 1994.

2. Z.Q.Xie and C.M.Lyneis, Proceedings of the 13th International Workshop on ECR Ion Sources, February 26-28, College Station, Texas USA, (1997), p. 16.

3. Z.Q.Xie and C.M.Lyneis, Rev. Sci. Instrum. **66**, 4218 (1995).

4. H. Koivisto, D. Cole, A. Fredell, C. Lyneis, P. Miller, J. Moskalik, B. Nurnberger, J. Ottarson, A. Zeller, J. DeKamp, R. Vondrasek, P.A. Zavodszky and Z.Q. Xie, in Proceedings of the Workshop on Production of Intense Beams of Highly Charged Ions, Catania, Italy, Sept. 24-27, 2000, to be published by the Italian Physical Society.

5. A. Shumshurov, [e-mail: sharkov@vitep5.itep.ru] private communication (2000).

Spatial Distribution of Ion Species in a Source Plasma for Broad Beams

N. Sakudo, K. Hayashi, Y. Nishiyama, K. Komatsu,
J. Miyamoto, M. Yutani and K. Awazu*

Kanazawa Institute of Technology, 7-1 Ohgigaoka, Nonoichi, Ishikawa 921-8501 JAPAN
*Hokuriku Industrial Advancement Center, Oyama-machi, Kanazawa, Ishikawa 920-0918 JAPAN

Abstract. When a chemical compound is used as the source material, the spatial distribution of each fragment-ion species in the source plasma is different. Therefore, if ions are extracted with multi-aperture electrodes as a broad beam, the ion-species abundance in each beamlet is also different. It is important to control the plasma parameters for keeping the ion-species uniformly distributed in the broad ion beam. We constructed a new apparatus equipped with a quadrupole mass spectrometer that can be moved in vacuum to measure local ratios of the fragment ions. The experiment was carried out with CF_4 as the source material. The results suggest that the ion-species distributions are not consistent with the electron-density distribution. The ion-species distributions are strongly dependent on the CF_4 molecule flow in the plasma chamber, while the electron-density distribution is not.

INTRODUCTION

Plasma processing for next-generation semiconductor devices requires high-density and good-uniformity plasmas. Several high-density plasma sources have been developed for various kinds of material processing [1]. Recently, high-density plasmas have been applied not only to semiconductor-device fabrication but also to non-semiconductor material processing using three-dimensional ion implantation, i.e. PSII (Plasma Source Ion Implantation) [2]. Plasma nitriding of steels by PSII has been intensively studied for applications in the automotive industry (e.g., engine parts such as gears, camshafts and turbocharger injector nozzles [3]). Since narrow ion sheaths are preferred in order to conformably cover the complex shapes of targets with the plasma, high-density plasmas have been chosen and studied. In order to control process uniformity, the plasma parameters, including EEDF (Electron Energy Distribution Functions) and IEDF (Ion Energy Distribution Functions) in plasma sources, have been studied [4]. As for the source-plasma uniformities in large area plasmas, only a limited number of plasma parameters, such as electron density and temperature, have been studied and discussed so far. In terms of further precise studies of large-volume plasma sources, besides other plasma parameters we are also focusing on the distributions of ion species. A quadrupole mass spectrometer that can be moved perpendicular to the beam extraction direction, is placed in a vacuum chamber. Ion-species distributions are measured for different types of gas-introduction systems. We suggest a new method to study and improve the source-plasma uniformity by pulsing the plasma with respect to pulsing the supply of gas.

EXPERIMENTALS

A schematic drawing of the experimental apparatus is shown in Figure 1. The ICP chamber consists of an

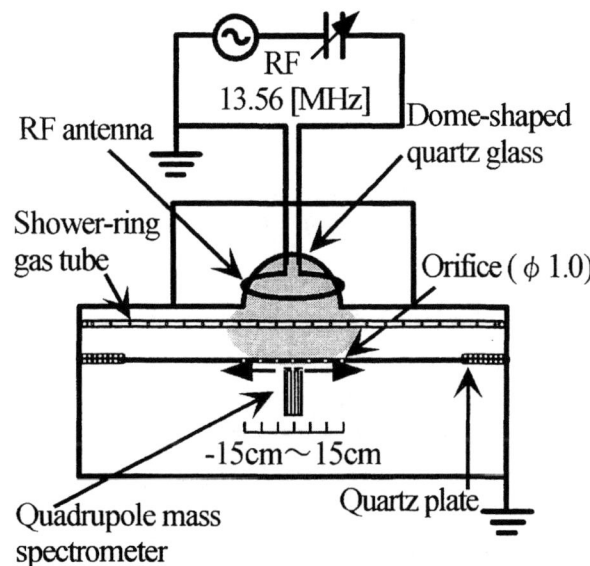

FIGURE 1. Schematic drawing of the ICP chamber.

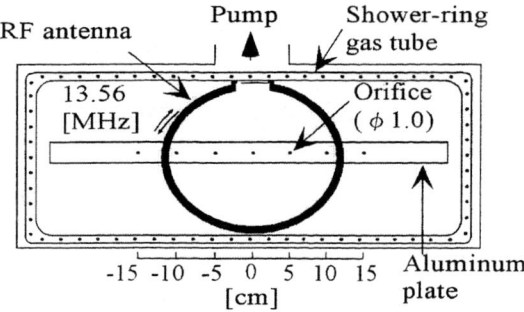

(a) Top view of the chamber equipped with the shower-ring gas tube.

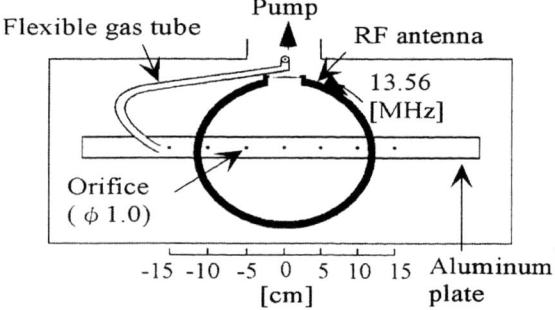

(b) Top view of the chamber equipped with the flexible gas tube.

FIGURE 2. Top views of the chamber with different gas tubes.

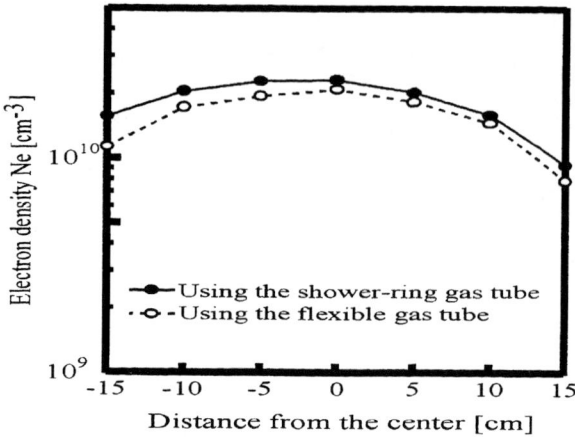

FIGURE 3. Electron-density distributions with different gas introduction methods. (RF power: 1000 W, Gas pressure:1.0 Pa, Gas flow rate: 80 sccm.)

aluminum vessel, a dome-shaped quartz glass of 300 mm in diameter and a quartz plate. Either a langmuir probe or a quadrupole mass spectrometer can be placed in the vacuum chamber and moved horizontally to measure the horizontal distributions of the electron density or the ion-species distributions in the source plasma, respectively. The antenna which has a loop of 280 mm in diameter is placed around the dome-shaped quartz glass. An RF power of 1000 W is fed to the RF antenna to generate plasma. CF_4 gas is introduced into the reaction room through a mass flow controller, and the gas pressure is set at 1.0 Pa with a flow rate of 80 sccm. Two types of gas-introducing tubes, utilized in this experiment, are shown in Figures 2 (a) and (b). The former shows a top view of the reaction chamber with a shower-ring gas tube which can feed gas uniformly from the circumference of the plasma and the latter shows the top view with a single-exit flexible tube which can feed gas from the left side of the plasma.

RESULTS

Electron-Density Distributions with Different Gas Introduction Methods

Electron densities in the plasma are measured at seven points (-15 cm to +15 cm) horizontally from the center of the chamber respectively for two different gas introduction methods. Figure 3 shows electron-density distributions. Plots in black in the figure show data derived by gas introducing from the circumference of the plasma towards the center of the chamber and plots in white show data when CF_4 is fed from one point on the left side of the plasma. Although the gas-introducing techniques are structurally different, the electron-density distributions are not dramatically changed.

Ion-species Distributions over the ICP Source

The Langmuir probe for measurement of plasma densities is exchanged for a quadrupole mass spectrometer for measuring the distributions of ion species in the CF_4 plasma. The quadrupole mass spectrometer can also be moved horizontally under the quartz plate. Because the operating pressure of the quadrupole mass spectrometer has to be less than 10^{-2} Pa, the space under the quartz plate is pumped differentially with a turbo molecular pump, while the reaction region above the quartz plate is kept at 1.0 Pa. Ions in the plasma are extracted through seven orifices in the aluminum plate and measured by the quadrupole mass spectrometer, as shown in Figure 1. First, CF_4 is introduced into the reaction region from the circumference of the plasma with the shower-ring gas tube. Figure 4 shows the distributions of ion species. The main positive ion components are CF^+, CF_2^+ and CF_3^+. By introducing CF_4 uniformly into the plasma, symmetrical distributions of the ion species are obtained.

Then the shower-ring gas tube is replaced with a flexible gas tube which can feed gas from one point on the left side of the plasma to the center of the reaction room. The results are shown in Figure 5. Ion-species distributions resulting by use of the flexible gas tube, especially those of

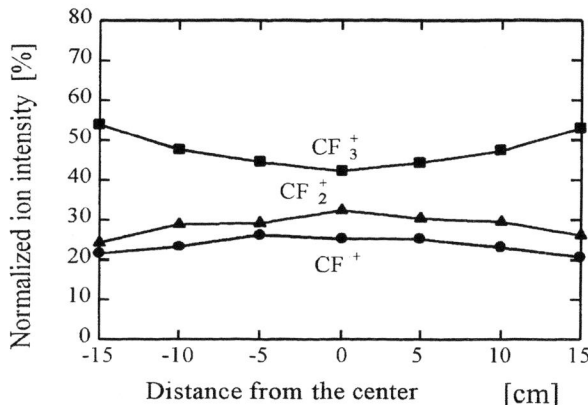

FIGURE 4. Normalized distributions of the dominant ion species, with the shower-ring gas tube. (RF power: 1000 W, CF_4 gas pressure: 1.0 Pa, CF_4 flow rate: 80 sccm.)

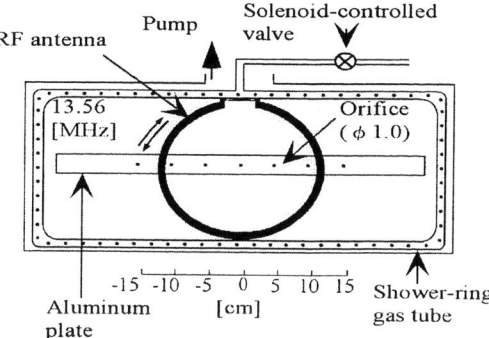

(a) Gas introduction controlled by the solenoid valve.

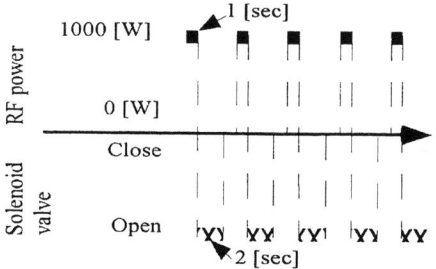

(b) Periodic gas introduction alternated with pulsed-RF cycle.

FIGURE 6. Pulsed gas introduction and its time interval with pulsed plasma

CF_3^+ species, are not symmetric. These results suggest that ion-species distributions depend on the gas introduction methods although the electron-density distribution seems to be symmetric independent of gas introduction methods, as shown in Figure 3.

In terms of uniformity of ion-species in a large-volume plasma, the technique used to introduce the gas will play one of the most important roles [5]. However, even if gas is introduced from the circumference of the plasma, the ion-species distribution is not sufficiently uniform as shown in Figure 4. Therefore, we suggest a new approach to generate uniform plasmas. The method used a solenoid-controlled valve that is installed at the front end of the shower-ring gas tube, as shown in Figure 6 (a). CF_4 is introduced into the reaction room by opening the valve for two seconds, and after closing the valve, the gas diffuses uniformly all over the reaction room. Then, pulsed plasma is generated subsequently with the time interval shown in Figure 6 (b). Figure 7 shows the ion-species distribution resulting from periodic gas introduction followed by pulsed plasma generation. Figure 8 shows the

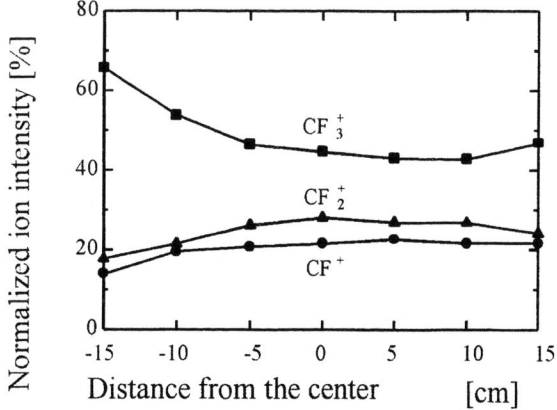

FIGURE 5. Normalized distributions of the dominant ion species, with the flexible gas tube.
(RF power: 1000 W, CF_4 gas pressure: 1.0 Pa, CF_4 flow rate: 80 sccm.)

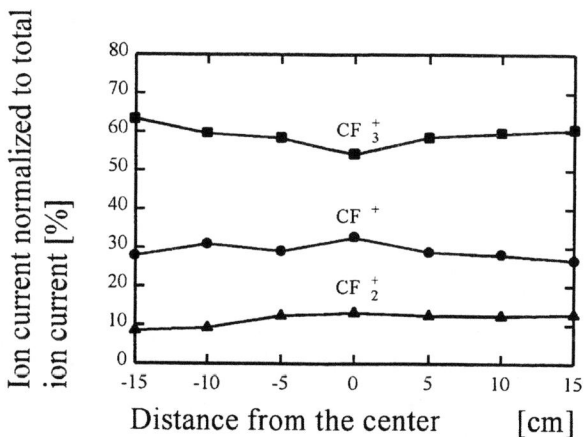

FIGURE 7. Mass spectra and the normalized distributions of the dominant ion specieswith pulsed gas introduction and plasma generation. (RF power: 1000 W, CF_4 gas pressure: 1.0 Pa, CF_4 flow rate: 80 sccm.)

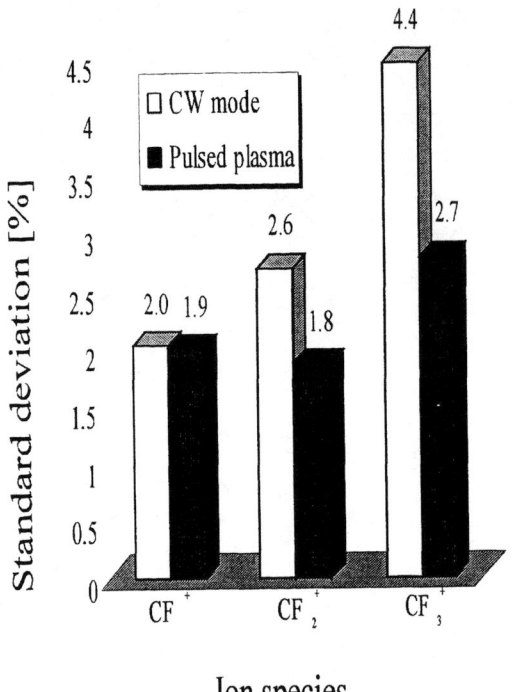

FIGURE 8. Comparison of the uniformities of ion species generated with CW and pulsed plasmas. (RF power: 1000 W, CF_4 gas pressure: 1.0 Pa, CF_4 flow rate: 80 scc)

uniformity of ion-species in comparison with the uniformities produced in CW mode. Operating both the plasma generation and the gas introduction in pulse modes improves the uniformity of the ion-species.

DISCUSSIONS AND CONCLUSIONS

When a chemical compound is used as the source material for broad ion beams, beam uniformity must be characterized not only by ion current-density distribution but also by the uniformity of each fragment-ion species.

From our study, it is evident that ion-species distributions do not always depend on the electron-density distributions. Distributions of ion-species change due to gas-introduction techniques, independent of the electron density distribution. Gas molecules are quickly dissociated after being introduced into the plasma and some of their fragments are ionized near the outlets of the gas tubes. Some fragments with high sticking coefficients, like metals, are deposited on the surface nearest to the place where they appear and they cannot reach the center of the plasma. This results in non-uniform distributions of ion-species. These phenomena in the source chamber cannot be ignored for realizing uniform broad ion beams. In this study, we suggest a new approach for improving the uniformity of the ion species in large-volume plasmas by optimizing the conditions for gas introduction and plasma generation. This technique can be effectively utilized in pulsed-broad ion beam applications.

ACKNOWLEDGMENTS

This research was partially financed by the New Energy and Industrial Technology Development Organization (NEDO).

REFERENCES

1. Timothy A. Grotjohn, *Rev. Sci. Instrum.* **65**, 1298 (1994).
2. N. Sakudo, K. Awazu, H. Yasui, E. Saji, K. Okazaki, Y. Hasegawa, N. Ikenaga, K. Kanda, Y. Nambo and K. Saitoh, "Development of Hybrid Pulse Plasma Coating System", to be published in this proceedings.
3. P. C. Johnson, *Plasma physics*, ed. Richard Dendy, p. 352. Cambridge Univ. Press, Cambridge (1996).
4. Junzo Ishikawa, *Rev. Sci. Instrum.* **69**, 863. (1998).
5. S. Okuji, N. Sakudo, K. Hayashi, M. Okada, T. Onogawa, T. Maesaka, Y. Nishiyama, K. Toyoda, S. Yashima and T. Ishida, *Rev. Sci. Instrum.*, **71**, 716 (2000).

The VERA Heavy Ion Program – Status and Prospects

R. Golser, G. Federmann, W. Kutschera, A. Priller, P. Steier, C. Vockenhuber

Vienna Environmental Research Accelerator, Institut für Isotopenforschung und Kernphysik, Universität Wien, Währinger Straße 17, A–1090 Wien, Austria / Europe

Abstract. Since the start of VERA operation in 1996 the long term goal has been to perform AMS up to the heaviest radionuclides. Last fall a major step forward was made with the installation of movable slits which float up to 13 kV inside the injector bouncer chamber. The injector mass resolution has been measured to be ≈900 at 80% transmission. Recently, we started to investigate isotopes of Pb, U, and Pu, with the goal to measure the radionuclides ^{210}Pb (22 y), ^{236}U (23 My), and ^{244}Pu (81 My). The detection is performed with a two-foil TOF system (450 ps resolution) followed by a Bragg ionization chamber. In addition we report on our first studies of doubly negative cluster ions, like $(C_{10})^{2-}$.

INTRODUCTION

Accelerator Mass Spectrometry (AMS) of very heavy radionuclides has been the long term goal of the Vienna Environmental Research Accelerator (VERA) since the beginning in 1996. The ion optical components are dimensioned to transport all nuclides up to Pu. Although this is, of course, a necessary condition, it is certainly not sufficient to actually perform heavy ion AMS with a 3 MV machine. In the following we present two examples of what we have reached so far, and where we want to go. In the first chapter we describe the general layout of VERA, with some emphasis on the initial problems with our injector magnet and their subsequent solution. Then, first results of AMS with ^{210}Pb will be shown. Finally, we describe our first experiment with doubly negative ions, where an AMS system is particularly useful to avoid some pitfalls of conventional mass spectrometry.

THE VERA AMS SYSTEM

VERA is based upon a 3 MV Pelletron tandem accelerator, Model 9SDH-2, manufactured by National Electrostatics Corp., Middleton, WI, USA. A recent description, including a schematic layout, may be found in [1]. The essential components at the low-energy side are a 40 sample source of negative ions by cesium sputtering (MC-SNICS by NEC) at a pre-acceleration voltage up to 70 kV, a 45° spherical electrostatic analyzer, and a 90° double-focusing injection magnet. The magnet is equipped with an insulating vacuum chamber that allows for fast sequential isotope injection (bouncing) by biasing the chamber rapidly up to +13 kV. At the high-energy side the main ion filters are a 90° double-focusing analyzing magnet, and an ExB velocity filter. The present setup proofed to be very satisfactory for high precision ^{14}C work, e.g. [1, 2]. AMS of ^{10}Be and ^{26}Al can also be performed without undue limitations [1]. For ^{129}I and more heavy ions however, the analyzing capabilities of the ExB-filter are not fully sufficient for low background AMS [1]. In 2001 we will therefore replace the velocity filter by a 90° spherical electrostatic analyzer (radius 2000 mm, gap 45 mm, max. voltage ±90 kV, max. ion electric rigidity E/Q = 4 MV).

For the ^{210}Pb experiment described below, a time-of-flight spectrometer was followed by a Bragg-type ionization chamber for energy analysis. The TOF system comprises two electrostatic mirror assemblies, where the ion-induced electrons from a thin self-supporting C-foil (≈4 µg/cm^2) hit a Multi-Channel-Plate to provide the start and the stop signal, respectively. The time resolution measured with Au^{5+} ions of 18 MeV is about 450 ps. The energy resolution of the Bragg-detector is between 3% and 5%. For the experiments with doubly negative ions only a surface barrier detector was used.

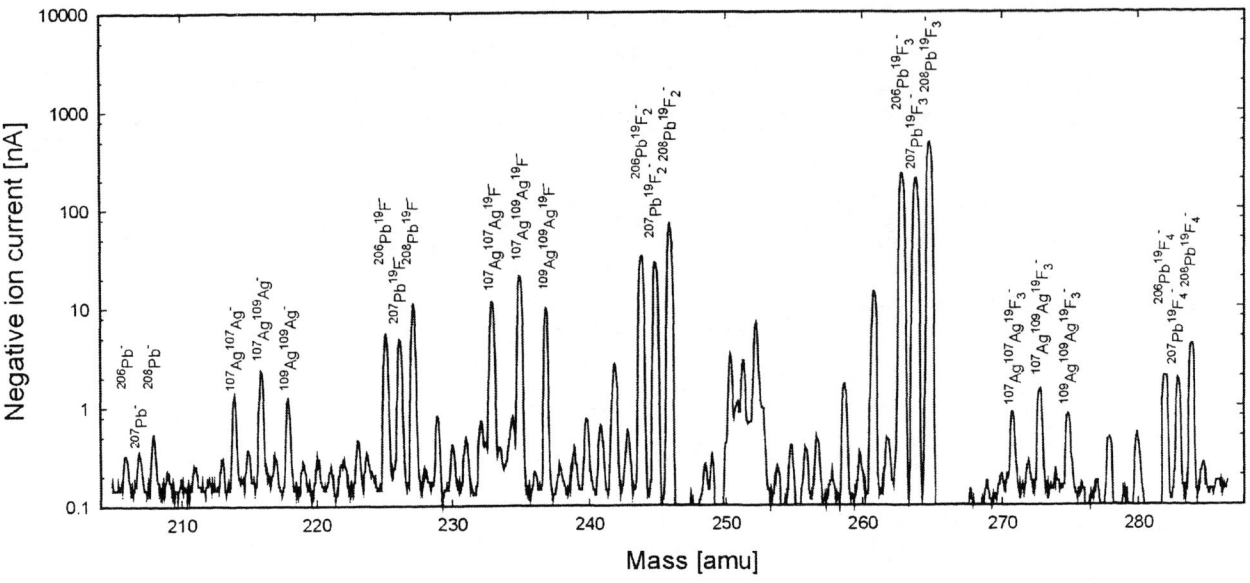

FIGURE 1. Injector magnet scan of negative ions obtained from PbF$_2$ mixed with Ag powder. The scan shows the high mass resolution ($\approx$900) achieved with the new injector slits (object and image slits are set to $\approx$1.0 mm), transmission through the slits and the accelerator is $\approx$80%.

The new injector slits

After the installation of VERA in 1996 an astigmatic focusing of the injection magnet became apparent. To correct for this deficiency an additional pair of magnetic quadrupoles were installed just before the magnet. Since the quadrupoles act on the ions after they have been accelerated by the bouncer voltage, fast sequential injection is still possible. This solution turned out to be fully adequate for ^{14}C, ^{10}Be and ^{26}Al which are measured with all slits wide open ($\pm$12 mm). For AMS of heavy nuclides like ^{129}I, ^{236}U, or ^{244}Pu a high resolution injection system is mandatory since the multitude of closely spaced stable ions - either atomic or molecular - becomes overwhelming (for a careful discussion of the problems with molecular fragments, the reader is referred to Ref. [3] by L. Kilius *et al.*). Here, the slits must be closed to about $\pm$1...$\pm$2 mm. Because quadrupoles can provide only net focussing, the original analyzing slits are too far from the magnet, about 25 cm downstream of the common waist for the x- and the y-direction (the beam propagates along the z direction). When the x waist is established at the original slit position, the beam becomes astigmatic and we get poor transmission through the accelerator due to the beam diverging in y. Moving the original slits closer to the magnet is impossible, so additional x-slits were installed inside the injector vacuum chamber. As mentioned before, the injector chamber is bounced up to 13 kV. Since the additional slits must follow this voltage rise, they need to be insulated from the mounting flange and from the manual controls. With the new slits, injector resolution is measured to be about 900 and transmission through the slits and the accelerator is still about 80%. To demonstrate the performance of the new slits, Fig. 1 shows a mass scan at the upper limit of our injector mass range.

^{210}Pb DILUTION SERIES

Fig. 1 serves the second purpose of illustrating the mass spectrum obtained from a sample of PbF$_2$ mixed with Ag powder. A dilution series of the radioisotope ^{210}Pb (half-life 22.3 y) in PbF$_2$ was intended to be a thorough test of the heavy ion measuring capabilities of VERA. For the details of this experiment the reader is referred to the thesis by P. Steier [4]. ^{210}Pb is the first heavy radioisotope without a stable isobar interference (^{209}Bi is the heaviest stable nuclide). Pb atomic ions give extremely low negative currents, so PbF$_x^-$ ions were used (fluorine has only one stable isotope ^{19}F). Unexpectedly, the highest negative ion currents came from PbF$_3^-$, cf. Fig. 1. ^{210}Pb^{19}F$_3^-$ was injected into the accelerator; the high energy side was tuned to ^{210}Pb^{5+}. Fig. 2 shows the time spectra (integrated over energy and normalized to ^{208}Pb^{5+} current) of three samples (out of seven) with nominal ^{210}Pb : ^{208}Pb ratios 5.4×10^{-10}, 3.1×10^{-11}, and blank, respectively. Also indicated in Fig. 2 is the ^{210}Pb region-of-interest. We wish to emphasize that all counts outside of this region must be considered background and must not be used to determine isotope ratios. A careful evaluation of all seven samples is given in [4] and can not be reproduced here. Suffice it to say that the linear relationship between the nominal and the measured values is perfectly established within counting statistics ($\approx 5\%$). ^{210}Pb background is of the order of 10^{-12}, overall efficiency (^{210}Pb^{5+} detected per ^{210}Pb atom in sample) is about 0.3%.

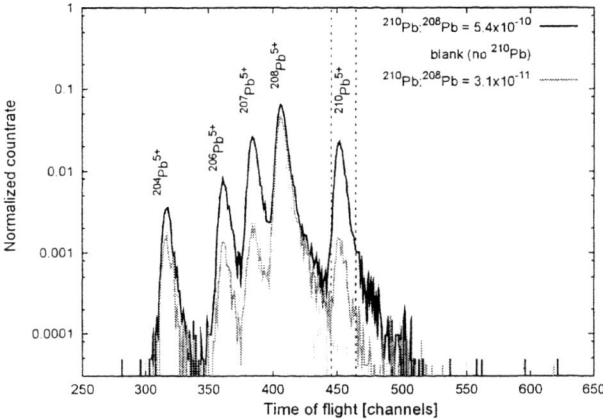

FIGURE 2. TOF spectra for two PbF$_2$ samples spiked with ^{210}Pb and a blank. The spectra are normalized to ^{208}Pb^{5+}.

DOUBLY NEGATIVE IONS

The fact that AMS is almost exclusively based on (singly) negative ions lead naturally to studies on their existence and their properties, e.g. [5]. Whereas the existence of doubly negative *atomic* ions has never been proved, doubly negative carbon clusters have first been reported 1990 by Schauer *et al.* [6] and later by Gnaser *et al.* [7]. Utilizing AMS they have been unambiguously identified 1996 by Middleton and Klein (M&K) [8]. Two exclusive properties of AMS facilitate an incontrovertible confirmation of doubly negative ions: the accelerator allows for the efficient break-up of molecular species in the stripping process and it provides sufficient energy to identify the ions by their nuclear charge.

Our attempts to study doubly negative species have so far been directed to test the principal suitability of VERA for this task. The maximum terminal voltage of 3 MV available with VERA is considerably lower than the 8 MV used by M&K. And up to now we do not have an ionization chamber for particle identification. If necessary we can, however, use a passive method based on the energy loss in a thin foil. The lower terminal voltage of VERA shifts the most probable charge state towards lower values. Often, charge state 1+ is the most prominent. For example, when the carbon cluster $(C_{10})^{2-}$ breaks up in the terminal stripper at 3 MV, only 600 keV are available per C, yielding about 50% C^+, about 35% C^{2+}, and about 5% C^{3+} [9]. In principle, charge state 1+ bears the risk of incomplete molecular fragmentation. However, recent ^{14}C measurements [9] by the Zurich group at sub MV terminal voltages demonstrate that the modest stripper densities used by VERA (about 1 μg/cm^2 Ar) suppress molecular components by many orders of magnitude.

We have repeated part of the work of M&K with the particular carbon cluster $(C_{10})^{2-}$ in the form $(^{13}C^{12}C_9)^{2-}$. It is injected at m/q = 60.5 and turns out to be the most intense doubly negative carbon cluster (1 pA order of magnitude). Very intense currents in the immediate neighborhood come from $^{12}C_5^-$ at m/q = 60, and both $^{12}C_4^{13}C^-$ and $^{12}C_5H^-$ at m/q = 61. Due to the improved mass resolution of our injector system mentioned before, we see the double negative ions at m/q = 60.5 clearly isolated. When a $(^{13}C^{12}C_9)^{2-}$ breaks up in the stripper, each of the 10 carbons begins to develop a history of its own. After 'sufficient' interactions with the stripper gas the fragments may be considered independent particles (see below). So, each ^{12}C has a certain probability P to exit the stripper in charge state 1+ and to become registered in the surface barrier detector. The MCA counts it in the peak corresponding to the particle energy E_1, i.e. near channel 100 in our case, cf. Fig. 3. Since there are up to nine $^{12}C^+$ from each $(^{13}C^{12}C_9)^{2-}$ it may frequently happen that two $^{12}C^+$ make it into the detector simultaneously, giving a signal with $2\times E_1$ and thus contributing one event to the channels near 200. As Fig. 3 shows, eventually even three, or four, or five,..., or all nine can make it. (Due to the low count rate, we can safely exclude pile-up, i.e. accidental coincidences from different clusters).

If all $^{12}C^+$ from a single cluster are statistically independent, the number of coincidences of M ^{12}C out of the total N within the cluster follow a Binomial distribution. Effects that may lead to deviations are: (a) The M = 2 intensity of $^{12}C^+$ might be increased due to interference with $(^{12}C_2)^{2+}$. We can exclude the existence of this molecule in significant amounts by the data from Zurich [9] and by our own: if many $(^{12}C_2)^{2+}$ would survive stripping, $(^{12}C_2)^+$ would be even more likely. When injecting $(^{12}C_5)^-$, which has the same velocity, the analyzed $(^{12}C_2)^+$ current was below our measuring limits, i.e. far below 1‰ of the $^{12}C^+$ current. (b) Another double negative cluster with mass 121 is conceivable, $(^{12}C_{10}H)^{2-}$. This cluster would be injected together with $(^{13}C^{12}C_9)^{2-}$. If $(^{12}C_{10}H)^{2-}$ breaks up in the stripper, it can yield up to 10 $^{12}C^+$ and would obviously increase the coincidence rate. Even with the ion detector we do not see any H^+ when injecting m/q = 60.5 and so we can exclude the existence of $(^{12}C_{10}H)^{2-}$ beyond 1ppm. After changing the injection magnet gradually towards m/q = 61, a tremendous amount of H^+ from the breakup of $^{12}C_5H^-$ begins to appear.

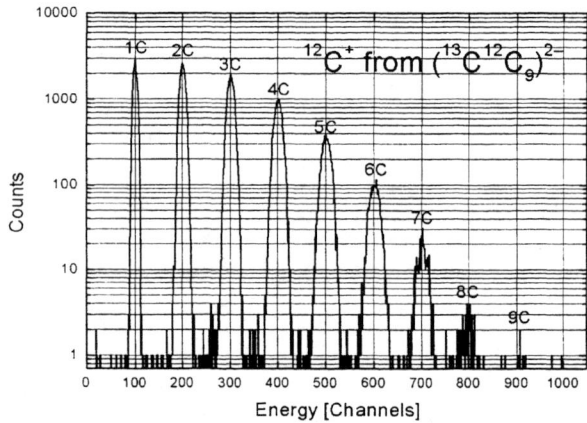

FIGURE 3. Energy spectrum of the $^{12}C^+$ registered in the detector after the break-up of $(^{13}C^{12}C_9)^{2-}$ in the terminal. The labels indicate the M-fold coincidences.

In conclusion, VERA seems to be well suited for studies on doubly negative ions. What may we speculate about the further role of AMS in this field? Principally, two different questions can be addressed: (a) Does a particular di-anion exist? And (b), what is its structure? The work by M&K leaves the impression that attempts to determine the structure by Coulomb explosion imaging are controversial. This should also be seen in the light of competitiveness. Photodetachment photoelectron spectroscopy [10] has implicit advantages to study the properties of multiply charged anions: direct measurements of excess electron binding energies are possible and even detailed insights into the electronic structure may be gained. An interesting first approach to combine photodetachment and the detection capabilities of AMS was made by Berkovits et al. [11].

Excluding the existence of a certain di-anion by AMS is possible only beyond rather crude limits: the di-anion has to exist for about 10 µs to reach the terminal stripper, it has to survive strong electric field gradients, etc. So, the real strength of AMS will be in *proofing* the existence of reasonably long-lived doubly negative ions. AMS has already brought new di-anions to light: Klein and Middleton have shown that $(BeC_n)^{2-}$, n ∈ {4, 6, 8, ..., 14} are surprisingly abundant [12]. And they have verified the existence of $(BeF_4)^{2-}$ and $(MgF_4)^{2-}$ [13] in agreement with theoretical predictions. However, their search for $(LiF_3)^{2-}$ and for $(Li_2F_4)^{2-}$, which were also predicted, did not yield positive results [13]. It is argued that a sputter ion source might produce these ions in vibrational states high enough to destabilize them. Thus it would be logical to substitute Li by Na, K, or Cs, and/or F by Cl, or I. This is exactly what we intend to do in the near future. As a recent paper shows [10], the quest for the smallest stable multiply charged anions in the gas phase has started. With AMS as a very powerful tool, VERA wants to take part in this quest.

REFERENCES

1. Priller, A., Brandl, T., Golser, R., Kutschera, W., Puchegger, S., Rom, W., Steier, P., Vockenhuber, C., Wallner, A., Wild, E., *Nucl. Instr. Meth.* B **172**, 100-106 (2000).

2. Rom, W., Golser, R., Kutschera, W., Priller, A., Steier, P., Wild, E., *Radiocarbon* **40/1**, 255-263 (1998), and *Radiocarbon* **41/2**, 183-197 (1999).

3. Kilius, L.R., Zhao, X.-L., Litherland, A.E., Purser, K.H., *Nucl. Instrum. Meth.* B **123**, 10-17 (1997).

4. Steier, P., "Exploring the limits of VERA: A universal facility for accelerator mass spectrometry", *PhD thesis*, University of Vienna, 2000.

5. Berkovits, D., Ghelberg, S., Heber, O., Paul, M., *Nucl. Instrum. Meth.* B **123**, 515-520 (1997).

6. Schauer, S.N., Williams, P., Compton, R.N., *Phys. Rev. Lett.* **65**, 625 (1990).

7) Gnaser, H., Oechsner, H., *Nucl. Instr. Meth.* B **82**, 518 (1993).

8. Middleton, R., Klein, J., *Nucl. Instr. Meth.* B **123**, 532-538 (1997).

9. Jacob, S.A.W., Suter, M., Synal, H.-A., ., *Nucl. Instr. Meth.* B **172**, 235-241 (2000).

10. Wang, Xue-Bin, Wang, Lai-Sheng, *Phys. Rev. Lett.* **83**, 3402-3405 (1999).

11. Berkovits, D., Heber, O., Klein, J., Mitnik, D., Paul, M., *Nucl. Instr. Meth.* B **172**, 350-354 (2000).

12. Klein, J., Middleton, R., *Nucl. Instr. Meth.* B **159**, 8-21 (1999).

13. Middleton, R., Klein, J., *Phys. Rev.* A **60**, 3515-3521 (1999).

Accelerator System at The Wakasa-wan Energy Research Center

S. Hatori, Y. Ito, R. Ishigami, K. Yasuda, T. Inomata, T. Maruyama, K. Ikezawa, K. Takagi, K. Yamamoto, S. Fukuda, K. Kume, G. Kagiya, T. Hasegawa, M. Hatashita, M. Yamada, H. Yamada, M. Dote, N. Ohtani, S. Kakiuchi, Y. Tominaga, S. Fukumoto and M. Kondo

The Wakasa-wan Energy Research Center, Fukui 914-0192, Japan

Abstract At Wakasa-wan Energy Research Center, Fukui, Japan, the accelerator system is now under construction. The system consists of a 200 keV ion implanter, a 5 MV Schenckel type tandem accelerator and a 200 MeV proton synchrotron. The ion implanter is for the improvement of the surface of the material. By using the tandem accelerator, we can perform the element analysis, the improvement of the bio-species and ion-implantation into various materials. The tandem accelerator works as an injector for the synchrotron and the synchrotron is for the cancer therapy and also bio and material science.

INTRODUCTION

The Wakasa-wan Energy Research Center (WERC) was constructed as the first institution close to local industry under the suggestion of Science Council of Japan. WERC is located at Tsuruga, Fukui Prefecture and the area is famous for the location of many nuclear plants. The researches at WERC are in very wide range, for example, the study of utilization of the heat emitted from the power stations and/or natural energy, reduction of the load to environment by synthesis of hydrocarbon by deoxidization of carbon oxide, safeguard research, environmental science and utilization of radiation. WERC are opened to the local industry and scientists. We perform the research of application of our machine to local industry and the training of technicians in the area.

Now, we are constructing an ion beam accelerator system. By using ion beams with wide range energy from 200 keV ion implanter, 5 MV tandem accelerator and 200 MeV proton synchrotron, we will study material science, development of ion beam analysis, cancer therapy or improvement of plants. FIGURE 1 shows our accelerator complex. For the application of ion beams, four irradiation rooms are prepared.

MICROWAVE ION SOURCE

A microwave ion source is for high intense implantation experiments. The maximum extraction voltage is 200 kV. Gas elements such as Kr, Ar, O and N and solid elements such as C, P, Si and Al are available for the ionization. The machine consists of a microwave ion source, an analyzer magnet, a post-acceleration tube, a bending magnet, a triplet quadrupole lenz and a target chamber. The frequency of the micro wave is 2.45 GHz and the magnetic field of 1.2 kG higher than ECR (875 G) is applied. The plasma density amounts to $10^{11} \sim 10^{12}/$cm^3.

TANDEM AND SYNCHROTRON SYSTEM

Ion sources

We have two ion sources. One is called "main ion source" and is used for the ionization of hydrogen and solid elements. The work function of the surface of the target adsorbing Cs vapor is reduced. By plasma sputtering the surface of the target and electrons penetrating through the potential barrier for binding electrons, plasma or target elements are converted to negative ion.

Another for the gas elements, especially He. "He-ion source" consists of heated cathode type positive ion source and charge exchange sell. The charge exchange is done by Li vapor. The ion source can generate 50 µA He negative ion.

When the tandem beam is injected to the synchrotron, we extract a pulsed beam by pulse-driving the arc electrode and applying the pulsed voltage to the electrostatic deflector.

Tandem accelerator

The maximum terminal voltage of 5 MV is generated by Schenkel rectifier. In order to enable transport and accelerate high intense beam (DC: 100 µA, pulse: 18 mA×250 µs×0.5 Hz), the conveyor current amounts to 500 µA. By using large capacitance for the terminal condenser, the ripple of the voltage when injection of pulse beam is reduced to 2 kV at the terminal voltage of 5 MV. It corresponds to energy dispersion of 4×10^{-4} and one tenth less than RFQ-DTL system.

For the charge exchange, an argon gas stripper is used. In order to transmit the high intense beam, the inside diameter of the stripper canal is comparative large 15 mm. The concentration of the gas into the stripper section is done by the recirculation by four turbomolecular pumps (50 l/s×4).

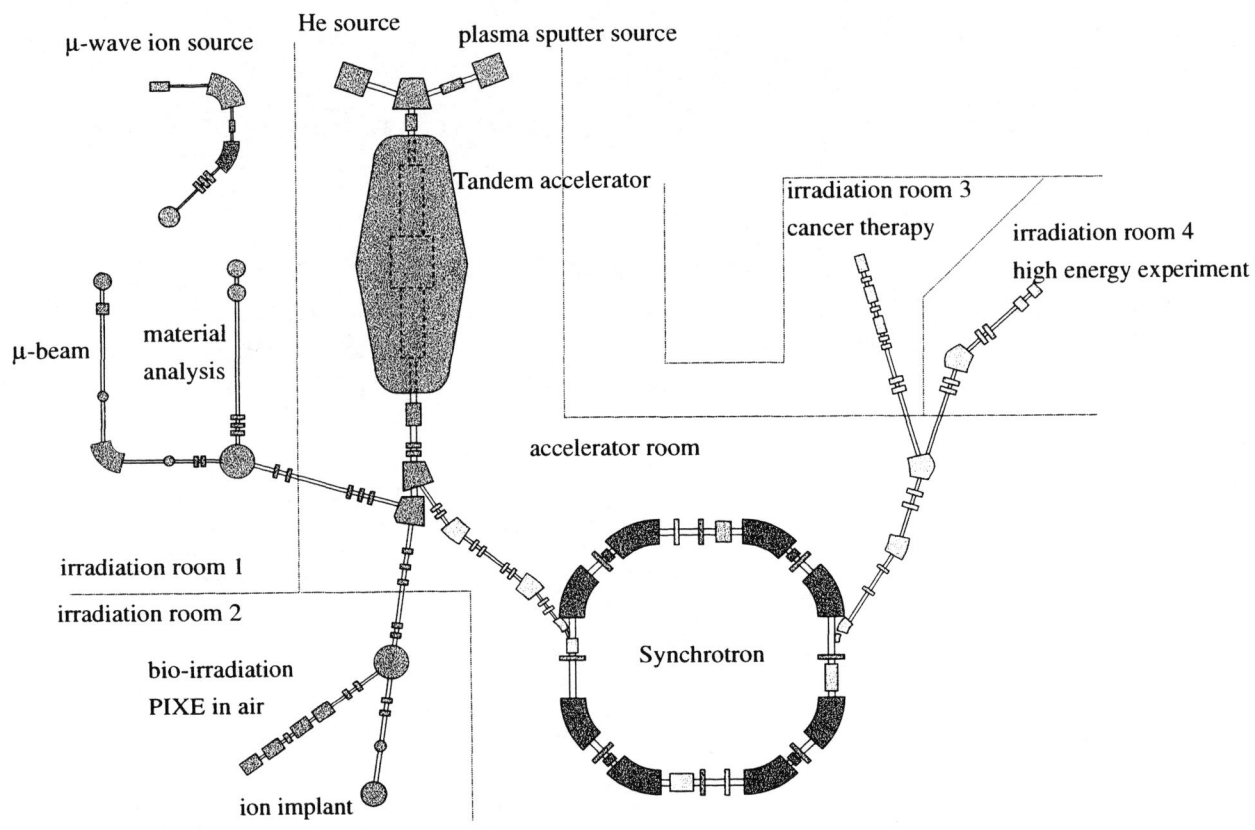

FIGURE 1 Schematic layout of WERC accelerator system

Synchrotron

The tandem beam is injected to the synchrotron and then accelerated to 200 MeV in the case of proton and 55 MeV/u for the heavy ion.

The circumfence of the synchrotron is 33.2 m. The superperiodicity is 4. Each lattice has QF-D-QD-D-QF and it is operated in separate function style. Horizontal and vertical tunes are 1.75 and 0.85, respectively.

The beam from tandem accelerator is injected by multi-turn injection method. The beam remains captured for about ten turns. The acceleration is done by an asynchronous RF cavity. In order to reduce the space charge effect in the early acceleration period, the higher harmonic wave is added to the cavity. After the acceleration, the beam is slow-extracted during 0.5 sec flat top period by a RF-knockout method. FIGURE 2 shows the beam intensity in the synchrotron as a function of time in the one capture-acceleration-strage-decceleration period. The design and achieved intensity of the tandem-synchrotron system is shown in TABLE 1.

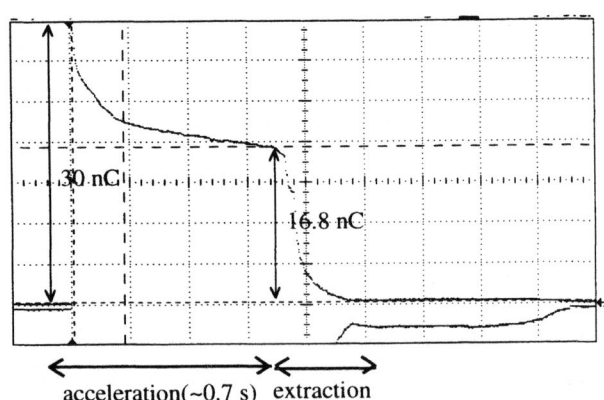

FIGURE 2 Beam intensity in the synchrotron during the acceleration period.

TABLE 1 Design and achieved beam intensities

	design	achieved
injection to tandem	18mA	9.8mA
extraction from tandem	6mA	4.7mA
injection to synchrotron	6mA	4.5mA
straged particle (200MeV)	1.3×10^{11}	1.0×10^{11}
extraction from synchrotron (200MeV)	10nA	8.0nA

BEAM LINES

The beam extracted from the tandem accelerator of which terminal voltage is set at maximum 1.7 MV is transported to the irradiation room 1. The room has two beam lines. One is for the microanalysis experiments. The beam is focused to the size of 2 μm by using a set of magnetic-quadrupole doublet. In the terminal chamber, PIXE, RBS and ERDA experiments are available. Another for the low energy implantation.

In the irradiation room 2, we can perform the experiments setting the terminal voltage at maximum 5 MV. We have two beam lines in the room. One is for the medium energy implantation or ion beam analysis such as transmission type ERDA. From the end of the another beam line, the beam can be extracted to the air. By using three apertures, the beam size is reduced to 10 μm diameter. We can apply the beam to the bio-irradiation or archeological matter.

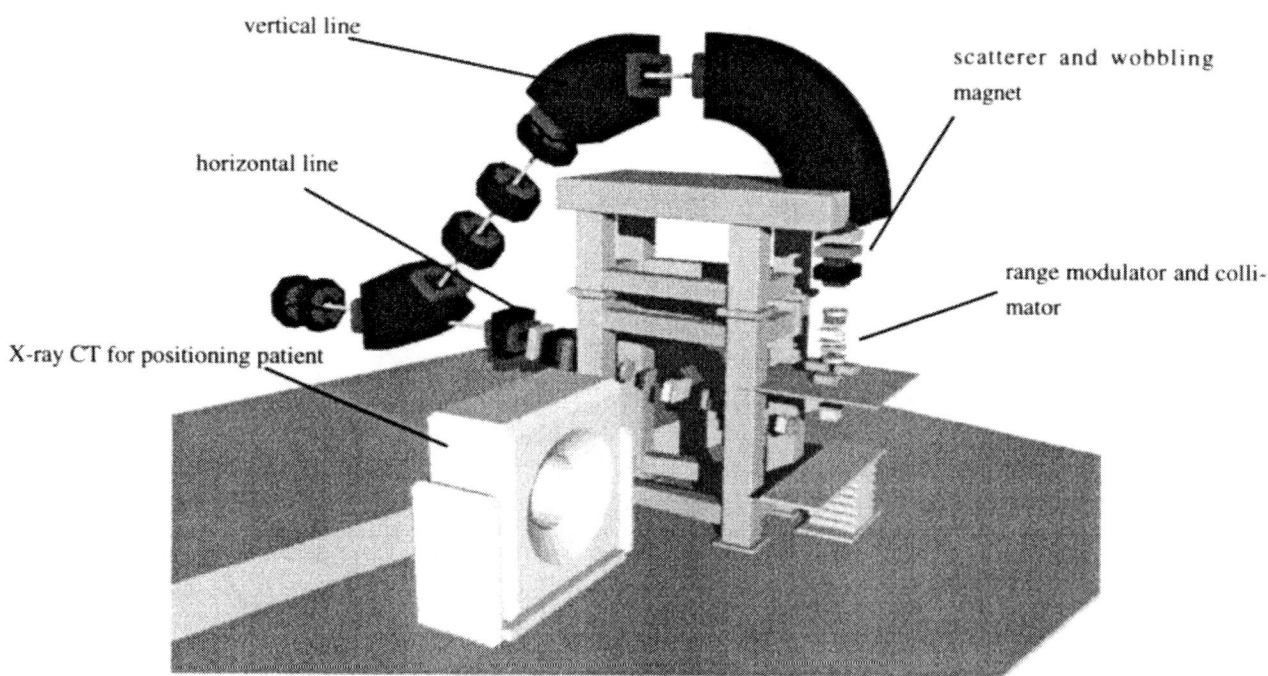

FIGURE 3 Conception of therapy beam lines

In the irradiation room 3 and 4, we can use the beam extracted from the synchrotron. In the room 3, we have two beam lines. Both lines are for the cancer therapy. The beam is extracted in the perpendicular direction from the one beam line and horizontally extracted from the other. Both lines have scatterers and wobbler magnets which are used for making the irradiation area with the diameter of 15 cm. The beam lines are now under construction and Figure 3 shows the conception. In the room 4, high energy irradiation experiments or developments for the therapy are performed. By using this beam line, the quality test of the beam for the therapy, i.e., check of the Bragg peak, the spread out Bragg peak (SOBP) and the beam profile in the water has been performed. Figure 4 shows the observed Bragg peak and SOBP of the 180 MeV proton in the water and that the beam is available for the cancer 5 cm long in the depth.

SUMMARY

The Wakasa wan Energy Research Center was born as an institute very close to the local industry and science. In order to contribute the local industry, we have been constructing accelerator system in the aim of the utilizaition of the radiation.

The energy of the beam from the accelerator complex is from sub MeV to sub GeV and its applications are in the very wide range such as material science, archeology, agriculture and fishery, cancer therapy and so on. In the next century, we start clinical irradiation.

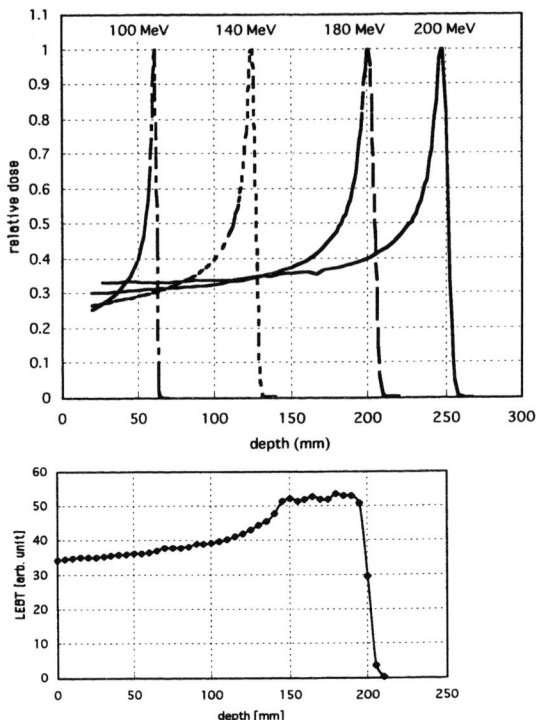

FIGURE 4   Observed Bragg peaks and SOBP

Bragg peaks (upper) were measured by using solid state detector in the water target and changing the incident energy of the proton beam. SOBP (lower) is for the 180 MeV proton beam in the water.

The NSCL coupled cyclotron project: status and latest news

T. Baumann

National Superconducting Cyclotron Laboratory, Michigan State University, East Lansing, MI 48824-1321, USA

Abstract. The upgrade of the National Superconducting Cyclotron Laboratory (NSCL) is well under way and nearing completion. With the coupled superconducting cyclotrons K500 and K1200, the NSCL will be able to deliver high-intensity heavy-ion beams for nuclear reaction studies at intermediate energies and for the production of rare isotope beams via projectile fragmentation. I am reporting on the planned facility and the current status of the upgrade project. Selected highlights of the physics opportunities at the upgraded facility are reviewed. Specific advantages of using fast beams for rare isotopes will be discussed.

INTRODUCTION

The National Superconducting Cyclotron Laboratory (NSCL) houses two cyclotrons, the world's first superconducting cyclotron K500, and its successor, the in terms of total beam energy most powerful cyclotron K1200. Since it started operation in 1989, the K1200 has delivered a large array of beams ranging from hydrogen to uranium with energies up to 200 MeV/nucleon that were used for basic nuclear physics research. With the increasing demand of intense rare isotope beams on the one hand, and the quest to study nuclear systems at extremes of temperature and density on the other, the need for higher beam intensities and energies arose. The coupled cyclotron project [1] addresses this need in a very efficient way, since it makes use of the two existing cyclotrons K500 and K1200, as well as existing ion sources and experimental facilities.

In order to accelerate heavy ions to high energies, high charge states are needed. Ion sources are, however, limited in intensity for high charge states. By splitting the acceleration into two steps, this limitation can be overcome. The K500 cyclotron will accelerate a low-charge-state beam, which the ion source can produce with high intensity, to moderate energies of not more than 20 MeV/nucleon. Upon injection into the K1200, this beam will be stripped to a higher charge state and can then be accelerated to the full energy.

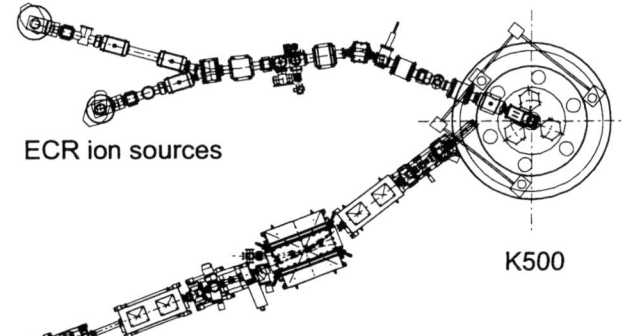
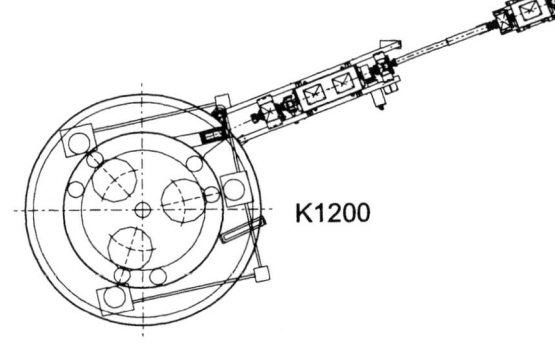

FIGURE 1. Top view of the two cyclotrons (K1200, to the left, and K500, above) together with the ion sources, injection and coupling line. The superconducting and the room-temperature ECR ion sources are located east of the K500 cyclotron. They feed their beams through a low-energy, high-intensity injection line into the K500 cyclotron. The direct injection line to the K1200 cyclotron is not shown on this drawing.

Some beams, e.g. low energy heavy ions and ^{4}He beams, are more efficiently accelerated by the K1200 in stand-alone mode, and this will still be possible after the upgrade.

---

* The NSCL is supported in part by the National Science Foundation grant number PHY95-28844.

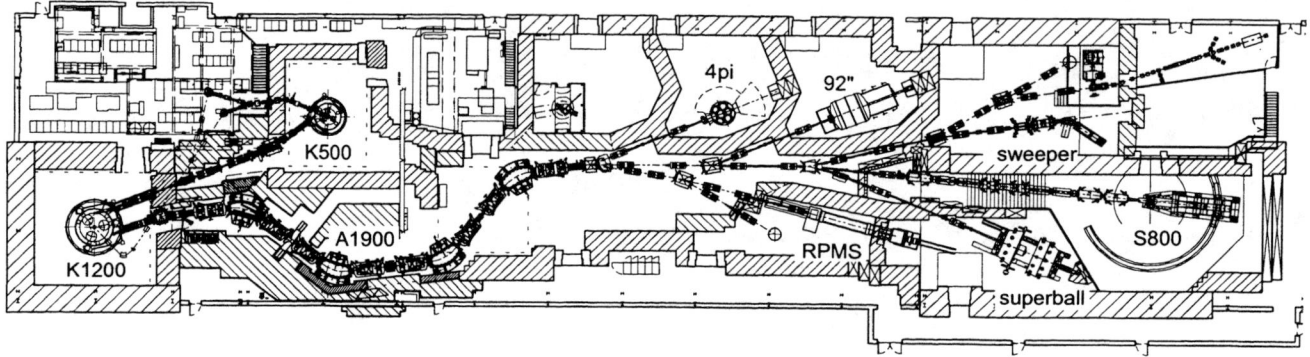

FIGURE 2. This is a top view of the coupled cyclotron facility after its completion. From the K1200 cyclotron (bottom left), the primary beam impinges onto a production target at the beginning of the A1900 beam analysis system. The reaction products are then dispersed by two large 6 Tm, 45° dipole magnets, enabling a selection by mass-to-charge ratio. A wedge-shaped degrader can be inserted at the intermediate focal plane (center) to achieve an isotopically clean separation of the reaction products. The two following dipoles match the dispersion at the intermediate focal plane and make the complete system achromatic. From the end of the A1900 (center), the rare isotope beam can be transported to any of the experiments described in the text.

To make optimum use of the beams delivered by the coupled cyclotrons, the new beam analysis system A1900 [1] for the separation of secondary beams produced by projectile fragmentation is being built. The A1900 has an improved acceptance and bending power compared to the previous beam analysis system. It will serve a dual function as beam transport system to make the primary beam available at any experimental station, and as the key element in the production and separation of rare isotope beams. The A1900 can cope with a magnetic rigidity higher than that of the primary beam, so that very neutron rich nuclei can be studied at the maximum energy of the K500 $\otimes$ K1200 cyclotron.

EXPERIMENTAL EQUIPMENT

A large array of experimental instruments will be available after the upgrade is completed. More information on the experimental devices and references can be found in the NSCL annual report [2]. In the following I give a brief overview.

The 4π Array is a charged particle detector array designed to study nucleus–nucleus collisions at intermediate energies. Using three different types of detectors, low pressure multiwire counters, Bragg curve counters, and fast/slow plastic scintillator telescopes, the 4π Array is able to detect and identify reaction products ranging from protons to fission fragments with a solid angle of close to 4π. The 4π Array is equipped with three high-resolution forward detectors, a high-rate forward array consisting of 45 phoswitch detectors covering angles up to 20°, a 16 element silicon/plastic array for angles between 1.5° and 2.9°, and a 16 element fast plastic detector for angles of 0.5°–1.5°.

The 92" scattering chamber is designed to house complex detector arrays in vacuum. The chamber is 234 cm (92") in diameter, about 300 cm long and contains a mounting platform, a turntable, a rotating arm, and a target positioning mechanism.

The Sweeper Magnet is a 4 T dipole magnet and a new development that will be ready after the cyclotron upgrade is completed. Many experiments involve neutron-rich, weakly bound nuclei, and often the final states will be neutron-unbound. To measure these states it is necessary to detect the decay products, i.e. a charged particle in coincidence with a fast neutron, both typically emitted close to the beam axis. The purpose of the sweeper magnet is to deflect the charged fragments from the axis to a shielded location at larger angles, where the charged particles are stopped or detected. For the detection of the charged particles, a detector array or the S800 spectrograph can be used. The neutrons will be detected under zero degree further down the beam axis in the *Neutron Walls*, a set of large area (2×2 m^2) liquid scintillator neutron detectors. Since the charged reaction products are stopped in a shielded location, the amount of background neutrons is greatly reduced.

The Reaction Products Mass Separator (RPMS) is a device to separate high energy heavy ion reaction products according to their mass-to-charge ratio. A velocity filter (Wien filter) is located directly in front of the RPMS, enabling a very clean separation of weak secondary beams.

The Superball is the University of Rochester's five-segment, 16,000 liter gadolinium-loaded liquid scintillator 4π neutron calorimeter surrounding a vacuum scattering chamber. With its 52 photo-multiplier tubes, it can detect multi-neutron events and is used as a multiplicity filter in nuclear reaction studies.

The S800 Spectrograph is a high-resolution superconducting magnetic spectrograph. With an energy resolution of 10^{-4}, an energy acceptance of 10%, and an angular acceptance of 20 msr, it is ideally suited for the analysis of products steming from direct reactions of secondary beams such as knock-out or charge-exchange reactions. In this type of experiments, a good particle identification of the reaction products is required.

The Segmented Germanium Detector Array is a new development for high-resolution in-beam γ-ray spectroscopy. Due to the 32-fold segmentation of the germanium crystal, this detector can be positioned close to the target where the γ rays are originating. This significantly improves the efficiency, while the segmentation keeps the effective opening angle of the detector small and the resolution at an optimum. A total of 18 detectors are ordered and can be used in different configurations. Besides using this detector array by itself, it can be combined in setups with the sweeper magnet or the S800 spectrograph.

In addition to the instruments described above a facility to investigate rare isotopes with short half-lives is currently being developed. In this facility, an energetic rare isotope beam that is separated by the A1900 beam analysis system will be stopped in a gas catcher cell and subsequently re-accelerated to low energies.

Two beam lines will be dedicated to accommodate experimental setups from outside users.

SCIENTIFIC OPPORTUNITIES

The significant intensity gains for primary beams of the coupled cyclotron, see Fig. 3, have a direct impact on the rare isotope beams that can be produced and used for experiments. With the intensities that will be available after the upgrade (see Fig. 4), we can study regions of the nuclear chart that were inaccessible in the past. Nuclear structure research and nuclear astrophysics will benefit from the availability of beams far from the valley of stability, since many of the most interesting problems are hidden in these regions.

One of the very interesting research topics will be to investigate isotopes that lie in the path of the astrophysical r-process (rapid neutron capture), which is responsible for the creation of about half of the heavier elements in our universe. The r-process is located in the very neutron rich region of the nuclear chart, and in the past only very few of the nuclei that take part in the r-process have been investigated experimentally. With the upgraded cyclotron

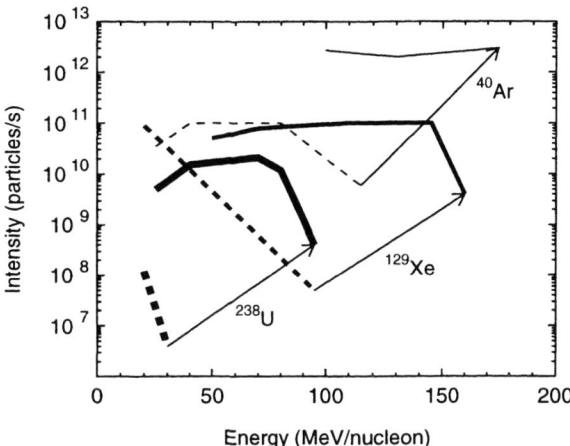

FIGURE 3. Estimated primary beam intensities and energies for the coupled cyclotron (solid lines) for isotopes of uranium, xenon, and argon (bold, medium, and thin lines). The dashed lines show the intensities that were available before the upgrade.

facility, a large number of these nuclei will be accessible for the first time. Figure 4 shows the estimated yields for some of the isotopes in the path of the r-process.

Another interesting question that is a major long-term experimental challenge for the research with rare isotope beams is the exploration of the very neutron-rich regions of the nuclear chart with respect to the limits of nuclear existence. Although the neutron drip-line, the limit of nuclear existence on the neutron-rich side of the nuclear chart, will remain unexplored to a large extent with even the most advanced proposed facilities, the detailed study of neutron-rich nuclei is a crucial step towards the understanding of the limits of nuclear binding. The question if the models of nuclear matter, like the shell model, do still apply to very neutron-rich nuclei, has to be investigted. Since some of the nuclear structure properties change slowly with neutron number, the exact position of the neutron drip-line is extremely sensitive to model parameters. The experimental determination of these properties, like the single particle energies, is very important for describing the neutron drip-line. With the rare isotope beams of the upgraded facility, many of these neutron-rich nuclei can be measured for the first time, and many others can be studied in more detail than possible with previous intensities.

Fast beams produced by projectile fragmentation or fission and separated in flight are the key method to reach neutron rich short-lived isotopes. Besides the fact that a large number of isotopes can be investigated only by this technique, fast beams offer an economic production of medium-energy beams of rare isotopes, an increased luminosity by using thick secondary targets, re-

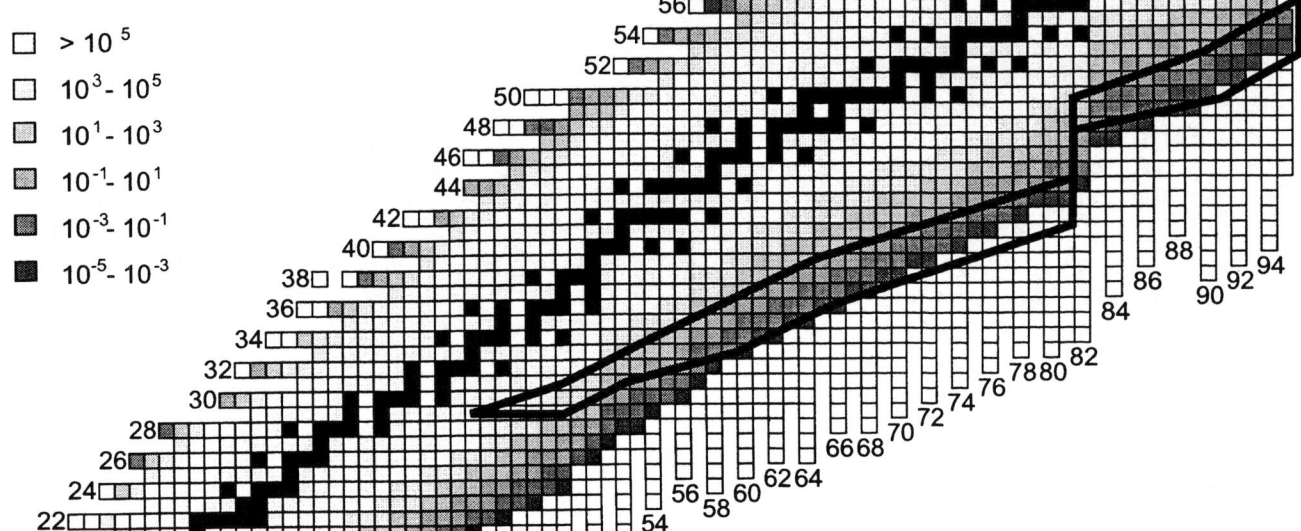

FIGURE 4. This figure shows a section of the nuclear chart, with the proton number on the vertical and the neutron number on the horizontal axis. The various shades of grey correspond to the estimated production yield of the respective isotope in particles per second, that the coupled cyclotron facility will be able to achieve. This figure also indicates the approximate path of the r-process (solid black line) that takes place when heavier elements are formed during the astrophysical nucleosynthesis. With the cyclotron upgrade, a large number of the nuclei in the path of the r-process will be accessible for the first time [3, 4].

duced background due to in-flight tracking identification on a particle-by-particle basis, efficient detection due to strong forward focusing, and low beam losses due to a chemistry-independent separation and fast transport to the experiment. Beam intensities down to the order of 10^{-5} particles per second, which translates to less than one particle per day, are already sufficient for some studies.

THE LATEST STATUS

Following is a brief list of the status (as of the time of writing) of major components of the upgrade project. A new room-temperature ECR ion source is completed and the beam development from this source is in progress. The ECR–K500 injection line is completed and being optimized. The K500 cyclotron is operational with an adequate performance. Injection and extraction studies are underway. The K500–K1200 coupling line is complete and ion optics test are being performed. The injection channel of the K1200 is installed and the stripper foil mechanism is being tested. All magnets of the A1900 fragment separator are installed and tested. The complete beam analysis system will be completed and ready for debugging by the end of the year 2000. The commissioning of the coupled cyclotron facility in planned to be finished by June 30, 2001.

CONCLUSION

The coupled cyclotron project will be completed within the next year and open up many new possibilities to the nuclear physics research community. With the anticipated intensities of rare isotope beams, many open questions in nuclear structure and nuclear astrophysics can be addressed. The coupled cyclotron facility will be a prime location for the production of rare isotope beams for the next decade and bridge the time until the arrival of the Rare Isotope Accelerator.

REFERENCES

1. NSCL White Paper, *The K500 ⊗ K1200*, MSUCL-939, 1994.

2. NSCL 1999 Annual Report [online], <http://www.nscl.msu.edu/research/1999_Annual_Report/>, 2000.

3. NSCL White Paper, *Scientific Opportunities with Fast Fragmentation Beams from the Rare Isotope Accelerator* [online], <http://www.nscl.msu.edu/research/ria/whitepaper.pdf>, March 2000.

4. Schatz, H., private communication.

Recycling and Recommissioning a Used Biomedical Cyclotron

L. R. Carroll and F. Ramsey
Carroll & Ramsey Associates

J. Armbruster
International Isotopes, Inc.

M. Montenero
Electronic Visions, Inc.

Abstract

Biomedical Cyclotrons have a very long life, but there eventually comes a time when any piece of equipment has to be retired from service. From time to time, we have the opportunity to help find new homes for used cyclotrons which, with relatively modest overhaul and refurbishment, can have many additional years of productive service, and thus represent a very valuable asset.

The reasons for retiring a cyclotron vary, of course, but in our experience it is often due to an institution's changing priorities or changing needs, rather than due to any fundamental age-related deficiency in the cyclotron itself. In this paper we'll report on the relocation and successful restoration of a used TCC CP-42 cyclotron, which was moved from M.D. Anderson Hospital in Houston to Denton, Texas in early 1998, where it is presently being used for R&D and commercial production of biomedical isotopes. Ownership of the machine has been transferred to the University of North Texas; facility, manpower, and operational resources are provided by International Isotopes, Inc.

ORIGINAL MOTIVATION AND GENESIS OF THE CP-42 CYCLOTRON

In a traditional, positive-ion cyclotron – exemplified by the CS-series of machines once built by The Cyclotron Corporation (TCC) of Berkeley, CA -- extraction of the beam is accomplished by means of an electrostatic channel. Inevitable beam loss at the entrance to the extraction channel, and consequent localized heating of extraction system components limits the amount of external beam that can be safely and reliably utilized.

Efficient, commercial-scale production of biomedical isotopes demands very high beam currents – much higher than can be easily extracted from a positive-ion cyclotron. Since *internal* (+) ion beam currents of the order of 100's of μA are easily achieved, commercial radiopharmaceutical producers often utilize internal targets, thus obviating the need to extract the beam.

However, operation of high-current internal targets carries a substantial price in that the cyclotron itself is subject to an intense, damaging neutron flux which can eventually lead to failure of major system components, as well as high-level, persistent, neutron-induced activation of the several tons of steel and copper comprising the main magnet. Neutron-induced activation also leads to persistent high levels of radiation in and around the cyclotron, making service and maintenance highly problematic.

Negative-ion Acceleration and Extraction

The initial concept and design of the CP-42 cyclotron was motivated largely by a desire to provide high beam current on targets which are external to -- and at a substantial distance from – the cyclotron itself, thereby mitigating the activation problems. This was accomplished by accelerating negative hydrogen (H-) ions and extracting the beam by means of charge-exchange in a thin foil. Utilizing this technique, an extraction efficiency of nearly 100% is achieved. Thus, the CP-42 was rated for an external proton beam current of 200 μA. over a range of energies from 10 MeV to 42 Mev, extracted from one of 9 separate beam ports.

In the early 1980's, TCC built five CP-42's: Four were intended for commercial-scale production of radiopharmaceuticals; one of the CP-42's was sold to the M. D. Anderson Hospital in Houston, TX for use in fast-neutron therapy for the treatment of cancer, funded by a research grant from the U. S. National Cancer Institute.

The Demise of TCC

Unfortunately, the scale and magnitude of the CP-42 development project severely strained the financial resources of the Cyclotron Corporation. The company went bankrupt in 1983, leaving all of the CP-42 installations in a state of partial completion. Fortunately, the projects were indemnified by an insurance company which immediately stepped forward to retain the pool of TCC technical personnel. One commercial customer elected to take a cash settlement instead of completing their project. That machine was subsequently shipped to Argentina and rebuilt under the supervision of the Cyclotron Group at KFK Karlsruhe. All of the other installations were eventually completed and accepted. All of the commercial CP-42 cyclotron installations are still operating to this day.

The CP-42 at Houston was intended primarily as a tool for the treatment of cancer using fast neutrons. In time, however, the outcome of clinical trials demonstrated results which were not substantially better than those achieved by well-managed photon therapy. Interest in the use of neutron therapy by researchers and funding agencies waned .....

In the 1980's positron-emission tomography (PET) began to emerge as an essential tool for biomedical research, with significant potential as a clinically-useful diagnostic imaging modality. During this period the Houston CP-42 had also been used for the production of PET / biomedical isotopes. Unfortunately, the CP-42 was just too big and expensive a machine to operate for cost-effective, hospital-based PET, and was eventually shut down and moth-balled.

TRANSFER OF OWNERSHIP

The administration at M.D. Anderson Hospital were looking for a way to remove the CP-42 from the premises, since the now-idle cyclotron was taking up valuable space in the hospital. They arranged for a transfer of ownership to another institution in the University of Texas System -- the University of North Texas (UNT) at Denton. Facility, manpower, and operational resources were to be supplied by International Isotopes, Inc. (I^3), a new start-up Company just entering the Radioisotope and radiopharmaceutical production business, who would be leasing the CP-42 system from the university. In the fall of 1997, Carroll & Ramsey Associates (CRA) were contacted by Dr. Joe Beaver, vice President for System Development at I^3.

CRA was asked to provide technical assistance and project supervision for dismantling the Houston CP-42 and moving it to the I^3 facilities in Denton, as well as helping to rebuild and re-commission the machine. The intention was to utilize the cyclotron for targetry and process development and validation while the primary accelerator facility at I^3 – the former injector Linac from the recently-canceled Super-Conducting Super-Collider Project -- was being refurbished and upgraded to adapt and "ruggedize" the injector Linac for eventual high-duty-cycle commercial service as I^3' s main isotope-producing accelerator.

The process of dismantling and moving the CP-42 was initiated under the direction of Mr. Robert Wetzel, senior project manager at I^3. We first conducted a careful review and reconnaissance of the M.D. Anderson facility with consideration for heavy lifting, maneuvering in close quarters, access, floor loadings, etc. A walk-through and planning session with the owners, riggers, and I^3 personnel was conducted in the weeks prior to the actual move.

Dismantling and loading the CP-42 onto five trucks required 10 days. A crane was brought in for 3 days to lift the cyclotron, power supplies and ancillary equipment from the basement through an access hatch-way in an adjoining parking lot with relatively minor disruption to the everyday hospital routine.

RE-ASSEMBLY AND SYSTEM TEST

When the CP-42 arrived at the I^3 facility in Denton, the vault intended to house the cyclotron hadn't yet been built, so the machine was assembled on the floor of a warehouse in order to evaluate and repair – as needed – all of the power supplies, mechanical subsystems, vacuum system, RF system, etc.

Since the Cyclotron Corporation had long-since gone bankrupt, the former owners of the CP-42 may have felt (albeit incorrectly) that they had nowhere to turn for outside technical assistance. During the previous years of service at M.D. Anderson there had been many ad hoc field changes and "repairs" which were well-intentioned, but not always optimum in their implementation. However, as a matter of CRA company policy, we proceeded under the premise that – with certain very specific exceptions – the original design concept of the CP-42 was sound, and that our primary goal would be to repair worn or damaged parts, and to restore the system to as close to the original factory configuration as possible.

The major exceptions were: 1) implementation of an improved high-voltage Dee insulator design; 2) changes to the ion-source water-cooling concept plus changes in various other aspects of the mechanical assembly and alignment of the ion source, and 3) a complete revision and replacement of the computer-aided operator-control interface, including substitution of a PC - based control console for the original DEC Model PDP-1103 computer, substitution of PLC's and other off-the-shelf components (such as stepper-motor control cards, etc.) for the original TCC-designed subsystem control modules.

Overhaul of Major System Components

The magnet and vacuum tank were dis-assembled in order to clean and polish all the O-ring surfaces which form the main vacuum seal between the upper and lower pole base and pole-tip plates. This required removal of the magnet hills, the vacuum tank, pole tip plates, plus the upper magnet coil, since that is held in place by the upper pole-tip plate and the upper portion of the vacuum tank. (The lower coil could stay in place.)

All four of the Varian VHS-10 vacuum diffusion pumps were overhauled and rebuilt using Vendor-supplied repair kits. Major repairs and replacement of worn parts were also implemented on the mechanical backing pumps. After refurbishment, repairs, and final leak-checking were completed, a base pressure in the 10^{-7} torr range was achieved.

All of the major power supplies and systems were 'stripped to the bone' to check for loose or corroded electrical wiring connections as well as any leaks in plumbing connections.

New Control System

By the time the system was decommissioned at M. D. Anderson Hospital, it had not been run for several years. The documentation was in disarray and many sections were missing. The control computer system had been modified from the original PDP 1103 system supplied by TCC and, although the computer system was still functional, the source and support disks were not readable. Inability to modify the control system for the new installation essentially rendered it useless.

When the CP-42 arrived in Denton the basic vacuum control system, comprising a control panel and a Texas Instruments "5TI" programmed-logic controller was quickly reassembled and placed into operation to maintain vacuum until a new control system concept was devised.

The control requirements were evaluated along with the existing system hardware components. The original PDP 1103 computer had already been replaced by an ensemble of microcomputers, but most of the original proprietary TCC distributed control bus was still in use. 5TI PLC's were still being used throughout the system for interlocking functions and vacuum controls.

It was decided that the new system would use a standard personal computer (PC) with LabView^{tm} as the graphical user interface. This would allow reasonable ease of programming and interfacing with other hardware. Moreover, I^3 was already using LabView for various other laboratory and control applications, so this would help maintain a degree of standardization and uniformity within the company.

Due to unavailability of parts, the 5TI PLC's were replaced with contemporary Modicon^{tm} PLC's linked to the PC with Modbus Plus^{tm} communications. The new PLC's would handle the interlocking functions, and also provide most of the I/O for LabView. Additional I/O cards were installed in the PC for interfacing to the control encoder knobs, beam current monitoring and Beam-line power supply control and monitoring.

In the Main Magnet supply, only the closed loop analog portion of the controls was re-used. A PLC was used to provide all control and monitoring signals. Another PLC was used for miscellaneous interlocks, beam-line interlocks and controls, variable extractor controls, and harmonic coil controls. Additional PLC's were used for a redundant Personnel Access and Safety System and for Target Transfer and Handling controls. The Magnet Supply itself was taken apart, components inspected and tested, and then reassembled. It was then connected to the Magnet and tested on the warehouse floor prior to permanent installation in the new facility. The Anode Power Supply and Ion Source Supply were treated in a similar manner, but could not be fully tested until all the systems were installed in an acceptable climate-controlled environment. After initial overhaul and after lengthy and thorough operational testing, the original power supplies (main magnet, RF, and ion source) have proven to be sound.

Only one major failure occurred -- a high-voltage line-isolation transformer in the RF power supply shorted out. Fortunately, a vendor was found who was able to re-build the transformer on a priority schedule. Most other problems were mechanical in nature such as worn motor gearing, control linkages, etc. Other problems occurred in the initial start-up of the RF system, but these proved to be largely due to lack of good environmental controls while construction of the facility was still underway. Once the room temperature and -- most importantly -- the humidity were brought under control, the system has been operating very reliably.

Vault Design and Construction

While repairs to the CP-42 were underway, designs for the vault were evaluated . After weighing the overall costs, advantages, and disadvantages of various designs against the mission at hand (this being intended more as a tool for "R&D" rather than a dedicated "production" facility), I^3 chose to build a single, large room to house the cyclotron, beamlines, and targets. Based upon the intended target products and estimated beam currents, it was determined that a wall thickness of six feet and a roof thickness of four feet would be sufficient to reduce the radiation fields outside the vault to less than 0.5 mR/hr in any uncontrolled area. It was also planned that supplementary shielding would be installed directly around the target stations to further reduce radiation fields. In any case, radiation measurements would be performed under actual operating conditions to verify that the shielding was sufficient and, if necessary, additional shielding would be installed.

The partially re-built CP-42 was then moved into place and the vault built around it. The inner wall of the vault was assembled using traditional wood frame construction techniques. In addition, polyethylene beads were placed between the wall studs to enhance neutron shielding. After pouring a concrete footer, the 6 ft. thick main shielding walls were constructed by dry-stacking concrete blocks. An interlocking pattern was used to help assure that there were no direct seams through the wall. The vault roof was constructed using "Glue Lam" beams, each 10 ½" X 24 ¾" X 34 ft. place on 18" centers over ¾" plywood sheets. Polyethylene beads were also placed between the roof beams to enhance neutron shielding. Concrete blocks were then stacked on top of the roof platform using an interlocking pattern. All penetrations (ducts, wire-ways, etc.) through the shield wall were constructed to eliminate line-of-sight paths. Inside the accelerator vault, all of the power and control cabling, plus hoses and plumbing for system cooling, were installed under a raised "computer" floor. A maze -- also constructed of concrete block -- provides access to the interior of the vault. The final "trimming" of the maze was established from radiation measurements during initial operation of the CP-42. All entrances to the vault and maze are controlled with the Personal Access Safety System (PASS).

OPERATIONAL RESULTS

First internal beam was achieved on October 22, 1998, approximately one year after the initial contact between I^3 and CRA, and approximately 8 months after arrival of the CP-42 at the Denton Facility. The first target irradiation was on Feb. 12, 1999. Target runs are now done routinely -- "on demand" -- to meet the needs of the radio-chemistry department. Until additional system upgrades are implemented, and until certain critical back-up spare parts fabricated or procured, beam currents are normally limited to 100 uA, with occasional excursions to higher levels.

The first product was to be ^{201}Pb / ^{201}Tl, so the first extractor and beamline were positioned for a nominal 28 MeV, based on equilibrium orbit data derived from original magnetic field measurements done at the factory. The final position for the extractor (a fixed "pop-up" extractor) was determined experimentally by measuring the energy of the extracted beam using the stacked foil method. Construction of a thallium target had proceeded concurrently with the project of rebuilding of the CP-42. Once the correct position for the extractor was found, the beam exit angle was established and the beamline and its target station fixed in place.

Next came rebuilding and installation of the Variable-Energy Extractor (VEA) and its beamline. This required a great deal of careful mechanical assembly, alignment, and testing to insure smooth and repeatable operation. A data base of extractor locations versus energy was developed from equilibrium orbit data. Once the VEA and its beamline were in place, beams of various energies (15, 18, 20, and 30 MeV) were extracted to a probe in the "Combo" steering magnet on the exit of the accelerator. Beams of 15 and 18 MeV were also transported to a beam-stop at the end the beamline. Two additional beamlines (16 and 18 MeV) were subsequently installed and are currently in use for isotope production. In calendar 1999, the CP-42 was operated for 16,161 uA-hours in 65 separate runs. In calendar 2000, the CP-42 has been operated (as of Nov 1, 2000) for 24,251 uA-hours in 136 separate target runs.

Acknowledgment

In addition to those named previously, we wish to acknowledge the invaluable help and encouragement from I^3 senior staff and, in particular, Dr. Homer Hupf.

The New IBA Laboratory to be installed at Universidad Autonoma de Madrid

A. Climent-Font, F. Agulló-López, O. Enguita, O. Espeso-Gil, G. García, and C. Pascual-Izarra

Instituto Universitario de Ciencia de Materiales "Nicolás Cabrera" C-XVI, Universidad Autónoma de Madrid, Campus de Cantoblanco, E-28049 Madrid, Spain

Abstract. The Universidad Autonoma de Madrid is installing a 5 MV tandem ion accelerator in its Campus which will be dedicated to ion beam analysis of materials. The voltage generating system will be a Cockcroft-Walton type. This is the second installation of this kind in Spain after the 3 MV Van de Graaff tandem installed in Seville in 1998. In this paper we describe the characteristics of the machine, the accelerator hall and the main lines of research foreseen in this accelerator laboratory. The tentative time schedule for the accelerator to be operative is estimated as fall 2001.

INTRODUCTION

The scientific community in Spain working in subjects related to Materials Science and Solid State Physics, has been steadily growing for the last 25 years. During this period many laboratories in the universities and from the Spanish Research Council (Consejo Superior de Investigaciones Científicas; CSIC) around the country, have acquired equipment for growing materials of interest for the new emerging technologies, mainly in the form of single crystals and thin films, and also a variety of analytical techniques for its characterization.

Until recently Ion Beam Analytical techniques did not benefit from this effort to strengthen the scientific quality and efficiency of the Spanish research. Only in 1998, after a long process, was the first IBA laboratory installed in the south of Spain in the University of Seville. It is equipped with a 3 MV Van de Graaff tandem accelerator from the firm National Electrostatics Corporation. The second IBA laboratory is now in the process of being installed in the campus of the Universidad Autónoma Madrid (UAM). The accelerator will be in this case a 5 MV Cockcroft-Walton tandem from High Voltage Engineering Europa (HVEE) in the Netherlands. We describe in this paper the technical aspects of the machine and the building housing it, and the scientific research purposes.

THE ACCELERATOR LABORATORY AT UAM

Since the UAM foundation in the late sixties, the Physics Division, under the direction of Prof. Nicolás Cabrera, had a strong orientation towards Solid State and Materials Science Physics. Nowadays, the scientific community in these research fields has a strong presence in the UAM campus as, besides the proper university laboratories, it is the location site of a Materials Research Institute from the CSIC. Not far from the campus there is a Microelectronics Institute with manifest interest in the IBA techniques. Besides the research institutes, in the surrounding of the campus there are electronic industries which might as well require services from the UAM accelerator laboratory. The installation of the accelerator laboratory in such an environment seems, therefore, quite adequate.

Other institutions in Madrid interested in IBA techniques include museums like El Prado, which would be interested in the analysis of paintings, the National Museum of Archaeology where curators and archaeologists are interested in studying metallic alloys like bronzes and gold artifacts, and the Institute of Spanish Cultural and Historical Patrimony. We have to mention that in the area of archaeometry, archaeologists from the Department of Archaeology and Ancient History of the UAM also share the interest in the new accelerator. The City Hall of Madrid has also manifested his interest in applications of PIXE to environmental studies.

The 5 MV Tandem and its housing

The decision for the choice of the machine, tandem versus single ended, was taken after several meetings held in Madrid with European experts. There were some doubts about the convenience of a single ended or a tandem machine, but finally the tandem was selected because of its versatility for providing an ion beam of almost any atomic element.

A maximum terminal voltage of 5 MV was thought adequate in order to obtain heavy ion beams like Si with energies in the range of 26 to 30 MeV to be able to perform heavy-ion ERDA, and Au and I beams for time of flight detection applications. As with such an accelerator protons can be accelerated up to 10 MeV and He ions up to 15 MeV, the accelerator hall will be shielded against radiation by concrete walls 1 meter thick, foreseeing future applications with high energetic H and He as well as allowing for the possibility of using deuterium beams for Nuclear Reaction Analysis. Figure 1 shows a drawing to scale of the accelerator hall with the accelerator and the foreseen extension lines.

The machine is now being built by HVEE, which will be supplying also the general purpose line and its chamber.

The high voltage is produced by a Cockcroft Walton voltage multiplier system. This kind of generator, being all made with solid state components, has the advantage over the Van de Graaff system for generating voltage of not having moving parts inducing vibrations and noise (electrical acoustical).

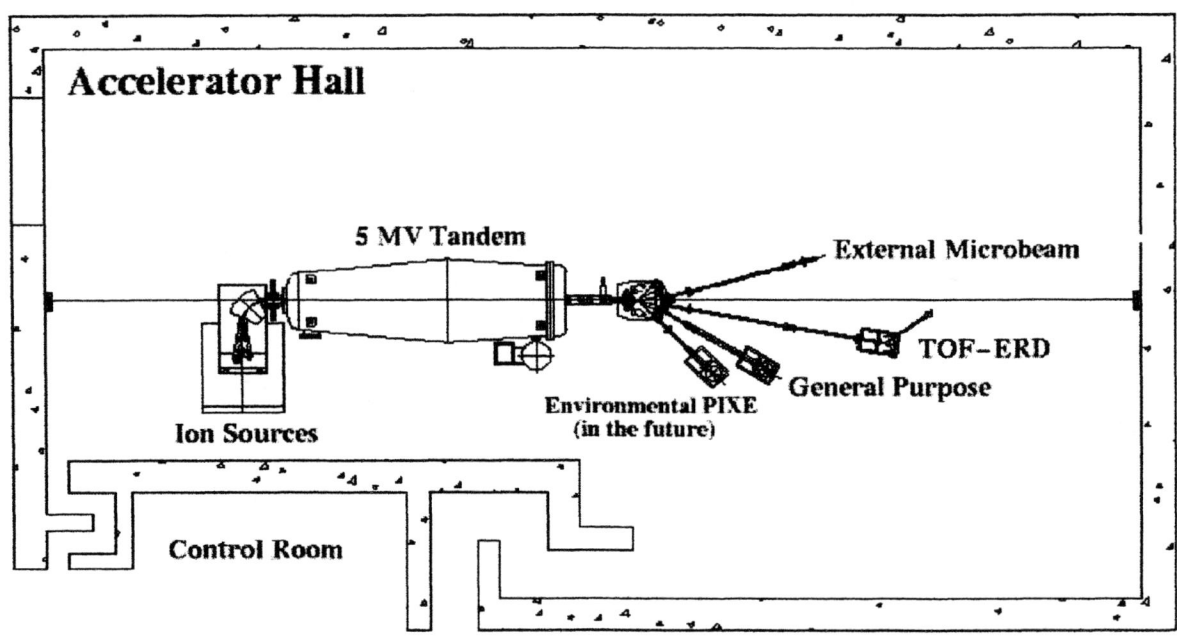

FIGURE 1. Drawing to scale of the accelerator hall housing the UAM 5 MV tandem accelerator showing its main elements and the foreseen extension lines. The annex building for auxiliary laboratories, offices and other services is adjacent to the control room side of the hall.

Reaching 5 MV with the Cockcroft-Walton voltage generator is an innovative project of HVEE who took the challenge of using the diode-capacitor voltage multiplier array to supply a voltage higher than the usual 2 or 3 MV in these systems.

The external microbeam set up is from Oxford Microbeams and we are already designing the extension lines for the microbeam and the complete setup for the time of flight in coincidence with energy elastic recoil detection measurements (TOF-ERD).

The hall is 32 meters long and in the area of the extension lines is 18 meters wide. It has two labyrinth shaped entrances; a small one to have direct access to the ion sources and a bigger one giving access to the extension lines and the experimental chambers. In that way the necessity of passing along the accelerator tank while the system is running is greatly reduced, avoiding exposure to the X-ray radiation produced in the high voltage terminal zone.

The whole building has, attached to the accelerator hall on the side of the control room, a two story annex with auxiliary laboratories, offices, meeting rooms, etc. The building is now under construction.

The research lines

From what has been said above about the laboratories around the campus, and the expectations of different and varied institutions, it is easy to guess that so far we consider at least three main lines of research for the accelerator laboratory: materials science, archaeometry and environmental studies. Other new lines may emerge in the future.

Specific applications for materials science research will be studies in thin films in general, single crystals, implantation studies, diffusion processes and interfaces phenomena. The techniques available will be RBS, Channeling, conventional ERD, TOF-ERD, NRA, PIXE and PIGE. These studies will be carried out using the general purpose chamber and the TOF-ERD setup. We hope that this will be a very important area of research due to the elevated number of research groups working in these fields in our campus.

Studies in archaeometry will be performed mainly using the external microbeam. We are already in touch with museum curators and archaeologists to carry out studies on metals and alloys, gems and stones, paintings, and papers and parchments. The techniques mainly used in this area will be PIXE and PIGE and, less frequently, RBS. Collaborations in this field with art historians and archaeologists have already started [1], [2].

The extension line for environmental studies is, so far, the least developed. We plan to build a new extension line (at 45° in Figure 1) with a dedicated chamber with automatic loading of samples and a high throughput. In this case the main technique to use will be PIXE. In this field, as well, we have previous collaborations with the Department of Environment of City Hall of Madrid [3].

Besides these main three lines, basic studies on stopping powers of the usual beam elements like H and He will be of interest to the UAM accelerator laboratory. We are also interested in a precise determination of the stopping powers of elements like H, B, C, N and O in materials normally used as absorber layers in conventional ERD.

Laboratory staff

We are now in the process of nucleating a research team. So far the staff working in the management, designing the extension lines and doing preliminary research in other well established laboratories are two senior professors from the UAM, a postdoctoral research associate, three PhD students and a technician. We aim, for the time that the accelerator will begin to be operative, to increase the team with another postdoctoral research associate, another PhD student, and two technicians.

Initial funding

The acquisition of the accelerator and the construction of the building has an initial budget of about 4 million US dollars. Funding has been provided by the UAM, the regional government of Madrid; CAM, and the European Community.

ACKNOWLEDGMENTS

We gratefully acknowledge the assistance in several meetings held to define philosophical and technical issues of what an accelerator laboratory should contain and offer, to G. Amsel (Univ. Paris VII), G.G. Bentini (LAMEL Inst. Bologna), J. Gyulai (Tech. Univ. Budapest), R. Hellborg (Univ. Lund), E. Hodgson (CIEMAT, Madrid), M.F. da Silva (Sacavem

Tech. Inst.), J.C. Soares (Univ. Lisbon), and P.D. Townsend and D. Hole (Univ. Sussex).

REFERENCES

1. Climent-Font, A., Demortier, G., Palacio, C., Montero, I., Ruvalcaba-Sil, J.L., and Díaz, D., *Nucl. Instr. and Meth.* **B 134**, 229-236 (1998).

2. Perea, A., Montero, I., Demortier, G., and Climent-Font, A., "Analysis of the Guarrazar treasure by PIXE," in *Ion beam study of art and archaeological objects*, edited by G. Demortier and A. Adriaens, Brussels: European Commission, 2000, pp.99-101.

3. Climent-Font, A. Swietlicki, E., and Revuelta, A., *Nucl. Instr. and Meth.* **B 85**, 830-835 (1994).

Radiation Effects Facilities, Dosimetry and Program at the Indiana University Cyclotron Facility

C. C. Foster, C. M. Berg, E. R. Hall, S. B. Klein, B. v. Przewoski, and K. M. Murray*

Indiana University Cyclotron Facility, Bloomington, Indiana 47408, USA
*KM Sciences, Bloomington, Indiana 47401, USA

Abstract. High energy protons are useful for ground based simulation of the effects of space radiation on electronic devices and systems. Present and planned facilities to support the ongoing IUCF radiation effects research program (RERP), which uses protons with energies as high as 200 MeV, are described and results of high energy proton dosimetry studies are presented.

INTRODUCTION

The use of accelerator produced high-energy protons to study the response of microelectronic devices and systems to simulated space radiation environments is well established[1,2]. IUCF has an active proton-based radiation effects research program, which has served many government, industrial and academic users since 1994. The facilities used to support this program are described in the proceedings of the 14th International Conference on the Application of Accelerators in Research and Industry [3]. The present paper will briefly review those facilities, report on calibration and validation studies for high-energy proton dosimetry and describe expanded radiation effects research facilities planned for the future.

IUCF is converting the 200 MeV proton cyclotron from use for nuclear physics research to a research facility for proton therapy and for radiation effects studies in biological, material and electronic systems. This conversion requires the reconstruction of the high energy beam transport system to convert it from the momentum dispersive system used for nuclear physics research to an achromatic system which is optimal for proton therapy. This reconstruction, which is in progress now and expected to be completed early in 2003, provides an opportunity to implement the improved facilities for radiation effects described in this paper. Throughout the reconstruction period, access to beam for radiation effects tests and research will be provided for an average of about two weeks out of every two months. After the reconstruction is complete, beam sharing on a fast time scale with proton therapy will allow access to beam for radiation effects studies 24 hours each day for about 300 days each year.

RADIATION EFFECTS RESEARCH STATION

One of the beam lines of the IUCF is configured to provide an accurately known radiation field. The configuration (the RERS) is shown schematically in Figure 1 and described in detail in reference [3]. It uses a spreading foil to initially scatter the proton beam so that a more uniform distribution of beam intensity may be obtained at the test site. This initial scattering also makes the intensity distribution less sensitive to any drift of the focus or position of the beam in the vacuum line.

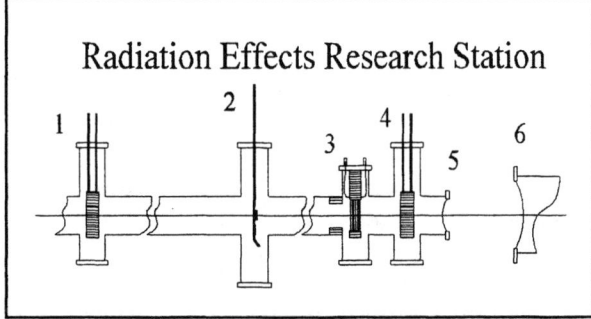

Figure 1. The RERS. Beam goes from left to right. Number 1 is a removable stop which turns the beam on and off, #2 is a target ladder with a scintillator for beam alignment and a Cu foil to spread the beam, #3 is the defining collimator and Secondary Emission Monitor (SEM), #4 is a removable Faraday cup used to calibrate the SEM, #5 is the exit window, and #6 is the entrance to the beam dump. Test devices are placed, in air, between 5 and 6.

A 5 cm inside diameter collimator is located 216 cm downstream from the scattering foil. This collimator serves to define the beam to a 7 cm diameter beam spot at the position of the device to be tested. Immediately following the collimator is the secondary emission monitor, (SEM) consisting of 15 half mil thick Cu foils alternately biased to collect the secondary electrons produced by the proton beam as it passes through. The Faraday cup is located 20 cm downstream from the SEM. It consists of a Cu block about 9 cm by 8 cm and 5 cm thick, (more than enough to stop 200 MeV protons). Two high strength, rare earth ceramic, permanent magnets are attached to the back of the Cu block to provide a trapping field for the secondary electrons produced at the face of the Faraday cup. Also, the outer foils of the SEM are positively biased at 100 Volts to return any secondary electrons back to them so that they would not be a source of secondary electrons that might get to the Faraday cup. This method of secondary electron control works well at these proton energies because there are few secondaries produced (.022 electrons per proton per copper surface are observed) and these secondaries are of low energy [4] (about 30 eV). We routinely measure less than 2 pA Faraday cup currents without beam. These leakage currents are measured and subtracted from the several nA beams used in these dosimetry comparisons. The Faraday cup is mounted so that it can be withdrawn to allow the beam to pass through an exit window and into the region, in air, where samples can be placed for irradiation.

The SEM measures the beam current. In order to determine the flux, a beam intensity profile is measured using GAFCHROMIC™ film [1], the change in optical density of which is proportional to the dose delivered to the film. This method is used for fluxes in the range from 10^6 p's/sec cm^2 to 10^{11} p's/sec cm^2. A proton energy range from 40 to 200 MeV is provided by degrading the proton beam with Cu plates placed directly up stream of the device under test. The reduction in flux due to multiple scattering in the degraders has been measured and corrections are made [3].

To confirm the accuracy of this Faraday cup/SEM dosimetry system, a careful comparison [5] was made with an independently calibrated Markus ion chamber [6] and a Shultz type [7] water calorimeter. The results of this comparison is presented later in this paper.

The need for extending the lower limit of the flux range to below 10^2 p's/sec cm^2 to serve some users, required the use of direct counting of the protons in the beam using a small scintillator and a high count rate photo-multiplier system. The highest fluxes measurable with such a scintillator system are limited by the onset of pulse pile up and sag in the gain of the photo-multiplier tube. The lower limit (about 3×10^6 p's/sec cm^2) of the fluxes measurable by the Faraday cup/SEM system, which has a gain of about 1.4, is due to the noise current of the system (about 2 pA). In order to compare the scintillator and the SEM systems in an overlapping range, operation of the SEM in air (as a parallel plate ion chamber with a gain of about 360) was explored. The results of this study also follows.

ION CHAMBER AND CALORIMETER COMPARISON

The calorimeter was placed in the beam 18.5 cm downstream of the exit window. The central part of the calorimeter is a cylinder of water 10 cm in diameter and 10 cm long in the beam direction with temperature sensors placed at the center, 5 cm deep. An acrylic phantom was constructed with dimensions similar to the Calorimeter. It is shown in Figure 2. It was designed with an insert in which an ion chamber could be placed such that it would see the same flux of protons as seen by the temperature probes of the Shulz type calorimeter.

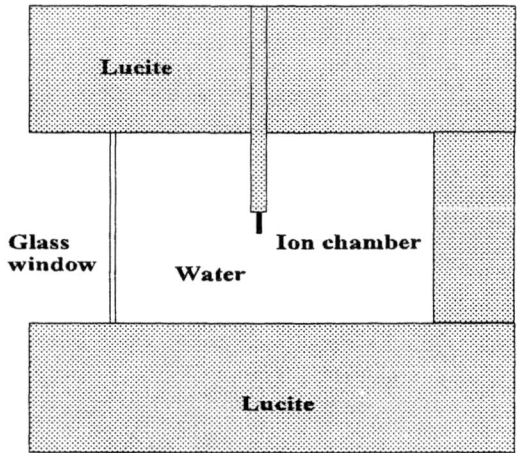

Figure 2. The phantom chamber which was identical to the calorimeter except that the ion chamber was located at the position of the thermistors of the calorimeter. Films were placed at the front face of the calorimeter and also just ahead of the ion chamber to measure the change in fluence due to 5 cm of water.

Films were placed just in front of the calorimeter and just in front of the phantom calorimeter. A small film was also placed just ahead of the ion chamber. The

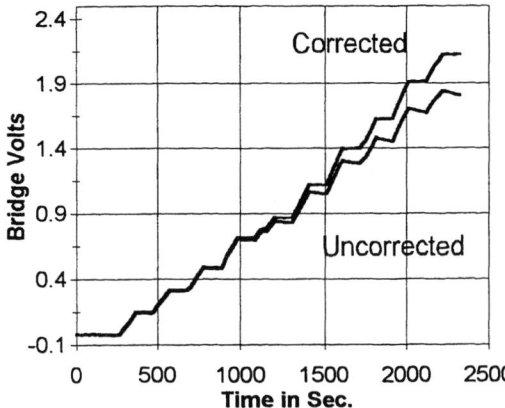

Figure 3. Bridge voltage versus time for the thermistors during a calorimeter measurement. The lower curve is the raw calorimeter response and the upper curve is the calorimeter response with the "cooling" removed.

TABLE 1. Comparison Results.

Test No.	Energy at the thermistor and ion chamber.	Average ratio of the calorimeter dose to Faraday Cup dose.	Average ratio of the ion chamber dose to Faraday cup dose.
69	154 MeV	0.987 +/-0.014	0.984 +/-0.002
76	165 MeV	0.987 +/-0.009	1.005 +/-0.005
78	163 MeV	0.996 +/-0.019	0.990 +/-0.005
81	131 MeV	0.997 +/-0.010	0.993 +/-0.012

ratio of the fluences at the two locations in the phantom determined from the film doses and the appropriate proton energies provided a factor to correct for the change in fluence of the proton beam while passing through 5 cm of water. It was assumed that this factor for the real calorimeter was identical to that for the phantom.

The calorimeter, being surrounded by a constant temperature water jacket at 4 degrees Celsius, does not make an adiabatic measurement. Heat is continually flowing in or out of the instrument. The Voltage across the probes, a quantity which is proportional to the temperature of the probes [5,7], was measured at one second intervals and digitally recorded throughout the calorimeter run. In order to determine the heating due to the protons, runs were made consisting of 12 to 15 cycles of beam on and beam off, using an "on" time of 5 minutes followed by an "off" time of 5 minutes. The "off" time temperatures were then fitted with a third order polynomial to provide a "cooling" function. This "cooling" function was then used to correct all the data to get the "true" heating due to the protons. The effect of this correction is evident from the fact that the slope of the corrected curve during the beam "off" times is zero. Similar run cycles were used when exposing the phantom to get the ion chamber responses. Figure 3 shows the raw calorimeter response together with the heating curve with the cooling removed.

Table I shows the results of these comparisons of the calorimeter and ion chamber to the Faraday cup.

In this table each figure is an average of 12 to 15 measurements together with their standard deviations. Measurements were made on each of 4 different days and at different energies.

The Markus ion chamber has a documented calibration [8] which can be traced to the National Institute of Standards. The Faraday cup system together with it's software, BeamMonster, was found to indicate a higher dose than the ion chamber by 0.7% with a standard deviation of 1.2%. The Faraday cup system also indicated a higher dose than the calorimeter data by 0.8% although it has a standard deviation of 1.9%. Hence, all three systems, the ion chamber, the calorimeter, and the Faraday cup system are in agreement to within 1% with an experimental uncertainty of about 2%.

LOW FLUX DOSIMETRY

In order to explore the use of the SEM in air, 5 mil thick kapton windows were inserted just upstream and just downstream of the SEM and the vacuum plumbing modified to allow operation of the SEM in air (as an "Air-SEM") while maintaining vacuum in the beam line and Faraday cup (see Figure 1). After aligning the beam and measuring the beam profile with the SEM in vacuum to verify that the system was operating normally, the SEM was operated in air and the average beam current, as measured on the Faraday cup, varied. The ratio of the charge measured by the Faraday cup to that measured by the SEM for a standard time interval was determined. The ratio was found to be constant (0.00270) to within +/-1% in the range of beam currents from 0.1 to 1.0 nA. For beam currents above 1.0 nA, the Air-SEM saturates because the 100 V bias is too low to collect all the charge. The ratio of the FC/SEM ratios for the Air-SEM to the vacuum SEM gives a gain of 530 +/-10, which makes the Air-SEM effective at low fluxes.

For fluxes in the range below about 10^5 p's/sec cm^2, a small scintillator (1/4"x1" and ½" thick) had been used for several experiments. This scintillator was placed just downstream of the exit window (Figure 1) and 1.2 cm away from the center of the beam profile. The fluence at the center of the beam profile was then calculated from the counts for a given time recorded by the scintillator/photo-multiplier system divided by the area of the scintillator (1.613 cm^2) and corrected for the offset from the center of the profile. This procedure was checked by direct comparison with the Air-SEM using the FC/Air-SEM ratio of 0.0027 +/-0.0003 and found to be accurate to within +/- 2%. This indicated that losses of counts in the scintillator due to edge effects was small. It was, therefore, decided to calibrate against the Air-SEM a scintillator of small enough area to over lap the flux range of usefulness of the vacuum SEM. This scintillator, which had an area of 0.4 cm^2 (1/4"x1/4" and ½" thick), was calibrated in the same manner and found to be accurate within +/- 2% as well. This work has shown that the combination of the vacuum SEM and the small scintillator provides accurate, convenient and continuous dosimetry over the entire range of fluxes from 10 to 10^{11} p's/sec cm^2.

FUTURE FACILITIES

The new beam line configuration being built at IUCF for proton therapy will incorporate several features of use for radiation effects studies. These are: an achromatic high energy proton beam, beam sharing and beam intensity modulation on a millisecond time scale, momentum analyzed energy degraders on the beam line to each end station, and the possibility of providing beams with a uniform beam profile through the use of non-linear magnetic fields. The technical basis of these features and their advantages for radiation effects are described in reference [9]. It is planned to build two beam line end stations for radiation effects research. One will be to the west and the other to the east of the main trunk line at the north end of the accelerator building. The end station on the east line will be identical to the present RERS and will not have a momentum analyzed energy degrader system. The end station to the west will have a new large area beam profile (up to 40 cm x 40 cm) either through the use of non-linear expansion of the beam or through multiple scattering in a foil and a large drift distance. This large area radiation effects research station will make available high energy proton fluxes as high as 10^9 p's/sec cm^2. This will be useful for radiation effects tests of entire commercial-off-the-shelf (COTS) systems [10]. Both the east and west end stations will exploit fast beam sharing and intensity modulation. It is planned to have the east station operational in the late summer of 2001 and the west station operational early in 2003.

CONCLUSION

The radiation effects research program at IUCF is an active program which is based on convenient access to high energy protons and careful dosimetry. The present facilities were described as were validating dosimetry studies. Planned facility development to support the program well into the future were outlined.

REFERENCES

1. Petersen, E. L., *IEEE Transactions in Nuclear Science*, **45**, 2550-2562 (1998).
2. Petersen, E. L., *IEEE Transactions in Nuclear Science*, **44**, 2174-2187 (1997).
3. Foster, C. C., et al., "Radiation Effects Test Facility at the Indiana University Cyclotron Facility," in *Application of Accelerators in Research and Industry-1996*, edited by J. L. Duggan and I. L. Morgan, AIP Conference Proceedings 392, New York, 1997, pp. 1131-1134.
4. Murray, K. M. Stapor, W. J., and Casteneda, C., *Nucl. Instr. And Meth.*, **A281**, 616-621 (1989).
5. Jones, A. Z., et al., *IEEE Transactions in Nuclear Science*, **46**, 1762-1765 (1999).
6. Markus, B., "Ionization Chambers, Free of Polarity Effects, Intended for Electron Dosimetry inAgriculture, Industry, Biology, and Medicine," *IEAE*, Wien, 1973.
7. Schulz, R. J., Wuu, C. S. , and Weinhous, M. S., *Medical Physics*, **14**, 790-796 (1987).
8. Accredited Dosimetry Laboratory Report # ION4461, *University of Wisconsin*, 1996.
9. Foster, C. C. and Hall, T., "Radiation Effects Research and Test Facilities at the Indiana University Cyclotron Facility," in *1999 Digest of Papers*, edited by J. P. Egan and E. D. Graham, Government Microcircuit Applications Conference - 1999 Proceedings 34, California, 1999, pp. 544-547.
10. Culpepper, W., "Radiation Susceptibility Assessment of Off-The-Shelf (OTS) Hardware," in *Workshop Notes*, 4th Annual International Workshop on Commercialization of Military and Space Electronics, California, 2000, pp. 311-317.

"GAFCHROMIC™ " is a trademark of GAF Chem. Corp.

A CW RFQ Injector for the IUCF Cyclotron

D.L. Friesel, V. Anferov, and R.W. Hamm[*]

Indiana University Cyclotron Facility, 2401 Milo B. Sampson Lane, Bloomington, IN 47408, USA
*AccSys Technology, Inc, 1177 Quarry Lane, Pleasanton, CA 94566-4757, USA

Abstract. Work has begun to upgrade the IUCF 210 MeV proton cyclotrons for use as a dedicated proton source for cancer treatment and other applications requiring medium energy dc proton beams. A major performance and reliability upgrade to replace the 600 kV Cockcroft Walton with a 700 keV proton CW RFQ as the pre-accelerator for the 15 MeV injector cyclotron is presented.

INTRODUCTION

The Indiana University Cyclotron Facility (IUCF), shown in Fig. 1, consists of a 600 kV Cockcroft Walton pre-accelerator, a 15 MeV proton Injector cyclotron and a 210 MeV proton Main cyclotron [1]. The separated sector cyclotrons, commissioned in 1975, were funded by the National Science Foundation (NSF) to perform intermediate energy nuclear physics research until 1998. NSF sponsored nuclear research at IUCF now continues with a new Cooler Injector Synchrotron (CIS) and the IUCF Cooler ring [2].

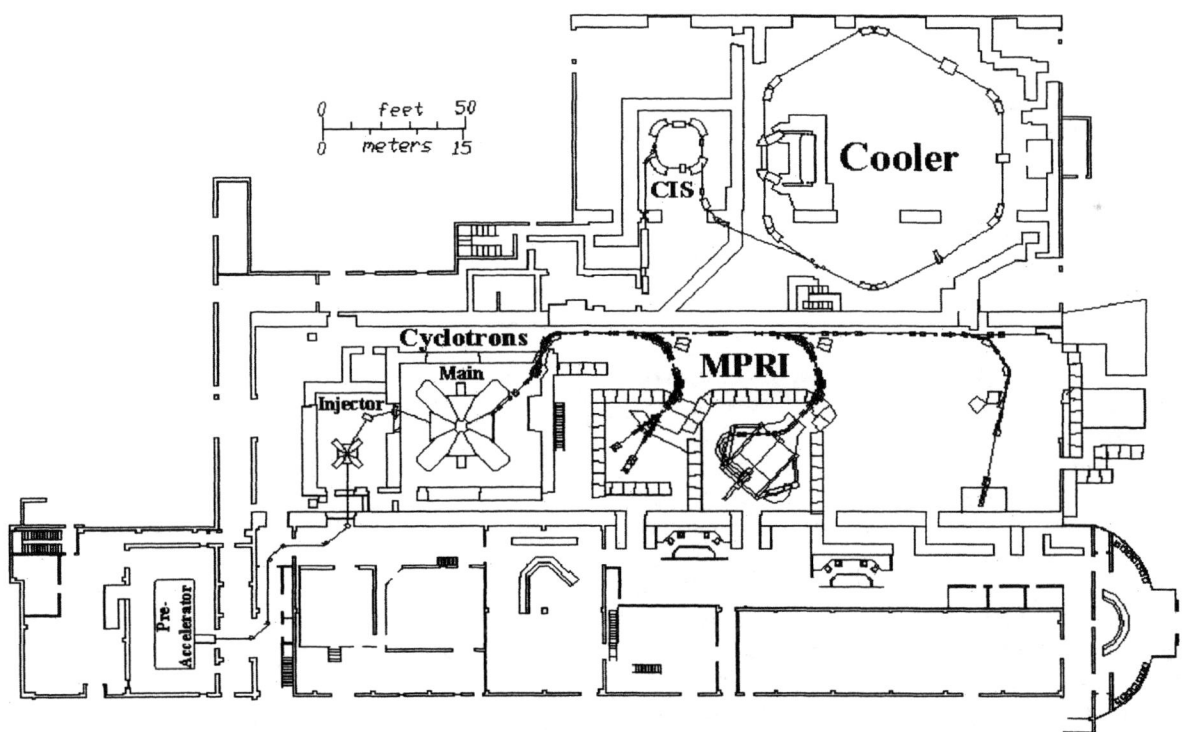

FIGURE 1. The IUCF synchrotron based physics research and planned cyclotron based MPRI medical facilities layout. Construction of the medical beam delivery and patient treatment facilities are now underway.

Cyclotron operations have continued, however, on a significantly reduced schedule for a variety of commercial, medical and radiation effects applications [3]. By virtue of their separated sector design, these machines have proven to be a reliable and energy efficient medium energy proton source. Consequently, IUCF, in conjunction with the Advanced Research Technology Institute (ARTI) at Indiana University, has developed a plan to convert these valuable cyclotron facility assets into a dedicated Proton Therapy cancer treatment facility, called the Midwest Proton Radiation Institute (MPRI) [4]. These plans are now being realized by virtue of recent (August, 2000) State and Federal funding appropriations totaling 12 million dollars. Addition funding from the National Institute of Health NIH is pending.

MPRI Facility Description

Construction has begun on the MPRI facilities shown in Fig. 1. The original cyclotron beam delivery and research hall, covering approximately 12,000 ft^2, is being reconfigured to house required oncology, medical physics, and nursing space, as well as the unique proton treatment facilities. In its initial construction phase, the new medical facility will include a re-designed matching beam transmission line to deliver achromatic beams to two medical treatment rooms, one room with two fixed treatment stations for eyes (macular degeneration), and/or head and neck tumors, and a second room with a full 360 degree rotating gantry. A third target room shown to the far right in Fig. 1 will be used for radiobiological and other radiation effects research activities. Space is provided in the design for a second gantry in the new MPRI Hall (not shown), but will require additional funding before its' construction can begin.

The MPRI beam transmission line begins at the main cyclotron exit with a winding achromatic section followed by a long, straight, trunk line to a beam dump at the far right of the hall. The trunk line incorporates unique beam intensity modulation and beam splitting tools to facilitate the efficient use of all treatment and research facilities. These beam manipulation tools, pioneered and used at IUCF for over 14 years to deliver beams to multiple scientific users [5], will permit the simultaneous delivery and monitoring of variable intensity beams to each treatment room. In addition, the beam line branch to each treatment room has a local energy selection system (ESS) to permit the independent delivery of the desired beam energy and distribution as well. Consequently, these tools will permit the IUCF cyclotrons to simultaneously deliver different beam energies, intensities and particle distributions to the three MPRI treatment rooms.

Beam development of the newly constructed achromat section begins in November 2000, the trunk line is expected to be completed and ready for beam by November of 2001, and the first use of the fixed beam room for patient treatment is planned for the end of 2002.

CYCLOTRON UPGRADES

A first step toward validating the use of the IUCF cyclotrons as a proton source for medical treatment was a historical review of the operational reliability of all related accelerator hardware systems. The IUCF cyclotrons operated for about 7000 hours annually during its peak operating years (1983-1993), and maintained a user availability of about 90%. The major recurrent sources of downtime for the cyclotrons during that operating period (and beyond) were the polarized ion source and 600 kV Cockcroft Walton pre-accelerator systems, which together accounted for about 50% of the total reported breakdown time. The remainder of the accelerator systems (dc power, water, vacuum, unpolarized ion sources, etc) had a reliability record consistent with the 95% beam delivery reliability required for medical applications. The complex polarized ion source, which alone accounted for 25% of the downtime, is not required for medical applications, and will be replaced with a reliable high intensity unpolarized proton source. This leaves the 600 kV ion source terminals as the only other high maintenance accelerator system requiring replacement.

Conceptual Design For A CW RFQ Pre-Accelerator

The IUCF main cyclotrons were designed to be variable particle and energy machines with a maximum proton energy of 220 MeV. Energy variability was a key feature of its use for scientific research, but was a feature that significantly complicated its daily operation and generally reduced user availability and reliability. We have therefore reconfigured the cyclotrons to operate at a fixed proton energy of 205 MeV. This has the added advantage of removing the variable energy requirement for the pre-accelerator and allows the possibility of using a Radio Frequency Quadrupole (RFQ) Linac for this purpose. IUCF recently gained experience with an RFQ Linac during the construction of CIS. A commercial 7 MeV H$^-$ RFQ/DTL Linac was purchased from AccSys Technology, Inc and found to be a reliable and reproducible source of protons for injection into the booster synchrotron [6]. We desire this beam phase space reproducibility to achieve routine quick turn-on and stable operation of the cyclotrons. In addition, high-energy operation of the cyclotrons has been compromised by the voltage capability of the ion source terminal (600 kV), which limits the proton beam energy for cyclotron injection to 620 keV. This low energy requires beam injection into the cyclotron at a radius where the ideal dipole field shape is distorted by saturation effects. Thus, IUCF has again enlisted the services of AccSys Technology, Inc. to design a 20 keV low energy beam transport system (LEBT), 700 keV Proton RFQ, and 700 keV transport system (HEBT) to replace the high voltage terminals as the cyclotron injector. While the higher injection energy permits injection into a good field region of the cyclotron, the RFQ exit beam properties are not well matched to the pulse structure and phase space require -ments for injection into the cyclotron acceleration spiral. The cyclotrons operate at an rf frequency of 35.2 MHz for 205 MeV protons, and have an rf phase acceptance of $6°$. The ideal injected beam pulse width for efficient acceleration is thus 0.5 nsec. The injector cyclotron also requires an injected beam energy spread of less than ± 600 keV and a beam width of less than 4 mm to pass through the injection inflectors.

A conceptual design for a 700 keV RFQ proton pre-accelerator for the cyclotrons, proposed by AccSys Technology, Inc, is shown in Fig. 2, and consists of a 20 keV, 1 mA proton ion source and LEBT, a 700 MeV RFQ and a short matching beam line to the injector cyclotron. Beam acceleration will be on the 2nd harmonic of the acceleration frequency, or 17.75 MHz. The ion source must produce a 1 mA peak intensity beam (6.24×10^{15} H^+/sec) into a total emittance of 0.5 π mm-mrad. The CW RFQ operates at 213 MHz, 12 times the cyclotron beam pulse frequency. A parallel plate chopper in the LEBT produces 4.9 nsec bunches (1 rf period at 213 MHz) at 17.75 MHz for the injection of 2.9×10^7 H^+/bunch into the RFQ at 17.75 MHz (82 µA average). IUCF will be responsible for the construction of the source, chopper, and LEBT electrostatic lens system for matching this beam into the RFQ. The 1.3 m RFQ transmission is predicted to be 84%, leaving an average beam intensity of 70 µA for Cyclotron injection. A short beam transport system from the RFQ exit to the injector has a quad triplet to focus the beam into a 6 mm diameter spot at the injector inflector in the center of the cyclotron. A 213 MHz, 14 kV re-buncher is used to reduce the phase width of the beam from the Linac to $6°$ of cyclotron rf phase. An average current of 21 µA with an energy spread of ± 600 keV is expected to be injected into the cyclotron acceleration spiral.

CW RFQ Performance Calculations

The 700 keV RFQ is 1.32 m long and has 113 cells for beam matching, bunching and acceleration. Beam transmission calculations through the RFQ were performed by AccSys Technology, Inc using their modified code *PARMQ* to determine the input requirements and to optimize resultant output beam properties. The calculation used 2930 macro-particles representing 10^4 H^+ ions each. The results, shown in Fig. 3, illustrate the 20 keV H^+ input and calculated 700 keV H^+ output beam phase space areas used for the above performance predictions. The total input and output emittances for the Linac are 0.5 πmm-mrad and is well matched to the acceptance of the injector cyclotron. The 20 keV input beam must be symmetrically focused into a 2.5 mm diameter spot at

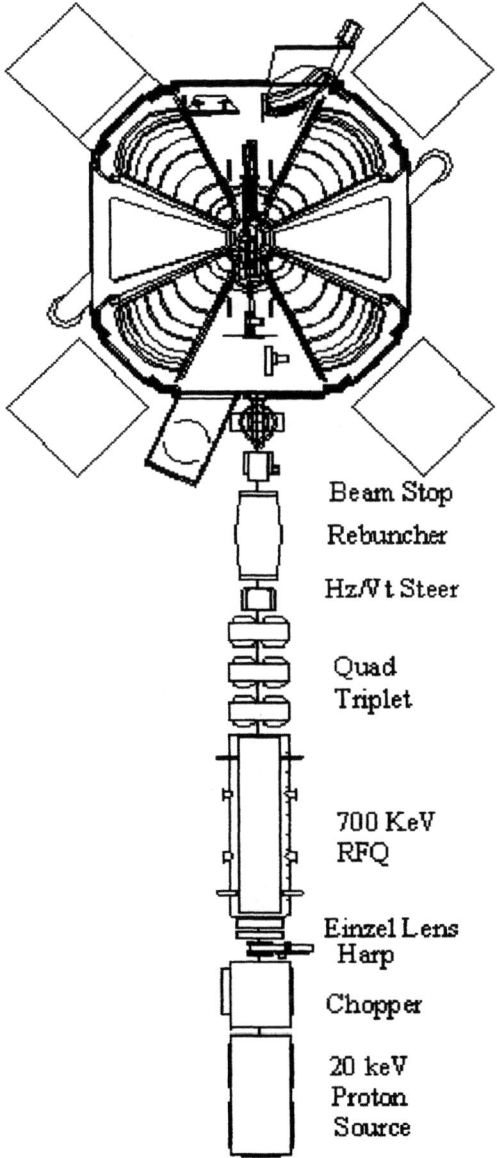

FIGURE 2. Preliminary design for a 700 keV RFQ pre-injector for the IUCF Injector Cyclotron.

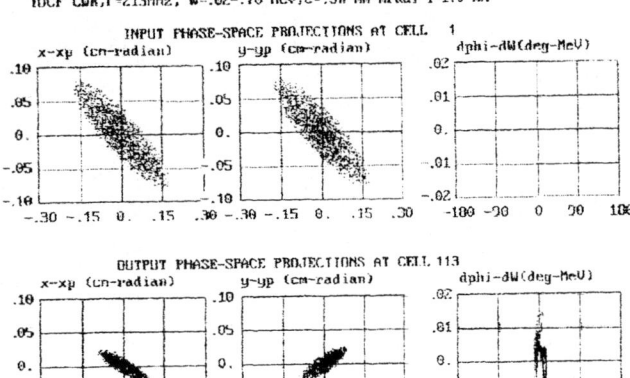

FIGURE 3. Input and calculated output phase space areas for the 700 keV CW RFQ.

the entrance of the RFQ, and slowly grows to a 3 mm diameter during acceleration. The transmission is calculated to be 84%. The RFQ phase width of the 700 keV beam is ± 16°, which is well within 6° of cyclotron rf phase acceptance. The natural energy spread of the resulting 70 μA (average) 700 keV RFQ exit beam is ± 1 keV FWHM, which is over twice the acceptance of the cyclotron.

Beam Transport calculations for transmitting this beam into the cyclotron were made using the code *TRACE3D*. A Quad triplet focuses the beam to a roughly parallel 6 mm diameter round beam at the entrance of the inflector 1.96 m downstream from the RFQ exit flange. The calculated properties of the 700 keV proton beam at the inflector using the 213 MHz

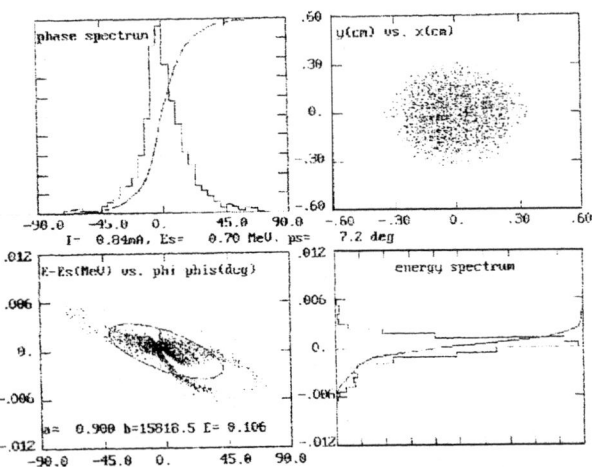

FIGURE 4. Calculated properties of the 700 keV proton beam at the cyclotron inflector entrance.

re-buncher is shown in Fig. 4. 21 μA (30%) of the 70 μA exiting the RFQ falls within the acceptance limits of the cyclotron specified above. Transmission into the cyclotron without using the re-buncher reduces the injected beam intensity by a factor of 2. Transmission of beam through the cyclotrons from this point has typically been 1%, which provides up to 2 μA of protons with a pulse period of 56 nsec. This is 4 times the minimum required intensity specified for medical operations beam operations. Injection without using the re-buncher reduces these numbers by a factor of 2, which is still more than required for medical operations. Hence the use of the re-buncher provides a comfortable performance safety margin for routine operations for applications.

ACKNOWLEDGEMENTS

I wish to thank the IUCF cyclotron operations group for their efforts in providing some of the cyclotron performance and operating history data and for helping to prepare some of the figures presented here.

REFERENCES

1. R.E. Pollock, PAC'89, IEEE 89CH2669-0, Chicago, 17 (1989).

2. D.L. Friesel *et al*, EPAC' 00, Vienna, 539 (2000), (http://accelconf.web.cern.ch/accelconf/e00/papers/mop5b05.pdf) to be published.

3. C.C Foster *et al*, AIP Conf. 392. (1996) pp. 1131

4. J.M. Cameron *et al*, Applications of Accelerators in Research and Industry, CP475, (1999) pp. 1026-1028.

5. D.L. Friesel *et al*, The 12th International Conference on Cyclotrons and their Applications, edited by B. Martin and K. Zeigler, World Scientific Publishing Co., Berlin, 1989, pp 380—384.

6. D.L. Friesel and R. Hamm, 19th Intl. Linac Conference, ANL-98/28 Vol. 1, Chicago, IL, 1998, pp 61-63.

Electrostatic focusing accelerator consisting of multiple coaxial cylinders

A. D. Dymnikov[*] and G. García[†]

[*]Radiological Research Accelerator Facility, Columbia University, 136, So. Broadway, Irvington, NY, 10533, USA
[†]Centro de Investigaciones Energéticas Medioambientales y Tecnológicas (CIEMAT), Avenida Complutense 22, Edificio 2, 28040 Madrid, Spain

Abstract. An electrostatic accelerating and focussing system for electron with energies up to 150 keV is presented in this paper. The energy and angular resolution reached by this systems give the ability to realize spectrometric studies of scattered electrons by molecular target in condensed phase. At 150 keV, the calculation of the electron trajectories, the aberration treatment and the optimization of the system have been performed by means of a relativistic matrix method.

INTRODUCTION

Electron collisions with atoms and molecules are important processes in many scientific and technological applications. Particularly, those connected with radiation dosimetry, radiation damage and biological effects are especially interested in the production of secondary electrons and their successive interaction with the target constituents (atoms, molecules, clusters, macromolecules,...). These applications require microscopic energy deposition models in which differential and integral cross section of the main processes that can take place, over a wide energy range, are needed. This motivated our systematic studies [1-3] on the cross section for electron scattering by atoms and molecules in the energy range 0.5-10 keV. In these studies we combined experimental techniques with calculations and simulations to obtain reliable total cross sections, integral elastic and inelastic data as well as differential elastic values. These results give a quite complete description of the electron scattering processes at the considered energy and are very useful in those application where the material of interest is in gas phase. However, an important part of the targets of interest in these fields are liquids and solids. In order to extend our studies to condensed phase targets, we require the design of new apparatus to operate at higher energies but maintaining the energy and spatial resolution needed in spectroscopic applications.

In this paper we present an accelerating and focusing system that provide an electron beam with an energy of 150 keV in conditions of small energy spread and minimum spot size. This system will be used to the study of multiple scattering processes and energy deposition in liquid and solid targets.

150 KEV ELECTRON ACCELERATOR SYSTEM

General Description

An schematic view of the apparatus is given in Fig. 1. Electrons, produced by an emitting filament, are selected in energy by a hemispherical electrostatic analyzer. The beam is formed at the exit of the analyzer by two slits of 0.5 mm in diameter. At this point, the energy of the electron beam is 100 eV with an energy spread of about 250 meV. The primary beam is then accelerated up to 5 keV by a two-slit system of 0.5 mm which also define its emittance for a given brightness. A five-cylinder electrostatic lens accelerates the beam up to 150 eV and it is focussed on the target with a spot size which has been minimized in accordance with the emittance of the beam.

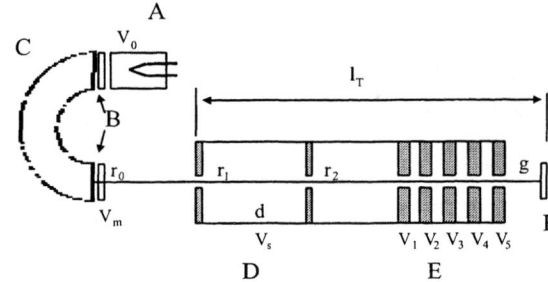

FIGURE 1. Sketch of the 150 keV electron accelerator. A, electron gun. B Monochromator apertures. C electrostatic energy analyser. D, acceptance slit system. E, five element lens. F, target. Parameters given in Table 1 are also shown.

Method of calculation and optimization of the beam parameters

We consider the focusing system, which consists of two round diaphragms and five cylinders, having equal diameter with thin walls and rotational symmetry about the central axis z. There is a charged particle beam with initial energy E_0 and with a given emittance, which is determined by two diaphragms separated by a distance d. The first diaphragm is the object diaphragm with radius r_1 and the second one is the aperture diaphragm with radius r_2. For a given brightness the emittance defines the beam current. We consider the differential equation of motion of the charged particles accurate to terms of third order inclusive. The following problem is solved: what are the values of V_1-V_5, r_1, d and the geometry of the system which provide the acceleration and the minimum beam spot size on the target?

It is convenient to choose a set of variables, for which the phase volume remains unchanged during the beam motion. For the electrostatic field these variables have the following form:

$$x_1 = x, \quad x_2 = \frac{p(z)}{p(0)} x', \quad y_1 = y, \quad y_2 = \frac{p(z)}{p(0)} y',$$

where $p(z) = \sqrt{\gamma^2(z) - 1}$,

$$\gamma(z) = \gamma(0) + \varphi(0) - \varphi(z), \quad \gamma(0) = 1 + \frac{E_0}{W_0}.$$

Here $p(z)$ is the dimensionless momentum of the axial particle, $\gamma(z)$ is its relative total energy, W_0 is the rest energy of an axial particle and $\varphi(z) = V(z)q/W_0$ is the dimensionless axial potential. In our study we take $E_0 = 5$ keV.

The analysis and calculation of the nonlinear systems of equations for monochromatic beam formation in the static field are considerably simplified by transforming from the nonlinear differential equations of motion in the phase space (x_1, x_2, y_1, y_2) to the system of linear equations in extended phase space - the phase-moment space. This is the essence of the method of embedding in phase-moment space [4]. For the differential equation of motion of the particles accurate to terms of k-order we have the phase-moment space of k-order. In the paraxial case (the equation of the first order) we usually obtain two linear equations for two phase-moment vectors of the first order $\tilde{x}[1] = \{x_1, x_2\}$ and $\tilde{y}[1] = \{y_1, y_2\}$. If the motion of the monochromatic beam is described by the third order equation, we have two linear equations for two phase-moment vectors of the third order:

$$\tilde{x}[3] = \{x_1, x_2, x_1^3, x_1^2 x_2, x_1 x_2^2, x_2^3, x_1 y_1^2, x_1 y_1 y_2,$$
$$x_1 y_2^2, x_2 y_1^2, x_2 y_1 y_2, x_2 y_2^2\}$$

and

$$\tilde{y}[3] = \{y_1, y_2, y_1^3, y_1^2 y_2, y_1 y_2^2, y_2^3, y_1 x_1^2, y_1 x_1 x_2,$$
$$y_1 x_2^2, y_2 x_1^2, y_2 x_1 x_2, y_2 x_2^2\}.$$

The writing of the nonlinear equation in a linearized form makes it possible to construct its solution using a matrizant, which is independent of the initial vector $\{x_{10}, x_{20}, y_{10}, y_{20}\}$, whereas the solution of the nonlinear equation is sought for each value $\{x_{10}, x_{20}, y_{10}, y_{20}\}$.

To the nonlinear equation of the third order we can associate a linear equation for the phase moments: $dx[3]/dz = P(z)x[3]$, where $P(z)$ depends on the axial potential $\varphi(z)$ and its first four derivatives. For the electrostatic axisymmetric field, the equation in y - plane is obtained from the equation in x - plane if $x \to y$, $y \to x$. The solution of this equation is written in terms of the matrizant $R(z/z_0)$ in the form:

$$x[3] = R(z/z_0) x_0[3], \quad R(z_0/z_0) = I$$

A continuous generalized analogue of Gauss brackets [5] is used to calculate the matrizant for the motion equation with the field coefficient matrix. In this method there is a rigorous conservation of the phase volume of the beam at each stage of the calculation.

For our numerical model of the matrizant we choose the analytical model of the axial potential in the form of piecewise-continuous function:

$$\varphi(z) = \varphi_{2j-1}(z), \quad \text{if} \quad z_{2j-2} \leq z \leq z_{2j-1}, \quad \text{and}$$
$$\varphi(z) = \varphi_{2j}(z), \quad \text{if} \quad z_{2j-1} \leq z \leq z_{2j};$$

where $\varphi_{2j-1}(z) = \text{const} = U_{2j-1}$, $j=1,2,\ldots$, $\varphi_{2j}(z)$ is changed from U_{2j-1} to U_{2j}, while z is changed from z_{2j-1} to z_{2j} and the four first derivatives of $\varphi_{2j}(z)$ are zero for $z = z_{2j-1}$ and for $z = z_{2j}$. With these conditions the function $\varphi_{2j}(z)$ has the following form:

$$\varphi_{2j}(z) = U_{2j-1} + (z - z_{2j-1})^5 (\frac{\Delta U_j}{\Delta z_j^5} +$$
$$(z - z_{2j})(-5\frac{\Delta U_j}{\Delta z_j^6} + (z - z_{2j})(15\frac{\Delta U_j}{\Delta z_j^7} +$$
$$35(2z - 3\Delta z_j - 2z_{2j-1})(z - z_{2j})\frac{\Delta U_j}{\Delta z_j^9}))),$$

where

$$U = \frac{qV}{W_0}, \quad \Delta U_j = U_{2j+1} - U_{2j-1}, \quad \Delta z_j = z_{2j} - z_{2j-1}.$$

It is known that the information about the averaged characteristics of a beam can be obtained by calculating the moments of the particle distribution function in phase space [6]. We consider the beam motion as a motion of the closed phase set. This allows us to introduce the matrix of the moments $\mathbf{M}$ of the distribution function over whole totality of the phase coordinates, where

$$M(z) = \int_\Omega f(x_1, x_2, y_1, y_2) x[3] \tilde{x}[3] dx_1 dx_2 dy_1 dy_2.$$

The integration is performed over the apertures of two diaphragms. The averaged radius $r(z)$ of the beam is determined by the matrix element $M_{11}(z)$. We suppose that $f(x_{10}, x_{20}, y_{10}, y_{20}) = 1$ for $(x_{20} + x_{10}/d)^2 + (y_{20} + y_{10}/d)^2 \leq (r_2/d)^2$, $x_{10}^2 + y_{10}^2 \leq r_1^2$, and $f(x_{10}, x_{20}, y_{10}, y_{20}) = 0$ for $(x_{20} + x_{10}/d)^2 + (y_{20} + y_{10}/d)^2 > (r_2/d)^2$, and $r(0) = r_1$. In this case we obtain $r(z) = \sqrt{M_{11}(z)}$,

$$M(z) = R(z/z_0) M(0) \tilde{R}(z/z_0),$$

$$M(0) = \int_\Omega x_0[3] \tilde{x}_0[3] dx_{10} dx_{20} dy_{10} dy_{20}$$

$M(0)$ is a function of r_1, of emittance $em = r_1 r_2 / d$, and of d.

To apply a numerical optimization to our system a merit function has to be defined. We choose merit function as $\rho = r(l_{tot})$ for a given emittance. Before the merit function can be evaluated the third order matrizant $R(z/0)$ has to be calculated. Since this matrizant depends on the particle trajectory, the first-order stigmatic property, which is described by the equation $R_{12}(l_{tot}) = 0$, must be satisfied before calculating this matrizant

The merit function is a function of r_1 and of d. All remaining parameters are fixed when we are seeking for the minimum value of ρ, and we find this minimum for different parameters. The radius of the object diaphragm has the strongest influence on the beam spot size for the given emittance. It is very important to use the optimal r_1 for obtaining the smallest beam spot size.

Results of calculations and spot size

The main geometrical parameters that define the characteristic of the primary beam are the emittance, which is defined by the aperture (r_1, r_2) and separation of the two crossed slits. The total length of the accelerating-focusing system is given by the distance between the entrance to the two-slit ensemble and the target and the position of the target. Other defining parameter is the distance between the last cylinder and the target (g). The values obtained for these parameters in our optimization procedure are shown in Table 1.

TABLE 1. Value of the parameters defining the geometry of the 150 keV electron accelerator

Description	Symbol	Length (m)
Monochromator slits	r_0	5.0×10^{-4}
Acceptance slit apertures	r_1, r_2	1.0×10^{-4}
Acceptance slit separation	d	1.12
Total accelerator length	l_T	3.0
Distance to the target	g	1.0×10^{-1}

The axial potential along the five-element electrostatic system are shown in Fig.2. This distribution has been adjusted to obtain a spot minimum on the target position. The bias potentials to be applied for each element of the system in order to reach the minimum spot condition are shown in Table 2.

TABLE 2. Applied potentials to the elements forming the entire focusing accelerator system

Element	Symbol	Potential (kV)
Filament (cathode)	V_0	-155
Monochromator slits	V_m	-154.9
Acceptance slit	V_s	-150
1st element lens	V_1	-150
2nd element lens	V_2	-75
3rd element lens	V_3	-127
4th element lens	V_4	-75
5th element lens	V_5	0
Target		0

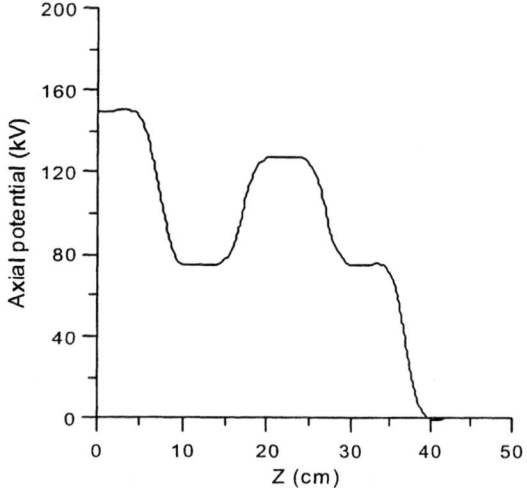

FIGURE 2. Variation of the axis potential along the axis coordinate (Z)

The shape of the spot obtained by calculating the trajectories of the electrons in the above geometrical and bias potential conditions is shown in Fig. 3. The diameter of the optimum spot is of the order of 5×10^{-6} m

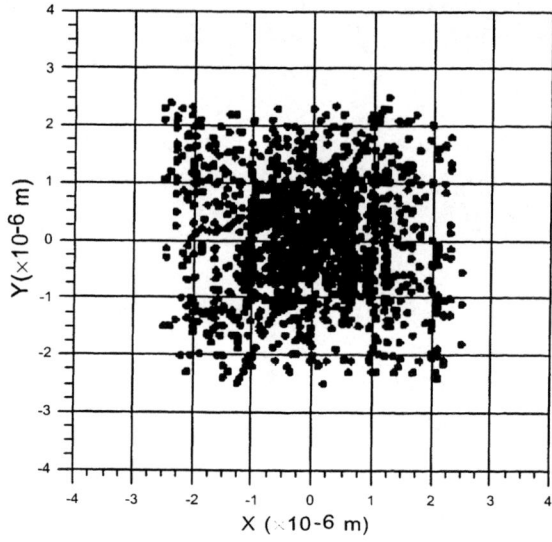

FIGURE 3. Simulated spot of the electron beam on the target region working at the optimum conditions.

ACKNOWLEDGEMENT

This work was partially supported by the Spanish *Programa Nacional de Promoción General del Conocimiento* (Project BFM2000-0012)

REFERENCES

1. G. García, M. Roteta, F. Manero, F. Blanco and A. Williart, J. Phys. B **32**, 1783 (1999).
2. G. García and F. Manero, Phys. Rev. A **57**, 1069 (1998).
3. G. García and F. Manero, Chem. Phys. Lett. **280**, 419 (1997).
4. H. Azbaid, A. D. Dymnikov and G. Martínez, Nucl. Instrum. and Meth. B **158**, 61 (1999).
5. Dymnikov and R. Hellborg, Nucl. Instr. and Meth., A 330 (1993), 323.
6. A.D. Dymnikov and E. A. Perelshtein, Nucl. Instr. and Meth., 148 (1978), 567.
7. A.D. Dymnikov, Nucl. Instr. and Meth., A 363 (1995), 435.

High-Intensity γ-ray Source

J. H. Kelley,[1,3] B. T. Crowley,[2,3] V. N. Litvinenko,[2,4] S.H. Park,[2,4] I.V. Pinayev,[2,4]
E. C. Schreiber,[2,3] W. Tornow,[2,3] Y. Wu,[2,4] and H. R. Weller[2,3]

[1]*Department of Physics, North Carolina State University, Raleigh, North Carolina 27695*
[2]*Department of Physics, Duke University, Durham, North Carolina 27708*
[3]*Triangle Universities Nuclear Laboratory, Durham, North Carolina 27708*
[4]*Duke Free-Electron Laser Laboratory, Durham, North Carolina 27708*

Abstract. A mono-energetic tunable source of 100% linearly polarized γ rays has been developed at the Duke Free-Electron Laser Laboratory in conjunction with Triangle Universities Nuclear Laboratory. The OK-4 FEL is coupled to a 1-GeV electron storage ring and generates intense beams of visible or UV photons. In γ-ray production mode, the OK-4 photons Compton scatter from high-energy electrons inside the optical cavity leading to backscattered γ rays. The strong correlation between scattering angle and γ-ray energy permits a selection of the energy spread of the γ-ray beam that depends on a simple geometrical aperture located along the optical axis. Results obtained/(design parameters) indicate γ-ray beams with energies of 2.2-58/(2.0-175) MeV, ΔE/E<1.0% and total fluxes greater than $10^7/(10^{10})$ γ rays/s.

INTRODUCTION

Electromagnetic probes have been used for many years to study properties of nuclei. Because the form of the electromagnetic interaction is well determined, the probe, i.e. γ rays or electrons, can be decoupled from the interaction being studied. To extend previous studies, γ-ray beams with high fluxes, tunable energies, a narrow energy spread and polarization are required.

Very high fluxes of γ rays ($>10^{10}$) have been commonly available from Bremsstrahlung facilities, where high-energy electrons are stopped in thin targets leading to the production of photons ranging from γ rays with the full electron-beam energy to visible photons with much lower energies. The energy of γ rays that are produced from Bremsstrahlung can be most efficiently determined by "tagging" the photons, that is, measuring recoiling electrons that produce γ rays and relating the electron energy to the energy of the γ ray it produced. However, the rates of photon "tagging" arrays are limited to around 10^4 γ rays/sec/MeV.

A solution to this dilemma appears to be available in the form of Compton scattering facilities that combine electron storage rings with intense laser light sources to generate high-energy γ rays. Because the electrons in a storage ring re-circulate millions of times per second, high electron currents can be stored in electron rings. When mono-energetic laser photons with energies of 2-12 eV, for example, are Compton scattered by these high-energy electrons, the result is a beam of γ rays that has a well-understood energy spread, whose peak energy is given by,

$$E_\gamma \approx 4\gamma_e^2 E_{\text{laser photon}}. \quad (1)$$

E_γ is the backscattered γ-ray energy and γ_e is the relativistic factor for the electron ($\gamma_e = E_{\text{electron}}/(0.511\ \text{MeV})$). For example, when 3.25 eV laser photons are Compton backscattered by 270 MeV electrons, a beam of 3.6-MeV γ rays is generated. Furthermore, because the laser photon polarization is maintained by the scattered γ-ray photon the beam is essentially 100% polarized.

Additional benefits are obtained when the laser photon source is a tunable Free-Electron Laser (FEL) that uses the electron storage ring to produce laser photons. In the region where laser photons are generated the path of circulating electrons must be aligned with the optical cavity of the FEL photons. When lasing is established, the electron beam axis and the laser photon axis are "self aligned", and the axis of the Compton backscattered γ-ray beam coincides with the laser beam axis. Because of the correlation between Compton scattering angle and γ-ray energy, a simple geometrical aperture that is located along the FEL optical axis can select the γ-ray beam energy spread.

FREE-ELECTRON LASER

A free-electron laser consists of a high-energy electron beam source, an optical cavity, and a magnetic undulator (a series of magnets that generate a region of alternating magnetic fields). As the electron beam passes through the undulator the electrons are "wiggled" and therefore emit radiation in the form of photons. The FEL equation,

$$\lambda_{laserphoton} = \frac{\lambda_{wiggler}}{2\gamma_e^2}\left[1 + \frac{K^2_{wiggler}}{2}\right], \quad (2)$$

gives the relation between $\lambda_{laser\,photon}$, the wavelength of generated laser photons, $\lambda_{wiggler}$, the wiggler period or the period of the oscillating magnetic field in the undulator, and $K_{wiggler}$, a dimensionless parameter that is proportional to the magnitude of the magnetic field. A FEL can provide laser light at shorter wavelengths (higher energy) than those available from conventional lasers by using high-energy electron beams.

DUKE FREE-ELECTRON LASER LABORATORY

The Duke Free-Electron Laser Laboratory (DFELL) comprises two FEL devices, the Mark III IR FEL and the OK-4 FEL. The most recent addition to the facility, the OK-4 FEL (see Fig. 1), is made up of an Optical Klystron (wiggler) and a buncher that are installed on the Duke storage ring. Injection of electrons into the storage ring is provided by a 240-to-275 MeV linac that is capable of adding 10^9 electrons/sec. A rf-cavity has been included in the storage ring and can either serve the purpose of replenishing the electron energy that is lost due to synchrotron radiation or boosting the circulating electron energies up to 1.1 GeV during a short ramping period.

Electrons in the storage ring pass through the OK-4 FEL, which has two planar "wiggler" segments that wiggle the electrons to produce linearly polarized photons. Photon beams ranging from 194 to 730 nm (2-6 eV) are presently available; though a broader range of photon wavelengths is anticipated with future upgrades, see table 1. The optical cavity consists of two mirrors whose surface is concave with a 27.27-meter radial curvature. The length of the optical cavity is 53.73 meters. In order to permit lasing, the circumference of the storage ring is twice the length of the optical cavity so that the period of oscillation of the FEL photons, ~360 ns, is identical to the circulation period of the relativistic electrons in the storage ring. Because the periods of the laser photons in the optical cavity and the circulating electrons are synchronized, new photons that are generated by the electrons, as they pass through the undulator, are added to the existing laser photon bunch.

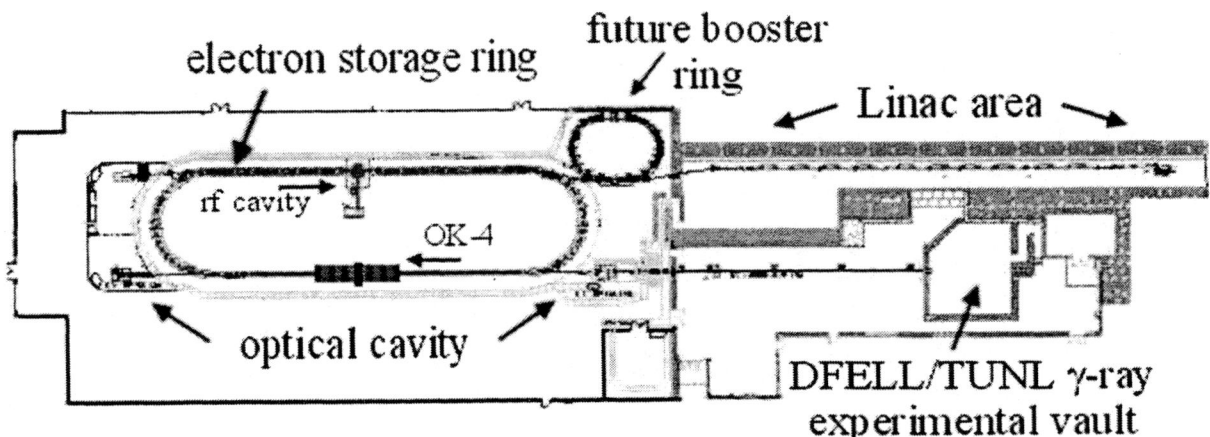

FIGURE 1. The general floor plan of the OK-4 FEL at the Duke Free-Electron Laser Laboratory.

DFELL/TUNL HIGH-INTENSITY GAMMA-RAY SOURCE

The OK-4 FEL can be operated in γ-ray production mode by Compton scattering FEL photons from the electrons that are circulating in the storage ring. Some Compton scattering facilities generate γ rays by backscattering "conventional" laser photons from the electrons in a storage ring. However, the photon flux available from conventional lasers is small when compared with those available from a storage ring FEL. For this reason, the γ-ray flux available from the HIGS facility exceeds the flux of other Compton scattering laser/storage ring facilities by several orders of magnitude. The HIGS facility generates γ rays by the addition of a second electron bunch, the scattering bunch, into the storage ring that is 1/2 the circumference of the storage ring away from the first electron bunch, the lasing bunch. Photons that were generated by the lasing bunch collide with the scattering electron bunch in a field free region, and as the laser photons interact with the relativistic electrons they are Compton backscattered as γ rays.

First results from the HIGS facility [1] measured the properties of a ~12.2 MeV γ-ray beam. In that study, a beam whose total flux was 2×10^5 γ rays/sec was collimated with a 3 mm aperture to produce a γ-ray beam whose energy spread was 1.2% and whose final flux was $\sim 2 \times 10^3$ γ rays/sec. Improvements in the electron optics have led to enhancements in the γ-ray beam properties. Using appropriate collimation, γ-ray beams with energy spreads of ΔE/E~0.5% have already been produced [2].

A benefit of using Compton backscattered laser photons to generate γ rays is that the photon polarization is maintained by the scattered γ rays. The γ-ray polarization was verified [1] by studying deuteron photodisintegration, $^2H(\gamma,n)p$, with the 12.2-MeV γ-rays. At this energy, deuteron photodisintegration proceeds entirely via electric dipole interactions, E1; this means that neutrons are emitted preferentially in the electric field plane of incident γ rays.

A measurement of the neutron analyzing power was used to measure the γ-ray beam polarization. The neutron analyzing power (Σ_n) is a relative measurement comparing the probability of having neutrons ejected in the horizontal plane (the electric polarization plane of the FEL photons) vs. the probability of having neutrons ejected in the vertical plane. By assuming that the γ-ray beam is 100% linearly polarized, a determination of the neutron analyzing power found $\Sigma_n=0.93\pm0.06$ and is in good agreement with a previous measurement, $\Sigma_n=0.95\pm0.02$ [3]. This indicates that the assumption of 100% polarization was justified.

Significant facility upgrades are planned. Presently, electrons are injected into the storage ring from the linac at energies of 240-280 MeV, which is the operating range of the linac. However, to allow high-energy γ-ray production, electrons in the storage ring are frequently ramped to energies above 280 MeV. In the operating mode above $E_e=280$ MeV, when electrons are lost from the storage ring over time they cannot be replenished from the linac on a continuous basis. Improved performance will be provided with the addition of a booster storage ring into the injection system that will permit injection up to electron energies of 1.2 GeV. When the booster injector ring is installed, it is expected that total γ-ray fluxes will be sustained at fluxes greater than those shown in Fig. 2.

TABLE 1. A summary of the operating parameters for the OK-4 FEL and the DFELL/TUNL HIGS facility

Facility Operating Parameters	Present	Design (with booster)
ring current	40 mA	200 mA
electron injection energy	0.24-0.28 GeV	0.2-1.2 GeV
electron energies (with ramping)	0.2-1.1 GeV	0.2-1.2 GeV
electron energy acceptance	6% ring, 2.5% rf	6% ring, 3% rf
revolution frequency	2.8×10^6 Hz	2.8×10^6 Hz
FEL photon energy	2-6 eV	2-12 eV
FEL photon wavelength	200-620 nm	100-620 nm
γ-ray energies	2-58 MeV	2-175 MeV
γ-ray intensities	$> 10^7$ γ rays/sec	$>10^{10}$ γ rays/sec
γ-ray beam energy spread	0.5%	0.4%

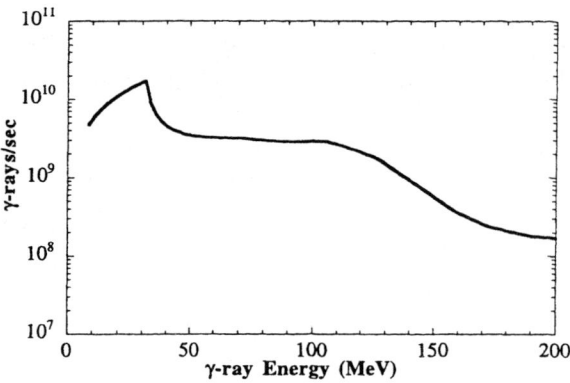

FIGURE 2. Expected total γ-ray flux as a function of peak γ-ray energy [4]. Assumptions included lasing bunch/(scattering bunch) currents of 10 mA/(100 mA).

HIGS RESEARCH PROGRAM

The developing research program at HIGS takes advantage of the narrow energy spread of the γ-ray beams and the 100% linear polarization of the beam.

First results from a study of the near-threshold ^{2}H(γ,n)p reaction have been published [5]. The inverse reaction p(n, γ)^{2}H is important for the synthesis of light ions in the early universe. However, because of difficulties with the p(n, γ)^{2}H measurement, there are substantial uncertainties in the near-threshold cross sections, and theoretical models are used to estimate cross section values. A determination of the ^{2}H(γ,n)p neutron analyzing power, Σ_n, can be used to determine the electric dipole (E1) and magnetic dipole (M1) strengths that contribute to the reaction and can be used to constrain predictions of the relevant cross sections. In the HIGS study [5], a beam of 3.58-MeV γ rays was used to measure Σ_n following deuteron photodisintegration. The observations, Σ_n (E_γ =3.58 MeV, φ=150)= 0.78±0.04, were found to be in good agreement with the predictions of Arenhövel, Σ_n (E_γ =3.58 MeV, φ=150)=0.81 [6]. A future effort is intended to measure Σ_n closer to the threshold energy.

A new study that is under development will measure the polarizabilities of the proton and the neutron. Nucleon polarizabilities are fundamental quantities that indicate the ability of external electric/magnetic fields to induce electric/magnetic dipole moments. The experiment involves measuring the cross sections and analyzing powers of intermediate energy γ rays (50-100 MeV) that are Compton scattered from the nucleon. Because γ rays at the HIGS facility are linearly polarized, the magnitude of the electric and the magnetic polarizabilities can be determined without any additional assumptions.

Finally, installation of a new Optical Klystron (OK-5) is planned. This upgrade will add the ability to produce left or right circularly polarized photons. Use of the OK-5 FEL will permit production of circularly polarized γ rays or linearly polarized γ rays and will greatly increase the capabilities of the facility.

SUMMARY

Progress at the DFELL/TUNL high-intensity γ-ray source has permitted the availability of γ-ray beams ranging in energy from 2-58 MeV with uncollimated, total fluxes above 10^7 γ rays/sec. With the aid of collimation, the energy spread of the γ-ray beam has been reduced to $\Delta E/E$~0.5%. The planned installation of the OK-5 FEL and a 1.2-GeV booster ring for the injection system will improve the capabilities and performance of the DFELL/TUNL high-intensity γ-ray source.

ACKNOWLEDGMENTS

Supported in part by the Department of Energy DE-FG02-97ER41042, DE-FG02-97ER41033 and the Office of Naval Research #N00014-94-1-0818.

REFERENCES

1. V.N. Litvinenko et al., *Phys. Rev. Lett.*, **78**, 4569 (1997).

2. S.H. Park. Ph.D. thesis, Duke University, 2000.

3. W. Del Bianco et al., *Phys. Rev. Lett.* **47**, 1118 (1981).

4. V.N. Litvinenko and J.M.J. Madey, Nucl. Insr. and Meth. A. **375**, 580 (1996); SPIE **2521**, 55 (1995).

5. E.C. Schreiber et al., *Phys. Rev. C*, **61**, 61604 (2000), and E.C. Schreiber. Ph.D. thesis, Duke University, 2000.

6. H. Arenhövel and M. Sanzone, *Photodisintegration of the Deuteron* (Springer-Verlag, Berlin, 1991).

Conceptual Design of a 100 MW Electron Beam Accelerator Module for the National Hypersonic Wind Tunnel Program*

Larry X Schneider

Applied Accelerator Technologies Department, Sandia National Laboratories
Albuquerque, New Mexico 87185, USA

Abstract. The National Hypersonic Wind Tunnel program requires an unprecedented electron beam source capable of 1-2 MeV at a total average beam power of 100-200 MW for several seconds. Although a 100 MW module is a two-order extrapolation from demonstrated average power levels, the scaling of accelerator components appears reasonable. This paper will present an evaluation of component and system issues involved in the design of a 100 MW electron beam accelerator module with precision beam transport into a high pressure flowing air environment.

INTRODUCTION

The design of a Medium Scale Hypersonic Wind Tunnel (MSHWT) facility is being explored to address deficiencies in present ground test capabilities above Mach 8. Conventional wind tunnel techniques involving isentropic expansion from a high pressure high-temperature source can not support this hypersonic parameter space and operation time without introduction of substantial high-temperature material challenges. Radiative energy addition, a concept proposed by Princeton University [1], provides a potential means to extend wind tunnel technology above Mach 8 while preserving prototypic flight conditions in the test section. For an output section that allows 50-100 cm diameter test objects, approximately 100-200 MW will need to be added to the airflow from an external power source. An electron beam guided into an air expansion nozzle is a potentially efficient means to add external energy to the flow [2]. Generating a continuous, multi-second, 1-2 MeV electron beam at this power level would represent a two order of magnitude increase in the power demonstrated by an electron beam accelerator in this class. The beam power of the MSHWT accelerator system would exceed the combined output power of the approximately 1000 industrial accelerators installed worldwide. An artist's concept of the MSHWT facility is shown in Fig. 1.

ACCELERATOR SYSTEM DESIGN

The MSHWT system is comprised of four subsystems: a high voltage DC power supply, electron injector and accelerating column, aerodynamic window, and a beam transport magnet system.

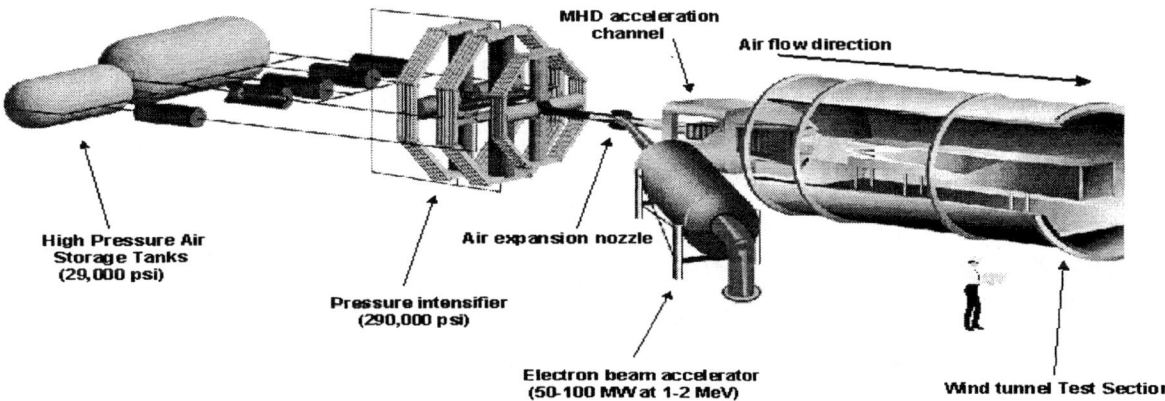

FIGURE 1. Artist's concept of a Medium Scale Hypersonic Wind Tunnel Facility. Electron beam accelerators play a key role for both energy addition into the nozzle and for enhancing the conductivity in a MagnetoHydroDynamic (MHD) afterburner.
* This work supported by the USAF and managed by the Arnold Engineering and Development Center.

Scaling electron accelerators to the 100-200 MW range will require a careful review of the physical limitations of each component from the birth of the electrons at the cathode through the acceleration phase and transport of the beam into the high pressure expansion nozzle.

There are three major classes of electron beam accelerator systems that have the capability to scale to megawatt outputs at beam energies in the few MeV range. They are: Radio Frequency (RF), Direct-Current (DC), and repetitive pulsed accelerator systems. Each technology base has intrinsic and practical limitations. The MSHWT application requires energy addition that is relatively constant with respect to the time scales involved in the dynamic air flow in the expansion nozzle. Conventional RF LINAC accelerator technologies can be designed to operate in a CW mode at a frequency from approximately 100-1000 MHz. Although this meets the temporal uniformity requirements, existing RF technologies are limited to an efficiency of approximately 50%. Pulsed accelerator technologies, such as Sandia's RHEPP technology [3], has a relatively weak cost scaling relationship with output power. However, the efficiency of this and other repetitive pulsed accelerators is presently limited to approximately 60%. DC accelerator technology is attractive from several standpoints. Inductively coupled rectifying transformer systems offer simplicity of design, robustness, and a significant capability to scale cost effectively to very high power levels. For these reasons, inductively coupled rectifying transformer systems were selected for further analysis in the MSHWT facility conceptual design

Several DC accelerator power supply concepts were studied under a Sandia contract by the Delta Division of the Efremov Institute in St. Petersburg, Russia. The Delta group was selected due to their industrial experience with inductively coupled transformer accelerator systems. Fig. 2 shows a three-phase, iron-core, rectifying transformer concept for a 2 MeV, 100 MW power supply. A 2 MeV accelerator column and differential pumping system for a foil-less window is shown to the left of the high voltage power supply. Although the high voltage output must be insulated from the grounded core in this design concept, the coupling of flux from the primary windings around each vertical core to the high voltage secondary is very high. Closed core concepts can have electrical efficiencies > 95%. The power supply and accelerator are insulated with SF_6 and N_2 and would use commercially available 25 kV diodes in the rectifier assembly.

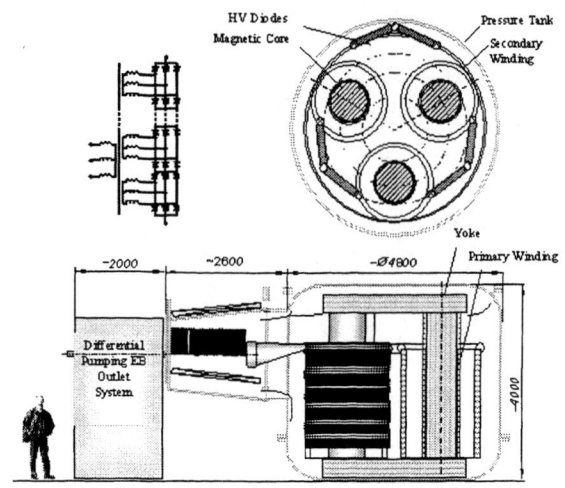

FIGURE 2. 2 MeV, 100 MW DC rectifying transformer power supply concept. Dimensions are in cm.

Injector Physics

A 2 MeV, 100 MW accelerator module will require 50 amps of beam current. At an accelerating potential of 50 keV, this injector will generate a 2.5 MW beam. A conventional Pierce-type injector can not deliver this current in a reasonable diameter beam due to space charge limitations. The high injector perveance (15.6 μperv) will require a numerically designed geometry to optimize the space charge limited flow. Fig. 3 shows an injector design operating at 50 keV and 50 A with a beam output diameter just under 3 cm. The injector is immersed in a 0.01 T solenoidal magnetic field in this simulation that was generated by a Poisson-solving code developed at Sandia. The electron source would be a standard thermionic cathode material such as Lanthanum hexaboride (LaB_6) that has ample current density and lifetime characteristics. The design shown in Fig. 3 operates at a conservative 2.5 A/cm². After initial acceleration to 50 keV in the injector, the next accelerating gap can also limit the total beam current due to space charge effects. Space charge limits in the first accelerating column gap can be estimated by modifying the planer Child-Langmuir equation to account for the injected 50 keV beam. Enhanced flow is described by Eq. (1).

$$j_{enh} = j_o \, F(\chi) \ \text{A/cm}^2 \quad (1)$$

where j_o is the Child-Langmuir current, $F(\chi) = \chi^{3/2}\,[(1-1/\chi)^{3/4} + 1]^2$ and $\chi = -\varphi_2 / (\varphi_1 - \varphi_2)$. φ_1 is the beam energy entering the gap and φ_2 is the beam exit energy (injection energy + gap energy).

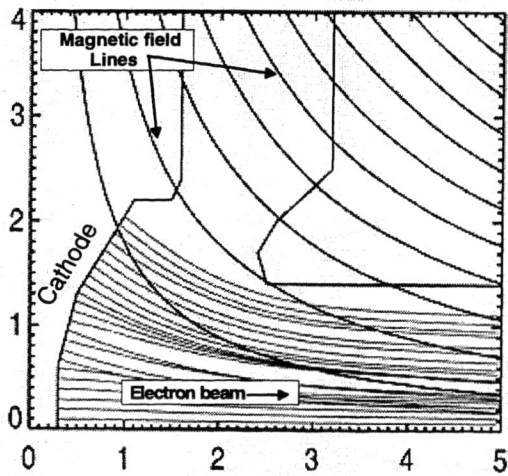

FIGURE 3. 50 keV, 50 A injector. This cross-section is symmetric around the lower horizontal axis. Dimensions are in cm.

Industrial accelerator columns routinely operate with electric field gradients of 1-1.5 MV/m on the ceramic insulators. For an accelerating gap in the column of $d_1 = 5$ cm, a gap voltage of $V_1 = 60$ kV (1.3 MV/m), and a beam diameter of 3 cm, the non-enhanced space charge limited current is only 9.7 A. For a 50 keV injected beam and a gap potential of $V_1 = 60$ kV, $\chi = 2.2$ and $F(\chi) = 8.5$. This allows a maximum current in the 3 cm diameter beam of $I_o = 82$ A. The injected 50 keV significantly enhances the flow in the first column gap. A similar gain is seen in the second gap d_2. For a beam entering at 110 keV, $\chi = 2.8$ and $F(\chi) = 14$. The maximum current space charge will allow in gap d_2 is $I_2 = (9.7 \text{ A})(14) = 136$ A. As the beam accelerates past 500 keV and becomes relativistic, the space charge limit raises significantly. Based on this space charge analysis, it is reasonable to envision a 100 MW module at 2 MeV or a 50 MW module at 1 MeV.

In practice, there will be beam optics issues that impact the peak current capability of this accelerator column. At these power levels, beam losses in the column must be extremely low to prevent damage to the accelerating gaps and ceramic insulators. The beam divergence must be very low and beam halo that could impact the gap electrodes must be extremely well controlled. These issues could limit a system of this size to several amperes instead of 10's amperes. However, the MSHWT accelerator system will require that the cathode, injector, and accelerating column be immersed in an externally supplied axial magnetic field to mitigate magnetic mirroring effects. This axial solenoidal field will also serve to confine the beam as it leaves the finite divergence injector and accelerates through the column. The axial magnetic field is a key requirement to extend the beam current into the 10's of ampere regime.

Beam transport to the high pressure nozzle

This application requires that the electron beam travel from its origin in the vacuum insulated accelerator to a high pressure expansion nozzle within the wind tunnel chamber. The beam will exit the high vacuum environment of the transport region through a aerodynamic window where it will then be guided and compressed to approximately 0.5 cm diameter in the Energy Injection Region (EIR) as shown in Fig. 4

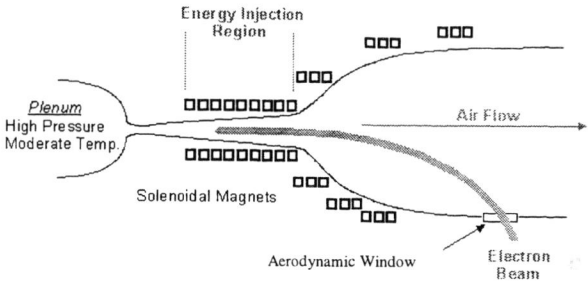

FIGURE 4. Cross-section of the expansion nozzle and energy injection region.

The region from the aerodynamic window to the entrance of the EIR will require modeling to evaluate the magnitude of gas breakdown in the approximate 1 atm environment that extends for about 0.3 m. As the beam enters the 100 MPa environment of the EIR, electron scattering, space charge, and charge migration issues will dominate the physics concerns in this region. At these high power levels, high energy electrons can not be allowed to impact the wall of the nozzle. The scattering beam must be compressed radially as it enters the nozzle using a series of multi-Tesla solenoidal magnets. Collisional effects in the high pressure nozzle have been modeled using Cyltran to set the magnitude of the solenoidal B-field in the EIR. Cyltran is a Monte-Carlo particle-in-cell code developed at Sandia that accounts for the collision of electrons with gas atoms in this region. A 20 Tesla magnetic field was required to minimize beam loss to the nozzle walls. Further modeling in the EIR will be required as Cyltran does not account for beam self-fields or the effects of electron induced gas chemistry.

Once the peak B-field strength is set in this region of the MSHWT system, the B-field profile back to the cathode can be designed. The 20 T magnetic field is sufficiently high to reflect back all the 1-2 MeV electrons entering this field if those electrons are born in a zero-field region at the injector. This mirroring effect can be estimated by considering the electron's gyrofrequency and conservation of energy as the particle converts azimuthal velocity to axial velocity. The maximum solenoidal B-field that an electron can propagate into (starting from a zero field) can be estimated as,

$$B_{max} = \frac{2c_o m_e \gamma \beta}{q_e r} \quad (2)$$

where B_{max} = B-field strength (T) and r = beam radius (m). For a 2 MeV electron beam starting with a 0.5 cm diameter outside the solenoid, mirroring will occur at a field strength above approximately 6.6 T. To overcome this effect the electrons must be immersed in a continuous solenoidal B-field back to cathode electron source. Determining the magnitude and optimum profile of the axial magnetic field back to the cathode will require numerical modeling techniques. The minimum field strength needed at the cathode can be estimated as,

$$B_o/B_m = \sin^2 \theta \quad (3)$$

where B_o = minimum field strength, B_m = maximum field strength, and θ = divergence angle. The high perveance injector configuration and imperfect electric fields through the accelerator column will produce a beam with 10's milli-radian divergence at the exit of the accelerator column. If we assume 40 mrad and use B_{max} = 20 T, B_o will need to be on the order of 0.03 T at the cathode and increase to 20 T at the EIR. Minimizing the equivalent divergence in the injector and accelerator column will be important to reducing the cost of the integrated magnet system.

Aerodynamic window

Conventional foil windows will not survive the heating from a multi-second beam much above a few mA/cm². A foil-less aerodynamic window will be required to transition the beam from the vacuum line into the approximately 1 atm environment just downstream of the expansion nozzle. An exit aperture of < 1 cm diameter will be required to inject beam into the wind tunnel chamber. This can be accomplished through conventional differential pumping techniques or potentially through other techniques that can reduce the overall high volume vacuum pumping requirements and system cost. A plasma porthole [4], a high temperature, low density, high viscosity plasma channel is being evaluated for the high pressure stage. A pressure reduction from approximately 1 atm to approximately 350 mTorr can occur in this stage, leading to a reduction by a factor of > 200 over standard differential pumping techniques.

SUMMARY

Demonstrated industrial DC accelerator technology forms the basis of the MSHWT accelerator concept described in this report. Scaling issues in extrapolating this technology base have been examined at a conceptual level. This work has not identified any fundamental physics that would prevent a system of this magnitude from being developed and fielded into a reliable hypersonic ground test facility. However, several areas of this system require further analysis. The next phase of this system design will include analysis and simulations in key risk areas such as the injector, accelerator column, and beam transport in the non-vacuum environment. Many of these issues will be explored in the next several months.

Proof-of-principal energy addition experiments, in conjunction with Princeton University, have already been completed at Sandia at a power level of 150 kW into a flowing gas nozzle [5]. The project team is currently preparing for a 1 MW energy addition demonstration at Sandia in the summer of 2001.

ACKNOWLEDGEMENTS

The National Hypersonic Wind Tunnel team consists of members from Princeton University, MSE Technology Applications, Inc., the USAF Arnold Engineering and Development Center (AEDC), Lawrence Livermore National Laboratories, and Sandia National Laboratories. The artists concept shown in Fig. 1 was generated by AEDC. The required beam conditions and magnetic field requirements were generated by Dr. Ron Lipinski of SNL based on thermodynamic conditions generated by Princeton University. The injector simulation shown in Fig. 3 was generated by Barry Marder, SNL.

REFERENCES

1. R. B. Miles, "Radiatively Driven Hypersonic Wind Tunnel", AIAA Journal, Vol. 33, No. 8, August 1995.
2. R. J. Lipinski, "Conceptual Design for an Electron-Beam Heated Hypersonic Wind Tunnel," Sandia National Laboratories report, SAND97-1595, July 1997.
3. L. X Schneider, "Repetitive high energy pulsed power technology development for industrial applications", Accelerators in Research and Industry., AIP, CP392, 1997.
4. A. Hershcovitch, "High-pressure arcs as vacuum-atmosphere interface and plasma lens for nonvacuum electron beam welding machines, …", J. Appl. Phys. 78 (9), Nov. 1, 1995.
5. P. Barker, "A 150 kW Electron Beam Heated Radiatively Driven Wind Tunnel Experiment", presented at the AIAA Conference, Jan. 2000, Reno, NV.

Laser-cooling of Ions and Ion Acceleration in the RF-Quadrupole Ring Trap PALLAS [†]

U. Schramm, T. Schätz, and D. Habs

Sektion Physik, LMU München, D-85748 Garching, Germany

Abstract. We report on the commissioning of the table-top RF-Quadrupole storage ring PALLAS*, dedicated to the experimental realization of crystalline ion beams. Images of large prolate ^{24}Mg$^+$-ion crystals are presented, operating PALLAS locally as a linear rf-trap and applying laser-cooling, beautifully confirming prior theoretical and experimental work. For few-ion strings, a radial oscillation of the crystal was observed, stabilized only by additional indirectly cooled impurity ions. On the challenging way towards crystalline ion beams, we demonstrate the acceleration of an ion distribution initially at rest, i.e., the very first operation of PALLAS as a storage ring, still to be heavily improved. Further applications, as the systematic investigation of residual gas damping, or sympathetic cooling for, e.g., the mass determination of heavy radioactive isotopes are briefly discussed.

INTRODUCTION

The idea to guide and store ion beams in RF-Quadrupole structures has many applications in modern accelerator physics. Besides the well-known mass analysis, ion guides gain importance as emittance improvers for, e.g., radioactive ion beams at ISOLDE (CERN) (1), especially combined with residual gas damping (2). RFQ-storage rings, i.e., ion guides bent to a circle, were suggested for the storage of intense low energy beams (3) and should principally allow the systematic investigation of ultra-cold ion beams down to the fully space-charge dominated regime, the crystalline ion beam (4, 5, 6).

Regarding the latter in heavy ion storage rings like Astrid, TSR or ESR (7), only a liquid-like Coulomb-order of very few heavy ions electron-cooled at the ESR (8) and a weak hint for beam ordering of dispersively laser-cooled ^{9}Be$^+$ beams at the TSR (9) were reported up to now. This is believed to be due to the too low symmetry and periodicity of the corresponding lattice functions (10). Nevertheless, low order crystalline structures were proposed to be reachable in the present machines (11) overcoming shear by well adapted gradient laser-cooling in combination with direct transverse laser-cooling (12).

We therefore constructed the RF-Quadrupole storage ring PALLAS to experimentally elucidate fundamental issues of crystalline ion beams and as a prototype for a versatile class of table top storage rings.

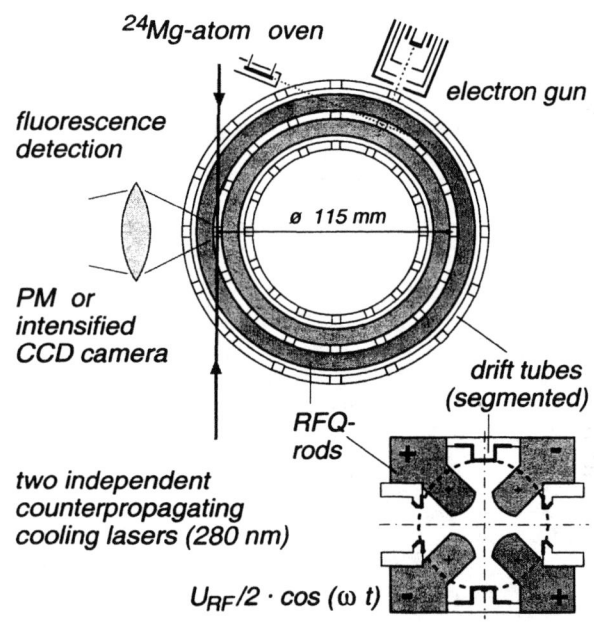

FIGURE 1. Axial and radial cut through the rf-quadrupole ring trap, emphasizing the drift tube structure. One section for laser-cooling and fluorescence detection and the location of the Mg-oven and the electron gun is sketched.

EXPERIMENTAL SETUP

The PALLAS storage ring is constructed like an rf-quadrupole trap bent to a closed circle with a diameter of 115 mm and an aperture radius of 2.5 mm, following (13). Typically the trap is operated at an rf-frequency of

---

[†] supported by the DFG under Contract No. Ha1101/8
* PAul Laser-cooLing Acceleration System

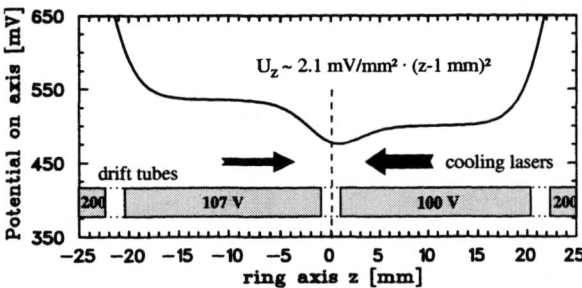

FIGURE 2. Typical axial potential in one of the laser-cooling sections, used to retain the position of a cold ion ensemble in front of the CCD-camera. The neighboring drift-tubes are biased to 107 V and 100 V respectively, causing the small potential dip due to the local ground penetration.

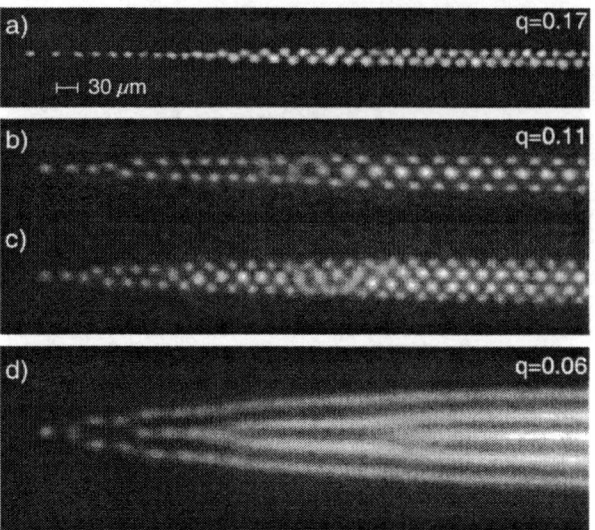

FIGURE 3. Series of images of very prolate ion crystals, axially confined only by the weak potential, shown in fig. 2. As indicated, the radial focusing strength (q) is decreased while the particle number going from a) to d) is increased, continuously raising the linear density λ. For image c), e.g., λ ranges starting with the string from 0.5 over 1.2 (zig-zag) to 2.2-3.2 for the single shell helical structures. The recording time of the CCD camera was 0.4 s (d) 3.2 s).

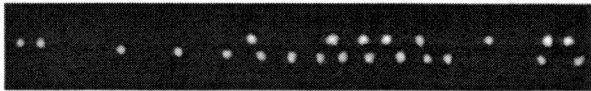

FIGURE 4. Ion crystal containing only about 25% directly laser-cooled ^{24}Mg$^+$-ions, but clearly maintaining the full crystalline order. The conditions are comparable to fig. 3b).

$\omega = 2\pi \cdot 6.3$ MHz and an amplitude of $U_{RF} = 140$ V. For ^{24}Mg$^+$-ions, the corresponding stability parameter $q = 2eU_{RF}/m\omega^2 r_0^2$ amounts to 0.12 and the single-particle secular frequency $\omega_{sec} = q\omega/\sqrt{8}$ in the radial pseudo-potential, equivalent to the betatron frequency in a synchrotron, to $2\pi \cdot 250$ kHz.

As pointed out in fig. 1, the quadrupole structure is surrounded by 16 segmented drift tubes, giving rise to an axial potential of 0.5 % of the voltage applied (5). These tubes can either be biased to manipulate the axial position of ions in the ring or independently switched to several 100 V at a present rate of 110 kHz, intended to accelerate stored ^{24}Mg$^+$-ions to a velocity of about 5000 m/s.

Inside the trap, a weak effusive beam of (isotope enriched) ^{24}Mg-atoms is ionized by a focused electron beam (initially 10 mA at an energy of 80 eV). Ions produced are simultaneously laser-cooled applying two counterpropagating laser beams detuned to 300 MHz below the $3s^2S_{1/2} - 3p^2P_{3/2}$ transition used for Doppler-cooling. Frequency doubled (14) dye lasers provide the required wavelength of 280 nm. The beams with a typical power of several mW are focused tangentially into the trap to a waist of ≈ 0.4 mm. While scanning the laser frequency to either probe or cool the ion distribution, the integrated fluorescence signal is recorded on a fast photomultiplier. Pictures of the ions are taken with an image intensified CCD camera at a constant laser detuning slightly below resonance to obtain maximum brightness.

To operate PALLAS as a linear ion trap, axial confinement has to be provided in the region where the laser is present. As demonstrated in fig. 2, this is achieved by biasing the neighboring drift tubes to about 100 V. The remaining ground penetration then generates a small potential dip between these tubes (giving rise to an axial frequency of only $2\pi \cdot 650$ Hz), ideally suited to confine an ensemble of cold ions, as described in the following section.

EXPERIMENTAL RESULTS

Ion Crystals

Having in mind the formation of crystalline ion beams, the structural development of ion Coulomb-order was theoretically treated for elongated crystals under parabolic axial confinement (4, 15). Introducing a dimensionless linear density $\lambda = a \cdot N/l$ in units of the Wiegner-Seitz radius a (N being the number of ions per unit length l), ion crystals were predicted to develop from a linear chain over a zig-zag band to single and multiple shell helical structures with increasing λ. Up to now, these special structures have only been observed in linear (ring) traps (13, 16, 17, 18, 19), being in excellent agreement with the upper MD simulations.

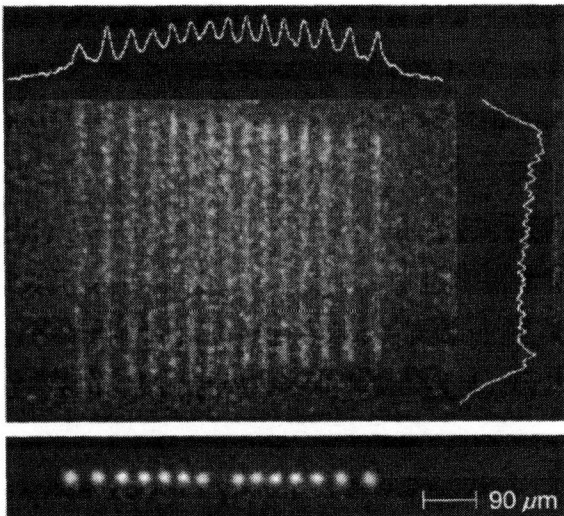

FIGURE 5. Image of a linear ion crystal, consisting of 14 ^{24}Mg$^+$-ions and one "dark" ion. In the upper figure, the crystal ions exhibit strong radial oscillations, while strictly keeping their axial positions (as elucidated by the projections also shown). In the lower figure one "dark" ion has jumped into the structure, immediately (time resolution 50 ms) stabilizing the ion string (The recording time of the sequence presented was 1.8 s to improve spatial resolution).

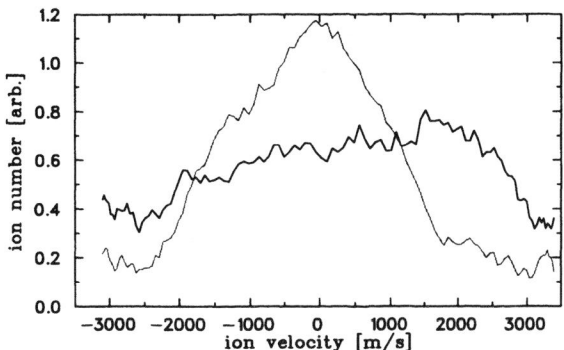

FIGURE 6. Velocity dependent resonance fluorescence of laser probed ^{24}Mg$^+$-ions, representing the ion velocity distribution after loading (grey) and with drift tubes switched at a synchronous velocity of 1200 m/s (black). The ions were cooled by residual gas damping (10^{-6} mbar He).

Nevertheless, controlling these structures is a key requirement aiming for the first generation of crystalline beams in PALLAS. Besides the fact, that the ring trap was designed to minimize axial potentials wells, cold ions can be retained in the potential dip, discussed above. This enables the formation of very prolate ion crystals (fig. 3), on the one hand beautifully confirming preceding work. On the other hand, the development of the structure with increasing λ can be nicely followed. Yet, the transition between two different, almost degenerate states of a single shell helical structure (15) in the range of the linear density $\lambda \approx 2$ to $\lambda \approx 3$ is resolved in the one crystal in fig. 3c). For even larger densities, the ions form multishell configurations, where the individual ions cannot be resolved any more.

A special feature of the axial potential (fig. 2), being slightly weaker on the right side, carefully balanced by the laser-cooling force for ^{24}Mg$^+$-ions, is its property to shift impurity ions, like N_2^+, that are not directly interacting with the laser to the right end of the crystal. Disabling this potential asymmetry and artificially loading additional "dark" ions into the trap still leads to well-shaped Coulomb crystals (fig. 4), although only few ^{24}Mg$^+$-ions are directly cooled. Obviously, this effect of sympathetic cooling can be exploited to cool ions not sensitive to laser-cooling (20) and will be exploited as discussed below.

Furthermore, a puzzling effect was often observed, reducing the ion number in a pure ^{24}Mg$^+$ crystal to only a few ions, expected to form a perfect linear chain. The ion crystal starts to oscillate radially while remaining axially perfectly cold. Inserting a single "dark" ion into the structure immediately stabilizes the ion string. Since laser-cooling predominantly influences the axial degree of freedom, the ion ensemble is sensitive to any small perturbance in the radial direction. Thus, it seems, that either the change in the collective mass of the crystal or the fact, that the dark ion enhances the energy transfer between the degrees of freedom is enough to suppress the immense radial oscillation of this few-ion crystal.

Ion Acceleration

To operate PALLAS as a storage ring, an ion ensemble initially at rest has to be accelerated and thereby overcome accidental potential wells, since presently the injection of ion beams (3) is not foreseen for technical reasons. Therefore, after loading the ring and eliminating impurity ions, the drift tubes have to be periodically switched (e.g., following a step like function suggested in (5)) with a synchronous frequency. After successful acceleration, continuous laser-cooling will maintain the beam velocity (21). The acceleration voltage will then be reduced to avoid unnecessary heating of the beam.

A very first test of this ion acceleration technique was undertaken using a low pressure (10^{-6} mbar) He buffer atmosphere to initially cool the ion distribution and provide some damping during continuous acceleration. The laser was only introduced to probe the distribution. In fig. 6, the resulting velocity distribution is presented compared to the initial distribution. The ions were subject to a synchronous tube switching velocity of 1200 m/s. The mean energy was shifted by 870 m/s, but the distribution was also strongly heated.

DISCUSSION AND OUTLOOK

Summarizing, laser cooling of ^{24}Mg$^+$ ions weakly confined axially leads to the well-known formation of prolate ion crystals. Their perfect appearance assures at least in the section of the ring observed the trap quality, crucial for the storage ring operation. Here, despite of the first acceleration of ions, the scheme has to be strongly improved.

Being able to store beams in PALLAS will then allow systematic cooling studies concerning the formation of crystalline ion beams and their stability, predicted to critically depend on the lattice function of a storage ring. In contrast to a synchrotron, the focusing properties of an RFQ storage ring can be chosen over a wide range, varying the rf-potential or frequency. More generally, the operating conditions for such table-top storage rings will be studied in detail, opening a field for new applications where a finite unidirectional velocity of the particles is desired.

Furthermore, sympathetic cooling of, e.g, radioactive ions would allow to include ions of interest into an ordered structure, thereby strongly reducing their kinetic energy and precisely defining their spatial position. Analyzing the mass dependent secular frequencies of the ion species in the trap directly allows to non-destructively identify the particles involved, since the excitation of the "dark" ions is immediately transferred to the ^{24}Mg$^+$ ions via the strong Coulomb-coupling inside the crystal. Thus, the resonance-fluorescence of the continuously laser-cooled ^{24}Mg$^+$ ions decreases as a function of the excitation amplitude of the "dark" ions. This scheme could be further improved, exciting Eigenfrequencies of the strongly coupled system.

REFERENCES

1. Herfurth, F., et al., *Nucl. Instr. Meth. A* in press (2000) and CERN-EP/2000-0562.
2. Lunney, M.D., et al., CSNSM 97-02 (1997) and Lunney, M.D., Moore, R.B., *Int. J. Mass Spec. Ion Proc.* **190/191**, 153 (1999).
3. Rugiero, A.G., "The Circular RFQ Storage Ring", in *Part. Acc. Conf. (1999)*, edited by Luccio, A., MacKay, W., New York, pp. 3731.
4. Habs, D., and Grimm, R., *Ann. Rev. Nucl. Part. Sci.* **45**, 391 (1995).
5. Schätz, T., Schramm, U., Habs, D., *Hyp. Int.* **115**, 29 (1998).
6. Schätz, T., Habs, D., Podlech, C., Wei, J., Schramm, U., "Towards crystalline ion beams - the PALLAS ring trap", in *Trapped Charged Particles and Fundamental Physics (1998)*, edited by Dubin, D., Schneider, D., AIP Conf. Proc. 457, Woodbury, 1999, pp. 269.
7. *Crystalline Beams (Erice 1995)*, edited by Maletic, D.M., and Ruggiero, A.G., World Scientific (1996).
8. Steck, M., et al., *Phys. Rev. Lett.* **77**, 3803 (1996) and Hasse, R.W., *Phys. Rev. Lett.* **83**, 3430 (1999).
9. Eisenbarth, U., et al., *Hyp. Int.* in press (2000).
10. Schiffer, J.P., in (7) pp. 217 and Wei, J., et al., in (7) pp. 229.
11. Wei, J., et al., *Phys. Rev. Lett.* **80**, 2606 (1998).
12. Lauer, I., et al., *Phys. Rev. Lett.* **81**, 2052 (1998).
13. Birkl, G., Kassner, S., and Walther. H., *Nature* **357**, 310 (1992).
14. Schramm, U., Peters, A., Habs, D., *Hyp. Int.* **115**, 57 (1998).
15. Rahman, A., and Schiffer, J.P., *Phys. Rev. Lett.* **57**, 1133 (1986) and Hasse, R.W., and Schiffer, J.P., *Annals of Phys.* **203**, 419 (1990).
16. Raizen, M.G., et al., *Phys. Rev. A* **45**, 6493 (1992).
17. Drewsen, M., et al., *Phys. Rev. Lett.* **81**, 2878 (1998).
18. Nägerl, H.C., et al., *Applied Phys. B* **66**, 603 (1998).
19. Block., M., et al., *J. Phys. B.* **33**, L375 (2000).
20. Bowe, P., et al., *Phys. Rev. Lett.* **82**, 2071 (1999) and Molhave, K., et al., *Phys. Rev. A* **62**, 011401(R) (2000).
21. The initial acceleration of the ion cloud only by the velocity dependent laser force is not possible, since a) the overlap with the ring is far too small and b) small potential wells could not be overcome by the maximum laser force of only several mV/mm.

The CERN-EU Radiation Facility for Dosimetry at Flight Altitude and in Space

A. Ferrari, A. Mitaroff and M. Silari

CERN, 1211 Geneva 23, Switzerland

Abstract. A reference facility for the inter-comparison of active and passive detectors in complex high-energy neutron fields is available at CERN since 1993. A positively charged hadron beam (a mixture of protons and pions) with momentum of 120 GeV/c hits a copper target, 50 cm thick and 7 cm in diameter. The secondary particles produced in the interaction traverse a shield made of either 80 cm of concrete or 40 cm of iron. Behind the iron shield, the resulting neutron spectrum has a maximum at about 1 MeV, with an additional high-energy component. Behind the concrete shield, the neutron spectrum has a second pronounced maximum at about 70 MeV and resembles the high-energy component of the radiation field at commercial flight altitudes created by cosmic rays. Recent Monte Carlo calculations are presented, performed for different beam conditions and shielding configurations in view of a possible upgrade of the facility for measurements related to the space program.

INTRODUCTION

A reference radiation facility for the calibration and inter-comparison of dosimetric devices in complex high-energy stray radiation fields is available at CERN since 1993 (1, 2). In addition to the interest for testing instrumentation and passive detectors used around high-energy particle accelerators, this program is partially supported by the European Commission in the framework of a research program for the assessment of radiation exposure at civil flight altitudes (3, 4). These reference fields are, in fact, sufficiently similar to the cosmic ray field encountered at 10-20 km altitude such that instrumentation is tested at CERN and subsequently used for in-flight measurements on aircraft (see, for example (5, 6)).

The aim of the present paper is to summarize briefly the essential features of the facility, which is discussed more extensively in (1, 2), address its possible application to measurements related to the space program and reports on recent calculations performed in view of extending its use to broader particle spectra.

THE CERN-EU HIGH-ENERGY REFERENCE FIELD FACILITY (CERF)

CERF is set up at one of the secondary beam lines (H6) from the Super Proton Synchrotron (SPS), in the North Experimental Area on the Prévessin site of CERN (*cerf* in French means deer; there is a fenced area where several deer live close to the experimental hall where CERF is operational). A positive hadron beam with momentum of usually 120 GeV/c is stopped in a copper target, 7 cm in diameter and 50 cm in length, which can be installed in two different positions inside an irradiation cave. The secondary particles produced in the target traverse a shielding, on top of these two positions, made up of either 80 cm concrete or 40 cm iron. These roof-shields produce almost uniform radiation fields (mostly neutrons) over two areas of 2x2 m^2, each of them divided into 16 squares of 50x50 cm^2. Each element of these "grids" represents a reference exposure location. Additional measurement positions are available behind the lateral-shielding of the irradiation cave, at the same angles with respect to the target as for the two roof positions. Shielding is either 80 cm or 160 cm concrete, and at both positions 8 additional exposure locations (arranged in 2x4 grids made up of the same 50x50 cm^2 elements) are provided. The nominal measurement points are at the center of each square at 25 cm above floor, i.e. at the center of a 50x50x50 cm^3 air volume, where the radiation field is calculated. The intensity of the primary beam is monitored by an air-filled ionization chamber at atmospheric pressure, placed in the beam just upstream of the target, connected to a current digitizing circuit. By adjusting the beam intensity on the target one can vary the dose equivalent rate at the reference positions, typically in the range from 25 μSv/h to 1 mSv/h on the iron-roof and from 5 to 600 μSv/h on the concrete-roof. The energy distributions of the various particles (mainly neutrons, but also photons, electrons, muons, pions and protons) at the various exposure locations are known by Monte Carlo simulations performed with the FLUKA code (7, 8). Details of the latest simulations are given in (2). The neutron energy distributions calculated for a primary beam of positive particles (35 % protons, 61 % pions and 4 % kaons, as determined experimentally) with 120 GeV/c momentum are shown in figure 1.

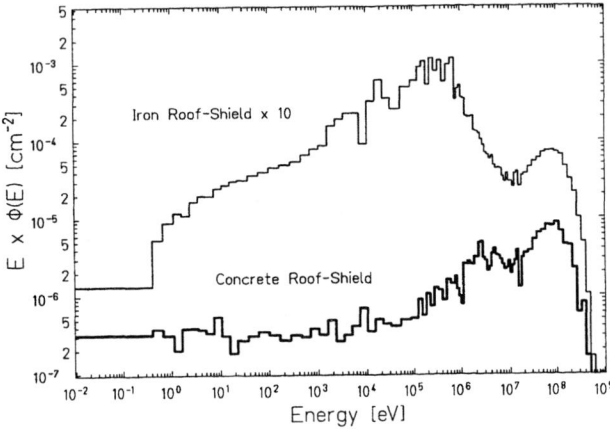

FIGURE 1. Neutron spectral fluences (lethargy) on the concrete and iron roof-shields (neutrons per primary beam particle incident on the copper target).

The spectrum outside the iron shield is dominated by neutrons in the 0.1- 1 MeV range, whilst the energy distribution outside the concrete shield shows a large relative contribution of 10-100 MeV neutrons. Therefore these exposure locations provide wide spectrum radiation fields well suited to test dosimetric instrumentation under different conditions. The fluence rate of other hadrons is much lower than that of neutrons. The photon fluence is almost one order of magnitude less than that of neutrons on the iron roof-shield, but almost a factor two higher than the neutron one on the concrete roof-shield, because of the contribution from (n,γ) reactions. The electron fluence is about one order of magnitude less than that of neutrons and the muon fluence almost three orders of magnitude less. However, an additional muon component is also present which directly comes from the upstream secondary production target in the H6 beam line and from pion decay in the beam line. These muons stream over the concrete and iron roof-shields. Their intensity depends on various factors which are not under direct control, as the angle under which secondary particles are guided from the production target into the H6 beam line, as well as the intensity of secondary beams in neighboring beam lines.

The accuracy of the calculated neutron spectral fluences was verified by extensive measurements made with a Bonner sphere spectrometer using an ^{3}He proportional counter, employed bare and within a set of five polyethylene spheres (83 mm, 108 mm, 133 mm, 178 mm and 233 mm in diameter), and with the spherical version of the LINUS rem counter (9, 10, 11, 12). The response function of LINUS extends to several hundreds MeV due to a lead converter added to the polyethylene moderator. There is excellent agreement between the FLUKA predictions and the experimental results (2, 13). The neutron spectrometry measurements were recently repeated with a Bonner sphere system which now includes two dedicated high-energy channels based on the LINUS design (14). The results are presently being analyzed.

RECENT DEVELOPMENTS

Several institutions from all over Europe, but also from the USA and Japan, use the facility since several years, testing various types of passive and active detectors: Tissue Equivalent Proportional Counters (TEPC), GM-counters, different types of rem counters, bubble detectors, scintillator based dose-rate meters, electronic pocket dosimeters, Si-diodes, track etch detectors, TLDs, films, nuclear track detectors, recombination chambers, multisphere systems, CR39 foils. In addition to dosimetric instrumentation, a group investigated the effect of radiation on computer memories, whilst another one performed several tests on a prototype of a beam loss monitor for the future CERN Large Hadron Collider (LHC). There is also a collaboration with the ATLAS-muon background group of CERN. CERF has also already been used to test instrumentation which was then flown in space (15, 16, 17) or on high-altitude flights (18). The neutron spectral distribution on the concrete roof-shield sufficiently resembles the spectral distribution encountered at civil flight altitude and in space. It was then decided to study whether a different shielding configuration can produce different particle fields.

Preliminary Monte-Carlo simulations were performed with the FLUKA code (7, 8) to investigate this possibility, particularly in view of further measurements in the framework of the space program. To avoid unnecessary time-consuming calculations, simulations were carried out in a simplified spherical geometry rather than modeling the complete facility. The aim was to understand if a given target/shielding combination and angular scoring region would indicate a promising situation which could subsequently be investigated more thoroughly. Eventually, the actual facility will have to be properly modeled. Calculations were performed for the "standard" 50 cm thick, 7 cm diameter copper target as well as for smaller targets, one 10 cm long and 7 cm in diameter, another 10 cm long and 1 cm in diameter. The study aims to produce a radiation field mainly rich in high-energy protons similar to that found in the stratosphere (19) or high-energy neutrons and other secondaries as found inside the space station or a spacecraft (20). The intention is also to increase the available dose equivalent rate at the exposure locations.

First, calculations were made for the "standard" primary beam incident on the target, i.e. a mixed beam composed of one third protons and two third positive pions of 120 GeV/c momentum. The target was at the centre of a spherical shield, with scoring done in conical regions 10° wide (0° to 90°). Neutrons, protons, pions, muons, photons and electrons were scored. A 10 cm thick aluminium shield was placed at either 1 m or 1.2 m distance from the target. Since in a practical situation the dose equivalent rate produced by such configuration will be much higher than that presently available at CERF, the reference exposure area will have to be shielded. An additional layer of material was therefore included in the simulations, made up of either 80 cm thick concrete placed at 2 m distance from the aluminium layer, or 40 cm iron backed by 40 cm concrete placed at 2.5 m distance from the aluminium. The most interesting results were obtained with the "standard" copper target and the latter shielding configuration described above. The particle spectra in the forward direction (0° to 10° angular region) are shown in figure 2.

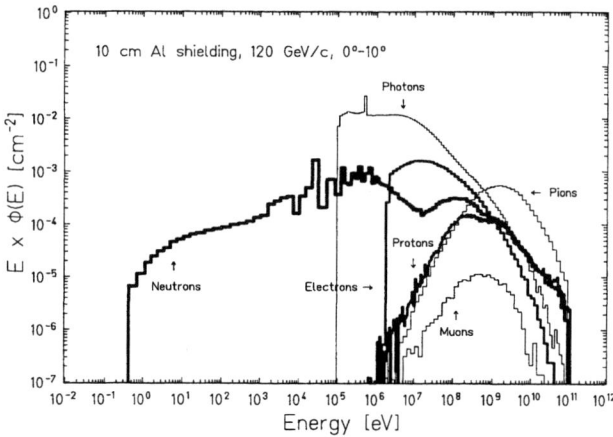

FIGURE 2. Particle spectral fluences (lethargy) in the 0°-10° angular region for a 120 GeV/c beam (particles per primary beam particle incident on the target). The target is copper, 50 cm thick and 7 cm in diameter; 10 cm thick aluminum shield placed at 1 m from the target; external shielding is 40 cm iron backed by 40 cm concrete placed at 2.5 m distance from the aluminum slab.

A simulation made without the back shield showed that the shield is responsible for the low energy neutron component shown in the figure. One should note, in particular, that the high-energy component (above 100 MeV) of the proton spectrum is similar to the energy spectrum of cosmic ray protons. Also, the neutron energy distribution extends up to about 100 GeV versus a few hundreds MeV of the present CERF configuration on the concrete roof-shield. The neutron fluence is about $6.5 \cdot 10^{-3}$ cm^{-2} and $4.3 \cdot 10^{-3}$ cm^{-2} per primary particle on the copper target, in the forward (0°-10°) and transverse (80°-90°) directions, respectively, as compared to the present figure of $3.5 \cdot 10^{-5}$ cm^{-2}. On the other hand, the pion component is too high in comparison to space. This configuration has the advantage that an independent exposure area could in principle be set-up and run in parallel to the present configuration (i.e. simultaneous data taking on the concrete roof-shield or iron-roof shield). Another series of simulations were performed with increasing aluminium thickness, with the aim of reducing the pion component relative to the other particles. Figure 3 shows the results with 40 cm aluminium at 30°-40° emission angle, where it is actually seen that the pion component decreases with respect to neutrons with increasing aluminium thickness and emission angle.

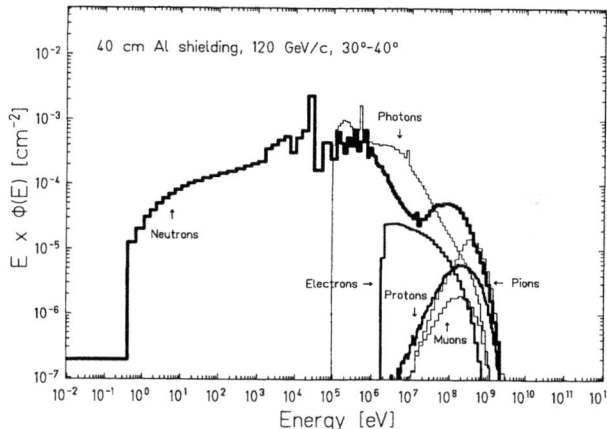

FIGURE 3. Particle spectral fluences (lethargy) in the 30°-40° angular region for a 120 GeV/c beam (particles per primary beam particle incident on the target). The target is copper, 50 cm thick and 7 cm in diameter; 40 cm thick aluminum shield placed at 1 m from the target; external shielding is 40 cm iron backed by 40 cm concrete placed at 2.5 m distance from the aluminum slab.

The influence of primary beam momentum on the produced secondary particles was investigated in the range 40-400 GeV/c. The simulations were performed with the actual particle composition of the beam impinging on the target: this is made up of about 85 % pions and 15 % protons at 40 GeV/c, 2/3 pions and 1/3 protons at 120 GeV/c, 1/3 pions and 2/3 protons at 205 GeV/c and protons only at 400 GeV/c. It turned out that the spectral shape and the secondary particle composition do not change much with beam momentum, but the absolute fluence per beam particle on target increases by about a factor of 2 by raising the momentum from 205 to 400 GeV/c and by a factor of 7 from 40 to 400 GeV/c. With increasing emission angle the neutron component becomes more and more dominant; Figure 4 shows the particle spectra for 205 GeV/c and 60°-70° emission angle.

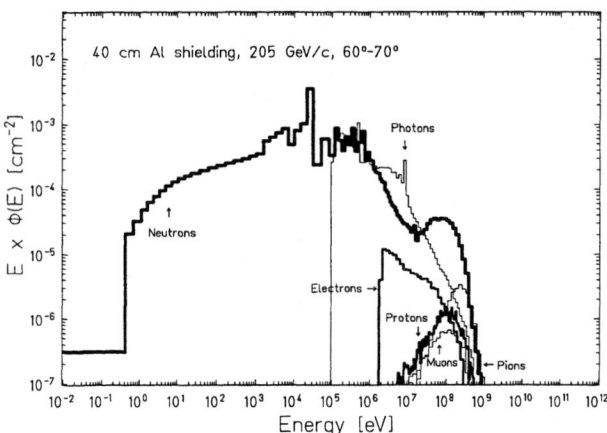

FIGURE 4. Particle spectral fluences (lethargy) in the 60°-70° angular region for a 205 GeV/c beam (particles per primary beam particle incident on the target). The target is copper, 50 cm thick and 7 cm in diameter; 40 cm thick aluminum shield placed at 1 m from the target; external shielding is 40 cm iron backed by 40 cm concrete placed at 2.5 m distance from the aluminum slab.

ACKNOWLEDGEMENTS

The operation of CERF is partially supported by the European Commission, Directorate General XII, contracts no. F13P-CT92-0026 (1992-1995) and F14P-CT95-0011 (1996-1999). A. Fassò, M. Höfert, T. Otto, G.R. Stevenson and L. Ulrici have contributed substantially to the set-up, start-up and operation of the facility over the past years.

REFERENCES

1. Höfert, M., Stevenson, G.R., "The CERN-CEC high-energy reference field facility," in *8th International Conference on Radiation Shielding*, American Nuclear Society, Inc., Illinois, USA, 1994, pp. 635-642.

2. Birattari, C., Ferrari, A., Höfert, M., Otto, T., Rancati, T., Silari, M., "Recent results at the CERN-EC high-energy reference field facility," in *Satif-3 Shielding Aspects of Accelerators, Targets and Irradiation Facilities*, NEA/OECD, Paris, 1998, pp. 219-234.

3. European Radiation Dosimetry Group, "Exposure of air crew to cosmic radiation," in *EURADOS Report 1996-01*, edited by McAulay, I.R., Bartlett, D.T., Dietze, G., Menzel, H.G., Schnuer, K., Schrewe, U.J., 1996.

4. "Study of radiation fields and dosimetry at aviation altitudes," Co-ordinated by the Dublin Institute for Advanced Studies, School of Cosmic Physics, Dublin (Ireland). Report DIAS-99-9-1.

5. Birattari, C., Esposito, A., Fassò, A., Ferrari, A., Festag, J.G., Höfert, M., Nielsen, M., Pelliccioni, M., Raffnsøe, C., Schmidt, P., Silari, M., Radiat Prot Dosim **51**, 87-94 (1994).

6. Alberts, W.G., Alevra, A.V., Ferrari, A., Otto, T., Schrewe, U.J., Silari, M., Radiat Prot Dosim **86**; 289-295 (1999).

7. Fassò, A., Ferrari, A., Ranft, J., Sala, P.R., "New developments in FLUKA modelling of hadronic and EM interactions," in *Simulating Accelerator Radiation Environments (SARE3)*, edited by Hirayama, KEK Proceedings 97-5, 1997, pp. 32-44.

8. Ferrari, A., Rancati, T., Sala, P.R., "FLUKA application in high-energy problems: from LHC to ICARUS and atmospheric showers," in *Simulating Accelerator Radiation Environments (SARE3)*, edited by Hirayama, KEK Proceedings 97-5, 1997, pp. 165-176.

9. Birattari, C., Ferrari, A., Nuccetelli, C., Pelliccioni, M., Silari, M., Nucl Instrum Meth **A 297**, 250-257 (1990).

10. Birattari, C., Esposito, A., Ferrari, A., Pelliccioni, M., Silari, M., Radiat Prot Dosim **44**, 193-197 (1992).

11. Birattari, C., Esposito, A., Ferrari, A., Pelliccioni, M., Silari, M., Nucl Instrum Meth **A 324**, 232-238 (1993).

12. Birattari, C., Esposito, A., Ferrari, A., Pelliccioni, M., Rancati, T., Silari, M., Radiat Prot Dosim **76**, 135-148 (1998).

13. Birattari, C., De Ponti, E., Esposito, A., Ferrari, A., Magugliani, M., Pelliccioni, M., Rancati, T., Silari, M., "Measurements and simulations in high-energy neutron fields," in *Satif-2 Shielding Aspects of Accelerators, Targets and Irradiation Facilities*, OECD/NEA, Paris, 1996, pp. 171-197.

14. Birattari, C., Cappellaro, P., Mitaroff, A., Silari, M., "Development of an extended range Bonner Sphere Spectrometer," in *Advanced Monte Carlo for Radiation Physics, Particle Transport Simulation and Applications*, Lisbon, Portugal, 2000 (in press)

15. Badhwar, G.D., Robbins, D.E., Gibbons, F., Braby, L.A., "Response of a Tissue Equivalent Proportional Counter to neutrons" (to be published).

16. Badhwar, G.D., Keith, J.E., Cleghorn, T., "Neutron dosimetric measurements onboard the space shuttle," in *Predictions and Measurements of Secondary Neutrons in Space*, edited by Badhwar, G.D., USRA, Houston, 1998.

17. Luszik-Bhadra, M., Matzke, M., Schuhmacher, H., "Developments of personal neutron dosemeters at PTB and first measurements in the space station MIR," ibid. ref. (16)

18. Bartlett, D.T., Hager, L.G., Tanner, R.G., Stelle, J.D., "Measurements of the high-energy neutron component of cosmic radiation fields in aircraft using etched track dosimeters," ibid. ref. (16)

19. Reitz, G., Radiat Prot Dosim **48**, 5-20 (1993).

20. Armstrong, T.W., Colborn, B.L., "Monte Carlo predictions of secondary neutron spectra inside the International Space Station and comparison with space measurements," ibid. ref. (16)

Simulation of Cosmic Neutron Flux Using an Electron Linac

T. J. Collens, A. P. Tonchev, R. M. Jaber, K. C. Kennedy, and J. F. Harmon

Idaho Accelerator Center, Idaho State University, Campus Box 8106, Pocatello, ID 83209, USA

Abstract. It has been shown that cosmic neutron flux can generate upsets in DRAM's and other electronics. A model simulating cosmic neutron flux has been developed using a 30 MeV Electron Linac. The slowing down of the electron beam (bremsstrahlung) in the target ultimately produces photoneutrons. MCNP4C is used to model the transport of neutrons through to the detector. Optimization of the model includes varying target thickness and placement. This simulation shows that cosmic neutron flux can be approximate up to 17 MeV.

INTRODUCTION

The possibility of cosmic rays causing semiconductor upsets has been suspected and studied for many years. More recently, a link between electronics upsets and cosmic neutrons has been shown. Collisions between high-energy neutrons and silicon atoms cause heavy ions to recoil, which deposits a disruptive charge in the silicon [1,2].

The fundamental form of cosmic radiation, as found in outer space, is called primary cosmic radiation. The content of primary cosmic radiation is approximately 87% protons, 12% alpha particles, and 1% heavier nuclei and electrons [3]. As primary cosmic rays enter the earth's atmosphere they collide with atmospheric atoms. Collisions not only attenuate the radiation, but also generate large numbers of secondary particles. The secondary particles consist of high-energy neutrons, protons, and short-lived subatomic particles.

Simulating the cosmic neutron spectrum may be useful for studying its effects on materials and devices. The very high energies of the atmospheric spectrum make traditional abundant neutron sources unsuitable as simulation sources. Reactor neutrons are somewhat limited by their Maxwell distribution, and radioactive neutron sources do not approach the high energies required. Accelerators are ideal sources to simulate such energetic phenomena.

Los Alamos National Laboratory Weapons Neutron Research (WNR) facility has an 800 MeV proton accelerator that was used to simulate the atmospheric neutron spectrum. Using such a highly energetic beam results in a spallation neutron source. The term spallation refers to the production of several neutrons from a single energetic particle. As many as thirteen neutrons are produced from each proton from the Los Alamos WNR beam, which enables it to match the cosmic neutron spectrum accurately up to 500 MeV [2].

Here, the possibility of simulating the cosmic neutron flux with a more accessible electron LINAC has been investigated. This alternative method was modeled with a 30 MeV electron accelerator to produce a non-spallation neutron source. The accelerator modeled in the simulations is based on the new LINAC accelerator located at Idaho Accelerator Center of Idaho State University. This accelerator is capable of outputting 2 to 30 MeV electrons at 360 pulses per second with an average beam current of 0.3 mA. Using a 30 MeV LINAC limits the simulation to the lower energies, but it may still be useful as the cosmic neutron flux falls off rapidly at higher energies.

The slowing down of the electrons in the target material generates bremsstrahlung radiation, which in turn generates neutrons primarily by way of (γ,n) and (γ,xn) reactions [4]. Monte Carlo computer models indicate that the cosmic neutron flux spectrum can be simulated in the 5-17 MeV range using the 30 MeV LINAC.

MODEL THEORY

The atmospheric neutron spectrum is relatively low in intensity but reaches extremely high energies. Figure 1. shows the approximate normalized atmospheric neutron spectrum. As can be seen from this figure, the atmospheric neutron flux is highest at

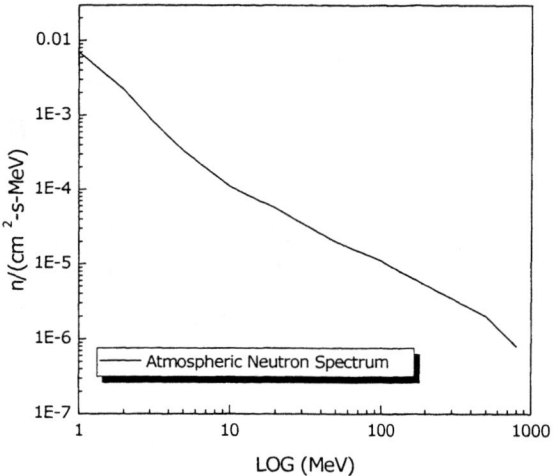

FIGURE 1. Atmospheric Neutron Flux Spectrum

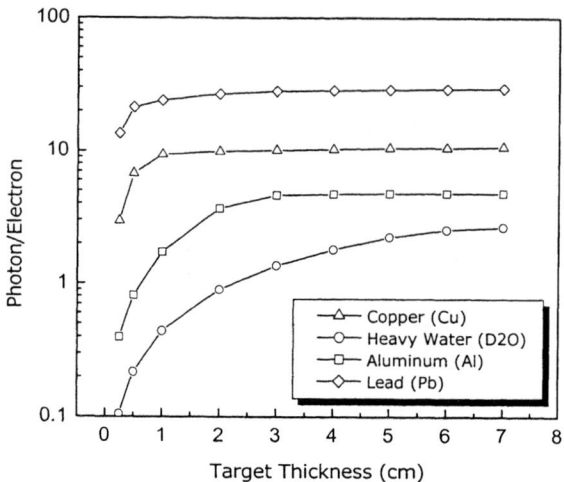

FIGURE 2. Bremsstrahlung Yield vs. Target Thickness

lower neutron energies and then slowly decreases as it approaches 1000 MeV.

Producing neutrons with an electron beam and target is primarily dependent upon bremsstrahlung production and on how readily the bremsstrahlung radiation is converted to neutrons. This results in the selection of a target or converter that possesses the qualities of both high bremsstrahlung production and a large neutron production cross-section.

The production of bremsstrahlung is dependent on the energy of the electrons being slowed down, target thickness, and the material or materials that make up the target [5]. Figure 2. is the estimated internal bremsstrahlung yield based on Monte Carlo code for several target materials and thicknesses at 15 MeV. An increase in the Z of the material relates to greater photon production within the target. This also illustrates the saturation behavior for photon production based on target thickness.

The photons produced interact with the target material to produce photoneutrons primarily by way of (γ,n) and (γ,2n) reactions. This type of reaction generally has a threshold energy around 7 MeV. Photoneutron cross-sections exhibit a Giant Resonance in which the cross section has a resonance centered about a few MeV, which allow (γ,n) and (γ,2n) reactions to account for the majority of a material's total photoneutron cross-section. Other neutron producing reactions such as (γ,3n) and (γ,pn) do contribute, but only slightly in most materials. Figure 3. illustrates the Giant Resonance in the photoneutron cross-section for lead as taken from Veyssiere [6].

Knowing the bremsstrahlung spectrum and the cumulative photoneutron cross section, the total yield may be calculated [7]. The total photoneutron yield is given by:

$$Y_n = \frac{\rho V N a}{A} \int_{E_{Th}}^{E_{Max}} \sigma(E) \cdot \phi_\gamma(E) dE \qquad (1)$$

where Na is Avogadro's Number, V is the volume of the target, and σ is the total photoneutron production cross-section.

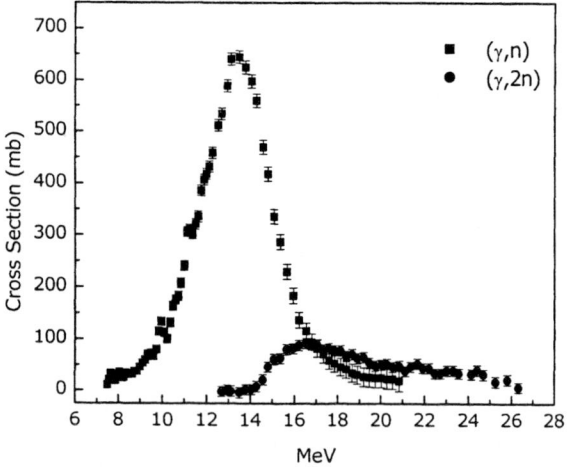

FIGURE 3. Photoneutron Cross Section of Pb-208

SIMULATION MODEL

The model consists of the target, and the shielding surrounding the detection point. The target is a lead circular cylinder, 2 cm in radius. A 5 cm thickness along the beam axis nears the saturation of bremsstrahlung production in most materials but does not serve as an excessively thick shield to internally generated neutrons. Lead was chosen for its accessibility, relatively high bremsstrahlung production and acceptable photoneutron production cross-section. The total neutrons emitted from a target exhibit a saturation behavior with respect to target thickness similar to bremsstrahlung production.

The total yield was calculated based on an MCNP4C estimation of the bremsstrahlung spectrum and the neutron production cross section [8]. The photoneutron spectrum was estimated using the pre-equilibrium code GNASH [9]. Given the maximum energy of photons, GNASH calculates the estimated neutron spectrum. Using the neutron spectrum from GNASH and the calculated neutron yield, a neutron source spectrum was generated and input into MCNP4C.

To help avoid inescapable gamma radiation the detector was shielded by 10 cm of lead. As another precaution the detector is located perpendicular to the beam and target to avoid the more intense gamma radiation exiting the rear of the target. The general arrangement of the model, including the target and the detector position, are shown in Figure 4.

RESULTS & CONCLUSION

The simulation was run at maximum endpoint bremsstrahlung energies of 16, 20, 25, and 27 MeV. This is a good sample of reasonably achievable accelerator energies. Results were calculated assuming an average 0.1 mA beam current. The simulation spectrum is much more intense than the actual atmospheric spectrum and therefore must be divided by a constant to allow direct comparison.

The simulations show that a better match to the atmospheric spectrum is possible with higher beam energies. Simulations at all beam energies are directly compared to the atmospheric neutron spectrum in Figure 5. The best fit to the atmospheric spectrum occurs with the 27 MeV electron beam simulation. Between neutron energies 5 and 17 MeV, the simulation achieves a reasonable concurrence. Figure 6 shows the 27 MeV electron beam simulation and the range in which there is an acceptable match with the atmospheric neutron spectrum.

Possible improvements on this LINAC and target neutron source model could be achieved by utilizing a more complex target design. Such target improvements may include multiple materials and asymmetrical target geometry. The simulation accuracy may also be improved by modeling the target cooling that will be necessary when obtaining experimental data.

One clear disadvantage to using a LINAC is that the simulation is limited to a range of lower energies dependent on the maximum beam energy. However, the effectiveness of this method may be enhanced by the fact that the atmospheric flux is highest at lower energies and that fewer electronics upsets result from

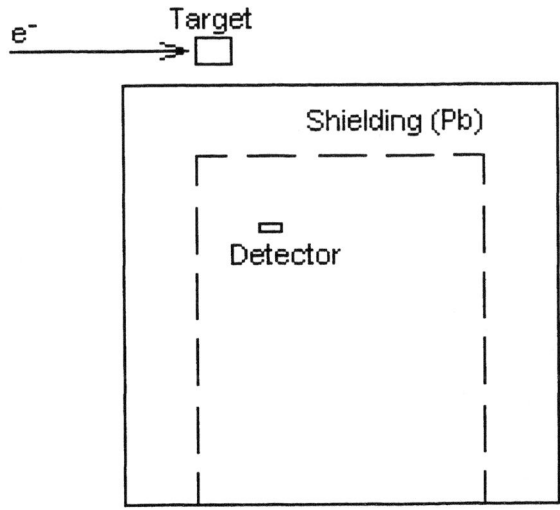

FIGURE 4. Simulation Model Geometry

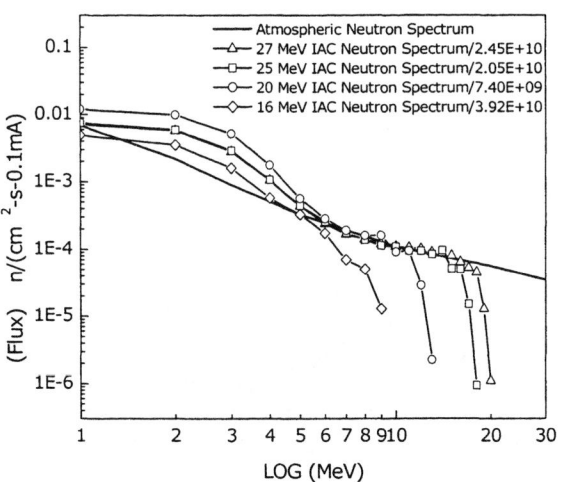

FIGURE 5. Simulation at All Beam Energies

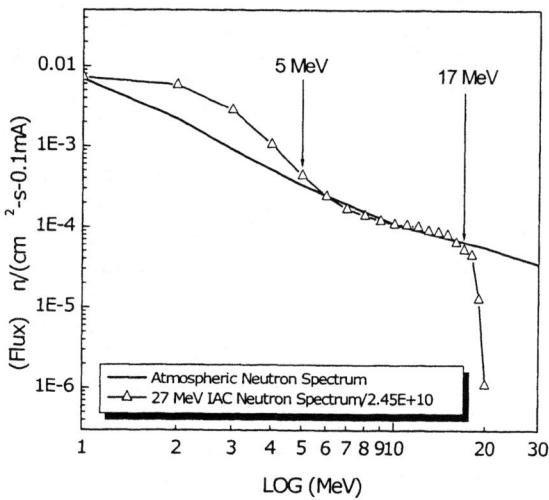

FIGURE 6. Simulated 27 MeV Beam Source/2.45E+10

the extreme high-energy side of the atmospheric neutron spectrum [2].

The use of a LINAC provides an intense neutron source that can be used to simulate the atmospheric neutron spectrum at lower energies. This type of simulation may not be as comprehensive as sources that reach much higher neutron energies, but they may prove useful as preliminary lower-cost testing once more experimental data is acquired.

ACKNOWLEDGMENTS

This research was supported by Micron Technology Inc. in conjuction with the Idaho Accelerator Center.

REFERENCES

1. Gosset C.A. et al., *IEEE Trans. Nucl. Sci.* **40**, 1845-1852 (1993).
2. McKee, W.R. et al., "Cosmic Ray Neutron Induced Upsets as a Major Contributor to the Soft Error Rate of Current and Future Generation DRAM's," in *IEEE International Reliability Physics Proceedings,* Texas, 1996.
3. Lamarsh, J.R., *Introduction to Nuclear Engineering*, 2nd Edition, Addison Wesley Publishing, Massachusetts, 1983, pp. 427-435.
4. Leo, W.R., *Techniques for Nuclear and Particle Physics Experiments*, Springer Verlag, Berlin, 1994, pp. 38-45.
5. Mayo, R.M., *Introduction to Nuclear Concepts for Engineers*, American Nuclear Society, Illinois, 1998, pp. 276-281.
6. Veyssiere, A. et al, *Nucl. Phys. A*, **159**, (1970).
7. Swanson, W.P., *Health Phys.* **35**, 353-367 (1978).
8. Briesmeister, J.F., MCNP – A General Monte Carlo Particle Transport Code, Version 4C, Los Alamos National Lab, 2000.
9. Young, P.G., Arthur, E.D., and Chadwick, M.B., "Comprehensive nuclear model calculations: Theory and use of the GNASH code." Pages 227-404 of: Gandini, A., Reffo, G. (eds), Proc. of the IAEA Workshop on Nuclear Reaction Data and Nuclear Reactors - Physics, Design, and Safety. Singapore: World Scientific Publishing,Ltd., Trieste, Italy, April 15 - May 17, 1996.

RIKEN RI Beam Factory Project

Yasushige Yano, Takeshi Katayama, Akira Goto, and Masayuki Kase

RIKEN, Wako, Saitama 351-0198, JAPAN

Abstract. The RI beam factory is under construction at RIKEN. In the first phase, a new multi-stage acceleration system will be started to operate in 2004. This system consists of three ring cyclotrons with K=520 MeV, 980 MeV and 2500 MeV in each. It will boost the heavy-ion energies obtained by the existing K540-MeV ring cyclotron up to 350 MeV/nucleon. The heavy-ion beams are converted into intense RI beams via the projectile fragmentation or fission. The construction of the second-phase building will be started in 2002. In the second phase, an accumulator cooler ring and an electron-RI beam collider will be built. The first electron scattering experiment on unstable nuclei is scheduled for 2009.

INTRODUCTION

In recent years the advent of radioisotope (RI) beams has opened up a number of fascinating new fields. To further develop the new fields of science, the RIKEN Accelerator Research Facility (RARF) has undertaken construction of the RI beam factory (RIBF) as a next generation facility that is capable of providing the world's most intense RI beams over the whole range of atomic masses.

Figure 1 shows the plan view of the RIBF. This new facility will add new dimensions to the RARF's existing capabilities. At present the RARF has the world-class heavy-ion accelerator complex consisting of a K540-MeV ring cyclotron (RRC) and a couple of different types of the injectors: a variable-frequency Wideröe linac (RILAC) and a K70-MeV AVF cyclotron (AVF). Moreover, its projectile-fragment separator (RIPS) provides the world's most intense light-atomic-mass RI beams. In the factory, a cascade of a K520-MeV fixed-frequency ring cyclotron (fRC), a K980-MeV intermediate-stage ring cyclotron (IRC) and a K2500-MeV superconducting ring cyclotron (SRC) will be a post-accelerator for the existing RRC. This new cyclotron system will be able to boost the RRC beam's output energy up to 400 MeV/nucleon for light ions and 350 MeV/nucleon for very heavy ions. The goal of the beam intensity is higher than 1pμA.

As in the existing RIPS, RI beams will be generated mostly by projectile fragmentation. In addition, fission of a uranium beam will be used for the production of very neutron-rich isotopes in the medium mass region. A BigRIPS will be installed to generate RI beams with much larger magnetic rigidity.

The RIBF includes a multi-use experimental storage rings (MUSES) consisting of an accumulator cooler ring (ACR) and an electron-RI beam collider (e-RI Collider). MUSES will enable us to conduct various types of advanced experiments: electron scattering on unstable nuclei, precision mass measurements, and atomic physics with cooler electron beams. In the original MUSES project, ion-ion merging or head-on collisions and X-ray spectroscopy of unstable nuclei are planned. We will request the budget for realizing these as an RI-beam application program.

CONSTRUCTION PHASE

The construction of the RIBF is scheduled in two phases: in the first phase, the new cyclotrons and the BigRIPS will be built, and the RI-beam experimental installation and the MUSES will be constructed in the second phase.

RILAC

The RILAC will be the initial accelerator for the RIBF. In order to upgrade the RILAC performance in the beam intensity, the pre-injector system consisting of a frequency-tunable folded-coaxial RFQ linac (FC-RFQ) equipped with an 18-GHz ECR ion source (ECRIS-18) has been developed. The FC-RFQ has

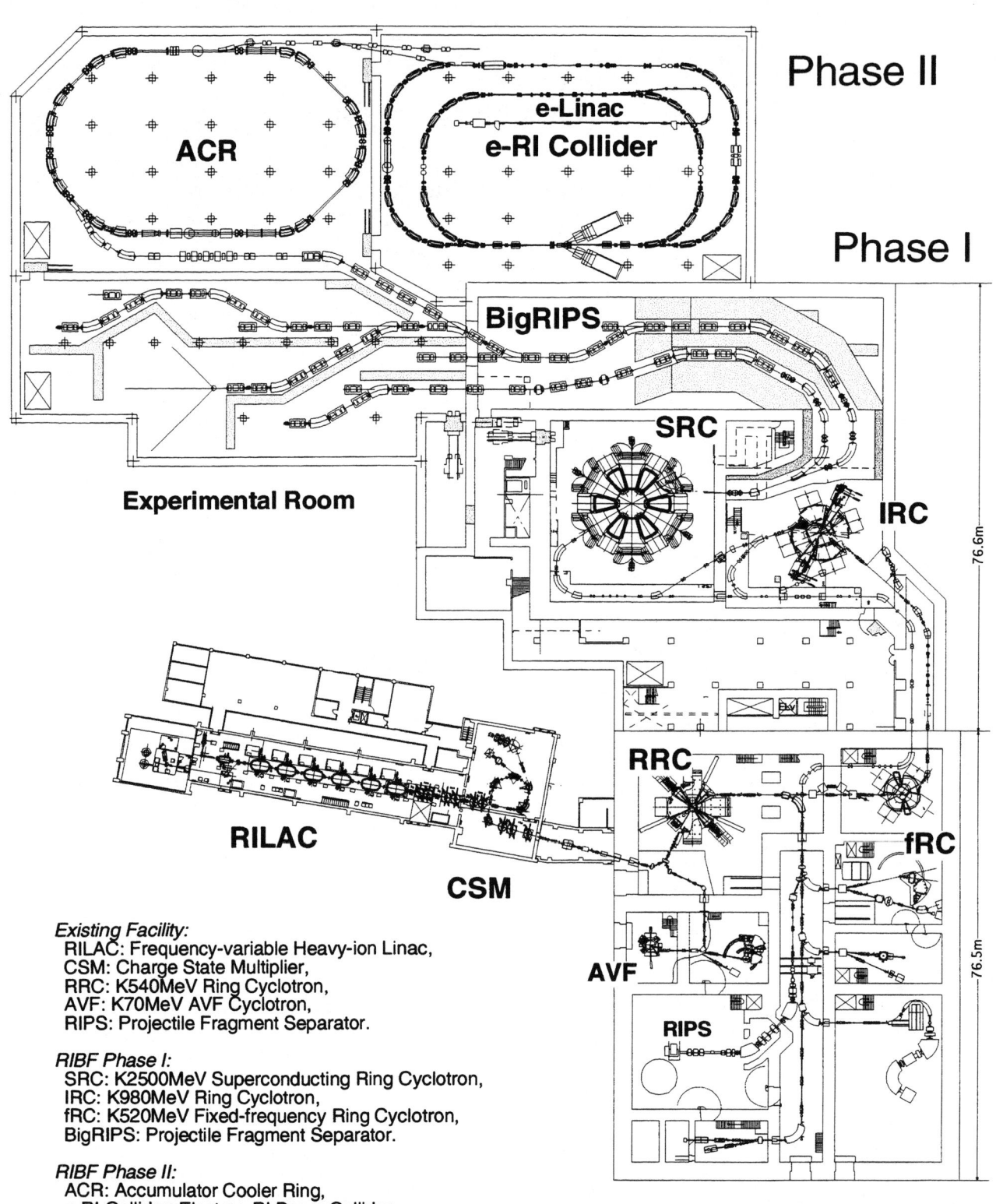

FIGURE 1. The layout of the RIKEN RI Beam Factory (RIBF). The RIBF is the extention of the existing heavy-ion accelerator facility. The first phase of the RIBF will be completed in 2003. The construction of the second-phase building will be started in 2002 and will be finished in 2005. The first electron-scattering experiment on unstable nuclei is scheduled for 2009.

successfully covered heavy-ion beams in the energy-mass region required, and the beam transmission efficiency of about 90 % at the maximum was obtained. In addition, high-intensity highly-charged ion beams have been produced by the ECRIS-18.

fRC-IRC-SRC

The velocity gains in the fRC, the IRC and the SRC are 2.0, 1.5 and 1.5, respectively. When bypassing the fRC, the combination of the IRC and the SRC accelerates light ions up to around Ar up to 400 MeV/nucleon, Kr^{30+} up to 300 MeV/nucleon, U^{58+} up to 150 MeV/nucleon and U^{49+} up to 100 MeV/nucleon. The beam intensity is expected to be 1 pµA for light ions, but to be less than 1 pµA for very heavy ions. With the fRC, which is operated at a fixed rf frequency, very heavy ions from Kr to U can be accelerated up to 350 MeV/nucleon with high intensities (1 pµA), because the low-frequency operation of the RILAC allows us to take full advantage of high-intensity, low-charge-state ions from the ion source. For example, U^{10+} ions are accelerated by the RILAC, and then U^{35+} through the RRC and U^{72+} through the fRC, and U^{88+} through the IRC and the SRC. Nearly 1% of U^{10+} ions from the ECRIS are accelerated to the final energy due to the charge stripping loss.

The IRC is a four-sector room-temperature ring cyclotron. The maximum sector field is 1.9 T and the mean extraction radius is 4.15 m. The rf frequency is ranged from 18 to 38 MHz.

The SRC is a six-sector superconducting ring cyclotron. Its design has recently been changed. In the new design, the cyclotron is almost completely covered with soft-iron slabs of about 1 m in thickness except in the central region in order to shield the stray magnetic field and the radiation. Owing to this design the maximum magnetomotive force has been significantly reduced to 3.7 MAT/sector. The maximum sector field is 3.8 T and the mean extraction radius is 5.36 m. The rf frequency is ranged from 18 to 38 MHz. The total weight amounts to about 9,000 tons.

The fRC is a four-sector room-temperature ring cyclotron. The mean extraction radius is 2.77 m. The rf frequency is fixed to be 36.7 MHz.

Two of the sector magnets of the IRC have been completed and the magnetic field is being mapped. The completion of the IRC is scheduled for the spring of 2001. The design of the SRC has been finalized. The completion of the SRC is scheduled for 2003. The fundamental design is being made for the fRC.

Big RIPS

The BigRIPS consists of a couple of achromatic spectrometers connected to each other in series. The first-stage spectrometer is used as a projectile fragment separator to produce RI beams, while the second one is used as a spectrometer to make particle identification of RI beams as well as to measure their momentum and angular dispersion. RI beams are tagged event by event with respect to momentum, angle and ion species. The reason why we adopt this tandem scheme is that purity of RI beams obtained by the first-stage is expected to be still poor due to the nature of energy loss in our energy domain. RI beams tagged are transported to various experimental set-ups placed downstream.

The BigRIPS is designed to have large acceptance, so that it can efficiently collect projectile fragments produced by in-flight fission of uranium beams. This in-flight fission has high capability of producing very neutron-rich medium-heavy RI beams, because it has large production cross section for those isotopes. The acceptance of BigRIPS has been taken to be 80 ~ 100 mrad for angle and 6 % for momentum. Such large acceptance allows to collect almost 50 % of fragments produced by the in-flight fission of uranium beams at 350 MeV/nucleon, when the symmetric fission fragments are selected. For producing RI beam species in other region of nuclear chart, such as proton-rich isotopes and lighter neutron-rich isotopes, projectile fragmentation of suitable heavy ions is to be used. In this case, the BigRIPS can collect almost 100 % of fragments produced.

The first-stage of BigRIPS is a mirror-symmetric and achromatic system with two dipole magnets. A wedge-shaped energy degrader is placed at its intermediate focus to make isotopic separation. The second-stage is a mirror-symmetric achromatic system with four dipole magnets. The use of four dipole magnets allows to obtain higher momentum resolution which is required for the tagging of RI beams. The maximum bending power of BigRIPS is about 8 Tm, when its optics is tuned for the high acceptance mode to collect fission fragments. On the other hand, when the optics is tuned for lower acceptance, which is nearly a half of the high acceptance, the bending power limit increases to 9.5 Tm, which corresponds to that of isotopes with A/Z=3 and E=400 MeV/nucleon. This mode is to be used for collecting relatively light neutron-rich RI beams produced by projectile fragmentation. The quadrupole magnets of BigRIPS are superconducting, so that the high acceptance and the large bending power such as 20 T/m with a warm bore of 24 cm in diameter can be achieved. Its prototype has been built recently and the design has been successfully confirmed by the test.

Huge radiation shielding made up of normal concrete slabs of 7000 tons in total weight surrounds the production target area. A space is reserved for an extra BigRIPS line.

ACR

The RI beams are transported to the injection point of the ACR along the length of 70 m. At the end point of transport line, a debuncher system will be installed to reduce the momentum spread of this beam.

The ACR of 134.8 m in circumference has asymmetric lattice structure to efficiently cool the accumulated RI beams. The maximum magnetic rigidity is 8.0 Tm, the horizontal/vertical acceptances are 125/40 mm*mrad and the momentum acceptance is $\pm 1\%$. The isochronous operation mode is also prepared.

RI-beam micro-pulses coming from the BigRIPS are injected into the ACR by means of the multi-turn injection. Then, the rf-stacking is immediately performed. After repeating the injection and the rf-stacking by 15~20 times, the beam is cooled by means of the stochastic cooling. One cycle of injection, rf-stacking and cooling is expected to take typically 200 msec. This cycle is repeated in the intrinsic lifetime of the RI or until the space charge limit is reached. Following this process the accumulated beam is finally cooled to less than 0.1% in momentum spread and 1π mm*mrad in transeverse emittance by combining the stochastic and the electron cooling techniques.

The number of RI ions stored in the ACR is determined by the balance of the supply rate and decay rate due to the intrinsic lifetime. This is limited by space charge effect rather than the lifetime for RI ions neighboring on the stability line, which has high production rate and long lifetime.

The accumulated RI beams in the ACR will be fast extracted and one-turn injected into the e-RI collider.

e-RI Collider

The e-RI collider for the electron scattering experiments on unstable nuclei consists of an electron and an ion storage rings that intercept at a colliding point with each other.

The electron-beam energy is 500~700 MeV which is obtained by an electron linac. This energy is determined by experimental requirement of resolution for electron scattering angle. To get high luminosity within small beam-beam effect the stored current needed is 500 mA and the beam emittance required is 1π mm*mrad, which is almost the same as that of the RI beam from the ACR.

The electron ring has a FODO structure in arcs to realise the required emittance. A 500-MHz rf cavity is installed to compensate radiation loss. Circumference of the electron ring is 120.00 m and the harmonics is 200.

The ion ring has maximum rigidity of 8 Tm, which is the same as that of the ACR. The lattice of the arc section has mirror symmetry to make dispersion suppression easy. An electron cooler is located in a short straight section to compensate emittance growth by beam-beam effect. An RF cavity is also located to keep bunch length. The circumference of the ion ring is 179.32 m and the harmonics is 32.

In the colliding section, where two rings intercept, a lattice of each ring is designed so as to get large luminosity with no mechanical interference between the magnets of the two rings. The β values determined are 0.02 m for the electron beam in both of horizontal and vertical directions and 0.1 m for the ion beam. The crossing angle is 20 mrad (1.145°), which reduces the luminosity to be half of that for head-on collision. A colliding space is 3.0 m and quadrupole doublets are located at both ends to focus both electron and ion beams. Additional septum-like quadrupole doublets are used for the ion beam because focusing power of the common doublets is not enough for it. Natural chromaticity in this section, which becomes large due to strong strength of the quadrupoles and large value of β, will be corrected locally. Devices for an experimental setup, a window to an electron detector, a shielding against synchrotron light and so on are being designed.

The goal of the luminosity is higher than $10^{27}/cm^2/s$ enough to clearly measure charge distributions in unstable nuclei.

SCHEDULE

The construction of the first-phase building, and the fabrication of the IRC, the SRC and the BigRIPS will be finished early in 2003. In 2003, the overall installation and final tuning of these machines will be done. We will submit the budget for the fRC in FY2002, and we expect to install this cyclotron during the above final tuning. The first RI beam from the BigRIPS is scheduled for the spring of 2004.

The construction of the second-phase building will be started in 2002 and will be finished in 2005. After its completion, major experimental installations, e.g. a large acceptance spectrometer and a gamma-ray detector, etc., will be installed. The ACR will be commissioned in 2006. The first electron scattering experiment on unstable nuclei is scheduled for 2009.

A Portable Neutron/Tunable X-Ray Source Based on Inertial Electrostatic Confinement

George H. Miley

University of Illinois, Fusion Studies Laboratory
102 NEL, 103 S. Goodwin Avenue
Urbana, IL 61801, U.S.A.
Ph: 217-333-3772, Fax: 217-333-2906, email: g-miley@uiuc.edu

Abstract. Inertial Electrostatic Confinement (IEC) offers a unique ion-beam–plasma-target configuration for production of neutrons via D-D or D-T fusion reactions. Research at the U. of IL has developed a unique "STAR" mode of operation where a basketball-shaped grid in the spherical (r ~ 15 cm) vacuum vessel creates intense ion beams focused at the center of the vessel, forming a dense fusing plasma core (target). Key advantages of this unique design are that grid sputtering is greatly reduced and good beam focusing is achieved. Commercial versions of this concept have been developed that offer 10^7 2.45-MeV D-D neutrons/sec (or 10^9/sec D-T). Such units are typically used to replace Cf-252 sources for industrial NAA. Next generation devices with rates above 10^9/sec D-D are currently under development. The IEC also provides a small tunable x-ray source (5-100 keV) for research applications by reversing the grid potential and also installing electron emitters. The changeover requires several hours down time, or, if needed, a separate dedicated IEC x-ray unit could be constructed.

INTRODUCTION

The U. of IL IEC uses a transparent cathode grid inside a vacuum vessel (anode) in a spherical configuration or a hollow cathode-anode

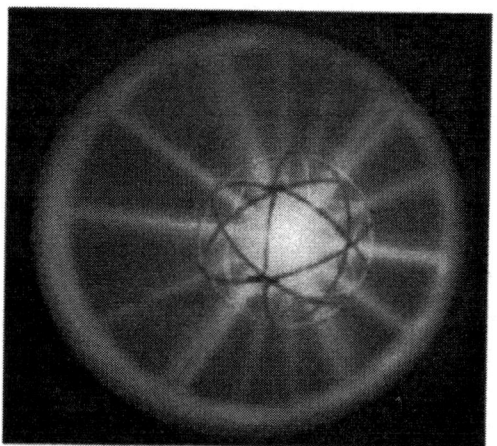

FIGURE 1 Photograph of a spherical unit in operation (above). The grid and dense core region, just visible in the port opening, are enlarged in the top right photograph. In the photograph of an operating cylindrical IEC (below, right), the dark ring structures are the electrodes

configuration in cylindrical geometry [1,2]. These two geometries are illustrated in the photographs in Fig. 1. Their construction is illustrated in Fig. 2. The current spherical unit employs a single grid for simplicity, although multigrid versions offer somewhat better efficiency. Likewise, the cylindrical unit is based on the simplest electrode design and arrangement for ease of construction.

Typical operating parameters are given in Table 1. A wide range of geometric dimensions are possible. However, once the electrode spacing is selected, the voltage and pressure are fixed by the Paschen breakdown relation [3].

TABLE 1. Typical Operational Parameters

Parameter	Value
Pressure	0.1-5.0 m Torr
Voltage	50-100 kV
Device Diameter	50-100 cm
Input Power	0.1-10 kW

The spherical unit offers a "point-type" source with an off-the-shelf steady-state rate of 10^7 2.5-MeV D-D neutrons/sec (n/s) and possibly up to 10^8 D-D n/s in 50-Hz pulsed versions. (With a D-T gas fill, these rates would increase by the cross-section ratio of about 10^2). The cylindrical version offers similar neutron rates, but provides a "line-like" source extending up to ~25 cm length. The latter configuration is most advantageous for applications requiring uniform coverage of broad surface areas, e.g. analysis of materials carried on a conveyor belt. Versions of the spherical unit are now produced commercially by Daimler-Benz Aerospace Corporation (DASA) [4].

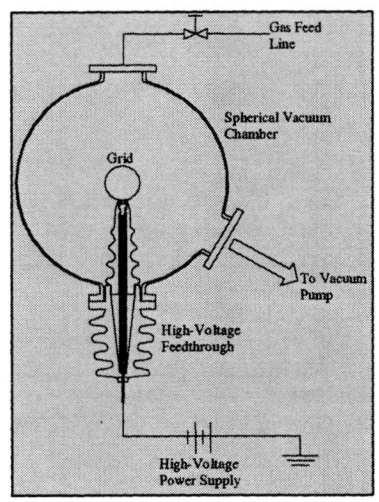

(a)

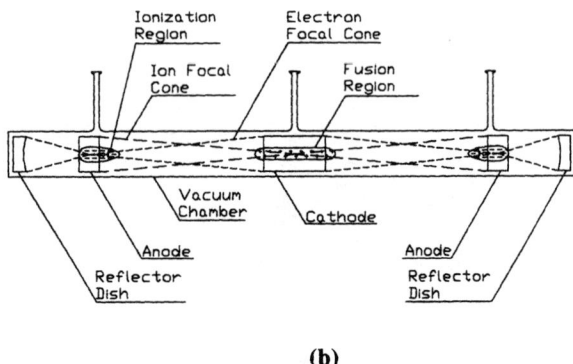

(b)

FIGURE 2. Schematic of the cross section of (a) a spherical IEC unit showing the grid and high voltage insulator structure and (b) a sketch of the cylindrical device and electrode structure.

TUNABLE X-RAY SOURCE

Another important advantage of the IEC, especially the smaller laboratory units, is that it can be converted to an attractive tunable x-ray source with minimum alteration of the apparatus [5]. X-ray operation essentially involves reversing electrode polarities and adding electron emitters.

The resulting electron Bremmstrahlung radiation has a broad energy spectrum extending up to the applied voltage (~10 – 60 kV). A typical measurement corresponding to an applied voltage of 30 kV is shown in Fig. 3. A FWHM of about 15 kV is obtained with a peak intensity at about 20 kV. This makes possible some small-scale laboratory x-ray experiments that would otherwise necessitate traveling to a synchrotron-type "light" source.

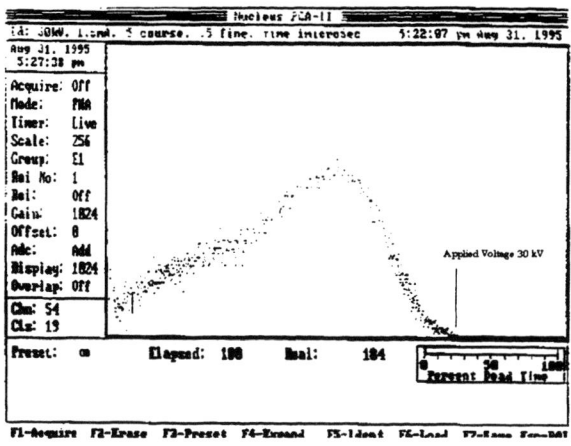

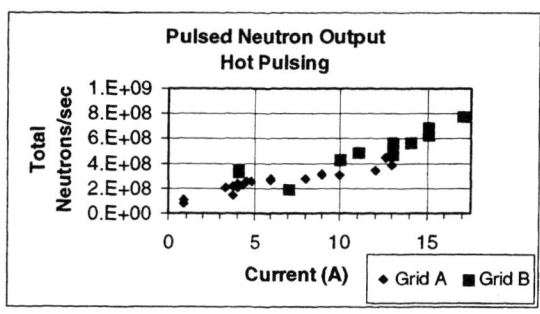

FIGURE 4. Pulsed neutron output at peak of pulse vs. peak cathode current.

Figure 3. Measured X-ray intensity energy spectrum. The y-axis shows the normalized intensity. The peak energy occurs at ~ 20kV (x-axis).

RECENT PULSED OPERATION STUDIES

Pulsed operation is attractive for certain NAA applications and also provides an attractive route to higher neutron yields since neutron production scales with ion current which can be very high using pulsed power technology [8]. The specialized pulser designed for such operation resembles a transmission line-type pulser. A hydrogen-thyratron switch is used to initiate the pulse and discharge the energy storage capacitors. Typical pulse lengths with this set-up are ~ 10 µs, consistent with requirements set by the ion recirculation time in the potential well.

Neutron production as a function of pulsed current for several different cathode-grid designs is presented in Figure 4. Data was collected for pulsed operation for cathode voltages of ~50 kV, and currents up to 17 A.

The data in Figure 4 was taken with two different grids. The first, a small reference grid was used (Grid A); a second, larger grid was also used (Grid B). Grid B produced a slightly higher fusion reaction rate. This can be explained since the larger grid allows a lower neutral background pressure cf. Paschen theory of breakdown [3], reducing charge exchange losses. In conclusion, these preliminary results suggest pulsed operation of the IEC can be very attractive and provides an important compliment to conventional steady-state operation.

SCALE-UP ISSUES

An understanding of the underlying plasma-electrodynamic physics of the IEC is needed to evaluate its potential for scale up to even higher neutron rates. The spherical version will be used here to explain the basic phenomena. In it, the cathode grid extracts and accelerates ions created in a plasma discharge created between it and the anode (vacuum vessel wall). As these ions converge in a small volume around the center of the sphere, a virtual anode is created which accelerates and focuses electrons into a yet smaller concentric volume.

These converging electrons develop a negative

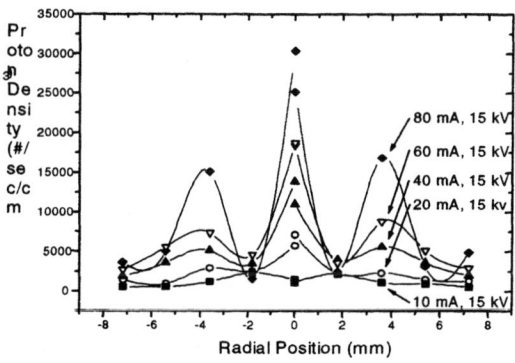

FIGURE 5. Measurement of the potential-well structure using a collimated D-D proton detector.

potential well within the positive "hill," which traps ions, greatly reducing energetic ion leakage and creating an intense region of beam-beam fusion reactions. Y. Gu recently achieved a measurement of this potential structure using a collimated proton detector to observe the source rate profile of D-D protons across the central core region [9]. Some of his results are shown in Figure 5. A triple peak profile, indicative of a potential "double well", occurs at higher currents (≥ 40 ms). The ability to scale-up to yet higher neutron rates requires generating a larger and deeper ion trap, which in turn involves controlling ion energies and angular momentum while increasing ion currents. Since such scale-up involves a velocity-space (vs. physical space) phenomena, higher neutron rates do not force larger unit sizes. However, a modest size increase of the total system dimensions is required to accommodate voltage holding, increased cooling and radiation shielding requirements.

CONCLUSIONS

The IEC devices described here represent a new class of neutron sources based on ion-beam–plasma-target fusion. They create a dense fusing region by focusing or concentrating the beam ions. In addition, energy efficiency is achieved by generation of a potential structure such that ions are trapped and recirculated. In contrast to conventional magnetic confinement fusion experimental devices, the IEC is unique in using electrostatic fields alone without the complication of magnetic field coils. The IEC uses ion inertia to obtain confinement without violating limits imposed by Earnshaw's theorem [10]. Key benefits of this approach are reflected in present devices through simplicity, compactness and long lifetime. Since the devices described here represent "first generation" designs, further improvements can be expected as research progresses.

ACKNOWLEDGMENTS

The IEC group at the U. of IL, Robert Stubbers, Brian Jurczyk, Mike Williams, and John DeMora, have been instrumental in this work. Helpful discussions with R. Nebel and D. Barnes (Los Alamos National Laboratory), R.W. Bussard (MC^2), M. Ohnishi (Kansai University), and R. Hirsch (Rand Corp), are gratefully acknowledged. This work was partially supported by DASA Contract No. 505013-D MIL.

REFERENCES

[1] Miley, G. H. "A Portable Neutron/Tunable X-Ray Source Based on Inertial Electrostatic Confinement," *Proceedings, Ninth Symposium on Radiation Measurements and Application*, edited by H.C. Griffin, W.L. Rogers, and K. Rengan, North Holland, 1999.

[2] Miley, G. H. "A Novel 2.5 MeV D-D Neutron Source," *Journal of Brachytherapy International*, Vol. 1, No. 1, 111-121, (1997).

[3] Miley, G. H. et al., "Discharge Characteristics of the Spherical Inertial Electrostatic Confinement (IEC) Device," *IEEE Trans. on Plasma Science*, Vol 25, No. 4, pp. 733-739, August 1997.

[4] Miley, G. H. and Sved, J. "The IEC Star-Mode Fusion Neutron Source for NAA—Status and Next-Step Designs", *Proceedings*, IRRMA '99, Raleigh, NC, 1999.

[5] Gu, Y. and Miley, G. H. "Spherical IEC Device as a Tunable X-ray Source," *Bult. APS*, 11, 1851 (1995).

[6] Gu, Y., Heck, P., and Miley, G. H. "Ion Focus Via Microchannels in Spherical Inertial Electrostatic Confinement and Its Pulsed Experimental Results," *1995 IEEE International Conference on Plasma Science*, IEEE Conf. Rec. 95CH35796, 266-267 (1995).

[7] Jurczyk, B., Gu, Y., and Miley, G. H. "Resonant Ion Driven Oscillation (RIDO) Concept," *Proceedings 39th Annual Meeting Division of Plasma Physics*, Pittsburgh, PA, Vol. 42, No. 10, p. 1818, November 17-21, 1997.

[8] Miley, G. H., Nadler, J. et al. "Issues for Development of Inertial Electrostatic Confinement (IEC) for Future Fusion Propulsion", *35th AIAA/ASME/SAE/ASEE Joint Propulsion Conference*, Los Angeles, CA, 1999, AIAA-99-2140.

[9] Gu, Y. and Miley, G. H. "Experimental Study of Potential Structure in a Spherical IEC Fusion Device," *IEEE Transaction of Plasma Science*, Vol 28, No. 1, 331-346, February 2000.

[10] Glasstone, S. & Lovberg, R., Controlled Thermonuclear Reactions: An Introduction to Theory and Experiment, D. Van Nostrand Co., NY, 1960, p.47.

A High Intensity Radiation Effects Facility

V.H.Rotberg, O.Toader and G.S.Was

Michigan Ion Beam Laboratory
Department of Nuclear Engineering and Radiological Sciences
The University of Michigan, Ann Arbor, MI 48109, USA

Abstract The facility of the Michigan Ion Beam Laboratory at the University of Michigan has been upgraded to conduct high intensity radiation effects studies on materials. This upgrade is necessary to pursue higher radiation damage levels than the studies previously conducted. To achieve this capability a new volume ion source was installed which can produce several times more H- current than the previous duoplasmatron. We will describe the objectives of the research and the facility as well as applications to a variety of radiation damage problems.

Introduction

The need for basic information regarding the microstructural effects of neutrons in components of aging light water reactors motivated the creation of the facility described in [1] (fig.1).

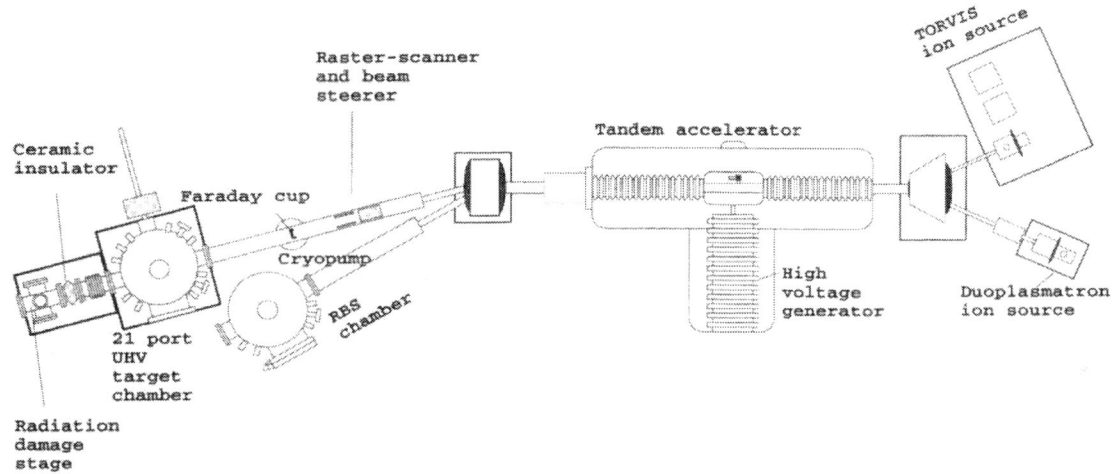

Fig.1 A schematic of the tandem accelerator and associated beam lines. The proton irradiation stage is located at the end of the leftmost beam line

Although it would seem preferable to study samples exposed to neutrons in-core, factors such as time, accessibility and expense make the search for alternatives attractive. If ion irradiation can generate microstructures comparable to those of interest, the radiation damage caused by neutrons can be studied in a more controlled and safer environment. The success of this approach has been demonstrated in a variety of studies [4-9] in which samples of stainless steel were irradiated for long periods of time with high intensity beams of protons at controlled temperatures. Doses of about 1 dpa (displacement per atom) were achieved in periods of about 40 hours with beam densities of about 10 $\mu A/cm^2$. Since the continuing interest in this work would require higher doses, it became apparent that the existing setup would be insufficient to carry out the irradiations in reasonable time periods due to frequent interruptions for equipment maintenance and personnel fatigue. A search was therefore initiated for an ion source that could provide larger intensities in order to replace the existing duoplasmatron source.

Ion sources

Due to transmission losses through the accelerator and the need to overscan the samples to assure beam uniformity it is necessary to have up to 60 μA of H- extracted from the ion source in order to achieve a 40 μA

of high-energy beam. This current is above the specifications for comfortable operation of the duoplasmatron ion source although intensities of up to 90 μA H⁻ were sometimes achieved. Long periods of adjustments were required together with frequent alignments and cleaning of internal components. Due to unrepeatable behavior of the source as it is pushed to operate above its specifications, it was sometimes necessary to allow for periods of up to 1 week to achieve a good intensity beam before an irradiation could begin. During this time the source had to be dismantled several times for alignments and filament coating also would need to be redone until a satisfactory source could be built. Once optimized it could maintain the beam through the current irradiation but it's other applications, such as providing He⁺ for surface analysis, would require further maintenance and dismantling.

After many consultations, we concluded that a practical replacement would be the TORVIS (TORoidal Volume Ion Source). Built commercially by National Electrostatics Corporation (Middleton, WI) it is a DC version of the source that was initially developed at Brookhaven National Laboratories by Pelec and Alessi [3]. The axial region of the source is separated from the outer region by a conical magnetic dipole field. This field prevents fast electrons from destroying the H⁻ ions that are formed in the axial region by the dissociative attachment of excited H_2 with slow electrons (fig.2).

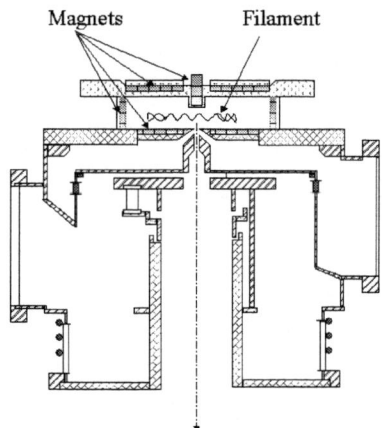

Fig.2 The TORVIS source. The plasma chamber containing the filament and surrounded by magnets is on top, followed by the extractor and lens assembly. The two side ports connect to high-speed turbomolecular pumps.

To control and monitor the parameters of the source such as gas pressure, filament current, arc current, etc. a software control program was developed. It is based on National Instruments Corp. (NI), Labview and controls the source parameters via serial ports. The interface between the computer and the source is an electronic device called an I/Oplexer. Three of these are used, which contain different analog and digital input/output modules. This setting allows the computer to be electrically isolated from the power supplies, which are on a high voltage platform. The Graphical User Interface is easy to understand and the source parameters can be logged into a file for future reference. The TORVIS source has proven to be extremely reliable requiring very low maintenance. We can easily obtain about 300 μA at the entrance of the accelerator, and in excess of 150 μA at the high-energy side. The only limitation on using very high currents is the power supply for the high voltage generator of the accelerator, which limits the total load to about 1 mA.

Description and capabilities

The irradiations are conducted at the Michigan Ion Beam Laboratory at the University of Michigan. This facility houses a 1.7 MV tandem accelerator built by General Ionex Corp. in which the high voltage is generated by a rectifier stack. Due to continuous use at high voltages, the oscillator tubes of the original push-pull circuit, which provides the high voltage radio frequency, needed frequent replacement at considerable expense and difficulty with unacceptable disruptions to the research programs. Due to this, the circuit was replaced by a solid-state power supply built by Accelerator Systems Inc., Atkinson NH. This has ensured continuous and reliable operation for long periods.

Presently there are 3 ion sources, a duoplasmatron which can provide about 1μA of He⁻, 50 μA H⁻, a sputter source with a capability of producing negative ions of heavier elements and the newly installed TORVIS with a capability of producing up to 400 μA of H⁻. The latter has replaced the duoplasmatron in its function of generating H⁻ beam. Depending on the charge state of the ion emerging from the gas stripper situated at the high voltage terminal, beams up to 5 MeV can be obtained. For protons, the maximum energy is 3.4 MeV. Irradiations have been conducted with beams of energy as low as 500 keV with adequate transmission through the accelerator tube. A raster-scanning system allows targets up to 5 cm. in diameter with the present beam line configuration. The beam quality is monitored and adjusted via a profile monitor manufactured by National Electrostatics Corp. Samples are mounted on a stage inside an electrically isolated chamber and maintained under high vacuum in the range low $10^{-7} - 10^{-9}$ torr. Temperature control is achieved by mounting the samples on a copper block with a liquid metal (indium or tin) coupling to facilitate heat conduction between the samples and the stage (fig.3).

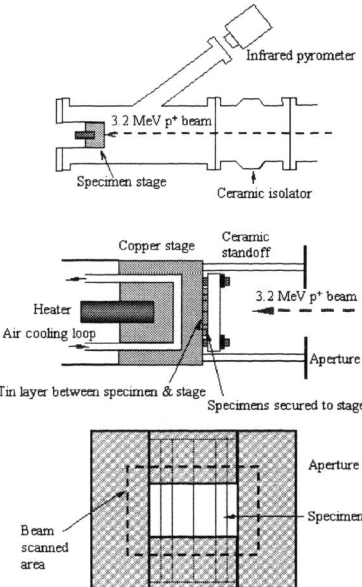

Fig. 3. Schematic of the irradiation stage for high intensity radiation studies. The middle drawing shows details of the specimen stage with the cooling loop and heater. At the bottom a detail of the aperture system for uniformity control.

The nominal surface temperature can be controlled to within +/- 5 ^{0}C of the desired goal temperature, which can range between 50^0 C and 500^0C. Higher sample temperatures (> 600^0 C) can be achieved by using a nickel stage identical in design to the copper stage and the same mounting procedures. Simultaneously heating the stage from the rear with an electric heater cartridge inserted into a cavity in the back of the stage and cooling the stage with air or water maintains the temperature during irradiation. With a power density of about 40 W/cm^2, a temperature difference of 20 - 100^0 C between the front of the samples and the back of the stage is typical during irradiation at any temperature. The sample temperature is monitored using an infrared pyrometer that can be remotely controlled to scan the irradiated region. Thermocouples, which are connected to each individual sample, are continuously monitored from a data acquisition program written in Labview. These are also used to calibrate the pyrometer via emissivity adjustment prior to applying the proton beam. The incident beam is focused down to a spot approximately 3 mm in diameter and then raster-scanned across the samples. About half of the total beam is scanned onto a 4-aperture system that completely surrounds the samples. The apertures, which are not cooled, including mounting screws, are completely constructed of tantalum, in order to withstand the temperature increase. They are directly supported on the stage by ceramic standoffs. Electrical feedthroughs carry the currents to the computer where balancing the current on each of these by adjustment of horizontal and vertical steering ensures uniform irradiation.

The computer monitors the irradiation process by reading the current on the samples and on the four apertures surrounding them. The thermocouples are connected to a specialized card from National Instruments Corp. (NI) for temperature monitoring, and then sent to the computer via a data acquisition card. This card also monitors the current on the apertures, stage and the signal coming from the pyrometer. A digital counter input is attached to the digitized output of a current integrator connected to the stage. An analog output is used to send a signal to an audible alarm that can be triggered when certain conditions are not met. The main data acquisition screen displays information about all the parameters of interest, which is simultaneously saved to a file for future reference. Both the computer that controls the source and the one that controls the data acquisition process are connected via a local computer network. The next goal would be to remotely access, control and view the parameters of both computers. Labview is a versatile language that allows this to be accomplished and we are in the process of implementing this capability. In view of the high current capabilities of the new ion source we expect that what used to be a 40 h irradiation to have about 1 dpa would take about 12h.

Results

The validity of the approach taken to simulate neutron damage by high-energy protons is demonstrated in comparison irradiations of 316 stainless steel with neutrons and protons. Figures 4 and 5 show grain boundary composition profiles, and the increase in yield strength due to neutron and proton irradiation of the same heats of material. Neutron irradiations were conducted at 274°C and dose rates around 5×10^{-8} dpa/s in the Barseback reactor in Sweden, and proton irradiations were conducted in MIBL at 360°C and at a dose rate of 7×10^{-6} dpa/s. As shown in Fig. 4, the composition profiles of Ni, Cr and Si for the respective irradiations are nearly identical in magnitude and spatial extent and capture the complicated "W" shaped chromium profile at the grain boundary. The same is true in Fig. 5 where the hardening of the alloy as a function of dose falls on nearly the same curve. Similar agreement occurs for microstructure and IASCC susceptibility. Figures 6 –8 provide examples of the range of irradiation capabilities that are accessible in the MIBL damage facility. Figure 6 shows the variation in grain boundary chromium content in one iron-base and two nickel-base austenitic alloys following proton irradiation to 0.5 dpa. Note that the irradiation temperature spans from 200°C to 600°C. Some experiments have also been conducted as high as 700°C. Figure 7 shows an example of a low temperature irradiation of an austenitic 304 stainless steel. All irradiations were done at 50°C and samples were then annealed at temperatures up to 500°C to remove the radiation damage, resulting in softening of the alloy. Irradiations with the new source have been conducted up to doses of 10 dpa and experiments are being planned with doses that exceed this value. However, the versatility of the source allows for extremely low dose irradiations as shown in Figure 8 for model reactor pressure vessel alloy (Fe-0.9Cu-1.0Mn) irradiated to doses as low as 0.001 dpa (10,000x lower than for austenitic alloys). Results show that the hardening obtained with proton irradiation is in excellent agreement with that obtained with neutron or electron irradiation. These results serve to show the wide range of applicability of the radiation damage facility (in temperature, dose, dose rate and target alloy) and the success in using proton irradiation to study neutron irradiation effects.

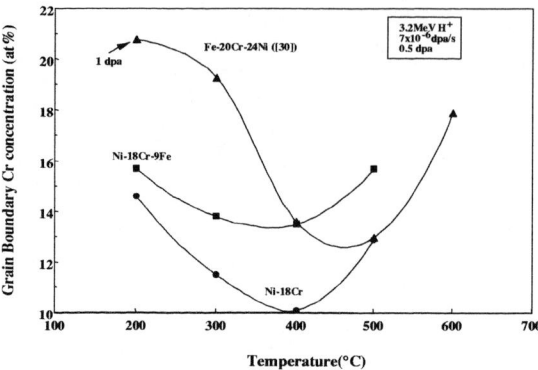

Figure 6: Measured grain boundary Cr concentration as a function of irradiation temperature for Ni-18Cr, Ni-18Cr-9Fe, and Fe-20Cr-24Ni irradiated with 3.2 MeV protons to 0.5 dpa. Ref. 8

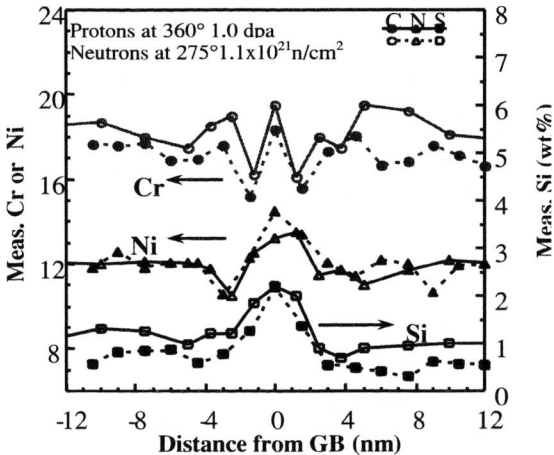

Figure 4: Comparison of Cr, Ni, and Si segregation profiles for proton and neutron irradiation of CP 316 SS to 1.0 dpa Ref 7.

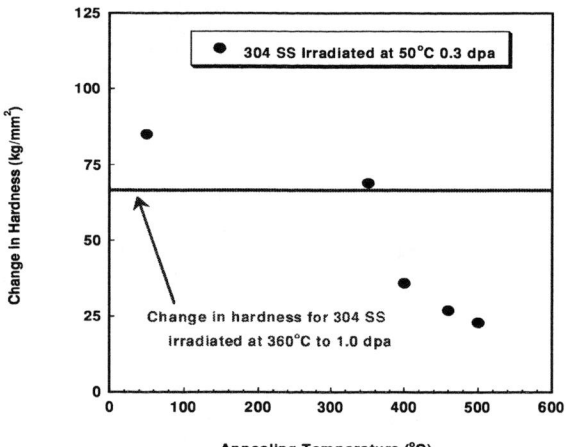

Figure 7: Results of microhardness for post-irradiation annealing of 304 SS irradiated at 50° C to 0.3 dpa. Annealing times were 0.5 hours except at 350° C where the annealing time was 3.5 hours. Ref. 9.

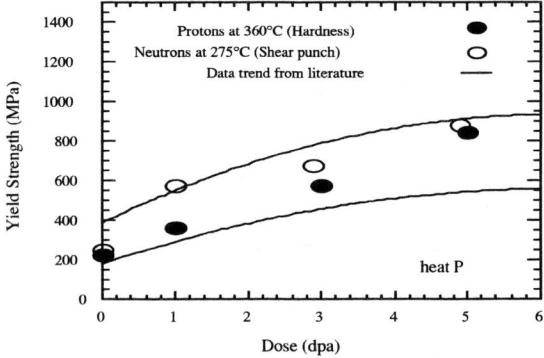

Figure 5: Change in yield stress under proton and neutron irradiation for the same heat of 316SS

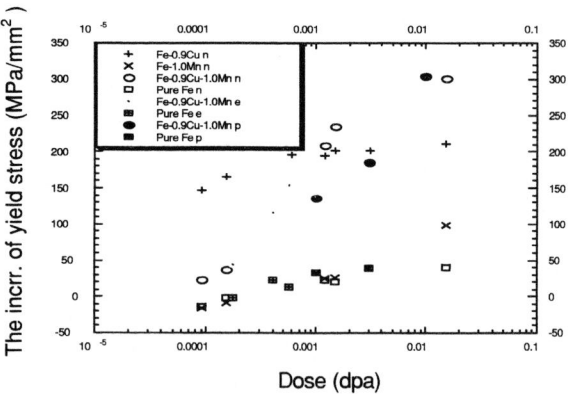

Figure 8: Comparison of yield strength increment on some model alloys versus dose for neutron, proton and electron studies (n, p, e respectively). Ref. 5 for n,e, Ref. 6 for p.

Summary

The upgraded facility of the Michigan Ion Beam Laboratory for Surface Modification and Analysis at the University of Michigan offers the capability of radiation damage studies using protons at high or low dose rates under practical time scales with precise temperature control in a computer controlled and monitored experiment. The commercial availability of the TORVIS ion source has been instrumental in this achievement.

References

1. D.L.Damcott, J.M.Cookson, V.H.Rotberg, G.S.Was, NIM B99 (1995) 780.
2. G.S.Was,T.R.Allen, J.T.Busby, J.Gan,, D.Damcott,D.Carter,M.Atzmon,E.A.Kenik, Journal of Nuclear Materials 270 (1999) 96.
3. J.G. Alessi and K. Prelec. IEEE Particle Accelerator Conference (1991, San Francisco, CA, USA).
4. J.M.Cookson,G.S.Was, P.L.Andresen, Corrosion-Vol.54,No.4.
5. D.E.Alexander et al. Int. Conf. On Environmental Degradation of Materials in Nuclear Power Systems (Newport Beach, CA,1999).
6. Qingkay Yu -private communication.
7. A. Jenssen, L. Ljungberg, J. Walmsley, and S. Fisher, Corrosion, 54, No.1 (1998) 48.
8. T.R.Allen et al. Journal of Nuclear Materials 244 (1997) 278, Damcott et al. J.Nucl Mat. 225 (1995) 97.
9. B. R. Grambau –private communication
10. Odette and Lucas, J. Nucl. Mater., 179-181 (1991) 572-576.
11. Bergenlid, U., Haag, Y. and Pettersson, K, The Studsvik MAT 1 Experiment. R2 Irradiations and Post-Irradiation Tensile Test. Studsvik Report STUDSVIK/NS-90/13, Studsvik Nuclear, 1990.
12. M. Kodama, Proceedings of the Eight International Symposium on Environment Degradation of Materials in Nuclear Power System-Water Reactors (NACE International, Amelia Island, FL, 1997) 831.
13. Jenssen, A., written communication, 1998.
14. G.R. Odette and G.E. Lucas, Fusion Reactor Materials-Semiannual Progress Report for Period Ending March 31, 1989, US Department of Energy, DOE/ER-0313/6 (1989) 313.

Applications for the RFD Linac Structure

Donald A. Swenson

Linac Systems, 1208 Marigold Drive NE, Albuquerque, NM 87122

Abstract. With the successful completion and operation of the "Proof-of-Principle" prototype of the Rf Focused Drift tube (RFD) linac structure, our attention has now turned to the identification of the first applications for this new compact and economical linac structure. The principal medical applications are for the production of short-lived radioisotopes for the positron emission tomography (PET and SPECT) application, epithermal neutron beams for the boron neutron capture therapy (BNCT) application, and nanoamperes of energetic (250 MeV) protons for proton therapy. The structure can be configured as a compact injector linac for proton synchrotrons. The structure can be configured as a pulsed cold neutron source to support cold neutron physics and its applications. The principal industrial applications include nondestructive testing (NDT), thermal neutron radiography (TNR), thermal neutron analysis (TNA), and pulsed fast neutron analysis (PFNA). Brief descriptions of these RFD-linac-based systems will be presented.

DESCRIPTION AND STATUS OF THE RFD LINAC STRUCTURE DEVELOPMENT

The revolutionary Rf Focused Drift tube (RFD) linac structure[1-3] is under development at Linac Systems. This structure resembles a drift tube linac (DTL) with radio frequency quadrupole (RFQ) focusing incorporated into each drift tube. The RFD drift tubes comprise two separate electrodes operating at different electrical potentials as excited by the TM_{010} drift tube linac cavity fields, each supporting two fingers pointing inwards towards the opposite end of the drift tube forming a four-finger geometry that produces an rf quadrupole field along the axis. Particles traveling along the axis traverse two distinct regions, namely, the gaps between the drift tubes where the acceleration takes place, and the regions inside the drift tubes where the rf quadrupole focusing takes place. This new structure could become the structure of choice to follow RFQ linacs in many scientific, medical, and industrial applications.

A 2.5-MeV prototype of the RFD linac structure has been constructed at Linac Systems to serve as the "proof of principle" (POP) for this new linac structure. The POP prototype[4,6,8] came into operation on June 19, 2000 in the Linac Systems laboratory in Waxahachie, TX[9]. This unit comprises a 25-keV proton ion source, an einzel-lens-based LEBT, a 0.65-m-long RFQ linac to 0.8 MeV, and a 0.35-m-long RFD linac to 2.5 MeV. Both the RFQ and RFD linac structures operate at 600 MHz. The two linac structures are resonantly coupled and powered by a collection of 12 planar triodes. The extreme simplicity of the interface between the two structures contributes to the practicality of this operational test on a limited budget. The overall length of the two linacs, including their interface, is only one meter. The entire assembly is evacuated by 2 turbomolecular pumps and 1 ion pump.

We did not have enough rf power to excite the structure to its design gradient[4]. To reduce the required power, we lowered the rf gradient by 24% and shortened the tank by 116 mm. Consequently, the energy and intensity of the beam from the POP was less than expected, but the performance[9] established the validity of the approach. Future RFD linac designs will employ a number of improvements that will rectify these problems. We are confident that, with these improvements and adequate rf power, the performance of the RFD linac structure will come up to our expectations.

FEATURES OF THE RFD LINAC STRUCTURE

RFD linac structures, which employ the same rf electric focusing as RFQ linacs, have the same small diameter beams that we find in RFQ linacs. These smaller diameter beams allow better coupling of the rf electric fields to the beam and higher frequency operation, resulting in smaller, lighter weight and more efficient linac structures, which in turn, mean less rf power to generate, less thermal load to cool, and less surface area to evacuate. The higher frequencies also offer the possibility of higher gradient operation (shorter structures). These compact linac structures will be more transportable than their predecessors, easier to enclose in radiation shielding, and less expensive to fabricate, maintain and operate.

ISOTOPE PRODUCTION FOR PET

A new source of PET isotopes[4] is being developed at Linac Systems, based on the RFD linac structure. This source has a beam energy of 12 MeV and an average beam current of 120 μA. It is packaged as two units, namely a linac unit and an rf power unit. The linac unit is similar in size and weight to a large Xerox machine, namely, 3.1-m long, 0.8-m wide, and 1.2-m high, with a weight of 1000 kg. The proton beam will be transported a short distance from the linac unit into a shielded, isotope-production target cell. Only minimal shielding is required around the linac unit and no shielding is required around the rf power unit. The electrical power and cooling requirements for the entire source are only 54 kVA. This system offers simultaneous irradiation of up to three targets. This compact and efficient unit promises to have a significant impact on the cost of acquiring, operating, and maintaining PET facilities. The parameters for the RFD portion of this structure are presented in the "PET" column of Table I.

EPITHERMAL NEUTRONS FOR BNCT

The job of producing enough epithermal neutrons for the BNCT application with an accelerator-based source is a formidable task – commonly understood to require 10 to 100 mA of 2.5 to 4 MeV protons on lithium or beryllium targets that can withstand the bombardment. The accelerator that had been identified as closest to meeting this challenge is the RFQ linac. However, the RFD linac structure, designed specifically to reduce the cost of the linacs in the few-MeV range, may win the role as the optimum structure for the BNCT application.

An epithermal neutron source, based on the RFD linac structure would have the following components: an ion source, a short low-energy transport system (LEBT), an RFQ linac section, an RFD linac section, an rf power system, a high-energy beam transport (HEBT) system, a beam target, and a neutron beam moderator/filter system.

The relatively modest beam current requirement of 10 mA at 25 keV can be obtained from a microwave ion source. A short LEBT, with Einzel lens focusing, will suffice to prepare the 10-mA proton beam for injection into the RFQ linac section. Because of the exceptional low-energy capabilities of the RFD structure, this RFQ linac section need only go to 0.75 MeV. We have chosen to operate the RFQ and RFD units at 460 MHz and at the very conservative excitation of only 1.6 Kilpatrick.

Three features of the RFD linac structure combine to suggest that it will have a much higher duty factor capability than the equivalent RFQ structure[5,7], namely its high shunt impedance, its large internal surface area, and its simple cylindrical shape. The power dissipation in the RFD structure of this system is only 90 kW. CW klystrodes, IOT amplifiers, and klystrons represent ideal rf power sources for this application. The parameters for the RFD portion of this structure are presented in the "BNCT" column of Table I.

THERMAL NEUTRON BEAMS

Intense neutron beams have applications for explosive detection in luggage, cargo, and land mines, corrosion detection in aircraft structures and bridges, void detection in munitions and fuel cells, non-destructive imaging of complex structures, and quantitative multi-element analyses of samples. Thermal Neutron Analysis (TNA) works by exposing materials to thermal neutrons and analyzing the prompt gamma rays resulting from the absorption of neutrons by the material. Computers then search for specific combinations of atomic elements that characterize explosives and drugs. Neutron Activation Analysis (NAA) works by irradiating objects with neutrons to produce radionuclides and subsequent detection of the delayed characteristic gamma rays emitted by decaying radionuclides. The gamma ray spectra yield the concentrations of various elements in the object. Hundreds of different types of materials can be detected. Thermal Neutron Radiography (TNR) is a powerful non-destructive imaging technique for the internal evaluation of complex structures. It involves attenuation of neutron beams by an object and recording the attenuation as images on film or video.

We promote a compact RFD-linac-based neutron source to support these techniques, allowing them to be applied at the location of the object, rather than transporting the object to the neutron source. This source has a flux of 3×10^{12} n/sec/4π. The parameters for the RFD portion of this structure are presented in the "TNA, NAA, & TNR" column of Table I.

HIGH INTENSITY PROTON BEAMS

In the past, we have promoted the RFD linac structure as a very compact and economical structure for acceleration of proton beams with relatively low peak beam currents (10 mA) in both the pulsed and CW formats. However, we are not surprised to find that the structure also has high intensity capabilities (100 mA) in both of these two temporal formats.

In fact, the RFD linac structure can handle any beam current that the better-known RFQ structure can handle. A linac structure that can accelerate 100-mA of protons from the output of a 1.5-MeV RFQ to an energy of 10 MeV is described here.

The advantage of the RFD approach over the RFQ/DTL approach is that it leads to a modest size and power requirement for this high intensity beam current. The length of the 10-MeV structure is only 3.75 m and the total rf power requirement for the structure is only 1.53 MW, which includes 0.85 MW of beam power. The beam power in this configuration is more than half of the total rf power requirement.

Because of the properties of the RFD structure, this unit would be capable of CW operation. Of the 0.68 MW of structure power, only 20% goes to the 44 drift tube assemblies (including stems), implying only 2.8 kW of power per drift tube assembly in the CW mode. Because the drift tubes have a relatively small effect on the resonant frequency of the structure, their temperature tolerance is several °C. A flow of 2.7 gpm of cooling water through the cooling channel in each half of the drift tube body will limit the temperature rise to 2 °C.

Another advantage of the RFD linac structure is that the entire structure is "radiation hard", unlike permanent-magnet focused linac structures.

The parameters for the RFD portion of this structure in the CW mode are presented in the "High Intensity" column of Table I.

ENERGY BOOSTER FOR PROTON THERAPY

One option for proton therapy is to boost nano-amperes of 10-12 MeV protons from some proton source to 250 MeV with compact S-band (3000 MHz) proton linac structures. We propose the RFD linac structure for the 12-70 MeV portion of this system. There are a number of suitable options for the 70-250 MeV portion of the system.

The 12-to-70 MeV RFD linac section is extremely compact. It would be only 4 meters long with an outer diameter of only 12 cm. It would be driven by a single 10-MW, S-Band klystron similar to those used for most medical electron linacs. A model of this structure has been fabricated. The parameters for this portion of the system are presented in the "Proton Therapy" column of Table I.

AVAILABILITY

All of the linac systems described in this report are based on our patented RFD linac structure, which has been under development at Linac Systems for eight years. The basic principles of the RFD linac structure are well established. The physics and mechanical designs for all of these systems are mature, and fabrication techniques for all components have been established.

These systems are available only through the mechanism of development contracts, where partial payment is made at the start of the contract, interim payments are made in the course of the contract, and the final payment is made upon acceptance of the system by the customer. Delivery would typically be two years after receipt of the order (ARO).

ACKNOWLEDGMENTS

The people who played a significant role in the development of the RFD linac structure and the POP prototype are: Frank Guy and Ken Crandall (accelerator physics), Joel Starling (mechanical engineering and commissioning), Jim Potter (rf power), John Lenz (thermal calculations), and Sylvia Revell (radiation safety).

TABLE 1. RFD Linac Portions of 5 Different Linac Systems

Model	PET Isotopes	BNCT Neutrons	TNA, NAA, & TNR	High Intensity	Proton Therapy
Energy (MeV)	12	2.5	4	10	70
Frequency (MHz)	600	460	600	350	3000
Pulse Structure	Pulsed	CW	Pulsed	CW	Pulsed
Beam Current, Peak (mA)	10	10	10	100	.01
Beam Current, Average (mA)	0.12	10	1	100	.001
Length (m)	2.28	1.5	1.3	3.75	4.0
Diameter, Inner (m)	0.38	0.49	0.38	0.66	0.08
Bore Hole Diameter (mm)	3.2	3.2	3.2	7.0	2.0
Number of Drift Tubes	45	41	40	44	153
Axial Electric Field (MV/m)	7.2	2.0	4.0	3.8	20.0
RF Power, Structure (MW)	1.250	0.090	0.280	0.690	10.0
RF Power, Beam (MW)	0.120	0.025	0.040	0.850	0.0
RF Power, Total (MW)	1.370	0.115	0.320	1.540	10.0
Facility Power and Cooling (kVA)	54	340	84	3,620	30

REFERENCES

1. D.A. Swenson, "RF-Focused Drift-Tube Linac Structure", LINAC'94, Tsukuba, 1994.

2. D.A. Swenson, Crandall, Guy, Lenz, Ringwall, & Walling, "Development of the RFD Linac Structure", PAC'95, Dallas, 1995.

3. D.A. Swenson, F.W. Guy, K.R. Crandall, "Merits of the RFD Linac Structure for Proton and Light-Ion Acceleration Systems", EPAC'96, Sitges, 1996.

4. D.A. Swenson, K.R. Crandall, F.W. Guy, J.M. Potter, T.A. Topolski, "Prototype of the RFD Linac Structure", LINAC'96, CERN, Geneva, 1996.

5. D.A. Swenson, "CW RFD Linacs for the BNCT Application", CAARI'96, Denton, 1996.

6. D.A. Swenson, K.R. Crandall, F.W. Guy, J.W. Lenz, W.J. Starling, "First Performance of the RFD Linac Structure", LINAC'98, Chicago, 1998.

7. D.A. Swenson, "Compact, Inexpensive, Epithermal Neutron Source for BNCT", CAARI'98, Den., 1998.

8. D.A. Swenson, F.W. Guy, and W.J. Starling, "Commissioning the 2.5-MeV RFD Linac Prototype", PAC'99, New York, 1999.

9. D.A. Swenson, "Status of the RFD Linac Structure Development", LINAC2000, Monterey, 2000.

Status report of the development of a multi-mA self-extracted H$^+$-beam cyclotron

S. Lucas, W. Kleeven, M. Abs, E. Poncelet, Y. Jongen

IBA, Chemin du cyclotron, B-1348 Louvain-La-Neuve, Belgium.

Abstract. In 1992, IBA developed a high intensity cyclotron for the production of Pd-103. Up to now 16 internal-target machines have been installed and are delivering 1 mA on average. Unfortunately such configuration suffers from two major drawbacks: i- little flexibility on the shape and size of the beam on target limiting the total power that target can tolerate, ii-activation of cyclotron components due to neutron production and primary-beam scattering. In 1995 IBA proposed a new method for the extraction of multi-mA positive ions [1] without the need of a deflector or a similar device. The extraction is obtained by a sudden and substantial reduction of the Lorentz force at the radial pole edge allowing the beam to escape from the machine (self-extraction principle). It was decided in 1998 to construct a prototype to test that extraction technology. This paper presents the status of the development. Focus is put on the RF system, ion source, magnetic configuration and final layout.

INTRODUCTION

A large percentage of radioisotopes used in nuclear medicine are produced by cyclotrons. Among them Pd-103 is used for curative purposes and is commercially produced by cyclotrons able to accelerate mA of proton beam to the kinetic energy of 14 MeV. Such cyclotrons can be divided into two categories: internal target and external target. Beam extraction is accomplished by either the use of a stripper or the use of an electrostatic deflector.

Stripping requires the acceleration of H$^-$ particles and the use a thin foil for beam interception. Despite of being very efficient this technology suffers the dissociation of the H- interaction with the background gas, a limited lifetime of the stripping foil and the high cost of the external ion source needed to produce a large amount of H-. All those limitations can be overcome with the use of a H$^+$ beam and an electrostatic deflector. But septum activation, high voltage instabilities and outgasing phenomena limit severely the overall performance of the machine.

The second category is made of H$^+$ with an internal target. This technology suffers from little flexibility as to the shape and distribution of the beam on target. Consequently the incident proton beam intercepts the target at grazing angles and a large fraction of the incoming beam is scattered in the cyclotron in an uncontrolled way. Activation of the machine results from that and from neutron generation.

In order to overcome the above limitations, IBA decided in 1998 to develop a multi-mA self-extraction cyclotron. This prototype will allow the extraction of several mA of proton beam in a non-intercepting way opening the door to new production capabilities never reached before. This paper presents a short description of different sub-systems, and the status of the development of the machine.

DESCRIPTION OF THE CYCLOTRON

Magnetism

As described in a previous paper [2] the design includes a few unconventional features such as i) a hill gap that is quasi-elliptical, ii) two opposite sectors that have an extended radius, iii) a groove machined on those extended poles which creates a sharp dip in the magnetic field and a region where the field index is smaller than −1 and iv) the presence of two opposed

Sm-Co harmonic kickers located at an azimuth of +-90° with respect to the entrance of the groove.

The elliptic gap creates a magnetic field which remains isochronous up to the radial pole edge and also allows a very fast transition from this isochronous region into the field dip. The harmonic kickers move the last internal orbit from the limit of the isochronous region into the groove, assuming a well-behaved extraction of the beam.

The extraction of the beam is tuned by two pair of centering coils that will adjust the beam to pass correctly above the extraction kickers. Finally, a gradient corrector located inside the machine, just after the extraction groove corrects the divergence of the beam. Additional details can be found in [2] and in [3].

Dee and Cavities

The specifications for the self-extraction cavities are classical, apart from some special requirements on the high energy gain per turn at the extraction radius and on the reduced voltage in the central region to try to reduce RF sparking as much as possible. Two stems per dee maintain a vertical symmetry in the cavity and reduce the risk of heating the magnet structure by vertical RF currents.

Figure 1 shows the dee shape, optimized by taking into account a high energy gain per turn in the central region (up to 1 MeV), a high impedance in the region associated with acceleration up to 13 MeV to get a high voltage ratio between the extraction and the central region, and an energy gain per turn of 200 kV at the extraction radius to have a high turn separation.

The stems have been located as close as possible to the central region to reduce the voltage in the center of the machine (43 kV peak-to-peak), and to get 55 kV peak-to-peak at the extraction radius. The gap between the dee and the poles is 5 cm at the extraction radius.

The calculated Q is about 4900.

Coupling is done capacitively in the median plane and tuning is achieved with a movable capacitor plate.

Harmonic 4 has been chosen to maximize the kinetic energy gain per turn.

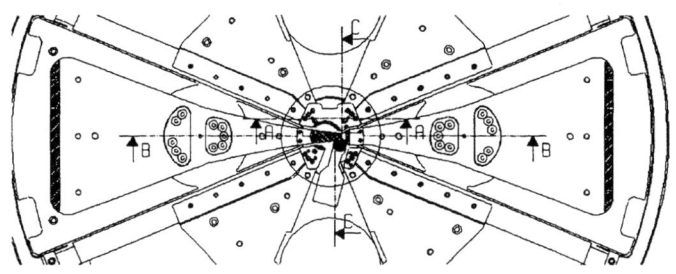

Figure 1: Dee and central region

RF system

A classical 3 stage amplifier chain is used: pre-driver (1-100W), driver (3-12 kW) and final power amplifier (200 kW).

Driver Amplifier

We use a recently-developed driver amplifier with the following features:

i) 3 different adjustable frequency ranges: 40-42 66-68 and 72-74 MHz, ii) air cooled single high gain power tetrode, iii) grid driven topology (grounded cathode) to insure high gain (20 dB), iv) RF power ranging from 3 to 12 kW under 50Ω load, v) class-AB biasing providing an average plate efficiency of 70% at full power, and vi) stable, easily tuned and packaged in a standard 19" rack mount

The schematic is illustrated in figure 2. Input and output are matched to standard 50Ω impedance with the use of the components CV1-C1-TL1-R1 for the input circuit and L-Cs for the output circuit. Input tuning is done by the neutralization inductance Ln.

The output circuit is formed by inductance L and capacitor Cs to match the output and also to reconstruct the missing alternance during the tetrode non-conduction period.

Biasing voltage applied to the grid (Ugrid) sets-up the quiescent point, and therefore the class. In this amplifier, class-AB2 has been chosen as a compromise between power dissipation and linearity, but the positive grid-cathode input voltages induces small grid current. The input signal is applied to the grid but also superimposed to the dc bias voltage. The amplitude of this signal must swing along the entire tube load curve.

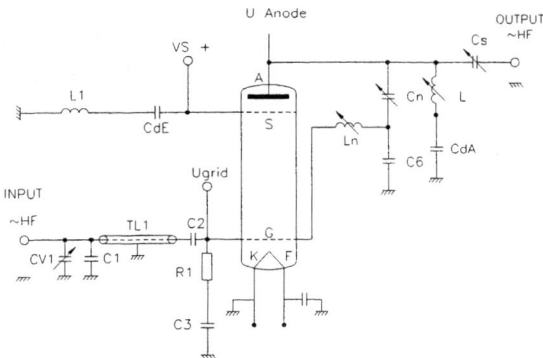

Figure 2: Schematic diagram of the universal driver amplifier

Because the amplifier is biased in class-AB, the anode current Ia is be pulsed and is heavily loaded with harmonics. The cavity's high Q assures a strong filtering of these undesired harmonic frequencies. The circuit made of Cn-C6-Ln acts as a neutralization circuit.

If a frequency change has to occur within one range as specified above (e.g. 40-42Mhz, central freq = 41 Mhz), only inductance Ln should be tuned, the value of Cn remains unchanged.

CdE and CdA are respectively the decoupling capacitors for screen and anode voltages. C2 and C3 are blocking capacitors to prevent the bias voltage to reach either the input or the ground.

Final Power Amplifier.

In order to test the capabilities of this prototype, the machine is equipped with a new final power amplifier (FPA) able to deliver up to 200kW of RF power to the accelerating cavity. The FPA is directly connected to the cavity via a short water-cooled 50 ohm line. Its own length maintains a constant voltage ratio between the RF tube and the cavity for the different cavity impedances met with high beam loadings. This also allows the tube to always operate at its best efficient working point because it is always working with the same anode voltage swing.

The tetrode used (TH681) is able to deliver up to 300 kW at the frequency involved (67 MHz). It is made of pyrolitic grids and exhibits a high gain in cathode driven topology (16 dB). The anode of the tube is fed with a 16 kV power supply. 17 amperes are required at full output power (200 kW). Plate efficiency is greater than 72% at full output power.

Thanks to its high gain, the tube needs a fairly low screen voltage below 1kV to draw the RF peak current without exceeding the grid current.

The input RF circuit consists of a three quarter wave resonator. The impedance transformation is done by the ratio between the line impedance and the internal tube impedance. The tuning and the matching adjustments are done by a movable short-circuit and a sliding sleeve, which serves as a line impedance reduction.

The space between the grid and the screen finds its first resonance around 100 MHz. This insures a very low reaction between the input and the output circuits giving a very good stability to the system.

The output tank consists on a folded quarter wave resonator where the output power is taken at a fixed 50 ohms point with respect to the line's end. The tuning of the circuit is done by a line impedance change close to the tube's anode. The output circuit is also equipped with a special design of micro-wave absorbers that kills the push-pull 1.2 GHz circulating resonance of the tube.

Diagnostic tools

Different diagnostic tools have been developed for this prototype. Probes equipped with Cu water-cooled multi-sleeves will be used to measure the vertical beam dimension and turn separation. Those probes are motorized and data recording is done automatically in a logger.

Optimization of the Ion source position will be made by monitoring the extracted current versus source-position on a water-cooled beam dump located at a radius of 7 cm (700 keV) from the central region.

Beam current measurement will be performed by non-intercepting devices placed on the beam line. Beam profile monitoring (BPM) is more problematic. We demonstrated at IBA that an interceptive BPM equipped with a simple 500 µm diameter W-wire can successfully measure the profile of a 300 µA 14 MeV proton beam spread over 350 mm^2. Unfortunately such a system is not suitable for higher beam currents. Therefore we are investigating the development of a device based either on optical pumping and absorption measurements [4] or ionization as described in [5].

Another diagnostic system used to tune the RF has also been developed. It pulls the active power out of the cavities without the need of the accelerated beam.

Full RF power is loaded through the cavities rather than loading the amplifier directly by a 50 ohms dummy load. Therefore both power-coupling subsystem and RF regulation are tested at the same time. The system is based on a capacitive coupling made underneath the dee, a coupling adjustment made by a variable stub and a power 50 ohms water cooled RF load. The choice of a capacitive coupling rather than an inductive one has been driven by the lack of space to introduce the probe. The capacitive probe is introduced in a 60 mm diameter hole made in the magnet yoke that emerges in the accelerating cavity. The stub is placed outside the machine and acts like a RF current bypass. This configuration allows the RF current pumped in the cavity to be quasi-constant and hence does not really influence the frequency tuning of the cavity.

Ion Source

The ion source is a classical internal PIG source [6] with cathodes made of Ta and anode (chimney) made of Cu-W alloy. The exit slit size is 0.6 mm x 12 mm. The beam extraction slit can be adjusted by exchange of the chimney. The source to puller gap is set around 1 mm. More than 5 mA of proton beam can be extracted from this type of source [7].

Final Layout

The final layout in shown in figure 3. Transport calculations show that the larger horizontal dimension of the beam is at the entrance of the quadrupole (9) and is 6 cm. The largest vertical beam dimension is 4 cm and is at the end of the quadrupole (9). Overall beam size on the beam dump will be 6 x 4 cm.

CONCLUSIONS

With the help of this prototype, we hope to demonstrate and validate the concept of self-extraction. The cyclotron is now assembled. The assembly of the RF amplifiers have still to be finished. We hope to produce beam by the end of 2000. Results will be presented in a future paper.

REFERENCES

1. Y. Jongen, D. Vanderplassche,, P. Cohilis, Proc. 14th Int. Conf on Cyclotrons and their applications, Cape Town, South Africa, 1995, World Scientific publisher, pp 115.

2. W. Kleeven et al, "Self-Extraction in a compact high intensity H+ cyclotron at IBA", Proc. 7th European Part. Accel. Conf. (EPAC 2000), Vienna, Austria, pp 2530-2532.

3. W. Kleeven, et al, "Self-Extraction in a small High current H+ cyclotron at IBA", this conference.

4. B. Pottin et Al, "Optical beam profiler for high current beams", Proc. 7th European Part. Accel. Conf. (EPAC 2000), Vienna, Austria, pp 960-962.

5. P. Cameron et Al, "The RHICS ionization beam profile monitor", Proceedings of the 1999 Particle Accelerator Conference, New-York, 1999, p.p 2114-2116

6. B.F. Gavin, in "The Physics of Ions Sources and technology", Wiley & Sons, NY, 1989, p. 167.

7. Y. Jongen et al, Proceeding of "Fourth European Particle Accelerator Conference, World Scientific-1994, p. 2627.

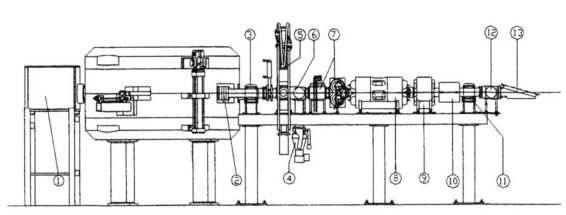 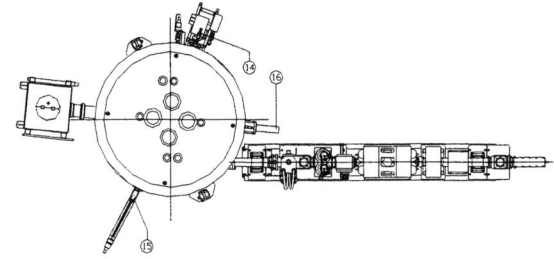

Figure 3: General description of the cyclotron and the beam line. 1: Final Amplifier – 2: Permanent Co-Sm Quadrupole magnet – 3: Steering – 4: Pumping unit – 5: Faraday & Beam viewer – 6: BPM – 7: X & Y collimators – 8: Quadrupole – 9: Sextupole – 10: Beam current measurement – 11: Scanning unit – 12: diagnostic tool – 13: Beam dump – 14: ion source – 15: Radial probe – 16: Cavity tuning.

SECTION VII

SYNCHROTRON EXPERIMENTS AND FACILITIES

Threshold Photoelectron Spectroscopy Using Synchrotron Radiation

G. C. King[1] A. J. Yencha[2] and M. C. A. Lopes[3]

[1]Department of Physics and Astronomy, The University of Manchester, Manchester M13 9PL, UK
[2]Department of Chemistry, State University of New York at Albany, Albany, New York 12222, USA
[3]Departmento de Fisica, ICE, Universidade Federal de Juiz de Fora, Juiz de Fora-MG, CEP 36036-330, Brazil

Abstract. Recent developments in the techniques of threshold photoelectron spectroscopy using synchrotron radiation are discussed. The types of information that such studies can provide about molecules are considered. Examples of current experimental techniques of threshold spectroscopy are given in the context of the high photon resolution now provided by synchrotron radiation sources. These techniques are illustrated by recent experimental results.

INTRODUCTION

Threshold photoelectron spectroscopy (TPES) offers the important characteristics of high sensitivity and high-energy resolution when combined with current synchrotron radiation sources. It also provides information that cannot be obtained with conventional photoelectron spectroscopy (PES). In TPES the collection energy of the electron spectrometer is fixed and is tuned to accept electrons of nominally zero kinetic energy. (In practice this means the collection of electrons of energies less than about 1 meV). The incident photon energy is varied and as the photon energy is scanned across an ion threshold a zero energy electron is produced so that the TPES spectrum maps out the energy levels of the ion. TPES can provide very high-energy resolution (~ 1meV) spectra that are essentially Doppler-free. It is characterised by very high detection efficiency, with a solid angle of collection of nominally 4πsr. Furthermore it can provide both of these simultaneously. The large collection angle does, however, mean that the technique cannot provide any angular information as all the threshold photoelectrons are collected regardless of their initial ejection angle. Vibrational intensities observed in conventional PES are determined by the Franck-Condon principle. In contrast the vibrational intensities in TPES are distinctly non-Franck-Condon. For example it is common to excite levels with vibrational quantum numbers of 20 or more. This behaviour is due to indirect ionisation processes that are more likely to occur close to the thresholds for ionisation. The information obtained from TPES measurements is mainly spectroscopic in nature. However dynamical information regarding the mechanisms for single and also double ionisation can be obtained [e.g. 1-5].

TECHNIQUES OF THRESHOLD PHOTOELECTRON SPECTROSCOPY

The Penetrating-field Technique

This technique depends upon the penetration of a weak electrostatic field into the interaction volume that preferentially draws out photoelectrons of near-zero energy over a solid angle of 4π sr [6]. The extraction system provides very high threshold energy resolution (< 2 meV) by virtue of the very fast fall off in collection efficiency with energy and simultaneously a very large solid angle of collection. Because a fraction of any energetic photoelectrons produced will be emitted into the solid angle subtended by the extracting electrode, these are removed by placing a conventional cylindrical electrostatic deflection analyser after the penetrating field stage. A TPES spectrum is obtained by scanning the photon energy and measuring the yield of threshold photoelectrons. Although resolutions better than a few meV are extremely difficult to achieve with conventional deflection analysers, such resolutions are readily achievable using the penetrating-field technique.

The Pulsed-field Ionisation Technique

This technique is used on the Chemical Dynamics Beamline at the Advanced Light Source (ALS). It consists of a steradiancy type analyser and a TOF spectrometer [7,8] and exploits the timing structure of the ALS. The synchrotron radiation is quasi-continuous except for a single dark gap of 112 ns in each beam cycle that allows timing experiments to be performed. After an elapsed time of ~ 20 ns from the start of the dark gap an electric field pulse of 1.5 V cm^{-1} is applied across the interaction region for a duration of 40 ns. This results in field ionisation of any long-lived Rydberg states of the target molecule and acts to accelerate the threshold electrons produced into the TOF electron spectrometer. Prior to the pulsed field a small dc field (0.04 – 0.2 V cm^{-1}) is applied across the interaction region to push any hot electrons that might have been present due to direct ionisation towards the electron detector as they are formed. Near the end of the pulsed-field period a gate is opened in the output of the electron detector for 10 ns and electrons collected in this period are accumulated. The opening of the electron collection gate is set to coincide with the flight time of electrons initially created in the synchrotron dark gap with zero kinetic energy. With this apparatus resolutions of 1.0 cm^{-1} have been demonstrated [8].

RESULTS AND DISCUSSION

TPES Studies

The penetrating-field technique has been used to perform TPES studies on a large number of molecules including the halogens [9-11], the halogen hydrides and deuterides [12-14] and most recently the interhalogen molecule ICl [15]. It is interesting to note that there are very few previous experimental investigations of ICl and these are of modest resolution. This illustrates the robustness of the TPES technique that performs with high performance even for aggressive gases. This TPES study of ICl was obtained with a penetrating-field electron spectrometer used in conjunction with the 5m McPherson monochromator at the Daresbury Laboratory SRS.

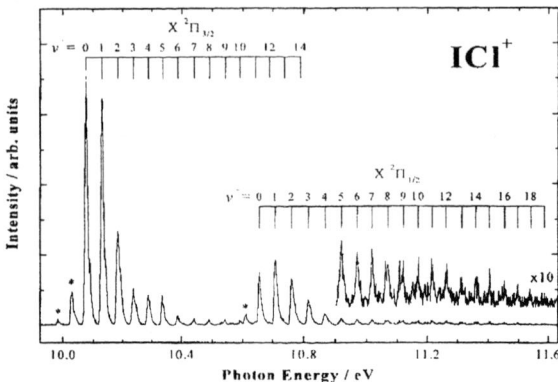

FIGURE 1. A TPES spectrum of ICl$^+$ covering the energy region of the X$^2\Pi_{3/2,1/2}$ band systems.

The ICl$^+$ (X $^2\Pi_i$) band system is shown in figure 1 at an energy resolution of 2 meV. Numerous well-resolved vibrational bands are observed in both spin-orbit components. In the first few vibrational bands in the (X $^2\Pi_{3/2}$) system one can clearly see partially resolved rotational branch profiles. Vibrational structure components were analysed using a second-order Dunham fit of the peak energy positions that yielded the following vibrational constants: for ICl$^+$ (X $^2\Pi_{3/2}$): ω_e = 0.0532 eV, $\omega_e x_e$ = 1.67 x 10^{-4} eV with a value of T$_e$ of 10.0505 eV and for ICl$^+$ (X $^2\Pi_{1/2}$): ω_e = 0.0542 eV, $\omega_e x_e$ = 2.58 x 10^{-4} eV with a value of T$_e$ of 10.6278 eV. The observed adiabatic ionisation potentials in the two components are found to be 10.076 ± 0.002 eV and 10.655 ± 0.002 eV respectively yielding a spin-orbit splitting of 0.579 ± 0.002 eV in excellent agreement with the PES results[16] and in good agreement with the calculated value of 0.561 eV [17]. Based on a simple Franck-Condon calculation using a Morse potential program together with the measured intensities of the vibrational peaks of ICl$^+$ an internuclear distance of 0.2224 ± 0.0001 nm was determined that can be compared with the known internuclear distance in the ground state of ICl of 0.2320878 nm. This shows the expected slight decrease in the internuclear distance upon removal of an antibonding electron.

PFI-PE Study of HF and Ion-pair Formation

A PFI-PE spectrum of HF over the photon energy range 15.9 – 16.5 eV is presented in the top panel of figure 2 [18]. This range encompasses the formation of the $v^+ = 0$ and 1 vibrational systems of the HF$^+$ (X $^2\Pi_{3/2,1/2}$) spin-orbit states. A resolution of 0.6 meV was determined from the width of the sharpest features in the spectrum. As can be seen extensive rotationally resolved structure is found over the entire energy region. Also shown, in the middle panel is the TPES spectrum of HF over the same photon energy range at a resolution of ~ 3 meV [19]. It shows partially resolved rotational structure in the $v^+ = 0$ band system of the $^+$(X $^2\Pi_{3/2,1/2}$) states of HF$^+$ that appears to be influenced by a strong resonance at about 16.06 eV. In the bottom panel of figure 2 is presented the F$^-$ photoexcitation function from ion-pair formation in HF at a resolution of ~ 13 meV [20]. All of the structure observed in the F$^-$ photoexcitation function has been attributed to ion-pair formation via photoexcited Rydberg states that are predissociated by the (V $^1\Sigma^+$) ion-pair potential.

The PFI-PE spectrum (top panel of figure 2) is dominated by two series of peak features at about 16.04 and 16.40 eV that are identified the $v^+ = 0$ and 1 bands of HF$^+$ (X $^2\Pi_{3/2,1/2}$). These structures are assigned to rotational lines of the two spin-orbit components F_1 and F_2. It should be noted that the relative intensities of the rotational profiles are essentially identical to those observed in the VUV laser PFI-PE spectrum of Mank et al. [21] except for small variations due to the different temperature conditions in the two experiments. Based on the spectral results of figure 2 the $R_1(0)$ transition energies in the $v^+ = 0$ and 1 bands are found to be at 16.0456(4) eV and 16.4056(4) eV respectively, in good agreement with the results of Mank et al. [21]. Thus, it can be concluded that the synchrotron based PFI-PE method is equivalent the VUV laser based method.

The highly structured region between the $v^+ = 0$ and 1 vibrational bands in the PFI-PE spectrum of HF is attributed to ion-pair formation with the detection of F$^-$ ions by the TOF spectrometer. This conclusion is supported by the large degree of correspondence between the features in the PFI-PE spectrum and the F$^-$ photoexcitation function, and by the energy positions of $nd\pi$ Rydberg series converging to the two spin-orbit components of the $v^+ = 1$ band system. The F$^-$ signal is explained as being due to a dc F$^-$ signal produced during the light-on portion of the synchrotron period *under field-free conditions*.

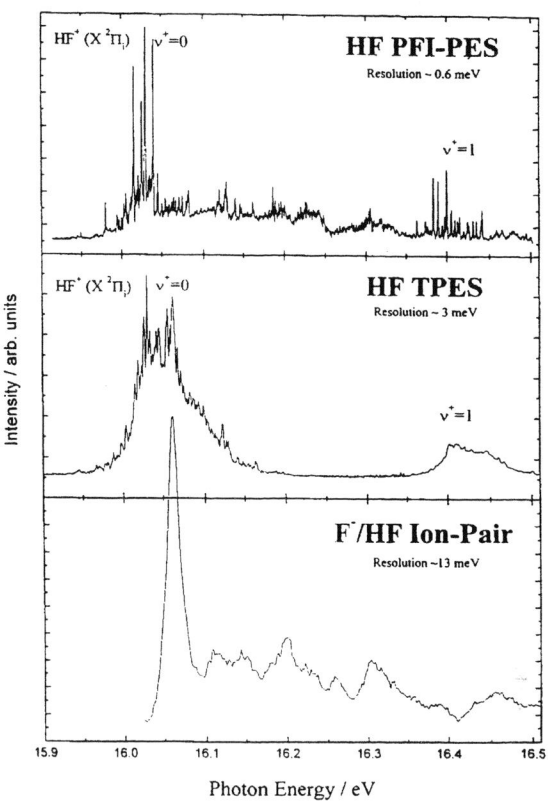

FIGURE 2. Comparison of the PFI-PE spectrum of HF (top), the TPES spectrum of HF (middle) and the F$^-$ excitation function of HF (bottom) over the photon energy encompassing the $v = 0$ and 1 vibrational levels of HF$^+$ (X$^2\Pi_{3/2,1/2}$).

CONCLUSIONS

TPES can provide very high sensitivity and very high-energy resolution when combined with current synchrotron radiation sources. This resolution is of the order of 1meV and enables the study of photoionisation processes at the individual rotational level. TPES is also very sensitive to the presence of indirect photo ionisation processes which gives access to very high vibrational levels. The very high sensitivity of the technique makes its use advantageous in the study of targets of low density and It may be expected that this will lead to its future use in the study of , for example, free-radicals and clusters.

REFERENCES

1. G.C. King, M. Zubek, P.M. Rutter, F.H. Read, A.A. MacDowell, J.B. West and D.M.P. Holland, J. Phys. B: At. Mol. Opt. Phys. 21 (1988) L403.
2. R.I. Hall, L. Avaldi, G. Dawber, M. Zubek, K. Ellis and G.C.King, J. Phys. B: At. Mol. Opt. Phys. 24 (1991) 115.
3. S. Cvejanovic, R.C. Shiell and T.J. Reddish, J. Phys. B: At. Mol. Opt. Phys. 28 (1995) L707.
4. P.A.Heimann, U. Becker, H.G. Kerkhoff, B. Langer, D. Szostak, R. Wehlitz, D.W.Lindle, T.A. Ferrett and D.A. Shirley, Phys. Rev. A 34 (1986) 3782.
5. D.B. Thompson, P. Bolognesi, M. Coreno, R. Camilloni, L. Avaldi, K.C. Prince, M. deSimone, J. Karvonen and G.C. King, J. Phys. B: At. Mol. Opt. Phys. 31 (1998) 2225.
6. S. Cvejanovic and F.H. Read, J. Phys. B: At. Mol. Phys.7, (1974) 1180.
7. C.-W. Hsu, M. Evans, P.A. Heimann and C.Y.Ng, Rev. Sci. Instrum. 68 (1997) 1694.
8. G.K. Jarvis, Y. Song and C.Y. Ng, Rev. Sci. Instrum. 70 (1999) 2615.
9. A.J. Yencha, M.C.R. Cockett, J.G. Goode, R Donovan, A Hopkirk and G C King. Chem. Phys. Letts. 229 (1994) 347.
10. A.J. Yencha, A. Hopkirk, A. Hiraya, R.J. Donovan, J.G. Goode, R.R.J. Maier, G.C King and A. Kvaran. J. Chem. Phys. 99 (1995) 7231.
11. A.J Cormack, A.J. Yencha, R.J. Donovan, K.P. Lawley, A. Hopkirk and G.C. King, Chem. Phys. 213 (1996) 439.
12. A.J. Yencha, A.G. McConkey, G. Dawer, L. Avaldi, M.A. MacDonald, G.C. King and R.I. Hall. J. Elec. Spectrosc. Rel. Phenom. 73 (1995) 217.
13. A.J. Cormack, A.J. Yencha, R.J. Donovan, A. Hopkirk and G.C. King, Chem. Phys. 221 (1997) 175.
14. A.J. Cormack, A.J. Yencha, R.J. Donovan, A. Hopkirk and G.C. King, Chem. Phys. 238 (1998) 109.
15. A.J. Yencha, M.C.A. Lopes and G.C. King, Chem Phys Letts. (2000) in press.
16. A.W. Potts and W.C. Price, Trans. Faraday Soc. 67 (1971) 1242.
17. J.M. Dyke, G.D. Josland, J.G. Snijders, P.M. Boerrigter, Chem. Phys. 91 (1984) 419.
18. A.J. Yencha, M.C.A. Lopes, G.C. King, M Hochlaf, Y. Song and C.-Y. Ng, Faraday Discuss. 115 (2000) in press.
19. A.J. Yencha, A.J. Cormack, R. Donovan, A. Hopkirk and G.C. King, J. Phys. B: At. Mol. Opt. Phys. 32 (1999) 2539.
20. A.J. Yencha, A. Hopkirk, J.R. Grover, B-M. Cheng, H. Lefebvre-Brion and F. Keller, J Chem Phys 103 (1995) 2882.
21. A. Mank, D. Rodgers and J.W. Hepburn, Chem. Phys. Letts. 219 (1994) 169.

Chemical State Analysis of Iron in Nerve Cells

S. Fujisawa*, A.M. Ektessabi*[+] and S. Yoshida[¶]

Department of Precision Engineering, Graduate School of Engineering, Kyoto University
Yoshida Honmachi, Sakyo-ku, Kyoto, 606-8501, Japan
¶*Department of Neurology, Wakayama Medical College*
27-9 Bancho, Wakayama City, 640-8511, Japan

Abstract. The chemical states of iron contained in tissues obtained from a patient with parkinsonism-dementia complex (PDC) were investigated using micro beams from a synchrotron radiation source. XRF analyses were performed at energies 7.160 keV, highly above the iron absorption edge, and 7.120 keV, slightly above the Fe^{2+} absorption edge to suppress the excitation of Fe^{3+}. Iron was detected in neuromelanin granules and one of glial cells in the PDC tissue. The results show differences in the chemical state of the iron in the neuromelanin granules and the glial cells. The Fe^{3+}/Fe^{2+} ratio of iron contained in the glial cell is considerably higher than that of the neuromelanin granules.

INTRODUCTION

Synchrotron radiation x-ray fluorescence (SRXRF) spectroscopy is a powerful method for trace element analysis. Recent developments in synchrotron radiation sources provide the impetus for XRF studies, with detection limits and spatial resolution being improved considerably [1]. XRF analyses have been used in the studies of biological specimens [2~4] because measurement can be performed in air and that wet samples can be analyzed [5]. A characteristic feature of SRXRF analysis is that incident x-ray energy is variable. When incident energy is above the absorption edge, the result is independent of the chemical state of the absorbing element, while when incident energy is near absorption edge, the result is dependent on the chemical state. Using these phenomena, we can obtain information about the density and the chemical states of the elements [5,6].

In this study, we applied SRXRF analysis to a pathological specimen which was obtained from the midbrain of a patient with a neurodegenerative disease. Some of studies of neurodegenerative diseases such as Parkinson's disease [2,7,8] and Alzheimer's disease [9,10] have reported that accumulations of metallic elements such as iron can be seen in the brain tissues of patients. The cause of neuronal degeneration is still unknown, but there are some indications that the accumulation of metallic elements or the possible changes in the chemical state of transition metals in brain tissues could be related to the neuronal degeneration [11,12]. It is important to have information on density and, especially, the chemical state of the iron in brain tissues from patients with neurodegenerative diseases. In this study, we focused on the density and the chemical state of iron contained in tissues of a patient with neurodegenerative disorder.

EXPERIMENTAL

Materials and Sample Preparation

Autopsy specimens from the midbrain, including substantia nigra, were obtained from a 56-year-old male patient who was diagnosed with Parkinsonism-dementia complex (PDC). PDC is a disorder unique to Guam, which is characterized by widespread Alzheimer-type neurofibrillary tangles [13,14]. The specimen was fixed in 10 % formalin, and embedded in paraffin. Sections 8 μm in thickness were cut and mounted on a Mylar film for x-ray analysis.

X-ray analysis

X-ray analysis using synchrotron radiation on autopsy specimens and reference samples were performed at beam line 39XU of SPring-8, at the Japan Synchrotron Radiation Research Institute (JASRI). Synchrotron radiation from the storage ring (8GeV, maximum current 100 mA) was monochromated with a

---

+ Contact author: A.M. Ektessabi. E-mail: h51167@sakura.kudpc.kyoto-u.ac.jp

Si(111) double crystal monochromator. Incident x-ray beams were restricted by an x-y slit and pinhole set. Incident beam size was about 10 μm in diameter. Measurements were performed in vacuum.

XRF analysis was performed when the incident x-ray energy was 7.160 keV and 7.120 keV. Incident photon flux was monitored with an air-filled ion chamber. Fluorescence x-rays were collected by a solid state detector (SSD). X-ray fluorescence energy was calibrated by metal thin films.

In order to measure the x-ray absorption coefficient curve of Fe^{2+} (FeO) and Fe^{3+} (Fe_2O_3), x-ray absorption fine structure (XAFS) analysis was performed with the energy resolution at 0.5eV.

Chemical state analysis

The x-ray absorption coefficient depends on the valence state and neighborhood structure of the absorbing elements. The x-ray absorption coefficient curves of FeO (Fe^{2+}) and Fe_2O_3 (Fe^{3+}) are shown in Figure 1. The absorption edge shift between Fe^{2+} and Fe^{3+} can be seen. At the incident x-ray energy of 7.160 keV, absorption coefficients have almost the same value in the case of Fe^{2+} and Fe^{3+}. On the other hand, the absorption coefficient is clearly different for Fe^{2+} and Fe^{3+} at an incident energy of 7.120 keV. Almost only Fe^{2+} can be selectively excited at this incident x-ray energy. As seen in this example, at an incident x-ray energy highly above absorption edge, elements can be excited indiscriminately. At an incident x-ray energy near the absorption edge, the absorption coefficient is sensitive to the chemical state of the element. Thus, XRF analysis with selected incident x-ray energies can provide information about both the density and the chemical states of elements.

TABLE 1. The x-ray absorption coefficient of FeO (Fe^{2+}) and Fe_2O_3 (Fe^{3+}) at the incident energies of 7.160 keV and 7.120 keV.

	7.160 keV	7.120 keV
FeO (Fe^{2+})	0.565	0.399
Fe_2O_3 (Fe^{3+})	0.498	0.075

RESULTS

As shown in Figure 1, the curves of the x-ray absorption coefficient have two different absorption edge in the cases of FeO (Fe^{2+}) and Fe2O3 ($Fe3^+$) with a certain shift in energy. The absorption edge of Fe^{3+} has shifted towards a higher energy than that of Fe^{2+}. Table 1 shows that the absorption coefficients of Fe^{2+} and Fe^{3+} at the incident x-ray energies of 7.160 keV and 7.120 keV. At the incident x-ray energy of 7.160 keV, both Fe^{2+} and Fe^{3+} can be excited. On the other hand, at the incident x-ray energy of 7.120 keV, Fe^{3+} is not excited selectively.

XRF analysis was performed at several points in the PDC nigral tissues with the incident x-ray energies of 7.160 keV and 7.120 keV. An optical microscopic photograph of the tissue is shown in Fig. 2. We can see some neuromelanin granules and glial cells in the

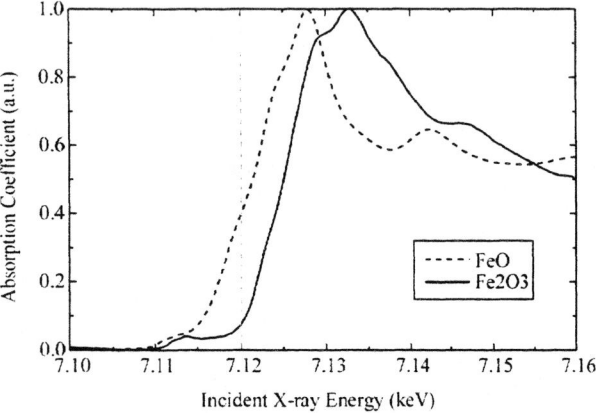

FIGURE 1. X-ray absorption fine structure (XAFS) spectra of FeO (Fe^{2+}) and Fe_2O_3 (Fe^{3+}).

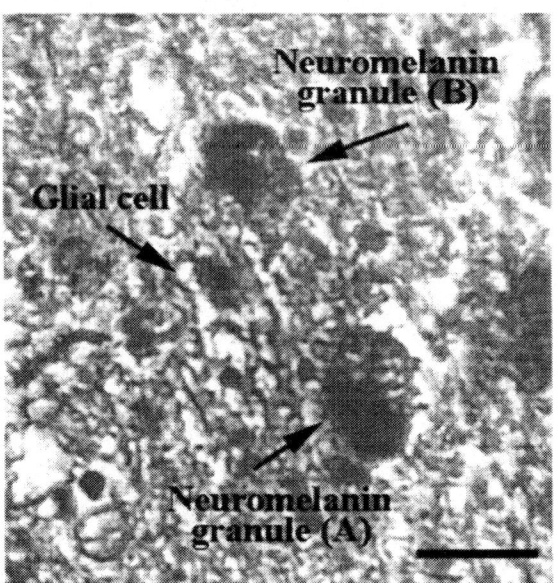

FIGURE 2. An optical microscopic photograph of the PDC tissue. Scale bar is 20 μm.

TABLE 2. Peak area of x-ray fluorescence spectra in the PDC tissues with the incident energy of 7.160 keV and 7.120 keV. Measurement time was 200 seconds for each point. XRF yields were normalized by incident x-ray intensity.

Target	XRF yield (7.160keV)	XRF yield (7.120keV)	Fe^{3+}/Fe^{2+}
Grial cell	3.98E-05	9.81E-06	5.46
Neuromelanin(A)	2.60E-05	1.42E-05	0.47
Neuromelanin(B)	1.03E-05	2.94E-06	3.55
Extra cellar tissue	5.17E-06	1.90E-06	1.77

tissue. Typical XRF spectra in the tissue are shown in Figure 3. The ordinate and abscissa represent XRF intensity and fluorescent energy, respectively. XRF intensity is normalized using fluorescent intensity divided by I_0. I_0 is the incident x-ray intensity, proportional to photons per second, as measured by the ionized chamber.

Table 2 shows the XRF yield of iron at several points in the tissue. Iron was detected at a high density in the neuromelanin granules and one of the glial cells. We obtained the Fe^{3+}/Fe^{2+} ratio using the XFR yields and the absorption coefficients (Table 1). We can see that the Fe^{3+}/Fe^{2+} ratio varies at these points in the tissues. For example, the Fe^{3+}/Fe^{2+} ratio of iron contained in the glial cell is about ten times higher than that in the neuromelanin granule (A). These results are based on the following assumptions: (i) the absorption coefficient curves of the different valence states of iron

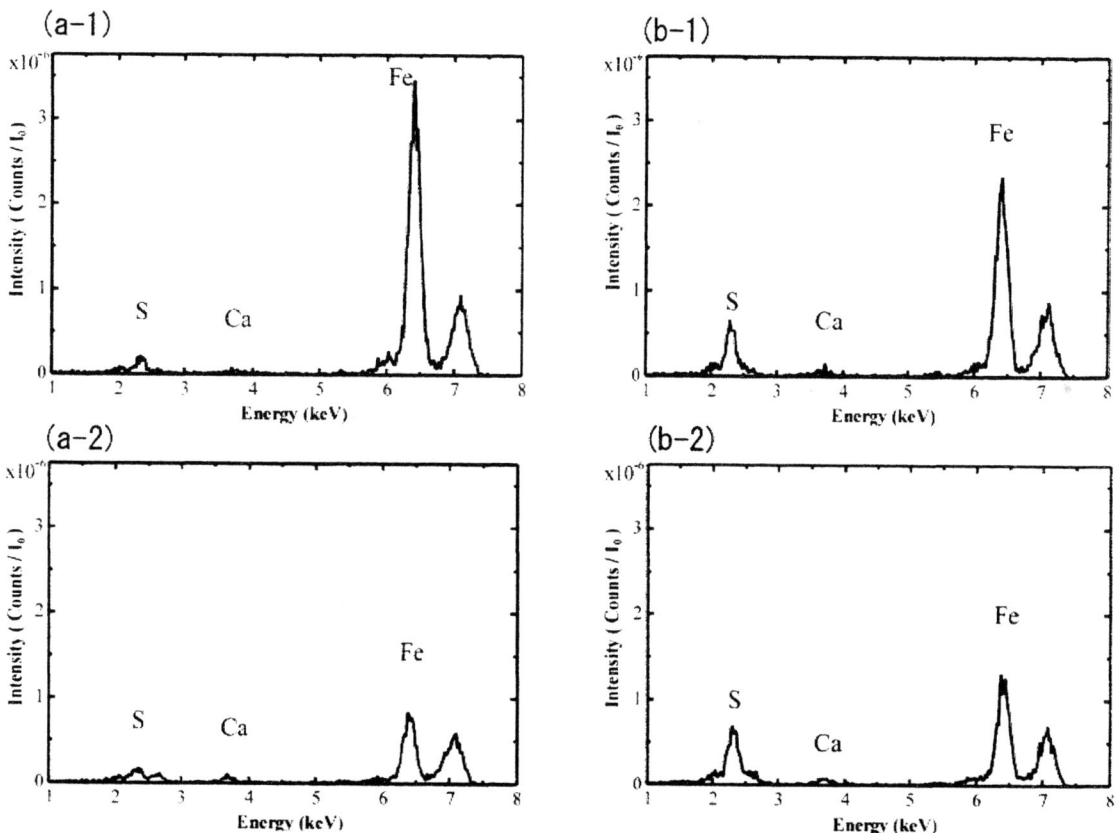

FIGURE 3. Typical XRF spectra in the tissue with PDC. Measurement points were inside of glial cell and neuromelanin granules. (a-1) glial cell, incident energy of 7.160 keV, (a-2) glial cell, 7.120 keV, (b-1) neuromelanin granules, 7.160 keV, (b-1) neuromelanin granules, 7.120 keV. Measurement time was 200 seconds for each point.

(Fe^{2+} and Fe^{3+}) were represented by only iron oxides (FeO and Fe_2O_3), (ii) iron contained in the tissues is a superposition of Fe^{2+} and Fe^{3+}.

Using these assumptions, the chemical state of iron can be well distinguished at the different points in the tissues.

CONCLUSIONS

In this study, we applied SRXRF analysis to the pathological specimen with two typical x-ray energies: one above the iron absorption edge, and the other near the absorption edge. We focused our attention on the sensitivity of the absorption coefficient near the absorption edge to the chemical state of iron, and we obtained information about the density and the chemical state of the iron.

The high brilliance of the SR micro beam makes it possible to analyze trace elements in a nondestructive state. Using SRXRF analysis, we demonstrated the semi- quantitative chemical state analysis of iron contained in pathological tissues at the singe cell level. This technique can be applied widely to investigations of biological samples.

In conclusion, iron was detected at a high density in neuromelanin granules and one of the glial cells in the tissue of a PDC patient. It is shown that there are differences in the chemical state of the iron in the neuromelanin granules and the glial cells. The Fe^{3+}/Fe^{2+} ratio of iron contained in the glial cell is one order of magnitude higher than that in the neuromelanin granules.

ACKNOWLEDGEMENT

The SR micro beam XRF analysis was done at the SPring-8, Japan Synchrotron Radiation Research Institute (JASRI) (project 1999B0059-NL-np, 2000A0104-CL-np and 2000A0105-CL-np).

REFERENCES

1. Saisho, H., Gohshi, Y., eds. Saisho H., Gohshi, Y., *Applications of Synchrotron Radiation to Material Analysis*, New York, Elsevier, 1996, chap. 2.
2. Ektessabi, A., Yoshida, S., Takada, K., *X-Ray Spectrom.*, **28**, 456-460 (1999).
3. Ektessabi, A. M., Rokkum, M., Johansson, C., Albrektsson, T., Sennerby, L., Saisho, H., Honda, S., *J. Synchrotron Rad.* **5**, 1136-1138 (1998).
4. Valkovic, V., Moschini, G., *Riv. Nuovo Cimento.* **16**, 1-55 (1993).
5. Sparks, Jr. C. J., eds. Winick, H. and Doniach, S., *Synchrotron Radiation Research*, New York, Plenum, 1980, chap. 14.
6. Sakurai, K., Iida, A., Gohshi, Y., *Adv. X-ray Anal.* **32**, 167-176 (1989).
7. Dexter, D. T., Wells, F. R., Lees, A. J., Agid, F., Agid, Y., Jenner, P., and Marsden, C. D., *J. Neurochem.* **52**, 1830-1836 (1989).
8. Hirsh, E. C., Brandel, J. P., Galle, P., Javoy-Agid, F., Agid, Y., *J. Neurochem.* **56**, 446-451 (1991).
9. Connor, J. R., Snyder, B. S., Beard, J. L., Fine, R. E., Mufson, E. J., *J. Neurosci. Res.* **31**, 327-335 (1992).
10. Good, P. F., Perl, D. P., Bierer, L. M., Schmeidler, J., *Ann. Neurol.* **31**, 286-292 (1992a).
11. Gerlach, M., Ben-Shachar, D., Riederer, P., Youdim, M. B. H., *J. Neurochem.* **63**, 793-807 (1994).
12. Robb-Gaspers, S. J., Connor, J. R., eds. Connor J. R., *Metals and Oxidative Damage in Neurological Disorders*, New York, Plenum, 1997, chap. 18.
13. Hirano, A., Kurland, L. T., Krooth, R., Lessell, S., *Brain* **84**, 642-661 (1961).
14. Hirano, A., Malamud, N., Kurland, L. T., *Brain* **84**, 662-679 (1961).

Recent Results in Photoionization of Atoms and Ions using Undulator Radiation

François J. Wuilleumier, D. Cubaynes, and J.-M. Bizau

Laboratoire Spectroscopie Atomique et Ionique, UMR CNRS 8624, Université Paris-Sud, 91405-Orsay, France

Abstract. Recent progress in the production of photon beams delivered by undulators allowed us to obtain new low- and high-resolution results in photoionization of atoms and ions. Using the 2nd generation Super ACO storage ring, we have measured cross sections for higher-order correlation satellites in lithium and single and double photoionization of multiply-charged ions. With the 3rd generation Advanced Light Source storage ring, we performed highly-resolved angle-integrated and angle-resolved experiments, including the study of correlation satellites and hollow states in alkali-atoms.

At the beginning of the 80's, new storage rings dedicated to producing only synchrotron radiation (SR) were built. On these so-called second generation SR sources, the flux available from undulators allows to obtain monochromatic photon beams in the 10^{13} photons/sec range within a relative spectral bandwidth of typically 10^{-3}, which is a gain of more than one order of magnitude as compared to the photon flux collectable from the bending magnets. Since the beginning of the 90's, we have entered a new phase with the construction of the low-emittance third-generation SR sources, delivering high-brightness photon beams, i. e.; allowing the increasing of the spectral resolution into the 10^4 range while keeping the same photon-flux level of 10^{13} photons/sec. Several of these rings are now in use. With the latest one, BESSY II, a spectral resolution of 600 µeV at 70 eV photon energy has been reached. Although these new sources have not been built primarily for atomic physics experiments, their wider availability is causing a new expansion of the research activities in the field.

All gas-phase experiments require high-intensity photon beams, because of the low density of the gaseous sample to be used in differential measurements (photoelectron, photoion, and fluorescence spectrometry), and because of the low cross sections of most of the photoionization processes to be investigated. Some experiments also need high-spectral resolution but can be carried out with a low resolution in the detection channel. Ultimately, highly differential measurements such as angle-integrated and angle-resolved photoelectron spectrometry may require simultaneously high-resolution in both excitation and detection channels, for instance when numerous final ionic states close in energy have to be resolved. In the past year we had the privilege to have access to undulator beam lines on a second generation storage ring, Super ACO in Orsay, and on a 3rd generation storage ring, the ALS in Berkeley. In what follows, we will illustrate how the use of both types of SR sources give valuable information on low-cross section photoionization processes and, especially, on low-density samples (atoms and multiply-charged ions).

LOW-RESOLUTION HIGH-INTENSITY EXPERIMENTS

Because of the relative large size of the electron beam (0.5 to 1 mm) and the divergence of the emitted photon beams in the 2nd generation SR sources, the spectral resolution cannot be better than 0.1-0.2 eV when 10^{13} photons/sec are needed in the 100 eV range. These characteristics are sufficient in a number of cases to perform measurements of partial photoionization cross sections using photoelectron/photoion spectrometries.

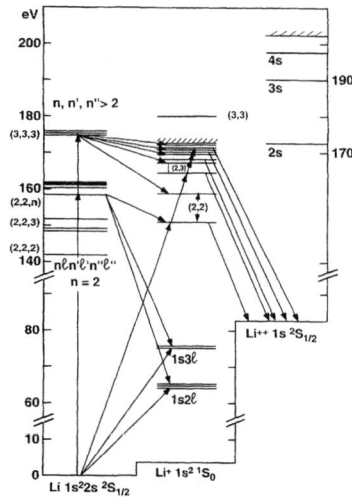

FIGURE 1. Energy level scheme of Li, Li$^+$, and Li^{2+}.

Let's take first the example of correlation satellites in low-Z elements, namely in lithium. Fig. 1 shows the energy level diagrams of Li, Li$^+$, and Li^{2+}. The lowest doubly-excited $2s^2$ ^{1}S Li$^+$ state has an energy of 151.7 eV. Between this energy and the 2s threshold of Li^{2+} (173 eV), the Li$^+$ doubly-excited states can be reached directly by photoionization into the continuum of one of the 1s electrons, accompanied by excitation of the outer 2s electron onto an nl orbital via correlation effects, or indirectly by one-electron autoionization decay of the highly excited (2l, n'l', n"l") hollow states. Most of these doubly-excited states of Li$^+$ undergo Auger decay to the ground state of Li^{2+}. Thus, the detection of the second step Auger decay lines provides direct information on the primary photoionization/ photoexcitation process, since fluorescence decay of these doubly-excited states is negligible.

The electron lines measured in a photoelectron spectrum can be seen in Fig. 2 where photons of 163 eV energy have been used to photoionize Li atoms.

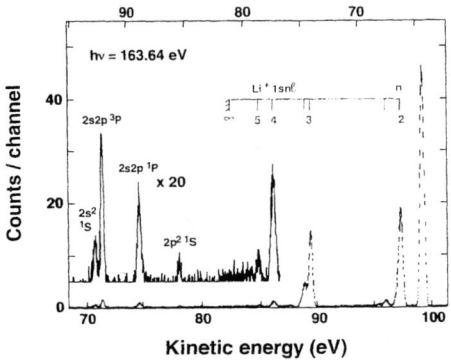

FIGURE 2. Electron spectrum following photoionization of Li by 163 eV photons.

The lines observable below the Li double-ionization threshold at 81.03 eV binding energy (BE, upper scale of the figure) are the 1snl correlation satellites, and the main lines (1s2s ^{1,3}S at 64.41 eV and 66.32 eV BE, respectively.) The lines which are visible on the left part of the figure above 81.03 eV BE results from the Auger decay of the (2l, 2l') doubly excited states. The corresponding photolines would appear at low kinetic energies with a poor intensity, because of the low transmission of the electron spectrometer used in these experiments (a cylindrical mirror analyzer, CMA, with a resolution of 0.4 eV at 80 eV kinetic energy). To measure the energy dependence of the intensity of these correlation satellites with two vacancies in the K-shell, we therefore decided to make the measurement on the well-resolved Auger lines (since we were using undulator radiation from Super ACO). For measurements of Auger lines, the low spectral resolution of the monochromatic photon beams does not affect the resolution of the observed electron lines. In fact, we increased considerably the intensity of the Auger lines by opening widely the slits of the monochromator. In this way, we measured the low relative cross sections to the 2l2l' Li$^+$ states over a wide energy range up to 450 eV. In Fig. 3, we show the branching ratios between the measured intensity of the 2s3s ^{3}S Li$^+$ line (upper panel) and 2s2p ^{1}P line (lower panel) and the 1s2s ^{3}S main line. The energy variation of these two branching ratios differs considerably, revealing the different mechanisms leading to their production. The 2s3s ^{3}S satellite results mainly from a double shake-up process, in which the remaining 1s electron is shaken up to the 2s orbital of the ion while the 2s electron is simultaneously shaken-up to a 3s orbital. The energy dependence shows the characteristic increase of the relative intensity of this satellite, starting with a low value close to threshold (0.4%) and increasing towards higher photon energies to reach a plateau of about 1.2% at high energy. The 2s2p ^{1}P satellite results mainly from interchannel coupling having the result that the remaining 1s-electron is excited to a 2p orbital. Like in the case of core-ionization in sodium, the relative intensity of this satellite shows a maximum at threshold (here, 1.3%) and decreases rapidly with increasing kinetic energy of the photoelectron, with an asymptotic value of 0.2% at high energy.

As a second example of low-resolution high-intensity measurement we show the photoionization of multiply-charged ions. Experimental studies of this process are still scarce due to the low-density of ions which can maintained in steady state conditions. Thus, high-intensity undulator radiation is absolutely needed for this kind of experiments.

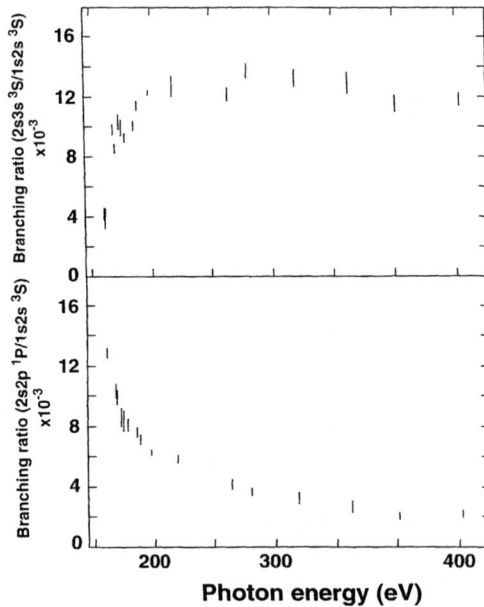

FIGURE 3. Branching ratios 2s3s^{3}S/1s2s ^{3}S (upper pannel) and 2s2p ^{1}P/1s2s ^{3}S (lower pannel) in Li.

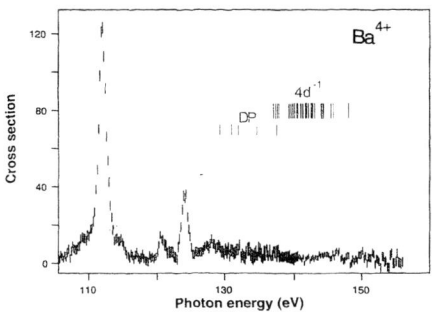

FIGURE 4. Photoionization spectrum of Ba^{4+} ions.

In a previous study [1], we analyzed photoionization of multiply charged ions of Xe combining the use of an ECR ion source with SR emitted from an undulator of Super ACO. For such ions, the natural lifetime of an inner vacancy in the 4d subshell is sufficiently high (above 100 meV) to allow measurements with a relatively low spectral resolution (0.1 to 0.5 eV in the 100-150 eV photon energy range). Here, we went one step further in adding an oven into the ECR source in order to produce multiply-charged ions of metallic vapors. In Fig. 4, we show the photoion spectrum resulting from single photoionization of Ba^{4+} ions to Ba^{5+} final states, according to: Ba^{4+} $4d^{10}5s^25p^4$ + $h\nu$ → Ba^{4+*} $4d^95s^25p^4nl$ → Ba^{5+} $4d^{10}5s^25p^3\varepsilon l$, the excited nl orbitals being mostly nf orbitals: 4f (close to 112 eV) and 5f (close to 124 eV). This dominant intensity of the 4d → nf transitions and the negligible intensity of Ba^{4+} double photoionization (which would result from single 4d-photoionization followed by Auger decay of the 4d-hole) confirms that the collapse of the nf orbitals is complete for Ba^{4+} (it occurs, in fact, for Ba^{3+}).

HIGH SPECTRAL RESOLUTION-HIGH INTENSITY EXPERIMENTS

In this paragraph we give two examples of electron spectrometry experiments for which a high-spectral resolution is needed, while the use of a low resolution spectrometer is still acceptable. We present first partial cross sections for resonant photoionization of lithium atoms [2] at 142.28 eV, i.e., at the excitation energy of the $2s^22p$ ^2P hollow Li state (see Fig. 1). The data were obtained using undulator radiation from the 3rd SR source ALS and the same CMA spectrometer as previously. We show in Fig. 5 (left pannels), partial cross sections for photoionization of Li into the 1s2s ^1S, 1s2p ^3P, and $1s^2$ ^1S continuum channels (from top to bottom). Here, the spectral resolution is 20 meV. The modest resolution (0.3 eV) of the CMA allows, however, resolving the corresponding electron lines. The profiles summarize the different behaviors which can be generally expected for autoionizing resonances [3]: in the ^1S channel, the shape of the cross section has a Fano-type profile, revaling a strong interference between the direct and resonant pathways; in the ^3P channel, one observes a symmetric Lorentzian profile, due to weak interferences between the strong resonant decay and the low direct photoionization cross section into this channel; and in the $1s^2$ ^1S channel one measures a window resonance, resulting from destructive interferences. The solid lines going through the points are fitted profiles using the formulae established by Starace [4], the dashed lines are the results of an R-matrix calculation. The lifetime of the $2s^22p$ ^2P state was measured to be 0.118 (5) eV.

Angle resolved studies [5] of the autoionization of the hollow excited states in lithium could also be measured in the same conditions, i.e., with a high-intensity, high spectral resolution photon beam and the low resolution CMA. In Fig. 5 (right pannels), we show the variation of the angular distribution parameter determined over the energy range of the $2s^22p$ ^2P resonance for [(1s2p ^3P)εl] and [(1s2p ^1P)εl] electrons. Both show a similar photon energy dependence, with β-values near 2 outside the resonance region and deep minima, of the order of 0.3–0.4, near the energy of the resonance. The full lines are here the results of R-matrix calculations after convolution with the spectral bandwidth (30 meV). Especially worthwhile to be noted is the fact that the photon energy at which minima occur into the β-values do not coincide with the energy of the resonance. This results from fast variations of the matrix elements and phase shift over the energy of the resonance. The occurence of this effect seem to be

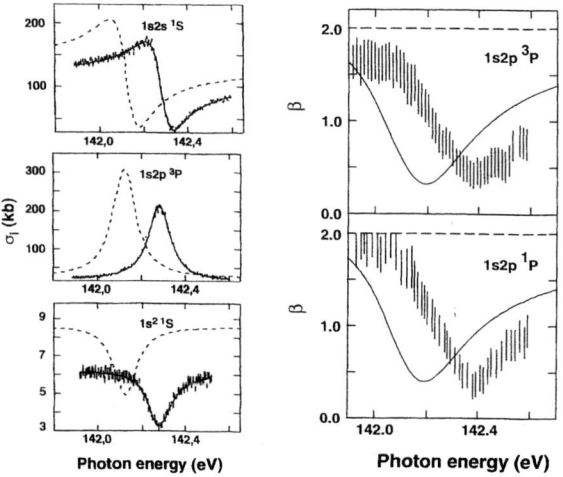

FIGURE 5. Partial cross sections (the left pannels, from ref. 2) and asymmetry parameters β (right pannels, from ref. 5) over the $2s^22p$ ^2P triply-excited resonance of Li.

quite general [6], although only high spectral resolution experiments allows an accurate determination of its value.

HIGH-RESOLUTION HIGH-INTENSITY EXPERIMENTS

Evidently, the level of the performances reached with a 3rd generation SR source is so high, that new apparatus need to be designed to make the best use of their high-resolution high-intensity photon beams. This means that the resolution in the detection channel must be similar to the resolution in the excitation channel. This is the case with the new SCIENTA-200 hemispherical analyzer whose ultimate resolution is 1 to 2 meV (when rare gases are used to calibrate the apparatus). We were able to use this spectrometer in collaboration at ALS crucial to get more insight into the dynamics of the photoionization process. Fig. 6 shows the Li$^+$ 1snl photoelectron spectrum [7] covering the binding energy from the 1s2l diagram lines up to and accross the double ionization threshold, measured at 100 eV photon energy. The region of $n \geq 4$ satellite transitions, shown on the left-upper part of the figure was measured with a resolution of 46 meV FWHM. A highest resolution of 37 meV FWHM was ultimately used to analyze in more details the 1s3l and 1s4l transition satellites, resulting from a spectral resolution of 30 meV, and a spectrometer resolution of 20 meV with a negligible contribution of the Doppler effect. Using such spectra, we were able to determine the relative intensities of the transitions with different values of Δl of the angular momentum transferred to the outer 2s electron (up to $\Delta l = 2$). Here again, the 1sns satellites results from a shake-up of the outer electron ($\Delta l = 0$), while interchannel coupling is responsible for the transitions with $\Delta l \geq 1$. The main result of this high-resolution experiment is that for 1snl satellite transitions with $n \geq 4$, the high-angular momentum lines ($\Delta l = 2$ and 3) contribute predominantly to the satellite cross sections [7].

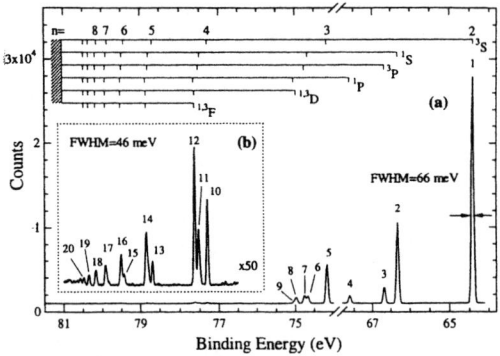

FIGURE 6. Photoelectron spectra of Li recorded with 100 eV photons (from Ref. 7).

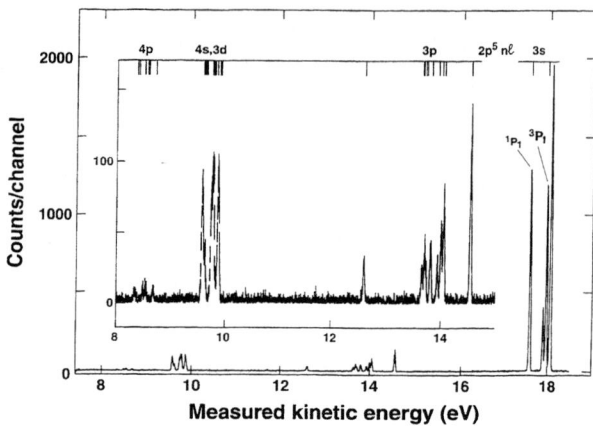

FIGURE 7. Photoelectron spectrum of Na recorded with 59 eV photons and a total resolution of 33 meV FWHM.

Finally, we show in Fig. 7 one of the highly-resolved photoelectron spectra of Na atoms we have measured at ALS. We observed many new satellite transitions and were able to resolve the various J components of several LS terms, within the 2p^{5}3s ^{1,3}P main lines as well as within several group of 2p^{5}nl satellites.

There are limitations in the use of a high-resolution electron spectrometer to study metal vapors. First, the density acceptable in the source volume of the analyzer must be kept significantly smaller than in the case of a lower resolution electron spectrometer. The very accurate adjustment of all components in a SCIENTA analyzer precludes a regular cleaning of the pieces inside of the spectrometer, as it can be done every day in a simpler spectrometer like the CMA. The limitation of the sample density makes also much more difficult two-color experiments involving the use of laser and synchrotron radiation, since the possibility to get a high density of atoms in an excited state is strongly governed by radiation trapping.

ACKNOWLEDGMENTS

The authors would like to thank S. Diehl, N. Berra, C. Blancard, J. Bozek, E. Kennedy, and J. Mosnier for their help in taking some of the data. They gratefully acknowledge the support of the CEA-DAM.

REFERENCES

1. Bizau, J.-M., et al., Phys. Rev. Lett. **84**, 435-38 (2000).
2. Diehl, S. et al., Phys. Rev. Lett. **76**, 3915-3918 (1996).
3. Fano, U,. Phys. Rev. **124**, 1866-1878 (1961).
4. Starace, A. F., Phys. Rev. A **16**, 231-242 (1977).
5. Diehl, S., et al., Phys. Rev. Lett. **84**, 1677-1680 (2000).
6. West, J. B., et al., J. Chem. Phys. **104**, 3923 (1996).
7. Cheng, W. T., et al. Phys. Rev. A **62**, 052501-9 (2000).

Synchrotron Radiation Total Reflection for Rainwater Analysis

Silvana Moreira Simabuco and Edson Matsumoto

Campinas State University, Civil Engineering Faculty - Water Resources Department
P. O. Box 6021 – Zip Code 13083-970 – Campinas - SP- Brazil
e-mail: silvana@fec.unicamp.br

Abstract. In this work Total Reflection X-Ray Fluorescence Analysis with excitation by Synchrotron Radiation (SR-TXRF) has been used for rainwater trace element analysis. The samples were collected in four different sites at Campinas City, SP, Brazil. Rainwater samples of 10 µl were added to Perspex reflector disks, dried under vacuum and analyzed for 100 s measuring time For the calibration system standard solutions with gallium as internal standard were prepared.. The detection limits obtained for K-shell lines varied from 29 ng.ml^{-1} for sulfur to 1.3 ng.ml^{-1} for zinc and copper, while for L-shell the values were 4.5 ng.ml^{-1} for mercury and 7.0 ng.ml^{-1} for lead.

INTRODUCTION

The total reflection X-ray fluorescence technique was introduced in 1971 by Yoneda and Horiuchi [1] and developed by Aiginger and Wobrauschek [2,3]. This method is based on the incidence of an X-ray beam at small angle (denoted critical angle) on the flat surface of a support or carrier (for example, quartz or Perspex) on which the sample to be analyzed is deposited. In this condition the scattering effect is minimized and thus a better peak-background ratio is obtained reducing in this way the detection limits.

The background intensity also can be reduced using linear polarization of the exciting radiati on.

The most intensive X-ray source available nowadays is the synchrotron, providing outstanding properties of brilliance, linear polarization and natural collimation.

Another advantage of this technique is the small volumes required for liquid sample analysis (microliters) or small masses (micrograms) for solid samples after chemical digestion.

TXRF is especially suitable for ultra-trace analysis of pure waters such as rain and drinking water and its advantages when compared to other methods such as atomic absorption spectrometry (AAS) and inductively coupled plasma techniques (ICP-AES and ICP-MS) are the multielement determination and low cost [4].

The aim of this work is apply the Total Reflection with Synchrotron Radiation (SR-XRF) to determine trace elements in rainwater sampling in four different sites in Campinas City, SP, Brazil.

MATERIALS AND METHODS

The quantitative analysis can be made through Equation 1, because the sample can be considered as thin film so that absorption and enhancement effects can be neglected.

$$I_i = S_i \, C_i \qquad (1)$$

where I_i represents the intensity (cps) for K or L X-ray line for the element i; C_i the concentration (in ppm or µg.ml^{-1}) and S_i the sensitivity for this element (cps.µg^{-1}.ml).

The thin film formed on the Perspex support does not have a regular geometry and therefore the X-ray intensities depend on the thin film position. This geometric effect [4,5] can be corrected computing the relative intensity for each element in relation to an internal standard added in every sample and standard.

Instrumentation

For the X-ray detection a hyperpure Ge detector with 140 eV resolution at 5.9 keV (Mn K_α line) was employed, while for the excitation a white beam of synchrotron radiation [6] with 2 mm width and 1 mm height under total reflection conditions was used.

Sample and Standard Preparations

To set up the calibration curve, five standard solutions containing the elements V, Cr, Mn, Fe, Co, Ni, Cu, Zn, As, Se, Tl and Pb were prepared at different and well-known concentrations (from 0.3 to 3.8 $\mu g.L^{-1}$) with Ga addition as internal standard.

For the sample preparation 1 ml of each sample was taken and 10 μl of Ga (1025 $\mu g.ml^{-1}$) was added resulting in a Ga concentration of 10.148 $\mu g.ml^{-1}$ as internal standard in the sample. Then an aliquot of 10 μl was pipetted onto a Perspex disk and dried in vacuum thus giving rise to a shaping thin layer of approximately 5 mm diameter (two replicates).

The samples were excited for 100 s and X-ray spectra obtained were evaluated by the software QXAS [7] in order to obtain the X-ray intensities.

Sampling Data

The site and date sampling for rainwater samples are shown in the Table 1.

TABLE 1. Site and date sampling for rainwater samples.

Site	Sample	Date
Water Treatment Plant 1 and 2	Ch-1	03/29/98
	Ch1-2	04/29/98
	Ch1-3	05/17/98
	Ch1-4	06/19/98
	Ch1-5	07/20/98
	Ch1-6	08/03/98
Water Treatment Plant 3 and 4	Ch2-1	03/29/98
	Ch2-2	08/03/98
Water Treatment Plant Capivari River	Ch3-1	03/29/98
	Ch3-2	08/03/98
Water Supply Atibaia River	Ch4-1	03/29/98

RESULTS AND DISCUSSION

On the bases of the measurements performed on standard solutions, the calibration curve for synchrotron radiation total reflection as shown the Figure 1 was obtained.

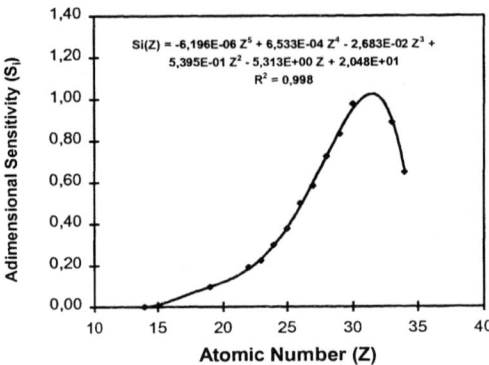

FIGURE 1. Adimensional sensitivity curve for Synchrotron Radiation X-ray Total Reflection

The detection limits were calculated by equation [8,9]:

$$LMD_i = 3 \cdot \sqrt{\frac{I_i(BG)}{t}} \cdot \frac{C_{Ga}}{I_{Ga} \cdot S_i} \qquad (2)$$

where $I_i(BG)$ is the background intensity for the element i; I_{Ga} the internal standard intensity (Ga); C_{Ga} the internal standard concentration (Ga), S_i the relative sensitivity for the element i and t the measuring time.

Detection limits (in $\mu g.L^{-1}$ or $ng.ml^{-1}$) for synchrotron radiation total reflection, calculated using the equation above, were extrapolated for 1000 seconds measuring time (figure 2).

As can be visualized in this figure, the detection limits obtained for K-shell changed from 29 $ng.ml^{-1}$ for sulfur to 1.3 $ng.ml^{-1}$ for zinc and copper. For L-shell the detection limits were 4.5 $ng.ml^{-1}$ for mercury and 7.0 $ng.ml^{-1}$ for lead.

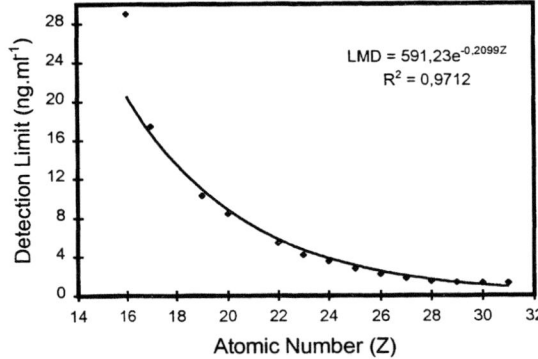

FIGURE 2. Detection Limit for Synchrotron Radiation Total Reflection (SR-TXRF) for 1000 s.

The concentrations of rainwater samples were calculated using the equation presented in Figure 1 and the results are presented in Figures 3 and 4.

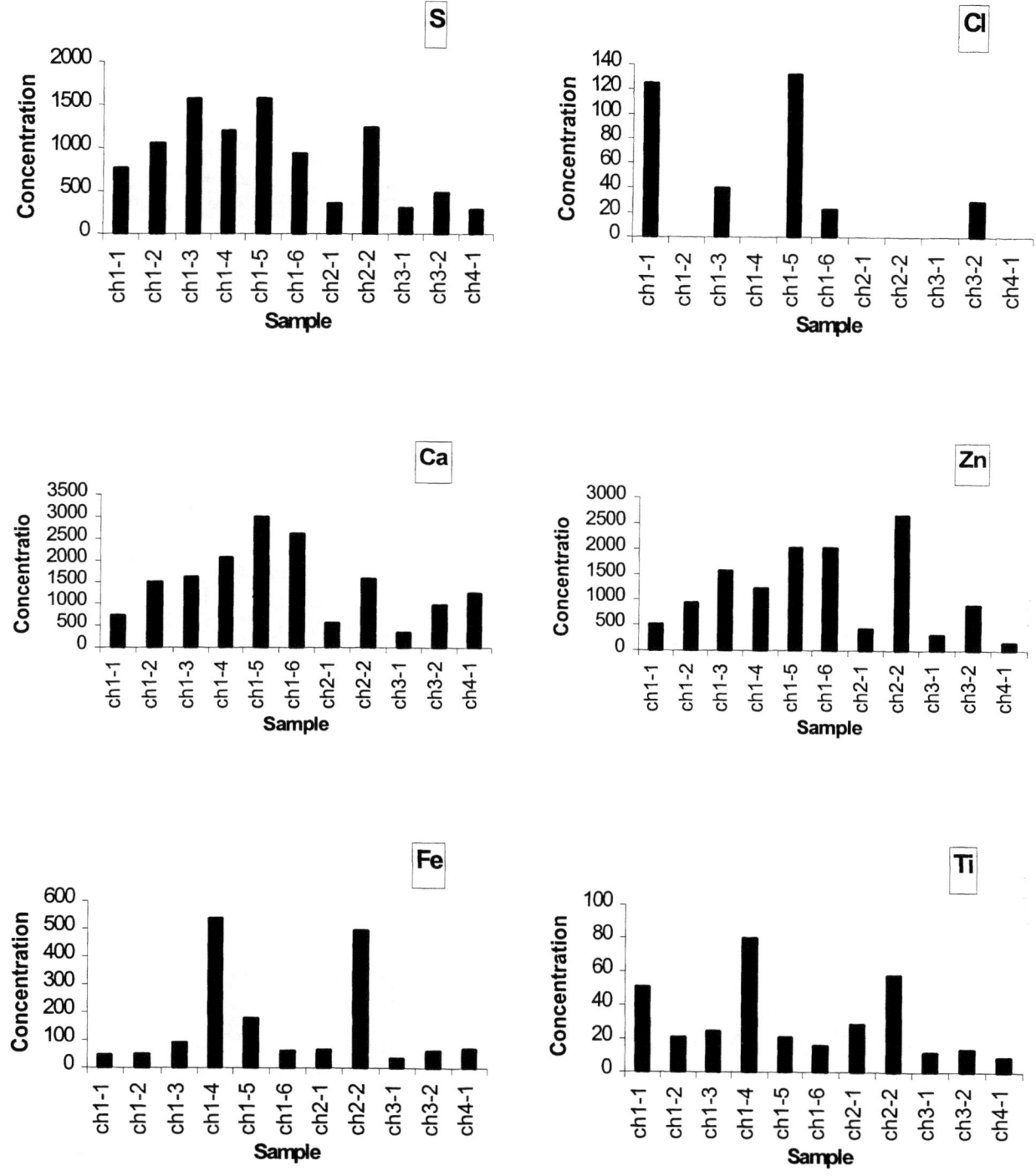

FIGURE 3. Concentration (ng.ml^{-1}) of S, Cl, Ca, Zn, Fe and Ti for rainwater samples by Synchrotron Radiation Total Reflection X-ray Fluorescence (SR-TXRF).

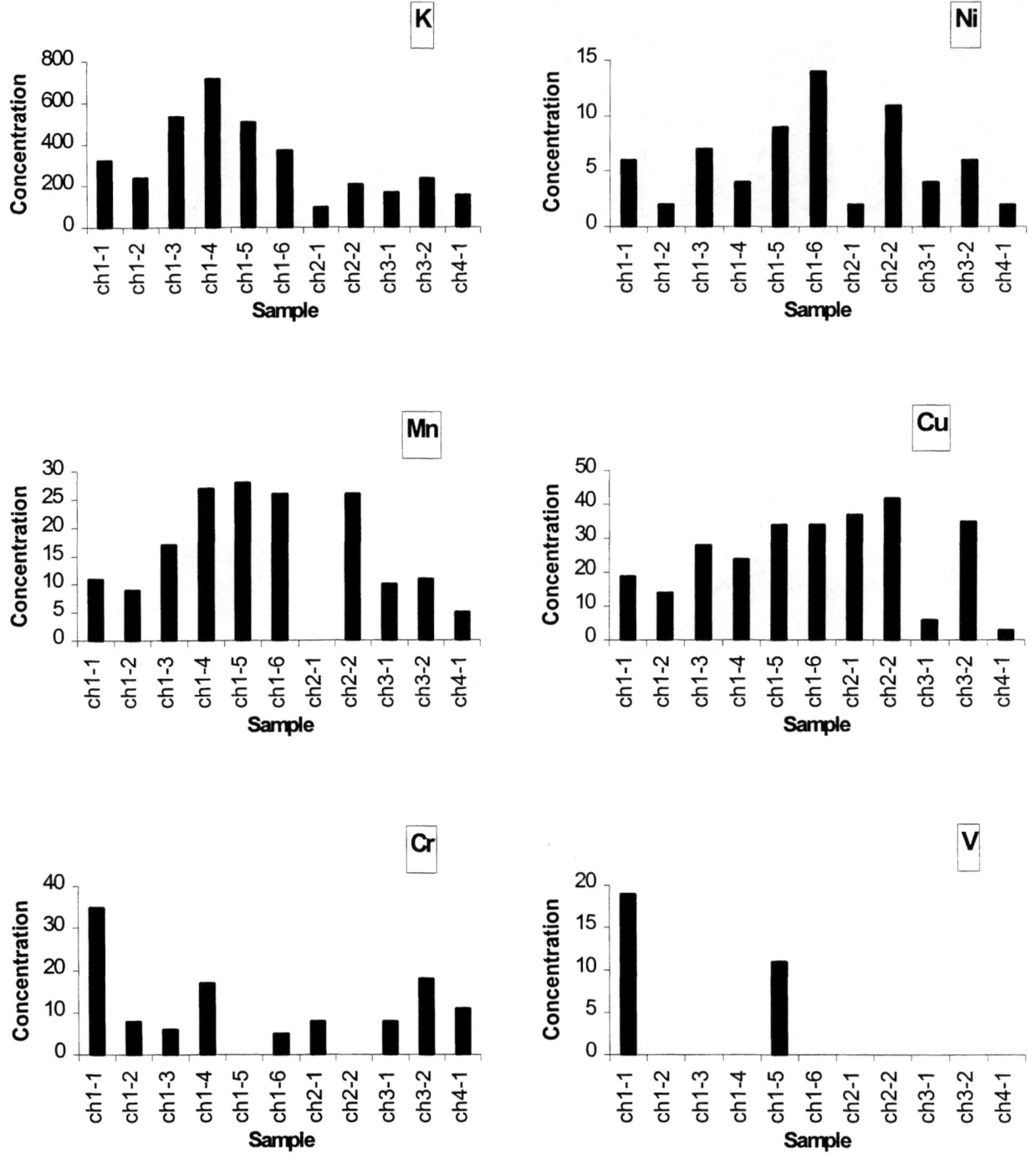

FIGURE 4. Concentration (ng.ml^{-1}) of K, Ni, Mn, Cu, Cr and V for rainwater samples by Synchrotron Radiation Total Reflection X-ray Fluorescence (SR-TXRF).

A standard reference material (Drinking Water Pollutants - Aldrich- 41,393-3) containing Cr, As, Se, Ba and Pb was analyzed in order to test the procedure. Table 2 shows the results and it can be see that the measured values agree well with the certified values.

TABLE 2. Measured and certified values in the standard reference material (Drinking Water Pollutants)

	Concentration ($\mu g.ml^{-1}$)	
Element	Certified Value	Measured Value
Cr	10.00±0.49	9.74±0.09
As	10.00±0.49	10.05±0.14
Se	5.00±0.05	5.00±0.25
Ba	100.00±0.53	99.98±0.53
Pb	10.00±0.49	9.86±0.61

As can be see by the table 1 the accuracy of this method is about 3% and the precision about 2%. In general, the detection limits values obtained are in good agreement with those reported by other workers in different synchrotron radiation facilities [4, 5, 10, 11].

CONCLUSIONS

The elements Ca, S, Fe and Zn presented the highest concentrations in every sample. Comparing the Ch1-1, Ch2-1, Ch3-1 and Ch4-1 samples collected in the same day (03/29/98) and different sites was not observed a significative variation for the same element but for the other three samples, Ch1-6, Ch2-2 and Ch3-2, sampling in 08/03/98, can be observed a small variation in the concentration for the same element.

When the Ch2-1 and Ch2-2 samples are compared, which were sampled at the same site but on different dates (03/29/98 and 08/03/98) it can be noted that the values for the Ch2-2 sample are higher than for the Ch2-1 sample. The same fact happened for the Ch3-1 and Ch3-2 samples, collected in the same date cited above.

This can be explained by the fact of August be a rainier month than March, thus, the particulate material present in the atmospheric is higher and when it rains this material is dragged together rainwater, resulting in a higher concentrations in the sample.

ACKNOWLEDGMENTS

Research (partially) performed at National Synchrotron Light Laboratory (LNLS) and financial supported by FAPESP.

REFERENCES

1. YONEDA, Y and HORIUCHI, T., *Rev. Sci. Instrum.*, **42**, 1069 (1971).

2. AIGINGER, H. and WOBRAUSCHEK, P., *Nucl. Instr. Meth.*, **114**, 157 - 158 (1974).

3. WOBRAUSCHEK, P. and AIGINGER, H., *Anal. Chem.*, **47**, 852 – 855 (1975).

4. KLOCKEMKÄMPER, R. and VON BOHLEN, A., *X-ray Spectrom.*, **25**, 156 – 162 (1996).

5. LADISICH, W., RIEDER, R., WOBRAUSCHEK, P., X-ray fluorescence analysis with monoenergetic excitation. *John Wiley & Sons Ltd., 1994.*

6. PÉREZ, C. A., RADTKE, M., SÁNCHEZ, H. J., TOLENTINO, H., NEUENSHWANDER, R., BARG, W.,RUBIO, M., BUENO, M. I. S., RAIMUNDO, I. M., and ROHWEDDER, J. R., *X-ray Spectrom.*, **28**: 320 – 326 (1999).

7. Quantitative X-ray Analysis System (QXAS) software package, IAEA, Vienna, (1970).

8. CURIE, L. A., *Anal. Chem.*, **40**(3): 586 - 593, (1968).

9. LADISICH, W., RIEDER, R., WOBRAUSCHEK, P., and AIGINGER, H., *Nucl. Instr. Meth. in Phys. Res.*, **330A**: 501 - 506, (1993).

10. MUIA, L. M., RAZAFINDRAMISA, F., VAN GRIEKEN, R., *Spec. Chim. Acta B*, **46B** (10): 1421 - 1427, (1991).

11. HOFFMANN, P., KARANDASHEV, V. K., SINNER, T., and ORTNER, H. M., *Fresenius J. Anal. Chem.*, **357**: 1142 - 1148, (1997).

Quantitative Analysis of Biomedical Samples Using Synchrotron Radiation Microbeams

Ali Ektessabi[*], Shunsuke Shikine[*], and Sohei Yoshida[†]

[*]*Department of Precision Engineering, Graduate school of Engineering, Kyoto University, Kyoto, 606-8501, Japan*
[†]*Division of Neurological Diseases, Wakayama Medical College, Wakayama, 640-8115, Japan*

Abstract. X-ray fluorescence (XRF) using a synchrotron radiation (SR) microbeam was applied to investigate distributions and concentrations of elements in single neurons of patients with neurodegenerative diseases. In this paper we introduce a computer code that has been developed to quantify the trace elements and matrix elements at the single cell level. This computer code has been used in studies of several important neurodegenerative diseases such as Alzheimer's disease (AD), Parkinson's disease (PD) and parkinsonism-dementia complex (PDC), as well as in basic biological experiments to determine the elemental changes in cells due to incorporation of foreign metal elements. The substantia nigra (SN) tissue obtained from the autopsy specimens of patients with Guamanian parkinsonism-dementia complex (PDC) and control cases were examined. Quantitative XRF analysis showed that neuromelanin granules of Parkinsonian SN contained higher levels of Fe than those of the control. The concentrations were in the ranges of 2300-3100 ppm and 2000-2400 ppm respectively. On the contrary, Zn and Ni in neuromelanin granules of SN tissue from the PDC case were lower than those of the control. Especially Zn was less than 40 ppm in SN tissue from the PDC case while it was 560-810 ppm in the control. These changes are considered to be closely related to the neuro-degeneration and cell death.

INTRODUCTION

Synchrotron radiation x-ray fluorescence (SR-XRF) spectrometry is a powerful and non-destructive method for the investigation of trace elements [1]. Distribution and the concentration of trace elements can be analyzed with high special resolution [2].

In this study we performed quantitative XRF analysis on substantia nigra (SN) tissues from a patient with Guamanian parkinsonism-dementia complex (PDC) and a control case. The metal elements are considered to play an important role in neurodegenerative disorders such as PDC, Parkinson's disease (PD), and Alzheimer's disease (AD) [3,4]. The interaction between proteins and transitional metals has a critical effect on the progress of these disorders through the generation of free radicals that cause oxidative stress [3,5]. The accumulation of transitional metal elements in the brain tissues has been reported in cases of PD and AD patients [5-7].

PDC is a neurodegenerative disorder in the western Pacific region [8-10]. In the present study the local concentrations of transitional metal elements that were accumulated in SN tissue of the PDC patient and a control case are quantified from the XRF spectra. The quantitative information about these elements is significant for the elucidation of the mechanisms of these diseases. The samples from each case were also investigated histologically by staining after XRF analysis.

E-mail (Ali Ektessabi): h51167@sakura.kudpc.kyoto-u.ac.jp

Basic Equations for Quantification

The intensity of the fluorescence x-ray $I_f(\lambda)$ that is generated from element i in the sample with the thickness of t is given by

$$I_f(\lambda) = QI_0(\lambda)\mu_i(\lambda)W_i \operatorname{cosec}\phi \{1-\exp(-\mu^*\rho t)\}\frac{1}{\mu^*} \quad (1)$$

where $I_0(\lambda)$ is the intensity of the incident x-ray with wavelength λ, $\mu_i(\lambda)$ is the effective mass attenuation coefficient of element i for incident wavelength λ, W_i is the weight fraction of element i, ϕ is the incidence angle, and ρ is the density of the sample. Q is the constant determined, mainly, by ionization cross section, fluorescence yield, the solid angle to the detector, and the path of x-rays. Q is given by

$$Q = CP_i \quad (2)$$

where C is the constant that is determined by the geometrical parameters of the set-up instruments, therefore C is constant under the same experimental condition. P_i is the probability that a characteristics line of element i is emitted. μ^* is given by

$$\mu^* = \mu(\lambda)\operatorname{cosec}\phi + \mu(\lambda_f)\operatorname{cosec}\varphi \quad (3)$$

where $\mu(\lambda)$ is the effective mass attenuation coefficient of the sample for incident wavelength λ, λ_f is the wavelength of the fluorescence x-ray, and φ is the take-off angle of the fluorescent x-ray. The derivations of these equations are given in detail in the literature [11].

In this study, the samples are thin sections and their major constituents are light elements. Therefore the magnitude of $\mu^* \rho t$ is considered to be small and Eqn (1) can be approximated as

$$I_f(\lambda) = CP_i I_0(\lambda)\mu_i(\lambda)W_i \rho t \operatorname{cosec}\phi \propto W_i \rho t. \quad (4)$$

Under this approximation, the fluorescent x-ray intensity from each element is proportional to the incident x-ray intensity and to the area density of the element $W_i \rho t$.

P_i and $\mu_i(\lambda)$ can be obtained from the existing database and handbooks [12]. $I_0(\lambda)$ was monitored by an ionization chamber. C is calculated by comparing the peak areas of XRF spectra obtained from thin pure metal film reference samples whose thicknesses are already-known. $I_f(\lambda)$ is obtained from peak areas of XRF spectra that are irradiated from the sample and are measured by Si detector.

Quantification was performed in the procedure as follows. The peak areas were calculated using a computer code developed by our group for the quantification of very low content elements in biomedical samples. In this program, the background is estimated from the untreated spectra, and the peak is obtained using Gaussians curve fitting and the least squares method. The absorption of fluorescent x-ray by the sample and the consequent excitation of other elements are not considered here. A typical example of the results that are calculated by this program is shown in figure 1.

The program is written in Visual Basic, and includes the following process: i) reduction of the background, ii) curve fitting of the peaks, iii) separation of peaks if essential, iv) smoothing of the spectra if essential, v) processing of the spectra and comparison of the reference material (thin film) vi) calculation of the possible errors in measurement process. Users can modify fitting parameters according to their needs.

INSTRUMENTS AND METHODS

XRF analysis in this investigation was performed at Photon Factory in beam line 4A. The incident x-ray energy was 14.3 keV and the beam size was approximately 7x5 μm^2. The synchrotron radiation was monochromatized by a multilayered reflecting mirror. Monochromatized x-rays were focused using slits and Kirkpartrick-Beaz optics [13]. Irradiating x-ray to the sample generated fluorescent x-ray and Si detector monitored its intensity. The incident and transmitted x-rays were monitored by ionization chambers that were set in front of and behind the sample. The sample was fixed on the holder on a stage that was moved by pulsemotors for elemental distribution imaging. A monitoring system equipped with a CCD camera was used to determine the measurement position. The analysis was carried out in air.

Using hematoxylin-eosin and Bodian stainings, combined stainings were performed for histological investigation after XRF analysis.

EXPERIMENTAL PROCEDURE

Brain tissues were obtained by autopsy from a 56-year-old male patient with a Guamanian PDC and were fixed in 10 % formalin. Then SN tissues were embedded in paraffin and were cut into 8 μm thin sections. The sample for XRF analysis was made by putting the section on a Mylar film. The SN tissues of an age-matched male patient without PDC were treated in the same way and were prepared as a control.

The optical microscope observation of the unstained PDC nigral section revealed that the section contained only a few neuromelanin granules much less than the control. The loss of neurons is characteristic of PDC. The elemental distribution images were obtained in the areas that contained surviving neuromelanin granules. From the results of the imaging, the measurement points A1-A3, B1-B2, and C1 were selected for further point-measurement: neuromelanin granules, nigral tissues, and glial cells respectively. The measured points are shown in Fig. 2(a and b), which are the optical microscopic photographs of the stained sample. XRF spectra were obtained at these points. The typical spectra that were measured in each section are shown in Fig. 3(a). The measurement time was 200 seconds. Quantitative analysis was then applied to the spectra and the calculated values for

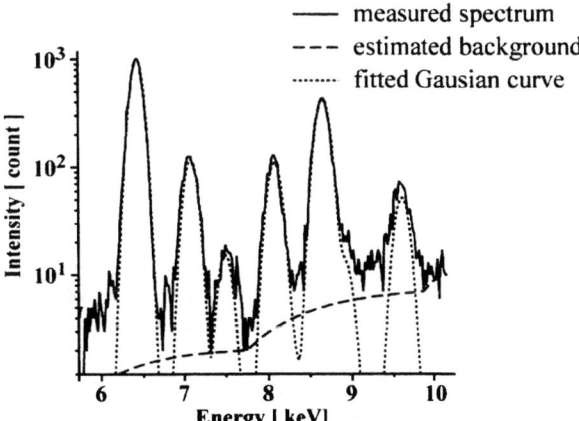

FIGURE 1. A typical example of the results calculated with a computer code that was made by our group.

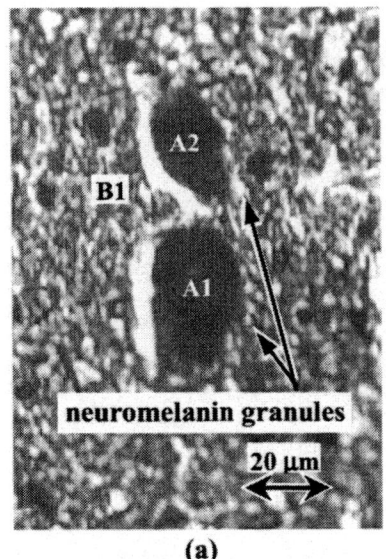
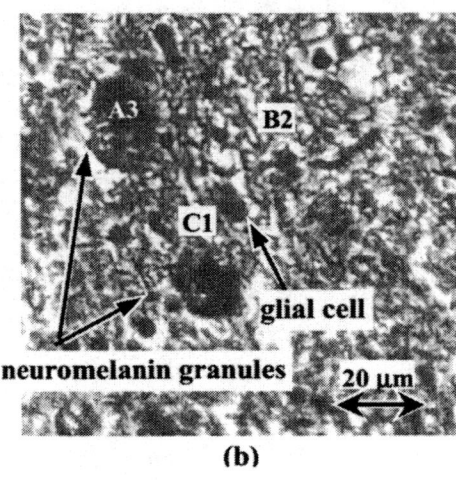

FIGURE 2. The optical microscopic photographs in the different areas of the sample from the parkinsonism-dementia complex (PDC) case. The neuromelanin granules, the nigral tissues, and the glial cell are observed at measurement points A1-A3, B1-B2, and C1 respectively.

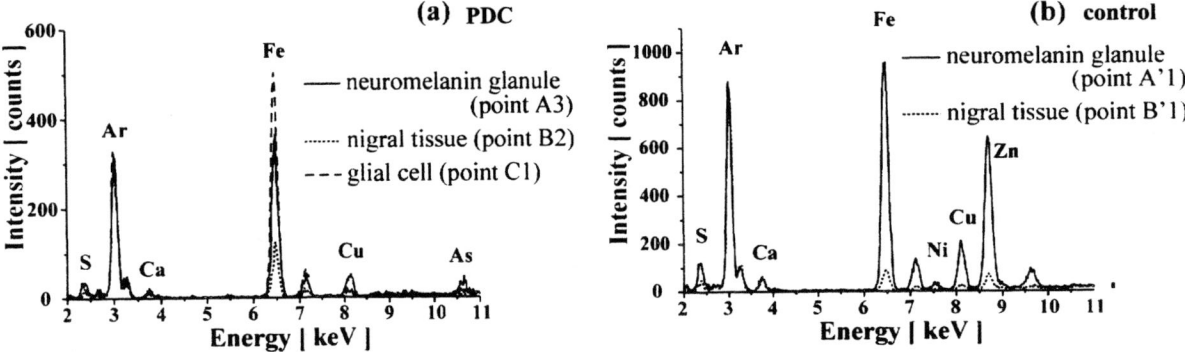

FIGURE 3. The spectra measured in the substantia nigra (SN) tissues of patients with (a) the Guamanian PDC and (b) the control cases. Measured points in (a) are shown in figure 2(b). Points A, B and C are located in the neuromelanin granules, in the nigral tissues, and in the glial cell respectively. The measurement time was 200 sec at each point. The incident beam energy was 14.3 keV.

concentrations of Fe, Cu, and Zn are shown in Table 1.

The analysis for the sample from the control case was performed according to the same process as the PDC case. Measurement points A'1-3 and B'1-2 were determined in neuromelanin granules and in nigral tissues. The typical XRF spectra obtained in each section from the control case are shown in Fig. 3(b). The concentrations of Fe, Cu, and Zn that were calculated from spectra and the results are also shown in Table 2.

EXPERIMENTAL RESULTS

The spectra (Fig. 3(a)) revealed that S, Fe, Cu and As were accumulated in neuromelanin granules more than in surrounding nigral tissues. In the spectrum that was obtained in the glial cell, the high peak of Fe can be seen but the peaks of S, Cu, and As are as low as those measured in nigral tissues.

The spectra obtained in the neuromelanin granules from the control subject (Fig. 3(b)) showed clear peaks of Zn and Ni, but in the spectra from the PDC case they are hardly detected. On contrary, the peak of As that was detected in the neuromelanin granules with the PDC case could not be seen in the spectra from the control case.

The results of the quantitative analysis showed that the neuromelanin granules of the PDC case contained Fe and Cu with concentration in the range of 2300-3100 ppm and 210-320 ppm respectively. In the control case, the concentrations of Fe, Cu, and Zn in

Table 1. The quantification results from the PDC case.

Measurement points	concentration [ppm]		
	Fe	Cu	Zn
Neuromelanin granule			
A1	3.1×10^3	3.2×10^2	$<4.0 \times 10^1$
A2	2.7×10^3	3.0×10^2	$<4.0 \times 10^1$
A3	2.3×10^3	2.1×10^2	$<4.0 \times 10^1$
Nigral tissue			
B1	7.3×10^2	6.0×10^1	$<3.0 \times 10^1$
B2	7.2×10^2	5.2×10^1	$<2.0 \times 10^1$
Glial cell			
C1	3.2×10^3	6.2×10^1	$<2.0 \times 10^1$

Table 2. The quantification results from the control case.

Measurement points	concentration [ppm]		
	Fe	Cu	Zn
Neuromelanin granule			
A'1	2.2×10^3	2.9×10^2	8.1×10^2
A'2	2.0×10^3	1.7×10^2	6.3×10^2
A'3	2.4×10^3	1.7×10^2	5.6×10^2
Nigral tissue			
B'1	3.1×10^2	9.3×10^1	1.4×10^2
B'2	2.0×10^2	3.6×10^1	8.0×10^1

the neuromelanin granules were 2000-2400 ppm, 170-290 ppm, and 560-810 ppm respectively.

CONCLUSION

The present study introduces a computer code that has been developed to quantify the trace elements at the single cell level. This code can be used to investigate several important neurodegenerative diseases such as AD, PD and PDC, as well as to investigate basic biological samples in order to study changes in cells due to the incorporation of foreign metal elements. As a typical case, concentrations of the metal elements in neuromelanin granules and nigral tissues from a patient with PDC were investigated and compared to a control case. It was also possible to observe the same samples histologically by staining after the elemental analysis.

The results of XRF spectroscopy showed that Fe and As had accumulated in neuromelanin granules, likely due to the progression of PDC. On the contrary, a reduction of Ni and Zn in the neuromelanin granules was found in the PDC case as compared to the control. Quantification to the XRF spectra that were obtained from neuromelanin granules showed that the concentrations of Fe were in the ranges of 2300-3100 ppm and 2000-2400 ppm in the PDC and the control cases respectively. The concentration of Cu was 210-320 ppm in the PDC case and 170-290 ppm in the control case. Zn was hardly detected in the PDC case though neuromelanin granules in the control case contained it at a concentration of 560-810 ppm. In the PDC case, the glial cell adjacent to neuromelanin granules contained Fe with a high concentration of 3200 ppm.

The quantitative information about the metal elements in bio samples that obtained using XRF analysis is considered to bring new insight into the mechanism of neurodegenerative disorders.

ACKNOWLEDGEMENTS

A copy of β-version of the program can be obtained in written request, free of charge, from Ali Ektessabi. The XRF analysis was carried out in Photon Factory, High Energy Physics Research Institute, Tsukuba. The authors would like to express their appreciation to Prof. A. Iida for his advice and support. The authors are grateful to K. Takada, for his work on development of the computer code during his presence in our laboratory.

REFERENCES

1. Sparks Jr. C.J., *Synchrotron Radiation Research*, Plenum, New York, 1980, chap. 14.
2. Saisho, H. and Gohshi, Y., *Application of Synchrotron Radiation to Material Analysis*, Elsevier, New York, 1996, chap. 2.
3. Bush, A.,*Current Opiniton in Chemical Biology*, 4, 184-191(2000).
4. Kienzl, E., Jellinger, K., Stachelberger, H., and Linert, W., *Life Sciences*, **65**, 1973-1976(1999).
5. Thong, P.S.P., Watt, F., Leong, S.K., He, Y., and Lee, T.K.Y., *Nuclear Instruments and Methods in Physics Reserch B*, **158**, 349-355(1999).
6. Ektessabi, A.M., Fujisawa, S., Takada, K., Yoshida, K., Maruyama, H., and Shin, R.W., *Int. J. of PIXE*, **9**, 297-303(1999).
7. Ektessabi, A.M., Yoshida, S., and Takada, K., *X-ray Spectrom.*, **28**, 456-460(1999).
8. Perl, D.P., Gaijdusek, D.C., Garruto, R.M., Yanagihara, R.T., and Gibbs, C.J., *Science*, **217**, 1053-1055(1982).
9. Garruto R.M., and Yase, Y., *Trends in Neuro Science*, **9**, 368-374(1986).
10. Piccardo, P., Yanagihara, R., Garruto, R.M., Gibbs, C.J., Jr., and Gajdusek, D.C., *Acta Neuropahol.*, **77**, 1-4(1988).
11. Lachance, G.R. and Claisse, F., *Quantitative x-ray fluorescence analysis*, John Wiley & Sons, New York, 1995, chap. 2.
12. Nastasi, M., Barbour, J.C., and Tesmer, F.R., *Handbook of Modern Ion Beam Material Analysis*, Materials Research Society, Pennsylvania, 1995, pp.375-382.
13. Iida, A. and Noma, T., *Nuclear Instruments and Methods in Physics Research*, **B82**, 129-138(1993).

LIGA SPINNERETS FOR MICROFIBERS

B.-Y. Shew, Y. Cheng

Synchrotron Radiation Research Center, #1, R&D Rd. VI, SBI Park, Hsichu, Taiwan, R.O.C.

Abstract. LIGA technology was used to fabricate innovative spinnerets with capillaries of high aspect ratio, any cross section, low surface roughness and various materials. A new technique, termed as " ultra-deep LIGA process", was developed to fabricate very thick microstructures (up to 2 mm) with high aspect ratio (more than 30). After die electroforming, polymer microstructures were mass replicated by injection molding. In stead of Ni depositing, NiCo alloy plating techniques were developed to improve the wear resistant ability of the die and the capillaries. In this report, comprehensive processes of the LIGA spinneret were demonstrated. New generation fibers with extra-fine size, new functionality yet in low cost could be expected after this exploration..

INTRODUCTION

Polyester is a synthetic compound in which small molecules are linked through the formation of an ester group. Though the polyester industry has existed since the 1950's, technological advances now create great rooms for this industry. One area with large potential is that of microfibers, which can breathe like natural silks and have a softer feel than traditional polyester fibers. Polyester fibers can be divided into two types: staple (fibers cut into length of 1 1/2' or longer) and filaments. Filaments are measured in denier. One denier is difined as one gram of the polyester at a length of 9,000 meters. The fine filaments under 0.3 denier are termed as extra-fine denier filaments (or microfibers).

PET raw material is fed into a spinneret with 24-144 capillaries to produce filaments at a speed over 3,000 m/min. The polyester fibers could have special functionality if they were extruded from the shaped capillaries. For example, the fibers with triangle cross section yield shining appearance; the hollow fibers are light in weight but it can keep body heat efficiently. The capillary width for microfiber is usually larger than 100μm to prevent polymer blocking in direct spinning process. Therefore, making the capillary deeper is much important than making the capillary smaller for microfiber production. From theoretical point of view, the aspect ratio of a shaped capillary should be larger than 6 to extend the fully developed flow in the capillary. The spinnerets are normally fabricated by micro-EDM technique. However, this method could not satisfy the requirements to produce complex shape, high aspect ratio spinnerets in low cost.

LIGA process is known as a powerful technique to fabricate high-aspect ratio microstructures (HARMS) with high precision. LIGA is the abbreviation of German words of "*Lithography*", "*Galvanik*" and "*Abformung*", which mean lithography, electroforming and molding in English [1]. Highly intense and collimated synchrotron X-rays is used as the light source for deep X-ray lithography. The lights near the hard X-rays are used to provide the depth of exposure. PMMA (polymethylmethacrylat) is commonly used as the X-ray resist due to its high contrast and low surface roughness under developing. In addition, LIGA process could provide microstructures with high uniformity, any lateral geometry, various materials and low surface roughness. All these advantages are necessary for a high value spinneret.

Fig.1 is the schematic diagram of the mass production processes for the LIGA spinnerets. The die is fabricated through deep X-ray lithography and electroforming. The stainless body (SUS 630) is machined by conventional drilling (Fig. 1a). As illustrated in Fig.1b, columnar microstructures are molded by polymer injecting. Capillaries are then electroformed direct from the surface of stainless body. Finally, the polymer is removed by thermal process. The plating surface is then ground until the orifice revealed. (Fig.1c).

In this paper, the manufacturing process of LIGA spinneret is presented. Shaped capillary with high aspect ratio (>30) was fabricated successfully by this process. The issues in X-ray lithography, electroforming and microinjection molding will be discussed in the following sections.

DEEP X-RAY LITHOGRAPHY

Substrate preparation

In LIGA process, X-ray resist is usually cast and polymerized directly on metallic substrate. Another method is to laminate PMMA and metal plate by hot pressing. However, the thermal expansion coefficient and elastic behavior are quite different between polymer and metal material. These properties mismatch will induce stress, which would degrade the adhesive strength and distort the laminated structure. This is especially serious when the PMMA resist is very thick.

A sandwich structure (PMMA/Al/PMMA) was proposed to balance the stress. The composite substrate was laminated with PMMA glue by cold pressing. Before that, the aluminum substrate was chemical etched appearing rough surface to promote mechanical interlocking between PMMA and Al substrate. Very thin titanium film (~800 Å) was sputtered on Al surface to encourage the initiating of metal deposition so that no chemical process was needed before die electroforming.

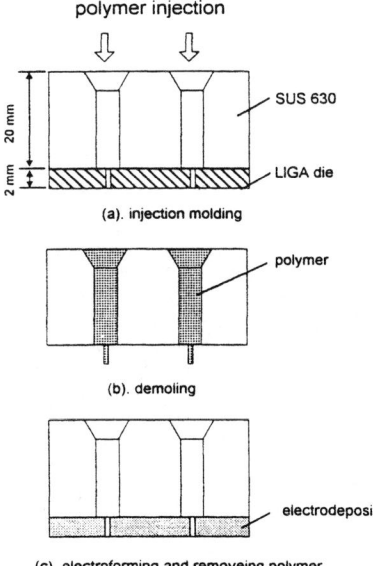

FIGURE 1. Duplication processes of the LIGA spinnerets.

Ultra-deep LIGA process

Deep X-ray lithography (DXL) utilizing a synchrotron light source is a powerful method to fabricate HARMS. Microstructure with an aspect ratio high than 100 has been reported by this technique. However, developing process become slowly in deep trench due to the difficulty in mass transfer. The phenomena will prolong the developing time and therefore damage the microstructure due to the micro-attack of the G. G. developer.

There are many ways, such as agitating, heating and dosage elevating, to increase the developing rate. Agitating is usually not useful for microtrenches. Heating could significantly enhance the diffusion rate and chemical reactions; however, the selectivity between exposed/unexposed area will be sacrificed. Increasing X-ray dosage could forward the developing process due to the reduction in polymer molecular weight. From our previous experiment, the developing rate of PMMA resist is almost linearly proportional to the X-ray dosage. In addition, there is not serious side effect with dosage elevation, which make it a more safe way to accelerate developing process.

As X-ray penetrating PMMA resist, the intensity attenuate because of the absorption effect. In order to develop the resist in a reasonable time, the bottom dosage should be high than ~ 4 kJ/cm^3. Whereas, the top dosage must typically low than ~ 20 kJ/cm^3 otherwise the induced gases upon X-ray irradiating will destroy the brittle PMMA structure. Under these two boundary conditions, it is difficult to modify the dosage distribution in PMMA resist and therefore, to fabricate very thick microstructure.

The common method to resolve this problem is using hard X-ray concerning its high penetration power. However, the induced photoelectrons will scatter and then degrade the lithography precision [2]. The difficulties in the light source access and mask fabrication, together with the safety concern for hard X-ray make this method inappropriate for making very thick microstructure.

A new strategy as "ultra-deep LIGA process" was developed in this program to fabricate couple-millimeters thick microstructure [3]. The principle is conducting DXL through successive exposing and developing with a conformal mask. After first lithography process, the dosage could be elevated beyond the constraint since the generated gases could escape through the porous surface. The dose distribution in the resist could be greatly modified and the developing rate could be substantially accelerated. The detail arguments about ultra-deep LIGA process are addressed elsewhere [3].

Conformal Mask

In order to eliminate the alignment complexity in multiple exposing process, the idea of the conformal mask was adopted in this experiment. Although the conformal mask could only be used for one time, the tool for mass production is the die but not the mask in LIGA process. The manufacturing procedure of the conformal mask is schematic illustrated in Fig.2. First, thin copper layer was sputtered or co-laminated on PMMA. Thick resist (JSR 137N, Japan) is then patterned by UV lithography directly on copper layer. The parameters of resist process, including spin coating, baking and developing condition, were optimized and a resolution of 1 µm was achieved for a thickness of 32 µm thick. The resist structure was further treated by RIE (Reactive Ion Etching) to trim the resist foot and remove the chemical residuals on the copper surface. As illustrated in Fig.2b, gold absorber was then plated through the resist. Non-cyanide gold bath (Technic 25-E) was used in this experiment considering the comparability with the resist material.

The necessary thickness of gold layer was calculated theoretically according to the character of X-ray source and the exposure depth. The electron energy of Taiwan Light Source (TLS) at the SRRC is 1.5 GeV. The critical photon energy and its critical wavelength are 2.3 keV and 0.54 nm respectively. According to our calculation, disregarding the effect of back-scattered electrons, a 10 µm gold absorber is required to make microstructures of 300 µm height [4].

After stripping the resist and copper layer, the conformal mask is now ready for subsequent DXL process (Fig. 2c). There are several advantages by successive exposing process combined with a conformal mask:
1. No alignment required.
2. No proximity gaps between resist and mask.
3. Thinner absorber and less exposure time.
4. High developing rate; and then very deep microstructure with high aspect ratio could survive after DXL process.

Fig.3 is the SEM photograph of PMMA structure fabricated through double exposing and developing process. As shown in Fig.3(a), no stepwise discontinuity was observed on the side wall. X-ray total reflection was believed to be the protection mechanism since the calculated critical angle (5.3 mrad) was much larger than the deviation angle (0.3 mrad) of the synchrotron X-ray. Stress balance by the sandwich structure also responds for the remarkable result.

This figure also shows that the Cu layer was slightly overetched to guarantee it will not affect the DXL accuracy. However, this process should be well controlled so that the absorber could survive after all the DXL process. Fig.3(b) is an arc PMMA structure (1mm thick and 100 µm wide) standing on the rough Al substrate. Shaped capillary for hollow fibers could be replicated from this mold by electroplating process.

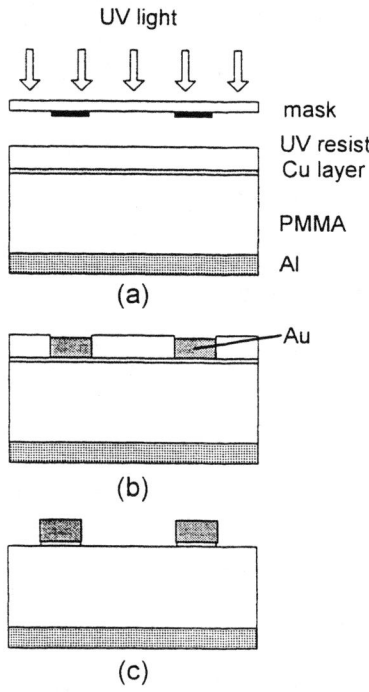

FIGURE 2. Manufacture process of the conformal mask for deep X-ray lithography. The geometric dimension in the schematic diagram is not in scale.

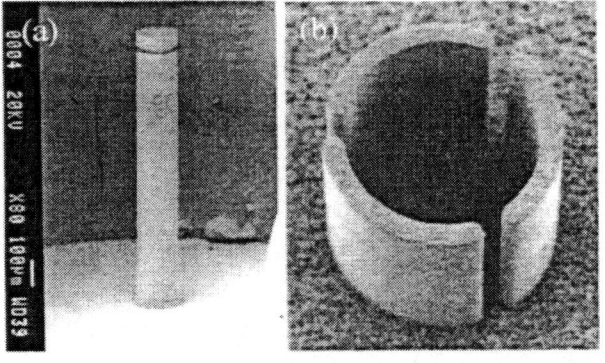

FIGURE 3. SEM photograph of the PMMA microstructure made by double exposures to achieve 1mm depth. The diameter of the column in 5(a) is 160 µm. The line width of the arc in 5(b) is 100 µm.

DIE ELECTROFORMING

Nickel Plating

Nickel is one of the most popular electrodeposit for protective and decorative applications. Among various electrolytes for Ni plating, sulfamate bath is the better choice for micro electroforming due to its low stress deposit. A commercial Ni sulfamate bath with a $[Ni^{2+}]$ concentration of 76 g/l was used in this study. The current density was range from 0.5~5 ASD (A/dm^2) dependent on the geometry and aspect ratio of the mold. Since the mass transfer is very slow inside deep trench, the organic additive might degrade under such a harsh environment. The chemical stability of these additives should be specially concerned for micro electroforming.

Fig.4 is the plated Ni capillaries with various cross sections. The polymer inside the orifice had been removed and the surface was ground without obvious damage. Comparing with EDM micromachining, LIGA process is especially suit for those spinneret with complex shaped capillaries as illustrated in this figure. The uniformity and the aspect ratio of the orifices, which is very important for high speed spinning, could be well controlled and fabricated by LIGA process without any difficulties.

The electrodeposit was further flattened by re-distribute the uneven current fluxes through a catholic shield. The results were evaluated by measuring the thickness across the deposit. As indicated from Fig.5, the thickness of the deposit was effectively smooth out by a PMMA shield (75% opening 6 mm ahead from substrate).

As we known, the wear resistance is dependent on the hardness of the material. The Vicker's hardness of Ni deposits range from 400 to 150 at current densities of 1 and 5 ASD respectively. However, high plating rate is more appreciated from the production aspect. In order to improved the duration of the LIGA die, NiCo alloy-plating technique in high deposition rate was also developed in this study.

Alloy Plating

Composition control is very critical in alloy plating because the metal ions have different tendency to catch the electrons and to deposit. Nickel and cobalt ion have the similar reduction potential, that mean the composition of the NiCo alloy is easier to be controlled than others, such as NiFe alloy. In this experiment, the cobalt ion was supplied by dissolving Co metal in the sulfamate electrolyte. The other compositions and plating parameter are similar as in Ni plating.

Fig.6 is the hardness and cobalt content of the alloy deposit as a function of cobalt sulfamate concentration in the electrolyte $[Co(SO_3NH_2)_2]$. The indentation load is carefully choused so that the measured hardness was not affected by the Al substrate underneath. As indicated from this figure, the cobalt contents increase consistently with $[Co(SO_3NH_2)_2]$. The hardness also increases at first but then saturate (Hv~400) as $[Co(SO_3NH_2)_2]$ larger than 20 g/l. The strengthening mechanism is primarily account for the lattice distortion by adding alloy element.

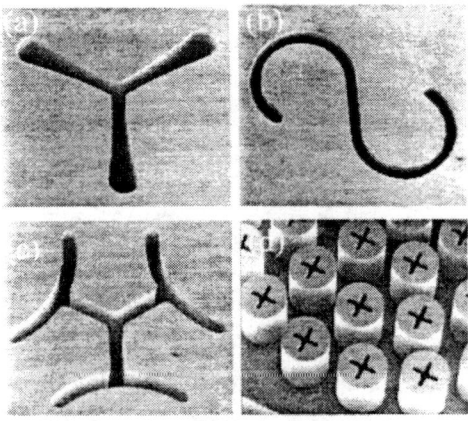

FIGURE 4. Ni plated capillaries for shaped (a, b, c) and conjugate (d) spinneret with a thickness of 1mm. The minimum line width in (a~c) and (d) is 100 μm and 50 μm respectively.

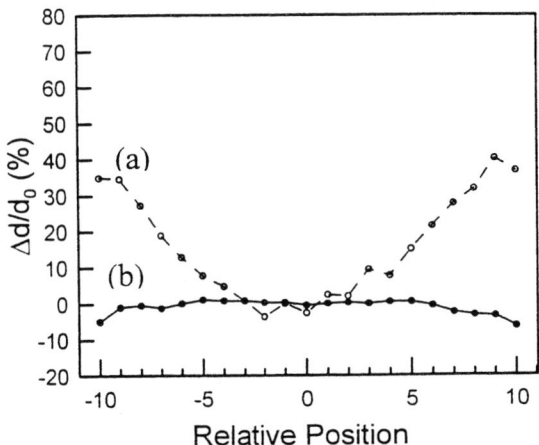

FIGURE 5.. Thickness distribution of the nickel deposits. (a). Without shielding (b). With shielding (75% opening, 6 mm ahead from the substrate)

MICRO INJECTION MOLDING

A conventional injecting machine was used to conduct the molding process and polypropylene (PP) was utilized as the molding material in this study. The injection parameters, including pressure, temperature, speed, holding and cooling time were optimized to control the uniformity and integrity of the molded microstructure. Fig.7 (a) is the photograph of molded result with 72 W-shaped polymer structure on the stainless body, which is now ready for subsequent plating process. As indicated from Fig.7(b), the polymer structure appear smooth side wall and well defied geometry. The debris on the top surface is acceptable because the complimentary electrodeposit will be ground to appear the orifice.

Polymer shrinkage is a serious problem in molding process. This is primarily accounted for the enormous thermal expansion mismatch between polymer and metallic materials. This problem was successively resolved by the special designs of shrinkage constrains and injecting runner. Systematic study about microinjection process will be presented elsewhere.

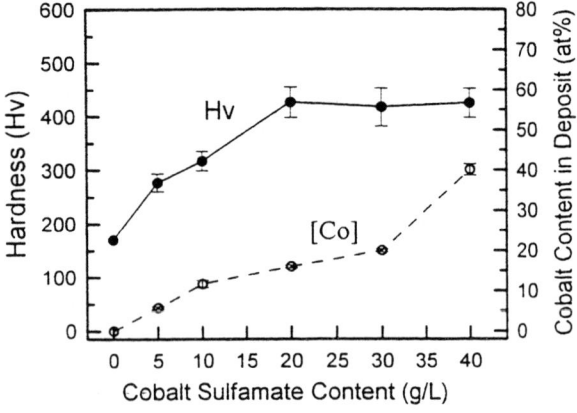

FIGURE 6. Hardness (Hv) and cobalt content (at. %) in the deposit as a function of cobalt sulfamate concentration in the electrolyte.

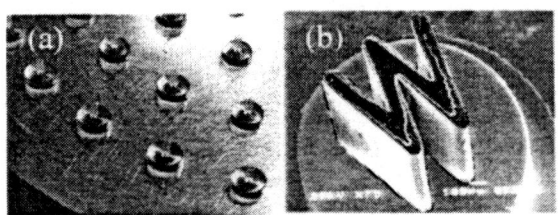

FIGURE 7. (a). Optical photo of the molding result, which reveal numerous polymer microstructures standing on the SUS substrate. (b). Magnified image of an individual polymer structure. The thickness and line width is 2mm and 100 µm respectively.

SPINNERET ELECTROFORMING

Surface Treatment

As illustrated in Fig.1(c), the capillaries will be replicated by electroforming from the stainless substrate. Since there is always a stable, dielectric oxide layer on the SUS surface, it must be cleaned prior to electroplating for guaranteeing the adhesive strength. The adhesion strength between Ni deposit/SUS304 was measured by pull-off test and the results were plotted versus temperature (Fig.8). As indicated from this figure, the adhesive strength decreases gradually with increasing temperature. However, it is much high than the applied pressure (80 kg/cm^2) in PET extrusion.

After final electroforming process, the top surface was ground to appear the capillaries. The machining process should be carefully control, otherwise the orifice edge will be damaged upon high shear stress. Fig.9 is the final product of the LIGA spinneret with 72 W-shaped capillaries in NiCo alloy materials. The depth and the minimum feature size of the capillaries are 2mm and 70 µm respectively.

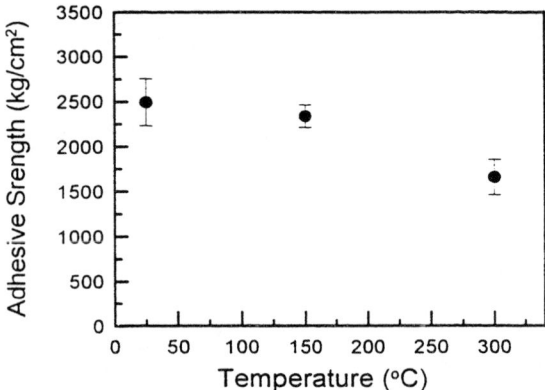

FIGURE 8. Adhesive strength between Ni/SUS630 interface at different temperature.

FIGURE 9. LIGA NiCo spinneret with 72 W-shaped capillaries.

The performance of the LIGA spinneret was co-evaluated with traditional ones by fiber spinning. Fig.10 is the cross section of the extruded fibers. The fibers inside the dashed circle were obtained from the LIGA spinneret. Although the true dimension is not the same, the photo reveals that the shape and the dimension of the "LIGA fiber" are much more uniform than the others. That means the "LIGA fibers" are not easy broken during spinning and the dyeing uniformity will be excellent. High quality and low cost fibers could then be expected using LIGA spinnerets.

Composite Plating

During fiber extruding, the spinneret suffer harsh condition of high temperature (~300 °C) and serious erosion. Some inorganic additives, such as TiO_2, will accelerate the wear behavior in some extension. Therefore, high temperature hardness is particularly concerned for the durability of the spinneret. As indicated in Fig.11, the hardness of Ni deposit is lower than that of the SUS 630 at any temperature; the hardness of NiCo alloy is similar with SUS 630 at room temperature, however it degrades as temperature increasing.

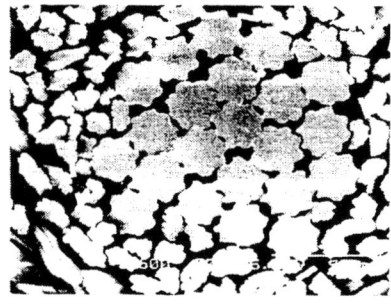

FIGURE 10. Cross-section of the extruded fibers.

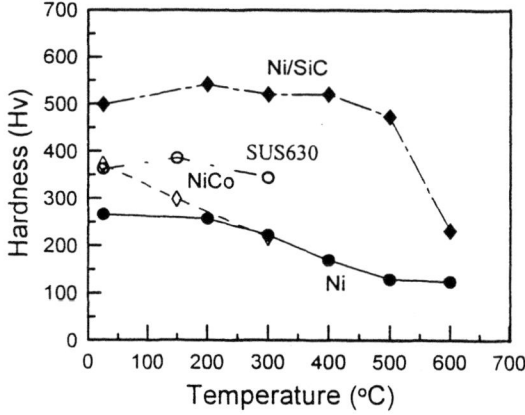

FIGURE 11. Hardness versus temperature of various depositsmaterials.

In order to improve the lifetime of the LIGA spinneret, Ni/SiC composite plating technique was also developed in this study. Preliminary results showed that the composite coating maintains its hardness (~ Hv 500) up to a temperature of 500°C. This makes the spinneret more durable over the conventional ones. In the future, this technique will be applied to make the capillaries to extend the lifetime of LIGA spinnerets.

SUMMARY

Spinneret is a key component in producing high quality fibers. The spinneret is normally fabricated by the micro-EDM technique. However, this method can not satisfied the requirements of capillaries with complex shape and high aspect ratio. Innovative technology is therefore urgently needed to facilitate the fabrication of high performance spinnerets.

In this paper, LIGA was used to make the spinneret. A precise die was first fabricated by deep X-ray lithography combined with electroforming technique. Spinnerets were duplicated through injection molding and plating. To meet the requirement in capillary depth, a new technique, termed the "ultra-deep LIGA process", was explored to make very thick microstructures with high aspect ratio. Instead of Ni deposit, NiCo alloy and Ni/SiC composite plating techniques were utilized to improve the wear resistant ability of the LIGA die and the spinneret. In this report, the manufacturing process for the LIGA spinneret was demonstrated. A spinneret with capillaries 2 mm deep and 70 mm wide was successfully fabricated by this method. New generation fibers can be expected as a result of this development.

ACKNOWLEDGEMENT

The authors thank the National Science Council of the Republic of China for its financial support under contract no. of NSC-85-2622-E007-010.

REFERENCES

1 E. W. Becker, W. Ehrfeld, P. Hagmann, A. Maner, and D. Munchmeyer, *Microelectron. Eng.*, vol. 4, 35-56, 1986.
2. G. Feiertag, W. Ehrfeld, H. Lehr, A. Schmid, M. Schmid, *Microelectronic Engineering*, 35, 557-560, 1997.
3 Y. Cheng, B. Y. Shew, C. Y. Lin, D. H. Wei and M. K. Chyu, *J. Micromech. Microeng.*, 9, 58-63, March 1999.
4. Y. Cheng, N.-Y. Kuo, and C. H. Su, *Review of Scientific Instrument*, 68(5), 2163-2166, May 1997.

The New Normal-Incidence-Monochromator Facility at CAMD

E. Morikawa and C.M. Evans and J.D. Scott

J. Bennett Johnston, Sr., Center for Advanced Microstructures and Devices
Louisiana State University, Baton Rouge, LA 70808

Abstract. Installation of the new low-energy, high-resolution beamline at CAMD is nearing completion. It is capable of covering the spectral range from the visible to ca. 50 eV. Coupled with a Scienta SES-200® hemispherical-analyzer electron spectrometer, the primary utilization of this beamline will be the investigation of material surfaces. Also of importance are valence-shell-electronic excitations and ionizations of atoms and molecules. The beamline is described, particular attention being given the optics. Optical calculations resulting in thorough simulation of the beamline capabilities are presented and throughput and spectral resolution are provided.

INTRODUCTION

A consortium of spectroscopists[1] interested in having an intense, highly monochromatic source of radiation covering the spectral range from ca. 5 eV to 50 eV joined together to acquire funding for a beamline to be installed at the Louisiana State University, J. Bennett Johnston, Sr., Center for Advanced microstructures and Devices, a 1.3 – 1.5 GeV synchrotron-radiation source. The impetus for acquiring such a beamline is the study of valence and first-inner-shell electronic structure of condensed-material surfaces and of atoms and molecules.

The properties of the CAMD electron storage ring have been published elsewhere.[2,3] Table 1 is a listing of the design parameters of the ring and Figure 1 contains plots of the radiation flux produced in a bending magnet.

TABLE 1. Design Parameters of CAMD Storage Ring.

Parameter	Value
Electron beam energy (GeV)	1.3/1.5
Maximum beam current (mA)	280/150
Bending radius (m)	2.928
Characteristic energy (keV)	1.66/2.56
Critical wavelength (Å)	7.47/4.84
Natural emittance @ 1.5 GeV (m-rad)	3.4×10^{-7}

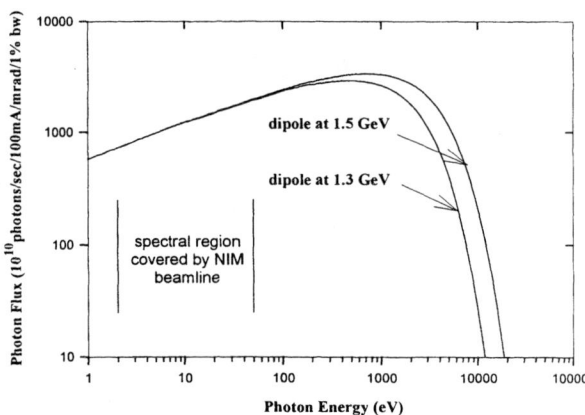

FIGURE 1. Photon-flux spectra from a bending magnet of the CAMD electron storage ring at 1.3 and 1.5 GeV.

BEAMLINE REQUIREMENTS

The user consortium initiated discussions regarding beamline design by stating a clear-cut set of requirements for the radiation at the sample position:

• High energy resolution of radiation over the range from ca. 2 eV to 50 eV.

- High-intensity photon beam delivered to the experiment station; along this line of thought, the entire 70 mrad horizontal fan available from the CAMD radiation port should be available.

- Beam should possess a high degree of plane polarization.

- The angular distribution of the radiation beam should lie within a cone of ±1°.

With these requirements and those imposed by the geometry of the proposed endstation, a Scienta SES-200® hemispherical-analyzer electron spectrometer, and the geometric restrictions of the storage ring and experiment hall, a beamline was designed and advertised to the commercial market in an invitation-to-bid construction of the entire beamline. McPherson Instruments, Inc. was awarded the contract. The details of this beamline are given in the next section.

RESULTING BEAMLINE; DESCRIPTION

The best monochromators, with regard to resolution, over the desired spectral region, use spherical gratings with incidence angles near normal and source and camera arms of equal length. This greatly reduces coma and other spherical aberrations introduced by the grating surface figure. There are several tried and proven mounts fitting these criteria; on-and-off-plane eagle mounts and the McPherson mount. It was decided to utilize the McPherson mount,[4] a configuration requiring the rotation of the grating while translating it along the bisector of the angle formed by the entrance and exit arms of the reflecting system when the grating is positioned to produce zero-order focus at the exit slit. Considering geometric constraints of the experiment hall along with resolving power, it was decided to select an optimal monochromator focal length of 3.00 meters. For a standard grating size of 50 mm wide by 100 mm long, to fill the grating, the vertical angle should be ca. 33 mrad (= 100/3) and the horizontal angle, ca. 17mrad (= 50/3); using the distance from the focus and the dimensions of the grating. Approximation of required angle from distance from slit and size of grating alone is exact only if the slit image is a point. Figure 2 is a plot of the synchrotron-radiation source of 70 mrad horizontal fan. The dimensions, vertical size ≈ 1 mm and horizontal size ≈ 4mm, will be used in the following discussion to estimate the size of the slit image and, thus, justify approximating it as a point.

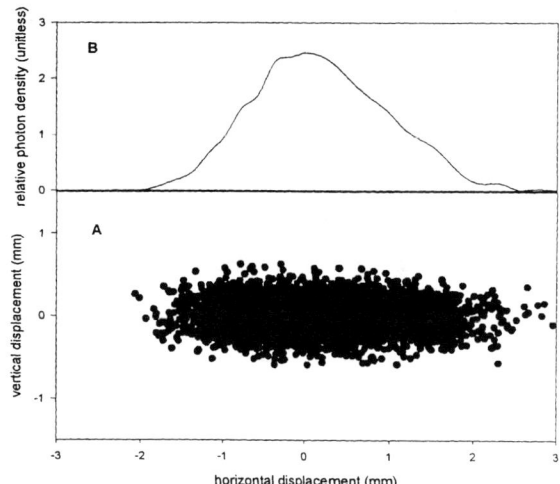

FIGURE 2. A. scattergram plot of ray-source points for a synchrotron-radiation source (2 to 50 eV photon energy) based on an electron beam having the characteristics of the CAMD ring operating at 1.3 GeV. B. Relative photon distribution of the source as a function of horizontal displacement

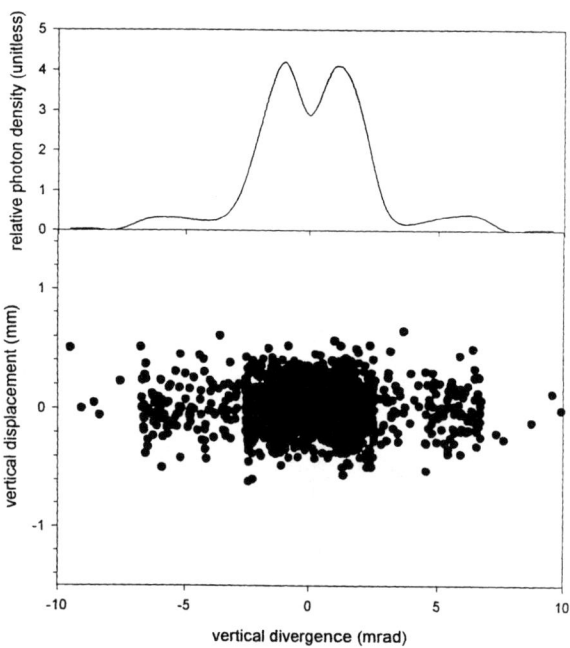

FIGURE 3. Vertical phase-space plot of CAMD synchrotron radiation; 5 eV photons, 1.3 GeV beam energy, at the source.

Optics between the source and the entrance slit of the monochromator must accomplish four tasks, which will be presented as two pairs of tasks:

- Focus the horizontal dimension of the beam at the monochromator entrance slit and transform the horizontal angle of the source (70 mrad) to that of the beam at the entrance slit (17 mrad).

- Focus the vertical dimension of the beam at the monochromator entrance slit and transform the vertical divergence of the source (ca. 7.5 mrad, see Figure 2) with the vertical divergence at the entrance slit (33 mrad).

The plots in Figures 2 and 3, as well as the other radiation density plots and ray-tracing results were created using the ray-tracing program, SHADOW.[5]

The angular magnification for the horizontal divergence is 0.24 (= 17/70) and of the vertical is 4.4 (= 33/7.5). Optical magnification generally refers to spatial magnification, the reciprocal of angular magnification; therefore, the optics between the source and the entrance slit must magnify vertically by 0.23 and horizontally by 4.17. From Figure 2, this would indicate an entrance-slit image of ca. 0.23 mm (= 1 mm × 0.23) vertical size and ca. 16.7 mm (=4 mm × 4.17) horizontal size. The vertical size is negligible with respect to the grating length; however, the horizontal size is ca. 30% of the stated grating width. The actual grating widths are 60-70 mm wide; therefore, calculation of angular divergence using grating size and distance between entrance slit and grating is considered satisfactory.

To accomplish these tasks, an ellipsoidal mirror is employed 2800 mm from the source and is made such that it focuses the beam at 11,760 mm from the mirror,

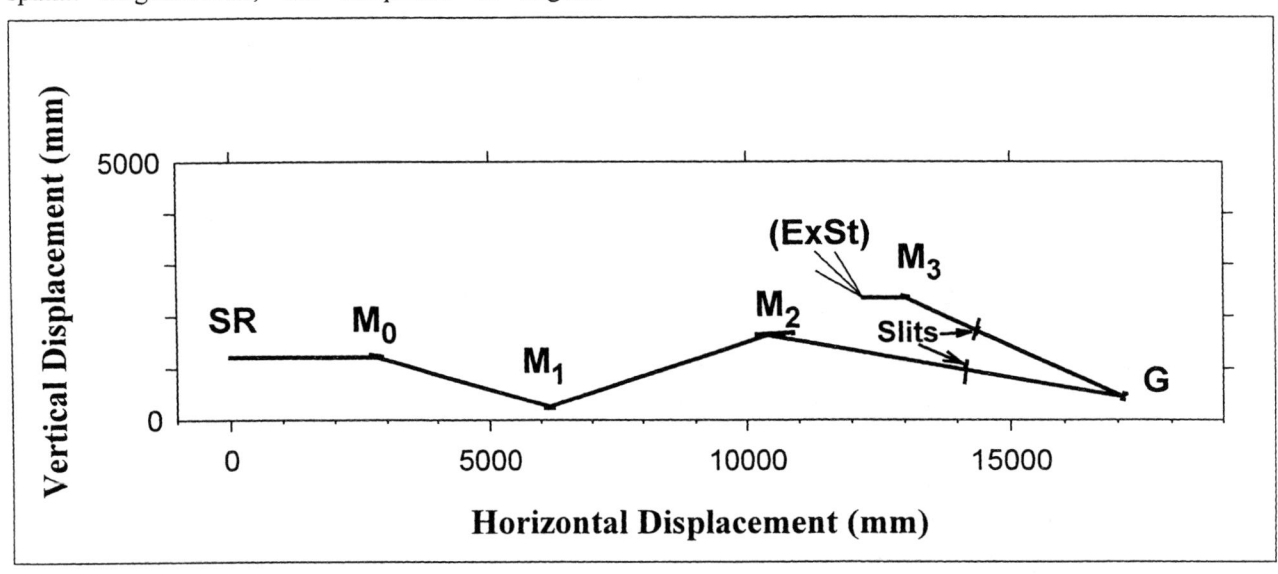

FIGURE 4. Elevation ray-trace drawing of the CAMD 3-meter normal-incidence-monochromator beamline at CAMD. Indicated beamline elements are described below:

SR   The synchrotron-radiation source

M_0   Copper-alloy, water-cooled, gold-coated ellipsoidal mirror. Object distance = 2800 mm and image distance = 11,760 mm. Incidence angle = 82.5°. Size, 200 mm wide x 220 mm long.

M_1   Zerodur® (special optical quality, heat-tempered glass), gold-coated cylindrical mirror (meridial focus). Radius = 3360 mm. Inc. angle = 73°. Size, 150 x 150 mm². Distance from M_0 = 3500 mm.

M_2   Float-glass, plane mirror that can be bent to form a cylinder to give a coma-corrected pair with M_1 and focus (vertical) beam at entrance slit. (Theoretical) radius = 16,047 mm. Incidence angle = 76°. Distance from M_1 = 4449 mm and from entrance slit = 3812 mm. Size, 90 mm wide and 660 mm long (pole is located 200 mm from upstream edge of mirror).

G   3-m radius spherical grating, ruled at 1200 grooves/mm, size, 60 mm wide and 110 mm long, Two gratings, one with Al plus MgF_2 overcoat for range 4.1 – 12 eV and one gold coated for 10 – 50 eV. A 600 groove/mm, Al/MgF_2-coated grating would cover into the visible region.

M_3   Zerodur® ellipsoid, object distance 1500 mm (also distance from exit slit) and image distance 800 mm [also distance to experiment station, (**ExSt**) or sample position]. Incidence angle = 78°.

for a magnification of 4.2; therefore, the horizontal focus criterion is met. The vertical magnification, however, now required is 0.054. Because the virtual source of the system used to accomplish this vertical magnification is the same as the focus, at least two vertically focusing mirrors are required. By using two cylindrical mirrors, each producing the same coma (but oppositely signed), a coma-free image at the entrance slit can be produced. Several simultaneous conditions must exist for the two cylinders; both mirrors must coexist within the 11,760 mm space between the ellipsoidal mirror and the monochromator entrance slit, the pair must focus the virtual source back upon itself, the magnification at the focus must be 0.054, the coma produced by each must be equal in magnitude but opposite in sign and the beamline must stay above but near the plane of the experiment-hall floor. The conditions were satisfied simultaneously by using an iterative numerical method. Figure 4 is an elevation view of the beamline. The results of the theoretical solution for the coma-corrected pair (M_1 and M_2) are given in the figure. (Actually, the exact solution for the mirror pair gave a calculated incidence angle for M_2 of 76.34°, with a radius is 16,507 mm. Ray-tracing results using both sets of parameters for M_2 gave practically identical results regarding the image at the entrance slit.)

SIMULATED CHARACTERISTICS OF MONOCHROMATIC RADIATION

Simulations were made using SHADOW ray-tracing calculations.[5] The results are given in Table 2. Three different monochromator-entrance-slit widths are used; 500 μm (low resolution), 100 μm (moderate resolution) and 15 μm (high resolution). The resolving power, $R = \lambda / \Delta\lambda$ (λ is the central wavelength), was estimated by measuring the spatial distribution of a monochromatic beam of the desired photon energy using a specific entrance-slit width. From this distribution and knowledge of the reciprocal linear dispersion of the grating for the particular wavelength at the exit slit plane, the spectrum of the radiation emerging through any exit-slit width was derived. This spectrum then was used to determine the resolving power for the given conditions. The exit slit was "matched" to the entrance slit by adjusting it to just accommodate the spatial displacement of the monochromatic beam. This, of course, would be impossible to do in a laboratory situation; there, the best match would be to adjust the exit slit to the same width as the entrance slit. For each of three photon energies, 5, 25 and 45 eV, and each of the afore mentioned entrance slits, the resolving power, the photon intensity (normalized to 100 mA of ring current at 1.3 GeV operation) and the plane polarization (per cent parallel to the exit slit) are given for the "matched" exit-slit width. Also given are the intensity and polarization at the sample position (focus of M_3); the full image and a masked 1×1 mm^2 are considered in the table.

REFERENCES

1. W. Plummer, Univ. TN, J. Allen, Univ. MI, P. Dowben, Univ. NB,. J. Erskine, Univ. TX and R. Gooden, Southern Univ.

2. Craft, B. C., Findley, A. M., Findley, G. L., McGlynn, S. P., Scott, J. D., Watson, F. H., *Rev. Sci. Instrum.* **60**, 2144-2147 (1989).

3. Bluem, H., Scott, J. D., *Nucl. Instr and Meth. In Phys. Res.* B **99**, 274-276 (1995).

4. Samson, J. A. R., *Techniques of Vacuum Ultraviolet Spectroscopy*, New York: John Wiley and Sons, Inc., 1967, pp. 62-64.

5. Cerrina, F.,, *SHADOW*, University of Wisconsin, Department of Electrical Engineering.

TABLE 2. Characteristics of Monochromatic Radiation.

Photon Energy	500 μm Entrance Slit	100 μm Entrance Slit	15 μm Entrance Slit	Sample Position 100 μm Ent. Slit	
				full image	1×1 mm^2
5 eV	1.0×10^{12} phot./sec R = 4000 Polarization = 99%	7.3×10^{10} phot./sec R = 16,500 Polarization = 99%	4.8×10^9 phot./sec R = 47,500 Polarization = 99%	3.8×10^{10} ph/s pol.= 99.7%	8.2×10^9 ph./s pol.= 99.8%
25 eV	2.2×10^{12} phot./sec R = 807 Polarization = 94.5%	1.7×10^{11} phot./sec R = 3340 Polarization = 94.5%	7.5×10^9 phot./sec R = 11,830 Polarization = 94.5%	8.2×10^{10} ph/s pol. = 96.8%	1.8×10^{10} ph/s pol. = 97.1%
45 eV	1.1×10^{12} phot./sec R = 460 Polarization = 90.5%	9.0×10^{10} phot./sec R = 1800 Polarization = 90.5%	4.2×10^9 phot./sec R = 6700 Polarization = 90.5%	4.3×10^{10} ph/s pol. = 92.7%	1.1×10^{10} ph/s pol. = 93.1%

The Development of the GCPCC Protein Crystallography Beamline at CAMD

Mitchell D. Miller[a], George N. Phillips, Jr.[a], Mark A. White[b], Robert O. Fox[b], and Benjamin C. Craft, III[c]

[a]Dept. Biochemistry & Cell Biology, Rice University, Houston, TX 77005
[b]Dept. Human Biological Chemistry & Genetics, University of Texas Medical Branch, Galveston, TX 77555-0647
[c]Center for Advanced Microstructures & Devices, Louisiana State University, Baton Rouge, LA 70806

Abstract. The Gulf Coast Protein Crystallography Consortium (GCPCC) is developing a beamline at the LSU/CAMD synchrotron. This beamline will be capable of standard macromolecular multiple-wavelength anomalous diffraction (MAD) phasing experiments over an energy range of 7-17.5 keV. The optical configuration uses a vertical collimating mirror, a channel-cut Si (111) monochromator and a focusing toroidal mirror. Built off of the CAMD 7 T superconducting, energy-shifting wiggler, this beamline will deliver a flux comparable to an NSLS bending magnet protein crystallography (PX) beamline. The beamline delivery and commissioning timetable calls for full 24/7 user operations by October 2001. The beamline design, with supporting calculations and ray tracing, is presented to substantiate the expected performance of this beamline for efficiently collecting accurate MAD data from macromolecules.

INTRODUCTION

The Gulf Coast Protein Crystallography Consortium (GCPCC) members include Louisiana State University, Rice University, University of Texas Medical Branch, Baylor College of Medicine, Oklahoma Medical Research Foundation, University of Texas – Austin, University of Houston, and Texas A & M University. In addition to providing access to the consortium laboratories, the GCPCC beamline will allocate 25 % of the beamline time to general users. The beamline is designed to collect protein crystallography MAD phasing data between 7-17.5 keV. This range includes K edges from iron to yttrium and L edges from neodymium to uranium.

The radiation source for this beamline is a 7 T superconducting, energy-shifting wiggler at the Center for Advanced Microstructures and Devices (CAMD) in Baton Rouge, LA [1]. The critical wavelength for the wiggler is 1.18 Å (10.5 keV) when operated at 7 Tesla with a 1.5 GeV electron beam. The X-rays produced by the CAMD wiggler at 1.5 GeV are harder than those from an NSLS bending magnet at 2.58 GeV. This results in a slightly higher flux/mA at the Se K edge for the CAMD source. The wiggler has five poles. The magnitude of the flux produced by the superconducting side poles relative to the central pole is 16% at 7 keV, 3 % at 12 keV, and 0.6% at 17 keV.

OPTICAL DESIGN & RAY TRACING

The optical design is typical of modern beamlines optimized for MAD. A qualitative anamorphic diagram is shown in Fig. 1. The first mirror collimates the beam vertically. Collimation increases the flux transmitted by the slits/monochromator for a given bandwidth. This maximizes the resolution of the monochromator. The mirror also reduces the thermal heat load on the first monochromator crystal by acting as a low pass filter. The wavelength is selected by a Si (111) channel-cut monochromator. Si (111) has a rocking width of 21 mrad at 12.7 keV and this results in an intrinsic energy resolution ($\Delta\lambda/\lambda$) of 1.33×10^{-4} [2]. The experimentally realized energy resolution is lower because of the effects of beam divergence and thermal deformation of the crystal surface. The second mirror focuses the beam horizontally and vertically on the experimental sample.

The program Shadow [3] was used to model the beamline and evaluate the optimal size and placement of the optical elements. Over the small acceptance of the beamline considered here, the variation of the radius of curvature of the electron orbit is negligible. For the purpose of modeling, the source is represented as a simple synchrotron with a bending radius of 0.71 meters. For heat-load calculations, all five wiggler poles are considered.

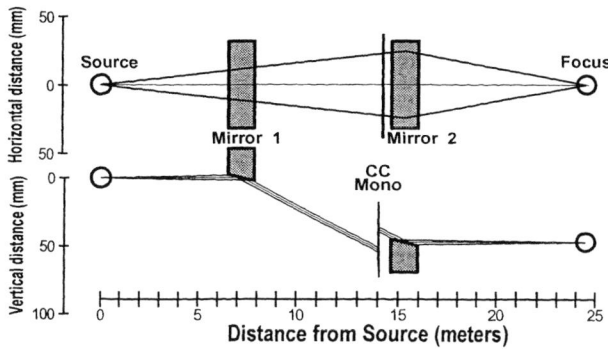

FIGURE 1. Anamorhic diagram of the PX beamline showing the location of the vertical collimating mirror (M1), the channel-cut monochromator, and the refocusing toroidal mirror (M2).

The location of the focusing mirror (M2) presents a trade-off between flux and the horizontal divergence at the sample. By de-magnifying the beam horizontally, the flux through a given aperture can be increased but the divergence of the beam transmitted through the aperture also increases. After comparing the flux and beam divergence at the sample for a several different demagnifications, the value of 0.55 was selected. The 0.55 demagnifying optics of the GCPCC beamline results in 2.3 horizontal milliradian divergence at the crystal with good flux. The divergence at the crystal can be reduced with the slits upstream of the monochromator at a cost in flux. This may be useful for some projects with long unit cells. When the horizontal crossfire is limited to 1.8 milliradians, the flux is 15 % less then that at 2.3 milliradians.

After ray tracing the complete beamline with "infinite" optics, that is optics which have unlimited length, rays which made it through the final aperture were back-traced through the system. This allows one to look at the distribution of rays that made it through the aperture at any point along the beamline. The backtraced rays are used to verify the correct dimensions for a mirror. The footprint of the rays on the surface of mirrors is shown in Fig. 2 with a box drawn to represent the size of the optics we will use (1350 mm x 20 mm for M1 and 1350 x 27 mm for M2). These optics are large enough that most of the rays strike the useable surface, although some rays are lost, especially at low energy where the opening angle of the source is larger.

The GCPCC beamline focal spot is shown in Fig. 3. The scatter plot shows rays over a very large area (8 mm x 4 mm). However, if one looks at the contour plot, the peak is centered over a much smaller area (0.4 mm x 1.5 mm). The fact that the spot size is somewhat larger than the sample aperture should make the beamline less sensitive to small fluctuations in beam position.

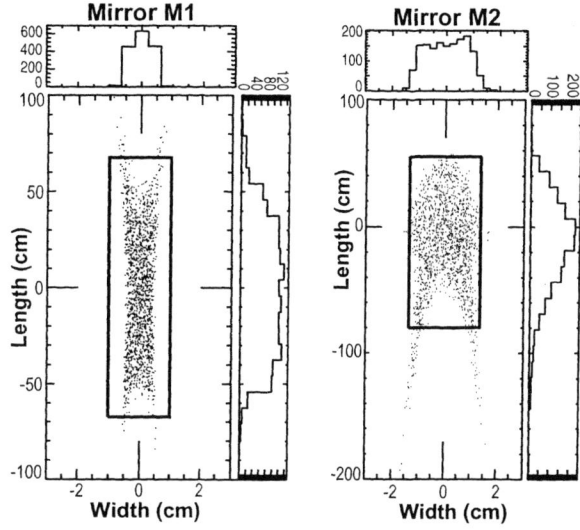

FIGURE 2. Scatter plot showing the location of the backtraced rays (those that made it through the 200 μm sample aperture) on the surface of the two mirrors at 12.7 keV. Note that the 1.4 m long mirrors capture most of the rays making it through the sample aperture. The pole of M2 is not at its center.

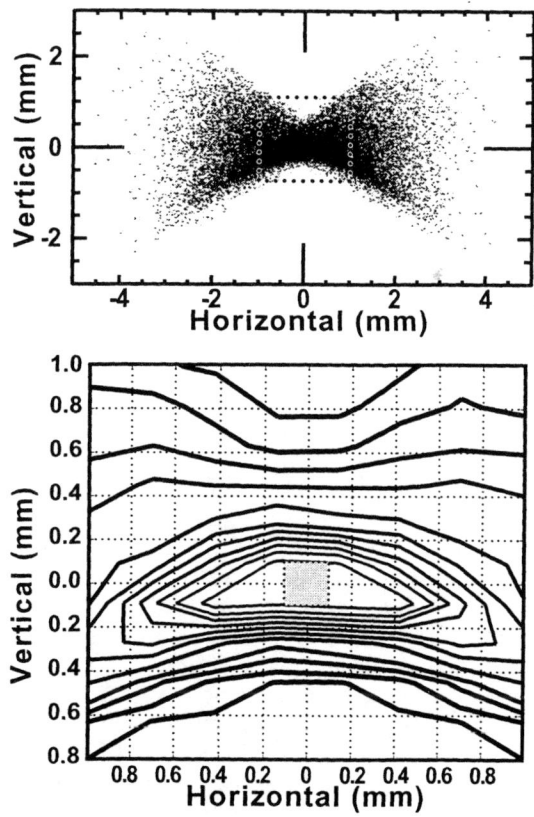

FIGURE 3. Focused spot at sample. Top scatter plot. A 200 μm reference aperture drawn on the lower contour plot.

To provide an estimate of the flux that the GCPCC beamline will deliver, we modeled the NSLS beamline X8C using the same methods as were used for the GCPCC beamline. X8C has optical components that are similar to the GCPCC design. There is a rhodium coated collimating mirror, a Si (111) double crystal monochromator, and a rhodium coated focusing mirror [4,5]. The flux ratio of the two model beamlines is GCPCC/X8C = 0.988. This represents the intensity obtained from the weighted sum of the rays that make it through the 200 micron aperture for the bandpass of the monochromator. At energies higher than 8.75 keV, the CAMD wiggler produces more photons/mA of circulating current at 1.5 GeV than an NSLS bending magnet at 2.58 GeV. Differences in flux will come down to differences in the average circulating current of the storage rings. Intensities measured in 1997 on NSLS bending magnet PX beamlines at 1 Å wavelength under 2.58 GeV operations normalized to 300 mA with a 200 μm square aperture show that the flux at different beamlines varies by a factor of 2-3 (P. Siddons, personal communication). Based on these measurements and calculations, the GCPCC beamline should deliver a flux within the range of existing NSLS bending magnet PX beamlines.

To estimate the heat load, we consider all five wiggler poles and assume that we accept 3 milliradians with an electron beam current of 400 mA. This is a reasonable current for estimating maximum heat load. Using these parameters, the total power output from the source is 160 Watts. The first mirror and beryllium window serve to reduce the heat load on the monochromator crystal. These two elements combine to remove approximately 50 % of the total power output in the 3 mrad fan allocated to the PX beamline. Cooling water will be used to remove heat from the fixed aperture, the collimating mirror, the first beryllium window, the adjustable slits, the first optical surface of the channel-cut monochromator crystal, the white beam stop and photon shutters.

IMPLEMENTATION

Beamline development on the CAMD wiggler is being carried out in two phases (TABLE 1). In the Phase 1 configuration, two beamlines are to be developed: a micromachining beamline and the protein crystallography beamline. The present vacuum chamber in the dipole bending magnet immediately downstream of the wiggler allows 25 milliradians of the radiation fan from the wiggler to be brought out of the accelerator. A new dipole vacuum chamber is being developed with two additional flanges allowing extraction of more wiggler radiation. In both Phase 1 and Phase 2, the two beamlines closest to the centerline of the wiggler fan have a common interface to the dipole vacuum chamber. Additionally, in both configurations, the extreme angular extents of wiggler radiation accepted by these two beamlines are plus and minus 12 milliradians. To limit disruption to the PX beamline during the Phase 2 dipole chamber upgrade, the front-end design was constrained to not require the repositioning of any of the PX beamline optical components. The M1 mirror tank is common to the two beamlines closest to the wiggler centerline.

TABLE 1. Utilization of the Wiggler Fan (Phase 1 & 2).

Fan	Phase 1 Use	Fan	Phase 2 Use
Outside		9 mrad	Lithography
		19 mrad	Gap
9 mrad	Lithography	3 mrad	Beamline #4 [a]
3 mrad	Gap	9 mrad	Gap
	Center of Wiggler Fan		
9 mrad	Gap	9 mrad	Gap
3 mrad	PX beamline	3 mrad	PX beamline
		25 mrad	Gap
Inside		3 mrad	DCM beamline

[a] This beamline is currently uncommitted.

Diagnostic screens are located between each optical element to aid in commissioning. This allows the effect of each optical element to be monitored independently without adjusting the other elements. A motor control system is being developed to allow the optical elements to be precisely aligned. The system uses Compumotor 6K series motion controllers and the MX beamline control software [6]. A graphical user interface is being developed to provide the necessary control for commissioning and routine user operations. The motor control system gives the beamline staff control over Mirror M1 – height, pitch, and roll, 4 independent blades of the adjustable aperture, Monochromator – pitch, yaw and detune, Mirror M2 – bend, height, pitch, yaw and horizontal position. In addition, there is a motorized slit unit inside the hutch, which serves to define the beam extents on the sample. The diagnostic screen actuators are also controlled through this system.

Mirrors

Both mirrors are 1.4 m long with useable lengths of 1350 mm. The mirrors are coated with rhodium, which is free of absorption structure over the energy range of the beamline. The grazing angle of 0.20 degrees (3.5 mrad) provides good reflectivity until 18 keV. The first mirror functions to vertically collimate the beam before the monochromator crystal. We are approximating the ideal parabolic figure with a ground cylinder with a radius of 4154 m. To reduce the vertical size of the collimated beam, the mirror is placed as close as possible to the wiggler source (7.25 m). This first optic is fabricated from single crystal silicon and is cooled using cooling blade. This mirror is installed in a "bounce down" configuration.

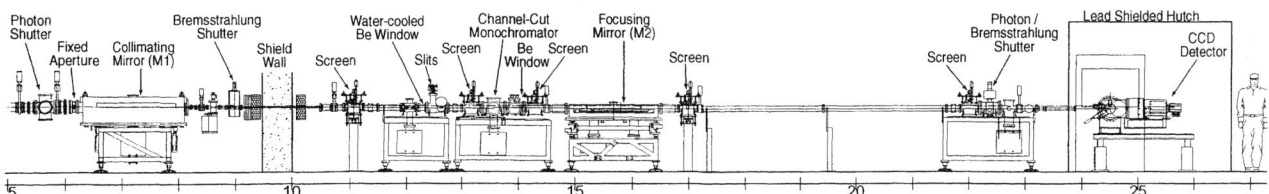

FIGURE 4. Elevation schematic of the GCPCC PX beamline.

The second mirror serves to focus the X-ray beam horizontally and vertically. Its ideal ellipsoid focusing figure is approximated with a toroid. The surface figure is realized by bending a long mirror into which the 3.92 cm sagittal radius had been ground. The tangential radius is adjustable between 4–6 km using a stepper motor driven U-bender. By adjusting the bending radius and the inclination angle, the beamline focus can be adjusted between 24.65 ± 0.15 m. The pole of M2 is located 15.8 m from the source. This mirror is in a "bounce up" configuration and returns the beam parallel to the plane of the ring.

Channel-Cut Monochromator

The geometry of the channel-cut crystal accepts the entire 4.7 mm high incident beam over the energy range from 7 to 17.5 KeV. The monochromator crystal has a flexure that allows the second surface of the crystal to be slightly de-tuned from the first. This provides a mechanism for rejecting higher order harmonics, which is important when working at low energies. At high energy, harmonic wavelengths are removed by the reflectivity cutoff of mirror M1. The small crystal gap (6.5 mm) minimizes the change in beam height upon changing the wavelengths. This is particularly important within a MAD data set where we would like to minimize the need to adjust mirror heights or realign the endstation. For example, a 200 eV change in wavelength at the Se K edge would only change the beam height by 5 microns. Changing between the extremes in the wavelength spectrum would result in a beam height change of 450 µm, which would necessitate changing the height of the second mirror and the endstation. The first crystal reflects the beam up and the second crystal brings the beam parallel to the incident beam but 12.5-12.9 mm higher. The first crystal surface of the monochromator is indirectly cooled via a water-cooled block. Cooling the crystal in this manner is to reduce thermal deformation of the crystal and help to maintain the energy resolution needed for MAD phasing.

Endstation design

We are currently finalizing the endstation design. The endstation will have a large format CCD based detector. The detector will have readout times of 1-4 sec. The goniostat will feature a 2θ offset, and will be mounted on a motorized kinematic experimental table will be with apertures and diagnostics necessary for aligning the system. There will be a fluorescence detector and cyrocolling system typical of modern MAD protein crystallography beamlines.

Manufacture and Installation

The first shipment (November 2000) comprises the collimating mirror system and components through the first screen (Fig. 4). This allows us to commission the first mirror while the remaining items are still in manufacture. The remaining beamline components will be installed in January 2001. The detector and endstation components are expected in Feburary 2001. The beamline delivery and commissioning timetable calls for full 24/7 user operations by October 2001.

ACKNOWLEDGMENTS

We would like to thank Dr. D. Peter Siddons for helpful discussions. Funding for the beamline is provided by the National Science Foundation through the biological infrastructure award DBI-9871464 with interagency matching funds from the NIGMS at the National Institutes of Health and the GCPCC member institutions.

REFERENCES

1. Borovikov, V. M., Craft, B., Fedurin, M. G., Jurba, V., Khlestov, V., Kulipanov, G. N., Li, Q., Mezentsev, N. A., Saile, V. and Shkaruba, V. A. J. Synchrotron Rad. 5, 440-442 (1998).
2. Helliwell, J. R., Macromolecular Crystallography with Synchrotron Radiation, Cambridge, Cambridge Univ. Press, 1992.
3. Welnak, C., Anderson, P., Khan, M., Singh, S. and Cerrina, F. Rev. Sci. Instrum. 63, 865-868 (1992).
4. Trela, W. J., Bartlett, R. J., Michaud, F. D. and Alkire, R. Nucl. Instr. and Meth. A 266, 234-237 (1988).
5. Alkire, R. W., Sagurton, M., Michaud, F. D., Trela, W. J., Bartlett, R. J. and Rothe, R. Nucl. Instr. and Meth. A 352, 535-541 (1994).
6. Lavender, W., MX – Data Acquisition and Control System, http://www.imca.aps.anl.gov/mx/, Sept. 2000.

SECTION VIII

POSITRON EXPERIMENTS

An Intense, Compact Fourth-Generation Positron Source Based on Using a 2 MeV Proton Accelerator

Noel A. Guardala, J. Paul Farrell* and Vadim Dudnikov*

*Naval Surface Warfare Center/Carderock Division, W. Bethesda, MD 20817-5700,
Brookhaven Technology Group, Inc. Nesconset, NY 11767*

Abstract. A method of producing an intense source of e^+'s that uses the internal pair production decay of the 6.05 MeV, 0^+ first excited state of O^{16} is described. The first excited state decays overwhelmingly by this process with the competing decays of double gamma-ray emission and atomic electron internal conversion contributing about 1 decay in 10^4. The most effective way to produce this excited state of ^{16}O in terms of compactness, power consumption and general ease of operation is to use the strong proton-induced resonance on F^{19} at 1.89 MeV. This resonance is the most favorable low-energy proton resonance in F^{19} for populating the 6.05 MeV, 0^+ state. We predict low-energy e^+ beams with intensity of 10^7-10^8/s by using off-the-shelf accelerator technology producing a 2 MeV proton beam with average currents of 1 to 5 mA.

INTRODUCTION

At a recently held workshop sponsored by the Institute for Theoretical Atomic and Molecular Physics, ITAMP[1], the need for new powerful and relatively inexpensive e^+ sources was discussed. These sources would be based on either large stored amounts of a long-lived e^+ emitting radioisotope (RI), or large, high-power accelerators (several hundred kW of power consumption). This paper describes a e^+ source that is:

- compact in terms of operation and power requirements
- inexpensive in terms of original cost and operating expenses
- versatile with ability to perform diverse experiments in areas of pure and applied physics.

This paper discusses a novel approach to producing high-flux e^+ beams which not only bypasses the need for large amounts of stored RI activities and the use of large, powerful electron accelerators but in addition, it does not produce radioactive materials nor are there any direct neutrons or high radiation fields near the operating area of the source. These very attractive features are a direct results of the proposed method of e^+ production. Namely e^+/e^- pairs are produced from the decay of an extremely short-lived (ca. 70 fs) excited nuclear state from a *stable* nucleus, ^{16}O.[2]

This new source can be used to produce low-energy, thermalized e^+ beams to perform "traditional" e^+ experiments such as positron annihilation (PALS) lifetime spectroscopy and correlated annihilation radiation spectroscopy (CARS). It can also produce nearly monoenergetic high-energy (keV – MeV) e^+'s and by the technique of in-flight annihilation of fast e^+'s, it can produce a tunable beam of high-energy (MeV) photons. The high energy positron beam can be used to do depth profiling, microprobe applications, channeling and atomic collisions studies. The nearly monochromatic MeV photons can be used in a variety of applications, such as, nuclear absorption studies for either pure or applied physics and radiography or imaging applications[1].

PHYSICAL BASIS OF POSITRON BEAM PRODUCTION

As a basis for positron beam production, we first consider the methods used by researchers in the field

---

[1] http://itamp.harvard.edu

[2] A patent has been filed by Brookhaven Technology Group, Inc. for this method of positron beam production.

of positron annihilation spectroscopy (PAS). PAS is commonly used to characterize the electronic structure of materials. The technique depends upon generating a bright, monoenergetic positron beam and directing it on the surface to be studied. There are three methods currently used to generate beams for PAS.

- Positron emitting radioisotope (RI) sources [2]
- Pair production from Bremsstrahlung produced by a high energy electron linac. [3]
- *In-situ* production of a positron emitting RI source, i.e. ^{13}N [4].

These represent respectively the first *three generations* of positron-emitting sources. All of these methods have significant drawbacks due to problems associated with the handling/production of highly radioactive materials in the first case, due to very high radiation fields and induced radioactivity generated by the high energy beams in the second and due to problems involving the production of fast neutrons inextricably linked to the method of producing the ^{13}N via a (d,n) reaction on a suitable ^{12}C – containing target such as high-purity graphite or diamond. The innovative method of generating a positron beam that is the subject of this paper is based on the process called *internal or nuclear pair production*. This electromagnetic decay process, in which an electron-positron pair is emitted from an excited nucleus, can only occur when the angular momentum of the excited state, with energy $> 2m_0c^2$, is populated in a nucleus that has the same angular momentum ground state. This is different than ordinary gamma decay followed by pair production. Ordinary de-excitation of the nucleus (in the absence of particle emission) takes place via emission of a single quanta of electromagnetic radiation (single gamma-ray emission) involving dipole, quadrupole and higher transitions downward eventually to the ground state of the nucleus. These processes require change in angular momentum of one unit or more. Therefore $0^+ \to 0^+$ transitions cannot be executed via single quanta (gamma ray) electromagnetic emission. These monopole transitions can only occur via three distinct processes:

- Multiple gamma ray emission, typically double or triple gamma ray de-excitation between the two 0^+ levels
- Nuclear internal conversion in which an atomic electron non-radiatively receives de-excitation energy
- The process of internal pair production or conversion in which an electron-positron pair is produced by electromagnetic transition inside the nucleus.

The probability of the first two de-excitation processes is very small in the case of the 6.05 MeV excited state of ^{16}O. Collectively they account for roughly 1 in 10^4 transitions between the first excited state and the ground state of ^{16}O. Thus, internal pair production is the overwhelmingly favored method of de-excitation in this instance.

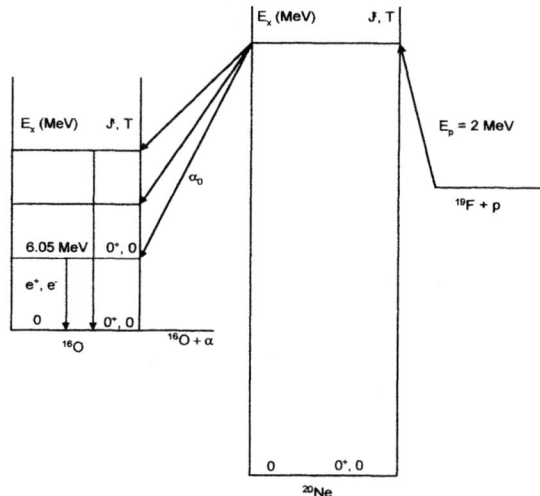

FIGURE 1 The method of generating positrons is shown in the energy level diagram for the reaction ^{19}F(p,αe$^+$e$^-$)^{16}O.

In internal nuclear pair production, the excited nucleus decays by direct emission of an electron and positron pair. The electron and positron share the excited state energy less the sum of the rest masses. The specific reaction we propose to exploit this method of positron production is the reaction ^{19}F(p, α e$^+$e$^-$)^{16}O. This reaction has many resonances in the energy range up through a proton energy of ~ 3 MeV and higher. The cross section, $\sigma(E = 2.2$ MeV$) = $ ~ 30 mb has a width of ~ $\Gamma = 0.5$ MeV.

The energy of the 0^+ first excited state in ^{16}O is 6.05 MeV. In the decay of this state, the electron and positron share the kinetic energy of ~ 5 MeV (= 6.05 – 2 m_0c^2) that remains after accounting for the rest mass of the electron positron pair. Figure 1 shows the energy level diagram for this reaction. This reaction can be exploited using standard accelerator technology including a standard proton ion source and standard tandem or radio frequency quadrupole (RFQ) accelerator. The energy of ~ 2 MeV is easily achieved with either type of accelerator system. The reaction produces no residual radioactivity and only small amounts of local shielding are required during operation.

The positron yield per incident proton from this reaction can be estimated from the relation,

$$Y = n \times \int \sigma \, dx = (n/\rho) \times \int \sigma(E) |S(E)|^{-1} \, dE, \quad (1)$$

where n is the number of fluorine nuclei per cm^3, σ is the cross section, ρ is the density of the target material, S(E) (MeV-cm^2/mg) is the stopping power and the integral is over the path length in the first expression and over the energy in the second. The cross section for this reaction was measured by Ranken, et al.[5]. The stopping power and the number of fluorine nuclei per milligram is determined by the choice of target. For an SF$_6$ target, the calculated thick target yield is ~ 5.2 x 10^{-7} positrons per proton at a proton energy of ~ 2 MeV. Thus, a 1 mA proton beam incident on a thick, SF$_6$ target produces about 3 x 10^9 positrons/sec.

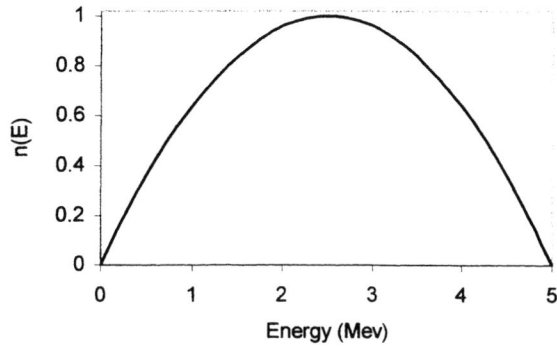

FIGURE 2. The energy spectrum of the emitted positrons from the internal pair conversion of the 6.05 MeV excited state of O^{16}

All positron sources evolve from a three body decay process. It is well known that three body decay results in a broad energy spectrum. due to the conservation of energy and momentum. In β$^+$ decay processes, the positron spectrum may be skewed to higher or lower energy depending on the type of transition responsible for the decay. In the case of *internal nuclear conversion* (IC), the decay energy is shared equally between the electron and positron since both particles have the same mass[3].

The energy spectrum that results from the first excited state of ^{16}O is shown in Figure 2. The spectrum peaks at ~ 2.5 MeV, which is ½ the center of mass energy available after taking account of the rest masses. Another factor which influences the shape of

---

[3] There is a small difference in the energy spectrum due to the difference in coulomb forces outside the nucleus.

the spectrum is the relatively long decay time for the 6.05 MeV first excited state. Indeed the lifetime broadening of this particular excited 0$^+$ state was of particular interest in nuclear structures studies many years ago [6].

To form the positrons into a mostly monoenergetic beam, the energy spread must be sharpened with minimum loss of intensity. There are several ways to accomplish this. The choice of method will depend on cost, the requirement for intensity and permissible energy spread. One option is to use a spectroscopic method to select a portion of the full spectrum. Another is to thermalize the spectrum by transporting the positrons through a material that has a negative work function for positrons. Thermalization reduces the energy spread to a few eV. It is the preferred method of generating highly monochromatic beams that are needed for PAS. About 1% of the initial positrons survive the selection process. That is, reducing the initial broad energy spectrum results in a factor of 10^{-2} in the number of positrons that survive to be formed into a beam. The number of monoenergetic positrons per second per mA of protons is obtained by multiplying the number of positrons per proton, the number of protons per mA and the thermalization factor. The result is:

(5 x 10^{-7} e$^+$/proton) * (6 x 10^{15} protons/mA-s) * (10^{-2} thermal e$^+$/total e$^+$) = 3 x 10^7 thermal e$^+$/mA-s.

The accelerator driven reaction ^{19}F(p, αe$^+$e$^-$)^{16}O, thus produces a thermal positron yield of ~ 3 x 10^7 e$^+$/mA-s of protons.

ACCELERATOR CONFIGURATIONS FOR GENERATING POSITRONS

Standard RF quadruple and tandem accelerators can be used to generate positrons using the technique described here. These are relatively low energy machines but high proton current is required to generate useful positron beams. Brookhaven Technology Group, Inc. has developed a compact high current negative hydrogen ion source that can be used in combination with a 1.1 MV Tandem to deliver up to 2 mA DC at 2.2 MeV to the target.

This surface plasma semiplanotron (SPS) with spherical focusing was first developed at the Budker. Institute of Nuclear Physics (BINP)[7-8]. The source, shown schematically in Fig. (3), consists of cathode, anode with emission aperture ~ 1 mm diameter, suppresser electrode, grounded extractor, magnet system, cathode insulators, and a DC high density cold hollow cathode glow discharge supported by voltage between cathode and anode. In operation, the source has produced beam current density of 500 mA/cm^2. In pulsed operation, the source produces beam current as high as 100 mA in millisecond pulses with 50 to 60 Hz repetition rate.

commercial high current tandem accelerator with approximately 1.1 MV terminal voltage. The high current, ~ 2.2 MeV proton beam is focused onto a ^{19}F target which may be a compound such as SF_6 in solid or liquid state.

CONCLUSION

A novel *fourth generation* e$^+$ source has been conceived based on a compact and efficient accelerator design. This source is capable of not only producing low-energy e$^{+'}$s desirable for condensed matter studies but will also be ideal for the production of monoenergetic beam of MeV-range positrons and gamma rays suitable for both pure and applied physics research.

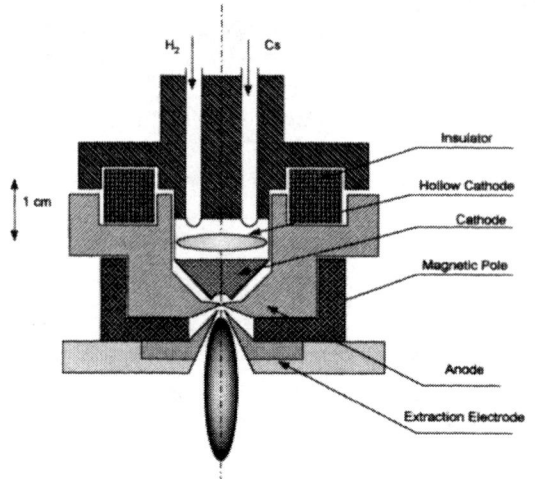

FIGURE 3. Surface plasma semiplanotron negative ion source.

Figure 4 is a block diagram of the positron beam system based on use of a high current tandem accelerator to produce the 2 MeV proton beam. It consists of an SPS ion source, deflection magnet and a

REFERENCES

[1] J.P. Farrell, et. al., Innovative method of using in flight annihilation of fast positrons to detect explosives, SPIE 13th Annual AeroSense Symposium, Orlando, FL, 1999 **SPIE 3710**, 446-453 (1999).
[2] C. D. Beling and M. Charlton, Contemp. Phys. **28**, 241, (1987).
[3] H. Tanaka and T. Nakanishi, Nucl. Instr. Meth., **B62**, 259, (1991)

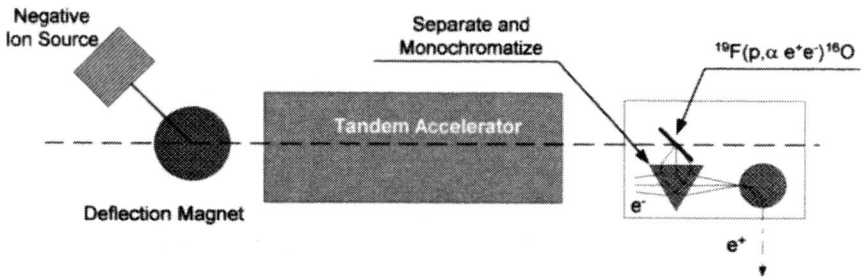

FIGURE 4. Block diagram of positron beam system consisting of SPS ion source and commercial 1 MV high current tandem accelerator.

[4] B. J. Hughey, R. E. Shefer, R. E. Klinkowstein and K. F. Canter, Conference Proceedings **392**, pp 455-459, Applications of Accelerators in Research and Industry, Ed. J. L. Duggan and I. L. Morgan, AIP Press, New York, 1997
[5] W. A. Ranken, T. W. Bonner and J. H. McCrary, Phys. Rev., **109**, 1646, (1958).
[6] M. Birk, J. S. Sokolowski and Y. Wolfson, Nuc. Phys., A216, 217, (1973).

[7] W.T.Diamond, Y.Imahori, J.W. MacKay, et al., Efficient negative ion sources for tandem injection, Rev. Sci. Istrum., **67** (3), p1404 (1996).
[8] A. Bashkeev, V. Dudnikov, Continuously operated negative ion surface plasma source, AIP Conference Proc. No. **210**, p.329, 1990, Fifth International Symp. On Production and neutralization of Negative Ions and Beams, BNL, NY, (1990).

Slow Positron Beams - A Versatile Tool for Studying Ion Implantation Defect Related Phenomena

B J Sealy, A P Knights, R M Gwilliam, C P Burrows* and P G Coleman*

University of Surrey Ion Beam Centre, SEEITM, University of Surrey, Guildford, Surrey, GU2 7XH, UK
*School of Physics, University of Bath, Claverton Down, Bath, BA2 7AY, UK

Abstract. Positron annihilation spectroscopy (PAS) with beams of controllable energy positrons has shown great promise as a technique for providing information on the concentration and distribution of vacancy-type, open-volume defects following the implantation of silicon. PAS is entirely non-destructive, requires no pre-measurement treatment of the sample and has a range of sensitivity of approximately 10^{15} - 10^{19} defects cm^{-3}. The Surrey Ion Beam Centre in collaboration with the University of Bath positron group has been investigating a number of novel applications of PAS such as ion beam dosimetry (including both low energy ion implantation and SIMOX production). Resulting from this work has been the development of a novel wafer-mapping tool compatible with a commercial environment. This paper will discuss applications of a prototype instrument which may make the use of PAS common in the fabrication plant.

INTRODUCTION

Ion implantation is one of the most important processing tools in semiconductor technology and implanters can be described as ubiquitous in the microelectronics industry. The current move to smaller dimensions (allowing more devices per cm^{-2}) is presenting ion implantation with a new set of problems which must be overcome if it is to maintain current size reduction trends. One such area in which advances are required is metrology. In particular there is a need for new, in-line, non-destructive technology which can provide a direct measurement of implantation dose and ion distribution [1].

Beam-based positron annihilation spectroscopy (PAS) is a potentially powerful tool in semiconductor process control and for general use in an R&D environment. Its desirable features are: (i) it is non-destructive; (ii) it is extremely sensitive to the vacancy-type defects created by ion implantation, and is consequently sensitive to ion doses as low as 10^9 - 10^{10} cm^{-2} in some circumstances; (iii) it is depth-tunable (by controlling the positron implantation energy), being especially suited to probing depths from 10 nm to 1 μm below the surface, and being sensitive to sub-keV ion implantation; (iv) it can be used for large-area mapping; (v) it does not require a post-implant annealing step before measurement, in contrast to some electrical characterization techniques used to monitor ion-implanted layers [2]); (vi) it requires no post-processing pre-measurement preparation: and (vii) it does not significantly power-load the substrate.

PAS MONITORING OF ION IMPLANTED MATERIAL

The three basic PAS techniques employed to study solids for many years are lifetime spectroscopy, angular correlation of annihilation radiation, and Doppler broadening spectroscopy [3]. The last of these is the preferred method for rapid monitoring of defect concentrations, and when performed with controllable-energy positron beams one can also extract information on the defect depth profile. Limited information can also be gained on the nature of defects - for example whether they are mono- or di-vacancies, or vacancy clusters/voids.

The principle of Doppler broadening spectroscopy has been described fully elsewhere [3,4]. Positrons of controllable energy E are implanted into the sample (mean depth in Si $<z>$ ~ $15.5E^{1.6}$ nm, E in keV, with a profile width ~ $<z>$) and after thermalisation in ~1 ps undergo diffusive motion until they are annihilated from either (a) the free state, (b) a trapped state (i.e., in a vacancy-type defect), or (c) at the surface. Because the mean electron momentum in each of these three states is different, so is the extent of the Doppler broadening of the annihilation gamma ray line at 511 keV. For example, in the vacancy-trapped state the absence of nearby core electrons reduces the Doppler broadening and the annihilation line is narrower. The broadening, measured using a high-resolution Ge detector, is characterized typically by the sharpness parameter S, defined as the ratio of the central part of the annihilation line (i.e., Ge detector photopeak) to its total area; the narrower the line (e.g., for a trapped positron), the higher is S. In summary, an increase in the measured linewidth parameter S is the signature of trapping in defects. Positrons are efficiently trapped by vacancy-type defects, so that concentrations as low as 10^{-7} per atom can be detected. Additionally, by measuring the mean value of S as a function of the positron implantation energy E, and by using appropriate fitting procedures, one can extract information about the depth profile of the defects - or, more simply, one can tune the positron probe to an energy at which the response to the subsurface damage is maximised. The current status of beam-based PAS to ion-implanted silicon has recently been reviewed in [5].

A practical positron beam system used for monitoring ion implanted material should incorporate features of research apparatus developed over the past fifteen years which create a user-friendly instrument. Important design criteria are: safe radioactive source handling, safe operation of high voltage supplies and vacuum system, robust design, efficient production of slow positrons with consequently high data collection rates and short run times, compact optics with source/detector shielding capability, facility for rapid sample changing and positioning, mm-diameter positron beam for wafer mapping, and computer control of system functions, data collection and analysis. The cost of such an instrument will be comparable to, or less than, that of other instruments currently used for monitoring implant uniformity and dose. All these features are being incorporated into a prototype system under construction in the authors' laboratories [6], and it is their hope that the instrument will be developed into the first compact, commercially-available positron beam for routine use in materials science.

ILLUSTRATIVE DATA

Experimental Technique

The data presented in this paper were accumulated on the University of Bath slow positron beam system, described by Chilton and Coleman [7]. Ion implantation was performed at the University of Surrey Centre for Research into Ion Beam Applications.

2, 10 and 25 keV B^+ in Cz Si

Figure 1 shows the raw data for Cz implanted with B^+ ions of energies 2, 10 and 25 keV at doses between 10^{12} and 10^{15} cm^{-2}. Incident positron energies have been converted to mean probed depth values using the expression stated above. The data have been chosen to demonstrate the sensitivity of PAS to both ion energy and dose. Uncertainties on individual S values are reflected in the scatter of the data points.

The peak in S is the signature of vacancy-type damage; at the highest positron energies used the parameters for all samples tend to the bulk value for undamaged Si; this is here normalised to unity. The maximum value of S here, ~1.04, is similar to that seen in all of our earlier data, and is consistent with trapping by open-volume defects of the size of a divacancy or divacancy-impurity complex.

Figure 1(a) illustrates the dependence of the positron parameter on ion dose. The reproducible nature of the shape of this dose dependence provides the foundation for the use of PAS as an ion dosimeter.

In Figure 1(b) data are shown for ions implanted with the same dose (~10^{13} cm^{-2}) at 2, 10 and 25 keV. This dose was chosen so that the saturation condition (S approaching 1.04) was not reached. Two features are evident; firstly, that the PAS response increases with increasing ion energy (indicating an increasing defect concentration), and secondly that the peak of the positron response moves to deeper depths below the

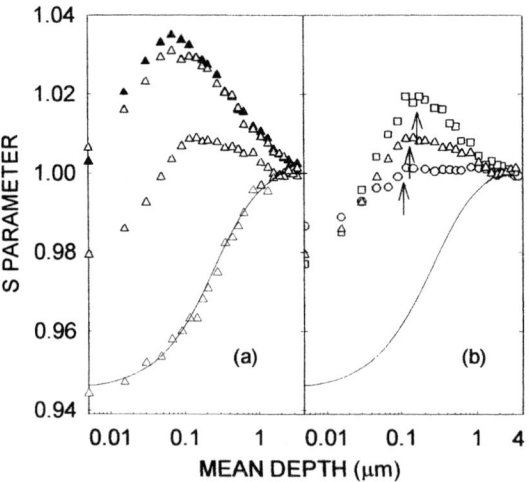

FIGURE 1. Normalised S parameter vs mean positron depth for B^+ ions implanted into Si. (a) 10 keV ions at nominal doses 10^{12}, 10^{13}, 10^{14} and 10^{15} cm^{-2} (increasingly dark symbols). (b) Ion dose = 10^{13} cm^{-2}: circles 2 keV, triangles 10 keV, squares 25 keV. Solid lines: fit to data for unimplanted Si.

sample surface as the ion energy increases, as indicated approximately by the three arrows. It is possible to arrive at these conclusions by direct inspection of the raw data because the data sets on each figure are directly comparable. It is important to exercise caution when drawing conclusions from raw data, however. For example, the apparent decrease in the PAS peak response with ion dose in Figure 1(a), suggesting that the damage is occurring at progressively shallower depths, is in fact a result of the low-dose data being affected by positron diffusion to, and annihilation at, the sample surface (with a lower S value). The data points at smaller depths are 'pulled down' by this effect and the peak consequently moves to the right.

It is therefore necessary to subject the raw data to straightforward analysis to allow for surface diffusion effects before extracting ion dose from the peak in the S parameter plots such as those in Figure 1(a). As implied above, similar care must be taken before extracting crude depth information from the positions of the arrows such as those in Figure 1(b). Whereas simple data reduction can be applied quickly, fuller analysis of the data can be performed using standard fitting programs. Thus, if the PAS device outlined above and described in more detail in [6] and [8] employs a mm-diameter positron beam which can be scanned across a wafer surface, the two-dimensional data common to traditional characterization tools such as photothermal spectroscopy is extended to three dimensions, providing data rich in information on defect structure.

The samples used in Figure 1 were also subjected to SIMS analysis after the positron data were taken; total ion doses and junction depths (i.e., depths at which the B concentration falls to the background doping level) were measured for three selected samples - at ion energies of 2, 10 and 25 keV. Total doses measured in this way will be used to calibrate the PAS-based dosimetry technique. Junction depths are seen to correlate with the trend seen in the raw PAS data, and the calibration of the latter to allow rapid measurement of the former would be a valuable extension of the information provided by the PAS instrument.

Sensitivity to Ion Dose

The precision of the positron technique is illustrated in Figure 2, in which the normalised peak PAS response as a function of ion dose, in this case designated R, is plotted for 125 keV Si^+ ions implanted into Si in the range $10^{11} - 10^{12}$ cm^{-2}. In the middle of

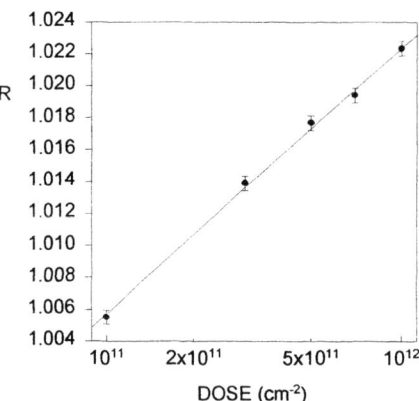

FIGURE 2. R (normalised S parameter for 7 keV positrons) vs ion dose for 125 keV Si^+ ions in Si at doses between 10^{11} and 10^{12} cm^{-2}. The solid line is a fit to the data.

this range a change of 2×10^{10} cm^{-2} in ion dose leads to a change of 3×10^{-4} in R. This precision in R is achieved if S is measured to $\pm 10^{-4}$, requiring ~10^7 total events to be recorded in the annihilation gamma photopeak. With current counting rates of ~10^3 s^{-1}, this is thus achieved in 10^4 s; with the adoption of sophisticated pile-up rejection and high counting rate electronics this time could be reduced by two orders of magnitude. (A precision of 10^{-11} cm^{-2} in the middle of the range in Figure 2 requires run times $\approx 10^2$ s at present count rates, therefore reduced to ~seconds with advanced techniques.) It is probable that the sensitivity and hence precision of the positron technique will be improved by (a) measuring a lineshape parameter other than S (b) using a dummy sample other than Si for which S is more sensitive to changes in ion dose.

PAS is a potentially powerful process control tool because it is applicable to any ion over a wide range of doses - and, importantly, is non-destructive. To illustrate the different ranges of dose sensitivity for low and high-energy ions we compare in Figure 3 the normalised peak PAS response as a function of ion dose for 1keV B^+ and 2MeV Si ions implanted into Cz Si. Error bars are approximately represented by the size of the points. The ordinate in the latter plot is the ratio of the peak S parameter value to that for undefected bulk Si, with allowance made for the effect of diffusion to the surface; this analysis procedure is described in detail in [8]. In the case of 1 keV B^+, however, the change in S parameter is predominantly due to a change in effective surface S value, and the normalised values shown in Figure 3 have been derived from simple ratios scaled to the same range as that for the high-energy Si data. The treatment of very low-energy ion implant PAS data is still being developed and will form the basis of a future publication.

Notwithstanding this inherent difference between the high- and low-energy ion raw data, it is still clear that the PAS response is non-linear. The range of PAS sensitivity to dose in the former is typically 10^{10} - 10^{15} cm^{-2}, whereas for low energies the range is ~10^{14} - 10^{16} cm^{-2}. These ranges fortuitously encompass the ranges of doses typically implanted at these energies. For example, source drain implants are performed at low ion energies at doses of typically 10^{14} - 10^{16} cm^{-2}: voltage threshold adjustment implants are at medium energy (10 - 100 keV) at 10^{11} - 10^{13} cm^{-2}: n well and p well implants are high (MeV) energies, in the 10^{12} - 10^{15} range.

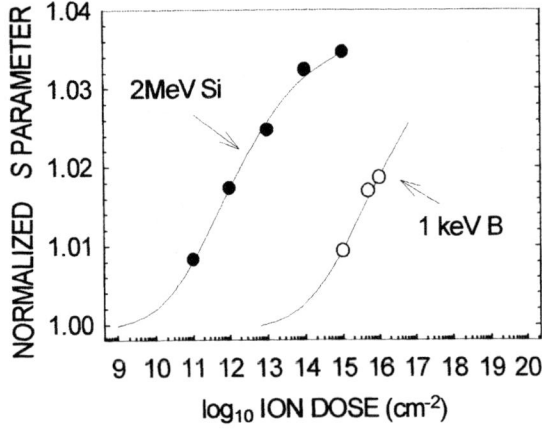

FIGURE 3. PAS response to high and low energy implants. The solid line fit to the 2 MeV Si data is shifted laterally to lie beneath the B data points. All data for 7 keV positrons.

CONCLUSIONS

An instrument employing beam-based positron annihilation spectroscopy is being developed in a collaborative project between the University of Bath and the Surrey Ion Beam Centre. The device will be capable of providing three-dimensional information on implanted wafers non-destructively, and has proven sensitivity to implants of all ion species and energies (from sub-keV to MeV) in commonly-used ion dose ranges. The PAS response to ~10^2 eV ions has been found to be fundamentally different from that to higher energy implants, being dominated by a change in the characteristic surface parameter rather than by the increase in the subsurface parameter associated with trapping at open-volume defect sites. New analysis procedures are being developed for sub-keV implanted samples.

The applicability of the device extends beyond implantation process control to general R&D assessment of a wide range of technologically-important films, layers and interfaces. The compact, automated instrument, which delivers a mm-diameter beam of controllable-energy positrons to the wafer being studied, is currently under construction.

ACKNOWLEDGEMENTS

This work was supported by EPSRC, UK, under grants numbers GR/M51895 and GR/M54001. The authors are grateful to F Malik for help with data collection.

REFERENCES

[1] The National Technology Roadmap for Semiconductors, available at http://notes.sematech.org/
[2] For example see Ziegler J, *Ion Implantation Science & Technology*, Academic Press, New York, 1984.
[3] Asoka-Kumar P, Lynn KG, Welch DO, *Journal of Applied Physics* **76**, 4935 (1994).
[4] Simpson PJ, Vos M, Mitchell IV, Wu C, Schultz PJ, *Physical Review B* **44**, 12180 (1991).
[5] Knights AP, Coleman PG, *Defect and Diffusion Forum* **183-5**, 41 (2000).
[6] UK patent file no. 9818330.4.
[7] Chilton NB, Coleman PG, *Measurement Science and Technology* **6**, 53-59 (1995).
[8] Coleman PG, Knights AP, Gwilliam RM, *Journal of Applied Physics* **86**, 5988-5992 (1999).

Positron Annihilation Studies on Stable and Undercooled Metal Melts at the Stuttgart Pelletron

H. Stoll, A. Siegle* and J. Major

Max-Planck-Institut für Metallforschung, Heisenbergstr. 1, D-70569 Stuttgart, Germany

Abstract. If the phase transition liquid → solid is suppressed on cooling, metal melts may be transformed into the metastable state of the 'undercooled melt'. An MeV positron (e^+) beam, which can easily be implanted into all kind of samples facilitates considerably positron-lifetime and Doppler-broadening measurements on liquids. By taking advantage of the high data-acquisition rate of the beam-based $\beta^+\gamma\Delta E_\gamma$-coincidence technique developed at the Stuttgart Pelletron measurements with good statistics in Ga, Bi, Sn, In, and Pb were performed not only in the solid states and in the 'stable' metal melts but also in the metastable undercooled melts at temperatures below the melting point T_m. As long as the metals stay in the liquid states, no sudden changes of electron density or electron momentum distribution at the positron site were observed at the melting point. Compared to the undisturbed solid, significantly longer positron lifetimes and much narrower Doppler broadening were observed in all metal melts investigated. These findings are explained by the formation of self-localized e^+ polaron states. Theory predicts two types of the 'polaronic' e^+ states which differ in their effective positron masses m^+. In Bi, Sn, In, and Pb 'low-m^+' polaron states showing effective masses of 1 to 4 proton masses are found. In Ga 'high-m^+' polaron states are observed with even much larger effective positron masses.

INTRODUCTION

One of the main structural differences between liquids like metal melts and crystalline solids becomes manifest in the pair correlation function $g(r)$ which describes the mean density of particles versus the distance r from a reference particle (1, 2). $g(r)$ can be obtained by X-ray or neutron scattering (2). The structures and pair correlation functions of a liquid and a crystal are sketched in Fig. 1: In contrast to the crystalline state (cf. Fig. 1 c,d) there is no long and medium range order in liquids (Fig. 1, a,b). At distances large compared to the mean distance of the particles the pair correlation function $g(r)$ in liquids therefore tends to a constant (which is 1 because of the normalization of $g(r)$). But different to completely disordered systems, e.g., gases, short range order still exists in liquids caused by the interaction of the atoms or molecules: hence $g(r)$ shows a strong peaking at short range (Fig. 1b).

$g(r)$ is defined as average in space and time. Thermally activated motions of atoms may however cause local minima in the density of atoms, the so-called 'free volumes', which may trap positrons. Indeed, experimental evidence is given in the section 'Results' of the present paper which shows that in metal melts the electron density at the positron site is significantly lower than the average electron density. However, the simple idea of positrons being trapped at local minima of the atomic density in metal melts analogous to the well known positron trapping at vacancy-like defects in crystals is misleading: The characteristic times of atomic movements in metal melts are in the range of 10^{-13} s to 10^{-11} s (2, 3) and thus at least one order of magnitude shorter than the positron lifetimes ($\approx 10^{-10}$ s) in those melts (in the solids the movements of atoms during the positron life can be neglected). The problem to explain the behavior of positrons in metal melts was solved by the comprehensive 'polaron model' given by Seeger (4). The application of this model to our experimental data will be discussed in the last section of this paper.

EXPERIMENTAL SET-UP

The experiments described in the present paper were performed at the Stuttgart MeV positron beam (5). A schematic drawing of the age-momentum-correlation (AMOC) set-up which was used is shown in Fig. 2. By taking advantage of the $\beta^+\gamma\Delta E_\gamma$-coincidence technique (6), a start detector with almost unity detection efficiency is applied which results in high count rates and virtually background-free lifetime spectra. Since only

---

* present address: Robert Bosch GmbH, D-70442, Stuttgart, Germany

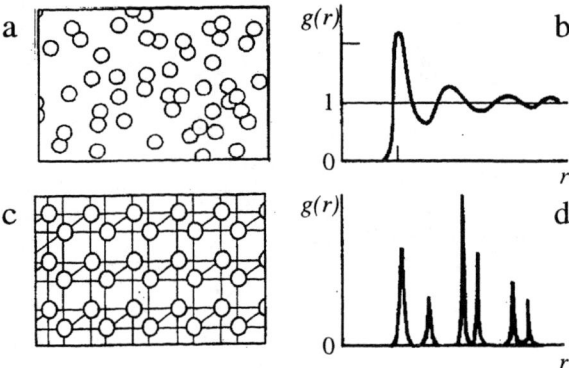

FIGURE 1. Structure and pair correlation function $g(r)$ of the liquid (a,b) and the crystalline solid state (c,d) according to (2). In contrast to the solid, there is no long and medium range order in the liquid. Short range order is evident in the liquid by a strong peaking of the pair correlation function at small r.

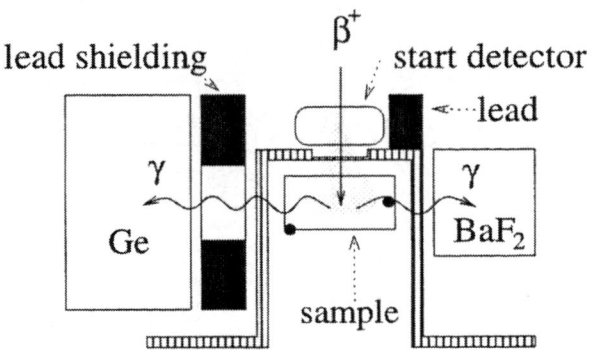

FIGURE 2. Schematic set-up of the experiments. The sample is placed in a cooled or heated crucible located in an evacuated chamber. Temperature sensors are marked by points (●). The MeV positron beam is implanted through a thin metal window. By letting the positrons pass through a 5 mm thick plastic scintillator (Pilot-U) before implantation into the sample, a start signal with 100% detection efficiency is generated. The corresponding stop signal for positron lifetime measurements is produced by one of the 511 keV annihilation γ photons in a BaF_2-detector. The energy of the second annihilation photon is measured in a high-purity germanium detector. Lead shields and the triple-coincidence measurement eliminate events from positrons annihilating outside the sample.

triple-coincidence events between the β^+-start detector, the stop detector, and the energy detector (cf. Fig. 2) were collected, events from positrons annihilating outside the sample could be eliminated effectively. About 10^6 triple coincidences were collected for each partial measurement.

RESULTS

Figure 3 shows the temperature dependence of the mean positron lifetime $\bar{\tau}$ and the Doppler broadening S parameter measured on gallium, bismuth, tin, indium, and lead in the solid and molten state and in the undercooled melt. The data were extracted from the corresponding two-dimensional AMOC reliefs. As long as the metals stay in the liquid states, no sudden changes of electron density or electron momentum distribution at the positron site were observed at the melting points. More details are given in (7).

In *gallium* the positron lifetime and the S parameter increase at the melting point in spite of the fact that Ga and also Bi contract upon melting (Ga by 3.2%, Bi by 3.35%). Gallium could easily be undercooled by 30 K in a fused quartz crucible.

In *bismuth* the change of the positron lifetime at the melting point is much stronger than the change in the S parameter.

In *tin* the changes at the phase transition are somewhat weaker than in gallium

In solid *indium* as well as in *lead* pronounced trapping of positrons at thermally generated vacancies is observed at higher temperatures. In both materials a decrease of the positron lifetime is found at the phase transition compared to the positron lifetime in monovacancies. The Doppler measurements show about the same S parameter at the melting points for the solids and the stable and undercooled melts. Compared to the undisturbed lattice, the positron lifetime as well as the S parameter is significantly higher in the melts. Thus beside the positron trapping at thermally generated vacancies at higher temperatures which concerns the solid states only, in principle there is no difference in positron annihilation in the melts of indium and lead in comparison to the other metals investigated.

THE POSITRON POLARON MODEL

A model of positron diffusion in solid and liquid metals proposed by Seeger (4) is based on the formation of 'acoustical e^+-polarons'. The electrostatic interaction energy between positrons and the neighbouring metal-ion cores may be lowered by displacing neighboring atoms away from the positrons (8, 9). In this way the effective positron mass m^+ is very much enhanced. The e^+ polaron state may be stable or metastable depending on the balance between the above gain in energy and the increase in elastic energy due to displacements of atoms and in kinetic energy due to the localization of the positrons. According to (4) two different kinds of e^+-polaron states exist: (i) A 'low-m^+ polaron state' with an effective positron mass m^+ of the order of magnitude of the mass

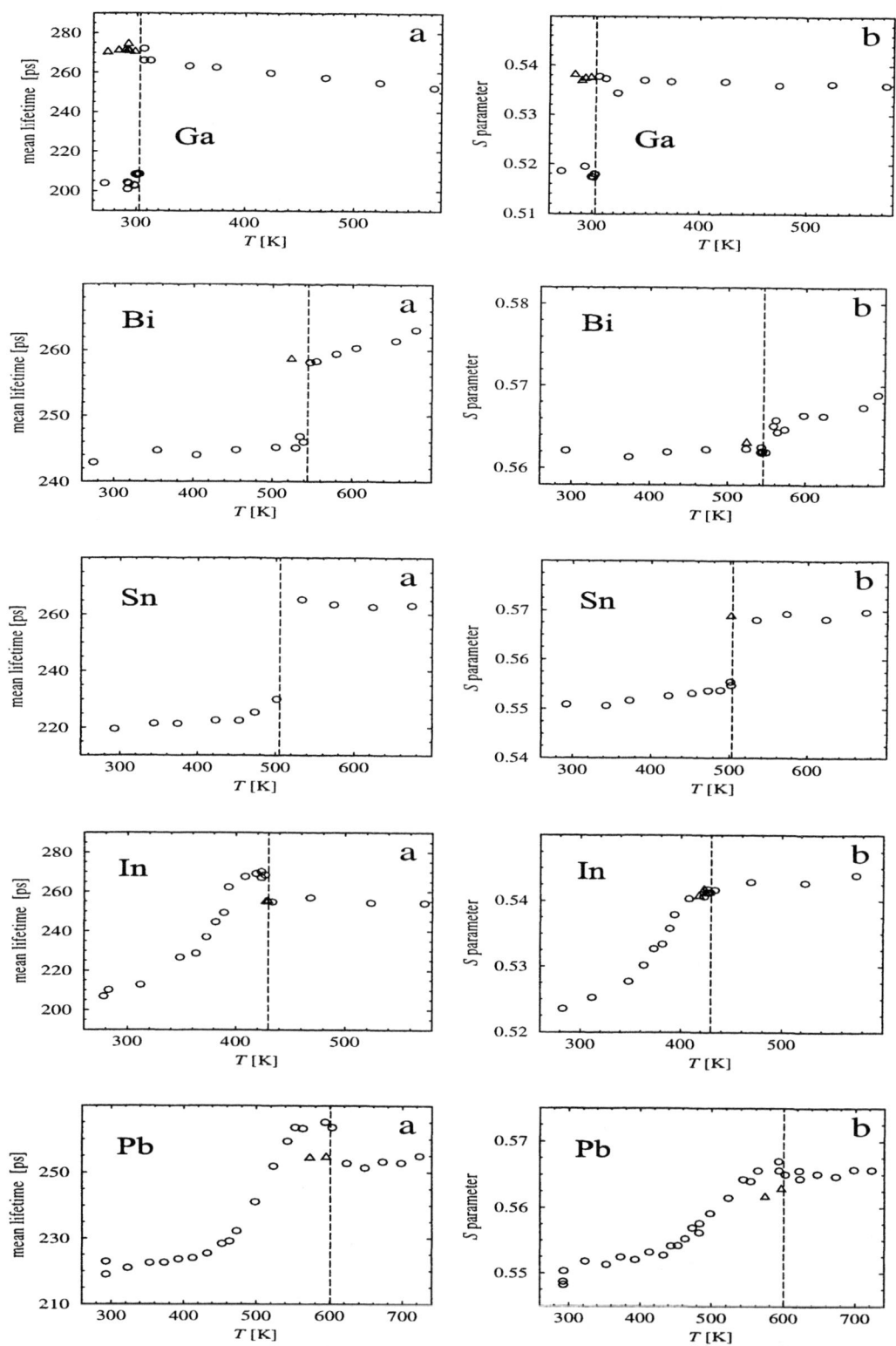

FIGURE 3. Temperature dependences of the mean positron lifetimes $\bar{\tau}$ (a) and the Doppler broadening S parameters (b) of Ga, Bi, Sn, In, and Pb. For the determination of the S parameter a central energy interval of 1.6 keV was chosen for Ga and In, and of 1.7 keV for Bi, Sn, and Pb. The dashed lines denote the melting points T_m (Ga: 302.9 K, Bi: 544.5 K, Sn: 505.2 K, In: 429.7 K, and Pb: 600,7 K). Measurements in the solids and stable melts are marked by circles. Measurements in the undercooled melts are indicated by triangles.

of a proton or less, which moves by mean free path motion and shows a temperature dependence of positron diffusivity $D^+ \sim T^{1/2}$ and (ii) a more heavy 'high-m^+ polaron state' moving in a thermally activated hopping process (4).

Compared to the solid state it is much easier to displace atoms in a metal melt. Thus the formation of e^+ polaron states is more likely. When stable e^+ polarons are formed, the electron density at the positron site and the annihilation of positrons with high-momentum core electrons is reduced. Thus the increase of positron lifetime and S parameter at the melting point reported in the present paper supports the formation of stable e^+ polaron states in the melts.

In addition, the e^+ polaron model (4) perfectly explains the temperature dependences of the diffusion lengths L^+ which were measured in liquid metals by means of a slow positron beam (10, 11, 12).

According to Seeger (4) the effective mass m^+ of a 'low m^+ polaron state' can be estimated from the positron diffusivity D^+, the mean free path of the positrons l, and the temperature T:

$$D^+ \approx l \left(\frac{3k_B T}{m^+} \right)^{1/2}, \quad (1)$$

where k_B denotes the Boltzman constant.

The diffusivity D^+ can by calculated from the diffusion length L^+ and the positron lifetime τ by the Einstein relation:

$$D^+ = \frac{L^{+2}}{2\tau}. \quad (2)$$

Values for liquid Sn, Bi, In, and Pb are summarized in Table 1. In all these metal melts the effective mass of the positron is between one and four times the mass of a proton. These values of m^+ in the order of the proton mass are well accounted to low-m^+ polaron states.

The low m^+ polaron state model cannot explain the gallium melt data. Eq. 1 and 2 would yield an effective positron mass greater than the mass of a gallium atom. Positrons in liquid gallium are in a high m^+ polaron state, which is also supported by the temperature dependence of the positron diffusivity D^+. The diffusivity shows an exponential temperature dependence with an activation energy of about 0.2 eV (4), i.e. the majority of positrons migrate by hopping.

Seeger suggested the following test of his model (4): The formation of a 'high m^+ polaron state' in the gallium melt should increase the Doppler broadening S parameter at the phase transition solid → liquid because of the reduced annihilation probability of the positrons with core electrons. It was predicted by Seeger (ref. (4), p. 13, lower part) that in bismuth this effect should be absent or, at least, much smaller. This prediction is indeed fulfilled (cf. Fig. 3, sub-figures on the right-hand side).

Table 1. Estimation of the positron diffusivity D^+ and the effective positron mass m^+ in units of proton mass m_p according to Eq. 1 and 2 in different metal melts just above the melting point $T \approx T_m$ (Bi: 544.5 K, Sn: 505.2 K, In: 429.7 K, Pb: 600.7 K). The diffusion lengths L^+ are taken from (12), the positron lifetimes τ from (7). The mean free path length l is approximated by the first maximum of the pair correlation function r_1 (ref. (2), p. 54).

metal	τ [ps]	L^+ [nm]	$l \approx r_1$ [nm]	D^+ [m^{2}s^{-1}]	m^+/m_p
Bi	258	19±2	0.34	$0.7 \cdot 10^{-6}$	3
Sn	265	26±2	0.32	$1.3 \cdot 10^{-6}$	1
In	256	23±2	0.32	$1.0 \cdot 10^{-6}$	1
Pb	253	18±2	0.33	$0.6 \cdot 10^{-6}$	4

ACKNOWLEDGMENTS

The authors would like to thank Prof. A. Seeger for his continuous support, Prof. H. D. Carstanjen for many fruitful discussions, and Dr. P. Bandžuch for his help in preparing the manuscript. Financial support by the Deutsche Forschungsgemeinschaft (DFG) within the scope of the Schwerpunktprogramm 'Unterkühlte Metallschmelzen' is gratefully appreciated.

REFERENCES

1. Shimoji, M., *Liquid Metals*, Academic Press, London, 1977.
2. Waseda, Y., *The Structure of Non-Crystalline Materials*, McGraw-Hill, New York, 1980.
3. March, N. H., and Tosi, M. P., *Atomic Dynamics in Liquids*, Macmillan, London, 1976.
4. Seeger, A., *Appl. Surface Sci.* **85** (1995) 8.
5. Carstanjen, H. D., Decker, W., and Stoll., H., *Z. Metallkd.* **84** (1993) 368.
6. Stoll, H., *MeV Positron Beams*, in: Positron Beams and Their Applications, ed. P. G. Coleman, World Scientific, Singapore, 2000 pp.237.
7. Siegle, A., Dr. rer. nat. thesis, Universität Stuttgart 1998, (Cuvillier, Göttingen, ISBN 3-89712-129-8).
8. Seeger, A., *Appl. Phys.* **7** (1975) 85.
9. Seeger, A., *Appl. Phys.* **7** (1975) 257.
10. Gramsch, E., Ph.D. thesis, The City University of New York, 1992.
11. Gramsch, E., Lynn, K. G., Throwe, J., and Kanazawa, I., *Phys. Rev. Lett.* **67** (1991) 1282.
12. Gramsch, E., Lynn, K. G., Throwe, J., and Kanazawa, I., *Phys. Rev. B* **59** (1999) 14282.

Low Energy Positrons at Semiconductor Surfaces

N. G. Fazleev [1,2], J. L. Fry [1], and A. H. Weiss [1]

[1] *Department of Physics, Box 19059, The University of Texas at Arlington, Arlington, Texas 76019, USA*
[2] *Department of Physics, Kazan State University, Kazan 420008, Russian Federation*

Abstract. Positron-annihilation-induced Auger spectra from the clean and exposed to hydrogen and oxygen Si(100)-(2×1) surface are analyzed by performing calculations of positron states and annihilation characteristics. Positron surface and bulk states are calculated for different hydrogen and oxygen coverages by solving Schrödinger's equation numerically using the finite-difference method and taking into account discrete lattice effects and the charge redistribution at the surface. The reconstructed Si(100)-(2×1) surface is described within the Dimer-Adatom-Stacking fault model. Calculations performed for the clean Si surface show that the positron surface state wave function is localized mostly on the vacuum side of the topmost layer of Si atoms. When hydrogen or oxygen is absorbed on the Si surface the positron wave function is displaced away from substrate atoms. As a result of this displacement, the overlap of the positron wave function with Si core electrons and, consequently, the annihilation probability of Si core electrons reduce, in agreement with experimental data.

INTRODUCTION

The adsorption of hydrogen and oxygen on a Si surface has attracted increased interest due to applications of these processes in device technology. Hydrogen adsorption is used mostly to lower the surface energy of Si and thus stabilize the Si surface prior to molecular beam epitaxy. However the difficulty of detecting hydrogen using standard surface spectroscopies leaves many open questions regarding the behavior of the absorbed hydrogen on a semiconductor surface. Oxygen adsorption is of interest due to its significance in Si device processing and in metal-oxide semiconductor device fabrication. Numerous studies of oxygen absorption have been directed towards better understanding of the formation of SiO_2 and a Si/SiO_2 interface. Nevertheless, the initial stages of oxidation of a semiconductor surface are still not well understood.

Recently the Si(100) surface with adsorbed hydrogen and oxygen has been studied using positron annihilation induced Auger electron spectroscopy (PAES) [1,2]. In PAES experiments, most of the low-energy positrons implanted into the sample under study diffuse back to the vacuum-solid interface and are trapped into a surface state. A certain fraction of surface-trapped positrons annihilate with neighboring core-level electrons, creating core-hole excitations and initiating Auger processes almost exclusively in atoms in the topmost layer. Since PAES intensities are sensitive to spatial distribution of the positron wave function at the surfaces of interest, the method has already been used to selectively obtain chemical information from the topmost atomic layer and to clarify the nature of the positron surface state [3].

PAES spectra taken from the clean Si(100) surface display a strong positron annihilation induced Auger signal in the measured energy range corresponding to the $L_{2,3}VV$ Auger transition for Si [4]. When the Si surface was exposed to hydrogen and oxygen the Si $L_{2,3}VV$ Auger intensities were observed to decrease approximately exponentially with increasing gas exposure in accordance with the Langmuir model of adsorption [2]. The purpose of this paper is to analyze from first principles these experimental results by performing calculations of positron surface and bulk states and positron annihilation characteristics for the Si(100) surface with different hydrogen and oxygen coverages.

THEORY

The potential due to the surface felt by a positron contains an electrostatic Hartree part $V_H(\mathbf{r})$ and a correlation part $V_{corr}(\mathbf{r})$. The Hartree potential $V_H(\mathbf{r})$ at the Si(100) surface is approximated by a superposition of atomic Coulomb potentials $V_{Coul}^{at}(|\mathbf{r}-\mathbf{R}|)$ from all atoms located within a predetermined radius of the evaluation point, where $\mathbf{R}$ defines the positions of the host nuclei. We perform atomic calculations within the local-spin-density approximation [5] using the exchange-correlation functional from Ref. 6. To account for the effects of the charge redistribution at the surface we use the method of Weinert and Watson [7]. Following this method we place atoms in a "compensating" potential well of magnitude 0.10 Ry extending from the atom center out to one Wigner-Seitz radius, R_{W-S}, and then linearly ramping to a value of 0.00 Ry at $2R_{W-S}$ and beyond. Schrödinger's equation is solved self consistently for each bound electron state of the Si atom with the inclusion of the "compensating" potential well. Electron wave functions then provide the modified atomic electron densities and corresponding atomic Coulomb potentials at the Si(100) surface via Poisson's equation. It has been shown that the superposition of the atomic electron densities provides an adequate description of the total electron density $n_-(\mathbf{r})$ in interstitial regions, where the positron wave function is mainly localized in the bulk, and of the rise of the electron density near atom chains, when compared with self consistent calculations [8].

In constructing $V_{corr}(\mathbf{r})$ at a surface we exploit the fact that the correlation component of the positron potential deep inside and far outside the semiconductor surface is well described by the local density approximation (LDA) and the image potential, respectively. It is possible then to divide space into two regions, namely, the bulk and image potential regions, where the two models are applied. Within the LDA, $V_{corr}(\mathbf{r})$ in the bulk region of a semiconductor is approximated by $V_{corr}^{EG}(n_-)[f(n_-,\varepsilon_g)]^{1/3}$, where $V_{corr}^{EG}(n_-)$ is the correlation energy for a positron in a homogeneous electron gas of density n_- [9]. This approximation is justified by the fact that inside the bulk the positron wave function mainly resides in interstitial regions between atoms where the electron density is slowly varying. The function, $f(n_-,\varepsilon_g)$, is a reduction factor that accounts for the diminished screening response of semiconductors to charged particles due to the existence of a band gap. The "gap parameter" ε_g describes the effect of the band gap on the electron-positron correlation. A reasonable fit to numerical results, obtained from screening calculations for point charges in a semiconductor, can be obtained using the interpolation formula [10]:
$f(n_-,\varepsilon_g) = 1 - 0.37\,\varepsilon_g/(1+0.18 r_s)$, where $r_s=(3/4\pi n_-)^{1/3}$. We consider ε_g to be a parameter, and use the value $\varepsilon_g = 0.2$ that was shown to reproduce well experimental annihilation rates for delocalized positron states in several IV- and III-V-type semiconductors. Outside the Si surface $V_{corr}(\mathbf{r})$ is described by the corrugated image potential [11,12]. Positron surface states in the present paper are calculated by solving Schrödinger's equation numerically using a modified relaxation technique [11,12] with boundary conditions that the positron wave function vanishes far into the bulk and the vacuum. The outermost plane of substrate atoms is taken to reside at $Z = 0$. The image plane position Z_0 is determined from Lang-Kohn theory [13]. The positions of the Si atoms are determined according to the modified dimer model for the Si(100)-2×1 reconstructed surface consisting of asymmetric and tilted dimmers [14].

RESULTS AND DISCUSSION

The results of calculations of the positron surface state wave functions show that the positron is trapped mainly in the image-correlation well just outside the clean Si(100)-(2×1) surface. The positron wave function has its maximum outside the top most layer of atoms, and experiences a rapid drop with distance into the Si lattice and overlaps very little with the second layer of Si atoms. It also follows from these calculations that the hydrogen or oxygen overlayer pushes the positron wave function away from the Si substrate, substantially reducing the overlap of the positron wave function with Si atoms. Computed binding energies, E_b, for a positron trapped at the Si(100) surface with different coverages of hydrogen and oxygen are shown in Table 1. These results for E_b show that adsorption of oxygen on the Si surface due to its larger atomic electronic charge causes larger changes in E_b compared to changes in binding energies due to the adsorption of hydrogen.

To clarify the behavior of positron bound states at the semiconductor surface the positron work function, Φ_p, is also calculated for the Si(100) surface with different coverages of hydrogen or oxygen. Calculations are performed using the same positron potential constructed for the surface by imposing periodic boundary conditions sufficiently

TABLE 1. Theoretical positron surface state binding energies, E_b, positron work functions, Φ_p, and positron annihilation probabilities with relevant core electrons, $p_{n,l}$, at the Si(100) surface, both clean and covered with hydrogen and oxygen.

System	E_b (eV)	Φ_p (eV)	$p_{n,l}$ (%)		
			Si 2s	Si 2p	Si 2s+2p
Si(100)	2.06	1.28	0.28	0.84	0.84
Si(100)+H(1/4ML)	3.63	1.47	0.11	0.32	0.43
Si(100)+H(1/2ML)	3.68	1.46	0.09	0.27	0.36
Si(100)+H(1 ML)	3.70	1.44	0.08	0.23	0.31
Si(100)+O(1/4ML)	3.96	1.91	0.14	0.41	0.41
Si(100)+O(1/2ML)	3.97	1.88	0.13	0.37	0.37
Si(100)+O(1 ML)	4.00	1.87	0.07	0.20	0.20

far into the bulk, and assuming **k**=0 to be the lowest Bloch state. Computed Φ_p are also shown in Table 1. Since in each case E_b is significantly larger than Φ_p it may be concluded that positron surface states are stable on the Si(100) surface for all coverages of hydrogen and oxygen. The increase of hydrogen and oxygen coverages on the Si(100) surface leads only to rather small changes of Φ_p.

Positron annihilation rates $\lambda_{n,l}$ with specific core-level electrons, described by quantum numbers n and l, are computed within the Independent Particle Model from the overlap of positron and core-level electron densities. The total annihilation rate, λ, (the inverse of the positron surface state lifetime, τ,) of a surface trapped positron is calculated from the overlap of positron and electron densities in the bulk region using the Local Density Approximation. Positron annihilation probabilities $p_{n,l}$ with specific core-level electrons, described by n and l, are obtained by dividing the positron annihilation rate $\lambda_{n,l}$ by the total positron annihilation rate λ: $p_{n,l} = \lambda_{n,l} / \lambda$. Results for τ and $p_{n,l}$ calculated for the Si(100) surface with different coverages of hydrogen and oxygen are shown in Table 1. It follows from these results that probabilities of a surface-trapped positron to annihilate with 2s and 2p core electrons of Si atoms that determine the Si $L_{2,3}$VV Auger intensity decrease with the increase of gas coverages. It also follows from Table 1 that the adsorption of oxygen on the Si(100) surface causes larger decrease of $p_{n,l}$ than the adsorption of hydrogen. This is consistent with theoretical calculations of positron wave functions at the Si(100) surface. The adsorbed oxygen atoms push the positron wave function further away from the Si atoms into the vacuum than the hydrogen atoms due to their larger electronic charge, thus reducing the probability of annihilation of the surface-trapped positron with the substrate core level electrons.

A good fit to the exponential-like decrease of the Si $L_{2,3}$VV Auger intensity measured by PAES with the increase in the hydrogen and oxygen exposure of the Si(100) surface was obtained in Ref. 2 using the Langmuir model of adsorption kinetics and assuming that the decrease in the clean surface PAES intensities by 60 and 71% after 4500 L hydrogen and 9000 L oxygen exposure, respectively, was linear with the increase in gas coverages. It follows from Table 1 that the obtained theoretical core annihilation probabilities are consistent with the decrease observed in the experimental Si $L_{2,3}$VV Auger intensity with the increase in hydrogen and oxygen coverages [2]. We note that the ordered overlayer systems were assumed even at coverages below saturation to facilitate calculations. The disorder that is likely to be present can be expected to lead to an averaging-effect, which would produce a relation between the adsorbate coverage and the PAES intensity closer to the linear relation assumed in Ref. 2.

CONCLUSIONS

In this work, techniques proven to work for metals have been extended to treat positrons at the adsorbate covered semiconductor surface. Appropriate screening factors have been introduced for semiconductors, and both positron bulk and surface states have been obtained from the same potentials for the Si(100) surface with adsorbed hydrogen and oxygen. Stable positron surface states have been found for all hydrogen and oxygen coverages on the (100) surface of Si. Annihilation probabilities of surface trapped positrons with Si 2s and 2p core-level electrons have been computed and found to decrease with the increase of gas coverages. The obtained theoretical results are consistent with the decrease of the positron annihilation induced Si $L_{2,3}VV$ Auger signal from the Si(100) surface with the increase in adsorbed hydrogen and oxygen coverages. Calculations performed in this paper have proven that PAES can be used to detect the presence of hydrogen and oxygen on the Si(100) surface.

ACKNOWLEDGMENTS

We would like to thank A.P. Mills, Jr. and K.G. Lynn for useful discussions. This work was supported in part by the National Science Foundation, the Robert A. Welch Foundation, and a University of Texas at Arlington Research Enhancement grant.

REFERENCES

1. Weiss, A. H., *Materials Sci. Forum* **105-110**, 511-520 (1992).
2. Kim, J. H., Yang, G., and Weiss, A. H., *Surf. Sci.* **396**, 388-393 (1998).
3. Weiss, A. H., *Solid State Phenomena* **28-29**, 317-340 (1993).
4. Fazleev, N. G., Kuttler, K. H., Fry, J. L., and Weiss, A. H., *Appl. Surf. Sci.* **116**, 304-310 (1996).
5. Gunnarsson, O. and Lundqvist, B. I., *Phys. Rev. B* **13**, 4276-4298 (1976).
6. Ceperly, D. M. and Adler, B. J., *Phys. Rev. Lett.* **45**, 566-569 (1980).
7. Weinert, M. and Watson, R. E., *Phys. Rev. B* **29**, 3001-3008 (1984).
8. Puska, M. J., Mäkinen, S., Manninen, M. and Nieminen, R. M., *Phys. Rev. B* **39**, 7666-7679 (1989).
9. Arponen, J. and Pajanne, E., *Ann. Phys.* **121**, 343-389 (1979).
10. Brandt, W. and Reinheimer, J., *Phys. Rev. B* **2**, 3104-3112 (1970).
11. Nieminen, R. M. and Puska, M. J., *Phys. Rev. Lett.* **50**, 281-284 (1983).
12. Fazleev, N. G., Fry, J. L., Kuttler, K., Koymen, A. R. and Weiss, A.H., *Phys. Rev. B* **52**, 5351-5363 (1995).
13. Lang, N. D. and Kohn, W., *Phys. Rev. B* **7**, 3541-3550 (1973).
14. Haneman, D., *Adv. Phys.* **31**, 165-194 (1982).

SECTION IX

FUSION EXPERIMENTS

Performance Enhancement of Negative Ion Sources for the JT-60U Tokamak

L. R. Grisham*, M. Kuriyama, M. Kawai, T. Itoh, N. Umeda, and JT-60U Team

Princeton Univ. Plasma Phys. Lab., P. O. Box 451, Princeton, N. J. USA 08543
Japan Atomic Energy Research Institute, Naka-Machi, Ibaraki 311-0193 Japan

Abstract. The negative ion based neutral beam system now operating for plasma heating and current drive on the large JT-60U tokamak marked the first application of negative ion source technology to the production of high current, high voltage beams for conversion to neutral atomic beams. This pioneering system has demonstrated the technical feasibility of negative ion based neutral beams for future fusion devices. Because this was a very large advance in the state of the art with respect to all system parameters, the principal physical processes governing the performance of the ion source and accelerator were somewhat different than had been the case with earlier much smaller negative ion sources. We have explored the physical mechanisms limiting the power and pulse length capability of these large sources, and have implemented ameliorating changes to reduce power loading on the accelerator grids and increase the transmitted power fraction.

INTRODUCTION

The JT-60U [1] tokamak in Naka, Japan is one of the largest tokamaks operating in the world today to conduct studies in magnetic confinement nuclear fusion for eventual application to fusion power plants. In the past, many fusion research devices have used beams of energetic neutral hydrogen isotope atoms injected across the confining magnetic field lines to heat the plasma, and also in some cases to drive part of the current circulating in the plasma, which in turn creates the poloidal component of the confining magnetic field. Previous generations [2,3] of neutral beam systems first produced positive ions of the desired hydrogen isotope, electrostatically accelerated the ions, and then converted a portion of the fast ions back to atoms by passage through a gas cell. However, for beam energies above 80 keV for deuterium, this neutralization process becomes rapidly less efficient.

Larger magnetic confinement devices require higher energy hydrogen isotope atomic beams for efficient current drive in the plasma core, and for heating of the central plasma. Negative ions of hydrogen can be converted to neutral atoms by passage through a gas cell with an acceptable efficiency (58-60%) over a wide range of beam energies. Unfortunately, because the electron affinity of hydrogen is only 0.75 eV, it is much more difficult to produce and extract negative hydrogen ions than their positive counterparts.

As the culmination to an extensive development program [4], JT-60U became the first device to use large high-current negative ion sources to produce beams of energetic neutral atoms [5]. Designed with the aim of eventually producing a total of 10 megawatts of neutral beams at 500 keV for 10 sec from two sources on one beamline, this system represented a very large advance in the state of the art in terms of negative hydrogen current, power, and source size. Such a large step is intrinsically challenging, and indeed, during early experiments, the beam transmitted power, pulse length capability, and usable voltage were less than planned. Although these limitations were engendered by a variety of processes, they were all manifest as excessive power deposition upon the accelerator and ground grids of the ion sources due to the interception of the grids by divergent beam particles. The grids were designed with water cooling sufficient to remove steady state heat loads equivalent to about 5% of the accelerated beam power, but individual grids were being struck by as much as 15% of the accelerated power. This then limited the beam current and power which could be achieved without striking high voltage breakdowns between the grids. Since the optics were of the Pierce weak-focusing variety ubiquitous among ion sources used for fusion neutral beams, the curent limitations translated into limitations upon the usable voltage as well. Several of the principal phenomena limiting the source performance arise from characteristics of large area sources, and thus were less important in their smaller predecessors. This paper addresses measures undertaken to address these phenomena.

BEAM STRIPPING

One source of the heavy grid interception was excessive stripping of the fragile negative ions while transiting the grid structure. Ions which were stripped to neutrals while still within the accelerator experienced only a subset of the electrostatic lenses, and thus were frozen onto generally divergent trajectories. Since beam ions were neutralized after passing through differing lengths of the accelerating field, some of them had energies lower than the nominal accelerating voltage, producing a lower energy continuum in the beam. The stripping was found to arise from a steeply increasing pressure in the source and accelerator as a function of time, which occurred as a consequence of the long vacuum time constant of the gas feed system. While it was not immediately feasible to change the geometric characteristics that gave rise to this behavior, it was possible to change the gas pulse timing to allow the gas to equilibrate before the arc, which in turn allowed a lower gas throughput to be used. After this improvement, a Doppler shift measurement showed that essentially all of the beam power transmitted to JT-60U was at the full acceleration energy.

SECULAR DEPENDENCE

After the sharp time dependence of the source and accelerator pressure was corrected, a strong secular dependence still persisted in nearly all of the source parameters, including the arc voltage and current, the extraction current, the extraction grid bias current, and the fraction of co-extracted electrons in the beam. In particular, the arc impedance was declining significantly, which in turn reflected a change in arc characteristics, and consequently a time dependence in the extractable negative ion current density. This led to time-varying divergence in the beam, which increased the average grid interception. It was found that the time required for the arc to equilibrate in negative ion sources of this sort is very long, and that the early operation had been done during the long turn-on transient. This problem was also corrected, so that the source plasma characteristics are equilibrated by the inception of beam extraction. This was accomplished by increasing the time the arc was on prior to beam extraction. The period required for equilibration was a decreasing function of arc and filament power, with 1.5 to 2.0 seconds being sufficient for most conditions, as opposed to the 0.5 second of arc prior to beam extraction which had been used during early operations.

Recently, further control over the secular dependence of the arc has been implemented with a filament control system which allows the programming of eight different values of the filament heating current at different times during the arc and beam pulse. This facilitates stable operation for longer pulses. It also results in lower average filament temperatures than was previously the case, reducing the evaporation of tungsten, and also the incidence of unipolar arcs which erode the filament. It is expected that this will increase filament lifetime, and reduce cesium burial by evaporated tungsten deposited on the source surfaces.

SPATIAL NON-UNIFORMITY

A strong spatial non-uniformity in the source plasma persisted even in the equilibrated arc, and this in turn led to a non-uniformity in the local negative ion current density extracted from different areas of the source grids. Since the same voltage gradients are applied over the whole area of the grids, local variations in current density result in mismatches between the radially outward force of the beam space charge and the radially inward electrostatic focussing field, and cause position-dependent divergence growth.

A number of diagnostic techniques were used to assess the plasma and negative ion current density non-uniformity. These included a beam-scanning movable calorimeter, the sharpness of reverse-accelerated electron beamlet burn marks on the back of the source, the relative temperatures of the five vertically arrayed sectors of the plasma grid, Langmuir probes, and most usefully, a technique in which the relative magnitudes of the arc currents flowing through each of the eight cathode groups were used as a measure of the relative plasma density in the vicinity of each group. All five of these techniques revealed a consistent view of plasma non-uniformity dominated by a top-down asymmetry, with higher density at the top of the source, declining in the lower third by several tens of percent. This led to a similarly-structured non-uniformity in the accelerated beam intensity, which in turn gave rise to increased deposition of power on the accelerator and ground grids, and thus to decreased voltage holding capability.

The most useful of these uniformity assessment techniques was the one we developed using the relative magnitudes of the arc currents flowing through the 8 cathode groups. This data could be easily gathered in a single shot and also had the capability to reveal any time dependence to the spatial non-uniformity. Unlike the Langmuir probes, this technique was not susceptible to contamination by the cesium used in the arc chamber to increase negative ion production.

We developed this assessment technique after realizing that, with the cathodes operating in the space charge limited regime, which had been the customary operating regime for these sources, the current flowing across each cathode sheath was a measure of the local plasma density. Thus, measurements of the arc currents flowing through each of the eight cathode groups yielded a measure of the plasma density in the vicinity of each group of filaments, and thus of the plasma uniformity. These were effectively functioning as reverse Langmuir probes, but without concerns about cesium contamination or extra biasing circuits.

In attempting to understand the mechanism driving the vertical inhomogeneity, all of the source parameters were scanned through their full accessible ranges, and those which could be reversed, such as the grid bias voltage across the plasma grid sheath and the direction of the longitudinal current flowing through the plasma grid to produce an electron filter field in the extraction region, were reversed. The non-uniformity was largely unchanged by any of these variations, and was also little altered by the presence or absence of cesium within the chamber. This leads to the supposition that it is probably driven at least in part by the one parameter that could not be reversed: the direction of arc current flow through the filaments and into the plasma. This gives rise to a continuous magnetic field flowing the length of the source. Unlike the magnetic field produced by the filament heating current, the magnetic field from the arc is not self-canceling. The potential for this causing appreciable non-uniformity is greater for sources with long extraction areas, such as this one (each source weighs 6.2 metric tons) than for its smaller predecessors.

A number of techniques were employed to reduce or compensate for the spatial inhomogeneity. Series resistors were added to the arc circuits for each of the eight filament groups. These resistors were independently adjustable to values between 25 and 150 milliohms, with the lowest values being installed in the regions exhibiting the lowest plasma densities. This produced some success, with decreases of the nonuniformity by a factor of two, but this is not a perfect solution because the required resistor balance tends to vary somewhat with arc and filament power and pulse length.

In order to reduce the power striking the grids due to edge effects, 19% of the plasma grid extraction area was masked, with equal amounts covered at the top and bottom of the source.

The heating current flowing through the filament cathodes was also varied to alter the operating regime of the sources, and to gain some control over the arc impedance. There are two potential disadvantages to operating the cathodes in the space charge limited mode, as had been the traditional case for these sources. One is that, as mentioned earlier, the current flow across the cathode sheath is proportional to the local plasma density. Thus, more primary ionizing electrons will be accelerated across the cathode sheaths located in high density regions, and less in low density regions. The resulting uneven distribution of ionizing electrons will probably tend to reinforce the intrinsic non-uniformity in the plasma distribution.

The second potential disadvantage to space charge limited cathode operation is that the discharge conditions determine the arc impedance, with the source operator having little independent control over the voltage drop across the cathode sheath, which determines the acceleration energy of the primary electrons. In any given set of discharge conditions, there is presumably some primary electron voltage which is most suitable for optimizing the extractable negative ion current. The freedom to tune for this optimum is partially constrained under space charge limited operation, since the voltage is strongly linked to the arc power.

We experimented with modest reductions in the cathode filament heating power to move the operating regime partly into the emission limited regime, where the current flowing across the sheath is determined by the emission from the thermionic filament, rather than by the space charge limit of the plasma sheath. Under these conditions, the acceleration energy of the primary ionizing electrons is partly controllable by the filament heating current.

Under some conditions, operation of the cathode filaments partly into the emission limited regime produced some increase in the beam power accelerated through the grids to the beam calorimeter, without increasing the power incident upon the accelerator and ground grids of the beam accelerator assembly. It is unclear at this time how much of this improvement was due to optimizing extractable negative ion current, and how much might have been due to subtle plasma uniformity improvements. Figure 1 shows the arc voltage as a function of filament heating voltage. In the space charge limited regime, the arc voltage is nearly independent of the filament heating voltage (and therefore filament heating power). In the emission limited regime, the arc voltage changes with the filament heating voltage and power. Figure 2 shows the power transmitted to the calorimeter and the power striking the grids as the discharge is moved from the space charge limited regime of cathode sheath operation towards emission limited operation.

when the discharge is not fully into the space charge limited regime. Diamonds show the power transmitted to the calorimeter, dark squares the power striking the ground grid, and circles the power hitting the second accelerator grid.

CONCLUSION

This study clarified the role of various physical processes which are important in large cesiated negative ion sources, and it thereby contributed to improved negative ion beam performance on the JT-60U tokamak. The acceleration efficiency of deuterium has increased from 55% before these improvements to as high as 72% afterwards, and to as much as 80% in the rarely run hydrogen beams. The fraction of the beam power intercepting the accelerator and ground grids has declined by amounts typically in the range of 30 – 40 %. The maximum injected deuterium neutral beam power achieved so far as a result of these improvements is 5.2 megawatts at 350 keV for 0.77 second [5], and the maximum pulse length is 2.0 seconds at 4.0 megawatts and 360 keV.

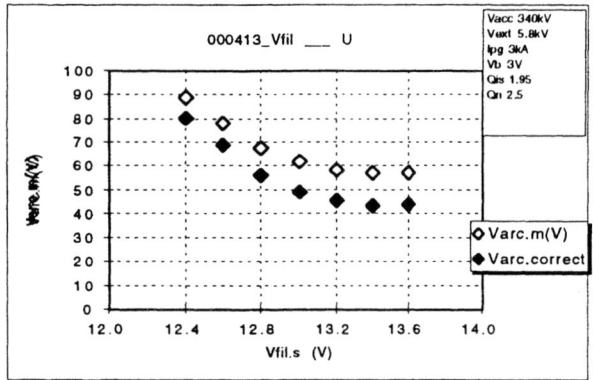

FIGURE 1. The arc enters the emission-limited regime, where the discharge voltage changes, at a filament heating voltage near 12.8 volts. The arc power is 120 kW throughout the scan.

REFERENCES

1. M. Kuriyama et. al., *Journal of Nuc. Sci. and Tech.*, 35, 739-742 (1998).

2. Y. Okumura et. al., *Rev. Sci. Instrum.* 67, 1018-1022 (1996).

3. T. Oikawa et. al., *Proc. 17th International Conf. Of Fusion Energy*, Yokohama Oct. 1998.

4. T. Oikawa at. al., to appear in *Proc. 18th International Conf. On Fusion Energy*, Sorrento Oct. 2000.

5. M. Kuriyama et. al., *Rev. Sci.Instrum.* 71, no. 2, pt II, 751-755 (2000).

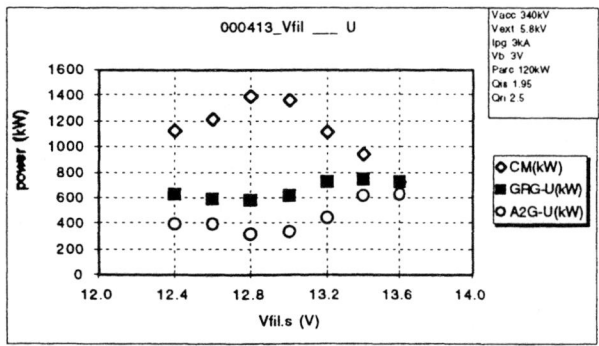

FIGURE 2. Under these arc conditions, some improvement is observed in the form of greater transmitted beam power and reduced grid interception

Fusion Neutronics – Streaming, Shielding, Heating, Activation

H. Freiesleben, D. Richter, K. Seidel, S. Unholzer

Institute for Nuclear and Particle Physics
Technische Universität Dresden, 01062 Dresden, Germany

Abstract. The International Thermonuclear Experimental Reactor (ITER) represents an important step towards a fusion power plant. Controlled fusion will be realized in a d-t-plasma magnetically confined by a Tokamak configuration. The first wall of the plasma chamber, blanket and vacuum vessel of ITER form a compact assembly for converting the kinetic energy of fusion neutrons into heat while simultaneously shielding the superconducting coils efficiently against neutron and accompanying photon radiation. This shielding system can be investigated with neutrons generated by low-energy accelerators. We report on experiments concerning shielding and streaming properties of a mock-up where energy spectra of both neutrons and protons were measured. They are compared with predictions of Monte Carlo calculations (code MCNP-4A) using various data libraries. The agreement justified the use of measured spectra as basis to calculate design parameters such as neutron and photon heating, radiation damage, gas production, and activation. Some of these parameters were also directly measured. The results validate the ITER design.

INTRODUCTION

A central point in the development of a fusion reactor like ITER is the proper design of the shielding system which, in practice, consists of the first wall of the plasma chamber, blanket and vacuum vessel. They form a compact assembly which prevents the superconducting magnetic coils to be heated by neutron and photon radiation, while simultaneously converting the kinetic energy of neutrons (14.1 MeV) originating from deuterium-tritium fusion into heat. However, the shielding capability is significantly reduced by neutrons streaming through unavoidable channels in the blanket and vacuum vessel. The nuclear part of the ITER shielding design rests on Monte Carlo simulations using the Monte Carlo code MCNP-4A [1], which treats the transport of coupled neutron and photon radiation in three dimensions, in connection with the Fusion Evaluated Nuclear Data Library (FENDL) [2,3].

To validate the design, a mock-up of the shielding system (with and without streaming channel) was assembled at the Frascati Neutron Generator (FNG) [4], and neutron and photon flux spectra were measured at several positions within the mock-up. They were compared with those resulting from MCNP-4A simulations. Since experimental and calculated spectra were in rather good agreement, other important design parameters could be calculated as well with good reliability, such as activation, nuclear heating, gas production and radiation damage, where the latter two are not experimentally accessible at presently existing neutron sources.

EXPERIMENTAL DETAILS

A sketch of the mock-up is shown in Fig. 1. It consists of a 1-cm-thick copper plate to replicate the first wall followed by a 94-cm-thick block made out of

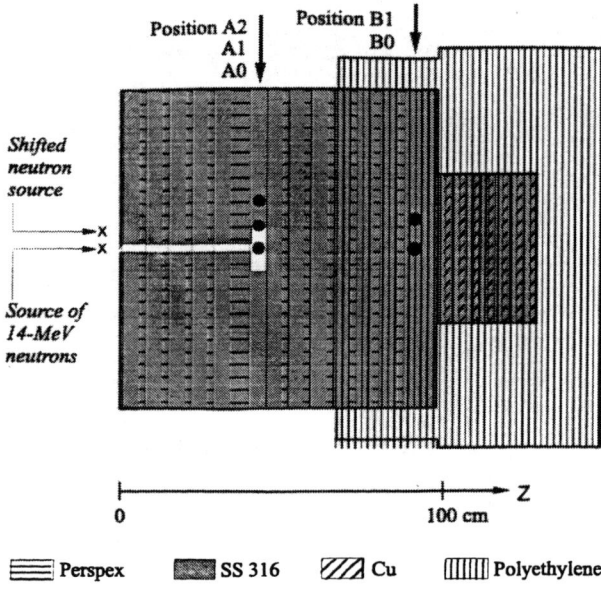

FIGURE 1. Horizontal cut of the mock-up, with neutron source and detector positions

alternating plates of 5-cm-thick stainless steel SS316 and of the water equivalent material Perspex with lateral dimensions of about 100 cm × 100 cm as a substitute for the water cooled blanket and vacuum vessel. It is backed with a 30-cm-thick block of alternating SS316 and copper plates to simulate the toroidal field coils. Polyethylene serves as a shield against background due to room-return. The 14.7 MeV neutron source was positioned at a distance of 5.3 cm from the copper plate for all measurements. Irradiations were carried out for the bulk shield assembly with source position on the central axis and detector positions for deep penetrations (outer blanket wall: z=41.4 cm and outer wall of the vacuum vessel: z=87.6 cm) and for an assembly with streaming channel (⌀ 2.8 cm) and cavity (14.8 cm × 4.8 cm × 5.2 cm in height) in order to study the reduced shielding efficiency in case of mechanically bolted blanket modules. Three irradiations were carried out: two with the neutron source either facing the closed or the open channel, one with the neutron source shifted by 5.3 cm, thus allowing neutrons to enter the open channel under 45°. Detectors were placed at the two z-positions through horizontal channels in SS316 plates in their center. A, B and A0, B0 were positions on the axis of the closed and open channel, respectively. A1, A2 and B1 were measurement positions shifted horizontally against the open channel by 7.5 cm, 15.0 cm and 9.0 cm, respectively. Measurements at off-axis detector positions will not be discussed in this contribution.

The neutron energy spectra $\Phi_n(E)$ were measured in the range from ~ 30 keV to 15 MeV. A set of gas filled proportional counters was used for measurements in the energy range up to 1 MeV, a Stilbene and a NE213 scintillator spectrometer was used above ~1 MeV. The NE213 scintillator spectrometer was simultaneously used to measure gamma energy spectra $\Phi_\gamma(E)$ for $E_\gamma > 0.2$ MeV, where pulse shape discrimination was used to separate events due to neutrons or gammas. Spectral fluences were determined from the pulse height distributions by methods of deconvolution which are described in detail in [5,6,7] for all detectors.

Calculations of fluence spectra at various detector positions in both the bulk shield and the streaming assembly were carried out applying the Monte Carlo code MCNP-4A [1]. The geometry of the mock-up and its surrounding (e.g. concrete walls of the building, metallic support of the assembly) were precisely accounted for; the radiation pattern of the uncollimated neutron source was considered in detail. For the simulation of deep penetrations of neutrons and photons with satisfying statistical accuracy, several variance-reducing techniques were applied of which geometry splitting with Russian roulette was the most effective one.

Cross-section data were taken from the Fusion Evaluated Nuclear Data Library (FENDL) [2,3]. It is a comprehensive, validated and extensively tested nuclear data library for fusion applications which comprises parts from ENDF/B-VI (USA), JENDL-3 and JENDL-FF (Japan), BROND-2 (RF), and EFF-3 (EU). FENDL-2 is the improved 1999 version which supersedes FENDL-1 from 1994. FENDL-2 is used here.

RESULTS AND DISCUSSION

Neutron and Photon Fluence Spectra

Examples for measured and calculated spectral neutron and photon fluences normalised to one source neutron are shown in Fig. 2 and Fig. 3. For the bulk shield experiment detector positions A an B apply. The same detector positions hold for the assembly with streaming channel and with shifted (A0S) or unshifted source (A0,B0).

Two observations could be readily made. 1. An open channel drastically enhanced the flux in comparison to the bulk shield assembly at positions A and B. For energies > 0.1 MeV (10 MeV) this factor is 9.7 (73.4) at position A and 6.0 (37.8) at position B. An

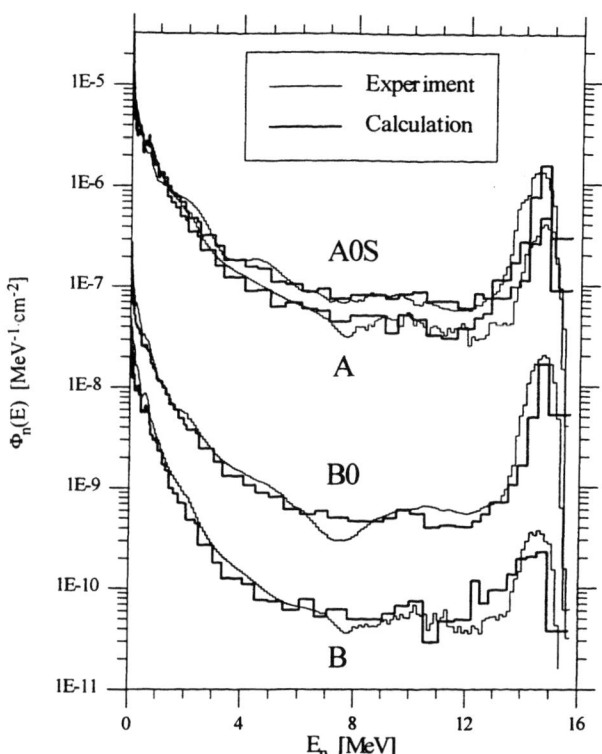

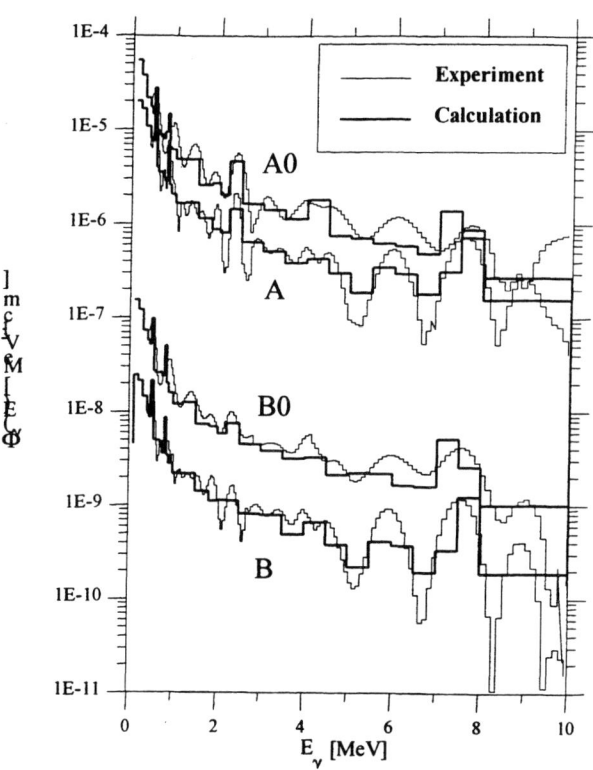

Figure 2: Spectral neutron fluences normalised to one source neutron for the bulk shield experiment (detector positions A and B) and for the same detector positions in the assembly with streaming channel (A0S and B0)

Figure 3: Spectral photon fluences normalised to one source neutron for the bulk shield experiment (detector positions A and B) and for the same detector positions in the assembly with streaming channel (A0 and B0)

open channel acted as a streaming path for neutrons and deteriorated the shielding capacity in comparison to the bulk where the flux of neutrons with energies > 0.1 MeV (10 MeV) was reduced between position A and B by a factor of 453 (936). Only 10^{-5} neutrons of those with normal incidence to the surface of the bulk reached position B, which guarantees the proper shielding of the toroidal field coils. 2. For both assemblies the agreement between experimental data and calculated results was better than 13 % at all energies at position A with a systematical underestimation of the data. This was as large as 30% for deep penetration at position B.

For any detector position, the distance to the neutron source corresponded to many mean-free-paths of fast neutrons in iron (4.5 cm), hence neutron transport was dominated by multiple interactions. Uncertainties in cross section data entered into fluence calculations multiplicatively. Therefore, an agreement of calculation and experiment within about 30% was quite satisfying.

Fig. 3 shows spectral photon fluences normalised to one source neutron for the bulk shield experiment (detector positions A and B) and for the corresponding detector positions in the assembly with streaming channel (A0 and B0). The photon spectrum is dominated by a broad background of photons following (n,n') reactions. The discrete lines in Fig. 3 stem from the decay of excited states in ^{56}Fe. These spectra prove the validity of the unfolding procedure.

Similar but far less drastic observations than in the neutron case could be made. 1. The open channel showed a higher photon flux than the bulk shield. For energies > 0.4 MeV, this factor was 3.0 at position A and 5.8 at position B. The overall attenuation between A and B amounted to 700 for the bulk shield. 2. There was a good agreement between experimental data and calculated results for position A. At position B an underestimation by the calculations of about 10% was found.

Photons are produced by inelastic processes of fast neutrons as well as by capture of thermal neutrons. The attenuation length for photons in steel (1.4 cm for E=0.4 MeV and 2.8 cm for E=1.5 MeV) is shorter than that of neutrons. Therefore, the photon fluxes at deep penetrations are determined by neutron transport and gamma production cross sections. The good agreement of calculated and measured photon fluences showes the production cross sections to be satisfyingly evaluated in the whole energy range. Also, the slowing-down

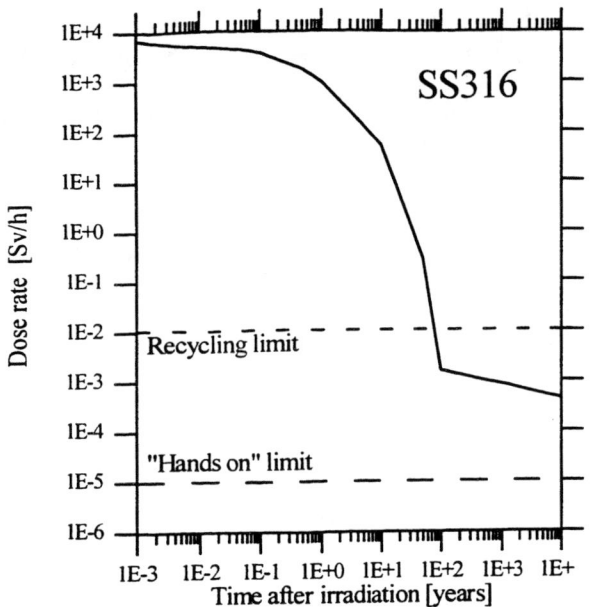

FIGURE 4. Contact dose rate of 1g of ITER steel SS316 after irradiation with a 14.9 MeV neutron flux representing a power of 1 MW/m² for the period of one year as function of decay time.

of fast neutrons to thermal energies appears to be well simulated by the transport calculation.

Reaction Rates and Nuclear Heating

The number of induced reactions N_R at given detector position per one source neutron on 10^{24} probe nuclei is

$$N_R = \int \Phi_n(E) \sigma(E) \, dE \qquad (1)$$

and was calculated using cross section data $\sigma(E)$ from the IRDF-90 File [8]. Results are given in Tab.1 for both experimental and calculated fluence spectra $\Phi_n(E)$ yielding $N_{R,exp}$ or $N_{R,calc}$, respectively.

During the bulk shield experiment activation of thin foils placed at various positions in the mock-up was carried out [7]. From their activity, the number of reactions taking place was determined. This number is also listed in Table 1 in the column $N_{R,activation}$. A very good agreement was found which encouraged to calculate activity and dose rate of ITER steel after irradiation. An irradiation of SS316 steel with a 14.9 MeV neutron flux representing a power of 1 MW/m² for one year was simulated and the cooling down was followed. The calculation was carried out with the most recent version of the European Activation System EASY-99 [9]. The decay curve shown in Fig. 4 reflects the contribution of nuclides with different half-lifes. The short term activity is mainly of interest for heat production and shut-down dose rates, whereas the long term activity is important for waste management. The long term dose rate is due to reactions on impurities, which prevents reaching the hands-on limit of 10µSv/h after 50 years of cooling, which is the ultimate design level of ITER.

Nuclear heating in steel at position A and B is dominated by gamma interaction processes according to calculations. With $\Phi_\gamma(E)$ one obtains

$$H_\gamma = \int \Phi_\gamma(E) \, h_\gamma(E) \, dE \qquad (2)$$

where $h_\gamma(E)$, the KERMA data (heating numbers) was taken from FENDL-1. Since $h_\gamma(E)$ strongly increases with E, the H_γ values are sensitive to the high-energy part of $\Phi_\gamma(E)$. For position A (B) a value of $H_\gamma = 4.61 \times 10^{-7}$ MeV/g (6.69×10^{-10} MeV/g) was calculated. These numbers are in good agreement with those measured at the same position with thermoluminescence detectors (TLD), namely $H_\gamma = 4.61 \times 10^{-7}$ MeV/g (7.44×10^{-10} MeV/g) per one source neutron [7]. These values are well within the ITER design limit [10].

TABLE 1. Number of reactions on 10^{24} probe nuclei per one source neutron derived from measured and calculated fluences and deduced from activation data.

Position	Reaction	$N_{R,experimental}$	$N_{R,calculated}$	$N_{R,activation}$
A	^{97}Nb(n,2n)	$(2.69\pm0.16)\times10^{-7}$	$(2.24\pm0.04)\times10^{-7}$	$(2.34\pm0.02)\times10^{-7}$
A	^{27}Al(n,α)	$(7.06\pm0.36)\times10^{-8}$	$(6.67\pm0.11)\times10^{-8}$	$(6.93\pm0.35)\times10^{-8}$
A	^{56}Fe(n,p)	$(6.61\pm0.30)\times10^{-8}$	$(6.21\pm0.07)\times10^{-8}$	$(6.47\pm0.32)\times10^{-8}$
A	^{58}Ni(n,p)	$(4.57\pm0.23)\times10^{-7}$	$(4.46\pm0.10)\times10^{-7}$	$(4.72\pm0.24)\times10^{-7}$
A	^{115}In(n,n')	$(6.06\pm0.36)\times10^{-7}$	$(4.35\pm0.07)\times10^{-7}$	$(4.71\pm0.28)\times10^{-7}$
B	^{97}Nb(n,2n)	$(2.27\pm0.03)\times10^{-10}$	$(2.00\pm0.12)\times10^{-10}$	$(2.46\pm0.12)\times10^{-10}$
B	^{56}Fe(n,p)			$(6.92\pm0.35)\times10^{-11}$
B	^{58}Ni(n,p)	$(6.01\pm0.91)\times10^{-11}$	$(4.91\pm0.30)\times10^{-11}$	$(5.46\pm0.27)\times10^{-11}$

TABLE 2. Number of produced gas atoms and displacements per atom (dpa) for 10^{24} probe nuclei normalised to one source neutron as derived from measured and calculated fluences.

Position, Fluence		He-atoms	H-atoms	Dpa
A	$\Phi_n(E)_{experimental}$	2.95×10^{-8}	1.60×10^{-7}	3.89×10^{-3}
A	$\Phi_n(E)_{calculated}$	3.00×10^{-8}	1.50×10^{-7}	3.95×10^{-3}
B	$\Phi_n(E)_{experimental}$	3.13×10^{-11}	1.71×10^{-10}	6.42×10^{-6}
B	$\Phi_n(E)_{calculated}$	2.58×10^{-11}	1.44×10^{-10}	5.50×10^{-6}

Gas Production Rate and Radiation Damage

The number of reactions with an alpha particle or a proton in the exit channel directly yields the number of produced He- and H-atoms. Tab. 2 shows these quantities as derived from $\Phi_n(E)_{exp}$ and $\Phi_n(E)_{calc}$. The agreement among these numbers reflects the agreement of the neutron fluences which were used to determine these numbers. Also included is the number of displacement per atom (dpa) which was determined by applying the damage cross sections from IRDF-90 [8]. These numbers were necessary to estimate the life time of components which are considered permanent in the ITER design (vacuum vessel, backplate, field coils).

SUMMARY

The experimental data obtained at a 14 MeV neutron generator on a bulk shield mock-up of ITER and their comparison with results from advanced transport codes like MCNP-4A using recent data libraries show an agreement within 30% at the most critical position, i.e. at the position of the superconducting field coils. The irradiation load of neutrons and photons is much smaller than the present design specifications allow for. However, streaming through channels, gaps or ducts penetrating the bulk shield may significantly worsen the situation as it reduces the effectiveness of the shield.

ACKNOWLEDGMENTS

This work was supported by the European Fusion Technology Programme. We gratefully acknowledge the contributions of P. Batistoni, M. Angelone, M. Pillon and L. Petrizzi from FNG, Frascati, Italy and of U. Fischer, Forschungszentrum Karlsruhe, Germany.

REFERENCES

1. Briesmeister, J. F., (ed.), *Report Los Alamos National Laboratory*, LA-12625 (1993).

2. Ganesan, S. and McLaughlin, P. K *International Atomic Energy Agency*, IAEA-NDS-128 (1994).

3. Herman, M., *International Atomic Energy Agency*, INDC(NDS)-395 (1999).

4. Martone, M. et al., *Journal of Nuclear Materials*, 212-215, p. 1661-1664 (1994).

5. Büermann, L. et al., *Nucl. Instrum. Methods* A332, 483 (1993).

6. Tichy, M., *PTB Laboratory Report PTB-7.2-1993-1*, Braunschweig, 1993

7. Batistoni, P. et al., *Fus. Eng. Design* **47**, 25-60 (1999)

8. Kocherov, N. P. and Mc Laughlin, P. K., *IAEA-NDS-141, Rev.2*, Vienna, 1993

9. Forrest, R. A. and Sublet, J.-Ch., FISPACT-99, *Reports Culham Scientific Centre, UKAEA FUS 407 and UKAEA FUS 409*, Culham December 1998

10. General Design Requirements Document (GDRD), S10 GDRD 2 95-02-10 F1.0, ITER EDA-JCT, June 1995

Integrated Neutral Beam Measurements in a Tokamak Environment

D.M. Thomas

Fusion Division
General Atomics

Tokamaks are, in many respects, the most promising avenue for the development of fusion power. The continual improvement in the performance of these devices and our understanding of them is due in great measure to the development of accurate plasma diagnostics. Many of the most crucial measurements required to assess our progress on these experiments are based in one way or another upon collisional interactions of injected neutral beams with the plasma. These measurements include such fundamental parameters as the ion temperature, rotation, and density profiles, electric and magnetic field structure, and local studies of the plasma turbulent transport. Maximizing the obtained information for a given geometry of plasma, beams, and possible viewchords represents an interesting challenge to the experimentalist. Advances in detector and analysis techniques allow us to take full advantage of the beam/plasma emission for these measurements.

INTRODUCTION

In the quest for fusion power, improving the performance of stable magnetic confinement schemes has been a key scientific issue. The most well developed scheme, the tokamak, has led the charge in this area. Recent progress on heating and controlling fusion-grade plasmas using the tokamak concept has led to the achievement of near scientific breakeven (Fusion power/input power~1) in several experiments worldwide, albeit on a transient basis. Our challenge for the future is to maintain and improve these performance levels on a steady-state (and ultimately economical) basis.

A crucial element of this progress has been the implementation of plasma diagnostics to measure various plasma parameters in situ. These measurements have become more and more complex as the plasma environment becomes more hostile (radiation, remote sightlines, larger dimensions, hotter plasmas). In particular, the recent development of the highest performance "advanced" tokamak modes has relied on the careful crafting of the plasma pressure and rotation profiles, requiring good internal measurements of these parameters with high spatial temporal resolution. Fortunately, the high intensity neutral beams typically used for heating these plasmas represent a powerful tool in this regard. By properly collecting and interpreting the optical emission resulting from beam-plasma collisions, in many cases we can infer the relevant plasma behavior without the need for material contact with the plasma. These techniques, and the relevant cross section work necessary to properly interpret the measurements, have been utilized on essentially all devices having heating beams or dedicated diagnostic neutral beams. In this paper, we will briefly review some of these techniques as they are employed on the DIII–D tokamak experiment.

MEASUREMENTS USING BEAM-PLASMA COLLISIONS ON DIII–D

The DIII–D program employs eight neutral deuterium beams for plasma heating, permitting over 20 MW of injected power per shot [1]. The beams are typically 40 keV/AMU $^2D°$, but can easily be run down to ~30 keV/AMU if necessary. A versatile beam control system permits modulation of each of the beams down to 5 ms pulse widths. The production of diatomic and triatomic deuterium ions in the ion source results in full, half, and third energy components in the neutral beam after acceleration and neutralization. The species mix injected into DIII–D at optimum perveance typically results in initial beam densities of 1×10^{15} atoms/m^3 in the beam footprint of 0.26×0.14 m (height x width). The typical 1/e divergence angle for the beams is 0.66° in the horizontal plane and 1.30° in the vertical plane at optimum perveance [2].

We may distinguish between possible measurements based on beam–plasma collisions as being due to either:

1. Beam excitation processes

$$D^o + (A^{z+}, D^+, e^-) \rightarrow D^* ,$$

where the subsequent decay emission gives information about the beam atoms and their environment; or

2. Charge transfer processes

$$D^o + A^{z+} \rightarrow D^+ + A^{(z-1)+*} ,$$

where the decay emission is now indicative of the parent ion population. Although not discussed in this paper, observation of the neutral particle emission from reactions of this type for the case z=1 can also be used to study the distribution function of the parent ion, using neutral particle analysis at the periphery of the plasma.

CHARGE EXCHANGE SPECTROSCOPY

On DIII–D, detailed measurements of impurity ion behavior (heating, confinement, and transport) are made using reactions of Type 2 above [3]. A rather extensive set of viewing chords (24 tangential and 16 vertical) have been installed and collimated to collect light from ~6–12 mm spot sizes. These views are arranged so as to be as tangent to the magnetic flux surfaces as possible at the point of beam intersection. This results in the maximum spatial resolution given the finite beam dimensions. The coverage in the edge region is especially good (chord spacing ~6 mm) in order to isolate fine scale structure in the edge. The progressive development of high quantum efficiency CCD detectors, efficient grating spectrometers, and high-speed readout electronics permits us to extract spectral information on every chord over the range 300–1100 nm with gating times down to 0.32 ms [4,5]. The resulting spectra are fit with a nonlinear least-squares fitting routine to identify the various peaks. Extensive spatial, wavelength, and intensity calibrations of all channels are conducted on a routine basis. These calibrations permit us to extract the impurity ion density, velocity, and temperature profiles from the amplitude, shift, and width of the spectral lines To obtain absolute ion density profiles, one also needs to know the local beam density, species mix, and energy-dependent charge transfer cross sections. These basic measurements then let us infer impurity ion pressure profiles. Since the inner wall of the DIII–D vacuum vessel is lined with carbon tiles, our workhorse transition for these measurements is the C VI ($\Delta n = 8 \rightarrow 7$) transition at 529.05 nm, resulting from the recombination of fully stripped carbon ions with the beam atoms.

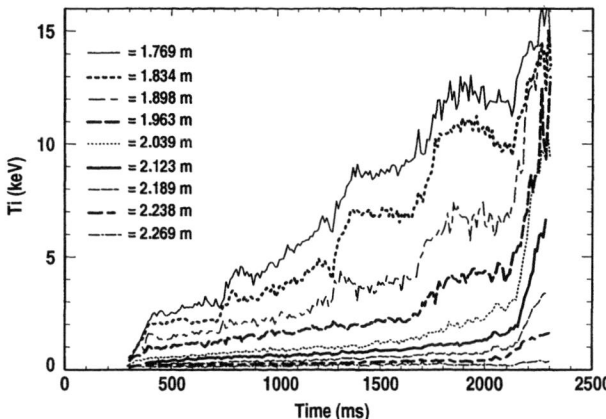

FIGURE 1. C VI ion temperature measurements on DIII–D, showing internal and external transport barriers.

Figure 1 demonstrates the quality of the present system, where we are able follow the formation of a transport barrier both in the core [internal transport barrier (ITB)] and edge [high confinement (H–mode)] of a high performance DIII–D discharge. The data quality is such that we can perform rigorous tests of theories dealing with formation of these enhanced confinement regimes. In this regard, the precise determination of the poloidal and toroidal velocity profiles is particularly valuable as it allows us to infer the intrinsic radial electric field structure using force balance arguments. We now believe that it is the shear in the electric field (or more properly, the shear in E×B flow) that plays a pivotal role in the suppression of turbulence and the improved confinement behavior [6].

BEAM EMISSION SPECTROSCOPY

Improvements in detector technology have also benefited observations of beam emission reactions of Type 1. Because the beam emission is proportional to excitation rate, and the rate itself is relatively insensitive to temperature in the energy range of interest, variations in the detected emission rate may be related directly to the underlying fluctuations in the plasma density [7–9]. By viewing the beam at a non-normal angle, the Doppler-shifted D_α beam emission may be distinguished spectroscopically from the large thermal D_α emission along the line of sight. Because of the large Doppler shifts obtained on DIII–D, we can use interference filters for this purpose rather than grating spectrometers, resulting in vastly improved light levels. Large collection optics and a two-dimensional array of fibers allow us to measure radially and poloidally localized density fluctuations with high time resolution (1 MHz digitization rate) and high spatial resolution (~1 cm resolution). This allows us to analyze the turbulent state of the plasma in great detail. An example is shown in Fig. 2, where neon was puffed into the periphery of the plasma in an attempt to suppress turbulence and improve confinement [10]. For this data, 32 spatial channels were deployed to examine the outer third of the plasma minor radius, with one 16-channel radial array and two poloidal arrays being used. Plotted is the relative density fluctuation power spectra for a neon puff shot and reference shot, before and after the neon injection occurs. The wavenumber listed is determined experimentally by the channel separation. We note two distinct effects of the neon injection: the overall power of the fluctuation spectrum is substantially reduced, with preferential lowering of the higher frequencies, corresponding to preferential suppression of the higher-k turbulent modes. These effects occur on the same timescale as do improvements in the global plasma confinement. These techniques have been extended to provide two-dimensional images of the turbulence [11].

MOTIONAL STARK EFFECT (MSE) SPECTROSCOPY

Performing more detailed spectroscopy on the beam emission yields additional information about the plasma. The neutral deuterium beam atoms are transiting a large magnetic field $\mathbf{B}=\mathbf{B}_{tor}+\mathbf{B}_{pol}$ as they enter the plasma. Because of the strong motional electric field ($\mathbf{E}=\mathbf{v}\times\mathbf{B}$) produced in the rest frame of the beam atoms, the D_α line is actually split into orthogonally polarized components (σ,π) by the Stark effect. When viewed in a direction perpendicular to E, the σ and π components are polarized perpendicular and parallel to the direction of the electric field, respectively. The emission of one of the components is then polarization analyzed to determine the pitch angle at an ($\mathbf{B}_{pol}/\mathbf{B}_{tor}$) of the local magnetic field. From this profile, one can then infer details about the internal current distribution, a critical element of plasma stability [12,13]. In order to utilize this effect, one needs sufficient perpendicular beam velocity $\mathbf{v}\times\mathbf{B}$ to separate the σ and π lines, as well as viewing tangent to **B** (to achieve good spatial resolution) and somewhat tangent to **v** (to achieve sufficient Doppler separation of the full, half, and third energy components from the thermal background). On DIII–D, we have three separate viewing geometries to attempt to satisfy these requirements [14].

The specifics of the current profile distribution are extremely important for the development and stabilization of high-performance plasmas. Coupled with a magnetic equilibrium reconstruction code, the MSE measurements have crucially improved both our understanding of plasma stability and confinement as

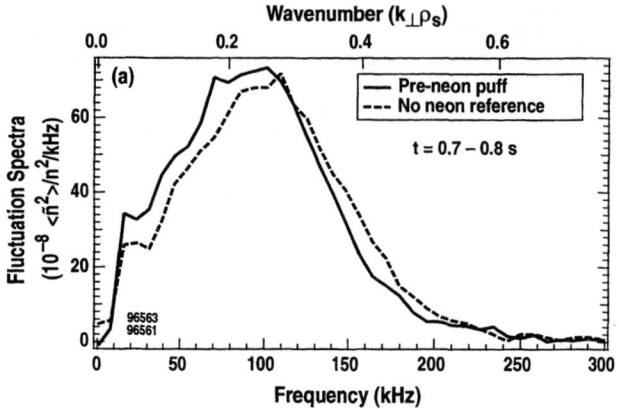
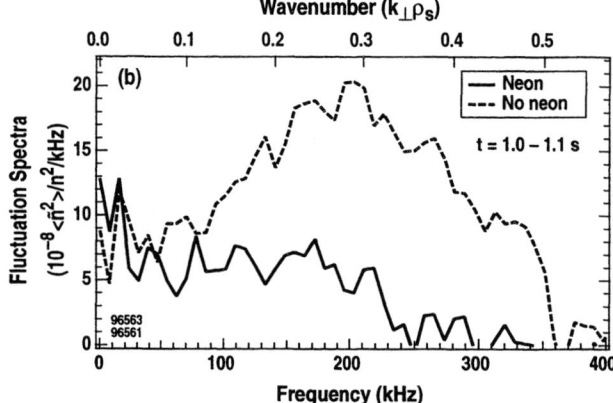

FIGURE 2. BES power spectra with and without N_e puff [10].

well as our ability to create new regimes of plasma performance.

LITHIUM BEAM ZEEMAN SPECTROSCOPY

One limitation of MSE becomes apparent in studies of H–mode edge plasmas. The large intrinsic electric fields found in these plasmas causes Stark splitting effects which are indistinguishable from the MSE, leading to a superposition of magnetic and electric field effects and precluding an unambiguous determination of B_{pol} from MSE alone [15,16]. As is shown in Fig. 3, a set of radial MSE views was implemented to mitigate this E_r effect, but the radial resolution of these channels is insufficient to examine fine scale structures in the edge. Simultaneous observation using CER to identify the E_r at the same location helps somewhat in this regard. To improve the precision of the edge magnetic field determination, we are deploying an atomic lithium beam on DIII–D [17].

The lithium 670 nm 2S-2P resonance line is split by the Zeeman effect, with the π and σ lines exhibiting similar polarization behavior as utilized in MSE, but

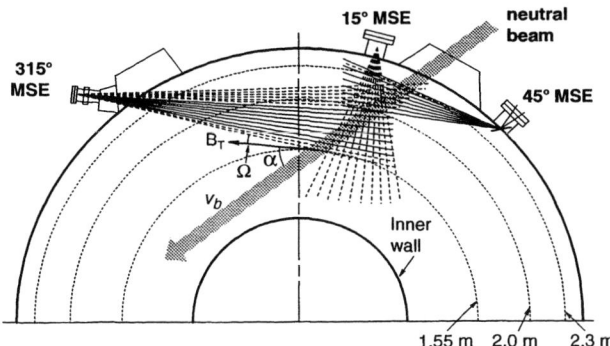

FIGURE 3. MSE views on DIII–D [14].

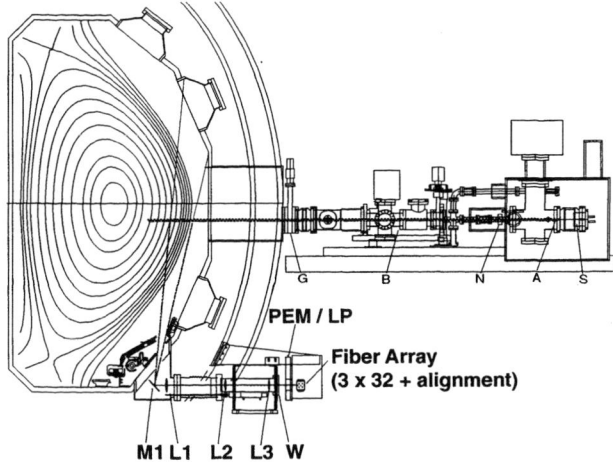

FIGURE 4. Lithium beam system on DIII–D showing accelerator and optical system [17].

referenced to the local B. Because this transition has negligible Stark splitting, the use of this line eliminates the electric/magnetic ambiguity. Also, the type of source utilized creates no half- and third-energy components in the beam. Finally, because the cross section for collisional excitation is so large ($\sim 500 \times D_\alpha$), we can utilize a very small beam, permitting very fine spatial resolution (Fig. 4). The key to successfully making this measurement is handling the somewhat smaller spectral splitting ($\sim 10\times$) relative to MSE measurement, requiring much better spectral resolution. This results in a conflict with light collection efficiency and the possible time resolution. Additionally, because of the enhanced ionization/charge transfer cross sections for Li^0, beam penetration is an issue and the measurement is only feasible in the edge plasma region.

CONCLUSION

Tokamak diagnostics based on beam-plasma collisions provide a variety of unique information about the interior of fusion grade plasma devices. They serve a vital role for improving our understanding of these devices and help in assessing the ultimate prospects of the tokamak as a potential power source.

ACKNOWLEDGMENTS

The measurements presented here represent just a small fraction of the substantial contributions made on beam-based diagnostics on DIII–D by General Atomics, the University of Wisconsin, and Lawrence Livermore National Laboratory plasma diagnostic groups. I would also like to thank the DIII–D Neutral Beams Group for providing the specific information on the operation and performance parameters for the DIII–D heating neutral beams. This work was supported by U.S. DOE Contract No. DE-AC03-99ER54463.

REFERENCES

1. Hong, R., et al., "Enhancement of DIII–D Neutral Beam System for Higher Performance," Proc. 17th Symp. on Fusion Technol., Rome, 1992.
2. Chiu, H., et al., "Measurement of Neutral Beam Profiles at DIII–D," Proc. 13th Top. Mtg. Technol. Fusion Energy, Nashville, 1998.
3. Gohil, P., et al., "The Charge Exchange Recombination Diagnostic System on the DIII–D Tokamak," Proc. 14th IEEE/NPSS Symp. on Fusion Engineering, California, 1991, Vol. 2 (IEEE, New Jersey, 1992) p. 1199.
4. Thomas, D.M., et al., Rev. Sci. Instrum. **68**, 1233 (1997).
5. Burrell, K.H., et al., "Improved CCD Detectors for the Charge Exchange Spectroscopy System on the DIII–D Tokamak," to be published in Rev. Sci. Instrum. (2000).
6. Burrell, K.H., et al., Phys. Plasmas **4**, 1499 (1997).
7. Fonck, R.J., Duperrex, P.A., Paul, S.F., Rev. Sci. Instrum. **61**, 3487 (1990).
8. Durst, R.D., Fonck, R.J., Cosby, G., et al., Rev. Sci. Instrum. **63**, 4907 (1992).
9. McKee, G., et al., Rev. Sci. Instrum. **70**, 913 (1999).
10. McKee, G., et al., Phys. Rev. Lett. **84**(9), 1922 (2000).
11. Fenzi, C., et al., "2D Turbulent Imaging in DIII–D via Beam Emission Spectroscopy," to be published in Rev. Sci. Instrum. (2000).
12. Levinton, F.M., et al., Phys. Rev. Lett. **63**, 2060 (1989).
13. Wroblewski, D., et al., Rev. Sci. Instrum. **61**, 3552 (1990).
14. Rice, B.W., et al., Rev. Sci. Instrum. **70**, 815 (1998).
15. Rice, B.W., et al., Nucl. Fusion **37**, 517 (1997).
16. Zarnstorff, M.C., et al., Phys. Plasmas **4**, 1097 (1997).
17. Thomas, D.M., et al., "Prospects for Edge Current Density Determination using LIBEAM on DIII–D," to be published in Rev. Sci. Instrum. (2000).

SECTION X

RADIATION PROCESSING:
FACILITIES AND APPLICATIONS

A New Electron Accelerator Facility for Commercial and Educational Uses

R. M. Uribe and C. Vargas-Aburto

Program on Electron Beam Technology, Kent State University, Kent OH 44242-0001, USA

Abstract. A 5 MeV 150 kW electron accelerator facility (NEO Beam Alliance Inc.) has recently initiated operations in Ohio. NEO Beam is the result of a "partnership" between Kent State University (KSU) and a local plastics company (Mercury Plastics, Inc.). The accelerator will be used for electron beam processing, and for educational activities. KSU has created a university-wide Program on Electron Beam Technology (EBT) to address both instructional (including workforce training and development) and research opportunities. In this work, a description is made of the facility and its genesis. Present curricular initiatives are described. Preliminary dosimetry measurements performed with radiochromic (RC) dye films, calorimeters, and alanine pellets are presented and discussed.

INTRODUCTION

According to recent statistics [1], the plastics industry in the U.S. employs 1.3 million people and provides $274 billion in annual shipments. It is the fourth-largest manufacturing industry in the United States. In Ohio this industry employs approximately 113,000 people thereby making the state second only to California in terms of employment. In 1996 Ohio had shipments of plastic-related products of more than $20 billion. A sizable part of the plastics industry in Ohio is concentrated in the North East part of the state from the city of Akron, where most of the rubber industry was developed, towards the city of Cleveland, where many small and medium size (with 200 and less employees) custom plastic processors are located. Some of these processors produce extruded thermoplastics for the appliance, automotive, plumbing, and other industries. Thermoplastics are often crosslinked to enhance their properties. For example, polyethylene becomes flexible, rubbery, and reaches a higher melting point when irradiated to doses from approximately 10^4 to 10^6 Grays [2]. Crosslinking can be achieved by chemical methods [3] and by electron irradiation [2]. However, the improved properties of irradiated plastic materials as well as the advantages that this technology has over the chemical methods have added to its increased popularity.

For the last decade or so, faculty at KSU have been actively involved in the study of radiation effects on materials. Specifically, efforts have focused on studying the effects of electron and proton radiation on space solar cells [4] and in the development and characterization of radiochromic dye films as radiation dosimeters [5]. An accelerator capable of providing an electron beam with suitable energies (in the MeV range) and good spatial homogeneity, is needed to carry out experimental work in these fields.

As a result, KSU and MPI decided to establish an electron accelerator facility. The Kent Regional Business Alliance (KRBA), a local non-profit organization interested in promoting university-industry partnerships and economic development for the region provided the mechanism required to finance and help to implement the initiative. Under this KSU-MPI "partnership", the accelerator is reserved for use by MPI and KRBA on an equal basis. KRBA's 50% allocation will be used for educational activities and support to industry. KSU plays a double role: as a user of beam time, and as technical support to the NEO Beam facility. All curricular and research (including industrial R&D) activities are the responsibility of KSU. Below, a description is given of the facility, the various academic activities (curricular and research), and a summary of preliminary dosimetry measurements.

THE FACILITY

The accelerator is housed in a 1,800 m^2 structure. The beam room (vault) has an area of approximately 57 m^2. Access to the vault is provided through a maze that allows product to be introduced continuously, and yet prevents radiation from exiting (See Fig. 1). The entrance to the maze is 1.8 m wide by 2.4 m high. Product can be introduced using a cart conveyor system, a reel-to-reel bulk handling system, or a series of openings through the shielding walls. Each cart carries a 1.8 x 1.2 m^2 aluminum tray to transport up to 375 kg of product. The speed of the conveyor system can be changed from 0.025 to 0.28 m/s. The bulk handling system is able to accommodate tubing or cable with outer diameter in the range (0.48 - 3.17) x 10^{-2} m (0.188 - 1.2 in), and is able to operate at speeds of up to 10 m/s.

The electron accelerator is a 5 MeV 150 kW Dynamitron manufactured by Radiation Dynamics Inc. (Edgewood NY).[1] The accelerator was designed to operate reliably at energies as low as 1 MeV. Electron currents range from 30 mA at the higher energies, down to hundreds of microamperes at the low energy end. A beam scanning system provides a uniform electron field 1.22 m in length at the exit window (located 1.5 m above the floor.

The facility also includes two laboratories; one for dosimetry measurements, and one for fabrication and characterization of plastic materials. A conference/teaching room with distance learning and video conferencing facilities is also available.

ACADEMIC ACTIVITIES

To take maximum advantage of the instructional and research opportunities offered by the NEO Beam facility, KSU has formally established a Program on Electron Beam Technology (EBT) that cuts across all academic units. The Director of the EBT program reports to both the Vice Provost for Research and the Vice Provost for Regional Campuses. A summary of the curricular activities related to this program is presented below.

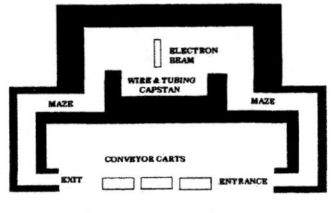

FIGURE 1. Schematic diagram of the electron accelerator vault and access maze.

[1] The mention of any commercial products is for the sake of clarity and does not constitute an endorsement by the authors for any of the products mentioned in the paper.

Curricular

Certificate Program and Associate of Applied Science Degree

To date a certificate program and an Associate of Applied Science (AAS) Degree in Radiation Polymer Engineering have been developed. The certificate program represents a subset of the Associate of Applied Science in Mechanical Engineering Technology (Integrated Manufacturing Option) Degree. Table 1 below shows a typical sequence of courses required for the AAS degree. These degrees are housed in the School of Technology (SOT), a college-level academic unit with a multitude of programs and degrees (undergraduate and graduate) in technology-related fields.

TABLE 1. Major courses in the Associate of Applied Science Degree, Radiation Polymer option.

Course Title	Credit hours
Engineering Drawing	3
Computer Aided Drafting	4
Properties of Materials	3
Microprocessors and Robotics	4
Robotics and Flexible Automation	3
Computer Integrated Manufacturing	3
Introduction to Plastics	4
Reinforced Plastics	3
Radiation Polymer Technology I	3
Radiation Polymer Technology II	3
Statistical Process Control	4

The courses in Radiation Polymer Technology introduce the students to the theoretical and practical aspects of radiation physics, radiation processing of polymers, radiation dosimetry, radiation safety, and characterization techniques for irradiated polymers. The practical portions of the program make heavy use of the NEO Beam facility and its associated laboratories. The distance learning capabilities will be available before the end of the year (2000), and will be used to enhance the curricular and research activities of the EBT program.

Curriculum Initiatives at the Undergraduate and Graduate Levels.

Additional courses are presently being developed within the SOT in Radiation Technology to address the future needs of technically prepared people in this field. These courses will be used as part of a new AAS degree, a new Bachelor of Science in Technology degree, and a Master of Technology (MTEC) degree. Selected courses are listed in Table 2.

TABLE 2. Proposed courses for a new program on Radiation Technology at the undergraduate and graduate levels.

Course Title	Level/Credit hours
Manufacturing Processes of Radiation Technology	Sophomore (3)
Radiation and the Environment	Sophomore (3)
Electron Beam Technology	Junior (3)
Industrial Radiation Safety	Junior (3)
Radiation Effects in Materials	Senior/Graduate (3)
Advanced Topics in Radiation Technology	Senior/Graduate (3)
Dosimetric Methods and Techniques	Senior/Graduate (3)
Applications of Electron Beam Technology	Graduate (3)
Industrial Radiation Sources and Systems	Graduate (3)
Dosimetry in Radiation Processing	Graduate (3)
Analysis Techniques in Radiation Technology	Graduate (3)

Dosimetry

Calibration of Dosimeters with Alanine Pellets

Since alanine is classified as a reference standard dosimetry system by ASTM [6], the decision was made to simultaneously irradiate alanine with other dosimeters. Therefore, five sets of alanine pellets supplied by NIST and three other dosimeter systems, were irradiated to measure their response to electron irradiation. The systems were: polystyrene calorimeters [7], and two different types of RC dye films (Far West Technology FWT-60-00, batches 2D1 and 7F7, Risø National Laboratory, B3 films).

Prior to irradiation the alanine pellets were placed in polystyrene disks to simulate the body of the calorimeters; the RC films were placed on top of the polystyrene disks. The disks were placed inside of polystyrene foam boxes identical to those used for the polystyrene calorimeters [7]. The polystyrene foam boxes were placed on a cart and irradiated with 4.5 MeV electrons at different beam currents and cart speeds, so as to provide several doses.

After irradiation, the alanine dosimeters were returned to NIST for dose evaluation using standard electron spin resonance methods. The temperature increase in the calorimeters and the absorbance change in the RC dye films were determined according to established techniques [7, 8]. In this manner the dose-response curves were obtained.

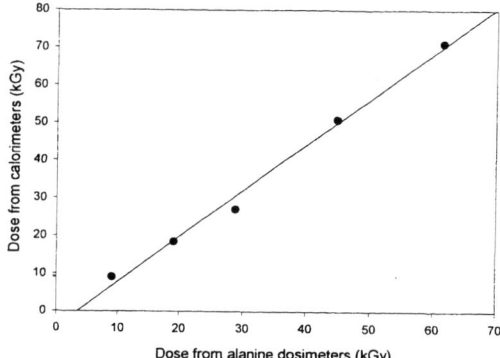

FIGURE 2. Calorimeter response versus dose as measured by alanine pellets. The line represents a linear fit to the data.

Dose Calibration Versus Beam Current and Conveyor Speed

Beam current and cart speed were calibrated in terms of dose using polystyrene calorimeters. The experiment was divided in two parts. First, five calorimeters were irradiated with a beam current of 20 mA at five evenly spaced cart speeds from 0.05 to 0.15 m/s. Second, five calorimeters were irradiated at five evenly spaced beam currents from 5 to 25 mA (cart speed: 0.1 m/s). The electron energy and the scan amplitude were fixed at 4.5 MeV and 1.22 m at the beam exit window, respectively. The dose in the calorimeters was determined as described in ref. [7].

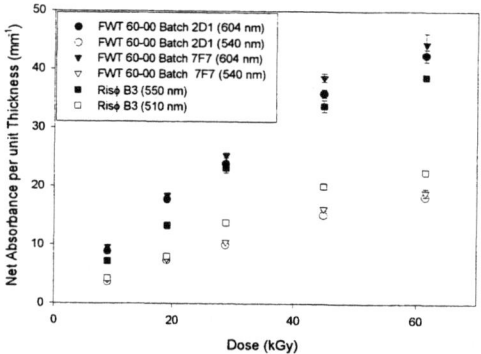

FIGURE 3. Dose-Response curves for the two radiochromic dye films discussed in this work. The absorbance values were obtained with a Lambda 18 Perkin Elmer spectrophotometer.

Results

Figs. 2 and 3 show the results of the calibration of the different dosimeter systems with alanine pellets. Fig. 2 shows that the dose recorded by the calorimeter is not the same as that measured by the alanine pellets. The difference between the two is still under investigation.

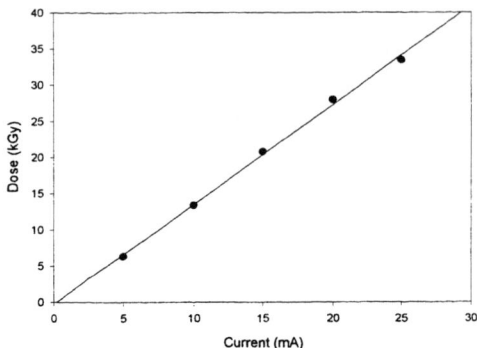

FIGURE 4. Variation of dose in a polystyrene calorimeter as a function of beam current.

Fig. 3 shows the dose-response graphs for the two RC films. Their response was obtained at 604 and 540 nm for the FWT-60-00 film and 550 and 510 nm for the Risø B3 film. The results indicate that the RC films can be used beyond the dose range to which they were irradiated, although at 60 kGy they show the onset of saturation. The response for the two different batches of the FWT-60-00 RC films is very similar, even though the 2D1 batch was obtained almost ten years before the 7F7 batch.

Figs. 4 and 5 show the dependence of dose on beam current and cart speed in a polystyrene calorimeter. As expected, the dose increases linearly with the beam current at a rate of 1.47 kGy/mA, and it is inversely proportional to the cart speed, at a rate of 2.86 kGym/s. These values are only valid for the beam energy and scanning amplitude used in these experiments.

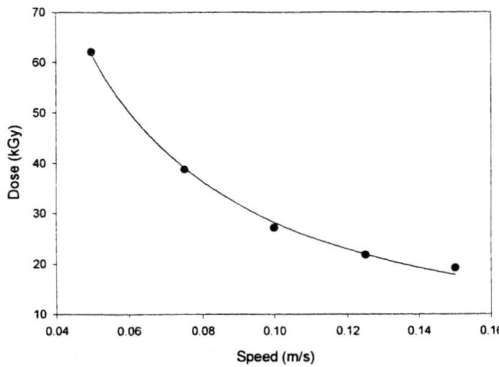

FIGURE 5. Variation of dose in a polystyrene calorimeter as a function of the conveyor cart speed.

ACKNOWLEDGEMENTS

The authors wish to thank Mr. Mike Shuman for technical assistance during the performance of the experimental part of this work and Mr. Michael Czayka for valuable discussions regarding the curriculum development. This work was partially supported by NASA Glenn Research Center under grant number NCC3-721.

REFERENCES

1. Limbach,B.,M., http://www.plasticsindustry.org/ about/news/99releases/ohio_summit.htm,(1999).
2. Woods, R.J., and Pikaev, A.K., *Applied Radiation Chemistry: Radiation Processing*, New York: John Wiley&Sons.,1994,p.346.
3. Mark, J.E., Eisneberg, A., Graessley, W.W., Mandelkern, L., and Koenig, J.L., *Physical Properties of Polymers*, Washington D.C.: ACS,1984,p.87.
4. Karlina, L. B., Blagnov, P. A., Boiko, M. E., Kozlovskii, V. V., Kudriatsev, Yu. A., Solov'ev, V. A., Vargas-Aburto, C., Uribe, R. M., and Brinker, D. J., "Radiation Hard $In_{0.53}Ga_{0.47}As$ Solar Cells with Zn Diffused Emitters" in *Twenty Sixth IEEE Photovoltaic Specialists Conference-1997*, IEEE: New Jersey, 1997, pp.1007-1010.
5. Uribe, R. M., Vargas-Aburto, C., McLaughlin, W. L., Walker, M. L., and Dick, C. E., "Electron and Proton Dosimetry with Custom-Developed Radiochromic Dye Films" in *Radiation Protection Dosimetry: Solid State Dosimetry-1993*, edited by E. P. Goldfinch et al., Radiation Protection Dosimetry, **47**,1-4,1993,pp.693-696.
6. Guide E 1261, "Guide for Selection and Calibration of Dosimetry Systems for Radiation Processing," in *Annual Book of ASTM Standards*, Vol. 12.02, American Society for Testing and Materials, West Conshohocken,PA,1999.
7. Miller, A., *Radiat. Phys. Chem.* **46**, 1243-1246 (1995).
8. E 1275-98, "Standard Practice for Use of a Radiochromic Film Dosimetry System," American Society for Testing and Materials, West Conshohocken PA: ASTM, 1998, 5 pp.

Performance of the Electron Beam Fluidized Bed Process For Disinfection and Disinfestation of Stored Products[1]

D. A. Cleghorn[2], S. V. Nablo[2] and David N. Ferro[3]

[2]Electron Processing Systems, Inc., 6 Executive Park Dr., North Billerica, MA 01862
[3]Dept. of Entomology, Fernald Hall, University of Massachusetts, Amherst, MA 01003

Abstract. A high velocity fluidized bed has been used to present product to a medium energy electron beam permitting effectively isotropic treatment of each (air) suspended particle. A pilot system is described based upon a 250 kV self-shielded Electrocurtain® accelerator which has been used for several years for surface disinfection (i.e. microorganism elimination) of a variety of agroproducts. Disinfestation studies (i.e. insect elimination) in winter wheat have used product velocities to 2000 m.min^{-1} for the evaluation of mortality of various life stages of "internal" feeders, such as the rice weevil (*Sitophilus oryzae*), which spends some 3 weeks at 30°C inside the grain kernel in its larval stage before emergence as an adult. Results are presented for the determination of the electron energies required to achieve effective larvicidal treatment for internal feeders in winter wheat.

INTRODUCTION

The treatment of aggregates such as powders, pellets, spices, seeds and cereal grains, with ionizing radiation presents the common problem of how best to present the product to the source. For gamma rays and X-rays, bulk palletized treatment, usually involving product rotation to improve treatment uniformity, is widely employed. There has been limited use of electron beams for this purpose, largely due to the high energies required for the useful penetration of conveyorized beds of the product, or of streams in ballistic flow (1).

The relatively high scattering cross-sections of energetic electrons in the 100-500 keV range provide an attractive route using a unidirectional source, to the isotropic treatment of particles supported in an air stream...a fluidized bed. The product to air loading factor of the bed can then be selected to provide good beam utilization at commercially acceptable throughput level (2).

The studies reported here have been directed to the application of shelf-shielded electron beam fluidized bed (ebfb) systems to both the disinfection and disinfestation of agricultural products. Although its use for the sterilization of powders is obvious, our current facilities do not provide the aseptic conditions required for this practice.

Product Handling Geometry

For the first of these applications for microorganism (usually pathogen) reduction in agroproducts, relatively high doses are required, up to 30 kGy for heavily contaminated spices. As a consequence, it is useful to maximize product residence time in the beam-affected volume, usually by a longitudinal presentation or with product flow along the electron beam. For the second application (disinfestation), the doses required for efficacious treatment are typically two orders of magnitude lower, so that even at high product velocities, a transverse presentation is used. Electron processors are typically characterized by a yield factor, k, in kGy.m.min^{-1}mA^{-1}. This is then used for process control via the relation:

$$D = k I / v \qquad (1)$$

where D is the delivered dose in kilograys, I is the beam current in milliamperes and v is the product(bed) velocity in meters per minute. Yield values for typical processors, for transverse and longitudinal product presentation, will perform roughly as the ratio of beam half-width to beam length. For example, the 225 kV Electrocurtain (3) used in this work with a 5 cm x 30 cm beam, has a measured longitudinal to transverse yield ratio of 6.8.

---

[1] This work has been supported under SBIR Grants from the U.S. Department of Agriculture (96-33610-3115 and 99-33610-7369)

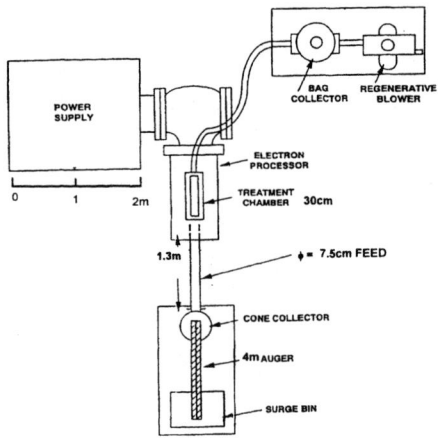

FIGURE 1. Longitudinal Electron Fluidized Bed System

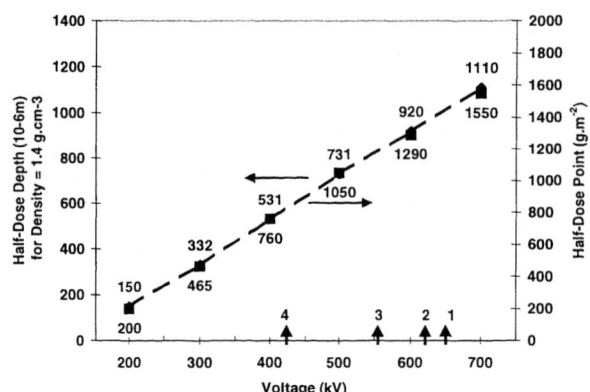

FIGURE 2. Penetration Performance for Fluidized Bed Process

The pneumatic product handling system utilized stainless steel sanitary tubing with Tri-Clamp® fittings. A 5-h.p. regenerative blower and Vac-U-Max (4) filter bag product recovery unit was used with variable speed feeders providing flow rates up to 300 g.s^{-1} for process evaluation. A schematic of the longitudinal geometry used in this work is shown in Fig. 1.

Dose Delivery Measurements

Radiochromic film dosimetry 50 g.m^{-2} thick (5) was used to determine the dose delivered to the surface of particles carried in the fluidized bed. The dosimeter films are loaded into the product prior to processing. Based upon the relatively large scatter (± 15%) shown in our film dosimetry work, we decided to run eight films (1 cm x 1 cm) during the course of a run, data points were rejected if they exceeded 3σ deviation from the mean. Some typical results from some fifty runs conducted with sesame seed are shown in Table 1 and illustrate the good reproducibility for this "in product" active dosimetry for the system. Using equation (1) the calculated velocities are shown in the last column. The average velocity calculated from the tabulated data is 761 m.min^{-1}.

The 5-h.p. regenerative blower and its exhaust connections to the filter recovery system remained fixed throughout these runs and all others during these studies. A real time monitor (6) developed for these systems is used for providing archival record of the electron beam machine output based upon the bremsstrahlung spectrum generated in the accelerator window. Techniques for validation of the performance of the monitor, for control and quality assurance of critical processing applications, are under development.

Product Penetration Considerations

Our studies have ranged from single surface treatment of an aggregate product (such as leafy herbs) to deeper penetration in cereal grains in order to affect burrowing larvae. Hence a detailed knowledge of effective product penetration in the bed in essential. Figure 2 shows some experimentally determined penetration performance data taken with the systems used in these studies. In each case the 50% point on the depth:dose curve is plotted and it is then converted in Fig. 2 (left hand ordinate) to depth in product of density 1.4 g.cm^{-3}, which is taken as typical for the endosperm of cereal grains. The four arrow markers on the abscissa are described in the disinfestation section.

It was possible to employ Monte Carlo code calculations (7) to examine the ability to treat particulates with these energy sources. These studies showed that with good control of feed rate and bed velocity, the bed thickness could be held well within the penetration capability of the electron source, providing uniform treatment of the product.

Lethality Studies: Microorganism Disinfection

A set of lethality curves for the natural bioburden in marjoram run on a 225 kV pilot is shown in Figure 3. These data illustrate a typical bioassay approach for this industry in which yeasts, molds and coliform bacteria levels are assayed, in addition to the total plate count. Although the current regulation would permit

TABLE 1. Dosimetry during Sesame Seed Runs (Feed Rate = 75g.s^{-1})

Run #	Dose (kGy)	Beam Current (mA)	Velocity (m.min^{-1})
149	6.8 ± 1.3	13.3	757
151	7.2 ± 0.6	13.3	714
185	7.2 ± 0.7	14.2	763
186	6.7 ± 1.1	13.4	774
187	6.9 ± 1.3	14.2	796

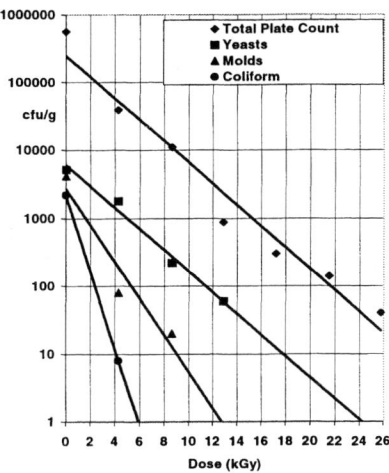

FIGURE 3. Lethality Results for ebfb treated Marjoram

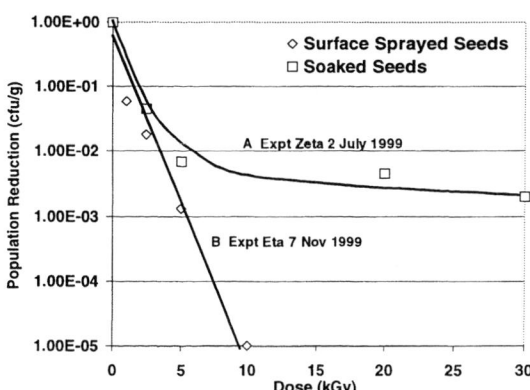

FIGURE 4. 225kV ebfb lethality Study for Alfalfa Seeds Soaked and Surface Sprayed with K12 *(E. coli* surrogate)

30 kGy treatment, in this case one would select a dose of < 15 kGy at which point a total plate count (TPC) of 1000 colony forming units per gram (cfu.g^{-1}) would be realized with no coliform bacteria present in the product.

This rather straightforward application to ground spices/leafy herbs was then extended to the disinfection of sprout seeds, where the elimination of pathogens (strains of *Salmonella* spp. and *Escherichia coli*) is required. Many chemical techniques (8) have been pursued; e.g. perchlorate soaks, for this purpose with varying degrees of success. The present process has been evaluated as a possible physical treatment of sprout seeds while maintaining acceptable embryo viability (for germination).

Because non-pathogenic inoculae must be used in our laboratory, we have employed an *E.coli* surrogate in these alfalfa sprout seed studies. Two types of inoculation have been utilized, spray inoculation and soaking inoculation. For the former, samples of 1.5 kg of seed are sprayed with 30-40 mL of the surrogate culture to a level of 10^6 cfu.g^{-1} in phosphate buffered saline, then air dried for 6 hours before treatment.

The differences in the inoculae distribution for these two cases are illustrated in the lethality curves of Fig. 4 taken at 225 kV. An initial rapid decrease in *E. coli* surrogate population results as the shell and epithelium of the seed's endosperm is treated to a depth of approximately 300 μm (See Fig. 2). The endosperm remains totally untreated beyond 400 micrometers depth and results in the surviving population shown in curve A. In curve B however, with surface inoculation, good lethality is shown with a 10^{-5} population decrease for a D_{10} value of 2 kilograys, even with the known penetration of the sprayed inoculum into the endosperm via cracks in the seed shell. The presence of deep-seated pathogens in naturally contaminated seed has been demonstrated (9) and is revealed in the hypochlorite work as well. The data of Fig. 4 and related studies with naturally contaminated alfalfa seed confirm the need for deep penetration for pathogen elimination in this product.

As pointed out earlier, the advantage of the fluidized bed process lies in its isotropic treatment. For alfalfa seed with a mean diameter of 1.3 mm and density of 1.2 g.cm^{-3}, a half dose treatment capability of 780 g.m^{-2} is adequate. Figure 2 yields a required voltage of 420 kV for "bulk" endosperm treatment of this product.

Mortality Studies: Insect Disinfestation

A great deal of effort has been devoted to the study of insect mortality at various life stages using both gamma ray and energetic electron sources. With the restrictions of the product presentations used in these studies no mortality information is available for electron energies under one megavolt. The goal of the work in this laboratory, suggested by entomologists at the USDA, was the evaluation of the process for internal feeding insects. We were reasonably confident of its efficacy for surface deposited eggs and (emergent) adults, so the focus in the first studies was the evaluation of mortality for the internal feeding larval stages of the rice weevil (*Sitophilus oryzae*) and the lesser grain borer (*Rhyzopertha dominica*).

With the assistance of the USDA's Laboratory in Manhattan, KS, colonies of both insects were established at U.MA, Amherst, MA and studies began using our 225 kV ebfb system at various life stages. Typical results are shown in Fig. 5 for *S. oryzae* in which it is clear that as the larvae burrow into the grain and develop (via 4 molts or instars) they become more vulnerable to treatment as the "protective" shell offered by the grain decreases e.g. at 21 days.

Sharifi and Mills (10), using X-ray analysis of infested wheat kernels, were able to measure larval tunnel dimensions for the various instars (molts) of *S. oryzae* over their three week "internal" life stages. Using hard winter wheat with an average minor kernel

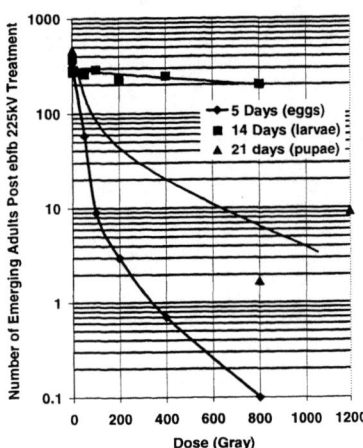

FIGURE 5. Survival of *Sitophilus oryzae* (Rice Weevil) ebfb Treated at 225kV at Different Developmental Stages.

radius of 1263 µm (density 1.4 g. cm^{-3}), the protective shell thicknesses are calculated in the third column of Table 2. The electron energies required for 50% penetration at each stage are shown in column 5 and are indicated along the abscissa of Fig. 2. In the second column, the age of each molt stage is shown as days after egg deposition (oviposit).

In order to test the validity of these range:energy considerations on adult mortality, trials were conducted using unilateral treatment on an accelerator (11) providing 400-700 kV performance. Data for these trials, conducted on *S. oryzae* larvae at 10, 15 and 22 days after oviposit, are presented in Table 3 at two treatment levels (200 and 800 gray) to confirm process efficacy dependence on energy and dose.

The tests reveal a much more sensitive dependence of insect mortality on energy than on dose as expected. Clearly, the fourth (22-day) instar is less well protected by the seed shell and shows very high mortality across this dose:energy range. The effects of low dose treatment on adult emergence and fecundity are being studied for an improved definition of these important treatment parameters.

TABLE 2. *S. oryzae* Protective Layers for Winter Wheat

Instar	Age (Days)	Shell Thickness (µm)	(g.m^{-2})	Voltage (kV)
1	7-8	1108	1551	650
2	12-13	1008	1411	625
3	16-18	848	1187	550
4	22-23	593	830	425

TABLE 3. Emergent Adult Mortality (in percent) as a Function of Energy for *S. oryzae* Larvae in Wheat

	400 kV		500 kV		700 kV	
Age days	200 Gray	800 Gray	200 Gray	800 Gray	200 Gray	800 Gray
10	55	66.2	52.4	88.6	92.9	100
15	74.1	60.4	80.4	92.1	99.5	99.6
22	97.6	97.3	99.6	100	100	100

CONCLUSIONS

These experiments have demonstrated the efficacy of using medium energy electrons (E ≤ 500 keV) for winter wheat and rice disinfestation using the fluidized bed for product presentation to the energy source. Mortality and emergent adult sterility data at all life stages are now needed to evaluate a fixed dose:energy treatment in product with mixed age populations.

ACKNOWLEDGEMENTS

The support of this work through USDA-SBIR grants is gratefully acknowledged. The advice and assistance of David Hagstrum of the Grain Marketing and Production Research Center, USDA, Manhattan, KS and the careful entomological work of Andy Slocombe at U.MA, Amherst are recognized. The microbiological work of Peter Slade at NCFST Summit-Argo, IL has been essential to these studies.

REFERENCES

1. Zakladnoy, G.A., Menshenin, A., Pertsovsky, E.S., Slimov, R.A., Cherepkov, V.G., Bogolyubov, B.F., *Radiat. Phys. Chem.* **34**, pp. 991-994, 1989.
2. Nablo, S.V. and Cleghorn, D.A., *Final Phase 2 report USDA Grant 96-33610-3115*, USDA, Box 2243, 1400 Independence Ave., Washington, D.C. 20250-2243, July 30, 1999.
3. Electrocurtain is a registered trademark of Energy Sciences Inc, 42 Industrial Way, Wilmington, MA
4. Vac-U-Max, 37 Rutgers St, Belleville, NJ. 07109.
5. Far West Technology, Inc, 330-D South Kellogg Ave., Goleta, CA. 93117.
6. Nablo, S.V., Kneeland, D.R. and McLaughlin, W.L., *Radiat. Phys. Clem.*, **46**, pp. 1377-1383, 1995.
7. Jenkins, T.M., Nelson, W.R. and Rindi, A., eds., *Monte Carlo Transport of Electrons and Photons*, New York: Plenum Press, 1988.
8. Taorima, P.J. and Beuchat, L.R., *Jour. Food Prot.* **62**, pp. 318-324, 1999.
9. Itoh, Y., Sugita-Konishi, Y., Kasuga, F., Masaki, I., Hara-Kudo, Y, Saito, N., Noguchi, Y., Konuma, H., Kumagai, S., *Appl. Environ. Microbiol.* **64**, pp. 1532-1535, 1998.
10. Sharifi, S. and Mills, R., *J. Econ. Entomol.* **64**, pp.1114-1118, 1977.
11. North Star Research Corp., 4421-A McLeod Rd, N.E. Albuquerque, NM, 87109.

Treatment of Foods with High-Energy X Rays

M. R. Cleland, J. Meissner, A. S. Herer and E. W. Beers

Ion Beam Applications, s.a. Chemin du Cyclotron, 3, B-1348 Louvain-la-Neuve, Belgium

Abstract. The treatment of foods with ionizing energy in the form of gamma rays, accelerated electrons and X rays can produce beneficial effects, such as inhibiting the sprouting in potatoes, onions and garlic, controlling insects in fruits, vegetables and grains, inhibiting the growth of fungi, pasteurizing fresh meat. poultry and seafood and sterilizing spices and food additives. After many years of research, these processes have been approved by regulatory authorities in many countries and commercial applications have been increasing. High-energy X rays are especially useful for treating large packages of food. The most attractive features are product penetration, absorbed dose uniformity, high utilization efficiency and short processing time. The ability to energize the X-ray source only when needed enhances the safety and convenience of this technique. The availability of high-energy, high-power electron accelerators, which can be used as X-ray generators, makes it feasible to process large quantities of food economically. Several industrial accelerator facilities already have X-ray conversion equipment and several more will soon be built with product conveying systems designed to take advantage of the unique characteristics of high-energy X rays. These concepts will be reviewed briefly in this paper.

INTRODUCTION

Ever since World War II, the beneficial effects of treating foods with ionizing energy have been investigated in many countries. These effects include inhibiting sprouting in potatoes, onions and garlic, delaying the ripening and senescence of fruits and vegetables, killing insects in fruits, vegetables and grains, inhibiting the growth of fungi, pasteurizing fresh meat, poultry and seafood and sterilizing spices and dried food additives [1]. Extensive animal feeding studies and chemical analyses have demonstrated the safety of consuming foods treated in this manner. As a result, regulations permitting the use of this technique for preserving fresh foods and enhancing their safety have been established in many countries. Public acceptance has been supported by favorable articles in newspapers and magazines, and commercial sales of such products are increasing.

Gamma rays, accelerated electrons and high-energy X rays are suitable energy sources for these applications. They all have similar biological and chemical effects, which are caused by ionizing critical molecules in living cells with low-energy electrons. These active agents are ejected from atoms and molecules within the treated material by the high-energy sources mentioned above. The selection of the energy source for a particular application is usually based on practical considerations, such as product size, shape and density, dose distribution, processing rate and cost.

Treatment with accelerated electrons provides the highest processing rate and the lowest unit cost. On the other hand, electrons have relatively short ranges in solid materials. With the regulatory energy limit of 10 MeV, they cannot be used to treat packages of food with thicknesses greater than 10 cm. Gamma rays from cobalt-60 sources are more penetrating, and can be used to treat packages with thicknesses of at least 30 cm. X rays, produced with high-energy electron beams, are even more penetrating than gamma rays. They also have some other attractive characteristics for food processing, which are discussed in this paper.

The development of X-ray conversion equipment for high-power, high-energy electron accelerators has made X-ray generators a practicable choice for treating products that cannot be done with electrons directly. Throughput rates are now sufficient to meet industrial requirements. Several X-ray processing facilities are equipped to use this relatively-new technology, and more will be built to meet the increasing demand for safer food products.

X-RAY PROPERTIES

Short-wavelength, high-frequency electromagnetic radiation is produced when energetic electrons strike any kind of material. The technical term for this form of energy is bremsstrahlung (braking radiation) and the common name is X rays. The energy distribution of the X-ray photons is continuous and extends up to the maximum energy of the incident electrons. The power conversion efficiency increases with the electron energy and with the atomic number of the material. For electron energies greater than 1 MeV, most of the X-ray power is emitted in the forward direction. The angular divergence of the photons decreases as the electron energy increases. The penetration of X rays in treated materials increases and the efficiency for using this energy also increases with the electron energy. These basic properties have been revealed by many studies, both theoretical and experimental [2,3]. An extensive list of publications on this topic can be found in Reference 3.

All of the characteristics mentioned above indicate that the highest available X-ray energy should be used for processing commercial products. Many years ago, a Joint Expert Committee of the United Nations (UN) Agencies, Food and Agriculture Organization (FAO), International Atomic Energy Agency (IAEA), World Health Organization (WHO), recommended an energy limit of 5 MeV to avoid inducing active isotopes in the food. This cautious limit was adopted by the Codex Alimentarius Commission and by many governmental agencies around the world. A recent review by a Consultants Meeting sponsored by FAO and IAEA has concluded that the energy limit for X-ray processing could be increased to 7.5 MeV, without compromising the safety of the food. On the basis of this review and other evidence, [4,5] a petition will be submitted to the US Food and Drug Administration requesting their approval of this higher energy limit. This will enhance the usefulness of this important process.

Theoretical simulations of basic X-ray properties have been done with the ITS Monte Carlo Code [2,6]. This method can provide accurate data about the efficiency for converting electron beam power to X-ray power, the photon energy spectrum, the angular dispersion, the penetration and the dose variations in complex product shapes. The data presented here were calculated for electron energies of 5, 7.5 and 10 MeV incident on a tantalum X-ray target to show the effects of increasing electron energy. Neutron emission from the target would be negligible at 5 and 7.5 MeV but it would be significant at 10 MeV [5].

The maximum X-ray emission from a target with a high atomic number is obtained with a thickness about 40% of the electron range. For the 5 and 7.5 MeV simulations, the thicknesses of the tantalum target plate, the cooling water channel and the stainless steel backing plate were assumed to be 1.2, 2.0 and 2.0 mm, respectively. For the 10 MeV simulations, the corresponding thicknesses were 1.6, 2.0 and 4.0 mm. The combined thicknesses were sufficient to prevent the incident electrons from penetrating the targets. These calculations confirmed earlier studies, which indicated that the efficiency for converting electron-beam power to useful X-ray power increases in proportion to the electron energy. The values obtained here were 8.2, 13.3 and 16.2 percent at 5, 7.5 and 10 MeV, resp.

The photon energy spectra are shown in Fig. 1. It is evident that the maximum photon energy is the same as the incident electron energy. The total number of photons and the proportion of high-energy photons increase with the electron energy. The most probable photon energy is about 0.3 MeV for each spectrum. The low-energy part of the 10 MeV curve is attenuated by the greater thickness of the target for this energy.

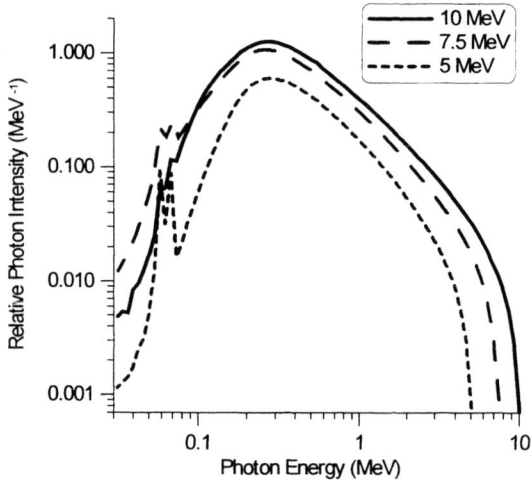

Figure 1. X-ray photon spectra for incident electron energies of 5, 7.5 and 10 MeV.

The angular dispersions of photons emitted in the forward direction are shown in Fig. 2. The emission is concentrated in the direction of the electron beam. This effect becomes more pronounced as the energy increases. The forward concentration increases the penetration in products passing by the target on a conveyor because most of the radiation is perpendicular to the product surface and relatively little is tangential. The small divergence of a high-energy X-ray beam is different from the isotropic emission of gamma rays. With that type of energy source, the larger angles of incidence increase the dose near the surface of the product. In effect, this reduces the penetration of the radiation.

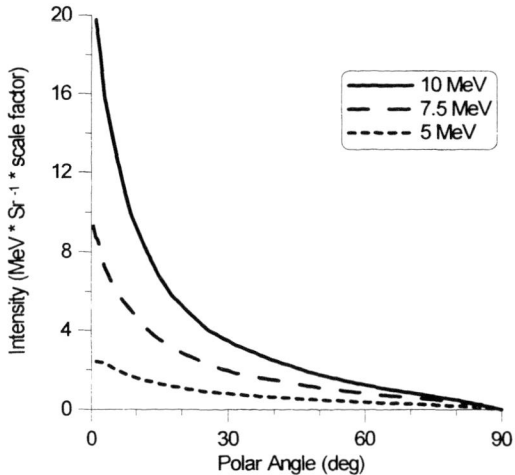

Figure 2. X-ray angular dispersions of X rays for incident electron energies of 5, 7.5 and 10 MeV.

A three-dimensional dose distribution in a moving package of unit-density material treated with 7.5 MeV X rays is shown in Fig. 3. The calculations were done by extending the scanning X-ray source beyond the package in the direction of conveyor motion. The data show a saddle-shaped pattern with maximum doses at the leading and trailing edges and minimum doses at the sides in the direction of beam scanning. The higher doses are caused by divergent rays reaching these zones while the package is moving toward and away from the X-ray target. The lower doses are the result of the imbalance of scattered photons at the edge of the package.

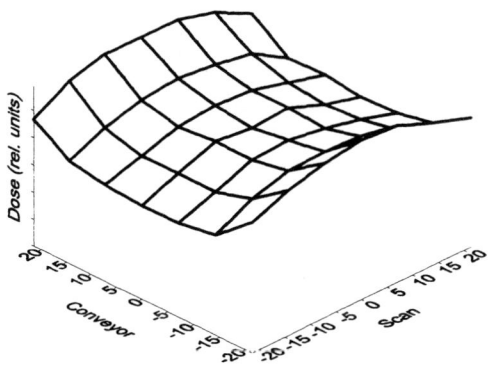

Figure 3. A saddle-shaped dose distribution in a package moving through a scanning 7.5 MeV X-ray beam.

Depth-dose distributions in a moving package of unit-density material treated with 7.5 MeV X rays are shown in Fig. 4. The upper curve is for the leading and trailing edges, the intermediate curve is for the middle zone and the lower curve is for the package sides in the direction of scanning. The apparent dose reduction near the surface is not realistic. It was a result of the method of calculation, which was done in two steps to reduce the computer time. First, the photon emission from the target was obtained and then this was used to simulate the depth-dose distribution. This procedure omitted the secondary electrons from the X-ray target, which would have contributed to the surface dose and eliminated this effect.

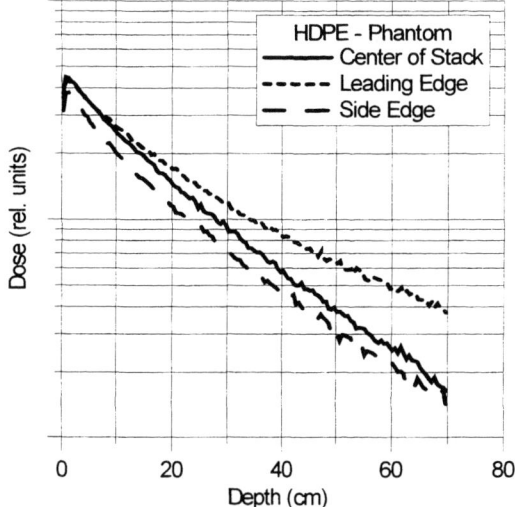

Figure 4. Depth-dose distributions showing the edge effects in a 50 x 50 x 70 cm package of material with unit-density treated with 7.5 MeV X rays.

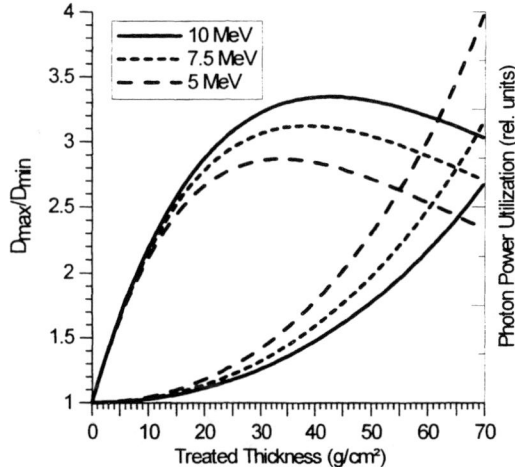

Figure 5. The upper curves show effective photon power utilization vs package thickness for 5, 7.5 and 10 MeV X-ray energies. The lower curves show max/min dose ratios for the same conditions.

Thick packages can be treated from opposite sides to improve the dose uniformity. Then the highest dose occurs on the entrance and exit surfaces and the lowest dose occurs midway between these surfaces. The effective X-ray energy utilization is usually defined as the minimum dose multiplied by the total mass of the package. With this definition, there is an optimum thickness for the maximum effective energy utilization. With thinner packages, the mass decreases more than the minimum dose increases, while with thicker packages, the mass increases less than the minimum dose decreases. These relationships are indicated by the effective energy or power utilization vs thickness curves shown in Fig. 5.

The maximum effective energy utilizations and the optimum package thicknesses increase with the incident electron energy. The optimum thicknesses are 34, 38 and 43 cm for 5, 7.5 and 10 MeV, respectively. In comparison, the optimum thickness for gamma rays from a large-area cobalt-60 source is about 28 cm [7].

Max/min dose ratios vs thickness are also shown in Fig. 5 for 5, 7.5 and 10 MeV. These ratios decrease with increasing electron energy for the same package thickness. Nevertheless, they all have about the same max/min ratio of 1.5 at the optimum thicknesses. In comparison, the max/min dose ratio for gamma rays from a large-area cobalt-60 source is about 1.75 at the optimum thickness of 28 cm [7].

X-RAY PROCESSING

The product conveying system within the treatment room must be designed to present opposite sides of the product carriers to the scanning X-ray beam to obtain acceptable dose uniformity. A simple, two-pass system would give satisfactory results, as is indicated by the data presented in this paper. A more complex, four-pass system with two conveyor tracks side-by-side could provide more uniform dose distributions and increase the X-ray power utilization efficiency. However, practical considerations of cost, reliability and transit time through the facility will influence the design of the product conveying system for particular applications. When treating refrigerated products, it will be important to have short transit times to keep the product temperature below the regulatory limit.

The processing rate in an X-ray treatment facility is directly proportional to the X-ray power and inversely proportional to the absorbed dose. The SI unit of dose is the gray (Gy), which is defined as the absorption of 1 joule of energy per kilogram of mass. This means that 1 kW of X-ray power could treat 1 kg/s with a dose of 1 kGy, if the dose distribution were uniform and the utilization efficiency were 100%.

Industrial electron accelerators can now provide up to 200 kW of beam power with electron energies up to 5 MeV, and more powerful equipment will soon be available. With these capabilities and 8% conversion efficiency, the emitted X-ray power would be 16 kW. Assuming an effective utilization efficiency of 40% and a minimum dose of 1.5 kGy, which would be sufficient to pasteurize fresh meat, the processing rate would be 16 x 0.40 / 1.5 = 4.27 kg/s or 15,360 kg/h or 33,800 lb/h. This treatment capacity will be more than enough for many commercial facilities.

REFERENCES

1. *Preservation of Food by Ionizing Radiation, Vols. I, II & III,* edited by E. S. Josephson and M. S. Peterson, CRC Press, Boca Raton, FL, 1982, 1983.

2. Meissner, J., Abs, M., Cleland, M. R., Herer, A. S., Jongen, Y., Kuntz, F. and Strasser, A., "X-Ray Treatment at 5 MeV and Above" in *Proceedings of the 11th International Meeting on Radiation Processing,* edited by J. H. Hubbell and A. Miller, *Rad. Phys. Chem.,* **57**, 647-651 (2000).

3. E 1608, "Standard Practice for Dosimetry in an X-Ray (bremsstrahlung) Facility for Radiation Processing", *Annual Book of ASTM Standards,* American Society for Testing and Materials, West Conshohocken, PA, 1999, pp. 810-820.

4. "Consultant's Meeting on the Development of X-Ray Machines for Food Irradiation", meeting chaired by A. Brynjolfsson, Food and Agriculture Organization, International Atomic Energy Agency, Vienna, Austria, 1995.

5. Brynjolfsson, A., "Natural and Induced Radioactivity in Food", in *Proceedings of the International Conference on Future Nuclear Energy Systems - Global '99,* American Nuclear Society, Chicago, Il, 1999.

6. CCC-467, "Integrated TIGER Series of Coupled Electron/Photon Monte Carlo Transport Codes System", *RSICC Software Directory,* Radiation Safety Information Computational Center, Oak Ridge National Laboratory, Oak Ridge, TN 1999.

7. Cleland, M. R. and Pageau, G. M., "Comparisons of X-Ray and Gamma-Ray Sources for Industrial Irradiation Processes", in *Proceedings of the Ninth International Conference on the Applications of Accelerators in Research and Industry,* Nuclear Instruments and Method in Physics Research, North Holland Physics Publishing, Amsterdam, The Netherlands, 1987. pp. 967-972.

A Sub-Picosecond Pulsed 5 MeV Electron Beam System

J. Paul Farrell[1], K. Batchelor[1]
I. Meshkovsky[2], I. Pavlishin[2], V. Lekomtsev[2], A. Dyublov[2], M. Inochkin[2],
and T. Srinivasan-Rao[3]

[1] *Brookhaven Technology Group, Inc,* [2] *Optoel Scientific Innovation Company,* [3] *Brookhaven National Laboratory*

Abstract. Laser excited pulsed, electron beam systems that operate at energies from 1 MeV up to 5 MeV and pulse width from 0.1 to 100 ps are described. The systems consist of a high voltage pulser and a coaxial laser triggered gas or liquid spark gap. The spark gap discharges into a pulse forming line designed to produce and maintain a flat voltage pulse for 1 ns duration on the cathode of a photodiode. A synchronized laser is used to illuminate the photocathode with a laser pulse to produce an electron beam with very high brightness, short duration and current at or near the space charge limit. Operation of the system is described and preliminary test measurements of voltages, synchronization and jitter are presented for a 5 MeV system. Applications in chemistry, and accelerator research are briefly discussed.

INTRODUCTION

Picosecond and sub-picosecond pulsed electron beams have current applications in accelerator research and in pulsed radiography for the study of transient phenomena using both direct electrons and x-rays produced by Bremsstrahlung. In accelerator research, fast pulsed high voltage systems are used to study field emission, photoemission in the presence of high fields and the formation and propagation of high brightness electron beams. Naturally, fast pulsed beams have practical use in pulse radiography where the requirements include precise synchronization between energy impulse and probe.

The first compact laser triggered sub-nanosecond high voltage generators were developed for Brookhaven National Laboratory by the Russian company, Optoel, in 1994[1]. They provided, respectively, a high voltage output of 0.2 - 0.5 MV on a 20 Ohm load, and 0.5 - 1 MV pulse on an 80 Ohm load with 0.1 to 0.15 ns rise time and pulse duration adjustable from 0.2 ns to 2 ns with 1 Hz repetition rate. The basic components of the generator are a low voltage LC circuit, a pulse transformer and a pulse forming line. There is a laser triggered SF_6 gas switch on the output of the transformer. Sensors are located at different distances along the pulse forming line to monitor the voltage wave form.

In this paper, we describe a new, short pulse, electron beam system that uses a pulse generator, a laser triggered spark gap and very high gradient photocathode to produce a compact, short pulse synchronizable electron beam. Preliminary test results are shown for a 5 MeV, 10 to 100 ps system.

THE BASIC SYSTEM

The basic components of the short pulsed electron beam system shown in Figure 1 are:

- master timer and laser system
- high voltage pulse power supply and laser triggered spark gap
- pulse forming system (PFL)
- photodiode electron gun and beam transport
- diagnostic and/or experimental test region.

The master timer and laser systems are standard commercial components. The high voltage power supply consists of a pre-pulser that provides an 80 kV pulse to the trigger electrode of a commercial spark gap. The spark gap discharges a capacitor to energize the primary coil of a high voltage resonant transformer. The output of this resonant transformer is developed across a coaxial high voltage spark gap. The spark gap is triggered by an axial laser pulse that discharges the high voltage output of the transformer into a pulse forming line (PFL). In the PFL, the high voltage pulse is shaped and transformed to produce a flat top high voltage of ~ 200 ps rise and fall time and 1 ns duration on the cathode of a photodiode.

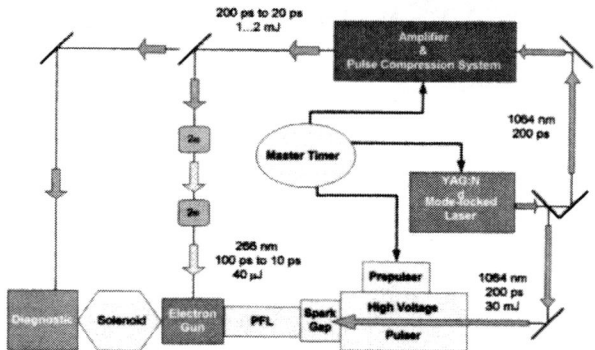

FIGURE 1. In this example, a mode locked YAG laser system (1064 nm) with 200 ps duration is used to trigger the spark gap and the same laser is used to provide a short pulse excitation to the photocathode. The photo pulse is frequency quadrupled to obtain the optimum uv frequency range (255 to 266 nm) for photo emission. Other laser arrangements are possible that can provide electron beam pulses down to 100 fs.

Laser System

The laser system performs three tasks:

- trigger high voltage spark gap
- illumination of the photo-cathode.
- synchronization of voltage, current and diagnostic pulse

Because the same laser provides illumination of the photocathode and the diagnostic signal, the electron beam and the probe are very highly synchronized.

High Voltage Pulse Power Supply

The high voltage pulse power supply consists of the low voltage pulse generator, a resonant pulse transformer, capacitance and a laser triggered spark gap switch. The purpose of the high voltage pulse power supply is to produce a voltage pulse on the high-voltage electrode of the laser triggered spark gap switch.

In operation, a low voltage pulse system generates voltage pulses with amplitude 25 - 100 kV on the primary coil of the resonant transformer. Standard commercial spark switches are used as switching elements in this part of the circuit. An 80 kV pre-pulser is used to trigger this spark gap. The duration of the first half wave voltage on the primary coil of the pulse transformer is ~ 700 ns. Transformer oil is used as insulating medium for both the low voltage pulser and the resonant pulse transformer.

On the 5 MV pulser, the resonant pulse transformer is designed to produce an output voltage of up to 2.5 MV (1st half wave) with pulse duration ~ 500 ns. The transformer consists of four main parts: the casing, the primary coil, high-voltage winding and high-voltage capacitance. The total capacitance of high-voltage electrode and high-voltage capacitance to ground is ~ 60 pF.

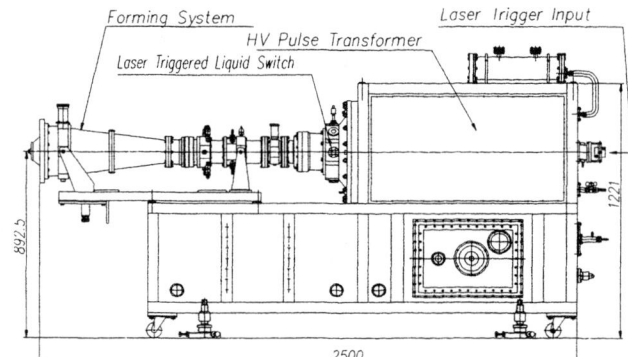

FIGURE 2. Side view of the 5 MV high voltage pulser.

The dielectric material in the output high voltage laser triggered spark gap is pressurized SF_6 gas for low voltage systems. A special dielectric liquid is used in the 5 MV system. The purpose of the laser triggered spark gap is to switch the energy accumulated at the output of the pulse transformer and to form a rectangular pulse with minimum time spread due to commutation of the spark gap. A coaxial tube with glass rod passes through the volume of the high-voltage pulse transformer and serves to transport the laser pulse to the inter-electrode gap.

Pulse Forming System (PFL)

The purpose of the pulse forming system is to produce an output pulse with rise time of ~ 150 ps and duration of ~ 1 ns at the photocathode. The pulse forming system of the 5 MV pulser consists of five parts: the charging inductance, the forming line, a self-breakdown switch, a transport line and the voltage transforming section. The impedance of the line seen at the triggered spark gap is 10 Ohms and the pulse length is 1 ns. This impedance determines the duration of the pulse at the output of the PFL. The rise time and jitter of the pulse at the cathode are dependent on reliable and consistent operation of a low-inductance, multi-channel, self-breakdown liquid switch that is incorporated in the PFL The final output voltage of 5 MV is achieved at the cathode by cylindrically symmetric transformation from an initial 10 Ohm line to final impedance of 160 Ohm. The output of the transforming line is terminated in a characteristic resistive impedance at the vacuum diode.

Preliminary tests of the operation of the 5 MV pulser are currently underway. First the liquid switch was tested with 25 mm gap. The liquid switch was triggered with use of the Nd-YAG laser at a wave length of 1060 nm, pulse duration of 200 ps and pulse energy of 10 - 30 mJ. The laser pulse was transported along the axis of the pulse transformer and into the dielectric switch from the side of the negative electrode. The applied voltage was 2.5 MV. At this voltage, the starting delay was measured to be 20 - 25 ns and the jitter was 0.5 ns.

Next, the 10 Ohm pulse forming line was connected to the high-voltage transformer and its performance was tested. The choice of recharging inductance enabled the line to fully charge in 3 - 5 ns with multiplication coefficient of 1.2 - 1.3 with respect to the transformer output voltage.

Finally, the tapered impedance transforming section of the PFL was tested. In this section, the impedance terminating impedance is 160 Ohms which results in a factor of four increase in voltage. Capacitive probes used to measure voltage along the PFL were calibrated by driving the sections at low voltage using a solid state pulser with ~ 150 ps rise time.

The Photo-Diode Electron Gun

The electron gun is a photodiode with a copper cathode and 1 mm stainless steel aperture[2-6]. The anode-cathode gap can be adjusted *in-situ* so the cathode can be conditioned without exposing the surface to atmosphere. To produce high brightness beams, the electric field at the cathode is ~ 1 GV/m. With a 1 mm diameter laser spot at the cathode, the steady state space charge limited emission current is ~ 500 A. After conditioning, the dark current is less than 1 % of the photocurrent.

Beam current higher than the space charge limit can be achieved with short pulse operation but internal pressure due to Coulomb forces causes longitudinal and transverse expansion of the beam.

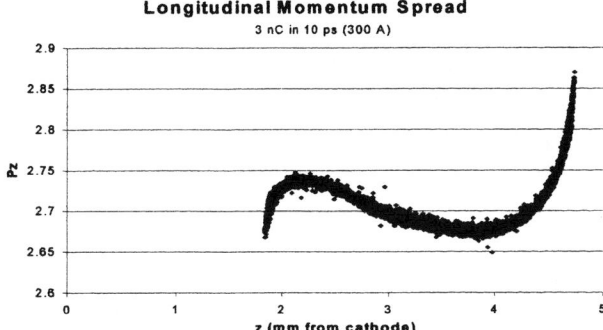

FIGURE 3. Longitudinal momentum spread for a 3 nC, 10 ps bunch calculated using the time dependent PIC code, MAFIA. P_z is in units of $\beta\gamma$.

Figure 3 shows the spatial pulse length and momentum spread of a 1 MeV beam at a distance of 3 mm from the cathode (~ 2 mm beyond the anode). Coulomb forces have pushed the leading edge of the bunch toward higher energy while edge is retarded in energy. As the beam progresses, the bunch width expands due to the difference in energy of the leading and trailing ends.

Beam Transport and Synchronization

The electron bunch is accelerated to relativistic speed in a distance of less than 1 mm in the photodiode. This rapid acceleration reduces effects of space charge and results in a beam with low divergence and small diameter at the anode. After exiting from the high field region of the diode, the beam diverges due to the lens effect of the transition from high field to no field. A solenoid lens is placed up close to the anode to collect the beam and focus it prior to injection into another high energy accelerator or for to the low energy diagnostic tests or experiments.

The 5 MeV system is designed to operate continuously at a pulse repetition rate of 0.1 Hz. The pulse rate is limited by the power of the charging system. Higher repetition rate is possible with larger power supplies and it may be necessary a heat exchanger to remove heat from the low voltage pulser.

FIGURE 4. Beam transport system at BNL used to characterize the beam from the sub-picosecond electron gun. The cathode is just above the stanchion on the right. It is followed by a solenoid focusing magnet and a dipole magnet that is used to measure energy spread.

When the high voltage laser trigger pulse, the laser pulse to the photodiode and the laser diagnostic (probe) pulse are derived from the same laser system synchronization depends primarily on the jitter in the high voltage spark gap. This jitter has been measured to be $< \sim 0.5$ ns. Jitter in the firing of the high voltage spark gap affects the energy of the pulse as well as the arrival time of the flat pulse at the cathode.

The synchronization of arrival of the electron bunch and the diagnostic pulse is simplified because it depends only on the difference in path length between the two pulses. This single laser method can be used down to 10 ps in duration. For applications that require electron bunches less than 10 ps, it will be necessary to employ a two laser system.

CONCLUSION

We have built and are currently testing a new 5 MeV pulsed electron beam system that is designed to produce electron bunches in the 10 ps to 100 ps range. The system can be configured to produce shorter bunches (~ 100 fs) using a more complex, and more expensive, two laser system.

Applications include accelerator research and use as an injector into higher voltage accelerators. It can also be used as a complete system for the study of transient phenomena using direct electrons or Bremsstrahlung.

ACKNOWLEDGMENTS

We would like to acknowledge the work of J. Smedley for doing the MAFIA calculations and for measurements of quantum efficiency in the presence of high fields and we would like to acknowledge Thomas Tsang, John Schill and Robert Conde for keeping the laser and high voltage pulser systems operating during the installation and characterization runs. This work is supported in part by DOE SBIR DE-FG02-97ER82336. The work at BNL is supported by DOE contract number DEACO2-98CH10886

REFERENCES

1. T. Srinivasan-Rao and J. Smedley; Table Top, Pulsed, Relativistic Electron Gun with GV/m Gradient; BNL 63517; pres. 7th Advanced Accelerator Concepts Workshop, Lake Tahoe, CA, 12-18 October (1996);AIP Conf. Proc. 398, eds. S. Chaltopadhyay, J. Mc Cullough, & P. Dahl, p. 730, AIP, Woodbury, NY (1997).

2. K. Batchelor, J. P. Farrell, G. Dudnikova, I. Ben-Zvi, T. Srinivasan-Rao, J. Smedley, and V. Yakimenko; A High Current, High Gradient, Laser Excited, Pulsed Electron Gun; BNL 65678 ; pres. EPAC'98, Stockholm, Sweden, 22-26 June (1998); Proc. 6th European Particle Accelerator Conf., S. Myers, L. Liljeby, Ch. Petit-Jean-Genaz, J. Peole, & K.-G. Rensfelt, eds., p. 791 (1998).

3. T. Srinivasan-Rao, J. Smedley, K. Batchelor, J. P. Farrell, and G. Dudnikova; Comparison of Electrostatic and Time Dependent Simulation Codes for Modeling a Pulsed Power Gun; BNL 65748 ; pres. 12th Int'l. Conf. on High Power Particle Beams (Beams'98), Haifa, Israel, 7-12 June (1998); M. Markovits and J. Shiloh, eds., IEEE Catalog #98EX103, p. 549 (1998).

4. T. Srinivasan-Rao, J. Smedley, J. Schill, K. Batchelor, and J. P. Farrell; Dark Current Measurements at Field Gradients Above 1 GM/m; BNL 65746; pres. 8th Workshop on Advanced Accelerator Concepts, Baltimore, MD, 6-11 July (1998).

5. T. Srinivasan-Rao, J. Smedley, K. Batchelor, J. P. Farrell, and G. Dudnikova; Optimization of Gun Parameters for a Pulsed Power Electron Gun; BNL 65748 ; pres. 8th Workshop on Advanced Accelerator Concepts, Baltimore, MD, 6-11 July (1998);

6. T. Srinivasan-Rao, J. Schill, I. Ben-Zvi, K. Batchelor, J.P. Farrell, J. Smedley, X.E. Lin, and A. Odian; Simulation, Generation, and Characterization of High Brightness Electron Source at 1 GV/m Gradient; BNL 66464; pres. PAC'99 Conf., New York, NY, 3/29-4/2/99; Proc. 1999 Particle Accelerator Conf., Eds. A. Luccio & W. Mackay, p. 75 (1999).

SECTION XI

MEDICAL APPLICATIONS:
POSITRON EMISSION TOMOGRAPHY AND TARGETS,
RADIOISOTOPE PRODUCTION,
PARTICLE THERAPY BNCT MEASUREMENTS

Characterization of Neutron and Photon Sources from a 10.5 MeV Proton Beam on [^{18}O] Enriched Water

L. F. Miller[*], L. W. Townsend[*], C.W. Alvord[**]

[*]*Nuclear Engineering Department, University of Tennessee, Knoxville, TN 37996*
[**]*CTI, inc., 810 Innovation Drive, Knoxville, TN 37932-2571*

Abstract. The production of F-18 from a 10.5 MeV proton beam on oxygen-18 results in significant yields of neutrons and photons. In order to optimize personnel shielding that satisfies regulatory requirements, it is essential that both the intensity of both neutrons and of photons be determined as a function of energy and angle, which was accomplished by combining results from measurements and from calculations. Energy dependence for neutrons was estimated as a function by unfolding Bonner ball measurements, a hyper-pure germanium detector was used to obtain measurements of the photon spectra, and a well established computer program was used to obtain the calculated values. The radiation intensity was determined from calibrated survey meters for neutrons and for photons. The energy and angular dependence obtained from measurements and calculations agree within the uncertainty of the measurements, but calculated results, scaled by measurements, were used for input to radiation shield design studies. The neutron yield is sufficiently high to be of interest for several applications.

INTRODUCTION

In order to obtain a more accurate neutron spectrum, measurements and calculations were performed for neutrons and for photons. The intensity, energy spectra, and angular distribution were needed for both neutrons and for photons to accurately specify the source term. Thus, the following measurements were made:

1. gamma spectroscopy measurements using a hyperpure germanium (HpGe) detector,
2. Bonner ball measurements for neutrons at ninety, fifty, and zero degrees relative to the beam centerline, and
3. field intensity measurement for neutrons on the beam centerline using a calibrated, NIST traceable, survey instrument.

GAMMA SPECTROSCOPY MEASUREMENTS

A HpGe detector, and associated instrumentation, were calibrated for energy dependence using a NIST traceable gamma source; although, the system was not calibrated for absolute efficiency. These data show qualitative features as expected, but they have not been analyzed in detail. Spectra were obtained at 90, 50 and 0 degrees off the beam axis, and an example spectrum for the first run at ninety degrees off the beam axis is shown in Figure 1.

BONNER BALL MEASUREMENTS

Six Bonner balls were used for these measurements. In order to evaluate detector response and to assure that responses from each Bonner ball were qualitatively correct, measurements were performed using a Cf-252 source. Results from this experiment, and from several cyclotron runs, are shown in Table 1.

You may note from the normalized measurements that count rates from the 10 and 12-inch balls are relatively higher for the Cf-252 measurement than for any of the three angular measurements from the cyclotron. This suggests that the energy spectrum from the cyclotron, at the point of measurement, has a relatively softer-shaped spectrum than does the Cf-252 spectrum.

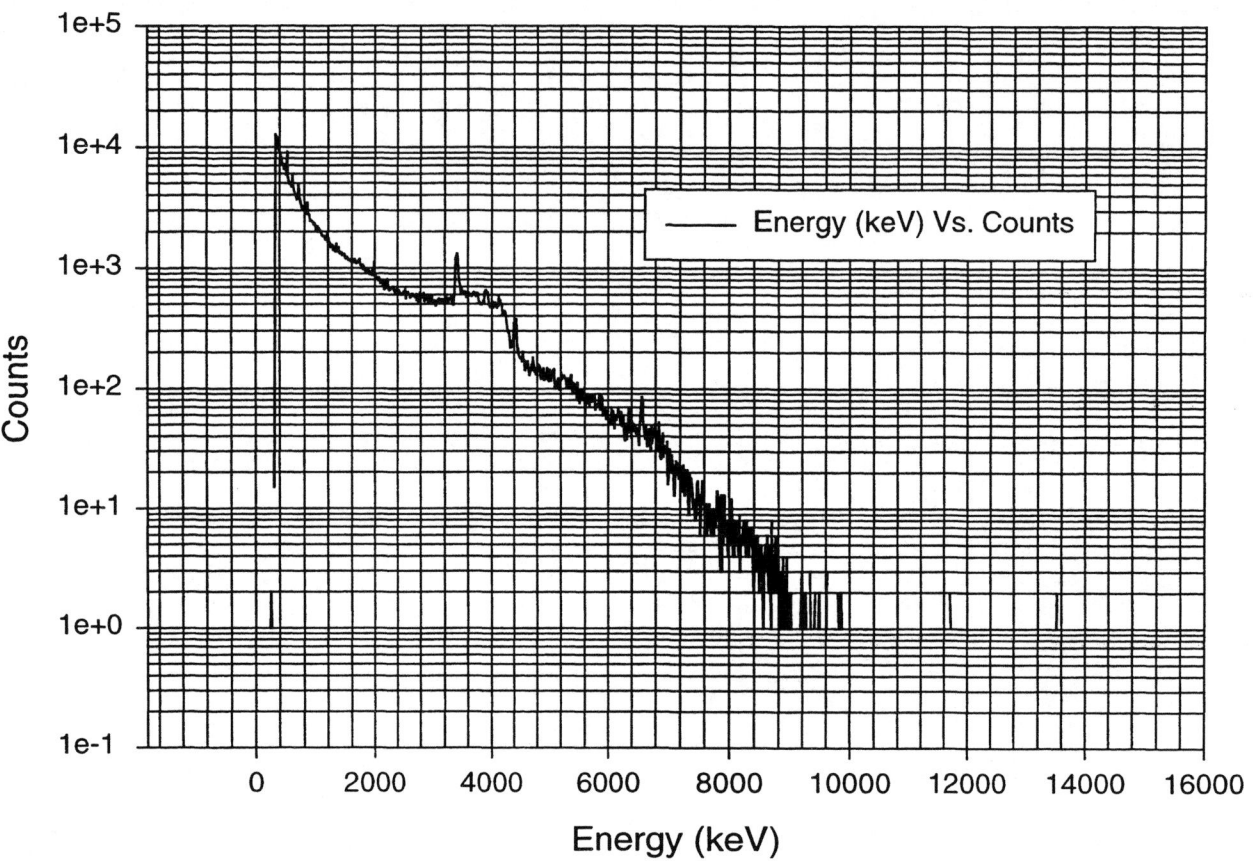

FIGURE 1. Gamma ray spectrum at 90 degrees off beam axis

TABLE 1. Results from Bonner ball measurements for ^{252}Cf and for three cyclotron runs.

	Cf-252	90 Deg	50 Deg	Zro Deg	90 Deg	50 Deg	Zro Deg	90 Deg	50 Deg	Zro Deg	Cf
	Cts/100 Sec	Cts/10 Sec	Cts/30 Sec	Cts/30 Sec	CPM	CPM	CPM	Norm	Norm	Norm	Norm
12" ball	158	3104	17174	24810	18624	34348	49620	0.052088	0.057083	0.05534	0.080695
10" ball	330	6177	30472	46947	37062	60944	93894	0.103655	0.101283	0.104717	0.168539
6" ball	597	11411	57604	87874	68466	115208	175748	0.191485	0.191464	0.196006	0.304903
4" ball	507	17947	92051	136973	107682	184102	273946	0.301165	0.30596	0.305524	0.258938
3" ball	224	12211	61840	89386	73266	123680	178772	0.20491	0.205544	0.199379	0.114402
2" ball	97	6661	32015	47989	39966	64030	95978	0.111777	0.106412	0.107041	0.04954
bare	45	2081	9704	14343	12486	19408	28686	0.034921	0.032254	0.031993	0.022983
Sum	1958				357552	601720	896644	1	1	1	1
					1	1.68288808	2.507730344				

The Bonner ball measurements are unfolded using the BUNKI (Hertel, Johnson) computer program based on a Cf-252 spectrum as an initial guess for the energy dependence, and the adjusted spectrum is not very different from a Cf-252 spectrum, except below about 10 keV. It can also be noted from the unfolded spectra illustrated below that the unfolded spectra contain more high-energy neutrons than does the Cf-252 spectrum. This suggests that the measured spectrum, at the point of the measurement, is represented by a Cf-252 spectrum with reasonable accuracy; however, the modal energy of the unfolded spectra appear to be about 1 MeV higher (about 2 MeV) than is the case for a Cf-252 spectrum (which is less than 1 MeV). It is concluded that the spectrum at the point of measurement is significantly degraded due to the increase of the unfolded spectra below 10 keV, which is probably due to scattering of higher energy neutrons. The measurement point was about fifteen feet from the source and many of the neutrons that interacted with the Bonner balls were probably scattered from the floor and other equipment in the area, as can be seen from the unfolded spectra relative to a Cf-252 spectrum shown below.

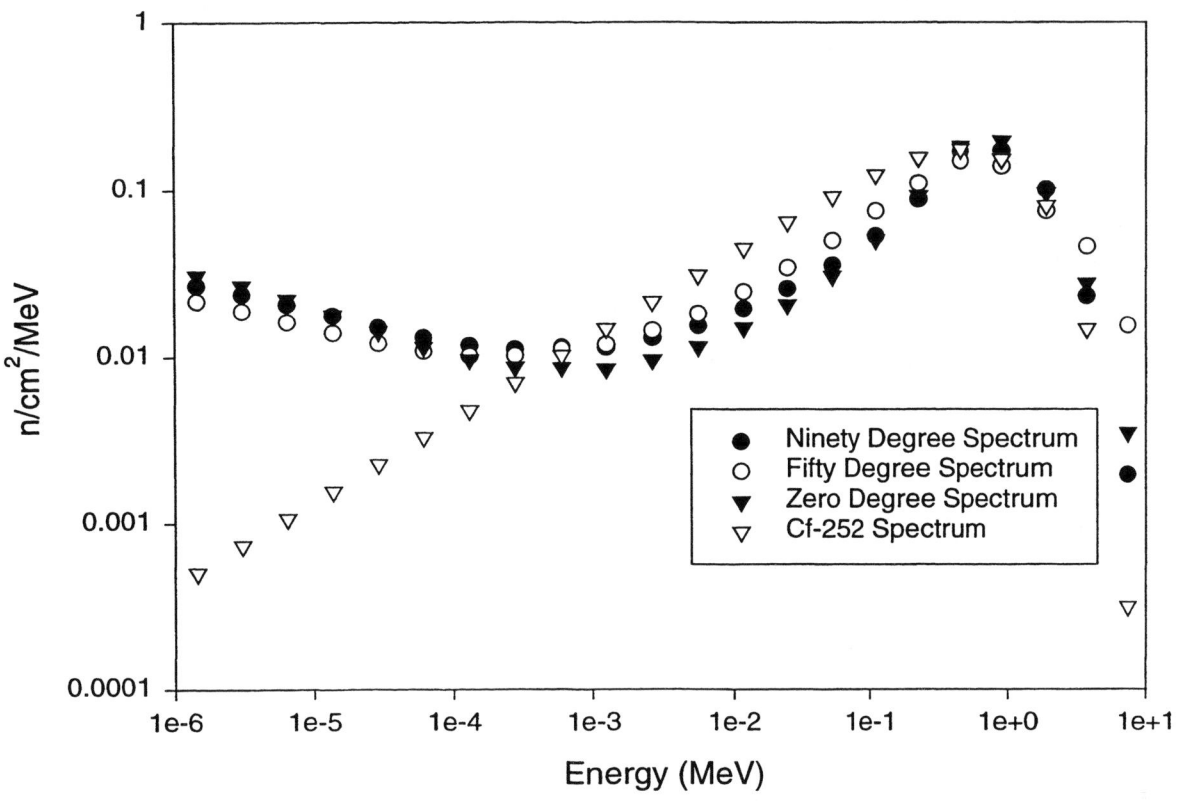

FIGURE 2. Comparison of unfolded Bonner ball spectra at 3 angles and Cf-252 spectrum

CALCULATIONS FOR THE SOURCE TERM

Calculations for the energy dependent source term at zero degrees off the beam path were performed using the Alice computer program (Townsend), and results are listed in Table 2. The columns labeled Egc and Enc are the energy group boundaries specified for the Alice calculation. Those labeled Eg_tr and En_tr are energy group boundaries that correspond to the 27 neutron, 18 gamma, neutron-photon coupled cross section library used in the discrete ordinates calculation. The calculated values are interpolated, integrated, and normalized to obtain the source by group for input to the coupled neutron-photon transport calculation.

A comparison between the Cf-252 neutron spectrum and the one calculated using Alice is shown in Figure 3.

TABLE 2. Results from calculations for photon and neutron spectra at zero degrees off the beam axis.

Egc(eV)	Gam_Src	Enc(eV)	Neut_Src	Eg_tr(eV)	Gam#/Src Interpolated	Int_gam	Gam_Fraction	En_tr(eV)	Neut#Src Interpolated	Int_neut	Neut_Fraction
1.75E+07	0	1.38E+07	0	1.00E+07	1.20E-03	2.40E+03	4.76E-03	2.00E+07	0.00E+00	0.00E+00	0.00E+00
1.70E+07	0	1.33E+07	0	8.00E+06	8.30E-03	1.25E+04	2.47E-02	6.43E+06	4.79E-02	1.64E+05	4.17E-01
1.65E+07	0	1.28E+07	0	6.50E+06	2.10E-02	3.15E+04	6.25E-02	3.00E+06	8.80E-02	1.01E+05	2.56E-01
1.60E+07	0	1.23E+07	0	5.00E+06	4.13E-02	4.13E+04	8.19E-02	1.85E+06	9.00E-02	4.05E+04	1.03E-01
1.55E+07	0	1.18E+07	0	4.00E+06	5.37E-02	5.37E+04	1.07E-01	1.40E+06	8.46E-02	4.23E+04	1.07E-01
1.50E+07	0	1.13E+07	0	3.00E+06	6.19E-02	3.10E+04	6.14E-02	9.00E+05	7.15E-02	3.58E+04	9.06E-02
1.45E+07	0	1.08E+07	0	2.50E+06	7.02E-02	3.51E+04	6.96E-02	4.00E+05	3.17E-02	9.51E+03	2.41E-02
1.40E+07	0	1.03E+07	0	2.00E+06	8.25E-02	2.81E+04	5.56E-02	1.00E+05	1.00E-02	8.30E+02	2.10E-03
1.35E+07	0	9.75E+06	0	1.66E+06	1.02E-01	3.36E+04	6.66E-02	1.70E+04	0.00E+00	0.00E+00	0.00E+00
1.30E+07	0	9.25E+06	0	1.33E+06	1.21E-01	3.98E+04	7.90E-02	3.00E+03	0.00E+00	0.00E+00	0.00E+00
1.25E+07	0	8.75E+06	0	1.00E+06	1.40E-01	2.81E+04	5.57E-02	5.50E+02	0.00E+00	0.00E+00	0.00E+00
1.20E+07	0	8.25E+06	0	8.00E+05	1.58E-01	3.15E+04	6.25E-02	1.00E+02	0.00E+00	0.00E+00	0.00E+00
1.15E+07	0	7.75E+06	0	6.00E+05	1.82E-01	3.64E+04	7.23E-02	3.00E+01	0.00E+00	0.00E+00	0.00E+00
1.10E+07	0	7.25E+06	0.038767916	4.00E+05	2.17E-01	2.17E+04	4.31E-02	1.00E+01	0.00E+00	0.00E+00	0.00E+00
1.05E+07	0.00038	6.75E+06	0.044211687	3.00E+05	2.39E-01	2.39E+04	4.74E-02	3.05	0.00E+00	0.00E+00	0.00E+00
1.00E+07	0.001197	6.25E+06	0.050027563	2.00E+05	2.65E-01	2.65E+04	5.25E-02	1.77	0.00E+00	0.00E+00	0.00E+00
9.50E+06	0.002476	5.75E+06	0.056160419	1.00E+05	2.94E-01	1.47E+04	2.92E-02	1.3	0.00E+00	0.00E+00	0.00E+00
9.00E+06	0.004127	5.25E+06	0.062513782	5.00E+04	3.11E-01	1.24E+04	2.47E-02	1.13	0.00E+00	0.00E+00	0.00E+00
8.50E+06	0.006191	4.75E+06	0.068936053	1.00E+04	0.3246			1	0.00E+00	0.00E+00	0.00E+00
8.00E+06	0.008255	4.25E+06	0.075220507			5.04E+05	1.00E+00	0.8	0.00E+00	0.00E+00	0.00E+00
7.50E+06	0.011144	3.75E+06	0.081105292					0.4	0.00E+00	0.00E+00	0.00E+00
7.00E+06	0.014859	3.25E+06	0.086176957					0.325	0.00E+00	0.00E+00	0.00E+00
6.50E+06	0.02105	2.75E+06	0.089884234					0.225	0.00E+00	0.00E+00	0.00E+00
6.00E+06	0.028066	2.25E+06	0.091441566					0.1	0.00E+00	0.00E+00	0.00E+00
5.50E+06	0.035496	1.75E+06	0.089581036					0.05	0.00E+00	0.00E+00	0.00E+00
5.00E+06	0.041274	1.25E+06	0.082524807					0.03	0.00E+00	0.00E+00	0.00E+00
4.50E+06	0.049529	7.50E+05	0.066786108					0.01	0.00E+00	0.00E+00	0.00E+00
4.00E+06	0.053656	2.50E+05	0.016662073					1.00E-05			
3.50E+06	0.057784									3.95E+05	1.00E+00
3.00E+06	0.061911										
2.50E+06	0.070166										
2.00E+06	0.082548										
1.50E+06	0.11144										
1.00E+06	0.140332										
5.00E+05	0.198116										

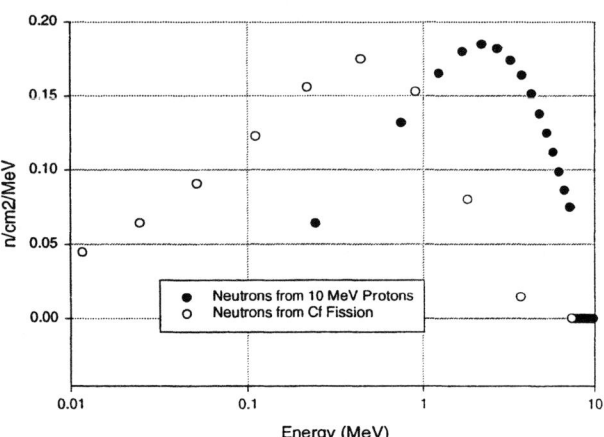

FIGURE 3. Comparison of calculation of ^{18}O(p,n)^{18}F spectrum from Alice and ^{252}Cf spectrum

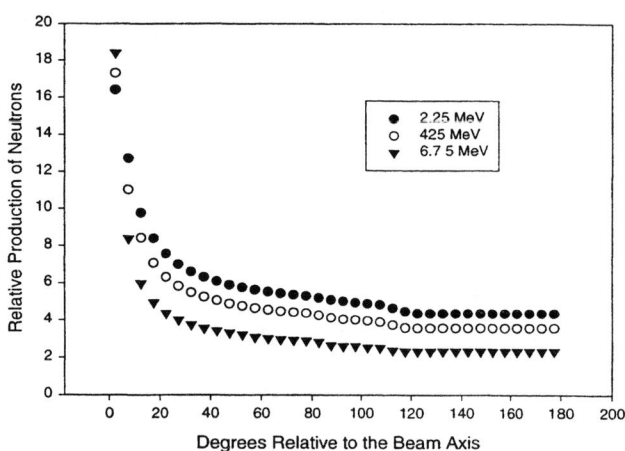

FIGURE 4. Neutron angular distribution at three neutron energies.

The angular distribution for several neutron energies, energy distribution of photons, and angular distribution of photons are illustrated in Figures 4-6 below.

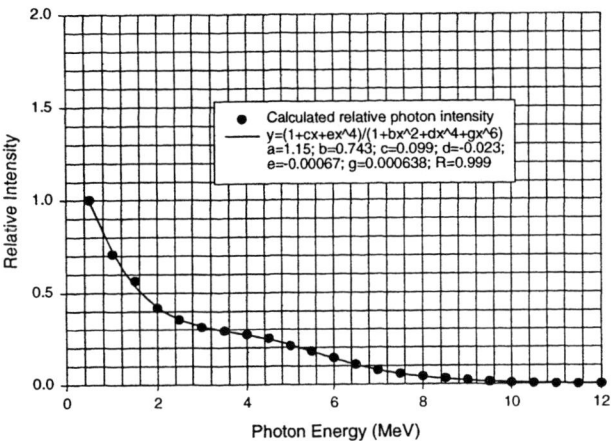

FIGURE 5. Calculated gamma energy spectrum

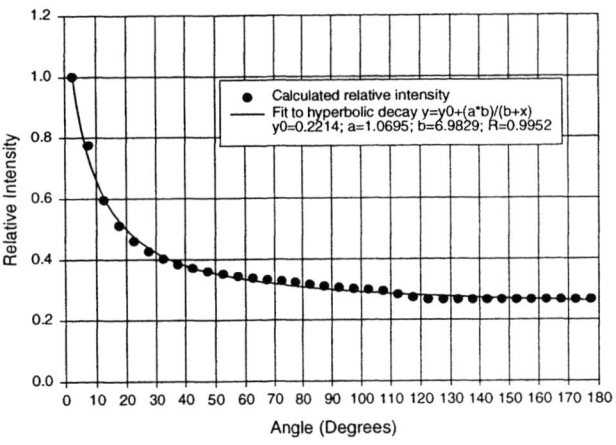

FIGURE 6. Calculated gamma angular distribution

SOURCE INTENSITY

The source intensity is based on neutron and gamma survey instrument measurements that are calibrated to NIST traceable sources. The results shown below are reported.

TABLE 3. Neutron and Gamma Survey Data

Run Number	Run Duration	mR/Hr Neutron	mR/Hr Gamma	Distance To Target	Angle To Target
1	5 min	400	18.5	9' 6"	135°
2	5 min	700	32	15' 6"	0°

Note that the dose equivalent rate due to neutrons is about a factor of twenty greater than due to gammas. However, the radiation weighting factor for neutrons is about a factor of ten; thus, these measurements suggest that the particle density of neutrons is about a factor of two greater than for photons. There are some complicating factors that need to be considered. One is that, according to the calculated energy spectra of photons and neutrons, there are about a factor of three fewer photons and neutrons at energies of several MeV. At an energy of one MeV, however, the Alice computer program predicts about a factor of four more photons than neutrons. Another is that there are several centimeters of brass shielding of the source. This would preferentially attenuate lower energy photons, but it would have little effect on the neutrons. Thus, the measurement should indicate relatively more neutrons than photons than would actually be the case. The total number of photons and neutrons emitted form the source is probably less important than their energy distribution for two reasons. One is that most of the photons produced are at about 0.5 MeV, or lower, which are rapidly attenuated. Another reason is that the photon dose at the surface of the shield is probably almost exclusively due to neutron capture.

The median energy of neutrons produced by 10 MeV protons on F-18 is about 2 MeV. At 10 mm in the ICRU sphere, the dose equivalent per unit fluence varies from about 135 to 176 x 10^{-12} Sv cm^2 [page 548 of the Radiological Health Handbook, Table 13.2]. For a dose equivalent rate of 1 mrem/hr, we obtain the following,

$$\phi = \left(\frac{1\,mrem/hr}{175\,x10^{-12}\,Sv\,cm^2}\right)\left(10^{-5}\frac{Sv}{mrem}\right)\left(\frac{1\,hr}{3600\,\sec}\right)$$

$$\phi \cong 15\frac{n/cm^2\,s}{mrem/hr}$$

Table 13.1.2 [page 514 of the Radiological Health Handbook, reports a fluence to dose equivalent conversion factor of 28 x 10^6 n/cm^2/rem, and this converts to about eight (n/cm^2/sec)/(mrem/hr). These average to about 12 (n/cm^2/sec)/(mrem/hr).

The reference measurement at 15.5 feet from the target on the beam axis resulted in a dose equivalent rate of 700 mrem/hr for a beam current of 2 microamps. The surface area of a sphere with a radius of 15.5 feet is 2.8 x10^6 cm^2. The estimated source strength is then given by,

$$S(n/s) = \left(\frac{12\,n/cm^2 s}{mrem/hr}\right)(2.8 \times 10^6 cm^2)(700\,mrem/hr)$$

$$S(n/s) \cong 2.4 \times 10^{10} n/s$$

Thus, the measured source term on the beam axis is about 1×10^{10} (n/s)/ microamp, and the calculated angular distribution of varies by about a factor of four. However, the neutron beam must pass through about 5 cm of brass to exit the source, and the detector response is composed of both scattered and uncollided neutrons. These factors are assumed to compensate and the angle-averaged neutron source strength is estimated to be about 1×10^{10} (n/s)/ microamp.

CONCLUSIONS AND RECOMMENDATIONS

Estimates of the source term for neutrons and for photons are based on measurements and on calculations with modest adjustments based on qualitative assessment. The estimate for the neutron source strength is believed to be more accurate than is the estimate for the gamma source. This is due to strong variability in the photon source strength as a function of energy and on preferential attenuation of the lower energy photons by the sample changer. A qualitative estimate for the uncertainty in the neutron source strength is about a factor of two and about a factor of four for the photons. The primary reason for the relatively large uncertainty is that the angle-averaged source is not accurately determined, scattering effects on the survey instrument are not known, and energy spectra effects on the fluence-to-dose equivalent conversion of the instrument is not known. In order to reduce this uncertainty, one should refine both the neutron and the photon angle- and energy-dependent measurements. However, the effort required to achieve this objective is probably not worth the cost for the improved result. The derived source term is sufficiently accurate to allow incorporation into 3D Monte Carlo simulation (MCNP) and pursue further validation using existing shielding assemblies and measurements.

REFERENCES

1. Hertel, Nolan, Personal Communication, Transfer of the Unfolding Code, February 2000

2. Johnson, T.L., et. al., "Recent Advances in Bonner Sphere Neutron Spectrometry," in *Theory and Practices in Radiation Protection and Shielding*, American Nuclear Society, April 1987

3. Carroll, L.R., et. at., "Radiation Measurements Related to the Design of a Self-Shielded Accelerator System," CTI, 1985

4. Carroll, L.R., "Radiation Safety Aspects of the RDS-112 Radioisotope Delivery System," Siemens, 1990

5. Alvord, Bill, "High-Power Fluoride Production System," CTI, September 1999

6. Townsend, L.W. Personal Communications, Calculations for p-n reactions on F-18 using the ALICE Computer Program, February, 2000

7. Shleien, Bernard, editor, The Health Physics and Radiological Health Handbook, Scinta, Inc, Silver Spring Md. 20902, 1992

Target and Accelerator Developments at CTI

C.W. Alvord, A.J. Mendez, D.E. Wittner

CTI Cyclotron Systems Inc., 810 Innovation Dr., Knoxville, TN 37932

Abstract. The accelerator products marketed by CTI have exclusively focused on proton-only, low energy (11 MeV) designs. This choice best suited the research customer, interested in producing several doses a day of a variety of positron emitting compounds. The PET cyclotron market has evolved into a high output, cost driven, competitive radiotracer production environment. A thoughtful analysis of the choices of energy and particle reveals that an 11 MeV proton accelerator outfitted with target changers and automated target loading and unloading equipment is still the best choice for FDG distribution. However technological innovations are required to face the challenges of the rapidly growing PET radiotracer business. Modifications to the CTI line of accelerators developed to face this evolving need will be presented.

INTRODUCTION

As the global PET market grows, we have continued to evaluate the shape of the current and future distribution businesses. Using existing data on the cost of operation of FDG manufacturing sites, a spreadsheet model (Microsoft Excel) that simulates cost per dose was generated. The results of input of many varying FDG sales scenarios reinforced the choice CTI has remained committed to; 11 MeV proton-only accelerators running on enriched materials. The relationship of cost per dose to number of doses produced daily is presented in Figure 1.

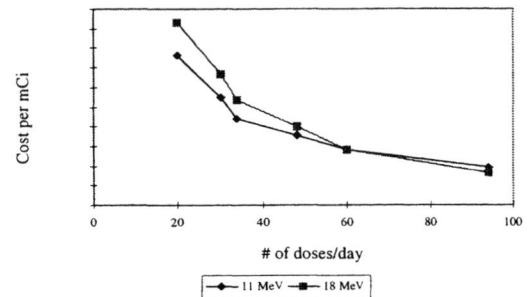

FIGURE 1. Cost per dose vs. # of doses/day

The model included many subtle factors including decommissioning costs, courier costs, and electrical power. The graph shows that at production levels from minimal to moderately high, the cost per dose from a center operating with an 11 MeV machine is more cost effective. The only case where the higher energy machine was more cost effective (approximately 97 doses per day) was a model where *two* RDS-111s were required to meet demand. The additional benefit from uptime due to the redundant accelerators was not included in cost analysis, but further contributes to the attractiveness of the 11 MeV machine.

The realization of this fact has driven CTI R&D efforts in the direction of reliability rather than alternate energies or particles. Recent developments include modifications to the loading and unloading equipment for the fluoride targets, major changes to the ion source to improve reliability and ease of service, and software enhancements that control costs and raise the level of service. These improvements will be summarized in the next three sections.

One-piece expansion chamber/manifold

The high-pressure enriched water target for fluoride production on the RDS-111 is outfitted with a combination pressure transducer/confinement valve/expansion chamber referred to collectively as the Shielded Delivery System. The schematic for the Shielded Delivery System is shown in Figure 2.

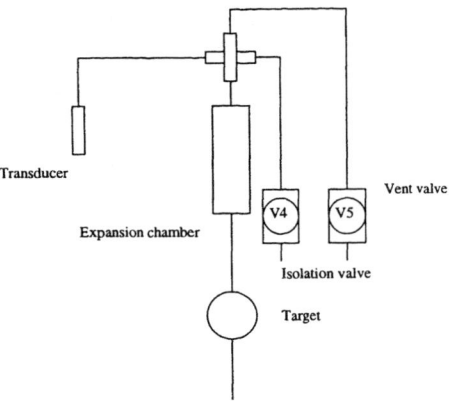

FIGURE 2. Shielded Delivery System schematic

The Shielded Delivery System is mounted near the target inside the shields to provide for control of any migration of aqueous fluoride activity away from the target. The expansion chamber is provided to control the pressure/power relationship of the target to maximize performance/reliability. All other components must minimize volume to make the expansion chamber function properly. The overall scheme is one of circuitous plumbing and multiple compression fitting connections. It was desirable to reconfigure the same components into a single assembly, minimizing connections and parts count.

Working with the General Valve Division of Parker, we developed an assembly that mimics this configuration but greatly reduces the part count. That assembly is pictured in Figure 3. This change has resulted in faster assembly times, minimized troubleshooting and improved yields. Impact on service cycles has not been measured, but is expected to improve.

Ion source reliability improvements

Background

The ion source for the RDS-111 cyclotron is of the Penning Ionization Gauge type, in which H⁻ is extracted from a plasma established in a low-pressure H_2 gas environment. The essential elements (shown in Figure 4) are a pair of tantalum cathodes (electron emitters), an anode containing the cylindrical discharge volume, and an axial magnetic field (provided by the cyclotron, ~1 T). The negative ions are extracted radially from a slit in the anode by a puller field of ~7 MV/m (source is biased to –15 kV). The mechanism for H⁻ production is outlined in Figure 4 below.

The discharge is established ("striking the arc") by applying a few kV to the cathodes (negative with respect to the anode), while flowing H_2 gas at low pressure (few tenths of torr) into the discharge volume. Electrons are accelerated to a few hundred eV from the cathodes by field emission and, constrained by the magnetic field, oscillate axially (reflex, or Penning, discharge) through the production volume. Collisions with the hot electrons ionize and vibrationally excite hydrogen molecules,

$$H_2 + e \rightarrow {}^*H_2^+ + 2e \rightarrow {}^*H_2 + e.$$

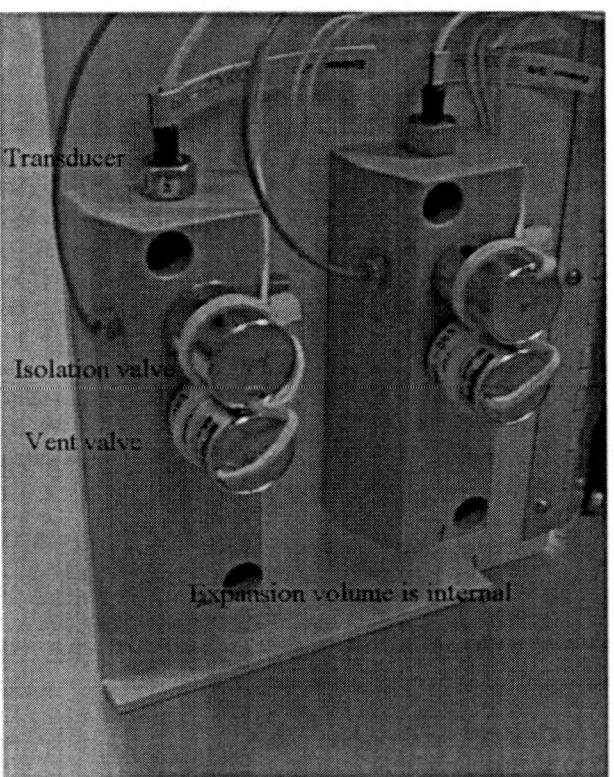

FIGURE 3. Manifold assembly

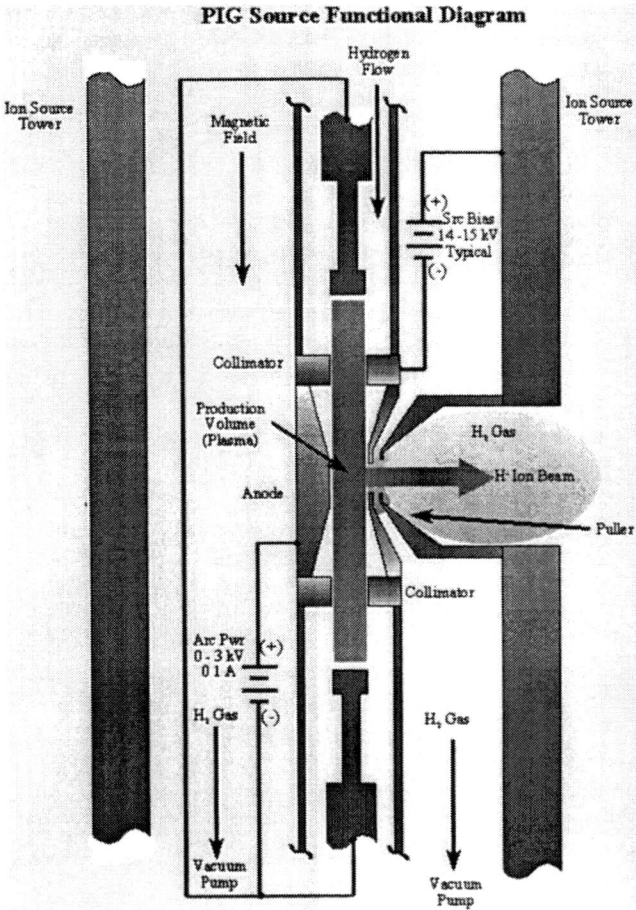

FIGURE 4. PIG source functional diagram

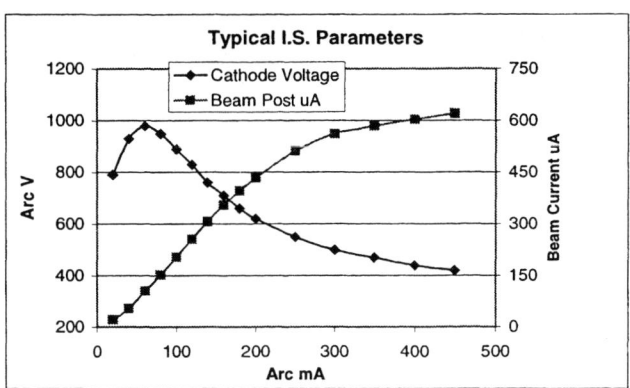

FIGURE. 5. Typical ion source operating parameters.

Back-bombardment by positive ions heats the cathodes to the point of thermionic electron emission.

The negative ions are formed outside the plasma predominantly by dissociative attachment of cold (< few eV) electrons and vibrationally-excited molecules, both of which can cross the magnetic field and diffuse out of the plasma:

$$*H_2 + e \rightarrow *H_2^- \rightarrow H^- + *H.$$

The arc discharge is characterized by two distinct regimes: when field emission dominates, arc currents are low and $dV/dI > 0$; when thermionic emission dominates (at higher arc currents) $dV/dI < 0$. For our source, the V – I curve peaks around 1 - 2 kV at an arc current of about 100 mA (see Figure 5 for typical operating curves).

Implementation

A number of features distinguish the new ion source from its predecessor. The ion source as it was could produce, when running well, sufficient beam currents for producing the desired PET radioisotopes. However, it was difficult to maintain the source in that condition. Typical failure modes included: failure to strike, arc shorting, and low or fluctuating output. There were two main culprits accelerating the source towards these failure modes: air leaks and organic contamination, both of which produced extremely high sputtering rates on the tantalum cathodes. The situation was further complicated by the difficulties in assembling and positioning the ion source, which led to another common failure mode—cathodes wearing off to one side (because of misalignment between the *B*-field and the cathode axes) and becoming "unstrikable". The remedies for these ailments are outlined below.

Contamination

To reduce the contamination from organics, plastic and other UHV incompatible materials were essentially eliminated from the ion source—the lower cathode insulator was changed to Al_2O_3 ceramic, and the upper cathode insulator was changed to PEEK. Additionally, the Teflon-insulated lower cathode wire was replaced with a UHV-rated KAPTON-insulated HV wire.

To improve gas cleanliness, the cathode-insulating ceramic tubes were replaced with small ceramic insulators, in appearance much like those on spark plugs. These were designed so that they formed gas seals with the cathode rods and outer stainless steel support tubes, and thus constrained the gas largely to the discharge region. This was accomplished by flowing the gas down the center of the upper cathode rod and having it emerge near the cathode, below the upper cathode ceramic.

As an added benefit, the above modification permitted the elimination of the internal o-rings in the lower cathode support assembly and the addition of pump-out slots on the stainless steel support tubes.

With these reductions in sources of contamination, we believe evaporation is beginning to compete with sputtering as a principal mode of cathode wear.

Structural integrity/serviceability

Wherever possible, the new source incorporates positive locating techniques to eliminate the need for user alignment. In its current form, the only permissible adjustment is the rotational position of the anode/emitter housing.

The new cathode ceramics, in addition to forming gas seals, also constrain the cathodes to proper concentricity with the anode and collimators, thus completely eliminating the premature, "off-the-edge" failure mode of the earlier source.

The location of cathode emitting surface relative to collimator face is now set by locating shoulders, preventing the gap from becoming too large (increasing the arc voltage) or too small (increasing the cathode wear rate and possibility of shorting).

The lower cathode rod is no longer pinned, but rather it is spring-loaded to provide correct location and positive gas seal at the cathode ceramic.

Results

The new ion source began shipping in early June of this year. Feedback from the field has been overwhelmingly positive. Although the main goal of the redesign was to improve reliability, serviceability and longevity, users are reporting modest improvements in output as well. Typically, users of the RDS-111 require 40 µA bombardments at 11 MeV for up to 2 hours at a time, often on two beamlines simultaneously (80 µA total beam). Modest arc powers of 0.2 A at 800 V for 40 µA, or 0.45 A at 370 V for 80 µA, are typically required to achieve these specifications. Under these conditions, time between rebuilds is now measured in months, rather than weeks. In factory testing, an early prototype of the new source ran for over 220 hours continuously at 0.25 A before a minor short (requiring intervention) ended the test. Cathode wear during the test was negligible. As more of the new sources are commissioned in the field, we will be better able to quantify the lifetime/reliability improvements.

Remote Diagnostics

Readers familiar with particle accelerators will recognize that a large number of dependent subsystems must function very well together to achieve desired daily production levels. The complex nature of the system makes diagnosis of problems difficult. Often, the field service engineer requires detailed assistance from the R&D engineers in the factory, occasionally requiring the R&D engineers to travel to a site on short notice for diagnosis of difficult problems. We initiated a project to study alternative methods of supporting the remote sites from the home office.

The purpose of the RDS-111 Remote Diagnostics Project is to improve the reliability and availability of CTI RDS-111 cyclotrons. This is accomplished by providing high speed real-time access to remote cyclotron sites, coupled with advanced diagnostic software functions to assist with automatic collection and analysis of critical performance parameters. The connection is usually made via an ISDN line and uses NetOp Remote Control (Danware) connectivity software. CTI has a policy in place to govern safe and ethical access of the remote computer, and obtains permission from the local responsible individuals before connecting. Remote dialup capability is protected from outside intervention by several layers of security. Advanced security features exist on the ISDN modem to prevent unauthorized access. The remote control software has additional security that can be turned on for non-ISDN access modes.

The process allows CTI to assess both immediate and long-term cyclotron performance, and to generate predictive maintenance schedules for a particular machine. It also allows CTI to watch trends over several machines. Finally, it is used to reduce the time required to manufacture and test cyclotrons on the plant floor.

There are several other benefits from use of this capability. Periodic software maintenance and/or upgrades can be performed without sending an expert to the site. Moreover, trending and data analysis for predictive maintenance recommendations can be developed. In the future it is expected that remote training will be possible, where one instructor in the factory can instruct many students at remote sites, or provide one-on-one specialized training at a single site. This occurs informally already, as operators at the site can watch while research engineers at the factory perform diagnostics or check parameters. The software is loaded on the audio-visual computer in the conference room at CTI, enabling the entire R&D staff to operate a machine remotely, while having the customer on the speaker phone to discuss alternatives.

Remote Diagnostics generates benefits to both the customer and CTI. For CTI, it enhances our ability to service our customers in a timely manner, saves travel costs for service engineers, and provides invaluable information for developing intelligent programs for preventive maintenance scheduling. For the customer site, it translates into quicker response, fewer technical problems, and dramatic increase in up-time.

Case studies

Two examples of how Remote Diagnostic capability has immediately proven beneficial follow. A site was exhibiting erratic beam tuning. Smith-Garren plots indicated that main magnet behavior was changing as the day progressed. Using remote diagnostics, we tracked cyclotron operational parameters over several weeks. We were able to correlate the behavior to the rise in temperature of the steel. We then found that the steel at the top of the cyclotron was being pushed up by the aluminum vacuum tank as it expanded with increased temperature, thus changing the magnetic characteristics of the machine. The steel top was shimmed appropriately to provide for thermal expansion of the vacuum tank, which fixed the magnet shift problem. This was an unexpected find, and would normally have required several trips made from the factory. Most of the work was performed remotely, and the shimming was a normal service call.

In a second instance, a customer called to report difficulty achieving dual beam. A research engineer contacted the remote site, compared operational parameters against those taken earlier and archived at CTI, and informed the customer that he had inadvertently changed the RF amplitude from the optimum value. This was corrected quickly and the customer was able to make his daily production run.

SUMMARY

Research continues in the direction of higher reliability at CTI. Ion source cathode lifetime, as well as target isolation foil lifetime, are issues of materials science that remain to be fully understood. Software engineering has much to contribute in terms of uptime, reliability, and machine intelligence. These technologies are of at least equal importance to the basic engineering and physics that contribute to accelerator development, and continue to be a focus of work at CTI.

GE PETtrace and Associated Systems, 4 years experience in Cambridge

J C Clark, F I Aigbirhio, P Burke and SPMJ Downey*

University of Cambridge UK(present address) The Mayo Clinic, Rochester MN*

Abstract. In 1994 The University of Cambridge selected a General Electric (GE) PETtrace cyclotron for installation at it's Clinical School site at Addenbrooke's as part of a Positron Emission Tomography (PET) Neuroimaging package.
The cyclotron was installed in a vault with labyrinth access. There is an associated Radiopharmaceutical laboratory with 3 hot cells. Equipment was delivered by GE for the production of [^{18}F]FDG, [^{11}C] Iodomethane and $^{11}C/^{15}O$ gas processing.
We report our experience with the cyclotron and associated equipment with reference to:
1. Cyclotron vault radiation shielding performance.
2. Post irradiation dose rates in the Cyclotron vault
3. Cyclotron and subsystems performance , problems and solutions.
4. Radiosynthetic equipment performance problems and solutions.
5. The Clinical Research program at The Wolfson Brain Imaging Center

INTRODUCTION

The use of Cyclotrons in Medicine has expanded enormously in the past few years. In 1952 The Cyclotron at Hammersmith Hospital London UK was the first to be installed on a hospital site. It was a large and complicated machine requiring a team of engineers and operators to run and maintain it. In the mid 60's the second medical cyclotron was installed at the Mallincrodt Institute of Radiology at Washington University St Louis. It was a much a more compact machine but still required significant manpower. The current commercially available medical cyclotrons designed for use in Positron Emmision Tomography (PET) are even more compact and user friendly. They are negative ion machines with an intelligent computerized control system which can be operated by the user who is usually the Radiopharmaceutical Chemist preparing the radiotracers for PET imaging studies. Maintenance of these machines can be contracted to the manufacturer or sometimes undertaken by local staff in a service partnership arrangement. In Cambridge we have chosen the latter and consequently have needed to learn at lot about the cyclotron and it's associated plant.

THE GE PETTRACE CYCLOTRON

The cyclotron is a dual particle negative ion isochronous sector focussed machine with proton and deuteron energies of 16.5 and 8.4MeV at beam currents up to 110 and 60 microamps respectively. It is equipped with 6 target ports with the possibility of extraction two beams to irradiate two targets simultaneously. Radionuclide productions are fully automated and the control system incorporates an interlock system that handles malfunctions and operational errors in order to protect personnel and equipment. An important service feature is the provision of a portable lap top computer (PETtrace Service System,PSS) which can be connected to the control network for full diagnostic access to all systems. The machine can also be operated in test mode from the PSS.

Cyclotron Vault and Radiopharmaceutical Chemistry Facility

The Cyclotron is installed in a low Sodium concrete vault to reduce neutron activation with access via a labyrinth. The radiopharmaceutical chemistry laboratories are close coupled with the vault as shown in figure1. In order to comply with radiation safety regulations for licensing, neutron and gamma ray doses were measured at the strategic positions around the facility shown in bold capitals on figure 1 and listed in table 1. The laboratory is equipped with 3 lead shielded "hot cells" where automated radiochemical synthetic equipment is housed. A adjoining laboratory houses the radiopharmaceutical quality control equipment and one of the two computer terminals which may be for operating the cyclotron the other being located in the PET scanner control room. All the cyclotron power supplies and the heat exchanger are housed in an adjacent room which has radiation tight penetrations through the labyrinth and vault walls for the interconnecting cables and pipes. The primary water cooling relies on hospital chilled water supplies which are shared with several other users. The facility has a built in radiation monitoring system that measures the gamma dose rates at strategic positions together with gas phase monitoring of radioactivity in the working environment and in the extracts from the vault and hot cells.

Cyclotron Operations

Overall the cyclotron has performed to specification and has met all our requirements for multi user access. As mentioned above we have contracted to maintain the machine ourselves which has worked out very well. However in the course of our maintenance duties we have had to cope with a wide variety of problems some routine and trivial some not so trivial! The vacuum system is very straightforward and has given little cause for concern., neither has the water cooling system. The magnet has been trouble free but the magnet power supply had to be upgraded as a GE factory Modification Instruction (FMI). Unfortunately before the FMI kit had been installed all the silicon controlled rectifiers (SCR's) failed taking the SCR firing boards with them. With new SCR's and drivers the supply has given no further problems. The RF power is generated in two stages before being transmitted to the cyclotron via a large coaxial cable. A driver power amplifier (DPA) feeds the main power amplifier which then drives the cyclotron dee structure. We suffered a rather expensive failure of the DPA which we hope will not happen again when GE have supplied an FMI to eliminate the "weak link" components. The ion source power supply is a standard Glassman unit which has only given us one small problem with the output load resistors which was probably due to accidental RF feedback when components in the ion source RF filter failed. The Accelerator Control Unit, ACU and Chemistry Control Unit CCU make use of Single Board Computers (SGC's) have given little cause for concern. The SUN Sparc Master control computer has been generally trouble free and the associated software has been kept up to date by GE. As we have gained operational experience we have fed back to GE information about specific desirable software enhancements as have other users. In the cyclotron the major items in need of service are the ion source and the targets. Access to the ion source is very straightforward after the vacuum has been turned off and the magnet door opened. The machine, having a vertical median plane, leaves the ion source readily accessible at about shoulder height. We have a complete spare ion source assembly and keep it "ready to go" with cleaned anodes, new Tantalum cathodes and insulators. The changeover takes about 10 minutes followed by a vacuum pump-down of about 30 minutes back to operating levels. The dee structure has been trouble free and any Centre region adjustments are carried after cleaning are carried out with due care and attention! The stripper foil extraction systems have also been generally trouble free with what sees to be almost infinite carbon foil lifetime. Two carousels of foils permit the dual extraction of protons or deuterons. On the target systems the Helium system for the beam exit and target entrance foil cooling has needed significant attention mainly with the plastic and rubber components in the pipes, valves and membrane pump. As with most cyclotron targets, preventative maintenance, which is essential to maintain optimal output and product quality, can give cause for radiation dose concerns for the operator. The main problem comes from the metal foil windows but with pre-planned cooling times we are not encountering any radiation dose limits on our staff. The design of the quick sealing target interconnects allows any target to be removed in less than 10seconds. Automatic beam tuning on target uses beam current readings from collimators above and below the beam exit port. As part of our development of safe systems of work we have measured the radiation doses around the cyclotron and have found little cause for concern other than the target foils mentioned above.

TABLE 1. Radiation dose survey data around the perimeter of the cyclotron vault

Location	Gamma dose rate micro Sv/hr	Neutron dose rate MicroSv/hr	Cyclotron proton beam current on nitrogen filled ^{11}C target microA
A Vault door to labyrith	1.5	0.25	40
B gas cylinber store	<0.3	0	40
C Door to service void	1	0	40
D Centre of vault roof (plant room)	4	0.2	40
E By Plant room steps	0.2	0	40
F At the side of plant room	0.2	0	40

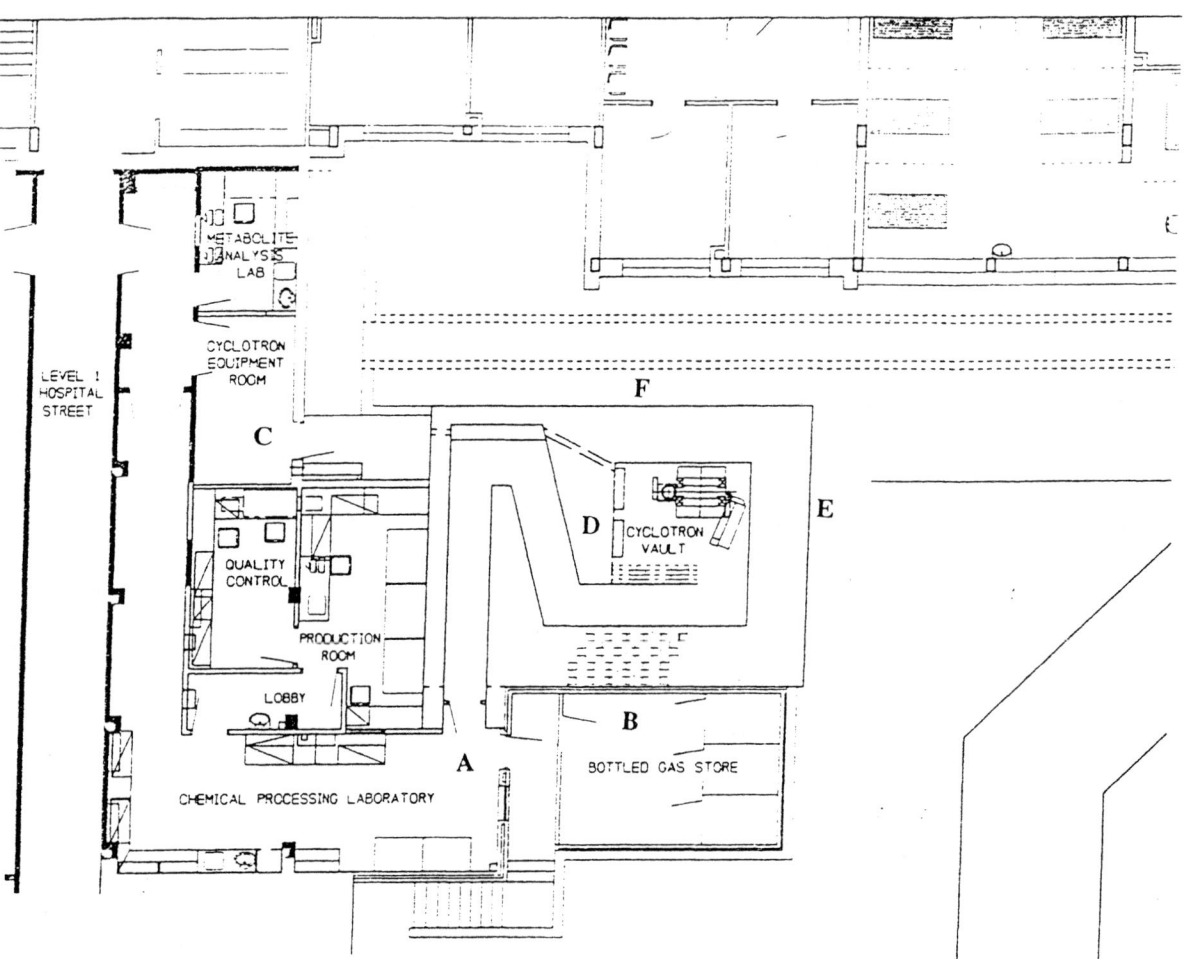

FIGURE 1. Wolfson Brain Imaging Center Cyclotron and Laboratory.

Chemistry Operations

We have three GE supplied pieces of chemistry equipment.

1. FDG Microlab is a device for the automated synthesis of 2-[^{18}F]fluoro-2-deoxyglucose from the ^{18}F (half life 110 mins)fluoride generated in the cyclotron target. The synthesis is based on a solid phase reaction[1] process which works reasonably well at intermediate levels of radioactivity but is not amenable to scale-up. GE have developed a new version of the device which uses the same hardware but can cope better with higher levels of radioactivity. Both systems have the pharmaceutically desirable feature of a single use disposable cassette.

2. Methyl Iodide Microlab is a novel device for converting $^{11}CO_2$ (^{11}C half life 20 mins) into the important radiolabeling precursor [^{11}C] methyl iodide[2]. It operates by catalytically converting $^{11}CO_2$ to [^{11}C]methane which then reacted with iodine vapour at high temperatures to generate the product at exceptionally high specific radioactivities (mCi/micromole) compared to the classical approach using wet chemistry.

3. The gas processing cabinet is a lead shielded unit which houses all the chemical conversion and purification devices that enable the various gaseous products that are recovered from the targets to be converted to other desired chemical forms. With the exception of the ^{15}O (^{15}O half life 2 mins) processing system we have found the systems adequate for our needs. As originally configured the ^{15}O processing system could only make one chemical form of labeled product at a time with an unacceptable time delay for the change between products. For our clinical research using ^{15}O we needed three forms $^{15}O_2$, $C^{15}O$ and $H_2^{15}O$ simultaneously. We have solved this problem by providing multiple take-offs from the target outlet and assembling additional gas flow controls and processing components in the process cabinet. In addition we have provided a device located in the PET scanner room that synthesizes $H_2^{15}O$ ready for automatic injection into the patient as and when required.

The CLINICAL IMAGING FACILITY

Most of our clinical research is directed towards studying patients with acute brain injury. Consequently our PET scanner is sited very close to the Neurointensive care ward. This ensures that transfer distances for patients are minimized. There is also a 3T MRI in the facility far enough away from the PET scanner to avoid magnetic interference but again close enough to make transfers distances short. We use the ^{15}O tracers mentioned above together with ^{18}F FDG to evaluate regional metabolism in the brain and the effects of clinical interventions and therapies. The MRI scanner is used for anatomical registration of the PET data as well as many functional methods eg.fMRI.

ACKNOWLEDGMENTS

The clinical research program at The Wolfson Brain Imaging Centre is supported by the UK Medical Research Council and Smith Kline Beecham through The Forsight Challenge Award Scheme.

REFERENCES

1. Toorongian SA Mullholland GK Jewett DM Batchelor MA and Kilbourn MR Nucl Med Biol (1990) 17(3) 273-9.

2. Larson P Ulin J and Dahlstrom K J LablComp Radiopharm (1995) 37 73-75.

A Multi-Run Chemistry Module For The Production Of [^{18}F]FDG

B. Sipe, M. Murphy, B. Best, S. Zigler, J. Lim,* E. Dorman,* T. Mangner,‡
M. Weichelt†

CTI, Inc., Knoxville, TN 37932
*The Queen's Medical Center, Honolulu, HI 96813
‡Children's Hospital, Detroit, MI 48201
†PETNet Pharmaceutical Services, Omaha, NE 68108

Abstract. We have developed a new chemistry module for the production of up to four batches of [^{18}F]FDG. Prior to starting a batch sequence, the module automatically performs a series of self-diagnostic tests, including a reagent detection sequence. The module then executes a user-defined production sequence followed by an automated process to rinse tubing, valves, and the reaction vessel prior to the next production sequence. Process feedback from the module is provided to a graphical user interface by mass flow controllers, radiation detectors, a pressure switch, a pressure transducer, and an IR temperature sensor. This paper will describe the module, the operating system, and the results of multi-site trials, including production data and quality control results.

INTRODUCTION

As the production of [^{18}F]FDG moves from the hospital setting into commercial distribution centers, the need to produce larger and larger quantities of [^{18}F]FDG is readily apparent. Distribution centers must accomplish this task with high reliability and without excessively increasing the radiation exposure of operators. Pursuant to these goals, we have developed a new chemistry module for the production of up to four batches of [^{18}F]FDG in a single set up. This paper describes the process, hardware and software utilized by the module. In addition, we discuss the initial data obtained from multi-site trials, including production and quality control results.

PROCESS DESCRIPTION

The multi-run [^{18}F]FDG module relies on the proven Hamacher method for the production of [^{18}F]FDG. [1,2] Prior to starting a batch sequence, the module automatically performs a series of self-diagnostic tests, including a reagent detection sequence. The module then executes a user-defined production sequence followed by an automated process to rinse tubing, valves, and the reaction vessel prior to the next production sequence. Process feedback from the module is provided to a graphical user interface by mass flow controllers, radiation detectors, a variety of switches, transducers, and temperature controllers.

HARDWARE DESCRIPTION

The [^{18}F]FDG module measures approximately 36 cm (14") wide x 49 cm (19") high x 41 cm (16") deep. The front of the module contains the reaction vessel heating/cooling assembly and miscellaneous gauges. The left side of the module houses the [^{18}F]fluoride ion trap-and-release columns, the [^{18}F]FDG purification columns, and waste collection vials. The back of the module has connections for compressed air, cables, inert gas, and traps for gaseous radioactive by-products. The [^{18}F]FDG module has a removable cover for the top and back of the module. Removal of this cover reveals another control panel for gauges, valves, and liquid detectors. This control panel and the right side panel have hinges that allow the module to be opened for service.

The [^{18}F]FDG module consists of several subsystems: (a) a reagent delivery subsystem, (b) a reaction vessel heating/cooling subsystem, (c) a radiation detector subsystem, (d) a pneumatic gas delivery subsystem, and (e) a pressure integrity test subsystem. Each subsystem is described below.

Reagent Delivery Subsystem

To accommodate multi-run capability, the reagent delivery system dispenses reagents and solutions from reservoirs, which consist of septum-sealed, glass reagent vials. The system dispenses different volumes by changing the depth of a needle used to the remove the liquid. In principle, this dispensing method has only two sources of error in the volume of delivered reagent: error in the diameter of the vial and error in the vertical position of the needle. The design minimizes the first source of error by specifying commercially-available vials with tight internal diameter tolerances. The second source of error is minimized by accurately controlling the needle position with a linear actuator mechanism (see below).

Liquids are transferred from the vial through a needle. A second needle supplies the gas over-pressure needed to transfer the liquid. In practice, the two needles are arranged concentrically. Gas enters the vial through the annulus between the inner and outer needles, while liquid exits through the inner needle (Figure 1).

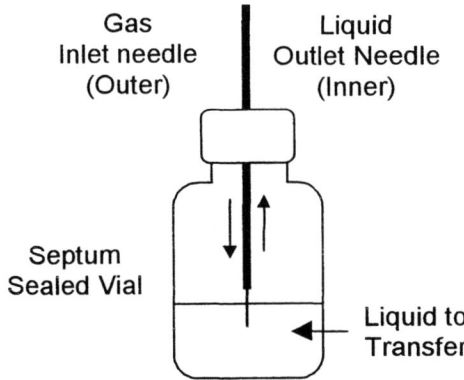

FIGURE 1. The concentric needle design for delivery of reagents and solutions.

The concentric needle design offers several advantages over a parallel needle design. The concentric design allows the use of a small gauge needle for the liquid delivery. This reduces the effect of sudden pressure changes and thus provides more control during the liquid delivery. The increased structural integrity provided by the larger gas needle dramatically reduces the possibility of bending the liquid needle. In addition, the smaller gauge needle forms a "pilot" hole for the larger gauge needle, which reduces the incidence of septum coring. Thus, the design simultaneously allows the use of a non-bending, small gauge liquid needle (to better control liquid delivery) and a reduced-coring, large gauge gas needle (to provide structural integrity).

Linear Actuator

The needles are mounted on a head which, in turn, mounts to a screw shaft linear actuator. The rotation of the shaft moves the head up or down to the desired vertical position. A belt-driven, rotating potentiometer is coupled to the screw shaft to provide vertical position feedback for the needle head. Power to the drive motor and the feedback from the potentiometer are interfaced to the computer/control system to allow accurate positioning of the needles.

Three locations are defined for the linear actuator: top, bottom and intermediate. The top position defines the upper range of motion for the arm, while the bottom position defines the lower range of motion. The intermediate position defines the approximate lowest position where the reagent carousel can rotate without obstruction by the needles. Limit switches are present at each position to provide feedback for the computer/control system. It is possible to position the arm at any location between the top and bottom position on the linear actuator.

Rotary Carousel

A variety of reagents and solutions are necessary for the production of [^{18}F]FDG. In addition, solvents are necessary for cleaning the apparatus between production runs. To accommodate this requirement with a single needle head, a rotating reagent carousel is employed to position the desired reagent under the needle head. The rotating carousel has ten reagent positions on its outer ring and five positions on its inner ring, thus providing space for up to fifteen reagent vials. Separate needles are used to access the reagents on the inner and outer rings. The vials may be large volume (27 ml) or small volume (10 ml), and are securely mounted in individual vial "slots" with a removable top plate. The carousel rotates on a pneumatically driven mechanism that has positional feedback for the computer/control system. A spring loaded detent assures accurate positioning of the carousel. The carousel is readily removable to allow access to the reagent vials.

Liquid Detectors

The liquid outlet needle is connected to small-bore flexible tubing (e.g., 1/16" o.d. PFA) to route the liquid during the transfer process. To ensure the successful transfer of liquid, the reagent delivery system employs ultrasonic detectors that sense the presence of liquid in the tubing. The detectors do not directly contact the liquid, which eliminates the possibility of reagent contamination and detector corrosion.

The detectors allow the operator to fill the reagent vials in the carousel with any volume of liquid, and then perform an "auto-detect" sequence to determine the quantity of liquid in the vials. This eliminates the need to accurately measure the volume when filling the reagent vials, thus expediting the set up process.

Reaction Vessel Heating/Cooling Subsystem

The heating/cooling block assembly has two temperature zones that separately cool and/or heat the upper and lower portions of the reaction vessel. This allows the execution of chemical processes over a temperature range up to 200 C. Heat is supplied to each temperature zone by the passage of compressed air over two resistively heated rods. The temperature of each heated zone is controlled by a temperature controller. Both zones may be independently cooled by compressed air vortex coolers.

The reaction vessel is a commercially available, borosilicate test tube with a threaded top. The vessel is screwed into a reactor head assembly, which seals against the top of the vessel with an o-ring. The reactor head assembly provides a conduit for the penetration of tubing in and out of the reaction vessel. The reaction vessel is closed to prevent the release of radioactive materials during processing. A retractable tube enters through the center of the reactor head. In the fully down position, the bottom of the tube reaches the bottom of the reaction vessel. This position is used to deliver solutions from the reaction vessel, and to bubble gas into the reaction vessel. In the fully up position, the bottom of the tube reaches the approximate middle of the reaction vessel. This position is used to deliver reagents to the reaction vessel. A small o-ring in the reactor head assembly creates a positive seal against the retractable tube.

An infrared temperature sensor measures the temperature of the bottom of the glass reaction vessel during processing. The sensor provides real-time temperature measurements of the reaction mixture, and is non-contact to prevent contamination of the reaction mixture. The sensor allows direct feedback of temperature shifts resulting from changes in process conditions (e.g., the end of an evaporation, etc.).

Radiation Detector Subsystem

The radiation detector subsystem has four detectors, one for each possible [^{18}F]fluoride ion resin cartridge. One resin cartridge is used for each batch of [^{18}F]FDG. Each detector is a photodiode with an integral cesium iodide scintillator that is housed in a 2 cm (0.8") diameter enclosure with a BNC connector. The analog output of the detector control board is 0 to 10 V proportional to 0 to 5 Ci (at 3 cm). Since only one of these four detectors is used at a time, the analog signals are multi-plexed so that only the signal of interest is available as feedback to the computer control system. A fifth radiation detector is located at the bottom of the reaction vessel. Each detector may be calibrated to provide output in mCi or GBq.

This subsystem has inserts that allow for the attachment of the anion exchange cartridges. The inserts are designed to match the cartridge and provide reproducible placement of the cartridge in front of the radiation detector.

Commercial[3] or homemade anion exchange resin cartridges may be used in the [^{18}F]FDG module. The configuration of the cartridge should be selected to maximize the [^{18}F]fluoride ion trapping efficiency. The release efficiency should be optimized with the solution used to release the [^{18}F]fluoride ion, especially the volume of the solution and the quantity of potassium carbonate. This solution must also be suitable for use later in the [^{18}F]FDG production process. The ionic form of the anion exchange resin plays a critical role in the ion exhange process, as well as subsequent [^{18}F]FDG processing. The carbonate (CO_3^{2-}) or bicarbonate (HCO_3^-) forms of the resins are acceptable, but the chloride ion form should be avoided because this leads to the production of 2-chloro-deoxyglucose.[4]

Pneumatic Gas Delivery Subsystem

The pneumatic gas delivery subsystem provides pressurized gas to transfer liquids and solutions throughout the [^{18}F]FDG module, as well as to perform bubbling during solvent evaporations. The nitrogen gas is delivered through electronic mass flow controllers, which are interfaced to the computer/control system to allow remote gas flow setpoints and feedback. The flow controllers are accurate to within less than 1 sccm.

A compressed air manifold supplies air to the reaction vessel heaters and coolers, to the pneumatically-actuated piston that drives the retractable tube in the reaction vessel, and to the pneumatically actuated reagent carousel. A normally-open valve controls the flow of air to the manifold, thereby allowing testing of the compressed air subsystem. A pressure switch attached to the manifold allows these tests to be performed automatically.

Pressure Integrity Test Subsystem

The final step in the production of [^{18}F]FDG is the membrane filtration of the product as it enters the final product vial. In order to ensure complete removal of bacteria, it is necessary to maintain the integrity of the membrane during the filtration process. The integrity of the filter may be assured by testing the wetted membrane filtration device after completing the filtration process. The pressure retention test is a commonly used integrity test for membrane filters.[5] In this test, the pressure of the gas on the wetted membrane is set to a point just below the bubble point. After initial pressurization, the supply of gas is removed, and the pressure monitored to determine if the membrane "holds" pressure.

The [^{18}F]FDG module utilizes an automatic pressure integrity test that reduces manipulation of the membrane filtration device, thereby reducing radiation exposure to the operator. The method, which is based on the pressure retention method, uses an electronic pressure transducer to measure the pressure on the membrane filter and to send this information to the computer control system. The transducer is attached to the gas inlet for the integrity test, which in turn is attached to the sidearm of a rotating 3-way stopcock. In one position, the stopcock allows the final product to flow through the membrane filtration device, while in the other position, the stopcock allows the pressurized gas to enter for the pressure integrity test. A vented filter is attached to the upstream side of the 3-way stopcock to vent air bubbles during the final product transfer. A non-vented filter is attached to the downstream side of the 3-way stopcock. The non-vented filter is tested by the pressure integrity test. Note that, if the test for the non-vented fails, the vented filter provides a back-up. A diagram of the apparatus appears in Figure 2.

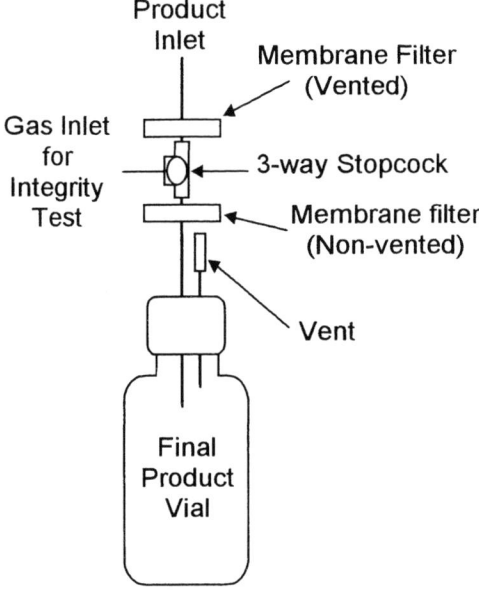

FIGURE 2. Apparatus used for automated filter hold test.

SOFTWARE DESCRIPTION

The multi-run [^{18}F]FDG module is controlled by a computer operating under the Windows® NT operating system. The control software consists of graphical pages that allow automated and manual operation of the [^{18}F]FDG module. In addition, the software provides access to synthesis parameters, hardware calibration values, historical data, and an electronic batch record for each batch.

The main page for the [^{18}F]FDG module appears in Figure 3. The various hardware components are represented graphically on this page. For example, solenoid valves appear as small circles and distribution valves appear as polygons. Animation effects change the appearance of hardware components depending on feedback from the linear actuator, flow controllers, etc. Navigation buttons for different chemistry modules appear on the left side of the page and automated task buttons appear on the right. The automated task buttons execute command files to initialize and set up the [^{18}F]FDG module. Similar command files are used in the production of [^{18}F]FDG. Individual valves, relays, and analog signals may be controlled from the main page or from a secondary text page. The secondary text page provides access to pages for synthesis parameters, hardware calibration factors, and a reagent definition table.

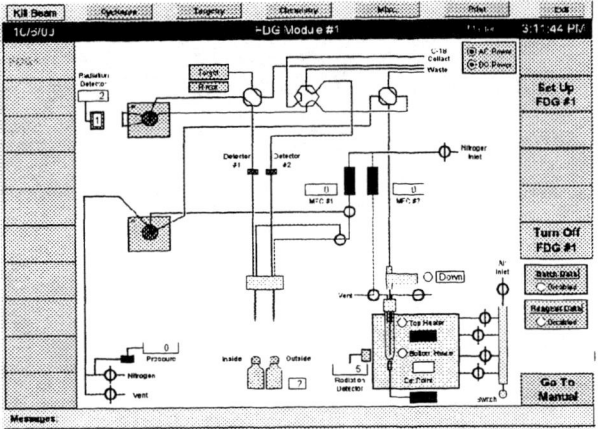

FIGURE 3. The main graphic page for the multi-run [^{18}F]FDG module.

RESULTS

Infrared Sensor Vs. Internal Thermocouple

To determine the sensitivity of the IR sensor to changes in the temperature of solutions in the reaction vessel, we performed experiments to compare temperature measurements from the sensor with those from an internal thermocouple. Figure 4 shows the temperature profile for the evaporation of 1 ml of acetonitrile. The top curve is the temperature measured by the IR sensor, and the bottom curve is the temperature measured by the internal thermocouple.

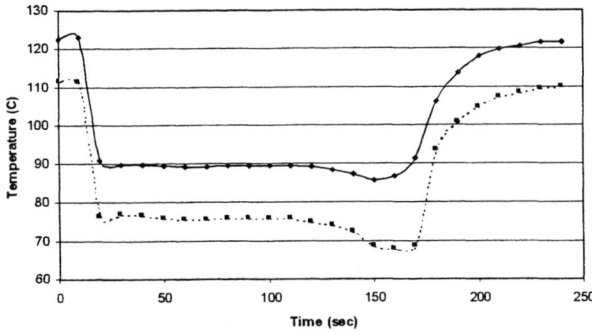

FIGURE 4. Temperature profile for the evaporation of acetonitrile measured by an internal thermocouple and the external IR sensor.

In Figure 4, a temperature increase indicative of complete evaporation is evident at approximately 175 sec. The similarity in the two profiles demonstrates that the sensitivity of the IR sensor is sufficient to establish the end of the solvent evaporation. This allows the control software to terminate an evaporation based on temperature instead of time.

Reproducibility of Evaporations

The reproducibility of solvent evaporations, especially in the preparation of [^{18}F]fluoride ion, is important to ensure consistent batch-to-batch results. Figure 5 illustrates a composite temperature profile of repeated acetonitrile evaporations. The error bars for each data point represent one standard deviation obtained from the average of three determinations. The region of greatest dispersion in the data occurs when the temperature rapidly rises at the point of complete evaporation. Although not illustrated in Figure 5, a similar region of variability occurs in experiments performed with the internal thermocouple, indicating the region is not an artifact resulting from use of the external IR sensor. Together, these results demonstrate the difficulty in predicting the exact moment of complete evaporation and highlight the utility of real-time temperature measurements of the solution.

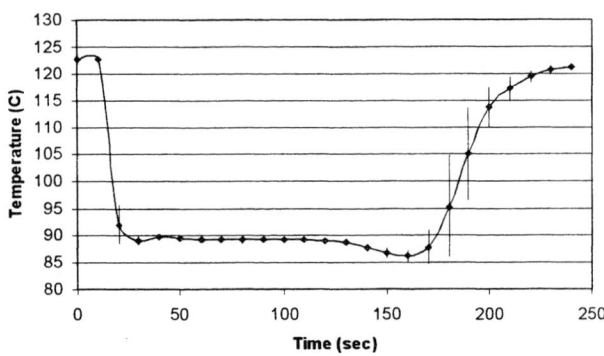

FIGURE 5. Composite temperature profile for repeated evaporations of acetonitrile. The vertical error bars represent one standard deviation.

Reproducibility of Reagent Additions

The reproducibility of reagent additions is also important to ensure consistent batch-to-batch results. The reproducibility was measured in two ways. First, the ability to accurately and reproducibly position the linear actuator was measured with a calibrated micrometer. The results of these measurements indicate that the position of the linear actuator arm may be reproducibly set and controlled to within less than one millimeter. Second, the reproducibility of liquid deliveries was tested with acetonitrile (AcN). Figure 6 is a plot of the amount of acetonitrile delivered as a function of the vertical position of the linear actuator arm. The vertical error bars represent the standard deviation in the average of all repetitions (n = 10). In all cases, the relative standard deviation (σ/average) is less than 5%, which demonstrates that this method of reagent addition is suitable for use in the production of [^{18}F]FDG.

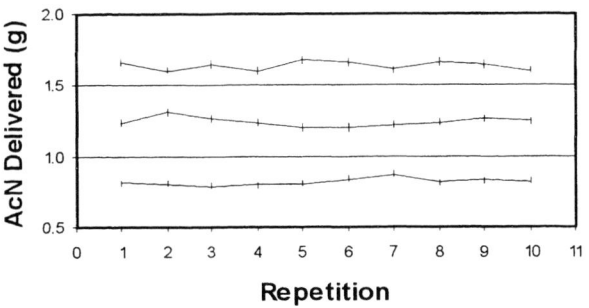

FIGURE 6. Repeated delivery of acetonitrile (AcN) for different linear actuator arm movements. The upper, middle and lower data sets correspond to a movement of 80, 60 and 40 analog counts, respectively. The vertical error bars represent one standard deviation in the average of all repetitions (n = 10).

Historical Data

The feedback built into the [^{18}F]FDG module provides real time access to a variety of process parameters, including gas flow rates, radiation levels, and temperature measured by the IR sensor. In addition, the control software logs the value of these parameters to permit a review of historical data. An example of the historical data available from the [^{18}F]FDG module is illustrated in Figure 7, which shows the changes in these parameters during a typical production run. This data is useful for monitoring historical production trends, as well as trouble shooting production problems.

Production and Quality Control Data

The initial production results from three different beta sites have given decay corrected yields of [^{18}F]FDG in the range of 55 to 85%. After optimizing production parameters, we expect the decay corrected yields to stabilize in the range of 70 to 80%. The synthesis time in the beta tests have ranged from 32 to 36 minutes. Quality control results have been encouraging, with product consistently meeting or exceeding the USP requirements for radiochemical purity, radionuclidic purity, chemical purity, bacterial endotoxins and sterility. Further work will focus on yield optimization, reducing the synthesis time, and developing validation data for regulatory authorities.

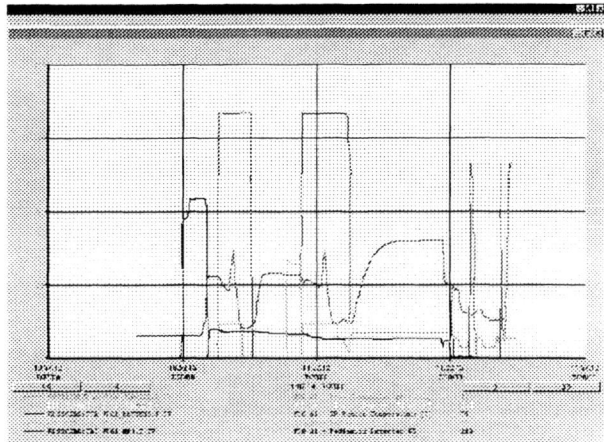

FIGURE 7. An example of historical data available from the multi-run [^{18}F]FDG module. The various lines represent values for radiation detectors, nitrogen gas flow rates and the temperature of the reaction vessel during a typical [^{18}F]FDG production run.

REFERENCES

1. Hamacher, K., Coenen, H.H., and Stöcklin, G., *J. Nucl. Med.* **27**, 235-238 (1986).

2. Padgett, H.C., Schmidt, D.G., Luxen, A., *et al.*, *Appl. Radiat. Isot.* **40**, 433-445 (1989).

3. Watkins, G. L., Richmond, J. C. W., Koeppel, J. A. Hichwa, R. D., "Off-the-Shelf Anion Exchange Devices: Are they Useful in Trapping/Releasing [^{18}F]Fluoride Ion?" *J. Nucl. Med. Suppl.* **39**, 239-240P (1998).

4. Alexoff, D.L., Casati, R., Fowler, J.S., *et al.*, *Appl. Radiat. Isot.* **43**, 1313-1322 (1992).

5. Sterilizing Filtration of Liquids (Technical Report #26), *J. Pharm. Sci. Technol.* **52**, 24-26 (1998).

The Plasma Separation Process As A Pre-Cursor For Large Scale Radioisotope Production

Nigel R. Stevenson

Theragenics Corporation
5203 Bristol Industrial Way, Buford, GA 30518

Abstract. Radioisotope production generally employs either accelerators or reactors to convert stable (usually enriched) isotopes into the desired product species. Radioisotopes have applications in industry, environmental sciences and most significantly in medicine. The production of many potentially useful radioisotopes is significantly hindered by the lack of availability or by the high cost of key enriched stable isotopes. To try and meet this demand, certain niche enrichment processes have been developed and commercialized. Calutrons, centrifuges and laser separation processes are some of the devices and techniques being employed to produce large quantities of selective enriched stable isotopes. Nevertheless, the list of enriched stable isotopes in sufficient quantities remains rather limited and this continues to restrict the availability of many radioisotopes that otherwise could have a significant impact on society. The Plasma Separation Process is a newly available commercial technique for producing large quantities of a wide range of enriched isotopes and thereby holds promise of being able to open the door to producing new and exciting applications of radioisotopes in the future.

INTRODUCTION

Stable enriched isotopes are frequently used in the production of radioisotopes. Typically, these materials are either bombarded with protons (or sometimes with deuterons, alphas, etc.) from accelerators or immersed in neutrons within reactor cores. The corresponding transmutation reactions produce the proton-rich or neutron-rich isotopes, respectively.

ISOTOPE PRODUCTION CONSIDERATIONS

Depending on the radioisotope production reaction the specific target isotope that is required may pose unique problems:

1. Target isotope is mono-isotopic (e.g., Be-9, F-19, Al-27, Rh-103, etc.) or very close to this (e.g., N-14 [99.6%] & N-15 [0.4%], V-51 [99.8%] & V-50 [0.2%], etc.). This is the most favorable scenario where isotope enrichment is unnecessary. Competing reactions from other isotopes that could result in the production of impurities are a non-issue. Competing *proton* reactions from the mono-isotopic target isotope may still cause difficulties e.g., (p,n) vs. (p,2n) etc.

2. Target isotope is one of several stable isotopes (e.g., Tl-203 [30%] & Tl-205 [70%], O-16 [99.76%] & O-17 [0.04%] & O-18 [0.2%], etc.) that usually requires separation or enrichment:

   - If the natural enrichment of the required target isotope is high and the competing reactions do not produce contaminants that cannot be removed – then further isotopic enrichment may not be required. For example, the production of ^{64}Cu (half-life = 12.7 hrs) via ^{63}Cu(n,γ)^{64}Cu can proceed in a reactor with natural Cu even though there are two (sizeable) natural isotopes - ^{63}Cu [69%] & ^{65}Cu [31%]. This is because the competing product (^{66}Cu) has a relatively short half-life (5 min).

- If competing reactions are a problem then enrichment may be required. For example, in the production of ^{123}I, one possible reaction is ^{123}Te(p,n)^{123}I. However, competing reactions such as ^{124}Te(p,n)^{124}I produce longer-lived, higher energy radioactive contaminants if natural Te were used as a target material. For this reason, highly enriched ^{123}Te must be used [1]. Since this requirement is expensive and the reaction is not convenient for most commercial accelerator energies - usually alternative reactions are selected [2] to produce this isotope.

3. The corresponding target isotope may not be stable or it may have such an extremely small natural abundance that it effectively prohibits its use for this purpose. The desired radioisotope may still be produced with spallation reactions or via fission chains. Extremely low yield reactions, such as double neutron absorption in reactors, are on rare occasions considered to produce the desired isotope if no other alternative is available.

ISOTOPIC ENRICHMENT

Several methods for isotopic enrichment exist that lend themselves to the specific characteristics of different element. These include calutrons, gaseous centrifuges, laser, diffusion, distillation and, most recently, plasma separation.

Commercial Enrichment Processes

Large scale production of enriched stable isotopes have been undertaken by governmental agencies and by commercial companies:

1. *Calutrons*: Large electromagnetic separators. These systems have been operated by governmental agencies in USA and Russia for several decades. These aging systems are expensive to operate and are being phased out in favor of modern efficient alternatives. Calutrons are capable of separating almost any element into its constituent isotopes.

2. *Gaseous Centrifuges*: Elements or compounds that are gaseous are candidates to be separated by this technique. Commercial facilities exist in the Netherlands and Russia. Low enriched ^{235}U for power reactor fuel is the main use of this technique – although several other isotopes can be produced on a commercial scale (Xe isotopes, ^{112}Cd, etc.).

3. *Others (Lasers, Cryogenics Distillation and Gaseous Diffusion)*: Techniques that may work well for (predominantly) light gaseous elements or compounds e.g. ^{18}O.

Plasma Separation Process

The Plasma Separation Process (PSP), invented several decades ago by John Dawson [3], employs the following steps (see Fig. 1):

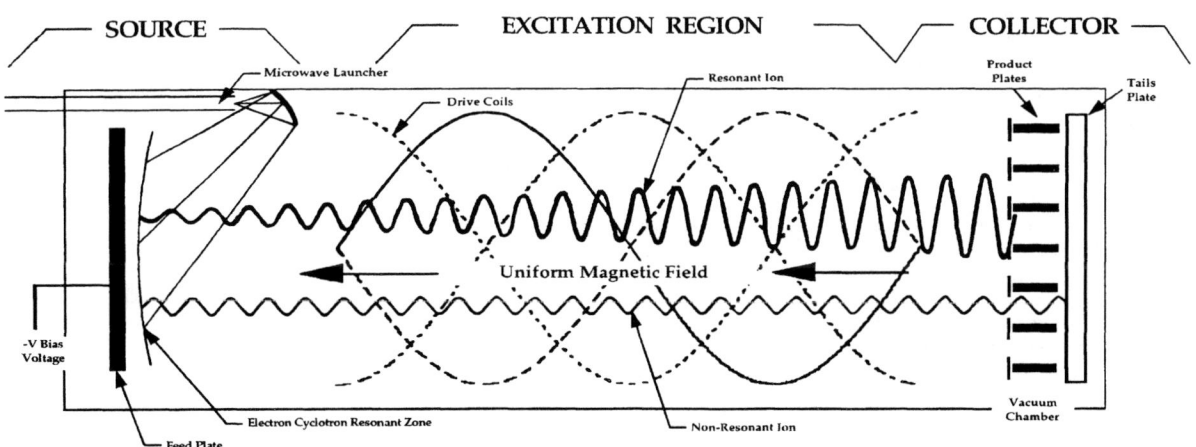

Figure 1. Plasma Separation Process

I. The feed target material (usually metallic) is vaporized by sputtering with microwaves.

II. A plasma region is produced by microwave heated electrons.

III. Ions enter the region of the solenoidal magnetic field where oscillating electric fields are applied.

IV. Resonant ions gain energy and travel in a helical path of increasing radius. Non-resonant ions also travel down the bore of the magnet but are generally out of phase and do not gain energy.

V. Specially designed collector plates are arranged to selectively catch the energetic (resonant) ions while allowing low-energy (non-resonant) ions to pass through to the tails plate.

The prototype device, constructed by governmental agencies several years ago, was recently leased to a commercial company (Theragenics Corporation) who are setting up a production facility in Oak Ridge, TN. The initial intent of this project is the production of ^{102}Pd as a pre-cursor to the production of ^{103}Pd in a reactor. However, this PSP can produce large quantities (typically several kg-tonnes/yr) of almost any enriched isotope between Ca and Bi.

DISCUSSION

The potential advantages of using stable isotopes in large quantities for industrial applications has been recognized [4] e.g., ^{10}B, ^{7}L and depleted ^{64}Zn (several tonnes/yr) in the nuclear industry; even numbered Cd isotopes for HeCd lasers; ^{28}Si to enhance semiconductor thermal conductivity, etc. With the PSP becoming available for commercial production - stable isotopes that are used for the above applications as well as feed materials for radioisotope production can now be produced economically in sizeable quantities. This fact may, in turn, open the way for more prolific use of these products in medicine, environmental sciences and industry.

The opportunities for beneficial uses of radioisotopes in society for medical and other applications is enhanced with the availability of the PSP to produce the feed materials needed to produce these. The availability of large quantities of enriched stable isotopes now opens the way for exciting applications that previously may have been overlooked or rejected as being impractical.

ACKNOWLEDGEMENTS

The author would like to acknowledge the assistance of Mr. Ron Warren in producing this document.

REFERENCES

1. Barrall, R. C. et. al., Eur. J. Med. **6**, 411-415 (1981).

2. Tarkanyi, F., et. al., Appl. Radiat. Isot. **42**, No. 3, 221-228 (1991).

3. Dawson, J. M., et al., Phys. Rev. Lett. **37**, 1547-1550 (1976).

2. "Beneficial Uses and Production of Isotopes", published by the Nuclear Energy Agency of the Organization for Economic Co-Operation and Development, OECD publications (1998).

Thermal Performance of CTI, Inc. Enriched Water Targets

Arthur E. Ruggles[*], Charles W. Alvord[**]

[*]*Nuclear Engineering Department, University of Tennessee, Knoxville, TN 37996 (aruggles@utk.edu)*
[**]*CTI, inc., 810 Innovation Drive, Knoxville, TN 37932-2571*

Abstract. Oxygen 18 enriched water targets are bombarded with 10.5 MeV protons to produce Fluorine 18 in cyclotron targets manufactured by CTI, inc. A thermal model of the target and target holder is developed that allows the relationship between the beam power applied to the target and the vapor volume produced in the target to be predicted. The model is compared with data on vapor volume production during bombardment for a range of beam power. The data agree well with predictions from the thermal model.

INTRODUCTION

The dynamics of heat transport in small volume, high-pressure (>15 bar) enriched water targets have been studied and presented previously. As the business of producing [^{18}F]Fluorodeoxyglucose (FDG) grows, the understanding of the thermal process in typical enriched water targets (~1 cc total volume, bombardment times of 1-3 hours, beam power of 400 to 1000 watts) is important to a growing number of production facilities and the manufacturers that provide equipment to them. Basic engineering equations have been applied to the low pressure (~ 1 bar) problem[1], yielding good correlation of predicted and observed temperatures vs. power. More recently computational fluid dynamics codes have been used to model the high pressure single and two-phase flow in steam and water targets[2,3]. This work represents a return to simpler methods, utilizing engineering formulae to predict behavior at elevated pressure.

Water in the target must change phase to accommodate the energy deposition from the proton beam. Efficient Fluorine production requires that liquid phase enriched water remain in the path of the proton beam throughout the target exposure time. The expense of enriched water encourages minimum target inventory. In the standard CTI target design, the operating pressure is maintained with Argon cover gas. Volume increase attributable to gas and vapor production during target bombardment is accommodated by the cover gas volume. Knowledge of the amount of vapor volume generated in the target during bombardment is important to sizing the target load volume and the target cover gas volume such that the desired operating conditions are sustained. Basic thermal performance attributes of the CTI Oxygen 18 enriched water target are simulated using integral control volume methods. The ultimate heat sink for the thermal power deposited in the target is process water circulated around the outside of the Silver target holder. The volume fraction of vapor in the target holder during bombardment is predicted as a function of the beam power. These results are in agreement with data taken from a target subjected to a range of beam powers. The simulation offers a tractable tool to support the design of targets for future CTI cyclotron systems.

THE TARGET VOLUME THERMAL MODEL

The target volume is modeled as a circular cylinder of radius 0.0059 meters and 0.008 meters in length, as shown in Fig. 1. The forward face of the cylinder faces the beam and is defined by a beam window. The beam window is cooled by Helium external to the target volume and this face is taken to be adiabatic in this simulation. The remaining surfaces of the target volume are adjacent to the silver target holder. The thermal performance of the target holder is considered separately in the next section.

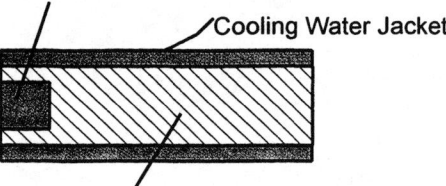

FIGURE 1. Schematic of Water Target Volume in Silver Target Holder

It is instructive to first examine some basic performance attributes of the target in order to appreciate relationships between beam power, target volume average volumetric heating, and target surface average heat flux. The target pressure is taken as 682 psia and the thermal power absorbed by the target is 400 watts. The target volumetric heating is evaluated by dividing the beam power by the target volume. The target average volumetric heating is 460 MW/m^3, which is just over half of the peak volumetric heating of 800 MW/m^3 expected in the Spallation Neuton Source currently being developed by the U. S. Department of Energy. The average surface heat flux is evaluated by dividing the beam power by the heat transfer area, where the heat transfer area is the area of the cylinder including just one end. The average surface flux at the interface between the target and the target holder is 1.0 MW/m^2. As a reference, the heat flux that causes a transition from nucleate boiling to film boiling (so-called critical heat flux) in water at one atmosphere is roughly 1.0 MW/m^2.

It is also useful to estimate the maximum power the target can absorb without creating vapor. A representative heat transfer coefficient for single phase natural convection in an enclosed volume similar to the target volume is 400 W/m^{2}C. The saturation temperature for the target water is 260 C, and the minimum temperature in the system is the heat sink process water which is taken as 40 C in this simulation. These assumptions allow for an estimation of the maximum power the target volume can reject without creating vapor as 33 Watts. Note that vapor production is expected at lower power due to the peaking of the beam energy deposition in the forward center of the target volume. The peak volumetric energy deposition causes rapid volumetric heating which would lead to vapor production prior to the power level predicted above.

The evaluation of the target thermal performance goes forward assuming the target volume is horizontally stratified, with the top volume of the target filled with vapor and the bottom filled with liquid. The entire target volume is assumed to be near saturation. The heat transfer between the portion of the target volume filled with vapor and the target holder is evaluated using models for laminar falling film condensation. A model for the laminar falling film condensation heat transfer coefficient, h_c, is taken as[1],

$$h_c = 1.13 \frac{\left[\rho_l(\rho_l - \rho_v)g \cdot h_{fg} \cdot k_l^3\right]^{1/4}}{\left[L \cdot \mu_l \cdot (T_{sat} - T_{wall})\right]^{1/4}} \quad (1)$$

where ρ is density, g is gravitational acceleration, h_{fg} is the latent heat of evaporation, k_l is the liquid conductivity, L is a characteristic length, μ_l is the liquid viscosity, T_{sat} is the saturation temperature, and T_{wall} is the wall temperature. All units are SI and properties are taken for standard (i.e., not enriched) water saturated at 260 Celsius. If the characteristic length is taken at 0.005 meters, the behavior of the condensation heat transfer coefficient is given in Fig. 2.

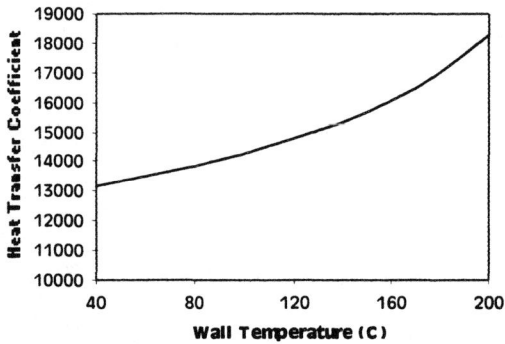

FIGURE 2. Condensation Heat Transfer Coefficient as a Function of Wall Temperature.

The power attributable to condensation heat transfer is given in Fig. 3 as a function of wall temperature for the case where half of the target volume is occupied by vapor. The characteristic length in the condensation heat transfer coefficient evaluation is taken as 0.005 meters.

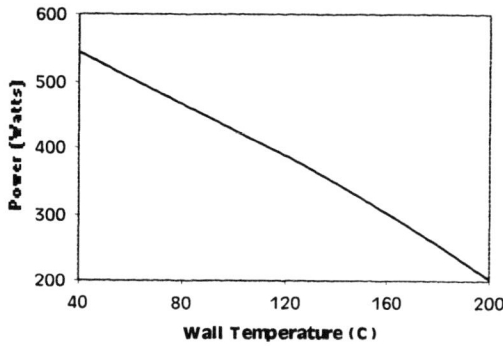

FIGURE 3. Power Attributable to Condensation Heat Transfer at 50% Vapor Volume.

The heat transfer between the water in the lower portion of the target and the target holder is estimated by considering the vapor volume generation rate for the nominal target power of 400 Watts. If all the beam power is used to generate vapor, then the vapor mass generation rate is given by the beam power divided by the latent heat of evaporation. The vapor volume generation rate is evaluated as the mass generation rate divided by the vapor density as 2.4x10^{-4} m^3/s. This indicates that vapor volume equal to just over eleven times the total target volume is produced each second if single phase heat transfer to the target holder is ignored. The vapor volume production rate is used to estimate the average liquid velocity in the target by assuming the vapor is produced in the forward region of the target and allowing the vapor to occupy half of the target volume. This produces a liquid velocity of 0.215 m/s and facilitates approximation of the heat transfer coefficient between the liquid portion of the target and the target holder. A single phase forced convection heat transfer coefficient, h_l, model is employed[2],

$$h_l = \frac{k_l}{L} 0.023 \left(\frac{\rho_l \cdot v_l}{\mu_l} \right)^{0.8} (\Pr_l)^4. \quad (2)$$

where $\Pr_l$ is the liquid Prandtl number. This model gives a heat transfer coefficient of 3,490 W/m^{2}C when the characteristic length is taken as 0.005 meters.

The power attributable to heat transfer between the liquid and the target holder when the liquid occupies half of the vapor volume is given in Fig. 4.

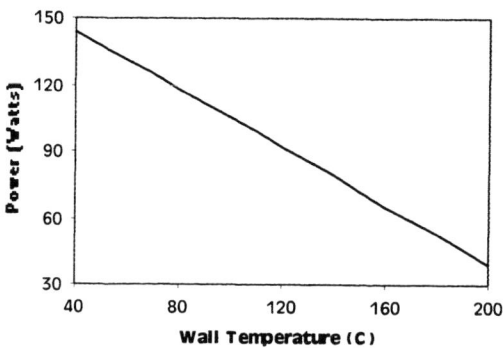

FIGURE 4. Power Attributable to Single Phase Heat Transfer at 50% Vapor Volume.

The inner wall temperature of the target holder, T$_{wall}$, is taken as uniform. This assumption is motivated by the silver conductivity of 400 W/m-C, which makes the target holder quite effective in conducting thermal power. The one dimensional temperature gradient in the silver target holder due to an applied flux of 1.0 MW/m^2 is 25 degrees Celsius per centimeter. This indicates the target inner wall temperature is not likely to vary by more than 25 C for the nominal operating condition where such flux is average. The total target power is presented as a function of the common inner wall temperature in Fig. 5. The ratio of the power attributable to condensation over the total power is offered in Fig 6.

The target power is evaluated as a function of the inner wall temperature for the case of 20% vapor volume in Fig. 7 and 70% vapor volume in Fig. 8. The ratio of power attributable to condensation over the total power is offered for 20% vapor volume in Fig. 9 and 70% vapor volume in Fig. 10.

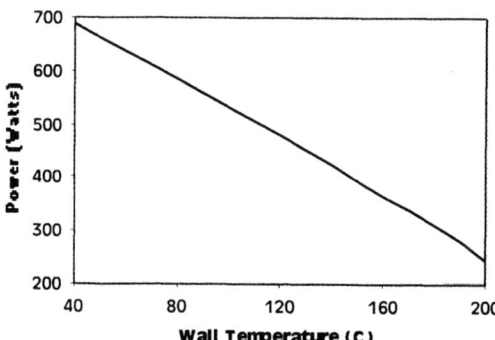

FIGURE 5. Target Power versus Inner Wall Temperature for 50% Vapor Volume.

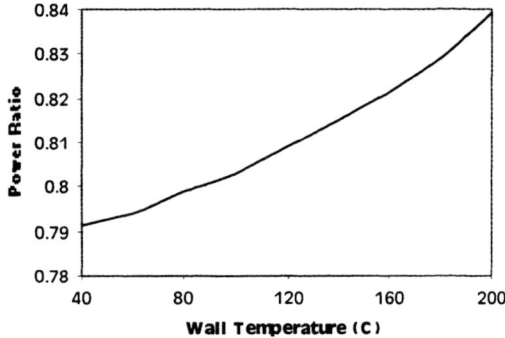

FIGURE 6. Target Power Attributable to Condensation Divided by the Total Power at 50% Vapor Volume.

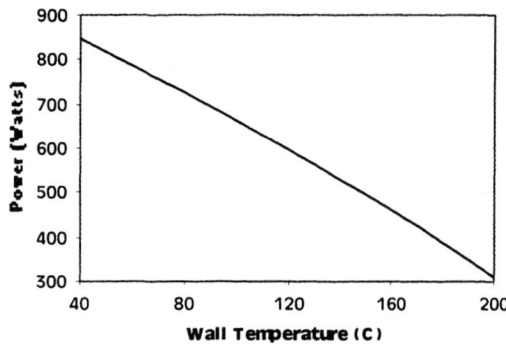

FIGURE 9. Target Power versus Inner Wall Temperature for 70% Vapor Volume.

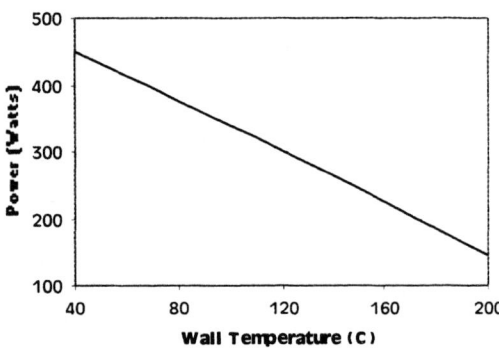

FIGURE 7. Target Power versus Inner Wall Temperature for 20% Vapor Volume.

It is noted that even for the 20% vapor volume case, the heat transfer attributable to condensation is important as evinced in Fig. 8.

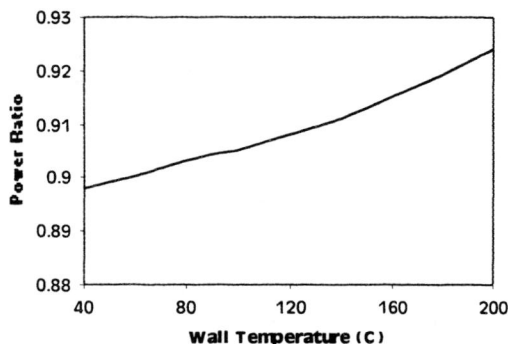

FIGURE 10. Target Power Attributable to Condensation Divided by the Total Power at 70% Vapor Volume.

Target Holder Thermal Model

The silver target holder is modeled as a single cylindrical fin with a base temperature equal to the inner wall temperature of the target. The outer circumference of the target holder is cooled with process water of temperature 40 Celsius. The power transferred from the fin, q_f, is given as,

$$q_f = m_a (T_{wall} - T_{out}) \tanh(m_f L_f) \quad (3)$$

where,

$$m_a = \left(h_{ex} P_h k_s A_{xs}\right)^{1/2}$$

and,

$$m_f = \left[\frac{h_{ex} P_h}{k_s A_{xs}}\right]^{1/2}.$$

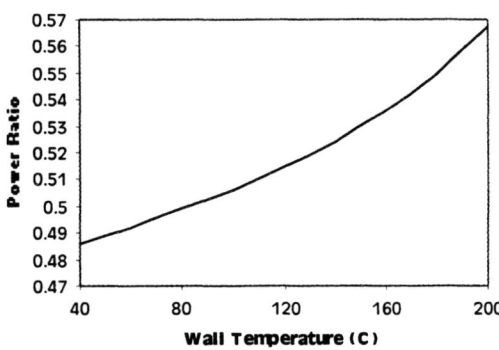

FIGURE 8. Target Power Attributable to Condensation Divided by the Total Power at 20% Vapor Volume.

Parameter P_h is the perimeter of the target holder, A_{xs} is the cross-sectional area of the target holder, h_{ex} is the external heat transfer coefficient and k_s is the conductivity of silver.

The heat transfer coefficient between the process water and the outer surface of the target holder, h_{ex}, is difficult to evaluate. The outer circumference of the target holder is threaded with eight threads per inch. The threads are nominally 2.1 mm deep and form the flow channel for the process water. The outer diameter of the target holder is 25.2 mm and slips into a smooth bore in the target holder carousel with clearance of around 0.25 mm. Significant flow will leak through the clearance between the outer diameter of the target and the inner diameter of the target carousel bore. Secondary flows are to be expected in the process water flow inside the spiral threads. All these complications motivated a bounding calculation, with the minimum external heat transfer coefficient, h_{ex}, taken as 5,000 W/m²C and the maximum value taken at 10,000 W/m²C. The target holder outer diameter is taken at 22 mm. The length of the target holder as a fin, L_f, is taken as 76 mm. Fig. 11 shows the target holder power as a function of target volume inner wall temperature for the two bounding external heat transfer coefficient values.

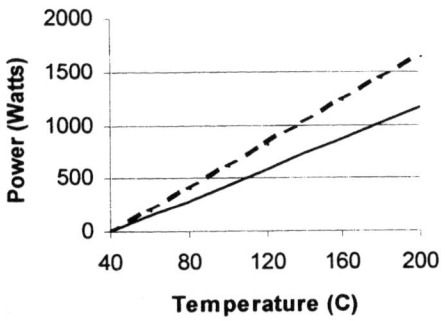

FIGURE 11. Target Holder Power Rejected to Process Water as a Function of Target Volume Inner Wall Temperature. Dashed line for 10,000 W/m²C, solid for 5,000 W/m²C.

Experimental

Measurements were undertaken to determine the volume of headspace in the standard target configuration and the amount of steam/water expansion encountered at a given beam intensity. The layout and calculated volumes are depicted in Figure 12.

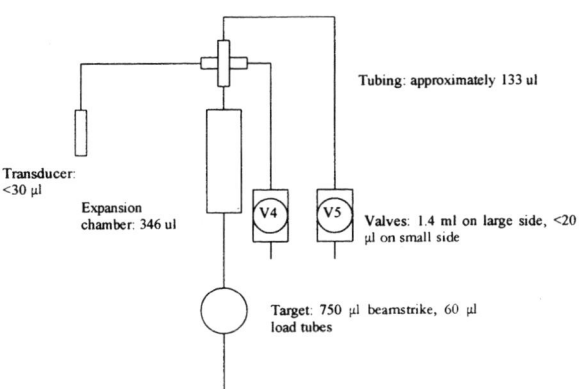

FIGURE 12. Target system layout.

From calculation, one expects (conservatively) 1359 µl of volume, including the beamstrike. The target is nominally loaded with 1100 µl by a Cavro syringe pump (1.6 µl repeatability). This results in a calculated headspace of 259 µl. This headspace can be increased by altering valve direction of valves V4 and V5. In the embodiment tested, V5 was reversed, giving a calculated 1.66 ml headspace

Cold Measurements

The actual headspace on an RDS-111 at the factory (SN DV27) was measured. The method for volume determination used was collection of expanded gas. In this method the target is pressurized with argon overpressure gas. The system is monitored with a transducer that has been calibrated with a separate test gauge. The fitting on the outboard side of valve V5 is replaced with a 140 ml syringe outfitted with a compression fitting nose. Valve V5 is opened and the syringe plunger allows equilibration with atmosphere. The overpressure was 512, 517, and 516 psig over three measurements. The syringe displacement was 95, 97 and 97 ml respectively. This calculates to 2.728, 2.763 and 2.755 ml of internal volume. Subtracting an 1100 µl load from the volume, the headspace when operating for this target configuration is 1.63 ml, which correlates well with calculation. A value of 1.63 ml was used for evaporated volume determinations based on beam-on pressure in the next section.

Bombardments

The target system with 1.63 ml headspace had been employed in an experimental bombardment using a hexagonal grid window support. During the bombardment the proton current was increased by 5 μA every 5 minutes up to 65 μA on target. This represents approximately 58.5 μA on the target water and the remainder intercepted by the aluminum grid, which is thicker than the proton range in aluminum at 11 MeV. Window deflection vs. pressure is minimized by this geometry, removing some potential sources of error in the measurement of the evolved gas/vapor void volume. The pressure and volume of evaporated water versus the beam current are displayed in Figure 13.

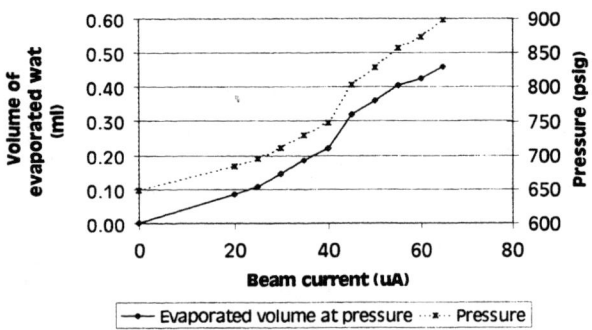

FIGURE 13. Target pressure and evolved volume vs. beam current.

Target Thermal Model Integration and Comparison With Data

The target holder thermal model and the target volume thermal model have the target volume inner wall temperature as a common parameter. These models are solved simultaneously for the power value that matches the target volume inner wall temperature. The target power is given as a function of the target vapor volume fraction in percent in Fig. 14 for the two bounding values of target holder external heat transfer coefficient. Data taken from a target subjected to a range of beam power values with provision for measurement of the displaced liquid volume are also presented on Fig. 14.

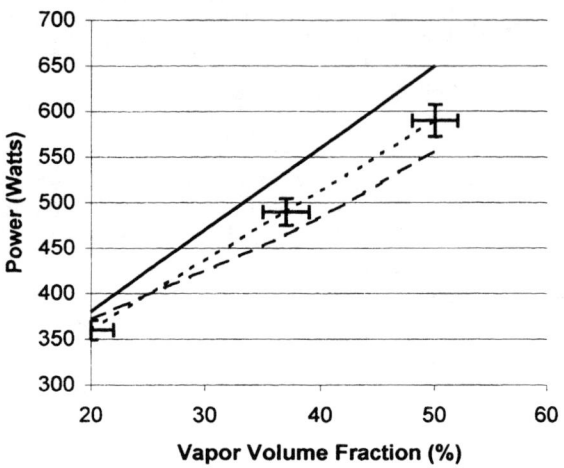

FIGURE 14. Target Power versus Vapor Volume Fraction in Percent. Solid line is model with external heat transfer coefficient equal 10,000 W/m^{2}C, dashed line is model with external heat transfer coefficient equal 5,000 W/m^{2}C, data line presented with error bars.

Conclusions

The target simulation uses simple heat transfer models from an undergraduate level heat transfer text. These models are representative of the heat transfer phenomena and are intended to provide a tractable simulation suited to supporting design improvement activities. Readers should note that the simulation was developed prior to the experiment that produced the data line presented in Fig. 14. The data validate the simulation performance better than expected and indicate that model refinements may not be required.

REFERENCES

1. Steinbach J., Guenther K., Loesel E., Grunwald G., Mikecz P., Ando L., Szelecsenyi F., and Beyer G.J. Temperature Course in Small Volume [^{18}O]Water Targets for [^{18}F]F- Production, *Appl. Radiat. Isot.*, **41**, No. 8, pp 753-756, 1990.

2. Alvord C.W., Computational Fluid Dynamics Study of Water Target Design and Operational Parameters, Talk given at Fifteenth International Conference on the Application of Accelerators In Research and Industry, 1998, Denton Texas

3. Lenz J.W., and Ruth T.J., An Enriched Water Heat Pipe Target for F-18 Production, Talk given at Fourteenth International Conference on the Application of Accelerators In Research and Industry, 1996, Denton Texas

4. Holman, J. P., *Heat Transfer*, McGraw-Hill: Fourth Edition, pp. 355-360, 1976.

On the production of radioactive stents

K. Schlösser, H. Schweickert

Forschungszentrum Karlsruhe GmbH, Hauptabteilung Zyklotron, Postfach 3640, D-76201 Karlsruhe, Germany

Abstract. In the last years radioactive stents proved to inhibit neointima formation. This paper describes the actual status of producing such radioactive stents. After a short discussion of the different radioisotopes suitable for radioactive stents, potential production methods are discussed. The ion beam implantation of P-32 applied at the Karlsruhe Research Center shall be described in more detail.

INTRODUCTION

In the last years radioactive stents proved to be able to inhibit neointima formation which is the main reason for restenosis after stent implantation[1]. In the following sections, present and future possibilities to produce such radioactive stents shall be described.

IDEAL ISOTOPES FOR A RADIOACTIVE STENT

As far as we know today, the conditions for an appropriate radioisotope are:

- The depth in tissue, where the dose of the radiation should be deposited, is believed to be in the order of several mm.
- The dose outside these several mm around the stent should be as low as possible.
- The optimum time for applying radiation to the vessel wall, while presently un known, is believed to be several days after the implantation of the stent such that the half-life should be at least several days.

From these conditions, it follows that γ-emitters with ranges of many centimetres in tissue have too long a range and that α-emitters with a range of a few μm only have too short a range for this application. The best suited radioisotopes are pure high-energy ß⁻-emitters or/and if the dose must be delivered further away from the stent pure low-energy x-ray emitters.

There are only a few radioisotopes (Table 1) which fulfill these conditions. At the moment, it seems that P-32 is the most ideal isotope, because it is a pure ß⁻-emitter and almost all of its energy is absorbed within the artery wall that has undergone dilatation. Furthermore, the 14.3 day half-life gives sufficient time for final assembly and sterilisation and allows for an useful shelf life. P-32 also has the advantage of being one of the most readily available and least expensive radioisotopes, which is routinely used in large quantities in many hospitals. The only other pure ß⁻-emitter with a sufficient half-life and high energy ß⁻ is Y-90. The disadvantages of Y-90 are a significant shorter half-life of 2,7 days and higher material costs relative to P-32.

There are two potentially interesting x-ray emitters. From our point of view, they are only of interest, if the depth where the dose has to be applied will be significantly deeper than the depth the ß⁻-emitters can reach. Pd-103 has the advantages of a long half-life and commercial availability. On the other hand, Pd-103 undergoes 100% decay into a metastable isomeric state in Rh-103 which is emitting conversion electrons of 36.8 keV energy. These low energy electrons have a range of 37 μm only and deposit a very high dose on the tissue at the surface of the stents, if they are not shielded. Cs-131 is a pure x-ray emitter and would therefore be better suited.

PRODUCTION OF RADIOACTIVE STENTS

In principle, there are three possibilities for the production of radioactive stents (Tab. 2), which will be discussed briefly.

ACTIVATION

This method (Fig. 1) uses the direct production of radioisotopes via a nuclear reaction by bombarding the stent material by a high-energy beam of light ions (protons, deuterons or α-particles) which is accelerated in a cyclotron or by neutron irradiation of the stent material in a reactor. As there is no appropriate target material in the usual stock material from which stents are made, thus "ideal" radioisotopes cannot be produced in this way. Nevertheless, the first radioactive stents with a significant activity for animal tests were activated by this method. The Cr Ni-steel of a PS-stent was activated by 17 MeV protons of the Karlsruhe compact cyclotron. Several nuclei were produced: Co-55, Tc-95, Mn-52, Tc-96, Mo-99, Co-56, Co-58 and Co-57. The predominant dose was created by electron-capture nuclei emitting soft x-rays and ß$^+$-decays. With these radioactive stents the first successful tests on rabbit iliac arteries were performed. A similar technique was used to produce the ß$^+$-emitter V-48 in a nickel-titanium stent by 8.5 MeV protons.[2]. In both cases, a number of high-energy γ-rays are produced in addition to the ß-particles and x-rays desired. This leads to a non-negligible whole body dose.

TABLE 1. The best suited isotopes for radioactive stents

Radioisotopes	Half-lives	Radiation	Tests
P-32	14.30 d	Pure ß$^-$-Emitter (1,7 MeV)	Animal, Human [1,3]
Y-90	2.67 d	Pure ß$^-$-Emitter (2,3 MeV)	Animal[4]
Pd-103	16.96 d	x-ray (21 keV) + conversion e$^-$	Animal (not published)
Cs-131	9.69 d	Pure x-ray (30.4 keV)	-

TABLE 2 Methods for the production of radioactive stents

Method	Advantages	Disadvantages
Activation	No wash-off No contamination problems	Unsuitable radioisotopes
Ion-Implantation	Any material can be implanted Very low wash-off No contamination problem Can easily vary activity amount along the stent	None
Chemical	In principle less expensive than the other methods	Wash-off is problematic Contamination possible New materials need new technology Change of the stent surface

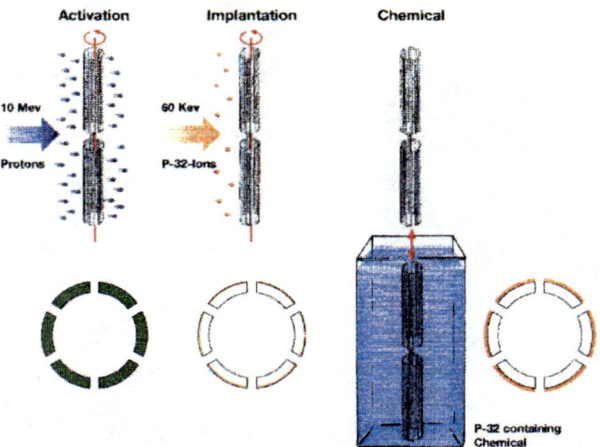

Figure 1. Schematic representations of the different production processes In the cases of activation and implantation, the radioisotopes are safely located inside the stent material. Chemical processes will deposit the activity on the surface of the stent.

ION IMPLANTATION

The ion implantation process (Fig. 1) is well-suited for all the "ideal" radioisotopes listed. Since we believe that P-32 is the most attractive radioisotope, the technology has been developed for this isotope. The basic technology consists of a very special high-efficiency ion source for P-32, an accelerating system and a 90°-mass separator in order to separate most impurities from the P-32 beam (Fig. 2).

The stents are implanted with P-32 on a special platform, where up to 30 stents can be irradiated simultaneously (Fig. 3). In order to achieve a homogeneous activity distribution on the surface, the stents are rotated in the P-32-beam which is swept horizontally and vertically. After the implantation process, the quality of each manufactured stent is tested extensively.

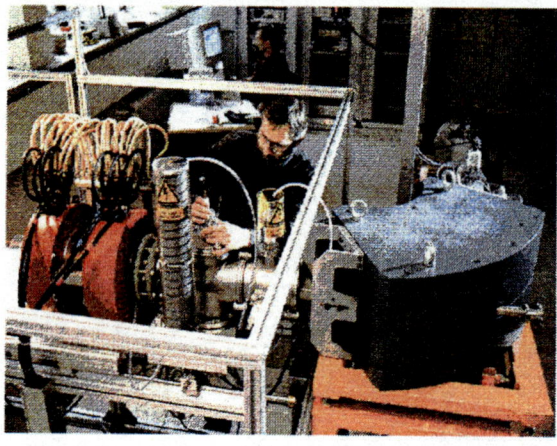
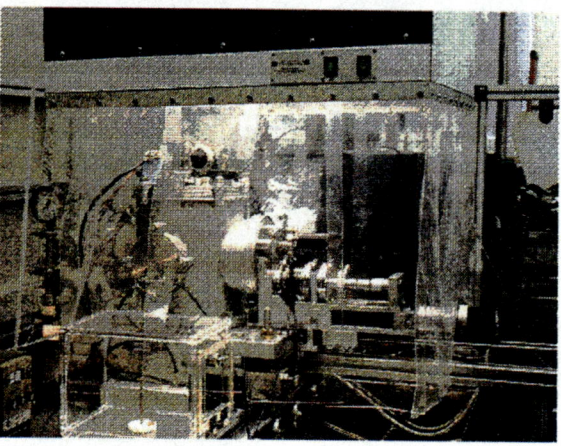

FIGURE 2. Photograph of the implanter during the commissioning phase. On the left, the special ECR ion source, the accelerating section and a 90° magnet are shown. On the right, the actual irradiation chamber in the laminar flow hood can be seen.

The absolute activity and the activity distribution on the stents are measured by a special scanning system (Fig. 4). The calibration of this system is traceable to NIST in the USA and to the Physikalische Technische Bundesanstalt in Braunschweig (German NIST equivalent).

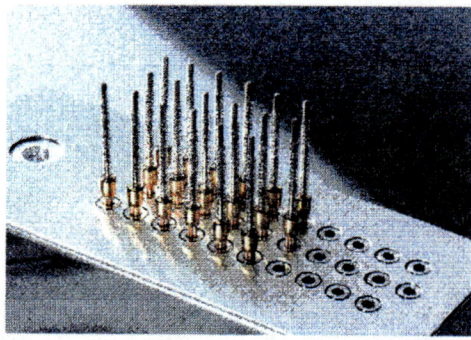

FIGURE 3. Photograph of the irradiation tool for BX stents by IsoStent

After the implantation process, the stents are washed in an ultrasonic bath for 15 min in saline (42°C) in order to remove all potential surface contaminations. The activity washed off is measured for each stent separately and typically lower than 1% of the total activity. Further washing in sonicated saline removes about another 0.3% of the activity of the stent. Since the combination of ultrasound and NaCl is a fairly aggressive treatment, the release of activity into the blood stream is very small, and no coating at all is required to encapsulate the activity. Although an uncoated ion-implanted stent cannot be considered a sealed source, a potential contamination of the catheter laboratory is excluded, because the P-32 atoms are embedded beneath the surface of any metal stent. We performed wipe tests and ultrasonic washing in alcohol to check for removable activity. The amount of activity) found there is < 0.02% of the stent activity. The stents are delivered in a Lucite shield mounted on a stent delivery system in order to ensure the radiation protection of the operator.[5] Thus, there is no need to touch the stents

The technology of the production of P-32 implanted stents has now been developed to the extent that a routine high-quality production of even large numbers can be achieved for a reasonable price.

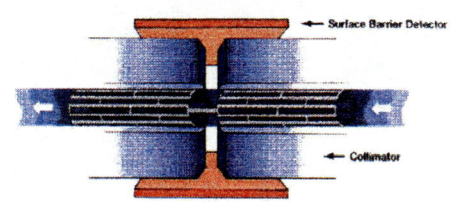
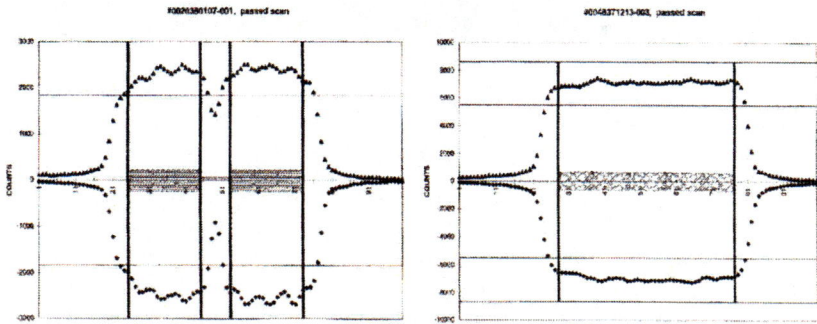

Figure. 4. Schematic representation of the stent activity and activity distribution measurement system. The measurement is done on two sides simultaneously. Typical results of activity distribution measurements for an PS- and BX-stent. This measurement guarantees a homogeneous activity distribution on the struts.

CHEMICAL

In this case, the stents are coated on the surface by an appropriate radioisotope. The major prerequisite for such stent coatings is an extremely fast bonding between the radioisotope and the stent material in order to prevent "leakage" of the radioisotope to other parts of the body. There are no publications up to now on such a procedure. Regarding P-32, we tested several methods for stainless steel samples like carrier-free P-32 direct deposition, direct deposition with subsequent heating, direct deposition with various pretreatments of the samples, P-32 deposition using Ag- and Au-plated stents, electroplating, etc. None of the stents produced by these methods were able to match our wash-off tests described above.

For Pd-103 a galvanic coating Process is under development[6]. Since this involves a high quality gold coating, excellent retention of radioactivity is achieved.

CONCLUSION

Up to now, the ion beam implantation of radioactive P-32 is the best and safest method to produce radioactive stents. At the moment, the ion source technology and the quality tests are being improved towards a higher productivity. We are sure that a sufficient number can be produced at a reasonable price, when radioactive stents will have shown their clinical usefulness.

REFERENCES

1. Albiero, R., Adamian, M., Kobayashi, N., Amato, A., Vaghetti, M., Di Mario, C., and Colombo, A., *Circulation* **101**, 18-26 (2000)

2. Eigler, N. L., Li, A. N., Whiting J. S., and DeFrance, A., A ^{48}Vanadium Brachytherapy Source for Treatment of Coronary Artery Restenosis; in: *Vascular Brachytherapy*, edited by R. Waksman, S.B. King, I.R. Crocker and R.F. Mould, Nucletron B.V. 1996, Chapter 23

3. Serruys, P.W., and Kay, I.P., *Circulation* **101**, 3-7 (2000)

4. Taylor, A.J., Gorman, P.D., Hudak, C., Tashko, G., Sweet, W., Farb, A., and Virmani, R., *Int J Radiat Oncol Biol Phys* **46(4)**, 1019-24 (2000)

5, IsoStent, Inc, Radioisotope Stents. in *Handbook of VascularBrachytherapy*, edited by R. Waksman, and P.W. Serruys, Martin Dunitz, Ltd 1998

6. Frey, A.W., Moeslang, A., and Przykutta, F., *European Heart Journal* **21**, 399 (2000)

Proposal For A New High-Energy Isotope Production Facility At LANSCE

E. J. Pitcher[+], M. W. Cappiello[+] and H. A. O'Brien[*]

[+]*Los Alamos National Laboratory*, [*]*O'Brien & Associates*
Los Alamos, New Mexico

Abstract. The original targeting system used for producing radionuclides via 800-MeV proton-induced spallation reactions at Los Alamos has been deactivated while a new 100-MeV targeting facility is being implemented. The feasibility of constructing another facility to retain the capability for 800-MeV proton bombardment has been examined. A new feature is the incorporation of tungsten targets to enhance energetic neutron production. A neutron flux of similar magnitude to the proton flux would result, aided by the placement of reflector shields to reduce escaping energetic particles. Interspersed among the tungsten targets are a total of 25 experimental target irradiation positions. A hydraulic target transfer system facilitates independent target insertion and retrieval. In many cases, this mixed neutron/proton irradiation environment significantly increases the radioisotope production rate above that of using protons alone. As a means of validating predicted isotope production rates for the new facility, predicted yields using the MCNPX and CINDER'90 codes were compared with measured yields obtained prior to the deactivation of the 800-MeV facility. Of the 10 isotopes for which measured data exist, activities at EOB are within 35% of the calculated yields.

INTRODUCTION

Radioisotopes are important tools in biological and medical research that have led to major advances in our understanding of biological functions and disease processes. With the development of more specific delivery agents such as monoclonal antibodies, antibody fragments, and peptides, major advances are projected in radiotherapy. Other growth areas foreseen in clinical practice include bone palliation treatment, brachytherapy applications in inoperable tumors, nuclear cardiology, and positron emission tomography procedures.

Research radionuclides and radionuclides proposed for incorporation into promising new nuclear medicine products are frequently unavailable or very expensive. Clinical trials, which are essential to the development of promising and exciting new therapies, often require large quantities of radionuclides that are not always readily available.

Isotope production at the high-energy beam stop area at Los Alamos Neutron Science Center (LANSCE) has been in operation since 1976. Production targets were placed directly in the high-current (1 mA), high-energy (800-MeV) proton beam. Over the years, a large variety of isotopes have been made for medical research using nuclear spallation reactions. These isotope production operations were run parasitically, taking whatever beam that remained from other funded projects. With the recent loss of beam activities for medium-energy physics and Accelerator Production of Tritium (APT) materials research, isotope production at 800 MeV ceased. Prospects for further operation are uncertain.

A new isotope production targeting facility is being built at the 100-MeV location along the linear accelerator at LANSCE. At 100 MeV, the energy level is well below the 200-300 MeV threshold required for spallation reactions. Without an 800-MeV target facility, the capability to produce spallation isotopes has been lost. These isotopes are needed to complement those that will be produced at 100 MeV.

Based on the perceived need for a spallation target facility at LANSCE, a study was commissioned to develop a preconceptual design of a Medical Isotope Target Station for operation at the high-energy end of LANSCE.

FACILITY

The 800 MeV Isotope Production facility (IP-800) design consists of a vacuum vessel, a spallation target, a rabbit targeting system, reflector, shielding, heat removal systems, coolant purification systems, beam transport, and control systems. The proposed facility would be located within the present thin target area (TTA) to allow the use of existing hot cells, to avoid interference with potential long-pulse spallation source (LPSS) operations, and to avoid the beam-line maintenance costs associated with a location near the present A-6 beam stop. By positioning the Target and Reflector Assembly away from the normal Line A beam path, the proposed facility would maintain compatibility with potential future operations, namely Advanced Hydrodynamics Facility (AHF), LPSS, and Accelerator Transmutation of Waste (ATW).

The Target and Reflector Assembly consists of two major subsystems: the Spallation Target/Rabbit Module (STRM) and the Reflector Modules. The STRM generates spallation neutrons, positions the target rabbits for irradiation, and stops the proton beam. The Reflector Modules reflect the target-escaping particles back into the STRM and act as a first layer of water-cooled shielding. Coolant and hydraulic rabbit pipes connect the Target/Reflector Assembly to the LANSCE chemistry hot cell to facilitate loading and removal.

The STRM contains a total of 45 tungsten spallation targets, which are placed into close-packed rung tubes connecting the coolant distribution manifolds. Twenty-five of these tubes contain the isotope target rabbits. The spallation targets consist of tungsten cylinders about 6 inches long, completely clad with a thin corrosion-resistant barrier. Surrounding the inner rabbit tubes are concentric, clad tungsten cylinder targets with varying wall thicknesses. Wall thicknesses vary from 0.080 inch in the front to solid 1.420-inch-diameter rods in the back. The overall length of the STRM is 36 inches.

Due to anticipated radiation-induced damage, the entire Spallation Target/Rabbit Module has been designed to permit periodic replacement. The Reflector Modules can also be replaced, but this event is expected to occur on a less frequent basis.

PHYSICS ANALYSIS

The primary goal of the physics calculations is to maximize the isotope production capability of the IP-800 by trying to maximize the proton and neutron fluxes (in the energy ranges of interest) seen by the rabbits. The MCNPX computer code (Ref. 1) was used to perform the physics analyses for fluxes, power densities, and reaction rates, and the CINDER '90 code (Ref. 2) was used to calculate isotopic compositions. All calculations assume a proton energy of 800 MeV. The beam was assumed to have a Gaussian distribution, with a 2-sigma of 3.0 cm. Isotope production and fluxes are normalized to a current of 0.75 mA, which is the expected load on the target. Powers and power densities are normalized to 1 mA to ensure that the target can handle the heat load if the full LANSCE beam becomes available.

Rabbit locations within the STRM are identified in Fig. 1. Table 1 shows the volume and total proton and neutron flux in each rabbit location. This table also contains an estimate of ^{82}Sr (produced via proton-induced reactions) and ^{32}P (produced via high-energy neutron-induced reactions) production in each location. To estimate isotope production, energy-dependent fluxes were tallied by MCNPX in each of the 25 rabbit locations. An auxiliary code was then run that used these fluxes and LAHET-generated cross sections to estimate direct yield isotope production. For these calculations, the irradiation time is assumed to be equal to the half-life of the produced isotope.

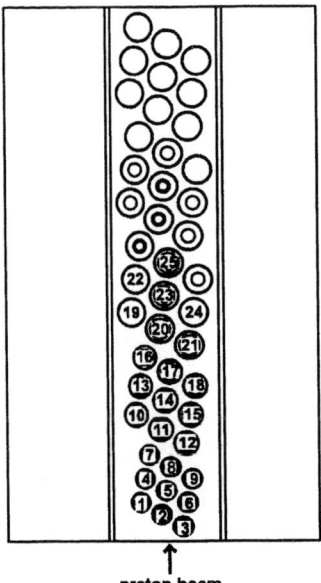

FIGURE 1. Identification of rabbit locations in IP-800 target.

Table 1. Summary Of Fluxes And Isotope Production Results For Baseline Design

Rabbit #	Volume (cm^3)	Proton Flux (p/cm^{2}sec)	Neutron Flux (n/cm^{2}sec)	^{82}Sr direct yield from Mo target (Ci)	^{32}P direct yield from S target (Ci)
1	2.04	5.46E+13	1.44E+14	1.41	0.57
2	2.04	1.51E+14	2.68E+14	3.90	1.29
3	2.04	1.19E+13	1.27E+14	0.30	0.34
4	2.04	2.60E+13	1.60E+14	0.70	0.49
5	2.04	1.46E+14	3.22E+14	3.98	1.44
6	2.04	2.67E+13	1.86E+14	0.72	0.57
7	2.04	1.28E+13	1.53E+14	0.35	0.39
8	2.04	1.36E+14	3.06E+14	3.77	1.31
9	2.04	5.41E+13	2.22E+14	1.49	0.76
10	4.73	2.43E+13	1.94E+14	1.58	1.26
11	4.73	1.02E+14	3.01E+14	6.60	2.61
12	4.73	7.68E+12	1.50E+14	0.51	0.79
13	4.73	1.21E+13	1.68E+14	0.82	0.91
14	4.73	7.92E+13	2.82E+14	5.26	2.25
15	4.73	1.19E+13	1.63E+14	0.80	0.90
16	4.73	6.81E+12	1.41E+14	0.46	0.72
17	4.73	6.77E+13	2.36E+14	4.53	1.86
18	4.73	2.27E+13	1.65E+14	1.53	1.02
19	8.53	1.15E+13	1.44E+14	1.41	1.33
20	8.53	4.59E+13	1.96E+14	5.51	2.46
21	8.53	3.23E+13	1.61E+14	3.69	1.88
22	8.53	2.22E+13	1.22E+14	2.37	1.32
23	8.53	1.01E+13	8.77E+13	1.07	0.82
24	8.53	1.40E+13	8.60E+13	1.29	0.86
25	8.53	8.64E+12	6.18E+13	0.53	0.58

ISOTOPE PRODUCTION ANALYSIS

Table 2 shows the irradiation setup used for the sample production analysis, while Table 3 shows the results of the calculation. The radiopurity and specific activity values are given at the cooling times listed after the radiopurity values. Irradiation times were assumed to be equal to the production isotope's half-life or 270 days, whichever was shorter. The number of cycles per year was determined from an estimated 250 days of operation per year. The selection of the target and its position in the target array shown in Table 3 is arbitrary. The main purpose of this table is to illustrate that a variety of targets and products can be irradiated and produced simultaneously.

To maximize the yields of a particular radionuclide, a yield profile for the complete irradiation facility must be calculated as illustrated in Table 1 for ^{82}Sr and ^{32}P. With this information, the highest yielding positions can be selected for irradiation, depending on the total amount of the radioisotope desired per irradiation.

An examination of the single-cycle production and annual production columns in Table 3 reveals that substantial amounts of the various radioisotopes can be made simultaneously. This table reflects the extraction of a single isotope from each target. In practice, multiple isotopes may be recovered from each target, as is done presently.

ACCURACY OF ISOTOPE PRODUCTION RATE CALCULATIONS

As a means of validating predicted isotope production rates for this new proposed facility, predicted radionuclide yields from the design codes were compared with measured yields from the existing isotope production facility at LANSCE. Of the 10 isotopes for which measured data exist, measured activities at the end-of-bombardment (EOB) were found to be within 35% of the calculated yields. These results lend confidence to the use of the MCNPX and CINDER '90 codes to predict isotope production rates for the new facility.

Table 2. Sample Irradiation Set-Up

Rabbit Number	Volume (cm³)	Production Isotope	Target Material	Irradiation Time (Cycle Time) (days)	Cycles Per Year
1	2.04	^{76}As	RbBr	1.1	228
2	2.04	^{72}Se	RbBr	8.5	29
3	2.04	^{73}As	RbBr	80.3	3
4	2.04	^{32}P	S	14.3	17
5	2.04	^{186}Re	W	3.78	66
6	2.04	^{47}Sc	Ti	3.35	74
7	2.04	^{109}Cd	Ag	270	1
8	2.04	^{127}Xe	CsI	36.4	6
9	2.04	^{188}W	Ir	69.4	3
10	4.73	^{117m}Sn	Sb	13.6	18
11	4.73	^{82}Sr	Mo	25.4	9
12	4.73	^{178}W	Re	21.6	11
13	4.73	^{22}Na	Al	270	1
14	4.73	^{103}Pd	Ag	17.0	14
15	4.73	^{179}Ta	W	270	1
16	4.73	^{88}Y	Zr	106.6	2
17	4.73	^{153}Sm	Gd	1.9	131
18	4.73	^{111}In	Sn	2.8	89
19	8.53	^{32}Si	KCl	270	1
20	8.53	^{68}Ge	RbBr	270	1
21	8.53	^{68}Ge	RbBr	270	1
22	8.53	^{68}Ge	RbBr	270	1
23	8.53	^{88}Zr	Mo	83.4	3
24	8.53	^{67}Cu	Zn	2.6	96
25	8.53	^{26}Al	SiC	270	1

Table 3. Summary Of Results From Example Irradiation Set-Up

Rabbit Number	Prod. Isotope	Single-Cycle Prod. (Ci)	Annual Prod. (Ci)	Radiopurity (%)	Specific Activity (kCi/g)
1	^{76}As	0.34	77	23 (10 hrs cooling)	118
2	^{72}Se	0.51	14.8	56 (3 days cooling)	1.6
3	^{73}As	0.09	0.28	78 (27 days cooling)	2.6
4	^{32}P	9.22	156	98 (10 hrs cooling)	103
5	^{186}Re	0.13	8.4	5 (3.78 days cooling)	0.08
6	^{47}Sc	4.22	312	67 (1 day cooling)	171
7	^{109}Cd	0.06	0.06	100 (10 days cooling)	0.0003
8	^{127}Xe	2.2	13.2	65 (10 days cooling)	0.03
9	^{188}W	0.06	0.19	0.5 (20 days cooling)	0.02
10	^{117m}Sn	0.51	9.1	44 (5 days cooling)	0.29
11	^{82}Sr	9.1	81.6	52 (9 days cooling)	3.8
12	^{178}W	3.8	41.7	50 (1 day cooling)	0.9
13	^{22}Na	0.54	0.54	100 (30 days cooling)	1.24
14	^{103}Pd	20.0	280.5	95 (17 days cooling)	2.5
15	^{179}Ta	6.25	6.25	59 (90 days cooling)	0.32
16	^{88}Y	4.4	8.8	84 (36 days cooling)	1.8
17	^{153}Sm	0.8	104	84 (15 hours cooling)	5.4
18	^{111}In	3.45	306	77 (1 day cooling)	73.8
19	^{32}Si	1.76E-7	1.76E-7	98 (3 days cooling)	4.2E-07
20	^{68}Ge	0.95	0.95	95 (90 days cooling)	0.21
21	^{68}Ge	0.76	0.76	96 (90 days cooling)	0.20
22	^{68}Ge	0.82	0.82	97 (90 days cooling)	0.37
23	^{88}Zr	5.9	17.6	96 (27 days cooling)	2.9
24	^{67}Cu	1.8	173	50 (2.6 days cooling)	17.3
25	^{26}Al	5.9E-6	5.9E-6	100 (30 days cooling)	5.1E-06

CONCLUSIONS

A pre-conceptual design of a versatile isotope production facility at the high-energy (800-MeV) end of LANSCE has been completed. Enhanced capabilities have been incorporated into the design. These include a mixed proton and neutron irradiation environment to enhance isotope production and a total of 25 individual target irradiation positions that use a hydraulic transfer system to allow rapid insertion and removal of individual production targets. The irradiation facility is linked to existing hot cells to facilitate recovery and packaging of irradiated targets. Location of the facility would maintain compatibility with existing LANSCE programs, as well as any potential new program or project.

Physics analyses were performed to determine particle fluxes, power densities, reaction rates, and isotopic compositions. Predicted radionuclidic yields from the analyses were validated by comparisons with actual measured yields from the existing isotope production facility. Of the 10 isotopes compared, measured activities at EOB were within 35% of the calculated yields. This is considered excellent agreement and demonstrates the accuracy of our predictions for isotope production rates. Radioisotope yield calculations show that substantial amounts of isotopes can be produced.

ACKNOWLEDGEMENT

This report summarizes the results of an extensive study performed at Los Alamos National Laboratory. In addition to the authors of this report, those that participated in the larger study included D. Poston, R. Barber, P. D. Ferguson, M. Wilson, L. Sanchez, J. D. Shelton, and T. Spatz. For copies of the study, entitled "LANSCE 800-MeV Isotope Production Facility: A Pre-conceptual Design", contact M. W. Cappiello at LANL, Los Alamos, NM 87545.

REFERENCES

1. Hughes, H. G., et al., "MCNPX™ – The LAHET™ /MCNP™ Code Merger," *Proceedings of the Third Workshop on Simulating Accelerator Radiation Environments (SARE3)*, May 7-9, 1997, KEK, Tskuba, Japan, p. 44; H. G. Hughes, et al., "Recent Developments in MCNPX™," *Second International Topical Meeting on Nuclear Applications of Accelerator Technology,* September 20-23, 1998, Gatlinburg, TN, p. 281.

2. Wilson, W.B., et al., "The Status of Nuclear Data for Transmutation Calculations," *ANS RP&S Topical Meeting on Advancements in Radiation Protection & Shielding*, April 21-25, 197, North Falmouth, MA; W. B. Wilson, et al., "Status of CINDER '90 Codes and Data," *Proceedings of the Third Workshop on Simulating Accelerator Radiation Environments (SARE4)*, September 14-16, 1998, Knoxville, TN, p. 69.

Computer Study of Isotope Production for Medical and Industrial Applications in High Power Accelerators

S. G. Mashnik and W.B. Wilson
T-16, Theoretical Division, LANL, Los Alamos, NM 87545
tel.: 505-667-9946, e-mail: mashnik@t2y.lanl.gov
K. A. Van Riper
White Rock Science, PO Box 4729, Los Alamos, NM 87544

ABSTRACT

Methods for radionuclide production calculation in a high power proton accelerator have been developed and applied to study production of 22 isotopes. These methods are readily applicable both to accelerator and reactor environments and to the production of other radioactive and stable isotopes. We have also developed methods for evaluating cross sections from a wide variety of sources into a single cross section set and have produced an evaluated library covering about a third of all natural elements that may be expanded to other reactions. A 684 page detailed report on this study, with 37 tables and 264 color figures is available on the Web at http://t2.lanl.gov/publications/.

The widespread use of radionuclides in medical and industrial applications is steadily increasing, leading suppliers to seek out new production facilities. A reliable supply chain is necessary to both encourage new applications and to replace aging production sources. The United States, in particular, faces a domestic production shortfall. It has been a policy of this country to import radioisotopes from Canada and other countries. The lack of a reliable supply has led to supply problems at times that could be ameliorated by a domestic production facility.

Among the possibilities for radionuclide production are high power accelerators, either purpose built or alongside existing applications. As an example of the latter, a recent study by the Medical University of South Carolina [1] discussed the production of medical radioisotopes at the proposed Accelerator Production of Tritium Facility (APT) [2].

We have undertaken a study to see to what extent existing nuclear data models are applicable to calculations of radionuclide production in a high energy, high power environment. We chose the APT target/blanket assembly as a typical environment in which to study isotope production. In a previous report [3], we considered the production of two radioisotopes – ^{18}F and ^{131}I – at two locations in the APT blanket. We have extended that study to look at the production rates of 22 isotopes in nearly 500 locations throughout the APT target and blanket. In addition to the 100 milliamp 1.7 GeV proton beam energy assumed in the previous study, we also treat beam energies of 1.0, 1.2, 1.4, 1.6, and 1.8 GeV (all at 100 milliamps).

It should be noted that we have chosen the APT accelerator for our study as an example with which to demonstrate the possibility of production of radioisotopes at such a facility. Radioisotopes can be produced also at other high power accelerators, projected for the Accelerator Transmutation of nuclear Wastes (ATW) or for Neutron Spallation Source (SNS) facilities, as well as at nuclear reactors. Our computational method is not limited to a particular facility and can be used to study production of radioactive and stable isotopes at any accelerator or reactor. For practical reasons, the emphasis of our present study is on radioisotopes.

We have prepared numerous figures and tables as part of this work. Space available in this paper prevents their inclusion here. Our 684 page detailed report on this study [4], with 37 tables and 264 color figures is available on the World Wide Web at http://t2.lanl.gov/publications/ and a limited number of hard copies may be available from the authors.

Figure 1 is a 3-dimensional rendering of a computer model of the target/blanket assembly. The model is based on the Todosow geometry with a 16 cm × 160 cm beam area. The beam enters the assembly through the window on the right. The beam strikes tungsten-dominated ladders in the center of the assembly; the position of these ladders correspond to the pipes extending from the manifolds. The lateral blankets extend to either side of the beam line and ladder area. The downstream blanket encompasses the region to the left of the ladder region, while the upstream blanket is the narrow area between the entrance window and the ladder region.

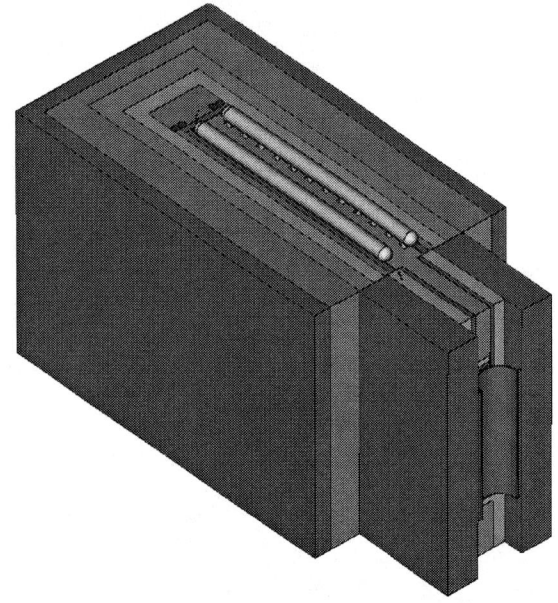

Figure 1: APT target/blanket assembly.

Tally Locations. Subdivision of cells in the upstream, downstream, and lateral blanket regions and in the beam cavity between ladders yielded approximately 183 cells in which the fluxes were tallied. Figure 2 shows the cell locations in a slice in the X-Z plane through the middle of the target model. We did not tally fluxes in the ladders, nor in the coolant pipes.

Vertical Segments. We segmented the tallies into five equally spaced vertical segments. Color figures with cross section through the upstream blanket, showing the segmenting and further details may by found in our comprehensive report [4].

The fluxes, and hence production rates, were greatest in the middle segment, decreasing towards the top and bottom. Unless otherwise stated, the results presented here are for the middle segment.

Locations for Detailed Results. We present detailed summaries of the production rates at four locations in our report [4].

For each reaction, we found the cell with the maximum rate over our entire set of tallies, and the cell with the maximum rate over all tallies excluding the

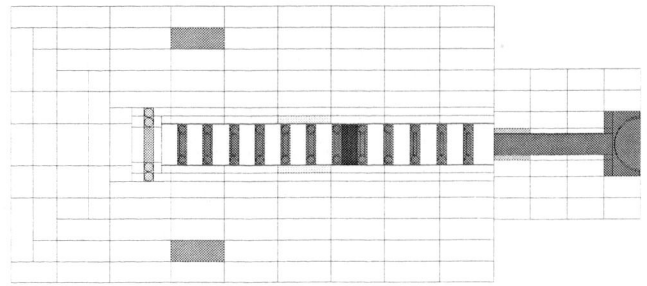

Figure 2: A slice through the middle of the APT target/blanket assembly showing cells where tallies were taken.

beam cavity cells.

We did not model any fixtures, such as irradiation tubes, that would be required to produce radioisotopes in any location. Our results thus assume any such fixtures would have no effect on the fluxes.

Spectrum plots of the neutron and proton fluxes in each segment of the four selected locations are given in our detailed report [4].

Flux Color Contours on Planes. We prepared a variation of the target geometry for display of the fluxes (and production rates) as color-coded contours on a plane [4]. We introduced 5 planes to represent the 5 vertical segments; the vertical position of the plane lies at the center of the corresponding segment. Each plane thus represents the flux averaged over a segment.

Figure 3 shows a contour plot of the neutron flux for a 1.0 GeV beam energy. The detailed Web report [4] includes similar (color) plots for neutron and proton flux contours at the 6 energies we consider.

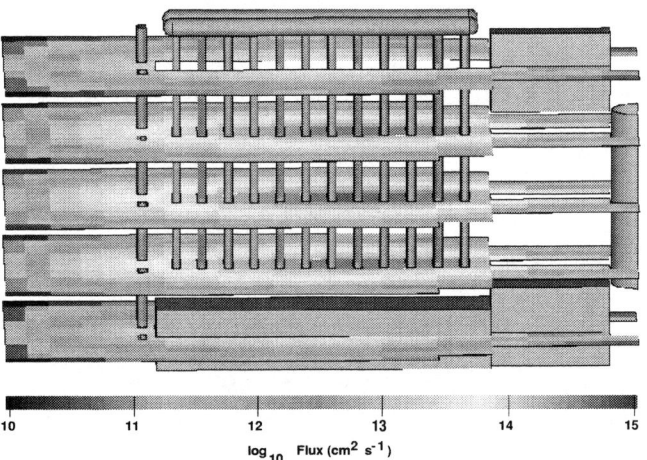

Figure 3: Contour plot of the neutron flux at a beam energy of 1.0 GeV.

At present, neither available experimental data nor any of the current models or phenomenological systematics can be used alone to produce a reliable evaluated activation cross section library covering a large range of target nuclides and incident energies. Therefore, we chose to create our evaluated library [5] by constructing excitation functions using all available experimental data along with calculations using some of the more reliable codes, employing each of these sources in the regions of targets and incident energies where they are most applicable. When we have reliable experimental data, they, rather than model results, are taken as the highest priority for our approximation. Wherever possible, we attempted to construct a smooth transition from one data source to another.

The recent *International Code Comparisons for Intermediate Energy Nuclear Data* organized by NEA/OECD at Paris [6], our own comprehensive benchmarks [3, 5, 4, 7, 8], several studies by Titarenko et al. [9] and the recent Ph.D. thesis by Batyaev [10], specially dedicated to benchmark currently available models and codes, have shown that a modified version of the Cascade-Exciton model (CEM) [11] as realized in the code CEM95 [12] and LAHET code system [13] generally have the best predictive powers for spallation reactions at energies above 100 MeV as compared to other available models. Therefore, we choose CEM95, the recently improved version of the CEM code [14], and LAHET (version 2.83) above 100 MeV to evaluate the required cross sections. The same benchmarks have shown that at lower energies, the HMS-ALICE code [15] most accurately reproduces experimental results as compared with other models. We therefore use the activation library calculated by M. B. Chadwick [8] with HMS-ALICE for protons below 100 MeV and neutrons between 20 and 100 MeV. In the overlapping region, between 100 and 150 MeV, we use both HMS-ALICE and CEM95 and/or LAHET results. For neutrons below 20 MeV, we consider the data of the European Activation File EAF-97, Rev. 1 [16, 17] with some recent improvements by M. Herman [18], to be the most accurate results available; therefore we use them here.

Measured cross-section data from our compilation described in [5], when available, are included together with theoretical results and are used to evaluate cross sections for study.

Results for 308 proton-induced and 342 neutron-induced evaluated cross sections are shown in 109 color figures on the Web, in our detailed report [4].

We consider production rates for the following 22 end product nuclides of medical importance [1]: ^{18}F, ^{35}S, ^{89}Sr, ^{133}Xe, ^{22}Na, ^{67}Cu, ^{89}Zr, ^{131}Cs, ^{32}Si, ^{32}P, ^{67}Ga, ^{95}Zr, ^{137}Cs, ^{32}P, ^{68}Ga, ^{95}Nb, ^{193m}Pt, ^{33}P, ^{68}Ge, ^{68}Ga, ^{131}I, and ^{195m}Pt. To produce these nuclides, we calculated neutron and proton reactions on 70 stable, naturally occurring isotopes of 25 elements in the neighborhood of the targets investigated (See details in [4]). ^{138,139}La, For each reaction, we 1) constructed a continuous energy representation of the cross section from the evaluation tables; 2) formed a flux-weighted average cross section for each particle flux at each location; 3) computed the one-hour irradiation end product P production rate per gram of target to each target nuclide/reaction product p radionuclide combination; 4) formed the one hour irradiation production rate per gram of target nuclide or naturally occurring element.

The flux-weighted cross section σ_{tp} for each target t and reaction product p is found by

$$\sigma_{tp} = \frac{\int_0^\infty \phi(E)\sigma_{tp}(E)dE}{\int_0^\infty \phi(E)dE} = \frac{\sigma_{tp}\Phi}{\Phi} .$$

The cross section σ_{tP} leading to end product P is taken as the sum of all cross sections for the direct production of P and products p decaying to P. The cross section σ_{ZP} for element Z leading to end product P is obtained as the natural-abundance-weighted sum of the cross sections σ_{tP} of the various naturally occurring target nuclides of the element.

The production rate R_{tP} (Ci/g-hr) for each target t – end product P combination is

$$R_{tP} = N_t \sigma_{tP} \Phi [1 - \exp(-\lambda_p T)] ,$$

where λ_p is the decay constant (s^{-1}) of end product P, $T = 3600$s corresponds to a one-hour irradiation, $N_t = N_0/A_t$ is the atom density (atoms/g) of the target material, N_0 is Avogadro's number (6.022×10^{23} atoms/mole), and A_t is the atomic weight of the target. A_t is taken as the integer mass number for isotopic targets and as the atomic weight for the elements.

Our Web report [4] includes complete tables for the production rates all isotopes studied, for all beam energies. Also, on the Web are tables that gives detail of these production rates, including the flux-averaged cross sections and production rates for intermediate products, and color contour plots of production rates for natural element target at a beam energy of 1.7 GeV.

We selected three groups of cells, all in the middle segment, to explore the dependence of the production rates on position in the target/blanket assembly and on the beam energy. Figures showing the variation of production rates with beam energy and distance from natural element targets are included in our Web report. The rates increase with beam energy, by between factors of 1.5 and 5 from beam energies of 1.0 to 1.8 GeV. The downstream cells closest to the beam cavity are less sensitive to beam energy than those further downstream. There is less spread in the beam energy dependence in the lateral cells than in the other groups. Within the beam cavity, the rates peak at a distance of 140 cm. The rates decrease exponentially with distance into the lateral blanket. Downstream of the first few cells in the downstream blanket, the rates also decrease exponentially with distance into the downstream blanket.

Commenting on the feasibility and economic viability of the production of any particular isotope is beyond the scope of this work. Such information, which must come from experts in the radionuclide arena, will be a vital ingredient in choosing which cross sections should be subjected to greater scrutiny.

This study was partially supported by the U. S. Department of Energy.

REFERENCES

[1] Spicer, K. M., Baron, S., Frey, G. D., et al., "Evaluation of Medical Radionuclide Production with the Accelerator Production of Tritium (APT) Facility," Medical University of South Carolina final report (July 15, 1997).

[2] Browne, J. C., Anderson, J. L., Cappiello, M. W., et al., in *The Savannah River Accelerator Project and Complementary Spallation Neutron Sources*, edited by F. T. Avignone and T. A. Gabriel, World Scientific, Singapore, 1998, pp. 14-36.

[3] Van Riper, K. A., Mashnik, S. G., Chadwick, M. B., et al., Los Alamos National Laboratory Report LA-UR-97-5068 (1997); hppt://t2.lanl.lanl.gov/publications/.

[4] Van Riper, K. A., Mashnik, S. G., and Wilson, W. B., Los Alamos National Laboratory report LA-UR-98-5379 (1998); hppt://t2.lanl.lanl.gov/publications/.

[5] Mashnik, S. G., Sierk, A. J., Van Riper, K. A., and Wilson, W. B., in *Simulating Accelerator Radiation Environments (SARE4)*, edited by T. A. Gabriel, ORNL, 1999, pp. 151-162; E-print *nucl-th/9812071*.

[6] Blann, M., Gruppelar, H., Nagel, P., and Rodens, J., *International Code Comparison for Intermediate Energy Nuclear Data*, NEA OECD, Paris, 1994; Michel R., and Nagel, P., *International Codes and Model Intercomparison for Intermediate Energy Activation Yields*, NSC/DOC(97)-1, OECD, Paris (1997); http://www.nea.fr/html/science/pt/ieay.

[7] Mashnik, S. G., Sierk, A. J., Bersillon, O., and Gabriel, T. A., Nucl. Instr. Meth. A **414**, 68-72 (1998).

[8] Koning, A. J., Chadwick, M. B., MacFarlane, R. E., Mashnik, S. G., and Wilson, W. B., ECN Report ECN-R-98-012, Petten (1998).

[9] Titarenko, Yu. E., Shvedov, O. V., Igumnov, M. M., et al., Nucl. Instr. Meth. A *414*, 73-99 (1998).

[10] Batyaev, V. F., Ph.D. thesis, ITEP, Moscow, 1999.

[11] Gudima, K. K., Mashnik, S. G., and Toneev, V. D., Nucl. Phys. A **401**, 329-361 (1983).

[12] Mashnik, S. G., "User Manual for the Code CEM95," JINR, Dubna (1995), http://www.nea.fr/abs/html/iaea1247.html.

[13] Prael, R. E., and Lichtenstein, H., Los Alamos National Laboratory report LA-UR-89-3014 (1989).

[14] Mashnik, S. G., and Sierk, A. J., in *Simulating Accelerator Radiation Environments (SARE4)*, edited by T. A. Gabriel, ORNL, 1999, pp. 29-51.

[15] Blann, M., Phys. Rev. C **54**, 1341-1349 (1996); Blann, M., and Chadwick, M. B., Phys. Rev. C **57**, 233-243 (1998).

[16] Muir, D. W., and Koning, A. J., in *Accelerator-Driven Transmutation Technologies and Applications*, edited by H. Condé, Uppsala University Press, 1997, vol. 1, pp. 469-475.

[17] Sublet, J.-Ch., Kopecky, J., Forrest, R. A., and Nierop, D., UKAEA, Culham, Abigdon, Oxfordshire OX 14 B, United Kingdom (1997).

[18] Herman, M., Los Alamos National Laboratory report LA-UR-96-4914 (1996).

Radio Indium and Gallium Labeled Porphyrins for Medical Imaging

P.V. Kulkarni, D. Jain and J. Narula

The University of Texas Southwestern Medical Center, Dallas, TX., Yale University School of Medicine, New Haven, CT, Hahnemann University Hospital, Philadelphia, PA

Abstract. In vitro and in vivo studies have shown that porphyrins-a family of closely related compounds localize in tumor tissue and proliferative smooth muscle cells. These compounds are photosensitive and are widely used in photodynamic therapy. These ring like molecules-macrocycles contain a central cavity and bind a variety of cations including transitional elements. Some of the related compounds have been labeled with In-111, Ga-67/68 and utilized for imaging tumor and atherosclerotic lesions. We labeled coproporphyrin, a well defined single molecule, with In-111 and used it to image atheromateous plaques in rabbits.

INTRODUCTION

Porphyrins are a large class of deeply colored red or purple fluorescent compounds having in common a substituted aromatic macrocyclic ring consisting of four pyrrole-type residues, linked together by four methine bridging groups. These molecules have certain distinguishing characteristics. The main function of porphyrins and related compounds are to bind metal atoms. Porphyrins are among the best ligands known in terms of thermodynamic stablity and kinetic non-lability. They have the ability to absorb visible light and convert it to other forms of chemical and physical energy. In vitro and in vivo fluorescent studies have shown that these molecules tend to accumulate in tumor tissue and proliferative smooth muscle cells (1). Thus, they are widely being used in photodynamic therapy. They are also being evaluated as sensitizers in chemotherapy and radiation therapy. Paramagnetic elements such as gadolinium can be incorporated into these molecules and may be used as contrast enhancement agents in magnetic resonance imaging (MRI). In-111 and Ga-67 are cyclotron produced radionuclides. They are widely used in nuclear medicine for imaging studies because of their favourable nuclear and chemical properties. Porphyrins can be labeled with radio gallium and indium and appear to be promising agents in oncologic and cardiovascular imaging studies.

Porphyrins are a family of closely related molecules. The studies described in the literature utilized compounds that are not pure single molecules but are a complex mixture of closely related compounds. Thus the results vary based upon the nature of the product used. Here we review the techniques used in radiolabeling, the nature of these molecules and their application. We labeled a well characterized single molecule-coproporphyrin (CP) with relatively high hydrophilicity (due to 4 carboxlic groups) and evaluated its potential to localize lipid laden cells in atheroscelrotic lesions in experimental animals (rabbits).

FLUORESCENCE AND PHARMACODYNAMIC STUDIES

Hematoporphyrin derivative (HPD), Photofrin II and other related compounds have distinct absorption spectra and are wisely used in photodynamic therapy. Some of these compounds are retained in the skin for an extended period after photodynamic therapy. Severe skin reactions have been observed after unintended light exposure for periods of a month or more after treatment. Activation of these agents by light at wavelength corresponding to the drugs absorbance band results in formation of highly toxic products: singlet oxygen molecular oxygen and oxygen radicals (2). Fluorescence of Porphyrins has been used clinically to localize malignant neoplasms because of their selective accumulation in these tissues (3,4).

It has been suggested that rapidly proliferating tissues in general, may preferentially concentrate some of these compounds. Proliferation of smooth muscle muscles, which are rich in lipids may contribute to the growth of atheromatous plaques. Spears et al. (5) showed that hematoporphyrin derivative (HPD) concentrated within atheromatous plaques of aorta of a variety of animal species (rabbits, monkeys). They examined the postmortem aorta for fluorescence in experimental moels of atherosclerosis 48 h after parenteral administration of HPD.

Radiolabeling of Porphyrin Related Compounds

Many porphyrin related compounds have slow blood clearance and localization of the agent in such target tissues as tumor and atherosclerotic plaques requires 48-72 hours. Because of their relatively longer half lives compared to Tc-99m (t1/2: 6h), and favorable coordination chemistry with porphyrins, In-111 and Ga-67 are ideally suited for labeling porphyrins for imaging studies.

Wong et al. (6) described a simple and efficient method of labeling hematoporphyrin derivative (HPD), with In-111 using indium chloride. Origitano et al. (7) reported that efficiency of radiolabeling using indium chloride was not consistent. They showed that rapid, consistent, high efficiency labeling of hematoporphyrin derivative was feasible by the transchelation technique using indium-oxine. Table I, describes the Rf of various components using solvent system: acetonitrile/water (50:50) and instant thin layer chromatographic (ITLC-SG) sheets.

TABLE 1. Rf values for ITLC-SG chromatogram with Acetonitrile/Water (50:50)

Reagent	Radioactive Peak Rf	Fluorescent Peak Rf
In-111 Oxine	0.0	-
In-111 Porphyrin	1.0	1.0
In-111 chloride	0.0	-
Porphyrin	-	1.0

Human Studies with In-111 Labeled Photofrin II

Photofrin II is a complex mixture of porphyrin monomers and dimers. It is most actively being utilized for photdynamic therapy of patients with brain tumors and other neoplasms. This agent was labeled with In-111 by a simple, rapid, high efficiency radiolabeling technique employing In-111 oxine (7). Origitano et al. studied the uptake and distribution of this agent in 20 patients with intracranial neoplasms, using single photon emission computed tomography (SPECT) with volume rendering in three dimension. Their studies showed that there was excellent correlation between SPECT images and contrast enhanced computed tomography (CT) or magnetic resonance (MR) images. Regions of focal uptake on SPECT images correlated with the surgical histopathological findings of the neoplasm. They observed that the kinetics of photosensitizer varied according to the tumor's histological findings, the patient's use of steroids, and among patients with similar types of tumor histology. Peak ratio of target-to-nontarget tissue varied from 24 to 72 hours after injection. SPECT images allowed repeated noninvasive, qualitative evaluation and the uptake and distribution of photosensitizing agents. It was concluded that imaging studies with radiolabeled photosensitizing agents would permit the scientific study of the factors affecting the efficacy of phtodynamic therapy in the treatment of intracranial neoplasms in humans. This would allow tailoring of photodynamic therapy to each patient.

Atherosclerotic Plaque Imaging

Fluorescent studies have demonstrated the preferential uptake of hematoporphyrin derivative (HPD) by atheromateous plaque in mammalian aorta (5). It is desirable to develop a technique to document the plaque radiographically. Thus it was proposed that a suitable porphyrin related compound may be labeled with a gamma emitter and a scintigraphic imaging technique may be used to detect the plaque. The plaques consist of many components such as macrophages and lipid laden cells. It is believed that presence of large number of lipid laden cells in a plaque may render it to be pathologically unstable and would make it susceptible for rupture leading to serious consequences. Thus, we hypothesized that radiolabeled porphyrin compounds would be suitable agents for localization of lipid rich atherosclerotic plaques, and thus would allow not only imaging the plaque but permit characterization of plaques (stable vs. unstable) by non-invasive means.

TABLE 2. Blood Clearance and Activation Wavelengths for some Porphyrins

Compound	Clearance	Activation Wavelength
Photofrin	24-48 h	630 nm
Benzoporphyrin	5 min.	690 nm
Texaphyrin	Rapid	732 nm
Tin ethyltiopurpurin	Rapid	664 nm

Wong et al. (6) labeled hematoporphyrin derivative (HPD) with In-111 by heating Indium chloride with HPD at pH 7.4, for 30 minute at 120°C. They evaluated the efficacy of In-111 labeled HPD in localizing and detecting atheromas in atherogenic New Zealand white rabbits. Biodistribution data of In-111 HPD in rabbits sacrificed 48 h post injection of the radiopharmaceutical indicated the significant amount of the tracer accumulated in the liver (36% ID), Spleen (1 % ID) and the kidneys (4% ID). Their studies showed that In-111 labeled HPD had increased uptake in atherosclerotic paque segments (~0.01 % ID/g tissue) compared to normal blood vessels (~0.0026 % ID/g). The blood clearance of the labeled agent was slow, thus the plaque could be imaged only after 24 h after the administration of the tracer. They observed that gastrointestinal and urinary radioactivity interfered in many cases resulting in equivocal images.

Coproporphyrin

It has been observed that the hydrophobic components in the porphyrins have better affinity for tumor tissue and less hydrophobic components may have beter affinty for proliferating smooth muscle cells. Thus, we hypothesized that the less hydrophobic well defined single component- coproporphyrin III hydrochloride (CP) may be labeled with In-111 and may be used for localizing atherosclerotic plaque in experimntal animals.

Coproporphyrin (CP) is relatively less hydrophobic compared to HPD due to the presence of 4 carboxylic groups. Thus, we chose this compound for labeling with In-111 and evaluated it for its potential in imaging experimental atherosclerotic lesions. We further characterized, by histologic and histochemical techniques, the type of cells the tracer accumulated in the plaques.

Purified coproporphyrin was obtained from Porphyrin Products, Inc. Logan, UT. We labeled coproporphyrin by transchelation technique described by Origitano et al. (7) by utilizing In-111 oxine. Briefly, In-111 oxine was diluted with 3 ml of sodium bicarbonate pH 7.4. Coproporphyrin solution (pH 7.4) 1.5 ml (2.5 mg/ml) was added. Mixed gently for 15 min. at room temperature. The vial was sealed, evacuated and then heated in an oil bath for 20 minutes at 120°C. The solution was cooled and diluted with pH 7.4 saline and filtered through a membrane filter.

Labeling efficiency was determined using ITLC strips and solvent system ethanol/water/ammonia (60:120:4). Another solvent system used was aetonitrile/water (50:50). The developed strips were dried and observed under a long UV wavelength light. The strips were cut into 1 cm. sections and were counted in a gamma well counter. The labeled product migrated to the solvent front, the free or unbound product remaind at the point of application. Radiochemical purity was always >95% in all experiments.

Animal Model of Atherosclerosis

New Zealand white rabbits were used in this study. Intradiaphragmatic aorta was de-endothelialized by balloon catheter. The animals were fed high fat, high cholesterol diet containing 2% cholesterol and 6% peanut oil for 12 weeks. Control animals were fed regular rabbit chow.

In-111 labeled CP, 1-1.5 mCi (~1 mg) was administered I.V. to anesthetized rabbits. Serial blood samples were obtained for blood clearance study. Planar whole body imaging was performed with a gamma camera in left lateral (LL) position at 5 min, 30 min., 1, 2, and 3 hours. Whole body SPECT imaging was performed at 2 hours in some animals. Animals were sacrificed after the last image. The aorta removed, opened and examined under UV light for fluorescence and imaged with a gamma camera.

RESULTS

Radiotracer (In-111 CP) cleared rapidly from circulation with t1/2 α of 15 min. and t1/2 β of 32 minutes. Abdominal aorta were well visualized in all experimental animals and none in control animals. Two hour images provided the earliest best visualization of the plaques. Maximum uptake (%ID/g$\pm$SD) was seen in liver (0.73$\pm$0.07), spleen (0.38$\pm$0.11), renal cortex ((0.13$\pm$0.05 and bone marrow (0.06$\pm$0.02). Atherosclerotioc lesion to control ratio was ~11.

Histologic and histochemical characterization of CP uptake in atherosclrotic lesion demonstrated that agent was localized in the regions stained by oil Red-O dye which stains the lipid laden cells but not on the regions of macrophages. These studies demonstrated that the lipid pool in atherometous lesions is the predominant site of radiotracer uptake. Unstable atherosclerotic plaques are pathologically characterized by attenuated fibrous cap, large lipid pool and intense macrophage infiltration. Due to rapid clearance and affinity for atherosclerotic lipid pool, radiolabeled CP should be an attractive agent for detection of unstable plaques.

SUMMARY

Our studies demonstrated that In-111 labeled coproporphyrin is taken up by atherosclerotic plaques and allows noninvasive imaging of experimental animal atherosclerosis . Blood clearance is rapid enough to enable imaging as early as 2 hours after injection of the tracer. It is an interesting novel radiotracer for noninvasive imaging of lipid rich atheromatous plaques with a potential for use in humans. Radiolabeled porphorin related compounds enable tailoring of photodynamic therapy of tumors in humans and experimental studies demonstrated the potential of these agents in cardiovascular imaging studies for localizing atherosclerotic lesions.

ACKNOWLEDGMENTS

Authors would like to thank Drs. Peter Antich and Robert Parkey for their encouragement and the technical support of Dr. Anca Constantinescu and staff at UT Southwestern Medical Center at Dallas, TX and Hahnemann University Hospital, Philadelphia, PA.

REFERENCES

1. Dougherty, T..J., *Porphyrin Localization and Treatment of Tumors*, New York: Alan R. Liss, 1984, pp. 75-87.

2. Henderson,B. W., and Dougherty, T.J., *Photochem Photobiol* **55**:145-157 (1992)

3. Lipson ., R. I., Baldes, E. J. and Olson, A. M., *Dis. Chest* **46**, 676-679 (1964)

4. Diamond, I., McDonagh.,A. F., Wilson., C. B., et.al., *Lancet* **II**, 1175-1177 (1972)

5. Spears, J. R., Serur, J., Shrophire, D., and Paulin, S., *J Clin Invest* **71**:395-399 (1983)

6. Wong., D. W., *Int. J. Appl. Radiat. Isot.* 35, 691-692 (1984)

7. Origitano., T. C., Karesh., S. M., Reichman., O. H., et. al., *J. Nucl. Med* **29** 927 (1988)

Non-Standard Isotope Production and Applications at Washington University

Timothy J. McCarthy, Deborah W. McCarthy, Richard Laforest, Heather M. Bigott, Frank Wüst, David E. Reichert, Michael R. Lewis and Michael J. Welch

Mallinckrodt Institute of Radiology, Washington University School of Medicine, St. Louis, MO 63110, USA

Abstract. The positron emitting radionuclides, oxygen-15, nitrogen-13, carbon-11 and fluorine-18 have been produced at Washington University for many years utilizing two biomedical cyclotrons; a Cyclotron Corporation CS15 and a Japan Steel Works 16/8 cyclotron. In recent years we have become interested in the production of non-standard PET isotopes. We were initially interested in copper-64 production using the $^{64}Ni(p,n)^{64}Cu$ nuclear reaction, but now apply this technique to other positron emitting copper isotopes, copper-60 and copper-61. Copper-64 is being produced routinely and made available to other institutions. In 1999 over ten Curies of copper-64 were produced, making copper available to thirteen institutions, as well as, research groups at Washington University. We are currently developing methods for the routine productions of other PET radioisotopes of interest, these include; bromine-76, bromine-77, iodine-124, gallium-66 and technetium-94m.

INTRODUCTION

At Washington University there are two biomedical cyclotrons available for the production of PET (positron emission tomography) radionuclides. These are a Cyclotron Corporation CS-15 cyclotron and a Japanese Steel Corporation 16/8. For many years, we have been producing ^{15}O, ^{11}C, ^{13}N and ^{18}F on these cyclotrons. Over the past several years we have developed techniques for the production of other medium half-life radionuclides. The characteristics of the isotopes produced are listed in Table 1.[1]

Solid Target Holder

A high-power solid target holder has been developed and utilized for the production of ^{64}Cu by the $^{64}Ni(p,n)^{64}Cu$ nuclear reaction (using ^{64}Ni electroplated onto a gold disk).[2,3] This patented target system was developed in collaboration with Newton Scientific Incorporated (Cambridge, MA) under SBIR and STTR funding. The target accommodates a 1.9 cm x 1.5 mm thick disk. During irradiation, the target is held in place by vacuum, and a pneumatically controlled air cylinder provides a water seal to the back of the disk (Figure 1). High purity water from a self-contained water chiller flows through the air cylinder head and across the back of the target disk during an irradiation providing, efficient cooling. After the irradiation, the water is purged from the cylinder head, the head is retracted and the disk is ejected with a small overpressure of dry nitrogen in the target chamber. The target is then transported via a pneumatic line to a hot cell for processing

Isotope Production

The solid target holder was initially developed for production of copper-64, we are now able to produce ^{60}Cu and ^{61}Cu using the same approach.[4] The target holder has also proven useful for the production of, ^{76}Br, ^{77}Br, ^{124}I, ^{66}Ga and ^{94m}Tc.[5]

Halogens

The halogens are produced from the corresponding copper selenide (for bromine) and copper telluride or tellurium dioxide (for iodine).[6] In both cases, the target material is pressed and then melted into a 6 mm diameter depression (1 mm deep) on a platinum coated

TABLE 1. Positron emitting radionuclides of interest at Washington University

Isotope	Half-life	Decay Modes/ % (maximal β^+ energy, MeV)	Reaction	Natural abundance of target isotope
Oxygen-15	123 s	β^+/100 (1.74)	^{14}N(d,n) ^{15}N(p,n)	99.6 %
Nitrogen-13	9.96 m	β^+/100 (1.20)	^{12}C(d,n) ^{16}O(p,α)	98.9 % 99.8 %
Carbon-11	20.34 m	β^+/>99 (0.97) EC/0.19	^{14}N(p,α)	99.6 %
Fluorine-18	109 m	β^+/97 (0.635) EC/3	^{18}O(p,n)	0.204 %
Copper-60	23.7 m	β^+/100 (3.92)	^{60}Ni(p,n)	26.1 %
Technetium-94m	52.0 m	β^+/72 (2.47) EC/28	^{94}Mo(p,n)	9.3 %
Copper-61	3.32 h	β^+/60 (1.22) EC/40	^{61}Ni(p,n) ^{60}Ni(d,n)	1.25 % 26.1 %
Gallium-66	9.49 h	β^+/56.5 (4.15) EC/43.5	^{66}Zn(p,n)	27.8 %
Copper-64	12.7 h	β^+/19 (0.66) EC/43 β^-/38	^{64}Ni(p,n)	1.16 %
Yttrium-86	14.74 h	β^+/34 (3.15) EC/66	^{86}Sr(p,n)	9.9 %
Bromine-76	16.2 h	β^+/57 (3.98) EC/43	^{76}Se(p,n)	9.1 %
Bromine-77	2.4 d	β^+/0.7 (3.15) EC/99.3	^{77}Se(p,n)	7.6 %
Iodine-124	4.18 d	β^+/25 (3.15) EC/75	^{124}Te(p,n)	4.8 %

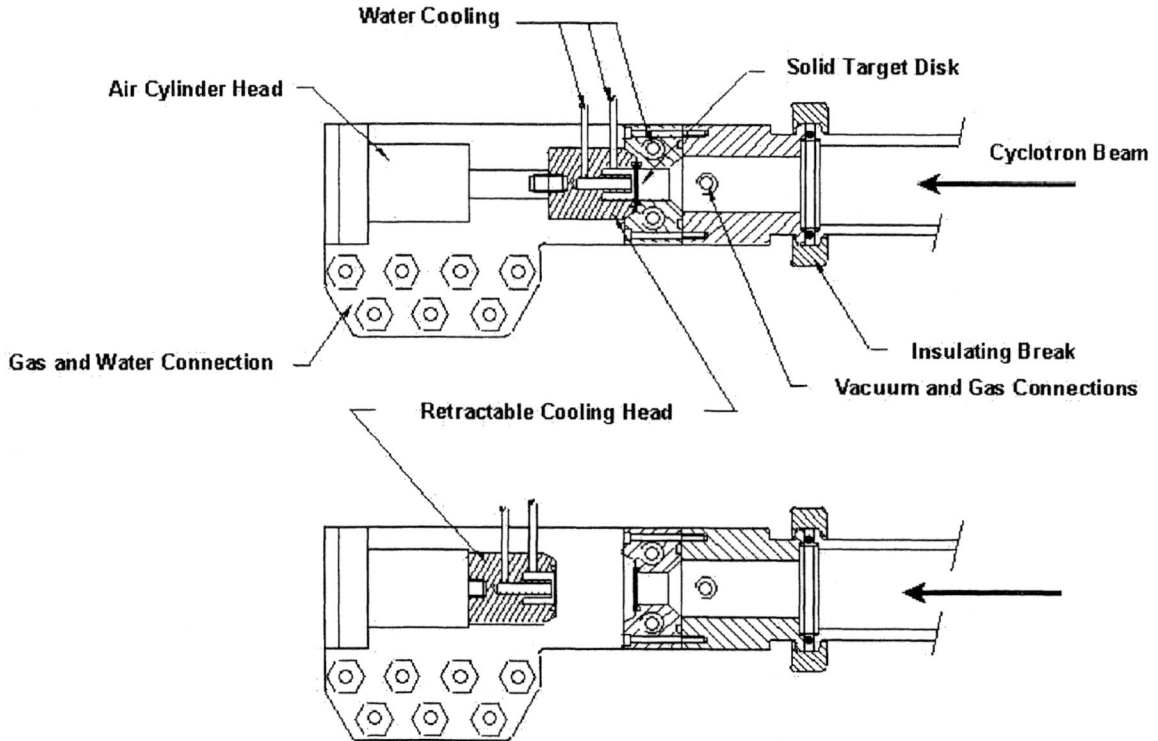

FIGURE 1. Solid target holder that is mounted to the CS-15 cyclotron.

Tungsten. The dimensions of the disk are identical to those of the gold disks used in the copper isotope productions. Due to the poor heat conductivity of Cu_2Se and Cu_2Te or TeO_2, a thick target will result in target melting even at low beam intensity. The target thickness is thus chosen so that the maximum production is achieved without stopping the beam in the target material, thus avoiding the Bragg peak where the power density is maximum. Cu_2Se target thickness' of approximately 150mg/cm^2 are routinely fabricated for a yield of 2.5 mCi/μA-h of ^{76}Br. The targets are then placed into the high power holder. For the production of ^{124}I, the beam is degraded from 14.5 MeV to 11 MeV, in order to eliminate the concurrent production of ^{123}I via the (p,2n) reaction. Beam intensities around 5 μA are commonly used without damage or loss of target material.

The radiohalogens are extracted from the target by a dry distillation technique similar to previously published work.[7-9] The target, placed inside a quartz tube, is heated by induction just below the melting point of the target material. A gentle stream of Ar displaces the activity and also prevents oxidation of the target and target material at high temperature. The activity is then recovered by rinsing the quartz tube with water. The use of inductive heating allows heating of the target to its melting temperature in less that one minute.

Gallium-66 and Technetium-94m

Gallium-66 has been produced via the $^{66}Zn(p,n)^{66}Ga$ nuclear reaction on enriched ^{66}Zn foils. We have designed an aluminum disk with a mounting system that allows for fast release of the activated foils. This assembly was bombarded in our solid target holder. The ^{66}Ga produced in these experiments has been used in several labeling and imaging studies.[10]

The processing of ^{94m}Tc was accomplished by a thermal distillation approach in a quartz apparatus according to Rösch et al.[11] The distillation approach allows the convenient separation ^{94m}Tc as $HTcO_4$ from the MoO_3 target in a temperature gradient within the designed quartz apparatus, employing wet air as the carrier gas. The entire process takes 20 minutes, and the recovery rate of MoO_3 was >85%.

NCI Funded Research Resource

Recently, the National Cancer Institute designated the Mallinckrodt Institute of Radiology at Washington University Medical School as a Research Resource in Radionuclide Research. The purpose of this resource is to provide novel radionuclides for collaborative research and other applications. The research component allows investigators from other institutions to visit Washington University, access the isotopes and our laboratory facilities. A second aspect of the resource allows us to ship radioisotopes to investigators at other institutions throughout the United States.

Copper-64 has been the first isotope to be produced and distributed thorough the resource, we anticipate that the other isotopes will all be available within the next six months. We have been routinely producing ^{64}Cu at Washington University and are evaluating it as a PET imaging isotope[12,13] and as a radiotherapeutic isotope.[14] Our collaborators at UCLA are evaluating ^{64}Cu-PTSM for PET imaging of in vivo cell trafficking.[15] Through a collaboration with Sam Gambhir at UCLA and Anna Wu and Jack Shively at City of Hope, we have shipped ^{64}Cu for labeling of a novel engineered antibody fragment administered to mice bearing carcinoembryonic antigen-positive tumor xenografts. High uptake in the CEA-positive tumor was observed using the microPET scanner (Concorde Microsystems, Knoxville, TN).[16] In addition, we are making ^{64}Cu available to 13 institutions across the country. In 1999, approximately 10 curies of ^{64}Cu was produced and made available to investigators at Washington University (Table 2), and 43 shipments of ^{64}Cu were made to other institutions.

TABLE 2. Copper-64 production at Washington University.

Year	Copper-64 yield (mCi)	Number of bombardments
1995	2,519	14
1996	4,242	29
1997	5,626	37
1998	7,924	41
1999	10,375	36
2000 (thru' Sept)	9,970	38
TOTAL	40,656	195

We have also been producing copper-60 on a routine basis (Table 3). Investigators at Washington University and collaborators are utilizing ^{60}Cu-labeled ATSM for delineating hypoxia in tissues.[17-19] Copper-61 has also been utilized for imaging and labeling studies. To date we have produced 1,784 mCi of ^{61}Cu in 18 bombardments.

TABLE 3. Copper-60 production at Washington University.

Year	Copper-60 yield (mCi)	Number of bombardments
1996	1,235	3
1997	1,810	5
1998	5,480	17
1999	8,165	28
2000 (thru' Sept)	9,400	28
TOTAL	26,090	81

ACKNOWLEDGMENTS

The authors are grateful to Bill Margenau for his invaluable technical assistance. The National Institutes of Health (R42 CA86307) and the Department of Energy (FG02-97ER82442) supported this research.

REFERENCES

1. Nickles, R.J., *J. Labelled Comp. Radiopharm.* **30**, 120-122 (1991).

2. McCarthy, D.W., Shefer, R.E., Klinkowstein, R.E., et al. *Nuc. Med. Biol.* **24**, 35-43 (1997).

3. Welch, M.J. McCarthy, D.W., Klinkowstein, R.E. *U.S. Patent* 6,011,825. (2000).

4. McCarthy, D.W., Bass, L.A., Cutler, P.D. et al. *Nuc. Med. Biol.* **26**, 351-358 (1999).

5. Welch M.J., Lewis, M.R., Downer J.B., et al. *Eur J. Nuc. Med.* **25**, 9798 (1998).

6. McCarthy, T.J., Laforest R., Downer, J.B., et al. "Investigation of ^{124}I, ^{76}Br and ^{77}Br production using a small biomedical cyclotron – can induction furnaces help in the preparation and separation of targets?" *Proceedings of the 8th Workshop on Targetry and Target Chemistry*, St. Louis, Mo June 23-27, 1999.

7. Tolmachev, V. Löqvist, A., Einarsson, L., et al. *Appl. Radiat. Isotopes* **49**, 1537-1540 (1998).

8. Kovacs, Z. Blessing, G., Qaim S.M., et al. *Appl. Radiat. Isotopes* **36**, 635-642 (1985).

9. Vallburg, W., Paans A.M., Terpestra J.W., et al. *Appl. Radiat. Isotopes* **36**, 961-964 (1985).

10. Mathias C.J., Lewis, M.R., Reichert, D.E., at al. *J. Nuc. Med.* **41**, 42P (2000).

11. Rösch, F., Novgorodov, A.F., and Qaim, S.M., *Radiochimica Acta* **64**, 113-120 (1994).

12. Lewis, J.S., Sharp, T.L., Jones, L.A. et al. *J. Nuc. Med.* **41**, 115P (2000).

13. Sharp T.L. Lewis, J.S. Herrero, P., et al. *J. Nuc. Med.* **41**, 175P (2000).

14. Lewis, J.S., Buettner, T.L., Connett, J.M., et al. *J. Nuc. Med.* **41**, 115P (2000).

15. Adonai, N., Nguyen, K. Walsh, J., et al. *J. Nuc. Med.* **41**, 256P (2000).

16. Wu, A.M., Yazaki, P.J., Tsai, S.W., et al. *Proc. Natl. Acad. Sci. USA* **97**, 8495-8500 (2000).

17. Mintun, M.A., Berger, K.L., Dehdashti, F., et al. *J. Nuc. Med.* **41**, 58P (2000).

18. Dehdashti, F., Mintun, M.A., Lewis, J.S., et al. *J. Nuc. Med.* **41**, 34P (2000).

19. Welch, M.J., and McCarthy, T.J., *J. Nuc. Med.* **41**, 315-317 (2000).

Production of Ultra-Pure I-123 from the ^{123}Te(p,n)^{123}I Reaction

H.B. Hupf, J.E. Beaver, J.M. Armbruster and J.P. Pendola

International Isotopes Incorporated, Denton, Texas

Abstract. The problems and potentials for producing I-123 (T1/2= 13.27 h) with high radionuclidic purity (RNP) in Curie quantities has been discussed in many publications for over 35 years. Depending upon the nuclear reaction selected, the isotopic enrichment of the target material used and the energy of the protons incident on the target, I-124 (T1/2 = 100 h), I-125 (T1/2= 59.4 d) and/or I-126 (T1/2 = 13 d) may be co-produced with the I-123 and significantly reduce the RNP at the time of clinical use. For the (p,n) reaction, the important factors which influence RNP are Te-123 enrichment, Te-123:Te-124 ratio and the incident proton energy. Using targets of 99.3% Te-123 enrichment with a Te-123/Te-124 greater than 990 and 15.7 MeV incident proton energy, we have produced I-123 with a RNP of 99.9+% at EOB. Thirty hours after EOB, the average RNP is 99.8+%. Production yields are being increased gradually, from near 500 mCi to 1 Ci per run at EOB.

INTRODUCTION

While the predominant medical use of I-123 is for thyroid studies as sodium iodide, there is an increasing demand for the radionuclide as a tracer label for various agents used for cancer, neurological and cardiac imaging. More than two dozen accelerator production methods for I-123 are theoretically possible, however the proton reactions are generally favored. Proton induced reactions on isotopically enriched xenon, tellurium and natural iodine (100% I-127) have been used to produce curie quantities of I-123 with varying percentages of co-produced radionuclidic impurities. The ^{127}I(p,5n)^{123}Xe (2.0 h) ---- ^{123}I reaction requires 60 to 70 MeV protons and co-produces ^{125}Xe (17 h) by the (p,3n) reaction and leads to ^{125}I as a radionuclidic impurity. I-123 produced from ^{124}Xe proceeds through three reactions at 20 to 30 MeV; (p,2n)^{123}Cs (5.8 m) ----- ^{123}Xe (2.0 h)---- ^{123}I (13.27h), (p,pn)^{123}Xe ---- ^{123}I and (p,2p)^{123}I. Some ^{125}Xe is co-produced through several possible nuclear reactions and leads to small quantities of I-125 in the I-123 product. The ^{124}Te(p,2n)^{123}I reaction at 27 MeV incident proton energy co-produces ^{124}I from the (p,n) reaction and from the (p,2n) reaction on traces of ^{125}Te in the enriched target material.

Pure I-123 could be produced from the ^{123}Te(p,n) reaction at 16 MeV incident proton energy if 100% enriched Te-123 was used as target material[1,2]. We have obtained 99.3% enriched ^{123}Te with 0.1% ^{124}Te to evaluate the yield and radionuclidic purity of the I-123 produced by the (p,n) reaction using incident protons at 15.8 +/- 0.1 MeV.

MATERIALS and METHODS

Bombardments were performed using a Cyclotron Corporation CP-42 variable energy cyclotron. The isotopic analysis of the Te-123 target material is shown in Table I.

TABLE I. Analysis of Te-123 Enrichment

Isotope	120	122	123	124	125	126	128	130
%	0.05	0.55	99.3	0.1	.05	.05	.05	.05
+/-		0.05	0.1	.05				

Radiometric analyses were performed using an intrinsic germanium detector coupled with an EG&G multichannel analyzer. Energy and efficiency calibrations were performed using NIST standards.

Initially, 49 mg of the enriched Te-123 was pressed into a cup target made from 99.9+% silicon, covered with a thin havar window and

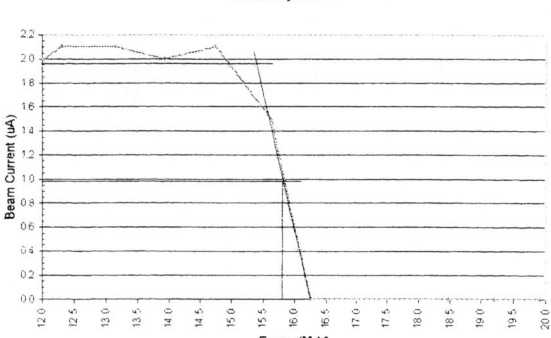

Figure 1: Beam Energy Measurement

fixed to the 15 MeV variable energy proton beam line. Two such targets were run for a total of 2.95 µA-hr and 0.43 µA-hr respectively for analysis of radionuclidic purity. Following 60 hours of radiation cooling, the targets were disassembled and a few mgs of the Te-123 were analyzed repeatedly for I-123, I-124, and I-126. Two weeks after EOB, when the I-123 had decayed, the samples were analyzed for Te-123m and I-125. Since I-123 and Te-123m both have principal gamma emissions at 159 keV, 83% and 84% abundant respectively, the Te-123 assay was used to correct the I-123 value, both decay corrected to EOB.

A dedicated beam line, target station and target transfer system to an adjacent hot cell was assembled and tested. The proton energy at the target station was measured at 15.8 MeV using a standard transmission foil energy method and the results are shown in Figure 1.

Twenty-two Te-123 electroplated copper flat plate targets were subsequently bombarded to assess target integrity, radionuclidic purity and I-123 production rate. The beam current on the target series was gradually increased from 30 to 90 uA, total bombardment time from 1 to 3.6 hours and the iodine was separated from the target material in the adjacent hot cell by distillation. Radiometric analyses for nuclide identification, radionuclidic and radiochemical purity was performed for each bombardment. Prior to the enriched Te-123 runs, several naturally occurring tellurium targets were run, chemically separated and analyzed, to validate the procedures. Data from the natural Te runs are not included below.

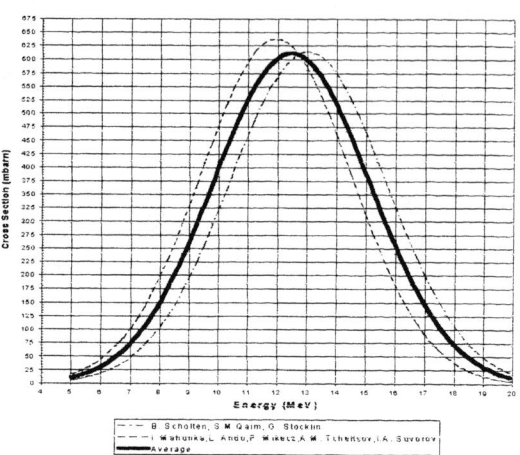

Figure 2: Smoothed Composite Excitation Function

RESULTS and DISCUSSION

The choice of incident proton energy of 15 to 16 MeV was based on a smoothed composite excitation function derived from previous work reported by others [2,3] and shown in Figure 2. Results of the powder cup targets, shown in Table II, demonstrate that very high purity I-123 is produced as predicted by the literature.

TABLE II. Powder Cup Target Results
*Te-123 contribution to the I-123 subtracted from the I-123 EOB yield.

Beam	Time,hr	% at EOB I-123	I-124	Te-123m*
1. 2.95 µA-hr	3	99.96	0.038	0.5
2. 0.43 µA-hr	1.3	99.97	0.026	0.3

Results for the thin target series of electroplated targets, where the produced iodine was separated from the tellurium target material, are shown in Table III. Column 2, Target,mg/cm^2, represents the actual plating thickness of the target. Since the targets were bombarded at a grazing angle, the actual thickness seen by the proton beam is thicker and varies with the actual plating thickness / SIN of the grazing angle.

TABLE III. Electroplated Target Results

I.D.	Target	Run	Beam	μA-hr	% at EOB			Total
					I-123	I-124	Te-121	I-123
	mg/cm²	hr	μA		%	%	%	mCi
0413	9.1	1.0	30	30	99.92	0.07	None Detected	18.73
0414	16.5	1.0	40	40	99.92	0.07	None Detected	23.57
0418	12.5	1.0	50	50	99.97	0.07	None Detected	34.60
0419	11.8	2.2	30	60	99.85	0.02	0.11	48.67
0420	11	2.2	40	85	99.96	0.03	None Detected	40.46
0425	12.0	1.3	50	65	99.96	0.03	None Detected	40.46
0609	22.6	1.3	50	65.7	99.96	0.04	None Detected	100.56
0616	15.6	2.0	50	100	99.96	0.03	None Detected	129.06
0317	14.0	3.1	50	151	99.98	0.02	None Detected	177.00
0503	13.1	2.2	60	120	99.96	0.03	None Detected	140.48
0509	20.8	3.0	60	147	99.96	0.03	None Detected	153.95
0218	14.0	1.6	60	62.2	99.96	0.03	None Detected	130.74
0629	17.1	2.1	80	160	99.93	0.07	None Detected	324.30
0302	19.6	3.0	80	240	99.96	0.07	None Detected	565.32
0306	18.5	3.2	85	255	99.98	0.02	None Detected	345.95
0334	10.2	3.0	5	255	99.97	0.03	None Detected	337.65
0335	18.0	3.5	85	300	99.96	0.04	None Detected	561.70
0336	10.2	3.5	85	300	99.97	0.03	None Detected	421.75
0349	10.0	3.6	90	300	99.95	0.05	None Detected	550.80
0350	11.3	3.4	90	305	99.97	0.03	None Detected	493.62
0370	14.2	3.4	90	300	99.98	0.02	None Detected	557.19
0387	10.0	3.4	90	300	99.98	0.02	None Detected	459.42

The average radionuclidic purity of the I-123 is 99.96% at EOB, 99.92% at 15 hours post EOB, 99.84 at 30 hours and 98.80% at 72 hours post EOB. The radiochemical purity for each lot is 95+% as iodide determined by chromatography in 70% methanol in water.

The above values meet or exceed current industry specifications. Iodine 125 and I-126 were not detected in any batches when counted at least 10 days post EOB. Note that lot 0419 in Table III contained 0.11% Te-121 (T1/2= 17 d). the Te-121 arises from the decay of I-121 (T1/2=2.1h) produced during bombardment. It is avoided, as evidenced in Table III, by allowing approximately 1 hour from EOB till the start of the iodine distillation for sufficient decay of the I-121 before distillation.

Based on the appearance of the targets following bombardments, beam currents up to 90 μA per hour were well tolerated on target plates up to 23 mg/cm^2.

The yields for the first six enriched targets, numbers 0413 thru 0425 in Table III, averaged 0.67 mCi/μA-hr and were run primarily to verify the radionuclidic purity and target integrity with increasing beam current. Following these runs, some improvements to the beam strike on target were made. The yields for the next 16 targets, numbers 0609 thru 0387, averaged 1.59 mCi/μA-hr. The apparent thickness to the beam ranged between 100 and 200 mg/cm^2. These yields are in reasonable agreement with published thick target yield data [2] shown in figure 3.

Additional work is currently underway to increase the target thickness to the proton beam and approach yields of 5 mCi/uA-hr.

CONCLUSION

Ultra-Pure I-123 can be produced from the ^{123}Te(p,n) ^{123}I reaction with 15 to 16 MeV protons using very highly enriched Te-123. Using an appropriate target plate thickness and beam to target grazing angle, curie quantities per run can be achieved. Economic considerations require an efficient chemical recovery and reuse of the Te-123 target material.

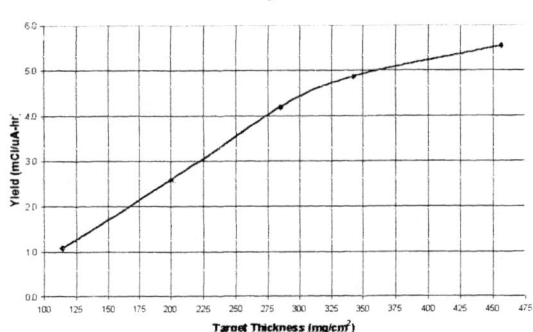

Figure 3: Calculated Yield as a Function of Target Thickness

ACKNOWLEDGEMENTS

We wish to acknowledge the assistance of Jamie Ellis for many helpful processing suggestions, Phil Wanek and the QC staff for the radioanalyses and contributions to the project.

REFERENCES

1. Barrall,R.C.,Beaver,J.E.,Hupf,H.B.,Rubio, F.R., Eur J Nucl Med (1981) 6:411-415; Production of High Purity I-123 with 15 MeV Protons.
2. Mahunka,I., Ando,I., Mikecz P., Tcheltsov, A.N., Suvorov,I.A., J Radioanal Nucl Chem Letters, 213 (2) 135-142 (1996); I-123 Production at a Small Cyclotron for Medical Use.
3. Schloten, B.,Qaim,S.M.,Stocklin,G., Appl Radiaiti Isot, 40(1987) 127, Excitation Functions of Proton Induced Nuclear Reactions on Natural Tellurium and Enriched Te-123 via the Te-123(p,n) I-123 at a Low Energy Cyclotron.
4. Mahunka, I.,Ando, I.,Kikecz,P.,Tcheltsov, A.N., Suvorov,I.A., Proc. 6th Symp Appl Cycl, June 1-4,1992, Turku, Finland.

Cyclotron Production and Potential Clinical Application of Iodine-124 Labeled Radiotracers

R. Finn, J. Balatoni, P. Kothari, K. Pentlow, Y. Sheh, C. Lom, J. Dahl, W. Eckelman[1], P. Plascjak[1], H.R. Adams[1], S.M. Larson

Memorial Sloan-Kettering Cancer Center, New York, NY and Clinical Center, National Institutes of Health, Bethesda, MD[1]

Abstract. Positron emission tomography (PET) is a dynamic molecular imaging technique applicable to clinical research, drug development as well as clinical diagnoses. The potential for PET is derived from specificity of the radiotracers and radioligands that are synthesized to monitor the biochemical or physiological processes. Further developments will depend on an increasing availability of unique radiotracers. Iodine-124, a radionuclide that has potential for both diagnostic and therapeutic applications, possesses a half-life of 4.18 days and decays by positron emission (23.3%) and electron capture (76.7%). The preparation of this radionuclide via the $^{124}Te(p,n)^{124}I$ nuclear reaction is described as well as chemistry associated with the preparation of specific radiotracers and radiopharmaceuticals incorporating iodine-124 at Memorial Sloan-Kettering Cancer Center.

INTRODUCTION

Treatment decisions in oncology are increasingly guided by information on the biologic characteristics of the tumors. In addition to size, location and extent of tumor, more specific tumor biologic properties, such as measurement of cellular proliferation and the expression of particular tumor proteins, can influence the treatment decision. Nuclear Medicine, a specialty of medical imaging which uses a variety of radionuclides incorporated into specifically designed pharmaceutical agents to achieve a time-dependent molecular image, is ideally suited to measure regional biology.

Positron Emission Tomography (PET) has become an increasingly important tool for nuclear medicine with most PET studies reported using the radiotracer [^{18}F]-FDG to trace glucose metabolism, a non-specific process essential for tumor growth but also necessary for inflammatory responses and tissue repair. The success of PET in oncology has stimulated a renewed demand among clinical investigators for short-lived radionuclides and suitable tracers for targeted studies of tumor biology, including cellular proliferation, protein and membrane synthesis, tissue hypoxia, tumor receptor and/or gene product expression. Iodine-124, a positron emitting radionuclide, is a nuclide which has both diagnostic and therapeutic applications.

The detailed preparation of iodine-124 via the $^{124}Te(p,n)^{124}I$ has recently been reported (1) and therefore, this manuscript will concentrate upon the synthetic route for the preparation of specific radiotracers and radiopharmaceutical development incorporating iodine-124 at Memorial Sloan-Kettering Cancer Center.

METHODS AND MATERIALS

General

Chemicals and solvents of the highest available quality were purchased from either Aldrich, J.T. Baker, or Fisher Scientific. All

HPLC solvents were filtered (0.45 μm, nylon or PTFE, Alltech) prior to use. Water (ultra-pure, ion free quality) was obtained from a Millipore Alpha-Q Ultra-pure water system.

Enriched tellurium dioxide/aluminum oxide solid targets were irradiated on either the CS-15 cyclotron at Memorial Sloan-Kettering Cancer Center or the CS-30 cyclotron at the Clinical Center, National Institutes of Health. The enriched granular elemental tellurium was purchased from NF Chemical, Port Chester, NY and had an initial isotopic composition of 99.7% with Te-125 accounting for 0.28%. The tellurium dioxide was synthesized as described in reference (2).

Multichannel analyses were performed using an end window HPGe detector (83.5 cm^3, FWHM 1.8 keV @ 1.33 MeV) and Canberra model 35 plus Analyzer. A BioScan model 200 imaging scanner and autochanger 1000 was used to analyze thin layer radio-chromatograms.

Production of Iodine-124

Iodine-124 was prepared using the ^{124}Te(p,n) and ^{124}Te(d,2n) nuclear reactions (the MSKCC CS-15 and NIH CS-30 cyclotrons respectively) upon the same target matrix. The target material consisted of an admixture of 250 mg of tellurium-124 dioxide with 6.7% by weight of aluminum oxide. The target matrix was formed by melting the tellurium dioxide and aluminum oxide admixture within the platinum cavity. For proton irradiations, the platinum target backing served as an internal monitor of target penetration by the beam. Target thinning or beam positional changes were signaled by concurrent production of gold-194 ($t_{1/2}$ = 39.5 h; $E_{threshold}$ = 3.31 MeV) and gold-196 ($t_{1/2}$ = 6.18d; $E_{threshold}$ = 2.28 MeV) via the (p,n) nuclear reactions on the corresponding stable platinum nuclide.

Radioiodine was recovered by a dry distillation process with efficient trapping of the volatile iodine species on the inner surface of a coated Pyrex tube. Coated Pyrex tubes were prepared by wetting their interior with 0.1 N NaOH prepared in 95% ethanol and vacuum drying. The concentration of NaOH per tube ranged from 50 to 80 micrograms. The trapped radioiodine within the tube was removed quantitatively by rinsing with 50 to 200 uL of water or appropriate buffer. Complete details of the procedure have been recently published (1).

CLINICAL APPLICATIONS OF IODINE-124

Nuclear medicine scans, particularly Positron Emission Tomography (PET), directly exploit the dissimilarity of tumor cell characteristics in comparison to normal cells. Malignant transformations apparently alter the enzymology of cells as evidenced by an increased rate of glycolysis, protein synthesis rate and DNA syntheses (3-5). Some examples of compounds labeled with iodine-124 at various stages of clinical investigation for specific studies within our program on improvement in cancer detection and treatment using cyclotron-produced radionuclides are outlined.

Thyroid Dosimetry

In thyroid cancer, the transport of iodide, the first step in thyroid hormonogenesis, is catalyzed by the Na+/I- symporter (6), an intrinsic membrane protein that is activated by thyrotropin (TSH) (7). The property of iodide trapping and organification is retained by differentiated thyroid cancer cells and forms the basis for the use of radioiodine diagnoses and therapies under endogenous TSH or exogenous recombinant TSH stimulation. The use of sodium iodide [I-124] dosimetry combines the superior resolution and quantitative potentials of PET for the detection of disease and determination of effective radioiodine therapy. Poorly differentiated thyroid cancer (ie, tall cell variants), anaplastic thyroid cancer and Hurthle cell carcinomas do not retain iodide.

Expression of Multi-drug Resistance

Colchicine, a naturally occurring alkaloid and a potent inhibitor of cellular mitosis, is a member of the multi-drug resistance family of drugs. As a potential indicator of resistance, the C-10 methoxy group of n-colchicine has been labeled using ^{11}C- and ^{13}C-iodomethane (8). However, the restrictions imposed by the short half-life of the carbon-11 compound prompted our investigation into the syntheses of colchicine analogues (9) labelled with radiohalogens. The preparation of no-carrier added ^{124}I labeled 10-desmethoxy-10-iododocolchicine has been accomplished by allowing 10-desmethoxy-10-bromocolchicine to react with iodine-124 labeled sodium iodide in freshly distilled 2-butanone. Separation of the labeled compound has been achieved using HPLC on silica columns with 89:10:1 parts dichloromethane: acetonitrile: methanol as solvent.

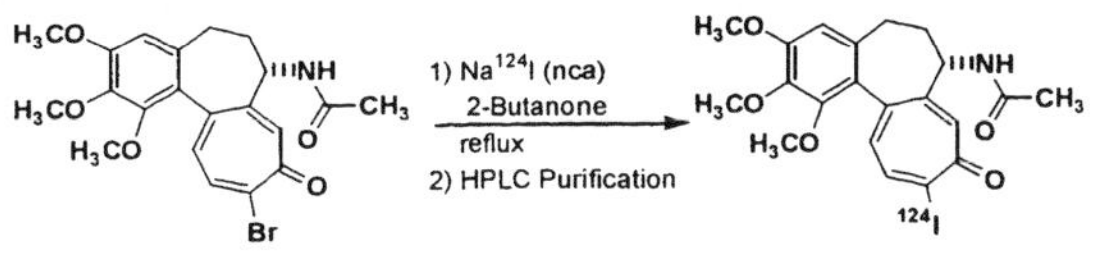

Figure 1. Synthetic scheme for Iodocolchicine.

Imaging Transgene Expression

A noninvasive, clinically applicable method for imaging the expression of successful gene transduction in target tissue or specific organs would be of immense value. Radioiodinated FIAU (2'-fluoro-2'deoxy'1-β-D-arabinofuranosyl-5-iodo-uracil) has been an effective probe for imaging the expression of the "marker/reporter gene", *HSV1-tk*. Moreover, FIAU labeled with iodine-124 has been used to obtain quantitative in vivo PET images of HSV1-tk gene expression (10) with superior sensitivity and resolution over that of SPECT iodine-131 radiolabeled FIAU images. An improved synthetic route has been investigated in which 2'-fluoro-2'-deoxy-1-β-D-arabinofuranosyl-5-(tri-n-butyltin)-uracil (11) is allowed to react with $Na^{124}I$ according to the scheme below. Typical radiochemical yields of the isolated product averaged 62% with radiochemical purities greater than 95%.

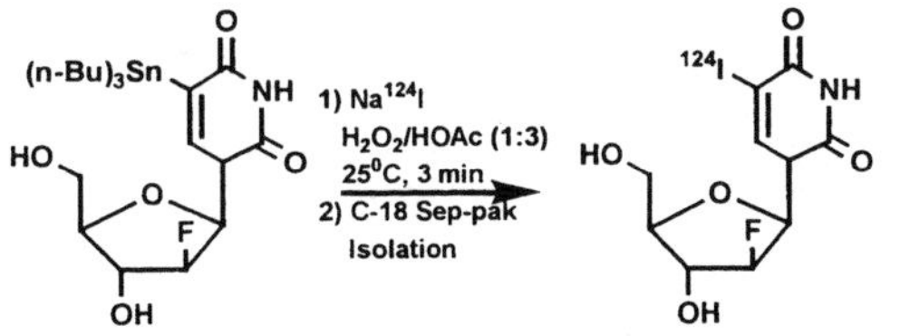

Figure 2. Synthetic scheme for FIAU

Radiolabeled Monoclonal Antibody Application

Radiolabeled monoclonal antibodies are showing increased promise for oncologic diagnosis as well as therapeutic application. The goal of radioimmunotherapy is to deliver a large radiation dose to the tumor over a finite time period but to minimize dose rate effects and radiation damage to normal tissue such as bone marrow. An increase in effectiveness of the particular radiolabeled monoclonal antibody could be achieved through the careful matching of the specific radionuclide with consideration of its biological half-life. Taking advantage of the unique decay characteristics of iodine-124 and the sensitivity and resolution of the PET camera to produce time dependent molecular images, several monoclonal antibodies have been chosen for clinical evaluation. The radiolabeled monoclonal antibody will allow an extension from functional process imaging in tissue to pathologic processes and more accurate nuclide directed treatments.

CONCLUSION

Concurrent with the technical improvements being made with the intrinsic resolution and reconstruction of images obtained with positron emission tomographs, is the increased availability of a variety of short-lived, radiolabeled substrates possessing the unique potential to serve as indicators of "in vivo" alteration of biochemical processes. Iodine-124 is such a radionuclide. Although the positron energy, positron abundance and other associated energetic emissions make the imaging characteristics of iodine-124 less ideal than fluorine-18, it has been shown that satisfactory imaging can be achieved (12). In the case of iodine-124, some photons are in cascade with the positron emission and therefore, in true time coincidence with the resulting annihilation photons. This situation results in the detection of spurious true coincidences between a gamma ray photon and a single annihilation photon (13, 14) but is in fact, correctable.

Despite of the complex decay scheme for iodine-124, the spatial resolution of images for this radionuclide is comparable to that of images obtained with the more conventional PET radionuclides and the half-life of 4.2 days is appropriate for slow physiological processes and for the clearance of nonspecific radioactivity.

ACKNOWLEDGEMENT

Supported in part by the Cancer Center Support Grant NCI-P30-CA08748, U.S.

Department of Energy DE-F02-86-E60407, and NIH Cooperative Research and Development Agreement.

REFERENCES

1. Sheh, Y., Koziorowski, J., Balatoni, J., Lom, C., et al., Low energy cyclotron production and chemical separation of "no carrier added" iodine-124 from a reusable, enriched tellurium-124 dioxide/aluminum oxide solid solution target, *Radiochim. Acta* **88**, 169-73 (2000).
2. *Handbook of Preparative Inorganic Chemistry*, G. Brauer (ed.) Vol. 1, Academic Press, 1963, pp. 447-48.
3. Weber, G., Enzymology of cancer cells, *N Engl. J. Med.*, **296**, 541-51 (1977).
4. Warburg, O., The metabolism of tumors, New York: Richard R. Smith, Inc., 1931, pp. 129-69.
5. Kallinowski, F., Schlenger, K. H., et al., Blood flow, metabolism, cellular microenvironment and growth rate of human tumor xenografts, *Cancer Research* **49**, 3759-64 (1989).
6. Dai, G., Levy, O., Carrasco, N., Cloning and characterization of the thyroid iodide transporter, *Nature* **379**, 458-60 (1996).
7. Kaminsky, S. M., Levy, O., Salvador, C., Dai, G., Carrasco, N., Na(+)-I- symport activity is present in membrane vesicles from thyrotropin-derived cells, *Proc. Natl. Acad. Sci. USA* **91**, 3789-93 (1994).
8. Kothari, P. J., Finn, R. D., Larson, S. M., Syntheses of colchicine and isocolchicine labelled with Carbon-11 and Carbon-13, *J. Label. Compd. Radiopharm.* **36**, 521-28 (1995).
9. Staretz, M. E., Hastie, S. B., Synthesis and tubulin binding of novel C-10 analogues of colchicine, *J. Med. Chem.* **36**, 758-64 (1993).
10. Blasberg, R., Tjuvajev, J. G., Herpes simplex virus thymidine kinase as a marker/reporter gene for PET imaging of gene therapy, *Q. J. Nucl. Med.* **43**, 163-9 (1999).
11. Balatoni, J., Finn, R., Blasberg, R., Tjuvajev, J., Larson, S., in: *Applications of Accelerators in Research and Industry*. Proceedings of the Fifteenth International Conference: Duggan JL, Morgan IL, (eds.); 1998 November; Denton, Texas. New York: AIP, 1999: AIP Conference Proceedings 475, Part 2, pp. 984-86.
12. Daghighian, F., Pentlow, K. S., Larson, S. M., Graham, M. C., DiResta, G. R., Yeh, S. D. J., Macapinlac, H., Finn, R. D., Arbit, E., Cheung, N. -K., In vivo kinetics of radiolabeled monoclonal antibody in human tumor: PET studies of I-124 labeled 3F8 Mab in glioma. *Eur. J. Nucl. Med.* **20**, 402-09 (1993).
13. Pentlow, K. S., Finn, R. D., Larson, S. M., Erdi, Y. E., Humm, J. L., Effects of cascade gamma rays in PET imaging and quantitation, *J. Nucl. Med.* **40**, 280P, (1999).
14. Pentlow, K. S., Finn, R. D., Larson, S. M., et al. Quantitative imaging of yttrium-86 with PET: The occurrence and correction of anomalous apparent activity in high density regions. *Clin. Pos. Imag.* (in press).

Industrial Production of ^{131}I by Neutron Irradiation and Melting of Sintered TeO$_2$

Jose Alanis* and Manuel Navarrete**

Nuclear Center of Mexico, ININ, Ocoyoacac, 50045 Mexico
**Faculty of Chemistry, Building D, CU, UNAM, 04510 Mexico

Abstract. Optimal conditions of temperature and reaction rate have been settled to produce high purity TeO$_2$ by the chemical reaction between Te and HNO$_3$. Also, heating and time conditions for sintering this product have been found, in order to create cavities in the crystal inside, where a gaseous element such as iodine can be adsorbed with minimal leaking. In this way it is fabricated a suitable target to be irradiated with thermal neutrons for obtaining ^{131}Te ($t_{1/2}$=24.8m) and ^{131m}Te ($t_{1/2}$=30h) by (n,γ) nuclear reactions. Irradiation time has been chosen to get ^{131}Te saturation activity (t_i=150m) because much longer irradiation times do not increase significantly total activity. Since parents ^{131}Te and ^{131m}Te have shorter half life than daughter ^{131}I ($t_{1/2}$=8.05d) optimal cooling time must permit daughter activity to grow up till a maximum (t_c=4d). Then, sintered cylinder shaped radioactive sample is manipulated in a hot cell, transported and put on a quartz tray, keeping Health Physics regulations. The quartz tray is inside a small electric oven enclosed in an airtight box with negative pressure (water 0.5cm). There, it is gradually heated till melting point (733°C). From 400°C on, vapors are pumped out and bubbled in two solutions: one is 0.1M NaOH, which retains nearly 99.9% of pumped ^{131}I. Other is 0.02M Na$_2$CO$_3$ (60%) plus 0.0025M NaHCO$_3$ (40%), which retains the remaining sample residue. Air filtering is accomplished by activated carbon and alumina filters in the inflow, glass wool fiber before bubbling and activated carbon again in the outflow.

INTRODUCTION

The production of ^{131}I is going to start soon at industrial scale in Mexico, by neutron irradiation followed by melting of TeO$_2$. This chemical must be produced in the optimum conditions of temperature and reaction rate during the oxidation of Te with HNO$_3$, in order to create crystals with maximum purity, suitable for the sintering process (1). This process consists in heating gradually the TeO$_2$ crystals in an oven till 700°C, and then for 5 minutes longer, with the aim to create cavities inside the crystalline structure, which permit a good adsortion of gaseous ^{131}I when it is released by decaying of ^{131}Te ($t_{1/2}$=25m and ^{131m}Te ($t_{1/2}$=30h), once the crystals have been irradiated with thermal neutrons (2). After an adequate cooling time which allows the maximum yielding of ^{131}I from decaying of both radioisotopes, the radioactive sample is melted at 733°C in an electric oven placed inside an airtight lucite box at negative pressure (0.5 cm of water) to avoid ^{131}I leaking to environment. From 400°C on, ^{131}I vapors are pumped out and bubbled in two alkaline solutions where 100% of them are dissolved. Air flowing through the oven is filtered al 3 points: input, before bubbling and output.

EXPERIMENTAL

In a previous paper (1) are described 7 experiments carried on with the purpose of finding the adequate temperature to induce the reaction:
Te + 4HNO$_3$ → TeO$_2$ + 4NO$_2$ + 2H$_2$O,
in order to obtain TeO$_2$ crystals pure enough to yield a maximum volume of gaseous ^{131}I once

Figure 1.- Experimental array to produce TeO_2

Figure 2.- Experimental array to melt radioactive $^{(131+131m)}TeO_2$ and to pump out ^{131}I

Figure 3.- Hot cell to melt radioactive sintered $^{(131+131m)}TeO_2$ and pump out ^{131}I

they are sintered, neutron irradiated and melted. The appropriate temperature to heat the reagents was 125°C. So, 10 g of metallic tellurium (Merck 1880) and 56 ml of 12N HNO_3 are put in the reaction chamber heated by an electric metallic blanket in presence of an airflow, used to carry away the NO_2 vapors, which are dissolved in three water containers, approximately 250 ml each, after being refrigerated by a water shirt at 5°C. The pumped airflow has a double function: to contribute to Te oxidation and to carry out the NO_2 vapors produced by the reaction. Suitable air pressure for the pumping was fixed at 1.2 cm of Hg by manometer, at a volume rate of 35 l/m. The reaction rate is a function of heating applied to reagents, and it can be measured by titrating the increasing HNO_3 concentration at the third bubbler water container at different times, by taking out aliquots to be titrated with 0.001N NaOH solution. Fig. 1 shows the experimental set used. When no vapors of NO_2 are released, reactor is drained out from the acid excess by opening a valve. Then, temperature is raised up to 330°C in order to dry the reaction product. Once cooled, TeO_2 is removed from the reactor and heated in an oven at 660°C during 8 hours, to remove nitrogen impurities present in it. The sintering process of TeO_2 crystals consists of putting them in a quartz tube, shaped as a cylinder, and once it is fixed in a laboratory support, it is placed in an oven and heated gradually till 700°C and then for 5 minutes longer. Once taken out from the oven and cooled down at room temperature, sintered cylinder shaped TeO_2 is easily detached from the quartz tube and ready to act as a neutron target. Thermal neutron irradiation for sintered TeO_2 sample was carried on at the following conditions: neutron flux of approximately 1.65×10^{12} n/cm^2-s in the Triga Mark III reactor of the Nuclear Center of Mexico. Irradiation time has been chosen to get the ^{131}Te saturation activity (t_i=150m) because much longer irradiation time does not increase significantly total activity. The suitable cooling time should be 4 days (t_c=4d), because this is the time taken to reach a maximum activity of ^{131}I by decaying from its parents ^{131}Te and ^{131m}Te (3). Radioactive, sintered, cylinder shaped sample, is manipulated in a hot cell, transported and put in a quartz tray inside a small electric oven enclosed in a lucite airtight box with negative pressure (0.5 cm of water), following the Health Physics regulations. There, it is gradually heated to the melting point (733°C). From 400°C on, melting vapors are pumped out and bubbled in two solutions: first one is 0.1M NaOH and second one is 0.02M Na_2CO_3 (60%) plus 2.5×10^{-3}M $NaHCO_3$ (40%). Airflow through the oven is filtered at entry by activated carbon and alumina. Before being bubbled in the alkaline solutions, ^{131}I vapors pass through one glass wool filter, and before being released, melting vapors pass again through an activated carbon filter. Fig. 2 shows the experimental set used.

RESULTS AND DISCUSSION

99.9% of the ^{131}I pumped out is dissolved in the first alkaline solution (0.1M NaOH), while the rest is caught by the second one and the activated carbon at the outflow. Shielding lead bricks all around the lucite airtight box allows a maximum dose of 5.66 Sv/h on the lead bricks surface when the experiment is running. However, the increasing background factor in the room at operating conditions has not been reduced from a value of 5, but this figure should be reduced by enlarging the lead shielding, since the whole system seems not to present any leak (pipes made from conventional PVC and screw fastened belts in every junction with glass). In our conditions of neutron flux, Te target mass (10 g), irradiation time (t_i=150m), and cooling time (t_c=4d), the ^{131}I activity collected in the first alkaline solution is approximately 400 μCi. This solution is ready to be conveniently diluted, to fix pH and distributed. Fig. 3 shows the hot cell to melt the radioactive, sintered $^{(131+131m)}TeO_2$ and pump out the ^{131}I.

REFERENCES

1.- Alanis J. and Navarrete M., Optimal conditions to obtain sintered TeO_2, useful as a thermal neutron target for the ^{131}I industrial production, *Part. Sc.and Tech.,an Int. J.* (in press).

2.- Alanis J. and Navarrete M. Optimal parameters to produce ^{131}I by neutron irradiation and melting of sintered tellurium dioxide, *Nucl. Instr. and Meth. in Phys. Res.* A422, 1999, pp 10-15.

3.- Choppin G. and Rydberg J. *Nuclear Chemistry, Theory and Applications*, Ed. Pergamon Press, Oxford, 1980, p. 74

The IBA State-Of-The-Art Proton Therapy System, Performances And Recent Results

D. Prieels, B. Marchand, B. Bauvir, P. De Crock, G. Gevers,
S. Schmidt, G. André, S. Ternier, Y. Jongen

Ion Beam Applications, S.A.,Chemin-du-Cyclotron 3, B-1348 Louvain-la-Neuve,Belgium

Abstract. In recent years IBA has continued its development of state-of-the-art systems for Proton Therapy. While the machine performance at the NPTC is such that all clinical specifications are met, IBA has continued to improve the proposed equipment to set even higher standards. Improvements in the ion source control, gantries and patient alignment systems will be addressed in the oral presentation and the first results obtained with the Pencil Beam Scanning algorithm will be presented.

INTRODUCTION

IBA has been installing its' state-of-the-art proton therapy system in the Northeast Proton Therapy Center since 1996. The machine performance is such that all clinical specifications [1] are being met. Full tests for the double scattering treatment mode and partial tests for the wobbling treatment mode have been performed.

Even during installation and now commissioning of the system, IBA has continued its design development to achieve even higher standards. The developments do not just involve improvements in the current design such as gantries or ion source control, but they led to developments that will lead proton therapy into a new era of active treatment capabilities.

IBA is developing an innovative pencil beam scanning (PBS) system. This new mode of treatment, in the vague today, was proven in the standard nozzle. It will now be implemented in a dedicated designed nozzle and will allow the therapists to perform the highest precision conformal treatments.

The IBA system is innovative in a sense that we will allow simultaneous control of the beam intensity and the speed of the beam spot.

CLINICAL RESULTS IN DOUBLE SCATTERING AND WOBBLING

Extensive tests of the double scattering treatment mode in the IBA standard nozzle at the NPTC were performed. These nozzles are designed to operate in two different treatment modes: double scattering and wobbling.

In order to summarize the clinical performances achieved in Treatment Room 1 at the NPTC, Table 1 presents the required specifications versus the achieved performances. Note that even if some of the achieved performances seem to be out of the specifications, it does not necessarily mean that the specification is not achieved. It merely indicates that that they are not achieved in TR1 with the actual options of the double scattering mode. for instance the maximum fields can only be obtained with the wobbling mode, and the minimum range is obtained with a particular passive option which is not currently implemented in TR1 as decided by MGH [1].

TABLE 1. Clinical Performances in Double Scattering [2]

	Preferred Specification	Minimum Specification	Achieved Performance
Range in patient	max = 32 g/cm² min = 3.5 g/cm²	max = 28 g/cm² min = 5.0 g/cm²	22.7 - 29.0 g/cm² (ϕ = 12 cm) 4.6 - 24.9 g/cm² (ϕ = 20 cm)
Range modulation	step of 0.5 g/cm² (R > 5 g/cm²) step of 0.2 g/cm² (R < 5 g/cm²)	step of 1.0 g/cm² (R > 5 g/cm²) step of 0.3 g/cm² (R < 5 g/cm²)	0.05 - 0.35 g/cm² (R>5.9 g/cm²) 0.05 - 0.1 g/cm² (R < 5.9 g/cm²)
Range adjustment	step of 0.1 g/cm² (R > 5 g/cm²) step of 0.05 g/cm² (R<5 g/cm²)	step of 0.1 g/cm²	< 0.1 g/cm² (R > 5 g/cm²) < 0.05 g/cm² (R < 5 g/cm²)
Average dose rate	2 Gy/min (F < 25 x 25 cm R > 9.8 g/cm²) 2 Gy/min (F < 10 x 10 cm R > 3.5 g/cm²)	0.5 Gy/min (F < 25 x 25 cm R > 9.8 g/cm²) 0.5 Gy/min (F < 10 x 10 cm R > 3.5 g/cm²)	1.9 - 140 Gy/min
Field size	40 x 30 cm	26 x 22 cm	ϕ = 20 - 24 cm
Dose Uniformity	+/- 2.5 %	+/- 4%	+/- 2.5 %
Effective SAD	> 3 m	> 2 m	$\approx$ 230 - 240 cm
Distal dose falloff (80%-20%)	< physical limit + 0.1 g/cm²	< 0.6 g/cm²	physical limit + 0.1 g/cm²
Lateral penumbra (80%-20%)	< 2 mm	< 4 mm	2.6 - 5.2 mm

The preferred specifications ask for a <u>range in patient</u> between 3.5 and 32 g/cm². We know that the maximum range of 32 g/cm² can only be achieved in wobbling. In double scattering, because of the energy loss in the scatterers, the maximum range we reached is 24.9 g/cm² for fields of 20 cm diameter and 29.0 g/cm² for fields of 12 cm diameter. The minimum range of 3.5 g/cm² can be achieved in double scattering but has not been measured since the minimum range of the passive options that the hospital chose for nozzle 1 is 4.6 g/cm².

A Bragg peak at the full energy was measured in the wobbling treatment mode. The range of this Bragg peak was 33.07 g/cm², which proves that the preferred specification will effectively be achieved in wobbling.

The preferred specification requires the <u>range modulation</u> to be adjustable by steps of 0.5 g/cm² over the full depth (0.2 g/cm² for ranges < 5 g/cm²). This specification is achieved. The adjustment is even better than the preferred specification.

The preferred specification requires the <u>range to be adjustable</u> by steps of 0.1 g/cm² (0.05 g/cm² for ranges < 5 g/cm²). This specification is achieved. The adjustment is even better than the preferred specification although the minimum adjustment was not measured.

The preferred specification requires a beam current sufficient to deliver 2 Gy/min <u>dose rate</u> over a 25 x 25 cm field size modulated to full depth for energies of about 115 MeV (= 9.8 g/cm²). The beam current must also be sufficient to deliver 2 Gy/min over a 10 x 10 cm field size modulated to full depth for all energies over 70 MeV.

Because the IBA system uses a fixed energy cyclotron with a degrader and a momentum-analyzing device, the efficiency decreases rapidly with the energy. This signifies the need for a higher beam current to deliver a dose of 2 Gy/min at the lowest ranges. For safety reasons the current at the cyclotron exit is limited to 300nA, although it could deliver much more than that. Assuming the maximum output current, the preferred dose rate can always be achieved unless for the very low ranges (< 4.5 g/cm²).

We know that the <u>maximum required field</u> of 30 x 40 cm will only be obtained in wobbling. In double scattering, we decided to limit the field size to 24 cm in diameter. The size-measured fields vary between 19 and 24 cm in diameter.

The preferred specification requires <u>dose uniformity</u> within +/- 2.5% unless in the case of passive scattering, precluded by physics limitations of an optimized beam spreading system. The dose non-uniformity due to aperture scattering is not part of the required specification and not IBA's responsibility.

The dose uniformity specification was checked by measuring lots of dose profiles in a water phantom under several conditions. These results showed that the uniformity stays within +/- 2.5% excepted sometimes at the extremity of some proximal transverse profiles, but this is due to edge effect of the aperture.

<u>The effective source of protons</u> in a double scattering system is located somewhere between the two scatterers. To measure exactly where the source is, two methods were used.

Although the two results are slightly different, both methods confirm that the SAD is about 230 cm.

The preferred specification constraints the momentum spread of the beam to contribute no more than 1 mm to the <u>distal fall-off</u> of a pristine Bragg peak. This specification is achieved provided that it deals with a non-scattered beam, i.e. if the achieved distal dose fall-off is plotted vs. the range at nozzle entrance, not the range in patient! Indeed, the scattering induces inherently important energy loss (especially for large fields) and energy spread and therefore increases the distal fall-off. The physical limit is therefore considered as the distal fall-off obtained with a mono-energetic beam at nozzle entrance.

In order to achieve the preferred specification, the slits need to be more closed. At low energies however, the slits should not be closed too much in order to obtain the preferred dose rate.

The preferred specifications constraints the beam delivery design to contribute no more than 2 mm to the 80-20% <u>lateral penumbra</u> above that which occurs from the natural penumbra increase due to multiple scattering effects in the patient. This penumbra is to be measured in air, at isocenter, with a beam defining collimator located 20 cm downstream.

The results showed that the preferred specification could only be achieved in single scattering, where in double scattering, the results vary between 2.6 and 5.2 mm.

RECENT RESULTS IN PENCIL BEAM SCANNING

Pencil beam scanning is a method in which a proton beam spot is moved by magnetic scanning while the beam intensity is adapted simultaneously, yielding finally to the desired dose distribution as aimed by a frame by frame treatment planning (each frame corresponds to one energy selection). The objectives will be to optimize time history evolution of the 3 manipulated variables (beam current, scanning magnet currents (X & Y)) along a pre-determined beam path in order to reach the prescribed dose in a very conformal way and within the minimum amount of time. The path planning is done with respect to the dynamic evolution and the associated constraints on the commanded variables while ensuring to reach the prescribed dose distribution within fixed tolerances.

IBA Pencil Beam Scanning Basics

The PBS nozzle itself will consist in different subsystems allowing (1) to control the beam (2) to monitor the beam and (3) to align the patient. position during the irradiation. All elements will be removable to facilitate the maintenance except the extra pair of quadrupoles and the two scanning magnets that will be considered as integral part of the nozzle frame.

The pair of quadrupoles allows to adjust the width of the beam spot at isocenter. A theoretical study showed that, at maximum energy, we could reach a spot of 2.5 mm (one sigma). In practice, the size of the pencil beam will be adjustable during the irradiation between σ = 2.5 mm and σ = 10 mm To allow rapid changes, the quadrupoles will be laminated.

A Scanning Magnets Power Supply of high performances is used to drive the scanning magnets that guide the beam spot into the target. This power supply consists in two fast IGBT's PWM (8kHz & 5kHz) inverters with magnet voltages "inner" regulation loops. The velocity settling time is 500 µsec, while the maximum speed obtained are: for the fast magnet 2000 cm/sec and for the slow magnet 200 cm/sec.

The beam intensity is regulated by a digital predictive controller, the Ion Source Electronic Unit (*ISEU*), driving the cyclotron source arc current and the feed-back is taken from an ionization chamber placed directly at cyclotron exit. The settling time for the ISEU is only 300 µsec.

Because of the small size of the pencil beam and thanks to the accurate control of the beam position, the IBA PBS does no longer need any patient specific device (aperture and range compensator) to conform the dose distribution to the target volume.

Using a range modulator inside the nozzle is not possible because of the scattering in the modulator would open the pencil beam. Therefore, the range is adjusted upstream in the ESS (Energy Selection System), which consists in a set of absorbers, followed by a pair of achromatic dipoles. Slits are used to limit the emittance and the energy spread of the beam.

The range modulation is therefore obtained by changing stepwise the whole beam line. For each step, the whole beam line can be tuned in less than 2 seconds. Accordingly, the degrader will also determine the minimum depth inside the patient. Taking into account the ESS efficiency for smallest energies will give the maximum dose rate for a given field size. The scanning surface is parallel to the body surface while irradiation will be done plane by plane.

First Experimental Results

The first version of the control algorithm [1] has been tested experimentally at the NPTC (Northeast Proton-Therapy Center) in Boston on the Nozzle developed by IBA for Scattering and Wobbling treatment mode. Indeed, this Nozzle is equipped with the same scanning magnets to be used in the *PBS* Nozzle and the proton therapy equipment also includes the *ISEU* to control the Beam Intensity Modulation. Unfortunately, this Nozzle is not equipped with the "extra" quadrupoles nor vacuum chamber. As a consequence, the beam spot size used for those first tests is quite large. However, the results demonstrate how accurate we can control simultaneously the beam intensity modulation and the beam spot movement to reach the prescription.

The first experimental tests were done with "raster" scanning path for the spot. The figure below shows what we mean by "raster" scanning:

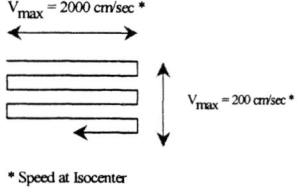

Figure 1: Raster scanning principle

Non-uniform distribution

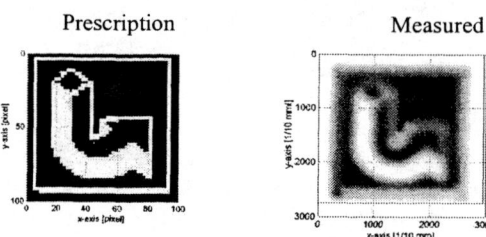

Figure 2: IBA Logo

This irradiation is shows how well the first version of the algorithm is able to deal very fast with dose gradient along the beam path and also shows the precision of the contour conformation. Note also that the time to irradiate this profile is very short, 2,1 second! The following table summarizes the characteristics of the irradiation.

TABLE 2. Settings for the Non Uniform Distribution

Beam spot maximum velocity:	500 cm/sec
Field size:	25 x 25 cm²
Number of passes:	1
Total irradiation time:	2,1 seconds

CONCLUSION

IBA proposes an innovative pencil beam scanning that allows truly conformal therapy with 3D-intensity modulation. It allows non-homogenous distributions and faster patient treatment. No apertures or compensator will result in easier operation and lower costs. Moreover, inherently to the principle, lowest activation can be achieved since almost all protons end up in the patient.

The first experimental results are very encouraging. Future work will consist in developing the dose monitoring system and optimizing the regulation loops.

REFERENCES

1. K. Gall et al, *NIM B* **79**, 881-884 (1993)

2. D. Prieels, "Clinical Performances in Double Scattering", NPTC Technical Report 88.17.53.011 Rev A

3. R. Sépulchre et al, "Dynamic Delivery Planning in IBA Proton Pencil Beam", EPAC '2000, Vienna, June 2000

Proton Synchrotrons for Cancer Therapy

George B. Coutrakon[†]

[†]*Loma Linda University Medical Center, 11234 Anderson Street, Loma Linda, CA 92354*

Abstract Synchrotrons have long been recognized for their superior capabilities in proton and heavy ion therapy. Their compactness and ease of beam energy control make them ideally suited to this application. The range of available intensities insures safety against high dose accidents such as have occurred with conventional electron accelerators. For heavy ion and heavy ion therapy, synchrotrons have been the exclusive choice among particle accelerators. In this paper, four synchrotrons designed for dedicated therapy facilities are reviewed and performance data are discussed.

INTRODUCTION

The past decade has witnessed a large growth in new proton facilities, which are either in construction or are already being used to treat patients. The advantages of proton therapy over conventional x-ray treatments for cancer and other diseases have been well documented[1,2].

In this paper we focus on the accelerators used at these facilities, particularly synchrotrons. Accelerator requirements for a new facility are discussed, followed by some examples of synchrotron's capable of fulfilling those requirements. In addition to the proton synchrotron at Loma Linda University Medical Center (LLUMC), two proton synchrotrons are nearing completion in Japan. Heavy ion facilities in Chiba, Japan (HIMAC) and Darmstadt, Germany (GSI) also use synchrotrons to accelerate carbon ions for patient treatments. In Heidelberg, Germany, a new heavy ion facility is being planned for the German Cancer Research Center that will use a synchrotron to accelerate protons and heavier particles. At University of Indiana (Bloomington) a 250 MeV weak focussing synchrotron has been recently commissioned for physics research but could be converted for medical use at a future date. Some of the highlights of these accelerators will be discussed here.

ACCELERATOR REQUIREMENTS FOR PROTON THERAPY

The general requirements for proton therapy accelerators have been addressed in other papers[3,4] and only a brief discussion will be presented here.

Accelerator requirements are, in many respects, coupled to the type of beam delivery system (nozzle) used in each treatment room. The function of a nozzle is to spread a small pencil beam that enters the treatment room into a large lateral area of uniform dose, and to modulate the beam energy to obtain a uniform dose over a predetermined thickness or depth in the patient. Many techniques to achieve this goal have been discussed in the literature[5,6].

In the future, proton and heavy ion nozzles will scan the beam across the tumor volume. In this approach, a mono-energetic beam of small lateral size (less than 1cm diameter) is swept by magnetic deflection across the tumor volume in a manner similar to a raster scan of an electron beam in a television screen. Every point or pixel in the target region may receive its prescribed dose by varying the sweep speed of the scanning magnets or by varying the intensity from the accelerator, or both. Also, varying the energy from the accelerator treats different depths in the patient. Given this scenario, which represents the most stringent demands for hadron therapy operations, typical requirements on energy and intensity can be characterized as follows.

For all therapy applications, the range of energies lies between 70 and 250 MeV. This corresponds to a depth penetration between 3cm (e.g. eye tumors) to 37cm for pelvic tumors. The energy accuracy for any beam delivered to the treatment room is $\pm 0.1\%$ (Ref. 4). This level of accuracy is important when adding Bragg Peaks of different energies to create a SOBP which has a uniform dose in the tumor region along the depth axis. It is necessary therefore to change the beam energy quickly (<2-3 sec) even within a single treatment.

Intensity levels (protons/minute or protons/pulse) from the accelerator are driven by the requirement to keep treatment times low. Typical treatment times should be less than 2 or 3 minutes for doses up to 2 Gray using the maximum field size desired. Fig. 1 shows the fluence required to achieve 1 Gray of dose as a function of the depth modulation. Intensity vs. dose rate can than be calculated for the desired maximum field size.

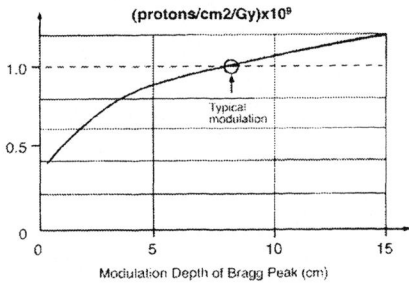

Fig. 1: Proton fluence per Gray versus modulation depth or width of spreadout Bragg Peak (SOBP). [7]

Beam Intensity Uniformity Requirements - To achieve uniform dose while sweeping a small beam across the tumor volume, the extracted intensity from the accelerator must be held constant or at least regulated with high precision. A typical requirement is +/-3% intensity fluctuation for frequencies below 1 kHz. The Lawrence Berkeley Laboratory heavy ion synchrotron produced the first raster scan beam for patient treatments in 1993. More recently, spill uniformity measurements at LLUMC have demonstrated that +/-2.4% intensity ripple is achievable with intensity feedback control to an accelerator quadrupole magnet[8].

PROTON THERAPY SYNCHROTRONS

All synchrotrons have several common features. All operate in pulsed mode with periodic cycles of beam injection, acceleration, extraction and magnet reset for the next injection. Typically the injection phase is quite short (less than one ms) and extraction may be as short as a single turn in the ring (less than 1 µs) or stretched over many seconds (slow resonant extraction). Slow extraction (0.25 second or longer) is desirable for therapy applications due to the requirements of the beam delivery systems in the treatment rooms. During each phase of the pulsed operation, the beam current is held at constant radius in the machine, which implies that the RF acceleration frequency must increase during acceleration. In the case of the Loma Linda machine, the RF cavity must accommodate a frequency swing of 1 to 9 Mhz.

Synchrotrons for proton therapy fall into two categories: strong focussing ($\nu > 1$) and weak focussing ($\nu < 1$) where ν is the horizontal betatron tune for the accelerator. Both types achieve adequate intensity for proton therapy. Strong focussing synchrotrons have the advantage of requiring smaller magnet apertures in the guide field, but require more space and complexity for additional focussing quadrupoles. The differences in achievable intensity between the two designs are small at these energies. Two of the synchrotron designs discussed in this paper are weak focussing and two are strong focussing. With the exception of the proposed GSI accelerator, which will also accelerate heavy ions, the sizes of these accelerators are between six and seven meters in diameter.

Synchrotrons require an injector consisting of an ion source, a low-energy linear accelerator, and focussing and bending magnets to direct the beam into the synchrotron's dipole guide field. All synchrotrons discussed here use RFQ linear accelerators because they can achieve proton beams between 2 and 4 MeV in a short length. When injection energies above 4 MeV are required, a drift tube linac following the RFQ is often used. The injector for the Loma Linda synchrotron uses a 1.5 meter long RFQ, which produces 2 MeV proton beam energy with 24 mA peak current for the one µsec injection time. A debuncher (sometimes called an energy compactor) reduces the energy spread after the RFQ to +/-0.5%. This is important for weak focussing machines with large dispersion, so that the beam size does not become too large in the synchrotron. Injection into the synchrotron can either be single turn or multi-turn. Since beam current from injectors is usually limited, multi-turn injection offers a means of increasing the intensity (or dose rate) by injecting the low energy beam over many turns in the synchrotron. Increasing the energy of the injected beam will increase the maximum intensity allowed by space charge forces in the synchrotron. Both of these improvements yield higher performance but also increase costs significantly. Three of the synchrotrons discussed here use RFQs built by AccSys Technology (Pleasanton, CA). The Loma Linda RFQ from AccSys Technology has worked reliably for ten years without problems.

Because of the pulsed nature of synchrotrons, the cycle time of the accelerator becomes an important factor to deliver higher dose rates, thus requiring a fast cycle time. At the same time, the nozzles in the treatment rooms require a long extraction time to accommodate the limitations of the beam spreading devices and dosimeters. In addition, the accelerator cycle time is limited by the ramping speed of the dipole power supplies. For synchrotrons with 250 MeV beam energy discussed here, the ramp up and ramp down times are each about 0.5 sec. The nozzles also require beam spills of 0.3 sec to 1.0 sec duration. Not surprisingly, the fastest cycle times of these accelerators that can accommodate the slow spill requirements of the nozzles are between 1.5 to 2.2 sec (see Tables 1 and 2).

The minimum diagnostic instrumentation for a synchrotron usually includes non-destructive beam position monitors which measure the horizontal and vertical centroid of the beam at multiple locations in the ring and a current toroid to monitor beam intensity throughout the acceleration cycle. A wire that scans across the synchrotron aperture at selected locations to measure beam profiles in the ring would be extremely useful as well. It is also helpful for the injection line to have beam profile monitors to examine the shape of the beam near the injection to the ring.

Weak Focussing Synchrotrons

Both the Mitsubishi (Fig. 2) and LLUMC synchrotron accelerators are weak focussing with zero gradient magnetic fields. Both use edge angles focussing on the ends of the dipole magnets to achieve vertical focussing, which eliminates the need for additional focussing quadrupoles. However, edge angle focussing in the vertical plane decreases the focussing in the horizontal plane which leads to larger magnet aperture requirements and larger "good field" regions inside the dipole magnets. Both machines have similar values for the betatron tune in the horizontal and vertical dimensions. Some of the beam characteristics of the Mitsubishi design and the Loma Linda design are listed in Table 1. The space charge limit for both of these machines is estimated at 1×10^{11} protons per pulse. The Mitsubishi accelerator is still under testing at the factory in Kobe, Japan. It will be shipped for use in a proton therapy center after commissioning.

The Loma Linda synchrotron has been in use at LLUMC since 1990 and has treated over 5000 patients. It has run very reliably six days per week for the past ten years. Over the past five years, fewer that 2% of all scheduled treatments were cancelled due to accelerator repairs. In most cases, lost treatment days could be rescheduled on weekends. Using four treatment rooms serviced by the one synchrotron, Loma Linda has demonstrated a capability of treating over 100 patients in a single day. The synchrotron intensity, approximately 1×10^{12} protons/minute, is capable of a dose rate of 1 to 1.5 Gray/minute for field diameters of 20cm and 10cm SOBP's. A detailed performance study of the synchrotron was published in 1994[9]. Of all the synchrotrons discussed here, only the Loma Linda machine has experimental measurements of maximum

TABLE 1. Accelerator Comparison

	Mitsubishi	LLUMC/ Optivus
Injection Energy	3 MeV	2 MeV
Maximum Repetition Rate	1.5 sec	2.2 sec
Maximum Intensity (Protons Per Pulse)	5×10^{10} ppp	3.5×10^{10} ppp
Maximum Intensity	2×10^{12} P/min	1×10^{12} P/min
Energy Range	70 – 250 MeV	20 – 300 MeV
Ring Diameter	6.8 meters	6.0 meters
Horizontal Dispersion (ave.)	7.0 meters	9 meters
Horizontal Tune	0.75	.58
Vertical Tune	1.34	1.3
Extraction Tune (horiz.)	2/3	½

intensity and energy capability. The other three synchrotrons still have not progressed to that stage and the numbers in Tables 1 and 2 represent design specifications rather than actual measurements for those machines. Optivus Technology, Inc. of San Bernardino, CA is responsible for marketing future facilities using the LLUMC accelerator design and may be contacted for further information.

Strong Focussing Synchrotrons

Hitachi of Japan is presently commissioning their 70-270 MeV synchrotron (Fig. 3) for the University of Tsukuba hospital. It consists of a 7 MeV injector using multi-turn injection with the machine cycle synchronized with the patient's respiration. This will be advantageous for treatments in the thoracic region of the body to reduce targeting errors due to patient motion. When respiratory gating is not required, machine cycle times as low as 2 seconds will be achievable.

The medical staff at GSI in Darmstadt, Germany has treated 68 patients with carbon beams since 1997, and wish to build a dedicated treatment accelerator for cancer therapy using protons, helium, carbon and oxygen ions at Heidelberg. Their proposed accelerator (Fig. 4) has many features in common with the Hitachi synchrotron including multi-turn injection, a variable

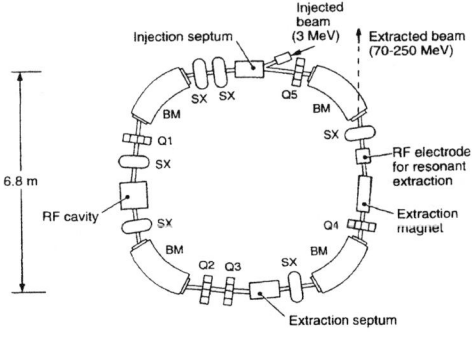

Fig. 2: The Mitsubishi proton synchrotron.
BM=Bending Magnet, Q=Quadrupole, SX=Sextupole

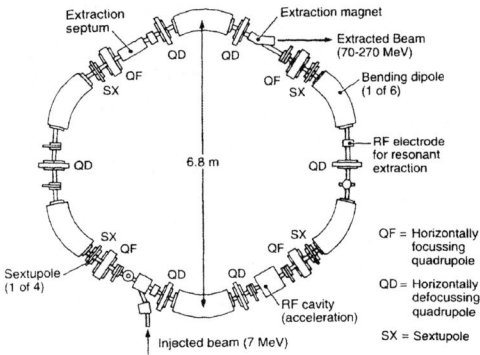

Fig. 3: Hitachi proton synchrotron.

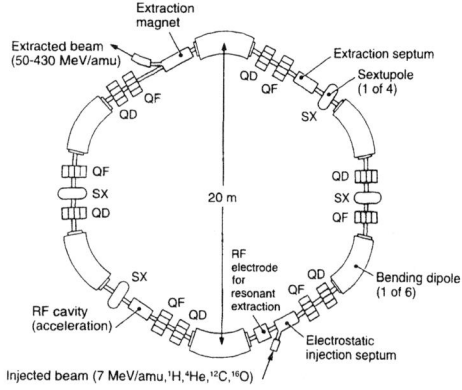

Fig. 4: The proposed synchrotron for Heidelberg, Germany for proton and light ion therapy.

TABLE 2. Accelerator Comparison

	Hitachi	GSI
Injection Energy	7 MeV	7 MeV/amu
Max. Repetition Rate	2 sec	2 sec
Maximum Intensity (Protons Per Pulse)	2×10^{11} ppp	4×10^{10} ppp
Maximum Intensity	6×10^{12} P/min	1.2×10^{12} P/min
Extraction Energy	70 –270 MeV	50 – 430 MeV/amu
Respiration Gated Option	yes	yes
Synchrotron Diameter	6.8 meters	20 meters
Horizontal Dispersion (ave.)	2.0 meters	2.9 meters
Horizontal Tune	1.7	1.7
Extraction Tune (horiz.)	5/3	5/3

repetition rate, and extraction on the horizontal betatron resonance using RF knockout technique. The cycle times of the accelerators with and without respiratory gating are nearly the same. The GSI accelerator has a diameter three times larger than the other three accelerators, to accommodate the higher rigidity (or particle momentum) needed for carbon beams relative to protons.

ACKNOWLEDGEMENTS

The author wishes to thank Sashi Harada of Mitsubishi Electric Corp., Yoshi Takada of Tsukuba University, and Hartmut Eickhoff of GSI for their valuable input concerning their synchrotron designs. Special thanks are given to Vivian Dillard for her excellent work in preparing this manuscript in its final form. The author takes responsibilities for any inaccuracies that may have been presented. The reader should contact the institutions directly for further details on the synchrotron designs

REFERENCES

1. Wilson, R. R., "Radiological Use of Fast Protons", *Radiology*, **47**, 487-491, 1946.
2. Slater, J. M., Archambeau, J. O., Miller, D. W., Notarus, M. I., Preston, W., Slater, J. D., "The Proton Treatment Center at Loma Linda University Medical Center: Rationale For And Description Of Its Development" *Int Journal of Radiation Oncology Biology & Physics*, **22**, Pp. 383-389, 1992.
3. Alonso, J. R., "Design Criteria for Medical Accelerators" *Ion Beams in Tumor Therapy* (Ed. Ute, L) Chapman & Hall, 1995, Chapter 19.
4. Coutrakon, G. B., Slater, J. M., Ghebremedhin, A., "Design Consideration for Medical Proton Accelerators", *Proc Of The 1999 Particles Accelerators Conference*, New York, New York, 1999, In Press.
5. Haberer, T., Becher, W., Schardt, D., Kraft, G., "Magnetic Scanning System for Heavy In Therapy" Nuc. Instr. and Methods., **A330**, pp 296-305 (1993)
6. Coutrakon, G. B., Bauman, D., Lesyna, D., Miller, J., Nusbaum, J., Slater, J., Johanning, J., Miranda, J., "A Prototype Beam Delivery System for the Proton Medical Accelerator at Loma Linda." *Journal of Med. Phys*, **18(6)**, pp. 1093-1099, 1991.
7. Private communications, Andy Koehler, Harvard Cyclotron Laboratory, Cambridge, MA, 1990.
8. Coutrakon, G. B., Ghebremedhin, A., Johanning, J., Koss, P., Jenkins, G., *Proc of The 14 CAARI[th]* "Spill Uniformity Measurements for a Raster Scanned Proton Beam" International Meeting Conference, AIP Proceedings No. 392, pp 1265-1272, 1996.
9. Coutrakon, G. B., Hubbard, J., Johanning, J., Maudsley, G., Slaton, T., Morton, B., "A Performance Study of The Loma Linda Proton Medical Accelerator" *Journal of Med. Phys.* **21(11)**, pp 1691-1701, 1994.

Ion beam therapy: overview of the world experience

J. M. Sisterson

Northeast Proton Therapy Center, Massachusetts General Hospital, 30 Fruit Street, Boston, MA 02114

Abstract. In 1946, R. R. Wilson first proposed the use of proton beams in radiation therapy. At 19 centers worldwide, over 26,000 patients have been treated with proton beams; ~64% of these patients have been treated since 1990. The good long term follow-up results that are available for selected treatment sites has led to an increased interest in having proton radiation therapy available in a hospital setting. The first such hospital-based facility began operation in 1990, the second in 1998 and several more are under construction. ~2500 patients were treated with helium and neon ions at the first heavy ion facility which closed in 1992. ~800 patients have been treated at the two currently operating heavy ion facilities built in the 1990s.

INTRODUCTION

In radiation therapy, we want to deliver the optimal dose to the target volume while giving as little dose as possible to all adjacent normal tissues and critical structures. Proton and heavy ion beams have similar properties, which can be exploited to achieve these objectives [1]. These beams have a finite range in material, with an increased rate of energy loss near the end of range (the "Bragg" peak), and because they are relatively heavy particles there is little scatter as the beams penetrate material. There is essentially no dose beyond the Bragg peak for proton beams, but due to fragmentation there is a low dose tail beyond the Bragg peak for ion beams.

Protons are low-LET (Linear Energy Transfer) particles so the Radio-Biological Effect (RBE) of proton beams is similar to that of x-rays or photons. Clinically for protons, a constant RBE of 1.1 is usually used, although experiments have shown that the RBE varies slightly through the Bragg peak to a maximum value of ~1.3 near the end of range [2]. The principal advantage of proton radiation therapy is the superior dose distributions designed to conform closely to a complex target volume, which can be achieved. In this way, minimal dose is given to adjacent normal tissue. Significant improvement in these dose distributions can be achieved using intensity modulated proton beams [3].

Heavy ions are high-LET particles, and have correspondingly higher RBEs. The penumbra and the width of the Bragg peak of an ion beam decrease with increasing Z, while the RBE in the entrance region and the distal fragmentation tail increase with increasing Z. Therefore, heavy ion beams have both the advantage of good dose distributions and an increased RBE. Helium ions (the 'lightest' heavy ion) have a biological action similar to that of protons. Treatment plans for all heavy ion beams, are more complex than those for protons because the plans have to account for the variable RBE through the Bragg peak and the distal fragmentation tail [4].

PROTON RADIATION THERAPY
Facilities

The first patients were treated with proton beams in the 1950s but it was almost 40 years before the first hospital based facility, designed specifically for proton therapy, was built at Loma Linda University Medical Center (LLUMC) in 1990. Before LLUMC's construction, all patients were treated at facilities using accelerators originally designed for some other application.

One such facility is the Harvard Cyclotron Laboratory (HCL), which has a 160 MeV synchrocyclotron built in 1949 for nuclear physics research. The first patient was treated at HCL in 1961. Since then, nearly one third of all patients in the world treated with proton beams have been treated at HCL in collaboration with physicians from Massachusetts General Hospital (MGH).

At some facilities, 60–70 MeV proton accelerators originally designed to produce neutron beams for therapy are used. The resulting proton beams have small depths of penetration, which can be used to treat eye tumors.

At several facilities, beam time for patient treatments is available only a few weeks each year, because the primary use of their accelerators is physics research. This time constraint limits both the number of patients that can be treated and the number of fractions that can be given to any one patient.

For the 18 operating proton therapy facilities in 1999 (the two programs at the Paul Scherrer Institute, Switzerland are considered as one facility), Table 1 shows the number of patients treated each year. At 7 facilities, the limited proton beam energy allows the treatment of eye tumors only. At one facility, only intracranial targets are treated using a single fraction. At the other ten facilities, a variety of sites are treated. The accelerators used in proton therapy include: at 10 facilities, cyclotrons; at 5 facilities, synchrocyclotrons; at 3 facilities, synchrotrons

TABLE 1. Patient statistics for 1999

# patients treated in 1999	All centers # centers	Multi treatment facilities # centers	Centers treating eyes only # centers
0 – 10	3	1	2
10 – 50	4	3	1
50 – 100	5	2	3
100 – 200	1	1	
200 – 300	3	2	1
300 – 400			
400 – 500	1	1	
500 – 600			
600 – 700			
700 – 800	1	1	
Total # centers	18	11	7
Total # of patients	2586	1382	1204

In 2000, there are two operating hospital-based dedicated proton therapy facilities, one at LLUMC and the other at the National Cancer Center Hospital East (NCCHE), Kashiwa, Japan. New hospital-based facilities nearing completion include the Northeast Proton Therapy Center (NPTC), MGH, Boston, USA and several in Japan.

To treat large tumor volumes with uniform dose distributions, proton beams are spread out both laterally and in depth. At most facilities, passive scattering techniques using combinations of scattering foils and brass plugs or annuli are used to spread the beam laterally. Range modulators - usually rotating lucite 'propellers' or ridge filters - are used to give uniform dose distributions in depth [5].

Scanning proton beams give improved conformal dose distributions, but more complex technology is required to produce the beams. A limited number of patients have been treated using scanning beams since the early 1980s at Chiba in Japan [6]. Scanned beams have been used to treat patients at PSI in Switzerland since 1996 [7]. At several facilities, scanning beam capability is under development.

To deliver the proton beam at any angle to the patient, proton gantries have been developed. At LLUMC, 'corkscrew' gantries are used, while at PSI a novel gantry design is used where the patient couch rotates with the gantry. At both NCCHE and NPTC, conventional gantry designs are used. A good review of the latest developments in proton therapy techniques is given in [8].

Patient Statistics and Follow-up results

As Figure 1 shows, the number of patients treated with proton beams each year is still increasing, indicating that the demand for proton therapy centers worldwide is not yet satisfied. Almost half of all the 26,603 patients treated worldwide through 1999 have been treated since 1993.

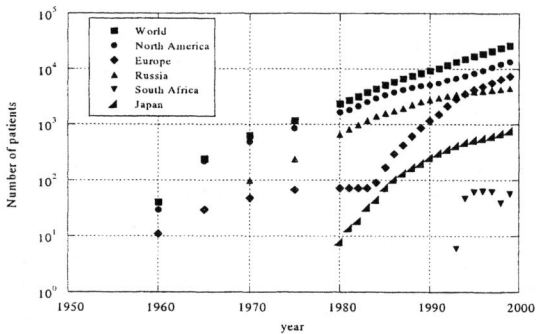

FIGURE 1. Proton therapy cumulative patient totals.

The continued growth in both the numbers of patients treated each year and the number of operating proton therapy facilities is shown in Figures 2 and 3, which give patient statistics for 1990 and 1999. In 1990, it is estimated that 866 patients were treated at 10 institutions.

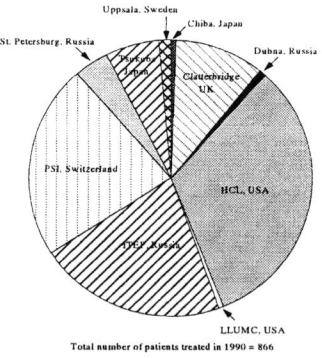

FIGURE 2. Patient statistics in 1990. An estimated 866 patients were treated worldwide.

By 1999, the number of patients treated tripled to 2586, and there were 18 operating institutions. Even so, almost half the patients were treated at two institutions, LLUMC and HCL.

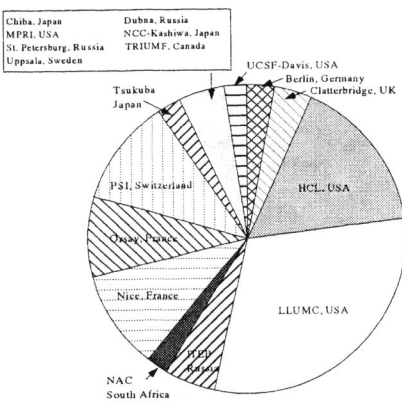

FIGURE 3. Patient statistics in 1999. An estimated 2586 patients were treated worldwide.

Detailed estimated patient data for selected treatment sites are presented in Table 2 for 1990, 1995, and 1999. For each year the number of patients treated for a particular site is expressed as a percent of that year's patient total. Note that by 1995 about twice, and by 1999 about three times as many patients were treated using proton beams as were treated in 1990.

TABLE 2: Estimated patient totals by treatment site

Year	1990	1995	1999
Site/ % of annual patient total	%	%	%
Uveal melanoma	46.7	41.6	33.4
Age related Macular Degeneration	0	2.7	9.3
All other eye	2.1	2.6	3.9
All eye patients	48.7	46.9	46.6
All chordoma & chondrosarcoma*	4.0	4.7	4.9
Prostate	0.8	16.8	18.4
Liver	1.4	1.1	1.5
AVM#	11.2	4.9	2.7
Head & Neck	0.5	1.7	3.4
Esophagus	0.6	0.3	0.1
Lung	0.6	0.4	0.5
All other sites	32.2	23.1	22.0
Patient totals	866	1841	2586
Number of facilities	10	15	18

* includes all treatment sites
# for 1990, data from ITEP, Moscow not available.

For several treatment sites, there are now good long-term follow up results available. The sites for which data are available include: choroidal melanomas [9]; chordomas and chondrosarcomas of the skull base [10]; prostate cancer [11]; arteriovenous malformations (AVMs) [12]; a summary of results for several sites is given in [13].

ION BEAM RADIATION THERAPY
Facilities

From 1957–1992, helium beams were used for radiation therapy at the University of California, Berkeley. Helium ions have a small high-LET component otherwise their biological properties are similar to protons. From 1977–1992 at the Bevelac, University of California, 433 patients were treated, most with 670 MeV/amu neon beams, but carbon, silicon and argon beams were also used [14].

Now in 2000, there are two operating heavy ion facilities. The first patient was treated at HIMAC, Chiba, Japan in 1994 and 829 patients have been treated by September 2000. Carbon ions of 290, 350 and 400 MeV/amu are usually used for therapy. The facility has three treatment rooms, one room with a horizontal beam, one with a vertical beam and one with both horizontal and vertical beams. The beam delivery system uses wobbler magnets to deliver a uniform treatment field in two dimensions.

The second operating facility is at GSI, Darmstadt, Germany, where patients are treated in collaboration with DFKZ, Heidelberg. This facility began operation in 1997, and 72 patients had been treated by July 2000. Horizontal beams of carbon ions and a sophisticated 'pencil beam' scanning system are used. In the near future, there will be three operating heavy ion facilities, because commissioning of the Hyogo Hadron Therapy Facility, Harima Garden City, Japan is underway. At this facility, both protons and carbon ions will be available and there will be 6 treatment rooms with 7 beam lines.

Besides the three facilities described here, there are several proposals for future heavy ion facilities in both Europe and Asia. A good review outlining the past, present and future heavy ion facilities can be found in reference [14].

Patient statistics and Follow-up

The cumulative number of patients treated with heavy ion beams (excluding helium beams) from 1992 is shown in Figure 4. By the end of 1999, ~1200 patients had been treated world wide with ions heavier than helium. Long term follow up data for selected sites is available for the patients treated in the programs at Berkeley [15], and will soon be available for the patients treated at HIMAC, Japan [16].

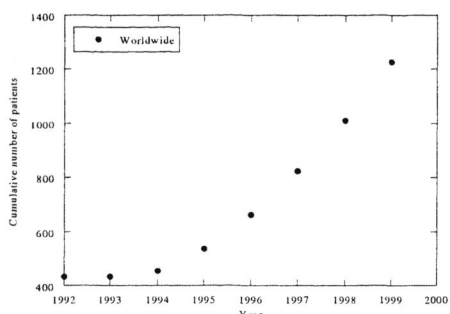

FIGURE 4: Heavy ion radiation therapy cumulative patient data.

CONCLUSIONS

The advantage of using proton beams in radiation therapy is to achieve superior dose localization. Dose distributions can be designed, which conform closely to target volumes of complex shapes, allowing the optimal dose to be given to the target volume while sparing adjacent normal tissue and critical structures. Significant improvement in these dose distributions can be achieved using intensity modulated proton beams.

Worldwide over the past ~50 years, more than 27,000 patients have been treated with proton beams and good long-term follow-up results are available for several treatment sites. There is still an increasing number of patients treated each year, so there is still considerable worldwide interest in having more proton therapy centers available. A conservative view of the future is shown in Figure 5, where it can be seen that the demand for proton therapy facilities is not yet satisfied.

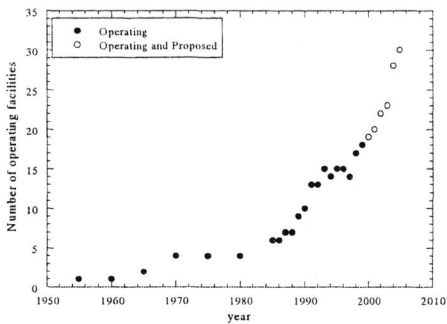

FIGURE 5: The future of proton therapy.

The role of heavy ions in radiation therapy has still to be determined. At this time, only ~3300 patients have been treated, worldwide (including helium beams), and there is limited long-term follow up data. The experience at the two operating facilities plus that expected from future new facilities will be essential to determine the treatment sites that will benefit by heavy ion radiation therapy.

Proton and heavy ion facilities that are designed to be used in the hospital setting demand accelerators and beam delivery systems that are very reliable, easy to maintain, easy to operate and are cost effective. The challenge to the accelerator community is to meet such specifications while at the same time achieving the superior dose distributions possible with these radiation modalities.

ACKNOWLEDGEMENTS

I am grateful to all my colleagues in proton and heavy ion therapy who were kind enough to share with me their patient statistics. Obviously without their help, this review could not have been prepared.

REFERENCES

1. Wilson R. R., *Radiology* **47**, 487 (1946).
2. Robertson J. B., *et al.*, *Cancer* **35**, 1664-77 (1975).
3. Lomax A., *Phys. Med. Biol.* **44**, 185-205 (1999).
4. Wambersie A., *Strahlenther. Onkol.* **175 Suppl 2**, 39-43 (2000).
5. Chu W. T., *et al.*, *Rev. Sci. Instr.* **64**, 2055-2122 (1993).
6. Kanai T., *et al.*, *Med. Phys.* **7**, 365-9 (1980).
7. Pedroni E., *et al.*, *Med. Phys.* **22**, 37-53 (1995).
8. Pedroni E., "Latest developments in proton therapy" in *Proc. of European Particle Accelerator Conference EPAC 2000*, Vienna, Austria, June 25-30, 2000.
9. Gragoudas E. S., *et al. Arch Ophthalmol.* **118**, 773-8 (2000).
10. Munzenrider J. E., *et al.*, "Skull base tumors: treatment with three-dimensional planning and fractionated x-ray and proton radiotherapy", in *Textbook of Radiation Oncology*, eds. S. A. Leibel and T. L. Phillips, Philadelphia, W. B. Saunders, 1998, pp. 347-56.
11. Slater J. D., *et al.*, *Urology* **53**, 978-84 (1999).
12. Harsh G., *et al.*, *Neurosurg. Clin. N. Am.* 10, 243-56 (1999).
13. Loeffler J. S. and Smith A. R., "Proton beam radiation therapy", in *Encyclopedia of Cancer*, ed. J. R. Bertino, 2nd edition, San Diego, Academic Press, (in press).
14. Alonso J. R., "Review of ion beam therapy: present and future" in *Proc. of European Particle Accelerator Conference EPAC 2000*, Vienna, Austria, June 25-30, 2000.
15. Castro J. R., et al., "Particle radiation therapy", in *Textbook of Radiation Oncology*, eds. S. A. Leibel and T. L. Phillips, Philadelphia, W. B. Saunders, 1998, pp. 1223-40.
16. H. Tsujii, private comm. (2000).

Progress of Particle Therapy in Japan

F. Soga

Division of the Accelerator Physics and Engineering, National Institute of Radiological Sciences(NIRS), 4-9-1 Anagawa, Inage-ku, Chiba 263-8555, Japan

The starting time in particle therapy in Japan was not so early compared to other western countries, but the recent development of medical applications using accelerator beams has been outstanding. After 15 years experience in proton therapy at University of Tsukuba and especially the promising results of recent years in heavy ion therapy at HIMAC, many projects of charged particle therapy are promoted at various places. In this paper, the present status of these projects and ongoing clinical results are described.

INTRODUCTION

In Japan, particle therapy began in 1975 at National Institute of Radiological Sciences (NIRS). By using the cyclotron imported from France, the fast neutron therapy started. Then, with use of the same cyclotron, the first proton therapy was started in 1979. Since then other institutes followed either neutron or proton therapy by using different accelerators. In 1994, the big project of heavy ion therapy called HIMAC project started its clinical trial in NIRS. Now in our country, several therapy projects, especially charged particle therapy are running or under construction. The paper will describe this increasing trend of therapy facilities and present status of cancer treatments with use of accelerator beams.

NEUTRON THERAPY

The fast neutron therapy using a cyclotron began at NIRS. The 30 MeV deuteron beam bombards thick Be metal and break-up neutrons of about 13 MeV in mean energy are used for the treatment. The depth-dose distribution of neutrons in the human body shows the exponential dumping along the depth which is similar to that of gamma rays, as is shown in Fig. 1. From 1975 to 1993 ; about 2200 patients were treated during 19 years at NIRS. At the Institute of Medical Science at the University of Tokyo, using 14 MeV deuteron beams they tried also the fast neutron therapy. It was started in 1976 and about 400 patients were treated during 10 years.

It turned out [1] that the results of treatments were not much better than photon treatment on the whole except for some kinds of tumor sites such as the salivary gland or other head and neck cancers, and lung cancer of Pancoast tumor. However, in NIRS the long experience of high LET(Linear Energy Transfer) radiation in neutron treatment gave a fundamental basis to set up and proceed with the ongoing heavy ion therapy project.

PROTON THERAPY

The clinical trial of proton therapy was started in 1979 using 70 MeV proton beam from the NIRS medical cyclotron which had been mainly used for the fast neutron therapy. As is well known, the depth-dose distributions of charged particles are completely different from photon and neutron which are seen in Fig. 1. The absorbed dose curves have a low flat plateau after particles impinge and penetrate into the body, and near the end of their range a sharp peak appears (Bragg Peak). By adjusting the energy of the protons to the depth of tumor, one can expect very good localization of the dose distribution around the tumor. Actual tumor in the patient has its own thickness along the beam path and certain volume in the three dimensions.
In order to irradiate the tumor volume with uniform dose distribution, various techniques have been developed for enlargement of the Bragg Peak with superposition and beam scanning in the transverse plane.

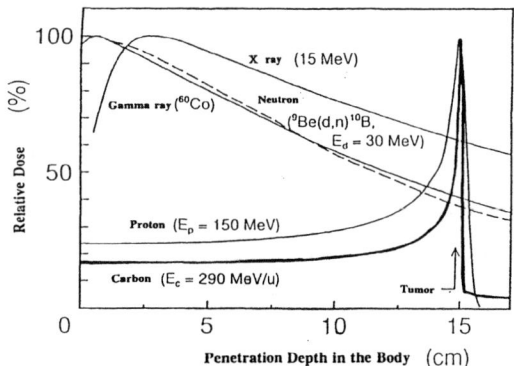

FIGURE 1. Depth-dose curves in tissue of various radiations

The energy of 70 MeV corresponds to only 38 mm penetration. Accordingly, the targets with use of this beam have been quite limited. In early years of the trial, the most of the treatments were performed on the superficial tumors using spot scanning technique to spread the irradiation field of the beam[2]. Recently, most of the proton treatments in NIRS are eye melanoma. But the occurrence rate of eye cancer in Japanese is much smaller compared to people in the western countries. The number of treated patients with protons in NIRS is so far about 140 during past 20 years. Rotating range modulator is used to spread the Bragg Peak. Local control rate of ocular melanoma during last 13 years is about 88 %[3].

The Proton Medical Research Center(PMRC) at University of Tsukuba started proton therapy in 1983. They used the degraded beam of 250 MeV from the 500 MeV booster synchrotron which is originally an injection synchrotron for 12 GeV main ring for study of nuclear or particle physics attached to High Energy Physics Laboratory. As the energy in PMRC is sufficiently high for many different tumors, they have placed main stress on deep-seated tumors in the body [4]. The sites are diversified ; liver, esophagus, lung, skull base, gynecology, bladder and head or neck cancer, etc. The total number of patients treated by March of 1999 is not much during 17 years ; about 600, because schedule of treatments has been affected by operation of main ring. But they have accumulated precious examples of above tumor sites for effectiveness of proton therapy, and are still ongoing in their treatments.

HEAVY ION THERAPY

As the first medically dedicated heavy ion facility in the world [5], HIMAC (Heavy Ion Medical Accelerator in Chiba) has performed clinical trial with use of carbon beam since 1994. Besides the characteristic of sharper dose-localization which is due to smaller multiple scattering and energy straggling than protons, the heavy ion has another strong advantage of biological aspect which can not be found in proton. That is higher Relative Biological Effectiveness and lower Oxygen Enhancement Ratio. The former means that in order to induce the same amount of eradication in biological cells, heavy ion shows stronger efficiency than photon. The latter means that irradiation of heavy ion to oxygen-deficient radio-resistant cells which lies at the central portion of tumor works more strongly than photon. Both effects in heavy ions such as carbon are 2 or 3 times as effective as protons and photons. Therefore, it is earnestly hoped that heavy ion therapy will cure patients of locally advanced cancer or hard-to-treat cancer which other modalities of radiotherapy cannot treat well.

The HIMAC is composed of three ion sources, RFQ linac, Alvarez linac, two synchrotrons, a beam transport system and three therapy rooms. Two synchrotrons are simultaneously operated with different energies. The maximum energy of the accelerator is 800 MeV/u which corresponds to 30 cm penetration in the body for silicon beams. However, in case of carbon, 400 MeV/u is sufficient for reaching 30 cm in tissue. Therefore, 290, 350 and 400 MeV/u have been used for carbon therapy depending on the position of tumors in the patients[6].

They have been conducting therapy for the following sites : Head and neck, Central Nervous System, Lung, Liver, Prostate, Cervical, Bone and Soft Tissue, Esophagus, Skull Base and Miscellaneous (not included above but individual adaptive objects being selected). Along each protocol for each site which is set up after careful discussion among committee members comprising medical doctors, variety of treatments have been performed[7]. These are selection of type of tumor, selection of patients, amounts of dose, number of fractionation, the toxicity study and investigation of the efficacy of carbon beams for various tumors. Also the dose escalation study has been performed.

FIGURE 2. The view of delivery system of the secondary beam port

By the time of February 2000, the number of treated patients amounted to 675. Concerning local control rate in various sites, it is still a bit premature to make definite conclusions, however, many doctors feel that the result of heavy ion irradiation turns out so far remarkably good especially with respect to deep-seated tumors and hard-to-treat tumors by other modalities. Those are head and neck, lung, liver and especially large bone/soft tissue cancer which is impossible to treat with other methods.

As long as the accelerator technology is concerned, one fruitful technique is the synchronized irradiation to the respiration of patients. In order to minimize the irradiation area in the tumor located in the abdomen such as lung and liver, attachment of the sensor for the movement associated with respiration and development of beam extraction from synchrotron by RF knock-out method are combined. Only at the timing of expiration of the patients is the beam turned on. The other stress of recent years has been placed on the utilization of secondary beams. By irradiation of radioactive beam such as ^{11}C(half life : 20 min.) in the tumor, the irradiation volume will be able to be observed with an aid of positron camera or positron emission tomography (PET) in near future[8]. This kind of beam may become useful both for diagnosis and treatment. Fig. 2 is a photograph of secondary beam port which is under the construction. The patient's bed will be added within this year.

Present irradiation system in HIMAC is a kind of static method. We adopt the wobbler magnets with a scatterer foil to enlarge the beam irradiation field in transverse plane, and multileaf collimator or patient collimator to define the cross-sectional area of irradiation. In longitudinal direction the bar ridge filters of various thickness has been used to spread the Bragg Peak. Now in order to improve the dose localization on the tumor, the development of three dimensionally conformed irradiation method has been investigated. The dynamic movement of range shifters is incorporated with that of multileaf collimator to automatically tailor the size and shape of the irradiation field corresponding to the target shape at each depth in a tumor volume of the patient. Also another subject is the research for spot scanning method. This technique is particularly important in case of weak intensity of the beam such as a secondary beam which will be realized in near future.

EXPANSION OF PARTICLE THERAPY

In addition to PMRC and HIMAC, there are 5 ongoing projects which are either running of treatment, or under planning or construction in Japan. These are : at the Kantou area, National Cancer Center at Kashiwa, and new medical proper machine at University of Tsukuba; at Middle area, Shizuoka prefecture; and at Kansai area, Fukui prefecture and Hyogo prefecture. The present status of these projects are individually described below.

(1) National Cancer Center East Hospital at Kashiwa

This institution is under direct control of the Ministry of Health and Welfare. The accelerator is 235 MeV cyclotron, which was delivered by Sumitomo Heavy Industries Ltd. in conjunction with Ion Beam Applications, (Belgium). There are three treatment rooms, two with rotational gantries and one with fixed horizontal beam port. The building was completed in 1997. The treatments started in November, 1998. They use the irradiation system composed of double scatterer foils and ridge filter. By summer of this year about 30 patients have been treated mainly on head and neck cancer and lung cancer.

In due course they want to treat 400 cancer patients every year.

(2) Hyogo Ion Beam Medical Center

Self-governing body of Hyogo prefecture started the feasibility study of 2nd heavy ion facility in Japan near the end of HIMAC construction. The project aims at proton therapy as well as carbon therapy. The maximum energy of accelerator is set to be 230 MeV proton and 320 MeV/u for carbon beam which correspond to 30 and 20 cm range in the body, respectively. Two gantries are for usage of protons. For carbon beam, there are three courses, a horizontal, a vertical and a 45 degree oblique beam line for irradiation to patients. The bird's eye view of the facility is shown in Fig. 3. Now the tests of acceleration are proceeding[9]. Already both proton and carbon beams have been extracted from the synchrotron with sufficient intensities.

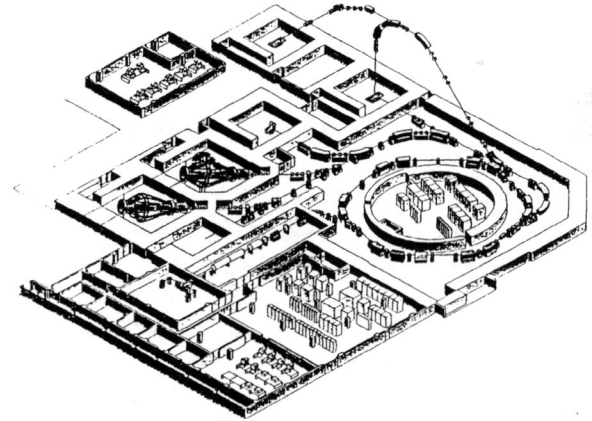

FIGURE 3. The bird's eye view of Hyogo Ion Beam Medical Center

The knowledge and experience in HIMAC are utilized as much as possible together with many improvements.

Probably the proton therapy will be expected to start earlier than heavy ion therapy and hopefully in 2001.

(3) University of Tsukuba

The Ministry of Education, Science and Culture has funded a proper proton therapy facility based on a hospital three years ago. This is renewal project after the long substantial achievement of proton therapy in PMRC which was constructed 18 years ago subsidiary to physics research. The construction is ongoing. The accelerator complex is composed of duoplasmatron ion source, 3 MeV RFQ linac, 7MeV DTL linac and 250 MeV synchrotron. They will use two gantries and the irradiation system composed of double scatterers, ridge filter, range shifter and so on. As this facility is attached to university, it is expected to take an important role in education, especially in the field of medical physics which has not been sufficiently developed in our country, producing many useful specialists for radiotherapy.

(4) The Energy Research Center, Wakasa Bay

Fukui prefecture opened the Energy Research Center at Wakasa Bay in 1994. Among many equipments and facilities for research and development for industry and technology in that area, there is an accelerator comprising tandem electrostatic accelerator plus synchrotron. Utilization of this accelerator is concerned mainly in research for technology, material science, agriculture, forestry and fishery. But they added therapeutic application to their program. The energy of proton beam from the synchrotron is 200 MeV at maximum and now the irradiation system is under construction. As this is not a medical proper machine, they want to develop various techniques for treatments. But from the viewpoint of social welfare in the local area, about 50 patients will be treated every year from 2001.

(5) Shizuoka Prefecture Cancer Center

Shizuoka prefecture decided in 1995 to newly establish a cancer center near the Mt. Fuji. They will construct a new hospital and proton therapy facility. The design of accelerator and irradiation system is ongoing and now almost finished. The planning view is shown in Fig. 4. In order to aim at the compact system of accelerator complex, the synchrotron is directly connected to the 3 MeV RFQ linac. Maximum energy of the synchrotron is 235 MeV. With the wobbler magnets, they will have three treatment rooms ; two gantries and a horizontal beam port. It is expected that treatments will be started in 2003 and the number of treated patients is expected to be about 400 per year.

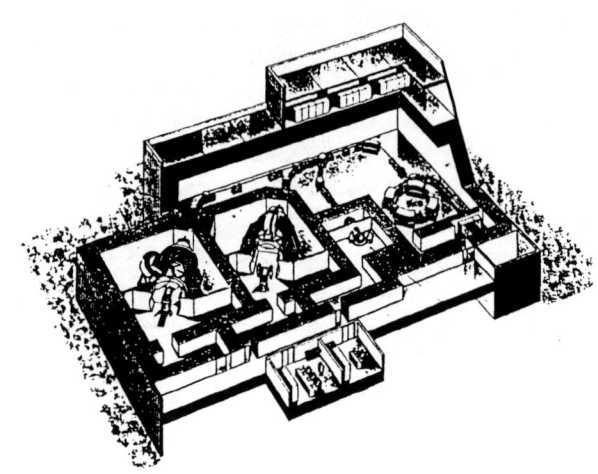

FIGURE 4. View of the facility in Shizuoka Prefecture Cancer Center

CONCLUSIONS

In Japan, the number of deaths due to malignant cancers are approximately 280 thousands per year and the trend shows that this rate is rapidly increasing partly because the average life time of Japanese ; 77 in men and 83 in women are still upward. Considering the promising results of ongoing charged particle therapy, it is intended to increase the facilities in various districts together with the research and development of medical accelerator and irradiation technology.

REFERENCES

1. Tsunemoto, H., and Morita, S., *Proceedings of the EULIMA Workshop*, Nice, 1989, pp 57-65.
2. Kanai, T., Kawachi, K., Kumamoto, Y., Ogawa, H., Yamada, T., Matsuzawa, H. and Inada, T. *Medical Physics* **7**, 365(1980)
3. Nakano, T. *et al.*, *J. Jpn. Soc. Ther. Radiol. Oncol.* **9**, Suppl. 2, 42 (1998)
4. Ohara, K., et al, *Int. J. Radiation Oncology Biol.Phys.* **38**, 367(1997)
5. Hirao, Y. *et al.*, *Nucl. Phys.* **A538**, 541c (1992)
6. Soga, F., *Rev. Sci. Instrum.* **71**, 1056 (2000)
7. Tsujii, H., *Proceedings of 6th Int. Meet. on Prog. in Radio-Oncology*, Salzburg, 1998, pp709-721
8. Kanazawa, M. *et al.*, *Proceedings of 6th European Particle Accelerator Conference*, Stockholm, 1998, pp2357-2359
9. Akagi T. *et al,. Proceedings of 11th Symposium on Accelerator Science and Technology*, Harima, 1997, pp116-118

Medical Applications of in vivo Neutron Inelastic Scattering and Neutron Activation Analysis: Technical Similarities to Detection of Explosives and Contraband

J. J. Kehayias

USDA Human Nutrition Research Center on Aging at Tufts University
Boston, MA 02111, USA

Abstract. Nutritional status of patients can be evaluated by monitoring changes in elemental body composition. Fast neutron activation (for N and P) and neutron inelastic scattering (for C and O) are used in vivo to assess elements characteristic of specific body compartments. There are similarities between the body composition techniques and the detection of hidden explosives and narcotics. All samples have to be examined in depth and the ratio of elements provides a "signature" of the chemical of interest. The N/H and C/O ratios measure protein and fat content in the body. Similarly, a high C/O ratio is characteristic of narcotics and a low C/O together with a strong presence of N is a signature of some explosives. The available time for medical applications is about 20 min – compared to a few seconds for the detection of explosives – but the permitted radiation exposure is limited. In vivo neutron analysis is used to measure H, O, C, N, P, Na, Cl and Ca for the study of the mechanisms of lean tissue depletion with aging and wasting diseases, and to investigate methods of preserving function and quality of life in the elderly.

INTRODUCTION

In vivo neutron activation analysis and inelastic neutron scattering have provided the body composition field with a basic tool for direct measurement of the elemental composition of tissue. The penetrability of neutron radiation allows for in vivo bulk analysis of the whole body at minimal radiation exposure. The elemental analysis is then used for the assessment of the body's major compartments (protein, bone, water, fat etc) and their physiological balance.

On December 21st 1988, Pan Am flight 103 exploded over Lockerbie, Scotland, killing 270 people as a result of a small hidden bomb planned by terrorists. This event gave birth to an intense search for effective non-destructive luggage inspection methods based on elemental analysis. A ^{252}Cf- based luggage inspection system designed to detect nitrogen was installed at JFK airport in January 1990. At about the same time, 90 km to the east, investigators were studying the wasting of end-stage AIDS patients at Brookhaven National Laboratory[1]. They were using ^{238}Pu-Be neutron sources to measure total body nitrogen (TBN). Most explosives have a uniquely high content of nitrogen. In the human body, nitrogen is a measure of our protein (> 98% of body's nitrogen is in protein). Two very different applications of bulk analysis became connected due to their focus on the same element. As elemental body composition methods improved to include carbon, oxygen and hydrogen, so did the applications for the detection of explosives and narcotics. Today, there is an impressive coincidence in the elements of common interest and, therefore, an overlap in technology.

IN VIVO ELEMENTAL ANALYSIS

We divide the human body mass into four main chemical compartments which change with age and health status. Water is the largest compartment followed by protein and fat (defined chemically as triglycerides). Bone mineral, the fourth compartment, includes 98% of the body's calcium. In vivo body composition methods, involving gamma ray spectroscopy, can be classified into the following categories: delayed neutron activation (used for Ca, Cl, Na by n capture and P and N by fast n reactions), neutron inelastic scattering (for C and O), prompt gamma neutron activation (for H, Cl and N) and natural gamma ray counting of ^{40}K. The major elements of the body are listed in Table 1 in order of contribution by weight.

TABLE 1. The major elements of the body for a 70 Kg "standard man" and method of assessment.
FNAA= delayed fast neutron activation analysis, NAA= delayed neutron activation analysis, NIS=neutron inelastic scattering, PGNA=prompt gamma neutron activation

Element	Amount (g)	% of body weight	Contributing compartment	Method
Oxygen	43,000	61	Water (mostly)	NIS
Carbon	16,000	23	Fat (mostly)	NIS
Hydrogen	7,000	10	All	PGNA
Nitrogen	1,800	2.6	Protein	PGNA,FNAA
Calcium	1,000	1.4	Bone	NAA
Phosphorus	780	1.1	Bone, muscle, others	FNAA
Sulfur	140	0.2	Several	
Potassium	140	0.2	Muscle (mostly)	K-40 counting
Sodium	100	0.1	Extra-cellular water	NAA
Chlorine	95	0.1	Extra-cellular water	NAA

Single Element Analysis

There are two compartments, protein and bone ash, which can be assessed directly by measuring only one element. This is because more than 98 % of the body's nitrogen and calcium (TBCa) are in protein and bone respectively:

Protein=TBN*6.25
Bone ash = TBCa *2.94

Nitrogen can be measured by prompt gamma neutron capture using moderated ^{252}Cf, ^{238}Pu-Be or ^{241}Am-Be neutron sources. The resulting 10.83 MeV gammas are recorded with large NaI(Tl) detectors shielded from the neutron beam[2]. An alternative method for TBN is the use of 14 MeV fast neutrons and the (n,2n) reaction resulting in a positron emitter[3]. The D-T neutron energy is below threshold for the other major positron emitting reactions (from O and C). The radiation exposure from either method is less than 0.80 mSv, but the fast neutron method carries a lower body-thickness-correction error.

One of the first extensive medical research applications, involving thousands of volunteers, took place at Brookhaven National Laboratory, Upton, NY. A ^{238}Pu–Be whole body neutron activation facility was built for the measurement of TBCa and the study and management of osteoporosis[4]. With this disease, even small changes in bone mineral (2% loss per year) could eventually result in significant reduction of bone strength and fracture. Delayed neutron activation analysis provided a method precise enough to monitor the management of bone loss and to evaluate the efficacy of treatments. The method used was the thermal neutron capture reaction on ^{48}Ca, which offered high sensitivity, high energy gamma rays to be detected (3.1 MeV), and a convenient half-life of ^{49}Ca (9 min) for the transport of the patient from the activation facility to a shielded whole body detector. Similarly, use of a fast neutron source and the reaction ^{31}P(n,α)^{28}Al makes possible the assessment of muscle phosphorus[3].

Finally, the large whole-body shielded gamma ray counters used as the detector for delayed neutron activation made possible the measurement of the natural radioactivity of the body due to the ^{40}K isotope[5]. Since ^{40}K is present in constant proportion (0.0118%) to the total mass of total body potassium (TBK), gamma ray activity constitutes a measure of TBK (after corrections for background and self-absorption). This provides us with a measurement of body cell mass, which is the metabolizing, oxygen-consuming portion of fat-free mass.

Measurement of Ratios

The Carbon-to-Oxygen Ratio

Based on the observation that the carbon-to-oxygen ratio (C/O) in tissue is a measure of fat content, we developed a model, which correlates this ratio to percent body fat[6]. Carbon and oxygen are measured by neutron inelastic scattering (Figure 1) at low radiation exposure (less than 0.08 mSv). A relationship between % body fat (F) and C/O (weight ratio) can be derived directly from the stoichiometry of the compartments contributing to total body carbon (TBC) and oxygen (TBO).

$$F = \frac{0.722 * C/O - 0.116}{0.610 * C/O + 0.653}$$

We found this method to be insensitive to assumptions about the composition of lean tissue, including the value of the hydration of lean tissue[6]. This approach to measuring body fat offers several technical and physiological advantages. Measurement errors (such as variation in neutron output) usually affect proportionally TBC and TBO leaving the C/O measurement unchanged. The method requires no empirical adjustments, it is not sensitive to ethnicity, age or health status, and it provides regional as well as total body measurements.

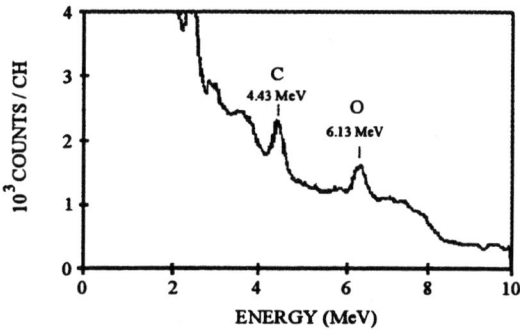

FIGURE 1. Gamma ray spectrum from a healthy volunteer acquired in coincidence with the fast neutron pulses.

The Carbon-to-Hydrogen Ratio

When using a pulsed D-T neutron generator to measure body carbon and oxygen, the data acquisition system operates in coincidence with the neutron pulses. Acquisition of spectra from the same detectors timed between neutron bursts give a strong hydrogen signal at 2.22 MeV from thermal neutron capture[7]. The C/H ratio is characteristic of tissue "dryness". This is because C/H for fat, glycogen and protein is high, ranging from 6.53 for fat to 7.61 for protein, and zero for water (Table 2). A simple C/H measurement can derive the level of hydration, even when the rest of the body composition is not known.

TABLE 2. Measurable ratios (by weight) using D-T neutrons.

Material (stoichiometry)	C/O	C/H
Fat ($C_{55}H_{102}O_6$)	6.88	6.42
Protein ($C_{100}H_{159}N_{26}O_{32}S_{0.7}$)	2.35	7.49
Glycogen ($C_6H_{10}O_5$)	0.90	7.15
"Average" human, whole body	0.50	2.54
Obese human, normal hydration	1.10	4.02
TNT ($C_7H_5N_3O_6$)	0.88	16.68
Cocaine ($C_{17}H_{21}NO_4$)	3.18	9.65

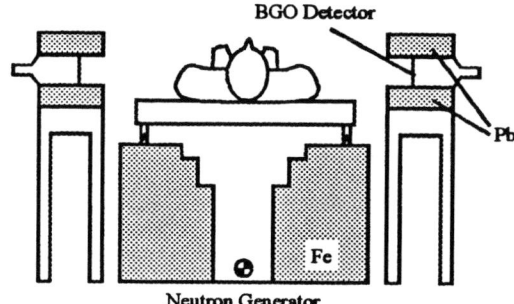

FIGURE 2. The neutron generator is positioned below the patient's scanning bed surrounded by steel shielding. Collimated BGO detectors are positioned at a 90° to the scanning.

The Neutron Generator Facility

The fast neutron source is an A-325 pulsed neutron generator, which utilizes a sealed D-T neutron tube (MF Physics Corp., Colorado Springs, CO). The tube consists of a Penning ion source, a focus electrode, an accelerator column and a zirconium tritide target. The target contains <1 Ci of tritium. The housing of the sealed tube assembly is filled with dielectric fluid for high voltage insulation. The ion source operates with crossed electric and magnetic fields generated by an anode potential at 5-7 kV and a permanent magnet. The neutron tube incorporates a reservoir element to supply deuterium gas for ion source operation and to control gas pressure and accelerator beam current, by adjusting electrically the temperature of the reservoir. Ions from the source are extracted and accelerated to the target, which is kept at a negative potential typically 60 to 120 KV. For the carbon and oxygen applications we use this generator at a pulsing mode (10 KHz with 10% duty factor) and at an output of 5×10^7 n/s/4π. The patient is scanned over the neutron generator (Figure 2) for 12 to 24 min and the gamma rays from the inelastic neutron scattering are recorded using collimated 127 × 76 mm BGO ($B_4Ge_3O_{12}$) crystal detectors[8].

The delayed 14 MeV neutron activation measurement of phosphorus and nitrogen requires a different arrangement. The patient is irradiated without scanning and with the head protected to eliminate the participation of brain phosphorus to the total P signal. The same neutron generator is used for the irradiation, although its pulsing rate is now irrelevant. After the 6 min irradiation, the patient is transported to a whole body gamma ray counter for data acquisition using NaI(Tl) detectors. To achieve the desired precision for P and N required for muscle and protein assessment, the delayed fast-neutron activation requires higher neutron flux (10^9 n/s/4π) and a total body radiation exposure of about 0.60 mSv.

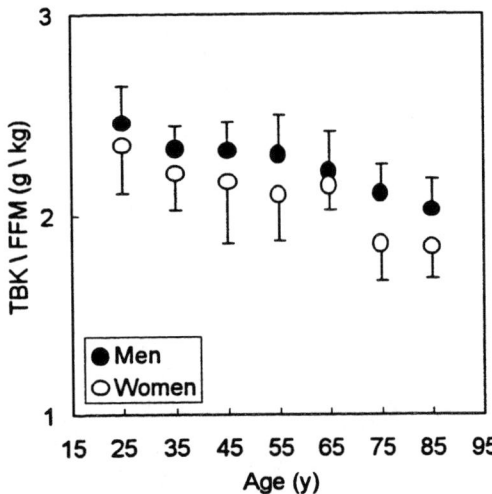

FIGURE 3. Potassium content of lean mass for 14 groups of volunteers plotted against the age of each group. Lean (fat-free) mass was measured by neutron inelastic scattering using the C/O ratio method. Adapted from reference 5.

DISCUSSION

The field of body composition is not new, but it has found new applications and new tools. There is a need to evaluate the efficacy of new anabolic clinical interventions, designed to manage catabolic diseases and to provide vital nutritional support to AIDS patients and the elderly. Change in body composition is usually the only early measurable outcome of these treatments. Any body composition method used to assess the outcome of treatment is required to measure, rather than assume, the composition of lean tissue. With the availability of small portable neutron generators, in vivo neutron activation has been expanded to include neutron inelastic scattering and fast neutron activation. The models used for analysis are simple, direct, and applicable universally to all types of patients. A recent example of the value of in vivo tissue analysis is the introduction of the "quality of lean" principle. Independent measurements of TBK (by counting) and lean mass (using fast neutrons) showed that the potassium content of lean is not constant as previously assumed[5]. TBK represents the portion of lean mass responsible for the metabolic function of the body, and is an indirect indicator of muscle mass and nutritional status. Figure 3 shows the decline of the "quality of lean" mass with age. We observe a similar decline in patients with wasting diseases. It is the goal of the anabolic treatment to preserve or increase the quality of lean mass.

There is a list of elements of common interest to body composition, detection of explosives, chemical weapons and narcotics. Nitrogen is the measure of protein in the body, or the indication of an explosive in a piece of luggage. C/O can measure fat and lean in vivo or serve as a signature for a large amount of hidden cocaine. Portable neutron generators can be used for extensive multi-center clinical trials on nutritional support or as instruments for identifying unexploded mines. The medical applications are limited by radiation exposure, but the available time for the assessment can be as long as 20 min, the environmental conditions are well controlled, and the elemental composition of the sample is always within the physiological requirements of living tissue.

REFERENCES

1. Kotler, D. P., Tierney, A. R., Wang, J., and Pierson, R. N., Am J Clin Nutr **50**, 444-447 (1989).
2. Vartsky, D., Ellis, K. J., and Cohn, S. H., J Nucl Med **20**, 1158-1165 (1979).
3. Kehayias, J. J., Smith, D., Roubenoff, R. et al., Appl Radiat Isotop **49** (5/6), 737-738 (1998).
4. Cohn, S. H., Ellis, K. J. and Wallach, S., Am J Med **57**, 683-686 (1974).
5. Kehayias, J. J., Fiatarone, M. A., Zhuang, H and Roubenoff, R., Am J Clin Nutr **66**, 904-910 (1997).
6. Kehayias, J. J., Zhuang, H., Hughes, V. and Dowling, L., Appl Radiat Isot **49** (5/6), 723-725 (1998).
7. Kehayias, J. J. and Zhuang, H., Nucl Instru and Methods in Phys Res **B79**, 555-559 (1993).
8. Kehayias, J. J., Zhuang, H., Dowling, L. et al., Nucl Instru and Methods in Phys Res (A353), 444-447 (1994).

Uranium Target for Electron Accelerator Based Neutron Source for BNCT

A.P. Tonchev, F. Harmon, T. J. Collens, K. Kennedy, A. Sabourov
Idaho Accelerator Center, Idaho State University, USA

Y.D. Harker, D.W. Nigg, J.L. Jones
Idaho National Engineering and Environmental Laboratory

Abstract. Calculations of the epithermal-neutron yield of photoneutrons from a uranium-beryllium converter using a 27 MeV electron linear accelerator of have been investigated. In this concept, relativistic electron beams from a 30 MeV LINAC impinge upon a small uranium sphere surrounded by a cylindrical tank of circulating heavy water (D_2O) nested in a beryllium cube. The photo-fission neutron spectrum from the uranium sphere is thermalized in deuterium and beryllium, filtered and moderated in special material (AlF_3/Al/LiF), and directed to the patient. The results of these calculations demonstrate that photoneutron devices could offer a promising alternative to nuclear reactors for the production of epithermal neutrons for Neutron Capture Therapy. The predicted parameter for the epithermal flux is more than 10^8 n.cm^{-2}.mA^{-1}.

INTRODUCTION

Boron Neutron Capture Therapy (BNCT) has been the subject of renewed worldwide research activity in recent years. When boron (^{10}B) is delivered to the patient, and is irradiated with thermal neutrons, the products of the ^{10}B(n,α)^{7}Li reaction deposit their energy within ~ 10μm of the reaction origin. Because of the enhanced uptake of the boron delivery agent in tumor cells, a higher radiation dose can be delivered to the tumor compared to normal tissue.

Until now all clinical trials of this BNCT concept utilized neutron beams produced by the cores of nuclear reactors [1]. However, accelerator-based neutron sources have enjoyed wide use and offer the advantages of long-term stability, ease of control, and absence of radioactive material [2,3]. The design of such a source, using near threshold charge particle reactions, was performed in a collaborative effort of the Idaho Accelerator Center and the Idaho National Engineering and Environmental Laboratory [4,5]. Recently, attention has been focused upon the development of epithermal neutron beams using electron accelerators [6]. The feasibility of an accelerator-based source of epithermal neutrons for BNCT has been investigated at energies around photo-neutron emission. It is the goal of this work to investigate an alternative epithermal neutron source for BNCT applications using depleted uranium and a 27 MeV electron beam from an electron linear accelerator.

PERFORMANCE ESTIMATE & DESIGN

Concept

The production of neutrons through the use of electron beams is well established and has played an important role in experimental neutron physics. The electron beam strikes a heavy-metal target producing bremsstrahlung, which, through the (γ,n) process, produces neutrons in subsequent layers of the target. If neutrons could be generated with comparable efficiency at a more modest electron energy (below 30 MeV for example), the size, operation cost, and operation staff could be significantly reduced. For production of neutrons with lower energy electrons, suitable for BNCT purposes, attention is immediately focused on beryllium and deuterium. This is due to their low reaction thresholds of 1.67 and 2.22 MeV, respectively. While these thresholds are quite low compared to heavier elements, the cross sections for the (γ,n) process are also smaller by about two orders of magnitude than the giant dipole resonance cross sections for heavier nuclei. For the heavy material like tungsten, the nuclear cross section is ~ 0.4 b at the peak of the giant dipole resonance near 14 MeV. In the case of fissile elements like thorium, uranium, and plutonium, the nuclear cross section around the maximum of the giant dipole resonance are double, owing to the photofission process. With greater photoneutron and photofission cross sections, fissile materials have an enhanced neutron yield. This makes them suitable for possible use as neutron sources in multiplying assemblies. The absolute percentage neutron yields per one incident electron are shown in Table 1. Our calculations are compared with the experimental data [7]. As can be seen from Table 1, plutonium is the most efficient neutron target followed by the uranium isotopes ^{235}U, ^{238}U, Pb (natural), and D_2O. The total photoneutron yield for fission materials is doubled compared to natural lead for the same energy and material thickness.

Photoneutron target

A bremsstrahlung beam was simulated on the basis of the parameters of the new linear accelerator (LINAC) at the Idaho Accelerator Center. This LINAC machine is capable of accelerating electrons to energies from 2 to 30 MeV and operates at 360 pulses per second with pulse width (full width at half maximum) from 4 μs to 10 ps. The average beam current extracted from the LINAC can reach 300 μA.

A 3.95 kg depleted U target was used to simulate photoneutron production along with a bremsstrahlung beam having a maximum end-point energy of 27 MeV. A uranium sphere with a radius of 3.81 cm was used as a target. The target is surrounded with a 0.7 mm steel shell (Fig. 1). Using the uranium sphere simultaneously as a bremsstrahlung target and a main neutron generator allows one to use almost 100% of the produced gamma-radiation in a 4π geometry for neutron production. Inspite of the angular distribution of the bremsstrahlung spectrum, the gamma production from every angle contributes to the total neutron production in the uranium sphere. Uranium has other advantages compared to many other materials. It has a very high Z number, which guarantees maximum bremsstrahlung production. In addition to the main photoneutron production reactions that occur in uranium such as (γ,n), $(\gamma,2n)$, and photofission (γ,f), neutron production is enhanced with (n,2n) and (n,3n) processes resulting from the initial "hard" photoneutron spectrum.

TABLE 1. Neutron yields in percent for one incident electron from different target materials

Element	Thickness, g/cm^2	Neutrons in % per one electron n_{tot} [n_{fiss}]			
		Calc. $E_{\gamma max}$ = 21.0 MeV	Exper. $E_{\gamma max}$ = 21.0 MeV	Calc. $E_{\gamma max}$ = 34.4 MeV	Exper. $E_{\gamma max}$ = 34.4 MeV
D_2O	8.5	0.056	---	0.10	---
Pb-natural	14.6	0.18	0.20 ± 0.02	0.41	0.46 ± 0.02
	29.1	0.26	0.28 ± 0.02	0.62	0.72 ± 0.04
	43.7	0.29	0.32 ± 0.03	0.71	0.79 ± 0.04
	58.2	0.30	0.32 ± 0.03	0.75	0.80 ± 0.03
^{238}U	24.3	0.59 [0.26]	0.61 ± 0.04	1.50 [0.70]	1.24 ± 0.06
	48.6	0.72 [0.32]	0.72 ± 0.05	1.87 [0.88]	1.71 ± 0.08
	72.9	0.75 [0.33]	0.71 ± 0.06	1.94 [0.91]	1.82 ± 0.07
	97.2	0.75 [0.33]	---	1.97 [0.92]	1.76 ± 0.07
^{235}U	24.2	0.70 [0.59]	0.71 ± 0.06	1.78 [1.56]	1.54 ± 0.06
	48.4	0.86 [0.73]	0.84 ± 0.06	2.24 [1.96]	2.12 ± 0.09
^{239}Pu	21.5	0.74 [0.56]	0.84 ± 0.06	1.81 [1.40]	1.75 ± 0.09
	43.0	0.94 [0.71]	0.92 ± 0.08	2.36 [1.82]	2.01 ± 0.12

Neutron Moderator

The electron energy in the uranium sphere is completely converted into radially-inward-directed bremsstrahlung radiation. Neutrons are subsequently generated in the uranium target from nuclear reactions: (γ,n), (γ,f), (γ,2n), (γ,3n), (n,2n), and (n,3n). The uranium cylinder is surrounded with D_2O, which serves as a cooler and the second neutron generator. The uranium-deuterium converter is placed into a beryllium cylinder, which thermalize the initial fast neutron component coming from the uranium target. Deuterium and beryllium also generate neutrons at the expense of the γ-rays scattered by uranium and serve as neutron moderators as shown in Figure 2. On the backside of the neutron moderator is an aluminum reflector (Al_2O_3). A key component of this design is the use of a highly efficient neutron moderator and filtering material. The neutron filter/moderator should have a high resonance scattering cross section in the fast energy range and low cross section in the epithermal range to provide the 1 eV to 10 keV epithermal window. In this study, based on the experience of epithermal beam development at INEEL, Fluental material was selected for a filter/moderator. This material is manufactured by hot isostatic pressing of a mixture of 69% (by weight) aluminum fluorite, 30 % aluminum, and 1 % lithium fluoride. A block of this material 0.40 cm length provides the neutron filtering and primary moderating. Downstream of the filtering and moderation region was a bismuth gamma-shield, followed by a conical neutron collimator.

Performance estimates and Discussion

Total neutron flux in the uranium sphere from 100μA of 27 MeV electrons equals 8.27E12 neutrons. The initial energy distribution of the neutron spectrum is continuous, and ranges from a few eV to the maximum end-point energy of the electron beam impinging the uranium target. Neutron spectra have a peak at 1.68 MeV. In comparison with the fission spectrum (242Cf for example), the low-energy neutron components in the photoneutron spectra are smaller than those in the fission spectra. Therefore, to produce the epithermal neutron flux suitable for BNCT, a lot of moderation has to be done. To calculate the photoneutron production in uranium, D_2O and beryllium volumes, ACCEPT an electron-photon-coupled transition transport code is used. The photoproduction from ACCEPT was converted to a photoneutron yield using standard photoneutron cross sections. The neutron energy spectrum was obtained with the help of the GNASH code [9]. Then, the neutron spectrum from every material was implemented in the MCNP code to perform the main particle tallies. The calculated neutron spectrum at the collimator exit port is shown in Figure 3.

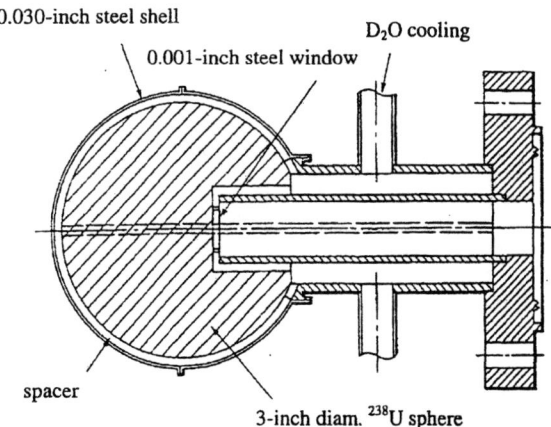

FIGURE 1. Uranium photoneutron target.

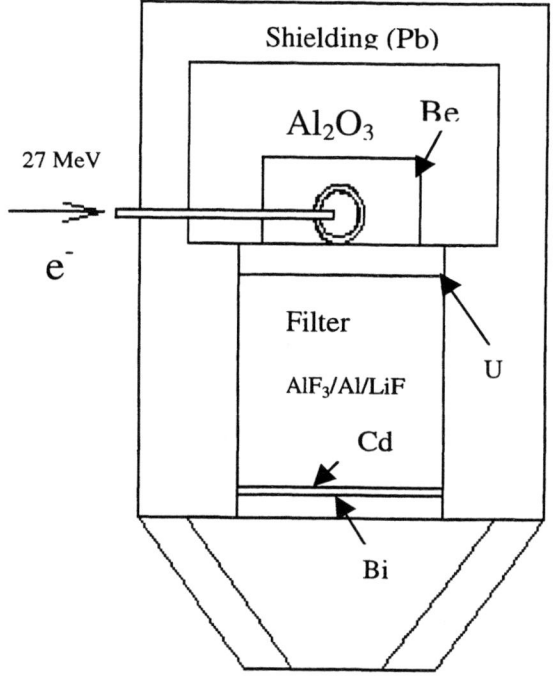

FIGURE 2. View of the electron accelerator based neutron source for BNCT with epithermal-neutron filter.

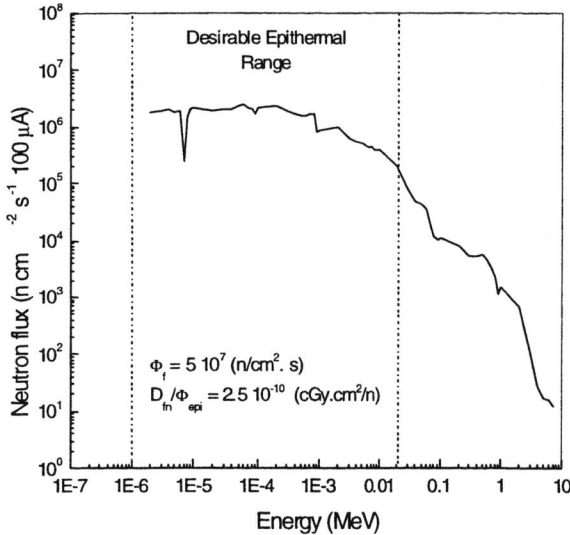

FIGURE 3. Calculated epithermal neutron spectrum at the collimator exit.

CONCLUSION

An accelerator-based bremsstrahlung photoneutron source is a logical choice for three main reasons: First, in contrast to a radioactive source (such as Sb-Be or actinide-Be), the accelerator-based source can be switched off without moving any part of the source into a heavy shield. Second, the accelerator-based sources are inherently more luminous than is conveniently achievable by radioactive sources. Third, a bremsstrahlung photoneutron source offers the possibility of adding x-ray interrogation of the waste drum as an aid in interpreting the results of neutron interrogation. A photoneutron source can control neutron flux more quickly than a neutron source using a reactor because it is possible to control neutron flux by switching on/off of an electron beam. Another advantage of a photoneutron source compared with a nuclear reactor is the small neutron production zone together with an epithermal flux that is competitive with an average power reactor. A main challenge for this concept may be gamma shielding.

REFERENCES

[1] Kiger III, W.S., Sakamoto, S., Harling, O.K., *Nucl. Science and Engineering* **131**, 1-22 (1999).

[2] Yanch, J.C., Shefer, R.E., Klinkowstein, R.E., Howard, W.B., Song, H., Blachburn, B., and Binello, E., "Research in Boron Neutron Capture Therapy at MIT LABA" in *Application of Accelerators in Research and Industry*-1996, edited by L.J. Duggan and I.L. Morgan, Eds., AIP Press, New York (1997) pp. 1281-1284.

[3] Allen, D.A., and Beynon, T.D., "A Design Study for an Accelerator-Based Epithermal Neutron Beam for BNCT", *Phys. Med. Biol.*, **40**, 807 (1995).

[4] Kudchadker, R.J., Lee, C.L., Harker, Y.D., Harmon, F., "Experimental Dosimetry and Beam Evaluation in a Phantom for Near Lithium Threshold Accelerator Based BNCT", in *Application of Accelerators in Research and Industry* -1998, edited by L.J. Duggan and I.L. Morgan, Eds., AIP Press, New York (1999) pp.1056-1059.

[5] Lee, C.L., Zhou, X.-L., Hamm, R.W., Harmon, F., Kudchadker, R.J., and Harker, Y.D., "Temperature rise in Lithium targets for Accelerator Based BNCT Using Multi-Fin Heat Removal", *Application of Accelerators in Research and Industry* -1998, edited by L.J. Duggan and I.L. Morgan, Eds., AIP Press, New York (1999) pp.1041-1044.

[6] Nigg, D.W., Michell, H.E., Harker, Y.D., Harmon, J.F., Advances in Neutron Capture Therapy, Volume 1, Elsevier, Amsterdam, p. 477.

[7] Groce, D.E., Alter, C.P., Herring, D.F. Trans. Amer. Nucl. Soc. p. 179.

[8] Hablieb, J.A., and Mehlhorn, T.A., "ITS – The Integrated TIGER Series of Coupled Eletron/Photon Monte Carlo Transport Codes". SAND91-1634, Sandia National Laboratory. March, 1992.

[9] Young, P.G., Arthur, E.D., and Chadwick, M.B. "Comprehensive nuclear model calculations: Theory and use of the GNASH code." Pages 227-404 of: Gandini, A., Reffo, G. (eds), Proc. of the IAEA Workshop on Nuclear Reaction Data and Nuclear Reactors - Physics, Design, and Safety. Singapore: World Scientific Publishing, Ltd.), Trieste, Italy, April 15 - May 17, 1996.

[10] Briesmeister, J.F., MCNP – Monte Carlo N-Particle Transport Code System, Version 4C, Los Alamos National Lab, 2000.

Neutron Capture Therapy (NCT) Enhancement of Fast Neutron Radiotherapy: Application to Non-Small Cell Lung Cancer

G.E. Laramore*, K.J. Stelzer*, R. Risler*, J.L. Schwartz*, J.J. Douglas*, J.P. Einck*, D.W. Nigg#, C.A. Wemple#, J.K. Hartwell#, Y.D. Harker#, P.R. Gavin+, and M.F. Hawthorne¶

*Department of Radiation Oncology, Box 356043, University of Washington Medical Center, Seattle, WA 98195-6043, U.S.A

#Idaho National Engineering and Environmental Laboratory, P.O. Box 1625, Idaho Falls, Idaho 83415-1575, U.S.A

+Department of Veterinary Clinical Sciences, Washington State University, Pullman, WA 99164-6610, U.S.A.

¶Department of Chemistry and Biochemistry, University of California / Los Angeles, Los Angeles, CA 90024-1569, U.S.A.

Abstract. Fast neutron radiotherapy utilizes neutrons in the energy range of several millions to several tens of millions of eV to treat human malignancies. These fast neutron beams produce a small cloud of "slow" neutrons as they penetrate the body. If one can selectively attach isotopes having large neutron capture cross sections (such as ^{10}B) to cancer cells, these "slow" neutrons can be used to enhance the killing of tumors. We describe a multidisciplinary effort to apply this technique to the treatment of patients with inoperable, non-small cell lung cancers. Problems in target design, compound development, beam optimization, and radiobiological experiments are discussed.

INTRODUCTION

Fast neutron radiotherapy had its beginnings shortly after the discovery of the neutron by Chadwick. The earliest clinical trials showed considerable toxicity and little efficacy [1] and the field languished until the 1950s, when mammalian cell culture techniques allowed a better understanding of the radiobiological differences between conventional, low linear energy transfer (LET) photon or electron radiation and high LET fast neutron radiation. Since the resumption of clinical trials at Hammersmith Hospital, London, England, in the 1960s, over 20,000 patients worldwide have received fast neutron radiation as all or part of their cancer therapy. A review of this clinical work has recently been given by Laramore [2]. While neutron radiotherapy has not proven to be the panacea that was originally hoped for, nevertheless it has been shown to be highly effective in the treatment of certain classes of tumors: malignant salivary gland tumors [3], locally-advanced prostate cancers [4,5], and sarcomas of bone and soft tissues [6]. For many other tumors it has shown equivalent results to conventional photon irradiation. The tumors where fast neutrons show a therapeutic advantage appear to be those which readily repair damage from conventional low LET radiation. This is manifested by a high, relative biological effectiveness (RBE) factor. Normal tissue side effects preclude simply escalating the neutron dose further [7]. What is needed is a way of selectively augmenting the radiation dose to tumor cells without increasing the radiation dose to normal tissue cells and a boron neutron capture therapy (NCT) boost offers a way of accomplishing this. To bring this concept into clinical use requires a coordinated effort to optimize the neutron beam and to develop an appropriate carrier agent for the tumor to be treated. While we will mention work on the latter topic, in this paper we will emphasize the physics related to beam optimizaton.

NCT ENHANCEMENT OF A FAST NEUTRON BEAM

The standard therapy beam at the University of Washington is produced by a 50.5 MeV p --> Be reaction resulting in a spectrum of neutron energies extending out to the maximum proton energy. The target itself consists of 10.5 mm of Be on a Cu backing. While effective in treating patients, this beam is suboptimal for the NCT reaction. Apart from resonance effects, the neutron capture cross section for ^{10}B and other elements scales approximately as $E^{-1/2}$, where E is the neutron energy. This makes the capture process less efficient at high energy; for example, the capture cross section of a 1 MeV neutron is approximately 6000 times smaller than for a thermal neutron. However, the fast neutrons are moderated as they penetrate tissue producing a small, thermalized component. This component has been measured in a water phantom and while dependent upon field size, would typically enhance the energy delivered to the tumor by about 0.1% per µg/g of ^{10}B in the tumor[8].

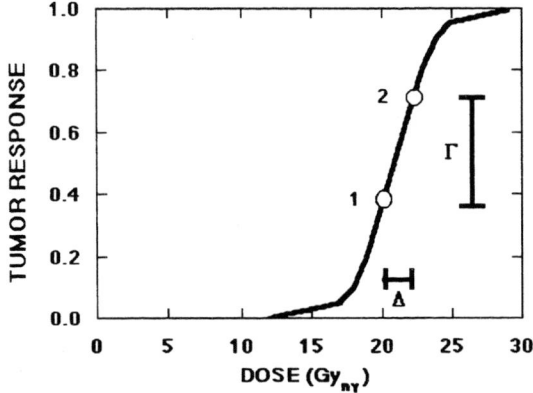

Figure 1. Schematic illustration of dose response curve for a tumor treated with fast neutron radiotherapy. Normal tissue tolerances limit the dose to corresponding to point "1" with the attendant tumor control probability. Adding a dose increment, Δ, to the dose via NCT shifts the tumor control to that indicated by point "2". The magnitude of the improvement, Γ, depends on the slope of the dose response curve. Reprinted with permision from Laramore et al.[8], © 2000, Elsevier Science.

We are working with a ^{10}B-compound, $(B_{10}H_{10})^{-2}$, carried as a sodium salt, which will produce ^{10}B-tumor levels of about 100 µg/g. The enhanced dose would thus be about 10% using our standard target. While this might seem to be a small gain, there are two important multiplicative factors in terms of the expected tumor response. The first relates to its high relative biological effectiveness (RBE) factor because of the very high linear energy transfer (LET) of the emitted ^{4}He and ^{7}Li particles which on the average deposit about 2.35 MeV within a 10 micron distance. The second multiplicative factor is due to the steepness of the dose response curve and the fact that fast neutron dose alone brings us to the steep portion of the curve. This is illustrated in Fig. 1 which shows that a small incremental increase, Δ, in tumor dose can result in a large change, Γ, in tumor control probability.

The expected enhancement has also been demonstrated in the V79 cell line [8], the Harding Passey melanoma system [9,10], the 36B10 rat glioma model [11], and in a human melanoma test system [12].

BEAM OPTIMIZATION

The net effect of the NCT enhancement over a course of fractionated radiotherapy can be described via a varient of the linear-quadratic model as given in Eq. (1)

$$S = \prod_{i=1}^{n} \{ [e^{-\alpha D_{N+\gamma} + \beta D_{N+\gamma}^2 + \Delta t_i \ln 2 / T_d}] [e^{-A P \sigma_B N_B D_B}] \} \quad (1)$$

where the first term in brackets represents the time-dependent, linear-quadratic formula for the cell killing from the primary beam and the second bracketed term represents the enhancement from the thermalized neutron component in the presence of ^{10}B. In Eq. (1), α and β are the usual linear-quadratic coefficients, Δt_i is the time interval between the i$_{th}$ and the i-1$_{th}$ radiation fractions, n is the total number of fractions, T_d is the effective tumor cell doubling time, A is a parameter of order unity that relates to the position of the ^{10}B atoms in the tumor cells, P is the probability of a neutron capture event fission process resulting in the killing of the cell, σ_B is the neutron capture cross section, N_B is the number of ^{10}B atoms on the cell, and $D_{N+\gamma}$ and D_B are, respectively, the "fast" and "slow" neutron components given at each fraction. By convention, the γ ray contaminant is included in the fast neutron dose, $D_{N+\gamma}$. The enhancement effects are contained entirely in the second term since we are dealing with ^{10}B concentrations sufficiently small so as not to deplete the primary neutron beam.

The point to note is that it is the product of the ^{10}B concentration in the tumor and the effective thermal neutron fluence that enters into the exponential describing the enhancement. While simply degrading the fast neutron beam would increase the effective thermal neutron component, we want to do this in such a manner as not to drastically reduce the penetration of the overall beam. Otherwise, it would not be possible to treat deeply-seated tumors as we do routinely with the standard therapy beam.

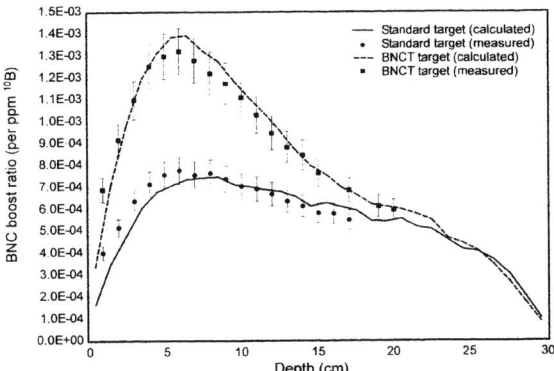

Figure 2. The BNCT boost ratio as a function of depth in a water phantom for 10 cm x 10 cm fields produced by the two targets with one ppm ^{10}B. The "boost ratio" is defined as the ^{10}B produced KERMA divided by the total neutron plus gamma dose produced by the incident beam. Reprinted with permission from Nigg et al. [16].

We have used the LAHET [13] code to model the proton transport in various neutron targets and MCNP [14] with the 100XS cross section library [15] to model the subsequent neutron transport to evaluate various target designs. Each set of model calculations included the target, the iron beam flattening filters just downstream from the target, and an approximate model of the iron collimator assembly. The measurements were validated using combinations of cadmium-covered foil dosimeter packages consisting of indium, gold, and aluminum and also cadmium-covered gold foils contained within hollow boron spheres. The details of the beam unfolding measurements and the comparison with the model calculations are described in detail by Nigg et al. [16]. A target consisting of 5 mm of Be overlying 2.5 mm of W on a Cu backing gives approximately twice the effective thermal neutron flux at a depth of 5-6 cm in a water phantom without significantly degrading the penetration of the overall beam. Detailed measurements show that energy has been shifted from the intermediate energy portion of the beam into the low energy portion of the beam without greatly affecting the high energy portion of the neutron spectrum. The cost for this is a reduction in KERMA factor by about 25% which simply means that it will take a few minutes longer to treat a patient. The BNCT boost ratios for the optimized target and the standard target are shown in Fig. (2).

The increased boost ratio means that approximately a 20% gain in physical dose can be achieved with ^{10}B-tumor levels of 100 μg/g. The downward shift in beam energy is expected to cause in increase in the RBE of the unenhanced beam by about 7-8% near the surface and about 3-5% at depths $\geq$ 10 cm. Prior to using this new beam on patients, baseline radiobiological studies must be performed since normal tissue tolerance factors determine the neutron dose that can be given in a particular clinical situation.

APPLICATION TO NON-SMALL CELL LUNG CANCER

The choice of a target tumor system represents an interplay of several factors. The tumor must be one where fast neutron radiotherapy alone reaches the steep portion of the dose response curve. This means that the NCT boost enhancement should yield a clinically-meaningful improvement in local control. The ^{10}B-carrier agent must be one that selectively incorporates into the tumor relative to the critical dose-limiting normal structures which vary with the particular tumor being treated. Finally, the tumor must be common enough to warrant the interest of a major pharmaceutical company since the ^{10}B-carrier agent must be made under "good manufacturing practice" standards and go through extensive testing prior to being utilized in a clinical trial. Non-small cell lung cancer fits all of these criteria.

In 2000 the American Cancer Society estimates that there will be 164,100 cases of lung cancer and 156,900 deaths due to this disease [17]. Of these, about 35,000-40,000 cases will be localized but surgically-unresectable non-small cell lung cancer. The last randomized trial of the Radiation Therapy Oncology Group testing the efficacy of fast neutron radiotherapy for inoperable non-small cell lung cancer showed a survival benefit compared to conventional radiotherapy for squamous cell tumors and for the good prognostic subgroup (all histologies) [18]. These are the subgroups of patients who have tumors less prone to developing early distant metastases and where improved local control might positively impact survival. An analysis of the failure pattern showed that patients tended to fail in areas adjacent to the spinal cord which was the dose-limiting structure in most cases. GB-10 is excluded from the spinal cord and so this critical structure would not receive the NCT boost enhancement seen by the tumor and other non-critical normal tissues in the treatment fields. A randomized, dose-searching study using GB-10 and fast neutrons will start in the fall of 2000 for patients with high grade gliomas of the brain and will then be extended to patients with non-small cell lung cancer.

SUMMARY

In spite of numerous technological advances in radiation therapy treatment planning and delivery, fast neutron radiotherapy appears to have approached its limits in terms of the tumor dose that can be safely given. This in turn restricts the tumor control that can be achieved. A NCT boost as described in this paper offers the possibility of enhancing the dose delivered to the tumor on a cell-by-cell basis. For this approach to develop into an effective form of treatment, there must be an interplay between three factors: (i) ^{10}B carriers that localize selectively in the tumor relative to the surrounding dose-limiting normal tissues must be developed, (ii) the therapy beam must be optimized to take advantage of both the fast neutron component and the thermalized neutron component developed as the beam penetrates tissue, and (iii) clinical trials must be performed to critically evaluate this treatment approach relative to "standard" therapeutic methodologies. All three factors are now coming together in the treatment of non-small cell lung cancer. An appropriate ^{10}B-carrier has been identified and taken through the necessary pharmacokinetic studies, a layered target has been developed that will increase the effect of the NCT boost without compromising the fast neutron treatment, and FDA approval for a phase I trial has been obtained. Given the pace of such clinical trials, it will be several years before we can critically evaluate the efficacy of this treatment in the clinical setting.

REFERENCES

1. Stone, R.S., *Am. J Roentgenol.* **5**, 771-785 (1940).
2. Laramore, G.E., *Sem. Oncol.* **24**, 672-685 (1997).
3. Laramore, G.E., Krall, J.M., Griffin, T.W., Duncan, W., Richter, M.P., Saroja, K.R., Maor, M.H., and Davis, L.W., *Int. J. Radiat. Oncol. Biol. Phys.* **27**, 235-240 (1993).
4. Laramore, G.E., Krall, J.M., Thomas, F.J., Russell, K.J., Maor, M.H., Hendrickson, F.R., Martz, K.L., Griffin, T.W., and Davis, L.W., *Am. J. Clin. Oncol. (CCT)* **16**, 164-167 (1993).
5. Russell, K.J., Caplan, R.J., Laramore, G.E., Burnison, C.M., Maor, M.H., Taylor, M.E., Zink, S., Davis, L.W., and T. W. Griffin, T.W., *Int. J. Radiat. Oncol. Biol. Phys.* **28**, 47-54 (1993).
6. Laramore, G.E., Griffith, J.T., Boesplflug, M., Pelton, J.G., Griffin, T.W., Griffin, B.R., Russell, K.J., Koh, W., Parker, R.G., and Davis, L.W., *Am. J. Clin. Oncol. (CCT)* **12**, 320-326 (1989).
7. Laramore, G.E. and Austin-Seymour, M., "Fast neutron radiotherapy in relation to the radiation sensitivity of human organ systems", in *Relative Radiosensitivities of Human Organ Systems, III. Advances in Radiation Biology, Vol. 15.*, edited by K. I. Altman and J. Lett, Academic Press, Orlando, 1992, pp. 153-193.
8. Laramore, G.E., Wootton, P., Livesey, J.C., Wilbur, D.S., Risler, R., Phillips, M., Jacky, J., Buchholz, T.A., Griffin, T.W., and Brossard, S., *Int. J. Radiat. Oncol. Biol. Phys.* **28**, 1135-1142 (1994).
9. Poller, F., Sauerwein, W., and Rassow, J, *Radiother. Oncol.* **21**, 179-182 (1991).
10. Sauerwein, W., Ziegler, W., Olthoff, K., Streffer, C., Rassow, J., and Sack, H, *Strahlenther. Onkol.* **165**, 208-210 (1989).
11. Buchholz, T.A., Rasey, J.S., Laramore, G.E., Livesey, J.C., Chin, L., Risler, R., Hamlin, D., Wootton, P., Wilbur, D.S., Phillips, M.H., Spence, A.M., and Griffin, T.W, *Radiology* **191**, 863-867 (1994).
12. Laramore, G.E., Risler, R., Griffin, T.W., Wootton, P., and Wilbur, D.S., *Bull. Cancer / Radiother. (Suppl. 1)* **83**, 191s-197s (1996).
13. Prael, R.E. and Lichtenstein, H. "The LAHET code system." LA-UR-89-3014, Los Alamos National Laboratory, 1989.
14. Briesmeitster, J.F. "MCNP--A general Monte Carlo N-particle transport code, Version 4A." LA-12625-M, Los Alamos National Laboratory, 1993.
15. Little, R.C. "Summary documentation for the 100XS cross section library (Release 1.0)." XTM:95-259, Los Alamos National Laboratory, October, 1995.
16. Nigg, D.W., Wemple, C.A., Risler, R., Hartwell, J.K., Harker, Y.D., and Laramore, G.E., *Med. Phys.* **27**, 359-367 (2000).
17. Greenlee, R.T., Murray, T., Bolden, S., and Wingo, P.A., *CA Cancer J. Clin.* **50**, 7-33 (2000).
18. Koh W.J., Krall, J.M., Peters, L.J., Maor, M.H., Laramore, G.E., Burnison, C.M., Davis, L.W., Zink, S., and Griffin, T.W., *Int. J. Radiat. Oncol. Biol. Phys.* **27**, 499-505 (1993).

SECTION XII

ION IMPLANTATION: SEMICONDUCTORS, MATERIALS MODIFICATION, CLUSTERS, NANOTUBES, ORGANIC MATERIALS, SOURCES FOR IMPLANTATION

Fermi-level dependent diffusion of ion-implanted arsenic in germanium

T. Ahlgren,[*] J. Likonen,[†] S. Lehto,[†] E. Vainonen-Ahlgren and J. Keinonen

Accelerator Laboratory, University of Helsinki, P.O. Box 43, FIN-00014 University of Helsinki, Finland

Abstract. The diffusion of arsenic has been studied in <100> Ge, implanted with $1 \cdot 10^{15}$, 120-keV $^{75}As^+$ ions/cm^2. The implanted samples were subjected to annealing in argon atmosphere in the temperature range of 450 - 550°C. The annealing times varied between 0.5 and 91 h. The As concentration profiles were measured by secondary ion mass spectrometry. A Fermi-level-dependent diffusion of As atoms was observed and quantitatively explained by diffusion via Ge vacancies with charge states 0 and 2-. The activation energies and pre-exponential factors for As diffusion through neutral Ge vacancies was found to be 3.4 ± 0.3 eV and 2.02×10^{20} nm^2/s, respectively, and through doubly negatively charged vacancies 2.9 ± 0.3 eV and 1.89×10^{16} nm^2/s, respectively. The solid solubility limit of As increases with temperature from about 1×10^{19} cm^{-3} at 450°C to about 5×10^{19} cm^{-3} at 550°C.

INTRODUCTION

The increasing importance of Ge in applications such as $Si_{1-x}Ge_x$ devices, multi-junction GaAs/Ge and GaInP/GaAs/Ge solar cells (1, 2) motivates studies on dopant diffusion in Ge. Wojtczuk et al. (1) noticed that during growth of a GaAs layer onto Ge, a p-n junction was created through the in-diffusion of Ga and As, resulting in a two-junction tandem cell. Later on this process has been used to create n-type layers in Ge. In a recent study on phosphorus diffusion into Ge, Söderval et al. (3) observed "box-type" penetration profiles which are an indication of concentration dependent diffusion. Mitha et al. (4, 5, 6) have investigated the effect of pressure on the diffusion of implanted As atoms in germanium. They concluded that the diffusion is not entirely mediated by vacancies. On the other hand, if the diffusion proceeds only through vacancies, the vacancy formation volume must be unexpectedly low or the migration energy high. Further studies giving information on atomic defects in Ge acting as diffusion vehicles, was conducted by Werner et al. (7). They studied the self-diffusion of Ge as a function of pressure, temperature and doping, and concluded that this process is mediated by neutral and singly negatively charged vacancies. In our recent study on As diffusion from a GaAs layer into Ge, we showed that Ge vacancies with the charge states 0 and 2- are the diffusion vehicles responsible for As migration in Ge (8). No contribution of the singly negatively charged vacancy was observed, indicating that Ge vacancy could be a negative U-center (9) which changes directly from neutral to double negative charge state without occupying the 1- state.

EXPERIMENTAL

Commercially prepared samples of p-type (Ga dopant concentration 2×10^{17} cm^{-3}), <100>-oriented single crystal Ge were implanted by using the 120-kV isotope separator at the University of Helsinki. The 120-keV room-temperature implantations to total fluences of $1 \cdot 10^{15}$ $^{75}As^+$ atoms/cm^2 were performed in vacuum (10^{-4} Pa), and the <100> crystal axis was tilted 7° off the beam direction.

The crystallinity of the implanted samples, confirmed with Rutherford backscattering spectrometry channeling technique (10), was restored in an initial annealing process at 450°C for 1 h. The samples were heat treated in a quartz-tube furnace in Ar atmosphere at a pressure of $\approx 10^5$ Pa. The annealing temperatures were measured with a calibrated Chromel-Alumel thermocouple in close contact with the samples.

Depth profiling of As was carried out using a double focusing magnetic sector SIMS (VG Ionex 1X70S). The negative secondary ions $^{151}AsGe^-$ and $^{152}Ge_2^-$ were analysed using 12-keV Cs^+ primary ions. The ion beam current was 100 nA and the rastered area was 330 x 350 μm^2. Crater wall effects were eliminated using a 10% electro-

---

[*] Corresponding author. Fax: +358 9 19140042; E-mail: tommy.ahlgren@helsinki.fi
[†] Permanent address: Technical Research Centre of Finland, Chemical Technology, P.O. Box 1404, FIN-02044 VTT, Finland

nic gate and 1 mm optical gate. The depth of the craters was measured by a Dektak 3030ST profilometer after the SIMS analyses. The uncertainty of the crater depth was estimated to be 5 %. The SIMS instrument was calibrated using ion implanted standard samples for As.

DIFFUSION MODEL

Arsenic is a substitutional n-type dopant in Ge and diffuses through neutral and charged Ge vacancies. Positively charged vacancies can be ruled out, because the substitutional As atom is a donor and thus has charge 1+ and repels positive vacancies. Our earlier study on As diffusion from a GaAs overlayer into Ge showed that only neutral and doubly negatively charged vacancies are needed to describe the As diffusion. Generally the substitutional diffusion coefficient can be written as

$$D = g \cdot f \cdot a^2 [V] \nu, \quad (1)$$

where g is the geometrical factor depending on the crystalline structure, f is the correlation factor, a is the jump length, $[V]$ and ν are the concentration and jump frequency for the vacancy through which the diffusion proceeds, respectively. The concentration of doubly negatively charged vacancies (in contrast to the neutral ones), increases with the increasing energy of the Fermi-level which depends on the electron concentration (12):

$$[V^{2-}] = [V^{2-}]_i (n/n_i)^2, \quad (2)$$

where $[V^{2-}]$ and $[V^{2-}]_i$ are the extrinsic and intrinsic doubly negatively charged vacancy concentrations, respectively, n and n_i are the extrinsic and intrinsic (13) electron concentrations, respectively. The extrinsic effective diffusion coefficient for As atoms can thus be written as

$$D_{As}^{eff} = D_{As}^0 + D_{As}^{2-}(n/n_i)^2, \quad (3)$$

where D_{As}^0 and D_{As}^{2-} are the diffusion coefficients of As diffusing through 0 and 2− charged Ge vacancies, respectively. The diffusion equation, where the change in the As concentration, $[As_{Ge}^+]$, is given as a function of depth x and time t, is the following one:

$$\frac{\partial [As_{Ge}^+]}{\partial t} = \frac{\partial}{\partial x}\left(D_{As}^{eff}\left[\frac{\partial [As_{Ge}^+]}{\partial x} + \frac{[As_{Ge}^+]}{n}\frac{\partial n}{\partial x}\right]\right). \quad (4)$$

The second term is due to the electric field produced by the electrons (12). By applying the charge neutrality condition and using the semiconductor equality $pn = n_i^2$, where p is the hole concentration, we get the electron concentration as a function of As and Ga concentrations

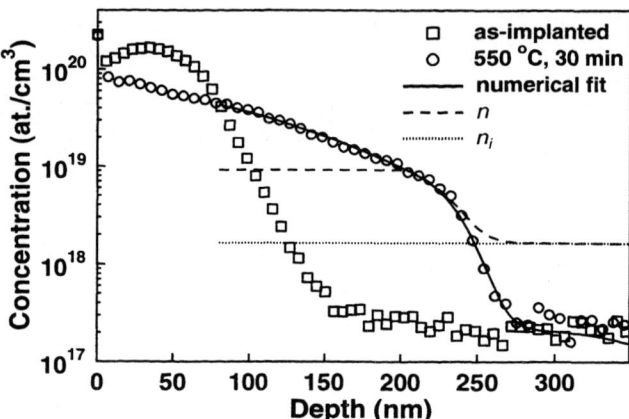

FIGURE 1. The numerical fit to the 550°C, 30 min annealed experimental profile. n and n_i are the extrinsic and intrinsic electron concentrations, respectively.

$$n = 0.5\left([As] - [Ga] + \sqrt{([As]-[Ga])^2 + 4n_i^2}\right). \quad (5)$$

Equation (5) is valid only when the concentration of As atoms is below some value, which is typically between 10^{18} and 10^{20} at./cm^3. When the As concentration exceeds this value the electron concentration does not increase anymore, but stays constant or even drops if electrical compensation occurs. Choosing the maximum electron concentration as a fitting parameter makes it possible to take this effect into account. Other fitting parameters are the diffusion coefficients D_{As}^0 and D_{As}^{2-}.

The concentration dependent diffusion equation (4) solved numerically with the effective diffusion coefficient calculated from Eqs. (3) and (5), resulted in a concentration distribution to be compared with an experimental profile. The diffusion coefficients D_{As}^0 and D_{As}^{2-} and the maximum electron concentration were obtained by least squares fitting.

RESULTS

Figure 1 shows the resulting fit to the 550 °C, 30 min annealed profile. The As solid solubility limit is seen to be about 5×10^{19} at./cm^3 and the maximum electron concentration is fitted to be about 1×10^{19} el./cm^3. The numerical fit is quite good with the diffusion coefficients $D_{As}^0 \approx 0.30$ and $D_{As}^{2-} \approx 0.029$ nm^2/s.

In Fig. 2 the fits are plotted for the As profiles observed after annealing at 450 °C for 91 h, 475 °C for 24 h, and 550 °C for 30 min. It can be noted that the calculated profiles are in a good agreement with the experimental ones.

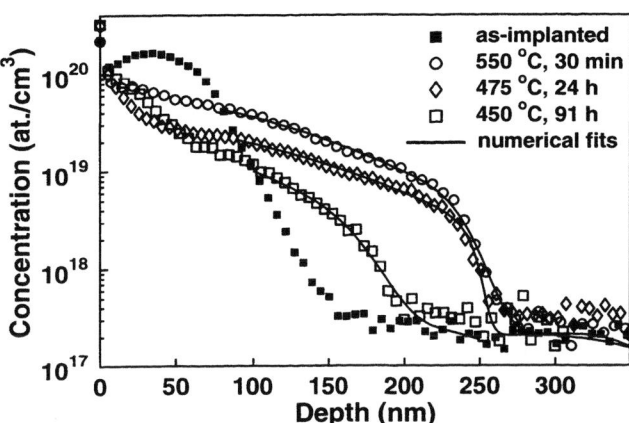

FIGURE 2. The numerical fits for the profiles observed after annealing at 450 °C for 91 h, 475 °C for 24 h, and 550 °C for 30 min.

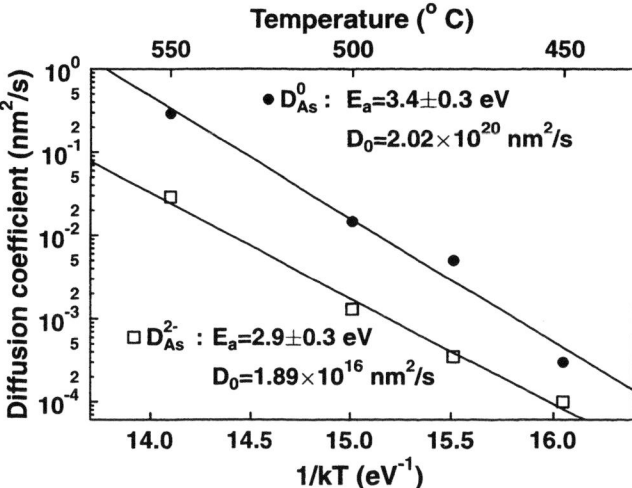

FIGURE 3. The Arrhenius plots with the corresponding activation energies and pre-exponential factors for As diffusion via Ge vacancies.

The solubility limit increases with temperature and the quite steep As profiles are nicely reproduced for the diffusion proceeding through the vacancy combination D_{As}^0 and D_{As}^{2-}.

Figure 3 shows the resulting Arrhenius plots, where the diffusion is well described by the equation $D = D_0 exp(-E_a/k_b T)$, where D_0 is the pre-exponential factor, E_a the activation energy, k_b is the Boltzmann's constant, and T is the absolute annealing temperature. The values for the pre-exponential factor and the activation energy are 2.02×10^{20} nm^2/s and 3.4 ± 0.3 eV, respectively, for diffusion through neutral vacancies and 1.89×10^{16} nm^2/s and 2.9 ± 0.3 eV, respectively, for diffusion through doubly negatively charged ones.

DISCUSSION AND CONCLUSIONS

As can be seen in Fig. 3, the pre-exponential factors for the diffusion through neutral and doubly negatively charged vacancies differ by four orders of magnitude. This difference can be due to the differences in the intrinsic vacancy concentrations, diffusion coefficients, or a difference in the correlation factors f^0 and f^{2-} (14) related to the As diffusion through the two different vacancy types, Eq. (1).

The activation energies obtained here are quite close to the total activation energy of 3.09 eV for the Ge self-diffusion reported by Werner, Mehrer and Hochheimer (7). They studied the self-diffusion of Ge by tracer-diffusion method. Their conclusion was that this process is mediated by neutral and singly negatively charged vacancies. Only neutral and doubly negatively charged vacancies are considered here, and no presence of singly negatively charged vacancies is observed in this study or our previous study (8). Hence, we suggest that the Ge vacancy cannot exist in the charge state 1- at any position of the Fermi level. The Ge vacancy could thus be a negative-U center, which changes directly from neutral to double negative charge state without occupying the 1- state. This kind of behavior has previously been observed for vacancies in Si (15), where the charge state 1+ is not thermodynamically stable.

The experimental steep penetration profiles, obtained for P diffusion in germanium by Söderval et al. (3), could according to the current observations also be explained by introducing diffusion through neutral and doubly negatively charged vacancies.

In summary, we have studied diffusion of implanted arsenic into germanium. The diffusion is quantitatively explained by the diffusion of substitutional As through Ge vacancies with the charge states 0 and 2-.

ACKNOWLEDGMENTS

This work has been supported by the Academy of Finland (Project N° 1018553).

REFERENCES

1. S. J. Wojtczuk, S. P. Tobin, C. J. Keavney, C. Bajgar, M. M. Sanfacon, J. D. Scofield, and D. S. Ruby, IEEE Trans. Electron Devices **37**, 455 (1990).

2. P. K. Chiang, J. H. Ermer, W. T. Nishikawa, D. D. Krut, D. E. Joslin, J. W. Eldredge, B. T. Cavicchi, and J. M. Ol-

son, Proc. of 25th IEEE Photovoltaic Specialists Conference, May 13-17, 1996, Washington D.C., p. 183-186.

3. U. Södervall and M. Friesel, Defect and Diffusion Forum, **143-147**, 1053 (1997).

4. S. Mitha, S. D. Theiss, M. J. Azis, D. Schiferl, and D. B. Poker, Materials Research Society Symposium Proceedings **325**, 189 (1994).

5. S. Mitha, M. J. Azis, D. Schiferl, and D. B. Poker, Appl. Phys. Lett. **69**, 922 (1996).

6. S. Mitha, M. J. Azis, D. Schiferl, and D. B. Poker, Defect and Diffusion Forum, **143-147**, 1041 (1997).

7. M. Werner, H. Mehrer, and H. D. Hochheimer, Phys. Rev. B **32**, 3930 (1985).

8. E. Vainonen-Ahlgren, T. Ahlgren, J. Likonen, S. Lehto, J. Keinonen, W. Li and J. Haapamaa, Appl. Phys. Lett. **77** 690 (2000).

9. N. A. Stolwijk in Landolt-Bernstein, New Series, **III/22b**, edited by M. Schultz (Springer-Berlin), 451 (1989).

10. E. Rauhala, T. Ahlgren, K. Väkeväinen, J. Räisänen, J. Keinonen, J. Likonen and K. Saarinen, J. Appl. Phys. **83** (1998) 738.

11. J. Likonen, E. Vainonen-Ahlgren, T. Ahlgren, S. Lehto, W. Li, and J. Haapamaa, to be published in Proc. of SIMS XII.

12. J. W. Mayer and S. S. Lau, *Electronic Materials Science For Integrated Circuits in Si and GaAs*, Macmillan Publishing Company, New York, 1990.

13. Semiconductors, Group IV Elements and III - V Compounds, ed. O. Madelung, Springer-Berlin, 28 (1991).

14. J. R. Manning, *Diffusion Kinetics for Atoms in Crystals*, D. Van Nostrand Company, 1968.

15. G. Watkins, *Deep centers in semiconductors : a state of the art approach*, New York Gordon & Breach 1986

The Alternative Ion Implantation Approaches for Ultra-Shallow Junction

Wei-Kan Chu[a], Jiarui Liu[a], Jianyue Jin[a], Xinming Lu[a], Lin Shao[a], Qingmian Li[a] and Peiching Ling[b]

[a]*Department of Physics, and Texas Center for Superconductivity, University of Houston, Houston, Texas 77204*
[b]*Advanced Materials Engineering Research, Inc., Sunnyvale, CA 94086*

Abstract. Ion implantation has been a traditional and well-developed technique for junction formation in semiconductor devices for more than 30 years. However, now a big challenge to this technique is the high implant current and throughput at very low energy for ultra-shallow junction formation due to the space-charge limit on the low energy beam. Molecular ions or clusters containing B atoms, such as SiB, SiB_2, GeB, and B_n, are used as implantation ion species. Due to the mass ratio of the ion to B atom, the higher ion energy can overcome the current limit, while the B constitute carries a lower energy for shallower junction formation. Additional benefits of the bombardment by accompanied particles, such as Si and Ge on the suppression of channeling effect and TED were studied. Recoil implantation and a technique of defect engineering for junction formation were also studied. Formation of a junction of less than 30nm was tested by both techniques.

INTRODUCTION

In the ultra-shallow junction implant process in the semiconductor industry, the implant beam energy for different ions, such as B, P or BF_2, is being reduced down to a few keV or a few hundreds eV. The ion beam transport, and the subsequent process throughput is declined dramatically due to the space charge limitation. In BF_2 implantation, the beam energy is higher for a lower energy of the B fragment atom, so the shallower junction can be formatted. In BF_2 implantation, the F contamination may cause an unwanted chemical effect to the substrate which could adversely affect material and device performance.[1] Extensive effort has been placed on the development of very low energy ion implantation equipment for the semiconductor industry. Up to date, a very low energy ion implanter for B implantation below 1 keV with the monomer ion beam current at ~ 1 mA has been developed. In this development, two major low energy transport concepts are used; one is based on the ion deceleration in front of the targets, and another one uses a low energy drift system. The deceleration concept can not avoid the energy contamination especially at very low energy due to the high charge exchange cross section. The low energy drift system is extremely difficult due to the space charge problem. The space charge effect, at low energy, results in a lateral acceleration of the ions causing divergence of the ion beam and the plasma instability. These are fundamental conflicts between the physical concepts and the requirements of the junction fabrication. It is clear that some new concepts of the ion beam transport and the surface doping process are needed. There are many alternative concepts in development. In this paper, we present some recent results on small cluster ion implantation, recoil implantation and defect engineering application in shallow junction formation.

MERITS AND PROBLEMS OF CLUSTER ION IMPLANTATION

Cluster ion implantation is a natural and logical extension of the present ion implantation technique. In ion beam transport, the cluster ion beam used the advantage of higher energy transport of the heavier ion species for a low mass fragment atom implant. In the beam-solid interaction, the cluster ion implantation will have some deviation from the traditional monomer ion-solid interaction. This deviation of the physics in beam-solid interaction may cause either problems or benefits in junction formation.

Shallow junction formation by an implantation of decaborane, $B_{10}H_{14}$, has been demonstrated.[2] Recently, GeB$^-$ Cluster ions have been used to effectively produce a 0.65-2keV boron implant for low energy ion implantation.[3] In our group, we have systematically studied the small cluster ion implantation for shallow junction formation; the potential benefits and the problems.

The principal potential benefit is the gain in beam transport. The beam transport at low energy is limited by the Chil-Lagmuir law and will increase proportional to $V^{3/2}$. Suppose we use a big cluster of mass Mc with the light fragment atom of mass Mi. The energy of the cluster can be increased by a factor of Mc/Mi for a given energy of fragment atom. In the case of cluster ion transport this enhancement is proportional to $(Mc/Mi)^{3/2}$. When a cluster contains n boron atoms, the total gain in beam transport is n x $(Mc/Mi)^{3/2}$. If the number of identical boron atoms is n in a cluster, the actual gain in beam transport is $n^{5/2}$. The beam transport gain for different small clusters is summarized in Tab. 1.

Another benefit on using cluster ion implantation is the possible suppression of the deep penetration of the implant due to channeling tail. A heavier fragment such as Ge from GeB cluster will facilitate the amorphization process during the ion implantation.

One of the problems in cluster ion implantation is the non-linear effect in the cluster-solid interaction. This non-linear effect, or collective effect, is in all aspects in the cluster-solid interaction, including projected range, straggling, defect formation, sputtering and others. How the non-linear effect will influence the TED (Transient Enhance Diffusion) after implantation is an open problem.

We have studied the cluster ion implantation for ultra-shallow junction formation with different small cluster ions, such as GeB, SiB, SiB$_2$, and Bn, The small clusters were employed because the non-linear effect is simpler and can be observed and measured experimentally in our lab.

SHALLOW JUNCTION FORMATION BY SMALL CLUSTERS

As mentioned above, space charge effect greatly limits the beam current, which is a barrier for practical use of low energy ion implantation. To overcome this problem we used a GeB$^-$ cluster ion beam to effectively reduce the boron energy while keeping the total cluster ion energy at a high level because the fractional energy of B is only 13% of the total energy in a GeB$^-$ ion cluster. The transport gain due to this enhanced beam energy is a factor of 21 (see Tab.1). In addition to alleviating the space charge effect, the use of a GeB$^-$ cluster is expected to combine Ge amorphization and B ion implantation into one process to reduce the channeling effect. This consideration was shown on a TRIM (Transport and Range of Ions in Matter) simulation from which we found that the projected range of Ge in the GeB cluster was larger than that of B. The amorphization process is very effective due to the heavy damage by Ge. It is almost equivalent to a preamorphization for the whole implantation process. Another advantage of the GeB$^-$ cluster ion implantation is that Ge is a group-IV and thus does not cause an unwanted chemical effect to the substrate as F does in BF$_2$ ion implantation. Other cluster ions, SiB, SiB$_2$, and Bn, have similar transport gain as shown in Tab.1, but the preamorphization effect is much weaker due to the lighter constitutes in the clusters. A typical depth distribution of Ge and B in GeB cluster ion implanted silicon at 15 keV is shown in Fig. 1 from TRIM simulation. The samples were prepared following a conventionally adopted cleaning procedure.[4] The GeB, SiBn and Bn cluster ions were obtained by using the SNICS source of a tandem accelerator. The cluster ions were carefully selected through an electromagnet and then they went through a scanner before it reached the target where uniform ion implantation took place. Then RBS (Rutherford Backscattering Spectrometry) was used to check chemical contamination and uniformity of the Ge implantation. The uniformity and dosage of the B-implanted samples were checked by NRA (Nuclear Reaction Analysis), based on the $^{11}B(p,\alpha)^8Be^*$ reaction, which at a proton energy of 660

TABLE 1. Energy reduction and Transport gain in small cluster ion transport.

Ion Species	E_B/E cluster	Transport Gain
B$_2$	1/2	5.7
B$_3$	1/3	16
GeB	13%	21
GeB$_6$	8%	272
SiB	28%	6.7
SiB$_6$	11%	150
LaB	7%	50
LaB$_6$	5%	483
B$_{10}$H$_{14}$	9%	378
B$_{18}$H$_{22}$	5%	1610

keV shows a broad resonance. The sensitivity of the NRA is adequate due to the large reaction cross section and the experiment arrangement. [5]

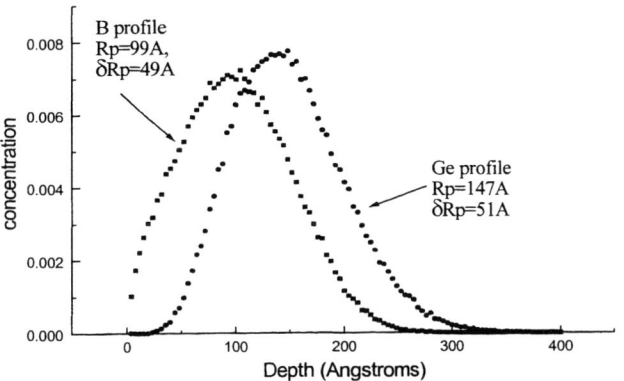

FIGURE 1. TRIM simulation profiles of Ge and B fragments from GeB cluster ions of 15 keV. The effective energy of Ge atom is 13 keV and the effective energy of B atom is only 2 keV.

The SIMS profiles of B in an as implanted silicon by $1 \times 10^{15}/cm^2$, 15keV GeB$^-$ cluster ions are shown in Fig. 2 along with the profiles after annealing. The implantation was done at room temperature with normal incidence. The effective B energy is 2 keV while the GeB cluster energy is 15 keV.

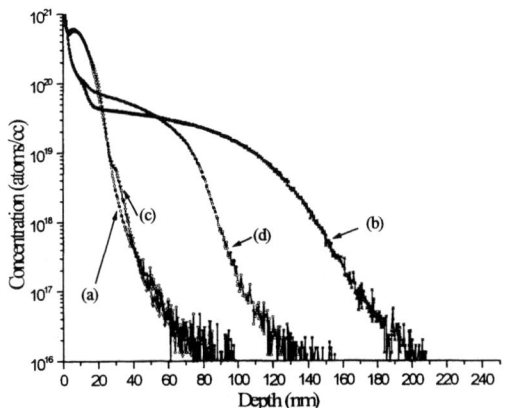

FIGURE 2. SIMS of samples of 15-keV, 1×10^{15} GeB$^-$ implanted into Si. Annealing conditions: (a) as-implanted, (b)1000°C/10s, (c) 550°C/300s, (d) 550°C/300s+1000°C/10s.

Fig. 2 gives the SIMS profiles for samples annealed under various procedures. One is rapid thermal annealing at 1000°C for 10 seconds and the other is two steps annealing with 550°C/300s and then 1000°C/10s. We believe that the GeB$^-$ cluster ion implantation has resulted in an amorphous layer, which reduces the channeling effect and thus the as-implanted profile depth. However, the co-implanted Ge also produced more interstitial than B did in the interstitial rich area, which is located beyond the a-c (amorphous-to-crystalline) interface. The excess interstitial have been reported to cause severe TED which neutralized the benefit of the reduction in channeling effect, and much is being done to reduce the excess interstitial. Worth noting in the profiles is the remarkably reduced junction depth of the sample with two-step annealing as compared to that of the one-step-annealed sample. Based on the previously reported work on the suppression of TED, it is conceivable that in our case the TED has been cut down by a factor of two because the amount of self-interstitial has been greatly reduced during the first step annealing at 550 °C, which converts the amorphous Si into single crystal Si with minimum boron disturbance.

RECOIL IMPLANTATION AND DEFECT ENGINEERING

Another alternative method for shallow junction formation investigated in our group is recoil implantation. [6] Recoil implantation from a solid dopant source had been investigated to make junction by irradiation Si or Ge ions. Dopant can be introduced into a substrate by ion-mixing to form a shallow junction. A generally mis-conception is that the recoiled dopant profile will become deeper while implantation energy is higher. Previous studies have shown that the central region of recoil distribution can be well approximated by an exponentially decaying function ($C(x) = Ae^{-x/L}$) with a decay length L linearly dependent on incident ion energy E_0. However, we have found that 500 keV high energy recoil implantation produces a shallower B profile than lower energy 50 and 10 keV implantation. This surprising discovery is presented in Fig. 3, where both as implanted B profile and the profile after rapid thermal annealing (1000oC/10s) confirmed that the junction depth due to 500 keV Si bombardment is much shallower than that by 50 keV. The experiment agrees well with our analytical calculation. [5] This surprising fact led to our better understanding of the role of the defects in Si ion irradiation, which is more important than the recoil process. In the Si ion irradiation process, the momentum imparted to the interstitial causes their distribution to be deeper than vacancies. Therefore, an excess vacancy rich region was formed close to the surface and excess interstitials were left in the deep range due to the spatially separated Frenkel Pairs. The separation between the vacancy distribution and interstitial distribution increases when implantation energy increases. TRIM calculation of the excess defect population in Figure 4 shows a net vacancy region that has formed deep around ½ Rp of Si projected range. The dashed lines are extra atoms introduced by implantation and the solid lines are defect distribution from spatial separated Frenkel Pairs, where a negative concentration means vacancy and positive concentration means Si self interstitial or excessive Si implants. Since the depth distribution of the vacancy and self interstitial of Si can be tailored by Si implantation in Si, B diffusion can be retarded in the vacancy rich region and enhanced in the interstitial rich region.

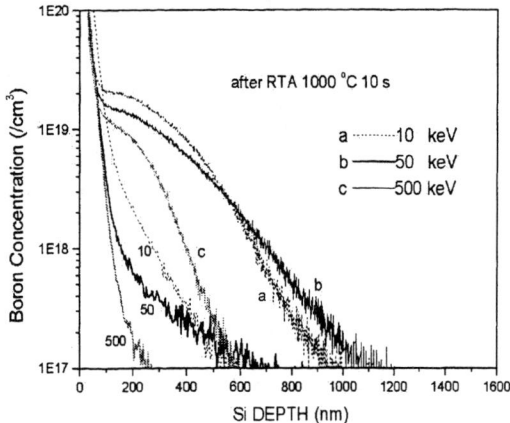

FIGURE 3. SIMS profiles of B recoils of as irradiated and after Rapid Thermal Annealing (1000°C/10s). The three Si ion irradiations to create recoils were 10keV, 50 keV and 500 keV for dosage of $5 \times 10^{14}/cm^2$.

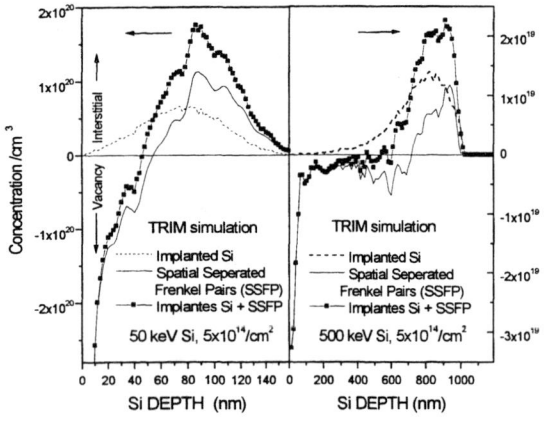

FIGURE 4. TRIM simulation of a) 50 keV and b) 500 keV, $5 \times 10^{14}/cm^2$ Si ions implanted into Si. The dashed lines are extra atoms introduced by implantation and the solid lines are defect distribution from spatial separated Frenkel Pairs (SSFP). Negative concentration corresponds to vacancy while a positive one corresponds to interstitial defects.

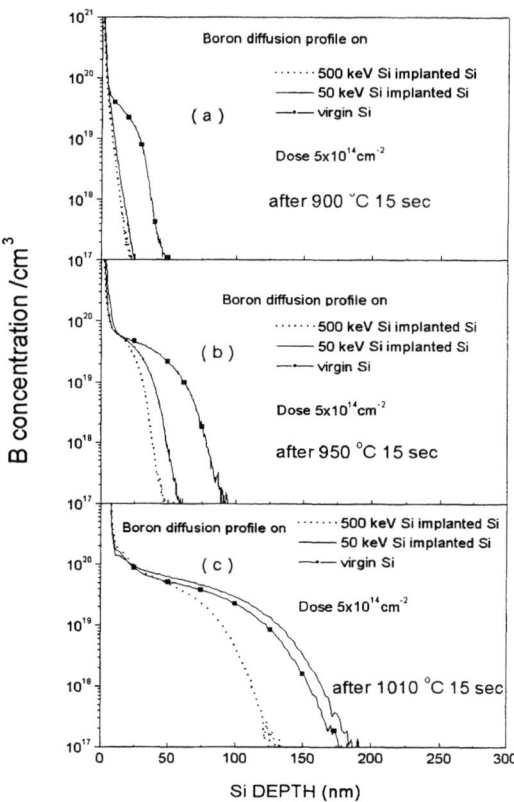

FIGURE 5. SIMS profiles of diffused boron from surface deposited layer into ion irradiation damaged substrate, samples were first irradiated with 50 keV or 500 keV Si ions, then 10 nm B layers were deposited and annealed under 900 °C (a) or 950 °C (b) or 1010 °C (c) for 15 seconds.

To verify this technique of defect engineering, an inverse experiment was performed. The silicon wafer was irradiated with Si ions of 500 and 50 keV to the dosage of $5 \times 10^{14}/cm^2$ and then deposed a thin B layer on the self-irradiated wafer. There is no recoil process. The effect of the Si ion created defect on the boron diffusion from the deposed top layer was investigated. Rapid thermal annealing of 900°C, 950°C and 1010°C for 15s was conducted for both irradiated samples. The results are shown in Fig. 5. B diffusion in 500 keV irradiated Si was reduced significantly and such suppression became more obvious when the incident Si dosage was increased. However, B diffusion in the 50 keV Si irradiated sample is higher. It was suggested that the vacancy rich region near the surface produced by the imbalance after irradiation suppress the B diffusion. By such judicious placing of vacancy and interstitial defects in different depth, B diffusion can be retarded or enhanced. In this experiment, a sub-10 nm deep junction was formed by this technique of point defect engineering. This technique opens up a new approach to the ultra-shallow junction fabrication.

CONCLUSION

Two alternative ion implantation approaches for ultra-shallow junction were successfully investigated. One used small B containing cluster ions, which is logical extension of traditional ion implantation. The other used co-implantation by either recoil implantation or by judiciously placing vacancy or interstitial in different depth. Both techniques can produce junctions shallower than 30 nm.

ACKNOWLEDGEMENT

This project was support by the State of Texas through the Advanced Technology Program and through the University of Houston, the facilities at the Texas Center for Superconductivity have been used.

REFERENCES

1. Jones, E. C., and Ishida, E. *Materials Science and Engineering,* R24, 1-80(1998).

2. Goto, K., Matsuo, J., Tada, Y., Tanaka, T., Mamiyama, Y., Sugii T., and Yamada, I., *IEEE, IEDM* 97-471.

3. Xinming Lu, Lin Shao, Q. Y. Chen, Jiarui Liu, Peiching Ling and Wei-Kan Chu, *MRS Spring meeting,* April 2000.

4. Wolf, S., and Tauber, R.N., *Silicon Processing for the VLSI Era, Vol.-Process Technology*, Sunset Beach, California:Lattice Press, 1986. p.516.

5. Jiarui Liu, Xinming Lu and Wei-kan Chu, to be published.

6. Lin Shao, Xinming Lu, Jian-Yue Jin, Qinmian Li, P. A. W. van de Heide, Jiarui Liu, and Wei-Kan Chu., App. Phys. Lett. 76(2000)3953

Slicing Dielectric Crystals With Ions: A New Material Processing Technique For Electronic And Optoelectronic Materials Integration

Richard M. Osgood, Jr., Antonije M. Radojevic, and Miguel Levy

Microelectronics Sciences Laboratories, Columbia University, New York, NY 10027
Tel: 212-854-4462. Fax:212-860-6182. E-mail: osgood@columbia.edu

Hassaram Bakhru

Department of Physics, State University of New York, Albany, NY 12222

Abstract. This paper describes a recently developed technique for accomplishing thin film transfer using a well-defined single crystal obtained from standard bulk growth. An external implantation exposure is used to prepare the material for the subsequent film transfer step. The films obtained with this method retain their single-crystal bulk properties.

INTRODUCTION

Since its invention in 1987, epitaxial liftoff has been proposed and used for achieving heterogeneous integration of many III-V and elemental semiconductor integrated systems.[1] For example, it has been shown to be an effective technique for integrating heterojunction bipolar transistors or diode lasers on Si substrates, thereby achieving the optimized performance of each of these separate materials. Despite this success, it has been heretofore impossible to use this technique to integrate devices of many other important material systems, chiefly non-semiconductors on these same substrates.

A good example of this problem is the need for integration of transition metal oxides on a semiconductor platform. In one particular case there has long been a need for on-chip thin film optical isolators. These devices are typically made from bismuth-substituted yttrium iron garnet (BiYIG) and thus would require epitaxial growth of a high temperature mixed oxide on a single crystal semiconductor surface, a growth technology that is challenged by a complex high-temperature chemistry, complex stoichiometry, along with the usual problems of lattice matching. Similar problems are seen with other technologically important metal oxides and ferroelectric materials, such as Lithium Niobate ($LiNbO_3$) and ferroelectric oxides.

In recent years a number of different techniques have been used to study the deposition of $LiNbO_3$ and other ferroelectric films, such as $BaTiO_3$, $PbTiO_3$, $Pb(Zr, Ti)O_3$, onto various substrates. Chemical vapor deposition,[2] sol–gel processing, RF sputtering, and pulsed-laser deposition are among the methods reported in the literature. Highly textured but polycrystalline films of $LiNbO_3$ on silicon have been produced both by sol–gel processing on MgO buffer layers and by RF sputtering onto silicon nitride layers.[3,4] Although high-quality films have been fabricated by these techniques, many of the electrical and electro-optical properties reported are generally not comparable to those of bulk single-crystal material. Finally, in many case growth is not possible, much less the formation of more complex material structures needed for lift-off.

THE CIS TECHNIQUE

Recently, a technique called Crystal Ion Slicing (CIS) has been applied to magnetic garnets and ferroelectrics, and made it possible to obtain high-quality freestanding single-crystal microns-thick films of these materials.[6-8] The CIS films are found to retain or closely approximate the single-crystal properties of the bulk material. The technique allows integration of thin films with otherwise incompatible yet technologically very important materials into planar hybrid systems.

The CIS process employs the formation and subsequent preferential etching of a buried damaged sacrificial layer obtained by implanting energetic (~MeV) He^+ particles into a single-crystal bulk material. As the ions penetrate into the target they scatter and lose energy. At higher energies the scattering is primarily electronic with no lattice distortion, while at lower energies, the dominant stopping mechanism is governed by Rutherford scattering. The host nuclei thus become significantly dislodged towards the end of the ionic range, leading to the formation of a narrow damaged sacrificial layer well beneath the surface, which exhibits preferential etching upon immersion into an etchant solution. The energy of the implantation can be adjusted to select the film thickness, while the total dose (~10^{16} cm^{-2}) will determine the amount of damage introduced in the sacrificial layer. Transport-of-ions-in-matter (TRIM) calculation is used to determine the ion range and implantation profile, and thus the thickness of the films to be obtained. For example, implanting at 3.8 MeV in $LiNbO_3$ will result in ~ 9.8 μm thick films, while at 2.25 MeV the film thickness is ~5.2 μm.

During the implantation process, the samples are mounted on a specially designed, 2"-diameter water-cooled target holder and the temperature of the substrate is maintained below 400 °C. As an added precaution, the beam current during the implantation is kept low (0.25 mA/cm^2), while the implantation uniformity is checked with four Faraday cups outside the target holder.

A deep undercut will form in the sacrificial layer during the etch process and uniformly (its fronts parallel to the facets) progress with time, resulting in a freestanding CIS film.[7, 8] The undercut length can be easily determined using Nomarski-prism optical microscopy. The etch selectivity, i.e. the ratio of the etch rates along and normal to the sacrificial layer plane, can be very high, and in the case of $LiNbO_3$ reaches 1000.[7]

FIGURE 1. Scanning electron micrograph of a CIS film of $LiNbO_3$ residing on glass.

By heat-treating the implanted bulk samples prior to etching, the etch rates and etch selectivity can be significantly enhanced by factors as large as 140.[8] We attribute this to the fact that upon implantation, He^+ does not form a solid solution within the host matrix; heat treatment thus drives the ions to regions of large damage where they coalesce, generating additional vertical stress. This intuitive picture is corroborated by the Secondary Ion Mass Spectroscopy studies where the additional stress buildup due to He^+ migration can be easily seen from the measured spectra.[8]

Consider the case of $LiNbO_3$. Etch rates as high as 4 μm/minute were obtained, resulting in ~10×10 mm^2, high-quality CIS films fabricated in a short period of time. The technique is particularly useful for fabrication of longer (>20 mm) and narrower (~3 mm) films, as it is the shorter film dimension that determines the time required for lift-off. A CIS film of $LiNbO_3$, fabricated in this manner and then placed on a glass platform, is shown in Fig. 1.

An alternate route to fabricating thin CIS films has also been applied to $LiNbO_3$.[9, 10] The implanted samples are placed on a doped (dummy) Si wafer with the implanted side facing the Si surface, and then heated to a temperature of ~100 °C, causing a pyro-electric charge buildup in $LiNbO_3$ and effective electrostatic bonding through a thin native SiO_2 layer at the interface. Rapid temperature increase is then applied, resulting in microns-thick CIS films obtained by thermal shock. These films have comparable qualities to those obtained by the wet-etch process.

Rapid thermal treatment was also found to be instrumental in removing the residual damage in the film region after the implantation. Rocking curve $\theta-2\theta$ measurements with Cu $K\alpha$ x-rays performed on $LiNbO_3$ have shown that the detached films have the same crystallographic structure as virgin bulk and little residual damage.[6] However, in order to remove the

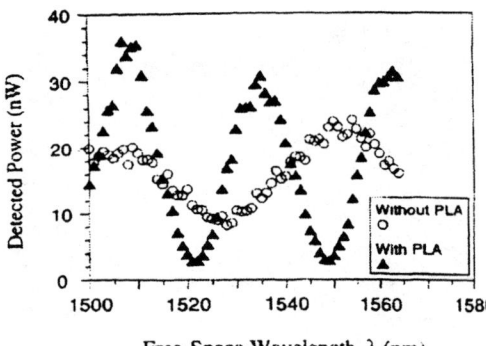

FIGURE 2. Effect of annealing on the optical LiNbO$_3$ modulator response: output-power variation before (○) and after (▲) applying post-lift-off annealing (PLA).

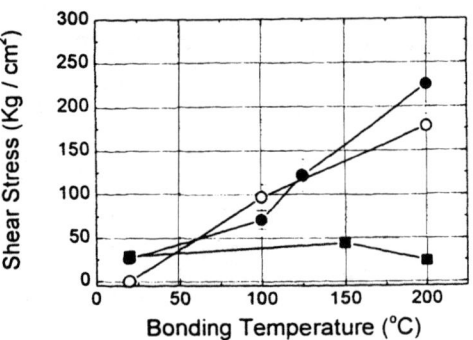

FIGURE 3. Bond strength of BiYIG to Si as a function of heating temperature: implanted (○) and virgin (●) BiYIG on hydrophilic Si, and virgin BiYIG on hydrophobic Si(■)

residual mechanical stress in the CIS films and preserve their bulk material properties, especially the excellent linear and nonlinear $\chi^{(2)}$ – related characteristics, additional annealing step is required upon lift-off, leading to a full recovery of linear and second-order nonlinear properties (see Fig. 3).[11-13]

Different methods have been studied in order to integrate the CIS films with important heterogeneous substrates, such as InP, Si and GaAs. Once fabricated, the thin films can easily be mounted on an arbitrary planar surface and held firm via the attractive Van der Waals force.[6] Bonding can also be achieved by applying an etchant-proof epoxy, wax, or polymer between the top film surface and the hetero-substrate prior to lift-off. Direct low-temperature (~200 °C) bonding has been used for integration of the liquid-phase-epitaxy-grown YIG films with various semiconductors, which in conjunction with the CIS process resulted in ~3-μm-thick BiYIG films on a 6-μm-thick film of GGG integrated with Si.[14] Shear-stress measurements were used to describe the bond and assess its strength by comparing its character under different surface conditions and semiconductor substrates. The measured bond strength for the BiYIG-Si interface has shown an increase with bonding temperature (Fig. 2). It was also found that the as-implanted and virgin samples had the same bonding strength with Si. Hydrophilic surface preparation is an important step in achieving direct bonding at low temperature. The bond strength for BiYIG on GaAs or InP heated to 200 °C was found comparable to that obtained for Si-to-Si bonding.[14] Similar results were obtained with LiNbO$_3$ films integrated with InP and Si, either by low-temperature direct bonding or pyroelectric bonding, as described above. Other methods such as anodic bonding, wafer bonding, metal or polymer bonding can also be applied in the hetero-integration process.

MATERIALS AND DEVICES

The technique of Crystal Ion Slicing has been successfully applied to a variety of materials that can be classified as metal-oxides, magnetic garnets and ferroelectrics; these include Y$_3$F$_5$O$_{12}$ (YIG) and Bi$_x$Y$_{3-x}$F$_5$O$_{12}$ (Bi-YIG), LiTaO$_3$, KTaO$_3$, SrBa$_x$Ti$_{1-x}$O$_3$, PbZr$_x$Ta$_{1-x}$O$_3$, etc. Due to its superior optical properties and ubiquity in various aspects of modern technology, much attention has been devoted to applying the CIS process on LiNbO$_3$. This review will thus concentrate on thin-film devices obtained with this material.

Thin films of LiNbO$_3$ have been fabricated with a 4-10 μm film thickness and different crystal orientation.[8, 11] The CIS process is also found to be transparent to the direction of the spontaneous material polarization; thus, an arbitrary polarization domain structure can be preserved upon ion implantation, subsequent heat treatment and wet etching. This fact was effectively used for fabrication of thin films of periodically poled LiNbO$_3$ for quasi phase-matched frequency conversion with bulk-like response at wavelengths of ~1.55 μm.[13]

One of the first practical applications of the CIS films of Z-cut LiNbO$_3$ and LiTaO$_3$ is the fabrication of a fully packaged freestanding thin-film pyroelectric detector.[10] Such a detector provides for better responsivity and a broader modulation frequency of the incoming signal compared to a detector mounted on a heat sink; thus, the CIS film detector response will depend only on the variation of the film thickness. Although CIS films generally have a thickness variation of <100 nm primarily due to roughness of the undercut surface,[8, 13] recent experiments have shown that this variation can be readily reduced to <20 nm.

More recently, zero-order λ/2–plates for integrated optics applications at ~1550 nm were fabricated from X-cut $LiNbO_3$.[9] The plates are characterized with conversion ratios in excess of 30 dB and negligible material loss. A polarization-independent performance of a hybrid-integrated device comprised of a CIS film retarder and straight single-mode silica waveguides was also demonstrated. The measured CIS wave-plate figures of merit and the superior material properties of $LiNbO_3$ make the CIS films excellent candidates for integrated polarization-insensitive devices.

In addition to the hybrid materials integration, the CIS films have the advantage of being only a few microns thick, thus allowing for low-voltage electrical tuning and utilization of the largest electro-optical coefficients in devices where such a control is desired. Low half-wave voltage of ~8 Vcm for full optical switching at ~1550 nm has been demonstrated in static modulation experiments on Z-cut $LiNbO_3$ films.[12] Furthermore, several schemes for high-speed, impedance-matched, low-voltage, low-microwave-loss, microwave-optical velocity matched modulators were theoretically studied,[17] and their experimental realization is currently underway.

CONCLUSION

The technique of Crystal Ion Slicing has emerged as a very effective tool for fabrication of microns-thick mesoscopic dielectric films with preserved bulk properties, and their subsequent integration with otherwise incompatible, technologically very important materials into hybrid optical and electronic systems. The relatively benign fabrication steps of the CIS process indicate that the technique would require only minor alterations in order to achieve full compatibility with conventional microfabrication techniques. Several devices based on CIS films have already been reported and have either shown potential for surpassing, or have already exceeded their commercially available counterparts. Additional work is in progress to extend the technique to other materials and develop tunable integrated devices for optical, microwave and electronic applications.

ACKNOWLEDGMENT

This work was made possible through the generous support of the U.S. AFOSR/DARPA program (contract No. F49620-99-1-0038), and the FAME program (contract No. N00173-98-1-6014).

REFERENCES

1. Yablonovitch, E., Gmitter, T., Harbison, J. P., and Bhat, R., *Appl. Phys. Lett.* **51**, 2222-2224 (1987).

2. Sakashita, Y., and Segawa, H., *J. Appl. Phys.* **77**, 5995-5999 (1995).

3. Yoon, J.-G., and Kim, K., *Appl. Phys. Lett.* **68**, 2523-2545 (1996).

4. Griffel, G., Ruschin, S., and Croitoru, N., *Appl. Phys. Lett.* **54**, 1385-1387 (1989).

5. Rost, T. A., He, L., Rabson, T. A., Baumann, R. C., and Callahan, D. L., *J. Appl. Phys.* **72**, 4336-4343 (1992).

6. Levy, M., Osgood Jr., R. M., Liu, R., Cross, L. E., Cargill III, G. S., Kumar, A., and Bakhru, H., *Appl. Phys. Lett.* **73**, 2293-2295 (1998).

7. Levy, M., Osgood Jr., R. M., Kumar, A., and Bakhru, H., *App. Phys. Lett.* **71**, 2617-2619 (1997).

8. Radojevic, A. M., Levy, M., Osgood Jr., R. M., Kumar, A., Bakhru, H., Tian, C., and Evans, C., *Appl. Phys. Lett.* **74**, 3197-3199 (1999).

9. Radojevic, A. M., Levy, M., Osgood Jr., R. M., Kumar, A., and Bakhru, H., *Photon. Tech. Lett.* **11**, (December 2000).

10. Lehman, J. H., Radojevic, A. M., Osgood Jr., R. M., Levy, M., and Pannel, C., *Opt. Lett.* **25**, 1657-1659 (2000).

11. Radojevic, A. M., Levy, M., and Osgood Jr., R. M., *Appl. Phys. Lett.* **75**, 2888-2890 (1999).

12. Ramadan, T., Levy, M., and Osgood Jr., R. M., *Appl. Phys. Lett.* **76**, 1407-1409 (2000).

13. Radojevic, A. M., Levy, M., Osgood Jr., R. M., Jundt, D. H., Kumar, A., and Bakhru, H., *Opt. Lett.* **25**, 1034-1036 (2000)

14. Izuhara, T., Levy, M., and Osgood Jr., R. M., *Appl. Phys. Lett.* **76**, 1261-1263 (2000).

15. Albaough K. B., and Rasmussen, D. H., *J. Am. Cerram. Soc.* **75**, 2644-2648 (1992).

16. Levy, M., Osgood Jr., R. M., Bhalla, A. S., Guo, R., Cross, L. E., Kumar, A., Sankaran, S., Bakhru, H., *Appl. Phys. Lett.* **77**, 2124-2126 (2000).

17. Georma, I., Savi, P., and Osgood Jr., R. M., *Photon. Tech. Lett.* **11** (December 2000)

Recoil implantation of boron into silicon by high energy Silicon ions

L. Shao,[a] X. M. Lu,[a] X. M. Wang,[a] I. Rusakova,[a] G. Mount,[b] L. H. Zhang,[b] J. R. Liu[a] and Wei-Kan Chu[a]

[a] *Department of Physics and Texas Center for Center for Superconductivity, University of Houston, Houston, TX 77204-5932*
[b] *Charles Evans & Associates, 810 Kifer Road, Sunnyvale, CA 94086-5203*

Abstract. A recoil implantation technique for shallow junction formation was investigated. After e-gun deposition of a B layer onto Si, 10, 50, or 500 keV Si ion beams were used to introduce surface deposited B atoms into Si by knock-on. It has been shown that recoil implantation with high energy incident ions like 500 keV produces a shallower B profile than lower energy implantation such as 10 keV and 50 keV. This is due to the fact that recoil probability at a given angle is a strong function of the energy of the primary projectile. Boron diffusion was showed to be suppressed in high energy recoil implantation and such suppression became more obvious at higher Si doses. It was suggested that vacancy rich region due to defect imbalance plays the role to suppress B diffusion. Sub-100 nm junction can be formed by this technique with the advantage of high throughput of high energy implanters.

INTRODUCTION

Fabrication of sub-100nm junction in the ultra-large scale integration technology becomes extremely difficult for traditional implantation. Currently available low energy implanters still face challenges like low beam current due to space charge effects. The relentless drive to find alternative doping methods has lasted two decades. Recoil implantation, the method of introducing dopant from the surface deposited layer into silicon by knock-on, had been studied by various authors, with focus on doping efficiency and recoil spectrum [1-3]. Conventional belief is that the higher the implantation energy, the deeper the recoil profile. In order to get shallow junction, much effort was focused on the low energy recoil implantation. It has been shown that the conventional belief is incorrect, and the higher the energy of the incident ions, the shallower the recoil profiles [4]. This opens a new technique to form ultra-shallow junction with the advantage of a high through-out of high energy implanter. Furthermore, defect imbalance by high energy implant is also helpful to suppress boron diffusion from the view point of defect engineering [5, 6]. Boron diffusion in the shallow junction formation by high energy recoil implantation was presently discussed in this literature.

EXPERIMENTAL

A boron layer (0.4 nm) was first deposited by e-beam evaporation onto an n-type (100) Czochralski-grown Si wafer, then 10, 50, and 500 keV Si ion beams were used to knock the boron atoms into the Si substrate by means of ion beam recoiling. The deposition process was immediately preceded by an HF etch to remove the native oxide. Deposition was performed at a rate of 0.1 nm/sec under a base pressure of 3×10^{-6} torr. A liquid nitrogen cooled substrate holder was used to decrease the mobility of the deposited boron atoms to avoid boron island formation. After deposition, nuclear reaction analysis (NRA) was used to check the amount of boron deposition and atomic force microscopy (AFM) was used to examine the film uniformity. The surface roughness for a bare Si sample was 0.3 nm based on the AFM measurements. For the 0.4 nm boron layers, the roughness was still 0.3 nm if using liquid nitrogen cooled substrates. This is compared to the 2 nm surface roughness when deposited at room temperature. There was no obvious boron island structure observed under AFM for cold deposition. After boron deposition under liquid nitrogen temperature, samples were irradiated separately with $5\times 10^{14}/cm^2$ 10 keV, 50 keV and 500 keV Si ions.

RESULT AND DISCUSSION

Figure 1 shows the SIMS profiles of boron distribution obtained using 5×10^{14} /cm^2 10 kev, 50 keV and 500 keV Si irradiation on 0.4 nm boron deposited Si substrate. The dashed lines are our calculated results. SIMS analysis was performed with 1 keV primary O$_2$ beam. A 25 nm boron profile was obtained by 5×10^{14} /cm^2, 500 keV Si irradiation on the Si sample overlaid with 0.4 nm boron layer, while deeper boron profiles were obtained from the 10 keV and 50 keV recoil implantation.

When Si ions pass through boron thin film, only elastic nuclear scattering generates boron recoils. During the rare nuclear collision between the incident ions and target atoms, target atoms can be permanently displaced from their original sites. The recoils may be divided into: (1) long range recoils created by primary ions with small impact parameters and short range primary recoils with large impact parameters, (2) short range secondary recoils produced by initially displaced target atoms, and (3) cascade recoils resulting from a series of uncorrelated low-energy atomic displacements. A typical recoil distribution has very high concentration near the surface, which decreases rapidly to a region of nearly constant slope and then falls off rapidly again at the deep region. It has been reported that in the central region of recoil distribution, the concentration profile $C(x) = Ae^{-x/L}$ has a characteristic decay length L (Å) $= 3.75\gamma E_0$ [7], where γ is the maximum fractional energy transfer to recoils given as

$$\gamma = \frac{4M_1 M_2}{(M_1 + M_2)^2}.$$

However, when the incident Si energy is increased, recoil boron distribution in the surface region would deviate significantly from an exponentially decaying function. The higher the incident energy, the more rapidly the recoil boron distribution falls off. This is consistent with the fact that the cross sections for large angle deflection become significant when the primary ion energy is increased, thus forming a sharp surface concentration within the first few hundred angstroms in the implantation direction.

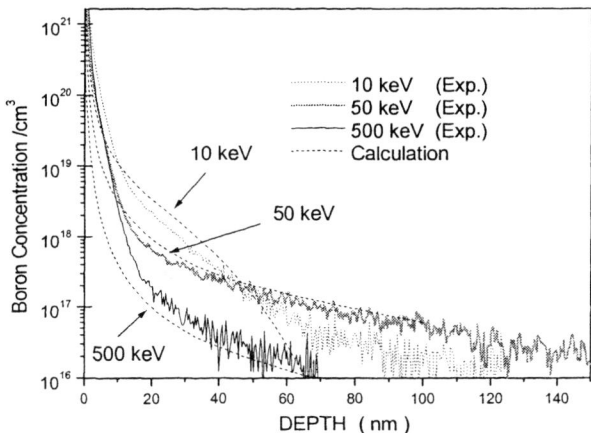

FIGURE 1. SIMS profile of boron distribution by 5×10^{14} /cm^2 10 kev, 50 keV or 500 keV Si implant through 0.4 nm thick B layer deposited Si.

The density of recoil as a function of recoil energy is known as the primary recoil spectrum, which is a well known concept in neutron radiation-damage theory where the thickness of the target is small compared with the range of the neutron. Similar condition also appeared in our experiment with ultra-thin e-gun deposited boron layer. Under such condition, the energy loss of the irradiation ion in the thin film is negligible, and interval $(T, T+dT)$ by an ion with energy E_0 can be described by

$$N_R(E_0,T)dT = \frac{dN_R(E_0)}{dT}dT = N\Delta x \frac{d\sigma(E_0)}{dT}dT$$

where Δx is the target thickness and T is the recoil energy. The function $N_R(E_0,T)dT$ defined above is the thin-target primary recoil spectrum. Based on the nuclear cross section suggested by Lindhard et al. [8,9]. For each incident ion, the number of recoiled boron atoms with transferred energy T can be taken to the form:

$$N\Delta x \frac{d\sigma(E_0)}{dT}dT = \frac{1}{2}N\Delta x \pi a^2 \left[f(t^{1/2})dt / t^{3/2} \right]$$

where $N\Delta x$ is the number of deposited boron atoms per unit area,

$$a = 0.8853 a_0 (Z_1^{2/3} + Z_2^{2/3})^{-1/2},$$

$$t = TE(M_2/4M_1)(a/Z_1 Z_2 e^2)^2,$$

a_0 is the Bohr radius. The scattering function $f(t^{1/2})$ depends on the inter-atomic potential. 20 eV was used for binding energy + interface potential in the calculation. Recoils were assumed stopped if transferred energy T fell below this threshold. Once T is known, the boron recoil angle can be calculated from the classical theory of two-body collisions. With transferred energy T and the scattering angle, the projected range and straggling tables from the Transport of Ions in Matter (TRIM) simulation [10] were used to get the spatial distribution of recoiled boron atoms in Si. In the calculation in figure 1, we choose an analytic form proposed by Wilson, Haagmark and Biersack [11] based on the LSS [8,9] approximation:

$$f = a(b^{1+c} - 1 - (1+c)lnb)/(b^{2+c} - 2b + b^{-c})$$

with constants a =0.56258, b =1.1776$t^{1/2}$, c =0.62680. Experimental and calculation results in figure 1 agree well except for some shifts near the surface region. Such shifts may be caused by the mixing between primary O_2 ions and boron atoms in the residual boron layer. The distribution in the tail region is caused by head on collisions or collisions with very small impact parameters. High energy recoils have higher energy and deeper range, which causes the cross-over of 50 keV recoil distribution and 10 keV recoil distribution. However, due to the low probability of hard collisions with small impact parameters, the concentration of recoil atoms merge into the noise level in 500 keV recoil implantation.

In addition to introducing boron into silicon, 500 keV implantation also introduces point defects, interstitial and vacancies. The momentum transferred to the target atoms results in the distribution of interstitials being a little deeper than vacancies. After dynamical recombination of local point defects, excess interstitials remain near the projected range of the ion in conjunction with excess vacancies close to the silicon surface. Many investigations have reported on the interaction of high energy implantation induced defect with defects from lower energy implants [12]. Implantation of high-energy Si ions has been shown to result in a reduction in the enhanced diffusion of a keV B implant during post-implant annealing [12]. With the ability to create a vacancy rich region, high energy recoil implantation has the advantage to suppress boron diffusion. This effect is demonstrated in Fig. 2. It shows the SIMS as-recoiled and annealed boron profile with 5×10^{14} cm^{-2} 10 keV, 50 keV or 500 keV Si implantation. The sample was annealed under 1000 °C for 10 seconds. The advantage of high energy recoil implantation can be seen from the comparison between

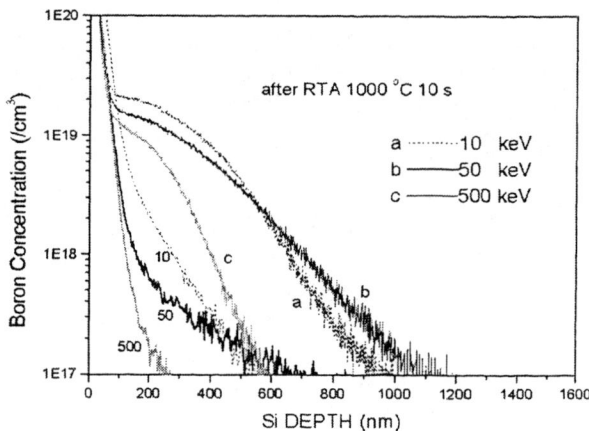

FIGURE 2. SIMS profiles of annealed samples with 5×10^{14} cm^{-2} 10 keV, 50 keV or 500 keV Si implantation. Samples were annealed under RTA 1000 °C for 10 seconds.

the profiles of the 10 keV low energy Si and the 500 keV high energy Si The annealed sample of 50 keV recoil implantation shows a deeper junction depth than that of 500 keV at the concentration 1×10^{17} cm^{-3}. This can be explained by a wider, or more deeply separated vacancy and interstitial defects in higher energy implantation. TRIM simulation of the excess defect population shows a net vacancy region formed and deep around ½ Rp of Si projected range.

Figure 3 shows the SIMS profiles of boron distribution in the sample obtained using 500 keV Si implantation and the profiles of annealed samples after 10 s, 1000 °C rapid thermal annealing (RTA) in N_2 or two-step annealing with 550 °C furnace annealing

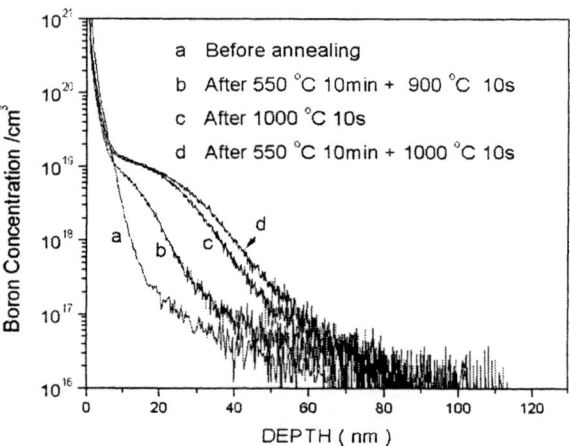

FIGURE 3. SIMS profiles of annealed samples with 5×10^{14}/cm^2 500 keV Si recoil implantation. Samples were annealed under RTA 1000 °C for 10 seconds, or two step annealing with 550 °C furnace annealing + RTA.

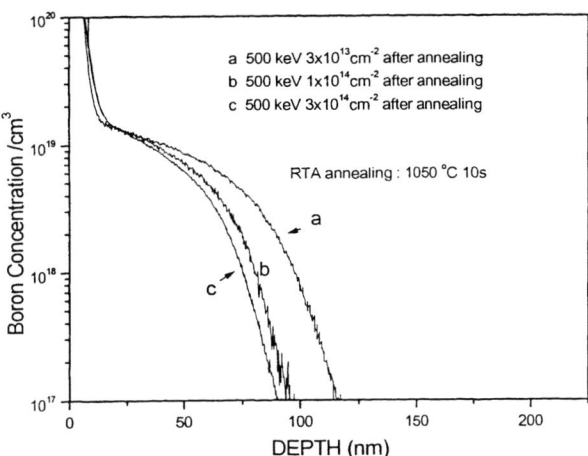

FIGURE 4. SIMS boron profile of samples obtained using 5×10^{14} / cm^2 irradiation on 0.4 nm boron deposited Si 500 keV incident energy. Si dose are 3×10^{13}, 1×10^{14}, or 3×10^{14}/cm^2 respectively.

followed by RTA. The junction depth after 10 s, 1000 °C RTA is 52 nm and sheet resistance is 920 Ω/square. A shallower 36 nm junction can be obtained via 10 min, 550 °C furnace annealing + 10 s, 900 °C with a higher sheet resistance 1620 Ω/square. It was recently demonstrated that the formation of the silicon boride phase can result in a Si interstitial supersaturating and cause boron enhanced diffusion (BED) [13]. During recoil implantation and RTA, boron precipitates and boron silicide may form and BED may play an important role to broaden boron profiles during annealing. It has been shown recently that the effect of BED can be minimized by superimposed high energy pre-implantation.

The higher the dose of 500 keV Si implantation, the more boron atoms introduced into silicon, this brings up a deeper junction after implantation. however, the defect imbalance, or vacancy population was increased also at a higher dose. The final junction depth after annealing may become shallower. Figure 4 shows that such TED suppression became more obvious when incident Si dose was increased. Samples were boron deposited and recoiled by 500 keV Si with the dose ranging from 3×10^{13} to 3×10^{14}/cm^2, then annealed under 1050 °C for 10 seconds. The boron diffusion depth after annealing became shallower when incident dose was increased.

In summary, we have shown that 500 keV high energy recoil implantation produces a shallower B profile than lower energy 10 keV and 50 keV implantation. This agrees well with the analytical calculated distribution of recoiled B atoms. Sub-100 nm junction can be formed by this technique with the advantage of high throughput of high energy implanters.

The high energy bombardment produce a surface vacancy-rich region with deep located interstitials, the higher the dose, the higher the vacancy concentration and the slower the boron diffusion. The demonstration indicates the possibility on diffusion control by point defect engineering which will be elaborated in a future paper.

ACKNOWLEDGEMENTS

This work was supported by the State of Texas through the Advanced Technology Program and through the Texas Center for Superconductivity at the University of Houston.

REFERENCES

1. Bruel, M., Floccari, M., and Gailliard, J. P., Nucl. Instr. and Methods Phys. Res. B **182/183**, 93(1981)

2. Grob, A., Grob, J. J., Mesli, D. Salles, D., and Siffert, P., Nucl. Instr. and Methods Phys. Res. B **182/183**, 85(1981)

3. Liu, H. L., Gearhart, S. S., Booske, J. H., and Wang, W., J. Vac. Sci. Technol. B **16**(1), 415 (1998)

4. Shao, L., Lu, X. M., Jing, J. Y., Li, Q. M., Liu, J. R., Heide, P. A. W., and Chu, Wei-Kan, Appl. Phys. Lett., **76**(26) (2000)3953

5. Shao, L., Lu, X. M., Wang, X. M., Rusakova, I., Liu, J. R., Chu, Wei-Kan, unpublished

6. Saito, S., Kumagai, M., Kondo, T., Appl. Phys. Lett. **63**(1993) 37

7. Christel, L. A., J. F. Gibbons, J. F. and S. Mylroie, Nucl. Instr. and Methods Phys. Res. B **182/183** (1981) 187-198

8. Lindhard, J. Nielsen, V., and Scharff, M., Mat. Fys. Medd. Dan. Vid. Selsk. **36** (14) (1970)

9. Lindhard, J., Scharff, M., and Schiott, H. E., Mat. Fys. Medd. Dan. Vid. Selsk. **37** (14) (1963)

10. Biersack, J. P., and Haggmark, L. G., Nucl. Instr. and Meth. Phys. Res. B **174**(1980) 257

11. Wilson, W. Haagmark, L. G. and Biersack, J. P., Phys. Rev. B. **15**(1977) 2458.

12. Venezia, V. C., Haynes, T. E., Agarwal, Aditya, Pelaz, L., Gossmann, H.-J., Jacobson, D. C. and Eaglesham, D. J., Appl. Phys. Lett.,**74**,1299(1999).

13. Libertino, S., Benton, J. L., Jacobson, D. J., Eaglesham, D. J., Poate, J. M., Coffa, S., Kringhøj, P., Fuochi, P. G. and Lavalle, M., Appl. Phys. Lett. **71**(3), 389 (1997)

Low Energy Implantation of Boron with Decaborane Ions

Marek Sosnowski

Department of Electrical and Computer Engineering, New Jersey Institute of Technology, Newark, NJ 07102

Abstract. Implantation of molecular ions of decaborane ($B_{10}H_{14}$) is an alternative path to ultra shallow doping of Si with B ions of very low energy (< 1 keV). Because of their mass, the molecular ions with an energy an order of magnitude larger than an energy of B^+ monomer ions achieve the same implantation depth. In addition, the molecular ions transport ten times more B per unit charge. To assess the feasibility of this approach, the properties of the decaborane ion beams with energies from 2 to 10 keV were examined. The ions were generated in an electron impact ionization source and transported to a sample chamber through a 2.5 m long beam line with an analyzing magnet. Experiments with electrostatic beam deflection show that the large ions survive the transport in the implanter environment and that neutralization is negligible. Si samples were implanted with decaborane ions and the implanted dose measured by current integration was compared with the amount of retained ^{11}B obtained by nuclear reaction analysis. The retained dose was found to be larger for decaborane ions, which may be attributed to a sputtering yield of Si, smaller than for low energy B^+ ions.

Development of ion sources capable of generating decaborane ion beams has reached the stage where batches of wafers can be implanted. The implanted B profiles and electrical characteristics of test MOS transistors fabricated using implantation with decaborane ions and B^+ and BF_2^+ ions of equivalent energy were found to be very similar. The results confirm the potential of decaborane ion beams as an alternative technology for manufacturing of ultra shallow p-type junctions in Si. More research is needed to fully understand the effects of cluster ions in semiconductors.

INTRODUCTION

Implantation of energetic ions into solids has played a critical role in the development of semiconductor technology over the last 25 years. It has made possible precise doping of the semiconductor layers in various parts of electronic devices, as well as fine adjustment of threshold voltage of MOS transistors. Fabrication of today's multi-layer silicon integrated circuits may utilize 20 implantation steps with ions of different species and energy up to a few MeV. The standard ion implantation technology, however, reaches its limit at the low end of the energy spectrum needed for the very shallow doping of next generation MOS transistors. The International Technology Roadmap for Semiconductors projects the need for 25 to 43 nm deep junctions in the 0.13 μm semiconductor devices, expected by the year 2002 and for 20 to 33 nm junctions in the 0.1 μm devices projected for the year 2005 [1]. The very small projected ion range requires low ion energy, particularly for light ions, which for example for B^+ ions needed for p-type ultra shallow junctions in silicon is of the order of 100 eV. Extraction from an ion source and transport of such low energy ions is hindered by the Coulomb forces (beam space charge), to the extent that standard ion implanters cannot deliver sufficient ion currents for commercial semiconductor implantation. New designs of implanters based, for example, on the deceleration of energetic beams in front of the target, have been introduced [2]. Another solution to the problem of low energy implantation may be plasma immersion [3]. This technique, however, does not discriminate among various ion species and its potential for precise dose control remains uncertain. Both methods are being evaluated by the industry. An alternative approach, discussed in this paper, is based on energetic beams of cluster ions, which produce implantation effects equivalent to those of monomer ions at a lower energy. This is due to partitioning of the ion kinetic energy among its constituent atoms proportionally to their mass. For example, each B atom in a B_{10}^+ cluster ion carries only one tenth of the beam energy. Moreover, the charge for a given atom fluence decreases by the same factor and the beam space charge as well as wafer charging are minimized accordingly.

Considerable interest has been generated by first reports of ultra shallow junctions formed by implantation of clusters formed by ionization of decaborane ($B_{10}H_{14}$). Experimental MOS devices with such junctions were made by Fujitsu in collaboration with Kyoto University [4,5]. Good device characteristics were reported even though the exact composition of the implanted species was not known, as the ion beam had not been mass analyzed. Subsequently, transient enhanced diffusion (TED) of boron in Si samples implanted with decaborane ions at Kyoto University was measured and compared with TED after implantation with boron ions of equivalent energy and dose (500 eV, 10^{15} cm^{-2}). No significant differences were found in the effects of both implantations [6] but the issue deserves further study with a well characterized beam and at still lower energy, where TED is expected to be negligible. Independently, Dirks et al at *Philips*, compared the effects of implantation of Si with decaborane ions and with B$^+$ ions of equivalent energy using microwave ion source and a high energy accelerator with deceleration of the ion beam [7]. The results showed that the effects of decaborane ions, such as B depth profiles, are essentially the same as those of monomer boron ions with ten times less energy, except for larger crystal damage in Si implanted with decaborane. More crystal damage with decaborane ions was also reported by the Kyoto group [8].

While the initial results of using decaborane ion beams for shallow B implantation were encouraging, it was understood that the acceptance of the technology by the semiconductor industry will depend on the feasibility of obtaining such beams of sufficient intensity in ion implanters. Attempts to use the compound in conventional implanter ion sources have not produced cluster ions. To evaluate the feasibility of generating such ion beams, an experimental ion implanter with an electron impact ion source was built at NJIT. The mass spectra of ions generated by electron bombardment of decaborane vapor were also measured using a quadrupole mass spectrometer. It was found that cluster ions containing 10 B atoms (B$_{10}$H$_x^+$) are the predominant component in the ion mass spectra (70-95%), even at elevated temperature (250 - 350°C) [9,10]. This finding was supported by our measurements of mass analyzed beams in the experimental ion implanter. The results were independently confirmed by more recent data obtained at Axcelis (formerly Eaton) with a different type of an ion source. Thus the B$_{10}$H$_x$ cluster appears to be more stable than had been expected, which is promising for the prospect of using decaborane for shallow boron implantation in silicon devices.

Here, we present data on properties of the B$_{10}$H$_x^+$ ion beams, in the energy range 2 - 10 keV, obtained with an experimental electron impact type ion source and a full-scale research ion implanter. They show that the beams of those ions can be effectively transported through an ion implanter without significant neutralization or breakup. B depth profiles measured by SIMS in Si wafers implanted with B$_{10}$H$_x^+$ and B$^+$ ions of equivalent energy are the same but it appears that the retained dose achieved with the molecular ions is higher than with the monomer ions for the same B fluence. The effect may be due to a different Si sputtering yield per impinging B atom with the two types of ions. Very little is known today about the details of interaction of energetic ionized clusters or large molecules, with a crystal lattice. The mechanisms involved are expected to be different than in the case of monomer ions because of the complex nature of the projectile. Important fundamental questions, regarding dopant clustering and defect production and diffusion are still to be answered before the full potential of this promising technology can be assessed. In the mean time the progress in ion source development, reported recently by Axcelis, resulted in sufficient beam currents for implantation of batches of wafers with test MOS devices. The most significant tests compare devices implanted with B$_{10}$H$_x^+$ ions and with the "standard" B$^+$ or BF$_2^+$ ions of equivalent B energy, while the rest of the wafer processing is the same. Two sets of such tests reported recently show very similar performance of devices implanted with these different ions [11,12]. This confirms that implantation of B into Si using decaborane ions has a high potential to become an important industrial process.

B$_{10}$H$_X^+$ ION BEAMS.

Properties of decaborane ion beams were investigated using an experimental implanter system with an electron impact ion source, resembling the design of a Bayard-Alpert ionization gage [13], analyzing magnet, beam line with deflection and scanning plates, and a sample chamber. Details of the system are described elsewhere [10]. The length of the

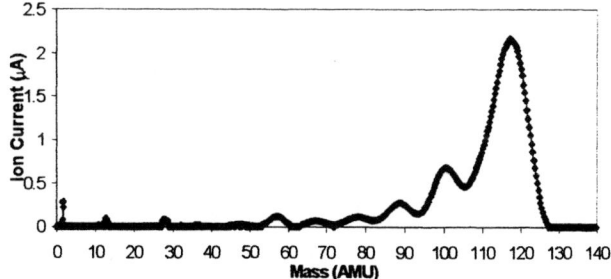

FIGURE 1. Ion current measured on the sample as a function of ion mass at the beam energy of 6 keV.

ion path from the source to the sample was 2.5 m in vacuum in the 10^{-6} Torr range. An example of the ion mass spectrum, measured at the beam energy of 6 keV, is shown in Fig. 1. The mass spectra were obtained by measuring the ion current on the sample block as a function of the magnet current. The relation between the magnet current and ion mass was established by a calibration procedure using the peaks of argon ions Ar^+ and Ar^{++}. The broad peaks correspond to cluster ions with different number of B atoms, from 1 to 10. Their width is related to the presence of boron natural isotopes (20% ^{10}B, 80% ^{11}B) and to dissociation of different numbers of hydrogen atoms ($14 - x$) from the original decaborane molecule. To evaluate a possible break-up and neutralization of the large cluster ions in collisions with residual gas molecules, we deflected the $B_{10}H_x^+$ beam using electrostatic plates in front of a silicon sample. If there was a neutral beam component, then some implanted boron would be found in the sample at the position of an undeflected beam. Break-up of the large ions into smaller fragments would also result in implantation at points on the sample between the deflected and undeflected beam positions. Implanted boron was detected in the sample by measurements of α-particles from the reaction of the predominant boron isotope, ^{11}B, with 650 keV protons:

$$^{11}_{5}B + p \rightarrow ^{8}_{4}Be + \alpha$$

Alpha particles were detected at a scattering angle of 170° by a silicon surface barrier detector, covered by a 10 micron mylar foil to block scattered protons. The yield of alpha particles for a given proton beam charge is directly proportional to the number of ^{11}B atoms in the sample. The proportionality factor was established experimentally by a measurement on a sample with a known boron concentration. The sample implanted with $B_{10}H_x^+$ ions was analyzed with the proton beam, 1 mm in diameter, at 1 mm intervals along the line of deflection. The results are shown in Fig. 2. The peak at 2.25 cm is exactly at the calculated position of the deflected beam. An arrow indicates the position of the undeflected beam (at 5.3 cm), where neutralized ions would be implanted. The number of alpha particles detected there is equal to the background and is below 1% of the peak intensity.

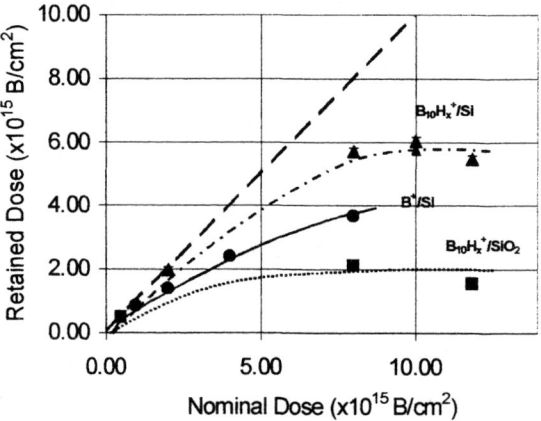

FIGURE 3. Retained B dose as a function of implanted dose for decaborane (5 keV) and boron (500 eV) ion implants. B^+ data from Axcelis-Lucent work [14].

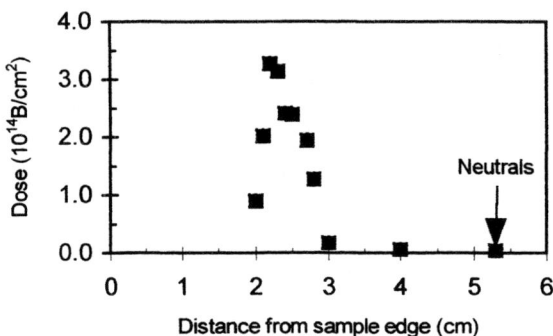

FIGURE 2. Boron dose measured in a Si sample implanted with a deflected ion beam. The arrow shows the position of the undeflected beam where neutralized ions would be implanted.

RETAINED BORON DOSE

Boron dose in Si and SiO_2 samples implanted with decaborane ions was measured by the NRA method, described in Section II, and compared with the dose calculated from the integrated ion beam current. Samples of Si (100) with a native oxide layer as well as samples treated by an HF dip, prior to implantation, were used. No discernible difference in the B dose implanted in those samples was found. The SiO_2 samples were cut from oxidized Si wafers with the oxide thickness of 10 nm. B doses retained in these samples were significantly lower than in the Si samples. The results obtained at the $B_{10}H_x^+$ energy of 5 keV are presented in Fig 3. The doses calculated from the ion beam fluence are compared with the results of the NRA analysis (the retained dose). When the two numbers are equal, a data point lies on the straight dashed line. This is observed for low implanted doses. At higher doses the retained dose tends to saturate, a behavior characteristic of sputtering limited implantation. The SiO$_2$ data indicate the expected higher sputtering yield for this material. Fig. 3 also shows data on implantation of B$^+$ ions in Si at the energy of 500 eV [11]. The retained dose also tends to saturate but at a lower level than in the case of decaborane ions. The data thus show that significantly more B atoms can be implanted into Si with decaborane ions than with B$^+$ ions of equivalent energy.

SUMMARY AND CONCLUSIONS

Beams of the B$_{10}$H$_x^+$ cluster ions can be generated in a carefully designed ion source, mass analyzed, transported through a beam line, without significant break-up or neutralization of the ions.

Measurements of the retained boron dose indicate that more boron can be implanted into silicon with decaborane ions than with boron ions of equivalent energy and fluence. The dose retained in silicon dioxide is significantly lower than in silicon. The difference between the retained dose and the dose calculated from the ion beam fluence is attributed to sputtering of the sample by the ion beam.

Development of ion sources capable of generating decaborane ion beams has reached the stage where batches of wafers can be implanted. The electrical characteristics of test MOS transistors fabricated using implantation with decaborane ions and ULE B$^+$ and BF2$^+$ ions of equivalent energy were found to be very similar. The results confirm the potential of decaborane ion beams as an alternative technology for manufacturing of ultra shallow p-type junctions in Si. More research is needed to fully understand the effects of cluster ions in semiconductors.

ACKNOWLEDGEMENTS

The above review summarizes contributions of a number of researchers, notably, Dale Jacobson of Lucent – Bell Laboratories, Aditya Agarwal and John Poate presently at Axcelis. The help of Tom Horsky, presently at Brooks Automation, in setting up an experimental ion implanter and of Matt Donatucci and Luping Wang of ATMI in handling decaborane is gratefully acknowledged. Finally, the research at NJIT would not be possible without the hard work and dedication of graduate students, Maria Albano and Vijay Babaram.

This research at NJIT was supported in part by grants from the National Science Foundation (GOALI program), Sematech, and SRC.

REFERENCES

1. "The International Technology Roadmap for Semiconductors", *Semiconductor Industry Association*, (1999).
2. Applied Materials Product Announcement (1996).
3. Cheung, N. W., En, W., Jones, E., and Yu, C., *Mat. Res. Soc. Symp. Proc.*, **279**, 297 (1993).
4. K. Goto, J. Matsuo, T. Sugii, I. Yamada and T. Hisatsugu, *IEDM-96, IEEE*, 768-771 (1997).
5. Goto, K., Matsuo, J., Tada, Y., Tanaka, T., Momiyama, Y., Sugii, T., and Yamada, I., *IEDM-97, IEEE*, 18.4.1-18.4.4 (1997).
6. Agarwal, A., Gossmann, H. J., Jacobson, D. C., Eaglesham, D. J., Sosnowski, M., Poate, J. M., Yamada, I.,. Matsuo, J., and Haynes, T. E., *Appl. Phys. Lett.*, **75**, 2015 (1998).
7. Dirks A. J., Bancken, P.H.L., Politiek, J. Cowern, N.E.B., Snijders, J.H.M., van Berkum, J.G.M., and Verheijen, M.A., *Ion Implantation Technology-IIT'98 Internat. Conf.* Kyoto, Japan, June 1998.
8. Matsuo, J., Seki, T., Aoki, T., Yamada, I., *Ion Implantation Technology-IIT'2000 Internat. Conf.* Alpbach, Austria, September 2000.
9. Sosnowski, M., Gurudath, R., Poate, J. M., Mujsce, A. and Jacobson, D. C., *Mat. Res. Soc. Symp. Proc.*, **568**, 49-54 (1999).
10. Sosnowski, M., Albano, M. A.,. Babaram, V., Gurudath, R., Poate, J.M., Jacobson, D.C., *Journal of Electrochemical Society* - to be published (2000).
11. Jacobson, . D.C., Gossmann, H-J., .Sosnowski, M., Albano, M.A., Babaram, V., Poate, J.M., Agarwal, A., Horsky, T., *Ion Implantation Technology-IIT'2000 Internat. Conf.* Alpbach, Austria, September 2000.
12. Perel, A.S., Krull, W., Hoglund, D., Jackson, K., *Ion Implantation Technology-IIT'2000 Internat. Conf.* Alpbach, Austria, September 2000.
13. Kirschner, J., *Rev. Sci. Instrum.*, **57**, 2640-2642 (1996).
14. Agsrwal, A., Gossmann, H. J., Jacobsen, D.C., private communication.

Ultra shallow Sb doped layer formation in Si (001) by the use of recoil implantation

K. E. Daley[1], D. T. Vonk[2], R. J. Culbertson[2]

[1] *Science and Engineering of Materials Program, Arizona State University, Tempe, AZ 85287-1704, USA*
[2] *Department of Physics and Astronomy, Arizona State University, Tempe, AZ 85287-1504, USA*

Abstract. Recoil implantation was performed to create ultra shallow Sb doped layers in an Si (001) substrate. The technique consists of the initial deposition of thin (40 to 140 nm) Sb layers followed by high energy Ar^+ ion irradiation and final chemical stripping of the residual Sb film. The resulting Sb atoms are recoil implanted into the underlying Si substrate. The results show a linear dependence of Sb concentration with Ar^+ ion dose. High resolution Medium Energy Ion Scattering (MEIS) measurements have shown the projected range to be 3.0 nm with a doped layer width of 2.8 nm. The deposited Sb layer thickness is also shown to be a weakly dependent parameter in determining the Sb concentration for a given ion energy and dose.

INTRODUCTION

The ever decreasing MOSFET feature size in modern day microprocessor fabrication has resulted in an intense interest to produce ultra shallow junction regions.[1] As the lateral surface dimensions of the devices decrease, the source and drain regions must also decrease to maintain normal transistor behavior. Conventional low energy implantation techniques have been the main focus of attention in achieving this goal, but do face many difficulties.[2]

An alternative technique to this low energy implantation is the process of recoil implantation. The technique consists of the initial deposition of dopant material onto substrate wafers followed by high energy ion irradiation of the layer. The resulting dopant atoms are recoil implanted into the underlying substrate material. The remaining deposited layer is chemically stripped leaving the implanted dopant atoms in place.

Previous research has shown the technique of recoil implantation to be very effective in producing thin Sb doped layers. Layers of 20 nm thickness were created by Kwok using recoil implantation.[3] Bruel and coworkers showed the recoil efficiency (number of atoms implanted per incident ion) of Sb irradiated with Ar+ ions.[4] Their results showed the recoil efficiency to be independent of ion energy and deposited layer thickness. This energy independence has also been shown by Baumvol.[5] There was no mention of doped layer thickness or concentration control in these research papers.

In this paper we expand upon this recoil implantation methodology to produced ultra shallow, abrupt and well controlled impurity layers. We show a quantitative relationship of concentration verse incident ion dose. We also show results of doped layer control under uniformity variations in deposited layer thickness. In our final results we have measured the Sb concentration depth profile and found a full width at half maximum (FWHM) of 2.8 nm.

EXPIREMENTAL PROCEEDURE

Sb was thermally evaporated onto 2.5 cm x 5.0 cm Si (001) substrate wafer pieces at a base pressure of 1 x 10^{-6} Torr. The samples were positioned at varying distances from the Sb source to give a range of deposited layer thickness. The layer thickness was chosen to vary from 40 nm to 140 nm to give a layer distribution for performing various experimental tests on layer thickness dependence. The pieces were further cut into 1 cm x 1 cm samples to be used as recoil implantation targets.

The implantation was performed using Ar^+ ions with a fixed energy controlled over the range of 100 keV to 200 keV. The beam energy and dose were varied as parameters to be investigated. The implantation was done by scanning the ion beam across a 1 cm x 1 cm aperture with the samples held on an electrically isolated base below it. Dose was determined by direct integration of the sample holder current. The current density was held between 1×10^{-5} to 2×10^{-5} A/cm^2. All implantation was performed at room temperature at a base pressure in the 8×10^{-6} to 1×10^{-5} Torr region.

The post implanted samples were chemically stripped of the deposited Sb layer with a solution of HCl:HNO$_3$:H$_2$O (1:1:1) held at room temperature. Samples were individually dipped into the solution with Teflon tweezers, then rinsed in 18 MΩ deionized water. The etch time was approximately 30 seconds, but the underlying Sb doped layer was found to be unaffected by longer etch times.

Rutherford backscattering spectrometry (RBS) was performed in the Facility for Ion Beam Analysis of Materials (IBeAM) laboratory at Arizona State University using 2 MeV He^{++} ions at normal incidence. The laboratory scattering angle was 170¡.

High resolution depth profiling using medium energy ion scattering (MEIS) analysis was also performed with 190 keV He$^+$ ions at an incident angle of 54.7° with the resulting 2D spectrum covering a scattering angle range from 57.6° to 83.6°. The MEIS system at Arizona State University was previously located at IBM Almaden Research Center and has been described elsewhere[6]. It includes a two-dimensional toroidal electrostatic analyzer[7].

RESULTS AND DISCUSSION

Figure 1 shows the results of a 140 nm thick film implanted with doubling 200 keV Ar$^+$ dose increments from 5×10^{15} cm^{-2} to 8×10^{16} cm^{-2}. The figure shows the RBS spectrum of the Sb peak. The peak areal density increases by a factor of two for each doubling of dose.

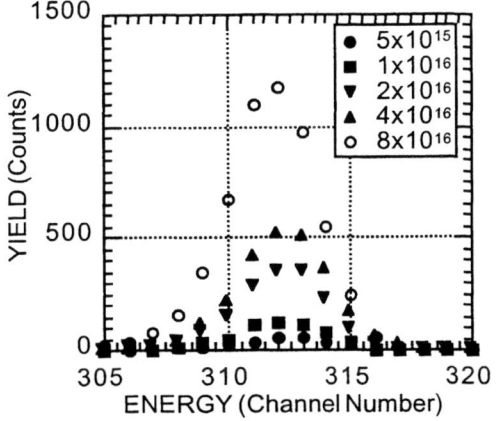

FIGURE 1. RBS spectra showing Sb counts with varied ion doses.

The computer simulation code RUMP[8] was used to quantify the Sb concentration with the resulting Sb concentration versus ion dose plotted in Figure 2. This demonstrates the linear relationship between Ar dose and Sb areal density.

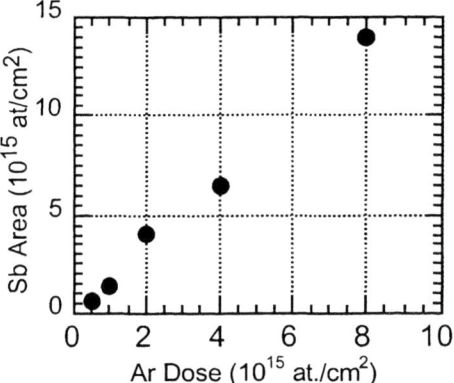

FIGURE 2. Sb concentration vs. dose.

The as-deposited Sb layer thickness was also tested to see if it would affect the Sb concentration for a given implant energy and dose. Figure 3 shows the results for a 5×10^{15} cm^{-2} Ar$^+$ ion dose at 150 keV on 43, 50 and 75 nm deposited films. The RBS spectra show nearly identical Sb peaks which indicates a deposited film thickness independence at least over a 30 nm window.

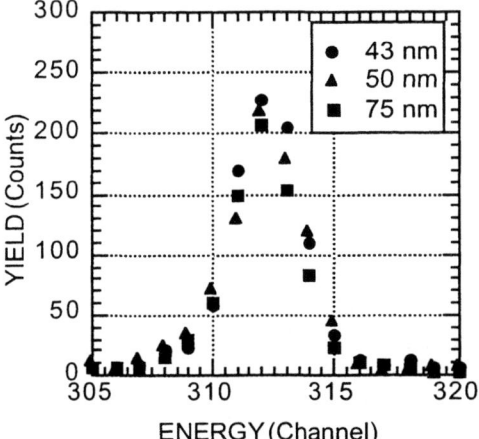

FIGURE 3. RBS spectrum of Sb peak showing layer thickness independence. The three films were 43, 50 and 75 nm of deposited Sb each implanted at 150 keV with a dose of 5×10^{15} cm^{-2} Ar$^+$.

The doped layer thickness obtained in all these experiments was not conclusively determined due to detector resolution limitations. The detector resolution of the RBS system used was limited to about 10 - 20 nm. High angle tilts to increase effective thickness were still inconclusive in obtaining a definitive thickness.

MEIS was used to determine the layer width to within a resolution of approximately 0.7 nm. The MEIS system yields a 2-dimensional array of detected particle counts at each scattering angle channel and each scattering energy channel. The energy channels were converted to depth values for each scattering angle channel in this array [9] using a surface energy approximation.[10] The resulting depth spectra for each scattering angle were recombined in bins with width 0.1 nm. These results shown in Figure 4 indicate a projected range of 3 nm and FWHM of 2.8 nm. . This result is supported quite well with computer simulations obtained using the Monte Carlo code Stopping and Range of Ions in Matter (SRIM)[11]. The data clearly shows that high impact parameter collisions with very low recoil energy transfers is the dominate mechanism at work for this process. This result was also conjectured by previous work done by Baumvol[5] even though their measurements were not sensitive enough to measure the final layer width.

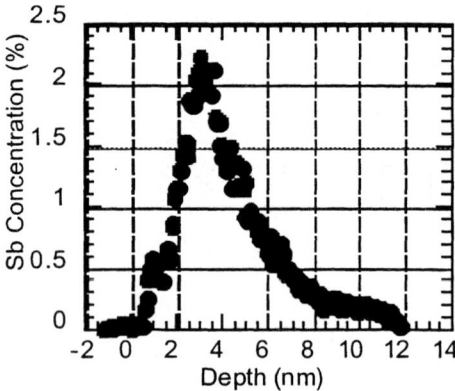

FIGURE 4. Sb concentration vs. depth, as determined by MEIS.

CONCLUSION

It has been shown that recoil implantation is effective in producing ultra thin Sb doped layers with a layer width of 2.8 nm. The Sb concentration is well controlled by the Ar $^+$ ion dose, showing a linear relationship. The concentration has been shown to be independent of deposited layer thickness variations at least over a 30 nm range.

ACKNOWLEDGEMENTS

The authors gratefully acknowledge technical support from Q. B. Hurst, T. J. Michael, J. Pucci, T.W. Karcher, and B.J. Wilkens, as well as assistance with data reduction and visualization by P.A. Scowen and B.A. Ashcroft.

REFERANCES

1. *The National Technology Roadmap For Semiconductors*, (1997).
2. R. Dejule, *Semiconductor International*, 50-56 (April, 1997).
3. H.L. Kwok, *Solid State Phenomena*, **1**, 195-201 (1988).
4. M. Bruel, M. Floccari, J. Labartino, J.F. Michaud, and A. Soubie, *Nuclear Instruments and Methods*, **189**, 135-140, (1981).
5. I.J.R. Baumvol, *Collision Processes of Ion, Positron, Electron and PhotonBeams with Matter*, Proceedings of the Latin American School of Physics, 4-24 Aug, 359-367, (1991).
6. D.E. Fowler, M. W. Hart, and J. V. Barth, *Vacuum* **46**, 1127-1131 (1995).
7. R.M. Tromp, M. Copel, M. C. Reuter, M. Horn, and J. Speidell, *Rev. Sci. Instrum.* **62**, 2679-2683 (1991).
8. L. A. Doolittle, Ph.D Thesis - Cornell University, 1985.
9. D.T. Vonk, to be published.
10. W.-K. Chu, J.W. Mayer, and M.-A. Nicolet, *Backscattering Spectrometry*, Academic Press, NY, NY (1978).
11. J.F. Ziegler, http://www.research.ibm.com/ionbeams/home.htm.

Low Energy Ion Beams for Surface Modification and Film Deposition

J. Wayne Rabalais

Department of Chemistry, University of Houston
Houston, Texas 77204-5641

Abstract. The chemical and physical interactions of ions with surfaces in the range 5 eV to several keV are reviewed. A general description of the interactions of hyperthermal ions with surfaces is presented. The phenomena involved are discussed as a function of the ion kinetic energy and a "transition energy region" is defined for hyperthermal ion – surface interactions. The uses of low energy ion beams, engineering of interfaces with specific properties, and current technologically important ion – surface processes are discussed.

INTRODUCTION

Hyperthermal reactive ions impinging on surfaces provide a method for deposition/growth/synthesis of materials within a nonequilibrium UHV environment. This allows independent control over ion energy and type, ion fluence and dose, substrate temperature, and background gases. The depth of penetration and/or interaction of the impinging ions is determined by the kinetic energy. The excellent control over ion dose allows deposition of thin films, e.g. < 30 Å, with sharp film-substrate interfaces. The UHV conditions allow deposition onto atomically clean, well-ordered surfaces.

ION – SURFACE INTERACTIONS

Ion beams provide a method of delivering unique atomic and molecular species to surfaces while controlling the interaction parameters by means of the ion kinetic energy. Consider the various phenomena that can occur when hyperthermal ions impinge on surfaces. These are illustrated in Fig. 1.
(i) electronic interactions – The incoming ions can be neutralized by electron capture from the surface. In the reverse process, electrons from incoming ions or atoms can be captured by the surface. Either process can result in excited electronic states. These processes depend on the relative energies of the filled and unfilled energy levels of the atoms and the surface.

(ii) photon emission – As a result of electron exchange and energy level crossings in the close encounters between atoms, electrons can be promoted to highly excited states from which they can emit photons.
(iii) scattering and recoiling – Ions can be scattered from the surface and atoms of the surface itself can be recoiled either into or out of the surface in positive, neutral, or negative charge states.
(iv) sputtering – The momentum imparted to surface atoms by impinging ions can result in "sputtering" of atoms, molecules, fragments, and clusters in various charge states.
(v) adsorption, desorption, & chemical reactions – The impinging ions can be adsorbed on the surface, they can cause desorption of materials from the surface, and chemical reactions can occur between the constituents.
(vi) interstitials, displacements, and replacements – The ions can be inserted into the lattice as interstitial atoms without displacement of host atoms or they can displace host atoms, thereby creating a "Frenkel pair". This results in "radiation damage".

THE "TRANSITION REGION"

A variety of ions with kinetic energies in the range $1 - 10^7$ eV are used for growth and modification of material surfaces and interfaces. Various phenomena are dominant or emphasized in different energy regions and thereby, the chemical and physical

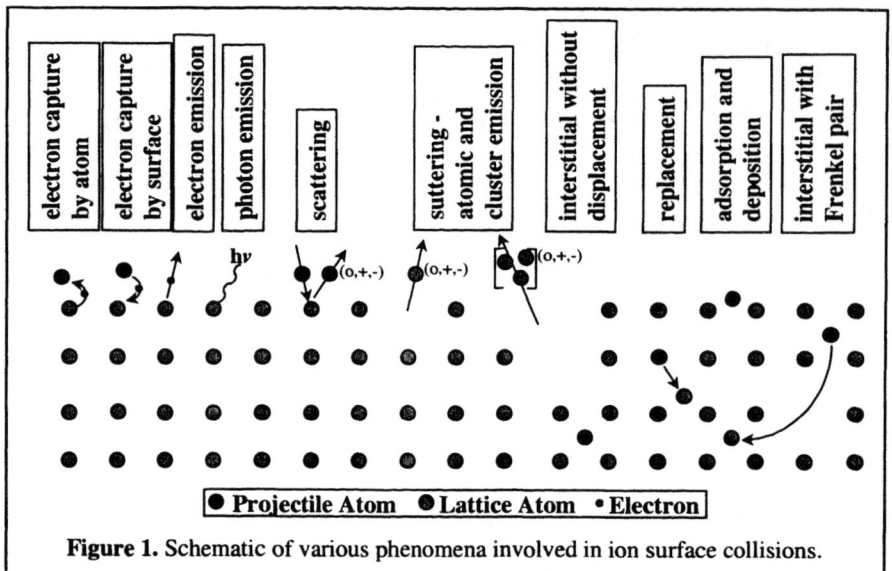

Figure 1. Schematic of various phenomena involved in ion surface collisions.

processes that are induced by the ion impacts are also controlled by this energy. The terminology that has evolved to define approximate energy ranges is as follows: < 1 eV – thermal, 1 to 500 eV – hyperthermal, 0.5 to 10 keV – low energy, 10 to 500 keV – medium energy, and > 0.5 keV – high energy. A schematic diagram of the various phenomena involved, along with an energy scale is shown in Fig. 2.

known as subplantation. In this hyperthermal region, the binary collision approximation[1] is not a valid assumption and simulations of the collision processes use molecular dynamics calculations[2]. The higher energy limit of this region is an ion energy of several keV. Here the chemical bond and electronic energies are much lower than the ion energy. The sputtering yield of surface atoms becomes equivalent to or higher

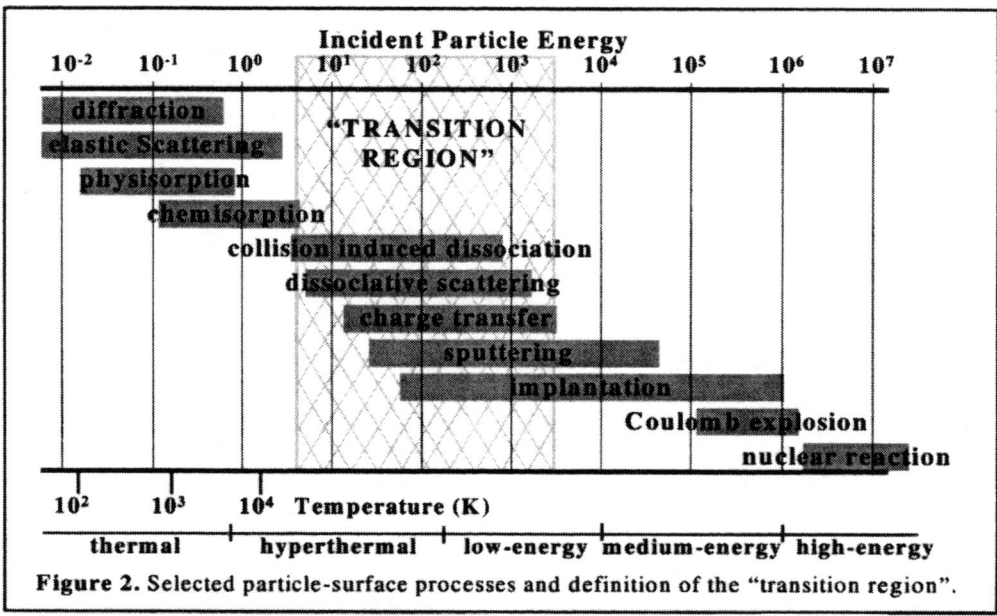

Figure 2. Selected particle-surface processes and definition of the "transition region".

The transition region is the region of interest for film deposition, surface modification, or inducing reactions. The lower energy limit of this region is an ion energy of several eV. This energy is equivalent to chemical bond energies and electronic excitation and ionization energies. The energy losses by the ion are primarily due to phonon excitation and electronic excitation and ionization. The ions have sufficient energy to penetrate into subsurface layers, commonly

than the ion beam flux. Processes deleterious to film growth, such as sputtering, defect formation, and roughening begin to dominate. In this region, the binary collision approximation is a good assumption and can be used to calculate ion penetration depths, sputtering yields, induced damage, etc.

Hyperthermal ions can physisorb or chemisorb on a surface, dissociate and chemisorb as fragments, abstract surface atoms, and dissociate if they are polyatomic. The ions may react with the substrate atoms, creating new chemical species that may either remain on the surface or be volatile or readily sputtered, leading to surface etching. Direct reactions can result in new chemical species, e.g. oxide and nitride growth. These low energy ions deposit their kinetic energy into the surface layers without affecting the bulk region. As the kinetic energy increases, phenomena such as sputtering, annealing, mixing, defect formation, and creation of unique surface topologies begin to occur.

FUNDAMENTAL ION – SURFACE RESEARCH

The development of new materials with specific surface and interface chemical, physical, and/or morphological characteristics is one of the great challenges of the next century. As trends in microelectronic devices move towards smaller, more densely packed components, it is necessary to develop new methods for fabrication and analysis of nanometer scale structures. The versatility of ion beams for surface modification has thrust this field to the forefront of fundamental research over the past few decades and where it will undoubtedly remain for some time.

Figure 3 illustrates some of the most significant areas of ongoing fundamental research in ion-surface interactions. In an article of this length, it is not possible to discuss all of these research areas. Instead, each area will be mentioned briefly and references to the relevant literature will be provided.

diamondlike carbon[4] that is free of hydrogen have been deposited in several laboratories. Attempts at deposition of carbon nitride[5] have produced mixed phases, varying C/N ratios, and amorphous morphologies. Recent advances[6] are moving closer to the goal of producing high quality β-C_3N_4. (3) Since mass-selected ion beams are composed of a single isotope, they naturally lend themselves to producing isotopically pure films[7]. Controlling the isotopic distribution affects the thermal conductivity, refractive index, electronic band gap, and mobilities of charge carriers. (4) Mixed interface layers can be formed by introducing two or more ions, e.g. mixed metal oxides can be produced from two ion beams in an oxygen environment[8]. (5) Chemical functionalities[9] can be imparted to surfaces by using molecular ions or ions of reactive elements, thereby altering the surface reactivities. (6) It has been demonstrated[7,10] that there is a synergism between the ion energy and substrate temperature. This synergism facilitates epitaxial growth at temperatures below those normally used in deposition of thermal atoms, e.g. molecular beam epitaxy. (7) Selective doping[11] of semiconductors with specific concentrations of dopant at specific levels below the surface is a well known process that is used to alter electrical characteristics. (8) The surface morphology[12] of materials can be altered by ion beam sputtering. In some cases, due to selective sputtering of one component of a compound surface, it is possible to create micron and nanometer scale structures with different chemical compositions. (9) Surface composition and structure analysis[13] using low energy ion beams has undergone rapid development due to its extreme surface sensitivity and ability to

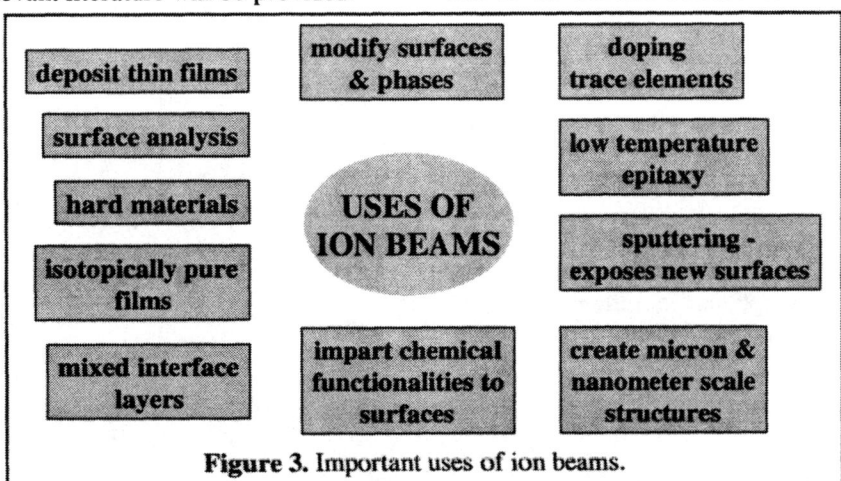

Figure 3. Important uses of ion beams.

(1) Since ion beams can be made from all of the elements, films of any of the nonvolatile elements can be deposited[3]. The excellent dosimetry and energy control of ion beams allows deposition of films with precise thicknesses. (2) Hard materials such as

directly detect surface hydrogen. Recent ion scattering imaging experiments[14] are developing the technique as a real-space surface crystallography.

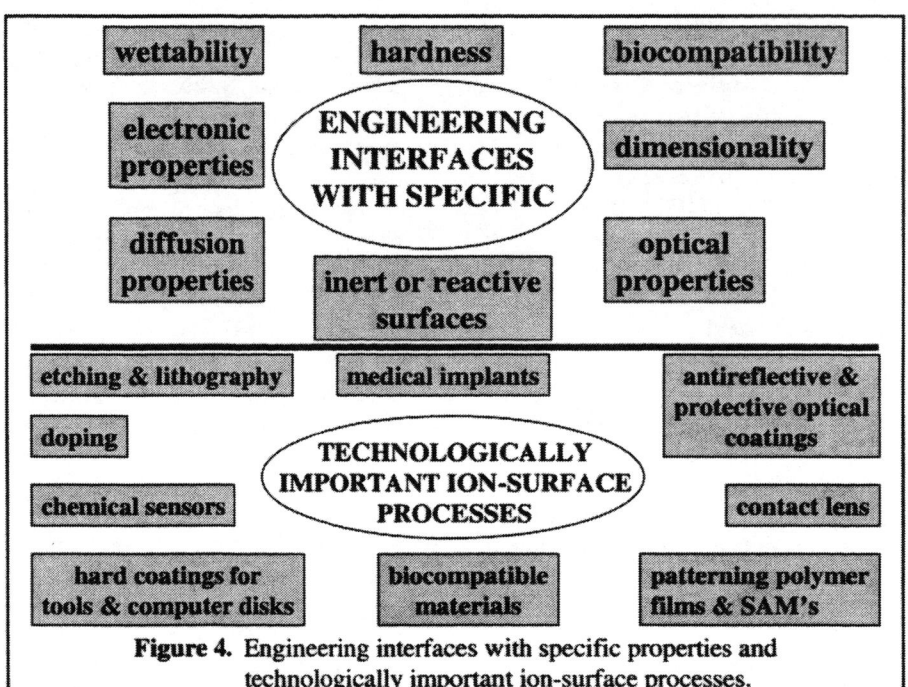

Figure 4. Engineering interfaces with specific properties and technologically important ion-surface processes.

APPLICATIONS OF ION – SURFACE PROCESSES

As a result of the intense fundamental research in ion beam – surface interactions, many technologically important applications have emerged as illustrated in Figure 4. The upper part of the figure depicts the engineering of interfaces with specific properties, such as wettability, hardness, biocompatibility, dimensionality, inert or reactive surfaces, and optical, electronic, and diffusion properties. The lower part of the figure depicts some of the current technologically important ion – surface processes. These include etching, lithography, doping, chemical sensors, hard coatings for tools and computer disks, medical implants, biocompatible materials, contact lens, patterning of polymer films and self-assembled monolayers, and antireflective and protective coatings.

In conclusion, future studies of ion – surface interactions and their applications promises to yield exciting advances in the synthesis of new materials with unique technological applications.

ACKNOWLEDGMENTS

This work was supported by the National Science Foundation under Award No. DMR-9616440.

REFEFRENCES

1. Parilis, E. S., Kishinevsky, L. M., Turaev, N., Baklitzky, B. E., Umarov, F. F., Verleger, V. Kh., and Bitensky, I. S., *Atomic Collisions on Solids*, North-Holland, New York, 1993.
2. Garrison, B. J., Kodali, P. D. S., and Srivastava, D., *Chem. Rev.* **96,** 1327 (1996).
3. Kasi, S. R., Kang, H. Sass, C. S., and Rabalais, J. W., *Surf. Sci. Rep.* **10,** 1 (1990).
4. Lifshitz, Y., Kasi, S. R., and Rabalais, J. W., *Phys. Rev. B,* **41,** 10468 (1990).
5. Marton, D., Boyd, K. J., Al-Bayati, Todorov, S. S., and Rabalais, J. W., *Phys. Rev. Lett.* **73,** 118 (1994).
6. Wu, M. L., Grurz, M. U., Dravid, V. P., Chung, Y. W., Anders, S., Freire Jr., F. L., and Mariotto, G., *Appl. Phys. Lett.* **76,** 2692 (2000).
7. Rabalais, J. W., Al Bayati, A. H., Boyd, K. J., Marton, D., Kulik, J., Zhang, Z., and Chu, W. K., *Phys. Rev. B,* **53,** 10781 (1996).
8. Lee, H., Lee, S. M., Ada, E. T., Kim, B., Weiss, M., Perry, S. S., and Rabalais, J. W., *Nucl Instrum Meth. Phys. Res. B,* **157,** 226 (1999). .
9. Lee, S. M., Ada, E. T., Lee, H., Marton, D., and Rabalais, J. W., *Nucl Instrum Meth. Phys. Res. B,* **157,** 220 (1999).
10. Lee, S. M., Fell, C. J., Marton, D, and Rabalais, J. W., *J. Appl. Phys.* **83,** 5217 (1998).
11. Chanson, E., Picraux, S. T., Poate, J. M., Borland, J. O., Current, M. I., Diaz de la Rubia, T., Eaglesham, D. J., Holland, O. W., Law, M. E., and Magee, C. W., *J. Appl. Phys.* **81,** 6513 (1997).
12. Springholz, G., Holy, V., Pinczolits, M., and Bauer, G., *Science* **282,** 734 (2000).
13. Rabalais, J. W., Science **250** 521 (1990). .
14. Bykov, V., Houssiau, L., and Rabalais, J. W., *J. Phys. Chem. B,* **104,** 6340 (2000).

Development Of Corrosion-Resistant Metal Nitride Coatings Via Ion Beam Assisted Deposition

J. Derek Demaree

U. S. Army Research Laboratory
Aberdeen Proving Ground, Maryland 21005-5069

Abstract. Hard nitride coatings, often considered candidates to replace electroplated chromium in tribological applications, can provide greater corrosion resistance than chromium through the development of a thin, highly bipolar passive film with significant incorporation of ammonium ions and cation-selective oxyanions. In this study, coatings of Cr-N, Mo-N, and Cr-Mo-N have been synthesized with ion beam assisted deposition (IBAD) in an attempt to understand the synergism involved in the production and incorporation of these protective oxyanions. In addition to providing resistance to anodic dissolution, it is thought that the enhanced production of inhibitive oxyanions in these coatings may provide some protection against pinholes and scratches, creating a self-healing coating.

INTRODUCTION

Ion beam assisted deposition (IBAD) has shown the ability to reproducibly tailor the adhesion, stress, and stoichiometry of protective metallurgical coatings for maximum tribological performance in specific applications of interest to the Army. [1] Hard metal nitride coatings produced by IBAD are being considered as replacements for electroplated hard chromium (EHC) in a number of Army systems, especially as environmental considerations make the once-common use of EHC more expensive and cumbersome. [2] The hardness and excellent wear properties of chromium nitride (Cr_xN_y) coatings deposited by IBAD and other physical vapor deposition techniques is well known [3-11], but the corrosion behavior of nitride coatings has not been extensively studied. The most detailed description of nitrogen and chromium interactions in aqueous corrosion, have come out of studies of the electrochemistry of nitrogen-bearing stainless steels [12,13]. Using highly sensitive surface analysis techniques, researchers have found that nitrogen tends to segregate at the surface of a metal undergoing anodic dissolution, so that the small amounts of nitrogen commonly added to steels can have a significant effect on the corrosion behavior. The primary kinetic barrier to corrosion in these steels is still a Cr_2O_3-based passive film, but the presence of N enhances some secondary barriers: namely, surface segregation of Cr and Mo underneath the passive film, and the presence of chromate and molybdate oxyanions at the passive film / solution interface. It is thought that nitrogen from the alloy is incorporated into the passive film as ammonia and then hydrated at the solution interface to form soluble ammonium ion [14]. This reaction raises the interfacial pH, increasing the incorporation of oxyanions like chromate or molybdate into the outer (cation-selective) portions of the passive film. According to the bipolar model of passive film first proposed by Sakashita and Sato [15], these oxyanions can alter the ionic transport through the passive film by enhancing ionic rectification. This enhanced bipolar film (outer cation-selective layer and inner anion-selective layer) slows anodic dissolution both through 1) the electrostatic repulsion of aggressive anions from the electrolyte by the fixed oxyanionic charge and 2) by enhancing the deprotonation of chromium hydroxide to form a highly stable and protective anhydrous Cr_2O_3.

In previous work, we have used XPS and electrochemical studies to understand the role of N more fully, especially its synergism with oxyanion formers like Cr and Mo [16-18]. We have reported [19-20] that the passive films formed on IBAD Cr nitride coatings are much thinner and more strongly bipolar than those formed on pure Cr, and contain large amounts of ammonia and chromate in the outer layers of the film. Since the Pourbaix diagram indicates that molybdate is more stable than chromate over a wide pH range [21], it is thought that this pH buffering mechanism will be even more effective in enhancing molybdate production. In this study, we have synthesized nitride coatings containing both Cr and Mo, and examined their corrosion behavior to see whether we can combine the protective properties of Cr nitride coatings with the inhibitive behavior of Mo-based coatings to produce a hard tribological film with enhanced corrosion-resistant properties.

EXPERIMENTAL

Chromium nitride coatings were deposited in at the National Defense Center for Environmental Excellence facility operated by CTC Corporation in Johnstown, PA. The chamber is cryogenically pumped with a base pressure of approximately 3×10^{-7} torr. The pressure during deposition was approximately 8×10^{-5} torr, and consisted almost entirely of N_2 from the ion source. Cr and Mo were evaporated in 10 kW e-beam evaporation sources, and the substrates were irradiated with 700 V nitrogen ions from a 10-cm diameter RF ion source at an ion current density of approximately 20-30 $\mu A/cm^2$. The deposition rate was varied from 0.20 to 0.60 nm/s to adjust the relative ion-to-atom arrival ratio. Ion arrival rates were chosen based on the results of earlier work in the deposition of IBAD Cr-N [20]. Compositions studied were: pure Mo (bulk sample), Mo nitride (50% N), Cr (IBAD coating with argon ions), three Cr nitride coatings (~7% N, ~30% N, and ~50% N), and two mixed Cr-Mo nitrides (Cr:Mo ratios of approximately 1:2 and 2:1, both with ~35% N). Coating thickness ranged from 150 nm to 500 nm. The composition and thickness of the coatings was measured using Rutherford backscattering spectrometry. All electrochemical tests were performed in 0.1 M hydrochloric acid (HCl) using a computer-controlled potentiostat, a $Ag/AgCl_2$ reference electrode, and a platinum counter electrode. The samples were briefly polarized to cathodic potentials (-1500 mV for 1 minute) to remove some of the air formed oxide immediately upon immersion. This was followed by argon deaeration for 1 hour while the open circuit potential (OCP) was monitored. After this, the samples were stepped to a potential of +600 mV (where Cr is passive and Mo is not) and held there for 5 minutes, while the anodic current was monitored. If the coating was not removed by this polarization, the sample was again allowed to rest for 30 minutes, while the OCP was monitored, and then potentiodynamically scanned from 30 mV below the OCP to 1500 mV above the reference potential.

RESULTS AND DISCUSSION

The corrosion potential of the coatings immediately upon immersion is shown in Figure 1. With only one exception, all nitride coatings had significantly more noble corrosion potentials than the base metals. Cr and Cr nitride coatings generally required 20-30 minutes to reach a stable potential, although the coatings containing nitrogen reached this much more quickly. While most coatings steadily rose to equilibrium potential as the surface passivated, the coating containing the most nitrogen (~50%) first quickly rose to a very high potential and then slowly dropped to potentials typical of lower-nitrogen coatings. This is likely due to a slow depletion of nitrogen in the coating just under the passive film, as noted previously [19]. Mo and Mo nitride coatings reached equilibrium potentials more quickly than Cr coatings, but the corrosion potential of Mo nitride was much more erratic than other coatings studied, indicating that the passive film formed was less stable than those on the other coatings. Adding Cr to Mo nitride coatings resulted in more stable passive films, although the corrosion potential of Cr1Mo2 nitride was more active than pure Mo or Cr. Adding still more Cr produced a passive film that was as stable as those on other Cr nitride coatings, but reached extremely high corrosion potentials.

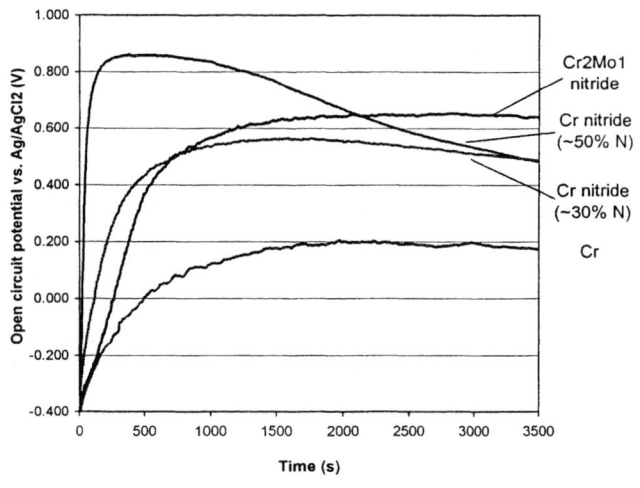

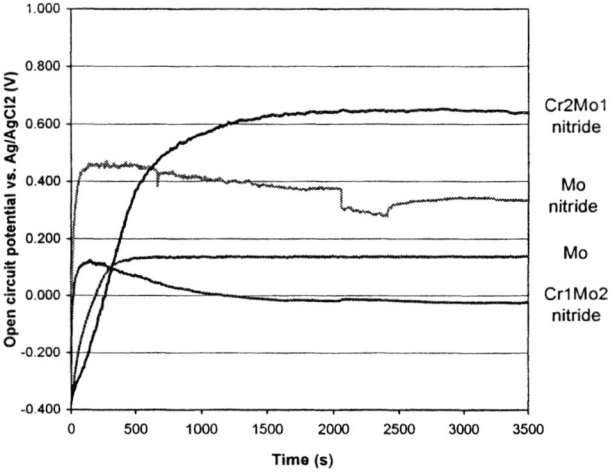

FIGURE 1. Open circuit potential of metals and metal nitrides immediately upon immersion into 0.1 M hydrochloric acid solution. Deaeration with argon gas was begun at t=0.

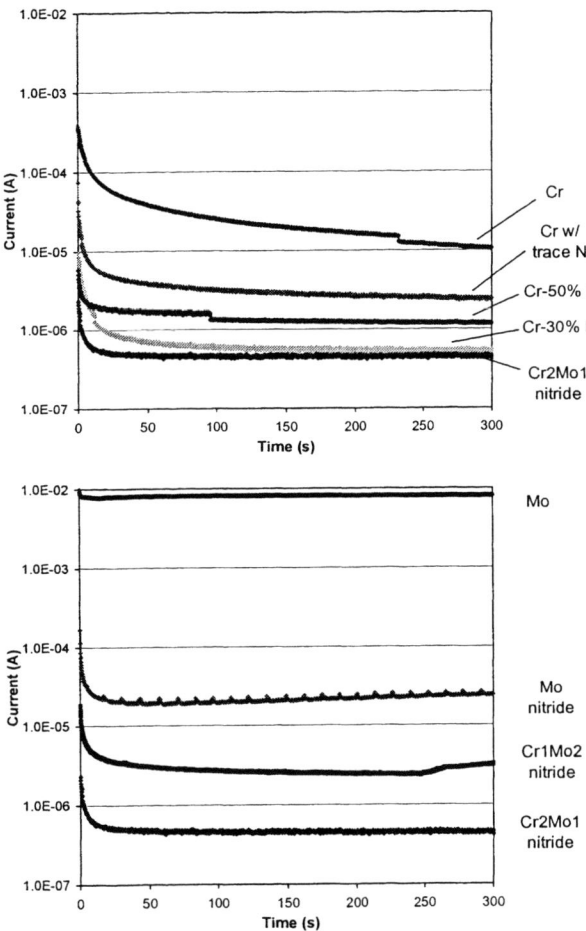

FIGURE 2. Passive current density at +600 mV vs. Ag/AgCl$_2$ in 0.1 M hydrochloric acid solution.

After reaching equilibrium corrosion potential, the samples were stepped to +600 mV, a potential where Cr is passive, but Mo is not. The passive current density is plotted in Figure 2. Cr nitride exhibits lower passive current densities than pure Cr, and reaches an equilibrium current density much more quickly. As expected, pure Mo corroded at a very high rate at this potential – almost three orders of magnitude more rapidly than Cr. Despite the unstable nature of the passive oxide formed on Mo nitride noted above, its passive current density was 100x less than that of pure Mo. As expected, adding Cr to Mo nitride further enhanced passivity to levels comparable to Cr nitride. When the coating consisted primarily of Cr nitride with some Mo present (Cr2Mo1 nitride), the coating passivated almost immediately, reaching extremely low current densities (5 x 10^{-7} A/cm^2).

Potentiodynamic scans of the coatings are shown in Figure 3. Note that an instrumentation artifact is present in all of the scans, introducing apparent jumps in the current as the potentiostat moves from one current range to another. Furthermore, the two coatings that corroded at a relatively high rate (Mo nitride and Cr1Mo2 nitride) were completely removed from the glass substrates before the scan was completed, and so exhibit sudden drops in current density at a potential around +1000 mV. The potentiodynamic scans of Cr and Cr nitride coatings show that all coatings are passive in this solution, up to the transpassive potential (~1000 mV) where the trivalent Cr is converted to hexavalent species. Molybdenum showed no evidence of passivity in this solution, but Mo nitride showed a more noble corrosion potential and a breakdown potential of ~ 600 mV vs. Ag/AgCl$_2$. Molybdate oxyanions are highly soluble in these solutions, and so do not form a highly protective passive film relative to Cr, but the effect of nitrogen is striking: inducing passivity and significantly lowering the corrosion rate.

As can be seen in the scans of the Cr-Mo nitrides, the relative amounts of Cr and Mo have a large effect on the nature of the passivity. Both coatings are passive up to the Cr transpassive potential, indicating that the passive film consists primarily of Cr species, although the limited amount of Cr available in the bulk of the Cr1Mo2 nitride clearly affects the ability of the passive film to repair itself and continue to remain protective under polarization. The Cr2Mo1 nitride coating has excellent corrosion resistance, with an open circuit potential almost as high as the transpassive potential for Cr, and an extremely low open circuit corrosion rate. It is hoped that molybdate species released at these high corrosion potentials may not only enhance the corrosion resistance of the coating, but also provide a source of mobile, soluble inhibitive oxyanions to protect against pinhole corrosion.

CONCLUSIONS

All coatings containing nitrogen are nobler than pure Cr, and passivate more quickly with passive current densities as much as 20x lower than Cr. Nitrogen induces passivity in Mo coatings, but the passive films are not very stable or protective unless the coating contains mostly Cr nitride. Cr2Mo1 nitride coatings exhibit extremely noble corrosion potentials, and low anodic dissolution rates. Further work is need to confirm the mechanism responsible for Mo passivation in the Mo nitride coatings, and to confirm the suspected enhanced production of molybdate on all nitride coatings containing molybdenum.

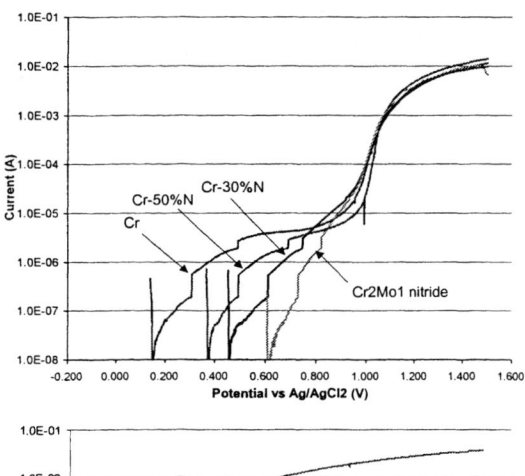

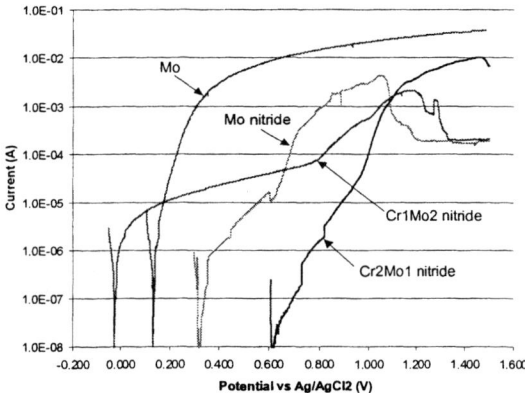

FIGURE 3. Potentiodynamic scans in 0.1 M hydrochloric acid solution, 1 mV/s, from −30 mV vs. open circuit potential, to +1500 mV vs. Ag/AgCl$_2$.

ACKNOWLEDGMENTS

The author would like to acknowledge the work of Melissa Klingenberg and Jerry Stem of CTC Corporation, and Clive Clayton and Gary Halada of the State University of New York at Stony Brook.

REFERENCES

1. J. K. Hirvonen, *Materials Science Reports*, **6** (1991) 745.

2. J. K. Hirvonen and J. D. Demaree, in *Advances in Coatings Technologies*, eds. C. R. Clayton, J. K. Hirvonen and A. R. Srivatsa (Warrendale, PA: The Metallurgical Society, 1996) 53.

3. S. Komiya, S. Ono, N. Umezu and T. Narusawa, Thin Solid Films, **45** (1977) 433.

4. T. Sato, M. Tada and Y. C. Huang, Thin Solid Films, **54** (1978) 61.

5. D. Wang and T. Oki, Thin Solid Films, **185** (1990) 219.

6. K. Kashiwagi, K. Kobayashi, A. Masuyama and Y. Murayama, J. Vac. Sci. Technol. A, **4**(2) (1986) 210.

7. P. M. Fabis, R. A. Cooke and S. McDonough, J. Vac. Sci. Technol. A, **8**(5) (1990) 3809.

8. K. K. Shih, D. B. Dove and J. R. Crowe, J. Vac. Sci. Technol. A, **4**(3) (1986) 564.

9. K. Sugiyama, K. Hayashi, J. Sasaki, O. Ichiko and Y. Hashiguchi, Surf. Coat. Technol., **66** (1994) 505.

10. W. Ensinger, M. Kiuchi, Y. Horino, A. Chayahara, K. Fujii and M. Satou, Nucl. Inst. Meth. B, **59/60** (1991) 259.

11. Milosev and B. Navinsek, Surf. Coat. Technol., **60** (1993) 545.

12. S. Suzuki, T. Nakazawa and Y. Waseda, ISIJ International, **36**(10) (1996) 1273.

13. C. R. Clayton, G. P. Halada and J. R. Kearns, Mat. Sci. Eng., A**198** (1995) 135.

14. D. Kim, C. R. Clayton and M. Oversluizen, Mat. Sci. Eng, A**186** (1994) 163.

15. M. Sakashita and N. Sato, in *Passivity of Metals*, eds. R. P. Frankenthal and J. Kruger (Princeton, NJ: The Electrochemical Society, 1978) 479.

16. G. P. Halada and C. R. Clayton, J. Vac. Sci. Technol. A, **11**(4) (1993) 2342.

17. G. P. Halada, D. Kim and C. R. Clayton, Corrosion, **52**(1) (1996) 36.

18. G. P. Halada, M. E. Monserrat, C. R. Clayton and J. D. Demaree, in *Advances in Coatings Technologies for Surface Engineering*, eds. C. R. Clayton, J. K. Hirvonen and A. R. Srivatsa (Warrendale, PA: The Metallurgical Society, 1996) pp. 339-351.

19. J. D. Demaree, W. E. Kosik, C. R. Clayton and G. P. Halada, in *Surface Engineering in Materials Science I*, eds. S. Seal, N. Dahotre, J. Moore and B. Mishra (Warrendale, PA: The Metallurgical Society, 2000) pp. 335-345.

20. J. D. Demaree, C. F. Fountzoulas, J. K. Hirvonen, M. E. Monserrat, G. P. Halada and C. R. Clayton, in *Atomistic Mechanisms in Beam Synthesis and Irradiation of Materials*, eds. J. C. Barbour, S. Roorda and D. Ila (Material Research Society Proceedings, vol. 504, 1998).

21. M. Pourbaix, *Atlas of Electrochemical Equilibria in Aqueous Solutions* (Houston, TX: National Association of Corrosion Engineers, 1974).

Ion Beam Induced Pore Structure Changes In Porous silicon

F. Pászti[a], A. Manuaba[a], Z.E. Horváth[b], E. Szilágyi[a], G. Battistig[b]

[a]*KFKI-Res. Inst. for Particle and Nuclear Physics, P.O.Box 49. H-1525 Budapest, Hungary*
[b]*MTA-Res. Inst. for Technical Phys. and Materials Science, P.O.Box 49. H-1525 Budapest, Hungary*

Abstract. In present work a new, accelerator based analytical method will be presented, which in favorable conditions is able to determine such morphological properties of porous samples, as porosity and average pore diameter. If the sample has columnar structure, it is also possible to tell the pore direction and check the ideality of the structure. The method relies on measuring the width of the resonance peak at 3045 keV in the energy spectra of He ions elastically backscattered from ^{16}O atoms. To demonstrate the feasibility of the method, measurements on porous silicon samples will be compared to TEM photos. Various effects of ion implantation on the porous structure will be clearly demonstrated.

INTRODUCTION

Ions neither lose energy nor suffer large angle scattering in cavities embedded in porous samples, hence RBS (Rutherford Backscattering Spectrometry)[1] seems to be insensitive for such features. Nevertheless, because the individual ions cross fluctuating amount of material on their way in and out, an extra, structure induced energy spread appears in the energy spectra of the backscattered ions[2-6]. The effect is especially well seen in resonant backscattering measurements[4-8], where a sharp resonance in the scattering cross-section results in a *resonance peak* in the spectra. A well known example is the 10 keV wide resonance in the ^{16}O$(\alpha,\alpha)^{16}$O elastic scattering at E_r = 3045 keV, where the cross-section 16 times enhances.

In present work this resonance will be applied to investigate ion-implantation caused changes in the pore-structure of columnar porous Si (PS) [9]. In as prepared columnar PS straight pores run parallel to each other perpendicular to the surface. In normal ambient the internal pore-walls soon become oxidized. The pore structure is never *ideal*. The walls are rippled, bent and have rough surfaces and/or fluctuations in their direction. Ion implantation changes significantly the pore structure: decreases this "ideality"[5], tilt and bend the walls[5-7] and decreases the porosity (volume fraction occupied by pores)[5, 7, 10-11]. After a brief introduction to the "*resonance method*", our results on the above effects will be shortly summarized.

THE RESONANCE METHOD

The resonance method bases on the mentioned resonance. It was introduced and detailed earlier[5, 7-8], we just summarize here its theory for the special case of columnar PS and our experimental setup. We routinely use the method. It is very sensitive, works also on untreated bulk samples and it takes just ≈2 h to take a width curve at ≈25 points.

Our detector is mounted at Θ = 165° in CORNELL geometry, (in the plane defined by the tilt axis and the beam). If the pores tilt by τ_0 from the surface normal, the sample is rotated so, that this tilt can be fully compensated by sample tilt. The incident energy of ions is set by ΔE = 100 keV excess energy above E_r.

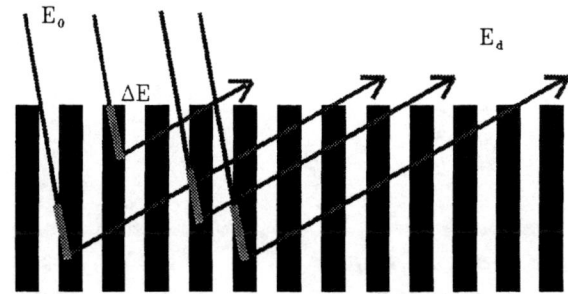

FIGURE 1. Schematic presentation of energy fluctuation in porous samples. The ions loose energy only in the pore walls, hence they reach the resonance energy at various depths.

For resonant scattering, the ions of $E_0 = E_r + \Delta E$ incident energy have to slow-down first to E_r. In porous samples the individual ions travel inward various distances in holes, hence, reach E_r at various depths (Fig. 1). This depth fluctuation transforms outward into a spread in the detected energy, ΔE_{in}. Due to the crossed holes outward, another energy spread arises, ΔE_{out}. The conventional energy spread, ΔE_{conv}, caused by detector resolution, resonance width, etc.[12] also contribute. All these spreads can be assumed Lorentzian and independent from each other, hence they combine linearly in the width of the resonance peak[8]:

$$W \approx \Delta E_{in} + \Delta E_{out} + \Delta E_{conv}. \tag{1}$$

ΔE_{in} is especially large in columnar PS when the beam is parallel to the pores. In this case the ions penetrate large distances remaining in the same wall or pore, hence reach E_r at rather different depths. Tilting the pores away from the beam, the ions cross more and more walls before reaching E_r and the depth fluctuation reduces. The average number of crossed walls inward is

$$N \approx \frac{\Delta E \sin|\tau - \tau_0|}{(1-P)S_{in}D}, \tag{2}$$

where τ and τ_0 are the tilt angles of the sample and the pores, P is the porosity and D is the average distance between the pore centers. S_{in} and S_d are average stopping powers in the walls inward and at the detected energy. They are 209 and 337 keV/μm in SiO_1 and increase just $\approx 10 \div 12\%$ from Si to SiO_2[13].

Assuming the energy fluctuation to be proportional to the average energy loss in each crossed wall, multiplied by the $\Delta N \approx \sqrt{N}$ uncertainty in N, one get

$$\Delta E_{in} \approx \Delta E_\perp \frac{1}{\sqrt{\sin|\tau - \tau_0|}}. \tag{3}$$

$$\Delta E_\perp = \sqrt{\frac{S_d^2 \Delta E D P(1-P)}{S_{in} \cos^2(\pi - \Theta)}} \tag{4}$$

is the energy spread for ions crossing the walls perpendicularly. At $\tau = \tau_0$ Eq. 3 has a singularity. In real cases $\Delta N_{min} = 1$, hence the $\Delta E_{in}(\tau)$ curve is truncated at

$$\Delta E_{in}^{max} \approx \frac{\sqrt{P} \Delta E S_d}{S_{in} \cos(\pi - \Theta)}, \tag{5}$$

therefore, it really resembles a Lorentzian curve.

ΔE_{out} is similar[8], just $\sin|\tau - \tau_0|$ changes to $\sqrt{1/\cos^2(\pi - \Theta) - \cos^2|\tau - \tau_0|}$. The maximum value changes correspondingly. $\Delta E_{conv} \approx 50$ keV in our case.

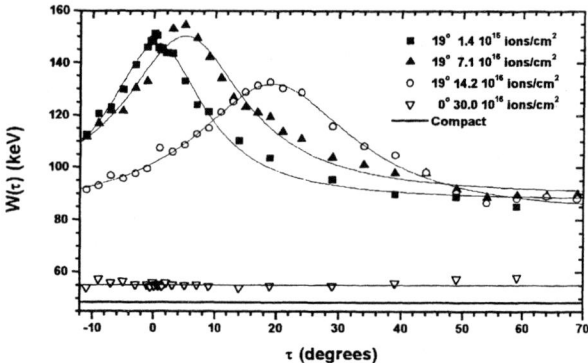

FIGURE 2. Experimental and fitted width curves taken on the PS sample at various spots implanted at the indicated conditions. The "compact" curve belongs to compact SiO_1.

The $W(\tau)$ "width curve" (width of resonance peak vs. sample tilt, e.g., Fig. 2) is also Lorentzian[7]

$$W(\tau) = W_0 + \frac{H}{(2(\tau - \tau_0)/\Delta\tau)^2 + 1}. \tag{6}$$

In columnar PS, the fitting parameters give direct information on the pore morphology as follows[7]. The position of maximum, τ_0, gives the average tilt of the pores. The height in ideal samples would depend just on P

$$H_{ideal} \approx \Delta E_{in}^{max} + \Delta E_{out}^{max} - \Delta E_{in}\left(\frac{\pi}{2}\right) - \Delta E_{out}\left(\frac{\pi}{2}\right) \approx$$
$$\Delta E_\perp \left(\sqrt{\frac{\Delta E}{S_{in}D(1-P)}} + \frac{1}{\sqrt{\tan(\pi - \Theta)}} - 2 \right) \approx \frac{S_d \Delta E}{S_{in}} \sqrt{P}. \tag{7a}$$

Due to not ideal walls (cf. Introduction) the measured height is always smaller by an "*ideality factor*", I

$$H \approx I \frac{S_d \Delta E}{S_{in}} \sqrt{P}. \tag{7}$$

The base level, W_0, independently of the ideality is

$$W_0 \approx 2 \frac{S_d \sqrt{\Delta E / S_{in}}}{\cos(\pi - \Theta)} \sqrt{DP(1-P)} + \Delta E_{conv} \tag{8}$$

$\Delta\tau$ is the width of $W(\tau)$. Expressing $|\tau - \tau_0|$ in Eq. 3 at $\Delta E_{in} = \Delta E_{in}^{max}/2$, one gets $\Delta\tau/2$ for the ideal case. In real samples $W(\tau)$ smears out, but keeps the peak area constant, hence the real $\Delta\tau$ is $\Delta\tau_{ideal}/I$

$$\Delta\tau \approx \frac{2}{I}\arcsin\left(\frac{4DS_{in}(1-P)}{\Delta E}\right) \approx \frac{8DS_{in}(1-P)}{I\Delta E}. \quad (9)$$

To avoid accumulation of errors, in practice, one might accept the P value obtained by gravimetry for the as prepared sample, and get I from Eq. 7, then D from Eq. 9 or 8. Treating the sample (e.g., by implantation or annealing) and supposing that D remains unchanged one can follow the changes in P and I from the evolution of W_{0exp} and H_{exp} (Eq. 8 and 7). Since W_0 does not depend on the pore regularity, Eq. 8 holds also for spongy samples where the pores run irregular.

Please note that according to Eq. 8 the $W_0(P)$ base level vs. porosity function describes a half ellipsoid which horizontal axis run from $P = 0$ to $P = 1$ at a height of ΔE_{conv}. Its vertical axis is proportional to $\sqrt{D}$. Therefore, the method is especially sensitive for small P values and its sensitivity just slowly decreases with decreasing D. In turn, W_0 reaches its maximum at $P = 0.5$ and it is almost insensitive for moderate porosity changes there.

EXPERIMENTAL RESULTS AND DISCUSSIONS

Let us see now some practical applications of the theory. About 15 μm thick columnar type PS layer was prepared by anodic etching of a p$^+$ <100> silicon wafer of 0.003 Ωcm resistivity at 38 mA/cm^2 current density in 1:1:2 mixture of HF (50%), water and ethanol. The obtained porosity and inter-pore distance was $P_v \approx 66\%$ and $D \approx 27$nm. The sample was then slightly oxidized (1 h, 300 °C in dry N_2 + ≈1% O_2 atmosphere). At almost unchanged morphology the porosity decreased to $P \approx 50\%$, since the measured composition was $SiO_{0.45}$.

Selected spots of the wafer were ion-implanted by 4 MeV $^{14}N^+$ ions in a two axis goniometer chamber of ≈1×10^{-4} Pa at various tilts, α_{imp}, and fluences, ψ (Table 1). The first 4 irradiations were aligned (ions fly along the <100> axis of the wafer). The beam of ≈200 nA current was swept through a 3×3 mm^2 diaphragm.

In-situ resonant backscattering measurements were carried out on the implanted spots and on a non-irradiated spot by $E_0 = 3145$ keV $^4He^+$ ions. The measuring beam was 1mm high and 0.5 mm wide.

TABLE 1. Implantation data, fitting parameters of the measured $W(\tau)$ functions and calculated porosity and ideality factors. Notes: 1) ψ is given in 10^{16} ions/cm^2 units; 2) the error in porosity emerges from the 2 keV error in W_0, it is ≈0.3% at $P = 0.5\%$ and ≈6% at $P = 40\%$; 3) the fitting was uncertain, only W_0 was fitted.

Name	α_{imp} [°]	ψ [1)]	W_0 [keV]	τ_0 [°]	$\Delta\tau$ [°]	H [keV]	P [2)] [%]	I [%]
Error	0.05	3%	2	0.3	1	3	.3÷6	8
V	-0	0	88	0.2	15	68	36	71
0M	-0.3	3.0	90	0.0	20	60	50	53
0H	-0.3	7.5	88	0.3	26	54	36	56
0S	-0.3	15	74	-1.4	42	34	10	66
0U 3)	-0.3	30	55	0.0	-	0	0.4	-
4L	4	1.5	85	0.3	19	60	26	72
4M	4	3.0	84	0.3	20	59	24	74
4H	4	7.5	85	1.7	27	57	26	69
4S	4	15	83	6.6	28	32	22	42
9L	9	1.5	88	0.5	18	61	36	63
9M	9	3.0	86	1.5	23	60	29	69
9H	9	7.4	89	2.9	24	55	41	53
19L	19	1.4	88	0.3	19	60	36	62
19M	19	2.8	86	1.4	22	61	29	70
19H	19	7.1	89	4.9	24	61	41	59
19S	19	14	82	19.2	31	50	20	69
19R	19	16	67	24.4	32	39	5	-
30L	30	1.3	92	1.5	17	63	50	55
30M	30	2.6	86	3.0	25	60	29	69
30H	30	6.5	87	9.9	26	57	32	63
50L	50	1.0	87	0.8	20	68	32	75
50M	50	1.9	87	2.4	23	62	32	68
50H	50	4.8	87	10.0	28	58	32	64
60L	60	0.8	87	3.0	17	50	32	55
60M	60	2.6	86	8.0	33	46	29	53
60H	60	3.8	89	12.6	31	49	41	48
60S	60	7.5	89	27.6	27	51	41	49
70L	70	0.5	90	1.7	18	69	50	61
70M	70	2.0	89	6.8	25	69	41	67
70H	70	3.8	92	18.9	29	53	50	47

Since $\Delta E = 100$ keV, E_r was reached at ≈1 μm average depth in the porous sample, hence the S_e electronic and S_n nuclear stopping powers of the $^{14}N^+$ ions did not change significantly in this region. This way, the results were affected mainly by α_{imp} and Ψ. The energy distribution of backscattered ions was measured by a surface barrier detector (15 keV resolution, 2.5 msr solid angle). The obtained spectra were analyzed by RBX code[14]. To obtain the $W(\tau)$ distributions, the measurements were carried out for each spots at ≈30 different tilt angles, from −10° to 70°. The measuring dose was 20 μC. Example spectra can be found in ref. 8.

The widths of resonance peaks were determined by fitting the spectra. The obtained $W(\tau)$ curves were then fitted by Lorentzian curves (see Fig.2 and Table 1).

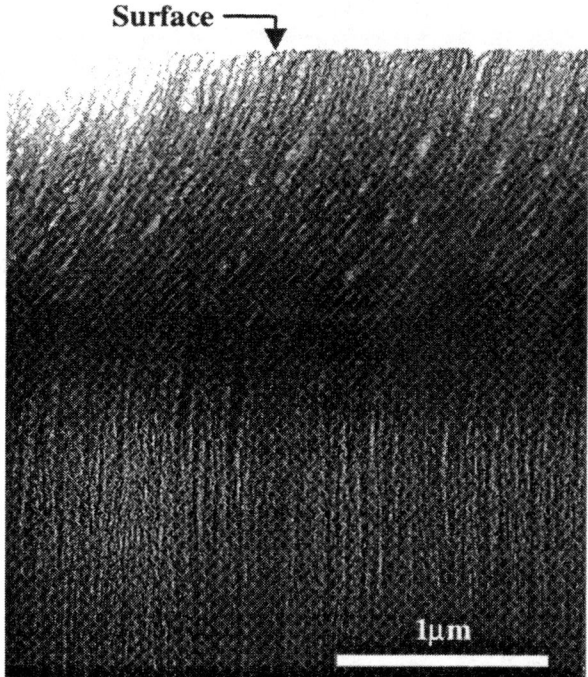

FIGURE 3. Cross-sectional TEM micrograph taken on sample "70H" (see table 1) From bottom to top: i) original structure: the walls have rough surfaces and are slightly bent, $D \approx 27$ nm. ii) near R_p P decreases (darker tone), all the walls still exists but, because of their tilting to ≈33°, D decreases to ≈22 nm. iii). close to the surface the walls became smoother, they are rippled, curved and tilt to ≈18°, but P and D have not changed significantly.

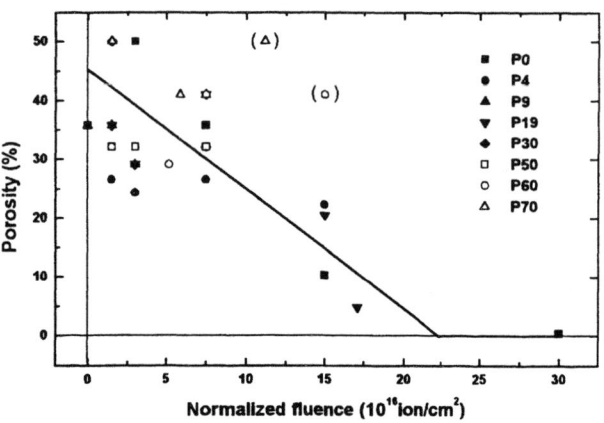

FIGURE 4. Porosity vs. normalized fluence at the indicated α_{imp} sample tilts. For the highest fluence implants at α_{imp} = 60° and 70° the surface might became rough and the tilt of the pore walls was too high to determine the base level precisely.

Let us see now first, how the method performs on the virgin sample, "V". Here $W_{0exp} = 88$ keV, $\Delta\tau_{exp} = 15° = 0.26$ rad and $H_{exp} = 68$ keV. Accepting the assumed porosity of 50%, one obtains $H_{ideal} = 114$ keV, hence $I = 60\%$, indicating a regular sample. D from Eq. 9 and 8 are 19 nm and 25 nm, both agree reasonably with the ≈27 nm determined by TEM (Transmission Electron Microscopy), (Fig. 3).

For the sample implanted by 4 MeV N ions, the following changes were observed. It was shown elsewhere[7] that the pore walls tilt as

$$\tau_0 \approx 3° \times 10^{-16} \frac{cm^2}{ion} \Phi \frac{\sin(\alpha_{imp} + \tau_0/2)}{\cos(\alpha_{imp})}, \quad (10)$$

i.e., the tilting seems to be driven by the moment transferred by ions to the pore walls perpendicularly and by the energy transferred to nuclear processes, E_{nucl}. (The $\cos^{-1}(\alpha_{imp})$ term takes into account that in tilted cases the ions cross larger distances till the same depth).

Let us now determine from Eq. 8 the porosity for each implanted sample (Table 1). It decreases linearly with the normalized fluence, $\Psi/\cos(\alpha_{imp})$, until it disappears at $\Psi \approx 22 \times 10^{16}$ ion/cm^2 (see Fig. 4).

Accepting the measured porosity, one can determine the ideality factor from Eq. 7. The result (Table 1 and Fig. 5) indicates that the ideality decreases just slightly, from ≈65% to ≈50%, i.e., the pore walls remain straight and smooth.

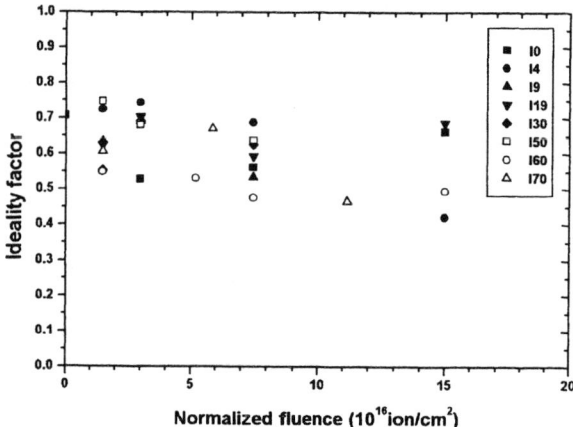

FIGURE 5. Ideality factor vs. normalized fluence at the indicated α_{imp} sample tilts.

TEM photos (see Figs. 3 and 6) confirm the above findings, i.e., near the surface the walls tilt in agreement to Table 1, the porosity does not decreases significantly and the walls remain nearly ideal (or they became even more smooth). Deeper, however, near to the R_p projected range of the ions where E_{nucl} is orders of magnitude higher, all these changes are much pronounced. At sample 19S (Fig. 6) the pores completely disappear there and the sample becomes compact. Meantime, the interpore distance seems to remain constant just the pores disappear at more and more places as one approaches the compact region. Interestingly, the pore walls does not tilt more than ≈33° anywhere.

FIGURE 6. Cross-sectional TEM micrograph taken on sample "19S" (see Table 1). From bottom to top are: i) compact crystalline Si; ii) intermediate layer between porous and crystalline Si; iii) completely compacted amorphous layer near R_p; iv) partially compacted region: the pores are still extinguishable, but disappear place to place, they tilt ≈30°; v) close to the surface the walls tilt ≈18°.

CONCLUSIONS

It was shown, that the resonance method can be successfully applied to determine the pore structure of PS samples and their changes under ion implantation. TEM pictures also confirm, that the pore walls tilt and the porosity decreases under ion implantation, while the interpore distance remains nearly constant and the ideality of the pore walls just slightly decreases. The porosity decrease is caused by complete collapse of pores at selected places rather than continuous thickening of pore walls or their complete tilting down and squeezing together, as it was suggested earlier[7]. The resonance method complements well TEM: it is faster, give quantitative results but cannot show fine details. It can be applied also on other porous samples containing oxygen. Measurements on zeolites are under way.

ACKNOWLEDGMENTS

The work was supported by Hungarian grants OTKA No. T-030327 and T-031756 and AKP No. 98-116 2.2/21. The porous samples were provided by the Technological Laboratory of MTA-MFA.

REFERENCES

1. *Handbook of modern ion beam materials analysis,* Eds. J.R. Tesmer and M. Nastasi, Pittsburgh: MRS, 1995.

2. Szilágyi E., Hajnal Z., Pászti F., Buiu O., Craciun G., Cobianu C., Savaniu C. and Vázsonyi É., *Mat. Sci. Forum* **248-249,** 373-376 (1997).

3. Hajnal Z., Szilágyi E., Pászti F. and Battistig G., *Nucl. Instr. and Meth.* **B 118,** 617-621 (1996).

4. Pászti F. and Szilágyi E., *Vacuum* **50,** 451-462 (1998).

5. Pászti F. and Battistig G., in print in *Phys. Stat. Sol.*

6. Pászti F., Szilágyi E., Manuaba A., Horváth Z.E., Battistig G., Hajnal Z. and Vázsonyi É., *Workshop on Ion and Slow Positron Beam Utilisation (Costa da Caparica, Portugal, 15-17 Sept. 1998)* OECD Proc., Paris: Nuclear Energy Agency, pp. 145-155.

7. Pászti F., Szilágyi E., Manuaba A. and Battistig G., *Nucl. Instr. and Meth.* **B 161-163,** 963-698 (2000).

8. Pászti F., Szilágyi E., Horváth Z.E., Manuaba A., Battistig G., Hajnal Z., Vázsonyi É., *Nucl. Instr. and Meth.* **B 136-138,** 533-539 (1998).

9. Canham L., *Properties of Porous Silicon,* London: INSPEC, The Inst. of Electrical Engineers, 1997.

10. Pászti F., Manuaba A., Szilágyi E., Vázsonyi É., Vértesy Z., *Nucl. Instr. and Meth.* **B 117,** 253-259 (1996).

11. Simon A., Pászti F., Manuaba A., Kiss Á.Z., *Nucl. Instr. and Meth.* **B 158,** 658-664 (1999).

12. Szilágyi E., Pászti F., Amsel G., *Nucl. Instr. Meth.* **B 100,** 103-121 (1995).

13. Ziegler J.F., *SRIM code,* http://www.research.ibm.com/ionbeams/home.htm#SRIM

14. Kótai E., *Nucl. Instr. Meth.* **B 85,** 588-596. (1994).

Comparison between two deposition methods for zirconia film

N. K. Huang, D. Z. Wang

Key Lab for Radiation Physics and Technology of Education Ministry of China
Institute of Nuclear Science and Technology, Sichuan University, Chengdu, 610064,
P. R. China

Abstract. Zirconia films prepared with two deposition methods, a reactive ion sputter method and a dual ion beam one, were characterized with RBS, XPS, TEM and XRD analyses. The results show that atomic ratio of [O]/[Zr] approximately constant throughout the layer, showing various zirconium oxides can be formed with the two methods by choosing suitable preparation parameters, and different phase structure can be obtained by selecting deposition method and substrate material. Comparison of optical properties for the zirconia films with these deposition methods is discussed in this paper.

1. INTRODUCTION

ZrO_2 is a high refractive index material with low loss and low scatter in the infrared region and has a large optical band gap and a high laser damage threshold, which makes it very useful in commercial application. There are many methods in use for depositing films such as evaporation, magnetron sputter, reactive sputter and ion beam assisted deposition.[1-3] In this paper, zirconia films with two methods are reported. One is a dual ion beam deposition technique, this one could produce stable films with high packing densities and have a favorable reproducibility in depositing films because each deposition parameter can be varied independently, therefore, it is relatively easy to identify and modify the critical parameters. Another is a reactive ion sputter deposition method. This one can be used for producing a variety of compound films. The compound films can be produced by which metal components can be obtained by sputtering with an inert gas ion beam extracted from an ion source and non-metal components can be introduced by leakage of reactive gas.

Our interest in deposition methods for zirconia films stems from our investigation on the optical properties. In this paper we try to compare these two deposition methods which is better to prepare zirconia films with good optical properties.

2. EXPERIMENTAL

Two deposition methods were used to prepare zirconia films. One is a dual ion beam deposition technique, another is a reactive ion beam sputter method. The dual ion beam deposition system is shown schematically in Fig.1. Two Kaufman ion sources are used to supply Zr metal atoms and O bombardment ions to the substrate respectively. The metal atom flux is sputtered by a 1keV Ar ion beam from the zirconium target with a purity of 99.98%, and the O bombardment ion flux, with an energy of 100eV, is supplied to the growing film. The base vacuum is 5×10^{-4}Pa, and during the deposition it was 4×10^{-2}Pa. The features of a reactive ion beam sputter system shown in Fig.2 differing from the dual ion beam deposition system are that oxygen, as a reactive gas, was leaked into the chamber through an adjusting needle valve and was stabilized at a desired pressure. And the base vacuum before deposition is about $(1\sim2)\times10^{-4}$Pa higher than that in the dual ion beam deposition system.

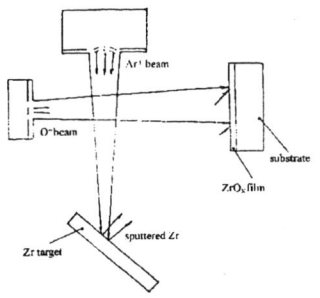

FIGURE.1 Schematic diagram of a dual ion beam deposition system

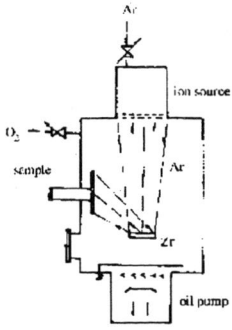

FIGURE 2. Schematic diagram of a reactive ion sputter deposition system

3. RESULTS AND DISCUSSION

Fig.3 shows the XPS depth profile of Zr-O films on Si(100) with a reactive ion sputter method under O_2 leakage to the pressure of 5×10^{-3} Pa. It is found that the film consists of three layers. Such a structure for the Zr-O films can also be found under O_2 leakage to the different pressure and with a dual ion beam method under different oxygen bombardment ion beam current densities. One layer of about $5\sim8$nm exists on the surface of the films, which was contaminated by carbon. XPS wide scan on the uppermost surface manifests (not shown here) that apart from the elements of the film Zr, O and auxiliary gas argon. Carbon from residual gas such as CO, CO_2 etc. in the chamber invaded into the film during deposition. Precise analysis with XPS showed that some dissociated carbon adsorbed on the top surface and some may form carbide with Zr due to Zr affinity.

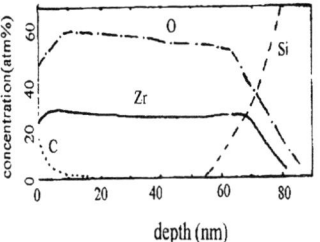

FIGURE 3. XPS depth profile of Zr-O film deposited by a reactive ion sputter method with a leakage of O_2 gas to the pressure of 5×10^{-3}Pa.

The second layer is the bulk of the film. It is found in Fig.3 that the atomic ratio of [O]/[Zr] throughout this layer is almost constant (here in Fig.3 about 2), which was also confirmed by RBS measurement [3]. With O_2 leakage to the different pressure in the chamber the atomic ratio of [O]/[Zr] is varied from zero to about slightly higher than 2. It seems various zirconium oxides including sub-oxides could be formed. In order to check them, XPS narrow scan of Zr3d was used. Fig.4 shows the Zr3d spectra for the Zr-O film deposited by a dual ion beam technique with different from oxygen ion beam current densites. It is found that different Zr oxidized states can be obtained according to the results of Morant et al[4], in which per oxidation state of zirconium is with an energy shift of about 1.06eV with respect to the metal Zr^0. Hence with an oxygen ion beam current density of $3\mu A/cm^2$, for example, the binding energy Zr3d at Zr3d3/2 180.9eV and Zr3d5/2 178.8eV manifests Zr^0, in comparison with the binding energy of metal zirconium of Zr3d3/2 at 181.1eV and Zr3d5/2 at 178.7eV, i.e. the formation of α-Zr solid solution, where oxygen in the bulk of the film solubilized into metalic zirconium. With an oxygen ion beam current density of 25 $\mu A/cm^2$, Zr^{+4} state in the bulk of the film has been formed. With different oxygen ion beam current densities, different zirconium oxidized states of Zr^{+1}, Zr^{+2}, Zr^{+3} and Zr^{+4} could be found in the bulk of the films.

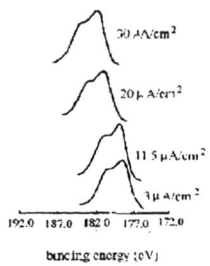

IGURE 4. Zr3d XPS spectra for the bulk of the films deposited by a dual ion beam method with different oxygen ion beam current ensities.

The third layer is a transition layer between film and substrate, in which some bridging configurations such as Zr-O-M depending on substrate materials could be formed as similar as that on the surface due to Contamination.[5] Further analyses on the transition layer will be made.

Fig.5 shows TEM micrographs for Zr-O film on NaCl substrate prepared by a reactive ion sputter method under the leakage of O_2 gas to the pressure of $5\times10^{-3}Pa$. It can be seen that by examining the selected area diffraction patterns that the film is mainly amorphous, where the diffraction rings shown in Fig. 5(b) are too wide to identify crystalline phases. By changing substrate material, for example, Si(100) wafer is used, some crystalline phase with preferred orientation could be found. Fig.6 shows TEM micrographs for Zr-O film on a NaCl substrate prepared by a dual ion beam technique with an oxygen ion beam current density of $30\mu A/cm^2$. It can be seen that the film contains a uniform fine-grained microstructure with crystallite size of about 50nm (Fig. 6(a)), and the film consists of a mixing of crystalline and amorphous material, the diffraction rings shown in Fig.6(b) seems a cubic phase in comparing with the calculation data based on the measurement and standard one. By selecting Al as substrate, monoclinic phase apart from cubic was also discovered by XRD analysis.

 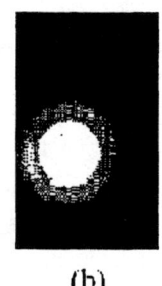

(a)   (b)

FIGURE 5. TEM image and corresponding diffraction pattern of the Zr-O film deposited by a reactive ion sputter method with a leakage of O_2 gas to the pressure of $5\times10^{-3}Pa$. (a) bright-field image (70000×) (b) electron diffraction pattern of (a), k=22.8

In order to examine the optical quality of the ZrO_2 films prepared with a dual ion beam and a reactive ion sputter methods. Refractive index measurement is made and its results are shown in Table1.

It can be seen that the maximum of refractive index n=2.2 for the ZrO_2 film prepared with the dual ion beam method is larger than that with the reactive ion sputter deposition technique, showing the density of the film prepared with a dual ion beam method is higher than that prepared with a reactive ion sputter one. Again, excessive oxygen by bombarding into film or leakage into chamber leads to decrease in their refractive index. The reason could be that excessive oxygen retained in the films could decrease the densities of the films.

Table 1. Refractive index measurement

Oxygen N	Oxygen ion current Density (μA/cm^2)		Oxygen leakage Pressure×10^{-3}Pa	
	25	30	5	7
	2.2	1.6	2.0	1.6

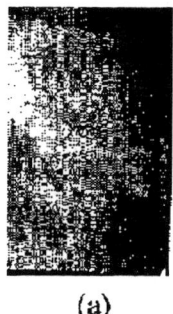 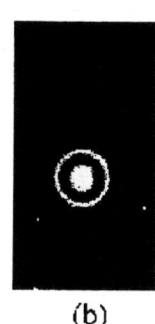

(a) (b)

FIGURE 6. TEM image and corresponding diffraction pattern of the Zr-O film deposited by a dual ion beam technique with an oxygen ion bean current density of 30μA/cm^2. (a) bright-field image (12000×) (b) electron diffraction pattern of (a), k=22.8

4. CONCLUSIONS

Microanalyses and optical measurement for the Zr-O films prepared with two deposition methods, a reactive ion sputter technique and a dual ion beam one, were reported above, the following conclusions can be made.

(1) With film preparation of a dual ion beam method or a reactive ion sputter technique. Various zirconium oxide with different chemical states of zirconium such as Zr^{+4}, Zr^{+3}, Zr^{+2}, Zr^{+1}, and Zr^0 could be formed by selecting different deposition parameters.

(2) The deposited Zr-O films were found to consist of three layers, one layer of about 5~8nm on the top surface was contaminated by residual gas such as CO, CO_2 etc in the chamber, the second layer is the bulk of the Zr-O film where the atomic ratio of [O]/[Zr] shows almost constant throughout the bulk of the film, and it is varies from zero to slight higher than 2, this means the various zirconium oxides were formed. The third layer is a transition layer between film and substrate, where some bridging configuration such as Zr-O-M could be formed.

(3) Phase characterization shows that amorphous and metastable cubic and normal monoclinic phases can be formed by these deposition methods.

(4) Optical measurement shows that the maximum of refractive index of n=2.2 can be obtained for the ZrO_2 film prepared with a dual ion beam technique, which is higher than that maximum of n=2.0 for the ZrO_2 film prepared with a reactive ion sputter method. It is suggested that the density of ZrO_2 film deposited by a dual ion beam technique is higher than that by a reactive ion sputter method.

Acknowledgements

The author gratefully acknowledges the support of K. C. Wong Education Foundation, Hong Kong

REFERENCES

1. P. J. Martin, J. Mater. Sci., 21(1986)1.
2. N. K. Huang, H. Kheyrandish and J. S. Colligon, Mater. Res. Bull., 27(1992)239.
3. N. K. Huang, H. Kheyrandish and J. S. Colligon, Phys. Status. Solid A 132(1992)405.
4. C. Morant, J. M. Sanz, L. Galan, L. Soriano and F. Rueda, Surf. Sci., 218(1989)331.
5. Y. S. Tang, N. K. Huang, Solid State Communications 77(5)(1991)341.

Effects of Proton Irradiation on the Critical Current Densities of YBa$_2$Cu$_3$O$_{7-\delta}$ Thin Films

Chong Wang[1], Hye-Won Seo[1], Chun-Jung Su[1,2], Yuh-Huah Wang[1,2], Q.Y. Chen[1], Xinming Lu[1], J.R. Liu[1], Tom Johansen[1,3], and W.K. Chu[1]

[1]*Texas Center for Superconductivity and Department of Physics, University of Houston, Houston, TX 77204, USA*
[2]*Department of Materials Science and Engineering, National Chiao-Tung University, Hsin-Chu, Taiwan, Republic of China*
[3]*Department of Physics, University of Oslo, P.O. Box 1048, Blindern, Oslo 3, Norway*

Abstract. Systematic quantitative measurements have been performed in order to study the effect of proton irradiation on the critical current density $J_c(B,T)$ for YBa$_2$Cu$_3$O$_{7-\delta}$ (YBCO) thin films using magneto-optic imaging (MOI) technique. Results for external magnetic fields up to 400 Oe and temperature ranging from 5 to 80 K are presented for a 600 nm thick YBCO film irradiated by proton up to a dosage of 10^{16} cm^{-2} at 500 keV. While no significant differences were observed for T>50 K, the J_c values consistently increased with the increasing external field for T<50 K.

INTRODUCTION

Proton irradiation has been previously used to effectively introduce point defects that serve as the magnetic flux-pinning centers in superconductors[1]. In so doing, the critical current density of a superconductor can be greatly increased. However, it remains unclear as to what roles that irradiation plays in the enhancement of J_c in terms of the dosages of irradiation. While magnetic susceptibility has been widely used for the determination of J_c, this quantity reflects the volume average without revealing the spatial variations of the physical properties of the material. In contrast, magneto-optical imaging (MOI) allows the direct observation of local magnetic properties of a superconductor. MOI is based on the large Faraday effect of Bi:YIG epitaxially grown on GGG substrates[2]. The ferrite garnet film (5 micron thick) is placed directly on top of the sample under study, and images are formed via a polarizing microscope with sample situated within the optical path of two crossed polarizers[2]. Perpendicular magnetic induction from the super-conductor causes the polarization of incoming light to rotate and the rotated component can thus pass through the analyzer, carrying the signature of local magnetization. The circulating current density, which is Jc, can then be measured without similar errors from magnetization and transport measurement methods. This technique has been applied successfully to YBCO thin films, BiSCCO tapes, single crystals and melt-processed high-Tc materials[3-6]. We believe that the MOI method would provide much more reliable J_c measurements.

EXPERIMENTALS

Lithography was used to define straight boundaries on the YBCO thin film strips essential in applying the MOI technique. The films under study were 600nm thick samples. One sample was irradiated up to 1×10^{16} protons/cm^2 using a 500 keV proton beam. The images of these samples, cooled to a temperature range from 5 K to 80 K and magnetic field from 143 Oe to 358 Oe, were then taken by the MOI. From these images the magnetic induction profiles within the samples were determined by the gray level of each pixel and, based on which, the critical current densities were calculated.

RESULTS AND DISCUSSIONS

Fig. 1(a) shows the magnetic image of an HTS strip of width 2w, with the gray level indicting the perpendicular field intensity B_z. The profile $B_z(x)$ along the indicated line traversing the sample is as given in Fig. 1(b), where the strip width (2w) and interval of the zero-B_z region (2a) are clearly demonstrated. Based on a previous calculation[7], the associated J_c is extracted from the equation: $a=w/cosh(B_a/B_f)$, where B_a is the applied field and $B_f=\mu_0/\pi J_c d$ is a characteristic field for the given film geometry with d being the film thickness. The obtained $J_c(B,T)$ data indicate that J_c essentially has no field dependence for T>50 K, but increases with field at lower temperatures. At low field, the $J_c(T)$ fits well with Ginzburg-Landau theory, giving $J_c(T)=J_c(0)\times(1-(T/T_c)^2)\times(1-(T/T_c)^4)^{1/2}$, where T_c is the critical temperature, but deviations emerge at higher field. In Fig. 2, we compare the $J_c(T)$ of non-irradiated and proton irradiated films at 358 Oe. It indicates no difference at higher temperature, but at lower temperature the J_c is higher for the irradiated one. Shown in Fig.3 are the zero temperature critical current densities, $J_c(0)$, as derived from such fitting, which appear to increase linearly with B for the range of fields studied on the irradiated sample. In comparison, the field enhancement levels off beyond 200 Oe for the non-irradiated sample. The origin of the field-enhancement in J_c is being investigated for various dosages of proton irradiation and will be reported elsewhere.

Note that at 500 keV, the projection range of proton is 8.43 μm, much larger than the film thickness and hence most of the protons are expected to fully penetrate the superconducting film. This leaves behind a uniformly distributed defects across the film thickness and avoids any effects of proton doping. The density of atom displacement is about 8×10^{-4} atoms/ion/nm. This gives 0.48 vacancies per incident ion. For the given dosage of 10^{16} cm^{-2}, there are 0.027 proton bombardments per unit cell of $YBa_2Cu_3O_{7-\delta}$, giving about 0.013 vacancies per unit cell. In effect, at this dosage, the area density of the vacancies can range from 5×10^{15} cm^{-2} for a completely random distribution of the vacancies to 2×10^{13} cm^{-2} for a completely ordered simple cubic array (emulating a columnar defect case). In terms of flux quantization, since a single fluxoid gives a magnetic flux of $\Phi_0=2.07\times10^{-7}$ Gauss-cm^2, such surface densities of vacancies would require 10^6-10^9 Gauss (or 10^5 Tesla) to saturate every vacancy with a quantum of flux line. With a field of 358 Gauss, the occupation of the vacancies by the flux quanta is thus, in essence, about 0.01 % for the orderly array in which a flux line links a linear chain of vacancies and 0.358 ppm for the case of completely randomly distributed vacancies.

At such a small fraction of occupancy, the flux quanta are in fact very diluted in the vacancy sea if the vacancies are indeed randomly distributed and, as a consequence, each can be treated as non-interacting with others. There then would be enough sites for them to hop to when the Lorentz force, $F_L \approx J_c \times \Phi_0/\xi^2$ ($\approx$ 6000 Dynes/cm at T=0 K), where ξ is the coherence length of super-electrons in the YBCO (typically ~O(1 nm)), exceeds the pinning force F_p provided by the pinning centers (i.e. vacancies). In this case, the enhancement of Jc by the external field should not be expected. It is thus possible that the vacancies have favored an arrangement of orderly array, most likely bent one way or another, which gives a much higher surface density of vacancy dictated pinning centers (for an extremely ordered state, ~0.01 % of sites being occupied).

In summary, we have successfully used the MOI technique to determine the J_c of proton-irradiated HTS thin films. Magnetic field enhancement of J_c has been observed.

(a)

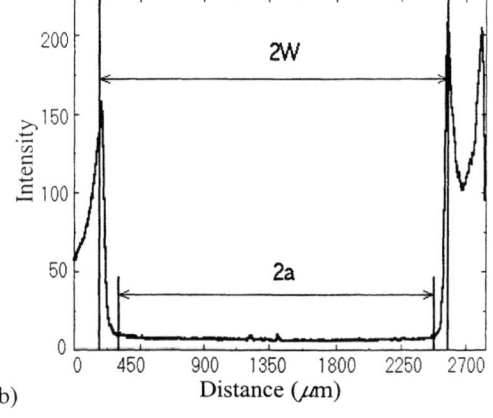

(b)

Figure 1. (a) MO Image of YBCO film (T=10K, B_a=143Oe). The traversing white line gives the cross-section profile of B_z shown in (b). (b) The peaks represent the boundary of YBCO strips

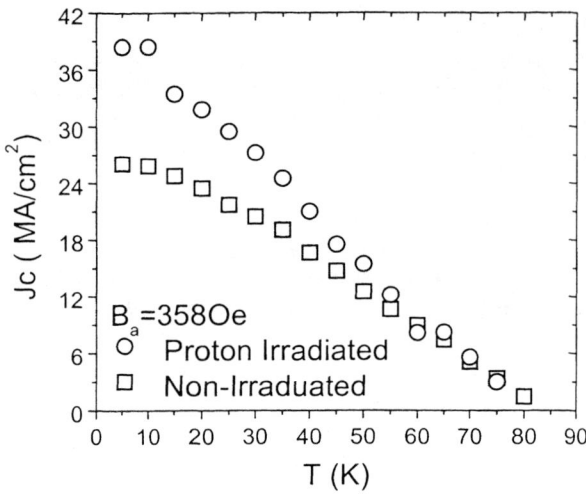

Figure 2. Effect of irradiation on Jc(T).

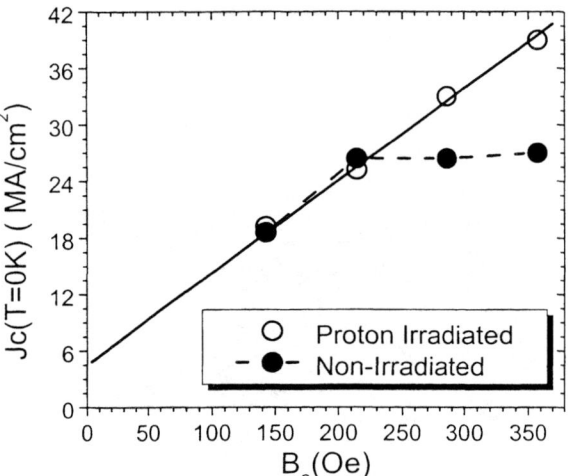

Figure 3. B-dependence of zero temperature Jc

This work was supported by the State of Texas through the Texas Center for Superconductivity at the University of Houston.

REFERENCES

1. Ignatiev, A., Zhong, Q., Chou, P. C., Zhang, X., Liu, J. R., and Chu, W. K., *Appl. Phys. Lett.* **70**, 1474 (1997).

2. Koblischka, M.R., and Wijngaarden, R.J., *Supercond. Sci. Technol.* **8**, 199(1995).

3. Johansen, T. H., Baziljevich, M., Bratsberg, H., Galperin, Y., Lindelof, P. E., Shen, Y., and Vase, P., *Phys. Rev. B* **54**, 16264(1996).

4. Uspenskaya, L. S., Vlasko-Vlasov, V. K., Nikitenko, V. I., and Johansen, T. H., *Phys. Rev. B* **56**, 11979(1997).

5. Frello, T., Baziljevich, M., Johansen, T. H., Andersen, N. H., Wolf, Th., and Koblischka, M. R., *Phys. Rev. B* **59**, R6639(1999).

6. Gaevski, M. E., Bobyl, A. V., Shantsev, D. V., Galperin, Y. M., Johansen, T. H., Baziljevich, M., Bratsberg, H., and Karmanenko, S. F., *Phys. Rev. B* **59**, 9655(1999).

7. Zeldov, E., Clem, J. R., McElfresh, M., and Darwin, M., *Phys. Rev. B* **49**, 9802(1994)

Measurement of Elastic Deformation of a Thin Foil by MeV-Energy Heavy Ion Irradiation

H. Tsuchida*, I. Katayama[†], S. C. Jeong[†], H. Ogawa*, N. Sakamoto*, and A. Itoh[‡]

*Department of Physics, Nara Women's University, Nara 630-8506, Japan
†Institute of Particle and Nuclear Studies, KEK, Tsukuba 308-0801, Japan
‡Quantum Science and Engineering Center, Kyoto University, Kyoto 606-8501, Japan

Abstract. Experimental results are presented for the first time on the deformation of thin aluminum foils induced by 8-MeV Si^{4+} ions irradiation. The deformation was measured systematically using a laser displacement meter with accuracy of 0.1 μm, as a function of the incident beam current ranging from 1 to 200 nA for two different foil thickness of 2 and 5 μm foils. Protrusive deformation profile was observed only within the beam spot size. The deformation was found to be completely elastic without hysteresis and was independent of the beam current. The deformation was increases linearly as the beam current increases up to 10 nA, above which it increases more rapidly. The surface expansion for a given foil is saturated at some beam current. The present method will be developed as a new method to study ion-solid interaction relating to high-density electronic excitation, thermal deformation of thin foil and ion track structure.

INTRODUCTION

When an energetic ion is incident on solid, the ion slows down due to a series of collisions with atoms and electrons in the target. In high-energy collisions, where electronic stopping is dominant, the electronic excitation leads to the motion of target atoms through the electron-phonon interaction. Since a highly ionized region is formed along the ion path in the target, the motion of target atoms is then also influenced by Coulomb repulsion between ionized target atoms. These processes are well known in study on electronic sputtering phenomenon, where various models related to the ion track have been proposed such as thermal, shock wave, pressure pulse models and so forth [1-4].

We can also consider the motion of target atoms from a macroscopic point of view. One of the quantities relating is an increase of the target size by ion collisions, which stems from a change of distance between lattice atoms of a target. These are known as ion-irradiation effects, and various studies have been performed. Klaumünzer and Schumacher [5] studied the growth of a metallic glass ($Pb_{80}Si_{20}$) irradiated with 285-MeV Kr ions and measured a change of the lengths and widths of the sample before and after the beam irradiation with an optical microscope. Their results have shown that the target size perpendicular to the beam axis grows where as the size parallel to the beam axis shrinks. In this case, the target irradiated with ions shows the permanent damage.

The change in the target dimension by a beam irradiation might be associated with a momentum transfer from incident ions to targets, that is, an incident beam force. We have performed an experiment to measure the beam force on a target using a torsion balance and a laser displacement meter [6]. In parallel we have also studied a deformation of thin target foils induced by ion irradiation, for the first time, with a high precision laser displacement meter. A beam of 8-MeV Si^{4+} ion and an aluminum foil target were used. We used thin foil for the targets with thickness comparable to stopping range of the projectile in order to expect a deformation of the foil as much as possible under a given beam force. The irradiation of high-energy ion beam may heat both surface and bulk of the target. This is contrasted with a laser irradiation, which mainly leads to surface heating. Furthermore, it is to be noticed, though it is obvious, that the force exerted by beams is microscopically so different from mechanical force. In the following, the beam current dependence of deformation and surface expansion are reported for two different thicknesses of aluminum foils. Although originally we had intended to apply the present method to investigate a material characteristic of a foil such as Young's modulus, the result becomes completely difference from what we expected.

EXPERIMENTAL METHOD

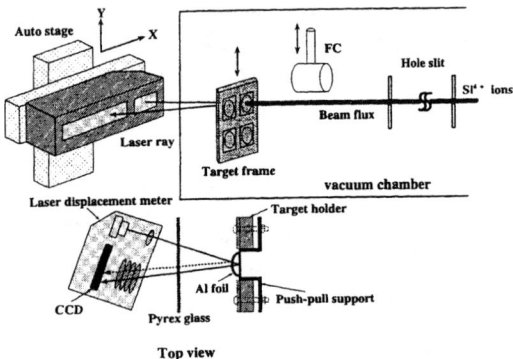

FIGURE 1. Schematic diagram of the experimental apparatus.

The experiment was performed with a 1.7 MV Peletron accelerator at Nara Women's University. A schematic diagram of the experimental apparatus is shown in Fig. 1. Targets of aluminum foil with two different foil thicknesses of 2 and 5 μm were irradiated with 8-MeV Si^{4+} ions. An aluminum foil was mounted on a holder equipped with a cylindrical push-pull support. Using this support, we carefully got rid of wrinkles in the foil. An extremely even plane of the foil was achieved. The target holder was fixed to a stable target frame. To reduce a mechanical vibration coming mainly from a vacuum pump, a rubber flange was used in a joint connecting between a scattering chamber and a beam duct. A beam was carefully collimated with two hole-slits separated by about 1 m from each other. The beam axis was perpendicular to the surface of target foil. The beam spot size at the target was about 3 mm in diameter, which was measured with a fluorescence of ZnS mounted on the target frame. The beam current was measured with a biased Faraday cup located just in the upstream of a target foil. The base pressure of the scattering chamber was kept below 7×10^{-7} Torr during the measurements.

A foil deformation induced by irradiating ion-beams was measured using a high precision laser displacement meter (LDM) of Keyence LE-4000, which has a precision of 0.1 μm accuracy. The measurement was carried out at a fixed projectile energy and with a fixed beam spot size. The principle for the measurements of deformation is as follows. A laser ray from LDM hits back of a target foil through a vacuum window of Pyrex glass. The spot size of laser ray at a target is less than 250 μm in diameter. A part of laser ray reflected from the foil passes through the other vacuum window. The laser ray reflected at angle of 135 degree with respect to the incident laser ray direction was detected with a CCD after an auto-focusing condenser lens. The focal position signal from a CCD was converted to the distance of reflecting point of laser ray from the displacement meter. The displacement of target foil can be determined by measuring the difference of two signals from LDM for on-irradiation and off-irradiation of beams. The laser ray used has a maximum power of 0.94 mW. As discussed in our previous paper [6], the irradiation of laser ray from LDM has negligible effect on deformation of the foil.

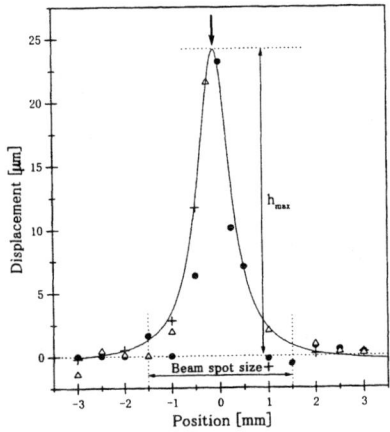

FIGURE 2. Deformation profile for aluminum foil of 5 μm irradiated with 8-MeV Si^{4+} ions. The incident beam current is 20 nA and the spot size of beams is about 3 mm in diameter. Marks indicate the experimental data.

A LDM was mounted on an auto-scanning XY-stage, where X corresponds to the horizontal and Y the vertical directions with respect to the beam axis, respectively. The system enables us to measure deformation profile of a target foil during an ion-beam irradiation. An example of the deformation profile for 5 μm aluminum foil at beam current of 20 nA was shown in Fig. 2. One can see that deformation of foil occurs only within the beam spot. The deformation profile in shape is almost normal distribution in this plot. The stopping range of 8-MeV Si ions in aluminum targets is 4.22 μm, calculated from SRIM code [7]. The 5 μm aluminum foil stops the projectiles completely. Therefore, the total kinetic energy of a projectile was deposited to the target foil. In contrast, the projectile can penetrate through a 2 μm aluminum foil. In this case, the energy deposition from the projectile to the target can be estimated to be about 2.25 MeV. This value is the same as the projectile energy of Si ions for aluminum targets corresponding to the stopping range of 2 μm.

RESULTS AND DISCUSSION

As shown in Fig. 2, the experimental results for 5 μm foil show that the deformation occurs within the size of beam spot. The same result is obtained for 2 μm foil, which is thinner than the range of projectile. In this case, electronic energy-loss plays a significant role. Thus, these results indicate that electronic excitation and ionization of target cause deformation of the foils. It is well known that a projectile penetrating a solid creates a cylindrical track. There are two regions in the ion track. One is infra-track created directly by incident beams. In this region most of incident energy is deposited and a region of high-density electronic excitation and ionization is formed. The other is ultra-track created by δ–electrons. The radii of cylindrical track can be calculated from simple formulae in [4]. In the present case of 8-MeV ^{28}Si in aluminum (target density; 2.7 g/cm^3), the radii for infra- and ultra-track are 3.6 Å and 87.8 Å respectively. This result indicates that energies of incident ions are deposited in very narrow region along the ion path.

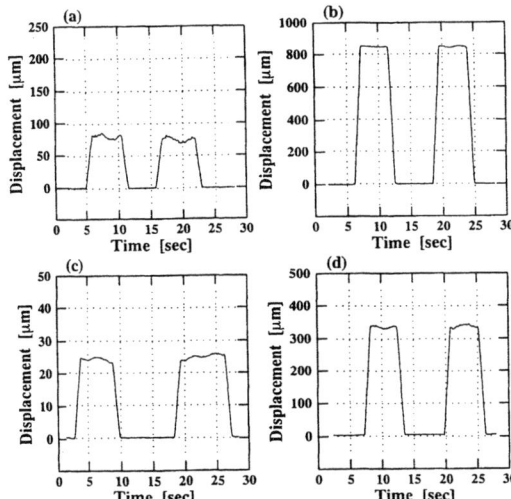

FIGURE 3. Time response of deformation for different thickness D and beam current I_0. (a); D=2 μm, I_0=20 nA, (b); D=2 μm, I_0=140 nA, (c); D=5 μm, I_0=20 nA, and (d); D=5 μm, I_0=140 nA.

Figure 3 demonstrates some examples of time response of deformation for a different beam current and a different thickness of target foil. The observation point for deformation (the point of hitting laser ray) is as indicated with an arrow in Fig. 2. Hereafter the observation is carried out at this position in the present experiment. The vertical axis in Fig. 3 indicates the values of displacement measured by the LDM. The value of zero in vertical scale is corresponding to no beam irradiation. A remarkable feature can be seen from this figure. The deformation occurs immediately after the beam irradiation without time lag (Note that there is 1 sec data-averaging time). The deformation is kept constant during beam irradiation. The slight variation of the deformations seen in Figs. a, c and d were checked and supposed to be caused by a beam instability. The foil unwinds completely itself just as the beam irradiation is stopped. The completely same response appears repeatedly. This means that the deformation is completely elastic and there is no hysteresis. These features are very peculiar and unexpected new phenomenon in beam irradiation experiments.

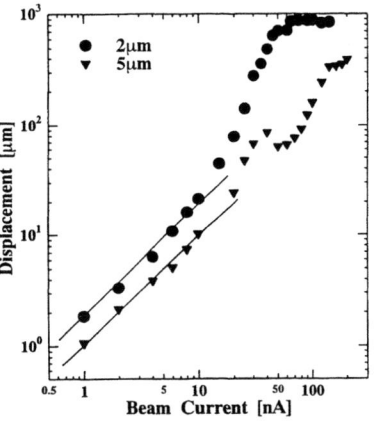

FIGURE 4. Incident beam current dependence of displacement h_{max} for two different foil thickness of aluminum by bombardment with 8-MeV Si^{4+} ions.

In Fig. 4, the maximum deformation, h_{max}, is plotted as a function of the beam current. The proportional relationship can be seen at beam current from 1 to 10 nA, which is irrespective of the foil thickness. The deformation for 2 μm foil is larger compared to that for 5 μm at the same beam current. At more than 10 nA, a different increase rate for deformation is evident. It is not clear whether there starts a different mechanism for deformation of the foil or a kind of pile-up effect. It might be relating to different thermal diffusion processes for lattice and electrons or complex structure of ion-track formation process. For 2 μm foil the deformation reaches the maximum of about 800 μm at around 70 nA beam current. From experimental results of the deformation shown in Fig. 4, the rate of surface expansion for given foils, S/S_0, is estimated by assuming that the deformation profile is a conical shape. Here, S_0 is the original surface area, which is the same as the size of beam spot (Φ3 mm). $S = S' - S_0$, and S' the side area of a cone. The h_{max} is corresponding to the height of a cone. In this assumption, S/S_0 can be replaced by a rate of linear expansion, L/L_0, where L_0 is the radius of the original surface area, $L = L' - L_0$, and L' the length of general line in a cone. Hence, $S/S_0 = L/L_0$ is given by

$$\frac{\Delta L}{L_0} = \sqrt{1 + \left(\frac{h_{max}}{L_0}\right)^2} - 1. \quad (1)$$

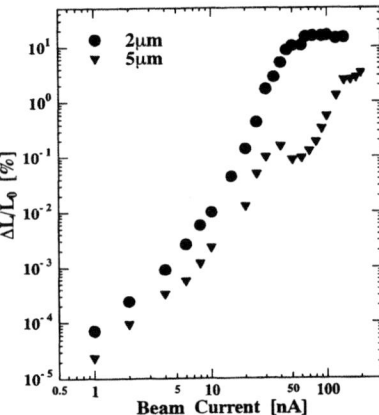

FIGURE 5. $\Delta L/L_0$ (%) as a function of incident beam current. (For details, see text.)

The results are plotted in Fig. 5. For 2 μm foil the maximum surface expansion is about 10 %. This value is considerably large in comparison with a linear expansion of about 2 % expected at the melting point for aluminum [8]. On the other hand, for 5 μm aluminum foils the maximum value of S/S_0 is about 3 %. The surface status of the foils after irradiation with a high beam current is extremely flat and has no change compared to that before the beam irradiation. The measurement was repeated several times and showed a quite good reproducibly. Thus, it was concluded that there is no permanent swelling and plasticity deformation for the foils irradiated by ion-beams within this experiment.

CONCLUSION

In summary, we have studied for the first time the deformation for aluminum foils of 2 and 5 μm in thickness by irradiation with ion-beam of 8-MeV Si^{4+} ions, using a high precision laser displacement meter. The main conclusions in this work are as follows. (1) The deformation of foils occurs only at area of beam irradiation and deformation profile has a conical shape. (2) The deformation of the irradiated foils is completely elastic without hysteresis. (3) No permanent swelling and plasticity deformation are observed for the foils after beam irradiation. (4) For a thin aluminum foil of 2 μm in thickness the rate of surface expansion reaches 10 % at more than 70 nA. This value is extraordinarily large compared to that for linear expansion of aluminum at the melting point. These results may suggest that the deformation is associated with a kind of heat effect, which is to be investigated further.

ACKNOWLEDGMENTS

We wish to express our sincere thanks to Dr. N. Yamada of National Research Laboratory of Metrology for useful discussion, and to M. Haba and J. Karimata for their technical supports during the present experiments. The authors of KEK would like to acknowledge the encouragement of the experiment by Prof. T. Nomura and E-group staffs of KEK. This work was supported by the Grant-in-Aid of the Japanese Ministry of Education, Science, Sports and Culture.

REFERENCES

1. Johnson, R. E., and Brown, W. L., *Nucl. Instr. and Meth.*, **198**, 103 (1982).

2. Reimann, C. T., *K. Dan Vidensk. Selsk. Mat. Fys. Medd* **43**, 351 (1993).

3. Håkansson, P., *K. Dan Vidensk. Selsk. Mat. Fys. Medd* **43**, 593 (1993).

4. Sundqvist, B. U. R., *Int. J. Mass Spect. Ion Proc.* **126**, 1 (1993).

5. Klaumünzer, S., and Schumacher, G., *Phys. Rev. Lett.* **51**, 1987 (1983).

6. Katayama, I., Jeong, S. C., Tsuchida, H., Ogawa, H., Sakamoto, N., and Itoh, A., *4th AISAMP*, invited paper, Taipei (2000).

7. Zieglar, J. F., Biersack, J. P., and Littmark, U., *The Stopping and Range of Ions in Matter (Computer Code SRIM Version 2000)*, Pergamon Press, New York (1985).

8. Simmons, R. O., and Balluffi, R. W., *Phys. Rev.* **117**, 52 (1960).

In situ imaging of highly charged ion irradiated mica

L.P. Ratliff and J.D. Gillaspy

Atomic Physics Division, National Institute of Standards and Technology, Gaithersburg, MD 20899 USA

Abstract. We have studied the modification of mica surfaces due to the impact of Xe^{44+} ions by imaging the ion-exposed surfaces with atomic force microscopy in vacuum. By incorporating the microscope into the vacuum chamber where the samples are exposed to the ions, we rule out posterior modification of these features in air. The features, raised bumps 19(2) nm in diameter, are similar to those imaged previously in air, however, their heights appear to be larger than previously reported.

There is a growing interest in the interactions of highly charged ions (HCIs) with surfaces, which is driven by a basic scientific interest in the physical processes involved and by an increasing number of applications that are being developed [1,2]. This interest stems from the fact that HCIs carry massive amounts of potential energy (for example, the Xe^{44+} ions used in the present work have 51 keV of potential energy). Because this potential energy is released at the surface [3], while the kinetic energy is released along the ion's path through the solid, the processes that occur at the surface are dominated by the potential energy. This strong influence of the potential energy has been demonstrated in studies of particle emission during HCI impact [4-6] (for example, x-ray emission, sputtering, secondary ion emission and electron emission), and in studies of surface damage after ion exposure [7,8].

With these experiments, as well as theoretical investigations [9,10], we have learned a great deal about the evolution of the projectile as it approaches and penetrates the surface. As described in the 'classical over-the-barrier model' [9], the ion begins to extract electrons from the surface even before impact. Many of these electrons become bound in high-lying states of the projectile, forming a 'hollow atom' with many empty core states. As the ion nears the surface, it can begin to decay, for instance by Auger ionization. Upon penetration of the surface, the electrons that are weakly bound are 'peeled off' and replaced with electrons from inside the surface, forming a secondary hollow atom.

While the formation and decay of the hollow atom is well described by models, at least in a general sense, the response of the surface is still poorly understood. Several models have been proposed to describe this situation. The first such model asserts that, if electrons are removed from the surface faster than they can be replaced from the bulk, there will be a localized charge imbalance which will cause a 'Coulomb explosion' [11]. This explosion can result in the sputtering of a large number of ions and neutral atoms or, in the case of mica, blistering [12]. A second model says that the ion leaves many excitations near its impact site [13]. These excitations promote atom pairs from bonding states to antibonding states. If the excitation density is above a threshold, they can lead to a destabilization of the crystal structure and permanent structural damage [14,15]. Materials that support self trapped excitons can be described by a third model. In this case, self-trapping can localize the excitations very near the surface until they decay into color centers, which migrate to the surface and result in sputtering [6].

One approach to studying the surface response to HCI impact is to image the impact sites after ion exposure in order to determine the size and morphology of the impact features. Such studies can give us information about the spatial extent of the ion-surface interaction and perhaps about the mechanism for their formation. In addition, for the various applications that are under development, it is useful, and in some cases necessary, to understand and characterize the damage left by the ion. Furthermore, if we can image these features, the morphology might suggest other applications that have not yet been considered. Because they are very small (approximately 20 nm in diameter and approximately 1 nm high), however, observation of the ion-induced features has been limited to a few materials on which atomically smooth surfaces can be prepared easily and reliably. One such surface is mica. Because of its layered structure, mica is easily cleaved with adhesive tape to reveal large, single crystalline terraces. Such a freshly cleaved surface can be imaged with contact mode atomic force microscopy (AFM) in air to reveal crystal structure. Because of these favorable characteristics, mica was used as the target for the most extensive, systematic studies of HCI-induced surface damage [7,8,16].

The studies of surface damage on mica from HCI impact have shown that each ion impact leaves a bump on the surface and that the size of these bumps is independent of kinetic energy in the range, 4.4 to 880 keV (Xe^{44+} projectiles) [7,16]. Furthermore, the volume of the bumps increases monotonically with charge state, and therefore, potential energy [7,8]. In one study, where HCIs of various species and kinetic energies were used, it was found that the volume increased linearly with projectile charge state from an apparent threshold at $q = 30$. Another study used only xenon ions with 100 keV of kinetic energy so that the charge state (and therefore, potential energy) was the only independent variable. In this case, the volume increased more rapidly than the charge state. In fact, the volume correlated more strongly with the potential energy, which has a non-linear relation to the charge state. Although the variation of feature size with charge state seems to be different in the two studies, they are not inconsistent. All of the data points in these two studies that correspond to the same ion species and charge state agree to within the stated uncertainty limits. The apparent difference in slope might arise from the fact that the former work involved ions of different mass (note that the relationship between charge state and potential energy varies with mass).

Because the variation in the size of the observed features is systematic and reproducible, we believe that the method of measuring the features with AFM in air is sound and useful. That is, the dimensions being measured are physically meaningful. In order to better understand the nature of these features, however, we need to look carefully at the imaging technique. There are some aspects of imaging in air that can change the apparent features from their original form. First, exposing the samples to air allows the features themselves to be modified [17]. For example, the impact sites are likely to contain many broken bonds, which would make them more reactive than the surrounding single-crystalline material. When such a surface is exposed to air, hydrocarbons or other contaminants might preferentially attach to the impact site, significantly changing its morphology from that of the original feature. Second, in air, there is a layer of water adsorbed on the surface. The capillary forces due to this water layer increase the force between the sample and the probe, possibly deforming the surface. The purpose of this study is to better understand the nature of the features as created by the HCIs. To this end, we have exposed mica surfaces with Xe^{44+} ions and imaged them in vacuum without allowing them to come into contact with air. We chose Xe^{44+} ions for our experiments because they were used in both of the previous experiments with which we compare our results. Because it has been established that the morphology of these features is independent of kinetic energy [7,16], we used a single kinetic energy, 350 keV, for all of our measurements.

Experimental

The ions are produced in the NIST electron beam ion trap (EBIT) and extracted into a beam line system that has been described previously [18]. The beam that reaches the target consists of approximately one million $^{136}Xe^{44+}$ ions per second at 350 keV in a diameter of 3 mm.

The mica samples were cleaved in air using adhesive tape and immediately inserted into a load lock chamber. From the load lock, they were passed between the various chambers of the apparatus while remaining under a vacuum of 10^{-7} Pa. This apparatus consists of a target chamber, where they are exposed to the ions and the imaging chamber, which is equipped with a scanning probe microscope. This microscope was used previously in scanning tunneling mode to image HOPG (highly oriented pyrolitic graphite) samples [19]. Because mica is a good insulator, all of the imaging was performed using AFM. We used contact mode AFM with an applied normal force of 1 nN. The samples were imaged, exposed to the HCIs, then imaged again without exposure to air. In order to study the effect of air exposure on the ion-induced features, the exposed samples were brought out of the vacuum, reinserted and imaged again.

Results and Discussion

Before exposure to the ions, the mica surfaces were featureless except for crystal structure observed in scans of small areas. In the exposed regions of the surfaces, we consistently found bumps that are 19(2) nm in diameter (the uncertainties given in this paper are single standard deviation and are based on repeated sampling) (see Figure 1) and each bump appears to correspond to a single ion impact; this is in agreement with previous measurements [7,8,16]. The fact that these features are convex, even before removal from the vacuum, means that this morphology is a direct result of the ion impact rather than, for

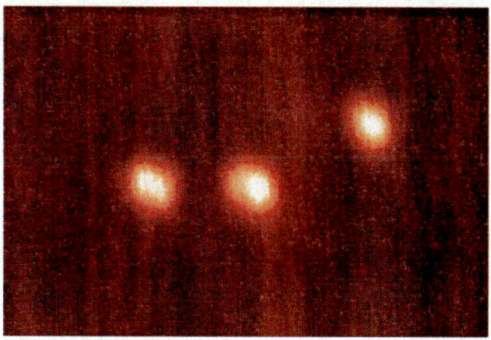

FIGURE 1. AFM image showing the topography of three Xe^{44+} impact sites on mica. The image is 50 nm x 80 nm and the features are approximately 20 nm in diameter and 1 nm high.

instance, contaminant adsorption at the impact sites [17]. Therefore, our observation is consistent with the interpretation of these features as blisters [7,16]. The heights measured in the current work, however, are somewhat larger than those previously reported. In the images taken before exposure to air, the features were 1.0(3) nm high, approximately double the heights measured previously with AFM in air [7,8,16]. The feature heights that were measured after the samples were exposed to air and returned to the vacuum for imaging, varied somewhat but remained in the range between the height of those not exposed to air (approximately 1 nm) and the height of those imaged in air (approximately 0.5 nm [7,8,16]). Quantitative measurement of feature height is difficult, however, because the topography is convoluted with other characteristics of the probe and of the surface, such as hardness and friction.

In order to correctly interpret the images, it is important to understand the lateral forces. The term lateral force refers to the torque exerted on the cantilever as it is scanned, for example, over a feature of elevated friction. When scanning in the forward direction (the direction perpendicular to the symmetry axis of the cantilever, referred to below as the 0° direction) this torque manifests itself as a lateral displacement of the beam of light in the 'bouncing beam' method of detection [20]. When imaging an area that is flat, but has a modified coefficient of friction, if the lateral force signal decreases when scanning in the forward direction it will increase when scanning in the reverse direction (this is because the frictional force is opposite to the direction of travel). If this surface is scanned in the direction along the axis of symmetry of the cantilever (90°), there will be no features in the lateral force image because the torque on the cantilever will tend to move it up and down, modifying the normal force, and thus mimicking topography. Such an increase in friction has been observed in tracks formed in mica by fast ions [21]. While the lateral and normal forces are nominally independent, there can be a coupling between the two that interferes with the measurement of feature height. Parks *et al.* [16] found that, in imaging mica exposed with Xe^{44+}, the lateral force played an important role in the measurement of the heights of the ion induced features. The heights measured when scanning in one direction were larger than those measured when scanning in the opposite direction; the quoted height values are averages of the two measurements. In order to minimize this effect, the normal force was kept to its minimum stable value (recall that the magnitude of the frictional force is proportional to the normal force). Interestingly, Ruehlicke *et al.* [7] did not observe this effect.

In the present work, we find that many of our lateral force images have an asymmetric character (see Figure 2) while others have the character of flat areas of modified friction (see Figure 3) that was seen by Parks *et al.* [16]. With respect to the asymmetric images, when the scanning direction is changed, their orientation does not rotate, as if the features are truly asymmetric. We interpret this phenomenon as resulting from the torque on the cantilever as it traverses the changing slope of the ion-induced bumps [20]. While many of our lateral force images show the type of frictional character seen in fig. 3, in addition to asymmetric character seen in fig. 2, the images taken before exposure to air are dominated by the later. With

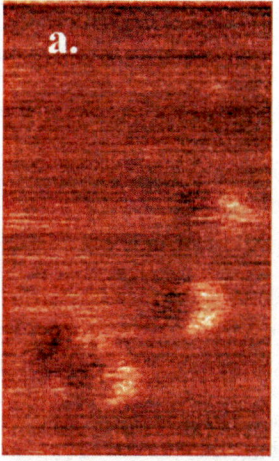 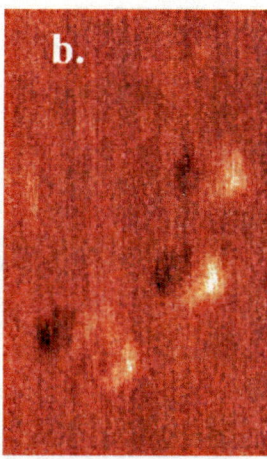

FIGURE 2. Lateral force images taken before the sample was exposed to air. The images are 100 nm across and were scanned at a) 0° and b) 90°. The asymmetric nature of these features is interpreted as resulting from a torque on the cantilever as it traverses the changing slope of the ion-induced bumps.

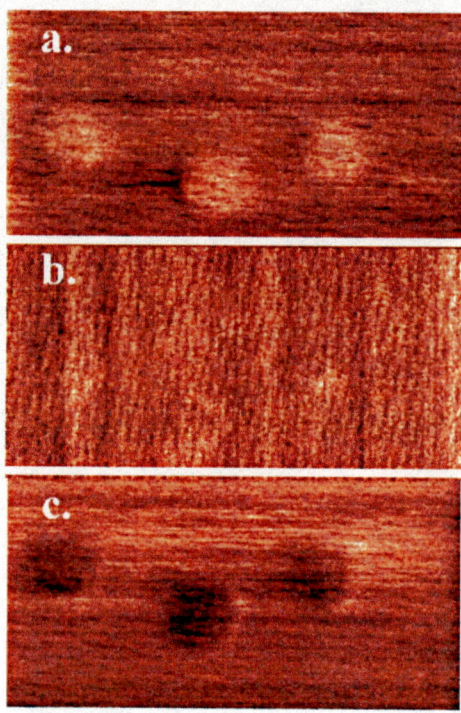

FIGURE 3. Lateral force images taken after the sample was exposed to air. The images are 100 nm across and were scanned at a) 0°, b) 90° and c) 180°. The features appear to have a modified coefficient of friction as compared to the background.

the possible exception of images taken very soon after exposure to air (less than one hour in the vacuum), we do not see the difference in forward and reverse heights observed by Parks *et al.* [16].

It was previously reported that repeated scanning tends to 'erase' the features; that is, the bumps flatten [16] or even peel open to reveal a crater [7]. While there were examples in the present work where slight modification in the features was observed after repeated scanning, generally the feature heights remained constant. One possible explanation for this discrepancy is that the adsorption of water and other contaminants in the air plays a role in the erasure. Alternatively, perhaps the imaging in the present work is done with a normal force that is sufficiently weak that the features are not significantly affected by the probe. It is not surprising that the effective normal force in vacuum is less than that in air because the capillary force due to adsorbed water is reduced. A weaker normal force could also explain the apparent increase in feature height over those measured in air; the tendency of the probe to compress the features during scanning is reduced, allowing the true topography to be measured.

Conclusion

We have confirmed that the features created by HCI impact on mica are protrusions as previously observed in air. This supports the assertion that these features are in fact blisters, and that they are caused by a delamination of the layers of the mica due to the deposition of the ion's large potential energy [7,8,12,16]. The heights of the features appear larger than those measured previously in air. This difference could reflect an actual modification of the surface when in air or an artifact of the imaging. In either case, because the vacuum environment allows more control over atmospheric conditions that are known to influence imaging with AFM, images taken in vacuum are more likely to reflect the true topography of the features created by the ions. Because of the similarity of features imaged in vacuum to those imaged in air, this study supports the general conclusions of the previous systematic studies [7,8,16].

Acknowledgements

We thank Doug Alderson for his technical assistance.

References

1. Hamza, A.V. *et al.*, J. Vac. Sci. Technol. A, **17**, 303-305 (1999).
2. Marrs, R.E. *et al.*, Rev. Sci. Instrum., **69**, 204-209 (1998).
3. Hattass, M. *et al.*, Phys. Rev. Lett., **82**, 4795-4798 (1999).
4. Schenkel, T. *et al.*, Prog. Surf. Sci., **61**, 23-84 (1999).
5. Arnau, A. *et al.*, Surf. Sci. Rep., **27**, 117-239 (1997).
6. Aumayr, F. *et al.*, Int. J. Mass Spectrom., **192**, 415-424 (1999).
7. Ruehlicke, C. *et al.*, Nucl. Instrum. Methods B, **99**, 528-531 (1995).
8. Parks, D.C. *et al.*, Nucl. Instrum. Methods B, **134**, 46-52 (1998).
9. Burgdorfer, J. *et al.*, Phys. Rev. A, **44**, 5674-5685 (1991).
10. Stolterfoht, N. *et al.*, Int. J. Mass Spectrom., **192**, 425-436 (1999).
11. Parilis, E.S., in *Proceedings of the International Conference on Phenomena in Ionized Gasses*, edited by 1969, 94.
12. Parilis, E., *et al.*, Nucl. Instrum. Methods B, **116**, 478-481 (1996).
13. Schenkel, T. *et al.*, Phys. Rev. Lett., **81**, 2590-2593 (1998).
14. Stampfli, P. and Bennemann, K.H., Appl. Phys. A, **60**, 191-196 (1995).
15. Herrmann, R.F.W. *et al.*, Appl. Phys. A, **66**, 35-42 (1998).
16. Parks, D.C. *et al.*, J. Vac. Sci. Technol. B, **13**, 941-948 (1995).
17. Neumann, R., Nucl. Instrum. Methods B, **151**, 42-55 (1999).
18. Ratliff, L.P. *et al.*, Rev. Sci. Instrum., **68**, 1998-2002 (1997).
19. Minniti, R. *et al.*, Physica Scripta, , In Press (2000).
20. Colton, R.J. *et al.*, *Procedures in Scanning Probe Microscopies*, John Wiley and Sons, Chichester, 1998.
21. Hagen, T. *et al.*, J. Vac. Sci. Technol. B, **12**, 1555-1558 (1994).

Functional Fabrication of MEMS by ion implantation

S.Nakano[1], H.Ogiso[2,1]

[1] Mechanical Engineering Laboratory, Agency of Industrial Science and Technology,
Ministry of International Trade and Industry, Tsukuba, Japan
[2] National Institute for Advanced Interdisciplinary Research, AIST, MITI, Tsukuba, Japan

Abstract. An ion-implantation material-modification technique was applied to a micro-electro mechanical systems (MEMS) fabrication technique in order to enhance the functionality of MEMS. Ion implantation, which is well known as *doping technology* in semiconductor and surface modification technology, can alter the characteristics of a substrate by the addition of ions. However, when the object is a microscale device such as MEMS, such implantation involves metallurgy of the micro material, because size, depth and area of the modified area are on the same order as the size of the microscale device. When the characteristics that can be controlled by ion implantation are combined with other properties, such mechanical, electrical, optical, and chemical, a wide range of characteristics can be easily controlled simply by changing the operating parameters, such as ion species, energy, dose, and substrate temperature. By effectively utilizing region selectivity, which is an advantage of ion implantation, the local physical properties of a micro device can be controlled. Consequently, in the design of MEMS devices, material properties can be controlled to enhance the functionality of the device. In this study, we used this ion implantation technique, which only involved injection of ions and etching that changed the chemical property of the substrate material, to fabricate a micro device, e.g., a microcantilever beam, that has low elasticity and electric conductance.

INTRODUCTION

When a mechanical device is designed, the material and its properties, as well as the structural characteristics, such as shape and movement of the device, must be considered. When both the material and structure are optimized, the device shows high performance.

However, only a few materials can be used for micro-electro mechanical systems (MEMS), because forming process dose not arrowed material variations. Currently, only basic materials are used, such as silicon and glass, which are deposited as thin films as silicon nitride, silicon carbide, etc. Such a small selection of materials limits the performance of MEMS, and consequently, wide-ranging practical applications of MEMS have not been realized. Therefore, to realize a more highly functional MEMS, we are developing advanced MEMS that more freely choose material characteristics, and that can easily be fabricated with existing lithography technology. The technique that we used to develop such advanced MEMS is the ion implantation material modification technique [1,2,3]. Ion implantation is known as a surface modification technique, but can also be used for bulk modification when applied to an MEMS that is similar in size to micro-meter orders. Ion implantation is also well known as a semiconductor process, and is as easily applicable to MEMS as is silicon lithography. In this report, we describe the concept of MEMS ion implantation and then show examples of the types of material properties that can be achieved.

Micromachining process technique

Ion implantation is applicable to general lithography technology in the processing of MEMS. However, we found that the micromachining of MEMS is more conveniently done by ion implantation process. This process is simple in that it only involves micropatterned ion implantation and etching (Figure 1). By using ion implantation, we were able to change the material properties of the substrate of the MEMS, such as mechanical, electrical, optical, and chemical properties. When the ion implantation decreases the etching rate of the substrate, the modified regions remains as microstructures after etching. The resulting microstructure is composed only of an ion-implanted layer. Furthermore, a wide range of characteristics for the modified material can be obtained simply by using different species of ions.

In this study, we made a microcantilever beam. We used a (100) silicon substrate, which is a common material for MEMS, and some kind of ions, which are gold, platinum, titanium, iron, carbon, boron and fluorine, and silicon ions. The substrate temperature during ion implantation was kept at 95 K. A micro-patterned metal stencil was applied during implantation. The advantages of the metal mask were long-term stability and repeated usage. Schmidt et al. [4], and van Kan et al. [5] showed that a narrow diameter beam can directly "draw" a micropattern. After ion implantation, the substrate was etched by a conventional anisotropic silicon etching solution, potassium hydroxide (KOH). The substrate was etched from only the backside which was also made the etch stop pattern by ion implantation. Etching rates depended on the ion dose, requiring at least 1×10^{17} cm^{-2} [1]. The resulting microcantilever showed good reproducibility in its shape, and showed no deflection, which originates from residual stress (Figure 2).

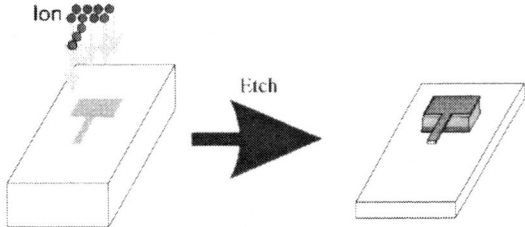

FIGURE 1. Fabrication of a microcantilever; the micro structure remained after micro-patterned ion implantation and substrate etching.

FIGURE 2. SEM image of fabricated microcantilever beam. 1×10^{17} cm^{-2} gold ions were implanted with a patterned mask.

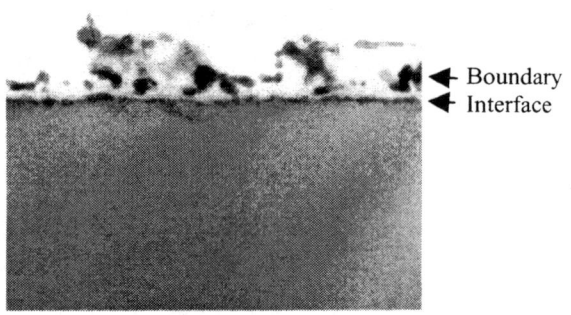

FIGURE 3. XTEM image of the gold ion-implanted silicon substrate after 30 minutes of etching. Precipitated gold grains were observed at the silicon interface and the surface boundary.

Cross-sectional TEM analysis

To evaluate rhe reasons of etch stop effect, we prepared silicon samples that were implanted with ions and then etched for a fixed time, The etching depth and the surface of the silicon was then examined by using a cross-sectional transmission electron microscope (XTEM). Small black grains (about 5 nm in diameter) composed of gold and iron were observed at the silicon interface (Figure 3). At the boundary, larger gold and iron grains (up to several tens of nanometers) were observed. This means that when etching dissolved the silicon atoms, the gold atoms that have lower solubility were precipitated as grains. As etching progressed, the grains grew. Furthermore, the etching rate was inversely proportional to the gold distribution (Figure 4). The gold grains that have lower solubility protected the silicon solvent, thus decreasing the etching rate. However, the number of larger gold grains on the boundary decreased with increasing etch time. This suggests that the grains on the boundary were separated the from surface due to the etching. On the other hand, the number of grains at the interface increased with increasing etch time. This suggests that the grains at the interface are related to the etching rate. We previously developed this etch stop model and described more details on other paper [6].

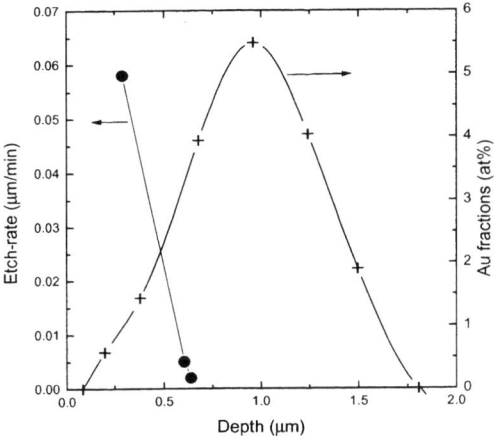

FIGURE 4. Average etching rate (•) calculated by dividing the etch depth by the etch time, and density of the implanted gold ions (+).

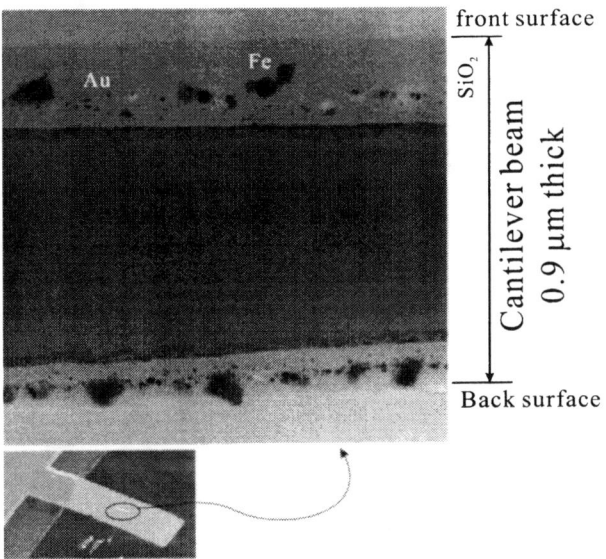

FIGURE 5. XTEM image of 1×10^{17} cm^{-2} gold-ion implanted silicon cantilever beam.

Figure 5 shows an XTEM photograph of the microcantilever beam. The center of the microcantilever was cut in the length direction. The photograph shows that the thickness of the microcantilever was 0.9μm, and that gold was distributed over the entire cantilever. The microcantilever was composed only of an ion-implanted layer. The surface of each side of the microcantilever showed both a gray contrast and black grains. Energy dispersive X-ray analysis (EDX) shows that the gray surface layers were SiO$_2$ and that the black grains were gold and iron. The iron grains are dissolved from the stainless-steel holder for single side etching and precipitated on the microcantilever.

Elastic properties of the microcantilever

Young's modulus, E, is a measure of elasticity of an object. For the microcantilever beams we determined E by using the resonant frequency method [7,8]. When a cantilever is subjected to two-dimensional deflection vibration, E can be expressed as

$$E = 48\pi^2 \frac{\rho L^4 f_i^2}{\alpha_i^4 h^2}, \qquad (1)$$

where ρ is the density of the cantilever, h is its thickness, and subscript i is the number of the mode. The parameter α_i is determined from the criterion for a beam fixed at one end and free at the other. The density of gold ions distributed in the silicon was not known, so we assumed $\rho = 2.3\times10^3$ Kg/m^3, which is the density of silicon. Figure 4 shows that E calculated by using Eq. (1) was relatively low, ranging from 13 and 85 GPa. Figure 4 also indicates that E decreased with increasing ion dose. Microstructures with such low E have not previously been achieved, and such structures will be useful for making high-sensitivity elastic sensors, such as micro accelerometers.

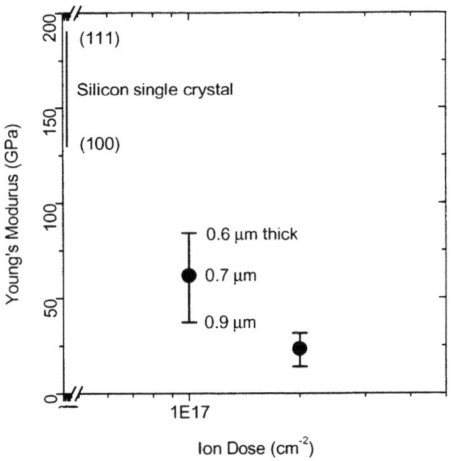

FIGURE 6. Young's modulus of the microcantilevers. Error bar shows the difference in the thickness.

Other ion species

Microcantilevers were also fabricated using other ion species: platinum, titanium, iron, carbon, boron and fluorine, and silicon. For each species, a silicon substrate was implanted with 1×10^{17} cm^{-2} ions at 95 K, and then etched by 30% KOH at 345 K. The microcantilevers implanted with the metallic ions, namely, platinum, titanium, and iron, were also constructed the microcantilevers. The similarity is due to the effect of the above-mentioned etching stop model. The cantilevers implanted with non-metallic ions, namely, carbon and boron, were successfully fabricated. For both microcantilevers, the solubility in KOH was decreased by the implantation. Furthermore, the carbon-implanted microcantilever showed formation of silicon carbide, and the boron-implanted microcantilever showed the etch stop effect of boron-implanted silicon[9]. However, cantilevers implanted with silicon and fluorine were not successfully fabricated, because the solubility in KOH was not affected by the implantation.

Discussion

We successfully prepared microstructures by using this ion implantation process except for examples the gold, platinum, titanium, iron, carbon, and boron. We were unsuccessful when using fluorine and silicon ions. If the model of etching stop that we proposed is correct, then the requirement for successful fabrication is that the ion species must only lower the solubility for the etch solution, and therefore various ion species can be used. Differences in the implanted ion species will allow microstructures of different characteristics to be fabricated. Further evaluation of this technique and the resulting microstructures will advance the design of materials that can be used in MEMS devices. In addition, the microstructures which are fabricated by this process, was separated from the substrate, and therefore the structure was not adversely affected as a leakage by the substrate. Therefore, such ion implanted microstructures can be used in the fabrication of supersensitive sensors.

Conclusion

We proposed and developed a micromachining technique involving ion implantation and etching process. This technique is suitable for use in conventional silicon lithography, resulting in simple, effective micromachining. A microcantilever beam was made using this technique. When gold ions were used, the resulting microcantilever was composed from only ion implantated layer, and showed no evidence of a deflection that is a common effect of residual stress. To explain the application of this working method, we proposed the etch stop model, which explains that the decrease in the etching rate of the ion-implanted microstructures was partly due to the precipitation of implanted gold at the interface. Elasticity of this cantilever (measured using the resonance frequency method) was relatively low, ranging from 13 to 85 GPa and was dependent on the ion dose. This ion implantation technique can be applied to micro devices that require elastic deformation, such as microaccelerometers and microgyroscopes, because such material has lower elastic properties can improve the performance of such microdevices.

REFERENCE

1. S.Nakano, H.Ogiso, and A.Yabe, Nucl. Instrum. Methods B **155** (1999) 79-84

2. S.Nakano, H.Ogiso, H.Sato, S.T.Nakagawa, Surf. Coat. Technol. **128-129** (2000) 71-75

3. S.Nakano, K.Yamanaka, H.Ogiso, and T.Koda, Proc. 3rd Int. Symp. Micromachine and Human Science, Nagoya, (1992), 51

4. B.Schmidt, L.Bischoff, and J.Teichert, Sensors and actuators A**61** (1997) 369-373

5. J.A. van Kan, J.L.Sanchez, B.Xu, T.Osipowicz, and F.Watt, Nucl. Instr. Meth. B**148** (1999) 1085-1089

6. S.Nakano, H.Ogiso, S.Nakagawa, H.Ishikawa, Nucl. Instr. Meth. B (in press)

7. H.J. Butt, and M. Jaschke, Nanotechnology **6** (1995) 1-7

8. S.Nakano, and K.Yamanaka, Jpn.J.Appl.Phys. **36** (1997) 3265-3266

9. H. Seidel, L.Cspregi, A.Heuberger, H.Baumgartel, *J. Electrochem. Soc.*, **137** (1990) 3626-3632

Experiments Using A 200 kV Implanter and A 5 MV Tandem Accelerator

Ryoya ISHIGAMI, Yoshifumi ITO, Keisuke YASUDA, and Satoshi HATORI

The Wakasa Wan Energy Research Center, 64-52-1, Nagatani, Tsuruga, Fukui 914-0192, Japan

Abstract. N$^+$ ions with an energy of 190 keV were implanted into an Al alloy (95% Al and 5% Mg) to a dose of 1.5×10^{19} ions/cm^2. A layer of AlN with 1.4 μm thickness was obtained. The amounts of InN deposited on GaAs or Al$_2$O$_3$ were measured by RBS using He^{2+} ions with an energy of 3.14 MeV generated by a tandem accelerator. The thickness was estimated to be 0.047 μm and 0.26 μm in each case. An experiment on transmission ERDA using He^{2+} ions with an energy of 15 MeV is proposed for the measurement of deuterons in thick Ti foil with good depth resolution.

INTRODUCTION

W-MAST (Wakasa-Wan Energy Research Center - Multi-purpose Accelerator with Synchrotron and Tandem) has been constructed in order to activate the local communities and industries through the development of science and technology[1]. The experimental facility consists of the accelerator complex of a synchrotron and a tandem accelerator, and a stand-alone ion-implanter.

The 200 kV ion implanter based on a microwave ion source will be used for ion beam modification of material surfaces and for the application to semiconductor materials such as III-V nitrides and to high-performance permanent magnets composed of rare earth and 3d elements. The tandem accelerator with a terminal voltage of 5 MV maximum will be utilized not only for the injector to the synchrotron accelerator but also for material modification and as a diagnostic beam for RBS, ERDA, PIXE, etc.

In W-MAST, four irradiation rooms are equipped for various research; in room 1 material diagnostics by tandem beams with low energy, in room 2 material modification and diagnostics by tandem beam with high energy, in room 3 proton cancer therapy by synchrotron beam and in room 4 experimental study using the synchrotron beam.

In this report, three kinds of items are given; 1) experimental results on nitrogen implantation into Al in a very high dose condition, 2) results on RBS analysis of III-V semiconductors, 3) proposal of a transmission ERDA for light element detection in the deeper position of the matter.

ION IMPLANTATION

In the 200 kV ion implantation device, nitrogen implantation into Al, Si, and Ti have been tried. At the present stage, Al is mainly used as the target sample, because AlN can be obtained from implantation at room temperature without subsequent annealing. Another reason is that AlN has the potential as a basic material in the microelectronic industry because of its high electrical resistivity, high thermal conductivity, and excellent thermal and chemical stability.

The detailed description of the ion implantation device and its performance is found elsewhere, so only the description related to this experiment is given. N$^+$ ions with an energy of 190 keV were implanted into an Al alloy sheet (polycrystalline of 95 % Al and 5 % Mg) in thickness of 1.2 mm. The surface of the sample was polished to a mirror finish. The implantation area was (8-9.5) mm × (5-8) mm and the ion current flowing to the sample was about 1 mA. The water-cooled sample holder made of copper prevented the sample temperature from increasing during the ion implantation. Thermal radiation from the sample surface was not observed during the ion implantation of the time interval of 30 minutes even in high current density of 3 mA/cm^2. The implantation chamber was evacuated to 1×10^{-5} Pa, and the residual vacuum was typically 0.8×10^{-4} Pa during the implantation.

The crystallography of the N implanted Al layer was investigated by X-ray diffraction (XRD). XRD study revealed that the wurtzite type hcp AlN structure with a c-axis preferred orientation was formed in the N implanted Al layer.

The depth profiles for the implanted layer were obtained by means of Auger electron spectroscopy (AES) combined with 3 keV Ar^+ ions for sputter etching. Figure 1 shows compositional depth profiles of the atomic concentration of Al, N, and O, in case of ion dose of 1.5×10^{19} ions/cm^2. The current density is 1.5 mA/cm^2. Figure 1 indicates that the nitrogen atomic ratio does not exceed the stoichiometric ratio of AlN as reported so far[2]. The thickness of the AlN layer is about 1.4 μm.

Lucas et al.[3] reported that the retained dose of nitrogen atoms was saturated beyond a certain incident dose in 100 keV N_2^+ implantation into Al, where they estimated that the saturation dose was $\sim 7.4 \times 10^{17}$ ions/cm^2. Miyagawa et al.[4] estimated from the comparison of the calculated results by dynamic-SASAMAL with the experimental results that the saturation dose was $\sim 8 \times 10^{17}$ ions/cm^2 in 50 keV N^+ implantation into Al. The existence of the saturation dose may lead that the thickness of the AlN layer has upper limit for a given implantation energy. We have observed that the width of the AlN layer increases with the incident dose of N^+ and that the thickness becomes almost constant (~ 0.7 μm) when the dose changes from 4×10^{18} to 8×10^{18} ions/cm^2. The saturation dose was estimated to be $\sim 1.7 \times 10^{18}$ ions/cm^2 in 190 keV N^+ implantation into Al on assumption that the atomic density of AlN was 4.8×10^{22} at./cm^3. We have also observed that the width becomes broader as the dose increases in the very high dose case more than 8×10^{18} ions/cm^2. Above results will be reported elsewhere in details. We only point out that the thickness of the AlN layer in the dose of 1.5×10^{19} ions/cm^2 observed here is twice that of the quasi-saturation (~ 0.7 μm).

RBS MEASUREMENT

Material analysis by means of RBS, ERDA and PIXE is possible at a beam line in the irradiation room 1, using the tandem accelerator beam, where H^+ (1-3 MeV, 5 μA in maximum), He^{2+} (1.5-5 MeV, 1 μA in maximum) and C^{3+} (2-6.8 MeV, 2.5 μA in maximum) are available. PIXE analysis for the Japanese paper "washi" and the cow bone have been done, which will be reported elsewhere. Here, RBS analysis of the InN thin film made by MOCVD method is described.

Figure 2 show RBS energy spectra, where He^{2+} with an energy of 3.14 MeV are used as incident beam and scattered angle is set to be 140°. Here, samples are (a) a thin film of InN deposited on GaAs and (b) a thin film of InN deposited on Al_2O_3. The backscattered

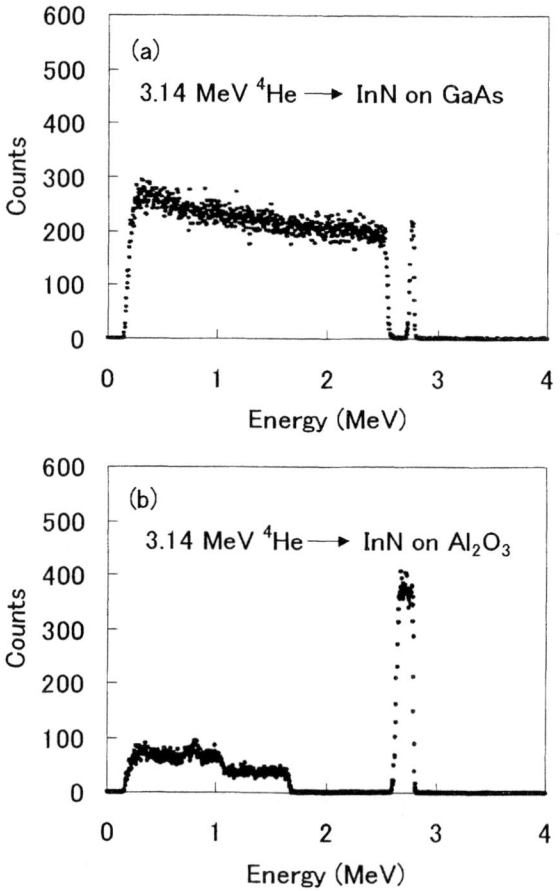

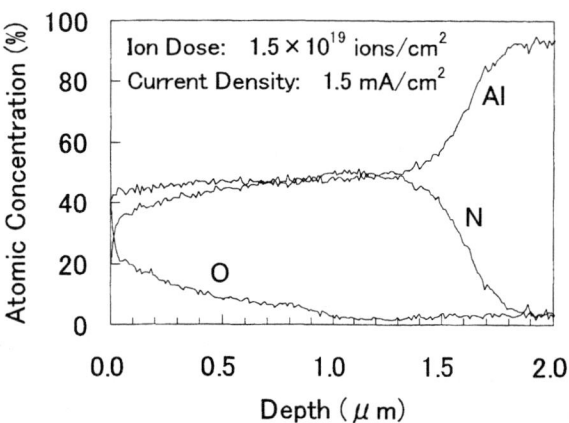

FIGURE 1. Compositional depth profiles of Al, N, and O for N-implanted Al.

FIGURE 2. RBS spectra of InN films deposited by means of MOCVD. Substrates are (a) GaAs and (b) Al_2O_3.

signals of In are separated from ones of the elements composed of the substrates (Ga, As, Al, etc), since the atomic number of In is high enough in comparison with that of those elements.

The thickness of the InN deposited on the substrates can be estimated from the RBS spectra shown in Fig. 2. The width of In signal in Fig. 2 (b) is 0.163 MeV. This value yields directly that the thickness of the InN is 0.22 μm with the stopping power of He from Ziegler[5]. The width of In signal in Fig. 2 (a) is 0.043 MeV, which is too small for the direct estimation of the thickness given above. The number of the atoms is found to be 1.5×10^{17} atoms/cm² from the In signal area in Fig. 2 (a). The thickness is estimated to be 0.047 μm from the value of the line density.

It is emphasized that the thickness of the thin films on the substrate composed of lower atomic number can be found non-destructively by using the RBS method.

TRANSMISSION ERDA

In the irradiation room 2, H⁺ (10 MeV, 100 μA), He²⁺ (15 MeV, 50 μA), and C⁴⁺ (25 MeV, 50 μA) generated by tandem accelerator will be available for various experimental studies. Here, we focus on the transmission ERDA method for the measurements of light elements in the metals by using 15 MeV He²⁺ beams. This can yield the measurement of the light elements with good depth resolution in such thick films. Let's consider the transmission ERDA experiment as shown in Figure 3. Deuterium is buried in a layer of 2 μm thickness within the Ti. Here, Ti is used as the metal film because Ti is a well-known hydrogen storage metal. The total width of Ti is assumed to be 72 μm. The range of He with an energy of 15 MeV is estimated to be 83 μm in Ti, so the incident He beam can penetrate the Ti in this arrangement. The incident angle of the He²⁺ is normal to the Ti target. Deuterons recoiled in the direction of recoil angle 17.8° are detected, where a detection angle of 17.8° is available in the scattering experiment in the room 2. An Al 27 μm thick is installed in front of the particle detector, because a lot of He²⁺ might be also injected in the particle detector without the Al foil.

The depth profile of D in Ti is derived from the energy spectrum of deuterium. Figure 4 shows the relation between the depth of deuterons and the energy of the recoil deuterons, where the stopping power of D and He used are reported ones [5, 6].

The depth resolution for deuterons δD is expressed as

$$\delta D = \Delta E / \frac{dE}{dl}, \qquad (1)$$

where E is the energy of the deuterons detected and ΔE is FWHM (full width at half maximum) of the energy spreading of E. The value ΔE depends on not only the energy straggling δE_s but also energy width δE_a due to the angular spreading by multi-scattering.

The energy straggling δE_s of the detected deuteron was calculated in two cases; first D is recoiled at the surface of the upstream side of the target and second, D at the surface of the downstream side. The values of δE_s are estimated to be 0.13 MeV in the first case and 0.23 MeV in the second case. The values of $(\delta E_s)^2$ are given by a summation in quadrature of the energy straggling of He in Ti, D in Ti, and D in Al.

The width of the angular spreading can be found from the film thickness, by use of the experimental results reported by many authors[7-16]. FWHM of the angular spreading of a deuteron is estimated to be

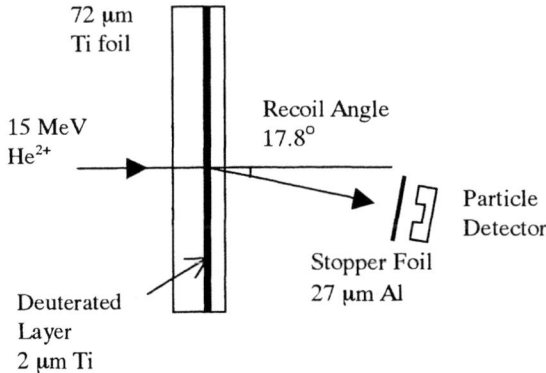

FIGURE 3. Schematic drawing of the transmission ERDA measurement.

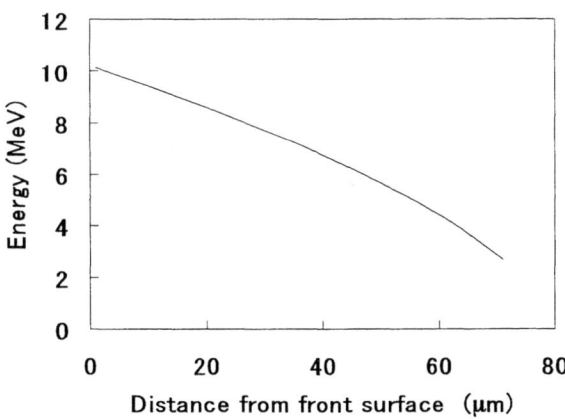

FIGURE 4. Calculated energy of a recoil deuteron detected as a function of depth in a Ti foil of 72 μm in thickness. An incident ion is He²⁺ with an energy of 15 MeV. The recoil angel is 17.8°.

TABLE 1. Calculated energy straggling δE_s, energy width caused by angular spreading δE_a, total energy width ΔE, and depth resolution δD at a front surface and a back surface of a Ti foil of 72 μm in thickness in transmission ERDA using He ions with an energy of 15 MeV. The recoil angle is 17.8°.

	the front surface	the back surface
Energy straggling δE_s	0.13 MeV	0.23 MeV
Angular spreading δE_a	0.48 MeV	0.26 MeV
Total energy width ΔE	0.50 MeV	0.35 MeV
Depth resolution δD	6.25 μm	1.95 μm

0.062 rad in the first case and 0.121 rad in the second case.

Let M_1 be the mass of the incident particle and M_2 be the mass of the recoil one. The recoil kinematic factor K' is expressed as a function of a recoil angle ϕ,

$$K' = \frac{4M_1M_2}{(M_1+M_2)^2}\cos^2\phi$$

The derivative of deuteron energy E_1 with respect to recoil angle ϕ is given by the following expression.

$$\frac{dE_1}{d\phi} = -\frac{4M_1M_2}{(M_1+M_2)^2}E\sin 2\phi . \qquad (2)$$

The energy width δE_a of a deuteron is derived from the angular spreading using eq. (2). The energy width δE_a is estimated to be 0.48 MeV in the first case and 0.26 MeV in the second case. Here the value of E used in the second case is 4.19 MeV, which is the mean energy of the He beam after penetration through the Ti foil.

The value ΔE in eq. (1) can be expressed as
$$\Delta E^2 = \delta E_s^2 + \delta E_a^2 . \qquad (3)$$
The ΔE is found to be 0.50 MeV and 0.35 MeV in the first and second cases, respectively.

CONCLUSION

N^+ ions with an energy of 190 keV generated by a 200 kV ion implanter were implanted into an Al alloy (95% Al and 5% Mg) to a dose of 1.5×10^{19} ions/cm². The N depth profile was a trapezoid with a flat-topped plateau almost equal to the stoichiometric ratio of AlN. The thickness of the AlN layer was 1.4 μm which was broader than the expected one.

The amounts of InN deposited on GaAs or Al_2O_3 were measured by RBS using He^{2+} ions with an energy of 3.14 MeV generated by a tandem accelerator. The thickness of InN on the substrate was found to be 0.26 μm and 0.047 μm. An experiment on transmission ERDA using 15 MeV He^{2+} was proposed. Deuterium is buried in a layer of 2 μm thickness within Ti, and the total width of Ti is 72 μm. The recoil angle of the deuterons is 17.8°, and Al with 27 μm thickness is installed in front of the particle detector. It was calculated that the depth resolution for the deuterons is 6.25 μm at the front surface of the Ti foil and 1.95 μm at the back surface.

ACKNOWLEDGEMENT

The authors would like to thank Prof. Akio Yamamoto of Fukui University for supplying the InN samples.

REFERENCE

1. Okada, O., Fukuda, S., Hatori, S., and Ito, Y., *Proceedings of the 13th International Conference on Higt-power Particle Beam (BEAMS 2000)*, Nagaoka, Japan, (2000), to be published.
2. Ohira, S,. and Iwaki, M., *Mater. Sci. and Eng.*, **A116**, 153 (1989).
3. Lucas, S., Terwagne, G., and Bodart, F., *Nucl. Instrum. Methods* **B50**, 401 (1990).
4. Miyagawa, Y., Nakao, S., Saitoh, K., Ikeyama, M., Tanemura, S., and Miyagawa, S., *Nucl. Instrum. Methods*, **B106**, 170 (1995).
5. Ziegler, J. F., *Helium Stopping Powers and Ranges in All Elemental Matter*, New York: Pergamon Press, 1977.
6. Andersen, H. H., and Ziegler, J. F., *Hydrogen Stopping Powers and Ranges in All Elements*, New York: Pergamon Press, 1977.
7. Tirira J., Serruys Y., and Trocellier, P., "Elastic Scattering: Cross-Section and Multiple Scattering" in *Forward Recoil Spectrometry Applications to Hydrogen Determination in Solids*, New York: Plenum Press, 1996 p. 74.
8. Schmaus, D. and L' Hoir, A., *Nucl. Instr. and Meth.* B4, 317-331 (1984).
9. Högberg, G., Nordén, H., and Berry, H. G., *Nucl. Instr. and Meth.* 90, 283-288 (1970).
10. Spahn, G. and Groeneveld, K. O., *Nucl. Instr. and Meth.* 123, 425-429 (1975).
11. Hooton, B. W., Freeman, J. M., and Kane, P. P., *Nucl. Instr. and Meth.* 124, 29-39 (1975).
12. Knudsen, H. and Andersen, H. H., *Nucl. Instr. and Meth.* 136, 199-201 (1976).
13. Anne, R., Herault, J., Bimbot, R., Gauvin, H., Bastin, G., and Hubert, F., *Nucl. Instr. and Meth.* B34, 295-308 (1988).
14. Andersen, H. H., Bøttiger, J., Knudsen, H., and Petersen, P. M., *Phys. Rev.* A10, 1568-1577 (1974).
15. Andersen H. H. and Bøttiger, J., *Phys. Rev.* B4, 2105-2111 (1971).
16. Schwabe, S. and Stolle, R., *Phys. Stat. Sol. (b)* 47, 111-118 (1971).

Film growth using mass-separated ion beams

H. Hofsäss, C. Ronning, and H. Feldermann

II. Physikalisches Institut, Universität Göttingen,
Bunsenstrasse 7-9, D-37073 Göttingen, Germany

Abstract. Mass-separated low energy ion beams provide clean, well defined and controllable deposition conditions for film growth and are thus particularly suited to study basic mechanisms of thin film growth. The recent advances and emerging applications of mass-separated ion beam deposition (MSIBD) are discussed. During the last years MSIBD has been extensively applied to study the growth of tetrahedral amorphous carbon (ta-C) and cubic boron nitride (c-BN) films. The nucleation and growth conditions are discussed and compared with model predictions. The microstructure and properties of F-, B- and N- containing carbon films are briefly discussed.

INTRODUCTION

The deposition of ions in the energy range of few tens of eV up to several keV is a thin film growth process far from thermodynamic equilibrium. Ion beam deposition has been extensively applied to synthesize a variety of diamondlike thin film materials [1] like tetrahedral amorphous carbon (ta-C) [2] and cubic boron nitride (c-BN) [3]. Mass selected ion beam deposition (MSIBD) provides the most clean and well defined deposition conditions [4]. Various deposition parameters can be independently controlled, which makes it possible (i) to grow isotopically pure thin films, (ii) compound film materials or layered film structures of variable composition, and (iii) to study the nucleation processes and the phase formation. MSIBD has been applied since the early eighties to study the growth of diamondlike carbon [5], carbon nitride [6], boron carbide [7] and c-BN [3]. Other studies include group III nitride growth [8] and combined ion beam and molecular beam epitaxy [9]. In this paper the experimental aspects and recent applications of MSIBD growth of diamondlike materials are discussed.

EXPERIMENTAL

Ions are produced in an ion source and accelerated to typically 30 keV. Mass separation is done by 90° sector magnet and electrostatic lenses guide the mass separated ion beam through differential pumping stages into the deposition chamber operated under ultrahigh vacuum (UHV) conditions. In order to get a homogeneous deposition over a larger area of the substrate, the beam is scanned across the substrate surface. Close to the substrate the ions are decelerated down to the desired energy by applying a bias voltage to the substrate relative to the voltage level of the ion source. The final energy is adjustable between a few eV and several keV.

Our system is equipped with a process control to select a specific mass and also to adjust the deceleration voltage [10]. A measurement of the deposited charge is used to periodically switch the mass separation magnet. During each switching cycle 10^{15} ions are typically deposited onto an area of 2 cm^2. The film composition can be varied by changing the relative amount of different ions deposited per switching cycle. Our MSIBD system provides a beam current up to 50 µA. The growth rates are low, typically between 10 and 100 nm/h. MSIBD is therefore not suitable for most industrial coating applications. However, it is ideal to study the growth process and to investigate the materials properties of films grown under specific growth conditions.

Several descriptions of MSIBD systems can be found in the literature [9,11,12]. Many of the MSIBD systems are modified ion implanters [5] connected to a high vacuum deposition chamber without sophisticated deceleration stages. Some systems are designed as dual-beam deposition systems [13,14] where two mass separated ion beams are merged together. Special

MSIBD systems with in-situ characterization tools like Auger electron spectroscopy (AES), X-ray and UV photoelectron spectroscopy (XPS,UPS) were developed [12,15].

GROWTH PROCESSES

Film growth processes taking place during ion beam deposition (IBD) are significantly different compared to most PVD and CVD methods and also compared to ion implantation. IBD is a sub-surface growth process where phase formation takes place below the surface in a depth of a few nm, determined by the range of the incoming ions. Lifshitz et al. have introduced the term 'subplantation' to characterize the ion beam deposition process and to distinguish it from surface dominated growth processes and from ion implantation [16]. The subplantation model assumes a three stage deposition process: an extremely rapid collisional stage, a fast thermalization stage, usually treated as a thermal spike, and a slow relaxation stage. For ion energies below a few keV the major fraction of the kinetic energy is converted into phonon excitations initially confined to a rather small volume. Thermal spike processes are therefore dominant, whereas at energies well above 10 keV it is electronic stopping, the generation of collision cascades, and defect production. In this way the subplantation and implantation regimes can be distinguished.

We have introduced the cylindrical spike model to describe ion deposition processes [17], taking into account energy loss along the ion track, collision cascade effects, and energy conversion to phonons and electronic excitations. The evolution of a diamondlike structure is a consequence of a large number n_T of atomic rearrangements of n_S atoms in the spike volume during a thermal spike. The resulting local structure is determined by the boundary conditions around the spike volume and the initial spatial distribution of the deposited energy. The cylindrical spike model is able to explain experimentally observed ion energy dependence for formation of ta-C and c-BN thin films. The substrate temperature is expected to play a minor role for phase formation. Indeed, nanocrystalline films of the high melting point materials c-BN [18], AlN, and GaN [8] can be grown by ion beam deposition even at room temperature. On the other hand, thermally driven diffusion processes may suppress the formation of a diamondlike phase as in the case a-C deposited at elevated substrate temperatures [19,20].

With increasing ion energy sputtering becomes a more and more important process. However, as long as no volatile components are produced during film growth, sputter coefficients are negligibly small at low ion energies and may reach values around 0.5 in the keV region [21]. Significant losses due to volatile components occur for example during carbon nitride and carbon fluoride film growth [22,23].

Ion Energy Dependence

The ion energy dependence of ta-C formation has been studied extensively using filtered cathodic vacuum arc deposition (FCVAD) and MSIBD [24]. For MSIBD grown films the sp^3-bond fraction, determined from the plasmon energy as a function of the carbon ion energy is shown in fig. 1. The plot shows a maximum sp^3-content around 100 eV and a slowly decreasing sp^3-content with further increasing energy. The best agreement with the experimental data for MSIBD grown films is obtained for the cylindrical spike model [17]. To illustrate this, we have plotted in fig.1 the parameter n_T/n_S, calculated using the cylindrical spike model and describing the probability for formation of a highly sp^3-bonded phase.

As-grown ta-C films exhibit an extreme compressive stress up to 10 GPa. The correlation between compressive film stress and sp^3-content for MSIBD grown films is shown in fig.1. The ion energy dependence of compressive stress and sp^3-bond fraction show an approximately linear correlation. However, a microscopic understanding of compressive stress generation in ta-C films is still lacking.

Characteristic for MSIBD grown c-BN films is an ion energy threshold of 125 eV and a substrate temperature threshold of about 150 °C to trigger c-BN nucleation (Fig.2). The nucleation of c-BN always requires the existence of an oriented h-BN intermediate layer. Once nucleated the ion energy can be reduced significantly down to about 60 eV and c-BN growth is maintained [25]. Our group has shown that, once nucleated, the substrate temperature can be reduced down to room temperature and c-BN growth is maintained [18]. Both experimental results clearly demonstrate that the threshold energy and temperature are only related to the nucleation but not to the continued growth of c-BN. The cylindrical spike model predicts about 30-50 eV ion energy as lower limit and about 3 keV as upper limit for c-BN formation.

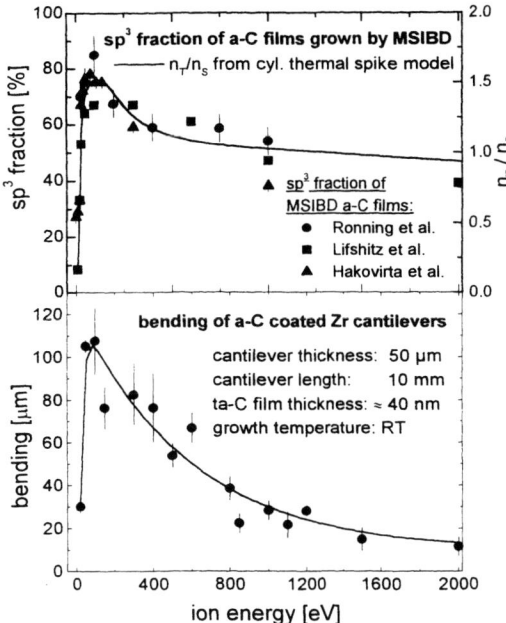

Figure 1: Top: Fraction of sp³-bonded atoms in a-C films grown at room temperature by MSIBD as a function of ion energy [24]. Also shown is the parameter n_T/n_S of the cylindrical thermal spike model, describing the probability for diamondlike phase formation [17]. Bottom: bending of Zr cantilever substrates due to compressive films stress in room temperature deposited a-C films as a function of ion energy.

The Role of Film Composition

It is possible to grow films doped with a given concentration of impurity atoms by selecting a certain ion charge ratio. This has been used to incorporate small amounts (0.2- 10 at.%) of B, N, and P impurities into ta-C films to study their electrical properties. B-, N- and P-incorporation resulted in a strong increase of the conductivity with increasing impurity concentration which was explained by an increased hopping and Frenkel-Poole-conduction [1].

The systematic variation of the N-content in a-C films was used to study the evolution of C-N bonding, starting from pure ta-C towards films with a composition of C_2N [26]. The XPS analysis of the C1s and N1s core level binding energies in these films has clearly shown the formation of graphitic CN films with increasing N-content, rather than a 2-phase system with one phase being C_3N_4, as proposed by other groups.

We also investigated the application of amorphous boron carbide films as a hard and low resistivity coating. The resistivity of B-doped ta-C films continuously decreases as the B-content increases. B_xC films were deposited at room temperature from 100 eV ions starting from pure ta-C up to a composition of B_4C by variation of the ion charge ratio $[^{11}B^+]:[^{12}C^+]$. The result of EELS and Stress measurements are plotted in Fig. 3. EELS indicates a continuous change from a dense ta-C phase towards a lower density B_4C compound. In contrast, the compressive stress has a significant minimum for a film composition near B_1C_1.

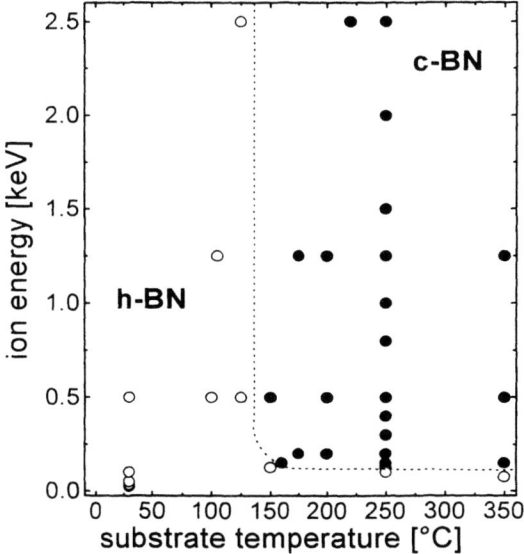

Figure 2: Deposition parameter dependence of the grown BN phase for MSIBD of $^{11}B^+$ and $^{14}N^+$ ions.

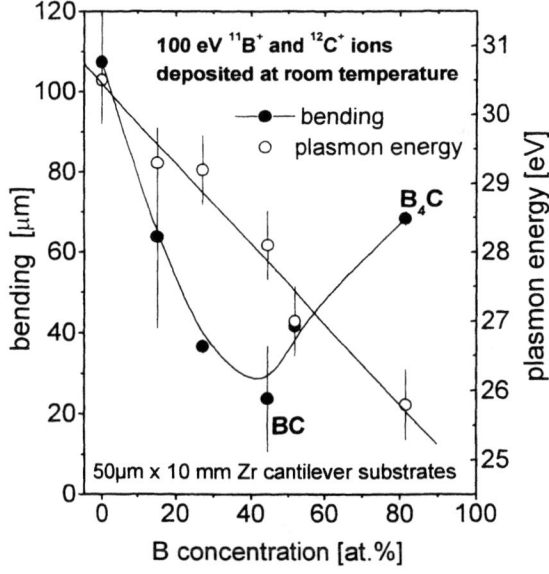

Figure 3: Compressive stress and plasmon energy of B_xC thin films grown at room temperature by MSIBD using 100 eV $^{11}B^+$ and $^{12}C^+$ ions. A cantilever bending of 110 μm corresponds to a compressive stress of about 10 GPa.

Finally we report on fluorinated amorphous carbon films (a-C:F) grown by room temperature deposition of 100 eV ^{12}C$^+$ and ^{19}F$^+$ ions on Si [23]. Similar to CN$_x$ film growth the formation of volatile components limits the maximum achievable F content to probably 30 at.%. Surprisingly, whereas the mass density and the compressive stress decreases continuously for increasing F-content, the plasmon energies are almost constant at around 30 eV. Thus, the valence electron density remains comparable to a F-free ta-C film. Most probably a-C:F films with higher F content have a porous microstructure. Compared to PVD or CVD fluorinated carbon films, the MSIBD films are thermally stable up to about 600 °C in vacuum.

APPLICATIONS

Apart from protective coating applications we want to mention to recent developments. A possible application of diamondlike films are emitters for field emission displays. To enhance the field emission, nm-size conducting ion tracks were generated in initially high resistivity ta-C films by heavy ion irradiation [27]. Amorphous B$_x$C films have a sufficiently high hardness and electrical conductivity and are therefore suited as hard conductive coating for Si tips used in atomic force microscopy [28]. B$_x$C coated AFM tips operated in the Kelvin mode are applied for simultaneous measurements of the surface topography and the work function [29].

ACKNOWLEDGEMENTS

This work was financially supported by the Deutsche Forschungsgemeinschaft under the auspices of the trinational D-A-CH cooperation on the synthesis of superhard materials and the german Bundesminister für Bildung und Forschung.

REFERENCES

1. Hofsäss, H. and Ronning, C., in Proc. 2nd Int. Conf. on Beam Processing of Advanced Materials, edited by J. Singh, S. M. Copley, J. Mazumder (ASM International, Materials Park, 1996), p. 29-56.
2. Silva, S.R.P., Robertson, J., Milne, W. I., Amaratunga, G.A.J., *Amorphous Carbon: State Of The Art; Proc. 1st International Specialist Meeting on Amorphous Carbon* (World Scientific, Singapore, 1998)
3. Ronning, C., Feldermann, H., and Hofsäss, H., Diam. Relat. Mater. 9 (2000) 1767
4. Ronning, C., Hofsäss, H., in *Diamond Materials IV, Proc. 4th Int. Symposium*, eds: K.V. Ravi and J.P. Dismukes (Electrochemical Society, Pennington, 1995), p. 359.
5. Koskinen, J., J. Appl. Phys. 63, 2094-2097 (1988).
6. Marton, D., K.J. Boyd, A.H. Albayati, S.S. Todorov, and J.W. Rabalais, Phys. Rev. Lett. **73**, 118-121 (1994).
7. Todorov, S.S., D. Marton, K.J. Boyd, A.H. Al-Bayati and J.W. Rabalais, J. Vac. Sci. Technol. A **12**, 3192 (1994).
8. Ronning, C., Dreher, E., Feldermann, H., Sebastian, M., Zweck, J., Fischer, R., Hofsäss, H., Mat. Res. Soc. Symp. Proc. Vol. **449**, 331 (1997).
9. Iida, T., Makita, Y., Kimura, S., Winter, S., Yamada, A., Shibata, H., Obara, A., Niki, S., Fons, P., and Tsai, Y., Appl. Phys. Lett. **63**, 1951 (1993).
10. Hofsäss, H., Biegel, J., Ronning, C., Downing, R.G., and Lamaze, G.P., Mater. Res. Soc. Symp. Proc. Vol. **316**, 881 (1994).
11. Miyazawa, T., S. Misawa, S. Yoshida and S. Gonda, J. Appl. Phys. **55**, 188 (1984).
12. Lifshitz, Y., Kasi, S.R., and Rabalais, J.W., Mat. Sci. For. Vol. **52 & 53**, 237 (1989).
13. Y. Horino, N. Tsubouchi, K. Fujii, T. Nakata, T. Tagaki, Nucl. Instr. Meth B **106**, 657 (1995)
14. Boyd, K. J., Marton, D., Todorov, S. S., Al-Bayati, A. H., Kulik, J., Zuhr, R. A., and Rabalais, J. W., J. Vac. Sci. Technol. A **13**, 2110 (1995).
15. Lau, W.M., Feng, X., Bello, I., Sant, S., Foo, K.K., Lawson, R.P.W., Nucl. Instr. Meth. B **59/60**, 316 (1991).
16. Lifshitz, Y., Kasi, S.R., Rabalais, J.W., Phys. Rev. Lett. **62**, 1290 (1989)
17. Hofsäss, H., Ronning, C., Sebastian, M., and Feldermann, H., Appl. Phys. A **66**, 153 (1998).
18. Feldermann H., et al. J. Appl. Phys. (2000) submitted
19. Lifshitz, Y., Kasi, S.R., Rabalais, J.W., Eckstein, W., Phys. Rev. B **41**, 10468 (1990).
20. Davis, C.A., Amaratunga, G.A.J., Knowles, K.M., Phys. Rev. Lett. **80**, 3280 (1998).
21. Hofsäss, H., Feldermann, H., Sebastian, M., and Ronning, C., Phys. Rev. B **55**, 13230 (1997).
22. Hofsäss, H., Ronning, C., Feldermann, H., Sebastian, M., Mat. Res. Soc. Symp. Proc. **438**, 575 (1997).
23. Ronning, C., Merk, R., Feldermann, H., Harbsmeier, F., Hofsäss, H., MRS Symp. Proc. 593 (2000) to be published
24. Ronning, C., Dreher, E., Thiele, J.-U., Oelhafen, P., Hofsäss, H., Diam. Relat. Mater. **6**, 830 (1997).
25. Hahn, J., Richter, F., Pintaske, R., Röder, M., Schneider, E., Welzel, T., Surf. Coat. Technol. **92**, 129 (1997).
26. Ronning, C., Feldermann, H., Merk, R., Hofsäss, H., Reinke, P., Thiele, J.-U., Phys. Rev. B **58**, 2207 (1998).
27. M. Waiblinger, Ch. Sommerhalter, B. Pietzak, J. Krauser, B. Mertesacker, M. Ch. Lux-Steiner, S. Klaumünzer, A. Weidinger, C. Ronning, H. Hofsäss, Appl. Phys. A. 69 (1999) 239
28. Hofsäss, H., Boneberg, J., Leiderer, P., German patent No. 197 52 202.5
29. Hofsäss, H., et al., Annual Report Solid State and Cluster Physics, University Konstanz, Germany (1997) p. 119, (http://www.ub.uni-konstanz.de/kops/volltexte/1999/145/pdf/145_1.pdf)

Diffusion and roughening during ion beam erosion of graphite surfaces

S. Habenicht and K. P. Lieb

II.Physikalisches Institut and Sonderforschungsbereich 345,
Universität Göttingen, Bunsenstr. 7-9, D-37073 Göttingen, Germany

Abstract. The ripple structure of graphite (HOPG) surfaces irradiated with 5-50 keV Xe-ions at inclined incidence was measured via scanning tunneling microscopy. In the temperature range 300 - 470 K, the deduced variations of the surface roughness and ripple wavelength were found to follow the predictions of the Bradley-Harper continuum theory in linear approximation. From both quantities, an activation energy of $\Delta E = 0.12(5)$ eV was deduced for radiation-enhanced surface diffusion. The variation of the ripple structure as function of all relevant parameters, such as ion mass, energy, fluence and impact angle and substrate temperature, therefore, confirm the validity of the Bradley-Harper approach for noble-gas irradiated graphite. In order to investigate the transition from linear to non-linear behaviour, the fluence dependence of the roughness was measured.

INTRODUCTION

Facetting or ripple formation of solid surfaces occurs, when beams of heavy ions hit the surface at tilted incidence to the surface normal [1]. The surface morphology is governed by the interplay between the angular-dependent sputter yield and the surface diffusion of atoms. While sputtering generally enhances the contrast of the surface structures, diffusion acts in the opposite direction and smoothens the surface. For ion-irradiated samples of Highly Oriented Pyrolytic Graphite (HOPG), scanning tunneling microscopy (STM) experiments under ambient conditions have proven to give easy access to the details of the ion-induced surface structures, down to atomic resolution [2,3]. In the case of HOPG irradiated with noble gas ions, recently systematic studies of the ripple structure as function of the tilting angle of the ion beam ($\theta = 0 - 80°$) and its mass (M_I = 40 - 130 amu), energy ($E = 2 - 50$ keV) and fluence ($\Phi = 4\times10^{16} - 5\times10^{18}$ ion/cm^2) [4-7] have been performed. As a general result of these investigations, it has been realized that both the continuum theory by Bradley and Harper [8] and Monte Carlo simulations performed by Koponen et al. [9] reproduce the observed orientation of the wave pattern with the impact angle θ and the scaling of the wavelength λ with the parameters M_I, E and Φ [4-6]. In particular, the experimentally observed energy scaling of the wavelength with the ion energy, $\lambda \propto E^p$, where the exponent p depends on the ion mass M_I, was explained by the spatial distribution of the radiation damage, which is produced in the collision cascade near the surface via the nuclear part of the stopping power [5,6,10].

Another important quantity which should affect the ripple structure is the atomic mobility near/on the surface. So far it has been approximated by the diffusion approach used by Wolf and Villain [11] and by Mullins [12]. These basic assumptions concerning the surface mobility and ist effects on the height evolution should be applicable to almost every kind of surface diffusion. One of the aims of the present study was to investigate the influence of atomic diffusion on the ripple structure by varying the substrate temperature T, and to compare it with the continuum theory. To this end, the temperature dependence of the wavelength λ and surface roughness w have been measured to get deeper insights into the role of surface mobility. Concerning the kind of diffusion process, one may hope to distinguish between radiation-enhanced diffusion during the collisional phase of the ion impact and thermal surface diffusion in the relaxation phase, i.e. after the collision phase has ended. Furthermore, the fluence dependence of the surface roughness w was measured in order to highlight the transition from the linear to the non-linear regime inferred in [4].

EXPERIMENTALS

The experiments were carried in a very similar way as described in our previous work [4-7]. Freshly cleaved

HOPG samples were irradiated with a 5-50 keV Xe^+ ion beam provided by the low-energy ion implanter IOSCHKA [13] and the heavy ion implanters ADONIS and IONAS and hitting the (0001) surface at an angle of $\theta = 30°$ to the normal direction. The ion beam was scanned across the sample to cover homogeneously a 10x10 mm^2 implantation area. At an ion fluence of $\Phi = 3 \times 10^{17}$ ions/cm^2, wave patterns develop, whose wavelength and roughness were measured at room temperature in air, by means of scanning tunneling microscopy (NANOSCOPE II, $Pt_{80}Ir_{20}$ tip). Fig. 1a and illustrate the height topographies of micrographs of samples irradiated at $T = 300°C$ and $430°C$. The temperature dependence of the deduced wavelength λ and roughness w are displayed in fig. 2a and b. Fig. 1c and 1d serve to illustrate surface topographies obtained at ion fluences of $\Phi = 2 \times 10^{18}$ and 5×10^{18} Xe-ions/cm^2 at $E = 5$ keV, $\theta = 60°$ and room temperature. Fig. 3 illustrates the fluence dependence of the roughness w in the range $\Phi = 4 \times 10^{16} - 5 \times 10^{18}$ Xe-ions/cm^2.

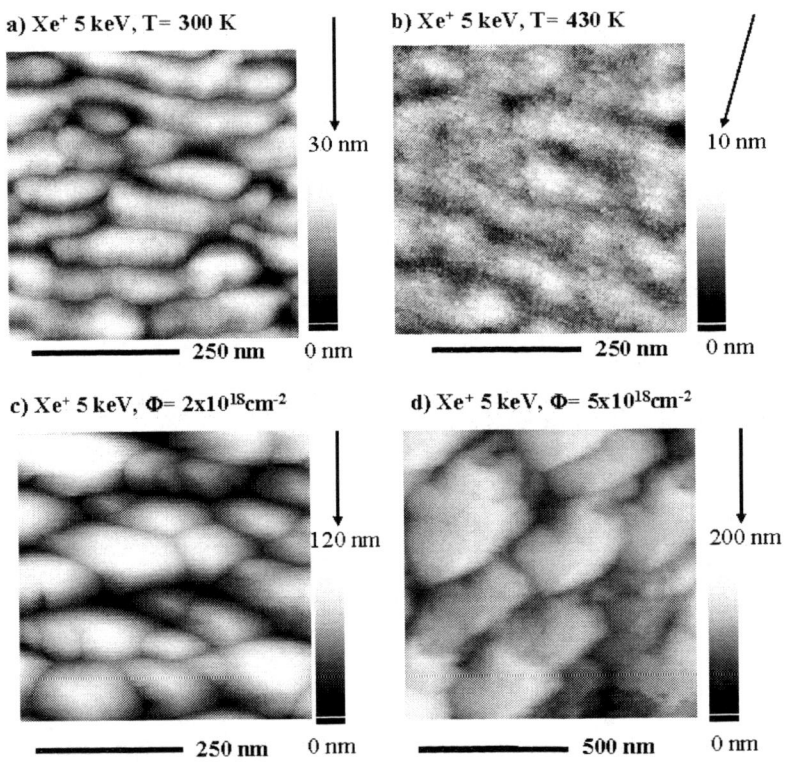

Figure 1: Micrographs of HOPG surfaces irradiated with 5 keV Xe ions at inclined incidence.
a) Ion fluence $\Phi = 3 \times 10^{17}$/cm^2, substrate temperature $T = 300$ K; b) $\Phi = 3 \times 10^{17}$/cm^2, $T = 430$ K;
b) c) $\Phi = 2 \times 10^{18}$/cm^2, $T = 300$ K; d) $\Phi = 5 \times 10^{18}$/cm^2, $T = 300$ K

RESULTS AND DISCUSSION

In the frame of the theory of Bradley and Harper [8], the evolution of the surface morphology $h(r,t)$ with the erosion time t is contained in the structure factor $S(k,t) = |h(k,t)|^2$, i.e. the Fourier transform of $h(r,t)$,

$$h(k,t) \propto \int_r \exp(ikr)[h(r,t) - \bar{h}(t)]dr \quad (1)$$

The function $h(r,t)$ is the solution of the differential equation of erosion, which can be described by the theory of Bradley/Harper and Barabasi [8,14].

$$\frac{\partial h}{\partial t} = -F_0 + C\frac{\partial h}{\partial x} + C_x\frac{\partial^2 h}{\partial x^2} + C_y\frac{\partial^2 h}{\partial y^2} \quad (2)$$

$$+ \Lambda_x\left(\frac{\partial h}{\partial x}\right)^2 + \Lambda_y\left(\frac{\partial h}{\partial y}\right)^2 - B\nabla^4 h + \eta$$

The first six term in Eq. (2) are related to the angular dependence of the sputtering yield, Y_0 denoting the sputtering yield at normal incidence ($\theta = 0°$) and the terms with $C_{x,y}$ the curvature dependence of the sputter yield $Y(\theta)$. The term containing B characterizes the smoothening due to surface diffusion. The xy-coordinates are defined along (x) and perpendicular to (y) the projection of the ion beam onto the surface, respectively. In the linear regime of Eq. (2) and under the influence of stochastic roughening η at the constant mean ion fluence Φ, the ripple wavelengths $\lambda_{x,y}$ and structure factor $S(k,t)$ are given by the expressions

$$\lambda_{x,y} = 2\pi \sqrt{\frac{2B}{C_{x,y}}},$$

$$S(k,t) = |h_0(k)|^2 \exp(R_k t) + \frac{\xi}{R_k}(\exp(R_k t)$$

(3)

where $k = (2\pi/\lambda)$ and $R_k(T) = C_x k^2 - B(T) k^4$. The fluence and temperature dependence of the roughness are contained in the exponential function $w(\Phi,T) \propto \exp[R_k(T)\Phi]$, where the temperature variation enters via the Boltzmann factor $B(T) \exp(-\Delta E/k_B T)$, ΔE being the activation energy of the diffusion process.

Hence, both the ripple wavelength $\lambda_{x,y}$ and roughness w are known functions of T and Φ and can be used to test the underlying theory. At the chosen tilting angle $\theta = 30°$ and ion fluence $\Phi = 3 \times 10^{17}/\text{cm}^2$, the wave pattern is oriented perpendicular to the ion beam direction (x-waves) [4]. The measured temperature dependence of the wavelength λ_x, indeed, follows the predicted trend, i. e. the product $\lambda_x(T) T^{1/2}$ varies proportionally to the Boltzmann factor $\exp(-\Delta E/k_B T)$, with the fitted activation energy $\Delta E = 0.10(7)$ eV. Similarly, the temperature variation of the roughness w at these values of θ and Φ is in agreement with the prediction of the Bradley-Harper approach and provides a consistent value of the activation energy, $\Delta E = 0.14(8)$ eV. Taking both values together, one arrives at $\Delta E = 0.12(5)$ eV. Combining these results with a determination of the surface mobility from the wavelength and the roughening evolution - free of external parameters - , one finds new and interesting insights into the

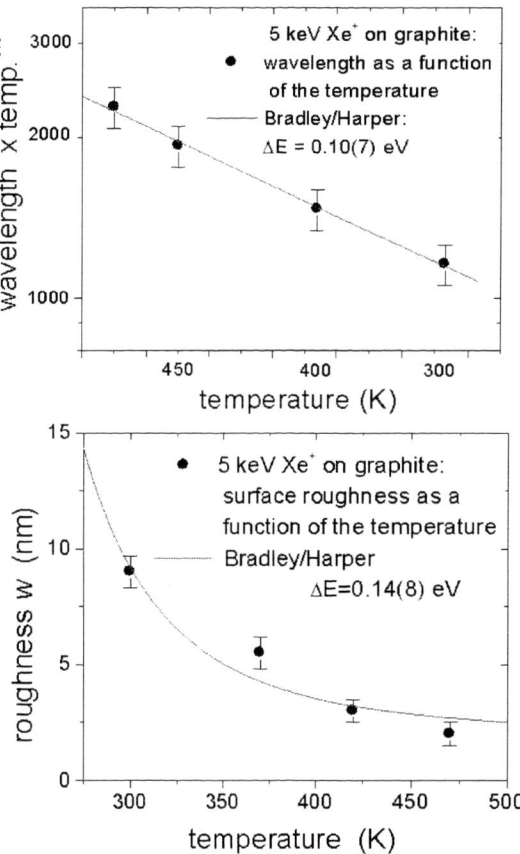

Figure 2: Temperature dependence of the ripple structure for 3×10^{17} Xe-ions/cm^2 hitting HOPG at 5 keV and $\theta = 30°$. The fits correspond to the predictions of the theory by Bradley and Harper [8]. Upper graph: Scaling of the expression $\lambda(T) T^{1/2} \propto \exp(-\Delta E/k_B T)$. Lower graph: Measured and fitted temperature dependence of the roughness $w(T)$.

topic of surface diffusion during ion bombardment, identifying radiation enhanced diffusion to play an important role in the surface evolution process [7]. The present results together with those of previous studies on Xe- and Ar-irradiated HOPG [4-7], lead to the interesting conclusion that the Bradley/Harper theory is able to reproduce the dependence of the ripple structure on all relevant ion parameters, such as the impact angle and the mass, energy and fluence of the beam, at least up to ion fluences of about $10^{18}/\text{cm}^2$. However, as shown in Figs. 1d, the ripple structure dissolves at ion fluences exceeding 2×10^{18} Xe/cm^2, where non-linear effects prevail. For 5 keV Xe ions hitting the surface at $\theta = 60°$ and $70°$ [4],

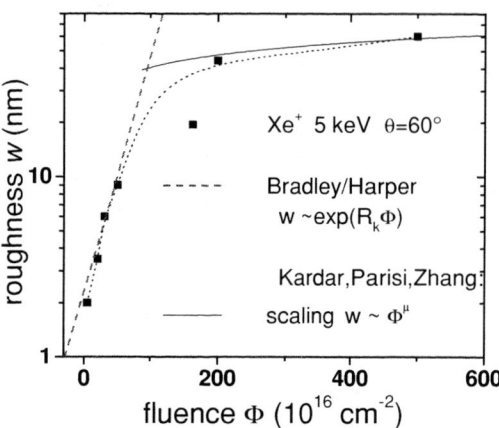

Figure 3: Fluence dependence of the surface roughness w measured at $\theta = 60°$, $E = 5$ keV and $T = 300$ K. Note the transition from the linear regime [$w \propto exp(R_k \Phi)$] to the non-linear regime [$w \propto \Phi^\mu$] which occurs around $\Phi = 2\times10^{18}$ Xe-ions/cm^2.

one finds, indeed, that the structure function can be parametrized by a power-law, $S(k) = k^{-\nu}$, $\nu = 2.7$, as predicted by Cuerno, Barabasi and Eklund and coworkers on the basis of the KPZ theory [14,15,16]. The exponent ν depends on the coupling parameters $\Lambda_{x,y}$. The transition from linear to non-linear behaviour should also be visible in the fluence dependence of the roughness parameter w. As shown in Fig. 3, the roughness increases steeply at low fluences, following the dependence $w \propto exp(R_k\Phi)$ predicted by the linear theory, but merges into a much slower increase at and above 2×10^{18} Xe$^+$-ions/cm^2, following a power-law relation $w \propto \Phi^\mu$ as predicted by the non-linear approach.

In conclusion, sputter erosion of HOPgraphite by heavy ions can be understood in terms of existing theories and has, indeed, been proven to show the predicted dependencies in all details. Further investigations will focus on the limitations of the erosion theory for rising energy regions, where bulk effects like segregation, nucleation and grain growth processes and also spike effects become important and start to superpose the erosion dominated surface morphology. Additionally, new experiments pointing on the dynamical behavior of the surface topography with rising erosion time are in progress. With the help of new and improved observation techniques it will be possible to observe effects like ripple propagation and its velocity dispersion, the role of correlation in surface height evolution and noise influence on surface morphologies in real-time on a microscopic scale [16].

ACKNOWLEDGMENTS

The authors are indebted to Dr. U. Geyer for his help with the STM analyses. The help of Dr. F. Roccaforte and Dr. F. Harbsmeier during the implantation procedure is acknowledged. Furthermore the authors want to thank Prof. W. Bolse for suggestions and discussions during the experiments. This work was funded by the Deutsche Forschungsgemeinschaft.

REFERENCES

1. S. Ghose and S. Karmohapatro, Adv. Electron Electron. Phys. **79** (1990) 73; A. L. Barabasi and H. E. Stanley, *Fractal concepts of surface growth* (Cambride University Press, Cambridge, 1995).
2. K. Reimann, W. Bolse, U. Geyer and K. P. Lieb, Europhys. Lett. **30** (1995), 463; Mat. Res. Soc. Symp. Proc. **354** (1995) 301.
3. W. Bolse, K. Reimann, U. Geyer and K. P. Lieb, Nucl. Instr. Meth. **B118** (1996) 488.
4. S. Habenicht, W. Bolse, K. P. Lieb, K. Reimann and U. Geyer, Phys. Rev. **B60** (1999) R2200.
5. S. Habenicht, U. Geyer, K. P. Lieb, F. Roccaforte and C. Ronning, Nucl. Instr. Meth. **B161-163** (2000) 962.
6. S. Habenicht, H. Feldermann, U. Geyer, H. Hofsäß, K.P. Lieb and F. Roccaforte, Europhys. Lett. **50** (2000) 209.
7. S. Habenicht, Phys. Rev. **B63** (2001), 12519
8. R. M. Bradley and J. M. E. Harper, J. Vac. Sci. Techn. **A6** (1988) 2390.
9. I. Koponen, M. Hautala and O.-P. Sievänen, Phys. Rev. Lett. **78** (1997) 2612; Nucl. Instr. Meth. **B129** (1997) 349.
10. K. B. Winterbon, P. Sigmund and J. B. Sanders, Mat. Fys. Medd. Dan. Vid. Selsk. **37** (14) (1970) 1.
11. D. E. Wolf and J. Villain, Europhys. Lett. **13** (1990) 389.
12. W. W. Mullins, J. Appl. Phys. **28** (1957) 333.
13. S. Habenicht, W. Bolse and K. P. Lieb, Rev. Sci. Instr. **69** (1998) 2120 and references therein.
14. M. Kardar, G. Parisi and Y. Z. Zhang, Phys. Rev. Lett. **56** (1986) 889.
15. R. Cuerno and A. L. Barabasi, Phys. Rev. Lett. **74** (1995) 4746; R. Cuerno, H. A. Makse, S. Tommassone, S. T. Harrington and H. E. Stanley, Phys. Rev. Lett. **75** (1995) 4464.
16. E. A. Eklund, R. Bruinsma, J. Rudnick and R. S. Williams, Phys. Rev. Lett. **67**, 1759 (1991).
17. S. Habenicht, K.P. Lieb, J. Koch and A. D. Wieck, to be published.

Films Formed by Hybrid Pulse Plasma Coating (HPPC) System

K. Awazu[a,c], N. Sakudo[b], H. Yasui[c], E. Saji[d], K. Okazaki[e], Y. Hasegawa[f],
N. Ikenaga[g], T. Sato[h], Y. Nambo[i], K. Saitoh[j]

[a]*The Hokuriku Industrial Advancement Center, Oyama-machi, Kanazawa, Ishikawa 920-0918 JAPAN*
[b]*Kanazawa Institute of Technology, 7-1 Ohgigaoka, Nonoichi, Ishikawa 921-8501 JAPAN*
[c]*Industrial Research Institute of Ishikawa, Ro-1 Tomizu-machi, Kanazawa, Ishikawa 920-0223 JAPAN*
[d]*Industrial Technology Center of Fukui, Kawai-washizuka-machi, Fukui, Fukui 910-0102, JAPAN*
[e]*Fujita Giken Co. Ltd, Terai-machi, Nomigun, Ishikawa 923-1115, JAPAN*
[f]*Onward Ceramic Coating Co. Ltd., Neagari-machi, Nomigun, Ishikawa 929-0111, JAPAN*
[g]*Shibuya Kogyo Co. Ltd., 2-232 Wakamiya, Kanazawa, Ishikawa 920-0054, JAPAN*
[h]*Nachi-Fujikoshi Corp., 1-1-1 Fujikoshi-honmachi, Toyama, Toyama 930-8511, JAPAN*
[i]*Nicca Chemical Co. Ltd., 4-23-1 Bunkyo, Fukui, Fukui 910-8670, JAPAN*
[j]*National Industrial Research Institute of Nagoya, 1-1 Hirate-cho, Kita-ku, Nagoya 462-8510, JAPAN*

Abstract. A new coating system was developed which consists fundamentally of plasma CVD and ion mixing. The system employs plasma-source ion implantation (PSII) combined with pulsed–gas introduction and pulsed-plasma generation. The Hybrid Pulse Plasma coating (HPPC) system is constructed from main three components, that is, a pulsed-gas introducing apparatus, a pulsed RF-microwave generator for high-density plasma and a negative high-voltage pulse power supply for PSII. The plasma densities were measured during the coating process using a Langmuir probe and/or a 10 GHz-microwave interferometer. Each process was monitored as a function of the plasma density as well as the change of the gas pressure. The mechanical properties of formed films were measured and the methods were discussed how to uniformly form films on complicated surfaces of three-dimensional workpieces.

INTRODUCTION

Ion implantation has been researched for improving the wear, corrosion and fatigue resistance and the friction properties of metals since 1970 [1]. Now ion implantation has been applied for extending the lives of forming and cutting tools such as punches, taps and drills [2]. However, the process of ion implantation is a line-of-sight process, so that a uniform implantation has to use beam scanning. This technique is difficult to implant uniformly for 3-dimennsional surfaces of workpieces.

Plasma-source ion implantation (PSII) modified the surface properties of materials by 3-dimensional ion implantation, which of the original technique was developed by Conrad and co-workers [3]. Plasma ions are accelerated across the sheath electric field and perpendicular implanted into the surface. For the practical application of this technique to industrial materials, very high-voltage pulses up to 250kV are required to modify the depth needed for engineering [4]. Furthermore, the technique of plasma-CVD (Chemical Vapor Deposition) has a problem that the molecules are dissociated at the plasma boundary before they reach the workpiece surface, resulting in poor thickness uniformity of deposited layers [5].

The more cost-effective technique for engineering parts with a complex shape, and 3-dimensional surface may be to combine PSII and deposition technique [6][7]. A new coating system was developed which consists fundamentally of plasma-CVD and ion mixing [8]. The system employs PSII combined with pulsed–gas introduction and pulsed-plasma generation. The Hybrid Pulse Plasma Coating (HPPC) system is constructed from main three components, that is, a pulsed-gas introducing apparatus, a pulsed RF-microwave generator for high-density plasma and a negative high-voltage pulse power supply for PSII.

In this paper, plasma densities are measure by using a Langmuir probe and a 10GHz microwave interferometer, and monitoring the plasma densities during the coating process. Diamond-like carbon (DLC) films were formed by the HPPC system. The films are measured with a micro-hardness tester, a scratch tester and the uniformity micro-hardness tester, a scratch tester and the uniformity of the films is tested by model dies developed for this research. The performances of the system were discussed.

EXPERIMENTAL

The hybrid-pulse plasma coating (HPPC) system, as shown in Fig.1, is constructed from main three components, that is, pulsed gas introduction, pulsed RF-microwave coupling to generate higher plasma-density and negative high-voltage pulse power supply to perform ion mixing by PSII. The principle of the HPPC system is presented in detail in ref.8, and the specifications are illustrated in table 1. Space distribution of plasma density was measured at first by a Langmuir probe [5]. During pulsed-plasma the plasma-densities were monitored by a 10GHz-microwave interferometer by timing of the ignition of plasma.

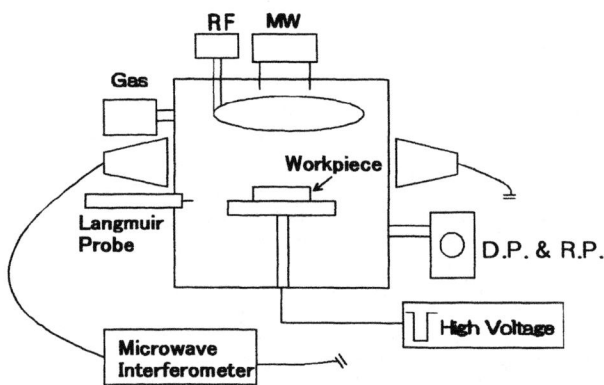

FIGURE 1. Schematic drawing of the HPPC system.

TABLE 1. Specifications of the HPPC system.

Pluse introduction of hydrocarbon gas
    Pulse repetition of gas: 1 Hz
    Cas pressure: 0.3-1.2Pa
Pulse plasma geration
    RF power: max 1.5kW, 13.56MHz
    MW power: max 3.0kW, 2.45GHz
Negative high-voltage generation
    Power: max 40kV-20A
    Pulse width: 5-500 μs
    Pulse repetition: max 1000Hz
Plasma density mesurement
    Langmuir probe: 0.5m
    Microwave interferometer: 10GHz
Vaccum chamber
    Size: 1m × 1m × 1m

Hydrocarbon gases of toluene (C_7H_8) were used as a source of surfaces of workpieces. Gases are introduced with an air valve at 1Hz, and then plasma is ignited and negative high-voltage pulses are applied to the workpieces. Their timings are shown in Fig.2. DLC films were formed on the workpieces by testing for one to two hours. Test conditions are standard following that gas pressure 0.3Pa, pulse width 10 μs and repetition 1000Hz of negative high-voltage pulse.

The hardness tests of DLC films were carried out using an ultra hardness tester (DUH, Shimadzu) with a triangle type indenter. The test load was 9.8 mN and the loading speed was 0.13 mNs^{-1}.

The scratch tests were carried out by a scratch tester with a tip radius of ϕ 0.2mm by applying constant loads of 40, 50, 60, 70 and 80N. After that, the test surfaces were observed by a laser microscope and the critical loads were decided by being or not being the detachments of the film by scratching.

The uniformity of the DLC films is tested using the workpieces and model dies developed for this research.

RESULTS

Plasma densities measured by a 10 GHz microwave interferometer

The distributions of plasma-densities are measured during a pulse of 500ms, in which argon or toluene gas as a carrier gas is introduced in a chamber and both of RF-power 1kW and MW-power 1kW are generated.

Figure 3 shows the relationships between the times of a trigger from the ignition of plasma and the plasma densities measured by the microwave interferometer. The density in argon gas is constant during a pulse of 500ms, but that in toluene gas tends to be smaller as the increase of passing times of a pulse time. This is due to dissolve the

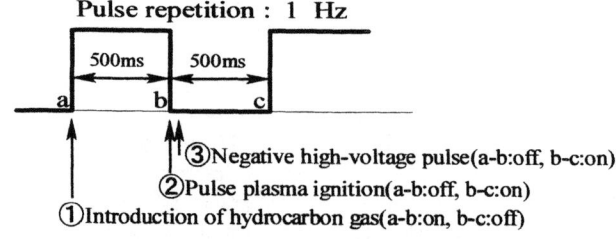

FIGURE 2. Pulse timing in the HPPC system.

toluene to some fragments and to deposit on the surface of the workpieces and/or the walls. So, the width of a pulse time is the smaller, the better is to have the constant density. As the result, the plasma density of pulse plasma is more constant than that of continuous plasma when the hydrocarbons or an organic metal is used as a carrier gas. Therefore, pulse plasma is useful in the case of the hydrocarbon gas as this research.

Figure 4 shows the hardness of DLC films measured by an ultra-micro hardness tester. This hardness evaluates from the depth of indentation with a triangle indenter. DLC films are formed in the conditions of pulse voltage 5.0, 7.5 and 10kV, and RF-power 1kW, toluene gas pressure 0.3Pa. The pulse voltage is the higher; the hardness is the larger, so DLC films are able to be harder when higher pulse voltages are applied on the workpieces. The modules of elasticity are also calculated from the indenting process of an indenter, and shows in the figure. As a result, it is found that high pulse-voltage needs to form hard DLC films

Mechanical properties

Figure 5 shows the relationship between critical loads of the detachment of DLC films by scratching and the pulse voltage for mixing (a), and the tracks observed with a laser microscope after scratching (b). The mixing is a coating process of DLC-film, which consists of the surface finishing, argon bombardment, ion mixing and DLC coating. After the ion mixing is carried out with the pulse voltage of 5, 10 and 15kV for 15 minutes, DLC films are formed on them for 60 minutes. The thickness of their films is approximately 1.5 μm. The critical loads increase to 50 to 70N when the negative pulse voltages for mixing become 5 to 15kV. These values are enough beyond 40 to 50N of critical loads which DLC films on punches and dies are required as an advantage tool. The results of scratch test indicate that the higher mixing voltage is, the more adhesive film forms.

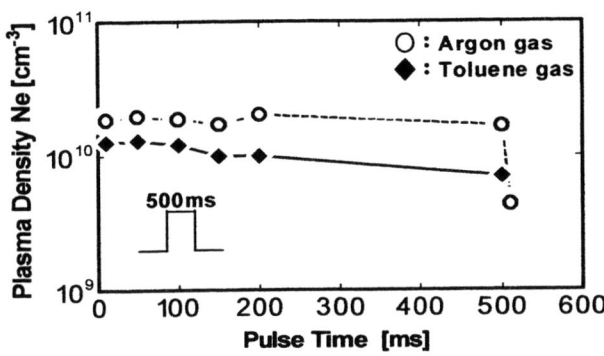

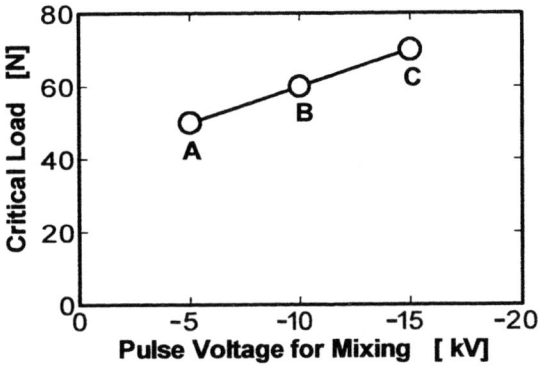

(a) Relationship between critical load and negative-high pulse voltages for mixing. Test duration: 1 hour.

FIGURE 3. Plasma densities during a pulse width of 500ms introducing argon gas and hydrocarbon gas of toluene. (Gas pressure: 0.3Pa, RF: 1kW, MW: 1kW)

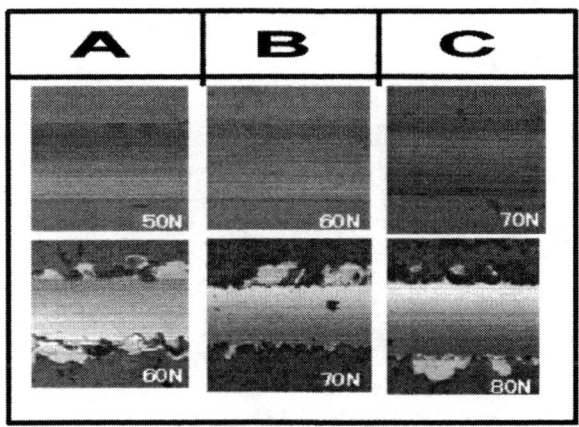

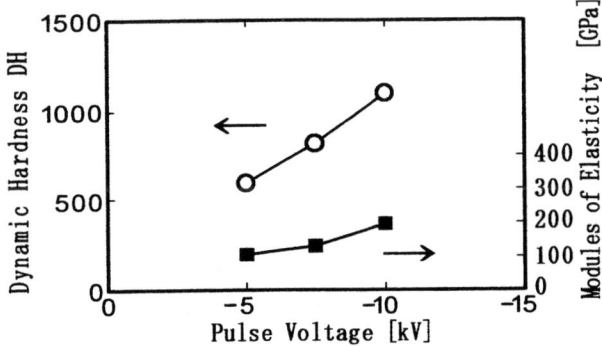

(b) Tracks observed with a laser microscope after scratch tests.

FIGURE 4. Relationships between pulse voltage and hardness and modules of elasticity of DLC films formed by HPPC system.

FIGURE 5. Results of scratch tests for DLC films formed by changing in negative-high pulse voltages at the mixing process.

Uniformity of DLC films

Figure 6 shows the schematic drawing (a) of a model die for testing DLC film-uniformity with aspect ratio (D/d), and the results measuring the thickness of workpieces at a and b. While aspect ratio is up to 4, DLC films are nearly 1.2 μm in thickness and is uniform at the bottom. DLC films at the side are also coated by approximately 0.2 μm for coating time of 1 hour. Therefore, it is found that DLC films are uniformly coated on the surface of the workpieces with 3-dimensinal shapes.

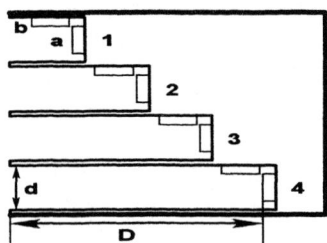

(a) Model dies for testing the aspect ratio D/d.

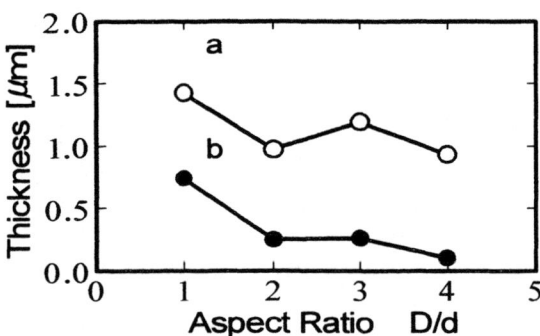

(b) Relationships between aspect ratio and thickness of DLC films at the places of a and b.

FIGURE 6. Uniformity of DLC films formed by the HPPC system.

CONCLUSION

The hybrid pulse plasma coating system, which fundamentally consists of plasma-CVD coating and ion-mixing, is effectively to form the coating films with uniformity and good adhesion on a complexly shaped surface of workpieces. The features of this system are the following 4 points. First is to generate highly dense and uniform plasma for shortening the ion sheath length. Second is to develop a new method of pulse gas introduction for uniformly coating metallic or ceramic layers on a three-dimensional surface. Third is to develop an ion-mixing technology by PSII for good adhesion of the CVD layer to the workpiece. Fourth is to develop a new technique for timing the three pulses for gas introduction, plasma ignition and ion implantation.

ACNOWLEDGEMENTS

This research was financially supported by New Energy and Industrial Technology Development Organization (NEDO).

REFERNCES

1. J. K. Hirvonen, "Ion Implantation, Treatise on Materials Science & Technology", Vol.18, Academic Press, New York (1980).
2. B. Torp, B. R. Nielsen, N. J. Mikkelsen, C. Astade, Surf. & Coat. Tech., **84**, 557 (1996).
3. R. Conrad, J. L. Radtke, R. A. Dodd, F. J. Worzala and N. C. Tran, J. Appl. Phys., **62**, 4591 (1987)
4. J. N. Matossian, R. Wei and J. D. Williams, Surf. & Coat. Tech., **96**, 58 (1997).
5. N. Sakudo, K. Hayashi, Y. Nishiyama, K. Komatsu, J. Miyamoto, M. Yutani and K. Awazu, "Special Distribution of Ion Species in a Source Plasma for a Broad Beam", to be published.
6. D. J. Rej and R. B. Alexander, J. Vac. Technol., B12, 2380 (1994).
7. J. V. Mantese, I. G. Brown, N. W. Cheung and G. A. Collins, MRS Bulletin, 52 (1996-8).
8. N. Sakudo, K. Awazu, H. Yasui, E. Saji, K. Okazaki, Y. Hasegawa, N. Ikenaga, K. Kanda, Y. Nambo and K. Saitoh, "Development of Hybrid Pulse Plasma Coating System", to be published.

Chemical Functionalization And Modification Of Carbon Nanotubes Through Ion Bombardment

Boris Ni and Susan B. Sinnott*

Department of Materials Science and Engineering, University of Florida, Gainesville, Florida 32611-6400

Abstract. Classical molecular dynamics simulations have been performed to investigate the chemical functionalization of single-walled (SWNT) and double-walled carbon nanotubes (DWNT) through CH_3^+ ion bombardment at 10, 45, and 80 eV. The simulations show that the process is highly efficient and that chemical functionalization occurs at every incident ion energy considered. However, significant differences in the response of the SWNTs and DWNTs are predicted from the simulations. At 45 and 80 eV defect formation and cross-linking between nearby nanotubes occurs. These new defect structures could substantially alter the mechanical and electrical properties of nanotubes

INTRODUCTION

Carbon nanotubes are widely considered to be promising materials for a wide variety of engineering applications due to their unique mechanical and electrical properties (1). However, to use nanotubes in engineering applications it is necessary to control their structure and in some cases modify this structure in a reproducible manner. These modifications include chemical functionalization with various chemical groups (2). Our previous computational investigation of single-walled nanotubes (SWNTs) functionalization by energetic ion collision showed that adhesion of CH_3^+ ions or heavy fragments such as CH_2 and CH takes place at all the energies considered (3). In addition, cross-linking between neighboring nanotubes was induced at incident ion energies of 80 eV (3). Other methods that have been shown to chemical functionalize carbon nanotubes make use of carbodiimide chemistry (4) or electrophilic reactants (5,6). However, Ref. 3 is the first to predict a mechanism for cross-link formation in a nanotube bundle.

The objective of the present work is to make a comparative study of the functionalization of SWNTs and double-walled nanotubes (DWNTs) through ion bombardment. We expect that since the elastic properties of SWNTs are different from those of DWNTs (7) the results of bombardment could potentially be quite different.

COMPUTATIONAL DETAILS

The approach used in this computational study is classical molecular dynamics simulations. Specifically, a third-order Nordsieck predictor corrector (8) is used with a time step of 0.2 fs. The forces are determined using a reactive empirical bond order potential for hydrocarbons (9-11) that has been widely used by ourselves and others to study carbon nanotubes (12-16) and ion-deposition processes (17-20). The long-range van der Waals interactions between the nanotube walls are characterized with a Lennard-Jones potential that is nonzero only after the covalent interactions have gone to zero. The combined expression for calculating the binding energy is therefore

$$V_b = \sum_i \sum_{j<I} [\, V_r(r_{ij}) - B_{ij}\, V_a(r_{ij}) + V_{vdw}(r_{ij})\,]$$

where V_b is the binding energy, r_{ij} is the distance between atoms i and j, V_r is a pair-additive term that takes into account the interatomic core-core repulsive interactions, V_a is a pair-additive term that models the attractive interaction due to the valence electrons. The term B_{ij} is a many-body empirical bond-order term that

*Corresponding author

modulates valance electron densities and depends on atomic coordination and bond angles. Finally, V_{vdw} is the Lennard-Jones potential. It should be pointed out here that this potential does not account for changes in the electronic structure of the atoms as a result of high-energy deposition. The assumption taken here and elsewhere (21,22) is that the impact energies are low enough that electronic excitations do not play a significant role in the chemical reactions that take place following the collisions.

The initial arrangement of carbon nanotubes is shown in Figure 1. They are arranged in a close-packed bundle and each nanotube in the bundle is 50 Å long. Therefore each SWNT consists of 800 atoms while each DWNT consists of 1202 atoms. The SWNTs considered in all of the reported simulations are (10,10) armchair nanotubes while the DWNTs are (5,5) within (10,10) armchair nanotubes.

About 5 Å at the edge of every nanotube had Langevin frictional forces (8) applied to mimic the heat dissipation properties of a real nanotube bundle. No other constrains have been set up on the system. Periodic boundary conditions were also applied to simulate nanotubes that are considerably longer than 50 Å.

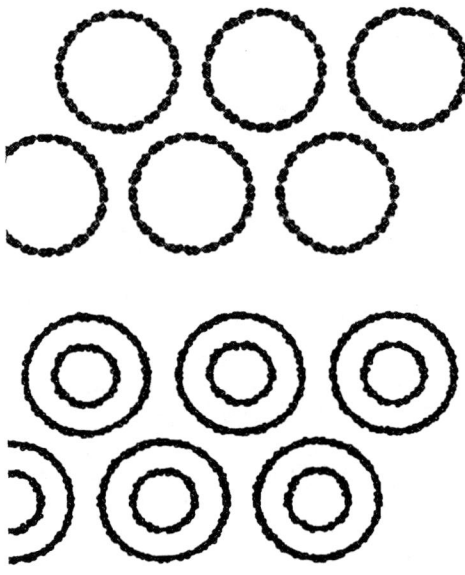

FIGURE 1. Initial structures of carbon nanotube bundles (SWNTs and DWNTs) used in the simulations.

RESULTS AND DISCUSSION

A comprehensive summary of the simulation results is presented in Table 1. The reported data is the averaged results of 35-40 trajectories for each system and incident energy. When the CH_3^+ ions are deposited at incident energies of 10 eV only two kinds of phenomena are predicted from the simulations. The first one is scattering of the intact CH_3^+ ions from nanotube without any reaction having taken place while the second event is the adsorption of CH_3 to the external nanotube wall. At this energy the simulations predict that the functionalization process is highly efficient with about 70% of impacting ions bonding to the walls of both SWNTs and DWNTs. A representative snapshot of the outcome of a typical trajectory is shown in Figure 2(a). The deposition process does not damage the nanotube walls for either the SWNTs or the DWNTs except for very slight local deformation due to the alteration of adsorptive site hybridization from sp^2 to sp^3 and subsequent local change of bondlengths from 1.42 Å to 1.55 Å. At such low energy the impacting ions do not probe the elasticity of the nanotubes which is why there is almost no difference in the statistical spread of the data in Table 1.

This is not the case at 45 eV of incident energy where most of the ions loose one or two of their hydrogen atoms on impact with the carbon nanotube bundle. In addition, about half of the time (52% for SW and 48% for DW) the larger ion fragments such as CH and CH_2 adsorbed on the outer surface. Some of these fragments bond to two or three nanotube carbon atoms which stabilizes the adsorbed functional group (see Figure 2(b)) at the surface. For both SWNTs and DWNTs less then 25% of the total number of fragments are reflected from the walls without forming a chemical bond. In about 19% of cases for SWNTs and only about 1% of time for DWNTs hydrocarbon fragments from the ion insert carbon atoms into the nanotube wall structure which creates usual defect structures (Figure 2(c) shows a common example where heptagons are created). In the case of the DWNTs 20% of time the ion deposition process deforms the outer wall of the nanotube so that one or two carbon atoms are pushed out of the outer wall to stick in between the two walls. Sometimes that atom that is pushed out is the carbon atom from the incident ion. This creates cross-links between the walls, as shown in Figure 2(d). Because the SWNTs consist of only one wall, similar processes result in adsorption on the outer wall or insertion into the wall rather than cross-link formation. These interwall covalent connections are of interest because it is known that electrical conductivity in multiwalled nanotubes is provided by the outer shell only (23). In addition it is generally accepted that the junctions of two different nanotubes of different helicity will exhibit diode conductivity (24). Therefore we suppose that cross-linking such as that shown here could substantially alter not only the mechanical properties of the nanotubes but also their electrical properties.

At 80 eV only 8% and 4% of all the fragments are scattered away after deposition for the SWNTs and DWNTs, respectively. The rest of the time the incident

TABLE 1. Percentage of events taking place in ionic CH_3^+ bombardment of a bundle of SWNTs and DWNTs.

	SWNT			DWNT		
	10eV	45eV	80eV	10eV	45eV	80eV
Scattering of CH_3	24.0	8.0	4.0	31.7	11.7	3.3
Scattering of CH_2		9.3	2.7			
Scattering of CH or C		6.7	1.3		12.5	0.8
Adsorption of CH_3 on outside wall	76.0	5.3	-	68.3	4.2	0.8
Adsorption of CH_2 on outside wall		24.0	4.0		23.3	1.7
Adsorption of CH on outside wall		28.0	10.7		25.0	9.2
Adsorption of C on outside wall			26.7		2.5	19.2
Adsorption of C on inside wall or between walls			26.7		20.0	46.7
Adsorption of C under 2nd wall						2.5
Incorporation of C into outer wall with defects		18.7	24.0		0.8	15.8

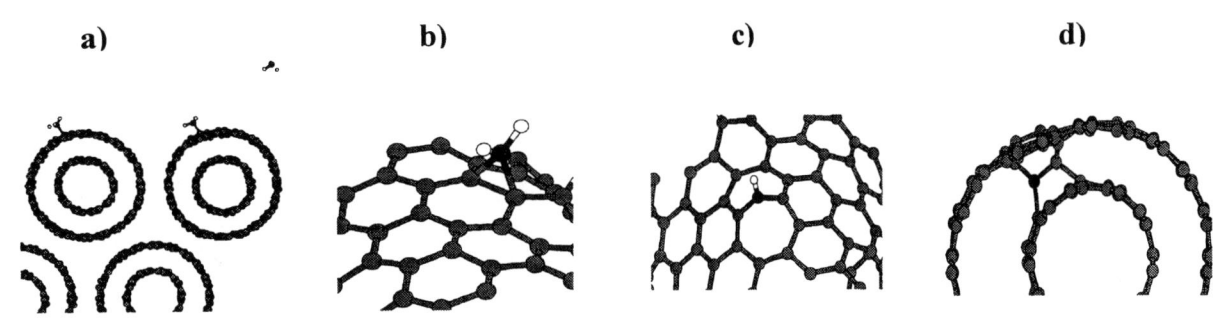

FIGURE 2: Snapshots of typical events taking place during ion deposition on carbon nanotube bundles a) DWNT at 10eV; b) SWNT at 45 eV, c) SWNT at 45 eV, and d) DWNT at 45 eV. The darker atoms are carbon atoms from the incident CH_3^+ ions.

ion or, much more likely, fragments of it adhered to the nanotube walls. However the most common occurrence is the insertion of single carbon atoms from the incident ion into the outer wall of the nanotube or their "knocking out" and exchange with an atom in the outer wall. This leads the knocked out atom to bond on the inside of the wall in the case of SWNTs and to bond between the walls, with possible cross-link formation, in the case of the DWNTs. The difference between the elastic properties of SWNTs and DWNTs is reflected in the percentage of carbon atoms that lead to carbon adsorbed on the inside wall or between the walls. The SWNTs deforms more easily on impact and, by the same token, returns to its original shape more easily. In contrast, the repulsive van der Waals forces between the inner and outer shells of the DWNTs prevent easy deformation of the outer shell. Therefore, carbon atom(s) knocked out of the outer shell usually stick in between walls the walls. In very rare cases, about 2.5% of the time, the carbon atom knocked out of the outer shell by the incoming ion is able to knock out a carbon atom out of inner wall and take its place. The carbon atom from inner wall is subsequently adsorbed on the interior of the inner shell of the DWNT. In addition to these events, at 80 eV the simulations predict that knockout events occur that lead to cross-link formation between neighboring nanotubes. These inter-nanotube cross-links could be used to stabilize the nanotube bundle with respect to shear, something that is not otherwise possible for a pure nanotube bundle.

CONCLUSIONS

In this paper we report on the results of classical molecular dynamics simulations that were used to study the chemical functionalization of SWNT and DWNT bundles through ion deposition. Incident ion energies of 10, 45 and 80 eV were considered in the study. The simulations predict that at 10 eV the CH_3^+ ion bonds chemically to the outer walls of the nanotubes the majority of the time. Only negligible differences in the results are predicted for the SWNTs and DWNTs. In contrast, at 45 eV the incident ion is fragmented on impact the majority of the time and these fragments form chemical bonds to the walls, form a variety of defect structures, and cause cross-linking between the inner and outer walls of the DWNTs. At 80 eV the incident fragments into even smaller fragments the overwhelming majority of the time. These ion fragments induce many of the same events that are predicted to occur at 45 eV with the additional formation of cross-links between the nanotubes in the bundle. These inter-nanotube cross-links are predicted to occur for both SWNTs and DWNTs.

ACKNOWLEDGMENTS

We gratefully acknowledge support from the National Science Foundation (CHE-9708049) and the Advanced Carbon Materials Center at the University of Kentucky which is supported by the NSF (DMR-9809686). This work was also supported in part by a grant from the NASA Ames Research Center (NAG 2-1121).

REFERENCES

1. Ebbesen, T.W., *Physics Today* **49**, number 6, 26, 1996.

2. See, for example, Yakobson, B.I., and Smalley, R.E., *American Scientist* **85**, 324, 1997.

3. Ni, B., and Sinnott, S.B., *Phys. Rev. B* **61**, R16343 (2000).

4. Liu, J., Rinzler, A.G., Dai, H., Hafner, J.H, Bradley, R.K., Boul, P.J., Lu, A., Iverson, T., Shelimov, K., Huffman, C.B., Rodriguez-Macias, F., Shon, Y.-S., Lee, T.R., Colbert, D.T., and Smalley., R.E., *Science* **280**, 1253, 1998.

5. Chen, J., Hamon, M.A., Hu, H., Chen, Y., Rao, A.M., Eklund, P.C., and Haddon, R.C., *Science* **282**, 95, 1998.

6. Wong, S.S., Joselevich, E., Woolley, A.T., Cheung, C.L., and Lieber, C.M., *Nature* **394**, 52, 1998.

7. Hertel, T., Martel R., and Avouris, P., *J. Chem. Phys. B* **102**, 910, 1998.

8. *Computer Simulation of Liquids*, M. P. Allen and D. J. Tildesley, Oxford University Press, New York, 1987.

9. Brenner, D.W., *Phys. Rev. B* **42**, 9458, 1990.

10. D.W. Brenner, O.A. Shenderova, S.B. Sinnott, J.A. Harrison, *J. Phys. C: Condensed Matter* (to be submitted).

11. Brenner, D.W., *Phys. Status Solidi B* **217**, 23, 2000.

12. Garg, A., Han, J., and Sinnott, S.B., *Phys. Rev. Lett.* **81**, 2260, 1998.

13. Yakobson, B.I., Brabec, C.J., and Bernholc, J., *Phys. Rev. Lett.* **76**, 2511, 1996.

14. Cornwell, C.F., and Wille, L.T., *Solid State Commun.* **101**, 555, 1997.

15. Harrison, J.A., Stuart, S.J., Robertson, D.H., and White, C.T., *J. Phys. Chem. B* **101**, 9682, 1997.

16. Yakobson, B.I., Campbell, M.P., Brabec, C.J., and Bernholc, J., *Comp. Mat. Sci.* **8**, 341, 1997.

17. Wijesundara, M.B.J., Hanley, L., Ni, B., and Sinnott, S.B., *Proceedings of the National Academy of Science, USA* **97**, 23, 2000.

18. Wijesundara, M.B.J., Ji, Y., Ni, B., Sinnott, S.B., and Hanley, L., *Journal of Applied Physics* (in press).

19. Brenner, D.W., Shendrova, O.A., and Parker, C.B., *Mat. Res. Soc. Symp. Proc.* **438**, 491, 1997.

20. Kerford, M., and Webb, R.P., *Nucl. Instrum. Meth. B* **153**, 270, 1999.

21. Barone, ME., and Graves, D.B., *J. Appl. Phys.* **77**, 1263, 1995.

22. Kwon, Y.-K., and Tomanek, D., *Phys. Rev. B* **58**, R16001, 1998.

23. Chico, L., Crespi, V.H., Benedict, L.X., Louie, S.G., and Cohen, M.L., *Phys. Rev. Lett.* **76**, 971, 1996.

DLC Film Formation by Ar Cluster Ion Beam Assisted Deposition

Teruyuki Kitagawa [a,b], Isao Yamada [b], Jiro Matsuo [c], Allen Kirkpatrick [d] and Gikan H. Takaoka [c]

[a] *Nomura Plating Co. Ltd., Nishiyodogawa, Osaka 555-0033 JAPAN*
[b] *Laboratory of Advanced Science and Technology for Industry, Himeji institute of Technology, Ako-gun, Hyogo 678-1205 JAPAN*
[c] *Ion beam engineering experimental laboratory, Kyoto University, Sakyo, Kyoto 606-8501 JAPAN*
[d] *Epion Corporation, Billerica MA 01821 USA*

Abstract. A novel method employing Gas Cluster Ion Beam (GCIB) bombardment during deposition of C_{60} is proposed for forming DLC film. This method, which offers low process temperatures and compatibility with large substrate areas, produces pure carbon films exhibiting outstanding physical characteristics. The impacts of energetic gas cluster ion result in deposition of extremely high energy densities into very localized and shallow atomic level regions of a target surface. These conditions are able to cause transformation of C_{60} into an ultra-hard form of carbon. Example DLC films have been deposited onto various substrates at room temperature using Ar gas cluster ions at 3 to 9 keV. Films deposited using Ar cluster ion energy of 7 keV exhibited Vickers hardness value of 50GPa, compared with typical values of 30 GPa for DLC films prepared by other methods.

INTRODUCTION

DLC (Diamond-like carbon) films have been widely used in the last few years for a number of applications, because of their superior properties, which include not only high hardness but also low friction coefficient, transparency and chemical stability. DLC films have been produced by numerous vapor phase methods [1, 2], but they present various problems for practical use. To obtain DLC film with high hardness, the substrate has to be heated to high temperature during the deposition. Consequently, adherence to substrate with high thermal expansion coefficient is poor, and intermediate layers are normally used. As well, it is difficult to obtain large area deposition of uniform quality by various CVD processes, which are the mainstream methods of DLC formation. On the other hand, in the case of PVD and ion beam assisted methods it is difficult to obtain a smooth surface and to achieve DLC with high hardness because of the damage from ion-surface interactions.

Thus, because of inherent problems in the deposition methods, DLC films produced by present technologies have a limited range of applications, despite the fact that in general DLC films have higher application potential than most of the other hard coating materials. In view of the remarkable recent progress in the field of electronics, hard coatings with good properties are required for various devices in the next generation [3], and thus improvements in DLC deposition technologies seem to be required.

As a breakthrough in DLC synthesis, a novel method employing gas cluster ion beam (GCIB) assisted deposition is proposed. In this method, energetic gas cluster ions deliver extremely high energy densities into very localized and shallow atomic level regions of a substrate surface [4]. Consequently, because the impact point of a cluster ion for an instant attains conditions of high pressure and high temperature [5], it is considered that the phase transition from sp^2 to sp^3 is enhanced even when the substrate is held at room temperature. As well, it is expected that films with quite smooth surfaces can be achieved by the lateral sputtering effect that is a unique characteristic of cluster ion bombardment [6,7]. Moreover, a large area deposition of uniform quality is easily obtained by scanning the cluster ion beam.

EXPERIMENT

Figure 1 shows a schematic diagram of the DLC deposition apparatus. Films were synthesized by simultaneously irradiating Ar cluster ion beams onto the room temperature substrates and evaporating fullerene as a carbon source.

A neutral Ar cluster beams is formed by adiabatic expansion of gases through a small nozzle into high vacuum [8], and the beam is ionized by electron bombardment. Monomer ions and small cluster ions are eliminated by the electrical fields of the ionizer and the retarding potential technique [9]. For these experiments the ionization voltage was 150 V, the electron current was 160 mA, and the Ar cluster ions were accelerated up to 9 kV. The mean size of Ar cluster ions was 1,000 atoms/ion, therefore the average energy per constituent atom was less than 9 eV. The current density of Ar cluster ion beam was more than 3 $\mu A/cm^2$ for substrate of 9 cm^2, with acceleration voltage above 7 kV.

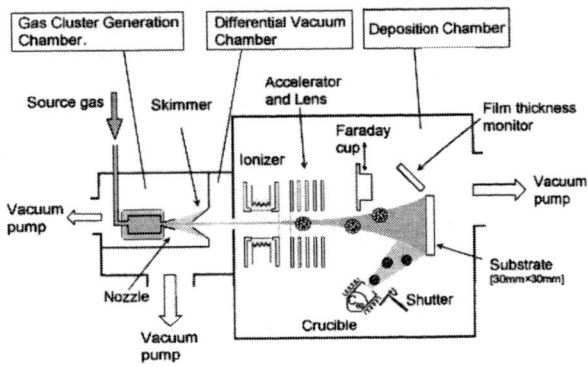

FIGURE 1. Schematic diagram of gas cluster ion beam assisted deposition apparatus.

Fullerene, consisting mainly of C_{60} was evaporated from a heated crucible. Fullerene presents some advantages for use in deposition. It doesn't consist of sp^2 orbitals that are typical to the structure of graphite [10]; also, the deposition is performed with hydrogen-free material. It has been reported that DLC films obtained with hydrogen-free material have higher thermal stability than those with hydrogen [11]. Because fullerene has lower sublimation temperature than that other carbon materials [12], it is also relatively easy to maintain a constant evaporation rate. Furthermore, due to recent improvements in fullerene production it is easy to obtain fullerene with a high content of C_{60} [13].

The films were deposited in a vacuum better than 1×10^{-5} torr and the substrate was kept at room temperature during the deposition. The substrates were silicon, steel and iron. The number of fullerene molecules was measured during the deposition using a film thickness monitor located outside beam area (see figure 1). In this study we used the parameter R, defined as the ratio of fullerene particles to Ar cluster ion directed to the substrate, which is represented by the current density at the substrate surface.

The hardness of the films was measured using a Micro-Vickers hardness tester with 15 g load and a holding time of 15 seconds, which gives a indentation small enough (0.2μm) not to be affected by the substrate, as the thickness of the films was above 1 μm. The film adhesion on the various substrates was measured using a scratch tester at 30 N load, 30 N/min loading speed, at a speed of 10 mm/min. The binding structure was evaluated using Raman spectroscopy. Average surface roughness (Ra) of the films was assessed using atomic force microscope (AFM).

RESULTS AND DISCUSSION

3.1. Optimization of Forming DLC Films

It has been experimentally demonstrated that the sputtering yields of various materials with Ar cluster ion were one order of magnitude higher than those with Ar monomer ion [14]. In the case of fullerene deposition under cluster ion bombardment, the phenomenon of high physical sputtering with Ar cluster ions was also taking place, and therefore, the ratio of Ar cluster ion to fullerene molecules had to be optimized. Ar cluster ion with 7 keV energy, which was considered appropriate energy, was used for the optimized condition. Figure 2 shows the dependence of R on the film growth rate. The film growth rate (vertical axis) represents the value of the film thickness actually measured on the substrate, divided by that measured by the film thickness monitor. In the

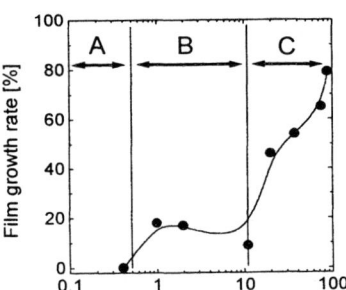

FIGURE 2. Fullerene/Ar cluster ion ratio dependence on film growth rate.

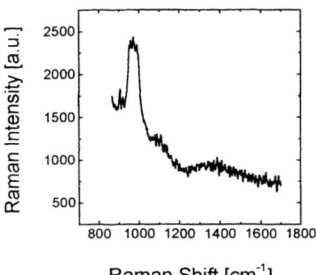 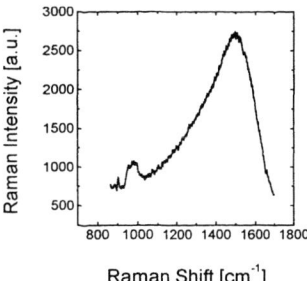 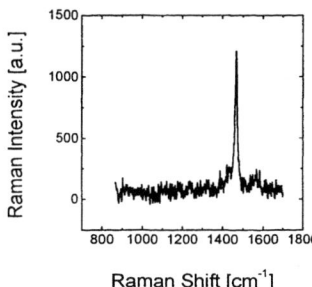

FIGURE 3. Raman spectra of the substrate treated under each condition of fullerene/Ar cluster ion ratio. From left to right: the Raman spectra under A, B and C conditions respectively.

region A of the graph, R was less than 0.5, and no film was deposited on the substrate. In the region B, R was between 0.5 and 10, and the maximal film growth rate was approximately 20 %. It can be observed that in region C of the graph, where R-value was above 10, the film growth rate increases with increasing R, finally reaching 80 % when R ratio is 100.

Figure 3 shows the Raman spectra of the substrate treated under each of the A, B and C conditions. The typical Raman spectrum for the substrates treated under condition A shows only a silicon peak of the substrate material, meaning that the deposited carbon was removed from the substrate by the sputtering of Ar cluster ions. The typical Raman spectrum of substrates treated under condition B shows the typical peak of DLC similar to that obtained by various other DLC deposition methods [1, 2], indicating that DLC was formed under these conditions. The typical Raman spectrum of substrates treated under conditions from region C shows only the C_{60} peak of fullerene [15], indicating that DLC film was not formed under these conditions.

From these results, the mechanism of interactions between energetic Ar cluster ion and fullerene molecule was considered as follows: Starting in condition C from a ratio of about 1 % Ar cluster (at a fullerene/Ar cluster ion ratio of 100), an increase in Ar cluster ion content vs. fullerene causes a decrease in fullerene film thickness because of the sputtering phenomenon. In condition B however, the interaction between the Ar cluster ion and the fullerene molecule results not only in sputtering but also in a phase transition from fullerene to DLC. If the Ar cluster ion ratio is increased further, it is considered that the DLC film, which is formed by irradiation with Ar cluster ion, is completely sputtered away. Because the binding energy of DLC is stronger than that of the fullerene molecule, it is considered that the sputtering yield of DLC is lower than that of fullerene. Consequently, in condition B, a steady state thickness is obtained for R-values between 0.5 and 10 under these experimental conditions.

3.2. Characteristics of DLC Films

Figure 4 shows the Vickers hardness of DLC films formed on silicon substrate with varied Ar cluster ion acceleration energy from 5 keV to 9 keV. The R-value was kept at the optimized conditions. When the DLC film was formed with 7 keV Ar cluster ion, the Vickers hardness reached a maximum of 50 GPa (5,000kg/mm^2). DLC films formed under other acceleration energy conditions exhibited Vickers hardness of over 30 GPa (3,000kg/mm^2).

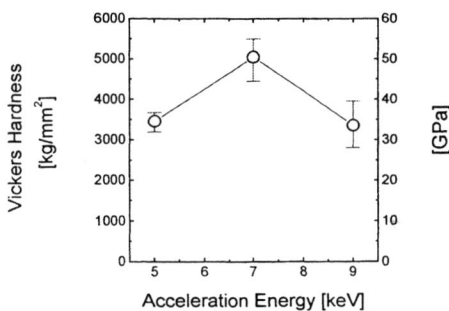

FIGURE 4. Ar cluster acceleration energy dependence on Vickers hardness of DLC films.

Figure 5 shows the critical adhesion values of the films. The adhesion values of DLC films formed by the cluster method on silicon substrate were approximately the same as those of prior-method DLC on the same substrate. In the case of steel and iron substrates, the values were half those of DLC films on silicon substrates. However, DLC films formed by the CVD process could not be directly adhered on metal substrates because of the large mismatch in thermal expansion coefficient.

Table 1 shows the average surface roughness (Ra) of films and the substrate. The value of DLC films formed at varied cluster acceleration energy was less than 5 Å. These results indicate that the substrate surface (Ra 4 Å) is not roughened by the deposition. In the case of non-optimized DLC conditions, films show an average roughness of 21 Å, which is significantly rougher than that of the DLC film.

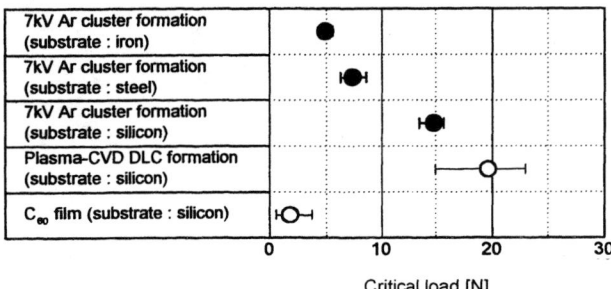

FIGURE 5. Film adhesion onto various substrates.

TABLE 1. Average surface roughness (Ra) of the DLC films and substrate and the film under non-optimized condition, measured using AFM.

sample	Average roughness (Ra)
DLC deposition with 5 keV Ar cluster ion	2 Å
DLC deposition with 7 keV Ar cluster ion	5 Å
DLC deposition with 9 keV Ar cluster ion	3 Å
Film deposition under non-optimized DLC forming condition	21 Å
Silicon used as a substrate	4 Å

CONCLUSIONS

DLC films with Hv of 50 GPa were deposited by a novel method of cluster ion beam assisted deposition. These films exhibit smooth surface and good properties of hardness and adhesion, combined with deposition at low temperature – a feature that enables deposition on substrates with high thermal expansion coefficients such as steel and other metals.

The high-density irradiation on a solid surface with a cluster ion was capable to produce the phase transition from fullerene to DLC films with high hardness under particular conditions. Further investigation will be carried out varying acceleration energy and cluster size to seek further improvement of the films.

ACKNOWLEDGMENTS

This paper has been edited for publication by http://editscience.com.

REFERENCES

1. J. Ullmann, Nucl. Instr. and Meth. in Phys. Res. B, **127**, 910 (1997).

2. H. Tsai and D.B. Bogy, J. Vac. Sci. Technol. A, **5**, 3287 (1987).

3. T.W. Scharf, R.D. Ott, D. Yang and J.A. Barnard, J. Appl. Phys., **85**, 3142 (1999).

4. T. Aoki, J. Matsuo, Z. Insepov and I. Yamada, Nucl. Instr. and Meth. in Phys. Res. B, **121**, 49 (1997).

5. Z. Insepov and I. Yamada, Nucl. Instr. and Meth. in Phys. Res. B, **112**, 16 (1996).

6. H. Kitani, N. Toyoda, J. Matsuo and I. Yamada, Instr. and Meth. in Phys. Res. B, **121**, 489 (1997).

7. N. Toyoda, H. Kitani, N. Hagiwara, T. Aoki, J. Matsuo and I. Yamada, Mater. Chem. And Phys., **54**, 262 (1998).

8. N. Toyoda, M. Saito, N. Hagiwara, J. Matsuo and I. Yamada, Proceedings of the 12-th International Conference on Ion Implantation Technology, Kyoto, Japan, IEEE, 1234 (1998).

9. H. Katsumata, J. Matsuo, T. Nishihara, T. Tachibana, E. Minami, K. Yamada, M. Adachi and I. Yamada, Proceedings of the 12-th International Conference on Ion Implantation Technology, Kyoto, Japan, IEEE, 1195 (1998).

10. R. C. Haddon, Acc. Chem. Res. **25**, 127 (1992).

11. A. A. Voevodin, S. D. Walck, J. S. Solomon, P. J. John, D. C. Ingram, M. S. Donley, and J. S. Zabinski, J. Vac. Sci. Technol. A **14**, 1927 (1996).

12. J. Abrefah, D. R. Olander, M. Balooch, and W. J. Siekhaus, Appl. Phys. Lett. **60**, 1313 (1992).

13. Y. Saito, M. Inagaki, H. Shinohara, H. Nagashima, M. Ohkohchi and Y. Ando, Chem. Phys. Lett., **200**, 643 (1992).

14. J. Matsuo, N. Toyoda, M. Akizuki, and I. Yamada, Nucl. Instr. and Meth. in Phys. Res. B, **121**, 459 (1997).

15. H.G. Busmann, H. Gaber, H. Strasser and I. V. Hertel, Appl. Phys. Lett., **64**, 43 (1994).

Molecular Dynamics Simulations of Cluster Ion Implantation for Shallow Junction Formation

Takaaki Aoki, Jiro Matsuo Gikan Takaoka and Isao Yamada [†]

Ion Beam Engineering Experimental Laboratory, Kyoto University
Sakyo, Kyoto 606-8501, Japan

Abstract. Cluster ion implantation is a useful technique for low energy implantation. In order to investigate the implant and damage formation by cluster ions, molecular dynamics simulations of boron monomer (B_1) and cluster with the size ranging from 2 (B_2) to 10 (B_{10}) impacting on silicon substrate were performed. When atom/clusters with the same energy per atom of 230eV/atom, all impacts show a similar implant profile, except for a vertical chain-like cluster with the size larger than 4. The latter finding is attributed to the clearing way effect. B_{10} impact induces displacements several times larger than B_1; this is because the high-density particle and energy deposition by B_{10} causes more knocked-on substrate atoms in the shallow substrate region and these knocked-on atoms tend to remain as displacements, an effect which is not observed in impacts of B_1. This higher displacement yield by B_{10} is believed to avoid the formation of point defects at low ion dose and to suppress transient enhanced diffusion.

INTRODUCTION

As the scale of LSI device decreases, the formation of high-quality shallow p-type junction becomes more important. In order to fabricate a sub-0.1μm p-MOS device, boron atoms are expected to be implanted with less than 1keV. However, as the energy of implant energy decreases, it becomes more difficult to obtain enough current for industrial fabrication because of the space charge effect [1]. Furthermore, the transient enhanced diffusion (TED) of dopant boron atoms becomes a serious problem as the incident energy decreases [2].

The boron cluster ion implantation technique using decaborane ($B_{10}H_{14}$) has been proposed as a solution for shallow junction formation [3]. It has been experimentally observed that the implant range of B atoms using the $B_{10}H_{14}$ ion implantation is equivalent to that of monomer B ions accelerated with 1/10 of the energy of $B_{10}H_{14}$. There are other advantages: cluster ion implantation is expected to have a nonlinear effect caused by the high-density irradiation of incident atoms. As shown in previous works of small carbon and cluster irradiation [4,5], $B_{10}H_{14}$ should be considered as a material on the border between cluster and monomer. It is important to examine the similarity and differences between $B_{10}H_{14}$ and B_1 ion implantation.

In this paper, molecular dynamics (MD) simulations of small boron cluster with the size less than 10, and boron monomer implantation into Si(001) substrate were performed. The dependencies of the implant depth, implant efficiency and damage formation on the size and structure of clusters are examined and the advantages of cluster ion implantation technique are discussed.

SIMULATION MODEL

In order to examine the implant process of small boron clusters, the MD simulations of B_1, B_2, B_4 and B_{10} monomer/cluster impacting on a Si(001) substrate were performed. The Stillinger-Weber potential model [6] was applied to the inter-atomic potential of Si-Si, and the Ziegler, Biersack, and Littmark (ZBL) model [7] was applied to B-B and B-Si potentials. A Si(001) substrate was prepared as a target material, which consists of 32768 atoms with a cube side of about 90Å

[†] Present Address: *Laboratory of Advanced Science and Technology for Industry, Himeji Institute of Technology, CAST, Kamigori, Ako, Hyogo, 678-1205, Japan*

TABLE 1. Structure, mean implant depth and implant efficiency of boron cluster.

	Structure of Cluster	Implant Depth [Å]
		Implant Efficiency

Num. of Stacks	B_1		B_2		B_4		B_{10}	
1	o	18.625	o-o	20.325	o-o-o-o	18.25	o-o o-o	18.2
		0.83		0.835		0.79		0.784
					⊙⊙⊙	23.125		
						0.85		
2			⊙/⊙	22.287	⊙⊙/⊙⊙	21.4	(cluster)	19.03
				0.885		0.89		
3								0.804
4					⊙/⊙/⊙/⊙	29.025		
						0.98		
10							⊙/⊙/⊙	32.1
								0.972

and the periodic boundary conditions were applied on this target.

In this work, B_1, B_2, B_4 and B_{10} are radiated on the Si substrate with an incident energy of 230eV/atom. As for B_2, B_4 and B_{10} impact, these clusters are radiated with different geometries, as shown in table 1. The reason for this is that, as the cluster size decreases, the structure and orientation of the cluster at impact becomes more significant, whereas large clusters have a spherical structure for which the orientation is undefined.

B_1 monomer and vertical B_2, B_4 and B_{10} chains are implanted with an incident angle of 7° to the surface normal and rotated 30° to the (001) direction to avoid channeling, and other clusters are implanted at normal direction. In order to obtain statistical properties, such as depth profile of implant atoms and displacements, 100 simulations for B_1 and B_2, and 25 simulations for both B_4 and B_{10} were performed respectively at different impact points.

RESULTS AND DISCUSSION

Implant Profile and Efficiency of Boron Cluster

The mean implant depths and implant efficiencies for various clusters are summarized in table 1. Each boron atom/cluster was irradiated on the target with the same incident energy of 230eV/atom, so that the total incident energy of B_2, B_4 and B_{10} are 0.46keV, 0.92keV and 2.3keV, respectively. The results are categorized by the properties of cluster, the cluster size and the number of stacked atoms in a line perpendicular to the surface. Table 1 indicates that, B_2, B_4 and B_{10} implantation, (except for the case of the vertical B_4 and B_{10} chains), shows the same implant profile and efficiency as B_1, about 82%.

To understand the enhancement mechanism of implantation depth and efficiency by chain-like cluster impact, the differences in the B implant profile were investigated according to the initial position in the cluster. Figure 1 shows the distribution of B atoms implanted by vertical B_{10} chain, according to the initial position in the cluster. As can be seen, the atoms -1 and -2, which impact on the substrate first, show shallower distribution than atoms -9 and -10, which impact last, and almost all of the backscattered atoms come from atoms -1 and -2. The model of enhancement in implant depth and efficiency by the

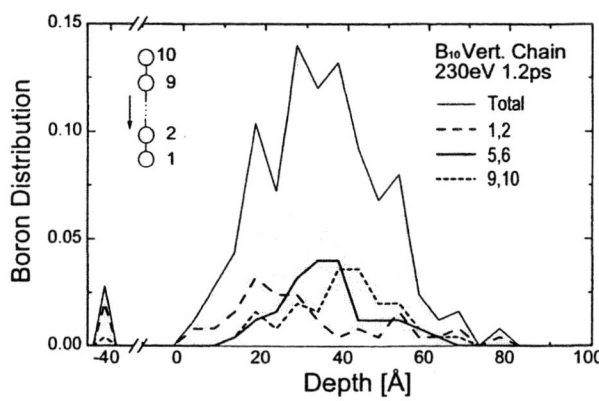

FIGURE 1. Profile of B atoms implanted by vertical B_{10} chain, according to the initial position in the cluster.

vertical-chain cluster is explained as follows: the first B atom of a chain cluster knocks-on a substrate Si atom and the following B atoms can thus penetrate to a deeper site. This collisional process is considered as the minimum model for the clearing-way effect [4].

Unlike vertical B_4 and B_{10} chain impact, the implant efficiencies of vertical B_2 and perpendicular square B_4 clusters show a similar value to that of B_1 and horizontal B_4 chain cluster. These results suggest that at this incident energy of 230eV/atom, the stack number of two in B_2 and in perpendicular B_4 square does not have enough density to cause an improvement in implant efficiency, but this changes at a stack number of four. This assumption is supported by the spherical B_{10} with the stack number of two or three and which shows similar implant depth and efficiency similar to that of the horizontal B_{10} chain rather than the vertical B_{10} chain.

As the number of stacks in the cluster increases, so does the mean implant depth. In the case of the vertical B_4 chain cluster impact, the cluster collapses immediately after it penetrates the first layer of the substrate. However, the B_{10} chain maintains coherency of velocity within the substrate and each B atom continues to penetrate deeply into the substrate. In these collisional processes of small B clusters, the interactions between B atoms are considered to be less probable and each B atom in the cluster acts in a way similar to monomer ions with the same energy per atom.

Damage Formation by Boron Cluster Impact

Figure 2 shows the time dependence of the number of displacements per a single B atom. The displacements are defined as those Si atoms, which have a potential energy of 0.4eV above the bulk state [8]. As shown in figure 2, the number of displacements reaches maximum at around 0.2ps and then decreases. The maximum number of displacements does not depend on either cluster size or cluster structure.

Figure 3 shows the time dependence of the mean kinetic energy of B atoms at the impact of B_1 and B_{10}. As can be seen in figure 3, the energy transition process of both shows similar profile. Therefore, it can be considered that each implanted B atom interacts with the substrate atoms individually and the kinetic energy of the incident atoms is transferred to the substrate without overlapping. This suggestion agrees with the aforementioned result that implant depth and implant efficiency do not depend on size and structure, except for the case of the vertical chain cluster.

The damage recovery process takes a longer time than the energy deposition from the projectile to substrate, and shows a different behavior depending on the cluster size. As shown in figure 2, the

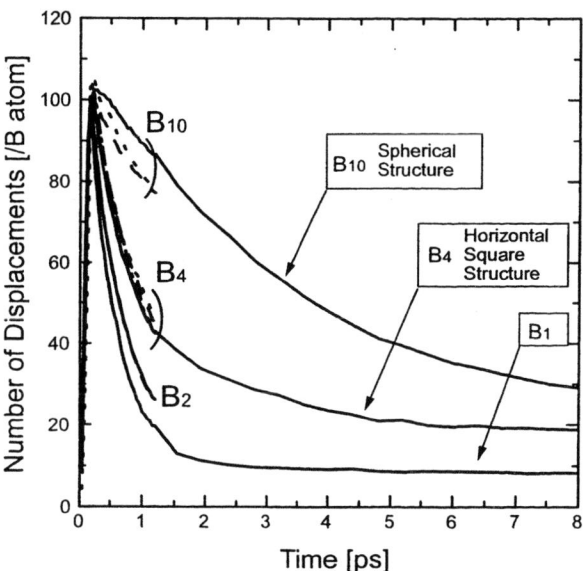

FIGURE 2. Time dependence of the number of displacements per B atoms by B clusters with various sizes.

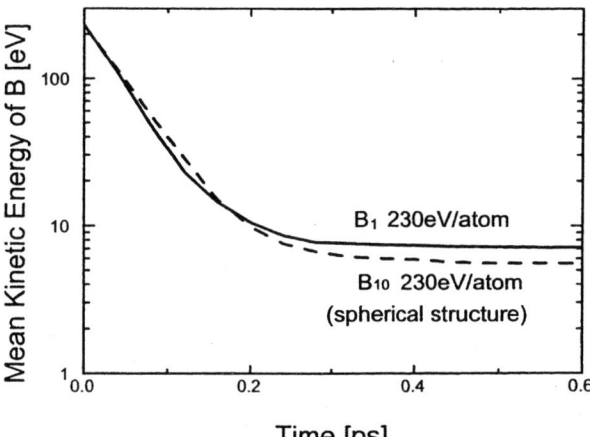

FIGURE 3. Time dependence of the mean kinetic energy of B atom at the impact of B_1 and spherical B_{10}

displacements induced by B_1 recover rapidly in 2ps and about eight displacements remain 8ps after the impact. However, the damage recovery rate slows down as the cluster size increases. In the case of spherical B_{10} impact, about 30 displacements, which is four times higher than B_1, still remain 8ps after impact. The high yield displacement by cluster ion impact is due to the high-density energy irradiation effect. When a B_{10} cluster impacts the substrate, B_{10} deposits its incident kinetic energy of 2.3keV in a finite region on the surface so that a large number of energetic knocked-on atoms are created. These knocked-on atoms interact with each other, and then it is considered that they remain in the form of lattice

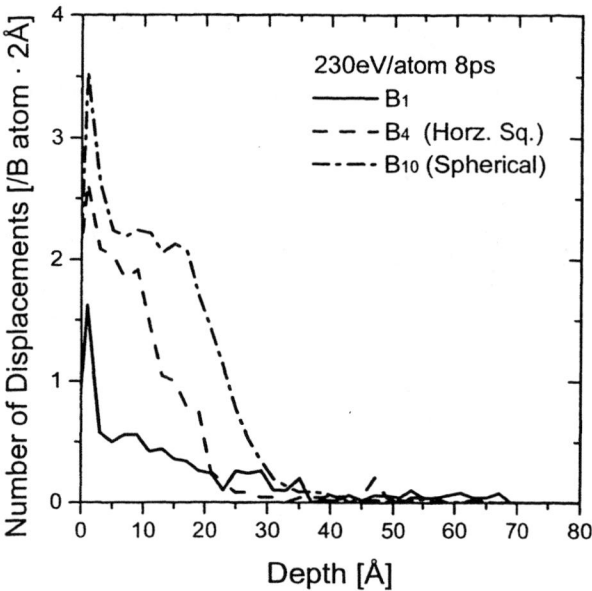

FIGURE 4. Depth profile of displaced Si atoms by the impact of B_1, B_4, and B_{10}.

deformation. Therefore, the yield of displacement by B_{10} remains several times higher than that of B_1.

Figure 4 shows the depth profiles of displacements induced by B_1, horizontal B_4 and spherical B_{10} cluster at 8ps after impact. For B_1 implantation, transient displacements are formed along the trajectory of the incident atom, however, these displacements easily recover because the energy deposited at each collision is small. In this case, knocked-on displacements reside around the incident B atom, and this is termed 'end-of-range' damage [9]. The end-of-range displacements are statistically observed in the region deeper than 30Å in figure 4, which shows a larger ratio of displacements for B_1 compared to boron clusters. It is reported that these displacements tend to cause transient enhanced diffusion (TED), which is serious problem in high-quality shallow junction formation using conventional monomer ion implantation [2,9].

On the other hand, B_{10} cluster creates a high density of displacements on the surface at the impact point because of the high-density energy irradiation effect. This damaged region is considered to be amorphized and appears as a box-like shape from the surface to a depth of 20Å, which is comparable with the mean implant depth of the B atoms. Because of the reduction in TED without pre-amorphization, this characteristic damage formation by B_{10} is expected to result in significant advantages in shallow junction formation. Through annealing, the reconstruction of a substrate irradiated with B_{10} clusters proceeds from the bottom of the amorphized layer to the surface of the substrate. Therefore, interstitial Si atoms tend to move to the top surface thus avoiding B atom diffusion into deeper regions of the substrate. It has been observed experimentally that low-energy $B_{10}H_{14}$ implantation into a Si substrate does not cause TED [3,10], similarly to the effect of B monomer implantation into a well pre-amorphized Si substrate.

SUMMARY AND CONCLUSIONS

MD simulations of B_1, B_2, B_4 and B_{10} impacting on Si (001) substrates were performed and the dependence of cluster size and structure on the impact process was examined. In each B cluster impact, the implant depth and implant efficiency obtained were similar to those of monomer ions, except for vertical chain-like cluster with size larger than 4. In this exceptional case, deeper implant depth and higher implant efficiency were observed due to the clearing-way effect.

The non-linearity of cluster impact is shown in the damage formation mechanism. The number of displacements produced by one B atom in one impact increases to the same maximum for both B cluster and B monomer. However, the damage recovery process is different, depending on cluster size. Damage induced by B_{10} recovers more slowly, with 4 times more displacements remaining after 8ps, compared to B_1. These displacements by B_{10} clusters concentrate in the near surface region of the impact point. This characteristic damage formation is expected to prevent transient-enhanced-diffusion of incident B atoms and thus achieve high-quality shallow p-type junctions.

REFERENCES

1. Moffatt, S., *Nulc. Instr. and Meth.* **B 96** 1 (1995).
2. Jager, H. U., *J. Appl. Phys.* **78** 176 (1995).
3. Goto, K., Matsuo, J., Tada, Y., Moriyama, Y., Sugii, T., and Yamada, I., *IEDM Tech. Digst.* (1997) p. 471.
4. Aoki, T., Seki, T., Matsuo, J., Insepov, Z., and Yamada, I., *Mat. Chem. and Phys.* **54** 139 (1998).
5. Seki, T., Tanomura, M., Aoki, T., Matsuo, J., and Yamada, I., *Mat. Res. Soc. Symp. Proc.* **504** 93 (1998).
6. Stillinger, F. H., and Weber, T. A., *Phys. Rev.* **B31** 5632 (1985).
7. Ziegler, J. P., Biersack, J. P., and Littmark, U., *The stopping and range of ions in solids;* New York: Pergamon Press (1985).
8. Caturla, M. J., de la Rubia, T. Diaz, and Gilmer, G. H., *Nucl. Instr. and Meth.* **B196** 1 (1995).
9. Jones, K. S., Elliman, P. G., Petravic, M. M., and Kringhoj, P., *Appl. Phys. Lett.* **68** 3111 (1996).
10. Kusaba, T., Shimada, N., Matsuo, J., and Yamada, I., *1998 International Conference on Ion Implantation Technology Proceedings* (1999) 1258.

Defects and Nanocluster Engineering in MgO

A. V. Fedorov[1], A. van Veen[1], M. A. van Huis[1], H. Schut[1], B. J. Kooi[2], J. Th. De Hosson[2], R. L. Zimmerman[3]

[1] *Interfaculty Reactor Institute, Delft University of Technology, NL-2629 JB Delft, The Netherlands*
[2] *Materials Science Centre, University of Groningen, Nijenborgh 4, 9747 AG Groningen, The Netherlands*
[3] *Alabama A&M University, Center of Irradiation of Materials, AL 35762-1447, USA*

Abstract. The optical properties of MgO crystals are known to change after introduction of nanosize metal precipitates. In this work the formation of metallic nanoclusters in the presence of nanosize rectangular shaped cavities was studied. The rectangular cavities were formed by 30 keV He^+ implantation followed by 1273 K annealing. The formation of cavities and their location was established by Positron Beam Analysis (PBA). The rectangular shape and their alignment in (100) direction was observed by X-TEM. Subsequently, the samples were implanted with 600 keV Ag and 1000 keV Au in order to introduce the metal ions in the vicinity of the cavities. The samples were then annealed to provide the formation of nanoclusters. The evolution of the implantation induced defects was monitored by PBA. The optical properties were studied by light absorption measurements.

INTRODUCTION

Metal nanoclusters in MgO are known to change the optical properties [1] in this metal oxide. An absorption band centered at 580-600 nm, known as Mie resonance, has been observed after Au/Ag implantation in a number of studies [2-3]. Conventionally metal nanoclusters are formed after 1-2 MeV implantation followed by high temperature anneals. An alternative way to form nanoclusters was studied in [4]. Prior to the metal ion implantation a chain of nanosize cavities was created as a result of 30 keV He^+ ion implantation followed by annealing. In the complementary TEM study [5] it was observed that the cavities are elongated in the [100] direction and have a rectangular form. Subsequently, Au ions were introduced by low energy, 30 keV implantation. Then the samples were annealed to provide diffusion of the metal ions to the cavity layer. However, only a small fraction of Au reached the cavities located at a depth of about 200 nm. However, most of the Au ions created a band of clusters close to the surface within the implantation range. In the present study MgO samples containing a band of nanocavities have been implanted with high-energy Au and Ag ions. The energies were chosen to introduce the metal ions directly in the vicinity of the cavity band.

EXPERIMENTAL

Single crystal MgO (100) samples with dimensions $10 \times 10 \times 1$ mm^3 were obtained from Kelpin GmbH (Hamburg, Germany). The samples were treated in the following way: The nanosize cavities were formed by 30 keV $^3He^+$ implantation followed by annealing in air at 1273 K for 0.5h. After this the samples were implanted with 30 keV Au at room temperature (RT) and 1000 K. Another set of four samples with cavities was implanted with 600 keV Ag and 1 MeV Au ions at RT and at 873 K. After the implantation the samples were step-wise annealed in air in order to form the nanoclusters. The optical properties of the samples were studied by light absorption. The structure of the defects introduced by implantation was studied by X-TEM and the Positron Beam Analysis (PBA). A detailed description of the PBA technique is presented in [6].

With PBA the sample is probed by a mono-energetic beam of positrons with energies ranging from 0.1 to 25 keV which correspond to typical depths from the surface up to 3000 nm. Annihilation of

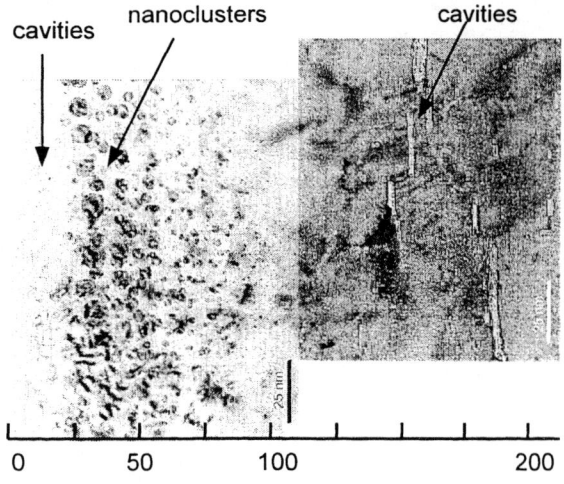

FIGURE 1. X-TEM picture of two bands of cavities and a band of nanoclusters obtained after a combination of 30 keV He and Au implantations and anneals discussed in the text. The depth location of the bands is shown in nanometers.

a positron with an electron at rest results in two photons of 511 keV each. If the electron has a non-zero momentum the energy of the photons is Doppler shifted resulting in a broader energy distribution. The energy spectrum of the 511 keV photon annihilation peak is characterized by two parameters: S and W. The S - parameter is defined as the contribution of the central part of the peak and is related to annihilations with valence or conduction electrons. The W - parameter is defined by the contribution of the "wings" in the spectrum and is related to annihilations with high momentum core electrons. Thus, high S - values are usually attributed to vacancy related defects where the chance for a positron to annihilate with a high-momentum core electron is low. Another contribution to high S - values is provided by the annihilation of positronium which can be formed in cavities of nanometer size.

Low - energy Au ion implantation

The bands of nanosize cavities and Au nanoclusters observed in X-TEM are shown in Fig. 1. In this work a MgO sample containing a band of cavities at a depth of about 200 nm was implanted with 30 keV Au ions to a dose of 10^{16} cm^{-2}. The sample was subsequently annealed in air at 1273 K. It is seen that a band of nanoclusters is formed within 25-75 nm of the surface. The cavity layer formed prior to the Au implantation had developed into a band of elongated internal cracks during the anneals. The cracks are presumably formed by merging of a few adjacent cavities into one. Additionally, the Au implantation has created another band of rectangular nanosize cavities within the first

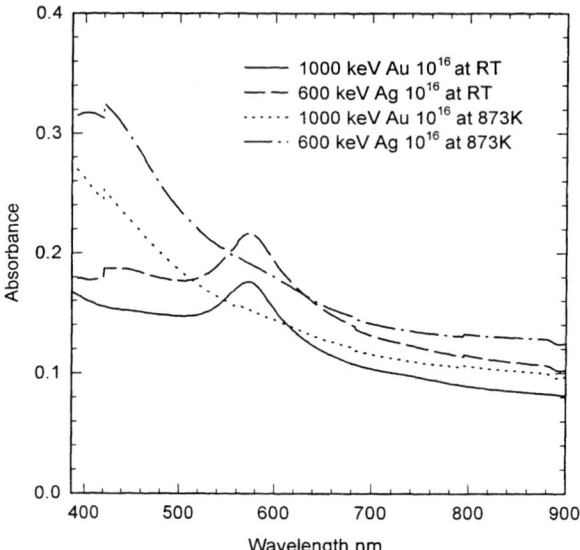

FIGURE 2. Light absorption measurements of the MgO samples with cavities after Au and Ag ion implantation at RT and 873 K with the indicated energy and dose.

25 nm. The fact that the implantation induced vacancies have agglomerated into cavities rather than escaping at the surface suggests a very short diffusion length within the damage region. The Au nanoclusters were later studied by X-Ray diffraction and X-TEM, and were found to be semi-coherent with the MgO host lattice [7].

High - energy Au and Ag ion implantation

In order to introduce the metal ions directly into the vicinity of the cavity band the implantations were carried out at energies which correspond to the depth of the original cavity band, i. e. 200 nm. Thus, the implantation energies for Au and Ag ions were 1 MeV and 600 keV, respectively. To enhance diffusion two samples were implanted at an elevated temperature of 873 K. The light absorption measurements performed directly after the implantations are shown in Fig. 2. The Mie resonance peak is observed only in the case of the RT implantations. In the case of the Ag implantation another peak, located at 420 nm, is observed both for RT and 873 K implantations.

Next, the samples were annealed in air and the evolution of the implantation induced defects was monitored with the help of PBA. The S-curves measured on the samples implanted with Au and Ag are shown in Figs. 3 and 4, respectively. The peak at 5-7 keV, which corresponds to a depth of 200-250 nm, is ascribed to the presence of the cavities created by the implantation. After the Au / Ag ion implantations the peak is split into two components. This effect is even better observed in the S- curves c) and d) measured

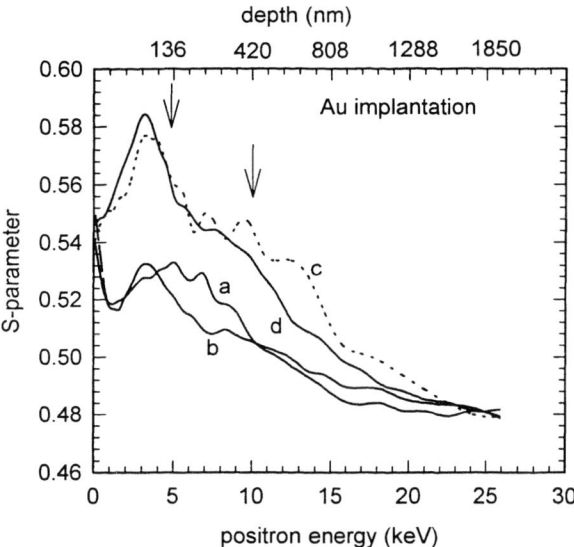
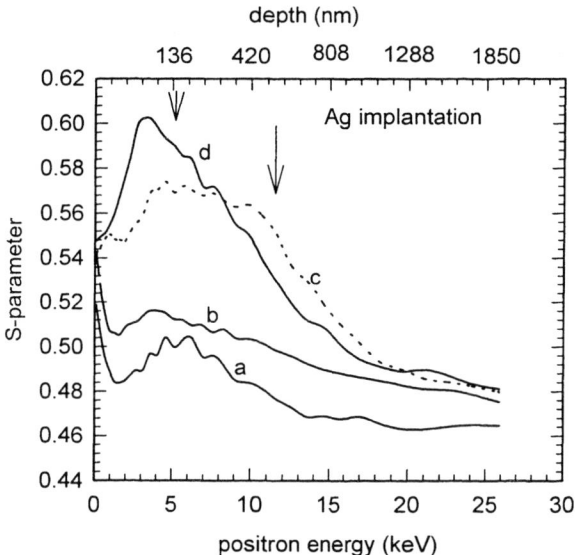

FIGURE 3. S- curves measured at different stages of the sample treatment: a) 30 keV He implantation and 1273 K annealing; b) followed by 1 MeV 10^{16} cm^{-2} Au implantation at RT; c) followed by 973 K annealing. Curve d) is measured on the sample without cavities after 1 MeV 10^{16} cm^{-2} Au implantation at RT followed by 973 K annealing.

FIGURE 4. S- curves measured at different stages of the sample treatment: a) 30 keV He implantation and 1273 K annealing; b) followed by 1 MeV 10^{16} cm^{-2} Ag implantation at RT; c) followed by 973 K annealing. Curve d) is measured on the sample without cavities after 1 MeV 10^{16} cm^{-2} Au implantation at RT followed by 973 K annealing.

after the high temperature treatment. Note that the S-curve d) is measured on the reference sample which did not contain cavities. Similar double peak curves were observed earlier and are reported in [8-9]. In the present X-TEM study and the one reported in Ref. [8] it was observed that the band of nanoclusters is confined between two bands of cavities. Moreover, in a Cu$^+$ ion implanted sample the central band contained both Cu precipitates and cavities [8]. Thus, two peaks in the measured S - curves are related to the bounding cavity bands while the dip in between is ascribed to the presence of metal precipitates.

To obtain the depth location of the defected layers the S - curves measured prior and after the metal ion implantations were analyzed with the program VEPFIT [6]. The S -curve a) ascribed to the original band of cavities was fitted with a model containing 3 layers: a layer of cavities between two layers of defect-free MgO matrix. The S - curve b) measured after the Au implantation was fitted with a 5 layer model: 3 defected layers (2 with cavities and one with metal precipitates) and 2 defect free layers. The results of the fit are shown in Fig. 5. The stopping ranges of ions obtained by the TRIM [10] simulations are also shown for comparison. The implantation energies of He, Au and Ag ions were chosen to yield the same implantation range. However, a discrepancy within 100 nm is observed between the predicted by TRIM ranges and the location of the defected layers obtained by VEPFIT. Assuming that the highest S - value is due to the formation of positronium (Ps) inside the cavity the theoretical limiting value for the S parameter in MgO, $S_{Ps} = 0.577$ [11]. The S - parameter measured in the defect-free MgO is $S_{bulk} = 0.476$ agrees with earlier measured values.

According to the previously performed studies, for example [3], the annealing temperature of 973 K is not high enough to provide formation of nanoclusters. From the PBA studies alone is difficult to say whether the metal ions have precipitated into a cluster or are still dissolved in the matrix. To determine this an additional X-TEM study is needed.

CONCLUSIONS

1. Nanosize cavities developed after He implantation were used as precursor for the precipitation of the metal nanoclusters.

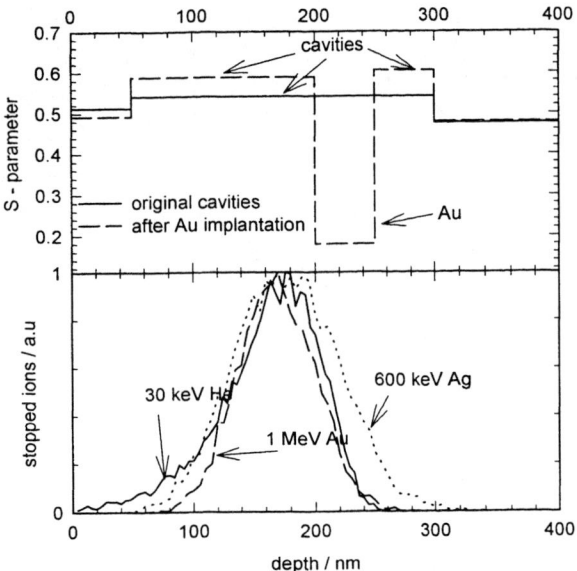

FIGURE 5. The results of the fitting performed with the VEPFIT program: Defect layers characterized by different S-values are shown as a function of depth. Ranges of the implanted ions with the indicated energies are also shown in the bottom graph..

2. Metal ions were introduced by low-energy (30 keV) Au implantation and high energy Au (1MeV) and Ag (600 keV) implantations.

- In the case of low-energy implantation nanoclusters are predominantly formed within the implantation range close to the surface.

- In the case of high-energy implantation only the samples implanted at room temperature show the Mie absorption band. The band disappears after annealing the samples at 973 K.

REFERENCES

1. P. Chakraborty, J. of Materials Science **33** (1998) 2235-2249.

2. C. W. White, J. D. Budai, S. P. Withrow, J. G. Zhu, E. Sonder, R. A. Zuhr, A. Meldrum, D. M. Hembree, Jr., D. O. Henderson, S. Prawer, Nucl. Instr. and Meth. B **141** (1998) 228-240.

3. R. L. Zimmerman, D. Ila, E. K. Williams, S. Sarkisov, D. B. Poker, D. K. Hensley, Nucl. Instr. and Meth. B **141** (1998) 308-311.

4. A. V. Fedorov, M. A. van Huis, A. van Veen, H. Schut, Nuclear Instr. and Meth. in Phys. Res., B **166-167** (2000) 215-219.

5. B. J. Kooi, A. van Veen, J. Th. M. de Hosson, H. Schut, A. V. Fedorov, F. Labohm, Appl. Phys. Lett. **76-9** (2000), 1110-1112.

6. A. van Veen, H. Schut, J. de Vries, R. A. Hakvoort and M. R. IJpma in: American Institute of Physics conference proceedings 218, *Positron beams for solids and surfaces* (eds. P.J. Shultz, G.R. Massoumi, P.J. Simpson) (1990) 171-196.

7. M. A. van Huis, A. van Veen, A. V. Fedorov, T. Himba, B. J. Kooi, J. Th. M. De Hosson, submitted to NIM B.

8. M. A. van Huis, A. V. Fedorov, A. van Veen, P. J. M. Smulders, B. J. Kooi, J. Th. M. De Hosson, Nuclear Instr. and Meth. in Phys. Res., B **166-167** (2000) 225-231.

9. J. Xu, A. P. Mills,Jr., A. Ueda, D. O. Henderson, R. Suzuki, and S. Ishibashi, Phys. Rev. Lett, **83-22** (1999) 4586-4589.

10. J. F. Ziegler, J. P. Biersack, V. Littmark, *The Stopping and Range of Ions in Solids*, ed. J. F. Ziegler, Pergamon Press (1985).

11. A. V. Fedorov, A. van Veen, H. Schut, MRS Meeting 1999, Mat. Res. Soc. Proc. Editors S. J. Zinkle, G. E. Lucas, R. C. Ewing, J. S. Williams, Vol. **540** (1999) 231-236.

Size-Specific Reactions of Transition Metal Clusters in Collision with Simple Molecules

M. Ichihashi*, T. Hanmura[†], R. T. Yadav[†], and T. Kondow*

*Cluster Research Laboratory, Toyota Technological Institute: in East Tokyo Laboratory, Genesis Research Institute, Inc., 717-86 Futamata, Ichikawa, Chiba 272-0001, Japan
[†]East Tokyo Laboratory, Genesis Research Institute, Inc., 717-86 Futamata, Ichikawa, Chiba 272-0001, Japan

Abstract. Reactions of nickel cluster ions, Ni_n^+ (n=3–11), with methanol molecules, CH_3OH, were studied at collision energies less than 1.0 eV by use of a tandem-type mass spectrometer. It was found that size-dependent chemisorption, demethanation and carbide formation of CH_3OH proceed on Ni_n^+; the demethanation proceeds on Ni_4^+, while the carbide formation on Ni_7^+ and Ni_8^+. The initial steps of the reactions are physisorption and chemisorption of methanol on Ni_n^+. The size dependence was explained in terms of kinematics on a reaction potential and the geometrical structure of Ni_n^+.

INTRODUCTION

Experimental investigations on reactions of simple molecules on clusters of transition metals in the gas phase have been performed intensively. Among these metals, nickel is one of the most frequently used elements in practical catalysis, in which the diameter distribution of nickel aggregates is closely related to their reactivity and selectivity. In this regard, reactivity of nickel cluster ions, Ni_n^+, has been investigated with a particular attention to their size-dependent reactivity [1-4]. For instance, Irion and his co-workers have investigated dehydrogenation of ethylene on Ni_n^+ (n=2–15) and revealed that Ni_n^+ with n=2, 5–15 react with ethylene into $Ni_n^+(C_2H_2)_m$, whereas $Ni_{3,4}^+$ is nonreactive [1]. Similar size-specific reactions between Ni_n^+ (n=4–31) and CO have been reported by Wöste and his collaborators [3]; $Ni_n(CO)_i^+$ is produced in the entire n-range studied, while $Ni_mC(CO)_p^+$ only from $Ni_{4,5,7}^+$. The studies of this kind have been undertaken under multiple collision conditions, which introduce undesired complexity in understanding the reaction mechanism. In order to elucidate a size-specific reactivity, we have investigated reactions of methanol, CH_3OH, on a size- and energy-selected nickel cluster ion, Ni_n^+ (n=3–11), under single collision conditions. In the present study, methanol was chosen as the reactant, because it is one of the basic chemicals and often emerges as a reaction intermediate and a precursor in synthesis, decomposition and oxidation reactions involving hydrocarbons, such as the Fischer-Tropsch process.

EXPERIMENTAL

A part of the apparatus employed in the present study has been reported in our previous papers [5-7], so that the equipment newly installed in the present experiment are described in detail. As shown in Fig. 1, a xenon-ion beam was produced from an ion source (CORDIS Ar25/35c, Rokion Ionenstahl-Technologie), accelerated up to 10 keV, and divided into four beams by a series of the acceleration plates. Each Xe+ beam was collimated and allowed to bombard one of four nickel metal targets. The ions sputtered from the targets were focused by conical electrodes into the first octopole ion beam guide (OPIG). The ions were decelerated in the OPIG and cooled in helium gas (> 10^{-3} Torr) at a given temperature of 77-300 K. The cluster ions thus cooled were transported in the second and the third OPIG's and mass-selected in the first quadrupole mass filter. A size-selected cluster ion was admitted into a collision region in the fourth OPIG surrounded by a collision chamber filled with methanol gas at a pressure of $\sim5\times10^{-5}$ Torr, which was measured by a spinning rotor gauge. Product ions in

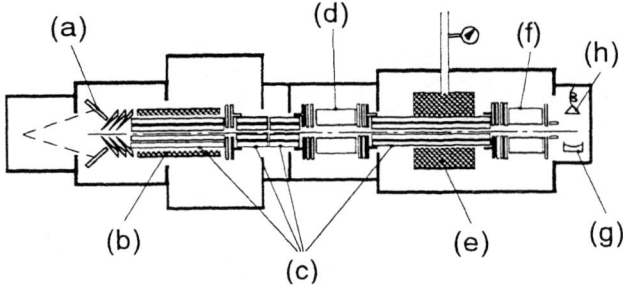

FIGURE 1. Schematic drawing of the apparatus used. (a) nickel target, (b) cooling chamber filled with helium gas, (c) octopole ion beam guide (OPIG), (d) and (f) quadrupole mass filters, (e) collision chamber filled with methanol gas, (g) ion-conversion dynode and (h) secondary electron multiplier.

the collision region, which were transported through the fourth OPIG, were mass-analyzed in the second quadrupole mass filter, and were detected by an ion-conversion dynode biased by –10 kV, which is followed with a secondary electron multiplier. The signal from the detector was amplified, discriminated and processed in electronic circuits based on a microcomputer.

The translational-energy spread of a parent cluster ion was measured to be less than 3 eV (laboratory frame), by changing a retarding potential of the collision region in the fourth OPIG. This energy spread gave rise to the collision-energy spread of 0.2 eV in the center-of-mass frame for the collision of Ni_8^+ with a methanol molecule.

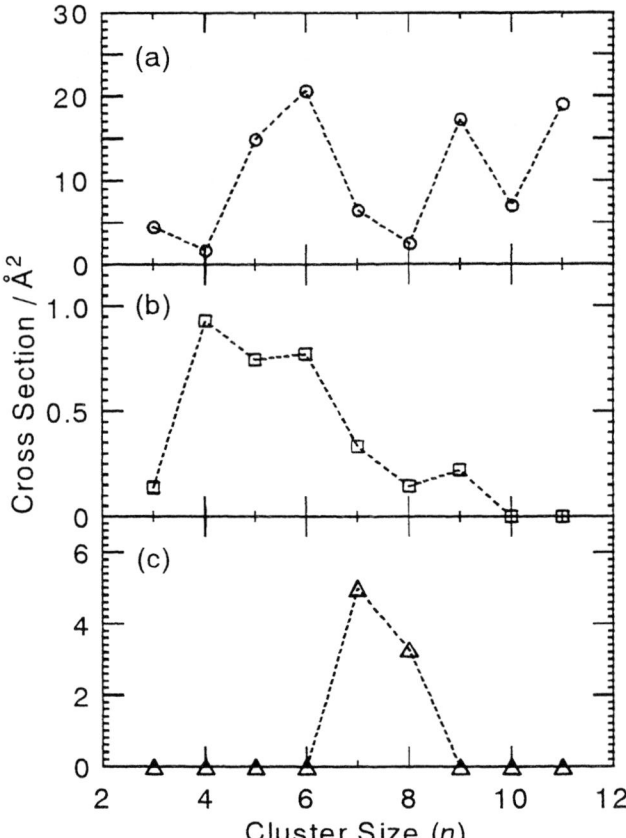

FIGURE 2. Cross sections for the production of $Ni_n^+(CH_3OH)$ [panel (a)], Ni_nO^+ [panel (b)] and $Ni_{n-1}C^+$ [panel (c)] as a function of the cluster size. The collision energy and the internal temperature are 0.1 eV and 300 K, respectively.

RESULTS AND DISCUSSION

Figure 2 shows the reaction cross sections as a function of the cluster size. Three types of product ions were observed: $Ni_n^+(CH_3OH)$, Ni_nO^+ and $Ni_{n-1}C^+$. These ions are considered to be produced by the following reaction pathways:

$$Ni_n^+ + CH_3OH \rightarrow Ni_n^+(CH_3OH), \quad (1)$$

$$Ni_n^+ + CH_3OH \rightarrow Ni_nO^+ + CH_4, \quad (2)$$

$$Ni_n^+ + CH_3OH \rightarrow Ni_{n-1}C^+ + NiO + 2H_2. \quad (3)$$

In $Ni_n^+(CH_3OH)$, CH_3OH should be chemically adsorbed onto Ni_n^+ (chemisorption), and otherwise (physisorption), it has to be detached readily. In some cases, the chemisorbed species, $Ni_n^+(CH_3OH)$, further reacts to form Ni_nO^+ (demethanation) or $Ni_{n-1}C^+$ (carbide formation). As shown in Fig. 2, the cross section for the chemisorption, the demethanation and the carbide formation are the largest at $n=6$, 4 and 7, respectively.

The cross sections for these reactions always decrease with the collision energy; reactions (1), (2) and (3) have no energy threshold. Figure 3(a) shows the collision energy dependence of reactions (1) and (2) on Ni_4^+. The cross sections obtained experimentally were compared with the Langevin cross section [8], which is almost equal to the cross section for the physisorption. As the Langevin cross section is much larger than the sum of the cross sections (total cross sections) for reactions (1), (2) and (3), CH_3OH which once physisorbs on Ni_n^+, $[Ni_n^+...CH_3OH]$, desorbs sizably, while the rest of the physisorbed species is converted to the chemisorbed one, $Ni_n^+(CH_3OH)$. Figure 3(b) shows the branching fractions of the chemisorption and the demethanation as a function of the collision energy. The branching fraction of the demethanation increases with the

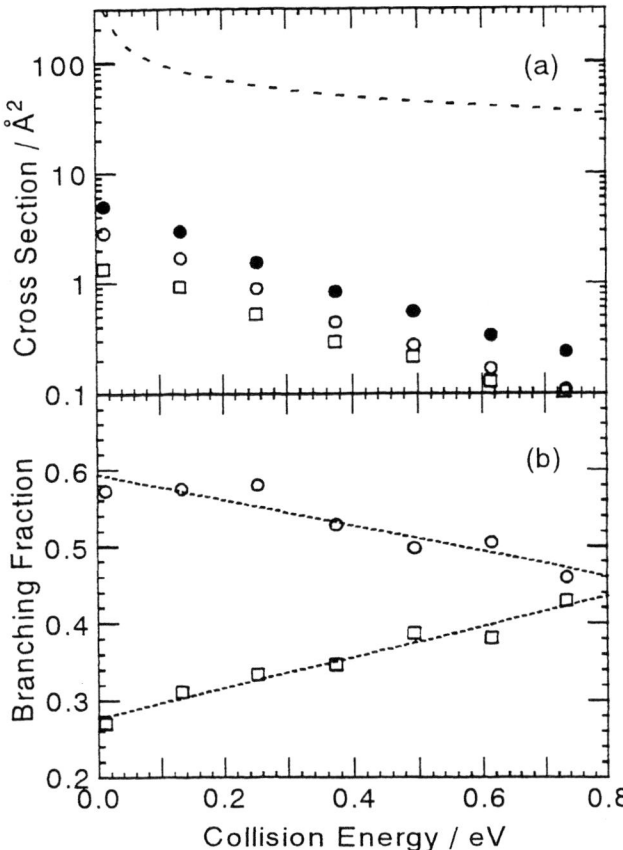

FIGURE 3. Reaction cross sections [panel (a)] and branching fractions [panel (b)]: total cross section (close circle), the production of $Ni_4^+(CH_3OH)$ (open circle), Ni_4O^+ (open square) as a function of the collision energy. The internal temperature of Ni_4^+ is 300 K. The dashed line in panel (a) shows the Langevin cross section, and the dotted lines in panel (b) show eye guides.

collision energy, and this indicates that there is an energy barrier between the chemisorption and the demethanation.

Let us consider a collision event between CH_3OH and Ni_n^+. Upon CH_3OH approaching to Ni_n^+, CH_3OH is at first trapped into a shallow potential well (physisorbed state) due to a charge–dipole interaction, and then it is transferred to a much deeper well (chemisorbed state) attributed to the superposition of the electron clouds between CH_3OH and Ni_n^+. To a sizable extent, the chemisorbed species, $Ni_n^+(CH_3OH)$, further reacts into Ni_nO^+ or $Ni_{n-1}C^+$.

There must be an energy barrier between the physisorbed and the chemisorbed state. The energy barrier is estimated from the dependence of the total cross section on the internal temperature of Ni_n^+. Figure 4 shows the total reaction cross sections as a

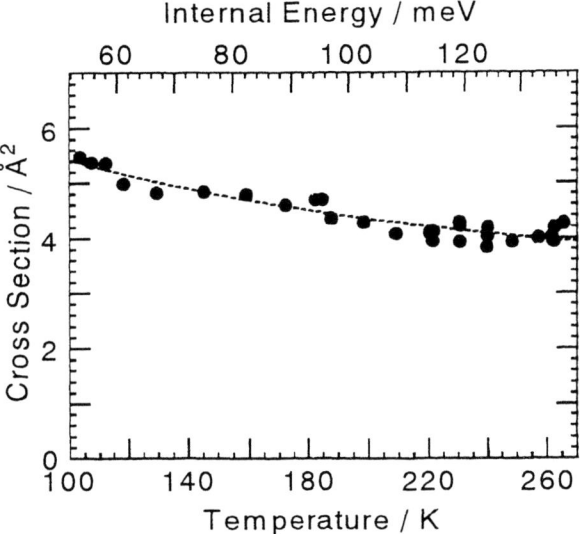

FIGURE 4. Total reaction cross section for Ni_4^+ as a function of the internal temperature of Ni_4^+. The collision energy is 0.1 eV. The dashed line shows the calculation with the best-fit value of ΔE.

function of the internal temperature (energy) of Ni_n^+. The cross section decreases with the increase of the internal temperature. In the framework of the kinematic model, the dependence on the internal energy is introduced from the rate constant, k_1, for the desorption from the physisorbed state, and the rate constant, k_2, for the transferring to the chemisorbed state. Then, the total reaction cross section, σ_r, is given as

$$\sigma_r = \sigma_L [k_2/(k_1+k_2) + \{k_1/(k_1+k_2)\} \exp(-(k_1+k_2) t)], \quad (4)$$

where σ_L represents the Langevin cross section. In the present experiment, the reaction time, t, given by the flight time from the collision region to the second quadrupole mass filter, turns out to be several hundred microseconds. The rate constants, k_1 and k_2, are estimated from the RRK theory [9], which depend on the energy barrier, ΔE, between the physisorbed and the chemisorbed states. The internal-energy dependence of the cross section was fitted to the theoretical dependence given by eq. (4) with leaving ΔE as a variable parameter. The size dependence of ΔE exhibits the minimum at $n=6$, and the two maxima at $n=4$ and 8. The size dependence of ΔE shows a clear correlation to the size dependence of the reaction cross section (see Fig. 2). Namely, the chemisorption is dominant at $n=6$, at which ΔE reaches the minimum, whereas the demethanation and the carbide formation are dominant at $n=4$ and 8, respectively, at which ΔE reaches the maxima. The sharp contrast indicates that

(i) the chemisorption proceeds more readily when the ΔE is low, and (ii) the demethanation and the carbide formation proceed more readily when ΔE is high.

On the reaction potential along the reaction coordinate, there are valleys corresponding to the physisorption, the chemisorption, and dissociative chemisorption leading to the demethanation (or the carbide formation); let us define the energies, E_1 and E_2, of the transition states between the physisorption and the chemisorption and between the chemisorption and dissociative chemisorption leading to the demethanation (or the carbide formation), respectively. On this potential, the demethanation cross section was estimated in the kinematic model. In the calculation of the demethanation cross section, the energy of the physisorbed state, E_{phys}, is approximated to be 0.4 eV by the energy of a charge–induced dipole interaction, while the energy of the chemisorbed state, E_{chem}, is calculated to be 1.5 eV by using a density functional theory. Throughout the calculation, E_{phys} and E_{chem} of Ni_n^+ is assumed to be independent of n, and only E_1 ($=E_{phys}-\Delta E$) was changed with ΔE. The demethanation cross sections thus calculated reproduce the measured cross sections in the $n \geq 4$ range. This agreement supports a mechanism that the size-dependent demethanation cross section is influenced by the barrier height, ΔE.

The carbide formation has a significant cross section only at $n=7$ and 8. As described above, the barrier height, ΔE, reaches the other maximum at $n=8$. The relation between the cross section and the barrier height seems to be explained similarly by the reaction scheme mentioned in the demethanation. However, this scheme cannot explain the result that the carbide formation proceeds at $n=7$ and 8, but not at $n=4$, at which the barrier heights are high. This difficulty can be solved by introducing a geometrical constraint to a reaction site on Ni_n^+, as is the case of methanol adsorption on a nickel surface, at which surface morphology plays an important role [10–12]. On the analogy of the surface reaction, one can explain the size-specific reactivity of methanol on Ni_n^+ in terms of matching of the Ni–Ni distance with the C–O distance. In the carbide formation, the reaction system passes through a transition state, in which CH_3OH occupies one nickel atom for its carbon atom and another nickel atom for its oxygen atom. The energy barrier leading to the carbide formation is lowered when the Ni–Ni distance of Ni_n^+ is the most comfortable to CH_3OH in the transition state, that is, the rate of the carbide formation changes sensitively with the Ni–Ni distance. In fact, the Ni–Ni distance of Ni_n^+ calculated by the density functional method is elongated at $n \simeq 6$ with increase in n. The comparison of this finding with the size-specific reactivity of Ni_n^+ leads us to conclude that the interatomic distance of Ni_n^+ with $n \geq 6$ matches well with the C–O bond length of CH_3OH in the transition state, whereas Ni_n^+ with $n \leq 5$ does not so that the carbide formation proceeds only on $Ni_{7,8}^+$ but not on Ni_4^+. On the other hand, the rate of the demethanation does not seem to be sensitive to the Ni–Ni distance because CH_3OH in the transition state responsible to the demethanation occupies a single site on Ni_n^+. Therefore, the demethanation could take place on Ni_4^+ and $Ni_{7,8}^+$. However, the demethanation does not proceed on $Ni_{7,8}^+$, probably because the Ni sites on Ni_n^+ with $n \geq 6$ are occupied more favorably by CH_3OH in the transition state which leads to the carbide formation.

REFERENCES

1. M.P. Irion and A. Selinger, Ber. Bunsen-Ges. Phys. Chem. **93**, 1408 (1989).

2. M.P. Irion, Int. J. Mass Spectrom. Ion Proc. **121**, 1 (1992).

3. Š. Vajda, S. Wolf, T. Leisner, U. Busolt and L.H. Wöste, J. Chem. Phys. **107**, 3492 (1997).

4. E.K. Parks, L. Zhu, J. Ho and S.J. Riley, J. Chem. Phys. **100**, 7206 (1994).

5. M. Ichihashi, S. Nonose, T. Nagata and T. Kondow, J. Chem. Phys. **100**, 6458 (1994).

6. J. Hirokawa, M. Ichihashi, S. Nonose, T. Tahara, T. Nagata and T. Kondow, J. Chem. Phys. **101**, 6625 (1994).

7. M. Ichihashi, T. Hanmura, R. T. Yadav and T. Kondow, J. Phys. Chem. A, in press.

8. R.D. Levine and R.B. Bernstein, Molecular Reaction Dynamics (Oxford Univ. Press, Oxford, 1974).

9. J.A. Draves, Z. Luthey-Schulten, W.-W. Liu and J.M. Lisy, J. Chem. Phys. **93**, 4589 (1990).

10. J.J. Vajo, J.H. Campbell and C.H. Becker, J. Phys. Chem. **95**, 9457 (1991).

11. J.E. Demuth and H. Ibach, Chem. Phys. Lett. **60**, 395 (1979).

12. L.J. Richter and W. Ho, J. Chem. Phys. **83**, 2569 (1985).

Low Energy Cluster Beam Deposition: a Novel Approach to the Synthesis of Nanostructured Materials

S. Iannotta*, P. Milani**

*CEFSA-CNR Centro per la Fisica degli Stati Aggregati, 38050 Povo di Trento (TN), Italy
**INFM-Dipartimento di Fisica, Università di Milano, Via Celoria 16, 20133 Milano, Italy

Abstract. Cluster beams produced by Pulsed Microplasma Cluster Sources (PMCS) and by hyperthermal supersonic seeded sources are very suitable to produce nanostructured materials with a fine control on their structure, morphology and functional properties. Their major advantage is the possibility to control the state of the precursor in terms of energy, momentum and state of aggregation by properly tuning the source parameters. The cluster deposition energy (up to several tens of eV) can be tuned by changing the seeding gas and the expansion conditions in order to produce films with tailored nanostructures. Nanostructured materials with very different physical-chemical properties: refractory metals, carbon, SiC and organic pi-conjugated systems can be produced.

INTRODUCTION

The structural and functional properties of thin films, assembled atom by atom, are largely determined by the kinetic energy of the ions impinging on the substrate during the film growth [1]. The use of clusters instead of atoms, as building blocks, can open new possibilities for the synthesis of materials where the structural and functional properties are also determined by the hierarchical organization of units with dimensions ranging from mesoscopic to nanoscopic scale [2]. This approach can be extended to aerodynamic accelerated molecules and oligomers achieving similar results [3]. Several different physical and chemical routes are currently being used for the synthesis of clusters and for assembling nanostructured materials. The general requisites are the control of mass distribution, structure and chemical reactivity [4]. Moreover one should be able to control the degree of coalescence of the clusters during the formation of the nanostructured material.

The realization of intense and stable cluster sources can make cluster beam deposition a viable technique for the synthesis of nanostructured films. The synthesis and processing of nanostructured materials where only the mean cluster dimensions influence properties of the system can benefit from the use of cluster beams. Of course this technique is not convenient for the production of bulk materials, however nanostructured thin films and composite materials consisting of clusters embedded in transparent and polymeric matrices can be produced efficiently. The properties of these systems can be varied by controlling the mass and energy distribution of the cluster with a precision that does not require the use of beams monochromatic in mass and energy.

In view of the use of clusters as building blocks of nanostructured thin films, intense and stable beams must be used and a good control on cluster mass and kinetic energies distribution must be achieved. These characteristics can be obtained with the use of beams produced by supersonic expansions. A supersonic beam can be schematically described as a gas stream expanding very rapidly from a high-pressure region (source) to a low-pressure region. The characteristics of the beam are mainly determined by the size and shape of the nozzle orifice and by the pressure difference between the two regions [5]. Compared to effusive beams used in Molecular Beam Epitaxy, supersonic beams provide higher intensity, and directionality allowing the deposition of epitaxial films with very high growth rates [6]. Custer beams

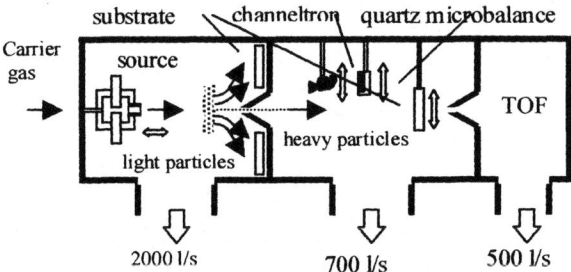

FIGURE 1: Schematic view (not to scale) of the apparatus for the production of supersonic cluster beams and for the deposition of nanostructured films.

can be used to grow nanostructured thin films where the original cluster structure is preserved after the deposition. The use of supersonic expansions may improve the deposition rate, and favor a better control on cluster mass distribution, thus making this technique competitive compared to other synthetic methods [2].

SETUP AND RESULTS

Figure 1 shows the principle of operation of the cluster beam apparatus that we have developed for the deposition of nanostructured thin films.

The first chamber hosts the cluster source and has a base pressure of typically 1×10^{-7} Torr, during source operation the average pressure is maintained in the range of $1-3 \times 10^{-5}$ Torr. The supersonic cluster beam enters the second chamber through an electroformed skimmer of 2 mm diameter. The second chamber is equipped with a sample holder which can intersect the beam, a quartz microbalance for beam intensity monitoring, and can alternatively host a beam-chopper or a fast ionization gauge for time of flight measurements of the velocity distribution of particles in the beam. During deposition the background pressure is typically 1×10^{-7} Torr. The third chamber hosts a linear time of flight mass-spectrometer (TOF/MS) that is placed collinear to the beam axis in order to achieve the best transmission. The detector of the TOF/MS, being sensitive to high-speed neutral clusters as well, can also be used for time of flight characterization [7].

The clusters are produced by a pulsed microplasma cluster source (PMCS) originally developed in our laboratory [7]. The PMCS is based on the following principle: a He pulse is directed against a target and it is ionized by a pulsed discharge fired between the target (cathode) and an electrode (anode). The target ablation is obtained by He plasma sputtering. A schematic representation of the PMCS is shown in Fig. 2: the source consists of a ceramic body with a channel drilled to intersect perpendicularly a larger cylindrical cavity. The channel hosts two rods of the material to be vaporized. A solenoid pulsed valve faces one side of the cavity. A removable nozzle closes the other side of the cavity. The discharge, driven by high voltage (between 500 and 1500 V), is very intense (~ 1000 A), lasting a few tens of microseconds, and produces the plasma for the ablation of the cathodic material. The ablated material is quenched by He and condenses in clusters that are carried out of the source in a seeded supersonic expansion through a nozzle of 2 mm diameter and 8 mm length. During typical operation the average pressure inside the cavity is of several hundreds Torr and the source body reaches a temperature of ~ 400 K. Stagnation temperature of the carrier gas is, however, a function of time, evolving very rapidly down to ~ 100 K as expansion takes place and the source gets empty. The velocity of the carrier gas is thus about 2000 m/s at the time the first clusters come out of the nozzle, but slows down to ~ 1000 m/s at the tail of the cluster pulse. A velocity slip of the clusters with respect to the carrier gas is also present but becomes of some relevance only for clusters exiting late from the source, when the stagnation pressure is reduced.

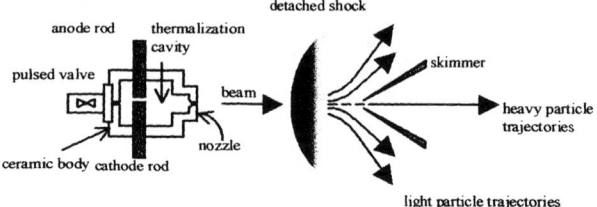

FIGURE 2: Expanded view of the pulsed microplasma cluster source (PMCS) and of the region near the skimmer where a shock wave is formed. The trajectories of the heavy and light particles are schematically shown.

In analogy with the case of laser vaporization sources [9], a cavity, where cluster aggregation occurs, is beneficial for several reasons. The erosion of the cathode does not affect significantly the dynamics of the gas during the expansion. The cavity itself decouples the cathode sputtering from the cluster formation process. The stability of the source is substantially improved and electrode erosion affects mainly the intensity but not the mass distribution of the clusters.

With typical discharge conditions, we obtain a lognormal cluster mass distribution in the range of 0-1500 atoms/cluster, with a maximum peaked at around 400 atoms/cluster and an average size at about 950 atoms/cluster. The kinetic energy of the clusters is lower than 0.5 eV/atom, well below the binding energy of carbons in the cluster. At cluster impact on the surface there is thus no substantial fragmentation of the aggregates and deposited films keep memory of the structure the clusters had in the gas phase [7].

Due to the long gas pulse exiting from the source, (i.e. high duty cycle regime), the source-skimmer distance D_{sk} and the background pressure strongly affect the expansion. Depending upon D_{sk}, a shock wave can be produced in front of the skimmer, causing mass separation effects and changing the final characteristics of the beam. In the reported experiments, the nozzle-skimmer distance has been varied from 40 mm to 16 mm. With D_{sk} = 40 mm deposition rates of 8 nm/min are routinely obtained on a substrate placed at 300 mm from the source. Circular films with a radius of 1 cm and uniform thickness can be deposited in the second chamber of the apparatus. Intersecting the beam in the first chamber films with an area of several cm^2 can be prepared.

If species with different weights are present in the gas to be expanded, the heavier constituents concentrate along the core of the beam [10, 11]. Different mechanisms have been proposed to account for these effects. Waterman and Stern have suggested that the flux of lighter species diverges radially more rapidly after the nozzle due to the greater thermal velocity components [12]. Reis and Fenn [10] have shown that mass separation can be obtained by exploiting the interaction of the beam with the shock wave detached from the skimmer. Due to their different inertia, light species follow diverging streamlines after the shock front, while heavy species are not diverted and can follow straight trajectories through the skimmer (see Fig. 2). Large clusters are then concentrated in the central portion of the beam, whereas the lighter ones are at the periphery. We have used separation effects to deposit thin films, with different cluster mass distributions, by intersecting different regions of the beam spot with a substrate.

Separation effects in front of the skimmer should enrich the periphery of the beam of small chain-like clusters, leaving large fullerene-like clusters in the beam center. Films grown using the periphery of the beam are then expected to show a very disordered structure; on the other hand films grown with the central region of the beam should be characterized by a disordered graphitic structure reminiscent of the fullerene-like character of the clusters [13]. Raman spectroscopy of films deposited with different portions of the cluster beam confirms the presence of a "memory" effect. Films grown with small clusters show a Raman spectrum typical of amorphous carbon, whereas films assembled with large fullerene-like clusters show a Raman spectrum typical of disordered graphite [7]. Fig. 3 shows a SEM micrograph of the cluster-assembled film. It is well known that the morphology of thin films shows, regardless of the material, universal characteristics [1]. These ubiquitous features consisting of arrays of columnar and conical structures are correlated with low adatom mobility. This is due to the deposition regime where the impinging particles have a low mobility. The general occurrence of morphological similarities on different length scales, as the growth takes place, suggests that a non-specific mechanism is responsible for the observed morphologies and that formation, growth and dynamics of film surfaces can be described in terms of scaling relations and universality classes. Cluster beam deposition allows to investigate how the precursor particle dimensions affect thin film morphologies and, more generally, if there is any influence of the particle dimensions on the scaling parameters. The film growth mechanisms with cluster beam deposition have been studied for the very initial stages at sub-monolayer coverages [14]. In cluster-assembled carbon films we observe the formation of microstructures typical of atom-assembled thick films deposited at glancing angle, where shadowing effects are dominant [15].

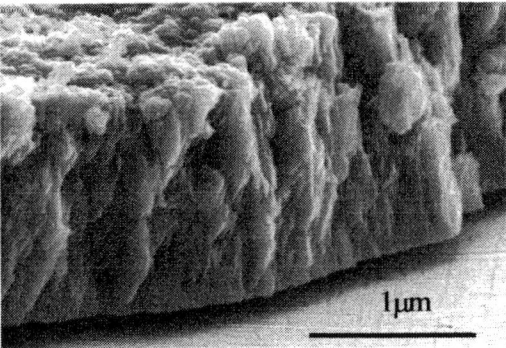

FIGURE 3: Section of a thick cluster-assembled carbon film. Conical and nodular structures are visible.

Cones and spherical nodules develop as the film grows. At higher magnification the cones are composed by dendritic structures (not shown). The large number of these defects indicates that the mobility of the clusters is very low. However, the role of the deposition rate must also be decoupled and separately studied. An AFM picture (Fig. 4) shows the granular structure based on clumps of spherical aggregates with typical diameter of few tens of nanometers.

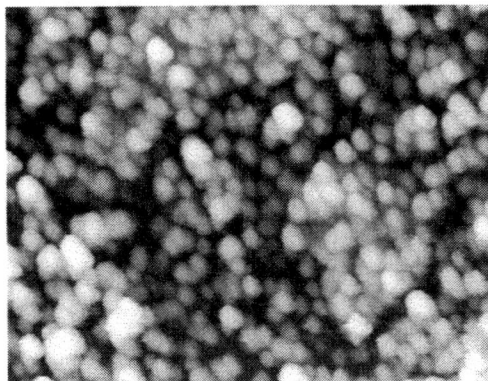

FIGURE 4: AFM micrograph of the surface of a nanostructured carbon film. The micrograph has been taken in tapping mode with a Digital Nanoscope IIIA instrument. The scan size is 10 μm x 10 μm.

This porous structure on different length scales has several consequences on the structural and functional properties of the carbon films. Stresses can be easily accommodated by this open structure [16], films with thickness of several microns can be grown without delamination on metallic and polymeric substrates at room temperature. The high surface area is also very attractive for electrochemical applications such as super capacitors [17]. Nanostructured carbon films are also promising candidates for the realization

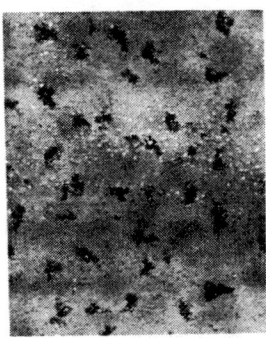 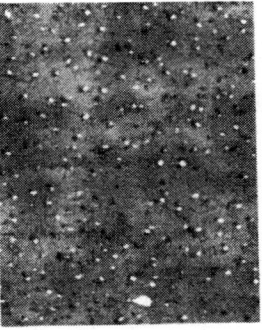

FIGURE 5: AFM micrographs (12 μm scan size) of SiC synthesis on Si(111)7x7. Left: thermal activation (900°C). Right: growth by supersonic beam of C_{60} (~5 eV).

of field emission devices in alternative to nanotube-based films [18].

Supersonic seeded beam of aerodynamically accelerated clusters and oligo-tiophenes can also be fruitfully used to prepare films of large interests for photonics and electronics. We have used a supersonic beam of pure C_{60} to activate kinetically the carbonization on a UHV atomically clean Si(111)7X7 surface. Here the major problems arise from the large number of defects basically due to the mismatch in lattice parameters (20% and to the high temperatures needed in the carbonization process [20]. We have proposed to activate the carbonization by the combined effect of temperature and kinetic energy of fullerene clusters aerodynamically accelerated, by seeding in He, up to 30-50 eV [21]. Fig. 5 shows the comparison of the difference in morphology obtained by AFM scans. The strong reduction in size and density of defects is clearly evident. In such samples LEED shows clearly a strong degree of crystallinity of the films [21]. We also made a supersonic seeded beam of oligothiphene [3], a prototype system for organic photonics and eletronics. Here the control of morphology, structure

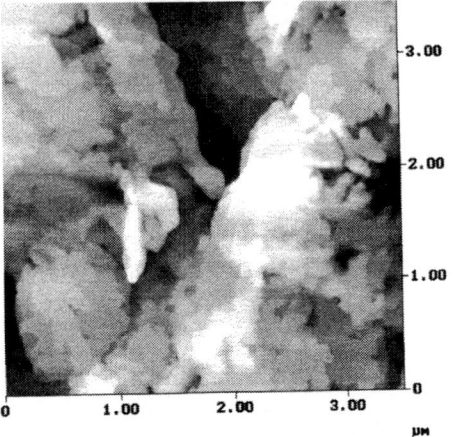

FIGURE 6: AFM micrograph of an oligothiophene film grown by seeded beams (8 eV). The "lamellar" structure can be controlled by the energy of the beam

(polymorphism) and density of defects is critical [23] to achieve functional performance. The control of beam parameters, in particular energy and momentum, allows an unprecedented control on the morphology and structure. Fig. 6 shows an AFM micrograph of typical sample prepared by a supersonic beam (8 eV). The structure and optical response of such films are very similar to those reported for single crystals [23]. The control achieved on morphology and structure is very promising to achieve the wanted charge and energy mobilities in these films.

CONCLUSIONS

In conclusion the use of supersonic cluster beams opens new perspectives for the synthesis of nanostructured and high quality films. The high deposition rates, the control on energy and cluster mass distribution makes this technique competitive with other synthetic routes for materials where a well-defined morphology and structure is required. Films grown with cluster beams show novel features of great interest both for pure and applied science.

This work is supported by CNR (MADESS II) and INFM (Advanced Research Project CLASS)

REFERENCES

1 J.E. Sundgren: in "Diamond and Diamond-like Films and Coatings", R.E. Clausing, L.L. Horton, J.C. Angus, P. Koidl (eds.), NATO ASI Series B 266, Plenum Press, New York (1991)
2 P. Milani, S. Iannotta, "Cluster Beam Synthesis of Nanostructured Materials", Springer Verlag, Berlin-Heidelberg (1999)
3 S. Iannotta, T. Toccoli, F. Biasioli, A. Boschetti, M. Ferrari, Appl. Phys. Lett. **76** (2000) 1845
4 A.S. Edelstein, R.C. Cammarata, in "Nanomaterials: Synthesis, Properties and Applications", IOP Publishing, Bristol (1996)
5 G. Scoles (ed.), "Atomic and Molecular Beam Methods", Oxford University Press, Oxford (1988)
6 D. Eres, D.H. Lowndes, J.Z. Tischler: Appl. Phys. Lett. **55** (1989) 1008
7 E. Barborini, et al., Chem. Phys. Lett. **300** (1999) 633
9 P. Milani, W.A. deHeer, Rev. Sci. Instrum. **61** (1990) 1835
10 V.H. Reis, J.B. Fenn, J. Chem. Phys. **39** (1963) 3240
11 E.W. Becker, K. Bier, Z. Naturforsch. **9a** (1959) 975
12 P.C. Waterman, S.A. Stern, J. Chem. Phys. **31** (1959) 405
13 R.O. Jones, G. Seifert, Phys. Rev. Lett. **79** (1997) 443
14 A. Perez et al., J. Phys. D: Appl. Phys. **30** (1997) 709
15 K. Robbie, et al., J. Vac. Sci. Technol. A **13** (1995) 1032
16 D. Donadio, L. Colombo, P. Milani, G. Benedek, Phys. Rev. Lett. **83** (1999) 776
18 L. Diederich, E. Barborini, P. Piseri, A. Podesta', P. Milani, A. Schneuwly, R. Gallay, Appl. Phys. Lett., in press
19 A.C. Ferrari, B. Satyanarayana, J. Robertson, W.I. Milne, E. Barborini, P. Piseri, P. Milani Europhys. Lett. **46** (1999) 245.
20 T. Fuyuki, T. Hatayama and H. Matsunami Phys. Stat. Sol. B **202** (1997) 359
21 G. Ciullo, M. Moratti, T. Toccoli and S. Iannotta, Philos. Mag., B **80** (2000) 635.
23 A. Podestà, T. Toccoli, M. Milani, A. Boschetti, S. Iannotta, Surface Sci. Lett (in press)

A two-step annealing of shallow junctions formed by GeB⁻ cluster ion implantation of Si

Xinming Lu[1], Lin Shao[1], Jianyue Jin[1], Xuemei Wang[1], Q. Y. Chen[1], Jiarui Liu[1],
Peiching Ling[2], and Wei-Kan Chu[1]

[1]*Department of Physics and Texas Center for Superconductivity at University of Houston, University of Houston, TX 77204 Advanced*
[2]*Materials Engineering Research, Inc., Sunnyvale, CA 94086*

Abstract. A two-step annealing was performed to study the diffusion of B in the GeB- cluster ion implanted Si. Samples implanted with 15-keV, 1x1015/cm2 GeB- ion cluster were rapid thermal annealed in dry N2 at (i) 1000°C/10sec (one-step annealing), (ii) 550°C/300sec, and (iii) 550°C/300sec+1000°C/10sec (two-step annealing) respectively for comparisons. We found that the junction depth of the two-step annealed sample was only half that of the one-step annealed sample. We argue that this is due to the reduction of Si self-interstitials during the first step annealing, which consequently suppressed transient enhance diffusion (TED) significantly. The samples were also evaluated by transmission electron microscopy, RBS/Channeling and SIMS profiling. The results showed that the two-step annealing procedure led to a defect free recrystalized region with Ge atoms mostly in the substitutional sites.

INTRODUCTION

One of the most difficult tasks in the formation of p-type shallow junction by ion implantation is the suppress of TED (Transient Enhance Diffusion) during the thermal annealing process. TED accounts for a substantial fraction of the dopant redistribution after annealing and may have caused the dopant diffusivity to be larger than the equilibrium diffusivity value by a factor of 10^2 to 10^4 [1]. TED is greatly believed to be caused by the interstitials generated during ion implantation. The excess self-interstitials drive the dopants out of the substitutional position during the thermal annealing process and force the dopants to go deeper into the silicon substrate. Many approaches have been taken to reduce the TED [1,2], such as C co-implantation, preamorphization using Si or Ge. Stolk et al [3] utilized a B delta-doped layer to show that without Si implantation there was only thermal equilibrium diffusion of B, but when Si was implanted into the wafer, the B delta-doped layer would then experience the TED. This suggests that in order to suppress TED, one should manage to reduce the silicon self-interstitials.

Two-step annealing was extensively used to reduce radiation damage. Previous studies showed that solid phase epitaxial growth (SPEG) of radiation damaged silicon occurred at temperatures between 500°C and 600°C, and this SPEG resulted in essentially damage-free crystals based on RBS channeling measurements [4,5,6]. In one-step annealing above 600°C, more defects are introduced. The rate of SPEG depends on the type of dopant, dopant concentration, substrate orientation and temperature. A TEM study on Arsenic ion implantation by Chu et al [7] showed that at 550°C the best epitaxial regrowth of Si with minimum defect occurred. This is because (i) During the 550°C first step annealing. The amorphous Si regrows into single crystal and any imbalance of self interstitials and vacancies are mostly annihilated, while boron atoms stay mostly undiffused. (ii) During the 1000°C RTA (Rapid Thermal Annealing), line defects such as dislocation will be removed and boron will be activated with limited diffusion. The separation of crystal regrowth, point defects removal with the dopant activation and diffusion is the key idea to control the boron TED process. Recently, GeB cluster ions have been used for low energy ion implantation study[8]. In this work, we have used GeB⁻ cluster ion for implantation of B. All as-implanted samples have been used to undergo various annealing procedures for comparative studies.

EXPERIMENTAL

In our experiment, the GeB⁻ cluster ions were generated by an SNICS (Source for Negative Ions by Cesium Sputtering) ion source of a tandem accelerator. Si substrates from Czochralski-grown <100>, 10-20Ω•cm, n-type wafers were cleaned by trichloroethylene, acetone, methanol, de-ion water and 5% HF and then implanted by $1 \times 10^{15}/cm^2$, 15-keV, GeB⁻ cluster ions at room temperature with a normal angle to the ion beam. The partial energy of B for 15-keV GeB cluster ion implantation is 1.94keV, determined by the mass ratio. This use of mass ratio in a cluster to effectively reduce the boron energy has been proposed earlier by Ling et al [9]. After implantation, the samples were rapid thermal annealed in dry N_2 at (i) 1000°C/10sec, (ii) 550°C/300sec, and (iii) 550°C/300sec+1000°C/10sec using an AG Associate Heatpulse 210T annealer. The annealing time in the first step was chosen by considering the solid state epitaxial regrowth rate at 550°C, which is about 1.2Å/sec from previous study. Hence an annealing time of 300-second is enough to regrow the crystal with minimal defects. After annealing, the samples were characterized by Secondary Ion Mass Spectrometry (SIMS), Cross-sectional Transmission Electron Microscopy (XTEM), four point probe sheet resistance measurements and Rutherford Backscattering Spectrometry (RBS) /Channeling.

RESULTS

The SIMS profiles were performed at Charles Evans & Associates with an O_2 primary ion beam at 1.5 keV and 500nA. Fig. 1 gives the SIMS profiles for samples annealed under various procedures. Worth noting in the profiles is the remarkably reduced junction depth of the sample with two-step annealing as compared to that of the one-step annealed sample. Based on the previously reported work on the suppression of TED[1,2], it is conceivable that in our case the TED has been cut down by a factor of two because the amount of self-interstitial has been greatly reduced during the first step annealing at 550°C. Note from the figure that the profiles also show a larger slope for the two-step annealing as compared to that of the one-step annealing and parallel to the as-implanted profile at the concentration below $1 \times 10^{19}/cm^3$, much like a horizontal translation of the as-implanted profile. While a more detailed analysis is being conducted to understand the significance, if any, of this observation, we believe that the mechanism of diffusion would be an important part in this parallel shift.

The Ge profile was almost unchanged after annealing suggesting that only limited thermal diffusion occured during the thermal process. The exact role that Ge played in the two step annealing is a subject currently under study in our laboratory. However, naively, one can argue that Ge atoms would stay stationary while the B diffused through the Ge populated region, because in the two-step annealing the Ge are already placed in proper lattice positions during the first stage annealing. But because of strain in the Ge region [10] when Ge is in substitutional sites, B atoms traversing through the Ge region may have a strain compensation effect that lowers the overall free energy during the first-step annealing and thus hampers the subsequent B diffusion.

All samples were subjected to RBS and channeling[11] to determine if Ge atoms were in substitutional sites or not. RBS and ion channeling measurements were carried out on a 1.7 MV tandem accelerator in our laboratory. The calculation of substitutional percentage was based on the following formula [12,13]:

$$S = (1-\chi_{Ge})/(1-\chi_{Si})$$

Where S is the fraction of substitute Ge, χ_{Ge} is the fraction of interstitial Ge as measured by RBS and channeling, and χ_{Si} is the fraction of interstitial Si. The results showed that 97.8% Ge occupied the substitutional sites for the 1000°C/10sec annealing, 97.6% for 550°C/300sec, and 98% for the two-step

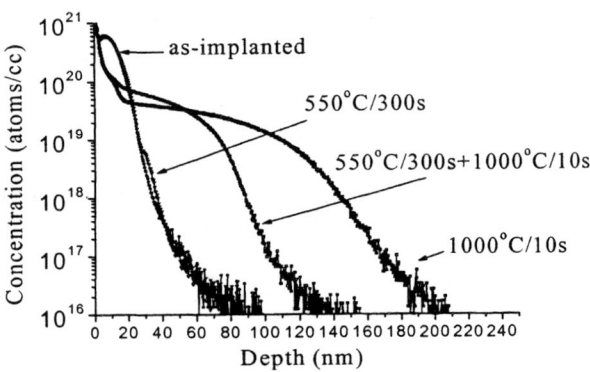

Figure 1. SIMS profiles of 15-keV, 1×10^{15} GeB⁻ implanted into Si with various annealing labeled.

annealing. The consistently similar S value led us to believe that the Ge atoms have all been placed in the proper substitutional sites in the Si lattice.

Figure 2. TEM of samples with 15-keV, 1×10^{15} GeB$^-$ implanted into Si. Annealing conditions: (a) as-implanted, (b) 1000°C/10s, (c) 550°C/300s, (d) 550°C/300s+1000°C/10s.

The SIMS profile of the 550°C/300sec sample was almost identical to that of the as-implanted sample except there was a kink around 280Å and this position was located around the interface between the amorphized and crystalline regions which is about 240Å according the cross sectional TEM. This may have been caused by the diffusing boron atoms interacting with the damage at the EOR (end of range) area[14,15], judged from two very similar characteristic depth.

TEM was done at Advanced Materials Engineering Research, Inc. (AMER) Sunnyvale, California. Fig.2 is the high-resolution cross-sectional TEM images for the four samples of 15 keV, 1×10^{15}cm^{-2}. Fig.2a shows the amorphized region with a clear a-c interface for the as-implanted sample. Fig. 2b-2d is the cross-section, respectively, of the 1000°C/10sec, 550°C/300sec, and two-step annealing samples. It is worth noting that there was no observable defects found in all samples except for some minor end of range lattice distortion possibly inflicted by Ge atoms. The exact cause of such distortion, however, must be further confirmed and is now being investigated in our laboratory. Note the difference between Fig. 2b and Fig. 2d in the surface roughness, with the two-step annealing sample being better than the 1000°C/10sec sample.

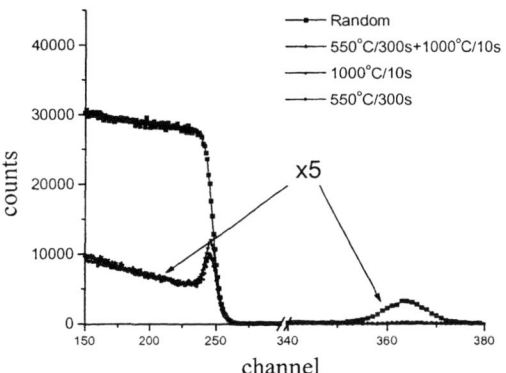

Figure 3. RBS/Channeling measurement for sample annealed with 1000°/10s, 550°/300s, 550°/300s+1000°/10s. The χ_m is 4.1%, 4.3%, and 4.1%.

RBS/Channeling were also carried out for these samples. Fig. 3 is the channeling measurements. The minimum yields of these three samples were 4.1% for the 1000°C/10sec sample, 4.3% for 550°C/300sec sample and 4.1% for the two-step annealing sample. The sheet resistance of these samples 440Ω/sq for the 1000°C/10sec annealing sample and 536Ω/sq for the two-step annealing sample. The dopant loss for the 1000°C/10sec annealing is 7%, while that for the two-step annealing is 28% based on Nuclear Reaction Analysis (NRA) measurements [17, 18].

CONCLUSION

In conclusion, we have utilized GeB$^-$ cluster ion implantation to study the two-step annealing for B shallow junction formation. The two-step annealing (550°C/300sec+1000°C/10sec) process gives a much shallower junction depth as compared to that of one step annealing (1000°C/10sec). We believe that the TED has been greatly reduced in the two-step annealing. The high-resolution TEM picture showed that the two-step annealing sample has defect free regrowth lattice structure, the consistent SIMS and TEM results suggests that Ge inflicted damage plays an observable role in the B diffusion.

ACKNOWLEDGEMENTS

This project was support by the State of Texas through the Advanced Technology Program and through the University of Houston, the facilities at the Texas Center for Superconductivity have been used.

REFERENCES

1. Jones, E. C., Ishida, E., Materials Science and Engineering, R**24** (1998) 1-80

2. Shao, L., Lu, X., Jin, J., Li, Q., Liu, J., Heide, P.A.W.van der and Chu, W.K. *Appl. Phys. Lett.* **26** (2000) 3953.

3. Stolk, P. A., Gossmann, H. J., Eaglesham, D. J., and Poate, J.M., *Nucl. Instrum. Methods Phys. Res.* **B96**,187 (1995)

4. Csepregi, L., Chu, W. K., Mueller, H., and Mayer, J. W., *Radiat. Eff.* **28,** 277(1976)

5. Csepregi, L., Kennedy, E. F., Gallagher, T. J., Mayer, J. W., and Sigmon, T.W., *J. Appl. Phys.* **48**, 4234(1977).

6. Kennedy, E. F., Csepregi, L., Mayer, J. W., and Sigmon, T. W., *J. Appl. Phys.* **48**, 4241(1977)

7. Alessandrini, E. I., Chu, W. K., and Poponiak, M. R., *J. Vac.Sci.Technol.,* **16**(2), Mar./Apr. 1979.

8. Lu, X. M., Shao, L., Jin, J. R., Li, Q., Rusakova, I., Chen, Q. Y., Liu, J., Ling, P., Chu, W.K., MRS meeting, San Francisco, April 2000

9. Ling, P. Strathman,M. D., Ling,C. H., *1998 International Conference on Ion Implantation Technology Proceeding,* Ed. J. Matsuo, G. Takaoka, I, Yamada, IEEE Catatog Number 98EX144, ISBN 0-780304538-X (softbound), and 0-7803-8 (microfiche). Also, see US Patents No. 5,863,831, issued January 26, 1999 and 5,763,319, issued June 9, 1998.

10. Eberl, K., Iyer, S. S., Zollner, S., Tsang, J. C., and Legoues,F. K., *Appl. Phys. Lett.* **60** (1992) 3033.

11. Chu, W. K., Mayer, J. W., and Nicolet, A., *Backscattering spectrometry* (Academic Press, New York 1982)

12. North, J. C., and Gibson,W. M., *Appl. Phys.Lett.* **16,** 126(1970)

13. Feldman, L. C., Mayer, J. W., Picraux, S. T., "*Materials Analysis by Ion Channealing*", Academic Press, Inc. New York,1982.

14. Barrett, C. R., Nix, W. D. and Tetelman, A. S., *The Principles of Engineering Materials* (Prentice-Hall, Englewood Cliffs, NJ, 1973), p. 260.

15. Roth, E. G., Holland, O. W., Venezia, V. C., and Nielsen, B., *J. Electronic Materials*, Vol. **26**, No. 11, 1997.

16. Liu, J. R., Lu, X. M., Chu, W. K., *Cross Section of $^{11}B(p,\alpha)$ ^{8}Be Reaction,* Submitted to Nucl. Instr. &Meth.. B

17. Liu, J. R., Lu, X. M., Chu, W. K., *Differential Cross Section of $^{11}B(p,\alpha)$ ^{8}Be Reaction* Submitted to Nucl. Instr. &Meth.. B

Nonlinear Effect of Carbon Cluster Induced Damage in Silicon

Zhaoxia Xie, Xuemei Wang, Xinming Lu, Lin Shao, Jiarui Liu and Wei-Kan Chu

Department of Physics and Texas Center for Superconductivity, University of Houston, Houston, Texas 77204

ABSTRACT: Carbon clusters C_n for n=1 up to 10 were obtained from a SNICS ion source. The clusters with the same velocity were implanted into silicon crystal. We have observed a nonlinear effect on radiation damage of silicon due to the size of the cluster ions. The larger the size of the cluster ions, the larger the radiation damage after normalized with the atomic dosage. The dependence of the damage on the implantation dosage was also studied. The quantitative characterization of the damage was performed by RBS/Channeling analysis and the nonlinear effect in the cluster-solid interaction will be presented.

INTRODUCTION

During the last few years, cluster ion beam was successfully applied in material processing. Cluster ion beam has been applied to produce ultra-shallow junctions in semiconductor device fabrication [1-3]. High rate sputtering has been observed with the sputtering yield of about 2 orders of magnitudes higher than the sputtering yield by monomers with the same energy per atom [4]. Atomic scale smoothing of surfaces by cluster ion bombardment has been applied to different materials to get the roughness of 0.2-0.8 nm [5,6]. Cluster ion beams also have been used for thin film formation at low substrate temperature [7,8]. Unfortunately, the interaction of a cluster of n atoms is not a result of the sum from n single-atomic ion interaction. This is the non-linear effect in cluster-solid interaction (or collective effects) [9]. The non-linear effect is dramatic and it should appear in all aspects in cluster-solid interaction, including projected range, straggling, defect formation and sputtering. The non-linear effect needs to be studied in details to benefit the cluster ion applications. In recent years, a few experiments on non-linear effect were performed with MeV polyatomic ion beams, where the electronic energy loss dominated. These papers were concentrated on defect production and energy loss [10-12]. The low energy range of keV or below, where the nuclear energy loss governed, is more important for material applications. Unfortunately, there is no systematic experiment of non-linear effect in this energy range. Aoki et al. studied the damage creation due to carbon clusters by molecular dynamic simulation [13]. This is basically due to the difficulties of getting the small clusters experimentally. Recently, we have obtained record large beam for small clusters from a SNICS source [14]. Clean carbon clusters from monomer up to C_{10} for the beam current from a few hundreds µA to 50 nA were obtained. These clusters allow us to do some systematic experimental studies of the cluster solid-interaction.

In this paper we concentrate on experimental measurement of radiation damage induced by C_1-C_{10} clusters in keV energy range. He^+ ion channeling in Si is used to measure the amount of radiation damage.

EXPERIMENTAL

The carbon cluster ions of different size (C_1-C_{10}) were extracted from a Source of Negtive Ion by Cesium Sputtering (SNICS) using a high density graphite target. In the source, a focused beam of Cs^+ ions of a few keV bombards the cathode to produce negative C_n ions. The negative C_n ions are extracted

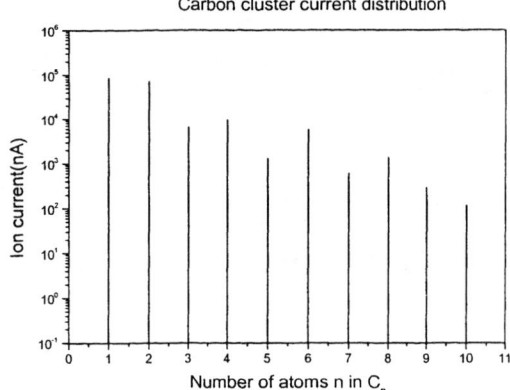

FIGURE 1. Mass Spectrum of carbon clusters from the SNICS ion source at typical sputtering voltage of 5keV.

in a few keV range and can be further accelerated up to 70keV. The carbon clusters were selected with an electromagnet of 30°. The electromagnet is able to deflect ions up to 350amu. An einzel lens and an adjustable double slit are used in front of the sample. The beam current was measured with a Faraday cup. A typical mass spectrum of carbon clusters is shown

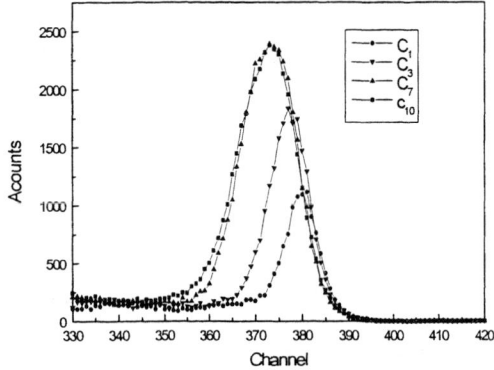

FIGURE 2. RBS/channeling spectra of C_1, C_3, C_7, C_{10} implanted Si at tilted angle of 15°.

in Fig.1. The cluster sizes larger than 10 were contaminated by Cs (C_{11}) and CsCn (C_{n+1}).

P-type Si wafers were implanted with the same energy per atom (6 keV) to the same atomic dosage (1×10^{15} atoms/cm^2). The dosage rate was kept approximately the same for all clusters. The samples were tilted for 15° to the incident beam. C_6 was implanted in four different dosages to study the damage-dosage dependence.

RESULTS AND DISCUSSION

The implanted samples were measured using glancing angle RBS/channeling for beam normal incidence and the detector at 98° with good depth resolution. Figure 2 shows the spectra of the surface-damage peaks measured by glancing angle RBS/channeling. The displaced Si atoms due to the cluster bombardment contribute to the surface-damage peaks. The peaks come from two parts: one is from the carbon cluster ion damage and the other is from the

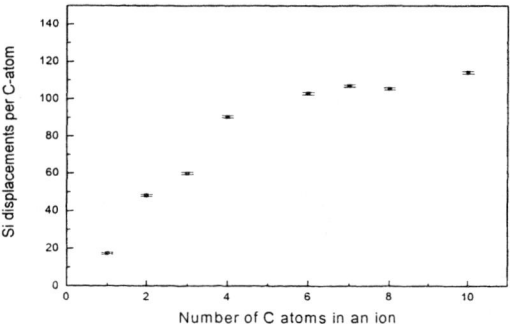

FIGURE 3. Number of displaced Si atoms induced by C_1-C_{10} Si samples tilted at 15°.

surface peak of the virgin wafer. Thus the number of displaced Si atoms is determined from the difference of the net counts of the surface peak for the bombarded sample and the virgin Si wafer. The damage in dependence on the cluster size is shown in Fig. 3. The vertical scale shows the number of displaced Si atoms per incident C atom. As we have expected, the implantation damage increases with the number of atoms in a cluster. This is a clear demonstration of the nonlinear effect of irradiation damage by clusters. Worth noting, that from C_1 to C_4 the increase is very fast, but from C_6 to C_{10} the damage increase declined indicating that there is a tendency that the irradiation damage will saturate as the number of atoms in an ion increase to around 8-10. There was contamination of Cu and Ag in C_5 and C_9 respectively, so the points were dropped out. They may get corrected with further experimentation.

Our results can be qualitatively understood by the "overlapping model" in the damage formation [13,15]. As a general picture, the "overlapping model" suggested that some temporal damage might convert into permanent damage due to the damaged area formed by single fragment- atoms from the cluster overlapped. Quantitatively, our experimental damage-cluster size dependence in Fig. 3 is quite different from the molecular dynamic simulation by Aoki et al. (see Fig. 3 in reference 13). First, the experiment shows a very fast increase of the damage per atom for the clusters from C_1 to C_4 by a factor of about 5, but the molecular dynamic simulation indicate an increase of about 20% only. Second, the saturation was observed around C_6–C_{10}, but the simulation shows a maximum damage at about C_{19}. The difference in the energy-per-atom (6 keV/atom to 2 keV/atom) would not give such a big difference in the cluster size dependence. However, due to the complicity of cluster-solid interaction, more research needs to be done. An experiment with a broad range of the cluster size up to C_{20} would be very helpful to clarify this problem.

Figure 2 shows the RBS/Channeling surface-damage profile for C_1, C_3, C_7, C_{10} clusters. As we can see from the shape of the peaks, the front of the peaks are the same but the damage peak moves deeper and deeper from C_1 to C_{10}. The depth profile of the damage is higher than the detection resolution, so it indicates a real deeper damage of the larger clusters. This also indicates that the projected range is increasing with the cluster size. The non-linear effect in the projected range seems very pronounced also, but a quantitative measurement should be done with better depth resolution. This may be explained by Coulomb explosion; the strong Coulomb interaction between naked nuclei after the electron is stripped may knock some fragment-atom forward with enough energy to form deeper damage.

Figure 4 shows the dosage dependence of the number of displaced Si atoms. The damage was induced by C_6 on tilted Si sample by 15°. The beam current in this measurement was kept almost constant. The damage is increasing with the dosage and tends to be saturated after $2 \times 10^{15}/cm^2$.

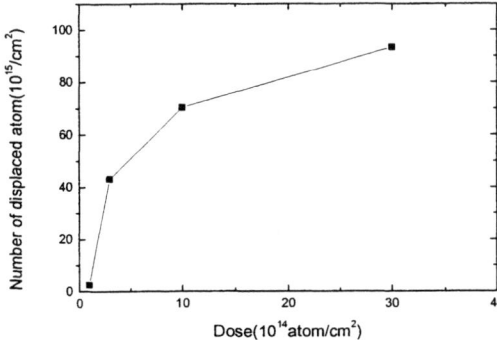

FIGURE 4. Number of displaced Si atoms induced by C_6 in dependence of the total dosage.

CONCLUSION

Carbon cluster ions (C_1-C_{10}) induced damage on silicon substrate was investigated by cluster implantation in Si at the same energy-per-atom and for the same dosage of atoms/cm^2. The damage was measured by glancing angle RBS/channeling. A dramatic non-linear effect was observed in damage-per-atom vs. cluster size dependence. For small cluster-size from 1 to 4, the damage-per-atom increases by a factor of 5 and the damage tends to saturate for C_8-C_{10}. This can be qualitatively understood in "overlapping model", but the quantitative disagreement is dramatic and has to be clarified.

The depth profile of the damage in dependence of the cluster size clearly indicates a non-linear effect of the projected range of the clusters. The non-linear effect on the projected range will be quantitatively studied with better depth resolution.

ACKNOWLEDGEMENT

This project was supported by The State of Texas through the Texas Center for Superconductivity, partially by the 003652-797 ARP and the National Science Foundation through the Materials Research Science and Engineering Center.

REFERENCE

1. Matsuo, J., Takeuchi ,D., Aoki T., Yamada, I., Ishida, In. E.,,Banerjee, S., Meta, S., Smith, T. C., Current, M., Larson, L., Tash(Eds.) A., *IEEE Proceedings of the 11th International Conference on Ion Implantation Technology.* Austin, TX, vol. 1, issue 1. 16-21 June 1996, IEEE Service center, Piscataway, NJ, 1997, P.768.

2. Takeuchi, D., Shimada, N., Matsuo, J., Yamada, I. Ishida, in: E., Banerjee, S., Meta, S., Smith, T.C., M., Current, Larson, L., A. Tash(Eds), *IEEE Proceedings of the 11th International Conference on Ion Implantation Technology.* Austin, TX, vol. 1, issue 1. 16-21 June 1996, IEEE Service center, Piscataway, NJ, 1997, P.772.

3. Chu, Wei-Kan, Liu, Jiarui., Jin, Jianyue., Lu, Xinming., Shao Lin and Li, Qingmian., *This proceedings.*

4. Yamada, I., Matsuo, J., Advanced metallization for future ULSI. In: K.N. Tu, J.M. Poate, J.W. Mayer, L.C. Chen(Eds.), *Material research Society,* Pittsburgh, 1997. P.265

5. Toyoda, N., Matsuo, J., Yamada, I., in: Duggan, J.,L., Morgan(Eds.), I.L. *Proceedings of the Application of Accelerators in Research and Industry' 96*. AIP Press. New York, 1997, P.483.

6. Chu,,W.,K., Li, Y.,P.,Liu, J.,R., Wu, J.,Z., Tidrow,,S.,C., Toyoda, N., Matsuo, J., Yamada, I., *Appl. Phys. Lett.* **72**(1997)246.

7. Yamada, I., *NEDO Consortium Program 1997*, Project Office: Management Office: Osaka Science and Technology Center, Utsubo, Osaka Japan: R&D for ultra-high quality transparent conductive film fabrication.

8. Yamada, I., *Japan Science and Technology Corporation, Agency of Science and Technology(JST), Innovative Research Application Program, 1998-2000.*

9. Sigmund, P., Bitensky, I.S., Jensen, J., *Nicl. Instr. Meth. B* **112**(1996)1-11

10. Döbeli, M., Ames, F., Ender, R. M., Suter, M., Synal, H. A., Vetter. D., *Nucl. Instr. and Meth. B* **106** (1995) 43-46

11. Baudin, K. , Brunelle, A., et al *Nucl. Instr. and Meth. B* **94**(1994)341-344

12. Tomaschko, Ch., Brandl, D., et al, *Nucl. Instr. and Meth. B* **103** (1995) 407-411.

13. Aoki, T., Seki, T., Matsuo, J., Insepov, Z., Yamada, I., *Nucl. Instr. and Meth. B* **153** (1999) 264-269

14. Wang, Xuemei., Lu, Xinming., Shao, Lin., Liu, Jiarui., Chu, Wei-Kan., *This proceedings*.

15. Aoki, T., Seki, T., Tanomure, M., Aoki, T., Matsuo, J., Yamada, I., *MRS Symp. Proc.* **504**(1999) 93.

Development of Gas Cluster Ion Beam Equipment

M. E. Mack,

Epion Corporation, Billerica MA USA

Abstract. The development of a fully mechanically scanned gas cluster ion beam tool is described. Average surface roughnesses of better than 10 Å can be achieved with this equipment

INTRODUCTION

A new and rapidly growing form of ion-beam processing is gas-cluster ion beams (GCIB) [1]. The ions utilized in GCIB are clusters of atoms, typically, argon, nitrogen or oxygen with 500 to 5000 or more atoms contained in the cluster [2]. GCIB is directed not toward electrical modification of a layer of the processed material but to physical and chemical modification of the substrate surface. Typical accelerated energies for these clusters are in the range of 15 to 25 keV so that the energy per atom is only 3 to 30 eV. Penetration depths are proportionately reduced, and the majority of the material interactions occur only in the top 0.5 to 3 nm. Moreover, collapse of the weakly bound cluster on impact with the strongly bound solid surface results in the conversion of much of the forward directed momentum of the cluster into lateral momentum [3]. As a result, the angular distribution of sputtered material from the surface is no longer Lambertian but is laterally lobed, reflecting the lateral exit momentum. This lateral momentum assists in smoothing of the surface, a principle application of GCIB [3,4]. Other applications include surface cleaning, cluster chemical etching and cluster assisted (reactive) film deposition [5,6]. In Sec. 2 the generation of ion cluster beams is discussed while section 3 discusses the end station necessary to process such substrates. Section 4 shows initial processing results with this equipment (the Epion *Ultra Smoother*™ 50M). Conclusions are summarized in section 5.

GENERATION OF ION CLUSTER BEAMS

Figure 1 illustrates a typically system for generating ion cluster beams. The clusters are generated by gas expansion through a sonic or supersonic nozzle into a vacuum. The adiabatic expansion results in a sharp decrease in gas temperature. For such an expansion

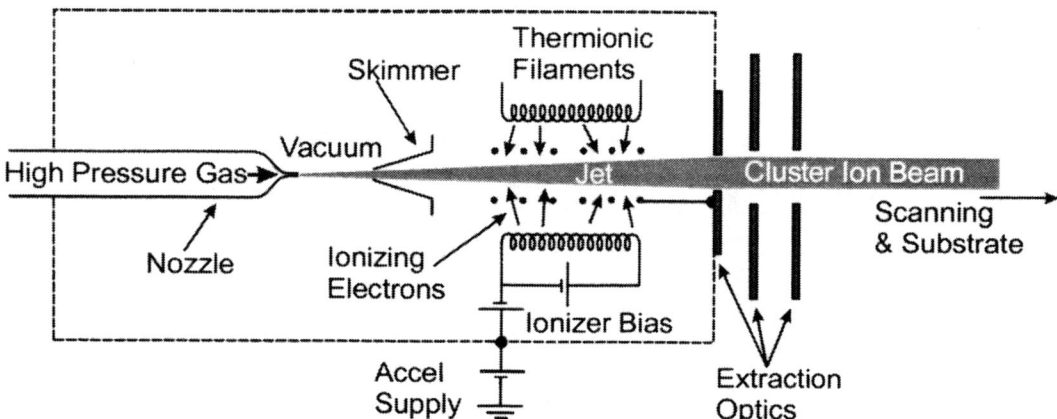

Figure 1. Cluster Generation and Ionization.

the temperature in the jet is

$$T = T_0 \left(P / P_0 \right)^{\frac{\gamma-1}{\gamma}} \quad (1)$$

where T_o and T are the initial and final gas temperatures, P and P_0 are the initial and final pressures, and γ is the specific heat ratio. For rare gases such as argon $\gamma=1.67$ while for diatomic gases like nitrogen and oxygen, $\gamma=1.4$. For the nozzles typically in use, throat diameters are in the range of about 0.1 mm. Upstream gas pressures are 4 to 8 atmospheres while ambient pressures downstream of the nozzle are 1 to 50 mTorr. Thus, gas temperatures in the jet of a few degrees Kelvin are expected resulting in condensation nucleation of the gas and formation of clusters. In practice, temperatures are slightly higher than predicted by equation (1), because of boundary layer effects in the flow and because of release of the heat of condensation on nucleation. Due to the lower specific heat ratio for the diatomic gases a higher required pressure ratio would be expected for cluster formation and this is borne out in practice. If the jet is left free in space a shock wave will be formed reheating the gas at the location of the first Mach disk [7]. This would destroy any clusters that had been formed. To prevent this occurrence the colder, central core region of the gas jet containing the clusters is passed through a skimmer as shown in Fig. 1.

To form the ion beam, the clusters within the jet are ionized by electron impact. The ionizing electrons, which are produced by thermionic emission from filaments, are then accelerated through a cylindrical screen by a bias potential of up to a few hundred volts. While this would seem to be a rather inefficient means of ionization, the large size of the clusters results in an increased ionization cross section [8] so that in excess of 70% of the clusters can be ionized by this means. Once ionized, the cluster beam is extracted to form an ion beam. Typically, extraction is coupled with a filter to separate out ionized monomers, which have not coalesced into clusters [5]. Extracted cluster ion beam currents of 50 μA or more can be achieved in this fashion. Note, that even at these currents, because of the high mass of these ions, the beams are of high perveance [9] and beam focussing and transport must be handled accordingly.

SUBSTRATE PROCESSING

Once extracted the cluster ion beam must be painted uniformly across the substrate in order to provide the smoothing desired. Early cluster processing equipment often utilized electrostatic scanning to cover the substrates. However, electrostatic scanning suffers several major drawbacks for high perveance beams. Of necessity the scanner voltages remove space charge

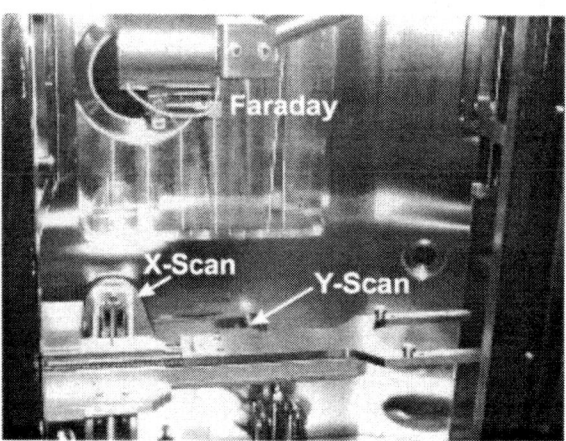

Figure 2 Mechanical Scanned End Station

neutralizing electrons from the beam. Mutual repulsion of the ions then causes an increase in the divergence of the beam. Scanner deflection angles are limited and so overall beamline path length must grow to cover 200 or 300 mm substrates. As a result of both the long path length and the increased divergence, spot size on the substrate may be comparable to the substrate size, reducing beam utilization and degrading uniformity.

Magnetic scanning has been used for high perveance beams [10] but the extreme magnetic rigidity of the cluster ion beams preclude magnetic scanning. Fortunately mechanical scanning provides a viable alternative. Figure 2 shows the two axis mechanical scan system used in the Epion *Ultra Smoother*™ processing systems. The horizontal (x-axis) and vertical (y-axis) scan motions are provided by commercial ball slides mounted together at right angles. The entire mechanism rotates about a horizontal axis so that substrates are loaded horizontally as shown in fig. 2 while processing occurs with the mechanism flipped up at an 85° angle. The ball screw mechanisms are lubricated with a very long life, and very low vapor pressure (10^{-10} Torr) lubricant. Typical base pressures for the end station are in the 10^{-8} Torr range and service life is expected to be in excess of six months. All mechanisms are located below the wafer during processing to minimize any possible particle contamination. The two axes shown are scanned at very different speeds with a speed ratio of roughly 40:1. The scan pattern used coupled with the fine pitch of the scans across the substrate and the large beam size (5 cm) relative to the pitch all ensure a uniformity well below the specified 2% one sigma. The Faraday shown in fig. 2 measures beam current each fast scan and adusts scan speed and scan pattern accordingly. This closed loop operation ensures that dose repeatability substrate to substrate and batch to batch is also well below 2% In many instances charging of substrates is a concern. For this reason the Epion *Ultra Smoother*™ includes a neutralization system in the beamline. The neutralizer provides up to 1000 μA of low energy electron emission to provide beam and substrate neutrality during processing. CHARM2 wafers attest to the effectiveness of this system with less than 3 volts average of charging positive and negative for isolated substrates and less than 5 volts for grounded substrates.

PROCESS RESULTS AND APPLICATIONS

Fig. 3 shows atomic force microscope photographs of smoothing with mechanical scanning in the case of a nickel iron alloy. These results were obtained using a Digital Instruments Nanoscope. Such materials are of interest for magnetic memories and read heads. Process dose was 1E15 cm^{-2} with 50 μA of beam current at 20 kV of acceleration energy. Both 3A and 3B show

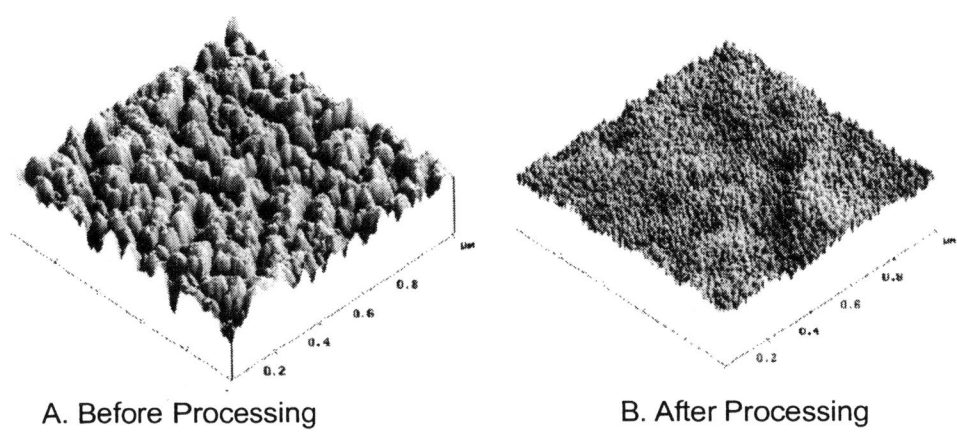

A. Before Processing B. After Processing

Figure 3. Typical Smoothing Results: Nickel Iron Alloy

a 1x1 μm section of the sample. Roughness at this dose was improved from a mean roughness of 20 Å before processing to 10 Å after processing. Typically smoothness follows a nearly exponential form with dose, with roughness diminishing rapidly at low dose (1E14 to 8E13) and eventually reaching a plateau at high dose (2E15 to 1E16 cm-2)[3].

CONCLUSIONS

Processing the surfaces of solids by bombardment with an argon ion cluster beams results in decontamination, etching and smoothing of those surfaces to very high figures of merit. This is especially attractive for applications in microelectronics and photonics where conventional ion processing often roughens surfaces or causes other forms of damage. The architecture of the machine, which has been developed, provides a stationary beam and minimizes beamline path length, ensuring maximum transmission to the processed substrate. This architecture is especially suitable to the high perveance ion cluster beams and is extendable to the much higher beam currents expected in the future.

ACKNOWLEDGEMENTS

The valuable and essential assistance of many coworkers, who participated in this development, is gratefully acknowledged. This work was supported in part by the U.S. DoC, through a NIST-ATP award (70NANB8H4011).

REFERENCES

[1]. I. Yamada, G. H. Takaoka, M. I. Current, Y. Yamashita and M. Ishi, Nucl. Inst. Methods B74, 341 (1993).

[2]. N. Toyoda, M. Saito, N. Hagiwara, J. Matsuo and I. Yamada, 1234, in *Ion Implantation Technology-98* edited by J. Matsuo, G. Takaoka and I. Yamada, IEEE, Piscataway, N. J. (1999).

[3]. N. Hagiwara, N. Toyoda, J. Matsuo and I. Yamada, 1230, in *Ion Implantation Technology-98* edited by J. Matsuo, G. Takaoka and I. Yamada, IEEE, Piscataway, N. J. (1999).

[4]. R. McEachern, J. Matsuo and I. Yamada, 1066, in *Ion Implantation Technology-98* edited by J. Matsuo, G. Takaoka and I. Yamada, IEEE, Piscataway, N. J. (1999).

[5]. E. Minami, W. Qin, M. Akizuki, H. Kastumata, J. Matsuo and I. Yamada, 1191, in *Ion Implantation Technology-98* edited by J. Matsuo, G. Takaoka and I. Yamada, IEEE, Piscataway, N. J. (1999).

[6]. H. Katsumata, J. Matsuo, T. Nishihara, T. Tachibana, E. Minami, K. Yamada, M. Adachi, and I. Yamada, 1195, in **Ion** *Implantation Technology-98* edited by J. Matsuo, G. Takaoka and I. Yamada, IEEE, Piscataway, N. J. (1999).

[7]. R. Campargue, J. Phys. Chem. 88, 4466 (1984).

[8]. § 5.6 Cluster Ion Sources in H. Zhang, *Ion Sources*, Springer-Verlag, Berlin (1999)

[9]. More correctly these are beams of high Poissonance. See A. T. Forrester, *Large Ion Beams*, Wiley & Sons, New York (1988)

[10]. G. Ryding, T. H. Smick, M. Farley, B. F. Cordts, R. P. Dolan, L. P. Allen, B. Mathews, W. Wray, B. Amundsen and M. J. Anc, 436, in *Ion Implant Technology-96* edited by E. Ishida, S. Banerjee, S. Mehta, T. C. Smith, M. Current, L. Larson, A. Tasch, and T. Romig, IEEE, Piscataway (1997)

O_2 Cluster Ion Assisted Deposition for Tin doped Indium Oxide (ITO) films

Jiro Matsuo, Gikan Takaoka and Isao Yamada*

*Ion Beam Engineering Experimental Laboratory., Kyoto University.
Sakyo, Kyoto, 606-8501, JAPAN*

あ B S T O_2 Gas Cluster Ion Beam assisted deposition technique has been developed to form ultra high quality tin doped indium-oxide(UHQ-ITO) films. This deposition process uses large cluster ions which can transport thousands of atoms in a ion with very low energy per constituent atom. Interactions between cluster ions and substrate atoms occur in the near-surface region and cluster ions can deposit their energy with a high density in a very localized surface region. The energetic oxygen clusters collapsed at the surface and reacted with the metal atoms and about 10% of them were incorporated, when the kinetic energy of the cluster ion was above 5 keV. Oxidation reaction can be enhanced by energetic cluster ion bombardment which offers a new technique for ion assisted thin film formation. Very smooth, highly transparent (>80 %) and low resistivity films, were obtained by using a 7 keV oxygen cluster ion beam. In order to realize high throughput for industrial application, a Multi-beam Gas Cluster Ion Beam equipment has been newly developed.

I. INTRODUCTION

There is a strong requirement to provide low temperature oxidation processing for semiconductor and optical device fabrication. Energetic beams are quite useful to achieve low temperature oxidation processing due to their excess kinetic energy [1-3]. Sn doped indium oxide (ITO) film, which is one of the most extensively used oxide, is transparent in the visible and electrically conductive. It has been applied to optoelectronic devices such as transparent electrodes of flat panel display(FPD) devices and photovoltaic devices, due to its relatively low resistivity ($<1 \times 10^{-3}$ $\Omega \cdot cm$) and high transparency to visible light (above 80% at 550nm).

As recent developments in technology, better electrical and optical properties and surface morphology are required for various optoelectronic device applications. For example, resistivity values lower than 1×10^{-4} $\Omega \cdot cm$ is required for transparent electrodes used in color liquid crystal digital (LCD) displays. Specifically, the transparent electrode used in color LCD needs to be formed on a heat-sensitive organic color filter. The temperature during film formation has to be kept below 150°C. However, it is difficult to form high quality films in this temperature range with current techniques such as magnetron sputtering or vacuum evaporation [4]. Furthermore, super-flat surfaces are required for the transparent electrodes used in organic electro-luminescence devices.

To meet these requirement for advanced devices, we have proposed O_2 cluster ion beam assisted deposition technique[5,6]. We have demonstrated direct oxidation of silicon surfaces at room temperature with a low energy oxygen cluster ion [6] and PbO_2 film formation under oxygen cluster ion irradiation during Pb evaporation [7]. Further, high quality tin doped indium oxide (ITO) film formation at room temperature has been reported recently [8].

Energetic cluster ion beams are very useful for surface processing, providing shallow implantation, high-rate sputtering, surface smoothing, cleaning and film formation as a consequence of the unique irradiation effects [5-14]. Clusters, aggregates of atoms or molecules, are interesting not only as a new state of matter but also as a new beam approach for material processing. When a cluster with the size of 1000 is accelerated with energy of 10 keV, each constituent atom has only 10 eV, and these 1000 atoms collide with the surface within a several nm^2. Thus, clusters impact a surface with a low equivalent energy (low velocity) but

*Present address: Laboratory of Advanced Science and Technology for Industry, Himeji Institute of Technology, CAST, Kamigori, Ako, Hyogo, 678-1205 Japan

with extremely high density.

It was found that energy control of the incident species (atoms or molecules) is a critical issue to reduce the resistivity of ITO films. When the energies of the incident species are high, resistivity of the film increases due to damage formed with ion bombardment. Obviously, when the energies are very low, enhancement of oxidation would be suppressed. Therefore, desirable energy of the incident species ranges from a few eV to a few tens of eV. The cluster ion beam has a suitable energy to meet this requirement. n this paper, oxygen cluster ion beams have been utilized in low resistivity ITO film formation.

II. THE GAS CLUSTER ION BEAM ASSISTED DEPOSITION EQUIPMENT

ITO film formation with O_2 cluster ion beam assisted deposition technique has been reported.[5-7] High quality ITO films can be formed at low temperatures. However in order to apply this technique industrially, it is necessary to increase the cluster ion beam current to deposit films on large area. A new multi-beam gas cluster Ion beam equipment is developed to obtain large cluster ion beam currents. Fig.1 shows a schematic diagram of multi-beam equipment. To operate this system easily and safely, control of the pumping system, the power supply, gate valves as well as the gas flow with computer is preferable.

This system equipped with two electron beam evaporators for evaporation of materials. ITO pellets or In , Sn metals are able to evaporate. The electron beam evaporation method is widely used for evaporation in industry, because of its wide range of control for evaporation rate. The deposition rate from the two electron beam evaporators are measured by crystal rate monitors and the rate can be kept at constant by feeding back mechanism. This help to stabilized the Sn composition ratio in the films, when co-evaporation technique is applied. While the cluster source chamber is evacuated by 880 l/s turbo molecular pump, and the deposition chamber is evacuated by 20" diffusion pump (17,500 l/s) down to 5×10^{-8} Torr.

The technique for the generation of gas cluster ion beams and for the cluster-size separation has been described previously[7,12] Oxygen cluster beams have been formed by adiabatic expansion through a Laval nozzle into the vacuum chamber. The clusters were ionized by electron bombardment, and the size selection of the cluster ions was performed using the retarding technique.

The cluster size distribution was measured using a Time Of Flight (TOF) technique. Cluster beams were ionized by electron bombardment with energy of 70eV in the Wiley-MacLaren type TOF system. The duration time of the extraction pulse was 10 ns and the repetition rate was 200Hz. Figure 2 shows typical TOF spectra of oxygen cluster beam. Oxygen cluster size is distributed up to several thousands and the mean cluster size is about 2000. Each size of oxygen cluster can be separated, as shown in the Figure 2-(b), when the cluster size is small (N<30). As shown in Figure 2-(a), mass resolution of the TOF system used in this study does not have sufficient mass-resolution to observe the peaks of each cluster size.

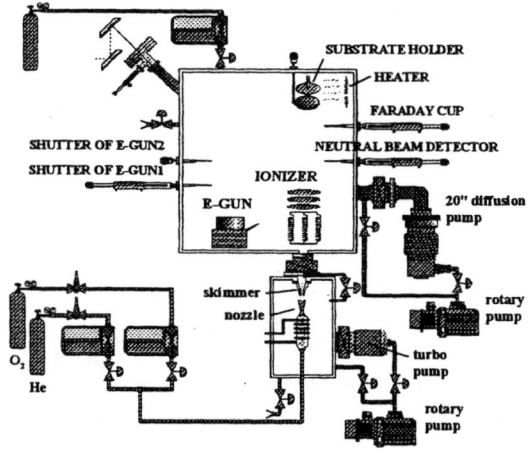

Fig.1 Schematic diagram of the Multi-beam equipment which was developed to obtain the large cluster ion beam current and to form high quality ITO films.

III. ITO FILM FORMATION

ITO films have been formed at room temperature by evaporating In and Sn in conjunction with the irradiation by O_2 cluster ions. It is quite difficult to obtain transparent films in oxygen ambient at room temperature. The irradiation area becomes atomically smooth and visibly transparent compared to the unirradiated area. According to Secondary electron Microscope(SEM) metallic Indium was precipitated in the unirradiated area. When unaccelerated cluster beams were used, transmission efficiency of the films deposited is not sufficient to optical applications. Highly transparent(>80%) films can be obtained, when oxygen clusters were ionized and accelerated at the energy of above 5 keV[7]. Therefore, oxidation reaction can be enhanced significantly with energetic oxygen cluster ion irradiation.

If mean cluster size is 2000, the average kinetic energy of the oxygen molecules in the cluster with the energy of 5 keV is only 2.5 eV. This energy is well below to the threshold energy of damage formation. As a consequence, the films with low resistivity were obtained. The deposition rate of In was about 15 Å/min. The In/O ratio of the films were measured as a function of acceleration voltage (Va). Stoichiometry of InO_x films was measured by

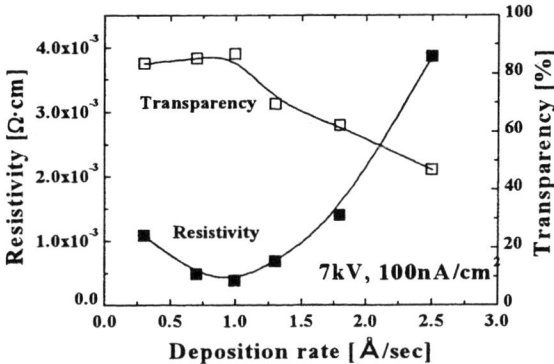

Fig. 3 Deposition rate dependence of resistivity and transparency

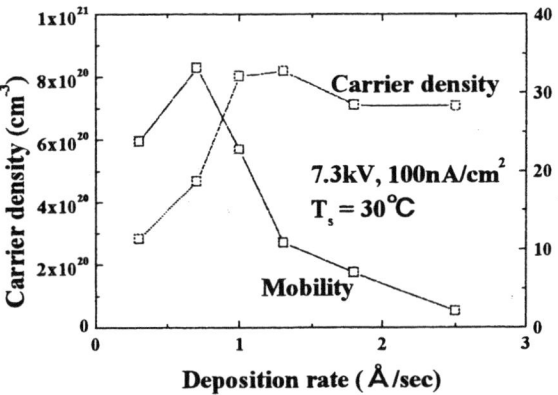

Fig. 4 Deposition rate dependence of carrier density and mobilty of undoped In_2O_3 films

Rutherhord Back Scattering (RBS) technique. Indium evaporation rate was kept at 20 nm/min. When O_2 cluster ion energy was above 5 keV, stoichiometric In_2O_3 films were grown at room temperature. The oxygen cluster beam current density was about 100 nA/cm^2 at 3 keV.

Transparent films could not be obtained at low acceleration voltage (<3 kV), even when deposition rate was as low as 2 nm/min. When ion current density was kept at 100 nA/cm^2, highly transparent films were easily obtained at 7keV. This result clearly demonstrates that energetic cluster ion has a great capability to oxidize the In surface. High acceleration voltage of above 5keV is necessary for cluster ions to form high quality ITO films as a result of enhanced chemical reactions. Approximately 10 % of oxygen atoms of the irradiated O_2 cluster ions have been incorporated in the films.

According to SEM observation of ITO films formed at various acceleration voltage, small particles, likely to be precipitates of metallic In were found below 3 keV. No particle was

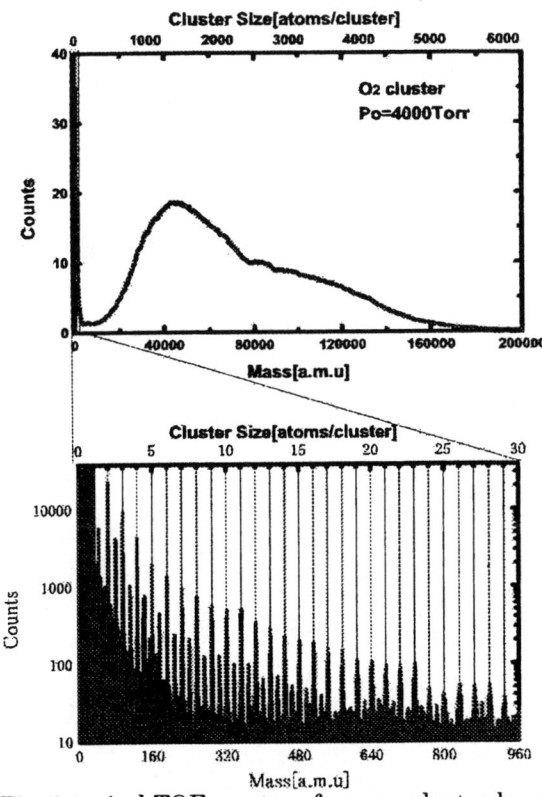

Fig. 2 typical TOF spectra of oxygen cluster beam

observed when stoichiometric ITO films were obtained. AFM measurements clearly show that very smooth films were formed above 5 keV. The surface roughness of the films was less than 2nm, which is sufficiently smooth to be utilized in optoelectronic devices.

In order to study bombardment effect of the cluster ion, un-doped In_2O_3 films were deposited with various conditions. As shown in Fig. 3, deposition rate was changed with fixed cluster ion current density. When deposition rate was low, more O_2 cluster ions were irradiated. Film resistivity decreased with deposition rate, because the deposited films were sufficiently oxidized by O_2 cluster ion bombardment. Stoichiometric In_2O_3 films were grown, when deposition rate was below 1 Å/s. Therefore, the transparency of these films was above 80 %.

The resistivity increased with decrease of the deposition rate below 1 Å/s. Hall measurements (Fig.4) of these films reveal that the carrier density decreases with the deposition rate. The Hall mobility did not decrease. Free carriers in ITO films are generated from both oxygen vacancies and tin atoms. In un-doped In_2O_3 films, free carriers are only from oxygen vacancies, which could be easily created by high energy ion bombardment. Nevertheless, the carrier density decreased with deposition rate, when cluster ions were irradiated. This result clearly indicates that the number of oxygen vacancies decreases by the bombardment of the cluster ions.

Recently, heteroepitaxial un-doped In_2O_3 film growth on single crystalline yttria stabilized zirconia(YSZ) were reported with Pulse Laser Deposition (PLD)[4] or Molecular Beam Epitaxy (MBE) technique [15]. The growth temperatures in these reports were about 300 °C. The carrier density of the un-doped In_2O_3 films deposited with the cluster ion assisted technique at 300 °C was quite close to the carrier density of the heteroepitaxial un-doped In_2O_3 film. Therefore, O_2 cluster ion assisted deposition technique has a great capability to form high quality oxide films without damage.

IV. SUMMARY

The Multi-beam Gas Cluster Ion Beam equipment was newly developed to obtain high quality ITO films for industrial applications. High intensity oxygen cluster ion beams were generated by supersonic expansion. The average size of the cluster ions was about 2000 according to TOF measurements. High quality ITO films were obtained by O_2 cluster ion assisted deposition at room temperature. About 10% of oxygen atoms in the cluster ions were incorporated into the films, when the kinetic energy is above 5 keV. Films with a transmission greater than 80 % have been deposited. The bombardment effect of cluster ions offers a new ion assisted thin film formation technique.

REFERENCES

1. S.Kimura, E. Murakami, T. Warabisako, H. Sunami and T. Tokuyama: IEEE Electron Device Lett., vol. **EDL-7** (1986) 38.

2. Y.Kawai, N. Konishi, J. Watanabe and T. Ohmi: Appl. Phys. Lett. **64** (1994) 2223.

3. S. Todorov and E. R. Fossum: J. Vac. Soc. Technol. **B6** (1988) 466.

4. R.B.H.Tahar, T.Ban, Y.Ohya and Y. Takahashi., J. Appl. Phys. **83**(1998), p2631

5. W. Qin, R.P. Howson, M. Akizuki, J. Maastuo, G. Takaoka, and I. Yamada, "Indium oxide film formation by O2 cluster ion-assisted deposition,". Material Chem.&Phys. **54** (1998) 258 (in press).

6. M. Akizuki, J. Matsuo, I. Yamada, M. Harada, S. Ogasawara and A. Doi, Jpn. J. Appl. Phys., **35** (1996) 1450.

7. M. Akizuki, J. Matsuo, M. Harada, S. Ogasawara, A. Doi, K. Yoneda, T. Yamaguchi, G. H. Takaoka, C. E. Ascheron and I. Yamada: Nucl. Instr. and Meth. B, **99**,(1995), p.229.

8. J. Matsuo, W. Qin, M. Akizuki, T. Yodoshi and I. Yamada. Mat. Res. Soc. Symp. Proc., in press (1997)

9. I. Yamada, W.L. Brown, J.A. Northby and M. Sosnowski, Nucl. Instr. And Meth. B,79 (1993) **223**.

10. H. Kitani, N. Toyoda, J. Matsuo and I. Yamada, Nucl. Instr. and Meth. B, **121**,(1997), p.489

11. T. Aoki, J. Matsuo, Z. Insepov and I. Yamada, Nucl. Instr. and Meth. B, **121**,(1997), p.49

12. Yamada, and J. Matsuo, Z. Insepov, D. Takeuchi, M. Akizuki and N. Toyoda, J.Vac. Sci. Technol A, **14**,(1996), p.781

13. R. Beuhler and L. Friedman, Chem.Rev., **86** (1986) 521

14. F. Hagena, Rev. Sci. Instrum. **63** (1992)2374

15. N. Taga,M. Maekawa, Y. Shigesato, I. Yasui, M. Kakei and T.E. Haynes: Jap. J. Appl. Phys. **37**(1998), p6524

Small clusters in strong laser fields

Christian Siedschlag and Jan M Rost

Max-Planck-Institute for the Physics of Complex Systems, Nöthnitzer Str. 38, D-01187 Dresden, Germany

Abstract. A theoretical model for the dynamics of rare gas atom clusters is developed which focuses on the motion of all particles and takes quantum effects into account only on the basis of effective rates. We demonstrate that similar characteristica of the energy absorption in metal and rare gas clusters can have a very different dynamical origin. The key quantity to distinguish the processes is the dependence on the laser frequency of the optimum size R of the cluster for energy absorption.

INTRODUCTION

The interaction of rare gas atom clusters as well as metal clusters with strong laser pulses has been studied experimentally and theoretically. Thereby, a large range of parameters has been covered, although not systematically. In particular the seize of the clusters, respectively the number N of constituent atoms, varies between 10 to 10^6. Depending on this number very different theoretical models have been formulated, from a hydrodynamic description for very large clusters (i) (1) to classical Monte Carlo calculations using electron ionization cross sections as input (ii) (2, 3) and for metal clusters, focusing on the valence electron cloud, time-dependent-local-density approaches (TDLDA) (4, 5), recently, including nuclear motion (iii) (6).

The following study is of category (ii). We are using classical mechanics to describe the motion in time of all charged free particles, electrons and atoms. The initial ionization of the electrons from individual atomic cores is modelled with a tunnelling rate in a static electric field, however including the contributions to an effective field at a time from all neighboring charged particles. This relatively simple approach allows us to treat quite a few particle individually, in the present case of a Ne_{16} cluster $16 \times 10 = 160$ electrons and 16 ions, i.e. roughly 170 particles. However, it is also easily possible to increase N considerably.

We expect the most interesting effects from a variation of the pulse duration T since it is of the same order of magnitude as the time scale of nuclear motion. There are at least four relevant time scales reflecting the complexity of the problem, see table 1

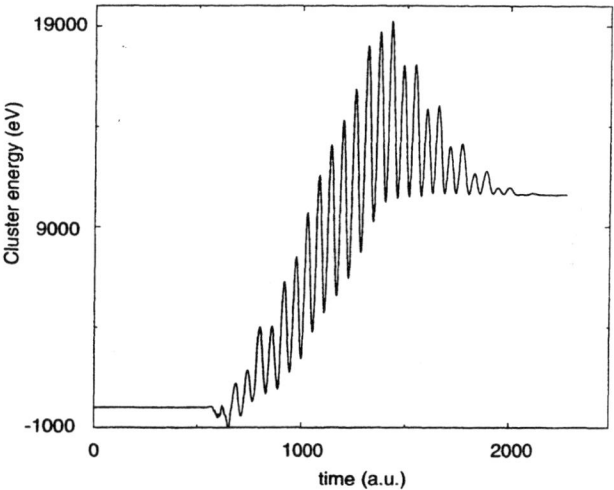

FIGURE 1. Absorbed energy of a Ne_{16} cluster during an intense laser pulse of 56 fs = 2300 a.u. duration with peak intensity $8.9 \times 10^{14} W/cm^2$ and frequency of $\omega = 0.055$ a.u..

ENERGY ABSORPTION IN RARE GAS ATOM CLUSTERS

Figure 1 shows the result of a typical run for Ne_{16}, the curve is drawn for the duration of the laser pulse which has a $\sin^2$ envelope. One sees that the energy absorption sets in with a delay. This is a consequence of the rare gas atoms: One needs a few mobile electrons first which can be driven by the light field. In a metal cluster the delocalized valence electrons supply such type of electrons from the beginning. Note, that the oscillations are due to the ponderomotive potential.

If one keeps the fluence (i.e. the energy content of the entire laser pulse) constant while varying the pulse duration T the necessary peak intensity I scales as $I \propto 1/T$ and

Table 1. Time scales in laser-cluster interaction for a light cluster of some 10 atoms

type of motion	origin	basic time unit
bound electronic motion	Kepler period in hydrogen ground state	10^0 a.u.
oscillation of the light field ω	inverse frequency $2\pi/\omega$	10^2 a.u.
laser pulse duration	100 fs	10^4 a.u.
ionic motion	$p_{ion} + p_{e^-} \approx 0 \Rightarrow t_{ion} \approx m_{ion}/m_{e^-} \cdot t_{e^-}$	10^4 a.u.

one obtains the energy absorption as a function of pulse duration as shown in Fig. 2. The most remarkable feature is the decreasing energy absorption with decreasing intensity (increasing pulse duration). Only for very small T, i.e. extremely strong intensities, a normal increase is observed. In this regime the cluster behaves like separate atoms.

ENERGY ABSORPTION IN METAL CLUSTERS

A similar behavior of the energy absorption as in Fig. 2 for a rare gas atom cluster was also observed experimentally in a metal cluster (7). The authors have proposed an interesting mechanism to explain this characteristic behavior, namely a plasmon induced resonance. Starting from a delocalized electron cloud for the cluster with density $\rho(t=0)$ before the beginning of the pulse they argue that with the subsequent ionization of the cluster and the expansion of the ions the density will decrease, and so will the characteristic eigenfrequency $\omega(\rho(t))$ as a function of time. At the time t_c when $\omega(\rho(t_c)) = \omega$, the laser frequency, the resonance condition is fullfilled and energy absorption is very efficient. If this optimum time t_c has good overlap with the laser pulse profile in time, a maximum in the energy absorption, similar as in Fig. 2 is reached.

The characteristic signature of the described process is a strong frequency dependence of the energy absorption *during* the pulse. It is controlled by the evolution of the cluster radius R as a function of time which determines the density $\rho(t)$.

FREQUENCY DEPENDENCE OF ENERGY ABSORPTION UPON THE CLUSTER RADIUS

To clarify if a similar mechanism can explain the energy absorption in rare gas clusters we have determined the energy absorption over the entire laser pulse starting from different initial cluster radii R. We find that in our

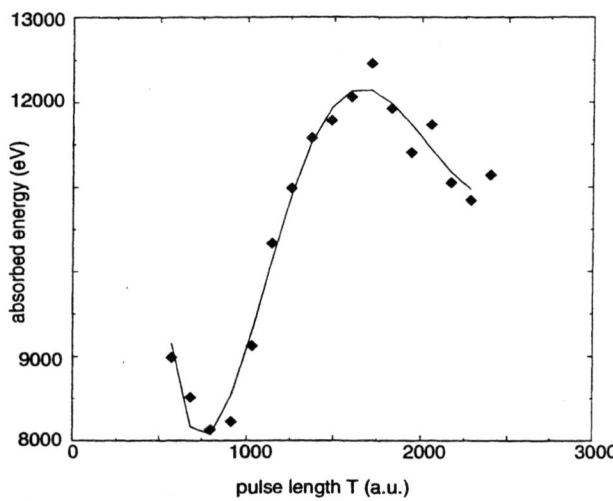

FIGURE 2. Absorbed energy of a Ne_{16} cluster in an intense laser pulse of frequency $\omega = 0.055$ a.u. for different pulse durations T.

case the energy absorption does *not* depend on the frequency, the maximum occurs at the same cluster radius (see Fig. 3).

This behavior can be explained with a mechanism known from molecules, namely "molecular enhanced ionization" (8). Briefly, the effect of another attractive charge center close by allows an electron to tunnel easier under the barrier of the potential from its own parent ion than in the so called separated and united atom limits, i.e., if the two positive charge centers are on top of each other or infinitely separated from each other.

One could argue that our findings originate in the ionization mechanism we use, namely the tunnelling ionization as described above. Hence, we have simulated the metal situation of a delocalized electron cloud.

ENERGY ABSORPTION WITH DELOCALIZED ELECTRONS

In our approach we can simulate a delocalized electron cloud by artificially making the individual binding poten-

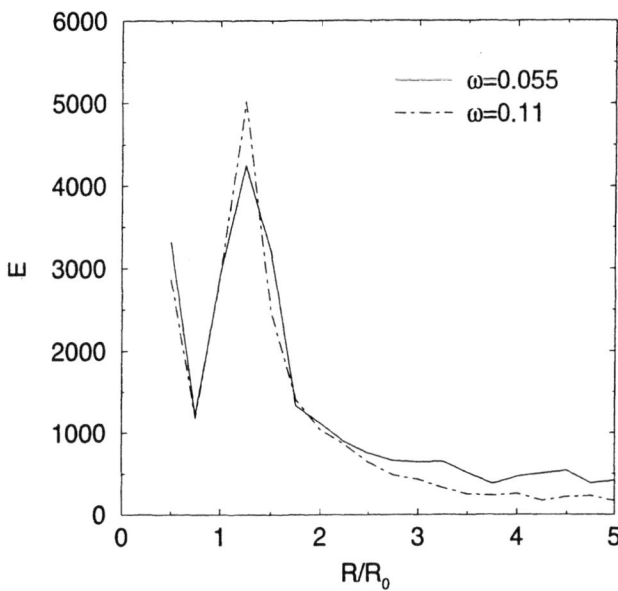

FIGURE 3. Absorbed energy of a Ne_{16} cluster in an intense laser pulse of frequency $\omega = 0.055$ a.u.(solid) and $\omega = 0.11$ a.u.(dashed), pulse duration $T = 20$ cycles, and peak intensity of $I = 0.89 \times 10^{14} W/cm^2$ for different initial cluster radii R, where R_0 is the optimum radius.

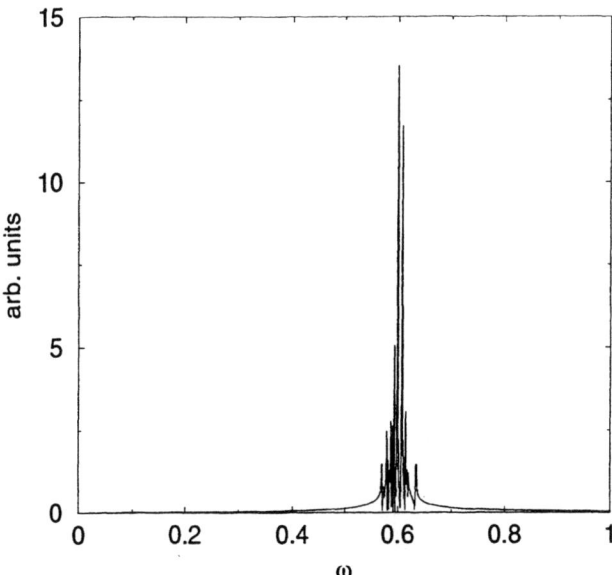

FIGURE 4. Electronic excitation spectrum in a Ne_{16} cluster for localized electrons (a=2, Z=1 in Eq. 1).

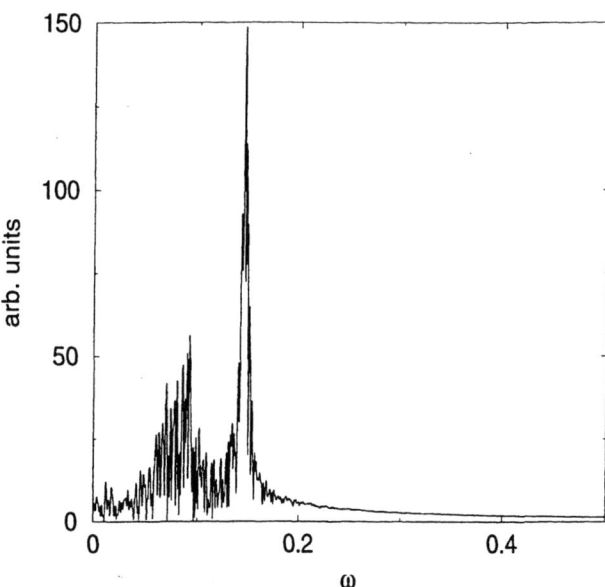

FIGURE 5. Same as Fig. 4 for delocalized electrons (a=30, Z=1 in Eq. 1).

tial of an electron to its parent ion with charge Z so broad that the individual potentials overlap for the characteristic interatomic distances in a cluster. For this purpose we have enlarged the parameter a in the soft-core potential

$$V(r) = -Z(r^2 + a^2)^{-1/2} \qquad (1)$$

from $a = 2$ to $a = 30$. As a proof that one deals in this case with delocalized electrons one can look at the excitation spectrum as a function of frequency, shown in Figs. 4 and 5.

One clearly sees the single peak at the energy for local excitations in Fig. 4 and the two peaks for local and collective excitation in Fig. 5. The absorption mechanism for the delocalized electrons (for a laser intensity which has been scaled down to match the small binding energy in the $a = 30$ case) is indeed laser frequency dependent, as one can see on Fig. 6.

SUMMARY

We have developed a model for treating atomic clusters in strong laser fields. The model predicts an energy absorption which has a maximum for a certain pulse duration and leads, therefore, to the same qualitative conclusions as they have been drawn in the case of metallic clusters. However, we have shown that very different mechanisms are behind this similar finding. In our case, we can identify a mechanism similar to the one found in

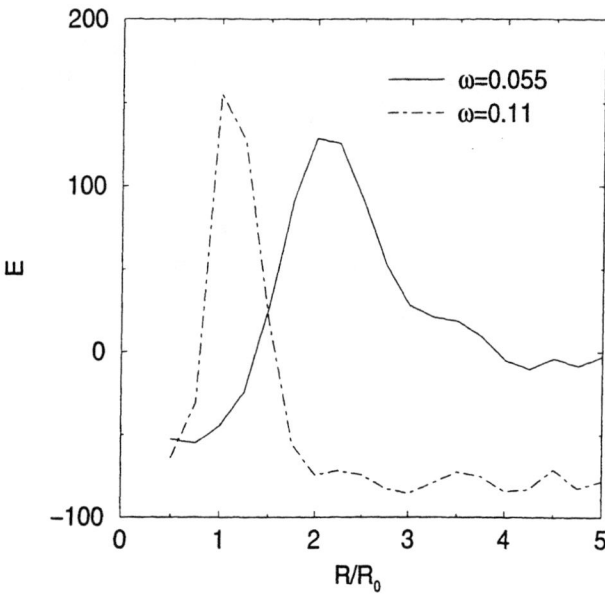

FIGURE 6. Absorbed energy of a "metallic" cluster with delocalized electrons (see text) in an intense laser pulse of frequency $\omega = 0.055$ a.u.(solid) and $\omega = 0.11$ a.u.(dashed), pulse duration $T = 20$ cycles, and peak intensity of $I = 3.51 \times 10^{12}$ W/cm^2 for different initial cluster radii R, where R_0 is the optimum radius for $\omega = 0.11$.

molecules with the characteristic signature that the optimum pulse duration is linked to an optimum seize of the molecule or cluster which is independent of the laser frequency. In the metallic case the proposed mechanism, relying on delocalized electrons, is strongly frequency dependent. We could confirm within our approach this feature modelling delocalized electrons. Hence, it appears that a similar behavior of the energy absorption in clusters can have a very different dynamical origin.

ACKNOWLEDGMENTS

We would like to thank P. Corkum, K.-H. Meiwes-Broer, and R. Schmidt for illuminating discussions.

REFERENCES

1. T. Ditmire et al, Phys. Rev. A 53, 3379,
2. I. Last and J. Jortner, Phys. Rev. A 62, 013201-1
3. C. Rose-Petruck et al, Phys. Rev. A 55, 1182
4. U. Saalmann and R. Schmidt, Phys. Rev. Lett. 80, 3213
5. P. G. Reinhard, to appear in Physics Reports
6. E. Suraud and P.G. Reinhard, Phys. Rev. Lett. 85, 2296
7. K.H. Meiwes-Broer, Phys. Rev. Lett. 82, 3783
8. PB Corkum et al, Phys. Rev. Lett. 75, 2819

STM Observation of a Si Surface Irradiated with a single Ar Cluster Ion

Toshio Seki and Gikan H. Takaoka

Ion Beam Engineering Experimental Laboratory, Kyoto University, Kyoto, Japan

Abstract. New surface modification processes, such as surface smoothing and shallow implantation, have been demonstrated using gas cluster ion irradiations because of the unique interaction between cluster ions and surface atoms. In order to reveal the cluster-surface interaction, the traces created by Ar cluster ion impact on Si and the annealing process of the traces were investigated using Variable Temperature Scanning Tunneling Microscope (VT-STM). In the STM image of a Si surface irradiated with cluster ions at 8keV, large craters were observed. When this irradiated sample was annealed at 600°C, the hole of the crater remained, but the outer rim disappeared. The hemispherical damage in the target was recovered.

INTRODUCTION

A cluster is an aggregate of a few to several thousands atoms. Because many atoms constituting a cluster ion bombard a local area, high-density energy deposition and multiple-collision processes are realized. Because of the interactions, cluster ion beam processes can produce unusual new surface modification effects, such as surface smoothing, high rate sputtering and very shallow implantation [1-4]. Various outstanding applications of the cluster ion beam have included so far: high quality tin doped indium oxide (ITO) films obtained by O_2 cluster ion assisted deposition at room temperature [5], smoothing of diamond films by Ar cluster beam [6,7], formation of an ultra shallow junction by using $B_{10}H_{14}$ ion implantation [8]. Moreover, the irradiation with cluster ions enhances the chemical reactions on the substrate surface. SiO_2 films are formed on Si substrates at room temperature by irradiation with either CO_2 cluster ions [9] or O_2 cluster ions [10]. The reactive sputtering occurs on Si surfaces at room temperature by irradiation with SF_6 cluster ion [11]. In order to reveal such cluster-surface interaction, a single trace formed by a cluster ion impact on a solid surface was investigated using STM.

EXPERIMENTAL

The traces created by Ar cluster ion impact on a Si(111) surface and the annealing process of the traces were investigated using an ion beam system combined with a Variable Temperature Scanning Tunneling Microscope (VT-STM) in Ultra High Vacuum (UHV). After irradiation with cluster ions, the surfaces can be observed with VT-STM from room temperature to 800°C in situ. Fig. 1 shows a schematic diagram of the ion beam system combined with VT-STM. The target chamber, the exchange chamber and the STM chamber are kept in UHV ($< 2\times10^{-9}$ Torr). The ion source can be used to generate either a cluster ion beam or a monomer ion beam. The ion beam enters the target chamber through a differential pumping chamber.

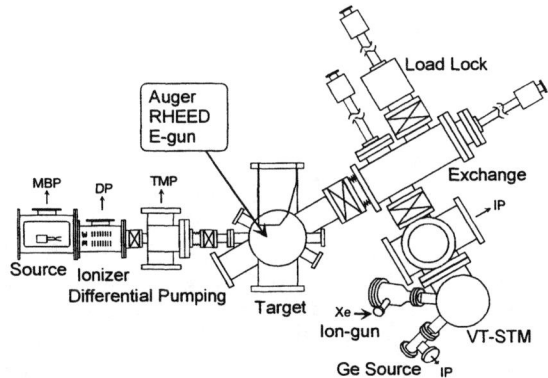

FIGURE 1. Schematic diagram of the ion beam system combined with VT-STM.

After irradiation the sample can be moved into the STM chamber without being exposed to air. STM observations can be carried out while heating the sample (< 800°C) by direct current.

The sample of Si(111) was initially cleaned chemically, and then transferred to the UHV chamber. The sample was flashed to 1250°C for a minute to eliminate the oxide layer in this chamber. We used a chemically polished W tip for the STM probe. The conditions used for these observations were: tunneling current - 0.5nA, bias voltage - +2.0V and the chamber pressure was lower than 5×10^{-10} Torr. Before irradiation, the 7×7-reconstructed surface was observed clearly at room temperature. A Si(111) substrate was irradiated with Ar cluster ions at acceleration voltage of 8 kV.

RESULTS AND DISCUSSION

Fig. 2(a) shows an STM image of the surfaces at room temperature. Because the mean size of the Ar cluster is about 1000, it is expected that the typical effect caused by large cluster impact would occur. The ion dose was about 6.3×10^{10} ions/cm^2, and the number of incident Ar cluster ions in the area of 200×200 nm^2 was about 25. About 12 doughnut shape traces can be observed in this image. The number of traces is of the same order with that calculated from the ion dose, indicating that these are single ion traces formed by individual Ar cluster ion impacts. The traces of cluster ion impact had a distinctively different shape from those of the monomer [12]: small hole shaped traces were observed in the case of monomer ion irradiation, whereas large doughnut shaped traces were observed in the case of cluster ion irradiation.

Fig. 2(b) shows a high-resolution image and a cross section diagram of a large doughnut-like trace with an outer diameter of about 80 Å and an inner diameter of about 25 Å. The cross-sectional image of the doughnut-like trace shows that the traces created by the cluster ion impacts were a crater shape with a big hole in center. The outer rim of the crater is 4.2 Å above the substrate surface level. This indicates again that cluster impact processes are quite different from the summation of the separate monomer impacts.

The process of cluster ion impact and crater formation has been discussed in case of cluster ion irradiation on Highly Oriented Pyrolitic Graphite (HOPG) surfaces [13]. The Ar cluster penetrated and created hemispherical damage in the target. After the Ar atoms escape as vapor from the target, a hole remains at the center of the damage region and a crater is created. This crater formation agrees with the result of STM observation as shown in Fig. 2(b). When a hemispherical damage is created in the target, the

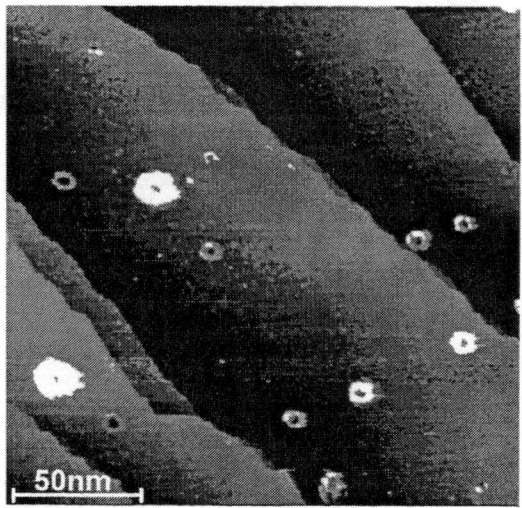

(a) Ar cluster irradiated surface

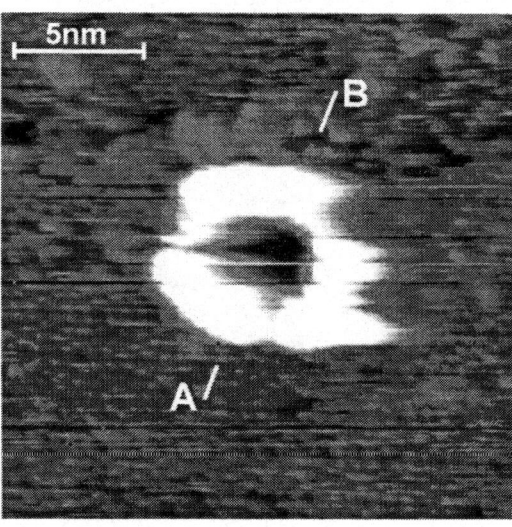

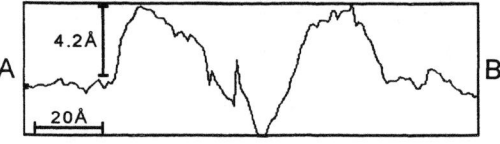

(b) High-resolution image and a cross section diagram of a large doughnut-like trace

FIGURE 2. STM images of Si(111) surface irradiated with Ar cluster ions at 8 kV.

number of Si atoms disordered by a cluster ion impact is several thousands. According to calculation with TRIM [14], the number of vacancies generated by an Ar ion is about 48. The number of atoms disordered by a cluster ion is hundred times larger than that displaced by a monomer ion. This indicates that the cluster ion beam can modify surfaces with lower ion dose than monomer ion beam.

Fig. 3 shows a distribution of the diameter of craters created by Ar cluster impacts at 8 kV. The crater diameter has a wide distribution, from 50 Å to 200 Å and the mean diameter is about 80 Å. This wide distribution can be caused by the size distribution of cluster ions and extremely large traces may be created by double charged cluster ions. If the size of clusters is 1000, the energy per one Ar atom is 8 eV. Although the energy per atoms is very low, large traces are created by Ar cluster impacts. This result may be caused by the multiple-collision effect.

The Si(111) surface was annealed after irradiation with Ar cluster ions at 8 kV. Fig. 4(a-c) shows STM images of the surface observed at various temperatures. The annealing time was about 30 minutes. At 300°C, the shape of the craters was still that of the initial craters. At higher temperature, part of the outer rim of craters appears chipped. Then, at 600°C, the outer rim of the craters disappeared, but the holes remained. The chipped area of the rim increased and the height of the rim decreased with annealing temperature. At 600°C, the outer rim of the crater disappeared and the 7×7-reconstructed structure was observed at the site of the rim. This result shows that the hemispherical damage created by the cluster impact was removed at this temperature.

Fig. 5 shows a schematic model of crater formation and annealing after cluster impact. It summarizes the results presented in Fig. 4. A crater is formed by an Ar cluster impact and the shape is maintained after annealing at 300°C. Because the atoms forming the

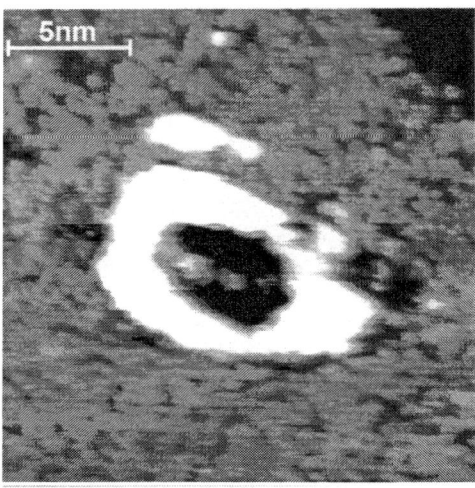

(a) At 300°C

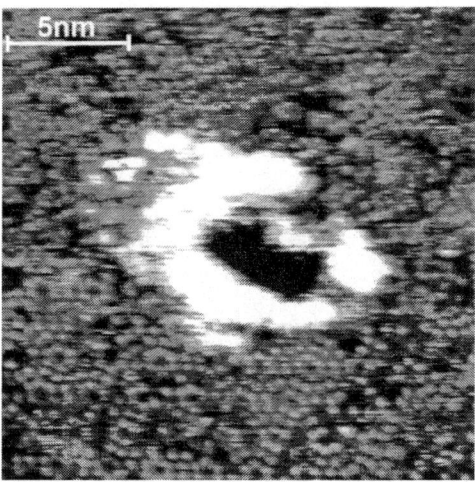

(b) At 400°C

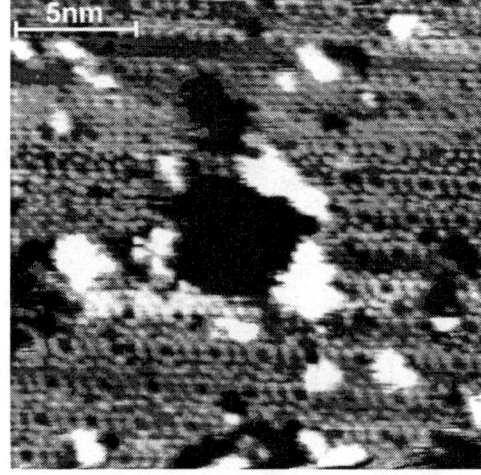

(c) At 600°C

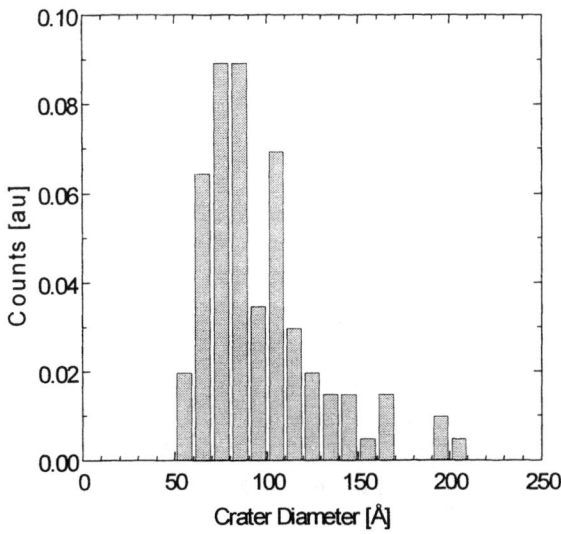

FIGURE 3. Distribution of the diameter of craters created by Ar cluster impacts at 8 kV.

FIGURE 4. STM images of craters observed at indicated temperature after irradiation with Ar cluster ions.

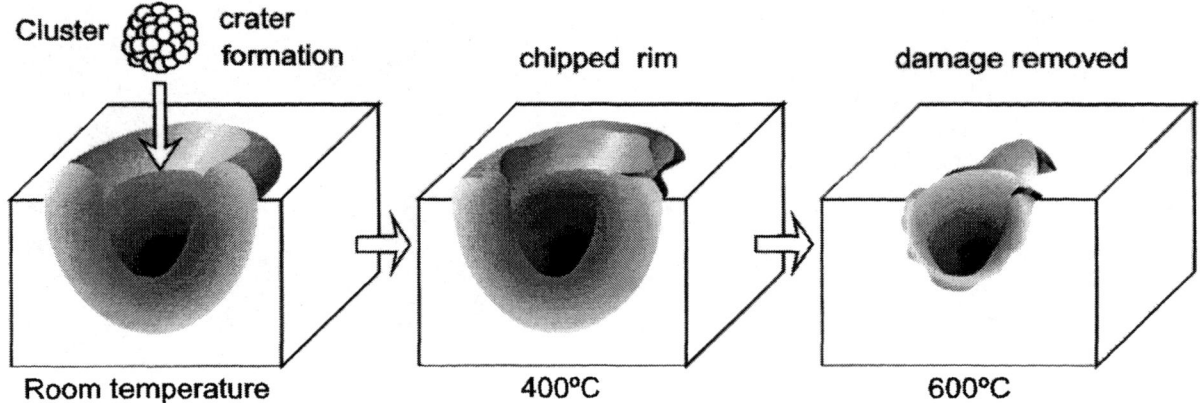

FIGURE 5. Schematic model of crater formation and annealing after cluster impact.

rim of crater can migrate above 400°C, the outer rim of crater is gradually chipped away above 400°C. The outer rim of crater disappears and the hemispherical damage in the target is removed at 600°C, but the hole remains.

CONCLUSION

The traces created by Ar cluster ion impact were observed and the annealing process of the traces was investigated. The trace of a cluster ion impact showed a crater shape, with a diameter of about 80Å. This indicates that cluster impact process is different from a summation of separate monomer impacts. The crater shape was maintained after annealing at 300°C. Because atoms forming the rim of the crater can migrate above 400°C, a part of the outer rim of the crater becomes chipped above 400°C. The outer rim of the crater disappears and the hemispherical damage in the target is removed at 600°C, but the hole remains.

ACKNOWLEDGMENTS

This paper has been edited for publication by http://editscience.com.

REFERENCES

1. I.Yamada, J.Matsuo, Z.Insepov and M.Akizuki, Nucl. Instr. and Meth. **B106**, 165 (1995).
2. I.Yamada, W.L.Brown, J,A,Northby and M.Sosnowski, Nucl. Instr. and Meth., **B79**, 223 (1993).
3. G.H.Takaoka, G.Sugawara, R.E.Hummel, J,A,Northby, M.Sosnowski and I.Yamada, Mat. Res. Soc. Symp. Proc., **316**, 1005 (1994).
4. Z.Insepov, M.Sosnowski and I.Yamada, Advanced Materials '93 IV/Laser and Ion Beam Modification of Materials, ed. I.Yamada et al, Trans. Mat. Res. Soc. Jpn. **17**, 1110 (1994).
5. W.Qin, R.P.Howson, M.Akizuki, J.Matsuo, G.Takaoka and I.Yamada, Mater. Chem. Phys., vol.54, no.1-3, 258 (1998).
6. N.Toyoda, N.Hagiwara, J.Matsuo and I.Yamada, Nucl. Instr. and Meth. **B148**, 639 (1999).
7. A.Nishiyama, M.Adachi, N.Toyoda, N.Hagiwara, J.Matsuo and I.Yamada, AIP conference proceedings (15-th International Conference on Application of Accelerators in Research and Industry) **475**, 421 (1998).
8. N.Shimada, T.Aoki, J.Matsuo, I.Yamada, K.Goto and T.Sugui, J. Mat. Chem. and Phys. **54**, 80 (1998).
9. M.Akizuki, J.Matsuo, M.Harada, S.Ogasawara, A.Doi, K.Yoneda, T.Yamaguchi, G.H.Takaoka, C.E.Asheron and I.Yamada, Nucl. Instr. and Meth. **B99**, 229 (1995).
10. M.Akizuki, J.Matsuo, S.Ogasawara, M.Harada, A.Doi and I.Yamada, Jpn. J. Appl. Phys., **35**, 1450 (1996).
11. N.Toyoda, H.Kitani, J.Matsuo and I.Yamada, Nucl. Instr. and Meth., **B121**, 484 (1997).
12. T.Seki, T.Aoki, J.Matsuo and I.Yamada, Nucl. Instr. and Meth., **B164-165**, 650 (2000).
13. T.Seki, T.Kaneko, D.Takeuchi, T.Aoki, J.Matsuo, Z.Insepov and I.Yamada, Nucl. Instr. and Meth., **B121**, 498 (1997).
14. J.P.Biersack and L.G.Haggmark, Nucl. Instr. and Meth., **174** 257 (1980).

The Extraction of Small Cluster Ions from Cesium Sputtering Ion Source

Xuemei Wang, Xinming Lu, Lin Shao, Jiarui Liu, and Wei-Kan Chu

Department of Physics, and Texas Center for Superconductivity, University of Houston, Houston, Texas 77204

Abstract. We found that the Cesium Sputtering Ion Source is suitable to deliver small cluster ions. We obtained the record beam current for carbon cluster ions C_n with n=1 up to 10. The beam current for C_2, C_5 and C_{10} was 70.7μA, 1.29μA and 116nA respectively. The negative cluster ions of extraction B_n, Si_n and Ge_n and their applications are also reported

INTRODUCTION

The Cluster ion beam opened up some new fields of its application in material processing. The Cluster ion beam has been applied to produce ultra-shallow junctions in semiconductor device fabrication [1,2]. High rate sputtering has been observed with the sputtering yield of about 2 orders of magnitudes higher than the sputtering yield by monomers with the same energy per atom [3]. Atomic scale smoothing of surfaces by cluster ion bombardment has been applied to different materials to get the roughness of 0.2-0.8 nm [4,5]. Cluster ion beams also have been used for thin film formation at low substrate temperature [6,7]. On the other hand, a dramatic non-linear effect (or collective effects) has been observed in cluster-solid interaction. The non-linear effect appeared in all aspects of cluster-solid interaction such as ion range, straggling, sputtering and defect formation. The non-linear effect needs to be studied in detail to fulfill the requirements for the cluster ion applications. In this paper, we report some results of extraction of small cluster ions from a cesium sputtering (SNICS) ion source. The sputtering target materials, sputtering voltage, current and the target geometry were optimized to get a higher cluster beam current. The optimized cluster ion beams of B_n, C_n, Si_n, Ge_n, SiB_n and GeB were applied for material modification and in non-linear effect study in cluster-solid interaction. We present the experiment arrangement to produce carbon clusters C_n (n=1-10), Boron clusters B_n (n=1-5), silicon clusters Si_n (n=1-6) and germanium clusters Ge_n (n=1-5). The energy of the cluster ions can be in a range from 1 keV up to 70 keV, and then can be accelerated up to several MeV in a tandem accelerator with good beam intensity for material research and cluster-solid interaction investigation.

EXPERIMENTAL

Small clusters can be formed by various techniques, such as pulsed arc cluster ion source (PACIS),[8] by laser vaporization,[9] sputter source with gas aggregation,[10] positive sputtering source and negative sputtering source [11, 12]. The cesium sputter ion source can deliver small cluster ion beams for nA to μA beam current and is the most versatile for different clusters formation from solid materials. It generates negative cluster ions by sputtering a solid material with positive cesium ions. The energy of the Cs beam is in a range of 1-10 kV. The sputtered negative ion beam is extracted by a potential of 5-30 kV. The expected cluster beam is selected and deflected 30° by a magnet. The magnet is able to deflect ions up to a mass of 350 amu. The ion current is measured with a Faraday cup, which is located at the low energy end of a 5SDH Tandem accelerator (from National Electrostatic Corp.). The cluster ion beam can be used at the low energy end or accelerated in the Tandem accelerator up to MeV energy. All results presented here were measured at the low energy end Faraday cup.

RESULT AND DISCUSSION

Boron cluster

Negative boron cluster ions were extracted with different sputtering targets: natural boron, enriched ^{10}B and enriched ^{11}B targets. For ^{10}B, enriched up to 99.82%, -140 mesh boron powder was used to produce boron clusters of $^{10}B_n$ (n=1-5). Stable and optimized ^{10}B cluster ion mass spectrum at a sputtering voltage of 5 kV is shown in Fig.1.

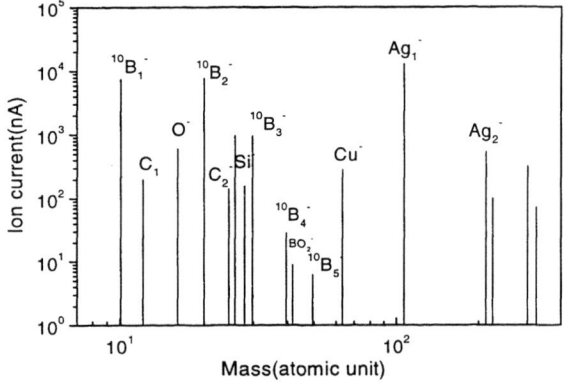

FIGURE 1. Mass spectrum of Boron clusters measured from the SNICS ion source at typical sputtering voltage of 5 kV.

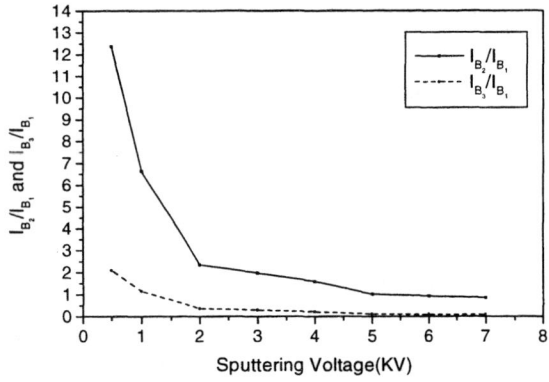

FIGURE 2. Cluster ion beam intensity ratio I_{B2}/I_{B1} and I_{B3}/I_{B1} in dependence of the sputtering voltage of the SNICS source.

The stable beam current was observed after several hours of operation. Well distinguished boron cluster peaks are observed from $^{10}B_1^-$ up to $^{10}B_2^-$, $^{10}B_3^-$, $^{10}B_4^-$, $^{10}B_5^-$. The monomer boron beam current is about 7.45 µA and the beam current for $^{10}B_5^-$ is about 6.23nA. A very interesting behavior of $^{10}B_2^-$ and $^{10}B_3^-$ was observed when the beam current was measured in dependence of the sputtering voltage. The I_{B2}/I_{B1} and I_{B3}/I_{B1} ratio are shown in Fig. 2. At the sputtering voltage between 4-8 kV, the $^{10}B_2^-$ beam intensity is at the level of $^{10}B_1^-$. At the sputtering voltage below 1 kV, the $^{10}B_2^-$ cluster beam is higher than the $^{10}B_1^-$ by a factor of more than 10. Natural boron and enriched ^{11}B targets show similar beam current for the clusters. The cluster beam current is increasing with the sputtering current and the sputtering voltage. Referring to the previous high power sputtering source in [11] and [13], we believe that the sputtering source, both positive and negative, can deliver mA of B_2 cluster ion beam.

Carbon cluster

Small carbon clusters are useful for the cluster-solid interaction investigation for several reasons. First, carbon clusters from SNICS source have a very high beam current; the beam current of C_1 is several hundreds µA and the beam current of C_{10} is about 50-100 nA. Second, the carbon clusters can be very clean with very low contamination, because the sputtering target can be from pure graphite. In our measurements, carbon cluster ions were produced by a high density graphite target. Carbon is one of the easiest negative ions to be extracted in the SNICS ion source. The full mass spectrum for carbon cluster ions is shown in Fig.3.

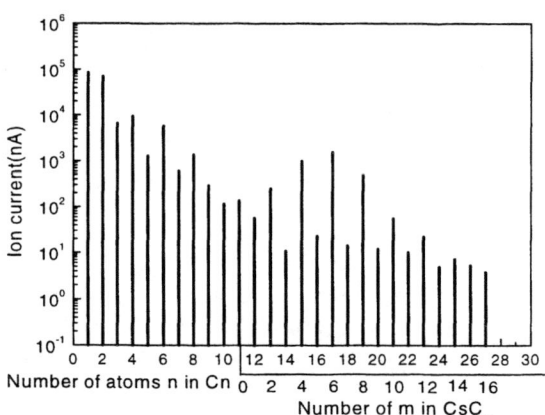

FIGURE 3. The full mass spectrum of carbon clusters from the SNICS ion source at the typical sputtering voltage of 5 kV.

A carbon target with solid graphite without a copper holder was used to avoid the possible spectrum contamination due to Cu. Carbon cluster ion current distribution can be measured up to C_{27}, but unfortunately, when the cluster size is larger than 11, the carbon clusters is contaminated by CsC_n clusters. If this contamination problem can be solved either by high-resolution magnet or by positive sputtering ion

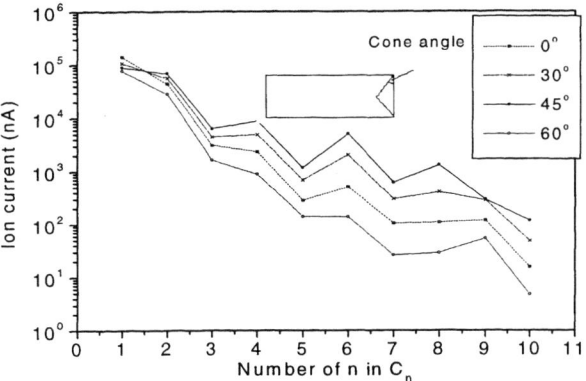

FIGURE 4. The intensity of the C-cluster ion beam with the front cone angle of 0°, 30°, 45°, 60° in dependence of the cluster size.

source (without Cs), the small carbon clusters from monomer up to C_{27} can be used as a powerful tool in cluster-solid interaction investigations. The intensities of even-n clusters through n=10 are stronger than odd-n clusters, while the cluster of odd-n for n>11 show higher intensities. In particular, the relative intensity beam of CsC_4^- (C_{15}) and CsC_6^- (C_{17}) showed pronounced enhancement. This kind of magic number is related to cluster structure itself. We have used this clean and easy graphite target to optimize the target geometry in this SNICS source with the reflecting sputtering. Similar to the early investigation in the transmit sputtering geometry [13], targets with different cone angle of 0°, 30°, 45°, and 60° were tested. The cluster beam current for different target shape is shown in Fig. 4. It was interesting that the highest monomer carbon beam current was obtained from 0° cone (flat head) target for more than 200μA, but the highest cluster beam current was obtained from the target with cone angle of 45°. The intensity of C_{10}^- is typically over 100 nA. The carbon cluster ion yield increases with the sputtering voltage in a range of 1kV—8kV.

Silicon and Germanium cluster

The cathode targets of Silicon and Germanium were made of high purity crystal. Like graphite, the silicon and germanium targets provide prolific negative ions. The mass spectrum of Si and Ge clusters is shown in Fig.5. The intensity of negative ion currents is also dependent on cathode voltage. At a typical sputtering voltage of 5 kV, the cluster ion beam current for Si_5 and Ge_5 is more than 5nA and 40nA respectively. This cluster ion intensity can be enhanced with the sputtering voltage and current.

It is worthwhile to notice that clusters with boron atoms in the form of molecules like GeB, SiB and SiB_2 have been obtained with good beam current. These boron containing small clusters were successfully applied for shallow junction study [14].

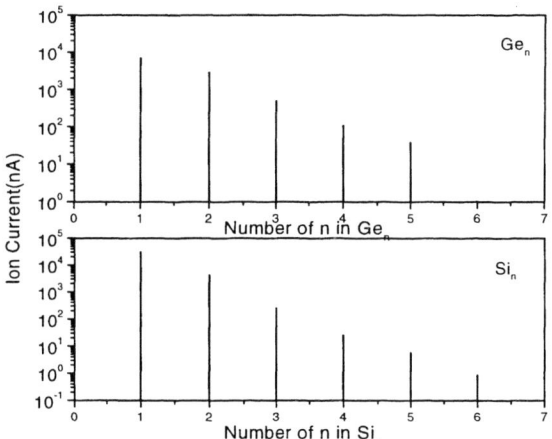

FIGURE 5. The full mass spectrum of Si_n and Ge_n from SNICS ion source at typical sputtering voltage of 5 kV.

CONCLUSION

Production of small cluster ions of B_n, C_n, Si_n and Ge_n has been investigated in SNICS ion source. Record high ion beam currents of these clusters were obtained due to the optimization in sputtering voltage, sputtering current, target materials and the target geometry. These cluster ions were successfully applied in a shallow junction study and in the basic research of the non-linear effect of cluster-solid interaction. Particular attention was paid to the boron and carbon cluster. The cluster ion yields from SNICS are adequate for a wide variety of experiments. It would contribute a lot to fundamental research in cluster-solid interaction and applied topics in material modification with cluster ions.

ACKNOWLEDGMENT

This project was supported by The State of Texas through the Texas Center for Superconductivity, partially by the 003652-797 ARP and the National Science Foundation through the Materials Research Science and Engineering Center.

REFERENCES

1. Matsuo, J., Takeuchi, D., Aoki, T., I. Yamada, I., In:.Ishida, E., Banerjee, S., Meta, S., Smith, T.C., Current, M., Larson, L., Tash(Eds.), A., IEEE Proceedings of the 11th International Conference on Ion Implantation Technology. Austin, TX, vol. 1, issue 1. 16-21 June 1996, IEEE Service center, Piscataway, NJ, 1997, P.768.

2. Takeuchi, D., Shimada, N., Matsuo, J., Yamada, I., in: Ishida, E., Banerjee, S., Meta, S., Smith, T.C., Current, M., Larson, L., Tash(Eds), A., IEEE Proceedings of the 11th International Conference on Ion Implantation Technology. Austin, TX, vol. 1, issue 1. 16-21 June 1996, IEEE Service center, Piscataway, NJ, 1997, P.772.

3. Yamada, I., Matsuo, J., Advanced metallization for future ULSI. In: Tu, K.N., Poate, J.M., Mayer, J.W., Chen(Eds.), L.C., *Material research Society Symposium Proceedings*, **Vol. 427** Materials Research Society, Pittsburgh, P.265(1997).

4. Toyoda, N., Matsuo, J., Yamada, I., in: Duggan, J.L., Morgan (Eds.), I.L., *Proceedings of the Application of Accelerators in Research and Industry' 96*. AIP Press. New York, P483 (1997).

5. Chu, W.K., Li, Y.P., Liu, J.R., Wu, J.Z., Tidrow, S.C., Toyoda, N., Matsuo, J., Yamada, I., *Appl. Phys. Lett.* **72**, 246 (1997).

6. Yamada, I., NEDO **Consortium Program 1997**, Project Office: Management Office: Osaka Science and Technology Center, Utsubo, Osaka Japan: R&D for ultra-high quality transparent conductive film fabrication.

7. Yamada, I., Japan Science and Technology Corporation, Agency of Science and Technology(JST), Innovative Research Application Program, 1998-2000.

8. Siekmann, H. R., Luder, Ch., Faehrmann, J., Lutz, H. O.,and Meiwes-Broer, K. H., *Phys.* **D. 20**, 417(1991).

9. Bondybey V. E., and English, J. H., *J. Chem Phys.* **74**, 1978(1981).

10. Sigmund, P., Bitensky, I.S., Jensen, J., *Nucl. Instr. and Meth* **B112**,1-30(1990).

11. Fayet, P., McGlinchey, M.J., and Woste, L.H., *J. Am. Chem. Soc.* **109**, 1733(1987).

12. Shen Dingyu, Jiang Dongxing, Wang Xuemei, Li Shuozhong, Qian Xing, *Nucl.Instr. and Meth* **A348**, 47-50 (1994).

13. Middleton, R., *Nucl. Instr. and Meth* **144**, 337(1977).

14. Chu, W.K., Liu, J.R., Jin, J.Y., Lu, X. M., Shao Lin., and Li, Q.M., this proceeding

Spatial Control of Nanoparticle Structures Using Dynamic Processes under High Flux Cu⁻ Implantation

N. Kishimoto*, Y. Takeda*, N. Umeda[†], N. Okubo[†], and C.G. Lee*

*National Research Institute for Metals, Tsukuba, Ibaraki 305-0047, Japan
[†]University of Tsukuba, Tsukuba, Ibaraki 305-8573, Japan

Abstract. Negative Cu ions of 60 keV, at high dose rates, have been applied for nanoparticle fabrication in insulators, using both conventional implantation and dynamic negative ion mixing (DNIM). Intense ion beams cause significant in-beam rearrangements of implants, i.e., spontaneous precipitation and depth-directional atomic migration. These dynamic processes are applicable for spatial control of nanoparticle structures. The spatial control is also possible by changing substrates, i.e., amorphous and crystalline SiO_2 and a spinel crystal of $MgAl_2O_4$. The other approach for the spatial control is to use a DNIM method with dynamic growth, which presents a thick uniform structure of nanoparticles.

INTRODUCTION

High-current technology of light ions has matured and is utilized for high-power applications, such as fusion reactors and pulsed power devices [1,2]. Pulsed proton devices have reached hundreds of kA up to MeV [2]. High-flux ions of medium-mass elements have been pursued to conduct shallow-impurity doping into semiconductors. Although gaseous or volatile elements are relatively easy to produce intense ion beams, high-flux heavy (metal) ions of 1 mA-class are not yet routinely available. Since metal elements require a large latent energy for ionization, an electric power supply near the earth potential is favorable. Therefore, a tandem-type accelerator with a negative ion source is suitable to attain high-flux ions. Recent progress on negative ion sources [3,4] has produced intense negative ion beams for material research.

For material applications, there are two practical implications of the high flux, i.e., efficient implantation over a wide area and high dose-rate implantation to utilize either high dose or high kinetic rate. Since material modification with metal precipitation requires larger doses (1 ~10%) than that with shallow-impurity doping in semiconductors (0.1 ~100 ppm), high-flux implantation is demanded for metal precipitation. Also, an atomic flux of 1 mA, e.g., over 10 cm², is comparable with that of gas-phase deposition such as vacuum evaporation (typically ~0.1 nm/s). In this context, we have applied high-flux implantation for metal nanoparticles embedded in insulators, which exhibit a nonlinear fast response, enhanced by a surface plasmon resonance [5]. To control nanoparticle structures, ordinary methods are to set an adequate ion-energy for the depth distribution and to anneal implanted specimens for the precipitation. Unlike the conventional methods, we have focused on kinetic phenomena induced by high-flux ions [6-8]. To perform ion implantation into insulators, surface charging is a serious problem, particularly at low energies $< 10^2$ keV. Application of negative ions suppresses surface charging down to several volts [4], balancing with the secondary electron emission. Therefore, an implantation system suitable for insulators is a combination of low-energy negative ions and high-energy (positive) ions. Use of negative ions provides us with the capability of a wide dose-rate range and with a kinetic tool to spatially control nanoparticle morphology. The merits due to negative and intense ions are also applicable to so-called dynamic ion mixing for insulators, combining low-energy negative-ion implantation and vacuum evaporation. Use of high-flux heavy ions causes ion-induced mixing and significant mass injection. We hereafter call this method Dynamic Negative Ion Mixing (DNIM), which is capable of film deposition with ion implantation over a wide range, irrespective of the ion penetration depth. In this paper, we first describe the high-flux accelerator system for nanoparticle formation in insulators. Next, experimental results of spatial control of nanoparticle structures are described from kinetic viewpoints.

HIGH-CURRENT HEAVY-ION ACCELERATOR SYSTEM

A heavy-ion accelerator system has been developed primarily to perform high-flux metal-ion implantation into insulators. The system, including a high-power YAG laser, is named the EPF (Extreme-Particle-Field) system, implying combined interactive fields of ions and photons. After the whole system is outlined, negative ion devices both for ion-alone implantation and for dynamic negative-ion mixing are detailed.

TABLE 1. Specifications of the EPF system.

Item	Specifications
Ion Source	Plasma-Sputter Type
Ion Species	B - Au
Accel. Voltage	60 kV or 2 MV
Ion Current	≤3 or ≤1 mA (ion species dependent)
Beam Transport	VPQM Triplet, Analyz.Magnet
Scanning	Vert. 5 kHz, Horiz. 890Hz
Pulsation	Multi-Strip-Line Chopper
Beam Detection	BPM, Wide-Band/high load FC, Micro-Channel Plate
YAG Laser	5 J/pulse@1.06 μm, 2.8J/pulse@ 532nm
EB Evaporator	4 kV×1 A, Quartz monitor
Temperature	RT - 1000 K, Th.C. and IR Pyrometer
In-situ Devices	CCD Spectrometer, EDS, SIMS, XPS

The main accelerator is capable of 1 mA-currents from B to Au, dependent on ion species. Characteristic features of the system are: i) intense negative ions of low (60 keV)- and high energy (~6 MeV), ii) a high-power YAG laser for co-irradiation with photons and iii) in-situ measuring devices. Specifications of the EPF system are summarized in Table 1. A high-power YAG laser is also installed into the implantation chamber. Details of the co-irradiation experiments were reported elsewhere [9].

A key device is an intense ion source for production of negative heavy ions, which is shown in Fig. 1. The negative ion source is a plasma-sputter type with a cusp magnetic field and a Cs supply [3]. A Xe plasma, ignited by a LaB_6 filament, confined by the cusp magnetic field, powerfully sputters the solid target, whose surface is covered with Cs atoms to promote negative ion production. The ion-source operation requires a delicate balance between the Xe arc and the Cs supply. The intense negative ions, extracted at 60 keV, are transported either directly to a low-energy implantation chamber or to the main acceleration tubes equipped with a Schenkel-type power-supply. Against the strong space-charge effects of currents at mA-levels, an optimized powerful lens system, including a VPQM (variable permanent quadrupole-magnets) triplet, attained total Cu^- current of 3 mA at the injector position. The low-energy negative beam-line utilizes the negative ions at 60 keV up to 3 mA, by either conventional implantation or dynamic negative-ion mixing. Negative ions enable us to conduct low-energy implantation to insulators. Since MeV positive ions are not subjected to the surface charging, the whole system provides ion implantation into insulators over a wide energy range from 60 keV to 6 MeV.

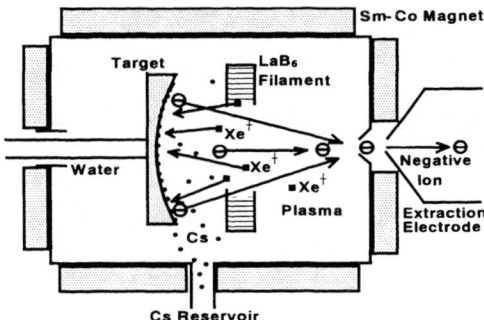

FIGURE 1. Negative ion source of plasma-sputter type.

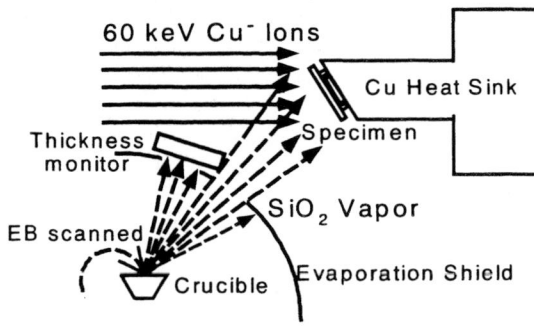

FIGURE 2. Schematic diagram of a Dynamic Negative-Ion Mixing method.

As for the DNIM method, film deposition of amorphous(a-)SiO_2 by electron-beam(EB) evaporation and ion implantation of 60 keV Cu^- were concurrently conducted onto an a-SiO_2 disk substrate. Figure 2 shows a schematic diagram of a dynamic negative-ion mixing method. To minimize interference between a negative-ion beam and molecules of SiO_2 evaporated, a 1/4-spherical shield was installed surrounding the EB-evaporator. The incident angles of Cu^- ions and the SiO_2-vapor flux were +19 and -15 degrees from the normal to the substrate, respectively. A Cu aperture mask with a 12 mmφ-hole clamped a substrate to the specimen stage via steel springs. Dose rate of Cu^- ranged from 5 to 50 μA/cm^2. The deposition rate of SiO_2 was maintained constant at 0.2 or 0.4 nm/s.

APPLICATIONS TO NANOPARTICLES EMBEDDED IN INSULATORS

Our spatial control of nanoparticles is associated with high energy-deposition rate of high-flux ions. Initial negative charge of the incident ions does not effect the intra-solid processes. Characteristic features of the kinetic control are discussed focusing on nanoparticle formation.

Spontaneous Nanocrystal Growth

With low-current ions, extensive work for insulators has been carried out [10,11]. Most of the past experiments employ positive ions at dose rate less than 5 $\mu A/cm^2$ and some of them (at the higher energy) require post-implantation annealing to promote metal precipitation and to annihilate defects.

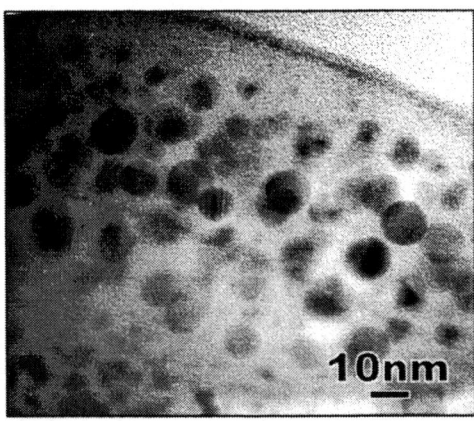

FIGURE 3. Cross-sectional TEM image of a-SiO$_2$ that was implanted with 60 keV Cu$^-$ at 260$\mu A/cm^2$ to 3 ×10^{16} ions/cm^2 (2 mmϕ-mask), tilted from the edge-on direction.

In our case, metal nanoparticles spontaneously grow during implantation. Figure 3 shows a cross-sectional TEM image that was implanted with 60 keV Cu$^-$ at 260 $\mu A/cm^2$ to 3 × 10^{16} ions/cm^2. Copious nanoparticles of 10 nm in diameter spontaneously form by the ion implantation. Each particle is round and is confirmed to be Cu single crystals by electron diffraction. The nanoparticle growth is explained by Ostwald ripening under radiation-induced diffusion. One of the merits of high-flux ions is thus the spontaneous growth of nanoparticles whose size is suitable for optical applications [6-8]. Optical absorption spectra are dependent on dose rate and the mask aperture, and mostly give a surface plasmon peak of 2.2 eV, which results in a large nonlinear susceptibility $\chi^{(3)}$ [8].

Material response to high-flux ions depends on the substrate species. To study crystallinity effects, we first conducted high-flux implantation into crystalline (c-)SiO$_2$ [6]. The c-SiO$_2$ showed a stronger depth rearrangement, but the Cu particle size is similar to a-SiO$_2$. The similarity between crystalline and amorphous substrates is reasonable, because c-SiO$_2$ is susceptible to amorphization under heavy ion irradiation.

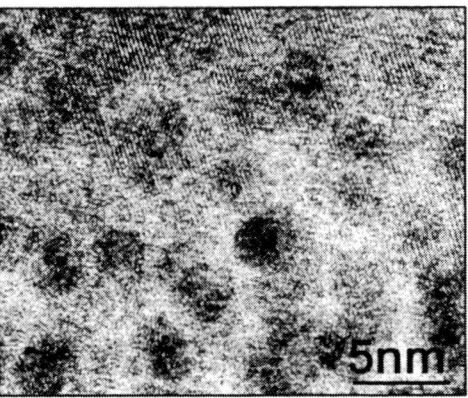

FIGURE 4. Cross-sectional TEM image of MgO·2.4(Al$_2$O$_3$) implanted with 60 keV Cu$^-$ at 10 $\mu A/cm^2$ to 3 ×10^{16} ions/cm^2.

Another crystalline substrate of Mg-Al spinel showed a contrasting precipitation behavior. Figure 4 shows a cross-sectional lattice image of Mg-Al spinel, MgO·2.4(Al$_2$O$_3$) that was implanted with 60 keV Cu$^-$ at 10 $\mu A/cm^2$ to 3 × 10^{16} ions/cm^2. Nanoparticles much smaller than in a-SiO$_2$ form in the non-stoichiometric spinel. Quantitative analysis of Cu nanoparticles indicated that full precipitation and no depth-oriented migration of implants occurred in the spinel. The spinel suppressed long-range atomic rearrangement of nanoparticles. It is thus demonstrated that nanoparticle growth and the depth distribution are controllable by using appropriate substrate species, as well as by adjusting dose rate.

Depth Control by In-beam Kinetics

Under ion implantation, particularly at high fluxes, in-beam rearrangements of nanoparticles occur with increasing dose rate. It is well-known that a bimodal depth-distribution of nanoparticles often occurs [10,11] in insulators. We have pointed out [8] that narrowing and the depth reduction of the atomic profile occur in a-SiO$_2$ with increasing dose rate. The narrowing tendency was somewhat stronger in c-SiO$_2$ than in a-SiO$_2$ [6]. The mechanism which we have proposed is a combined process of energy-gradient-induced diffusion and enhanced sputtering at high dose rates [7]. The narrowing process leads to a sharp distribution of implants.

Figure 5 shows a cross-sectional TEM image of a-SiO$_2$ implanted with 60 keV Cu$^-$ at 45 $\mu A/cm^2$ to 3 ×

10^{16} cm². At the critical dose rate, a two-dimensional distribution of Cu nanocrystals is obtained. During the narrowing and the depth reduction of the implant profile, the Ostwald ripening causes a further centripetal force. This rearrangement process can be called a self-assembly induced by high-flux ions. The small amount of nanoparticles gives a locally large susceptibility $\chi^{(3)}$ of $10^{-(7-8)}$ esu.

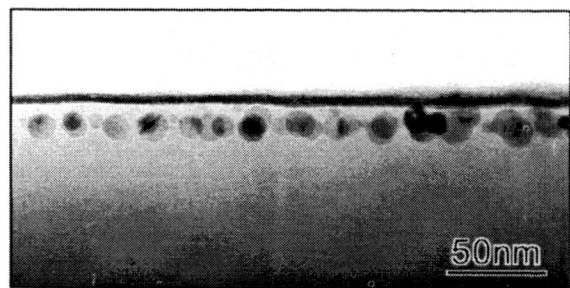

FIGURE 5. Cross-sectional TEM image of a-SiO₂ implanted with 60 keV Cu⁻ at 45 µA/cm² to 3 ×10¹⁶ ions/cm².

Thick Film Formation of Nanoparticles

The DNIM method succeeded in fabricating thick and uniform films, dispersed with Cu nanocrystals.

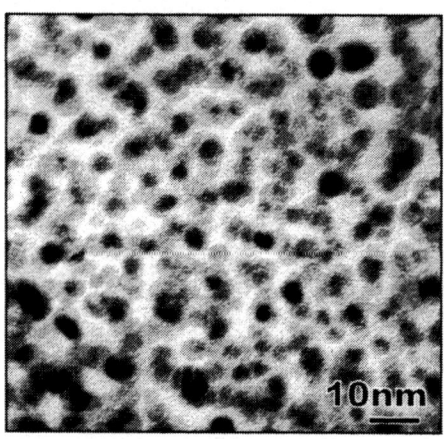

FIGURE 6. Cross-sectional TEM image of a specimen fabricated by DNIM method. The evaporation rate and dose rate are 0.4 nm/s and 20 µA/cm², respectively.

Figure 6 shows a cross-sectional TEM image of a specimen fabricated with the DNIM method, where evaporation rate and dose rate are 0.4 nm/s and 20 µA/cm², respectively. Nanoparticles of 5 nm in diameter uniformly distributed throughout the thickness of 500 nm. In the conventional implantation without film deposition, nanoparticles were located, more or less, in the vicinity of the projected range (~45 nm) as seen in Fig. 3. The thickness is virtually unlimited, until film exfoliation begins. Distributions of particle size and number density are uniform along the thickness. The average particle size of 5 nm in this case is smaller than that in the conventional implantation at the same dose rate (~10-15 nm). The smaller size in the DNIM process may be ascribed to efficient energy dissipation because of the moving surface. The particle size in the DNIM processing was controllable, to some extent, by changing dose rate.

SUMMARY

High-flux negative Cu ions have been applied for nanoparticle fabrication in insulators, using both conventional implantation and dynamic negative ion mixing. Intense ions caused spontaneous precipitation and depth-directional atomic migration of nanoparticles, including the 2D-assembly. Spatial control of nanoparticles with the kinetic processes was demonstrated by changing dose rate and substrates. The DNIM method also controlled spatial distribution in a thick uniform structure. The kinetic aspects develop unique tools to control the spatial structures.

REFERENCES

1. D.J. Rej, R.R. Bartsch, H.A. Davis, R.J. Faehl, J.B. Greenly and W.J. Waganaar, Rev. Sci. Instrum. **64**, 2753-2760 (1993).

2. Y. Yatsui, X.D. Kang, T. Sonegawa, T. Matsuoka, K. Masugata, Y. Shimotori, T. Satoh, S. Furuuchi, Y. Ohuchi, T. Takeshita and H. Yamamoto, Physics of Plasmas **1**, 1730-1735 (1994).

3. Y. Mori, G.D. Alton, A. Takagi, A. Ueno and S. Fukumoto, Nucl. Instrum. & Methods Phys.Res. **A273** 5-12 (1988).

4. J. Ishikawa, H. Tsuji, Y. Toyota, Y. Gotoh, K. Matsuda, M. Tanjyo and S. Sakaki, Nucl. Instrum. & Methods Phys. Res. **B96**, 7-12 (1995).

5. R.F. Haglund Jr., L.Yang, R.H. Magruder III, C.W. White, R.A. Zuhr, L. Yang, R. Dorsinville and R.R. Alfano, Nucl. Instrum.& Methods Phys.Res.**B91**, 493-498 (1994).

6. N. Kishimoto, N. Umeda, Y. Takeda, C.G. Lee and V.T. Gritsyna, Mater. Res. Soc. Symp. Proc. **540**, 153-158 (1999).

7. N. Kishimoto, N. Umeda, Y. Takeda, C.G. Lee, V.T. Gritsyna,. Nucl. Instrum.& Methods Phys. Res. **B148**, 1017-1022 (1999).

8. N. Kishimoto, Y. Takeda, N. Umeda, V.T. Gritsyna, C.G. Lee, T. Saito. Nucl. Instrum. & Methods **B166/167**, 840-844 (2000).

9. N. Kishimoto, N. Okubo, C.G. Lee, N. Umeda and Y. Takeda, Nucl. Instrum. & Methods Phys. Res. in press (2001).

10. H. Hosono, H. Fukushima,Y. Abe, R.A. Weeks and R.A. Zuhr, J. Non-Cryst. Solids, **143**, 157-161 (1992).

11. R.H. Magruder III, R.F. Haglund, Jr, L.Yang, J.E. Wittig and R.A. Zuhr, J. Appl. Phys., **76**, 708-715 (1994).

Application of Ionizing Radiation for Nano-Cluster Engineering

D. Ila*, R. L. Zimmerman, and C. I. Muntele

Center for Irradiation of Materials, Alabama A&M University
Normal. AL 35762-1447 USA
*Corresponding author: Tel. (256) 851-5866, Fax (256) 851-5868, e-mail ila@cim.aamu.edu

D. B. Poker and D. K. Hensley

Solid State Division, Oak Ridge National Laboratory, Oak Ridge, TN 37831 USA

Abstract. We report the results of our investigation of producing nanoclusters of Au in silica using post bombardment by MeV silicon. This technique has resulted in producing Au-nanoclusters at fluences of two orders of magnitude less than what is traditionally used. This is accomplished by first implanting Au into silica, then subsequently bombarding the silica with MeV Si ions. The size of the nanoclusters, ranging from 1 to 10 nanometers, is controlled by the implantation dose and total electronic energy deposited by post implantation-bombarding ions in the implanted layer. With the use of an indirect measurement method, such as optical absorption spectrophotometry (non-destructive), and a direct method, such as transmission electron microscopy (destructive), we show how, and at what concentrations, metallic ions nucleate to form nanoclusters by irradiation assisted nucleation at a dose below that needed for spontaneous nanocluster formation.

INTRODUCTION

Fabrication of a variety of optical devices relies mostly on changes in both the linear and nonlinear properties of glasses. The traditional technique used to change the linear properties of glasses mostly involved melting selected metals with glass, cooling the melt to form homogeneous glass, and then forming the metal colloids by a reheating process. Recently, metallic ion implantation and thermal annealing have been used to introduce similar effects and nonlinear optical properties near the surface [1-8]. The techniques used to form nanoclusters may be categorized as follows: A) room temperature implantation, followed by high temperature annealing; B) room temperature implantation at doses above the threshold for spontaneous nanocluster formation; C) ion implantation at elevated temperatures. An attractive property of ion implantation is that ions can be focused in a well-defined space in an optical device, to induce local changes in its linear and nonlinear properties.

It has long been known that small metallic particles or colloids embedded in dielectrics produce colors associated with optical absorption at the surface plasmon resonance frequency [6,7], which depends on the index of refraction of the host substrate and the electronic properties of the colloids formed in the host material. For clusters with diameters much smaller than the wavelength of light, the theories of Mie [8] can be used to calculate the absorption coefficient (cm^{-1}) of the composite:

$$\alpha = \frac{18 \cdot \pi \cdot Q \cdot n_0^3}{\lambda} \times \frac{\varepsilon_2}{\left(\varepsilon_1 + 2 \cdot n_0^2\right)^2 + \varepsilon_2^2} \quad (1)$$

where Q is the volume fraction occupied by the metallic particles, n_o, is the refractive index of the host medium, and ε_1 and ε_2 are the real and imaginary parts of the frequency-dependent dielectric constant of the bulk metal. Equation (1) is a Lorentzian function with a maximum value at the surface plasmon resonance frequency (v_p). Values of ε_1 for the metals, as a function of wavelength, are tabulated in [9]. Using the measured n_o for various wavelengths, one can predict from equation (1) the photon wavelengths for the surface plasmon resonance frequencies of metallic colloids in the photorefractive host materials, as shown in Figure 1. This figure shows that for silica host material, the absorption band for Au should be near 530 nm.

To produce nanoclusters such as gold in a silica medium, the implantations are done at high doses (1 to 2×10^{17} ions/cm^2) to overcome the solubility of the implanted species in the substrate. This is also why there is a wide distribution of cluster sizes produced by the above techniques (Fig. 2a). To overcome this problem,

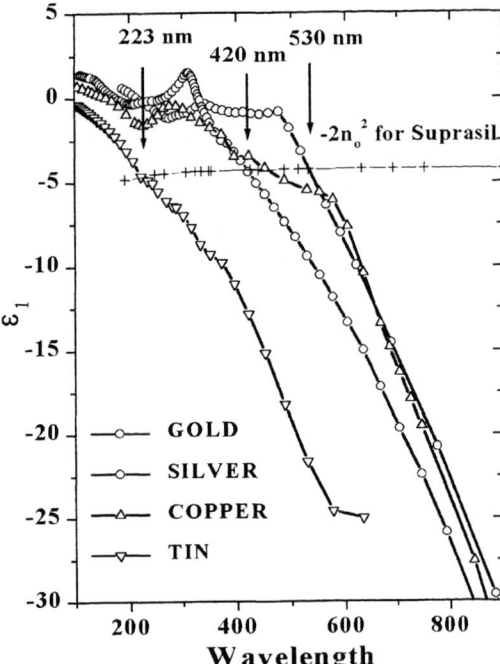

FIGURE 1. A plot of the real part of the frequency-dependent dielectric constant of the bulk metal (Au, Sn, Ag, Cu). It also shows the value of the dielectric constant ($\varepsilon_1 = -2 n_0^2$) in the silica host, for optimal extinction coefficient.

since 1994, we initiated a series of investigations into how to confine the implanted ions in a narrow layer. This was done by generating chemical barriers, engineering defect barriers on both sides of the implanted layer, as well as combining two non-equilibrium processes: ion implantation and post-implantation irradiation [9, 10]. The latter technique, the focus of this paper, is a combination of metallic ion implantation, followed by MeV ion bombardment, all at room temperature. The result was the production of gold nanoclusters in silica at fluences as low as 5×10^{15} ions/cm^2, with a fine control of the size and uniformity of the nanoclusters (see Fig. 2b).

EXPERIMENTAL PROCEDURES

The $5 \times 5 \times 0.5$ mm silica glass used, Suprasil-300, is a commercial product of Heraeus Amersil, Inc. and is of known purity. 2.0 MeV gold ions were implanted at a low current density of 2 μA to produce a layer with gold concentrations between 5×10^{15} ions/cm^2 and 1.2×10^{17} ions/cm^2. The beam current was kept low to avoid the premature formation of gold clusters due to ion beam heating. The depth of the implanted layer (0.48 μm), as well as the penetration range of the post-implantation bombarding particles, was calculated using the SRIM computer code [11] and measured using Rutherford backscattering spectrometry (RBS). The energy of the bombarding Si particles (5.0 MeV) was selected such that the Si ions stopped beyond the range of the Au implanted layer, so that the energy deposited in the gold-containing layer was due mainly to the electronic energy loss. Fluences of silicon between 5×10^{15} ions/cm^2 and 2×10^{17} ions/cm^2 were investigated, in order to establish optimum conditions for gold nanocluster formation. In all of the above post-implantation bombardments the temperature of the silica host was kept at 300K.

The optical absorption photospectrometry (OAP) results given by the small nanoclusters formed were obtained using two separate methods. The first method was to subtract the OAP result acquired after the Au-implanted silica (but before Si-bombardment) from the results acquired after the Si-bombardment. The other method was to subtract OAP results of an Au-implanted sample and a Si-bombarded one from the OAP result of the sample that was first Au-implanted and then Si-bombarded. Both of these techniques produced similar results.

RESULTS AND DISCUSSION

Higher ion beam fluences and higher atomic numbers of the bombarding ions resulted in more damage to the silica glass, and a subsequent change of the optical properties [7, 13]. This was observed by OAP during all bombardments. These effects are reduced with heat treatments above 970K [7, 8]. Fig. 3 shows the optical absorption spectra for the 2.0 MeV Au-implanted silica at 1.2×10^{17} ions/cm^2 after annealing at temperatures of 873K, 1273K and 1473K, each for one hour. As shown in this figure, the higher the post-implantation heat treatment, the more pronounced is the observed optical absorption band that appears around the wavelength calculated using the Mie theory (526 nm) [14].

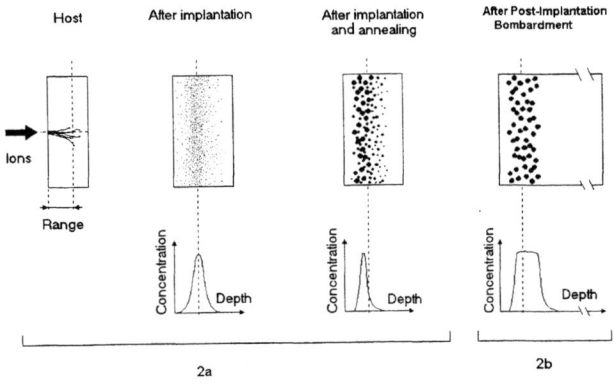

FIGURE 2. Silica host after Au implantation followed by: a) annealing; b) MeV Si ion post-implantation bombardment.

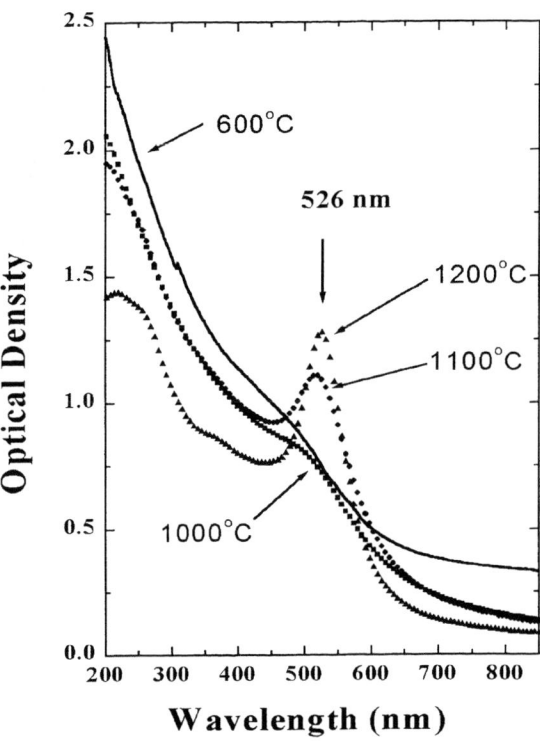

FIGURE 3. Optical absorption spectra bands for the 2.0 MeV Au-implanted silica at 1.2×10^{17} ions/cm² after annealing at temperatures between 873K, 1273K and 1473K each for one hour.

Figure 4 shows the optical absorption spectra from silica: (i) implanted with 2.0 MeV Au at 1.2×10^{17} ions/cm², at room temperature; (ii) implanted with 6×10^{16} ions/cm² Au at 2.0 MeV and post bombarded by 5.0 MeV Si at 6×10^{16} ions/cm², 1×10^{17} ions/cm², and 2×10^{17} ions/cm². A comparison of the absorption bands (526 nm) obtained from heat-treated Au-implanted silica, and the absorption bands (520 nm) obtained from post-implantation Si-bombarded silica, indicates a 6 nm shift. The only difference is that the position of the absorption band is slightly red-shifted, which is due to thermal relaxation of the Si-bombarded silica, thus recovering its original index of refraction value. As we increased the post-implantation bombardment fluence to 2×10^{17} ions/cm² the damage to the substrate caused the gold ions to migrate. The damage is shown by the widening of the absorption band and a general increase in the absorption for all wavelengths.

Rutherford Backscattering Spectrometry (RBS) spectra of the Au-implanted silica at 1.2×10^{17} ions/cm², before and after bombardment by 5 MeV Si, at various fluences are shown in Figure 5, along with a RBS spectrum from Au film on silica used as a reference.

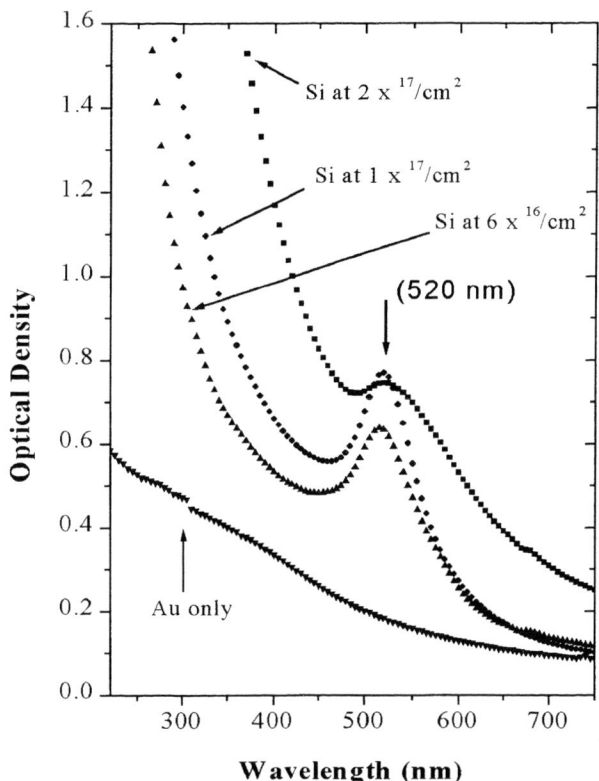

FIGURE 4. The optical absorption spectra band from Au implanted Suprasil-300.

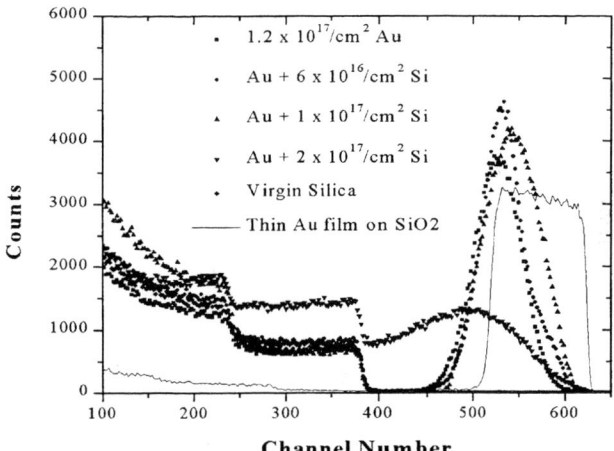

FIGURE 5. RBS spectra of 2.0 MeV Au-implanted Suprasil-300 at 1.2×10^{17} ions/cm², before and after bombardment by 5 MeV Si at various fluences.

This figure indicates that as the post-implantation bombardment fluence increases above 6×10^{16} ions/cm², the efficiency of producing well-defined layers is reduced. The increase to 2×10^{17} ions/cm² causes irreversible damage to the silica substrate.

Fig. 6 shows the optical absorption spectra from silica:

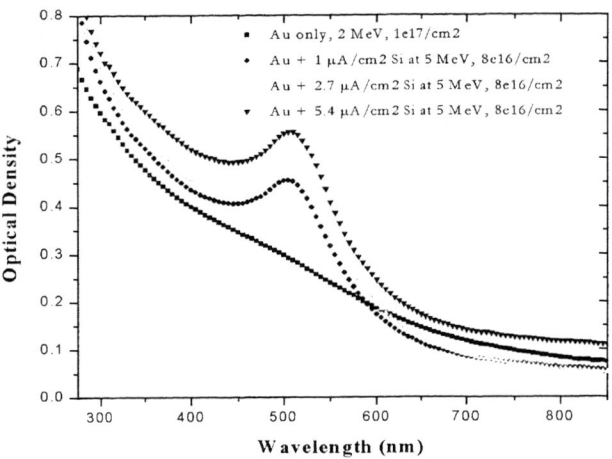

FIGURE 6. Optical absorption spectra of Au implanted silica.

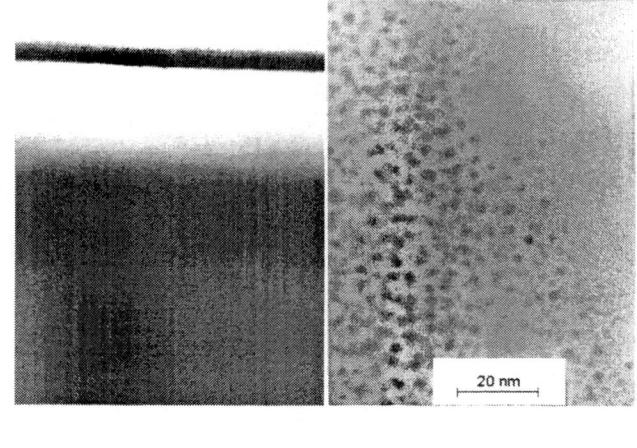

(a)   (b)

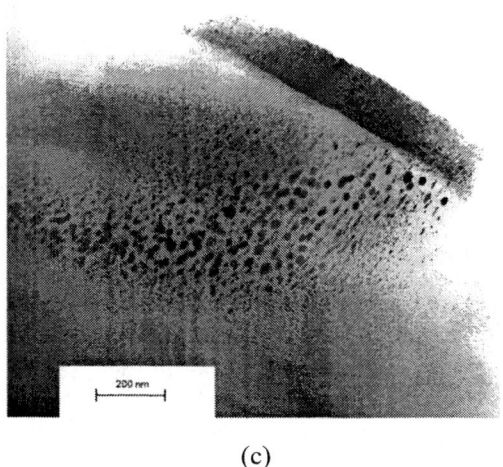

(c)

FIGURE 7. TEM micrographs of silica, taken after each step performed in the process of gold nanocluster formation.

(i) implanted with 2.0 MeV Au at 1.2×10^{17} ions/cm^2, at room temperature; (ii) implanted with 2.0 MeV Au at 1×10^{17} ions/cm^2 and then bombarded by 5.0 MeV Si at 8×10^{16} ions/cm^2 at current densities of 1 µA/cm^2, 2.7 µA/cm^2, and 5.4 µA/cm^2 1×10^{17} ions/cm^2, and 2×10^{17} ions/cm^2. A comparison of the absorption bands and their FWHM, as well as their TEM results, indicates that the increase in the current density doesn't help the nanoclusters formation or their confinement. In fact, an increase in the current density increases the substrate temperature, which results in migration of Au ions deeper into the substrate, as well as increasing the absorption base for all wavelengths.

Fig. 7 shows the comparison of TEM micrographs of silica, implanted at room temperature: (a) 2.0 MeV Au at 1.2×10^{17} ions/cm^2; (b) 2.0 MeV Au at 6×10^{16} ions/cm^2, and then bombarded by 5.0 MeV Si at 6×10^{16} ions/cm^2; and (c) 2.0 MeV Au at 5×10^{15} ions/cm^2, and then bombarded by 5.0 MeV Si at 2×10^{17} ions/cm^2. The micrograph 7(b) shows almost a uniform size of Au nanoclusters, whereas the micrograph 7(c) shows Au ions occupying the migrating voids generated by the damage to the silica substrate.

Table I shows the energy deposited by the electronic (ε_e) and nuclear stopping (ε_n), and the integral of the product of the extinction coefficient and the wavelength [10, 14] at the absorption band after post-implantation bombardment with silicon ions of 1.2, 2.0, and 5.0 MeV.

The energy deposited at the location of the Au-implanted layer is calculated using the SRIM computer code [11]. In this table the extinction coefficient (α) is normalized and has arbitrary units.

To evaluate the mechanism by which the nanoclusters are formed using post-implantation bombardment, we implanted gold and silicon at energies such that their implantation ranges were equal. We implanted 5.5 MeV Au ions (at similar concentrations as above), in the 1.4 µm implantation range, followed by 1.0 MeV Si ion bombardment at various fluences. The result was that no Au nanoclusters were produced, thus re-confirming that the process of nanocluster formation is due to the electronic energy lost by the MeV Si ions in the gold-implanted layer of silica.

TABLE 1. Deposited energies.

Si Beam Energy (MeV)	ΔE (eV) due to ε_e	ΔE (eV) due to ε_n	α·λ (nm)
1.2	107,017	6,976	4.1
2	176,868	4,814	8.6
5	373,018	2,980	27

CONCLUSIONS

A comparison of the area under the absorption band for the Si-bombarded samples at constant fluence and different energies with the electronic energy deposited at the Au-implanted depth indicates a similar increasing trend. Meanwhile, the energy deposited in the Au-implanted layer, due to the nuclear stopping power of Si ions at different energies, shows the opposite trend.

The energy deposited by electronic excitations contributes to both formation of large size nanoclusters and an increase of their volume fraction. That is why the integral under the absorption band is increasing faster than the increase in the electronic energy deposited by Si ions at higher energies.

RECOMMENDATIONS

To check the validity and precision of our technique, we are producing SiO_2 films with uniform Au concentrations on SiO_2 substrates, using ion beam assisted deposition (IBAD). This will reduce the uncertainty due to the energy straggling in the implanted top layer of the samples described herein. In addition, we have a square shaped Au concentration, which makes tracing of the Au diffusion into the substrate easier.

The solution to produce a more confined metallic nanocluster in any host material, with better control over size distribution at much lower implantation doses, is to combine post-bombardment with MeV ions with ion implantation. The combined two non-equilibrium techniques can induce the formation of Au nanoclusters with fine control on their sizes, ranging from 0.5 nm to several nanometers, depending on the combination of implanted dose and the electronic energy deposited per post-implantation bombarding ions, thus generating an optical absorption band at about 520 nm and a third order nonlinearity in a more confined volume.

ACKNOWLEDGEMENTS

This project was supported by the Center for Irradiation of Materials at Alabama A&M University and Alabama EPSCoR-NSF Grant No. OSR-9559480. The work at ORNL was sponsored by the Division of Materials Science, U.S. Department of Energy, under Contract DE-AC05-96OR22464 with Lockheed Martin Energy Research Corp.

REFERENCES

1. G. W. Arnold, *J. Appl. Phys.* **46**, 4466 (1975)
2. G. W. Arnold and J. A. Bordes, *J. Appl. Phy.* **48**, 1488 (1977).
3. R. H. Magruder III, R. A. Zuhr, D. H. Osborne, Jr., *Nucl. Inst. & Meth. in Phys. Res.* **B99**, 590 (1995).
4. R. F. Haglund, Jr., D. H. Osborne, Jr., R. H. Magruder, III, C. W. White, R. A., Zuhr, P. D. Townsend, D. E. Hole, and R. E. Leuchtner, *Mat. Res. Soc. Symp. Proc.* **Vol. 354**, 629 (1995).
5. C. W. White, D. S. Zhou, J. D. Budai, R. A. Zuhr, R. H. Magruder and D. H. Osborne, *Mat. Res. Soc. Symp. Proc.* **Vol. 316**, 499 (1994).
6. K. Fukumi, A. Chayahara, M. Adachi, K. Kadono, T. Sakaguchi, M. Miya, Y. Horino, N. Kitamura, J. Hayakawa, H. Yamashita, K. Fujii and M. Satou, *Mat. Res. Soc. Symp. Proc.* **Vol. 235**, 389 (1992).
7. D. Ila, Z. Wu, R. L. Zimmerman, S. Sarkisov, C. C. Smith, D. B. Poker, and D. K. Hensley, *Mat. Res. Soc. Symp. Proc.* **457**, 143 (1997).
8. D. Ila, E. K. Williams, S. Sarkisov, C. C. Smith, D. B. Poker, and D. K. Hensley, *Nucl. Instr. Meth. in Phys. Res.* **B141**, 289 (1998).
9. D. Ila, Z. Wu, C. C. Smith, D. B. Poker, D. K. Hensley, C. Klatt and S. Kalbitzer, *Nucl. Instr. Meth. in Phys. Res.* **B127**, 570 (1996).
10. D. Ila, E. K. Williams, C. C. Smith, D. B. Poker, D. K. Hensley, C. Klatt and S. Kalbitzer, *Nucl. Instr. Meth. in Phys. Res. B*, (1999).
11. J. F. Ziegler, J. P. Biersack and U. Littmark, *The Stopping and Range of Ions in Solids* (Pergamon Press Inc., New York, 1985).
12. E. R. Schineller, R. P. Flam and D. W. Wilmot, *J. Opt. Soc. Am.* **58**, 1171 (1968).
13. P. D. Townsend, *Nucl. Instr. Meth. in Phys. Res.* **B46**, 18 (1990).
14. G. Mie, *Ann. Physik* **25**, 377 (1908).
15. M. Verhaegen, L.B. Allard, J. L. Brebner, M. Essid, S. Roorda, J. Albert, *Nucl. Instr. Meth. in Phys. Res.* **B106**, 438 (1995).

Formation and Application of Metal Nanoclusters in SiC

Robert L. Zimmerman[a], D. Ila[a], C. Muntele[a], I. Muntele[a], A. L. Evelyn[a],
D. H. Hensley[b] and D. B. Poker[b],

[a] *Center for Irradiation of Materials, Alabama A&M University, Normal, AL 35762, USA*
[b] *Solid State Division, Oak Ridge National Laboratory, Oak Ridge, TN 37831, USA*

Abstract. Using high dose and energy ion implantation, followed by thermal annealing, we have formed nanoclusters of various ion species in Si-face 6H-SiC. The implantation of ions into any semiconductor material followed by thermal annealing leads to a change of the electrical and optical properties at the implanted layer owing to the formation of defects and to the formation of nano-clusters with diameters less than 30 nm. In this paper, we will present the results from keV and MeV ion implantation into SiC and propose applications for optical and electrical device fabrication.

INTRODUCTION

As a wide bandgap semiconductor (3.23 eV) with high index of refraction (2.7), high fracture toughness (3.1 MPa-m$^{1/2}$), high thermal conductivity (~5.0 W/cm), high saturated electron drift velocity (~2.7 x 10^7 cm/s) and high breakdown electric field strength (~3 MV/cm), SiC is a material of choice for high temperature, high voltage, high frequency and high power applications [1-3].

Increasingly, bombardment and implantation with high-energy ions are being used to provide the surface topology and material modification of SiC microelectromechanical systems. We report the introduction of metal colloids into optical materials by metal ion implantation and the possibility of fabricating optical devices. Recently, it has been shown that because of its outstanding thermal stability SiC can be employed both as an oxygen and as a hydrogen sensor that operates in a temperature regime considerably higher (up to 1000°C [4-9]) than conventional sensors such as tin oxide (SnO$_2$) or silicon.

A: Formation of Nanoclusters in SiC

Introducing metal colloids into an optical material will change the color of the material [10, 11]. Similar effects near the surface as well as changing the nonlinear optical properties can be induced by ion implantation followed by thermal annealing [12-16]. For clusters with diameters much smaller than the wavelength of light λ, the theories of Mie [17] can be used to calculate the absorption coefficient of the composite:

$$\sigma = \frac{18\pi Q n_0^3}{\lambda} \cdot \frac{\varepsilon_2}{\left(\varepsilon_1 + 2n_0^2\right)^2 + \varepsilon_2} \quad (1)$$

where **Q** is the volume fraction occupied by the metallic particles, n_0 is the refractive index of the host medium, and ε_1 and ε_2 are the real and imaginary parts of the dielectric constant of the bulk metal. Equation 1 is a Lorentzian function with a maximum value when the light frequency equals the surface plasmon resonance frequency (ω = ω$_p$), where:

$$\varepsilon_1(\omega_p) + 2n_0^2 = 0 \quad (2)$$

Values of ε_1 for the metals as a function of wavelength are tabulated in [18]. The index of refraction measured by prism coupling for the SiC used in this work was 2.26 (the published index is 2.655). From Equation (2) one can predict the photon wavelengths for the surface plasmon resonance frequencies of metallic colloids in photorefractive host materials (Figure 1).

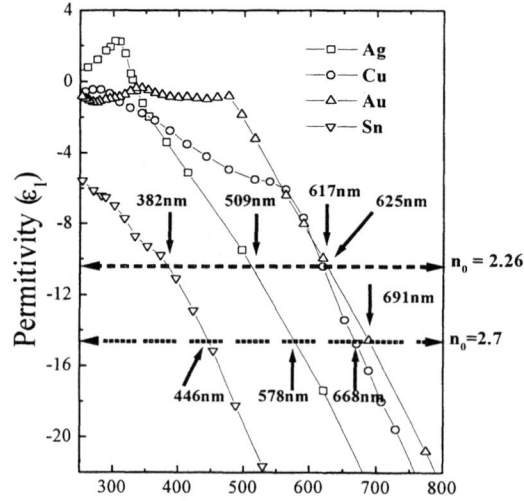

FIGURE 1. Permitivity vs. wavelength for Ag, Cu, Au and Sn. The surface plasmon resonance occurs when Eq. 2 is satisfied. Measured and expected refractive indices of SiC are shown at the horizontal dashed lines.

We have implanted SiC samples from CREE Research Inc., (6H-N type, 3.5° off axis, doped with nitrogen at $9.2 \times 10^{17}/cm^3$, Si face) with ions such as 1.5 MeV Si, 1.0 MeV Au, 2.0 MeV Ag, 2.0 MeV Cu, and 160 keV Sn at fluences between $8 \times 10^{15}/cm^2$ to $5 \times 10^{17}/cm^2$ both at room temperature and at 500°C.

In previous work [13, 14] on other optical materials, we used inert ions such as He or Ar, or an ion that is one of the constituents of the host crystal, such as Si or O, to observe the change in the optical properties of the bombarded material as a function of the implantation parameters. Figure 2 shows the optical absorption spectra for SiC bombarded by 1.5 MeV Si ions at various fluences, at room temperature and at 500°C. The room temperature bombardment caused severe darkening, even at fluences as low as $4 \times 10^{16}/cm^2$, shifting the absorption edge from 350 nm to 650 nm. Implantation at 500°C allows the damage to recover. If Si nanocrystals form, they absorb light in the far IR region and do not interfere with our observations. These spectra from the Si bombardment were used to evaluate the optical absorption due to ion bombardment damage.

Figure 3 shows optical absorption spectrum of a sample implanted at 500°C with 3×10^{17} Au/cm^2 at 1.0 MeV and annealed at 1000°C in argon for one hour. The nitrogen impurity center absorbs red light in a reference sample and a difference spectrum is shown in Fig. 3. At Au fluences less than $3 \times 10^{17}/cm^2$ no peaks due to plasmon resonance were discernible.

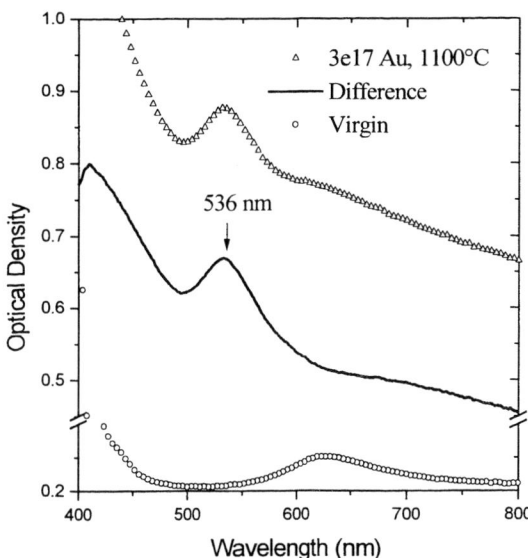

FIGURE 3. Optical Density vs. Wavelength for 1 MeV Au implanted at 500°C into 6H, N-doped SiC heated to 1000°C in Ar. Au ion fluence was $3 \times 10^{17}/cm^2$.

The size of the Au clusters can be estimated from the peak position and the width of the resonance through the relation $r = A_m v_f / \Delta \omega_{\frac{1}{2}}$ [19] where A_m is a constant, taken to be 1.5 for Au, v_f is the Fermi velocity of electrons in gold and $\Delta \omega_{\frac{1}{2}}$ is the full width at half maximum of the absorption peak. From Fig. 3 the Au cluster diameter is estimated at 5 nm.

Figure 4 shows typical optical absorption spectra for SiC implanted with 160 keV Sn at room temperature and then annealed at 200°C in argon for one hour. This figure also shows that as implantation fluence increases the absorption baseline increases. Using the measured absorption band, 406 nm, and equation 2 the calculated index of refraction for SiC at the implanted volume is 2.22. Increasing the Sn fluence produced an increase in the height of the optical absorption band due to formation of Sn nanoclusters. Observation of the formation of Sn nanoclusters in SiC using visible optical spectrometry was made possible because of the high index of refraction of the host material. In other hosts the absorption band from Sn nanoclusters would have been located in the UV region.

Implantation at room temperature results in a large increase in optical absorption that can mask the surface plasmon resonance absorption band. This difficulty is alleviated by implanting at elevated temperature (500°C), which inhibits the formation of defects. The broad plasmon resonance spectra indicate that the nanoclusters are

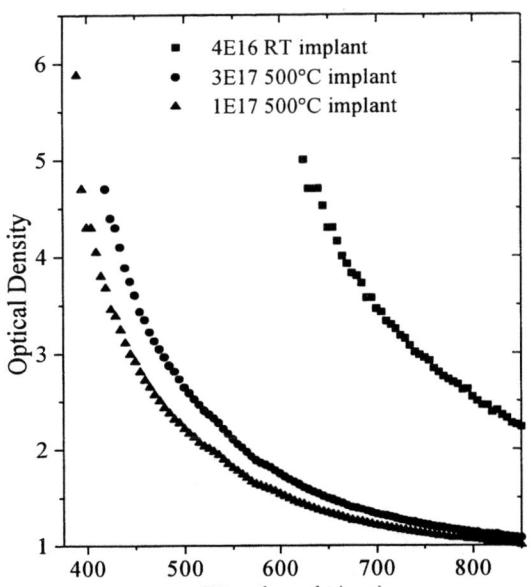

FIGURE 2. Optical density vs. wavelength for 1.5 MeV Si implanted into 4H, p-type SiC at 28°C and 500°C.

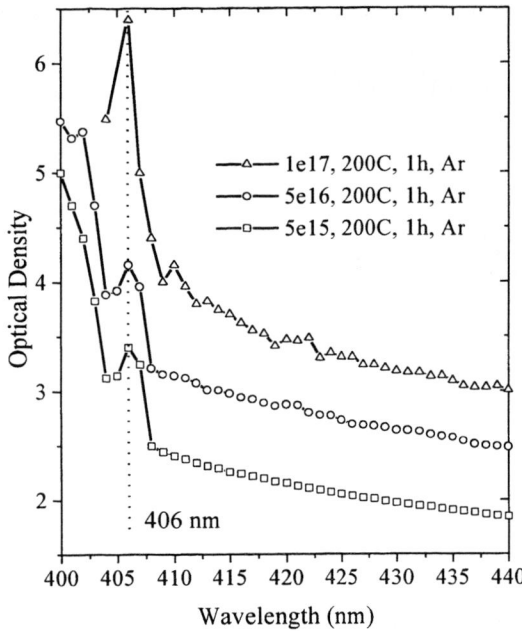

FIGURE 4. 160 keV Sn implanted into 6H, n-type SiC at 28°C and annealed at 200°C for 1 hour.

very small. Further work is necessary at higher implantation and annealing temperatures.

B: SiC Gas Sensor

Depositing a catalytic metal such as Pd onto SiC results in a Schottky diode behavior. The adsorbed gas changes the space charge region under the metal clusters, which in turn affects the conductivity of the crystal. This change in conductivity is measured and can be correlated to surface concentrations and to the levels of the sampled gas in the ambient [8]. Rather than applying a palladium film [9], we have implanted palladium ions into 6H, n-type SiC samples. The implantation was performed at 500°C, in order to minimize the induced implantation damage, to fluences between 3×10^{14} at/cm^2 and 3.2×10^{16} at/cm^2, and implantation energies of 70 keV and 130 keV. The energies were chosen using the SRIM code [12] in order to get high near surface concentrations of Pd ions. Also, from the same code, the approximate implantation range was established as being 40 nm and 65 nm.

A potential difference of ±1V was applied across the sample, and the current was measured using a probe on the implanted side (Figure 5). Measurements were taken for temperatures near 23, 70, 145, and 215 °C in a controlled environment. Air was supplied alternately with an H$_2$-Ar gas mixture with 4% hydrogen.

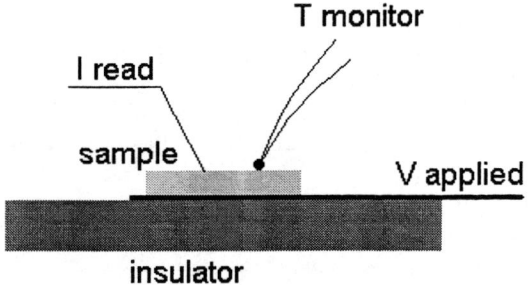

FIGURE 5. Schematic of the I-V setup.

RESULTS

Figure 6 shows that at room temperature (23°C) the current increases in the presence of the H$_2$-Ar mixture. This was the behavior shown for each of the SiC samples independent of the implanted Pd fluence, although the current differential increases monotonically with Pd fluence.

Measurements were performed on pristine SiC samples and revealed no current differential with the gas change for temperatures from 23°C to 240°C. When the H$_2$-Ar environment is replaced by air, the current reverts to its initial value. At higher substrate temperatures, the differential current for H$_2$-Ar ambient and air ambient decreases as the substrate temperature approaches

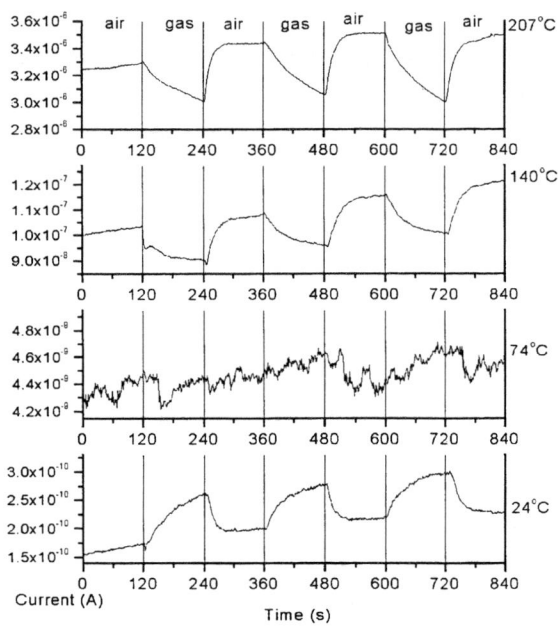

FIGURE 6. Current response for a sample implanted with 3.2 x 10^{16}/cm^2, 130 keV Pd. A constant potential difference of 1 volt was applied as shown in Figure 5.

60°C to 84°C, where the current differential is essentially nonexistent. At temperatures higher than 84°C the substrate current in the presence of H_2–Ar gas is smaller than when air is present. The relative sensitivity to hydrogen inverts at about 84°C.

These observations differ from those reported in the literature [4, 6, 9] for SiC with palladium deposited on the surface for which the current in the presence of hydrogen is more at every temperature, and reverts to the initial value when air is supplied instead of hydrogen. No sense inversion has been reported. We observe similar behavior both for positive and negative applied voltages with the inverted current differential at the higher temperatures. For absolute voltages above 1.2 V, the junction breaks down, and no more sensing behavior can be observed.

The fact that p-n junction breakdown destroys sensing behavior makes us believe that the sensing properties are related to the rectifying properties of the device rather than to the modification in the conductivity of palladium, or palladium hydride. The similar behavior for both positive and negative applied voltages may indicate a structure more like n-p-n than p-n junction. Unlike surface films of Pd on SiC, Pd ion implantation may produce an almost symmetrical structure.

The slow response time observed for all samples and shown in figure 6 is owing to the fact that the hydrogen must diffuse to palladium layer. Similarly, when air is introduced, the diffusion process is reversed. The recovery process depends on the removal of the hydrogen that was adsorbed in the device. Two processes occur: the chemical affinity for hydrogen by the oxygen in the air, and the outgassing due to the device's operation at elevated temperatures. At room temperature the first process is dominant, while the outgassing begins to contribute at temperatures above 100°C.

CONCLUSIONS

In this manuscript we discussed two of the applications of SiC crystals: optical devices and electronic sensors. The first application is new. The high index of refraction in SiC, 2.655, allowed the observation of the absorption spectra of Sn nanoclusters. The strength of Si-C bond as well as the diffusivity of metals in SiC should be used to study the possibility of formation of an epi-layer of nanoclusters for various materials with a high volume fraction, which will enable us to use much lower implantation doses and lower power lasers to operate optical devices.

ACKNOWLEDGMENTS

This project was supported by the Center for Irradiation of Materials at Alabama A&M University, NASA-GRC Contract No. NAG3-2123, and the Division of Material Sciences, U. S. Department of Energy, under contract DE-AC05-00OR22725 with the Oak Ridge National Laboratory, managed by UT-Battelle, LLC.

REFERENCES

1. J. R. O'Connor, J. Smiltens, *Silicon Carbide, A High-Temperature Semiconductor* (Pergamon, NY, 1960).
2. G. L. Harris, *Properties of Silicon Carbide* (INSPEC, London, 1995).
3. K. Shenai, R. S. Scott and B. J. Baliga, IEEE Trans. Elec. Dev. **36,** (1989) 1811.
4. G.W. Hunter, P.G. Neudeck, G.D. Jefferson, G.C. Madzsar, C.C. Liu and Q.H. Wu, *Report E-7773 NASA*, (1993).
5. G.W. Hunter, P.G. Neudeck, C.C. Liu and Q.H. Wu, Conference on Advanced Earth-To-Orbit Propulsion Technology, (1994).
6. L–Y Chen, G.W. Hunter, P.G. Neudeck, D. Knight, C.C. Liu and Q.H. Wu, *Proc. 190th Meeting of Electrochemical Society* (1996).
7. L. A. Spetz, A. Baranzahi, P. Tobias, I. Lundstrom, Physica Status Solidi (A) Applied Research **162**, 1, (1997) 493-511.
8. G. Muller, G. Krotz, E. Niemann, Sensors and Actuators **A43**, 1-3, (1994) 259-268
9. M. A. George, M. A. Ayoub, D. Ila, D. J. Larkin, Mat. Res. Soc. Symp. Proc. **572,** (1999) 123-128.
10. G. Fuchs, G. Abouchacra, M. Treilleux, P. Thévenard, and J. Serughetti, Nucl. Instr. and Meth. in Phys. Res. **B32,** (1988) 100.
11. G. Abouchacra, G. Chassagne, and J. Serughetti, Radiation Effects **64** (1982) 189.
12. J. F. Ziegler, J. P. Biersack, U. Littmark, *The Stopping and Range of Ions in Solids*, Pergamon Press, NY, 1985.
13. G. W. Arnold, J. Appl. Phys. **46,** (1975) 4466.
14. C. W. White, D. S. Zhou, J. D. Budai, R. A. Zuhr, R. H. Magruder and D. H. Osborne, Mat. Res. Soc. Symp. Proc. **316,** (1994) 499.
15. E. K. Williams, D. Ila, A. Darwish, D. B. Poker, S. S. Sarkisov, M. J. Curley, J-C. Wang, V. L. Svetchnikov, H. W. Zandbergen, Nucl. Instr. and Meth. in Phys. Res. **B148,** (1998)1074 .
16. D. Ila, E. K. Williams, S. Sarkisov, C. C. Smith, D. B. Poker, and D. K. Hensley, Nucl. Instr. and Meth. in Phys. Res. **B141,** (1998) 289.
17. G. Mie, Ann. Physik **25,** (1908) 377.
18. D. R. Lide, Ed., *CRC Handbook of Chemistry and Physics*, 76th Edition (CRC Press, Boca Raton, 1987).
19. W. T. Doyle, Phys. Rev. **111,** (1958) 1067.

Application of Ion Beams for Polymeric Carbon Based Biomaterials

A. L. Evelyn[*]

Center for Irradiation of Materials, Alabama A&M University, Normal, AL 35762 USA

Abstract. Ion beams have been shown to be quite suitable for the modification and analysis of carbon based biomaterials. Glassy polymeric carbon (GPC), made from cured phenolic resins, has a high chemical inertness that makes it useful as a biomaterial in medicine for drug delivery systems and for the manufacture of heart valves and other prosthetic devices. Low and high-energy ion beams have been used, with both partially and fully cured phenolic resins, to enhance biological cell/tissue growth on, and to increase tissue adhesion to GPC surfaces. Samples bombarded with energetic ion beams in the keV to MeV range exhibited increased surface roughness, measured using optical microscopy and atomic force microscopy. Ion beams were also used to perform nuclear reaction analyses of GPC encapsulated drugs for use in internal drug delivery systems. The results from the high energy bombardment were more dramatic and are shown in this paper. The interaction of energetic ions has demonstrated the useful application of ion beams to enhance the properties of carbon-based biomaterials.

[*]In collaboration with D. Ila, R. L. Zimmerman and K. Bhat, Center for Irradiation of Materials, Alabama A&M University, David B. Poker and Dale K. Hensley, Oak Ridge National Laboratory

INTRODUCTION

Historically, polymeric carbon-based materials have been subjected to ion beams for analysis and modification. Current research on biocompatible materials still use ion beams in this regard. The application of both high and low energy ions in the modification and analysis of polymeric carbon for use as biocompatible materials are well documented. One form of polymeric carbon biomaterial, glassy polymeric carbon (GPC), is chosen to illustrate the applications of ion beams. GPC, an amorphous organic polymeric carbon material that is widely used as a biomaterial, is used in the manufacture of prosthetic heart-valves and other percutaneous devices. It possesses high chemical inertness and biocompatibility. In many instances, the mismatch in compatibilities of these devices results in device failure over a relatively short time. In spite of its characteristics, GPC mechanical heart valves still present thromboembolic problems, which are related to the interface between its surface and the biological tissue. This is due mainly to the natural cicatrization processes that occur with implanted devices. Thromboembolism, which can result in device dysfunction, is the most important problem to be solved in cardiac prostheses research today. Thromboresistence can be improved by treatment of the surface of the GPC devices with ion beams. GPC material is being considered for use in certain drug delivery systems, because of its structure and its chemical characteristics described above. The GPC material is tailored with ion beams for such applications.

Ion Beam Effects on Polymeric Materials

Ion beams are essentially ionizing radiation, which causes chemical changes to occur in polymeric carbon materials. Several processes can occur separately, or in combination with one another, during ion bombardment of the material. Transmitted phonons or excitons may break chemical bonds, producing scission and polymer chain cross-linking. Dissociated atoms, small molecules and other radicals may diffuse through the material. Dehydrogenation can result in dangling bonds, which may eventually saturate and results in chain cross-linking [1]. Other molecular emission processes, double-bond formation, triple-bond formation, dipole formation and precipitate formation by self-clustering of the injected species can also occur.

Ion beam radiation induces greater property changes in polymeric materials than comparative sources, such as UV-light, electron beams, γ-rays and x-rays. This is due partly to the lower cross sections for these sources, compared with those for the heavier energetic ions [1].

Induced effects in ion-irradiated polymers depend on the amount of energy deposited in the material as a function of time and space. Energy deposited per unit volume as a function of the ion fluence can be expressed as the energy density $D_e = \phi S_e$ (eV/mm^3), given by the product of the ion fluence (ϕ=ions/mm^2) and energy loss, where S_e = dE/dx [2]. Various types of alterations in the polymer structure take place, corresponding to ion fluence ranges. At low energy densities (~ 10^{20} eV/mm^3), cross-linking may result in new bond structure formation. Conversely, chain scission can result from broken bonds. This results in changes in the solubility and molecular weight distributions as well as modifications to transport properties and polymer morphology. Desorption of simple molecules occurs as weaker bonds are broken and certain elements are liberated. In this fluence range, there is no overlapping of the ion tracks [2].

At increased fluences, and greater energy densities (~ 10^{21} eV/mm^3), the ion tracks are much closer together and begin to overlap. This results in a more widespread rearrangement of the molecular bonds, leading to a complete change in the polymer structure. The material at this stage has different electronic structures and the physical properties change as new functional groups form. The polymer backbone may also change at these greater energy densities. The chemical composition changes as new molecular species are liberated [2].

At progressively higher fluences and energy densities (~ 10^{22} eV/mm^3) there is complete overlap of the ion tracks and a much more stable polymer is produced. The properties of the material at this stage resemble that of hydrogenated amorphous carbon, in terms of the electronic and optical properties, hardness and chemical inertness [2]. Studies have shown that the presence of atoms (oxygen, nitrogen, fluorine, etc.), other than carbon and hydrogen, in the unirradiated polymer materials stabilizes the final amorphous structure differently than that obtained from hydrocarbon polymers [3].

Linear energy transfer (LET) is the energy deposited per unit path length of the ion's range in the irradiated material [4-9]. Ions have shorter penetration ranges in polymers than do other forms of radiation, such as UV, x-rays, γ-rays or electron beams, of comparable energy. Consequently, the energy loss per unit path length or LET is greater for ions [1].

Structure of GPC

GPC is made from cured phenolic resins (resol), in an inert environment. After curing at 60 °C, the resin is pyrolyzed at low temperature rates to avoid changing shape or disruption due to volatile decomposition products. Heat treatment to 550 °C produces an electrically conducting material due to hydrogen release and conjugation of the aromatic rings forming graphene planes in random arrangement. For heat treatment at 650 °C, the material still presents open porosity due to space between the ribbons. Further heat treatment to 1000 °C the GPC pores remain but progressively close, reducing permeability as the graphitic planes aggregate themselves to form the final structure of the GPC. The final GPC structure is as random graphene planes and the material appears dark, hard and vitreous. Its density (1.45) is significantly lower than that of graphite from

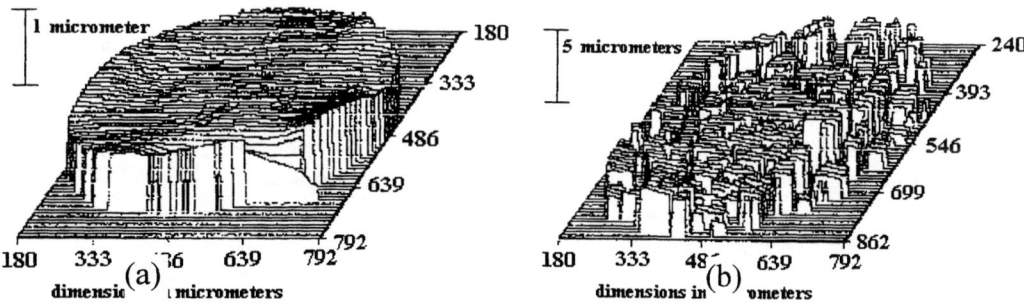

Figure 1. Surface profile of 700 °C GPC (a) before ion bombardment, and (b) after bombardment with 8 MeV oxygen ions at a fluence of 2.8x10^{13} ions/cm^2.

which one may deduce a relative pore volume of about 35%.

ION BEAM APPLICATIONS

Treating the surface of the GPC biomaterial with ion beam irradiation can improve its biocompatibility, by increasing its roughness and subsequently increasing the adherence of the endothelialized tissue that is in contact with the implanted prosthetic component's surface. The texture desired for this medical purpose should be a profile of cavities in the surface whose dimensions are sufficient (about 10-20 µm) to allow formation of tissue that adheres to the material surface [8].

Figure 1 compares the surface roughness (measured with an optical profiler) for oxygen bombarded and non-bombarded phenolic resin samples, pyrolyzed at 700 °C. The sample shown in Figure 1b was bombarded with high-energy ions at the Surface Modification and Characterization (SMAC) facility at Oak Ridge National Laboratory in Oak Ridge, TN. Table 1 shows the irradiation parameters for the ion beam treatment [8]. There is a threshold fluence for each ion and energy, over which the roughness tends to decrease as the ion fluence is increased. Ion bombardment beyond this critical fluence causes densification and polishing of the initially rough, porous surface. Samples prepared at 700 °C showed greater RMS roughness than those prepared at 1500 °C, which is probably due to GPC's hardness and purity, both of which increase with heat treatment temperatures. For these samples, atomic force microscopy (AFM) [8] showed a specific texture (Fig.2), which presented 40 µm width regions, separated by 10-30 µm gaps. For fluences higher than 2.8×10^{13} oxygen ions/cm^2 the roughness decreased. The ions' stopping range varies to some degree with fluence, owing to the change in the density of the material. However, for this work this change was assumed negligible.

Ion beams may also be used to increase the porosity of GPC. Ion bombardment improves the permeability by opening the existing pores and making them interconnected. However, if the fluence is higher than the pore density of the material near the surface, ion bombardment makes the pores unavailable (9, 10). The resulting pores or voids can then be used as capsules for drug delivery systems. The availability porosity after oxygen bombardment has been measured by lithium absorption. The lithium absorption was carried out at 700 °C in a molten 99.6% lithium chloride bath for duration of one hour. The near surface concentration profile of lithium in these samples was measured by Li(p, 2α) nuclear reaction analysis (NRA) [11]. Alpha particles from the reaction were observed from the surface to a depth of about 10 µm below the surface of the GPC, beyond which the energy of the protons is insufficient to react with lithium. A 100 nanoampere 1.03 MeV proton beam and a silicon surface detector, at a back angle of 170° were used to perform the NRA.

The surface texture of the oxygen bombarded GPC was observed with AFM and the results correlated with the previously measured available porosity.

CONCLUSION

The results show increased surface roughness, as well as increased porosity in the ion bombarded GPC. This is direct evidence of the efficacy of the applications of ion beams to the production of carbon-based biomaterials. The textured GPC that results from ion bombardment has enhanced biocompatibility. Bombardment by fast heavy ions transfers energy to recoil carbon atoms, as well as produces residual positive charge near the primary ion track. The damage along the ion track apparently causes neighboring pores, near surface of GPC, to connect. The ion beam bombardment enhanced available porosity measured by lithium absorption correlates with the enhanced roughness of the surface of GPC.

These studies have demonstrated the usefulness of ion beams in modifying carbon-based polymeric materials.

ACKNOWLEDGMENTS

The author acknowledges the support of the Center

TABLE 1. Ions, energies, fluences and stopping ranges used in ion beam treatment.

Ion	Energy (MeV)	Fluence (ions/cm^2)	Stopping Range of Ion
Oxygen	8	1.0, 2.8 and 10 $\times 10^{13}$	6.77 µm
Carbon	6	3.0, 10 and 30 $\times 10^{13}$	6.75 µm
Silicon	5	0.5, 3.4 and 10 $\times 10^{13}$	4.14 µm
Gold	10	1.0, 10 and 100 $\times 10^{14}$	2.84 µm

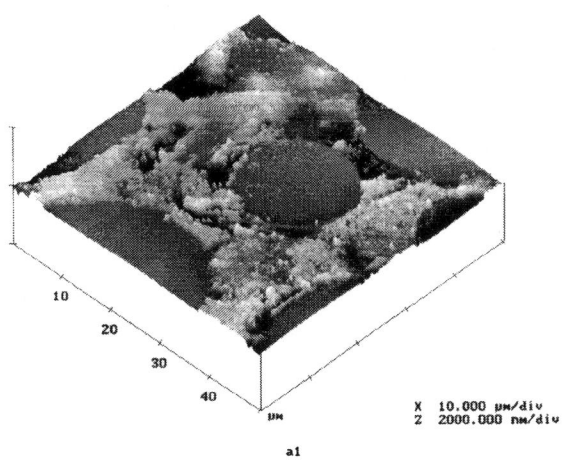

Figure 2. AFM image of a smaller region of the same sample described in Figure 1.

for Irradiation of Materials at Alabama A&M University and the Division of Materials Sciences, U.S. Department of Energy, under contract DE-AC05-96OR22464 with Lockheed Martin Energy Research Corp.

REFERENCES

1. E. H. Lee, G. R. Rao, M. B. Lewis, and L. K. Mansur, *J. Mater. Res.* **9**, 1043 (1994).

2. L. Calcagno, *Nucl. Instr.& Meth.* **B105**, 63 (1995).

3. G. Marletta, *Nucl. Instr.& Meth.* **B46**, 295 (1990).

4. E. Balanzat, N. Betz and Buford, *Nucl. Instr. &. Meth. B* **105**, 46 (1995).

5. Chapiro, *Nucl. Instr.& Meth.* **B105**, 5 (1995).

6. D. Ila, A. L. Evelyn and Y. Qian, *Mat. Res. Soc. Symp Proc.* **338**, 613 (1994).

7. G. M. Jenkins and K. Kawamura, *Polymeric Carbons-Carbons Fiber*, Cambridge, Cambridge University Press, 1976.

8. M. G. Rodrigues, A. L. Evelyn, D. Ila, , R. L. Zimmerman, D. B. Poker, and, D. K. Hensley, "Radiation Enhanced Porosity and *Roughness of Biomaterials*" in *Interfaces, Adhesion, & Processing in Polymer Systems*, edited by S. H. Anastasiadis et al., MRS Symposium Proceedings, Spring 2000 (In Print).

9. D. Ila, G. M. Jenkins, R. L. Zimmerman, A. L. Evelyn, *Mat. Res. Soc. Symp. Proc.*, **331**, (1994) 281.

10. R. L. Zimmerman, D. Ila, D. B. Poker, S. P. Withrow, *Application of Accelerators in Research and Industry*, Duggan & Morgan, New York, 1996, p957.

11. R. L. Zimmerman, D. Ila, G. M. Jenkins and D. B. Poker, *Nucl. Instr.& Meth.* **B106**, 550-554 (1995).

A Negative Ion Beam Application for Improving Biocompatibility of Polystyrene Surface

Junzo Ishikawa[*], Hiroshi Tsuji[*], Hiroko Sato[†], Hitoshi Sasaki[*], and Yasuhito Gotoh[*]

[*] *Department of Electronic Science and Engineering, Kyoto University, Sakyo-ku, 606-8501 Kyoto, Japan*
[†] *Department of Polymer Chemistry, Kyoto University, Sakyo-ku, 606-8501 Kyoto, Japan*

Abstract. Improving biocompatibility of polystyrene surface by negative-ion implantation was investigated in respect to two kinds of cells: human umbilical vascular endothelial cells (HUVEC) and neurons (rat adrenal phechromocytoma: PC-12h). Negative ions of silver were implanted to non-treated polystyrene dishes (NTPS) at conditions of 20 keV and 3×10^{15} ions/cm^2 with or without a pattering mask with narrow slits array of 60 μm in width. Negative ions were used to avoid charge-up problem. The contact angle of pure water on ion-implanted NTPS surface was about 75°, which was lower than 86° for the original surface of NTPS. It was found from XPS surface analysis that this change to hydrophilic surface was caused by introduced functional groups of C-O, C=O and O-C=O, during and after the ion implantation. After cell culture experiment, HUVEC cells attached only on the ion-implanted region of NTPS. On the ion-implanted narrow region with a width of 60 μm, attached HUVEC cells aligned with a longitudinal direction. As for PC-12h, the neuron cells also showed the same selective attachment property on the patterned NTPS surface by ion implantation. Besides, neurites of neural protrusion extended from cell body were also attached only on the ion-implanted region.

INTRODUCTION

Improvement of biocompatibility of polymers has been desired in biological and medical fields. For artificial hybrid-type blood tube to be attached vascular cells, improving cell-attachment property is required since most polymers have poor biocompatibility. Precise and fine control of cell-attachment property on polymer surface is also required to form an artificial neuron network in culture. It is desired in the neuroscience for investigation of the detailed mechanism of information transport and for study of the influence of heavy metals in neurons in the medical fields. The control of neurites-extension properties is also required. Besides, such artificial neuron network with in/out-put electrodes will serve as an interface between human nerve system and external silicon circuit. This "bio-interface" is required in near future for the development of advanced artificial hand, leg and foot those might be controlled by human nerve system.

Ion implantation is a useful method of the surface modification of polymers because it can modify only surface layer and/or dope desired elements. Suzuki et al. showed that polymers had cell-attachment property of bovine vascular cells by the ion implantation of various positive noble-gas ions at relatively high implantation energy of 150 keV [1,2]. In the conventional ion implantation using positive ions, however, a charge-up problem is inevitable because insulators of most polymers used in the medical field are insulators. It makes control of both implantation energy and dose amount inaccurate. On the contrary, an ion implantation using negative ions has an advantage of "charge-up free" even in insulators and isolated electrodes [3,4]. We applied this negative-ion implantation with silver-negative ions for surface modification of polymer surface [5,6] and showed improvement of cell-attachment properties for polystyrene.

NEGATIVE-ION IMPLANTATION

Silver-negative ions were produced and extracted from a cesium-sputter type negative ion source (Neutral and Ionized Alkaline metal Bombardment-type Negative Ion Source, NIABNIS) [7] with a cone shaped sputtering target made from a pure material to be ionized. The extracted negative ion beam was transported to an implantation chamber after mass-separated with an electromagnetic sector magnet at bending angle of 30 degrees. Polystyrene dishes were placed on a base plate of the collector cup and a mask was placed over the sample surface at a distance of 1 mm. The mask was a nickel thin plate of 40 μm in thickness and had aperture slits of 60 μm-wide and 4 mm-long with a spacing of 60 μm in a region of 4 x 8 mm. The ion beam passed through a limiter with a hole of 11.28 mm diameter and entered the sample surface through the mask in the collector cup. As for implantation conditions for evaluation of surface properties, ion energy and dose were varied in ranges of 5 - 20 keV and in 10^{14} - 10^{16} ions/cm^2, respectively. Current density was less than 300 nA/cm^2 in order to avoid melting of polystyrene. Residual and background gas pressures were about 1.0×10^{-3} Pa and 1.4×10^{-4} Pa, respectively.

Polystyrene dishes (NTPS, untreated polystyrene dish, No. 25060-60, Corning) and spin coated polystyrene thin films (SCPS) on glass were used in this experiment. NTPS and SCPS have no cell-attachment properties on their surfaces.

For evaluation of negative-ion implanted surface, contact angle of pure water was measured for surface wettability. Oxygen atoms and functional groups introduced during ion implantation were evaluated by X-ray photoelectron spectroscope. Surface morphology was measured by atomic force microscopy.

In cell culture experiment, we examined two kinds of cells, human vascular cells and rat nerve cells, for evaluation of biocompatibility for negative-ion-implanted polystyrene dishes.

SURFACE PROPERTIES OF ION-IMPLANTED POLYSTYRENES

For biocompatibility of materials, wettability is an important factor. We measured contact angle of pure water on the Ag-implanted polystyrene samples. Contact angles of Ag-implanted NTPS without the mask (Ag/NTPS) are shown as a function of the ion dose and the ion energy in Figs. 1(a) and 1(b), respectively. The contact angle on Ag/NTPS decreased with an increase in ion dose and energy. At conditions of 20 keV and 3×10^{15} ions/cm^2, the contact angle came down to 74° from 86° for the original surface. But the contact angle at 30 keV was high. The reason is considered that the most surface layer was changed to hard amorphous carbon layer from polystyrene by ion bombardment. Ag-implanted spin-coated polystyrene (Ag/SCPS) also showed a decreasing dependence of contact angle similar to Ag/NTPS. The contact angle came down 78 degrees at the same condition from the original contact angle of 90°, which was a little larger than Ag/NTPS.

Oxygen atoms and functional groups introduced during ion implantation were analyzed by the apparatus of AXIS-165S, Shimazu/Kratos for XPS. From the ratio O_{1s} and C_{1s} spectra in XPS measurement, we obtained an atomic ratio of introduced oxygen to carbon in polystyrene, O/C, of the sample surface. Fig. 2 shows the O/C ratio for Ag/NTPS (20 keV) at various ion doses. The O/C ratio for original NTPS is almost zero. Considerable oxygen atoms were introduced even at a low dose of 1.0×10^{14} ions/cm^2, and gradually increased with an increase in ion dose from 1.0×10^{14} to 1.0×10^{16} ions/cm^2. Detailed C_{1s} spectra of Ag/NTPS (20keV, 3×10^{15} ions/cm^2) and composition of functional groups are shown in Figs. 3(a) and 3(b), respectively. The C_{1s} spectra included photoelectrons from of bonding states in functional groups such as C-O and C=O. C_{1s} spectra were decomposed into four distributions based on the chemical shifts [8] and shown by various lines in Fig. 3(a). Thus, the sample included atomic bonds of about 81 % C-C(C=C), 10% C-O, 7 % C=O, and 2 % O-C=O. These functional groups such as hydrophilic carbonyl and carboxyl ones are considered to contribute to lowering contact angle of Ag/NTPS. Ion implantation would result in many voids and defects in the surface layers. Then, these voids absorbed oxygen atoms from residual gas in the vacuum or/and from the air after taking out of the chamber. And functional groups were formed and they gave hydrophilic property to the surface.

As for the change of surface shape of ion implanted polystyrene, sputtered depth and surface roughness were measured by an atomic force microscopy (AFM). An AFM image of Ag/NTPS (20keV, 3×10^{15} ions/cm^2) is shown in Fig. 4. The line-shape ditches on the surface, which were traced on the mold surface, were remained after the ion implantation. The

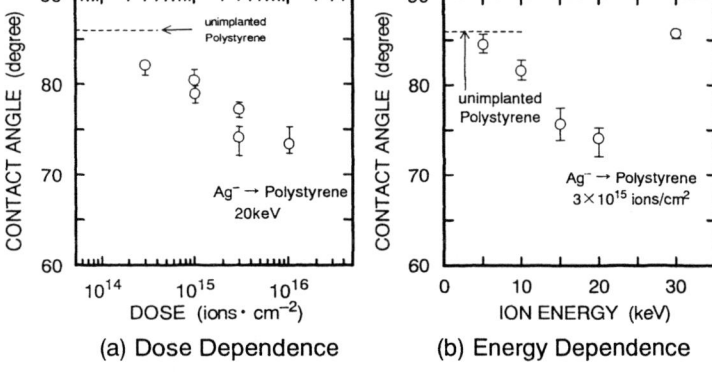

(a) Dose Dependence (b) Energy Dependence

FIGURE 1. Contact angle of Ag/NTPS as a function of (a) the ion dose at ion energy of 20 keV, and (b) the ion energy at ion dose of 3×10^{15} ions/cm^2. Dashed lines indicate a contact angle of the original NTPS.

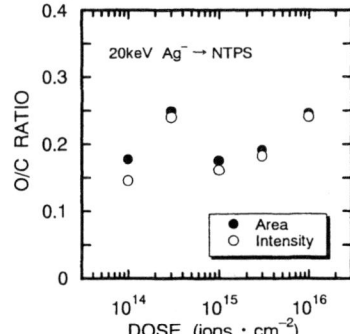

FIGURE 2. Atomic ratio of O/C in the surface layer of Ag/NTPS at 20 keV. Two kinds of evaluation with spectra area of XPS O_{1s} and C_{1s}, and their peak intensities are shown by open and closed circles, respectively.

sputtered depth of Ag/NTSP surface was about 15 nm for conditions of 20keV and 3×10^{15} ions/cm^2. The average surface roughness (Ra) of dishes after negative-ion implantation is shown in Fig. 5 for large area by solid marks and for small area avoiding line ditches of 1×1 μm^2 by open marks. In the small area, Ra was 1.2 nm for Ag/NTPS(20keV, 3×10^{15} ions/cm^2) while 1.7 nm for unimplanted surface.

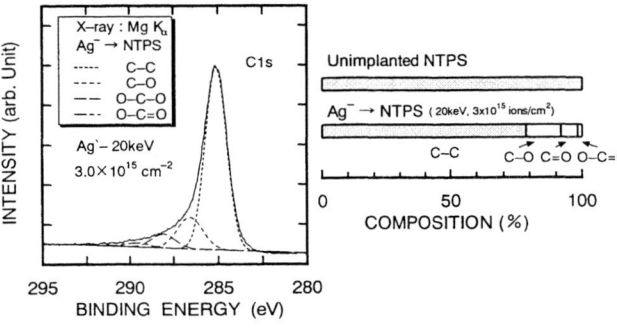

FIGURE 3. Introduced functional groups. (a) XPS C$_{1s}$ spectra of Ag/NTPS at 20keV, 3×10^{15} ions/cm^2, with peak separation by fitting with four gaussian distributions, and (b) composition of the functional groups.

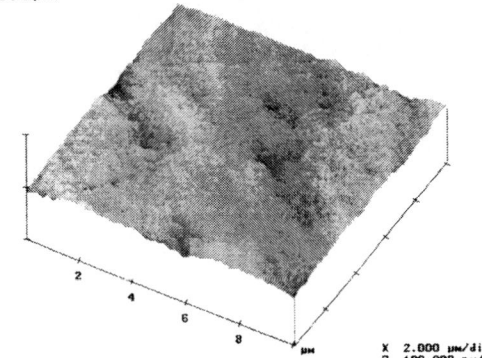

FIGURE 4. AFM image of the Ag/NTPS dish (20keV, 3×10^{15} ions/cm^2) in an area of 100×100 μm^2.

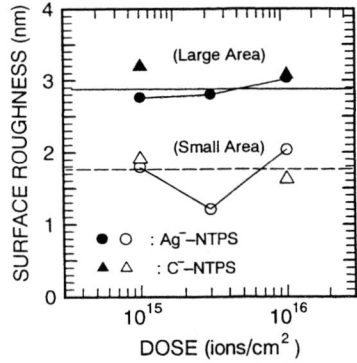

FIGURE 5. Surface roughness (Ra) of ion-implanted NTPS for large area (100×100 μm^2) by solid marks and for small (1×1 μm^2) by open ones.

BIOCOMPATIBILITY OF ION-IMPLANTED POLYSTYRENES

Two kinds of cells; human umbilical vascular endothelial cell (HUVEC) and rat adrenal phechromocytoma (PC-12h) were cultured on Ag-implanted polystyrene dishes for evaluation of cell-attachment property.

Vascular Cell Attachment Properties

HUVEC cells were seeded on the sample dishes after sterilized for one day in 70% ethanol, and were cultured in a culture medium containing some supplements in an incubator at 37°C in air flow with 5% CO$_2$ for 2 weeks. The culture media were changed every three days. HUVEC growth and attachment to the dish surfaces were observed by a phase contrast microscope (Olympus, CK-2).

Phase contrast micrograph of human umbilical vascular endothelial cells on Ag/NTPS dishes (20keV, 3×10^{15} ions/cm^2) after 25 hours in culture is shown in Fig. 6, where a dark area in shape of a part circle is the ion-implanted region. HUVEC cells attached on only the Ag-implanted area with spreading pseudopodia, but the unimplanted area showed no cell-attachment property. Figure 7 shows the number of attached HUVEC cells in an area of 0.636 mm^2 on the various

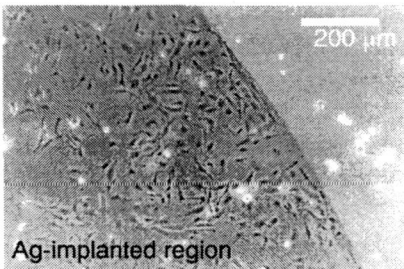

FIGURE 6. Phase-contrast micrograph of HUVEC cells on the Ag/NTPS at 20 keV with 3×10^{15} ions/cm^2.

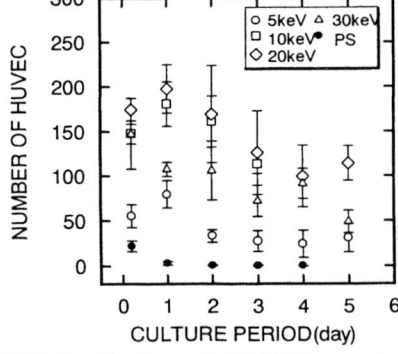

FIGURE 7. Number of HUVEC attached on the surface of various Ag-implanted NTPS as a function of the cultured days.

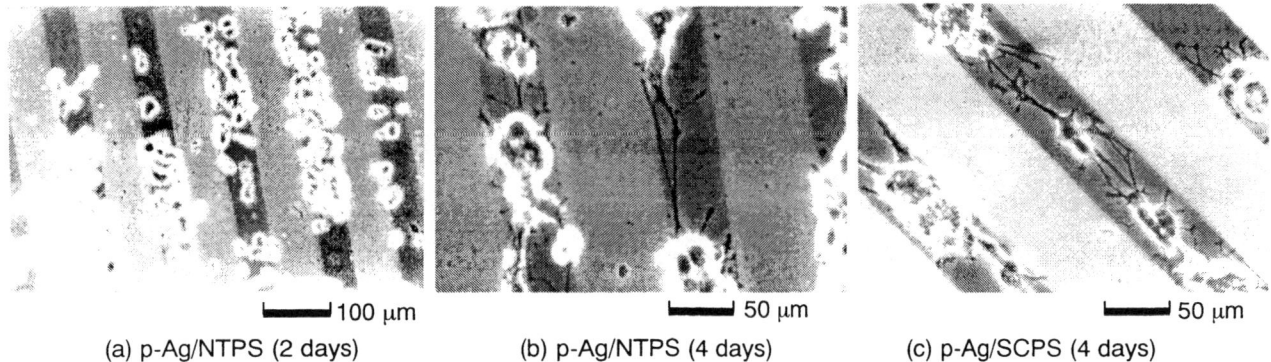

FIGURE 8. Phase-contrast micrographs of PC-12h cells cultured on Ag-implanted polystyrenes; (a) p-Ag/NTPS dish after 2-day culture without NGF, (b) p-Ag/NTPS dish after 4 days (2 days with NGF), and (c) p-Ag/SCPS film for 4 days (2 days with NGF). Note that dark stripe parts are implanted regions.

Ag/NTPS at 3×10^{15} ions/cm^2. On the unimplanted polystyrene, the number of HUVEC rapidly decreased to zero after 1 day. The Ag/NTPS at 20 keV showed the better attachment property, and there remained more than 120 cells after 5 days and this number was comparable to the seeded number of about 150 cells.

Neuron and Neurites Attachment Properties

Nerve cells of PC-12h were seeded on samples of patterned Ag-implanted polystyrenes of dishes and thin film (p-Ag/NTPS and p-Ag/SCPS, 20keV, 3×10^{15} ions/cm^2) after sterilized. In the first 2 days for growth in number, the cells were cultured in a high-glucose culture medium (DMEM, Nissui), 5% fatal bovine serum (FBS, Bio-Whittker) and 5% heat-inactivated horse serum (HS, Gibo) with antibiotics in a humid, 5% CO_2 incubator at 37 °C. For another 2 days (4 days in total) for extension of neurites, cells were cultured in a serum-free DMEM with a nerve growth factor (NGF, 50 ng/ml). As a control sample in the experiment of cell culture, cells were cultured on a collagen (40 µg/ml) coated tissue-culture polystyrene dishes (TCPS, No. 25010-60, Coring) at the same time.

Appearance of PC-12h cells cultured on each sample was observed by a phase-contrast optical microscope. The observed micrographs of cultured PC-12h cells are shown in Fig. 8; (a) on p-Ag/NTPS after the first 2 days in culture without NGF, (b) on p-Ag/NTPS after 4 days (2 days with NGF), and (c) on p-SCPS after 4 days. In the micrographs, each dark striped area is the Ag-implanted region with about 60 µm in width. PC-12h cells are appeared in round stone-shape and they attached only on the implanted region for all cases. The black and fine lines in Figs. 8(b) and 8(c) are neurites expanded from cell bodies. The neurites also outgrew only on the ion implanted regions. The cells were normally grown on ion implanted samples as similar as on the control sample.

CONCLUSION

Negative-ion implantation was found to be an effective method for the surface treatment of cell-attachment and neurites extension properties on the polymer. The improvement of the property was contributed by the surface hydrophilic property due to functional groups introduced by ion bombardment. By using negative ion implantation, we are able to form an artificially designed neural network.

REFRENCES

1. Y. Suzuki, M. Kusakabe, K. Kusakabe, H. Akiba and M. Iwaki, *Nucl. Instrum. and Methods in Phys. Res. B*, **B59/60**, 698-701 (1991).
2. Y. Suzuki, M. Kusakabe, J.S. Lee, M, Kusakabe, M. Iwaki and H. Sasabe, *Nucl. Instrum. and Methods in Phys. Res. B*, **65**, 142-147 (1992).
3. H. Tsuji, Y. Toyota, J. Ishikawa, S. Sakai, Y. Okayama, S. Nagumo, Y. Gotoh, and K. Matsuda, "Charging Voltage Measurement of An Isolated and Insulators during Negative-Ion Implantation" in *Proc. of the 10th Int. Conf. on Ion Implantation Technololgy-94*, Elsevier, 1995, pp.612-615.
4. H. Tsuji, J. Ishikawa, S. Ikeda, and Y. Gotoh, *Nucl. Instrum. and Methods in Phys. Res. B*, **B127/128**, 278-281 (1997).
5. H. Tsuji, H. Satoh, S. Ikeda, Y. Gotoh and J. Ishikawa, *Nucl. Instrum. and Methods in Phys. Res. B*, **B141**, 197-201 (1998).
6. H. Sato, H. Tsuji, S. Ikeda, N. Ikemoto, J. Ishikawa and S. Nishimoto, *J. Biomed. Materials Res.*, **44**, 22-30 (1999).
7. J. Ishikawa, Y. Takeiri, H. Tsuji, T. Taya and T. Takagi, *Nucl. Instrum. and Methods in Phys. Res. B*, **B232**, 186-195 (1984).
8. G. Beamson and D. Briggs: *High Resolution XPS of Organic Polymers - The SCINTA ESCA300 Database*, John Wiley & Sons, Chichester, UK, 1992.

Plasma Source Ion Implantation Technology for Engineering Surfaces of Materials

E.H. Wilson, D.F. Lawrence, K. Sridharan, and P.W. Sandstrom

University of Wisconsin
Madison, WI 53706, USA

Abstract. Plasma Source Ion Implantation* (PSII) is a non-line-of-sight technique for energetic ion surface modification of materials. At the University of Wisconsin there are presently three PSII systems two of which measure about $1m^3$ and a third that measures $0.1m^3$. Plasma generation is achieved in vacuum through filamentary, RF, DC-pulsed, or glow discharge. High voltage pulsing is achieved using a tetrode modulator that pulses at up to 60kV or by a solid-state pulser that can supply 20kV. Recently, a crossatron modulator capable of 40kV and 1kA peak anode current was built in-house. Surface properties of a wide range of materials have been beneficially modified using PSII in ion implantation, film deposition, energetic ion mixing, and sputtering modes. Industrial field testing of PSII-treated parts has yielded promising results but successful commercialization requires judicious selection of applications which effectively exploit the unique aspects of PSII as a surface modification tool. *J.R.Conrad U.S. Patent# 4764394, 1988

INTRODUCTION

Ion implantation alters the near surface regions of a material by the introduction of atomic or molecular species. In conventional beam-line implantation, ions produced away from the target are directed as a beam onto a target surface. Alternatively, in Plasma Source Ion Implantation (PSII) a plasma created around the target surface provides the source of ions for implantation. A high voltage pulse applied to the target accelerates ions into the target surface with potentials typically ranging from -1 to -100 kV. This process does not have inherent species selectivity, as any positive ion in the sheath will be accelerated towards the surface during the voltage pulse. Both beamline and PSII can result in the formation of non-equilibrium phases in the materials' surface, that are not dictated by thermodynamic constraints. PSII has been shown to enhance the wear and corrosion resistance of surfaces and has advantages in ultra-shallow semiconductor doping. In the lower energy regime, thin films of diamond-like carbon can be successfully synthesized. The low process temperature, its non-line-of-sight nature and the ability to treat many large surfaces at once are advantages of PSII as a surface modification treatment.

PLASMA SOURCE ION IMPLANTATION PROCESS

A PSII apparatus consists of an electrically isolated conductive stage which is pulsed to a negative potential in a chamber under high vacuum. When the stage is pulsed in the presence of a plasma all non-insulated surfaces which are biased experience ion bombardment. Ions are extracted from the bulk of the plasma as the plasma sheath edge rapidly expands away from the negatively biased surface. After the sheath has reached equilibrium, the small potential gradient across the presheath extracts the necessary ions to the sheath edge.[1] Thus the presheath potential replenishes ions after those initially in the sheath during expansion have been implanted into the target. The presheath is necessarily asymmetric due to the large potential differences between the system walls at ground and the high negative bias on the stage. The resulting ion bombardment conforms to the surface on a length scale dependent on the sheath thickness. This process yields non-line-of-sight implantation capable of treating complex geometries. The process is schematically illustrated in Figure 1.

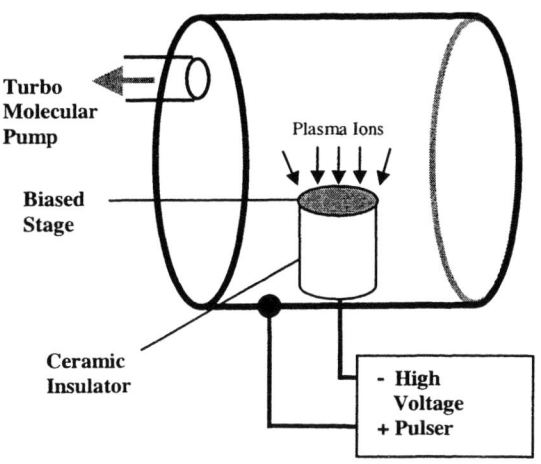

FIGURE 1. Schematic representation of PSII apparatus.

Plasma Producing Apparatus

Before producing the plasma, the system is typically pumped to a background pressure of 10^{-6} Torr to reduce the production of ions from contaminant gases in the chamber. The base pressure is achieved with a turbomolecular pump attached to a stainless steel chamber through a butterfly valve which limits feed gas throughput during operation. The butterfly valve position also affects the residence time of gas in the chamber which has implications on the population of various species produced from the breakdown of complex molecular gases. Although PSII does not have mass separation capabilities, with careful control of vacuum vessel materials and cleanliness, it is possible to achieve implantation purities suitable for use in semiconductor devices. The ability to process semiconductor materials is facilitated by a gas handling system that includes negative pressure exhausted gas cabinets, safety interlocks with reactive gas sensors, a remote shut-off, and pumps equipped with corrosive gas resistant components. In addition a liquid-vapor pressure delivery system is available in a symmetric dual-injection configuration.

In order to produce the plasma, process gases are introduced to pressures in the range of 0.1 - 50 mTorr depending on the plasma discharge method. The plasma can be produced by various methods. The systems at the University of Wisconsin can generate plasma through filamentary, radio frequency (RF), direct current (DC)-pulsed, or glow discharge. Filamentary discharge is produced by running a heating current through banks of thoriated tungsten filaments which are biased to repel thermionically emitted electrons. In-vacuum replacement banks are available in the event of filament failure. RF discharges can be ignited using capacitively or inductively coupled antennae. An inductively coupled plasma (ICP) can be produced using a 13.56 MHz generator and matching network capable of delivering 3 kW of absorbed power. A DC-pulsed power supply is used in PSII-assisted deposition to power a sputter cathode. This power supply enables the sputter cathode to be run in a non-poisoning mode for reactive systems such as tantalum and nitrogen. Finally, a glow discharge plasma can be produced with no external source using the high voltage applied to the target. The resulting ion cascade typically produces a lower density plasma for the duration of the pulse.

High Voltage Pulsing Apparatus

PSII modifies surfaces using target pulsing which negatively biases the target with a wave train of high voltage pulses. The alternative steady state biasing of a target is made less practicable by several phenomena. Steady state biasing of a target can reduce ion production as the plasma sheath expansion decreases bulk plasma volume.[2] Equipment power limitations also commonly set pulse constraints. Ordinarily a square wave train is generated to turn on a high voltage modulator for the duration of the pulse width. Pulse widths ranging from 1 - 200 μs are used in the systems at The University of Wisconsin. Repetition rates of 25 - 250 Hz allow the pulsing system to recharge and dissipate heat. A typical voltage pulse is shown in Figure 2.

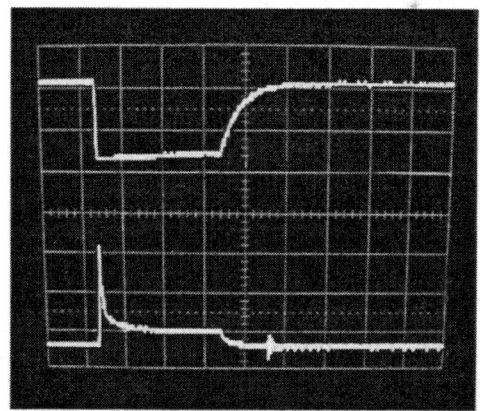

FIGURE 2. Oscilloscope trace of a −20 kV pulse (upper) and stage current (lower).

The duty cycle is 1/100 for a 100 μs pulse width at 100 Hz pulse repetition rate. In addition to the voltage pulse, the oscilloscope trace shows the current drawn by the stage. The stage current gives a direct measure of implanted ion flux when corrected for secondary electron emission from the stage.[3] The dose may be calculated simply from the product of the time averaged current density to the stage $\dot{j}$, and the duration of the implant $t_{implant}$, shown in equation (1). The dose calculation may be written as,

$$Dose = \dot{j} \cdot t_{implant} = \frac{\bar{i}_{pulse} \cdot t_p \cdot f}{(\gamma + 1) \cdot A} \cdot t_{implant} \quad (1),$$

where the average current of one pulse $\bar{i}_{pulse}$, pulse width t_p, repetition rate f, and implanted area A are measured to determine current density. A correction is made for the number of secondary electrons ejected from the stage for each impacting ion, γ since this contributes to current in the same direction as collected ions. A correction may also be made for species which have a significant fraction of multiply charged ions by accounting for the charge difference.

High voltage pulsing systems can have various design configurations. Each of the three PSII chambers at the University of Wisconsin has a pulsing system which provides different capabilities. One system uses a tetrode modulator to negatively bias the stage up to –60 kV. This 1 m³ system uses a 15 V wave generator that is optically isolated for safety from a grid pulser which biases the control grid within the tetrode modulator. When the wave generator pulses, the grid pulser swings from -500 V to a positive bias allowing conduction through the tetrode. The modulator and stage are oil cooled which keeps implant targets at or near room temperature.

A second 0.1 m³ system has a solid-state Insulated Gate Bipolar Transistor (IGBT) pulse modulator capable of a 20 keV implantation. At 10 kV the modulator can produce up to 75 A peak current. This system has an inverted wafer stage and is designed such that no ferrous materials contact the plasma during semiconductor processing. Due to much higher dose rate accessible with this modulator, the stage may be cryogenically cooled with liquid nitrogen.

A third modulator with a 1 kA pulsed-current capability has also been developed. This modulator uses a Hughes Crossatron™ as the high voltage switching element, and is usable with implantation energies up to 40 kV. The Crossatron™ is a gridded hydrogen-discharge device that combines the best features of thyratron and tetrode switches. This modulator employs a fast IGBT driver to provide turn-on and turn-off pulses to the Crossatron™ at pulse-repetition rates greater than 1 kHz. The high current capability of this modulator enables implantation with larger target surfaces and higher dose rates. The high average-current capability (500mA) of this modulator may also be exploited for target heating during ion implantation.

RESULTS

The PSII process has been used in a number of modes to modify surfaces of materials.[7] High energies (upwards of ~10 kV) result in ion implantation into the surface of materials. In the low energy regime (up to ~5 kV), thin film deposition can be made to occur on the surface of materials. For example, diamond-like carbon (DLC) films can be deposited at low energies using acetylene as a precursor gas.[4] Ion-implantation prior to film deposition has been shown to improve film-substrate adhesion.[9] Mixing of multilayer thin films can be achieved by energetic ion bombardment while differential sputtering between elements can be used to change the surface chemistry of alloys.

Most work on ion implantation using PSII has been conducted using implant species from plasmas with relatively simple gas phase chemistries (e.g., N_2, CH_4, O_2). Depth of the modified region is on the order of fractions of a micron. Laboratory tests using flat substrates have shown that surface hardening and improvement in wear resistance can be achieved by PSII using doses in the range of 10^{17} to 10^{18} ions/cm² and energies upwards of about 30 kV.[7]

Figure 3 illustrates an example of surface hardening for a PSII nitrogen ion implanted AISI A-2 tool steel (3×10^{17} ions/cm² @ 50 kV). Because the implanted layer is very thin, improvements in hardness are observed only at low test loads. Figure 4 shows

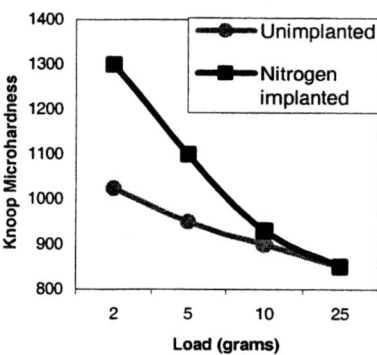

FIGURE 3. Hardness profiles of A-2 tool steel in the unimplanted and nitrogen ion implanted conditions.

wear track profiles after fretting wear testing of unimplanted and PSII nitrogen ion implanted (3×10^{17} ions/cm^2 @ 50 kV) Ti-6Al-4V surgical alloy test flats against a ruby ball stylus.[5] The smaller dimension of the wear scar on the nitrogen ion implanted sample is indicative of superior wear resistance. Use of softer material stylus such as low carbon or stainless steel has shown that the change in surface chemistry due to PSII treatment significantly reduces the propensity for

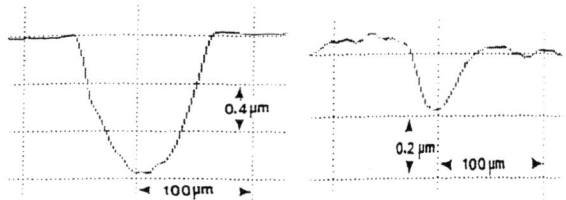

FIGURE 4. Profile of wear scars on surgical alloy Ti-6Al-4V in the unimplanted (left) and PSII nitrogen ion implanted (right) conditions after fretting wear against a ruby stylus.

transfer of material from the stylus to test flat. This transfer of material is generally indicative of galling which is known to frequently occur in industrial metal forming operations, where it manifests as transfer of material from the work-piece to the tool.[7]

PSII provides a non-line-of-sight approach for the deposition of DLC films, a few microns in thickness. DLC films modified with Si and O incorporation have recently been produced at the University of Wisconsin. These films have a combination of high hardness and low surface energy, making them attractive candidate coatings for applications requiring improved release properties. Table 1 shows a comparison of these properties for steel and DLC and modified DLC films produced by PSII.

TABLE 1. A comparison of hardness and surface energies of DLC and modified DLC with steel.

Property	Alloy Steel	DLC	(Si-O-C) DLC
$H_{k(5g)}$ (kg/mm^2)*	350	1750	1900
γ (dynes/cm)	43	48	31

*Strongly dependent on prior heat treatment for steels and deposition parameters for films.

Industrial appeal for PSII process has come from its non-line-of-sight nature, low process temperature, and retention of dimensions and surface finish. Results of industrial field testing of PSII treated parts for a variety of applications has been summarized in reference 7. The tests generally indicate that PSII treatment provides best results under adhesive or mildly abrasive wear conditions.

PSII processing has not been successful in applications involving high temperatures, primarily due the shallow thickness of the modified layer which gets consumed quickly by oxidation. However, the use of organometallic precursor gases may allow deposition of high temperature-resistant carbides, nitrides, and oxides of transition metals in a non-line-of-sight manner.[9]

Semiconductor applications of PSII have received attention because the high dose rates achievable, which allow for the processing of large areas. The process is also being considered for the production of shallow junctions because of its ability to effectively operate in the low energy regime. Surface amorphization in Si has been observed after implantation at high dose rates in PSII.[10]

ACKNOWLEDGEMENTS

We acknowledge the technical assistance of N. Hershkowitz, S. Yan, and K. Kriewaldt. This work was supported by Wisconsin Plasma Processing and Technology Consortium grant # 133-J808.

REFERENCES

1. Hala, A.M., and Hershkowitz, N., "Two species presheath measurements in a multi-dipole plasma" IEEE International Conf. on Plasma Science 1999, pp. 142.

2. Lieberman, M.A., *Principles of Plasma Discharges*, New York: Wiley, 1994, pp. 526-538.

3. Shamim, M.M., Scheuer J.T., Fetherston R.P. and Conrad J.R., *J. Appl. Phys.* **70**, 4756-4759 (1991).

4. Nastasi, M., Elmoursi, A.A., Faehl, R.J. et al., *MRS Symposium Proc.* Vol. 396, 455-466 (1996)

5. Sandstrom, P.W., Sridharan, K., and Conrad, J.R., *Wear*, 166, 163-168 (1993).

6. Chun, M. "*High Dose Rate Effects in Silicon using Plasma Source Ion Implantation*", Ph.D. Thesis, University of Wisconsin, Madison, 1999.

7. Andre, A., ed., "*Handbook of Plasma Immersion Ion Implantation and Deposition*", John Wiley & Sons, New York, NY, September 2000.

8. Rej, D.J., *Handbook of Thin Film Process Technology*, IOP Publishing Ltd., E.2.3:1 – E.2.3:25, (1996).

9. Mantese, J.V., Brown, I.G., Cheung, N.W., and Collins, G.A., MRS Bulletin, 21. No. 8, 52 – 56, (1996).

Resonance Ultrasonic Vibrations in Cz-Si wafers as a Possible Diagnostic Technique in Ion Implantation

Z.Y. Zhao[*], S. Ostapenko[**], R. Anundson[*], M. Tvinnereim[*], A. Belyaev[**], M. Anthony[**]

Advanced Micro Devices, Inc., Fab25, 5204 East Ben White Blvd., M/S 608, Austin, Texas 78741, USA
**University of South Florida, Center for Microelectronics Research, 4202 E Fowler Ave., Tampa, FL 33620*

Abstract. The semiconductor industry does not have effective metrology for well implants. The ability to measure such deep level implants will become increasingly important as we progress along the technology road map. This work explores the possibility of using the acoustic whistle effect on ion implanted silicon wafers. The technique detects the elastic stress and defects in silicon wafers by measuring the sub-harmonic f/2 resonant vibrations on a wafer induced via backside contact to create standing waves, which are measured by a non-contact ultrasonic probe. Preliminary data demonstrates that it is sensitive to implant damage, and there is a direct correlation between this sub-harmonic acoustic mode and some of the implant and anneal conditions. This work presents the results of a feasibility study to assess and quantify the correspondent whistle effect to implant damage, residual damage after annealing and intrinsic defects.

INTRODUCTION

The need for high energy implants increases as the ULSI devices get smaller and the structure becomes more complex, especially in making retrograde wells and buried layers. Due to the limitation of current popular metrology techniques, the industry does not have an effective technique that can measure the variation of the implant energy and dose in this area. For instance, both Thermal wave (TW) and sheet resistance (Rs) do not have the sensitivity with such implants, since the implants are too deep to be probed. We face the challenge to find a technique that has the sensitivity and meets the stringent requirements of a clean room environment.

It has been reported recently that in response to injected ultrasonic waves, silicon wafers generate sub-harmonic resonance that are sensitive to defects and elastic stress.[1,2] By employing the w-mode of sub-harmonic resonance, the whistle effect as it is defined, the characteristics – amplitude dependence, frequency scan and spatial distribution, allow clear distinction versus harmonic vibrations of the same wafer. The origin of the sub-harmonic vibrations observed on 200mm silicon wafers is attributed to a parametric resonance of flexural vibrations in thin silicon circular plates, which is strongly related to defects in the silicon crystalline material.[2] Since ion implantation generates significant damage inside the silicon wafer, it is natural to relate the acoustic wave with its application in ion implantation. Since the measurement is fast, non-contact and nondestructive, it makes this technique a potential candidate for the semiconductor industry.

EXPERIMENT

Application of whistle effect from nonlinear resonance ultrasonic vibrations in silicon wafers is explored in this work.. Ultrasonic vibrations were excited in 200mm wafers using an external ultrasonic transducer. The vibration amplitude was recorded in a non-contact mode using a scanning acoustic probe. Whistle effect(generation of f/2 sub-harmonic mode) was observed in all wafers including control, implanted, and annealed. The measured quantities include whistle amplitude, frequency gate of the whistle excitation (f_{ex}), amplitude threshold (V_{th}) and its dependence versus f_{ex}. According to a general theory of parametric resonance in vibrating systems,[3] the elastic quality of a wafer in terms of whistle damping is higher when the maximum amplitude and width of frequency gate are increased, and opposite, when a threshold and the slope in V_{th} (f_{ex}) dependence are decreased.

Low Dose Implant Experiment

Two sets of 200mm Si wafers were used in this group. Effort was taken to ensure the wafers in the same set came from the same ingot. The implants were with phosphorus at the dose of 1e12/cm^2 at three different energies. In one set of wafers, pre-implant annealing was performed as an attempt to relax any possible initial elastic stress inside the wafers, hereafter referred as "pre-annealed". The other group is referred as "normal". The pre-implant annealing condition was 950 °C for 20 seconds in pure nitrogen ambient. Half of the wafers from the set that was annealed pre-implant and half of the wafers from the set that was not annealed pre-implant were annealed post implant. The post-implant annealing condition was 1000 °C for 10 seconds in pure nitrogen ambient. Control wafers (without implantation) were also included to compare the whistle effect of implantation in the silicon material. The wafers were measured using the W-mode of the applied acoustic wave.

Before the whistle measurement, TW and Rs were measured on the wafers. The control wafers all had ~50 TW units. After pre-implant annealing, the wafers without implantation had about 180 TW unit. This is puzzling, since there was no implant damage introduced in the silicon wafers. Possible oxide growth was first suspected, but the oxide thickness was measured to be 10 – 11Å, which is within the limit of the thickness of native oxide. Although it was not possible to identify the cause of this small variation, it is believed the intrinsic stress in the silicon wafers should have been relaxed due to this thermal treatment. This has been demonstrated in Figure 1, in which one sees the pre-annealed control wafers have higher whistle amplitude and wider windows. Post implant TW readings for the 0.1MeV, 1MeV and 2MeV implants were 630, 580 and 520 respectively. This indicates TW gets less sensitive as the ion energy increases. Sheet resistance was also attempted after post-implantation anneal. However, the four point probe could not pick up sensible signals from all the wafers. This observation demonstrates that neither of these techniques can accurately measure implants in these dose and energy ranges.

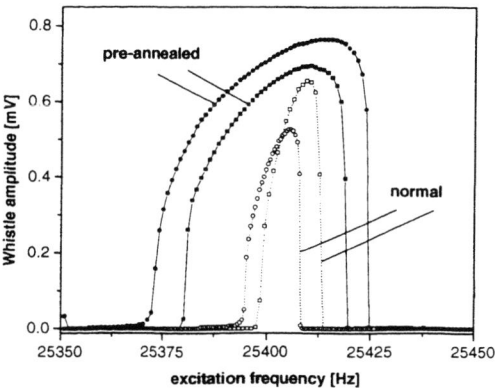

FIGURE 1. Excitation frequency scans of whistle amplitude in four control wafers. It is worth to mention that another control wafer from a different manufacturer exhibits much lower maximum w-amplitude (~0.2mV) and completely off the range of f_{ex}-windows.

Figure 2 shows whistle amplitudes of normal wafers subjected to implantation and post-implant annealing. The whistle amplitude for the as-implanted wafers seems to increase with implant energy. However, the response from the post-annealed wafers does not have a definite trend. The pre-annealed wafers all have similar readings and do not seem to correspond to the variation of the conditions. It is noted that the values of whistle amplitude correspond well to the total energy deposited through nuclear stopping of phosphorus ions in silicon but not to the energy deposited by electronic stopping (See Figure 3). It is reasonable to suggest that the whistle effect is a direct reflection of the lattice damage caused by the energy deposited in the silicon by nuclear stopping.

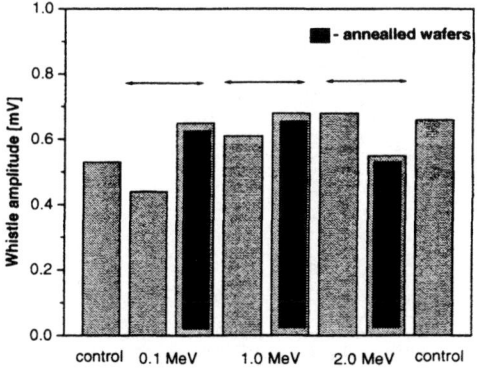

FIGURE 2. Maximum whistle amplitude for the normal wafers. The wafers are paired to check the effect of post-implant annealing. The implantation energies are depicted.

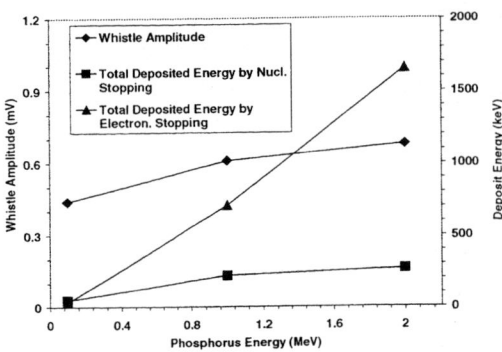

FIGURE 3. The comparison of whistle amplitude and the total energy deposited through nuclear stopping of Phosphorus ions in silicon. The stopping powers were calculated with TRIM95 [4].

A comparison of normal and pre-annealed wafers showed that all pre-annealed wafers have wider gate widths compared to normal wafers. This matches the differences of w-amplitudes of both sets in Figure 2. Table 1 presents the net changes of the frequency gate by comparing to the values of the average of the control wafers. It can be seen that the values are positive except for one, which is 0.1MeV wafer. It suggests that the whistle gate frequency responds to the implant damage and the implanted dopant positively. Although a conclusion is difficult to draw due to the limited set of tested conditions, this will be a direction worth pursuing in future work.

TABLE 1. The net whistle gate change after ion implantation for the two sets of wafers. Values are presented in Hertz.

E (MeV)	Normal as implanted	normal + post anneal	pre-annealed as implanted	pre-annealed + post anneal
0.1	-3	1	10.5	4.5
1	12	13	10.5	18.5
2	12	6	2.5	0.5

Figure 4 shows the V_{th} versus f_{ex} curves in two normal and two pre-annealed wafers. The picture illustrates that normal wafers have higher threshold voltage, ΔV_{pp} = 0.6 to 0.9 mV, compared to pre-annealed ones (at a specific frequency). They also have a slope difference by a factor of ~ 3 (averaged across all normal and pre-annealed wafers). This seems to suggest that normal wafers have larger damping of acoustic vibrations than the pre-annealed wafers. Again, this is consistent with previous data on maximum whistle amplitude and frequency gate.

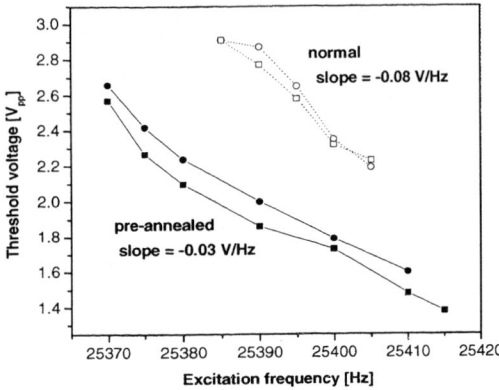

FIGURE 4. Frequency dependencies of threshold voltage for two types of Si wafers. Arrow indicates a decrease of the threshold in pre-annealed wafers.

Whistle Amplitude Response to Implants

A group of six wafers was implanted and annealed to study the whistle amplitude response to ion implantation. The measured whistle amplitudes and the implant conditions are depicted in Figure 5. The whistle amplitude clearly responded to the dose change, but not to annealing. It is interesting that the lower dose implant dramatically damped the amplitude. Although this is in agreement with the observations in Figure 2, it is opposite to the understanding of acoustic resonance mechanism and the observation in Figure 3. The common belief is that acoustic resonance is enhanced when the crystal quality is better. It is expected that the damage to the crystalline quality of the Cz-Si wafers decreases when the implant dose is decreased. This, in turn, should give less damping for the whistle amplitude. However, the observation did not support this argument.

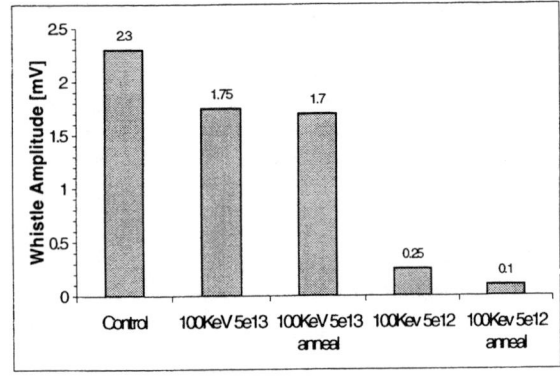

FIGURE 5. Whistle amplitude correspondence to implant dose and annealing. The wafers were not pre-annealed. The post implant anneal condition was 1000 °C for 10 seconds.

Whistle Amplitude Response to Annealing

Figure 6 shows yet another attempt to understand the effect of different annealing conditions on the whistle amplitude. In this case, the implant was 1MeV P$^+$ at 5e13/cm^2. Two anneals were applied to the implanted wafers. One was a 800 °C 10 second anneal intended to just repair the implant damage. The other went through a typical activation anneal of 1100 °C for 10 seconds in addition to the damage repair annealing. Consistent with the result in Figure 4, this group of wafers also showed strong whistle amplitude damping after thermal annealing, but it seems that the thermal budget of the damage repair annealing is sufficient to get the full scale of amplitude damping. The additional anneal did not contribute more damping.

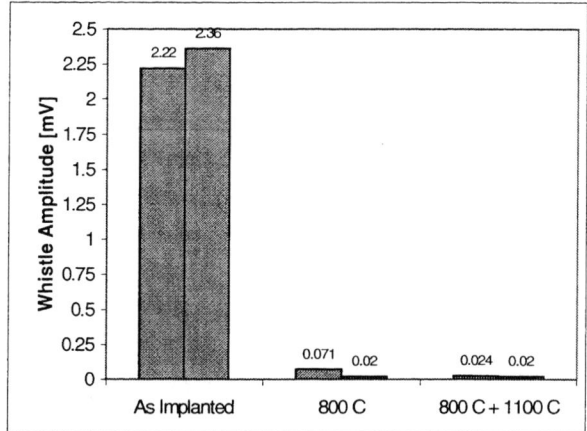

FIGURE 6. Whistle amplitude correspondence to implant and different annealing conditions. The implant was 1MeV P$^+$ at 5e13/cm^2.

CONCLUSIONS

Based on the data collected in this work, the pre-implant history of the wafer is important for acoustic quality achieved after the implantation and post-implant annealing. The wafers with pre-implant annealing have higher acoustic quality than the wafers without it when they are compared after identical conditions of implantation and annealing. It is possible that the pre-implant annealing relaxes the intrinsic stress in the silicon wafers which causes a higher acoustic response. The 100keV implanted wafers showed the lowest w-amplitude and frequency gate, and highest threshold. This is very clear in normal non-annealed samples before and after implantation. This trend is partially compensated after post-implant annealing, again indicating improvement of quality with post-implant processing. We may suggest that 0.1MeV phosphorus implantation, even at 10^{12} cm^{-2} dose, creates not only near-surface damage due to shallower implant depth (~0.1µm), but also causes a stress effect to a wafer. This is similar to the damping effect on w-mode delivered by SiO$_2$ layer. [2] The acoustic technique in this study would respond to the integral effect of damage and stress, which can be used for diagnostics. We will follow-up with theoretical analyses of the stress effect to w-mode characteristics.

In some cases, the measured variables showed response to the changes in experimental conditions. However, the data sometimes contradicts the common beliefs of general acoustic resonance theory. Although this is the first step, it is appropriate to suggest that there is much to be explored in the application of the acoustic wave to ion implantation in the semiconductor industry. While the technology progresses rapidly, the search for better metrology techniques must be accelerated as well.

ACKNOWLEDGMENTS

The authors would like to express their gratitude to the Diffusion/Implant module of Fab25, AMD, for the support they have received in the wafer implantation and annealing process. They also would like to thank their individual employers for granting them this academic experiment.

REFERENCES

1. S. Ostapenko and I. Tarasov, *Appl. Phys. Lett.*, **76** (16), 2217 (2000).

2. A. Belyaev, I. Tarasov, S. Ostapenko, S. Koveshnikov, V. A. Kochelap and A. E. Belyaev, ECS meeting (Phoenix, 2000), in press.

3. L. D. Landau, *Mechanics,* Pergamon Press (1960). P. 80.

4. J. F. Ziegler, Instruction Manual, TRIM version 95.4.

Ultrasonic Pulse from Fast Heavy-Ion Irradiation on Solids

Tadashi Kambara[a], Yasuyuki Kanai[a], Takao M. Kojima[a], Yoichi Nakai[a],
Akira Yoneda[a], Yasunori Yamazaki[a], and Kensuke Kageyama[b]

[a] *RIKEN (The Institute of Physical and Chemical Research), Wako, 351-0198 Japan*
[b] *Department of Mechanical Engineering, Saitama University, Urawa, 338-8570 Japan*

Abstract. Short pulse ultrasonic signals were observed from metallic samples irradiated with fast heavy ions. A pulsed beam of 26 MeV/u Xe ions impinged samples of polycrystalline Al and Cu plates with thickness of 5 mm and 10 mm. All the ions were stopped in the target. Three piezoelectric sensors were attached to the surface of the sample and detected the ultrasonic signals. The waveforms from the sensors were recorded as well as the time structure and intensity of the beam pulse. For both the Al and Cu samples, a sensor placed at the opposite side of the beam spot detected a very short pulse of longitudinal wave at the onset of the ultrasonic oscillation. The waveform of the pulse does not depend on the material and thickness of the sample much, but depends more on the position of the irradiation.

INTRODUCTION

When a fast heavy ions pass through a solid material, it deposits large amount of energy to the electronic system of the material. It is known that the high-density electronic excitations thus produced may result in a columnar defect which is deformation or change of the lattice structure along the trajectory of the ion. The formation of the columnar defects depends on the ion species, velocity and material. Many studies have been reported on the microscopic observation of the columnar defect as well as the resultant change of the electromagnetic properties of the material(1, 2, 3).

On the other hand, the dynamical processes leading to the change of the lattice structure are not well understood yet. The first stage of the process is described as a electronic excitation and ionization by the two-body collisions between ion and individual atom in the material. The following processes are determined by the interaction between the high-density electron excitations and the lattice and are expected to depend on the characteristics of the material. Such processes have been studies by real-time measurements like time-resolved photon spectroscopy(4).

The dynamical processes which result in the change and deformation of the lattice structure may cause temporary lattice deformation which penetrates in the material as elastic waves. If the frequency of the wave is below a few MHz, it can be detected by conventional ultrasonic sensors. Several works have been reported on ultrasonic waves by irradiation of electrons, positrons(5, 6), and slow ions(7, 8).

Recently we reported the first observation of ultrasonic waves generated by fast heavy ions(9). In the experiments, we irradiate short pulses of 26 MeV/u Xe ions on various materials as polycrystalline Al, single crystal of KCl and AL_2O_3 and observed ultrasonic waveforms in correlation with arrival time and intensity of the beam pulse. In the case of Al we found at the first edge of the waveform an interesting structure which changed its shape quickly with the distance between the sensor and the source. This structure was prominent only in Al and not observed for the single crystal of KCl and Al_2O_3. It was speculated to be from very high frequency ultrasonic waves.

Here we report new measurements of ultrasonic waves from metals to study the structure of the waveform more closely. We set another sensor on the opposite side of the sample with the beam spot, to catch the direct ultrasonic wave from the beam spot without reflections by the sample surfaces.

EXPERIMENTAL PROCEDURES

The experiments have been performed at the accelerator research facility of RIKEN. A Xe-ion beam was produced by an ECR ion source and was chopped to short pulses by a pulse-operated electric deflector and a subharmonic RF buncher. The beam was then accelerated up to energy of 26 MeV/u (velocity of 6.95×10^7 m/s) by a linear accelerator (RILAC) and a ring cyclotron (RRC) with an RF frequency of $f = 28$ MHz and was guided to an experimental beam line.

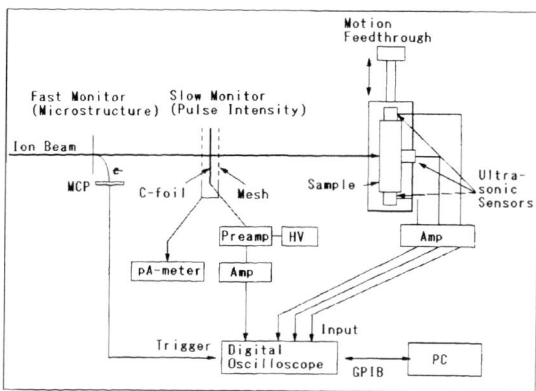

FIGURE 1. Experimental setup

The beam pulse consisted of one or two micro bunches with a width of 3ns and separated by 179 ns which was the period of the sub-harmonic buncher. The number of micro bunches in a pulse depended on the accelerator conditions which could not be well controlled. The interval between adjacent pulses was set to about 10 ms so that it was longer than the decay time of the observed ultrasonic vibrations.

Figure 1 shows the experimental setup at the beam line. Before hitting the sample, the ions passed through two secondary-electron monitors; one for the measurement of the microscopic time structure of the pulse (fast monitor) and the other for the measurement of relative ion intensity in the pulse (slow monitor).

In the fast monitor, the ions passed through an aluminum foil and the secondary electrons were isochronally guided by an electrostatic field and multiplied by a microchannel plate (MCP). No amplifiers were used for the fast-monitor signal to avoid a time delay. The time resolution of this system was about 1 ns, which was fast enough to resolve the structure of the micro bunches of the beam pulse.

In the slow monitor, the beam passed through a $30\mu g/cm^2$ carbon foil. The foil was biased to -190V so that the secondary electrons were extracted from the foil. The signal from the foil was amplified by a charge-sensitive preamplifier and a slow amplifier with a shaping time of $6\mu s$ so that the output pulse height was proportional to the number of the ions in the beam pulse.

The beam was then guided to a vacuum chamber where the sample was irradiated. The distance from the fast monitor and the sample was about 4m and the ion transit time was about 60ns. The samples were metallic polycrystalline Al and Cu plate with square-shaped faces of 35mm × 35mm and a thickness of 5 or 10 mm. The ion beam was incident on one face at right angles with a spot size of about 3mm × 3mm. According to calculations with TRIM code the range of the ions is about 280 μm in Al and 110 μm in Cu, which are much shorter than the thickness of the samples. Therefore all the ions were stopped in the material near the incident surface.

For the detection of ultrasonic signals, we have used piezoelectric sensors widely applied in Acoustic Emission (AE) method for non-destructive inspection of materials(10). The sensor (Fuji Ceramics, M304A) is equipped with a head amplifier and is sensitive to oscillation perpendicular to the surface up to about 2 MHz with the highest sensitivity at about 300kHz. Three sensors were attached to a sample: Two sensors (referred hereafter as right and left sensors) were on the both side planes, and one (referred as back sensor) at the center of the back plane opposite to the face plane where the beam hit. The sensors were pressed on the surface by a spring at about 1kgf with vacuum grease between them. The sample with the sensors was fixed on a frame at the end of a linear motion feedthrough. Moving the frame perpendicular to the beam direction, we could change the position of the beam spot on the sample.

In the measurements, a digital oscilloscope acquired the ultrasonic waveforms and the signals of the slow and fast monitors for each pulse of the incident beam. The signal of the fast monitor triggered the oscilloscope so that the time relation between the beam pulse arrival and ultrasonic detection were determined. Data acquisition was repeated for 20 to 50 shots in the same condition.

RESULTS

Here we show ultrasonic waveforms observed for samples of different material and thickness, and discuss the characteristics of the waves. The waveforms were sorted shot by shot according to the micro-bunch structure of the beam pulse and the single-bunch events were selected. The waveforms displayed below are averaged over 7 to 39 single-bunch shots to reduce the noise. It is important to notice that the most part of the waveform is affected by the elastic vibration of the sample and the intrinsic resonance of the sensors.

Figure 2 shows the waveform from the back and left sensors on the 5mm thick Al sample, along with the micro structure of the pulse from the fast beam monitor. The beam spot was at the center of the face of the sample, just opposite to the back sensor. The displayed waveforms are an average of 34 shots.

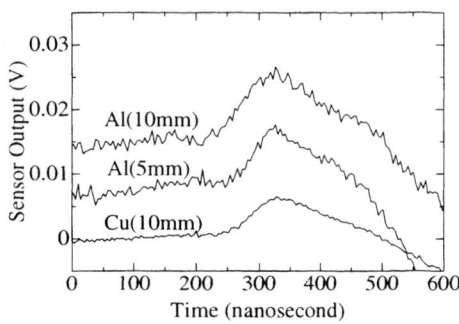

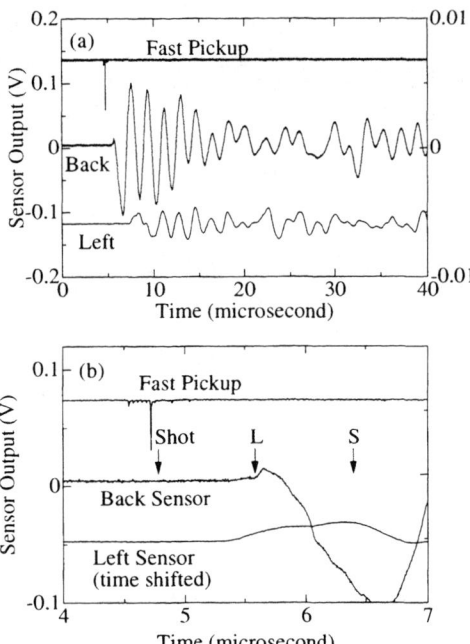

FIGURE 2. Ultrasonic waveforms from aluminum plate observed by the back and left sensors. Time structure of the beam pulse is also shown. Waveforms expanded in time are shown in (b) where Shot denotes the arrival time of the beam pulse, and L and S denote the estimated arrival times of longitudinal and shear waves to the back sensor.

FIGURE 3. Expanded waveforms of ultrasonic short pulse from 5 and 10 mm thick aluminum plate and 10 mm thick Cu plates observed by the back sensor. In order to align the short pulses from different samples, the waveform is shifted to the left by 1.92 μs for Cu (10mm), 0.55 μs for Al (5mm) and 1.43 μs for Al (10 mm).

The ultrasonic waveforms from the back and left sensors have oscillatory structures as shown in Fig. 2(a). The first rise of each waveform is delayed relative to the beam pulse due to sound propagation of the distance from the beam spot and the sensor which are 5mm for the back sensor and about 18 mm for the left sensor. Figure 2(b) shows the same waveforms but the part of the onset of the ultrasonic oscillation is expanded. The arrival time of the beam pulse on the sample, denoted as Shot, is delayed from the fast monitor signal by 60 ns which corresponds to the flight time from the monitor to the sample. The left-sensor waveform is shifted by 2 μs to the left in the figure in order to show the structure at the onset. The waveform at the onset is different between the back and left sensors: The back sensor waveform has a low and sharp pulse with shorter rise time than 100 ns but the left sensor waveform has a much broader structure. The pulse in the back sensor is 0.8 μs after the arrival of the beam pulse. The tabulated sound velocities in bulk aluminum are 6260 m/s for the longitudinal and 3080 m/s for the shear wave and the propagation time over the sample thickness of 5 mm is 0.80 μs and 1.62 μs respectively. The arrival times thus estimated are denoted as L and S in Fig 2(b). Obviously the onset time of the short pulse observed at the back sensor is consistent with the arrival time of the longitudinal wave. Therefore we conclude that the observed ultrasonic pulse is a very short longitudinal wave generated at the irradiation. On the other hand, there is no such structure in the waveform at the expected arrival time of the shear wave.

A similar short pulse is observed for 10 mm thick Al and Cu at the expected arrival time of longitudinal wave for each sample. Their waveforms are compared in Fig 3.

The waveforms look similar for Al and Cu, with rise time of 70-80 ns and a longer decay time. The similarity between the observed waveforms from different samples does not necessarily mean the similarity in the real waveforms since the waveform measured by the piezoelectric sensor is strongly affected by the characteristics of the sensor.

Figure 4 compares the waveforms from the back sensor on 10-mm thick Al sample at different beam spot positions; the beam spot at the center of the face (0 mm) and that shifted from the center by 3 mm, 6 mm and 12 mm. At 0 mm, the beam spot was just at the opposite side of the sensor. The shape of the first wave depends on the position of the beam spot and it gradually gets broader when the distance increases. The onset time matches the estimated arrival time of the longitudinal wave denoted by arrows in the figure. We speculate that the increase of

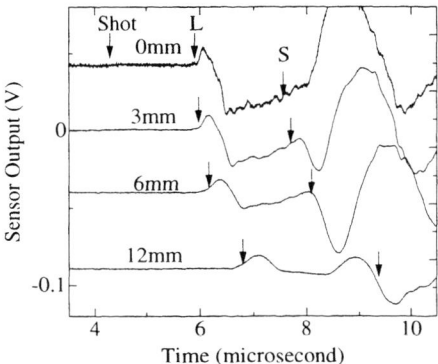

FIGURE 4. Beam-spot position dependence of the waveform for 10-mm thick Al sample. Arrows on the curves indicate the arrival times of longitudinal (L) and shear (S) waves estimated with the tabulated sound velocities in bulk aluminum.

the pulse width is due to the finite size of both the beam spot and the ultrasonic sensor.

In addition to the longitudinal wave, a direct shear wave can be observed by the sensor if it arrives at the sensor with a non-zero incident angle. On the other hand, if the sensor and the beam spot is on axis, the direct shear wave cannot be observed. Actually, when the beam spot was shifted by 3-12 mm, the observed waveforms has a structure nearly at the estimated arrival time of the shear wave.

DISCUSSIONS

Comparing the present and previous experimental results, we speculate that the structure previously observed at the onset of the ultrasonic wave was result of the short pulse waves, longitudinal and possibly shear waves, observed here. The short pulse wave is produced at the beam spot near the surface of the incident plane and propagates to any direction. When the sample is a rectangular bar and sensors are at its ends, like the setup in the previous experiment, part of the wave is deflected by the boundary of the sample and reaches the sensor simultaneously with the direct wave. The interference between the direct and reflected waves may cause the structure observed previously.

The irradiation dynamics which generates such short pulse wave is not known. However, from the previous measurements, appearance of the structure in the waveform depend on the material: Such structures were observed in metals but absent in single crystal KCl. If it is also the case for short longitudinal pulse, it may be some process other than the stress due to thermal expansion by deposited energy. Further experiments are planned to study the dynamical processes of the irradiation-induced ultrasonic wave.

REFERENCES

1. M. Toulemonde, S. Bouffard, and F. Studer: Nucl. Instr. and Meth. **B 91,** (1994) 108.
2. A. Iwase, N. Ishikawa, Y. Chimi, K. Tsuru, H. Wakana, O. Michikami, and T. Kambara: Nucl. Instr. and Meth. **B 146,** (1998) 557.
3. N. Nishida, S-I Kaneko, H. Sakata, H. Kajiwara, M. Matsumoto, T. Mochiku, K. Hirata, and T. Kambara: Physica **B 284-288** (2000) 967.
4. K. Kimura and W. Hong: Phys. Rev. B **58,** (1998) 6081.
5. I. A. Borshkovsky, V. D> Volovik, I. A. Grishaev, I. I. Zalyubovsky, V. V. Petrenko, G. A. Chekhutsky, and G. L. Fursov: Phys. Lett. **40A,** (1972) 97.
6. F. P. Denisov, S. I. Ilyin, B. N. Kalinin, V. M. Kuznetsov, A. P. Potylitsin, V. K. Tomchakov, V. N. Zabaev, and S. A. Vorobiev: Phys. Lett. **77A,** (1980) 266.
7. D. Adliene, L. Pranevicius, and A. Ragauskas: Nucl. Instr. and Meth. **209/210,** (1983) 357.
8. J. Teichert, L. Bischoff, and B. Köhler: Nucl. Instr. and Meth. **B 120,** (1996) 311.
9. T. Kambara, Y. Kanai, T. M. Kojima, Y. Nakai, A. Yoneda, K. Kageyama, and Y. Yamazaki: Nucl. Instr. and Meth. **B 164-165,** (2000) 415.
10. M. Shiwa, H. Inaba, and T. Kishi: Journal of JSNDI **39,** (1990) 374.

Ion Beam Enhanced Emission of Charged Particles from Hot Graphite

J. Lozano, Q.C. Kessel, E. Pollack and W.W. Smith

Department of Physics and the Institute of Materials Science, University of Connecticut, Storrs, CT 06269

Abstract. Thermal desorption spectroscopy of ions from positively biased graphite (grafoil) has been investigated by measuring the energies of the emitted ions with a hemispherical electrostatic analyzer and the masses with a residual gas analyzer under ultra-high vacuum conditions. Potassium is one of the ions emitted at temperatures above 800 °C. The present data show that under near threshold conditions (4V), ions appear with well-defined energies equal approximately to the bias voltage minus 4V. This phenomenon can be greatly enhanced by prior bombardment with an ion beam. It is not clear whether these energies are the result of resonant process on the hot surface or simply due to a process attributable to surface chemistry. At higher biases the peaks broaden in energy and the energy deficit increases.

INTRODUCTION

The emission of charged particles from a hot graphite surface is reported. Although the emission of "positive and negative electricity" is a well known phenomena (1), these new data determine the energies and masses of ions emitted from a positively biased sample. These energies do not correspond to the expected energy, in eV, for a singly charged ion released from a surface biased at V volts. Energy storage and energy transfer to and from graphite has long been of interest (2,3), and has been related to the concentration of defects, such as vacancies, interstitials and impurities. The excitation of phonons and plasmons in various forms of carbon are also familiar phenomena (4). Energy related effects are important to take into account when graphite is used at high temperatures, such as in fission reactors (as a moderator) and fusion reactors (as divertors). Because of this, the emission of particles from graphite surfaces has also been investigated by a number of techniques, including thermal desorption spectroscopy (TDS) (4), direct recoil spectroscopy, and elastic recoil detection (5). Penetration, trapping and the self-sputtering behaviors of ions impinging on graphite surfaces have also been investigated and show that impurity ions may be held strongly between the basal planes of graphite (6).

The technique used to obtain some of the present data is similar to TDS, except that the desorbed ions are either mass analyzed with a residual gas analyzer (RGA) or energy analyzed by a 150 degree, 100 mm radius, hemispherical electrostatic analyzer and detected with a channeltron electron multiplier. The

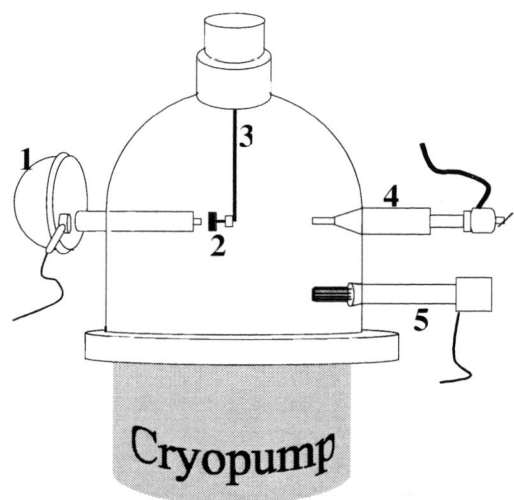

FIGURE 1 Experimental setup.
1) Electrostatic energy analyzer. 2) Sample and heater assembly. 3) Manipulator arm. 4) Ion gun. 5) Residual gas analyzer (quadrupole mass spectrometer).

hot graphite surface emitting the charged particles may be biased and consists of a piece of grafoil (7), a form of graphitic carbon, mounted either over a hot tungsten filament or on a 0.320 mm "button heater"(8). In both configurations the 0.532 mm thick grafoil is heated from one side and the ions detected are being emitted from the other side. The temperature of this side is measured with an optical pyrometer. The vacuum is maintained in the 10^{-10} torr range with a cryopump. Figure 1 shows an outline of the apparatus. The sample is placed on a carrier which fits, in turn, on the manipulator. The heated sample may be moved to face either the electrostatic analyzer or the residual gas analyzer. The application of a voltage to a wire placed near the path of the ions and neutral atoms entering the RGA permits the determination of the fraction of ions entering the RGA.

THE EXPERIMENT

Figure 2 shows two energy spectra of the emitted positive ions obtained at a temperature of 840 °C. Below 600 °C, peaks are not observed in the spectra, but as the temperature increases emission is observed and rapidly increases as the temperature is increased in the range studied. The electrostatic analyzer measures only the ionized particles' energies and gives no information about the corresponding masses, so the RGA is used for the determination of the masses. Only ions with energies greater than 7 eV may be mass analyzed at the present time, but the RGA is presently being altered to enable the observation of lower energy ions. The correspondence of phenomena observed by these two methods of detection is made through a determination of the counting rates as a function of the sample temperature and the bias voltage being applied to the sample. The data in Fig. 2a differ significantly from those in Fig. 2b. A well defined peak is observed which tracks with the bias voltage in a way that corresponds to the creation of singly charged ions with energy deficits of 4 eV; i.e. they appear at energies corresponding to the applied voltage minus 4V. The emission intensity of these ions falls rapidly with increasing bias voltage and approaches zero for a bias of 8 volts. The masses of these ions are not known at the present time. The data in Fig. 2b display a very different behavior. They have been determined to be potassium ions with the RGA and their energies do not directly correspond to the bias voltage. For example, curve 10 peaks at an energy of only 20 eV when the applied bias is 40 eV. Both sets of data may be significantly affected by prior ion bombardment. Implantation of approximately 10^{16} 2 keV Ar ions per cm^2 at room temperature can greatly enhance emission, especially for the data presented in Fig. 2a. Implantation of 10^{21} ions per cm^2 destroys the emission. For this latter dosage, the graphite surface loses its crystalline sheen and appears to have been amorphized. This aspect of the experiment suggests that the emission of these ions depends on the crystalline nature, perhaps the crystal size, of the hot graphite. For this reason experiments in the energy range of Fig. 2b have also been carried out with highly oriented pyrolytic graphite (HOPG).

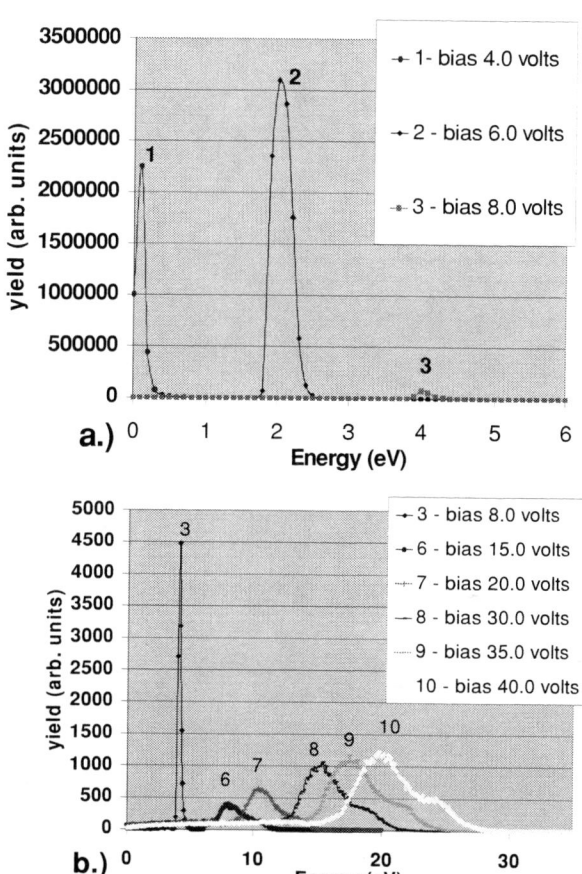

FIGURE 2 The ion yield is plotted versus the ion energy, with the bias voltage indicated for each curve. a.) low energy spectra indicating the emission of ions with an energy deficit of approximately 4 eV. b.) higher energy spectra for which the ions arrive in the detector with energies less than what would correspond to the bias voltage. Curve 3 from a) is shown for comparison.

The higher energy data obtained with HOPG were very similar to those obtained with grafoil (Fig. 2b) and it was determined that the ions emitted from the

HOPG were potassium. Alkali ions are known to become ionized on hot surfaces, e.g. beta-eucryptite ($Li_2O \cdot Al_2O_3 \cdot 2SiO_2$) (9) and Holmlid and coworkers have observed not only the emission of K ions, but also the emission of K_n ionic molecules for n values of 1 to 61 (magic numbers) (10,11,12). They attribute much of their data to the creation of Rydberg ions on the surface. For the present HOPG data, the source of the K ions appears to be the tungsten filament used to indirectly heat the HOPG. Potassium is often utilized in the processing of tungsten wire. Without the HOPG in place, the RGA was lined up with the filament and the K, mostly neutral, was observed. Placing the HOPG, approximately a mm thick, between the filament and the RGA resulted in a reduced neutral signal and a greatly enhanced ion signal. The button heater was installed to eliminate the K, but as Fig. 2b, obtained using grafoil shows, K is still observed. It is unclear whether the observed K ions are the result of prior contamination of the vacuum chamber or whether the grafoil contains trace amounts of K.

Grafoil is not a well characterized material in terms of its crystal structure and may be considered to be a collection of oriented graphite crystallites. Graphite is a unique layered crystal with very different electrical and thermal conductivities parallel and perpendicular to its basal planes. The foil does have a thermal conductivity of 140 W/m·K along its width and length and only 5 W/m·K through its thickness (7) (to be compared with values of 190 -390 W/m·K and 1-3 W/m·K, respectively for (13)). The premium grade of grafoil has a carbon content of 98% and may contain up to 450 PPM of sulfur. It probably contains forms of carbon, other than graphite, as well. Atoms, including C, are easily intercalated between the basal planes (14). The fast particles observed might be ejected from between the basal planes, from a surface or a crystal boundary and might be from the sample itself, or might be from a molecule or atom adsorbed onto the surface. In a review of fundamentals of coal combustion, Essenhigh (15) notes that the adsorption of molecules, such as O_2, H_2O and H_2, is non-uniform, with preferential adsorption on the edge atoms of the graphite crystallites. As the defects introduced by ion implantation may provide additional "edge atoms", the data of Fig. 2a may be due to chemical effects. On the other hand, the correspondence of the peaks in Fig. 2a to a well-defined energy deficit, together with their narrow widths, suggest they might be due to a resonant process at the surface that depends upon the defects in graphite.

ACKNOWLEDGMENTS

We have greatly benefited from conversations with our colleagues, Professors Best, Budnick, Fernando, and Sinkovic. Michael Newman and Ryan Sears assisted with the assembly of the apparatus and participated in the preliminary measurements. This research was supported by awards from the Research Corporation (Q.C.K.), the Connecticut Space Grant Consortium under NASA grant NCC5-390 (J.L.), The University of Connecticut Research Foundation, and Connecticut Innovations, Inc. This last grant was made possible through a collaboration with Advanced Technology Materials, Inc., of Danbury CT.

REFERENCES

1. Richardson, O.W., *The Emission of Electricity from Hot Bodies*, London, Longmans, Green and Co. 1916.
2. Prosen, E. J. and F. R. Rossini, *J. Res. Natl. Bur. Standards* **33**, 439 (1944)
3. Dienes, G. J. and Vineyard, G. H. *Radiation Effects in Solids*, Interscience publishers, Inc., New York 1957.
4. Siegle, R., Davies, J.A., Forester, J.S. and Andrews, H.R., *Nucl. Instr. and Meth. D* **90**, 606 (1994).
5. Ahmad, S., Akhtar, M.N., Qayyum, A., Ahmad, B., Babar, K. and Arshed, W., *Nucl. Inst. and Meth. in Phys. Res. B* **122**, 19 (1997).
6. Choi, W., Kim, C. and Kang, H. *Surf. Sci* **281**, 323 (1993).
7. Grafoil is the tradename of graphite based paper manufactured by UCAR Carbon Company, Inc.
8. Manufactured by HeatWave, Watsonville, CA.
9. Blewett, J. P. and Jones, E. J. *Phys. Rev.* **50**, 464 (1936).
10. Wang, J., Engvall, K., Holmlid, L., *J. Chem. Phys.* **110** (1999) 1212-1220.
11. Wang, J., Andersson, R., Holmlid, L., *Surface Sci,* **399** (1998) L337-341.
12. Wang, J., Holmlid, L., *Chem. Phys. Letters,* **295** (1998) 500-508.
13. Pierson, H. O., *Handbook of Carbon, Graphite, Diamond and Fullerenes*, Park Ridge, NJ, Noyes Publications, 1993, p. 157.
14. Dresselhaus, M. S., and Dresselhaus, G. *Adv. Phys.* **30**, 139 (1981).
15. Essenhigh, R.H., *Chemistry of Coal Utilization* (Second Supplementary Volume), Ch. 19, Elliot, M.A., Ed., John Wiley & Sons, New York, (1981).

High Power Photoconductive Semiconductor Switches Treated with Amorphic Diamond Coatings

Farzin Davanloo[*], Mugurel C. Iosif[*], Tiberius Camase[*], Carl B. Collins[*], and Forrest J. Agee[+]

[*]Center for Quantum Electronics, University of Texas at Dallas, P.O. Box 830688, Richardson, TX 75083-0688
[+]Air Force Office of Scientific Research, AFOSR/NE, 801 N. Randolph St., Arlington, VA 22203-1977

Abstract. Our recent efforts have resulted in implementation and demonstration of several intense photoconductively switched stacked Blumlein pulsers producing high power output pulses with risetimes as fast as 200 ps. A single GaAs photoconductive switch triggered with a low power laser diode array commutates these devices. During the avalanche-mode photoconductive switching of these pulsers at high powers, current filamentation associated with the high gain GaAs switches produces such high current density that switches are damaged near the metal-semiconductor interface and the lifetime is limited. This report presents progress toward improving the switch operation and lifetime by advanced treatments with the amorphic diamond coatings.

INTRODUCTION

Our recent efforts [1] have concentrated on development and demonstration of high power stacked Blumlein pulsers commutated by a single photoconductive semiconductor switch (PCSS). Presently, these devices operate with a switch peak power in the range of 50-80 MW and switch triggering laser pulse energies as low as 300 nJ. During the avalanche-mode photoconductive switching of these pulsers, the current is concentrated in filaments that extend from the cathode to the anode across the insulating region of the PCSS. Carrier recombination results in the emission of characteristic band gap photons in the near infrared region, which can be seen by an infrared viewer [2].

We have observed these effects in experiments where laser diodes provided trigger photons for the avalanche commutation of stacked Blumlein pulsers [1,3]. As soon as the avalanche is initiated, a single filament can be observed which approximately follows the collimated laser beam that was focused in a line from the cathode to the anode. Multiple branching from the avalanche initiation point has also been observed but with much less probability. Filamentary currents with densities of several MA/cm^2 and diameters of 15-300 μm passing through a narrow channel can cause switch damage, especially at the contacts points. A greater number of filaments during each cycle of commutation reduce the stress on the switch, thereby increasing its lifetime.

Our current research has been directed to study and implement the broadening of the current channels in the avalanche photoconductive switch in order to improve lifetime and increase switching peak power. Two approaches are being used. The first is control of laser diode beam delivery to the switch, and improvements of switch contacts, pulser configuration and charging mechanism. The second approach is application of amorphic diamond coatings to the PCSS switch electrodes to enhance operation and lifetime in Blumlein pulse generators. In this report we present the progress made in study and use of amorphic diamond films to treat high power PCSS.

AMORPHIC DIAMOND FILMS ON PHOTOCONDUCTIVE SWITCH MATERIALS

Basic research in our laboratory has described a conformal coating that has hardness of natural diamond and exceptionally high values of electron emissivity. Discovered at UTD, it is made from graphite and laser light without catalyst, noxious byproduct, or toxic wastes. This material was termed amorphous ceramic diamond and later shortened to "amorphic diamond" for convenience [4,5]. Deposited at room temperatures it forms a strong bond to any material onto which it is applied. Such a favorable combination of hardness, chemical bonding and an elastic modulus of 850 GPa should translate directly into an increased resistance to abrasive wear of components coated with amorphic diamond. It has been demonstrated that only a 1-3 μm coating of amorphic diamond could protect fragile substrates against erosive environments [4,5]. Analytical techniques have shown amorphic diamond to consist of nodules of tens of nanometers in diameters that are composed of sp3 (diamond) bonded carbon in a matrix of other carbons. The nodules seem to be disordered mixtures of the cubic and the rare hexagonal polytypes of diamond that have no extensive crystalline planes along which to fracture. Since it is condensed from laser plasmas produced under conditions that are also optimal for the growth of interfacial layers, the films of amorphic diamond are strongly bonded to the substrates onto which they are condensed [4,5].

In this work, amorphic diamond coatings were deposited on one side of highly resistive Si and GaAs substrates, the types used in our Blumlein pulsers as the photoconductive switch materials. Electrodes were attached to either side using conducting silver paint and epoxy. The grids like conducting electrodes were painted on the coating sides. For these measurements a Keithley 237 high voltage source-measure unit was used to provide constant voltage bias while monitoring the current through the sample. To avoid heating the sample the measurements were carried out with pulsed voltage sweep having on and off times of 50 and 450 ms, respectively. For measurements of forward diode characteristics, diamond coating was biased negative with respect to the substrate.

Typical current-voltage characteristics measured for a 0.3 μm nominal amorphic diamond on the p-type Si (100) with a resistivity in the range > 2000 Ω cm is shown in Fig. 1. For comparison, the I-V plot for an uncoated similar Si substrate is included in this figure. The rectifying behavior for the coated sample seen in Fig. 1 was attributed to the amorphic diamond /Si junction because the I-V plot for the uncoated sample showed symmetrical and ohmic character with change in the voltage polarity.

Current-voltage properties were also studied for the amorphic diamond on semi-insulating GaAs with resistivity of about 1.0×10^7 Ω cm. Results are presented in Fig. 2 for a bare GaAs sample and a 0.57 μm amorphic diamond film on GaAs. Again a symmetrical and ohmic current-voltage behavior is seen. The I-V characteristics differ considerably for the coated sample, especially for the forward current region where the coating side was biased low.

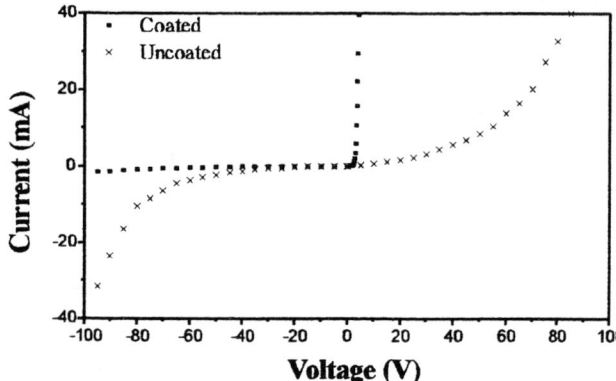

FIGURE 1. Current –Voltage characteristics measured in the dark for an uncoated and diamond coated Si with a high resistivity rating.

The rectifying character under reverse bias is clearly seen with conductivity approaching that of bare GaAs at higher bias voltages. This may be due to the release of trapped negative charges when the amorphic diamond is slowly depleted at reverse bias voltages above 200 V. This behavior was found completely reversible as tests were repeated.

Of importance to this study, is the rapid current increase in forward direction as seen in Figs. 1 and 2. This is attributed to the tunneling of electrons from amorphic diamond to the conduction band of the GaAs or Si. This process is similar to the Fowler-Nordheim tunneling. It is expected to provide pre-avalanche sites for operation of diamond coated PCSS and thereby diffuse the conduction current.

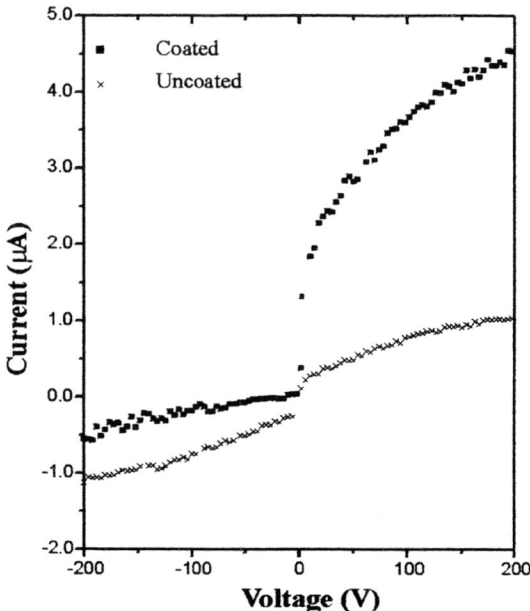

FIGURE 2. Current–Voltage characteristics measured in the dark for an uncoated and diamond coated Si with a high resistivity rating.

To measure the electrical breakdown strength of our amorphic diamond coatings, the I-V measurements were performed under the reversed biased conditions at higher voltages until a breakdown voltage was reached. Figure 3 plots the I-V characteristics, in the reverse biased conditions, for a 0.13 μm-thick amorphic diamond coating on a p-type Si substrate. The reverse breakdown voltage in excess of 425 V gives an electrical breakdown strength greater than 3×10^9 V/m if the entire diamond film is assumed to be depleted. The high electrical breakdown strength for amorphic diamond suggests that carrier velocity saturation in high field strength is allowed and the electron-electron scattering is of the same order as that in crystalline diamond [6].

SWITCH DESIGN CONSIDERATIONS

In our earlier work, improvements in the PCSS switch operation and lifetime have been examined by coating the triggered face of GaAs switch cathodes with strips of highly adhesive films of amorphic diamond. With the application of amorphic diamond, not only the switch lifetime was increased but also the damage at the cathode contact was found to be less than that found for the anode contact [3,7]. This indicated that diamond coating protected and hardened the cathode side. In addition, in experiments where the switch had no diamond coating, the majority of the time, a single current filament commutated the switches. The filament was initiated near the cathode and followed, approximately, one of the laser beams to the anode. Multiple branching was rarely seen [1]. In the case of switch with the diamond coating, the multiple branching was observed more often indicating an increase of pre-avalanche sites [3]. The amorphic diamond / PCSS junction properties discussed briefly in the previous section may have aided longevity of the PCSS in these earlier studies.

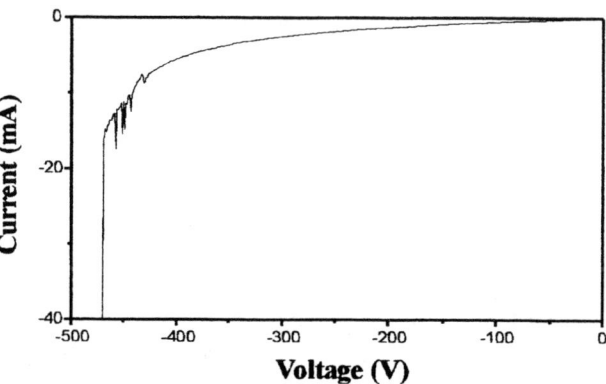

FIGURE 3. Current-Voltage characteristics showing the breakdown voltage for a nominal amorphic diamond film on p-type Si substrate. Measurements were performed in the dark under reverse bias conditions.

The switch/electrode configuration used in the PCSS lifetime studies was a part of a low profile switch assembly [1,7] that facilitated the use of a single photoconductive switch in the pulser, as can be seen in Fig. 4. The electrode assembly allowed for operation of PCSS in either lateral or opposed configurations. Layers of Kapton insulator were placed between the switch and the base electrode to restrict the current path through the top electrode holders in the lateral configuration. In the opposed configuration, switch was placed to contact one of the base electrodes on one side and the other top electrode holder on the other side. Layers of Kapton were used to isolate the switch from the base electrode on one side and the top electrode holder on the other side. The current path was through one of the top electrode holders into the GaAs bulk and through the other base electrode.

For both configurations, top copper electrode holders were connected to the base electrodes by means of several screws as shown in Fig. 4. The pulse forming lines from a prototype pulser were connected to the bottom of base copper electrodes, which were cast in a G-10 plastic plate. Each switch was fabricated from one half of a semi-insulating LEC grown GaAs wafer with a diameter and thickness of 5 cm and 0.5 mm, respectively. It was held in place by means of two copper holders screwed to the electrodes. Commutation of the switch was triggered at 905 nm by focusing the LD-220 laser diode array beam in two straight lines across the switch gap from cathode to anode.

It is anticipated that, by depositing films of amorphic diamond near the PCSS cathode electrode contact, the number of carriers and avalanche sites increase thereby improving the switch lifetime and performance. As described earlier, there is a rapid current increase in forward direction of an amorphic diamond/ GaAs heterojunction where the diamond coating is biased negative with respect to the GaAs substrate. This case is schematically presented in Fig. 5 for the switch/electrode in the opposed configuration, with the switch cathode coated with a strip of thin film amorphic diamond.

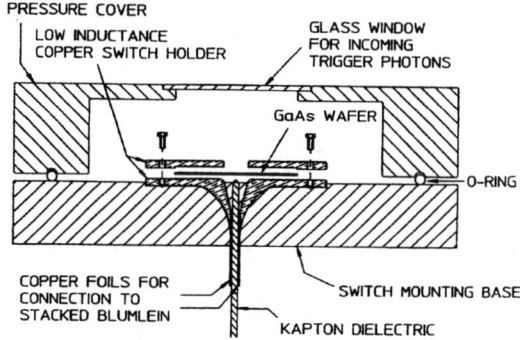

FIGURE 4. Schematic drawing of the switch assembly used in our photoconductively switched stacked Blumlein pulsers.

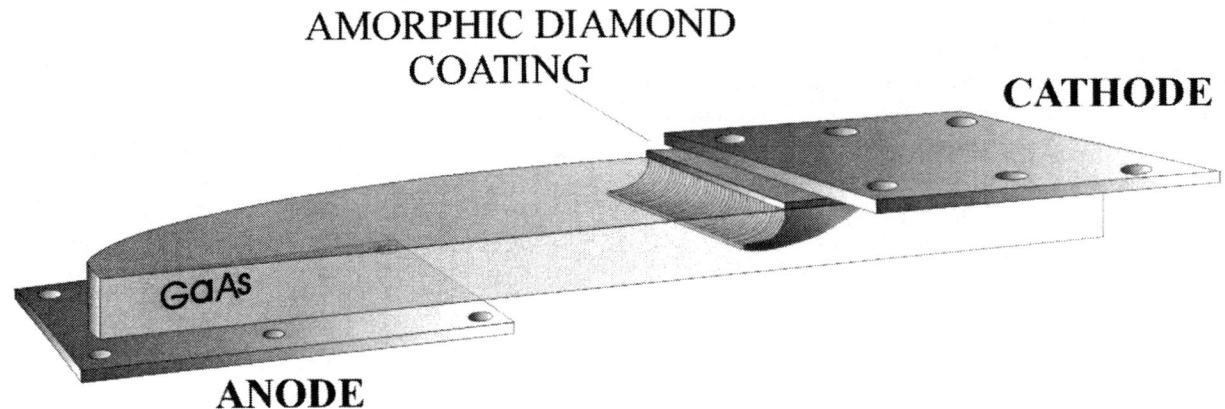

FIGURE 5. Schematic drawing of the switch/electrode assembly in the opposed configuration. Off-state carrier movement due to the tunneling of electrons from amorphic diamond to GaAs is also presented schematically.

Encouraging or inhibiting the current conduction flow at the interface between amorphic diamond and PCSS material may have pronounced effect upon off-state switch hold-off and switch performance. For example the tunneling of electrons from amorphic diamond to GaAs during the off-state stage of PCSS operation provides pre-avalanche sites that may diffuse conduction current upon switch activation. However, this may also increase leakage current at high fields causing switch shorting and failure. To avoid such problems for a particular charging voltage, it may be necessary to limit the current injection by controlling the switch gap, diamond film thickness and the laser diode beam delivery to the switch.

Diamond coating of the switch anode area may result in increased hold-off characteristics of the PCSS in the off-state stage of operation leading to longer switch lifetimes. In this case the amorphic diamond inhibits the flow of electrons at the interface until very high fields are reached. This is due to rectifying behavior of the amorphic diamond/ GaAs heterojunction operating under reverse bias condition as discussed earlier in this paper.

The semiconductor properties of amorphic diamond can be employed to improve the PCSS longevity by coating the switch cathode or anode areas or both. However, the critical issue to resolve is the switch design options that make optimal use of amorphic diamond coatings for long life operations of avalanche PCSS in stacked Blumlein pulsers. Design options include: switch configuration, switch gap setting, diamond film thickness, exact locations of diamond coatings and film qualities. Further research and lifetime studies are underway to characterize the elementary processes involved in conduction of PCSS devices coated with amorphic diamond during on- and off-states of operation.

ACKNOWLEDGMENTS

This work was supported by the Air Force Office of Scientific Research (AFOSR) under Grant F49620-00-1-0296. We would like to acknowledge the initial contributions of W. R. Osborn, R. K. Krause and D. L. Borovina in electrical characterizations of the amorphic diamond.

REFERENCES

1. F. Davanloo, C.B. Collins, and F.J. Agee, IEEE Trans. Plasma Sciences, 26, 1463-1475 (1998).
2. F.J. Zutavern, G.M. Loubriel, W.D. Helgeson, M.W. O'Malley, R.R. Gallegos, A.G. Baca, T.A. Plut, and H.P. Hjalmarson, "Fiber-optic Control of Current Filaments in High Gain Photoconductive Semiconductor Switches," in *Conference Record of the 1994 Twenty-First Power Modulator Symposium*, 1994, pp. 116-119.
3. F. Davanloo, R. Dussart, M.C. Iosif, C.B. Collins and F.J. Agee, "Photoconductive Switch Enhancements and Lifetime Studies for Use in Stacked Blumlein Pulsers," in *Proceedings of the 12th IEEE International Pulsed Power Conference*, Edited by C. Stallings and H. Kirbie, 1999, p. 320-323.
4. C. B. Collins, F. Davanloo, T. J. Lee, H. Park, and J. H.You, J. Vac. Sci & Technol B11, 1936-1941 (1993).
5. F. Davanloo, H. Park and C.B. Collins, J. Mat. Res., 11, 2042-2050 (1996).
6. E. A. Koronova and S. A. Sherchenco, Sov. Phys. Semicond. 1, 299 (1967).
7. F. Davanloo, H. Park, C. B. Collins and F. J. Agee, "Photoconductive Switch Enhancements for Use in Blumlein Pulse Generators," edited by J. L. Duggan and I. L. Morgan, AIP Conference Proceedings 475, New York, 1999, pp. 918-921.

SECTION XIII

NONDESTRUCTIVE ANALYSIS: DETECTION OF DRUGS, NUCLEAR MATERIALS, EXPLOSIVES, CEMENT ANALYSIS

FIGARO: Detecting Nuclear Material using High-Energy Gamma Rays from Oxygen

B. J. Micklich[a], D. L. Smith[a], T. N. Massey[b], D. Ingram[b], and A. Fessler[a]

[a]*Technology Development Division, Argonne National Laboratory, Argonne, IL 60439 USA*
[b]*John E. Edwards Accelerator Laboratory, Ohio University, Athens, OH 45701 USA*

Abstract. Potential diversion of nuclear materials is a major international concern. Fissile (e.g., U, Pu) and other nuclear materials (e.g., D, Be) can be detected using 6-7 MeV gamma rays produced in the $^{19}F(p,\alpha\gamma)^{16}O$ reaction. These gamma rays will induce neutron emission via the photoneutron and photofission processes in nuclear materials. However, they are not energetic enough to generate significant numbers of neutrons from most common benign materials, thereby reducing the false alarm rate. Neutrons are counted using an array of BF_3 counters in a polyethylene moderator. Experiments have shown a strong increase in neutron count rates for depleted uranium, Be, D_2O, and ^6Li, and little or no increase for other materials (e.g., H_2O, SS, Cu, Al, C, ^7Li). Gamma source measurements using solid targets of CaF_2 and MgF_2 and a SF_6 gas target show that proton accelerators of 3 MeV and 10-100 microamperes average current could lead to acceptable detection sensitivity.

INTRODUCTION

Significant quantities of special nuclear material (SNM) have been smuggled across borders or facility boundaries. This poses a major national and international concern. The danger could be reduced substantially by developing an effective, compact, transportable, and affordable inspection system that could be deployed easily at critical locations, e.g., border crossings or sensitive nuclear facilities.

The FIGARO concept uses a compact, low-energy proton accelerator to generate gamma rays that can interrogate objects such as packages, luggage, or containers for the presence of all SNM, including U-235, as well as other nuclear materials such as beryllium and deuterium. FIGARO gives an unambiguous signal for the presence of nuclear material, independent of size, shape, or chemical form. FIGARO is designed to provide high sensitivity and good signal-to-noise performance, and is highly resistant to counter-measures. Since photons are used as the probing radiation, there is little to no residual activity in inspected items. When the accelerator is turned off, there is minimal residual radioactivity to impede handling and transport. The system hardware could be made sufficiently compact to put in a small truck or transport aircraft for rapid and flexible field deployment in response to changing threats, and a single accelerator could serve multiple interrogation portals at a given location.

PHYSICAL PRINCIPLES

The $^{19}F(p,\alpha\gamma)^{16}O$ reaction is exothermic with a Q-value of 8.115 MeV. The reaction proceeds via the population of excited levels of the compound nucleus ^{20}Ne which have large α-particle decay widths to excited states of ^{16}O. The transitions to the ground state and four excited states of ^{16}O take place through the emission of five alpha groups. The second, third, and fourth excited states of ^{16}O de-excite almost exclusively by the emission of gamma rays with energies of 6.129, 6.917, and 7.116 MeV, respectively. Higher excited states also decay through these levels. Thus the $^{19}F(p,\alpha\gamma)^{16}O$ reaction produces nearly mono-energetic, high-energy gamma rays of 6-7 MeV.

These gamma rays are sufficiently energetic to produce neutrons by photo-fission and/or photo-neutron reactions in fissionable material (e.g., uranium and plutonium), as the reaction thresholds are in the range 5.5-6 MeV. In addition, these gamma rays will induce neutron emission from non-fissionable nuclear materials (e.g., deuterium, lithium-6, and beryllium). The (γ,n) thresholds are about 2.22, 5.73, and 1.85 MeV for these three nuclides, respectively. However, these 6-7 MeV gamma rays do not have sufficient energy to produce photoneutrons from common benign materials. The low resulting neutron background means that a DC accelerator can be used rather than a pulsed one, since we are not required to count delayed fission neutrons between pulses.

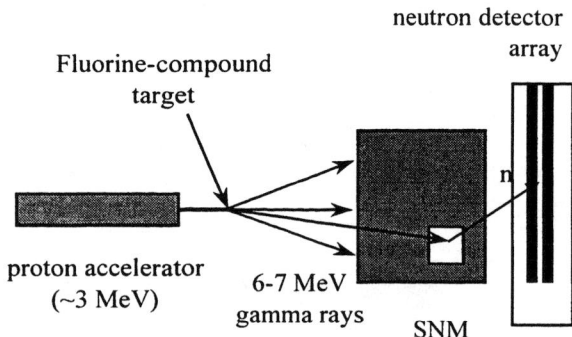

FIGURE 1. Schematic diagram of FIGARO operation.

Because the system would operate in a DC mode, and energy and time resolution are not required, one can use a very simple neutron detection system such as an array of moderated BF_3 or ^3He counters. We have chosen to use BF_3 because the energy released in the $^{10}B(n,\alpha)^7Li$ reaction is greater than that released in the $^3He(n,p)^3H$ reaction, allowing better rejection of low-energy gamma events and noise. Because very efficient use is made of both the accelerator-produced gamma rays and the neutrons generated by the interaction of those gammas, FIGARO has the potential to be a very efficient and effective detection system for all nuclear materials.

PROTON TARGET OPTIMIZATION

The requirements of a practical target for the production of energetic 6-7 MeV gamma rays from the $^{19}F(p,\alpha\gamma)^{16}O$ reaction are: (i) good mechanical strength and thermal conductivity to enable it to sustain significant beam currents at typical operating proton energies (up to 100 microamperes at 3 MeV, equivalent to 300 W of beam power) for an extended period of time with no significant deterioration; and (ii) minimal production of unwanted background neutrons from (p,n) reactions. For the latter consideration, reference is made to Table 1 (1). Although not shown in this table, the (p,n) threshold for mono-isotopic fluorine is 4.234 MeV.

Our earlier work in the development of this technology investigated the use of solid-compound fluorine targets (2). Both CaF_2 and MgF_2 in crystalline form were considered. These targets were found to be quite limited in their ability to survive even modest proton beam currents over reasonable time periods. Later a gas-target system employing SF_6 was tested and was found to be far superior to the solid targets for the present application. Initially, an existing target cell fabricated entirely from stainless steel was employed without modification. The gas-cell window selected was 5-micron-thick W foil. The tungsten windows thinned during use, most probably from being attacked by free fluorine radicals produced by dissociation of the target gas. This led to frequent catastrophic loss of the cell gas and failure of the accelerator vacuum system. Next, a 7-micron-thick Ni window was tested. While this arrangement was mechanically stable and contained the target gas reliably at pressures up to 35 psia and beam currents up to 2 µA, the yield of (p,n) neutrons was found to be unacceptably high above E_p = 3 MeV. Mono-isotopic Al has a relatively high (p,n) threshold (5.804 MeV) and also possesses desirable thermal and mechanical properties. Aluminum foil windows of both 15 and 20 micron thickness have been used with considerable success, with low neutron background and stable operation for extended periods at 13-15 psia and proton beam currents up to 4 µA. These are not fundamental physical limitations, so a target system capable of handling the more severe operating conditions expected of a fieldable system can and will be developed in the future.

TABLE 1. (p,n) Thresholds for Selected Elements*

Element	Isotope	Abundance (%)	p,n threshold (MeV)
Lithium	^7Li	92.41	1.880
Carbon	^{13}C	1.11	3.236
Oxygen	^{18}O	0.20	2.574
Sulfur	^{36}S	0.02	1.978
Chlorine	^{37}Cl	24.23	1.639
Chromium	^{53}Cr	9.501	1.406
	^{54}Cr	2.365	2.220
Manganese	^{55}Mn	100	1.032
Iron	^{57}Fe	2.119	1.647
	^{58}Fe	0.282	3.144
Cobalt	^{59}Co	100	1.887
Nickel	^{61}Ni	1.140	3.070
	^{64}Ni	0.926	2.496
Copper	^{65}Cu	30.83	2.167
Gold	^{197}Au	100	1.389

*Only isotopes with (p,n) thresholds below 3.5 MeV are listed

High-purity, 0.1-mm-thick Al was used to line the gas cell walls, beam apertures, and collimators to reduce the neutron background since Al has a high (p,n) threshold. SRIM (3) calculations indicated that this thickness was sufficient to stop the most energetic protons used (about 4.25 MeV). The only region of the gas target assembly which was not protected from incident protons by an Al layer was the stainless steel beam stop at the end of the cell. However, SRIM calculations showed that the range of protons in SF_6 at 13 psia and 4.25-MeV proton energy is significantly smaller than the actual cell length. Therefore it is highly unlikely that any protons could actually reached the unprotected beam stop. SRIM calculations also indicated a small probability of scattered protons reaching the side wall of the cell, but it was (as mentioned above) lined with Al sufficiently thick to stop these scattered charged particles.

This approach reduced the (p,n) neutron background considerably although it could not be eliminated completely due to unavoidable residual neutron yield from (p,n) reactions at proton energies above their respective thresholds on the minor isotopes ^{13}C, ^{18}O, and ^{36}S found in the target environment. These neutrons appear to be produced primarily at the gas cell entrance. However, we found that the effect of these neutrons on the FIGARO detector system could be reduced by about a factor of 2-3 by placing hydrogeneous shielding in the vicinity of the gas cell entrance region in such a way that neutrons from the interrogated sample were not obstructed. Background reduction is important for improving system sensitivity.

NEUTRON PRODUCTION MEASUREMENTS

A benchmark setup was used to detect neutrons produced by (γ,n) and (γ,f) reactions using the gamma rays generated by the $^{19}F(p,\alpha\gamma)$ source. The measurements were performed using the 4.5-MV Tandem Van de Graaff accelerator at the Ohio University John E. Edwards Accelerator Laboratory. A new beamline section was fabricated to minimize the effects of (p,n) reactions from elements deposited along the beamline in other experiments. A gas cell containing SF_6 was mounted on the end of this beamline, with the neutron detection array centered on the entrance of the gas cell and about 16 inches away horizontally. Protons of energy 3.0 MeV were used with a 15-micron Al window, so that the proton energy entering the SF_6 was about 2.65 MeV. The neutron detector array consisted of 19 BF_3 tubes mounted in three rows in a polyethylene moderating assembly. The outputs from all the tubes in a given row were summed and recorded as a pulse height spectrum for later analysis. A simpler detection system suitable for field application could consist of just a lower-level discriminator and a scalar.

A large number of samples, both pure elements and compounds, were examined to determine whether γ-induced neutrons were produced. Figure 2 shows the neutron count rates for selected materials that were examined, including several nuclear materials and common benign materials. The results have been corrected for background and have been normalized for integrated beam current and target mass. The nuclear materials DU (approximately 4.1 kg of depleted uranium was used as a substitute for fissile materials), beryllium, and heavy water show a clear neutron signal. Lead, which might be used to shield gamma rays from the target, also produces neutrons as expected. No other common benign materials tested to date (a total of about 20) have shown any significant increase in neutron detection rate above background.

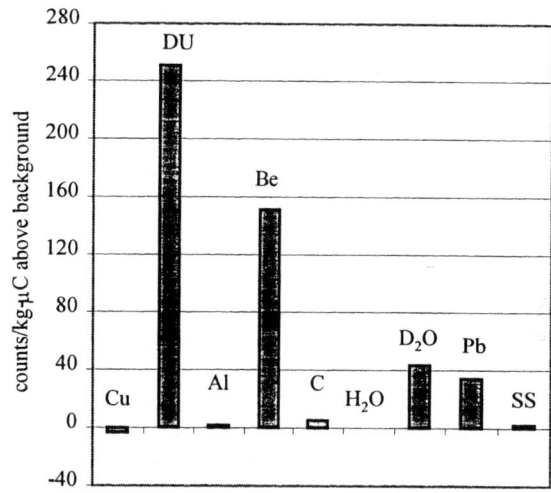

FIGURE 2. Normalized neutron count rates for selected nuclear and common materials.

In addition, studies have been performed to assess the resistance of the FIGARO technique to various countermeasures. Counter-measures to FIGARO could be medium- to high-Z gamma shielding to reduce γ-induced reactions, hydrogenous neutron shielding to prevent neutrons from reaching the detector, or some combination of the two. Both copper and lead gamma shields were investigated, along with neutron shields of polyethylene, borated polyethylene (BPE), lucite, and borax.

The results of some of these countermeasure studies are shown in Figure 3. Count rates of at least 10 counts/μC-kg above background are seen for even the case of one inch of lead in front of the DU with 3 inches of lucite behind. A more realistic geometry would move the gamma source, target material, and neutron detector further away from each other, but with increased detector coverage in solid angle. A reduction in count rate of no more than ten is expected. The detection of one kilogram of uranium with 1000 counts above background would require a proton beam charge of 1000 μC, which could be delivered by a 100 μA accelerator in ten seconds.

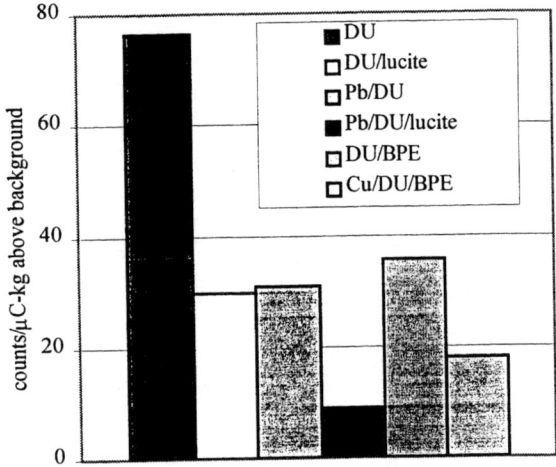

FIGURE 3. Normalized neutron count rate for trials employing selected countermeasures.

In a separate set of measurements, we examined the effect of placing the DU sample in an office wastebasket containing water. The DU sample was placed in the center of approximately 3-4 gallons of water. Results corrected for background and normalized to integrated beam current and target mass are given in Table 2. Even in this extreme case one can still see an increase in counts above background. It is important to arrange the test items so that the water minimally shields the detector array from the background neutrons produced near the gas cell entrance. If these are shielded, then the increase in neutrons counted due to the presence of the DU is offset by a reduction in the count of background neutrons. This points out another reason that the neutron background must be kept as low as possible.

Several measurements were also performed with the DU sample located inside a standard 9-inch×14-inch×22-inch carry-on suitcase filled with an assortment of clothes and personal articles. In each of several cases there was an excess of neutron counts above background sufficient to determine the presence of the sample (1000 counts above background) with an irradiation of approximately 50-250 μC.

TABLE 2. Results of Countermeasures Studies.

Case	Counts/μC	Counts/μC-kg above background
background avg.	35.3	-
DU	63.3	6.85 ± 0.02
DU in water	44.7	2.33 ± 0.02
water	36.5	-
background avg.	30.7	-
suitcase	15.3	-
suitcase + DU #1	118.4	21.27 ± 0.06
suitcase + DU #2	48.6	4.34 ± 0.04
suitcase + DU #3	122.6	22.30 ± 0.06

CONCLUSIONS

Gas targets of SF_6 have proven to be strong sources of gamma rays from the $^{19}F(p,\alpha\gamma)^{16}O$ reaction. Aluminum windows have been used at $E_p = 3$ MeV currents up to 4 μA (12 W power) without failure. Benchmark measurements have shown neutron production from a range of nuclear materials with insignificant production from benign materials. The basic technique also appears to be resistant to countermeasures. Further directions involve the design of a high-power gas target and investigation of ways to use the technique to discriminate various types of nuclear materials.

ACKNOWLEDGMENTS

This work was funded by the U.S. Department of Energy under contract W-31-109-ENG-38. The authors would like to thank the staff of the Ohio University John E. Edwards Accelerator Laboratory for their continued helpfulness during this research.

REFERENCES

1. QCALC code, available online from the National Nuclear Data Center. URL: http://www.nndc.bnl.gov.

2. Fessler, A., Massey, T. N., Micklich, B. J., and Smith, D. L., *Nucl. Inst. Meth. Phys. Res.* **A450**, 353 (2000).

3. Ziegler, J. F., Biersack, J. P. et al., SRIM 2000 code. URL:http://www.research.ibm.com/ionbeams/home.htm.

Relocatable Cargo X-Ray Inspection Systems Utilizing Compact Linacs

W. Wade Sapp[*], Andrey V. Mishin[#], William L. Adams[*], Joseph Callerame[*], Lee Grodzins[*], Peter J. Rothschild[*], Richard Schueller[*], Gerald J. Smith[*&]

[*]*American Science & Engineering, 829 Middlesex Turnpike, Billerica, MA 01821, USA*
[#]*AS&E High Energy Systems Division, 3300 Keller Street, #101, Santa Clara, CA 95054, USA*
[&]*Current Address: FEI Company, One Corporation Highway, Peabody, MA 01960-7990, USA*

Abstract. Magnetron-powered, X-band linacs with 3 – 4 MeV capability are compact enough to be readily utilized in relocatable high energy cargo inspection systems. Just such a system is currently under development at AS&E™ using the commercially available ISOSearch™ cargo inspection system as the base platform. The architecture permits the retention of backscatter imaging, which has proven to be an extremely valuable complement to the more usual transmission images. The linac and its associated segmented detector will provide an additional view with superior penetration and spatial resolution. The complete system, which is housed in two standard 40' ISO containers, is briefly described with emphasis on the installation and operating characteristics of the portable linac. The average rf power delivered by the magnetron to the accelerator section can be varied up to the maximum of about 1 kW. The projected system performance, including radiation dose to the environment, will be discussed and compared with other high energy systems.

INTRODUCTION

Cargo x-ray inspection systems utilize source energies between several hundred keV (useful for palletized cargo, cars, moderately loaded trucks and containers, etc.) and about 10 MeV. These higher energies are deemed[1] necessary to achieve a fully penetrated inspection of heavily loaded ISO shipping containers and railway cars. Unfortunately, complex cargo often produce such cluttered transmission x-ray images that a thorough inspection is impossible even though the load is fully penetrated.

While improved spatial resolution and contrast sensitivity are important attributes in dealing with clutter, it is the additional information obtained from backscatter images that often permits a successful inspection without opening the container. As a consequence, backscatter imaging is a prominent feature of all AS&E inspection systems. This applies to the commercial product line[2] as well as to new capabilities currently under development, such as the system described in this paper.

SELECTION OF X-RAY ENERGY

Penetration vs. X-Ray Energy

Penetration capability of a realizable inspection system involves much more than mass attenuation coefficients. For example, a system designed to inspect a tractor-trailer rig in a single pass needs acceptable uniformity of the x-ray beam intensity over a height of almost fourteen feet. However, the angular

---

[1] An independent study published in 1995 [1] measured the inspection effectiveness of detecting bulk quantities of drugs in a Cargo Laden Intermodal Container Scenario (Seaport Entry Scenario). There was surprisingly little difference between the effectiveness of a dual-source 8-10MeV system and a dual-source 450 keV system with backscatter imaging capability.

[2] Descriptions more detailed than that provided in sales brochures have been published in References 2 -- 5.

distribution of x-rays from the production target becomes more forward-peaked at higher energies, and this effect is most pronounced for the highest energy x-rays which are the most penetrating. As a consequence, higher energy x-ray sources must be moved further away from the cargo to achieve acceptable coverage. The greater distance results in less flux reaching the cargo, and this reduces the effective penetrating power. A recently published semi-empirical calculation [6] shows that penetration through steel increases only slightly between 3 MeV and 6 MeV and actually decreases at higher energies.

Stream-of-Commerce

Two common cargo inspection challenges are tanker trucks containing fuel, alcohol, or water and tractor-trailer rigs transporting fruits or vegetables. In both cases the inspected object has a maximum radiological thickness equivalent to very nearly eight feet of water. Also, these loads are not easily sub-divided for inspection by systems with lower penetrating power.

In typical streams-of-commerce at ports of entry in both the US and most foreign countries suspect loads with radiological thickness greater than eight feet of water are relatively unusual.

It is fortuitous that a well-designed 3 MeV cargo inspection system can penetrate[3] approximately eight feet of water.

X-Ray Energy

Radiation protection and collimation both become more difficult (expensive) at higher energies. This is yet one more reason to choose the lowest energy consistent with our inspection objectives. In the preceding paragraph 3 MeV was mentioned as a logical choice. We have decided on 3.5 MeV to give us flexibility in filtration design and a contingency margin for inevitable imperfections. It should be emphasized that 3.5 MeV is not an upper limit. Experience operating the current system will be the best guide to determining the upper limit for this compact, relocatable configuration.

---

[3] We have adopted the *de facto* definition of penetration as being that thickness of absorber behind which a large, radiologically thick object can just be discerned by a competent operator when the x-ray inspection system is operating in a standard mode.

ISOSearch

ISOSearch (Figure 1.) is an AS&E cargo inspection system featuring two 450 keV sources, one for each side of the cargo[4]. It produces four independent x-ray images of the inspected conveyance: two transmission images and two backscatter images. It is constructed inside two 40' ISO containers complete with x-ray detectors, an air-conditioned control room, and an optional motor-generator set (not shown). The end user can relocate this compact system and set it up in about three days (site preparation excluded). It is available with a variety of cargo transport systems from simple tuggers to a permanently installed rail system. It complies fully with all regulations for "cabinet" x-ray systems[5].

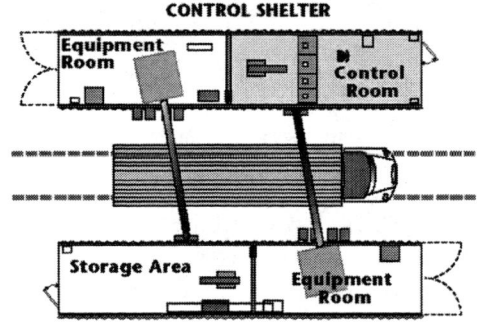

FIGURE 1. AS&E ISOSearch with dual 450 keV systems.

One facility recouped for the customer a third of its cost in about six months of operation in spite of the fact that significant areas of some of the containers could not be penetrated with 450 keV.systems. Significantly improved penetration was needed for

---

[4] The primary reason for dual sources is to view both sides of the conveyance with backscattered x-rays which have limited penetration in dense cargo.
[5] U.S Bureau of Radiological Health Standards for Cabinet X-Ray Systems (21 CFR 1020.40). Cabinet surfaces are defined as the exterior walls of the installed system. Portals are open (unshielded). The operator receives less than 1.0μG/hr (0.1mR/hr).

more thorough inspections of densely loaded conveyances.

ISOSearch PLUS 3.5 MeV

AS&E is currently developing a 3.5 MeV inspection subsystem using a very compact pulsed linac and a 1024-channel segmented detector array, both designed specifically for this application. This system is designed to be installed midway between the 450 keV beam planes in ISOSearch and provide a fifth, highly penetrating view of the conveyance to supplement the four images obtained with the 450 keV beams. An artist's conception of the new system is shown in Figure 2. Detailed computer simulations have shown that with judicious shielding and operating procedures this enhanced system will still satisfy "cabinet" regulations.

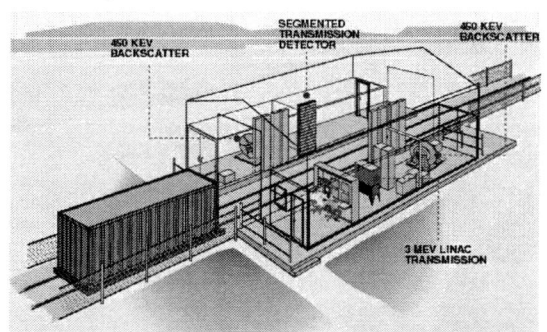

3.5 MeV LINAC

This enhanced ISOSearch is realizable only because of the very compact X-band linac and ancillary equipment[6]. In order to fit the linac system into the standard shipping container it was essential that its length be minimized so that the production target could be placed far enough away from the inspected cargo to have good beam coverage. The small size also significantly reduces the local shielding volume and weight. Only modest floor reinforcements will be necessary. To attain low environmental radiation to qualify as a cabinet x-ray system, we will need a local shielding thickness of about 14" of Pb. With such thickness, penetrations through the shield for power and water are generally the greatest sources of leakage radiation. While still a significant challenge, designing an effective labyrinth for the X-band rectangular wave guide is considerably simpler than for the much larger S-band guides. A system block diagram which is drawn to scale (except for cable lengths and diameters) is shown in Figure 3. An enlarged view of the linac alone is shown in Figure 4.

FIGURE 2. (at left) ISOSearch (two 450 keV sources with both backscatter and transmission detectors) supplemented by a single 3.5 MeV linac source and segmented detector for the transmitted x-rays. In practice the segmented detector will be shaped like an inverted "L".

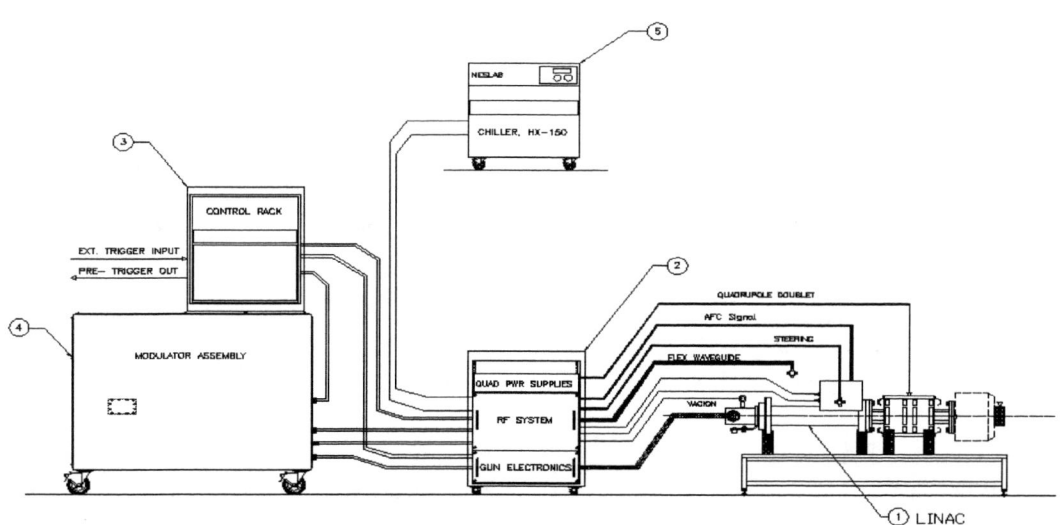

FIGURE 3. System block diagram for 3.5 MeV, magnetron-powered X-band linac. Approximately to scale except for cable and hose lengths. Linac rf length is 40 cm.

---

[6] Fortunately it is also less expensive than competing S-band systems.

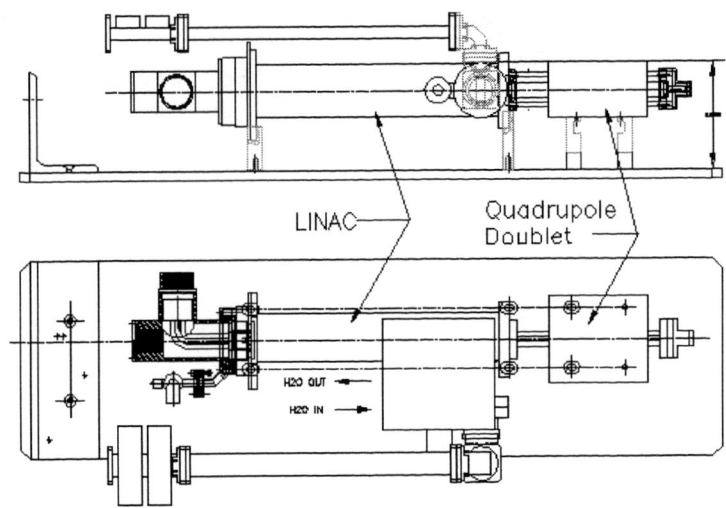

FIGURE 4. 3.5 MeV linac detail. Water cooled target is in the tip at the far right. Cable, water connections, and flexible rectangular wave guide are not shown.

The quadrupole doublet is designed to reshape the intrinsically circular beam into an ellipse that more nearly matches the acceptance of the x-ray collimator (not shown). The design parameters are given in Table 1. The entire system has been designed and is being fabricated by the AS&E High Energy Systems Division in Santa Clara, CA. Delivery is expected in October, 2000.

TABLE 1. 3.5 MeV Linac Design Parameters

Maximum Energy (fixed)	3.5 MeV
Dose Rate 1m From Target	100 R/min
Beam Spot on Target	0.5mm x 4mm
rf Frequency (X-band)	9303 MHz
Length of Accelerator (rf)	40 cm
Accel. Peak Current (max)	50 mA
Accel. Avg. Current (max)	50 µA
Avg. Elec.-Beam Pwr. (max)	200 W
Magnetron Peak Pwr. (nom)	1.0 MW
Pulse Voltage	35 kV
P. R. F. (nom.)	15 – 200 pps
Pulse Width	4 µs
Injection Voltage	17 kV
Power Required (wall-plug)	11 kVA

SUMMARY

A very compact X-band 3.5 MeV linac has been designed and built and will soon undergo acceptance tests. Its intrinsically small size and very small rectangular waveguide permit its integration into a relocatable cargo inspection system with very low leakage radiation.

REFERENCES

1. Leach, E. R. and Spradling, M. L., "The Systematic Testing of X-Ray Non-Intrusive Inspection Systems" in *Proceedings of the ONDCP International Technology Symposium (Nashua, 1995)*, pp. 6-17 – 6-28.

2. Sapp, W. W., Smith, G. J., Swift, R. D., "A Mobile X-Ray System for Non-Intrusive Inspection of Vehicles" in *Proceedings: Harnessing Technology to Support the National Drug Control Strategy, ONDCP International Technology Symposium (Chicago, 1997)*, pp. 14-19 – 14-25.

3. Smith, G. J., "X-Ray Inspection of Palletized Cargo for Security and Drug Interdiction" in *Proceedings: 13th Annual Security Technology Symposium and Exhibition (Virginia Beach, 1997)*, sponsored by ADPA/NSIA Security Division.

4. Swift, R. D., Lindquist, R. P., "Medium Energy X-Ray Examination of Commercial Trucks" in *Cargo Inspection Technologies*, edited by A. H. Lawrence, SPIE 2276 (San Diego, 1994).

5. Sapp, W. W., Rothschild, P. J., Schueller, R., Mishin, A. V., "New, Low-Dose 1 MeV Cargo Inspection System with Backscatter Imaging" in *Penetrating Radiation Systems and Applications II*, (to be published), SPIE 4142 (San Diego, 2000).

6. Sapp, W. W., Huang, S., "X-Ray Penetration for Cargo Inspection: A Semi-Empirical Formula" in *Proceedings of the ONDCP International Technology Symposium (Washington, 1999)*, pp. 22-18 – 22-26.

NELIS – An Illicit Drug Detection System

P. A. Dokhale[*], J. Csikai[+], P. C. Womble, and G. Vourvopoulos

Applied Physics Institute, Western Kentucky University, 1 Big Red Way, Bowling Green, KY 42101

Abstract. NELIS (Neutron Elemental Inspection System) is currently being developed to inspect pallets laden with various commodities for contraband (drugs, etc.). NELIS analyzes the characteristic gamma rays from the elements in drugs such as C, O, H, Cl, N, etc. that are produced by nuclear reactions from fast and thermal neutrons (Pulsed Fast/Thermal Neutron Analysis). Hidden drugs are identified through the measurement of the elemental content of the object, and the comparison of expected and measured elemental ratios. NELIS can be coupled with conventional X-ray imaging system to optimize the inspection capabilities at ports of entry.

INTRODUCTION

In principle, neutrons impinging upon an object can induce any of several nuclear reactions in the elements found in the object. These reactions, which include elastic scattering, (n,γ), $(n,p\gamma)$, $(n,n'\gamma)$, and neutron activation, can be utilized to identify the elements of interest by examining either the energy of the scattered neutrons or the characteristic γ-rays emitted. The penetrating ability of neutrons and γ-rays provides an effective way for measuring the elemental content of an interrogated material. Oil exploration[1], coal bulk analysis[2], cement analysis[3], and mineral exploration[4] have extensively used neutrons for elemental characterization. In the last few years, neutron based systems for the detection of contraband materials such as explosives[5,6] and drugs[7,8] have been under development.

The Pulsed Fast/Thermal Neutron Analysis (PFTNA) technique is a bulk analysis technique for elemental characterization[5]. The PFTNA method utilizes a pulsing d-T neutron generator, which has a neutron pulse duration of a few microseconds and a pulsing frequency of a few kHz. The 14 MeV neutrons produced by the generator can initiate fast reactions with such elements as C and O contained within the object under interrogation. The elements are identified through characteristic γ-rays (4.43 MeV for C and 6.13 MeV for O). By allowing the fast neutrons to moderate to thermal energies, other elements such as Cl, S, and Ca can be measured via the (n,γ) reaction. Finally, by shutting off the generator for a few minutes, neutron activation analysis can be performed to quantify elements such as Na, Si, etc. It is thus possible to identify a large number of elements by utilizing the reaction that will measure a particular element most effectively.

In 1999, we developed a pulsed neutron system called PELAN (Pulsed ELemental Analysis with Neutrons[5] for the detection of explosives. PELAN successfully passed its field trials in August 1999, detecting the presence of several explosives. Based on these results, and after several discussions with ONDCP, DEA, and US Customs, it was proposed that, drawing on the experience gained from PELAN, we should concentrate our efforts in detecting illicit drugs hidden in bulk media or in items shipped on pallets. In the Fall of 1999 period, we concentrated in taking data using various commodities. Drug simulants (having the same elemental composition as drugs) were hidden in various locations in these commodities. The amount of simulant varied between 2 and 3 kg, and in several instances, was dispersed in smaller amounts within the pallet. The figures below show the configuration used in the various experiments. In the cement case, (Figure 1), the neutron generator is located on top of the cement bags and the gamma-ray

---

[*] Permanent Address: Ward Center for Nuclear Sciences, Cornell University, Ithaca, NY-14853
[+] Permanent Address: Institute of Experimental Physics, University of Debrecen, H-4001 Pf. 105. Bem ter 18/A, Debrecen, Hungary

detector below the bags. In the other three cases, (Figures 2,3,4) the neutron generator is on the right, while the gamma-ray detector is on the left. In all cases, we were able to detect the changes in the elemental content of the commodity whenever part of the bulk material was replaced with a drug simulant. In the case of the soft drinks, the drug simulant was in small packages, and several of these were placed at various, random locations within the interrogated volume. Irrespective of the position, the elemental content (or the C/O ratio) changes appreciably whenever drugs are included in the commodity.

FIGURE 1. Drug simulants hidden in cement.

FIGURE 2. Drug simulants hidden in clay pottery.

FIGURE 3. Drug simulants hidden in CocaCola™.

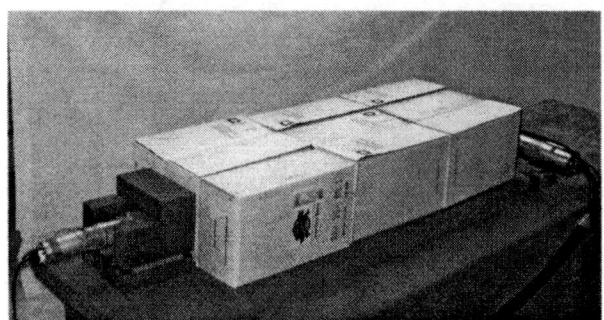

FIGURE 4. Drug simulants hidden in beans.

These results were shown to US Customs, and they suggested that we develop a prototype elemental pallet inspection system, tied to the prototype X-ray imaging system under construction.

NELIS

Based on the work that we have already performed and the needs of US Customs, a system called NELIS (Neutron ELemental Inspection System) is being developed. NELIS can inspect the contents of the pallet, looking for elemental compositions that are different from the "background" elemental composition. As "background" elemental composition, we define the elemental composition of the commodities on the pallet in the absence of hidden drugs within the commodity.

Science Applications International Corporation (SAIC) is developing a Pallet Inspection System for U. S. Customs. The system utilizes gamma rays to create a density image of the contents of a 1.2m x 1.2m x 2.4m pallet. The image is based on the VACIS system that SAIC has developed and scheduled to be deployed at several US entry points. The gamma-ray system is primarily looking for density anomalies (if the average atomic number Z is low), but is not able to identify whether the anomaly is due to the presence of drugs hidden within the pallet. In the case of high Z materials, if the bulk density is high, no effective imaging can be produced.

Figure 5 shows an artist's sketch of a Pallet Elemental Composition and Imaging System, composed of a VACIS gamma-ray system for imaging, and a NELIS neutron system for the elemental composition. The system is proposed to be transportable, with the two components being developed independently of each other. VACIS and NELIS are linked together at a Unified Workstation, allowing the operator to switch from a VACIS screen

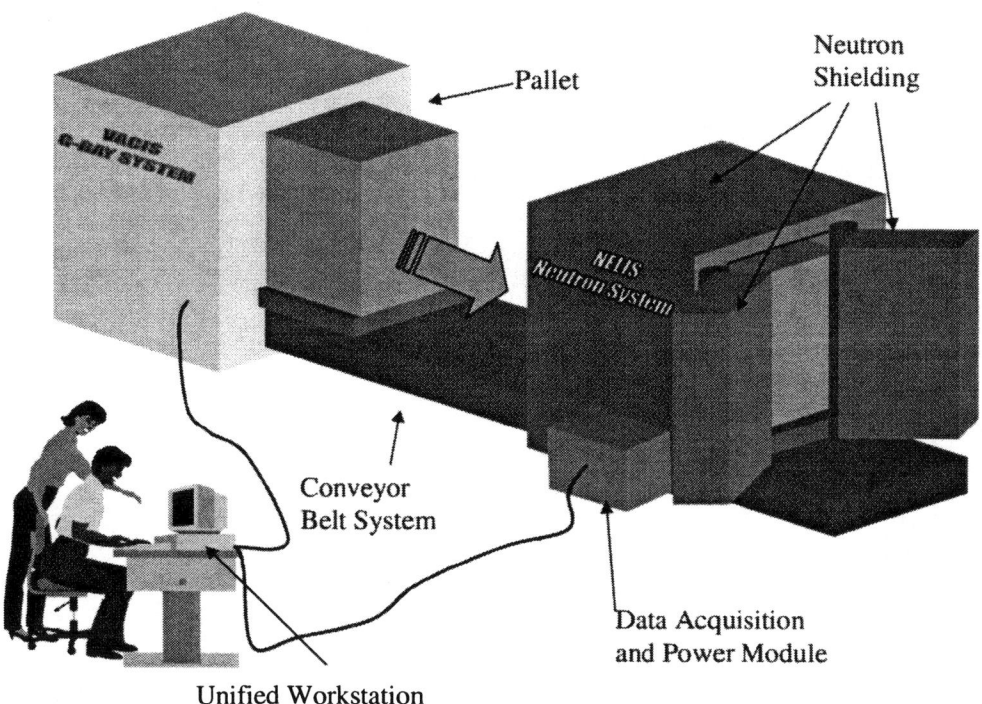

FIGURE 5. The Pallet Inspection System composed of a VACIS and a NELIS system.

to a NELIS screen and vice versa. This paper addresses only the development, construction, and testing of NELIS.

In order to limit the radiation exposure of personnel working around NELIS, an aluminum water tank 75-cm thick comprises NELIS's outer shell. The water tank serves also as a neutron moderator, creating thermal neutrons within the volume containing the pallet (see Figure 5). If NELIS is to be moved to another location, the water can be drained to minimize the weight of the system.

Upon the completion of the gamma-ray imaging by VACIS, the pallet moves to NELIS for elemental inspection. A pallet interrogation proceeds as follows:

1. A pallet enters NELIS and the shielding doors (Figure 5) close.

2. The neutron generator (see Figure 6) within NELIS is energized for a short period (up to 5 minutes). The neutrons are emitted in microsecond wide pulses, interacting primarily with the carbon and oxygen atoms in the commodity, emitting characteristic gamma rays. These gamma rays are detected with BGO detectors placed on the opposite side from the neutron generator (Figure 6). Between pulses, the fast neutrons emitted from the generator suffer multiple collisions, lose their energy and they slow down. The slowed down neutrons interact primarily with hydrogen, sulfur, chlorine, iron and various other chemical elements within the volume. The gamma rays emitted from these interactions are also measured with the same BGO detectors. This procedure of emitting fast neutrons and then slowing them down proceeds at a rate of 10,000 times/second. At the end of the interrogation period, the neutron generator is automatically switched off, stopping the production of neutrons. The acquired data are automatically analyzed within a few seconds in the Data Acquisition and Power Module (Figure 6). Upon the completion of the analysis, the results are fed to a computer code that contains a decision making tree. Based on a library already stored in the computer, the NELIS operator at the Unified Workstation (Figure 5) is informed automatically whether the elemental composition of the pallet conforms to the "background" elemental composition. If the analysis indicates that the elemental composition deviates

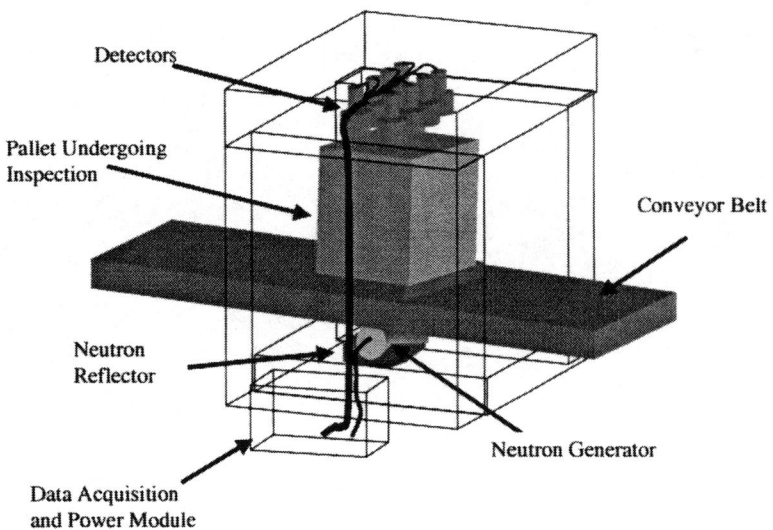

FIGURE 6. The NELIS System.

beyond the 95% confidence limit, the operator is warned accordingly.

SUMMARY

The Neutron ELemental Inspection System (NELIS) is being developed. NELIS is being built specifically to inspect pallets of commodities. The system can be used as a stand-alone device or can be utilized in conjunction with an x-ray inspection system. The system will compare the elemental content of the pallet under inspection to that of a well-established innocuous pallet of the same commodity. Any anomalies will be reported to the operator for further inspection.

ACKNOWLEDGMENTS

This research is supported by the Department of the Defense under contract DAAD07-98-C-0116.

REFERENCES

1. Scot, H. D., Stoller, C., Roscoe, B. A., Plasek, R. E., and Adolph, R., "A New Compensated Through-Tubing Carbon/Oxygen Tool for Use in Flowing Wells", Trans. Of the SPWLA *32nd Annual Logging Symposium*, Midland TX, June 1991, paper MM.

2. Kirchner, A. T., "On-line Analysis of Coal", *IAE Coal Research 40*, London, UK, Sept. 1991.

3. Baron, J. P., and Debray, L., "Potential of Nuclear Techniques for On-line Bulk Analysis in the Mineral Industry", *Applications of Nuclear Techniques*, Eds. G. Vourvopoulos and T. Paradellis, World Scientific Press, Singapore, 1991, p. 268.

4. Holmes, R. J, *Nucl. Geophys.* 1 (1), 41, 1987.

5. Womble, P. C., Vourvopoulos, G., and Paschal, J., "Multi-Element Analysis Utilizing Pulsed Fast/Thermal Neutron Analysis for Contraband Detection", SPIE Conference, Denver, on "Penetrating Radiation Systems and Applications", SPIE Vol. 3769 (1999) 189

6. Womble, P. C., Vourvopoulos, G., and Paschal, J., "PELAN: A Pulsed Neutron Portable Probe for UXO and Landmine Identification", SPIE Conference on "Penetrating Radiation Systems and Applications", SPIE Conference, San Diego, in press.

7. Vourvopoulos, G., and Thornton, J., "A Transportable, Neutron-Based Contraband Detection System" in Proc. of Counterdrug Law Enforcement: Applied Technology for Improved Operational Effectiveness, Nashua, NH, p.2-39, 1995.

8. Womble, P. C., Vourvopoulos, G., Dokhale, P. A., Ball Howard, J., and Paschal, J., "Neutron-Based Portable Drug Probe" in *Proc. Of Counterdrug Law Enforcement: Applied Technology for Improved Operational Effectiveness*, Washington, D.C., 6-1, 1999.

A Commercial Elemental On-line Coal Analyzer Using Pulsed Neutrons

Michael Belbot, George Vouvopoulos, and Jonathan Paschal

Applied Physics Institute, Western Kentucky University, 1 Big Red Way, Bowling Green, KY 42101, USA

Abstract. Because of its heterogeneity and the delay involved, traditional laboratory analysis of coal samples does not allow real time control of coal bulk parameters. Large excursions in important parameters (such as sulfur or calorific content) can be expensive and can be avoided with an on-line coal analyzer. The system that we developed utilizes nuclear reactions produced from fast and thermal neutrons and from neutron activation producing isotopes with half-lives longer than a few seconds. Characteristic gamma rays detected with BGO (bismuth germanate) detectors are used for the identification of the various chemical elements. The main features of the analyzer are elemental self-calibration independent of the coal seam; better accuracy in the determination of elements such as carbon, oxygen, and sodium; and diminished radiation risk. A prototype coal analyzer has been built and the first commercial model is currently being developed.

INTRODUCTION

A fast, accurate, real-time method of determining the elemental composition of coal is important to the coal industry for pricing, quality control, and reduction of SO_2 emissions. The calorific value (BTU/lb) coal is not measured directly, but determined through an algorithm that uses elements such as C, O, H, and S. Other elements such as Si, Ca, Al, Fe can be measured to determine the ash composition. Measurement of S is dictated by the Clean Air Act Amendments [1] that requires control of SO_2 emissions by coal-fired power plants. Sodium and Cl cause fouling and slagging that reduce boiler efficiency.

In a previous publication [2] we stated that blending was the method chosen by nearly 50% of the coal-fired power plants affected by the Clean Air Act. Continuous control of coal quality by blending different types of coal can only be done through a real-time determination of coal composition. This is unobtainable through chemical ASTM (American Society for Testing and Materials [3]) analyses, because of the time required in obtaining results.

Blending is often done to control fuel cost. For example, coal-fired plants operate at lowest fuel cost at a specified target BTU/lb value. Another example is compliance to environmental regulations, most notably SO_2 emissions. The sulfur and calorific content of coal vary depending on the region where coal is mined. By adding controlled quantities of various seams, both the BTU/lb and S content can be controlled at minimum cost. A real-time method of determining the elemental coal composition can allow the user to operate with a much smaller safety margin in the blend and thus use less coal from the more expensive seam. Nuclear techniques using interactions of neutrons with coal fulfill this need, because the coal is analyzed within minutes as it travels on a conveyor belt or as it falls in a chute. These neutron interactions produce gamma rays that have energies unique to each element (for example, 1633 keV for Na and 5420 keV for S).

Currently industrial coal analyzers use a radio-isotopic neutron source such as ^{252}Cf that employs the Prompt Gamma Neutron Activation Analysis technique (PGNAA). We have used the Pulsed Fast Thermal Neutron Analysis (PFTNA) technique [4, 5, 6] using a pulsed deuterium-tritium (d-T) sealed tube generator that produces 14 MeV neutrons. We have chosen this type of neutron source, because it enables one to use both fast and thermal neutron reactions. By separating the fast from thermal reactions by using a pulsed source [6], one can measure more elements and to measure them more accurately.

In this paper, we discuss a commercial on-line coal analyzer that is being developed and built at Western Kentucky University that employs the PFTNA technique. The commercial analyzer is based on a prototype that has already been built at Western Kentucky University (see [2]). In particular, the prototype demonstrated that a working model of the elemental analyzer using the PFTNA method could determine with good precision elements such as C, O, and Na. These elements are difficult to measure with a radioisotopic source.

PFTNA-BASED ELEMENTAL ANALYSIS

Coal continuously flows in a vertical chute that is irradiated with pulsed 14 MeV neutrons. During the neutron pulse, high-energy neutrons interact with the elements such as C and O emitting characteristic gamma rays. In between pulses, neutrons scatter off light elements in the coal and slow down to thermal energies. These lower-energy neutrons initiate thermal capture reactions with elements such as H, S, and Cl emitting gamma rays characteristic of these elements. Neutron activation is used for the measurement of Na, producing isotopes that have longer half-lives (on the order of seconds) than the fast and thermal capture reactions. The gamma rays produced from each type of nuclear reaction (fast neutron, thermal neutron, and activation) are acquired and stored in different spectra. This reduces the background as compared with the spectra taken with a radioisotopic source. The analysis of the experimental data was performed using SPIDER [7] a de-convolution computer code developed for the automatic extraction of the intensities of the characteristic gamma rays. See [2] and [6] for more details on SPIDER and PFTNA elemental analysis.

COMMERCIAL COAL ANALYZER

The commercial model of the elemental coal analyzer is housed in a temperature controlled land-sea container that is about 2.4 m (8 feet) high, 2.6 m (8.5 feet) wide and 4.0 m (13 feet) long. Figure 1 below is a cutaway view that shows the outside and internal structure of the coal analyzer. The analyzer is sufficiently shielded so that the emitted neutron radiation is below occupational limit immediately outside the container in areas where personnel will be present. Other areas will be exclusion zones while the analyzer is in operation. This permits maintenance and other work to be performed near the analyzer while it is in operation. A sampler and conveyor belt are used to divert coal from the main conveyor belt and feed it to the top of a vertical chute (main chute) inside the analyzer. An exit conveyor below the analyzer returns the analyzed coal back to the main conveyor belt. The speeds of the conveyor belt below the analyzer is adjusted so that the main chute is always filled to the top with coal. The analyzer can handle coal up to 10 cm (4 inches) in size at a maximum flow rate of 300 tons/hr.

The main chute is surrounded by a neutron generator, several gamma ray detectors, and a neutron detector for monitoring the output of the neutron generator. All aspects of the operation of the analyzer, such as data collection, analysis and display of data, movement of coal, diagnostics, and so forth, are computerized and controlled by a single master computer program. This program has an extensive set of diagnostics that monitor all devices associated with the analyzer and alert the operator of any errors with the appropriate corrective action. The analyzer is controlled through a touch screen that displays the results of the data analysis and the diagnostics. Details regarding the above software have been previously discussed in [2]. The master program allows remote control of the analyzer so that the operator need not be located near it. The analyzer is made more compact by eliminating a control room.

FIGURE 1. Cutaway view of commercial coal analyzer.

PERFORMANCE OF PROTOTYPE COAL ANALYZER

One of the reasons for building the prototype coal analyzer [2] was to determine the performance of an on-line analyzer using the PFTNA technique. Figure 2 below shows a plot of coal content for various parameters as determined by ASTM methods versus that measured by the analyzer. These plots demonstrate the performance of the prototype. The diagonal line in each plot indicates one to one correspondence.

It is important to note that the calibration of the analyzer [2] does not depend on prior knowledge of coal composition, because it is derived from measured parameters. Thus for all the points in figure 2, no recalibration was done. The points of figure 2 span a wide range of values, and their linearity support that the calibration is universal for all types of coal.

Sulfur and Na points in figure 2 indicate they are measured accurately. Calorific value has a larger standard deviation than the ASTM method. It should be noted, however, that accuracy of all parameters is the same regardless of whether a single seam or a blend is analyzed. This is because the calibration does not depend on prior knowledge of coal composition. Although the accuracy of moisture content is limited, it is roughly the same as that obtained with commercially available microwave meters. The on-line analyzer has the advantage over the commonly used microwave-based moisture meters that it can measure moisture even when the coal is frozen. This is because moisture is measured through a nuclear reaction with the hydrogen nuclei that is unaffected by whether the water is frozen.

EXAMPLE OF COAL ANALYZER USE

As stated in the introduction, coal blending is a widely used procedure for quality control in coal-fired power plants. An on-line elemental analyzer can be used to control such blending within narrow limits, so that substantial saving can be realized. An example using actual data is presented below. Figure 3 shows data from a power plant for about 125 days of operation. The data points are the calorific values in BTU/lb obtained by sampling the blended coal and analyzing it in a laboratory using the ASTM method.

The two blended coals shall be called "coal A" and "coal B". The calorific value and cost of coal A is about 11900 BTU/lb and $33.35/ton. That of coal B is about 8750 BTU/lb and $17.50/ton. The ideal blend where the plant runs efficiently at lowest fuel cost is 70% coal A and 30% coal B corresponding to a calorific value of 10950 BTU/lb for the blend. The average calorific value for the points shown in figure 3 is about 11400 BTU/lb. This corresponds to a blend of 84% coal A and 16% coal B with a composite cost of $30.81/ton. The precision of the calorific measurement of the on-line analyzer is 150 BTU/lb. If the on-line analyzer were used to control the blend, the worst case is where the calorific value of the blend is always 150 BTU/lb. higher than the desired value (10950 BTU/lb). The calorific value of that blend would be 11100 BTU/lb, which corresponds to a blend of 75% coal A and 25% coal B at a cost of $29.39/ton. Based on the data of figure 3, for every million tons of

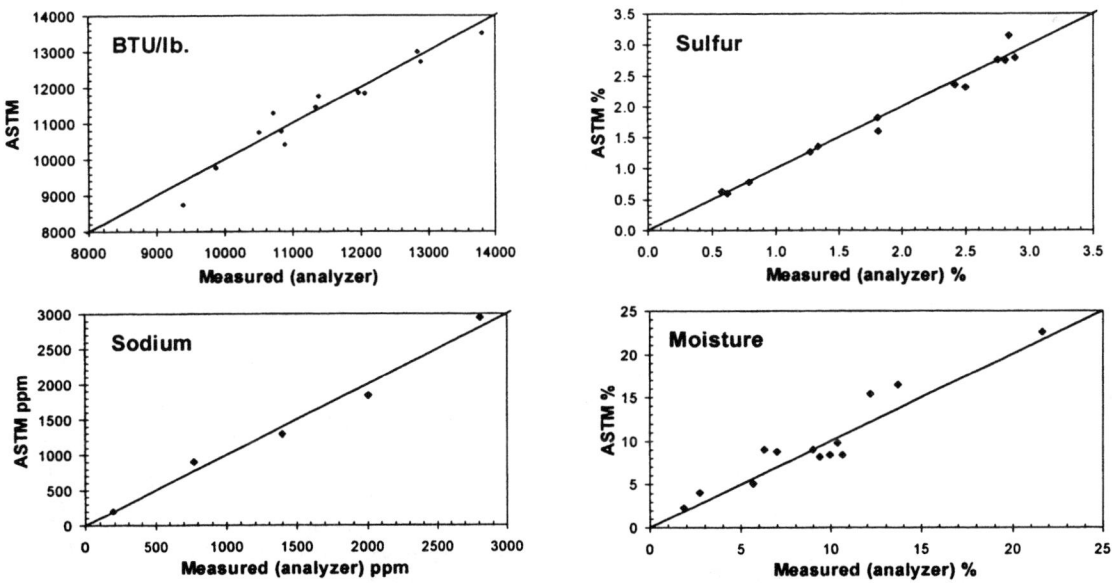

FIGURE 2. Plots of coal content for selected parameters measured by ASTM methods versus that measured by the prototype coal analyzer. Diagonal lines represent one to one correspondence.

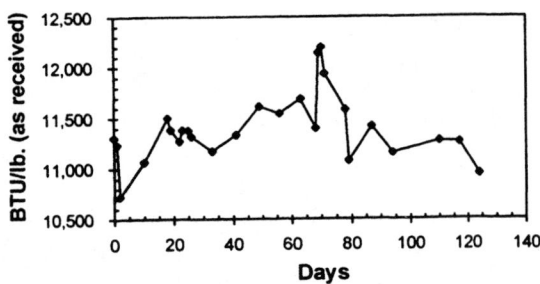

FIGURE 3. Data of a blend burned by a coal-fired power plant showing BTU/lb. (as received) versus days of operation.

CONCLUSION

Although commercial analyzers using ^{252}Cf based systems already exist, there many advantages to using pulsed neutrons. Carbon and O content can be directly measured over their complete ranges in coal without the need for recalibration. Sodium in coal, which causes fouling and slagging, can be measured with an on-line analyzer that uses PFTNA, which cannot be done with ^{252}Cf-based analyzers. Concerning the neutron generator itself, when it needs replacement, a new one can be shipped by common carrier without special shielding requirements. When de-energized, the radiation emitted from the beta decay of the tritium target is entirely stopped by the rugged stainless steel container in which the (D-T) neutron generator tube is housed. Unlike radioisotopic sources, the output of the neutron generator decays only when used. A ^{252}Cf source, on the other hand, has a half-life of 2 ½ years whether it is used or not.

The direct measurement of C and O means that calorific values can be measured without recalibrating for each particular seam analyzed. This cannot be done with ^{252}Cf based analyzers. For these types of on-line analyzers, an accurate analysis of BTU/lb content of blends of coal is impossible, unless the composition of the seams and the ratio of the blend are already known. Requiring this precludes the possibility of accurately blending different types of coal to control its calorific value.

We have shown that an on-line elemental analyzer can produce substantial financial savings to a coal-fired power plant. Because of the large amount of coal consumed each year by coal-fired power plants, a reduction of a few percent in fuel cost can result in savings that can easily justify the installation of the analyzer from a financial standpoint. The analyzer can also be utilized to save comparable amounts of money in compliance of SO_2 emissions, in optimization of boiler efficiency by controlling Na content, and in a number of other uses in coal quality and control.

ACKNOWLEDGEMENTS

We thank P. Huffine for his valuable help in sample preparation and operation and maintenance of the prototype. We also thank C. Campbell, S. Adelson, C. Simpson, and M. Young for their help in writing the acquisition code for both the prototype and commercial analyzer. The help of J. Riley and his staff with the ASTM analysis is appreciated. Research supported by the US Department of Energy Grant # DE-FC02-91ER75661, and the National Science Foundation Grant # 9760056 and #9901852.

REFERENCES

1. Dept. of Energy, Energy Information Administration publication, "The Effects of the Title IV Clean Air Act Amendments of the 1990 on Electric Utilities: an Update", 1997; see web address: "http://www.eia.doe.gov/ environment.html" for a summary and complete text.
2. Belbot, M.. Vourvopoulos, G., Womble, P., Paschal, J., "Elemental online coal analysis using pulsed neutrons", *Proceedings of the International Society of Optical Engineers*, edited by F. Patrick Doty, **3769**, 168-177 (1999).
3. *Annual Book of ASTM Standards, Section 5: Petroleum Products, Lubricants, and Fossil Fuels*, West Conshohocken, PA, 1998, **Vol. 5.01 – 5.05**.
4. Vourvopoulos, G., "Industrial On-Line Bulk Analysis Using Nuclear Technologies", *Nucl. Instr. Methods*, **B56/57**, 917-920 (1991).
5. Vourvopoulos, G., "Multi-parameter On-line Coal Bulk Analysis", *J. Coal Quality*, **12(2-3)**, 96-101 (1993).
6. Dep, L., Belbot, M., Vourvopoulos, G., and Sudar, S., "Pulsed neutron-based on-line coal analysis", *J. Radioanalytical Nucl. Chem.*, **234** 107-112 (1998).
7. Computer Code SPIDER (SPectral Interpolation and DEconvolution Routine), written by Sudar, S. and Paschal, J., Western Kentucky University, version 3.1, 1997.
8. *Keystone Coal Industrial Manual*, edited by A. P. Sandra, Intertec Publishing, Chicago, 1997, pp. 135-179.

Detection of Explosives With the PELAN System

P. C. Womble, C. Campbell, G. Vourvopoulos, J. Paschal, Z. Gácsi, and S. Hui[*]

Applied Physics Institute, Western Kentucky University, 1 Big Red Way, Bowling Green, KY 42101

Abstract. PELAN (Pulsed ELemental Analysis with Neutrons) is a small portable system for the detection of explosives. PELAN weighs less than 45 kg and is man portable. It is based on the principle that explosives and other contraband contain various chemical elements such as H, C, N, O, etc. in quantities and ratios that differentiate them from other innocuous substances. The pulsed neutrons are produced with a 14 MeV (d-T) neutron generator. Separate gamma-ray spectra from fast neutron, thermal neutron and activation reactions are accumulated and analyzed to determine elemental content. Data analysis is performed in an automatic manner and a final result of whether a threat is present is returned to the operator. PELAN has successfully undergone field demonstrations for explosive detection.

INTRODUCTION

Explosives have been shown[1] to be differentiated from innocuous materials and other contraband materials by the utilization of ratios of chemical elements e.g. C/O ratio and N/O ratios. The problem of identifying explosives is thus reduced to the problem of elemental identification.

Depending on the chemical elements that one wishes to measure, one might have to use neutrons of several energies. In many of the neutron-based applications for explosives currently in use, radioisotopic sources (Am-Be, ^{252}Cf) are utilized for neutron production. These sources can excite a host of chemical elements (H, C, S, Fe, Cl, etc.) through neutron capture reactions. However, there are other elements such as C and O which need neutron energies several MeV higher than those available from the radioactive sources. To satisfy this, a neutron source is required that can produce the high energy neutrons for measurement of elements such as C and O, and low energy (0.025 eV) for elements such as H and Cl. This can be accomplished with a pulsed neutron generator. This technique is called Pulsed Fast/Thermal Neutron Analysis (PFTNA).

The basis of PFTNA is a pulsed neutron generator utilizing the deuterium-tritium (d-T) reaction. The pulsed d-T neutron generator provides 14 MeV neutrons which in turn initiate several types of nuclear reactions ((n,n'γ), (n,pγ), (n,γ) etc.) on the object under scrutiny. The γ rays from these reactions are detected by a suitable set of detectors (usually bismuth germanate (BGO) scintillators). During the neutron pulse, the γ-ray spectrum is primarily composed of γ rays from the (n,n'γ) and (n,pγ) reactions on elements such as C and O, and is stored at a particular memory location within the data acquisition system. Between pulses, some of the fast neutrons that are still within the object lose energy by collisions with light elements composing the object. When the neutrons have energy less than 1 eV, they are captured by such elements as H, N, and Fe through (n, γ) reactions. The γ rays from this set of reactions are detected by the same set of detectors but stored at a different memory address within the data acquisition system. This procedure is repeated with a frequency of approximately 10 kHz. After a predetermined number of pulses, there is a longer pause that allows the detection of γ rays emitted from elements such as Si and P that have been activated. Therefore, by utilizing fast neutron reactions, neutron capture reactions, and activation analysis, a large number of elements contained in an

---

[*] Present Address: Department of Physics, 1396 Physics Building, Purdue University, W. Lafayette, IN 47907

object can be identified. Figure 1 shows the time sequence of the nuclear reactions taking place.

PFTNA uses low resolution, high Z detectors such as bismuth germanate (BGO) or gadolinium ortho-silicate (GSO). Data analysis of the resulting γ-ray spectra is performed with the computer code SPIDER, a spectrum deconvolution code[2] developed for the Windows 95/98/NT platforms.

THE PELAN SYSTEM FOR EXPLOSIVES

A PFTNA-based device called PELAN (Pulsed ELemental Analysis with Neutrons) was built for the characterization of explosives. PELAN is a small man-portable device, composed of a suitcase that contains the necessary power supplies for the neutron generator and the data acquisition system, and of a probe which is placed next to the object under interrogation (see Figure 2). The probe contains the neutron generator tube (upper horizontal tube in Fig. 2), the BGO γ-ray detector (lower horizontal tube), and the necessary material to shield the detector from the neutrons (vertical tube). The total mass of the probe and suitcase is less than 45 kg.

PELAN is controlled with a palm-top computer (Figure 2), connected to the PELAN suitcase with a 15-m data cable. The palm top provides fully automatic operation of PELAN. With a single touch command, all necessary power supplies are energized, neutrons are produced, and data accumulated for a

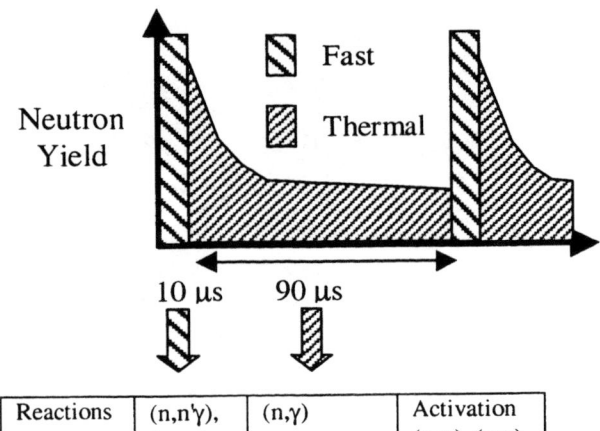

Reactions	(n,n'γ), (n,pγ)	(n,γ)	Activation (n,α), (n,p)
Gamma	Prompt	Prompt	Delayed
Elements	C, O	H, S, Cl, Fe	O, P, F

FIGURE 1. Pulsed neutron generator time sequence.

predetermined time. Upon the completion of data acquisition, the data are automatically reduced, analyzed, and the results of the interrogation are displayed on the palm-top screen.

Double-blind field trials with high explosives proved PELAN's ability to detect explosives in different scenarios. PELAN was operated in an automatic mode, without interference from the operator. The data acquisition and analysis were performed automatically by the PELAN software and a result was then returned to the operator.

FIGURE 2. The PELAN System.

The decision tree for explosives detection with the PELAN was based on previous measurements of elemental ratios, shown in Table 1. It should be emphasized that decision-making trees are not the same for all the various conditions under which explosives can be found. For example, a decision making tree for unexploded ordnance on the ground is different than one for unexploded ordnance in water.

TABLE 1. Measurements of elemental ratios of explosives with PELAN.

Material	Expected		Measured	
	C/O	N/O	C/O	N/O
C-4	0.71	1.0	0.74 ± 0.03	1.0 ± 0.1
TNT	1.2	0.5	1.11 ± 0.07	0.5 ± 0.2
RDX	0.53	1.0	0.56 ± 0.02	1.2 ± 0.1

RECENT ADVANCES IN PELAN

Temperature Stabilization of the BGO Detector

When a BGO crystal's temperature increases its light output decreases. In a practical sense, this decrease in light output effectively changes the gain of the amplification system. Therefore, the position of the various peaks in a spectrum will change. This process is reversible i.e. cooling the detector to its original temperature will return the peaks to their original position. A common method to compensate for this temperature-dependency is to utilize refrigerators to maintain a constant temperature on the crystal. For a portable system, this solution is not an option due to the increased bulk and weight.

An automatic system to track the temperature changes and adjust the gain of the amplification system to compensate for the current temperature conditions of the BGO crystal has been designed. The system consists of a temperature probe that is connected to a data acquisition card. An automated computer program monitors the change in temperature of the temperature probe.

The change in the light output of the BGO crystal as a function of temperature is well-documented. The exact function is given in Melcher et al.[3] and can also be found at the web-site for Bicron.[4] At room temperature, the light output (and by necessity, the gain of the amplification system) changes approximately 1% for every 1°C. Therefore, if the temperature of the detector is known, the gain of the amplifier can be adjusted to compensate for any changes in temperature. Figure 3 shows typical gain variations as the automatic system responds to a wide range of temperatures. The temperature of the detector was varied between −5° C to +45° C. In Figure 3, we can see that despite the wide fluctuations in temperature, the automatic system keeps the position of the centroid of a 662 keV ^{137}Cs γ-ray within 7%.

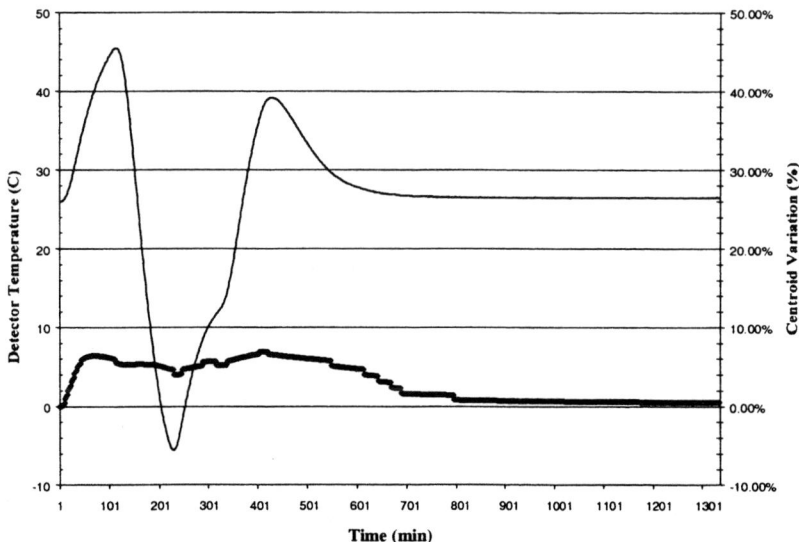

FIGURE 3. The thin line and left axis indicate the temperature of the detector. The dark data points and right axis indicate the variation of centroid position after the gain has been corrected for temperature.

Decision Making Algorithms

As mentioned above, individual substances can be differentiated by the ratios of chemical elements (C/O, C/N, etc.). However, most objects encountered are combinations of chemicals, not single chemicals. Also, the uncollimated detector in PELAN measures both the sample and its environment. The question is which decision-making algorithms can best determine whether a threat is present when innocuous materials surround the threat.

A large number of measurements have been taken of innocuous clutter with and without an explosive simulant present. The innocuous clutter was composed of clothes, tools, food-stuffs, and other common materials. The ratio, N/H, for a particular measurement is plotted versus C/O for that measurement in Figure 4. The various types of explosive simulant (C4, TNT, black-powder, dynamite) hidden within the innocuous clutter form groups in this figure along with a region that contains only innocuous material. Although for most of the cases the explosive simulants could be identified even in the presence of substantial clutter, the graph shows that there is a region where combinations of innocuous materials lie within these explosive groups. Thus, any set of Boolean logic statements must be complex and also must draw upon other ratios such as C/H, N/O, etc.

Currently, we are examining other methods such as fuzzy logic and neural networks for the identification of explosives. In some preliminary results, the fuzzy logic algorithm is more successful at identifying the explosive simulant present but is more likely to identify innocuous materials as explosives. Our work with neural networks is in its initial stages.

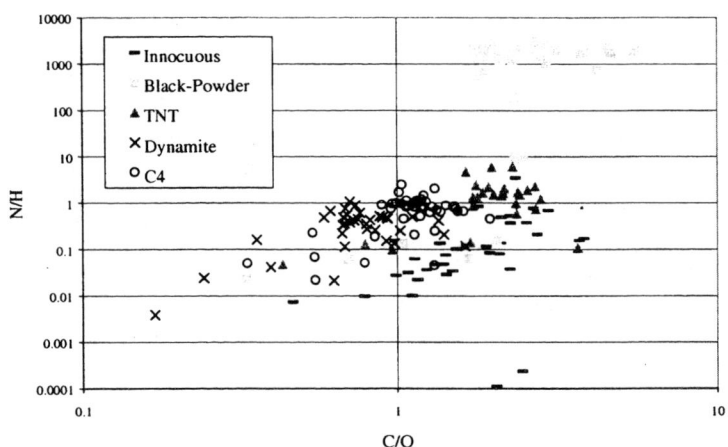

FIGURE 4. N/H versus C/O for a large number of measurements of innocuous materials with and without explosives.

ACKNOWLEDGMENTS

We would like to thank S. Adelson, J. James and G. Begtrup for their assistance in data collection.

This work was supported in part under by the Department of Defense under contracts DAAD07-98-C-0116 and DAAD05-98-C-022.

REFERENCES

1. Vourvopoulos G., Chemistry and Industry, 18 April 1994, p. 297-300.

2. Vourvopoulos, G., Dep, L., Sudar, S., Womble, P. C., and Schultz, F. J., "Neutron-Generator Based On-Line Coal Analysis: A Progress Report" in *Proc. 8th Int'l Conf. On Coal Science*, Eds. J.A. Pajares and J.M.D. Tascon, Elsevier Science B. V., Oviedo, Spain, 1995.

3. Melcher, C. L., Schweitzer, J. S., Manente, R. A., and Peterson, C. A., *IEEE Trans. On Nuc. Sci.*, **38**, 1991, 506-509.

4. Bicron web-site at http://www.bicron.com/bgo.htm.

A Portable System for Nuclear, Chemical Agent and Explosives Identification

W. E. Parker,[*] W.M. Buckley,[*] S.A. Kreek,[*] A. J. Caffrey,[‡] G.J. Mauger,[*] A. D. Lavietes,[*] and A. D. Dougan[*]

[*] *Lawrence Livermore National Laboratory, P. O. Box 808 Livermore, CA 94550*
[‡] *Idaho National Engineering and Environmental Laboratory, Idaho Falls, Idaho*

Abstract. The FRIS/PINS hybrid integrates the LLNL-developed Field Radionuclide Identification System (FRIS) with the INEEL-developed Portable Isotopic Neutron Spectroscopy (PINS) chemical assay system to yield a combined general radioisotope, special nuclear material, and chemical weapons/explosives detection and identification system. The PINS system uses a neutron source and a high-purity germanium γ-ray detector. The FRIS system uses an electromechanically cooled germanium detector and its own analysis software to detect and identify special nuclear material and other radioisotopes. The FRIS/PINS combined system also uses the electromechanically-cooled germanium detector. There is no other currently available integrated technology that can combine a prompt-gamma neutron-activation analysis capability for CWE with a passive radioisotope measurement and identification capability for special nuclear material.

INTRODUCTION

The Portable Isotopic Neutron Spectroscopy System (PINS) was developed for non-destructive evaluation of suspect chemical warfare material.[1,2] PINS performs its evaluation by neutron activation analysis. A high purity germanium (HPGe) crystal is used to detect characteristic gamma rays. A software package makes a determination about what chemical weapon or explosive material is present.

The Field Radionuclide Identification System (FRIS) was developed independently to identify any radioactive material.[3] FRIS uses an electromechanically cooled germanium detector (EMC HPGe) to detect passive γ-ray emissions. The EMC HPGe detector makes the system completely portable for long periods of time and significantly easier to deploy than systems which use liquid-nitrogen cooled germanium detector technology. The software associated with FRIS is a general purpose software identification package that has been tested extensively with liquid-nitrogen cooled germanium detectors.[4] Both the PINS and FRIS systems require the high-resolution γ-ray spectroscopy that is possible with high purity germanium detectors.

Both systems have been successfully field tested, and PINS has been extensively used for munitions identification.[5,6] The two systems can be combined to provide a complete general radioisotope, special nuclear material, and chemical weapons/explosives detection and identification system. The final system will consist of the EMC HPGe detector and one software package for data analysis.

DESCRIPTION OF THE TWO SYSTEMS: A. PINS

Neutrons from a radioisotopic ^{252}Cf source or from a sealed-tube neutron generator[7] are used to interrogate the contents of chemical weapons, explosives or a munition. Neutrons pass through the container material and are captured or scattered by the nucleus of one of the chemical elements within the munition.

Capture reactions (n,γ) and inelastic scattering reactions (n,n'γ) take place within the nuclei of the atoms that make up the chemical weapons material. The nuclei then emit high-energy gamma rays which are characteristic of the chemical element. A high-purity germanium (HPGe) detector detects gamma rays. Fig. 1 shows a good measurement setup, including the ^{252}Cf source, tungsten shadow blocks, the γ-ray collimator, and the HPGe detector.

The PINS system can identify the fills of most munitions in 1000 - 300 seconds or less, depending on the fill and size of the munition. PINS sensitivity is the highest for the various smoke fills, followed by the CW agents that contain chlorine, and then by the CW agents that contain phosphorus. PINS is least sensitive to explosive-filled items because of the relatively low neutron cross section of nitrogen.

The PINS software rechecks the energy calibration, identifies all of the γ-ray peaks in the

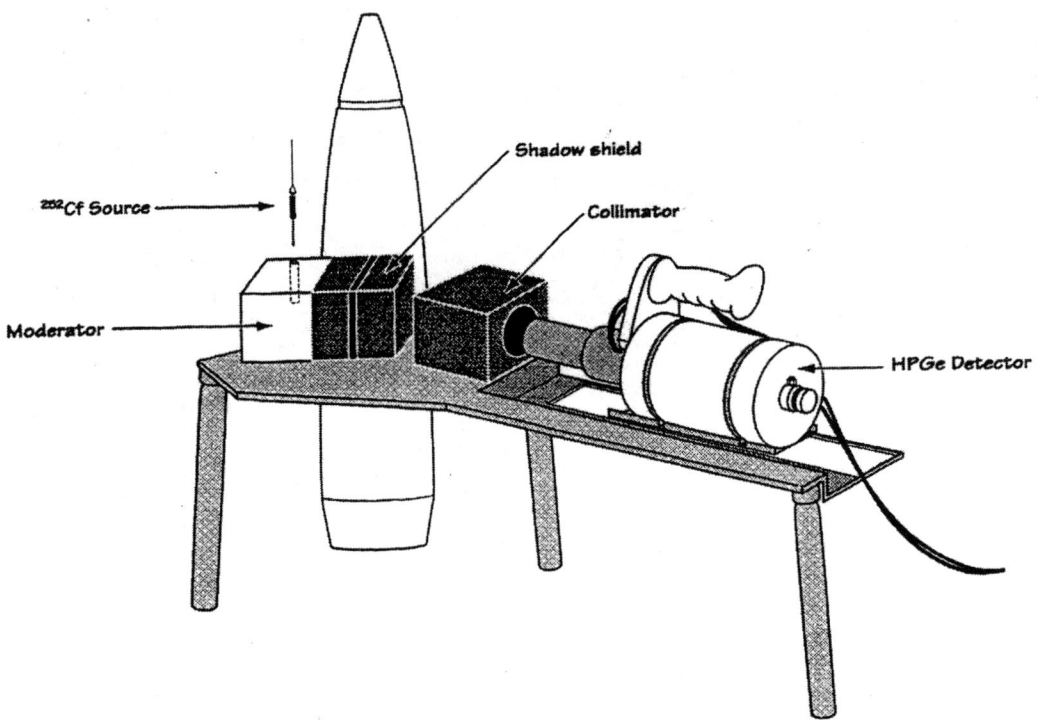

Figure 1. Experimental measurement setup for the PINS system. The ^{252}Cf source emits neutrons, which penetrate the container to illuminate the chemical weapon material. Tungsten shadow blocks protect the germanium detector from direct exposure to neutrons.

spectrum, and determines which chemical elements are present. The relative abundances of the chemical elements in the material are determined by the relative peak intensities associated with each γ-ray line. The PINS software then executes a decision–tree logic pathway to determine the item's fill. For field recovered suspect chemical weapons, the US Army Meteriel Assessment Review Board reviews the PINS data before making decisions regarding final disposition.

B. FRIS

The FRIS system, which couples the electromechanically cooled detector with general purpose γ-ray analysis software, was originally developed for US Customs to provide the ability to rapidly differentiate between special nuclear material and other radioactive materials being transported through U.S. ports of entry. The system is compact, portable, and capable of battery operation.

The electromechanically cooled detector is a self-calibrating, active-vibration-controlled electromechanical cooler coupled to a high purity germanium detector. It is the active-vibration control system that reduces the vibration to tolerable levels.[8]

The control system for FRIS operates on a Windows platform with a Visual Basic user interface. The time to process the data and to identify the radioisotopes is dependent on the strength of the source, distance between source and detector, and whether any shielding is present. For average sized sources found in standard field scenarios, 30 seconds is sufficient for data collection followed by a short time (5 - 10 seconds) to perform the analysis. The user interface reports either special nuclear material, industrial, medical radioisotopes, or "nothing to report." The user can click on any of these for further information regarding specific isotopes and confidence of reporting.

DATA

The first step in the integration of the PINS/FRIS systems was to verify that the electromechanically cooled detector would take data with sufficient resolution so that the PINS software could analyze that data. Introduction of the electromechanically cooled detector into the PINS system allows freedom from the requirements of liquid nitrogen. Measurements were made with mustard gas (HD) simulant. The experimental setup was the same as shown in Fig. 1 but with the EMC HPGe detector. For this test, the EMC HPGe detector was a 74% detector with resolution of 2.5 keV at 1332 keV (^{60}Co). Measurements with a LN cooled detector were made on the same day to insure a fair comparison. The LN cooled detector used was a 52% detector and had a 2.4 keV energy resolution at 1332 keV. The measurements were made with about 40% dead time, and data were collected for a total live time of 1000 seconds.

Figs. 2 and 3 and show the results of the γ-ray energy spectra for the mustard gas simulant collected with both detectors. Fig. 2 shows the 900 – 1400 keV energy range and Fig. 3 shows the 1900 – 2300 keV energy range. The only noticeable difference between the spectra collected with the two detectors is that the peak-to-background ratio is better for the LN cooled detector. The PINS software package correctly identified mustard gas for both tests.

As described in Figs. 2 and 3, the spectra clearly show the presence of chlorine and hydrogen, as well as aluminum, iron and silicon. Of particular interest in the identification of mustard gas are the chlorine peaks at 1164, 1950 and 1959 keV, a sulfur peak at 2230 keV, and the hydrogen peak at 2223 keV.

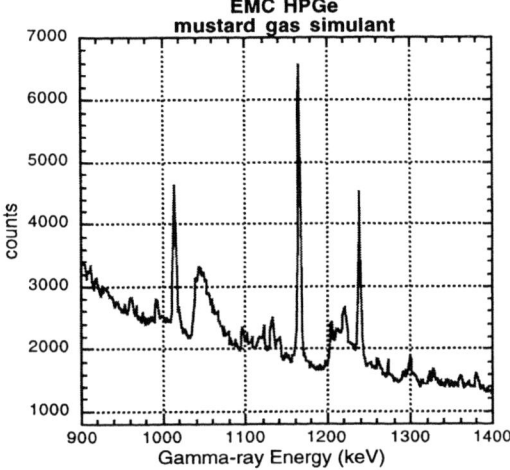 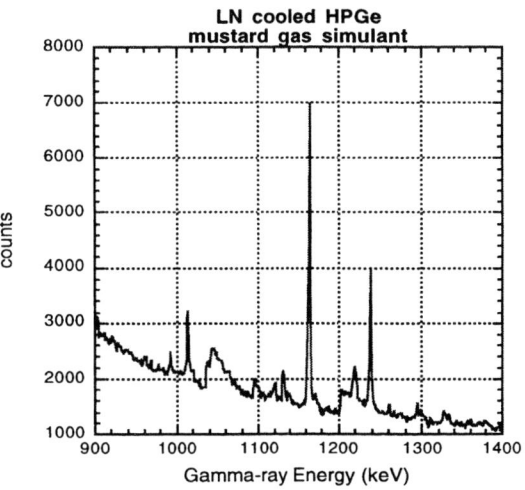

Figure 2. Mustard gas simulant from 900 – 1400 keV detected by the electromechanically cooled HPGe detector (left) and the LN cooled HPGe detector (right). The aluminum peak at 1014 keV is clearly seen, as is the germanium peak at 1039 keV, chlorine at 1164 keV, and iron at 1238 keV.

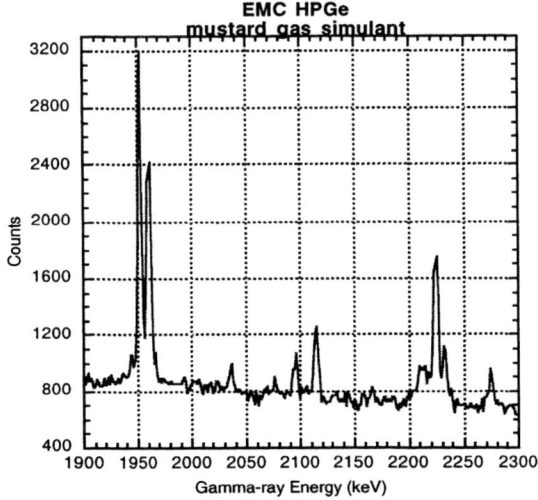 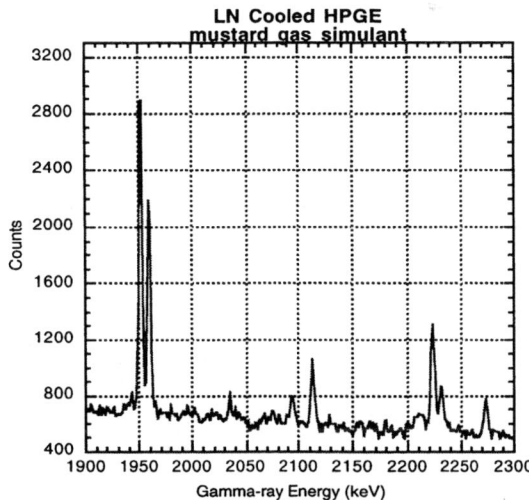

Figure 3. Mustard gas simulant in the 1900 – 2300 keV energy range detected by the EMC HPGe detector (left) and the liquid-nitrogen cooled HPGe detector (right). Two chlorine peaks can be seen at 1950 and 1959 keV, iron at 2113 keV, and hydrogen at 2223 keV. A sulfur peak can be seen at 2230 keV, just to the right of the H peak.

DISCUSSION

Our preliminary measurements indicate that the energy resolution of the electromechanically cooled germanium detector is sufficient for the PINS software to correctly analyze data. Our next set of tests will include explosives, and actual chemical agents instead of simulants.

We will then evaluate further integration of FRIS and PINS by selecting a software package that will analyze both the γ-ray data from radioactive material as well as the neutron induced radioactivity in PINS. One possibility is to use the current FRIS software for both chemical weapons/explosives and radioactive material identification. This will involve the development of an appropriate library and verification of the proper and adequate analysis of the (n,γ) and (n,n'γ) lines used in the PINS analysis.

Some fills, including nerve agent VX, can be reliably identified in as little as 100 seconds. However, in the case of high-explosive (HE) filled 75-mm munitions, 3,000 live-second data acquisition times are indeed necessary. Some military units therefore require 3,000 second counts for every single projectile, as the cost of repeating the assay is high. PINS will soon have improved analysis software that will inform the operator when the data are sufficient to identify the fill.

There are several additional features which would be desirable for the PINS/FRIS hybrid instrument. These modifications would improve its applicability for counter-terrorism applications. They are not required for the instrument to work, but would be most helpful for operation by personnel who do not have extensive detector technology training. Additional modifications would include automated hardware setup and simplified operation of the software, including automatic energy calibration for FRIS. For the simplified operation of the software, we need to develop a software shell with a graphical user interface to facilitate PINS/FRIS selection and to perform a set of operational checks and verifications. These operational checks can include: absorber, collimator, and shielding verification, as well as a check of neutron source presence or absence.

ACKNOWLEDGEMENTS

This work was performed under the auspices of the U.S. Department of Energy by Lawrence Livermore National Laboratory under Contract No. W-7405-Eng-48.

REFERENCES

1. A. J. Caffrey, J. D. Cole, R. J. Greenwood, *IEEE Transactions on Nuclear Science*, **39**, 1422 (1992).
2. A. J. Caffrey, B. D. Harlow, J. K. Hartwell, K. M. Krebs, G. D. McLaughlin, and A. L. Siedenstrang, "PINS Chemical Assay System Accuracy for Field-Recovered Munitions and Containers," Idaho National Engineering and Environmental Laboratory, INEEL/EXT-98-00218, April 1998.
3. G. J. Mauger, "Electromechanically Cooled High Purity Germanium Spectrometer System," p. 8 in the Lawrence Livermore National Laboratory Safeguards and Security Quarterly Progress Report to the U.S. Department of Energy, UCRL-ID-106454-99-1, January 1999.
4. R. Gunnink and J. B. Niday, "Computerized Quantitative Analysis By Gamma-Ray Spectrometry, Vol. I. Description of the GAMANAL Program," Lawrence Livermore National Laboratory, UCRL-51061, 1971.
5. A. L. Siedenstrang, G. D. McLaughlin, K. M. Krebs, J. K. Hartwell, R. J. Gehrke, and A. J. Caffrey, "PINS Chemical Assay System User's Manual," Idaho National Engineering and Environmental Laboratory, EG&G-PHY-10389, April 1994.
6. A. J. Caffrey, R. J. Gehrke, R. C. Greenwood, J. K. Hartwell, K. M. Krebs, G. D. McLaughlin, A. L. Siedenstrang, and K. D. Watts, "U.S. Army Experience with the PINS Chemical Assay System," Idaho National Engineering and Environmental Laboratory, EGG-NRP-11443, September 1994.
7. R. A. Alvarez, A. D. Dougan, M. R. Rowland, T.-F. Wang, *J. of Radioanal. and Nucl. Chem., Articles*, **192**, 73 (1995).
8. G. J. Mauger, W. E. Parker, B. B. Bandong, R. G. Lanier, and A. D. Lavietes, Proceedings of SPIE, "Penetrating Systems and Applications," F. P. Doty, Ed., pp. 43-50, Denver, Colorado, July, 1999.

Study of Cement Chemistry with Nuclear Resonant Reaction Analysis

J. S. Schweitzer*, R. A. Livingston[¶], C. Rolfs[§], H.-W. Becker[§] and S. Kubsky[§]

Department of Physics, University of Connecticut, Storrs, Connecticut 06269-3046
[¶]*Office of Infrastructure R&D, Federal Highway Administration, McLean, VA 22101*
[§]*Institut für Physik mit Ionenstrahlen, Ruhr-Universität Bochum, Bochum, Germany*

Abstract. Nuclear Resonance Reaction Analysis (NRRA) has been applied for the first time to measure the development of the hydrogen depth profile in the early stages of hydration of tricalcium silicate, the major constituent of Portland cement. To obtain the best spatial resolution, it is necessary to have good beam energy resolution. Three regions were observed in the profile, whose H concentrations were obtained by using the $^1H(^{15}N,\alpha\gamma)^{12}C$ reaction. By analogy with the hydration of alkali silica glasses, these are identified as the reaction-controlling surface layer, the gel layer, and the calcium-leached layer. The surface layer has an H concentration and thickness consistent with a few unit cells (1.1 nm) of tobermorite-like material. The inner regions exhibit diffusion-controlled growth with time until the hydrogen concentration approaches that of the surface layer at 4.25 ± 0.07 hrs. This event marks the end of the induction period.

INTRODUCTION

Le Chatelier discovered the basic process of Portland cement hydration over a century ago [1], but many details, particularly those about the kinetics of the reaction, remain uncertain. We report here the first use of nuclear resonance reaction analysis (NRRA) to measure the hydration depth profile and its development over the induction period.

The cementitious properties come from a reaction between water and Portland cement, yielding a calcium-silicate-hydrate gel (C-S-H), the binder in concrete, and calcium hydroxide [2]. The reaction equation is usually given in terms of tricalcium silicate, the main reactive constituent: $Ca_3SiO_5 + zH_2O \rightarrow Ca_xSi(OH)_y \bullet nH_2O + (3-x)Ca(OH)_2$. The Ca/Si ratio in the gel, x; hydroxyl content, y; and interlayer water content, n; all vary during the reaction. The values of y and n remain uncertain.

Thermal analysis data show that the kinetics of this reaction are very non-linear. After a brief rapid reaction period on the order of minutes, the rate slows down significantly for several hours, known as the induction period, during which the cement/water mixture stays relatively fluid. At the end of the induction period, the reaction rate accelerates rapidly and peaks at about 13 hours. By then, the cement has hardened and can no longer be worked. Recently the application of quasi-elastic neutron scattering has proven useful for investigating hydration kinetics [3], producing a detailed kinetic model of the post-induction period [4]. But, the exact mechanism that determines the induction period remains controversial.

Le Chatelier [1] proposed a through-solution process that depends on reaching supersaturation to end the induction period. Modern research favors instead a topochemical mechanism based on a surface-controlled rate-limiting step. Taylor compiled at least four different models for the induction period [2]. In general, it is proposed that a surface layer initially forms, inhibiting the reaction between the porewater solution and tricalcium silicate. After a period of time, the film layer breaks down, allowing the reaction to proceed at its maximum rate. The mechanism for the film breakdown may either be changes in the surface layer itself or sufficient alteration of the silicate substrate to cause a breakthrough.

The surface layer on hydrating tricalcium silicate has been examined by surface analysis methods including scanning electron microscopy, transmission electron microscopy, atomic force microscopy, and ESCA. These methods provide images of surface layer morphology. ESCA gives the calcium/silicon ratio.

As a complementary technique we have applied NRRA because of its inherent capability to provide in-situ measurements of the total hydrogen quantity and its depth distribution in a sample, as the reaction cross-section is significant only over a very narrow energy range. We used the E_R = 6.400 MeV resonance in $^1H(^{15}N,\alpha)\gamma^{12}C$ (width Γ_R = 1.44 keV, strength $\omega\gamma$ = 21.1 eV) [5]. This provides a H-detection sensitivity to about 10 ppm and a H-depth resolution of a few nm at the surface. While this technique has been applied to measure hydration reactions of several silicate-based materials including nuclear waste glass [6] and of medieval stained glass [7], this is the first attempt to measure the hydrogen depth profile in highly reactive cementitious materials to study the time evolution of the hydrogen distribution during the induction period.

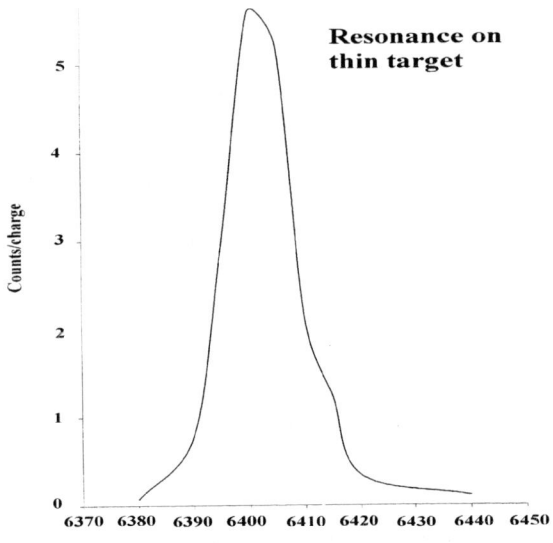

FIGURE 1. Measured total width from beam interaction with a monolayer of hydrogen.

EXPERIMENTAL PROCEDURE AND ANALYTICAL TECHNIQUE

The pellets were pure tricalcium silicate rather than Portland cement, which would contain several phases each with different hydration kinetics. The tricalcium silicate was fired, ground, and refired until only the triclinic phase was detected by XRD. The composition from XRF was 99.1% C_3S with 0.18% Fe_2O_3, 0.18%, SrO 0.17% Al_2O_3 and 0.14% MgO, with others <0.09%. Usually, tricalcium silicate is ground to a fine powder, but here, to provide a simple one-dimensional geometry, the powder was pressed into cylindrical pellets and sintered at ~1600 °C yielding a flat, smooth surface for hydration and ion beam irradiation. The 12.7-mm diameter allowed a relatively broad (3 mm) ion beam, to reduce localized heating in the pellet. The 5 mm pellet thickness was an infinite depth.

Several pellets were placed in a saturated calcium hydroxide solution and single pellets were removed at specific times for analysis. The pellets were stored and handled under inert nitrogen or argon until placed in the hydration bath to prevent premature hydration of the surface. To avoid reaction with carbon dioxide, the calcium hydroxide solution was made up with distilled deionized water. During the hydration period, the bath was purged with and kept under nitrogen gas at positive pressure. The bath was kept at 20 °C to compare with isothermal hydration kinetics data from other experiments such as quasi-elastic neutron scattering [4]. The hydration reaction is quenched so the samples can be placed in the vacuum system for analysis without the reactions continuing. We removed free water with an acetone rinse. Residual acetone was removed by drying for several hours in a vacuum at 10^{-6} Torr until no further outgassing was observed.

The 4 MeV Dynamitron tandem accelerator at the Ruhr-Universität Bochum [5] provided the ^{15}N ion beam. An RF system rather than mechanical belts generates the accelerating voltage, with very low ripple, and good energy resolution, giving nanometer scale spatial resolution for our sample measurements.

The beam setup for measuring H-depth profiles is given in Mehrhoff et al. [8]. Typical particle currents at the target were about 15 nA. The 4.44 MeV γ-ray emitted in the $^1H(^{15}N,\alpha)\gamma^{12}C$ reaction was observed in a 12-inch diameter x 12-inch long NaI(Tl) crystal (49 ± 3% photopeak efficiency). The beam energy resolution, determined by measurement of a hydrogen monolayer standard, was 13 keV (see Figure 1).

Each pellet represented a single point in the hydration history of tricalcium silicate. To scan the H depth profile in the specimen, the beam energy was increased stepwise starting just below the resonance energy of 6.400 MeV. At each step, a γ-ray spectrum was acquired and the 4.44 MeV peak area determined, typically 10,000 cts for 1 minute live time. The beam energy was increased typically in 10 keV steps up to 7 MeV to resolve thin surface layers. Above 7 MeV coarser steps (100-500 keV) were used as the profile changed more slowly with depth. The maximum beam energy was limited to 12 MeV to avoid interference from the next higher energy resonance.

RESULTS

A typical hydrogen depth profile (Figure 2) has several significant features: a surface Gaussian peak; a gradient that decreases non-linearly with depth, and a plateau region that extends into the material beyond the range of measurement (~2.5 μm). The NRRA raw data are in units of gamma-ray counts/nCoulomb of ^{15}N current vs. beam energy in MeV. The conversion factor from beam energy to depth, from Monte Carlo simulation of ^{15}N transport, is $d = (E_b - E_r) / 2.37 \times 10^3$ where E_b and E_r are beam and resonance energy. d is in mm for E in MeV. Ideal tricalcium silicate stoichiometry and density is assumed. The Gaussian peak can be associated with a surface layer that regulates penetration of hydroxyls into the tricalcium silicate and the reverse transport of Ca^{2+} into solution.

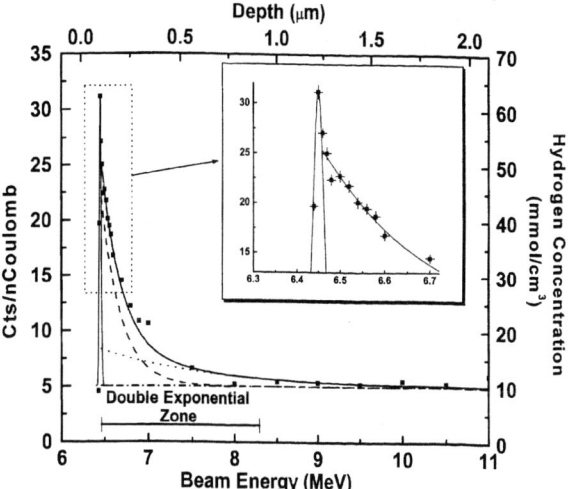
FIGURE 2. Typical hydrogen depth profile.

Henderson and Bailey [9], from TEM image analysis, proposed the surface layer is a monolayer tobermorite-like sheet, with an orthorhombic unit cell, (a = 1.12, b = 0.74 and c=2.28 nm). Tobermorite is a layer silicate that is the crystalline phase closest in formula to C-S-H gel, though there are significant differences between the two structures. The measured hydrogen concentration (53 ± 5 mmol/cm^3) agrees closely with that of ideal tobermorite, as contrasted to a water layer or calcium hydroxide. Tobermorite is used here only to represent the class of calcium silicate hydrates that could be in C-S-H gel, such as jennite, having similar H concentrations and densities. These data can't confirm the tobermorite-like structure. However, the surface layer thickness must be about 2-5 nm, a few unit cells (1.1 nm) of the postulated sheet. Future research is planned to refine this estimate.

The next region of the profile has a non-linear gradient suggesting a diffusion-controlled process, but it was not possible to fit the data to a simple Fickian diffusion model or a more complicated diffusion with reaction model. This is not surprising because the assumptions underlying these models are violated here. Water is diffusing into the substrate, but calcium is moving out simultaneously. Also, the structure itself changes over time from a crystal to a porous gel through which the water can move more freely. The best fitting purely empirical model used two exponential decay functions, suggesting this region may consist of two distinct zones. This is similar to the model proposed for reactions of alkali-silicate glasses with water. As shown in Figure 3, this is an outer zone where the gel is produced and an inner zone of silica hydration and calcium depletion. To quantify this region, a characteristic length scale, x_c, was defined such that for $x < x_c$, $A_1 \exp(-\mu_1 \xi) > 0.01 * A_2 \exp(-\mu_2 \xi)$.

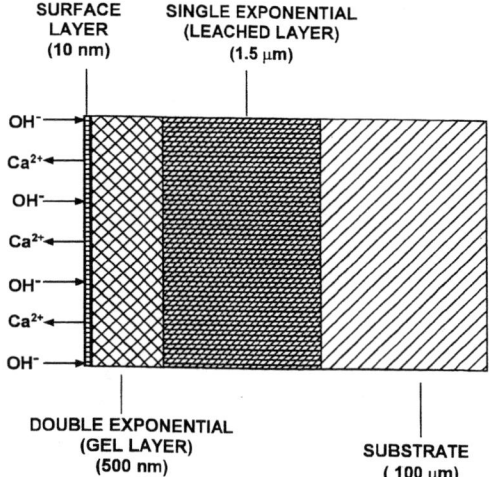
FIGURE 3. Schematic drawing of the arrangement of layers on tricalcium silicate grain surface. The outer layer controls ion exchange with the solution.

The plateau is fitted as a constant that varied between samples, and did not correlate with hydration time. It may be from residual acetone. Such problems in cement pastes have been reported. But, a control sample with just an acetone rinse and vacuum treatment had no plateau. As it forms in a few minutes, it may be due to a more rapid transport mechanism than crystal diffusion, such as through residual pores left after the pellet sintering, or through crystal defects.

As a function of time the Gaussian peak broadens and the double exponential shows a definite movement into the material. These effects are seen (Figure 4) in plots of Gaussian peak position and width and the characteristic length of the double exponential zone as functions of time. There is a distinct jump just past 4 hours. This event at 4.25 ± 0.07 hrs can be regarded as ending the induction period, consistent with the 3.5 ± 0.5 hrs of FitzGerald et al. [4] from quasi-elastic neutron scattering in samples hydrated at 20 °C.

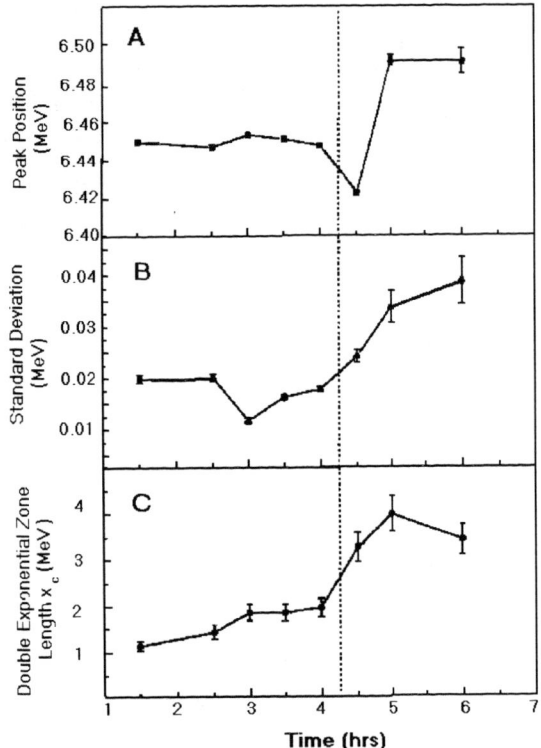

FIGURE 4. Plots of model parameters versus time. Figure 4A is the Gaussian peak centroid; Fig. 4B is the Gaussian width and Fig, 4C is the width of the mixed zone. The dashed vertical line indicates the effective end of the induction period at 4.25 ± 0.07 hours.

Within the limits of uncertainty, the Gaussian peak height is constant during the induction period. The hydrogen concentrations for the inner regions tend to approach this limiting value. This implies the presence of a surface layer where available hydrogen sites are saturated. The mean value of Gaussian peak height for the samples up to the end of the induction period is 26 ± 2 cts/nCoulomb or 53 ± 5 mmol/cm^3.

The surface layer appears to develop a well-ordered structure through topochemical growth from solution on a silicate template. Defects in this layer allow OH$^-$ ions to penetrate into the solid, and Ca^{2+} to pass into solution, leaving behind the silica tetrahedra. This reaction produces a zone of C-S-H gel much less dense than the tricalcium silicate (2.2-2.4 g/cm^3 vs. 3.12 g/cm^3) creating a swelling pressure. Pressure increases as the gel volume grows until the surface layer stresses exceed its strength and it ruptures, as has been seen with TEM [9]. The pressure extrudes whiskers of gel through the cracks into solution. For our configuration, it would show up as a surface layer thickening, and a widening of the Gaussian peak. This model for ending the induction period relies on mechanical aspects (swelling pressure and strength of surface layer) of the system rather than chemical phase transformation.

CONCLUSIONS

We have shown for the first time that NRRA can be used to measure the change of hydrogen profile in the early hydration of tricalcium silicate. A time-resolved method has been used to determine the end of the induction period with a precision of 1-2%. The hydrogen profile shape indicates that a nanometer scale surface layer controls the reaction during the induction period. The effects of temperature, minor elements in the cement and retarders or accelerators on the rate controlling steps of the induction period process including the permeability and strength of the surface layer and the growth of the reaction zone will be investigated. The NRRA method provides a powerful new quantitative tool for this research.

REFERENCES

1. Le Chatelier, H. L., *Experimental Researches on the Constitution of Hydraulic Mortars*, New York: McGraw Publishing Company, 1887.

2. Taylor, H. F. W., *Cement Chemistry, 2nd Edition*, London: Thomas Telford, 1997.

3. Livingston, R. A., Neumann, D., Allen, A. J., and Rush, J. J., in *Neutron Scattering in Materials Science II*, edited by B. Wuensch, D. Neumann and T. Russell, Pittsburgh: Materials Research Society, 1996, Vol. 376, p. 459.

4. FitzGerald, S., Neumann, D., Rush, J., and Livingston, R., *Chem. Mater.* **10**, 397-402 (1998).

5. Becker, H. W., Bahr, M., Berheide, M., Borucki, L., Buschmann, M. Rolfs, C., Roters, G., Schmidt, S., Schulte, W. H., Mitchell, G. E., and Schweitzer, J. S., *Z. Phys.* **A351**, 453 (1995).

6. Doremus, R. H. *Glass Science, Second Edition*, New York: John Wiley & sons, 1994.

7. Schreiner, M., Grasserbauer, M., and March, P., *Fresenius Z. Anal. Chem.* **331**, 428 (1988).

8. Mehrhoff, M., Aliotta, M., Baumvol, I. J. R., Becker, H. W., Berheide, M., Borucki, L., Domke, J., Gorris, F., Kubsky, S., Piel, N., Roters, G., Rolfs, C., and Schulte, W. H., *Nucl. Instr. Meth* **B132**, 671 (1997).

9. Henderson, E., and Bailey, J. E., *J. Mat. Sci* **28**, 3681 (1993).

A New On-Belt Elemental Analyser for the Cement Industry

B.D. Sowerby*, C.S. Lim*, J.R. Tickner*, C. Manias** and D. Retallack**

*Division of Minerals, Commonwealth Scientific and Industrial Research Organisation,
Private Mail Bag 5, Menai NSW 2234 Australia*
** Fuel & Combustion Technology International (Aust.) Ltd, 20 Stirling Street, Thebarton SA 5031 Australia

Abstract. On-line control of raw mill feed composition is a key to the improved control of cement plants. Elements of primary importance to the industry are calcium, silicon, aluminium and iron. Direct on-conveyor belt analysis of raw mill feed is required, independent of changes in belt loading, moisture content and both horizontal and vertical segregation. A new and improved on-conveyor belt elemental analyser for cement raw mill feed has been developed and tested successfully in Adelaide Brighton's Birkenhead cement plant. The analyser utilises two ^{241}Am-Be neutron sources and multiple BGO detectors to measure both neutron inelastic scatter and thermal neutron capture gamma rays. Dynamic tests in the plant on highly segregated material having depths in the range 100 to 200 mm have shown analyser total RMS errors of 0.49, 0.52, 0.38 and 0.23 wt.% (on a loss free basis) for CaO, SiO_2, Al_2O_3 and Fe_2O_3 respectively, when 10-minute counting periods are used.

INTRODUCTION

In a cement plant, the various components in cement raw mill feed (limestone, shale, iron ore, etc.) are typically fed onto a conveyor belt in sequence prior to milling and subsequent firing in a kiln. The on-belt elemental analysis of this raw mill feed would enable improved control of raw mix chemistry in cement plants. Compared to sampling and subsequent laboratory analysis, plant control is improved as timely on-belt measurements allow real time control action.

In the cement industry plant control is based on parameters such as Lime Saturation Factor, Silica Ratio, Alumina Ratio as well as hypothetical compounds such as C_3S, C_2S, C_3A and C_4AF [1]. These parameters are calculated from the proportions of calcium, silicon, aluminium and iron in the raw mill feed. Elements of secondary importance are magnesium, sulphur, chlorine, sodium, titanium and potassium.

In 1997 Fuel and Combustion Technology (FCT) commenced work on *Mastermind*, a cement plant control system integrating a series of novel on-line analysers with sophisticated control software. CSIRO Minerals was chosen to develop a new and improved on-belt elemental analyser for cement raw materials which is capable of accurately measuring key elements independent of both horizontal and vertical segregation and independent of changes in belt loading. Earlier commercial on-belt cement analysers (the Gamma-Metrics Cross-Belt Analyser and Scantech's Geoscan) utilise the thermal neutron capture (TNC) technique with a ^{252}Cf source and NaI(Tl) detectors. More recently a neutron generator-based on-belt cement analyser has been developed [2].

The present paper describes a new and improved on-belt elemental analyser for cement raw mill feed and the results of testing both in the laboratory and in service on a moving conveyor belt in the Adelaide Brighton Cement plant at Birkenhead, South Australia. The new analyser is called XENA (X-belt Elemental Nuclear Analyser) and it is marketed by FCT.

METHOD

Both neutron inelastic scatter (NIS) and TNC techniques involve bombarding the material of interest with neutrons and measuring the characteristic gamma rays that elements in the sample produce. In NIS, the neutrons promote the nucleus they scatter from into an excited state, which produces prompt gamma rays of specific energies as it decays back to the ground state. In TNC, a neutron is absorbed by a nucleus, with the neutron's binding energy being released as one or more prompt gamma rays again having specific energies. Higher-energy neutron sources such as ^{241}Am-Be or a D-T neutron generator can be used for both TNC and NIS applications.

After considering the relative advantages and disadvantages of ^{241}Am-Be and neutron generators in industrial applications, ^{241}Am-Be sources were selected for XENA. Compared to neutron generators, ^{241}Am-Be sources have the following advantages: significantly lower cost and longer life (^{241}Am-Be source encapsulation is guaranteed for 15 years); much less equipment complexity; minimal maintenance; and stable neutron output. The big advantage of neutron generators is that they can be turned off, therefore simplifying source transportation and shielding. In addition they can be pulsed (therefore enabling activation and detection of short-lived radioisotopes and the separation of NIS and TNC gamma rays) and ultimate source disposal is less expensive.

XENA also employs a novel source/detector configuration to achieve improved spatial uniformity compared to previous methods [3,4]. The Monte Carlo code MCNP was used to estimate the spatial sensitivity of various gauge configurations and to optimise the position of the sources and detectors [5]. The geometry best suited to on-belt cement raw meal analysis combines two transmission measurements with sources on opposite sides of the sample.

LABORATORY TESTS

Prior to designing a gauge for plant testing, laboratory tests were performed using a prototype gauge to test concepts and to compare results with Monte Carlo predictions. The laboratory tests were performed using 100 synthetic raw meal samples and 25 plant samples from the Adelaide Brighton Cement plant at Birkenhead, South Australia. The samples were measured over a range of depths between 80 and 200 mm. The results of the laboratory tests showed that the gauge was capable of accurately measuring the proportions of CaO, SiO_2, Al_2O_3, Fe_2O_3, MgO, Cl, K_2O and TiO_2 in both the synthetic and plant samples [3].

PLANT GAUGE DESIGN

The design of XENA was based on both experimental measurements in the laboratory and on Monte Carlo models of the gauge configuration [3,5]. The overall mechanical design is shown in Figure 1. The gauge is in two halves, each containing a 370 GBq (10 Ci) ^{241}Am-Be neutron source and two 76 mm ⌀ x 76 mm long (BGO) detectors in a transmission configuration. The use of multiple detectors and sources provides much reduced sensitivity to non-uniformity of composition both vertically and across the conveyor belt. A third BGO detector exposed to the 4.43 MeV source gamma rays is used to determine the material depth on the belt.

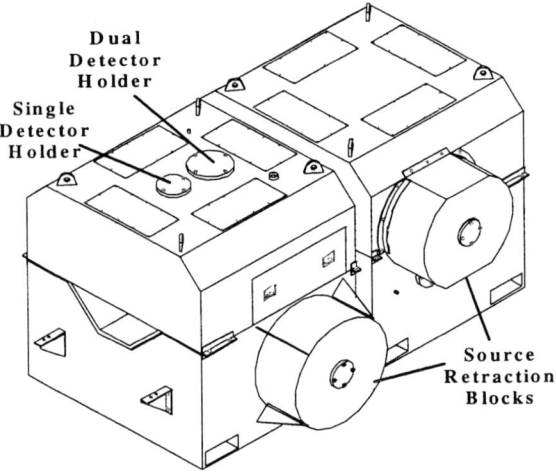

FIGURE 1. CAD drawing of the external view of the on-belt neutron-gamma elemental analyser XENA.

The two halves of the gauge are separated by a gap to allow the inclusion of a belt support roller. The gauge has been constructed to allow one half of the gauge to be moved horizontally by a small distance when access is required to the area between the two halves for maintenance. Within the gauge, a Teflon slider plate supports the belt. The two sealed cavities in which the detectors are mounted are cooled to about -20°C using recirculating refrigerant systems, as cooling of BGO detectors results in a significant improvement in detector resolution. For example the resolution of one BGO detector improved from 10.3% (FWHM resolution at 662 keV) at +15°C to 8.2% at -20°C.

PLANT RESULTS

XENA was successfully installed on the cement raw meal feed belt at the Adelaide Brighton Cement plant at Birkenhead, South Australia in March 1999. Material on the belt at the XENA location is segregated both horizontally and vertically. The preliminary calibration measurements performed on XENA at Lucas Heights were repeated at Birkenhead. These measurements involved seven plant raw meal samples in plastic sample boats [3].

The long-term stability of the energy scale of the gamma-ray spectra used in the composition measurements is critical and a software stabilisation code was developed. Tests on the 4.438 MeV carbon peak over a 12 hour period showed that the RMS variation in peak position was 0.019% for hardware stabilisation alone, and 0.004% after software stabilisation was included.

Reference spectra were collected for each detector while an empty belt moved through the analyser. All spectra used in composition measurements are scaled by these reference spectra, which are updated regularly. In this way, changes in the spectra produced by changes in the analyser (eg belt or slider plate wear) are automatically corrected for.

During normal plant operation, the composition ranges are tightly restricted to achieve maximum production rates and product quality. These ranges are too limited for an adequate calibration of the analyser. Instead, a series of on-line step change tests was implemented to finalise the plant calibration of XENA. Each step test consisted of two consecutive 45-minute changes to the normal proportions of the feed materials such that the changes during the second period exactly counterbalanced the changes during the first. This made it possible to increase significantly the calibration range without affecting product quality. Sixteen of the step change tests were performed when two mills were in operation (mean depth ~165 mm) and four were carried out when only one mill was in operation (mean depth ~115 mm). Plant factors prevented more tests.

During each step change test, samples were manually collected from the transfer point at the end of the conveyor belt (referred to as BELT) passing through XENA as well as from the nearest plant semi-automated sampling point (referred to as MILL). Between XENA and the MILL sampling points the raw meal is ground in a mill and some recycled material from the kiln is added to the mill. This recycled material potentially affects the correlation between the MILL and the BELT measurements.

The means and estimated errors of the BELT and MILL analyses showed that there were some discrepancies between the mean values for CaO and Al_2O_3 measured at the BELT and MILL, suggesting that the recycled material fed into the mills is calcium poor and aluminium rich. Mean values for the other elements agreed fairly closely. However, the sampling errors were not small, being of similar order to the total errors anticipated for XENA on the basis of the laboratory measurements [3]. As the plant control is configured to use the MILL results, the BELT results were corrected by adding the mean difference between the two measurement sets for each element.

It was found that Ca and Fe were best determined using the TNC gamma rays at 1.942 and (7.631, 7.646) MeV respectively whereas Si and Al were best determined from NIS gamma rays.

Comparisons of XENA and MILL for samples of depth 160-200 mm are shown in Table 1 and Figure 2. The total RMS errors shown in Table 1 were estimated by performing a 3-way Grubbs' analysis [6] between the XENA, BELT and MILL composition measurements for each sample. The correlation coefficients and errors for the 4 major elements are generally good; aluminium has the lowest correlation coefficient (0.77) due to the fact that it is present in small quantities and has fairly weak spectral lines.

Table 1. Summary of RMS errors and correlation coefficients for plant measurements (10 minute counting time) on cement raw meal of depth 160-200 mm.

	Repeatability Error (wt.%)	Total RMS Error (wt.%)	Correlation Coefficient
CaO	0.32	0.49	0.94
SiO_2	0.29	0.52	0.88
Al_2O_3	0.12	0.38	0.77
Fe_2O_3	0.15	0.23	0.85

In the present paper, the terms *total RMS error* and *repeatability error* are used to describe the accuracy of the composition measurements. The total RMS error is defined as the RMS uncertainty on the measurement by the gauge of the composition of a previously unseen sample. The repeatability error is the standard deviation of a set of repeated measurements of the same sample. Experimentally, the repeatability error was dominated by the statistical errors on the measured spectrum.

Note that chemical compositions are quoted in the oxide forms assumed in XRF analysis and on a loss-free basis (that is, excluding CO_2, moisture and any other material lost during firing in the kiln), in accordance with cement industry convention.

The calibration of XENA for lower depth samples was performed using the four step tests carried out with only one mill running (mean depth 115 mm) supplemented by plant data collected during normal plant operation. Subsequently the evaluation of the performance of XENA was carried out at all depths over a period of a month. The results showed that the composition measurements showed very little dependence on depth within the range 100-200 mm.

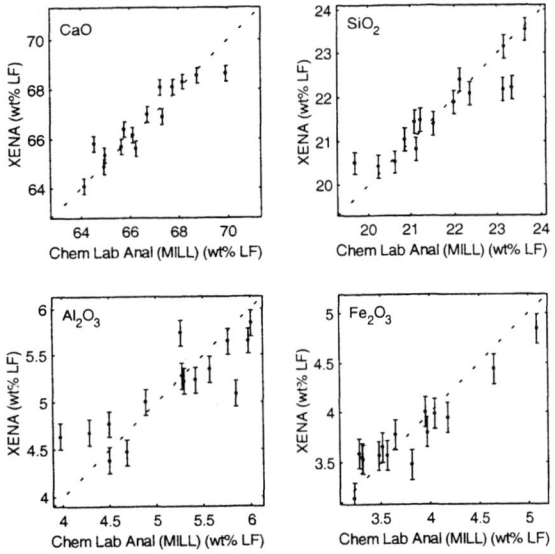

FIGURE 2. On conveyor belt XENA measurements of CaO, SiO_2, Al_2O_3 and Fe_2O_3 versus the chemical laboratory analyses of the MILL samples (expressed as wt.% loss-free (LF)). The centre points are the means for each of the step tests and the error bars show the 10-minute live-time repeatability errors. Reprinted from [3], © 2000, with permission from Elsevier Science.

SUMMARY

In summary, XENA has the following features:

- The choice of both neutron inelastic scatter and thermal neutron capture techniques. Compared to instruments that use only thermal neutron capture gamma rays, improved performance can be obtained for those elements such as Al, Si and Mg where NIS gives better results.

- Good spatial uniformity

- Calibration demonstrated over a wide depth range (80-200 mm at least)

- Relatively low source costs and more stable neutron output over long periods due to the use of long life radioisotope sources.

Since installation, the analyser has performed well during on-line tests. Stability has been excellent and the measurements from XENA have closely tracked the post-mill sampler XRF results.

ACKNOWLEDGMENTS

The authors wish to thank the staff of the Adelaide Brighton Cement plant at Birkenhead for their assistance and advice during the plant installation and testing of the analyser.

REFERENCES

1. Bye, G. C., *Portland cement: composition, production and properties.* Pergamon Press, Oxford, 1983.

2. Lebrun, P., Le Tourneur. P., Poumarede, B., Moller, H. and Bach, P., 15th International Conference "Applications of Accelerators in Research and Industry", ed. J.L. Duggan and I.L. Morgan, American Institute of Physics CP475, 695-698 (1999).

3. Lim, C.S., Tickner, J.R., Sowerby, B.D., Abernethy, D.A., McEwan, A.J., Rainey, S., Stevens, R., Manias, C. and Retallack, D., *Applied Radiation and Isotopes* **54**, 11-19 (2000).

4. Sowerby, B. D., Lim, C. S., Tickner, J. R., An improved bulk material analyser for on-conveyor belt analysis. Int. Patent Application No. PCT/AU98/01026 (1997).

5. Tickner J.R., *Applied Radiation and Isotopes* **53**, 507-513 (2000).

6. Grubbs, F.E., *J. Amer. Statistical Assoc.* **43**, 243-264 (1948).

SECTION XIV

TOMOGRAPHY AND RADIOGRAPHY

The Detector Problem in Fast Neutron Radiography

John I.W. Watterson, Richard M. Ambrosi and Heidar Rahmanian

Schonland Research Centre for Nuclear Sciences, University of the Witwatersrand, Private Bag 3
PO WITS, Johannesburg 2050, South Africa

Abstract. Fast neutron radiography has several advantages including the inherent brightness of an accelerator based neutron source in comparison to a reactor. In addition it can be used for the selective imaging of certain elements using fast neutron resonances (resonance imaging). One of the greatest problems in fast neutron radiography is the efficient detection of the fast neutrons. Such a system must consist of a neutron sensitive screen and a way of converting the image into an electrical signal. In most cases the method uses a screen that is based on the elastic scattering of neutrons by protons followed by the conversion of the proton energy into light in a scintillator material. This light subsequently produces an electrical signal using a CCD, or an amorphous silicon or other semiconductor screen. All such techniques involve a trade-off between image quality and efficiency. As the screen thickness increases, its efficiency also increases, linearly for small screen thicknesses, but saturating exponentially. On the other hand the effects of secondary neutron scattering and light spread also increase. The effect of neutron scattering in thick scintillators has been evaluated using Monte Carlo methods and is found to be significant for thick detectors.

INTRODUCTION

Fast neutron radiography is a relatively new technique [1]. It has a number of significant advantages over the thermal technique. These include the inherent brightness of an accelerator target as compared with a reactor moderator, the ability to distinguish between different elements through the use of resonance imaging and the fact that fast neutrons are in general more penetrating and can image larger specimens. The method of resonance neutron radiography uses the structure of the total neutron cross-sections to provide a method that is sensitive to a particular element or elements.

There are two important challenges that must be overcome before the promise of this fast neutron technique can be fully realised. These are the challenge of producing a neutron source of high intensity with a defined energy spread [2,3] and good geometry, and the development of imaging detectors that are both sensitive to fast neutrons and that also have a good spatial resolution [4]. In addition it could be an advantage if the detectors have some selectivity for the neutron energy [5]. This is because the scattered neutrons that degrade the images have lower energies than the unscattered neutrons. The process of neutron detection depends on the production of a charged particle (in the converter) and the subsequent detection of that particle. In the case of fast neutrons the cross-sections are relatively small and the converter should be thick, in order to be efficient. In such a detector there is an appreciable contribution from neutrons scattered in the detector itself.

THE DETECTION OF FAST NEUTRONS

Most significant methods for the detection of fast neutrons depend on the elastic scattering of protons and the subsequent conversion of the proton energy into a number of electrons. In turn these are often converted into light in a scintillator material. This light subsequently produces an electrical signal using a CCD, or an amorphous silicon or other semiconductor screen

The cross-section for the scattering of protons is very much lower than that for the absorption reactions ^{7}Li(n,α) or ^{10}B(n,α) used for detection in thermal neutron radiography.

Figure 1 shows the total cross section for neutron scattering in hydrogen as a function of energy. This cross-section drops rapidly with energy from about 11 barns at 150 keV to about 1 barn at 10 MeV [6]. In spite of this it is significantly larger than the cross-section for most other reactions that could be considered. This paper will be confined to the use of this reaction for the primary detection of the neutrons.

The primary efficiency of the detector can be defined as the ratio of the total number of neutrons that interact with the screen, to the number of neutrons that are incident on it. Defined in this way the primary efficiency ε_p is just $\varepsilon_p = \left(1 - e^{-N\sigma d}\right)$ where N is the density of the hydrogen atoms and d is the thickness of the screen.

Figure 2 shows the efficiency of a screen as a function of the thickness for a number of common hydrogen containing materials at a cross-section of 1.1 b. This is the cross-section for 8 MeV neutrons.

In the case of a converter in the form of a slab, the maximum thickness is limited by the range of spreading of light within the screen and its effect on the resolution and the transparency of the detector to its own light. Polyethylene, which has two hydrogen atoms for each carbon atom, is the most efficient, but it has the disadvantage of not being transparent when a scintillator is used (although it is translucent). Polypropylene is transparent although its primary efficiency is some 17% lower than for polyethylene.

A particularly effective form of scintillator is a block of polypropylene containing ZnS as the scintillator. The ZnS has several advantages including a very high light output for heavy charged particles, such as protons, and a low sensitivity for electrons. This detector is commercially available as the "PP Converter."

Thick Detectors

The problem of light spreading in thick scintillators could be addressed by using scintillating fibres. The geometrical problem of each of these fibres intersecting a number of different neutron trajectories could be addressed by having a "hedge-hog" type of detector with each fibre pointing at the source (at least for a point source).

In any thick detector, even of this type, there will

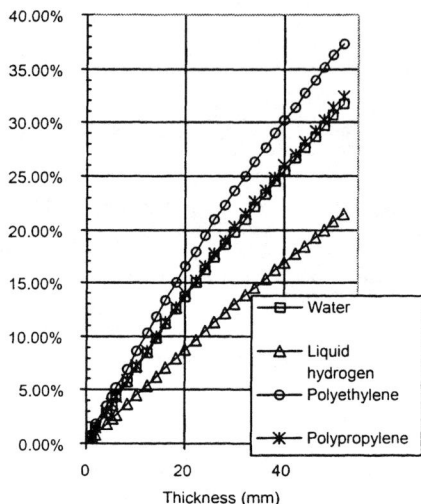

FIGURE 2 Primary efficiency for neutron detection for candidate materials for neutron screens in fast neutron radiography. Calculated for 8 MeV neutrons (Total cross-section 1.1b) as a function of the converter thickness

be scattering within the body of the converter. This effect has been modelled using a Monte Carlo method.

METHOD

The point scattering function for the scintillator material was evaluated with the Monte Carlo computer code MCNP-4A using a cylindrical geometry [7]. A line source of neutrons was directed down the axis of a hypothetical polyethylene cylinder and the tallies were collected in a series of annular concentric volumes with a pitch of 1mm. On this basis, a point scattering function was evaluated. This is the point scattering function for the detector and not, as is more usual, for the specimen.

The implication of the scattering for image formation was modelled by using the same Monte Carlo method as for the point scatter, but with a planar source extending half-way across the detector. The neutrons were in a parallel beam along the axis of the detector. This mimicked an infinite absorber covering half of the detector with a neutron source at a large distance so that parallax could be ignored. The geometry for this is shown in Figure 3.

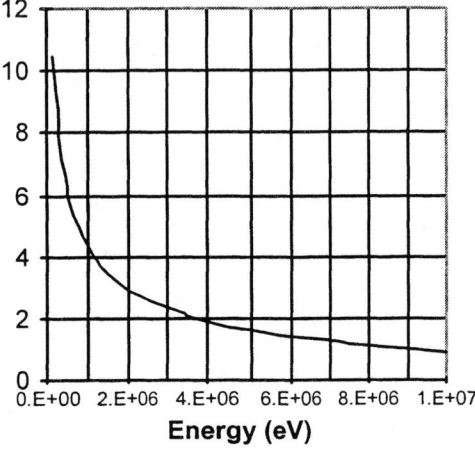

FIGURE 1 Total neutron cross-section for hydrogen as a function of neutron energy over the range 150 keV to 10 MeV [6].

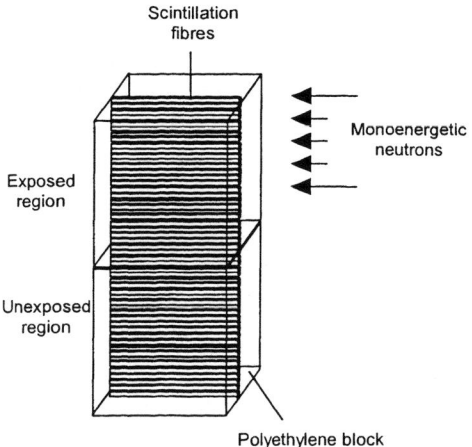

FIGURE 3. Geometry for the determination of the edge function using a Monte Carlo technique

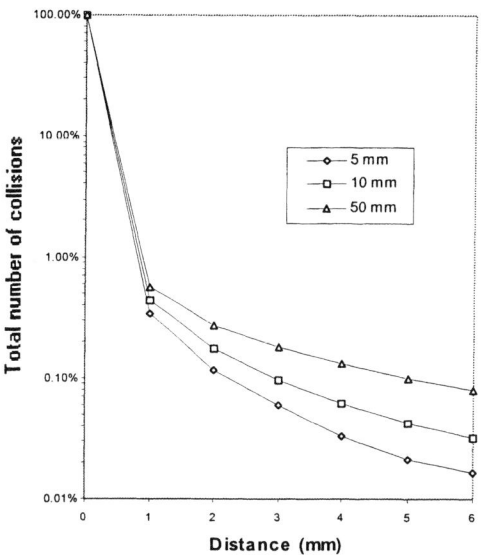

FIGURE 4. Point scatter function for a line source of neutrons incident on detectors (converters) of different thickness'.

RESULTS AND DISCUSSION

Point Scatter Function

Figure 4 shows the results obtained for the point scatter function [7] for three different thicknesses of converter, i.e. 5 mm, 10 mm and 50 mm. This is a point scatter function for the detector and not, as is more usual, for the specimen that is being imaged.

The amount of scatter into the shadow region increases significantly as the thickness of the converter increases. The ratio of the scattering increases significantly for the thicker detectors at larger distances from the line source.

The scatter at 1 mm from the line beam is less than 1% of the value in the beam. On the other hand the

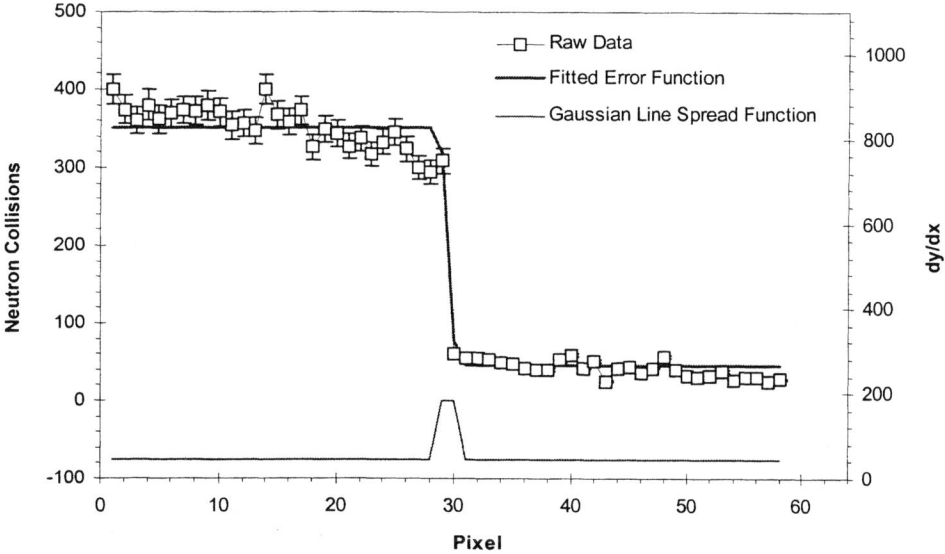

FIGURE 5. A cross section through an edge generated with MCNP-4A. The scintillator was 50 mm in length and the tally cells had a cross sectional area of 6.25×10^{-2} mm^2 (0.5 mm x 0.5 mm) The edge shown is fitted with an error function and the corresponding line spread function is also shown.

point scattering function must be integrated over the entire beam to assess the effect of scattering on the image of a feature from an extended object. In the case of an edge, the effect of this integration can be seen using the model shown in Figure 3, above [7].

The Edge Function

The edge function that was obtained for the geometry of Figure 3 is shown in Figure 5 for the case of 50 mm thick converter. The scattering kernels add to give a diffuse background inside the shadow that is some 15 % of the value near the edge of the direct beam. In addition there appears to be some structure in the response to the direct beam, that tends to reduce the contrast at the edge.

In order to obtain quantitative results for the resultant reduction in the contrast and the effect on the resolution an error function was fitted to the data to give the curve shown by the solid line through the data in Figure 5. This was then differentiated to give the solid line at the bottom of the figure (a Gaussian). In this case, with a 50 mm thick detector, the contrast was reduced from 100% in the ideal case to 88%,

TABLE 1 Contrast and Resolution as a Function of Detector thickness

Scintillator thickness (mm)	Contrast (%)	Resolution from Edge Function (mm)
5	99.8	0.5
10	98.9	0.6
50	88.2	0.8
100	76.7	1

while the FWHM of the Gaussian was 0.8 mm. The FWHM value is approximate since 0.5 mm cells were used for the tallies in the MCNP analysis of Figure 3.

Table 1 shows how the contrast and the resolution, determined in the same way as in Figure 5, varied with the thickness of the detector. With a 5 mm thick screen there is a negligible degradation in the image quality. At 10 mm the effect of scattering becomes noticeable and increases rapidly so that there is a large effect with a 100 mm thick screen.

CONCLUSIONS

Neutron sources for fast-neutron radiography, and in particular for resonance neutron radiography are accelerator based. Because of the difficulties associated with the development of small, high intensity accelerator based sources, and the low efficiency of fast neutron imaging detectors, images are often limited by the Poisson statistics of the events registered.

While work continues on the development of accelerators with high beam currents and target designs to address the first part of this problem, it is of particular interest to consider the physics of detector design. All neutron detectors are based on the production of energetic charged particles. In the case of fast neutrons the most efficient process is the elastic scattering of neutrons and protons. As can be seen from Figure 1, such detectors should be thick to increase the efficiency.

Several factors must be considered in the design these detectors. If they are based on a scintillator, the light must be transported out of the detector. This can probably be overcome using scintillator fibres in a suitable geometry. Thick detectors, particularly if they are not based on ZnS as a scintillator, are sensitive to gamma-rays. This effect can be reduced, for example by using a low gamma-ray yield reaction such as $D(d,n)^3He$ for neutron production.

All such detectors will scatter neutrons. This will lead to a degradation of the image. This effect has been modelled using a Monte Carlo technique with MCNP-4A and, as can be seen from Figure 5 and Table 1, the effect is significant for thick detectors. Since the figure of merit of an image depends directly on the contrast but only on the square root of the counts [8], the flux must be doubled to compensate for a reduction in the contrast from 100% to 80%.

REFERENCES

[1] Watterson, J.I.W., "The Development of a Computational Model for Fast Neutron Radiography", *SPIE Proceedings Series,* **2867**, 1997, pp. 358-361.

[2] Guzek, J., "Elemental Radiography Using Fast Neutron Beams", Unpublished PhD Thesis, University of the Witwatersrand, Johannesburg, 1999.

[3] Iverson, E.B., Lanza, R.C. and Lidsky, L.M., "Wndowless Gas Target for Neutron Production", SPIE Proceedings, **2867**, 1997, pp 513-516.

[4] Ambrosi, R.M. and Watterson, J.I.W, Nucl. Instrum. and Meth. **139B**, 279-285 (1998).

[5] Rahmanian. H. and Watterson, J.I.W., Nucl. Instrum. and Meth. **139B**, 466-4705 (1998).

[6] ENDF/B-VI, 1991

[7] Ambrosi, R.M., "A Model for the Physics of Image Formation in Fast Neutron Radiography", Unpublished PhD Thesis. University of the Witwatersrand, Johannesburg, 2000.

[8] Watterson, J.I.W., University of the Witwatersrand, Johannesburg, Schonland Research Centre, Report SRCNS 97/02

Computed Tomography Investigation of Microgravity-Tested Sand Samples

SUSAN N. BATISTE[1], KHALID A. ALSHIBLI[2], MARK R. LANKTON[1], STEIN STURE[1], ROY A. SWANSON[1], and NICHOLAS C. COSTES[1]

[1]Univeristy of Colorado at Boulder, [2]University of Alabama in Huntsville

Abstract. Computed Tomography (CT) is being used to investigate the complex internal structure of axisymmetric (triaxial) sand specimens. A series of triaxial experiments was conducted on dry Ottawa sand specimens at very low effective confining stresses in a microgravity environment aboard the Space Shuttle during two missions. Post-flight analysis includes studying the internal fabric and failure patterns using CT. In addition ground-tested specimens subjected to different compression levels are scanned to investigate the evolution of instability patterns, quantify void ratio variation, and provide a direct comparison with microgravity specimens. For an upcoming Shuttle mission, trial specimens are scanned to investigate an experimental reforming method for flight and evaluate techniques for reconstituting specimens. The CT technique demonstrates good ability to detect specimen inhomogeneities and localization patterns, and quantify void ratio variation within sand specimens.

INTRODUCTION

The mechanical behavior of granular materials is highly dependent on the arrangement of particles, particle groups and associated pore space. These geometric properties comprise the so-called *material fabric*. In the literature, fabric analysis techniques are mainly classified as destructive (e.g., specimen stabilization and thin-sectioning), and nondestructive techniques (e.g., magnetic resonance imaging, ultrasonic testing, x-ray radiography, and x-ray computed tomography) X-ray radiography has been used to trace density change of soil samples (e.g., Roscoe 1970, Vardoulakis and Graf 1985, Vardoulakis *et al.* 1985). However, it suffers from the limitation of not providing a three-dimensional (3-D) radiograph. Consequently, most of x-ray radiograph techniques have been performed on specimens tested under plane strain (two-dimensional strain) conditions. High resolution 3-D density images can be obtained by computed tomography. The first applications of tomography were in radio astronomy (Bracewell 1956) and today it is widely used in the areas of medical imaging, seismology, petroleum engineering, acoustic imaging, soil science, and powder industry (e.g., Coshell *et al.* 1994, Amos *et al.* 1996, Zeng *et al.* 1996, Denison and Carlson 1997, Denison *et al.* 1997, Phillips and Lannutti 1997).

In this paper, the use of CT analysis for sand specimens tested under triaxial conditions at very low effective confining pressures in a microgravity environment and on 1-g reformed specimens will be presented. We explain the density calibration techniques, showing and describing the failure patterns, and quantifying void ratio variations within the specimens.

MATERIALS AND METHODS

A series of conventional triaxial compression experiments were performed in the SPACEHAB module of the Shuttle Orbiter during the STS-89 mission to Mir in January, 1998. The experiments were conducted on cylindrical dry specimens (75 mm in diameter and 150 mm long, compressed 38 mm vertically) at extremely low effective confining pressures. More details about the experiments hardware and results can be found in Alshibli *et al.* (1996) and Sture *et al.* (1998).

CT SCANNING

The CT scans reported in this study were performed at the NASA/ Kennedy Space Center (KSC), Industrial Computed Tomography System (ICTS) using a CITA-201 scanner, manufactured by Scientific Measurement Systems, Inc. A Cobalt-60 gamma ray source was used, with an energy level of approximately 1.25 MeV and source spot size of 2.5 mm. The source aperture was set as 36 degree by 4 mm high. The 125 detector apertures were set at 2 mm wide by 4 mm high, which samples from approximately a 2-mm thick CT data region. The data object acquisition circle was 600 mm at 1581.85 mm from source. Ray spacing data acquisition was every 0.5 mm. Second generation data acquisition was used, with time per datum of 0.2 seconds, thus taking approximately 1 hour to acquire data per complete slice. Slices were computed at 1 mm intervals over the height of each specimen. For further information on the tomography system the reader is referred to Engel (1998).

CALIBRATION OF COMPUTED TOMOGRAPHY DATA

The CT data is used for two purposes: measuring internal structure geometry and density (or void ratio). The data has been calibrated for both, and measures of accuracy are presented below.

Spatial resolution was determined using ASTM standards (ASTM-E1695, 1995). The interface between an aluminum rod and air was examined and it was determined that data shows 50% modulation at 0.275 line pair per millimeter (lpm) and 10% at 0.53 lpm. Considering the average grain size of the material is 0.22 mm, one cannot resolve below 5.3 grain diameters at 10% modulation, or 8 grain size diameters at 50% modulation.

Calibration of density was also performed according to ASTM standards (ASTM-E1935, 1997). Water and Lexan regions included in scans of the specimens were used as standards. An Aluminum standard was also scanned by ICTS and the data were used in the calibration. To verify the calibration, the average CTN in a specimen was compared to its bulk density for 11 different specimens. The average CTN was measured using all CTN from each plane over the entire height and diameter of a specimen. The bulk density was calculated from initial mass and volume measurements, and volume change during testing. The void ratio standard deviation was found to be less than 0.017.

RESULTS AND ANALYSIS

Specimen Deformation Patterns

Three-dimensional reconstructions of the specimens using CT data have been prepared using the image processing software IDL (Interactive Data Language). Sides of volume renderings have been cut away to view the internal structure and axial planes through the center of a specimen have be generated and displayed in Figure 1.

A second use is to examine the development and progress of shear bands during sample compression. Researchers are interested in the strain levels at which shear bands develop, and density, orientation and size of shear bands. This type of information is obtained by preparing several specimens, and compressing them to different strain levels. Then, each specimen is scanned and a picture of internal structure formation is achieved. A sample of the specimens tested is shown in Figure 2.

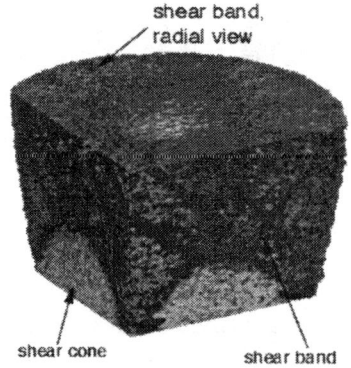

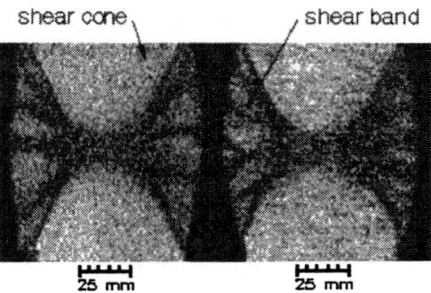

Figure 1. 3-D Volume Rendering and Two Orthogonal Axial Sections of the C25-Specimen

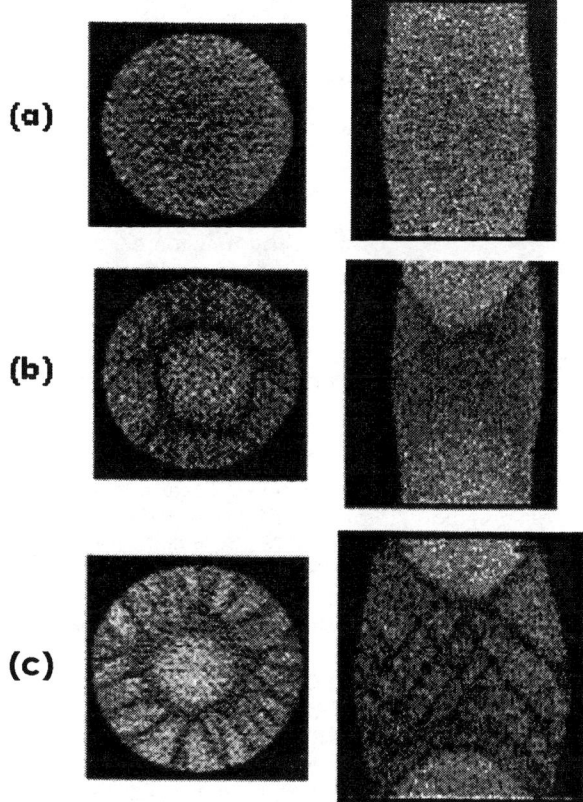

Figure 2. CT images of specimens compressed to (a-c) 5, 9, and 25% axial strain.

A third use of CT scanning is evaluating a new reforming method intended to take a specimen that has been compressed and is in a state similar to Figure 1 or 2c, and essentially erase the internal structure to prepare the specimen to be tested again. This has two purposes: first, to re-use samples while still on-orbit. This saves both time and money by compressing several missions worth of tests into one mission. Secondly, it allows researchers to lower the density of the specimen below what would hold up under the g-forces and vibrations of launch. The development program has included performing this reformation on the ground, and using CT scans to look for remnants of internal structure. Figure 3 demonstrates this program. On the far left is an undeformed, untested specimen, a model for what researches want a reformed specimen to look like. The middle and right specimens have both been reformed and scanned. It was expected that the sides of the specimen would not be straight due to gravitational force effects, so the analysis is concentrated on the internal density distribution. The middle specimen was untouched during the reformation process, and in general looks fairly homogeneous. However, a triangular area is visible in the very lower section. This is recognized as a shear cone structure which apparently was not broken up during reformation. The far right specimen was slightly disturbed during reformation, and shows no internal structure. This indicates that slight disturbance during reformation process is necessary in order to fully break up end structures.

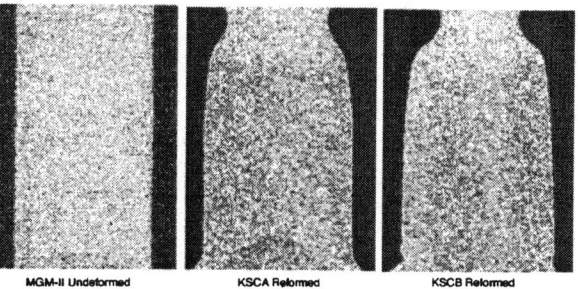

Figure 3. Comparison of an untested specimen and two reformed specimens.

Void Ratio Variations

While 3-D volumes and axial sections are useful for qualitatively understanding the internal structure of a triaxial specimen, other methods are used to examine the features quantitatively. The relationship between the CTN and specimens densities, described earlier, was used to obtain quantitative observations. The data are displayed as a contour map (Figure 4.)

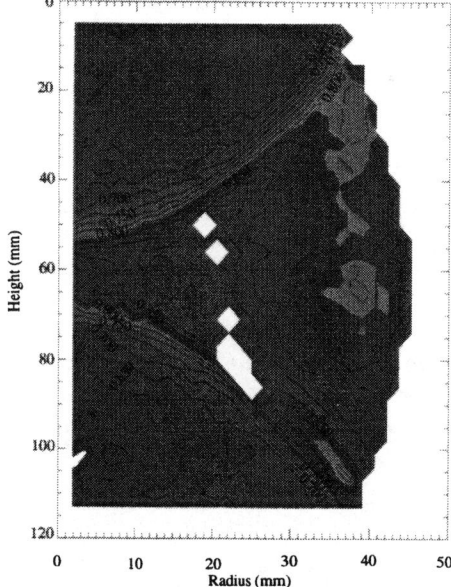

Figure 4. Void Ratio Distribution of C25 Specimen Represented by Contour Map

The contour maps were produced from a grid of average void ratios. The void ratios were averaged over small annuli within the specimen, centered around the central long axis. The annuli were at evenly spaced radii and heights throughout the specimen. This method reports the average void ratio in a circumferential region, appropriate to axisymmetric shearing structures. Two features of the contour maps should be noted. First, the shear bands illustrated earlier (Figures 1 & 2c) are not seen in contour mapping, as both shear bands and areas outside of shear bands were combined in each average void ratio calculation. Second, the apparent increase in void ratio seen at the outer edge of the specimen is not a physical property of the specimen, but is related to the edge effects of the CT scan at the specimen-water interface. Figure 4 shows an example of a fully compressed specimen (25% axial strain.) Shear bands are fully developed, and the quarter-circle areas of low void ratio at both ends of the map are indicative of fully developed shear cones. On the border of these shear cones is a sharp transition to higher void ratios. In the area outside of the shear cones, shear banding is present, and indicated by the large void ratios throughout the remainder of the specimen. At the smaller radii, the void ratio is high where several shear bands cross and the shear cones merge. Images of this area do not reveal individual shear bands, but a mass of low density material.

CONCLUSIONS

The CT technique demonstrated the ability to detect specimens inhomogeneities, localization patterns, and quantify void ratio variation within sand specimens.

ACKNOWLEDGMENTS

The authors gratefully acknowledge the financial support provided by NASA/ George Marshall Space Flight Center under contract No. NAS8-38779. Thanks are also due to Peter Engel of NASA/ Kennedy Space Center for help in performing the CT scans.

REFERENCES

Alshibli, K., Costes, N., and Porter, R. (1996). Mechanics of Granular Materials, Space Processing of Materials, *International Symposium on Optical Science, Engineering, and Instrumentation, SPIE-The International Society for Optical Engineering*, Denver, CO, Vol. 2809, Aug 1996, 303-310.

Amos, C. L., Sutherland, T. F., Radzijewski, B., and Doucette, M. (1996). A Rapid Technique to Determine Bulk Density of Fine-grained Sediments by X-ray Computed Tomography, *Journal of Sedimentary Research*, 66 (5), 1023-1039.

Bracewell, R. N. (1956). Strip Integration in Radio Astronomy, *Australian Journal of Physics*, 9, 198-217.

Coshell, L., McIver, R. G., and Chang, R. (1994). X-Ray Computed Tomography of Australian Oil Shales: Non-Destructive Visualization and Density Determination, *Fuel*, 73 (8), 1317-1321.

Denison, C., Carlson, W. D., and Ketcham, R. A. (1997). Three-dimensional Quantitative Texture Analysis of Metamorphic Rocks Using High-Resolution Computed X-ray Tomography: Part I. Methods and Techniques, *Journal of Metamorphic Geology*, 15(1), 29-44.

Denison, C., and Carlson, W. D. (1997). Three-dimensional Quantitative Texture Analysis of Metamorphic Rocks Using High-Resolution Computed X-ray Tomography: Part II. Application to Natural Materials, *Journal of Metamorphic Geology*, 15 (1), 45-57.

Engel, P. (1998). "NASA/KSC Industrial Computed Tomography (CT) System", Unpublished Report.

Phillips, D. H., and Lannutti, J. J. (1997). Measuring Physical Density with X-ray Computed Tomography, *NDT & E International*, 30 (6), 339-350.

Roscoe, K. (1970). The Influence of Strains in Soil Mechanics, *Geotechnique*, 20 (2), 129-170.

Sture, S., Costes, N., Batiste, S., Lankton, M., Alshibli, K., Jeremic, B, Swanson, R., and Frank, M. (1998). "Mechanics of Granular Materials at Low Effective Stresses," *ASCE, Journal of Aerospace Engineering*, 11 (3), 67-72.

Vardoulakis, I. and Graf, B. (1985). Calibration of Constitutive Models for Granular Materials Using Data from Biaxial Experiments, *Geotechnique*, 35(3), 299-317.

Vardoulakis, I., Graf, B., and Hettler, A. (1985). Shear Band Formation in a Fine Sand, *Proceedings of the 5th International Conference on Numerical Methods in Geomechanics*, Nagoya, 517-522.

Zeng, Y., Gantzer, C. J., Payton, R. L., and Anderson, S. H. (1996) Fractal Dimension and Lacunarity of Bulk Density Determined with X-ray Computed Tomography, *Soil Science Society of America Journal*, 60, 1718-1724.

Cold neutron and monochromatic X-ray micro-tomography

B. Masschaele[*], S. Baechler[#], P. Cauwels[*], J. Jolie[#], W. Mondelaers[*], M. Dierick[*]

[*]Department of subatomic and radiation physics, Ghent University, Proeftuinstraat 86, B-9000 Gent, Belgium
[#]Institute of physics, University of Fribourg, Pérolles, CH-1700 Fribourg, Switzerland

Abstract. Both neutrons and X-rays have very interesting properties for tomography. Tunable monochromatic X-ray beams were used for element sensitive tomography of samples containing heavy elements using the K-edge dichromatic scanning technique. This research was done at the high-energy beam-line of the ESRF (France). A new set-up has been build for cold neutron tomography at the spallation source SINQ at PSI (Switzerland). The high flux cold neutron beam transmissions are observed with a CCD camera via a neutron-to-visible-light converter. Many samples have been investigated with parallel beam geometry. At the moment the conversion-screen resolution is about 230 µm. A neutron focussing lens was installed to make a cone-beam geometry. With this set-up, the resolution will increase by the magnification. Software for computer reconstruction for parallel neutron and X-ray beams and neutron cone beams have been developed.

INTRODUCTION

Since a couple of years our research team has been working in the field of micro tomography [1],[2],[3]. The use of intense monochromatic X-ray beams led to the possibility of element sensitive 3D tomography. This research was done at the high-energy beam-line of the ESRF (France). Secondly, using an intense cold neutron beam for micro tomography led to the possibility of non-destructive testing of new samples. A new set-up has been build for cold neutron tomography at the spallation source SINQ at PSI (Switzerland).

X-ray tomography has been used for nondestructive analysis of samples for a long time. The same procedure can be applied using neutrons. By looking at the attenuation of the neutron beam in the sample from different angles, it is possible to do a neutron tomography. Because the properties of neutrons are so much different from X-rays, a complementary range of applications can be found. Due to the different nature of interaction and the complementary Z-dependent cross sections, X-ray are mainly used for heavy element tomography whereas neutrons are much more suitable for light element tomography. There are some disadvantages with the use of neutrons like the activation of samples and equipment due to nuclear reactions.

In the first part of this paper we will discuss monochromatic X-ray tomography and the second part gives an overview of our latest result with cold neutron micro-tomography.

MONOCHROMATIC X-RAYS FOR ELEMENT SENSITIVE TOMOGRAPHY

The standard procedure for tomography is to measure the attenuation of the radiation through a sample for several orientations. The recorded patterns are used to reconstruct the three-dimensional attenuation distribution for the whole sample. By scanning samples using photons having two different energies, one just below and one just above the K-edge of the element under investigation, the tomography becomes element sensitive.

We built a completely automated transportable measurement setup, also applicable for neutron tomography measurements. A schematic presentation can be seen in figure 1. The X-rays or neutrons are converted to visible light after passing the sample. The light is reflected by a surface mirror towards a low-light-level CCD camera. The samples are positioned by high precision stepping motors. At the moment the whole set-up is controlled by LabVIEW®-programs.

The X-ray conversion screen is made out of a YAG crystal.

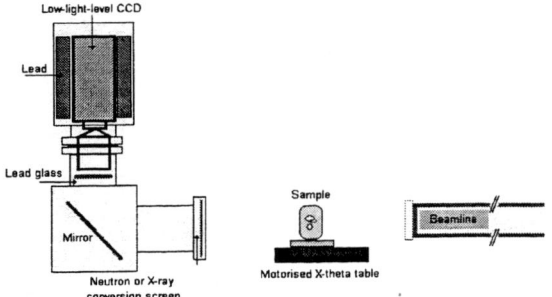

FIGURE 1. A schematic representation of the tomography set-up. In the case of neutrons a lens can be placed between the sample and the beam-line.

The tomography set-up is completely automated which allows the operator to leave the beam-line. When the storage ring needs to be refilled or if there is an accelerator problem the tomography is automatically paused and restarted when the beam is back.

At the high-energy beam-line ID15, there are two insertion devices: an asymmetrical multipole wiggler (AMPW) and a superconducting wavelength shifter (SCWS). We use the SCWS. There are permanent filters in the beam (0.7 mm C, 4.0 mm Be, 4.1 mm Al). The beam intensity is $5.13 \cdot 10^{13}$ ph s^{-1} mrad^{-2}, 0.1% bw, 0.1A at 95 keV. The energy resolution is 1keV.

Up to now we reached an image resolution of 15μm. This value can go down to 7μm depending on which camera lens is used. With this value samples of about 2cm in width can be scanned. A typical scan for one chemical element takes about two hours. One advantage of the monochromatic beam in comparison with a white or polychromatic beam is the absence of beam hardening. Hardening of the beam leads to artifacts that are very difficult to correct. An advantage of using a crystal monochromator is the possibility of changing the photon energy in a continuous way. For our set-up we use a Si 311 crystal with an asymmetric cut of 63°.

The normalization of the acquired projections is very important. It can be done in two ways. During the acquisition of each image the beam intensity is integrated. Afterwards every projection is normalized by a corresponding reference image I_o. t_{acq} is the camera acquisition time.

$$\text{Pix}(u,v)_{\beta,\text{norm}} = \frac{\text{Pix}(u,v)_\beta}{t_{acq}(\beta).\text{counts}(\beta)} \times \frac{t_{acq,o}.\text{counts}_o}{\text{Pix}(u,v)_o} \quad (1)$$

"Pix" is the pixel value at position u, v. β is the projection angle. t_{acq} is the camera acquisition time. "Counts" is a counter value for the integration of the beam intensity during the acquisition of the projection. A second, but less accurate, method is normalization on a region of the image. This method gives better results for neutron tomography because of the low counter values in this case.

Here below is an example of the possibilities of tunable monochromatic X-ray tomography (figure 2). First experiments were mainly done on artificial samples to investigate the technique. After a thorough analysis we started 3D tomography on biological, industrial and geological samples. With this technique we studied the metal diffusion from amalgam tooth fillings into the teeth. A number of teeth was scanned for silver, tin, mercury and gold. Within the precision of the set-up we could only detect a diffusion of tin around the filling. A new project is scheduled for this year where glass encapsulated nuclear waste will be subjected to X-ray and cold neutron micro-tomography.

FIGURE 2. Mercury containing cinnabar stone from Toscana. Left: reconstruction of stone with single energy. Right: the dual energy reconstruction only shows the mercury.

COLD NEUTRON MICRO-TOMOGRAPHY

First cold neutron micro-tomography experiments were performed at the cold neutron source of SINQ at PSI (Villigen CH). The neutrons are produced by spallation reaction of protons in a lead target. Neutron guides extract cold neutrons to the experiments from a cold moderator of about 20 liters of liquid deuterium, at the very low temperature of 25 Kelvin. The beam intensity at the sample is 10E8 neutrons/cm^2.s.

The same set-up is used as for X-ray tomography. The neutron beam (2 by 5 cm) is observed on a "neutron-to-visible-light"- convertor. This screen consists of three layers. First layer is an aluminum plate which serves as substrate and reflector. The second layer is Li. Alpha's are produced by the n-Li nuclear reaction. The alpha's are then converted to visible light in a ZnS layer. The screen determines the 230 µm resolution. Figure. 2 shows a paper tissue in a water filled lead container. The lead container has a wall thickness of 1.5cm. X-rays could never penetrate the lead cylinder. Neutrons almost don't notice the lead. On the other hand, the water in the tissue easily stops the neutron beam.

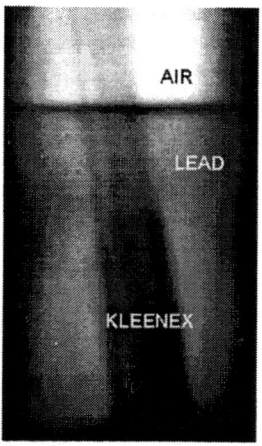

FIGURE 3. Neutron radiograph of water being absorbed by a paper tissue. The tissue is placed in a lead container with 1.5 cm thick walls.

The resolution of the neutron radiograph is determined with a strongly absorbing peace of Li. The edge between the Li and air can be used to determine the image resolution. As a definition for resolution we determine the number of pixels between ¼ and ¾ of the edge. This number is then multiplied with the pixel size. The resolution gets worse if the distance between the Li or sample and the screen becomes bigger. See figure 3. The reason is the 0.4° beam divergence. One can easily calculate the divergence of the beam from the fit to the data.

$$\mathrm{Div} = \mathrm{Arc\,tan}(\frac{7.3192}{1000}) = 0.42° \quad (2)$$

The cold neutrons are transported from the spallation target to the experiment room by means of supermirror coated neutron guides. The cold neutrons are reflected by the walls and have consequently a certain angle range. The divergence of the beam is also a function of the energy (higher energies have smaller scattering angles, also meaning less divergence):

$$\mathrm{Div}(°) = 0.093 \times \lambda(\text{Å}) \quad (3)$$

The maximum of the spectrum is 4Å. The beam spectrum begins at 1.8 Å has a tail up to 10 Å. The samples are therefore to be placed as close as possible to the screen. The set-up allows coming as close to the screen as 3cm.

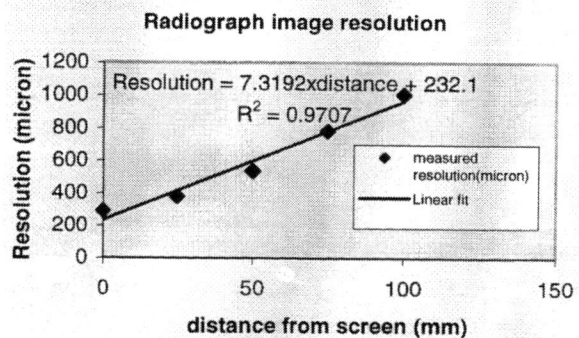

FIGURE 4. Neutron radiograph resolution as function of the distance from the conversion screen.

Below are two examples of neutron micro-tomography. Both samples are biological. The first picture is a cut through the reconstruction of a chestnut and the second picture is the reconstruction of a coral. Both samples are low Z material and are therefore ideal for neutron tomography. The cross-section as a function of Z, for neutrons, cannot be expressed with a simple equation like for X-rays. Since they have completely different attenuation coefficients, neutron tomography is complementary to X-ray tomography.

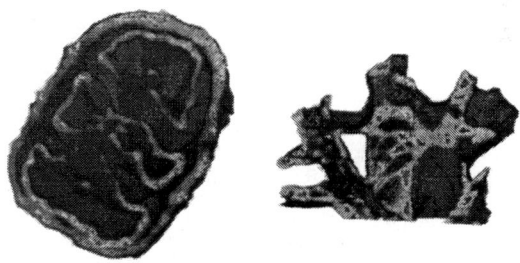

FIGURE 5. Neutron micro-tomography of a chestnut (left) and a coral (right)

Neutron micro-tomography with a neutron lens set-up

By using a neutron-focusing lens, it is possible to change the beam geometry from parallel beam to cone-beam. The lens is composed of a large number of poly-capillary fibers, parallel at the lens entrance and bent in such a way that all fibers converge towards a focal point. The diameter of the focused beam at the waist is smaller than 0.5 mm and the flux gain is greater than 30. At the waist a 5mm thick Li pinhole collimator stops background neutrons. The collimator opening defines the source and the diameter of the opening the source size. Normally, this will determine the image resolution. The advantage of the lens set-up is the possibility to magnify the sample projection.

$$\text{Magnification } M = \frac{d}{D} \quad (4)$$

D is source to screen distance, d the source to sample distance. The resolution of the screen is fixed, but when M is bigger than 2, the resolution becomes better than 250μm. A second property of the lens is the possibility of increasing the beam size to investigate larger samples. A disadvantage of the lens is the lower neutron flux. The acquisition time, with the Sensicam 12bit cooled CCD camera is more than 2 minutes compared to 200ms. For cone-beam tomography one has to scan the samples for more than 180°. Normally the sample projections are acquired over 360° degrees.

The software for cone-beam reconstruction from CCD images is far more complex than for parallel beam. The simple rotation over 360° is not sufficient to make an accurate reconstruction. Nowadays, most people make use of helix scanning. In our case, the sample is moving step-wise in the vertical direction while the rotation takes place. Two methods are studied at the moment. The first method is based on rebinning the pixels. The second is based on a generalized FDK algorithm.

OUTLOOK

Neutron micro-tomography is relatively new. Some aspects should need further investigation. The neutron beam is not monochromatic and beam hardening occurs. This effect will be studied with monte carlo simulations (MCNP4B) and we will write an algorithm to correct this effect. Neutron conversion screens will also be studied with MCNP. Helix tomography for accurate cone beam reconstruction will be applied.

CONCLUSION

Both, monochromatic X-rays and neutrons, have interesting properties for micro-tomography. Much research still has to be done in the field of neutron tomography. The resolution of neutron conversion screens has to improve. This and other papers have made comparisons between X-rays and neutrons [4]. The techniques are highly complimentary.

ACKNOWLEDGMENTS

We would like to thank M. Defrise of the VUB for helping us with the software for cone-beam reconstruction. This project is supported by the Institute for the Promotion of Innovation by Science and Technology in Flanders (IWT).

REFERENCES

1. M.Bertschy, J.Jolie, W.Mondelaers " Heavy element tomography using tunable gamma-ray beams " Applied Physics A 62 (1996) 437 - 44.

2. J. Jolie, T. Materna, B. Masschaele, W. Mondelaers, V. Honkimaki, A. Koch, T. Tschentscher "Heavy element sensitive tomography using synchrotron radiation above 100 keV" in 'Speciation, Techniques and Facilities for Radioactive Materials at Synchrotron Light Sources' Eds. Nuclear Energy Agency (1999) 249

3. T. Materna, J. Jolie, W. Mondelaers, B. Masschaele, V. Honkimaki, A. Koch, T. Tschentscher "Uranium sensitive tomographies with synchrotron radiation" J. Synch. Rad. 6 (1999) 1059

4. H. Kobayashi, M. Satoh "Basic performance of a neutron sensitive photostimulated luminescence device for neutron radiography" in NIM A 424 (1999) 1-8

Neutron Radiography Activity in the European Program Cost 524: Neutron Imaging Techniques

P. Chirco (1), P. Bach (2), E. Lehmann (3), M. Balasko (4)

(1) I.N.F.N., Bologna, Italy
(2) SODERN, 20 avenue Descartes, 94451 Limeil-Brévannes Cedex, France
(3) Paul Scherrer Institute, CH 5232 Villigen PSI, Switzerland
(4) Hungarian Academy of Sciences Atomic Energy Research, Budapest, Hungary

Abstract

COST is a framework for scientific and technical cooperation, allowing the coordination of national research on a European level, including 32 member countries. Participation of institutes from non-COST countries is possible. From an initial 7 Actions in 1971, COST has grown to 200 Actions at the beginning of 2000. COST Action 524 is under materials domain, the title of which being "Neutron Imaging Techniques for the Detection of Defects in Materials", under the Chairmanship of Dr P. Chirco (I.N.F.N.). The following countries are represented in the Management Committee of Action 524 : Italy, France, Austria, Germany, United Kingdom, Hungary, Switzerland, Spain, Czech Republic, Slovenia and Russia. The six working groups of this Action are working respectively on standardisation of neutron radiography techniques, on aerospace application, on civil engineering applications, on comparison and integration of neutron imaging techniques with other NDT, on neutron tomography, and on non radiographic techniques such as neutron scattering techniques. A specific effort is devoted to standardisation issues, with respect to other non European standards. Results of work performed in the COST frame are published or will be published in the review INSIGHT, edited by the British Institute of Non Destructive Testing.

INTRODUCTION

COST is a framework of scientific and technical cooperation, allowing the coordination of national research on a European level. COST Actions consist of basic and precompetitive research as well as activities of public utility.

There are 32 COST member countries : the fifteen EU Member States plus Iceland, Norway, Switzerland, Czech Republic, Slovakia, Hungary, Poland, Turkey, Slovenia, Croatia, Malta, Estonia, Bulgaria, Cyprus, Latvia, Lithuania and Romania.

COST cooperation was set up in 1971 by a ministerial conference attended by Ministers for Science and Technology of the 19 original COST countries.

This cooperation was widened two times to include Iceland, Hungary, Czechoslovakia and Poland at the second Ministerial Conference held on COST in Vienna in November 1991 and Malta, Estonia and Romania at the third Ministerial Conference held in Prague in May 1997.

This organisation, which laid the foundations for scientific cooperation at the European level, is based upon a flexible set of arrangements enabling different national organisations, institutes, universities and industry to join forces and make concerted efforts in a broad range of scientific and technical areas.

From the beginning, the European institutions (Commission and Council Secretariats) have played a particularly important role in the COST framework.

Since 1989, organisations and institutes from non-COST countries and especially from other Central and Eastern European countries may also participate in individual COST Actions, if there is a justified mutual interest.

From an initial 7 Actions in 1971, COST has grown to 200 Actions at the beginning of year 2000.

A special feature of all COST Actions is the complete freedom of participation by each country involved, according to national research priorities.

COST funding covers only the coordination expenses of each Action (scientific secretariat, contribution to workshops and conferences, publications, short term scientific mission, etc…).

COST Actions exist in the 17 following domains at present : informatics, food technology, oceanography, urban civil engineering, transport, medical research, environment, forests and forestry products, materials, chemistry, agriculture and biotechnology, miscellaneous, meteorology, fluid dynamics, social sciences, physics, telecommunications.

COST Action 524 is under materials domain.

THE EUROPEAN PROGRAM COST 524

The COST Action 524 is entitled "Neutron imaging techniques for the detection of defects in materials". This action was started on 23 March 1998 and will be terminated on 22 September 2002, under the chairmanship of 3 representatives :
- a chairman, Dr Piero Chirco (INFN/Italy),
- a vice-chairman, Dr Marton Balasko (KFKI/Hungary),
- and an E.C. representative, Dr Oliver Pfaffenzeller (European Commission DG XII/B1, Belgium).

Following countries are represented in the Management Committee of this Action : Italy, France, Austria, Germany, United Kingdom, Hungary, Switzerland, Spain, Czech Republic, Slovenia and Russia.

The six working groups (WG) of this Action are working respectively on :

WG1 – Standardisation of neutron radiography techniques,
WG2 – Aerospace applications
WG3 – Civil engineering applications
WG4 – Comparison and integration of non imaging techniques with other NDT,
WG5 – Neutron tomography
WG6 – Non radiographic techniques (such as neutron scattering techniques)

The present paper will give some highlights of the COST 524 Action, mainly related to the works of some of the Working Groups.

COMPARISON OF EUROPEAN NEUTRON RADIOGRAPHY FACILITIES

European neutron radiography facilities were systematically compared, in the frame of the working group WG1 called "standardisation". A recent paper (1) underlines the possibilities presently available in Europe for practical applications : quantitative non destructive studies, computed neutron tomography, or real time imaging.

There are 11 European member countries of the COST collaboration 524, whereas Switzerland and Russia play a special role being not explicitly funded as COST members.

Different neutron sources are used for neutron radiography. Some countries are operating large size research reactors, like France (3 reactors : ORPHEE, OSIRIS and ISIS), Germany (2 reactors : FRM-1 and BER-2, and FRM-2 in year 2001), Hungary (1 of the Russian WWS type), or TRIGA research reactors, like Austria, Italy and Slovenia.

A spallation neutron source SINQ of the PSI in Switzerland is also in operation, with two neutron radiography beam lines.

A mobile neutron source, based on a high power sealed D-T tube, is in operation at SODERN (France).

For a worldwide comparison of neutron radiography stations to other stations, the leading Japanese facility at the JAERI research reactor JRR-3M (2) is included here.

Neutron flux level

Most of the beam lines have a thermal spectrum, while cold neutrons are provided at the CEA-ORPHEE station, at the PSI-NCR facility, and partly by the TUM-channel 2.

At the HMI-beam line, the cold neutrons are further monochromatised and the beam intensity thus becomes very low. The cold neutron guides at CEA and PSI have the highest intensity but limited beam sizes. They are used for practical applications with extended samples in a horizontal scanning mode.

Figure 1 presents the compared neutron fluxes in a logarithmic scale.

Beam collimation

Beam collimation is described by the L/D ratio, which limit the obtainable spatial resolution by the inherent blurring. Unsharpness U_{beam} can be easily related to the distance between the object and the detector plane d :

$$U_{beam} = \frac{d}{L/D}$$

Figure 2 presents the spatial resolution given by the beam collimation at the different beam lines. The inherent detector resolution is given for reference, for X ray films, for image plate, and for CCD camera with scintillator. Most of the facilities cannot use completely the detector performances due to the unsharpness of the beam.

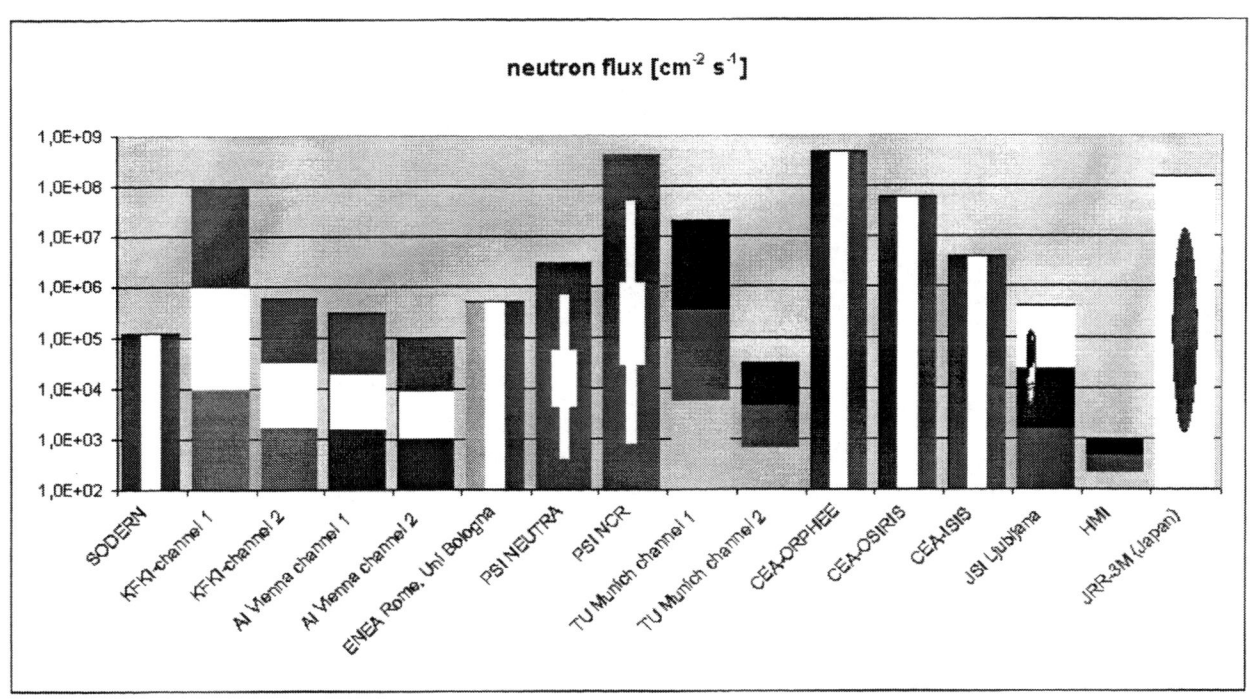

Fig. 1: Neutron beam intensities of European NR beam lines (a Japanese one is given for comparison)

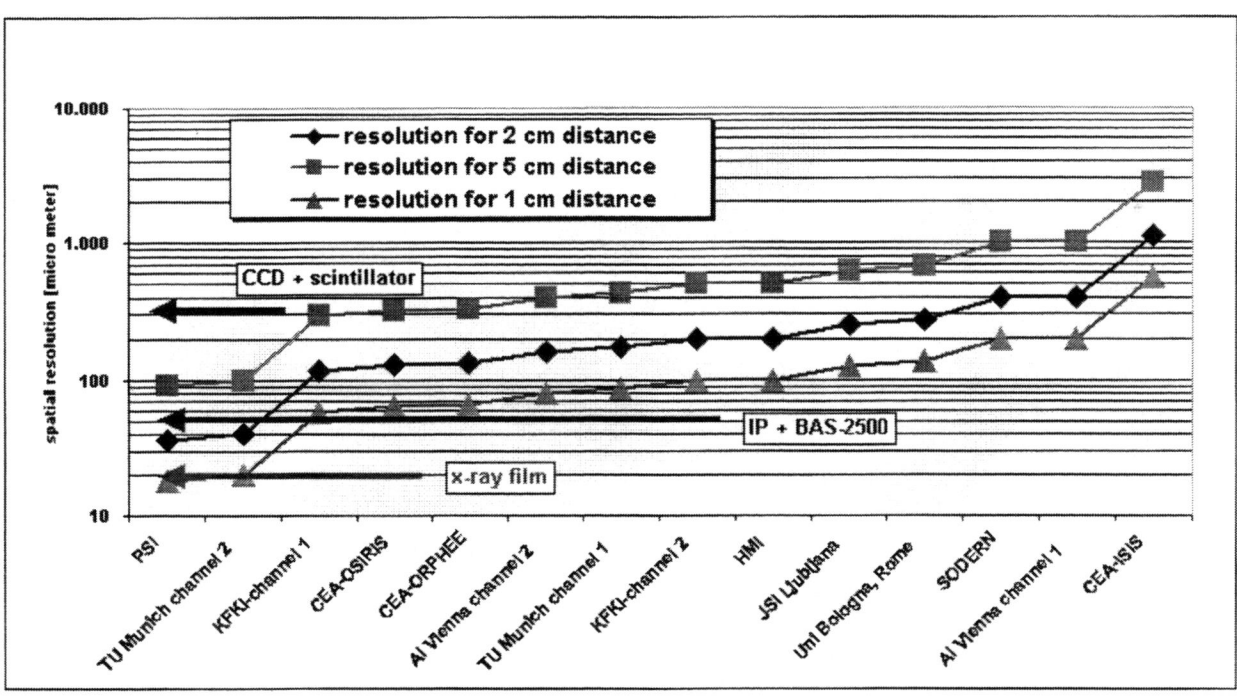

Fig. 2: The spatial resolution given by the beam collimation at the different beam lines is related to the inherent detector resolution of three radiography methods. Most of the facilities cannot use completely the detector performances due to the inherent unsharpness of the beam.

	TU Munich	ENEA Casaccia	KFKI Budapest	AI Vienna	Uni Fribourg	CEA Saclay	PSI	HMI	Lebedev Institute Moscov
country	Germany	Italy	Hungary	Austria	Switzerland	France	Switzerland	Germany	Russia
neutron source	FRM-I	reactor @ Cassacia	KFKI-reactor	AI-reactor	SINQ	ORPHEE	SINQ	BER-II	RRT
source type	MTR-reactor	TRIGA-reactor	WWS-reactor	TRIGA-reactor	spallation source	MTR-reactor	spallation source	MTR-reactor	reactor
neutron energy	thermal	thermal	thermal	thermal	cold	cold	thermal	cold+monoenergetic	monoenergetic
L/D	500	54 (estimated)	170	125			550	340	100
flux level [cm-2 s-1]	1.00E+05	5.00E+05	1.00E+08	1.00E+05	1.00E+08		3.00E+06	1.00E+04	2.00E+05
detector type	CCD-camera	CCD-camera	CCD-camera	CCD-camera	CCD-camera	not yet decided	CCD-camera	CCD camera	CCD camera
number of pixel x	512	192	512	512	1248		512	512	1024
number of pixel y	512	165	512	512	1024		512	512	1152
field of view in x [cm] (min to max)	6 to 28	1,5 to 12	1 to 10	4 to 21	2		2 to 30	0.1 - 2	1 to 35
field of view in y [cm] (min to max)	6 to 28	1,2 to 10	1 to 10	4 to 21	5		2 to 30	0.1 -2	1 to 35
exposure time / frame	20 - 120 s		40 ms	20 - 180 s	750 ms		10 - 60 s	200 -2000 s	variable
number of projections	200	200	360	200	400		200	30-100	up to 180
max. weight [kg]	20	1	1	1			20	5	up to 17
method for reconstruction	filtered backprojection	filtered backprojection	filtered backprojection		filtered backprojection (Shepp-Logan)		filtered backprojection	filtered backprojection, ART	filted backprojection, maximum likelyhood
tools for visualisation	MIRA, VG Studio	AC3D	own software	IDL			IDL, VG Studio	Sclicer, VG Studio	Own+Slicer
planned improvements	reactor FRM-II, new facility and setup	New CCD at new collimator	start of operation after set up		neutron lens for enlargement		higher resolution, lower exposure	new math. Tools larger CCD - camera	tomography on portable neutron sources
applications	adhesives, moisture, defects, reverse engineering	R&D	hydrogen in Zr tubes		tooth, batteries	explosites, initiators	turbine blades, moisture distribution, ceramics	defects, new imaging signals , R&D	Investigation of perspective application

Table 1: Comparison of the different approaches to neutron tomography within the COST-524 member countries

Fig. 3: Neutron tomography image of an aircraft turbine blade (taken at the P.S.I. station NEUTRA)

Gamma radiation background

The neutron/gamma ratio is an important factor : a low value creates a disadvantageous fog in the images. This ratio is very high, more than 10^8 cm^{-2} s^{-1} Sv^{-1} h, at KFKI/Hungary, at JSI/Slovenia, and especially at PSI/Switzerland on the two beam lines.

Beam size

Most of the beams have a circular shape, but neutron guide lines have commonly a rectangular cross section. The guides for cold neutrons have relatively small fields of view but high intensity. They are often used in a scanning mode, using a moving trolley for sample and detector.

This overview of existing NR facilities operated by COST action 524 members helps to understand the abilities for neutron radiography in Europe.

MOBILE NEUTRON RADIOGRAPHY EQUIPMENT FOR AEROSPACE APPLICATIONS

The need for mobile neutron radiography equipment comes mainly from aerospace applications, such as inspection of airplane structures for early corrosion detection, or inspection of turbine blades, where neutrons are well absorbed by organic materials, by water, etc…, even behind metallic housings.

Mobile neutron radiography equipment are either robotic systems moveable in a specific area, or smaller systems to install in a specific plant and to move after use to another plant. The purpose of such equipment is to put neutrons in suitable locations and only at the appropriate time, and to avoid sending numerous objects in front of a reactor or accelerator based neutron radiography station.

Different concepts for mobile equipment are related to the neutron energy (fast neutrons for a high penetration depth, or thermal neutrons for a good resolution), to the size and weight of the equipment (from 0.5 to 3 metric tons), and to the switchable source capability (switchable for neutron tube based equipments like DIANE, non switchable for isotopic neutron emitters).

A comparison of different systems is described in ref. 2. Effective performances are difficult to compare, while optimisations for different applications are numerous. An important criterion is the beam divergence L/D, which is between 10 and 70 for mobile equipments. Main imaging devices are neutron converters with specific vacuum cassette films, CCD cameras (cooled or not) looking at a scintillating screen such as Gd2O2S based screen, and imaging plates like photostimulable luminescent phosphors.

Taking the example of the DIANE mobile system designed by SODERN, the thermal neutron flux on the object is adjustable between 5.10^3 and 10^6 n/cm².s, and L/D is adjustable between 10 and 60. Exposure time is between 10 Minutes and a few hours, depending on sample type and expected resolution.

Some aerospace applications are the following :
- inspection of cracks in fighter ailerons,
- water and corrosion detection in structure parts,
- quality control inspection of turbine blades at the end of the manufacturing process, using a gadolinium based solution as a marker,
- delamination of composite materials after an impact, using a similar marker,
- inspection of pyrotechnic devices,
- inspection of lubrication films.

NEUTRON TOMOGRAPHY DEVELOPMENTS

For tomography, it is required to have many two-dimensional projections in digital form of an observed object rotating over 180° angle range with respect to its central axis. The detector position and the sample manipulation must be adjusted with the accuracy of the size of one pixel in the projections. Usually a stationary digital neutron detector and a turntable are installed in a quasi-parallel neutron beam with a high L/D ratio.

A summary of the most important properties of set-ups in the different institutions of the COST-524 countries dealing with neutron tomography is given in table 1. The 9 existing systems are based on a CCD-camera set-up with a neutron sensitive scintillator as detector.

Objects from a few cm to about 30 cm can be investigated. The limitation for higher spatial resolution with camera base systems is given mainly by the properties of the scintillator. The spread of light in the scintillating material limits the resolution to about 0.2 mm. A more complete comparison is given in ref. 3.

Visualisation of tomographic results is much more difficult, compared with radiographic imaging. This is due to the amount of data and to the need for adequate presentation in 3D with respect to object and defects shapes to investigate. An example of image of an aircraft turbine blade is shown in figure 3, supplied by the team of PSI in Switzerland.

Two tasks were started for all participants of the neutron tomography COST-524 Action :
- to compare the software tools applied for volume reconstruction by delivering a complete set of projections of one observed object,

- to compare detector and beam performance by a set of test specimens with well defined geometry and material composition.

All potential partners outside this collaboration are kindly invited for participation in the exchange of knowledge and experience in the future.

INTEGRATION OF NEUTRON IMAGING TECHNIQUES WITH OTHER NDT METHODS

For a number of applications, complementarity of neutron imaging and X-rays imaging is evident. Some interesting examples are the investigation of ageing defects in airplane wings, or the quality inspection of pyrotechnic devices. It is less evident to use complementarity of neutron imaging and ultrasonic or acoustic techniques.

Some participants in the COST-524 Action are working under the leadership of the group of KFKI in Budapest (Hungary), on dynamic neutron radiography associated with acoustic techniques for dynamic inspection of cooling devices. Study of new cooling agents in compressor type refrigerators was performed simultaneously by neutron radiography at the Budapest reactor, by vibration diagnostics in the 20 Hz – 20 kHz frequency range, and by acoustic emission technique up to 100 kHz for a better sensitivity (see ref. 4 and 5). Results are encouraging, and these techniques will be used soon in the quality control line for industrial products.

CONCLUSION

A specific effort was devoted to standardisation issues, and comparison of performances of different equipments available today in the countries involved in the COST-524 Action. All potential partners outside this collaboration are invited for participation in the exchange of knowledge and experience in the future. Results of work performed in the COST frame are published or will be published in the review INSIGHT, edited by the British Institute of Non Destructive Testing.

ACKNOWLEDGEMENTS

The present paper was performed in the frame of the COST-524 Action, under the auspices of the European Commission/DG XII. The authors wish to thank this organisation for its support.

REFERENCES

1. E.H. Lehmann
   15th World Conference on Non Destructive Testing (WCNDT), Roma, 15-21 October 2000
   Facilities for neutron radiography in Europe.
2. P. Bach
   INSIGHT Vol. 42, n° 4, April 2000
   Mobile neutron radiography systems and applications.
3. E.H. Lehmann et al.
   15th WCNDT, Roma, 15-21 October 2000-09-29
   Status and prospects of neutron tomography in Europe
4. M. Balasko et al.
   15th WCNDT, Roma, 15-21 October 2000
   Combined dynamic neutron radiography and vibration diagnostics for industrial applications.
5. M. Balasko et al.
   Proc. 6th WCNR, Osaka, May 1999
   Dynamic neutron radiography as a promising method for modern NDT techniques.

An Accelerator System for Neutron Radiography

Brian Rusnak, James Hall

Lawrence Livermore National Laboratory, Livermore, CA 94550

Abstract. The field of x-ray radiography is well established for performing non-destructive evaluation of a vast array of components, assemblies, and objects. While x-rays excel in many radiography applications, their effectiveness diminishes rapidly if the objects of interest are surrounded by thick, high-density materials that strongly attenuate photons. Due to the differences in interaction mechanisms, neutron radiography is highly effective in imaging details inside such objects. To obtain a high intensity neutron source suitable for neutron imaging, a 9-MeV linear accelerator is being evaluated for delivering a deuteron beam into a high-pressure deuterium gas cell. Since a windowless aperture is needed to transport the beam into the gas cell, a low emittance is needed to minimize losses along the high-energy beam transport (HEBT) and the end station. A description of the HEBT, the transport optics into the gas cell, and the requirements for the linac will be presented.

INTRODUCTION

The concept of using 10 – 15 MeV neutrons to image heavily-shielded, low-Z (atomic number) objects has been demonstrated [1]. To significantly increase the resolution and decrease the imaging time for neutron radiography, a neutron source with an intensity in excess of 10^{12} n/sec is required.

Present-day techniques for accelerator-driven neutron generation that use thin metal windows to separate a pressurized deuterium gas cell from the accelerator vacuum are limited in how much beam intensity can be delivered to the gas due to window heating, thus typically limiting the overall intensity of the neutron source to approximately 10^{11} n/sec.

Recent development efforts at Brookhaven National Laboratory (BNL) and Massachusetts Institute of Technology (MIT) [2] have advanced two "windowless aperture" systems to the point where they provide attractive alternatives to metal windows for pressurized gas cell applications. These systems are enabling technologies for achieving much higher deuteron beam intensities.

Utilizing this capability, an effort is underway at Lawrence Livermore National Laboratory (LLNL) to develop an integrated HEBT and end station design to specify the beam requirements for a high-intensity deuteron particle accelerator intended for neutron radiography.

Two primary constraints exist in transporting the particle beam to the gas cell. The first and more important constraint is that the beam must be focused into a narrow channel in the high-pressure deuterium gas cell to achieve maximum resolution by the imaging system. The second constraint concerns minimizing unnecessary induced and operational radioactivity in the overall machine. While the radiation from the pressurized gas cell will be large, if it is localized in only the gas cell, it is more easily shielded and controlled. To satisfy this constraint, the beam needs to be transported to the gas cell with minimal beam loss in the accelerator and the HEBT.

DIFFERENTIAL PUMPING AND PRESSURIZED GAS CELL

The present approach is to use a plasma porthole to isolate the 2 – 3 atmospheres of deuterium gas in the gas cell from the accelerator vacuum. Testing at MIT has shown that a plasma porthole running with argon can effectively plug a 5-mm-diameter channel and maintain a 10^{-4} Torr vacuum while holding over 2 atmospheres

---

* Work performed under the auspices of US Department of Energy by Lawrence Livermore National Laboratory under contract W-7405-Eng-48.

in the gas cell 125 mm away [3]. To reduce the gas load to the accelerator further, a series of three apertures tubes from 5 to 8 mm in diameter and 115 mm long each provide conductance-isolation between pumpout chambers in the differential pumping section. The combination of the small diameters of the gas cell and the aperture tubes with an overall longitudinal spacing of 800 mm requires the incoming particle beam be of a sufficiently small emittance so it can be transported cleanly through the system. By using the aperture system in combination with the plasma porthole, a vacuum on the order of 10^{-7} Torr is anticipated in the HEBT. Figure 1 shows the design concept of the end station used to develop the accelerator beam requirements.

Constraining the beam emittance so it can be cleanly transported through the aperture system allows a rotating aperture gas cell design to be used as a backup technology for isolating the high pressure gas from the accelerator vacuum. In a rotating aperture system, the beam holes must be as small as possible, and high levels of beam loss can adversely affect the lifetime of the rotating aperture system.

BEAM STOP

After the deuteron beam passes through the gas cell, it needs to be stopped in such a way that the spatial distribution of neutrons generated in the gas cell is minimally impacted. Conventional approaches for stopping a 9-MeV, 100 – 300 µA average-current deuteron beam involve impinging the beam either on a rotating disk or on a thin water-cooled metal target sloped to decrease the power density. Both of these approaches require significant hardware be placed in the flight path of the neutrons generated in the gas cell which would adversely affect the distribution of the neutron beam.

An alternative beam stop approach is proposed that should be simple, inexpensive, and cause minimal disruption to the neutron beam. The beam would be stopped in a high-Z, high-pressure gas behind the deuterium gas cell and use a parallel-flow, pressure-balanced interface that would allow the deuteron beam to go into the high-Z gas and be stopped there with limited gas mixing, as shown in Figure 1. This approach allows the deuterium gas cell to be a short length, which is needed for high resolution by keeping the interaction region small, and it helps keep the neutrons nearly monoenergetic, as a small energy spread is imparted on the beam as it passes through the deuterium gas cell.

By using a high-Z stopping gas, cleaning cross-contaminant gases from the other stream relies on the properties of the respective gases, that is, deuterium in the high-Z gas could be removed by gettering, where high-Z atoms in the deuterium could be removed by cryo-trapping. The 1 – 3 kW of average beam power would be removed by a heat exchanger in the high-Z gas recirculation loop.

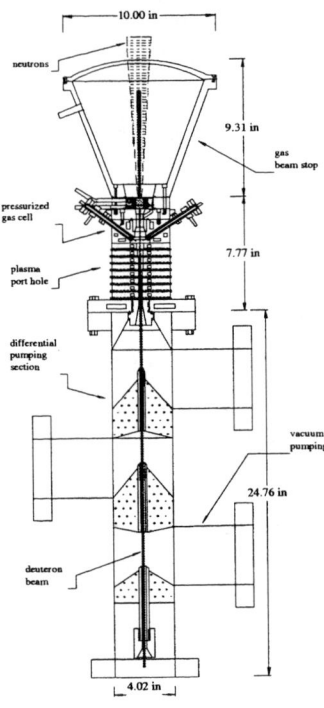

FIGURE 1. The end station drawing shows the differential pumping section, plasma porthole, pressurized gas cell, and beam stop concept.

END STATION ACCEPTANCE

The three apertures in the differential pumping section and the aperture of the plasma porthole, combined with the small beam channel needed in the gas cell, place a significant constraint on the beam emittance from the accelerator.

To generate a maximum acceptable emittance value for the end station, the procedure was to generate x-x' and y-y' acceptance parallelograms based on the length, spacing, and diameter of the apertures in the reference design. Using basic trigonometry, acceptance parallelograms can be determined both for the geometric acceptance to just transport the beam through the apertures and for the acceptance needed to geometrically transport the beam through a thin (1.5-

mm diameter) channel in the gas cell. The beam channel acceptance is the more stringent constraint and is what needs to be met to achieve the desired resolution for the radiography system. The beam channel acceptance as an area in phase space can be determined by:

$$A_0 = \frac{4 r_{ap1} r_{sp}}{L + \lambda_{sp}} - \frac{2 r_{ap1} \lambda_{sp} (r_{ap1} - r_{sp})}{L(L + \lambda_{sp})}$$

where r_{ap1} is the radius of the entrance aperture, r_{sp} is the radius of the beam spot in the channel, λ_{sp} is the length of the spot in the channel, and L is the distance from the first aperture to the beginning of the beam channel.

While this expression gives the area for the channel acceptance in x and y phase spaces, it is not entirely useful as a particle beam distribution rarely appears as a parallelogram in phase space. To obtain the maximum elliptical area (α_0) that corresponds to the acceptance parallelogram, it is observed that:

$$\frac{\alpha_0}{A_0} = \frac{\pi}{4}$$

for all angles of rotation of the largest ellipse that can be enclosed within a parallelogram. Applying this relation to parallelogram area, A_0, gives an expression for the maximum elliptical area in phase space that corresponds to the largest emittance that can be tolerated from the accelerator and still make the desired spot size without scraping the beam on the transport elements:

$$\alpha_0 = \frac{r_{ap1} r_{sp}}{L + \lambda_{sp}} \left\{ 1 - \frac{\lambda_{sp}}{2L} \left(\frac{r_{ap1}}{r_{sp}} - 1 \right) \right\} \pi .$$

Both areas are in millimeter-milliradians if the variables are input as millimeters. Figure 2 shows a plot of the acceptance parallelogram and the maximum elliptical area. The plot applies for both x-x' and y-y' since the apertures are round.

The maximum elliptical area calculated is an approximation since the plasma porthole will have some effect on the beam; however, an estimate of the lens effect due to the plasma shows it is less than 5%.

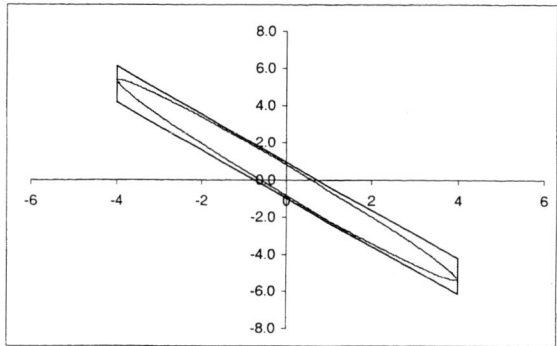

FIGURE 2. The plot shows the beam channel acceptance parallelogram and the maximum elliptical area that can be used to achieve the desired beam channel size in the gas cell. The x axis is millimeters and the y is milliradians.

BEAM FROM AN ACCELERATOR

To this point, the beam spot-size acceptance has been expressed as the largest elliptical area that can be enclosed by the acceptance parallelogram that comes from the beam channel geometry. To relate this to an actual accelerator beam emittance, comparisons will be made between the maximum elliptical area (α_0) divided by π, and the unnormalized, 5*RMS emittance from the accelerator, which contains at least 92% of the density for a Gaussian beam. To cleanly transport the beam through the end station,

$$\varepsilon_{unormalized, 5RMS} < \alpha_0 / \pi .$$

As shown in Figure 3, the longer the beam channel is in the gas cell, the smaller the beam emittance for the accelerator needs to be. The present gas cell design would have a 40-mm-long beam channel.

Conveniently, the transport code TRACE3D works in unnormalized 5-RMS emittances and allows straightforward assessments of potential accelerator designs, at least where ready output beam emittance simulations or data are available. Utilizing output parameters generated by PARMILA for a recent linac scoping study [4], TRACE3D was used to design a HEBT, comprised of a magnetic quadrupole and a quadrupole triplet, to transport the beam to the end station. The envelop for a beam having an emittance less than 3.5 mm-mrad was evaluated along the transport line and compared to the apertures in the HEBT and the end station for an acceptable beam transport tune that met the beam channel focusing requirement. The resulting plot, shown in Figure 4, shows the 5-RMS beam envelop as it is transported from the accelerator to the gas cell. From this plot, it is clear that a beam of sufficient quality to meet the beam channel requirement is readily transported through the HEBT and into the end station.

CONCLUSION

To use a particle accelerator for intense neutron radiography, a significant beam-quality constraint is encountered in focusing the beam to a very tight beam channel in the high pressure deuterium gas cell where the neutrons are generated. For the 40-mm-reference-design gas cell, the maximum emittance allowable for high-resolution radiography is less than 3.3 mm-mrad unnormalized, 5-RMS.

ACKNOWLEDGEMENTS

The authors would like to thank Prof. Richard Lanza of MIT and his students for their support in the ongoing development of the plasma porthole and differential pumping system.

REFERENCES

1. J. Hall *et al*, "Recent Results in the Development of Fast Neutron Imaging Techniques," to be published in these proceedings.
2. R. Lanza *et al*, "The Plasma Porthole: a Windowless Vacuum-Pressure Interface with Various Accelerator Applications," *Application of Accelerators in Research and Industry* (AIP **CP475**), J. Duggan and I. Morgan (eds.), New York, NY: AIP Press, 1999, pp. 932–935.
3. R. Lanza, private communication.
4. R. Hamm, private communication.

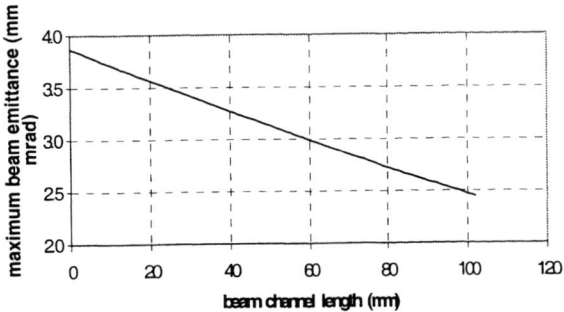

FIGURE 3. This plot shows the maximum unnormalized, 5-RMS beam emittance values needed to obtain 1.5-mm-diameter beam channels of different lengths in the gas cell. Plot is for an end station 775 mm long, with an entrance aperture radius of $r_{ap1}=4$ mm and a beam channel radius of $r_{sp}=0.75$ mm.

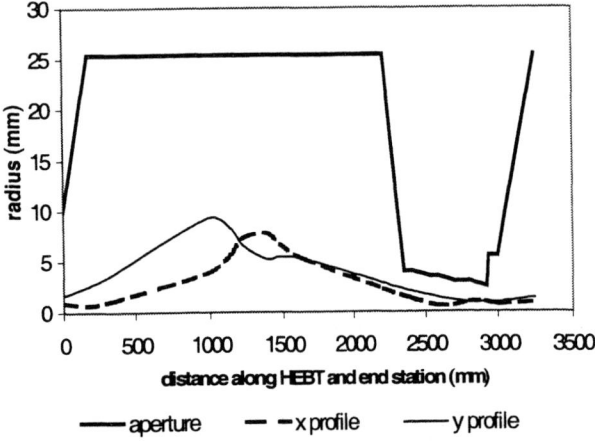

FIGURE 4. This plot shows both the 5-RMS beam envelop in x and y as a function of position in the HEBT, in the end station, and the gas cell where the beam channel requirement needs to be met. The average aperture to beam RMS ratios are 7.94 in x and 4.47 in y.

Fast Neutron Resonance Radiography for Security Applications

Gongyin Chen and Richard C. Lanza

Department of Nuclear Engineering, Massachusetts Institute of Technology, Cambridge, MA 02139

James Hall

Lawrence Livermore National Laboratory, Livermore, CA 94550

Abstract. Fast Neutron Resonance Radiography (NRR) has been suggested to detect explosives and drugs in passenger suitcases. In the NRR method, the 2-D elemental mapping of hydrogen, carbon, nitrogen, oxygen and the sum of other elements are calculated using fast neutron radiographic images taken at different neutron energies chosen to cover the resonance features of one or more elements. A radiographic image provides the 2-D mapping of the sum of elemental contents (weighted by the attenuation coefficients) and images taken at different neutron energies form a set of linear equations which can be solved for the mapping of individual elemental content. Explosives and drugs can be identified by their characteristic elemental composition. Different energy (2-6 MeV) neutrons can be obtained at different angles from a DD neutron source. A fixed-energy RFQ with a thick target can be used as the neutron source in NRR. Simulation results are presented in the paper.

INTRODUCTION

Fast Neutron Resonance Radiography (NRR) is a nuclear technique for the determination of 2-D elemental content distribution. It has the major advantages of nuclear techniques, providing elemental information for explosive and drug detection and the ability to penetrate thick objects [1]. NRR is based on fast neutron radiography. An accelerator is used to produce source neutrons (2-6 MeV) and a hydrogen-rich scintillator (typically plastic) is used to detect the transmitted neutrons. Light emitted from the scintillator is recorded using a CCD camera to form a radiographic image. For each image, the energy spectrum of the source neutrons is chosen to cover resonance features of one or more elements of interest.

A radiographic image is a 2-D map of projected attenuation. It can also be thought of as a 2-D map of the sum of content of all existing elements, weighted by their attenuation coefficients. For each pixel in the image, there exists a linear equation stating that the total attenuation equals the weighted sum of projected elemental contents. When we take another radiographic image at a different energy spectrum, the resulting linear equation has different attenuation coefficients (weighting factors) and total attenuation, but has the same projected contents, as the object is the same. In principle, when there are more equations than the number of existing elements, the set of linear equations can be solved for a definite Least-Squares solution of projected elemental contents.

The attenuation coefficients are related to the total neutron cross-sections over the energy spectrum of the source neutrons. Unlike x-ray attenuation, fast neutron attenuation coefficients vary only moderately from energy to energy and from element to element. The difference is typically within a factor of 2 to 3 and is neither too small nor too large. As a result, the linear equation set is a "good" problem. Elements other than hydrogen, carbon, nitrogen, and oxygen are modeled as one component to reduce the number of equations required. We will use H, C, N, O, and "other" to represent these five components. This is generally appropriate to luggage scanning.

FAST NEUTRON RESONANCE RADIOGRAPHY

The simplest case of NRR is to map one element at a time. We look for an energy region with a resonance peak/valley for one element while the cross sections of other elements are flat over the same energy range. For example, we might choose the sharp resonance peak at 2.077 MeV for carbon (see Figure 1). A radiographic image is taken at the "on-resonance" neutron energy, and another taken at an "off-resonance" neutron energy. The difference of the two images gives a 2-D map of the corresponding element.

It becomes increasingly difficult to find a single peak when more elements are considered. Resonance peaks found as described above are generally very narrow. This requires a variable energy charged particle accelerator and a thin target to generate a monoenergetic neutron source. When a radiographic image is taken with neutrons having a broad energy range, (e.g. 4.487-5.525 MeV as shown by the 25° stripe in Figure 1) resonance features of more than one element are covered. It turns out that two such images are not enough to give us the 2-D mapping of any element. However, when more images are taken, the 2-D mapping of all elements involved can be solved.

Using a broad-spectrum neutron source exploits the characteristically broad giant resonance features of elements. This improves resonance contrast (compared to choosing narrow peaks). In addition, differences among the images (seven or more) are added up, which is very important for low content elements such as nitrogen.

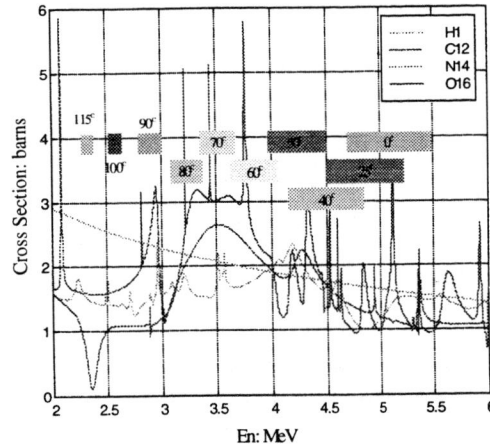

FIGURE 1. Total Neutron Cross Section. The horizontal stripes show the effective energy of a DD neutron source at $E_d = 2.3$ MeV, and $\Delta E_d = 0.8$ MeV (thick target).

Linear Attenuation Model

Images taken at different energies are normalized with corresponding open field views. The negative logarithm of the normalized image is a 2-D projection of total attenuation. There is a linear equation set for each pixel in the image:

$$a_{11} x_{11} + a_{12} x_{12} + ... + a_{1n} x_{1n} = b_1$$
$$a_{21} x_{21} + a_{22} x_{22} + ... + a_{2n} x_{2n} = b_2$$
$$...$$
$$a_{m1} x_{m1} + a_{m2} x_{m2} + ... + a_{mn} x_{mn} = b_m$$

In the equations, a_{ij} is the attenuation coefficient, x_{ij} is the unknown projected element content and b_i is the measured total attenuation. a_{ij} is related to the total cross section over the source neutron energy range which can be determined in advance. There are m equations with n unknowns. When $m>n$, there exists a least squares solution. In order to reduce the number of equations (images), all elements other than H, C, N, and O are represented with one attenuation component which is flat over the energy range. In the solution, the content of most other elements is split between "H" and "other". Thus, a larger relative error is expected for hydrogen, especially when a lot of aluminum is present. The flat component also compensates for internal scattering effects within the object.

The linear attenuation model assumes the following two facts: (1) the attenuation by one element is linearly dependent on its projected content and (2) the existence of one element does not affect the attenuation of other elements.

Setup and Neutron Source

Using the angular dependence of the DD neutron energy spectrum allows us to use a fixed energy accelerator such as an RFQ and a thick target to generate fast neutrons. RFQ ion accelerators are compact, high beam current devices usually designed to accelerate a specific ion to a fixed energy. Deuteron is the ion to be accelerated in this application and the target can be either high-pressure deuterium gas or a solid (deuterated) target.

Different energies are achieved by rotating the aligned object-detector assembly around the neutron source. This is different from tomography in which the detector and source are aligned and object rotates around a fixed axis. Suppose $E_d = 2.3$ MeV and $\Delta E_d = 0.8$ MeV (thick target). The energy range changes from 4.70-5.55 MeV at 0° to 2.27-2.33 MeV at 115° (see Figures 1 and 2). At each nominal angle, the neutron energy spectrum varies over the image, so different attenuation coefficients must be used for different parts in the images.

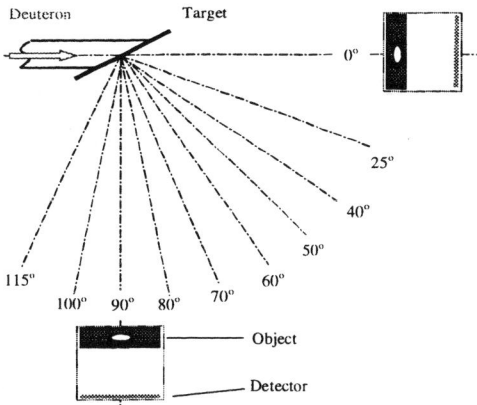

FIGURE 2. Neutrons of different energies are obtained by rotating the object-detector assembly around the source.

SIMULATION

The attenuation coefficients are generally smaller than the average cross section over the corresponding neutron energy range. It is not desirable to calculate these coefficients directly from cross section data. MCNP [2] is used to simulate the attenuation. A slab of pseudo single-element material is placed between a source and a detector, with the same geometry as in image simulation. The attenuation coefficients are calibrated for $0-1.0 \times 10^{24}$ atoms/cm^2 for each element.

Radiographic images are simulated using the radiation transport code "COG" developed at LLNL [3, 4]. The source-to-object and object-to-image distances used in the simulations were ~150 cm and 50 cm, respectively, giving an imaging angle of ± 10° and a magnification factor of 1.33. The energy spectrum of the neutron source was adjusted downward from 5.55 - 4.70 MeV to 2.33 – 2.27 MeV in a series of 10 runs to simulate scanning the object through angles ranging from 0° to 115° relative to the axis of a real DD neutron source. Images of the direct (unscattered) neutron flux from the source and scattered neutrons in several bins at lower energies were generated using 200 X 200 pixel imaging detectors with appropriate energy and scattered-particle masks. These direct/scattered neutrons are summed together (weighted by energy) to form real images. The simulations were each run with 5×10^8 test particles on the LLNL ASCI Blue Pacific supercomputer using 256 processors operating in parallel.

Test Object

The object modeled in the simulations is a sort of "Terrorist Overnight Bag." The bag itself consists of a thin aluminum shell (40 X 30 X 10 cm) with a wood handle, thick cloth covering and steel fittings. It contains a newspaper, a bag of sugar (105g), a stash of cocaine-HCl (105g), a travel umbrella, a 4" switchblade knife, a paperback book (presumably the "*Anarchist's Handbook*"), a block of plastic explosive (270 g; 50/50 wt.% mix of RDX and PETN), a pen and pencil set, a small camera, an automatic pistol with extra ammo clip, a flat paper notebook and a selection of cotton, wool and nylon clothing items. The bag is heavily loaded and has an average density of around 0.5g/cm^3. Figure 3 shows an image of the bag and marks the locations of hidden explosives and drugs. It is not possible to identify these items directly from the simulated images.

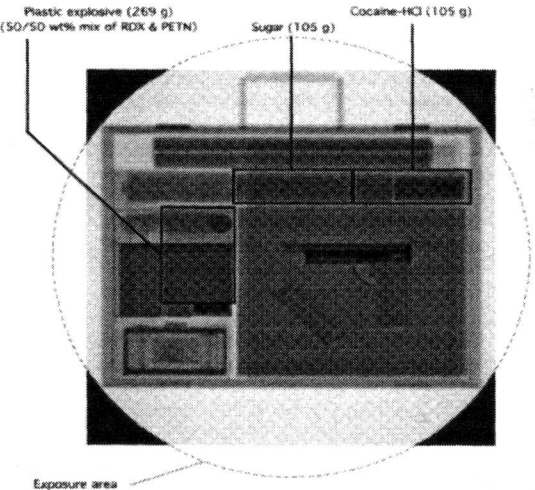

FIGURE 3. Neutron Radiographic Image Simulated at 0°.

NRR Result

The calculated elemental mappings are given in Figure 4. The plastic explosive can be identified by its high nitrogen/oxygen content and low hydrogen/ carbon content. The stash of cocaine looks different from sugar in that the drug has about equal amounts of hydrogen and carbon but very little oxygen. The bright bar overlapped on the drug in the carbon image is the polystyrene handle of an umbrella. A pistol, a knife and a battery are visible in the "other" picture. A glass lens in the camera is also very clear in the oxygen picture. Two aluminum buckles can be seen on the top in the Hydrogen picture. As we have mentioned, the calculation splits aluminum content into hydrogen and "other" results.

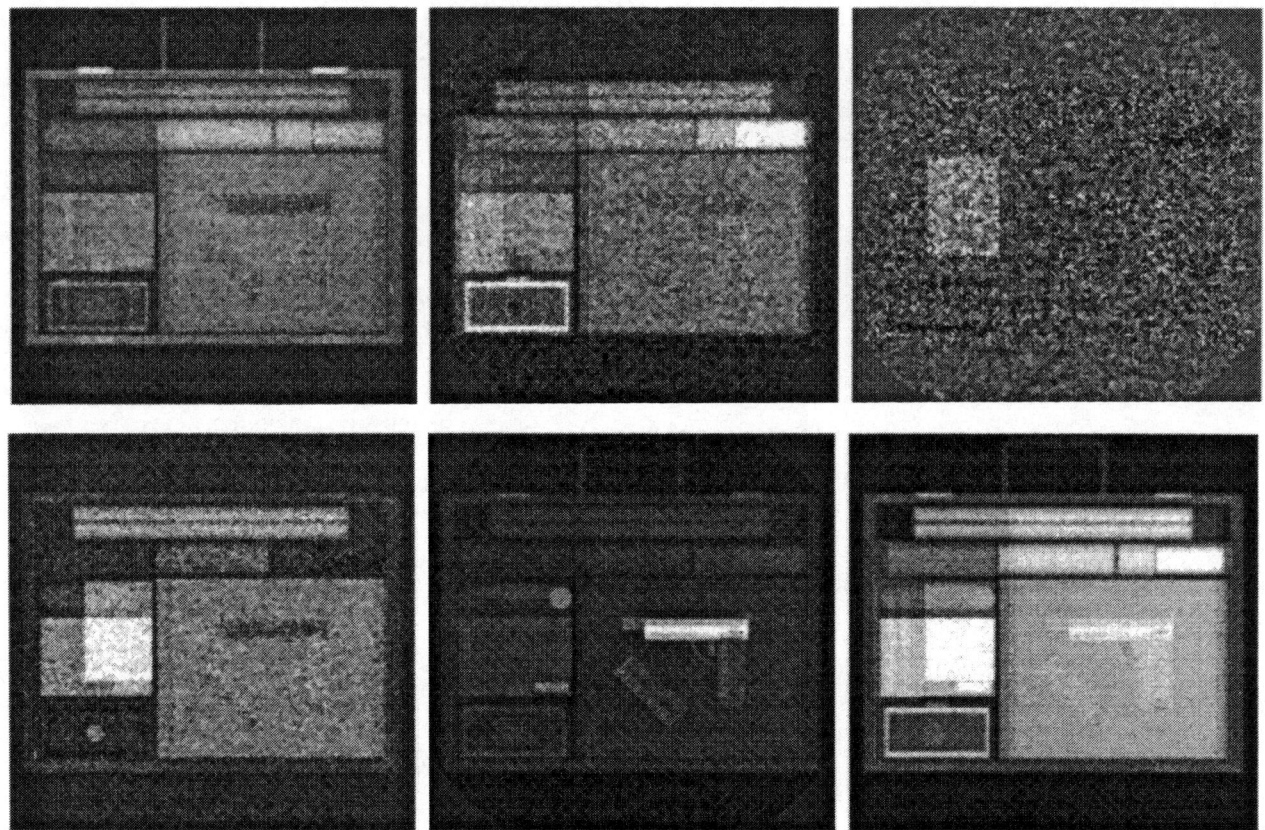

FIGURE 4. Projected elemental contents (atoms/cm^2) of the bag. Top line from left to right: hydrogen, carbon and nitrogen. Bottom line from left to right: oxygen, sum of other elements and sum of all elements.

SUMMARY

NRR is a promising nuclear technique for drug and explosive detection. It calculates elemental mappings of hydrogen, carbon, nitrogen, oxygen and other elements from a series of radiographic images taken at different energies chosen to cover resonance features of one or more elements of interest. Explosives and drugs can be identified by the signatures in their elemental composition.

A compact RFQ running at fixed energy can be used to generate fast neutrons. Neutrons of different energies are obtained by rotating the object-detector assembly around the DD neutron source. We estimate that a practical neutron flux can be achieved with 100μA average deuteron current.

Although our primary interest is in drug and explosive detection, the NRR principle is not restricted to such systems (*i.e.* H, C, N, O, "other"). Other examples might be in the area of on-site mineral analysis.

ACKNOWLEDGMENTS

This work is supported by the Office of National Drug Control Policy (ONDCP) under Contract DAAD07-98-C-0117 and the Federal Aviation Administration under Grant 93-G-053.

REFERENCES

1. T. Gozani, "Advances in accelerator based explosives detection systems", NIM **B79** (1993), pp. 601-604.

2. Judith F. Briesmeister, "MCNP—A General Monte Carlo N-Particle Transport Code", LANL, March 1997.

3. T. Wilcox and E. Lent, COG - A Particle Transport Code Designed to Solve the Boltzman Equation for Deep-Penetration (Shielding) Problems, Vol. 1 Users Manual, LLNL Rept. # M-221-1 (1989).

4. J. Hall, "Monte Carlo Modeling of Neutron and Gamma-Ray Imaging Systems," SPIE **2867**, 465 (1996).

Recent Results in the Development of Fast Neutron Imaging Techniques

James Hall, Frank Dietrich, Clint Logan and Brian Rusnak

Lawrence Livermore National Laboratory, P.O. Box 808, M/S L-050, Livermore, CA 94551-9900

Abstract. We are proceeding with the development of fast ($\approx$ 12 MeV) neutron imaging techniques for use in NDE applications. Our goal is to develop a neutron imaging system capable of detecting sub-mm-scale cracks, cubic-mm-scale voids and other structural defects in heavily-shielded low-Z materials within thick sealed objects. The final system will be relatively compact (suitable for use in a small laboratory) and capable of acquiring both radiographic and full tomographic image sets. The design of a prototype imaging detector will be briefly reviewed and results from several recent imaging experiments will be presented. The concurrent development of an intense, accelerator-driven neutron source suitable for use with the final production imaging system will also be briefly discussed.

INTRODUCTION

We are proceeding with the development of fast neutron imaging techniques for use in NDE applications involving the inspection of thick sealed objects ($\rho x \gtrsim 100$ g/cm^2). Our goal is to develop a high-energy neutron imaging system capable of detecting sub-mm-scale cracks, cubic-mm-scale voids and other structural defects in heavily-shielded low-Z materials within such objects. The final production system will be relatively compact (suitable for use in a small laboratory) and capable of acquiring both radiographic and full tomographic image sets. In order to expedite the development process and minimize associated technical risks, we are using commercially-available system components and proven neutron imaging techniques wherever possible. As currently envisioned, the final production system will consist of an intense, accelerator-driven D(d,n)^{3}He neutron source ($E_n \approx 12$ MeV @ 0°) with an effective yield of $\approx 8.5 \times 10^{10}$ n/sec/sr along the beam axis and an effective spot size ≈ 1.25 mm (FWHM), a multi-axis staging system to support and manipulate objects under inspection and an imaging detector (*cf.* Figure 1). The imaging detector itself will consist of a plastic scintillator viewed indirectly by a cryogenically-cooled, high-resolution (2048 X 2048; 24 μm pixel) CCD camera. The ultimate spatial resolution of the system should be $\approx 0.50 - 1.00$ mm (FWHM) at the object position.

The conceptual design of our imaging system [1, 2] and the results of early imaging experiments [3] have already been published elsewhere. In this paper, we will briefly review the design of a prototype imaging detector and present results from several recent imaging experiments conducted at Ohio University. The parallel development of an intense, accelerator-driven neutron source suitable for use with the final production imaging system will also be discussed.

IMAGING DETECTOR DESIGN

Our prototype for the imaging detector consists of a rigid plastic scintillator viewed indirectly by a single CCD camera. The camera assembly consists of a fast (f/1.00; 50 mm) photographic lens coupled through a remotely-controlled mechanical shutter housing to a thinned, back-illuminated, high-resolution (1024 X 1024; 24 μm pixel) CCD imaging chip with an anti-reflective (UVAR) coating on its active area to improve its sensitivity. The chip is cryogenically cooled and operates at a controller-stabilized temperature of -120 °C which effectively eliminates thermal electronic noise buildup in the pixel wells. This allows for extended image integration times ($\approx$ 1 hr) and greatly enhances the sensitivity of the camera in low light situations. Data is downloaded to the camera controller using a 16-bit A/D converter running at 50 kHz which imposes an average read-out noise on the image of $\approx$ 5 electrons/pixel. When compared to the full well capacity of the CCD chip ($\approx$ 325,000 electrons/pixel), this gives an effective dynamic range for the camera in excess of $\approx$ 50,000. A thin (0.125"), front-surfaced mirror fabricated from aluminized Pyrex glass is used to reflect light from the 30 cm X 30 cm X 4 cm-thick BC-400 plastic scintillator into the camera assembly which is mounted on dual-axis optical rails adjacent to (but well out of) the neutron beam path. The entire detector assembly is housed in a light-tight plywood enclosure with thin (0.125") aluminum entrance and exit apertures for the neutron beam.

RECENT IMAGING EXPERIMENTS

Experimental setup

In the experiments described here, a nearly monoenergetic, 10-MeV neutron beam was generated by focussing 6.85-MeV D^+ ions extracted from the Ohio University Accelerator Laboratory (OUAL) tandem van de Graaff accelerator into a cylindrical, 1-cm-diameter, 7.5-cm-long D_2 gas cell attached to the end of a beam line. The gas cell was capped with thin (≈ 5 μm) W entrance and exit windows and maintained at a static pressure of ≈ 45 psia (≈ 3.1 atm). Average D^+ ion currents measured at the gas cell were ≈ 8 μA during the runs and the diameter of the beam focal spot at the entrance window to the cell was ≈ 3.5 mm. This provided an effective neutron yield of $\approx 4.2 \times 10^9$ n/sec/sr along the beam axis (*i.e.* $\approx 1/20^{th}$ the intensity of the source proposed for our final production imaging system).

The test objects imaged in the experiments were mounted on a multi-axis staging system which was located on the beam axis ≈ 2 m downstream from the neutron source. The prototype imaging detector was located in a shielded detector cave ≈ 2 m further downstream behind a thick (≈ 1.5 m) concrete with a tapered polyethylene collimator. This provided a 2:1 image magnification factor with a clear field-of-view (FOV) $\approx$ 12" in diameter at the detector position (FOV diameter ≈ 6" at the object position).

Imaging results

A series of seven neutron imaging experiments have thus far been carried out at OUAL. These experiments focussed on radiographic (single view) imaging of conventional step wedges made of various materials and "slab" assemblies composed of blocks of low-Z materials (*e.g.* polyethylene) shielded by various thicknesses (≤ 4") of Pb or depleted uranium (D-38) and tomographic (multiple view) imaging of cylindrical test objects composed of nested shells of high- and low-Z materials (see [3] for details). In this paper we present recent radiographic imaging data on fractured ceramics shielded by D-38 and "mock" tomographic reconstructions of the British Test Object.

The ceramic/poly/D-38 test object consisted of a 4" X 2" X 1"-thick slab of borated-ceramic set atop a polyethylene slab of similar size and shielded by 1" of D-38 (areal density ≈ 50.2 g/cm^2). The ceramic piece featured two sets of 4- and 2-mm-diameter holes machined to depths of 0.160", 0.120", 0.080" and 0.040" (the smallest hole corresponds to a volume defect of ≈ 3 mm^3 and an areal density defect of ≈ 0.20 g/cm^2) and a narrow slot cut down from the top to a depth of 1" (*cf.* Figure 2a). This slot was used to facilitate cracking the ceramic along its centerline. The poly piece featured the same set of 4- and 2-mm-diameter holes but no crack. The ceramic was carefully reassembled with the fracture being barely visible to the naked eye and the shielded assembly was imaged in a series of 48 30 minute exposures (time integrated flux at object position $\approx 1.1 \times 10^{10}$ n/cm^2) with each image yielding ≈ 1400 counts/pixel above the CCD camera's built-in DC offset level. The final processed image and associated lineouts (*cf.* Figure 2b) clearly show the crack in the ceramic slab and all of the machined features including even the smallest, 2-mm-diameter, 0.040"-deep hole.

The British Test Object ("BTO"), on loan to the Department of Energy from AWE, Aldermaston for the express purpose of evaluating the performance of proposed NDE systems, consisted of a set of six nested cylindrical shells of (respectively) graphite, polyethylene, Al, W, polyethylene and W with a solid polyethylene core (*cf.* Figure 3a). The cylindrical shells were each segmented to provide an axially-symmetric joint structure. The full assembly had an OD of 18.9 cm and its areal density ranged from 125.93 g/cm^2 (along the centerline) to 176.40 g/cm^2 (along the limb of the poly core). Since the diameter of the BTO was somewhat too large to fit into our restricted FOV at OUAL, only a portion ($\approx 60\%$) of the object was imaged. A series of 12 30 minute exposures (time integrated flux at object position $\approx 2.4 \times 10^9$ n/cm^2) were taken of the assembly and, since it was axially symmetric, only a single viewing angle was used. The final processed image was then cropped at the center of the object and reflected about that line to produce an artificial image of the full assembly. This was then replicated 180 times to produce a "mock" tomographic data set. Subsequent reconstructions done using LLNL's Constrained Conjugate Gradient (CCG) algorithm (*cf.* Figure 3b) clearly show the gross structure of the BTO and also reveal the detailed joint structure in the outer shells. Note that, in spite of the fact that the tomographic data set used for reconstruction was derived in a somewhat artificial way, it *does* still provide an accurate demonstration of the capability of neutron imaging even when relatively short exposures are used.

NEUTRON SOURCE DEVELOPMENT

The development of an intense, high-energy neutron source suitable for use in a full-scale imaging system is proceeding in parallel with our work on the prototype detector and staging system. As noted above, we propose to use an accelerator-driven $D(d,n)^3He$ neutron source operating at ≈ 12 MeV. In order to meet our performance goals, the source will need to have an effective yield of ≈ 8.5×10^{10} n/sec/sr along the beam axis and an effective spot size ≈ 1.25 mm (FWHM). The most promising system evaluated thus far is a compact radio-frequency quadruplole (RFQ) accelerator coupled to a drift-tube linac (DTL) [4]. The accelerator system will need to deliver an average D^+ ion current ≈ 300 μA at the gas cell in order to produce a neutron flux sufficient to image objects of interest. While this is certainly achievable, it does pose another, in some ways more challenging, problem: the simultaneous requirements of a high deuterium ion current and relatively small focal spot size effectively preclude the use of conventional ("windowed") D_2 gas cell designs due to their inability to handle the large heat loads involved (≈ 250 kW/cm^2 (ave)). We are therefore collaborating with R. Lanza at the Massachusetts Institute of Technology (MIT) on the development of two types of "windowless" D_2 gas target assemblies - a "rotating aperture" design [5] and a "plasma window" design [6] - which could be coupled to a high-current D^+ accelerator and used as a neutron source. Operational testing of prototype targets is currently underway at MIT.

CONCLUSIONS

The experiments carried out thus far have demonstrated the feasibility of using high-energy neutron imaging in applications involving the inspection of thick sealed objects and we are poised to commit to the construction of a full-scale facility. Additional tests of the prototype imaging detector are planned at OUAL and key decisions on gas target and accelerator technologies will be made during early 2001.

ACKNOWLEDGMENTS

We would like to thank Prof. David Ingram of Ohio University and the staff of OUAL for their continued support in the testing of our prototype imaging detector and Jessie Jackson and Dr. Harry Martz of LLNL for their assistance in tomographic image reconstructions. We would also like to thank Prof. Richard Lanza of MIT and his students for their support in the ongoing development of the D_2 gas targets.

This work was performed at the University of California, Lawrence Livermore National Laboratory, under the auspices of the U.S. Department of Energy (contract # W-7405-Eng-48).

REFERENCES

1. F. Dietrich and J. Hall, "Detector concept for neutron tomography in the 10 - 15 MeV energy range", LLNL report # UCRL-ID-123490, 4 pp. (1996).
2. F. Dietrich, J. Hall and C. Logan, "Conceptual design for a neutron imaging system for thick target analysis operating in the 10 - 15 MeV energy range," published in J. Duggan and I. Morgan (eds.), *Application of Accelerators in Research and Industry* (AIP **CP392**), New York, NY: AIP Press, 1997, p. 837-840.
3. J. Hall, F. Dietrich, C. Logan and G Schmid, "Development of high-energy neutron imaging for use in NDE applications", SPIE **3769**, 31-42 (1999) (UCRL-JC-134562); abridged version also published in AIP **CP497**, 693-698 (1999).
4. B. Rusnak and J. Hall, "An accelerator system for neutron radiography," published in these proceedings.
5. E. Empey, Master's thesis (unpublished), Massachusetts Institute of Technology, Cambridge, MA, January 2000.
6. W. Gerber, R. Lanza, A. Hershcovitch, P. Stephan, C. Castle and E. Johnson, "The plasma porthole: a windowless vacuum-pressure interface with various accelerator applications," published in J. Duggan and I. Morgan (eds.), *Application of Accelerators in Research and Industry* (AIP **CP475**), New York, NY: AIP Press, 1999, p. 932-935.

COLLECTED FIGURES

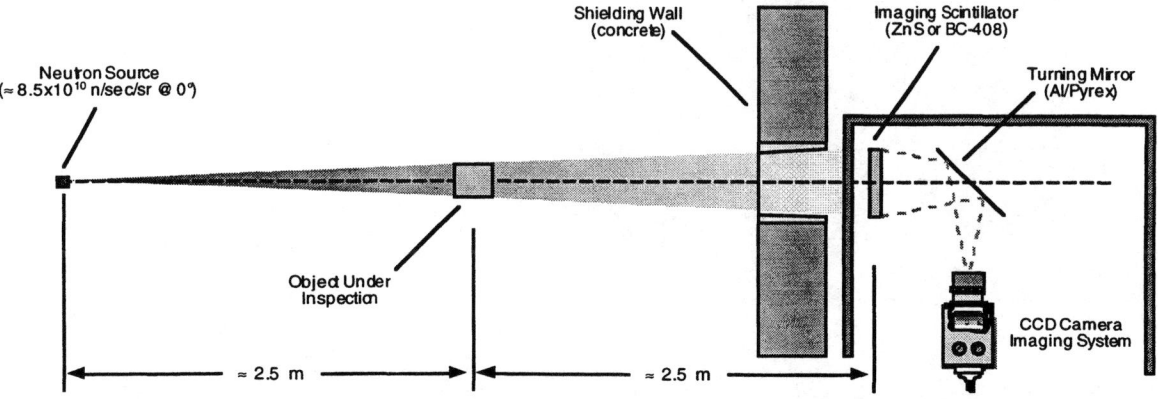

Figure 1: Conceptual design of our proposed high-energy neutron imaging system. The use of a 2:1 image magnification factor will minimize backgrounds at the imaging plane caused by internal scattering within the object under inspection.

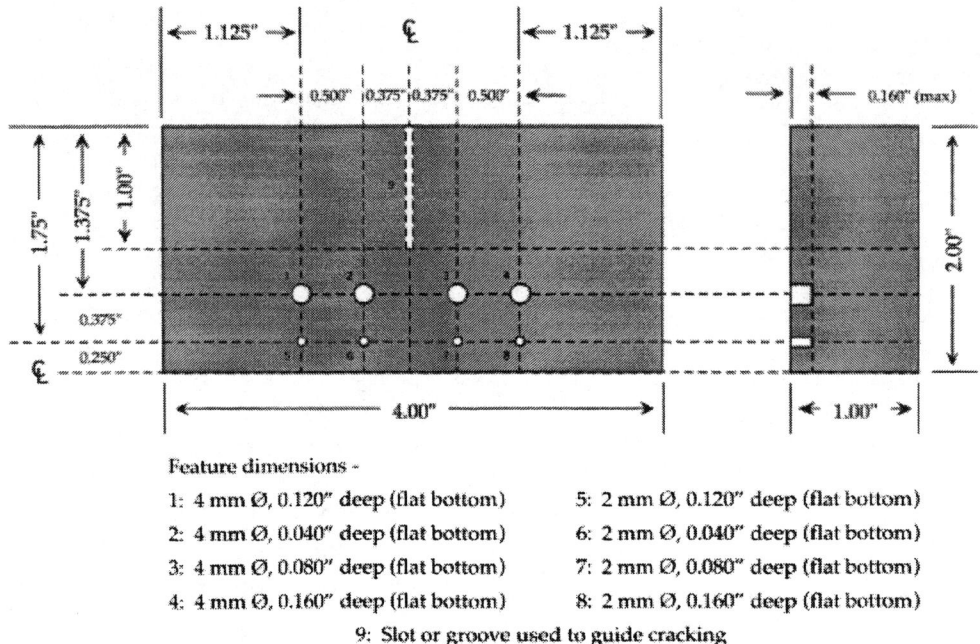

Feature dimensions –

1: 4 mm Ø, 0.120" deep (flat bottom)   5: 2 mm Ø, 0.120" deep (flat bottom)
2: 4 mm Ø, 0.040" deep (flat bottom)   6: 2 mm Ø, 0.040" deep (flat bottom)
3: 4 mm Ø, 0.080" deep (flat bottom)   7: 2 mm Ø, 0.080" deep (flat bottom)
4: 4 mm Ø, 0.160" deep (flat bottom)   8: 2 mm Ø, 0.160" deep (flat bottom)
9: Slot or groove used to guide cracking

Figure 2a: Design drawing of ceramic test object showing parallel sets of 4- and 2-mm-diameter holes machined to depths of 0.160", 0.120", 0.080" and 0.040" and a narrow slot cut down from the top to a depth of 1". This slot was used to facilitate cracking the ceramic along its centerline.

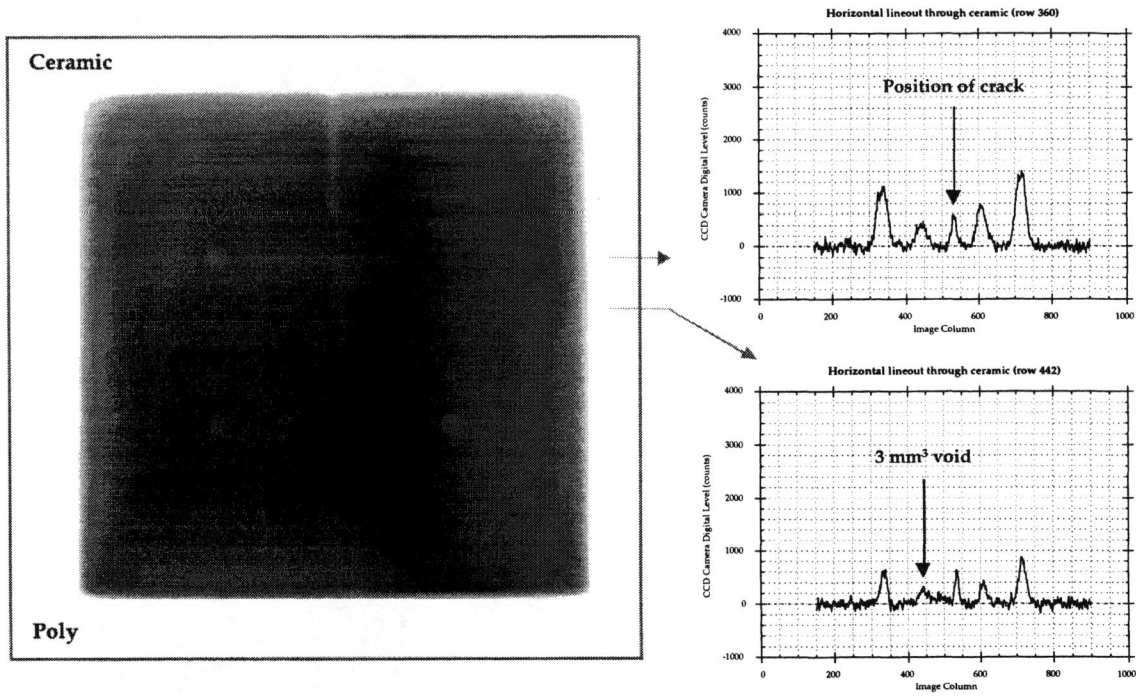

Figure 2b: Neutron radiograph of fractured ceramic and polyethylene test object shielded by 1" of D-38 ($\rho x \approx 50.2$ g/cm^2). The ceramic piece was carefully reassembled prior to imaging with the crack being barely visible to the naked eye. The processed image and associated lineouts clearly show the crack in the ceramic slab and all of the machined features in both the ceramic and polyethylene including even the smallest, 2-mm-diameter, 0.040"-deep hole.

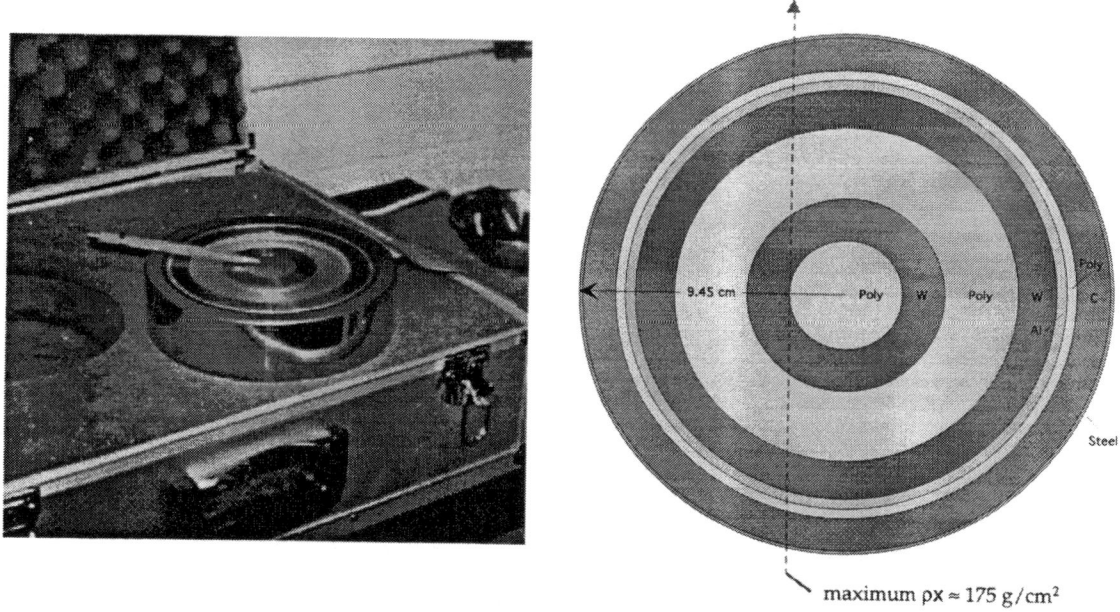

Figure 3a: Photograph and schematic drawing of the British Test Object (BTO) indicating structure and composition. The BTO is on loan to the DOE from AWE, Aldermaston for the express purpose of evaluating the performance of proposed NDE systems.

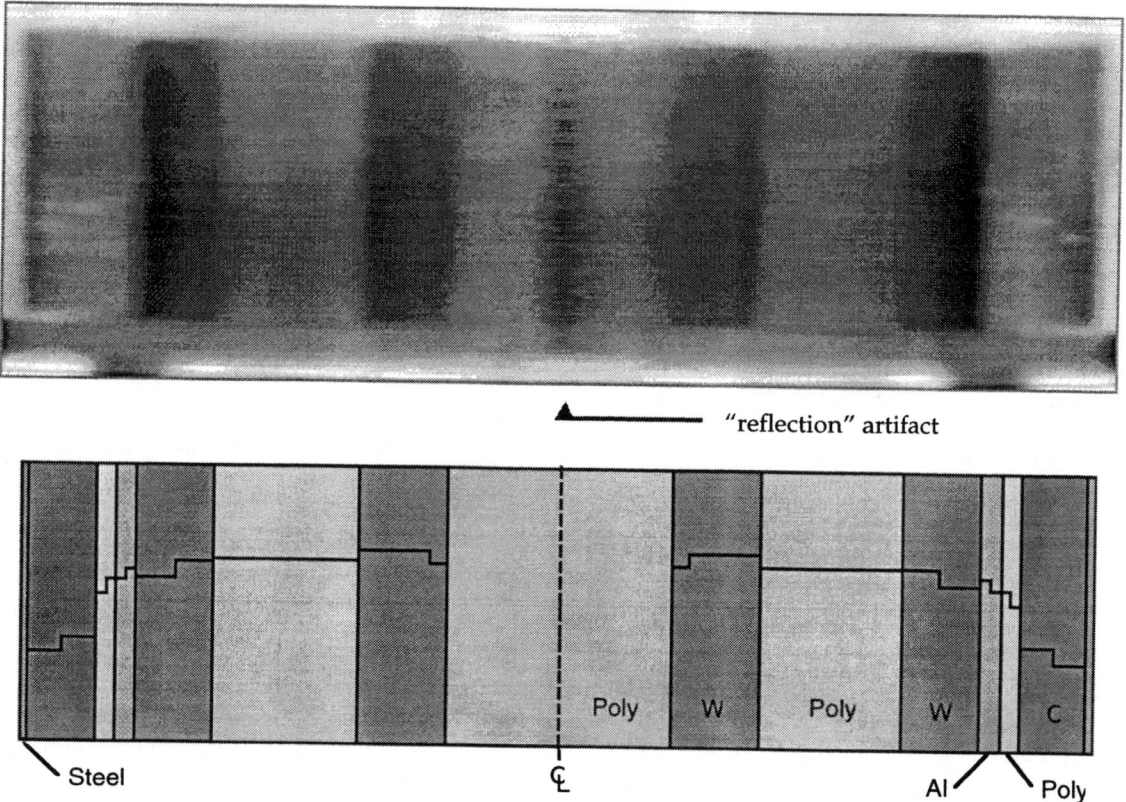

Figure 3b: "Mock" tomographic reconstruction of the BTO. The slice shown here was produced using a CCG reconstruction algorithm which takes the conical shape of the beam into account. Note that the gross structure of the BTO and the detailed joint structure in the outer shells is clearly visible in the reconstruction. The dark band at the center of the reconstruction is an artifact caused by reflecting the original radiograph about its centerline to form a complete image of the object.

Development of Semiconductor Detectors for Fast Neutron Radiography

R. T. Klann[a], C. L. Fink[a], D. S. McGregor[b], and H. K. Gersch[b]

[a]*Technology Development Division, Argonne National Laboratory, Argonne, IL 60439 USA*
[b]*Department of Nuclear Engineering and Radiological Sciences, University of Michigan, Ann Arbor, MI 48109 USA*

Abstract. A high-energy neutron detector has been developed using a semiconductor diode fabricated from bulk gallium arsenide wafers with a polyethylene neutron converter layer. Typical thickness of the diode layer is 250 to 300 μm with bias voltages of 30 to 150 volts. Converter thicknesses up to 2030 μm have been tested. GaAs neutron detectors offer many advantages over existing detectors including positional information, directional dependence, gamma discrimination, radiation hardness, and spectral tailoring. Polyethylene-coated detectors have been shown to detect 14 MeV neutrons directly from a D-T neutron generator without interference from gamma rays or scattered neutrons. An array of small diode detectors can be assembled to perform fast neutron radiography with direct digital readout and real-time display of the image produced. In addition, because the detectors are insensitive to gamma rays and low energy neutrons, highly radioactive samples (such as spent nuclear fuel or transuranic waste drums) could be radiographed.

INTRODUCTION

Neutron radiography is a well developed and commercially viable technique that is used throughout the world for non-destructive examinations. The technique employs a neutron source and collimators to produce a neutron beam. A sample is then placed in the neutron beam and recording media are placed directly behind the sample. The attenuation of the beam by the sample produces a two-dimensional "shadow image" of the sample. A more detailed description of neutron radiography can be found in the literature [1,2].

Thermal and epithermal neutron radiography are the most widely used forms of neutron radiography. These utilize low-energy neutrons which have a limited range in most materials of interest. Fast neutron radiography, i.e. utilizing neutrons with energies around 14 MeV produced from the D-T reaction, has been explored as a non-destructive examination tool for larger samples because of the greater penetrability of the neutrons [3-7]. The recording media at these energies has been almost exclusively a proton-producing plastic coupled with X-ray scintillation screens. A sheet of light-sensitive film is placed in contact with the scintillation screens to record a latent image on the film. The film is then removed and chemically processed to produce a radiograph. This process has been shown to produce reasonable radiographs, but it has serious limitations. The technique is slow because of the time it takes to produce the image (exposure time) and the time it takes to develop the film. The technique is inconsistent because of the chemical processing. In addition, the processing requires specialized equipment and chemicals which introduce environmental and industrial hazards for use and disposal. The imaging screens and film are sensitive to gamma radiation and X rays which means that the screens and film must be shielded (with some neutron loss) and samples cannot be radioactive. Another drawback is that the images produced are not digital. The radiograph must be scanned to produce a digital image with a loss in resolution and contrast due to limitations in scanning technology.

This paper discusses the development of a new type of coated semiconductor detector for the detection of fast neutrons. Small contacts are deposited on semi-insulating bulk gallium arsenide (GaAs) wafers and then coated with a hydrogen-rich material, e.g. polyethylene. A direct read-out of the detector count rate is obtained with standard electronics. This configuration allows multiple diodes to be created using a single wafer such that positional information can be obtained and used to produce an image [8,9]. The resolution of the radiography is limited by the size of the individual diodes.

THEORY OF OPERATION

When a voltage is applied across the GaAs wafer, a truncated high field or active region is produced near the contact [10]. The incident neutrons are converted into charged particles in the coating, i.e. recoil protons, which excite free charge carriers in the GaAs detector active region. The charge carriers are drifted to the detector contacts, and a preamplifier circuit measures the induced charge. Charges excited in the low field or substrate region are not collected. In addition, because the active region is so thin (10-20 µm), background gamma-ray interactions are reduced. Gamma rays that are absorbed in the active region are easily discriminated from the recoil protons. Figure 1 is a schematic representation.

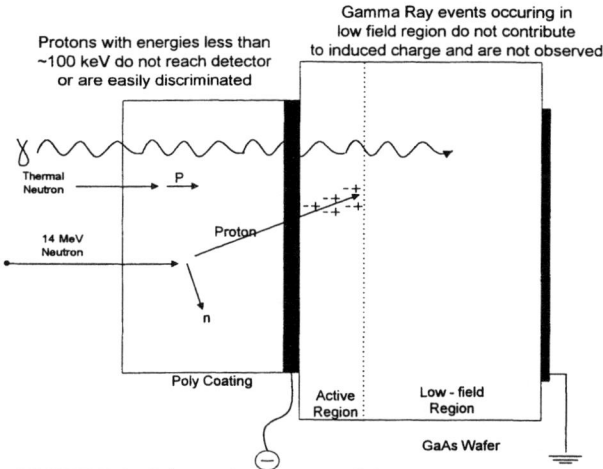

FIGURE 1. Schematic diagram of detector operation

DETECTOR FABRICATION

Schottky barrier bulk GaAs diodes were fabricated with high-density polyethylene (HDP) coatings. The GaAs diodes were manufactured in the following manner. Commercial bulk semi-insulating (SI) GaAs wafers were used for the devices. The back surfaces were lapped at 30 rpm with a 3 µm calcined aluminum oxide powder/ deionized water solution over an optically flat glass plate until 100 µm of GaAs material was removed. Afterwards, the backsides were polished with 0.3 µm calcined aluminum oxide powder mixed in a sodium hypochlorite solution over a chemically resistant polishing cloth at 65 rpm for 10 minutes. A final polish was performed with a 50:50:1 methyl alcohol:glycerol:bromine solution for 10 minutes over a chemically resistant polishing pad at 70 rpm. The wafers were cleaned in a series of solvents and etched in a 1:1:320 H_2SO_4:H_2O_2:deionized water solution for 5 minutes followed by a 2 minute etch in a 1:1 HCl: deionized water solution. The wafers were then cleaned in a deionized water cascade and blown dry with N_2.

The backsides were implanted at an angle of 7° from normal with ^{29}Si ions at an average energy of 100 keV at a dose amounting to 5×10^{13} ions per cm^2. The implants were activated with a rapid thermal anneal in Ar for 30 seconds at 800°C. Afterwards, a stacked layer of Ge (500 Å): Pd (1300 Å) was vacuum evaporated over the backsides, followed by a low temperature anneal of 250°C in N_2 for 30 minutes. Vacuum evaporation of a stacked layer of Ti (150 Å): Au (700 Å) completed the backside processing of the devices.

Front-side processing of the devices included lapping and polishing of the samples, in which the initial lapping with 3 µm calcined aluminum oxide powder was used to thin the wafers to 250 µm total thickness. Afterwards, the wafers were polished using the 0.3 µm calcined aluminum oxide powder mixed in a sodium hypochlorite solution followed by the methyl alcohol:glycerol:bromine solution. Again, the wafers were cleaned in a series of solvents and etched in a 1:1:320 H_2SO_4:H_2O_2:deionized water solution for 5 minutes followed by a 2 minutes etch in a 1:1 HCl:deionized water solution. Afterwards, the wafers were cleaned in a deionized water cascade and blown dry with N_2.

The basic pad area designs were patterned onto the surfaces with photoresist. A final etch in the patterns was performed with the H_2SO_4: H_2O_2:deionized water solution followed by the HCl:deionized water solution. The wafers were washed in a deionized water cascade and blown dry with N_2. A stacked layer of Ti:Au was evaporated over the wafer and lifted off in acetone. Other variations used a system of Ti:Pt:Au contacts.

Polyethylene-coated devices were manufactured by adhering various thicknesses of HDP sheets to the bare Schottky contacts. The HDP sample thicknesses included "Humiseal" only, 50 µm (0.002 inches), 75 µm (0.003 inches), 125 µm (0.005 inches), 250 µm (0.010 inches), 450 µm (0.018 inches), 900 µm (0.035 inches), and 2030 µm (0.080 inches). The individual devices were cleaved from the GaAs wafers, and fastened with silver-based epoxy to 1 mm thick aluminum oxide mounts. Figure 2 shows several detectors with different coatings. The detector area, defined by the diode size, is a 25 mm² circle with one diode on each detector.

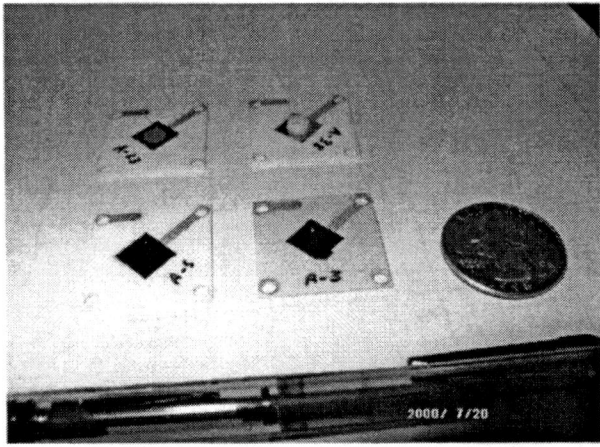

FIGURE 2: GaAs Detectors on Alumina mounts

TEST RESULTS

The GaAs Schottky barrier detectors were mounted in light-impenetrable Al boxes. The enclosed devices were then placed at a distance of 10 cm from the target cooling cap of an MF Physics A-711 Neutron Generator. The detectors were operated with reverse bias voltages between 30 and 150 volts.

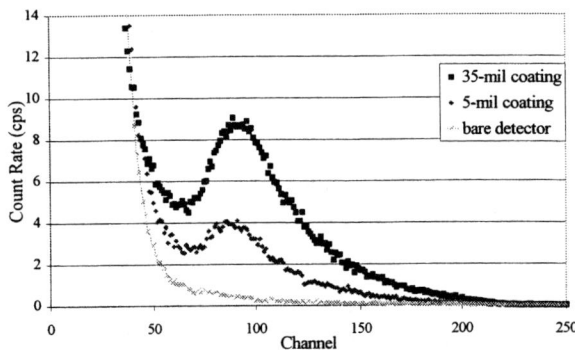

FIGURE 3: Fast neutron response for two different coating thicknesses (detector bias of 120 volts).

Figure 3 shows the results for coating thicknesses of 0.005 (5 mils) and 0.035 inches (35 mils) compared to the results from an uncoated detector for a bias of 120 volts. The peak around channel 90 is a result of the recoil proton energy deposited in the active region of the detector. The low energy tail is due to detector noise and background from gamma rays, X rays, and scattered neutrons. In practice, the lower level discriminator would be set in the valley of the spectrum to maximize the neutron signal to noise ratio. The relative efficiency of the 35-mil coating is 2.7 times that of the 5-mil coating. This is less than predicted from modeling studies [11] and is likely the result of energy straggling, edge effects, and effects from the contact materials, which were not accounted for in the modeling studies.

The range of the recoil protons exceeds the high field region of the detector. By increasing the bias on the detector, the high field region is increased, leading to higher energy deposition and a larger pulse from the detector. This can be observed in Figure 4 and Table 1. Figure 4 shows the results of increasing the detector bias while maintaining all other parameters constant. The entire curve is amplified due to the increase in the depth of the active region of the detector. The effect becomes apparent in Table 1. The neutron count rate is the total count rate of all channels above the valley. It is relatively constant (12% change) over the entire range of bias voltages while the total count rate has increased by a factor of 15. The noise is being amplified by the higher voltage and the gamma response is being increased because of the increase in active region.

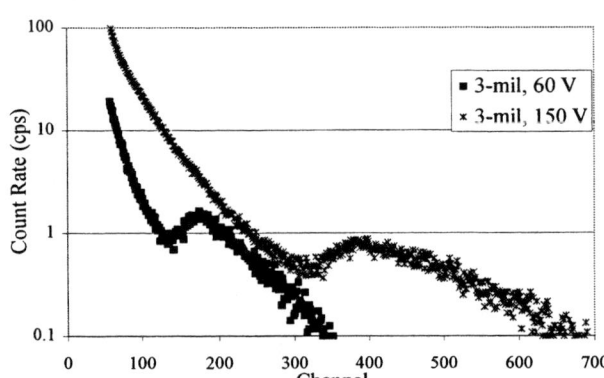

FIGURE 4: Fast neutron response for two different detector bias voltages.

Table 1 Effect of Bias Voltage (3-mil poly coating)				
Bias Voltage (volts)	Total Rate (cps)	Neutron Rate (cps)	Valley Channel	Peak Channel
30	211.6	138.7	85	100
60	502.7	141.0	140	175
90	1211.4	147.2	195	255
120	2103.8	154.2	250	330
150	3182.0	157.8	320	380

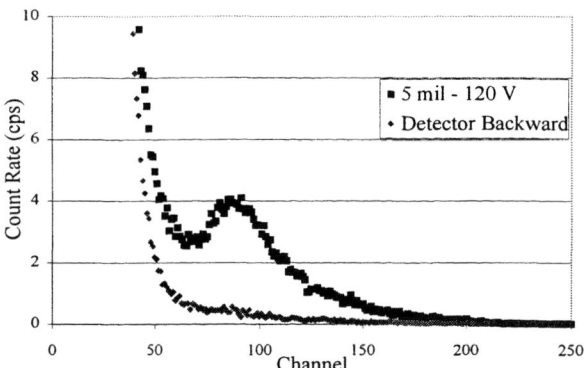

FIGURE 5: Directional response of 5-mil coated GaAs detectors to fast neutrons

Figure 5 compares the response of a detector facing the neutron source to that of a detector facing away from the neutron source. Figure 5 establishes that the device is directionally sensitive, in which the forward scattered recoil protons can enter the GaAs detector only if it is facing the general direction of the fast neutron source. If the response were due to any other reaction or radiation incident on the detector, the spectrum for the backward facing detector would show a response. Instead, the response of the backward facing detector is statistically identical to the uncoated detector response shown in Figure 3. This also demonstrates that there is a negligible response from lower energy neutrons due to scattering in the room. For radiography purposes, this helps to reduce background and eliminates the need for collimation or shielding of the detectors. In addition, it helps to improve image resolution and clarity because scattered neutrons simply do not contribute to the neutron signal.

CONCLUSIONS

Semiconductor diode detectors using bulk GaAs wafers have been fabricated and tested. The detectors have been shown to respond to fast neutrons with minimal response from scattered neutrons and gamma rays. With energy discrimination these effects can be easily reduced such that a clean signal is obtained for the fast neutron response. In addition, the detectors have been shown to be directionally dependent such that the response from background neutrons is significantly reduced.

The detectors are digital, in that the total count rate is summed to produce a numeric value for a given detector. With an array of these types of diodes, a digital fast neutron radiograph could be produced. The resolution of such a system would be dependent on the size of the individual diodes. Diodes have been produced with areas much smaller than 1 mm^2, which means that this approach is viable for fast neutron radiography.

ACKNOWLEDGMENTS

The submitted manuscript has been authored by a contractor of the U.S. Government under contract No. W-31-109-ENG-38. Accordingly, the U.S. Government retains a nonexclusive, royalty-free license to publish or reproduce the published form of this contribution, or allow others to do so, for U.S. Government purposes. Research at the University of Michigan was performed under an appointment to the U.S. Department of Energy Nuclear Engineering and Health Physics Fellowship Program sponsored by DOE Office of Nuclear Energy, Science, and Technology.

REFERENCES

1. Von Der Hardt, P., Rottger, H., *Neutron Radiography Handbook*, D. Reidel Publishing Company, London, 1981.
2. Berger, H., Iddings, F., *Neutron Radiography: A State-of-the-Art Report*, Non-Destructive Testing Information Analysis Center Report No. NTIAC-SR-98-01, TRI/Austin, Inc., 1998.
3. Klann, R. and Natale, M., "Fast Neutron Radiography Research at ANL-W," *Proceedings from the Fifth World Conference on Neutron Radiography*, edited by C.O. Fischer, et al., DGZfP BB53, Berlin, Germany, 1997, pp.382-390.
4. Klann, R., "Fast Neutron (14.5 MeV) Radiography: A Comparative Study," *Proceedings from the Fifth World Conference on Neutron Radiography*, edited by C.O. Fischer, et al., DGZfP BB53, Berlin, Germany, 1997, pp.469-483.
5. Richardson, A., *Materials Evaluation* **35**, 52-58 (1977).
6. Brzosko, J., et al., *Nucl. Inst. Meth. Phys. Res.* **B72**, 119-131 (1992).
7. Kim, K., et al., *Nucl. Inst. Meth. Phys. Res.* **A422**, 929-932 (1999).
8. McGregor, D., et al., *IEEE Trans. Nucl. Sci.* **43**, 1357-1364 (1996).
9. Manolopoulos, S., et al., *IEEE Trans. Nucl. Sci.* **45**, 394-400 (1998).
10. McGregor, D., and Kammeraad, J., *Semiconductors and Semimetals*, **43**, 383-442, (1995).
11. Klann, R., and McGregor. D., "Development of Coated Gallium Arsenide Neutron Detectors," *Proceedings from the Eighth International Conference on Nuclear Engineering (ICONE-8)*, edited by S. Anghaie, et al., ASTM, New York, New York, ICONE-8110, pp. 1-6.

SECTION XV

TEACHING UNDERGRADUATES WITH ACCELERATORS

Upper-Division Student Laboratory Experiments With a 2.5 MV Van de Graaff Accelerator

D. Bradbury*, W. Dukes, E. Gerber[†], T. Terry**, and R. S. Peterson
Physics Department, The University of the South, Sewanee, TN 37383
and
P. Sangsingkeow, PerkinElmer Instruments-ORTEC, Oak Ridge, TN 37830

*Present address: School of Engineering, Washington University in St. Louis
[†]Present address: Program in Applied and Computational Mathematics, Princeton University
**Present address: Department of Chemistry, University of North Carolina

Abstract. A modern physics laboratory that traditionally emphasized accelerated ion techniques was modified to direct student attention toward the application of these techniques to achieve specific research goals. Students were given an ORTEC surface-barrier detector and access to an accelerator, which they used to determine the elemental identity of materials in the detector and measure their physical properties. Familiarity with the accelerator techniques and the analysis of the data were quickly achieved in the effort to provide more time to meet the research goals of this experiment. The emphasis on the research goals instead of the accelerator-based techniques stimulated a more engaging lab experience.

INTRODUCTION

As physics faculty we continually search for pedagogy that will interest more students in becoming or remaining physics majors. Our traditional, undergraduate laboratory experiences may not contribute as much to the recruitment and retention of majors as we would like. This paper reports an attempt to modify a traditional lab to enhance the students' lab experience and still achieve the traditional teaching goals.

Over the past few years my students have traveled to the University of North Texas to use the 2.5 MV Van de Graaff accelerator to study Rutherford scattering and (p, γ) nuclear reactions. A major goal of this laboratory work was to give the students experience with new types of equipment in order to explore a more challenging and exciting array of physical phenomena. In the past these students have rated the experience highly in formal evaluations. However, it did not engender excitement and anticipation in younger students looking ahead to the same class.

The emphasis of this lab experience was changed to focus the students more upon experimental design and the achievement of the laboratory goals. Instead of studying nuclear scattering and interactions with accelerated beams, these techniques were used to analyze the physical characteristics of an unknown sample.

LABORATORY ASSIGNMENT

The students were given a sample and measurement goals that could be achieved by use of accelerated ion beams. The focus changed from the study of instruments and simple physical phenomena to understanding and using these same instruments and phenomena in themselves to achieve a research goal.

A surface barrier detector was chosen as the material to study. The students were asked to measure the composition and thickness of the thin-film coatings, to verify that the semiconductor was silicon, and to determine whether the silicon was n-type or p-type and the identity of the dopant element. In a semester with thirteen 3-hour labs, this consumed eight of those lab periods. Each student wrote four short lab reports on the results and a class-prepared presentation was given by one of the students at a regional SPS meeting. The students are co-authors of this paper.

Surface-barrier detectors are used in backscattering experiments and their calibration and operation have been part of previous years' lab experiments. Most students do not gain an understanding the physical principles of these diode detectors by simply operating and calibrating the devices. The details of diode construction can provide more opportunity to explore their operation as diodes. Preparation of the sample is the ideal, and these students hoped to make their own surface-barrier detector. As the resources necessary to make surface-barrier detectors are not available this small university, production of detectors was observed in Oak Ridge, Tennessee. PerkinElmer Ortec hosted the students, walking them through the complete manufacturing process. Although more than a day is required to produce a detector, technicians at PE Ortec were able to demonstrate all the steps by using several detectors at different stages of production. At the end of the tour, three unmounted detectors that had not passed quality assurance measurements were provided. Two of these became the samples that the students would study.

A surface-barrier detector is made from n-type silicon with a gold film forming the diode and an aluminum film coating on the side opposite the gold for electrical contact. Although a surface-barrier detector is atypical of the p-n junction diodes normally discussed in an electronics class or a modern physics textbook, students readily accepted the fundamentals of this diode's operation.

Measurement techniques were suggested that could be used to attack the research goals. Those that could be used in an accelerator experiment were Rutherford scattering, (p, γ) reactions, and PIXE (Particle-Induced X-ray Emission spectroscopy). In addition, x-ray fluorescence and neutron activation were discussed as ancillary measurement techniques using the same detectors.

The students performed several lab experiments in preparation for the accelerator experiments. These included γ-ray spectroscopy with NaI(Tl) detectors, a coincidence experiment using NIM electronics, and x-ray spectroscopy with an electron microscope and Si(Li) detector. In addition class time was used to review simple two-body mechanics of Rutherford scattering, the idea of a cross section, and the differential cross section.

EXPERIMENTAL RESULTS

The accelerator laboratory in the Physics Department at the University of North Texas has an array of accelerators, including the 2.5 MV Van de Graaff which was used for this set of experiments. The students arrived on a Friday afternoon and left the following Sunday evening. The weekend included standard Rutherford scattering experiments to observe the kinematic factors and the scattering cross section [1], shown in Figure 1. Thin film targets of gold on carbon and germanium on carbon were used for Rutherford scattering and PIXE. The experimental results are in excellent agreement with theoretical expectations. In the past these scattering experiments have been the focus of the students' efforts. These are excellent experiments because they bring many new concepts to the students' attention. These include beam current integration, measurement of the solid angle, calibration of the beam energy, operation of a vacuum system, energy loss in a thin film, and the use of a normalization detector.

Following the study of Rutherford scattering, the surface barrier detectors were introduced into the ion beam. Backscattering spectra were taken from both the aluminum and the gold sides. The gold and the silicon were identified directly from the kinematics of

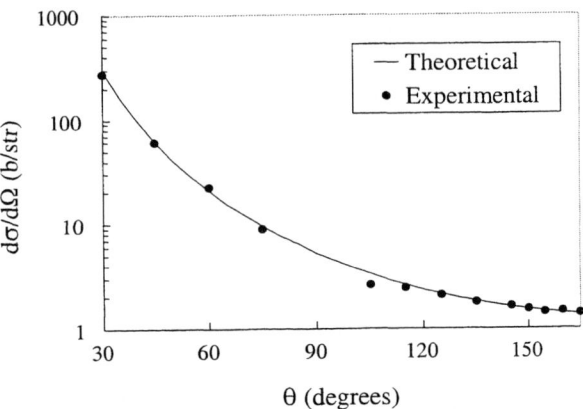

FIGURE 1. Differential scattering cross section measurements of 2 MeV alpha particles scattering from a thin germanium foil.

alpha particles scattered from the gold side of the sample. Scattering from the aluminum side produced overlapping aluminum and silicon peaks, preventing a direct identification of the aluminum. This spectrum is shown in Figure 2. There was no evidence of scattering from the dopant nuclei in either spectrum.

The experimental setup was then altered to include a Si(Li) detector for PIXE measurements. The x-ray emission from each side of the sample was observed and the x-ray yields were normalized to a fixed-angle measurement of Rutherford-scattered helium ions. Although the concept of normalization was obvious to the students once discussed, it was not anticipated in their discussions of the changes to the experimental setup. Absolute x-ray yields were determined by comparison to PIXE measurements of known samples of aluminum and gold.

The PIXE results gave enough data to identify the coating elements and the silicon, but there was not sufficient evidence to identify the dopant element. The measured x-ray yields gave film thicknesses for both the gold and aluminum.

The Rutherford scattering from the aluminum had initially given the students the false impression that they had experimental evidence for the dopant element from the peak centered at about 1,080 keV in Figure 2. However, their analysis of the mass associated with the peak gave wildly improbable answers for the dopant. Additional thought about the scattering from the gold side of the target and the lack of the same peak changed the speculation from an elemental source to an "interference" from overlapping silicon and aluminum spectral contributions. This was found to be the best explanation when the data were more carefully analyzed. A simple model for scattering from a thin

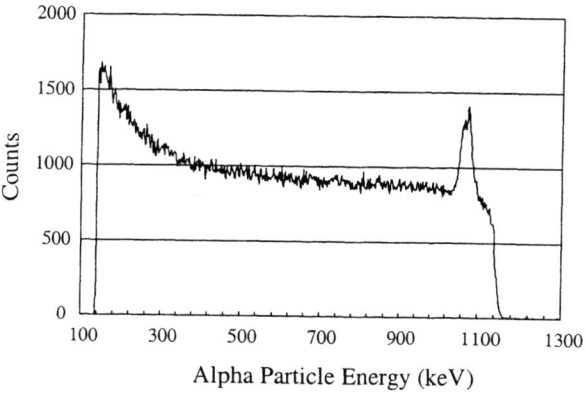

FIGURE 2. The backscattering spectrum from the silicon-aluminum side of the surface barrier detector gives an "interference" peak between the alphas scattered from silicon and those scattered from aluminum.

layer of aluminum on thick silicon was used with the scattering measurements to extract the thickness of the aluminum layer. The results are presented in Table 1, without an assigned error range.

TABLE 1. Thin-film Measurements.

Material	Rutherford Backscattering	PIXE
Gold	34 ± 2 μg/cm^2	37 ± 3 μg/cm^2
Aluminum	65 μg/cm^2	53 ± 2 μg/cm^2

Neutron activation and x-ray fluorescence were used to search for any signature of the dopant element. This involved the use of an intrinsic germanium detector for the observation of gamma spectra from the neutron activated sample. There was no clear evidence for dopant elements from either measurement.

Before finishing the visit, a 1.4 MeV proton beam was used to observe (p, γ) reactions in aluminum targets. This data was used to calibrate the magnet and to verify the energy calibration of the surface-barrier detectors used in the measurement of the alpha particle energies in the Rutherford scattering experiments. This was the most challenging of the experiments that we completed. The students easily accepted the concept of a nuclear resonance, but the concept of calibrating the magnet proved to be elusive.

The final analysis of the data was completed over the next five weeks. The measurements of the gold thicknesses on two of the samples were well within the range of values, 30-40 μg/cm^2 (see Table 1) that PerkinElmer suggested and the measured aluminum thicknesses were within the broader range of PerkinElmer values, 40-80 μg/cm^2. Although the x-ray and Rutherford-scattering results showed that the semiconductor is silicon, there was not sufficient evidence to identify the dopant element. Simple I-V measurements of the diode behavior of the samples showed the silicon to be n-type [2].

CONCLUSIONS

A traditional accelerator lab to study Rutherford scattering and x-ray production was altered to change the student focus towards research goals in the study of the physical properties of a prepared sample. For this lab the sample was a surface-barrier detector. Rutherford scattering and x-ray production were studied as techniques for measuring thin film thicknesses and making elemental identifications. The experimental results were within the expected values given by the manufacturer of the sample.

With such a small group of students, it is not statistically proper to use the results of the student evaluations for a global conclusion on the value of this lab exercise. However, three of the four students in the course rated the lab as the best experience of their college careers.

ACKNOWLEDGMENTS

RSP and students would like to extend our gratitude to Dr. Jerome Duggan for hosting our visit to the University of North Texas, to Dr. Mohamed El Bouanani and Dana Necsoiu for assisting in the accelerator experiments, and to Dr. Pat Sangsingkeow's staff of technicians at PerkinElmer Ortec for their demonstrations of surface-barrier detector production. Support for travel was provided by the Dean of the College of Arts and Sciences at the University of the South.

REFERENCES

1. Paul A. Tipler, *Modern Physics,* New York: Worth Publishers, Inc., 1978, pp. 133-43.
2. Glenn F. Knoll, *Radiation Detection and Measurement*, 3rd Edition, New York: John Wiley & Sons, Inc., 2000, pp. 377-81.

PIXE and Moseley's Law

James R. Huddle

Physics Department, U.S. Naval Academy, Annapolis, Maryland 21402-5026

Abstract. In an undergraduate atomic physics experiment, 2.8 MeV protons are used to bombard thick elemental targets. X rays emitted from the targets are detected with a Si(Li) detector, and their energies are measured with a multichannel analyser. The $K\alpha$ and $K\beta$ x-ray energies are each plotted as functions of $(Z-1)^2$, and the Rydberg energy, Rch, is extracted from the slopes of the resulting lines. The $K\alpha$ plot gives $Rch = 14.099 \pm 0.025$ eV, while the $K\beta$ plot gives 13.557 ± 0.027 eV. The experiment extends the Bohr Theory as discussed in class, serves to introduce PIXE as a quantitative technique for elemental analysis, gives the students experience in using modern instruments which may be used in later experiments, and experience in using uncertainty analysis to decide whether their measurements agree or disagree with other measurements or theoretical calculations. Although the experimental values differ from each other and from the accurately-known value of 13.606 eV by less than 4%, they do not agree to within the experimental uncertainties, forcing students to consider a systematic error arising from their use of the Bohr Theory to derive Moseley's Law.

INTRODUCTION

In two papers [1] published in 1913 and 1914, H. G. J. Moseley reported the results of his studies of the dependence of the frequency of characteristic x rays on atomic number. Since photon frequency is proportional to photon energy, we may express Moseley's Law [2] as:

$$E = Rch\left(\frac{1}{n_f^2} - \frac{1}{n_i^2}\right)(Z-b)^2.$$

Here, R is the Rydberg constant, Rch is the Rydberg energy = 13.60569172(40) eV [3], n_f is the electron's shell number after the x-ray transition, n_i is the shell number before the transition, and b is the electron screening constant = 1 (K x rays) or 7.4 (L x rays).

Moseley's results are the basis for our modern understanding that the atomic number, Z, which determines an atom's place on the periodic table, represents the number of protons in an atom's nucleus, and that it is Z, not the mass number A, which determines an element's chemical properties. As a bonus, Moseley was able to guess the existence of three elements previously unknown (Tc, Pm, and Re, with $Z = 43$, 61, and 75, respectively). Also, Moseley was able to detect impurities in his elemental samples, and suggested in his first paper that analysis of characteristic x-ray emission "may prove [to be] a powerful method of chemical analysis." Today, three methods for performing such studies are in routine use: X-ray Fluorescence (XRF), Proton-Induced X-ray Emission (PIXE), and the energy-dispersive x-ray analysis (EDX) often used by electron microscopists to determine the elemental compositions of the targets they image.

This paper describes an experiment in which undergraduate students use Moseley's Law to measure the Rydberg energy, Rch. Attractive features of this exercise include: (1) It yields a measurement of a quantity with an experimental uncertainty, allowing a quantitative comparison with theory, (2) It introduces students to solid state detectors, multichannel analysers and to PIXE as a method for materials analysis, all of which can be used in successive experiments, and (3) It forces students to consider the validity of Moseley's Law and the Bohr theory from which it is derived when they discover that their experimental value of Rch does not agree with the accurately known value within the experimental uncertainty.

EXPERIMENT

The general approach taken in this exercise is as follows: $K\alpha$ and $K\beta$ x-ray energies were measured for a number of thick samples of pure elements and simple compounds under proton bombardment. These energies were then plotted as functions of $(Z-1)^2$, and lines were fit to the data. For $K\alpha$ x-rays, $n_i = 2$ and $n_f = 1$, so Moseley's law says that the slope should be equal to $(3/4) Rch$. For $K\beta$, $n_i = 3$ and $n_f = 1$, so the slope should be $(8/9) Rch$. Hence, measured values for Rch can be determined, with uncertainties obtained from the uncertainty in the slope.

Undergraduate experiments in which the $(Z-1)^2$ dependence of K x-ray energies is studied have been reported previously [4-8], although only in the last of these is any consideration given to uncertainty analysis or comparison of the experimental results with theory.

None of these experiments uses PIXE. The present experiment was performed in the Naval Academy Tandem Accelerator Lab using the the external-beam milliprobe [9] beamline. The Lab is equipped with a National Electrostatics Corp. 5SDH Pelletron tandem accelerator. The milliprobe beamline is usually used for in-air PIXE in connection with materials analytical research; it was used for this experiment for the ease with which targets can be changed. A beam of protons is accelerated to an energy in the range 2.0 to 3.0 MeV, focussed and momentum-analyzed, then brought out of the accelerator's beamline vacuum system through a thin Kapton foil, and focussed onto a target. The resulting x rays were detected with an Ortec SLP Si(Li) detector, amplified with an Ortec 572 spectroscopy amp, and a spectrum was acquired with a Nucleus PCA-II personal computer analyser card in a Zenith 286 personal computer. The detection electronics are very simple; a schematic diagram can be found in the instruction manual for the detector or in Ref. 10. We have found that we need only a few nA of beam on the target; sometimes even this is enough to overload the detector, so when necessary, we used a filter of one or two layers of 25-μm-thick Al to cut the count rate to an acceptable level. The samples were placed inside a Plexiglas sample enclosure to protect laboratory users from the proton beam. The enclosure was interlocked in such a manner that opening the enclosure interrupts the beam by closing an electropneumatic valve in the beamline. (We have also interlocked our accelerator such that abnormal radiation levels, open lab doors, or failed warning signs shut down the charging chain [11].) Targets of "pure" elements including V, Cr, Fe, Co, Ni, Cu, Zn, Nb, Mo, Ag, Cd, In, Sn, and compound targets of CsI, NaCl, GaAs and Nichrome were used; one spectrum for each target was acquired on the multichannel analyser. Impurities detected in some of the samples also gave x-ray data for the elements Ti, Y, and Zr. After each spectrum was acquired, a Faraday cup was inserted to stop the beam, the beamline valve (down-beam of the Faraday cup) was closed, and the target was changed in preparation for the next spectrum. While the target was being changed, the x-ray energies obtained from the spectrum were entered into a spreadsheet for analysis after class.

LABORATORY CLASS MANAGEMENT

The experiment was run as part of the laboratory associated with our senior-level quantum mechanics course, which numbered about 35 students broken into four lab sections, two sections meeting on Tuesdays, and two on Thursdays. The experiment was conducted in two weeks. During the first week, students were given a tour of the lab and instrumentation, and were taught to calibrate the Si(Li) detector, acquire spectra, determine centroids and widths of x-ray peaks, and dump spectra and peak information to hard copy. When students reported to lab the following week, a technician had calibrated the detector and brought a beam of 2.8 MeV protons on target. Each of the four sections took data from four or five elemental targets and CsI and one or two objects of special interest to the students, such as a class ring or a coin. CsI is interesting because it demonstrates clearly that the x-ray yield is Z-dependent. After the last class, the data were pooled and made available to all students.

DATA ANALYSIS

The data are plotted in Fig. 1. While the plotting application we used will fit the data to a straight line, it

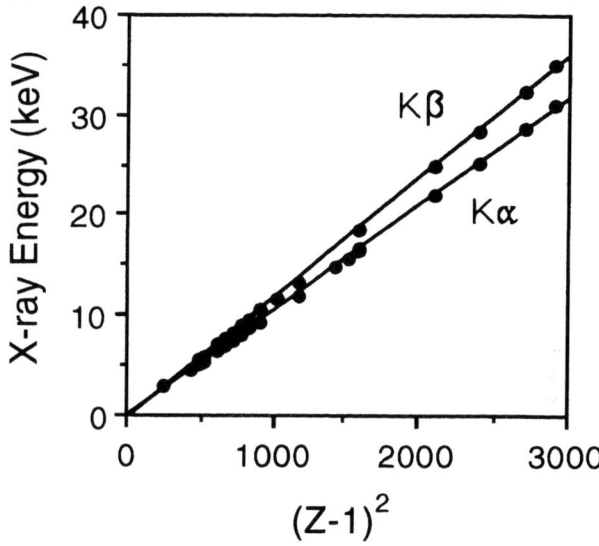

FIGURE 1. Moseley plot of the experimental data for $17 \leq Z \leq 55$. The slope of the $K\alpha$ line is 10.574 ± 0.019 eV, giving $Rch = 14.099 \pm 0.025$ eV, while the $K\beta$ slope is 12.051 ± 0.024 eV, giving $Rch = 13.557 \pm 0.027$ eV. Neither value agrees with the value of 13.606 eV from Ref. 3.

will not compute uncertainties in the slope and intercept. Therefore, the fit is done in a spreadsheet, using the least squares method [12]. The advent of powerful spreadsheet applications for personal computers has made fitting data to a straight line extremely simple; our students did this in just a few minutes using a spreadsheet template they had developed in a previous lab exercise. The slope of the $K\alpha$ line is 10.574 ± 0.019 eV, which gives a value $Rch = 14.099 \pm 0.025$ eV. The slope of the $K\beta$ line is 12.051 ± 0.024 eV which gives a value $Rch = 13.557 \pm 0.027$ eV.

DISCUSSION

Note that the experimental values for Rch do not agree with each other within the uncertainties, nor do they agree with the accurately known value [3] of 13.606 eV, although they differ from that value and from each other by less than 4%. The reason, of course, is that Moseley's Law is based on the Bohr Theory; here, we have experimental evidence which contradicts that theory. A plot of the accurately known x-ray energies tabulated by Bearden [13] as a function of $(Z-1)^2$ is shown for $3 \leq Z \leq 92$ in Fig. 2. Note that the plot is not linear, in <u>disagreement</u> with Moseley's Law. A box shows the range of elements used in this work.

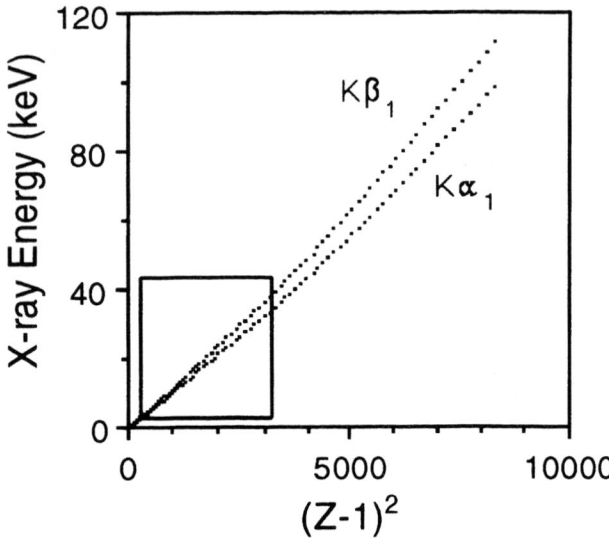

FIGURE 2. Plot of $K\alpha_1$ and $K\beta_1$ data for $3 \leq Z \leq 92$ from Ref. 13. Note that neither plot is linear, in disagreement with Moseley's Law. The box shows the region that includes the data taken in this work.

Application of quantum mechanics, and inclusion of the fine structure corrections leads to the following expression [14-16] for the energy of an electron in an inner shell, described in terms of the quantum numbers $|n,\ell,j>$:

$$E_{n,\ell,j} = Rch\frac{(Z-\sigma_1)^2}{n^2} - Rch\alpha^2\frac{(Z-\sigma_2)^4}{n^3}$$
$$\left(\frac{3}{4n} - \frac{1}{\ell+1} + \frac{j(j+1) - \ell(\ell+1) - s(s+1)}{2\ell(\ell+\frac{1}{2})(\ell+1)}\right)$$

In this expression, α is the fine structure constant, and σ_1 and σ_2 are "screening constants". The term in $(Z-\sigma_1)^2$ is the same as in the Bohr theory, while the term in $(Z-\sigma_2)^4$ includes the relativistic, spin-orbit and Darwin perturbation corrections, which are collectively known as the fine structure. Pauling and Goudsmit [15] state that σ_2 is "remarkably constant" in Z, although it varies from level to level, while σ_1 varies slightly and linearly with Z. This is what causes the the plots in Fig. 2 to curve slightly. The curved shape also explains why the authors of Ref. 8 are able to report agreement with the Bohr theory: They used a range of atomic numbers over which the slope happens to approximate the prediction of the Bohr theory more closely.

CONCLUSION

Student response to this experiment has been animated. The students enjoy working in a modern research lab, and they appreciate that, for once, when they "get the wrong answer", the systematic error is in the theory, not in the instrumentation nor in their lab technique. We have not yet tried quantitative materials analysis using in-air PIXE as a follow-on experiment, simply because we have several other experiments we like to do in this course. Students who take an elective course in nuclear physics do a Rutherford Backscattering experiment using the same accelerator and multichannel analyser.

Acknowledgements

The author gratefully acknowledges useful discussions with F.D Correll, J.R. Vanhoy and D.J. Treacy. This lab exercise would not have run smoothly without D.M. Moore's efforts to set the lab up and run the accelerator.

References

1. H.G.J. Moseley, Philos. Mag. **26**, 1024 (1913); **27**, 703 (1914). These two classic papers have been combined, edited and reprinted with editorial commentary in H.A. Boorse and L. Motz, *The World of the Atom* (Basic Books, New York, 1966) Vol. II, pp. 866-883.
2. See, for example, P.A. Tipler, *Modern Physics*, (Worth, New York, 1978) pp. 152-155.
3. P.J. Mohr and B.N. Taylor, Physics Today **53**, No. 8, Part 2, p. BG6 (August 2000).
4. C. Hohenemser and I.M. Asher, Am. J. Phys. **36**, 882 (1968).
5. W. Kiszenick and N. Wainfan, Am. J. Phys. **42**, 161 (1974).
6. P.J. Ouseph, K.D. Hoskins, J.I. Berman and A.J. Bolander, Am. J. Phys. **50**, 275 (1982); P.J. Ouseph and K.D. Hoskins, *ibid.*, 276 (1982).
7. D. Sway and D.A. Wells, Am. J. Phys. **6**, 208 (1938). This experiment is also described in T.B. Brown, ed., *The Taylor Manual of Advanced Undergraduate Experiments*, (Addison-Wesley, Reading, MA, 1959), pp. 440-442.
8. C.W.S. Conver and J. Dudek, Am. J. Phys. **64**, 335 (1996); see also the comment by K.R. Naqvi in Am. J. Phys **64**, 1332 (1996).

9. S.A. Maclaren, F.D. Correll, J.R. Huddle, J. Vanhoy, and W.D. Kulp, III, Nucl. Instrum. Methods **B56/57**, 708 (1991); F.D. Correll, J.R. Huddle and J. Vanhoy, *ibid.*, p. 1180.
10. J.L. Duggan, *Experiments in Nuclear Science*, Applications Note AN34, (EG&G Ortec, Oak Ridge, TN, 1984); *Laboratory Investigations in Nuclear Science*, (Tennelec/Nucleus, Oak Ridge, 1988).
11. J.R. Huddle, J.R. Vanhoy and F.D Correll, Nucl. Instrum. Methods **B56/57**, 1173 (1991).
12. See, for example, P.R. Bevington, *Data Reduction and Error Analysis for the Physical Sciences*, (McGraw-Hill, New York, 1969) pp. 92-113, or J.R. Taylor, *An Introduction to Error Analysis*, (University Science Books, Mill Valley, CA, 1982) pp.153-162.
13. J.A. Bearden, Rev. Mod. Phys. **39**, 78 (1967), reprinted in *CRC Handbook of Chemistry and Physics*, 51st ed., edited by R.C. Weast (Chemical Rubber Co., Cleveland, 1970), pp. E144-E182.
14. G.P. Harnwell and J.J. Livingood, *Experimental Atomic Physics*, (McGraw-Hill, New York, 1933), pp. 362-368.
15. L. Pauling and S. Goudsmit, *The Structure of Line Spectra*, McGraw-Hill, New York, 1930), pp. Chap. X, pp.170-191.
16. H.G. Kuhn, *Atomic Spectra*, (Academic, New York, 1962), pp. 232-243.

Accelerator-Based Techniques for the Support of Senior-Level Undergraduate Physics Laboratories

J.R. Williams, J.C. Clark and T. Isaacs-Smith

Physics Department, Auburn University, AL 36849

Abstract. Approximately three years ago, Auburn University replaced its aging Dynamitron accelerator with a new 2MV tandem machine (Pelletron) manufactured by the National Electrostatics Corporation (NEC). This new machine is maintained and operated for the University by Physics Department personnel, and the accelerator supports a wide variety of materials modification / analysis studies. Computer software is available that allows the NEC Pelletron to be operated from a remote location, and an Internet link has been established between the Accelerator Laboratory and the Upper-Level Undergraduate Teaching Laboratory in the Physics Department. Additional software supplied by Canberra Industries has also been used to create a second Internet link that allows live-time data acquisition in the Teaching Laboratory. Our senior-level undergraduates and first-year graduate students perform a number of experiments related to radiation detection and measurement as well as several standard accelerator-based experiments that have been added recently. These laboratory exercises will be described, and the procedures used to establish the Internet links between our Teaching Laboratory and the Accelerator Laboratory will be discussed.

INTRODUCTION

The Undergraduate Teaching Laboratory -- Since approximately 1980, the Auburn University Physics Department has offered upper-level undergraduate laboratory instruction in three one-quarter laboratory courses – one at the junior-level and two at the senior-level. The laboratory exercises taught in these three courses are related to material covered in lecture courses that contain no laboratory components per se. The senior-level courses are taken by 8-12 students each year (senior undergraduates and first-year graduate students). One sequence of laboratory exercises taught in one of these courses is based on radiation detection and the interactions of radiation with matter. The students complete five laboratory exercises based on information contained in references [1,2] that are provided for them:

(1) gamma ray detection using sodium iodide (NaI)
(2) gamma ray detection using high purity germanium (HPGe),
(3) gamma ray attenuation using NaI,
(4) charged particle detection using silicon surface barrier detectors (SBDs), and
(5) nuclear lifetime measurements using the method of delayed coincidence.

These laboratory exercises are reasonably sophisticated, and the students (working in groups of 2-3) are generally pushed to finish all five in an eight week period. For example, to complete exercise (5), students must simultaneously set up NaI and surface barrier detectors (with all their associated electronics) in order to detect gamma rays and charged particles in coincidence. The students are simply given a reference from the American Journal of Physics [3] and asked to repeat the measurement of the lifetime of the second excited state in ^{237}Np which decays by gamma ray emission following the alpha decay of ^{241}Am.

The students who carry out the exercises described above are also given a tour of the Accelerator Laboratory where they see practical applications for the radiation detection techniques they will learn. They see that the detectors and techniques that they are studying are identical to those employed in the Accelerator Laboratory by research faculty from a number of departments who use the accelerator for ion implantation and ion beam materials analysis (primarily Rutherford backscattering spectroscopy and light ion channeling). Graduate students who take the senior-level undergraduate laboratory courses usually do so in order to strengthen their backgrounds before they begin working in the Accelerator Laboratory or in a related area of materials research in some other department.

COURSE UPGRADES FOR THE SEMESTER SYSTEM

Starting just this fall, Auburn has switched from the quarter system to the semester system. The three upper-level laboratory courses formally taught over a period of three quarters have now been converted to two coursers offered during two semesters – one in the junior year and one in the senior year. In anticipation of this changeover, we have expanded the senior-level radiation detection and measurement course by developing exercises that introduce our students to accelerator-based materials analysis techniques. After completing the five radiation detection and measurement experiments described previously, the students will use the accelerator as a neutron source to produce ^{28}Al with the ^{27}Al(n, γ) reaction.[3] They will measure the 2.26 min half-life of ^{28}Al by detecting gamma rays from the decay of the first excited state in ^{28}Si. The students will also perform a classic experiment in which they measure the Rutherford scattering cross section for several materials. Subsequently, the students will then learn the fundamentals of thin film materials analysis using the techniques of Rutherford backscattering spectroscopy (RBS). The analysis program RUMP will be available, and the students will simulate numerous thin film RBS spectra for samples that they either prepare themselves in a using thermal evaporation system or that other faculty and students generate in the Accelerator Laboratory for thin film characterization.

The Accelerator Laboratory -- The charged particle accelerator facility is part of the Auburn University Leach Science Center. The heart of the facility is a National Electrostatics 2MV tandem accelerator that was installed during the fall of 1997. This accelerator is equipped with an RF charge-exchange ion source for alpha particle beams and a cesium sputter source for the production of proton beams and a wide range of heavy ion beams (e.g., carbon, silicon, gold). Depending on the final charge state of the accelerated ion, energies of up to 8 MeV can be obtained, with typical beam currents ranging from a few nanoamperes to several microamperes. Ion beams are routinely used in the accelerator laboratory for materials modification and analysis using the techniques of Rutherford backscattering spectrometry (RBS), ion implantation, light ion channeling, nuclear reaction analysis and proton-induced x-ray emission.

The Accelerator Laboratory is a unique facility on our campus. Physics Department faculty and staff maintain the accelerator, and our graduate students are the primary users of the facility. Beam time is also provided for a number of other departments including Electrical Engineering, Chemistry, Mechanical Engineering, Chemical Engineering and the Space Power Institute.

Two Internet links have been developed between the Accelerator Laboratory and the Teaching Laboratory in the Physics Department (Fig. 1). Our accelerator is automated for computer-controlled operation, and as a result, the machine can be operated remotely from a personal computer (PC3) located in the Teaching Laboratory. Remote operation is accomplished using *pcAnywhere* – a software package that we already have in the Accelerator Laboratory.[5] *pcAnywhere* captures the accelerator control system software running on PC1 and shares this software with PC3. Complete operational control of the accelerator is available from either PC. The tandem was manufactured by National Electrostatics Corporation located in Middlebury, WI, and we routinely use *pcAnywhere* between Auburn and Middlebury to download changes to the operating system software, run diagnostics, test accelerator operation, etc. A research associate in the Accelerator Laboratory is available to change targets in the analysis chamber when the accelerator is operated from the Teaching Laboratory. Currently, students are limited to putting beam on and off a sample by controlling a Faraday cup just in front of the analysis chamber. However, we are working on a video connection that will let the students view an analog beam current meter in the accelerator control room. This arrangement will permit the students tune the beam in the Faraday cup.

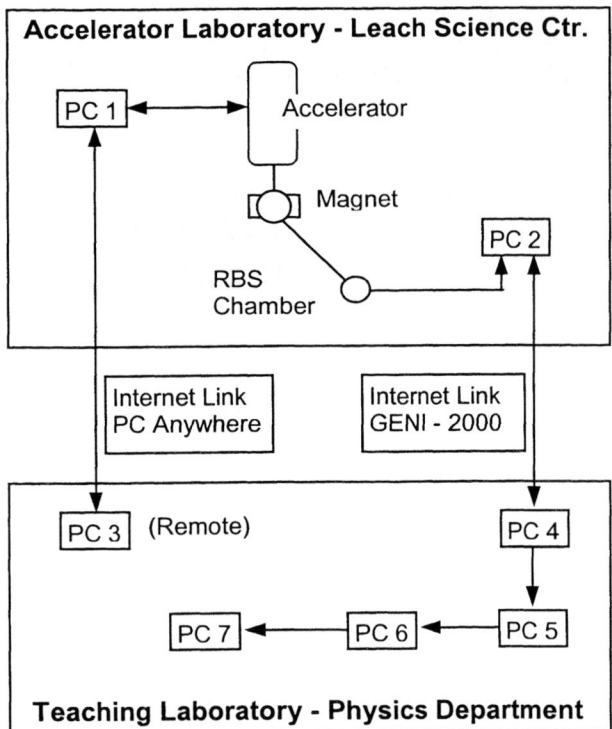

FIGURE 1. Internet links between the Accelerator Laboratory and the Upper-Level Teaching Laboratory in the Physics Department. Typically, four groups of two to three students each would work at PCs 4, 5, 6 and 7.

We have also acquired a second software package (*GENI-2000*)[6] that controls data acquisition and display when an ion beam is used to irradiate a sample in the RBS chamber. This software operates with a Canberra Industries multichannel analyzer (MCA) card in PC2. This card acquires and displays data live-time on PC2 using its own MCA software. The *GENI-2000* software contains all of the MCA software for the local operation using PC2, as well as additional features that allow the MCA card to be controlled remotely from the Teaching Laboratory. The MCA card in PC2 can be controlled remotely by only one PC – in this case PC4. However, when the *GENI* software is installed in PCs 5, 6 and 7 through the purchase of a right to copy option, live-time data acquisition can be viewed on these three computers as well. Spectra can be stored and subsequently analyzed on PCs 4 - 7.

SUMMARY

We have increased the number of laboratory exercises that are available for senior undergraduate and first-year graduate students in the Physics Department. These new experiments are accelerator-based and supplement basic laboratory exercises performed by our students for radiation detection and measurement. Two Internet links have been established between the University's Accelerator Laboratory and the Upper Level Teaching Laboratory in the Department – one for remote operation of the Pelletron accelerator, and one for remote, live-time data acquisition. Initially, new experiments will be conducted using the accelerator for neutron production and for the application of Rutherford backscattering techniques for thin film analysis. However, the new accelerator and the Internet links provide great flexibility for our senior-level undergraduate laboratory course. For example, new experiments will be added in the future in the areas of nuclear reaction analysis and ion implantation.

Short of actually working in the Accelerator Laboratory for one or two semesters, students will be exposed as realistically as possible to ion beam analysis techniques that support state-of-the-art materials research programs. These techniques are applied not only for semiconductor materials analysis, but also in almost any other research and development activity where thin film systems are fabricated and analyzed.

REFERENCES

1. G.F. Knoll, *Radiation Detection and Measurement*, New York, John Wiley and Sons, 2000.

2. J.L. Duggan, Laboratory Investigations in Nuclear Science, Tennelec, Inc. 1988.

3. A.A. Rollefson and R.M. Prior, *Am. J. Phys.* 46(10) 1007-1008 (1978).

4. A.A. Rollefson and R.M. Prior, *Am. J. Phys.* 46(10) 1077-1078 (1978).

5. Symantec, Inc. (www.symantec.com).

6. Canberra Industries, Inc. (www.canberra.com)

Undergraduate Participation in the Crystal Ball Baryon Spectroscopy Program at Brookhaven National Laboratory

Michael E. Sadler and L. Donald Isenhower*

Department of Physics, Abilene Christian University, Abilene, Texas 79699

Abstract. ACU undergraduates have played significant roles in the experimental program in baryon spectroscopy at the Alternating Gradient Synchrotron (AGS) at Brookhaven National Laboratory (BNL). The most recent data consist primarily of cross sections for $\pi^- p \to$ Neutrals at 130–750 MeV/c obtained with the Crystal Ball detector. An overview of the undergraduate involvement will be presented, culminating with the analysis of the data to obtain differential cross sections for $\pi^- p \to \pi^0 n$.

INTRODUCTION

A new Crystal Ball Collaboration [1] has been formed to utilize the SLAC Crystal Ball for measurements in baryon spectroscopy at BNL. E913 at the AGS is an experiment designed to study the formation of baryon resonances from $\pi^- p$ interactions and their decay into neutral final states, e.g. $\pi^- p \to \gamma n, \pi^0 n, \eta n, \pi^0 \pi^0 n$... E914 is a similar experiment for $K^- p$ interactions. The data reported here, differential cross sections for $\pi^- p \to \pi^0 n$ at 130–300 MeV/c, were taken with the Crystal Ball (CB) multiphoton spectrometer during Fall, 1998, in the C6 beam line at the Alternating Gradient Synchrotron at Brookhaven National Laboratory.

The physics motives for making more accurate measurements of $\pi^- p \to \pi^0 n$ in this region are to determine better the isospin-odd s-wave scattering length, to extrapolate scattering amplitudes to the non-physical region (e.g., for determinations of the πN σ term), to evaluate the πNN coupling constant and the mass splitting of the Δ resonance, to improve the determination of the mass, width and decay of the $P_{11}(1440)$ resonance and to investigate isospin invariance and charge splitting of the P_{33} scattering amplitude.

ACU undergraduates participated in the move of the CB from SLAC to BNL, in the reassembly of the detector after it had been dormant for eight years, in cabling, testing, and, finally, in preparing for the experiment and the data acquisition. Summers of 1999 and 2000 were spent analyzing the data at ACU. The students have been integrally involved in several aspects of this analysis.

THE CRYSTAL BALL

The CB detector, designed and built at SLAC, is a highly-segmented, total-energy electromagnetic calorimeter and spectrometer that covers over 90% of 4π steradians. A schematic is shown in Figure 1. The ball proper is a sphere with an entrance and exit opening for the beam and an inside cavity with radius of 25 cm for the liquid hydrogen target. It is constructed of 672 optically isolated NaI(Tl) crystals that detect individual γ's. Electromagnetic showers in the CB are measured with an energy resolution of $2.7\%/E^{1/4}$, where E is in GeV. Directions of the γ rays are measured with a resolution of 2-3 degrees in the polar angle. An electromagnetic shower from a single γ ray deposits energy in several crystals, called a cluster. The present cluster algorithm sums the energy from the crystal with the highest energy with that from the twelve nearest neighbors.

---

* Supported in part by U.S.D.O.E. by Grant DE-FG03-94ER40860
[1] The new Crystal Ball Collaboration consists of B. Draper, S. Hayden, J. Huddleston, D. Isenhower, C. Robinson and M. Sadler, *Abilene Christian University*, C. Allgower and H. Spinka, *Argonne National Laboratory*, J. Comfort, K. Craig and A. Ramirez, *Arizona State University*, T. Kycia, *Brookhaven National Laboratory*, M. Clajus, A. Marusic, S. McDonald, B. M. K. Nefkens, N. Phaisangittisakul and W. B. Tippens, *University of California at Los Angeles*, J. Peterson, *University of Colorado*, W. Briscoe, A. Shafi and I. Strakovsky *George Washington University*, H. Staudenmaier, *Universität Karlsruhe*, D. M. Manley and J. Olmsted, *Kent State University*, D. Peaslee, *University of Maryland*, V. Abaev, V. Bekrenev, N. Kozlenko, S. Kruglov, A. Kulbardis, I. Lopatin and A. Starostin, *Petersburg Nuclear Physics Institute*, N. Knecht, G. Lolos and Z. Papandreou, *University of Regina*, I. Supek, *Rudjer Boskovic Institute* and A. Gibson, D. Grosnick, D. D. Koetke, R. Manweiler and S. Stanislaus, *Valparaiso University*.

Crystal Ball
Multi-photon Spectrometer

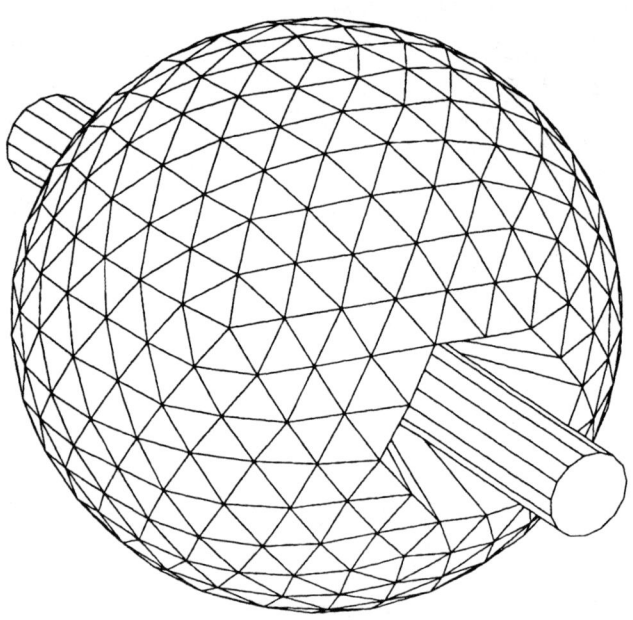

672 separate NaI(Tl) crystals
94% solid angle; with endcaps ~98%

$\sigma/E = 2.7\% \times E^{-1/4}$ (E in GeV)

$\emptyset_{ext} = 1.32$ m, $\emptyset_{cavity} = 0.50$ m, $\sigma_\theta \approx 2° - 3°$, $\sigma_\phi = 2°/\sin\theta$, $t = 16 X_0$

FIGURE 1. Schematic view of the Crystal Ball.

EXPERIMENT AND ANALYSIS

A 10 cm long liquid hydrogen target was installed inside the CB at the C6 beam line at the AGS. The beam phase space was measured by one drift chamber upstream and six drift chambers downstream of the last bending magnet. A data acquisition system was designed that utilizes the CEBAF Online Data Acquisition software (CODA). The trigger consisted of a disappearing beam requirement, determined by three scintillation counters upstream and veto counters downstream of the target, and a minimum energy in the CB. A veto barrel consisting of four plastic scintillation counters surrounding the hydrogen target was used to identify neutral (no signal in the veto barrel) and charged (at least one veto barrel signal) triggers. The charged triggers were prescaled by a factor of 10. Only the neutral triggers have received significant attention in the analysis thus far. An interface with the CERN Physics Analysis Workstation (PAW) software was implemented for on-line monitoring of the beam, detector and physics events.

The detector acceptance as a function of scattering angle was determined by using a GEANT Monte Carlo (MC) simulation. The measured beam phase space was used as input. An angular distribution of the outgoing π^0's at a given momentum was obtained from the VPI (now GW) SP99 partial-wave analysis (PWA).

The MC program propagated the electromagnetic processes from the $\pi^0 \to \gamma\gamma$ decay through the hydrogen target and container, the beam pipe, veto barrel scintillator and the CB elements. Surprisingly, on the order of 25% of the events produced a charged particle in the veto system according to the MC. The acceptance was corrected for this factor, which is one of the significant sources of systematic error at this point in the analysis.

In the present analysis the NaI crystals that border the entrance and exit tunnels for the beam were used as guard crystals, meaning that events were rejected if the highest energy in a cluster of crystals occurred in one of these crystals. Two-cluster events consistent with a single π^0 and three-cluster events consistent with a π^0 and a neutron were used in the present analysis.

The electron and muon contamination of the pion beam was measured from time-of-flight (TOF) at low momenta (below 300 MeV/c). The electron measurements were used to determine the efficiency of a differential gas Čerenkov counter in the beam downstream of the target. This efficiency (92.3%) was used to determine the electron contamination from the Čerenkov counter up to the highest momenta in the experiment. The on-momentum muon contamination was extrapolated from the measurements at low momentum. (On-momentum muons originate from pion decay near the production target.) The decay muons (muons that are registered in the beam counters which come from pion decay toward the end of the beam channel) must be calculated by Monte Carlo simulation and have only been estimated. As an example, the μ/π fraction was evaluated to be 3.5% and e/π was 5.9% at 299 MeV/c from these techniques.

Beam momentum calibrations for the C6 beam line were performed utilizing time-of-flight and stopping-range techniques. The ADC's for each crystal were calibrated by a combination of data from a ^{137}Cs source and monochromatic 129.4 MeV photons from the $\pi^-_{stopped} p \to \gamma n$ reaction. Cross comparisons between the calibrations for the beam momenta and the ADC's were also made from kinematic relationships. For example, at a given beam momentum the energy of the π^0 from $\pi^- p \to \pi^0 n$ (or an η from $\pi^- p \to \eta n$) is determined by the scattering angle. This self-consistency check between the momentum calibration of the beam and the energy response of the detector is a feature that has been lacking in previous measurements of neutral-particle final states.

Data were analyzed this past Summer to obtain absolute differential cross sections for $\pi^- p \to \pi^0 n$ at momenta

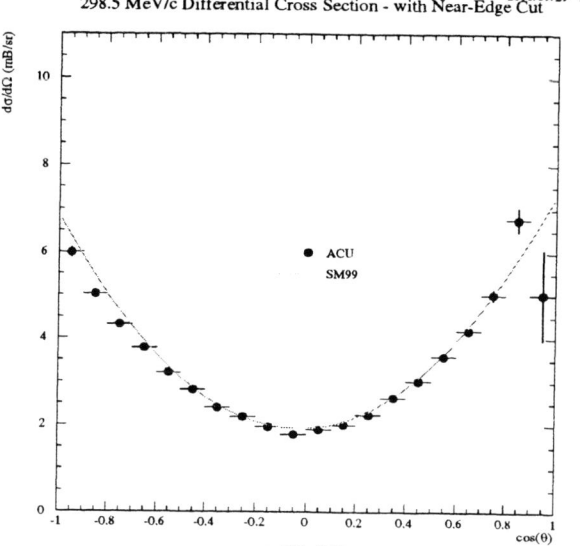

FIGURE 2. Preliminary differential cross section at $P_\pi = 299$ MeV/c.

from 130 to 300 MeV/c. An example of the results is shown in Figure 2 at 299 MeV/c, which is near the peak of the Δ, or P_{33}, resonance. Here the data are dominated by the P_{33} scattering amplitude which has an angular dependence given by $1 + 3\cos\theta$ as shown by the curve in Figure 2, taken from the GW SM99 phase shift analysis. Reproduction of this angular dependence, as well as the absolute normalization, gives confidence that the operation of the Crystal Ball and the acceptance calculation are reasonably well understood. These results were obtained by undergraduates who analyzed the data at ACU during Summer 2000.

UNDERGRADUATE PARTICIPATION

The complexities of modern experiments such as the one described above are foreboding to the uninitiated, particularly an undergraduate physics major. The Crystal Ball detector in the configuration used for the data presented here is one that can be understood by any undergraduate who understands the operation of a simple scintillation counter, usually encountered in the laboratory of a modern physics course taken in the second year. The processing of information from approximately 700 scintillators (including the beam counters amd veto counters) presents a logistical programming challenge involving computer programming and interactive graphics. These skills are quickly acquired by undergraduate physics students and they are often able to make meaningful contributions in data analysis after a week or two of concentrated effort.

Undergraduates obviously need initial guidance for most tasks since they lack the overall knowledge of how the different parts of an experiment fit together and may need to learn how to use test equipment. Many experiments are similar to the Crystal Ball program in that the detector is a large device comprised of subassemblies that are duplicated repeatedly. Once a subassembly is understood, then an undergraduate can repair, calibrate or re-assemble them with minimal supervision. Often the students become more proficient in the these tasks than their supervisor.

Below is a list of accomplishments and contributions made by undergraduates to experiments in which ACU has been involved over the past three years. A synergy has developed among the principal investigators, ACU administration, U.S.D.O.E., capable and hard-working undergraduates and accomplished collaborators, all motivated by good physics.

- Accompanying the cross-country move of the Crystal Ball from SLAC to BNL in order to monitor the environment-controlling apparatus (air conditioners and dehumidifers). This task was done by Ben Draper, an ACU undergraduate.

- Building the support stand for the Crystal Ball detector (with monetary support and in cooperation with Argonne National Laboratory). Draper, who had excellent mechanical skills, worked on the project with the ACU Industrial Technology Department and a local steel company to deliver the support stand on budget and ahead of schedule.

- Helping to install and test the 672 photomultiplier tubes on the NaI crystals on the CB. This task was done by Jeremy Huddleston and Zack Mulkey under the supervision of Vladimir Bekrenev. a Russian collaborator.

- Running the signal and high voltage cables for the CB, again done by Huddleston and Mulkey during a visit to BNL during their Christmas break in 1996.

- Repairing all of the integrate-and-hold (I&H) modules on the CB in order to decrease spread in pedestals by factors of 4-10 and eliminate excessive noise. Isenhower made the prototype and students duplicated the changes in the other 80 electronics modules before the 1998 run. In the process, the students (Matt Shaw and Robert Hance) acquired experience in acquiring the parts and materials for repair of the I&H modules and worked on other parts of CB front-end electronics.

- Setting up the hardware for data acquisition, including a Sun workstation and a VME crate for the front-end of the data acquisition hardware.

- Working closely with our Russian and other foreign colleagues to maximize their extensive contributions to the CB project. The students often help in purchasing materials or other dealings about which the foreign scientists have little knowledge in the U.S. The students often learn skills that they continue to perform when the foreign scientists return to their home institutions.

- Writing software to provide monitoring of operation of the CB and various rates during data acquisition. This task was accomplished by Cory Robinson, an ACU undergraduate, as a summer project at BNL in 1998.

- Working on campus during the summer and academic year, including building various support stands, assembling high-voltage distribution boxes, testing scintillators and plateauing PMT's for experiments at BNL and FNAL. Again, these tasks are often done in connection with the ACU Industrial Technology Department.

- Running Monte Carlo simulations for CB analysis.

- Running Monte Carlo simulations to aid in the design of future experiments.

- Fabricating scintillation counters for beam monitoring or other tasks, often done in conjunction with our Russian colleagues.

- Utilizing our campus computing facilities (Sun workstations and various Linux platforms) to analyze the CB data obtained in 1997 and 1998, as described above.

- Constructing a computer farm by upgrading old PC's to analyze CB data, particularly in the CPU-intensive Monte Carlo calculations. Shawn Hayden, an undergraduate proficient in computer hardware and software, oversaw the upgrade and has served as system manager of our local computers.

- Setting up a web page for ACU research work.

ACU has been involved in other experiments with different collaborations, most recently E789, E866 and P906 at Fermilab. We have joined the PHENIX collaboration at BNL in the past year. Significant contributions by ACU undergraduates on these experiments include:

- Maintaining and repairing all four hodoscope arrays for E866.

- Maintaining TDC's, Segmenters, and various other components of the Nevis Transport system used for DAQ in Fermilab E866.

- Repairing preamps and other such items for E866.

- Measuring, repairing, and reinstalling signal cables for E866 including a large number of Ansley cables.

- GEANT modeling of backgrounds for Fermilab P906 to help determine optimal absorber placement and provide a strong case for Fermilab to consider a fixed-target program to extend the E866 measurements.

- Participated in the construction of the world's largest cathode strip chambers for muon tracking in PHENIX at RHIC. Matt Shaw and Chris Kuberg spent a full year and Robert Hance spent one semester during the academic year at BNL assembling these chambers. Three other students, Austin Barker, Andy Brown, Wayne Bland and Jeb Qualls, spent all or part of this past summer at BNL on this task.

The Conference Experience for Undergraduates (CEU) program is in its third year as part of the APS meeting of the Division of Nuclear Physics. The Fall 2000 meeting in Williamsburg, Virginia had nine ACU undergraduates presenting their research projects. The students and the title of their projects follow:

- Andy Brown, *Very Low Energy Measurements of Pion-Nucleon Charge Exchange Scattering*

- Michael Daugherity, *Muon Tracking and Calibration in PHENIX*

- Shawn Hayden, *Crystal Ball Beam Contamination Studies*

- Chris Kuberg, *Construction of the World's Largest Cathode Strip Chambers for Muon Tracking in PHENIX at RHIC*

- Nathan Longbotham, *GEANT Background Modeling of Fermilab P906*

- Jebidiah Qualls, *The Muon Tracking Arms of the PHENIX Detector*

- Cory Robinson, *Correction of Crystal Ball Monte Carlo Geometry and Analysis of Lowest Momentum Runs*

- Matthew Shaw, *Construction of Cathode Strip chambers for PHENIX*

- Preston Willis, *Comprehensive Test Program for Multi-Chip Modules*

SECTION XVI

TARGETS FOR NUCLEAR RESEARCH

Tritium Target Manufacturing For Use in Accelerators

P. Bach, C. Monnin, M. Van Rompay *(1)*, A. Ballanger *(2)*

(1) SODERN, 20 avenue Descartes, 94451 Limeil-Brévannes Cedex, France
(2) CEA/DAM/DTMN-Valduc, 21120 Is sur Tille, France

ABSTRACT

As a neutron tube manufacturer, SODERN is now in charge of manufacturing tritium targets for accelerators, in cooperation with CEA/DAM/DTMN in Valduc. Specific deuterium and tritium targets are manufactured on request, according to the requirements of the users, starting from titanium target on copper substrate, and going to more sophisticated devices. A wide range of possible uses is covered, including thin targets for neutron calibration, thick targets with controlled loading of deuterium and tritium, rotating targets for higher lifetimes, or large size rotating targets for accelerators used in boron neutron therapy. Activity of targets lies in the 1 to 1000 Curie, diameter of targets being up to 30 cm. Special targets are also considered, including surface layer targets for lowering tritium desorption under irradiation, or those made from different kinds of occluders such as titanium, zirconium, erbium, scandium, with different substrates. It is then possible to optimise either neutron output, or lifetime and stability, or thermal behaviour.

INTRODUCTION

For more than 35 years, SODERN has been developing and manufacturing sealed neutron tubes, including small deuterated or tritiated targets. The CEA center of Valduc does a similar activity, concerning large deuterated or tritiated targets for accelerators.

SODERN and CEA/Valduc center have decided to join their efforts in this field for a better efficiency. Presently, SODERN does the commercialisation of the targets, and is operating all the manufacturing equipment in its premises or in the CEA laboratories. Targets are usually made on request, according to the specification of the users. Research and development of new targets or new related technologies are performed at SODERN, with the cooperation of CEA/Valduc center and other institutions.

The purpose of this paper is to present the technical capabilities of our group in this field, and some of the new developments in deuterated or tritiated targets technology, related to different applications.

TARGET MANUFACTURING PROCESS

Targets used in 14 MeV or 2.5 MeV neutron generators are usually made of metallic thin film deposited on a mechanical substrate, and loaded with deuterium and/or tritium. Such a target will emit neutrons when bombarded by a beam of accelerated deuterium and/or tritium ions, currently between 100 and 350 keV. In most cases, i.e. in accelerators, a beam of deuterons is directed toward a tritiated target. Titanium based targets are often used, as adherence of titanium after loading remains good.

The expected fusion reactions are the following:
Ti tritide targets : D-T reaction :

$$_1^2H + _1^3H \rightarrow _0^1 + _2^4He + Q$$

En = 14.1 MeV and Eα = 3.5 MeV Q = 17.6 MeV
Ti deuteride targets : D-D reaction :

$$_1^2H + _1^2H \rightarrow _0^1 + _2^3He + Q$$

En = 2.45 MeV and E$^3_{He}$ = 0.82 MeV Q = 3.26 MeV.
The working conditions of the targets depend on the desired neutron applications. Thus, it is fundamental to know the target and accelerator characteristics in order to establish the relations between the titanium hybrid thin film, the incident beam and the generated neutrons.

METAL DEPOSITION

Main parameters to control for metal deposition are the following :
- cleanliness and the surface state of substrate,
- effective area to metallize,
- type of metal,
- energy of atoms and molecules during deposition, related to the technology for deposition,
- substrate temperature during deposition,
- deposition rate,
- thickness of the film.

Substrates with a diameter up to 30 cm are manufactured. We are currently using copper or stainless steel substrates.

In a first step, the substrate is chemically etched and cleaned according to the material and the expected surface state. The substrate is then installed behind a metallization mask, and installed inside the metallization equipment. Different kind of metals are used, chosen from their hydrogen occlusion properties : titanium, zirconium, erbium, scandium, yttrium. Titanium is often selected, for its excellent adherence properties, and its loading ratio ranging from 0 to 2.

We are using either thermal evaporation, or electron gun evaporation, or cathodic sputtering technique, and the substrate holder is heated at the suitable temperature for an efficient outgassing. Temperature is then adjusted to the expected level, and metallization is started at the suitable rate. Temperatures are either room temperature or 200°C, sometimes more. Deposition rate lies in the 20 to 500 Å range, according to specifications. Deposition time results in the suitable thickness; this parameter may be controlled using a quartz microbalance. Typical thicknesses are 0.1 µm for thin targets, or between 1 and 8 µm for other targets such as those used for neutron irradiation.

GAS LOADING

Metallized substrates are loaded with hydrogen isotopes, either deuterium, or tritium, or a mixture of these two gases, using specific loading equipments with pumps, manifolds, gas reservoirs, etc.

In order to manufacture very clean targets with controlled loading rate, each target is placed in its own small vessel, outgased in ultra high vacuum conditions, then gas loaded individually. In that way, loading rate and loading conditions (temperature, pressure, maximum rate) are well controlled, for instance for an extended lifetime. Another possible process is to load some targets at the same time and at their maximum loading rate : in that case, loading conditions are not well controlled, and effective loading ratio is not known.

Evaluation of effective tritium activity has to be made using other methods and equipments. Very large targets, up to 950 Curie of tritium, are made according to this second process, but in an individual way : the loading rate is then known from loading conditions.

TARGET CHARACTERIZATION

Development of targets as well as characterisation of targets on request is made using a number of analytical methods, such as :
- laser profilometry or mechanical roughness measurements, for evaluating the roughness of the surface,
- scanning electron microscopy on deuterated films,
- X-ray diffraction for structural and phase analysis,
- Thermogravimetry, to study absorption/desorption kinetics of deuterium from titanium thin films,
- Neutron output measurement.

Results of characterisation are extensively described in the ref. 1.

Initial morphology of the titanium thin film deposited at 200°C and at 8 nm.s^{-1} growth rate shows like a multilayer overgrowth. At 400°C, small nodules are visible, and the film seems to be less porous.

From X-ray diffraction, titanium film is in its α form (H.C. structure), plane 001 being parallel to the surface.

After loading, titanium deuteride film is in the γ phase (C.F.C. structure), without preferential orientation. When loading is not complete, a mix of α and γ phase is observed (see ref. 2).

From thermal gravimetry measurements, increases of mass of titanium powder begins with a parabolic curve, showing a diffusion mechanism in a single phase medium, and continues with a sigmoidal shape, showing the growth of the γ phase. It is interesting to note a relatively long delay before the mass increase of the sample, which is very fast. This delay depends on superficial oxide layer and morphology of the layer. We observe a similar behaviour on thin films, but time constants are shorter.

Sputtering of some targets was studied after deuteron bombardment. The less altered targets are those showing the lower microstructural porosity, while substrate temperature during the deposition step was in the 400°C range and deposition range in the 8 nm.s^{-1}. The other targets showed superficial craters.

Lifetime of deuterated targets is extended and neutron output is generally very stable, when using a beam without impurities. Lifetime of tritiated targets is shorter; neutron output decreases during the 10 to 100 h of operation. Lifetime of deuterium tritium (50 % each) targets may be very extended, when using a mixed beam without impurities.

CONCLUSION

SODERN and CEA/Valduc center have developed different technologies for tritium targets, allowing the selection of the best configuration for each kind of use. SODERN made available targets for neutron energy calibration, targets for research accelerators, and targets for neutron irradiation, in sizes up to 30 cm diameter with a possible tritium activity as high as 950 Curie, compatible with a type A packing.

Lifetime of deuterated targets may be very extended with a stable output, using a beam without impurities. Lifetime of tritiated targets is shorter, and output decreases over the current 10 to 100 hours of use, depending on beam intensity. It is possible to use a deuterium and tritium loading process. Lifetime of such targets may be very long, in the 100 to 1000 hours, when using a deuterium and tritium beam without impurities.

With improvement of existing technologies one can foresee targets capable of use at temperature between 200°C and 350°C without significant loss of gas, and targets to be used up to 80°C with a very low tritium exhaust during ion bombardment.

REFERENCES

1. C. Monnin et al.
   20th World Conference of the INTDS : "Target for particle beams : preparation and use"
   Antwerp, Belgium, Oct. 2-6, 2000
   (to be published in NIM-A)
2. P. Bach et al.
   Le vide – Les couches minces – Vol. 2.12 – Mai 1982

Preparation of Actinide Targets by Electrodeposition for Heavy-Ion Studies and Laserspectroscopic Investigations

K. Eberhardt, P. Thörle, A. Nähler, N. Trautmann

Institut für Kernchemie, Johannes Gutenberg-Universität Mainz, D-55099 Mainz, Germany

Abstract. For the preparation of actinide targets electrochemical methods have been used. Electrolytic depostion in form of the actinide hydroxide on a Ta- or Ti-backing is performed from an aqueous solution, applying current densities up to several A/cm^2. In the case of Be or Al as backing material, molecular plating was applied. Here, the deposition from an organic solution (usually isopropanol) with current densities of only a few mA/cm^2 and voltages up to 1000 V occurs. With these techniques target densities up to 1 mg/cm^2 are possible. In most cases, prior to deposition, chemical separation procedures are required to ensure high purity of the target material. Targets with thicknesses ranging from fg/cm^2 up to mg/cm^2 for the actinides Ac to Es have been produced. A brief survey for the applications of actinide targets in chemical and physical studies is given.

INTRODUCTION

Electrodeposition is widely used for the preparation of actinide targets on metallic and non-metallic backing materials [1-7]. Very often deposition is accomplished from aqueous acidic solutions applying current densities up to several A/cm^2. In this case, due to their high negative electrode potential, hydrogen is evolved and the actinide elements are deposited in the form of their hydroxides. This technique is referred to as "electrolytic deposition" (ED). Deposition from organic media (isopropanol, isobutanol, acetone or ethylalcohol) is also possible. Here, the actinide compound, normally the nitrate, is dissolved in a small volume (5-20 µl) hydrochloric or nitric acid and the aqueous phase is mixed with the organic solvent. Under these conditions no electrolytic dissociation occurs while an electric current is applied. With this technique, called "molecular plating" (MP), highly uniform actinide layers with thicknesses up to 1 mg/cm^2 can be achieved and probably the same compound as originally dissolved in the aqueous phase is deposited at current densities in the range of mA/cm^2.

Compared to alternative techniques - evaporation, electrospraying or precipitation, respectively - electrodeposition offers high yields, simple handling and easy recovery of the target material. The present paper describes the electrodeposition of actinides from Ac to Es on various backing materials with target thicknesses ranging from fg/cm^2 for applications in the field of nuclear- and atomic physics up to 1 mg/cm^2 for heavy-ion experiments. Procedures for purification and recycling of target material are also discussed.

EXPERIMENTAL

Fig. 1 shows a cross section view of the cell used for electrodeposition. A cylindrical funnel made of polyether-etherketone (PEEK) confines the area to be plated and acts as a solution container with a volume of 10-12 ml. A Viton O-ring serves as seal to prevent losses during the deposition process. An additional glass plate (see Fig. 1) is used beneath the funnel if the diameter of the target spot is on the order of 4 millimeters or below. The backing foil is sealed to the funnel and the glass plate by means of a polyethylene cap and a titanium (Ti) cylinder that serves as the cathode. A platinum (Pt) or rhodium (Rh) wire is used as anode material. Molecular plating is carried out by applying a high voltage of 700 to 1000 V and a low current, typically several mA/cm^2, whereas electrolytic deposition is performed at 10-12 V, yielding current densities on the order of several A/cm^2 from an aqueous solution of pH 2-3 and containing ammonium chloride or ammonium sulphate as electrolyte. For cooling, the cell should be placed in an ice-water bath during deposition.

The choice of a proper backing material is a crucial point in many heavy-ion (HI) reaction studies. Very often a material with a low Z-number like beryllium (Be; Z=4) is used in order to prevent the production of interfering transfer products obtained in reactions with the beam. Here, molecular plating must be applied, since Be dissolves rather quickly in aqueous acidic solutions. For the same reason, Rh (Z=45) instead of Pt (Z=78) should be employed as anode material in order to prevent contamination of the target creating Po-isotopes. For the deposition of actinides onto other backing materials such as Ti, molybdenum (Mo) or tantalum (Ta), respectively, electrolytic deposition is a suitable method. Normally, deposition is performed on commercially available self-supporting foils with thicknesses ranging from 5 μm up to 50 μm. The foils should be pinhole-free and pre-cleaned with acid, water and rinsed with isopropanol before use.

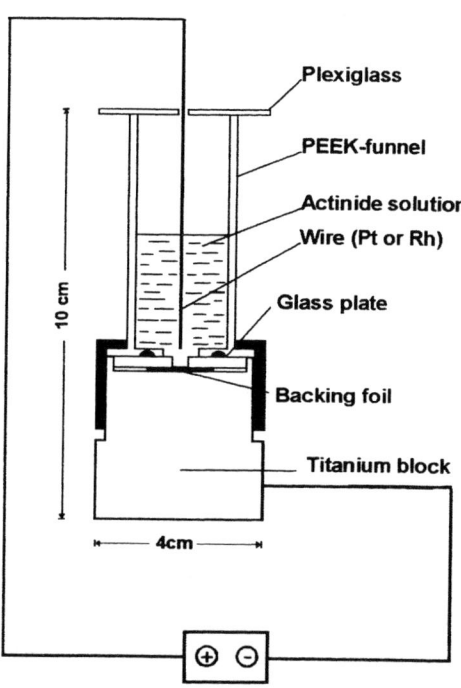

FIGURE 1. Cell for the preparation of actinide targets by electrodeposition from organic and aqueous solutions

In order to prevent excessive heating of a stationary target at high beam currents as delivered from heavy-ion accelerators, a multi-target device mounted on a rotating wheel is under development for actinide targets. Here, the rotation speed of the wheel is adapted to the pulse structure of the ion beam. Such an arrangement will be applied in experiments investigating the chemical properties of the heaviest elements at the Gesellschaft für Schwerionenforschung (GSI) in Darmstadt. One segment of such a target wheel is shown schematically in Fig. 2. Here, the target area is 1.93 cm^2 per segment, with three segments forming a complete circle. The banana-shaped backing is pre-mounted onto a frame that fits into a deposition cell similar to that shown in Fig 1.

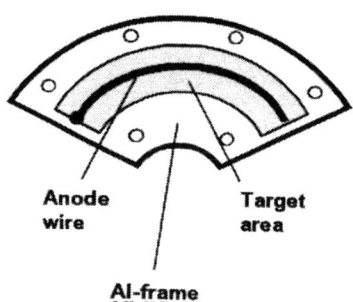

FIGURE 2. Schematic view of one of three segments for a rotating-wheel target arrangement. The banana-shaped target covers an area of 1.93 cm^2 (per segment).

Table 1 comprises some of the actinide targets that have been made at the Institut für Kernchemie in the last few years mainly for use in heavy-ion (HI) reaction experiments at accelerators and as fission targets in research reactors. In Table 2 targets for applications in the field of atomic physics and for ultratrace analysis of plutonium (Pu) in various environmental samples are listed. In addition to that, thorium (Th), uranium (U), and neptunium (Np) targets have been prepared for different applications.

Very often, prior to deposition, chemical separation procedures are required to ensure high purity of the target material. This is of special importance in many HI-experiments, since HI-reactions of impurities such as Pt or lead (Pb) with the beam have a much higher cross section compared to the HI-reaction of the target material itself. Furthermore, traces of Be after a recovery process prevent an effective deposition process in molecular plating. Thus, after recycling of an irradiated target used in an HI-accelerator, Be must be separated prior to further use.

For the separation of Pb from the trivalent actinides, a cation-exchanger (CIX; Dowex 50WX8) is used. The CIX has to be pretreated with pure water, 8 N HCl and 0.5 N HCl-solution at least five times. The actinide solution to be purified is evaporated to dryness and the material is dissolved in 3 ml 0.5 N HCl. This solution is transferred to the CIX-column. The column is first washed with 8 x 2 ml 0.5 N HCl. Then Pb is selectively eluted with 15 ml 1.5 N HCl, whereas the trivalent actinides remain on the column. In a sub-sequent step, the actinides are eluted with 15 ml 8 N HCl. The separation procedure is performed at a temperature of 55°C.

Beryllium is separated on an anion-exchanger (AIX; BioRad AG1X8) from 1 M HNO_3 in a methanol-water mixture (90 Vol% methanol) at room temperature. The original solution is evaporated to dryness and 2 ml of the HNO_3/methanol-mixture is added. The solution is then transferred to the AIX-column and the column is washed with 4 x 2 ml of the methanolic solution. Under these conditions, Be is eluted completely. Subsequently, the trivalent actinides are eluted with 5 x 2 ml 1 N HNO_3. ^{7}Be is used as a tracer to check for complete separation.

TABLE 1. Targets for use in HI-reaction experiments and as fission targets in research reactors

Isotope	Backing	Mass (total)	Method
Pu-240/242/244	Be/Mo	100-200 μg	MP/ED
Am-242[a]	Ti	20 μg	ED
Cm-245[a]	Ti	15 μg	ED
Cm-248	Be	300 μg	MP
Cm-248/Gd-152	Be	300 μg	MP
Cf-249	Be	200 μg	MP
Cf-251[a]	Ti	5 μg	ED
Cf-252[b]	Ti	3 μg	ED
U(nat)/Nd(nat)[c]	Be	800 μg	MP

[a] Serve as fission targets for use in a research reactor.
[b] Cf-252 serves as a spontaneous fission source.
[c] A total of 3 targets as part of a rotating wheel.

TABLE 2. Targets for use in atomic physics studies and for ultratrace analysis of Pu

Isotope	Backing	Number of Atoms	Method
Ac-227	Ta	10^{12}	ED
Pu-236/239	Ta	10^6-10^{12}	ED
Pu-244	Ti	10^{10} - 10^{13}	ED
Am-243	Ta	10^{14}	ED
Cm-248	Ta	10^{12}	ED
Bk-249	Ta	10^{11}	ED
Cf-249	Ta	10^{11}	ED
Es-249	Ta/Ti	10^{10} – 10^{12}	ED

APPLICATIONS

The Cm-targets listed in Table 1 have mainly been used in experiments to study the chemical properties of the heaviest elements, e.g. element 104, rutherfordium (Rf) [8]. For this, Rf is produced in the reaction ^{248}Cm(^{18}O,5n)^{261}Rf. After chemical separation, ^{261}Rf and the ^{257}No daughter are detected via α-α-correlation measurements at 8.28 MeV and 8.22 MeV, respectively. With ^{18}O projectiles, and especially lead impurities in the target, the α-emitting nuclides $^{211-214}$Po are produced with α-energies ranging from 7 up to 12 MeV that make the detection of the Rf-No decay chain more difficult. In Fig. 3 cumulated α-spectra of the Rf fraction after separation with the chemical apparatus ARCA [9] are shown. The upper spectrum was obtained with a target containing lead impurities, the lower spectrum corresponds to a target where lead has been separated prior to target production.

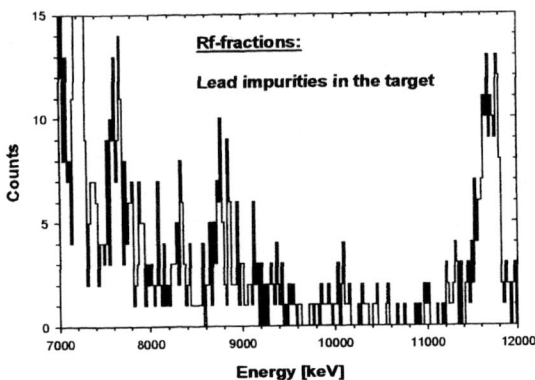

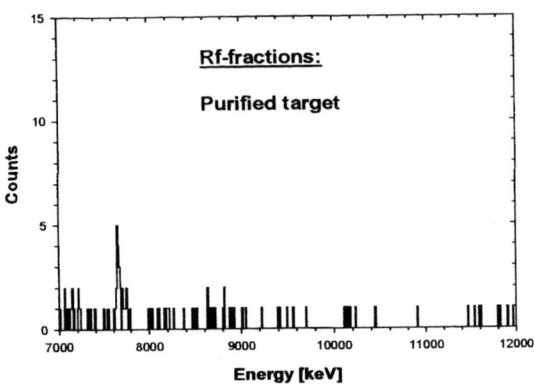

FIGURE 3. α-spectra of chemically separated Rf-fractions. Upper spectrum with lead impurities in the target. Lower spectrum results from purified target material.

Most of the targets compiled in Table 2 have been used for ultratrace analysis and for atomic physics measurements by means of Resonance Ionization Mass Spectrometry (RIMS). Due to its high sensitivity and isotopic selectivity, RIMS is a powerful tool for ultra-trace analysis of radiotoxic elements like Pu in the environment having a detection limit of 10^6 atoms [10]. Furthermore, RIMS enables the precise determination of the first ionization potential of the actinide elements with a sample size of $\leq 10^{12}$ atoms. For this, the actinide atoms under investigation are ionized in the presence of an electric field after multiple resonant laser excitation. Subsequently, the ions are detected in a time-of-flight mass spectrometer (TOF-MS). The first ionization potential is obtained by scanning the wavelength of the laser used for the last excitation step across the ionization threshold W_{th} at various electric field strengths F. Fig. 4 shows a plot of the ionization thresholds W_{th} (in cm^{-1}) of various actinide elements versus the electric field strength F. Extrapolation of W_{th} to F=0 by means of weighted least squares fits leads directly to the first ionization potential. With RIMS, the ionization potentials of Am, Cm, Bk, Cf and Es were determined for the first time and the ionization potentials of Th, U, Np and Pu were re-measured with high precision to demonstrate the capabilities of this method [11].

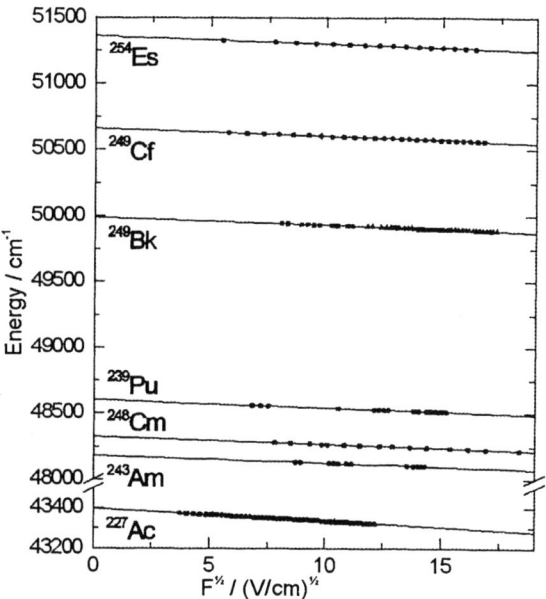

FIGURE 4. Plot of ionization thresholds versus square root of the electric field strength F for various actinide elements. The first ionization potential is obtained by extrapolation to zero field strength.

For the determination of the first ionization potential, the element under investigation is electrodeposited in form of its hydroxide onto a rectangular Ta filament (3.5 mm x 12 mm, 50 μm thick - see Table 2) and covered with a 1 μm Ti layer obtained by sputtering. By resistive heating of such a sandwich filament in the source region of the TOF-MS, the hydroxide is converted to the oxide, which is reduced to the elemental state by diffusion through the Ti layer. By evaporation from the Ti surface an atomic beam is created for interaction with the laser beams. The combination Ta-Ti has proven to be most efficient for the production of an atomic beam for the transprotactinium elements, whereas for RIMS on Ac the combination Ta-Zr is used, with Zr as the covering and reducing layer [12]. Filaments of this type are also used for measurements of the adsorption entropies and -enthalpies of, e.g., Es in the elemental state on Ta and Ti surfaces.

ACKNOWLEDGEMENTS

The authors are indebted to the European Commission Joint Research Center, Institute for Transuranium Elements, Karlsruhe, for long-term storage of an intense ^{252}Cf source and the subsequent chemical separation of the ^{248}Cm target material. One of us (N.T.) acknowledges fundings by the Gesellschaft für Schwerionenforschung, Darmstadt.

REFERENCES

1. Trautmann, N., and Folger, H., *Nucl. Instr. Meth.* **A282**, 102-106 (1989).
2. Ingelbrecht, C., Moens, A., Eykens, R., and Dean, A., *Nucl. Instr. Meth. Phys. Res.* **A377**, 34-38 (1994).
3. Evans, E.J., Lougheed, R.W., Coops, M.S., Hoff, R.W., and Hulet, E.K., *Nucl. Instr. Meth.* **102**, 389-401 (1972).
4. Parker, W., Bildstein, H., and Getoff, N., *Nucl. Instr. Meth.* **26**, 55-60 (1964).
5. Getoff, N., and Bildstein, H., *Nucl. Instr. Meth.* **36**, 173-175 (1965).
6. Getoff, N., and Bildstein, H., *Nucl. Instr. Meth.* **70**, 352-354 (1969).
7. Getoff, N., and Bildstein, H., *Nucl. Instr. Meth.* **200**, 151-160 (1982).
8. Kratz, J.V., "Chemical Properties of the Transactinide Elements" in *Heavy Elements and Related New Phenomena*, edited by W. Greiner and R.K. Gupta, New Jersey: World Scientific Publishing Co. Inc. 1999, pp. 129-193.
9. Schädel, M., Brüchle, W., Jäger, E., Schimpf, E., Kratz, J.V., Scherer, U.W., and Zimmermann, P., *Radiochim. Acta* **48**, 171-176 (1989).
10. Nunnemann, M., Erdmann, N., Hasse, H.-U., Huber, G., Kratz, J.V., Kunz, P., Mansel, A., Passler, G., Stetzer, O., Trautmann, N., and Waldek, A., *J. Alloys Comp.* **271-273**, 45-48 (1998).
11. Erdmann, N., Nunnemann, M., Eberhardt, K., Herrmann, G., Huber, G., Köhler, S, Kratz, J.V., Passler, G., Peterson, J. R., Trautmann, N., and Waldeck, A., *J. Alloys Comp.* **271-273**, 837-840 (1998).
12. Eichler, B., Hübener, S., Eberhardt, K., Erdmann, N., Funk, H., Herrmann, G., Köhler, S., Trautmann, N., Passler, G., and Urban, F.-J., *Radiochim. Acta* **79**, 221-233 (1997).

Status of the Target Development for the Heavy Element Program

B. Kindler[a], S. Antalic[b], H.-G. Burkhard[a], P. Cagarda[b], D. Gembalies-Datz[a], W. Hartmann[a], S. Hofmann[a], J. Kojouharova[a], J. Klemm[a], B. Lommel[a], R. Mann[a], S. Saro[b], H.-J. Schött[a], J. Steiner[a]

[a] *Gesellschaft für Schwerionenphysik (GSI), D-64291 Darmstadt, Germany*
[b] *Comenius University, SK-84248 Bratislava*

Abstract. We report on the status of target development for the heavy-element program at GSI, namely for the SHIP experiment. We present some recent results in enhancing the durability of the targets irradiated with the more and more intense heavy-ion beams. The enhancement of the beam intensity is necessary in order to keep beam times at reasonable length despite cross sections in the picobarn region. We synthesised new compound targets on ^{208}Pb -basis with significantly higher melting temperatures and built up a target monitoring system. To investigate the high-current targets and develop an active cooling system we built up a test bench for offline experiments.

INTRODUCTION

The search for new heavy elements is still an exciting and red hot scientific task. GSI was extremely successful in this field over the last decades and to stay on the ball the development has to move on. Since the cross sections of the heavy elements are decreasing with increasing mass, increasing beam currents are needed from the accelerator to keep the beam times at reasonable length. For element 112 for example the cross section for the synthesis was ~ 1 pb, that means one event per week on average.

But a higher beam current leads to new problems especially concerning the targets. Already now for some projectile-target combinations the limiting factor for the possible intensity is not the accelerator or the ion source, respectively, but the temperature the targets can stand. The heating through the energy loss of the heavy-ions in the lead or bismuth targets then leads to a partial melting of the target material.

SHIP

For the production of heavy elements at GSI the velocity filter SHIP (Separator for Heavy Ion reaction Products) is applied, which is shown in principle in Fig.1 [1]. The heavy-ion beam hits the target which is located on a rotating target wheel. The projectile nucleus interacts with the target nucleus and a shower of fusion products is produced. The fusion products are then electromagnetically separated from the primary beam and are finally implemented in a position sensitive Si-detector which is surrounded by additional detectors for the registration of escaping alphas, fission products, x-rays, and gammas.

THE TARGETS

On the target wheel eight banana-shaped targets are mounted [2]. On a carbon backing of 35 µm/cm² there is a layer of ~450 µg/cm² of the actual target material – in most cases lead or bismuth – and then comes a covering of 15 µg/cm² carbon. As mentioned above the main target materials for these experiments are ^{208}Pb with a melting temperature of 327.5 °C and ^{209}Bi with 271.3 °C, which is fairly low.

In this range the cooling through radiation is small. Since the target wheel runs under vacuum there is also no heat conduction via the surrounding atmosphere possible. The third mechanism for temperature reduction is conduction along the target itself. But since the target layers are very thin and the amorphous carbon layers are very bad thermal conductors, this process is not very effective either.

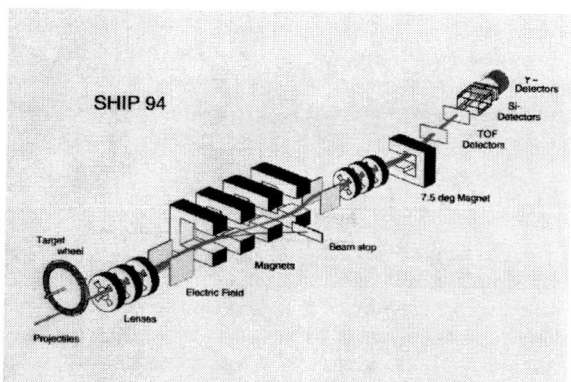

FIGURE 1. The velocity filter SHIP [3], with permission from IOP Publishing Limited. The drawing is approximately to scale, but the target wheel and the detectors are enlarged by a factor of 2. The wheel has a radius of 155 mm and the total length from wheel to detector is 11 m.

To enhance the durability and the survival rate of the targets at high currents we therefore followed three different possibilities in parallel that could possibly be combined later on.

Compound Targets

The first idea we had concerning the target material directly was to find some compound of lead and bismuth which is suitable and has a higher melting temperature compared to that of the pure element [4]. We decided to concentrate in the beginning on lead compounds since this is the target material that is needed more often. Besides there were more candidates for lead than for bismuth compounds when we investigated the phase diagrams and vapor pressures of possible compound partners.

The compounds or the other component had to fulfil several conditions. At first the compound as well as the additional component itself should be nontoxic. And the melting temperature should be considerably higher than 400 °C in order to profit from cooling through radiation losses. On the other hand the evaporation temperature should not be too high, because in the first place we get in trouble with the survival of the carbon backing. In the second place we then have to evaporate the target material with an electron gun and this process involves much higher material consumption than thermal evaporation. For highly enriched material that could be very cost-intensive.

Other preconditions for the compound partner in order to enhance the probability to really synthesize and evaporate a stable compound is that the vapor pressure and the melting or sublimation temperature should not differ too much from that of lead. Additionally the proportion of lead relative to the other component should be as high as possible and the reaction products of the additional component should have cross sections as low as possible in order to keep the cross sections of the fusion products of lead as high and undisturbed as possible.

Up to now we have two lead compounds already synthesized and evaporated, namely ^{208}PbS and Tm^{208}Pb$_3$.

^{208}PbS

Leadsulfide has a melting temperature of about 1130 °C but at approximately 950 °C it already has a vapor pressure of 10^{-3} bar. The evaporated targets look very homogeneous and they have a deep black color what should be advantageous concerning the radiation cooling. The targets do not show any signs of aging, oxidization or other visible alterations and they are mechanically stable. A disadvantage could be that the compound is nonmetallic and therefore electrostatic charging could become a problem.

Tm^{208}Pb$_3$

Tm^{208}Pb$_3$ is a metallic compound. Since there is no phase diagram of this element combination known we can only estimate the melting temperature from two neighboring phase diagrams of rare earth elements with lead, namely DyPb$_3$ that has a melting temperature of ~ 900 °C and YbPb$_3$ of 740 °C. Presumably the melting temperature of the thulium compound should be somewhere in-between. We choose thulium despite these unknown quantities since thulium is the most insensitive one of the rare earth elements that usually are easily oxidized. The material as evaporated looks a bit brittle but in principle homogeneous and mattly metallic. But this material shows an obvious aging after one or two weeks. There is a sort of whisker growth of metallic lead at the surface. Besides, the surface begins to look inhomogeneous and oxidized after some time.

From both compounds targets where produced but since beam time is rare and precious they could not be tested with the heavy ion beam until now. That is one reason to construct a set up especially for testing components for a high-current production target independently of the heavy-ion beam as it will be described below.

Target Cooling

Another possibility to avoid the melting of the targets is an active cooling with a He-jet. A He-jet will be blown at the target in the moment of irradiation thus transporting heat load away from the target surface. The distance of the reaction products flying through the atmosphere must not be too long because otherwise the charge states are changed and the intensity is reduced significantly. Therefore the whole target chamber of SHIP will have to be reconstructed in order to fit in the equipment for a differential pumping system. To get the right measures and distances for this new target chamber design test experiments are inevitable.

Target Monitoring

Another item we work on is a new target monitoring system. On the one side we will use a fast infrared camera to observe the thermal distribution over the target during irradiation. It is also helpful for the controlling of the experiments concerning the new target materials and the target cooling. On the other side we work on the implementation of an online thickness measurement of the targets by scanning them with an electron beam and analyzing the energy loss behind the target with a position sensitive detector. This will be done in a position at the wheel opposite to the irradiation spot.

TEST BENCH FOR HIGH-CURRENT PRODUCTION TARGET

Since beam time is the most expensive part of heavy-ion experiments we had to create the possibility to test all the improvements described above offline. We therefore constructed a set up where all the components developed can be built in to test and optimize the dimensions and the overall performance.

The test bench basically consists of three vacuum chambers arranged in a line as is shown in Fig.2. Each chamber is connected with a high vacuum pumping system with turbomolecular pumps. The target is situated in chamber 1. For the beginning we use a resistive heating for a standing target.

The target can be observed from the backside with the infrared camera (IRC) which is situated behind chamber 1. Through a germanium window (GW) the temperature distribution of the beam spot on the target as well as the melting of the target material can be recorded. Also depicted in Fig.2 is the first stage of a target cooling with a He-jet that is transported right to the target surface via a thin tube and guided along the surface by a plate-like flange.

Additionally the heating of the heavy-ion beam can be simulated by an electron beam. The electron gun is mounted in chamber 3 and the electron beam is guided and focussed by electromagnetic lenses (Q1,Q2,D1) on the target. By holding the power of the electron beam constant the durability of up to five targets that can be mounted on one ladder can be compared directly with each other.

In future the fixed target ladder will be replaced by a target wheel that could be driven in the same way as in the real SHIP-experiment. Further on the online thickness monitor will be implemented in the test bench with a second smaller electron gun mounted in front of chamber 1.

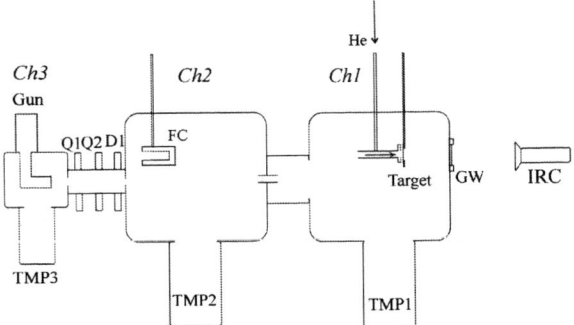

FIGURE 2. Test bench for high-current production target at ship. The target in chamber 1 (CH1) can be heated either by an electron beam (Gun) or by resistive heating. The electron beam is guided and focussed with electromagnetic lenses (Q1,Q2,D1) and can be measured by a Farraday cup (FC). The target can be cooled with a He-jet guided though a tube. The temperature distribution is measured via the infrared camera (IRC) through a Ge-window (GW).

SUMMARY

We described the design of a set up with which it is possible to test the new ideas and developments for a high-current target independently of the heavy-ion beam thus sparing the rare beam time for real production experiments. We showed different directions of investigations to enhance the durability of the targets as there were compound targets with higher melting temperatures, active target cooling with a He-jet and an effective target monitoring with an infrared camera and energy loss evaluation of electrons passing through the target.

REFERENCES

1. S. Hofmann and G. Münzenberg, *Rev. Mod. Phys.* **72**, 733 – 767 (2000).

2. H. Folger, W. Hartmann, F. P. Heßberger, S. Hofmann, J. Klemm, G. Münzenberg, V. Ninov, W. Thalheimer, P. Armbruster, *Nucl. Instr. and Meth.* **A 362**, 64-69 (1995).

3. S. Hofmann, Rep. Prog. Phys. **61**, 639 – 689 (1998).

4. D. Gembalies-Datz, W.i Hartmann, S. Hofmann, B. Kindler, J. Klemm, J. Kojouharova, B. Lommel, J. Steiner, to be published in *Nucl. Instr. and Meth.* **A**

Isotopic Germanium Targets for High Beam Current Applications at GAMMASPHERE*

J. P. Greene and T. Lauritsen

Physics Division, Argonne National Laboratory, 9700 S. Cass Avenue, Argonne, IL 60439 USA

Abstract. The creation of a specific heavy ion residue via heavy ion fusion can usually be achieved through a number of beam and target combinations. Sometimes it is necessary to choose combinations with rare beams and/or difficult targets in order to achieve the physics goals of an experiment. A case in point was a recent experiment to produce ^{152}Dy at very high spins and low excitation energy with detection of the residue in a recoil mass analyzer. Both to create the nucleus cold and with a small recoil-cone so that the efficiency of the mass analyzer would be high, it was necessary to use the ^{80}Se on ^{76}Ge reaction rather than the standard ^{48}Ca on ^{108}Pd reaction. Because the recoil velocity of the ^{152}Dy residues was very high using this symmetric reaction (5% v/c), it was furthermore necessary to use a stack of two thin targets to reduce the Doppler broadening. Germanium targets are fragile and do not withstand high beam currents, therefore the ^{76}Ge target stacks were mounted on a rotating target wheel. A description of the ^{76}Ge target stack preparation will be presented and the target performance described.

1. Introduction and Motivation

In order to search for hyperdeformation as well as linking transitions from superdeformed bands in the mass 150 region, an investigation of the decay quasi-continuum γ rays in the nuclei ^{151,152}Dy was needed. Sufficient statistics were required to extract and determine the character of the decay out as well as γ rays emitted while the nucleus is potentially hyperdeformed. This was accomplished using the reaction ^{76}Ge(^{80}Se,5n|4n)$^{151|152}$Dy and GAMMA-SPHERE [1] with the Fragment Mass Analyzer (FMA) [2]. The ATLAS accelerator was used to provide as much beam on target as allowed by the counting rates in GAMMASPHERE. Earlier experiments with fixed targets showed severe target damage due to re-crystallization which prompted the use of a rotating ^{76}Ge target wheel. Thin targets were needed to reduce the amount of Doppler broadening observed in the emitted γ rays. The 400 μg/cm^2 thickness of the ^{76}Ge target was chosen to reduce the number of normal decay γ rays in the decay out region in ^{151,152}Dy by taking advantage of isomers in the nuclei. To further optimize the experiment, a double stack of 400 μg/cm^2 ^{76}Ge targets was employed. This stacking of targets is a common experimental technique to reduce Doppler broadening without reducing yield, however it has never before been attempted with target wheels rotating at 600 RPM.

The crystalline nature of elemental germanium is a challenge for the production of freestanding foils for use in experiments with heavy-ion beams. Many techniques are available including centrifugation, vaporization using electron bombardment, and deposition employing electron beam or focused ion beam sources. A detailed listing of the various methods for the preparation of germanium films has been given by Meens and Ehret [3]. For our purposes, we employed vacuum deposition using a multi-pocket electron beam source of 270° geometry [4].

2. Germanium Targets

The ^{76}Ge separated isotope needed for the targets was obtained as an oxide from Oak Ridge National Laboratory (ORNL) and had an enrichment of 92.82%. The oxide was reduced to the metallic form using a hydrogen furnace [5,6]. The deposition was carried out using an electron beam source onto standard microscope slides, as described by Meens and Ehret [3]. The slides were first coated with NaCl as a parting agent immediately beforehand, using the same source. Although Ramsay [7], recommends BaCl as the optimum substrate for germanium film growth, we experienced difficulty with release of the foils using this salt. The source to substrate distance was 10 cm. The glass slides were heated to approximately 215 °C using a quartz lamp. This temperature was arrived at empirically from previous preparations of Ge targets. The pressure within the evaporator was 2×10^{-6} torr, provided by a cryopump. The ^{76}Ge films were then

---

*Work supported by the U.S. Department of Energy, Nuclear Physics Division, contract No. W-31-109-ENG-38.

floated off and eight quadrant targets were prepared with thicknesses of 300-400 µg/cm^2, enough for one double stacked target wheel.

3. GAMMASPHERE Target Wheel

In order to withstand the high beam currents necessary for the experiment, the targets were prepared as a rotating target wheel. The GAMMA-SPHERE target wheel was developed for use with volatile or low melting point target materials and has been described previously [8,9]. The targets were mounted on four quadrant frames, each with an open area of 2.62 cm^2. This allows for the higher beam power to be dissipated over a larger area. With the addition of beam wobbling in the vertical direction, so as not to degrade the mass resolution of the FMA, the power per unit area deposited in the target is substantially reduced, thus lowering the temperature within the target. As can be shown from previous calculations [10], the calculated power per unit area deposited in the rotating wheel target for the 314 MeV ^{80}Se beam with a current of 5 pnA was 7.15 mW/cm^2. This translates to a temperature within the target of about 99° C. This is to be compared with a calculated temperature of 488° C for a non-rotating target. In Figure 1, a plot of the time dependence of the heating within the target is given for the first 10 revolutions of the wheel. This heating from the beam would remain well below the melting point of 938.3° C for germanium, thus avoiding loss of target material and increasing the target lifetime.

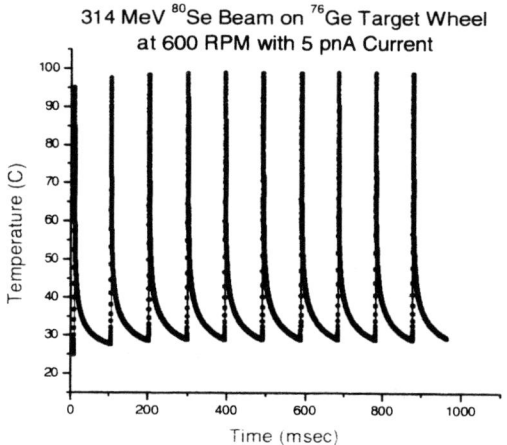

FIGURE 1. Plot of Temperature vs. Time showing the time dependence of the target heating over the first 10 revolutions of the target wheel.

4. Results and Conclusion

In conclusion, the preparation of isotopic germanium target wheels for GAMMASPHERE proved crucial to the success of the experiment. A double stack of 300-400 µg/cm^2 ^{76}Ge foils, prepared for a rotating target wheel, provided sufficient ^{151,152}Dy reactions for the experiment and withstood 314 MeV ^{80}Se beam currents of 5 pnA for six days of running. Examination of the target wheels after irradiation revealed severe damage due to re-crystallization within the foil, particularly for the front foil stack, facing the beam. The re-crystallization temperature for germanium occurs somewhere between 90° and 454° C which would indicate that the target was exposed to a deposited beam power greater than that calculated. This suggests that the focused beam spot may be smaller than expected. A photograph is given in Figure 2 showing target quadrants before and after bombardment by the heavy ion beam.

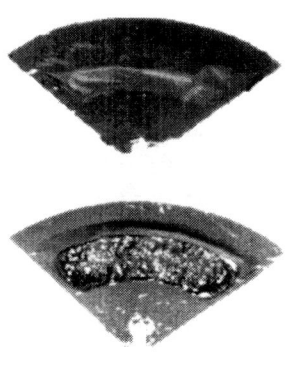

Figure 2. Photograph of ^{76}Ge target wheel quadrants before and after bombardment by 5 pnA 314 MeV ^{80}Se beam showing damage due to recrystallization in the target.

Acknowledgments

The authors would like to thank Dr. Donald Geesaman, the Physics Division Director, and Dr. Irshad Ahmad, the Target Facility Group Leader, for their continuing encouragement and support of these efforts. This work is supported by the U.S. Department of Energy, Nuclear Physics Division, under Contract No. W-31-109-Eng-38.

References

[1] I.Y. Lee, *et al.*, Nucl. Instr. and Meth. **A520** (1990) p. 641c
[2] C.N. Davids, *et al.*, Nucl. Instr. and Meth. **B70** (1992) p. 358
[3] A. Meens and G. Ehret, Nucl. Instr. and Meth. in Phys. Res. **A362** (1995) 53-59
[4] G.E. Thomas, J.P. Greene, P. Maier-Komor and R.H. Leonard, Nucl. Instr. and Meth. in Phys. Res. **A303** (1991) 162-164
[5] J.M. Heagney and J.S. Heagney, Proc. Conf. INTDS, **Report LA-6850-C** Los Alamos, NM (1977) p. 92
[6] H.U. Friebel, D. Frischke, R. Grossman and H.J. Maier, Nucl. Instr. and Meth. **167** (1979) 9-11
[7] D. Ramsay, Proc. Conf. INTDS, **AECL-5503**, Chalk River, Canada (1974) p. 151
[8] J. P. Greene, G.E. Thomas and R.H. Leonard, Nucl. Instr. and Meth. in Phys. Res. **A362** (1995) 81-89
[9] J.P. Greene, *et al.*, *Applications of Accelerators in Research and Industry*, J.L. Duggan and I.L. Morgan (eds.), The American Institute of Physics, **CP475** (1999) 929-931
[10] J.P. Greene, R. Gabor and J. Neubauer, in preparation

Temperature Calculations of Heat Loads in Rotating Target Wheels Exposed to High Beam Currents*

John P. Greene, Rachel Gabor[†] and Janelle Neubauer[‡]

Physics Division, Argonne National Laboratory, 9700 S. Cass Avenue, Argonne, IL 60439 USA

Abstract. In heavy-ion physics, high beam currents can eventually melt or destroy the target. Tightly focused beams on stationary targets of modest melting point will exhibit short lifetimes. Defocused or "wobbled" beams are employed to enhance target survival. Rotating targets using large diameter wheels can help overcome target melting and allow for higher beam currents to be used in experiments. The purpose of the calculations in this work is to try and predict the safe maximum beam currents which produce heat loads below the melting point of the target material.

1. Introduction and Motivation

In an effort to predict the lifetimes of targets under heavy-ion bombardment due to melting of the target material, calculations have been performed of the heat dissipation occurring within the target. For our research involving the production of heavy elements, the reaction cross-sections are small. Many days of beam time at high beam currents ($\gtrsim$ 10 pnA) on target are encountered. The targets produced for these experiments must be robust and capable of withstanding the intense currents involved. Calculations of the heating within the target can give a guideline to the safe limits to which they may be exposed without melting. However, melting is not the only factor involved in target longevity. Sputtering, stress due to radiation damage, re-crystallization, changes in emissivity and plastic deformation all can play destructive roles. In this work we will concentrate on the more straight forward target failure due to melting, and attempt to predict when this mode of failure occurs.

2. Description of the Calculations

In heavy-ion physics, high beam currents will eventually destroy targets through melting, sputtering or radiation damage. Tightly focused beams impinging on stationary targets of modest melting point material lead to short target lifetimes as beam currents are raised. Defocused or "wobbled" beams enhance target survival. However, rotating target wheels can often overcome target melting and allow for high beam currents to be used in experiments. The calculations in this work are aimed at predicting the safe range of beam currents which produce heat loads below the melting point of the target material. For our research involving reactions with small cross-sections, many days of experimental beam time using high beam currents are normal. The targets produced for these experiments must be robust and capable of withstanding the beam currents involved. Calculations of the heating within these targets give a guideline to the safe maximum beam currents they may be exposed to before melting.

The calculation must take into account how much heat the beam produces in the target, and how that heat is dissipated as the target rotates. The balanced equation of "heat in" = "heat out" needs to be satisfied. Heat is brought into the target by the energy loss of the beam inside the foil. This is known as the energy loss and may be calculated using a stopping power model such as TRIM [1]. Under fixed conditions, the heat produced is proportional to the beam current. The heat in the target decreases over time by conduction through the foil away from the beam spot given by the equation:

$$Q = (T - T_0)\lambda A/d$$

where λ is the thermal conductivity, A is the cross-sectional area and d the distance through the foil away from beam spot. In addition the target heating is dissipated by radiation as given by:

---

*Work supported by the U.S. Department of Energy, Nuclear Physics Division, Contract No. W-31-109-ENG-38.
[†]Harvey Mudd College, Claremont, California
[‡]North Central College, Naperville, Illinois

$$E = \varepsilon\sigma S(T^4 T_0^4)$$

where σ is the Stephan-Boltzman constant, S is the surface area irradiated by the gaussian shaped beam and ε is the emissivity of the material. A FORTRAN program [2] was developed to calculate the temperature distribution for the target material in the form of a wheel. To simulate wheel rotation, the power produced by the beam is applied to a slab of target material over a short time interval (beam on) and then removed. The target slab is then left to cool (beam off) until it again rotates into the beam. The basic equation is given as follows:

$$WI = mC_v \, dT/dt + (T-T_0)\lambda D/\rho + 2\varepsilon\sigma S(T^4-T_0^4)$$

with W equal to the energy loss, I the beam intensity, m is the amount of mass in the beam and C_v is the specific heat. In the next term, D is the target thickness (mass per unit area) and ρ is the density of the target material. In the last term, the factor 2 takes into account that energy is radiated from both sides of the foil. The resulting time-dependent partial differential equation for temperature was solved using the finite difference method [3].

The program code was then executed for our beam/target systems and the maximum target temperature was calculated as a function of beam current. The results clearly show the maximum beam current that keeps the target temperature below the melting point. In addition, the time dependence of the temperature within a slab of target material following the wheel rotation (beam on/beam off) was studied. Figure 1 shows plots of the target temperature vs. time for a 449 MeV, 2 pnA ^{86}Kr beam irradiating 0.5 mg/cm^2 Pb and Bi target wheels rotating at 600 RPM. The contrasting behavior of their temperature dependence may be attributed to differences in the emissivity and thermal conductivity of these two metals.

3. Input Parameters

The program requires as input the radius and speed of the rotating target. For the target material, several thermal properties are needed. The specific heat, C_v, gives a measure of the energy required to raise the temperature within the target, while the values for thermal conductivity and emissivity provide the means of lowering the target temperature. Table 1 gives the input parameters used for the calculations. The value for the emissivity of PbS was estimated to be 0.3 based on the spectral emissivity [4] and from similar values for PbO.

For the beam, the energy, current and beam spot size are needed as input. The energy loss was calculated using the thickness and material comprising the target. The power per unit area deposited in the target was determined using the energy loss and beam current but was also highly dependent on the size of the beam spot on the target.

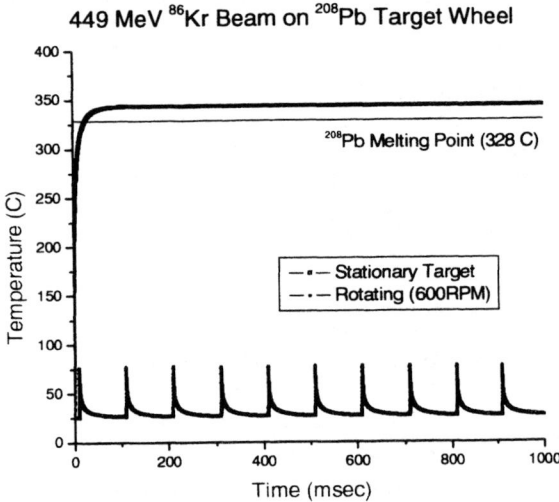

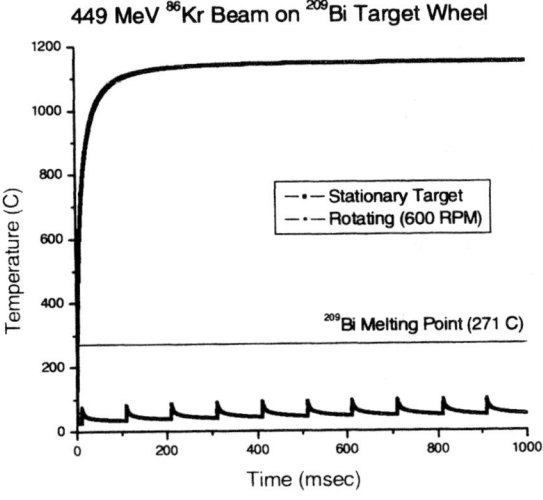

Figure 1. Plots of the target temperature vs. time for a 449 MeV ^{86}Kr beam of current 2 pnA irradiating 0.5 mg/cm^2 Pb and Bi target wheels rotating at 600 RPM.

Table 1.

Target	Emissivity [6]	Specific Heat (J/g-K) [7]	Density (g/cm³)	Thermal Conductivity (W/m-K) [7]	Energy Loss for 449 MeV ^{86}Kr	Melting Point (°C)
Pb	0.43	0.1288672	11.35	34.4	17.7	328
Bi	0.048 [7]	0.1221728	9.74	7.22	18.0	271
PbO	0.28	0.2050760	9.53	2.77	19.2	886
PbS	0.3 (est.)	0.2068396	7.50	2.30	19.5	1114

The program assumes a Gaussian shape to the beam with the highest current density contained in the peak of the distribution. From our previous example, for a 449 MeV ^{86}Kr beam with a current of 2 pnA and a focused spot size of 0.5 mm, the peak power deposited in a 0.5 mg/cm² ^{208}Pb stationary target was calculated to be 2.3 watts/cm². This translates to a maximum temperature within the target of 343°C, which is reached rather quickly (~ 50 msec). For a high current, tightly focused beam, the energy deposited would quickly vaporize the target material, and so the beam may be intentionally defocused. This may produce unwanted effects in the experiment.

A more practical approach used in many laboratories is to rotate the target. Using rotation, the small target section heated for a brief time interval by the beam spot, is allowed to cool until it once again rotates into the beam. The radius of the wheel and its speed of rotation determine this heating/cooling cycle. The limitations then become the preparation of large, segmented, target wheels and the availability and performance of motor drives and linkages for high speed rotation. In Figure 2, these effects are shown in a plot of maximum target temperature vs. beam current for the 449 MeV ^{86}Kr beam on ^{208}Pb targets. For the target wheel (r = 22 mm) rotating at 600 RPM, the melting point for the lead target (m.p. 328°C) is reached at a calculated beam current of about 12 pnA.

A further method employed involves "wobbling" the beam using a triangular waveform at approximately 5 Hz, using magnetic steerers upstream from the target position. The "wobbling" produces an amplitude about the vertical dimension, effectively increasing the area the beam illuminates on the target as it rotates. In essence the focused beam is traveling with a higher velocity (rotation plus "wobbling") across the target foil. Moreover, because of these non-coupled motions, the beam does not return to the same point on the target after one complete rotation of the wheel. This allows even further cooling to occur. Further work is needed to refine the calculation to include these effects.

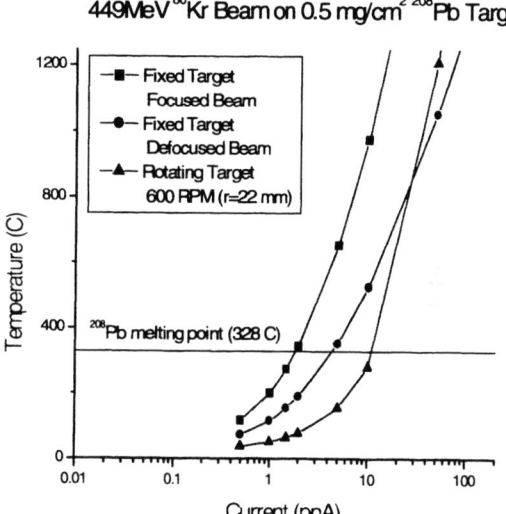

Figure 2. Plot of Temperature vs. Beam Current for an ^{86}Kr beam on a ^{208}Pb target wheel showing the effect of the beam dimensions and rotation.

4. Results of the Calculations

Calculations were carried out which explored the effects of rotation speed and wheel radius on target heating. For rotation speed, increasing the speed allows higher beam currents to be used before target melting. One can observe the rise and fall in temperature following a point on the target as it encounters the beam and then rotates away. Higher speed decreases this amplitude, shortening the cooling period before the next beam encounter. Ultimately we are constrained by the maximum speed of the motor drive systems employed (~1000 RPM). The temperature rise within the target as a function of wheel radius was investigated for two cases; small (r=22 mm) and large (r=88 mm) wheels already in use for experiments. By rotating the target and employing beam wobbling, using the beam parameters as before, the power deposited in the Pb target is reduced to 2.6 mW/cm^2 for the small wheel and 0.64 mW/cm^2 for the large wheel as compared to a stationary target. The large target wheel (r=88 mm) was designed to operate at 1000 RPM. This allows an order of magnitude increase in beam current before the onset of target melting.

Calculations were then performed for several target systems currently under study. A beam of 449 MeV ^{86}Kr was used to bombard four targets; Pb, Bi, PbO and PbS, and employing the GAMMASPHERE target wheel (r=22 mm) [5]. The results, plotted in Figure 3 show the limits of beam current the targets are calculated to withstand with respect to their melting points. In practice, target damage occurs before this limit of target melting is reached. For example, in Pb targets, island formation and migration of the lead material away from the beam heating is observed, leading to limited useful lifetimes.

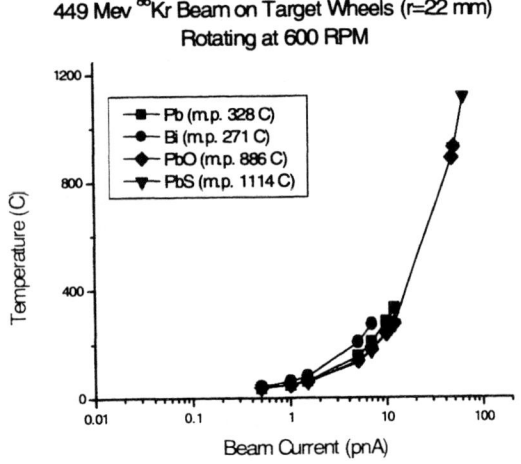

Figure 3. Plot of Temperature vs. Beam Current for Pb, Bi, PbO and PbS target wheels.

5. Conclusion and Future Considerations

In conclusion, the calculations performed showed the variation in target heating due to rotation speed and radius of the target wheel. The calculations also dramatically show the sensitivity of the target heating to the beam shape, which is why the beam is wobbled in addition to the target rotation. Preliminary experimental results are supported by the calculations. The initial tests, although intended primarily to explore accelerator capabilities and detector performance, irradiated targets at beam currents up to 15 pnA. The Pb and Bi targets were able to survive at this current using beam wobbling. The less robust PbO and PbS targets showed signs of rupture, although it is not clear that the damage was beam induced. Higher beam currents are expected in proposed heavy element synthesis experiments and calculations of this type will prove valuable in predicting target behavior under actual conditions. Further modifications to the calculations will include the contribution from the carbon backing foils used and for gas cooling of the target wheel.

Acknowledgments

The authors would like to acknowledge the previous research of Dr. Birger Back and James P. Done, a summer student working for him, upon which the present work is based. The experimental work was carried out under the direction of Dr. Robert Janssens.

We would also like to thank Dr. Donald Geesaman, the Physics Division Director, and Dr. Irshad Ahmad, the Target Facility Group Leader, for their continuing encouragement and support of these efforts. This work is supported by the U.S. Department of Energy, Nuclear Physics Division, under Contract No.W-31-109-Eng-38.

References

[1] J.F. Ziegler, J.P. Biersack and U. Littmark, The Stopping and Range of Ions in Solids, Pergamon Press, New York, USA 1985.
[2] B. Back, priv. comm.
[3] W.H. Press, S.A. Teukolski, W.T. Vetterling and B.P. Flannery, *Numerical Recipes in FORTRAN: The Art of Scientific Computing*, Cambridge Univ. press, Cambridge, MA, (1992)
[4] A. Goldsmith, T.E. Waterman and H.J. Hirschborn, Handbook of Solid Materials, The MacMillan Co., New York, USA 1961.
[5] *Applications of Accelerators in Research and Industry*, J.L. Duggan and I.L. Morgan (eds.), The American Institute of Physics, **CP475** (1999) 929-931
[6] *Radiant Properties of Materials*, Aleksander Sala, PWN – Polish Scientific Publishers, Warsaw, Poland (1986)
[7] Handbook of Chemistry and Physics, D.R. Lide (ed.), CRC Press, Inc. (1990)

APPENDIX

CAARI 2000 Participants

Accorsi, Roberto
MIT
Nuclear Engineering
19 Glendale Avenue
Sumerville MA 02139
617 258 8857
raccorsi@mit.edu

Agosteo, Stefano
Politecnico di Milano
Dipartimento Ingegneria Nucleare
Via Ponzio 34/3
I-20133 Milano Italy
39 02 2399 6318
39 02 2399 6309
stefano.agosteo@polimi.it

Aksoy, Abdulkadir
King Fahd Univ of Petrol & Mins
Energy Res Lab Div 11/RI
Research Institute Box416
Dhahran 31261 Saudi Arabia
966 3 860 3558
966 3 860 4281
aksoy@kfupm.edu.sa

Alanis, Jose
Nuclear Ctr of Mexico ININ
Balanco #9
Toluca Mexico 52042
+5 3 29 72 00
+5 3 29 73 06
jam@nuclear.inin.mx

Alanko, Tommi
University of Jyvaskyla
Department of Physics
P O Box 35
Fin 40351 Jyvaskyla FINLAND
358 14 2602390
358 14 2602351
tommi.alanko@phys.jyu.fi

Alarcon, Ricardo
Arizona State University
MIT/Bates Linear Accelerator
21 Manning Avenue
Middleton MA 01949 2846
617 253 9508
617 253 9599
alarcon@bates.mit.edu

Alford, Terry L
Arizona State University
Chemical & Materials Engineering
Tempe AZ 85287 6006
480 965 7471
480 965 3534
alford@asu.edu

Ali, Rami M.
University of Nevada, Reno
Department of Physics
MS 220
Reno NV 89557 0058
775 784 6798
775 784 1398
ali@physics.unr.edu

Aliabadi, Habib
Kansas State Univ
James R. Macdonald Lab
Dept of Physics CW 182
Manhattan KS 66506
785 532 2657
785 532 6806
habib@phys.ksu.edu

Allen, Leslie H
Univ of Illinois-Urbana
Material Science
1304 West Green Street
Urbana IL 61801
217 333 7918
217 244 1631
l-allen9@uiuc.edu

Allred, Anna
Cyclotron Institute
Texas A&M University
College Station, TX 77843
979-845-1411
979-845-1899
allred@comp.tamu.edu

Alton, Gerald D
Oak Ridge National Laboratory
Physics Division
POBox 2008 Bldg 6000 MS-6368
Oak Ridge TN 37831 6368
865 574 4751
865 574 1268
gda@ornl.gov

Alvord, C. William
CTI Cyclotron Systems
Targetry
810 Innovation Drive
Knoxville TN 37932
865 218 2454
865 218 3000
bill.alvord@cti-pet.com

Ammirati, Thomas
Connecticut College
Physics, Astronomy & Geophysics
270 Mohegan Ave
New London CT 06320
860 439 2349
860 439 5011
tfamm@conncoll.edu

Andrade, Eduardo Ibarra
UNAM
Instituto de Fisica
Apdo Postal 20 364
Mexico D.F. 0100 Mexico
52 5 622 50 35
52 5 622 5046
andrade@ifisicacu.unam.mx

Anthony, J Mark
University of South Florida
Director Ctr of Microelectromics
Research
Director Ctr of Microelectronics Res
4202 E Fowler Ave ENB 118
Tampa FL 33620 5350
813 974 2096
813 974 3610
manthony@eng.usf.edu

Anundson, Rick
Advanced Micro Devices
2900 South First Street
Austin TX 78704
512 602 5776

Aoki, Takaaki
Kyoto University
Ion Beam Engr Experi Lab
Yoshida-honmachi Sakyo
Kyoto 606-8501 Japan
81 75 753 5951
81 75 751 6774
t-aoki@kuee.kyoto-u.ac.jp

Ariyasinghe, Wickramasinghe
Baylor University
Department of Physics
P O Box 97316
Waco TX 76798 7316
254 710 2511
254 710 3878
w_ariyasinghe@baylor.edu

Arntz, Floyd
Diversified Technologies Inc
35 Wiggins Avenue
Bedford MA 01730
781 275 9444
781 275 6081
newhouse@divtecs.com

Arp, Uwe
SURF
National Institute of Standards & Tech
100 Bureau Drive Stop 8410
Gaithersburg MD 20899-8410
301 975 3233
301 208 6937
uwe.arp@nist.gov

Assmann, Walter
Universitaet Muenchen
Sektion Physik
Beschleunigerlabor Coulombwall 6
D 85748 Garching GERMANY
49 89 2891 4283
49 89 2891 4280
walter.assmann@physik.uni-muenchen.de

Avaldi, Lorenzo
IMAI del CNR
Area della Ricerca di Roma
Via Salaria Km 29.300 CP10
00016 Monterotondo Scalo Italy
39 6 90672238
39 6 90672235
avaldi@mlib.cnr.it

Avdonina, Nina
University of Pittsburgh
Astronomy & Physics
O'Hara Street
Pittsburgh PA 15260
412 624 9050
412 624 9360

nba@vms.cis.pitt.edu

Awazu, Kaoru
Industrial Research Instute of
Ishikawa Prefecture
Ro 1 Tomizu machi
Kanazawa 920 0223 JAPAN
81 76 267 8084
81 76 867 8090
awazu@irii.go.jp

Bach, Pierre H
SODERN
DAN
20 Avenue Descartes BP23
F94450 Limeil Brevannes FRANCE
33 1 45 95 70 04
33 1 45 95 70 70
pierre_bach@sodern.fr

Bakhru, Hassaram
University of Albany
Physics Department 216
1400 Washington Avenue
Albany NY 12222
518 442 4505
518 442 5260
hb694@cas.albany.edu

Ball, Gordon C
TRIUMF
Science Division
4004 Wesbrook Mall
VANCOUVER BC V6T 2A3
CANADA
604 222 7334
604 222 1074
ball@triumf.ca

Barbour, J Charles
Sandia National Laboratory
Nanostructure & Semiconductor
Dept 1112 MS 1415
Albuquerque NM 87185 1415
505 844 5517
505 844 1179
jcbarbo@sandia.gov

Barclay, Andy
Mallinckrodt
Cyclotron
2703 Wagner Okace
Maryland Heights MO 63043
314 654 7826
314 654 7456
ginnie.lingo@mkg.com

Barkyoumb, John H.
Naval Surface Warfare Ctr
Carderock Div, Code 0112
9500 MacArthur Blvd
West Bethesda MD 20895-5700
301227 1275
301 227 1150
barkyoumbjh@nswccd.navy.mil

Batiste, Susan
University of Colorado
Lab for Atmospheric & Space
Physics
1234 Innovation Drive
Boulder CO 80303
303 492 5052
303 492 6444
batiste@colorado.edu

Bauer, Rudolf W.
Lawrence Livermore Natl Lab
Physics
P O Box 808 - Mail Stop L-056
Livermore CA 94551 0808
925 422 4527
925 423 8086
bauer2@llnl.gov

Baumann, Thomas
Michigan State University
NSCL/Cyclotron Lab
South Shaw Lane
East Lansing MI 48824 1321
517 333 6437
517 353 5967
baumann@nscl.msu.edu

Baumer, Christian
University Munster
Wilhelm Klemm Str. 9
Munster, GERMANY 48149
49 251 833 4966
baumer@uni-muenster.de

Beaver, Joe E.
International Isotopes Inc
Radioiosotope Production
301 Jim Christal Road
Denton, TX 76207
940 484 9492
940 484 0877

Becchetti, Fred D.
University of Michigan
Dept of Physics, Room 1049
Randall Lab-500 E. University
Ann Arbor MI 48109 1120
734 764 1598
734 764 6843
fdb@umich.edu

Beebe, Edward N.
Brookhaven National Laboratory
Collider Accelerator Department
Bldg 930
Upton NY 11973
631 344 4849
631 344 5011
beebe@bnl.gov

Belbot, Michael D
Western Kentucky University
Physics & Astronomy
1 Big Red Way
Bowling Green KY 42103-3576
270 781 3859
270 781 1104
michael.belbot@wiu.edu

Bennett, Jonathan A
Vanderbilt University
Physics and Astronomy
6301 Stevenson Center
Nashville TN 37235
615 343 7698
615 343 1708
jonathan.a.bennett@vanderbilt.edu

Benveniste, Victor M
EATON Corp
SED
108 Cherry Hill Drive
Beverly MA 01930
978 921 9610
978 927 3652
vbenveni@bev.eten.com

Bergstrom, Paul M.
Lawrence Livermore National
Laboratory
Physics Directorate
7000 East Avenue, L-59
Livermore CA 94550
925 424 5775
925 422 9560
palko@llnl.gov

Berrah, Nora
Western Michigan University
Physics Department
1903 West Michigan Ave
Kalamazoo MI 49008 5151
616 387 4939
616 387 4939
berrah@wmich.edu

Berry, Henry Gordon
University of Notre Dame
Department of Physics
NSH 223
Notre Dame IN 46556
219 631 4012
219 631 5952
berry.20@nd.edu

Berryhill, Adam
PerkinElmer ORTEC
Detectors for Structure Research
801 south Illinois Avenue
Oak ridge TN 37831
865 481 2403
865 423 1306
adam_berryhill@perkinelmer.com

Bertholom, Rene
Thomson Tubes Electroniques
18 Ave di Marechal Juin
Mevdon La Foret 92366
France
33 1 30 703595
33 1 3070 3670

Bhalla, Chander P
Kansas State University
Department of Physics
Cardwell Hall
Manhattan, KS 66506 2601
785 532 1632
785 532 6806
bhalla@phys.ksu.edu

Bigelow, Alan W
RARAF Nevis Laroratories

136 S Broadway
Irvington NY 10533-2500
914 591 9244
ab1260@columbia.edu

Bixler, David L
Angelo State University
Physics Department
2601 W Avenue N
San Angelo TX 76909
915 942 2524
915 942 2188
david.bixler@angelo edu

Boatner, Lynn A.
Oak Ridge Nat'l Laboratory
Solid State Division
PO Box 2008 MS-6056
Oak Ridge TN 37831 6056
865 574 5492
865 574 4814
lb4@ornl.gov

Bolognese, Teresa
IPSN
DPHD/SDOS rue Auguste Lemaire
FRANCE F 92265
33 1 46 54 80 79
33 1 47 46 97 77
teresa.bolognese@ipsn.fr

Bolomey, Leonard A.
23 Samoset Drive
Salem NH 03079-1507
603 893 9011
lbolomey@lx.netcom.com

Boston, Andrew
University of Liverpool
Oliver Lodge Lab Oxford St
Liverpool MERSEYSIDE
L69 7ZE UK
44 151 794 6776
44 151 794 3348
ajb@ns.ph.liv.ac.uk

Botting, Tye
Texas A&M University/TEES
1095 Nucelar Science Rd, 3575
TAMU
College Station, TX 77843-3575
979-862-3660
979-862-2667
botting@trinity.tamu.edu

Bowyer, Sonya
Pacific Nowthwest National
Laboratory
Radiological and Chemical Sciences
POBox 999 P8-08
Richland WA 99352
509 376 3546
509 372 0672
sonya.bowyer@pnl.gov

Boyce, James R
Thomas Jefferson Lab
Free Electron Laser Div Rm
001/704-06
12000 Jefferson Ave MS 7A
Newport News VA 23606
757 269 7513
757 269 5024
boyce@jlab.org

Boyd, Richard N.
Ohio State University
Physics and Astronomy
174 W 18th Ave
Columbus OH 43210 1106
614 292 2875
614 292 7557
boyd@mps.ohio-state.edu

Braby, Leslie A.
Texas A&M University
Nuclear Engineering
TAMU 3133
College Station TX 77843-3133
979 862 1798
979 845 6443
labraby@tamu.edu

Brauer, Gerhard
Forschungszentrum Rossendorf
Ion Beam Physics/Materials
Research
Postfach 510119
Dresden 01314 GERMANY
49 351 260 2117
49 351 260 3285
g.brauer@fz-rossendorf.de

Braunstein, Gabriel H.
University of Central Florida
Physics
POBox 162385
Orlando FL 32816-2385
407 823 4192
407 823 5112
gbraunst@mail.ucf.edu

Brede, Hein
Physikalisch-Tecnische
Bundesanstalt
Johndosimetry and Accelerator
Mgmt
Bundesallee 100
D-38116 BS Germany
++49 531 592 6310
hein.brede@ptb.de

Bricault, Pierre
TRIUMF
4004 Wesbrook Mall
Vancouver V6T 2A3
British Columbia CANADA
604 222 7417
604 222 1074
pierre_bricault@triumf.ca

Brijs, Bert
IMEC
STDI
Kapeldreef 75
LEUVEN B-3001 BELGIUM
32 16 281306 desk
32 16 281608 lab
32 16 281501
BRIJS@IMEC.BE

Brown, John
Mallinckrodt
Chemistry
2703 Wagner Place
Maryland Heights MO 63043
314 654 7826
314 654 7456
ginnie.lingo@mkg.com

Browning, James F
Sandia National Laboratories
PO Box 5800
Albuquerque NM 87185 0871
505 284 2700 office
505 284 3399 lab
505 845 7536
jfbrown@sandia.gov

Brownridge, James D.
State Univ of New York-Binghamton
Physics Veatal Parkway East
PO Box 6016
Binghamton NY 13902-6000
607 777 4370
607 777 2546
jdbjdb@binghamton.edu

Brune, Carl R
Univ of North Carolina Chapel Hill
Physics
CB#3255 Phillips Hall
Chapel Hill, NC 27514-3255
919 660 2608
919 660 2634
carlb@physics.unc.edu

Brzosko, Jan S.
DIANA Hi-tech
152 Harrison Ave #2
Jersey City NJ 07304
201 332 2962
201 332 2962
brunoN50507@aol.com

Buckley, Kenneth
TRIUMF
4004 Westbrook Mall
Vancouver V6T 2A3
604 222 1047 x6308
604 222 1074
ken.buckley@triumf.ca

Budnar, Milos S
University of Ljubljana
Jozef Stefan Institute
61111 Ljubljana, POB 3000
Jamova 39 Ljubljana SI-1001
Slovenia
386 61 1773900
386 61 219385 or 1232120
milos.budnar@ijs.si

Budner, Greg
Varian Medical Systems
Industrial Products
26265 Hilliard Blvd
Westlake OH 44145
440 835 8860
440 871 7658

grag.budner@varian.com

Busch, Brett
Rutgers University
Physics and Astronomy
136 Freling Huysen Road
Piscataway NJ 08854-8019
732 445 3926
732 445 4991
bwb@physics.rutgers.edu

Butz, Tilman
University of Leipzig
Physics & Geosciences
Linnestr.5
04103 Leipzig GERMANY 04103
49 341 9732 701
49 341 973 2748
butz@physik.uni-leipzig.de

Cabrera-Trujillo, Remigio
University of Florida
Quantum Theory Project
PO Box 118435
Gainesville FL 32611-8435
352 392 8113
352 392 8722
trujillo@qtp.ufl.edu

Caffee, Marc W
Lawrence Livermore Natl Lab
Center for Accelerator Mass Spectro.
PO Box 808 L-237
Livermore CA 94550
510 423 8395
510 422 1002
caffee1@llnl.gov

Caines, Helen
Ohio State University
Van de Graaff Laboratory
1302 Kinnear Road
Columbus OH 43212
614 292 4775
614 292 4833
caines@mps.ohio-state.edu

Camilloni, Rossana
CNR IMAI
Area della Ricerche di Roma
Via Salaria Km.29.300 CP 10
1 00016 Monterotondo Scalo ITALY
39 06 906 72224
39 06 90672 238
camilloni@elettra.trieste.it

Campbell, John L.
University of Guelph
Physics
Guelph ONTARIO NIG 2W1
CANADA
519 824 4120 x3846
519 836 9967
jlc@physics.uoguelph.ca

Carroll, Lewis R
Carroll - Ramsey Associates
950 Gilman Street
Berkeley CA 94710
510 559 8153

510 559 8158
cra@carroll-ramsey.com

Chabal, Yves J
Lucent Technologies
Bell Laboratories 1C 462
600 Mountain Ave
Murray Hill NJ 07974 0636
908 582 4193
908 817 0640
yves@lucent.com

Chaput, Ernest
Economic Development Partnership
108 Cherry Hills Drive
Aiken, SC 29803
803-648-5402
803-649-5774
esandc@prodigy,net

Chartas, Aristotelis
Aristotelis
elis
Hellenic Military Academy
Physics Department
Vari 16673 Greece
30 1 6845571
30 1 6826608
hartasa@otenet.gr

Chen, Hui
University of Houston
5415 Scott 43
Houston, TX 77204
713 743 8261
chen2002@yahoo.com

Chen, Gongyin
MIT
Nuclear engineering
13 Audrey Street B-1
Cambridge MA 02139
617 253 6613
617 253 2343
gchen@mit.edu

Chen, Zhifan
Clark Atlanta University
James P Brawley Drive at Fair Street SW
Atlanta GA 30314
404 880 8631
404 880 8360
zchen@cau.edu OR
zchen@ctsps.cau.edu

Chen, Li
University of Houston
TcSUH
3201 Cullen Blvd
Houston TX 77204 5932
713 743 8259
713 743 8201
chenli@hotmail.com

Chesnel, Jean-Yves
CIRIL ISMRA
Centre Interdisciplinaire de
Recherche Ions Lasers
6 Boulevard Marechal Juin

F-14050 Caen Cedex FRANCE
33 2 31 45 25 75
33 2 31 45 25 57
chesnel@spalp255.ismra.fr

Cheung, Nathan W.
Univ of California-Berkeley
Dept of Elec Eng & Comp Sci
513 Cory Hall
Berkeley CA 94720
510 642 1615
510 642 2739
cheung@eecs.berkeley.edu

Chichester, David
Thermo Gamma-Metrics
5788 Pacific Center Blvd
San Diego CA 92121
858 450 9811
858 452 9750
chichester@gammametrics.com

Chmara, Frank
Peabody Scientific
PO Box 2009
Peabody MA 01960
978 535 0444
978 535 5827
fchmara@channel1.com

Chu, Wei-Kan
University of Houston
TX Ctr for Superconductivity
4800 Calhoun Road
Houston TX 77204 5932
713 743 8252
713 743 8201
wkchu@uh.edu

Chujo, Tatsuya
Brookhaven National Laboratory
Box 5000 Building 510C
Upton NY 11973-5000
631 344 5152
631 344 3253
chujo@bnl.gov

Church, David A
Texas A&M University
Physics Department MS-4242
408 Engr Physics Bldg Box 3578
College Station TX 77843 4242
979 845 2841
979 845 2590
church@phys.tamu.edu

Cipolla, Sam J
Creighton University
Physics Department
2500 California Street
Omaha NE 68178
402 280 2133
402 280 2140
samcip@creighton.edu

Clark, John C
University of Cambridge
Wolfson Brain Imaging Centre
Box 65 Addenbrooke's Hospital, Hills Road

CAMBRIDGE CB2 2QQ UNITED KINGDOM
44 1223 33 1815
44 1223 33 1826
jcc24@wbic.cam.ac.uk

Cleland, Marshall R.
Ion Beam Applications (IBA)
Rhodotron
20 Little Lane
Hauppage NY 11788
631 979 8718
631 979 8718
mrcleland@msn.com

Climent-Font, Aurelio
Universidad Autonomia Madrid
Campus De Canto Blanco
Dept de Fisica Aplicada C-12
Cantoblaco, E-28049, Spain
34 91 397 5264
34 1 397 3969
acf@uaw.es

Clough, Anthony S
University of Surrey
Department of Physics
Stag Hill GU2 5XH
Guildford Surrey ENGLAND
44 1483 259 407
44 1483 876 781
a.clough@surrey.ac.uk

Coban, Ali
King Fahd Univ of Pet and Min
Research Institute
CAPS PO Box 2044
Dhahran 31261 Saudi Arabia
966 3 860 3546, 3319
966 3 860 4281, 2266
coban@kfupm.edu.sa

Cohen, Scott M
Duquesne University
Physics Department
600 Forbes Avenue
Pittsburgh PA 15282-0321
cohensm@duq.edu

Coldwell, Robert L
Constellation Technology Corp
7887 Bryan Dairy Rd Ste 100
Largo FL 33777-1498
352 392 0793
352 392 0524
ufbobc@ufl.edu

Conard, E. Milo
Particle Accelerator Consultants
Ave des Moissonneurs 43
B-1325 Dion-Valmont
Belgium
32 10 22 77 06
32 10 22 77 06
milo.conard@ping.be

Courtney, Bill
International Isotopes Incorporated
301 Jim Christal Road
Denton TX 76207
940 484 9492

Coussement, Romain
Katholieke Universiteit Leuven
Instituut voor Kern-en Stralingsfysica
Celestijnenlaan 200 D
B 3001 Leuven (Heverlee)
BELGIUM
32 16 327 260
32 16 327 985
romain.coussement@fys.kuleuven.ac.be

Coutrakon, George
Loma Linda University Medical Ctr
Radiation Medicine
11234 Anderson Street RM B121
Loma Linda, CA 92354
909 824 4378 x42678
909 824 4083
coutrak@proton.llumc.edu

Craft, Benjamin
Louisiana State University
CAMD
6980 Jefferson Hwy
Baton Rouge LA 70806
225-334-6666
bcraft@unix1.sncc.lsu.edu

Cruz, Salvador A
Universidad Autonoma Metropolitana
Iztapalapa
Apartado Postal 55 534
09340 Mexico City Distrito Federal
525 804 4988
525 804 4611
cruz@xanum.uam.mx

Culbertson, Robert J
Arizona State University
Dept of Physics & Astronomy
Box 871504
Tempe AZ 85287 1504
480 965 0945
480 965 7954
robert.culbertson@asu.edu

Culbertson, Nicole
Arizona State University
Dept of Physics and Astronomy
Box 871504
Temple AZ 85287-1504
480 965 0581
480 965 7954
herbots@asu.edu

Culp, Randy
Arkansas Tech University
Department of Physics
50 Longcole Place
Russellville AR 72802
501 964 0833
501 964 0882
randy.culp@mail.atu.edu

Current, Michael I
Silicon Genesis Corporation
590 Division St
Campbell CA 95008
408 871 3082
408 871 8607
mcurrent@sigen.com

Dance, William E
Neutron Radiology-Consultant
3306 Whitehall Drive
Dallas TX 75229 2556
214 352 6956
same as work # - need to call before sending fax
wednr@aol.com

Davanloo, Farzin
University of Texas-Dallas
Center for Quantum Electronics
2601 N Floyd Rd NB11
2601 N Floyd Rd NB11
Richardson TX 75080 0688
972 883 2863
972 690 1167
fdavan@utdallas.edu

De Corte, Frans
University of Gent
Laboratory of Analytical Chemistry
Laboratory of Analytical Chemistry
Proeftuinstraat 86
B-9000 Gent, Belgium
32 9 2646627
32 9 2646699
frans.decorte@rug.ac.be

Demaree, John Derek
US Army Reserch Laboratory
Weapons & Materials Directorate
AMERL-WM-MC
Aberdeen Proving Ground MD 21005 5069
410 306 0840
410 306 0829
jdemaree@arl.army.mil

Denker, Andrea
Hahn-Meitner-Institut
Bereich F
P O Box 390128
D 14109 Berlin Germany
49 30 8062 2498
49 30 8062 2097
denker@hmi.de

Desta, Yohannes
LSU CAMD
6980 Jefferson HWY
Baton Rouge LA 70806
225 578 4618

Devlin, Matt
LANL
LANSCE 3
MS H855
Los Alamos NM 87545
505 665 0421
505 665 3705
devlin@lanl.gov

Dobrosavlevich, Alexander
Institute of Nuclear Sciences VINCA
TESLA Project

P O Box 522
11001 Beograd Yugoslavia
381 11 454 965
381 11 446 2226

Donahue, Douglas J
University of Arizona
Department of Physics
Building 81
Tucson AZ 85721
520 621 2480
520 621 9619
djd@physics.arizona.edu

Dousse, Jean-Claude
University of Fribourg
Dept of Physics
Chemin du Musee 3
CH-1700 Fribourg Switzerland
41 26 300 9073
41 26 300 9747
jean-claude.dousse@unifr.ch

Doyle, Barney L
Sandia National Laboratories
1111 Rad Solid Interac & Proces
PO Box 5800 MS 1056
Albuquerque NM 87185 1056
505 844 7568
505 844 7775
bldoyle@sandia.gov
micro96@somnet.sandia.gov

Drosg, Manfred
University of Vienna
Institute of Experimental Physics
Boltzmanng 5
A-1090 Wien Austria
43 1 4277 51112
43 1 4277 9511
drosg@exp.univie.ac.at

DuBois, Robert D
University of Missouri Rolla
Department of Physics
Rolla MO 65409
573 341 4708
573 341 4715
dubois@umr.edu

Duggan, Jerome L
University of North Texas
Department of Physics
PO Box 311427
Denton TX 76203-1427
940 565 3252
940 565 2227
jduggan@unt.edu

Dulloo, Abdul R
Westinghouse Electric Company LLC
Science & Technology Dept
1310 Beulah Road Bldg 302
Pittsburgh PA 15235-5081
412 256 2140
412 256 1007
dullooar@westinghouse.com

Dunlop, James C
Yale University
AW Wright Nuclear Structure Lab
PO Box 208124
New Haven CT 06520-8124
631 344 7781
631 344 7566
jcdunlop@star.physics.yale.edu

Dunnam, F Eugene
University of Florida
Physics Dept New Bldg #2364
P O Box 118440
Gainesville FL 32611 8440
352 392 1444
352 466 3538
dunnam@phys.ufl.edu

Dymnikov, Alexander D
Columbia University
RARAF-Health Sciences
136 S Broadway PO Box 21
Irvington NY 10533
914 591 9244
914 591 9405
ad455@columbia.edu

Eberhardt, Klaus
Johannes Gutenberg Universitat Mainz
Institut fur Kernchemie
Fritz Strassmann Weg 2
D 55128 MAINZ GERMANY
49 6131 3925321
49 6131 3924510
eberhard@mail.kernchemie.uni-mainz.de

Ederer, David L.
Tulane University
Physics Department
2001 Stern Science Center
New Orleans LA 70118
504 865 5520
504 862 8702
dlederer@mailhost.tcs.tulane.edu

Eguiluz, Adolfo G
University of Tennessee
Dept of Physics and Astronomy
1408 Circle Dr Nielsen Physics Bldg
Knoxville TN 37996 1200
865 974 9642
865 974 7843
eguiluz@utk.edu

Eguskiza, Mikel
SCKCEN
Kernspectrometrie (GKD)
Boeretang 200
B 2400 Mol BELGIUM
32 143 32 721
32 143 210 56
meguskiz@sckcen.be

Einfeld, Dieter
Forschungzeutnum Karlsruhe Gurbtf PEA
Postfach 3640
Karlsruhe D 76021 GERMANY
49 7247 82 6180
49 7247 82 6172
einfeld@anka.fzk.de

El Bouanani, Mohamed
University of North Texas
Department of Physics
PO Box 311427
Denton TX 76203 1427
940 565 3336
940 565 2227
me0008@unt.edu

Ellegaard, C
Niels Bohr Institute
Blegdamsvej 17
DK 2100 COPENHAGEN DENMARK 2100
45 35 325 209
45 353 254 25
ellegaard@nbi.dk

Escue, R. B.
UNT retired faculty
2400 Southridge
Denton TX 76205
rbe0001@unt.edu

Ethridge, D. Ray
Halliburton Energy Services
P O Box 42800
Mail Stop 8017
Houston TX 77242
281 496 8850
281 596 4279
ray.ethridge@halliburton.com

Evelyn, Arthur Leslie
Alabama A&M University
Center for Irradiation of Matls
POBox 1447 Natural & Physical Sci
Normal AL 35762 1447
256 851 5866
256 851 5868
leslie@cim.aamu.edu

Farrell, J Paul
Brookhaven Technology Group Inc
120 Lake Ave. South/Suite 15
Nesconset, NY 11767
631 751 7515
pfarrell@btg.cc

Fassbender, M
Los Alamos National Lab
Los Alamos NM 87545
505 667 8358
505 665 4955
mifa@lanl.gov

Fazleev, Neal G
University of Texas-Arlington
Physics
UTA Box 19059
Arlington TX 76019 0059
817 272 2469
817 272 3637
fazleev@uta.edu

Fedorov, Alexander V
Delft Univ of Technology

Interfaculty Reactor Inst
MekelWeg 15 NL
Delft 2629 JB the Netherlands
31 15 278 1876
31 15 278 6422
fedor@iri.tudelft.nl

Fehrenbach, Charles
Kansas State University
Department of Physics
116 Cardwell Hall
Manhattan KS 66506
785 532 3461
785 532 6806
macf@phys.ksu.edu

Feldman, Leonard C
Vanderbilt University
Dept of Physics & Astronomy
Box 1807 Station B
Nashville TN 37235
615 343 7273
615 343 1708
leonard.c.feldman@vanderbilt.edu

Feng, Huan
Brookhaven National Laboratory
Environmental Sciences
Bldg 901A
Upton NY 11973 5000
631 344 2081
631 344 5271
hfeng@sun2.bnl.gov

Fink, Dietmar
Hahn-Meitner Institute
SF4
Glienicker Str 100
D-14109 Berlin GERMANY
49 30 8062 3029
49 30 8062 2793
fink@hmi.de

Finn, Ronald D.
Memorial Sloan-Kettering Cancer Center
Radiology/Cyclotron Core Facility
1275 York Avenue
New York NY 10021
212 639 2458
212 717 3101
finnr@mskcc.org

Flanz, Jacob B.
Massachusetts General Hospital
Radiation Oncology
30 Fruit St
Boston MA 02114
617 724 9528
617 724 9532
flanz@hadron.mgh.harvard.edu

Flechard, Xavier
J R Macdonald Laboratory
Department of Physics
Kansas State University
Manhattan KS 66506 2604
785 532 2669
785 532 6806

Foster, Charles C.
Indiana University
Cyclotron Facility, Phys Dept.
2401 Milo B Sampson Lane
Bloomington IN 47408
812 855 2931
812 855 6645
foster@iucf.indiana.edu

Fournet, Justin
ULL
148 1/2 Catherine St
Lafayette LA 70503
fournetmicro@aol.com

Franz, Achim
Brookhaven National Laboratory
Physics Dept
Bldg 510C
Upton NY 11973-5000
631 344 8414
631 344 4592
achim@bnl.gov

Frawley, Anthony Dennis
Florida State University
Physics Department B-159
PO Box 3016
Tallahassee FL 32306 3016
850 644 4034
850 644 4478
frawley@fsuhip.physics.fsu.edu

Freeman, Charles G
SUNY Geneseo
Physics & Astronomy
1 College Cir
Geneseo NY 14454 1401
716 245 5286
716 245 5288
freeman@geneseo.edu

Freiesleben, Hartwig
Technische Universitaet Dresden
Institute for Nuclear & Particle Physics
Mommsenstrasse 13
Dresden 01069 GERMANY
49 351 463 5461
49 351 463 7292
freiesleben@physik.tu-dresden.de

Freitas, Maria do Carmo
ITN-QUIMICA
Est. Nac. 10
Sacavern 2685-053 Portugal
351 - 9550021
351 21 994 1039
cfreitas@itn1.itn.pt

Friesel, Dennis L
Indiana University
Cyclotron Facility
2401 Milo B Sampson Lane
Bloomington IN 47408
812 855 2944
812 855 6645
friesel@iucf.indiana.edu

Fujisawa, Shigeyoshi
Kyoto University
Precision Engineering
Yoshida Homachi Sakyoku
Kyoto 606-8501 JAPAN
81 75 753 5257
81 75 753 5259
n50250@sakura.kudpc.kyoto-u.ac.jp

Gabella, Bill
Vanderbilt University
WM Keck Free-electron Laser Center
Box 1816B
Nashville TN 37235
615 343 2713
615 343 1103
b.gabella@vanderbilt.edu

Gaelens, Michel
Universite' Catholique de Louvain
Centre de Recheeches du Cyclotron
Chemin du Cyclotron, 2
Louvain-la-Neuve B1348 Belgium
+32 10 47 32 75
+32 10 45 21 83
gaelens@cyc.ucl.ac.be

Galindo-Uribarri, Alfredo
Oak Ridge National Lab
Physics Bldg 6000
PO Box 2008 MS 6371
Oak Ridge TN 37831
865 574 6124
865 574 1268
uribarri@mail.phy.ornl.gov

Garcia, Gustavo
CIEMAT
Edificio 2
Avenida Complubense, 22
Madrid, E-28040, Spain
34 91 346 6583
34 91 346 6442
gustavo@ciemat.es

Gataullin, Marat
California Institute of Technology
CERN-EP Division
Geneva 23
CH 1211 SWITZERLAND
41 22 767 9258
41 22 767 8530
marat.gataoulline@cern.ch

Gelbke, Konrad
Michigan State University
Natl Superconduct Cyclo Lab
East Lansing MI 48824-1321
517 333 6300
517 333 6411
gelbke@nscl.msu.edu

Gianfranco, Ranieri
ALARA S.A.
125 Este de la Municipalidad de Tibas
Tibas, San Jose   COSTA RICA
+506 235 4001
+506 235 1359
alarasa@sol.racsa.co.cr

Gilliam, David M.
Natl Inst of Standards & Tech
Ionizing Radiation Division
MS 8461
Gaithersburg MD 20899 8461
301 975 6206
301 926 1604
david.gilliam@nist.gov

Glass, Gary A
University of Louisiana at Lafayette
Acadiana Research Lab
POBox 44210
Lafayette LA 70504 4210
337 482 6184
337 482 6190
glass@usl.edu

Goains, Chris
Baylor University
Physics Department
P.O. Box 97316
Waco TX 76798 7316
254 710 2511
254 710 3878

Golden, David E.
University of North Texas
Department of Physics
P O Box 311427
Denton TX 76203 1427
817 565 3260
817 565 4824
golden@bob.unt.edu

Golser, Robin
Universitaet Wien
Insti fuer Isotopenforschung
Waehringerstrasse Str 17
Wein A-1090 Austria
43 1 4277 51701
43 1 4277 9517
golser@ap.univie.ac.at

Gonzalez, Carlos
University of Texas-Houston
6431 Fannin
Houston TX 77030
713-500-7755
713-500-7771
carlos.gonzalez-lepera@uth.tmc.edu

Gonzalez, Alejandro D
Centro Atomico Bariloche
Com Nac'l de Energia Atomica
Av E Bustillo 9500
San Carlos de Bariloche
ARGENTINA
54 2944 445 233
54 2944 445 299
gonzalez@cab.cnea.gov.ar

Gonzalez Lepera, Carlos
University of Texas
Cyclotron Facility
6431 Fannin Street
Houston TX 77030
713-500-7755
713-500-7771
carlos.gonzalez-lepera@uth.tmc.edu

Good, Roger
Denton Regional Medical Hospital
3537 South I35E Ste 120
Denton TX 76205
940 243 2288 or
898-6206
cancerline@aol.com

Govil, Indra Mani
Panjab University
Department of Physics
Sector 14
Chandigarh 160014 INDIA
91 172 541 741
91 172 541741
imgovil@panjabuniv.chd.nic.in

Gozani, Tsahi
Ancore Corporation
Advanced Nucleonics Operation
2950 Patrick Henry Drive
Santa Clara CA 95054 1837
408 727 0607
408 727 8748
tsahi@ancore.com
Irene Nakasone's E-Mail:
irene@ancore.com

Gray, Thomas J
Kansas State University
James R McDonald Lab
Physics Dept Cardwell Hall
Manhattan KS 66506 2601
785 532 2663
785 537 9601
tgray@phys.ksu.edu

Greco, Richard
ULL
112 Castle Row
Lafayette LA 70506
337 784 0163
merit_2@excutc.com

Greene, John P
Argonne National Laboratory
Physics 203
9700 S Cass Ave
Argonne IL 60439
630 252 5364
630 252 6247
greene@anl.gov

Greenwood, Jason. B
Queen's University Belfast
Dept of Pure & Applied Phy
Univ Road, Belfast, BT7 1NN
N. Ireland, UK
44 28 902 73935
44 28 903 10785
j.greenwood@qub.ac.uk

Greife, Uwe
Colorado School of Mines
Department of Physics
1017 20th Street
Golden CO 80401
303 273 3618
303 273 3919
ugreife@mines.edu

Grime, Geoffrey W
University of Oxford
Department of Materials
Parks Road
Oxford OX1 3PH UK ENGLAND
44 1865 273 367
44 1865 273 418
geoff.grime@materials.ox.ac.uk

Grisham, Larry R.
Princeton University
Plasma Physics Laboratory
P O Box 451
Princeton NJ 08540-0451
609 243 3168
609 243 3248
lgrisham@pppl.gov

Grodzins, Lee
Miton Corp
800 Middlesex Turnpike Bldg 8
Billerica MA 01840
978 262 8655
978 262 8801
lgrodzins@miton.com

Groeneveld, Karl-Ontjes E
J. W. Goethe University
Institute für Kernphysik
August-Eulerstr 6
D-60486 Frankfurt am Main
GERMANY
49 69 7982 4251
49 69 7982 4212
groeneveld@em.uni-frankfurt.de

Groenewold, Gary S
Idaho National Engineering
and Environmental Laboratory
P O Box 1625
Idaho Falls ID 83415-2208
208 526 2803
208 526 8541
gsg@inel.gov

Grossmann, Rainer
University of Munich
Sektion Physik
Am Coulombwall 1
D-85748 GARCHING GERMANY
49 89 2891 4023
49 89 2891 4034
rainer.grossmann@physik.uni-munich.de

Guardala, Noel A
Naval Surface Warfare Ctr
Carderock Division Code 682
9500 MacArthur Blvd
West Bethesda MD 20817 5700
301 227 0592
301 227 2252
guardalana@nswccd.navy.mil

Guo, Baonian N.
Varian Semiconductor Equipment
Assoc Inc

Implant Process & Equipment Characterization
35 Dory Road MS GL 12
Gloucester MA 01930
978 282 2469 Gloucester OR
978 463 5026 Newburyport
978 281 1897
baonian.guo@vsea.com

Habenicht, Soenke
George-August Universitaet
Goettingen Leinestr.25
Bunsenstr.7-9
D37073 Goettingen Germany
49 40 5613 2653
49 551 394493
habenicht@physik2.uni-goettingen.de

Haberer, Thomas
GSI
Biophysics
Planckstrasse 1
64291 DARMSTRADT GERMANY
49 6159 712 627
49 6159 712 106
t.haberer@gsi.de

Hagel, Kris
Texas A&M University
Cyclotron Institute
M/S 3366
College Station TX 77843
979 845 1411
979 845 1899
hagel@comp.tamu.edu

Haight, Robert C
Los Alamos National Laboratory/LANSCE-3
PO Box 1663 MS-H855
Los Alamos NM 80545
505 667 2829
505 665 3705
haight@lanl.gov

Halka, Monica
Portland State University
12800 NW Springfield Rd
Portland OR 97229
503 725 4226
503 725 3888
halkam@pdx.edu

Hall, James M.
Lawrence Livermore National Laboratory
Physics / N-Division
P O Box 808 M/S L 050
Livermore CA 94551 0808
925 422 4468
925 423 3371
jmhall@llnl.gov

Hamm, Marianne E
AccSys Technology Inc
1177A Quarry Lane
Pleasanton CA 94566
925 462 6949 x103
925 462 6993
mhamm@linacs.com

Hamm, Robert W
AccSys Technology Inc
President
1177 A Quarry Lane
Pleasanton CA 94566
925 462 6949 x104
925 462 6993
rhamm@linacs.com

Harker, Yale D.
Idaho Nat. Engineering &Env. Lab.
Radiation Physics
PO Box 1625 MS2114
Idaho Falls ID 83415 2114
208 526 0707 & 533 4251
208 526 5208
ydh@inel.gov

Harmon, J. Frank
Idaho State University
Physics Department
Campus Box 8106
Pocatello ID 83209
208 282 5877
208 282 5878
harmon@physics.isu.edu

Hasan, Asad
Kansas State University
American University of Sharjah
Sharjah UAE
+9716 505 5524
971 655 85066
ahasan@aus.ac.ae

Hatori, Satoshi
The Wakasa-wan Energy Research Ctr
64 52 1 Nagatani Tsuruga
Fukui 914 0192 JAPAN
81 770 24 5625
hatori@werc.or.jp

Haynes, Tony E
Oak Ridge National Lab
Solid State Division Bldg 3003
PO Box 2008, MS-6048
Oak Ridge TN 37831 6048
865 576 2858
865 576 6720
hayneste@ornl.gov

Heinz, Andreas
Argonne National Laboratory
Physics Division
9700 S Cass Avenue
Argonne IL 60439
630 252 1925
630 252 2864
heinz@anlphy.phy.anl.gov

Hellborg, Ragnar
Lund Institute of Technology
Nuclear Physics Department
Solvegatan 14
LUND SE 223 62 SWEDEN
46 46 222 76 44
46 46 222 4709
ragnar.hellborg@nuclear.lu.se

Helmer, Richard G
Idaho Accelerator Center
Idaho Natl Engineering & Env Lab
PO Box 1625
Idaho Falls ID 83415-2114
208 526 4157
208 526 9267
helmerr@pcif.net

Hemmers, Oliver
Univ of Nevada Las Vegas
Department of Chemistry
4505 S Maryland Parkway
Las Vegas, NV 89154-4003
702 895 2691
702 895 4072
hemmers@nevada.edu

Hershcovitch, Ady
Brookhaven National Lab
CAD
Bldg 911C
Upton, NY 11973
631 344 4531
631 344 5954
hershcovitch@bnl.gov

Hichwa, Richard D
University of Iowa
Pet Imaging Center
Radiology
Iowa City IA 52242 1009
319 356 4104
319 353 6512
richard-hichwa@uiowa.edu

Hicks, Sally F.
University of Kentucky
Dept of Physics & Astronomy
Lexington KY 40506 0055
606 257 5845

Hirose, Masafumi
Sumitomo Heavy Industries Ltd
Research & Dev Center
2-1-1 Yato-cho
Tanashi-city Tokyo 188-8585 Japan
81 424 68 44 76
81 424 68 4477
msf_hirose@shi.co.jp

Hirsch, Helmut V B
The University at Albany SUNY
Biology Department
1400 Washington Avenue
Albany NY 12222
518 442 4311
518 442 4767
hirsch@albany.edu

Hitchcock, Adam P
McMaster University
BIMR
Hamilton ONT Canada L8S 4M1
905 525 9140
905 521 2773
aph@mcmaster.ca

Hjelm, Rex P
Los Alamos National Laboratory
LANSCE-12
PO Box 1663 MS H805
Los Alamos NM 87545
505 665 2372
505 665 2676
hjelm@lanl.gov

Hoelzle, Rainer
Inst fuer Festkoerperforschung
Forschungszentrum Juelich
Juelich 1 Germany
49 2461 61 3151, private 49 2461 53594
49 2461 61 2410
R.Hoelzle@kfa-juelich.de

Hoffmann, Volker
Hahn-Meitner Institut
SF4
Glienicker Str 100
D-14109 Berlin GERMANY
49 30 8062 2408
49 30 8062 2293
hoffmann-v@hmi.de

Hofsass, Hans
Universitaet Goettingen
II Physikalisches Institut
Bunsenstrase 7-9
D-37073 Goettingen GERMANY
49 551 397669
49 551 394493
hhofsae@uni-goettingen.de

Holder, Joe P
Texas A&M University
Lawrence Livermore National Lab
L414
Livermore CA 94550
925 422 2276
925 422 5940
holder4@llnl.gov

Holland, Orin Wayne
Oak Ridge National Laboratory
Solid State Division
PO Box 2008 MS 6048
Oak Ridge TN 37831-6048
865 576 2502
865 576 6720
hn2@ornl.gov

Hollerman, William A
University of Louisiana at Lafayette
Acadiana Research Laboratory
PO Box 44210
Lafayette LA 70504
337 482 1011
337 482 6190
hollerman@lousiana.edu

Horn, Kevin M
Sandia National Laboratories
15343 Rad Effects Experimentation
1515 Eubank SE
Albuquerque NM 87185 1167
505 845 7944
505 845 3471
kmhorn@sandia.gov

Horvat, Vladimir
Texas A&M University
Cyclotron Institute
MS3366
College Station TX 77843 3366
979 845 1411
979 845 1899
v-horvat@tamu.edu

Hossain, Tim Z
AMD Incorporated
PCAL
5204 E Ben White Blvd MS 613
Austin TX 78741
512 602 2542
512 602 7470
tim.hossain@amd.com

Huang, N. K.
Sichuan University
Inst of Nuclear Science & Tech
610064 Chengdu
Sichuan Province P. R. China
86 28 541 2230
86 28 541 0252
succ@pridns.scu.edu.cn

Hubler, Graham K.
Naval Research Laboratory
Surface Modification Code 6370
4555 Overlook Avenue S W
Washington DC 20375 5320
202 767 4786
Sec 202 767 4800
202 767 5301
hubler@ccs.nrd.navy.mil

Huddle, James R.
U. S. Naval Academy
Physics
572 M Holloway Rd
Annapolis MD 21402 5026
410 293 6672
410 293 3729
huddle@nadn.navy.mil

Hughey, Barbara J.
Newton Scientific Inc
245 Bent Street
Cambridge MA 02141
617 354 9469
617 354 9479
bhughey@world.std.com

Humes, Fred E
Economic Development Partnership
PO Box 1708
Aiken SC 29801
803 648 3362
803 641 3369

Hupf, Homer B
International Isotopes Inc
3100 Jim Christal Road
Denton TX 76207
940 484 9492 main #
940 380 7324 his office
940 484 0877
homerh@aol.com

Hutchinson, Donald
Oak Ridge National Laboratory
Measurement Science
PO Box 2008 Bethel Valley Rd
Oak Ridge TN 37831 6004
865 574 4730
865 574 1249
hutchinsondp@ornl.gov

Hynes, Shelly
University of Louisiana at Lafayette
138 Fern Street
Natchitoches, LA 71457
318 356 9337
337 482 6190
shelly_finley@hotmail.com

Iannotta, Salvatore
Italian National Research Council
Gas Surface Dynamics
Via Sommarive 18
38050 Povo di Trento ITALY
39 461 314 251
39 0461 810 628
iannotta@cefsa.itc.it

Ichihashi, Masahiko
Toyota Technological Inst
Cluster Ressearch Lab
717-86 Futamata
Ichikawa Chiba 272-0001 JAPAN
81 47 320 5910
81 47 327 8030
ichihashi@toyota-ti.ac.jp

Ikushima, Tatsushi
Osaka Science & Technology Center
1-8-4 Utsubo Hommachi nishiku
Osaka JAPAN 550-0004
81 6 6443 5322
81 6 6443 5319
ikushima@ostec.or.jp

Ila, Daryush
Alabama A&M University
Ctr for Irradiation of Materials
PO Box 1447
Normal AL 35762 1447
256 851 5866
256 851 5868
ila@cim.aamu.edu

Imanishi, Nobutsugu
Kyoto University
Dept of Nuclear Engineering
Sakyo
Kyoto 606-8501 Japan
81 75 753 5821same as fax
81 75 753 5821same as work
imanishi@nucleng.kyoto-u.ac.jp

Ingram, David C.
Ohio University
Dept of Phys & Astronomy
Clippinger Research Lab
Athens OH 45701
740 593 1705
614 593 1436

ingram@ohio.edu

Ishigami, Ryoya
Nagoya University
Dept of Crystalline Material Sci
Furo-cho, Chikusa-ku
Nagoya 464-01 Japan
81 770 24 5619
81 770 24 5605
rishigami@werc.or.jp

Ishii, Keizo
Tohoku University
Quantum Science and Energy Engineering
Aoba-ku Aramaki, Aza-Aoba
Sendai 980-8579 JAPAN
81 22 217 7931
81 22 217 7931
keizo.ishii@qse.tohoku.ac.jp

Ishikawa, Junzo
Kyoto University
Electronic Science & Engr
Yoshida-honmachi, Sakyo-ku
Kyoto 606-8501 Japan
81 75 753 5325
81 75 753 5324
ishikawa@kuee.kyoto-u.ac.jp

Ito, Yoshiaki
Kyoto University
Institute for Chemical Research
Uji, Kyoto 611-0011 Japan
81 774 38 3044
81 774 38 3045
yosi@elec.kuicr.kyoto-u.ac.jp

Jacobs, Dennis
University of Notre Dame
Department of Chemistry and Biochemistry
251 Nieuwland Science Hall
Notre Dame IN 46556 5670
219 631 8023
219 631 6652
jacobs.2@nd.edu

Jacobson, Dale C
Lucent Technologies
Silicon Processing Research
700 Mountain Ave Rm 1E 365
Murray Hill NJ 07974
908 582 6557
908 582 4228
dcj@lucent.com

Jakubassa-Amundsen, Doris H
University of Munich
Physics Section
Am Coulombwall 1
85748 Garching Bavaria GERMANY
49 069 798 4251
49 069 798 4212
doris.jaku@lrz.uni-muenchen.de

James, William Dennis
Texas A&M University
Ctr for Chemical Characterization
Teague Building
College Station TX 77843 3144
409 845 7630/2341
409 845 1655
wd-james@tamu.edu

Jankuhn, Steffen
University of Leipzig
Fakultät für Physik/ u Geowissen
Linnestr 5
D-04103 Leipzig Germany
49 341 9732 706
49 341 9732 497
jankuhn@rz.uni-leipzig.de
http://www.uni-leipzig.de/~nfp/

Javorsek, Dan
Purdue University
Physics
1820 Abnaki Drive
W Lafayette IN 47906
765 494 5381
765 494 0706
javorsek@hotmail.com

Jenkins, Greg S.
LOMA LINDA
17411 Crestlake LN
Riverside CA 92503
909 687 3706

Jeynes, Christopher
University of Surrey
The Ion Beam Centre
Guildford GU2 5XH
ENGLAND
44 1483 259 829
44 1483 259 391
c.jeynes@surrey.ac.uk

Jiang, Weilin
Pacific Northwest Natl Lab
Material Sciences
PO Box 999 MSIN K8-93
Richland WA 99352
509 376 5471
509 376 5106
weilin.jiang@pnl.gov

Jin, Mingji
University of Houston
3201 Cullen Blvd
Houston TX 77204
713 743 8218
713 743 8201
mjjin@hotmail.com

Jin, Jian-Yue
Varian Semiconductor Equipment
35 Dory Road
Gloucester, MA 01930
978 282 7521
978 281 1897
jianyue.jin@vsea.com

Johnson, Brant M
Brookhaven National Lab
Physics
PO Box 5000 Bldg 510C
Upton NY 11973 5000
631 344 4552
631 344 3253
brant@bnl.gov

Johnson, James H
Varian Associates
Radiation Division, Marketing
611 Hansen Way M/S H240
Palo Alto CA 94303

Jones, James L
Idaho National Engineering Lab
National Security Program
1181 Kortnee Drive
Idaho Falls ID 83415 2802
208 526 1730
208 526 5208
jlj@inel.gov

Jones, Keith W
Brookhaven National Laboratory
Environmental Sciences
Building 901A
Upton NY 11973 5000
631 344 4588
Sec Lori Barbier: 516 344 5125
631 344 5271
kwj@bnl.gov

Jongen, Yves
Ion Beam Applications S.A.
Chemin du Cyclotron, 3
B1348 Louvain-La-Neuve
Belgium
32 10 47 58 54
32 10 47 59 52
jongen@iba.be

Junker, Matthias
Laboratori Nazionali del Gran Sasso
Collab LUNA
S.S.17bis, km 18.910
67010 Assergi (AQ) ITALY
39 0862 437 298
39 0862 437 570
junker@lngs.infn.it

Kalyanaraman, Ramki
Oak Ridge National Lab
Si Processing Research Lab
1E 203 700 Mountain Ave
Murray Hill NJ 07974
908 582 2402
908 582 4228
ramkik@lucent.com

Kambara, Tadashi
RIKEN
Inst. of Physical & Chem.Res.Atomic Physics Laboratory
Wako, Saitama 351-0198
JAPAN
81 484 62 1111
81 48 462 4644
kambara@rarfaxp.riken.go.jp

Kasahara, Mikio
Kyoto University
Graduate School of Energy Science
Gokanosho
Uji Kyoto 611-0011 JAPAN

81 774 38 4408
81 774 38 4411
kasahara@energy.kyoto-u.ac.jp

Kato, Masahiko
Nagoya University
Crystalline Materials Science
Furo-Machi
Nagoya 464-8603 JAPAN
81 52 789 3604
81 52 789 5155
m-kato@nucl.nagoya-u.ac.jp

Kavanagh, Karen
Simon Fraser University
Physics Department
8888 University Drive
Burnaby BC V5A 1S6 CANADA
604 291 4244
604 291 3592
kavanagh@sfu.ca

Kavcic, Matjaz
J Stefan Institute
Department for Low and Medium
Energy Physics
PO Box 3000
LJUBLJANA SI-1001 SLOVENIA
386 61 1885266
386 61 1766500
matjaz.kavcic@ijs.si

Kehayias, Joseph John
Tufts University
USDA - HNRC
711 Washington Street
Boston MA 02111
617 556 3162
617 556 3344
kehayias@hnrc.tufts.edu

Keller, Roderich
E O Lawrence Berkeley Natl Lab
AFRD IBT
1 Cyclotron Road MS 71-259
Berkeley CA 94720
510 486 5223
510 486 5788
r_keller@lbl.gov

Kelley, John H
Duke University
Triangle Universities Nuclear
Laboratory
Science Drive
Durham NC 27708 0308
919 660 2631
919 660 2634
Kelley@.tunl.duke.edu

Kessel, Quentin C.
The University of Connecticut
Department of Physics and Institute
of Materias
2152 Hillside Rd
Storrs CT 06269 3136
860 486 4118
860 486 3346 or 4745
kessel@uconnvm.uconn.edu

Khan, Siraj M
U S Customs Service
Applied Technology Division
1300 Pennsylvania Ave NW Ste 1575
Washington DC 20229
202 927 2025
202 927 1418
siraj.m.khan@customs.treas.gov

Khiari, Fatah Zouhir
King Fahd Univ of Pet&Min
Ctr for Applied Phys Sci
KFUPM Box 1693
Dhahran 31261 Saudi Arabia
966 3 860 2989
966 3 860 4281
khiari@kfupm.edu.sa

Kieser, William (Liam) E
University of Toronto
IsoTrace Laboratory
60 Saint George Street
Toronto ONT M5S 1A7 Canada
416 978 2241
416 978 4711
liam.kieser@utoronto.ca

Kim, Sang-Wook
Woods Hole Oceanographic
Institution
Geology and Geophysics
McLean Lab MS 8
Woods Hole MA 02543
508 289 3701
508 457 2183
swkim@whoi.edu

Kim, Su
University of North Texas
PO Box 308470
Denton TX 76203
940 369 6433
su_kim8883@hotmail.com

Kindler, Birgit
Gesellschaft fur
Schwerionenforschung
GSI Targetlaboratory
Planckstrasse 1 Hessen
64291 DARMSTADT GERMANY
49 6159 71 2523
49 5159 71 2166
b.kindler@gsi.de

King, George C.
University of Manchester
Physics & Astronomy
Schuster Laboratory
Manchester M13 9PL England UK
44 161 275 4134
44 161 275 4259
george.king@man.ac.uk

Kinross-Wright, John
International Isotopes
1030 Dallas Drive #214
Denton TX 76205
940 390 3762
jkinross@intiso.com

Kirch, Klaus
Los Alamos National Laboratory
H 803
Los Alamos NM 87545
505 665 3821
505 665 4121
kirch@lanl.gov

Kiselev, Maxim Y
Eastern Isotopes
13 Crescent Court
Sterling VA 20164
202 297 3032
703 787 3032
maxim.kiselev@usa.net

Kishimoto, Naoki
National Research Institute for
Metals
Hybrid Resolution Beam Station
1 2 1 Sengen Tsukuba
Ibaraki 305 0047 JAPAN
81 298 59 5059
Lab: 59 5011
81 298 59 5010
kishin@nrim.go.jp

Kistenev, Edouard
Brookhaven National Laboratory
Physics Department
Upton NY 11973
631 344 7502
631 344 3253
kistenev@bnl.gov

Kitagawa, Teruyuki
Nomura Plating Co Ltd
Dept of Research & Development
12-20 5chome Himejima
Nishiyodogawa
OSAKA 555 JAPAN
81 66473 1357
81 6471 1308
terukita@nishiki.kuee.kyoto-u.ac.jp

Kitamura, Norio
Kyoto University
Precision Engineering
Yoshida Honmachi Sakyoku
Kyoto 606-8501 JAPAN
81 75 753 5259
81 75 753 5259
m52888@sakura.kudpc.kyoto-u.ac.jp

Klann, Raymond T
Argonne National Laboratory
Technology Development Div
975 Pennwood Lane
Argonne IL 60440
630 252 4305
630 252 1885
klann@anl.gov

Klody, George M.
National Electrostatics Corp
PO Box 620310
7540 Graber Road
Middleton WI 53562 0310

608 831 7600
608 256 4103
nec@pelletron.com

Knies, David L.
Naval Research Laboratory
Code 6370
4555 Overlook Av SW
Washington DC 20735
202 767 5659
202 767 5301
knies@nrl.navy.mil

Knox, John M.
Idaho State University
Physics Department
P O Box 8106
Pocatello ID 83209
208 236 2616
208 236 4649
knoxj@physics.isu.edu

Kobe, Donald H.
University of North Texas
Department of Physics
P.O. Box 305370
Denton TX 76203 5370
940 565 3272
940 565 2515
kobe@unt.edu

Kocbach, Ladislav
University of Bergen
Department of Physics
Allegaten 55
BERGEN N 5007 NORWAY
47 55 58 28 71
47 55 58 94 40
ladi@post.fi.uib.no

Kocharovskaya, Olga
Texas A & M University
Physics Department
MS 4242
College Station TX 77843-4242
979 845 2012
979 458 1235
kochar@atlantic.tamu.edu

Koehler, Paul E
Oak Ridge National Laboratory
Physics Division
MS6354 Bldg 6010
Oak Ridge TN 37831
865 574 6133
865 576 8746
koehlerpe@ornl.gov

Korol, Andrei V.
Universitaet Frankfurt am Main
Inst fur Theoretische Physik
Postsach 111932
Frankfurt am Main GERMANY
49 69 798 22634
49 69 798 28350
korol@th.physik.uni-frankfurt.de

Kotler, Jiri
MDS Nordion
447 March Rd

Kanata K2K1X8 Ontario Canada
613 592 2790
eho@mds.nordion.com

Krause, Herbert F
Oak Ridge National Laboratory
Physics Division
PO Box 2008 MS 6377
Oak Ridge TN 37831 6377
865 574 5049
865 574 1118
krause@mail.phy.ornl.gov

Kravchenko, Ivan
University of Florida
Dept of Physics
PO Box 118440
Gainesville FL 32653
352 392 9229
kravch@phys.ufl.edu

Kravchuk, Leonid V
Institute for Nuclear Research
Russian Academy of Sciences
60-th October Pr 7A
117312 Moscow RUSSIA
7 095 334 0061
7 095 135 2268
kravchuk@al20.inr.troitsk.ru

Krawchuk, Robert
Rutgers University
Lab Surface Modification
9 Carter Brook Lane
Princeton NJ 08540
609 921 7060
krawxhukrb@earthlink.net

Kruecken, Reiner
Yale University
Wright Nuclear Structure Laboratory
PO Box 208124 272 Whitney Ave
New Haven CT 06520
203 432 5616
203 432 3522
reiner.kruecken@yale.edu

Kudryavtsev, Yuri
The Leuven University
Instituut voor Kern- en
Stralingsfysica
Celestijnenlaan 200D
LEUVEN 3001 BELGIUM
32 16 327270
32 16 327985
yuri.kudriavtsev@fys.kuleuven.ac.be

Kulkarni, Padmakar V
University of Texas
Southwestern Medical Ctr
5323 Harry Hines Blvd
Dallas TX 75390-9058
214 648 2957
214 648 2991
padmakar.kulkarni@utsouthwestern.edu

Kuo, Thomas Y.T.
The Cyclotron Corporation
950 Gilman Street

Berkeley CA 94710
510 524 6769

Kuo, Thomas Y.
4654 N Larwin Avenue
Concord CA 94521

Kupchishin, Anatoly
Almaty State University
Physical-Technological Ctr
13, Lenin St
Almaty 480100 Kazakhstan
denvoronov@usa.net

Kvale, Thomas J
University of Toledo
Physics and Astronomy Dept.
2801 W Bancroft Street
Toledo OH 43606
419 530 2980
419 530 2723
tjk@physics.utoledo.edu

Lajoie, John
Iowa State University
Physics and Astronomy
12 Physics
Ames IA 50011
515 294 6952
515 294 6027
lajoie@iastate.edu

Lamm, A J
Silicon Genesis Corporation
590 Division Street
Campbell CA 95008
408 871 4353
408 871 8607
alamm@sigen.com

Lamoureux, Michele
University P & M Curie
L-DIAM C75 Physics
4 Place Jussieu
Paris 75252 Cedex 05 FRANCE
33 1 44 27 63 03
33 1 44 27 70 82
mla@ccr.jussieu.fr

Lamy, Thierry
CNRS/ISN/IN2P3
Institute des Sciences Nucleoires
F-38041 Grenoble Cedex
France 38000
33 4 76 28 41 33
33 4 76 28 41 43
lamy@isn.in2p3.fr

Landers, Allen
Western Michigan University
Physics Department
Kalamazoo MI 49008
616 387 5360
616 387 4939
allen.landers@wmich.edu

Lanford, William A
University at Albany
Department of Physics
1400 Washington Avenue

Albany NY 12222
518 442 2561
518 442 4486
lanford@thor.albany.edu

Lanza, Richard C.
Massachusetts Institute of Technology
Department of Nuclear Engineering
77 Massachusetts Av Rm NW13-221
Cambridge MA 02139
617 253 2399
617 253 2343
lanza@mit.edu

Laramore, George E
Univ of Washington Med Ctr
Radiation Oncology
PO Box 356043
Seattle WA 98195 6043
206 548 4100
206 598 3498
george@.radonc.washington.edu

Larson, Larry
SEMATECH
Doping Program Manager
2706 Montopolis Drive
Austin TX 78741 6499
512 356 7145
512 356 7640
Larry.Larson@SEMATECH.org

Larsson, Mats
Stockholm University
Department of Physics
P O Box 6730
Stockholm, S-11385 Sweden
46 8 1646 25
46 8 34 7817
mats.larsson@physto.se

Latornell, Doug
MDS Nordion Inc.
Vancouver Operations
4004 Wesbrook Mall
Vancouver BC V6K IV9 Canada
604 228 8952 X161
604 222 4627
dlatornell@mds.nordion.com

Lau, S. S.
University of California
ECE Dept 0407
9500 Gilman Drive
San Diego CA 92093 0407
858 534 3097
858 534 0556
lau@ece.ucsd.edu

Lee
Lee
Hang, Hang
University of Houston
3201 Cullen Blvd
Houston, TX 77204
713 743 8227

Lennard, William N
University of Western Ontario
Physics & Astronomy
London Ontario
N6A3K7 CANADA
519 661 2111 x86461
519 661 2033
wlennard@uwo.ca

Leonard, Douglas S
Univ of North Carolina Chapel Hill
Physics & Astronomy
145 Hamilton Rd
Chapel Hill NC 27514 3255
919 932 1280
919 660 2634
doug@tunl.duke.edu

Levin, Jon C
University of Tennessee
Dept of Physics & Astronomy
401 Nielson Physics
Knoxville TN 37996-1200
865 974 8705
865 974 7843
jlevin@utk.edu

Levine, Zachery
NIST
Photon Physics Group Bldg 245/B102
100 Bureau Dr Stop 8410
Gaithersburg MD 20899 8410
301 975 5453
301 208 6937
zachary.levine@nist.gov

Liang, Felix
Oak Ridge National Laboratory
Bethel Valley Rd MS 6368
PO Box 2008 Bldg 6000
Oak Ridge TN 37831 6368
865 574 4109
865 574 1268
liang@mail.phy.ornl.gov

Liao, Changgeng
Lanzhou University
P. O. Box 44
Lanzhou, Gansu 730001
P. R. China
931 8822342
931 8881996
wangxl@bepc2.ihep.ac.cn

Litherland, Albert E. (Ted)
University of Toronto
IsoTrace Laboratory
60 Saint George Street
Toronto M5S 1A7 Canada
416 978 3785
416 978 4711
ted.litherland@utoronto.ca

Liu, JiaRui
University of Houston
Texas Ctr for Superconductivity
3201 Cullen Blvd
Houston TX 77204 5932
713 743 8255
713 743 8201
jrliu@uh.edu

Liu, Yuan
Oak Ridge National Laboratory
Physics Division
POBox 2008 MS-6368 Bldg 6000
Oak Ridge TN 37831 6368
865 574 4761
865 574 1268
liuy@ornl.gov

Loiselet, Marc
Universite Catholique de Louvain
Centre de Recherches du Cyclotron
Chemin du Cyclotron 2
B-1348 Louvain-la-Neuve BELGIUM
32 10 47 36 45
32 10 45 21 83
loiselet@cyc.ucl.ac.be

Lommel, Bettina
Geselischaft fur Schwerionenforschung
Targetlabor
Planckstrabe 1
D 64291 Darmstadt GERMANY
49 6159 71 2691
49 6159 71 2166
b.lommel@gsi.de

Longo, Tony
Therangenic Corporation
5203 Bristol Industry Way
Buford GA 30518
770 271 0233
longot@theragenics.com

Lu, Xin-Ming
University of Houston
TcSUH
3201 Cullen Blvd.
Houston, TX 77204
713 743 8259
713 743 8201
xinlu@bayou.uh.edu

Lucas, Stephane
Ion Beam Applications (IBA)
Av du Cyclotron 3
B-1348 Louvain-La-Neuve Belgium
32 10 47 59 72
32 10 47 58 10
lucas@iba.be

Maas, A J H
Eindhoven Univ of Tech
Calipso b v
PO Box 513
NL-5600MB Eindhoven The Netherlands
31 40 247 4238
31 40 245 3587
a.j.h.maas@tue.nl

MacAdam, Keith B
University of Kentucky
Physics & Astronomy
177 Chemistry-Physics Bldg
Lexington KY 40506 0055
859 257 6101
859 323 2846

macadam@pop.uky.edu

Macek, Joseph H
University of Tennessee
Theoretical Physics
200 S College
Knoxville TN 37996 1501
865 974 0770
ORNL: 865 576 0510
865 974 6378
jmacek@utk.edu

Mack, Michael E
Epion Corporation
Director Beamline Technology
37 Manning Road
Billerica MA 01821
978 670 1910 x206
978 670 9119
mmack@epion.com

Maden, Colin
ETH Zurich
Institute of Particle Physics
ETH Hoenggerberg HPK H25
CH 8093 Zurich SWITZERLAND
41 1 633 2014
41 1 633 1067
maden@particle.phys.ethz.ch

Madison, Don Harvey
University of Missouri
Department of Physics
1870 Miner Circle
Rolla MO 65409 0640
573 341 4703
573 341 4715
madison@umr.edu

Maier, Hans Joerg
Universitaet Munich
Technological Laboratory
Am Coulombwall 1
D-85748 Garching GERMANY
49 89 289 14027
49 89 289 14034
hans-joerg.maier@physik.uni-muenchen.de

Mandler, John
INEEL
PO Box 1625
Idaho Falls Idaho 83415-2114
208 526 0355
208 526 9267
mwj@inel.gov

Manson, Steven T
Georgia State University
Physics and Astronomy
Atlanta GA 30303
404 6513082
404 651 1427
smanson@gsu.edu

Mantica, Paul
Michigan State University
Nat'l Superconducting Cyclotron Lab.
164 S Shaw Lane
East Lansing MI 48824-1321
517 333 6456
517 333 6562
517 353 5967
mantica@msu.edu

Marble, Daniel K
Tarleton State University
Math and Physics
Box T 470
Stephenville TX 76402
254 968 9880
254 968 9534
marble@tarleton.edu

Margetis, Spyridon (Spiros)
Kent State University
Physics Department
Smith Hall
Kent OH 44242
330 672 9739
330 672 2959
margetis@star.physics.kent.edu

Markoff, Diane
Duke University
Triangle Universities Nuclear Laboratory
POBox 90308
Durham NC 27708 0308
919 660 2624
919 660 2634
markoff@tunl.duke.edu

Marmar, Earl S
Massachusetts Institute of Tech
NW17 119 175 Albany Street
Cambridge MA 02139
617 965 1358
617 253 0627
marmar@psfc.mit.edu

Martinez, Horacio
UNAM Instituto de Fisica
Lab de Cuernavaca Colisiones
Apartado Postal 48-3
Cuernavaca Morelos 62251 Mexico
73 29 17 59
73 17 30 77
hm@ce.ifisicam.unam.mx

Martinez, Trinidad
National University of Mexico
Bldg D Chemistry Faculty
Ciudad Duniversitaria
Mexico D.F. 04510 Mexico
525 5622 5232
525 550 6083 or 56225232
tmc@servidor.unam.mx

Mashnik, Stepan G
Los Alamos National Laboratory
Group T16 Theoretical Div
3000 Trinity Dr MS B283
Los Alamos NM 87545
505 667 9946
505 667 1931
mashnik@t2y.lanl.gov

Masschaele, Bert
University Gent
Subatomic & Radiation Physics
Preftuinstraat 86
Gent 9000 Belgium
32 9 264 6532
32 9 264 6697
bert.masschaele@rug.ac.be

Mathot, Gilles
Facultes Universitaires
Laboratoire d'Analyses par Reactions Nucleaires
Notre-Dame de la Paix
rue de Bruxelles 61 B-5000 Namur
BELGIUM
32 817 254 79
32 817 254 74
gilles.mathot@fundp.ac.be

Matsunami, Noriaki
Nagoya University
School of Eng: Energy Eng & Sci
Furo-Cho, Chikusa-ku
Nagoya 464-8603 JAPAN
81 52 789 3777
81 52 789 3847
n-matsunami@nucl.nagoya-u.ac.jp

Matsuo, Jiro
Kyoto University
Ion Beam Eng Experimental Lab
Yoshida-Honmachi
Sakyo Kyoto 606 8501 Japan
75 753 5953
75 751 6774
jmatsuo@kuee.kyoto-u.ac.jp

Matteson, Samuel E.
University of North Texas
Department of Physics
PO Box 311427
Denton TX 76203 1427
940 565 2630
940 565 2515
matteson@unt.edu

Mattsson, Soren
Lund University
Radiation Physics Malmo
Malmo University Hospital
SE-20502 Malmo SWEDEN
46 40 331374
46 40 963 185
soren.mattsson@rfa.mas.lu.se

Mauron, Olivier
University of Friboug
Institut de Physique
Ch Du Musee 3
Fribourg CH 1700 SWITZERLAND
41 26 300 9075
41 26 300 9031
olivier.mauron@unifr.ch

Maxson, Donald R.
Brown University
Physics Department
Box 1843 Barus and Holley Bldg
Providence RI 02912
401 863 1442
401 863 2024

drmbarri@webtv.net

Mayer, James W.
Arizona State University
Center for Solid State Studies
P O Box 871704
Tempe AZ 85287 1704
480 965 9601 Nancy#4546
480 965 9004
james.mayer@asu.edu

McAninch, Jeffrey E
Lawrence Livermore Natl Lab
Center for AMS
PO Box 808 L 397
Livermore CA 94550
925 423 8506
925 423 7884
mcaninch1@llnl.gov

McCarthy, Timothy J
Washington University School of Medicine
Radiology Sciences
510 S Kingshighway Blvd Box 8225
St Louis MO 63110-1076
314 362 8429
314 362 9940
mccarthyt@mir.wustl.edu

McCullough, Robert W
The Queen's University
Dept of Pure & Applied Physics
University Road
Belfast BT7 1NN N Ireland UK
44 2890 273 541
44 2890 310785
rw.mccullough@qub.ac.uk

McDaniel, Floyd Del
University of North Texas
Department of Physics
P O Box 305370
Denton TX 76203-5370
940 565 3251
Cell Phone # 940 390 7330
940 565 2227
McDaniel@UNT.EDU

McDevitt, Daniel
North Carolina State University
PO Box 3233
Durham NC 27715
919 660 2535
919 660 2636
mcdevitt@tunl.duke.edu

McIntyre, Justin I
Pacific Northwest National Laboratory
Radiological & Chemical Science Group
POBox 999 Battelle Blvd R
Richland WA 99353
509 376 0085
509 372 0672
justin.mcintyre@pnl.gov

McIntyre, Laurence C
University of Arizona
Physics
1118 E 4th St
Tucson AZ 85721
520 621 6813
520 621 6813
mcintyre@physics.arizona.edu

McSherry, Donna
The Queen's University of Belfast
N Ireland
d.mcsherry@am.qub.ac.uk

Mehta, Rahul
University of Central Arkansas
Physics & Astronomy
Lewis Science Ctr Rm 171
201 Donaghey Avenue
Conway AR 72035 0001
501 450 5906
501 450 5914
rahulm@mail.uca.edu

Meister, J David
Technology Commericalization Int., Inc.
1650 Research Blvd. NE, Suite 200
Albuquerque, NM 87102
505-247-4100
505-247-1899
tcintl@nm.net

Mendenhall, Marcus H
Vanderbilt University
Reck Free Electron Laser Ctr
PO Box 351816 Station B
Nashville TN 37235-1816
615 343 6439
615 343 1103
marcus.h.mendenhall@vanderbilt.edu

Mendez, A.J.
University of North Carolina
Dept of Physics & Astronomy
Chapel Hill NC 27599 3255

Merabet, Hocine
University of Nevada Reno
Department of Physics
MS 220
Reno NV 89557-0058
775 784 1335
775 784 1398
hocine@physics.unr.edu

Micklich, Bradley J
Argonne National Laboratory
Technology Development Bldg 362
9700 South Cass Avenue
Argonne IL 60439
630 252 4849
630 252 5287
bjmicklich@anl.gov

Mikhailov, Serguei
CAFI-EICN
7, av. de l'Hotel de Ville
Le Locle, Switzerland 2400
+41329303690
+41329303691
mikhailov@eicn.ch

Miley, George H.
University of Illinois
214 Nuclear Engineering Lab
103 S Goodwin Av
Urbana IL 61801 2984
217 333 3772
217 222 2906
g-miley@uiuc.edu

Miller, Mitchell D
Rice University
Biochem & Cell Biol MS 140
6100 Main St
Houston TX 77005-1892
713 348 3051
713 348 5154
mitchm@bioc.rice.edu

Miller, Thomas Gill
Tensor Technology Inc
254 Brentwood Lane
Madison AL 35758
256 772 3737
256 772 3126
tgmiller@hiwaay.net

Miranda, Javier
Univ Nac Autonoma de Mexico
Instituto de Fisica
Apartado Postal 20-364
MEXICO DF 01000 MEXICO
52 5 622 5005
52 5 622 5009
miranda@fenix.ifisicacu.unam.mx

Mitchell, Ian V
University of Western Ontario
Physics and Astronomy
PAB 100 London ONTARIO
N6H 3K7 CANADA
519 661 3393
519 850 2422
j.mitchell@uwo.ca

Mitchell, Joseph H
Thomas Jefferson National Accelerator Facility
Accelerator Division-FEL
12000 Jefferson Ave, MS 12A1
Newport News VA 23606
757 269 7851
757 269 5703
mitchell@jlab.org

Mitchell, Gary E.
North Carolina State University
Physics
Box 8202
Raleigh NC 27695 8202
919 660 2638
919 660 2634
mitchell@laser.tunl.duke.edu

Moehs, Douglas Paul
Argonne National Laboratory
Physics Division Blgd 203
9700 S Cass Avenue

Argonne IL 60439
630 252 5643
630 252 8647
moehs@anlphy.phy.anl.gov

Monce, Michael N.
Connecticut College
Physics. Astronomy & Geophysics
270 Mohegan Av
New London CT 06320 4196
860 439 2348
860 439 5011
mnmon@conncoll.edu

Montenegro, Eduardo C.
Pontificia Universidade Catolica
Depto de Fisica
Rue Marques de S. Vicente 225
Gavea
Rio De Janeiro 22453-900 Brazil
55 21 529 9360
55 21 259 9397
ecmo@vdg.fis.puc-rio.br

Moog, Elizabeth R
Argonne National Laboratory
Advanced Photon Source
9700 S Cass Ave XFD 401
Argonne IL 60439
630 252 5926
630 252 9303
moog@aps.anl.gov

Moore, Ken
Mallinckrodt
Cyclotron
2703 Wagner Place
Maryland Heights MO 63043
314 654 7826
314 654 7456
ginnie.lingo@mkg.com

Moran, Jean E
Lawrence Livermore National Lab
Analytical & Nuclear Chemistry Div
PO Box 808 L-231
Livermore CA 94550
925 423 1478
925 422 3160
moran10@llnl.gov

Morgan, Ira Lon
International Isotopes Incorporated
3100 Jim Christal Road
Denton TX 76207
940 484 0877
IMorgan826@aol.com

Morrissey, David J.
Michigan State University
NSCL
South Show Lane
East Lansing MI 48824
517 333 6321
517 353 5967
morrissey@nscl.msu.edu

Mous, Dirk J. W.
High Voltage Eng Europa B. V.
P O Box 99
3800 AB Amersfoort
The Netherlands
31 33 461 9741
31 33 461 5291
dmous@highvolteng.com

Mukoyama, Takeshi
Knasai Gaidai University
16-1 Kitakatahoko-Cho
Hiraka 573-1001 Osaka JAPAN
81 72 856 1721
81 72 855 5507
mukoyama@khc.kansai-gaidai-u.ac.jp

Munson, Carter P
Los Alamos National Lab
MS E-526 P 24
Los Alamos NM 87545
505 667 7509
beeper:505 996 3738
505 665 3552
cmunson@lanl.gov

Muntele, Iulia
Alabama A&M University
PO Box 1447
Normal AL 35762
256 851 5866
256 851 5868
iulia@cim.aamu.edu

Muntele, Claudiu I.
Alabama A&M University
Natural and Physical Sciences
PO Box 1447
Normal, Alabama 35762
256 851 5866
256 851 5868
claudiu@cim.aamu.ede

Munzenrider, John E
Massachusetts General Hosp
Radiation Oncology
100 Blossom Street
Boston MA 02062
617 724 3655
617 724 9532

Nablo, Sam V
Electron Processing Systems Inc
6 Executive Park Dr
North Billerica MA 01862
978 667 6366
978 671 0122
eps@tiac.net

Nadeau, Marie-Josee
Christian-Albrechts Universitat
Leibniz-Labor fuer Altersbest & Iso
Max-Eyth Str. 11-13
24118 Kiel Germany
49 431 880 7390
49 431 880 3356
mnadeau@leibniz.uni-kiel.de

Nagadi, Mahmoud M.
King Fahd University KFUPM
Physics Department
PO Box 388
Dhahran 31261 Saudi Arabia
966 3 860 2255
966 3 860 2293
mmnagadi@kfupm.edu.sa

Nagaitsev, Sergei
Fermi Nat'l Accelerator Lab
Box 500 MS 345
Batavia IL 60510
630 840 4397
630 840 4552
nsergei@fnal.gov

Naidu, Seetala V.
Grambling State University
Department of Physics
RWE Jones Drive GSU
Grambling LA 71245
318 274 2574
318 274 3281
naidusv@alpha0.gram.edu

Nakamura, Takashi
Tokyo Institute of Technology
Physics
2-12-1 O-okayama
Meguro 152-8551 Tokyo JAPAN
81 3 5734 2652
81 3 5734 2751
nakamura@yap.nucl.ap.titech.ac.jp

Nakamura, Toshio
Nagoya University
Tandetron AMS 14C Dating Laboratory
Dating & Materials Research Center
Chikusa, NAGOYA 464-8602 JAPAN
81 52 789 2728
81 52 789 3092
nakamura@nendai.nagoya-u.ac.jp

Nakano, Shizuka
Mechanical Engineering Laboratory
AIST MITI
Advanced Machinery
1-2 Namiki
Tsukuba Ibaraki 305 8564 JAPAN
81 298 61 7163
81 298 61 7007
shizuka@mel.go.jp

Nakata, Jyoji
Kanagawa University
Science Department
2946 Tsuchiya, Hiratsuka-shi
Kanagawa-ken 259-1293 JAPAN
81 463 59 4111 ext 2708
81 463 58 9684
jyojin@info.kanagawa-u.ac.jp

Naqvi, Akhtar Abbas
King Fahd University KFUPM
Ctr for Applied Physical Sciences
PO Box 1815
Dhahran 31261 Saudi Arabia
966 3 860 4362
966 3 860 4281
aanaqvi@kfupm.edu.sa

Nash, David H.

Mallinckrodt Group
2703 Wagner Place
Maryland Heights MO 63043
314 770 7871
314 770 7456
dhnash@mkg.com

Nastasi, Michael
Los Alamos National Lab
BldgSM32 MS-K765
PO Box 1663
Los Alamos NM 87545
505 667 7007
505 665 2992
nasty@lanl.gov

Navarrete, Manuel
National University of Mexico
Inorganic & Nuclear Chemistry
Building D CU
04510 Mexico City MEXICO
525 622 52 32
525 622 52 32 or 525 584 51 80
jmnat33@servidor.unam.mx

Necsoiu, Daniela
University of North Texas
Department of Physics
P O Box 311427
Denton TX 76203 1427
940 565 3336
940 565 2227
dn0002@jove.acs.unt.edu

Nguyen, Hai
Kansas State University
116 Cardwell Hall
Manhattan KS 66506
785 532 2665
nguyht@phys.ksu.edu

Ni, Boris
University of Florida
Material Science & Engineering
PO Box 116400
Gainesville FL 32611-6400
352 846 3373
352 846 3355
borni@mail.mse.ufl.edu

Nigam, Mohit
Univeristy of North Texas
Department of Physics
PO Box 311427
Denton TX 76203 1427
940 565 3336
940 565 2227
mn0007@jove.acs.unt.edu OR
mohitnigam@hotmail.com

Nilsson, Thomas
CERN
ISOLDE
EP division GENEVA
SWITZERLAND
49 22 767 3809
49 22 767 8990
thomas.nilsson@cern.ch

Ningkang, Huang
Sichuan University
Institute of Nuclear Science
and Technology
Chengdu, P. R. of China

Ninomiya, Satoshi
Kyoto University
Department of Nuclear Engineering
Sakyo
Kyoto 606 8501 JAPAN
81 75 753 5821
81 75 753 5821
ninomiya@nucleng.kyoto-u.ac.jp

Norton, Gregory A
National Electrostatics Corp
VP Marketing
PO Box 620310
Middleton WI 53562 0310
608 831 7600
608 256 4103
nec@pelletron.com

Nurmela, Arto
University of Helsinki
PO Box 9 (Siltavourenpenger 20 M)
Helsinki FINLAND 00014
358 9 191 8368
358 9 191 8378
arto.nurmela@helsinki.fi

Obolensky, Oleg
University of Pittsburgh
Dept of Physics & Astronomy
Pittsburgh PA 15260
oleg@stribor.phyast.pitt.edu

O'Brien, Harold A
Los Alamos National Lab.
Physics Div.(P-3)
107 La Senda Rd
Los Alamos NM 87545
505 665 4179 or?0250
505 672 1685
hobrien@mesatop.com

Ognibene, Ted
CAMS/LLNL
L-397
Livermore CA 94551
925 424 6266
ognibene@llnl.gov

O'Kelly, Donna
University of Texas
Nuclear Engineering Teaching Lab
JJPickle Research Campus Bldg 159
Austin TX 78712
512 232 4174
512 471 4589
djokelly@mail.utexas.edu

Olson, Ronald E.
University of Missouri-Rolla
Department of Physics
1870 Miner Circle
Rolla MO 65409 0640
573 341 4933
573 341 4715
olson@umr.edu

O'Neil, James P.
Lawrence Berkeley National
Laboratory
Center for Functional Imaging
1 Cyclotron Rd MS 55-121
Berkeley CA 94720
510 486 5276
510 486 6208
jponeil@lbl.gov

Ordonez, Carlos A.
University of North Texas
Physics Department
PO Box 311427
Denton TX 76203 5370
940 565 4860
940 565 2515
fi26@vaxb.acs.unt.edu

Orphan, Victor J.
Science Applications Intl Corp
Adv. Tech. & Analysis
16701 West Bernardo Dr
San Diego CA 92127
858 826 9102
858 826 9224
victor.j.orphan@saic.com

Ozaki, Satoshi
Brookhaven National Laboratory
RHIC
Bldg 1005
Upton NY 11973 5000
631 344 5590
631 344 2166
ozaki@bnl.gov

Pacheco, Adriano M G
Technical University of Lisbon
Department of Chemical Engineering
(DEQ) and Centre for GeoSystems
(CVRM)
Av Rovisco Pais, 1 1049-001
Lisboa
Portugal
351 21 841 7442
351 21 841 7998
apacheco@ist.utl.pt

Parker, Winifred
Lawrence Livermore Natl Lab
Analytical & Nuclear Chemistry
PO Box 808 L-231
Livermore CA 94551
925 422 1215
925 422 3160
parker18@llnl.gov

Paszti, Ferenc
KFKI Research Institute
Hungarian Academy of Sciences
PO Box 49
H 1525 Budapest 114 HUNGARY
36 1 392 2744
36 1 395 9151
paszti@rmki.kfki.hu

Pathak, Anand P
University of Hyderabad

School of Physics
Central University P.O.
Hyderabad 500 046 India
91 40 30 10 500 ext 4316
91 40 3010181
app8p@uohyd.ernet.in

Peaslee, Graham F
Hope College
Chemistry Department
35 East 12th Street
Holland MI 49423
616 395 7117
616 395 7118
peaslee@hope.edu

Pedroni, Eros
Paul Scherrer Institute
Division of Radiation Medicine
Villigen-PSI
CH-5232 SWITZERLAND
41 56 310 3518
41 56 310 35 15
eros.pedroni@psi.ch

Perez, James A
Luther College
Department of Physics
700 college Drive
Decorah IA 52101
319 387 1629
perezjam@luther.edu

Perkins, Luke T.
EMR Photoelectric
20 Wallace Road
Princeton Junction, NJ 08550
609 897 8578
609 799 5788
luke.perkins@slb.com

Peterson, Randolph S
University of the South
Department of Physics
SPO 1193
Sewanee TN 37383 1000
931 598 1550
931 598 1145
rpeterso@sewanee.edu

Petra, Maria
Argonne National Laboratory
9700 S Cass Avenue
Argonne IL 60439
630 252 4039
630 252 4039
petra@anlphy.phy.anl.gov

Phaneuf, Ronald A
University of Nevada,-Reno
Department of Physics
225 Leifson Phys MS 220
Reno NV 89557 0058
775 784 6818
775 784 1398
phaneuf@physics.unr.edu

Pincosy, Phil
Lawrence Livermore National Lab
B Division L99
7000 East Ave
Livermore CA 94606
925 423 7118
925 422 3389
pincosy1@llnl.gov

Plasil, Franz
Oak Ridge National Laboratory
PO Box 2008 MS 6372
Oak Ridge TN 37831 6372
865 574 4711
865 576 2822
plasil@mail.phy.ornl.gov

Poate, John M
Axcelis Corporation
55 Cherry Hill Drive
Beverly MA 01915
978 232 4249
978 232 4221
john.poate@axcelis.com

Pochat, Jean-luc
IPSN/DPHD/SDOS
Bat 159 Cadarache Center
ST Paul Lez Durance FRANCE
13108
33 4 42 25 32 81
33 4 42 25 49 48
jean-luc.pochat@ipsn.fr

Porter, Len E
Washington State University
Radiation Safety Office
Nuclear Radiation Center Roundtop
Pullman WA 99164 1302
509 335 7057
509 335 1615
porter1@mail.wsu.edu>

Portillo, Salvador
Texas Christian University
Department of Physics and
Astronomy
TCU Box 298840
Fort Worth TX 76129
817 257 6394
817 257 7742
s.portillo@mail.tcu.edu

Potter, James M
JP Accelerator Works Inc
2245 47th Street
Los Alamos NM 87544 1604
505 661 8155
505 661 8156
jpotter@jpaw.com

Powers, Darden
Baylor University
Physics Department
P O Box 97316
Waco TX 76798 7316
254 710 2511
540 710 3878
darden_powers@baylor.edu

Pratt, Richard H
University of Pittsburgh
Physics & Astronomy
100 Allen Hall
Pittsburgh PA 15260
412 624 9052
412 624 9163
rpratt@pitt.edu

Price, Hywel
Rutherford Appleton Laboratory
Chilton Didcot
Oxfordshire OX11 OQX
Oxon United Kingdom ENGLAND
44 01235 445593
44 01235 446221
h.g.price@rl.ac.uk

Price, Jack Lewis
Naval Surface Warfare Center
Code 682Applied Matls Sci Dept
9500 MacArthur Blvd
West Bethesda MD 20817 5700
301 227 4163
Lab: 301 394 4995
301 227 2252
pricejl@nswccd.navy.mil

Pusa, Petteri
University of Helsinki
Siltavuorenpenger 20K
Helsinki 00014 FINLAND
358 9 191 8390
358 50 345 8559
petteri.pusa@helsinki.fi

Qaim, Syed M
Forschungszentrum Julich
GmbH(FZJ)
Institut fur Nuklearchemie
PO Box 1913 D 52425
JULICH GERMANY
49 2461 613 282
49 2461 612 535
s.m.qaim@fz-juelich.de

Quarles, Carroll A
Texas Christian University
Dept of Physics & Astronomy
P O Box 298840
Fort Worth TX 76129
817 257 7375
817 257 7742
c.quarles@tcu.edu

Radojevic, Antonije Tony
Columbia University
Microelectronics Sciences Lab
530 West 120th Street Rm 1001
New York NY 10027
212 854 8449
212 860 6182
tony@cumsl.ctr.columbia.edu

Rakers, Sven
University of Muenster
Nuclear Physics Institute
D 48149 Muenster
GERMANY
49 251 83 34966
49 251 83 34962
rakers@uni-muenster.de

Randers-Pehrson, Gerhard
Columbia University
RARAF 136 S. Broadway
P O Box 21
Irvington NY 10533
914 591 9244
914 591 9405
gerhard@r-p.net

Rathmell, Robert D
Eaton Corporation
Semiconductor Equipment Division
108 Cherry Hill Drive
Beverly MA 01915
978 524 9045
978 927 3652
rrathmel@bev.etn.com

Ratliff Peterson, Laura
National Institute of Standards & Technology
Atomic Physics Division
100 Bureau Dr Stop 8421
Gaithersburg MD 20890 8420
301 975 6580
301 975 5485
Laura.Ratliff@NIST.gov

Reading, John F
Texas A&M University
Physics Department
College Station TX 77843
409 845 5073
409 845 2590
reading@.physics.tamu.edu

Rehm, Ernst
Argonne National Laboratory
Physics Division Bldg 203
9700 South Cass Av
Argonne IL 60439
630 252 4073
630 252 9210
rehm@anlphy.phy.anl.gov

Revesz, Peter
Cornell University
Materials Science/Engineering
126 Bard Hall
Ithaca NY 14853
607 255 7179
607 255 2365
revesz@msc.cornell.edu

Reyes, Pedro G
Universidad Nacional de Mexico
Instituto de Fisica
Circuito Extenion S/11
Mexico D.F. Mexico
56 22 98 95
56 16 03 26
pedro@fis.unam.mx

Richard, Patrick
Kansas State University
Dept of Physics
106A Cardwell Hall
Manhattan KS 66502
785 532 6783
785 532 6806

richard@phys.ksu.edu

Richter-Sand, Robert J.
North Star Research Corp
4421 McLeod NE Ste A
Albuquerque NM 87109
505 888 4908
505 888 0072
rrsand@aol.com

Rickards, Jorge
Univ Nac Autonoma de Mexico
Instituto de Fisica
Apartado Postal 20-364
Mexico 01000 DF Mexico
525 622 5065
525 622 5009
rickards@fenix.ifisicacu.unam.mx

Rimini, Emanuele
Universitita di Catania
Dipartimento di Fisica
C so Italia 57
I-95129 Catania ITALY
39 095 7195418
39 095 383 023
rimini@infnct.ct.infn.it

Roberts, Mark L.
University of Georgia
Ctr for Applied Isotope Sciences
120 Riverbend Road
Athens GA 30602
706 542 1395
706 542 6106
roberts5@llnl.gov

Roberts, Andrew D
University of Wisconsin
Medical Physics and Psychiatry
1300 University Ave
Madison WI 53706
608 263 4369
608 262 2413
arobert5@facstaff.wisc.edu

Rogachev, Grigory
University of Notre Dame
4012 Parkwood Circle #2A
Mishawaka IN 46545
219 277 7152
grogache@nd.edu

Rost, Jan Michael
Max Planck Institute
fuer Physik komplexer Systeme
Noethnitzer Str 38
D 01187 DRESDEN GERMANY
49 351 871 2204
sec: 49 351 871 2202
49 351 871 2299
rost@mpipks-dresden.mpg.de

Rotberg, Victor Hanin
University of Michigan
Nuclear Engineering Department
2600 Draper Rd
Ann Arbor MI 48109 2104
313 936 0166
313 936 8820

victor.rotberg@um.cc.umich.edu

Rozsa, Csaba M.
Bicron
Advanced Technology Group
12345 Kinsman Road
Newbury OH 44141
216 564 2251 Voice#953-5674
216 564 8047

Rubensson, Jan Erik
Uppsala University
Physics Department
Box 530
SE 75121 Uppsala SWEDEN
46 18 471 3562
46 18 471 3524
jan_erik.rubenssom@fysik.uu.se

Rüch, Dorothea
Gesellschaft für

Schwerionenforschung
Planckstrasse 1, Postfach 110552
D-6100 Darmstadt Germany
49 6151 359 320
49 6151 359 988

Ruck, Dorothee M
Gesellschaft
fur Scherionenforschung
PF 110552, D-6100 mbH
DARMSTADT GERMANY
49 6159 712630
49 6159 712905
d.rueck@gsi.de

Ruggles, Arthur E
University of Tennessee
Nuclear Engineering
Pasqua Engineering
Knoxville TN 37996 2300
865 974 7563
865 974 0668
aruggles@utk.edu

Rusnak, Brian
Lawrence Livermore National Lab
Accelerator Tech Engineering
POBox 808 L 287
Livermore CA 94550
925 422 0435
925 424 3532
rusnak1@llnl.gov

Ruth, Thomas J.
TRIUMF
PET Program
4004 Wesbrook Mall
VANCOUVER BC V6T 2A3
CANADA
1 604 822 7753
1 604 222 1074
truth@triumf.ca

Ryan, James M.
University of New Hampshire
Space Science Center
39 College Rd 310 Morse Hall
Durham NH 03824 3525

603 862 3510
603 862 4685
james.ryan@unh.edu

Sabin, John R
University of Florida
Physics Dep Quantum Theory
P O Box 118435
Gainesville, FL 32611 8435
352 392 1597
352 392 8722
sabin@qtp.ufl.edu

Sadler, Michael E
Abilene Christian University
Physics 320B Foster Science Bldg
ACU Box 27963
Abilene TX 79699 7963
915 674 2189
1 888 227 7497
915 674 2146
sadler@physics.acu.edu

Sakudo, Noriyuki
Kanazawa Institute of Technology
7 1 Ohgigaoka Nonoichi
Ishikawa 921 8501 JAPAN
81 76 274 9262
81 76 274 9251
sakudo@neptune.kanazawa-it.ac.jp

Sale, Kenneth E.
Lawrence Livermore National
Laboratory
L 59 N Division
PO Box 808 7000 East Ave
Livermore CA 94550
925 423 0686
kesale@llnl.gov

Saleh, Adli A
Charles Evans & Associates
SINS Group
810 Kifer Road
Sunnyvale CA 94086
408 530 3782
408 530 3501
asaleh@cea.com

Sanders, Justin M.
University of South Alabama
Department of Physics
ILB 103
Mobile AL 36688 0002
334 460 6224 x2134
334 460 6800
jsanders@jaguar1.usouthal.edu

Sangsingkeow, Pat
Perkin Elmer
Production & Dev: Detector Tech
100 Midland Road
Oak Ridge TN 37830
865 481 2416
865 483 2133
pat_sangsingkeow@egginc.com

Sapp, William Wade
American Science & Engineering
829 Middlesex Turnpike
Billerica MA 01821
978 262 8634
978 262 8805
wsapp@as-e.com

Sato, Susumu
University of Tsukuba
Institute of Physics
Tenno-dai 1-1-1
Tsukuba 305 Ibaraki JAPAN
81 298 53 4249
81 298 53 6618
ssato@tac.tsukuba.ac.jp

Satou, Takahiro
Tohoku University
Dept of Quantum Science and
Energy Engineering
Aramaki Aza Apba 01, Aobako
Sendai 980 8579 JAPAN
81 22 217 7933
81 27 346 9690
takas@stein.qse.tohoku.ac.jp

Schempp, Alwin
University Frankfurt-Germany
Inst für Angewandte Physik
Robert Mayer Str 2-4
D-60054 Frankfurt/M Germany
49 69 7982 2802
49 69 7982 8510
a.schempp@em.uni-frankfurt.de

Schenkel, Thomas
Lawrence Livermore National Lab
Accelerator & Fusion Research
1 Cyclotron Road
Berkeley CA 94720
510 486 6674
tschenkel@lbl.gov

Schiebl, Christian
MCS Consult
Prof Leopold Hauer
Gasse 7 A-3552 Lengenfeld
AUSTRIA
43 2719 8783
43 2719 8783 4
schiebl@aon.at

Schiettekatte, Francois
INRS-Energie et Materiaux Conf
1650, Boul. Lionel-Boulet
C.P. 1020 Varennes
Quebec J3X 1S2

Schlyer, David J
Brookhaven National Lab
Cyclotron Lab, Chemistry
Bldg 901
Upton NY 11973
631 344 4587
631344 7350
Schlyer@bnl.gov

Schmidt-Boecking, Horst
Universität Frankfurt
Institut für Kernphysik
August-Euler-Str 6
Frankfurt/Main 90, D-60486
GERMANY
49 69 798 4252
49 69 798 24212
schmidtb@ikf.uni-frankfurt.de

Schneider, Dieter H.G.
Lawrence Livermore Natl Lab
Physical Sciences
P O Box 808 L-421
Livermore CA 94550-9900
925 423 5940
925 422 5940
Schneider2@llnl.gov

Schneider, Robert J.
Woods Hole Oceanographic
Institution
National Ocean Sciences
Accelerator Mass Spectrometry
Facility
Mail Stop 8
Woods Hole MA 02543
508 289 2756
508 457 2183
rschneider@whoi.edu

Schneider, Larry
Sandia National Laboratories
Applied Acceerator Tech
POBox 5800 MS 1152 Dept 1643
Albuquerque NM 87111
505 845 7135
505 284 6078
lxschne@sandia.gov

Schramm, Ulrich
Universitaet Munchen (LMU Munich)
Sektion Physik
Am Coulombwall 1
GARCHING D-85748 GERMANY
49 89 289 14077
49 89 289 14072
ulrich.schramm@physik.uni-muenchen.de

Schulz, Michael E.
University of Missouri at Rolla
Department of Physics
1870 Miner Circle
Rolla MO 65409-0640
573 341 4712
573 341 4715
schulz@.umr.edu

Schweickert, Hermann
Forschungszentrum karlsruhe GmbH
Cyclotron Laboratory
POBox 3640
D76021 KARLSRUHE GERMANY
49 7247 822 433
49 7247 823 156
schweick@hzy.fzk.de

Schweikert, Emile A.
Texas A&M University
Dept of Chemistry
College Station, TX 77843 3144
409 845 2341
409 845 1655

schweikert@mail.chem.tamu.edu

Schweitzer, Jeffrey S
University of Connecticut
Department of Physics
2152 Hillside Road U-46
Storrs CT 06269-3046
860 486 4978
860 486 3346
schweitz@phys.uconn.edu

Scott, John D
LSU
CAMD
3980 Jefferson Hwy
Baton Rouge LA 70806-8109
228 388 4605
225 388 6954
scott@lsu.edu

Seabury, Edward H
Los Alamos National Laboratory
LANSCE 3
MS H855
Los Alamos NM 87545
505 665 2797
505 665 3705
seabury@lanl.gov

Sealy, Brian J
University of Surrey
Electronic Engineering
Information Tech and Math
Guildford Surrey GU2 5XH UK
44 1483 879 139
44 1483 534 139
b.sealy@eim.surrey.ac.uk

Seidel, Carl W.
Carl W Seidel & Associates
Radioisotopes/Radiopharmaceuticals
208 Royal Oaks Place
Denton, TX 76205
940 387 3004
cell phone:940 300 3261
940 381 3036
carl.w.seidel@mindspring.com

Seki, Toshio
Kyoto University
Ion Beam Eng Experimental Lab
Sakyo
KYOTO 606 8501 JAPAN
81 75 753 4994
81 75 751 6774
seki@kuee.kyoto-u.ac.jp

Seo, Hye-Won
University of Houston
Physics - TCSUH
3201 Cullen Blvd
Houston TX 77204
713 743 8258
713 743 8201
fireflyshw@yahoo.com

Shafroth, Stephen M.
Univ of North Carolina-Chapel Hill
Physics & Astronomy
CB# 3255 Phillips Hall
Chapel Hill NC 27599 3255
919 962 3015
919 962 0480
shafroth@physics.unc.edu

Shao, Lin
University of Houston
Physics Department
3201 Cullen Blvd
Houston TX 77204
713 743 8259
713 743 8201
shaolin99@hotmail.com

Shapiro, Mark Howard
California State University at Fullerton
Physics Department
PO Box 6866
Fullerton CA 92834 6866
714 278 3884
714 278 5810
mshapiro@fullerton.edu
http://chaos.fullerton.edu/physiocs.html

Shepard, Kenneth W
Argonne National Laboratory
Physics Division
9700 S Cass Ave Bldg 203
Argonne IL 60540
630 252 4029
630 252 9647
kwshepard@anl.gov

Sherrill, Bradley M.
Michigan State University
Cyclotron Lab
South Shaw Lane
East Lansing MI 48824
517 333 6322
517 353 5967
sherrill@nscl.msu.edu

Shevchenko, Valeriy A
Taras Shevchenko Kiev University
Nuclear Physics Laboratory
Glushkova av 6a
Kiev 03022 Ukraine
380 44 261 34 97
380 44 220 82 85
shevfis@carrier.kiev.ua

Shew, Bor-Yuan
Synchrotron Radiation Research Ctr
No 1 R&D Rd VI
Science Based Industrial Park
70101 Hsinchu Taiwan
886 3 578 0281 7329
886 3 578 9816
yuan@alpha1.srrc.gov.tw

Shibata, Yasuyuki
Natl Inst for Environmental Studies
Environmental Chemodynamics Section
16-2 Onogawa
Tsukuba IBARAKI 305-0053 JAPAN
81 298 50 2450
81 298 50 2574
yshibata@nies.go.jp

Shikine, Shunsuke
Kyoto University
Precision Engineering
Yoshida Honmachi Sakyoku
Kyoto 606-8501 JAPAN
81 75 753 5259
81 75 753 5259
h50252@sakura.kudpc.kyoto-u.ac.jp

Shiner, David
University of North Texas
Physics Department
P O Box 305370
Denton TX 76203 5370
940 565 3824
940 565 2515
shiner@unt.edu

Shinn, Michelle
Jefferson Lab
Room 001/602-19 Mail Stop 6A
12000 Jefferson
Newport News VA 23606
757 269 7565
757 269 5519
shinn@jlab.org

Shutthanandan, Shuttha V
Pacific Northwest Natl Lab
Environmental Molecular Science Lab
PO Box 999, MSIN K8-93
Richland WA 99352
509 376 1928 office
509 376 2708 lab
509 376 5106
shuttha@pnl.gov

Sie, Soey H.
CSIRO: Exploration & Mining
HIAF Laboratory
PO Box 136, 51 Delhi Rd
North Ryde NSW 2113 AUSTRALIA
61 2 9490 8648
61 2 9490 8921
s.sie@dem.csiro.au

Silari, Marco
CERN
1211 Geneva 23
SWITZERLAND
41 22 7673937
41 22 767 5700
marco.silari@cern.ch

Simabuco, Silvana Moreira
Campinas State University
Water Resources
PO Box 6021 CEP 13083-970
Campinas City Sao Paulo Brazil
13083-970
55 19 788 2354
55 19 788 2411
silvana@fec.unicamp.br

Sisterson, Janet M.
Massachusetts General Hospital

Northeast Proton Therapy Center
30 Fruit Street
Boston MA 02114
617 724 1942
617 724 9532
jsisterson@partners.org

Smit, Ziga
University of Ljubljana
Physics
Jadranska 19
LJUBLJANASI 1000 SLOVENIA
386 1 1766 589
386 1 217 281
ziga.smit@fmf.uni-lj.si

Smith, David M
University of California Berkeley
Space Sciences Laboratory
Berkeley CA 94720
510 643 1585
510 643 8302
dsmith@ssl.berkeley.edu

Smith, Richard J.
Montana State University
Physics
EPS Bldg Room 214
Bozeman MT 59717
406 994 6152
406 994 4452
smith@physics.montana.edu

Smith, Paul A
Los Alamos National Laboratory
Los Alamos Nuclear Medicine
Initiative
3491 Trinity Drive Suite B
Los Alamos NM 87544
505 661 6830
505 661 6827
pas@pascd.com

Soga, Fuminori
National Institute of Radiological
Sciences
Division of Accelerator Physics and
Engineering
4-9-1 Anagawa Inage-ku
Chiba-shi 263 JAPAN
81 43 206 3170
81 43 251 1840
soga_f@nirs.go.jp

Solovyov, Andrey
A F Iokfe Physical - Technical
Institute
Russian Academy of Sciences
Politechnicheskaya 26
194021 St Petersburg RUSSIA
7 812 247 9191
7 812 247 1017
solovyov@rpro.ioffe.rssi.ru

Sone, Hayato
RIKEN-Institute of Physical and
Chemical Research
Surface & Interface Lab
2-1 Hirosawa, Wako,
Saitama 351-0198 Japan

81 48 462 1111 x3127
81 48 462 4656
sone@postman.riken.go.jp

Sosnowski, Marek
New Jersey Institute of Technology
Electrical & Computer Engineering
University Heights
Newark NJ 07102
973 596 3541
973 596 5680
sosnowski@njit.edu

Sowerby, Brian D.
CSIRO
Division of Minerals
PMB 5
MENAI NSW 2234 AUSTRALIA
612 9710 6719
612 9710 6789
brian.sowerby@minerals.csiro.au

Spemann, Daniel
Universitaet Leipzig
Fakultaet fuer Physik
Linnestr.5
D-04103 Leipzig GERMANY
49 341 973 2706
49 341 973 2497
spemann@nfp1.exphysik.uni-leipzig.de

Stark, Diane T
State Univ of New York at Albany
Biology
1400 Washington Ave
Albany NY 12222
518 442 4312
ds6144@cnsvax.albany.edu

Steinman, Don
Martin Marietta
Specialty Component
P O Box 2908
Largo FL 34649 2908
813 541 8385
813 541 8778

Stephenson, Edward James
Indiana University
Cyclotron Facility
2401 Milo B Sampson Lane
Bloomington IN 47408
812 855 5469
812 855 6645
stephens@iucf.indiana.edu

Stevenson, Nigel R
Theragenics Corporation
VP Isotope Production & Research
5203 Bristol Industrial Way
Buford GA 30518
770 271 0233
678 482 4909
stevensn@theragenics.com

Stewart, Larry D.
General Atomics
P.O. Box 85608
San Diego CA 92186 5608

858 455 4281
858 455 2494
larry.stewart@gat.com

Stirm, Chuck
Varian Medical Systems
559 877 3066

Stockli, Martin P
Kansas State University
Physics
116 Cardwell Hall
Manhattan KS 66506 2604
785 532 2661 = Office -6777 (lab)
785 532 6806
stockli@phys.ksu.edu

Stoll, Hermann
Max Planck Inst fur Metallforschung
Institute für Physik
Heisenbergstr 1 D-70569
STUTTGART 80 GERMANY
49 711 689 1848
49 711 689 1932
stoll@mpi-stuttgart.de

Stoudenmire, Sterling
WDRC Inc
953 Gondolier
Gulf Breeze FL 32361
850 934 5680
603 943 5926
sstouden@thelinks.com

Stracener, Daniel W.
Oak Ridge National Laboratory
MS 6368 Bldg 6000
Bethel Valley Road
Oak Ridge TN 37830
865 574 4725
865 574 1268
strace@mail.phy.ornl.gov

Strangis, Saverio R
CNEA
Prebistero Juan Gonzalez y aragon
No 15
Ezeiza Buenos Aires ARGENTINA
B1802AYA
54 11 4379 8250
54 11 4480 0615
srstrangis@yahoo.com

Strathman, Michael D
Thin Film Analysis Inc
250 Santa Ana Court
Sunnyvale CA 94086
408 238 6351
408 238 3466
michael@tfainc.com

Strieder, Frank
Ruhr-Universitat Bochum
Institut Fur ExperimentalPhysik III
Universithssstr.750
D-4478Bochum Germany
Universithssstr.150
D-44780 Bochum Germany
49 234 3223597
49 234 3214172

strieder@ep3.ruhr-uni-bochum.de

Sved, John
FusionStar
Space Infrastructure - Ctr Trauen
Eugene-Saenger Strasse 52
Fassberg D-29328 GERMANY
49 5055 598 203 or
49 421 539 5262
49 5055 598 206 or
49 421 539 5262
john.sved@astrium-space.com

Swann, Charles P
University of Delaware
Bartol Research Institute
217 Sharp Lab
Newark DE 19716
302 831 1279
302 831 1843
swann@bartol.udel.edu

Swenson, Donald A
Linac Systems
1208 Marigold Drive NE
Albuquerque NM 87122
505 798 1904
505 798 1902
daswenson@aol.com

Syed, Rashid H.
VA Medical Center
PET/Imaging Service
1 Veterans Drive 11-P
Minneapolis MN 55417 2300
612 725 1948 pager 612 660 7781
612 725 2068
rhs@pet.med.va.gov

Synal, Hans-Arno
Paul Scherrer Institut
c/o ETH Hoenggerberg
ETH-Honggerberg
CH-8093 Zurich, Switzerland
411 633 2027
41 1 633 1067
synal@particle.phys.ethz.ch

Takada, Masashi
National Inst of Radiological
Sciences
International Space Radiation Lab
4 9 1 Anagawa Inage-ku
Chiba 263 8555 JAPAN
81 43 206 3239
81 43 251 4531
m_takada@nirs.go.jp

Talib, Magda
Temple University
Mathematics
1805 N Broad St Office #638
Philadelphia PA 19122
215 204 5872
215 214 6433
magda@math.temple.edu

Tawara, Hiroyuki
Kansas State University
Manhattan KS 66506-2601

785 532 2653
785 532 6806
tawara@phys.ksu.edu

Templon, J A
University of Georgia
Physics and Astronomy
Athens GA 30602
706 542 2843
706 542 2492
templon@jlab.org

Tesmer, Joseph R.
Los Alamos National Lab
Materials Science & Tech
POBox 1663 MS K765
Los Alamos NM 87545
505 667 6370
505 665 2992
tesmer@mst.lanl.gov OR
joe.tesmer@lanl.gov

Theron, Chris
National Accelerator Centre
Materials Research Group
Box 72
FAURE 7131 SOUTH AFRICA
27 21 843 3820
27 21 843 3543
ctheron@nac.ac.za

Thevuthasan, Theva
Theva Suntharampillai
Pacific Northwest Natl Lab
Environmental Molecular Sci Lab
POBox 999 MSIN K8-93
Richland WA 99352
509 376 1375
509 376 5106
theva@pnl.gov

Thomas, Dan M
General Atomics
Fusion Group
3550 General Atomics Court
San Diego CA 92121 1194
858 455 2403
858 455 4156
dan.thomas@gav.com OR
thomas@fusion.gat.com

Thomas, Ross T
Virginia Military Institute
366 Mallory Hall
Lexington VA 24450
540 464 7503
thomast@mail.vmi.edu

Thompson, Jeffrey S
University of Nevada Reno
Department of Physics
Mail Stop 220
Reno NV 89557 0058
775 784 6821
775 784 1398
thompson@physics.unr.edu

Thompson, Michael
Cornell University
Department of MS&E

329 Bard Hall
Ithaca NY 14853
607 255 4714
607 255 2365
mot1@cornell.edu

Timmers, Heiko
Australian National University
Electronic Materials Eng
Canberra ACT 0200
AUSTRALIA
61 2 6279 8887
61 2 6249 0511
heiko.timmers@anu.edu.au

Toader, Ovidiu
University of Michigan
Michigan Ion Beam Lab
2600 Draper Road
Ann Arbor MI 48109
734 936 0166
734 936 8820
ovidiu@engin.umich.edu

Todd, Alan M
Advanced Energy Systems Inc
VP Accelerator & Special Projects
29 Airpark Road
Princeton NJ 08540
609 430 2125
cell: 609 841 5607
609 430 1460
todd@grump.com

Tombrello, Thomas A.
California Inst of Technology
Physics, Mathematics & Astronomy
103-33 Caltech
Pasadena CA 91125
626 395 4241
626 564 0267
tat@cco.caltech.edu

Toyoda, Noriaki
Massachusetts Institute of Tech
Materials Science & Engineering
77 Massachusetts Ave 13-4126
Cambridge MA 02139
617 253 5302
617 253 6782
ntoyoda@mit.edu

Trail, Carroll C
2304 North Lake Trail
Denton TX 76201
565-1336
cctrail@yahoo.com

Trickey, Samuel B
University of Florida
QTP 2324 New Physics Bldg
PO Box 118435
Gainesville FL 32611 8435
352 392 6978
Sec 352 392 1597
352 392 8722
trickey@qtp.ufl.edu

Triftshauser, Werner
Universität Der Bundeswehr

Insti fuer Nukleare Festkorperphy
Munchen, Werner Heisenberg Weg 39
D-85579 Neubiberg, Germany
49 89 6004 3505 /4
49 89 6004 3295

Tsuchida, Hidetsugu
Nara Women's University
Department of Physics
Nara 630 8506
JAPAN
81 742 20 3378
81 742 20 3378
t-hide@phys.nara-wu.ac.jp

Tsuji, Kouichi
Tohoku University
Materials Research
2-1-1 Katahira Aoba ku
Sendia 980-8577 JAPAN
81 22 215 2133
81 22 215 2131
tsuji@imr.tohoku.ac.jp

Tuniz, Claudio
Australian Permanent Mission to the United Nations
Australian Embassy
Mattiellistrasse 2 4/111
VIENNA A 1040 AUSTRIA
43 1 512 8580 119
43 1 504 1178
tuniz@anstagov.au

Uda, Masayuki
Waseda University
Laboratory of Materials Science and Technology
2 8 26 Nishi-Waseda Shinjuku-ku
Tokyo 169 JAPAN
81 3 5286 3308
81 3 5272 9799
muda@mn.waseda.ac.jp

Unal, Ridvan
J R Macdonald Laboratory
Department of Physics
Kansas State University
Manhattan KS 66506
ridvan@phys.ksu.edu

Uribe, Roberto M
Kent State University
Electron Beam Technology
117 Van Deusen Hall
Kent OH 44242-0001
330 672 2770
330 672 2894
ruribe@kent.edu

Uritani, Akira
Nagoya University
Dept of Nuclear Engineering
Furo-cho, Nagoya, Aichi 4648603
JAPAN
81 52 789 4695
81 52 789 5127
uritani@avocet.nucl.nagoya-u.ac.jp

Valkovic, Vlado
Institute Rudjer Boskovic
Experimental Physics
Bijenicka c.54
10000 Zagreb CROATIA
385 1 468 0101
Home Office 385 1 2341 190
385 1 468 0239
valkovic@rudjer.irb.hr

Van Brocklin, Henry F
Lawrence Berkeley National Laboratory
Center for Functional Imaging
1 Cyclotron Rd MS 55 121
Berkeley CA 94720
510 486 4083
510 486 4768
hfvanbrocklin@lbl.gov

Van den Bergh, Paul
Katholieke Universiteit Leuven
Instituut voor Kern-en Stralingsfysica
Celestijnenlaan 200D
B-3001 Leuven Belgium
32 16 32 72 70
32 16 32 79 85
paul.vandenbergh@fys.kuleuven.ac.be

van Nieuwenhuizen, Gerrit J
MIT
LNS MIT Group Brookhaven Natl Lab
46 Dawn Drive
Shirly NY 11967-5000
631 344 4342
631 344 5815
nieuwhzn@mit.edu

Van Oosterhout, Henri
High Voltage
Engineering Europa B V
PO Box 99
3800 AB Amersfoort The Netherlands
31 33 461 9741
31 33 461 5291
hoosterhout@highvolteng.com

Vanderberg, Bo H
Axcelis Technologies Inc
108 Cherry Hill Drive
Beverly MA 01915
978 524 9054
978 927 3652
bvandebe@bev.etn.com

Vanhoy, Jeffery R.
US Naval Academy
Department of Physics
572 Holloway Road
Annapolis MD 21402-5026
410 293 6671
410 293 3729
vanhoy@nadn/navy.mil

Varghese, S. L.
University of South Alabama
Physics Department
ILB 115
Mobile AL 36688 0002
334 460 6224
334 460 6800
svarghes@jaguar1.usouthal.edu

Velkovska, Julia
State Univ of New York Stony Brook
Physics & Astronomy
SUNY at Stony Brook
Stony Brook NY 11794
631 632 3273
631 632 8573
julia.velkovska@sunysb.edu

Verda, Raymond D
Los Alamos National Laboratory
Materials Science/Tech Div
MST 8 MS G755
Los Alamos NM 87545
505 665 6685
505 667 8021
verda@lanl.gov

Verkhoturov, S V
Texas A&M University
Chemistry
PO Box 30012
College Station TX 77843-3012
979 845 2344
979 845 2338
verkhoturov@mail.chem.tamu.edu

Vetter, Kai
LBNL
kvetter@lbl.gov

Videbaek, Flemming
Brookhaven National Laboratory
Physics Department
510D
Upton NY 11973
631 344 4106
631 344 1334
videbaek@bnl.gov

Viesti, Giuseppe
Istituto Nazionale de Fisica Nucleare
Sezione de Padova
Via Marzolo 8
I-35100 Padova ITALY
39 49 827 7124
39 49 876 2641
viesti@pd.infn.it

Villari, Antonio
GANIL
Bvd Henri Becquerel
C P 5027
Caen Cedex 5 14076 FRANCE
14076
33 231 454 458
33 231 454 720
villari@ganil.fr

Vincent, Laetitia
CNRS Orleans/ CERI
#A rue de la Ferdleree
Orleans FRANCE 45071
02 38 85 76 46

vincent@cnrs-orleans.fr

Vizkelethy, Gyorgy
Sandia National Laboratories
Dept.1111
1515 Eubank SE/ POB 5800-1056
Albuquerque NM 87185-1056
505 284 3120
505 844 7775
gvizkel@sandia.gov

Vlastou, Rosa
Natl Technical Univ of Athens
Physics Department
Zografou Campus
GR 112 53 Athens, Greece
003 01 7723008
30 1 7723025
vlastou@central.ntua.gr

Vogel, John S
Lawrence Livermore Natl Lab
Center for AMS L397
PO Box 808 7000 E Ave
Livermore CA 94551
925 423 4232
925 423 7884
jsvogel@llnl.gov

Von Reden, Karl F
Woods Hole Oceanographic Inst
Geology and Geophysics
360 Woods Hole Road MS#8
Woods Hole MA 02543
508 289 3384
508 457 2183
kvonreden@whoi.edu

Vourvopoulos, George
Western Kentucky University
Applied Physics Institute
1 Big Red Way
Bowling Green KY 42101
270 781 3859
270 781 1104
vour@wku.edu

Walter, Richard L
Duke Universtiy
Department of Physics
P O Box 90305
Durham NC 27708 0305
Work# 919 660 2629 & physics
dept# 919 660 2501
919 660 2634
walter@tunl.duke.edu

Walter, Kevin C
Technanogy
Senior Scientist
1601 Alton Parkway Suite B
Irvine CA 92606-4801
949 261 1420 x101
949 261 0311
kwalter@technanogy.net

Wang, Chong
University of Houston
Physics
3201Cullen Blvd.

Houston TX 77204
713-743-8259
713-743-8201
wangscott@hotmail.com

Wang, Xuemei
University of Houston
3201 Cullen Blvd

3201 Cullen Blvd
Houston TX 77204 5932
713 747 8256
woll9@yahoo.com

Wang, Yongqiang
University of Minnesota
IT Characterization Facility
100 Union St. SE
Minneapolis MN 55455
612 626 1019
612 625 5368
yqwang@tc.umn.edu

Ward, Sandra J
University of North Texas
Dept of Physics
PO Box 311427
Denton TX 76203-1427
940 565 4739
940 565 2515
sward@unt.edu

Warkentien, L.S.
Brookhaven Natl Lab
Dept Adv Tech,Eng Res App Div
PO Box 5000
Upton NY 11973-5000
516 344 2929

Watson, Rand L
Texas A&M University
Cyclotron Institute
MS 3366
College Station TX 77843 3366
979 845 1411
979 845 1899
watson@comp.tamu.edu

Watterson, John I W
University of Witwatersrand
Schonland Center Physics Dept
Private Bag 3 WITS
2050 JOHANNESBURG SOUTH AFRICA
27 11 787 0252
27 11 339 2144
watterson@src.wits.ac.za

Weathers, Duncan L.
University of North Texas
Department of Physics
PO Box 311427
Denton TX 76203 1427

940 565 2079
940 565 2227
weathers@unt.edu

Webb, Roger P
University of Surrey
Electronic Engineering
Info Technology & Mathematics
Guildford GU2 5XH Surrey UK
44 1483 259 830 Ext 2291
44 1483 534 139
r.webb@eim.surrey.ac.uk

Weber, Thorsten
University Frankfurt
Ins
Institut fuer Kernphysik
August Euler Str 6
Frankfurt 65191 GERMANY
49 69 798 24218
49 69 798 24212
weber@hsbpc1.ikf.physik.uni-frankfurt.de

Wehlitz, Ralf
Univ of Wisconsin-Madison
Synchrotron Radiation Ctr
3731 Schneider Dr Rt 4
Stoughton WI 53589-3097
608 877 2164
608 877 2001
wehlitz@src.wisc.edu

Weiss, Alex
University of Texas-Arlington
Center for Positron Studies
Physics Dept Box 19059
Arlington TX 76019 0059
817-272-2266
817 272 3637
B093a4w@utarlg.uta.edu or weiss@uta.edu

Wells, Eric
Kansas State University
J R Macdonald Lab
116 Cardwell Hall
Manhattan KS 66506 2604
785 532 2669
785 532 6806
ewells@phys.ksu.edu

Wells, Douglas
Idaho State University
Physics
Campus Box 8106
Pocatello ID 83209
208 282 3986
208 282 4649
wells@physics.isu.edu

Welton, Robert F
Oak Ridge National Lab
Physics Division, Bldg 6000
PO Box 2008 MS 6368
Oak Ridge TN 37831 6368
865 574 4753
865 574 1268
weltonrf@ornl.gov

Wender, Stephen A.
Los Alamos National Lab
LANSCE MS H803
P O Box 1663
Los Alamos NM 87545
505 667 1344
505 665 3705
wender@lanl.gov

Wenninger, Horst
CERN
DSU/ETT Division
CH 1211 GENEVA 23
Geneva CH 1211 SWITZERLAND
41 22 767 4097
41 22 767 8666
horst.wenninger@cern.ch

West, Thomas A
Thergenics Corporation
5203 Bristol Industrial Way
Buford, GA 30542
770 831 4331
770 271 6842
westt@theragenics.com

Wetteland, C. J.
Los Alamos National Lab
MSK765
Los Alamos, NM 87545
505 667 6133
505 665 2992
wetteland@lanl.gov

White, Gary Dane
Northwestern State University of Louisiana
Dept of Chemistry & Physics
Fournet Hall   College Av
Natchitoches LA 71497
318 357 5214
318 357 4219
white@alpha.nsula.edu

White, James T
Texas A&M University
Physics Department
College Station TX 77843
979 845 5490
979 845 2590
white@physics.tamu.edu

Wieman, Howard
LBNL
Nuclear Science Division
Berkeley CA 94720
631 344 7762
631 344 4206
hhwieman@lbl.gov

Wilkens, Barry J.
Arizona State University
Ctr. for Solid State Science
Tempe AZ 85287 1704
602 965 9613
602 727 6205
barry.wilkens@asu.edu

William, Ian
Queen's University Belfast
Belfast UK BT7 INN
44 28 90 273699
i.williams@gub.oc.uk

Williams, John Robert
Auburn University
Dept of Physics, 310 Nuclear Sci Ctr
106 Allison Lab
Auburn AL 36849 5318
334 844 4678
334 844 6917
williams@physics.auburn.edu

Williams, James M
Retired, now a consultant
7824 Beckett Ridge Court
Powell, TN 37849
865 574 6265
865 576 8135
notleck1b@aol.com

Wilson, Erik
University of Wisconsin
Plasma Source Ion Implantation Program
1500 Engineering Dr Rm 841 ERB
Madison WI 53706
608 265 4056
608 263 3632
wilsone@cae.wisc.edu

Winter, Helmut
Humboldt Universitaet
Physics
Invalidenstrasse 110
Berlin D 10115 GERMANY
49 30 7891
49 30 7899
winter@physik.hu-berlin.de

Wohrer-Beroff, Karine
Groupe De Physique Des Solides
Univ of Paris Jussieu 7-GPS
2 place Jussieu
75005 Paris FRANCE
33 1 4427 7309
33 1 4427 7309
wohrer@gps.jussieu.fr

Womble, Phillip C.
Western Kentucky University
Applied Physics Institute
1 Big Red Way
Bowling Green KY 42101
270 781 2518
270 781 1104
womble@wku.edu

Wong, Alfred Y
UCLA Plasma Physics Laboratory
Physics & Astronomy
1-129 Knudsen Hall MS 154705
Los Angeles CA 90025 1547
310 825 1642
310 206 2173
awong@physics.ucla.edu

Wood, Blake P
LANL
Plasma Physics Group P 24
MS E526
Los Alamos NM 87545
505 665 6524
505 665 3552
bwood@lanl.gov

Wu, Shiu-Chin
National Tsing Hua University
Department of Physics
101 Sec 2, Kuang Fu Rd
Hsinchu 30043 Taiwan R.O.C.
886 35 722 418
886 35 717 160
scwu@phys.nthu.edu.tw

Wuilleumier, Francois
Universite Paris Sud
Lab de Spectr Atom et Ion
Campus Orsay B.350
F-91405 Orsay Cedex FRANCE
33 1 69 15 65 36
33 1 69 15 58 11
francois.wuilleumier@lsai.u-psud.fr

Wuosmaa, Alan H
Argonne National Laboratory
Physics Division
9700 S Cass Ave
Argonne IL 60439
630 252 6545
630 252 6210
wuosmaa@anl.gov

Xie, Zhaoxia
University of Houston
3201 Cullen Blvd.
Houston, TX 77204
713 743 8218
713 743 8201
zhxxie66@yahoo.com

Xue, Jianming
Kyoto University, Sakyo
QSEC, Engineering Faculty
Kyoto 606-01
Japan
81 75 753 5846
81 75 753 5845
jmxu@nucleng.kyoto-u.as.jp

Yagishita, Akira
Institute of Material Structure Science
Photon Factory
Oho 1-1
Tsukuba Ibaraki 305-0801 JAPAN
81 298 64 5660
81 298 64 2801
akira.yagishita@kek.jp

Yamada, Yuichi
NEDO
Sunshine 60 29th F 3-1-1
Higashi-Ikebukuro, Toshinma-ku
Tokyo, JAPAN 170-0028
81-3-3987-9498
81-3-3987-9396
yamadayic@nedo.go.jp

Yamada, Isao

Himeji Institute of Technology
Laboratory of Advanced Science and Technology
Cast Kamigori Ako Hyogo
Japan 678-1205
81 791 58 002
i-yamada@kuee.kyoto-u.ac.jp

Yamamoto, Shunya
JAERI
Dept of Materials Development
1233 Watanuki-
Takasaki 370-1292 Gunma Japan
81 027 346 9422
81 27 346 9690
yamamoto@taka.jaeri.go.jp

Yang, Changyi
University of Melbourne
Microanalytical Research Centre
Parkville VIC 3010 AUSTRALIA
76203-1427
613 8344 5376
613 9347 4783
changyiy@hotmail.com

Yano, Yasushige
RIKEN
Cyclotron Laboratory
2-1 Hirosawa Wako-shi
Saitama 351-0198 Japan
81 48 462 1111
81 48 461 5301
yano@rikaxp.riken.go.jp

Yu, Yueh-Chung
Academia Sinica:Physics
128 Yen Chiou Yuan
Rd, Sec 2, Nankang
Taipei 11529 Taiwan 11529
886 2 2789 9669
control room no.:886-2-2789-9613
886 2 2783 4187
phycyu@ccvax.sinica.edu.tw

Yuichi, Yamada
NEDO
Sunshine 60 29th F 3-1-1
Higashi-Ikebukuno, Toshima-ku
Tokyo, Japan 170-0028
81 3 3987 9498
81 3 3987 9396
yamadayic@nedo.go.jp

Zaharakis, Konstantinos
Western Michigan University
Department of Physics
Kalamazoo MI 49008
616 387 5360
616 387 4939
k.zaharakis@wmich.edu

Zamkov, Mikhail
Kansas State University
1604 Roof Dr. 11
Manhattan, KS 66502
785 770 9828
mikhail@phys.ksu.edu

Zanini, Luca
Los Alamos National Laboratory
LANSCE 3
MS H855
Los Alamos NM 87545
505 665 3019
505 667 5377
zanini@lanl.gov

Zavodszky, Peter A.
Michigan State University
National Superconducting Cyclotron Laboratory
South Shaw Lane
East Lansing MI 48824
517 333 6460
517 353 5967
zavodszky@nscl.msu edu

Zeman, Herbert D.
Univ of Tennessee-Memphis
School of Biomedical Engr
899 Madison Ave Room 801
Memphis TN 38163
901 448 8454
901 448 7387
hdzeman@bellsouth.net

Zhang, Zuhua
University of Houston
TCSUH
3201 Cullen Blvd
Houston TX 77204 5932
713 743 8256
713 743 8201
tcsuh5@jetson.uh.edu

Zhao, Zhiyong
Advanced Micro Devices
FAB 25
5204 East Ben White Blvd MS 608
Austin TX 78741
512 602 2208
512 602 2209
zhiyong.zhao@amd.com

Zhuikov, Boris L
Institute for Nuclear Research
60th October Prospect, FA
117312 Moscow RUSSIA
7 095 334 0785
7 095 334 0184
bz@al20.inr.troitsk.ru

Zigler, Steve
CTI Cyclotron Systems
810 Innovation Drive
Knoxville TN 37932
865 218 2000
865 218 3000
steve.zigler@cti-pet.com

Zimmerman, Robert Lee
Alabama A&M University
Ctr for Irradiation of Materials
P O Box 1447 Dept of Physics
Normal AL 35762 1447
256 851 5844
256 851 5868
rlzimm@cim2.aamu.edu

Zouros, Theo J M
University of Crete
Dept of Physics
PO Box 2208
71003 Heraklion Crete GREECE
30 81 394 117
30 81 394 101
tzouros@physics.uoc.gr

Zwicker, Robert
Medical College of Virginia
Department of Rad Oncology
Box 980058
Richmond VA 23298 0058
804 828 4440
978 777 3594
rzwicker@hsc.vcu.edu

Author Index

A

Åberg, M., 394
Abs, M., 696
Accorsi, R., 491
Adams, E. M., 432, 454
Adams, H. R., 849
Adams, W. L., 1057
Adoui, L., 114
Agee, F. J., 1047
Agosteo, S., 555
Agulló-López, F., 643
Ahlgren, T., 887
Ahmad, I., 261, 559
Aigbirhio, F. I., 804
Akdogan, T., 339
Alanis, J., 853
Alarcon, R., 293
Alford, T. L., 443
Aliabadi, H., 36, 149, 172
Alshibli, K. A., 1091
Alton, G. D., 244, 277, 285
Alvord, C. W., 793, 799, 817
Ambrosi, R. M., 1087
Ames, F., 265
Andrade, E., 440
Andre, G., 857
Anferov, V., 651
Angélique, J. C., 254
Antalic, S., 1148
Anthony, M., 1036
Anundson, R., 1036
Aoki, T., 967
Aprahamian, A., 309, 346
Aramaki, T., 403
Armbruster, J. M., 335, 639, 845
Aslanoglou, X., 7
Avila-Perez, P., 512
Awazu, K., 623, 955
Äystö, J., 265

B

Bach, P., 1099, 1141
Back, B. B., 231
Baechler, S., 1095
Baer, D. R., 454
Bailey, A. M., 48, 89, 137, 522
Bailey, M.,
Bair, A. E., 443
Bajeat, O., 254
Baker, M. D., 231
Bakhru, H., 896
Balasko, M., 1099
Balatoni, J., 849
Ballanger, A., 1141
Ball, G. C., 297
Barrera, E., 440
Barros, L. I. C., 508
Barstis, T. L. O., 122
Barton, D. S., 231
Bartschat, K., 89, 137
Barué, C., 254
Basilev, S., 231
Batchelor, K., 787
Bates, B. D., 231
Batiste, S. N., 1091
Battistig, G., 919
Baum, R., 231
Baumann, T., 635
Bauvir, B., 857
Beaver, J. E., 845
Beck, B. R., 118
Becker, H.-W., 1077
Becker, J. A., 309
Beene, J. R., 277
Beers, E. W., 783
Belbot, M., 1065
Belyaev, A., 1036
Benis, E. P., 76, 149
Ben-Itzhak, I., 33, 36
Ben-Zvi, I., 615
Berg, C. M., 647
Bernardi, G., 164
Bernstein, L. A., 309
Berrah, N., 177
Berry, H. G., 80
Bertrand, F. E., 277
Best, B., 808
Betts, R. R., 231
Beukens, R. P., 390
Bhalla, C. P., 172, 201
Bhattacharyya, P., 559
Bialas, A., 231
Bibić, N., 599
Bigelow, A. W., 108
Bigott, H. M., 841
Bilheux, J.-C., 285
Bindel, R., 231
Bizau, J.-M., 711
Blackmon, J., 261
Bliman, S., 137, 213, 217
Blohm, G., 458
Bluem, H., 615
Bodart, F., 480

Bogucki, W., 231
Bohn, C., 615
Bollen, G., 265
Bongers, H., 265
Borasi, F., 261
Bouffard, S., 526
Bouly, J. L., 281
Bowles, T. J., 289
Bowyer, S. M., 563
Boyce, D., 335
Boyd, R. N., 358
Bozek, J. D., 177
Bradbury, D., 1125
Bray, I., 89, 137
Bremer, J. H., 526
Bricault, P., 239
Brijs, B., 470
Brown, C. M., 567
Brown, T. B., 315
Bruch, R., 48, 89, 137, 213, 217
Brune, C. R., 305
Bruyneel, B., 611
Bryan, W. A., 44
Brzosko, J. S., 277
Buckley, W. M., 1073
Budzanowski, A., 231
Burke, P., 804
Burkhard, H.-G., 1148
Burrows, C. P., 745
Busza, W., 231
Butz, T., 428

C

Cabrera, L., 512
Cabrera-Trujillo, R., 3
Caffrey, A. J., 1073
Cagarda, P., 1148
Caggiano, J., 261, 559
Callcott, T. A., 548
Callerame, J., 1057
Camase, T., 1047
Campbell, C., 1069
Campbell, J. L., 413
Cappiello, M. W., 828
Carnes, K. D., 33
Carney, J. P. J., 52
Carpenter, M. P., 559
Carroll, A., 231
Carroll, L. R., 301, 639
Carstanjen, H. D., 458
Casey, J. A., 607
Cassimi, A., 114
Castillo, F., 161
Cauwels, P., 1095

Cederkäll, J., 265
Ceglia, M., 231
Chabalier, B., 555
Chadwick, M. B., 346
Chang, Y. H., 231
Chauvin, N., 281
Chen, A. E., 231, 261
Chen, G., 1109
Chen, J., 72
Chen, Q. Y., 928, 983
Chen, Z., 112
Cheng, Y., 724
Chesnel, J.-Y., 114
Chirco, P., 1099
Chtangeev, M., 339
Chu, W.-K., 323, 891, 900, 928, 983, 987, 1007
Church, D. A., 118
Chutjian, A., 157
Cipolla, S. J., 56, 84
Cizewski, J. A., 559
Clark, J. C., 804, 1132
Cleghorn, D. A., 779
Cleland, M. R., 783
Climent-Font, A., 643
Cocke, C. L., 36
Coghen, T., 231
Cogneau, M., 269
Cohen, S. M., 29
Coldwell, R. L., 587
Cole, M. D., 615, 619
Coleman, P. G., 745
Collens, T. J., 675, 877
Collins, C. B., 1047
Colson, J. M., 269
Conner, C., 231
Conton-Rogan, S. E., 177
Costes, N. C., 1091
Coutrakon, G. B., 861
Craft, III, B. C., 734
Crothers, D. S. F., 168
Crowley, B. T., 659
Csikai, J., 1061
Cubaynes, D., 711
Culbertson, R. J., 908
Curdy, J. C., 281
Czyz, W., 231

D

Dabrowski, B., 231
Dahl, J., 849
Daley, K. E., 908
Datar, S. A., 25
Datz, S., 126

Davanloo, F., 1047
Davids, C. N., 261, 559
Davidson, J. R., 366
Davinson, T., 265
De Corte, F., 495, 504
Decowski, M. P., 231
De Crock, P., 857
De Hosson, J. T., 971
Deleu, J., 470
Demaree, J. D., 915
Denker, A., 417
Despet, M., 231
Deumens, E., 3
Devlin, M., 309, 346
DeWitt, H. E., 118
Dierick, M., 1095
Dietrich, F., 1113
Dobrosavljević, A., 599
Dodd, P. E., 516
Dokhale, P. A., 1061
Dorman, E., 808
Dörner, R., 185
Dote, M., 631
Dougan, A. D., 1073
Douglas, J. J., 881
Doupe, J. P., 390
Downey, S. P. M. J., 804
Doyle, B. L., 516, 531, 535
Dreizler, R. M., 185
DuBois, R. D., 181
Dudnikov, V., 741
Duggan, J. L., 25, 531, 535, 539
Dukes, W., 1125
Dulloo, A. R., 499
Dunnam, F. E., 421
Dymnikov, A. D., 655
Dyublov, A., 787

E

Eberhardt, K., 1144
Eckelman, W., 849
Ederer, D. L., 548
Eguskiza, M., 504
Einck, J. P., 881
Ektessabi, A. M., 707, 720
El Bouanani, M., 25, 335, 531, 535, 539
Elliman, R. G., 487
Elmore, D., 382
El-Zein, A., 44
Emhofer, S., 265
Engelhard, M. H., 454
Enguita, O., 643
Erlandsson, B., 377, 394
Espeso-Gil, O., 643

Evans, C. M., 730
Evelyn, A. L., 1020, 1024

F

Faarinen, M., 377, 394
Falsetti, A. B., 421
Farrell, J. P., 741, 787
Fazleev, N. G., 753
Federmann, G., 627
Fedorov, A. V., 971
Fehrenbach, C. W., 603
Feldermann, H., 947
Ferrari, A., 671
Ferro, D. N., 779
Fessler, A., 1053
Fillipone, B., 289
Fink, C. L., 1118
Finn, R., 849
Fiol, J., 164
Fischbach, E., 382
Fischer, T., 458
Fita, P., 231
Fitch, J., 231
Fleming, C. H., 209
Florescu, A., 60
Focke, P., 164
Foglio Para, A., 555
Fokitis, E., 436
Forstner, O., 265
Foster, C. C., 647
Fotiades, N., 309
Fournet, J., 522
Fox, R. O., 734
France III, R. H., 277
Franklin, W. A., 339
Fregenal, D., 164
Freiesleben, H., 763
Freitas, M. C., 508
Fremont, F., 114
Friedl, M., 231
Friesel, D. L., 651
Frolov, D. O., 591
Frolov, O. S., 591
Fry, J. L., 753
Fujisawa, S., 707
Fujitaka, K., 579
Fukuda, S., 631
Fukumoto, S., 631
Fursa, D. V., 137
Fursa, F. V., 89

G

Gabor, R., 1155
Gácsi, Z., 1069
Gaelens, M., 269
Gai, M., 277
Galuska, K., 231
Ganz, R., 231
Garcia, E., 231
García, G., 643, 655
Garrett, P. E., 309
Gaubert, G., 254
Gaudreau, M. P. J., 607
Gavin, P. R., 881
Gealy, M., 172
Gehrke, R. J., 366
Geller, R., 281
Geltenbort, P., 289
Gembalies-Datz, D., 1148
Gentens, J., 611
George, N., 231
Gerber, E., 1125
Gersch, H. K., 1118
Gevers, G., 857
Gibouin, S., 254
Gilbody, H. B., 205
Gillaspy, J. D., 935
Gillen, D. R., 205
Glass, G. A., 193, 425, 522, 548
Godlewski, J., 231
Godunov, A. L., 185
Golovkina, V., 213
Golser, R., 627
Gomes, C., 231
Gomi, S., 145, 543
González, A. D., 164
Gonzalez Lepera, C., 313
Goodworth, T. R. J., 44
Gorczyca, T. W., 177
Goto, A., 679
Gotoh, Y., 1028
Gottdang, A., 403
Graf, U., 555
Greene, J. P., 261, 1152, 1155
Greenwood, J. B., 157
Greiner, W., 17
Griesmayer, E., 231
Grisham, L. R., 759
Grodzins, L., 1057
Grötzschel, R., 7, 436
Gruber, L., 118
Guardala, N. A., 193, 741
Gulbrandsen, K., 231
Gund, C., 265
Gunnarsson, M., 394
Guo, B. N., 531, 539
Gushue, S., 231
Gwilliam, R. M., 745

H

Habenicht, S., 951
Habs, D., 265, 667
Hagel, K., 223
Haight, R. C., 309, 339, 346
Håkansson, H., 377
Håkansson, K., 377
Halik, J., 231
Hall, E. R., 647
Hall, J. M., 567, 1105, 1109, 1113
Halliwell, C., 231
Hamm, R. W., 651
Hanmura, T., 975
Hanni, J., 48, 89, 137, 213, 217
Haridas, P., 231
Harker, Y. D., 877, 881
Harmon, J. F., 342, 675, 877
Harss, B., 261
Hartmann, W., 1148
Hartwell, J. K., 881
Hasan, A. T., 36, 172
Hasegawa, T., 631
Hasegawa, Y., 955
Hatashita, M., 631
Hatori, S., 631, 943
Hawkey, T. J., 607
Hawthorne, M. F., 881
Hayashi, K., 623
Hayes, A., 231
Hayes, J. M., 407
Heintzelman, G. A., 231
Heinz, A., 261, 559
Hellborg, R., 377, 394
Helmer, R. G., 366
Hemmers, O. A., 189
Henderson, C., 231
Henderson, D., 261
Henning, W. F., 559
Hensley, D. H., 1020
Hensley, D. K., 1015
Herer, A. S., 783
Hicks, N., 72
Hicks, S. F., 315
Hill, R., 289
Hino, M., 289
Hinojosa, G., 213
Hoedl, S., 289
Hoffmann, V., 526
Hofmann, S., 1148
Hofsäss, H., 947
Hogan, G., 289

Holder, J. P., 118
Hollerman, W. A., 193, 425, 522, 548
Hollis, R., 231
Holynski, R., 231
Holzman, B., 231
Horn, K. M., 516
Horvat, V., 56, 93
Horváth, Z. E., 919
Hough, J. B., 339
Huang, N. K., 924
Huber, G., 265
Huck, A., 265
Huddle, J. R., 1128
Huerta, L., 440
Huguet, Y., 254
Hui, S., 1069
Huot, N., 555
Hupf, H. B., 335, 845
Husson, X., 114
Huygebaert, C., 470
Huyse, M., 235, 265, 611
Hynes, S. F., 522

I

Iannotta, S., 979
Icenhower, J. P., 454
Ichihashi, M., 975
Ikenaga, N., 955
Ikezawa, K., 631
Ila, D., 451, 484, 1015, 1020
Imai, M., 543
Imanishi, N., 145, 543
Ingram, D., 1053
Inochkin, M., 787
Inomata, T., 631
Iosif, M. C., 1047
Isaacs-Smith, T., 1132
Isenhower, L. D., 1135
Ishigami, R., 631, 943
Ishikawa, J., 1028
Ito, T. M., 289
Ito, Y., 40, 631, 943
Itoh, A., 931
Itoh, T., 759

J

Jaber, R. M., 675
Jacobs, D. C., 122
Jadczyk, A., 587
Jain, D., 837
Jakuba, D. H., 197
Jankuhn, S., 428

Janney, M. A., 250
Janssens, R. V. F., 261, 559
Jarin, P., 254
Javorsek II, D., 382
Jeong, S. C., 931
Jiang, C. L., 261
Jiang, W., 447
Jin, J., 323, 891, 983
Johansen, T., 928
Johns, G. D., 309
Johnson, E., 231
Johnston, I. M. G., 44
Jolie, J., 1095
Jones, G. D., 559
Jones, J. L., 877
Jongen, Y., 696, 857
Jonson, B., 265
Julin, R., 559

K

Kabuto, S., 403
Kageyama, K., 1040
Kagiya, G., 631
Kakiuchi, S., 631
Kambara, T., 1040
Kanai, Y., 1040
Kane, J., 231
Kang, K. D., 72
Karfidov, D., 72
Karim, K. R., 201
Kase, M., 679
Katayama, I., 931
Katayama, T., 679
Katzy, J., 231
Kawai, M., 759
Kawai, T., 289
Kearns, D. M., 205
Kehayias, J. J., 873
Keinonen, J., 887
Kekki, T., 555
Keller, R., 362
Kelley, J. H., 659
Kempkes, M. A., 607
Kennedy, K. C., 675, 877
Kern, W. G. E., 407
Kessel, Q. C., 1044
Kester, O., 265
Khellaf, A., 458
Khoo, T. L., 559
Kieser, W. E., 386, 390
Kiisk, M., 377, 394
Kim, S.-W., 407
Kimura, K., 470
Kindler, B., 1148

King, G. C., 703
Kirch, K., 289
Kirkpatrick, A., 963
Kishimoto, N., 1011
Kita, W., 231
Kitagawa, T., 963
Kitamura, T., 403
Klann, R. T., 1118
Kleeven, W., 696
Klein, M., 403
Klein, S. B., 647
Klemm, J., 1148
Knights, A. P., 745
Kocbach, L., 68
Koehler, P., 350
Koivisto, H., 619
Kojima, T. M., 1040
Kojouharova, J., 1148
Kokkoris, M., 7, 436
Komatsu, K., 623
Kondev, F. G., 559
Kondo, M., 631
Kondow, T., 975
Kooi, B. J., 971
Korol, A. V., 17, 64
Kossionides, S., 7, 436
Kothari, P., 849
Kotula, J., 231
Koubouras, G., 436
Kraner, H., 231
Krause, H. F., 126
Kravchenko, I. I., 421
Kreek, S. A., 1073
Krestow, J., 390
Krouglov, K., 235
Krücken, R., 319
Kruglov, K., 265
Kruse, O., 458
Kubsky, S., 1077
Kucewicz, W., 231
Kudryavtsev, Y., 611
Kuhn, M., 548
Kühnel, K.-U., 265
Kukk, E., 177
Kulinich, P., 231
Kulkarni, P. V., 837
Kume, K., 631
Kuriyama, M., 759
Kutschera, W., 627

L

Lacoste, A., 281
Ladadwa, I., 68
Laforest, R., 841

Lamoreaux, S. K., 289
Lamy, T., 281
Land, D. J., 193
Langley, A. J., 44
Lankton, M. R., 1091
Lanza, R. C., 491, 1109
Laramore, G. E., 881
Larson, S. M., 849
Lartigue, J., 512
Lasche, G. P., 587
Lauritsen, T., 559, 1152
Lavietes, A. D., 1073
Law, C., 231
Lawrence, D. F., 1032
Lecesne, N., 254
Lee, C. G., 1011
Lehmann, E., 1099
Lehto, S., 887
Lekomtsev, V., 787
Lemler, M., 231
Leroy, R., 254
Levy, M., 896
Lewis, M. R., 841
Lewitowicz, M., 254
Li, Q. M., 323, 891
Li, S. L., 108
Liang, J. F., 273
Liao, C., 425, 522
Lieb, K. P., 951
Ligocki, J., 231
Likonen, J., 887
Liljeby, L., 265
Lim, C. S., 1081
Lim, J., 808
Lin, C. D., 48, 89, 137
Lin, W. T., 231
Lindle, D. W., 189
Ling, P., 891, 983
Lister, C. J., 559
Litherland, A. E., 386, 390
Litvinenko, V. N., 659
Liu, C. N., 177
Liu, C.-Y., 289
Liu, J., 323, 891, 983, 987, 1007
Liu, J. R., 900, 928
Liu, Y., 244
Livingston, R. A., 1077
Logan, C., 1113
Loiselet, M., 269
Lom, C., 849
Lommel, B., 1148
Lopes, M. C. A., 703
López, C., 512
Lozano, J., 1044
Lu, X., 323, 891, 900, 928, 983, 987, 1007

Lu, X. M., 900
Lucas, S., 696
Lugo-Licona, M. F., 100
Lyalin, A. G., 64

M

Maazouz, M., 122
Maazouz, P. L., 122
Macek, J. H., 130
Mack, M. E., 991
Macri, J. R., 571
Magnusson, C.-E., 377, 394
Maheswaran, S., 432, 454
Maier, K. H., 417
Major, J., 749
Mandler, J. W., 366
Mangner, T., 808
Manias, C., 1081
Manly, S., 231
Mann, R., 1148
Manuaba, A., 919
Marchand, B., 857
Markoff, D. M., 327
Marques, A. P., 508
Marry, C., 254
Martin, J. W., 289
Martínez, H., 161
Martínez, T., 512
Maruyama, T., 631
Mashnik, S. G., 833
Masschaele, B., 1095
Massey, T. N., 1053
Mathot, G., 480
Matsumoto, E., 715
Matsuo, J., 963, 967, 995
Matteson, S., 25, 108
Matthews, J. L., 339
Mattsson, S., 377, 394
Mauger, G. J., 1073
Maunoury, L., 254
Maxwell, J. A., 413
Mayer, J. W., 443
McCarthy, D. W., 841
McCarthy, T. J., 841
McCartney, P. C. E., 153
McCleod, D., 231
McConkey, J. W., 153
McConnell, M. L., 571
McCullough, R. W., 153, 205
McDaniel, F. D., 25, 335, 531, 535, 539
McDonald, J. E., 277
McGrail, B. P., 454
McGrath, C. A., 153, 168, 309
McGregor, D. S., 1118

McGuire, J. H., 185
McKenna, P., 44
McNabb, D. P., 309
McSherry, D. M., 133, 168
Meissner, J., 783
Melzacki, K., 277
Mendez, A. J., 799
Merabet, H., 48, 89, 137, 213, 217
Mertens, A., 21
Meshkovsky, I., 787
Michałowski, J., 231
Micklich, B. J., 1053
Mignerey, A., 231
Milani, P., 979
Miley, G. H., 683
Miller, L. F., 793
Miller, M. D., 734
Miller, P., 619
Miller, R. S., 571
Miller, T., 382
Miranda, J., 100
Mishin, A. V., 1057
Mitaroff, A., 671
Mitchell, G. E., 331
Miyamoto, J., 623
Mizutani, Y., 403
Moewes, A., 548
Mondelaers, W., 1095
Monnin, C., 1141
Montenegro, E. C., 96
Montenero, M., 639
Morgan, I. L., 335
Morikawa, E., 730
Morris, C., 289
Moshammer, R., 133
Moukha, I., 265
Mount, G., 900
Mous, D. J. W., 403
Msezane, A. Z., 112
Mueller, P. E., 250
Mukoyama, T., 104, 141
Mülmenstädt, J., 231
Mulvaney, J. M., 607
Muntele, C. I., 451, 484, 1015, 1020
Muntele, I. C., 451, 1020
Murphy, M., 808
Murray, K. M., 647
Murray, S. N., 244

N

Nablo, S. V., 779
Nageswara Rao, S. V. S., 476
Nähler, A., 1144
Nakai, Y., 1040

Nakajima, K., 470
Nakamura, T., 579
Nakano, S., 939
Nambo, Y., 955
Naramoto, H., 466
Narula, J., 837
Nauwelaerts, S., 470
Navarrete, M., 512, 853
Nayandin, O., 177
Neal, M., 231
Necsoiu, D., 335
Neil, G., 615
Nelson, R. O., 309
Nešković, N., 599
Neubauer, J., 1155
Newell, W. R., 44
Ni, B., 959
Nigam, M., 25, 535, 539
Nigg, D. W., 877, 881
Nilsson, L. E., 394
Nilsson, T., 265
Nimets, O. F., 591
Ninomiya, S., 145, 543
Nishiyama, Y., 623
Nolen, J., 261
Nosslin, B., 394
Nouicer, R., 231
Nyman, G., 265

O

Obolensky, O. I., 60, 64
O'Brien, H. A., 828
O'Donnell, J. M., 339
Ogawa, H., 931
Ogiso, H., 939
Öhrn, Y., 3
Ohtani, N., 631
Oinonen, M., 265
Okazaki, K., 955
Okubo, N., 1011
Olszewski, A., 231
Ophel, T. R., 487
O'Rourke, S. F. C., 133, 168
Orr, N. A., 254
Ortman, W. K., 250
Osgood, Jr., R. M., 896
Ostapenko, S., 1036
Östberg, H., 394
Ostrowski, A. N., 265

P

Pacheco, A. M. G., 508
Pacquet, J. Y., 254
Pak, R., 231

Paradellis, T., 7
Pardo, R. C., 261
Park, I. C., 231
Park, S. H., 659
Parker, P., 261
Parker, W. E., 1073
Paschal, J., 1065, 1069
Pascual-Izarra, C., 643
Paskalov, G., 72
Pászti, F., 919
Patel, M., 231
Pathak, A. P., 476
Paul, M., 261
Pavlishin, I., 787
Pellemoine, F. L., 254
Pendola, J. P., 845
Pentlow, K., 849
Pernegger, H., 231
Persson, P., 377, 394
Peterson, R. S., 1125
Petrović, B., 499
Phaneuf, R., 213
Phillips, Jr., G. N., 734
Pichlmaier, A., 289
Pinayev, I. V., 659
Pineda, J. C., 440
Pitcher, E. J., 828
Plachke, D., 458
Plascjak, P., 849
Plesko, M., 231
Podlech, H., 265
Poker, D. B., 1015, 1020
Pollack, E., 1044
Pommé, S., 504
Poncelet, E., 696
Porter, L. E., 11
Postiau, N., 269
Powell, C., 277
Pratt, R. H., 52, 60
Price, J. L., 193
Prieels, D., 857
Priller, A., 627
Przewoski, B. v., 647
Purser, K. H., 386

Q

Qin, X., 122
Quarles, C. A., 56
Quinteros, C. L., 122

R

Rabalais, J. W., 911
Radojevic, A. M., 896
Rahmanian, H., 1087

Ramirez, A., 512
Ramsey, F., 639
Rangama, J., 114
Rataud, J. P., 254
Rathke, J. R., 615
Ratliff, L. P., 935
Ratzinger, U., 265
Rauniyar, R., 250
Ravazzani, A., 555
Ravn, H. L., 265
Reed, C., 231
Reeder, P. L., 563
Rehm, K. E., 261
Reichert, D. E., 841
Reis, M. A., 508
Reiter, P., 265, 559
Remsberg, L. P., 231
Renfrow, S. N., 531
Rensfelt, K. G., 265
Repnow, R., 265
Retallack, D., 1081
Reuter, M., 231
Reyes, P. G., 161
Richard, P., 36, 149, 172
Richter, A., 265
Richter, D., 763
Risler, R., 881
Robouch, P., 504
Rocha, M. F., 440
Rody, S. K., 254
Roland, C., 231
Roland, G., 231
Rolfs, C., 1077
Ronning, C., 947
Rosenberg, L., 231
Rosenthal, G., 72
Ross, D., 231
Rost, J. M., 999
Rotberg, V. H., 687
Rothschild, P. J., 1057
Ruddy, F. H., 499
Ruggles, A. E., 817
Rusakova, I., 900
Rusnak, B., 1105, 1113
Ruzycki, N., 548
Ryan, J. M., 231, 571
Ryckewaert, G., 269

S

Sabin, J. R., 3
Sabourov, A., 877
Sadler, M. E., 1135
Sadovnickiy, A. A., 591
Safkan, Y., 339

Saint Laurent, M. G., 254
Saitoh, K., 955
Saji, E., 955
Sakamoto, N., 931
Sakudo, N., 623, 955
Saladin, J. X., 309, 346
Sale, K. E., 567
Sanders, J. M., 209
Sanderson, J. H., 44
Sandstrom, P. W., 1032
Sangsingkeow, P., 1125
Santos, A. C. F., 96
Sanzgiri, A., 231
Sapp, W. W., 1057
Sarin, P., 231
Sasaki, H., 1028
Sato, H., 1028
Sato, T., 955
Saunders, A., 289
Savukov, I. M., 80
Sawicki, P., 231
Sayler, A. M., 33
Scaduto, J., 231
Scates, W. W., 342
Schätz, T., 667
Schauer, J. E., 76
Schempp, A., 265
Schiffer, J. P., 261
Schill, J., 615
Schillebeeckx, P., 555
Schlösser, K., 824
Schmidt, P., 265
Schmidt, S., 857
Schmidt-Böcking, H., 185
Schneider, D., 118
Schneider, L. X., 663
Schneider, R. J., 407
Schött, H.-J., 1148
Schramm, U., 667
Schreiber, E. C., 659
Schrieder, G., 265
Schueller, R., 1057
Schuler, T. M., 548
Schut, H., 971
Schwalm, D., 265
Schwartz, J. L., 881
Schweickert, H., 824
Schweitzer, J. S., 1077
Scott, J. D., 730
Seabury, E. H., 339
Sealy, B. J., 745
Seestrom, S., 289
Segel, R. E., 261
Seidel, J. G., 499
Seidel, K., 763
Seki, T., 1003

Seo, H.-W., 928
Serebrov, A., 289
Seweryniak, D., 261, 559
Shaffer, C. D., 60
Shah, M. B., 153, 168
Shakov, K. K., 185
Shao, L., 891, 900, 983, 987, 1007
Shea, J., 231
Sheh, Y., 849
Shevchenko, V. A., 591
Shew, B.-Y., 724
Shikine, S., 720
Shotter, A. C., 265
Shriner, Jr., J. F., 331
Shutthanandan, V., 432, 447, 454
Siddiqui, A. M., 476
Sie, S. H., 399
Sieber, T., 265
Siedschlag, C., 999
Siegle, A., 749
Siem, S., 559
Siems, A., 48, 89, 213, 217
Siemssen, R. H., 261
Sigaud, G. M., 96
Silari, M., 671
Simabuco, S. M., 715
Simon, H., 265
Sims, D. A., 399
Sinacore, J., 231
Singh, M. J., 36
Sinnott, S. B., 959
Sipe, B., 808
Sisterson, J. M., 865
Skeppstedt, Ö., 265
Skog, G., 377, 394
Skulski, W., 231
Smith, D. L., 289, 1053
Smith, G. J., 1057
Smith, M. S., 261
Smith, S. J., 157
Smith, W. W., 1044
Snell, G., 177
Sobocinski, P., 114
Soga, F., 869
Sole, P., 281
Solovjev, I. A., 64
Solov'yov, A. V., 17, 64
Sonzogni, A. A., 559
Sortais, P., 281
Sosnowski, M., 904
Sowerby, B. D., 1081
Spaulding, R., 342
Spemann, D., 428
Sridharan, K., 1032
Srigengan, B., 44
Srinivasan-Rao, T., 615, 787

Steadman, S. G., 231
Steier, P., 627
Steinberg, P., 231
Steiner, J., 1148
Stelzer, K. J., 881
Stenström, K., 377, 394
Stephans, G. S. F., 231
Stephenson, E. J., 575
Stevenson, N. R., 814
Stockli, M. P., 603
Stodel, C., 254
Stodulski, M., 231
Stoll, H., 458, 749
Stolterfoht, N., 526
Stopa, Z., 231
Stracener, D. W., 250, 257
Straczek, A., 231
Strangis, S. R., 313
Strathman, M. D., 463
Strek, M., 231
Strieder, F., 354
Sture, S., 1091
Su, C.-J., 928
Suárez, S., 164
Sukhanov, A., 231
Surowiecka, K., 231
Suter, G. F., 399
Suto, K., 403
Suzuki, T., 403
Svegborn, S. L., 394
Swanson, R. A., 1091
Swenson, D. A., 692
Szilágyi, E., 919

T

Taday, P. F., 44
Takada, M., 579
Takagi, K., 631
Takaoka, G. H., 963, 967, 995, 1003
Takeda, Y., 1011
Tang, J. L., 231
Taniguchi, S., 579
Tanner, V., 555
Tarisien, M., 114
Tawara, H., 36, 149, 172
Teng, R., 231
Ternier, S., 857
Terry, T., 1125
Terwagne, G., 480
Thevuthasan, S., 432, 447, 454
Thirolf, P. G., 265
Thomas, D. M., 768
Thörle, P., 1144
Thornberg, C., 377

Thorsson, O., 394
Thuillier, T., 281
Tickner, J. R., 1081
Tiitta, A., 555
Timmers, H., 487
Tipton, B., 289
Toader, O., 687
Tochio, T., 40
Todd, A. M. M., 615
Togawa, O., 403
Tolmanov, S. G., 185
Tominaga, Y., 631
Tomski, I., 390
Tonchev, A. P., 675, 877
Tornow, W., 659
Townsend, L. W., 793
Trautmann, N., 1144
Trigueiros, A. G., 48, 89, 213
Trzupek, A., 231
Tseng, H. C., 48, 89, 137
Tsuchida, H., 931
Tsuji, H., 1028
Tvinnereim, M., 1036
Tzvetkov, T., 122

U

Uda, M., 141
Ullmann, J. L., 339
Ullrich, J., 133
Umeda, N., 759, 1011
Ünal, R., 36, 172
Unholzer, S., 763
Uribe, R. M., 775
Utsuro, M., 289
Uusitalo, J., 261, 559

V

Vainonen-Ahlgren, E., 887
Vale, C., 231
Valind, S., 394
Van den Bergh, P., 235, 265, 611
Vandervorst, W., 470
Van Duppen, P., 235, 265, 611
Vane, C. R., 126
Vanhoy, J. R., 315
van Huis, M. A., 971
van Nieuwenhuizen, G. J., 231
Van Rinsvelt, H. A., 421
Van Riper, K. A., 833
Van Rompay, M., 1141
van Veen, A., 971
Vargas, C. A., 440

Vargas-Aburto, C., 775
Varghese, S. L., 209
Velkovska, J., 227
Verdier, R., 231
Vetter, K., 583
Vieux-Rochaz, J. L., 281
Villari, A. C. C., 254
Vizkelethy, G., 516
Vlaicu, A. M., 40
Vlastou, R., 7, 436
Vockenhuber, C., 627
Vogt, J., 428
von Hahn, R., 265
Vonk, D. T., 908
von Reden, K. F., 373, 407
Voulot, D., 205
Vourvopoulos, G., 1061, 1065, 1069

W

Wadsworth, B., 231
Walsh, D. S., 516, 531
Walter, G., 265
Wang, C., 928
Wang, D. Z., 924
Wang, T. F., 261
Wang, X., 900, 983, 987, 1007
Wang, Y., 443
Wang, Y.-H., 928
Warr, N., 315
Warren, M. W., 421
Was, G. S., 687
Watson, R. L., 93
Watterson, J. I. W., 1087
Weathers, D. L., 108
Weber, W. J., 447
Weichelt, M., 808
Weijers, T. D. M., 487
Weiss, A. H., 753
Weissman, L., 235, 265
Welch, M. J., 841
Weller, H. R., 659
Wells, D. P., 342
Wells, E., 33
Welsch, C., 265
Welton, R. F., 250
Wemple, C. A., 881
Wenander, F., 265
Wender, S. A., 339
White, M. A., 734
Wiedenhoeft, M., 177
Wiedenhöver, I., 261, 559
Williams, C. L., 250
Williams, I. D., 44, 157
Williams, J. R., 1132

Wills, A. A., 177
Wills, J. S. C., 407
Wilson, E. H., 1032
Wilson, W. B., 833
Winter, H., 21
Wittner, D. E., 799
Wolfs, F. L. H., 231
Womble, P. C., 1061, 1069
Wong, A. Y., 72
Woods, P. J., 265
Wosiek, B., 231
Wozniak, K., 231
Wu, Y., 659
Wuilleumier, F. J., 711
Wuosmaa, A. H., 231, 231
Wüst, F., 841
Wyslouch, B., 231

X

Xie, Z., 987
Xue, J., 145, 543

Y

Yadav, R. T., 975
Yamada, H., 631
Yamada, I., 963, 967, 995
Yamada, M., 631
Yamamoto, K., 631
Yamamoto, S., 466
Yamazaki, Y., 1040
Yang, C., 25, 535, 539
Yano, Y., 679
Yasuda, K., 631, 943
Yasui, H., 955
Yates, S. W., 315
Yencha, A. J., 703
Yoneda, A., 1040
Yoshida, S., 707, 720
Younes, W., 309
Young, A. R., 289
Young, P. G., 346
Yuan, J., 289
Yutani, M., 623

Z

Zaharakis, K. E., 93
Zaim, H., 244
Zalewski, K., 231
Zamkov, M., 149, 172
Zanini, L., 309, 346
Zarazúa, G., 512
Zavala, E. P., 440
Zavodszky, P. A., 619
Zhang, L. H., 900
Zhao, X.-L., 386, 390
Zhao, Z. Y., 1036
Zigler, S., 808
Zimmerman, R. L., 451, 484, 971, 1015, 1020
Zouros, T. J. M., 76, 149
Zychowski, P., 231